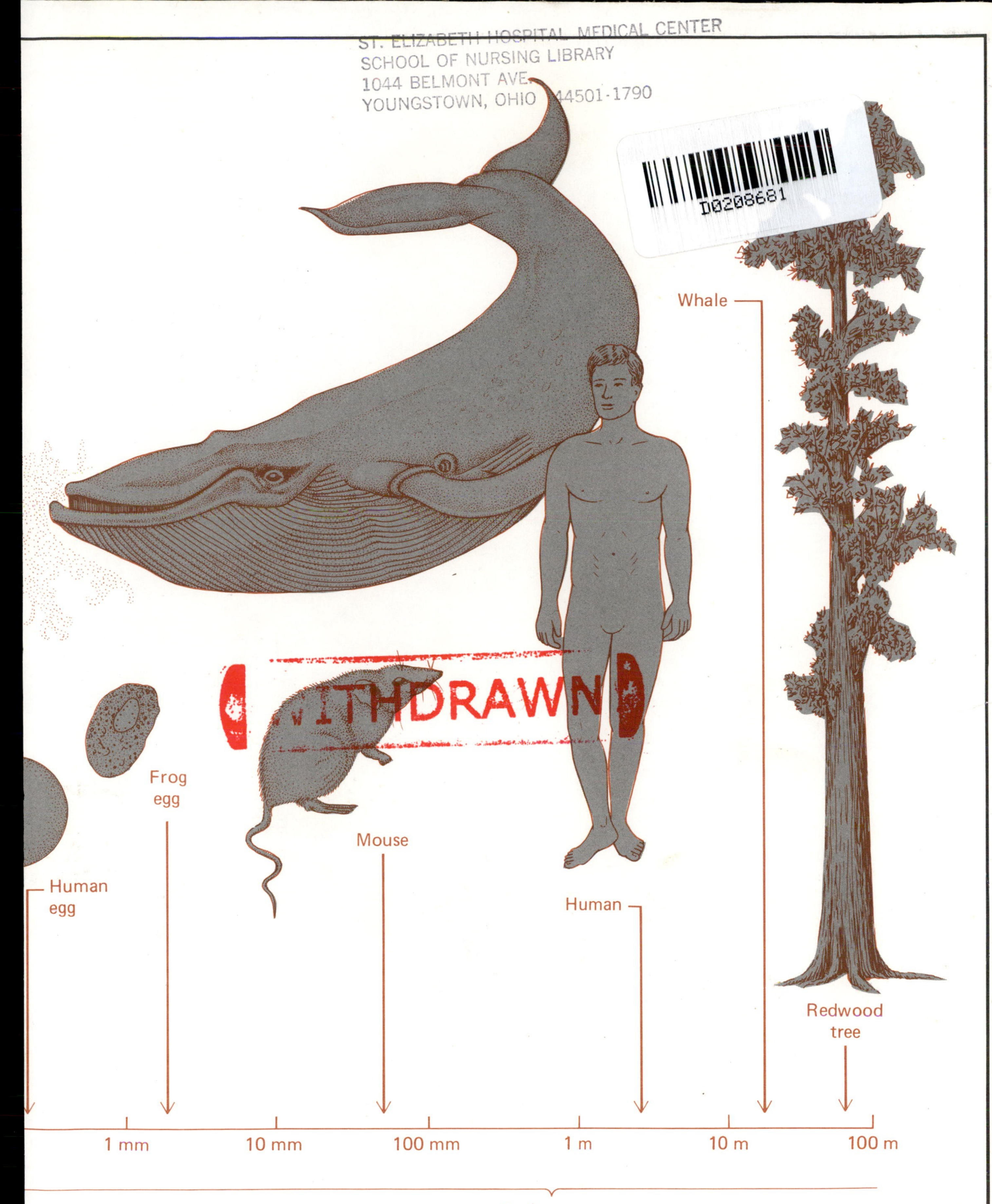
Whale
Frog egg
Mouse
Human egg
Human
Redwood tree
1 mm
10 mm
100 mm
1 m
10 m
100 m
Radio waves

BIOLOGY

CLAUDE A. VILLEE
HARVARD UNIVERSITY

ELDRA PEARL SOLOMON
HILLSBOROUGH COMMUNITY COLLEGE

P. WILLIAM DAVIS
HILLSBOROUGH COMMUNITY COLLEGE

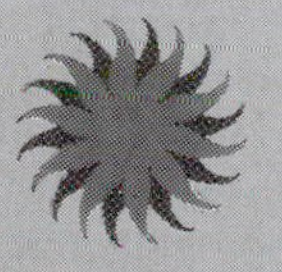

SAUNDERS COLLEGE PUBLISHING
Philadelphia New York Chicago
San Francisco Montreal Toronto
London Sydney Tokyo Mexico City
Rio de Janeiro Madrid

Address orders to:
383 Madison Avenue
New York, NY 10017

Address editorial correspondence to:
West Washington Square
Philadelphia, PA 19105

Text Typeface: Palatino
Compositor: York Graphic Services
Acquisitions Editors: Michael Brown and Ed Murphy
Developmental Editor: Don Reisman
Project Editor: Carol Field
Copy Editor: Elizabeth Galbraith
Art Director: Carol Bleistine
Art/Design Assistant: Virginia A. Bollard
Text Design: Emily Harste
Cover Design: Lawrence R. Didona
Text Artwork: J & R Technical Services
Production Manager: Tim Frelick
Assistant Production Manager: Maureen Iannuzzi

Cover: The brooding anemone, *Epiactis prolifera,* from Washington State, © Charles Seaborn.

Library of Congress Cataloging in Publication Data

Villee, Claude A. (Claude Alvin), 1917–
Biology.

Includes index and bibliography.

1. Biology. I. Solomon, Eldra Pearl.
I. Solomon, Eldra Pearl. II. Davis, P. William.
III. Title.
QH308.2.V56 1985 574 84-22234

ISBN 0-03-058477-9

67 071 9876543

CBS COLLEGE PUBLISHING
Saunders College Publishing
Holt, Rinehart and Winston
The Dryden Press

To
Dorothy Villee
Karen Davis
Mical, Amy, & Belicia

Preface

The goal of BIOLOGY is to give the reader, the beginning biology student, an understanding and appreciation of the vast diversity of living things, their special adaptations to their environment, and their evolutionary and ecological relationships. We have emphasized the basic unity of life and the fundamental similarities of the problems that are faced by all living organisms. Along with this, we have been very conscious of our responsibility to impress upon students that we are not alone on earth. We share our home with many thousands of varieties of living things. Indeed we are dependent upon countless organisms for our very survival, and such is our position of ecological domination of the biosphere that they in turn depend on us. This interdependence is stressed throughout the text.

FOCUS. The principles of biology can be learned using as a model the frog, dogfish, daisy, and even the colon bacillus. We have chosen a comparative approach. A large number of students have a special interest in human biology—in the structure, function and development of the human body—generated perhaps by their plans for a career in medicine, dentistry, or one of the allied health sciences, or simply by an interest in how their body is put together and how it works. For this reason, we have, as have other textbook authors, made frequent use of the human being as a biological model, and we have given attention to the human aspects of biology. These very same students, however, are those who stand to benefit most from our comparative, principles-oriented approach; for as they continue their professional education, they may have little additional exposure to such subjects as plant biology, invertebrate biology, ecology, and evolution.

This book attempts neither to be encyclopedic nor cursory, but presents the concepts of biology and their relevance to human beings in interesting and understandable fashion. There is no general agreement among biologists as to the sequence in which the several major topics in a general biology course should be taught. This is understandable, for reasonable arguments can be advanced for each of the many possible combinations and permutations. The various aspects of biology are intimately related, and each could be grasped much more readily if all the other aspects had been learned previously. Since this cannot be done (except perhaps by a student repeating the course!), each instructor must choose the sequence that seems optimal to him or her. Because of this, we have taken special pains to write each chapter and each part so that they do not depend heavily upon preceding chapters and parts. The various parts and chapters can be taken up in any of a number of sequences with pedagogic success.

ORGANIZATION. An appreciation of science requires not only a grasp of the content of science but also an insight into the processes by which scientific knowledge is acquired. An introduction to the methods of science is given in Chapter 1 of Part I, and throughout the text examples of experimental work are presented to illustrate modern methods of biological research.

Part I continues with a discussion of the molecular basis of biology, and the architecture of cells and tissues. The chemistry chapters (Chapters 2 and 3) have been designed to be clear and biologically relevant. Too often a student turns to the first chapters of a biology book and develops the notion that he or she has actually enrolled in an introductory chemistry course. Chapters 4 and 5 (Cells and Cell Membranes) include the highest quality electron and light micrographs available; many of these are accompanied by detailed line drawings that clarify the fine points of cell organization. From Chapter 2 (Atoms and Molecules), through Chapter 3 (Macromolecules), on to Chapters 4 and 5 (Cells and Organelles), we have introduced the student to increasing large levels of organization in biological systems. In Chapter 6, we discuss multicellularity, tissues, and organ systems. However, since not all instructors will want to begin a discussion of these subjects early in the course, the chapter has been conveniently divided along the lines of plants and animals, so that the separate sections can be used immediately preceding the units on The Structures and Life Processes of Plants (Part V) and The Structures and Life Processes of Animals (Part VI).

Part II discusses the properties and constituents of enzyme systems, the flow of energy through the world of life, and the grand metabolic adaptations by which living systems obtain and utilize energy by photosynthesis and cellular respiration. Both in Part II and elsewhere in the book we have attempted to integrate the fundamental details of cellular energetics with the broad patterns of energy flow throughout the world of life.

Part III, Genetics, begins with a discussion of mitosis (first introduced in Chapter 4) and meiosis, and then presents the principles of classical Mendelian genetics. Chapter 12 emphasizes research and clinical applications in human genetics and genetic disease. The chapter includes information on population genetics that helps the student better understand how genetic variation is maintained through generations. This information, however, can be adapted easily to a curriculum that covers population genetics along with evolution (the fundamentals of the Hardy-Weinberg law are summarized at the beginning of Chapter 45).

Chapter 13 discusses in depth the genetic code and the transfer of biological information in DNA molecules from generation to generation. Chapter 14 is devoted to RNA and protein synthesis; Chapter 15 discusses gene control, with attention to possible regulatory mechanisms in eukaryotes. Chapter 16 emphasizes research in molecular genetics and its possible applications. New advances in this rapidly expanding field are discussed throughout the genetics unit. Every chapter of BIOLOGY has been subjected to review by leading specialists; however, the chapters on molecular genetics (Chapters 13 to 16) are unique in the contributions made to the art and text by Drs. Roger McMacken and Jeffrey Corden of The Johns Hopkins University. Our knowledge in the subject of molecular genetics seems to be growing exponentially; the input of these two researchers has proven invaluable in making the chapters accurate, up to date, and exciting for students to read.

Part IV is devoted to the diversity of living organisms. It begins with a discussion of how and why living things are classified. Separate chapters are devoted to the viruses, monerans, protists, and fungi. The plant kingdom is given a comprehensive survey in two chapters: Chapter 22 presents the primitive land plants and discusses the evolution of mosses and ferns, and Chapter 23 describes the gymnosperms and angiosperms. The last three chapters of Part IV survey the invertebrate and vertebrate animals living today, their structural and functional adaptations, and their evolution. These are not, however, the only comparatively organized chapters in the book. The range of adaptations present in a variety of organisms is summarized in succeeding chapters in Parts V and VI on the systems and life processes of plants and animals.

Part V is devoted to a discussion of structure and life processes in plants. Separate chapters discuss in detail the physiological and morphological attributes of plants, plant growth and development, plant hormones, plant nutrition, and plant reproduction.

Part VI describes structure and life processes in animals. Each chapter begins with a comparative study of the particular adaptations of representatives of various animal phyla. Both structure and physiology are discussed in an evolutionary framework (that is, how different attributes might have evolved in response to stresses from the environment). Accompanying the discussions of human beings as biological models are a large number of high quality medical illustrations. These drawings are a unique learning aid in an introductory biology textbook.

The final section of this book, Part VII, explores the biology of populations. Evolution, animal behavior and ecology are the subjects of this unit. Chapters 45 and 46 discuss the general principles of evolution and the evidence for evolution. Chapter 47 introduces the student to the possible mechanisms of the origin of life itself on this planet. This chapter also summarizes some of the information in Part IV by giving a chronological history of the evolution of life on earth through an examination of the fossil record. Appropriately, since much of this knowledge is inexact, the chapter closes by presenting some of the important controversies in evolutionary theory today (such as the debate on punctuated equilibrium).

Behavior is discussed as a complex of adaptations in Chapters 48 and 49. Chapter 48 discusses behavior at the level of the individual organism, while Chapter 49 discusses the adaptive value of social behavior. Both these chapters are necessarily selective in their focus; their emphasis on adaptation, evolution, and the biological basis of behavior is designed to complement knowledge the student has acquired from other sections of the book.

Chapters 50 and 51 present principles of ecology from the standpoint of populations and communities and include discussion of adaptations, ecosystems, and the various types of biomes. Chapter 52, Human Ecology, emphasizes the impact of human beings upon the biosphere. Since, however, ecological themes are so much an integral part of the textbook, these chapters are, in part, a synthesis of the information the student has encountered elsewhere.

LEARNING AIDS. Many pedagogic aids have been included to help the student with the challenging task of mastering the principles of biology. Both **learning objectives** and a **chapter outline** are included at the beginning of each chapter. These help students as they begin to read a chapter and are useful later as a way to organize knowledge and study for an examination. Important new terms are set in **boldface** for emphasis throughout the text. Illustrations have been carefully designed to support and clarify concepts presented in the text, and tables are frequently employed to organize and summarize information.

Focus boxes present subjects in greater depth, introduce applications of material presented in the text, or integrate knowledge from the various subdisciplines of biology and related sciences. At the end of each chapter a **summary** in outline form is provided. There is also a **Post-test** (with answers at the back of the book) so that the student can evaluate his or her mastery of the factual material in the chapter. **Review questions** help the student to focus upon important concepts. At the end of each part there is a list of **supplementary readings**. These readings are selected specifically for the undergraduate biology major, and include numerous articles written at a level that the student can readily understand. A **glossary** giving the definitions of many important biological terms is integrated with the very comprehensive **index**. This feature is useful in enhancing student recall and allows the student to conveniently find references and examples in the text. Preceding the glossary/index is an **appendix summarizing the classification system** used in the text; this appendix also outlines minor phyla not covered in Part IV. Also included is an **appendix on common prefixes, suffixes, and word roots used in biology**.

SUPPLEMENTS. Many supporting materials have been provided to accompany this text. These include a comprehensive set of supplements for the use of both instructor and student: an **instructor's manual**, a **laboratory manual**, a **study guide** that includes additional review and self-tests, **overhead color transparencies** for classroom use, slides of **electron micrographs**, and a **test bank**.

CLAUDE A. VILLEE
ELDRA PEARL SOLOMON
P. WILLIAM DAVIS

Acknowledgments

We would like to express our deep appreciation and gratitude to the members of the editorial staff of Saunders College Publishing who have helped so much in developing this book and guiding it through the maze of details involved in transforming the initial manuscript to the final text. Most important was our Developmental Editor Don Reisman who pushed us—and himself—to what seemed like the very limits of our endurance (many over-fourteen-hour days and seven-day working weeks) in meeting an almost interminable series of tough deadlines and in striving for the highest possible quality. Don labored with us over every paragraph and every illustration. He is responsible in large part for the extensive, complex photograph program that supports and beautifully illustrates the text. His sharp eye and keen aesthetic instinct have spared the reader many a "slightly out of focus" photograph (which otherwise would surely have been included since we had taken them). We are most grateful for Don's deep involvement in every aspect of this project. We also thank Don's Developmental Assistant Amy Leary for her patient tracking of the many details of illustrations and permissions.

We are much indebted to Carol Field, our Project Editor, for her care and expertise in the difficult task of guiding the book through the production phase. We also want to thank Art Director Carol Bleistine and Manager of Editing, Design, and Production, Tim Frelick. We appreciate the continued support of our Publisher, Don Jackson, and our Acquisitions Editors Michael Brown and Ed Murphy. We are grateful to all of these dedicated experts who believed in this project and gave us the support necessary to make BIOLOGY a reality.

We want to thank Kathleen Callinan for typing portions of the manuscript. We are grateful to Phala Pesano for her dedication and hard work in word processing, rendering some of the preliminary drawings, and helping us prepare the glossary/index. We thank Harold Levin for the use of several of his illustrations in the chapters on evolution. We also want to acknowledge the help of Forrest Hearst of Microcomputer Systems, Inc., who patiently and generously helped us join the computer age.

Lastly, we acknowledge the patience and support of our families. Amy Solomon's and Belicia Efros's help in word processing, proofreading, bursting endless reams of computer paper, and in many other aspects of the project was important to us. We especially appreciate Mical Solomon's role in persuading us to make the big leap from typewriter to word processor and his continual help in keeping our computers' systems working.

REVIEWERS OF BIOLOGY. We want to express our thanks to the instructors and researchers who have helped shape BIOLOGY. They have shared with us both their talent and the pressures of meeting deadlines. Their suggestions have been invaluable.

John H. Adler, Michigan Technological University
Lester Bazinet, Community College of Philadelphia
William L. Bischoff, University of Toledo
George Bowes, University of Florida
W. H. Breazeale, Jr., Francis Marion College
Jeffrey Corden, The Johns Hopkins University
Harry O. Corwin, University of Pittsburgh
Stephen J. Dina, St. Louis University
Lee C. Drickamer, Williams College
Milton Fingerman, Tulane University
Elizabeth A. Godrick, Boston University
Peter Gregory, Cornell University
Thomas Hanson, Temple University
Mark Jacobs, Swarthmore College
Robert W. Korn, Bellarmine College
Victor Lotrich, University of Delaware
Arthur Mange, University of Massachusetts, Amherst
Roger McMacken, The Johns Hopkins University
Robert E. Moore, Montana State University
Thomas L. Naples, Delaware Community College
David O. Norris, University of Colorado, Boulder
Frank G. Nordlie, University of Florida
Jeanne S. Poindexter, The Public Health Research Institute of The City of New York, Inc.
Susan Pross, University of South Florida
Florence Ricciuti, Albertus Magnus College
Martin Roeder, Florida State University
Rodney A. Rogers, Drake University
Marvin J. Rosenberg, California State University, Fullerton
John A. Schmitt, Ohio State University
Richard C. Snyder, University of Washington
Charles L. Stevens, University of Pittsburgh
Daryl Sweeney, University of Illinois at Urbana-Champaign
Joseph W. Vanable, Purdue University
Jack Waber, West Chester State University
Lawrence Winship, Hampshire College
Drew H. Wolfe, Hillsborough Community College
John L. Zimmerman, Kansas State University

Contents Overview

Contents

part V
STRUCTURES AND LIFE PROCESSES IN PLANTS
569

part VI
STRUCTURES AND LIFE PROCESSES IN ANIMALS
655

part I

THE ORGANIZATION OF LIFE

Microscopic jewels, several species of algae glitter in this photomicrograph by Tom Adams. The globular colonies are *Volvox,* one of which contains several daughter colonies about to be released to start independent life on their own. Filamentous algae, *Spirogyra,* curve in the foreground and background. Strands of such algae consist of numerous cylindrical cells joined end-to-end. Also seen are two small crustaceans, a copepod (left) and *Chydorus* (right). (Magnification ×60 and enlarged photographically.)

1
A View of Life

OUTLINE

I. What is life?
 A. Specific organization
 B. Metabolism
 C. Homeostasis
 D. Growth
 E. Movement
 F. Responsiveness
 G. Reproduction
 H. Adaptation
II. The organization of life
 A. Organization of the organism
 B. Ecological organization
III. The variety of organisms
 A. Kingdom Monera
 B. Kingdom Protista
 C. Kingdom Fungi
 D. Kingdom Plantae (plants)
 E. Kingdom Animalia (animals)
IV. How biology is studied
 A. Systematic thought processes
 B. Designing an experiment
 C. How a hypothesis becomes a theory
 D. The ethics of science

LEARNING OBJECTIVES

After you have read this chapter you should be able to:

1. Distinguish between living and nonliving things, describing the features that characterize living things.
2. Define *metabolism* and *homeostasis,* and give examples of these processes.
3. Define *adaptation,* and describe its function in promoting perpetuation of a species.
4. List in sequence and briefly describe each of the levels of biological organization.
5. Describe the roles and interdependence of producers, consumers, and decomposers.
6. Identify the five kingdoms of living organisms, and give examples for each group.
7. Design an experiment to test a given hypothesis using the procedure and terminology of the scientific method.
8. Outline the ethical dimensions of the scientific method, and give examples of possible ethical problems that may arise in the course of scientific investigation.

If you have read anything about DNA, viruses, or how a vaccine works, you are already familiar with some aspect of biology unknown in the days of Aristotle, or even Darwin and Pasteur. During the past century the science of biology has undergone rapid change and has had a significant impact on the way we live. We are able to produce antibiotics, transplant hearts, and manipulate genes. Along with such exciting new developments, biology has, as always, been concerned with the study of interrelationships of the diverse forms of living things that inhabit our planet. It has enhanced our awareness of the exquisite complexity that characterizes all living things, and has helped us better appreciate our own impact on other living things and on the environment (Fig. 1–1).

This book is a starting point for your exploration of **biology,** the study of life. It will provide you with the tools that will enable you to become a part of this fascinating science. In this chapter we will examine some of our everyday assumptions about living things, and start to formulate a structure for organizing our knowledge. To begin with, as we attempt to study life, we need to develop a deeper understanding of what life is.

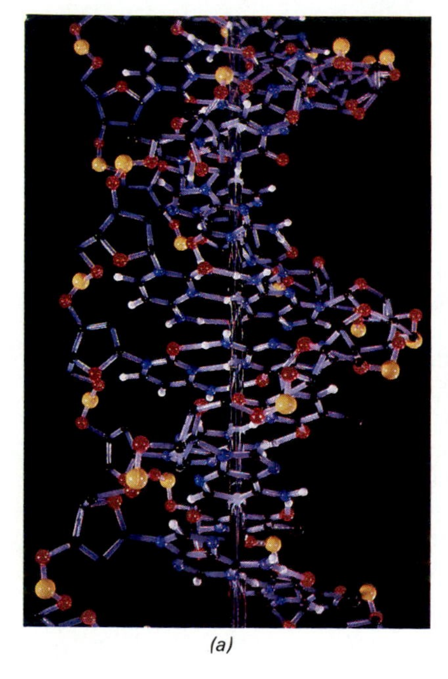

(a)

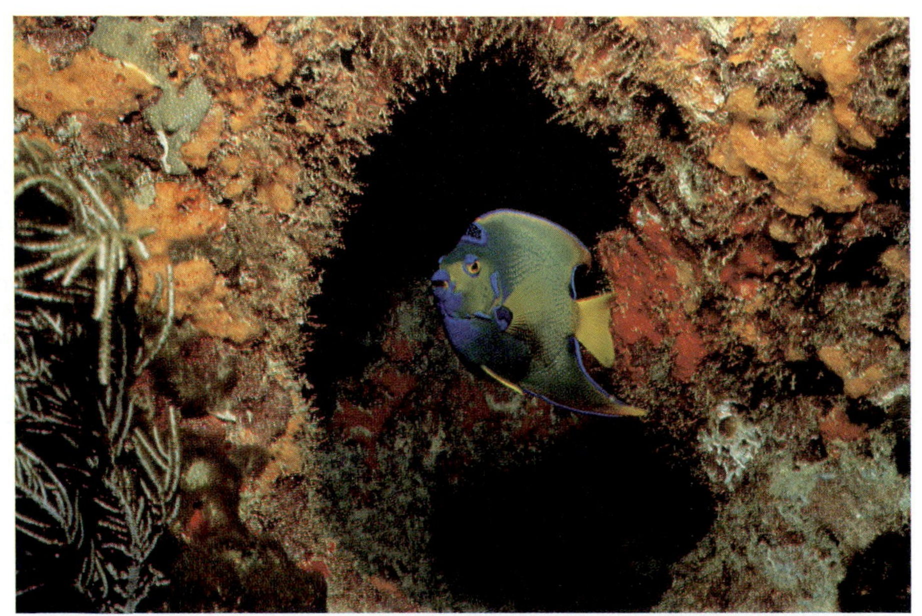

(b)

(c)

Figure 1–1 Some subjects of research in modern biology. (*a*) Essential to the continuity of life is an organism's ability to reproduce itself. Shown here is a model of DNA, the basic hereditary material of life. The different colored balls represent the different atoms that make up DNA: black = carbon; red = oxygen; white = hydrogen; yellow = phosphorus; blue = nitrogen. (*b*) An angelfish guarding a section of coral reef. Biologists are concerned with the physical characteristics of the fish, how its body functions, its behavior, and its interaction with the other living things that inhabit its environment. (*c*) Biologists also study the effects of human activities on the environment. Runoff of fertilizers from nearby farmland has caused a buildup of nutrients in this pond. As a consequence, there has been an explosive growth of algae on the surface of the water. Among its many effects, this layer of algae prevents sunlight from reaching plants and other algae below, causing them to die. Their decomposition reduces the concentration of oxygen in the water, which leads to fish kills. ((*a*), Phil Degginger; (*b*), Charles Seaborn; (*c*) Tom Adams.)

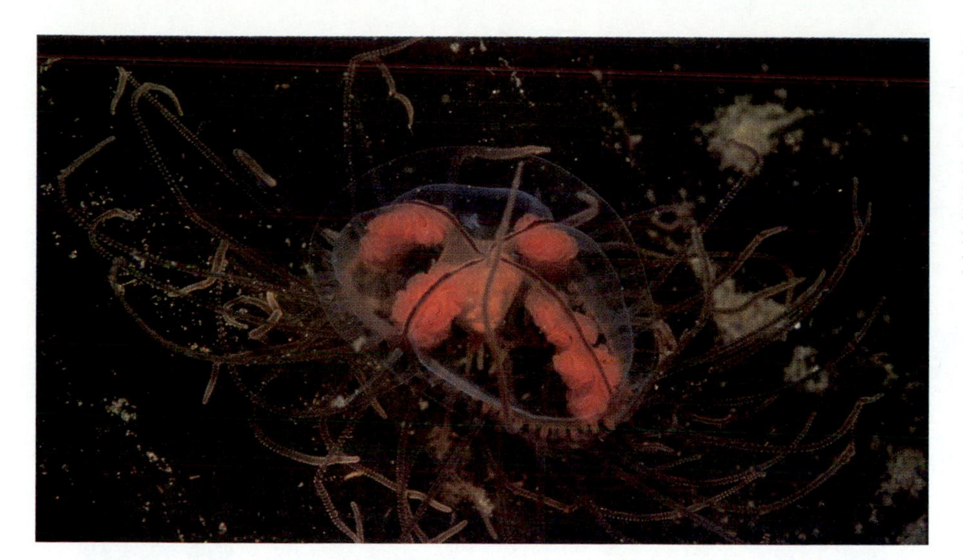

Figure 1–2 This jellyfish is a relatively simple animal; yet it is composed of millions of cells that are organized to perform specialized functions. For example, the cells that make up its numerous tentacles enable the animal to gather information from its environment and trap food. Cells in the orange region are specialized to carry out reproductive functions. (Charles Seaborn.)

What Is Life?

It is relatively easy to determine that a human being, an oak tree, and a grasshopper are living whereas rocks are not. Yet it remains very difficult to formally define life. At one time it was believed that a living system could be distinguished from a nonliving system by its possession of a special "vital force." Now, after centuries of searching, we understand that there is no single substance or force that is unique to living things. The best we can do toward defining life, therefore, is to reexamine our ideas about what life is, and to list the features that living things have in common. When we do this, we find that most living things do share certain characteristics—a precise kind of organization, metabolism, homeostasis, movement, responsiveness, growth, reproduction, and adaptation to environmental change—that distinguish them from nonliving things. We will consider each of these characteristics in the following sections.

SPECIFIC ORGANIZATION

Each kind of organism (living thing) is recognized by its characteristic appearance and structure. Living things are not homogeneous but are made of different parts, each with special functions. Although organisms vary greatly in size and appearance, all (except the viruses[1]) are composed of basic units called **cells.** The cell is the simplest part of living matter that can carry on all of the activities necessary for life. Some of the simplest organisms, such as bacteria, are unicellular; that is, they consist of a single cell. In contrast, the body of a human or an oak tree is made of billions of cells. In such complex multicellular organisms the processes of the entire organism depend upon the coordinated functions of the constituent cells (Fig. 1–2).

METABOLISM

In all living organisms chemical reactions take place that are essential to nutrition, growth and repair of cells, and conversion of energy into usable forms. The sum of all the chemical activities of the organism is called **metabolism.** Metabolic reactions occur continuously in every living organism (Fig. 1–3); when they cease, the organism dies.

Each individual cell of an organism constantly takes in new substances, alters them chemically in a variety of ways, and builds new cellular components. Some nutrients are used as "fuel" for cellular respiration, a process during which some of their stored energy is captured for use by the cells. Life on earth involves a neverending flow of energy within cells, from one cell to another, and from one organism to another.

[1] As we will see in Chapter 17, viruses can carry on metabolism and reproduce only by utilizing the metabolic machinery of *the cells they parasitize,* and so are said to be on the borderline between living and nonliving things.

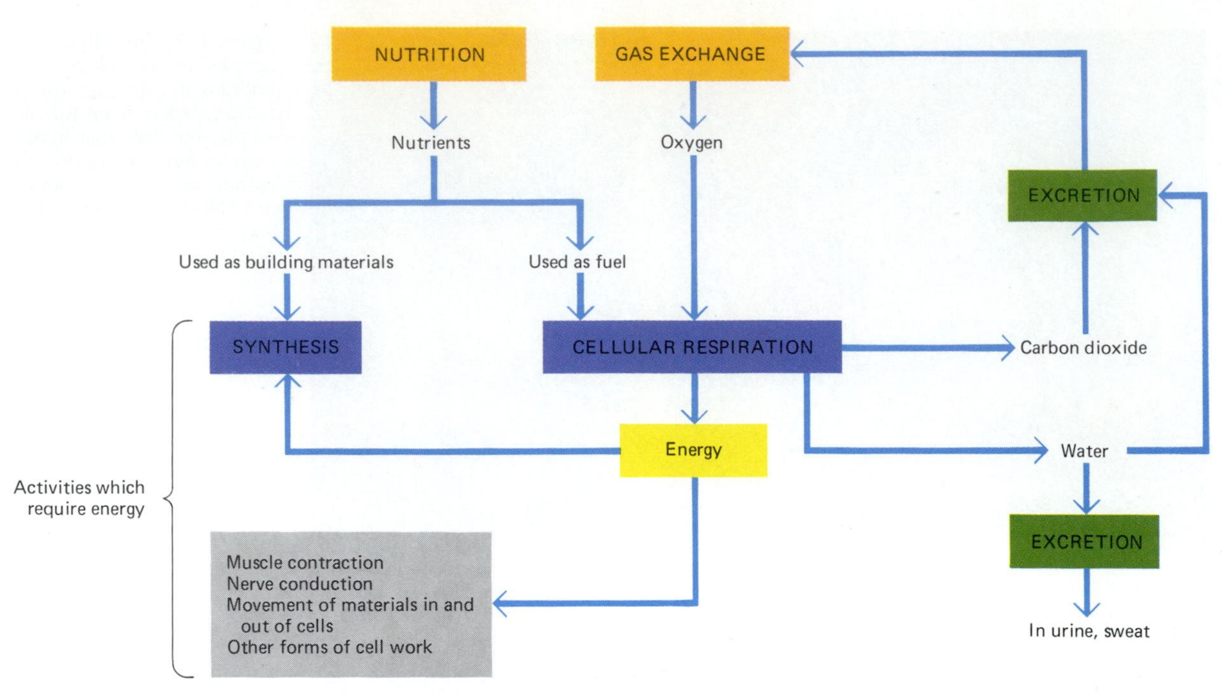

Figure 1–3 Relationships of some metabolic activities. Some of the nutrients provided by proper nutrition are used to synthesize needed materials and cell parts; other nutrients are used as fuel for cellular respiration, a process that captures energy stored in food. This energy is needed for synthesis and for other forms of cellular work. Cellular respiration also requires oxygen, which is provided by the process of gas exchange. Wastes from the cells such as carbon dioxide and water must be excreted from the body.

HOMEOSTASIS

In all organisms the various metabolic processes must be carefully and constantly regulated to maintain a balanced state. When enough of some cellular component has been made, its production must be decreased or turned off. When the supply of energy in a cell declines, appropriate processes for making more energy available must be turned on. These self-regulating control systems are remarkably sensitive and efficient. The tendency of organisms to maintain a constant internal environment is termed **homeostasis,** and the mechanisms that accomplish the task are known as **homeostatic mechanisms.**

The regulation of body temperature in the human being is an example of the operation of homeostatic mechanisms: When body temperature rises above the normal 37° C, the temperature of the blood is sensed by special cells in the brain that function like a thermostat. These cells send nerve impulses to the sweat glands to increase the secretion of sweat (Fig. 1–4). The evaporation of the sweat from the body surface lowers body temperature. Other nerve impulses cause the dilation of small blood vessels (capillaries) in the skin, making it appear flushed. The increased blood flow brings more heat to the body surface to be radiated away.

When body temperature falls below normal, the sensor in the brain initiates nerve impulses to constrict the blood vessels in the skin, reducing heat loss. If body temperature falls lower, nerve impulses may be sent to the muscles of the body, stimulating the rapid muscular contractions we call shivering, a process that generates heat.

GROWTH

Some nonliving things appear to grow. Crystals may form in a supersaturated solution of a salt; as more of the salt comes out of solution, the crystals may enlarge. However, this is not growth in the biological sense. Biologists restrict the term **growth** to those processes that increase the amount of living substance in the organism. Growth, therefore, is an increase in cellular mass that is brought about by an increase in the *size* of the individual cells, by an increase in the *number* of cells, or both (Fig. 1–5). Growth may be uniform in the several parts of an organism, or it may be greater in some parts than in others so that the body proportions change as growth occurs.

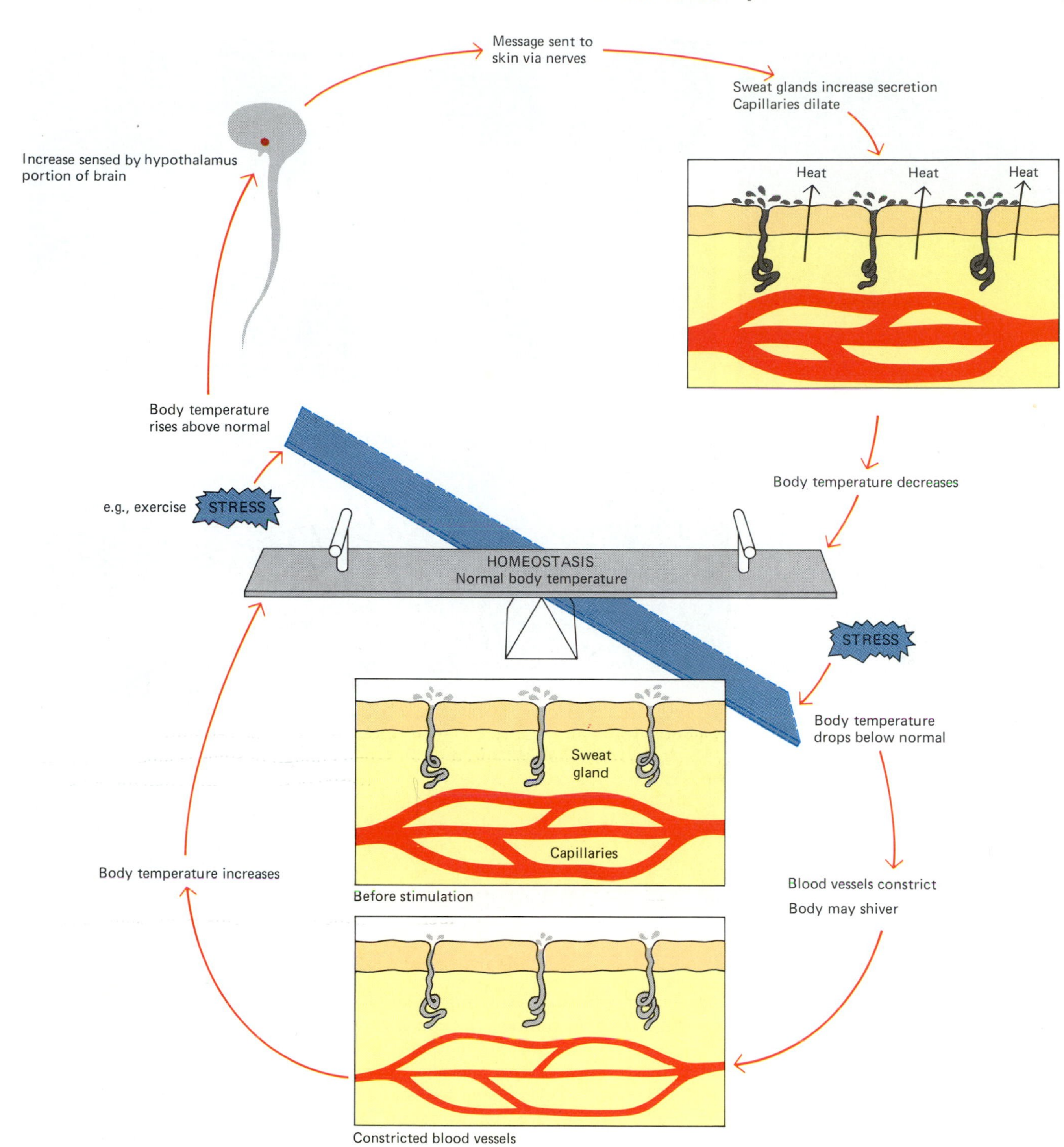

Figure 1–4 Regulation of body temperature in the human by homeostatic mechanisms. An increase in body temperature above the normal range stimulates special cells in the brain to send messages to sweat glands and capillaries in the skin. Increased circulation of blood in the skin and increased sweating are mechanisms that help the body to get rid of excess heat. When body temperature falls below the normal range, blood vessels in the skin constrict so that less heat is carried to the body surface. Shivering, in which muscle contractions generate heat, may also occur.

Some organisms—most trees, for example—continue to grow indefinitely. Many animals have a defined growth period that terminates when a characteristic size is reached in adulthood. One of the remarkable aspects of the growth process is that each part of the organism continues to function as it grows.

MOVEMENT

Movement, though not necessarily locomotion (moving from one place to another), is another characteristic of living things. The movement of most animals is quite obvious—they wiggle, crawl, swim, run, or fly. The movements of plants are much slower and less obvious but occur nonetheless. The streaming motion of the living material in the cells of the leaves of plants is known as **cyclosis**.

(a)

(b)

Figure 1–5 As living organisms grow, they take in substances and alter them to fit their own needs for internal organization and energy. Crystals of copper sulfate, shown in (*a*), can "grow" in a supersaturated solution, but their internal structure is undifferentiated and the crystals remain unresponsive to the environment. On the other hand, the scorpionfish in (*b*), which on the outside looks much like a rock, is waiting to gulp the next small organism that unwarily swims by. The scorpionfish will then refashion the raw materials provided by the prey to make more of its own cells and body parts—in other words, it will manufacture more scorpionfish. ((*b*), Charles Seaborn.)

Locomotion may result from the beating of tiny hairlike extensions called cilia or flagella, from the contraction of muscles, or from the slow oozing of a mass of cell substances termed **amoeboid motion** (Fig. 1–6). A few animals, such as sponges, corals, oysters, and certain parasites, do not move from place to place as adults. Most of these, however, have free-swimming larval stages. Even in the sessile (firmly attached, not free to move about) adults, however, cilia or flagella may beat rhythmically, moving the surrounding water past the organism; in this way food and other necessities of life are brought to the organism.

Figure 1–6 Several types of cellular movement. (Arrows indicate direction of motion of flagella or cytoplasm.)

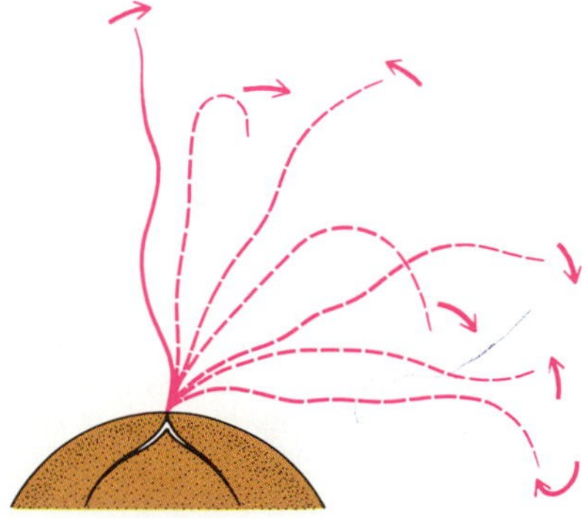

Beating of a single flagellum

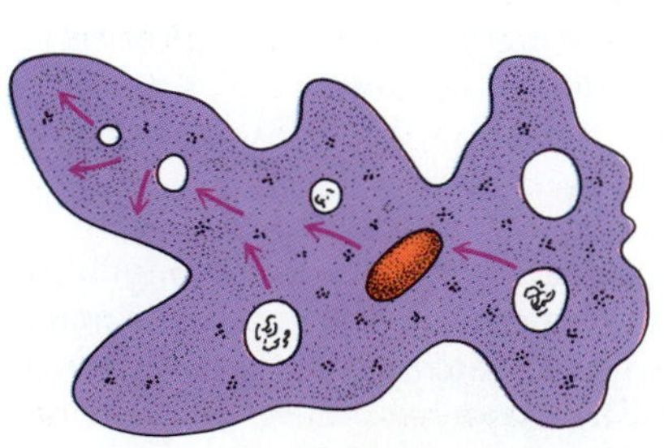

Amoeboid motion

Cyclosis

Figure 1–7 A leaf of the Venus's-flytrap attracting and capturing a lacewing. (E. R. Degginger.)

RESPONSIVENESS

Living things respond to **stimuli**, physical or chemical changes in their internal or external environment. Stimuli that evoke a response in most organisms are changes in the color, intensity, or direction of light; changes in temperature, pressure, or sound; and changes in the chemical composition of the surrounding soil, air, or water. In complex animals such as the human, certain cells of the body are highly specialized to respond to certain types of stimuli; for example, cells in the retina of the eye respond to light. In simpler organisms, such specialized cells may be absent, but the whole organism may respond to stimuli. Certain single-celled organisms respond to bright light by retreating.

The responsiveness of plants may not be as obvious as that of animals, but plants do respond to light, gravity, water, and other stimuli principally by the growth of different parts of their bodies. The streaming motion of the cytoplasm in plant cells may be speeded up or stopped by changes in the amount of light. A few plants, such as the Venus's-flytrap of the Carolina swamps (Fig. 1–7), are remarkably sensitive to touch and can catch insects. Their leaves are hinged along the midrib and possess a scent that attracts insects. The presence of an insect on the leaf, detected by trigger hairs on the leaf surface, stimulates the leaf to fold. The edges come together and the hairs interlock to prevent the escape of the prey. The leaf then secretes enzymes that kill and digest the insect. These plants are usually found in soil that is deficient in nitrogen. Fly-trapping enables these plants to obtain part of the nitrogen they require for growth from the prey they "eat."

REPRODUCTION

Although at one time worms were believed to arise from horsehairs in a water trough, maggots from decaying meat, and frogs from the mud of the Nile, we now know that each can come only from previously existing organisms. One of the fundamental tenets of biology is that "all life comes only from living things." If there is any one characteristic that can be said to be the very essence of life, it is the ability of an organism to reproduce its kind.

In simpler organisms such as the amoeba, reproduction may be **asexual** (Fig. 1–8), that is, without sex. When an amoeba has grown to a certain size it reproduces by splitting into two to form two new amoebas. Before it divides, an amoeba makes a duplicate copy of its hereditary material (genes), and one complete set is distributed to each new cell. Except for size, each new amoeba is identical with the parent cell. Unless eaten by another organism or destroyed by adverse environmental conditions such as pollution, an amoeba does not die.

In most plants and animals, **sexual reproduction** is carried out by the production of specialized egg and sperm cells that unite to form the fertilized egg, from which the new organism develops. With sexual reproduction, each offspring is not a duplicate of a single parent but is the product of the interaction of various genes contributed by both the mother and the father. Genetic variation is the raw material for the vital processes of evolution and adaptation.

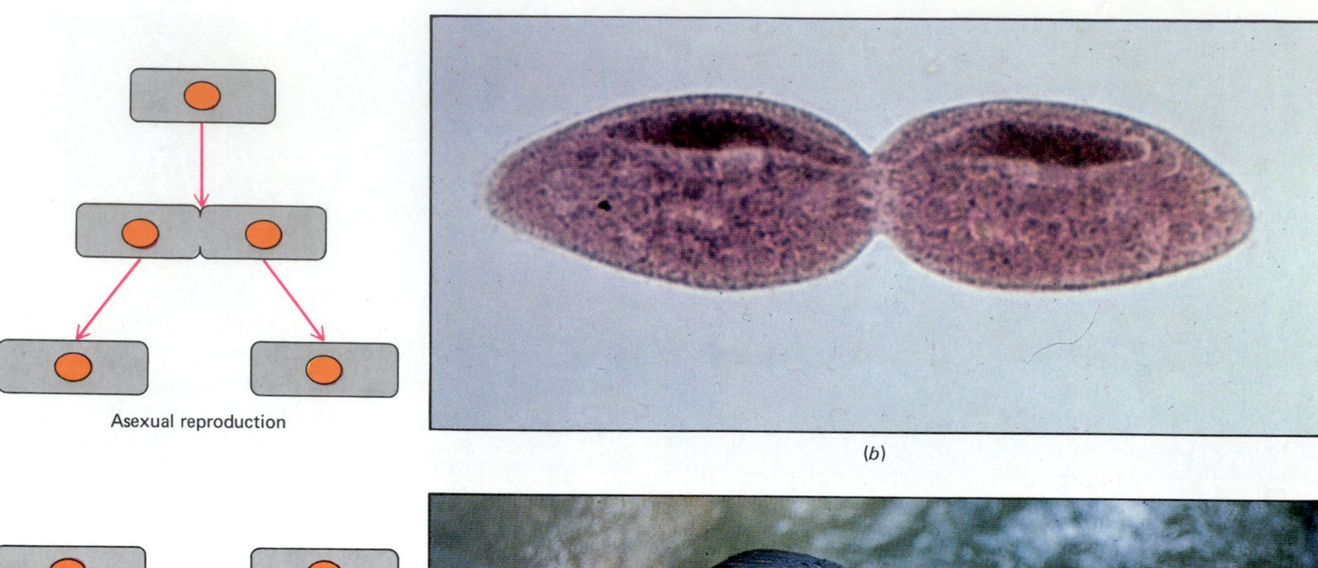

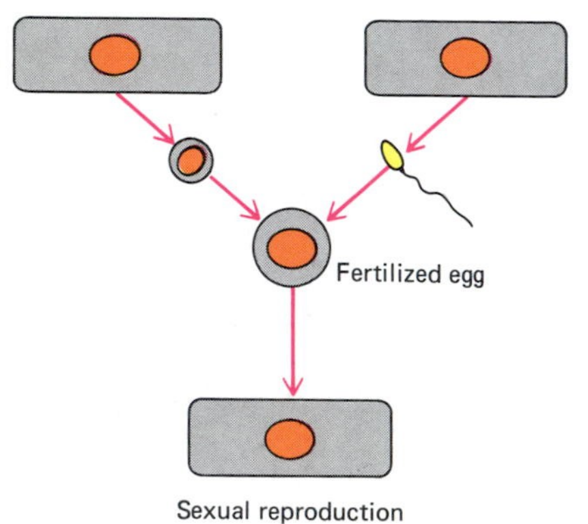

Figure 1–8 Approaches to reproduction. (*a*) In asexual reproduction, one individual gives rise to two or more offspring—all identical with the parent. In sexual reproduction, two parents each contribute a sex cell; these join to give rise to the offspring, which is a combination of the traits of both parents. (*b*) Asexual reproduction in the unicellular *Paramecium.* (*c*) Mating in river otters. ((*b*), Carolina Biological Supply Company; (*c*), courtesy of Busch Gardens.)

ADAPTATION

The ability of a species[1] to **adapt** to its environment is the characteristic that enables it to survive in a changing world. Adaptations are traits that enhance an organism's ability to survive in a particular environment. They may be structural, physiological, or behavioral, or a combination of all of these. The long necks of giraffes are an adaptation for reaching the leaves of trees, and the thick fur coats of polar bears are an adaptation for surviving frigid temperatures. Every biologically successful organism is actually a complex collection of coordinated adaptations.

Adaptation involves changes in species rather than in individual organisms. Many adaptations occur over long periods of time and involve many generations. They are the result of processes such as mutations (permanent chemical changes in genes) and natural selection (discussed in Focus on the Evolutionary Perspective). If every organism of a species were exactly like every other, any change in the environment might be disastrous to all, and the species would become extinct. Differences among individuals initiated by random mutation and disseminated by sexual reproduction provide for a differential in the ability of organisms to cope with changes in their surroundings; those best suited to cope with any specific change live to pass on their genetic recipe for survival. A particularly well-studied example of adaptation to a changing environment is the color of the peppered moth, *Biston betularia* (Fig. 1–9).

[1]A species is a group of similar organisms that freely interbreed and produce fertile offspring.

FOCUS ON

The Evolutionary Perspective

In his book *The Origin of Species*, published in 1859, **Charles Darwin** synthesized many new findings in geology and biology and delineated a comprehensive theory that has helped shape the nature of biological science to the present day. Darwin presented a wealth of evidence that the present forms of life on earth descended with modifications from previously existing forms. His book raised a storm of controversy in both religion and science, some of which still lingers. It also generated a great wave of scientific research and observation that has provided much additional evidence that evolution is responsible for the great diversity of organisms present on our planet.

Darwin observed that individual members of a species show some variation from one another. He also observed that many more organisms are born than survive into adulthood and reproduce. In simple terms, his theory of **natural selection** states that organisms with traits that best enable them to cope with pressures exerted by the environment are the ones most likely to survive and produce offspring. As a result, these traits become more widely distributed in the population. Over long periods of time, as organisms continue to change (and as the environment itself changes, bringing different selective pressures), the members of the population begin to look increasingly unlike their ancestors.

Eggs of the midshipman fish. Random events might be largely responsible for determining which of these developing organisms will reach adulthood and reproduce. However, certain desirable or undesirable traits that each organism might have will also contribute to its probability for success in its environment. Although not all organisms are as prolific as the midshipman fish, the generalization that more organisms are born than survive is true throughout the living world. (Charles Seaborn.)

The theory of evolution will be discussed in depth in Chapters 45 through 47. However, in this first chapter of a book that is designed to introduce you to modern biology, it is important to emphasize that, while evolution is itself a subdiscipline of biology, some element of an evolutionary perspective is present in almost every specialized field within biology. Darwin's theory of evolution has proved to be one of the great unifying concepts of biology. Though not all biologists accord natural selection the primary role that Darwin gave it, most do consider it to be one of the most important mechanisms of evolution. Biologists in almost every subdiscipline try to understand the features and functions of organisms and their constituent cells and parts by considering them in light of the long, continuing process of evolution. Additionally, biologists are constantly checking for verification of the evolutionary relationships among different organisms. Biology today is more than a science of describing and naming organisms and life processes. Biologists are concerned not only with the existence of structural similarities but with what these similarities (and differences) may tell us about how organisms are related to one another, and how things have come to be as they are.

(a)

(b)

Figure 1–9 Peppered moths illustrate the advantage of protective coloration and the ability of organisms to adapt to changes in their environment. (*a*) In England until the mid-19th century, dark moths were very rare. Until that time tree trunks were covered with white lichens, and dark moths were at a definite disadvantage because predaceous birds quickly spotted and devoured them. (*b*) When pollution that came with the Industrial Revolution killed the lichens on the tree trunks and covered the tree trunks with soot, the light-colored moths fell easy prey to birds. Now the dark-colored moths had the advantage, and their numbers increased dramatically. With recent efforts to control pollution, this trend has begun to reverse. ((*a*), (*b*), Michael Tweedie, Photo Researchers, Inc.)

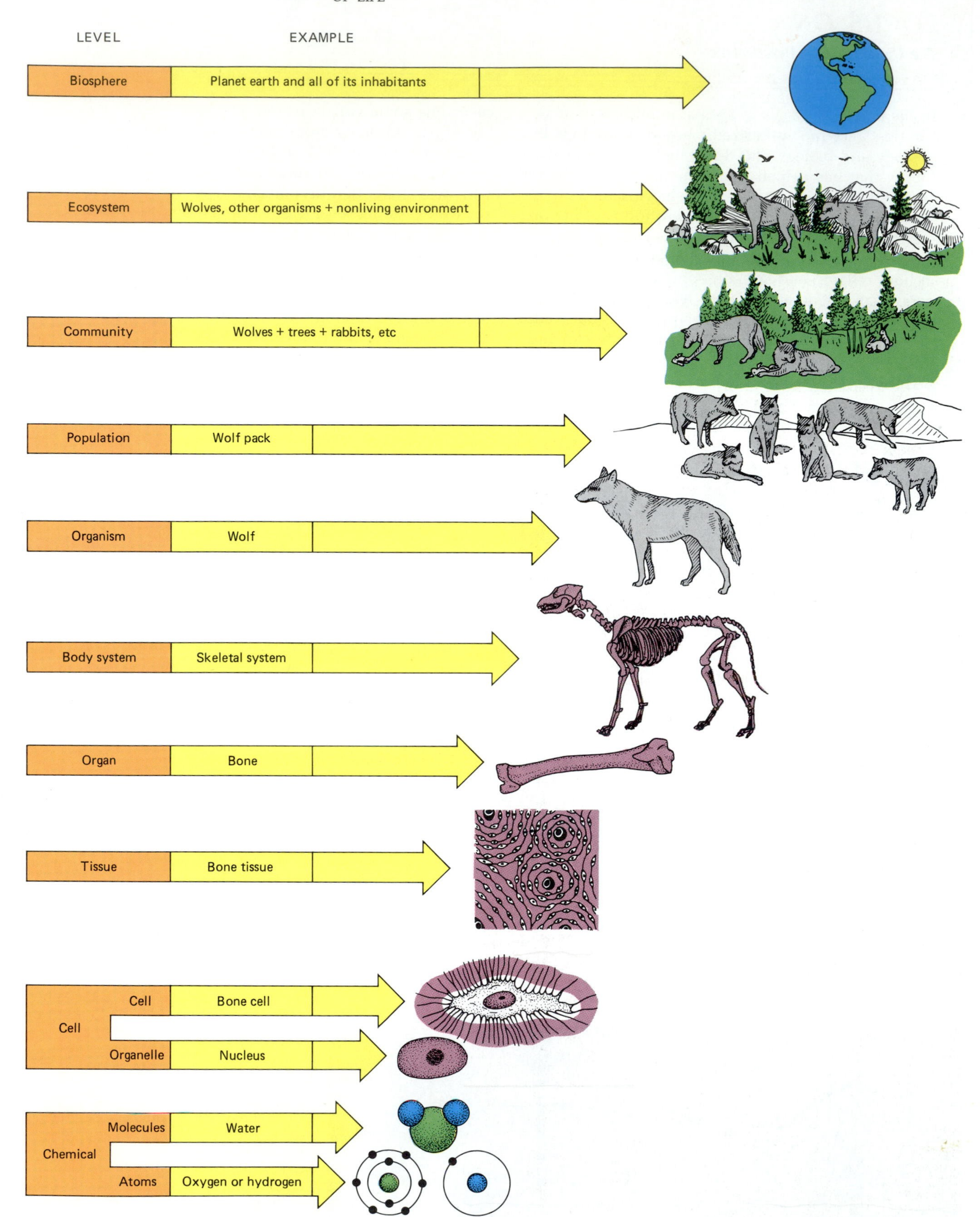

Figure 1–10 Levels of biological organization.

The Organization of Life

One of the striking features of life is its organization. We have already noted the cellular level of organization, but within each individual organism we can identify several other levels. Even when considering the interactions within and between groups of organisms, we can recognize a hierarchy of increasing complexity, as shown in Figure 1–10.

THE ORGANIZATION OF THE ORGANISM

The **chemical level** is the simplest level of organization. It includes the basic particles of all matter, **atoms,** and combinations of atoms called **molecules.** An atom is the smallest unit of a chemical element (fundamental substance) that retains the characteristic properties of that element. For example, an atom of iron is the smallest possible amount of iron. Atoms combine chemically to form molecules. For example, two atoms of hydrogen combine with one atom of oxygen to form one molecule of water.

At the **cellular level** we find that many diverse molecules may associate with one another to form complex and highly specialized structures within cells called **organelles.** The cell membrane that surrounds the cell, and the nucleus that contains the hereditary material, are examples of organelles. The cell itself is the basic structural and functional unit of life. Each cell consists of a discrete body of jelly-like cytoplasm surrounded by a cell membrane. The organelles are suspended within the cytoplasm (Fig. 1–11).

In most multicellular organisms cells associate to form **tissues,** such as muscle tissue or nervous tissue. Tissues, in turn, are arranged into functional structures called **organs,** such as the heart or the stomach. Each major group of biological functions is performed by a coordinated group of tissues and organs, called an **organ system.** The circulatory and digestive systems are examples of organ systems. Functioning together with great precision, the organ systems make up the complex multicellular organism.

ECOLOGICAL ORGANIZATION

Organisms interact to form still more complex levels of biological organization. All of the members of one species that live in the same area make up a **population.** The environment in which an organism or population lives is known as its **habitat.** The populations of organisms that inhabit a particular area and interact with one another form a **community.** Thus, a community can be composed of hundreds of different types of life forms. The study of how organisms of a community relate to one another and with their nonliving environment is called **ecology.** A community together with its nonliving environment is referred to as an **ecosystem.**

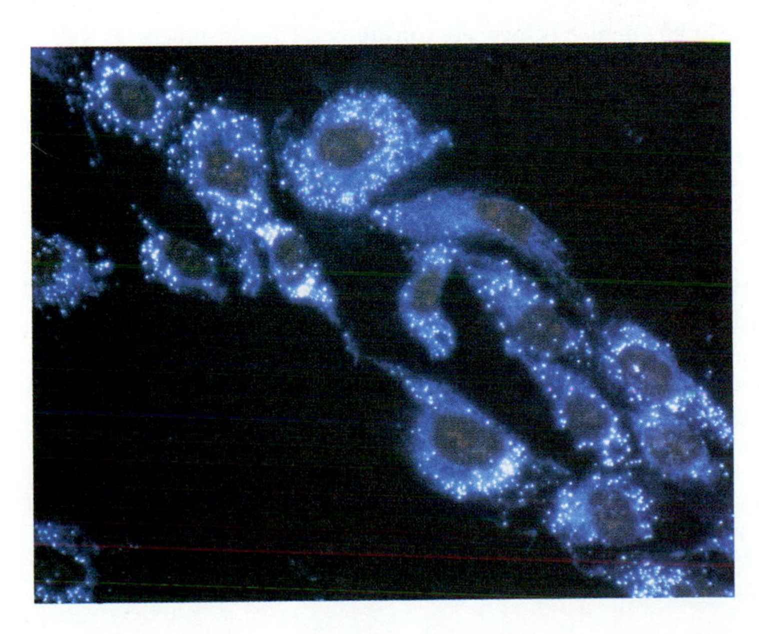

Figure 1–11 Human fibroblasts (cells from connective tissue) stained with fluorescent dye. Nuclei can be seen in red. (Courtesy of Dr. Paul Gallup.)

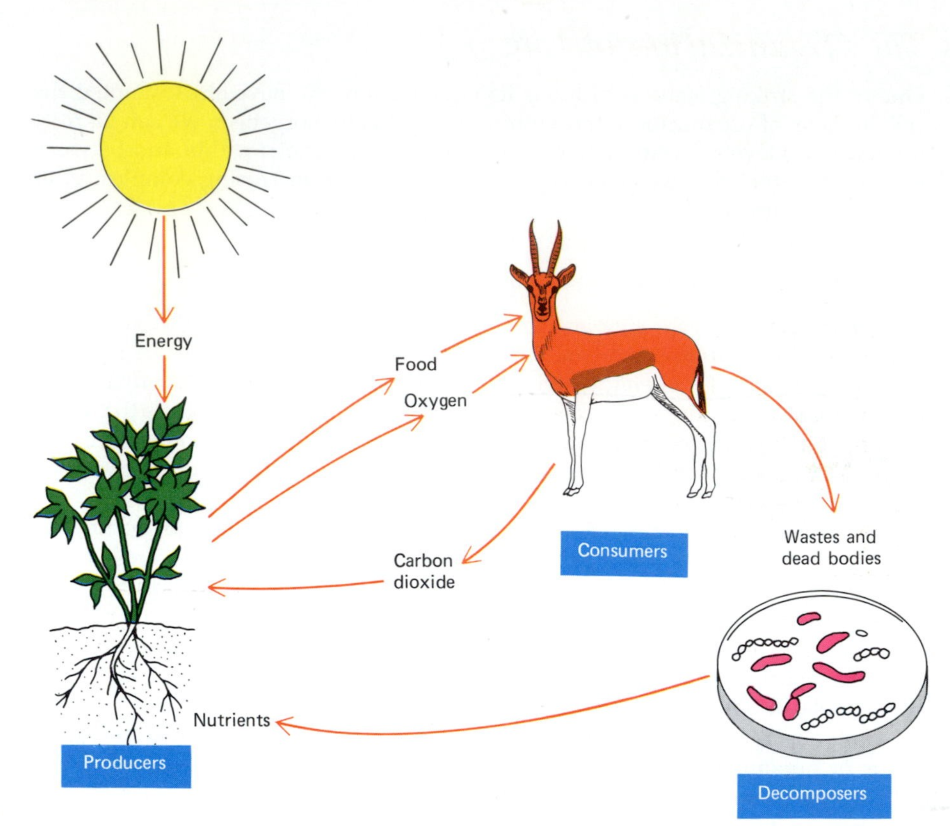

Figure 1–12 Interdependence of producers, consumers, and decomposers. Producers provide oxygen and food containing energy and nutrients for consumers. In turn, the consumers provide carbon dioxide needed for photosynthesis by the producers. The decomposers break down wastes and dead organisms so that minerals are recycled.

A self-sufficient ecosystem contains three types of organisms—producers, consumers, and decomposers—and has a physical environment appropriate for their survival (Fig. 1–12). **Producers,** or **autotrophs,** are algae, plants, and certain bacteria that can produce their own food from simple raw materials. Most of these organisms use sunlight as an energy source and carry out photosynthesis. During photosynthesis the energy from sunlight is used to synthesize complex molecules from carbon dioxide and water. The light energy is transformed into chemical energy, which is stored within the chemical bonds of the food molecules produced. Oxygen, which is required not only by plant cells but also by the cells of most other organisms, is produced as a by-product of photosynthesis.

Carbon dioxide + Water + Energy ⟶ Food + Oxygen

Animals, including human beings, are **consumers.** The consumers, as well as the decomposers, are **heterotrophs,** organisms that are dependent upon producers for food, energy, and oxygen. However, these organisms also contribute to the balance of the ecosystem. Like all living things (including producers), they obtain energy by breaking down food molecules originally produced during photosynthesis. The biological process of breaking down these fuel molecules is known as **cellular respiration.** When chemical bonds are broken during cellular respiration, their stored energy is made available for life processes (Fig. 1–13).

Food + Oxygen ⟶ Carbon dioxide + Water + Energy

Gas exchange between producers and heterotrophs by way of the nonliving environment helps to maintain the life-sustaining mixture of gases in the atmosphere.

Decomposers—the bacteria and fungi—are an important component of an ecosystem because they break down the wastes and the bodies of dead organisms, making their components available for reuse. If decomposers (and scavengers, such as vultures) did not exist, nutrients would become locked up in the dead bodies of plants and animals, and the supply of elements required by living systems would soon be exhausted.

A familiar example of an ecosystem is a balanced aquarium. Green plants or algae serve as producers, providing food, energy, and oxygen for the consumers,

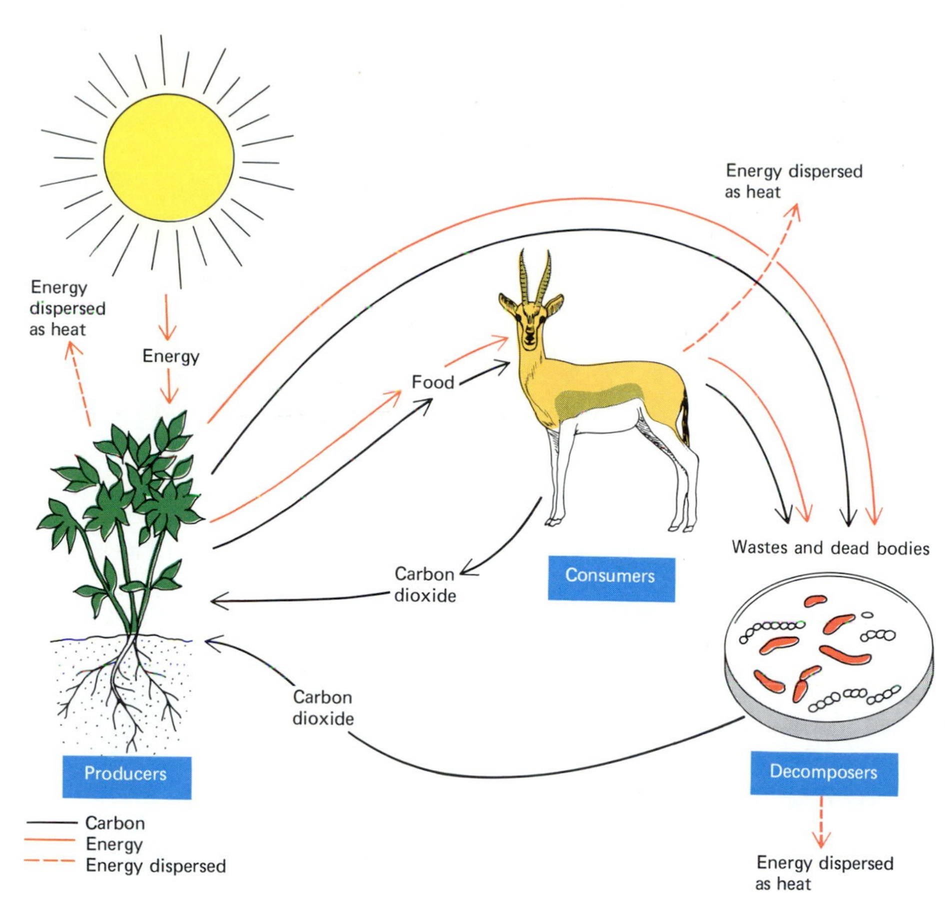

Figure 1–13 Flow of energy and carbon through the biosphere. Carbon and many other chemical elements are continuously recycled. Carbon is converted from carbon dioxide gas to food by producers. Consumers obtain carbon by eating producers. Carbon leaves the consumers and producers in the form of wastes and dead material, which is broken down by decomposers. Consumers break down carbon-containing compounds during cellular respiration, converting them in part back to carbon dioxide, and the cycle begins anew. Energy, on the other hand, cannot be recycled. Some of it is dispersed as heat during every energy transaction. For this reason a constant energy input from the sun is required to keep the biosphere in operation.

the fish. Bacteria and fungi are the decomposers that break down waste products and dead bodies, permitting nutrients to be recycled. As long as there is a continuing input of light energy for the plants, such a system may continue to thrive for months or even years. Eventually, though, it will collapse. A larger, natural ecosystem such as a pond is far more stable and is likely to last much longer. Its stability is greater, for unlike the balanced aquarium, the natural ecosystem consists of a great diversity of organisms. Chemicals and energy have a multitude of alternative pathways within such an ecosystem. Should one type of organism die out, blocking one pathway, other organisms take its place and the system as a whole is little disturbed.

An ecosystem can be as small as a pond (or even a puddle) or as vast as the Great Plains of North America or the Arctic Tundra. The largest ecosystem is the planet earth with all its inhabitants—the **biosphere.** (The term ecosphere is sometimes used instead of biosphere.)

The Variety of Organisms

The subject of biology is life, but how would it be possible to study life without a system for naming and classifying its myriad forms? The basic unit biologists have agreed upon in their classification of organisms is the **species.** It is difficult to give a definition of this term that can be applied uniformly throughout the living world, but we will define a species as a population of similar individuals, alike in their structural and functional characteristics, that in nature freely interbreed and produce fertile offspring.

Closely related species are grouped together in the next higher unit of classification, the **genus** (plural, genera). Each organism is given a scientific name consisting of two words, the genus and the species, both in Latin. The scientific name of the American white oak is *Quercus alba,* whereas the name of the European white oak is *Quercus robur.* Another tree, the white willow—*Salix alba*—belongs to a different genus. Our own scientific name is *Homo sapiens.*

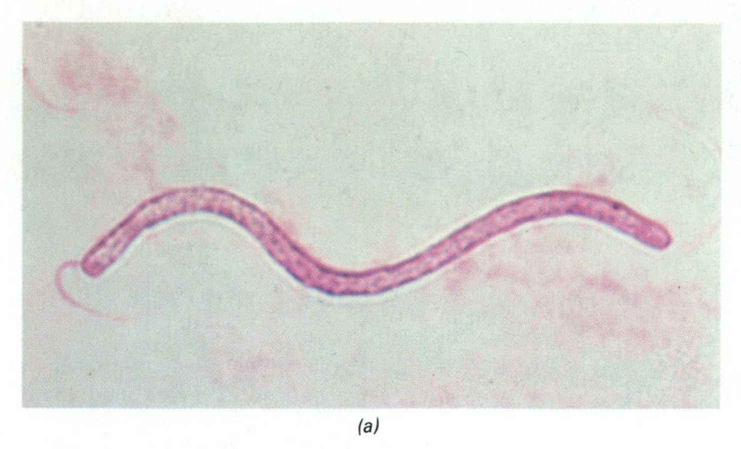

(a)

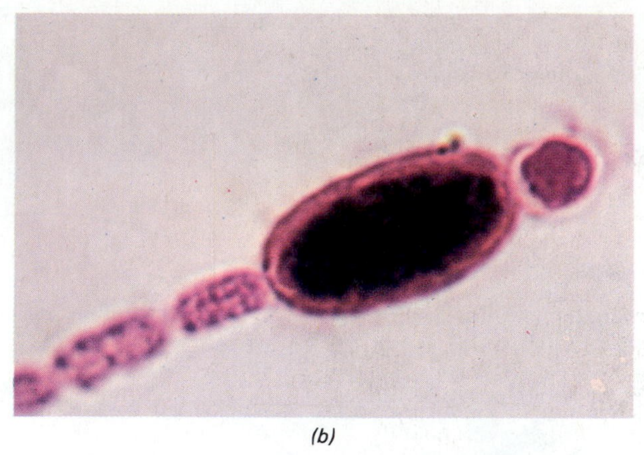

(b)

Figure 1–14 Members of kingdom Monera. (*a*) This bacterium (*Spirillum volutans*), which has been magnified several hundred times in this photomicrograph, propels itself by means of the whiplike flagellum at each of its ends. (*b*) This cyanobacterial colony (genus *Cylindrospermum*) shows some specialization of its cells. The cell at the end fixes nitrogen (converts nitrogen gas to forms usable by plants and other organisms). The greatly enlarged cell is a spore, which can survive long periods of unfavorable environmental conditions such as dryness. ((*a*), (*b*), Carolina Biological Supply Company.)

Organisms are assigned to increasingly broad categories in which they have fewer characteristics in common (Chapter 17). The broadest category commonly used is the **kingdom.** In this book, five kingdoms are recognized: Monera, Protista, Fungi, Plantae, and Animalia. A review of the kingdoms can be found in Chapter 17 (Table 17–1), and a more detailed discussion of them, in Chapters 18 through 26. We will refer to these groups repeatedly throughout the text as we consider the many kinds of problems faced by all living things and the various adaptations that have evolved in response to these problems.

KINGDOM MONERA

The single-celled **bacteria** and **cyanobacteria** (blue-green algae) of the kingdom **Monera** differ from all other organisms in that they lack a nuclear membrane as well as other membrane-bound organelles (Fig. 1–14). They are referred to as **prokaryotes.** All other organisms are **eukaryotes,** organisms with cells that have distinct nuclei surrounded by nuclear membranes, as well as a variety of other membranous organelles. Bacteria are microscopic organisms that serve as decomposers in ecosystems. Some bacteria cause diseases in human beings or in other organisms. The cyanobacteria are structurally similar to bacteria but contain the green pigment chlorophyll (as well as other pigments), which traps the energy of sunlight and enables these organisms to carry on photosynthesis. (Some true bacteria are also photosynthetic but employ alternative versions of the process.)

KINGDOM PROTISTA

Members of the kingdom **Protista,** known as protists, are primarily solitary, single-celled eukaryotes, but some species form loose aggregations of cells called **colonies** (see Part I opener). The animal-like protists, the **protozoa,** are generally larger than bacteria and are mobile. The plantlike protists include several divisions of algae; these contain chlorophyll and carry on photosynthesis (Fig. 1–15).

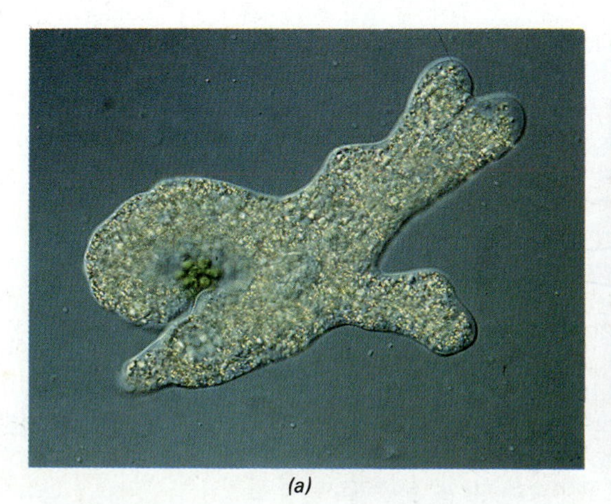

(a)

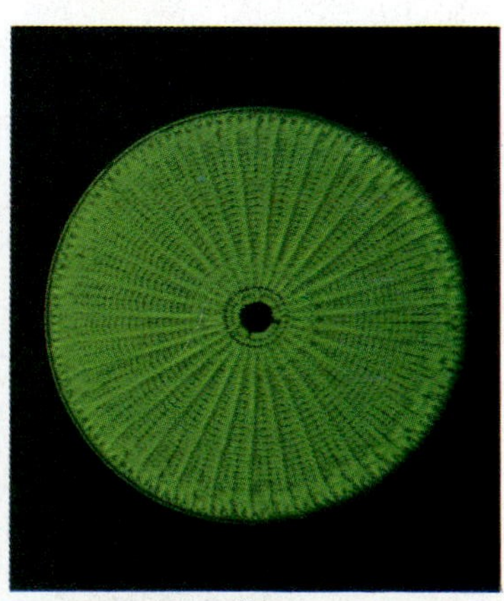

(b)

Figure 1–15 (*a*) A giant amoeba (a protozoan), *Chaos carolinense,* ingesting *Pandorina,* a multicellular alga. Though unicellularity imposes a limit on the size an organism can obtain, a few such organisms, such as *Chaos,* are actually larger than some of the simple multicellular organisms. (*b*) Shown here is a marine diatom (a plantlike protist). There are thousands of species of diatoms, which are important producers in marine ecosystems; a liter of seawater may contain up to half a million individual diatoms. ((*a*), (*b*), Michael Abbey, Photo Researchers, Inc.)

(a)

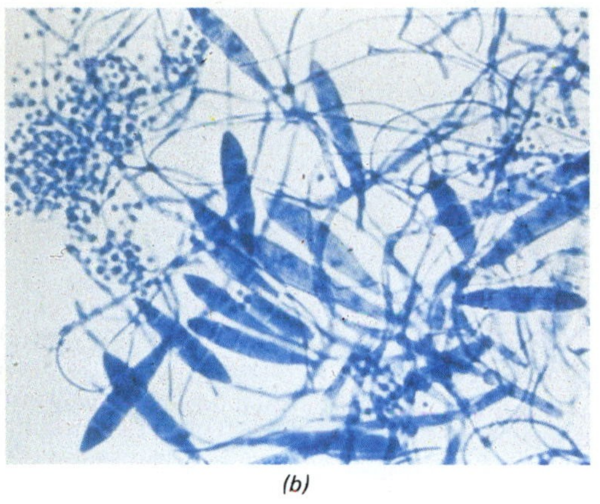

(b)

Figure 1–16 Kingdom Fungi. (*a*) Sulfur tufts, *Hypholoma fasciculare*. (*b*) A fungus (*Trichophyton mentagrophytes*) that can cause athlete's foot. ((*a*), courtesy of Leo Frandzel; (*b*), courtesy of Dr. Wilfred Little.)

KINGDOM FUNGI

The **fungi** are a diverse group of eukaryotes that differ widely in structure and in modes of reproduction (Fig. 1–16). They serve as decomposers, absorbing nutrients from dead leaves or other organic matter in the soil. Fungi produce spores during the reproductive process. There are two main groups of fungi, the **slime molds** and the **true fungi.** Slime molds are organisms found as slimy masses on decaying leaves and wood. During their life cycle they progress through a variety of very different forms. The true fungi include the unicellular yeasts, and the multicellular molds, mushrooms, and bracket fungi.

KINGDOM PLANTAE (PLANTS)

Plants are multicellular organisms adapted to carry out photosynthesis. Their photosynthetic pigments, such as chlorophyll, are located within membranous organelles called **chloroplasts.** Plant cells are surrounded by rigid cell walls containing cellulose and typically have large fluid-filled sacs called vacuoles. The kingdom Plantae includes the red and brown algae, the green algae, bryophytes, and vascular plants (Fig. 1–17).

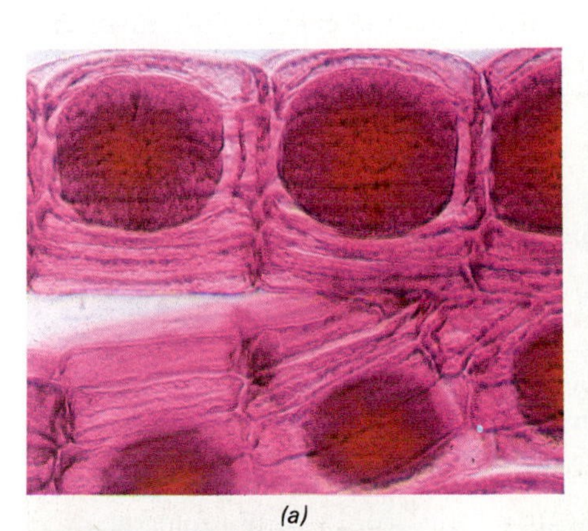

(a)

(b)

(c)

(d)

(e)

Figure 1–17 The plant kingdom claims many beautiful and diverse forms. (*a*) Red algae. (*b*) Liverwort, a bryophyte. Like other bryophytes, most liverworts are small plants that live in moist areas. (*c*) A cycad, a palmlike gymnosperm. The fossil record shows that cycads were very abundant during the Mesozoic era, more than 100 million years ago. (*d*) The bull thistle, *Cirsium Vulgare*, is an angiosperm with composite flowers; its flowers are very small and are grouped into compact structures that may be mistaken for a single flower. (*e*) The water lily, *Nymphaea*, is an angiosperm with a large flower consisting of six petal-like parts. ((*a*), (*b*), (*c*), Carolina Biological Supply Company; (*d*), (*e*), courtesy of Leo Frandzel.)

Red and brown algae are multicellular, primarily marine organisms, commonly called seaweeds. Green algae are simple plants that live in habitats ranging from salt and fresh water to moist environments on land. Most botanists believe that the complex plants evolved from organisms similar to the green algae living today.

Mosses, liverworts, and their relatives are **bryophytes.** These terrestrial (land) plants require a moist environment to complete their reproductive cycle. Because they lack an efficient system of internal transport, bryophytes do not become very large.

The vascular plants, or **tracheophytes,** include the oldest land plants as well as today's dominant land plants—the ferns, conifers **(gymnosperms),** and the flowering plants **(angiosperms).** Their efficient internal transport system moves water and nutrients from one part of the plant to another and enables them to grow to very large sizes.

KINGDOM ANIMALIA (ANIMALS)

All animals are multicellular. Their cells lack photosynthetic pigments, so animals must obtain nutrients by eating other organisms. Complex animals have a high degree of tissue specialization and body organization, which have evolved along with motility, complex sense organs, nervous systems, and muscular systems. Some ten major groups, or **phyla** (singular, phylum), of animals are recognized (Fig. 1–18). Among these are the following:

Sponges. The sponges are the simplest animals. They are aquatic and sessile. The sponge body is perforated with many pores, and food particles are strained from the water passing through.

Cnidarians. The cnidarians include jellyfish, sea anemones, and corals. These aquatic, mostly marine animals typically have stinging cells. The body is basically a simple sac with only one opening to the digestive cavity. The mouth (which must also serve as an anus) is usually surrounded by a circle of tentacles bearing stinging cells.

Flatworms. Like the cnidarians, flatworms have a digestive cavity connected to the outside by a single opening. They live in fresh or salt water, or may be terrestrial. Flatworms are **bilaterally symmetrical,** which means the body may be divided into roughly similar right and left halves. There is a concentration of nervous tissue and sense organs in the anterior (front) end of the animal, a definite advantage to an animal which moves in a forward direction.

Mollusks. Mollusks include the oysters, clams, scallops, octopods, snails, slugs, and squid. These animals have a complex body plan that is quite different from that of other animals. Most have a hard, calcareous (calcium-containing) shell, which serves as protection but makes locomotion difficult. They typically have a broad, muscular foot used in locomotion.

Annelids. The segmented worms, or annelids, are found in the oceans, in fresh water, or in moist, shady habitats. This group includes the earthworm, leeches, and a variety of marine worms. The annelid body is composed of a series of rings or segments; both the body wall and the internal organs are segmented.

Arthropods. Spiders, lobsters, insects, centipedes, and millipedes are among the more familiar arthropods. There are more arthropods in terms of both number and species—there are about a million species, mainly insects—than organisms in any other phylum. They live in a greater variety of habitats and can eat a greater variety of foods than the members of any other phylum. The term arthropod (jointed foot) refers to their paired jointed appendages.

Echinoderms. The spiny-skinned echinoderms include the sea stars, sea urchins, and sea cucumbers. These marine animals are radically different from all other animals but appear to be related to the chordates. The skin of echinoderms contains calcareous spine-bearing plates.

Chordates. The chordates have a skeletal stiffening rod (notochord), a tubular nerve cord, and paired gill slits. These structures (or their rudiments) are present in all chordate embryos but may be lost or transformed during development. The major chordate subphylum, the vertebrates, are characterized by a cartilaginous or bony vertebral column that surrounds and usually replaces the notochord. Vertebrates include sharks, bony fish, amphibians (frogs and salamanders), reptiles (snakes, lizards, turtles, alligators), birds, and mammals. Chordates are less diverse and much less numerous than insects but rival them in their adaptations to many life styles.

Figure 1–18 Some diverse life forms in the animal kingdom. (*a*) This tube coral, *Dendrophyllia gracilis*, a cnidarian, is a marine animal that looks like a flower. (*b*) The file clam is a mollusk, an animal whose soft body is covered by a protective shell. (*c*) The painted lady butterfly, an arthropod, migrates great distances; millions cross the Mediterranean Sea between Europe and Africa each year.

The chordates are adapted to many life styles: (*d*) The lion fish (*Pterois volitans*) is a native of coral reefs; its unusual fins are armed with glands that secrete a powerful toxin poisonous to humans as well as to other animals. (*e*) The roseate spoonbill (*Ajaia ajaja*) a wading bird, uses its spoonlike beak to gather shellfish and aquatic insects from tidal areas. (*f*) African lions (*Panthero leo*), among the fiercest of animals, are also among the most sociable; they live peaceably in prides (groups) of as many as 35. ((*a*), (*d*), Charles Seaborn; (*b*), (*c*), (*e*), James H. Carmichael, Coastal Creations; (*f*), courtesy of Busch Gardens.)

How Biology Is Studied

This book is about the systematic study of life—the science of biology. Because it is a science, biology offers you knowledge about life that is different from what you have gained through casual intuition or insight from your daily experiences. What distinguishes science is its insistence on a rigorous method to examine a problem and its attempts to devise experiments and tests of reason to validate its findings. The essence of the scientific method is the posing of questions and the search for answers to those questions. But the questions must arise from observations and experiments, and the answers must be potentially testable by further observation and experiment.

What makes science systematic is the attention it gives to organizing knowledge so it is readily accessible to all those who wish to build upon its foundation. In this way science is both a personal and a social endeavor. Science is not mysterious; by its sets of rules and procedures it makes itself open to all who wish to take on its challenges. Science seeks to give us precise knowledge about those aspects of the world that are accessible to its methods of inquiry. It is not a replacement for philosophy, religion, or art; and being a scientist does not mean excluding oneself from participation in these other fields of human endeavor.

SYSTEMATIC THOUGHT PROCESSES

The scientific method helps scientists to use thought processes in a rational, systematic, and error-minimizing way in order to discover, define, or delineate truth. We can define scientific truth as agreement with the facts. What scientists seek is *demonstrable* truth that can be presented in a logically compelling way. They try to distinguish between truth as it exists and belief or opinion about truth that cannot be demonstrated, or which may change as knowledge increases. Truth is objective; opinion about truth is subjective. Although pure objectivity may be impossible, scientists strive to be as objective as they can be.

It is out of the question, of course, for us to define truth in a way that is completely independent of our own experiences and our perceptions of the universe. What is most basic to the concept of scientific truth is the assumption that there is, in actuality, an external universe that is independent of the perceiving subject (us). The universe as we see it, in other words, is not merely a figment of our imagination; the world is not a dream despite the fact that our perceptions of it may be only a rough analogy to what is there. This belief, this agreement, as Einstein said, is the necessary basis of all science.[1]

The systematic thought processes on which science is based can for the most part usually be broken down into two categories: deduction and induction. With **deductive reasoning,** we begin with supplied information, called premises, and draw conclusions on the basis of that information. Deduction proceeds from general principles to specific conclusions.

Induction is almost the opposite of deduction. With **inductive reasoning,** we begin with specific observations from which we seek to draw a conclusion or discover a unifying rule or general principle. The inductive method can be used to organize raw data into manageable categories by answering the question, what do all these facts have in common? A weakness of this method of reasoning is that conclusions contain *more* information than the reported facts on which they are based. We go from many observed examples to all possible examples when we formulate the general principle. This is known as the **inductive leap.** Without it, we could not arrive at generalizations. However, we must be sensitive to the possibility that the conclusion is not valid. The extra information that inductive conclusions contain can come only from the creative insight of a human mind, and creativity, however admirable, is not infallible.

Even if a conclusion is based on thousands of observations, it is still possible that new observations can challenge the conclusion. However, the greater the number of cases that are employed, the more likely we are to draw accurate scientific conclusions. The scientist seeks to be able to state that any specific conclusion has a certain statistical probability of being correct.

---

[1]Science is sciencing. From Gonzalez, N. L.: *Science* 219:20, 28 Jan., 1983.

DESIGNING AN EXPERIMENT

The ultimate sources of all the facts of science are careful close observations and experiments, made free of bias and with suitable controls, and carried out in as quantitative a fashion as possible. The data collected may then be analyzed so that the observed phenomena may be brought into some sort of order. The data can be synthesized or reassembled, and whatever relationships that may exist can be discovered. On the basis of these initial observations the scientist makes a generalization, or constructs a hypothesis. A **hypothesis** is a trial idea about the nature of the collected data, or possibly about the connection between a chain of events, or even about cause-and-effect relationships between events. Predictions made on the basis of the hypothesis can then be further tested by controlled experiments.

It is in the construction of hypotheses and the ability to discern relationships between seemingly disparate events that scientists differ most. The ability to look at a mass of data and suggest a reason for their interrelations is rare. Science does not advance by the mere accumulation of facts, nor by the mere postulation of hypotheses. The two must go hand in hand in scientific investigations: observation, hypothesis, more observations, revised hypothesis, further observations, refining of the hypothesis, and so on, in a continuous process of intellectual feedback.

Let us follow the process of setting up an experiment. Suppose a pharmaceutical company wants to test a new drug to determine whether it will improve memory in elderly patients with memory problems. To test the drug, the company solicits the cooperation of physicians who work with such patients. The physicians administer a memory test and then prescribe the drug to 500 patients for a period of 2 months. They then administer another memory test, and find that the patients demonstrate a 20% increase in their ability to remember things. Can the drug company legitimately conclude that its hypothesis is correct, that the drug does indeed improve memory in elderly patients? Alternative explanations might be possible. The attention paid to the patients might in itself be a stimulus to their mental powers, for instance. Therefore, the conclusion cannot be considered valid.

To avoid such objections, a properly designed experiment must have a **control;** that is, a second experiment must be performed under the same conditions as those in the first, except that the one factor being tested should be varied. Thus, another similar group of patients must be given a **placebo,** a harmless starch pill similar in size, shape, color, and taste to the pill being tested. Neither group of patients should be told which pill—the drug or the placebo—has been given. In fact, to prevent bias, most medical experiments today are carried out in "double-blind" fashion: Neither the patient nor the physician know who is getting the experimental compound and who is getting the placebo. The pills or treatments are coded in some way unknown to physician or patient. Only after the experiment is over and the results are in is the code broken to identify the control and experimental patients. Another example of a controlled experiment is shown in Figure 1–19.

Not all experiments can be so neatly designed; for one thing, it is often difficult to establish appropriate controls. For example, we know that the carbon dioxide content of the earth's atmosphere is increasing because of the combustion of fossil fuels by our industrial civilization and because of widespread clearing and burning of forests. This increased carbon dioxide in the atmosphere produces a

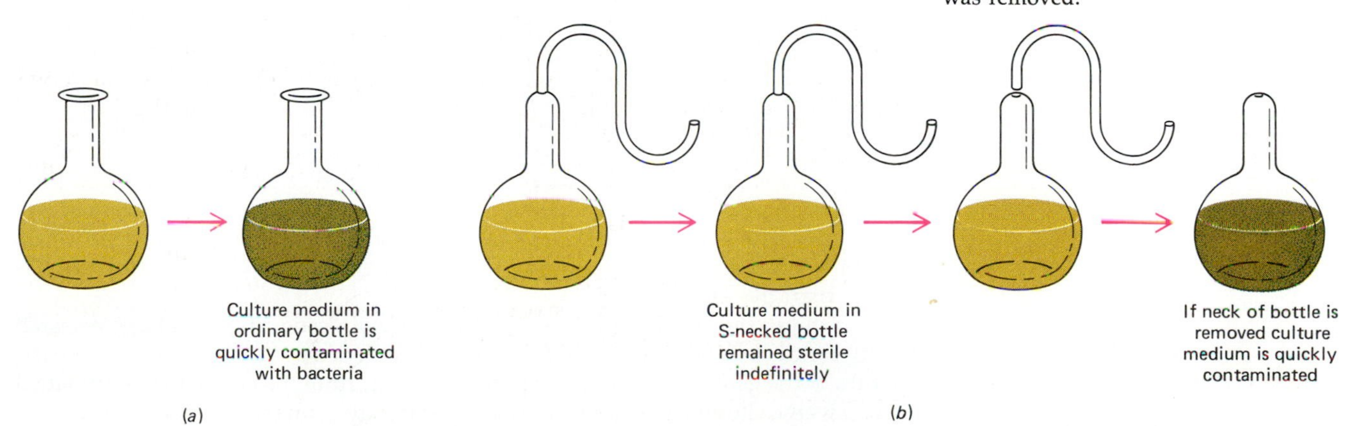
Culture medium in
ordinary bottle is
quickly contaminated
with bacteria
(a)
Culture medium in
S-necked bottle
remained sterile
indefinitely
(b)
If neck of bottle is
removed culture
medium is quickly
contaminated

Figure 1–19 Pasteur's experiments disproving the spontaneous generation of microorganisms. Nutrient broth (sugar and yeast) was placed in two types of flasks and boiled to kill any bacteria present. (*a*) Pasteur used as his control flasks with straight necks that permitted bacteria to settle into the broth, and in these flasks the broth was soon teeming with bacteria. However, Pasteur's experimental flasks, shown in (*b*), had long S-shaped necks that did not permit bacteria to enter, even though the flask was open to the air. Bacteria did not grow in such a flask unless its neck was removed.

"greenhouse effect," trapping heat from solar radiation. Some scientists have warned that this thermal blanket around the globe may increase the average temperature of the earth and ultimately alter its climate. Yet in the recent past, scientists have also argued that the accumulation of particulates in the atmosphere (soot) would moderate or cancel the effect of carbon dioxide increase. Even if the temperature of the earth does increase, however, given other variables in the earth's atmosphere, how could we be certain that the temperature change resulted from human activities?

This raises an important practical question. Obviously we do not have a second unindustrialized earth whose climate could be compared with our own. Without such a control, scientists have had to base their predictions of the future climate on mathematical modeling techniques that fall short of perfection. Should we postpone action pending the development of a perfectly predictive model? Clearly, that would involve a long wait, and by then it might be impossible to act effectively.

HOW A HYPOTHESIS BECOMES A THEORY

A hypothesis supported by a large body of observations and experiments becomes a **theory,** defined in the dictionary as "a scientifically acceptable general principle offered to explain phenomena; the analysis of a set of facts in their ideal relations to one another." A good theory relates facts that previously appeared to be unrelated and that could not be explained on common ground. A good theory grows; it relates additional facts as they become known; it may even suggest practical applications. It predicts new facts and suggests new relationships between phenomena.

A good theory, by showing the relationships between classes of facts, simplifies and clarifies our understanding of natural phenomena. Einstein wrote, "In the whole history of science from Greek philosophy to modern physics, there have been constant attempts to reduce the apparent complexity of natural phenomena to simple, fundamental ideas and relations." A theory that, over a long period of time, has yielded true predictions with unvarying uniformity, and is thus almost universally accepted, is referred to as a scientific **principle** or **law.**

THE ETHICS OF SCIENCE

Science and technology have a great impact on how we live. Yet scientists are not passive machines that record and explain phenomena in the world around them. They are part of that world, and as members of society, they participate in the same subjective cultural and political experience as people in other occupations. Science, therefore, is very much a social activity, and as such, the conduct of science is subject to ethical principles. Ethics may be applied to the scientific method itself, to the conduct of experiments, or to the public release and use of scientific findings. For instance, honesty and objectivity at every stage are indispensable if the results are to have any value. Deliberately or subconsciously doctored data may mislead scientists for generations. Great damage can be done by withholding results of a scientific investigation. Such a temptation is especially great among those who are directed to find "scientific" evidence that will support the claims made by their employers. In fact, a scientist's promotion or continued employment may even be dependent upon willingness to suppress "unfavorable" information.

There are other categories of ethical concern in science. Many questions surround the use of human subjects in medical research. Is it ethical to use prisoners who volunteer as subjects for experiments? Participation in such experiments is often linked to parole or to needed funds. Is it ethical to place patients in control groups when they might be benefited by a treatment being tested? The medical literature is full of reports on treatments now known to be useless, or even harmful, which were used for years but finally were abandoned as experience showed their ineffectiveness. Many scientists think that there is a time in the development of any new treatment when it is not only morally justified but morally required to do carefully controlled tests on human beings to be sure that the new treatment is better than the former one.

Ethical concerns have also been raised regarding some types of basic research being conducted today, most notably research in genetics. We understand the basic nature of the genetic code, and are perhaps not far from the ability to manipulate it extensively, altering and remaking living creatures. In recombinant DNA tech-

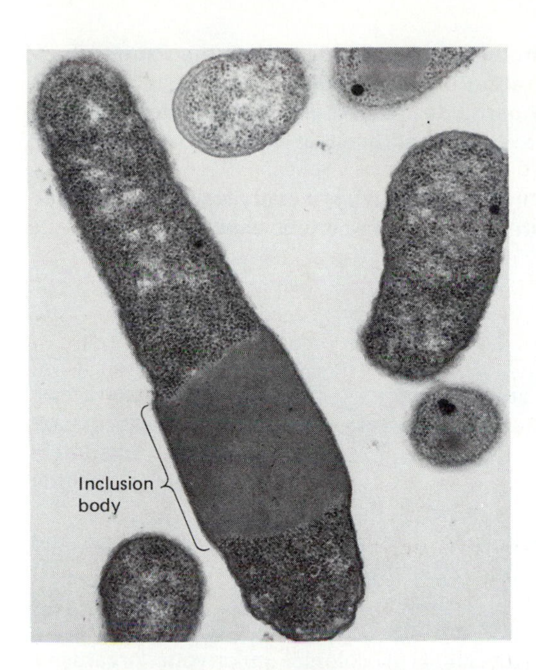

Figure 1–20 A genetically engineered *Escherichia coli* bacterium (magnified 68,000 times). Its hereditary material has been modified so that the bacterium produces large quantities of the hormone human insulin. Individuals with the disease diabetes mellitus have insufficient amounts of this hormone, which is necessary for the normal metabolism of sugar. Since the bacterium itself has no use for the insulin, and has no way of excreting it, instead the hormone accumulates in an "inclusion body" within the bacterial cell. The insulin can be harvested by destroying the bacterium. (Courtesy of Dr. Daniel C. Williams and the Lilly Electron Microscope Laboratory.)

niques, genes are transplanted from one organism into another, usually a bacterium, enabling the recipient to produce new substances. In this way bacteria can be used to produce needed therapeutic products such as insulin, which can then be marketed commercially (Fig. 1–20). Further research in this area could lead to cures for certain genetic diseases. What are the ethics, however, of working on a project to develop lethal bacteria that are resistant to antibiotics for use in biological warfare? Or what risk is there that some newly fashioned microbe might be released from a laboratory and infect human beings? Such an accident has not yet occurred, but how do we weigh the risks of this happening against the benefits of research?

All of us stand to benefit or lose from the discoveries of biological research. Biologists have an ethical responsibility to try to prevent the misuse of their discoveries. However, society, which includes nonscientists, influences scientific discoveries by contributing or withholding funds for research and by imposing regulations and standards on research. This means that much of the responsibility for scientific ethics is shared by society at large. For their part, therefore, it seems that scientists have an interest in keeping the general public informed about their work.

SUMMARY

I. A living organism is able to maintain metabolic homeostasis, move, grow, respond to stimuli, adapt, and reproduce its kind.
  - *A.* All living things (except the viruses) are composed of cells.
  - *B.* Metabolism is the sum of all the chemical activities that take place in the organism, which include the chemical reactions essential to nutrition, growth and repair, and conversion of energy to useful forms.
  - *C.* Homeostasis is the tendency of organisms to maintain a constant internal environment.
  - *D.* Living things grow by increasing the size and number of their cells.
  - *E.* Movement, though not necessarily locomotion, is characteristic of living things.
  - *F.* Living things respond to stimuli and adapt to their environment.
  - *G.* Reproduction may be asexual, in which the offspring are identical with the parent, or sexual, in which the offspring reflect the characteristics of two parents.

II. There is a hierarchy of biological organization.
  - *A.* A complex organism is organized at the chemical, cellular, tissue, organ, and organ system levels.
  - *B.* The basic unit of ecological organization is the population. Various populations form communities; a community and its physical environment is referred to as an ecosystem. The planet earth and all of its inhabitants may be regarded as a giant ecosystem, the biosphere.

III. Living organisms may be classified in five kingdoms. Each kind of organism is assigned to a genus and a species.
  - *A.* Kingdom Monera includes the bacteria and cyanobacteria.
  - *B.* Kingdom Protista includes several phyla of protozoa and several divisions of algae.
  - *C.* Kingdom Fungi consists of slime molds and true fungi, including yeasts and molds.
  - *D.* Kingdom Plantae includes red and brown algae, green algae, bryophytes, and vascular plants.
  - *E.* Kingdom Animalia consists of about ten major phyla including sponges, cnidarians, flatworms,

mollusks, annelids, arthropods, echinoderms, and chordates.

IV. The scientific method is a system of observation, hypothesis, experiment, more observation, and revised hypothesis.

A. Deductive reasoning and inductive reasoning are two categories of systematic thought process used in the scientific method.

B. A hypothesis is a trial idea about the nature of an observation or relationship.

C. A properly designed scientific experiment must have a control and must be as free as possible from bias.

D. Honesty and objectivity at every stage of investigation are essential in scientific experimentation.

POST-TEST

1. The sum of all the chemical activities of the organism is termed ______________.
2. The tendency of organisms to maintain a constant internal environment is termed ______________.
3. A ______________ is a physical or chemical change in the internal or external environment that evokes a response in an organism.
4. The streaming motion of the living material in the cells in the leaves of plants is termed ______________.
5. The splitting of an amoeba into two is an example of ______________ ______________.
6. An organism must be able to ______________ to stimuli in the environment in order to survive.

Match the terms in Column A with their descriptions in Column B.

Column A	*Column B*
7. Atom	a. Group of tissues arranged into functional structures
8. Cell	b. Combination of two or more atoms
9. Molecule	c. Smallest particle of an element that retains the characteristic properties of that element
10. Organ	d. Specialized structures within cells
11. Organ system	e. Association of similar cells to carry out a specific function
12. Organism	f. Structural and functional unit of life
13. Organelles	g. Groups of organs that function together to carry out one or more of the major life functions
14. Tissues	h. Consists of a group of coordinated organ systems

15. In an ecosystem we can distinguish producers, consumers, and decomposers. The principal producers are ______________ and ______________. Consumers include ______________. Decomposers include ______________ and ______________.
16. Organisms that lack a nuclear membrane and membrane-bound organelles are termed ______________.
17. The kingdom Monera includes the ______________ and the ______________.
18. Single-celled or colonial eukaryotic organisms make up the kingdom ______________.
19. A trial idea about the nature or connections between a chain of events is termed a ______________.
20. A ______________ is a scientifically accepted, well-tested hypothesis or group of related hypotheses offered to explain phenomena.

REVIEW QUESTIONS

1. Contrast a living organism with a nonliving object.

2. In what ways might the metabolism of an earthworm and a tiger be similar? What would be the metabolic consequences if an organism's homeostatic mechanisms failed?

3. What are the necessary components of a balanced ecosystem? In what ways are consumers dependent upon producers? upon decomposers?

4. To which kingdom would each of the following belong?
- *a.* an *Escherichia coli* bacterium
- *b.* a mold
- *c.* an amoeba
- *d.* a rose bush

5. Contrast a hypothesis and a law.

6. How would you go about testing the hypothesis that beriberi is caused by a deficiency of the vitamin thiamine? What would you consider to be proof that beriberi is caused by thiamine deficiency?

7. How would you describe the mode of operation of the scientific method?

8. What is meant by a *controlled experiment?*

9. Devise a suitably controlled experiment to show:
- *a.* whether a strain of mold found in your garden produces an effective antibiotic.
- *b.* whether the rate of growth of a bean seedling is affected by temperature.

2

Atoms and Molecules: The Physical Basis of Life

OUTLINE

I. Chemical elements
II. Atomic structure
  A. Isotopes
  B. Atomic mass
  C. Electrons and orbitals
III. Chemical compounds
IV. Chemical equations
V. Chemical bonds
  A. Covalent bonds
  B. Polar covalent bonds
  C. Ionic bonds
  D. Hydrogen bonds
  E. Other interactions between atoms
VI. Molecular mass
VII. Oxidation–reduction
VIII. Water
  A. Cohesive and adhesive forces
  B. Temperature stabilization
  C. Ionization
IX. Acids and bases
  A. pH
  B. Buffers
X. Salts

LEARNING OBJECTIVES

After you have read this chapter you should be able to:

1. Identify the chemical elements important in living things.
2. Describe the properties and roles of electrons, protons, and neutrons determining atomic structure.
3. Distinguish between the terms *atomic number, mass number, atomic mass,* and *molecular mass.*
4. Define the term *electron orbital,* and relate orbitals to energy levels; relate the number of valence electrons to the chemical properties of the elements.
5. Distinguish between the types of chemical bonds that join atoms to form ionic and covalent compounds, and give the characteristics of each type.
6. Define and use the terms *cation* and *anion.*
7. Define the terms *oxidation* and *reduction.*
8. Discuss the properties of water molecules and their importance in living things.
9. Define the terms *acid* and *base,* and discuss their properties.
10. Use the pH scale in describing the hydrogen ion concentration in living systems, and describe how buffers help to minimize change in pH.
11. Describe the composition of a salt, and tell why salts are important in living organisms.

How might we understand what makes a certain painting a great work of art? We could begin by studying lines and color, for every painting is a collection of these components. Yet it is only if the arrangement of these components strikes the observer in an especially compelling way that the painting may be considered a work of art. We begin our study of life in a similar way, by studying its simplest components—electrically charged atoms and molecules. Yet we remember that it is the specific organization and precise interaction of these components that define life.

During recent years, biologists have focused on analyses of the structure and function of cells, on subcellular structures and, finally, on experiments with isolated enzyme systems and genes. Molecular biologists are now exploring the very energy transformations and enzymatic reactions that underlie the manifestations of life. From all of these studies have emerged the following generalizations, which we can apply as we begin our own inquiry into the nature of life:

1. The chemical composition and metabolic processes of all living things are remarkably similar despite their great diversity in form.
2. The physical and chemical principles governing living systems are the same as those that govern nonliving systems.

Chemical Elements

The term **element** is applied to substances that cannot be decomposed into simpler substances by chemical reactions. The matter of the universe (see Focus on Matter and Energy) is composed of 92 naturally occurring elements, ranging from hydrogen, the lightest, to uranium, the heaviest. In addition to the naturally occurring elements, about 17 elements heavier than uranium have been made by bombarding elements with subatomic particles in devices known as particle accelerators.

About 98% of an organism's mass is composed of just six elements: oxygen, carbon, hydrogen, nitrogen, calcium, and phosphorus. Approximately 14 other elements are consistently present in living things, but in smaller quantities. Some of these, such as iodine and copper, are known as **trace elements** because they are present in such minute amounts.

Scientists have assigned each element a **chemical symbol**—usually the first letter or first and second letters of the English or Latin name of the element. For example, O is the symbol for oxygen, C for carbon, H for hydrogen, N for nitrogen, and Na for sodium (the Latin name is natrium). Table 2–1 lists the elements that make up a living organism and explains why each is important.

Whatever physical state matter may assume—gas, liquid, or solid—it is composed of units termed atoms. An **atom** is the smallest subdivision of an element that retains its chemical properties. The subdivision of any kind of matter ultimately yields atoms; these units cannot be divided by chemical means. Atoms are much smaller than the tiniest particle visible under a light microscope. By special scanning electron microscopy (see Chapter 4), with magnification as much as 5 million times, researchers have been able to photograph some of the larger atoms such as uranium and thorium.

Atomic Structure

Physicists have discovered a considerable number of subatomic particles, but for our purposes we need consider only three: protons, neutrons, and electrons. **Protons** have a positive electric charge; **neutrons** are uncharged particles with about the same mass as protons. Protons and neutrons make up almost all of the mass of an atom and are concentrated in the **atomic nucleus.** **Electrons** have a negative electrical charge and an extremely small mass (only about 1/1800 of the mass of a proton). The electrons, as we will see, behave as though they were spinning about in the empty space surrounding the atomic nucleus.

Each kind of element has a fixed number of protons in the atomic nucleus. This number, called the **atomic number,** is written as a subscript to the left of the chemical symbol. Thus ${}_{1}H$ indicates that the hydrogen nucleus contains one proton, and ${}_{8}O$, that the oxygen nucleus contains eight protons. It is the atomic num-

TABLE 2–1
Elements that Make up the Human Body

Name	*Chemical Symbol*	*Approximate Composition of Human Body by Mass (%)*	*Importance or Function*
Oxygen	O	65	Required for cellular respiration; present in most organic compounds; component of water
Carbon	C	18	Forms backbone of organic molecules; can form four bonds with other atoms
Hydrogen	H	10	Present in most organic compounds; component of water
Nitrogen	N	3	Component of all proteins and nucleic acids
Calcium	Ca	1.5	Structural component of bones and teeth; important in muscle contraction, conduction of nerve impulses, and blood clotting
Phosphorus	P	1	Component of nucleic acids; structural component of bone; important in energy transfer
Potassium	K	0.4	Principal positive ion (cation) within cells; important in nerve function; affects muscle contraction
Sulfur	S	0.3	A component of most proteins
Sodium	Na	0.2	Principal positive ion in interstitial (tissue) fluid; important in fluid balance; essential for conduction of nerve impulses
Magnesium	Mg	0.1	Needed in blood and body tissues; a part of many important enzymes
Chlorine	Cl	0.1	Principal negative ion (anion) of interstitial fluid; important in fluid balance
Iron	Fe	Trace amount	Component of hemoglobin and myoglobin; component of certain enzymes
Iodine	I	Trace amount	Component of thyroid hormones

Other elements, found in very small amounts in the body (the trace elements), include manganese (Mn), copper (Cu), zinc (Zn), cobalt (Co), fluorine (F), molybdenum (Mo), selenium (Se), and a few others.

ber, the number of protons in the nucleus, that determines the chemical identity of the atom. The total number of protons plus neutrons in the nucleus is termed the **mass number** and is indicated by a superscript to the left of the chemical symbol. The common form of oxygen atom, with eight protons and eight neutrons in its nucleus, has an atomic number of 8 and a mass number of 16. It is indicated by the symbol ${}^{16}_{8}O$.

ISOTOPES

Atoms of the same element containing the same number of protons but different numbers of neutrons, and hence having different mass numbers, are called **isotopes.** The three isotopes of hydrogen, ${}^{1}_{1}H$, ${}^{2}_{1}H$, and ${}^{3}_{1}H$, contain zero, one, and two neutrons, respectively. Elements usually occur in nature as a mixture of isotopes.

All of the isotopes of a given element have essentially the same chemical characteristics. However, some isotopes with excess neutrons are unstable and tend to break down, or decay, to a more stable isotope (usually becoming a different element). Such isotopes are termed **radionuclides** (or **radioisotopes**) since they emit high-energy radiation when they decay.

FOCUS ON

Matter and Energy

Living things, like nonliving objects, are composed of matter and energy. Matter occupies space and has mass (weight); energy is the ability to produce a change in matter (in its motion or state). Energy may take the form of light, heat, mechanical energy, electrical energy, or chemical energy.

At one time, matter and energy were considered to be distinct from one another. In the early part of this century, however, Albert Einstein pointed out that both matter and energy must have mass. His idea has been confirmed by astronomers, who have observed that a ray of light traveling from a distant star toward the earth bends as it passes the sun. This phenomenon demonstrates that the light is influenced by the gravitational influence of the sun and, therefore, has mass. The actual relationship between energy and a given mass is summarized in Einstein's famous equation $E = mc^2$, where E is energy, m is mass, and c is the velocity of light (a constant).

Before Einstein's time it was also believed that matter could not be created or destroyed but could only be converted from one form to another. (When you eat a steak and break down its molecules in cellular respiration, you do not actually transform its matter into energy; you are releasing the energy stored in the chemical bonds of the molecules that made up the steak.) Under appropriate conditions, however, as in the sun or a nuclear reactor, matter can be converted into energy. By applying Einstein's equation, you can see that, since c^2 is a very large number (3×10^8 m/sec), a great amount of energy can be obtained from the conversion of a small amount (mass) of matter. However, a living system that converts matter into the energy it needs for its metabolic reactions would have to be the product of science fiction. For one thing, these reactions take place at such high temperatures that the structure of the cell (and its immediate surroundings!) would be destroyed. Therefore, although physicists and some chemists might speak of matter and energy as though they could be interchanged, biologists, for purposes of expediency and practicality, discuss the matter and energy of living systems as being separate, distinct features.

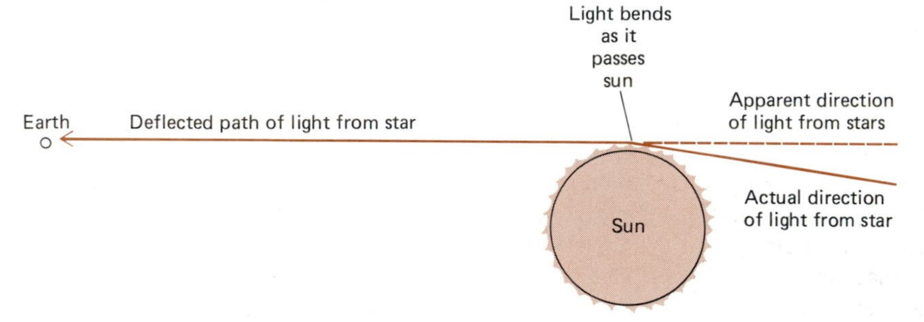

The deflection of light as it passes near the sun. The gravitational attraction of the sun "bends" the light, making it appear (to an observer on earth) that the light is coing from a different direction than it actually is.

Radionuclides such as ^3H (tritium) and ^{14}C have been extremely valuable research tools in biology and are useful in medicine for both diagnosis and treatment. Despite the difference in the number of neutrons, the body treats all isotopes of a given element the same chemically. The reactions of a fat, hormone, or drug can be followed in the body by labeling the substance with a radionuclide such as carbon-14 or tritium. For example, the active component in marijuana (tetrahydrocannabinol) has been labeled and administered intravenously. By measuring the amount of radioactivity in the blood and urine at successive intervals, it was determined that this compound remains in the blood for more than 3 days; products of the metabolism of this substance can be detected in the urine for more than 8 days.

Because radiation from radionuclides can interfere with cell division, such isotopes have been used in the treatment of cancer (a disease characterized by rapidly dividing cells). Radionuclides are also used to test thyroid gland function, to measure the rate of red blood cell production, and to study many other aspects of body function and chemistry.

ATOMIC MASS

The **atomic mass**[1] of an element is a number that tells us how heavy, on the average, an atom of that element is compared with an atom of another element. The mass of any single atom or molecule is exceedingly small, much too small to be conveniently expressed in terms of grams or micrograms. Such masses are expressed in terms of the **atomic mass unit (amu),** equal to the *approximate* mass of a proton or neutron. (The standard for comparing elements is based on assigning an atomic mass of exactly 12 to $^{12}_{6}C$, the most common isotope of carbon.) The atomic mass for an element reflects the masses of the mixtures of isotopes that occur in nature. For example, although more than 99% of the hydrogen atoms in a naturally occurring sample have an atomic mass of 1 amu (to be precise, on the carbon = 12 scale it is 1.0000078 amu), the atomic mass of hydrogen is 1.0079 amu. This reflects the fact that a small amount of deuterium, $^{2}_{1}H$ (mass number 2), and an even smaller amount of tritium, $^{3}_{1}H$ (mass number 3), occur along with the common form of hydrogen, $^{1}_{1}H$.

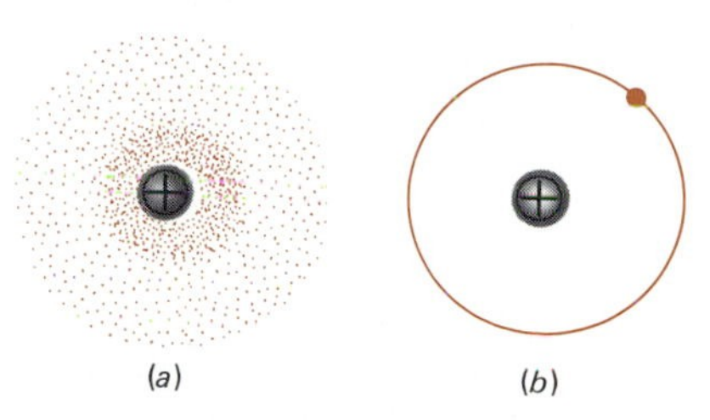

Figure 2–1 Two ways of representing the hydrogen atom. (*a*) An electron cloud. Dots represent the probability of the electron's being in that particular location at any given moment. (*b*) Bohr model of a hydrogen atom. Although the Bohr model is a less accurate way to depict electron configuration, it is commonly used because of its simplicity and convenience.

ELECTRONS AND ORBITALS

The space outside the atomic nucleus contains the electrons. Each electron bears a charge of −1, exactly equal but opposite to the charge on a proton. The electrons are attracted by the positive charge of the protons. Each type of atom has a characteristic number of protons and neutrons within the nucleus and a characteristic number of electrons around it, although the number and relative position of the electrons in the atom may change during chemical reactions. In a neutral atom the number of protons in the nucleus equals the number of electrons around it. The atom as a whole is in a state of electrical neutrality; that is, it has no net charge. The positive charges of the protons equal the negative charges of the electrons.

Electrons occur in characteristic regions of space termed **orbitals.** The lowest energy orbital is nearest the nucleus and is spherical in shape (see Fig. 2–2(*a*)). Other electron orbitals farther from the nucleus are either spherical or dumbbell-shaped or are represented by more complex three-dimensional coordinates. Orbitals represent the places where electrons are most probably found. Electrons whirl around the nucleus, now close to it, now farther away, so that an electron cloud surrounds the nucleus. One way of illustrating an atom is to show its electron orbitals as clouds, as in Figure 2–1(*a*). The density of the shaded areas is proportional to the probability that an electron is present.

Several electrons may have similar energies; these make up a **shell** and are said to be in the same **energy level.** The number of electrons in the outer energy level determines the chemical properties of atoms. The energy levels or shells of electrons in an atom can be represented by a series of concentric circles around the nucleus, as in Figure 2–1(*b*). It is important to keep in mind, however, that electrons do not circle the nucleus in fixed concentric pathways. Although each orbital may contain no more than two electrons, there may be several orbitals within a given energy level or electron shell.

The way that electrons are arranged around an atom is referred to as the **electron configuration** of that atom. Electrons always fill the orbitals nearest to the nucleus before occupying those farther away. The maximum number of electrons in the innermost shell (which is a single spherical orbital) is two; the second shell has four orbitals (one spherical and three dumbbell-shaped) and thus can contain a maximum of eight electrons (Fig. 2–2). The third shell has a maximum of 18 electrons arranged in nine orbitals, and the fourth, 32 electrons in 16 orbitals. Although the third and outer shells can each contain more than eight electrons, they are most stable when only eight are present. We may consider the first shell to be complete when it contains two electrons, and the other shells to be complete when they each contain eight electrons. The atomic structures of some elements important in biological systems—carbon, hydrogen, oxygen, nitrogen, sodium, and chlorine—are shown in Figure 2–3.

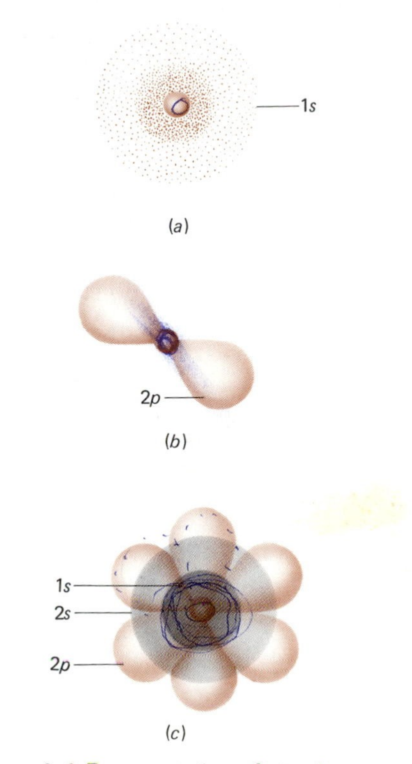

Figure 2–2 Representation of atomic orbitals. (*a*) The first energy level is a single spherical orbital (designated 1*s*) that can hold a maximum of two electrons. The electrons depicted in the diagram can actually be found anywhere within the dotted area. (*b*) One of the dumbbell-shaped (2*p*) orbitals of the second energy level. (*c*) The second energy level has four orbitals, one spherical (2*s*) and three dumbbell-shaped (2*p*). The six electrons of a carbon atom are distributed in orbitals as depicted here. Remember that the higher the energy level, the greater the average distance that the electron is located from the nucleus.

[1] For convenience we will consider mass and weight as equal, although this is not always true. Mass does not depend upon the force of gravity, but weight does. Thus, a person on the moon has the same mass on earth, but because of the moon's lower gravity, body weight is less.

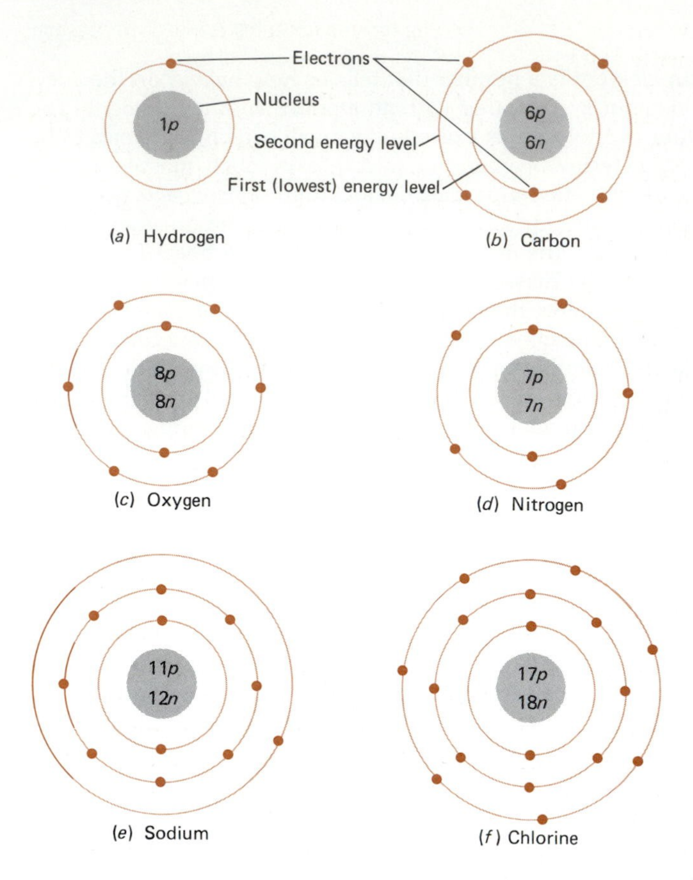

Figure 2–3 Bohr models of some biologically important atoms. (*a*) Hydrogen. (*b*) Carbon. (*c*) Oxygen. (*d*) Nitrogen. (*e*) Sodium. (*f*) Chlorine. Each circle represents an energy level, or electron shell. Dots on the circles represent electrons. *p*, proton; *n*, neutron.

Keep in mind that each atom is largely empty space. The distance from an electron to the protons and neutrons in the central nucleus may be 1000 times greater than the diameter of the nucleus itself. The tendency of the negatively charged electrons to fly off in space is countered by their attraction to the atomic nucleus due to the positive charge of the protons in the nucleus.

The more distant the energy level is from the nucleus, the greater is the energy of the electrons in that level. An electron can be moved to an orbital farther from the nucleus by providing it with more energy, or an electron can give up energy and sink back to a lower energy level in an orbital nearer the central nucleus. Energy is required to move a negatively charged electron farther away from the positively charged nucleus.

When energy is added to the system, an electron can jump from one level to the next, *but it cannot stop in the space in between.* To move an electron from one level to the next the atom must absorb a discrete packet of energy known as a **quantum,** which contains just the right amount of energy needed for the transition—no more and no less. The term quantum jump is used in everyday language to indicate a sudden discontinuous move from one level to another.

Chemical Compounds

A **chemical compound** is a substance that consists of two or more different elements combined in a fixed ratio. For example, water is a chemical compound consisting of two atoms of hydrogen chemically combined with one atom of oxygen. The properties of a chemical compound can be quite different from those of its constituent elements: At room temperature, water is usually a liquid; hydrogen and oxygen are gases.

A **chemical formula** is a shorthand method for describing the chemical composition of a compound. Chemical symbols are used to indicate the types of atoms in the molecule, and subscript numbers are used to indicate the number of each type of atom present. The chemical formula for molecular oxygen, O_2, tells us that this molecule consists of two atoms of oxygen. The chemical formula for water, H_2O, indicates that each molecule consists of two atoms of hydrogen and one atom of oxygen. (Note that when a single atom of one type is present, it is not necessary to write 1; we do *not* write H_2O_1.)

Another type of formula is the **structural formula,** which shows not only the types and numbers of atoms in a compound but also their arrangement. In any type of chemical compound the atoms are always arranged in the same way. From the chemical formula for water, H_2O, you could only guess whether the atoms were arranged H—H—O or H—O—H. The structural formula, H—O—H settles the matter, indicating that the two hydrogen atoms are attached to the oxygen atom.

Chemical Equations

During any moment in the life of an organism, be it a mushroom or a housefly, many complex chemical reactions are taking place. The chemical reactions that occur between atoms and compounds—for example, between methane (natural gas) and oxygen—can be described by means of chemical equations:

$$\underset{\text{Methane}}{CH_4} + \underset{\text{Oxygen}}{2\,O_2} \longrightarrow \underset{\text{Carbon dioxide}}{CO_2} + \underset{\text{Water}}{2\,H_2O} + \text{Energy}$$

In a **chemical equation,** the **reactants** (the substances that participate in the reaction) are written on the left side of the equation, and the **products** (the substances formed by the reaction) are written on the right side. The arrow means "yields" and indicates the direction in which the reaction tends to proceed.

The number preceding a chemical symbol or formula indicates the number of atoms or molecules reacting. Thus, $2\,O_2$ means two molecules of oxygen, and $2\,H_2O$ means two molecules of water. The absence of a number indicates that only one atom or molecule is present.

In some cases the reaction will proceed in the reverse direction (to the left) as well as forward (to the right); at **equilibrium** the rates of the forward and reverse reactions are equal. Reversible reactions are indicated by double arrows:

$$\underset{\text{Nitrogen}}{N_2} + \underset{\text{Hydrogen}}{3\,H_2} \rightleftharpoons \underset{\text{Ammonia}}{2\,NH_3}$$

Chemical Bonds

The chemical properties of an element are determined primarily by the number and arrangement of electrons in the *outermost* energy level (electron shell). In a few elements, called the "noble gases," the outermost shell is filled. These elements are chemically inert, meaning that they will not readily combine with other elements. Two such elements are helium, with two electrons (a complete inner shell), and neon, with ten electrons (a complete inner shell of two and a complete second shell of eight).

The electrons in the outermost energy level of an atom are sometimes referred to as **valence electrons.** When the outer shell of an atom contains fewer than eight electrons, the atom tends to lose, gain, or share electrons to achieve an outer shell of eight (zero or two in the case of the lightest elements).

The elements in a given compound are always present in a certain proportion by mass. This reflects the fact that atoms are combined by chemical bonds in a precise way to form the compound. A **chemical bond** is the attractive force that holds two atoms together. Each bond represents a certain amount of potential chemical energy. The atoms of each element form a specific number of bonds with the atoms of other elements—a number dictated by the valence electrons. The two principal types of chemical bonds are covalent bonds and ionic bonds.

COVALENT BONDS

Covalent bonds involve the sharing of electrons between atoms. A compound consisting primarily of covalent bonds is called a **covalent compound.** The smallest part of a covalent compound that retains the properties of the compound is a **molecule.**

A simple example of a covalent bond is the one joining two hydrogen atoms in a molecule of hydrogen gas, H_2 (Fig. 2–4(*a*)). Each atom of hydrogen has one

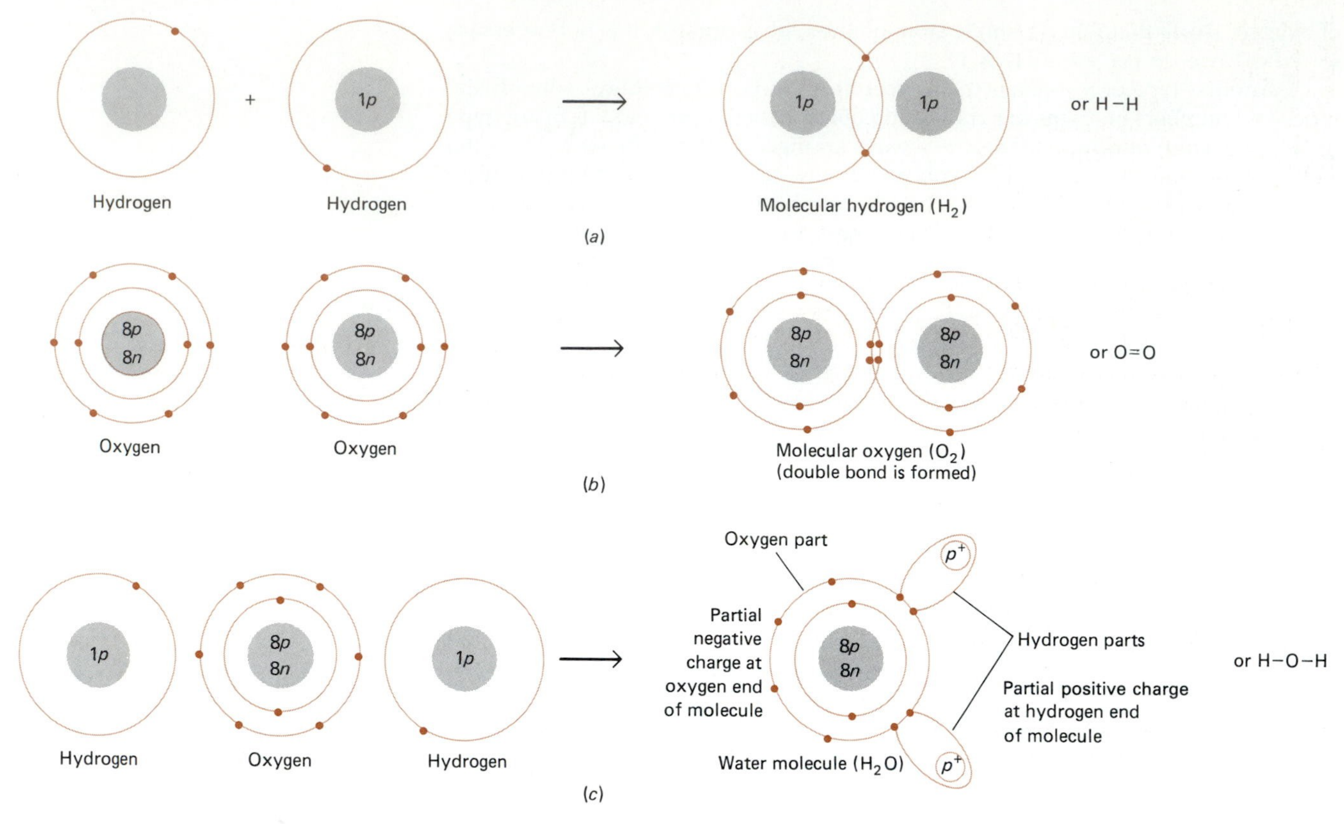

Figure 2–4 Formation of covalent compounds. (*a*) Two hydrogen atoms achieve stability by sharing electrons, thereby forming a molecule of hydrogen. The structural formula shown on the right is a simpler way of representing molecular hydrogen. The straight line between the hydrogen atoms represents a single covalent bond. (*b*) Two oxygen atoms share two pairs of electrons to form molecular oxygen. Note the double bond. (*c*) When two hydrogen atoms share electrons with an oxygen atom, the result is a molecule of water. Note that the electrons tend to stay closer to the nucleus of the oxygen atom than to the hydrogen nuclei. This results in a partial negative charge on the oxygen portion of the molecule and in a partial positive charge at the hydrogen end of the molecule. Although the water molecule as a whole is electrically neutral, it is a polar covalent compound.

electron, but two electrons are required to complete the first energy level. The hydrogen atoms have equal capacities to attract electrons, so neither donates an electron to the other. Instead, the two hydrogen atoms share their single electrons so that each of the two electrons is attracted simultaneously to the two protons in the two hydrogen nuclei. The two electrons thus whirl around *both* atomic nuclei and join the two atoms together.

A simple way of representing the electrons in the outer shell of an atom is to use dots placed around the chemical symbol of the element to represent the electrons. In a water molecule two hydrogen atoms are covalently bonded to an oxygen atom:

$$\mathrm{H}\cdot + \mathrm{H}\cdot + \cdot\ddot{\underset{\cdot\cdot}{\mathrm{O}}}\cdot \longrightarrow \mathrm{H}:\ddot{\underset{\cdot\cdot}{\mathrm{O}}}:\mathrm{H}$$

Oxygen has six valence electrons; by sharing electrons with two hydrogen atoms, it completes its outer level of eight. Each hydrogen atom obtains a complete outer level of two. The atoms of each element have a characteristic affinity for electrons. In a covalent bond between two different elements, such as oxygen and hydrogen, the electrons are pulled closer to the atomic nucleus of the element with the greater electron affinity (in this case, oxygen). (Note that in the structural formula H—O—H, each pair of shared electrons is represented by a single line. Unshared electrons are usually omitted in a structural formula.)

The carbon atom has four electrons in its outer energy level. These four electrons are available for covalent bonding:

$$\cdot\dot{\underset{\cdot}{\mathrm{C}}}\cdot$$

When one carbon and four hydrogen atoms share electrons, a molecule of methane, CH_4, is formed:

$$\begin{array}{ccc} & \mathrm{H} & \\ \mathrm{H} & :\ddot{\underset{\cdot\cdot}{\mathrm{C}}}: & \mathrm{H} \\ & \mathrm{H} & \end{array} \quad \text{or} \quad \begin{array}{ccc} & \mathrm{H} & \\ & | & \\ \mathrm{H}— & \mathrm{C} & —\mathrm{H} \\ & | & \\ & \mathrm{H} & \end{array}$$

Each atom shares its outer-level electrons with the other, thereby completing the first energy level of each hydrogen atom and the energy level of the carbon atom.

The nitrogen atom has five electrons in its outer shell:

$$\cdot\ddot{\text{N}}\cdot$$

When a nitrogen atom shares electrons with three hydrogen atoms, a molecule of ammonia, NH_3, is formed:

```
     ..
H  :N:  H        or        H—N—H
    ..                       |
    H                        H
```

When an electron pair is shared between two atoms, the covalent bond is termed a **single bond.** Two oxygen atoms may achieve stability by forming covalent bonds with one another. Each oxygen atom has six electrons in its outer shell. To become stable, the two atoms share two pairs of electrons, forming molecular oxygen (Fig. 2–4(*b*)). When two pairs of electrons are shared in this way, the covalent bond is referred to as a **double bond.** Some atoms form **triple bonds** with one another, sharing three pairs of electrons.

POLAR COVALENT BONDS

Electronegativity is a measure of an atom's attraction for electrons in chemical bonds. A covalent bond between atoms of different electronegativity is called a **polar covalent bond** (in contrast to the **nonpolar covalent bond** in a hydrogen molecule). In a water molecule the electrons tend to be closer to the nucleus of the oxygen atom than to the nuclei of the hydrogen atoms. The water molecule as a whole is electrically neutral, but it is polar, since the two hydrogen atoms have a partial positive charge and the oxygen atom has a partial negative charge (Fig. 2–4(*c*)).

The polarity of compounds is of prime importance in understanding the structure of biological membranes and their properties. Covalent bonds may have all degrees of polarity, from ones in which the electrons are exactly shared (as in the hydrogen molecule) to ones in which the electrons are much closer to one atom than to the other and the bond is quite polar. An ionic bond is one extreme of polarity in which the electrons are pulled completely from one atom to the other.

IONIC BONDS

When an atom gains or loses electrons it becomes a charged particle called an **ion.** Atoms with one, two, or three electrons in their outer shell tend to lose electrons to other atoms. When these atoms lose electrons they become positively charged as a result of the excess of protons in their nucleus. Positively charged ions are termed **cations.** Atoms with five, six, or seven valence electrons tend to gain electrons from other atoms and become negatively charged **anions.** Cations and anions play essential roles in transmission of nerve impulses, muscle contraction, and many other life processes (Fig. 2–5).

An **ionic compound** is a substance consisting of anions and cations that are bonded together by their opposite charges. The bonds between these ions are

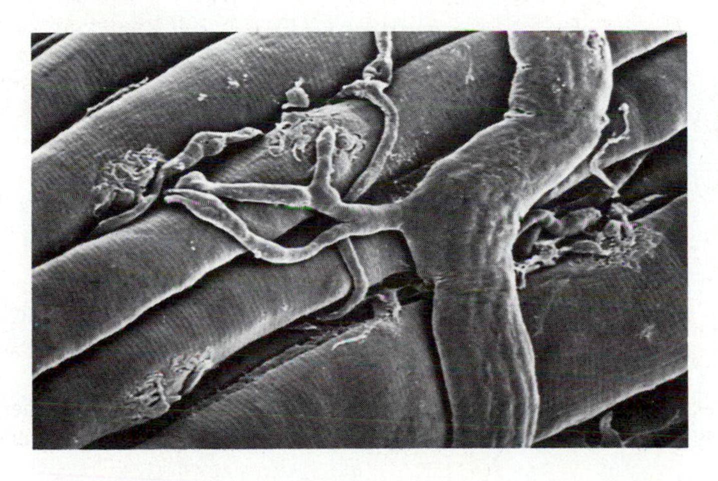

Figure 2–5 Sodium, potassium, and chlorine ions are among the ions essential in the conduction of a nerve impulse. This scanning electron micrograph shows a nerve fiber communicating with several muscle cells (approximately ×900). The nerve fiber transmits impulses to the muscle cells, stimulating them to contract. The muscle cells are rich in calcium ions, which are essential for muscle contraction. (From Desaki, J.: Vascular autonomic plexuses and skeletal neuromuscular junctions: A scanning electron microscopic study. *Biomedical Research Supplement*, 139–143, 1981.)

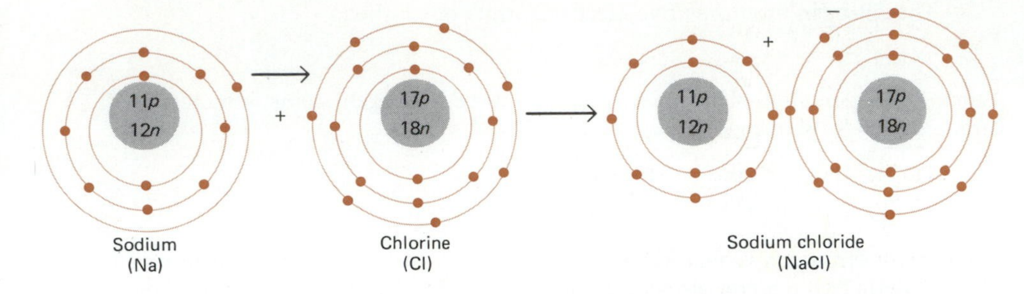

Figure 2–6 Formation of an ionic compound. Sodium donates its single valence electron to chlorine, which has seven electrons in its outer energy level. With this additional electron chlorine completes its outer energy level. The two atoms are now electrically charged ions. They are attracted to one another by their unlike electric charges, forming sodium chloride. The force of attraction holding these ions together is called an ionic bond.

termed **ionic bonds.** Such bonds differ from covalent bonds in that ionic bonds do not involve a sharing of electrons between atoms. Instead, they result from the complete transfer of the outermost electrons from one atom to another.

A good example of how ionic bonds are formed is the reaction that occurs between sodium and chlorine. A sodium atom, with atomic number 11, has two electrons in its inner shell, eight in the second, and one in the third shell. A sodium atom cannot fill its third shell by obtaining seven electrons from other atoms, for it would then have a vast excess of negative charge. Instead, it gives up the single electron in its third shell to some electron acceptor, leaving the second shell as the complete outer shell (Fig. 2–6). A chlorine atom, with atomic number 17, has 17 protons in its nucleus, two electrons in its inner shell, eight in the second, and seven in the third shell. The chlorine atom achieves a complete outer shell not by losing the seven electrons in its third shell (for it would then have a vast positive charge) but by accepting an electron from an electron donor such as sodium to complete its outer third shell.

When sodium reacts with chlorine, its outermost electron is transferred completely to chlorine. The sodium ion now has 11 protons in its nucleus, 10 electrons circling the nucleus, and a net charge of 1^+. The chlorine ion has 17 protons in its nucleus, 18 electrons circling the nucleus, and a net charge of 1^-. These ions attract each other as a result of their opposite charges. They are held together by this electrical attraction in ionic bonds to form sodium chloride,[1] common table salt.

Compounds joined by ionic bonds, such as sodium chloride, have a tendency to **dissociate** (separate) into their individual ions when placed in water. In the solid form of an ionic compound, the constituent ions require considerable energy to be pulled apart. Water, however, is an excellent **solvent;** as a liquid it is capable of dissolving many substances. This is because of the polarity of the water molecules. The localized partial positive charges (on the hydrogen atom) and partial negative charges (on the oxygen atom) on each water molecule attract the anions and cations on the surface of an ionic solid. As a result, the solid dissolves. In solution, each cation and anion of the ionic compound is surrounded by oppositely charged ends of the water molecules (Fig. 2–7(*a*)). This process is known as **hydration.**

[1] In both covalent and ionic binary compounds (binary denotes compounds consisting of two elements), the element having the greater attraction for the shared electrons is named second, and an -ide ending is added to the stem name—e.g., sodium chloride, hydrogen fluoride. The -ide ending is also used to indicate an anion, such as in chloride (Cl^-) and hydroxide (OH^-).

Figure 2–7 (*a*) Hydration of an ionic compound. The crystal of NaCl consists of regularly spaced ionic bonds between the Na^+ and Cl^-. When NaCl is added to water, the partial negative ends of the water molecules are attracted to the positive sodium ions and tend to pull them away from the chlorine ions. At the same time, the partial positive ends of the water molecules are attracted to the negative chlorine ions, separating them from the sodium ions. When the NaCl is dissolved, each of the sodium and chlorine ions is surrounded by water molecules electrically attracted to it. (*b*) A model of a crystal of NaCl. The actual crystal contains billions of ions.

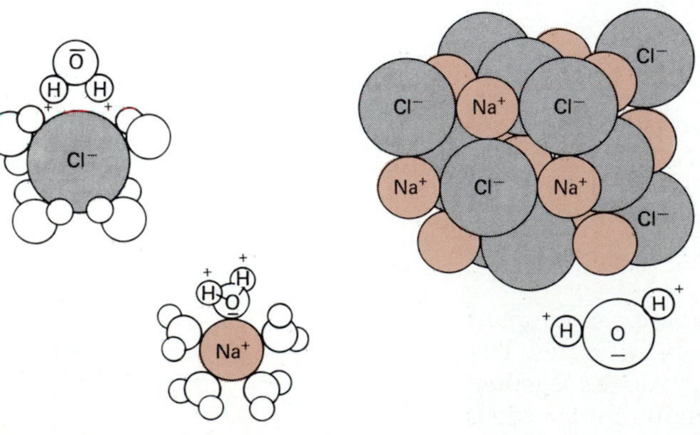

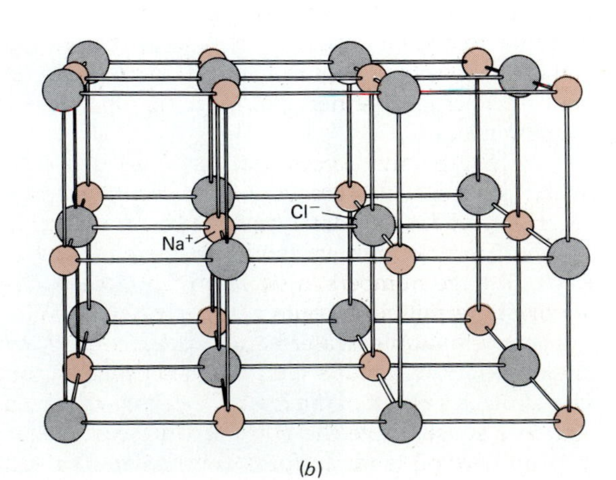

$$\underset{\text{Sodium chloride}}{NaCl} \xrightarrow{\text{in } H_2O} \underset{\text{Sodium ion}}{Na^+} + \underset{\text{Chloride ion}}{Cl^-}$$

$$\underset{\text{Calcium chloride}}{CaCl_2} \xrightarrow{\text{in } H_2O} \underset{\text{Calcium ion}}{Ca^{2+}} + \underset{\text{Chloride ions}}{2\ Cl^-}$$

$$\underset{\text{Sodium sulfate}}{Na_2SO_4} \xrightarrow{\text{in } H_2O} \underset{\text{Sodium ions}}{2\ Na^+} + \underset{\text{Sulfate ion}}{SO_4^{2-}}$$

The term molecule does not adequately explain the properties of ionic compounds such as NaCl. Hydrated sodium and chlorine ions do not interact with each other to the extent that "molecules" of sodium chloride can be said to exist. Likewise, if you observe a model of NaCl in its solid crystal state (Fig. 2–7(*b*)), you will see that each ion is actually surrounded by six ions of opposite charges. The molecular formula NaCl indicates that sodium and chlorine ions are present in a one-to-one ratio, but in the actual crystal, no discrete molecules composed of one Na^+ ion and one Cl^- ion are present.

HYDROGEN BONDS

Another type of bond that is extremely important in biological systems is the **hydrogen bond.** When hydrogen is combined with oxygen (or with another electronegative atom), it has a partial positive charge because its electron is positioned closer to the oxygen atom. Hydrogen bonds tend to form between a hydrogen atom that is covalently bonded to oxygen or nitrogen and some other electronegative atom, usually oxygen or nitrogen (Fig. 2–8). The atoms involved may be in two parts of the same molecule or in two different molecules.

Hydrogen bonds are weak and are readily formed and broken. They have a specific length and orientation; this feature is very important in their role in helping to determine the three-dimensional structure of large molecules such as nucleic acids and proteins. Hydrogen bonds, though relatively weak individually, occur in large numbers in the double helix of DNA and in the alpha helix of proteins (Chapter 3). The large number of bonds present compensates for the relative weakness of the individual bonds.

The water molecules in liquid water and in ice are held together in part by hydrogen bonds. The hydrogen atom of one water molecule, with its partial positive charge, is attracted to the oxygen atom of a neighboring water molecule, with its partial negative charge, forming a hydrogen bond. Each water molecule can form hydrogen bonds with a maximum of four neighboring water molecules.

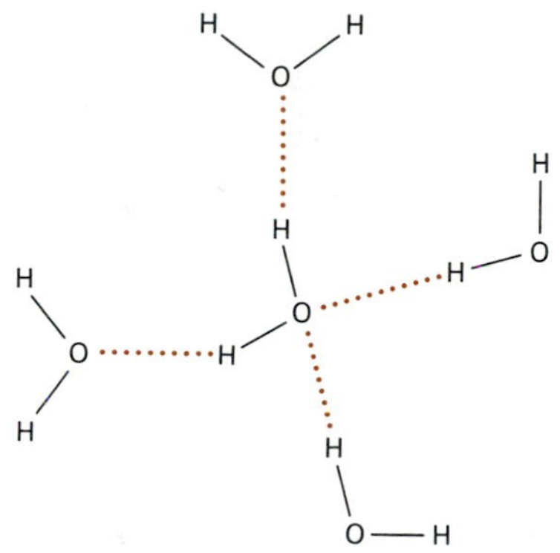

Figure 2–8 Hydrogen bonding of water molecules. Each water molecule tends to form hydrogen bonds with four neighboring water molecules. The hydrogen bonds are indicated by dotted lines. The covalent bonds between the hydrogen and oxygen atoms are represented by solid lines.

OTHER INTERACTIONS BETWEEN ATOMS

Two other types of interactions between molecules are van der Waals forces and hydrophobic interactions. These bonds, though weak, are important in maintaining the shape of many of the complex molecules in living cells. They are also important in holding together groups of nonpolar molecules, as, for example, in cell membranes.

The attractive forces between molecules called **van der Waals forces** occur when the molecules are very close together and are due to interaction of their electron clouds. Van der Waals forces are weaker and less specific than the other types of interactions we have considered. They are most important when they occur in large numbers, a situation that occurs when the shapes of the molecules involved permit close contact between the atoms.

Hydrophobic (water-hating) **interactions** occur between groups of nonpolar molecules. Such groups tend to cluster together and are insoluble in water. This clustering is a result of the hydrogen bonds that hold the water molecules together, and in a sense, drive the nonpolar molecules together. Hydrophobic interactions explain why oil tends to form globs when it is added to water.

Molecular Mass

The molecular mass of a compound is the sum of the atomic masses of its constituent atoms; thus, the molecular mass of water, H_2O, is (2 × 1 amu) + (16 amu), or 18 atomic mass units. (Owing to the presence of isotopes, atomic mass units are not whole numbers. However, for our purposes each atomic mass value has been rounded off to a whole number.) The molecular mass of the simple sugar glucose, $C_6H_{12}O_6$, which occurs very widely in plants and animals, is (6 × 12 amu) + (12 × 1 amu) + (6 × 16 amu), or 180 atomic mass units.

The amount of a compound whose mass in grams is equivalent to its molecular mass is termed one **mole.** (The mole can also refer to elements.) Thus one mole of glucose has a mass of 180 grams. A one-molar solution, represented by 1 *M*, contains one mole of the substance (e.g., 180 grams of glucose) in one liter of solution. Chemical compounds react with each other in quantitatively precise ways. For example, when glucose is burned in a fire or metabolized in a cell, one mole of glucose reacts with six moles of oxygen to form six moles of carbon dioxide (CO_2) and six moles of water.

$$\underset{\text{Glucose}}{C_6H_{12}O_6} + \underset{\text{Oxygen}}{6\ O_2} \longrightarrow \underset{\text{Carbon dioxide}}{6\ CO_2} + \underset{\text{Water}}{6\ H_2O} + \text{Energy}$$

The mole is a very useful unit because we cannot do experiments with individual atoms or molecules. A mole represents a very large number of units (6.02×10^{23}) that chemists can weigh and work with. Thus, in considering chemical reactions, it is important to think in terms of these units—moles.

Oxidation–Reduction

Rusting—the combination of iron with oxygen—is a familiar example of oxidation and reduction.

$$4\ Fe + 3\ O_2 \longrightarrow 2\ Fe_2O_3$$

Oxidation is defined as a chemical process in which an atom, ion, or molecule loses electrons. In rusting, iron is changed from its metallic state to its iron(III) (Fe^{3+}) state; it is being oxidized.

$$4\ Fe \longrightarrow 4\ Fe^{3+} + 12\ e^-$$

The e^- stands for electron. At the same time, oxygen is changed from its molecular state to its charged state:

$$3\ O_2 + 12\ e^- \longrightarrow 6\ O^{2-}$$

When oxygen accepts the electrons removed from the iron, it is being reduced. **Reduction** is a chemical process in which an atom, ion, or molecule gains electrons. Oxidation and reduction reactions occur simultaneously because one substance must accept the electrons that are removed from the other. Oxidation–reduction reactions are sometimes referred to as **redox reactions.**

Electrons are not easily removed from covalent compounds unless an entire atom is removed. In living cells, oxidation almost always involves the removal of a hydrogen atom from a compound; reduction often involves the addition of hydrogen.

Water and Its Properties

A large part of the mass of most organisms is simply water. In human tissues the percentage of water ranges from 20% in bones to 85% in brain cells. The water content is greater in embryonic and young cells and decreases as aging occurs. About 70% of our total body weight is water; as much as 95% of a jellyfish or certain plants is water. Water is not only the major component of organisms but also one of the principal environmental factors affecting them. Many organisms live within the sea or in freshwater rivers, lakes, and puddles. The physical and chemical properties of water have permitted living things to appear, to survive, and to evolve on this planet.

Water will dissolve many different kinds and great quantities of compounds. Because of its solvent properties and the tendency of the atoms in certain compounds to form ions when in solution, water plays an important role in facilitating chemical reactions. Water itself is a reactant or product in many chemical reactions that occur in living tissue. Water is also the source, through plant metabolism, of the oxygen in the air we breathe, and its hydrogen atoms are incorporated into the many organic compounds in the bodies of living things. Water is also an important lubricant. It is present in body fluids wherever one organ rubs against another and in joints where one bone moves on another.

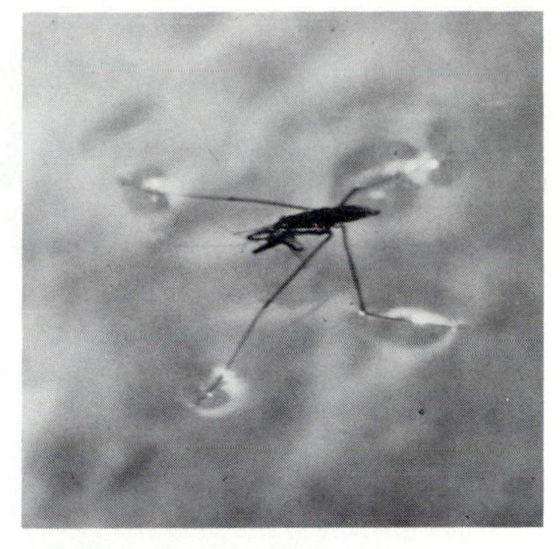

Figure 2–9 A water strider on the surface of a pond. Fine hairs at the ends of its legs spread its weight over a large area, allowing the body of the animal to be supported by the surface tension of the water.

COHESIVE AND ADHESIVE FORCES

Water exhibits both cohesive and adhesive forces. Water molecules have a very strong tendency to stick to each other; that is, they are **cohesive.** This is due to the hydrogen bonds among the molecules. Water molecules also stick to many other kinds of substances (i.e., those substances that have charged groups of atoms or molecules on their surfaces). These **adhesive** forces explain how water makes things wet. Water has a high degree of **surface tension** because of the cohesiveness of its molecules; its molecules have a much greater attraction for each other than for molecules in the air. Thus, water molecules at the surface crowd together, producing a strong layer as they are pulled downward by the attraction of other water molecules beneath them (Fig. 2–9).

Adhesive and cohesive forces account for the tendency, termed **capillary action,** of water to rise in very-fine-bore tubes (Fig. 2–10). Water also moves through the microscopic spaces between soil particles to the roots of plants by capillary action; and capillary action plays some part in the rise of water through the stems of plants to their leaves.

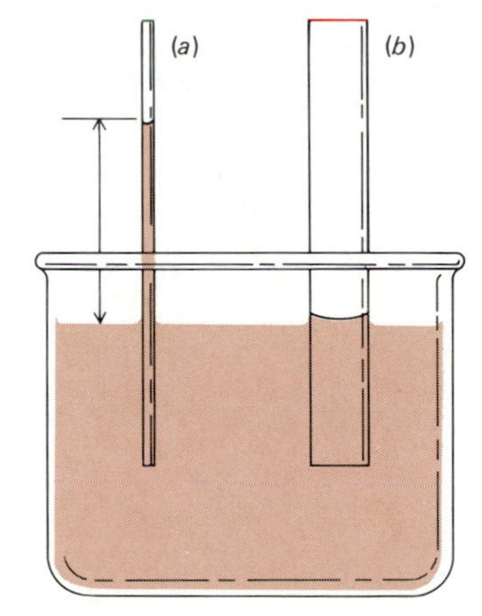

Figure 2–10 Capillary action in glass tubes of different diameters. (*a*) In the smaller tube, adhesive forces attract water molecules to charged groups on the surfaces of the tube. Other water molecules inside the tube are then "pulled along" by cohesive forces (hydrogen bonds between the water molecules). (*b*) In the large-diameter tube, a smaller percentage of the water molecules line the glass. Because of this the adhesive forces are not strong enough to overcome the cohesive forces of the water beneath the surface level of the container, and water in the tube rises only slightly.

TEMPERATURE STABILIZATION

Water has a high **specific heat;** that is, the amount of energy required to raise the temperature of water by one degree Celsius is quite large. The high specific heat of water results from the hydrogen bonding of its molecules. Raising the temperature of a substance involves adding heat energy to make its molecules move faster—to increase the kinetic energy of the molecules. Some of the hydrogen bonds holding the water molecules together must first be broken to permit the molecules to move more freely. Much of the energy added to the system is used up in breaking the hydrogen bonds, and only a portion of the heat energy is available to speed the movement of the water molecules (increase the temperature of the water).

Because so much heat loss or heat input is required to lower or raise the temperature of water, the oceans and other large bodies of water have relatively constant temperatures. Thus, many organisms living in the oceans are provided with a relatively constant environmental temperature. The high water content of plants and animals living on land helps them to maintain a relatively constant internal temperature. The rates of chemical reactions are greatly affected by temperature, generally doubling for each 10°C increase in temperature. The reactions of biological importance can take place only within a relatively narrow temperature range, and water helps to minimize temperature fluctuations.

Hydrogen bonds contribute another important property of water. Whereas most substances become more dense as the temperature decreases, water is most dense at 4°C and then begins to expand again (becoming less dense) as the temperature decreases further. This expansion occurs because its hydrogen bonds become more rigid and ordered. As a result, frozen water (ice) floats upon the denser cold water (Fig. 2–11). Note that the expansion of water takes place even before it actually freezes; this explains why a pond freezes from the surface down, rather than from the bottom up: As water temperature drops, the colder water—in the narrow temperature range of 0°C to 4°C, where it is less dense—rises to the pond surface; it then freezes to form a lid of ice. This ice insulates the water below from the wintry chill so that it is less likely to freeze. Organisms that inhabit the pond are able to survive the frigid winter below the icy surface.

Because its molecules are held together by hydrogen bonds, water has a high **heat of vaporization.** More than 500 calories are required to change a gram of liquid water into a gram of water vapor. A **calorie** is a unit of heat energy (defined as 4.184

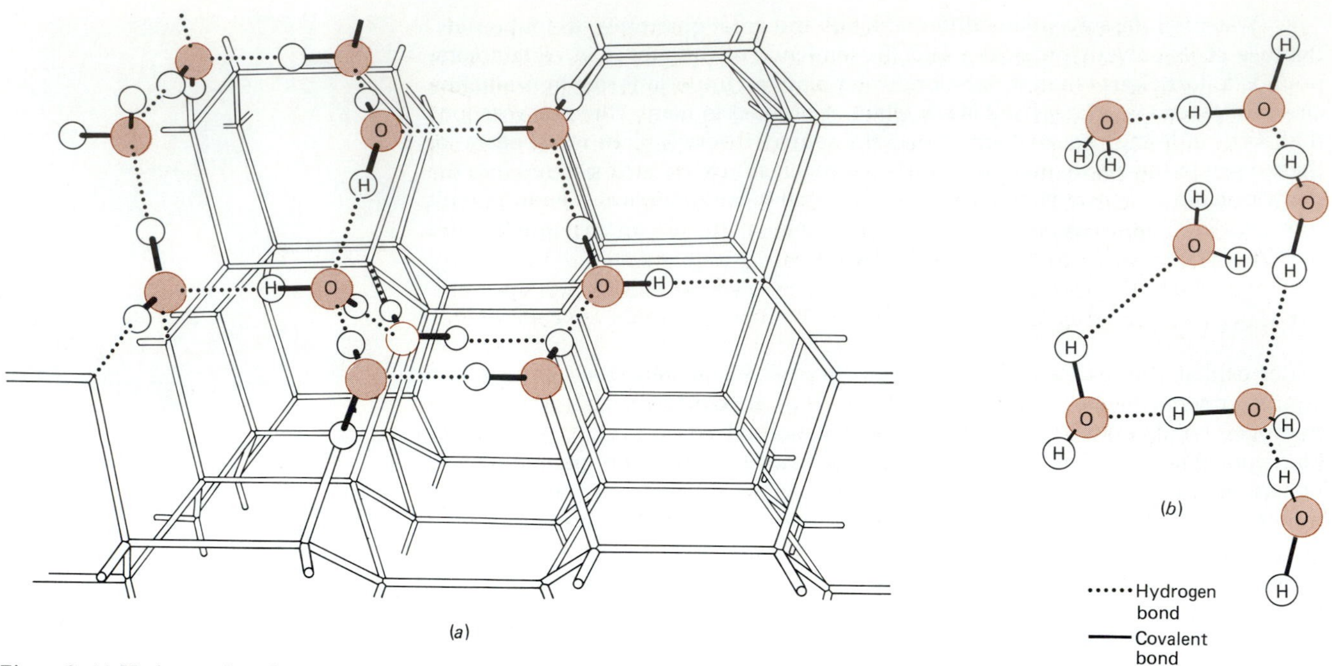

Figure 2–11 Hydrogen bonding in ice compared with that in liquid water. (*a*) Note the regular, evenly distanced hydrogen bonds in the hexagonal superstructure of ice. (*b*) When ice melts to produce liquid water, the crystal structure collapses. The hydrogen bonds occur less consistently and are of unequal length. (Redrawn permission of Arthur Geis)

joules) that equals the amount of heat required to raise the temperature of one gram of water one degree Celsius. It may surprise you to learn that a great deal of heat is liberated into the environment when liquid water changes to ice. These properties of water are crucial in stabilizing temperatures on earth. The quantity of water on the earth's surface is enormous; this large mass resists the warming effect of heat and cooling effect of low temperatures. Since water absorbs heat as it changes from a liquid to a gas, the human body can dissipate excess heat by the evaporation of sweat, and a leaf can keep cool in the bright sunlight by evaporating water from its surface. Water's high heat conductivity makes possible the even distribution of heat throughout the body.

IONIZATION

A further characteristic of water molecules is their slight tendency to **ionize,** that is to dissociate into ions: hydrogen ions (H^+) and hydroxide ions (OH^-). In pure water a very small number of water molecules form ions in this way. The tendency of water to dissociate is balanced by the tendency of hydrogen ions and hydroxide ions to reunite to form water:

$$HOH \rightleftharpoons H^+ + OH^-$$

Since water splits into one hydrogen ion and one hydroxide ion, the concentrations of hydrogen and hydroxide ions in pure water are exactly equal. Such a solution is said to be **neutral,** neither acidic nor basic (alkaline). The slight tendency of water molecules to form ions results in a concentration of hydrogen ions and of hydroxide ions of 0.0000001 (10^{-7}) mole per liter.

Acids and Bases

An **acid** is a substance that dissociates in solution to yield hydrogen ions (H^+)[1] and an anion.

$$\text{Acid} \longrightarrow H^+ + \text{Anion}$$

An acid is a proton *donor.* A **base** is defined as a proton *acceptor.* Most bases are

[1]The H^+ immediately combines with water, forming a hydronium ion (H_3O^+). However, by convention H^+, rather than the more accurate H_3O^+, is used.

substances that dissociate to yield a hydroxide ion (OH^-) and a cation when dissolved in water. The part of an acid remaining after the dissociation of the H^+ is termed the **conjugate base.** The addition of a proton to a base yields its **conjugate acid.** Acids turn blue litmus paper red and have a sour taste. Hydrochloric acid (HCl) and sulfuric acid (H_2SO_4) are inorganic acids (relatively small compounds that generally do not contain carbon). Lactic acid ($CH_3CHOHCOOH$) from sour milk and acetic acid (CH_3COOH) from vinegar are two common organic acids (more complex compounds containing carbon atoms). Bases turn red litmus paper blue and feel slippery to the touch. Sodium hydroxide (NaOH) and ammonium hydroxide (NH_4OH) are inorganic bases. In later chapters we will encounter a number of organic bases such as the purine and pyrimidine bases that are components of nucleic acids.

Acids and bases dissociate when dissolved in water, releasing H^+ ions and OH^- ions, respectively. When the concentration of hydrogen ions in the solution is greater than 0.0000001 *M* (10^{-7} *M*) the solution is acidic. When the concentration of hydrogen ions is less than 10^{-7} *M* (and the concentration of hydroxide ions is greater than 10^{-7} *M*) the solution is basic, or alkaline. In an aqueous solution, the concentration of hydrogen ions multiplied by the concentration of hydroxide ions equals 10^{-14}. In pure water at neutrality, both H^+ and OH^- ions are present in a concentration of 10^{-7} *M*, and the product $10^{-7} \times 10^{-7} = 10^{-14}$. In a solution containing 0.1 *M* HCl, the concentration of H^+ is 0.1 *M* or 10^{-1} *M*, and the concentration of OH^- is 10^{-13} *M*; $10^{-1} \times 10^{-13} = 10^{-14}$.

pH

Since the concentration of hydrogen ions is usually small, it is much more convenient to express the degree of acidity or alkalinity in a fluid in terms of **pH,** defined as the logarithm of the reciprocal of the hydrogen ion concentration, log $[1/(H^+)]$. The pH scale is thus a logarithmic one, extending from 0, the pH of a 1 *M* acid such as HCl, to 14, the pH of a 1 *M* base such as NaOH (Fig. 2–12). The hydrogen ion concentration of pure water is 10^{-7} *M*; its pH is 7.0. At pH 7.0 the concentrations of H^+ ions and OH^- ions are exactly equal. Since the scale is logarithmic, a solution with a pH of 6 has a hydrogen ion concentration 10 times greater than a solution with a pH of 7 and is much more acidic. A pH of 5 represents another tenfold increase, so a solution with a pH of 4 is 10 × 10 or 100 times more acidic than a solution with a pH of 6. Solutions with a pH of less than 7 are acidic and contain more H^+ ions than OH^- ions. Solutions with a pH greater than 7 are alkaline, or basic, and contain more OH^- ions than H^+ ions. The contents of most animal and plant cells are neither strongly acidic nor alkaline but are an essentially neutral mixture of acidic and basic substances. Any considerable change in the pH of the cell is incompatible with life (Fig. 2–13).

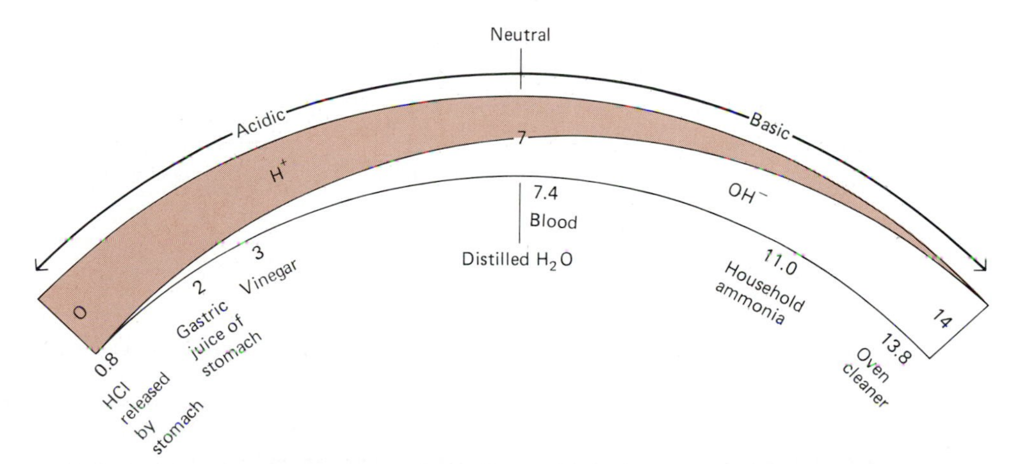

Figure 2–12 The pH scale. A solution with a pH of 7 is neutral because the concentrations of H^+ and OH^- are equal. The lower the pH below 7, the more H^+ ions are present, and the more acidic the solution is. As the pH increases above 7, the concentration of H^+ ions decreases and the concentration of OH^- increases, making the solution more alkaline (basic).

Figure 2–13 The effects of "acid rain." Sulfur oxides emitted from fossil fuel plants and industry, and nitrogen oxides, mainly from automobile exhaust, are converted in the moist atmosphere into acids of, respectively, sulfur and nitrogen (e.g., sulfurous and nitrous acid). These acids are dispersed over wide areas by air flow patterns in the atmosphere. Whereas the pH of unpolluted rain averages 5.6, in some parts of the United States and Canada, the pH of rain has been measured at 4.2 and *lower*. Most fish species die at a pH of 4.5 to 5.0. Acid rain also affects vegetation. The roots of this spruce tree have withered and died, while the rest of the plant has been deprived of nutrients and has suffered greatly reduced efficiency in its photosynthesis. (Townsend P. Dickinson, Photo Researchers, Inc.)

BUFFERS

Many homeostatic mechanisms operate to maintain appropriate pH values. For example, the pH of human blood is about 7.4 and must be maintained within very narrow limits. Should the blood become too acidic, coma and death may result; excessive alkalinity can result in overexcitability of the nervous system and even convulsions.

A **buffer** is a substance or combination of substances that resists changes in pH when acids or bases are added. The buffer either accepts or donates hydrogen ions. A buffer consists of a weak acid and its conjugate base, or a weak base and its conjugate acid. One of the most common buffering systems, and one that is important in human blood, is carbonic acid and the bicarbonate ion. Bicarbonate ions are formed in the body as follows:

$$\underset{\text{Carbon dioxide}}{CO_2} + \underset{\text{Water}}{H_2O} \rightleftharpoons \underset{\text{Carbonic acid}}{H_2CO_3} \rightleftharpoons H^+ + \underset{\text{Bicarbonate ion}}{HCO_3^-}$$

As indicated by the arrows, the reactions are reversible.

When excess hydrogen ions are present in blood or other body fluids, bicarbonate ions combine with them to form carbonic acid, a weak acid.

$$H^+ + HCO_3^- \rightleftharpoons \underset{\text{Carbonic acid}}{H_2CO_3}$$

The carbonic acid is unstable and quickly breaks down into carbon dioxide and water.

Buffers also work to maintain pH when hydroxide ions are added. A buffer may release hydrogen ions, which combine with the hydroxide ions to form water.

$$OH^- + H_2CO_3 \longrightarrow HCO_3^- + H_2O$$

Salts

When an acid and a base are mixed together, the H^+ of the acid unites with the OH^- of the base to form a molecule of water. The remainder of the acid (anion) combines with the remainder of the base (cation) to form a salt. Hydrochloric acid reacts with sodium hydroxide to form water and sodium chloride:

$$HCl + NaOH \longrightarrow H_2O + NaCl$$

A **salt** may be defined as a compound in which the hydrogen atom of an acid is replaced by some other cation. A salt contains a cation other than H^+ and an anion

TABLE 2–2
Some Biologically Important Ions

Name	*Formula*	*Charge*
Sodium	Na^+	1+
Potassium	K^+	1+
Hydrogen	H^+	1+
Magnesium	Mg^{2+}	2+
Calcium	Ca^{2+}	2+
Iron	Fe^{2+} or Fe^{3+}	2+ [iron(II)] or 3+ [iron(III)]
Ammonium	NH_4^+	1+
Chloride	Cl^-	1−
Iodide	I^-	1−
Carbonate	CO_3^{2-}	2−
Bicarbonate	HCO_3^-	1−
Phosphate	PO_4^{3-}	3−
Acetate	CH_3COO^-	1−
Sulfate	SO_4^{2-}	2−
Hydroxide	OH^-	1−
Nitrate	NO_3^-	1−
Nitrite	NO_2^-	1−

other than OH^-. Sodium chloride, NaCl, is a compound in which the hydrogen ion of HCl has been replaced by the cation Na^+.

When a salt, an acid, or a base is dissolved in water, its dissociated charged particles can conduct an electric current; these substances are called **electrolytes.** Sugars, alcohols, and many other substances do not form ions when dissolved in water; they do not conduct an electrical current and are termed **nonelectrolytes.**

Cells and extracellular fluids (such as blood) of plants and animals contain a variety of dissolved salts, which include many important mineral ions. Such ions are essential for fluid balance, acid–base balance, and, in animals, nerve and muscle function, blood clotting, bone formation, and many other aspects of body function. Sodium, potassium, calcium, and magnesium are the chief cations present, and chloride, bicarbonate (HCO_3^-), phosphate (PO_4^{3-}), and sulfate (SO_4^{2-}) are important anions (Table 2–2).

The body fluids of terrestrial animals differ considerably from sea water in their total salt content. However, they resemble sea water in the kinds of salts present and in their relative abundance. The total concentration of salts in the body fluids of most invertebrate marine animals is equivalent to that in sea water, about 3.4%. Vertebrates, whether terrestrial, freshwater, or marine, have less than 1% salt in their body fluids.

Most biologists believe that life originated in the sea. The cells of those early organisms became adapted to function optimally in the presence of this pattern of salts. As larger animals evolved and developed body fluids, this pattern of salts was retained, even when some of their descendants migrated into fresh water or onto land. Some animals have evolved kidneys and other organs, such as salt glands, that selectively retain or secrete certain ions, thus resulting in body fluids with somewhat different relative concentrations of salts. The concentration of each ion is determined by the relative rates of its uptake and excretion by the organism.

Although the concentration of salts in cells and body fluids of plants and animals is small, this amount is of great importance for normal cell function. The concentrations of the respective cations and anions are kept remarkably constant under normal conditions. Any marked change results in impaired cellular functions and ultimately in death.

SUMMARY

I. The chemical composition and metabolic processes of all living things are very similar; the physical and chemical principles that govern nonliving things also govern living systems.

II. An element is a substance that cannot be decomposed into simpler substances by chemical reactions.

A. The matter of the universe is composed of 92 elements ranging from hydrogen, the lightest, to uranium, the heaviest.

B. Six elements—carbon, hydrogen, oxygen, nitrogen, phosphorus, and calcium—make up about 98% of an organism's content by weight.

III. Atoms are composed of a nucleus containing protons and neutrons and a cloud of electrons around the nucleus in characteristic energy levels and orbitals.
  A. Atoms of the same element that contain the same number of protons but different numbers of neutrons, and therefore have different mass numbers, are called isotopes.
  B. In a neutral atom, the number of protons equals the number of electrons, so the atom has no net electrical charge.

IV. Atoms are joined by chemical bonds to form larger, more complex structures called compounds.
  A. Covalent bonds are strong, stable chemical bonds formed when atoms share electrons forming molecules.
    1. Covalent bonds are nonpolar if the electrons are shared equally between the two atoms.
    2. Covalent bonds are polar if one atom has a greater affinity for electrons than that of the other.
  B. An ionic bond is formed when one atom donates electrons to another. An ionic compound is made up of positively charged ions (cations) and negatively charged ions (anions).
  C. Hydrogen bonds are relatively weak bonds formed when a hydrogen atom in one molecule is attracted to a highly electronegative element such as oxygen or nitrogen in another molecule or in another part of the same molecule.

V. The molecular mass of a compound is the sum of the atomic masses of its constituent atoms.

VI. Oxidation is a chemical process in which a substance loses electrons; reduction is a chemical process in which a substance gains electrons.

VII. Water accounts for a large part of the mass of most organisms.
  A. Water molecules are cohesive owing to the hydrogen bonding between the molecules, and they also adhere to many kinds of substances.
  B. Water has a high degree of surface tension because of the cohesiveness of its molecules.
  C. Water has a high specific heat, which helps organisms to maintain a relatively constant internal temperature; this property also helps to keep the oceans and other large bodies of water at a constant environmental temperature.
  D. Other important properties of water include its high heat of vaporization, its slight tendency to form ions, and its ability to dissolve many different kinds of compounds.

VIII. An acid is a substance that dissociates in solution to yield hydrogen ions and an anion. Acids are proton donors. Bases are proton acceptors. A base generally dissociates in solution to yield hydroxide ions.
  A. The pH scale extends from 0 to 14, with 7 indicating neutrality. As the pH decreases below 7, the solution is more acidic. As a solution becomes more basic (alkaline), its pH increases from 7 toward 14.
  B. An acid and its conjugate base or a base and its conjugate acid can act as a buffer to resist changes in the pH of a solution when acids or bases are added.

IX. A salt is a compound in which the hydrogen atom of an acid is replaced by some other cation. Salts are the source of many mineral ions that are essential for fluid balance, nerve and muscle function, and many other body functions.

POST-TEST

1. The six elements that make up some 98% of the mass of most organisms are ______________, ______________, ______________, ______________, ______________, and ______________.
2. The chemical symbol for carbon is ______________; for hydrogen, ______________; and for oxygen, ______________.
3. Elements such as cobalt, present in minute amounts in living things, are referred to as ______________ ______________.
4. The three major types of subatomic particles are ______________, ______________, and ______________.
5. Particles with a negative electric charge and an extremely small mass are termed ______________.
6. The number of protons in the nucleus, called the ______________ ______________, is written as a subscript to the left of the chemical symbol.
7. The sum of the protons and the neutrons in the nucleus of the atom, termed the ______________ ______________, is indicated by a superscript to the left of the chemical symbol.
8. Atoms of the same element containing the same number of protons but different numbers of neutrons are termed ______________.
9. Electrons move about the central nucleus of the atom in characteristic regions termed ______________.
10. Each orbital may contain at most ______________ electrons.
11. The tendency of the negatively charged electrons to fly off into space is countered by their attraction to the atomic nucleus due to the ______________ charge of the protons in the nucleus.
12. The atoms of a few elements, such as ______________ and ______________, have a complete outermost shell of electrons; these are called ______________ ______________.
13. The attraction holding two atoms together is called a ______________ ______________.
14. Electrically charged atoms are called ______________.
15. Positively charged atoms are termed ______________, and negatively charged atoms are termed ______________.

Match the terms in Column A with their definitions in Column B.

Column A	*Column B*
16. Covalent bond	a. Electrons in the outer orbit that determine how many electrons an atom can donate, receive, or share
17. Hydrogen bond	b. The combination of two or more atoms joined by covalent chemical bonds
18. Ionic bond	c. Substances participating in a reaction
19. Molecular mass	d. Substances produced in a chemical reaction
20. Molecule	e. Transfer of an electron from an electron donor to an acceptor and the binding together of two particles of opposite charge
21. Products	f. Atoms joined by the sharing of electrons between them
22. Reactants	g. Weak bond that holds water molecules together
23. Valence electrons	h. Sum of the atomic masses of the constituent atoms

24. Water tends to rise in very-fine-bore tubes, a phenomenon termed ______________ ______________.

25. The amount of energy required to change one gram of liquid water to one gram of water vapor is termed the ______________ ______________ of water.

26. The logarithm of the reciprocal of the hydrogen ion concentration is termed ______________.

27. An acid is a proton ______________; a base is a proton ______________.

28. A solution of a weak acid and its conjugate base is called a ______________.

REVIEW QUESTIONS

1. Distinguish between:
   a. an atom and an element
   b. a molecule and a compound
   c. an atom and an ion
2. How do isotopes of the same element differ? What is a radioisotope?
3. Contrast electrons with protons and with neutrons.
4. Compare ionic and covalent bonds, and give specific examples of each.
5. Write a chemical equation depicting the hydration of:
   a. sodium chloride
   b. calcium chloride
6. What properties of water make it an essential component of living matter?
7. How would a solution with a pH of 5 differ from one with a pH of 9? from one with a pH of 7?
8. Why are buffers important in living organisms? Give a specific example of how a buffer system works.
9. Differentiate clearly between acids, bases, and salts. What are the functions of salts in living organisms?
10. Why must oxidation and reduction occur simultaneously?
11. Describe a reversible reaction that is at equilibrium.
12. What are valence electrons? What is their significance?
13. What are hydrogen bonds? What is their significance?
14. How is each of the following determined?
    a. atomic number
    b. molecular mass

3
Biologically Important Molecules

OUTLINE

I. Carbohydrates
- *A.* Monosaccharides
- *B.* Disaccharides
- *C.* Polysaccharides
- *D.* Modified and complex carbohydrates

II. Lipids
- *A.* Neutral fats
- *B.* Phospholipids
- *C.* Carotenoids
- *D.* Steroids

III. Proteins
- *A.* Subunit structure
- *B.* Protein structure: levels of organization
- *C.* Functions of proteins

IV. Nucleic acids
- *A.* Subunit structure
- *B.* Related nucleotides

LEARNING OBJECTIVES

After you have read this chapter you should be able to:

1. List the principal groups of organic compounds present in organisms, and describe the characteristics and biological roles of each.
2. Distinguish among monosaccharides, disaccharides, and polysaccharides, and discuss the monosaccharides and polysaccharides of major importance in living systems.
3. List the several kinds of lipids found in living organisms, and describe their composition, characteristics, and biological roles.
4. Describe the chemical structures, properties, and functions of proteins and amino acids.
5. Describe the levels of organization of protein molecules.
6. Describe the structure of the nucleic acids, DNA and RNA, and distinguish between nucleotides and nucleosides.
7. Describe the roles in metabolism of the nucleotides ATP and NAD.

The types of substances present, and even their relative proportions, are strikingly similar in cells from various parts of the body and in cells from different organisms. Both a human liver cell and a unicellular amoeba contain about 80% water, 12% protein, 2% nucleic acid, 5% lipid, 1% carbohydrate, and a fraction of other substances. We have already considered the importance of water and other **inorganic compounds** in Chapter 2. Proteins, nucleic acids, lipids, and carbohydrates are all **organic compounds,** complex compounds that contain the element carbon[1] (Table 3–1). In this chapter we will focus on these major groups of organic compounds and their importance in living organisms. (Vitamins, another class of organic compounds, will be discussed in Chapter 34.)

Some organic compounds serve as structural components of cells or tissues; others are metabolized to supply energy for the cell's functions; still others regulate the rates and directions of metabolic processes within the cell. Because of the properties of carbon an immense variety of organic compounds is possible. With four electrons in its outer energy level, the carbon atom can share electrons with other atoms and form four covalent bonds. Carbon atoms share electrons not only with other elements (typically, hydrogen, oxygen, and nitrogen) but also with other carbon atoms to form chains of varying lengths. The chains may be unbranched or branched, or the carbon atoms may join to form rings. Adjacent carbon atoms may form single bonds, or by sharing additional pairs of electrons, they may form double (—C=C—) bonds or triple (—C≡C—) bonds.

Carbohydrates

Sugars, starches, and celluloses are typical **carbohydrates.** Sugars and starches are important sources of fuel for living cells, whereas celluloses are structural components of plants. Carbohydrates contain carbon, hydrogen, and oxygen atoms in a ratio of approximately one carbon : two hydrogens : one oxygen. Carbohydrates may be classified as monosaccharides, disaccharides, or polysaccharides.

MONOSACCHARIDES

Monosaccharides are simple sugars that usually contain from three to six carbon atoms. **Glucose** (also called dextrose) and **fructose** (also called levulose) are examples of monosaccharides. Each is composed of a single hexose unit (which consists of six carbon atoms) having the formula $C_6H_{12}O_6$. Compounds that have identical molecular formulas but different arrangements of atoms, such as glucose and fructose, are termed **isomers.** Their different atom arrangements give the two sugars different chemical properties. Glucose is the most abundant hexose in the bodies of humans and other animals. Other carbohydrates that we eat are converted to glucose in the liver. The concentration of glucose in the blood is kept at a constant level even though the cells continuously remove glucose from the blood for use as an energy source. The pentoses (five-carbon sugars) ribose and deoxyribose are components of nucleotides and nucleic acids.

The arrangement of atoms in molecules may be shown by structural formulas (Fig. 3–1) in which the atoms are represented by their symbols—C, H, O, and so forth—and the chemical bonds are indicated by connecting lines. Recall that hydrogen can form one bond (corresponding to its single electron) with other atoms, oxygen can form two bonds, and carbon, four bonds.

Actually, molecules are not the simple two-dimensional structures depicted on a printed page. In fact, the properties of a compound depend in part on its specific three-dimensional structure. Molecules of glucose and other monosaccharides in solution do not exist predominantly as the extended straight carbon chains shown in Figure 3–1; instead, they exist mainly as rings formed when a chemical bond connects carbon-1 to the oxygen attached to carbon-5. For example, glucose and fructose in solution typically have the ring configurations shown in Figure 3–2. In solution these sugars are found as mixtures of open-chain forms, α (alpha)

[1] Some simple carbon compounds, including those related to carbonic acid and its salts, to the oxides of carbon (such as CO_2), or to the cyanides, are considered inorganic.

TABLE 3–1
Some of the Major Groups of Organic Compounds Important in Body Function

Class of Compound	*Component Elements*	*Description*	*How to Recognize*	*Principal Functions in Body*
Carbohydrates	C, H, O	Contain approximately 1 C:2 H:1 O (but make allowance for loss of oxygen and hydrogen when sugar units are linked).	Count the carbons, hydrogens, and oxygens.	Cellular fuel; energy storage; structural component of plant cell walls; component of other compounds such as nucleic acids and glycoproteins.
		1. Monosaccharides (simple sugars)—mainly 5-carbon (pentose) molecules like ribose or 6-carbon (hexose) molecules such as glucose and fructose.	Look for the ring shapes:  hexose or pentose	Cellular fuel; components of other compounds
		2. Disaccharides—2 sugar units linked by a glycosidic bond, e.g., maltose, sucrose.	Count sugar units.	Components of other compounds
		3. Polysaccharides—many sugar units linked by glycosidic bonds, e.g., glycogen, cellulose.	Count sugar units.	Energy storage; structural components of plant cell walls.
Lipids	C, H, O	Contain relatively less oxygen relative to carbon and hydrogen than do carbohydrates.		Energy storage; cellular fuel, structural components of cells; thermal insulation.
		1. Neutral fats. Combination of glycerol with from 1–3 fatty acids. Monoacylglycerol contains 1 fatty acid; diacylglycerol contains 2 fatty acids; triacylglycerol contains 3 fatty acids. If fatty acids contain double carbon-to-carbon linkages (C=C) they are unsaturated; otherwise they are saturated.	Look for glycerol at one end of molecule: H H—C— H—C— H—C— H	Cellular fuel; energy storage.
		2. Phospholipids. Composed of glycerol attached to one or two fatty acids and to an organic base containing phosphorus.	Look for glycerol and side chain containing phosphorus and nitrogen.	Components of cell membranes
		3. Steroids. Complex molecules containing carbon atoms arranged in four interlocking rings (three rings contain six carbon atoms each and the fourth ring contains five)	Look for four interlocking rings: 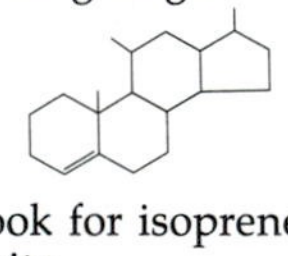	Some are hormones; others include cholesterol, bile salts, vitamin D.
		4. Carotenoids. Red and yellow pigments; consist of isoprene units.	Look for isoprene units.	Retinal (important in photoreception) and vitamin A are formed from carotenoids.
Proteins	C, H, O, N, usually S	One or more polypeptides (chains of amino acids) coiled or folded in characteristic shapes.	Look for amino acid units joined by C—N bonds.	Serve as enzymes; structural components; muscle proteins; hemoglobin.
Nucleic acids	C, H, O, N, P	Backbone composed of alternating pentose and phosphate groups, from which nitrogenous bases project. DNA contains the sugar deoxyribose and the bases, guanine, cytosine, adenine, and thymine. RNA contains the sugar ribose, and the bases guanine, cytosine, adenine, and uracil. Each molecular subunit, called a nucleotide, consists of a pentose, a phosphate, and a nitrogenous base.	Look for a pentose-phosphate backbone. DNA forms a double helix.	Storage, transmission, and expression of genetic information.

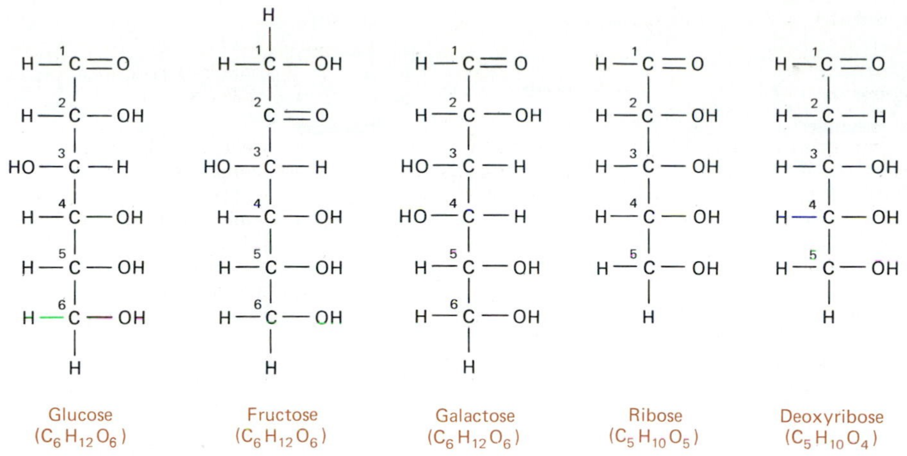

Figure 3–1 Structural formulas of some important monosaccharides (simple sugars). The monosaccharides are represented here as straight chains, called stick formulas. Although it is convenient to show monosaccharides in this form, they are more accurately depicted as ring structures, as shown in Figure 3–2. Corresponding carbon atoms have been numbered in the stick formulas and ring structures. Note that glucose, fructose, and galactose are structural isomers—that is, they have the same chemical formula, $C_6H_{12}O_6$, but their atoms are arranged differently.

forms, and β (beta) forms. In the α form the hydroxyl group of carbon-1 of glucose (C-2 of fructose) is *below* the plane of the ring. In the β form the hydroxyl group is *above* the plane of the ring.

DISACCHARIDES

A **disaccharide** (two sugars) is a carbohydrate that can be split into two monosaccharides. The disaccharide **maltose** (malt sugar) consists of two chemically combined glucose molecules. **Sucrose,** the sugar we use to sweeten our foods, consists of glucose combined with fructose. **Lactose,** the sugar present in milk, is composed of one molecule of glucose and one of galactose, another hexose monosaccharide. The covalent bond that joins the two monosaccharide units in a disaccharide is called a **glycosidic bond.** An α-1,4-glycosidic bond forms when the hydroxyl group on carbon-1 is in the α position. A β-glycosidic bond forms when the hydroxyl group is in the β-position on C-1 as the linkage forms.

During digestion, maltose is degraded to form two molecules of glucose:

Maltose + Water ⟶ Glucose + Glucose

Similarly, sucrose is degraded during digestion to form glucose and fructose:

Sucrose + Water ⟶ Glucose + Fructose

Structural formulas for the compounds in these reactions are shown in Figure 3–3. Because water is added during the breakdown of a disaccharide, this type of reaction is referred to as a **hydrolysis** reaction.

POLYSACCHARIDES

The most abundant carbohydrates are polysaccharides such as starches, glycogen, and celluloses. A **polysaccharide** is a single long chain, or a branched chain, consisting of repeating units of a simple sugar, usually glucose (Fig. 3–4). The precise

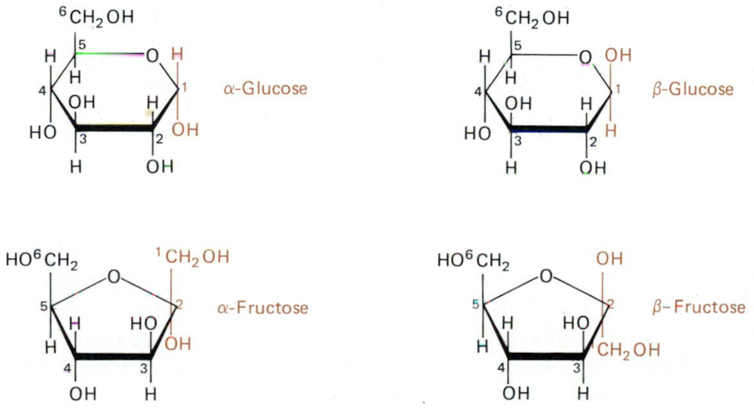

Figure 3–2 Two monosaccharides, glucose and fructose, drawn to represent their ring structures. At each angle in the ring is a carbon atom; its presence is understood by convention. Each number on the ring or on an attached group corresponds to the numbered carbons in the stick diagrams of glucose and fructose shown in Figure 3–1. The thick, tapered bonds in the lower portion of each ring indicate that the molecule is a three-dimensional structure. The thickest portion of the bond is interpreted as being the part of the molecule "nearest" the viewer. A solution of a sugar such as glucose or fructose may be regarded as a mixture of α, β, and straight chain forms. (Note also that $HOCH_2$ is sometimes used for convenience instead of CH_2OH.)

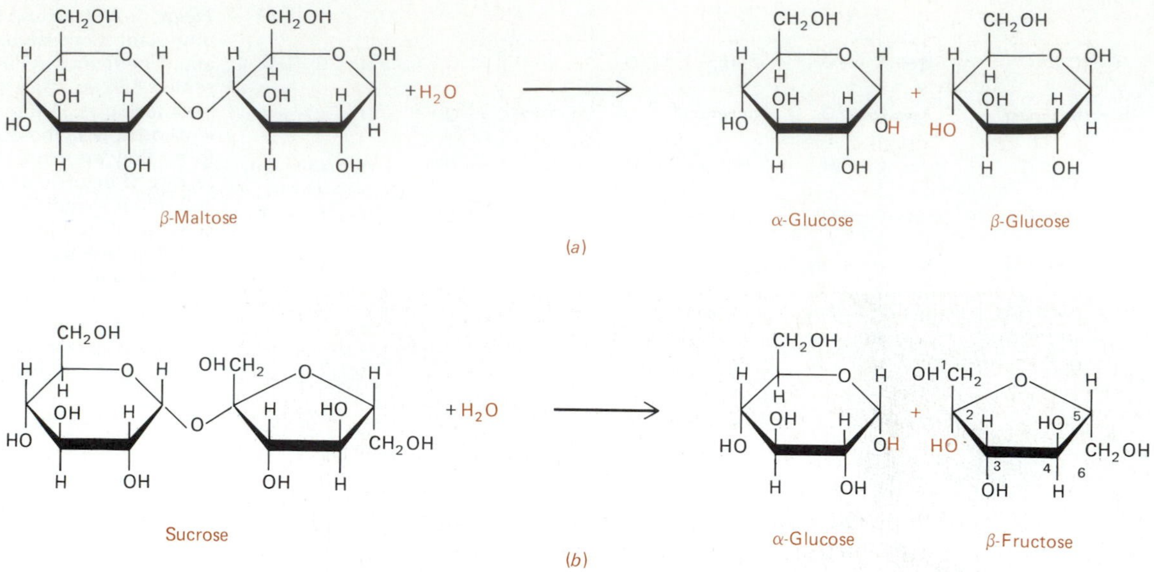

β-Maltose $+ H_2O \rightarrow$ α-Glucose + β-Glucose

(a)

Sucrose $+ H_2O \rightarrow$ α-Glucose + β-Fructose

(b)

Figure 3–3 A disaccharide can be degraded to yield two monosaccharide units. (*a*) Maltose may be broken down to form two molecules of glucose. This is a hydrolysis reaction that requires the addition of water. (*b*) Sucrose can be split to yield a molecule of glucose and a molecule of fructose. The carbons in the β-fructose molecule are numbered to show that, although "reversed," this is the same molecule as in Figure 3–2.

number of sugar units present varies, but typically thousands of units are present in a single molecule of a polysaccharide.

Starch is the storage form of carbohydrate in plants, whereas **glycogen** (sometimes referred to as animal starch) is the form in which glucose is stored in animals. Glycogen is a highly branched polysaccharide that is more water-soluble than plant starch. Glucose cannot be stored as such, for its small, readily soluble molecules would leak out of the cells. The larger, less soluble starch and glycogen molecules do not readily pass through the cell membrane. Thus, instead of storing simple sugars, cells store the more complex polysaccharides such as glycogen, which can be readily broken down into simple sugars as needed. In humans and other mammals, glycogen is stored especially in liver and muscle cells. Liver glycogen can readily be converted back to glucose; between meals, the liver releases this glucose into the blood, thus maintaining the concentration of glucose in the blood relatively constant.

Carbohydrates are the most abundant group of organic compounds on earth, and **cellulose** (Fig. 3–4(*c*)) is the most abundant carbohydrate, accounting for 50% or more of all the carbon in plants. Wood is about half cellulose; cotton is at least 90% cellulose. Plant cells are surrounded by a strong supporting cell wall consisting mainly of cellulose. As the principal structural component of the plant, cellulose is responsible for the rigidity of the plant body. Cellulose is an insoluble polysaccharide composed of many glucose molecules joined together. The β-glycosidic bonds joining these sugar units are different from the α-glycosidic bonds in starch and are not split by the enzyme that cleaves the bonds in starch.

Another important polysaccharide is **chitin,** the main component of the external skeletons of insects, crayfish, and other arthropods. It is also found in the cell walls of fungi. Chitin is a tough modified polysaccharide.

MODIFIED AND COMPLEX CARBOHYDRATES

Many *derivatives* of monosaccharides are important biological compounds. The amino sugars glucosamine and galactosamine are compounds in which an alcohol group (—OH) is replaced by an amino group (—NH_2) (see the section on proteins; Fig. 3–5; and Focus on Some Functional Groups Important in Biochemistry). Glucosamine is the molecular unit found in chitin; galactosamine is present in cartilage. Carbohydrates may also be combined with proteins to form glycoproteins or with lipids to form glycolipids. Such complex carbohydrates serve as structural components of cells and cell walls.

Cellulose

(a)

(b)

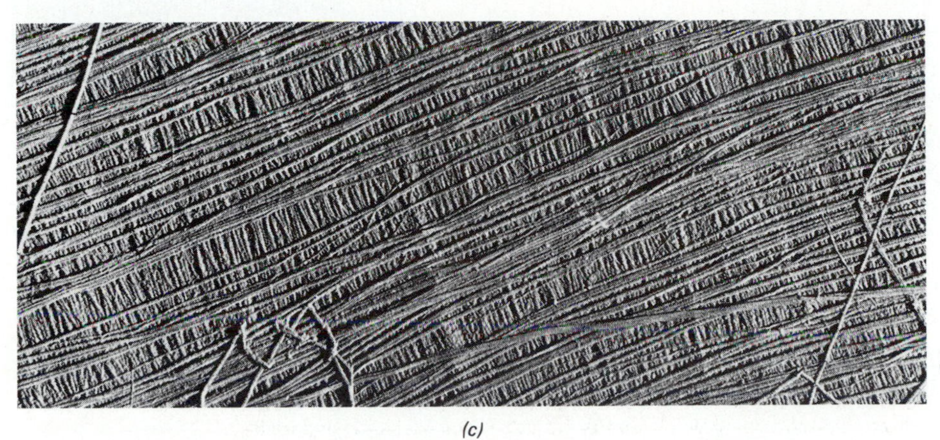

(c)

Glycogen or starch

(d)

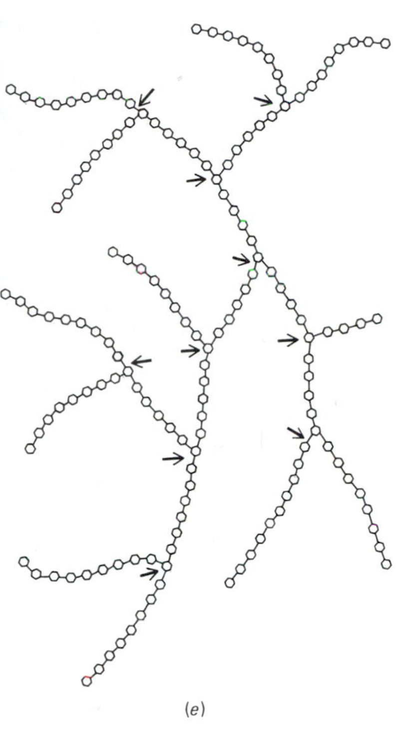

(e)

(f)

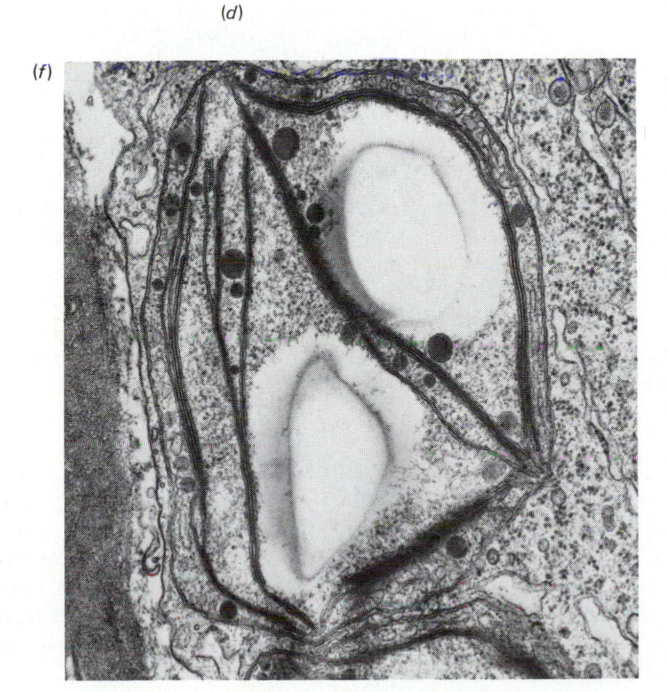

Figure 3–4 Some common polysaccharides. (*a*) The structure of cellulose. The cellulose molecule is an unbranched polysaccharide composed of approximately 10,000 glucose units joined by glycosidic bonds. (*b*) A more diagrammatic representation of cellulose structure. Each hexagon represents a glucose molecule bound by a glycosidic bond to the adjacent glucose molecule. (*c*) An electron micrograph of cellulose fibers from the cell wall of a marine alga (approximately ×24,000). (*d*) Molecular structure of glycogen or starch. These molecules are branched polysaccharides composed of glucose molecules joined by a slightly different type of glycosidic bonds. At the branch points there are bonds between carbon 6 of a glucose in the straight chain and carbon 1 of the glucose in the branching chain. Glycogen is more highly branched than starch. (*e*) Diagrammatic representation of starch or glycogen. The arrows represent the branch points. (*f*) An electron micrograph showing starch stored in a specialized organelle, a leukoplast, of an algal cell (approximately ×50,000). ((*c*), Omikron, Photo Researchers, Inc.); (*f*), Biophoto Associates, Photo Researchers, Inc.)

N-acetyl glucosamine

Chitin

(*a*)

(*b*)

Figure 3–5 (*a*) The amino sugar *N*-acetyl glucosamine (NAG) is found in bacterial cell walls and in chitin. Chitin consists of NAG units joined by glycosidic bonds. (*b*) Chitin is an important component of the exoskeleton of arthropods, such as the cicada. After spending 17 years in a subterranean larval stage, this insect crawls out upon a tree trunk or other support and then sheds its old exoskeleton, shown here, and flies off.

FOCUS ON

Some Functional Groups Important in Biochemistry

A **functional group** is a grouping of atoms that occurs commonly and that behaves the same no matter what molecule it is part of. On the basis of their functional groups, organic compounds can be assigned to families. In the following formulas, R represents the remainder of the molecule of which the functional group is a part.

Name	*Structural Formula*	*Example*
Alcohol	R—OH	H—C(H)(H)—C(H)(H)—OH Ethanol (the alcohol contained in beverages)
Amino group	R—NH_2	R—C(NH_2)—COOH Amino acid
Carboxyl group	R—COOH	R—C(NH_2)—COOH Amino acid
Methyl group	R—CH_3	H—C(H)(H)—OH Methanol (wood alcohol)
Phosphate group	R—O—P(=O)—OH	HO—P(=O)(OH)—O—R Phosphate ester (as found in ATP)
Ketone group	R—C(=O)—R	H—C(H)(H)—C(=O)—C(H)(H)—H Acetone
Aldehyde	R—C(=O)—H	H—C(H)=O Formaldehyde
Ester	R—C(=O)—O—R	H—C(H)(H)—C(=O)—O—C(H)(H)—H Methyl acetate
Sulfhydryl	R—SH	Found in proteins

Lipids

Lipids are a heterogeneous group of compounds that are soluble in nonpolar organic solvents, such as ether or benzene, but are relatively insoluble in water. They have a greasy or oily consistency. Like carbohydrates, lipids are composed of carbon, hydrogen, and oxygen atoms, but they have relatively less oxygen in proportion to the carbon and hydrogen than carbohydrates do. Among the groups of lipids especially important biologically are the neutral fats, phospholipids, carotenoids, steroids, and waxes. Some lipids are important biological fuels; others serve as structural components of cell membranes.

NEUTRAL FATS

The most abundant lipids in living things are the **neutral fats.** These compounds yield more than twice as much energy per gram as do carbohydrates and are an economical form for the storage of food reserves. Carbohydrates and proteins can be transformed by enzymes into fats and stored within the cells of adipose tissue. The layer of adipose tissue just under the skin serves as an insulator against the loss of body heat. An important adaptation of whales, which live in cold water and have no insulating hair, is an especially thick layer of fat (blubber) under the skin. In humans and other mammals the subcutaneous (that is, beneath the skin) fat keeps the skin firm in addition to restricting the loss of body heat.

A neutral fat consists of glycerol joined to one, two, or three molecules of a fatty acid. **Glycerol** is a three-carbon alcohol that contains three —OH groups (Fig. 3–6). A **fatty acid** is a long straight chain of carbon atoms with a carboxyl group (—COOH) at one end. About 30 different varieties of fatty acids are commonly found in animal lipids. They typically have an even number of carbon atoms. For example, butyric acid, present in rancid butter, has four carbon atoms, and oleic acid, the most widely distributed fatty acid in nature, has 18 carbon atoms.

Saturated fatty acids contain the maximum possible number of hydrogen atoms, while **unsaturated fatty acids** contain carbon atoms that are doubly bonded with one another and are not fully saturated with hydrogens (Fig. 3–6). Fatty acids with several double bonds are called **polyunsaturated fatty acids.** Fats containing unsaturated fatty acids are the **oils,** most of which are liquid at room temperature. Fats containing saturated fatty acids are solids at room temperature. For mammals, at least two fatty acids (linoleic acid and linolenic acid) are **essential** nutrients. **Essential nutrients** are nutrients that are necessary for metabolism but cannot be manufactured in the body from other substances; they must therefore be included in the diet.

When a glycerol molecule combines chemically with one fatty acid, a **monoacylglycerol** (sometimes called a monoglyceride) is formed. When two fatty acids combine with a glycerol, a **diacylglycerol** (or diglyceride) is formed; when three fatty acids combine with one glycerol molecule, a **triacylglycerol** (or triglyceride) is formed. In combining with glycerol, the carboxyl end of the fatty acid attaches to one of the —OH groups. (In the overall reaction that produces a fat, the *equivalent* of a molecule of water is removed from the glycerol and the fatty acid. However, the H^+ and the OH^- are removed from the reactants in separate steps and do not necessarily combine as H_2O when the reaction is complete. During digestion, the neutral fats are hydrolyzed to their constituent fatty acids and glycerol.)

PHOSPHOLIPIDS

Phospholipids are important constituents of cell membranes. A **phospholipid** consists of a glycerol molecule attached to one or two fatty acids, but the glycerol is also bonded to a phosphate group and to an organic base such as choline. Phospholipids contain nitrogen, present in the organic base (Fig. 3–7). (Note that phosphorus and nitrogen are absent in the neutral fats.)

The two ends of the phospholipid molecule differ physically as well as chemically. The fatty acid portion of the molecule is **hydrophobic** (water-hating) and is not soluble in water. However, the portion composed of glycerol and the organic base is ionized and readily water-soluble. This end of the molecule is said to be **hydrophilic** (water-loving). The polarity of these lipid molecules causes them to

Glycerol (a) Fatty acid

A triacylglycerol $+ 3H_2O$ $\xrightarrow{\text{Enzyme}}$

Products

Glycerol + Oleic acid + Linoleic acid + Palmitic acid

(b)

(c)

Figure 3–6 Neutral fats. (*a*) Structural formulas of glycerol and a fatty acid. (*b*) Hydrolysis of a triacylglycerol yields glycerol plus three fatty acids. Note that the triacylglycerol is an unsaturated fat—two of its fatty acid components contain double bonds between carbon atoms. (*c*) Honeybees on a brood comb. The comb is composed of wax secreted by special abdominal glands of the bees. It is a compound consisting of fatty acids and alcohols, and though it is classified as a lipid, it can be digested by very few animals.

Phosphatidic acid

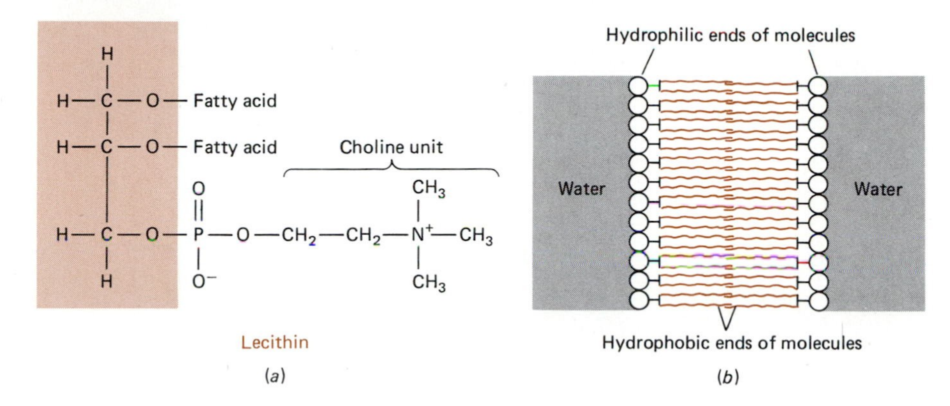

Figure 3–7 Phospholipids. (*a*) Many phospholipids are derivatives of phosphatidic acid, a compound consisting of glycerol bonded to two fatty acids and a phosphate group. Lecithin is a phospholipid found in cell membranes. It forms when phosphatidic acid combines with choline. (*b*) A lipid bilayer such as is found in cell membranes. The hydrophobic fatty acid ends are in color.

take up a certain configuration in the presence of water, with their hydrophilic water-soluble ends facing outward, interacting with the surrounding water. The hydrophobic tails face in the opposite direction. The cell membrane (Fig. 3–7) is a lipid bilayer composed of two layers of phospholipid molecules with their hydrophobic tails meeting in the middle and their hydrophilic heads oriented toward the outside of the cell membrane.

CAROTENOIDS

The red and yellow plant pigments called **carotenoids** are classified with the lipids because they are insoluble in water and have an oily consistency. These pigments, found in the cells of all plants, play some role in photosynthesis and in **phototropism,** the orientation of plants toward light. They consist of five carbon subunits known as **isoprene** units. Splitting in half a molecule of the yellow plant pigment carotene yields a molecule of **vitamin A, retinol** (Fig. 3–8). The light-sensitive chemical present in the cells in the retina of the eye is retinal, a derivative of vitamin A. In the presence of light, retinal undergoes a chemical reaction by which light stimuli are received.

It is interesting to note that photoreceptors or eyes have evolved independently in three different lines of animals—mollusks, insects, and vertebrates. These organisms have no common evolutionary ancestor equipped with eyes, yet the eyes of all have the same compound, retinal, involved in the process of light reception. That retinal is present in each of these types of eyes is not the result of a common evolutionary ancestry but rather of some unique fitness of this kind of molecule for the process of light reception.

STEROIDS

Although steroids are classified as lipids, their structure is quite different from that of other lipids. A **steroid** molecule contains carbon atoms arranged in four interlocking rings; three of the rings contain six carbon atoms and the fourth contains five (Fig. 3–9). The length and structure of the side chains that extend from these rings distinguish one kind of steroid from another.

Among the steroids of biological importance are cholesterol, bile salts, the male and female sex hormones, and the hormones secreted by the adrenal cortex.

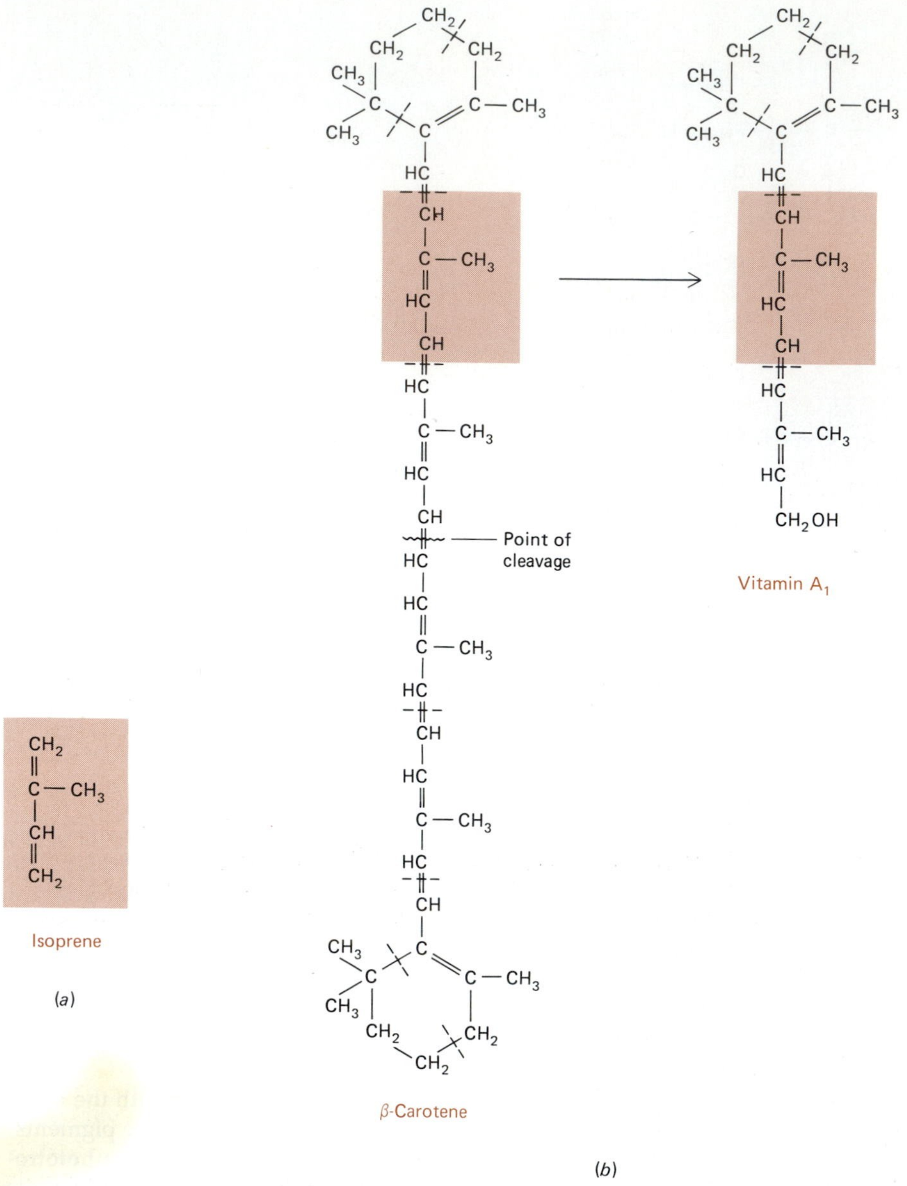

Figure 3–8 Carotenoids. (*a*) Isoprene, the molecular subunit found in carotenoids. (*b*) β-Carotene, the yellow pigment found in some plants that gives carrots, sweet potatoes, and other orange vegetables their color. Most animals can convert carotenoids to vitamin A.

Steroid hormones regulate certain phases of metabolism in a variety of animals, including vertebrates, insects, and crabs.

Steroids that have an —OH group are referred to as **sterols.** The sterol **cholesterol** is an important structural component of animal cell membranes. Plant cell membranes contain other types of sterols.

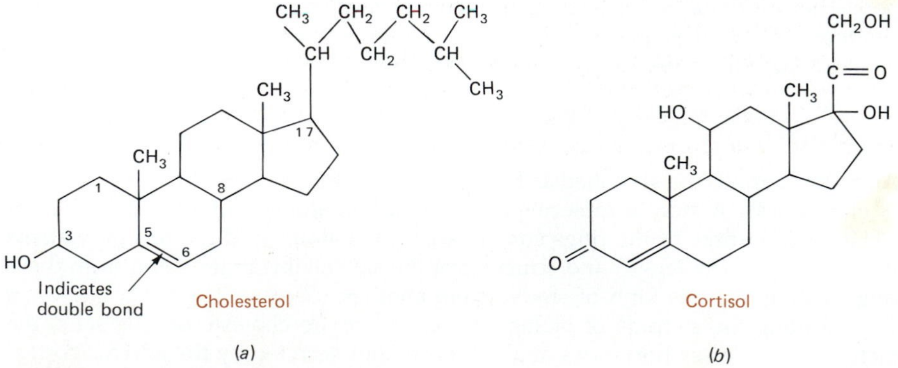

Figure 3–9 Steroids. All steroids have the basic skeleton of four interlocking rings of carbon atoms. Note that a carbon atom is present at each point in each ring. The first three rings each contain six carbon atoms, and the fourth ring contains five. For simplicity, hydrogen atoms have not been drawn within the ring structures.

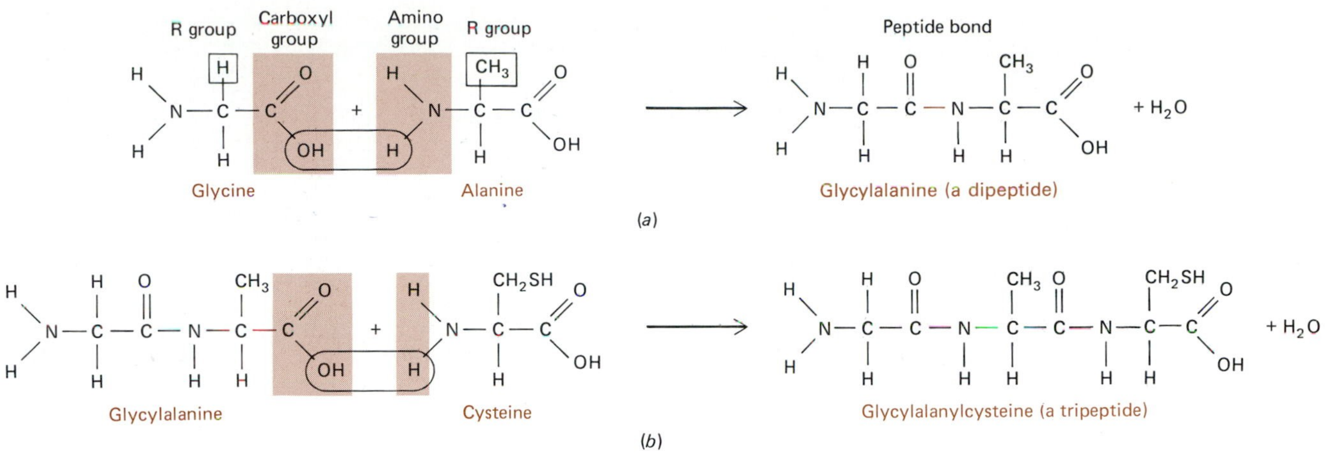

R group Carboxyl group Amino group R group

Glycine Alanine Peptide bond Glycylalanine (a dipeptide) $+ H_2O$

(a)

Glycylalanine Cysteine Glycylalanylcysteine (a tripeptide) $+ H_2O$

(b)

Figure 3–10 (*a*) Formation of a dipeptide. Two amino acids combine chemically to form a dipeptide. (*b*) A third amino acid is added to the dipeptide to form a chain of three amino acids (a tripeptide). The bond formed between adjacent amino acids is a peptide bond.

Proteins

All enzymes, certain hormones, and many of the important structural components of the cell are **proteins.** In addition to carbon, oxygen, and hydrogen, proteins contain nitrogen and, usually, sulfur.

SUBUNIT STRUCTURE

Proteins are among the largest molecules present in cells and share with nucleic acids the distinction of having great complexity and variety. Proteins are made of simpler components called **amino acids.** All of the more than 20 kinds of amino acids commonly found in proteins contain an **amino group (—NH_2)** and a **carboxyl group (—COOH)** bonded to the same carbon atom, but they differ in their side chains, abbreviated as **R groups** (see Focus on Amino Acids Present in Proteins). Glycine, the simplest amino acid, has a hydrogen atom as its R group or side chain; alanine has a methyl (—CH_3) group (Fig. 3–10).

Depending on the pH of the solution, the amino group, —NH_2, may accept a proton and become —NH_3^+, or the carboxyl group, —COOH, may donate a proton and become —COO^- (Fig. 3–11). As we saw earlier, a solution of an acid and its conjugate base serves as a buffer and resists changes in pH when an acid or base is added. Because of their amino and carboxyl groups, amino acids and proteins in solution resist changes in acidity or alkalinity, thereby serving as important biological buffers.

Amino acids are linked by **peptide bonds** between the amino group of one amino acid and the carboxyl group of the next. The formation of peptide bonds within a cell is a complex process. As with the reaction that produces a fat, water is shown as a product of the overall process that results in the formation of a peptide bond; however, the OH^- and H^+ are removed from the reactants in separate steps and are not necessarily present as H_2O at the completion of the reaction. When two amino acids combine, a **dipeptide** is formed (Fig. 3–10); a longer chain of amino acids is a **polypeptide.**

Each protein may contain hundreds of amino acids joined in a specific linear order. An almost infinite variety of protein molecules is possible. By analytic methods developed in the early 1950s investigators can determine the exact sequence of amino acids in a protein molecule. **Insulin,** a hormone secreted by the pancreas and

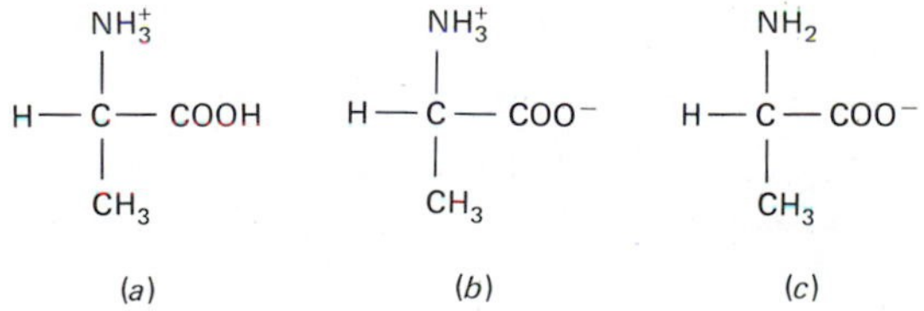

(a) (b) (c)

Figure 3–11 Change in the structure of an amino acid as the pH of the surrounding medium changes. Alanine is used here as an example. (*a*) Predominant structure of alanine at an acid pH such as 1. (*b*) Predominant structure of alanine at neutrality. (*c*) Predominant structure of alanine at an alkaline pH such as 12.

FOCUS ON

Amino Acids Present in Proteins

Common Name	*Symbol*	*Simplified Structural Formula*
Glycine	Gly	$H—CH—COOH$ $\quad\; \vert$ $\quad NH_2$
Alanine	Ala	$CH_3—CH—COOH$ $\qquad\quad \vert$ $\qquad\; NH_2$
Valine	Val	H_3C $\quad \diagdown$ $\quad CH—CH—COOH$ $\quad \diagup \qquad \vert$ $H_3C \qquad NH_2$
Leucine	Leu	H_3C $\quad \diagdown$ $\quad CH—CH_2—CH—COOH$ $\quad \diagup \qquad\qquad \vert$ $H_3C \qquad\qquad NH_2$
Isoleucine	Ile	CH_3 $\quad \diagdown$ $\quad CH_2$ $\qquad \diagdown$ $\qquad CH—CH—COOH$ $\qquad \diagup \qquad \vert$ $\quad CH_3 \qquad NH_2$
Serine	Ser	$CH_2—CH—COOH$ $\vert \qquad\quad \vert$ $OH \qquad NH_2$
Threonine	Thr	$CH_3—CH—CH—COOH$ $\qquad\quad \vert \qquad \vert$ $\qquad\; OH \quad NH_2$
Cysteine	Cys	$CH_2—CH—COOH$ $\vert \qquad\quad \vert$ $SH \qquad NH_2$
Methionine	Met	$CH_2—CH_2—CH—COOH$ $\vert \qquad\qquad\quad \vert$ $S—CH_3 \qquad NH_2$
Aspartic acid	Asp	$HOOC—CH_2—CH—COOH$ $\qquad\qquad\qquad \vert$ $\qquad\qquad\quad NH_2$
Asparagine	Asn	$H_2N—C—CH_2—CH—COOH$ $\qquad\; \Vert \qquad\qquad \vert$ $\qquad\; O \qquad\quad NH_2$
Glutamic acid	Glu	$HOOC—CH_2—CH_2—CH—COOH$ $\qquad\qquad\qquad\qquad\quad \vert$ $\qquad\qquad\qquad\qquad NH_2$
Glutamine	Gln	$H_2N—C—CH_2—CH_2—CH—COOH$ $\qquad\; \Vert \qquad\qquad\qquad\quad \vert$ $\qquad\; O \qquad\qquad\qquad NH_2$

Common Name	*Symbol*	*Simplified Structural Formula*
Arginine	Arg	$H-N(-C(=NH)-NH_2)-CH_2-CH_2-CH_2-CH(NH_2)-COOH$
Lysine	Lys	$CH_2(NH_2)-CH_2-CH_2-CH_2-CH(NH_2)-COOH$
Hydroxylysine*	Hyl	$CH_2(NH_2)-CH(OH)-CH_2-CH_2-CH(NH_2)-COOH$
Histidine	His	(imidazole ring: HN, N)$-CH_2-CH(NH_2)-COOH$
Phenylalanine	Phe	(benzene ring)$-CH_2-CH(NH_2)-COOH$
Tyrosine	Tyr	$HO-$(benzene ring)$-CH_2-CH(NH_2)-COOH$
Tryptophan	Trp	(indole ring: N–H)$-CH_2-CH(NH_2)-COOH$
Proline	Pro	(pyrrolidine ring: N–H)$-COOH$
4-Hydroxyproline	Hyp	$HO-$(pyrrolidine ring: N–H)$-COOH$

*Thus far, found only in collagen and in gelatin.

used in the treatment of diabetes, was the first protein for which the exact sequence of amino acids in its peptide chains was ascertained (Fig. 3–12). The first enzyme for which the amino acid sequence was determined was ribonuclease, a relatively small enzyme secreted by the pancreas (only 125 amino acids long!).

Hemoglobin, the protein in red blood cells that is responsible for oxygen transport, consists of 574 amino acids arranged in four polypeptide chains. Its chemical formula is $C_{3032}H_{4816}O_{872}N_{780}S_8Fe_4$. Various types of proteins differ from one another in the *number, types,* and *arrangement* of the amino acids they contain.

With some exceptions, plants can synthesize all of their needed amino acids from simpler substances. The cells of humans and animals in general can manufacture some, but not all, of the various kinds of biologically significant amino acids if the proper raw materials are available. Those that animals cannot synthesize but must obtain in the diet are known as **essential amino acids.** These amino acids are no more important as components of proteins than are other amino acids, but they are essential in the diet because they cannot be synthesized. Animals differ in their biosynthetic capacities; what is an essential amino acid for one species may not be essential in the diet of another.

PROTEIN STRUCTURE: LEVELS OF ORGANIZATION

Several different levels of organization can be distinguished in the protein molecule. The **sequence** of amino acids in a polypeptide chain constitutes its **primary structure.** This sequence, as we shall see later, is specified by the sequence of nucleotides in the DNA and RNA of the cell.

The **secondary structure** of protein molecules involves the coiling of the peptide chain into a helix or some other regular conformation (shape). Peptide chains ordinarily do not lie out flat, or coil randomly, but undergo coiling to yield a specific three-dimensional structure. A common secondary structure in protein molecules is known as the **α-helix** (alpha-helix). This involves the formation of spiral coils of the polypeptide chain. The α-helix is a very uniform geometric structure with 3.6 amino acids occupying each turn of the helix. The helical structure is determined and maintained by the formation of hydrogen bonds between amino acids in successive turns of the spiral coil (see Fig. 3–12). Another common type of secondary structure is the β-pleated (beta-pleated) sheet, present in fibroin, the protein of silk. This is composed of many strands of extended, zigzag chains of amino acids lying side by side and joined by hydrogen bonds (Fig. 3–12). The structure of the pleated sheet makes it flexible rather than elastic. Hydrophobic interactions play a role in the proper folding of macromolecules.

The **tertiary structure** of a protein molecule involves the folding of the α-helix upon itself; this folding imparts a specific overall structure to the protein molecule. Hydrogen bonding and ionic and hydrophobic interactions between one part of the peptide chain and another part hold the folds in place. **Disulfide bonds** (—S—S—) between certain amino acids and other covalent bonds may also be important in maintaining the tertiary structure of proteins. Most proteins assume a globular shape (see Focus on Classifying Proteins). The biological activity of the protein depends in large part on the specific tertiary structure of the molecule held together by these bonds. When a protein is heated or treated with any of a number of chemicals, the tertiary structure becomes disordered and the coiled peptide chains unfold to give a more random conformation. This unfolding is accompanied by a loss of the biological activity of the protein—for example, its ability to act as an enzyme. This change is termed **denaturation** of the protein.

Proteins composed of two or more subunits have a **quaternary** structure. This refers to the combination of two or more like or unlike peptide chain subunits, each with its own primary, secondary, and tertiary structures, to form the biologically active protein molecule. Hemoglobin is composed of four subunits, two identical α and two identical β subunits (Fig. 3–12(*d*)). These subunits are held together by noncovalent interactions.

FUNCTIONS OF PROTEINS

Some proteins serve as enzymes, some as hormones, and still others as important structural components of cells and tissues (see Focus on Classifying Proteins). The protein collagen, present in connective tissue, accounts for about one third of all the

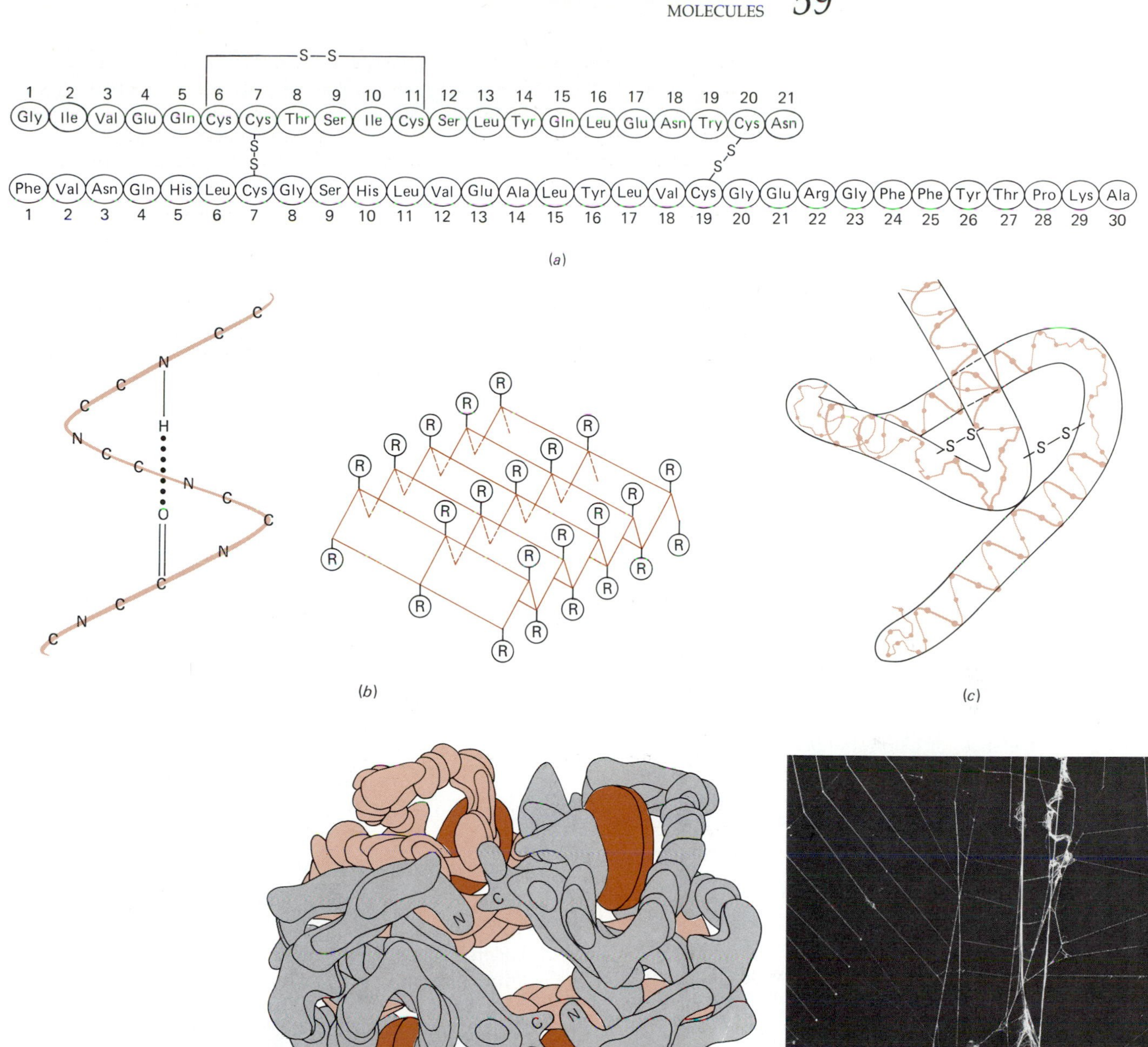

Figure 3–12 Protein structure. (*a*) The primary structure of the two polypeptide chains that make up the protein insulin. The primary structure is the linear sequence of amino acids. Each oval in the diagram represents an amino acid; the letters inside the ovals are symbols for the names of the amino acids (see Focus on Amino Acids Present in Proteins). Insulin is a very small protein. (*b*) The secondary structure of proteins is commonly an α-helix. The folds in the helix are held together mainly by hydrogen bonds between oxygen and hydrogen atoms. In some proteins, such as the silk protein fibroin, the backbone of the polypeptide chain is stretched out into a zigzag structure, rather than a helical one. (*c*) The tertiary structure results from the coiling and folding of the α-helix (or other secondary structure) into an overall globular or other shape. Hydrogen bonds, sulfhydryl bonds, and van der Waals forces are among the forces that hold the parts of the molecule in the designated shape. (*d*) Proteins that consist of more than one polypeptide subunit assume a final quaternary shape. Hemoglobin, a globular protein composed of four polypeptide subunits, is illustrated here. Its quaternary structure consists of the final shape in which the subunits combine. In hemoglobin each polypeptide encloses an iron-containing heme group (shown as colored disks). (*e*) The silk used by this garden spider to wrap its prey is a protein composed of several different amino acids. The silk is stored in liquid form in special glands within the spider's abdomen. As it is released, it hardens into extremely strong and flexible fibers. Each individual thread is composed of numerous fibers. ((*e*), Walker Van Riper, Courtesy of University of Colorado Museum.)

FOCUS ON

Classifying Proteins

There are many ways to classify proteins. Three common methods are by function, by composition, and according to their solubility.

I. Proteins grouped according to biological function

Enzymes	All enzymes are proteins.
Structural proteins	Proteins that are part of cells and tissues, such as collagen and elastin in connective tissues and keratin in skin, hair, and nails
Contractile proteins	The muscle proteins actin and myosin
Hormones	Insulin, growth hormone, and several other hormones are proteins.
Transport proteins	Hemoglobin and myoglobin both transport oxygen; a type of albumin in the blood transports fatty acids.
Defense proteins	Immunoglobulins (antibodies) protect the body against disease; fibrinogen in the blood is important in the clotting process.

II. Proteins grouped according to solubility

Globular proteins	Tend to be soluble in water because of their polar surface. Globular proteins are the most numerous group of proteins; they include all of the enzymes, the plasma proteins, and proteins found in cell membranes.
Fibrous proteins	Insoluble in water; elongated to form strong fibers; function as structural and supporting proteins. They include collagen, elastin, keratin, myosin, and fibrin (the protein in blood clots).

III. Proteins grouped according to composition—conjugated proteins can be classified according to the nonprotein component.

Lipoproteins	Proteins that contain fat and other lipids such as the lipoproteins in the blood
Glycoproteins	Proteins that contain sugars such as immunoglobulins that function in defense against microorganisms. Glycoproteins include many membrane proteins and collagen and other proteins found in connective tissues.
Nucleoproteins	Proteins that are bound to nucleic acids; found in chromosomes and viruses
Hemeproteins (or chromoproteins)	Proteins that contain a heme group: hemoglobin, myoglobin, and certain enzymes (cytochromes)
Metalloproteins	Proteins that contain one or more metallic ions; some enzymes

protein in the bodies of humans and other mammals. Proteins may also be broken down into their component amino acids and used in cellular respiration.

Each cell contains hundreds of different proteins, and each kind of cell contains some proteins that are unique to it. In fact, the characteristic types, distributions, and amounts of protein determine what a cell looks like and how it functions. A muscle cell contains a large quantity of the proteins myosin and actin, which are responsible for its appearance and its ability to contract. Red blood cells, in contrast, are characterized not by actin and myosin but by hemoglobin, which is responsible for the special function of oxygen transport.

Most proteins are species-specific—that is, they vary slightly in each species. Thus, the protein complement (as determined by the instructions in the genes) is also largely responsible for differences among species. The degree of difference in the proteins of two species depends upon their evolutionary relationship. Organisms less closely related have proteins that differ more markedly in amino acid sequence than do those of closely related forms. Investigations of the similarities of proteins such as hemoglobins in different species have added biochemical evidence to the ideas of evolutionary relationships derived from other types of evidence.

Some proteins differ slightly even among individuals of the same species, so that each individual is biochemically unique. Only genetically identical organisms—identical twins or members of closely inbred strains of animals—have identical proteins.

Nucleic Acids

Nucleic acids, like proteins, are large, complex molecules. They were first isolated by Friedrich Miescher in 1870 from the nuclei of pus cells and gained their name from the fact that they are acidic and were first identified in nuclei. There are two types of nucleic acids, **ribonucleic acid (RNA)** and **deoxyribonucleic acid (DNA);** the different kinds of RNA and DNA vary in some of their structural components and in their metabolic functions. DNA, present in the chromosomes in the cell nucleus, is the primary repository for genetic information. RNA is also present in the nucleus, as well as in cell structures called ribosomes, and is found in lesser amounts in other parts of the cell. The various kinds of RNA have specific functions in the process of protein synthesis.

SUBUNIT STRUCTURE

Nucleic acids are composed of **nucleotides,** molecular units that consist of a **nitrogenous base;** a five-carbon sugar; and a **phosphate group** (Fig. 3–13). **Purines** (with a double ring) and **pyrimidines** (with a single ring) are two types of nitrogenous bases present in nucleic acids. DNA contains the purines **adenine** (A) and **guanine** (G) and the pyrimidines **cystosine** (C) and **thymine** (T), together with the sugar **deoxyribose** and phosphate. RNA contains the purines adenine and guanine and the pyrimidines cystosine and uracil (U), together with the sugar ribose and phosphate. The removal of the phosphate group from a nucleotide yields a compound termed a **nucleoside,** composed of the base and ribose.

The molecules of nucleic acids are made of linear chains of nucleotides, each attached to the next by ester bonds between the sugar molecule of one and the phosphate group of the next. As we will see in the discussion of the genetic code (Chapter 14), the specific information of the nucleic acid is coded in the unique sequences of the four kinds of nucleotides present in the chain. An enormous amount of evidence affirms that DNA is responsible for the specificity and chemical properties of the **genes,** the units of heredity. There are several classes of RNA, each of which plays a particular role in the biosynthesis of proteins by the cell.

Figure 3–13 (*a*) The three major pyrimidine bases found in nucleotides. (*b*) The two major purine bases found in nucleotides. (*c*) A nucleotide (adenosine monophosphate). (*d*) A very diagrammatic representation of part of a nucleic acid molecule.

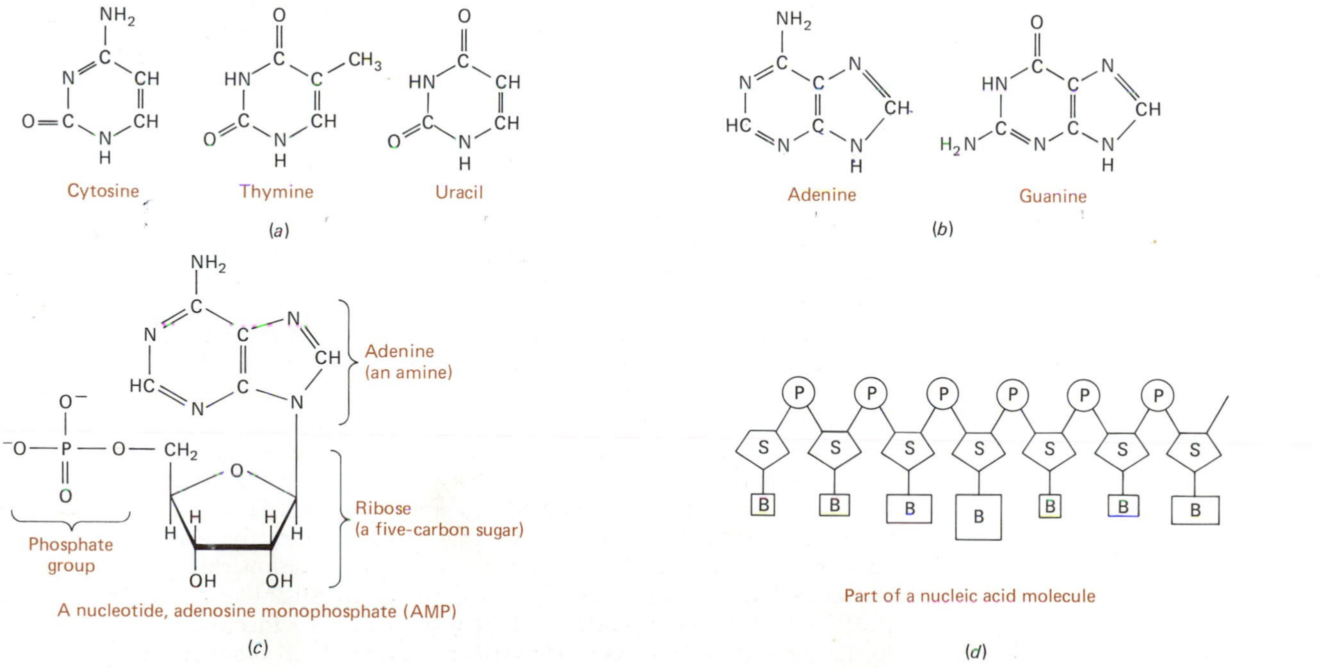

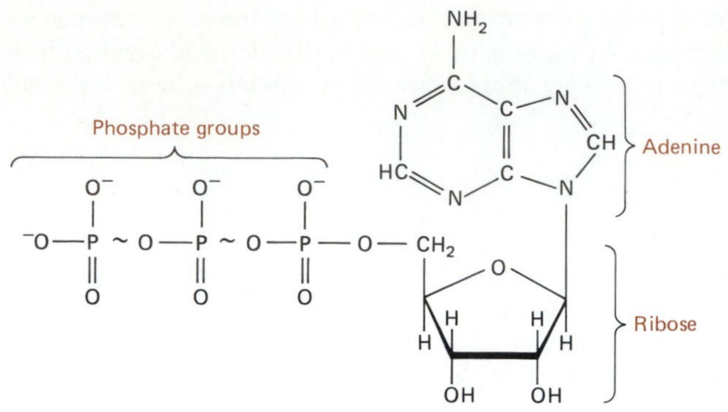

Figure 3–14 The structure of ATP, a nucleotide that stores energy in its energy-rich phosphate bonds.

RELATED NUCLEOTIDES

Besides their importance as subunits of nucleic acids, a number of nucleotides serve other vital functions in living cells. **Adenosine triphosphate,** or **ATP,** composed of adenine, ribose, and three phosphates (Fig. 3–14), is of major importance as the energy currency of all cells. The two terminal phosphate groups are joined to the nucleotide by special "energy-rich" bonds, indicated by the ~ symbol. These are energy-rich bonds in the sense that much free energy is released when the bonds are hydrolyzed. The biologically useful energy of these bonds can be transferred to other molecules. Most of the chemical energy of the cell is stored in these high-energy phosphate bonds of ATP, ready to be released when the phosphate group is transferred to another molecule.

Guanosine triphosphate, or GTP, is required in the synthesis of proteins; uridine triphosphate, or UTP, is required in the synthesis of glycogen; and cytidine triphosphate, or CTP, is required in the synthesis of fats and phospholipids. All

Figure 3–15 Formation of cyclic AMP from ATP.

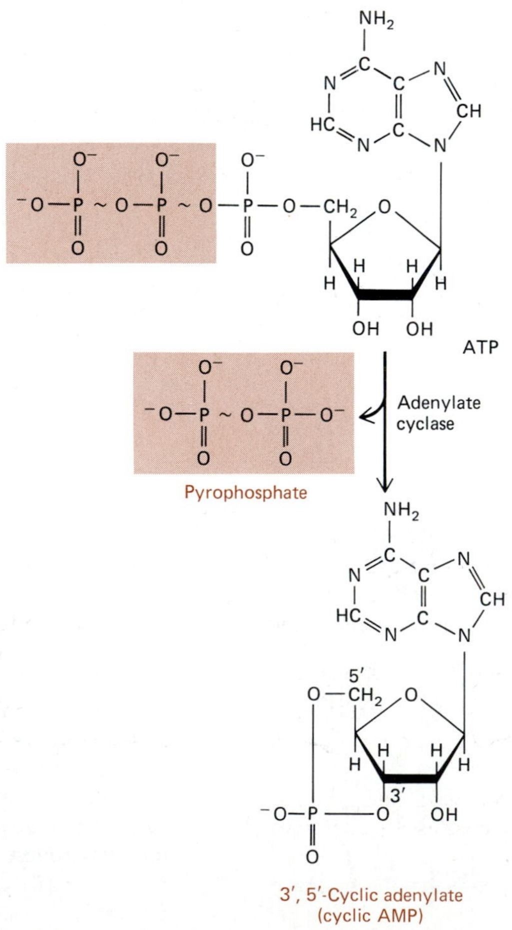

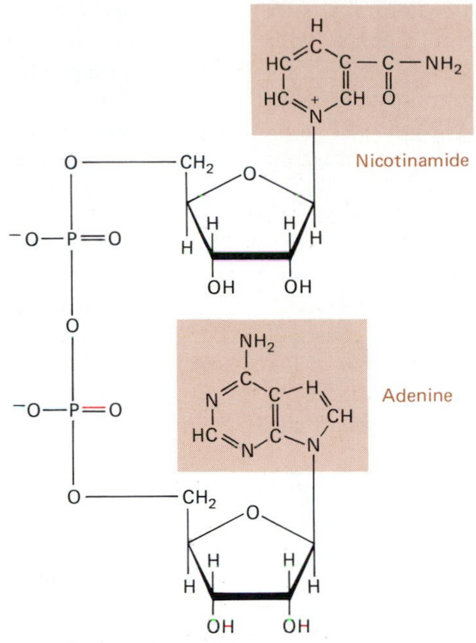

Figure 3–16 Structure of nicotinamide adenine dinucleotide, NAD^+.

four deoxyribose nucleoside triphosphates, abbreviated dATP, dGTP, dTTP, and dCTP, are required for the synthesis of DNA.

Each nucleotide may be converted by enzymes called **cyclases** to the cyclic monophosphate form (Fig. 3–15). ATP, for example, is converted to cyclic adenosine monophosphate, or **cyclic AMP,** by the enzyme adenylate cyclase. Cyclic nucleotides play important roles in mediating the effects of hormones and in regulating various aspects of cellular function.

Cells contain several dinucleotides that are of great importance in metabolic processes. **Nicotinamide adenine dinucleotide,** abbreviated **NAD^+**, consists of one nucleotide composed of nicotinamide (derived from niacin, a B vitamin), ribose, and phosphate attached to a second nucleotide composed of adenine, ribose, and phosphate (Fig. 3–16). NAD^+ is very important as a primary electron and hydrogen acceptor in biological oxidations within cells (Chapter 9). Nicotinamide adenine dinucleotide phosphate, NADP, serves as electron and hydrogen acceptor in certain other reactions. Its structure is similar to that of NAD^+ except that it has a third phosphate group attached to the ribose of the adenine nucleotide. Flavin adenine dinucleotide, FAD, contains riboflavin, a B vitamin, and serves as a hydrogen and electron acceptor in other dehydrogenation reactions. Our daily need for vitamins reflects the fact that dinucleotides are constantly breaking down and being resynthesized.

SUMMARY

I. The chemical components of cells and tissues include, in addition to water and inorganic salts, a great variety of organic compounds. The major groups of organic compounds are carbohydrates, lipids, proteins, and nucleic acids.

II. Carbohydrates contain carbon, hydrogen, and oxygen in a ratio of approximately one carbon : two hydrogens : one oxygen. Sugars, starches, and celluloses are typical carbohydrates.

- *A.* Monosaccharides are simple sugars such as glucose, fructose, or ribose. Glucose is an important fuel molecule in living cells.
- *B.* A disaccharide (e.g., sucrose) consists of two monosaccharide units.
- *C.* Most carbohydrates are polysaccharides (e.g., glycogen and starch), long chains of repeating units of a simple sugar.
  1. Carbohydrates are typically stored in plants as starch and in animals as glycogen.
  2. The walls of the plant cells are composed mainly of the polysaccharide cellulose.
- *D.* Glycolipids and glycoproteins serve as structural components of cells and cell walls.

III. Lipids are composed of carbon, hydrogen, and oxygen but have relatively less oxygen in proportion to carbon and hydrogen than do carbohydrates. Lipids have a greasy or oily consistency and are relatively insoluble in water.

A. The body stores fuel in the form of neutral fats. A fat consists of a molecule of glycerol combined with one to three fatty acids.

1. Three types of neutral fats are monoacylglycerols, diacylglycerols, and triacylglycerols.

2. Fatty acids also can be saturated or unsaturated (contain one or more double bonds).

B. Phospholipids are structural components of cell membranes.

C. Carotenoids are red and yellow pigments that play a role in photosynthesis and phototropism.

D. Steroid molecules contain carbon atoms arranged in four interlocking rings. Cholesterol, bile salts, vitamin D, male and female sex hormones, and the hormones of the adrenal cortex are important steroids.

IV. Proteins are large, complex molecules made of simpler components termed amino acids that are joined by peptide bonds. They are composed of carbon, hydrogen, oxygen, nitrogen, and sulfur.

A. Four levels of organization can be distinguished in protein molecules: (1) primary structure, the sequence of amino acids in the peptide chain; (2) secondary structure, the coiling of the peptide chains in a helix or some other regular conformation; (3) tertiary structure, the folding of the chain upon itself; and (4) quaternary structure, the spatial relationship of the combination of two or more subunits, each composed of a peptide chain.

B. Proteins are important structural components of cells and tissues. Many serve as enzymes or as hormones, and proteins may also be used as fuel. Some proteins are species-specific, and some vary slightly even among individuals of the same species, conferring biochemical individuality upon each organism.

V. The nucleic acids DNA and RNA store information that governs the structure and function of the organism. Nucleic acids are composed of carbon, hydrogen, oxygen, nitrogen, and phosphorus.

A. Nucleic acids are made of long chains of nucleotide units, each composed of a nitrogenous base (a purine or a pyrimidine), a five-carbon sugar (ribose or deoxyribose), and a phosphate group.

B. ATP is a nucleotide of special significance in energy metabolism. NAD, NADP, and FAD are electron and hydrogen acceptors in biological oxidations.

POST-TEST

1. Molecules (other than carbonates) containing the element carbon are termed ________________ ________________.

2. Some typical carbohydrates are ________________ and ________________.

3. Glucose and fructose are examples of ________________, composed of a single ________________ unit.

4. Compounds with identical molecular formulas but different arrangements of atoms are termed ________________.

5. The pentoses ________________ and ________________ are components of nucleotides and nucleic acids.

6. Glycogen is a storage form of ________________.

7. Glucosamine and galactosamine, sugars in which the hydroxyl group is replaced by an amino group, are examples of ________________ ________________.

8. Splitting a molecule of carotene in half yields a molecule of ________________.

9. An important component of animal cell membranes is the sterol ________________.

Match the compounds in Column A with their descriptions in Column B.

Column A	*Column B*
10. Carbohydrates	a. Compounds composed of carbon, hydrogen, and oxygen in a ratio of approximately 1:2:1
11. Carotenoids	b. Composed of a single hexose unit
12. Neutral fats	c. Composed of a long chain of single sugar units joined by glycosidic bonds
13. Nucleic acids	d. Composed of glycerol, phosphate, an organic base, and two fatty acid molecules
14. Nucleotides	e. Composed of glycerol, phosphate, an organic base, and two fatty acid molecules
15. Monosaccharides	f. Red or yellow pigments composed of five-carbon isoprene units
16. Phospholipids	g. Composed of carbon atoms arranged in four interlocking rings
17. Polysaccharides	h. Large, complex molecules composed of amino acids
18. Proteins	i. Large, complex molecules composed of nucleotide subunits
19. Steroids	j. Composed of a nitrogenous base, a five-carbon sugar, and phosphoric acid

20. Because of their amino and carboxyl groups, amino acids and proteins in solution serve as ________________.

21. The amino acids that an animal cannot synthesize but must obtain in the diet are termed ________________ ________________ ________________.

22. The sequence of amino acids in a polypeptide chain constitutes its ________________ ________________.

23. The coiling of a peptide chain into a helix or some other regular conformation is termed its ________________ ________________.

24. The structure of an α-helix is maintained by the formation of ________________ ________________ between amino acids in successive turns of the spiral coil.

25. The ________________ ________________ of a protein molecule involves the folding of the α-helix upon itself, with the folds held together by hydrogen, ionic, and hydrophobic interactions between one part of the chain and another.

26. The unfolding of a protein molecule induced by heat or by treatment with certain chemicals, and the accompanying loss of biological activity, is termed ________________.

27. Proteins serve in the body as ________________, ________________, or ________________ ________________.

28. The primary repository of genetic information is ________________.

29. The nitrogenous bases present in deoxyribonucleic acid include ________________, ________________, ________________, and ________________.

30. ________________ ________________, composed of adenine, ribose, and three molecules of phosphate, plays an important role in the storage and transfer of biologically useful energy.

31. An important primary electron and hydrogen acceptor in biological oxidations is ________________ ________________ ________________, composed of nicotinamide, ribose, and phosphate joined to adenine, ribose, and phosphate.

REVIEW QUESTIONS

1. Distinguish between inorganic and organic compounds.
2. Why is each of the following biologically important?
   a. proteins
   b. fats
   c. steroids
   d. nucleic acids
   e. carbohydrates
3. Distinguish among monosaccharides, disaccharides, and polysaccharides, and give specific examples of each.
4. Distinguish between monoacylglycerols and triacylglycerols; between saturated and unsaturated fatty acids.
5. What is the biological importance of each of the following?
   a. carotenoids
   b. phospholipids
   c. amino acids
   d. cellulose
6. Draw a structural formula of a simple amino acid such as glycine, and identify the carboxyl and amino groups.
7. There are thousands of different types of proteins in cells. How does one protein differ structurally from another?
8. Distinguish among primary, secondary, and tertiary protein structure.
9. Discuss the biological importance of
   a. ATP
   b. NAD
10. Describe the subunit structure of a nucleic acid.

4
The Life of Cells

OUTLINE

I. The cell theory
II. Viewing the cell
III. General characteristics of cells
IV. Inside the cell
  A. Endoplasmic reticulum and ribosomes
  B. The Golgi complex
  C. Lysosomes
  D. Mitochondria
  E. Plastids
  F. Peroxisomes
  G. Microtubules and microfilaments
  H. The microtrabecular lattice
  I. Centrioles
  J. Cilia and flagella
  K. Vacuoles
  L. The cell nucleus
    1. Nuclear envelope
    2. Chromosomes
    3. Nucleolus
V. Differences in some major cell types
  A. Plant and animal cells
  B. Prokaryotic and eukaryotic cells
VI. The cell cycle
  A. Interphase
  B. Mitosis
    1. Prophase
    2. Metaphase
    3. Anaphase
    4. Telophase
    5. The significance of mitosis
  C. Factors that affect the cell cycle

LEARNING OBJECTIVES

After you have read this chapter you should be able to:

1. Explain why the cell is considered the basic unit of life, and discuss the cell theory.
2. Compare the resolving power of the electron microscope and that of the light microscope, and distinguish between transmission and scanning electron microscopes.
3. Describe the general characteristics of cells.
4. Describe, locate, and list the functions of the principal organelles, and label them on a diagram or photomicrograph.
5. Distinguish between smooth and rough endoplasmic reticulum, and describe the functional relationship between ribosomes and endoplasmic reticulum.
6. Describe how the Golgi complex packages secretions and manufactures lysosomes.
7. Describe functions of lysosomes, and explain what happens when they leak.
8. Explain why mitochondria are referred to as the power plants of the cell.
9. Describe the nature and roles of microtubules, microfilaments, and the microtrabecular lattice.
10. Describe the structure of flagella and cilia and how they function in locomotion.
11. Describe the structure of the cell nucleus and its biological roles.
12. Distinguish between plant and animal cells and between prokaryotic and eukaryotic cells.
13. Identify the stages in the cell cycle, and describe the principal events characteristic of each.
14. Explain the significance of mitosis, and describe the process.

The fundamental unit that defines life is not an atom, ion, or molecule, for these particles make up nonliving as well as living matter. To truly understand the living world as something different from the nonliving requires a knowledge of the cell. Within the cellular structure of an organism, ions and molecules are precisely arranged, providing the machinery and environment that permits life processes to take place. Although it consists of many parts, the cell itself is considered the basic structural and functional unit of life because it is the smallest unit of living material capable of carrying on all the activities necessary for life.

We can think of the cell as a complete metabolic unit, because it has all of the chemical and physical components needed for its own maintenance and growth. When provided with essential nutrients and an appropriate environment, many kinds of cells can be kept alive in laboratory glassware for years. No cell part is capable of such survival.

The Cell Theory

The **cell theory** is one of the broadest and most basic generalizations of biology. This theory states that (1) all living things are composed of cells and cell products, (2) new cells are formed only by the division of preexisting cells, (3) there are fundamental similarities in the chemical constituents and metabolic activities of all cells, and (4) the activity of an organism as a whole can be understood as the collective activities and interactions of its interdependent cellular units.

Although several researchers made earlier contributions to our knowledge of the cell theory, two German investigators, the botanist Matthias Schleiden and the zoologist Theodor Schwann, are credited with presenting the first concise, yet comprehensive, statements of the cell theory. Schleiden and Schwann, in papers published in 1838 and 1839, respectively, pointed out that plants and animals are aggregates of cells arranged according to definite laws. It was largely through these papers that the concept of the cell as the "unit of life" was able to gain acceptance among biologists of the early 19th century.

An important extension of the cell theory, stated by Rudolf Virchow in 1855, is that new cells come into existence only by the division of previously existing cells. Cells do not arise by spontaneous generation from nonliving matter. The corollary to this, that all the cells living today can trace their ancestry back to ancient times, was pointed out by August Weissman about 1880.

Viewing the Cell

One of the biologist's most important tools for studying the structure of the cell is the microscope. In fact, cells were first described by Robert Hooke in 1665 when he examined a piece of cork using one of the crude microscopes of the 17th century. What Hooke saw (Fig. 4–1) were the cell walls of dead cork cells. In other observations Hooke described cell contents, but only some two centuries later was it realized that the important part of the cell is its contents, not its walls. The ordinary **light microscope** (greatly refined since Hooke's time) enabled biologists to discover most cell structures. The light microscope, the type of microscope used by students in most college laboratories, relies on visible light as the source of illumination. (An improved, more sophisticated kind of light microscope, the **phase contrast microscope,** has special lenses that can detect small differences in the refractive index for parts of a cell.)

During the past three decades the development of the **electron microscope** has enabled researchers to study the fine detail (ultrastructure) of cells. The electron microscope uses a beam of electrons (wavelength about 0.5 nm[1]) so that the object being studied is flooded with electrons rather than light waves (wavelength about 500 nm) (Fig. 4–2). **Magnification** is the ratio of the size of the image viewed under the microscope to the actual size of the object. Whereas the ordinary light microscope can magnify a structure about 1000 times, the electron microscope can magnify it 250,000 times or more.

[1] One nanometer is one *billion*th of a meter.

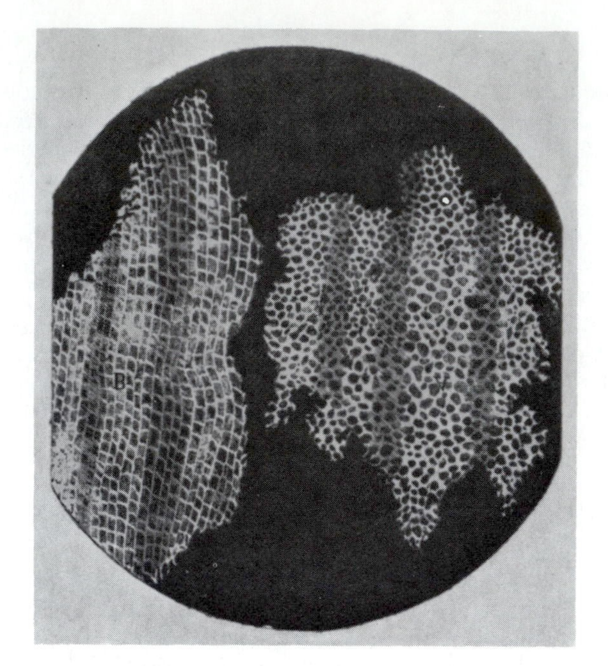

Figure 4–1 Drawing by Robert Hooke of the microscopic structure of a thin slice of cork. Hooke was the first to describe cells—his observations were based on the cell walls of these dead cork cells. (From the book *Micrographia,* published in 1665, in which Hooke described many of the objects he had viewed using the compound microscope he constructed.)

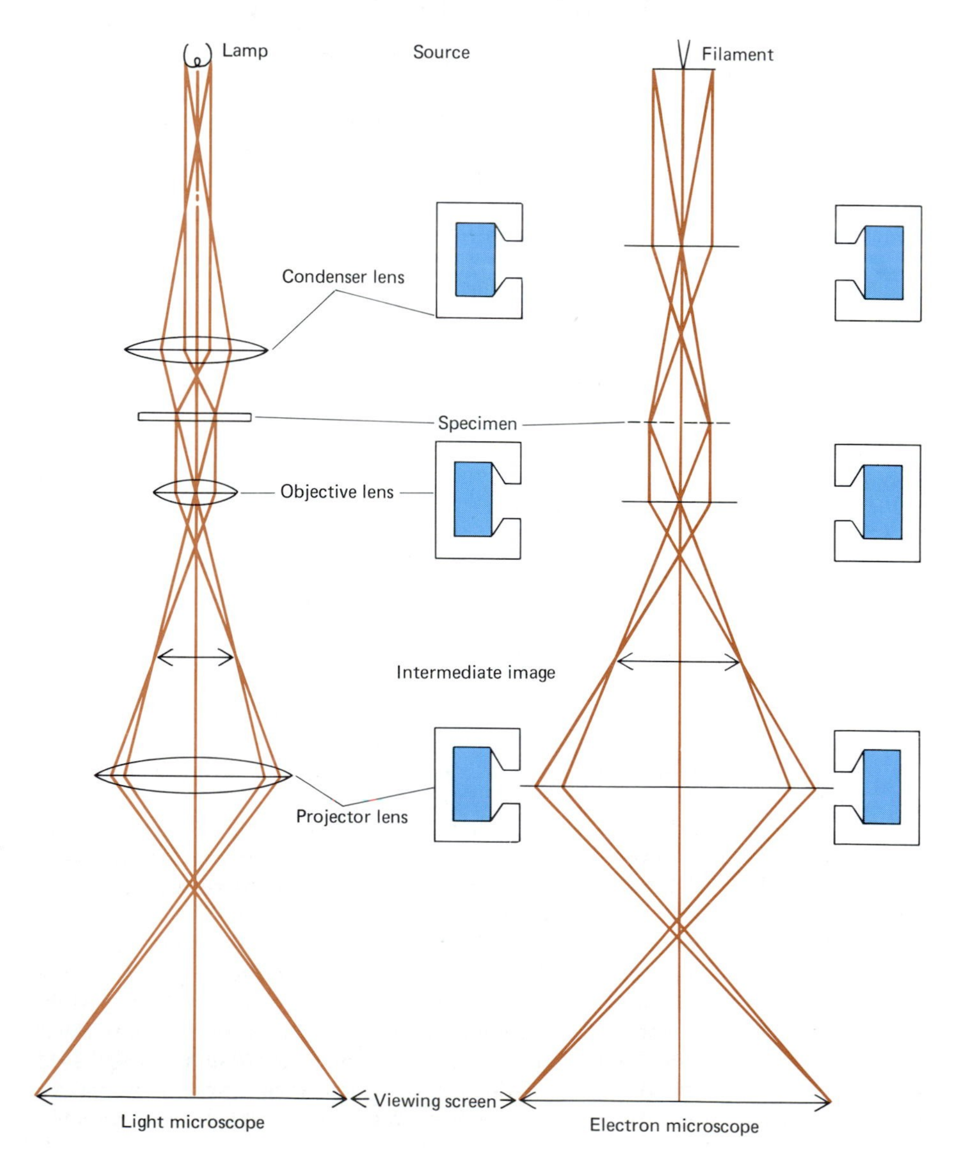

Figure 4–2 Comparison of a light microscope and an electron microscope. The light microscope is shown upside down so that the parts correspond to those of the electron microscope. In each kind of instrument, light rays or an electron beam is focused by the condenser lens onto the specimen. The objective lens forms a first magnified image of the specimen, which is further magnified by the projector lens onto a ground glass screen in the light microscope or onto a fluorescent screen in the electron microscope. The lenses in the electron microscope are actually magnets that bend the beam of electrons.

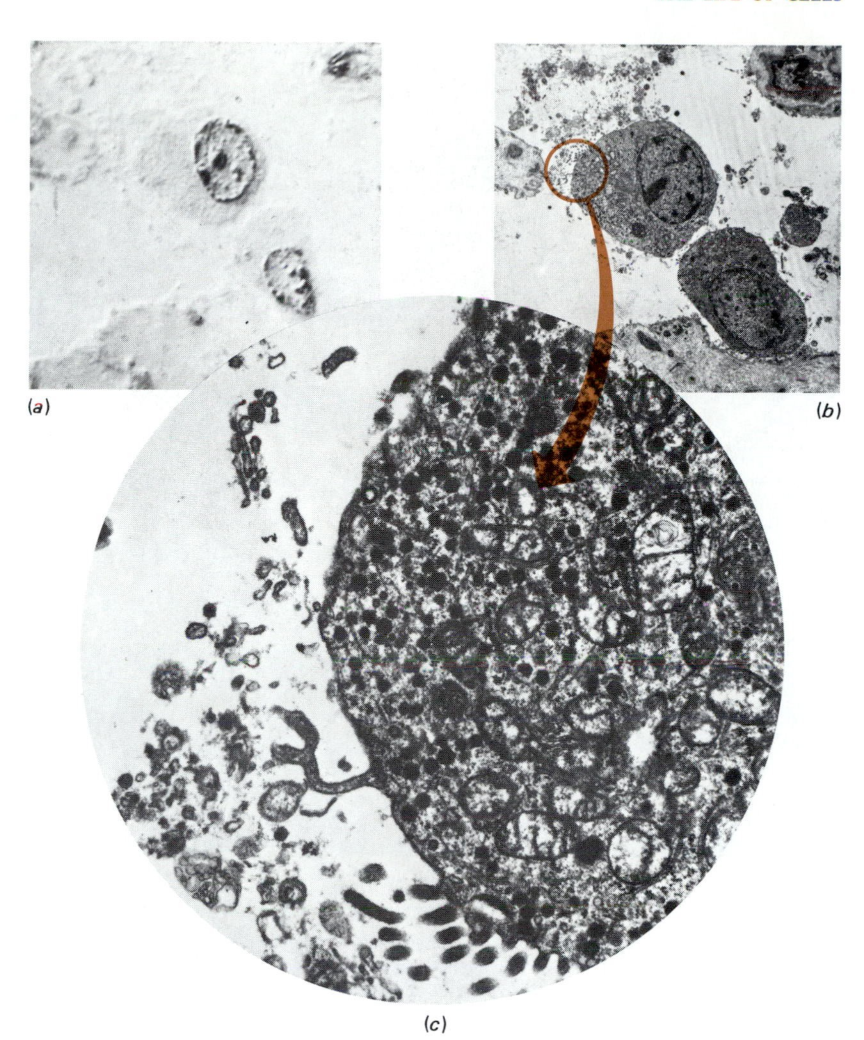

Figure 4–3 Comparison of a photograph taken with a modern light microscope and two taken with an electron microscope. (*a*) Lung cancer cells magnified 1800 times as seen through a light microscope. (*b*) The same cells, at the same magnification, seen through an electron microscope. The clearer detail is a result of the greater resolving power of the electron microscope. (*c*) A portion of one of the cells seen in (*b*) magnified about 17,000 times by the electron microscope. Note the black granules and other detail not visible at the lower magnification. (Courtesy of Zeiss. The electron micrographs were taken with the Zeiss EM 9S-2 by Dr. Harry Carter.)

A great advantage of the electron microscope is its superior **resolving power.** Even more important than magnification, resolving power is the ability to perceive fine detail and is expressed as the minimum distance between two points that can be distinguished as separate, distinct points rather than viewed as one blurred single point. Resolving power depends on the quality of the lenses and upon the wavelength used. Whereas the light microscope with the best lenses can resolve objects about 500 times better than the unaided human eye, the electron microscope has a resolving power of more than 10,000 times that of the human eye (Fig. 4–3). This is possible because the electron beams used as the source of illumination have much shorter wavelengths than those of visible light. The image formed by the electron microscope cannot be viewed directly. It first must be projected onto a television screen or photographic plate. A photograph taken with an electron microscope is called an **electron micrograph,** or **EM.**

In the **transmission electron microscope,** a beam of electrons passes through the specimen and strikes a photographic plate (Fig. 4–2) or a fluorescent screen. The specimen, embedded in plastic, must be cut in ultra-thin sections so that the beam of electrons can pass through it. This type of electron microscope has been most valuable for studying details of internal cell structure. In the **scanning electron microscope,** the electron beam does not pass through the specimen but causes secondary electrons to be emitted from the surface, which has been coated with a metal (gold). The contour of a specimen causes variations in the angle with which the beam strikes the various points in the specimen; this leads to variations in the intensity with which secondary electrons are emitted. A recording of the emission from the specimen provides a picture of its three-dimensional nature (Fig. 4–4). This special kind of micrograph provides information about the shape and the surface of the specimen that is not available from one obtained with the transmis-

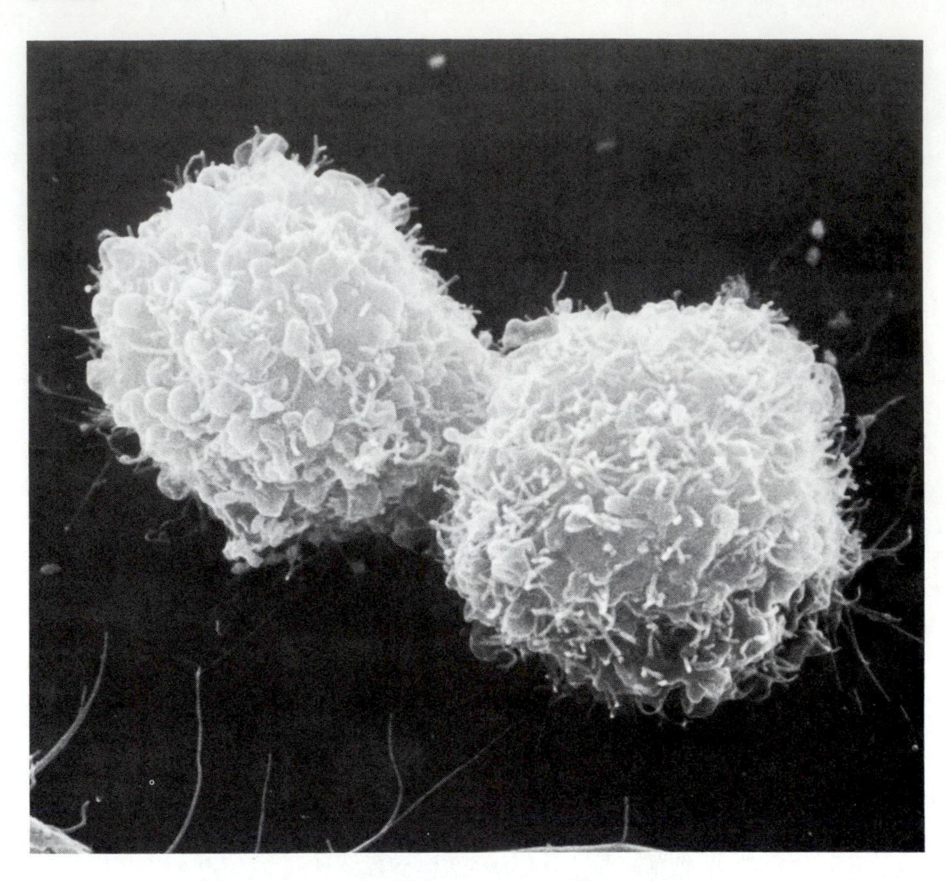

Figure 4–4 Scanning electron micrograph of a dividing lymphocyte (a white blood cell). (Courtesy of Dr. I. ap Gwynn.)

sion electron microscope. A *disadvantage* of electron microscopy is that, with present technology, live specimens cannot be observed (to avoid their scattering by air molecules, the electrons must be directed through a vacuum).

General Characteristics of Cells

The cells of plants and animals present a bewildering variety of sizes, shapes, colors, and internal structures, but all have certain features in common. Each cell consists of a tiny bit of jelly-like material surrounded by a cell membrane. Most cells contain a nucleus and other internal structures referred to as **organelles** that perform specific functions.

Most cells are microscopic in size. An average animal cell measures about 15 μm (micrometers[1]) in diameter; an average plant cell about 40 μm. The largest cells are birds' eggs, but these are highly specialized and are not really typical cells: Most of the egg cell consists of yolk, which is necessary for the nourishment of the developing bird but is not part of the functioning structure of the cell. Neither the shell nor the white of the bird's egg is considered part of the cell because these structures consist of nonliving material secreted by the mother bird's oviduct. The smallest cells known, those of certain microorganisms, are less than 0.3 μm in diameter.

The size and shape of a cell are related to the specific functions it must perform. Although cells tend to be spherical, many cell types have other characteristic shapes (Fig. 4–5). Some cells, like the amoeba and white blood cell, have the ability to change shape as they move about. Sperm cells have long whiplike tails used in locomotion. Nerve cells possess long extensions that permit them to transmit messages over great distances in the body. Epithelial cells, which are specialized to cover body surfaces, look like tiny building blocks.

[1] One micrometer is one *millionth* of a meter.

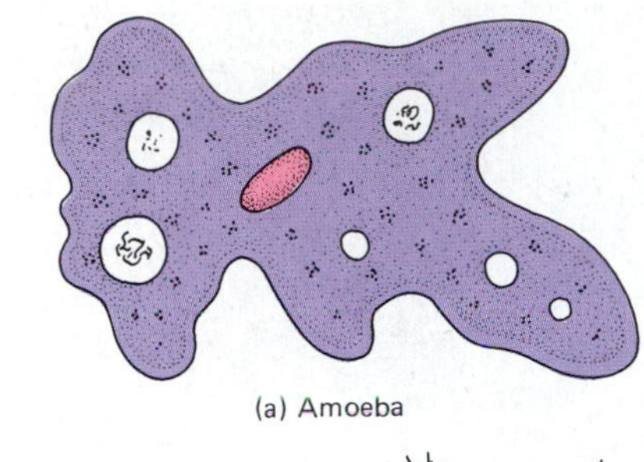

(a) Amoeba

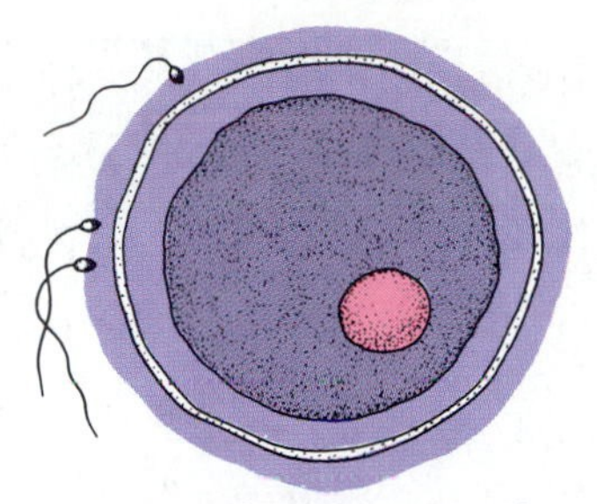

(b) Ovum (egg) and sperm cells

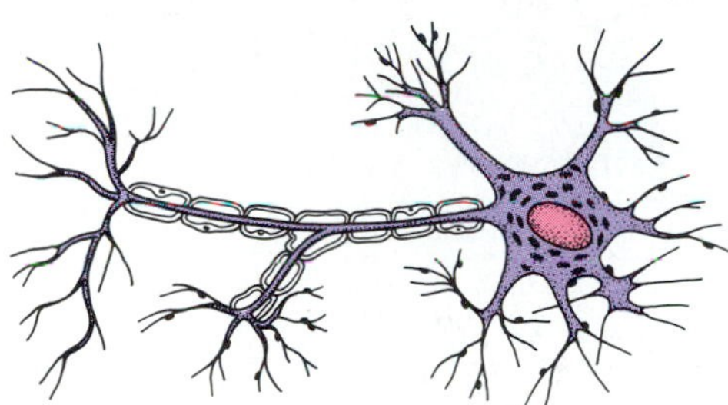

(c) Nerve cell

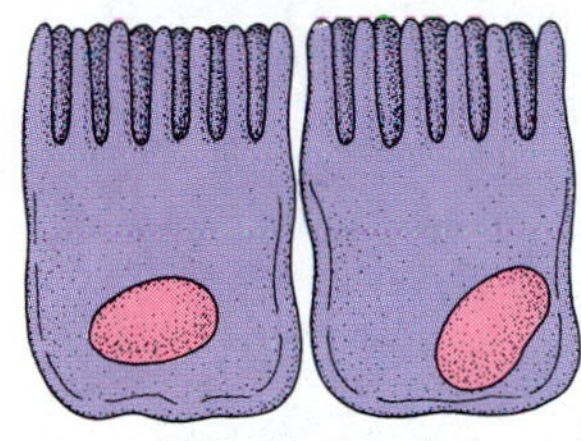

(d) Epithelial cells

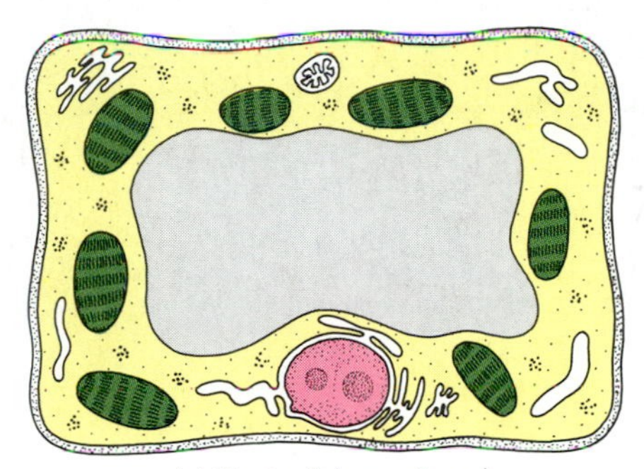

(e) Plant cell (parenchyma)

Figure 4–5 The size and the shape of cells are related to their functions. (*a*) The amoeba changes its shape as it moves from place to place. (*b*) An ovum (egg cell) and sperm cells. Ova are among the largest cells; sperm cells are comparatively tiny. Note the long tail (flagellum) used by the sperm cell in locomotion. By whipping the flagellum, the sperm can move toward the egg. (*c*) Nerve cells are specialized to transmit messages from one part of the body to another. (*d*) Epithelial cells join to form tissues that cover body surfaces and line body cavities. (*e*) The bulk of the organs of most young plants consists of parenchymal cells.

Inside the Cell

Early biologists thought that the cell interior consisted of a homogeneous jelly that they called **protoplasm.** With their limited tools they were able to recognize only a few internal structures such as the nucleus. With the electron microscope and other modern research tools, perception of the world within the cell has been greatly expanded. We now know that the cell is a highly organized, complex structure with its own control center, internal transportation system, powerplants, factories for making needed materials, packaging plants, and even a self-destruct system.

Today, the word *protoplasm* (if used at all) is used only in a very general way. The portion of the cell outside the nucleus is called **cytoplasm,** and the corresponding material within the nucleus is called **nucleoplasm.** Suspended within the fluid component of the cytoplasm and nucleoplasm are the various subcellular organelles (Fig. 4–6). The intracellular fluid, called **cytosol,** consists mainly of water, in which are dissolved amino acids, sugars, and other substances needed to manufacture larger molecules. Also present are structural proteins, enzymes used in cellular metabolism, and ions that help maintain an appropriate biochemical environment.

Most of the subcellular organelles are enclosed by **membranes.** These membrane-bound organelles effectively partition the cytoplasm into different compartments. The membrane acts as a barrier, making it possible for the chemical contents of an organelle to be different from the chemical environment in the general cytoplasm or in other organelles. Such differences in content permit metabolic processes to proceed in an orderly, effective manner.

The characteristics of the various organelles are summarized in Table 4–1. This table may be useful as an introduction to the study of cell structure and later as a checklist for review. The cell membrane is described in Chapter 5.

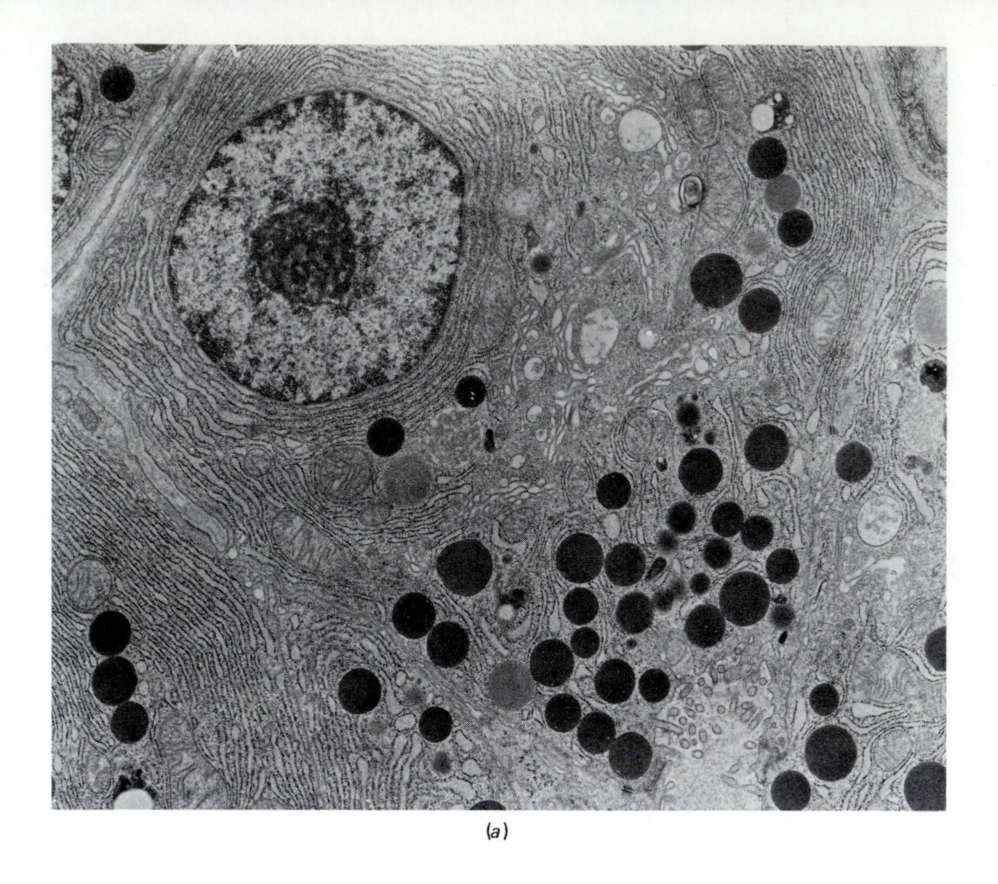

(a)

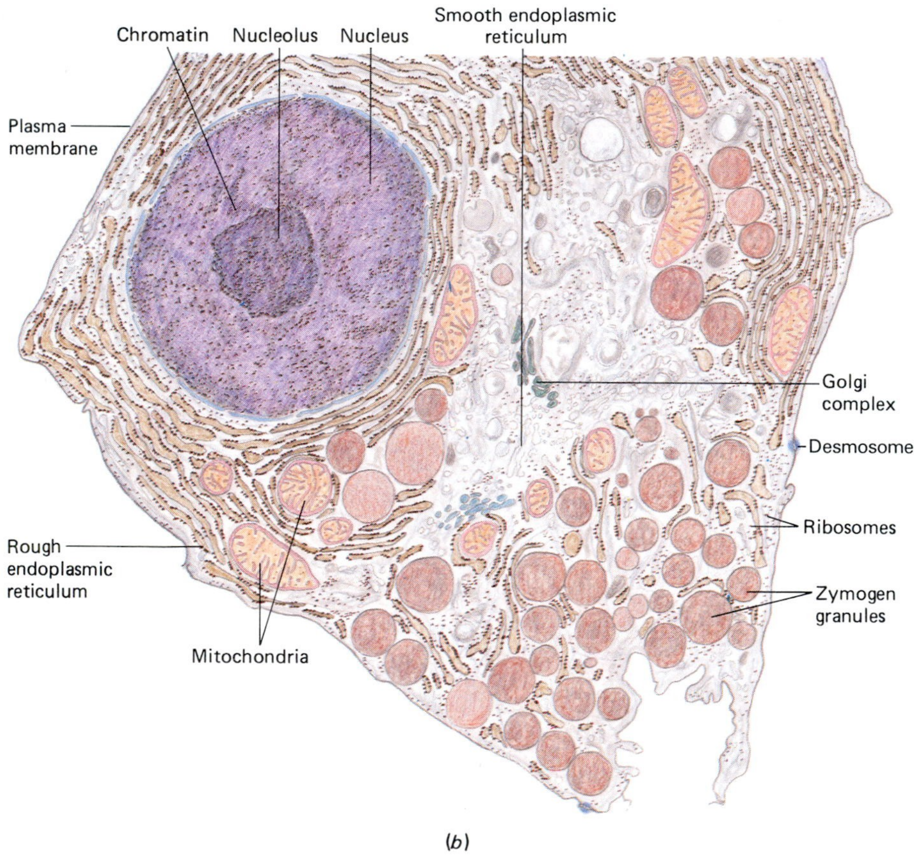

(b)

Figure 4–6 The structure of an animal cell. (*a*) Electron micrograph of a human pancreas cell, magnified ×16,000. Most of the structures of a typical animal cell are present here, however, like most of the cells of a complex, multicellular organism, this cell has certain features that permit it to carry out a specialized function. The large, circular dark bodies within the cell are zymogen granules containing inactive enzymes. Released from their storage cells and activated, these enzymes facilitate reactions such as breaking down peptide bonds during the digestion of proteins. (*b*) A drawing based on the electron micrograph, emphasizing the important structures of the cell. All these structures will be discussed in this chapter. Desmosomes, important structures in maintaining adhesion between cells, are discussed in Chapter 5. ((*a*), courtesy of Dr. Susumu Ito, Harvard Medical School.)

TABLE 4–1
Eukaryotic Cell Structures and Their Functions

Structure	*Description*	*Function*
Cell membrane	Lipid bilayer throughout which a variety of proteins are distributed in a mosaic pattern	Protection; regulates passage of materials in and out of cell; helps maintain cell shape; communicates with other cells
Endoplasmic reticulum (ER)	Network of internal membranes extending through cytoplasm; forms system of tubes and vesicles	Intracellular transport of materials
Smooth	Lacks ribosomes on outer surfaces	Produces steroids in certain cells; conduction of impulses in muscle cells
Rough	Ribosomes stud outer surfaces	Manufactures and transports proteins
Ribosomes	Nonmembranous granules composed of RNA and protein; some attached to ER	Manufacture protein
Golgi complex	Stacks of flattened membranous sacs	Packages secretions; manufactures lysosomes
Lysosomes	Membranous sacs containing hydrolytic enzymes	Release enzymes to hydrolyze proteins and other materials, including ingested bacteria; play a role in cell death
Vacuoles	Membranous sacs	Contain ingested materials or cellular secretions or wastes
Mitochondria	Sacs consisting of two membranes; inner membrane is folded to form cristae.	Site of most of the reactions of cellular respiration; power plants of cell
Plastids	Membranous; chloroplast contains disklike thylakoids.	Chloroplasts contain chlorophyll, which traps energy in photosynthesis.
Microtubules	Nonmembranous, hollow, spirally arranged tubes with walls of tubulin protein	Provide structural support; may have role in cell movement; components of centrioles, cilia, and flagella
Peroxisomes	Membranous sacs containing oxidative enzymes	Peroxisomes carry on metabolic reactions and split H_2O_2; glyoxysomes are sites of the glyoxylate cycle and take part in photorespiration
Microfilaments	Nonmembranous, rodlike, structures consisting of contractile protein	Provide structural support; may play a role in cell movement
Centrioles	Nonmembranous; a pair of hollow cylinders located within a region called the centrosome; each centriole consists of nine triple microtubules.	Mitotic spindle forms between these structures in animal cell.
Cilia	Nonmembranous; hollow tubes made of two central and nine peripheral microtubules; extend outside of cell	Movement of material outside the cell. Ciliated cells that line the respiratory tract beat to move mucus away from the lungs; not present in all cells
Flagella	Nonmembranous hollow tubes made of two central and nine peripheral microtubules; extend outside of cell; longer than cilia	Cellular locomotion; in the human body found only in sperm cells
Nucleus	Large spherical structure surrounded by a double nuclear membrane; contains nucleolus and chromosomes	Control center of the cell; contains the chromosomes
Nucleolus	Nonmembranous; rounded granular body within nucleus; consists of RNA and protein	Assembles ribosomes; may have other functions
Chromosomes	Nonmembranous; long threadlike structures composed of DNA and proteins	Contain the genes (hereditary units) that govern the structure and activity of the cell

ENDOPLASMIC RETICULUM AND RIBOSOMES

Electron microscopy reveals that the cytoplasm of a typical cell contains a complex maze of membranes with the appearance of spaghetti-like tubular strands. This complex of membranes is the **endoplasmic reticulum,** or **ER** (Fig. 4–7). Although the ER always constitutes a system of fluid-filled spaces enclosed by membranes, its appearance and extent may vary greatly in different cells. The tightly packed sheets of the ER may form tubules, some 50 to 100 nm in diameter. In certain regions of the cell the cavities of the ER may be expanded, forming flattened sacs called **cisternae.** In most cells the spaces of the ER appear to be interconnected. The cell membrane, nuclear membrane, and membranes of the ER all appear to be con-

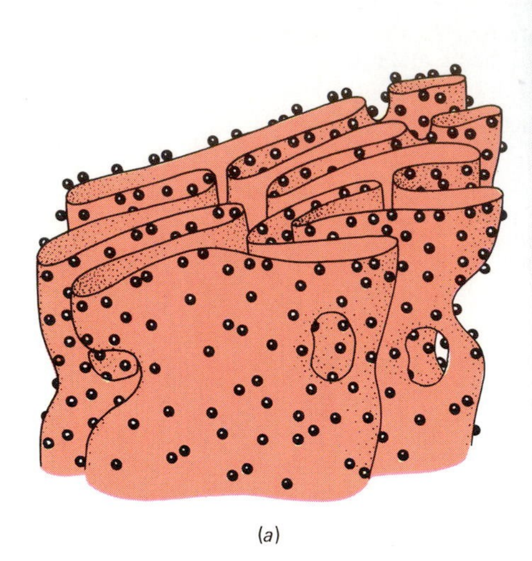

(a)

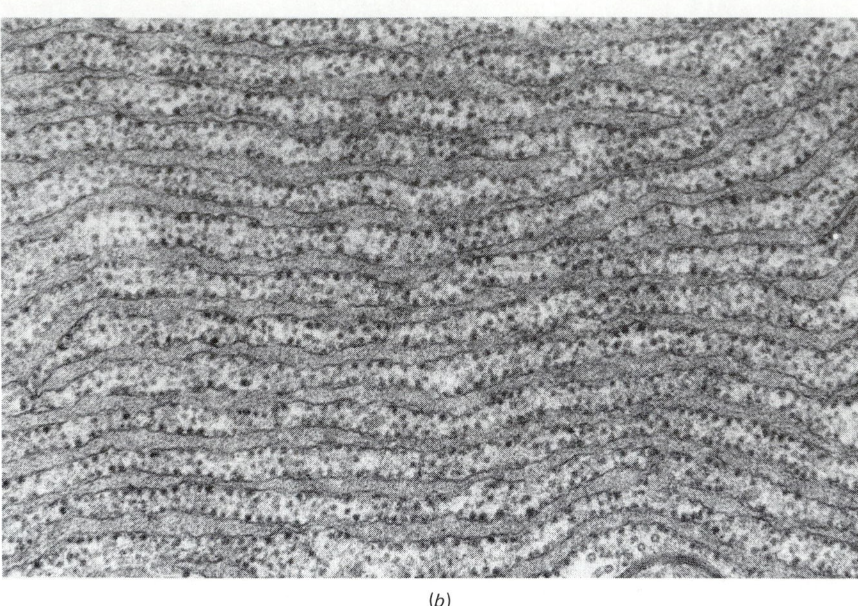

(b)

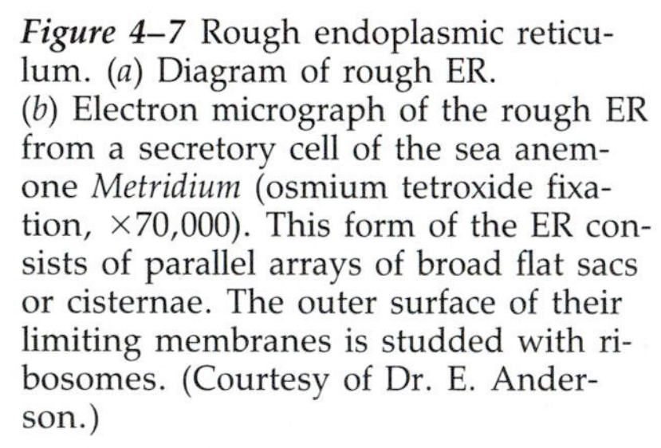

Figure 4–7 Rough endoplasmic reticulum. (*a*) Diagram of rough ER. (*b*) Electron micrograph of the rough ER from a secretory cell of the sea anemone *Metridium* (osmium tetroxide fixation, ×70,000). This form of the ER consists of parallel arrays of broad flat sacs or cisternae. The outer surface of their limiting membranes is studded with ribosomes. (Courtesy of Dr. E. Anderson.)

nected in some electron micrographs; some investigators think that all of these membranes have a common origin.

The membranes of the ER divide the cytoplasm into a multitude of compartments in which different groups of enzymatic reactions may occur. In fact, the membranes of the ER contain a variety of enzymes and serve as a framework of surfaces on which some of the biochemical reactions of the cell take place. Many enzymes work most effectively when they are anchored to membranes, and the extensive and complex folding of the ER provides an enormous surface area for enzymatic activities. The ER also functions as a system for transporting a variety of chemical substances from one part of the cell to another, and perhaps also to the exterior of the cell and into the nucleus. The cisternae of the ER may serve as temporary storage areas for certain substances. Still another important function of the ER is synthesis of certain types of compounds (including lipids, proteins, and complex carbohydrates).

Two types of ER can be distinguished, smooth and rough. **Smooth ER (SER),** so called because the outer surfaces of its membranes have a smooth appearance, produces steroids in certain cells. In muscle cells it plays a role in calcium storage and affects muscle contractions. In some cells it functions in cellular secretion. The smooth ER of liver cells functions in lipid metabolism and also in detoxifying certain poisons, as well as drugs. When an experimental animal is injected with the barbiturate phenobarbital, the amount of SER in the liver cells increases dramatically over a period of a few days and enzymes known to break down phenobarbital increase in concentration within the smooth ER membranes.

Rough ER (RER) has a granular appearance due to the presence of tiny organelles called **ribosomes** that stud its outer walls. Ribosomes are the site of protein synthesis. Rough ER is especially extensive in cells that synthesize proteins for export from the cell, such as those that secrete digestive enzymes. Rough and smooth ER are continuous with one another, and the cell can change the relative amounts of smooth and rough ER to meet its changing needs.

Ribosomes occur in all kinds of cells, from bacteria to complex plant and animal cells. A bacterium such as *Escherichia coli* contains some 6000 ribosomes in its single cell; a rabbit reticulocyte (the precursor of a red blood cell) contains some 100,000 ribosomes. Composed of RNA and protein, a ribosome consists of two subunits, shaped like overstuffed chairs, which are combined to form the active protein-synthesizing unit (see Fig. 14–9). Ribosomes are assembled in the nucleus and then pass out into the cytoplasm. In many cells, clusters of five or six ribosomes termed **polyribosomes** (see Fig. 14–7) appear to be the functional unit of protein synthesis. Not all ribosomes are attached to the ER; some float freely in the cytoplasm.

The **Golgi complex,** first described in 1898 by the Italian microscopist Camillo Golgi, is present in almost all cells except mature red blood cells. By electron microscopy the Golgi complex is seen to consist of layers of platelike membranes, which may be distended in certain regions to form **vesicles,** or **sacs,** filled with cell products (Fig. 4–8). In animal cells the Golgi complex is often located at one side of the nucleus.

The Golgi complex functions principally as a processing and packaging plant and is most highly developed in cells that are specialized to secrete products. As protein is manufactured along the rough ER, it is sealed off in little packets of membrane, forming tiny vesicles. These vesicles pass through the ER to the Golgi complex, with which they fuse to form new Golgi complex membranes. Within the Golgi complex the protein secretion may be concentrated by the actions of its membranes. During storage, proteins may be modified. Often some carbohydrate component is added to form a glycoprotein. The protein is then packaged within a larger sac made of membranes from the Golgi complex. Such **secretory vesicles** (called **secretory granules** when the secretion within them is dense) are released from the Golgi complex and move to the cell membrane. The secretory vesicle fuses with the cell membrane and releases its contents to the exterior of the cell. Cells lining the digestive tract produce and secrete digestive enzymes in this manner, and cells known as goblet cells produce and secrete mucus in this way. In an actively secreting cell such as a goblet cell, it is thought that the Golgi complex completely renews its membranes every 30 minutes!

The Golgi complex of plant cells produces a variety of extracellular polysaccharides used as components of the cell wall. (However, cellulose is produced by the cell membrane in most plants.) The Golgi complex in plant cells usually consists of separate stacks of membranes dispersed throughout the cell.

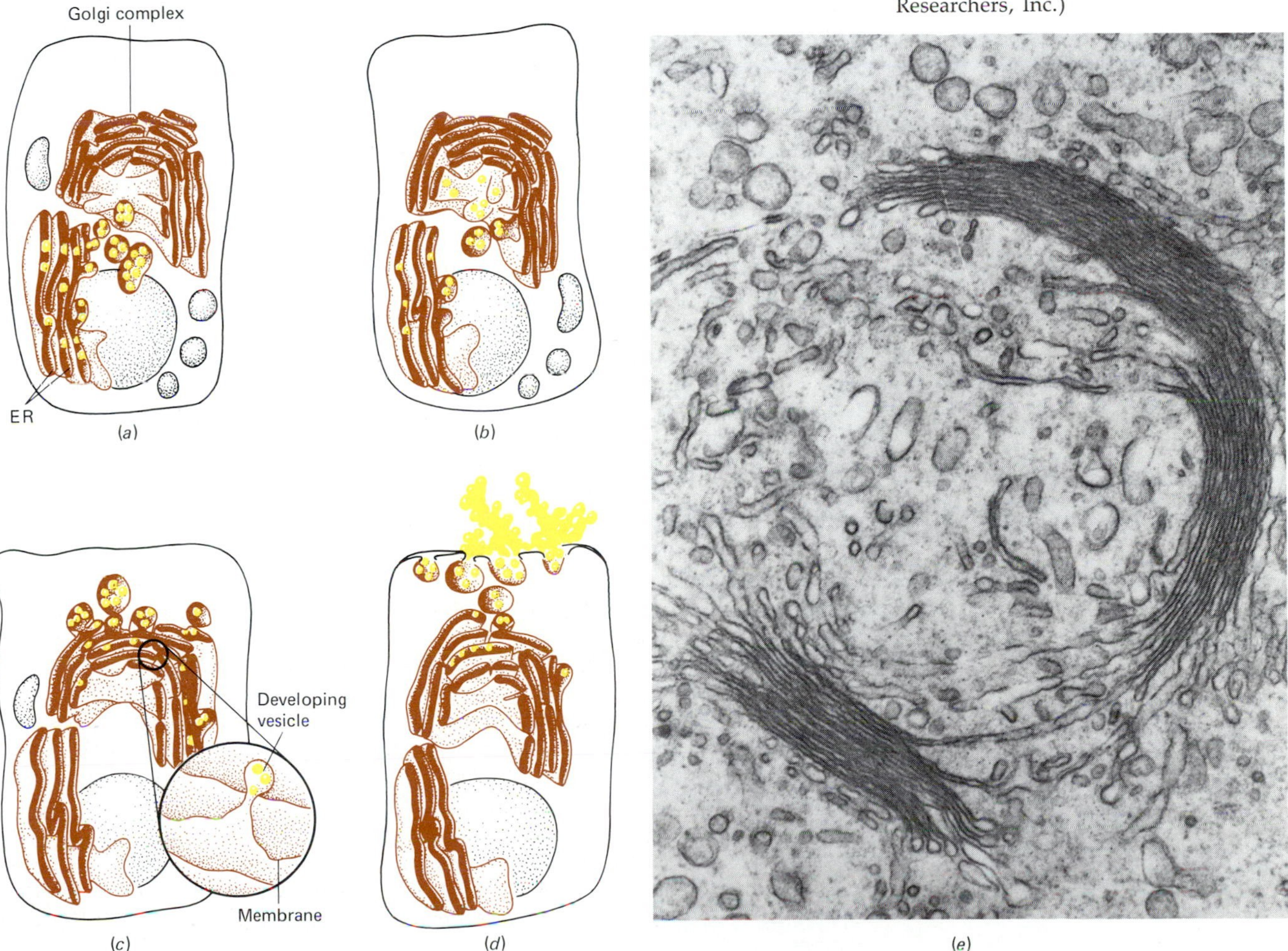

Figure 4–8 The Golgi complex. (*a*)–(*d*) The Golgi complex during a secretory cycle in a goblet cell. The tiny mucus droplets join to form larger drops, which are then released from the cell. (*e*) Electron micrograph of a section through a Golgi complex from a sperm cell of a ram. ((*e*), Don Fawcett, Photo Researchers, Inc.)

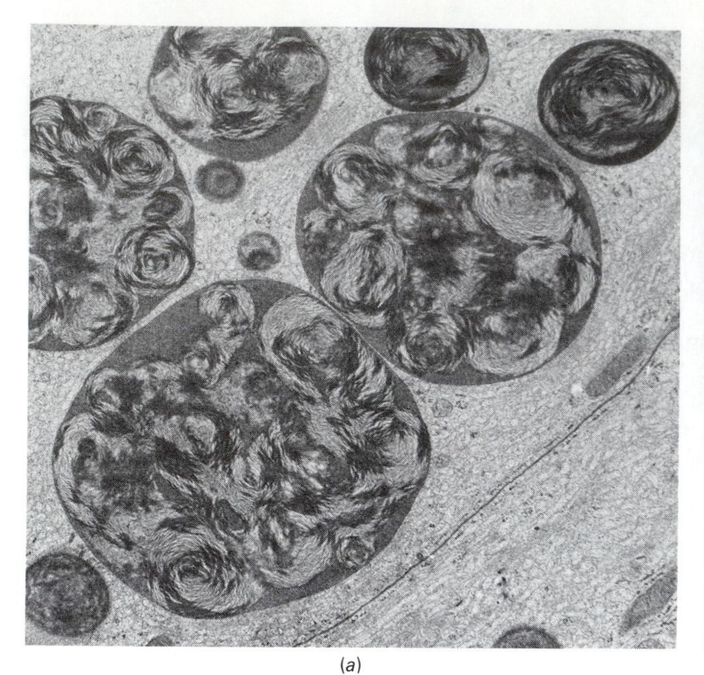
(a)

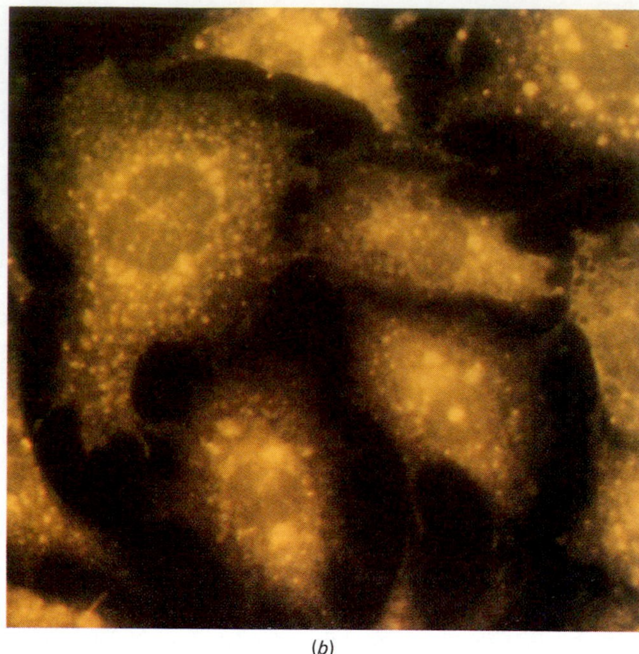
(b)

Figure 4–9 Lysosomes. (*a*) Electron micrograph showing different stages of lysosome formation. Primary lysosomes bud off from the Golgi complex. After a lysosome encounters material to be digested it is known as a secondary lysosome. The secondary lysosomes shown here contain various materials being digested. (*b*) Distribution of lysosomes in a cell. ((*a*), Don Fawcett, Photo Researchers, Inc.; (*b*), courtesy of Dr. Paul Gallup.)

LYSOSOMES

Intracellular digestive enzymes are manufactured along the rough ER and then transported to the Golgi complex. There the enzymes are enclosed in a bit of Golgi complex membrane, which is then pinched off and separated from the Golgi. Each little vesicle containing digestive enzymes is a **lysosome.** About 40 different enzymes have been identified within the lysosome. These enzymes are most active near the pH of 5.

The lysosomes are dispersed throughout the cytoplasm (Fig. 4–9). When a white blood cell ingests bacteria or some other scavenger cell (e.g., a macrophage) ingests debris or dead cells, this foreign matter is surrounded by a vesicle consisting of part of the cell membrane (see Fig. 5–11). One or more of the cell's lysosomes then fuse with the vesicle containing the foreign matter. The hydrolytic enzymes[1] can digest proteins, polysaccharides, and nucleic acids. Lysosomes that have encountered materials for digestion are known as secondary lysosomes. The lysosomal membrane itself is able to resist the digestive action of these powerful enzymes and prevent them from digesting the contents of the cell.

In a cell that is short of fuel, lysosomes may break down organelles so that their component molecules may be used as fuel. This sort of self-cannibalizing is termed **autophagy** (eating one's self). When a cell dies, lysosomes release their enzymes into the cytoplasm, where they break down the cell itself. This self-destruct system accounts for the rapid deterioration of many cells following death. Christian deDuve, the Belgian biochemist who discovered lysosomes, referred to them as "suicide bags."

During development of the embryo, some cells must die as structures are sculpted out of formless cellular masses. For example, in the early human embryo the hand develops first as a solid stump. As the individual fingers are fashioned, the cells forming the tissue between them must be destroyed. The destruction of these cells is carried out selectively and precisely by lysosomal enzymes.

Studies suggest that some forms of tissue damage and the aging process itself may be related to leaky lysosomes. Rheumatoid arthritis is thought to result in part from damage done to cartilage cells in the joints by enzymes that have been released from lysosomes. Cortisone-type drugs, which are used as antiinflammatory agents, stabilize lysosome membranes so that leakage of damaging enzymes is reduced.

[1] A hydrolytic enzyme splits a larger compound (such as a fat or polypeptide) into smaller units with the addition of water.

Lysosomes have been identified in almost all kinds of eukaryotic cells. Their occurrence in plant cells, however, is a matter of some debate. Some investigators have suggested that other specialized organelles found in plant cell cytoplasm may have functions similar to those of lysosomes.

MITOCHONDRIA

Cells contain tiny power plants called **mitochondria** (singular, **mitochondrion**), in which most of the reactions of cellular respiration (fuel breakdown with the release of energy) take place (Chapter 9). Mitochondria are most numerous in cells that are very active. A liver cell may contain more than 1000 mitochondria, and a muscle cell even more. In addition to their role in cellular respiration, mitochondria are thought to regulate calcium concentration within the cell. The size of mitochondria may vary, ranging from 2 to 8 μm in length (and 0.2 to 1.0 μm in diameter). They may appear as spheres, rods, or threads. A typical sausage-shaped mitochondrion is shown in Figure 4–10.

When living cells are examined by phase contrast microscopy, their mitochondria can be seen to move, to change size and shape, and to fuse with other mitochondria to form bigger structures or to cleave to form smaller ones. Mitochondria are usually concentrated in the region of the cell with the highest rate of metabolism.

Each mitochondrion is bounded by a double membrane. Both the outer and inner membranes consist of **lipid bilayers,** in which are embedded a variety of protein molecules. The outer membrane forms a smooth outer boundary. The inner membrane is folded repeatedly into parallel plates, called **cristae,** that extend into the center of the mitochondrial cavity (Fig. 4–10). These plates may meet and fuse with folds extending in from the opposite side. The shelflike cristae contain many of the enzymes necessary for cellular respiration. The semifluid material within the inner compartment, termed the **matrix,** contains other enzymes needed for cellular respiration.

PLASTIDS

The cells of plants and algae contain **plastids,** cytoplasmic organelles involved in the synthesis or storage of food. **Chloroplasts,** the most important plastids, contain the green pigment **chlorophyll,** which imparts the green color to plants and is of paramount importance in trapping the energy of sunlight for photosynthesis. In addition to chlorophyll, chloroplasts contain a variety of yellow and orange pigments termed **carotenoids.** A one-celled alga may have only a single large chloro-

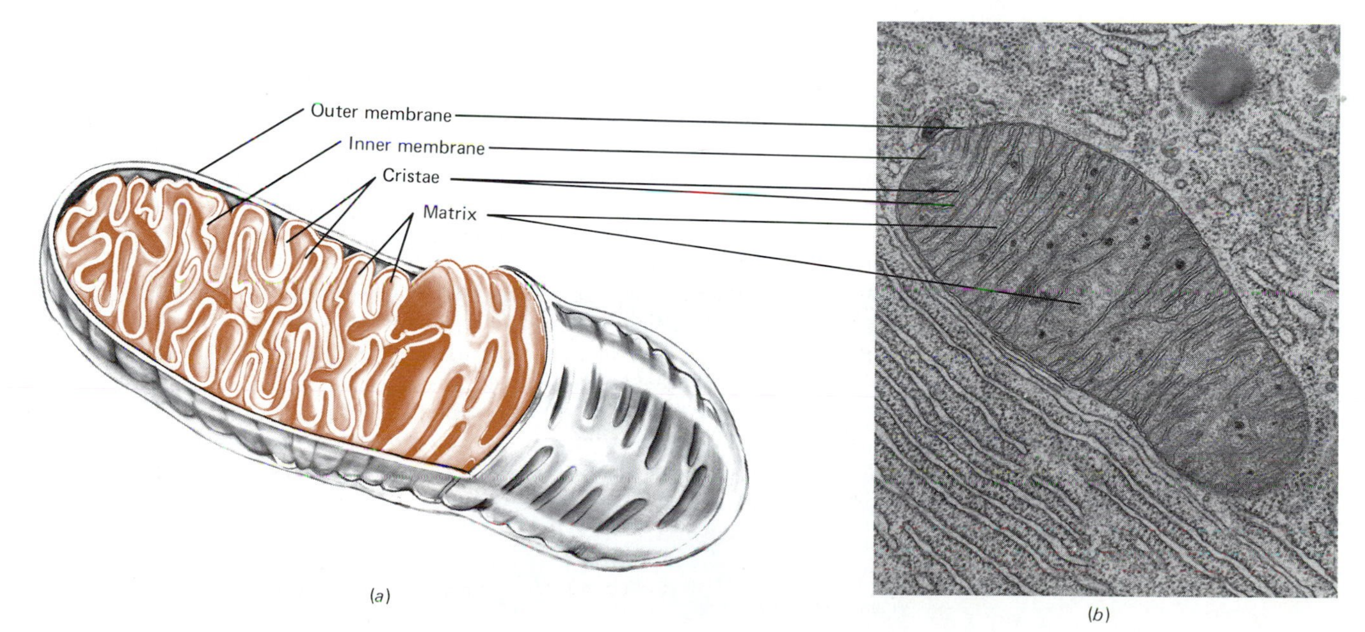

Figure 4–10 The mitochondrion. (*a*) Diagram of a mitochondrion cut open to show the cristae. (*b*) Electron micrograph of a typical mitochondrion from the pancreas of a bat showing the cristae and matrix (approximately ×80,000). Note the extensive rough ER at the lower left and some lysosomes at the upper right. (Courtesy of Dr. Keith R. Porter.)

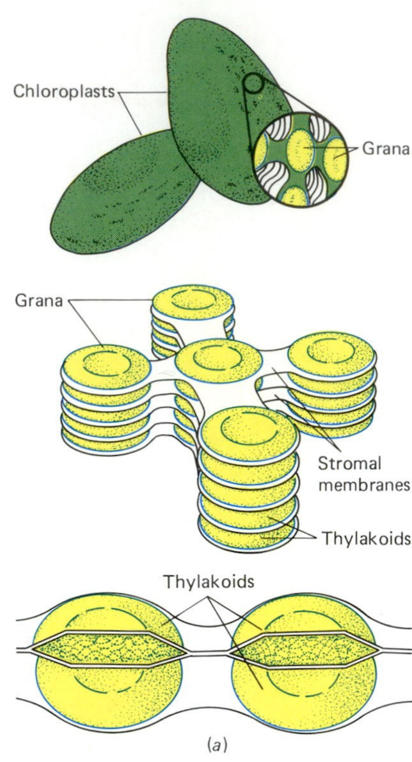

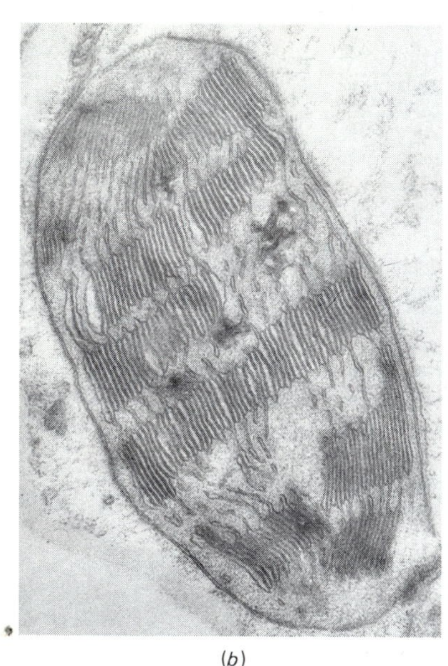

Figure 4–11 The chloroplast. (*a*) A small portion of a chloroplast is shown magnified to illustrate the arrangement of the thylakoids. (*b*) Electron micrograph of a chloroplast from a leaf of the tobacco plant, *Nicotiana rustica,* showing the fine structure of the grana (approximately ×30,000). Note the alternate layers of protein and lipid in the grana within the chloroplast, and the membrane separating the chloroplast from the surrounding cytoplasm. Chloroplast structure is shown in more detail in Figure 8–16. (Courtesy of Dr. E. T. Weier.)

plast, but cells of the flowering plants often possess 20 to 100 of these interesting organelles.

The chloroplast is not a simple bag of chlorophyll and carotenoids. Indeed, the capacity of chlorophyll to capture light energy depends upon its precise distribution within the chloroplast. The chloroplasts of plant cells are typically disk-shaped structures bounded by inner and outer membranes. Inside the chloroplast, a third distinct membrane is organized into disklike sacs called **thylakoids** (Fig. 4–11). Chlorophyll is present within the thylakoids, and the light-dependent reactions of photosynthesis (those involved in capturing the energy of sunlight) take place within these structures. Chloroplast structure and function will be discussed in greater detail in Chapter 8.

Like mitochondria, chloroplasts contain DNA, RNA, and ribosomes and have the ability to synthesize some proteins. Both types of organelles are also able to grow and to divide, forming daughter organelles.

Plant cells also contain colorless plastids termed **leukoplasts** (see Fig. 3–4), which serve as centers for the storage of starch and other materials. Starch is commonly deposited in leukoplasts present in storage roots and stems and in seeds. **Chromoplasts,** a third type of plastid, contain pigments that give their characteristic colors to flowers, ripe fruits, and autumn leaves.

PEROXISOMES

Peroxisomes are membrane-bound organelles that contain oxidative enzymes. They use oxygen to carry out metabolic reactions. Perhaps 20% of fatty acid oxidation (breakdown) takes place within peroxisomes. This reaction, as well as other reactions that take place within peroxisomes, produce hydrogen peroxide, a lethal oxidant within cells. However, before the cell is damaged, the hydrogen peroxide (H_2O_2) is split to yield H_2O and O_2. This is promoted by enzymes such as catalase, which are also present in the peroxisome. The hydrogen peroxide is used by catalase to oxidize various substrates, including alcohol. Peroxisomes in liver and kidney cells may be important in detoxifying certain compounds such as ethanol (the alcohol in alcoholic beverages). About 50% of the ethanol we drink is thought to be oxidized to a compound called acetaldehyde within the peroxisomes.

One type of peroxisome in plant cells cooperates with chloroplasts and mitochondria in the light-dependent process of photorespiration. In this process oxygen is consumed and carbon dioxide is released (discussed in Chapter 8). Photorespiration takes place in the leaves of many kinds of plants. Another type of plant cell peroxisome, referred to as a **glyoxysome,** is important in young plants before they are able to carry on photosynthesis. The glyoxysome contains the enzymes necessary for the conversion of fatty acids stored in the seed to sugars by way of a series of reactions known as the glyoxylate cycle. The sugars are used by the young plant as an energy source and as a component needed to synthesize other compounds. Animal cells lack glyoxysomes and are unable to convert fats into carbohydrates.

MICROTUBULES AND MICROFILAMENTS

Most cells have hollow cylindrical cytoplasmic subunits (about 25 nm in diameter) termed **microtubules** (Fig. 4–13). They appear to be important in maintaining and controlling the shape of the cell. Microtubules play a role in such cellular move-

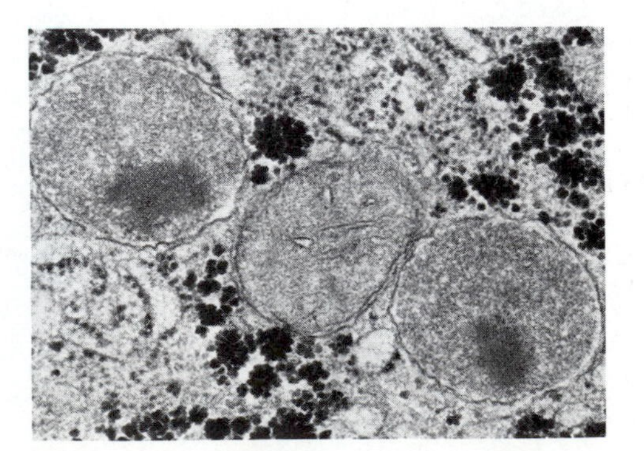

Figure 4–12 Transmission electron micrograph of a rat liver cell showing two circular peroxisomes with a section of a mitochondrion (center). The peroxisomes each show a region of crystalline material that may consist of oxidizing enzymes (magnification ×66,000). (Courtesy of Drs. Christian deDuve and Helen Shio, Rockefeller University.)

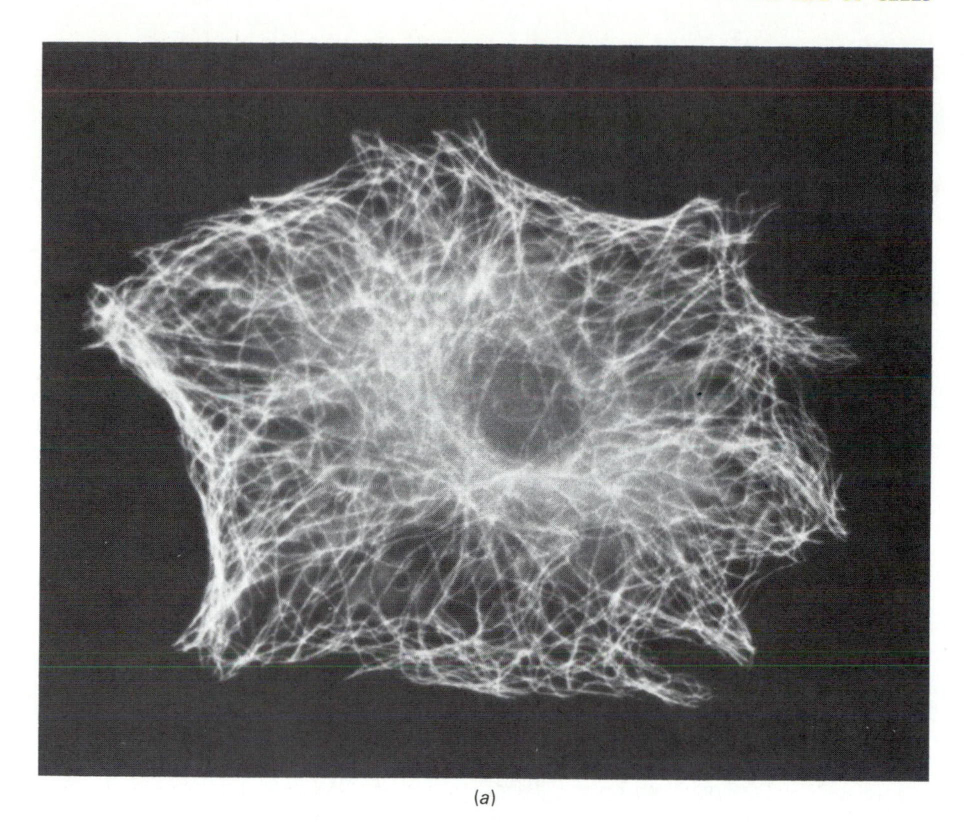
(a)

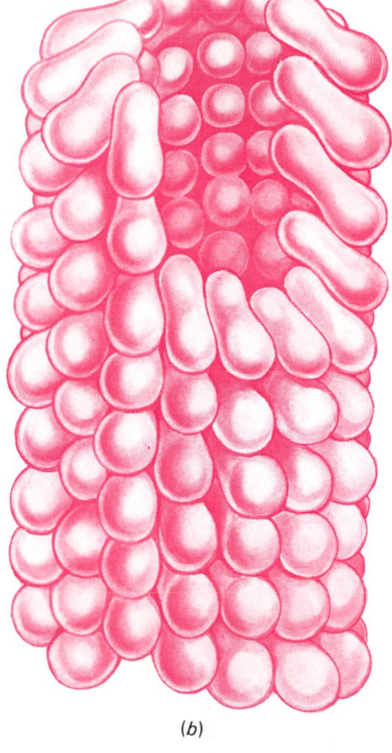
(b)

Figure 4–13 Microtubules. (*a*) Electron micrograph of a human cell showing the extensive distribution of microtubules throughout the cell. This cell was stained with fluorescent antitubulin, permitting the tubulin that makes up the microtubules to be viewed. (*b*) Structure of a microtubule. Part of the microtubule has been split in half lengthwise. Note the spiral arrangement of the component units and their two-lobed "barbell" shape. ((*a*), courtesy of Drs. Keigi Fujiwara, Hugh Randolph Byers, and Elena McBeath, Department of Anatomy, Harvard Medical School.)

ments as the movement of chromosomes on the spindle formed during mitosis (nuclear division) and are the major structural components of cilia and flagella.

Microtubules are composed largely of the protein **tubulin,** a dimer[1] of two very similar α and β subunits, each with a molecular weight of about 55,000. The walls of these hollow cylinders are about 4.5 to 7 nm thick. (This corresponds to the dimensions of the glycoprotein subunits and suggests that the walls are one molecule thick.) Microtubules can grow by the addition of more α and β subunits or can shorten by the disassembly of subunits. Microtubules within axons (extensions of nerve cells) play a role in the rapid transport of proteins and other molecules (such as the hormones of the posterior pituitary) down the axon to its tip (Chapter 39).

Solid cytoplasmic **microfilaments** are present in most cells in addition to the hollow microtubules. Composed of strings of protein molecules, these filaments play additional roles in cell structure and movement. The cytoplasm of skeletal muscle fibers contains many long, thin filaments termed **myofibrils.** Very thin filaments, composed of the protein actin, interact with thicker filaments composed of the protein myosin in the process of muscle contraction (Chapter 32). The actin filaments are identical with microfilaments found in other types of cells. Microfilaments composed of actin are associated with cellular movement such as the flowing of cytoplasm in amoeboid motion and in cytoplasmic streaming in certain plant cells. In many types of cells the microfilaments and microtubules form a flexible framework referred to as the **cytoskeleton.**

THE MICROTRABECULAR LATTICE

Even the **ground substance** (the formless background of cytoplasm) of the cell, which appears to be entirely homogeneous and structureless by light microscopy or by conventional electron microscopy, has been shown to have an internal structure when examined by high-voltage (1 million volts) electron microscopy. The cytoplasmic ground substance contains an irregular three-dimensional lattice of very slender protein threads that extend throughout the cytoplasm and are attached to the cell membrane. The interlinked filaments form a three-dimensional "spider web" in which are suspended the various intracellular organelles. The microtu-

[1] A dimer is a compound formed by the combination of two similar simpler units.

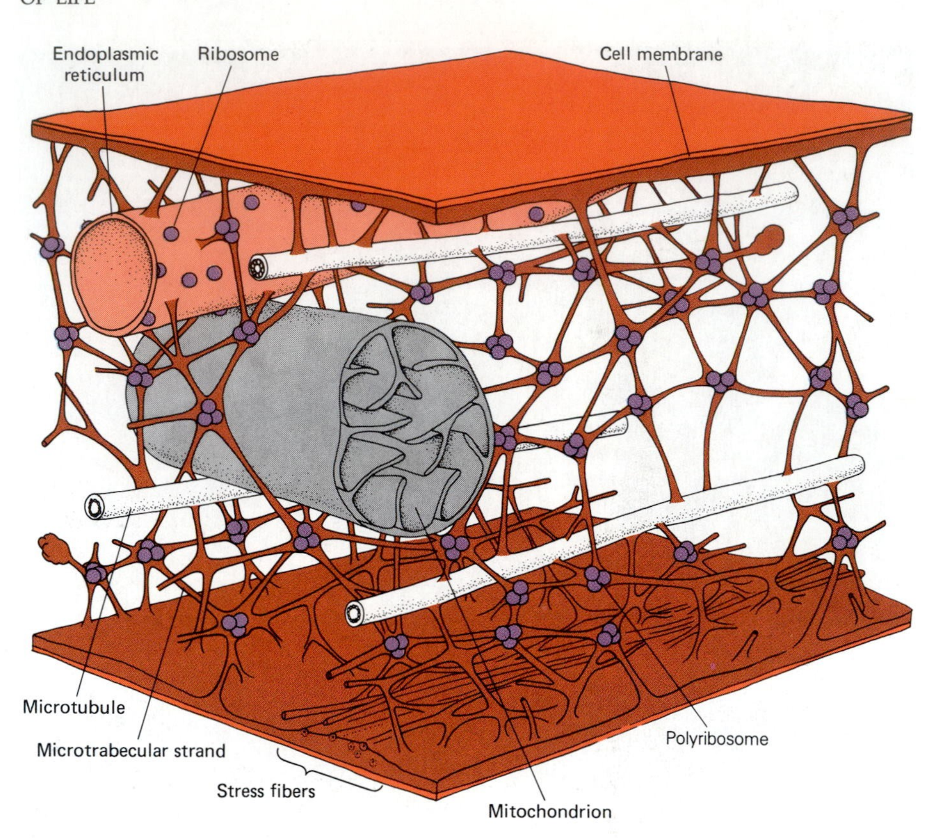

Figure 4–14 Model of the microtrabecular lattice drawn about 200,000 times its actual size. The lattice is traversed by microtubules, which appear to be attached to the trabeculae of the lattice. (Drawing based on work of Keith Porter.)

bules and microfilaments are coated with a material very similar to that composing the individual lattice filaments. Microtubules, microfilaments, endoplasmic reticulum , and mitochondria are integrated with the lattice and are suspended in it. This network has been named the **microtrabecular lattice** by Keith Porter, the American cell biologist who first observed it (Fig. 4–14).

The microtrabecular lattice extends throughout the cytoplasm and links the subcellular organelles, microtubules, and microfilaments into a structural and functional unit. The individual fibers of the lattice are 6 to 16 nm thick. In cells exposed to the drug cytochalasin B, which inhibits many kinds of cellular movement, the microtrabecular lattices become thickened and coarse, and the spaces between the lattice fibers become enlarged. This finding suggests that the microtrabecular lattice plays an important role in cell movement.

Many of the enzymes previously believed to be "soluble" in the cytoplasm may be bound to this microtrabecular lattice. Indeed, they may be bound in a nonrandom orientation; enzymes that act in sequence may be located on the lattice in the correct spatial orientation so that they can pass the substrate molecules (the molecules they act upon) from one to the next.

The microtrabecular lattice serves as a sort of intracellular musculature, undergoing local contractions and changes in shape that continually redistribute and reorient the intracellular organelles as the cell goes about its various functions. For example, the movement of granules within chromatophores (pigment cells) in the skin of certain invertebrates, fishes, frogs, and reptiles is under hormonal or nervous control. These movements enable the animal to lighten or darken its skin color rapidly to match the color of its surroundings or to present a color display that might frighten a predator (Fig. 4–15). The pigment granules can be moved quite rapidly by the microtrabecular lattice. The movement of the pigment granules is accompanied by marked changes in the structure of the lattice. The aggregation of the pigment granules into the center of the cell is accompanied by a shortening and thickening of the microtrabecular threads. This draws the granules into clusters and moves them along paths defined by microtubules to the central region of the cell. In the reverse process, by which the pigment granules are dispersed through the cell, the microtrabecular fibers elongate and move the pigment granules back to the peripheral region of the cell along the microtubules.

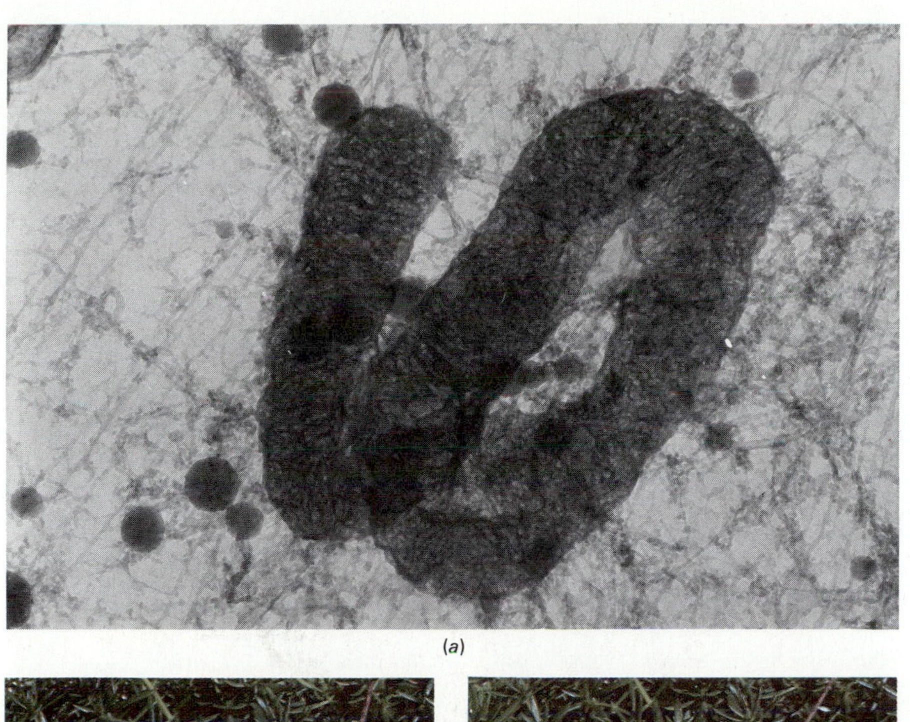

(a)

(b)

(c)

Figure 4–15 Chromatophores (pigment cells). (*a*) Photomicrograph of a chromatophore (approximately ×85,000). A large, S-shaped mitochondrion is suspended in microfilaments, the very thin, wirelike structures. The dark circular structures are pigment granules. (*b*), (*c*) The green anole lizard, *Anolis carolinesis*, changes color not, as one might guess for the purpose of camouflage, but according to its activity. Basking in the sun, the lizard is usually brown colored. When threatened or aggressive, the animal turns green. The color change can also be the result of other environmental cues, including stimuli such as changes in temperature and moisture. Any of these factors act to release hormones that in turn influence the movement of pigment in the microtrabecular lattice. ((*a*), courtesy of Drs. Keigi Fujiwara, Hugh Randolph Byers, and Elena McBeath, Department of Anatomy, Harvard Medical School; (*b*), (*c*) Luci Giglio.)

CENTRIOLES

Each animal cell possesses two tiny **centrioles,** organelles that function in cell division. A centriole is a hollow cylinder composed of nine triple microtubules (Figs. 4–16 and 4–27). Centrioles are located within a dense area of cytoplasm, the **centrosome,** which is usually near the nucleus.

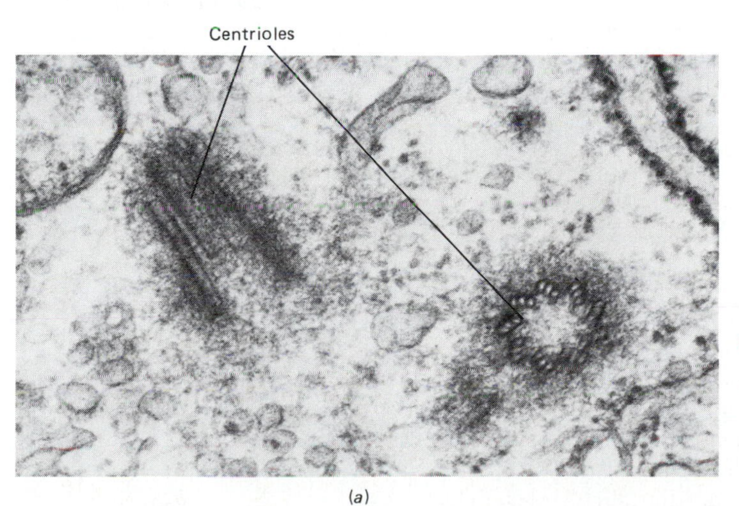

(a)

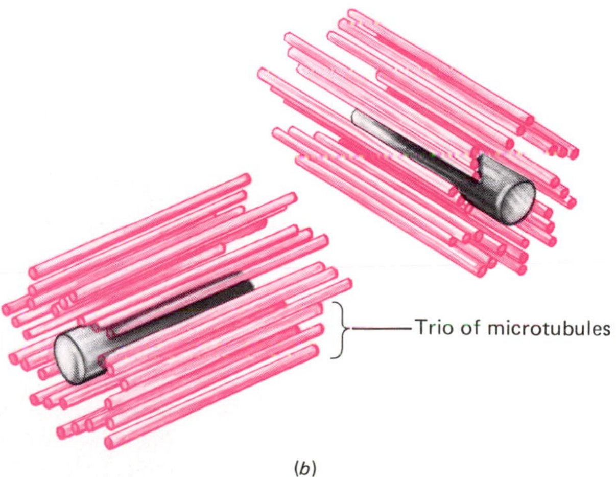

(b)

Figure 4–16 Centrioles. (*a*) Electron micrograph of a pair of centrioles from monkey endothelial cells (×73,000). (*b*) A line drawing of the centrioles. Note that one centriole has been cut longitudinally and one transversely. See also Figure 4–27. ((*a*), B.F. King, School of Medicine, University of California/BPS.)

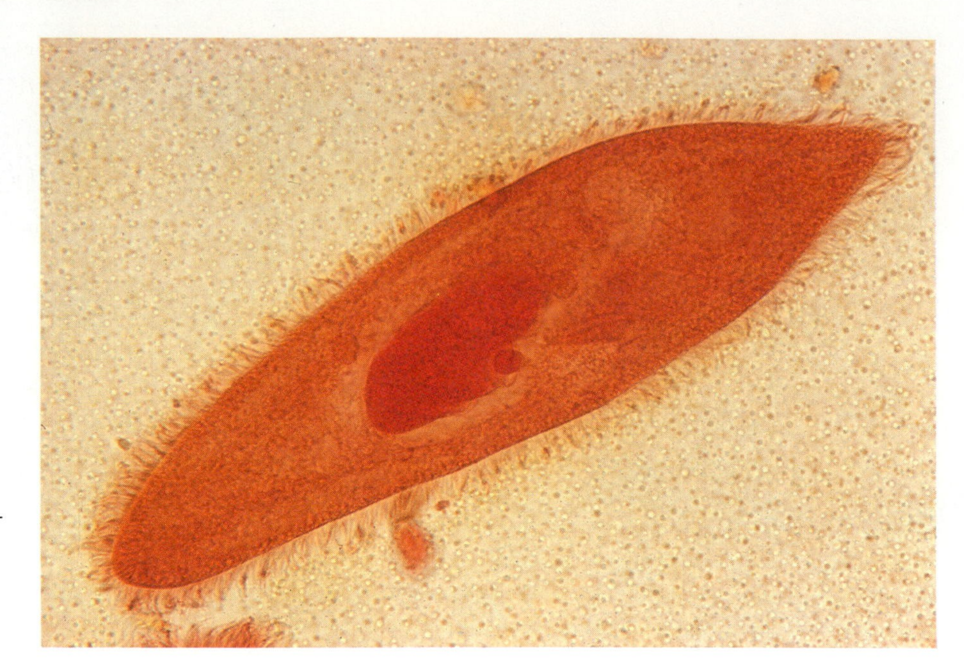

Figure 4–17 Photomicrograph of a paramecium showing numerous cilia along its surface. Paramecia are protists that propel themselves by means of their cilia. (Michael Abbey, Photo Researchers, Inc.)

CILIA AND FLAGELLA

Certain cells have movable whiplike structures projecting from the free surface. If a cell has one, or only a few, of these appendages that are relatively long in proportion to the size of the cell, they are termed **flagella** (singular, **flagellum**). If a cell has many short appendages, they are termed **cilia** (singular, **cilium**) (Fig. 4–17). Flagella and cilia are very similar in structure, and both function either in moving the cell through the surrounding liquid or in moving liquids and particles across the surface of the cell. Flagella or cilia are commonly found on unicellular and small multicellular organisms (Fig. 4–17) and on the sperm cells of animals. They may be the principal means of locomotion of such cells. Cilia also occur on the cells lining the internal ducts of many animals; their beating assists in moving material through these passageways.

The cilia and flagella of eukaryotic cells are remarkably uniform in their internal structure, whether they are present in a protist cell or in a large organism. Each cilium or flagellum consists of a slender, cylindrical stalk covered by an extension of the cell membrane and containing a cytoplasmic matrix with 11 groups of microtubules embedded in it (Fig. 4–18). Nine pairs of microtubules are arranged around two central microtubules. This is often referred to as the **9 + 2 arrangement** of microtubules. At the base of the stalk, within the main portion of the cell, is a **basal body,** whose structure is similar to that of a centriole. It is composed of nine triple microtubules. The basal body is essential to the function of the cilium or flagellum. It is the structure from which the stalk arises; in some cells, there are visible tufts of fibers that extend from the basal bodies down into the cytoplasm of the cell. The microtubules of which cilia and flagella are composed seem to play an important part in their movement, perhaps by complex sliding movements along one another that bend the whole organelle.

VACUOLES

A **vacuole** is a bubble-like space filled with watery fluid and bordered by a membrane. Vacuoles are present in many types of cells but are most common in the cells of plants and of certain protists (Fig. 4–19). Most protozoa have food vacuoles containing food undergoing digestion, and many have contractile vacuoles that remove excess water from the cell. More than half the volume of a plant cell may be occupied by a large central vacuole containing stored food, salts, pigments, and wastes. Plants lack waste disposal systems and often utilize vacuoles as a storage place for toxic materials. Such waste products are often aggregated as small crystals.

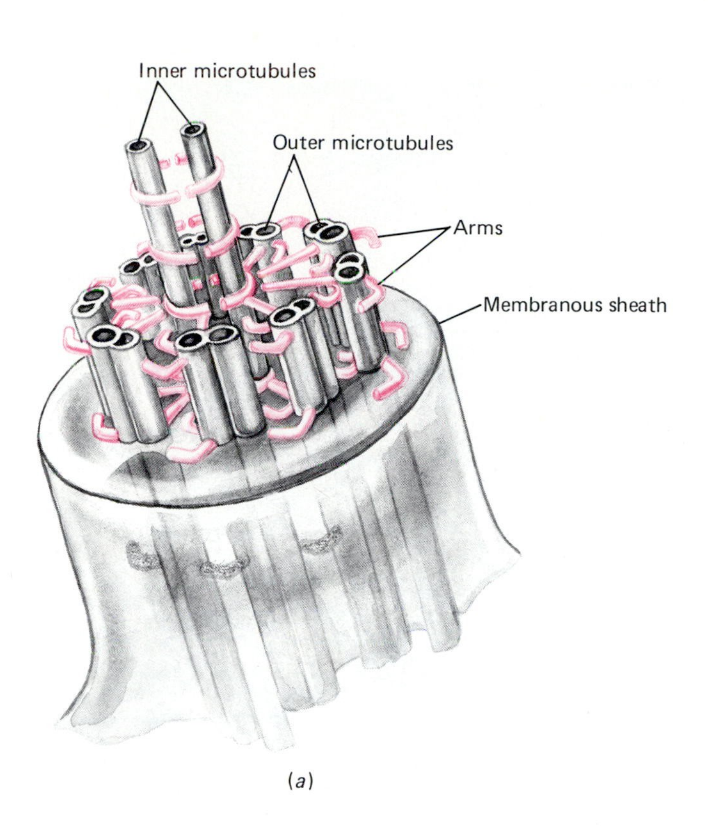

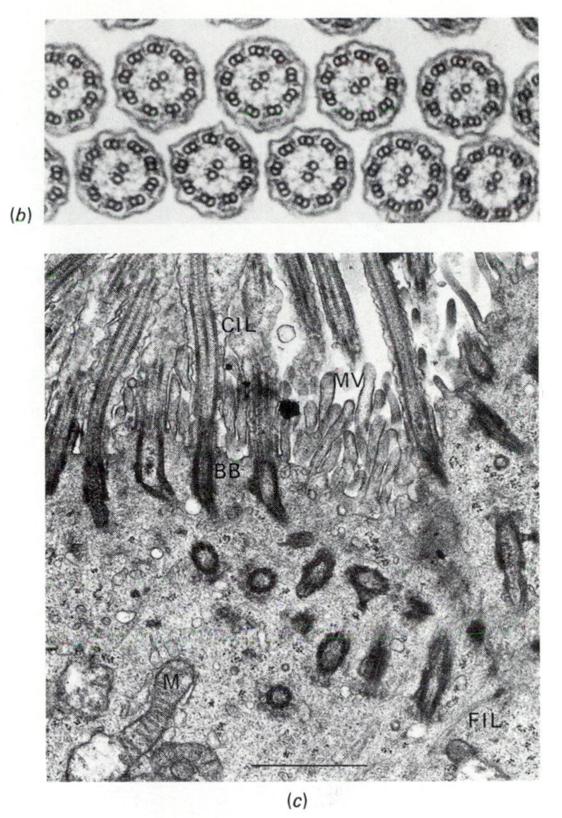

Figure 4–18 Cilia. (*a*) Structure of a cilium. (*b*) Electron micrograph of cross sections through cilia showing the 9 + 2 arrangement of their microtubules. (*c*) Electron micrograph of epithelial cells with cilia (approximately ×10,000). *CIL*, cilia; *BB*, basal body of cilium; *CM*, cell membrane; *MV*, microvilli (tiny extensions of the cytoplasm at the cell surface); *M*, mitochondrion. Black line indicates length of one micrometer. ((*b*), courtesy of Omikron, Photo Researchers, Inc.; (*c*), courtesy of Dr. Lyle C. Dearden.)

THE CELL NUCLEUS

Each cell contains a spherical or oval organelle known as the **nucleus** (Fig. 4–20). In some cells the nucleus has a relatively fixed position, somewhere near the center. In others it may move around freely and be found almost anywhere in the cell. The nucleus, as we will see, is of prime importance in the control of cellular processes. It contains the genes responsible for determining the chemical and physical traits of the cell and of the organism (see Focus on *Acetabularia*). Most cells have one nucleus, although there are a few exceptions. For example, the mammalian red blood cell loses its nucleus in the process of maturation. Skeletal muscles have several nuclei per cell. Other exceptions to the general rule of one nucleus per cell include the slime molds (Chapter 20) and the pollen tubes of seed plants (Chapter 21)

(Text continued on page 86.)

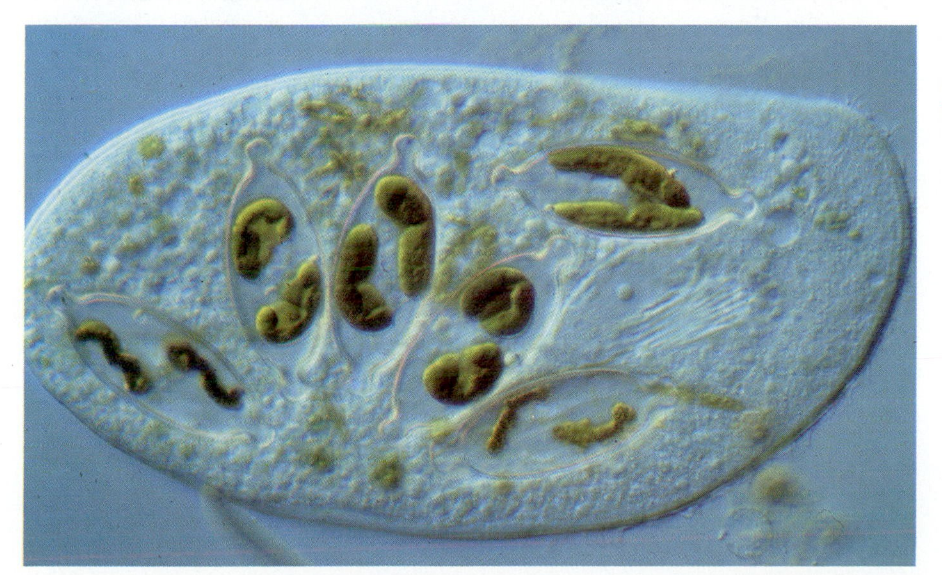

Figure 4–19 The protozoan *Chilodonella*. Inside its body are vacuoles that contain ingested diatoms (diatoms are small, photosynthetic plantlike protists). From the number of diatoms scattered about its insides, one might judge that *Chilodonella* has a rather voracious appetite (magnification ×150). (Walker England, Photo Researchers, Inc.)

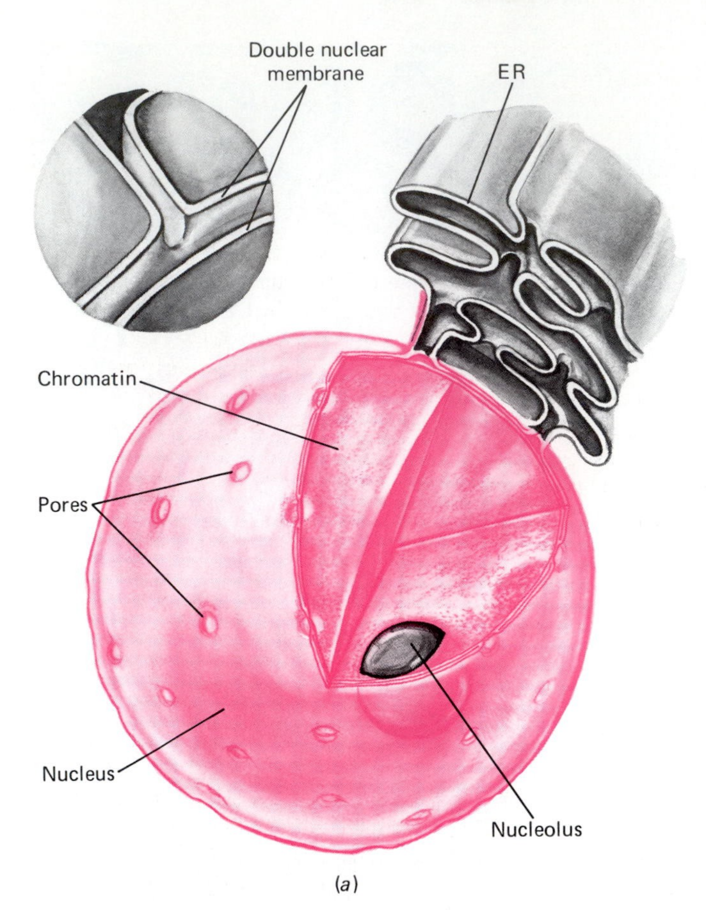

(a)

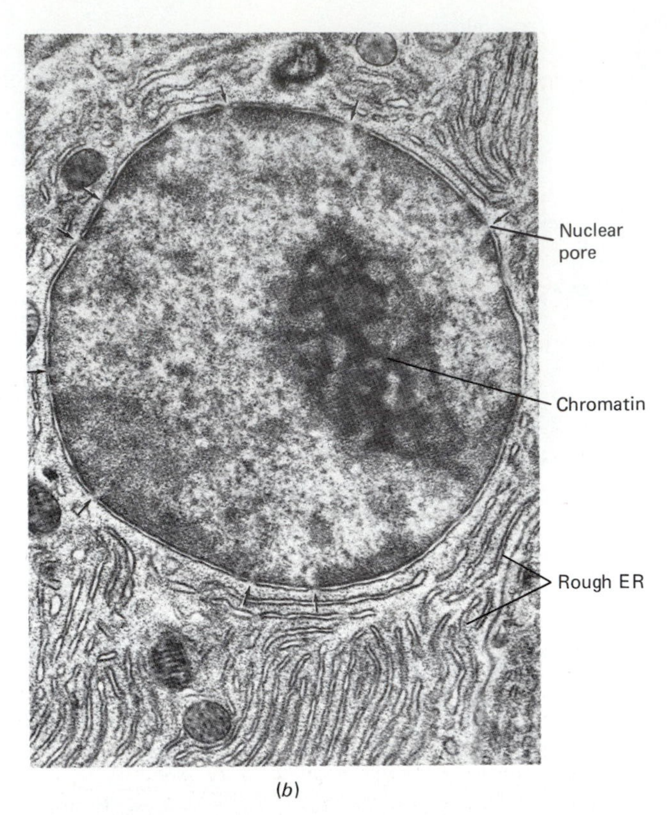

(b)

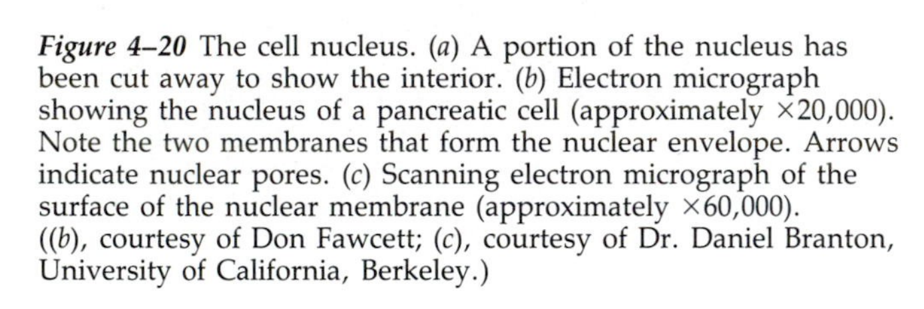

Figure 4–20 The cell nucleus. (*a*) A portion of the nucleus has been cut away to show the interior. (*b*) Electron micrograph showing the nucleus of a pancreatic cell (approximately ×20,000). Note the two membranes that form the nuclear envelope. Arrows indicate nuclear pores. (*c*) Scanning electron micrograph of the surface of the nuclear membrane (approximately ×60,000). ((*b*), courtesy of Don Fawcett; (*c*), courtesy of Dr. Daniel Branton, University of California, Berkeley.)

(c)

FOCUS ON

Acetabularia: The Mermaid's Wineglass and the Secret of Life

The role of the nucleus can be inferred from experiments in which it is removed from the cell and the consequences are examined. When the nucleus of a single-celled amoeba is removed surgically with a microneedle, the amoeba continues to live and move, but it does not grow and dies after a few days. We conclude that the nucleus is necessary for the metabolic processes (primarily the synthesis of nucleic acids and proteins) that provide for growth and cell reproduction.

But, you may object, what if the operation itself and not the loss of the nucleus caused the ensuing death? This can be decided by a controlled experiment in which two groups of amoebas are subjected to the same operative trauma, but in one group the nucleus is removed, while in the other it is not. We can insert a microneedle into some of the amoebas and push the needle around inside the cell to simulate the operation of removing the nucleus, but then withdraw the needle, leaving the nucleus inside. Amoebas treated with such a sham operation recover and subsequently grow and divide, but the amoebas without nuclei die. This permits the inference that it is the removal of the nucleus and not simply the operation that causes the death of the amoebas.

More details of nuclear function have been supplied by experiments involving *Acetabularia,* one of the largest known kinds of unicellular algae.

Introducing Acetabularia

In the imaginations of the more romantically inclined biologists, the little seaweed *Acetabularia* resembles a mermaid's wineglass. Less imaginatively, it has been described as looking like a little green toadstool measuring, at most, 5 to 7 cm in length (see photograph). Although it is a typical alga of tropical seas, it also occurs in some subtropical waters that are both shallow and somewhat rocky.

Acetabularia. (Courtesy of Charles Seaborn.)

In the 19th century, biologists discovered that this insignificant underwater plant consists of a single giant cell. Small for a seaweed, *Acetabularia* is gigantic for a cell. It consists of (1) a rootlike **holdfast,** (2) a long cylindrical **stalk,** and (3) at sexual maturity, a cuplike **cap.** The nucleus is found in the holdfast, about as far away from the cap as it can be. In due course, the nucleus divides by **meiosis** (a type of cell division in which the chromosome number is reduced to half the normal number), and its progeny of pronuclei swim up the stalk into the cap, where they become the pronuclei of sex cells. These are released upon maturity to swim away in search of partners. Although there are several species of *Acetabularia,* with caps of different shapes, all species function similarly.

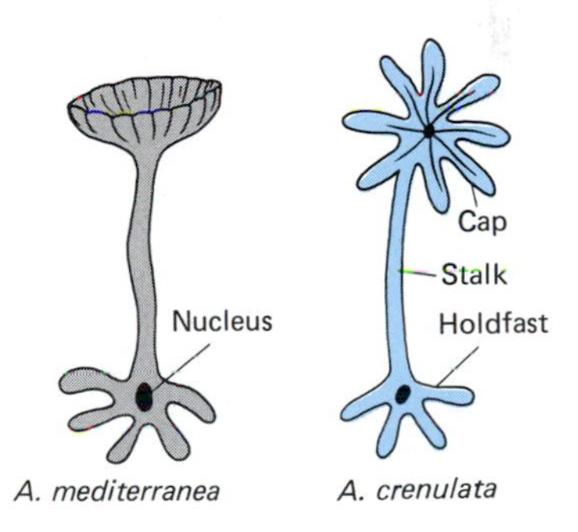

Hämmerling's and Brachet's Experiments

If the cap of *Acetabularia* is removed experimentally just before reproduction, another one will grow after a few weeks. Such behavior, common among lower organisms, is called **regeneration.** This fact attracted the attention of investigators, especially Hämmerling and Brachet, who became interested in the relationship that might exist between the nucleus and the physical characteristics of the plant. Because of its great size, *Acetabularia* could be subjected to surgery

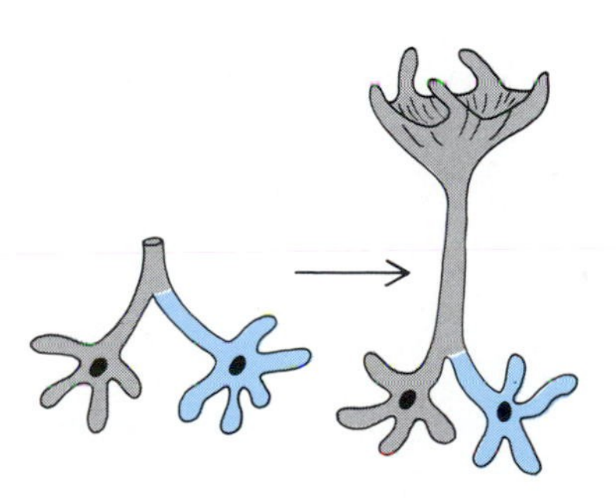

Continued on next page

impossible with smaller cells. These investigators and their colleagues performed a brilliant series of experiments that in many ways laid the foundation for much of our modern knowledge of the nucleus. In most of these experiments they employed two species of *Acetabularia, A. mediterranea,* which has a smooth cap, and *A. crenulata,* with a cap broken up into a series of finger-like projections.

The kind of cap that is regenerated depends upon the species of *Acetabularia* used in the experiment. As you might expect, *A. cren* will regenerate a "cren" cap, and *A. med* will regenerate a "med" cap. But it is possible to graft together two capless algae of different species. Through this union, they will regenerate a common cap that has characteristics intermediate between those of the two species involved. Thus, there is something about the lower part of the cell that controls cap shape.

Stalk Exchange

It is possible to attach a section of *Acetabularia* to a holdfast that is not its own by telescoping the cell walls of the two into one another. In this way the stalks and holdfasts of different species may be intermixed.

First, we take *A. mediterranea* and *A. crenulata* and remove their caps. Then, we sever the stalks from the holdfasts. Finally, we exchange the parts.

What happens? Not, perhaps, what you would expect! The caps that regenerate are characteristic *not* of the species donating the holdfasts but of that donating the stalks!

However, if the caps are removed once again, this time the caps that regenerate will be characteristic of the species that donated the holdfasts. This will continue to be the case no matter how many more times the regenerated caps are removed.

From all this we may deduce that the ultimate control of the cell is vested in the holdfast, because from now on, no matter how often the caps of these grafted plants are removed, they are always regenerated according to the species of the holdfast. However, there is a time lag before the holdfast gains the upper hand. The simplest explanation for this delay is that the holdfast produces some cytoplasmic temporary messenger substance whereby it exerts its con-

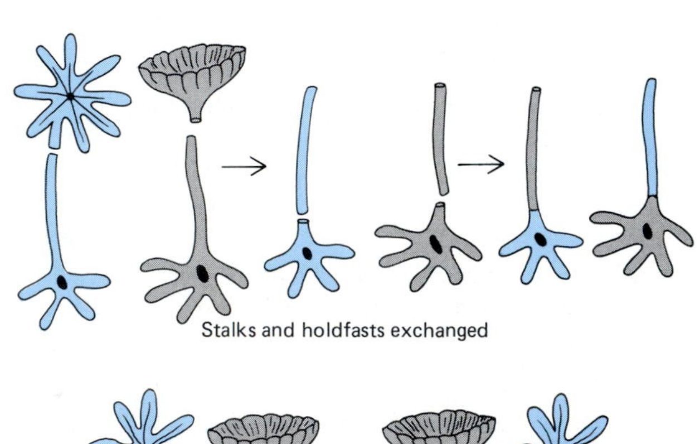

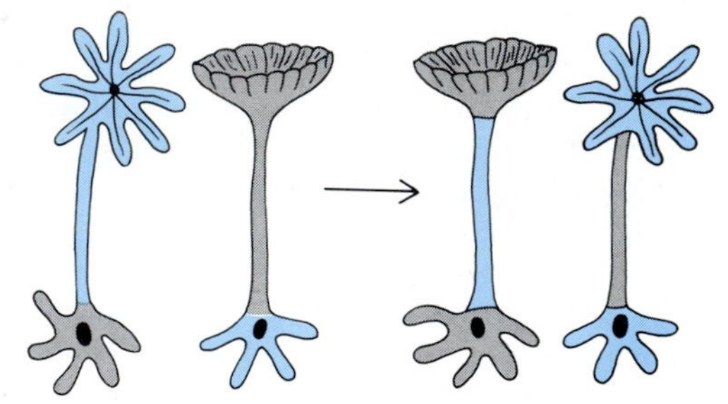

Nuclear Envelope

The nucleus is separated from the surrounding cytoplasm by a **nuclear envelope,** which regulates the flow of materials into and out of the nucleus. Electron microscopy reveals that the nuclear envelope is a flattened sac composed of two nuclear membranes separated by a very small space. The two membranes of the nuclear envelope are fused at intervals to form **nuclear pores**—channels through which the interior of the nucleus is in communication with the cytoplasm (Fig. 4–20). These

trol, and that initially the grafted stems still contain enough of that substance from their former holdfasts to regenerate a cap of the former shape. But this still leaves us with the question of what it is about the holdfast that accounts for its dictatorship. An obvious suspect is the nucleus.

Nuclear Exchange

If the nucleus is removed and the cap cut off, a new cap will regenerate. *Acetabularia,* however, is usually able to regenerate only once without a nucleus. If the nucleus of an alien species is now inserted, and the cap is cut off once again, a new cap will be regenerated characteristic of the species of the nucleus! If more than one kind of nucleus is inserted, the regenerated cap will be intermediate in shape between those of the species that donated the nuclei.

There is only one reasonable explanation for these observations: The control of the cell exerted by the holdfast is attributable to the nucleus that is located there. This information helped provide a starting point for research on the role of the nucleic acids in the control of all cells and all cellular life.

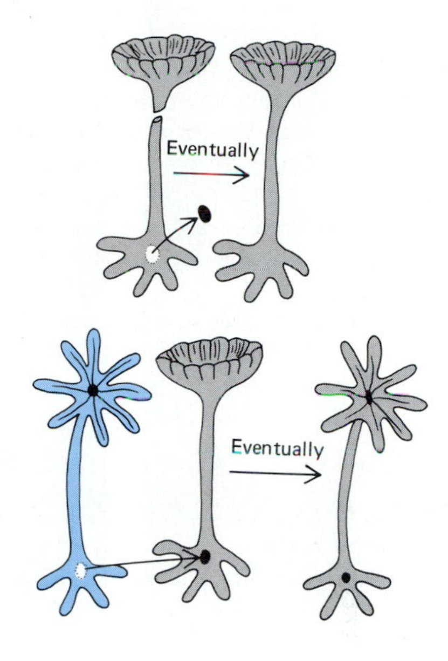
Eventually
Eventually

The Control of the Cell

The following sums up what is known of nuclear function based on findings in the experiments with *Acetabularia:*

1. Ultimate control of the cell is exercised by the nucleus. In the end, the form of the cap is determined by the kind of nucleus present in the holdfast. In the long run, the only thing that can successfully compete for control with a nucleus is another nucleus.
2. Some control of the form of the cap is exercised by the nonnuclear parts of the cell, presumably the cytoplasm or something in it. But since that substance can exercise control for just one regeneration, it must be limited in quantity and unable to reproduce itself without the nucleus. Perhaps it is perishable as well.
3. The source of the messenger substance must be the nucleus.

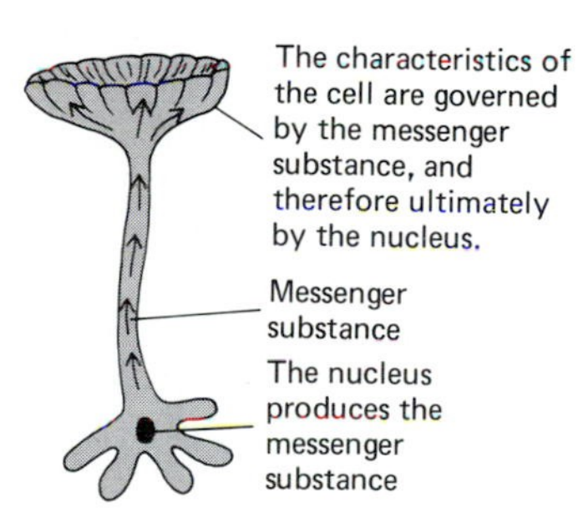
The characteristics of the cell are governed by the messenger substance, and therefore ultimately by the nucleus.
Messenger substance
The nucleus produces the messenger substance

pores, however, do not allow unrestricted diffusion between the nuclear and cytoplasmic compartments. Each pore appears to be closed by a delicate membrane somewhat thinner than other cellular membranes. This undoubtedly plays a role in controlling the passage of material through the pores. The outer of the two nuclear membranes appears to be continuous with the membranes of the ER and the Golgi complex described previously, and, by way of the latter structures, with the cell membrane.

(a)

(b)

Figure 4–21 Views of chromosomes and DNA. (*a*) Scanning electron micrograph of a chromosome from a hamster cell. Just prior to division, the loose threads of DNA that make up chromosomes shrivel into the knotted coils you see here. (*b*) Chromosomes consist of DNA and protein. This unusual photograph is a computer-generated view of the DNA molecule. It simulates a view down along the axis of the molecule. The bonds appear as solid white lines, and the atoms as hollow spheres of various colors; red = oxygen; blue = nitrogen; green = carbon; yellow = phosphorus; white = hydrogen. ((*a*) Drs. Susanne M. Gollin and Wayne Wray, Kleberg Cytogenetics Laboratory, Department of Medicine, Baylor College, and Biology Department, The Johns Hopkins School of Medicine. (*b*), courtesy of Computer Graphics Laboratory, University of California, San Francisco.)

Chromosomes

Within the semifluid ground substance inside the nucleus, termed **nucleoplasm,** are suspended a fixed number of threadlike bodies called **chromosomes** (Fig. 4–21). These are composed of tightly coiled DNA and protein and contain the hereditary material, the **genes**.

When a cell is killed by fixation with the proper chemicals and stained with appropriate dyes, chromosomes become visible under the light microscope. (These and other intranuclear structures can also be observed in a *living* cell by the use of phase contrast microscopy.) In a cell that is not dividing, the chromosomes usually appear as an irregular network of strands and granules termed **chromatin.** Just prior to nuclear division these strands condense into compact, rod-shaped chromosomes, which are subsequently distributed to the two daughter cells in exactly equal numbers. Each type of organism has a characteristic number of chromosomes present in each of its component cells. The fruit fly has 8 chromosomes, the garden pea 14, corn 20, the potato 48, the toad 22, the rat 42, the human 46, the duck 80, and the crayfish over 100.

The cells of complex plants and animals contain *two* of each kind of chromosome. For example, the 46 chromosomes in a human cell include 23 pairs, two of each of 23 different kinds. The chromosomes differ in their size and shape and in the presence of knobs or constrictions along their length. In most species the structural features of the different chromosomes are distinctive and permit the cytologist to distinguish the different pairs when they are condensed during cell division.

A cell with two complete sets of chromosomes is termed **diploid.** Sperm and

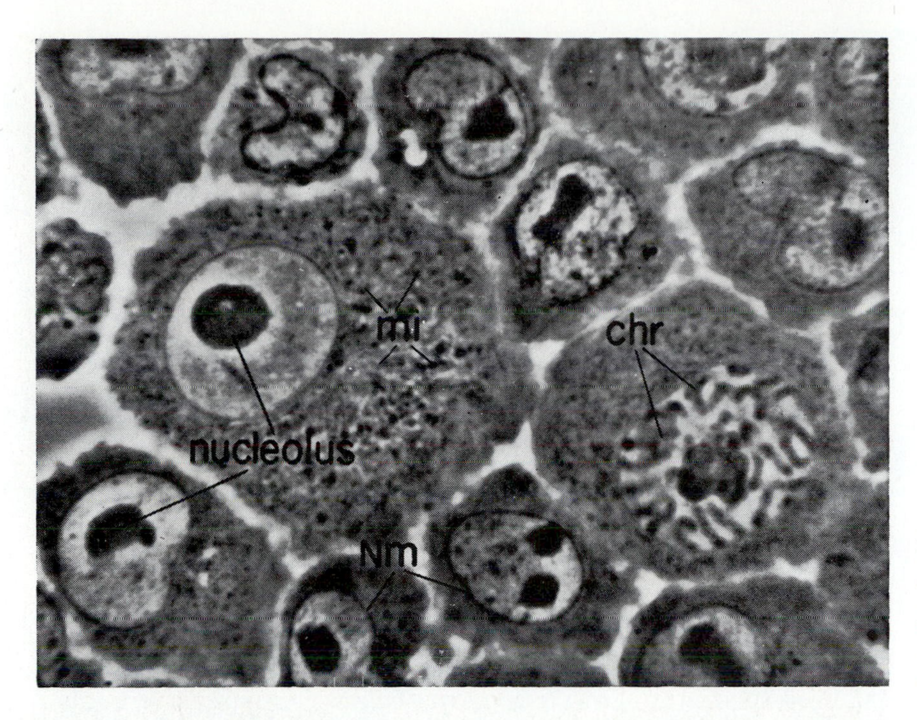

Figure 4–22 A photomicrograph taken by phase contrast microscopy of living cells from a tumor; *chr*, chromosomes; *mi*, mitochondria; *nm*, nuclear membrane. Chromatin is visible in most of the nuclei shown. Chromosomes are evident in the cell at the lower right.

egg cells, which have only one of each kind of chromosome—one full set of chromosomes—are termed **haploid.** They have just half as many chromosomes as the cells of the body of the members of that species. When the egg is fertilized by the sperm, the two haploid sets of chromosomes are combined and the diploid number is restored in the fertilized egg, or **zygote.** The zygote divides repeatedly, giving rise to diploid cells, and eventually develops into a diploid embryo and then an adult.

Nucleolus

Another structure within the nucleus, visible under the light microscope in a properly prepared cell, is the **nucleolus,** a compact, typically spherical body that is rich in RNA (Fig. 4–22). There may be more than one nucleolus in a nucleus, but the cells of any given species of plant or animal usually have a fixed number of nucleoli.

The nucleoli become disorganized when a cell is about to divide. After cell division is complete, they reappear at well-defined regions of the chromosomes, called **nucleolar organizers.**

The RNA incorporated into the ribosomes, termed **ribosomal RNA,** or **rRNA,** is synthesized in the nucleolus. Recall that the ribosomes are the structures in the cytoplasm on which proteins are synthesized. Their subunits, composed of rRNA and protein, are assembled within the nucleolus.

Differences in Some Major Cell Types

Even in their diversity the thousands of different kinds of cells share many common features of structure and function. There are, however, some important differences that we should note. (See Table 4–2.)

PLANT AND ANIMAL CELLS

An electron micrograph of a "typical" plant cell is shown in Fig. 4–23. Animal and plant cells differ in certain features: (1) although all cells are limited by cell membranes, plant cells are also surrounded by stiff cell walls of cellulose, which prevent their changing position or shape; (2) plant cells contain plastids such as chloroplasts, which are absent in animal cells; (3) plant cells have several or one large central water vacuole; (4) centrioles (and probably lysosomes) are absent in the cells of complex plants.

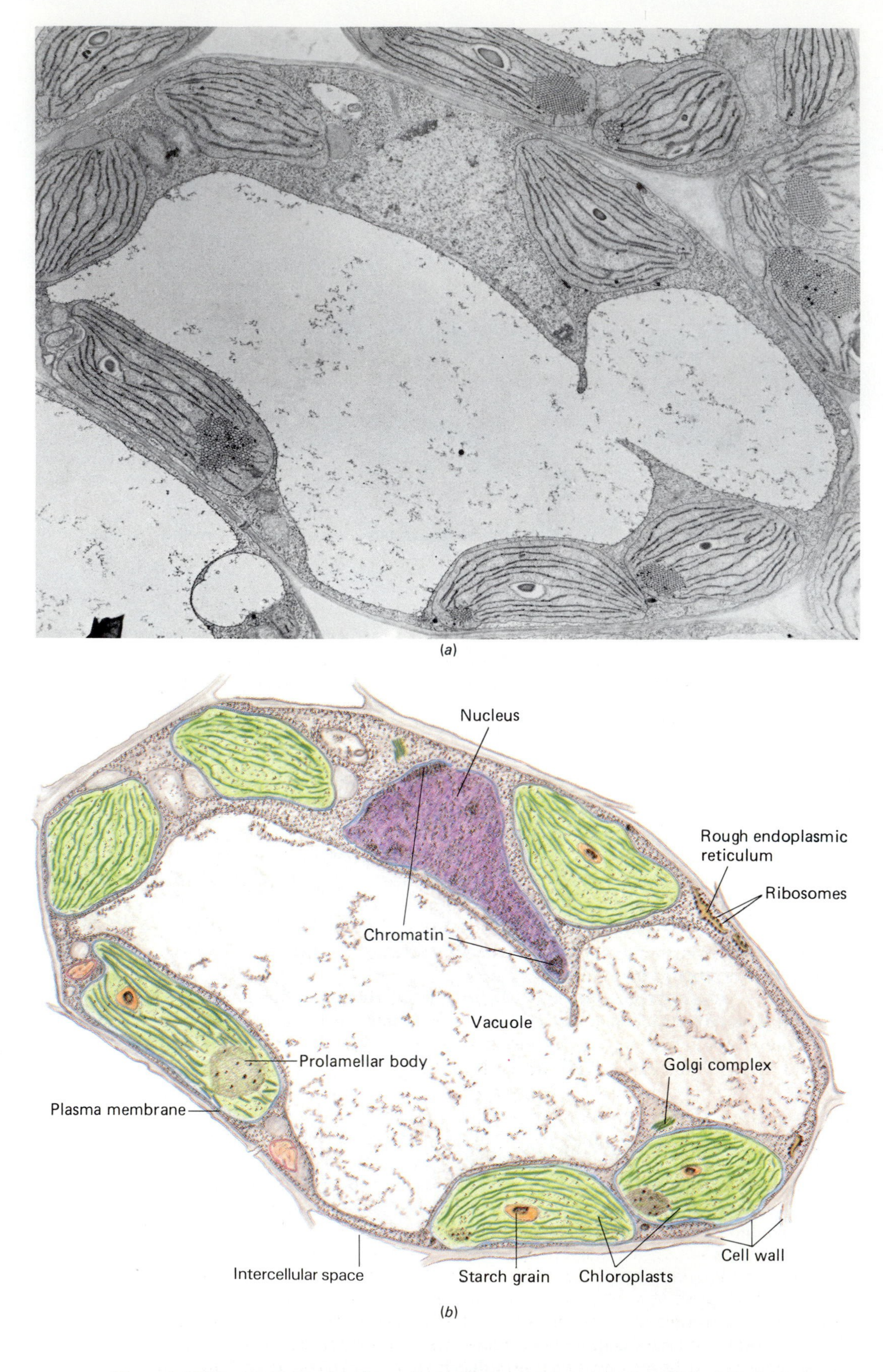

Figure 4–23 The structure of a plant cell. (*a*) Electron micrograph of a cell from the leaf of a young bean plant, *Phaseolus vulgaris,* magnified ×14,000. (*b*) A drawing highlighting the structures of the plant cell. Prolamellar bodies are membranous regions typically seen in developing chloroplasts. The structure of this cell would be different were it taken from some other portion of the plant, such as the stem or root. ((*a*), courtesy of Kenneth Miller, Brown University.)

TABLE 4–2
Major Features of the Cells of Prokaryotes, Plants, and Animals

	Prokaryotic Cell	*Plant Cell*	*Animal Cell*
Cell membrane	Present	Present	Present
Cell wall	Present; contains peptidoglycan	Present; contains cellulose	Absent
Nuclear membrane	Absent	Present	Present
Chromosomes	Composed only of nucleic acid; circular	Composed of DNA and protein; linear	Composed of DNA and protein; linear
Mitochondria	Absent	Present	Present
Endoplasmic reticulum	Absent	Present	Present
Golgi complex	Absent	Present	Present
Plastids	Absent	Usually present; chloroplasts containing chlorophyll	Absent
Ribosomes	Present	Present	Present
Vacuoles	Absent	Usually present	Small or absent
Centrioles	Absent	Absent in complex plants	Present
Lysosomes	Absent	Usually absent(?)	Often present
9 + 2 Cilia	Absent (simple, single-filament flagella may be present)	Absent in complex plants	Often present

PROKARYOTIC AND EUKARYOTIC CELLS

The cells described previously were all **eukaryotic** cells, cells that possess a nuclear membrane and a variety of complex membranous organelles such as mitochondria. Only bacteria and cyanobacteria have **prokaryotic** cells. These cells are generally smaller than eukaryotic cells, have no nuclear membrane, and lack distinct membranous organelles such as mitochondria.

Instead of a nucleus surrounded by a nuclear membrane, prokaryotic cells have a nuclear region in which lies a single large DNA molecule. The DNA has none of the associated histone proteins found in eukaryotic cells. Prokaryotic chromosomes are usually circular rather than rod-shaped, as are those of eukaryotic cells. The prokaryotic chromosome is a simple DNA chain composed of two linear sequences of nucleotides twisted together. Proteins are synthesized on ribosomes in prokaryotes just as they are in eukaryotes, but the ribosomes are smaller than and somewhat different in structure from those of eukaryotic cells.

The cell wall of prokaryotic cells is composed not of cellulose but of polysaccharide chains linked to one another by chains of amino acids (a complex called peptidoglycan). In some prokaryotic cells, the cell membrane is folded inward to form a complex of internal membranes called the **mesosome** (Fig. 4–24). Reactions of cellular respiration are thought to take place along these membranes. The cyanobacteria and the few bacteria that can carry out photosynthesis contain chlorophyll associated with flat, sheetlike structures termed lamellae, but the lamellae are not located within membrane-bound plastids. Some bacterial cells have flagella, but these contain a single microfilament and are structurally quite different from the flagella present in eukaryotic organisms.

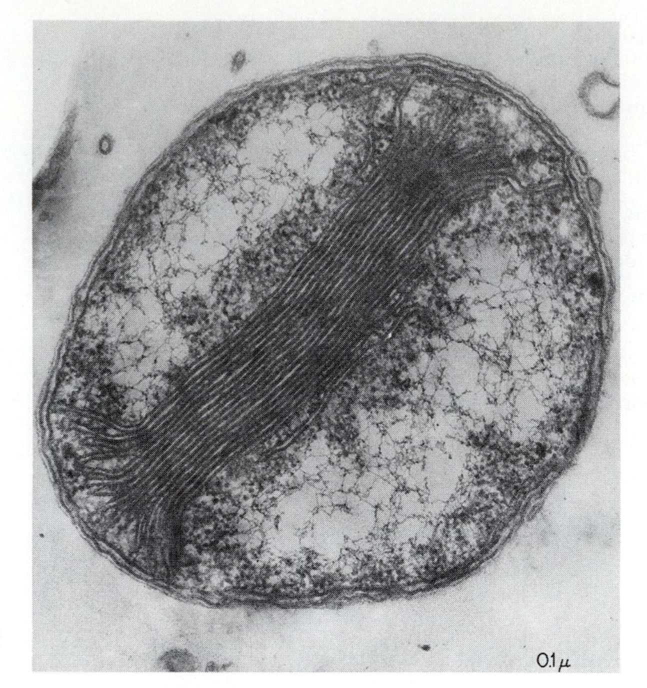

Figure 4–24 An electron micrograph of a thin section of a prokaryotic cell—the marine nitrifying bacterium, *Nitrosocystis oceanus.* A mesosome extends across the cell. On either side are light areas containing strands of DNA. Near the periphery of the cell are ribosomes, which appear as dark dots. Outside the plasma membrane is a cell wall composed of four dense layers. (Courtesy of Dr. S. W. Watson.)

The Cell Cycle

In cells that are capable of dividing, the **cell cycle** is the period from the beginning of one division to the beginning of the next division. The cell cycle is customarily represented in diagrams as a circle (Fig. 4–25). The length of time between two successive divisions, represented by a complete revolution of the circle, is termed the generation time, *T*. The generation time can vary over wide ranges but is usually several hours long in plant and animal cells.

Cell multiplication involves two main processes, mitosis and cytokinesis. The actual division of the cell to form two cells is **cytokinesis.** However, before a eukaryotic cell can divide, its nucleus must undergo **mitosis,** a complex division that precisely distributes a complete set of chromosomes to each daughter nucleus. Mitosis ensures that each new cell contains the identical number and types of chromosomes present in the original mother cell.

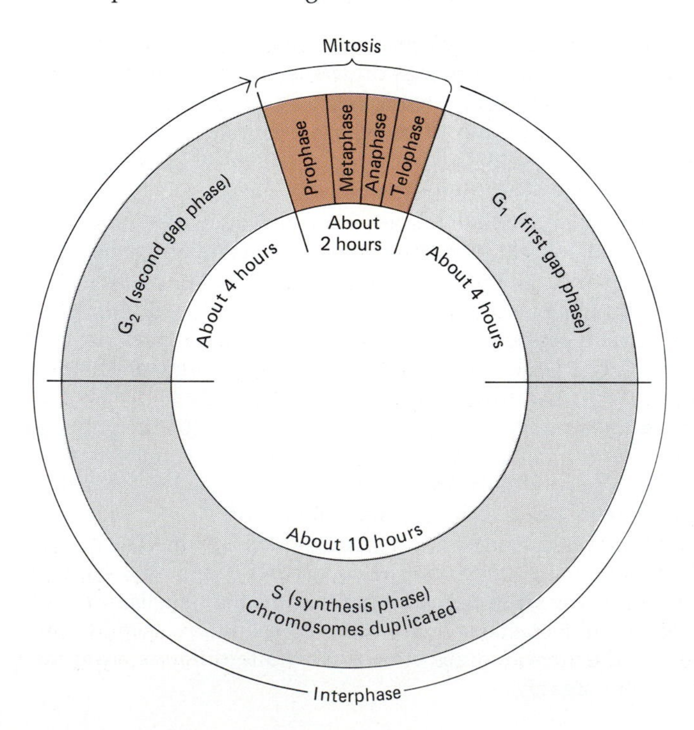

Figure 4–25 The cell cycle. Time relations are illustrative only; actual time varies with the cell type.

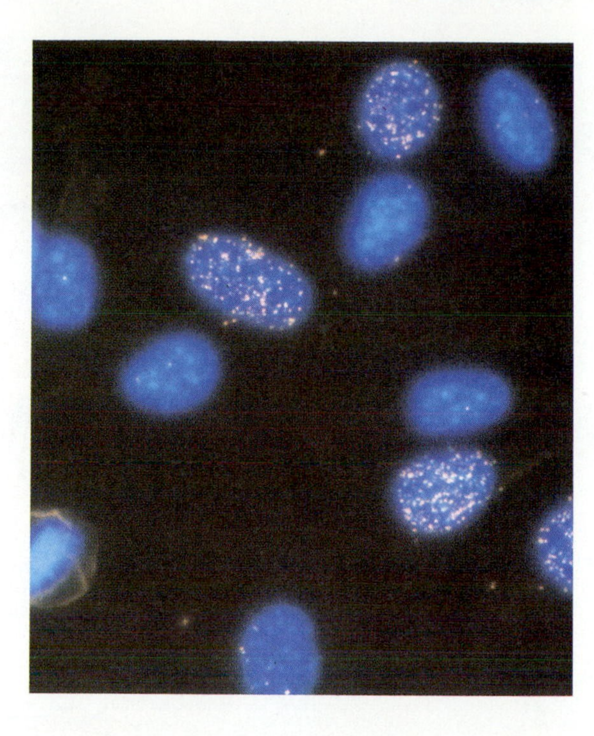

Figure 4–26 Photomicrograph of mouse cells in which thymidine (a DNA precursor) has been labeled with tritium. The orange grains indicate replicated DNA in interphase cells. Only the nuclei are clearly visible. (Jonathan G. Izant.)

INTERPHASE

Most of the life of the cell is spent in **interphase,** actively synthesizing needed materials and growing. The term *interphase* (between phases) refers to the fact that this stage of the cell's life cycle occurs between the phases of successive mitoses. In about 1950 it was recognized that chromosomes undergo replication during the interphase, and simply separate and are distributed to the daughter nuclei during mitosis. Experiments using tritium (^{3}H)-labeled thymidine, one of the precursors of DNA, showed that DNA is synthesized during interphase (Fig. 4–26). The period of DNA replication during interphase is termed the **synthesis phase,** or **S phase.** The time between mitosis and the beginning of DNA replication is termed the **G_1 phase,** or **first gap phase.** During the G_1 phase, the cell grows, and certain processes, such as increased activity of enzymes involved in DNA synthesis, occur that make it possible for the cell to enter the S phase and become committed to a future cell division.

After completion of the S phase, the cell enters a **second gap phase, G_2.** At this time there is an increase in protein synthesis as the final steps in the cell's preparation for division take place. The completion of the G_2 phase is marked by the beginning of mitosis.

A typical human body cell contains 46 threads of DNA (46 chromosomes) with a total length of 2 m or more. All of these threads are stuffed into a nucleus about 5 μm in diameter. During the complex replication process an exact copy is made of each of these 46 threads. Replication does not simply begin at one end of each thread and travel along to the other end; rather, each thread undergoes replication in many segments according to a definite program. The segments do not replicate in tandem in any one chromosome, nor does any one chromosome complete its replication before the next chromosome begins to replicate. When the last segments have been duplicated, DNA synthesis ceases and does not resume until the S phase of the next cycle.

MITOSIS

Each mitotic division is a continuous process, with each stage merging imperceptibly into the next one. However, for descriptive purposes mitosis, or the **M phase,** has been divided into four stages: prophase, metaphase, anaphase, and telophase (Figs. 4–27 and 4–28). Please refer to these figures as you read the description of each phase of mitosis, and to the individual figures that accompany the description of each stage of mitosis.

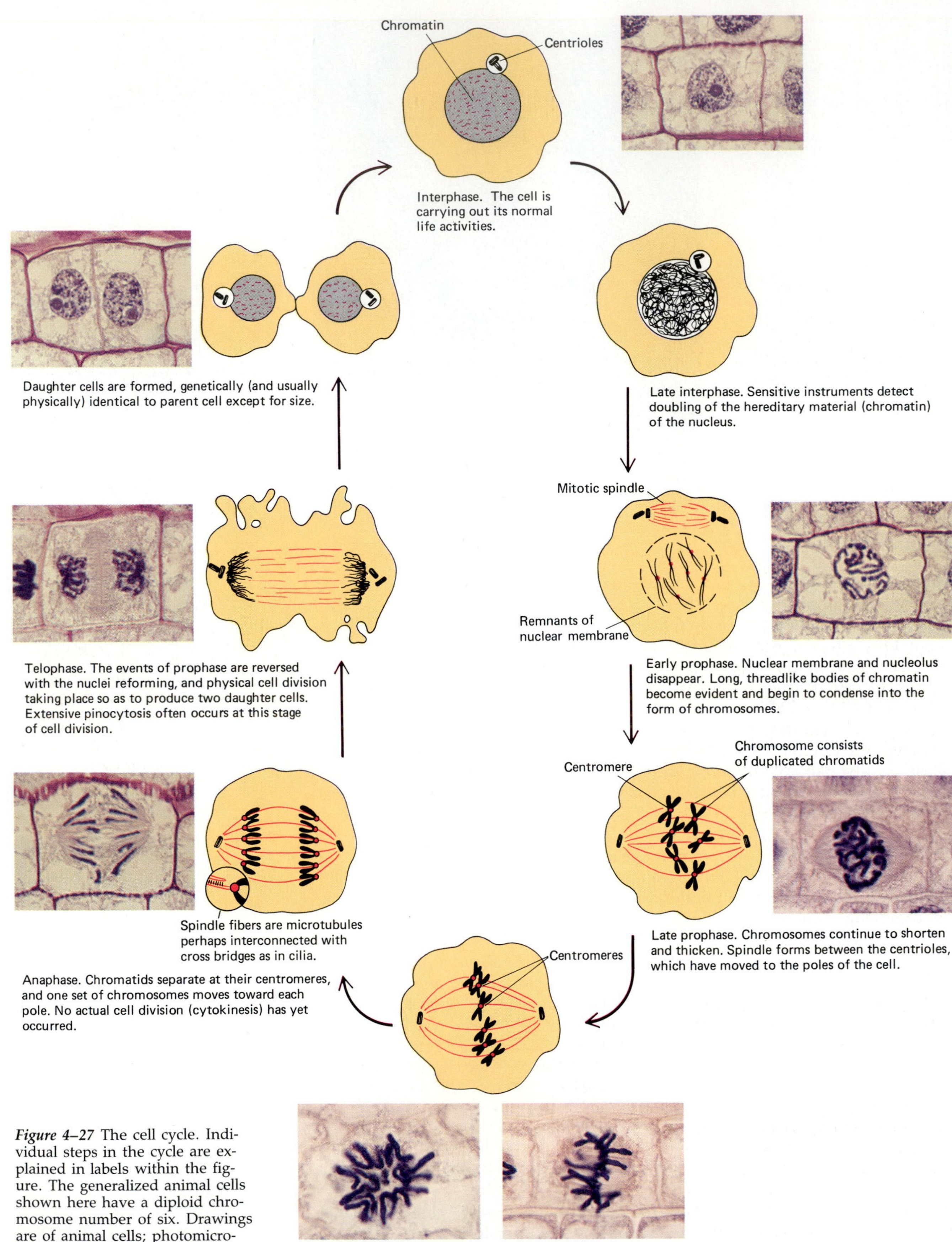

Chromatin
Centrioles
Interphase. The cell is carrying out its normal life activities.
Late interphase. Sensitive instruments detect doubling of the hereditary material (chromatin) of the nucleus.
Mitotic spindle
Remnants of nuclear membrane
Early prophase. Nuclear membrane and nucleolus disappear. Long, threadlike bodies of chromatin become evident and begin to condense into the form of chromosomes.
Centromere
Chromosome consists of duplicated chromatids
Late prophase. Chromosomes continue to shorten and thicken. Spindle forms between the centrioles, which have moved to the poles of the cell.
Centromeres
Early metaphase. Spindle fibers attach to the centromeres of the chromosomes. Chromosomes line up along the equatorial plane of the cell.
Spindle fibers are microtubules perhaps interconnected with cross bridges as in cilia.
Anaphase. Chromatids separate at their centromeres, and one set of chromosomes moves toward each pole. No actual cell division (cytokinesis) has yet occurred.
Telophase. The events of prophase are reversed with the nuclei reforming, and physical cell division taking place so as to produce two daughter cells. Extensive pinocytosis often occurs at this stage of cell division.
Daughter cells are formed, genetically (and usually physically) identical to parent cell except for size.

Figure 4–27 The cell cycle. Individual steps in the cycle are explained in labels within the figure. The generalized animal cells shown here have a diploid chromosome number of six. Drawings are of animal cells; photomicrographs are of plant cells (onion root tip, *Allium cepa*) (Carolina Biological Supply Company.)

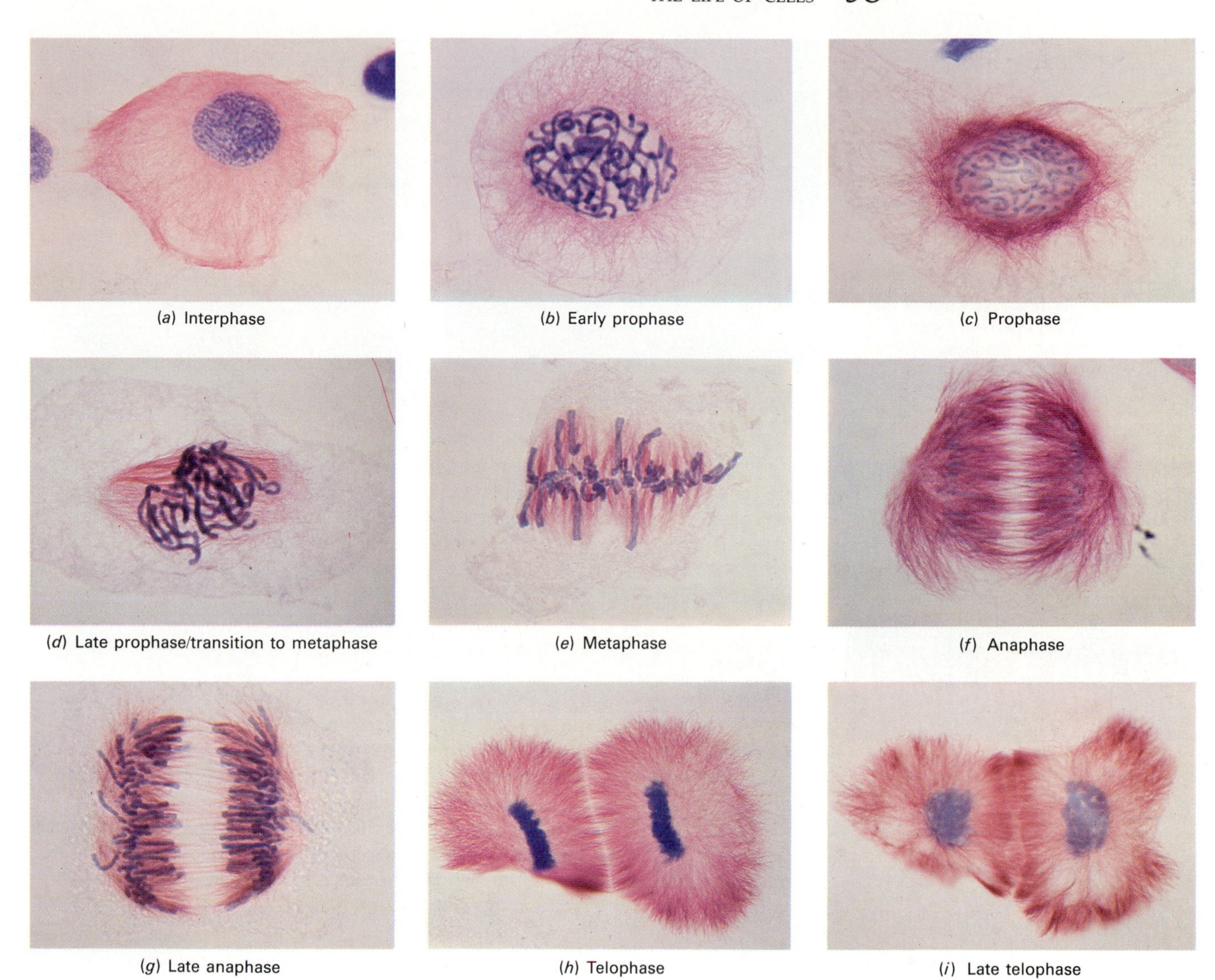

Figure 4–28 Interphase and the stages of mitosis in plant cells (*Haemanthus*) prepared with stains. (Andrew S. Bajer, University of Oregon.)

Prophase

The first stage of mitosis, **prophase,** begins when the chromatin threads begin to condense and appear as chromosomes. In forming the mitotic chromosomes, the long threads of DNA are condensed and coiled into much shorter bundles, permitting the chromosomes to separate and pass into the daughter cells without tangling.

As prophase proceeds, the chromosomes become shorter and thicker and are visible as dark, rod-shaped bodies under the light microscope. The chromosomes have not yet separated completely from their duplicates and are referred to as **chromatids.** Twin chromatids are joined at special structures called **centromeres.**

Early in prophase, the members of the two pairs of centrioles separate and migrate toward opposite poles of the cell. Centrioles are characteristic of animal cells but are not found in the cells of complex plants. Microtubules, composed mainly of protein, form and begin to organize into a **mitotic spindle.** In animal cells, similar microtubules extend outward in all directions from the centrioles; these clusters of microtubules are called **asters** (Fig. 4–29).

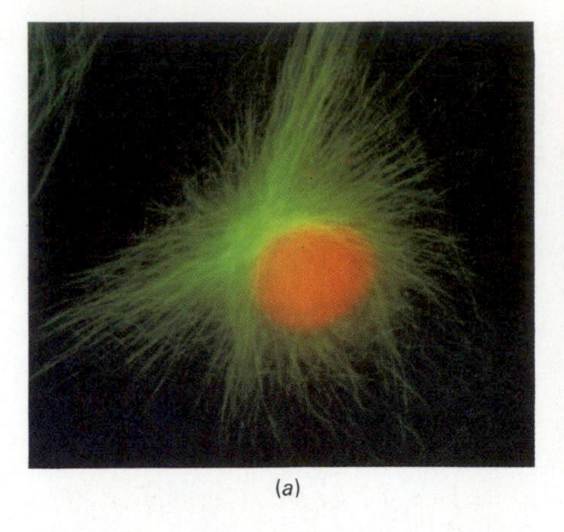

(a)

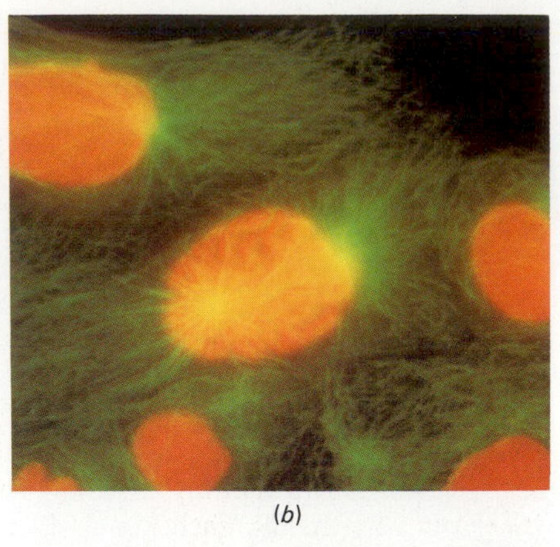

(b)

Figure 4–29 The transition from interphase to prophase in human cells grown in culture. The photographs in this figure and in Figures 4–30 to 4–32 are of cells stained with fluorescent dyes. Chromosomes are stained orange, and the microtubules are stained yellow-green. (*a*) Interphase. The microtubules are not yet organized into a mitotic spindle. (*b*) Prophase. Asters are moving toward opposite poles of the cell. (All photographs in Figures 4–29 to 4–32 by Jonathan G. Izant.)

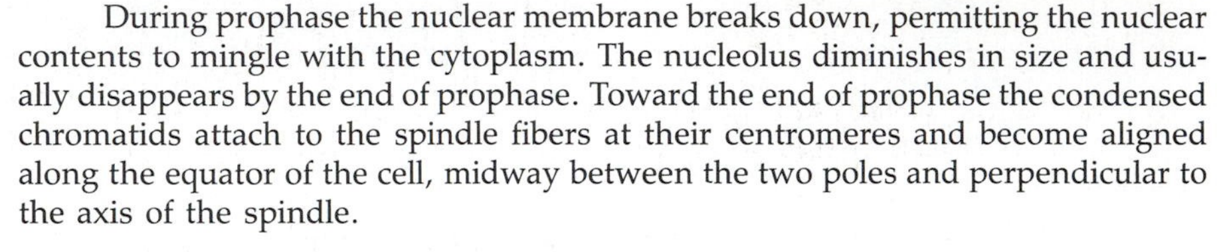

During prophase the nuclear membrane breaks down, permitting the nuclear contents to mingle with the cytoplasm. The nucleolus diminishes in size and usually disappears by the end of prophase. Toward the end of prophase the condensed chromatids attach to the spindle fibers at their centromeres and become aligned along the equator of the cell, midway between the two poles and perpendicular to the axis of the spindle.

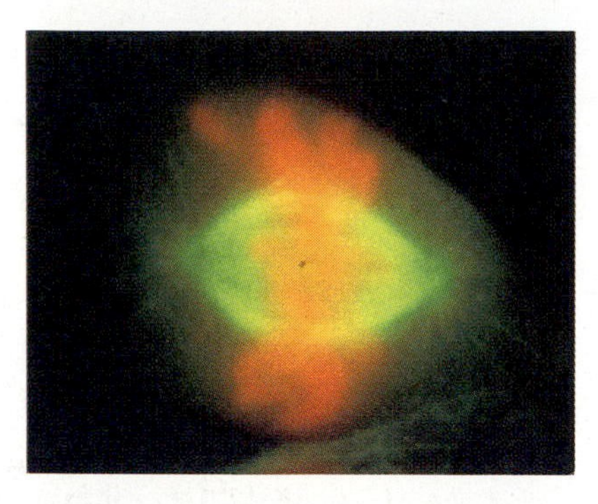

Figure 4–30 Metaphase. The mitotic spindle is well defined; chromosomes are lined up along the equatorial plane.

Metaphase

The period during which the chromatids are lined up along the equatorial plane of the cell constitutes **metaphase.** The mitotic spindle is complete; it is composed of numerous microtubules that extend from each pole to the equator. They end near the centrioles but do not actually touch them. The spindle is gel-like in consistency and is more viscous than the surrounding cytoplasm (Fig. 4–30).

During metaphase each chromatid is completely condensed and appears quite thick and discrete. Because metaphase chromosomes can be seen more clearly than those at any other stage, they may be photographed and studied to determine possible chromosome abnormalities (Fig. 12–3).

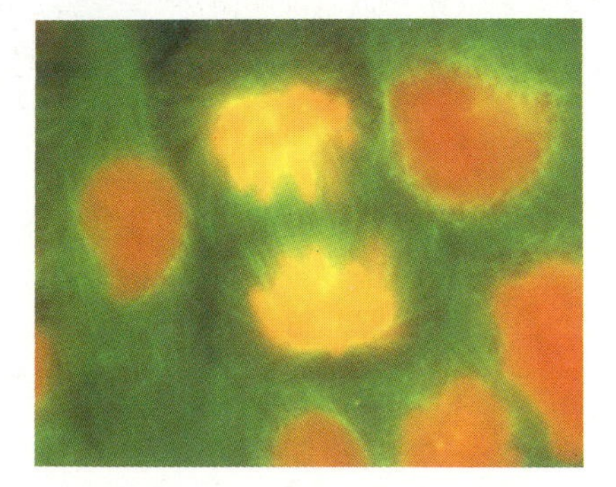

Figure 4–31 Late anaphase. A complete set of chromosomes is moving toward each end of the cell.

Anaphase

Anaphase begins with the separation of the centromeres of the sister chromatids of all the chromosomes. Each chromatid is now an independent chromosome. The separated chromosomes are slowly pulled toward opposite poles. The chromosomes move toward the poles with the centromeres (attached to the spindle fibers) leading the way and with the arms of the chromosomes trailing behind. Anaphase ends when a complete set of chromosomes has arrived at opposite ends of the cell (Fig. 4–31).

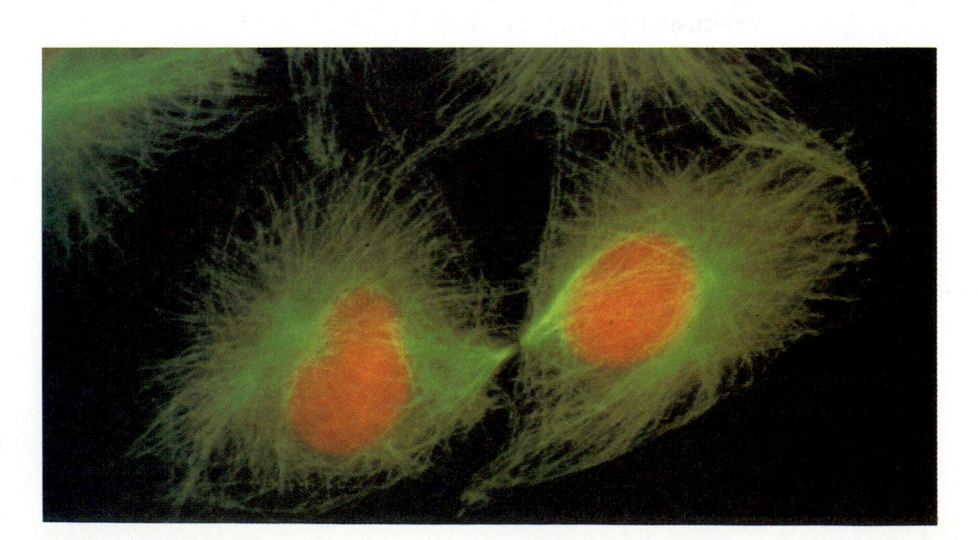

Figure 4–32 Late telophase. There is now a complete set of chromosomes at each end of the cell. Cytokinesis has begun, evidenced by the split between the groups of microtubules.

Telophase

The final stage of mitosis, **telophase,** is characterized by a return to interphase conditions. The chromosomes elongate by uncoiling. A new nuclear membrane forms around each set of chromosomes, produced at least in part from lipid components of the old nuclear membrane. Nucleoli reappear, while spindle fibers disappear (Fig. 4–32).

Cytokinesis, the actual division of the cytoplasm to yield two daughter cells, generally takes place during telophase (Fig. 4–28h). The division of an animal cell is accomplished by a furrow that encircles the surface of the cell in the plane of the equator. The furrow gradually deepens and separates the cytoplasm into two daughter cells, each with a complete nucleus. In plant cells, division occurs by the formation of a **cell plate,** a partition that forms in the equatorial region of the spindle and grows laterally to the cell wall. The cell plate forms from vesicles that break off from the ER. Each daughter cell then forms a cell membrane on its side of the cell plate, and the cellulose cell walls are added on either side of the cell plate.

The Significance of Mitosis

The remarkable regularity of the process of cell division ensures that each daughter cell will receive exactly the same number and kind of chromosomes that the parent cell had. Thus, each cell of a multicellular organism has exactly the same number and kinds of chromosomes as those in every other cell. If a cell should receive more or less than the proper number of chromosomes by some malfunctioning of the cell division process, the resulting cell may show marked abnormalities and may be unable to survive. The fact that a cell contains the genetic information needed for every characteristic of the organism explains why a single cell taken from a fully differentiated adult plant, for example, has the potential, under suitable conditions in cell culture, to develop into an entire new plant.

FACTORS THAT AFFECT THE CELL CYCLE

The frequency of mitosis varies widely in different tissues and in different species. Under optimal conditions of nutrition, temperature, and pH, the length of the cell cycle—the **generation time**—for any given kind of cell is constant. Under less favorable conditions, the cell cycle may be slowed; that is, the generation time is greater. However, it has not been possible experimentally to speed up the cell cycle and make cells grow faster. It appears that the length of the cell cycle is the time required for the cell to carry out some precise program that has been built into each kind of cell. The program has two parts: one having to do with replication of the genetic material in the chromosomes and the other involving the doubling of all of the constituents of the cell involved in growth.

When conditions are optimal, a bacterium can divide every 20 minutes. In the red bone marrow of human beings, an astonishing 10 million red blood cells are produced each second; that is, 10 million mitoses must occur per second. Cells that line the digestive tract, and cells in the reproductive layer of the skin, divide rapidly throughout life. In contrast, cell divisions in the central nervous system usually cease in the first few months of life.

In nearly all animal cells the production of substances that control the entrance of the cell into the S phase or the M phase depends on stimulation by growth-promoting substances present in the blood. These **growth factors** are small proteins and appear to act specifically on some kinds of cells but not on others. For example, a nerve growth factor is essential for the mitosis of sympathetic nerve cells.

Substances that inhibit mitosis, called **chalones,** counter the action of the growth factors. The various chalones are also very specific and affect only the type of tissue in which they are produced. For example, the chalone produced by skin cells inhibits mitosis by neighboring skin cells. Damaged skin cells are thought to synthesize less chalone, so that cells in the vicinity of a wound are released from this inhibition. They begin dividing, producing new tissue to provide for the healing of the wound. When enough healthy cells have been produced they synthesize enough chalone to inhibit further mitotic division, thus turning off the wound-healing process.

The cell cycle can also be affected by certain drugs. **Colchicine** is a drug used to block cell division in eukaryotic cells. This substance binds with a microtubule protein and interferes with the normal function of the mitotic spindle. The chromosomes cannot separate appropriately and move to the opposite ends of the cell. As a result, the cell may end up with an extra set of chromosomes. Plant cells can survive treatment with colchicine; in fact, plants consisting of cells with extra sets of chromosomes tend to be larger and more vigorous than normal plants.

Antibiotics such as streptomycin and the tetracyclines prevent mitosis indirectly by inhibiting protein synthesis in prokaryotic cells. This prolongs the G_1 phase of the cell cycle. Some drugs used in cancer therapy block one or more of the enzymes involved in DNA synthesis and cell division. Because cancer cells divide much more rapidly than most normal body cells, they are most affected by these drugs.

SUMMARY

I. The cell is considered the basic unit of life because it is the smallest self-sufficient unit of living material, and because organisms are made up of cells and their products.

II. The cell theory states that all living things are composed of cells and cell products; that new cells are formed only by the division of preexisting ones; that all cells share similar chemical components and have similar metabolic activities; and that the activity of an organism can be understood as the collective activities and interactions of its cells.

III. The resolving power of the electron microscope is far greater than that of the light microscope, enabling investigators to study the fine structure of the cell.

IV. Most cells are microscopic, but their size and shape vary according to their function.

V. The cell is bounded by a cell membrane, and most eukaryotic cells have a nucleus and other types of complex organelles that perform specific functions.
- *A.* The endoplasmic reticulum (ER) is a system of internal membranes that transport and store materials within the cell and that divide the cytoplasm into compartments.
  - *1.* The smooth ER produces steroids in certain cells.
  - *2.* The rough ER is studded along its outer walls with ribosomes that manufacture proteins.
- *B.* The Golgi complex concentrates secretions that are produced in the ER and packages them for export from the cell. It adds carbohydrate components to some compounds and also produces lysosomes.
- *C.* Lysosomes function in intracellular digestion.
- *D.* Mitochondria are bounded by a smooth outer membrane and a folded inner membrane. The cristae of the inner membrane contain the enzymes of cellular respiration.
- *E.* Cells of algae and plants contain plastids: chloroplasts, which contain chlorophyll; pigmented-filled chromoplasts; and colorless leukoplasts.
- *F.* Peroxisomes are membranous sacs containing oxidative enzymes.
- *G.* The shape of the cell is maintained by microtubules, microfilaments, and the microtrabecular lattice; these also play a role in cell movement.
- *H.* Centrioles, cilia, and flagella are composed of microtubules; cilia and flagella move cells through the surrounding medium or move the surrounding medium past the cells.
- *I.* The nucleus, the control center of the cell, is bounded by a double-layered nuclear membrane.
  - *1.* The long, threadlike chromosomes are suspended in the nucleoplasm. They are composed of DNA and proteins and contain the hereditary units, the genes.
  - *2.* The nucleolus functions in the synthesis of ribosomal RNA and in the assembly of ribosomes.

VI. Plant cells differ from animal cells in that they possess a rigid cell wall, plastids, and large water vacuoles; they lack centrioles.

VII. Prokaryotic cells of bacteria and cyanobacteria lack a nuclear membrane and other membranous organelles.

VIII. The cell cycle is the period from the beginning of one division to the beginning of the next division.
- *A.* Interphase can be divided into the first gap phase, the synthesis phase, and the second gap phase.
  - *1.* During the G_1 phase the cell grows and prepares for the S phase.
  - *2.* DNA is replicated during the S phase.
  - *3.* During the G_2 phase there is an increase in protein synthesis.
- *B.* During mitosis, a complete set of chromosomes is distributed to each end of the cell, and a daughter nucleus is formed around each set.
  - *1.* During prophase, chromatids condense, the nucleolus disappears, the nuclear membrane breaks down, and the mitotic spindle begins to form.
  - *2.* During metaphase, the chromatids line up along the equator of the cell; the mitotic spindle is complete.
  - *3.* During anaphase, the chromosomes separate and move toward opposite poles of the cell.
  - *4.* During telophase, a nuclear membrane forms around each set of chromosomes, nucleoli reappear, the chromosomes elongate, the spindle disappears, and cytokinesis generally takes place.

POST-TEST

Match the subcellular organelles in Column A with their functions in Column B.

Column A	*Column B*
1. Peroxisomes	a. Tiny sacs that contain oxidative enzymes
2. Centrioles	b. Intracellular transport of materials
3. Chromosomes	c. System of internal membranes that manufacture and transport proteins
4. Cilia	d. Granules involved in protein synthesis
5. Flagella	e. Packages secretory products of the cell
6. Golgi complex	f. Packets of hydrolytic enzymes
7. Lysosomes	g. Move materials along the outer surface of the cell
8. Microfilaments	h. Move the cell along
9. Microtubules	i. Cell's control center; contains chromosomes
10. Mitochondria	j. Involved in assembly of ribosomes
11. Nucleolus	k. Contain the genes
12. Nucleus	l. Bubble-like spaces within the cell containing ingested materials
13. Plastids	m. Site of most reactions of cellular respiration
14. Ribosomes	n. Provide structural support and plays role in cell movement
15. Rough endoplasmic reticulum	o. Component of cilia, flagella, and centrioles
16. Smooth endoplasmic reticulum	p. Function in cell division
17. Vacuoles	q. Membranous structures containing pigments

18. The complex maze of spaghetti-like tubular strands within the cytoplasm is termed the ______________ ______________.

19. Proteins to be exported from the cell are synthesized on ribosomes on the ______________ ______________ ______________.

20. Protein synthesized in the rough endoplasmic reticulum is transported in little packets to the ______________ ______________, where it is concentrated and repackaged for secretion from the cell.

21. The powerful hydrolytic enzymes contained in the ______________ are released when the cell dies and digest the cellular remains.

22. Membrane-bound organelles containing the enzyme catalase are termed ______________.

23. The shelflike folds of the inner mitochondrial membrane are termed ______________.

24. The cytoplasmic organelles of plants involved in the synthesis and storage of food are called ______________.

25. The cylindrical, hollow cytoplasmic subunits, ______________, play a role in controlling the shape and movement of cells.

26. The flexible framework of the cell by microfilaments and microtubules is termed the ______________.

27. The three-dimensional lattice of slender protein threads extending through the cytoplasm and attached to the cell membrane in which are suspended the various subcellular organelles has been named the ______________ ______________.

28. ______________ and ______________ are movable whiplike structures projecting from the cell surface, which move the cell through the surrounding liquid or move the liquid across the surface of the cell. Each is composed of 11 groups of ______________, arranged with ______________ in the center and ______________ around the circumference.

29. The large, spherical or oval ______________ contains the chromosomes.

30. Within the semifluid ground substance of the nucleus, the ______________, are suspended a fixed number of linear threadlike structures called ______________.

31. A cell with two complete sets of chromosomes is termed ______________; sperm and egg cells, which have only one complete set of chromosomes, are termed ______________.

32. The fertilization of a haploid egg by a haploid sperm produces a diploid fertilized egg or ______________.

33. In addition to a cell membrane, plant cells are surrounded by a stiff cellulose ______________ ______________.

34. The period from the beginning of one cell division to the beginning of the next is termed the ______________ ______________.

35. DNA for the new sets of chromosomes is synthesized during the ______________.

36. To facilitate the description of the process, mitosis has been divided into four stages: ______________, ______________, ______________, and ______________.

37. The period during which the chromosomes are lined up on the equator of the cell constitutes the ______________.

38. The division of the cytoplasm to yield two daughter cells is called ______________.

39. The drug ______________ binds to a microtubule protein, interferes with the mitotic spindle, and hence blocks cell division.

REVIEW QUESTIONS

1. Trace the development of the cell theory, and explain the theory.
2. What are the main differences between plant and animal cells? between prokaryotic and eukaryotic cells?

3. Draw a diagram of a cell, and label ten important organelles.
4. What are the functions of the nucleus? What evidence can you cite indicating the role of the nucleus in cell metabolism?
5. What are the functions in the cell of each of the following?
   a. ribosomes
   b. mitochondria
   c. Golgi complex
   d. endoplasmic reticulum
6. What are the functions of microfilaments and microtubules?
7. Why are lysosomes sometimes referred to as the self-destruct system of the cell?
8. What are the functions of plastids? Compare the structures of a chloroplast and mitochondrion.
9. What are the features of the S phase of the cell cycle? of the M phase?
10. What are the functions of
    a. cilia
    b. chromosomes
    c. centromeres
    d. chalones

5
The Cell Membrane

OUTLINE

I. Functions of the cell membrane
II. The structure of the cell membrane
  A. The lipid bilayer
  B. Membrane proteins
  C. Membrane receptors
  D. Microvilli
III. Cell walls
IV. How materials pass through membranes
  A. Diffusion
    1. Dialysis
    2. Facilitated diffusion
    3. Osmosis
    4. Isotonic, hypertonic, and hypotonic solutions
    5. Turgor pressure
  B. Active transport
    1. Characteristics of mediated transport systems
    2. Transport proteins
  C. Endocytosis and exocytosis
V. Cell junctions and communications
  A. Desmosomes
  B. Tight junctions
  C. Gap junctions

LEARNING OBJECTIVES

After you have read this chapter you should be able to:

1. Discuss the importance of the cell membrane to the cell, describing the various functions it serves.
2. Diagram the lipid bilayer of the cell membrane and its associated proteins, and describe the structure of the membrane.
3. Characterize membrane-bound receptors, and give their function.
4. Describe microvilli, and give their function.
5. Compare the structures surrounding animal cells with those surrounding plant and bacterial cells.
6. Contrast the physical with the physiological processes by which materials are transported across cell membranes.
7. Given concentrations and membrane characteristics, predict the direction of diffusion of solutes and solvents across differentially permeable membranes.
8. Solve simple problems involving osmosis. For example, predict whether cells will swell or shrink under various osmotic conditions.
9. Summarize the currently accepted hypothesis of how small hydrophilic molecules are transported across the cell membrane.
10. List the characteristics of mediated transport, and compare active transport with facilitated diffusion.
11. Compare exocytosis and endocytosis.
12. Describe and contrast desmosomes, gap junctions, and tight junctions.

In order to carry out the many chemical reactions necessary to sustain life, the cell must maintain an appropriate internal environment. It must regulate its own composition, providing constant conditions, despite changes in the outside world. This is possible because all cells, even the simplest ones, are physically separated from the external environment by a limiting **cell membrane,** also referred to as a **plasma membrane.**

Functions of the Cell Membrane

The cell membrane is not an inanimate wall. Quite the contrary, it is a complex mechanism that permits many selective interactions between the cell and its environment. Among its many functions are the following:

1. The cell membrane regulates the passage of materials into and out of the cell. If substances could pass freely into and out of the cell, its contents would then reflect the chemical composition of the surrounding medium, with disastrous results. This does not happen because the cell membrane can prevent the passage of certain substances while permitting, or facilitating, the passage of others. For example, the cell membrane facilitates the entrance of needed nutrients, and the exit of specific products and wastes, but prevents the loss of other substances from the cell. Several ways in which the cell membrane regulates the passage of materials will be discussed later in this chapter.
2. The cell membrane receives information that permits the cell to sense changes in its environment and to respond to them. Cell surfaces are equipped with **receptor proteins** that receive chemical messages from other cells. Among the substances that combine with such receptors are hormones, growth factors, and neurotransmitters (chemicals released by nerve cells). Some of these substances combine with the receptors in the membrane. Some hormone-receptor complexes pass through the membrane; others remain on the outer surface of the cell. Their combination with the receptor may stimulate the membrane to send a signal into the cell resulting in some type of response.
3. The cell membrane maintains structural and chemical relationships with neighboring cells. Certain proteins in the cell membrane permit cells to recognize one another, to communicate with one another, to adhere to one another when appropriate, and to exchange materials.
4. The cell membrane also protects the cell, may be involved in movement of the cell, may function in secretion, and, in some cells, is important in transmitting impulses.

The Structure of the Cell Membrane

The cell membrane is so thin—about 6 to 10 nm—that it can be seen only with the electron microscope. According to the currently accepted model of the cell membrane, the **fluid mosaic model,** the membrane consists of a rather fluid lipid bilayer (a double layer of lipid) in which a variety of globular proteins are embedded.

THE LIPID BILAYER

The lipid components of the cell membrane include phospholipids, glycolipids, and cholesterol. All of these have an important structural feature in common: They are asymmetrical, elongated molecules and have one highly polar, **hydrophilic** (water-loving) portion and one nonpolar, **hydrophobic** (water-hating) portion. The term **amphipathic** is used to describe a molecule that has both a hydrophilic portion and a hydrophobic portion.

Each of the amphipathic lipids in the cell membrane is composed of a hydrophilic head group, such as choline, and two hydrophobic tails, which are usually the fatty acid chains (see Fig. 3–7). When viewed with the electron microscope, the cell membrane is seen as two dark lines separated by an intermediate light zone;

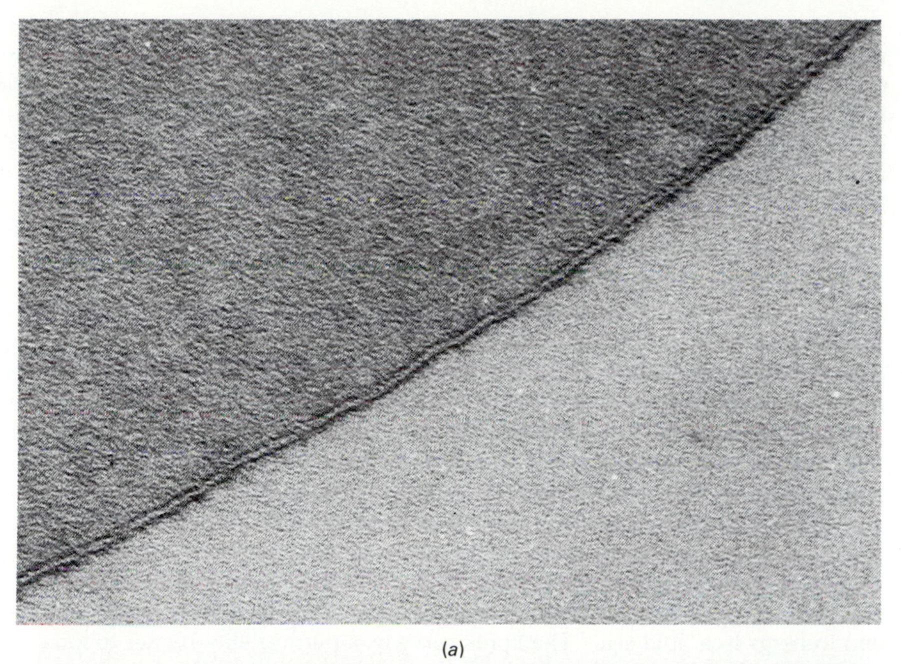

(a)

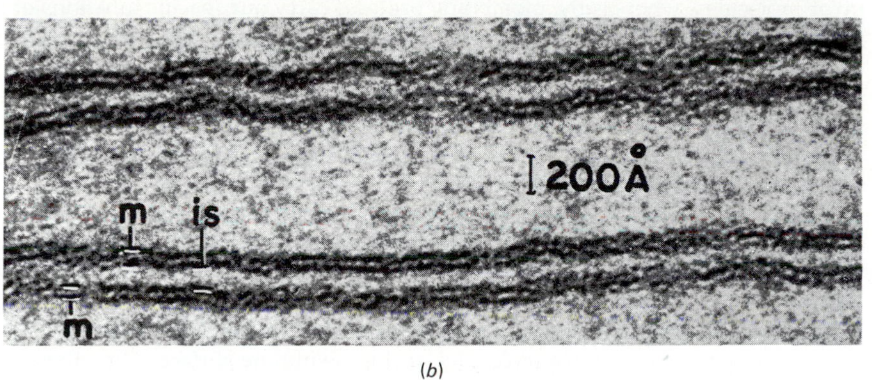

(b)

Figure 5–1 (a) The cell membrane is the interface between the cell and its environment. (b) Electron micrograph of a portion of a cell membrane (approximately ×240,000). The dark lines represent the hydrophilic heads of the lipids, while the light zone represents the hydrophobic tails. *m*, membrane; *is*, intercellular space. ((a), Omikron, Photo Researchers, Inc.)

apparently, the dark lines represent the hydrophilic heads of the lipids, and the light zone, the hydrophobic tails (Fig. 5–1). In diagrams the hydrophilic head groups are represented by circles and the hydrophobic tails are represented by two wavy lines (Fig. 5–2). In the cell membrane the nonpolar, hydrophobic fatty acid chains of the phospholipids meet and overlap in the middle, while the polar, hydrophilic heads are directed toward the outside of the membrane.

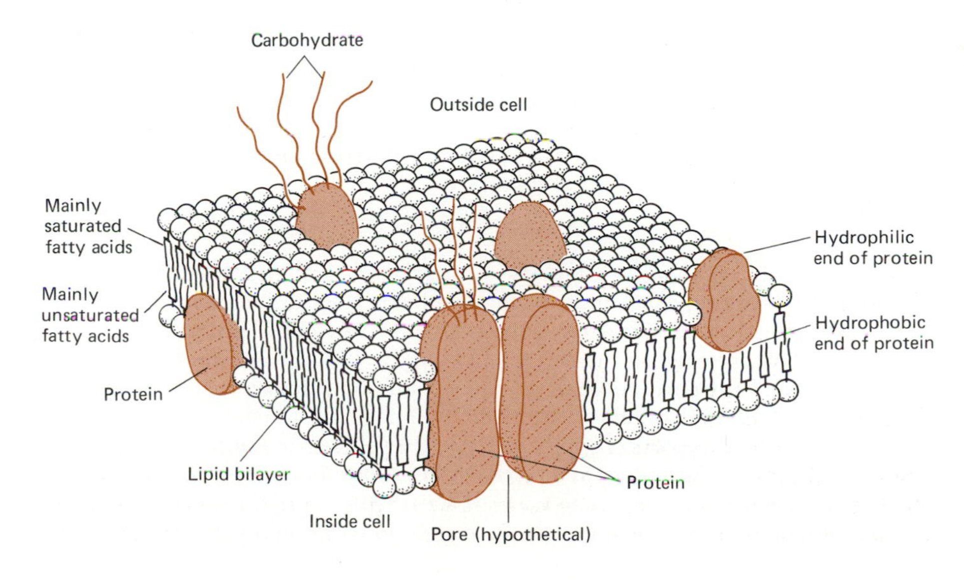

Figure 5–2 The architecture of the cell membrane drawn according to the current fluid mosaic model.

Analyses show that the two sheets of the bilayer differ in chemical composition, but the functional significance of the lipid asymmetry is not understood. The outer layer is especially rich in choline phospholipids and glycolipids, whereas the internal lipid layer is rich in other types of phospholipids. The fatty acids of the outer, choline-rich layer of lipids are primarily saturated fatty acids; the inner layer is rich in polyunsaturated fatty acids. There is effectively no exchange of lipids between the inner and outer layers of the bilayer.

The structure of a lipid bilayer is inherent in the amphipathic nature of the constituent lipids, and the formation of a lipid bilayer from phospholipids and glycolipids is a rapid and spontaneous process. The driving force for the formation of the lipid bilayer is the hydrophobic interactions of the hydrocarbon chains. When amphipathic lipids are placed in water, the polar heads show an affinity for water, whereas the hydrocarbon tails avoid it. Lipid bilayers are stabilized by the hydrophobic interactions of the fatty acid chains and by the ionic and hydrogen bonding between the polar head groups and the surrounding water molecules. Lipid bilayers not only assemble themselves but are self-sealing; that is, if a hole is made in it the bilayer will seal itself, thus repairing the tear.

MEMBRANE PROTEINS

The lipid bilayer is a fluid matrix in which the proteins can move around like "protein icebergs in a lipid sea." The lipids serve as a permeability barrier to ions and polar molecules, whereas the membrane proteins carry out specific functions of the membrane such as chemical transport and the transmission of messages. Cell membranes are dynamic structures, and their chemical composition and molecular arrangement may change with changing conditions. This very plastic quality of the lipid bilayer permits the cell to respond to a wide variety of external stimuli. Membranes may differ from one another in the numbers and kinds of proteins present; that is, membranes with different functions contain different proteins.

Based on their location in the membrane there are thought to be two categories of protein: integral proteins and peripheral proteins. **Integral proteins** penetrate deeply into the lipid layer. As a result of their amino acid composition, they have hydrophobic regions that are largely buried in the hydrophobic interior of the lipid bilayer, and hydrophilic regions that protrude from the membrane, permitting interactions with water-soluble molecules at the membrane surface. Thus these proteins are amphipathic, like the lipid components of the membrane. Integral proteins may be located primarily in the outer sheet of the membrane, protruding out into the exterior, or may be mainly associated with the inner sheet of the membrane, extending into the interior of the cell. A few proteins extend all the way through both sheets of the lipid bilayer and have a nonpolar middle portion embedded in the lipid bilayer and polar regions at each end extending to the membrane surfaces. **Peripheral proteins** are water-soluble and are thought to be weakly bound to the surface of the membrane.

Most of the membrane proteins are associated with the inner, cytoplasmic surface of the membrane (see Focus on Splitting the Lipid Bilayer). The membrane proteins that protrude from the outer surface (away from the cytoplasm) are largely glycoproteins, that is, proteins to which sugar residues are attached. Thus nearly all cells are "sugar-coated," like breakfast cereal. Little or no sugar is attached to the inner surface of the cell membrane or to any of the intracellular membranes. Membrane proteins have some freedom of lateral movement within the membrane and can change their position on the cell surface.

MEMBRANE RECEPTORS

In addition to keeping substances inside or outside the cell, membranes must receive and translate messages from outside the cell into some sort of intracellular response. Glycoproteins on the cell membrane are involved in interactions with other cells. They appear to serve as the cell's communication system with its outer environment, both with other cells and with free molecules such as hormones. The membrane-bound receptors, such as the type that binds the hormone insulin, are glycoproteins. Some hormones and certain other compounds are able to transmit a message across the cell membrane without actually crossing the membrane themselves. The binding of the hormone to its glycoprotein receptor triggers some

FOCUS ON

Splitting the Lipid Bilayer

High-resolution electron micrographs of cell membranes typically show a three-layered structure of lines in a dense–light–dense pattern (see Fig. 5–1). The two dense lines correspond to the polar heads of the lipid bilayer, and the light area between them corresponds to the hydrophobic region of the fatty acid chains. The cell membranes of all cells and the membranes of a great many subcellular organelles all appear to have this three-layered structure. Very-high-resolution electron micrographs indicate that the membrane has some sort of beaded substructure and that it is made of hexagonal subunits with a central dense granule 2.5 nm in diameter.

Cells can be rapidly frozen in liquid nitrogen and then fractured with a microtome knife. A small amount of ice is evaporated from the fracture surface ("etched"), and then a small amount of platinum or gold is deposited on the fracture surface, forming a replica of this surface. When the metallic coating is examined in the transmission electron microscope, it can be seen that the fracture typically splits each cell membrane into two half-membranes in the middle of the lipid bilayer. By this means, known as **freeze fracture** or **freeze etch,** the interior of the split membrane can be examined. The two faces are not identical: One, the face nearer the cytoplasm, contains many particles; and the other, the face near the outside of the cell, contains many pits (see accompanying figure). The particles are integral proteins, while the pits are spaces where the proteins had been.

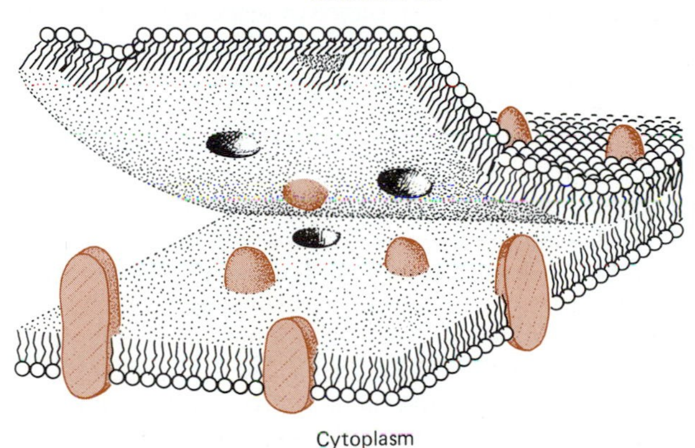

In the freeze-fracture method, the path of membrane cleavage is along the hydrophobic interior of the lipid bilayer, resulting in two complementary fracture faces: (1) an outwardly directed inner half-membrane presenting the P-face from which the majority of the globular proteins project, and (2) an inwardly directed outer half-membrane presenting the E-face, which is relatively smooth but shows occasional protein particles.

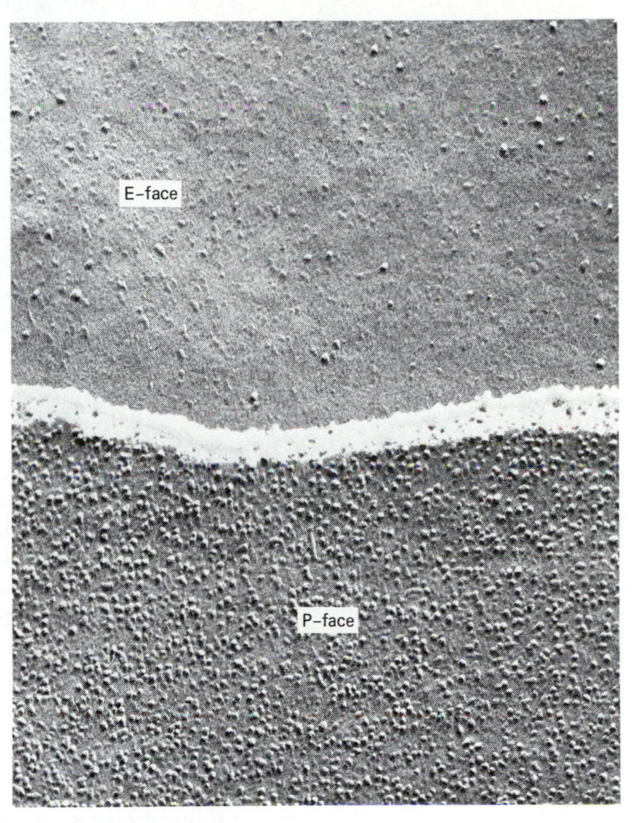

A freeze fracture made of a membrane from a cell of the eye of a monkey. Notice the greater number of proteins present in the P-face of the membrane. (Don Fawcett, Photo Researchers, Inc.)

change in shape in another glycoprotein molecule, one that extends all the way through the membrane. The change in this protein activates some enzyme or causes release of some substance that induces the biological response. Glycoproteins on the cell's surface also appear to play an important role in the ability of cells to recognize other similar cells during the cell movements that occur during development.

The protein molecule seems beautifully adapted to transmitting messages across the membrane: It has a receiving terminal on the outside of the membrane and a sending terminal in contact with the inner cytoplasm. The lipid bilayer serves as an insulating layer, preventing the transfer of unwanted signals.

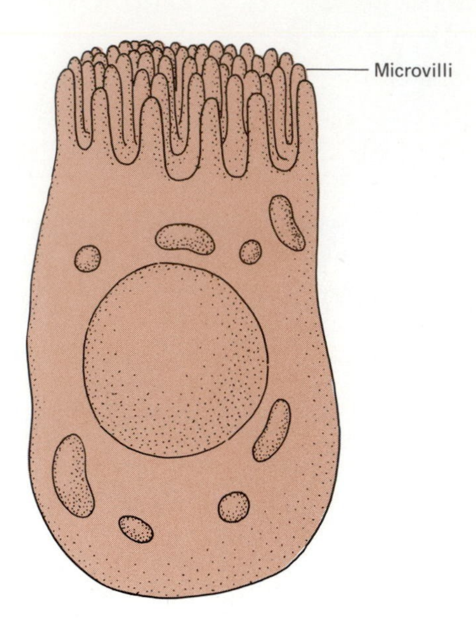

(a)

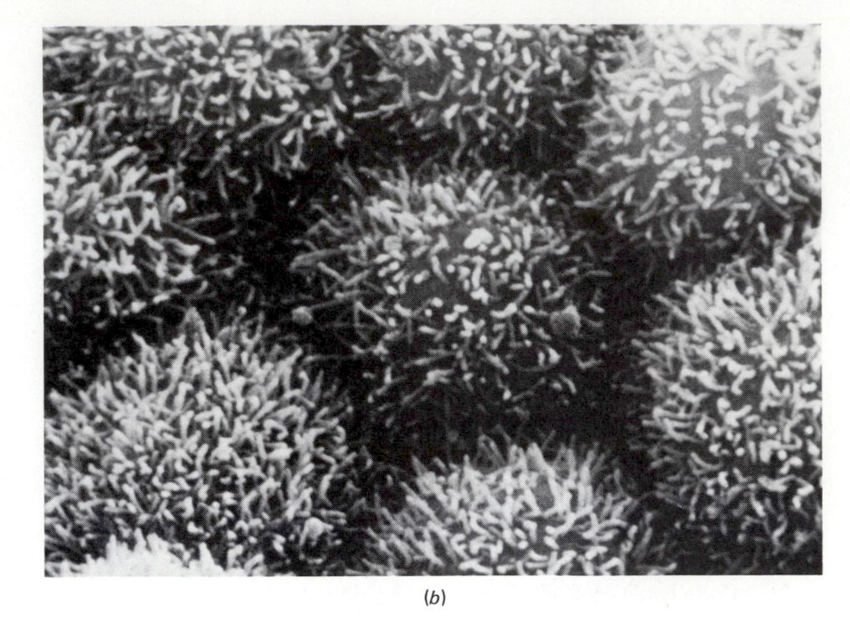

(b)

Figure 5–3 Microvilli. (*a*) Microvilli, present on the free surfaces of many types of cells, greatly increase the surface area for absorption of materials. (*b*) Scanning electron micrograph of ovarian epithelial cells from a mouse, showing microvilli (approximately ×65,000). ((*b*), courtesy of Dr. E. Anderson, Harvard Medical School.)

MICROVILLI

Multiple small evaginations (out-pocketings) of the cell membrane, termed **microvilli,** are present on the free surface of many types of cells, particularly those specialized to absorb materials (Fig. 5–3). Microvilli can enormously increase the surface area available for absorption of materials from the cell's environment. The number of microvilli can rapidly increase or decrease in response to environmental conditions or to changes in the metabolic needs of the cell.

Cell Walls

Cells of plants, algae, bacteria, and fungi have thick **cell walls** that lie just outside their cell membranes. The cell wall, which consists of carbohydrates secreted by the cell, protects and supports the cell and gives it a characteristic shape. The rigid plant cell wall is pierced in many places by tiny pores through which water and dissolved materials can pass. **Cellulose,** the main component of the plant cell wall, is usually present in the form of long threadlike fibers. The plant cell wall has been compared to reinforced concrete in which the cellulose fibers are like the steel rods and the matrix material around them, the concrete.

In addition to cellulose, the plant cell wall contains pectins and lignin. **Pectin** forms gels with sugar in the plant fluids, which are then less likely to flow out of a wound in a stem or leaf. Pectins obtained from plants are used in the preparation of jams or jellies. **Lignin** provides rigidity. Cells whose main function is support contain a great deal of lignin in their walls. In fact, wood consists primarily of cell walls in which the cellulose has been reinforced by lignin.

The cell walls of prokaryotes are not composed of cellulose. They contain a unique complex carbohydrate known as peptidoglycan, which will be discussed in Chapter 19. Many bacteria have a slimy polysaccharide capsule that lies just outside the cell wall and serves as an additional protective layer.

How Materials Pass Through Membranes

Whether a membrane will permit the molecules of any given substance to pass through it depends on the structure of the membrane and the size and charge of the molecules. A membrane is said to be **permeable** to a given substance if it will permit the substance to pass through, and **impermeable** if it will not permit the substance to pass. A **selectively permeable** membrane will allow some but not other substances to pass through it. Permeability is primarily a property of the membrane. All of the biological membranes surrounding cells, nuclei, vacuoles, mitochondria, chloroplasts, and the other subcellular structures are selectively permeable.

TABLE 5–1
Mechanisms for Moving Materials Through Cell Membranes

Process	*How It Works*	*Energy Source*	*Example*
Physical processes			
Diffusion	Net movement of molecules (or ions) from a region of greater concentration to one of lower concentration	Random molecular motion	Movement of oxygen in tissue fluid
Dialysis	Passage of small solute molecules through a selectively permeable membrane	Random molecular motion	Kidney dialysis
Facilitated diffusion	Carrier protein in cell membrane accelerates movement of molecules from region of higher to region of lower concentration.	Random molecular motion	Movement of glucose into some cells
Osmosis	Water molecules diffuse from a region of higher to a region of lower concentration through a selectively permeable membrane.	Random molecular motion	Water enters red blood cell placed in distilled water
Physiological processes			
Active transport	Protein molecules in cell membrane transport ions or molecules through membrane; movement may be against a concentration gradient (i.e., from a region of lower to a region of higher concentration).	Cellular energy	Movement of sodium out of cell against a concentration gradient (the "sodium pump")
Phagocytosis	Cell membrane encircles particle and brings it into cell by forming a vesicle around it.	Cellular energy	White blood cells ingest bacteria
Pinocytosis	Cell membrane takes in fluid droplets by forming vesicles around them.	Cellular energy	Cell takes in needed solute dissolved in tissue fluid.

Responding to varying environmental conditions or cellular needs, the cell membrane may present a barrier to a particular substance at one time and then actively promote its passage at another. By regulating chemical traffic in this way, the cell controls its own composition. The distribution of ions and molecules inside the cell can be very different from that outside the cell. In the nonliving world, materials move passively by physical processes such as diffusion. In living organisms, materials can also be moved actively by physiological processes such as active transport, phagocytosis, or pinocytosis (Table 5–1). Such active physiological processes require the expenditure of energy by the cell.

DIFFUSION

Some substances pass into or out of cells by simple physical diffusion. All atoms and molecules in liquids and gases tend to move in all directions until they are spread evenly throughout the available space (Fig. 5–4). **Diffusion** may be defined as the movement of molecules (or other particles) from a region of high concentration to one of lower concentration brought about by the kinetic energy of the molecules. The rate of diffusion is a function of the size and shape of the molecules, their electric charges, and the temperature.

The three states of matter—*solid, liquid,* and *gas*—differ in the freedom of movement of the constituent molecules. The molecules of a solid are closely

Figure 5–4 The process of diffusion. When a small lump of sugar is dropped into a beaker of water, its molecules dissolve, as shown in (*a*), and begin to diffuse throughout the water in the container, as seen in (*b*) and (*c*). Eventually, diffusion results in an even distribution of sugar molecules throughout the water in the beaker, as shown in (*d*).

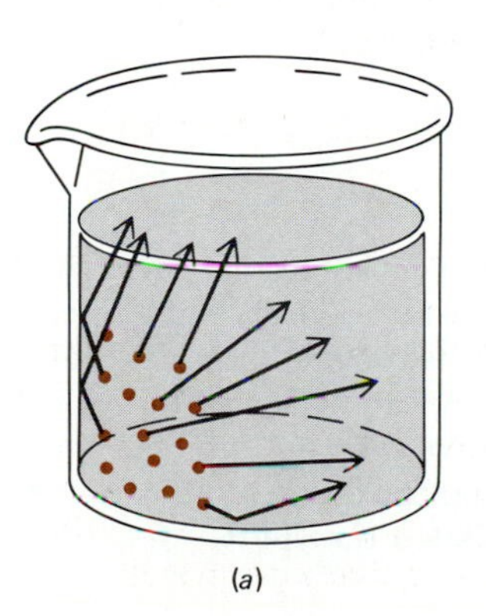

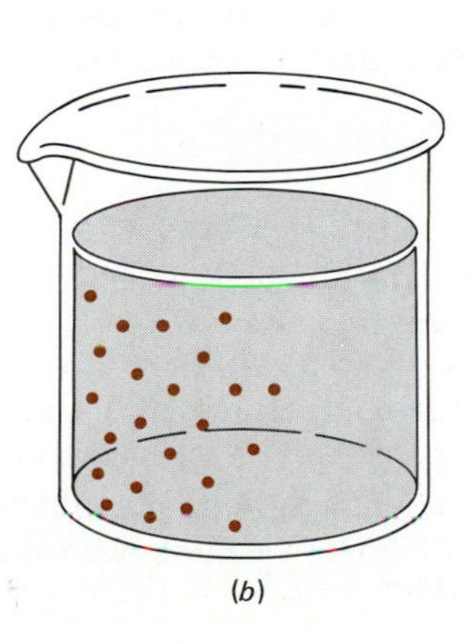

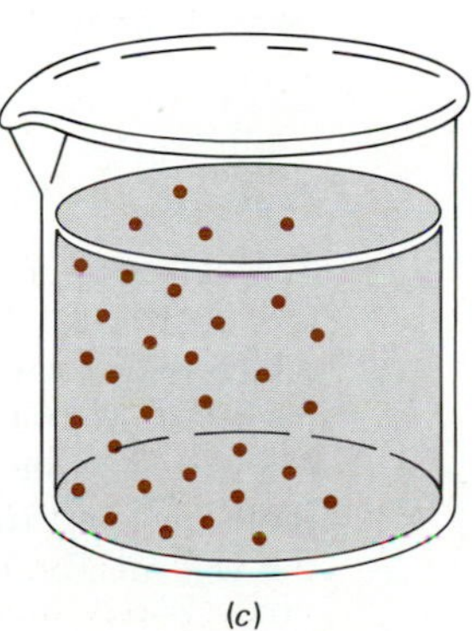

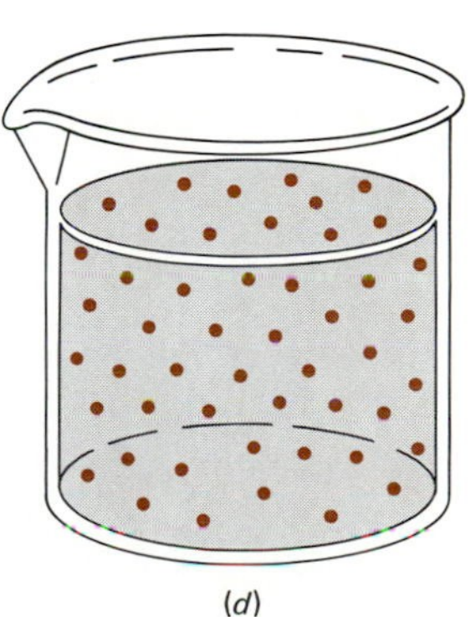

packed, and the forces of attraction between them allow them to vibrate but not to move around. In the liquid state, the molecules are farther apart; the intermolecular forces are weaker, and the molecules move about with considerable freedom. In the gaseous state, molecules are so far apart that intermolecular forces are negligible; molecular movement is restricted only by the walls of the container that encloses the gas.

Examining a drop of water under the microscope does not reveal the motion of the water molecules; they are much too small to be seen. When a drop of India ink, which contains fine particles of carbon, is added to the water, the motion of the carbon particles is visible under the microscope. Each carbon particle is constantly being bumped by water molecules. The recoil from these bumps moves the carbon particle in an aimless, zigzag path. This motion of small particles—termed **Brownian movement** after Robert Brown, an English botanist who first observed it when he looked through the microscope at pollen grains in a drop of water—provides a model of how diffusing molecules move.

As diffusion occurs, each individual molecule moves in a straight line until it bumps into something—another molecule or the side of the container. Then it rebounds and moves in another direction. An individual molecule may move as fast as several hundred meters per second, but each molecule can go only a fraction of a nanometer before it bumps into another molecule and rebounds. Thus the progress of any given molecule in a straight line is quite slow. Molecules continue to move even when they have become uniformly distributed throughout a given space. However, as fast as some molecules move in one direction, others move in the opposite direction, so that they remain uniformly distributed. The molecules of any number of different substances will diffuse independently of each other within the same solution; ultimately all become uniformly distributed.

Although diffusion occurs rapidly over microdistances, it takes a long time for a molecule to travel distances measured in centimeters. This overall slow rate of diffusion has important biological implications, for it limits the number of molecules of nutrients and oxygen that can reach an organism by simple diffusion. Only a very small organism that needs relatively few molecules per second can survive if it sits in one place and lets molecules come to it by diffusion. A larger organism must have some means either of moving to a new region or of stirring up its environment to bring new molecules to it. Alternatively, some organisms live where the environment is constantly passing by—in a river or in the intertidal zone on the seashore. Trees and other land plants have solved the problem by developing a tremendously branched root system, thus obtaining their water, salts, and other nutrients from a large area of the surrounding environment.

Dialysis

The diffusion of a **solute** (a dissolved substance) through a differentially permeable membrane is termed **dialysis.** To demonstrate dialysis, a cellophane bag can be filled with a sugar solution and immersed in a beaker of pure water (Fig. 5–5). If the cellophane membrane is permeable to sugar as well as to water, the sugar molecules will pass through it, and the concentration of sugar molecules in the water on the two sides of the membrane will eventually become equal. Subsequently, solute molecules will continue to pass through the membrane, but there will be no net change in concentration, for the rate of movement will be equal in the two directions. Kidney dialysis is a practical application of this process.

Facilitated Diffusion

The cell membrane is relatively impermeable to most polar molecules. This is of great biological advantage to the cell, because most of the compounds being metabolized within the cell are polar. This impermeability of the cell membrane prevents their loss. To transport polar nutrients such as glucose and amino acids through the lipid bilayer into the cell, systems of carrier proteins have evolved that bind these molecules and transfer them across the membrane. The transfer of solutes across the cell membrane by a carrier system is termed **mediated transport.**

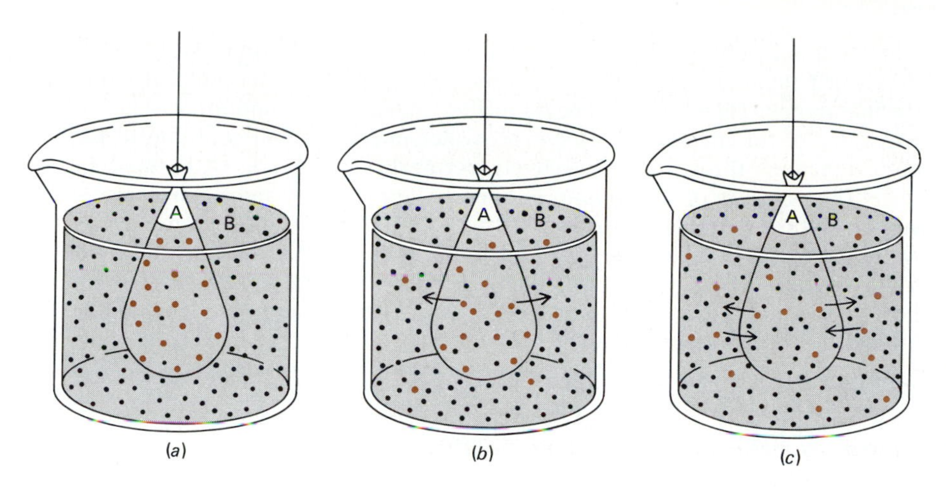

Figure 5–5 Dialysis. (*a*) A cellophane bag filled with a sugar solution is immersed into a beaker of water. The cellophane acts as a selectively permeable membrane, permitting passage of the sugar and water molecules but preventing passage of larger molecules. (*b*) Sugar molecules move through the membrane into the water in the beaker. (*c*) Eventually the sugar becomes distributed equally between the two compartments. Although sugar molecules continue to diffuse back and forth, the net movement is zero. The same is true for the water molecules. Black dots represent water molecules; colored dots represent sugar molecules.

In **facilitated diffusion** the carrier molecules simply serve as a passive conveyor belt that permits the substance to pass in either direction down a chemical gradient (a continuous change in concentration through a given space). The carrier proteins, called **permeases,** are thought to combine temporarily with the solute molecule and accelerate its movement through the membrane. The carrier protein is not changed by this action, and after it transports a solute molecule, it is free to bind with another solute molecule.

Osmosis

Osmosis is a special kind of diffusion in which molecules of water pass through a selectively permeable membrane, moving from the region where water molecules are more concentrated to where they are less concentrated. Most solute molecules cannot diffuse freely through the selectively permeable cell membrane, but water (solvent) molecules can pass freely through it.

When a 5% sugar solution is placed in a bag made of a selectively permeable membrane such a cellophane and suspended in water, the water molecules diffuse into the bag. This occurs because the larger sugar molecules are unable to penetrate the membrane and thus remain in the bag. If the liquid in the bag is 5% sugar, it is only 95% water. The liquid in the beaker is 100% water. Therefore, water molecules move from the container, where they are more concentrated, into the bag, where they are less concentrated. If a glass tube is inserted into the bag as shown in Figure 5–6, the column of water in the tube will rise. Eventually the liquid within the tube

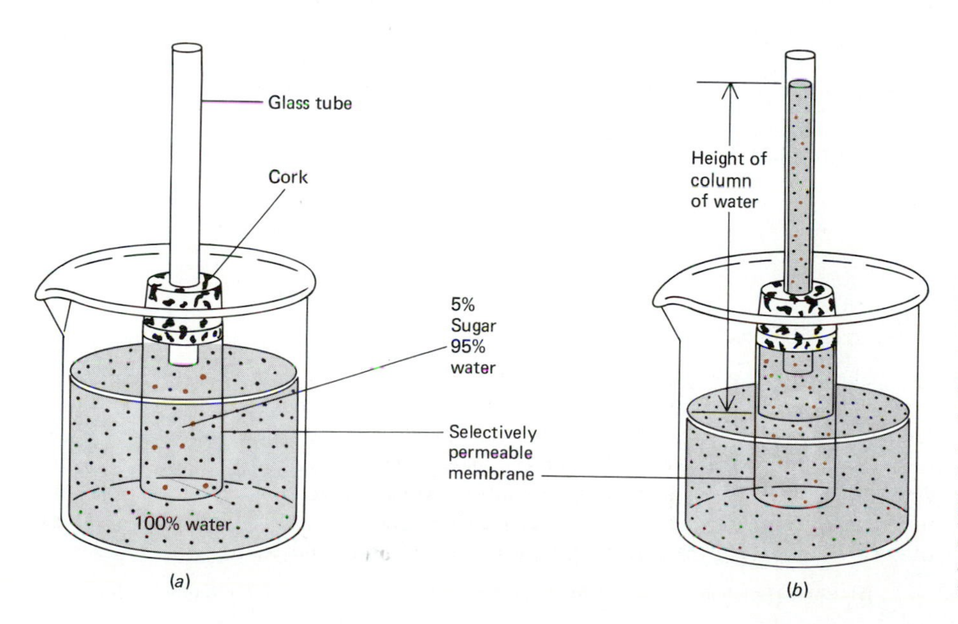

Figure 5–6 Osmosis. (*a*) A 5% sugar solution is placed in a bag made of a selectively permeable membrane and suspended in water. This membrane permits passage of the water molecules but not of the larger sugar molecules. Therefore, the water molecules pass across the membrane into the bag, causing the column of water in the glass tube to rise. (*b*) When equilibrium is reached, the pressure of the column of water in the tube just equals, and is a measure of, the osmotic pressure of the sugar solution.

will rise to a height at which the weight of the water in the tube exerts a pressure that is just equal to that resulting from the passage of the water molecules through the membrane. Then there will be no further change in the amount of water on the two sides of the membrane. However, water molecules will continue to pass in both directions through the selectively permeable membrane with equal speed.

The pressure of the column of water is termed the **osmotic pressure** of the sugar solution. Osmotic pressure results from **osmotic flow,** the tendency of the water molecules to pass through the membrane and equalize their concentration on the two sides. A more concentrated sugar solution would cause a greater osmotic flow and would cause water to rise to a higher level in the funnel. It would have a greater osmotic pressure, the force that must be applied to oppose the osmotic flow.

Isotonic, Hypertonic, and Hypotonic Solutions

Dissolved in the fluid compartments of every living cell are salts, sugars, and other substances that give that fluid a certain osmotic pressure. When such a cell is placed in a fluid with exactly the same osmotic pressure, there is no net movement of water molecules either into or out of the cell; the cell neither swells nor shrinks. Such a fluid is said to be **isotonic** (that is, of equal osmotic pressure to the fluid within the cell). Normally our blood plasma and all of our body fluids are isotonic to our cells; they contain a concentration of dissolved materials equal to that in the cells. A solution of 0.9% sodium chloride (sometimes called physiologic saline) is isotonic to the cells of humans and other mammals. Human red blood cells placed in 0.9% sodium chloride will neither shrink nor swell.

If the concentration of dissolved substances in the surrounding fluid is greater than the concentration within the cell, water tends to pass out of the cell, and the cell shrinks (Fig. 5–7). Such a fluid is said to be **hypertonic** to the cell. Human red blood cells placed in a solution of 1.3% sodium chloride will shrink. If the surrounding fluid contains less dissolved materials than the cell, it is said to be **hypotonic;** water then enters the cell and causes it to swell. Red blood cells placed in a solution of 0.6% sodium chloride will take up water, swell, and burst.

Turgor Pressure

The rigid cell walls of plant cells, algae, bacteria, and fungi enable these cells to withstand, without bursting, an external medium that is very dilute, containing only a very low concentration of solutes. Because of the substances dissolved in the cell's cytoplasm the cells are hypertonic to the outside medium. (The outside medium is hypotonic to the cell's cytoplasm.) Water tends to diffuse into the cells, filling their central vacuoles, and distending the cells. The cell swells, building up a pressure, termed **turgor pressure,** against the rigid cellulose cell walls (see Fig. 5–8). The cell wall can be stretched only very slightly, and a steady state is reached when the resistance of the cell wall to stretch prevents any further increase in cell size. There is no further net movement of water molecules into the cell. Turgor pressure is an important factor in providing support for the body of herbaceous plants. Thus, a flower wilts when the turgor pressure in its cells has decreased owing to the lack of water.

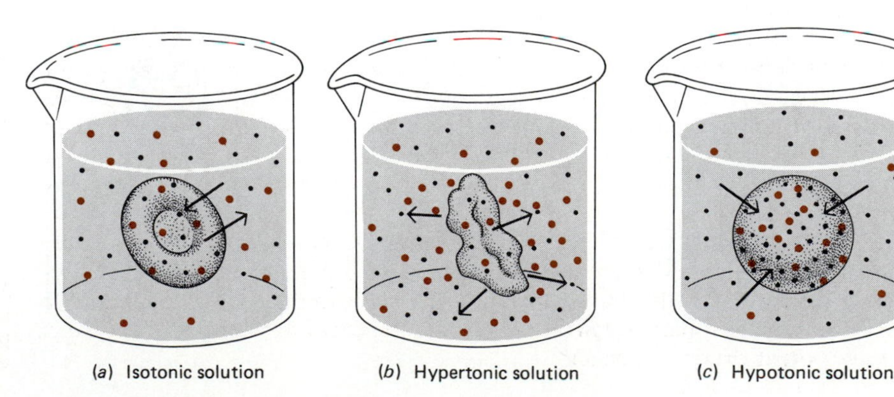

Figure 5–7 Osmosis and the living cell. (*a*) A cell is placed in an isotonic solution. Because the concentration of solutes (and thus of water molecules) is the same in the solution as in the cell, the net movement of the water molecules is zero. (*b*) A cell is placed in a hypertonic solution. This solution has a greater solute concentration (and thus a lower water concentration) than that in the cell. Therefore, it has greater osmotic pressure than the cell. This results in a net movement of water molecules out of the cell, and the cell becomes dehydrated, shrinks, and may die. (*c*) A cell is placed in a hypotonic solution. The solution has a lower solute (and thus a greater water) concentration than that in the cell. The cell contents therefore have higher osmotic pressure than the solution. There is a net movement of water molecules into the cell, causing the cell to swell. The cell may even burst.

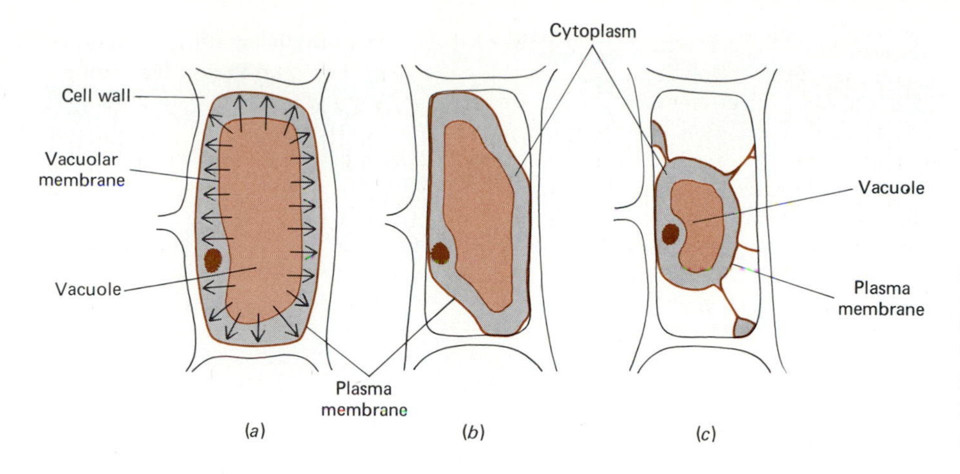

Figure 5–8 Turgor pressure in a plant cell. (*a*) In hypotonic surroundings, the contents of the cell fill the space within the wall. The arrows indicate the turgor pressure of the cell. (*b*), (*c*) If the cell is placed in a hypertonic medium, it loses water and its contents shrink. In (*c*) the cell is said to be plasmolyzed.

If a plant cell is placed in a hypertonic medium, it loses water to its surroundings, and its contents shrink. This process is called **plasmolysis** (Fig. 5–8). Plasmolysis can occur when large amounts of salts or fertilizers are contained in the soil or water around the plant.

ACTIVE TRANSPORT

The chemical composition of the cytoplasm is not identical with that outside the cell. For example, the concentration of sodium in the tissue fluid outside a human cell is about 10 times greater than inside, and the potassium concentration is about 35 times greater inside the cell than outside. Such differences are essential to the well-being of the cell. If the cell membrane were freely permeable, diffusion would quickly eliminate these differences in solute concentration.

The lipid bilayer prevents the diffusion of certain ions and molecules into or out of the cell but permits the passage of other ions and molecules. In addition, the cell can move materials from a region of low concentration to a region of higher concentration by **active transport,** a form of mediated transport. Working "uphill" against such a concentration gradient requires energy. Expenditure of energy in active transport is an example of how cells must work just to remain alive.

Characteristics of Mediated Transport Systems

Mediated transport systems, whether active transport or facilitated diffusion, have several characteristics. First, as the concentration of solute is increased, the amount transported per unit time increases—the rate of transfer increases in a parabolic curve that approaches a maximum (Fig. 5–9). It has been suggested that the transport system has specific sites to which the substance being transported is bound and that there is a finite number of such sites in the cell membrane. When all of these sites are filled, transport proceeds at its maximum rate. Increasing further the number of solute molecules available does not increase the rate of transfer. In contrast, the rate of simple physical diffusion of a solute across a membrane is directly proportional to the concentration gradient of the solute.

A second feature of mediated transport systems is the **specificity** of the system for the substance being transported. A transport system that mediates the transfer of glucose into the cell will have little or no effect on the transport of fructose or other monosaccharides or of disaccharides such as lactose. Experiments with many such systems have led to the conclusion that the transport process involves a protein with a binding site that is complementary to the substance being transported.

A third feature of mediated transport systems is that they may be specifically inhibited by substances that are structurally related to the compound being transported. Like enzyme inhibitors (Chapter 7), substances that compete for binding to the specific carrier protein decrease the rate of transport of the compound. Some of these are potent drugs or poisons.

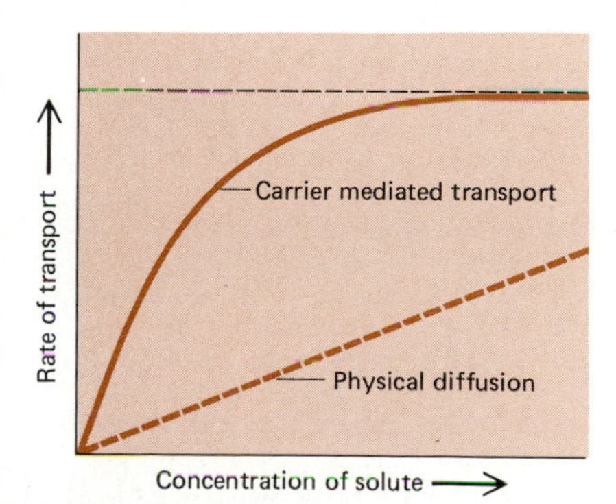

Figure 5–9 A comparison of the rates of transport across membranes by simple physical diffusion and by carrier-mediated transport, which shows saturation.

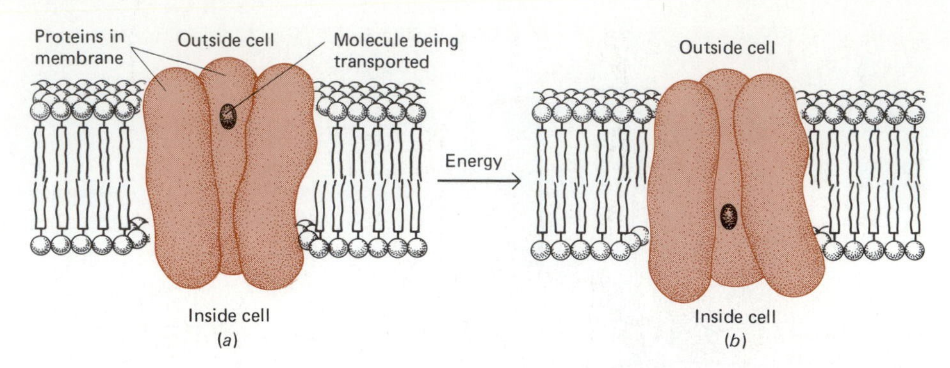

Figure 5–10 Active transport. (*a*) According to current theory, a group of transport proteins form a fluid-filled pore through which hydrophilic molecules can pass. When the molecule being transported binds to an active site on one of these proteins, it is thought that the transport protein changes shape. (*b*) The molecule is then forced through the channel and ejected on the other side.

Transport Proteins

The question of how hydrophilic compounds such as glucose and amino acids can pass through a hydrophobic lipid bilayer has puzzled biologists. An early guess suggested that the polar compound is bound to the membrane protein, which acts like a revolving door: It flips 180 degrees, unloads the polar compound to the interior of the cell, flips another 180 degrees, and reloads, so that the process continues. However, as more has been learned of the nature of the membrane proteins and the structure of the membrane, this hypothesis has lost favor. The membrane proteins have hydrophilic heads sticking out of the membrane and hydrophobic tails buried in the lipid bilayer. Flipping such a molecule would require a great deal of energy, and indeed, sophisticated experiments labeling one side of the membrane protein have shown that they do *not* flip.

The currently accepted hypothesis suggests that the transport proteins extend entirely through the membrane. Four or more transport protein molecules grouped in contact with each other (Fig. 5–10) provide a water-filled pore about 1 nm in diameter through which hydrophilic molecules can pass. Other calculations of the sizes of the pores in cell membranes suggest that they are about this order of magnitude. At some point on the interior of the membrane, there is an active site on the protein that specifically binds to one kind of molecule—glucose, for example. The binding of the transported molecule to the transport protein is postulated to trigger a change in the shape of the transport protein. As a result of this change in shape, the ion or molecule is forced through the channel and ejected on the other side.

ENDOCYTOSIS AND EXOCYTOSIS

Cells can also actively transport materials by exocytosis and by endocytosis. In **exocytosis,** a cell ejects waste products or specific secretion products such as hormones (Fig. 5–11). A membrane-enclosed vesicle within the cytoplasm fuses with the cell membrane. The vesicle then opens at the point of fusion, permitting release of the enclosed material to the exterior without the loss of other cell contents.

In **endocytosis,** materials are taken into the cell. Two types of endocytosis are phagocytosis and pinocytosis. In **phagocytosis,** which literally means "cell eating," the cell ingests large solid particles such as bacteria or food (Fig. 5–12). Folds of the cell membrane enclose materials outside the cell, forming a vesicle (also called vacuole) around them. The vesicle, still attached to the cell membrane, bulges into the cell interior, carrying with it a portion of the cell membrane and whatever medium it enclosed. The membrane then tightens like a drawstring purse and fuses together, leaving the invaginated portion as a free vesicle in the cytoplasm.

In **pinocytosis,** the cell takes in small drops of extracellular fluid. These are trapped by folds of the cell membrane and then pinch off into the cytoplasm as tiny fluid vesicles (Fig. 5–13). The contents of these vesicles are slowly transferred into the cytoplasm, and the vesicles themselves become smaller and eventually disappear. In a third type of endocytosis, **receptor-mediated endocytosis,** membrane proteins combine with specific proteins or small particles. The membrane invaginates, bringing materials into the cell (Fig. 5–13(*b*)).

Endocytosis can be induced in many kinds of cells by the presence in the surrounding medium of certain proteins or amino acids. For example, an amoeba exposed to such a protein solution develops short cytoplasmic projections, each

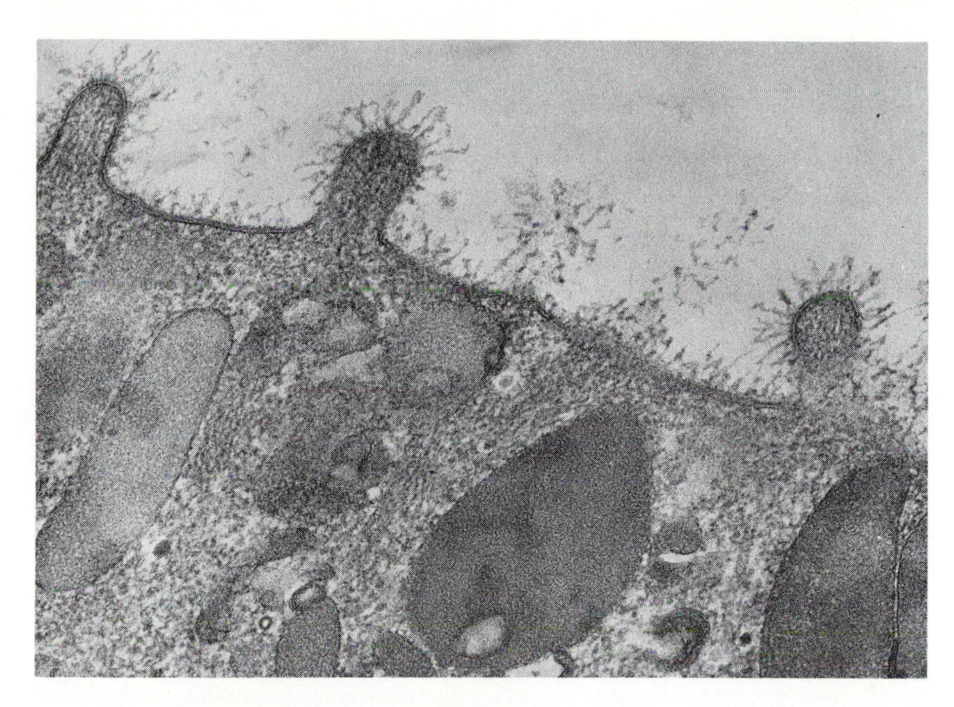

Figure 5–11 Exocytosis. A high magnification electron micrograph of the upper surface of a secreting cell. Secretion granules can be seen in the cytoplasm, approaching the cell membrane. The filaments projecting diffusely from the cell surface are probably hydrated proteins (magnification ×125,000). (J.F. Gennaro, Photo Researchers, Inc.)

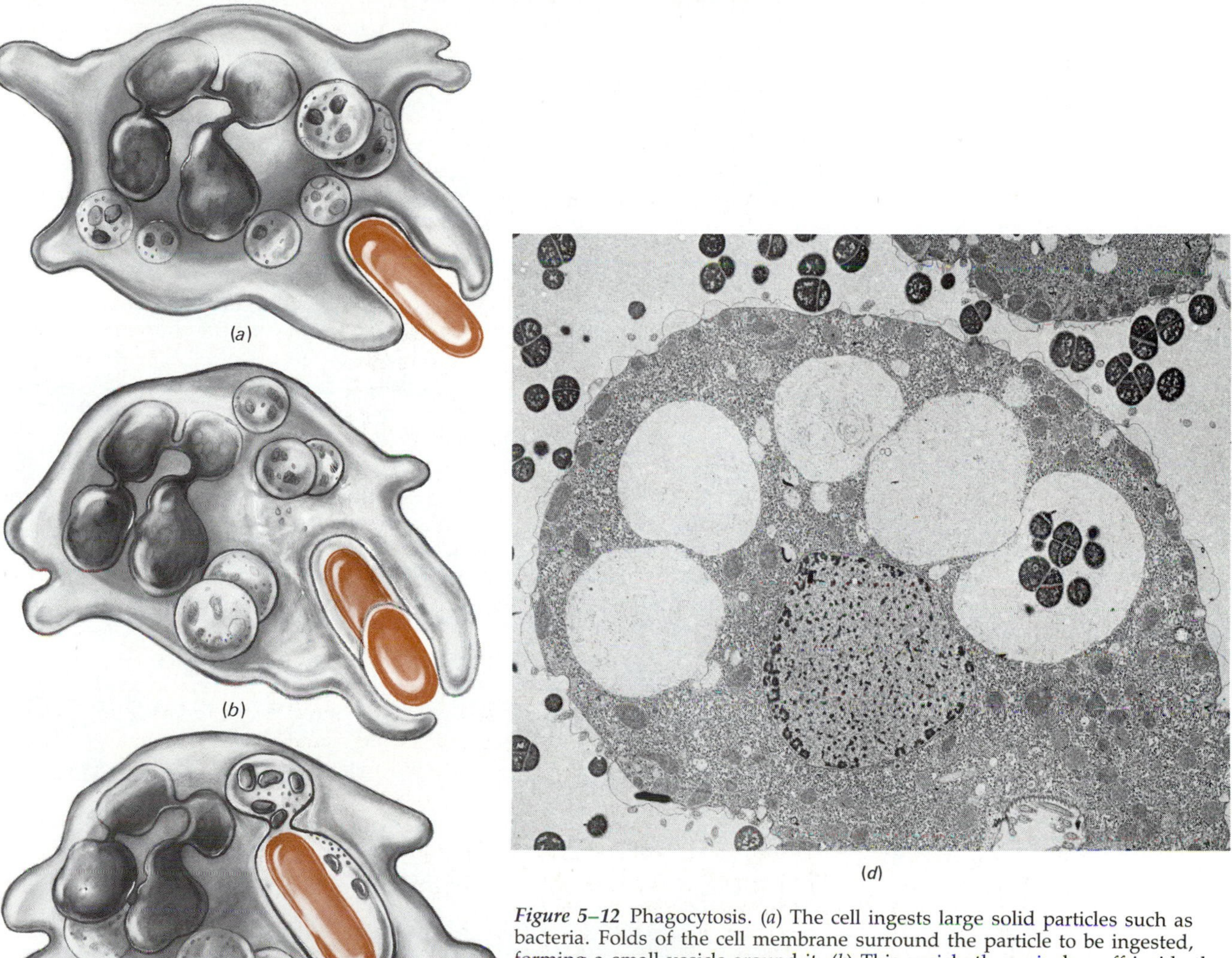

Figure 5–12 Phagocytosis. (*a*) The cell ingests large solid particles such as bacteria. Folds of the cell membrane surround the particle to be ingested, forming a small vesicle around it. (*b*) This vesicle then pinches off inside the cell. (*c*) Lysosomes may fuse with the vesicle and pour their potent digestive enzymes onto the ingested material. (*d*) Ingested bacteria held in vesicles (called phagosomes) of a *Tetrahymena* cell (a protozoan). The central vesicle (vacuole) contains particles of digested bacteria. ((*d*), T.J. Beveridge, Univ. of Guelph/BPS.)

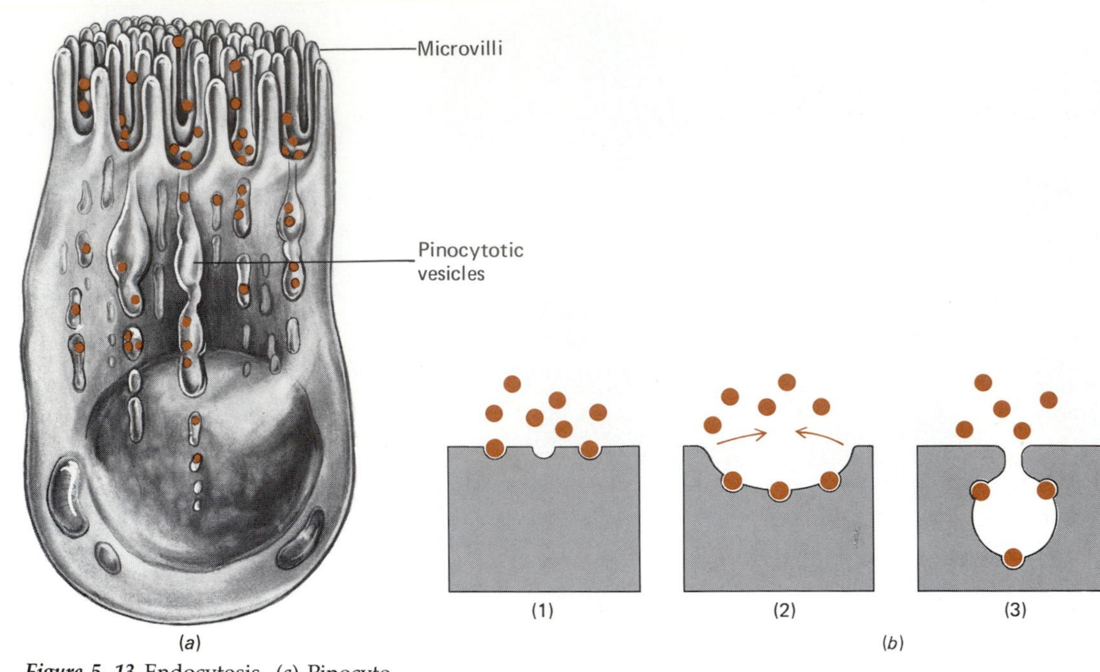

Figure 5–13 Endocytosis. (*a*) Pinocytosis. Tiny droplets of fluid are trapped by folds of the cell membrane, which then pinch off into the cytoplasm as little vesicles of fluid. The content of these vesicles is slowly transferred into the cytoplasm. Note the numerous microvilli. (*b*) Receptor-mediated endocytosis. Protein molecules in solution are bound to specific sites on the cell membrane. The membrane folds in to bring the protein molecules into a vesicle within the cytoplasm. (*c*) Endocytosis in a developing red blood cell in the bone marrow of a guinea pig. In this sequence, particles of ferritin are being taken inside the cell, where its iron is used in the synthesis of hemoglobin. (Don Fawcett, Photo Researchers, Inc.)

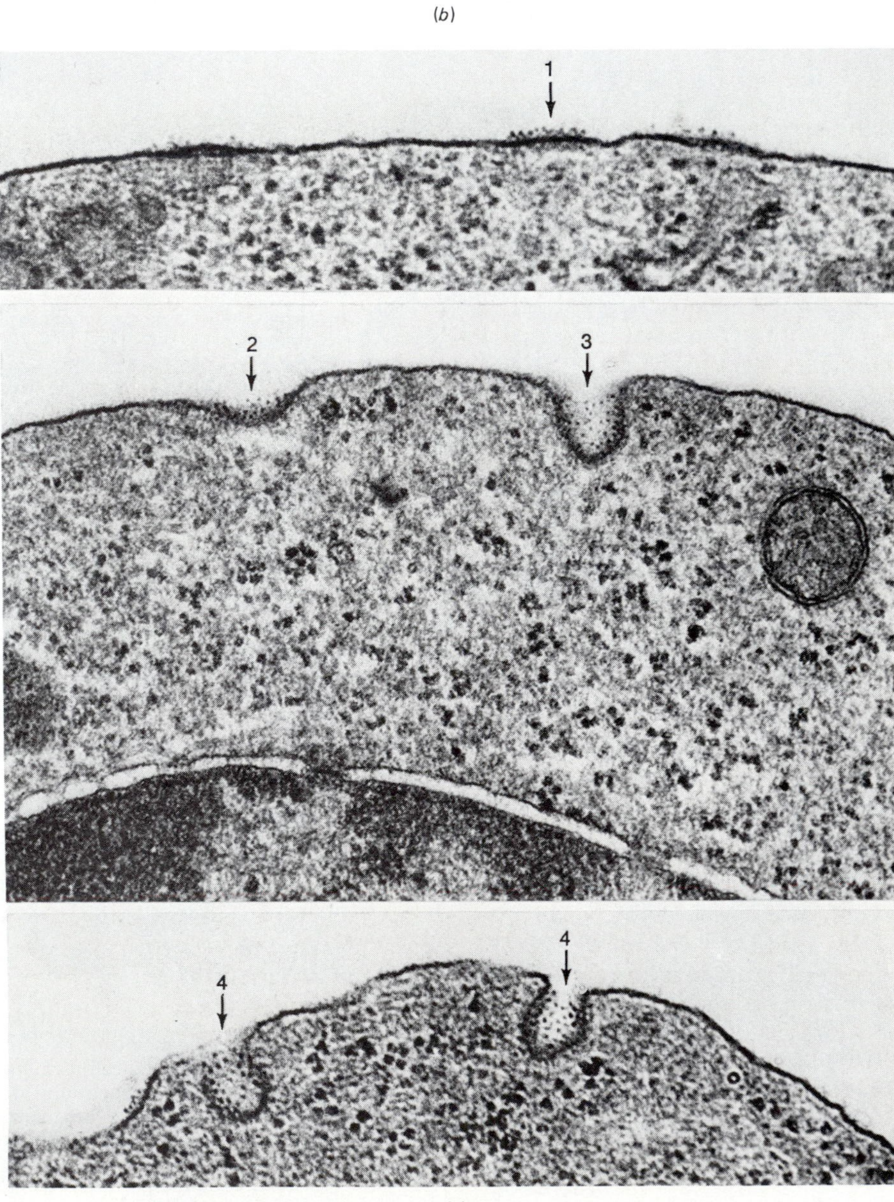

TABLE 5–2
Types of Intercellular Connections

	Characteristic Specializations in Cell Membrane	*Width of Intercellular Space*	*Function*
Desmosomes	Button-like electron-dense regions in each of opposing cell membranes; tonofilaments extend from cell membrane into cytoplasm, attached to cytoskeleton.	Normal size 24 nm	Hold cells tightly together
Tight junctions	Adjacent cell membranes fuse.	Intercellular space disappears as the two adjacent membranes fuse.	Form a continuous barrier; separate certain body cavities from others
Gap junctions	Channels or pores through the two cell membranes and across the intercellular space; membranes are specialized with highly ordered patterns.	Intercellular space greatly narrowed, to 2 nm	Provide for electrical communication between cells and for flow of ions

with a narrow undulating channel extending from its tip to its base. Vesicles form at the inner end of the channel, separate from it, and pass deeply into the cytoplasm. Endocytosis is an active, energy-requiring process, and mitochondria are usually oriented near these active undulating surfaces. In a single cycle of endocytosis an amoeba takes up a volume of fluid equal to 1% to 10% of its total volume. The process requires the synthesis of an area of new membrane equivalent to as much as 6% of the initial surface area of the amoeba. In endocytosis, the cell does not simply take up drops of the surrounding fluid but can take up protein molecules selectively. An amoeba can take up an amount of protein equivalent to that present in 50 times its own volume from the surrounding medium in 5 minutes. Endocytosis is a membrane activity that brings about the *selective* uptake of materials from the surrounding environment. It can account for the rapid transport of large, water-soluble molecules, such as proteins and nucleic acids, across the cell membrane. Endocytosis also occurs during cell division, producing rapid swelling of the daughter cells as they come into being in telophase.

Cell Junctions and Communications

In areas where cells are in close contact with one another, specializations of the cell membrane may provide for the adhesion of adjacent cells or for cell-to-cell communication (Table 5–2). A classic example of the latter is the synaptic junction between neurons.

DESMOSOMES

Adjacent epithelial cells, especially those of the epidermis (the upper layer of the skin), are so tightly adherent to each other that a strong mechanical force must be applied to separate them. The question of what holds together the cells of our skin was answered by electron microscopy. Discontinuous, button-like plaques called **desmosomes** are present on the two opposing cell surfaces, separated by an intercellular space about 24 nm wide (Fig. 5–14). Each desmosome appears as a region of dense material consisting of local differentiations of each of the opposing cell membranes. Desmosomes are made of protein and can be broken down by enzymes such as trypsin.

The function of the desmosome is purely mechanical—that of holding cell surfaces together. Desmosomes act like rivets or spot welds, points holding together two structures not connected otherwise. The cell membrane of the desmosome is of normal dimensions but appears to be thickened because of the presence

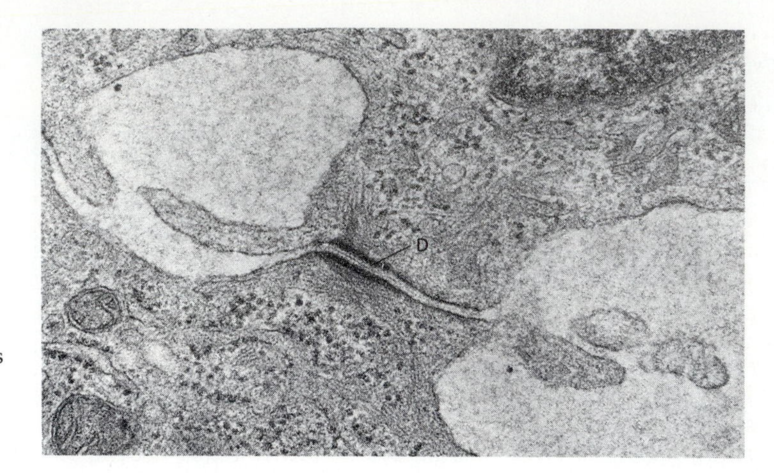

Figure 5–14 Electron micrograph showing a desmosome (*D*) between two cells of the ovarian epithelium of a rabbit (approximately ×70,000). (From Anderson, E.: *Journal of Morphology* 150:135–166, 1976.)

of a thin layer of electron-dense material on its inner cytoplasmic surface. Next to this is a somewhat thicker layer of fine filaments called **tonofilaments** embedded in the cytoplasm. Bundles of tonofilaments in the cytoplasm typically converge upon the desmosomes and appear to terminate in them; these may serve to anchor the ends of the desmosome within the cells it joins.

TIGHT JUNCTIONS

A different type of connection between cells, with physiological rather than simply mechanical functions, is the so-called **tight junction.** This is an area of tight connection between two adjacent cell membranes, a connection so tight that the junction becomes essentially impermeable. Unlike the desmosomes, which bridge an intercellular space of normal width, the tight junction involves the complete disappearance of that space. Each cell is right up against the adjacent cell membrane. Electron micrographs and freeze-etch studies have shown that, in the tight junction, the pulling together of cell membranes does not simply occlude the intercellular space but actually involves the *fusion* of the two membranes. Electron micrographs of the region show not four dark lines (indicating the presence of two normal bilayers) but only *three.* In other words, the outer surfaces of the two lipid bilayers have in some fashion melted together in a single layer. This zone of fusion completely encircles each cell (Fig. 5–15). A layer of such cells is in effect embedded in one essentially continuous cell membrane with no gaps or intercellular spaces.

Cells connected by tight junctions constitute a continuous barrier; such junctions are found in the apical portions of cells that separate certain body cavities from others or from connective tissue. In these situations a sharp physical separation between two compartments is essential. A classic example is the blood–brain barrier (Chapter 40), which involves the lining of the cerebral blood vessels, the cells of which are joined by tight junctions. Here the role of the tight junction is

Figure 5–15 Freeze-fracture replica illustrating a tight junction. P-face (*P*) and complementary E-face (*E*). Rabbit ovarian epithelium (×50,000). (Courtesy of Dr. E. Anderson, Harvard Medical School.)

essentially protective, for it blocks the passage of certain substances from the blood into the brain tissue. The complex and delicate operations of the brain would be disrupted if these substances could enter. The function of the tight junction barrier in other organs may be to keep substances *in* rather than out of a given physiological compartment. For example, the tight junctions in the epithelial lining cells of the ciliary body of the eye prevent the sodium ions and ascorbic acid (vitamin C) molecules present in very high concentrations in the anterior chamber of the eye from leaking back into the circulation.

GAP JUNCTIONS

A third type of intercellular connection, the **gap junction,** is like the desmosome in bridging the space between two cells, but the space is narrowed from 24 nm to 2 to 4 nm. It differs in that it connects not only the exterior membranes of the cells but also their cytoplasmic compartments. When a marker substance is injected into one of the cells connected by gap junctions, the marker passes rapidly into the adjacent cell, but it does not enter the space between the cells. A heavy metal such as lanthanum, which cannot cross the cell membrane, can penetrate the gap junction, entering the extracellular space or gap of 2 to 4 nm. However, when a thin section is made across the gap junction, perpendicular to the membranes that are connected, the electron microscope shows a hexagonally arranged mosaic of circular areas into which the lanthanum has not penetrated. The inference is that the gap junction consists of an array of channels or pores passing through the cell membrane, across the narrow intercellular space, and through the membrane of the adjacent cell (Fig. 5–16).

Passing along these channels where they cross the gap between the cells is another complementary set of channels through which extracellular substances can

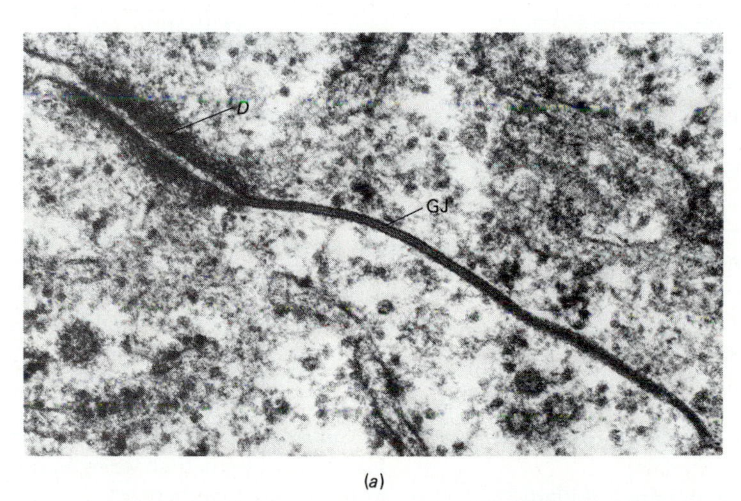

(a)

(b)

Figure 5–16 (*a*) A gap junction (*GJ*) and a desmosome (*D*) between ovarian cells of a rabbit (approximately ×180,000). (*b*) A freeze-fracture replica of the P-face (*P*) gap junction between two ovarian cells of a mouse (approximately ×70,000). ((*a*), from Albertini, D. F., and Anderson, E.: *Journal of Cell Biology* 63: 234–250, 1974; (*b*), from Anderson, E.: *Journal of Morphology* 156:339–366, 1978.)

move from one part of the extracellular space to another. The system is similar to a pair of metal tanks (the cells) connected by a close-set array of short pipes (the channels). Any substance outside the tanks can pass through the spaces between the pipes but will be unable to enter either the pipes or the tanks. The channels from cell to cell permit the passage of certain substances from one cell to the next. The narrowness of the connecting channels, about 1.5 nm, prevents the passage of large molecules from one cell to the next. However, the channels are large enough to permit the passage of molecules such as sucrose and a variety of dyes.

Gap junctions may also provide for electrical communication between cells. In some species the gap junction acts as a special sort of synapse (junction between nerve cells) where the impulse is transmitted electrically rather than chemically from one cell to the next. These electronic synapses are electrically coupled by a gap junction through which the signal can be transmitted as a surge of potassium ions. Such electronic synapses are found in the electrical organs of the electric eel, the torpedo, and the electric catfish. The electric organ consists of modified skeletal muscle cells. To generate an effective shock, all of these cells must "fire" precisely at the same moment. Chemical transmission across the synapse is too slow, requiring perhaps a millisecond, to permit the precise synchronous firing of all the muscle cells. However, this is readily achieved by the gap junctions of the electronic synapses, which provide a sort of short-circuit connection between the cells. The intercalated disks in cardiac muscle (Chapter 6) contain gap junctions that rapidly transmit impulses from one cell to the next so that all the muscle fibers in the ventricles contract simultaneously.

SUMMARY

I. The cell membrane functions (1) to regulate the passage of materials into and out of the cell; (2) to receive information that permits the cell to sense changes in its environment and to respond to them; (3) to communicate with other cells; and (4) to protect the cell.

II. The cell membrane surrounding each cell is composed of a lipid bilayer in which are embedded specific membrane proteins.

- *A.* Membrane proteins are typically glycoproteins with the carbohydrate portion on the exterior surface of the membrane. These are important in determining the properties of the membrane and in regulating cell–cell communication and interaction.
- *B.* Some membrane proteins are receptors, binding some specific hormones and transmitting specific information across the membrane to the cell's interior.
- *C.* Many types of cells have microvilli that increase the surface area for absorption of materials.

III. Thick cell walls surround the cell membranes of plant, algae, bacteria, and fungal cells.

IV. The cell membrane facilitates the entrance of certain molecules and prevents the entrance of others. It is an active structure that has carrier proteins to transfer specific ions or molecules in or out, with or against, a concentration gradient.

- *A.* Some substances pass through the cell membrane by simple diffusion. Osmosis is a special kind of diffusion in which molecules of water pass through a selectively permeable membrane, moving from the region where water molecules are more concentrated to where they are less concentrated.
- *B.* The cell membrane can actively transport materials against a concentration gradient. Small groups of transport proteins surround a fluid-filled channel that extends through the membrane. The molecule to be transported binds to one of the proteins and is forced through the channel.
- *C.* Cells "eat" and "drink" by endocytosis. Two types of endocytosis are phagocytosis and pinocytosis, processes that involve the infolding of a portion of the cell membrane and the synthesis of additional cell membrane material. Cells may also eject waste products or secrete substances such as hormones by the reverse process, exocytosis.

V. Adjacent cells are connected and held together by specializations of the cell membrane.

- *A.* Desmosomes are discontinuous button-like plaques made of proteins that serve as rivets or spot welds to hold cells together.
- *B.* Tight junctions are areas where the lipid bilayers are partially fused. Cells joined by tight junctions form a continuous barrier and are found where a sharp physical separation between two compartments is required.
- *C.* Gap junctions are cell-to-cell connections consisting of an array of small channels passing through the cell membranes by means of which small molecules can pass from one cell to the next.

POST-TEST

1. The cell is separated from the outside environment by the ______________ ______________, which regulates the passage of materials into and out of the cell.

2. Cell surfaces are equipped with ______________ ______________ that receive chemical messages from other cells.

3. In the currently accepted model of the cell membrane, the membrane consists of a rather fluid ________________ ________________ in which a variety of globular ________________ are embedded.

4. The lipid components of the cell membrane include ________________ and ________________.

5. All of these have a highly polar ________________ portion and a nonpolar ________________ portion.

6. The formation of the lipid bilayer from its constituents occurs ________________.

7. Lipid bilayers are self-________________ and self-________________.

8. Specific functions of the membranes such as chemical transport and the transmission of messages are carried out by the ________________.

9. The membrane proteins that extend through both sheets of the lipid bilayer and to the membrane surfaces are ________________.

10. The free surface of many kinds of cells have multiple small outpocketings called ________________ that greatly increase the surface available for absorption of material from the cell's environment.

11. The cells of plants have rigid cell walls outside of the cell membrane composed mainly of ________________.

12. The cell walls of prokaryotes contain a different complex carbohydrate called ________________.

13. A membrane that permits some but not other substances to pass through it is said to be ________________ ________________.

14. The movement of molecules from a region of high concentration to one of lower concentration, brought about by the kinetic energy of the molecules, is termed ________________.

15. The diffusion of a solute through a differentially permeable membrane is termed ________________.

16. A solution that has an equivalent concentration of solutes and hence the same osmotic pressure as that in the fluid within the cell is said to be ________________.

17. Because the fluid surrounding a plant cell is hypotonic to it, the cell swells against the rigid plant cell wall, developing a ________________ ________________ that provides support for the body of the plant.

18. The ejection by a cell of waste products or specific secretory products is termed ________________.

19. Materials are taken into a cell by ________________; in ________________, folds of the cell membrane enclose material outside the cell, forming a vesicle that bulges into the interior of the cell.

20. Cells may be held together by discontinuous, button-like plaques called ________________ on the two opposing cell surfaces.

21. Connections between cells in which the two membranes actually fuse and the intercellular space disappears are termed ________________ ________________.

22. A third type of intercellular connection, in which an array of channels or pores passes through the cell membrane, across a narrow intercellular space, and through the membrane of the adjacent cell, is known as a ________________ ________________.

REVIEW QUESTIONS

1. Differentiate clearly between diffusion and osmosis.
2. In what ways is the phenomenon of diffusion important to living things?
3. Make a diagram of the cell membrane showing the lipid bilayer, the various types of membrane proteins, and the carbohydrates on the outer surface of the membrane.
4. Compare the structure and composition of the membranes and walls surrounding plant, bacterial, and animal cells.
5. List and discuss the various functions served by the cell membrane.
6. Compare the chemical nature of the inner and outer layers of the lipid bilayer. How are these related to the function of the cell membrane?
7. Discuss the nature of the membrane proteins, and explain how their properties make them especially adapted for their functions.
8. Compare the transport of compounds across the cell membrane by physical diffusion with the transport by the process of mediated transport. What additional factor is present in the process of active transport?
9. Compare phagocytosis and pinocytosis. Give an example of exocytosis.
10. What are microvilli? What function do they serve in the activities of cells?
11. Explain what is meant by freeze-fracturing a membrane.
12. Define osmotic pressure. What components of a solution determine its osmotic pressure?
13. Distinguish among isotonic, hypertonic, and hypotonic solutions.
14. What is meant by the turgor pressure of a plant cell? What is responsible for this phenomenon?
15. Contrast gap junctions, tight junctions, and desmosomes.

6

Tissues, Organs, and Organ Systems

OUTLINE

I. Why are so many living things multicellular?
II. Animal tissues
  A. Epithelial tissues
  B. Connective tissues
    1. Fibers and cells of connective tissue
    2. Loose and dense connective tissues
    3. Elastic and reticular connective tissues
    4. Adipose tissue
    5. Cartilage and bone
    6. Blood
  C. Muscle tissue
  D. Nervous tissue
III. Animal organs and organ systems
IV. Plant tissues
  A. Meristematic tissues
  B. Permanent tissues
    1. Surface tissues
    2. Vascular tissues
    3. Fundamental tissues
V. Plant organs and plant systems

LEARNING OBJECTIVES

After you have read this chapter you should be able to:

1. Discuss the advantages and disadvantages of the multicellular condition.
2. Define tissue, organ, and organ system.
3. Describe the four principal kinds of animal tissue—epithelial, connective, muscular, and nervous tissues—and give their respective functions.
4. Describe the main types of epithelial tissue, and locate each in the body.
5. Describe the main types of connective tissue, and give their functions.
6. Compare the three types of muscle tissue and their functions.
7. Give the function of nervous tissue, and distinguish between neurons and glial cells.
8. List all the organ systems characteristic of complex animals, describe their functions, and discuss how each system is involved in maintaining the constancy of the internal environment.
9. Distinguish between meristematic and permanent tissues in plants.
10. Describe and give the functions of the following types of permanent tissues: surface tissues, fundamental tissues, and vascular tissues.
11. List and describe the component parts of the roots and shoots of plants.

Most one-celled organisms are so small that they are invisible to the unaided eye. Why are there no giant amoebas slithering around? And for that matter, why is a dog or a maple tree multicellular, instead of being composed of just one large cell?

Why Are So Many Living Things Multicellular?

One of the main reasons that cells do not get large is that it is inefficient for them to do so. All materials must pass into or out of the cell through the cell membrane, so the size of the membrane in comparison with that of the rest of the cell is critical. As a cell increases in size, its volume increases at a greater rate than that of its surface. (The surface of a sphere increases in proportion to the square of its radius, while the volume increases as the cube of its radius.) A large cell, then, has proportionately much less cell membrane for its volume than a smaller cell (Fig. 6–1). Perhaps a familiar example will help to make this concept clear. Consider potatoes. Would you rather peel one very large potato or eight small ones? The amount of mashed potatoes prepared would be the same, but those eight small potatoes have a lot more surface to peel!

As a cell grows, its surface (that is, the cell membrane) becomes unable to provide sufficient oxygen and nutrients for all regions of the cell. Wastes produced within the cell must move longer distances to reach the cell membrane and exit from the cell. Because diffusion is not effective over long distances, a large increase in cell volume would threaten the well-being of the cell. There seems also to be a limit to the amount of cytoplasm that a nucleus can control. In fact, very large cells may have more than one nucleus.

When its size approaches the limits of efficiency, a cell divides to form two cells. Each new cell is half the size of the mother cell, but the relative size of both the cell and nuclear membranes is greatly increased in proportion to the volume of the cell.

In unicellular organisms, cell division results in the production of two new individuals, whereas in multicellular organisms, the two new cells may remain associated to form part of the whole organism. The number of cells, not their individual size, is responsible for the different sizes of various organisms. The cells of an earthworm, a human being, or an elephant correspond in size; the elephant is larger because its genes are programmed to provide for a larger number of cells. In order to achieve a large size an organism must be multicellular.

With multicellularity also comes the specialization of cells. In a unicellular organism, such as a moneran or a protist, the single cell must carry on all the activities necessary for the life of the organism. An organism composed of many cells can assign specific tasks to different cells. Animals, most plants, and many fungi are multicellular. Their cells are specialized to perform certain functions, permitting the efficient division of labor among different groups of cells. For example, some groups of cells are specialized to transport materials, whereas other groups of cells provide for the energy needs of the organism. How do these cells associate and perform such specialized functions? To answer this we will examine the tissues, organs, and organ systems of animals and plants.

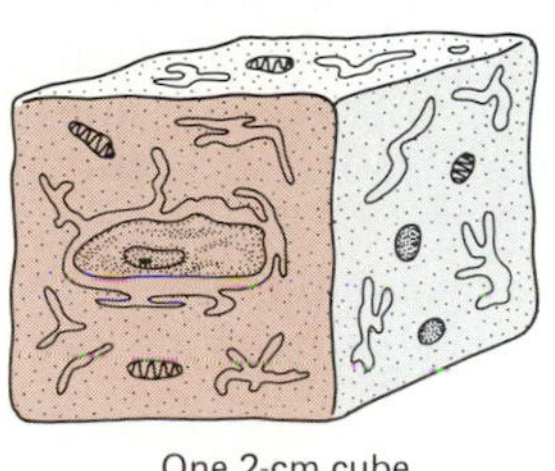

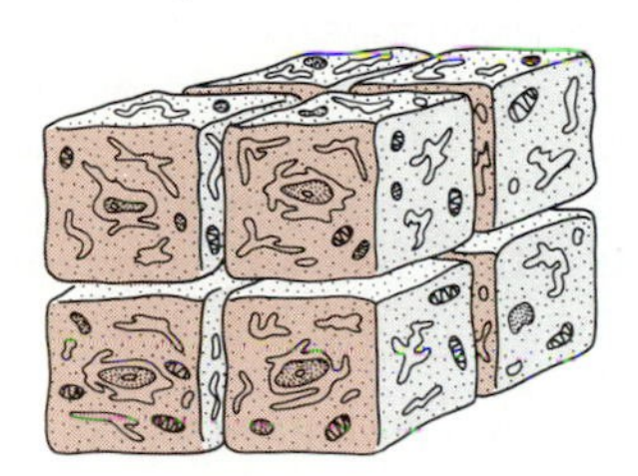

Figure 6–1 Eight small cells have a much greater surface (cell membrane) area in relation to their total volume than that of one large cell. This concept might be easier to grasp if you imagine that each of these cells is a potato. The amount of mashed potatoes you could prepare from eight small potatoes would be the same as from one large one, but which would you rather peel?

Animal Tissues

A **tissue** consists of a few types of closely associated cells that are adapted to carry out specific functions. Animal tissues are customarily classified as epithelial, connective, muscular, or nervous. Each kind of tissue is composed of cells with a characteristic size, shape, and arrangement.

EPITHELIAL TISSUES

Epithelial tissue (also called **epithelium**) consists of cells fitted tightly together, forming a continuous layer or sheet of cells covering a body surface or lining a cavity within the body. One surface of the sheet is attached to the underlying tissue by a **basement membrane** composed of tiny fibers and of nonliving polysaccharide material produced by the epithelial cells. In addition to the outer layer of the skin, the linings of the digestive and respiratory tracts and the lining of the kidney tubules are examples of epithelial tissues.

Epithelial tissues function in protection, absorption, secretion, or sensation. The epithelial layer of the skin covers the entire body and protects it from a variety of deleterious effects of the environment including mechanical injury, harmful chemicals, bacteria, and fluid loss. The epithelial tissue lining the digestive tract absorbs nutrients and water into the body. Other epithelial cells are organized into glands, adapted for the secretion of cell products.

Everything that enters the body or leaves it must cross one or more layers of epithelium. Food that is taken into the mouth and swallowed is not really "inside" the body. This occurs only when the substance is absorbed through the epithelium of the gut and enters the blood. The permeability properties of the various epithelia regulate to a large extent the exchange of substances between the different parts of the body and between the organism and the external environment. Finally, all sensory stimuli must penetrate the outer epithelium to be received.

Many epithelial membranes are subjected to continuous wear and tear. As outer cells are sloughed off they must be replaced by new ones from below. Such epithelial tissues generally have a rapid rate of cell division so that new cells are continuously produced to take the place of those lost (Fig. 32–6).

Three types of epithelial cells can be distinguished on the basis of their shape (see Table 6–1): **Squamous** epithelial cells are thin, flattened cells shaped like pancakes or flagstones. **Cuboidal** epithelial cells are short cylinders that in side view are cube-shaped, resembling dice. Actually each cell has a complex shape, usually that of an eight-sided polyhedron. **Columnar** epithelial cells look like tiny columns or cylinders when viewed from the side. The nucleus is usually located near the base of the cell. Viewed from above, or in cross section, these cells appear hexagonal in shape. Columnar epithelial cells may have cilia on their free surface that beat in a coordinated way, moving materials in one direction. Most of the respiratory tract is lined with ciliated epithelium; the ciliary beating helps to move particles of dust and other foreign material out of the lungs.

Epithelial tissue may be **simple**—that is, composed of one layer of cells— or **stratified**—composed of two or more layers (see Table 6–1). Simple epithelium is usually located in areas where materials must diffuse through the tissue or where substances are secreted, excreted, or absorbed. Stratified epithelial tissue is located in regions where protection is the main function. Stratified squamous epithelium is found in the skin, and lining the mouth and esophagus, of humans and other vertebrates. A third arrangement of epithelial cells is **pseudostratified** epithelium, so named because its cells falsely appear to be layered. All of the cells really do rest on a basement membrane, but not every cell is tall enough to reach the free surface of the tissue. This may give the false impression that there are two or more cell layers. Some of the respiratory passageways are lined with pseudostratified epithelium equipped with cilia.

Table 6–1 illustrates the main types of epithelial tissue and indicates where they are located in the body, as well as describing their functions.

The linings of the body cavities and the linings of the blood and lymph vessels are derivatives of **mesenchyme**, a generalized embryonic tissue that gives rise to connective tissues rather than epithelial germ layers. Structurally, however, they are in all respects typical epithelial cells. To distinguish them from the true epithe-

TABLE 6–1
Some Epithelial Tissues

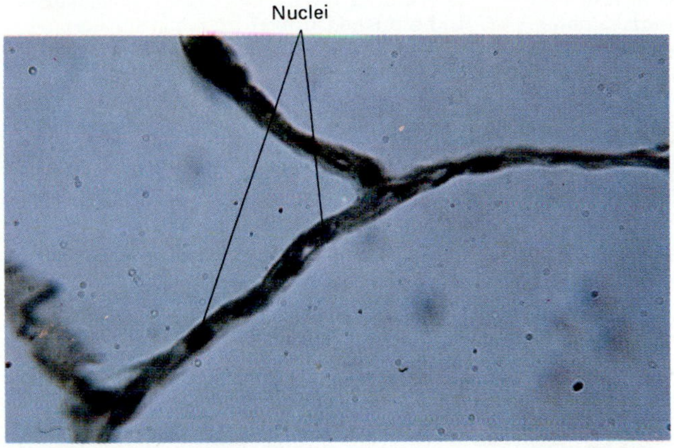

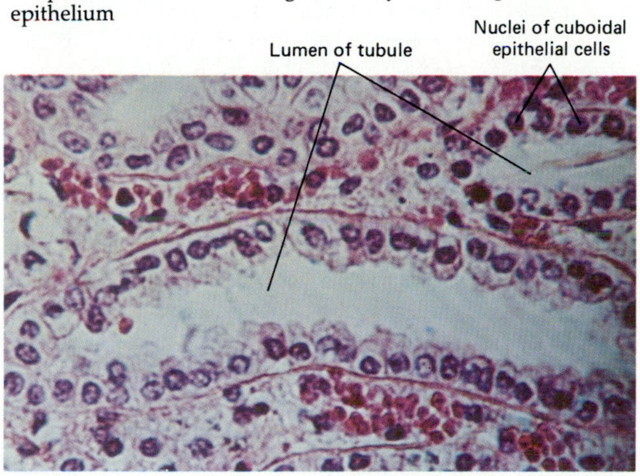

Type of Tissue	*Main Locations*	*Functions*	*Description and Comments*
Simple squamous epithelium	Air sacs of lungs, lining of blood vessels	Passage of material where little or no protection is needed	Cells are flat (often so flat that cytoplasm cannot be discerned) and arranged as single layer.
Simple cuboidal epithelium	Lining of kidney tubules, gland ducts	Secretion and absorption	A single layer of cells. From the side each cell looks like a short cylinder. Sometimes microvilli for absorption are present.
Simple columnar epithelium	Lining of much of the digestive tract; ciliated columnar epithelium lines the upper part of respiratory tract	Secretion, especially mucus secretion; absorption, protection, movement of mucous layer	Single layer of columnar cells, often with nuclei located in base of each cell almost in a row; sometimes with enclosed secretory vesicles (goblet cells), highly developed Golgi apparatus, and cilia

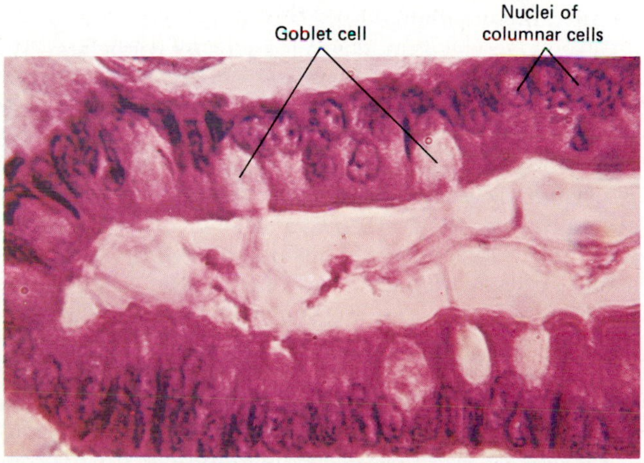

(Continued)

TABLE 6–1 (continued)

Type of Tissue	*Main Locations*	*Functions*	*Description and Comments*
Stratified squamous epithelium	Skin, mouth lining, vaginal lining	Protection only; little or no absorption or transit of materials; outer layer cells are continuously sloughed off and replaced from below.	Several layers of cells, with only the lower ones columnar and metabolically active. Division of lower cells causes older ones to be pushed upward toward surface.
Pseudostratified epithelium	Some respiratory passages, ducts of many glands, sometimes ciliated	Secretion, protection, movement of mucus	Comparable in many ways to columnar epithelium, except that not all cells are the same height. Thus, though all cells contact the same basement membrane, the tissue appears stratified. Nuclei not in line. Ciliated, mucus-secreting, or with microvilli

lia, the linings of blood and lymph vessels are termed **endothelium**, and the linings of the body cavities are termed **mesothelium**.

A layer of cells specialized to receive stimuli is called **sensory** epithelium. The olfactory epithelium in the lining of the nose, for example, contains cells that respond to the presence of certain chemicals in the air breathed in. These are responsible for the sense of smell.

A **gland** consists of one or more epithelial cells specialized to produce and secrete a product such as sweat, milk, mucus, wax, saliva, hormones, or enzymes (Fig. 6–2). The epithelial tissue lining the cavities and passageways of the body typically contain specialized mucus-secreting cells called **goblet cells**. The mucus lubricates these surfaces and facilitates the movement of materials.

CONNECTIVE TISSUES

The main function of **connective tissues** is to join together the other tissues of the body. Connective tissues also support the body and its structures and protect underlying organs. In addition, almost every organ in the body has a supporting

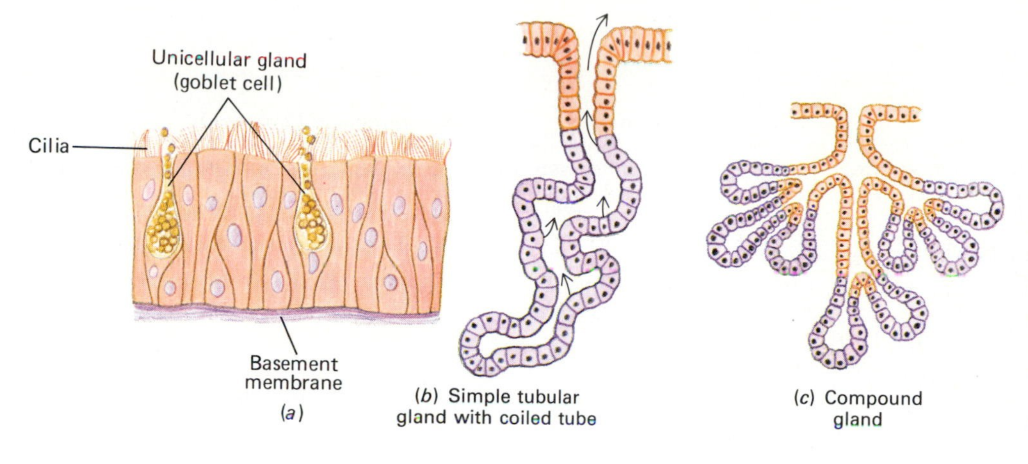

Figure 6–2 A gland consists of one or more epithelial cells. (*a*) Goblet cells are unicellular glands that secrete mucus. (*b*) Sweat glands are simple tubular glands with coiled tubes similar to the one shown here. (*c*) The parotid salivary glands are compound glands like the one shown here.

framework of connective tissue, called **stroma**. The epithelial components of the organ are supported and cushioned by the stroma.

There are many kinds of connective tissues and many systems for classifying them. Some of the main types of connective tissue are (1) loose and dense connective tissues, (2) elastic connective tissue, (3) reticular connective tissue, (4) adipose tissue, (5) cartilage, (6) bone, and (7) blood, lymph, and tissues that produce blood cells. These tissues vary widely in the details of their structure and in the specific functions they perform (Table 6–2).

Connective tissues contain relatively few cells embedded in an extensive **intercellular substance** consisting of threadlike microscopic **fibers** scattered throughout a matrix secreted by the cells. The **matrix** is a thin gel composed of polysaccharides. The cells of different kinds of connective tissues differ in their shape and structure and in the kind of matrix they secrete. The nature and function of each kind of connective tissue are determined in part by the structure and properties of the intercellular substance. Thus, to some extent, connective tissue cells perform their respective functions indirectly by secreting the matrix, which does the actual connecting and supporting.

Fibers and Cells of Connective Tissue

Connective tissue contains three types of fibers: collagen, elastic, and reticular. **Collagen fibers**, the most numerous type, extend in all directions and are composed of bundles of smaller parallel **fibrils** (Fig. 6–3). They are flexible but resist

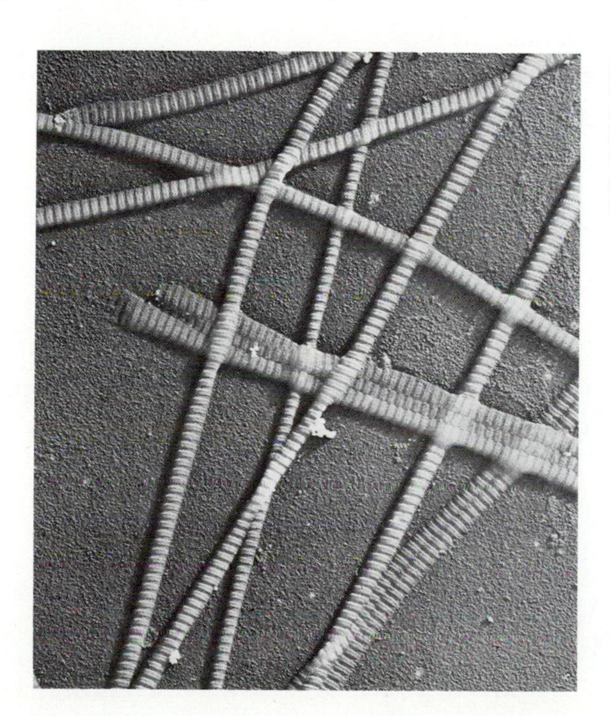

Figure 6–3 Electron micrograph of collagen fibrils teased from a preparation of calf skin (approximately ×33,000). Note the regular periodic striations of the fibrils, which indicate its repeating structural unit. (Courtesy of Dr. Jerome Gross.)

TABLE 6–2
Connective Tissues

Type of Tissue	*Main Locations*	*Functions*	*Description and Comments*
Loose (areolar) connective tissue 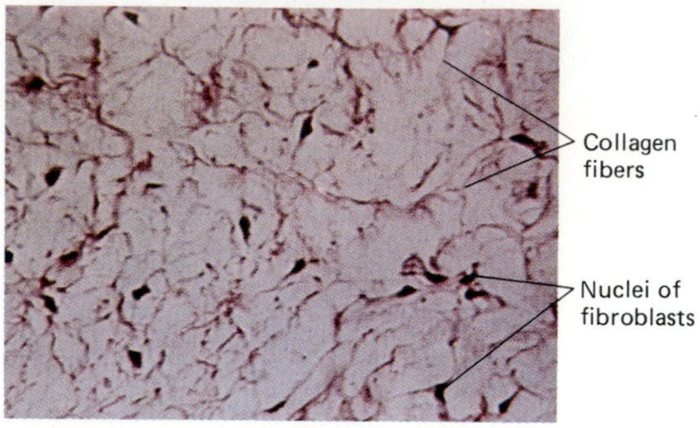	Every place where support must be combined with elasticity; e.g., subcutaneous layer	Support; reservoir for fluid and salts	Fibers produced by fibroblasts embedded in a semifluid matrix and mixed with an assortment of other cells
Dense connective tissue 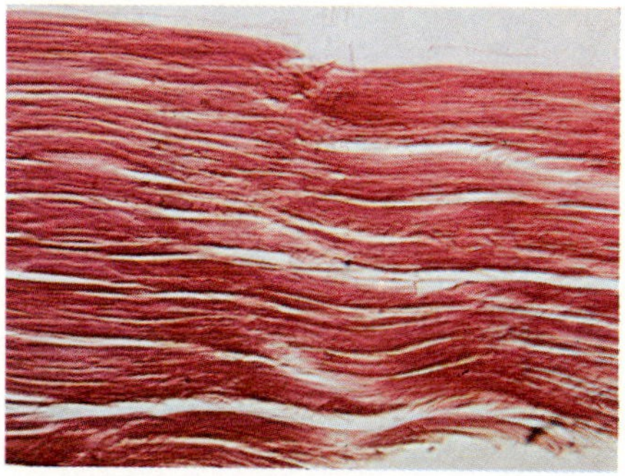	Tendons, strong attachments between organs; dermis of skin	Support; transmission of mechanical forces	Bundles of interwoven collagen fibers interdigitated with rows of fibroblast cells
Elastic connective tissue 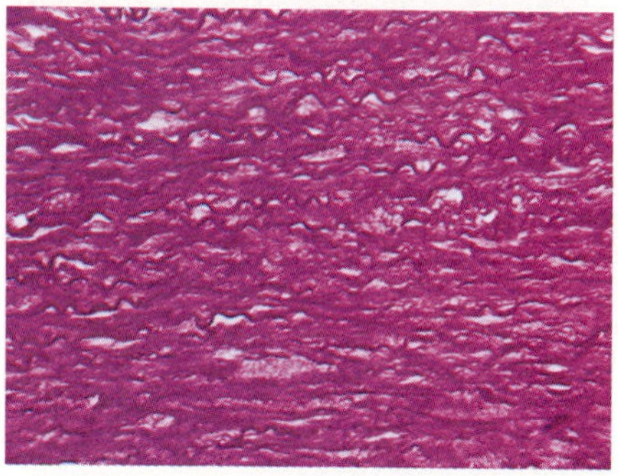	Structures that must both expand and return to their orginal size, such as lung tissue, large arteries; ligaments	Confers elasticity	Branching elastic fibers interspersed with fibroblasts
Reticular connective tissue 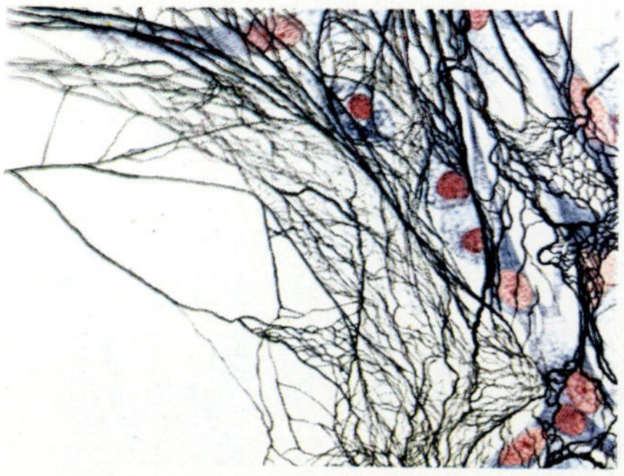	Framework of liver, lymph nodes, spleen	Support	Consists of interlacing reticular fibers

Type of Tissue	*Main Locations*	*Functions*	*Description and Comments*
Adipose tissue	Subcutaneous layer; pads around certain internal organs 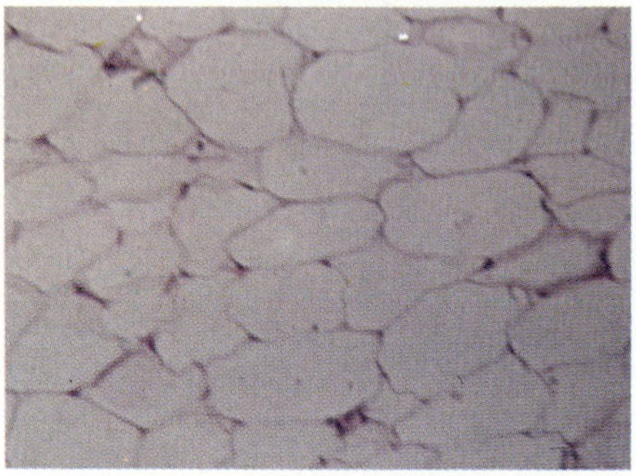	Food storage; insulation; support of such organs as mammary glands, kidneys	Fat cells are star-shaped at first; fat droplets accumulate until typical ring-shaped cells are produced.
Cartilage	Supporting skeleton in sharks, rays, and some other vertebrates; in other vertebrates, forms ends of bones; supporting rings in walls of some respiratory tubes; tip of nose; external ear 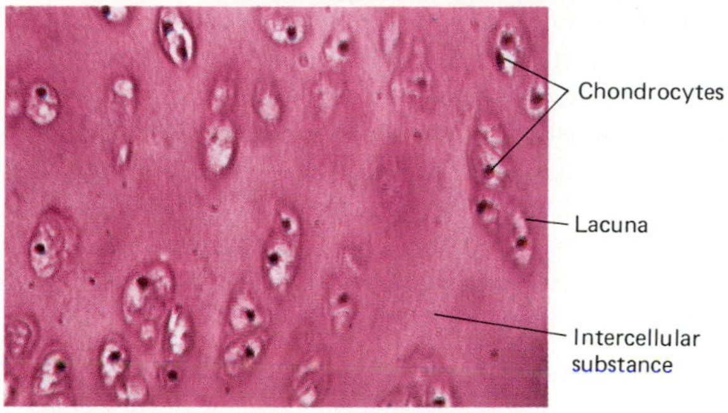	Flexible support and reduction of friction in bearing surfaces	Cells (chondrocytes) separated from one another by the gristly intercellular substance; occupy little spaces in it
Bone	Forms skeletal structure in most vertebrates 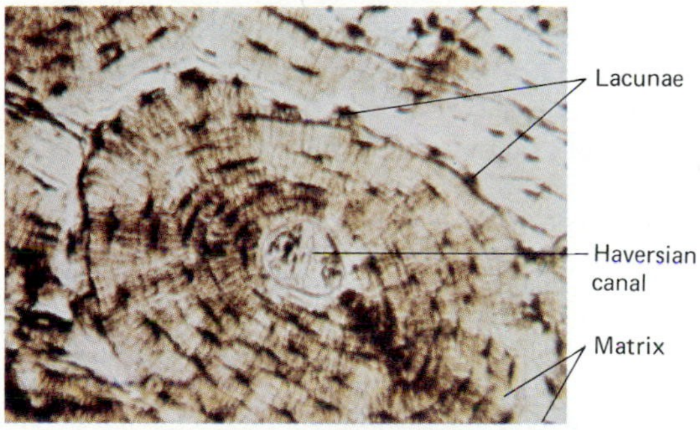	Support, protection of internal organs; calcium reservoir; skeletal muscles attach to bones.	Osteocytes located in lacunae; in compact bone, lacunae arranged in concentric circles about haversian canals
Blood	Within heart and blood vessels of circulatory system 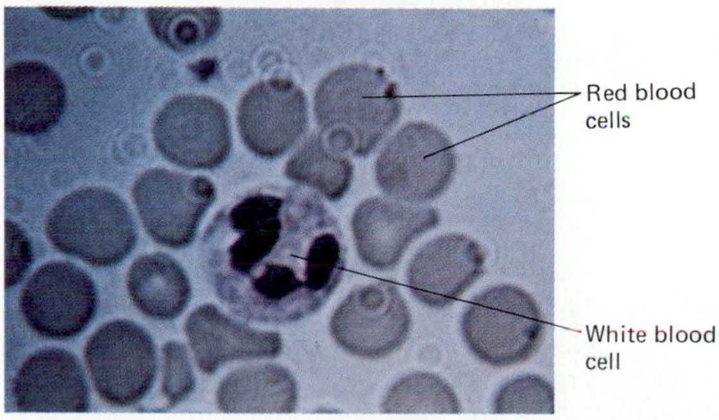	Transports oxygen, nutrients, wastes, and other materials	Consists of cells dispersed in a fluid intercellular substance

tension, stretching only very slightly in response to a pull. With a great-enough force they will break. **Elastic fibers** branch and fuse to form networks; they can be stretched by a force and then will return to their original size and shape when the force is removed. They are not composed of bundles of fibrils but do contain microfibrils evident by electron microscopy. **Reticular fibers** are very small, branched fibers that form delicate networks not visible in ordinary stained slides. The reticular fibers become apparent when a tissue is stained with silver.

Collagen and reticular fibers contain a unique protein, **collagen** (rich in the amino acids glycine, proline, and hydroxyproline). Collagen is a very tough material, and these fibers impart great strength to structures in which they occur. (Meat is tough because of its collagen content.) The tensile strength of collagen fibers has been compared to that of steel. When treated with hot water, collagen is converted into the soluble protein gelatin. Because there is so much connective tissue in the body, about one third of all mammalian protein is collagen.

Connective tissue also contains several types of cells. **Fibroblasts** are connective tissue cells that produce the protein and carbohydrate complexes of the matrix as well as the fibers. The fibroblasts release specific protein components that arrange themselves to form the characteristic fibers. Fibroblasts are especially active in developing tissue and in healing wounds. As tissues mature, the numbers of fibroblasts decrease, and they become less active. Inactive fibroblasts are referred to as **fibrocytes**.

Pericytes are undifferentiated (unspecialized) cells located along the outer walls of small blood vessels (capillaries) that run through connective tissues. It is thought that pericytes give rise to other types of cells when necessary. For example, when an injury occurs, pericytes are thought to multiply and give rise to fibroblasts that can produce the components needed to heal the wound.

Macrophages, the scavenger cells of the body, are also common in connective tissues. They wander through the tissues cleaning up cellular debris and phagocytizing foreign matter including bacteria (Fig. 6–4). Among the other types of cells seen in connective tissues are mast cells, which release histamine during allergic reactions; adipose (fat) cells; and plasma cells, which produce antibodies.

Loose and Dense Connective Tissues

Loose connective tissue (also called **areolar** tissue) is the most widely distributed connective tissue in the body. It is found as a thin filling between body parts and serves as a reservoir for fluid and salts. Nerves, blood vessels, and muscles are wrapped in this tissue. Together with adipose tissue, loose connective tissue forms the subcutaneous (that is, below the skin) layer, the layer that attaches skin to the muscles and other structures beneath. Loose connective tissue consists of fibers strewn in all directions through a semifluid matrix. Its flexibility permits the parts it connects to move.

Dense connective tissue is very strong, though somewhat less flexible than loose connective tissue. Collagen fibers predominate. In **irregular** dense connective tissue, the collagen fibers are arranged in bundles distributed in all directions

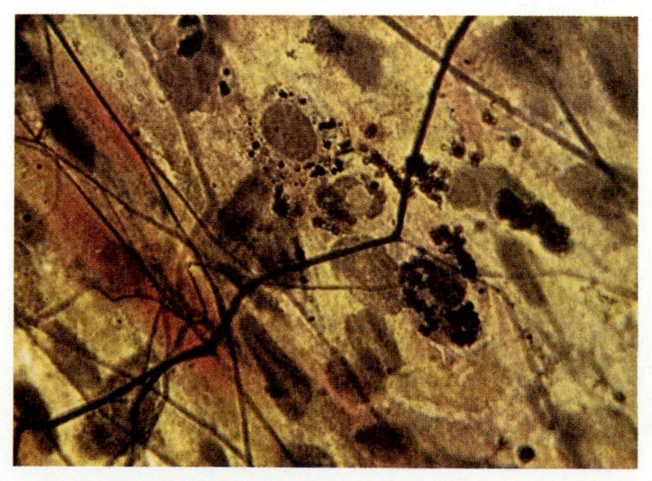

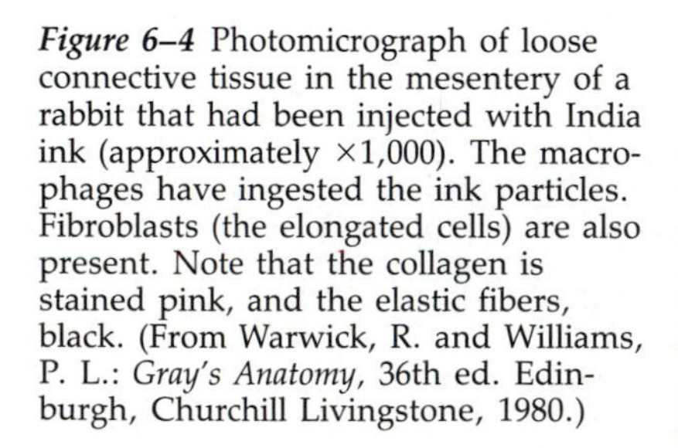
Figure 6–4 Photomicrograph of loose connective tissue in the mesentery of a rabbit that had been injected with India ink (approximately ×1,000). The macrophages have ingested the ink particles. Fibroblasts (the elongated cells) are also present. Note that the collagen is stained pink, and the elastic fibers, black. (From Warwick, R. and Williams, P. L.: *Gray's Anatomy*, 36th ed. Edinburgh, Churchill Livingstone, 1980.)

through the tissue. This type of tissue is found in the lower layer (dermis) of the skin. In **regular** dense connective tissue, the collagen bundles are arranged in a definite pattern, making the tissue greatly resistant to stress. Tendons, the cable-like cords that connect muscles to bones, consist of this tissue.

Elastic and Reticular Connective Tissues

Elastic connective tissue consists mainly of bundles of parallel elastic fibers. It is found in ligaments, the bands of tissue that connect bones to one another. Structures that must expand and then return to their original size, like the walls of the large arteries and lung tissue, contain elastic connective tissue. **Reticular connective tissue** is composed mainly of interlacing reticular fibers. It forms a supporting stroma in many organs, including the liver, spleen, and lymph nodes.

Adipose Tissue

Adipose tissue is rich in fat cells, which store fat and release it when fuel is needed for cellular respiration. It is found in the subcutaneous layer and in tissue that cushions internal organs. An immature fat cell is somewhat star-shaped. As fat droplets accumulate within the cytoplasm, the cell assumes a more rounded appearance (Fig. 6–5(*a*)). Fat droplets eventually merge with one another until finally a single large drop of fat is present. This large drop occupies most of the volume of the mature fat-storing cell. The cytoplasm and its organelles are pushed to the cell edges, where a bulge is typically created by the nucleus. A cross section of such a fat cell looks like a ring with a single stone and is sometimes called a **signet ring cell**. (Cytoplasm forms the ring, and the nucleus, the stone.)

When you study a section of adipose tissue through a microscope, it may remind you of chicken wire (Fig. 6–5(*b*)). The "wire" is represented by the rings of cytoplasm, and the large spaces indicate where fat drops existed before they were dissolved by chemicals used to prepare the tissue. The empty spaces may cause the cells to collapse, resulting in a wrinkled appearance.

Cartilage and Bone

The supporting skeleton of vertebrates is composed of cartilage or bone. **Cartilage** is the supporting skeleton in the embryonic stages in all vertebrates, but it is largely replaced in the adult by bone in all but the sharks and rays. The supporting structure of the external ear, the supporting rings in the walls of the respiratory passageways, and the tip of the nose in humans are examples of structures composed of cartilage. Cartilage is firm, yet elastic. Cartilage cells called **chondrocytes** secrete this hard, rubbery matrix around themselves and also secrete collagen fibers, which become embedded in the matrix and strengthen it. Chondrocytes eventually come to lie singly or in groups of two or four in small cavities called **lacunae** in the matrix (Fig. 6–6). The cartilage cells in the matrix remain alive. Cartilage tissue lacks nerves, lymph vessels, and blood vessels. Chondrocytes are nourished by diffusion of nutrients and oxygen through the matrix.

Bone is the major vertebrate skeletal tissue. It is similar to cartilage in that it consists mostly of matrix material containing lacunae and inhabited by the cells that secrete and maintain the matrix (Fig. 6–7). Unlike cartilage, however, bone is a highly vascular tissue with a substantial blood supply. Diffusion alone would never suffice for the nourishment of the bone cells, called **osteocytes**, because the matrix consists not only of collagen, mucopolysaccharides, and other organic materials but also of the mineral apatite, a complex calcium phosphate. Diffusion through such a substance would be impracticably slow. Thus the osteocytes of bone communicate with one another and with capillaries by tiny channels, **canaliculi**, which contain fine extensions of the cells themselves. Because it is important that no bone cell be located very far from the nearest blood vessel, in much bone the osteocytes are arranged around central capillaries in concentric layers called **lamellae**, which form spindle-shaped units known as **osteons**. The capillaries, as well as nerves, run through central microscopic channels known as **haversian canals**.

Bone also contains large multinucleated cells called **osteoclasts**, which can dissolve and remove the bony substance, as can the osteocytes themselves. The shape and internal architecture of the bone can gradually change in response to

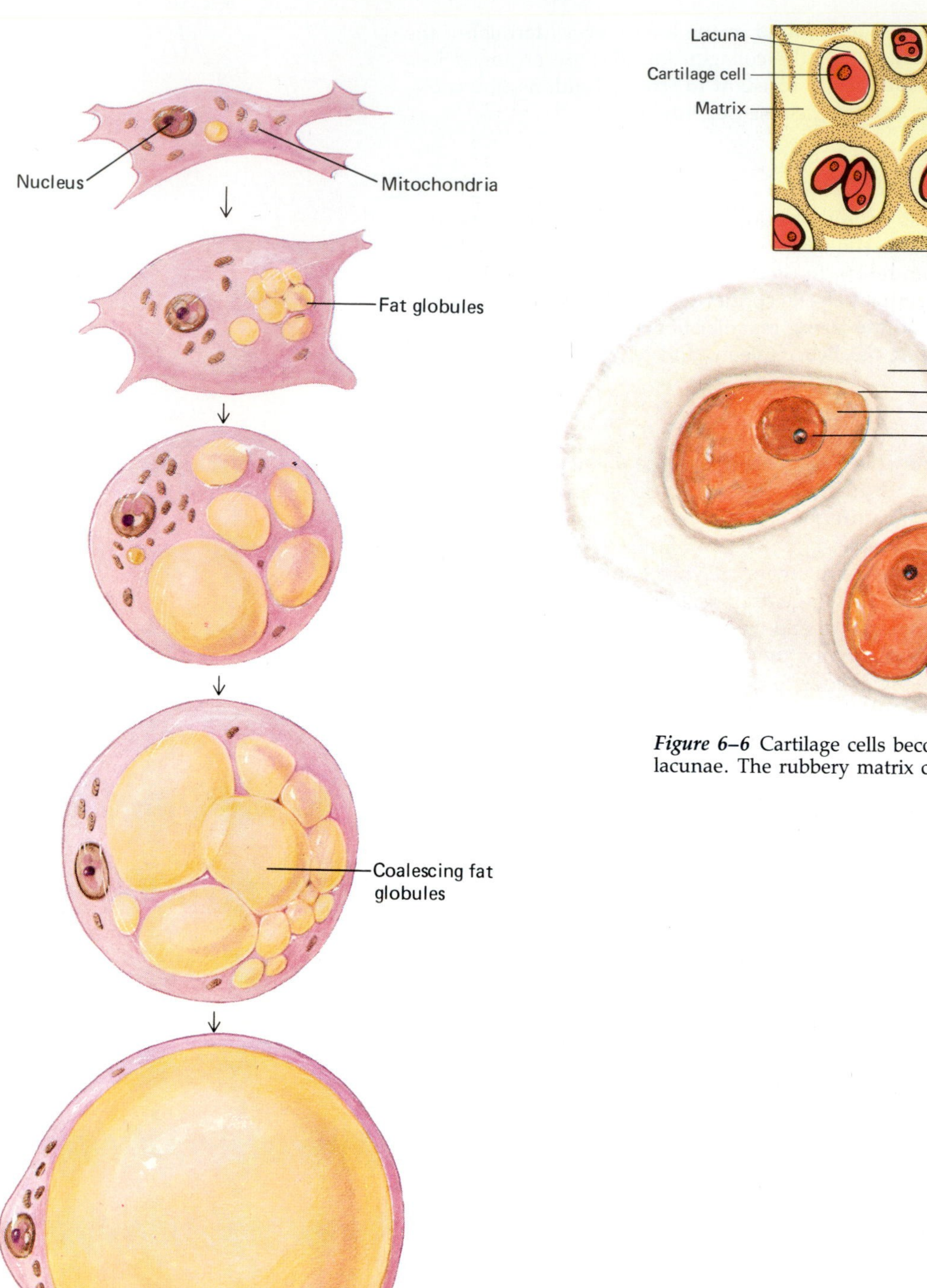

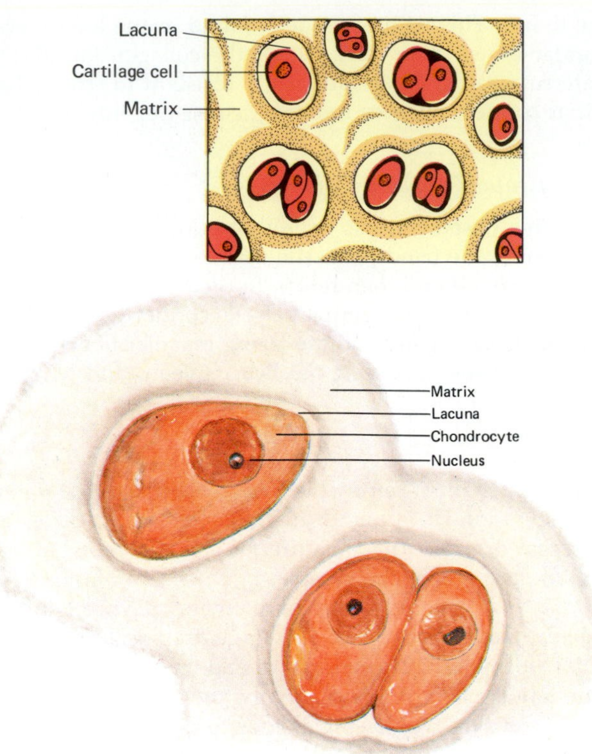

Figure 6–6 Cartilage cells become trapped in small spaces called lacunae. The rubbery matrix contains collagen fibers.

Figure 6–5 Storage of fat in a fat cell. (*a*) As more and more fat droplets accumulate in the cytoplasm, they coalesce to form a very large globule of fat. Such a fat globule may occupy most of the cell, pushing the cytoplasm and the organelles to the periphery. (*b*) Photomicrograph of adipose tissue. The fat droplets were dissolved by chemicals used to prepare the tissue, leaving large spaces. Because of these spaces, the cells tend to collapse and no longer appear round.

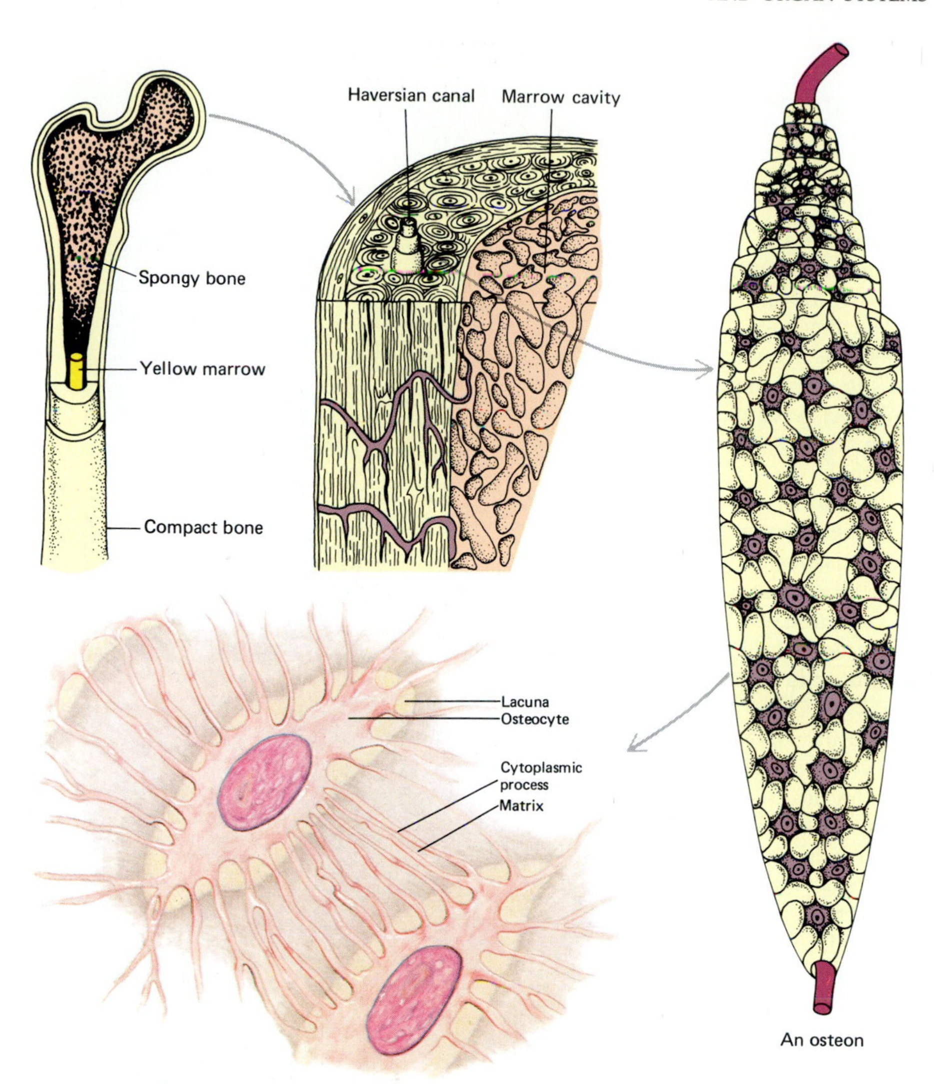

Figure 6–7 Compact bone is made up of units called osteons. Blood vessels and nerves run through the Haversian canal within each osteon. In bone the matrix is rigid and hard. Bone cells become trapped within lacunae, but communicate with one another by way of cytoplasmic processes that extend through tiny canals.

normal growth processes and to physical stress. The calcium salts of bone render the matrix very hard, and the collagen prevents the bony matrix from being overly brittle. Bones are amazingly light and strong. Most bones have a large **marrow cavity** in the center; this may contain yellow marrow, which is mostly fat, or red marrow, the connective tissue in which red and some white blood cells are made.

Blood

Blood, the fluid within our arteries and veins, is composed of red and white cells and platelets, suspended within **plasma**, the liquid, noncellular part of the blood. Plasma transports many kinds of substances from one part of the body to another. Some of these substances are simply dissolved in the plasma, whereas others are bound to proteins such as albumins. Most biologists classify the blood with connective tissues. Other biologists, however, consider blood to be a separate class of tissues, because the connective tissue cells secrete their surrounding matrix, whereas blood cells do not secrete the plasma.

The **red blood cells** (erythrocytes) of humans and other vertebrates contain the red respiratory pigment **hemoglobin**, which can combine easily and reversibly with oxygen. Oxygen, combined as oxyhemoglobin, is transported to the cells of the body by the red blood cells. The red cells of most mammals are flattened biconcave disks that lack a nucleus (Table 6–2); those of other vertebrates are oval and have a nucleus. In many invertebrate animals the oxygen-carrying pigments are not located within a cell but are dissolved in the plasma, coloring it red or blue.

Human blood contains five different kinds of **white blood cells**, each with a distinct size, shape, structure, and functions. None of the white blood cells contain hemoglobin, but some can move around by amoeboid motion and slip through the walls of the blood vessels, entering the tissues of the body to engulf bacteria and

FOCUS ON

Neoplasms—Unwelcome Tissues

A **neoplasm** (new growth) is an abnormal mass of cells. Neoplasms, or **tumors**, can develop in many species of animals and plants. A benign ("kind") tumor tends to grow slowly, and its cells stay together. Because benign tumors form discrete masses, often surrounded by connective tissue capsules, they can usually be removed surgically. Unless a benign neoplasm develops in a place where it interferes with the function of a vital organ, it is not lethal.

A malignant ("wicked") neoplasm, or **cancer**, usually grows much more rapidly than a benign tumor. Neoplasms that develop from connective tissues or muscle are referred to as sarcomas, and those that originate in epithelial tissue are called carcinomas. Unlike the cells of benign tumors, cancer cells do not retain the typical structural features of the cells from which they originate.

Cancer is thought to be triggered when the DNA of a cell is mutated (altered) by radiation, certain chemicals or irritants, or viruses. (See discussion of oncogenes in Chapter 16.) When the mutated cell multiplies, all the cells derived from it bear the identical mutation. Should the mutation interfere with the cells' control mechanisms, the cells begin to behave abnormally. Two basic defects in behavior that characterize most cancer cells are rapid multiplication and abnormal relations with neighboring cells. Though normal cells respect one another's boundaries and form tissues in an orderly, organized manner, cancer cells grow helter-skelter upon one another and infiltrate normal tissues (see figure). Apparently they are no longer able to receive or respond to signals from surrounding cells.

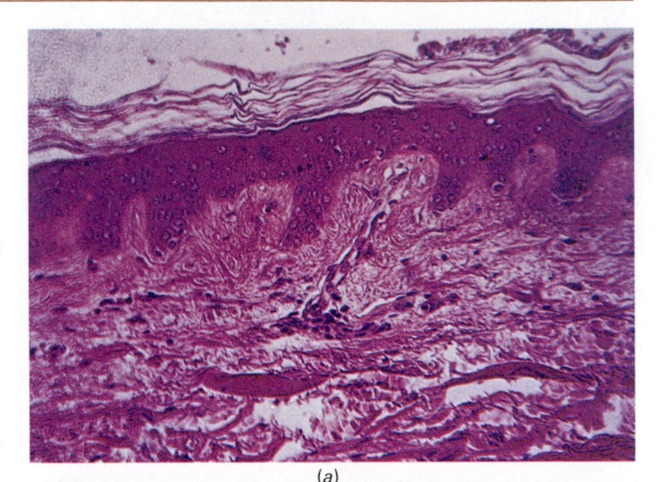

(*a*)

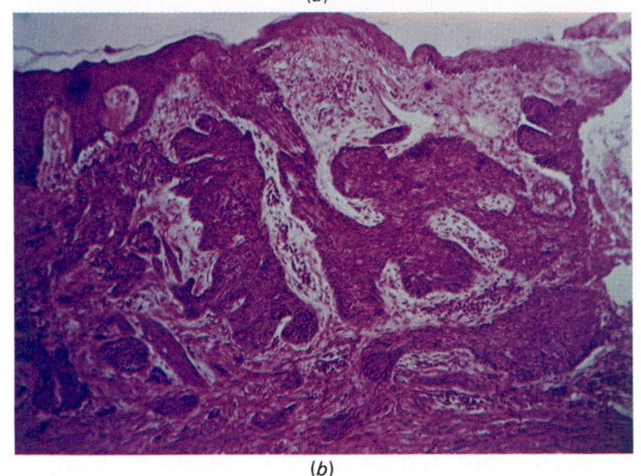

(*b*)

Normal skin tissue (*a*) compared with cancerous tissue (*b*).

Studies indicate that many neoplasms grow to only a few millimeters in diameter and then enter a dormant stage, which may last for months or even years. At some point, cells of the neoplasm release a chemical substance that stimulates nearby blood vessels to develop new capillaries that grow out toward the neoplasm and infiltrate it. Once a blood supply is ensured, the neoplasm grows rapidly and soon becomes life-threatening.

Death from cancer is almost always caused by **metastasis,** which is a migration of cancer cells through blood or lymph channels to distant parts of the body. Once there, they multiply, forming new malignant neoplasms, which may interfere with normal function in the tissues being invaded. Cancer often spreads so rapidly and extensively that surgeons are unable to locate all the malignant masses.

Why some persons are more susceptible to cancer than others remains a mystery. Some researchers think that cancer cells form daily in everyone but that in most persons the immune system (the system that provides protection from disease organisms and other foreign invaders) is capable of destroying them. According to this theory cancer is a failure of the immune system. Another suggestion is that different persons have different levels of tolerance to environmental irritants. As many as 90% of cancer cases are thought to be triggered by environmental factors.

A cancer patient's survival depends upon early diagnosis and treatment with a combination of surgery, radiation therapy, and drugs that suppress mitosis (chemotherapy). Since cancer is actually an entire family of closely related diseases (there are more than 100 distinct varieties), it may be that there is no single cure. Most investigators agree, however, that a greater understanding of basic control mechanisms and communication systems of cells is necessary before effective cures can be developed.

other foreign particles. The white cells constitute an important line of defense against disease bacteria.

Platelets are not whole cells but are small fragments broken off from large cells located in the bone marrow. They play a key role in the clotting of blood (Chapter 33).

MUSCLE TISSUE

The movements of most animals result from the contraction of the elongated, cylindrical, or spindle-shaped cells of **muscle tissue.** Each muscle cell, usually referred to as a **fiber**, contains many small, longitudinal, parallel contractile fibers called **myofibrils**. The proteins **myosin** and **actin** are the chief components of myofibrils. Muscle cells perform mechanical work only by contracting, getting shorter or thicker; they cannot exert a push.

Three types of muscle tissue are found in vertebrates (Fig. 6–8): **Cardiac muscle** is present in the walls of the heart. **Smooth muscle** occurs in the walls of the digestive tract, uterus, blood vessels, and certain other internal organs. **Skeletal muscle** makes up the large muscle masses attached to the bones of the body. Skeletal muscle fibers are among the exceptions to the rule that cells have only one nucleus; each skeletal muscle fiber has many nuclei. The nuclei of skeletal muscle fibers are also unusual in their position: They lie peripherally, just under the cell membrane. This is thought to be an adaptation to increase the efficiency of contraction. The entire central part of the skeletal muscle fiber is occupied by the contractile units, the myofibrils. Skeletal muscle cells may be as long as 2 or 3 cm.

By light microscopy, both skeletal and cardiac fibers are seen to have alternate light and dark transverse stripes, or **striations**. These microscopic stripes are involved in the contraction process, for they change their relative sizes during contraction (see Fig. 6–12): The dark stripes remain essentially constant, but the light stripes decrease in width. Striated muscle fibers can contract rapidly but cannot remain contracted. A striated muscle fiber must relax and rest momentarily before it

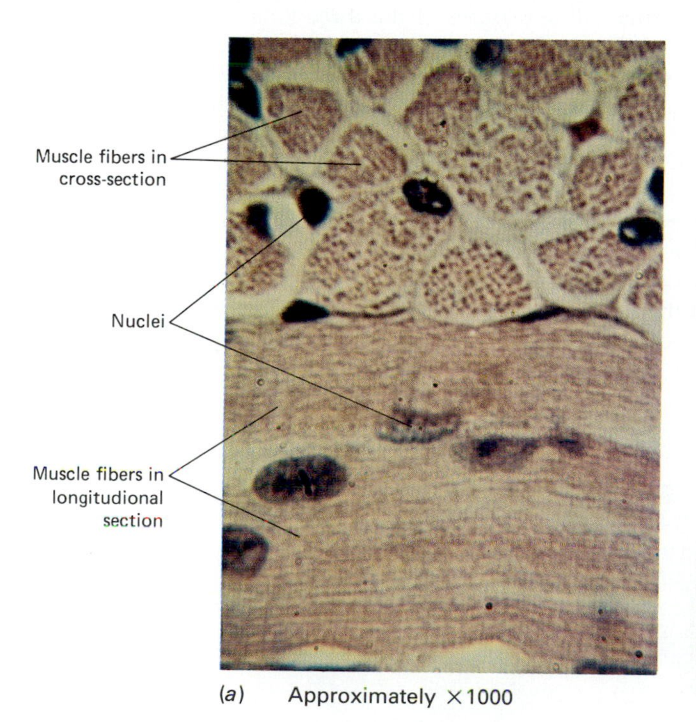

(*a*) Approximately ×1000

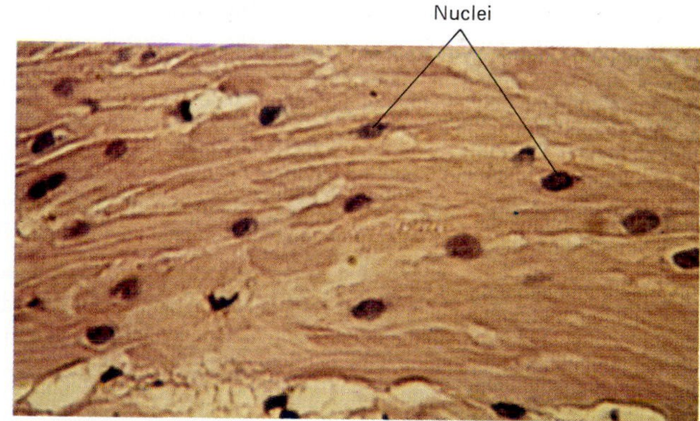

(*b*) Approximately ×450

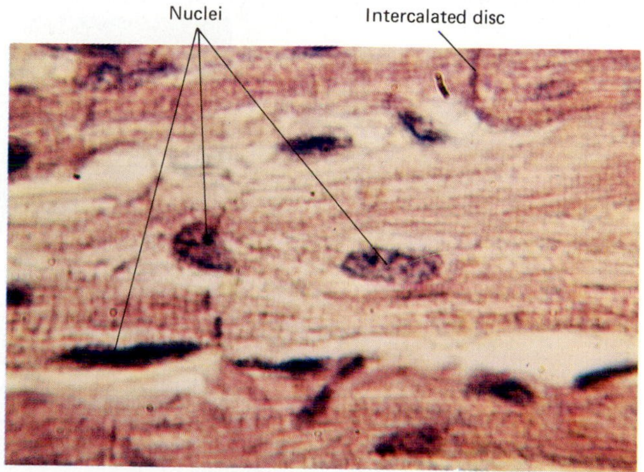

(*c*) Approximately ×900

Figure 6–8 Muscle tissue. (*a*) Skeletal muscle is striated and is under voluntary control. (*b*) Smooth muscle tissue lacks striations; its contraction is involuntary. (*c*) Cardiac muscle tissue is striated and has branched fibers; its contraction is also involuntary. The special junctions between cardiac muscle cells are called intercalated disks.

TABLE 6–3
Types of Muscle Tissues

	Skeletal	*Smooth*	*Cardiac*
Location	Attached to skeleton	Walls of stomach, intestines, etc.	Walls of heart
Type of control	Voluntary	Involuntary	Involuntary
Shape of fibers	Elongated, cylindrical, blunt ends	Elongated, spindle-shaped, pointed ends	Elongated, cylindrical fibers that branch and fuse
Striations	Present	Absent	Present
Number of nuclei per fiber	Many	One	One
Position of nuclei	Peripheral	Central	Central
Speed of contraction	Most rapid	Slowest	Intermediate
Ability to remain contracted	Least	Greatest	Intermediate

Nuclei; Cross striations
(*a*) Skeletal muscle fibers

Nuclei
(*b*) Smooth muscle fibers

Nuclei; Intercalated disks
(*c*) Cardiac muscle fibers

can contract again. Skeletal muscle fibers are generally under voluntary control, whereas cardiac and smooth muscle fibers are not usually regulated at will. Table 6–3 summarizes the distinguishing features of the three kinds of muscle. In the bodies of invertebrates, the distribution of muscle types may be quite different from that in vertebrates, and one or more types may be missing completely. Most of the muscles of arthropods are striated, whereas the muscles of mollusks are smooth.

NERVOUS TISSUE

Nervous tissue is composed of **neurons**, cells specialized for conducting electrochemical nerve impulses, and **glial cells**, cells that support and nourish the neurons (Fig. 6–9). Certain neurons receive signals from the external or internal environment and transmit them to the spinal cord and brain; other nerve cells process and store the information. This is the cellular basis for the complex functions of consciousness, memory, and thought.

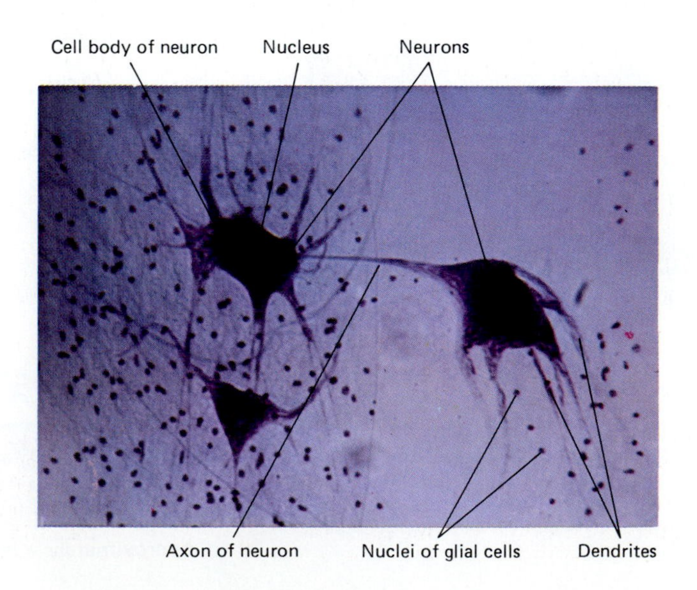

Figure 6–9 Nervous tissue consists of neurons and glial cells.

Neurons come in many shapes and sizes, but typically each has an enlarged **cell body**, which contains the nucleus, and from which two kinds of thin hairlike extensions project. **Dendrites** are fibers specialized for receiving impulses either from environmental stimuli or from another cell. The single **axon** is specialized to conduct impulses away from the cell body. Axons usually are long and smooth but may give off an occasional branch; they typically end in a group of fine branches. Axons range in length from a millimeter or two to more than a meter. Those extending from the spinal cord down the arm or leg in the human may be a meter or more in length. Neurons communicate at junctions called **synapses**; thus they are functionally connected and can pass impulses for long distances through the body. A **nerve** consists of a great many fibers bound together by connective tissue.

Animal Organs and Organ Systems

Complex animals have a great variety of organs. Although an animal organ may be composed mainly of one type of tissue, other types are needed to provide support, protection, and a blood supply and to allow transmission of nerve impulses. For example, the heart consists mainly of cardiac muscle tissue, but it is lined and covered by epithelium, contains blood vessels composed of connective tissue, and is regulated by nervous tissue.

The organ systems of complex animals include the integumentary, skeletal, muscle, nervous, circulatory, digestive, respiratory, urinary (excretory), endocrine, and reproductive systems (Fig. 6–10). See Table 6–4 for a summary of their principal organs and functions. In the digestive system, for example, organs include the mouth, esophagus, stomach, small and large intestines, liver, pancreas, and salivary glands. This system functions to process food, reducing it to its simple components. The digestive system transfers the products of digestion into the blood for transport to all of the cells of the body.

TABLE 6–4
Organ Systems of a Mammal and Their Functions

System	*Components*	*Functions*	*Homeostatic Abilities*
Integumentary	Skin, hair, nails, sweat glands	Covers and protects body	Sweat glands help control body temperature; as a barrier the skin helps maintain steady state.
Skeletal	Bones, cartilage, ligaments	Supports body; protects; provides for movement and locomotion; calcium depot	Helps maintain constant calcium level in blood
Muscular	Masses of skeletal muscle; cardiac muscle; smooth muscle	Moves parts of skeleton, locomotion; movement of internal materials	Ensures such vital functions as nutrition through body movements. Smooth muscle maintains blood pressure; cardiac muscle circulates the blood.
Digestive	Mouth, esophagus, stomach, intestines, liver, pancreas	Ingests and digests foods; absorbs them into the blood	Maintains adequate supplies of fuel molecules and building materials
Circulatory	Heart, blood vessels, blood; lymph and lymph structures	Transports materials from one part of body to another; defends body against disease	Transports oxygen, nutrients, hormones; removes wastes; maintains water and ionic balance of tissues
Respiratory	Lungs, trachea, and other air passageways	Exchange of gases between blood and external environment	Maintains adequate blood oxygen content and helps regulate blood pH; eliminates carbon dioxide
Urinary	Kidney, bladder, and associated ducts	Eliminates metabolic wastes; removes substances present in excess from the blood	Regulates blood chemistry in conjunction with endocrine system
Nervous	Nerves and sense organs, brain and spinal cord	Receives stimuli from external and internal environment, conducts impulses, integrates activities of other systems	Principal regulatory system
Endocrine	Pituitary, adrenal, thyroid, and other ductless glands	Regulates body chemistry and many body functions	In conjunction with nervous system, regulates metabolic activities and blood levels of various substances
Reproductive	Testes, ovaries, and associated structures	Provides for continuation of the species	Passes on genetic endowment of individual; maintains secondary sexual characteristics

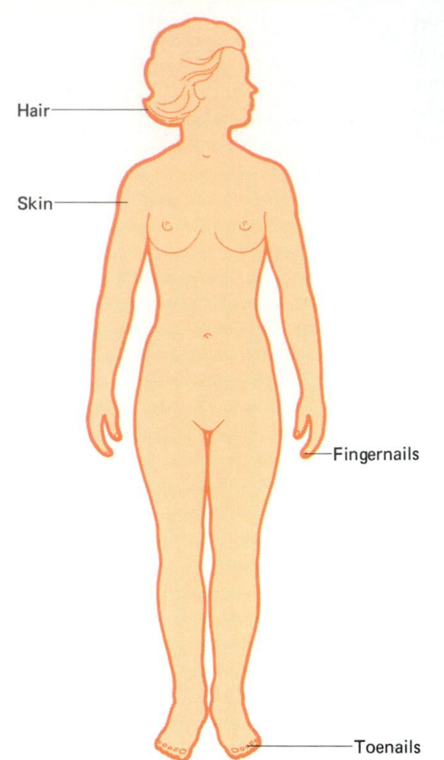

(1) The integumentary system consists of the skin and the structures such as nails and hair that are derived from it. This system protects the body, helps to regulate body temperature, and receives stimuli such as pressure, pain, and temperature.

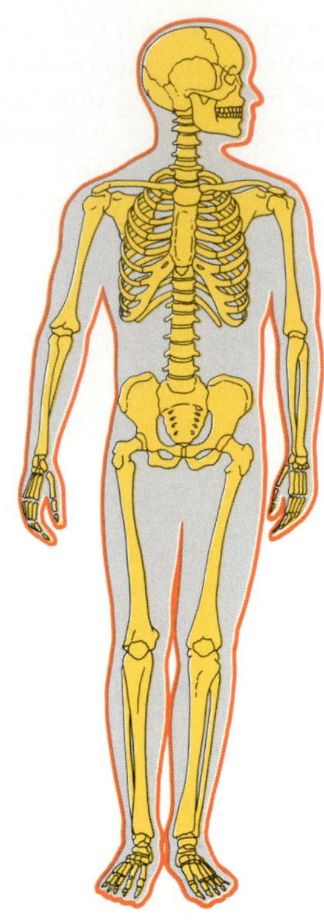

(2) The skeletal system consists of bones and cartilage. This system helps to support and protect the body.

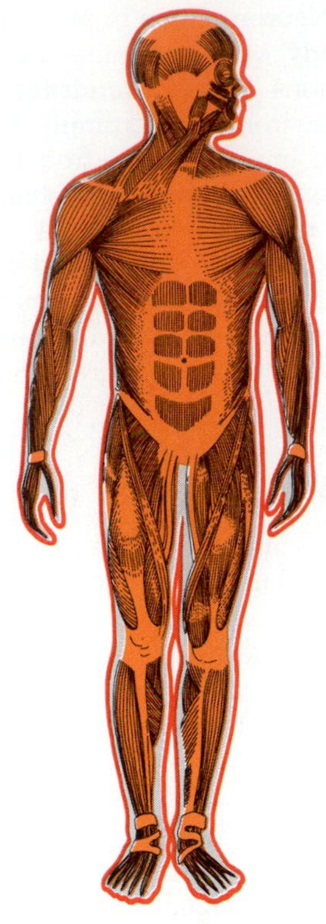

(3) The muscular system consists of the large skeletal muscles that enable us to move, as well as the cardiac muscle of the heart and the smooth muscle of the internal organs.

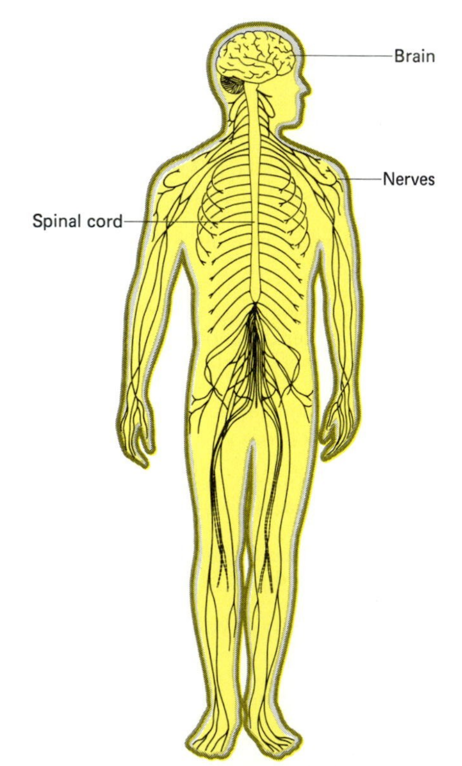

(4) The nervous system consists of the brain, spinal cord, sense organs, and nerves. This is the principal regulatory system.

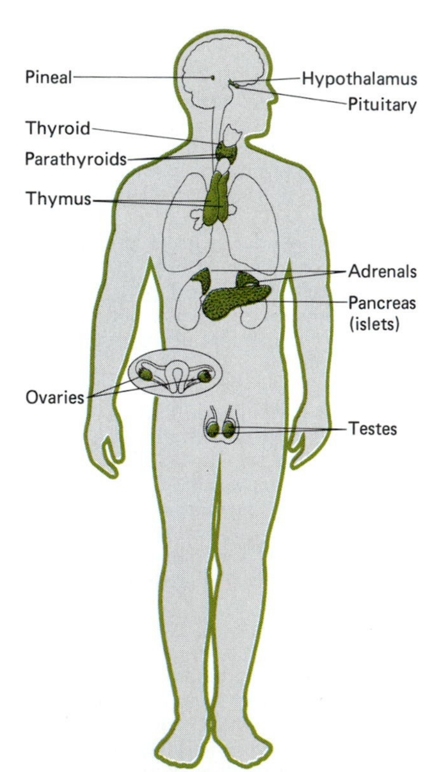

(5) The endocrine system consists of the ductless glands that release hormones. It works with the nervous system in regulating metabolic activities.

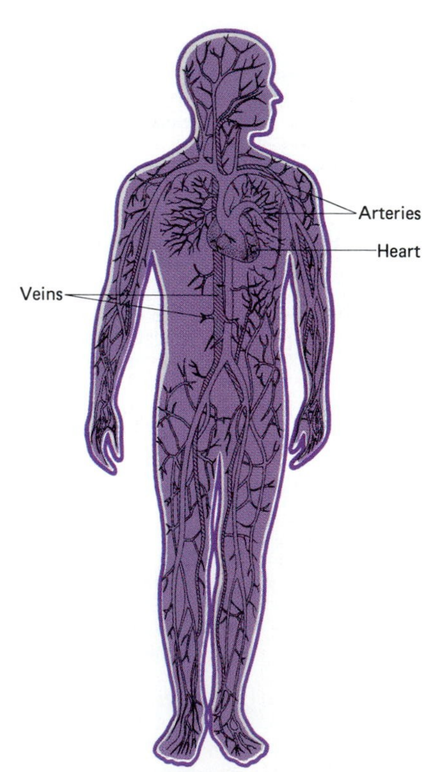

(6a) The circulatory system includes the heart and blood vessels. This system serves as the transportation system of the body.

Figure 6–10 The principal organ systems of the human body.

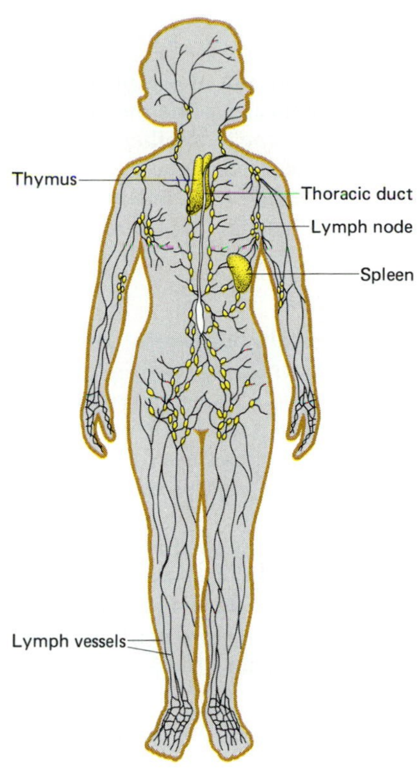

(*6b*) The lymphatic system is a subsystem of the circulatory system; it returns excess tissue fluid to the blood and defends the body against disease.

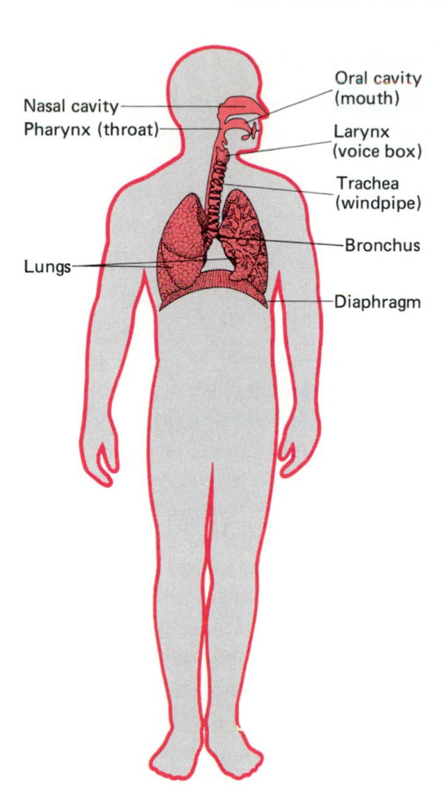

(*7*) The respiratory system. Consisting of the lungs and air passageways, this system supplies oxygen to the blood and excretes carbon dioxide.

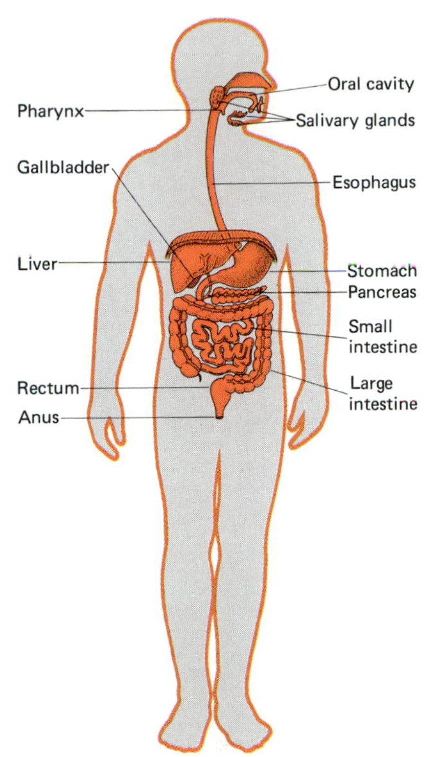

(*8*) The digestive system consists of the digestive tract and glands that secrete digestive juices into the digestive tract. This system mechanically and enzymatically breaks down food and eliminates wastes.

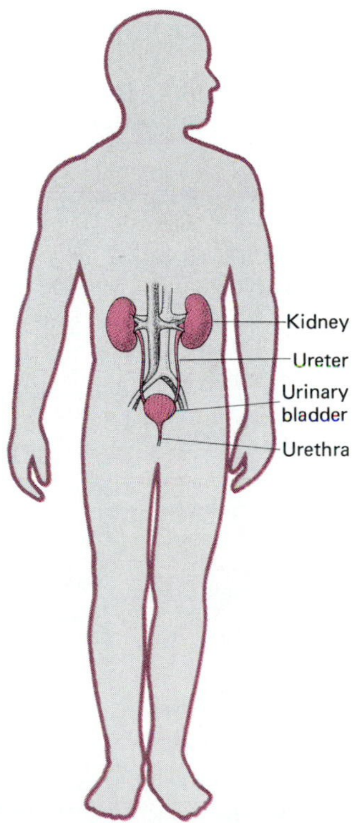

(*9*) The urinary system is the main excretory system of the body, and helps to regulate blood chemistry. The kidneys remove wastes and excess materials from the blood and produce urine.

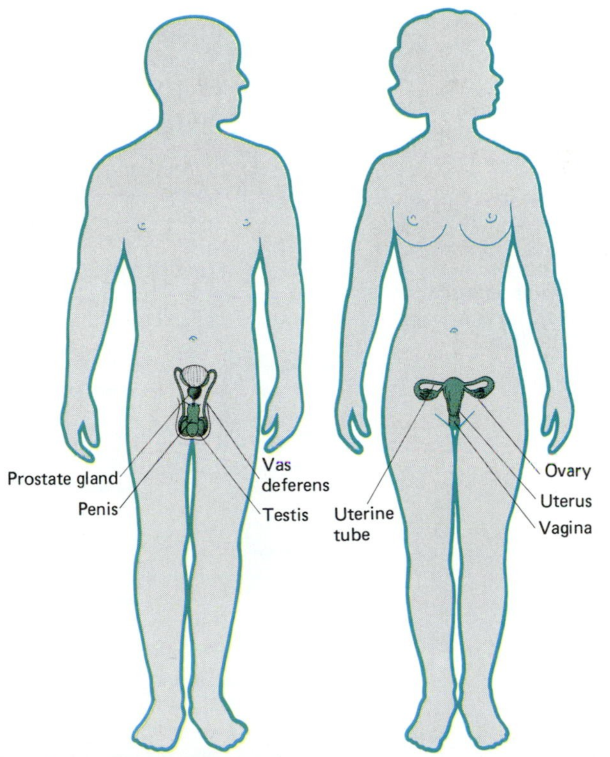

(*10*) Male and female reproductive systems. Each reproductive system consists of gonads and associated structures. The reproductive system maintains the sexual characteristics, and perpetuates the species.

Plant Tissues

It is difficult to categorize plant tissues, and biologists are not in complete agreement regarding their classification. Some cells may appear as intermediates between cell types, and a given cell may change from one type to another during its life. Plant tissues can be assigned to one of two major categories: meristematic tissues, composed of immature cells that undergo cell division, and permanent tissues, composed of mature, differentiated cells.

MERISTEMATIC TISSUES

The **meristematic tissues** of plants are composed of small, thin-walled cells with large nuclei and few or no vacuoles (Fig. 6–11). Their chief function is to grow, divide, and differentiate into all the other types of tissues. As the embryonic plant begins development, it is composed entirely of such tissues, collectively called **meristem**. As development proceeds, most of the meristem becomes differentiated into other tissues, but even in an adult tree there are regions of meristem that provide for continued growth. Meristematic tissues are found in the rapidly growing parts of the plant—the tips of the roots and stem and in the **cambium,** a layer of tissue that produces lateral growth, especially in the stem. The meristem in the tips of the roots and stems, called **apical meristem**, is responsible for the increase in the length of these structures. The meristem in the cambium, the **lateral meristem**, makes possible the increase in diameter of growing stems and roots.

PERMANENT TISSUES

Permanent, or mature, **tissues** are composed of mature, specialized cells. They include surface tissues, vascular tissues, and fundamental tissues (Table 6–5).

Surface Tissues

Surface tissues make up the protective outer covering of the body of the plant. The surface tissue of roots and stems in young plants and herbaceous plants is the **epidermis**. This is also the surface tissue of leaves. Surface tissues protect the underlying cells from drying, from mechanical abrasion, and from invasion by parasitic fungi, bacteria, and protists.

The **epidermis** of leaves and the cork layers of stems and roots are examples of surface tissues specialized to protect underlying tissues. The epidermis is often

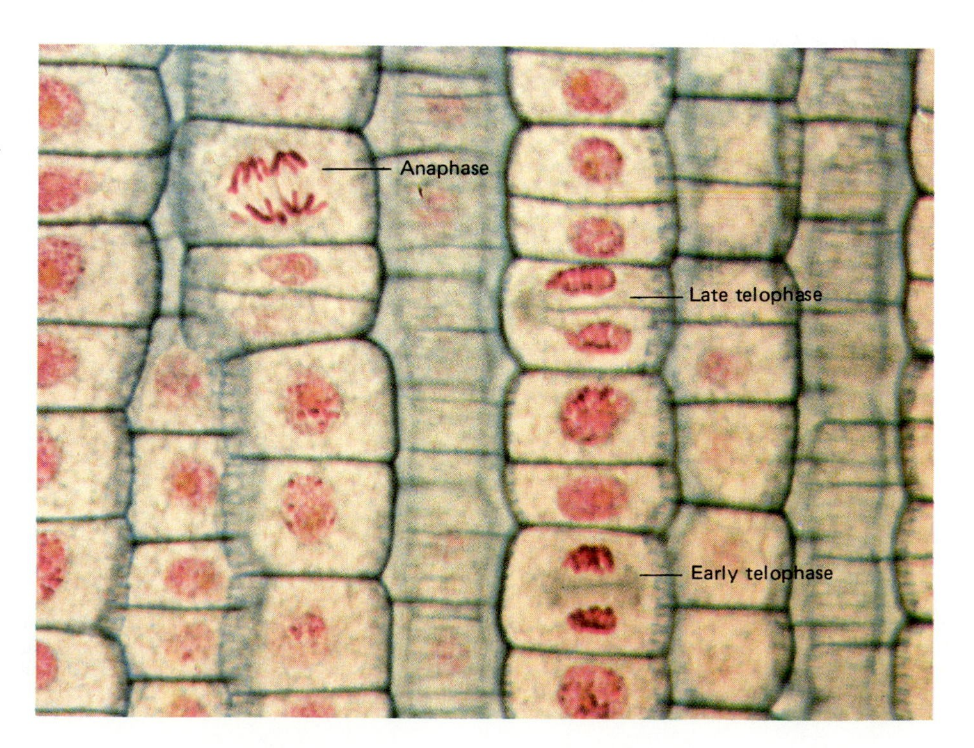

Figure 6–11 Meristematic tissue is composed of immature cells that grow, divide, and differentiate, giving rise to the other types of plant tissues. This tissue from the root tip of an onion (*Allium cepa*) shows cells in various stages of mitosis. (Courtesy of Triarch.)

TABLE 6–5
Plant Tissues

Type of Tissue	*Structure*	*Location*	*Function*
Meristematic	Small, thin-walled cells with large nuclei, few vacuoles	Tips of roots and stems, cambium	Growth, division, differentiation into other tissues

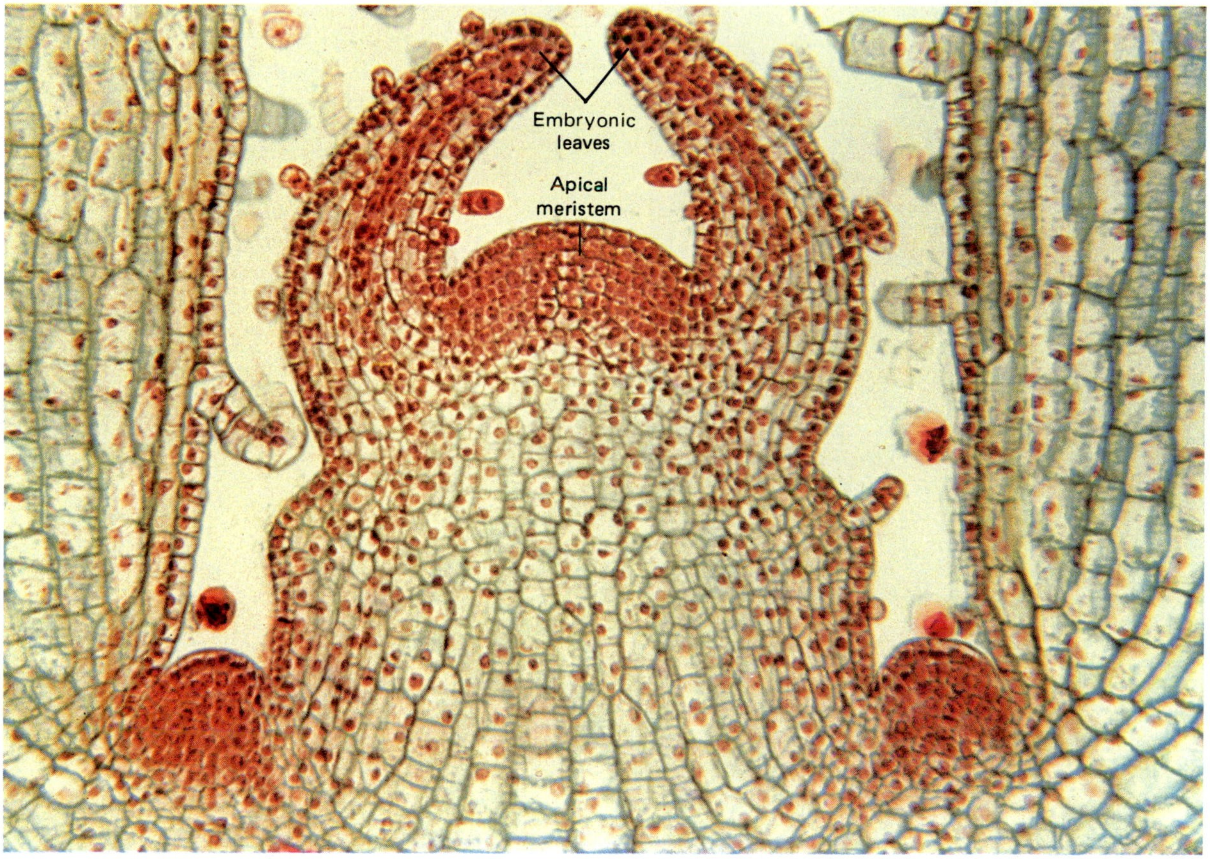

Apical meristem from *Coleus* stem, longitudinal section. (Courtesy of Triarch.)

Type of Tissue	*Structure*	*Location*	*Function*
Permanent (mature)	Mature, specialized cells		
Surface	Cells usually flattened, may secrete cutin or suberin	Epidermis of leaves, cork layers of stems and roots	Protection; in some (e.g., root hairs), absorption of water and nutrients

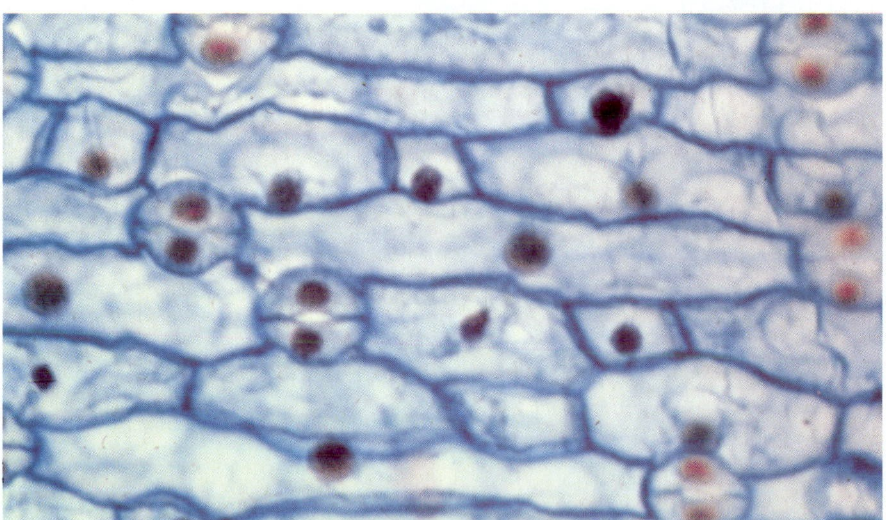

Onion leaf epidermis. (Carolina Biological Supply Company.)

(*Continued*)

TABLE 6–5
(continued)

Type of Tissue	*Structure*	*Location*	*Function*
Vascular			
Xylem	Long tubes composed of tracheids and xylem vessels	Roots, stems, and leaves	Transport of water and dissolved salts; support
Phloem	Long tubes composed of sieve tubes and companion cells	Roots, stems, and leaves	Transport of nutrients; support

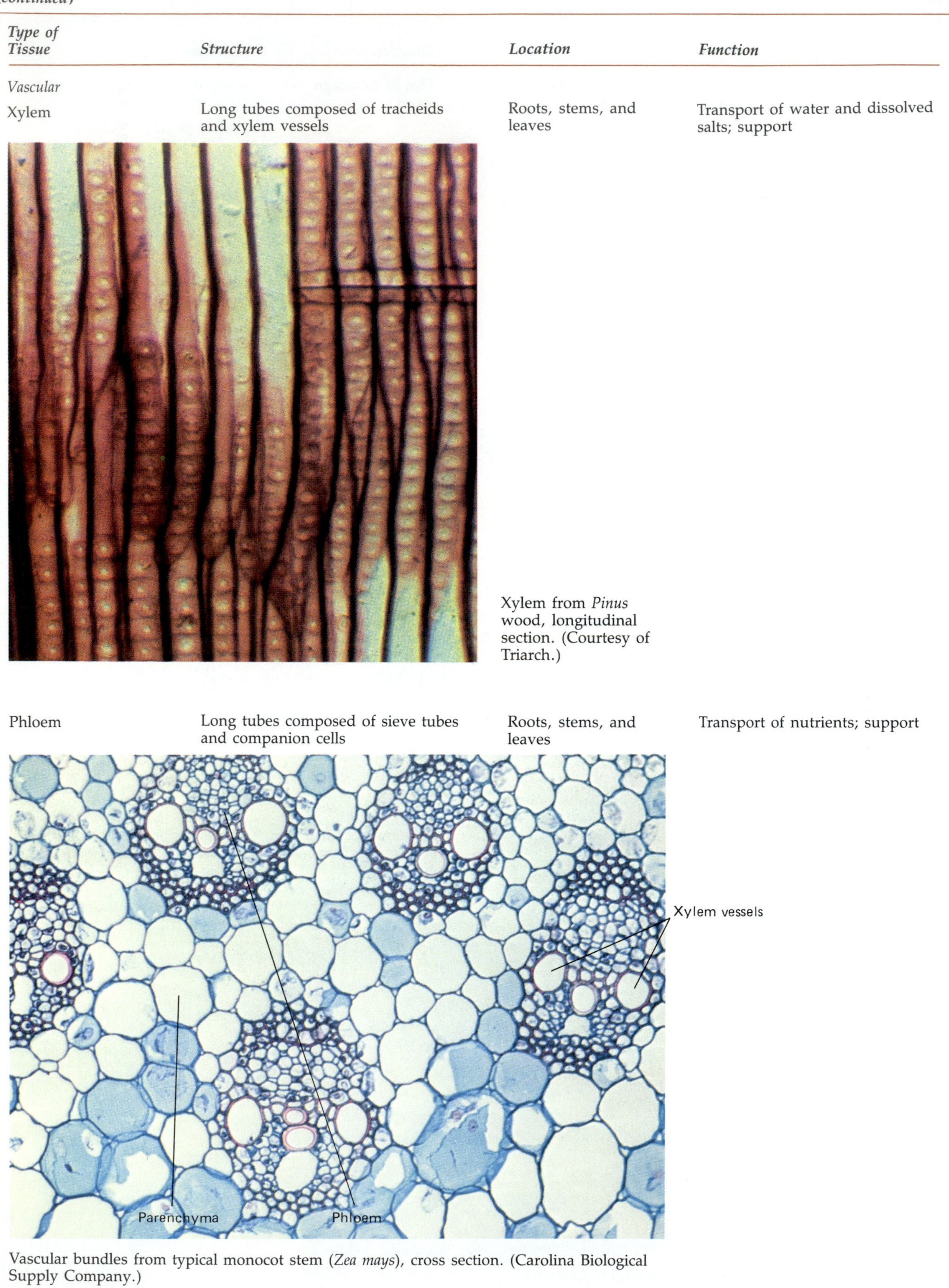

Xylem from *Pinus* wood, longitudinal section. (Courtesy of Triarch.)

Vascular bundles from typical monocot stem (*Zea mays*), cross section. (Carolina Biological Supply Company.)

Type of Tissue	*Structure*	*Location*	*Function*
Fundamental	Simple tissue; usually composed of one type of cell	Make up bulk of plant body	
Parenchyma	Cells with thin walls and central vacuoles	Roots, stems, and leaves	Photosynthesis; food storage

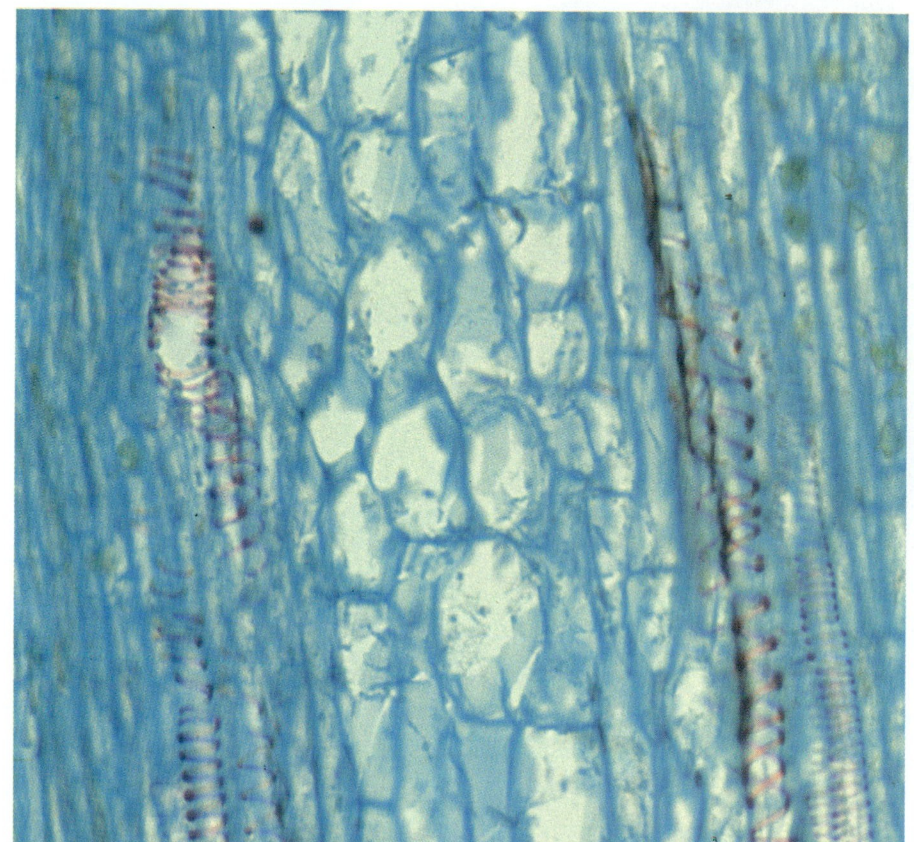

Cells from parenchyma (xylem vessels are stained purple). (Courtesy of John Dwyer.)

Type of Tissue	*Structure*	*Location*	*Function*
Chlorenchyma	Thin-walled, closely packed cells with large vacuoles and chloroplasts	Leaves, some stems	Photosynthesis

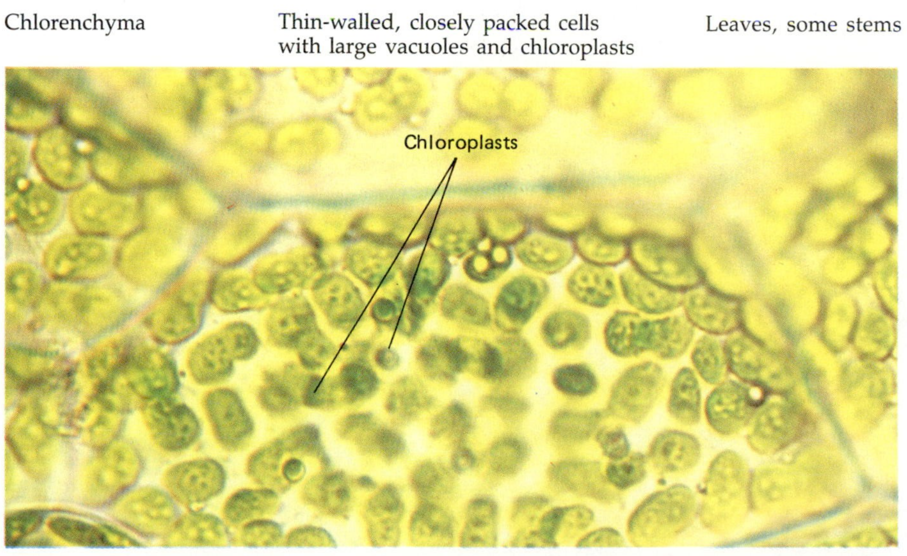

Chlorenchymal cells containing numerous chloroplasts. (Carolina Biological Supply Company.)

(*Continued*)

TABLE 6–5 (continued)

Type of Tissue	*Structure*	*Location*	*Function*
Collenchyma	Cells with walls thickened in the corners	Stems, leaf stalks	Support
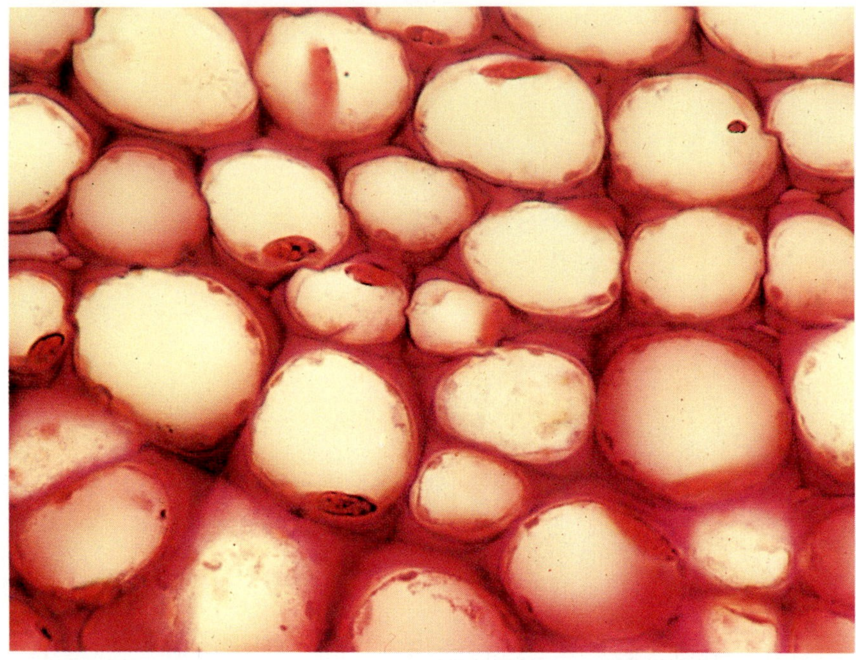 Cross section through part of a stem showing collenchyma. (Biophoto Associates, Photo Researchers, Inc.)			
Sclerenchyma	Cells with greatly thickened walls	Shells of nuts, hard parts of seeds	Support
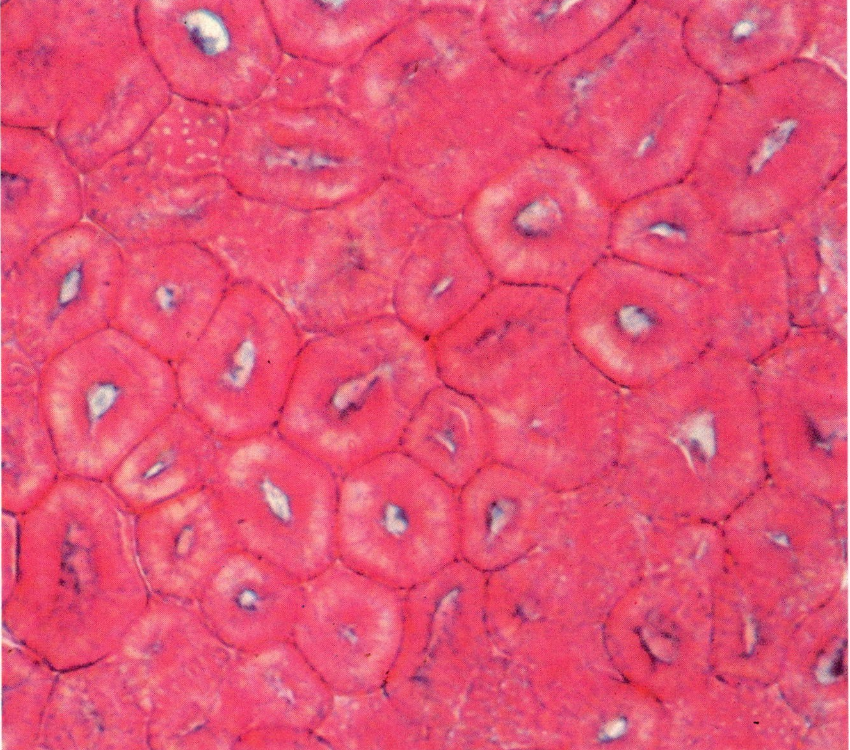 Sclerenchymal cells (stone cells) from a cherry pit. (Courtesy of John Dwyer.)			

only one cell thick, and epidermal cells are usually flattened. The epidermis of leaves secretes a waxy, waterproof material known as **cutin** that decreases the loss of water from the surface of the leaf. Specialized epidermal cells called **guard cells** are present on the surface of leaves. They occur in pairs around each tiny opening, called a **stoma**, into the interior of the leaf. The turgor pressure in the guard cells regulates the size of the stomatal aperture and hence the rate at which oxygen, carbon dioxide, and water vapor can pass into or out of the leaf.

Some of the epidermal cells in the roots have outgrowths called **root hairs**. These increase the absorptive surface of the root for the intake of water and dissolved minerals from the soil. Woody stems are covered by layers of **cork cells** produced by the cork cambium, another lateral meristem lying near the surface. Cork cells are closely packed; their cell walls contain a waterproof material, **suberin**. Because the suberin prevents the entrance of water into the cork cells themselves, they are short-lived. All mature cork cells are dead, like the cells in the outer layer of the epidermis of vertebrates. However, the epidermal layer of leaves, photosynthetic stems, and young roots is alive.

Vascular Tissues

Two types of vascular tissues are found in plants: **xylem**, which conducts water and dissolved salts (Fig. 6–12), and **phloem**, which conducts dissolved nutrients such as sugars and amino acids. In all higher plants the first xylem cells to develop are long **tracheids** with pointed ends and circular, spiral, or pitted thickenings in the walls. Subsequently other cells join end to end to form **xylem vessels**, which provide a continuous pathway from root tip to the uppermost leaves. As the vessels develop, the end walls dissolve and the side walls thicken, leaving a slender cellulose tube as much as 3 m long for the conduction of water. The cytoplasm and

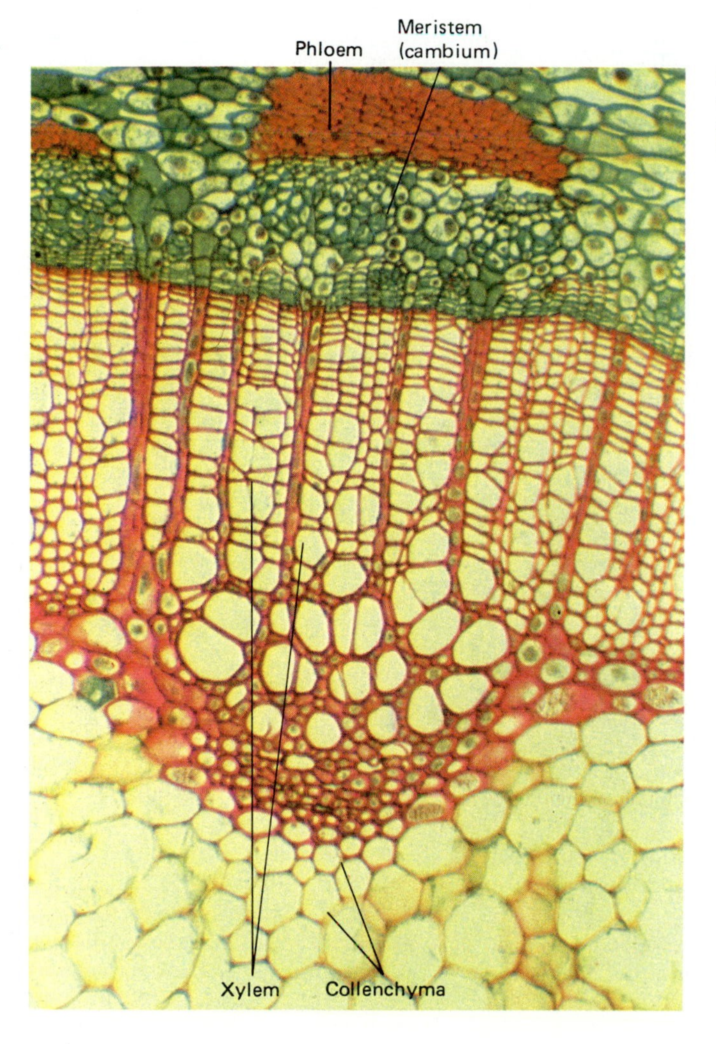

Figure 6–12 Cross section through part of a magnolia stem showing meristematic tissue, xylem, and collenchyma. (Courtesy of Triarch.)

nucleus of both tracheids and vessels eventually die, leaving tubes that continue to function to conduct water. The thickening, which involves the deposition of **lignin** (the substance responsible for the hard, woody nature of plant stems and roots), enables the xylem to act as a supportive as well as a vascular tissue.

The **sieve tubes** of the phloem are also formed by an end-to-end fusion of cells. However, the ends of these cells do not disappear but remain as a perforated plate, the **sieve plate**. Unlike the tracheids and vessels of the xylem, mature phloem sieve tubes remain alive and have an abundance of cytoplasm; however, they lose their nuclei. Adjacent to the sieve tubes are nucleated **companion cells**, which are thought to regulate the functions of the sieve tubes. Sieve tubes are found in the woody stems of plants, in the soft part just outside of the cambium layer.

Fundamental Tissues

The great mass of the plant body, including the soft parts of the leaf, both the central portion (the **pith**) and the peripheral portion (the **cortex**) of stems and roots, and the soft parts of flowers and fruits, is composed of **fundamental tissues**. The chief functions of the fundamental tissues are production and storage of food and support. The simplest of the fundamental tissues, called **parenchyma**, is found in roots, stems, and leaves and consists of cells with a thin wall and a thin layer of cytoplasm surrounding a central vacuole. Relatively unspecialized, parenchymal cells retain the ability to undergo cell division and to differentiate into other cell types. **Chlorenchyma** is modified parenchyma containing chloroplasts, in which photosynthesis occurs. Chlorenchymal cells are loosely packed and make up most of the interior of leaves and may occur in some stems. The cells have thin walls, large vacuoles, and many chloroplasts.

In some fundamental tissues, the corners of the cells are thickened and provide the plant with support. **Collenchyma**, which is found just beneath the epidermis of the stems and leaf stalks, has cells with such thickenings in the corners. Even more support and mechanical strength, however, is provided to stems and roots by **sclerenchyma**. In sclerenchymal cells, the entire cell wall is greatly thickened; these cells may take the form of long thin fibers with thickened walls. Sclerenchyma is found in the shells of nuts and in the hard parts of seeds.

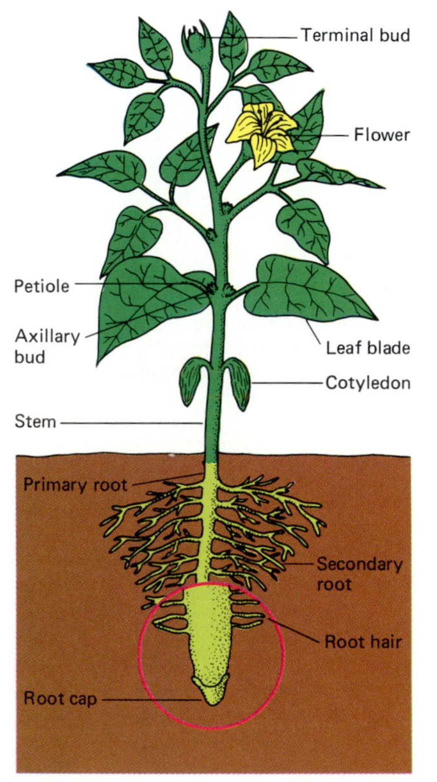

Figure 6–13 A young bean plant showing some of the plant organs. The circled portion of the root cap and root hairs indicates a magnified view.

Plant Organs and Plant Systems

Plants have fewer types of organs and organ systems than those of animals. The bodies of complex plants are composed of two major parts, the **root** and the **shoot**. These are generally considered the organ systems of the plant body. They can be distinguished by the arrangement of the vascular tissues, by the way lateral shoots and branch stems are formed, and by the presence of leaves on the shoot. The two, of course, are intimately connected, and many of the tissues, such as xylem and phloem, are essentially continuous from root tip to shoot tip. The roots function in anchoring the plant in the soil and in obtaining water and nutrients from the soil and then conducting them to the shoot.

The shoot consists of several organs: the stem, the foliage leaves, and the reproductive organs—the flowers and fruits (Fig. 6–13). The stem functions in support and transport of nutrients; the leaves function in photosynthesis, and the flowers, fruits, and seeds, in reproduction.

SUMMARY

I. Multicellular organisms are composed of a number of cell types, each specialized and adapted to carry out specific functions.
  A. A tissue is an aggregation of similarly specialized cells that associate to perform a specific function or group of functions.
  B. Several types of tissue may be united to form an organ; organs may be organized as an organ system.

II. Multicellular organisms are able to achieve a much larger size than that possible in single-celled ones.
  A. It is not efficient for a single cell to grow very large because its volume increases at a greater rate than its surface, making it difficult for the cell membrane to meet the cell's needs for oxygen and nutrients.
  B. In a multicellular organism, cells can specialize to perform specific functions.

III. Animal tissues are classified as epithelial, connective, muscular, or nervous.

A. Epithelial tissue may form a continuous layer or sheet of cells covering a body surface or lining a body cavity; some epithelial tissue is specialized to form glands.

1. Epithelial tissue functions in protection, absorption, secretion, or sensation.
2. Epithelial cells may be squamous, cuboidal, or columnar in shape.
3. Epithelial tissue may be simple, stratified, or pseudostratified. (Features of each are summarized in Table 6–1.)

B. Connective tissue joins other tissues of the body together, supports the body and its organs, and protects underlying organs.

1. Connective tissue consists of cells, such as fibroblasts and macrophages, and the intercellular substance secreted by the cells.
2. Some types of connective tissue are loose and dense connective tissue, elastic connective tissue, reticular connective tissue, adipose tissue, cartilage, bone, and blood. (Features of each are summarized in Table 6–2.)

C. Muscle tissue is composed of cells specialized to contract. Each cell is an elongated fiber containing many small longitudinal, parallel contractile units called myofibrils. The chief components of myofibrils are the proteins actin and myosin.

1. Skeletal muscle is striated and is under voluntary control.
2. Cardiac muscle is striated; its contraction is involuntary.
3. Smooth muscle, in which contraction is also involuntary, is responsible for movement of food through the digestive tract and for other forms of movement within body organs.

D. Nervous tissue is composed of neurons, which are cells specialized for conducting impulses, and glial cells, which are supporting cells.

V. Complex animals have a great variety of organs and ten principal organ systems. (See Table 6–4.)

VI. Plant tissues are classified as meristematic or permanent tissues.

A. Meristematic tissue is composed of immature dividing cells and can differentiate into other types of tissue.

B. Permanent tissues include surface tissues, fundamental tissues, and vascular tissues.

1. Surface tissues make up the outer covering of roots, stems, and leaves.
2. Fundamental tissues make up the great mass of the plant body: the soft parts of the leaf, most of the stem and roots, and the soft parts of flowers and fruits. Fundamental tissues provide support and produce and store food.
3. Two types of vascular tissues are xylem, which conducts water and dissolved salts, and phloem, which conducts dissolved nutrients such as sugars and amino acids.

VII. The plant body consists of two major parts, root and shoot. The shoot includes the stem, leaves, and the reproductive organs (flowers and fruits).

POST-TEST

1. A group of cells fitting tightly together to form a continuous sheet covering a body surface or lining a cavity of the body is termed an ______________.

2. The functions of epithelial tissues include ______________, ______________, ______________, and ______________.

3. On the basis of their shape we can distinguish ______________, ______________, and ______________ epithelia.

4. The mammalian respiratory tract is lined with ______________ ______________ ______________.

5. The outer layer of the skin is composed of ______________ ______________ ______________.

6. Epithelial cells specialized to produce and secrete a product are called ______________.

7. The supporting connective tissue framework of an organ is called ______________.

8. ______________ ______________ are small branched fibers forming delicate networks in tissues.

9. Match the terms in Column A with their definitions in Column B.

Column A	*Column B*
1. Canaliculi	a. Undifferentiated cells on outer walls of capillaries
2. Chondrocytes	b. Scavenger cells that clean up cellular debris
3. Collagen	c. The ground substance of connective tissue; part of the intercellular substance
4. Fibroblasts	d. A unique protein present in connective tissues, secreted by fibroblasts
5. Glial cells	e. Connective tissue cells that produce and secrete the proteins and other components of the matrix
6. Macrophages	f. Cartilage cells that secrete a flexible rubbery matrix
7. Matrix	g. Tiny channels in bone containing extensions of bone cells
8. Myofibrils	h. Fragments of cells that play a role in blood clotting
9. Pericytes	i. Small, longitudinal parallel contractile fibers in muscle cells
10. Platelets	j. Supporting cells present in nervous tissue

10. Erythrocytes contain the respiratory pigment ______________, which combines readily and reversibly with oxygen.

11. The liquid, noncellular part of the blood is termed ______________.

12. Match the terms in Column A with their definitions in Column B.

Column A	*Column B*
1. Cambium	a. Closely packed protective cells containing suberin
2. Chlorenchyma	b. Specialized epidermal cells occurring in pairs around each stoma
3. Cork cells	c. Fundamental tissue containing chloroplasts
4. Guard cells	d. Structures in phloem formed by end-to-end fusion of cells
5. Meristem	e. Vascular tissue that transports dissolved nutrients
6. Phloem	f. Vascular tissue that transports water and dissolved salts
7. Sieve tubes	g. Small opening into the interior of a leaf
8. Stoma	h. Composed of immature cells capable of cell division
9. Xylem	i. Layer of tissue that produces lateral growth in plants

13. The bodies of complex plants are composed of two major parts, the ______________ and the ______________.

14. The first xylem cells to develop are long ______________ with pointed ends and pitted thickenings in the walls.

15. The permanent tissues of the plant, composed of mature differentiated cells, include ______________, ______________, and ______________.

REVIEW QUESTIONS

1. What sort of arguments can you muster for the position that multicellular organisms have an advantage over unicellular organisms?
2. What are the functions of epithelial tissues? How are the cells adapted to carry out these functions?
3. What is the structure of bone? of adipose tissue? of loose connective tissue? How are each of these adapted to carry out its special functions?
4. Compare the properties of the three types of muscle.
5. Discuss the structure of a nerve cell and how this adapts it for its function.
6. Which tissues of plants and animals are comparable? Which are peculiar to one or the other?
7. Compare the structure and functions of xylem with those of phloem.
8. Name the kinds of tissues you would expect to find in the following organs: the lung, the heart, the intestines, and the salivary glands.
9. Compare meristematic tissues with permanent tissues.
10. List the principal organ systems found in a complex animal, and give the functions of each.

Recommended Readings for Part I

Books

Alberts, B., et al: *Molecular Biology of the Cell.* New York, Garland Publishing, 1983. A complete, but well-written and easy-to-read reference textbook.

Avers, C. J.: *Cell Biology,* 2nd ed. New York, D. Van Nostrand Co., 1981. A fine presentation of the details of cell structure and function.

Baker, J. J. W., and G. E. Allen: *Matter, Energy and Life:* An Introduction to Chemical Concepts. Reading, MA, Addison-Wesley Publishing Co., 1981. A presentation of the principles of thermodynamics and their application to studies of living systems. A difficult subject clarified.

Baum, S. J., and C. W. Scaife: *Chemistry: A Life Science Approach,* 2nd ed. New York, Macmillan Publishing Co., 1980. A chemistry text emphasizing those subjects of special interest to students of biology.

Bettelheim, F. A., and J. March: *Introduction to General, Organic and Biochemistry.* Philadelphia, Saunders College Publishing, 1984. A readable reference text for those who would like to know more about basic chemistry.

Dewitt, W.: *Biology of the Cell.* Philadelphia, Saunders College Publishing, 1977. An evolutionary approach to cell structure and function and the principles of cell biology.

Fawcett, D. W.: *The Cell,* 2nd ed. Philadelphia, W. B. Saunders Co., 1981. A study of cell structure through an exciting collection of electron micrographs.

Fessenden, R. J., and J. S. Fessenden: *Chemical Principles for the Life Sciences,* 2nd ed. New York, Allyn & Bacon, 1979. A chemistry text written primarily for students of biology, emphasizing the chemical aspects of life processes.

Hoagland, M.: *The Roots of Life: A Layman's Guide to Genes, Evolution and the Ways of Cells.* New York, Avon Books, 1979. A paperback written for the general public by a working scientist.

Holtzman, E., and A. B. Novikoff: *Cells and Organelles,* 3rd ed. Philadelphia, Saunders College Publishing, 1984. An integrated approach to the structural, biochemical, and physiological aspects of the cell.

Holum, J. R.: *Fundamentals of General, Organic, and Biological Chemistry,* New York, John Wiley and Sons, 1982. An excellent reference book for those who would like to learn more about basic chemistry and its applications to biology.

Margulis, L.: *Symbiosis in Cell Evolution.* San Francisco, W. H. Freeman & Co., 1981. A fascinating summary of the hypothesis that the cell organelles of eukaryotic organisms evolved from prokaryotes.

Toner, P. G., and K. Carr: *Cell Structure,* 2nd ed. London, Churchill Livingstone, 1979. A well-written text on cell structure and function.

Tribe, M. A.: *Metabolism and Mitochondria.* New York, Cambridge University Press, 1976. A brief text on the key role of mitochondria in intermediary metabolism.

Journal Articles

Albersheim, P.: The walls of growing plant cells. *Scientific American* 232: 80–96, 1975. Information about the special properties of cell walls, obtained in part from the study of the polysaccharides present.

Capaldi, R. A.: A dynamic model of cell membranes. *Scientific American* 230: 26–42, 1974. A well-illustrated description of the fluid mosaic model of the structure of the cell membrane.

Cloud, P.: The biosphere. *Scientific American* 249:176–189, 1983. A fascinating discussion of the relationship between microbial, animal, and plant life on earth and the physical environment.

DeDuve, C.: The lysosome. *Scientific American* 208: 64–78, 1963. An account of lysosome structure and function by the man who discovered them.

Grivell, L. A.: Mitochondrial DNA. *Scientific American* 248: 78–88, 1983. Mitochondria have their own genetic system.

Gross, J.: Collagen. *Scientific American* 204: 120–138, 1961. The remarkable features of the molecular structure of the most abundant protein in the human body.

Kornberg, R., and A. Klug: The nucleosome. *Scientific American* 244: 52–64, 1981. The primary subunit of chromosome structure is helical DNA wound around a core of histone proteins.

Lazarides, E., and J. P. Revel: The molecular basis of cell movements. *Scientific American* 240: 100–112, 1978. The role of microfilaments and microfibrils in all kinds of cellular movement.

Lodish, H., and J. E. Rothman: The assembly of cell membranes. *Scientific American* 240: 48–62, 1979. How a cell membrane grows but preserves the difference between its two sides.

Luria, S. E.: Colicins and the energetics of cell membranes. *Scientific American* 233: 30–46, 1975. A study of the energy involved in the movement of substances across the membranes of cells.

Marx, J. L.: Organizing the cytoplasm. *Science,* 222: 1109–1111, 1983. A discussion of the microtrabecular lattice and cytoskeleton, and their components.

Neutra, M., and C. P. Leblond: The Golgi apparatus. *Scientific American* 220: 100–122, 1969. Describes studies of the structure and function of the Golgi apparatus made using autoradiographic techniques.

Rothman, J. E., and J. Lenard: Membrane asymmetry. *Science* 195: 743–753, 1977. The asymmetric cell membrane has different proteins on its inner and outer surfaces. This article describes some of the cell functions that depend on this asymmetry.

Satir, B.: The final steps in secretion. *Scientific American* 233: 28–44, 1975. The interaction of the membrane of the secretory vesicle with the cell membrane in the process of exocytosis.

Sharon, N.: Glycoproteins. *Scientific American* 230: 78–92, 1974. A well-illustrated account of glycoproteins and their importance in the economy of the cell and especially of the cell membrane.

Shulman, R. G.: NMR spectroscopy of living cells. *Scientific American* 248: 86–93, 1983. Spectroscopy makes it possible to study metabolic processes in intact living cells.

Singer, S. J., and G. Nicolson: The fluid mosaic model of the structure of cell membranes. *Science* 175: 720–731, 1972. The first publication of Singer's theory and model of the cell membrane.

Unwin, N., and R. Henderson: The structure of proteins in biological membranes. *Scientific American* 250: 78–94, 1984. A discussion of the configurations of membrane proteins that permit them to become embedded in lipid, yet extend into the watery medium that surrounds that membrane.

Wessells, N. K.: How living cells change shape. *Scientific American* 225: 76–88, 1971. A fascinating description of the function of microtubules and microfilaments in the movements of cells.

part II

ENERGY IN LIVING SYSTEMS

You would not think that this praying mantis is consuming sunlight, but that is the ultimate source of energy in the biosphere. Plants, algae, and certain types of bacteria use the energy of sunlight to build complex organic molecules. Heterotrophes such as the cucumber beetle, shown being eaten by a praying mantis, obtain their energy directly from these producers. Others, such as the mantis, obtain their energy from other heterotrophes. In the end, all these organisms depend on sunlight for their energy. (Peter J. Bryant, UC-Irvine/BPS.)

7
Energy Flow Through the World of Life

OUTLINE

I. Energy transformations
II. Chemical directions
- *A.* The first law of thermodynamics
- *B.* The second law of thermodynamics
- *C.* Free energy
- *D.* Entropy and living systems
- *E.* Reversibility of reactions in living systems
- *F.* Equilibrium

III. Coupled reactions
IV. Factors that affect chemical reactions
- *A.* Activation energy
- *B.* Temperature
- *C.* Concentration
- *D.* Catalysts: introducing enzymes

V. How enzymes work
- *A.* Cofactors
- *B.* The role of enzymes in living cells

VI. Factors affecting enzyme activity
- *A.* Concentrations of enzyme and substrate
- *B.* Enzyme inhibition
- *C.* Temperature
- *D.* pH

VII. Energy flow through the biosphere

LEARNING OBJECTIVES

After you have read this chapter you should be able to:

1. Compare and contrast potential and kinetic energy, and give examples of each.
2. State the first and second laws of thermodynamics, and discuss their applications to biology.
3. Define and use the terms *energy*, *entropy*, and *free energy*.
4. Distinguish between exergonic and endergonic reactions, and describe how they may be coupled.
5. Describe the general effects of heat, substrate concentration, and catalysts on chemical reactions.
6. Describe the function and characteristics of enzymes, and explain how they work. (Include a description of cofactors and their action.)
7. Explain how enzymatic activity is affected by such factors as the amount of enzyme and substrate present, change in temperature, pH, and the presence of inhibitors.
8. Trace the flow of energy from the sun through the biological world.

The never-ending flow of energy within a cell, from one cell to another, and from one organism to another organism is the very essence of life. Living things are highly organized systems and must expend energy constantly to maintain this order and to counteract the tendency of all systems to become disordered. The cells that compose the bodies of living things have efficient systems for the transformation of one kind of energy into another. This energy is used in building complex chemicals or in the production of structures such as chloroplasts, which themselves transform one kind of energy into another.

The myriad chemical reactions of cells, which enable them to grow, move, maintain, and repair themselves, to reproduce, and to respond to stimuli, together make up the process of metabolism. In fact, **metabolism** can be defined as all the chemical and energy transformations that take place within the living organism. The metabolic reactions of all cells are remarkably similar despite the enormous differences in the appearance and adaptations of organisms. In most kinds of cells, glucose and other simple sugars are metabolized to carbon dioxide and water by cellular respiration in a stepwise series of reactions. During these conversions a part of the energy of the glucose molecule is conserved and made available to the cell to drive other processes. In this chapter we will see how these reactions are controlled and what factors affect their rates. Then we will briefly consider the overall flow of energy from the sun to the various kinds of organisms on the earth.

Energy Transformations

Energy, the ability to produce a change in the state or motion of matter, can exist in several different forms. These include heat, electrical, mechanical, chemical, and radiant energy (the energy of electromagnetic waves, such as radio waves, visible light, x rays, and gamma rays). **Potential energy**—that is, stored energy—is the capacity to do work owing to the position or state of a particle or mass. (A simple definition of work is the movement of a particle or mass through a distance.) In contrast, energy that is being expressed—that is, the actual motion of a particle of mass—is referred to as **kinetic energy.** An example of the conversion of potential energy to kinetic energy is the release of a drawn bow (Fig. 7–1): The tension in the bow and string represents stored energy; when the string is released, this potential energy is released so that the motion of the string propels the arrow. It would require the input of additional energy to draw the bow once again and restore the potential energy.

The study of energy transformations in living organisms is termed **bioenergetics.** Three major processes that transform energy can be distinguished in the biological world: photosynthesis, cellular respiration, and cellular work (Fig. 7–2). In **photosynthesis,** the radiant energy of sunlight is captured by the green pigment chlorophyll that is present in the chloroplasts of algae and plants. The radiant energy is then transformed into chemical energy, which is used in turn to synthesize carbohydrates and a variety of other complex molecules from carbon dioxide and water. The radiant energy of sunlight (which is a form of kinetic energy) is transformed into **chemical energy** (a type of potential energy) and stored in the bonds of carbohydrates and other complex molecules.

In order to become available to drive the many functions of the cell, the chemical energy of carbohydrates and other molecules undergoes a second type of energy transformation during cellular respiration. The reactions of **cellular respiration,** involving the step-by-step breakdown of glucose and other molecules, result in the production of energy-rich phosphate bonds.

The third major type of energy transformation, **cellular work,** occurs when the chemical energy of the energy-rich phosphate bonds is utilized by cells to do work (Table 7–1). This may be the mechanical work of muscular contraction, the electrical work of conducting a nerve impulse, the osmotic work of moving molecules against a gradient, or the chemical work of synthesizing new molecules for growth or for the storage of energy. As these transformations occur, some energy is lost to the environment and is dissipated as heat. Cells contain remarkably effective control mechanisms to regulate energy transformations.

Heat is a convenient form in which energy may be measured. This is why the study of energy has been termed **thermodynamics,** i.e., heat dynamics. Other

POTENTIAL KINETIC

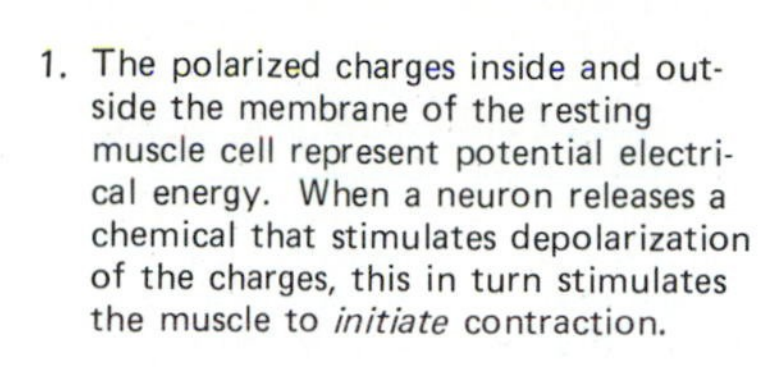

1. The polarized charges inside and outside the membrane of the resting muscle cell represent potential electrical energy. When a neuron releases a chemical that stimulates depolarization of the charges, this in turn stimulates the muscle to *initiate* contraction.

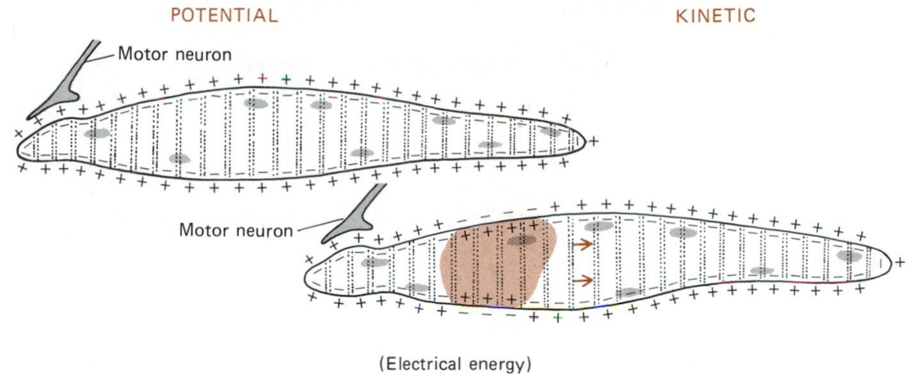

(Electrical energy)

2. Chemical energy is stored in the high energy phosphate bonds of a molecule called ATP. Breaking these bonds releases energy to carry out muscle contraction.

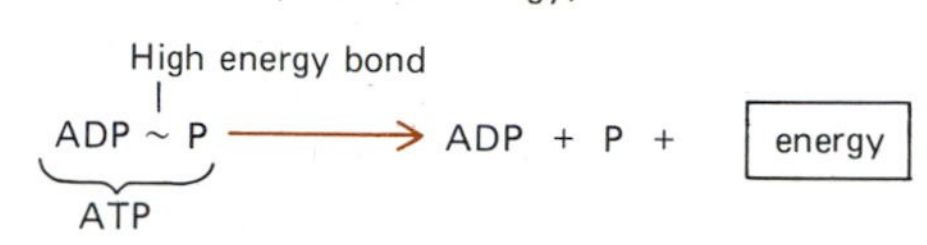

(Chemical energy)

3. This energy released inside each muscle cell causes submicroscopic thick filaments to pull on thin filaments by means of cross bridges. The thin filaments move past the thick ones, causing the cell to shorten.

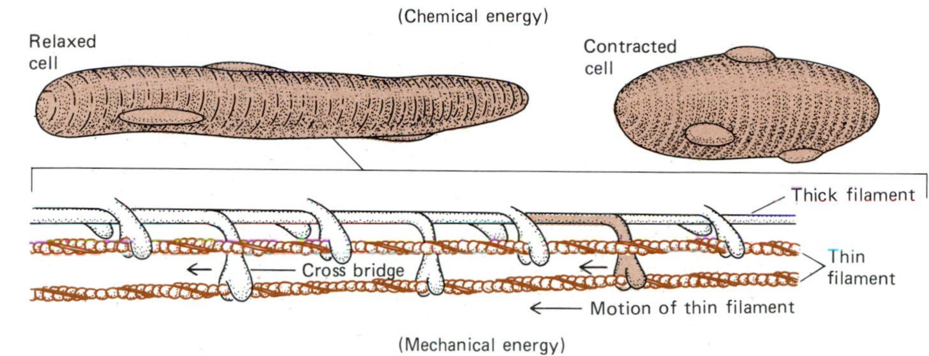

(Mechanical energy)

4. The archer's muscles contract, pulling on the bones, which then move and transmit the muscular force to the bow.

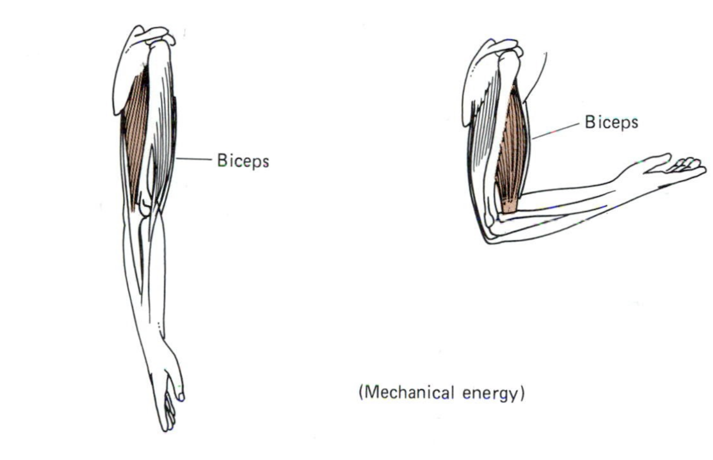

(Mechanical energy)

5. The archer has drawn the bow. The energy stored in the drawn bow is potential mechanical energy. When the archer releases her fingers the string moves forward and the arrow is propelled toward a target.

Figure 7–1 Shooting a bow and arrow illustrates some of the energy transformations that take place in a living system. For more information on ATP, see Focus on ATP: The Energy Currency of Cells; the actions of muscle cells are described in Chapter 32.

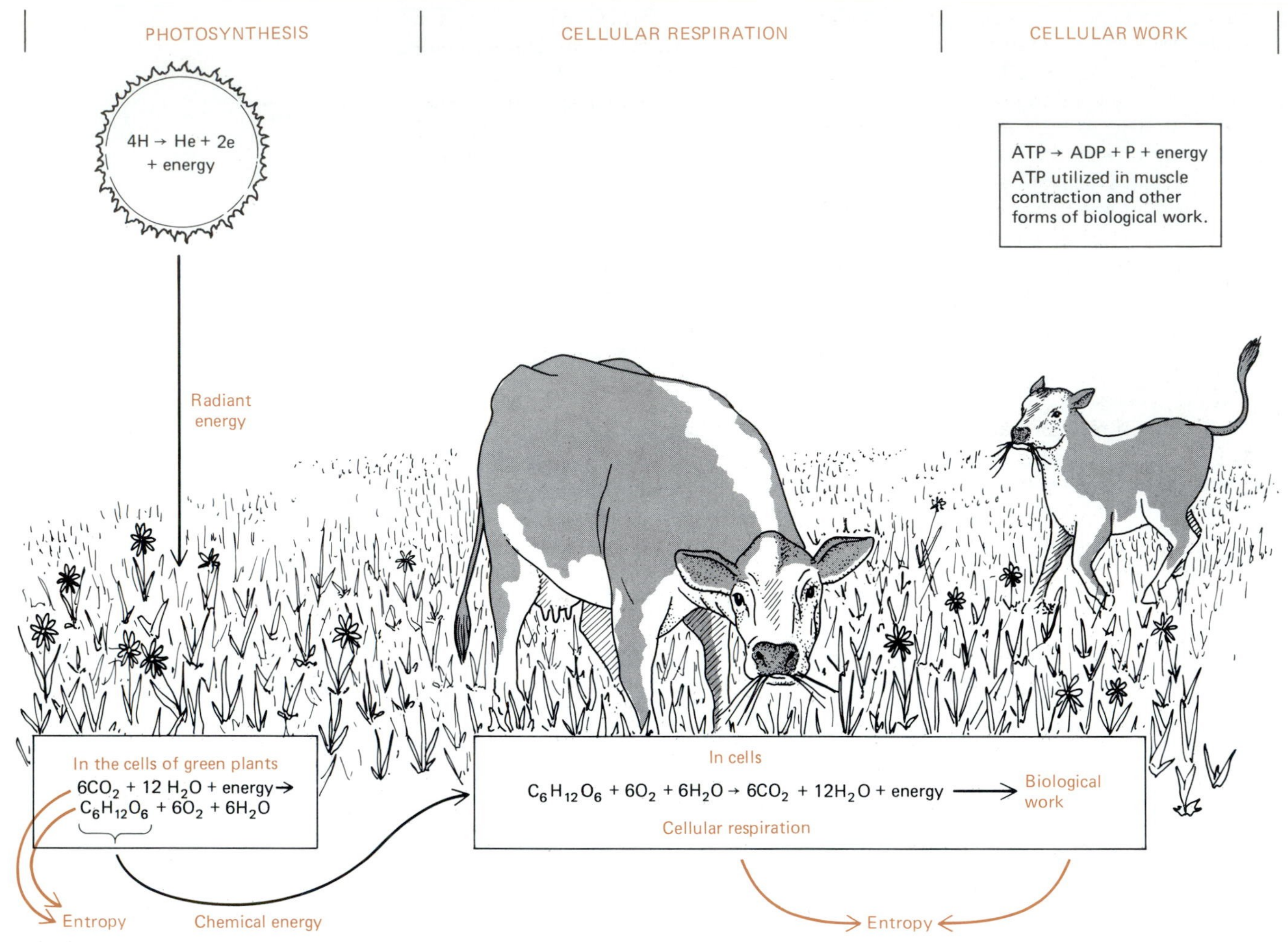

Figure 7–2 Three major types of processes that transform energy: photosynthesis, cellular respiration, and biological work. (The chemical reactions shown are simplified. See Chapters 8 and 9 for more detail.)

TABLE 7–1
Energy Transformations in Cells

Transformation	*Type of Cell*
Chemical energy to electrical energy	Nerve, brain
Sound to electrical energy	Inner ear
Light to chemical energy	Chloroplast
Light to electrical energy	Retina of eye
Chemical energy to osmotic energy	Kidney
Chemical energy to mechanical energy	Muscle cell, ciliated epithelium
Chemical energy to radiant energy	Luminescent organ of firefly
Chemical energy to electrical energy	Sense organs of taste and smell

forms of energy—radiant, chemical, electrical, or the energy of motion or position—can be converted to heat and measured by their effect in raising the temperature of water. Several units may be used in measuring energy; the unit most widely used in biological systems is the **kilocalorie[1], or kcal.**

Nearly every physical or chemical event in both living and nonliving systems is accompanied by a transfer of heat to the surroundings or by the absorption of heat from the surroundings. A process in which heat is transferred to the surroundings is said to be **exothermic.** A process in which heat is absorbed from the surroundings is termed **endothermic.** Many of the machines used in industry are heat engines. An engine driven by steam produced by the burning of coal in a boiler is a familiar example of a heat engine. However, heat is not a useful way of transferring or storing energy in biological systems. Under conditions of constant pressure,

[1] The kilocalorie is used by nutritionists in measuring the potential energy of foods, but they generally refer to this unit as a Calorie (with a capital C). Thus, a Calorie equals 1000 calories.

heat can do work only when it can flow from a region of higher temperature to a region of lower temperature. Living organisms are basically isothermal (equal-temperature) systems; there is no significant temperature gradient—that is, difference in temperature—among the various parts of the cell or the various cells in a tissue. Cells cannot act as heat engines, for they have no means of permitting heat to flow from a warmer to a cooler body. Furthermore, temperatures above 50°C would denature and inactivate enzymes and other cellular constitutents.

Cells utilize the energy stored in chemical bonds in complex organic molecules. However, this energy must be properly channeled if it is to perform work. Random energy release or the sudden release of large amounts of heat would be more likely to produce chaos than the order essential to living things.

Chemical Directions

Most chemical reactions are reversible, but within the cell a variety of control systems guide the reactions so that they occur primarily in one direction. This chemical directionality assures that organisms are able to carry out the orderly sequence of reactions that constitute metabolism and allow the safe storage of energy inside cell membranes. The factors that govern energy transformations in living systems correspond to the same laws that you might study in a physics or chemistry course. Let us therefore first review some important principles of thermodynamics.

THE FIRST LAW OF THERMODYNAMICS

The **first law of thermodynamics** states that energy can neither be created nor destroyed but can be changed from one form to another. For example, consider a car traveling down the highway: As the car moves forward, it has mechanical energy. If the engine is turned off, the car will continue forward for a distance, but eventually it will come to a halt. Where has all the energy gone?

All the time the car is traveling, some of its mechanical energy is being transferred to the molecules that make up the road surface. (You may have noticed that even at night, without the heat contributed by sunlight, the surface of a heavily traveled road can become quite hot.) When the engine is turned off, the additional mechanical energy needed to make up for the loss is no longer being supplied, and the transformation is allowed to proceed to completion. It is friction, largely with the road surface, that converts the mechanical energy to heat and stops the car (Fig. 7–3). In the absence of friction, a vehicle would travel indefinitely. This is what occurs in outer space.

What has happened is that the highly ordered, unidirectional energy of the motion of the car has been converted to a disordered, multidirectional form of energy. This is the energy of random molecular motion, or **heat.** None of the energy that propelled the car has been destroyed; it has just been transformed into a different form. Can that energy be recovered and put to further use? The following section discusses the difficulty of accomplishing this.

THE SECOND LAW OF THERMODYNAMICS

The difficulty in recovering, in a useful form, all the mechanical energy that has been transformed into heat has to do with a universal trend. This trend is actually an overwhelming probability that can be stated as a law, the **second law of thermodynamics:** Disorder in the universe is always increasing.

In your own day-to-day experiences you might have observed that creating order requires an input of energy—you must do work. For example, if you spill the many pieces of a jigsaw puzzle onto the floor it is very unlikely that the pieces will spontaneously reassemble themselves into the original picture. Likewise, if you drop a glass vase on the floor it will be difficult to gather all the shards of glass and fit them together to form the vase. The reason for this is that out of a multitude of possible ways in which the pieces can be assembled, only one represents the highly ordered form that is the vase or the finished puzzle. Given all possible forms of arrangement, order is an extremely unlikely state of existence.

Some of the concepts introduced here are best understood using mathematical models. However, it can be readily appreciated that disorder is more likely than

1. An automobile travels down the highway. Its energy is highly directional.

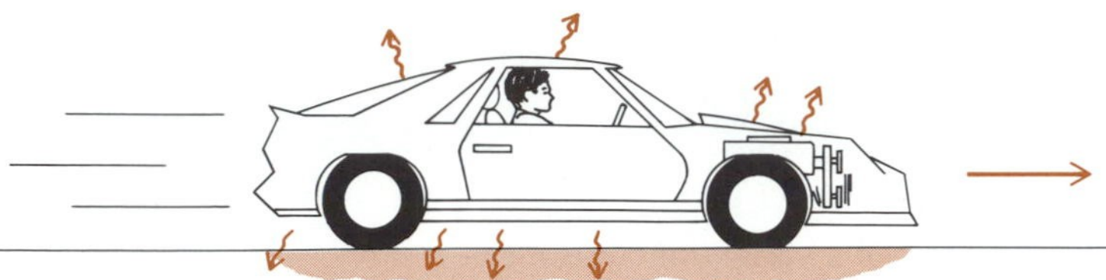

2. As the car travels, the surface of the road heats up, along with the surrounding air and the car's tires.

3. The engine of the car is turned off and the car begins to slow.

4. All the energy of the car's forward motion has been converted into heat and the car is now at a halt.

Figure 7–3 The car does not continue to travel forever because it is stopped by frictional forces. The mechanical energy of the car is converted to heat energy—with an increase in disorder.

order. From this we might surmise something about the ultimate fate of energy in the universe: that all energy will be degraded into its most disordered form, which is heat. We have considered the conditions under which heat can be made to perform work. In every transfer of energy, however, a certain amount of it cannot be recovered. This is the energy that has gone toward increasing the overall disorder of the system. **Entropy** is the measure of the disorder of a system—of the energy that has become randomized and uniform throughout a system. (Anything less than a uniform distribution of energy implies that some energy is still ordered; unevenly distributed energy represents less than total randomness.)

According to the second law of thermodynamics, entropy in the universe is always increasing. Eventually all energy will be random and uniform in distribution. There will be no differences in energy potential. Consequently, no energy will be able to flow, and no work can be done. All energy will become useless (which is also a way of stating the second law[1]).

It is important to understand that the second law of thermodynamics is consistent with the first law. The total amount of energy in the universe is not decreasing with time; but the energy available to do work is being degraded to random molecular motion.

[1] Another definition is one often used by engineers: All forms of energy are eventually degraded into heat, which is dissipated into the environment.

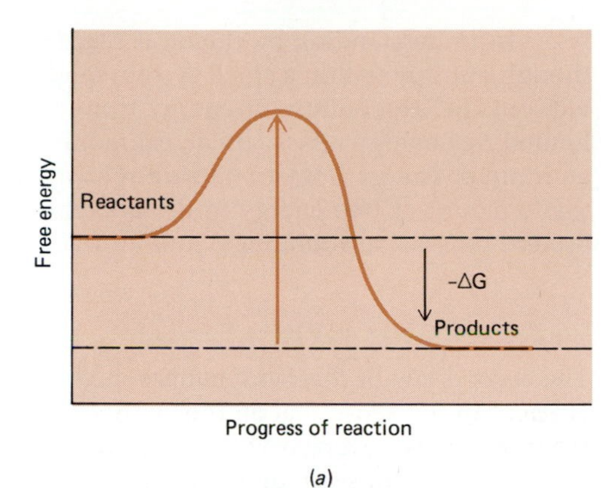

(a)

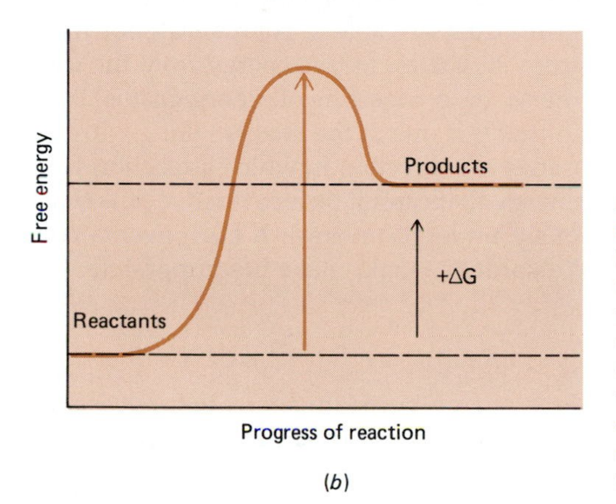

(b)

Figure 7–4 Free energy diagrams. (*a*) In an exergonic reaction, the products have less free energy than the reactants. In other words, ΔG has a negative value. (*b*) Endergonic reactions require a net input of free energy; hence the products have more free energy than the reactants. Note that even the exergonic reaction requires some input of energy to get started. This energy is termed activation energy.

FREE ENERGY

Free energy is that part of the total energy of a system that is available to do work under conditions of constant pressure and temperature. **Spontaneous reactions,** reactions that occur without outside intervention, release free energy and can therefore perform work; these are also called **exergonic reactions.** Nonspontaneous reactions require a net input of free energy and are said to be **endergonic** (Fig. 7–4). Most exothermic (heat-releasing) reactions are also exergonic, but if the disorder of the reacting system has increased, then more free energy is released than would be indicated by the amount of heat released. This increase in entropy can sometimes be sufficient to make even an endothermic (heat-absorbing) reaction or process occur spontaneously. Such is the case with the melting of ice and the dissolving of certain solids in liquids.

The maximum amount of biologically useful energy we can get from a reaction is limited by several factors: the change in heat (ΔH) of the reactants, which under conditions of constant pressure (as in a biological system) is roughly equal to the change in their internal energy; the temperature at which the reaction occurs; and the amount of energy that is involved in increasing or decreasing the disorder in the system. This amount of work is represented by the free energy change, ΔG. The equation used to calculate ΔG also serves to determine spontaneity:

$$\Delta G = \Delta H - T\,\Delta S$$

where ΔG = change in free energy; a minus sign (−) indicates an exergonic reaction
ΔH = change in heat, or, for our purposes, internal energy; exothermic reactions have a negative value and endothermic reactions have a positive value
T = temperature in kelvins[1]
ΔS = change in entropy

[1]A kelvin, abbreviated K, is a unit of measurement on a temperature scale called the Kelvin scale. A kelvin has the same magnitude as a Celsius (centigrade) degree; on the Kelvin scale, however, the freezing point of water is at 273 K, because kelvins are measured from absolute zero (0 K, or −273°C), the temperature at which all molecular motion stops. To convert from Celsius degrees to kelvins, use the equation K = °C + 273.

In understanding biochemical reactions it is important to realize that even though entropy within a small system might decrease, such ordering cannot occur without the expenditure of energy from somewhere. This is reflected in either a limited free energy release (in an exergonic reaction) or (in an endergonic reaction) an input of energy from an outside source. The net result is still a reduction in the total amount of free energy in the universe (the universe consists of the reacting system plus its surroundings) and an overall increase in entropy.

ENTROPY AND LIVING SYSTEMS

The second law of thermodynamics means that no energy transformation is 100% efficient in the sense that all the energy released can be made to work, or that the reaction can be reversed without an input of energy from an outside source. The second law might seem ultimately pessimistic (for eventually the universe will run out of useful energy); however, its implications for living systems are not altogether grim. Cells are able to compensate for the loss of their internal free energy to entropy by taking in free energy from the outside (through the process of photosynthesis, or by ingesting other organisms). Additionally, entropy is biologically useful in that it is one of the factors (along with the change in internal energy) that determines the direction in which a reaction proceeds. The fact that entropy represents energy that cannot be recovered means that certain key reactions will not spontaneously undergo reversal. If such events could happen, then this different kind of "disorder" would make life impossible.

REVERSIBILITY OF REACTIONS IN LIVING SYSTEMS

Free energy transformations that release a large amount of heat are inefficient because, as discussed earlier, heat cannot be stored in living cells or anywhere in living organisms. Therefore heat must be radiated to the surrounding environment; this depletes the supply of energy without performing work for the organism. Usually when free energy transformations take place in living systems, only a small amount of free energy is released as heat at one time. The unused free energy continues to be stored in chemical bonds.

Actually, almost any sudden, large release of free energy in a living organism is inefficient. In most biochemical reactions, there is little free energy difference between reactants and products. As a result, as long as there is the continued availability of external energy inputs, most of the reactions that occur within living cells—including even the important overall reactions of metabolism—are theoretically reversible; in fact, such **reversibility** is characteristic of many biochemical reactions. Reversibility allows cells to control their release of free energy in accordance with their needs, and it permits many of their large biological molecules to be rebuilt or otherwise recycled for continued use in metabolic processes.

EQUILIBRIUM

When the free energy difference between the products and reactants is zero, the reaction is said to be at **equilibrium.** At equilibrium it is impossible to say which are the products and which are reactants because the reaction can proceed in both directions at once. As a result of a change imposed on the reacting system (e.g., a change in temperature or pressure), equilibrium will shift and the reaction will proceed in a specific direction until once again the free energy difference is zero.

Large free energy changes that are not readily reversible will disrupt metabolic cycles within an organism. Any condition (such as a change in temperature or pressure) that might shift the equilibrium of important biochemical reactions too far in one direction is incompatible with life.

Coupled Reactions

The development and growth of an organism is marked by increasing organization. Accompanying this increasing order, and indeed essential to the maintenance of order, is an enormous number of chemical reactions, the *majority* of which are endergonic. In a sense, then, life (which consists of highly ordered systems) is a constant struggle against the second law of thermodynamics (the inexorable tendency toward disorder). How does a living organism, from the time it is "born" until the time it dies, drive its endergonic reactions and maintain a high degree of complexity?

We have already seen how cells compensate for their continuous loss of free energy by employing outside energy inputs. Two factors make these inputs available for use by the cell. First, organisms are part of a large universe with a vast reserve of free energy; and second, they have within themselves special structures, enzymes, and genetic information needed to direct their negatively entropic life processes.

To see how the chemical machinery of the cell is able to supply the energy to direct an endergonic reaction, consider the following free energy change, ΔG, in a hypothetical chemical reaction:

(1) $$A \longrightarrow B + C \quad \Delta G = +5 \text{ kcal/mole}$$

Since ΔG is positive, free energy is absorbed rather than released and the reaction is not spontaneous.

However, consider the following reaction:

(2) $$C \longrightarrow D \quad \Delta G = -8 \text{ kcal/mole}$$

This reaction can proceed spontaneously, since it *loses* free energy and therefore has a *negative* ΔG.

Note that although the reaction $A \rightarrow B + C$ has a positive free energy change (+5 kcal/mole), the reaction $C \rightarrow D$ has a larger and *negative* free energy change (−8 kcal/mole). Because the free energies of reactions are additive, cells can utilize exergonic reactions to drive endergonic reactions. When reactions (1) and (2) occur together, they form a system that has an overall negative free energy change (−3 kcal/mole). After all, if $A \rightarrow B + C$ and $C \rightarrow D$, ultimately $A \rightarrow B + D$.

To sum it up:

(1)	$A \longrightarrow B + C$	$\Delta G = +5$ kcal/mole
(2)	$C \longrightarrow D$	$\Delta G = -8$ kcal/mole
(3)	$A \longrightarrow B + D$	$\Delta G = -3$ kcal/mole

If D is the desirable product, then this is a thermodynamically feasible way to produce it. The second reaction pulls the first one along.

There are many examples of such related reactions in biology. They are called **coupled reactions,** because in them a thermodynamically favored exergonic reaction provides the energy needed to drive a thermodynamically unfavorable (endergonic) reaction. Thus the two are coupled together. We could predict, then, that for every endergonic reaction occurring in a living cell there must be a coupled exergonic reaction to drive it, and indeed this is true. Often, the exergonic chemical reaction involves the breakdown of adenosine triphosphate, or ATP. The ATP is manufactured in its turn by a series of strongly exergonic reactions of cellular respiration (Chapter 9). ATP is discussed in more detail in Focus on ATP: The Energy Currency of Cells.

Factors That Affect Chemical Reactions

The laws of thermodynamics tell us the factors that govern the direction of a chemical reaction, but they do not tell us anything about the rate of a reaction. For a reaction to actually take place three main requirements must be met:

1. The involved atoms and molecules must get close enough to one another for their electrons to interact.
2. They must approach one another in the proper orientation in order to interact. (Imagine two amino acids approaching one another in such a way that their amino groups were adjacent. No reaction would occur, because among other things, the amino end of one amino acid must unite with the carboxyl end of the other to produce a dipeptide.)
3. Existing chemical bonds between these atoms and molecules must be broken before new ones can be formed.

FOCUS ON

ATP: The Energy Currency of Cells

Adenosine triphosphate, better known as ATP, is an "energy-rich" molecule by virtue of the phosphoanhydride bonds between its first and second phosphates (*A* in the figure) and between its second and third phosphate groups (*B*).

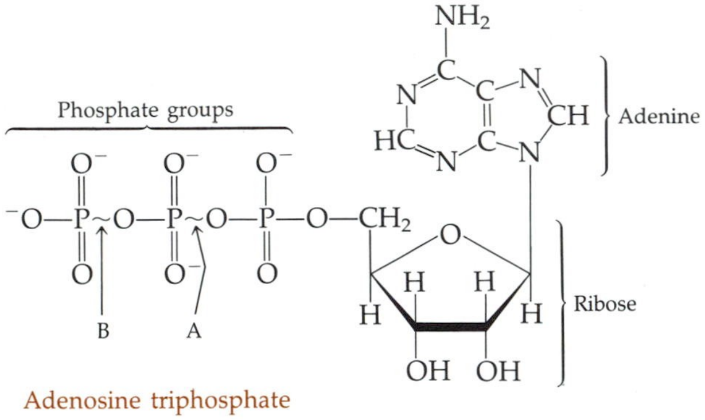

Phosphate groups

Adenine

Ribose

Adenosine triphosphate

A large amount of free energy is released when these bonds are hydrolyzed. We can think of these compounds as having high phosphate-group-transfer potential; that is, they have a high potential to transfer the phosphate group, together with some of the energy of the bond, to an acceptor molecule such as glucose.

$$\text{Glucose} + \text{ATP} \longrightarrow \text{Glucose 6-phosphate} + \text{ADP}$$

(ADP: Adenosine diphosphate)

ATP is formed from ADP and inorganic phosphate, P_i, when foodstuffs are oxidized or when light energy is trapped in photosynthesis. The cycle of ATP $\rightleftharpoons$ ADP, as shown, is the primary means by which energy is exchanged in biological systems. ATP is not a long-term storage form of energy but rather serves as the immediate donor of free energy. ATP molecules are formed and consumed rapidly, with a high rate of turnover and a short half-life. A person at rest uses up about 45 kg (90 lb) of ATP each day, but the amount present in the body at any given moment is less than 1 g. A person who exercises strenuously during the day uses up several times this amount.

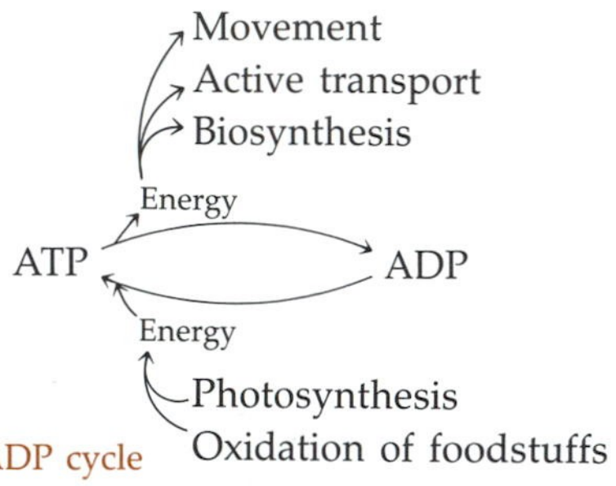

Movement
Active transport
Biosynthesis
Energy
ATP
ADP
Energy
Photosynthesis
Oxidation of foodstuffs

The ATP-ADP cycle

In other reactions ATP is split at the second phosphate group to yield adenosine monophosphate, AMP, and inorganic pyrophosphate, PP_i:

$$\text{ATP} \longrightarrow \text{AMP} + PP_i$$

The hydrolysis of ATP, ATP + $H_2O \rightarrow$ ADP + P_i, yields about -7.3 kilocal per mole of free energy, whereas the hydrolysis of glucose 6-phosphate, glucose 6-phosphate + $H_2O \rightarrow$ glucose + P_i, yields only about -3.3 kilocal per mole. Phosphoenolpyruvate (one of the intermediates in glycolysis) has an even higher phosphate-group-transfer potential than ATP and can transfer its phosphate group to ADP to form ATP. This is one of the reactions by which ATP is generated during the metabolism of sugars. The group-transfer potential of ATP is intermediate among the various biologically important molecules with phosphate groups. This permits ATP to donate phosphates (and energy) to some molecules and ADP to accept phosphates from others.

To increase the rate of a reaction, conditions under which these requirements are met must occur with greater frequency (Fig. 7–5). As we will see, several factors, including temperature, concentration of the reactants, and the action of catalysts, control the reaction rate. But how do reactions get started in the first place?

ACTIVATION ENERGY

If hydrogen and oxygen are mixed in a 2:1 ratio, and a spark is applied, the two react with one another to form water with an explosive release of energy.

$$2\,H_2 + O_2 \longrightarrow 2\,H_2O + \text{Energy}$$

No detectable hydrogen or oxygen remains. Clearly, the energy relationships of the substances involved favor a spontaneous reaction. Yet if the gas mixture is left in a

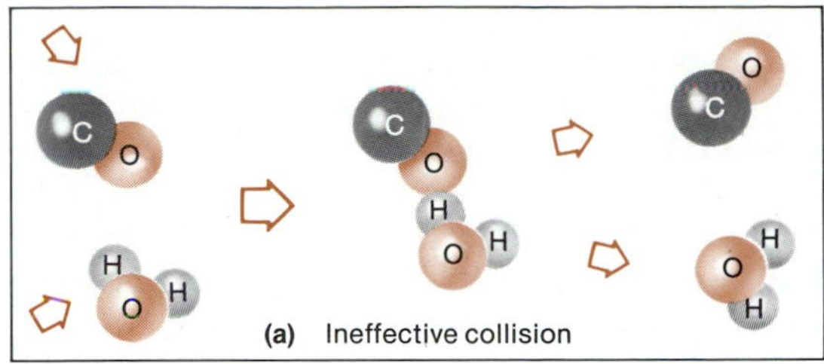

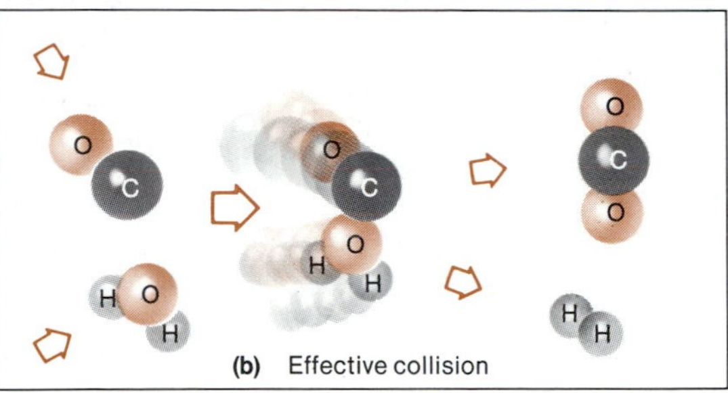

Figure 7–5 Ineffective and effective collisions between gaseous carbon monoxide (CO) and water. (*a*) The carbon monoxide and water molecules do not approach each other in the proper orientation for a reaction to occur. (*b*) A carbon monoxide molecule collides with a water molecule with sufficient force and in such a position as to remove an oxygen atom, so that carbon dioxide is formed. (Courtesy of R.E. Davis, K.O. Gailey, and K.W. Whitten.)

spark-proof container, 30 years later it might still be a mixture of hydrogen and oxygen, without a single molecule of water. In biochemistry the term "spontaneous" does not mean that such a reaction will proceed at an observable rate. Even a strongly exergonic reaction that releases more than enough energy as it proceeds is prevented from beginning by an energy barrier; if new bonds are to be formed, existing bonds must first be broken. The energy required to overcome this barrier and initiate the reaction is called **activation energy.**

In the reaction between H_2 and O_2 a spark can provide this activation energy. The spark produces an intense, localized heating of the gases, making it more likely that hydrogen and oxygen molecules will collide and that their existing covalent bonds will be strained or broken. When the atoms do interact to form water, the release of energy is so great that it produces heat. That heat serves to trigger the interaction of nearby molecules of hydrogen and oxygen. They, in turn, release a large amount of heat that provides activation energy for molecules in adjacent areas, and so on, in a steadily widening chain reaction that rapidly goes to completion.

Most strongly exergonic reactions proceed to completion very quickly—once they are started. Other reactions occur more slowly. The rates of these reactions are greatly influenced by temperature and by concentration of the reactants.

TEMPERATURE

Any change in temperature, if other conditions do not also change, will have an effect on the overall kinetic energy of the molecules in the reacting system . Not all molecules in a system at a given temperature have the same kinetic energy. Some have an amount above the average for the system; others have an amount below the average, and a few have exactly the average amount. Some molecules therefore might collide with each other with sufficient force to react, but the effect of their output of energy on nearby molecules might not be enough to cause the rapid chain of events that brings about the fast completion of the type of reaction that occurs between hydrogen and oxygen. Raising the temperature of the reaction system, however, will increase the average kinetic energy of the molecules and therefore the number that will have sufficient energy to overcome the energy barrier. Finally, the increased frequency of collisions of all kinds increases the number of collisions that take place in the right orientation. Although this is a haphazard effect at best, for every increase in temperature of 10°C the rate of most chemical reactions is doubled.

CONCENTRATION

Increasing the concentration of the reactants will also increase the rate of most reactions. Up to a point at least, the more concentrated reactants are in a solution, the more likely their molecules are to collide. In a strongly exergonic reaction two reactants, A and B, will react to form C and D until very little A and B is left. Many reactions reach a point, however, at which there is little free energy difference between the reactants and products. As this point is reached, the concentration of A and B is decreasing (but at a slower rate because there are fewer molecules that can collide) and the concentration of C and D is increasing (but also at a slower rate than when the reaction began). At equilibrium, A and B are colliding to form C and D at the same rate that C and D are colliding to form A and B. The concentration of all the reactants is constant at equilibrium; the forward and reverse reactions occur at equal rates. (See Focus on The Law of Mass Action.)

CATALYSTS: INTRODUCING ENZYMES

We have seen that heat can be used to increase the rate of chemical reactions. However, recall that cells cannot store heat as their means of storing energy. Additionally, the sudden release of heat that accompanies strongly exergonic reactions would damage cells, tissues, and complex organic compounds. While a science fiction robot might perhaps operate on exploding hydrogen, in living things the energy of chemical reactions is released much more slowly and smoothly. How is hydrogen induced to interact chemically in such a way that the energy released can be transformed into biologically usable form?

FOCUS ON

The Law of Mass Action

The **law of mass action** states that, when all other conditions are constant, the rate of the reaction is proportional to the concentrations of the reactants. Consider a bottle of club soda. It is essentially a solution of carbon dioxide dissolved in water under high pressure (for which you can easily pay more than a dollar). Yet that bottle of club soda is not really just dissolved carbon dioxide; it is a solution of carbonic acid. To make it, carbon dioxide had to react with water in this fashion:

$$H_2O + CO_2 \longrightarrow H_2CO_3$$

Yet as everyone knows, if the cap is left off, the opposite occurs:

$$H_2CO_3 \longrightarrow H_2O + CO_2$$

Bubbles of gas fizz to the surface; if the bottle is left open long enough, eventually no CO_2 is left. This is clearly an example of a reversible reaction. Notice that its direction depends upon the pressure of the carbon dioxide gas. However, this is just another way of saying that its direction depends upon the concentration of carbon dioxide gas.

When the club soda is manufactured, it is treated with high-pressure carbon dioxide. This causes large amounts of carbon dioxide to dissolve in water, producing a high concentration of the dissolved gas. The only way this concentration can be reduced is for it to react with the water to become carbonic acid. Yet this in turn produces increasing concentrations of carbonic acid, which increases the likelihood that some of those carbonic acid molecules will break down. Ultimately an equilibrium is reached, whereby just as many carbonic acid molecules break down as those forming. At that point the solution will contain all three molecular species: carbon dioxide, water, and carbonic acid at varying concentrations. So it will continue as long as the bottle remains unopened.

As soon as the cap is removed, carbon dioxide rushes out, and further amounts are lost by diffusion into the air, which has a very low carbon dioxide content. This reduces the concentration of carbon dioxide in the solution, rendering it less probable that molecules of carbon dioxide will collide with water molecules to make carbonic acid. In fact, the opposite occurs: The decrease in carbon dioxide means that carbonic acid already present will be converted to carbon dioxide and water, with the carbon dioxide steadily leaving the system. When equilibrium is reached, the concentrations of both carbon dioxide and carbonic acid will be considerably less than they were in the unopened bottle.

What this example illustrates is that when there is little free energy difference on the two sides of a chemical equation, the direction of a reaction will be governed mostly by the *concentrations* attained by products and reactants, with the reaction tending to proceed so as to minimize the difference in concentration between the chemical species on the two sides of the equation. If one of them is continually removed, for instance, by gaseous diffusion, evaporation, or precipitation, then even despite minimal energy differences, the reaction will nevertheless proceed to completion.

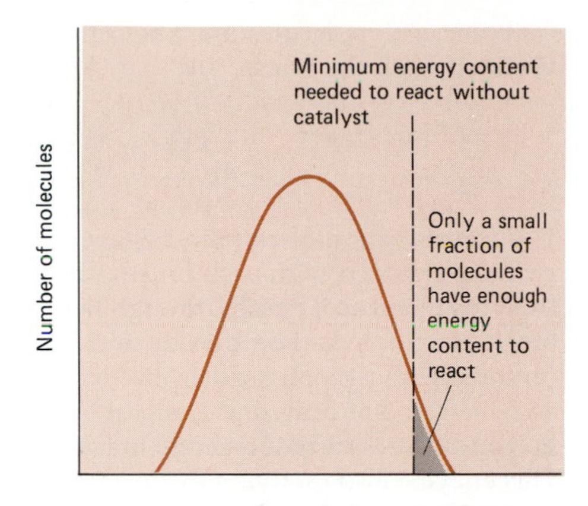

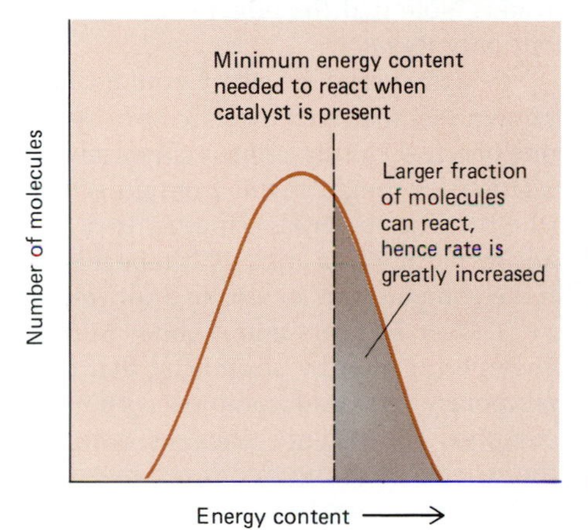

Figure 7–6 A catalyst such as an enzyme lowers the activation energy of a reaction and increases the fraction of the population of molecules with an energy content great enough to react and form the product. The catalyst does this by forming an unstable intermediate complex with the substrate, which then dissociates to form the product, freeing the catalyst to react with another molecule of substrate.

We saw earlier how adding a spark to a mixture of hydrogen and oxygen would yield a large amount of energy. How might this same reaction be made to occur at a slower rate with no explosion? One method of inducing hydrogen to react with oxygen is by the addition of platinum. Although this rare metal is very expensive, so little is needed that this would be an economical way of controlling the reaction rate. Tiny amounts of platinum dust added to the mixture of hydrogen and oxygen causes them to combine; although the same amount of heat is liberated, it is released more gradually so that no explosion takes place. When the reaction is completed, the platinum can be recovered unchanged and used again. The platinum serves as a **catalyst,** a substance that increases the rate of a chemical reaction without itself being permanently changed. In this reaction platinum attracts the molecules of hydrogen and oxygen to its surface, where they form a temporary complex with the platinum. The molecules are brought into close proximity to one another and can react with one another at a very modest temperature. As the catalyst, platinum greatly reduces the activation energy necessary to initiate the reaction (Fig. 7–6).

The catalysts typical of living organisms are proteins known as **enzymes.** As the properties of enzymes that make them such excellent catalysts have become more thoroughly understood, the design of industrial and laboratory catalysts with these same features has become possible.

How Enzymes Work

An enzyme can promote only a chemical reaction that would be able to proceed in its absence. There is nothing in the action of any catalyst that can change the operation of the second law of thermodynamics, so enzymes do not influence the direction of a chemical reaction or the final concentrations of the molecular species involved—but they surely can speed up the reaction. For example, consider the role of the enzyme carbonic anhydrase in the excretion of carbon dioxide from the body.

Carbonic acid molecules must be converted to water and carbon dioxide, which is liberated from the blood.

$$\underset{\text{Carbonic acid}}{H_2CO_3} \rightleftharpoons \underset{\text{Water}}{H_2O} + \underset{\text{Carbon dioxide}}{CO_2}$$

The concentration of carbon dioxide in the air in the lungs is low, and the air is partially changed with each breath; thus, carbon dioxide is continuously removed from the blood as it passes through the lungs. However, the carbonic acid does not break down into carbon dioxide and water rapidly enough to prevent a toxic concentration of carbonic acid in the blood. Why, then, are we still breathing? The enzyme carbonic anhydrase, present in tissues and blood cells, catalyzes the reaction and causes carbonic acid to break down fast enough to be effectively excreted. This enzyme also catalyzes the formation of carbonic acid from carbon dioxide and water. Note that the enzyme influences not the direction of the chemical reaction but only its *rate*.

Many years ago, the German chemist Emil Fischer pointed out the now well-known fact that enzymes are very **specific.** That is, an enzyme will catalyze only one or a few closely-related chemical reactions. Fischer suggested that this specific relationship might result from an actual "fit" between the enzyme and **substrate** (the reacting materials), that is, that the substrate fits the enzyme molecule like a key in a lock. (This analogy can be misleading, though; a key operates on a lock but if we compare the lock to the enzyme, in this case the lock operates on the key.)

Over 50 years ago, Leonor Michaelis elaborated the lock-and-key model of enzyme function by supposing that the reacting materials, the substrates, form a temporary chemical complex with the enzyme molecule, an **enzyme–substrate complex.** The enzyme–substrate complex then breaks down to yield the *products* of the reaction and the free, unchanged enzyme (Fig. 7–7). Michaelis calculated the reaction rate expected for his proposed model; he then carried out experiments measuring the reaction rate in terms of the amount of product generated at various concentrations of enzyme and substrate. The experimental reaction rates observed verified his prediction.

More direct evidence of the existence of enzyme–substrate complexes was obtained in separate experiments by David Keilin of Cambridge University and by Britton Chance of the University of Pennsylvania. Chance took a brown-colored peroxidase enzyme from horseradishes that can catalyze the breakdown of hydrogen peroxide to water and oxygen. When the brown enzyme was mixed with hydrogen peroxide, a green-colored enzyme–substrate complex was formed. This changed to a second, pale red enzyme–substrate complex, which finally split to give the original brown enzyme plus water and oxygen, the breakdown products of hydrogen peroxide. Only when oxygen is actually given off does the enzyme regain its original brown color. Since the color of a substance is strongly dependent on its chemical makeup, this "traffic light" behavior lent credence to the idea that an enzyme–substrate complex formed just before the products were released.

We now know beyond reasonable doubt that enzyme–substrate complexes do indeed form, and we even know how. The enzyme–substrate complex is a temporary chemical compound, joined by weak bonds (e.g., hydrogen bonds) between the enzyme and substrate. These enzyme–substrate bonds will form only at a specific place on the enzyme, called the **active site,** where the substrate can attach.

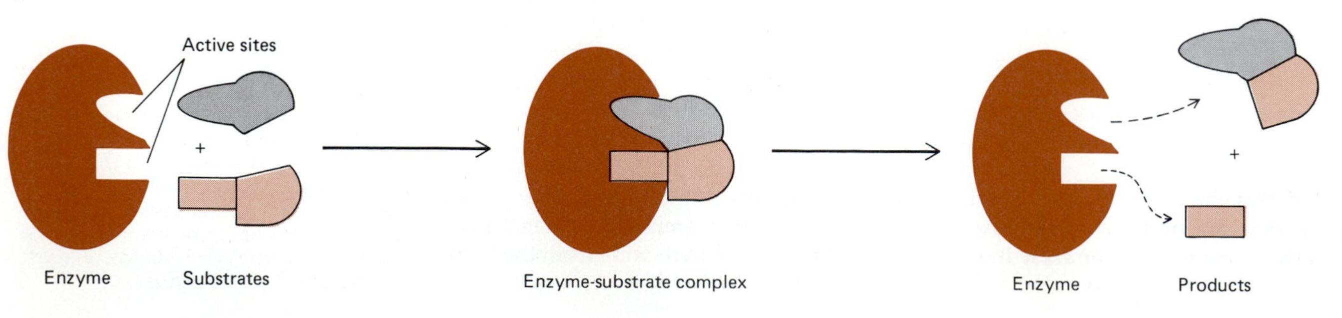

Figure 7–7 Lock-and-key mechanism of enzyme action. The substrates fit the active sites much as keys fit locks. Note that when the products separate from the enzyme, it is free to catalyze production of additional products; that the enzyme is not changed by the reaction; and that the active site occupies only a small fraction of the surface of the enzyme.

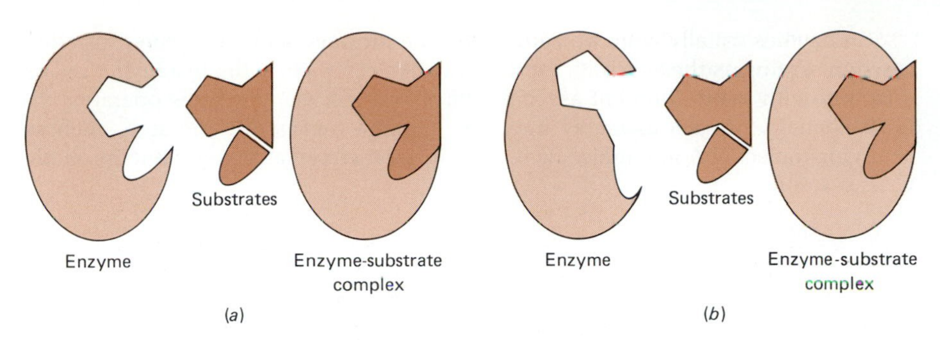

Figure 7–8 Comparison of models of the enzyme–substrate complex. (*a*) The lock-and-key model. (*b*) The induced-fit model. According to the induced-fit hypothesis, some change takes place in the shape of the enzyme as the substrate binds with the active site.

If there are two substrates, then two adjacent active sites will bring them together and hold them in the proper orientation for the reaction to occur. Some biochemists also suspect that the shape of the active site and the particular arrangement of its points of attachment produce strain in the critical bonds already existing within the substrate molecule. The active sites of enzymes are apparently not as rigid as was once thought. When the substrate binds to the enzyme it may induce a change in shape in the enzyme molecules (Fig. 7–8). This change is referred to as the **induced fit** of the enzyme to the substrate. The change in shape of the enzyme molecule can put strain on the substrate. This stress may cause bonds to break, thus promoting the reaction.

COFACTORS

Some enzymes (such as pepsin, secreted by the stomach) consist solely of protein. Other enzymes consist of two parts, a protein component referred to as the **apoenzyme** and an additional chemical component called a **cofactor.** Of the enzymes involved in the breakdown of glucose, several require a magnesium ion, Mg^{2+}, for activity; amylase, secreted by the salivary glands, requires chloride ions, Cl^-. Many of the biochemically necessary trace elements—those required by plants and animals in very small amounts—function as cofactors, usually as an integral part of the enzyme molecule. In fact, the metal is often a functional part of the active site itself.

In enzymes that have a *metal* cofactor, called **metalloenzymes,** catalysis is really a property of the metal but it is greatly enhanced by the attached apoenzyme. Recall that ordinary catalysts seem to operate mainly by bringing reactants together. Enzymes not only perform this action but also precisely orient the substrate molecules and stress their internal bonds. Together, the cofactor and apoenzyme perform very effectively. For example, hydrogen peroxide can be split by inorganic catalysts such as free iron atoms. However, iron accomplishes this only at a very slow rate. It would take *300 years* for an iron atom to split the same number of molecules of H_2O_2 cleaved in *one second* by a molecule of the enzyme catalase, which contains a single iron atom!

Organic cofactors are called **coenzymes.** A coenzyme contains some simple organic compound as part of the molecule. Some of the vitamins are used to make various coenzymes. Some coenzymes are an integral part of the active site; in such systems the coenzyme is probably mainly responsible for the catalytic properties of the enzyme. Many of the electron transport molecules that we will study in connection with cellular respiration (Chapter 9) are really coenzymes that could not operate without their associated enzymes.

In other reactions, the coenzyme functions more like an accessory substrate. Such a coenzyme is loosely bound to the active site and probably serves as a mediator between substrates in the following way: One substrate combines with the enzyme and coenzyme to form a complex. This complex then combines with the second substrate, to yield the products plus the enzyme and coenzyme (Fig. 7–9).

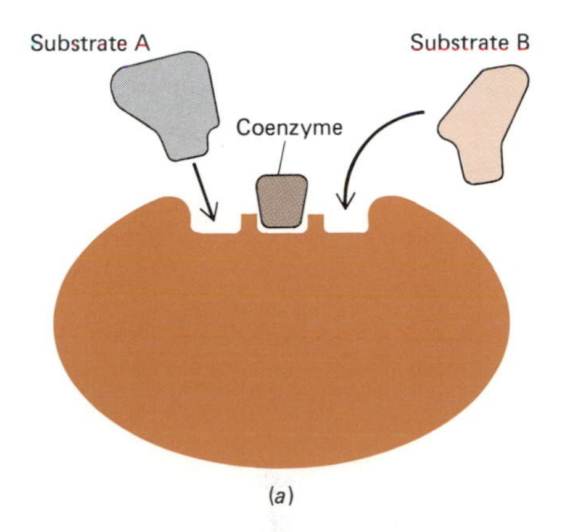

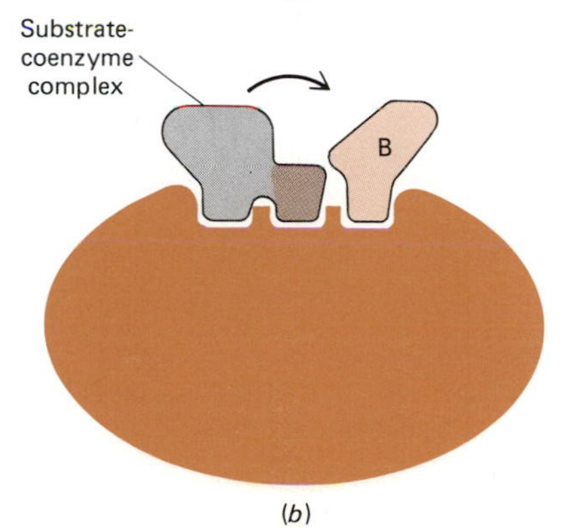

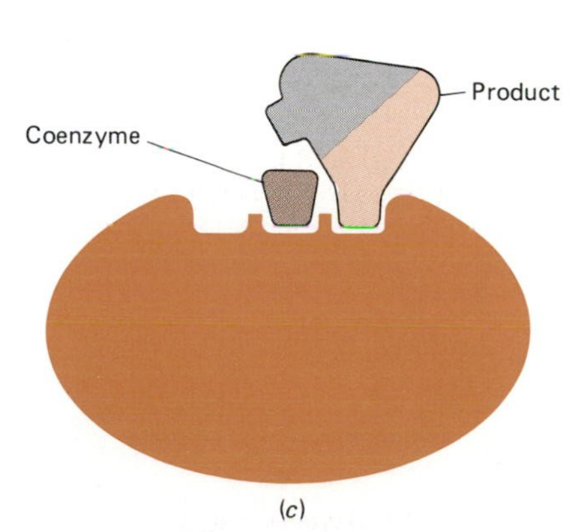

Figure 7–9 Some coenzymes are integral parts of the active site of the enzyme. Others, like the one shown here, are loosely bound to the active site. One substrate combines first with the coenzyme to form a coenzyme–substrate complex. Then the coenzyme–substrate complex combines with the second substrate. This complex yields the products and releases the coenzyme.

THE ROLE OF ENZYMES IN LIVING CELLS

Enzymes determine the order of chemical reactions within the cell and ensure that chemical reactions take place in the appropriate cellular locations and compartments. If a particular enzyme is located in the inner membrane of a mitochondrion, for instance, the reaction it catalyzes can take place *there* and nowhere else.

Enzymes usually work in teams, with the product of one enzyme-controlled reaction serving as the substrate for the next. We can picture the inside of a cell as a factory with many different assembly lines and disassembly lines operating simultaneously. In each assembly line, one enzyme carries out one step, such as changing molecule A into molecule B, and the next enzyme converts B to molecule C, and so on.

$$A \xrightarrow{\text{Enzyme 1}} B \xrightarrow{\text{Enzyme 2}} C$$

Germinating barley seeds contain two enzymes that will convert starch to glucose: amylase, which hydrolyzes starch to maltose, and maltase, which splits maltose to glucose. Altogether, 11 different enzymes working consecutively are required to convert glucose to lactate. The same series of 11 enzymes is found in human cells, in green leaves, and in bacteria.

Enzymes determine biochemical pathways. Consider the following branched pathway:

$$A \xrightarrow{\text{ABase}} B \xrightarrow{\text{BCase}} C \quad \begin{matrix} \xrightarrow{\text{CDase}} D \xrightarrow{\text{DEase}} E \\ \xrightarrow{\text{CFase}} F \xrightarrow{\text{FGase}} G \end{matrix}$$

Enzymes are used to regulate and to route these reactions. If E is needed some of the time and G at other times, then when E is needed the cell can produce CDase, and likewise, when G is needed the cell can produce CFase. Within some biochemical systems, one of the possible products of a reaction is never needed; in such cases, enzymes that can catalyze reactions leading to it need not be produced at all.

As you progress through your study of biology you will increasingly appreciate just how complex the biochemistry of any cell really is. If the various pathways could not be precisely controlled both for direction and for rate, nothing but hopeless chaos would result. This is the real significance of enzymes—their role in biochemical control.

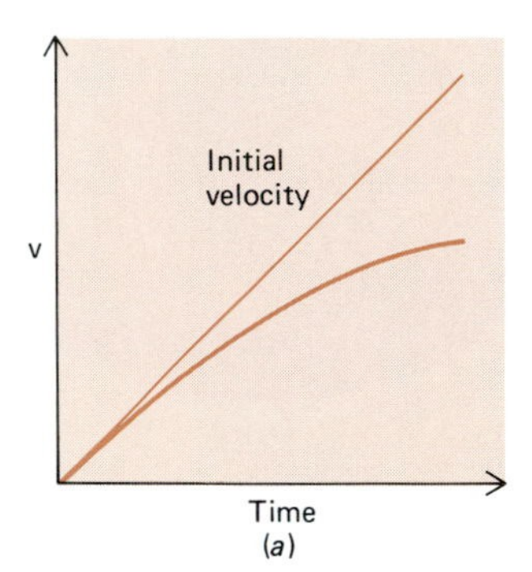

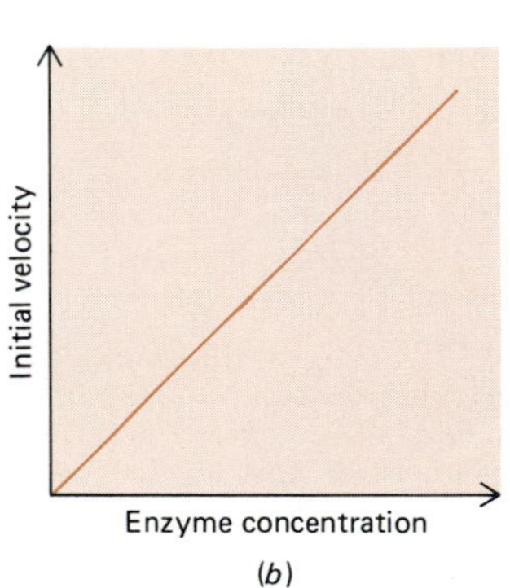

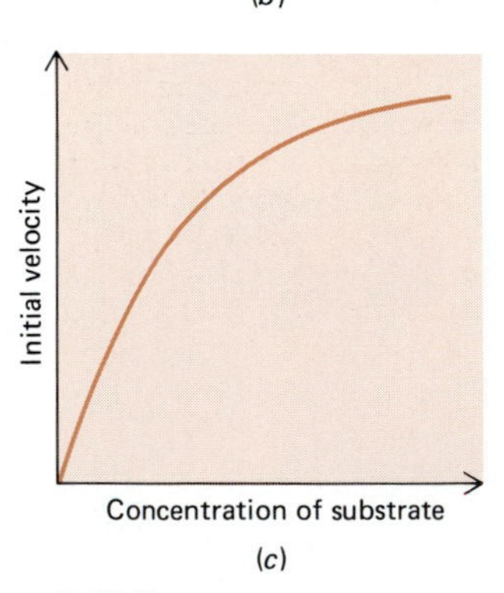

Figure 7–10 Enzyme reaction rate is influenced by several factors. (*a*) Reaction rate as a function of time. (*b*) Reaction rate as a function of the amount of enzyme added. Enough substrate and cofactors are added so that these do not limit the rate. (*c*) Reaction rate as a function of the amount of substrate. Enough enzyme and cofactors are added so that these do not limit the rate of the reaction.

Factors Affecting Enzyme Activity

The amount of enzyme and substrate present affects the rate of a reaction. In addition, the activity and specificity of an enzyme depend upon its molecular conformation, that is, upon its tertiary and quaternary structure. Hence enzyme activity is reduced or abolished by factors that change the molecular conformation: the presence of inhibitors; heat; and changes in pH.

CONCENTRATIONS OF ENZYME AND SUBSTRATE

If the pH and the temperature of an enzyme system are kept constant and if an excess of substrate is present, the rate of the reaction is directly proportional to the concentration of enzyme present (Fig. 7–10). Hence the *concentration* of an enzyme in a tissue extract can be determined by measuring the *rate* of the enzyme reaction. If the pH, temperature, and enzyme concentration of the system are kept constant, the initial rate of reaction is proportional to the concentration of substrate present, up to a limiting value. (If the enzyme system requires a cofactor, the concentration of *this* substance may, under certain circumstances, determine the overall rate of the reaction.)

Since enzymes may catalyze the attaining of equilibrium from either direction, what is usually considered to be the *product* of the reaction may also be a substrate for the reverse reaction. It, too, may bind to the enzyme at its active site. Thus, as a reaction proceeds and substrate is converted to product, the accumulating product may bind to the enzyme (thereby decreasing the number of enzyme molecules available to bind with substrate) and decrease the rate of the forward reaction.

The total number of enzyme molecules in a cell at any given moment is regulated by the rate at which the enzyme is produced on the ribosomes and by the rate that it is broken down within the cell. The process by which proteins are synthesized and its genetic control will be discussed in Chapter 14. Like all intracellular proteins, enzyme molecules are broken down by intracellular proteolytic enzymes.

When the rate of enzyme production exceeds the rate of enzyme breakdown, the amount of enzyme activity is increased. Cells can respond to a signal from the environment, such as a hormone, by increasing the rate of production or of breakdown of an enzyme.

ENZYME INHIBITION

Most enzymes can be **inhibited** (so that their activity is decreased) or even destroyed by certain chemical agents. Enzyme inhibition may be reversible or irreversible. **Reversible** inhibitors can be competitive or noncompetitive. In **competitive** inhibition, the inhibitor competes with the normal substrate for binding to the active site of the enzyme (Fig. 7–11). A competitive inhibitor usually is structurally similar to the normal substrate and so fits into the active site and combines with the enzyme; yet it is not similar enough to substitute fully for the normal substrate in the chemical reaction, and the enzyme cannot attack it to form reaction products. A competitive inhibitor occupies the active site only temporarily and does not permanently damage the enzyme. In fact, competitive inhibition can be reversed by increasing the substrate concentration.

In **noncompetitive** inhibition, the inhibitor binds with the enzyme at a site other than the active site. Such an inhibitor renders the enzyme inactive by altering its shape. Many important noncompetitive inhibitors are metabolic substances that regulate enzyme activity by combining reversibly with the enzyme.

Irreversible inhibitors combine with a functional group on an enzyme and permanently inactivate or even destroy the enzyme. Many poisons are irreversible inhibitors. Nerve gases, for example, poison the enzyme acetylcholinesterase, important in the function of nerves and muscles. Cytochrome oxidase, one of the enzymes of the electron transport system (part of cellular respiration), is especially sensitive to cyanide. Death results from cyanide poisoning because cytochrome oxidase is irreversibly inhibited and can no longer transfer electrons from substrate to oxygen.

A number of insecticides and drugs are irreversible enzyme inhibitors. Penicillin is a good example of such a drug. This antibiotic and its chemical relatives inhibit a bacterial enzyme, transpeptidase, which is responsible for establishing some of the chemical linkages in the material of which the bacterial cell wall is

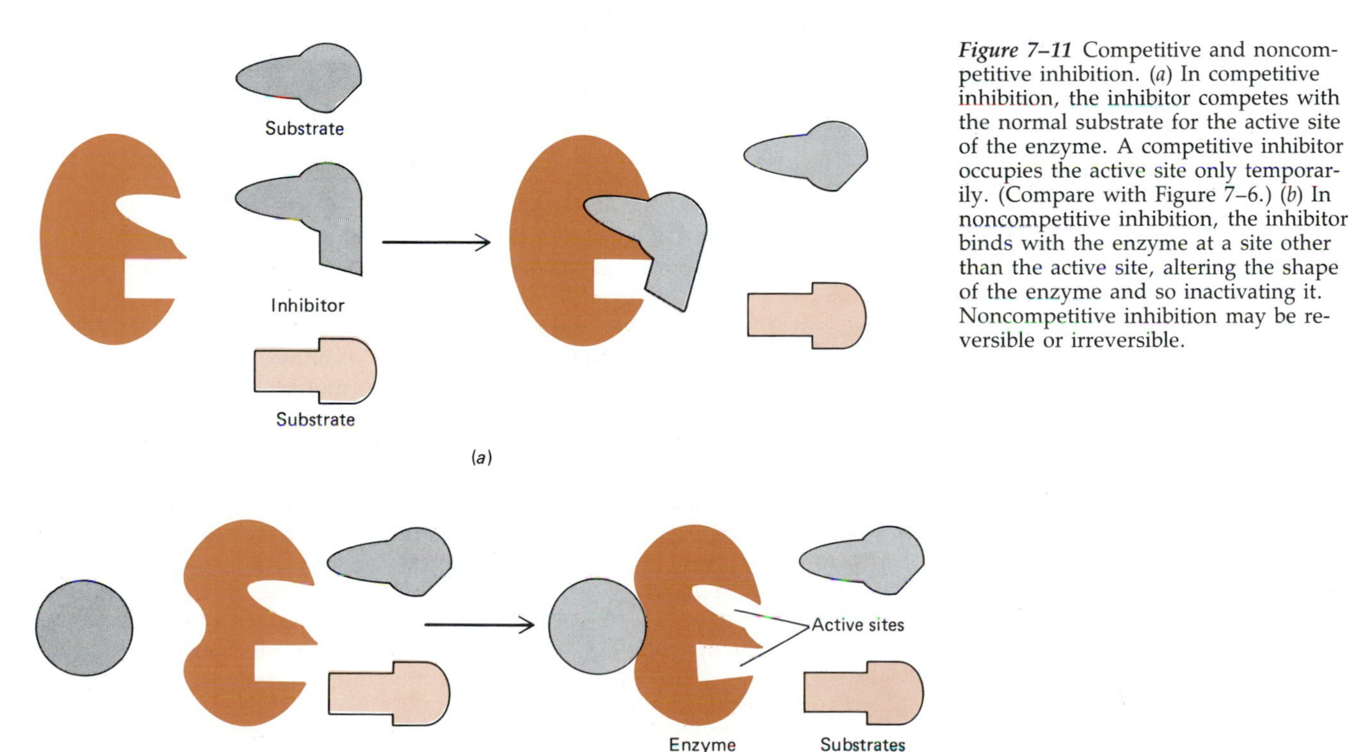

Figure 7–11 Competitive and noncompetitive inhibition. (*a*) In competitive inhibition, the inhibitor competes with the normal substrate for the active site of the enzyme. A competitive inhibitor occupies the active site only temporarily. (Compare with Figure 7–6.) (*b*) In noncompetitive inhibition, the inhibitor binds with the enzyme at a site other than the active site, altering the shape of the enzyme and so inactivating it. Noncompetitive inhibition may be reversible or irreversible.

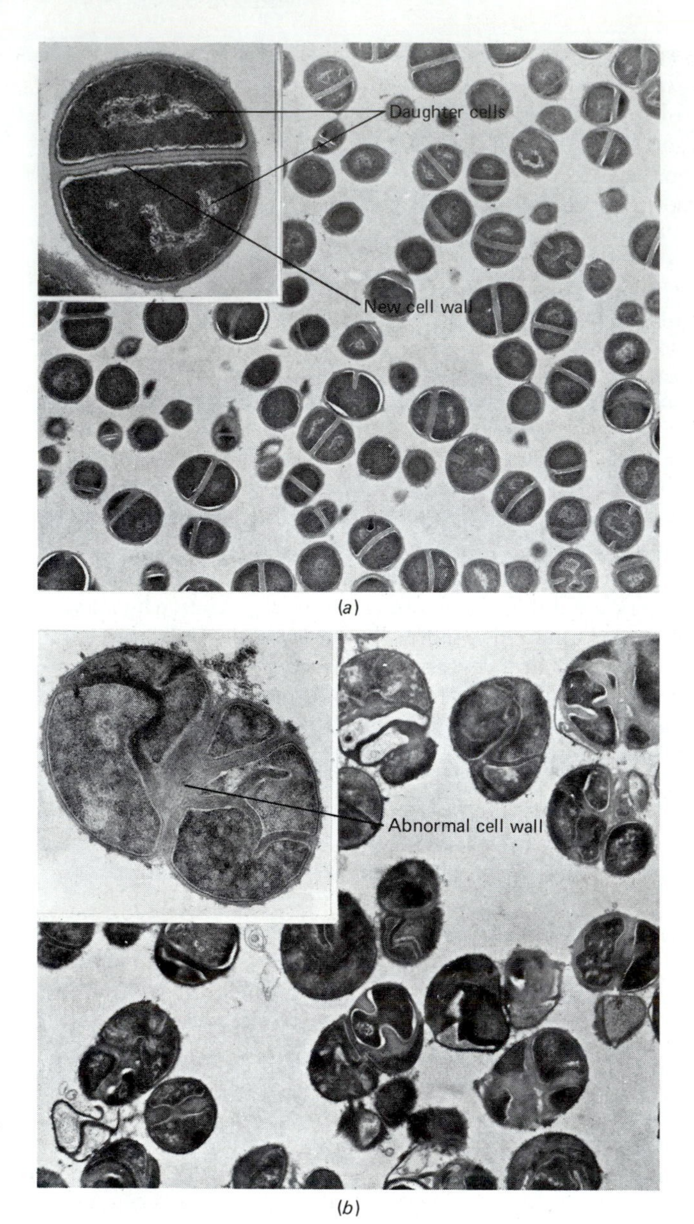

Figure 7–12 Antibiotic damage to bacterial cell walls. (*a*) Normal bacteria. Inset shows the new cell wall laid down between the daughter cells of a dividing bacterium. (*b*) Damaged bacteria. (Insets, approximately ×54,000.) (Courtesy of Drs. Victor Lorian and Barbara Atkinson, with permission of The American Journal of Clinical Pathology.)

composed. Unable to produce new cell walls, susceptible bacteria are prevented from multiplying effectively, as shown in Figure 7–12. Since human body cells do not possess cell walls and do not employ this enzyme, penicillin is harmless to humans, except for the occasional allergic patient.

Enzymes themselves can act as poisons if they get into the wrong compartment of the body. As little as 1 mg of crystalline trypsin injected intravenously will kill a rat. Several types of snake, bee, and scorpion venom are harmful because they contain enzymes that destroy blood cells or other tissues. The proteolytic enzymes of the pancreas, trypsin and chymotrypsin, are synthesized in the form of inactive enzyme **precursors,** molecules that are somewhat larger than the active enzyme. These are packaged in granules and secreted into the duct of the pancreas, thus protecting the pancreas itself from being digested by the enzymes it synthesizes. The enzymes are made active by other enzymes that cleave off a portion of the precursor molecule to yield the active enzyme. Acute pancreatitis, a serious, even lethal disease, occurs when the proteolytic enzymes become active while still within the pancreas and digest the cells of the pancreas and its blood vessels.

TEMPERATURE

Enzymes are inactivated by high temperatures. Enzymatic reactions occur slowly or not at all at low temperatures, but the catalytic activity reappears when the temperature is raised to normal. The rates of most enzyme-controlled reactions increase with increasing temperature but within limits. Temperatures greater than 50°C to 60°C rapidly inactivate most enzymes by denaturing the protein, altering its molecular conformation by causing the secondary and tertiary structure of the protein to unwind. This inactivation is usually not reversible; that is, activity is not regained when the enzyme is cooled. Most organisms are killed by even short exposure to high temperature; their enzymes are inactivated and they are unable to continue metabolism. There are a few remarkable exceptions to this rule: Certain species of bacteria can survive in the waters of hot springs, such as the ones in Yellowstone Park where the temperature is almost 100°C. These organisms are responsible for the brilliant colors in the terraces of the hot springs. Still other bacteria live at temperatures much above that of boiling water in undersea hot springs, where the extreme pressure keeps water liquified.

pH

The activity of an enzyme is markedly changed by alteration in the acidity or alkalinity of the reaction medium. Full enzyme activity requires a specific number of positive and negative charges on the enzyme. Changes in pH add or remove hydrogen ions from the protein, thereby changing the number of positive and negative charges on the protein molecule and affecting its activity. Pepsin, a protein-digesting enzyme secreted by cells lining the stomach, is remarkable in that it will work only in a very acid medium, optimally at pH 2. In contrast, the pH optimum of trypsin, the protein-splitting enzyme secreted by the pancreas, is 8.5, on the alkaline side of neutrality. Many of the enzymes that operate within cells become inactive when the medium is made very acid or very alkaline. Strong acids or bases irreversibly inactivate enzymes by permanently changing their molecular conformation. Most enzymes are active only over a narrow range of pH.

Energy Flow Through the Biosphere

The transfer of energy through the living world begins with the capture of the radiant energy of sunlight in the process of photosynthesis. Humans and other animals derive their energy from the foodstuffs that they eat. Our fruits and vegetables are obtained directly from plants, while our meat, fish, and shellfish are products of animals. These animals derive their energy supply from plants, algae, or other animals. Ultimately all of the food and energy of the animal world comes from the plants and algae.

Plants and algae require for growth only water, carbon dioxide, nutrient salts, and an abundant supply of radiant energy. This energy originates from the sun. At the very high temperatures (about 10 million degrees C) that occur in the interior of the sun, hydrogen atoms are transformed by thermonuclear reactions to helium atoms, with the release of energy in the form of gamma rays. The gamma rays react with electrons, and the energy is ultimately emitted as photons of light energy that pass out of the sun.

The total amount of energy arriving at the earth from the sun is about 13×10^{23} kcal per year. (The amount of solar energy reaching the earth each day is roughly equivalent to the energy of a million atom bombs of the size used in World War II.) About one third of this solar energy is reflected back into space from the surface of the earth as light. Much of the remaining two thirds is absorbed as heat by the ground, water, or vegetation. Some of this absorbed heat is utilized to evaporate the waters of seas and lakes, which condense as clouds that fall as rain.

Ecology is the science that deals directly with the transfer of energy among organisms and also the transfer of energy between the nonliving world and the living world. In the next two chapters, we will be studying only those energy transformations that take place within the living cell. However, the principles that apply to cellular energetics also apply to the flow of energy within an ecosystem. From the moment energy is liberated in the thermonuclear reactions that take place in the sun, energy becomes dispersed. Plants do not use all available sunlight for photosynthesis, and plants do not cover our entire planet. As a result, of the small fraction of solar energy that does reach the earth, less than 1% is transformed into biologically useful energy (Fig. 7–13). Additionally, when a plant is eaten by an

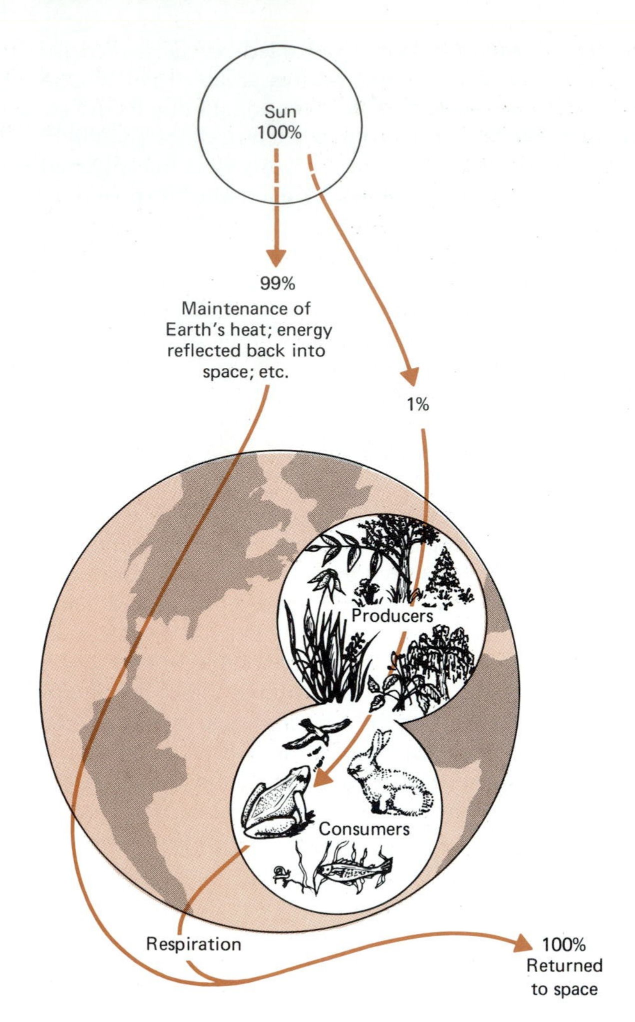

Figure 7–13 The flow of energy from the sun to the earth. Of the energy from the sun that passes in the earth's direction, 99% is reflected back into space or ends up heating the surface of the planet and the atmosphere. Photosynthetic organisms (plants, algae, and cyanobacteria) absorb approximately 1% of the energy that reaches the earth; even less than that is eventually used to form the complex organic molecules in which energy is stored for cellular respiration.

animal, or when bacteria decompose it, some of the energy is lost as heat, and the entropy of the system increases. When the animal in turn is eaten by another animal, a further decrease in free energy occurs as the second animal oxidizes the organic substances of the first animal, liberates energy, and synthesizes its own cellular constituents. Eventually all the radiant energy originally trapped by plants in photosynthesis is converted to heat, which is dissipated to the environment. The loss of all this energy might seem terribly inefficient; yet as long as there is enough hydrogen in the sun to guarantee a continuous supply of sunlight reaching the earth, there will be an adequate input of usable energy to sustain life on our planet.

SUMMARY

I. Life is characterized by the flow of energy from cell to cell and from organism to organism.
- *A.* Energy is the ability to produce a change in the state or motion of matter.
- *B.* Living things carry on three types of energy transformations.
  1. In photosynthesis algae and plant cells capture the radiant energy of sunlight and use it to synthesize organic compounds.
  2. Cells of living organisms metabolize these organic compounds by the reactions of cellular respiration, producing energy-rich phosphate bonds.
  3. The chemical energy of the energy-rich phosphate bonds is used by cells to do work such as synthesizing new molecules, contracting muscles, and conducting nerve impulses.

II. The study of energy and its transformations is thermodynamics.
- *A.* Energy can be changed from one form to another but can be neither created nor destroyed (first law of thermodynamics).
- *B.* Physical and chemical processes proceed in such a way that the entropy—the randomness or disorder—of the universe becomes maximal (second law of thermodynamics).
- *C.* As entropy increases during an irreversible process, the amount of free energy decreases.
- *D.* Chemical reactions that release energy to the system are termed exergonic. Reactions that require energy to drive them are termed endergonic.

III. The term coupled reactions is applied to two reactions that must take place together so that one can furnish energy or one of the reactants needed by the other.

A. The changes in free energy of a series of reactions are additive.
B. Thus, a thermodynamically unfavorable reaction (with a positive ΔG) can be driven by a thermodynamically favorable one (with a negative ΔG).

IV. There are three requirements for a chemical reaction to proceed.
A. The reactant molecules must get close enough to one another for their electrons to interact.
B. The reactant molecules must approach one another in the proper orientation in order to interact.
C. The reactant molecules must leave their current associations, necessitating the breaking of chemical bonds.
D. These conditions are promoted by increasing the temperature, increasing the concentration of reactants, and by catalysts.

V. Enzymes are organic catalysts that regulate the speed and specificity (but not the direction) of chemical reactions in living cells.
A. Enzymes greatly reduce the activation energy necessary to initiate a chemical reaction.
B. An enzyme combines with its substrate to form an intermediate enzyme–substrate complex, which then breaks down, releasing the enzyme and the products. The part of the enzyme that combines with the substrate is the active site.
C. Some enzymes require a cofactor such as a metal ion to function effectively; organic cofactors are called coenzymes.
D. The rates of enzyme reactions are affected by temperature, pH, and the concentrations of enzyme, substrate, and cofactors.
E. Most enzymes can be inhibited by certain chemical substances. Reversible inhibition can be competitive or noncompetitive.

VI. The ultimate source of energy for the earth's organisms is sunlight.

POST-TEST

1. The ability to produce a change in the state or motion of matter is known as ________.
2. The energy of a particle in motion is termed ________ ________.
3. The three major types of energy transformations in living things are ________, ________, and ________.
4. ________ is the branch of physics that deals with energy and its transformations.
5. "Energy may be changed from one form to another but is neither created nor destroyed" is a statement of the ________ ________ ________ ________.
6. A process in which heat is delivered to the surroundings is ________.
7. A process in which heat is absorbed from the surroundings is ________.
8. In thermodynamics the term ________ is applied to a disordered state of the system.
9. "Physical and chemical processes proceed in such a way that the entropy of the system becomes maximal" is a statement of the ________ ________ ________ ________.
10. The ________ energy, ΔG, of a system is that part of the total energy of the system that is available to do work under conditions of constant ________ and ________.
11. A reaction can occur spontaneously only if ΔG is ________.
12. A reaction that releases energy to the system is ________.
13. To drive a reaction that requires an input of energy, some reaction that yields energy must be ________ to it.
14. The free energies, ΔGs, of the reactions are ________.
15. The energy that is required to initiate a reaction is called ________ ________.
16. A substance that affects the rate of a chemical reaction without affecting its equilibrium point is a ________.
17. ________ are protein catalysts produced by cells.
18. Enzymes and their substrates combine temporarily to form an ________–________ ________.
19. The small portion of an enzyme molecule that combines with the substrate is the ________ ________.
20. The rate of an enzyme reaction is regulated by ________ and by the amounts of ________, ________, and ________.

REVIEW QUESTIONS

1. Trace the sequence of energy transformations from sunlight to the heat released in muscle contraction.
2. Contrast potential and kinetic energy, and give examples of each.
3. Contrast exergonic and endergonic reactions.
4. Life is sometimes described as a constant struggle against the second law of thermodynamics. Explain why this is true. How do organisms succeed in this struggle?
5. Why are coupled reactions biologically important?
6. What is activation energy? What is the relationship of a catalyst to activation energy?
7. Give the function of each of the following:
   a. active site of an enzyme;
   b. coenzyme.
8. Describe three factors that influence enzymatic activity.
9. Contrast competitive and noncompetitive inhibition.
10. Discuss the flow of energy through the biosphere.

8
Photosynthesis

OUTLINE

I. The hydrogen economy of the biosphere
II. Light and atomic excitation
III. Introducing chlorophyll
IV. The action spectrum of photosynthesis
V. Photosynthetic membranes
VI. The chemistry of photosynthesis
  A. The Hill reaction
  B. The reactions of photosynthesis
    1. The light reactions
      a. Antenna molecules
      b. Photosystem I
      c. Photosystem II
      d. How the photosystems interact
      e. How proton gradients form ATP: chemiosmosis
      f. The products of the light reactions
      g. Cyclic photophosphorylation
    2. The light-independent reactions: carbon dioxide fixation
      a. The C_3 pathway (the Calvin cycle)
      b. Energy considerations
      c. Regulation of the C_3 pathway
    3. The C_4 pathway
VII. Photorespiration

LEARNING OBJECTIVES

After you have read this chapter you should be able to:

1. Describe the physical properties of light and the nature of photochemical processes.
2. Describe how the absorption of photons can activate a compound such as chlorophyll, which can then transfer the energy to drive a reaction in another system.
3. Distinguish between the light-dependent and light-independent reactions of photosynthesis, and summarize the events that occur in each phase.
4. Draw a diagram of the internal structure of a chloroplast, and discuss how this structure facilitates the process of photosynthesis.
5. Describe the nature of a photosystem and the functions of antenna pigments and the reaction center.
6. Discuss the properties, functions, and constituents of the P700 and P680 photosystems in the light reaction.
7. Distinguish between cyclic and noncyclic photophosphorylation.
8. Describe how a gradient of protons is established across the thylakoid membranes and how this gradient functions in the synthesis of ATP.
9. Describe the chemical reactions involved in the conversion of CO_2 to glucose in photosynthesis, and indicate how many molecules of ATP and NADPH are required for the process.
10. Discuss the nature of the C_4 pathway and how this increases the effectiveness of the C_3 pathway in photosynthesis in certain types of plants.
11. Outline the reactions involved in photorespiration.

Consumers must depend upon other living things for the source of their energy. Ultimately, the vast amount of this energy comes from plants and algae. These producers convert solar energy into chemical energy, by a process called **photosynthesis.** The chemical energy from photosynthesis is stored in their cells in the form of carbohydrates and other organic molecules and is the fuel for the chemical reactions that sustain almost all forms of life on our planet. Photosynthesis is the process where ecology and biochemistry meet.

The Hydrogen Economy of the Biosphere

When the world runs out of petroleum, our civilization can keep going a while longer on coal as a fuel, and perhaps in time we will be able to employ solar power or thermonuclear power as a permanent solution to our energy needs. However, almost all power options yield energy in the form of electricity, which is not well suited for portable applications. If really efficient storage batteries are not developed, it is hard to see how automobiles or airplanes could be electrically powered. It would be hard to find extension cords that are long enough!

The problems that face us in supplying energy to fuel our industrial world are not new to the biological world. Much energy, for instance, can be yielded by combining hydrogen and oxygen to form water:

$$2\,H_2 + O_2 \longrightarrow 2\,H_2O + \text{Energy}$$

Hydrogen, however, is extremely reactive; not much of it exists in our atmosphere in uncombined form. Therefore, we must first find a source of energy to dissociate water.

Theoretically, the output of a power plant could be used to dissociate water electrolytically into hydrogen and oxygen:

$$2\,H_2O + 4\,e^- \longrightarrow 2\,H_2 + O_2$$

This is a strongly endothermic reaction. The energy input would be stored, in effect, in the hydrogen and oxygen gases. If they were allowed to react together, for example, in an automobile engine, their energy would be released.

The practical difficulty in all of this is storage and transportation of the hydrogen fuel, which would require clumsy means such as heavy tanks or extreme cold. To solve this problem, the hydrogen could be combined with carbon to form artificial hydrocarbon fuels similar to gasoline or propane. Such further processing would involve some sacrifice in energy content, but since these fuels do have a very high hydrogen content, they would still serve very well, as indeed they do now. Even if we had no carbon in elemental form with which to combine the hydrogen, hydrogen-carrying fuels could still be produced, at some further energy sacrifice, from such common carbon sources as carbon dioxide. Modification of the process would permit the production of still other energy-rich substances, such as alcohol or glycerine, which are important in industry.

Although the controlled manufacture of hydrogen fuel has not yet become a reality in human society,[1] the biological world has long made use of this energy source. The vast chemical industry of the biosphere has for millennia captured solar energy by the photolysis of water, has released the resulting oxygen gas, and has incorporated the hydrogen into energy-rich carrier compounds containing carbon and oxygen as well as hydrogen. These compounds (carbohydrates, for the most part) form the matter of biological commerce, for they also can be incorporated into the body structure of organisms as structural components. The energy of these carbohydrates and their derivatives is released in living cells by a long sequence of

[1]There is a possibility that hydrogen gas could actually be produced by algae under special artificial conditions. Not all plants seem capable of producing hydrogen as such, but when certain kinds of green algae are maintained under anaerobic conditions (without oxygen), they respond by synthesizing the enzyme hydrogenase. This enzyme enables them to utilize the energy of sunlight to split water into molecular hydrogen and oxygen. The hydrogenase accepts electrons from excited chlorophyll and unites them with protons to form hydrogen gas, H_2. However, the enzyme hydrogenase is rapidly inhibited by the very molecular oxygen that is generated as a by-product of this reaction. Therefore, in order to maintain the necessary anaerobic conditions, the oxygen must be continuously flushed out by a stream of inert carrier gas. Unfortunately, the rate of hydrogen production achieved so far is much too low to be a practical way of making hydrogen gas in commercial quantities.

enzymatically catalyzed reactions that ultimately unite the protons and electrons of hydrogen with oxygen to form water, completing the cycle. The rest is detail, but that detail is important enough to form the subject of this as well as the following chapter.

Light and Atomic Excitation

We are not used to seeing, touching, or tasting such things as atoms, neutrons, and electrons, and the images of those things reported to us by scientific instruments may be difficult to visualize. One of the most difficult concepts to grasp is the nature of light. How can something have characteristics of both a particle and a wave? Yet it can easily (though indirectly) be demonstrated that light behaves (we do not say *is*!) as both particles and waves. Light is electromagnetic radiation in a specific band of wavelengths. (A wavelength is the distance from one wave peak to the next.) Within the spectrum of visible light, violet has the shortest wavelength, and red light, the longest wavelength. (Ultraviolet light has a shorter wavelength than visible light, and infrared light has a longer wavelength. However, we must leave it to insects to perceive the former and pit vipers to perceive the latter.)

Light is composed of particles of energy called **photons.** The energy of a photon is different for light of different wavelengths. The shorter the wavelength, the more energy light has, and the longer the wavelength, the lower the energy. In other words, the energy of the photon is inversely proportional to the wavelength.

Light can excite certain types of molecules—thereby moving electrons into higher energy levels. Photons interact with atoms in a variety of ways, but all of them depend on the electronic structure of the atom. Recall that an atom consists of a nucleus surrounded by electrons in one or more energy levels. In the hydrogen atom a single electron occupies the first energy level. When the electron is in this first (lowest) energy level, the hydrogen atom is in its **ground state.** However, energy can be added to the electron so that it will attain a higher energy level. An important concept to understand is that these energy levels are discrete (Fig. 8–1). If insufficient energy is supplied, the electron will not move to a higher energy level (recall the discussion of a quantum in Chapter 2). If an electron is raised to a higher energy level than its ground level, the atom is said to be *excited.* Such excitation can result from the absorption of any kind of energy—often electrical or chemical. In photosynthesis, of course, the energy causing the excitation comes from light.

If a photon is energetic enough to raise one of the electrons of an atom to a higher energy state, one of two things may happen, depending on the atom and its surroundings:

1. The electron may soon return to its ground level. Energy is usually dissipated as heat or as light of a longer wavelength.
2. The electron may be lost, leaving the atom with a net positive charge. The electron may be accepted by a reducing agent (see discussion of oxidation–reduction in Chapter 2), which leaves the formerly excited atom with a net positive charge.

Introducing Chlorophyll

Virtually all living things are ultimately dependent upon a chemical substance that is specifically adapted to undergo excitation by light—**chlorophyll.** Chlorophyll is what makes green plants green, but it is also present in all of the variously colored plants and algae that engage in photosynthesis. Chlorophyll becomes excited either by light or by energy passed to it from other substances that have become excited by light.

Several kinds of chlorophyll exist. The most important of these is probably **chlorophyll *a*,** which is illustrated along with the similar but not identical **chlorophyll *b*** in Figure 8–2. Notice the complex structure of the "head" of the molecule. This is a **porphyrin ring,** another version of which occurs in the iron-containing hemoglobin of blood. However, unlike hemoglobin, chlorophyll contains a central atom of magnesium. The long phytol "tail" of the molecule is really a hydrocarbon whose solubility in lipids determines the orientation of the chlorophyll molecules in the internal membranes of the chloroplast. The molecular shape of chlorophyll is

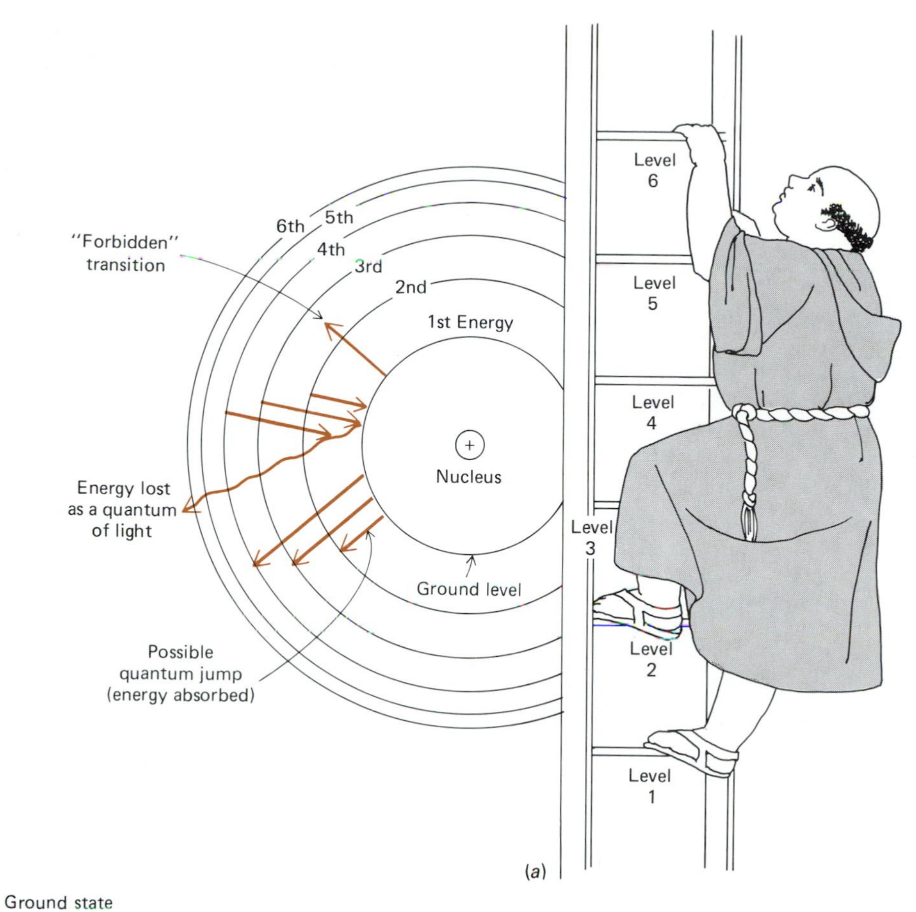

Figure 8–1 (*a*) Energy level transitions in a hydrogen atom. An electron is located in a discrete energy level and has a definite amount of energy. Note that if enough energy is absorbed, the electron can jump beyond the second energy level, but that electrons cannot occupy regions *between* energy levels. Sometimes the situation is compared to climbing a ladder: Going up or down requires stepping on a rung; it does no good to try to step in between them. (*b*) Model of a photochemical reaction. A quantum of light energy strikes an atom or the molecule of which the atom is a part. The energy of the photon may push the electron to an orbital farther from the nucleus. If the electron "falls" back to the next lower energy level, a less energetic photon is re-emitted. If the appropriate electron acceptors are available, the electron may leave the atom. In photosynthesis a chain of such acceptors bleeds the electron of the energy transferred from a photon and begins the light-dependent reactions.

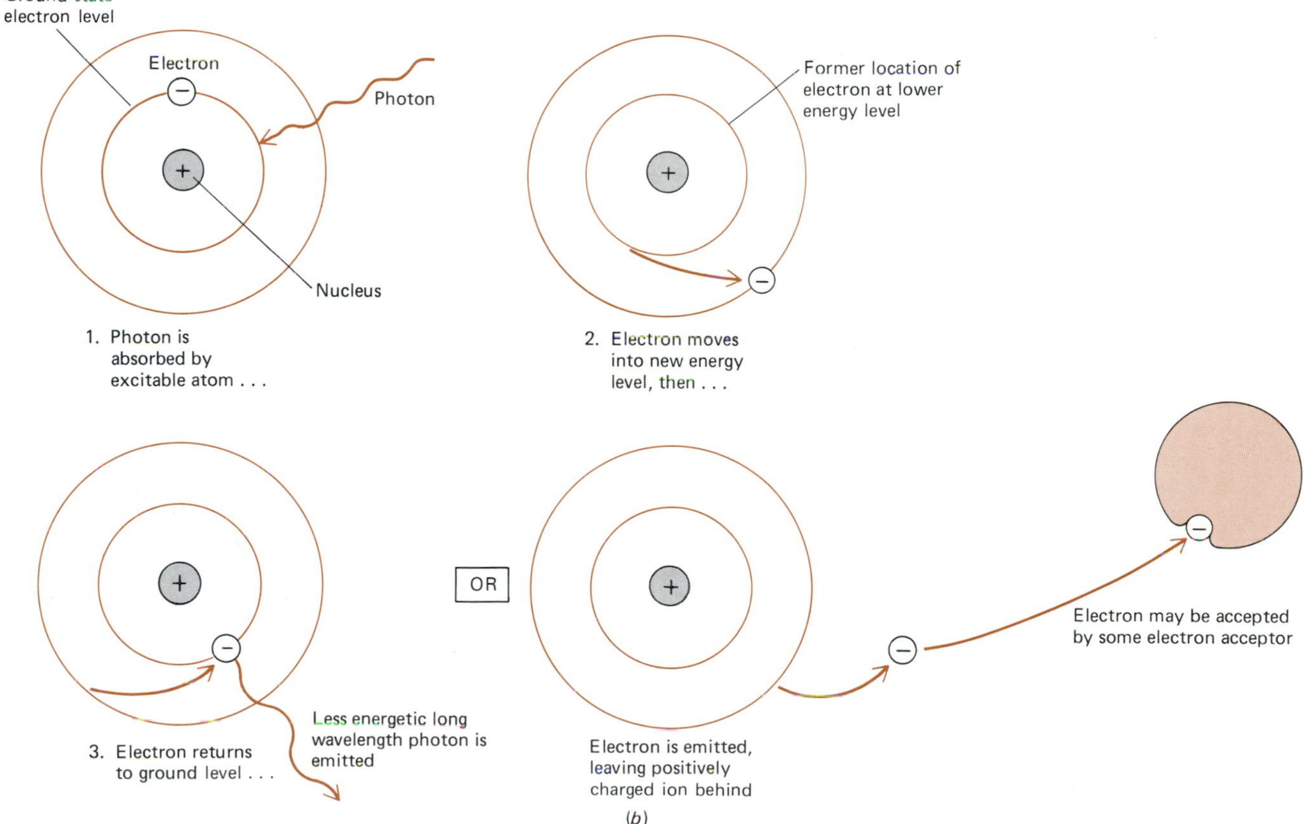

thought to make close-order packing of the molecules possible. This is important because, as we will see, chlorophyll molecules cooperate in the absorption and conversion of light.

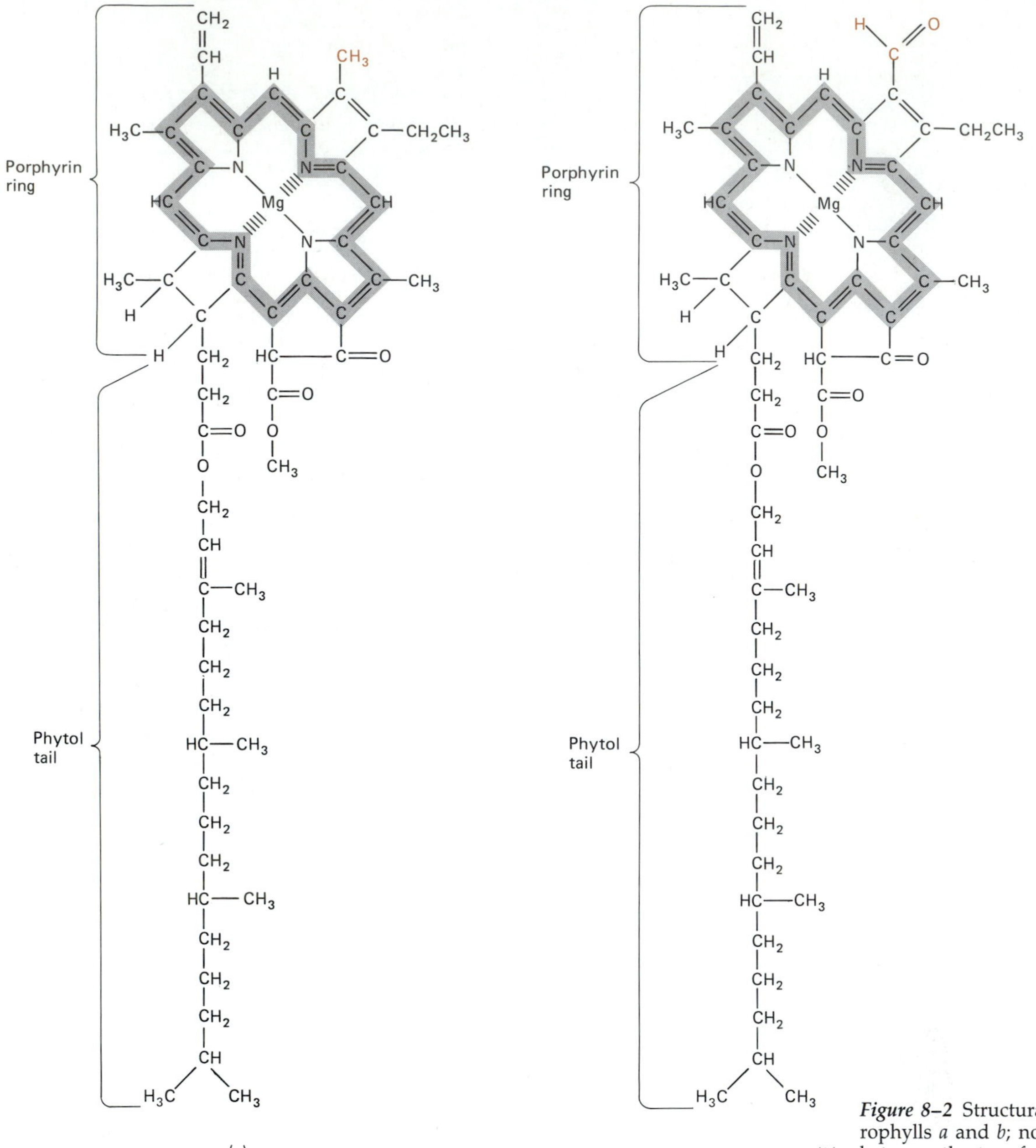

Porphyrin ring

Phytol tail

(a)

(b)

Figure 8–2 Structural formulas for chlorophylls *a* and *b*; note the differences between the two forms.

Photosynthesis depends entirely upon light in the visible part of the spectrum; however, some of these wavelengths are far more effective in exciting chlorophyll than are others. Plants are green because their leaves reflect most of the green light that falls upon them. If green light is reflected, then certainly most of it is not being absorbed or used. The light wavelengths that chlorophyll *does* absorb lie principally in the blue, violet, and red regions of the spectrum (Fig. 8–3). But this kind of **absorption spectrum** does not tell us exactly which wavelengths of light are actually most effective in photosynthesis.

The Action Spectrum of Photosynthesis

The relative effectiveness of different wavelengths of light in photosynthesis is given by the **action spectrum** of photosynthesis (Fig. 8–3). This action spectrum was determined a long time ago in one of the classic experiments in biology. In 1883 the German biologist T. W. Engelmann took advantage of the shape of the chloroplast in *Spirogyra,* a green alga. *Spirogyra* occurs as slimy strings in freshwater habitats, especially slow-moving or still waters (Fig. 8–4(*a*)). *Spirogyra* wins no beauty

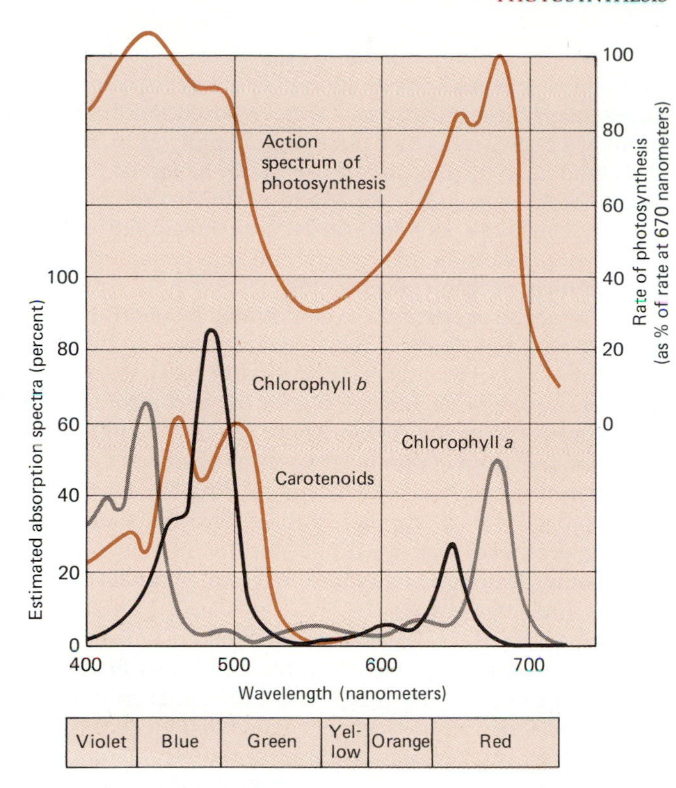

Figure 8–3 Action and absorption spectra of chlorophyll *a* and chlorophyll *b* and the pigment carotene. Note that the action spectrum (that is, the production of oxygen with light of different wavelengths) corresponds roughly to the combined absorption spectra of the chlorophylls and carotene, showing that even though carotene is not directly involved in photolysis, it enhances the ability of chlorophyll to become excited at regions of the spectrum that would otherwise lie outside its range of excitation.

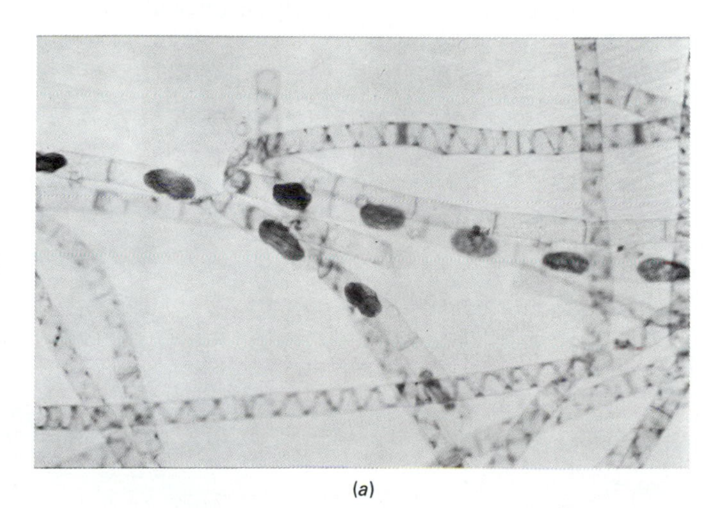

(*a*)

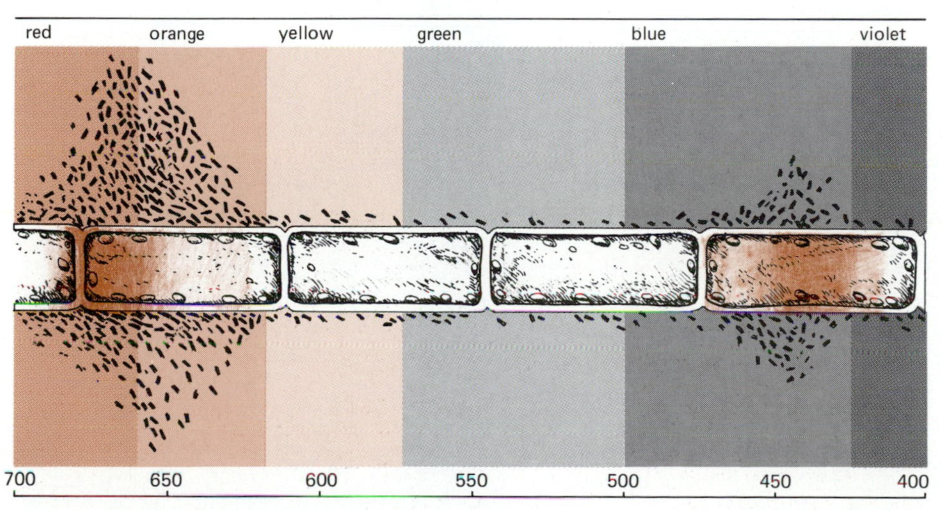

(*b*)

Figure 8–4 (*a*) Filaments of *Spirogyra*. (*b*) Engelmann's experiment to determine the wavelengths of light most effective for photosynthesis. Engelmann projected a spectrum onto a filament of the alga *Spirogyra*. He watched under a microscope as bacteria swam toward the algal cells emitting the most oxygen; the bacteria aggregated along the cells in the blue and red portions of the spectrum. ((*a*), E. R. Degginger.)

prize in bulk, but the individual cells are exquisitely beautiful, each containing a long, spiral, emerald-green chloroplast embedded in cytoplasm. Engelmann reasoned that if these chloroplasts were exposed to a spectrum produced by a prism, photosynthesis would take place most rapidly at the points where the chloroplast was illuminated by the colors most readily absorbed by chlorophyll—if chlorophyll were indeed responsible for photosynthesis. Yet how could photosynthesis be measured in those technologically unsophisticated days? Engelmann knew that photosynthesis produced oxygen, and that certain motile bacteria are attracted to areas of high oxygen concentration.

The action spectrum was determined by observing which strands of *Spirogyra* the bacteria swam toward: the strands located in the red and blue regions of the spectrum. The fact that the bacteria did not move toward such areas when *Spirogyra* was not present served as an experimental control showing that bacteria are not merely attracted to any region where red or blue light is being absorbed. Because the action spectrum of photosynthesis, as observed by Engelmann, closely matched the absorption spectrum of chlorophyll, Engelmann was also able to conclude that it was indeed the chlorophyll in the chloroplasts (and not another compound in another organelle) that is responsible for photosynthesis (Fig. 8–4). Numerous more sophisticated (but no more ingenious) studies have since amply confirmed Engelmann's conclusions.

However, the action spectrum of photosynthesis can be somewhat different from the action spectrum of pure chlorophyll, particularly in such strongly colored vegetation as red marine algae. The explanation is twofold: First, the chloroplasts contain accessory photosynthetic pigments (Fig. 8–5) such as **carotenoids** and **phycobilins** (accessory pigments found in red algae and cyanobacteria) in such large amounts that they mask the color of the chlorophyll, and for that very reason they absorb the green light that chlorophyll itself would reflect. Second, and more to the point, the accessory pigments transfer the energy of excitation that this green light would produce to chlorophyll molecules. The presence of such accessory pigments lets the alga utilize light in the green area of the spectrum more efficiently than

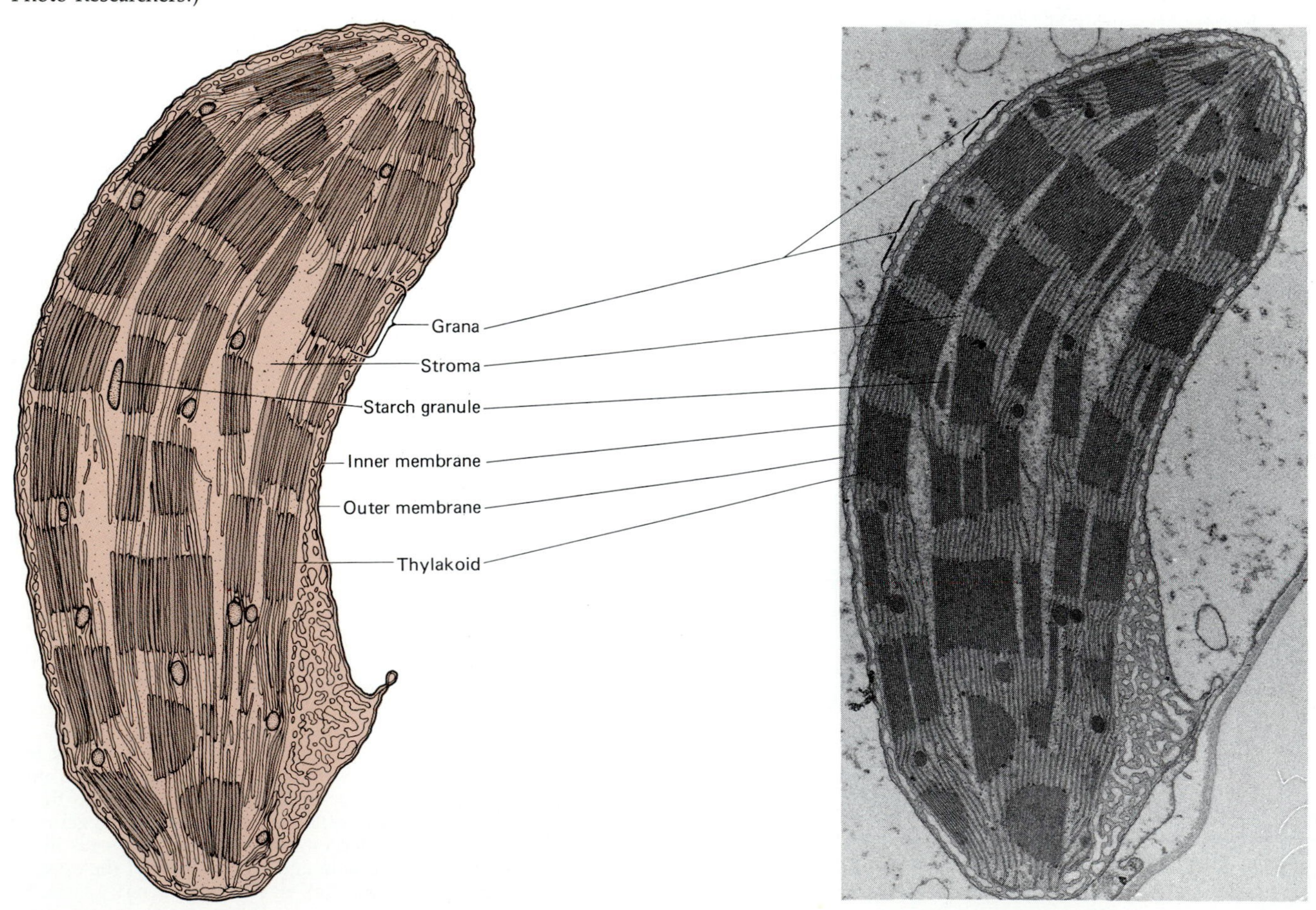

Figure 8–5 An electron micrograph and diagram of the chloroplast from a maize leaf cell. Note grana which consist of stacks of thylakoids. (D.K. Shumway, Photo Researchers.)

could, for example, a chrysanthemum plant. This is an important adaptation, for it permits these algae to live in deeper aquatic habitats where the red light most effective in photosynthesis has been filtered out by its passage through the water.

Terrestrial plants also contain some accessory photosynthetic pigments, obvious in something like a Japanese maple or a copper beech, but visibly present in most trees only when their leaves turn color in the fall. Toward the end of the growing season, the chlorophyll is metabolized and its magnesium is stored in the permanent tissues of the tree, leaving only the accessory pigments in the leaves. The presence of accessory photosynthetic pigments can be demonstrated by chemical analysis in almost any leaf, and it is well for us that leafy green vegetables contain carotenes, for some of these substances by a simple biochemical transformation become vitamin A in our bodies.

Photosynthetic Membranes

Chlorophyll and other photosynthetic pigments are found within the membranes of **thylakoids,** which are tiny, flattened sacs. All of the enzymes necessary for the light-dependent reactions are also associated with the thylakoid membranes. In the prokaryotes that photosynthesize, thylakoids often occur as extensions of the plasma membrane and may be arranged around the periphery of the cell. In photosynthetic eukaryotes, the thylakoids are located within the chloroplasts.

The thylakoid is the basic unit of photosynthesis. Yet the day is not far past when biologists tried—it now seems naively—to use solutions of chlorophyll to study photosynthesis in vitro. It didn't work, not even when all known accessory pigments and cofactors were added to the solution. It did not work for the same reason that a storage battery that has been run through a grinder will not work: The way the components are organized is just as important as their chemical properties.

Let us examine a **chloroplast** and its component thylakoids. Surrounded by a double membrane, the chloroplast has an interior packed with stacks of thylakoids (Fig. 8–6). These stacks are known as **grana** (singular, **granum**). The thylakoids look something like collapsed beach balls, all stacked into neat piles. Each thylakoid has a clearly distinguishable "inside" and "outside," though we must bear in mind that the outside of the thylakoid is still *inside* the chloroplast. Each thylakoid membrane is a phospholipid bilayer with embedded lumps of protein, very much like a plasma membrane.

Some thylakoid membranes extend from one granum to another. The fluid-filled region of the chloroplast outside the thylakoids between the grana is called the **stroma.** Most of the enzymes required for the light-independent reactions are found in the stroma.

The Chemistry of Photosynthesis

That photosynthesis produces oxygen has been known since the time of Joseph Priestley, who in the 18th century placed a mouse in a bell jar of carbon dioxide that had first been treated with a sprig of living mint for several days. The mouse was able to breathe the rejuvenated air and live. Early scientists thought that the overall reaction for photosynthesis was

$$6\ CO_2 + 6\ H_2O \longrightarrow C_6H_{12}O_6 + 6\ O_2$$

Some beginning chemistry students would infer from this that the oxygen was displaced from the carbon dioxide by the water. Many biologists originally believed the same thing, but an inconvenient fact suggested otherwise: A class of photosynthetic bacteria that usually employ hydrogen sulfide rather than water as a source of hydrogen produce elemental sulfur, but no oxygen, by the following reaction:

$$12\ H_2S + 6\ CO_2 \longrightarrow C_6H_{12}O_6 + 12\ S + 6\ H_2O$$

C. B. van Neil pointed out in the early 1930s that since these bacteria incorporate carbon dioxide in their organic compounds with no release of oxygen, the interpretation of plant photosynthesis might be incorrect. If the oxygen in plant photosynthesis did not come from the carbon dioxide it must have come from the water. This idea was confirmed when the heavy isotope of oxygen, ^{18}O, became available: In a classic experiment performed by Ruben and Kamen in 1941, water

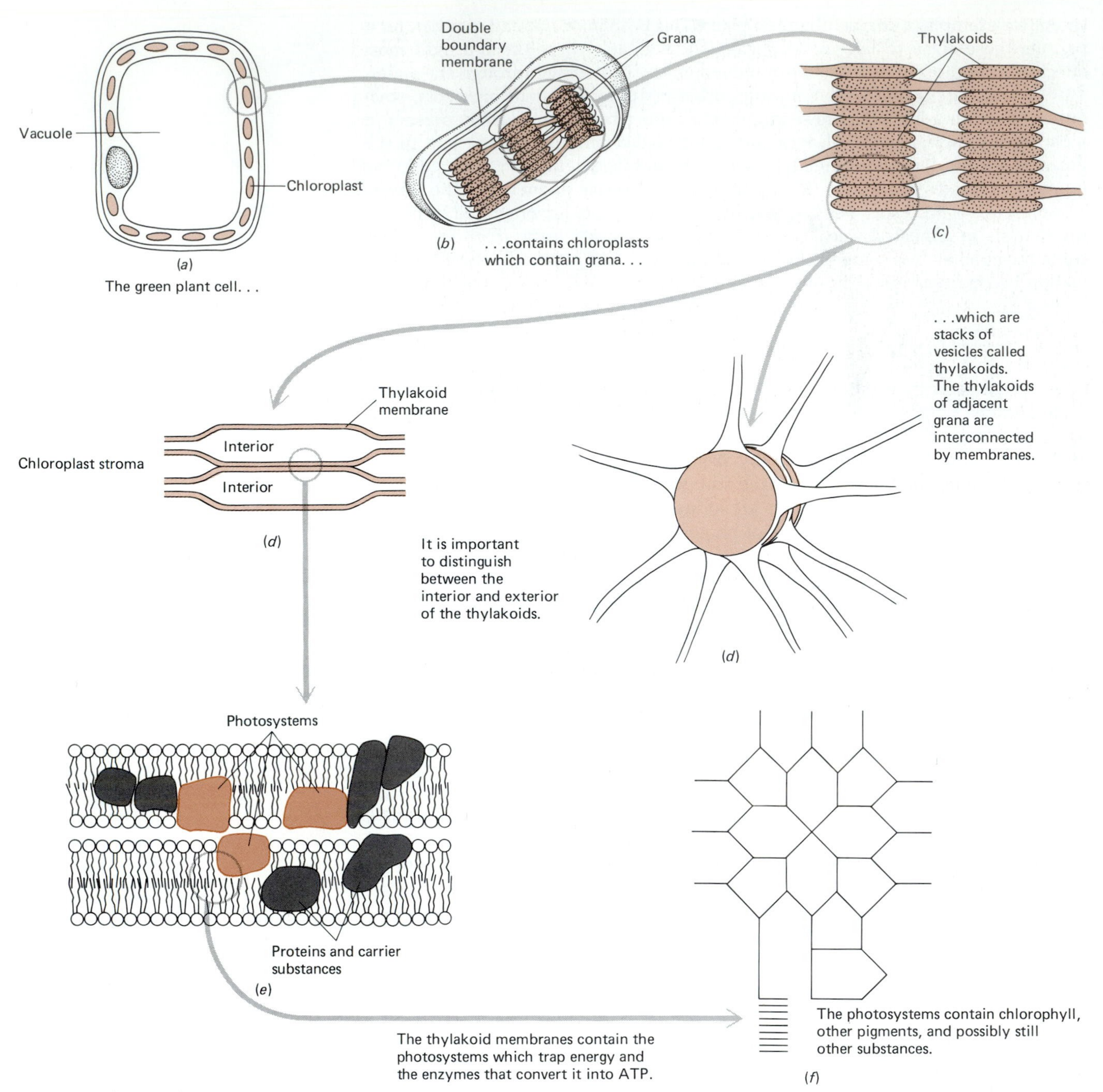
Vacuole
Chloroplast
(a)
The green plant cell. . .
Double boundary membrane
Grana
(b) . . .contains chloroplasts which contain grana. . .
Thylakoids
(c)
. . .which are stacks of vesicles called thylakoids. The thylakoids of adjacent grana are interconnected by membranes.
Thylakoid membrane
Chloroplast stroma
Interior
Interior
(d)
It is important to distinguish between the interior and exterior of the thylakoids.
(d)
Photosystems
Proteins and carrier substances
(e)
The thylakoid membranes contain the photosystems which trap energy and the enzymes that convert it into ATP.
The photosystems contain chlorophyll, other pigments, and possibly still other substances.
(f)

Figure 8–6 Levels of structure in the chloroplast.

was labeled with ^{18}O; the ^{18}O was liberated as $^{18}O_2$ during photosynthesis. When carbon dioxide was labeled with ^{18}O, the isotopic oxygen appeared in the carbohydrate produced, but not in the oxygen. However, this meant that more than 6 molecules of water must be involved in the overall reaction for photosynthesis. Based on these studies, the correct complete overall equation for plant photosynthesis is

$$6\ CO_2 + 12\ H_2O \longrightarrow C_6H_{12}O_6 + 6\ O_2 + 6\ H_2O$$

THE HILL REACTION

An even earlier indication that the oxygen evolved in photosynthesis comes from water molecules rather than carbon dioxide came from the classic experiments carried out by Robin Hill in 1937. Hill obtained oxygen when he exposed isolated chloroplasts to light in the presence of an electron acceptor, ferricyanide, but in the absence of CO_2 and $NADP^+$. This demonstrated that oxygen can be evolved with-

Figure 8–7 On sunny days the oxygen released by aquatic plants may sometimes be visible as bubbles in the water. This plant (*Elodea*) is actively carrying on photosynthesis, as evidenced by the oxygen bubbles. (E. R. Degginger.)

out the reduction of carbon dioxide if an electron acceptor such as ferricyanide is available. The absence of carbon dioxide in the mixture confirmed that the oxygen released comes from water molecules. In addition, the experiments demonstrated that the primary event in photosynthesis is the light-induced transfer of electrons from one compound to another. The electron donor is water in complex plants. For in vivo photosynthesis, the electron acceptor is **nicotinamide adenine dinucleotide phosphate, $NADP^+$.** The production of oxygen and hydrogen atoms during photosynthesis is known as the **Hill reaction** after its discoverer (Fig. 8–7).

THE REACTIONS OF PHOTOSYNTHESIS

As a result of the experiments of Melvin Calvin and many others, it has become clear that photosynthesis proceeds by a large number of chemical reactions that can be divided into the **light reactions,** or **light-dependent reactions,** and the **dark reactions,** or **light-independent reactions.** (The light-independent reactions are ultimately dependent upon light despite their name, because they are driven by the products of the light-dependent reactions; see Table 8–1 and Focus on Summary Equations for Photosynthesis.)

TABLE 8–1
Summary of Principal Reactions of Photosynthesis

Reaction Series	*Summary of Process*	*Needed Materials*	*End-products*
Light reactions (take place in thylakoid membranes)	Energy from sunlight used to split water, manufacture ATP, and reduce NADP		
Photochemical reactions	Chlorophyll energized; reaction center gives up energized electron to electron acceptor.	Light energy; pigments such as chlorophyll	Electrons
Electron transport	Electrons are transported along a chain of electron acceptors in the thylakoid membranes; electrons eventually reduce $NADP^+$; the splitting of water provides some of the H^+ that accumulates inside the thylakoids.	Electrons; $NADP^+$, H_2O	$NADPH + H^+ + O_2$; H^+
Chemiosmosis	H^+ are pumped across the thylakoid membrane, forming a proton gradient; they return across the membrane through special channels formed by the protein complex CF_0–CF_1; ATP is produced.	A proton gradient and membrane potential; $ADP + P_i$ (inorganic phosphate)	ATP
Light-independent reactions (take place in stroma)	Carbon dioxide fixation; carbon dioxide is combined with an organic compound.	Ribulose bisphosphate, CO_2, ATP, $NADPH + H^+$	Carbohydrates $ADP + P_i$ $NADP^+$

FOCUS ON

Summary Equations for Photosynthesis

Summary equation for the light-dependent reactions:

$$12\ H_2O + 12\ NADP^+ + 18\ ADP + 18\ P_i \longrightarrow 6\ O_2 + 12\ NADPH + 12\ H^+ + 18\ ATP$$

Summary equation for the light-independent reactions:

$$12\ NADPH + 12\ H^+ + 18\ ATP + 6\ CO_2 \longrightarrow C_6H_{12}O_6 + 12\ NADP^+ + 18\ ADP + 18\ P_i + 6\ H_2O$$

By canceling out the common items on opposite sides of the arrows in these two coupled equations, we obtain the simplified overall equation for photosynthesis:

$$6\ CO_2 + 12\ H_2O \longrightarrow C_6H_{12}O_6 + 6\ O_2 + 6\ H_2O$$

The Light Reactions

The light reactions of photosynthesis begin with chlorophyll. According to current theory, chlorophyll molecules and associated electron acceptors are located in units known as **photosystems.** Each photosystem contains several hundred chlorophyll molecules. There are two different types of photosystem, designated **photosystem I** and **photosystem II.** Each photosystem has a different chlorophyll composition.

The presence of two photosystems was suggested by an important series of experiments in which plant cells were exposed to monochromatic light (light of a single wavelength) and the rate of photosynthesis was measured. The rate of photosynthesis decreased abruptly when the light used had a wavelength longer than 680 nm, despite the fact that chlorophyll absorbs light over the range of 680 to 700 nm. This puzzling observation was clarified when it was found that the rate of photosynthesis with long-wavelength light can be enhanced by exposing the plant cells simultaneously to light of about 600-nm wavelength. The rate of photosynthesis in the presence of both 600-nm and 700-nm light was greater than the sum of the rates when the two kinds of light were used separately. These observations lead to the hypothesis that photosynthesis involves the interaction of two photosystems. Both can be driven by light with a wavelength shorter than 680 nm but only one of the photosystems can be driven by light of a longer wavelength. Photosystem I contains a reactive pigment (perhaps a special form of chlorophyll *a*) known as **P700,** because one of the peaks of its light absorption spectrum is at 700 nm. Photosystem II utilizes the pigment **P680,** with an absorption maximum at a wavelength of 680 nm.

Before going further let us review a few of the basic concepts of oxidation and reduction, because the light reactions of photosynthesis are based almost entirely upon oxidation–reduction chemical reactions. First, oxidation can be defined as either the loss of electrons or the loss of hydrogen from some chemical compound. Even water can be oxidized—by losing electrons or hydrogen. The dehydrogenation of water yields oxygen as a product. The opposite and complement of oxidation is reduction. Reduction involves the acceptance of either electrons or hydrogen by some chemical. Oxidation and reduction are coupled and complementary reactions. If one takes place, the other must also occur. If an electron is given up by a substance, it must be accepted by another substance.

Second, electrons are donated by compounds with a high redox potential to those with a lower redox potential, so that ordinarily electron flow is one-way, in much the same way that water flows downhill. Certain substances that are capable of oxidation and reduction can form cascades so that electrons are passed from one to another like pails of water in a bucket brigade. As each accepts and then releases an electron or electrons, it is alternately reduced, then oxidized, then reduced once again, and so on. In most biological redox systems of this sort, protons (H^+) follow the negatively charged electrons.

Third, electrons possess energy (as can be demonstrated by flipping a light switch). Every time an electron is passed from one acceptor to another, there is an opportunity to extract some of that energy; some energy must be given off in each such reaction. In other words, each oxidation–reduction is a small exergonic reaction that could contribute to the formation of ATP. In this way the electron energy is liberated slowly—a slow burn, rather than a fire.

ANTENNA MOLECULES. As can be seen in Figure 8–8, in most plants the efficiency of photosynthesis increases as light intensity increases, up to a point. Eventually, a **saturation intensity** is reached that represents a maximum rate of photosynthesis. Increases in light intensity past this point have no effect except, perhaps, eventually to "cook" the experimental system. Even at the saturation intensity only one molecule of oxygen is being produced per 2500 chlorophyll molecules. It seems that only one chlorophyll molecule in this tremendous aggregation is active; all the others serve only as antennae to gather solar energy. Their excitation energy is then funneled into the single chlorophyll molecule, the **reaction center,** that is able to give up its energized electron to an electron acceptor (Fig. 8–9).

Studies of the chlorophyll in photosynthetic bacteria indicate that for this transfer of energy to occur, a protein matrix holds the chlorophyll molecules in the photosystem in the proper orientation. Accessory pigments also function as antenna molecules in the photosystems. The time required for the transfer of energy from the various antenna molecules to the reaction center is less than 10^{-10} second.

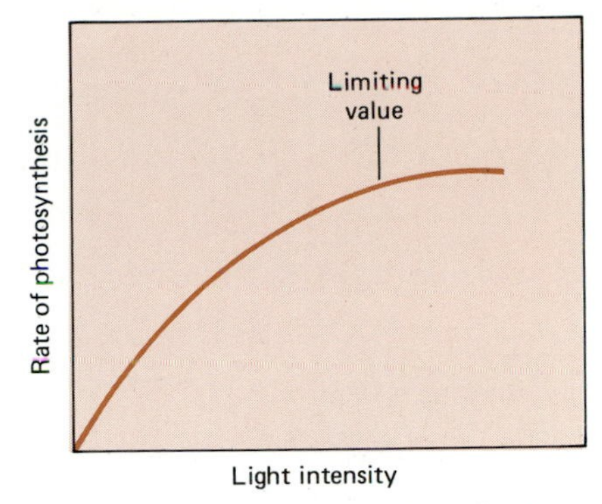

Figure 8–8 Even though only a small portion of the chlorophyll molecules in a thylakoid is excited at any one time, once this has occurred, further increases in light intensity do not increase the rate of photosynthesis above a limiting value.

PHOTOSYSTEM I. When photosystem I absorbs a photon, it emits an electron that is thought to be accepted by a protein known as **bound ferredoxin** (Fig. 8–10). Bound ferredoxin, a protein that contains iron and sulfur, is attached to the thylakoid membranes. Notice that when the electron is transferred from photosystem I to ferredoxin, photosystem I becomes positively charged or electron-deficient, a fact that we will consider further presently.

Next, bound ferredoxin transfers its electron first to soluble ferredoxin, and then to the electron acceptor $NADP^+$. When NADP is in an oxidized condition, it is strongly and positively charged. Actually $NADP^+$ is able to accept *two* electrons from soluble ferredoxin, although each molecule of ferredoxin can carry only one electron at a time and each excitation of photosystem I produces just one electron. Thus two photons of light are required to convert $NADP^+$ to the hypothetical reduced form, $NADP^-$. Next, $NADP^-$ combines with one proton, H^+, to form

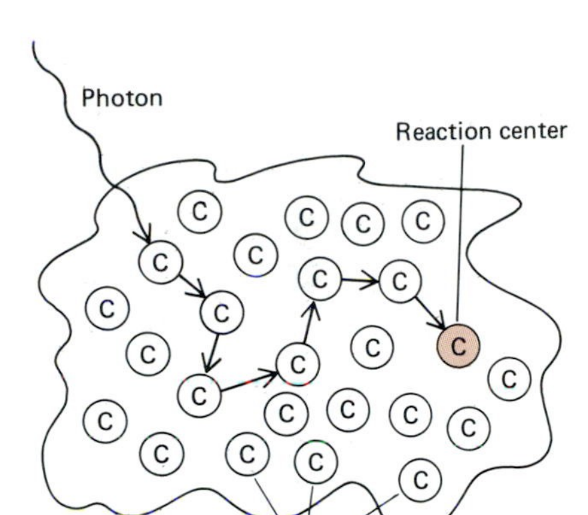

Figure 8–9 A photosynthetic unit. The many chlorophyll molecules (*c*) in the unit are excited by photons and transfer their excitation energy to a specially positioned pigment molecule (color), the reaction center.

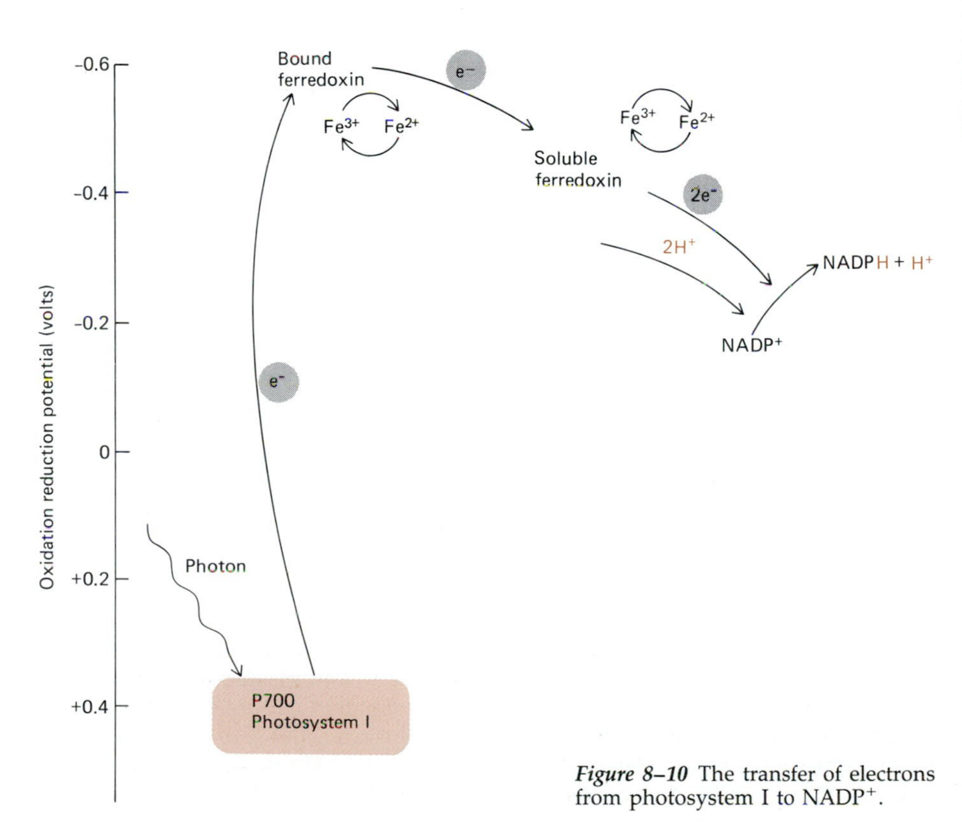

Figure 8–10 The transfer of electrons from photosystem I to $NADP^+$.

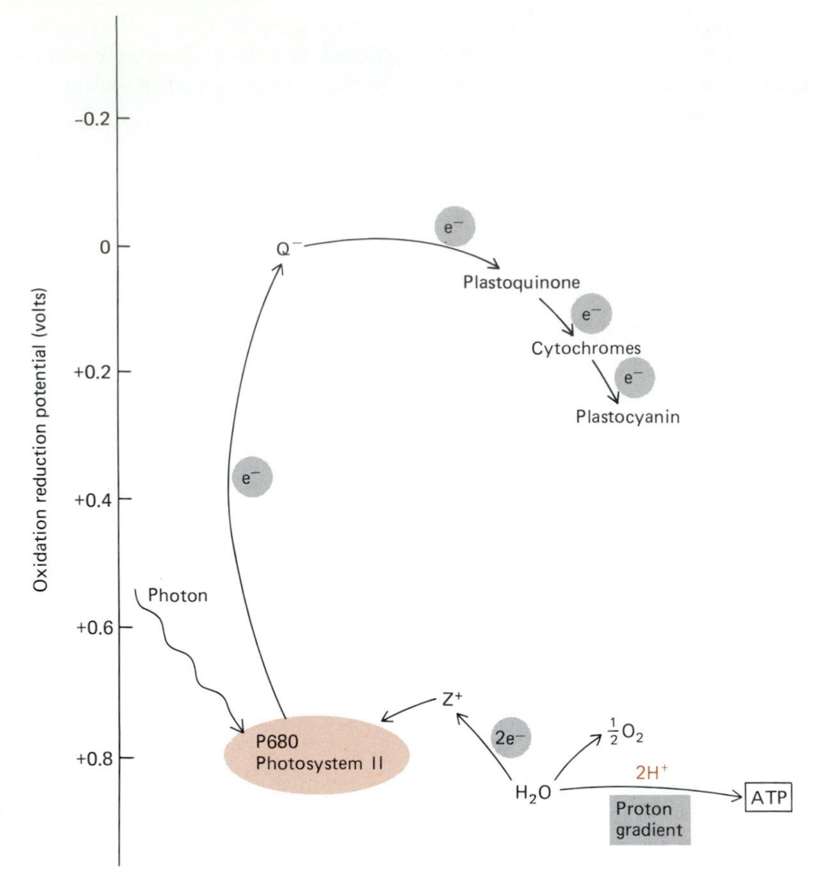

Figure 8–11 When photosystem II is activated by absorbing a photon, a strong oxidant, Z^+ (exact structure unknown), is produced. Its potential is so great that it is able to extract electrons from water, forming O_2 and contributing to the formation of a proton gradient that generates ATP. The electrons then flow to the reductant Q^-, whose structure is also as yet unknown but which is tightly bound to the thylakoid membranes. Then the electron passes to soluble plastoquinone and on to yet other acceptors.

NADPH. The proton originates, as we will see, in another stage of photosynthesis. The reaction is:

$$NADP^+ + H^+ + 2 \text{ ferredoxin (reduced)} \longrightarrow NADPH + 2 \text{ ferredoxin (oxidized)}$$

PHOTOSYSTEM II. Photosystem II is activated by a photon much as is photosystem I. Like photosystem I, it gives up an electron to a chain of electron acceptors (Fig. 8–11). Moreover, like photosystem I, photosystem II becomes positively charged as a result. However, this positive charge is immediately neutralized by electrons derived from surrounding water molecules. Consider water. Since hydrogen, the simplest possible atom, consists of a proton plus an electron, H_2O can be thought of as $p_2^+e_2^-O$. If the electrons, e^-, are removed, that leaves only protons and oxygen. The oxygen is released into the atmosphere.

$$2\,H_2O \longrightarrow 4\,H^+ + 4\,e^- + O_2$$

Please note that photosystem II is responsible for the photolytic dissociation of water and the production of atmospheric oxygen. Can you guess what happens to the protons?

HOW THE PHOTOSYSTEMS INTERACT. We left photosystem I a few paragraphs back with a positive charge and a mysterious input of protons. It cannot emit another electron, whether or not it becomes excited, until one is restored to it somehow. We have also seen that photosystem II produces an electron as a by-product of photolysis. *That electron is donated to photosystem I,* which retires its electron deficit. But it doesn't get there in one step. This electron is passed from one acceptor to another in a long chain of easily oxidized and reduced compounds. Among these are plastoquinone, a coenzyme related to vitamin K; several cytochromes, iron-containing proteins with a heme group like that in hemoglobin; and plastocyanin, a copper-containing protein similar to the oxygen-carrying pigment in horseshoe crab blood (Fig. 8–12). It was formerly thought that each of these reactions, or at least some of them, were linked to coupled reactions that produce ATP. It was a beautiful hypothesis that turned out to be untrue.

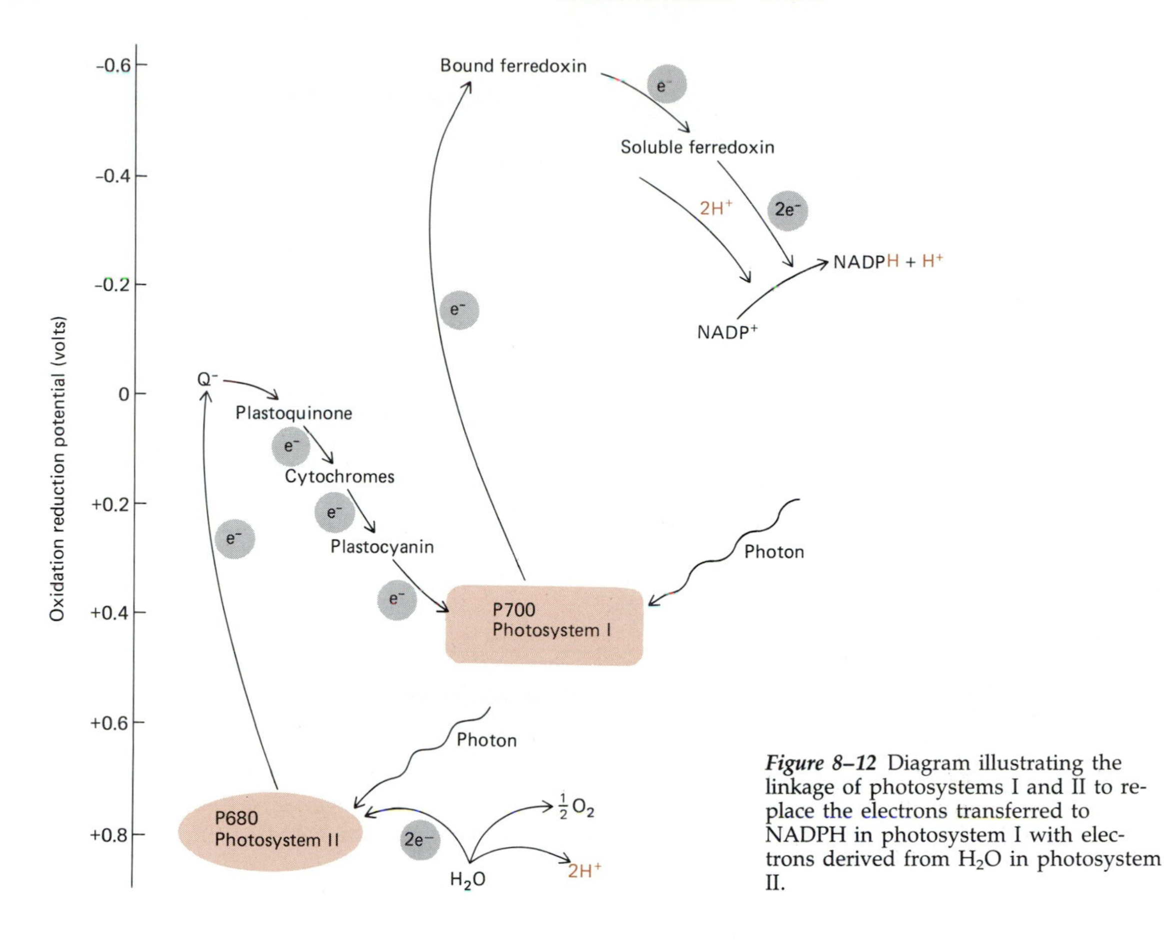

Figure 8–12 Diagram illustrating the linkage of photosystems I and II to replace the electrons transferred to NADPH in photosystem I with electrons derived from H_2O in photosystem II.

What actually happens is that, as electrons are passed from one acceptor to another, energy is released and is used to pump protons across the thylakoid membrane to the interior of the thylakoid. The protons return from the interior to the stroma through special channels in the membrane. It is this **proton flow,** rather than the electron flow responsible for it, that is used to produce ATP in photosynthesis. Recall, moreover, the protons that are formed by photolysis. They too possess energy. The chloroplast is constructed so as to milk a large part of the energy from the protons that have come from all sources, and to convert a large part of that energy into ATP.

HOW PROTON GRADIENTS FORM ATP: CHEMIOSMOSIS. According to the chemiosmotic theory (summarized here but discussed in more detail in Chapter 9), such energy-using organelles as chloroplasts accumulate protons in special membranous compartments. The protons then are allowed to diffuse across the compartment membranes. The membrane potential energy, created by the difference in proton concentration across the membrane, is harvested by a special protein complex. This complex uses a portion of that energy to synthesize ATP, the cell's universal energy exchange medium.

The photosystems and electron acceptors are embedded in the thylakoid membrane. It appears that as electrons travel through the chain of acceptors, protons are pumped into the thylakoid and accumulate there. Recalling that protons and hydrogen ions are the same, we can appreciate that the pH of the thylakoid interior will fall, and in fact it approaches a pH of 4 in bright light. This produces a difference of about 3.5 pH units across the thylakoid membrane—about a 5000-fold difference in H^+ concentration.

The initial evidence that there is a substantial pH difference across the thylakoid membrane came from some remarkable experiments by Andre Jagendorf in 1966. These experiments revealed that chloroplasts are able to synthesize ATP in the dark when a pH gradient was imposed across their thylakoid membranes. Jagendorf soaked some isolated chloroplasts in the dark in a medium with a pH of 4 for several hours. The chloroplasts were then rapidly mixed with a medium of pH 8 that contained ADP and inorganic phosphate. In these experiments, the pH of the

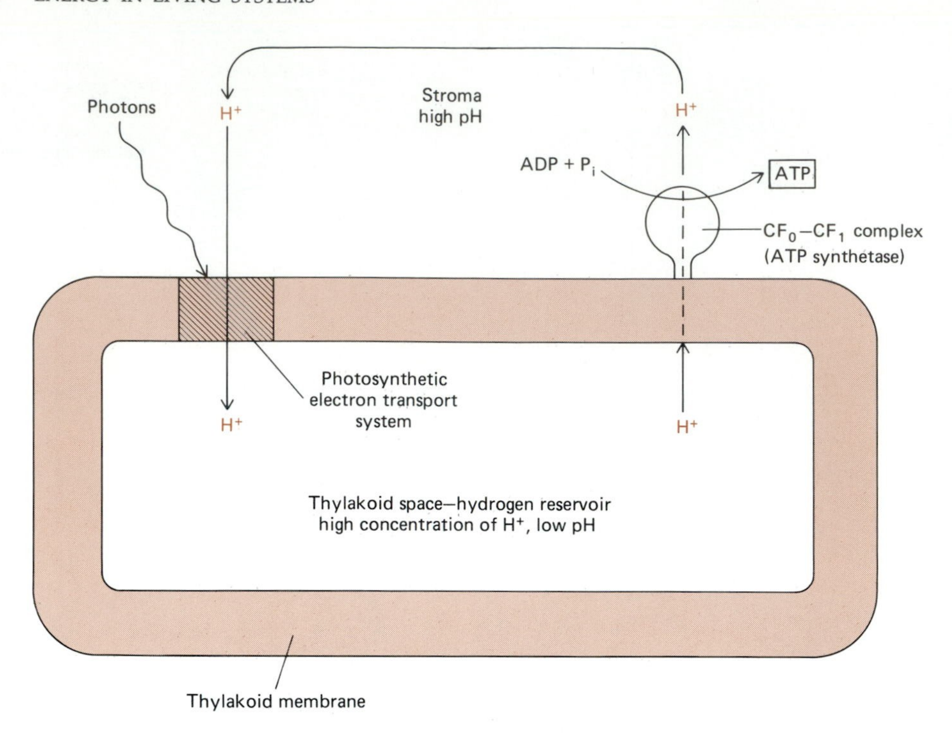

Figure 8–13 The proton gradient across the thylakoid membrane.

stroma increased to about 8, whereas the pH within the thylakoid space remained at 4. Then, as the gradient of hydrogen ions (protons) across the thylakoid membrane disappeared (as the protons moved from the thylakoid space to the stroma), ATP was synthesized.

In accordance with the general principles of diffusion, the highly concentrated hydrogen ions inside the thylakoid tend to diffuse out. However, they are prevented from doing so because the thylakoid membrane is impermeable to them except at certain points bridged by a remarkable protein, the CF_0–CF_1 complex. This complex extends across both monolayers of the thylakoid membrane, projecting from the membrane surface both inside and outside. These transmembrane particles constitute channels whereby protons *can* leak out of the thylakoids (Fig. 8–13). However, note that the concentrated proton solution inside the thylakoids represents a low-entropy state. If the protons can move out of the thylakoids into the surrounding space, the resulting random and more-or-less even distribution of protons within the chloroplast would be a high-entropy state. The second law of thermodynamics identifies this situation as one in which there is a potential for useful work resulting from the change in entropy.

As the protons pass through the CF_0–CF_1 complex, energy is released, and that energy is used to synthesize ATP. Just how this is accomplished is now the subject of very active investigation.

THE PRODUCTS OF THE LIGHT REACTIONS. The result of all the activity involved in the light reactions is that the energy of sunlight has been captured to photolyze (split) water. Protons and electrons liberated by this process produce ATP and NADPH (Fig. 8–14). The ATP is a source of energy to drive the light-independent reactions that we will next consider. The NADPH is needed as a source of reducing power or hydrogen—an essential ingredient in the organic compounds that those reactions will manufacture. Molecular oxygen is also produced by the photolysis reaction but is not used in the light-independent reactions. This process is referred to as **noncyclic photophosphorylation.**

CYCLIC PHOTOPHOSPHORYLATION. Electrons released from the P700 of photosystem I and transferred to bound ferredoxin, P430, can be transferred to yet another cytochrome, cytochrome b_{563} instead of to $NADP^+$. These electrons can then pass via cytochromes and plastocyanin back to the oxidized form of P700. This cyclic flow of electrons can generate ATP, and the process is termed **cyclic photophosphorylation** (Fig. 8–15). Photosystem II is not involved in the process of cyclic photophosphorylation, and oxygen is not produced. Cyclic photophosphorylation occurs when there is too little $NADP^+$ to accept electrons from reduced ferredoxin.

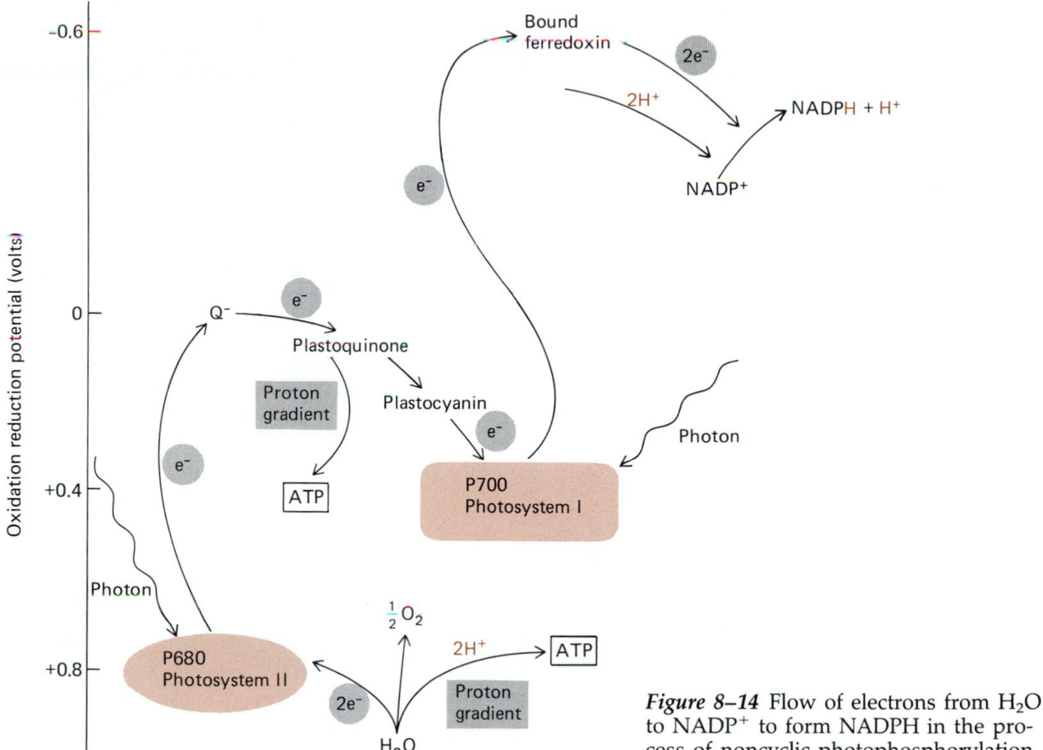

Figure 8–14 Flow of electrons from H_2O to $NADP^+$ to form NADPH in the process of noncyclic photophosphorylation.

Two energy-rich phosphates (ATPs) are formed as the energy of the electrons is transferred via phosphorylation. As the electrons flow through a series of acceptors, some of the radiant energy taken in is conserved in a chemical form as ATPs rather than being lost as fluorescence and heat. Cyclic photophosphorylation may account for some of the ATP needed to drive the light-independent reactions, but it does not produce NADPH, nor does it split water molecules to yield oxygen. By itself, cyclic photophosphorylation could never serve as the basis of photosynthesis because, as we will see shortly, NADPH is necessary in order for CO_2 to be reduced to carbohydrate. (Photosynthetic bacteria, however, which do not produce oxygen, contain only a single photosystem, which is analogous to photosystem I. The reactions that occur are analogous to cyclic photophosphorylation in plants.)

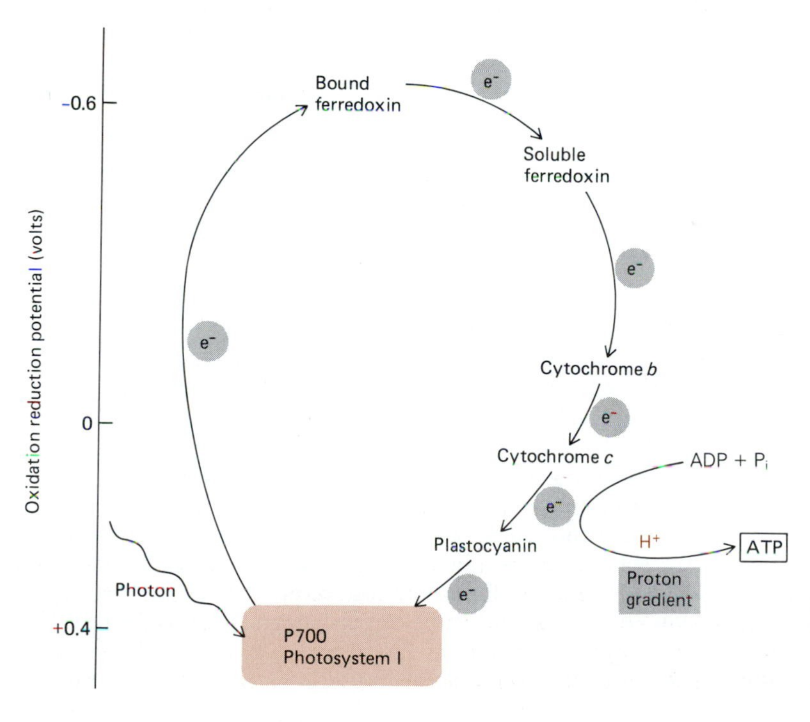

Figure 8–15 Flow of electrons from photosystem I in cyclic photophosphorylation to ferredoxin. The electrons then return to photosystem I.

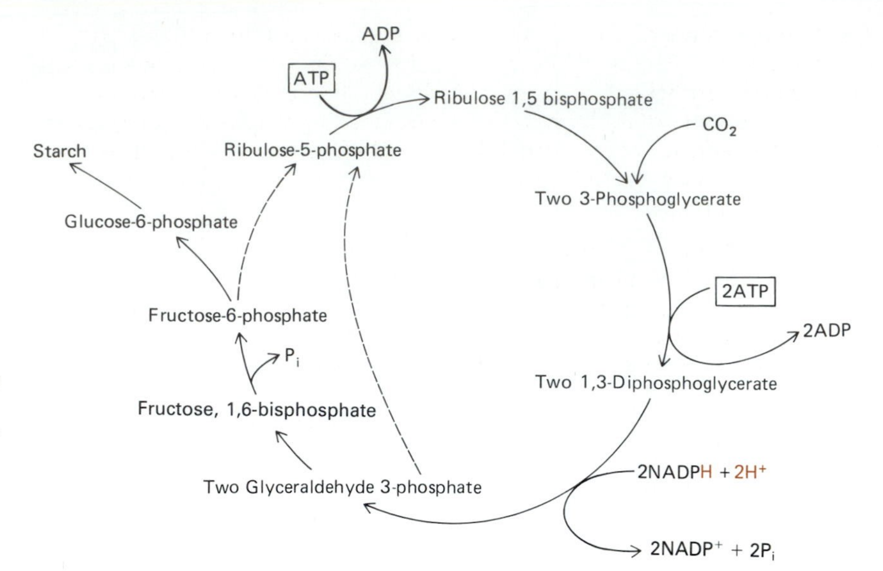

Figure 8–16 The Calvin cycle or C_3 pathway of light-independent reactions. Carbon dioxide is fixed, that is, brought into chemical combination by these reactions, of which are shown only a few. For example, ribulose 5-phosphate is formed from fructose 6-phosphate and glyceraldehyde 3-phosphate by a series of reactions indicated here only by dotted lines.

The Light-Independent Reactions: Carbon Dioxide Fixation

The light-dependent reactions are associated with the thylakoid membranes, whereas the light-independent reactions occur in the stroma. The essential feature of the light-independent reactions is the assimilation of carbon dioxide into organic molecules—that is, **carbon dioxide fixation.** New carbon-to-carbon bonds are formed, joining carbon dioxide to a preexisting organic molecule, and the product is then reduced by the addition of electrons.

The nature of these light-independent reactions was elucidated by an elegant series of investigations by Melvin Calvin and his colleagues. They incubated suspensions of the unicellular green alga *Chlorella* with $^{14}CO_2$, illuminated the suspension for a brief period of time, and then stopped the reaction by dropping the algal cells into hot alcohol. Then they attempted to identify the substances made by the cells that contained the ^{14}C. They chose algae for their experiments because the cells can be easily cultured. Subsequent experiments with a wide variety of organisms from photosynthetic bacteria to spinach and other complex plants have shown that the same reactions are involved.

THE C_3 PATHWAY (THE CALVIN CYCLE). The reactions by which carbon dioxide is assimilated occur very rapidly, and Calvin found that after as short an incubation as 60 seconds, many compounds contained radioactive carbon, ^{14}C. When cells were exposed to the $^{14}CO_2$ for only *5 seconds,* the only compound with a substantial amount of radioactivity was **3-phosphoglycerate.** Subsequent research revealed the nature of the molecule with which carbon dioxide combines initially; it proved to be **ribulose bisphosphate,**[1] formed by the transfer of a phosphate group from ATP to ribulose 5-phosphate (Fig. 8–16). Carbon dioxide is added onto ribulose bisphosphate, forming a transient six-carbon intermediate compound, which is rapidly cleaved to form two molecules of phosphoglycerate. The addition of carbon dioxide is catalyzed by an enzyme, **ribulose 1,5-bisphosphate carboxylase,** located on the stromal surface of the thylakoid membranes. This is an exceptionally abundant enzyme, making up more than 16% of the total protein of the chloroplasts. It is probably the single most abundant protein in plants, and possibly in living things.

The 3-phosphoglycerate formed is then transformed by a reaction that requires another ATP (from the light reactions) to drive it, to give diphosphoglycerate. An input of NADPH is needed to reduce the diphosphoglycerate to **glyceraldehyde 3-phosphate,** a three-carbon sugar. Two of these trioses (three-carbon sugars) condense to form one hexose (six-carbon sugar), **fructose 1,6-bisphosphate;** one of the phosphate groups is removed enzymatically to yield fructose 6-phosphate, and this undergoes a molecular rearrangement to become **glucose 6-phosphate.** The glucose phosphate can be added onto a molecule of starch and stored.

For the cycle to continue, ribulose bisphosphate must be regenerated from some of the six-carbon and three-carbon sugars. This is achieved by a series of

[1]Formerly called ribulose diphosphate.

transfers of two- or three-carbon units so that one hexose and three trioses end up as three pentoses [6 + (3 × 3) → (3 × 5)]. The first product of these transfer reactions, **ribulose 5-phosphate,** is phosphorylated by ATP to regenerate ribulose 1,5-bisphosphate. The series of reactions by which fructose 6-phosphate is formed and ribulose 1,5-bisphosphate is regenerated is termed the **Calvin cycle.** It is also called the **C_3 pathway,** because the first stable compound into which carbon dioxide is assimilated, 3-phosphoglycerate, contains three carbons.

ENERGY CONSIDERATIONS. Let us consider next how much energy is expended in synthesizing a molecule of six-carbon sugar. Six turns of the Calvin cycle are required, since only one carbon atom is reduced from the carbon dioxide state to the carbohydrate state by each turn. Twelve molecules of ATP (by this we really mean the free energy of these molecules) are spent in phosphorylating 12 molecules of phosphoglycerate to 1,3-diphosphoglycerate, and 12 molecules of NADPH are used in reducing 12 molecules of 1,3-diphosphoglycerate to glyceraldehyde 3-phosphate. An additional six molecules of ATP are used in regenerating ribulose 1,5-bisphosphate. The overall equation for the light-independent reactions of photosynthesis is

$$6\ CO_2 + 18\ ATP + 12\ NADPH + 12\ H^+ \longrightarrow C_6H_{12}O_6 + 18\ ADP + 18\ P_i + 12\ NADP^+ + 6\ H_2O$$

The energy of three molecules of ATP and two molecules of NADPH is required to convert one molecule of carbon dioxide to carbohydrate.

The formation of the two molecules of NADPH requires the capture of four photons by the P700 reaction center of photosystem I and the capture of an additional four photons by the P680 reaction center of photosystem II. One mole of photons of light with a wavelength of 600 nm has an energy content of 47.6 kcal, so the energy put into the system by eight moles of photons is 381 kcal. The change in free energy for the reduction of carbon dioxide to a hexose is 114 kcal per mole. Thus the overall efficiency of photosynthesis can be calculated as 114/381 or about 30%. Actual values, however, fall far short of this ideal, for a variety of reasons, and rarely exceed 1% to 3%. Sugar cane is among the most efficient photosynthesizers, sometimes reaching an 8% efficiency.

REGULATION OF THE C_3 PATHWAY. The rate-limiting step in the Calvin cycle is the assimilation of carbon dioxide into ribulose 1,5-bisphosphate to form two molecules of phosphoglycerate. The activity of the enzyme catalyzing this reaction, ribulose 1,5-bisphosphate carboxylase, is increased in three ways when the system is illuminated: First, the enzyme is activated by the NADPH formed when photosystem I absorbs light. Second, the rate of the enzyme reaction is increased when the system becomes more alkaline. The light-induced proton gradient that leads to the acidification of the thylakoid space makes the stromal space, where the carboxylase is located, more alkaline. Finally, the carboxylase is further activated by Mg^{2+}, and magnesium ions are released into the stromal space as protons are being pumped into the thylakoid space by light.

The C_4 Pathway

In laboratory studies of certain tropical grasses such as sugar cane, corn, and crabgrass, the first compounds to be labeled after the cells are incubated with $^{14}CO_2$ are four-carbon compounds with two carboxyl groups: **oxaloacetate, malate,** and **aspartate** (Fig. 8–17). This **C_4 pathway** is present in addition to the Calvin cycle in a number of orders of plants and apparently has evolved independently several times. Plants with the C_4 pathway evidently appeared first in geographical areas with high temperatures, high light intensities, and limited amounts of water. These plants have a higher temperature optimum, a higher light optimum, less loss of water by transpiration, and higher rates of photosynthesis and growth than those in plants having only the Calvin cycle. In fact, a major difference between C_3 and C_4 plants is that C_4 plants do not light-saturate even at the highest light intensities they encounter in nature. The rapid growth of crabgrass in your lawn on hot summer days can be blamed on the C_4 pathway.

The C_4 crop plants have yields that are two to three times larger than those of C_3 plants, provided that an abundance of light and water is available. In the United States, the average annual yield per acre in 1976 was 4872 pounds for corn and 2744

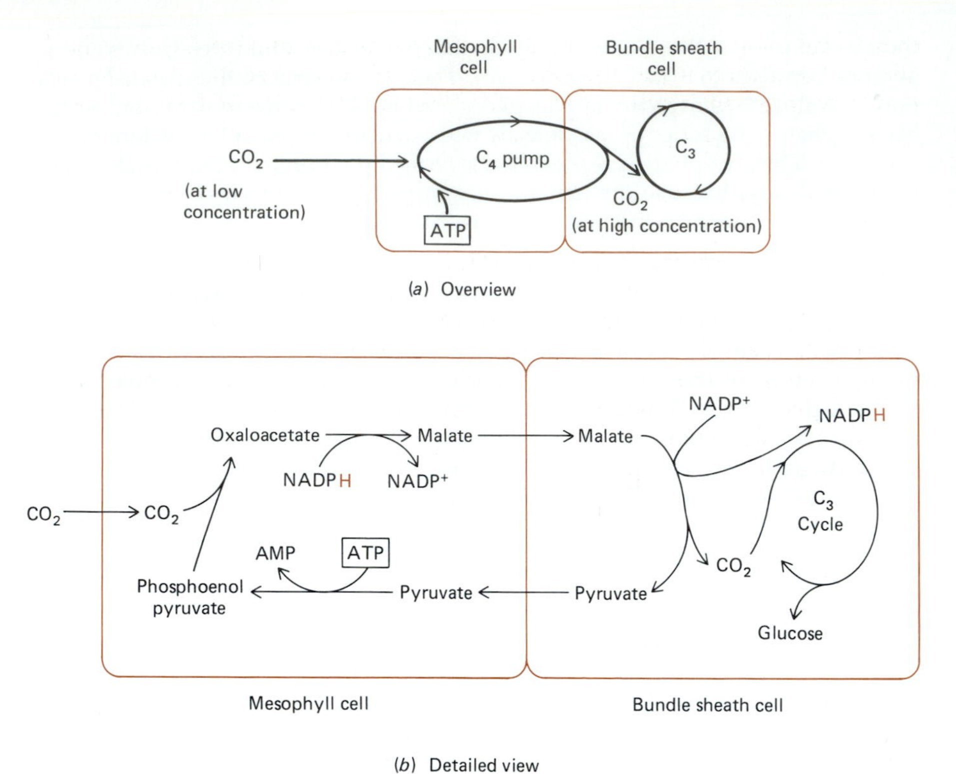

Figure 8–17 The C_4 series of reactions in photosynthesis. (*a*) If, as is usually the case, CO_2 is present in low concentration, the C_4 system readily absorbs it and in effect concentrates it for use by the C_3 system to which it is "pumped." Since the C_4 system does consume some energy, ultimately made available only by photosynthesis, this system is "worthwhile" to the plant only at high light intensities when the stomata can be kept open despite these conditions. Thus, to be effective, the C_4 system requires an abundance of both water and light. Yet under these conditions it can fix more carbon than the C_3 system can fix by itself. (*b*) A more detailed diagram, showing how some of the reactions catalyzed by enzymes in the C_4 plants add carbon dioxide to phosphoenol pyruvate (three carbons) to form oxaloacetate (four carbons).

pounds for sorghum (both C_4 plants), compared with 1560 pounds for soybeans and 1800 pounds for wheat (both C_3 plants).

The essential feature of the C_4 pathway is that it concentrates carbon dioxide in the cells that carry on photosynthesis by the C_3 pathway. Most C_4 plants have a layer of **bundle sheath cells** surrounding the vascular bundles of xylem and phloem (Fig. 8–18). A layer of **mesophyll cells** surrounds the bundle sheath cells. The enzymes of the Calvin cycle are present largely or entirely in the bundle sheath cells. The mesophyll cells (and many other plant cells) contain the enzyme phosphoenol pyruvate carboxylase that catalyzes the assimilation of carbon dioxide into **phosphoenol pyruvate,** or **PEP** (three carbons), to yield **oxaloacetate** (four carbons). The oxaloacetate may be converted to either malate or aspartate. The malate or aspartate then passes to the bundle sheath cells, where a different enzyme catalyzes the decarboxylation of malate; pyruvate, a 3-carbon compound is formed.

$$\text{Malate} + \text{NADP}^+ \longrightarrow \text{Pyruvate} + CO_2 + \text{NADPH} + H^+$$

The CO_2 released condenses with ribulose 1,5-bisphosphate and enters the Calvin

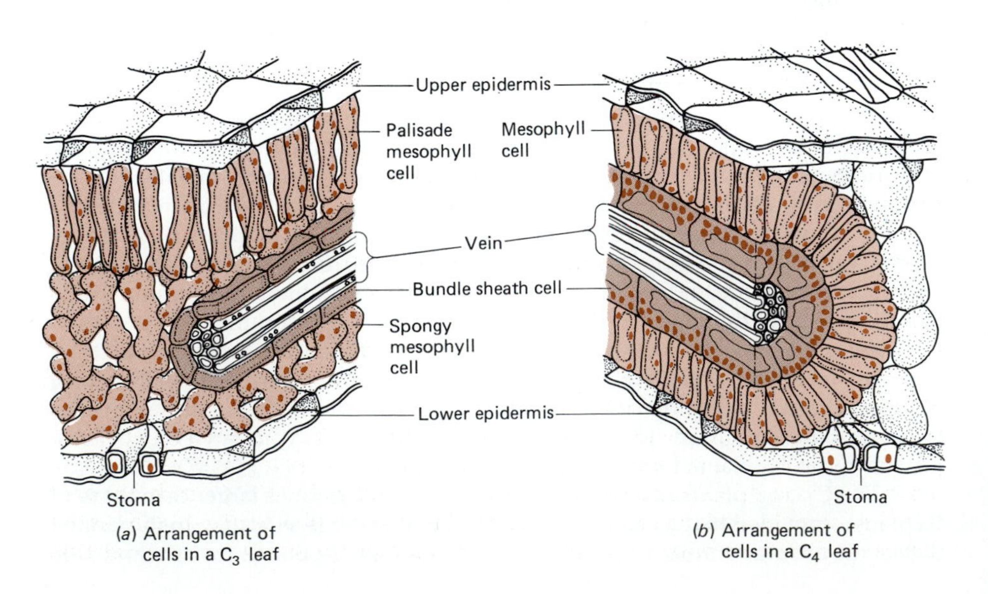

Figure 8–18 Diagram of bundle sheath cells, often found in C_4 plants.

cycle in the usual manner (see Fig. 8–17). The pyruvate formed in the decarboxylation reaction returns to the mesophyll cell where phosphoenol pyruvate is regenerated by the reaction of pyruvate with ATP. The overall reaction of the C_4 pathway is

$$\underset{\text{within the mesophyll cells}}{CO_2} + ATP \longrightarrow AMP + 2\ P_i + \underset{\text{within the bundle sheath cells}}{CO_2}$$

This equation reveals that two energy-rich phosphates are used up in transporting carbon dioxide to the chloroplasts of the bundle sheath cells. The role of the C_4 cycle is simply to increase the concentration of carbon dioxide within the bundle sheath cells so as to drive the C_3 cycle there. The operation of the C_4 cycle serves to increase the concentration of carbon dioxide within the bundle sheath cells some 10- to 60-fold over that in the cells of plants having only the C_3 pathway. The net reaction for the combination of the C_4 and C_3 pathways is

$$6\ CO_2 + 30\ ATP + 12\ NADPH + 12\ H^+ \longrightarrow C_6H_{12}O_6 + 30\ ADP + 30\ P_i + 12\ NADP^+ + 6\ H_2O$$

The combined pathway involves the expenditure of 30 ATPs per hexose, rather than 18 ATPs used in the absence of the C_4 pathway. The expenditure of the extra ATPs ensures a high concentration of carbon dioxide in the bundle sheath cells and permits them to carry on photosynthesis at a rapid rate. When light is abundant, the rate of photosynthesis is limited by the concentration of carbon dioxide available; this is never very high. However, at lower light intensities and temperatures, it is C_3 plants that have the advantage. For example, winter rye, a C_3 plant, grows lavishly when crabgrass cannot. The reason is that C_4 metabolism requires 5 ATPs to fix 1 mole of CO_2, whereas C_3 uses only 3 ATPs.

Photorespiration

Like animal cells, plant cells carry out the reactions of **cellular respiration,** primarily in their mitochondria, using substrates such as glucose and producing carbon dioxide. These reactions produce ATP, utilized to drive the metabolic processes in the plant cells. In addition, many plants, when illuminated, use O_2 and produce CO_2 by a different process termed **photorespiration.** The enzyme that catalyzes the combining of carbon dioxide with ribulose 1,5-bisphosphate is ribulose 1,5-bisphosphate carboxylase. This enzyme is an **oxygenase** as well as a carboxylase; in fact, oxygen and carbon dioxide compete for binding to the same active site on that enzyme. The oxygenase function of the enzyme causes ribulose 1,5-bisphosphate to react with molecular oxygen to form the three-carbon 3-phosphoglycerate and the two-carbon **2-phosphoglycolate** (Fig. 8–19). The phosphoglycolate is hydrolyzed to **glycolate** and inorganic phosphate, P_i. Then the glycolate is metabolized further in glyoxysomes (Chapter 4) to **glyoxylate** with the production of hydrogen peroxide, H_2O_2. The glyoxylate is metabolized further to carbon dioxide in the mitochondria.

When the concentration of CO_2 is high and that of O_2 is low, as in plants with the C_4 pathway, the *carboxylation* of ribulose 1,5-bisphosphate is favored, and the synthesis of carbohydrates by the C_3 pathway proceeds. However, when the concentration of carbon dioxide is low and that of oxygen is high, the *oxidation* of ribulose 1,5-bisphosphate is favored, and carbon dioxide is released.

The biological role of photorespiration is a mystery, for it appears to be a wasteful process in which organic compounds are converted to carbon dioxide without the production of ATP, or NADPH. (However, glyoxylate may function in the production of the amino acids glycine and serine.) In some C_3 plants, photorespiration may break down as much as half of the carbohydrates formed by photosynthesis. Photorespiration is probably an unavoidable consequence of the design of the ribulose 1,5-bisphosphate carboxylase enzyme, which must occur when this enzyme obtains its CO_2 in an oxygen-rich environment. If it were possible to block photorespiration, the plant would have a much greater net photosynthetic incorporation of carbon dioxide and would have a greater net yield. The C_4 pathway does this very thing by making carbon dioxide available to the ribulose 1,5-bisphosphate carboxylase enzyme without a great deal of accompanying oxygen. In ecological

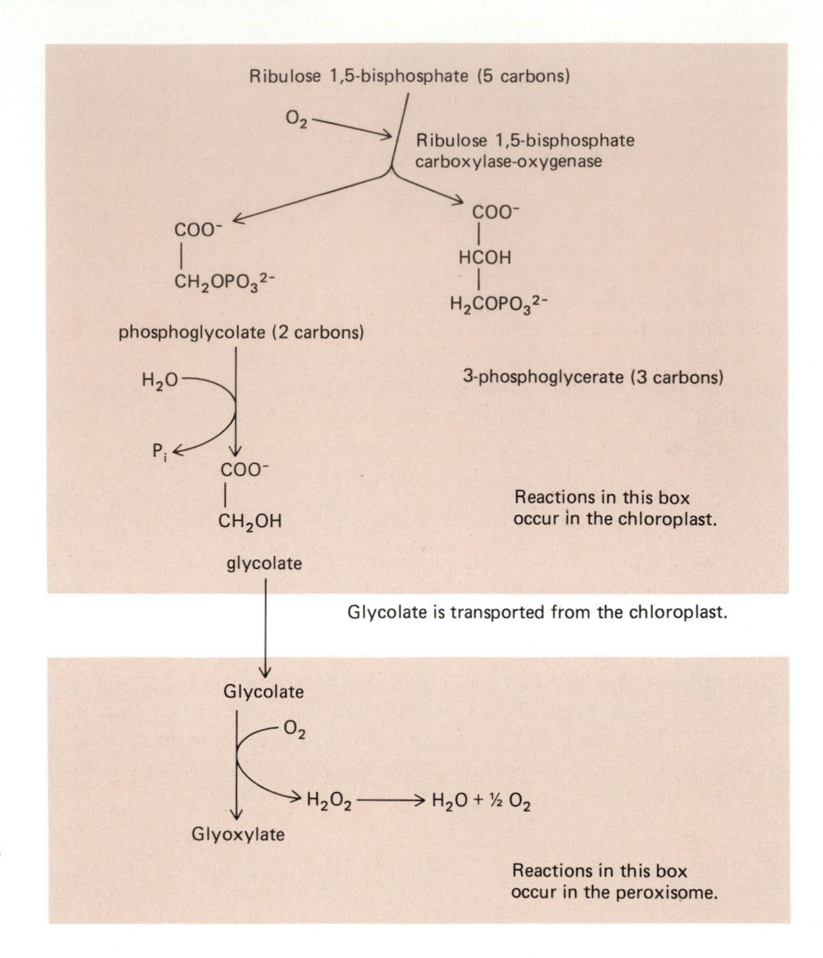

Figure 8–19 Formation of glyoxylate in photorespiration. Glyoxylate is produced within a glyoxysome, a type of peroxisome found in plant cells.

situations where light intensity is not a limiting factor for photosynthesis, the C_4 pathway gives its possessors a competitive advantage over those that do not use this pathway. (A somewhat comparable system, the CAM pathway, occurs in many desert plants as a specialization and will be discussed in Chapter 27.)

SUMMARY

I. Chlorophyll or accessory pigments are excited by photons. The resulting high-energy electrons are accepted by special compounds.

II. Chlorophyll is the substance that captures light energy; its absorption spectrum is closely similar to the action spectrum of photosynthesis.

III. The oxygen released by photosynthesis originates from water. The hydrogen liberated by this photolysis is used to reduce carbon dioxide with the formation of carbohydrate.
  A. Most molecules of chlorophyll do not yield electrons directly but function as antennae to pass their energy of excitation to central molecules that do yield high-energy electrons.
  B. Units called photosystems, containing chlorophyll molecules and electron acceptors, are the site of the light reactions. (Table 8–1 summarizes the reactions of the two photosystems.)
  C. Proton gradients formed directly or indirectly by the photosystems result in proton diffusion across thylakoid membranes. The proton-gradient energy is harvested by the CF_0–CF_1 complex, which uses it to synthesize ATP.

IV. The NADPH and ATP produced in the light reactions are used to convert CO_2 into hexoses and other organic compounds in light-independent reactions, by a series of reactions called the Calvin cycle or C_3 pathway.
  A. The light-independent reactions begin with the reaction of CO_2 with ribulose bisphosphate to form two molecules of the three-carbon 3-phosphoglycerate.
  B. The phosphoglycerate molecules form hexoses.
  C. Ribulose bisphosphate is regenerated by a complex series of reactions.
  D. Three ATP and two NADPH molecules are consumed in converting one CO_2 molecule into carbohydrate.

V. The C_4 pathway of photosynthesis enables tropical plants to take advantage of high light intensity. CO_2 is fixed into a four-carbon compound with the expenditure of ATP. The CO_2 is later removed and refixed by the C_3 pathway.

POST-TEST

1. Light is composed of particles of energy called ________________.

2. The relative effectiveness of different wavelengths of light in photosynthesis is given by the ________________ ________________.

3. Blue light has ________________ energy than red light.

4. When an electron absorbs a light quantum and moves to another orbital the molecule is said to be in an ________________ state.

5. In complex plants the electron donor in the light-dependent reactions is ________________.

6. The first stable, chemically defined products formed as a result of the light reactions are ________________ and ________________.

7. A process in which electrons are removed from an atom or molecule is termed ________________.

8. Chloroplasts contain many stacks of flattened, disk-shaped membranous sacs, termed ________________.

9. Each individual sac is called a ________________.

10. The chloroplast has three distinct membranes: ________________, ________________, and ________________.

11. These define three separate spaces: ________________ ________________, ________________, and ________________ ________________.

12. In addition to chlorophyll, most plants contain other photosynthetic pigments such as ________________ and ________________.

13. Chlorophyll and other pigments are organized into groups of several hundred pigment molecules termed ________________.

14. Each of these has one special pigment molecule, the ________________ ________________; the other ________________ chlorophylls absorb light and transfer the energy to it.

15. The oxygen molecules released in photosynthesis are derived from ________________.

16. Photosystem I, excited by light of 700-nm wavelength, generates ________________.

17. Photosystem II, excited by light of 680-nm wavelength, is very powerful and can remove ________________ and ________________ from water, forming ________________.

18. Photosystems I and II are linked by a series of ________________ ________________.

19. The cyclic flow of electrons from the P700 of photosystem I to electron carriers and back to P700, generating ATP, is termed ________________.

20. The transfer of electrons through the two photosystems produces a gradient of ________________ across the membrane.

21. As protons flow through the complex of enzymes in the thylakoid membrane, ________________ is generated and released into the ________________ space.

22. The light-independent reactions of photosynthesis begin with the condensation of CO_2 with ________________ to form ________________.

23. The reactions by which fructose 6-phosphate is formed and ________________ is regenerated is termed the ________________ ________________.

24. In the C_4 pathway of photosynthesis carbon dioxide condenses with ________________ ________________ to form oxaloacetate. This is converted to malate or aspartate and passes to the bundle sheath cells where malate is decarboxylated, yielding ________________ for photosynthesis.

25. In photorespiration, ribulose 1,5-bisphosphate reacts with ________________ instead of with CO_2 and yields phosphoglycerate and ________________.

REVIEW QUESTIONS

1. What is meant by the action spectrum of photosynthesis?
2. What properties of chlorophyll are most significant in photosynthesis?
3. What is the function of antenna chlorophylls? of the reaction center of the photosynthetic unit?
4. By what reaction is molecular oxygen produced from water?
5. Compare and contrast the components and functions of photosystems I and II.
6. How is a proton gradient established across the thylakoid membrane? How does this result in the synthesis of ATP?
7. What features distinguish cyclic from noncyclic photophosphorylation?
8. What features of the structure of a chloroplast are especially important in enabling the process of photosynthesis to proceed?
9. How are ATP and NADPH produced and utilized in the process of photosynthesis?
10. What strategies may be employed in the future to increase the world's supply of food?
11. Compare and contrast the C_3 and C_4 pathways with respect to fixation of carbon dioxide.
12. What is photorespiration? Why does this seem to be a wasteful process?

9
Cellular Respiration and Biosynthesis

OUTLINE

I. An overview of cellular respiration
II. Glycolysis
III. Formation of acetyl coenzyme A
IV. The citric acid cycle
V. The electron transport system and chemiosmotic phosphorylation
  A. Electron transport
  B. Oxidative phosphorylation
  C. Regulation of energy production
  D. The molecular organization of mitochondria
  E. The chemiosmotic theory of oxidative phosphorylation
VI. Energy yield from glucose
VII. Anaerobic pathways
VIII. Oxidation of other nutrients
  A. Oxidation of amino acids
  B. Oxidation of fatty acids
IX. Biosynthetic reactions

LEARNING OBJECTIVES

After you have read this chapter you should be able to:

1. Write a summary reaction for cellular respiration, and give the origin and fate of each substance involved.
2. Indicate where the various reactions of cellular respiration occur and why special carriers are needed to transfer certain intermediates across membranes.
3. Summarize the events of glycolysis, giving the key organic compounds formed and the number of carbon atoms in each. (Your instructor will specify the degree of detail required.) Indicate the number of ATP molecules consumed and produced, and the reactions in which dehydrogenation occurs.
4. Describe the conversion of pyruvate to acetyl coenzyme A.
5. Summarize the events of the citric acid cycle, and describe the fate of acetyl coenzyme A.
6. Examine a sequence of reactions such as glycolysis or the citric acid cycle, and identify the dehydrogenations, decarboxylations, and make-ready reactions.
7. Summarize the operation of the electron transport system including the process by which a gradient of protons is established across the inner mitochondrial membrane, and explain how this proton gradient drives the synthesis of ATP.
8. Compare aerobic respiration with fermentation in terms of ATP formation, final hydrogen acceptor, and end-products; give two specific examples of fermentation.
9. Indicate how the products of amino acid and fat metabolism are oxidized in the same metabolic pathways that oxidize glucose.
10. List and discuss five basic principles of cellular biosynthesis.

Every living cell must in some way extract free energy from the nutrients it captures from the environment. The enzymatic mechanisms utilized to extract free energy from compounds such as glucose, and to conserve a portion in a biologically useful form such as ATP, are very similar in all cells. Many of these reactions are catalyzed by enzymes located in the mitochondria, either bound to the mitochondrial membranes or present in the fluid matrix within the mitochondria.

An Overview of Cellular Respiration

The process by which cells extract energy from glucose, fatty acids, and other organic compounds, utilizing oxygen and producing carbon dioxide and water, is called **cellular respiration.** One of the principal pathways of cellular respiration involves the breakdown of the common nutrient glucose. The overall reaction for the metabolism of glucose can be summarized as follows:

$$C_6H_{12}O_6 + 6\ O_2 + 6\ H_2O \longrightarrow 6\ CO_2 + 12\ H_2O + \text{Energy}$$

You will recognize this as essentially the reverse of the process of photosynthesis described in the previous chapter.

The oxidation of glucose to carbon dioxide and water in cells cannot occur in a single reaction; there are no enzymes that can catalyze the direct attack by oxygen molecules on glucose. Instead, the oxidation of glucose occurs in a sequence of reactions that can be grouped into four phases (Fig. 9–1; see also Focus on Summary Reactions for Cellular Respiration):

1. **glycolysis,** the conversion of glucose to two molecules of pyruvate, a three-carbon compound, with the formation of some ATP;
2. formation of the compound acetyl coenzyme A and CO_2 from pyruvate and coenzyme A;

Figure 9–1 An overview of cellular respiration. The four main phases in cellular respiration are glycolysis, formation of acetyl coenzyme A from pyruvate, the citric acid cycle, and the electron transport system. Details of each of these processes are given in later illustrations in this chapter. This is a simplified diagram. For example, the hydrogens shown coming off a fuel molecule are immediately combined with a hydrogen carrier such as NAD⁺. More details are given in the Focus on Summary Reactions for Cellular Respiration.

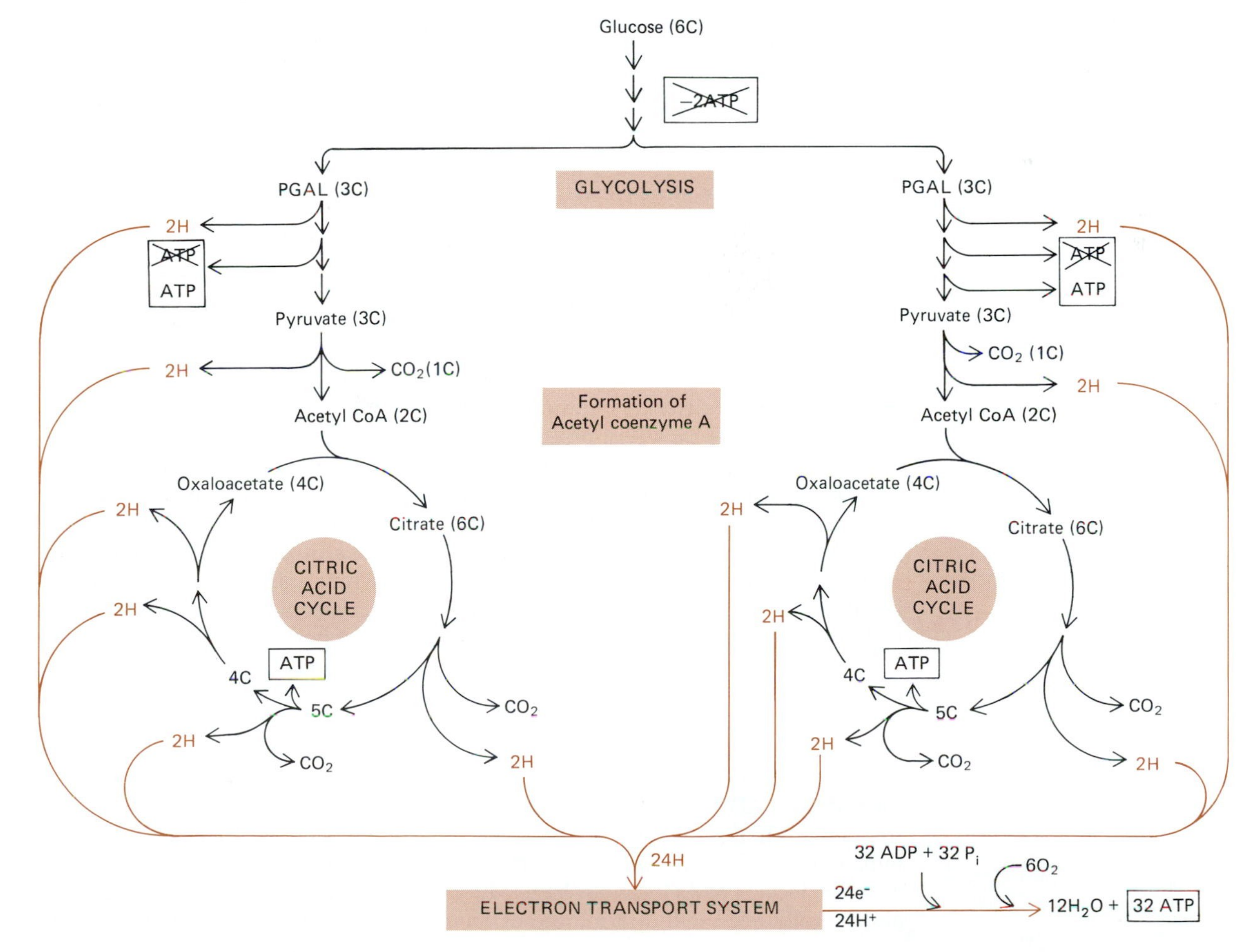

FOCUS ON

Summary Reactions for Cellular Respiration

Summary reaction for the complete oxidation of glucose:

$$C_6H_{12}O_6 + 6\ O_2 + 6\ H_2O \longrightarrow 6\ CO_2 + 12\ H_2O + \text{Energy}$$

Summary reaction for glycolysis:

$$\text{Glucose} + 2\ \text{ADP} + 2\ P_i + 2\ \text{NAD}^+ \longrightarrow 2\ \text{Pyruvate} + 2\ \text{ATP} + 2\ \text{NADH} + 2\ H^+ + 2\ H_2O$$

Summary reaction for the conversion of pyruvate to acetyl coenzyme A:

$$2\ \text{Pyruvate} + 2\ \text{Coenzyme A} + 2\ \text{NAD}^+ \longrightarrow 2\ \text{Acetyl CoA} + 2\ CO_2 + 2\ \text{NADH} + 2\ H^+$$

Summary reaction for the citric acid cycle:

$$2\ \text{Acetyl CoA} + 6\ \text{NAD}^+ + 2\ \text{FAD} + 2\ \text{GDP} + 2\ P_i + 2\ H_2O \longrightarrow 4\ CO_2 + 6\ \text{NADH} + 6\ H^+ + 2\ \text{FADH}_2 + 2\ \text{GTP} + 2\ \text{Coenzyme A}$$

Summary reaction for the processing of hydrogens (electrons and protons) in the electron transport system:

$$\text{NADH} + H^+ + 3\ \text{ADP} + 3\ P_i + \tfrac{1}{2}\ O_2 \longrightarrow \text{NAD}^+ + 3\ \text{ATP} + H_2O$$

3. the **citric acid cycle,** which converts the two-carbon acetyl group attached to coenzyme A to carbon dioxide and removes electrons and protons;
4. the **electron transport system** and chemiosmotic phosphorylation, which process the electrons and protons removed from the fuel molecule during the preceding phases. When enzymes known as dehydrogenases remove hydrogens from glucose or some other substrate, the hydrogens immediately combine with either the coenzyme **NAD⁺** (nicotinamide adenine dinucleotide) or the coenzyme **FAD** (flavin adenine dinucleotide). The coenzyme is reduced when hydrogen combines with it. In the electron transport system, hydrogens (or their electrons) are passed from NADH through a series of electron carrier molecules and ultimately react with oxygen to form water. As the hydrogens (or electrons) are transferred from one carrier molecule to the next, protons are pumped to the outside of the inner mitochondrial membrane. This establishes an electrochemical gradient of H^+ across the inner mitochondrial membrane. Then as protons flow back across the membrane, free energy is released and used to synthesize ATP from ADP and inorganic phosphate (P_i).

The many reactions involved in cellular respiration can be assigned to one of three types: First, **dehydrogenations** are reactions in which two hydrogens (actually, two electrons plus two protons) are removed from the substrate and transferred to a coenzyme such as NAD^+ or FAD, which acts as a primary acceptor. Second, **decarboxylations** are reactions in which a carboxyl group (—COOH) is removed as a molecule of CO_2. The carbon dioxide we exhale each day is derived from decarboxylations. Third, **"make-ready" reactions** are ones in which molecules undergo rearrangements so that they can subsequently undergo further dehydrogenations or decarboxylations. As we trace the reactions of cellular respiration we will encounter many examples of these three basic types.

In Chapter 7 we considered the important basic concept of the coupling of an exergonic (energy-yielding) reaction with an endergonic (energy-requiring) reaction. In living cells, the exergonic reactions usually involve the breakdown of ATP. The free energy liberated in the hydrolysis of the high-energy phosphate bonds of ATP is used to drive the reactions that require an input of free energy. ATP is formed from ADP and P_i when fuel molecules are oxidized or when light energy is trapped in photosynthesis. The basic mechanism of energy exchange in living systems is the cyclic conversion of ATP to ADP and back.

Molecules of ATP cannot be used for long-term energy storage. In fact, they are typically used within a minute of their formation; thus, the turnover of ATP molecules is very rapid. A resting human being consumes about 40 kg of ATP every 24 hours and as much as 0.5 kg per minute during strenuous exercise; yet the amount present in the human body at any given moment is less than 1 g.

Glycolysis

The sequence of reactions that convert glucose and other hexoses to pyruvate with the production of ATP is termed **glycolysis.** During this series of reactions a six-carbon sugar is enzymatically degraded to two molecules of **pyruvate,** each of which contains three carbon atoms. Each reaction is catalyzed by a specific enzyme, and there is a net gain of two ATP molecules per glucose molecule. The reactions of glycolysis take place in the cytoplasm. These reactions can proceed **anaerobically**—that is, in the absence of oxygen.

The first four steps of glycolysis serve to **phosphorylate** (add phosphate to) glucose and convert it to two molecules of the three-carbon compound **glyceraldehyde 3-phosphate,** often referred to as **PGAL.**[1] Two molecules of ATP must be invested in these reactions to activate the glucose molecule and prepare it to be split.

As we trace the steps of glycolysis, refer frequently to Fig. 9–2.

Figure 9–2 Glycolysis: The series of reactions by which glucose is metabolized to pyruvate. Each step is catalyzed by a specific enzyme. Although not indicated here, many of these reactions are reversible. Note that there is a net gain of 2 ATPs from glycolysis.

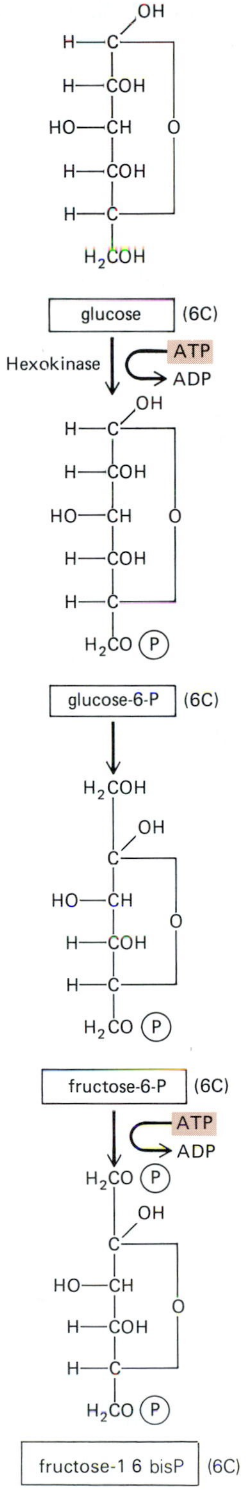

Step 1. The series of glycolytic reactions begins with the activation of glucose.

Glucose + ATP ⟶ Glucose 6-phosphate + ADP

Step 2. Glucose 6-phosphate undergoes a rearrangement reaction catalyzed by an isomerase to yield **fructose 6-phosphate.**

Step 3. Fructose 6-phosphate accepts a second phosphate group from ATP to form **fructose 1,6-bisphosphate.** The enzyme catalyzing this reaction, called phosphofructokinase, regulates the overall rate of glycolysis. It is inhibited by ATP and citrate and stimulated by AMP.

[1]PGAL stands for phosphoglyceraldehyde, another name for glyceraldehyde 3-phosphate.

Step 4. Fructose 1,6-bisphosphate is then split into two three-carbon sugars, **glyceraldehyde 3-phosphate (PGAL)** and **dihydroxyacetone phosphate.** Dihydroxyacetone phosphate is enzymatically converted to glyceraldehyde 3-phosphate for further metabolism in glycolysis.

fructose-1 6-bisP (6C)

Aldolase

CHO / CHOH / CH_2O (P) ⇌ CH_2O (P) / CO / CH_2OH

Glucose has been converted by the reactions discussed so far into two molecules of glyceraldehyde 3-phosphate. No biologically useful energy has been extracted from these reactions; indeed, there is a loss, because the high-energy phosphates from two molecules of ATP have been invested in the reactions. In the ensuing reactions the cell harvests some of the energy contained in glyceraldehyde 3-phosphate.

Step 5. Glyceraldehyde 3-phosphate reacts with an SH (sulfhydryl) group in the enzyme, glyceraldehyde 3-phosphate dehydrogenase, forming an H—C—OH group that can undergo dehydrogenation with NAD^+ as hydrogen acceptor. The product of the reaction is **phosphoglycerate,** still bound to the enzyme. Phosphoglycerate reacts with inorganic phosphate to yield **1,3-diphosphoglycerate** and free enzyme. Overall, the reaction is

$$\text{Glyceraldehyde 3-phosphate} + NAD^+ + P_i \longrightarrow \text{1,3-Diphosphoglycerate} + NADH + H^+$$

In this oxidation–reduction reaction, glyceraldehyde 3-phosphate has been converted to a diphosphoglycerate, and the newly added phosphate group is attached with an energy-rich bond.

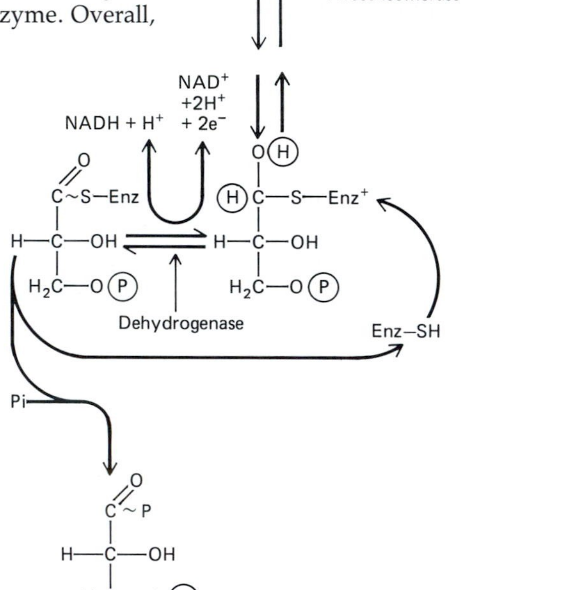

Step 6. The energy-rich phosphate reacts with ADP to form ATP. This transfer of energy from a compound with an energy-rich phosphate is referred to as a **substrate-level phosphorylation.**

$$\text{1,3-Diphosphoglycerate} + ADP \longrightarrow ATP + \text{3-Phosphoglycerate}$$

Two Diphospho-glycerate (2 × 3C)

2 ADP

2 ATP

Step 7. The 3-phosphoglycerate is rearranged to 2-phosphoglycerate by the enzymatic shift of the position of the phosphate group.

Two 3-Phospho-glycerate (2 × 3C)

Step 8. Next, an energy-rich phosphate is generated by the removal of *water* rather than by the removal of hydrogen atoms. The product is **phosphoenol pyruvate** (sometimes abbreviated **PEP**).

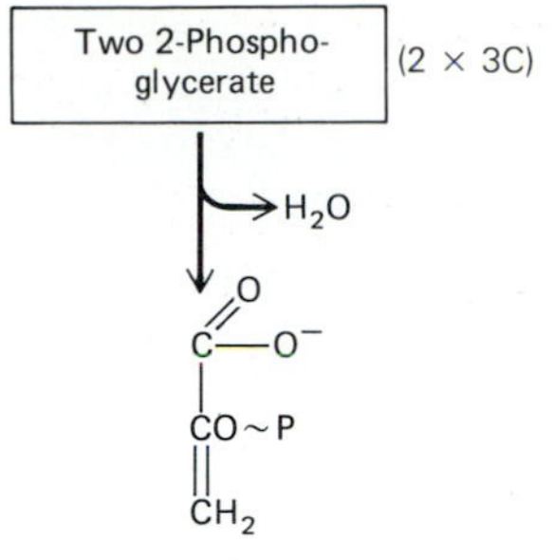

Step 9. Phosphoenol pyruvate can transfer its phosphate group to ADP to yield ATP and pyruvate.[1]

Phosphoenol pyruvate + ADP ⟶ ATP + Pyruvate

This is the second energy-rich phosphate group generated at the substrate level in the metabolism of glyceraldehyde 3-phosphate to pyruvate.

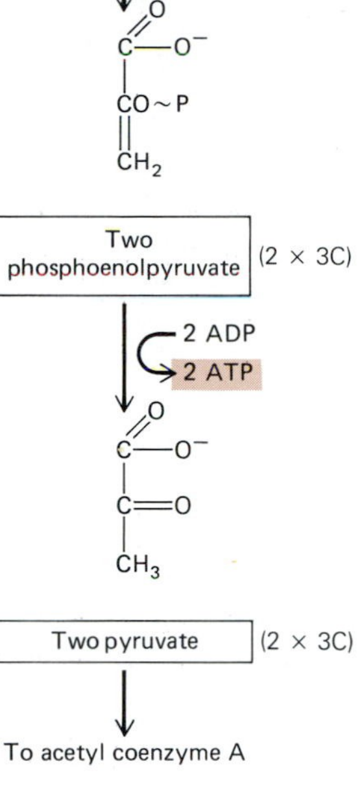

Each glucose metabolized yields *two* molecules of glyceraldehyde 3-phosphate. A total of *four* ATPs is produced as one molecule of glucose is metabolized to pyruvate (Fig. 9–2). However, two ATP molecules were utilized at the beginning of the process, one to convert glucose to glucose 6-phosphate and the second to convert fructose 6-phosphate to fructose 1,6-bisphosphate. The *net* yield in the process is 2 ATPs (4 ATPs produced minus 2 ATPs used up in the reaction). The overall reaction for glycolysis may be summarized as follows:

$$\text{Glucose} + 2\ \text{ADP} + 2\ P_i + 2\ \text{NAD}^+ \longrightarrow 2\ \text{Pyruvate} + 2\ \text{ATP} + 2\ \text{NADH} + 2\ H^+ + 2\ H_2O$$

Formation of Acetyl Coenzyme A

Pyruvate must be converted to a compound called **acetyl coenzyme A,** or simply, **acetyl CoA,** because the citric acid cycle accepts most of its fuel molecules in this form. Acetyl CoA consists of a two-carbon segment (from the original fuel molecule) bound to a large, complex compound, coenzyme A. This coenzyme contains pantothenic acid, one of the B complex vitamins.

In a complex set of reactions, pyruvate undergoes oxidative decarboxylation. First, a carboxyl group is removed as carbon dioxide (Fig. 9–3). Then, the two-carbon group remaining undergoes a make-ready reaction with coenzyme A. The compound formed has an H—C—OH group that can be dehydrogenated to yield acetyl CoA. The hydrogen removed is accepted by NAD^+. The overall reaction for the formation of coenzyme A is

$$2\ \text{Pyruvate} + 2\ \text{NAD}^+ + 2\ \text{Coenzyme A} \longrightarrow 2\ \text{Acetyl CoA} + 2\ \text{NADH} + 2\ H^+ + 2\ CO_2$$

Note that the original glucose molecule has now been oxidized to two acetyl (two-carbon) groups and two CO_2 molecules. The hydrogens removed have reduced NAD^+ to NADH, forming four NADH molecules (two during glycolysis and two during the oxidation of pyruvate).

NAD^+ → $NADH + H^+$; Coenzyme A

Pyruvate → Acetyl CoA + Carbon dioxide

Figure 9–3 The enzymatic conversion of pyruvate to acetyl coenzyme A.

[1] Pyruvate and many other compounds in glycolysis and the citric acid cycle exist as ions at the pH found in the cell. At lower pH values they associate with H^+ to form acids, such as pyruvic acid. In some textbooks they are represented in the acid form.

The Citric Acid Cycle

The **citric acid cycle** (also known as the tricarboxylic acid or TCA cycle, or the **Krebs cycle**) is the final common pathway for the oxidation of pyruvate, fatty acids, and the carbon chains of amino acids. The reactions of the citric acid cycle, which take place in the mitochondria, are described in Figure 9–4. The first reaction of the cycle occurs when coenzyme A transfers its two-carbon acetyl group to the four-carbon compound oxaloacetate, forming citrate, a six-carbon compound.

$$\underset{\text{2C compound}}{\text{Acetyl CoA}} + \underset{\text{4C compound}}{\text{Oxaloacetate}} \longrightarrow \underset{\text{6C compound}}{\text{Citrate}} + \text{CoA}$$

During the sequence of reactions that follows, the fuel molecule loses first one and then another carboxyl group (carbon–oxygen segment) as carbon dioxide. Hydrogen atoms are removed in certain reactions and transferred to the electron acceptors NAD^+ and FAD. One molecule of ATP is formed by a substrate level phosphorylation. Because two pyruvates are formed from each glucose molecule, the cycle must turn twice to process each glucose molecule. A summary reaction for the citric acid cycle is given in the Focus on Summary Reactions for Cellular Respiration.

The citric acid cycle consists of eight main steps (see Fig. 9–4):

Step 1. Acetyl CoA combines with oxaloacetate to form citrate.

Step 2. Citrate has neither an H—C—OH nor a $—CH_2—CH_2—$ group and cannot undergo dehydrogenation. It is converted by two make-ready reactions, the removal and addition of a molecule of water, to isocitrate.

Step 3. Isocitrate does have an H—C—OH group and undergoes dehydrogenation, and then decarboxylation to yield the five-carbon compound **α-ketoglutarate** and carbon dioxide.

Step 4. In the next step α-ketoglutarate undergoes oxidative decarboxylation to form **succinyl coenzyme A** and CO_2. Just as pyruvate is converted to the coenzyme A derivative of an acid with one less carbon atom (acetyl coenzyme A), α-ketoglutarate (five carbons) is converted to succinyl CoA (four carbons). The reactions involved include a dehydrogenation and a decarboxylation and, like the oxidative decarboxylation of pyruvate, require other cofactors in addition to NAD^+ and coenzyme A.

Step 5. The bond joining coenzyme A to succinate, like that in acetyl coenzyme A, is an energy-rich one and can be written ~S. The energy of the ~S bond of acetyl coenzyme A is used to bring about the addition of the acetyl group to oxaloacetate to yield citrate. The energy of the ~S bond of succinyl coenzyme A can be converted to an energy-rich phosphate bond in the form of **guanosine triphosphate, GTP** (a compound similar to ATP but containing guanine rather than adenine). The GTP is converted to ATP. This, like those we examined in glycolysis, is an example of an energy-rich bond formed at the substrate level by reactions not involving the electron transport system. In this step succinyl CoA is converted to succinate.

Step 6. Succinate is dehydrogenated to fumarate. The hydrogens are accepted by FAD, rather than by NAD^+.

Step 7. Fumarate is hydrated to form malate.

Step 8. Malate is dehydrogenated to form oxaloacetate, which can then combine with another molecule of acetyl coenzyme A. This completes the citric acid cycle.

In the course of the cycle, two molecules of CO_2 and eight hydrogen atoms (eight protons and eight electrons) are removed. The CO_2 produced accounts for the two carbon atoms that entered the citric acid cycle. You may have wondered why more hydrogen is generated by these reactions than entered the cycle with the fuel molecule. These hydrogens come from water molecules that are added during the course of the cycle. Extracting hydrogen from water requires energy, which is provided by the disruption of the chemical bonds of the fuel molecule. Some of this energy is in effect stored in the hydrogen thus generated. During the reactions of the citric acid cycle, one high-energy phosphate is synthesized at the substrate level.

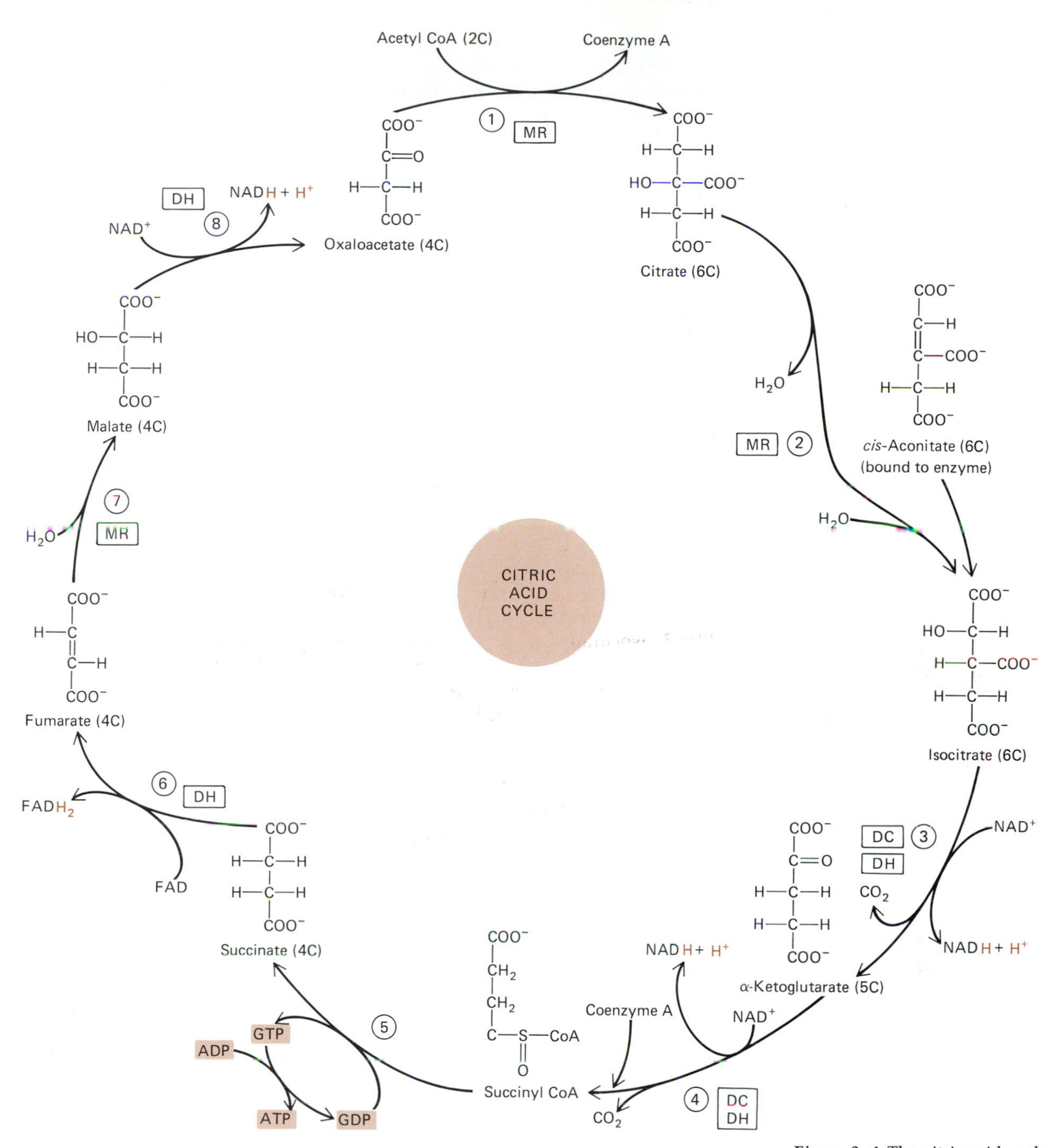

Figure 9–4 The citric acid cycle. During this series of reactions, acetyl coenzyme A, produced from glucose and other fuel molecules, is metabolized to yield carbon dioxide and hydrogen. The hydrogen is immediately combined with NAD$^+$ or FAD and is fed into the electron transport system. The reactions are designated *DH*, dehydrogenation; *DC*, decarboxylation; or *MR*, make-ready.

The Electron Transport System and Chemiosmotic Phosphorylation

At several points in glycolysis, the formation of acetyl coenzyme A, and the citric acid cycle, we have noted reactions in which hydrogens are removed from substrates and transferred to primary hydrogen acceptors—the pyridine nucleotide, NAD$^+$, or the flavin nucleotide, FAD. What becomes of these hydrogens? The hydrogen protons become separated from their electrons as the electrons are transferred from one to another of a series of electron acceptors. The protons (H$^+$) are pumped across the inner mitochondrial membrane into the intermembranous space. The inner mitochondrial membrane is impermeable to protons, and thus, an electrochemical gradient is set up across this membrane. This proton gradient provides energy for ATP synthesis. The final electron acceptor in the chain is molecular oxygen; when two electrons and two protons combine with oxygen, water is produced. This explains why we must breathe oxygen—so that it will be available to accept hydrogen and electrons passing through the electron transport system.

ELECTRON TRANSPORT

The main function of the flow of electrons along the electron chain appears to be that of pumping protons across the inner mitochondrial membrane to the intermembranous space. Electrons entering the electron transport system from NADH have a relatively high energy content. As they pass along the chain of electron carriers, such as cytochromes, they lose much of their energy. Some energy is conserved in the form of ATP, and the rest is released as heat. (The way in which this is accomplished will be discussed presently.) The passage of each pair of electrons from NADH to oxygen yields three ATPs.

Most of the electron carriers in the electron transport system are **cytochromes,** compounds with porphyrin rings somewhat similar to hemoglobin and chlorophyll. A cytochrome consists of a protein and a porphyrin ring enclosing an iron atom. It is the iron atom that combines with the electrons. There are several types of cytochromes, and each holds electrons at slightly different energy levels. Electrons are passed along from one cytochrome to the next in the chain, losing energy as they go. Finally, they combine with protons and oxygen, forming water (Fig. 9–5).

OXIDATIVE PHOSPHORYLATION

The flow of electrons is tightly coupled to the phosphorylation process and will not occur unless phosphorylation can proceed also. This, in a sense, prevents waste, for electrons will not flow unless energy-rich phosphate can be formed. If electron flow were uncoupled from phosphorylation, there would be no production of ATP and the energy of the electrons would be wasted as heat. Because the phosphorylation of ADP to form ATP is coupled with the oxidation of electron transport components, this entire process is sometimes referred to as **oxidative phosphorylation.**

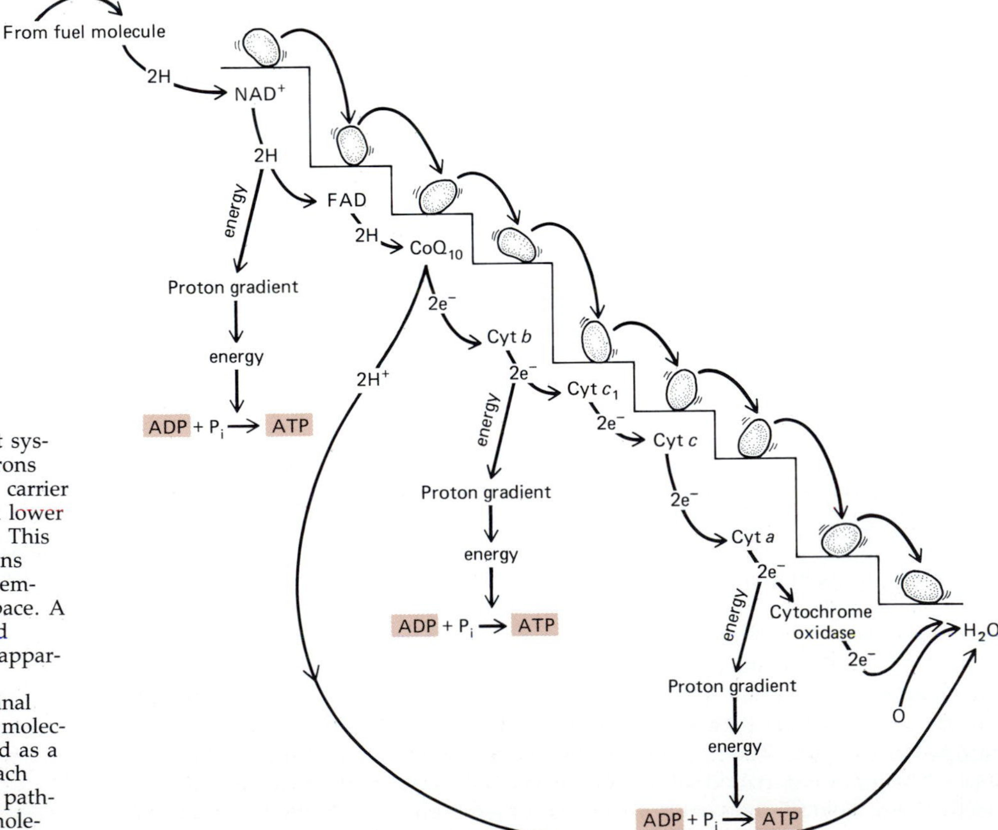

Figure 9–5 The electron transport system. As hydrogens or their electrons are transferred from one electron carrier molecule to another at lower and lower energy levels, energy is released. This energy is used to transport protons across the inner mitochondrial membrane to the intermembranous space. A membrane potential is established across the inner membrane; this apparently is the source of the energy needed to synthesize ATP. The final electron (and proton) acceptor is molecular oxygen, so water is produced as a product of these reactions. For each pair of hydrogens that enter this pathway, a maximum of three ATP molecules is produced.

In the process of oxidative phosphorylation the energy of the electron transfer potential of NADH (that is, its ability to transfer electrons to other compounds) is converted into the energy of the phosphate transfer potential of ATP. This is, in turn, the ability of ATP to transfer phosphate groups to other compounds. The electron transfer potential of a compound is reflected in its oxidation–reduction potential, or redox potential, which is expressed in volts. The redox potential of hydrogen gas, H_2, is defined as 0 volts.

A compound with a negative redox potential has a lesser electron affinity and therefore more readily loses electrons than does H_2. A compound with a positive redox potential has a greater electron affinity and more readily accepts electrons than does H_2. A strong reducing agent such as NADH has a negative redox potential; a strong oxidizing agent such as O_2 has a positive redox potential. The driving force of oxidative phosphorylation is the electron transfer potential of NADH.

REGULATION OF ENERGY PRODUCTION

The fact that phosphorylation is tightly coupled to electron flow provides a system of control that can regulate the rate of energy production and adjust it to the momentary rate of energy utilization. In a resting muscle, for example, oxidative phosphorylation continues until all of the ADP has been converted to ATP. Then, since there are no more acceptors of ATP, phosphorylation must stop. Since electron flow is tightly coupled to phosphorylation, the flow of electrons will also cease. When a biological process such as muscle contraction occurs, the energy required is obtained from the splitting of ATP to yield ADP plus inorganic phosphate, P_i, and energy. The ADP formed can then serve as an acceptor of phosphate and energy to become ATP once again. Oxidative phosphorylation continues until all of the ADP has again been converted to ATP. The electron transport system can produce ATP only when ADP is available. This serves as a kind of biochemical governor. Electric generating systems have an analogous control device that adjusts the rate of production of electricity to the rate of utilization of electricity.

THE MOLECULAR ORGANIZATION OF MITOCHONDRIA

Recall from Chapter 4 that each mitochondrion has two membranes, an outer smooth one and an inner one that is folded repeatedly to form the shelflike projections called cristae (Figs. 9–6 and 9–8). The enzymes of the citric acid cycle are found in the soluble matrix, in the internal region bounded by the inner mitochondrial membrane. The electron carriers of the electron transport system are tightly bound to the inner membrane (Fig. 9–7). This poses a problem: Most of the cell's ATP is generated by enzyme systems located on the inner membrane of the mitochondrion. However, ATP is *utilized* in the synthesis of proteins, fats, carbohydrates, nucleic acids and other molecules, in the transport of substances across cell membranes, in the conduction of nerve impulses, and in the contraction of muscle cells. All of these reactions occur largely or entirely *outside of* the mitochondria, in other parts of the cell. Yet large charged molecules such as ATP cannot readily diffuse across the lipid bilayer of the inner mitochondrial membrane. A special carrier protein for adenine nucleotides enables ADP and ATP to cross this permeability barrier. The flow of ATP is coupled to the flow of ADP; ATP can exit from the mitochondrial matrix only if ADP enters it. In any case, ATP can be manufactured only if the raw material ADP is continuously supplied. The coupled flow of ADP and ATP is mediated by an enzyme called **ATP–ADP translocase,** which makes up about 6% of the protein in the inner mitochondrial membrane.

THE CHEMIOSMOTIC THEORY OF OXIDATIVE PHOSPHORYLATION

Although it had long been known that oxidative phosphorylation occurs in the mitochondria, and many experiments showed that the transfer of electrons from NADH to oxygen resulted in the production of three ATP molecules, just how the ATPs were synthesized remained a mystery. From the redox potentials of the various members of the electron transport system it was possible to infer where the phosphorylations were coupled. The mitochondrial enzyme that converts ADP and P_i to ATP was isolated and characterized. Several theories suggested that some sort of energy-rich intermediate was formed and transferred its energy to drive the

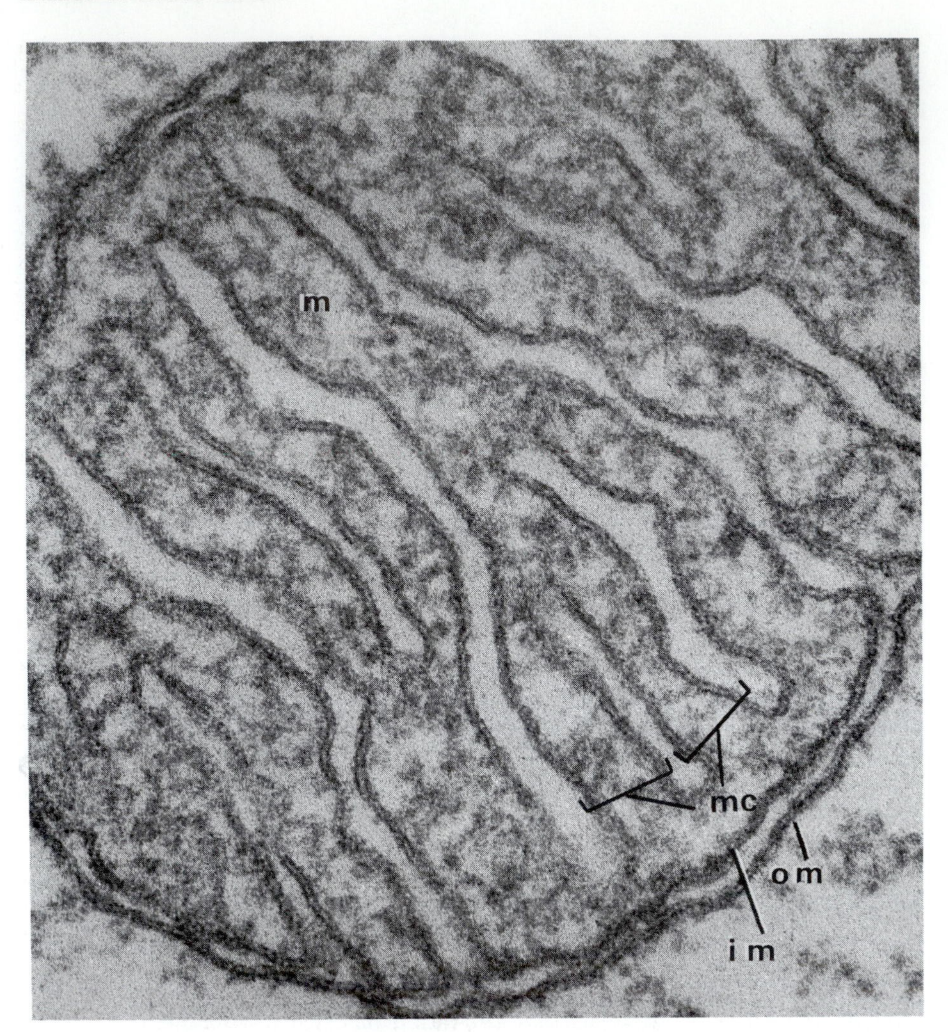

Figure 9–6 Electron micrograph of a mitochondrion from a pancreatic acinar cell (approximately ×207,000). The outer membrane (*om*) is smooth, but the inner membrane (*im*) is folded to form the cristae (*mc*); *m*, matrix. (Courtesy of G. E. Palade.)

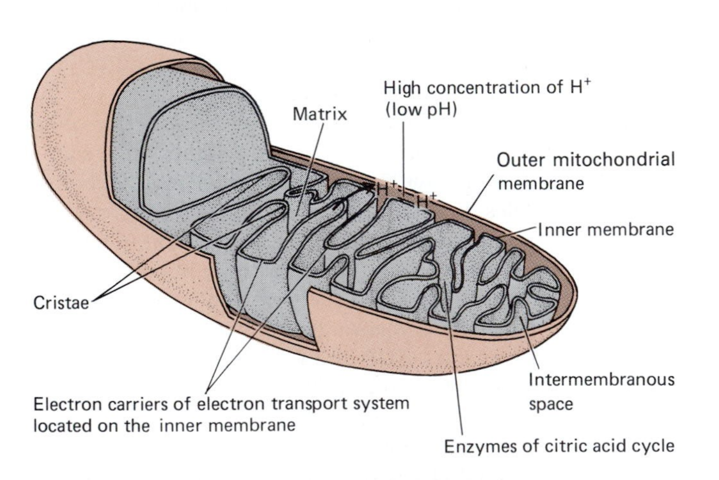

Figure 9–7 The organization of the mitochondrion.

synthesis of ATP, but as was the case with photosynthesis, no such intermediate could be discovered despite an intensive search in many laboratories.

A very different mechanism, the **chemiosmotic hypothesis,** was suggested by Peter Mitchell in 1961. This has been supported by experimental evidence from many laboratories, and Mitchell was awarded a Nobel Prize in 1978. Mitchell proposed that electron transport and ATP synthesis are coupled by a proton gradient across the mitochondrial membrane. According to this model, the stepwise transfer of electrons from NADH or $FADH_2$ through the electron carriers to oxygen results in the pumping of protons across the inner mitochondrial membrane into the space between the inner and outer mitochondrial membranes. Protons are pumped out of the matrix by three kinds of electron transfer complexes, each associated with a

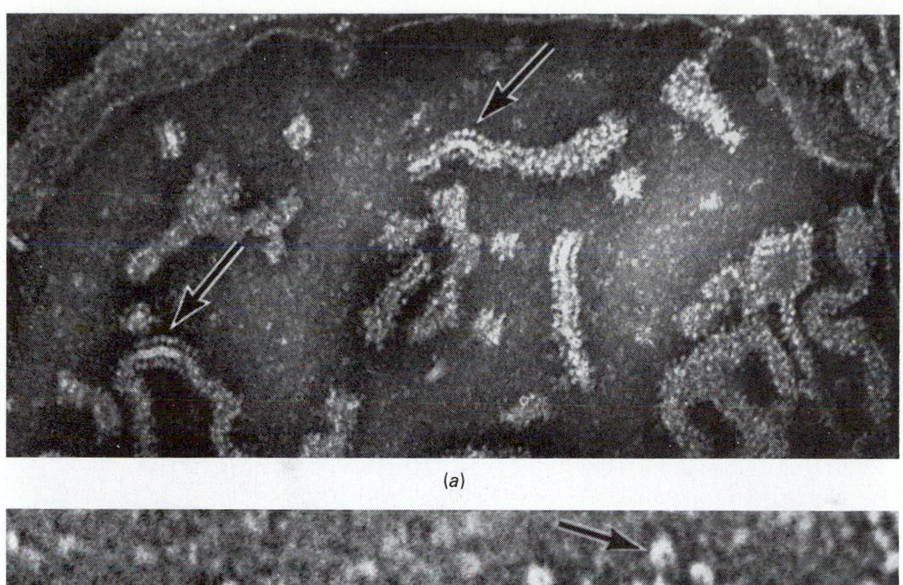

(a)

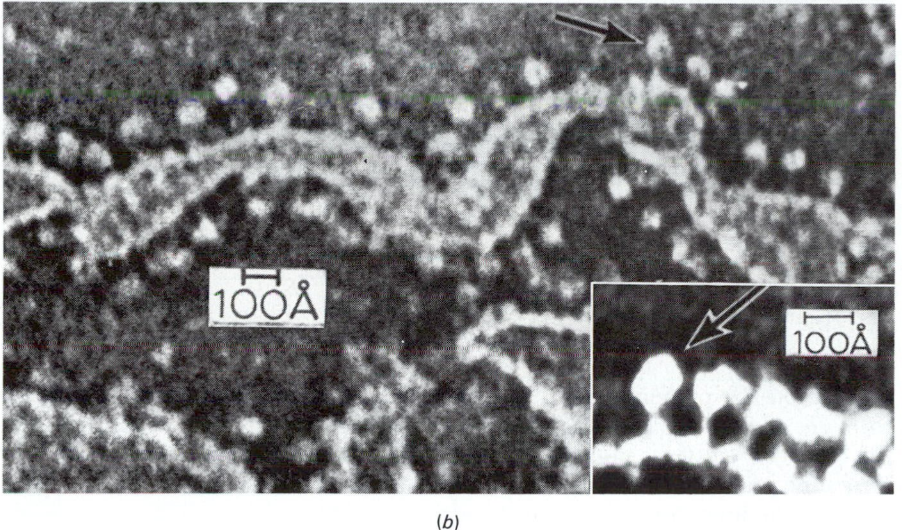

(b)

Figure 9–8 Electron micrograph of a mitochondrion swollen in a hypotonic solution and negatively stained with phosphotungstate. (*a*) Isolated cristae (×85,000). (*b*) At higher magnification (×500,000), the particles attached by a stalk to the surface of the cristae are evident. These may be the F_0–F_1 ATPase complex. (Courtesy of H. Fernandez-Moran.)

particular step in the electron transport system. This process generates a membrane potential across the inner mitochondrial membrane, with the medium in the intermembrane space positively charged (Fig. 9–9).

The difference in concentration of hydrogen ions between the matrix and the intermembrane space represents potential energy. This potential energy results in part from the difference in pH and in part from the difference in electric charge on the two sides of the membrane. When the protons are permitted to flow back down the proton gradient into the matrix, energy is released that can be used to convert ADP and P_i to ATP.

The protons flow back to the matrix of the mitochondria through special channels in the inner membrane. These channels occur in the enzyme **F_0–F_1 ATPase** that synthesizes ATP from ADP and P_i. (The F_0–F_1 ATPase of the mitochondria is similar to the CF_0–CF_1 complex that serves a corresponding function in the thylakoids during photosynthesis.) As the protons move down the energy gradient, the energy released is used to drive the synthesis of ATP. Thus the proton gradient across the inner mitochondrial membrane couples phosphorylation with oxidation.

The chemiosomotic model of the mitochondrion requires that there be a sharp topological and functional distinction between the inside and the outside of this organelle. This in turn implies that there must be an exact spatial organization of the various electron transport enzymes so that they have definite locations with respect to the inner and outer lipid monolayers of which the inner membrane is composed. However, some of these proteins are able to pass electrons to one another, and since the proteins have a variety of positions within the membrane, the electrons transferred in this fashion necessarily follow a circuitous path within the confines of the inner membrane of the mitochondrion. Although the details are still somewhat uncertain and controversial, it is clear that these electron transfers some-

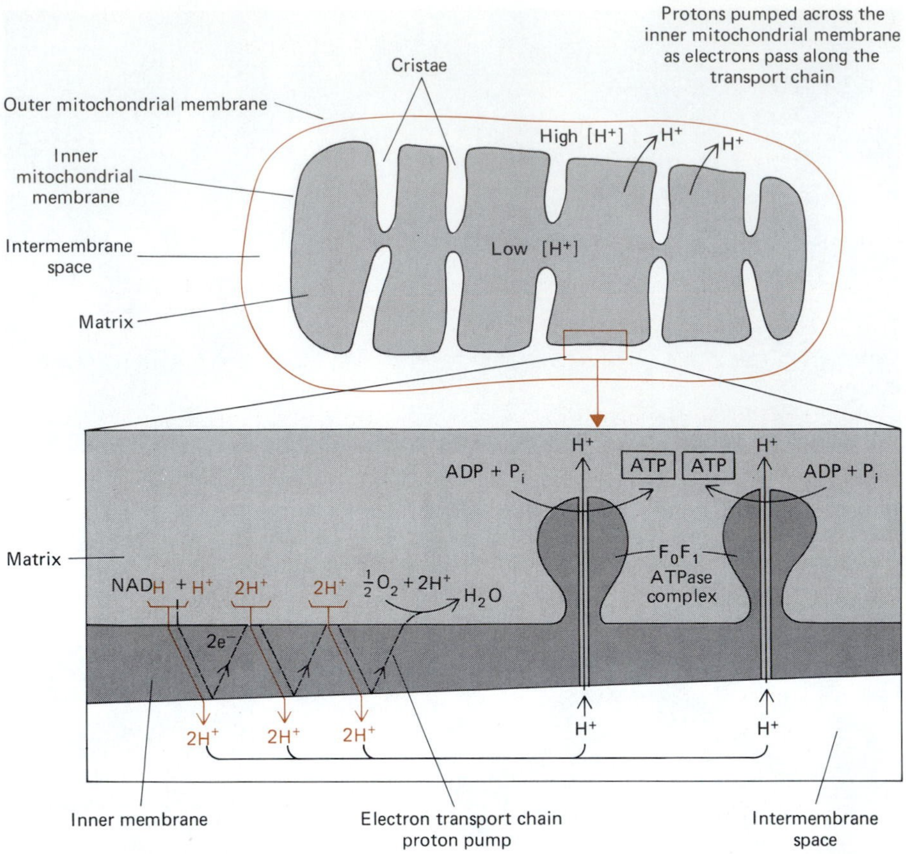

Figure 9–9 Chemiosmosis. According to the chemiosmotic theory, the electron transport chain in the inner mitochondrial membrane is a proton pump. The energy released during electron transport from substrate to oxygen is used to transport protons, H^+, from the mitochondrial matrix to the intermembranous space, where the protons become concentrated. There are three sites in the electron transport chain in which protons are transported. The protons are prevented from diffusing back into the matrix through the inner membrane everywhere except through special pores in ATPase molecule complexes in the membrane. The flow of the protons through the ATPase generates ATP at the expense of the free energy released as the protons pass from a region of high concentration (outside) to a region of low concentration (inside).

how enable the protein carriers to extract energy from the electrons as they pass through, and to use this energy to drive a proton pumping mechanism. Probably, the carriers themselves pump the protons—in a definite direction, against a concentration gradient, for storage in a definite location, the intermembranous space between the inner and outer mitochondrial membranes.

From the intermembranous space protons are permitted to leak gradually back into the matrix. However, the leakage of these protons is possible only through the F_0–F_1 protein complexes located in the membrane, because the inner membrane is otherwise impermeable to the passage of protons. Exactly how the F_0–F_1 complex manages to synthesize ATP is not yet known.

Energy Yield from Glucose

Let us now review where biologically useful energy is released and calculate the total energy yield from the complete oxidation of glucose. Table 9–1 summarizes the arithmetic involved. In glycolysis—(1) in the table—glucose is activated by the addition of two ATP molecules and converted ultimately to 2 pyruvate + 2 NADH + 2 H^+ + 4 ATP, yielding a direct net profit of two ATPs. The two pyruvates are metabolized, at (2), to 2 CO_2 + 2 NADH + 2 H^+ + 2 acetyl CoA. In the citric acid cycle, (3), the two acetyl coenzyme A molecules are metabolized to 4 CO_2 + 6 NADH + 6 H^+ + 2 $FADH_2$ + 2 ATP; and these are the two ATP molecules formed at the substrate level from succinyl coenzyme A.

Since the oxidation of NADH + H^+ in the electron transport system yields 3 ATPs per mole, the 10 NADH molecules can yield up to 30 ATPs. In skeletal muscle and some other types of cells, the 2 NADH molecules from glycolysis yield only 2 ATPs each (see Focus on Shuttles Across the Mitochondrial Membrane), so in those cells, the total number of ATPs from NADH is only 28. The oxidation of the reduced flavin, $FADH_2$, yields 2 ATPs per mole, so the two $FADH_2$ molecules yield 4 ATPs. Summing these at (4) in Table 9–1, we see that the complete aerobic metabolism of one mole (180 g) of glucose yields a maximum of 36 to 38 ATPs. Note that all but two ATP molecules were generated by reactions taking place in the mitochondria, and all but two of those generated in the mitochondria (the two formed at the

TABLE 9–1
Energy Yield from the Complete Oxidation of One Mole of Glucose

(1) Net ATP yield from glycolysis			2 ATP* (substrate level)
Also from glycolysis	2 NADH + H^+	⟶	4–6 ATP
(2) Two pyruvate molecules to two acetyl CoA molecules	2 NADH + H^+	⟶	6 ATP
(3) Two acetyl CoA molecules, through citric acid cycle			2 ATP (substrate level)
	6 NADH + 6 H^+	⟶	18 ATP
	2 $FADH_2$	⟶	4 ATP
(4) Total ATP yield			36–38 ATP

*These are the only two ATPs that can be generated anaerobically. Production of all the other ATPs depends upon the presence of oxygen.

substrate level in the citric acid cycle) were produced by the electron transport system.

When a mole of glucose is burned in a calorimeter, some 686 kcal are released as heat. The free energy stored in one energy-rich bond of ATP is about 7.3 kcal. When 36 ATPs are generated during the complete biological oxidation of glucose, the total energy stored in ATP amounts to 7.3 kcal × 36, or about 263 kcal. When energy-rich phosphate bonds are cleaved, this energy becomes available for cellular work. Thus, the efficiency of cellular respiration is 263/686, or about 38%. This compares very favorably with the efficiency of the finest machines. The remainder of the energy in the glucose is released as heat, utilized by some animals to help maintain body temperature.

FOCUS ON
Shuttles Across the Mitochondrial Membrane

The inner mitochondrial membrane is not permeable to NADH, which is a large molecule. The NADH produced in the cytoplasm from the dehydrogenations in glycolysis cannot diffuse into the mitochondria to transfer their electrons to the electron transport system. Unlike ATP and ADP, NADH does not have a carrier protein to transport it across the membrane. Instead, several systems have evolved to transfer the *electrons* of NADH (though not the NADH molecules themselves) into the mitochondria.

In liver, kidney, and heart cells, a special shuttle system known as the **malate–aspartate shuttle** transfers the electrons from NADH through the inner mitochondrial membrane. Once inside the matrix, malate passes the electrons to a different NAD^+, one already in the matrix. These electrons can then be passed along the electron transport system in the inner membrane, and three molecules of ATP can be produced.

In skeletal muscle, brain, and some other types of cells, another type of shuttle, the **glycerol phosphate shuttle,** operates. This shuttle requires more energy than the malate–aspartate shuttle. As a result, the electrons are at a lower energy level when they enter the electron transport chain. They are accepted by coenzyme Q, rather than by NAD^+, and so generate only two ATP molecules per pair of electrons. This is why the number of ATPs produced by the complete oxidation of a mole of glucose in skeletal muscle cells is 36 rather than 38. The NADH and $FADH_2$ produced in the oxidation of fatty acids are formed within the mitochondria and can pass electrons directly to the electron transport system.

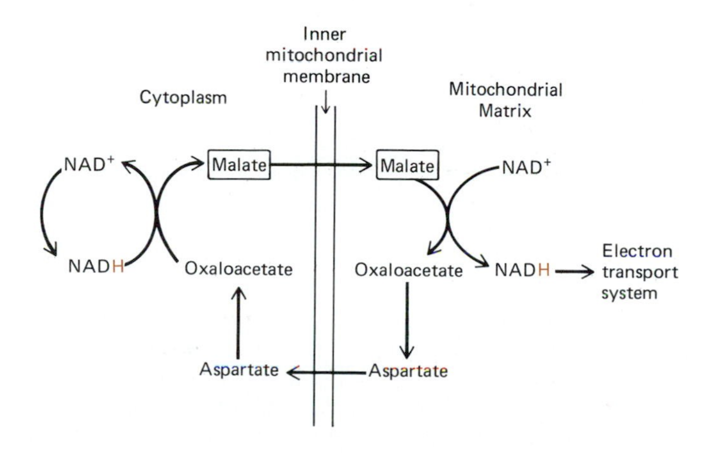

Anaerobic Pathways

The preceding discussion focused upon aerobic respiration, a reaction pathway that requires the presence of oxygen. In plant and animal cells neither the citric acid cycle nor the electron transport system can operate without oxygen. Oxygen is the final hydrogen and electron acceptor in the electron transport chain. When oxygen is not available, the last cytochrome in the chain cannot transfer its electrons. The preceding electron acceptor then has no acceptor to give its hydrogen electrons to, and so on down the chain. The entire system becomes blocked all the way back to NADH, and no further ATPs can be produced by way of the electron transport system.

Some bacteria carry on **anaerobic respiration,** that is, respiration that proceeds without oxygen. These organisms utilize nitrite (NO_2^-), sulfate (SO_4^-), or some other inorganic compound as electron acceptors.

In **fermentation,** which is also anaerobic, ATPs are generated by a process that utilizes organic compounds both as donors and as acceptors of electrons. Certain types of bacteria engage solely in fermentation; others shift to this process when deprived of oxygen. Production of alcohol from sugar is an example of fermentation. When deprived of oxygen, yeast cells convert pyruvate to **ethyl alcohol:**

$$\text{Pyruvate} + \text{NADH} + \text{H}^+ \longrightarrow \text{Ethyl alcohol} + \text{CO}_2 + \text{NAD}^+$$

If we had similar enzymes we could get drunk by running up several flights of stairs!

Certain bacteria and some animal cells, most notably muscle cells, can release energy anaerobically (as in fermentation) for a time by producing lactate (Fig. 9–10). The hydrogen atoms removed from the fuel molecule during glycolysis are transferred to pyruvate. The enzyme lactate dehydrogenase present in these cells converts pyruvate to **lactate** (the ionic form of **lactic acid**).

$$\text{Pyruvate} + \text{NADH} + \text{H}^+ \longrightarrow \text{Lactate} + \text{NAD}$$

This occurs during strenuous exercise when the amount of oxygen delivered to the muscle cells may be insufficient to keep pace with the rapid rate of glycolysis. Not all of the hydrogen atoms accepted by NAD^+ can be processed in the usual manner

Figure 9–10 Comparison of anaerobic with aerobic pathways. When oxygen is not available, fermentation occurs in certain bacteria, in yeasts, and in muscle cells.

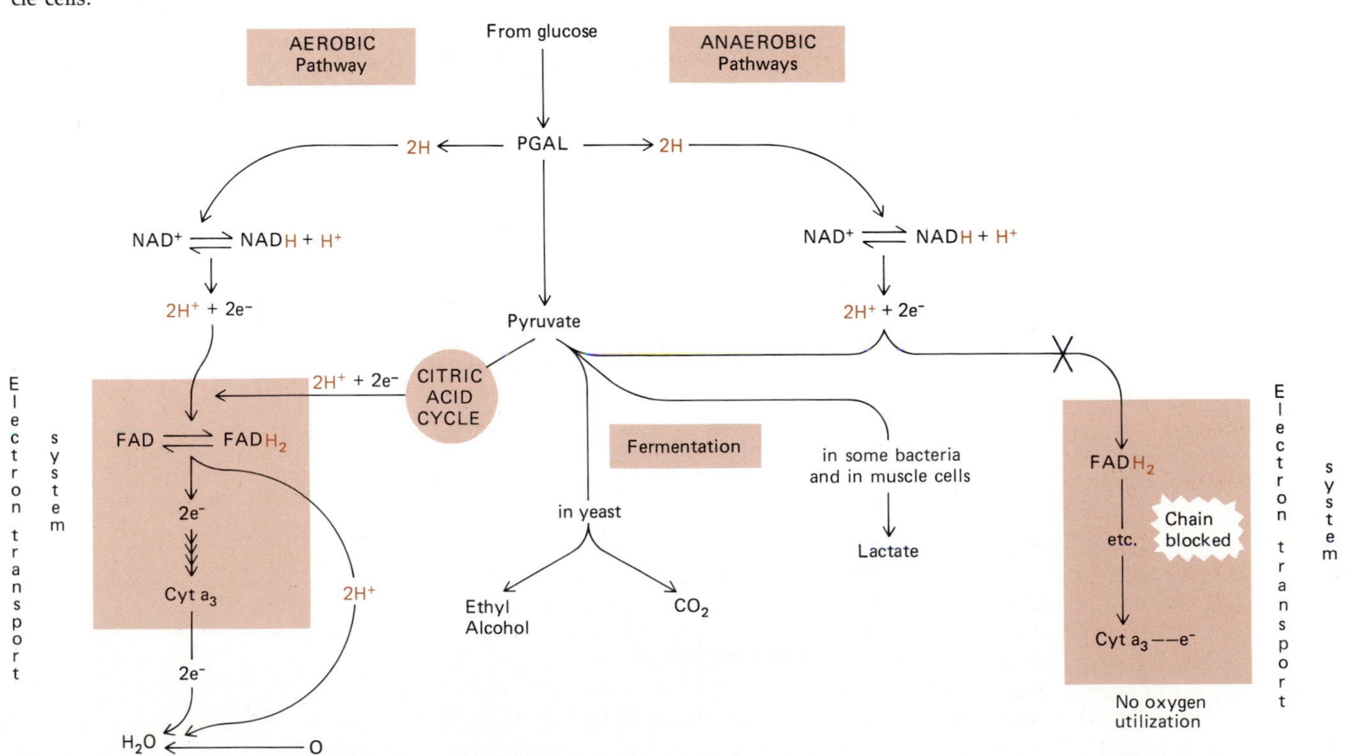

because there is a shortage of oxygen. Instead, these hydrogens are transferred to pyruvate to form lactate.

Lactate acidifies the blood, which stimulates respiration, causing rapid breathing. However, the additional oxygen supplied may not be sufficient to meet the needs of the rapidly metabolizing muscle cells. As lactate accumulates in muscle cells, it lowers the pH and contributes to **muscle fatigue.** Some of the lactate diffuses into the blood and is transported to the liver.

After strenuous exercise, there is generally a period of rapid breathing (panting). It is as though by running faster than the circulatory system can supply oxygen to the muscles, the athlete has incurred an **oxygen debt** that must be repaid after the exertion is over. The extra oxygen taken in is needed to oxidize the lactate in the liver cells to pyruvate, which can then be converted back to glucose or oxidized aerobically in the citric acid cycle.

Alcohol, the end-product of fermentation by yeast cells, can be burned; it could even be used as an automobile fuel. Obviously it contains a great deal of energy that the yeast cells were unable to utilize by anaerobic methods. Lactate, a three-carbon compound, contains even more energy than that in alcohol (a two-carbon compound). During aerobic respiration all available energy is released by completely oxidizing glucose or other fuel molecules. A net yield of only two ATP molecules can be produced anaerobically from a molecule of glucose, compared with 36 more produced when oxygen is available. The two ATP molecules produced during glycolysis represent less than 5% of the total energy in a molecule of glucose.

The inefficiency of anaerobic processes (fermentation) necessitates the use of a large amount of fuel. By rapidly degrading many fuel molecules, a cell can compensate somewhat for the small amount of energy that can be gained from each. To perform the same amount of work, a cell functioning anaerobically must consume up to 20 times as much glucose or other carbohydrate as an aerobic cell. For this reason skeletal muscle cells, which often function anaerobically for short periods, store large quantities of glucose in the form of glycogen.

Oxidation of Other Nutrients

Many organisms depend upon nutrients other than glucose (or in addition to glucose) as a source of energy. Human beings and many other animals usually obtain more of their energy by oxidizing fatty acids than by oxidizing glucose. Amino acids are also used as fuel molecules. Such nutrients can be transformed into one of the metabolic intermediates that can be fed into glycolysis or the citric acid cycle.

OXIDATION OF AMINO ACIDS

Amino acids are metabolized by reactions in which the amino group is first removed, a process called **deamination.** The amino group is converted to urea and excreted, but the carbon chain is metabolized and eventually enters the citric acid cycle. The sequence of reactions varies somewhat with different amino acids, but for each, a series of reactions modifies the carbon skeleton to produce a substance that is part of the citric acid cycle. Alanine, for example, undergoes deamination to become pyruvate, glutamate is converted to α-ketoglutarate, and aspartate yields oxaloacetate. Other amino acids may require several enzyme-catalyzed reactions in addition to deamination to yield a substance that is a member of the citric acid cycle. Ultimately the carbon chains of all the amino acids are metabolized in this way.

OXIDATION OF FATTY ACIDS

Each gram of triacylglycerol contains more than twice as many kilocalories as that in a gram of glucose or amino acids. When completely metabolized in cellular respiration, a molecule of a six-carbon fatty acid can generate up to 44 ATPs (compared with 36 for glucose, which also has six carbons). Both the glycerol and fatty acid components of a neutral fat can be used as fuel. Phosphate is added to glycerol, converting it ultimately to PGAL or another compound that undergoes glycolysis.

Fatty acids are oxidized and split enzymatically into two-carbon compounds (acetyl groups) bound to coenzyme A; that is, they are converted to acetyl coen-

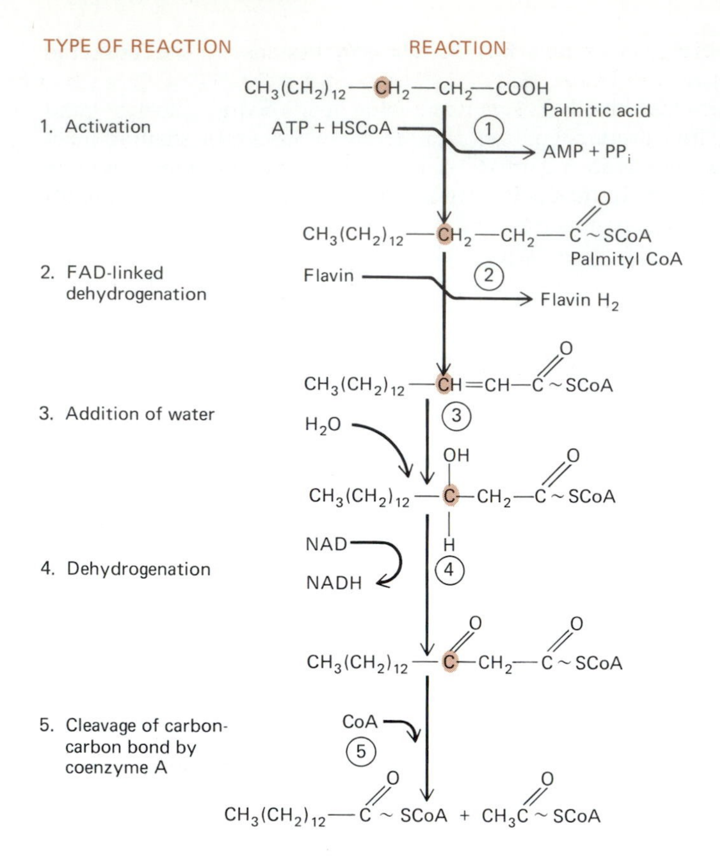

Figure 9–11 Chemical reactions in β-oxidation of fatty acids.

zyme A. This process, which occurs in the matrix of the mitochondria, is known as **β-oxidation** (beta oxidation). The acetyl CoA molecules enter the citric acid cycle and are converted to CO_2 with the release of hydrogen.

The long chain of a fatty acid is split into two-carbon acetyl CoA units by a cyclic sequence of reactions that include dehydrogenations and make-ready reactions but not decarboxylations. The overall process is termed β-oxidation because the oxygen atom is added to the β carbon, the second carbon from the carboxyl group. For example, palmitate, a fatty acid with a chain of 16 carbon atoms, is first activated by enzyme-catalyzed reactions with ATP and coenzyme A to form palmityl CoA (Fig. 9–11). This undergoes a dehydrogenation between the second and third carbons of the chain. The group undergoing dehydrogenation is a $—CH_2—CH_2—$ group; the primary acceptor of the electrons and protons liberated is a flavin. The product, a long-chain molecule with a double bond between carbons 2 and 3, undergoes a make-ready reaction, the addition of a molecule of water across the double bond. The resulting molecule has an H—C—OH group at the β carbon (Fig. 9–11) that can undergo dehydrogenation with NAD^+ as acceptor.

The product of the dehydrogenation, a long-chain fatty acid with a C=O group at the β carbon (carbon 3) and with coenzyme A still attached to its carboxyl group, reacts with a second molecule of coenzyme A. The coenzyme A attacks at the C=O group and attaches to it, cleaving off carbons 1 and 2 with the original coenzyme A group attached as acetyl CoA and leaving a carbon chain that is two carbons shorter. This product has coenzyme A attached to its carboxyl group and is ready, without any further activation, to be dehydrogenated by the enzyme that uses flavin as the primary acceptor.

Four enzymes working in succession catalyze these four reactions and yield an acetyl CoA molecule plus the remaining carbon chain with a coenzyme A attached. Seven such series of dehydrogenations and make-ready reactions will split the 16-carbon chain of palmitate into eight two-carbon fragments, each with a coenzyme A group attached—eight acetyl CoA molecules. These molecules can be metabolized in the citric acid cycle.

Each cycle of β-oxidation—dehydrogenation, hydration, dehydrogenation—produces one NADH molecule and one $FADH_2$ molecule. When their electrons are passed through the electron transport system, the NADH produces 3 and the

$FADH_2$ produces 2 ATPs, for a total of 5 ATPs. The seven cycles of β-oxidation that split palmitate to eight acetyl CoA molecules thus yield $7 \times 5 = 35$ ATPs. Each of the acetyl CoA groups yields 12 ATPs as it is metabolized in the citric acid cycle: $8 \times 12 = 96$ ATPs. Thus the yield of ATPs, from the complete oxidation of one palmitate to CO_2 and H_2O, is $96 + 35 = 131$ ATPs. This is why fat is such a superlative energy storage molecule (and why it takes so long to lose fat by dieting!).

Biosynthetic Reactions

In addition to the reactions that break down molecules and conserve their energy as ATP, all cells have a remarkable array of enzymes for synthesizing molecules of proteins, fats, carbohydrates, nucleic acids, steroids, and other compounds from simple precursors. This **anabolic** (synthetic) aspect of metabolism is quite complex, but a few basic principles of cellular biosynthesis can be listed here:

1. Each cell, in general, synthesizes its own proteins, nucleic acids, lipids, polysaccharides, and other complex molecules and does not receive them preformed from some other cell. Muscle glycogen, for example, is synthesized within the muscle cell and is not derived from liver glycogen.
2. Each step in a biosynthetic process is catalyzed by a separate enzyme. Each enzyme is genetically determined by a separate gene.
3. Although certain steps in a biosynthetic sequence may proceed without the use of ATP, the overall synthesis of these complex molecules is strongly endergonic and requires chemical energy. Why should this be true? Can you relate this to your understanding of the concepts of energy and entropy?
4. These synthetic processes use only a few substances as raw materials. These include acetyl coenzyme A, glycine, succinyl coenzyme A, ribose, pyruvate, and glycerol.
5. The synthetic processes are, in general, not simply the reverse of the processes by which the molecule is broken down. Each includes one or more steps that differ from any step in the degradation process. These steps are catalyzed by different enzymes, permitting separate control mechanisms for the synthesis and breakdown of the molecule.
6. The biosynthetic process includes not only the formation of the various macromolecular components but their assembly into the several kinds of membranes that compose the cell's outer boundary and its intracellular organelles. Each cell's constituent molecules are in a dynamic state—they are constantly being degraded and synthesized.

Even a cell that is not growing, not increasing in mass, uses a considerable portion of its total energy for the chemical work of biosynthesis. A rapidly growing cell must allocate a correspondingly larger fraction of its total energy output to biosynthetic processes. A rapidly growing bacterial cell may use as much as 90% of its total energy for the synthesis of proteins!

In biosynthetic reactions in which peptide, glycosidic, or ester bonds are formed, the bond is formed not merely by the simple removal of a molecule of water. The synthesis of sucrose in the sugar cane plant, for example, does not proceed via

$$\text{Glucose} + \text{Fructose} \longrightarrow \text{Sucrose} + H_2O$$

This reaction would require the input of energy, some 5.5 kcal per mole, to go to the right if all reactants were present in a concentration of 1 mole per liter. However, the concentrations of glucose and fructose in the plant cell are probably less than 0.01 mole per liter, and the concentration of water is very high, about 55 moles per liter. With all this water available, the equilibrium point of the reaction would be very far to the left, toward the splitting of sucrose into the two simple sugars, rather than toward sucrose synthesis.

Instead, in biosynthetic reactions one of the reactants is typically activated by a reaction with ATP. The terminal phosphate of ATP is transferred by an enzyme to glucose, with the conservation of some of the energy of the high energy phosphate bond. Glucose phosphate, with a higher energy content than that of free glucose, can react with fructose (in a reaction catalyzed by another enzyme) to yield sucrose and inorganic phosphate:

$$\text{ATP} + \text{Glucose} \longrightarrow \text{ADP} + \text{Glucose 1-phosphate}$$

$$\text{Glucose 1-phosphate} + \text{Fructose} \longrightarrow \text{Sucrose} + \text{Phosphate}$$

$$\text{Sum: ATP} + \text{Glucose} + \text{Fructose} \longrightarrow \text{Sucrose} + \text{ADP} + \text{P}_i$$

The overall reaction proceeds to the right because there is a net decrease in energy (and a net increase in entropy). Comparable reactions, with one of the reactants activated by forming its phosphate, adenylate, or coenzyme A derivative, occur in the synthesis of proteins, nucleic acids, and fatty acids.

SUMMARY

I. The pattern of cellular respiration is basically similar in all forms of life. The process by which glucose is oxidized to carbon dioxide and water with the release of energy can be divided into four phases: glycolysis, formation of acetyl coenzyme A, the citric acid cycle, and the electron transport system.

II. Glycolysis is the sequence of reactions that convert glucose to pyruvate with the net production of two ATP molecules. The glycolytic enzymes are dissolved in the cytoplasm.

III. Pyruvate moves into the mitochondria and is decarboxylated; the remaining two-carbon molecule undergoes dehydrogenation and combines with coenzyme A to form acetyl coenzyme A.

IV. Acetyl coenzyme A enters the citric acid cycle, where the two-carbon compound is completely degraded to carbon dioxide.

V. Hydrogens (protons and electrons) removed from the fuel molecule during glycolysis and the citric acid cycle are transferred to NAD^+ or FAD. These protons and electrons are then transferred to oxygen via the various electron acceptors that make up the electron transport system.

- *A.* In the electron transport system, phosphorylation is tightly coupled to electron flow; this coupling regulates the rate of energy production to the rate of energy utilization.
- *B.* As electrons are passed through the electron transport chain from primary acceptors to cytochromes and finally to oxygen, some energy is conserved as ATP.
- *C.* According to the chemiosmotic theory, the spatially oriented flow of electrons generates a proton gradient across the inner mitochondrial membrane, and the flow of protons back through the membrane is coupled to the generation of ATP.
- *D.* Special carrier proteins and shuttles are present to transfer ADP, ATP, protons, and the electrons from NADH across the inner mitochondrial membrane, which is relatively impermeable to large, charged molecules.

VI. Organic nutrients other than glucose can be converted into appropriate compounds and fed into the glycolytic or citric acid pathway.

- *A.* Amino acids can be deaminated and the carbon skeleton converted to a metabolic intermediate such as pyruvate.
- *B.* Both the glycerol and fatty acid components of fats can be oxidized as fuel. Fatty acids are converted to acetyl coenzyme A molecules by the process of β-oxidation.

VII. The cells of living things exist in a dynamic state and are continuously building up and breaking down the many different cell constituents.

- *A.* In general, each cell synthesizes its own complex macromolecules, and each step in the process is catalyzed by a separate enzyme.
- *B.* These biosynthetic reactions are strongly endergonic and require ATP to drive them.

POST-TEST

1. The reactions by which cells extract energy from fuel molecules, utilize oxygen, and produce carbon dioxide and water are termed ______ ______.

2. The conversion of glucose to pyruvate with the formation of ATP is termed ______.

3. The cyclic sequence of enzymatic reactions that converts acetyl coenzyme A to carbon dioxide and removes electrons and protons is called the ______ ______ ______.

4. The passage of electrons from substrate molecules to oxygen in the mitochondria involves a series of reactions that make up the ______ ______ ______.

5. Dehydrogenase molecules remove electrons from fuel molecules and transfer them to primary acceptors such as ______ and ______.

6. Both glucose and fatty acids must be ______ before they can be metabolized; this requires the investment of energy from ______.

7. Amino acids are oxidized by reactions in which the amino group is removed, termed ______, and the carbon chain eventually enters the ______ ______ ______.

8. In ______ reactions two electrons and two protons are removed from the substrate and transferred to a primary acceptor.

9. The carbon dioxide we exhale each day is derived from ______ reactions.

10. According to the chemiosmotic hypothesis, a ________ gradient is set up across the inner mitochondrial membrane; then, when ________ flow back to the mitochondrial matrix through special channels in the membrane, ________ is synthesized.

11. In the conversion of glyceraldehyde 3-phosphate to 1,3-diphosphoglycerate, ________ is synthesized.

12. The enzymes of the electron transport system are located on the ________ ________ of the ________.

13. The flow of electrons in the electron transport system is tightly coupled to ________.

14. The complete aerobic metabolism of one mole of glucose yields ________ ATPs or ________ kcal.

15. Although most of the cell's ATP is generated by enzymes located inside the ________, most of the processes in which ATP is utilized occur in the ________.

16. Production of alcohol from sugar is an example of ________.

17. During strenuous exercise, muscle cells convert pyruvate to ________.

REVIEW QUESTIONS

1. What arguments would you use to justify calling mitochondria the power plants of the cell?
2. Why are ATP molecules consumed during the first steps of glycolysis?
3. Examine a sequence of reactions such as glycolysis or the citric acid cycle, and identify the dehydrogenations, decarboxylations, and make-ready reactions.
4. Explain the roles in cellular respiration of coenzyme A, cytochromes, NAD^+, and FAD.
5. Calculate how much energy is made available to the cell by the operation of the citric acid cycle, by glycolysis, and by the electron transport system, and estimate the efficiency of each phase and of the overall process of cellular respiration.
6. Draw a diagram of the mitochondrion and indicate:
   a. the site of enzymes of the citric acid cycle;
   b. the site of the enzymes of the electron transport system; and
   c. the site of the proton gradient that drives the production of ATP from ADP and P_i.
7. Compare the conversion of pyruvate to acetyl coenzyme A with the conversion of α-ketoglutarate to succinyl coenzyme A.
8. The metabolism of a gram of fat (triacylglycerol) to CO_2 yields about 9 kcal, whereas the metabolism of a gram of glucose to CO_2 yields about 4 kcal. From what you know about the production of ATP in various reactions, how would you account for this?
9. Describe the reactions by means of which ATP is produced at the substrate level in glycolysis and in the citric acid cycle. In what ways do they differ?
10. What is the specific role of oxygen in a cell? What happens when cells are deprived of oxygen?
11. What is the end-product of fermentation in human muscle? in yeast cells?
12. What must be done to amino acids before they can be utilized as fuel?
13. Summarize the process of β-oxidation of fatty acids.
14. Why is it advantageous that synthetic reactions are generally not the reverse of the reactions in which molecules are broken down?
15. Summarize the chemiosmotic theory of ATP production.

Recommended Readings for Part II

Books

Alberts, B., et al. *Molecular Biology of the Cell.* New York, Garland Publishing, Inc., 1983. A detailed, very well-written account of energy conversions that take place within chloroplasts and mitochondria.

Barber, J. (ed.). *Topics in Photosynthesis,* vols. 1–5. New York, Elsevier, 1976–1981 (a continuing series). Carefully edited research papers reporting the latest developments in our understanding of the photosynthetic process.

Becker, W. M. *Energy and the Living Cell: An Introduction to Bioenergetics.* New York, Harper and Row, 1977. A paperback book containing a brief discussion of thermodynamics and the production and utilization of energy by living cells.

Clayton, R. K. *Photosynthesis: Physical Mechanisms and Chemical Patterns.* Cambridge, England, Cambridge University Press, 1980. An important resource for information about current concepts and experiments dealing with the photosynthetic process.

Clayton, R. K., and W. R. Sistron (eds.). *The Photosynthetic Bacteria.* New York, Plenum Press, 1978. A multiauthor book summarizing research on many aspects of the photosynthetic process in bacteria.

Hatch, M. D., and N. K. Boardman. *Photosynthesis.* New York, Academic Press, 1981. A detailed description of the chemistry and physics underlying the photosynthetic process; a valuable source book.

Lehninger, A. L. *Bioenergetics: The Molecular Basis of Biological Energy Transformations,* 2d ed. Menlo Park, Calif., Benjamin/Cummings, 1971. A classic presentation of energy transformations in living systems; an unusually clear exposition of a somewhat difficult subject.

Lehninger, A. L. *Principles of Biochemistry.* New York, Worth Publishers, Inc., 1982. A standard biochemistry text with a detailed, very clear account of bioenergetics, photosynthesis, and cellular respiration. A fine source book for students.

Racker, E. *A New Look at Mechanisms in Bioenergetics.* New York, Academic Press, 1976. A well-written exposition of the mechanisms involved in the process of bioenergetics by one of the major contributors to concepts in this field.

Stryer, L. *Biochemistry,* 2d ed., San Francisco, W. H. Freeman, 1981. A well-illustrated, readable text that covers the concepts of cellular energetics from the ground up.

Articles

Bjorkman, O., and J. Berry. High-efficiency photosynthesis. *Scientific American,* October 1973, 80–93. Description of the C_4 pathway, the structure of the leaf cells, and the biochemistry of the pathway of carbon dioxide.

Dickerson, R. E. Cytochrome c and the evolution of energy metabolism. *Scientific American,* March 1980, 136–153.

Hinckle, P. C., and R. E. McCarty. How cells make ATP. *Scientific American,* March 1978, 104–123. An interesting presentation of the chemiosmotic theory and how it may explain both photosynthesis and oxidative phosphorylation.

Miller, K. R. The photosynthetic membrane, *Scientific American.* October 1979, 102–113. Describes the structure of the thylakoid membrane and how it is adapted to convert light energy to chemical energy.

Stoeckenius, W. The purple membrane of salt-loving bacteria, *Scientific American,* June 1976, 38–50. A curious photosynthetic mechanism that uses a pigment rather like visual purple of animals instead of chlorophyll in capturing the energy of light.

part III

GENETICS

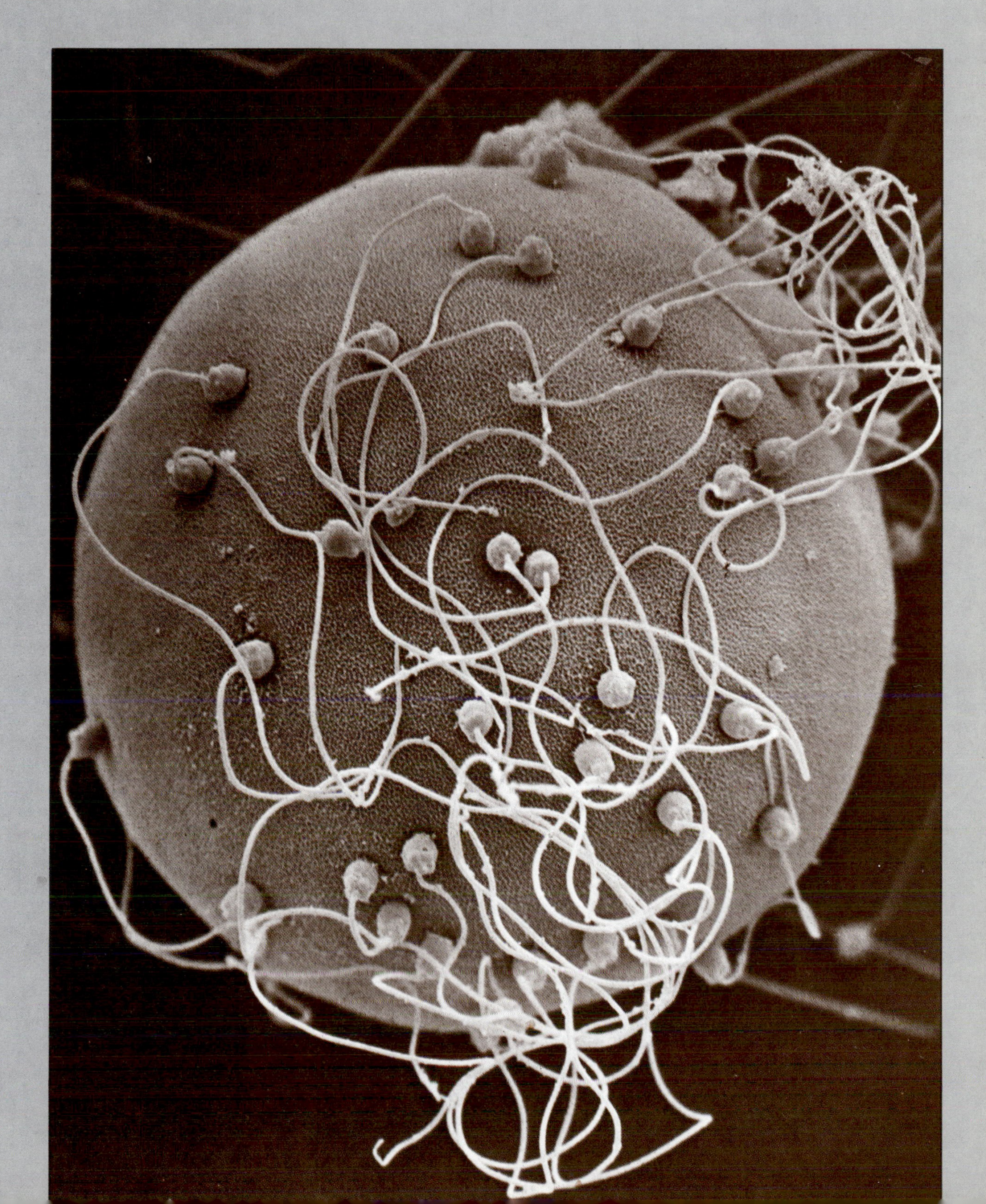

Scanning electron micrograph of sperm on the egg of a surf clam (×2,100). (David M. Phillips, The Population Council.)

10
Producing a New Generation: Meiosis

OUTLINE

I. Eukaryotic chromosomes
  A. Genes
  B. How many chromosomes?
II. Meiosis
  A. How does meiosis differ from mitosis?
  B. The process of meiosis
III. Spermatogenesis
IV. Oogenesis

LEARNING OBJECTIVES

After you have read this chapter you should be able to:

1. Distinguish between asexual and sexual reproduction.
2. Describe the structure of a chromosome, and give the function of a gene.
3. Distinguish between haploid and diploid, and define homologous chromosomes.
4. Contrast the events of mitosis and meiosis.
5. Compare spermatogenesis with oogenesis.
6. Describe the structure of a mature sperm.
7. Explain the role of the polar bodies in oogenesis.

The essence of the reproductive process is the production of a new generation of offspring that resembles the parental generation. This process involves the transfer of biological information to the new organism. Human beings have been aware for centuries that "like begets like" and that one of the prime characteristics of living things is their ability to reproduce their kind. This genetic (that is, via the genes) transmission of traits from parent to offspring is called **heredity.** The branch of biology concerned with the structure, transmission, and expression of genes is called **genetics.** Since its inception at the beginning of this century, the science of genetics has advanced with great speed and currently is developing at an even more accelerated pace, largely as a result of the science of molecular genetics, established in the 1950s.

Resemblances between parents and offspring are close but are usually not exact. The offspring of a particular set of parents differ from each other and from their parents in many respects and to different degrees. Such differences, termed **variations,** are characteristic of all living things. Some variations are inherited, whereas others may be due to the effects of temperature, moisture, food, light, and many other environmental factors on the development of the organism. Even the expression of inherited characteristics may be strongly influenced by the environment in which the individual develops.

The details of the reproductive process vary tremendously for different kinds of organisms, but we can distinguish two basic types of reproduction: asexual and sexual. In **asexual** reproduction a *single* parent splits, buds, or fragments to give rise to two or more offspring that have identical genes and therefore inherited traits that are very closely similar to those of the parent (Fig. 10–1). Asexual reproduction is usually a rapid process; it permits organisms that may be well adapted to their environment to produce new generations of similarly adapted organisms.

In contrast, **sexual** reproduction generally involves *two* parents: Each contributes a specialized gamete, an egg or a sperm; these fuse to form the fertilized egg, or **zygote.** The egg is typically large and nonmotile, with a store of nutrients to support the development of the embryo that results when the egg is fertilized. The

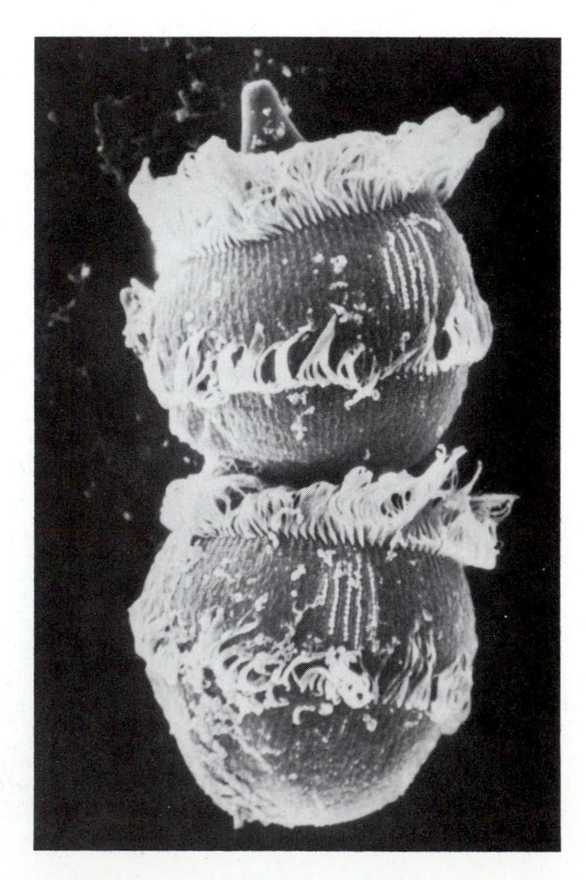

Figure 10–1 A scanning electron micrograph of an early stage of binary fission in the protozoan *Didinium nastutium.* This type of asexual reproduction is common among protozoa. (Courtesy of Dr. Eugene B. Small.)

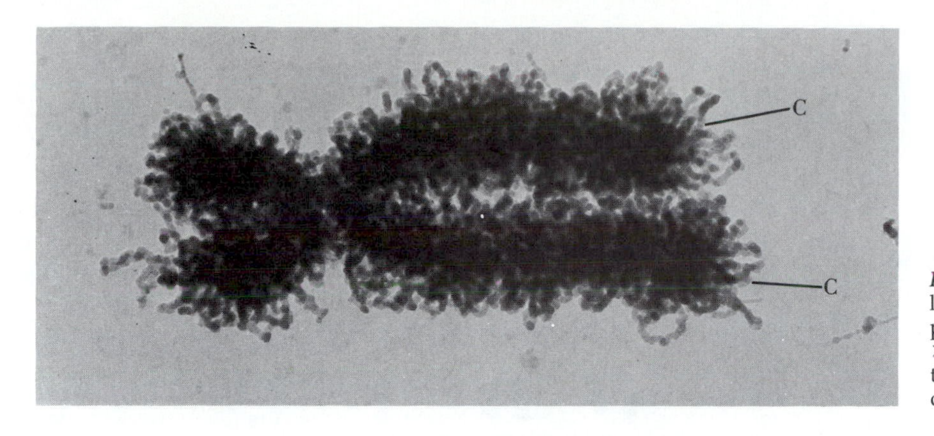

Figure 10–2 A human chromosome isolated and photographed at the metaphase stage of mitosis (approximately ×50,000). Note the two chromatids attached at their centromeres. (Courtesy of E. J. DuPraw.)

sperm is usually small and motile, adapted to swim actively to the egg by beating its long, whiplike tail. Sexual reproduction has the biological advantage of making possible the recombination of the inherited traits of the two parents, so that the offspring may be able to survive environmental changes or other stress better than either parent.

Eukaryotic Chromosomes

The nucleus of a eukaryotic cell contains many rodlike **chromosomes.** Except during cell division the chromosomes appear as long, thin, dark-staining threads called **chromatin.** These threads are fibers that contain about 60% protein, about 35% DNA, and about 5% RNA. The DNA is tightly bound to small, basic proteins called **histones.**

Just before and during mitosis (Chapter 4), the chromatin condenses dramatically and is visible as elongated, discrete chromosomes (see Figs. 4–23 and 10–2). Each chromosome contains a highly coiled and compacted single DNA molecule wound around groups of histone molecules (Fig. 10–3). The units of histone surrounded by DNA are called **nucleosomes;** they are repeating beadlike structures joined together by connecting DNA and histone.

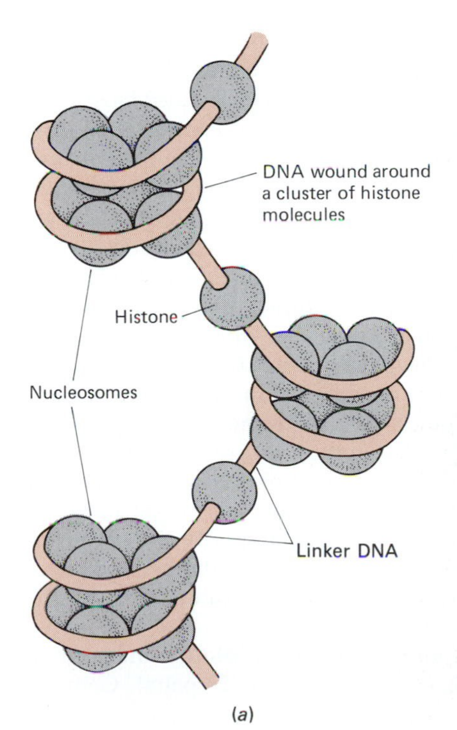

(*a*)

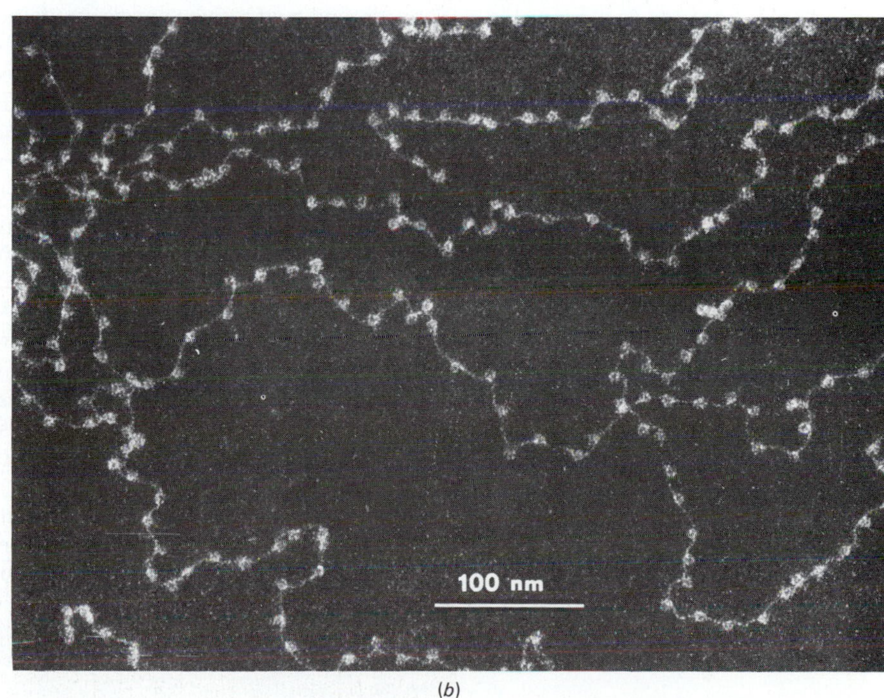

(*b*)

Figure 10–3 The units of histone surrounded by DNA in the chromosome are called nucleosomes. (*a*) The hypothetical structure of a nucleosome. Each nucleosome bead is thought to contain a set of eight histone molecules that form a protein core around which the double-stranded DNA is wound. The DNA wound around the histone consists of 146 nucleotide pairs; another segment of DNA, 60 nucleotide pairs long, links nucleosome beads. This linker DNA and the nucleosome bead make up the nucleosome. It is thought that one type of histone (H1) extends to cover the linker DNA and to contact beads of adjacent nucleosomes. This histone appears to be responsible for packing nucleosomes and may help link them to one another. (*b*) Nucleosomes from the nucleus of a chicken red blood cell. Each spherical structure is a nucleosome. Normally nucleosomes are packed more closely together but preparation procedure has spread them apart revealing DNA that links one nucleosome to the next. (Courtesy of D. E. Olins and A. L. Olins.)

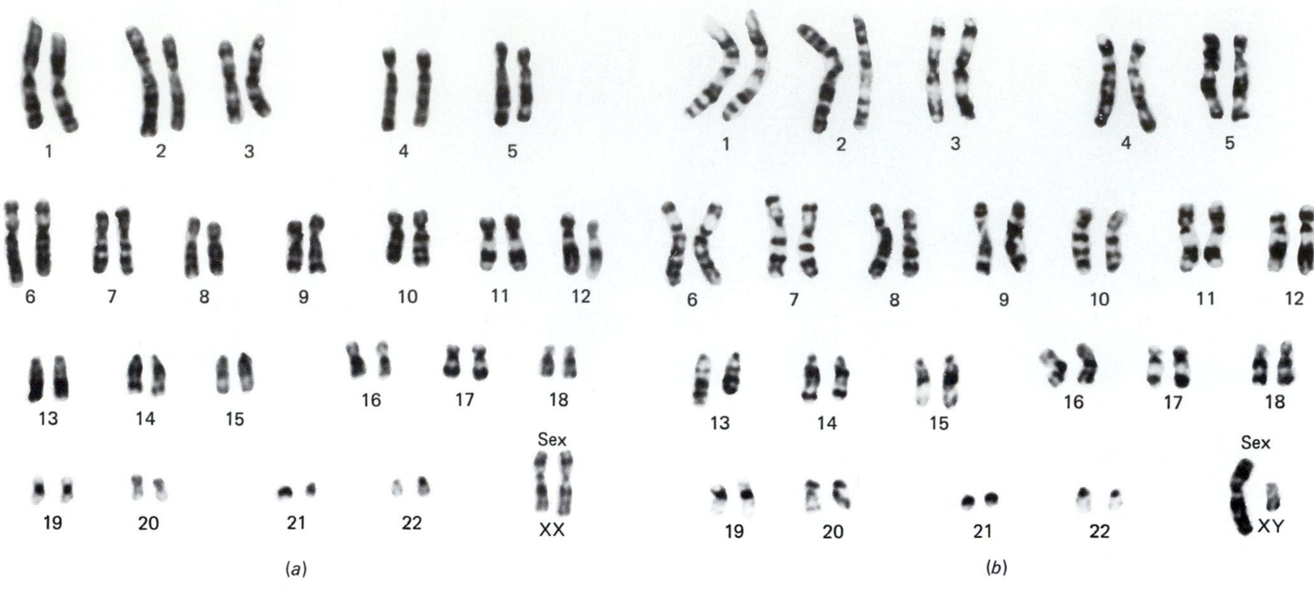

Figure 10–4 Photomicrographs, known as karyotypes, of normal human female and male chromosomes.

Each chromosome has, at a fixed point along its length, a small, light-staining zone called a **centromere** that controls the movement of the chromosome during cell division. As the chromosome becomes shorter and thicker just before cell division occurs, the centromere region becomes accentuated and appears as a constriction.

GENES

Each chromosome may contain hundreds, or even thousands, of genes. A **gene** can be roughly defined as a sequence of DNA nucleotides that codes for a specific RNA molecule. Many of the RNA molecules then code for a specific polypeptide chain. Each gene is thought to be made up of about six nucleosomes. A typical mammalian cell may have about 50,000 genes, but the actual number is not known.

HOW MANY CHROMOSOMES?

Every organism of a given species contains a characteristic number of chromosomes in each of its cells. Each cell in the body of every normal human being has exactly 46 chromosomes (Fig. 10–4). Many other species of animals and plants also have 46. It is not the *number* of chromosomes that differentiates the various species of animals but rather the information specified by the genes. A certain species of roundworm has only two chromosomes in each cell; some crabs have as many as 200 per cell. The highest chromosome number reported so far is about 1600, found in a radiolarian, a marine protist. Most species of animals and plants have chromosome numbers between 10 and 50. Numbers above and below this are comparatively rare.

Chromosomes normally exist in pairs; there are typically two of each kind in the somatic (body) cells of complex plants and animals. Thus the 46 chromosomes in human cells consists of 23 different pairs. The pairs differ in banding patterns, in length and shape, and in the presence of knobs or constrictions along their length. The members of a pair are referred to as **homologous chromosomes** (Fig. 10–5); they carry information for the same traits, though not necessarily the same information. For example, members of a pair of homologous chromosomes might carry genes that specify hemoglobin structure; however, one member might have the information for normal hemoglobin while the other might specify the abnormal hemoglobin that results in sickle cell anemia. In most species, the chromosomes vary enough in their morphological features so that cytologists can distinguish the different homologous pairs.

A set of two of each kind of chromosome is called the **diploid** number or **2n** number. In humans the diploid number is 46 chromosomes (23 pairs). Gametes

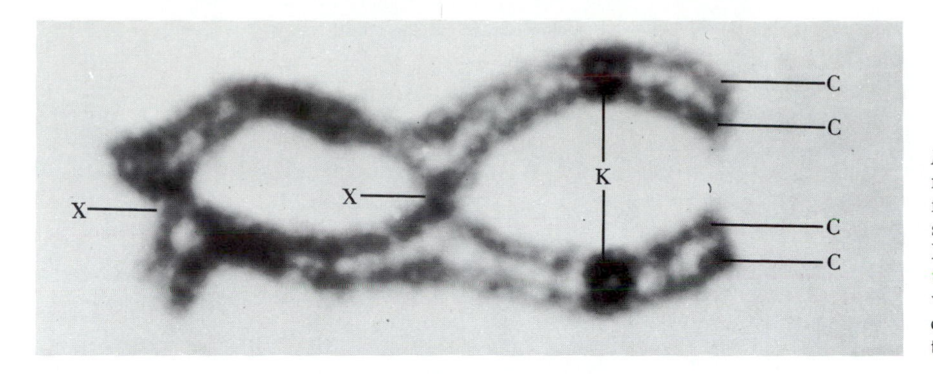

Figure 10–5 A pair of homologous chromosomes during late prophase of the first meiotic division of a salamander spermatocyte (developing sperm cell). Note the four chromatids that make up the tetrad (at C). The centromeres are visible at *K*. Crossing over produces the configurations shown at each *X*. (Courtesy of J. Kezer.)

have only one of each kind of chromosome, the **haploid,** or **n,** number. When sperm and egg fuse in fertilization, each gamete contributes its haploid set of chromosomes so that the fertilized egg (zygote) contains the diploid number. When the zygote divides by mitosis to form the first two cells of the embryo, each cell receives a complete diploid set of chromosomes. As mitosis and cell division continue, each body cell receives a diploid set of chromosomes. Thus, most body cells are diploid. However, there are exceptions. For example, some liver cells are **polyploid,** containing four complete haploid sets of chromosomes (tetraploid or 4n). Some plants are also polyploids that contain three or more sets of chromosomes per cell. In fact, about one third of all flowering plants and almost three fourths of all grasses are polyploid. Such plants are often larger and more hardy than diploid members of the same group, and they are important commercially. Modern bread wheat (*Triticum aestivum*) is a hexaploid (6n) developed from three different diploid (2n) species.

Meiosis

In Chapter 4 we examined the process of mitosis, which ensures that each daughter cell will receive exactly the same number and kind of chromosomes that the parent cell had, together with the features of the cell cycle. The constancy of the chromosome number in the cells of successive generations of organisms is ensured by the process of **meiosis,** a special type of division that occurs during the formation of eggs and sperm in animals and of spores in plants. The process of meiosis involves a pair of cell divisions during which the chromosome number is reduced to one half. Each gamete receives only half as many chromosomes as those contained in the body cells of the parent. When two gametes unite in fertilization, the fusion of their nuclei reconstitutes the normal 2n number of chromosomes.

HOW DOES MEIOSIS DIFFER FROM MITOSIS?

The events of meiosis are somewhat similar to the events of mitosis, but there are several important differences (Fig. 10–6): (1) In meiosis, there are two successive nuclear and cell divisions, yielding a total of four cells. (2) Each of the four cells produced in meiosis contains the haploid number of chromosomes—that is, only one of each homologous pair. (3) During meiosis, the homologous chromosomes containing genetic information from each parent are thoroughly shuffled, and one of each pair is randomly distributed to each new cell; the resulting gametes possess many different combinations of chromosomes, some of which may not have occurred in any prior generation.

THE PROCESS OF MEIOSIS

The process of meiosis consists of two nuclear and cell divisions designated the first and second meiotic divisions or, simply, **meiosis I** and **meiosis II.** Each of these includes prophase, metaphase, anaphase, and telophase. During the first meiotic division the members of each homologous pair of chromosomes separate and are

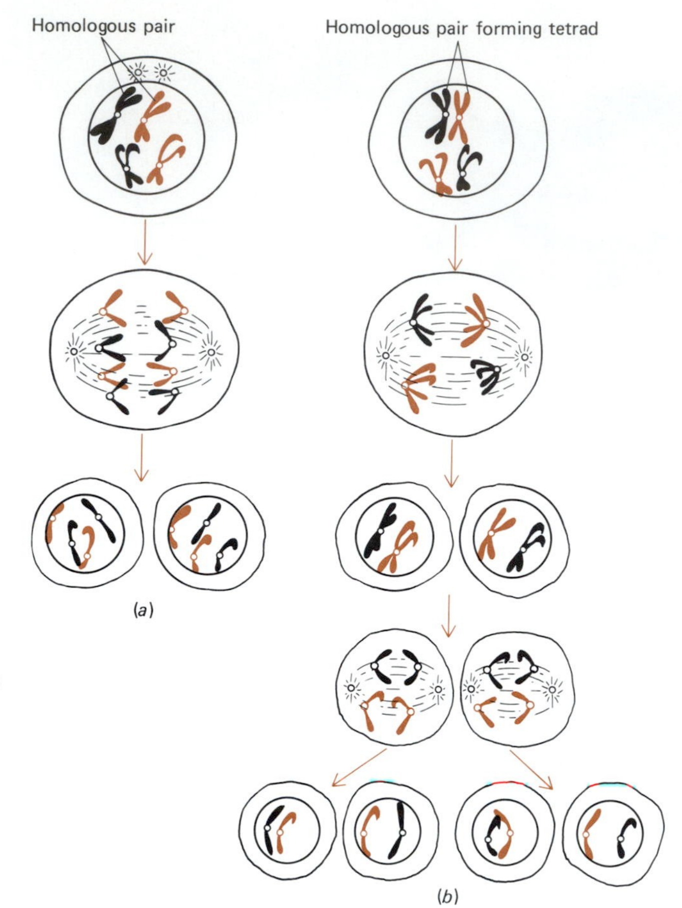

Figure 10–6 Meiosis compared with mitosis. The diploid number for the cell shown here is four. (*a*) Mitosis. Note that each daughter cell has an identical set of four chromosomes (two pairs), which is the diploid number. (*b*) Meiosis. Two divisions take place, giving rise to four daughter cells. Each daughter cell has only two chromosomes, one of each pair. The chromosomes shown in black originally came from one parent; those shown in color came from the other parent. Note that in the prophase of the first meiotic division, homologous chromosomes come together, forming tetrads.

distributed into separate cells. In the second meiotic division the chromatids that comprise each chromosome separate and are distributed to the daughter cells. The following discussion describes meiosis in an animal with a diploid chromosome number of four. Refer to Figures 10–7 and 10–8 as you read.

As in mitosis the chromosomes duplicate themselves during the S phase before meiosis actually begins. Recall from Chapter 4 that when a chromosome is duplicated, it consists for a time of two chromatids joined by their centromeres. In order to be considered a complete chromosome, the structure must possess an unshared centromere. During prophase of the first meiotic division, while the joined chromatids are still elongated and thin, the homologous chromosomes come to lie close together side by side along their entire length, by a process called **synapsis.** How the homologous chromosomes find one another is not known. In our example, since the diploid number is four, there are two homologous pairs. One of each pair is a **maternal chromosome,** originally inherited from the organism's mother, whereas the other member of each pair, a **paternal chromosome,** was contributed by the father. Since each chromosome is doubled at this time and actually consists of two chromatids, synapsis results in the coming together of four chromatids, forming a complex known as a **tetrad.** The number of tetrads equals the haploid number of chromosomes. In human cells there are 23 tetrads (and a total of 92 chromatids) at this stage.

During synapsis, homologous chromosomes become wrapped about one another, and genetic material may be exchanged between homologous chromosomes. This process is known as **crossing over.** The resulting **genetic recombination** greatly enhances the prospects for variety among offspring of sexual partners.

In many species, the prophase of the first meiotic division is an extremely extended phase during which the cell grows and synthesizes nutrients for the future embryo. In many types of gametes, the chromosomes assume unusual configurations during this phase. Hundreds of pairs of loops project from the chromatid axis, giving a lampbrush appearance from which comes their name—**lampbrush chromosomes** (Fig. 10–9). The loops are sites of intense RNA synthesis.

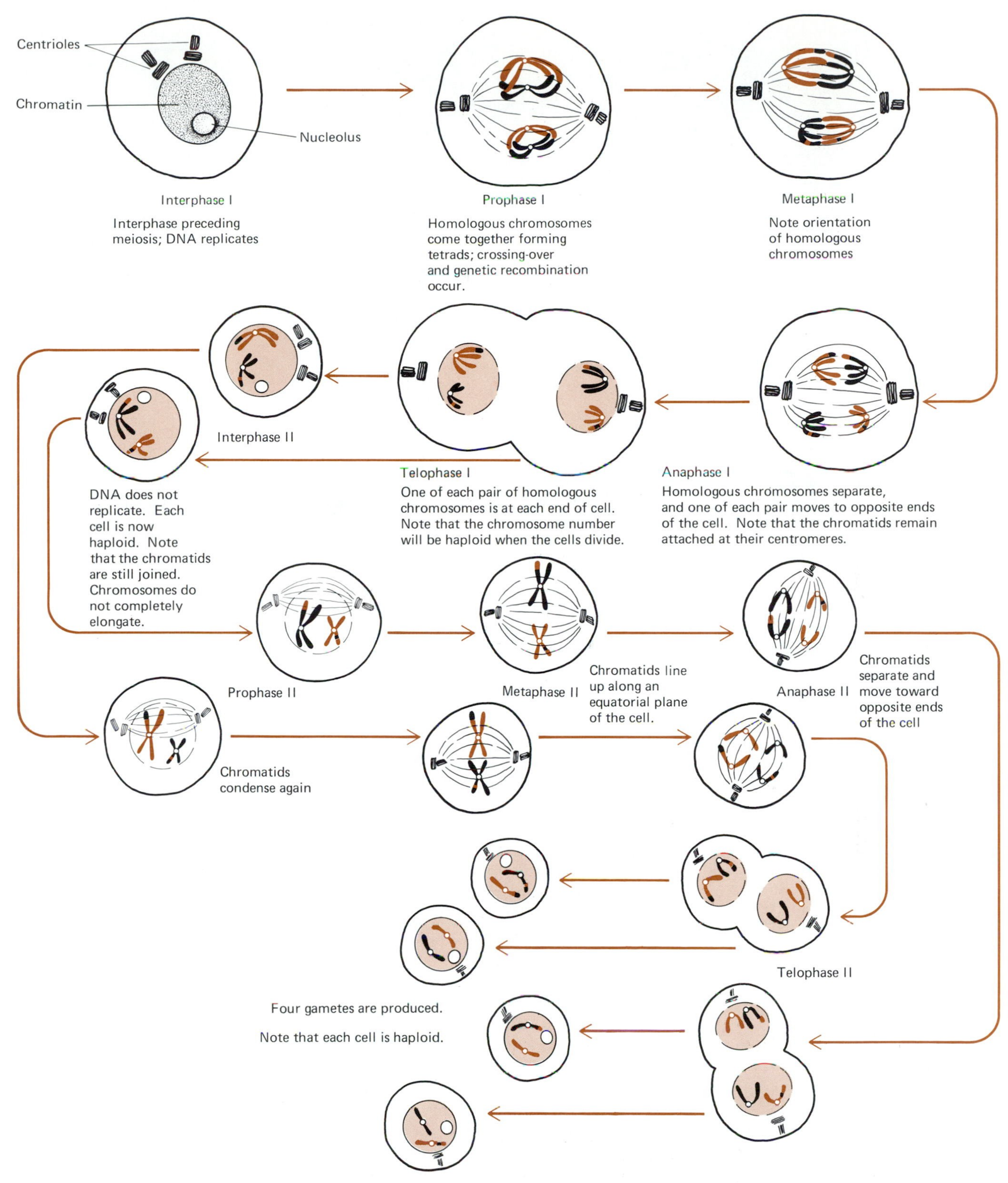
Centrioles
Chromatin
Nucleolus
Interphase I
Interphase preceding meiosis; DNA replicates
Prophase I
Homologous chromosomes come together forming tetrads; crossing-over and genetic recombination occur.
Metaphase I
Note orientation of homologous chromosomes
Interphase II
DNA does not replicate. Each cell is now haploid. Note that the chromatids are still joined. Chromosomes do not completely elongate.
Telophase I
One of each pair of homologous chromosomes is at each end of cell. Note that the chromosome number will be haploid when the cells divide.
Anaphase I
Homologous chromosomes separate, and one of each pair moves to opposite ends of the cell. Note that the chromatids remain attached at their centromeres.
Prophase II
Chromatids condense again
Metaphase II
Chromatids line up along an equatorial plane of the cell.
Anaphase II
Chromatids separate and move toward opposite ends of the cell
Telophase II
Four gametes are produced.
Note that each cell is haploid.

Figure 10–7 The stages of meiosis. The diploid number for the cell shown here is four.

While the unusual events characteristic of prophase I of meiosis are occurring, other events that are also characteristic of mitotic prophase take place. The centrioles move to opposite poles, a spindle forms between the centrioles, and the nuclear membrane dissolves.

The tetrads then line up along the equator of the spindle, and the cell is said to be in metaphase. Both chromatids of one chromosome are oriented toward the same pole. Their joined centromeres are attached to the spindle fibers of only one of the two poles. The pairs of centromeres of homologous chromosomes are separated

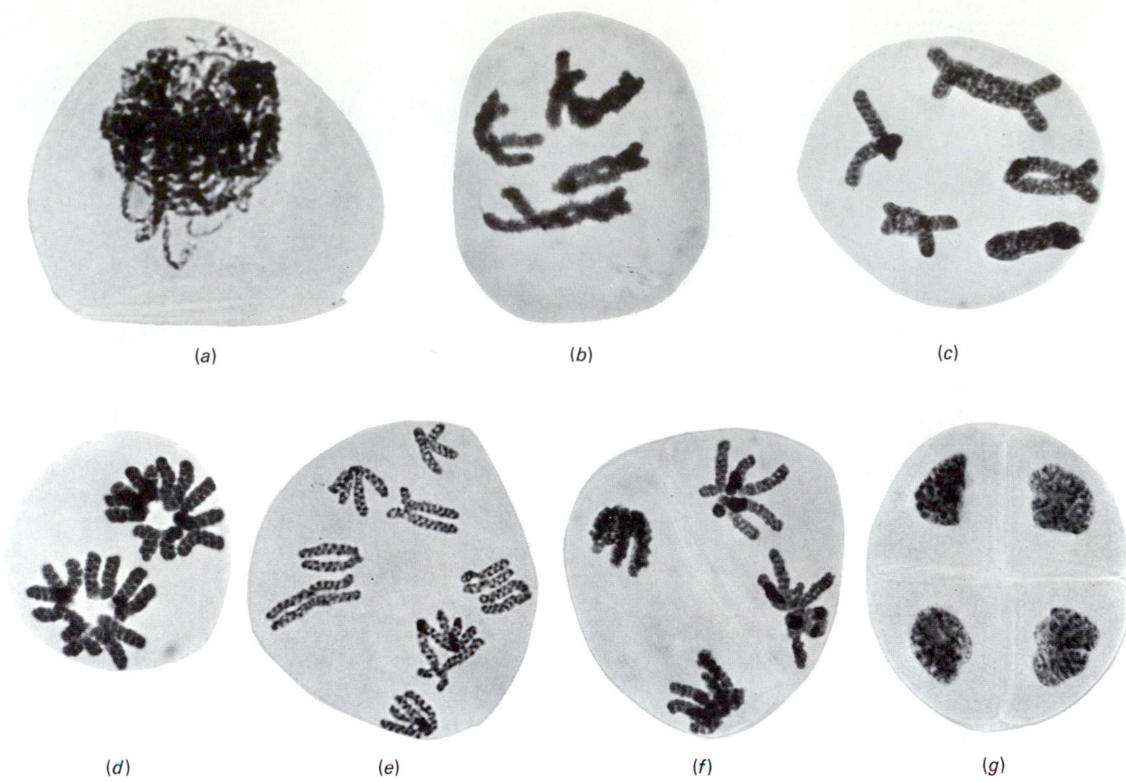

Figure 10–8 Photomicrographs of meiosis in the plant *Trillium erectum* (×2000). (*a*) Early prophase of the first meiotic division. (*b*) Later prophase of the first meiotic division. (*c*) Metaphase I. (*d*) Anaphase I. (*e*) Metaphase II. (*f*) Anaphase II. (*g*) Four daughter cells. (Courtesy of A. H. Sparrow and R. F. Smith.)

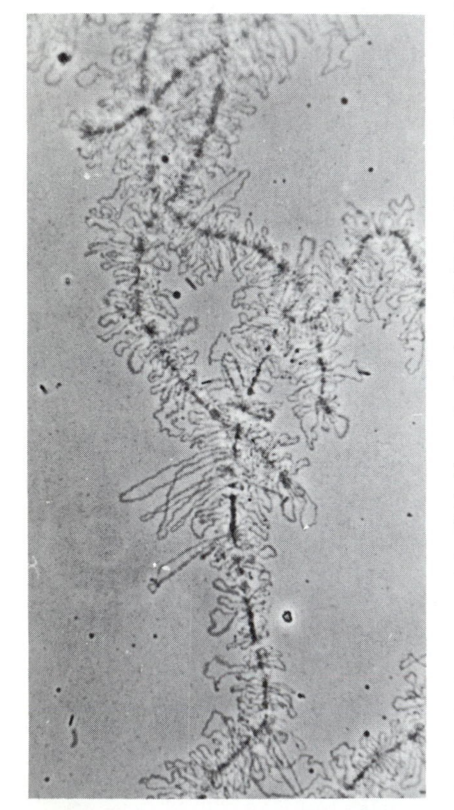

Figure 10–9 Photomicrograph of chromosome from the oocyte of the newt *Triturus viridescens,* showing the loops radiating from the central thread (approximately ×1100). The appearance of these loops inspired their name, lampbrush chromosomes. (Courtesy of Dr. Dennis Gould.)

so that only the ends of the homologous chromatids contact one another (Fig. 10–10).

During anaphase of the first meiotic division, the homologous chromosomes of each pair, but not the daughter chromatids, separate and randomly move toward opposite poles. The chromatids are still united by their centromeres. This differs from mitotic anaphase, in which the centromeres separate and the daughter chromosomes pass to opposite poles. In the telophase of the first meiotic division in our example, there are two duplicated chromosomes at each pole, that is, four chromatids (see Fig. 10–7). In humans there are 23 duplicated chromosomes (46 chromatids) at each pole. During telophase, the nuclei reorganize, the chromatids begin to elongate, and cytokinesis (cell division) generally takes place.

During the interphase that follows, there is no S phase, for no further chromosome replication takes place. In most organisms meiotic interphase is very brief; in some organisms it is absent. Since the chromatids do not completely elongate between divisions, the prophase of the second meiotic division is also brief. Prophase II is similar to a mitotic prophase; there is no pairing of homologous chromosomes (indeed, only one of each pair remains in the cell), and there is no genetic recombination.

During metaphase II, the chromatids again line up on the equator. These metaphases can be distinguished because in the first, the chromatids are arranged in bundles of four (tetrads), and in the second, the chromatids are arranged in groups of two. During anaphase II, the centromeres split and the daughter chromatids, now complete chromosomes, separate and move to opposite poles. Thus in the telophase of the second meiotic division there is one of each kind of chromosome, the haploid number at each pole. Nuclear membranes then form, the chromosomes gradually elongate forming chromatin threads, and cytokinesis occurs.

The two successive meiotic divisions yield four haploid nuclei, each containing one and only one of each kind of chromosome (see Fig. 10–7). Each of these cells has a different combination of genes. In plants meiosis generally occurs in structures called sporangia and gives rise to reproductive cells called spores. Later, plants that develop from the spores eventually produce eggs and sperm by mitosis.

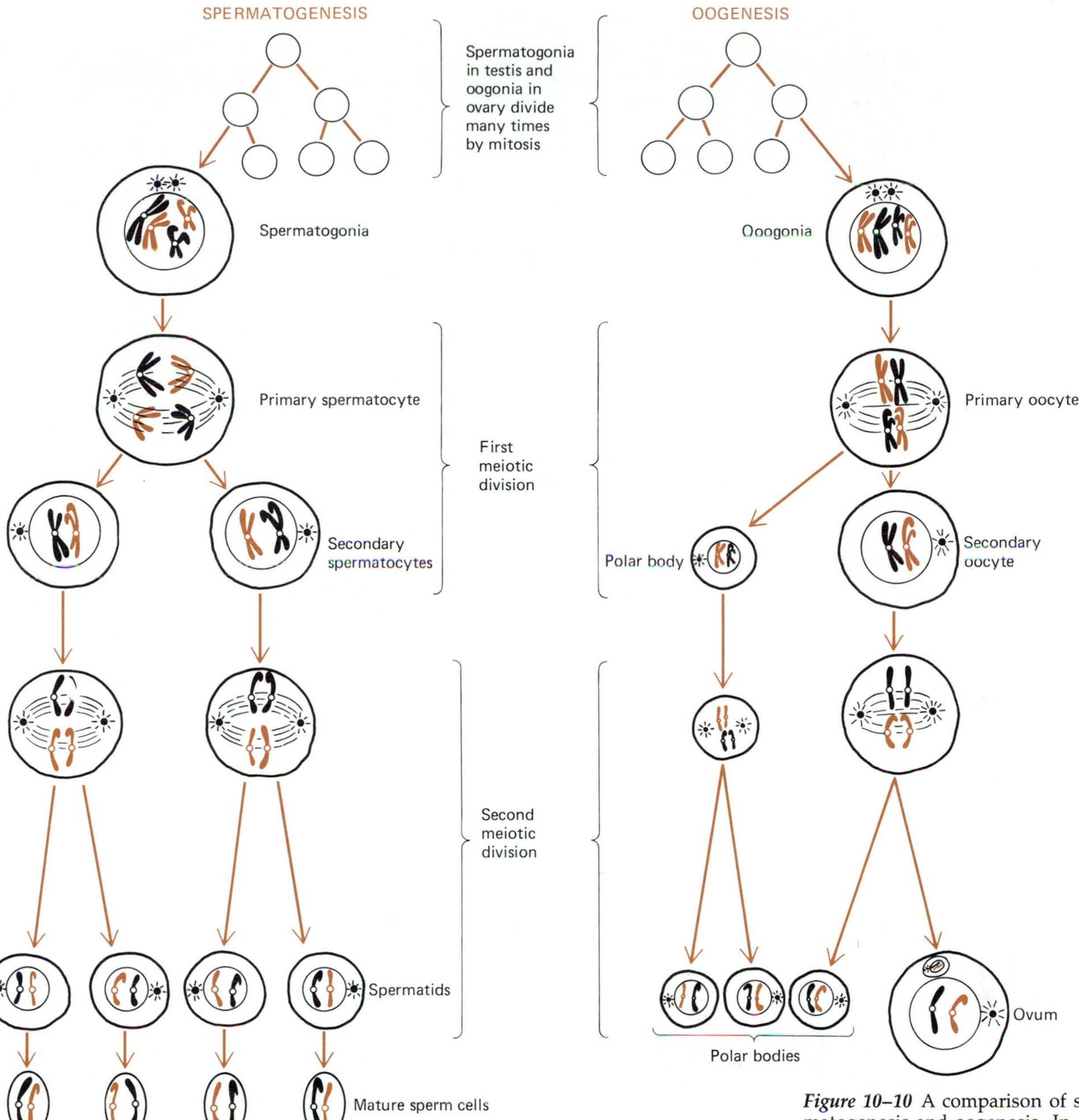

Figure 10–10 A comparison of spermatogenesis and oogenesis. In spermatogenesis, the primary spermatocyte divides by meiosis, giving rise to four spermatids; the spermatids differentiate, becoming mature sperm cells. In oogenesis, only one functional ovum is produced from each primary oocyte; the other three cells produced are polar bodies that degenerate.

(Plant reproduction will be discussed in Chapter 31.) In animals meiosis takes place in specialized reproductive structures and results in formation of sperm and eggs. Fundamentally, **gametogenesis,** or gamete formation, is similar in all organisms, but there are some differences in detail.

Spermatogenesis

In all vertebrates, including humans, **spermatogenesis,** or sperm production, takes place in the testes. Each human testis is made up of thousands of cylindrical **seminiferous tubules,** in each of which millions of sperm cells develop. The walls of these tubules are lined with primitive, unspecialized germ cells, **spermatogonia** (singular, spermatogonium). Throughout embryonic development and during childhood the spermatogonia divide mitotically, giving rise to additional spermatogonia. After sexual maturity, some of the spermatogonia undergo spermatogenesis; others continue to divide mitotically and produce more spermatogonia for later spermatogenesis. In most wild animals, there is a definite breeding season, typically in either spring or fall, during which the testes increase in size and spermatogenesis occurs; between breeding seasons the testes are small and contain only spermatogonia. In humans and most domestic animals, spermatogenesis occurs throughout the year once sexual maturity is reached.

Spermatogenesis begins with the growth of the spermatogonia into larger cells known as **primary spermatocytes** (see Fig. 10–10). These divide (in the first meiotic division) into two equal-sized **secondary spermatocytes,** which in turn divide (in the second meiotic division) to form four equal-sized spermatids. The **spermatid,** a spherical cell with a generous amount of cytoplasm, is a gamete with the haploid number of chromosomes. A complicated process of change (but not cell division) converts the spermatid into a functional **spermatozoon** (plural, spermatozoa), usually called simply a **sperm.** The nucleus shrinks in size and becomes the head of the sperm (Fig. 10–11), while the cell sheds most of its cytoplasm. Some of the Golgi bodies congregate at the front end of the sperm and form a point, the **acrosome,** which may aid the sperm in fertilization.

The two centrioles of the spermatid move to a position just in back of the nucleus. A small depression appears on the surface of the nucleus, and one of the centrioles, the **proximal centriole,** takes up a position in the depression at right angles to the axis of the sperm. The second or **distal centriole,** just behind the proximal centriole, gives rise to the axial filament of the sperm tail, or flagellum (Fig. 10–11). The axial filament has a 9 + 2 arrangement of microtubules like that in eukaryotic flagella (two central microtubules surrounded by a ring of nine pairs of microtubules).

The mitochondria move to the point at which head and tail meet, and form a small **middle piece** that provides energy for the beating of the tail. Most of the cytoplasm of the spermatid is discarded as residual bodies, which are taken up by phagocytosis by supporting cells called **Sertoli cells.** The mature sperm retains only a thin sheath surrounding the mitochondria in the middle piece and the axial filament of the tail.

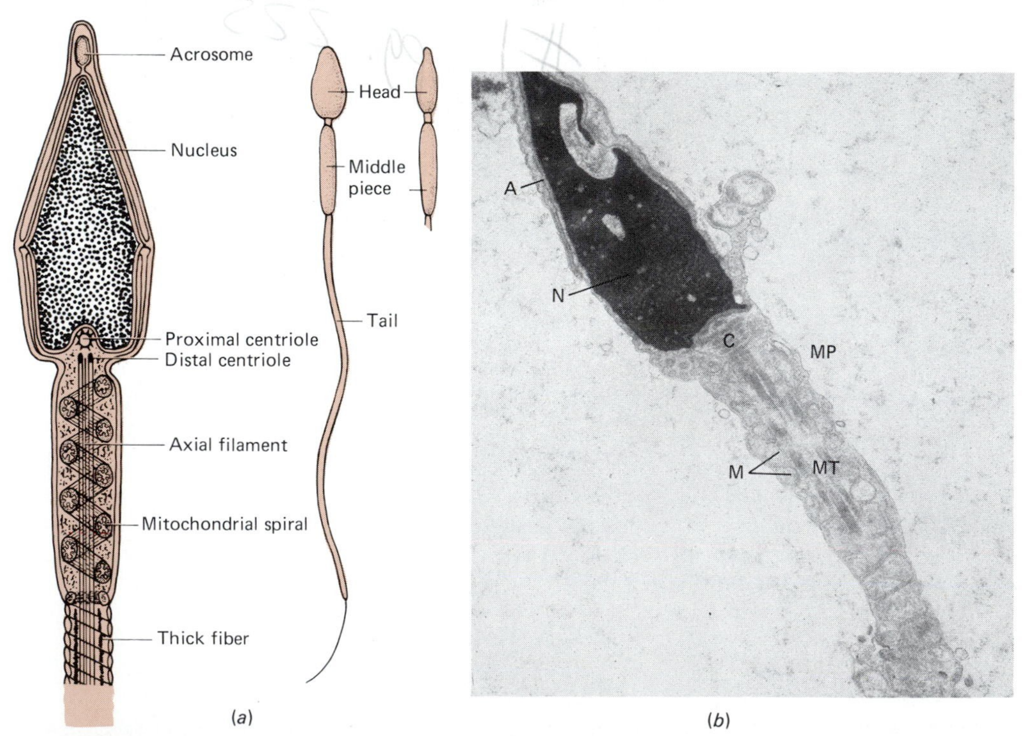

Figure 10–11 Sperm cell structures. (*a*) A mammalian sperm. *Left,* Head and middle piece, greatly enlarged, as seen in the electron microscope. *Middle, right,* Top and side views of a sperm as seen by light microscopy. (*b*) Electron micrograph of a mature sperm cell showing the head, midpiece, and first portion of the tail (approximately ×37,500). *A,* acrosome; *N,* nucleus; *MP,* midpiece; *M,* mitochondria; *MT,* microtubules; *C,* centrioles. (Courtesy of Dr. Lyle C. Dearden.)

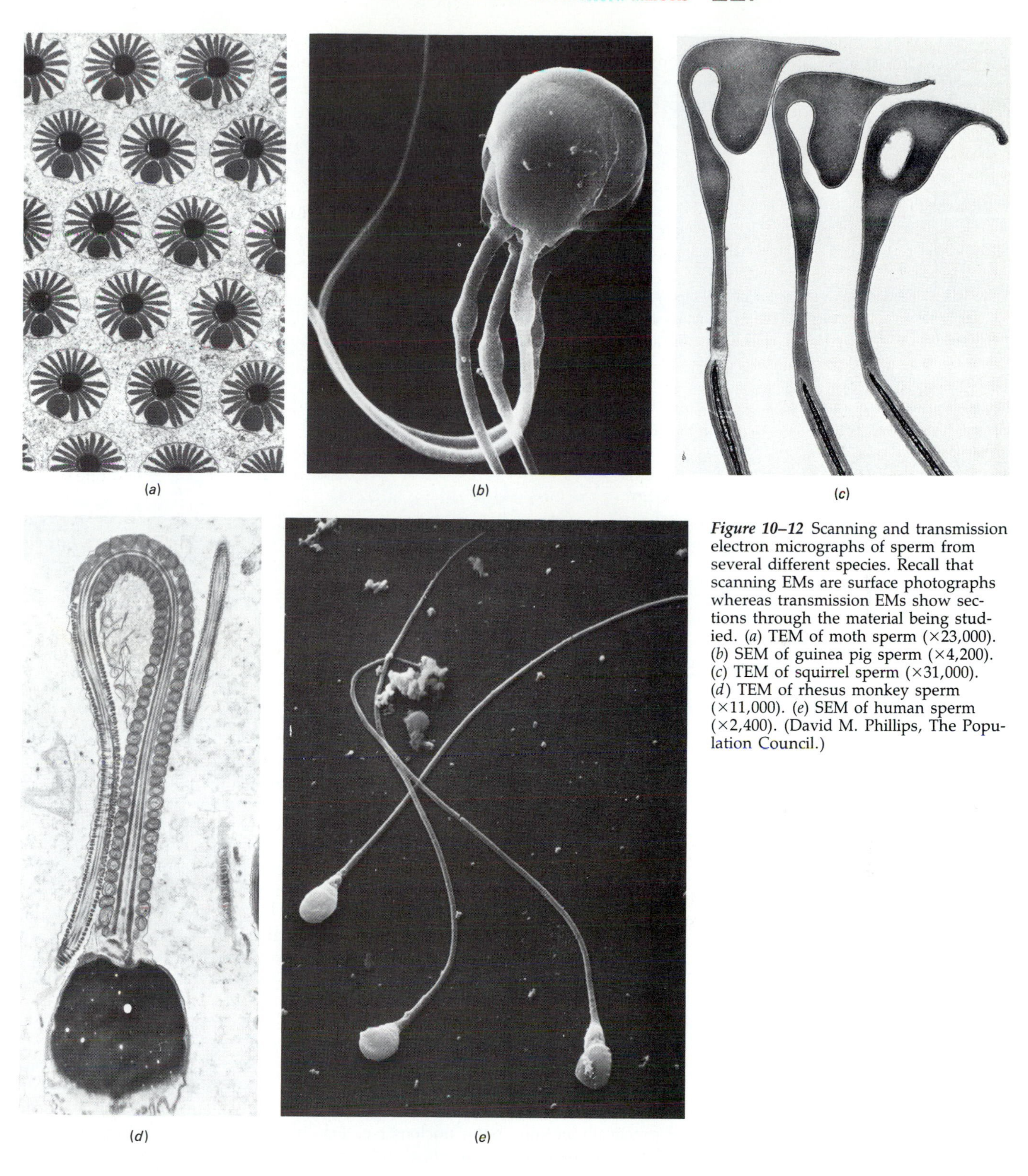

Figure 10–12 Scanning and transmission electron micrographs of sperm from several different species. Recall that scanning EMs are surface photographs whereas transmission EMs show sections through the material being studied. (*a*) TEM of moth sperm (×23,000). (*b*) SEM of guinea pig sperm (×4,200). (*c*) TEM of squirrel sperm (×31,000). (*d*) TEM of rhesus monkey sperm (×11,000). (*e*) SEM of human sperm (×2,400). (David M. Phillips, The Population Council.)

The sperm of various animal species may be quite different. There are great variations in the size and shape of the tail and in the characteristics of the head and middle piece (Fig. 10–12). The sperm of a few animals (e.g., the parasitic roundworm *Ascaris*) have no tail and move instead by amoeboid motion. Crabs and lobsters have a curious tailless sperm with three pointed projections on the head, which stick to the surface of the egg, holding the sperm securely in place. The middle piece uncoils like a spring, and pushes the nucleus of the sperm into the egg cytoplasm, thus accomplishing fertilization. Some plant sperm are ciliated, and in many vascular plants, the sperm is virtually reduced to a nucleus.

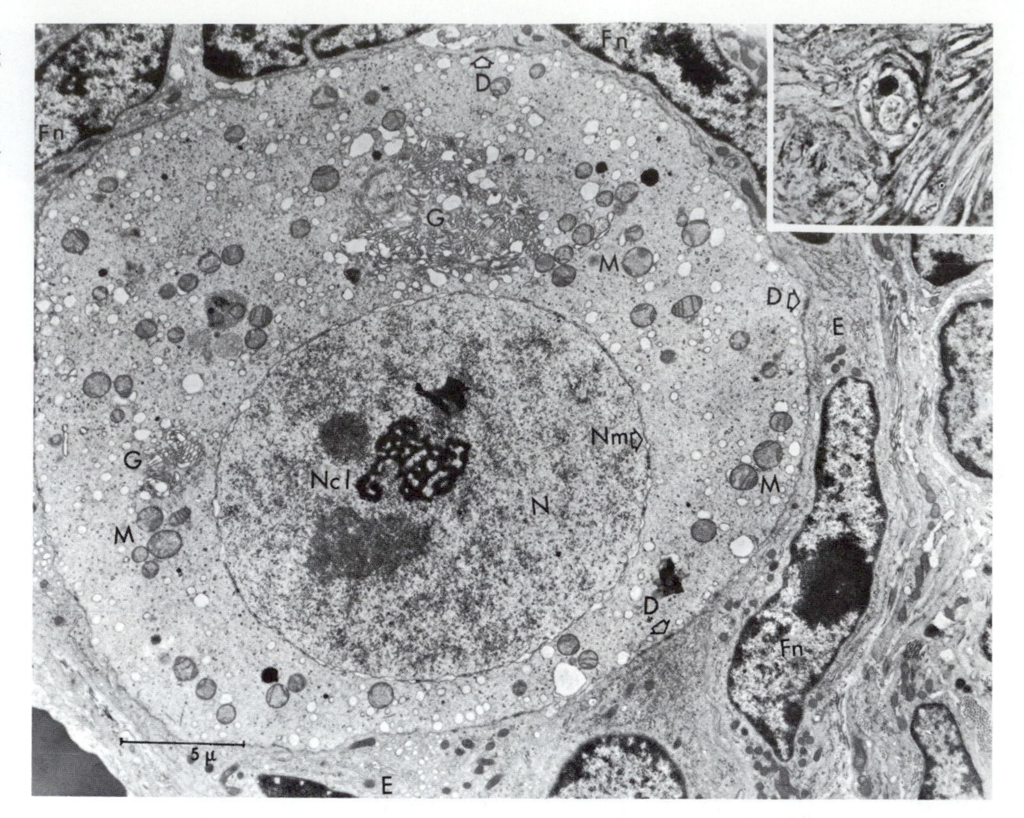

Figure 10–13 Electron micrograph of a young oocyte of a guinea pig showing Golgi material. *Inset,* Oocyte in a similar stage as seen by light microscopy. *N,* nucleus; *Ncl,* nucleolus; *Nm,* nuclear membrane; *G,* Golgi material; *M,* mitochondria; *D,* desmosomes between oocytes and follicle cells; *Fn,* nuclei of follicle cells; *E,* endoplasmic reticulum in follicle cells.

Oogenesis

The **ova** (singular, ovum), or eggs, develop in the ovary from immature sex cells, **oogonia.** Early in development the oogonia undergo many successive mitotic divisions forming additional oogonia, all of which have the diploid number of chromosomes. In many animals, notably the vertebrates, the oogonia and oocytes are surrounded by a layer of follicle cells. The developing ovum surrounded by its follicle cells forms a spherical structure referred to as a **follicle.** By the third month of human fetal development, the oogonia begin to develop into **primary oocytes** (Fig. 10–13). When a human female is born, her ovaries contain some 400,000 primary oocytes, which have attained the prophase of the first meiotic division. These primary oocytes remain in prophase until sexual maturity is reached. Then as each follicle matures, the first meiotic division resumes and is completed at about the time of ovulation (the release of the oocyte from the ovary), some 15 to 45 years after meiosis began!

The nuclear events of **oogenesis** (development of ova)—synapsis, the formation of tetrads, and the separation of the homologous chromosomes—are the same as those occurring in spermatogenesis. However, the division of the cytoplasm is unequal, resulting in one large cell, the secondary oocyte, which contains the yolk and nearly all the cytoplasm, and one small cell, the first **polar body,** which consists of practically nothing but a nucleus (see Fig. 10–10). (It was named a polar body before its significance was understood because it appeared as a small speck at the animal pole of the egg.)

The second meiotic division is arrested in the metaphase and is not completed until after fertilization. When meiosis is completed, the secondary oocyte again divides unequally into a large ovum and a small second polar body, both of which have the haploid chromosome number. The first polar body may divide into two additional second polar bodies, but generally it merely deteriorates. By the time the ovum is produced, it is already fertilized and is technically a diploid zygote. The small polar bodies soon disintegrate, so that each primary oocyte gives rise to just one ovum, in contrast to the four sperm formed from each primary spermatocyte. The unequal cytoplasmic division ensures that the mature egg will have enough cytoplasm and (in many species) stored yolk for the zygote to begin development and survive. The primary oocyte in a sense puts all its yolk in one ovum; the egg has (figuratively speaking!) neatly solved the problem of reducing its chromosome number without losing the cytoplasm and yolk needed for development after fertilization.

In fertilization the union of one haploid set of chromosomes from the sperm with another haploid set from the egg reestablishes the diploid chromosome number. Thus the fertilized egg, or zygote, as well as all the body cells developing from it by mitosis, has the diploid number of chromosomes. In each individual, exactly half of the chromosomes and half of the genes come from the mother and the other half from the father. Because of the nature of gene interaction, the offspring may resemble one parent more than the other, but the two parents make roughly equal contributions to its inheritance. (An exception is the inheritance of mitochondrial and chloroplast DNA, which is of maternal origin.)

SUMMARY

I. In the production of a new generation, genetic information is transferred from parent to offspring; this process is termed heredity. Genetics is the study of the structure, transmission, and expression of genes.
- *A.* Offspring produced by asexual reproduction have hereditary traits identical with those of the single parent.
- *B.* Sexual reproduction makes possible the recombination of traits from both parents.

II. Each chromosome consists of nucleosomes joined by connecting DNA and histone.
- *A.* A gene is that portion of chromosomal DNA that codes for a specific RNA molecule, which may in turn code for a specific polypeptide.
- *B.* Each somatic cell of every diploid (2n) organism of a given species has a characteristic number of pairs of chromosomes. The chromosome pairs differ in length and shape and other structural features.

III. In the formation of gametes—eggs and sperm—a special kind of cell division termed meiosis occurs. This consists of a sequence of two cell divisions in which the chromosome number is reduced to one half, the haploid (n) number. Each gamete contains one of each kind of chromosome, i.e., one complete set of chromosomes.
- *A.* During the prophase of the first meiotic division, the members of a pair of chromosomes undergo synapsis and crossing over. During crossing over genetic material is exchanged.
- *B.* The members of each pair of homologous chromosomes separate during the anaphase of the first meiotic division and are distributed to different daughter cells.
- *C.* During the second meiotic division, the two chromatids comprising each homologous chromosome separate, and each is distributed to a daughter cell.

IV. In spermatogenesis, the meiotic divisions convert each spermatocyte to four spermatids, which then undergo a process of growth and change and become the functional sperm. A sperm consists of a head capped with an acrosome, a middle piece containing mitochondria, and a long flagellum.

V. In oogenesis, meiosis divides the cytoplasm of the oocyte unequally so that the mature ovum receives nearly all of the cytoplasm and yolk and three small polar bodies receive only nuclear material.

VI. In fertilization, a haploid set of chromosomes from the sperm and a haploid set of chromosomes from the egg unite, forming the diploid number in the zygote.
- *A.* Half of the chromosomes of each individual are derived from the mother and the other half from the father.
- *B.* Because of gene interactions, the offspring may resemble one parent more than the other, but the two parents make equal chromosome contributions to the inheritance of the offspring.

POST-TEST

1. The tendency of individuals to resemble their parents is termed ________________.

2. The splitting, budding, or fragmenting of a single parent to give rise to two or more offspring with hereditary traits identical with those of the parent is termed ________________ ________________.

3. In sexual reproduction each of two parents contributes an egg or a sperm, which fuse to form a fertilized egg or ________________.

4. Between mitotic divisions the chromosomes of eukaryotic organisms appear as long, thin, dark-staining threads called ________________.

5. The chromosomes of eukaryotic organisms are composed of ________________, ________________, and ________________.

6. The unit of chromosomal structure, composed of a histone group wrapped with a coil of DNA, is termed a ________________.

7. The movement of the chromosomes during cell division is controlled by a small, clear region on the chromosome called the ________________.

8. The members of a pair of chromosomes are referred to as ________________ ________________.

9. That portion of a DNA chain that codes for a specific polypeptide is termed a ________________.

10. Eggs and sperm contain the ________________

number of chromosomes (____________), whereas the somatic cells contain the ____________ number of chromosomes (____________).

11. Cells containing more than two complete sets of chromosomes are termed ____________.

12. The pairing of homologous chromosomes during prophase I is known as ____________.

13. The exchange of segments of homologous chromosomes during synapsis is known as ____________ ____________.

14. In an organism with a diploid chromosome number of 12, each sperm will have ____________ chromosomes.

15. The production of sperm is known as ____________.

Match the terms in Column A with their definitions in Column B.

Column A	*Column B*
16. Acrosome	a. Group of four homologous chromatids seen in first meiotic prophase
17. Oocyte	b. Unspecialized germ cells from which sperm develop
18. Ovum	c. Haploid spherical cell that develops into mature sperm
19. Polar body	d. Tip of sperm head; plays role in fertilization
20. Seminiferous tubule	e. Supporting cell adjacent to spermatocytes
21. Sertoli cell	f. Precursor of egg
22. Spermatid	g. Small, nucleated cell formed by unequal division of oocyte
23. Spermatogonia	h. Mature female gamete
24. Tetrad	i. Fertilized egg
25. Zygote	j. Microscopic cylindrical structure in testis in which sperm develop

REVIEW QUESTIONS

1. How does meiosis differ from mitosis?
2. Define the following terms:
   a. diploid
   b. haploid
   c. homologous chromosomes
3. Describe the stages in meiosis, and indicate when and how synapsis and separation of homologous chromosomes occur.
4. How do you suppose certain liver cells become tetraploid?
5. Describe the structure of a chromosome. What are nucleosomes?
6. What is the relationship between genes and chromosomes? What are the functions of genes?
7. Two species may have the same diploid chromosome number and yet be very different. How can this be explained?
8. Compare the essential features of spermatogenesis and oogenesis.
9. Describe the process by which a spermatid differentiates into a mature sperm.
10. How is the diploid number reestablished in fertilization?
11. Why is cytoplasmic division unequal in oogenesis?
12. An organism has a diploid chromosome number of 10.
    a. How many pairs of chromosomes would it have in a typical body cell, such as a muscle cell?
    b. How many tetrads would form in the prophase of the first meiotic division?
    c. How many chromatids would be present in the prophase of the first meiotic division?
    d. How many chromosomes would be present in a mature sperm cell?

11
The Basic Principles of Heredity

OUTLINE

I. Genes and alleles
II. A monohybrid cross
- *A.* Homozygous and heterozygous
- *B.* Phenotype and genotype
- *C.* Test crosses

III. Calculating the probability of genetic events
- *A.* Types of probability
- *B.* The product law
- *C.* The sum law
- *D.* Applying the laws of probability

IV. Incomplete dominance
V. Deducing genotypes
VI. Dihybrid and trihybrid crosses
VII. Mendel's laws
VIII. Gene interactions
IX. Polygenic inheritance
X. Multiple alleles
- *A.* ABO blood types
- *B.* The Rh system

XI. Linkage and crossing over
XII. The genetic determination of sex
- *A.* The Y chromosome and maleness
- *B.* Nuclear sexing

XIII. Sex-linked and sex-influenced traits
XIV. Inbreeding, outbreeding, and hybrid vigor
XV. The mathematical basis of population genetics
- *A.* The gene pool and the Hardy-Weinberg law
- *B.* Estimating the frequency of genetic carriers
  - *1.* Albinism
  - *2.* Phenylketonuria

LEARNING OBJECTIVES

After you have read this chapter you should be able to:

1. Define and use correctly the terms *gene, allele, locus, genotype, phenotype, dominant, recessive, homozygous, heterozygous,* and *test cross.*
2. Solve problems in genetics involving monohybrid and dihybrid crosses, by applying Mendel's laws.
3. Explain how the method of progeny selection is used by commercial breeders in establishing a genetic strain that will breed true for a given trait.
4. Summarize and apply the concept of probability using the product and sum laws.
5. Solve problems in genetics involving incomplete dominance, polygenes, multiple alleles, and sex-linked traits.
6. Discuss ways in which genes may interact to affect the appearance of a single trait.
7. Discuss the phenomena of linkage and crossing over, and solve problems involving linked genes.
8. Discuss the genetic determination of sex and the role of the Y chromosomes in determining maleness.
9. Compare inbreeding and outbreeding, and discuss the genetic basis of hybrid vigor.
10. Summarize the mathematical basis of population genetics, using the Hardy-Weinberg principle to solve genetics problems such as estimating the frequency of genetic carriers.

Within the brief span of a decade the foundations of our present concepts of heredity and evolution were laid. Charles Darwin's *The Origin of Species* (1859) presented evidence supporting the concept of organic evolution and the theory of natural selection. Gregor Mendel discovered the basic laws of heredity in 1866, and Fredrich Miescher discovered nucleic acids in 1869. Although Darwin's views were widely discussed and accepted by most biologists within a few years, Mendel's and Miescher's contributions were unappreciated. Our present understanding of heredity really began in 1900 with the rediscovery of Mendel's work and the subsequent development of the science of genetics (see Focus on the Work of Gregor Mendel).

Interest in the nucleic acids lagged until the 1940s, when new chemical and crystallographic techniques yielded data that provided the basis for the research of James Watson and Francis Crick. These biologists provided evidence that the DNA molecule is a double helix. This was followed by Crick's suggestion (1961) that three adjacent nucleotides in a DNA chain compose a triplet code, specifying a particular amino acid, and by Marshall Nirenberg's "cracking" of the genetic code. Evidence accumulated since then indicates that the genetic code is virtually universal, that the same nucleotide triplet specifies a given amino acid in both bacteria and human beings.

Studies in genetics and evolution have continued to develop together, and our increasing understanding of the genetics of populations and of differential reproduction has modernized Darwin's concept of natural selection. Molecular biologists, comparing the details of the molecular structure of complex proteins in different species, are rediscovering and revalidating and sometimes reinterpreting the evolutionary relationships postulated decades ago on the basis of gross structural similarities and differences.

Genes and Alleles

From his experiments on inheritance in pea plants, Mendel made three key inferences, which we may restate as follows: (1) The units of heredity, now called genes, exist in pairs in individuals, but gametes have only one of each kind of gene. A member of a pair of genes that expresses itself is dominant, while one that does not express itself in the presence of a dominant gene is recessive. (2) During the formation of eggs and sperm each pair of genes separates independently of the members of other pairs of genes. (3) The members of the pairs of genes are assorted at random in the gametes.

Figure 11–1 This female captive bengal tiger is normally colored, but her offspring is white—not a true albino, but somewhat albinistic. The white tiger lacks the ability to produce normal pigments. Only about 50 white tigers are known to exist in the world, all descending from one white male captured in India in 1951. From their colors and relationship we can deduce that the mother is a genetically mixed carrier of the white trait. However, in her the normal allele codes for the production of pigments and so is dominant to the abnormal allele. Sexual reproduction makes it possible for offspring to differ genetically, and therefore in appearance, from their parents. The white animal is genetically pure for this trait. (Photographed at Busch Gardens, Tampa.)

The laws of heredity follow directly from the behavior of the chromosomes in mitosis, meiosis, and fertilization (discussed in the preceding chapter). Within each chromosome are many genes, each different from the others and each controlling the inheritance of one or more characteristics. The regularity of the mitotic process ensures that each daughter cell will have two of each kind of chromosome and therefore two of each kind of gene. As the chromosomes separate during meiosis and recombine in fertilization, so, of course, must the paired genes separate and recombine. All of the phenomena of Mendelian genetics depend on these simple facts. Each chromosome behaves genetically as though it were composed of a string of genes arranged in a linear order. The members of a homologous pair of chromosomes have similar genes arranged in similar order. The gene for each trait occurs at a particular point in the chromosome called a **locus** (plural, loci).

The inheritance of any trait can be studied only when there are two contrasting conditions, such as the yellow and green colors of Mendel's peas, normal pigmentation versus absence of pigmentation (**albinism**) in humans or other mammals (Fig. 11–1), or brown versus black coat color in guinea pigs. In simple cases, an individual expresses one or the other but not both of such contrasting conditions. Genes governing variations of the same trait that occupy corresponding loci on homologous chromosomes are termed **alleles** (Fig. 11–2).

The usage of the term allele emphasizes that there are two or more alternative forms of the gene at a specific locus in homologous chromosomes. Each of these forms can be assigned a single letter (or group of letters) as its symbol, with a capital letter designating one allele and the corresponding lowercase letter the other. It is customary to indicate the dominant allele with a capital letter and the recessive

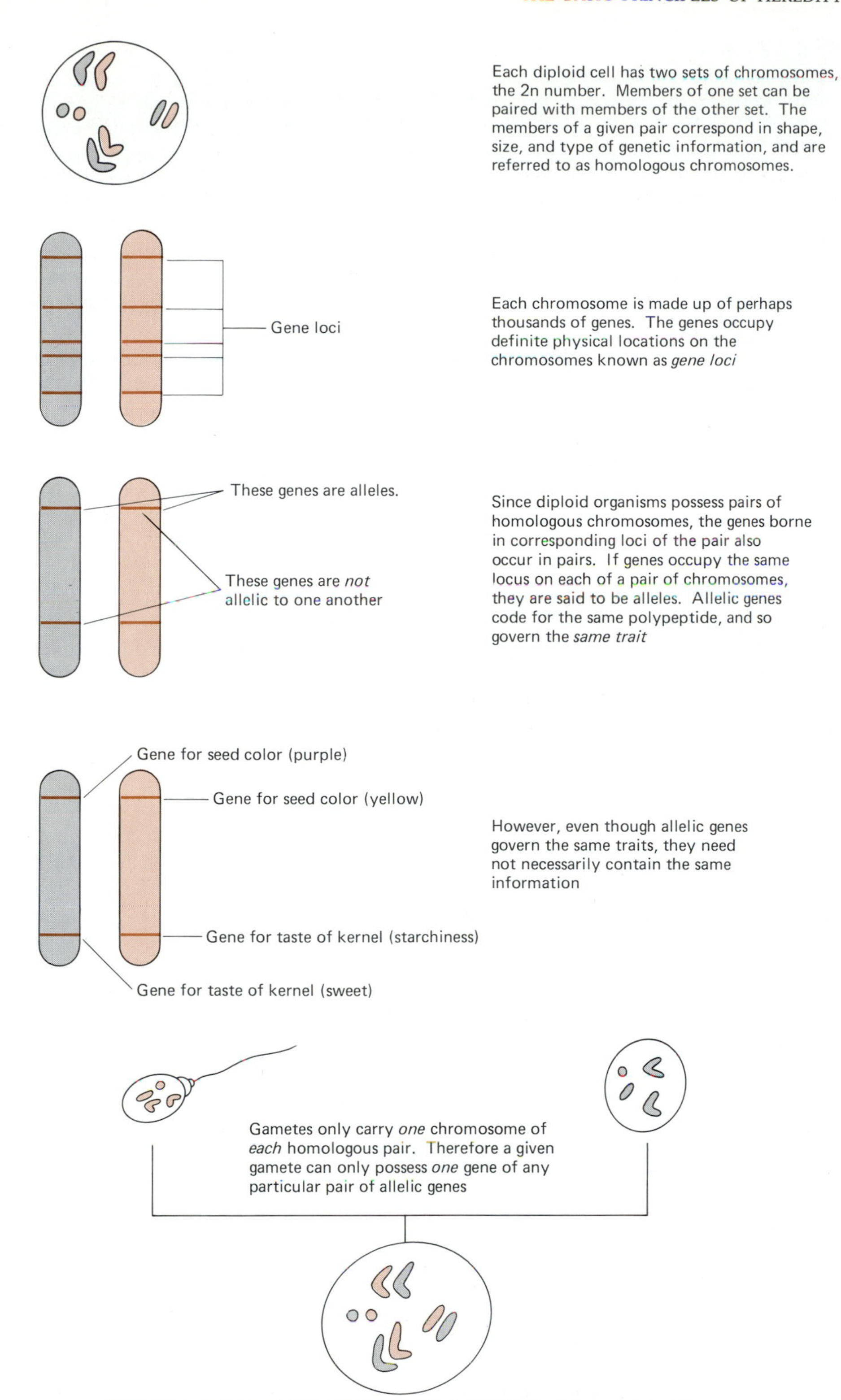

Each diploid cell has two sets of chromosomes, the 2n number. Members of one set can be paired with members of the other set. The members of a given pair correspond in shape, size, and type of genetic information, and are referred to as homologous chromosomes.
Gene loci
Each chromosome is made up of perhaps thousands of genes. The genes occupy definite physical locations on the chromosomes known as *gene loci*
These genes are alleles.
These genes are *not* allelic to one another
Since diploid organisms possess pairs of homologous chromosomes, the genes borne in corresponding loci of the pair also occur in pairs. If genes occupy the same locus on each of a pair of chromosomes, they are said to be alleles. Allelic genes code for the same polypeptide, and so govern the *same trait*
Gene for seed color (purple)
Gene for seed color (yellow)
However, even though allelic genes govern the same traits, they need not necessarily contain the same information
Gene for taste of kernel (starchiness)
Gene for taste of kernel (sweet)
Gametes only carry *one* chromosome of *each* homologous pair. Therefore a given gamete can only possess *one* gene of any particular pair of allelic genes
When the gametes combine to form a zygote, it, and the resulting embryo, will have homologous pairs of chromosomes, but of each pair, one member will be *maternal* in origin, and one *paternal* in origin. Each pair will bear allelic genes.

Figure 11–2 Homologous chromosomes and alleles.

allele with a lowercase letter. Thus, the letter *B* for black coat color and the letter *b* for brown represent alleles of the specific gene pair that determines coat color in guinea pigs.

The relationship between a given gene and the trait that it controls may be simple: A single gene pair may regulate the appearance of a single trait. Or the relationship may be more complex: One gene may participate in the control of several traits, or many genes may cooperate to regulate the appearance of a single trait. As you will learn in a later chapter, each gene is a portion of DNA in which biological information is stored as a triplet code in the sequence of nucleotides that compose the double helix of the DNA molecule. The information in each gene is "read out" and a specific protein is synthesized. The presence of the specific protein—an enzyme, for example—provides the chemical basis for the genetic trait.

A Monohybrid Cross

The usage of genetic terms and some of the basic principles of genetics can be illustrated by considering a simple **monohybrid cross,** that is, a cross between two individuals that differ with respect to a single characteristic. The mating of a "pure" brown male guinea pig with a "pure" black female guinea pig is illustrated in Figure 11–3. During meiosis in the spermatocytes in the male, the two *bb* alleles separate so that each sperm has only one *b* allele. In the formation of ova in the female, the *BB* alleles separate so that each ovum has only one *B* allele. The fertilization of this egg by a *b* sperm results in an animal with the alleles *Bb,* that is, one allele for brown coat and one for black coat. What color would you expect them to be—dark brown, gray, or perhaps spotted? In this instance they are just as black as the mother! The allele for black coat color is said to be *dominant* to the allele for brown coat color: It produces black coat color even when only one of the black alleles is present (*Bb*). The brown allele is said to be *recessive* to the black one: It produces brown coat color only when two alleles for brown color are present (*bb*).

The phenomenon of dominance explains, in part, why an individual may resemble one parent more than the other despite the fact that both make equal contributions to their offspring's genetic constitution. In one species of animal, black coat color may be dominant to brown, while in another species, brown may be dominant to black. Dominance is not completely predictable and can be determined only by experiment.

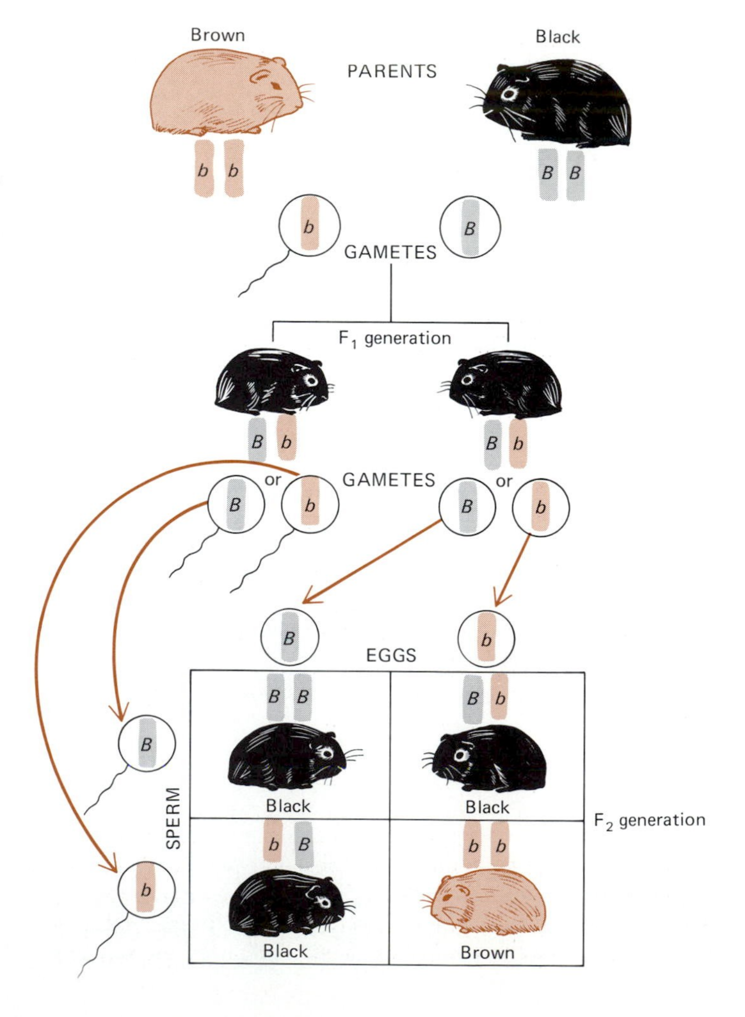

Figure 11–3 A monohybrid cross. The cross between a homozygous brown guinea pig and a homozygous black guinea pig. The F_1 generation includes only black individuals. However, the mating of two of these offspring yields F_2 generation offspring in the ideal ratio of 3 black : 1 brown, indicating that they are heterozygous.

HOMOZYGOUS AND HETEROZYGOUS ORGANISMS

An organism with two alleles exactly alike, such as two black coat alleles (*BB*) or two brown coat alleles (*bb*), is said to be **homozygous** for the trait. An organism with one dominant allele and one recessive allele (*Bb*) is said to be **hybrid,** or **heterozygous.** Using these terms we can now formulate better definitions for dominant and recessive alleles (or genes): A **recessive** allele is one that will produce its effect only in a homozygous individual; a **dominant** allele is one that will produce its effect in either a homozygous or a heterozygous individual.

During meiosis in the heterozygous black guinea pig, the chromosome containing the *B* allele becomes separated from the chromosome containing the *b* allele, so that each sperm or egg contains either a *B* allele or a *b* allele but never both. It follows that sperm (or eggs) containing *B* alleles and those with *b* alleles are formed in equal numbers by heterozygous *Bb* individuals. Since there are two types of eggs and two types of sperm, four combinations are possible in fertilization. There is no special attraction or repulsion between an egg and sperm containing the same kind of gene; hence these four possible combinations are equally probable. The combinations of eggs and sperm can be determined by algebraic multiplication:

$$(1/2\ B + 1/2\ b)\text{ eggs} \times (1/2\ B + 1/2\ b)\text{ sperm} = 1/4\ BB + 1/2\ Bb + 1/4\ bb$$

The possible combinations of eggs and sperm may also be represented in a "checkerboard" or **Punnett square** (as illustrated in Fig. 11–3). The types of eggs are represented along the top, the types of sperm are indicated along the left side, and the squares are filled in with the resulting zygote combinations. Three fourths of all offspring will be *BB* or *Bb* and have black coat color; one fourth will be *bb* and have brown coat color. The genetic mechanism responsible for the approximately 3:1 ratios obtained by Mendel in his pea breeding experiments is now evident. The generation with which a particular genetic experiment is begun is called the P_1 or parental generation. Offspring of this generation are called the F_1 or first filial generation. Those resulting when two F_1 individuals are bred constitute the F_2 or second filial generation, the grandchildren. Those resulting from the mating of F_2 individuals make up the F_3 generation, and so on.

PHENOTYPE AND GENOTYPE

The appearance of an individual with respect to a certain inherited trait in a given environment is known as its **phenotype.** The organism's genetic constitution (for that trait), usually expressed in symbols, is called its **genotype.** In the cross we have been considering, the ideal phenotypic ratio of the F_2 generation is 3 black-coated guinea pigs:1 brown-coated guinea pig, and the ideal genotypic ratio is 1 *BB*:2 *Bb*:1 *bb*. The phenotype may be a morphological characteristic—shape, size, color—or a physiological characteristic or even a biochemical trait such as the presence or absence of a specific enzyme required for the metabolism of a specific substrate. The phenotypic expression of genes may be altered by changes in the environmental conditions under which the organism develops.

In working genetics problems it is well to use the following procedure to avoid errors:

1. Write down the symbols you are using for each gene.
2. Determine the genotype of the parents, deducing them from the phenotypes of the offspring if this is necessary.
3. Indicate the possible kinds of gametes formed by each of the parents.
4. Set up a Punnett square, putting the possible types of sperm along its side and the possible types of eggs across its top.
5. Fill in the Punnett square and read off the genotypic and phenotypic ratios of the offspring.

TEST CROSSES

One third of the black guinea pigs in the F_2 generation of the mating of homozygous black × brown are themselves homozygous, *BB*; the other two thirds are heterozygous, *Bb*. Guinea pigs with the genotypes *BB* and *Bb* are alike phenotypically; they both have black coats. Since animals cannot, in general, be self-fertilized, how do you think a geneticist can distinguish the homozygous (*BB*) and heterozygous

FOCUS ON

The Work of Gregor Mendel

Gregor Mendel, an Austrian abbot who bred pea plants in his monastery garden at Brünn, in what is now Czechoslovakia, discovered the basic laws of genetics. Mendel had several types of pea plants in his garden and kept records of the inheritance of seven clearly contrasting pairs of traits, such as yellow versus green seeds, round versus wrinkled seeds, and green versus yellow pods.

When Mendel crossed plants with two different characteristics, such as yellow and green seeds, the plants in the next generation, the first filial or F_1 generation, resembled one of the two parents. The second filial or F_2 generation included individuals of both parental types. When Mendel counted these, he found that the two types of individuals were present in the F_2 generation in a ratio of approximately 3:1. For example, when he crossed tall plants with short plants, all the members of the F_1 generation were tall (see accompanying figure). When two of these F_1 generation tall plants were crossed, the F_2 generation included some tall and some short plants—787 tall and 277 short. Clearly, in the first generation the genetic factor (gene) for shortness was hidden or overcome by the gene for tallness. Mendel termed the gene for tallness *dominant* and the gene for shortness, *recessive*.

After he discovered that the crossing of the F_1 generation plants led to offspring in the second generation in an approximate ratio of three with the dominant characteristic to one with the recessive characteristic, it occurred to Mendel that each plant must have two genetic factors, whereas each gamete has only one. The F_1 generation tall plants also had two genetic factors—one for tallness and one for shortness—but because the tall gene was dominant, these plants were tall. However, when these F_1 plants formed gametes, the gene for tallness separated from the gene for shortness so that half of the gametes contained a gene for tallness, and the other half a gene for shortness. The random fertilization of eggs by sperm led to four possible combinations of genes: one with two genes for tallness, designated *TT*; one with two genes for shortness, *tt*; and two with one gene for tallness and one gene for shortness, *Tt* and *tT*. Because the tall gene is dominant to the short, three of the four kinds of offspring were tall plants and only one was short. Mendel's mathematical abilities enabled him to recognize that a 3:1 ratio would be expected among the offspring if each plant had two factors for any given characteristic rather than a single one.

By crossbreeding and counting the types of offspring, Mendel was able to detect regularities in the pattern of inheritance that had escaped earlier breeders. Mendel's conclusions were supported when chromosomes were finally seen and the details of mitosis, meiosis, and fertilization became known.

Mendel reported his findings at a meeting of the Brünn Society for the Study of Natural Science and published his results in the transactions of that society in 1866. The importance of his findings was not appreciated by other biologists of the time, and they were neglected for nearly 35 years.

In 1900, Hugo DeVries in Holland, Karl Correns in Germany, and Erich von Tschermak in Austria independently rediscovered the laws of inheritance that had been described by Mendel. The literature search that is a part of the conduct of scientific research led them to Mendel's paper in which these laws had been clearly stated 35 years before. They gave credit to Mendel by naming the basic laws of inheritance after him.

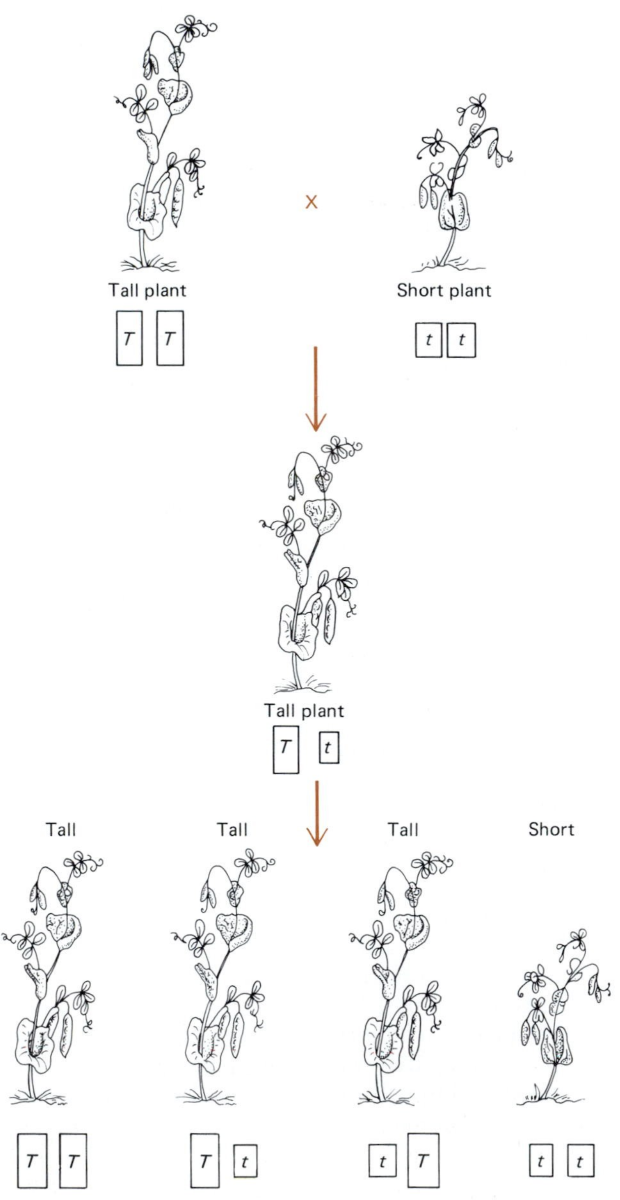

A diagram illustrating one of the crosses carried out by Gregor Mendel. Crossing a tall pea plant with a short pea plant yielded offspring all of which were tall. However, when these offspring were self-pollinated, the next generation included tall and short plants in a ratio of about 3:1.

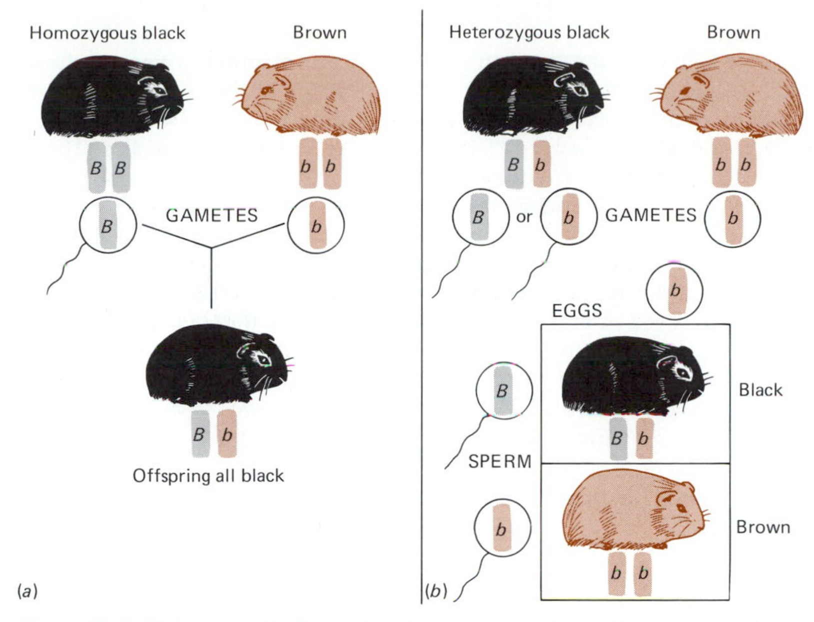

Figure 11–4 Test crosses to determine the genotype of a black guinea pig. (*a*) If a black guinea pig is mated with a brown and all of the offspring are black, the guinea pig probably has a homozygous genotype. (*b*) If, on the other hand, any of the offspring is brown, the black guinea pig must be heterozygous for this trait.

(*Bb*) black-coated guinea pigs? This is done by a **test cross,** by mating each black guinea pig with a homozygous brown (*bb*) guinea pig (Fig. 11–4). If all of the offspring are black, what inference would you make about the genotype of the black parent? If any of the offspring are brown, what conclusion would you draw regarding the genotype of the black parent? Can you be more certain about one of these two inferences than the other?

Mendel did just these sorts of experiments and bred heterozygous tall (*Tt*) pea plants with homozygous (*tt*) short ones. He predicted that the heterozygous parent would produce equal numbers of *T* and *t* gametes, whereas the homozygous short parent would produce only gametes containing *t*, and that this should lead to equal numbers of tall (*Tt*) and short (*tt*) individuals among the progeny. Thus, as a good hypothesis should, Mendel's hypothesis not only explained the known facts but enabled him to predict the results of other experiments.

This sort of testing is of great importance in the commercial breeding of animals or plants when the breeder is trying to establish a strain that will breed true for a certain characteristic. Two bulls, for example, may look equally healthy and vigorous; yet one will have daughters with qualities of milk production that are distinctly superior to those of daughters of the other bull. A breeder tests the genotypes of the breeding stock by making test matings and observing the offspring; if the offspring are superior with respect to the desired trait, the parents are thereafter used regularly for breeding.

Calculating the Probability of Genetic Events

All genetic ratios are properly expressed in terms of probabilities. In the examples just discussed, we saw that the offspring of the mating of two individuals heterozygous for the same gene pair would appear in the ratio of three with the dominant trait to one with the recessive trait. If the number of offspring is large enough, this ratio will be very closely approximated, as Mendel's experiments demonstrated. However, if the number of offspring is small, the ratio of the two types may be quite different from the expected 3:1. Why should this be? If there are only four offspring, any distribution—from all four with the dominant trait to all four with the recessive trait—might be found, although the latter would occur only very rarely. A better statement is that there are three chances in four (3/4) that any particular offspring of two heterozygous individuals will show the dominant trait and one chance in four (1/4) that it will show the recessive trait.

TYPES OF PROBABILITY

Three types of probabilities can be distinguished. **A priori probabilities** are those which can be specified *in advance* (*a priori*) from the nature of the event. For example, in flipping a coin the probability of obtaining heads is 1 in 2, and in casting dice the probability of obtaining a two is 1 in 6 (there are six sides to a die). These probabilities are independent of whether or not the event actually occurs.

In contrast, **empirical probabilities** are obtained by counting the number of times a given event occurs in a certain number of trials. For example, if a surgeon performs a certain type of operation on 500 people and 40 of them die, then the probability of death in this type of operation is 40 out of 500, or 0.08. Such empirical probabilities must be used in many fields of research where there is no theoretical basis—no *a priori* basis—for predicting the outcome. This type of probability is used in setting up the "risk tables" used widely by insurance firms.

If a scientist collects data about the number of individuals with certain traits in a given population and wants to know the extent to which these numbers agree with the ratio expected on the basis of some genetic theory (1:1, 3:1, and so on), he or she uses the methods of **sampling probability.**

THE PRODUCT LAW

If two events are independent of each other, the probability of their coinciding is the *product* of their individual probabilities. For example, the probability of obtaining heads on the first toss of a coin is 1/2, and the probability of obtaining heads on the second toss of a coin (an independent event) is also 1/2. The probability of obtaining heads twice on successive tosses of the coin is the product of their probabilities, $1/2 \times 1/2$, or 1/4: There is one chance in four of obtaining heads twice on two successive tosses of a coin. This **product law** of probability also holds for three or more independent events. For example, the probability of choosing at random a person who is male, has blood type A, and was born in June is $0.5 \times 0.4 \times 0.084 = 0.0168$. (The chance of being male is 1/2 or 0.5; of having blood type A is about 40% or 0.4; and of being born in June is 1/12 or 0.084.)

THE SUM LAW

According to the **sum law,** the probability that one or another of two independent events will occur is the *sum* of their separate probabilities. For example, in flipping a coin the probability that it will come up heads or tails is $1/2 + 1/2 = 1$. In rolling dice the probability that the die will come up *either* two or five is $1/6 + 1/6 = 1/3$.

APPLYING THE LAWS OF PROBABILITY

Using the product law, we can calculate how frequently in a large series of matings, if each mating results in exactly four offspring, all four would show the recessive trait. How often in the mating of two heterozygous black guinea pigs, *Bb,* should we expect to get a litter of four brown guinea pigs? The probability that the first one will be brown is 1/4. The probability that the second one will be brown is also 1/4. The fertilization of each egg by a sperm is an independent event, and we use the product law to calculate the combined probability of two (or more) events occurring together. The probability that all four of the offspring of any given mating will be brown is

$$1/4 \times 1/4 \times 1/4 \times 1/4 = 1/256$$

In other words, there is 1 chance in 256 that all four guinea pigs will have brown coat color.

Do not assume that if we actually mated 256 pairs of heterozygous guinea pigs and if each had four offspring, we would be *guaranteed* of getting one litter of four brown-coated offspring. Many people have lost vast sums of money in gambling by making similar mistakes! If we have made 255 such matings without getting a set of four brown offspring, what is the probability of getting four brown guinea pigs on the 256th try? We might be misled into thinking it is bound to happen, but in fact there is still only 1 chance in 256 that it will occur, since each of these matings is an independent event. The probability that any given offspring

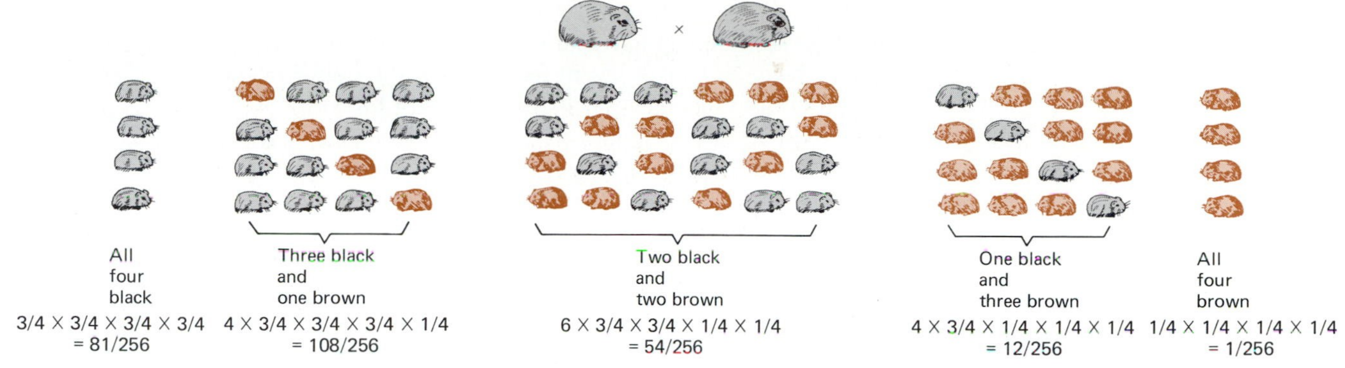

Figure 11–5 The possible distributions of black and brown offspring from heterozygous parents. Note the various ways in which a particular distribution can be obtained, for example, three black guinea pigs and one brown guinea pig in a litter of four. The probabilities for each general case are calculated.

will show the dominant trait, black coat color, is 3/4. We can, in a similar fashion, calculate the probability that all four guinea pigs in the litter will be black:

$$3/4 \times 3/4 \times 3/4 \times 3/4 = 81/256$$

There are 81 chances in 256 that all four guinea pigs will have black coat color.

How many different ways are there of getting three black guinea pigs and one brown guinea pig in a given mating? A look at Figure 11–5 shows that there are four ways: The first to be born could be brown and the next three would be black; or the second one born could be brown and the first, third, and fourth black. The other two possibilities are that the brown one would be the third or the fourth to be born. To calculate the probability that three of the four offspring will be black and the fourth will be brown we must multiply the number of possible combinations by the probability of each type:

$$4 \times 3/4 \times 3/4 \times 3/4 \times 1/4 = 108/256$$

This is the probability that there will be three black guinea pigs and one brown guinea pig in a litter of four. It may be surprising at first that the ideal 3:1 ratio is actually obtained in less than half of the total number of litters of four. However, when we add up the total numbers of black and brown offspring from a large number of matings, the 3:1 ratio is more and more closely approximated.

How many different ways are there of getting two black and two brown guinea pigs in a litter of four? There are six ways, as shown in Figure 11–5. From this we can calculate the probability that two of the four offspring will be black and two brown:

$$6 \times 3/4 \times 3/4 \times 1/4 \times 1/4 = 54/256$$

We can also see that there are four different combinations that give one black and three brown: The first born may be black and the rest brown, or the second will be black and all others brown, or the third or the fourth will be black. The probability of getting one black guinea pig and three brown guinea pigs is

$$4 \times 3/4 \times 1/4 \times 1/4 \times 1/4 = 12/256$$

All probabilities are expressed as fractions (or decimal fractions) ranging from *zero*, expressing an impossibility, to *one*, expressing total certainty. Probabilities can be multiplied or added just like any other fractions. In this example there are no other possibilities; the four guinea pigs must be either four black, or three black and one brown, or two black and two brown, or one black and three brown, or four brown. If we have made no arithmetical errors, the sum of the five probabilities will add up to one:

$$81/256 + 108/256 + 54/256 \times 12/256 + 1/256 = 256/256 = 1$$

The color of the iris of the human eye is inherited through several pairs of genes, but one pair is the primary factor differentiating brownish eye color from bluish. The allele for brown eye color, B, is dominant to the allele for blue, b. If two heterozygous brown-eyed people marry, what is the probability that they will have a blue-eyed child? Clearly, there is 1 chance in 4 that any child of theirs will have blue eyes. Each mating is a separate, independent event; its result is not affected by the results of any previous matings. If these two brown-eyed parents have had three brown-eyed children and are expecting their fourth child, what is the probability that the child will have blue eyes? Again, the unwary might guess that this one *must* have blue eyes, but in fact there is still only 1 chance in 4 of having a blue-eyed child and 3 chances in 4 of having a brown-eyed child.

Incomplete Dominance

From studies of the inheritance of many traits in a wide variety of organisms it is clear that one member of a pair of alleles may not be completely dominant to the other. Indeed, it may be improper to use the terms dominant and recessive in such instances. For example, red and white are common flower colors in Japanese four-o'clocks. Each color breeds true when these plants are self-pollinated. What flower color might we expect in the offspring of a cross between a red flowering plant and one that bears white flowers? Without knowing which is dominant we might predict that all would have red flowers or all would have white flowers. This cross was first made by the German botanist Karl Correns, who found that all the F_1 offspring have pink flowers! How can we explain that? Does this result in any way prove that Mendel's assumptions about inheritance are wrong? Quite the contrary, for when two of these pink-flowered plants were crossed, offspring appeared in the ratio of 1 red-flowered : 2 pink-flowered : 1 white-flowered (Fig. 11–6). In this instance, as in

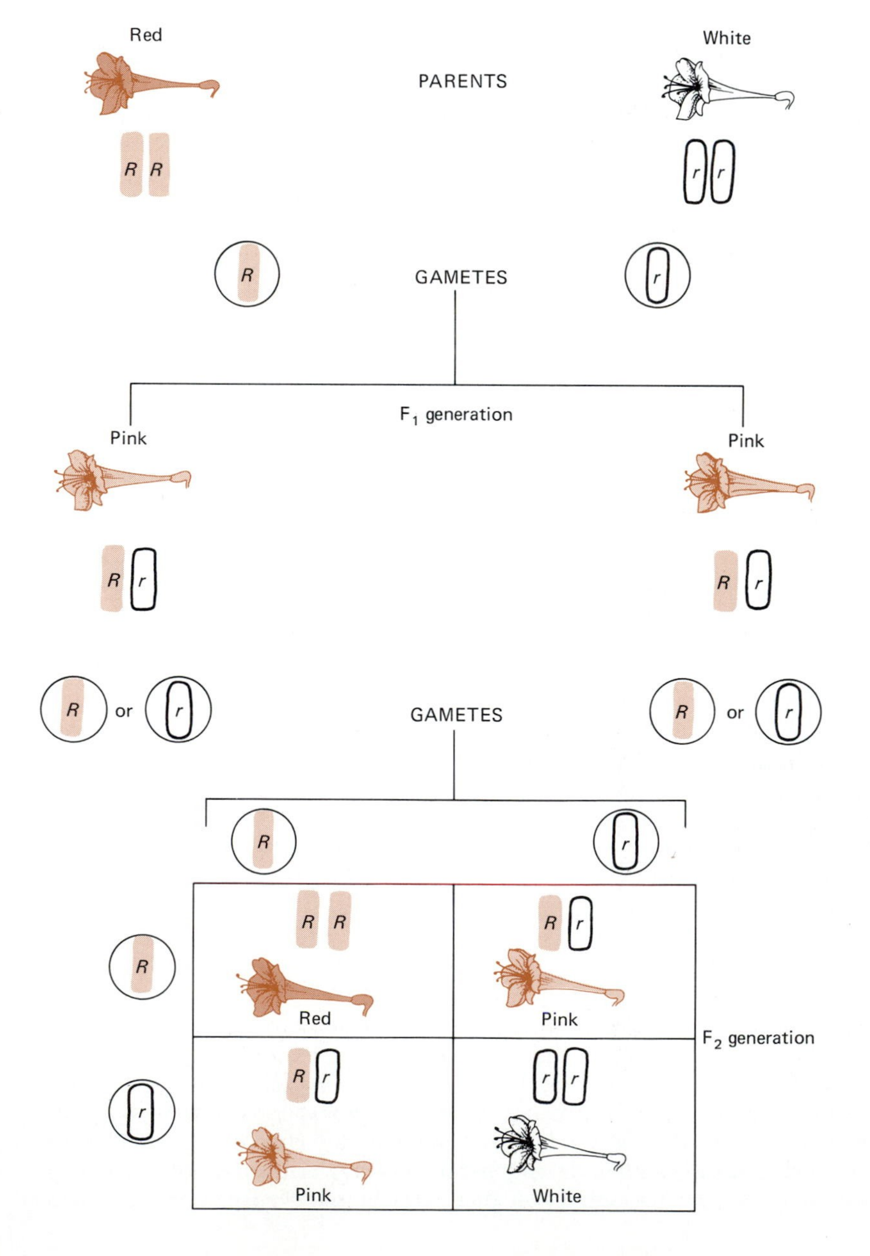

Figure 11–6 Incomplete dominance in Japanese four-o'clocks. Red is incompletely dominant to white in some types of flowers. A plant with the genotype *Rr* has pink flowers.

other aspects of science, the finding of results that differ from those predicted simply prompts scientists to reexamine and modify their assumptions to account for the new exceptional results. The pink-flowered plants are clearly the heterozygous individuals, and neither the red allele nor the white allele is completely dominant. When the heterozygote has a phenotype that is intermediate between those of its two parents, the genes are said to show **incomplete dominance.** In these crosses the genotypic and phenotypic ratios are identical.

Incomplete dominance is not unique to Japanese four-o'clocks. Red- and white-flowered sweet pea plants also produce pink-flowered plants when crossed. In both cattle and horses, reddish coat color is incompletely dominant to white coat color. The heterozygous individuals have roan-colored (that is, reddish with white spots) coats. If you saw a white mare nursing a roan-colored colt, what would you guess was the coat color of the colt's father? Is there more than one possible answer?

Deducing Genotypes

The science of genetics resembles mathematics in that it consists of a few basic principles, which, once grasped, enable the student to solve a wide variety of problems. The genotypes of the parents can be deduced from the phenotypes of their offspring. In chickens, for example, the allele for rose comb (*R*) is dominant to the allele for single comb (*r*). Suppose that a cock is mated to three different hens, as shown in Figure 11–7. The cock and hens A and C have rose comb; hen B has a single comb. Breeding the cock with hen A produces a rose-combed chick; with hen B, a single-combed chick, and with hen C, a single-combed chick. What type of offspring can be expected from further matings of the cock with these hens?

Since the allele for single comb, *r*, is recessive, all of the hens and chicks that are phenotypically single-combed must be homozygous, *rr*. We can deduce, then, that hen B and the offspring of hens B and C are genotypically *rr*. All of those individuals that are phenotypically rose-combed must have at least one *R* allele, and the cock and hen C are therefore *Rr*. The fact that the offspring of the cock and hen B was single-combed proves that the cock is heterozygous, *Rr*. The single-

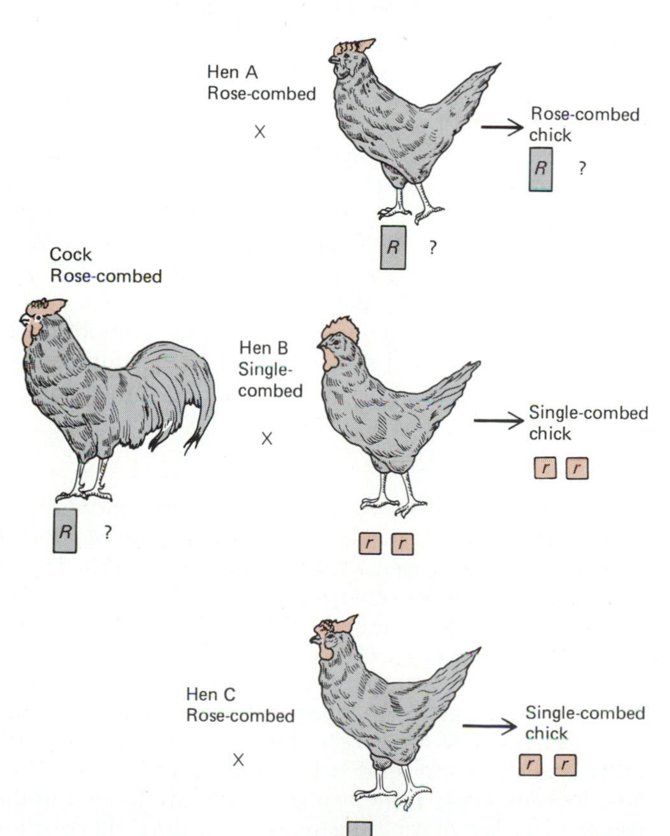

Figure 11–7 Deducing the parental genotypes from the phenotypes of the offspring. In chickens, the allele for rose comb (*R*) is dominant to the allele for single comb (*r*). Determine the unknown parental genotypes. See text for discussion.

combed chick received one *r* allele from its mother but must have received the other from its father. The fact that the offspring of the cock and hen C had a single comb proves that hen C is heterozygous, *Rr*. It is impossible to decide from the data given whether hen A is homozygous, *RR*, or heterozygous, *Rr*; further breeding would be necessary to determine this. Additional matings of the cock with hen B would result in one half rose-combed and one half single-combed individuals; additional matings of the cock with hen C would produce three fourths rose-combed and one fourth single-combed chicks.

Dihybrid and Trihybrid Crosses

Frequently geneticists must analyze the inheritance of two or more traits in the same group of individuals. A mating that involves individuals differing in two traits is called a **dihybrid cross.** The principles involved and the procedure of solving problems are exactly the same for monohybrid and for dihybrid (or polyhybrid) crosses. In the latter, of course, the number of types of gametes is greater, and the number of types of zygotes is correspondingly larger.

When two pairs of alleles are located in different (nonhomologous) chromosomes, each pair is inherited independently of the other; that is, each pair separates during meiosis independently of the other. When a homozygous black, short-haired guinea pig (*BBSS*, since short hair is dominant to long hair) and a brown, long-haired guinea pig (*bbss*) are mated, the *BBSS* animal produces gametes that are all *BS*, and the *bbss* animal produces only *bs* gametes. Each gamete contains one and only one of each kind of gene. The union of *BS* gametes and *bs* gametes yields only individuals with the genotype *BbSs*. All the offspring are heterozygous for hair color and for hair length, and all are phenotypically black and short-haired.

When two of the F_1 individuals are mated, each produces four kinds of gametes with equal probability: *BS*, *Bs*, *bS*, and *bs*. Hence, 16 combinations are possible among the zygotes (Fig. 11–8). There are 9 chances in 16 of obtaining a black, short-haired individual; 3 chances in 16 of obtaining a black, long-haired individual; 3 chances in 16 of obtaining a brown, short-haired individual; and 1 chance in 16 of obtaining a brown long-haired individual.

In a similar fashion, problems involving three pairs of genes may be solved. An individual heterozygous for three pairs of alleles located in different pairs of chromosomes will yield eight types of gametes in equal numbers. The union of these eight types of eggs and eight types of sperm will yield 64 possible zygotes in the F_2 generation. In peas, as Mendel demonstrated, yellow seed color (*Y*) is dominant to green (*y*), smooth seeds (*S*) are dominant to wrinkled (*s*), and tall plants (*T*) are dominant to dwarf (*t*). The mating of a homozygous yellow, smooth, tall plant (*YYSSTT*) with a homozygous green, wrinkled, dwarf plant (*yysstt*) will produce offspring that are all yellow, smooth, and tall (*YySsTt*). When two of these F_1 plants are mated, F_2 offspring are produced in the ratio of 27 yellow, smooth, tall:9 yellow, smooth, dwarf:9 yellow, wrinkled, tall:9 green, smooth, tall:3 yellow, wrinkled, dwarf:3 green, wrinkled, tall:3 green, smooth, dwarf:1 green, wrinkled, dwarf. Draw a Punnett square to verify these numbers.

Mendel's Laws

Now that we have learned about monohybrid and dihybrid crosses, we can state more formally how Mendel's laws apply to them. Mendel's **law of segregation** may be stated as follows: Genes exist in individuals in allelic pairs, and in the formation of gametes the alleles separate, or segregate, and pass into different gametes, so that each gamete has one and only one of each allele. This law can be illustrated by mating two heterozygous black guinea pigs. A brown guinea pig may be produced because the alleles of each parent segregate into separate gametes during meiosis.

Mendel's **law of independent assortment** states that the members of one gene pair separate from each other in meiosis independently of the members of other gene pairs and come to be assorted at random in the resulting gamete. (This law does not apply if the two gene pairs are located in the same pair of chromosomes.) This law may be illustrated by the dihybrid cross just discussed. The segregation of the *Bb* alleles is independent of the segregation of the *Ss* alleles.

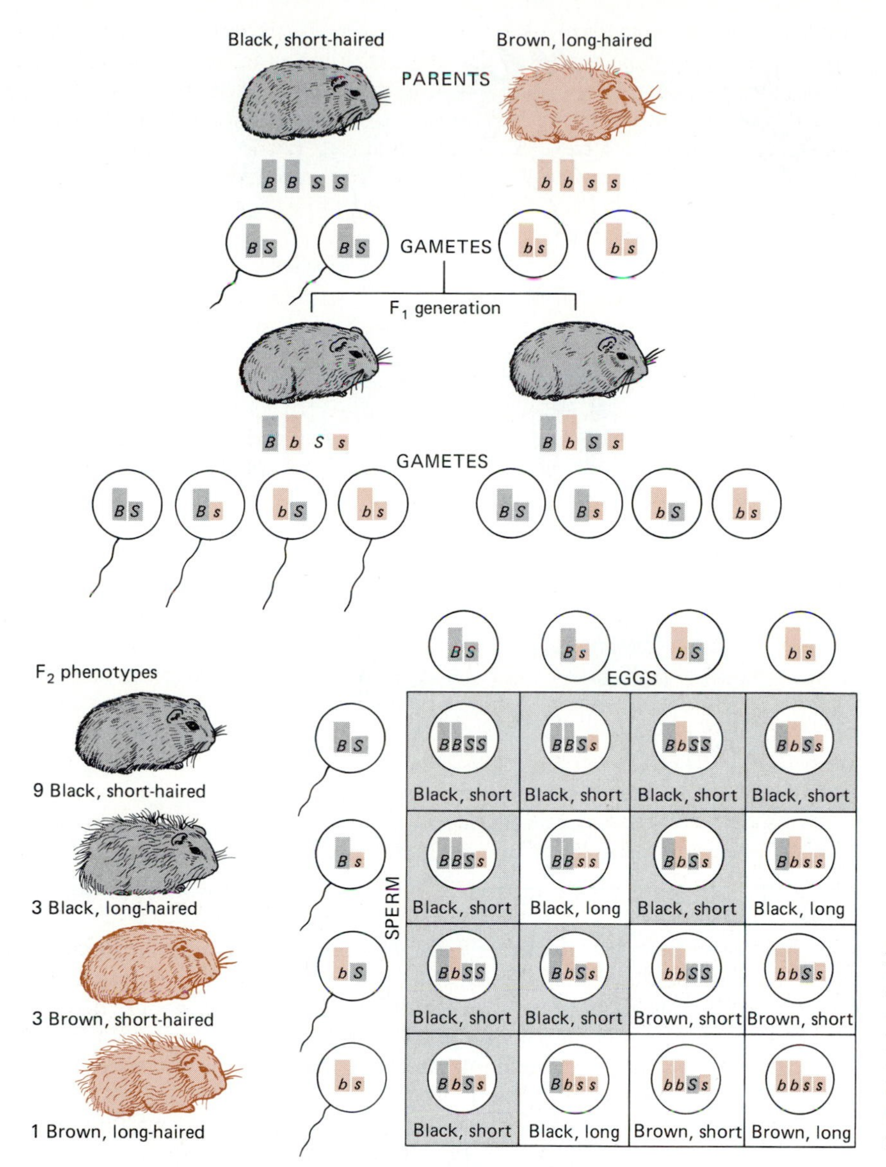

Figure 11–8 A dihybrid cross. When a black, short-haired guinea pig is crossed with a brown, long-haired one, all of the offspring are black and have short hair. However, when two members of the F_1 generation are crossed, the ratio of phenotypes is 9 : 3 : 3 : 1. Note that the two gene pairs considered here segregate independently.

Gene Interactions

In the examples presented so far, the relationship between a gene and its phenotype has been direct, precise, and exact. Each gene controls the appearance of a single trait. However, the relationship of gene to characteristic may be quite complex. Most genes probably have many different effects, a quality referred to as **pleiotropy.** This is dramatically evident in many genetic diseases in which multiple symptoms can be traced to a single pair of alleles. **Epistasis** is a common type of gene interaction in which the presence of a particular allele of one gene pair determines whether alleles of another gene pair will be expressed. By such mechanisms several pairs of genes may interact to affect a single trait, or one pair may inhibit or reverse the effect of another pair of genes. More than 12 pairs of alleles interact in various ways to produce the coat color of rabbits, and more than 100 pairs of alleles are concerned with the color and shape of the eyes in fruit flies.

One of the simpler types of gene interaction is illustrated by the inheritance of combs in poultry. The allele for rose comb, *R*, is dominant to that for single comb, *r*. Another gene pair governs the inheritance of pea comb, *P*, versus single comb, *p*. A single-combed fowl must have the genotype *pprr;* a pea-combed fowl is either *PPrr* or *Pprr;* and a rose-combed fowl is either *ppRR* or *ppRr* (Fig. 11–9). When a homozygous pea-combed fowl is mated to a homozygous rose-combed one, the offspring have neither pea nor rose comb but a completely different type called walnut comb. The phenotype of walnut comb is produced whenever a fowl has one or two *R* alleles plus one or two *P* alleles. What would you predict about the types of combs

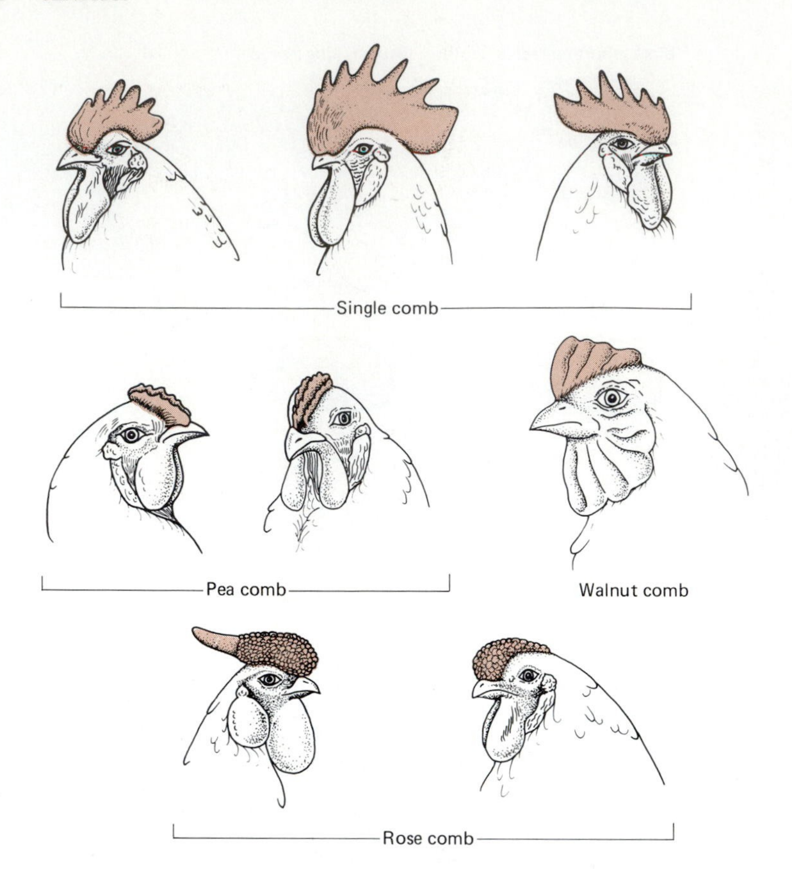

Figure 11–9 The different types of genetically determined combs in roosters. Two gene pairs govern the inheritance of these types of combs.

among the offspring of the mating of two heterozygous walnut-combed fowls, *PpRr*?

Two gene pairs that are inherited independently may interact in such a way that neither of the dominant alleles alone can produce its effect unless the other is present. Such gene pairs have been termed **complementary genes;** the action of each one "complements" the action of the other in the production of the phenotype. The presence of both dominants produces one trait; the alternate trait is seen when either one or both are absent.

In the course of experiments with sweet peas, Bateson and Punnett found that crossing two white-flowered races of sweet peas gave offspring that all had purple flowers! When two of the F_1 purple-flowered plants were crossed, the F_2 generation had offspring in the ratio of 9 purple-flowered:7 white-flowered (Fig. 11–10). Two gene pairs, designated *C–c* and *E–e*, proved to be involved: The dominant allele *C* codes for an enzyme needed in the production of a white compound from which purple pigment can be made by another enzyme coded for by the dominant allele *E*. The homozygous recessive *cc* is unable to synthesize the white compound, and the homozygous recessive *ee* lacks the enzyme to convert the raw material into a purple pigment. A race of sweet peas breeding true for purple flowers could be established by selecting two *CCEE* plants and mating them.

We have already seen that coat color in guinea pigs is determined by the *B–b* gene pair, with the *B* allele for black coat dominant to the *b* allele for brown coat. The expression of either phenotype, however, depends on the presence of the dominant allele of yet another gene pair. This allele, *C*, codes for the enzyme tyrosinase, which converts a colorless precursor into the pigment melanin and hence is required for the production of any kind of pigment. Thus the homozygous recessive *cc* lacks the enzyme and produces no melanin, and the animal is a white-coated, pink-eyed albino, no matter what combination of *B* and *b* alleles may be present. When an albino with genotype *ccBB* is mated to a brown guinea pig with genotype *CCbb*, the F_1 generation will be black-coated, *CcBb*. When two such animals are mated, their offspring will appear in the ratio of 9 black-coated:3 brown-coated:4 albino. Make a Punnett square to verify this. In this example of epistasis, two independent pairs of genes interact in such a way that one dominant will produce its effect whether or not the other is present, but the second will produce its effect only in the presence of the first.

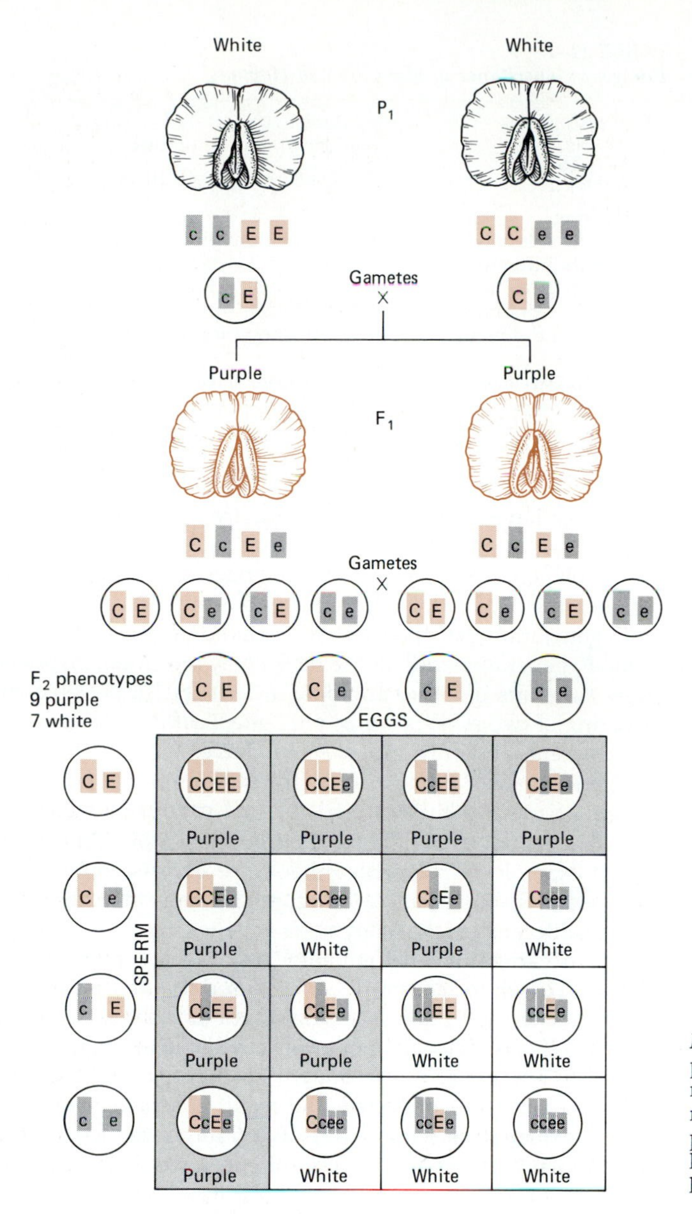

Figure 11–10 A cross between sweet peas illustrating the action of complementary genes. Two pairs of complementary genes regulate flower color. To produce purple flowers, the plant must have at least one *C* allele and one *E* allele.

Polygenic Inheritance

Many human characteristics, such as height, body form, and skin color, and many commercially important characteristics in animals, such as milk or egg production, the size of fruits, and so on, cannot be separated into distinct alternate classes and are not inherited through a single pair of genes. Several, perhaps many, different pairs of genes affect each characteristic. The term **polygenic inheritance** is applied when two or more independent pairs of genes have similar and additive effects on the same characteristic. Skin color inheritance in humans was studied by Davenport in Jamaica. He found that two major pairs of genes are involved, which he designated *A–a* and *B–b*.[1] The capital letters represent alleles producing dark skin—the more capital letters, the darker the skin. The alleles affect the skin character in an additive fashion. A full black has the genotype *AABB*, while a white has the genotype *aabb*. The F_1 offspring of an *aabb* person and an *AABB* person are all *AaBb* and have an intermediate skin color termed mulatto. The offspring of two such

[1] It is now thought that there are three or four pairs of genes involved in determining skin color, but two pairs are used here to illustrate the principle of polygenic inheritance.

TABLE 11–1
Polygenic Inheritance of Skin Color in Humans

Parents	*AaBb* (mulatto) × *AaBb* (mulatto)
Gametes	*AB Ab aB ab* *AB Ab aB ab*
Offspring	
1 with 4 dominants	*AABB*—phenotypically black
4 with 3 dominants	2 *AaBB* and 2 *AABb*—phenotypically "dark"
6 with 2 dominants	4 *AaBb*, 1 *AAbb*, 1 *aaBB*—phenotypically mulatto
4 with 1 dominant	2 *Aabb*, 2 *aaBb*—phenotypically "light"
1 with no dominants	*aabb*—phenotypically white

Based on Davenport's work. There are actually three or four pairs of polygenes involved.

mulattoes have skin colors ranging from the black to the white phenotype (Table 11–1).

Polygenic inheritance is characterized by an F_1 generation that is intermediate between the two completely homozygous parents and shows little variation and by an F_2 generation that shows a wide variation between the two parental types. Most of the F_2 generation individuals have some intermediate phenotype, and only a few show the traits of either grandparent. Of the 16 possible zygote combinations in offspring of two *AaBb* persons, only one, *AABB*, will have skin as dark as the black grandparent, and only one, *aabb*, will have skin as light as the white grandparent. The genes *A* and *B* produce about the same amount of darkening of the skin; hence, the genotypes *AaBb*, *AAbb*, and *aaBB* all produce similar mulatto phenotypes.

The model used here for the inheritance of skin color in humans is a rather simple example of polygenic inheritance because only two major pairs of genes were used. The inheritance of height in humans involves perhaps ten or more pairs of genes. Because of the many pairs of genes involved, and because height is modified by a variety of environmental conditions, the heights of adults range from perhaps 125 cm to 215 cm. If we measured the heights of a thousand adult American men taken at random, we would find that only a few are as tall as 215 cm or as short as 125 cm. The height of most would cluster around the mean, about 170 cm. When the number of people of each height is plotted against height in centimeters and the points are connected, the result is a bell-shaped curve called a **curve of normal distribution** (Fig. 11–11). If you measured the heights of a thousand men and women combined, what sort of curve would be generated by the data?

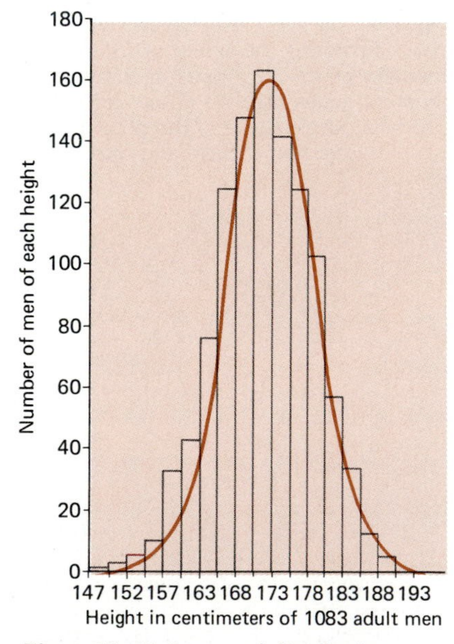

Figure 11–11 A normal distribution curve showing the distribution of the heights of 1083 adult white males. The blocks indicate the actual number of men whose heights were within the unit range. For example, there were 163 men whose heights were between 170 and 173 cm. The smooth curve is a normal curve based on the mean and standard deviation of the data. Note the bell shape of the curve.

Multiple Alleles

In the examples so far, we have dealt with situations in which at any locus—that is, any given position on the chromosome—there is one of only two alleles, the dominant and recessive alleles. At many, if not most, loci there may be additional possibilities, with genes producing phenotypes different from those of both the dominant and the recessive. The term **multiple alleles** is applied to three or more alternative forms of a gene that can occupy a single locus. Each of the alleles can produce a distinctive phenotype. Among the members of a population, any given diploid individual may have any two of the alleles but never more than two, and any gamete, of course, can have only one of them. But in the population as a whole there will be distributed three or more different alleles.

In rabbits, for example, a *C* allele causes fully colored coats. The homozygous recessive, *cc*, causes albino coat color. There are two other alleles at the same locus, c^h and c^{ch}. The allele c^h, in a homozygous individual, causes the "Himalayan pattern," in which the body color is white but the tips of the ears, nose, tail, and legs are colored. The c^{ch} allele, in a homozygous individual, produces the "chinchilla pattern," in which the entire body has a light gray color. These alleles can be arranged in a series—*C*, c^{ch}, c^h, and *c*—in which each is dominant to those following it and recessive to those preceding it. In other series of multiple alleles, some may be incompletely dominant so that the heterozygotes have a phenotype intermediate between those of their parents.

TABLE 11–2
*ABO Blood Types**

Phenotype (Blood Type)	*Genotypes*	*Antigen on RBC*	*Antibodies in Plasma*	*Frequency in U.S. Population (%)*	
				Western European Descent	*African Descent*
A	I^AI^A, I^Ai	A	Anti-B	45	29
B	I^BI^B, I^Bi	B	Anti-A	8	17
AB	I^AI^B	A, B	None	4	4
O	ii	None	Anti-A, anti-B	43	50

*This table and the discussion of the ABO system have been simplified somewhat. Actually, some type A persons have two type A antigens and are designated type A_1, while those with only one antigen are termed type A_2.

HUMAN ABO BLOOD TYPES

The human blood types O, A, B, and AB are inherited through multiple alleles. Allele I^A provides the code for the synthesis of a specific glycoprotein, antigen A, in the red blood cell. (Immunity will be discussed in Chapter 36, but for now we will define antigens as compounds capable of stimulating an immune response.) Allele I^B leads to the production of a different glycoprotein, antigen B. These glycoproteins are found on the surfaces of the blood cells that manufacture them. Allele i produces no antigen. The antibodies anti-A and anti-B are proteins that appear in the plasma of persons lacking the corresponding antigens on their red blood cells. (Antibodies are proteins that combine with specific antigens.) Allele i is recessive to the other two, but neither allele I^A nor allele I^B is dominant to the other. Rather, they are **codominant,** for both are expressed phenotypically. The symbols I^A, I^B, and i are used to emphasize that all three are alleles at the same locus. Persons with genotype I^AI^A or I^Ai have **blood type A** (Table 11–2); those with genotype i^Bi^B or I^Bi have **blood type B;** and those with genotype ii have **blood type O.** When both the I^A and I^B alleles are present, both antigen A and antigen B are produced in the red blood cells; persons with this I^AI^B genotype have **blood type AB** (Fig. 11–12).

Determining the blood types of the persons involved may be helpful in settling cases of disputed parentage. Such blood tests can never prove that a certain man is the father of a certain child; they determine only whether or not he *could* be the father. They may definitely prove that he could not be the father of a certain child. Could a man with type AB blood be the father of a child with blood type O? Could a man with blood type O be the father of a child with type AB blood? Could a type B child with a type A mother have a type A father or a type O father?

More than a dozen other sets of blood types, including the Rh and MN groups, are inherited through other genes, independently of the ABO blood types. Determining some of these types in a given person may be useful in establishing relationships that could not be made certain by the ABO blood type alone.

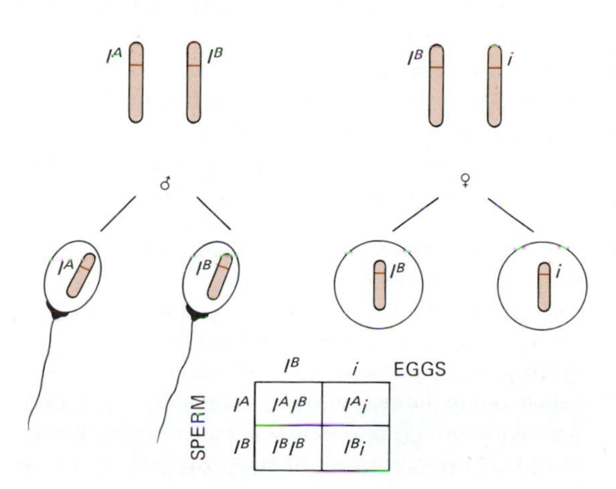

Figure 11–12 Multiple alleles for blood types. Three different alleles exist for ABO blood type. In this cross each parent produces only two kinds of gamete, but between them both, there are three different types of gametes with respect to ABO blood type. Four possible genotypes occur among their offspring. How many possible blood types can be found among the offspring?

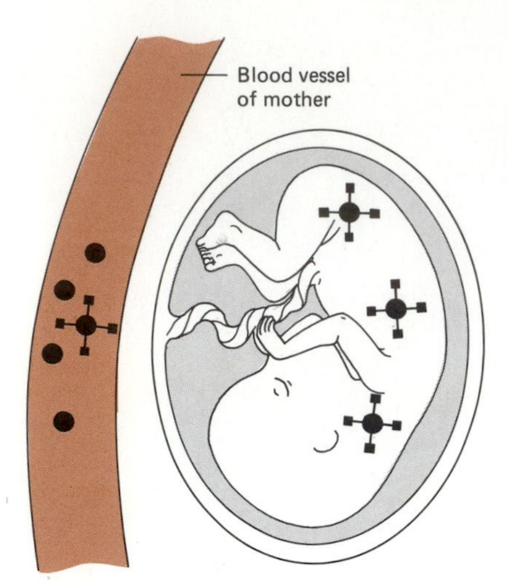

A few Rh+ RBCs leak across the placenta from the fetus into the mother's blood

(a)

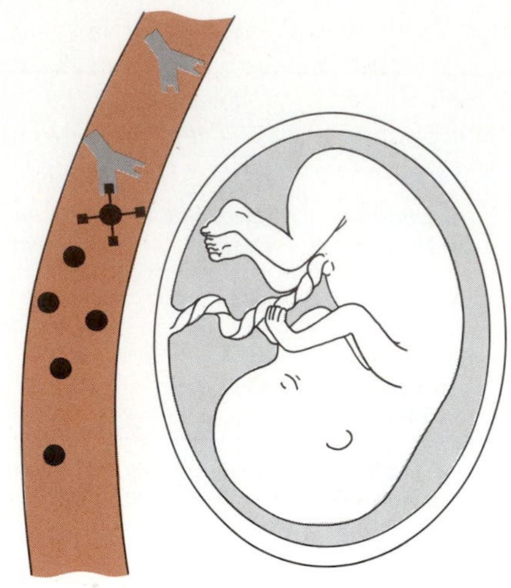

The mother produces D antibodies in response to D antigen on Rh+ RBCs

(b)

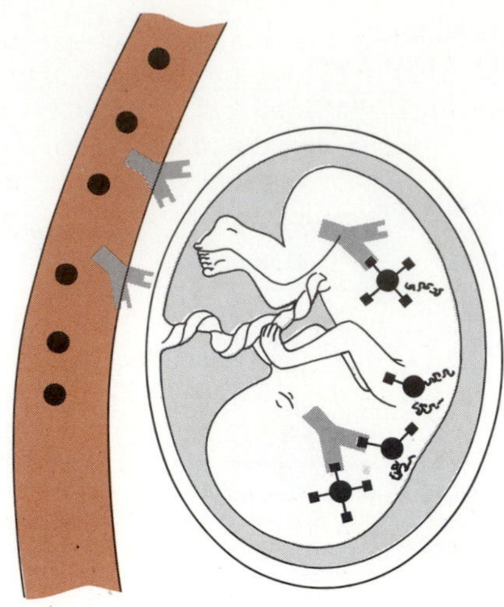

D antibodies cross the placenta and enter the blood of the fetus. Hemolysis of Rh+ blood occurs. The fetus may develop erythroblastosis fetalis.

(c)

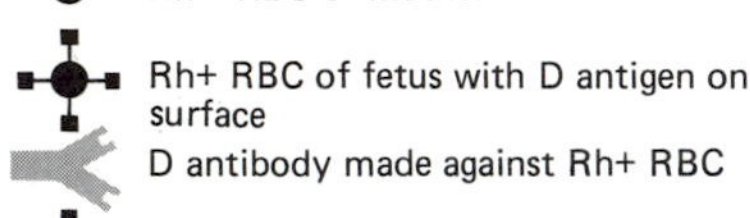

Figure 11–13 Rh incompatibility can cause serious problems when an Rh-negative woman and an Rh-positive man produce Rh-positive offspring. (*a*) Some Rh-sensitized red blood cells (RBCs) leak across the placenta from the fetus into the mother's blood. (*b*) The woman produces D antibodies in response to the D antigens on the fetal red blood cells. (*c*) Some of the D antibodies cross the placenta and enter the blood of the fetus, causing red blood cells to rupture and release hemoglobin into the circulation. The fetus may develop erythroblastosis fetalis.

THE Rh SYSTEM

Named for the rhesus monkeys in whose blood it was first found, the Rh system consists of at least eight different kinds of Rh antigens, each referred to as an Rh factor. By far the most important is **antigen D** (anti-D). About 85% of persons of western European descent are Rh-positive, which means that they have antigen D on the surfaces of their red blood cells. (Of course, this is in addition to the antigens of the ABO system.) The 15% or so of this population who are Rh-negative have no antigen D and will produce antibodies against that antigen when exposed to Rh-positive blood.

Although several kinds of maternal–fetal blood type incompatibilities are known, *Rh incompatibility* is probably the most important (Fig. 11–13). If a woman is Rh-negative and her husband is Rh-positive, the fetus may be Rh-positive, having inherited the D allele from the father. A small quantity of blood from the fetus may pass through some defect in the placenta (especially during the birth process) and into the mother's blood, sensitizing her white blood cells, which then produce antibodies to the Rh factor. When this woman becomes pregnant again, sensitized white blood cells produce antibodies that may pass through the placenta into the fetal blood and cause clumping of the red blood cells. Breakdown products of the hemoglobin released into the circulation damage many organs, including the brain. This disease is known as **erythroblastosis fetalis.** In extreme cases so many fetal red blood cells are destroyed that the fetus dies before birth.

When Rh-incompatibility problems are suspected, blood can be exchanged while the baby is still within the mother's uterus, but this is a risky procedure. Rh-negative women are now treated just after childbirth (or at termination of pregnancy by miscarriage or abortion) with an anti-Rh preparation. This drug apparently clears the Rh-positive cells from the mother's blood very quickly, thus minimizing the chance of sensitizing her own white blood cells. As a result, her cells do not produce the anti-D that could harm her next baby.

Linkage and Crossing Over

The chromosomes are inherited as units—they pair and separate during meiosis as units; thus, all the genes in any given chromosome tend to be inherited together. If the chromosomal units never changed, the genes on any one chromosome would always be inherited together. However, during meiosis when the chromosomes are pairing and undergoing synapsis, crossing over and recombination occur. During this process, homologous chromosomes may exchange entire segments of chromosomal material (Fig. 11–14). This exchange of segments occurs at random along the length of the chromosome. Several exchanges may occur at different points along the length of the same chromosome during a single meiotic division. The greater the distance between any two genes in the chromosome, the greater will be the likelihood that they will be separated by crossing over.

In fruit flies, the gene pairs coding for wing shape—allele *V* for normal wings and allele *v* for vestigial wings—and for body color—allele *B* for gray and allele *b* for black—are located in the same chromosomes (Fig. 11–15). They tend to be

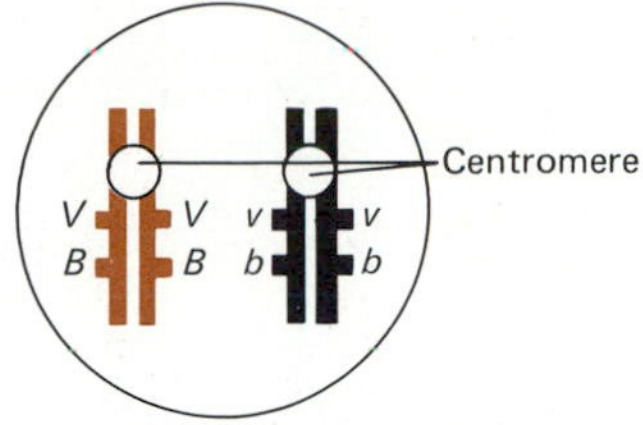

Two homologous chromosomes undergo synapsis in meiosis.

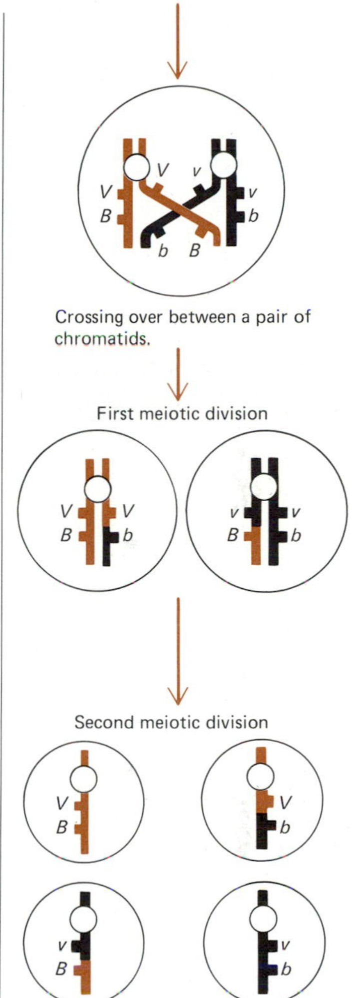

Four haploid gametes produced; here two crossover and two noncrossover gametes.

Figure 11–14 Crossing over and genetic recombination. Crossing over, the exchange of segments between chromatids of homologous chromosomes, permits recombination of genes (for example, as *vB* and *Vb*). The farther apart genes are located on a chromosome, the greater is the probability that an exchange of segments between them will occur.

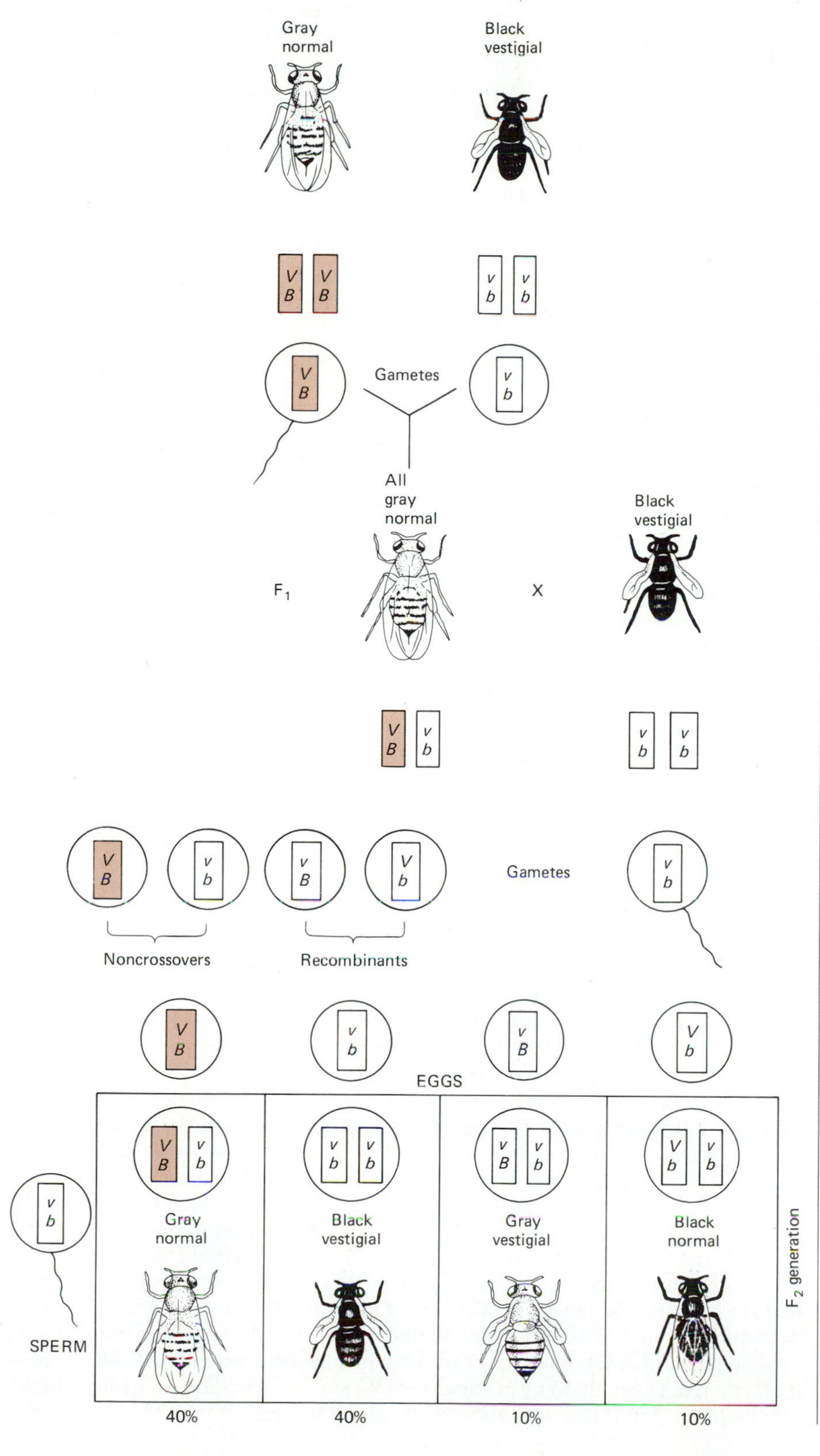

◀

Figure 11–15 A cross involving linkage and crossing over. In fruit flies, the genes for vestigial versus normal wings and for black versus gray body are linked; they are located in the same chromosome.

inherited together and are said to be **linked.** What characteristics would you predict in the offspring from a cross of a homozygous *VVBB* fly with a homozygous *vvbb* fly? They will all have gray bodies and normal wings and the genotype *VvBb*.

When one of the F_1 heterozygotes is crossed with a homozygous *vvbb* fly, the offspring appear in a ratio that differs from that of the ordinary test cross for a dihybrid. If the two genes governing this trait were not linked but were in different chromosomes, the offspring would appear in the ratio of 1/4 gray-bodied, normal-winged : 1/4 black-bodied, normal-winged : 1/4 gray-bodied, vestigial-winged : 1/4 black-bodied, vestigial-winged flies. If the genes were completely linked and no exchange of chromosomal segments occurred, then only the parental types—flies with gray bodies and normal wings and flies with black bodies and vestigial wings—would appear among the offspring, and these would be present in equal numbers (Fig. 11–14). However, there is an exchange of segments between the locus of allele *V* and the locus of allele *B*. Because of this crossing over of part of the chromosomes, some gray-bodied, vestigial-winged flies and some black-bodied, normal-winged flies—which together represent the crossover types, or recombinants—appear among the offspring. Most of the offspring resemble the parents and are either gray-normal or black-vestigial. In this particular instance, crossing over occurs between these two points in this chromosome at a frequency of about 20%. In such crosses, about 40% of the offspring are gray flies with normal wings. Another 40% are black flies with vestigial wings. These two make up the "noncrossover" class. Of the recombinants, 10% are gray flies with vestigial wings, and 10% are black flies with normal wings.

The distance between two genes in a chromosome is measured in **map units,** or crossover units, which represent the percentage of crossing over that occurs between them. The percentage of crossing over is calculated by dividing the total number of instances of each kind of crossing over (10 + 10) by the total number of offspring (40 + 40 + 10 + 10). The answer must be multiplied by 100 since a percentage is needed. Thus, *V* and *B* are said to be 20 map units apart. A 1% crossover value between two loci equals 1 map unit between them.

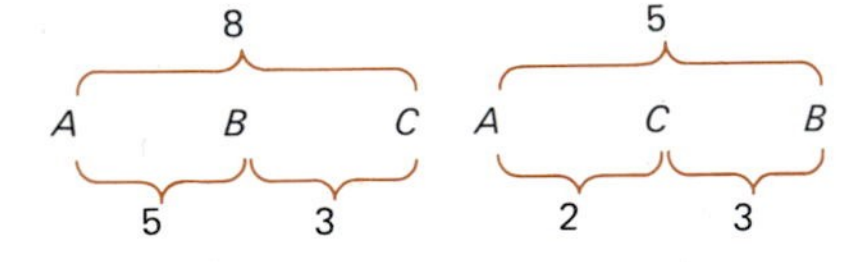

Figure 11–16 Genetic mapping. This diagram illustrates the means of determining whether gene *C* lies between or to the right of genes *A* and *B* from the percentage of crossing over between each of the possible pairs.

In a number of species, the frequency of crossing over between specific genes has been measured. All of the experimental results are consistent with the hypothesis that genes are present in a linear order in the chromosomes. If the three genes *A*, *B*, and *C* occur in a single chromosome, the amount of crossing over between *A* and *C* is either the sum of, or the difference between, the amounts of crossing over between *A* and *B* and between *B* and *C*. For example, as shown in Figure 11–16, if the crossing over between *A* and *B* is 5 map units and between *B* and *C* is 3 map units, the crossing over between *A* and *C* will be found to be 8 map units (if *C* lies to the right of *B*) or 2 map units (if *C* lies between *A* and *B*). By putting together the results of a great many such crosses, detailed chromosome maps of the location of specific genes on specific chromosomes in a number of species have been made (see Focus on Genetic Mapping).

Crossing over occurs at random, and more than one crossover between two loci in a single chromosome may occur at a given time. We can observe among the offspring only the frequency of recombinations, not the frequency of crossovers. The frequency of crossing over will be slightly larger than the observed frequency of recombination, because the simultaneous occurrence of two crossovers between two particular genes will lead to the reconstitution of the original combination of genes in a particular chromosome.

All the genes in a particular chromosome tend to be inherited together and comprise a **linkage group.** The number of linkage groups determined by genetic tests is always equal to the number of pairs of chromosomes. At present, the most detailed chromosome maps available are those for the bacterium *Escherichia coli,* which has one circular chromosome, and for the fruit fly, which has four pairs of chromosomes. The chromosomes of corn, mice, *Neurospora* (a mold), and certain other species of bacteria and viruses have also been mapped in considerable detail.

Linkage provides an explanation for the common observation that certain traits in humans and other organisms tend to be inherited together. Such traits are determined by genes that are located rather close together in a given chromosome. Crossing over provides another means by which genetic recombinations may arise. It plays a role in evolution by making possible new combinations of genetic units in the offspring, but it can also undo some useful linkages.

FOCUS ON

Genetic Mapping

A genetic map provides a summary of the genetic information about a species. The map indicates the loci (positions) of specific genes along a chromosome. Genetic maps have been at least partially worked out for many organisms including some bacteria, some fungi, corn (*Zea mays*), the fruitfly (*Drosophila melanogaster*), the mouse, and the human. More than 1000 gene loci have been identified on the map of *Drosophila's* four chromosomes. A partial map showing some of the gene loci is illustrated here. Note that both the normal and the common mutant phenotypes are indicated.

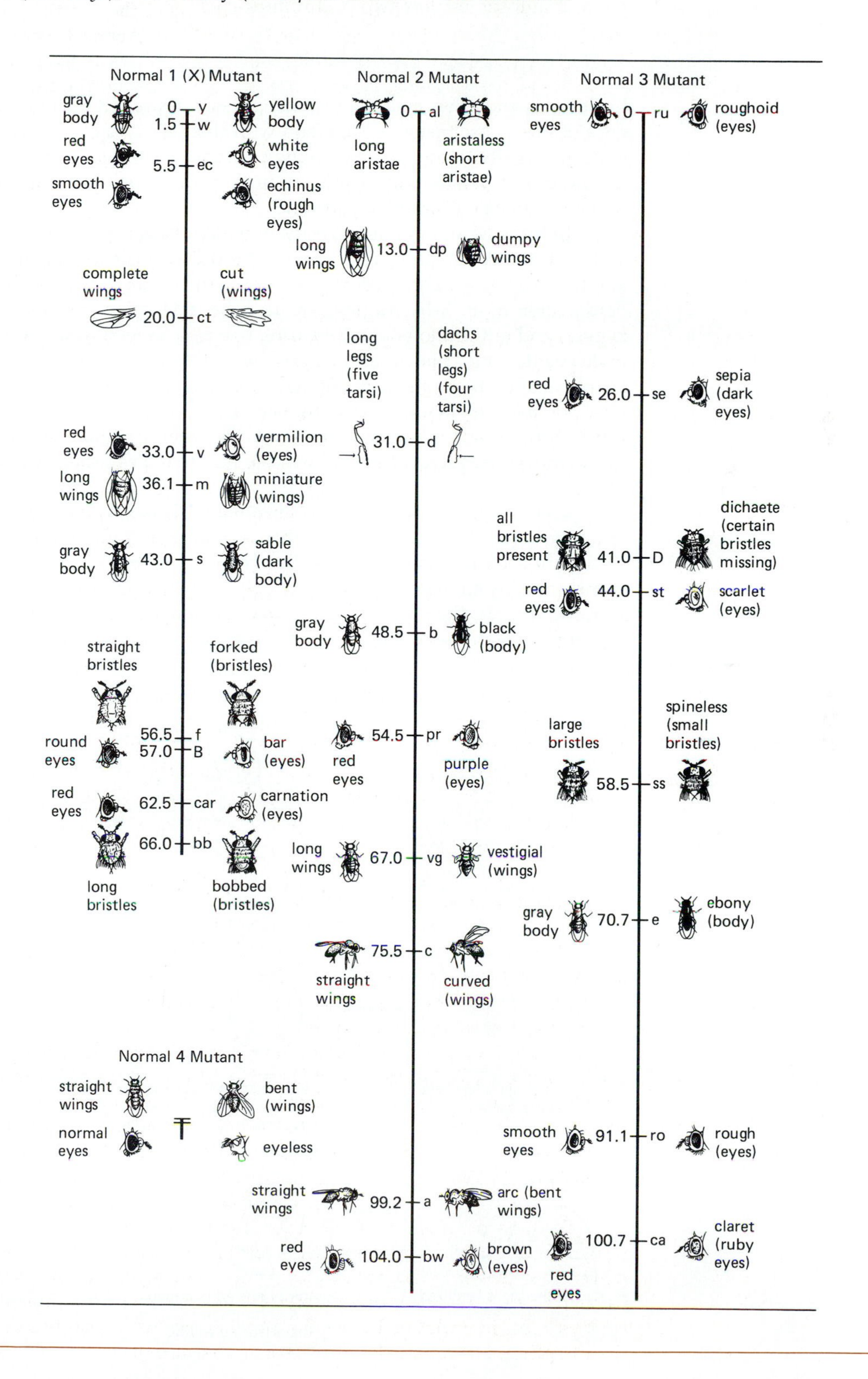

The Genetic Determination of Sex

An exception to the general rule that all homologous pairs of chromosomes are identical in size and shape is the sex chromosomes. The cells of the females of most species contain two identical sex chromosomes, called **X chromosomes;** in males, however, the set of two sex chromosomes is composed of a single X chromosome and a smaller **Y chromosome** with which it undergoes partial synapsis during meiosis. Human males have 22 pairs of **autosomes,** chromosomes other than the sex chromosomes, plus one X chromosome and one Y chromosome; females have 22 pairs of autosomes plus two X chromosomes.

THE Y CHROMOSOME

In humans and other mammals, maleness is determined in large part by the presence of the Y chromosome. A person with the XXY constitution is a nearly normal male in his external appearance, though with underdeveloped gonads (Klinefelter's syndrome). A person with one X but no Y chromosome has the appearance of an immature female (Turner's syndrome).

In humans and in other species in which the normal male has one X and one Y chromosome, two kinds of sperm are produced: Half contain an X chromosome, and half contain a Y chromosome. All eggs contain one X chromosome (Fig. 11–17). Fertilization of an X-bearing egg by an X-bearing sperm results in an XX, female zygote, and fertilization of an X-bearing egg by a Y-bearing sperm results in an XY, male zygote. Since there are equal numbers of X- and Y-bearing sperm, about equal numbers of each sex are born. Actually, some 106 boys are born to every 100 girls. One possible explanation of this numerical difference is that the sperm containing the Y chromosome might be lighter and able to move a little faster than an X-bearing sperm; consequently it would win the race to the egg slightly more than half the time.

This XY mechanism of sex determination is believed to operate in most species of animals. In birds and butterflies (Lepidoptera), the mechanism is reversed: Males are XX and females are XY. Sex chromosomes have been detected in some plants, notably strawberries, and probably exist in other plants with separate sexes. In some species, the organs of both sexes are present in each individual; sex chromosomes have not been found in such organisms.

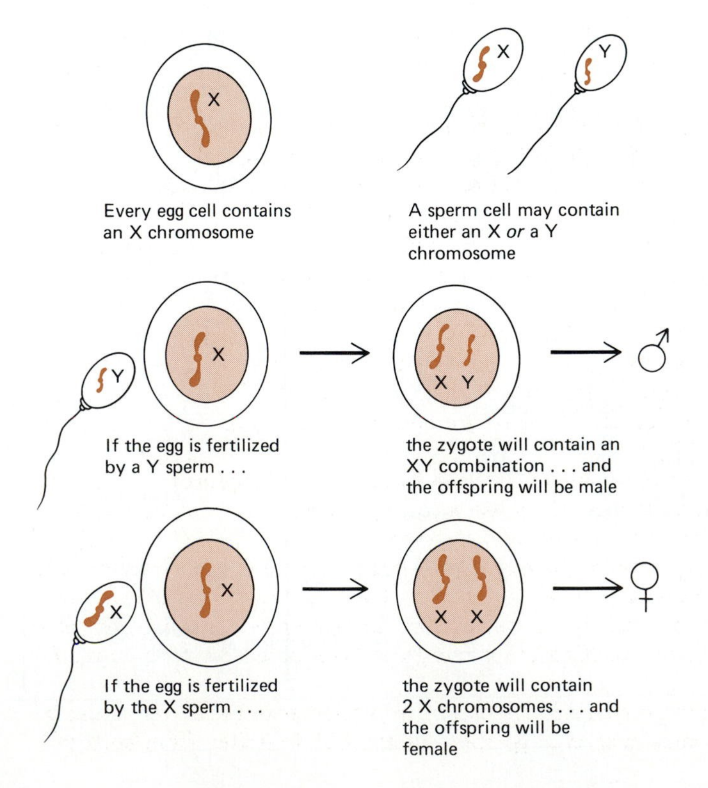

Figure 11–17 Sex is determined at the moment of fertilization by the sperm. An X-bearing sperm produces a female; a Y-bearing sperm produces a male.

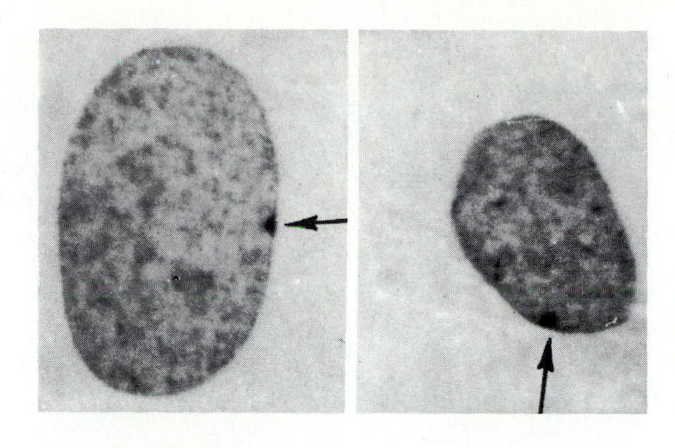

Figure 11–18 Barr body in human fibroblasts cultured from skin of a female (approximately ×2200). (Courtesy of Dr. Ursula Mittwoch, Galton Laboratory, University College, London.)

NUCLEAR SEXING

A dark spot of chromatin, called a **Barr body,** is visible at the edge of the nucleus of female mammalian cells during interphase (Fig. 11–18). This characteristic makes it possible to carry out **nuclear sexing,** to determine whether an individual is genetically female or male. The Barr body has been found to represent one of the two X chromosomes, which becomes dense and dark-staining; the other X chromosome resembles the autosomes and, during interphase, is a fully extended thread not evident by light microscopy. From this and other evidence, the British geneticist Mary Lyon has suggested that only one of the two X chromosomes in the female is active; the other is inactive. Which of the two becomes inactive in any given cell is a matter of chance, and the cells of a woman's body are of two kinds in which one or the other X chromosome is inactive. Since the two X chromosomes may have different genetic complements, the cells in a woman's body may differ in the effective genes present.

In mice and cats, which have several sex-linked genes for certain coat colors, the female heterozygous for such genes may show patches of one coat color in the midst of areas of the other color. This phenomenon, termed **variegation,** is evident in calico and tortoise-shell cats. The inactivation of one X chromosome apparently occurs early in development; thereafter all the progeny of that cell have the same inactive X chromosome. Although one X chromosome appears to be inactive, there are marked abnormalities when either X chromosome is completely missing from the chromosome complement of the cell (e.g., the XO condition of Turner's syndrome; see Chapter 12).

SEX-LINKED AND SEX-INFLUENCED TRAITS

The human X chromosome contains many genes, whereas the Y chromosome contains only a few, principally the genes for maleness. Traits controlled by genes located in the X chromosome, such as color blindness and hemophilia, are called **sex-linked** because their inheritance is linked with that of the sex of an offspring. A male offspring receives from his mother a single X chromosome and therefore all his genes for sex-linked characters. A female receives one X from her mother and one from her father. Males, having but one X chromosome, have only one of each kind of gene located in the X chromosome.

In the male, any gene that lies on the X chromosome will be expressed, whether or not such a gene is dominant or recessive in the XX female. A male with such an X-linked gene is neither homozygous nor heterozygous; he is said to be **hemizygous.**

The phenomenon of sex linkage involves only the X chromosome. For most sex-linked genes, the dominant allele leads to the normal trait and the recessive allele to the abnormal trait. In a female, two recessive sex-linked alleles must be present for the abnormal trait to be expressed, but in a male the presence of only one recessive allele can lead to the abnormal trait. One practical consequence is that while females may carry these traits, they usually find expression only in their male offspring. (A son receives his X chromosome from his mother, never from his father.)

In order to be expressed in a female, a recessive sex-linked trait must be present on both chromosomes; that is, the trait must be inherited from both par-

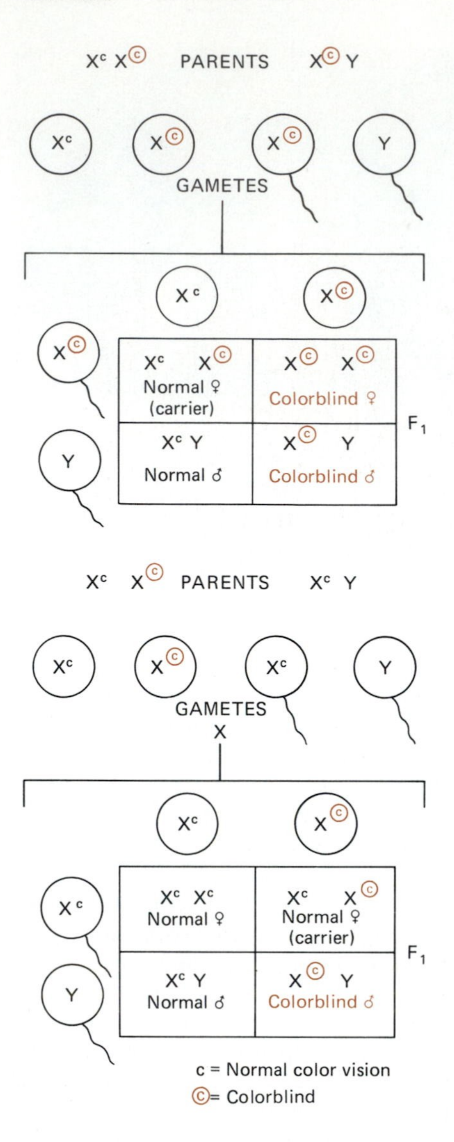

Figure 11–19 Sex linkage. Two crosses involving color blindness, a sex-linked characteristic. Note that the Y chromosome does not carry a gene for color vision.

ents. A color-blind girl, for example, must have a color-blind father and a mother who is at least heterozygous for color-blindness (Fig. 11–19). Such a combination is unusual, to say the least. Yet a color-blind boy need have only a mother who is heterozygous for the trait; his father can be normal.

Not all the characteristics that are different in the two sexes are sex-linked. For certain traits, known as **sex-influenced,** inheritance is through genes located in autosomes, but expression is altered or influenced by the sex of the animal. Males and females with identical genotypes may have different phenotypes. In sheep, for example, a single gene pair determines the presence or absence of horns: The *H* allele for the presence of horns is dominant in males but recessive in females, and the *h* allele for hornlessness is recessive in males but dominant in females. The genotype *HH* produces a horned animal regardless of sex. *Hh* produces the horned phenotype if the animal is male and a hornless phenotype if the animal is female, and *hh* produces a hornless animal whether it is a ram or a ewe.

Inbreeding, Outbreeding, and Hybrid Vigor

How do geneticists go about establishing a breed of cow that will give more milk or a strain of hens that will lay bigger eggs, or a variety of corn with more kernels per ear? By selection of the organisms that approach the desired phenotype, which are then used in further matings, true breeding strains with the commercially advantageous trait are gradually developed; that is, eventually a strain homozygous for all of the dominant (or recessive) polygenes involved is selected. It is clear that there is a limit to the effectiveness of breeding by selection. When a strain becomes homo-

Figure 11–20 Inbreeding and outbreeding. Corn plants from left to right represent successive generations of inbreeding (self-pollination). There is an obvious reduction in vigor for several generations, but eventually plants become homozygous at most genetic loci, and new generations are more uniform. (Courtesy of Connecticut Agricultural Experiment Station.)

zygous for all the polygenes involved, further selective breeding cannot increase the desired quality. Moreover, because of **inbreeding**—the mating of two closely related individuals—the strain may become homozygous for multiple deleterious traits as well! Certain dog breeds, for instance, are known for their susceptibility to congenital dislocation of the femur.

Inbreeding procedures are used widely by geneticists to improve strains of cattle, corn, or cantaloupes. A stock heterozygous for many recessive desirable traits may be improved by inbreeding. However, because inbreeding also enables recessive undesirable traits to appear phenotypically, it must be used with caution. Another problem is that traits considered desirable by a breeder might actually be handicaps under natural conditions. Human inbreeding increases the frequency of defects present at birth; this is why the marriage of first cousins or closer relatives is forbidden by law in some states.

The mating of individuals of totally unrelated strains, termed **outbreeding,** frequently leads to offspring that are much better adapted for survival than either parent; such improvement reflects a phenomenon termed **hybrid vigor.** Mongrel dogs are generally hardier than highly inbred purebreds. The mule, a hybrid resulting from the mating of a horse and a donkey, is a strong, sturdy beast, better suited for many tasks than either parent. However, like many animals that are the product of two different species, the mule cannot reproduce. (The slightly different horse and donkey chromosomes do not undergo proper synapsis during meiosis.) A large part of the corn, wheat, and other crops grown in the United States are hybrid strains. Each year the seed to grow these crops must be obtained by mating the original strains. The hybrid is quite heterozygous and gives rise, when mated, to a wide variety of forms, none of which is as good as the original hybrid. (Hybrid corn seeds are not normally planted, but are eaten instead!)

Hybrid vigor may be explained as follows: Each of the parental strains is homozygous for certain undesirable recessive genes, but any two strains are homozygous for different undesirable genes. Each strain contains dominant genes to mask the recessive undesirable genes of the other strain. One strain then might have the genotype *AAbbCCdd*, and another strain, the genotype *aaBBccDD*. The capital letters represent dominant genes for desirable traits, and the lowercase letters represent recessive genes for undesirable traits. The hybrid offspring, with the genotype *AaBbCcDd*, would combine all of the desirable and none of the undesirable traits of the two parental strains.

The Mathematical Basis of Population Genetics

A **panmictic** population is one in which organisms breed randomly so that genes combine essentially at random. Human populations are panmictic for certain genes; that is, mating is at random with respect to them. People are not greatly concerned

Figure 11–21 Hybrid vigor is exhibited by the hybrid corn at center. The inbred parent strains are shown at left and right. (Courtesy of Connecticut Agricultural Experiment Station.)

about blood types, for example, and therefore usually marry without taking them into account. Humans, therefore, are panmictic with respect to blood types. However, mating among human beings is not panmictic with respect to some genes—those governing skin color or height, for example.

The genetic composition of a population (all of the genes of all of the individuals in the population) is termed the **gene pool.** A panmictic population may be considered as a **pool of allelic genes,** any one of which has a chance of combining with any other, at least when borne by an individual of the opposite sex. What that chance is depends upon the frequencies of the alleles involved and upon no other factor.

THE GENE POOL AND THE HARDY-WEINBERG LAW

If brown eye genes are dominant to blue eye genes, why haven't all the blue eye genes disappeared? The answer lies partly in the fact that a recessive gene, such as the one for blue eyes, is not changed by having existed for a generation next to a brown eye gene in a heterozygous individual, *Bb*. The remainder of the explanation lies in the fact that as long as there is no selection for either eye color (that is, as long as people with blue eyes are just as likely to marry and to have as many children as people with brown eyes), successive generations will have the same proportions of blue- and brown-eyed people as those in the initial one. The composition of the gene pool will remain the same.

The reason for this may not be immediately obvious, but a brief exercise in mathematics will show the underlying principle. This principle, first proposed in the early 20th century independently by at least two scholars, is known as the **Hardy-Weinberg law** in honor of its discoverers.

Imagine a population of organisms that is panmictic for two allelic genes: *A*, whose frequency of occurrence in the population is p, and a, whose frequency is q. Their combined frequency is 1, so we may write

$$p + q = 1 \quad \text{so} \quad p = 1 - q \quad \text{or} \quad q = 1 - p \tag{1}$$

A gamete bearing gene *A* may combine with a similar one to form an *AA* zygote. Similarly, $A \times a \rightarrow Aa$; $a \times A \rightarrow aA$; and $a \times a \rightarrow aa$. The frequency of each combination is the product of the frequencies of its component genes, as follows:

	$p\ (A)$	$q\ (a)$
$p\ (A)$	$p^2\ (AA)$	$pq\ (aA)$
$q\ (a)$	$pq\ (Aa)$	$q^2\ (aa)$

Adding these terms gives the total of all genotypes:

$$p^2 + 2pq + q^2 = 1 \qquad (2)$$

where p^2 = frequency of AA
$2pq$ = frequency of Aa
q^2 = frequency of aa

In the next generation the present Aa organisms will be able to produce *either* A or a gametes; that is, half the gametes will be A and the other half a from this source. The AA organisms will, of course, yield only A gametes; the aa, only a gametes. Thus the total frequency of all A and a may be expressed as follows (where p' stands for the new frequency of A; q' for the new frequency of a):

$$\begin{aligned} p' &= p^2 + 1/2\ (2pq) && (3) \\ &= p^2 + pq \\ &= p^2 + p(1-p) && \text{from Equation (1)} \\ &= p^2 + p - p^2 = p \\ q' &= q^2 + 1/2\ (2pq) && (4) \\ &= q^2 + pq \\ &= q^2 + q(1-q) && \text{from Equation (1)} \\ &= q^2 + q - q^2 = q \end{aligned}$$

Therefore, if left undisturbed, gene frequencies in the gene pool of a panmictic population do not change from generation to generation, regardless of whether these genes are dominant or recessive.

The Hardy-Weinberg law holds true, however, only in a panmictic population and in the absence of mutation, natural selection, genetic drift, or exchange of genes between the population and any other population (i.e., gene flow). (These evolutionary mechanisms will be discussed in Chapter 45.) Without those perturbing forces, genetic frequencies in a freely interbreeding population will not change from generation to generation. Thus the Hardy-Weinberg law establishes a baseline from which evolutionary departures must take place.

ESTIMATING THE FREQUENCY OF GENETIC CARRIERS

Neither the value of p nor that of q, which are gene frequencies, can be measured directly. However, since the recessive phenotype can be distinguished, the value of q^2—the frequency of genotype aa—can be determined. From this the gene frequencies q (which is the square root of q^2) and p (which is $1 - q$) can be calculated. Finally, the frequencies of the other genotypes, p^2 (AA) and $2pq$ (Aa), can be calculated. To calculate the number of individuals in a population who are genetic carriers for a given trait (i.e., are heterozygotes, Aa), we need to know only that it is inherited by a single pair of genes and the frequency with which the homozygous recessive individuals appear in the population (Table 11–3). Consideration of some

TABLE 11–3
The Offspring of the Random Mating of a Population Composed of 1/4 **AA**, *1/2* **Aa** *and 1/4* **aa** *Individuals*

Mating			
Male	*Female*	*Frequency*	*Offspring*
AA ×	AA	1/4 × 1/4	1/16 AA
AA ×	Aa	1/4 × 1/2	1/16 AA + 1/16 Aa
AA ×	aa	1/4 × 1/4	1/16 Aa
Aa ×	AA	1/2 × 1/4	1/16 AA + 1/16 Aa
Aa ×	Aa	1/2 × 1/2	1/16 AA + 1/8 Aa + 1/16 aa
Aa ×	aa	1/2 × 1/4	1/16 Aa + 1/16 aa
aa ×	AA	1/4 × 1/4	1/16 Aa
aa ×	Aa	1/4 × 1/2	1/16 Aa + 1/16 aa
aa ×	aa	1/4 × 1/4	1/16 aa
			Sum: 4/16 AA + 8/16 Aa + 4/16 aa

specific examples of the genetics of populations should help in understanding these principles.

Albinism

Albinos are individuals with no pigment at all in their skin, hair, or eyes. Albinism is an inherited trait; in most forms of the disorder a specific enzyme, tyrosinase, is lacking. Tyrosinase catalyzes one of the reactions involved in the production of the melanin pigments. Albinism is inherited through a single pair of genes; it occurs about once in 20,000 births in individuals who are homozygous recessive for this trait. From this fact we can calculate that the frequency of *aa* individuals (q^2) is 1/20,000. The value of q can be calculated by taking the square root of q^2. The square root of 1/20,000 is about 1/141. Since $p = 1 - q$ or $1 - 1/141$, $p = 140/141$. From these values for p and q, we can calculate the value of $2pq$, which represents the frequency of the genetic "carrier" *Aa* individuals: $2 \times 140/141 \times 1/141 = 1/70$. Surprising as it may seem, one person in 70 is a carrier of albinism, although only one person in 20,000 is homozygous and displays the trait. At first glance it may seem odd that there are so many carriers in a population that contains so few homozygous recessives. Reflecting on the mathematical relations involved, however, should lead you to realize that this must be true. When q is small (such as 1/141), then q^2 will be very small, but $2pq$ will be much larger.

Phenylketonuria

Many human diseases that are known to be inherited are determined by recessive genes. The undesirable trait—the disease condition—is expressed in the homozygous recessive individual. When a man and a woman ask a geneticist whether they should have children, of primary concern is whether both are heterozygous for the same unwanted recessive trait. If they are, then there is 1 chance in 4 that any of their offspring will show this inherited disease.

Phenylketonuria (PKU) is an inherited disease in which there is a deficiency of the enzyme that in normal individuals converts phenylalanine to tyrosine. Phenylalanine and phenylpyruvic acid accumulate in the blood and tissues and are excreted in the urine. The accumulation of phenylalanine in the nervous tissues interferes with their function; hence individuals with unmanaged phenylketonuria usually develop a mental deficiency associated with the disease. (PKU can be successfully treated when the affected infant is promptly placed on a low phenylalanine diet.) The gene frequency q for this trait in the United States population is about 0.005, and the gene frequency p of the normal allele is 0.995. The incidence of the heterozygous state is about 1 in every 100 persons ($2 \times 0.995 \times 0.005 = 0.01$). Though physically and mentally normal, the heterozygous individual has a somewhat lower-than-normal ability to metabolize phenylalanine; however, the results of phenylalanine metabolism tests given to heterozygotes overlap those given to normal people, so that it is not possible to distinguish between the two with confidence. If a combination of family history and the phenylalanine metabolism test results makes it likely that both husband and wife are heterozygous carriers for phenylketonuria (or for any other recessive trait), it becomes possible to predict the probability (1/4) that they will have a child showing the abnormality.

SUMMARY

I. Mendel's inferences from his experiments with the breeding of garden peas have been tested repeatedly in all kinds of diploid organisms and found to be true.
   A. The units of heredity exist in pairs in individuals, but gametes have only one of each kind of gene.
   B. During the formation of eggs and sperm, each pair of genes separates independently of the members of other pairs of genes.
   C. Each chromosome behaves genetically as though it were composed of genes arranged in a linear order; the members of a homologous pair of chromosomes have genes arranged in a similar order.

II. In a cross of two homozygous individuals that differ in a single gene pair, a monohybrid cross, the offspring are heterozygous; they have one of each allele. Because one of the alleles may be dominant, the offspring usually resemble one of the parents more than the other.
   A. In the mating of two of these F_1 offspring there are 3 chances in 4 that one of the F_2 generation will resemble that parent with the dominant genotype and 1 chance in 4 that it will resemble the other parent.
   B. Some important basic genetic terms that reflect

important genetic concepts are locus, allele, dominant, recessive, monohybrid, homozygous, heterozygous, phenotype, and genotype.

III. Genetic ratios can be expressed in terms of probabilities.
   A. Probabilities are expressed as a fraction, the number of favorable events divided by the total number of events. This can range from zero (an impossible event) to one (a certain event).
   B. Probabilities can be multiplied and added like any other fraction.
   C. The probability of two independent events occurring together is the product of the probabilities of each occurring separately.
   D. The probability that one or another of two or more independent events will occur is the sum of their separate probabilities.

IV. In incomplete dominance, one allele may not be completely dominant to the other, and the offspring may be intermediate between the two parental types. For example, pink-flowered plants may be obtained by breeding red- and white-flowered sweet pea plants.

V. In a dihybrid cross, two individuals, each heterozygous for each of two traits, may produce offspring that exhibit the phenotypes in a 9 : 3 : 3 : 1 ratio.

VI. The relationship between a gene and its phenotype may be quite complex.
   A. Most genes have many different effects; this is known as pleiotropy.
   B. The presence of a particular allele of one gene pair may determine whether alleles of another gene pair will be expressed; this is called epistasis.

VII. In polygenic inheritance, two or more independent pairs of genes may have similar and additive effects on the phenotype.
   A. Many human characteristics such as height, body form, and skin color, as well as many traits in plants and animals, are inherited by polygenes.
   B. In polygenic inheritance, the F_1 generation is intermediate between the two parental types and shows little variation, whereas the F_2 generation shows wide variation between the two parental types.

VIII. The term multiple alleles is applied to three or more alleles that can occupy a single locus on the chromosome; each gene produces a specific phenotype.
   A. Any individual has any two of the alleles; any gamete has only one.
   B. The inheritance of the ABO human blood types is governed by three alleles.

IX. All the genes in any given chromosome tend to be inherited together—they are said to be linked. By measuring the frequency of crossing over between various genes it is possible to construct a genetic map of the chromosome.

X. The sex of humans and many other animals is determined by the X and Y chromosomes.
   A. Females have two X chromosomes; males have one X and one Y.
   B. In many species maleness is determined in large part by the presence of the Y chromosome.
   C. The fertilization of an X-bearing egg by an X-bearing sperm results in a female (XX) zygote. The fertilization of an X-bearing egg by a Y-bearing sperm results in a male (XY) zygote.
   D. A Barr body is present at the edge of the nucleus of female mammalian cells during interphase. This permits nuclear sexing of individuals.

XI. Traits controlled by genes located in the X chromosomes are referred to as sex-linked because their inheritance is linked with that of the sex of an offspring. A male receives all of his genes for sex-linked traits from his mother.

XII. Inbreeding, the mating of two closely related individuals, greatly increases the probability for an individual to become homozygous for recessive genes. Outbreeding, the mating of individuals of totally unrelated strains, increases the probability that the offspring will be heterozygous for many traits. These heterozygous individuals may be stronger and better able to survive than either parent, a phenomenon termed hybrid vigor.

XIII. According to the Hardy-Weinberg law, gene frequencies in a panmictic population do not change from generation to generation, regardless of dominance or recessivity, unless mutation, natural selection, or some other disturbing force is present. Gene frequency in a population can be calculated from the frequency of the recessive phenotype.

POST-TEST

1. The specific site in the chromosome of a given gene is termed its ______________.

2. Genes governing variations of the same trait and occupying corresponding loci on homologous chromosomes are termed ______________.

3. A cross between two organisms differing with respect to a single characteristic is a ______________ ______________.

4. An allele that has the ability to produce its phenotype in a heterozygous individual is a ______________ ______________.

5. A ______________ ______________ is an allele that can produce its phenotype only when present in a homozygous individual.

6. An organism with two identical genes is said to be ______________ for the trait; an organism with one dominant gene and one recessive gene is said to be ______________ for the trait.

7. The offspring of the parental generation are called the ______________ ______________ ______________ or the ______________ ______________.

8. The appearance of an individual with respect to a given inherited trait is known as its ______________.

9. The organism's genetic constitution, expressed in symbols, is called its ______________.

10. A probability that can be specified in advance is termed an ______________ ______________ ______________.

11. The probability that two independent events will coincide is the ______________ of their individual probabilities.

12. The probability that one or another of two independent events will occur is the ______________ of their individual probabilities.

13. A probability of ______________ expresses an impossibility; a probability of ______________ expresses a certainty.

14. In crosses of individuals with genes showing incomplete dominance, the genotypic and phenotypic ratios of the offspring are ______________.

15. A mating of individuals differing in two traits is called a ______________ ______________.

16. Two pairs of genes that are inherited independently and interact in such a way that neither dominant allele can produce its effect unless the other is present are termed ______________ ______________.

17. The term ______________ refers to two or more independent pairs of genes with similar and additive effects on a given trait.

18. Three or more alleles that may occupy a single locus are termed ______________ ______________.

19. The genes in a given chromosome tend to be inherited together and are said to be ______________.

20. By observing the presence or absence of a ______________ ______________ at the edge of the nucleus, scientists can determine the ______________ sex of the individual.

21. The mating of two closely related individuals such as brother and sister is termed ______________.

22. The offspring of totally unrelated parents may be better adapted for survival than either parent, a phenomenon called ______________ ______________.

23. An inherited lack of the enzyme tyrosinase results in the inability to produce the pigment melanin and the phenotype of ______________.

24. An inherited lack of the enzyme phenylalanine hydroxylase results in the accumulation of phenylalanine in tissues, which if untreated leads to physical and mental retardation; this disease is called ______________.

REVIEW QUESTIONS

1. Show by diagrams how the alleles in gene pairs located in different pairs of chromosomes segregate independently in meiosis.
2. In peas, yellow color in seeds is dominant to green. State the colors of the offspring of the following crosses:
   a. homozygous yellow × green
   b. heterozygous yellow × green; heterozygous yellow × homozygous yellow
   c. heterozygous yellow × heterozygous yellow
3. Could two albino parents have a normally pigmented child? Could two normally pigmented parents have an albino child?
4. If two animals heterozygous for a single pair of genes are mated and have 200 offspring, about how many are expected to have the dominant phenotype?
5. When two long-winged flies were mated, the offspring included 77 with long wings and 24 with short wings. Is the short-winged condition dominant or recessive? What are the genotypes of the parents?
6. A blue-eyed man, both of whose parents were brown-eyed, married a brown-eyed woman whose father was blue-eyed and whose mother was brown-eyed. This man and woman have a blue-eyed child. What are the genotypes of all the individuals mentioned?
7. Outline a breeding procedure whereby a true breeding strain of red cattle could be established from a roan bull and a white cow. (Roan is *Rr*.)
8. What is the probability of rolling a seven with a pair of dice?
9. What are the implications of the Hardy-Weinberg law? How is this applied to studies of human genetics?
10. Contrast the meanings of the terms genotype and gene pool.
11. In rabbits, spotted coat (*S*) is dominant to solid color (*s*), and black (*B*) is dominant to brown (*b*). A brown spotted rabbit is mated to a solid black one, and all the offspring are black spotted. What are the genotypes of the parents? What would be the appearance of the F_2 generation if two of these F_1 black spotted rabbits were mated?
12. The long hair of Persian cats is recessive to the short hair of Siamese cats, but the black coat color of Persians is dominant to the black-and-tan coat of Siamese. If a pure black, long-haired Persian were mated to a pure black-and-tan, short-haired Siamese, what will be the appearance of the F_1 offspring? If two of these F_1 cats are mated, what is the chance of obtaining a long-haired, black-and-tan cat in the F_2 generation?
13. In peas, tall plants (*T*) are dominant to dwarf (*t*), yellow color (*Y*) is dominant to green (*y*), and smooth seed (*S*) is dominant to wrinkled seed (*s*). What would be the phenotypes of the offspring of the following matings?
    a. *TtYySs* × *ttyyss*
    b. *TtyySs* × *ttYySs*
14. What is pleiotropy? What is epistasis?
15. A walnut-combed rooster is mated to three hens. Hen A, which is walnut-combed, has offspring in the ratio of 3 walnut : 1 rose. Hen B, which is pea-combed, has offspring in the ratio of 3 walnut : 3 pea : 1 rose : 1 single. Hen C, which is walnut-combed, has only walnut-combed offspring. What are the genotypes of the rooster and the three hens?
16. What kinds of matings result in the following phenotypic ratios?
    a. 3 : 1
    b. 1 : 2 : 1
    c. 9 : 3 : 3 : 1
    d. 9 : 7
    e. 1 : 4 : 6 : 4 : 1
    f. 2 : 1
    g. 47 : 47 : 3 : 3
17. The weight of the fruit in a certain variety of squash is determined by three pairs of genes, with *AABBCC* pro-

ducing 6-pound squashes and *aabbcc* producing 3-pound squashes. Each dominant gene adds 1/2 pound to the weight. When a 6-pound squash is crossed with a 3-pound squash, all the offspring weigh 4.5 pounds. What would be the weights of the F_2 fruits if two of these F_1 plants were crossed?

18. Mrs. Doe and Mrs. Roe had babies at the same hospital and at the same time. Mrs. Doe took home a girl and named her Nancy. Mrs. Roe took home a boy and named him Richard. However, she was sure she had had a girl and brought suit against the hospital. Blood tests showed that Mr. Roe was type O, Mrs. Roe was type AB, and Mr. and Mrs. Doe were both type B. Nancy was type A, and Richard was type O. Had an exchange occurred?

19. Explain the mechanism of genetic determination of sex in humans.

20. The barred pattern of chicken feathers is inherited by a pair of sex-linked genes, *B* for barred and *b* for no bars. If a barred female is mated to a non-barred male, what will be the appearance of the progeny? What commercial usefulness does this have?

21. One pair of genes for coat color in cats is sex-linked. The gene *B* produces yellow coat, *b* produces black coat, and the heterozygote, *Bb*, produces the tortoise-shell pattern of coat color. What kind of offspring result from the mating of a black male and a tortoise-shell female?

12
Human Genetics

OUTLINE

LEARNING OBJECTIVES

After you have read this chapter you should be able to:

1. Discuss the various reasons that human beings are not very favorable subjects for the study of inheritance.
2. Discuss the inheritance of mental abilities.
3. Distinguish between environmentally induced and inherited abnormalities, and between chromosome abnormalities and gene defects.
4. Describe the phenomenon of nondisjunction, and discuss its role in the production of Down syndrome, Klinefelter syndrome, and Turner syndrome.
5. Describe how amniocentesis is used in the diagnosis of human genetic abnormalities.
6. Discuss the scope and implications of genetic counseling.
7. Define eugenics, and distinguish between positive and negative eugenics.
8. Discuss the problems associated with the practice of eugenics including its possible impact on balanced polymorphism.

Quite naturally, geneticists have great interest in the study of human genetics. The trouble is that humans do not serve well as the subjects of certain types of genetic research. For the study of the mode of inheritance in any species, geneticists would prefer (1) to have standard stocks of genetically identical individuals, that is, **isogenic strains;** (2) to mate members of different isogenic strains; and (3) to raise the offspring under carefully controlled conditions. The organisms favored in genetic studies are bacteria, molds, fruit flies, mice, and corn. These organisms produce many offspring and have only a short time between successive generations.

In comparison, human beings are not ideal subjects for these types of studies of inheritance. Members of the human species are genetically very diverse; that is, they are heterozygous for many genes. Few human beings would be willing to select a mate according to the needs of genetic researchers, and in no human culture is genotype a conscious factor in choosing a mate. Furthermore, human families are small in number; more than 20 to 30 years elapse between generations; and understandably, few people would be willing to release their children to be raised in a controlled laboratory environment.

Despite these difficulties, a great deal has been learned about human inheritance, and the field is progressing rapidly. This results in part from the medical attention given to genetic disease in human beings. The extensive medical records of disease serve as a fine data pool upon which hypotheses may be based and against which they may be tested. Furthermore, some of the phenomena in human inheritance, originally quite puzzling, have been clarified by the solutions of analogous problems in the inheritance of bacteria, molds, flies, and mice. In the field of chemical genetics, many important discoveries made originally with human material have been confirmed by experiments using bacteria or molds. For example, the earliest evidence that genes determine the sequence of amino acids in proteins came from studies of the different types of human hemoglobin.

Geneticists studying human inheritance were forced to devise methods of measuring the relative frequency of contrasting traits in an entire population of individuals and of calculating the frequency of specific alleles. Each pair of genes, such as A and a, is distributed in the population in such a way that any member may have the genotype AA, the genotype Aa, or the genotype aa. If there is no selective advantage for any of these three genotypes (that is, if individuals with any of the three genotypes have equal probabilities of surviving), the frequency of these genes in successive generations of individuals will remain unchanged. As long as individuals with each of these three genotypes are just as likely to mate and have offspring as are individuals with the other genotypes, the three genotypes will be present in succeeding generations in the same proportion as in the initial generation. This concept of **genetic equilibrium** is fundamental to all studies of population genetics.

The early studies of human heredity usually dealt with readily identified single traits and their distribution among the members of a family, as illustrated by the **pedigree** shown in Figure 12–1. The development of methods of making inferences about the mode of inheritance of a trait from studies of its distribution in an entire population, based on the laws of probability, has been of great usefulness in studies of human genetics.

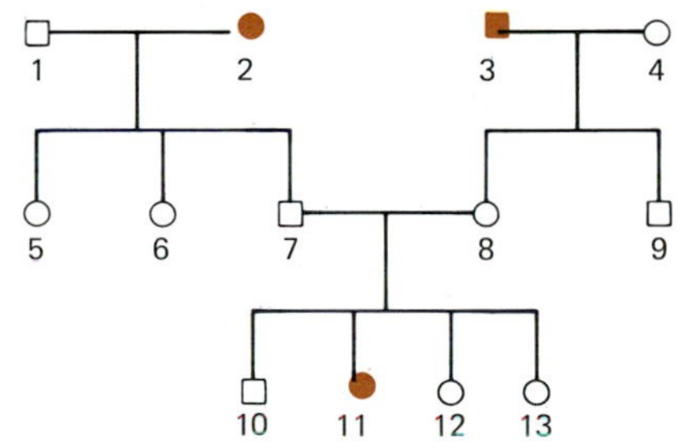

Figure 12–1 A pedigree for albinism. Males are indicated by squares and females by circles. Individuals showing the trait under study are indicated by color symbols and those not showing the trait under study are indicated by white symbols. Relationships are indicated by connecting lines; all members of the same generation are placed in the same row. Thus, *11* is an albino girl whose sisters, *12* and *13*, and brother, *10*, are normal. Her father, *7*, and mother, *8*, as well as her aunts, *5* and *6*, uncle, *9*, paternal grandfather, *1*, and maternal grandmother, *4*, are normal. Her paternal grandmother, *2*, and maternal grandfather, *3*, are albino.

Inherited Human Traits

The development of each organ of the body is regulated by a large number of genes. The mechanisms of inheritance of many physical traits and hundreds of specific enzymes are now known. In fact, the loci of many genes have been identified, and chromosome maps, although incomplete, have been worked out for each human chromosome.

The age at which a particular gene expresses itself phenotypically may vary widely. Most characteristics develop long before birth, but some, such as hair and eye color, are not fully expressed until shortly after birth. Some, such as amaurotic idiocy, become evident in early childhood while still others, such as glaucoma and Huntington's chorea, develop only after the individual has reached maturity.

The Inheritance of Mental Abilities

Human genetics concerns itself with the inheritance of not only disease states, but also of many normal human traits and abilities—for example, intelligence. The inheritance of mental ability or intelligence is one of the most difficult problems of human genetics. Psychological tests, such as the Stanford-Binet test, are prepared by setting arbitrary tasks for children and then determining what capacity can normally be expected of children of each age. In taking the test a child is given progressively more difficult problems until he or she is finally unable to solve them. The child's performance is then compared to those of a sample of children of all ages, and he or she is assigned a "mental age" accordingly.

It is difficult—to say the least—to set up a test that measures *innate* intellectual capacity. At least two factors affect the objectivity of any intelligence test:

1. the prior training of the person taking the test (some people have done poorly on certain types of tests given by the U.S. Army because not only were they unable to read or to conceptualize from flat, two-dimensional drawings, but they were even unaccustomed to holding a pencil).
2. the political, sociological, psychological, or educational theories of the *test maker*, who may consider certain abilities to be of higher intellectual status than that of others.

When such tests are given, a wide range of mental ability, from complete incompetence to excellent comprehension, has been found for each age group. A child of 6 who can solve problems ordinarily solved by children 8 years old is considered superior to one of 6 who can do only those normally done by 6-year-olds. The "mental age" as determined by this test is divided by the actual chronological age, and the quotient is multiplied by 100 to give the **intelligence quotient,** or **IQ.** When the intelligence quotients of a large number of people are measured they form a curve of normal distribution from zero to over 140, with the largest number of scores in the class defined as normal, and progressively fewer scores in the classes further from normal.

It is assumed that the kinds of tasks performed on IQ tests are representative of problem-solving behavior in general, and that the scores achieved represent general mental ability. Whether this assumption is justified or not, the distribution of IQs does conform to a normal or bell-shaped curve of random probability. Lower mental abilities may be caused by diseases such as syphilis or meningitis, by injuries sustained during birth or other environmental factors, or by some untreated congenital disorders. Some cases appear to result from direct inheritance, although the evidence for this is no longer considered as unassailable as it was in past years. (Some genes, such as those for the disease phenylketonuria, do indirectly produce lower mental abilities but are not involved in the determination of intelligence in physically healthy individuals.)

More recently developed intelligence tests have provided measures of primary abilities, such as the ability to reason inductively, the ability to memorize, and the ability to visualize objects in three dimensions. Special abilities—musical, artistic, mechanical, and mathematical—also seem to have some hereditary basis, but their inheritance is evidently separate from that of IQ. Since musical ability, for instance, is a complex function of pitch discrimination, tone memory, and a sense of rhythm, melody, and harmony, it is not surprising that inheritance of such ability should be complex. Total intelligence, if there is such a thing, would therefore have a very complex hereditary basis indeed.

Some investigators conclude from these studies that the upper limit of a person's mental ability is determined genetically (perhaps by a system of polygenes), but that how these inherited abilities are developed is determined by environmental influences, by training, and by experience. This is no trivial reservation. Malnutrition alone could make a potential Charles Atlas a 90-pound weakling. The same sort of thing almost certainly holds for mental abilities. Physical and mental traits are the result of the interplay of both genetic and environmental factors.

Human Cytogenetics

Cytogenetics is the branch of biology that uses the methods of cytology to study genetics. Many of the basic principles of genetics were discovered by experiments with lower organisms in which it was possible to relate genetic data with cytological events. These experiments involved making smears of cells and examining them with the microscope to see the number and structure of the chromosomes present. Some of the organisms used in genetics, such as the fruit fly *Drosophila,* have very few chromosomes (only four pairs). In *Drosophila* salivary glands and certain other tissues, the chromosomes are quite large, so that their structural details are readily evident.

KARYOTYPES

The normal human karyotypes for males and females are shown in Figure 10–4. The term **karyotype** refers both to the chromosome composition of an individual and to the photomicrograph showing that composition. Cells from the bone marrow, blood, or skin are incubated with plant lectins, which stimulate mitosis, and are then treated with colchicine to stop the process in the metaphase. The cells are placed in hypotonic solution, causing them to swell and enabling the chromosomes to spread out so that they can be visualized more readily. The preparation is fixed and the chromosomes are stained to reveal banding patterns. Then the chromosomes are photographed. Each chromosome is cut out of the photographic print and aligned so that the homologous pairs are placed together. Chromosomes are identified by their length, by the position of the centromere, by banding patterns, and by the presence of knobs or satellites. The gradation in size between several of the pairs is quite fine, and although the largest chromosome is some five times as long as the smallest one, there are only slight differences between some of the intermediate-sized ones (Figs. 10–4, 12–3, and 12–5).

BIRTH DEFECTS

A birth defect, or **congenital defect,** is simply one that is present at birth; it may or may not be inherited. Some congenital abnormalities are inherited, while others are produced by environmental factors that affect the developmental process. For example, if a woman contracts German measles during the first three months of pregnancy, there is a substantial risk that her offspring will show congenital malformations. Environmental factors that have been linked with birth defects are discussed in Chapter 44.

Certain abnormalities are the result of mutations involving a single gene; sickle cell anemia and albinism are examples of this type of defect. Other abnormalities, such as Down syndrome (mongolism), result from chromosome aberrations in which an extra chromosome may be present or a chromosome may be absent.

CHROMOSOME ABNORMALITIES

Although common in plants, the presence of multiple members of complete chromosome sets, **polyploidy,** is rare in animals. In fact, when present in all of the cells of the body, polyploidy is lethal in humans and many other animals. Triploidy is sometimes found in embryos that have been spontaneously aborted in early pregnancy. The few triploids or tetraploids that have been born alive and lived for a few days were found to be mosaics of cells, some of which were diploid.

More common in humans are chromosome abnormalities called **aneuploidies,** which involve the presence of an extra chromosome or the absence of one chromosome. An individual with an extra chromosome—that is, with three of one kind—is said to be **trisomic.** An individual lacking one of a pair of chromosomes is **monosomic.** These aneuploidies generally arise as a result of abnormal meiotic (or mitotic) division in which homologous chromosomes fail to separate. This phenomenon is called **nondisjunction.** For example, two X chromosomes that fail to separate might enter the egg nucleus, leaving the polar body with no X chromosome (Fig. 12–2). Alternatively, the two joined X chromosomes might go into the polar body, leaving the ovum with no X chromosome. Nondisjunction of the XY chromosomes in the male might lead to the formation of sperm that have both an X and a Y chromosome or of sperm with neither an X nor a Y chromosome. Chromosomal nondisjunction may occur during either the first or the second meiotic division; it may also occur during mitotic divisions and lead to the establishing

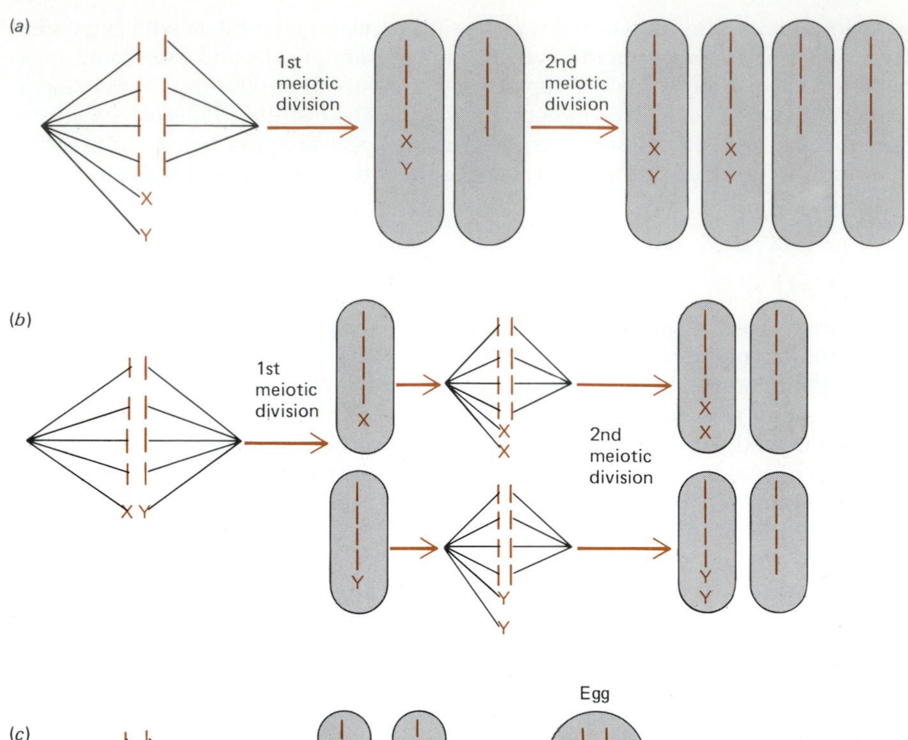

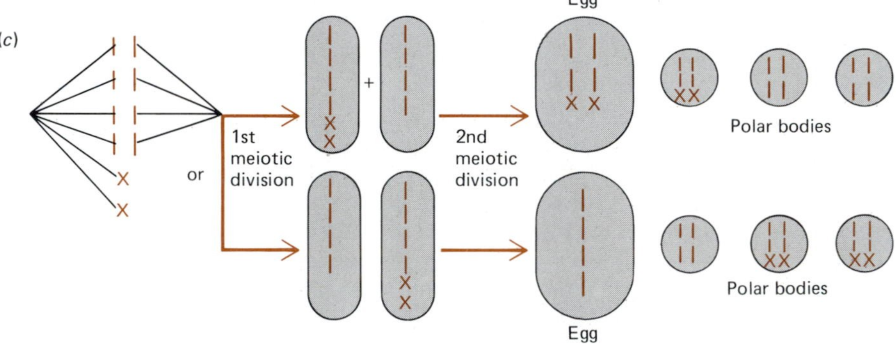

Figure 12–2 Nondisjunction of sex chromosomes. *(a)* Nondisjunction in the first meiotic division in the male results in two XY sperm and two sperm with neither an X nor a Y chromosome. *(b)* Nondisjunction in the second meiotic division in the male results in one sperm with two X chromosomes and one with no sex chromosomes, or in one sperm with two Y chromosomes and one with no sex chromosomes. *(c)* Nondisjunction in the female results in the formation XX or O eggs whether it occurs in the first or second meiotic division.

of a clone of abnormal cells in an otherwise normal individual. When an abnormal gamete unites with a normal one, the diploid zygote has the chromosome abnormality.

In some aneuploidies, a part of a chromosome may break off and attach to another chromosome. Such a **translocation** may result in an abnormally long or short chromosome. Table 12–1 summarizes some disorders that are produced by aneuploidies. (Also see Fig. 12–3.)

Down Syndrome

Cytogenetic studies have clarified the origin of one of the more distressing abnormal conditions in humans, that of **Down syndrome,** caused by the trisomy 21 aneuploidy (Fig. 12–4). Persons suffering from this syndrome have abnormalities of the face, eyelids, tongue, hand, and other parts of the body, and are retarded (in varying degrees that can be affected by environmental influences) in both their physical and mental development. They are also unusually susceptible to certain diseases, for example, leukemia. The term **mongolism** was originally applied to this condition because affected persons often have a fold of the eyelid similar to that typical of members of the Mongolian peoples. Down syndrome is a relatively common congenital malformation, occurring in 0.15% of all births, and is 100-fold more likely in the offspring of women 45 years or older than in the offspring of mothers aged less than 19. The occurrence of Down syndrome, however, is affected much less, if at all, by the age of the father.

Cytogenetic studies have revealed that most persons with Down syndrome are trisomics and have one extra chromosome 21, making a total of 47. The presence of this extra small chromosome is believed to arise by nondisjunction, usually in the

TABLE 12–1
Some Chromosome Abnormalities

Karyotype	*Common Name*	*Clinical Description*
Trisomy 13		Multiple defects, with death by age 1 to 3 months
Trisomy 15		Multiple defects, with death by age 1 to 3 months
Trisomy 18		Ear deformities, heart defects, spasticity, and other damage; Death by age 1 year
Trisomy 21	Down syndrome	Overall frequency is about 1 in 700 live births. True trisomy usually found among children of older (age 40+) mothers, but translocation resulting in the equivalent of trisomy may occur in children of younger women. A 35-year-old mother has a 1:200 chance of producing a child with this syndrome. A 40-year-old mother has a 1:50 chance of doing so, and at 44 the risk is 1:20! A similar, though less marked influence is exerted by the age of the father. Epicanthic skin fold (i.e., a fold of skin above the eye), although not the same as that in the Mongolian race, produces an oriental appearance—hence the former name "mongolism" for this syndrome. Varying degrees of mental retardation (usually an IQ of 70 or below), although more intelligent exceptions are known. Short stature, protruding furrowed tongue, transverse palmar crease, and cardiac deformities are common. Patients usually die by age 30–35; 50% die by age 3 or 4. Unusually susceptible to respiratory infections and to leukemia. Females are fertile, if they live to sexual maturity and, if able to reproduce, produce Down syndrome in 50% of their offspring.
Trisomy 22		Similar to Down syndrome but with more skeletal deformities
XO	Turner syndrome (gonadal dysgenesis)	Short stature, webbed neck, sometimes slight mental retardation. Ovaries degenerate in late embryonic life, leading to rudimentary sexual characteristics. Gender is female.
XXY	Klinefelter syndrome	Male with slowly degenerating testes, enlarged breasts
XYY		Unusually tall male, heavy acne of skin; some tendency to mild mental retardation
XXX		Despite triploid X chromosomes, usually fertile, fairly normal females
Short 5 (deletion of short arm of chromosome 5)	Cri-du-chat syndrome	Microcephaly, severe mental retardation. In infancy, cry resembles that of a cat. Defective chromosome is heterozygous.
Deletion of one arm of chromosome 21	Philadelphia chromosome	Chronic granulocytic leukemia

oocyte (hence the correlation between rate of occurrence and maternal age). In about 4% of patients with Down syndrome, only 46 chromosomes are present, but owing to a translocation, one is abnormal and contains extra genetic material.

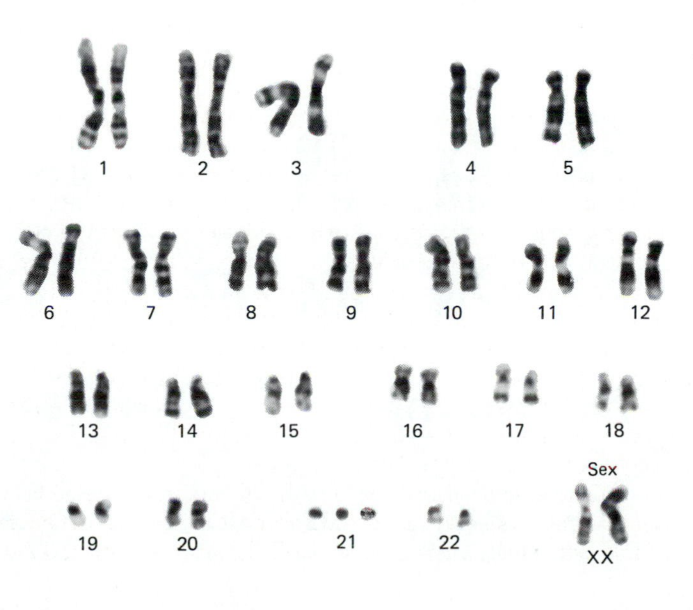

Figure 12–3 Karyotype of an individual with Down syndrome. Note the presence of an extra chromosome 21.

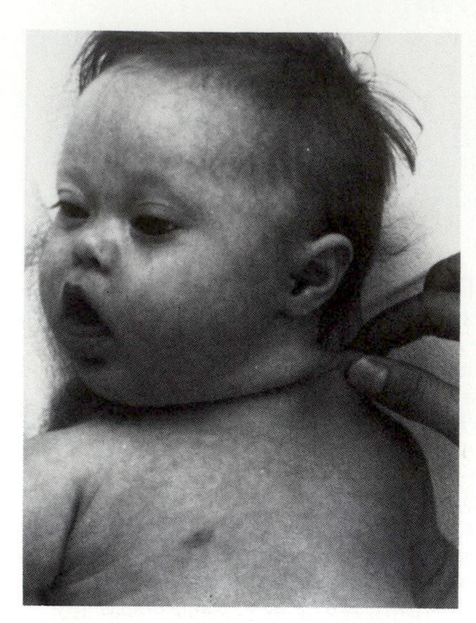

Figure 12–4 A child with Down syndrome.

The presence of this extra chromosome leads to the complex physical and mental abnormalities that characterize Down syndrome. Whether the extra genes in the third chromosome 21 leads to the production of an extra amount of certain enzymes and whether this is the basis for the abnormal physical and mental development are not known. When a certain chromosome or part of a chromosome has been added or deleted, the genetic imbalance results in defects in many types of organisms.

Down syndrome should be inherited as though it were a dominant gene, since an afflicted person would form gametes of which half would have the normal complement of 23 chromosomes and half of which would have 24 chromosomes. In the rare cases in which women with Down syndrome have had offspring by chromosomally normal men, they have produced normal and afflicted children in about equal proportions.

Klinefelter Syndrome

In another condition caused by an altered chromosome number, affected persons are outwardly nearly normal males but have small testes and produce few or no sperm (Fig. 12–5). The seminiferous tubules are very aberrant in microscopic appearance, and gynecomastia (a tendency for formation of female-like breasts) may also be present. This condition, called **Klinefelter syndrome,** usually becomes apparent only after puberty, when the small testes and gynecomastia may cause the affected individual to seek medical attention.

Because Barr bodies are present in the cells of such persons, these males were at one time thought to be XX individuals—that is, genetic females. However, when their chromosomes were examined cytologically and counted, it was found that they actually are trisomics and have 47 chromosomes; their cells have two X chromosomes and one Y chromosome. The fact that they are nearly normal males in external appearance emphasizes the strong, male-determining effect of the Y chromosome in humans.

Turner Syndrome

In persons with Turner syndrome, the external genital structures, though feminine, are those of an *immature* female. The internal reproductive tract is present but is also immature in development; the uterus is present but small, and gonads may be absent. The cells of these individuals lack Barr bodies, suggesting that they are males. However, they have only 45 chromosomes; they have one X chromosome but *no* Y chromosome. This type of disorder again emphasizes the importance of the Y chromosome in determining male characteristics.

Genes and Disease

More than 150 human disorders involving enzyme defects have been linked with genetic mutations (see Table 12–2). These disorders are sometimes referred to as inborn errors of metabolism. Phenylketonuria (PKU), discussed in the preceding

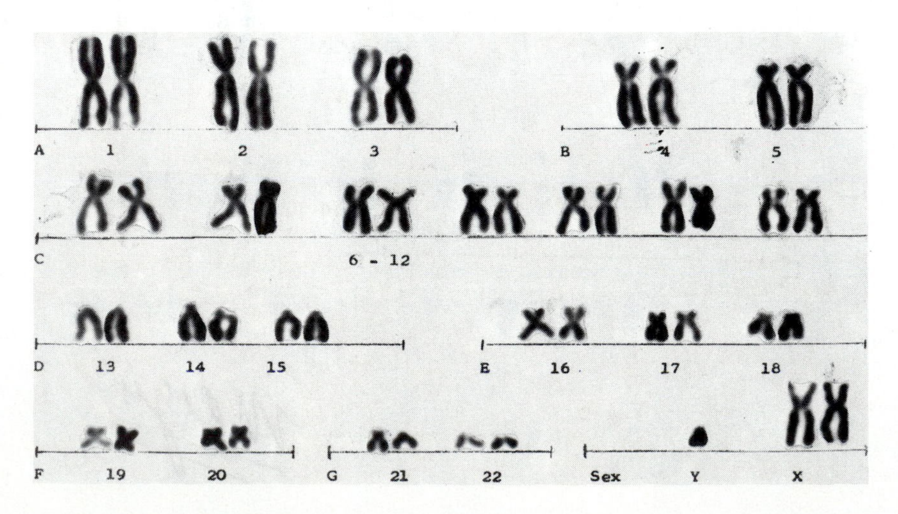

Figure 12–5 Karyogram of an individual with Klinefelter syndrome. This individual has an XXY karyotype, which is phenotypically male but sexually somewhat undeveloped. (Courtesy of Dr. Gilbert Echelman.)

TABLE 12–2
Some Important Genetic Disorders

Name of Disorder	*Mode of Inheritance*	*Clinical Description*	*Treatment, If Any*	*Comments*
Alkaptonuria	Autosomal recessive	Pigmentation of cartilage and fibrous tissue, with eventual development of arthritis. Presence of homogentisic acid causes urine to darken when it stands.	Arthritis may be treated.	Deficiency of enzyme, homogentisic acid dehydrogenase
Childhood pseudohypertrophic muscular dystrophy	X-linked recessive	Begins in the first 3 years of life. Muscles swell, then undergo fatty degeneration. Progressive muscular deterioration leads to confinement, then to death in the early twenties.	Symptomatic	Also known as Duchenne-type muscular dystrophy; extremely rare in females but heterozygotes sometimes exhibit minor muscle function defects
Cystic fibrosis	Autosomal recessive	High level of sweat electrolytes, pulmonary disease, cirrhosis of the liver, pancreatic malfunction, and, especially, nonsecretion of digestive enzymes. No spermatogenesis in males, but females sometimes reproduce. Life expectancy, 12–16 years, with some living into thirties and forties; commonest in persons of Northern European extraction	Symptomatic, with emphasis on digestive enzyme replacement and control of respiratory infections	CF kills more children than diabetes, rheumatic fever, and poliomyelitis combined. Exists in different degrees of severity. Thick mucus interferes with lung clearance.
Gangliosidosis (e.g., Tay-Sachs disease)	Autosomal recessive	Several types exist. One, Tay-Sachs disease, results from the deficiency of hexoseaminidase A. All variants involve the abnormal accumulation of sphingolipids, ordinarily released from nerve and other cells by action of whatever enzyme is deficient. Evidently several different enzymes are required. Blindness, paralysis, death in first few years of life in most cases		Tay-Sachs disease is especially prevalent among Jews of Eastern European ancestry.
Hemoglobinopathic disease (e.g., sickle cell anemia)	Group of autosomal recessive or incompletely dominant traits	Abnormalities of red blood cells caused by the presence of certain inappropriate amino acids at crucial locations in the hemoglobin molecule. In sickle cell anemia, for instance, one of the β-chain amino acids is the "wrong" one, resulting in decreased solubility of hemoglobin molecules in low-oxygen environments such as tissue capillaries. This causes extreme shape distortions such as sickling, which leads in turn to the premature destruction of the cell.	Varies with type of disease. Some (e.g., hereditary methemoglobinemia) may require no treatment. Some cannot be treated at all. Sickle cell anemia (SCA) can be treated to some degree.	These traits are similar and related but not allelic. Microcytic anemia is commonest in Mediterranean populations, sicle cell anemia in some black populations. In heterozygotes, SCA offers some protection against malaria.
Hemophilia	X-linked recessive	Chronic bleeding, including bleeding into joints, with resultant arthritis. More than one variety of hemophilia exists.	Treated with clotting factors	Even heterozygotes have some clotting factor deficiencies.

(continued)

TABLE 12–2
(continued)

Name of Disorder	*Mode of Inheritance*	*Clinical Description*	*Treatment, If Any*	*Comments*
Lesch-Nyhan syndrome	X-linked recessive	Slowly developing paralysis accompanied by mental deficiency and self-mutilation, with patients persistently biting themselves. Gout usually develops because of deficiency of the enzyme involved in purine metabolism. The heterozygous condition is detectable (see comments).	Gout may be treated. Neurological symptoms are not treatable, and early death is inevitable.	Deficiency of a specific enzyme is to blame. Half the cells of the female carrier are enzyme-deficient. Her hair follicles can be biopsied and studied. If this is done, they will be found to be enzyme-negative, enzyme-positive, or mixed, but not all of them contain the enzyme. Normally all of them would.
Phenylketonuria (PKU)	Autosomal recessive	Deficiency of liver phenylalanine hydroxylase leads to a chain of events beginning with excessive phenylalanine in the blood. This leads to a depression of the levels of other amino acids, leading in turn to excessively light coloration and mental deficiency.	A low-phenylalanine diet minimizes symptoms. Most states have extensive PKU screening programs, in which newborns are tested for excessive blood phenylalanine, or for presence of metabolic products in the urine.	Since melanin is synthesized from tyrosine, tyrosine deficiency caused by phenylalanine hydroxylase deficiency results in light coloring of skin and hair.
Red-green colorblindness				
Deutan variety	X-linked recessive	Patient can distinguish only 5 to 25 hues, as against the normal ability to see 150+. Though visual acuity is normal, the "green" cone pigment is deficient. Subjectively, all colors are perceived as hues of blue and yellow.		Actually a series of alleles of differing degrees of severity, with the more normal dominant over the more deficient varieties. In both protan and deutan forms, heterozygous females show some color vision defects.
Protan variety	X-linked recessive, not allelic to deutan	Similar to deutan variety except that here the "red" cone pigment is missing. NOTE: Other defects of color vision associated with one or another cone deficiency also exist. Most of these are sex-linked.		
Tyrosinase-negative oculocutaneous albinism (T− albinism)	Autosomal recessive	Absence of pigmentation due to functional absence of tyrosinase. Visual acuity 20/200 or less. Marked susceptibility to skin cancer.	Avoidance of sunlight.	Somewhat more common among blacks than whites
Tyrosinase-positive oculocutaneous albinism (T+ albinism)	Autosomal recessive	Reduction of pigmentation due to malabsorption of tyrosine by body cells. If heavy pigmentation is genetically specified, some pigmentation will survive, though in some cases phenotype is virtually identical with T−. Pigmentation and visual acuity improve with age.		Highest incidence in American Indians, less in blacks, least in whites. Hybrid T+/T− persons appear normal.

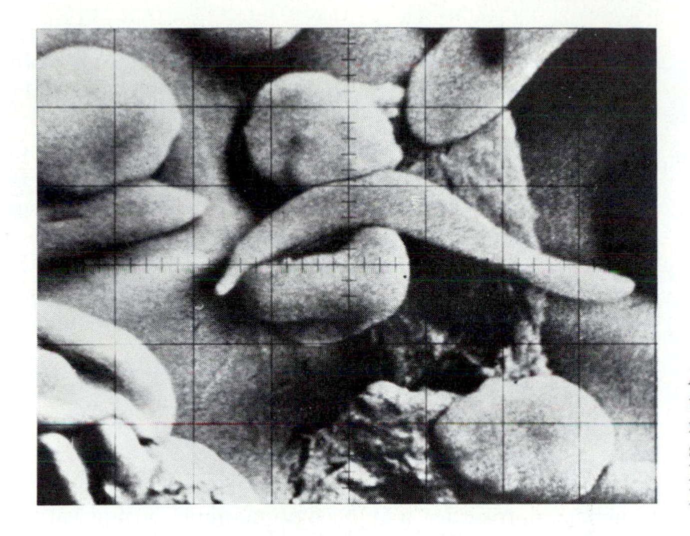

Figure 12–6 Red blood cells from a patient with sickle cell anemia (approximately ×4000). Note the abnormal shape of some of the cells. (Courtesy of Irene Piscipe-Rodgers, Philips Electronic Instruments.)

chapter, cystic fibrosis, and sickle cell anemia are well-known diseases that have been linked to gene defects. Most human genetic diseases are transmitted as autosomal recessive traits, and so are expressed only in the homozygous state.

SICKLE CELL ANEMIA

Sickle cell anemia is inherited as an autosomal recessive trait. The disease is most common in persons of African descent. The blood cells of a person with sickle cell anemia are shaped like a sickle cell or a half moon, whereas normal red blood cells are biconcave discs. The sickle cell contains hemoglobin molecules with a slightly different molecular structure from that found in normal red blood cells. The hemoglobin molecules in a person with sickle cell anemia have the amino acid valine instead of glutamic acid at position 6, the sixth amino acid from the amino terminal end, in the β chain. The substitution of an amino acid with an uncharged side chain (valine) for one with a charged side chain (glutamate) makes the hemoglobin less soluble, and it tends to form crystal-like structures that change the shape of the red blood cell (Fig. 12–6). This occurs in the veins after oxygen has been released from the hemoglobin. Their abnormal sickle shape slows blood flow and blocks small blood vessels, with resulting tissue damage and painful episodes. The abnormal blood cells have shortened life spans, leading to anemia in the victims of this disease. Other aspects of sickle cell anemia are discussed later in the chapter.

CYSTIC FIBROSIS

The most frequent autosomal recessive disorder in white children is **cystic fibrosis (CF)** (Fig. 12–7). One out of 20 persons in the United States is heterozygous for the gene for CF. Every five hours a homozygous baby who has active disease is born. Until the 1940s such babies soon died. By the 1950s, however, owing to advances in antibiotic therapy, half of them lived to be 3 years old, with a few living much longer. Today a victim of CF has a 50% chance of living into his or her twenties, but will spend four years of that time in a hospital. For each year of life, health care will cost thousands of dollars. CF sufferers usually die in what would be, for almost anyone else, the prime of life.

The most severe effect of CF is upon the respiratory system. CF causes the respiratory system to produce abnormally viscous and tenacious mucus. This heavy mucus also occurs elsewhere in the body (ducts of the pancreas and liver and the intestine, for example), but it is particularly serious in its effects on the respiratory system because of its action on the self-cleansing mechanisms of the lungs. The viscous mucus of the lung cannot be moved easily by the ciliary lining of the bronchi, and the action of those cilia may be abnormal too. Consequently, mucus forms pools in the bronchi rather than moving out normally. Thus, instead of serving its normal housekeeping function, the mucus serves as a culture medium for dangerous bacteria. These bacteria or their toxins then attack the surrounding tissues, leading to the development of recurring pneumonia, obstruction of the bronchial

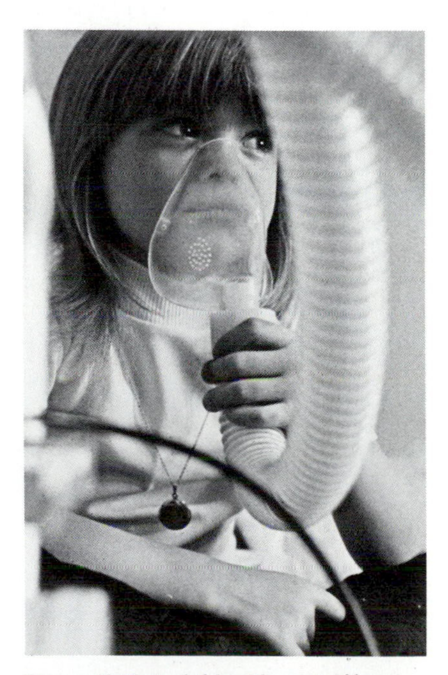

Figure 12–7 A child with cystic fibrosis using a nebulizer which disperses medications into a fine mist that can then be inhaled. (Courtesy of the Cystic Fibrosis Foundation.)

tubes by stretching, bronchitis, emphysema, and even bronchial asthma. Although the disease process of CF basically is known, how the abnormal gene produces these effects or how to cure the disease is not known–only the symptoms can be treated.

In sum, the CF sufferer is the victim of progressive chronic obstructive pulmonary disease that, together with malnutrition resulting from digestive malfunction, eventually is fatal—although it may take years to progress to that point. During those years, the afflicted person may suffer continuous and miserable physical illness and marked emotional distress. A combination of judicious but vigorous use of antibiotics to control bacteria, plus daily physical therapy to help clear mucus from the bronchial tree, nevertheless allows many victims of CF to live productive lives for many years.

Diagnosis of Genetic Abnormalities

Genetic abnormalities may become manifest during early intrauterine life or not until late in adult life. Because the possibility of preventing or alleviating the effects of a genetic abnormality is obviously greater if it can be detected as early as possible in life, efforts have been made over the years to detect such abnormalities at birth. For example, an infant proved to have PKU can be placed on a diet containing the least possible amount of phenylalanine in order to minimize the damage to the central nervous system. In the past 20 years, as physicians have become bolder in approaching the fetus while it is still in the uterus, diagnosis of a number of genetic abnormalities during fetal life has become possible.

In one diagnostic technique, known as **amniocentesis,** a sample of the amniotic fluid, together with the fetal cells present in it, is obtained by inserting a needle through the lower abdomen of the pregnant woman and through the wall of the uterus into the uterine cavity (Fig. 12–8). A syringe is attached to the needle and amniotic fluid is withdrawn. The amniotic fluid is centrifuged to sediment the cells, a sample of which is transferred to slides, fixed, and stained. The remaining cells are placed in a culture medium and grown for 2 to 3 weeks. Some of these cells are collected and assayed for the presence or absence of certain enzymes.

Cells obtained from amniotic fluid can be analyzed for the type of sex chromosomes, permitting a determination of the sex of the fetus. This can be useful in diagnosing the presence of severe sex-linked hereditary diseases such as hemophilia. Chromosomes can also be counted to identify trisomy, monosomy, and other chromosome abnormalities.

The number of genetic abnormalities that can be detected by amniocentesis has increased remarkably in the last few years and now includes many mutations as well as chromosomal aberrations detectable by their karyotype. Enzyme deficiencies can be detected by incubating cells recovered from amniotic fluid with the appropriate substrate and measuring the product. An increased concentration of α-fetoprotein in amniotic fluid has been shown to be associated with defects in the development of the neural tube such as spina bifida; this finding is proving useful as a diagnostic tool.

Genetic Counseling

Couples who have had one abnormal child or who have some other member of the family affected with a hereditary disease and hence are concerned about the risk of abnormality in a first or subsequent child may seek genetic counseling. Genetics clinics are available in most metropolitan centers to provide such advice.

Advice, of course, can be given only in terms of the *probability* that any given offspring will have a particular condition. The geneticist needs a carefully taken family history of both the man and the woman and may use tests for the detection of heterozygous carriers of certain conditions. Some diseases are inherited by a single pair of genes, and the probabilities are then easily calculated. For example, if one prospective parent is affected with a trait that is inherited as an autosomal dominant, such as that for Huntington's chorea, the probability that any given child will have the trait is 0.5.

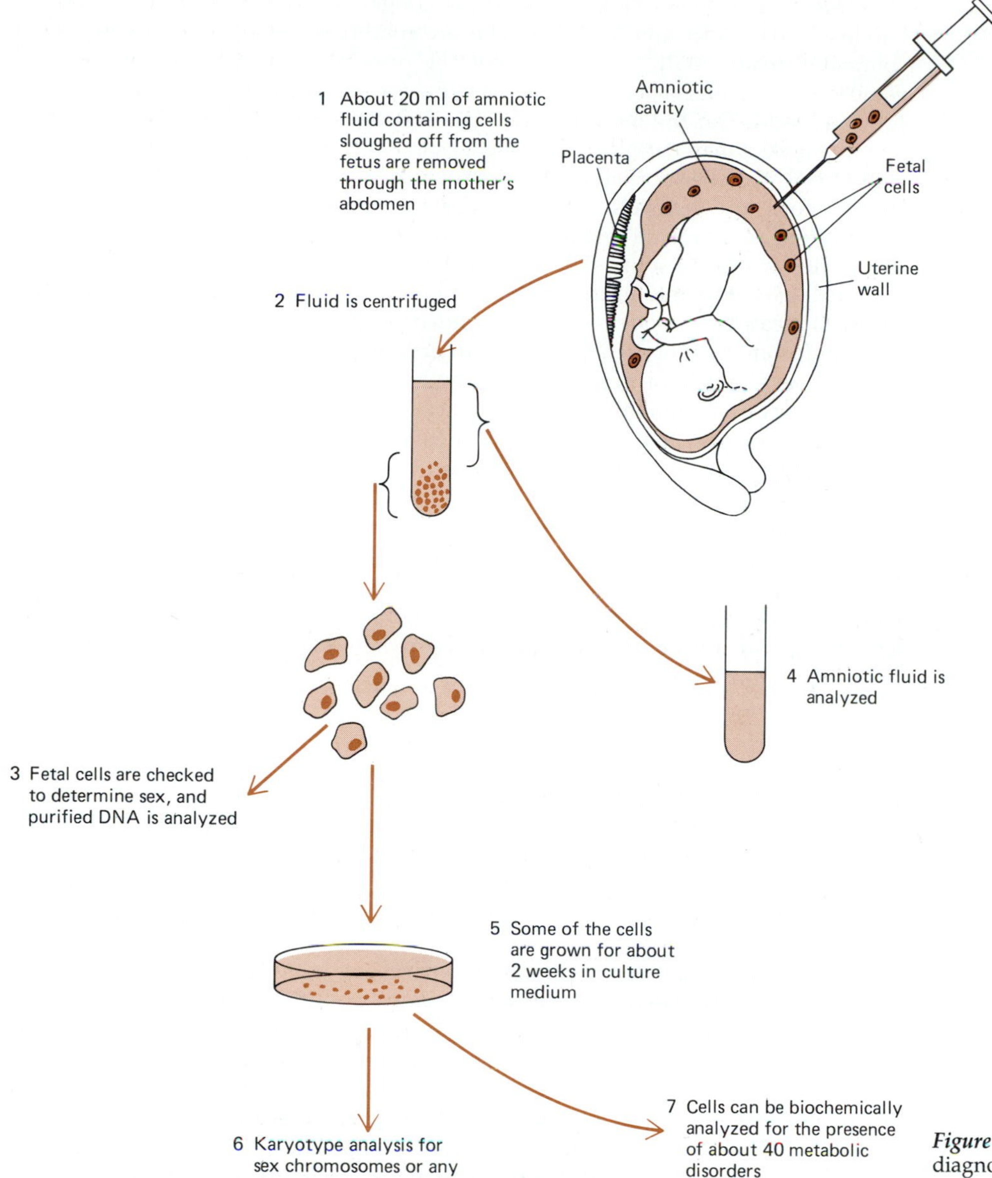
1 About 20 ml of amniotic fluid containing cells sloughed off from the fetus are removed through the mother's abdomen
Amniotic cavity
Placenta
Fetal cells
Uterine wall
2 Fluid is centrifuged
4 Amniotic fluid is analyzed
3 Fetal cells are checked to determine sex, and purified DNA is analyzed
5 Some of the cells are grown for about 2 weeks in culture medium
6 Karyotype analysis for sex chromosomes or any chromosome abnormality
7 Cells can be biochemically analyzed for the presence of about 40 metabolic disorders

Figure 12–8 The process of prenatal diagnosis of genetic disease by amniocentesis.

The birth of one child with a trait such as that for albinism or phenylketonuria to normal parents establishes that both parents are heterozygous carriers, and the probability that a subsequent child will be affected is 0.25. For a disease inherited by a recessive gene on the X chromosome, such as hemophilia, the probability depends on whether the father or the mother has the disease or is a carrier for it: A normal woman and an affected man will have daughters who are carriers and sons who are normal. The probability that a carrier woman and a normal man will have an affected son is 0.5, and the probability that they will have a carrier daughter is also 0.5.

In conditions in which the method of inheritance is unknown or is doubtful, an estimate of the probability of appearance of a given trait can be obtained from a table of empirical risk.[1] It is difficult to give a precise probability, because the trait may be inherited in different ways in different pedigrees. For example, the disease retinitis pigmentosum appears to be inherited as a sex-linked trait in some pedigrees and as an autosomal trait in others.

[1]Empirical risk simply means observed risk. By recording the instances of appearance of the trait in the offspring of affected parents geneticists can estimate the likelihood that children of specified crosses will have the trait. Naturally it is best to *calculate* the risk if possible, but this is practical only in cases of known simple inheritance of particular diseases.

Mental deficiency, epilepsy, deafness, congenital heart disease, anencephaly, harelip, spina bifida, and hydrocephalus are conditions about which inquiries are commonly made. With such conditions the possibility must be considered that some environmental factor played a role in the appearance of the abnormality in the previous child. Did the mother have some sort of infectious disease during pregnancy (e.g., German measles)? Was she on some sort of drug therapy (e.g., diethylstilbestrol)? Was she subjected to any sort of radiation? By dissecting environmental contributions, the geneticist can make a better estimate of the probability that genetic factors are involved and a better estimate of the probability of recurrence of the particular trait in some subsequent offspring.

It is now possible to detect the carriers of several genetic diseases, so that counseling can be provided when both husband and wife are heterozygous. For many autosomally inherited diseases, carriers show only half the level of enzyme activity characteristic of normal homozygotes. For example, screening programs have been set up in more than 50 major cities for Tay-Sachs disease, a degenerative neurological disease that causes death in early childhood. About 1 in 28 Jews of Eastern European descent have been found to be heterozygous for Tay-Sachs.

Persons heterozygous for sickle cell anemia have a mixture of normal and abnormal hemoglobins in their red blood cells; about 45% of the total hemoglobin has the abnormal chemical constitution. Such persons are said to have **sickle cell trait.** Their blood cells do not usually undergo sickling but can be made to do so when the amount of oxygen in the blood is reduced. This provides a very simple test, using only a drop of blood, by which the heterozygote can be distinguished from the homozygous normal.

Eugenics

Eugenics (literally, "well-born") is the process of genetically improving the human species over the course of generations. The concept is not new, but some of the mechanisms for manipulating genes, such as genetic engineering (discussed in Chapter 16) are very new indeed. Negative eugenics deals with the prevention of spreading undesirable genes, whereas positive eugenics encourages the spread of "good" genes.

It has been estimated that each of us carries as many as 15 deleterious recessive genes that fail to express themselves only because their alleles have not found similar alleles. (The chances of expression of such harmful genes are much enhanced in the offspring of relatives who, since they have ancestors in common, may very well also have many deleterious genes in common.) Furthermore, there are many genes that, while they do not produce frank disease, may reduce vitality to some degree or may produce some undesirable trait.

The totality of all harmful genes in a gene pool is called **genetic load,** and in all species that have been investigated it is substantial indeed. Obviously, the origin of genetic load lies in mutation, and anything that promotes mutation can be expected to increase genetic load. On the other hand, deleterious genes tend to be removed from a population by natural selection, an example of which is the death of an organism that such a gene might produce. The equilibrium that any one such gene actually reaches in a population probably results mostly from the balance attained by these two forces. In human terms, however, the selective process is a very costly one, both socially and in terms of human suffering.

NEGATIVE EUGENICS

There is nothing that can be done now to repair a mutation that produces a genetic disease such as cystic fibrosis. Could it be removed from the gene pool by a selection process instead? Such a proposal need not conjure up mental pictures of storm troopers leading people off to gas chambers. With modern methods of contraception the transmission of genes could be prevented by preventing reproduction. As the geneticist Theodosius Dobzhanzky used to say, "A genetic death need not produce a cadaver." Perhaps, then, we should practice **negative eugenics**—the deliberate manipulation of the human gene pool to reduce the frequency of harmful genes.

Yet this still raises many ethical problems. How can we ensure that those carrying deleterious genes refrain from reproducing? Would there be criminal penalties or compulsory sterilization? A further complication is that many genetic diseases cannot be detected in the heterozygous state. Relatively few genetic diseases are dominant. If they were, they would be rare, since they would tend to be eliminated as soon as the mutant gene that produced them came into being. To prevent carriers of genetic disease from reproducing, therefore, restrictions would need to apply to all close blood relatives of those having disease. Such restrictions would unfairly prevent many people of normal genotype from reproducing.

This leads to a practical objection. Those exhibiting recessive genetic disease represent only the tip of a much vaster iceberg—a reservoir of heterozygous genes. Although improved biochemical diagnostic techniques have rendered some genetic diseases detectable in healthy carriers, many others remain undetectable in heterozygotes. The most stringent selection against such a disease will reduce its incidence only a little in each succeeding generation. Using cystic fibrosis (CF) as an example, let us see just how little that could be expected to be.

With the CF gene assumed to be presently in a state of equilibrium (not a completely justified assumption, as we will see), its frequency can be calculated by the Hardy-Weinberg law. The incidence of CF in the United States population and in Northern European countries is 0.0004. This means that $q^2 = 0.0004$, and therefore $q = 0.02$. Let us suppose that no one with CF could reproduce (an assumption, by the way, that is almost true; affected females are reproductively handicapped so that most do not become pregnant, and males are almost always sterile). If the mutation rate is negligible (as it probably is), then complete selection against CF should reduce the frequency of the gene in each generation by the value of q^2.

Generation	*Incidence of CF*	*Gene Frequency in Population*
0	0.000400	0.02
1	0.000384	0.0196
2	0.000369	0.0192
3	0.000355	0.01883
4	0.000342	0.01848
5	0.000329	0.01814

That corresponds to about an 18% reduction in the incidence of the disease—but it would take more than 100 years to accomplish it. To cut the incidence of CF in half (to a rate of 0.0002) would take far longer than most government-sponsored programs endure. To reduce it to .0001 would take longer than the government itself would probably last!

This raises another question. Since those with CF are highly handicapped in reproduction and until recently died long before reproductive age anyway, we have a eugenics program administered by nature that has probably continued for thousands of years. Yet CF is still with us. How can that possibly be explained?

BALANCED POLYMORPHISM

We have already met the genetic disease sickle cell anemia, which results from an inappropriate amino acid substitution in the β chain of the hemoglobin molecule. The heterozygote has a mixture of the two hemoglobins, both normal and abnormal. It is thought that red blood cells with this mixture sickle when invaded by the malaria parasite and are selectively destroyed by the liver and spleen, along with their parasites. This would give heterozygotes an advantage if they lived in an area like central Africa where malaria is widespread. Under such circumstances natural selection will tend to maintain even a lethal gene in the gene pool. To be sure, there is little need for the sickling gene in Chicago or Manhattan; yet people who have it do live there. There was a need for it, however, in the part of Africa from which their ancestors came (see Figure 45–4). Will we ever need this gene again? Let us hope not, but it would be rash to rule out the possibility.

Such maintenance of genetic variety by differing or opposed selective pressures is called **balanced polymorphism,** the stable equilibrium of many allelic genes by natural selection. Could balanced polymorphism also explain CF? It is estimated that if heterozygous persons produced just 2% more offspring on the average than those completely without the trait, it would account for the observed incidence of this gene in our population. Persons known to be heterozygous for CF do, in fact, seem to have more children than others do. Why this should be the case is not known, but the fact itself, if confirmed by further studies, is instructive and admonitory. Even if eliminating the CF gene were possible, would it be wise?

The existence of variation in itself may actually be of adaptive value to a population for several reasons. For one thing, a completely homozygous population would have no genetic substratum on which natural selection could act. How can we predict what challenges the future may hold for the human species? A population that has a good prognosis for survival in the future is one that has maintained sufficient variation to permit further adaptive changes. Some rapidly breeding organisms adapt genetically even to seasonal crises. Observations on wild populations of fruit flies and other organisms with very short generation times have shown that their genetic pools do change adaptively, even in response to such changes in the environment as the alterations of the seasons. Winter to a fruit fly is like an Ice Age to us. If we were to eliminate the appropriate genes from our population (and how do we know which ones they are?) we might not survive the next crisis. Fortunately, the Neanderthal people (who were probably specifically adapted to Ice Age conditions) were not eugenics enthusiasts. If they were, we, their descendants, might not even exist.

POSITIVE EUGENICS

Positive eugenics is the attempt to promote human characteristics held to be desirable and to increase their representation in the population. The negative eugenics we have considered is essentially an attack on genetic disease. Positive eugenics is more like the selective breeding that has produced superior varieties of domestic animals and plants. It is not always possible to distinguish reliably between positive and negative eugenics; the line can be hard to draw. A good example involves proposals to improve the general intelligence of the population, which, so goes the argument, can be expected to deteriorate under the conditions of modern life, in which persons of low intelligence are as likely to reproduce as those of high intelligence.

Positive eugenics, at least in the area of intelligence, might appear to be justified, but closer examination suggests caution. In the first place, selection *for* a gene is the same as selection *against* its alternative. Viewed in this light, even if intelligence were governed by a single gene (an absurd assumption in view of the vast complexity of the brain), it would take the same kinds of ruthless measures and long periods of time to accomplish anything noteworthy as those required for the elimination of a genetic disease. In fact, if average intelligence were to be considered a genetic disease, then it is an extremely common one! The span of time needed to raise the human population substantially in intelligence would probably compare in order of magnitude to the total time span in which humanity has existed. Since "intelligence" is certainly governed by a multitude of genes whose actions and interactions are for the most part totally unguessed at, it is likely that such a eugenics program would require a geological epoch to implement and might have many unforseen consequences.

Additionally, there may be intellectual configurations that, while not measurable by currently fashionable standardized tests, could be highly desirable even today, to say nothing about the unpredictable world of the future. Standardized tests in current use concern themselves principally with the prediction of probable success in the kinds of academic tasks imposed in our schools, or in the kinds of ability emphasized in the industrial societies of the recent past. But the ability to program computers, for instance, is not closely allied to any of the old industrial tasks, including many mathematical tasks that are still a traditional part of the school and college curriculum. Intelligence tests of the future might well concern themselves with such postindustrial skills and might identify quite a different segment of the population as intellectually elite.

It is as if we were to develop a measurement of physical fitness that depended upon the average of biceps diameter, strength of grip, and total force exerted when the arm is flexed. Utter weaklings or magnificent physical specimens would score consistently low or high on the test, which would seem to confirm its reliability; such a test, however, would fail utterly to predict success in track events, though it might be very useful in weight lifting. We might even be able to prove that AQ (arm quotient) is inheritable and, from that, jump to the conclusion that exercise is unimportant as a determinant of athletic success! As Sir Peter Medawar, the English biologist and Nobel Prize winner, has said, "The IQ boys have tried to convince us that a person's IQ *is* his intelligence, which is on all fours with saying that a patient's temperature is his health."

Intelligence is a highly polymorphic trait, in many cases not realistically described in quantitative terms. Our society requires a *variety* of intellectual and personality traits. Do we really need millions of Nobel Prize winners? Still more, will we need them in the unguessable world of the 30th century? That is probably when we will get them, at the soonest, if we rely on eugenic breeding programs. We can't seem to plan realistically for the energy crisis over a 20-year period. It is absurd to anticipate what would make us genetically fit for the next millennium.

SUMMARY

I. The geneticist investigating human inheritance cannot make specific crosses of pure genetic strains but must rely on studies of pedigrees and on the laws of probability, studies of gene pools, and the principles of population genetics to discover methods of inheritance of specific traits.

II. A great variety of physical and biochemical traits are inherited.

III. Studies of the karyotype, of the number and kinds of chromosomes present in the nucleus, permit detection of individuals that are polyploid or have other chromosome abnormalities.
  - *A.* Such studies can detect individuals who are trisomic, having one extra chromosome, or who are monosomic, lacking one member of a pair of chromosomes.
  - *B.* Down syndrome, Klinefelter syndrome and Turner syndrome are examples of trisomy or monosomy.

IV. Mutations in genes are responsible for many inherited diseases such as phenylketonuria, sickle cell anemia, and cystic fibrosis. Most human genetic diseases are transmitted as autosomal recessive traits.

V. Some genetic diseases can be diagnosed long before birth by amniocentesis. Both chromosome abnormalities and some gene defects can be diagnosed by studying the fetal cells obtained.

VI. Genetic counselors can advise prospective parents who have a history of genetic disease regarding the probabilities of giving birth to affected offspring. It is now possible to detect the presence of a number of harmful recessive genes.

VII. Eugenics is the process of genetically improving the human species over the course of generations.
  - *A.* Negative eugenics is the deliberate manipulation of the human gene pool to reduce the frequency of harmful or "undesirable" genes.
  - *B.* Positive eugenics is the attempt to increase the frequency of genes held to be desirable.
  - *C.* Deliberate manipulation of the human gene pool could threaten balanced polymorphism, or could possibly remove from the gene pool a gene for a variation that may someday be necessary for the survival of the species.

POST-TEST

1. Standard stocks of genetically identical individuals are called ______________ ______________.

2. One hundred times the mental age divided by the chronological age yields the ______________ ______________.

3. The array of chromosomes present in a given cell is called the ______________.

4. An abnormality or defect present and evident at birth is termed a ______________ ______________.

5. The abnormality in which there is one more or less than the normal number of chromosomes is an ______________.

6. A person with an extra chromosome, with three of one kind, is termed ______________.

7. A person with one less than the normal number of chromosomes, having only one member of a pair, is termed ______________.

8. The failure of two homologous chromosomes to separate normally during cell division is called ______________.

9. The transfer of a part of one chromosome to a nonhomologous chromosome is termed ______________.

10. A person with ______________ ______________ is retarded in physical and mental development, has abnormalities of face, tongue, and eyelids, and is ______________ for chromosome ______________.

11. The XXY individual has the disorder known as ______________ ______________; the XO individual has ______________ ______________.

12. Inherited disorders due to defective or absent enzymes are called inborn errors of ______________.

13. The gene for sickle cell anemia produces an altered ______________ molecule, which is less soluble and is more likely to crystallize and deform the shape of the red blood cell.

14. In a person with ______________ ______________, the mucus is abnormally viscous and tends to plug the ducts of the pancreas and liver and to accumulate as pools in the lungs.

15. In the process of ______________ a sample of amniotic fluid is obtained by inserting a needle through the walls of the abdomen and uterus into the uterine cavity.

16. ______________ ______________ is the process of genetically improving the human species by preventing the spreading of undesirable genes.

REVIEW QUESTIONS

1. What means have been devised for overcoming the difficulties of studying human inheritance?
2. What is meant by nondisjunction? What human abnormalities appear to be the result of nondisjunction?
3. Distinguish between congenital and hereditary traits.
4. What factors might result in a child more intelligent than either of its parents?
5. Why is amniocentesis an important clinical tool?
6. How can carriers of certain genetic diseases be identified?
7. What is meant by inborn errors in metabolism? Give an example.
8. Imagine that you are a genetic counselor. A couple comes to you for advice because the woman had a sister who died of Tay-Sachs disease. What suggestions would you give them?
9. Distinguish between positive and negative eugenics, and give arguments for and against the use of each.

13
DNA: The Secret of Life

OUTLINE

I. Unraveling the genetic code
  A. Early clues
  B. DNA and the transfer of genetic information
II. DNA: a macromolecule
  A. Nucleotides
  B. The double helix
    1. Base pairing and hydrogen bonds
    2. Forms of DNA
III. DNA replication
IV. DNA information density

LEARNING OBJECTIVES

After you have read this chapter you should be able to:

1. Give reasons for identifying DNA as the fundamental genetic material of the cell, and describe the experimental evidence for this that is summarized in this chapter.
2. Describe the basic structure of the DNA molecule.
3. Summarize the process of DNA replication.

"History often hinges upon . . . an accident or other unforeseen whim of fate," wrote the historians Goldstein and Dillon. "Therefore, it cannot be reduced to a scientific formula or factored like a mathematical problem."[1] This is true even when the history of which we speak is the history of science.

At a faculty tea on an early summer midafternoon in 1951, Francis Crick learned from a colleague that " . . . a semirigorous argument hinted that adenine and thymine should stick to each other . . . A similar argument [Crick learned] could be put forward for attractive forces between guanine and cytosine."[2] This bit of shoptalk sparked the studies by Crick and James Watson that revealed the structure of DNA—the now-famous double helix.

Unraveling the Genetic Code

If a complete description of the structure and function of the human body were possible, how long would it have to be? While no one is in a position to say, 600 books of average length might not be an unreasonable estimate. Yet, amazingly, all of these directions are contained within the microscopic nucleus of the zygote. The only way so much information could be encoded in so small a space is for it to be contained in the basic pattern of organization of some chemical substance. A molecule that can bear information must be internally variable and large; in other words, it must be complex. Not many substances are capable of such complexity. Until Crick and Watson's discovery, most informed biologists and chemists could see no possible candidates other than proteins.

Neither fats, carbohydrates, nor steroids could serve for such miniaturization of information. The medium carrying the nuclear genetic information could not have been one upon which a message was simply *impressed*. The message had to be a *part of the molecule itself*, and that would be possible only by varying the combinations of the atoms of which the molecule was composed. In the simple molecules there are neither enough atoms nor sufficient variability in organization to carry messages.

Proteins were favored as possible information carriers because they were known to be composed of a large number of amino acid subunits, arranged in sequences so complex that even today relatively few have been mapped in detail. (When such mapping has been carried out, the printout of the amino acid sequence may fill several journal pages of fine print.) Remember that there are just 20 amino acids that occur in proteins; these are combined in different arrangements, each of which characterizes a specific protein. To get an idea of the variation involved, note that there are only 26 letters in the English alphabet; these 26 letters nonetheless would be sufficient for writing those 600 volumes mentioned earlier.

The cell nucleus certainly does contain protein. Presumably this protein was the specialized carrier of the genetic message. To be sure, the nucleus also contained a kind of nucleic acid (discovered in the 19th century by Friedrich Miescher), but the structure of this substance seemed far too simple for it to be capable of carrying information. It was composed of a simple pentose sugar (deoxyribose), phosphate, and four nitrogenous bases known as purines—that is, adenine and guanine—and pyrimidines—that is, thymine and cytosine. Erwin Chargaff of Columbia University had demonstrated by the early 1950s that there was a simple relationship among these bases—always, the amount of cytosine equaled the amount of guanine, and the amount of thymine equaled the amount of adenine, in the DNA extracted from any tissue derived from any organism whatever. No such simple relationship occurred among the amino acids of proteins! At that time DNA was thought to have no more important function than to hold the nuclear proteins in place, or perhaps to strengthen the chromosomes during cell division.

EARLY CLUES

Even earlier, two highly respected researchers had made observations that did not fit these prevailing conceptions about DNA. In 1928 the British investigator Freder-

[1]From Pranger, G. W.: *At Dawn We Slept*, Preface. New York, McGraw-Hill Book Co., 1980.
[2]From Watson, J.: *The Double Helix*. New York, Atheneum Press, 1968, p. 128.

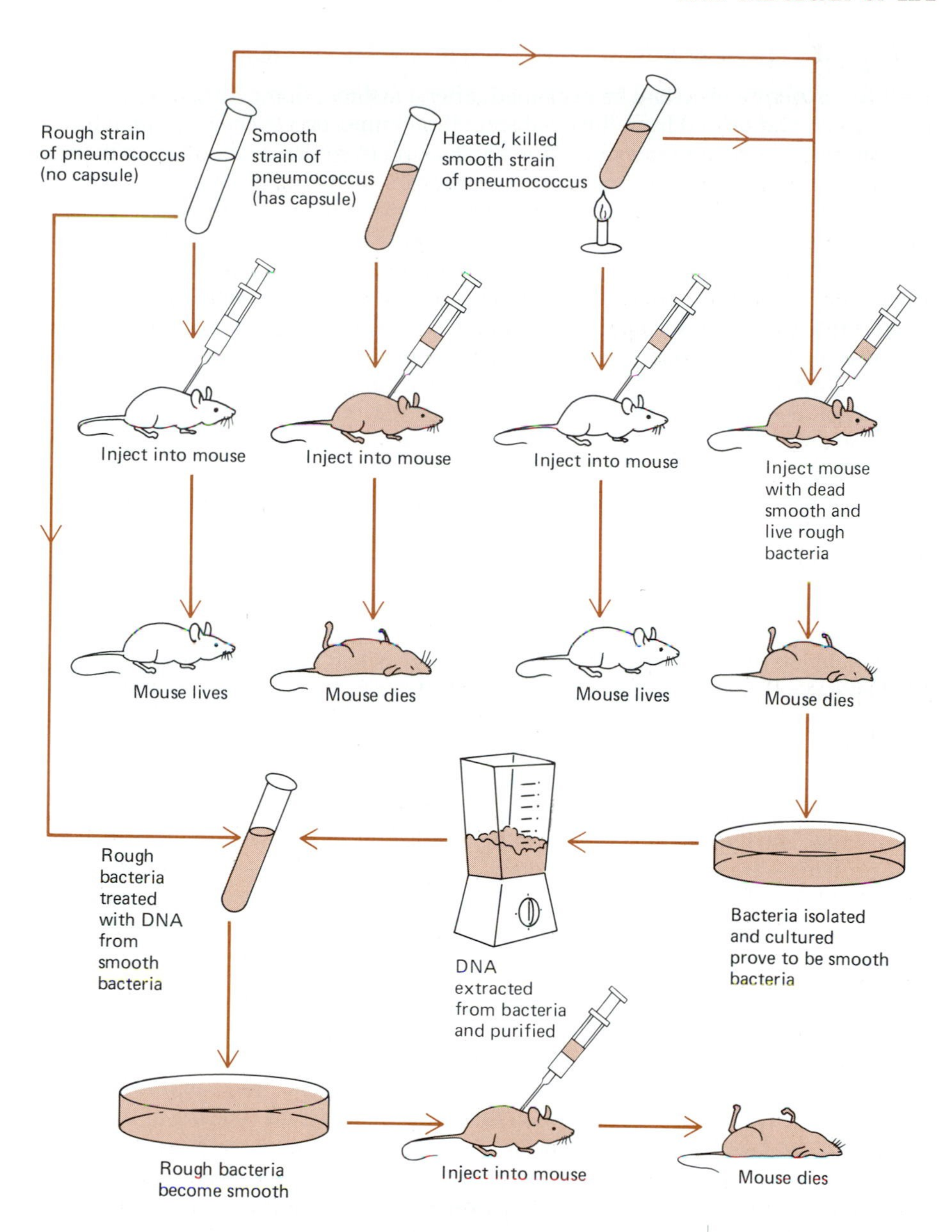

Rough strain of pneumococcus (no capsule)
Smooth strain of pneumococcus (has capsule)
Heated, killed smooth strain of pneumococcus
Inject into mouse
Inject into mouse
Inject into mouse
Inject mouse with dead smooth and live rough bacteria
Mouse lives
Mouse dies
Mouse lives
Mouse dies
Rough bacteria treated with DNA from smooth bacteria
DNA extracted from bacteria and purified
Bacteria isolated and cultured prove to be smooth bacteria
Rough bacteria become smooth
Inject into mouse
Mouse dies

Figure 13–1 The experiments of Fred Griffith and Oswald Avery. Griffith demonstrated the transfer of genetic information from dead, heat-killed bacteria to living bacteria of a different strain. Although neither the rough strain of *Pneumococcus* nor the heat-killed smooth strain could kill a mouse, a combination of the two did. Autopsy of the dead mouse showed the presence of living, smooth strain pneumococci. Later, Avery demonstrated the restoration of virulence to rough bacteria by treating them with DNA derived from smooth bacteria—proving that it was the DNA that carried the genetic information necessary for the bacterial transformation.

ick Griffith reported the results of what then impressed most biologists as a very eccentric experiment. Griffith has been described as one who " . . . seemed to live only to differentiate and describe all the pneumococci and hemolytic streptococci in the world."[1]

Bacteriologists had long known of the existence of nonvirulent strains of otherwise pathogenic bacteria that can be raised on laboratory media but that are so quickly destroyed by a host's immunological defenses that they are incapable of causing disease. One of these nonvirulent strains of *Pneumococcus* lacks a protective polysaccharide capsule, and forms *rough* colonies on agar in the laboratory. In contrast, the normal virulent strain produces glistening, *smooth* colonies with a polysaccharide capsule. Injected into mice, virulent smooth pneumococci kill the animals in short order, but the nonvirulent rough ones do not. Pursuing an immunological investigation with no obvious relationship to inheritance, Griffith injected dead, smooth pneumococci into mice *along with* live rough microbes. The mixture should have been harmless. Instead, many of the mice died. Griffith then cultured the blood of the dying mice and recovered live smooth pneumococci from it; these were fully virulent when injected into other mice. Even the descendants of such pneumococci could kill still *other* mice! (See Fig. 13–1.)

[1] Bendiner, E.: Avery: Making Sense of a "Stupid" Nucleotide. *Hospital Practice,* October 1982, pp. 195–219.

DNA AND THE TRANSFER OF GENETIC INFORMATION

Griffith's experiments could be explained, after a fashion. Some form of inheritable information had passed from the dead smooth pneumococci to the live rough ones. The information needed to make a capsule, lacking from those laboratory mutants, had somehow passed to them from the cadavers of the dead smooth microbes.

In the early 1940s, Griffith's remarkable experiments were repeated with modifications by Oswald Avery, a researcher at the Rockefeller Institute of New York City. Avery reasoned that some transforming principle, some chemical substance, had conveyed genetic information from the dead bacteria to the living. He and his research team first suspected that the substance was part of the polysaccharide coat of the pneumococcus—a reasonable belief, but wrong. Further experiments by these Rockefeller Institute scientists indicated that it was something from the *interior* of the dead cells that produced capsular transformation. Having eliminated all else by a lengthy series of experiments, they were left with DNA. They found that they could produce fibers of smooth pneumococcal DNA capable of transforming nonvirulent strains to virulent ones. In some manner, DNA carried genetic information (see Focus on Studies with Bacterial Viruses). But how?

DNA: A Macromolecule

The infallible eye of hindsight tells us that biologists had been misdirected for years by their entirely false impression of the simplicity of the DNA molecule. But what is DNA really like? What kind of picture did Watson, Crick, and their colleagues eventually develop?

Estimates of how much DNA a typical cell contains vary, ranging between 1 and 2 meters of the long, threadlike molecules. On a cellular scale that is astonishingly large. The longest molecules of other kinds that are known—polysaccharides and proteins—are composed of as many as thousands of repeating simple units, but DNA is much larger. All of the large biologically active molecules are composed of many small repeating subunits such as amino acids or hexose sugars. The fundamental units of DNA are the nucleotides (described in Chapter 3).

NUCLEOTIDES

To review, each nucleotide in DNA consists of a pentose sugar (deoxyribose), a phosphate, and a nitrogenous base—thymine (T), adenine (A), guanine (G), or cytosine (C). As shown in Figure 13–2, the deoxyribose forms the core of the molecule, with the base attached to it at the 1′ position and the phosphate attached at the 5′ position. The OH group at the 3′ position serves as a potential coupler to other nucleotides. (A number with a "prime" symbol indicates an atom on the pentose; a number without a prime indicates an atom of the base.)

It may have occurred to you that AMP is a nucleotide. Recall that AMP is what is left after ATP has donated its two terminal energy-rich phosphate groups in some coupled chemical reaction. The deoxyribose analogue[1] of AMP, dAMP, is one of the DNA nucleotides; that of ATP, dATP, also occurs in the cell. The remaining three bases of DNA also form triphosphates analogous to ATP: Guanine forms dGTP; thymine, dTTP; and cytosine, dCTP.

To form DNA, all the nucleotides that comprise it are joined together to form a **polymer,** a repeating-unit molecule of indefinite length. A nucleo*side* consists of a purine or pyrimidine base linked to a pentose; a nucleo*tide* is a phosphate ester of a nucleoside. A nucleoside triphosphate consists of a nucleoside with three phosphate groups attached. Additional nucleotides are added to a DNA chain by reactions involving nucleoside triphosphates. A nucleoside triphosphate is added to the last nucleotide in the chain, and pyrophosphate, PP_i (two phosphates linked together), is released. The resulting polynucleotide possesses a free 5′ phosphate and a free 3′ OH to which other nucleoside triphosphates can be linked to extend the chain further.

The formation of polynucleotides proceeds in one direction only, with new nucleotides being added only to the end of the growing chain that has a free 3′ OH available for attachment so that replication proceeds from the 5′ end of the DNA

[1] An analogue is structurally related to another compound, but its chemical effects may be different.

FOCUS ON

Studies with Bacterial Viruses

Further evidence that DNA carries genetic information came from studies with bacteria and **bacteriophages,** also called **phages,** which are the viruses that infect bacteria. A. D. Hershey and Martha Chase, working at Cold Spring Harbor, New York, showed that only the DNA and not the protein of the virus enters the bacterial cell. Hershey and Chase carried out an experiment to determine whether the phage, as it infects the bacterium, injects DNA, proteins, or both into the bacterial cell (see accompanying figure). They took advantage of the fact that DNA contains phosphorus, whereas protein does not, and that protein contains some sulfur atoms, whereas DNA does not.

By culturing bacteriophage on bacteria grown in a medium containing the radioactive isotopes ^{32}P and ^{35}S, Hershey and Chase were able to grow phage that contained ^{32}P in its DNA and ^{35}S in its protein. The radioactive phage particles were recovered, purified, and used to inoculate new, unlabeled bacteria. After the infection had begun, the bacterial cells were agitated in a blender to remove the extra virus and then were broken apart and analyzed. Outside the cell, the remains of the virus contained ^{35}S, but the bacterial cells contained radioactive phosphorus and very little, if any, radioactive sulfur. This is evidence that the DNA of the phage entered the cell, whereas the protein coat of the phage remained outside. This viral DNA, injected into the bacterial cell by the phage, in some way commandeers the machinery of the bacterial cell that ordinarily makes new bacteria and programs it to make new bacteriophage material instead. When the bacteriophage was allowed to multiply within the bacteria and then to escape, the new generation of bacteriophages contained ^{32}P but no ^{35}S.

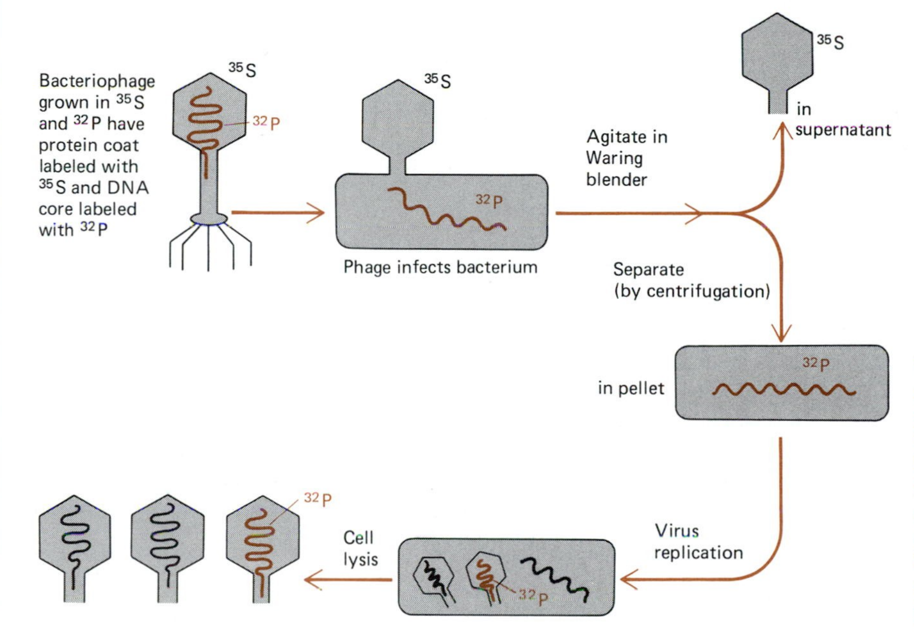

The Hershey–Chase experiment, which demonstrated that only the DNA component of bacteriophage is injected into bacteria, while the protein coat of the virus remains outside. All of the genetic information needed for the synthesis of both new protein coat and new viral DNA is provided by the parental viral DNA.

chain to the 3′ end. This occurs under the control of a specific enzyme system, using a preexisting strand of DNA as a **template** or mold. Thus the sequence of nucleotides in DNA is not random but *depends on a complex predetermined plan.*

THE DOUBLE HELIX

The primary clues about the structure of the DNA molecule came from studies using x-ray diffraction, carried out by Rosalind Franklin in the laboratory of M. H. F. Wilkins. The phenomenon of x-ray diffraction is quite difficult to understand, and a full description is outside the scope of this textbook. However, almost everyone has had occasion to observe the diffuse blur that forms around a street light on a foggy night. The myriad, randomly arranged droplets of water scatter the light in every conceivable direction, interfering with its ability to carry information to the eye.

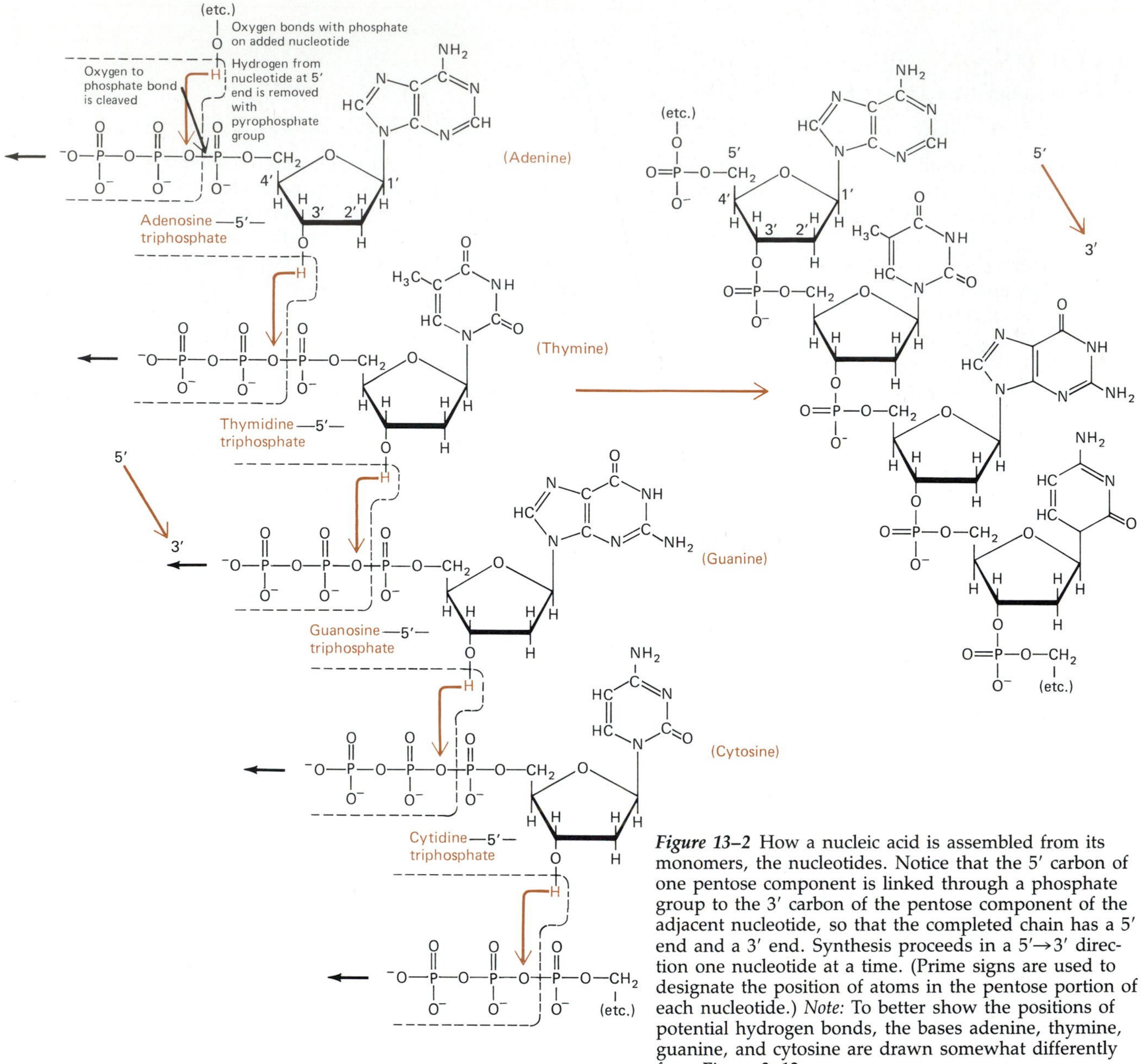

Figure 13–2 How a nucleic acid is assembled from its monomers, the nucleotides. Notice that the 5′ carbon of one pentose component is linked through a phosphate group to the 3′ carbon of the pentose component of the adjacent nucleotide, so that the completed chain has a 5′ end and a 3′ end. Synthesis proceeds in a 5′→3′ direction one nucleotide at a time. (Prime signs are used to designate the position of atoms in the pentose portion of each nucleotide.) *Note:* To better show the positions of potential hydrogen bonds, the bases adenine, thymine, guanine, and cytosine are drawn somewhat differently from Figure 3–13.

Light waves are not noticeably scattered by objects as small as atoms, however, because of the relatively long wavelengths of visible light. The wavelength of x-radiation, however, is so short that the x rays *are* scattered by atoms and molecules. Because of the regularity of atomic arrangements in crystals, when a pure crystal of DNA is bombarded with x rays the x rays are diffracted or bent in specific directions as they pass through the substance. The amount and nature of the bending and mutual interference of the x rays depends upon the structure of the molecule itself. The pattern of x-ray diffraction (Fig. 13–3) provides to the experienced eye a number of clues about the structure of the molecule. From such x-ray diffraction pictures Franklin and Wilkins inferred that the nucleotide bases of DNA (which are flat molecules) are stacked one on top of the other like a group of saucers. These x-ray diffraction patterns showed three major regularities or periodicities in crystalline DNA: one of 0.34 nm, one of 2.0 nm, and one of 3.4 nm.

On the basis of Chargaff's analytic results and Franklin and Wilkin's x-ray diffraction patterns, Watson and Crick proposed in 1953 a model of the DNA molecule (Fig.13–4) that ranks among the great intellectual achievements of history. The studies by chemists such as Linus Pauling had provided a great deal of information about the exact distance between the atoms that are bonded together in the DNA

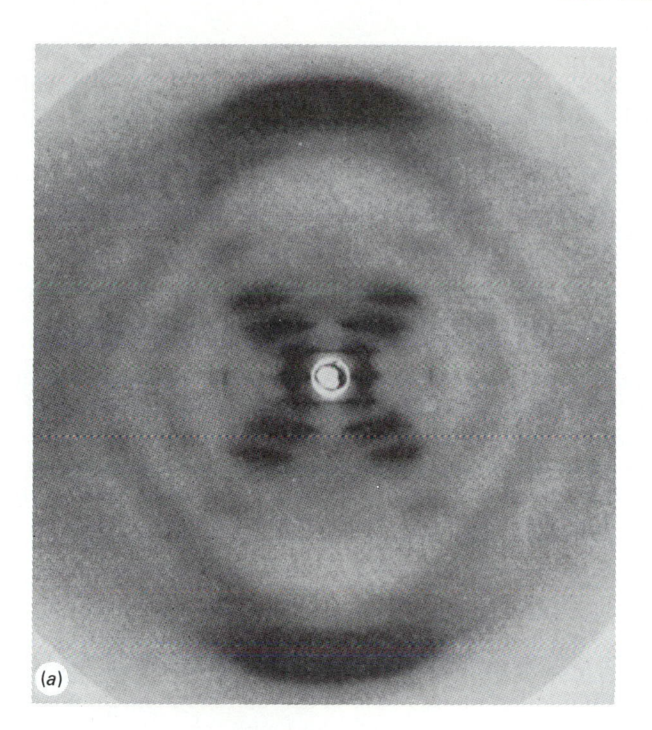

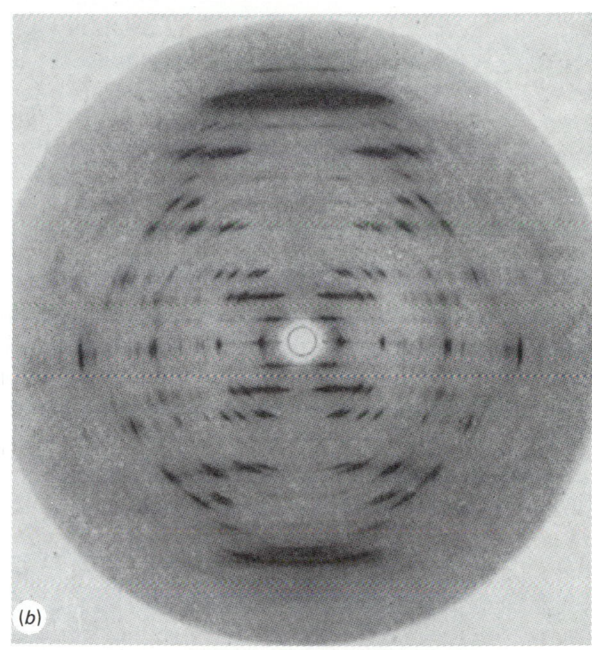

Figure 13–3 These x-ray diffraction photographs of suitably hydrated fibers of DNA show the so-called B configuration. *(a)* Pattern obtained using the sodium salt of DNA. *(b)* Pattern obtained using the lithium salt of DNA. This pattern permits a most thorough analysis of DNA. The diagonal pattern of spots (reflections) stretching from 11 o'clock to 5 o'clock and from 1 o'clock to 7 o'clock provides evidence for the helical structure of DNA. The elongated horizontal reflections at the top and bottom of the photographs provide evidence that the purine and pyrimidine bases are stacked 0.34 nm apart and are perpendicular to the axis of the DNA molecule. (Courtesy of Biophysics Research Unit, Medical Research Council, King's College, London.)

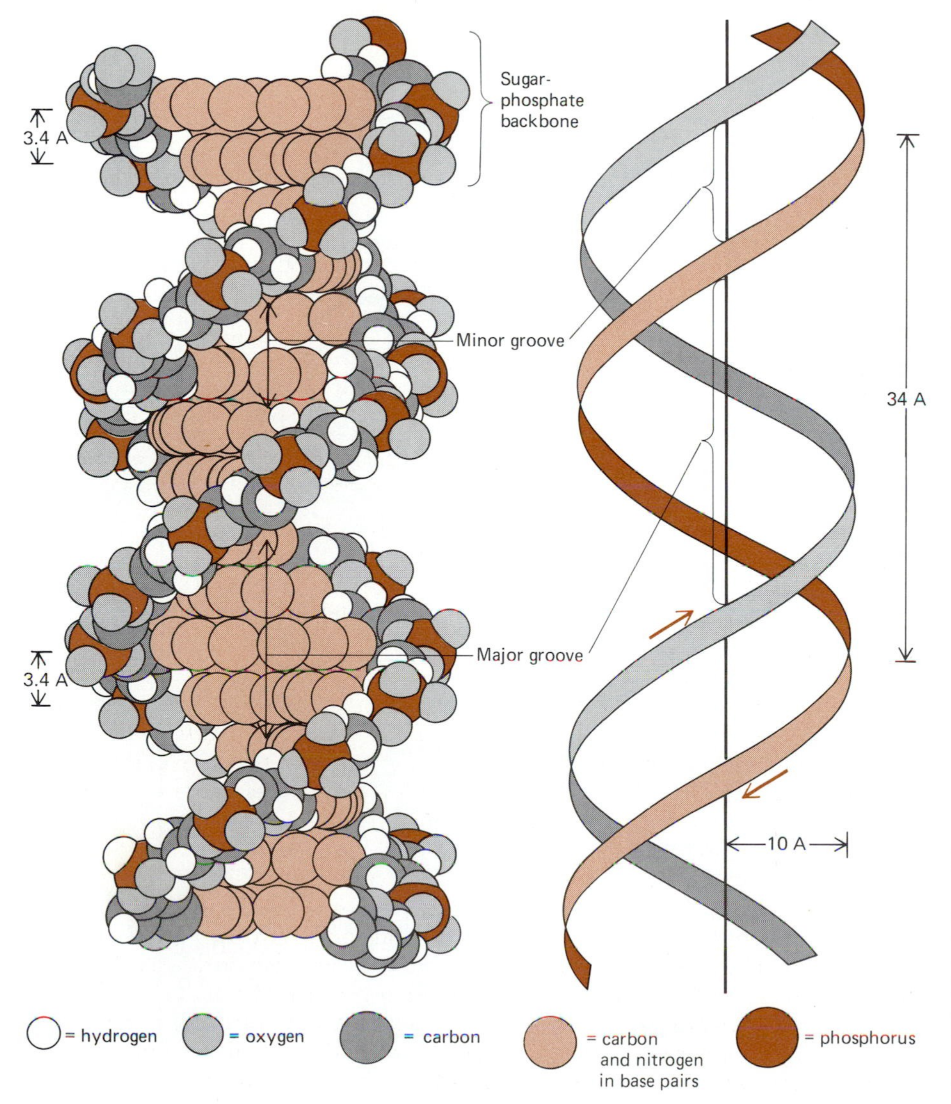

Figure 13–4 Molecular models of deoxyribonucleic acid (B form). On the left is a space-filling model of the DNA double helix. On the right is a diagrammatic model of the DNA double helix with certain of its dimensions shown in angstrom units. The ribbons represent the sugar-phosphate backbone of each strand, while the arrows indicate that the two strands extend in opposite directions. An angstrom unit is equal to an nm.

molecule, the angles between the bonds of a given atom, and the sizes of the atoms. Using this information, Watson and Crick began to build scale models of the component parts of DNA and then fit them together to agree with the experimental data.

It seemed clear to Watson and Crick that the 0.34-nm periodicity found by Franklin and Wilkins corresponded to the distance *between successive nucleotides* in the DNA chain. Furthermore, it was a reasonable guess that the 2.0-nm periodicity corresponded to the *width* of the chain. To explain the 3.4-nm periodicity, they postulated that the chain was coiled in a **helix,** or spiral. This 3.4-nm periodicity corresponded to the distance between successive turns of the helix. A chain can be wound around a cylinder either loosely or tightly; this corresponds to the steepness of the pitch of the coils in the helix. Since 3.4 is just 10 times the 0.34-nm distance between the successive nucleotides, it was clear that each full turn of the helix contained 10 nucleotides. From these facts Watson and Crick could calculate the density of a chain of nucleotides coiled in a helix 2 nm wide, with turns that were 3.4 nm long. Such a chain would have a density only *half* as great as the known density of DNA. To resolve this seeming contradiction, they postulated that there were *two* chains—a *double helix* of nucleotides—that made up a DNA molecule.

The next problem in the investigation of DNA structure was to determine the spatial relationships between the two chains that make up the double helix. Having tried a number of arrangements with their scale model, Watson and Crick found that the model that best fit all data was one in which the two nucleotide helices were wound together in *opposite directions* (Figs. 13–4 and 13–5), with the sugar phosphate chains on the outside of the helix and with the purines and pyrimidines on the inside, held together by hydrogen bonds between bases on the opposite chains. Though individually very weak, *collectively* hydrogen bonds are strong

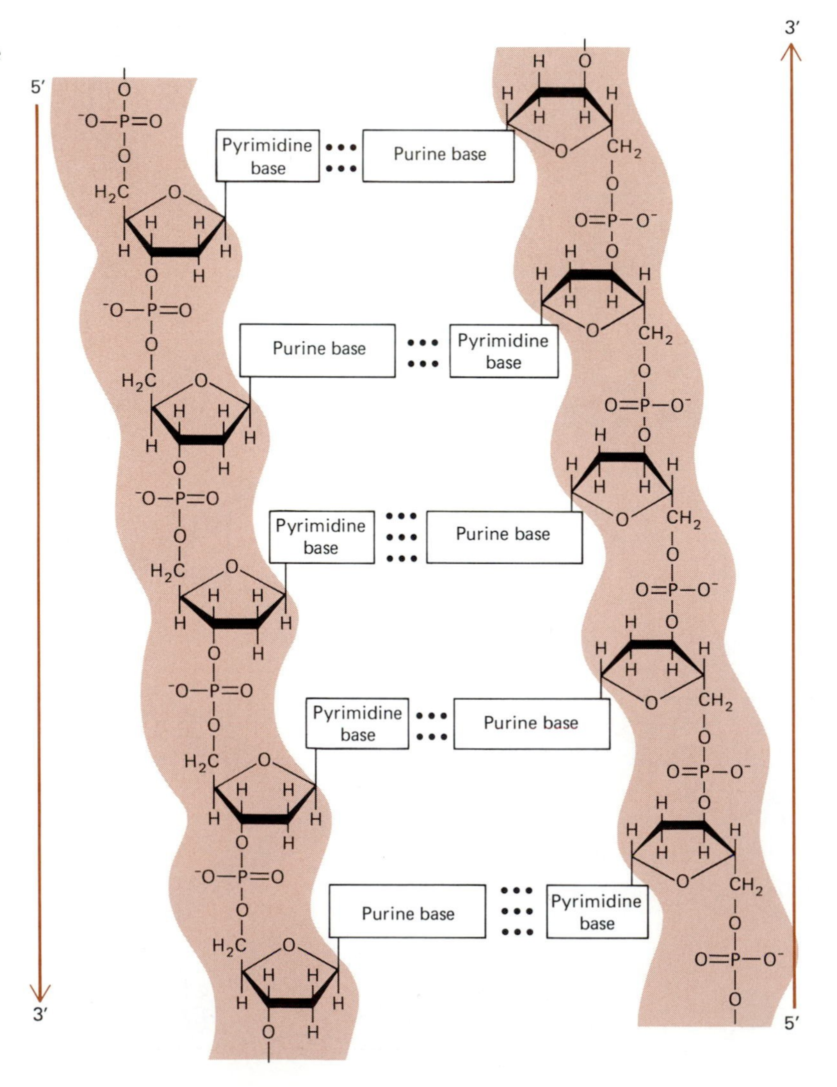

Figure 13–5 Schematic diagram of a portion of a DNA molecule showing the two polynucleotide chains joined by hydrogen bonds. The chains are not flat, as represented here, but are coiled around each other in helices (see Fig. 13–4). The two strands extend in opposite directions, as indicated by the arrows.

enough to hold the chains together and to maintain the helix. A double helix can be visualized by imagining the form that would be obtained by taking a ladder and twisting it into a helical shape, keeping the rungs of the ladder parallel. The sugar and phosphate molecules of the nucleotide chains make up the sides of the ladder, and the rungs are formed by the nitrogenous bases joined by hydrogen bonds.

Base Pairing and Hydrogen Bonds

Look again at Figure 13–2. Notice that the **pyrimidines** thymine and cytosine each contain *one* ring of atoms. The **purines** adenine and guanine, however, each contain two rings. Further study of the possible models made it clear to Watson and Crick that each crossrung must contain one purine and one pyrimidine, for the space available with the 2.0-nm periodicity would just accommodate one purine and one pyrimidine. Two purines (1.2 nm + 1.2 nm) would be too large for the space available; two pyrimidines (0.8 nm + 0.8 nm) would not come close enough together to form hydrogen bonds. Further examination of the detailed model showed that, although a combination of the purine adenine and the pyrimidine cytosine was the proper size to fit as a rung on the ladder, they could not be arranged in such a way that would form the correct hydrogen bonds. A similar consideration ruled out the pairing of the purine guanine and the pyrimidine thymine. However, adenine and thymine are joined by two hydrogen bonds, and guanine and cytosine by three hydrogen bonds.

The nature of the hydrogen bonds requires that adenine pair with thymine and guanine with cytosine. Two hydrogen bonds can form between adenine and thymine, and three hydrogen bonds between guanine and cytosine (Fig. 13–6). This concept of **specific base pairing** provided an explanation for Chargaff's finding that the molar amounts of adenine and thymine in any DNA molecule are always equal, and that the molar amounts of guanine and cytosine are always equal. The specificity of the kind of hydrogen bond that can be formed normally

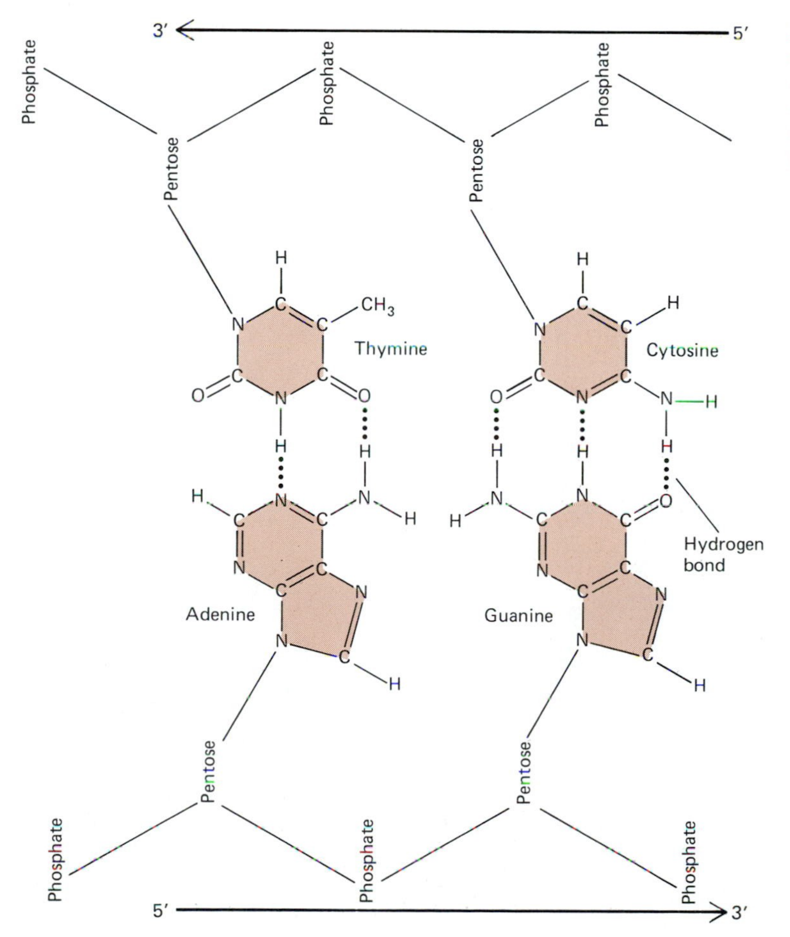

Figure 13–6 Diagram of the hydrogen bonding between the base pairs thymine and adenine *(left)* and cytosine and guanine *(right)* in DNA. The AT pair has two hydrogen bonds; the GC pair has three.

assures that for every adenine in one chain, there will be a thymine in the other chain. Similarly, for every guanine in the first chain, there will be a cytosine in the second chain. Thus the two chains are **complementary** to each other; that is, *the sequence of nucleotides in one chain dictates a complementary sequence of nucleotides in the other.* Moreover, the two strands extend in opposite directions and have their terminal phosphate groups at opposite ends of the double helix; because of this they are said to be **antiparallel.**

Forms of DNA

Several alternative forms of DNA have been discovered since Watson and Crick proposed their model of what is now known as **B-DNA.** Most of the alternative forms involve relatively minor variations in the pitch and tilt of the helix, but recently a radically different form has come to light, **Z-DNA** (Fig. 13–7). This form of DNA forms a left-handed helix rather than the "standard" right-handed Watson–Crick variety. Moreover, its groove zigzags, which is one of the reasons it is called Z-DNA. The significance of this form of DNA is beginning to be appreciated. Z-DNA is, for example, common in the light bands of *Drosophila* chromosomes. It is entirely possible that Z-DNA is involved in the regulation of eukaryotic genes by blocking transcription of the chromosomal regions in which it occurs (see Focus on Chromosomes and Genetic Expression).

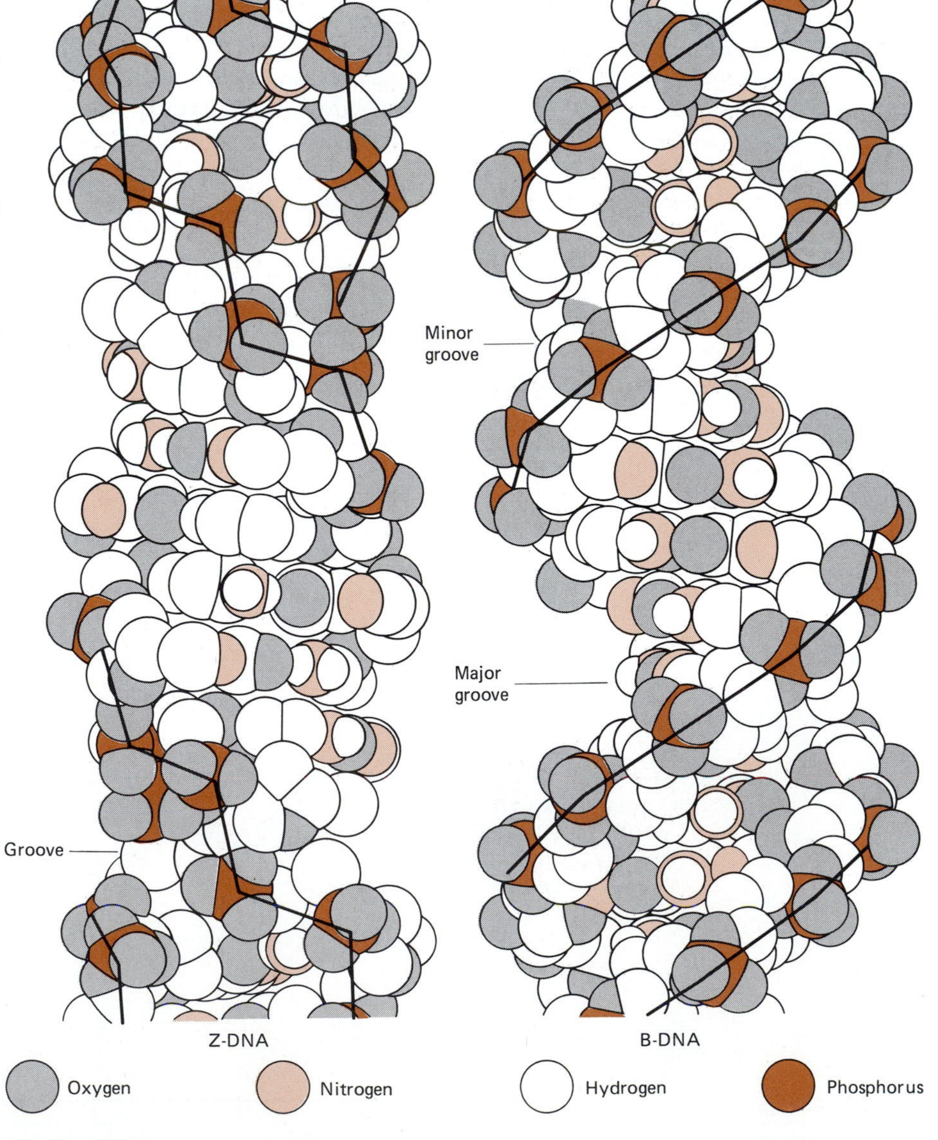

Figure 13–7 Z-DNA compared to B-DNA. Z-DNA forms a left-handed helix and its major groove zigzags. Because of the zigzag and because it is so different from all previously known forms of DNA, Z-DNA was named from the other end of the alphabet! Z-DNA is genetically inactive, and is most likely to form only when certain sequences of bases are present in the DNA strand. It may function as part of genetic control mechanisms, or as a spacer affecting the function of moveable genes. (From Quigley, G. J.: Left-hand and Right-handed helical DNA. *Science* 211, cover, January 1981. Copyright 1981 by the American Association for the Advancement of Science.)

FOCUS ON

Chromosomes and Genetic Expression

What do chromosomes do during interphase? Perhaps you have gained the impression that they exist only during actual cell division, but though they are greatly elongated, interphase chromosomes do exist, and indeed that is when they express the genetic information encoded in their DNA.

The best-studied example of this involves the chromosome puffs of salivary gland chromosomes from dipteran (fly) larvae. It might seem strange to single out dipterans for special study, but the chromosomes of their larvae have long been known to be among the largest in the entire world of life and therefore most easily visualized with the light microscope. Their great size results from a repeated process of DNA replication without accompanying cell division. It is easy to see their pattern of bands, to correlate various genetic defects with changes in that pattern, and perhaps most important, to notice sequential changes in gene activity that accompany different stages in development. The steroid hormone **ecdysone,** normally associated with molting in these insects, causes characteristic unraveling of certain specialized regions of the giant chromosome—the **chromosome puffs.** *Production of mRNA occurs at these unraveled regions.* As we will see in Chapter 14, mRNA is an intermediate substance that ultimately produces the proteins governing all life processes in cells.

The way in which chromosomes are packed into the interphase cell's nucleus may also be important. As described in Chapter 15, Barbara McClintock has shown that gene function in eukaryotes is often related to the position of the gene on the chromosome. It is believed that this position-determined function results for the most part from associations with other, regulatory genes on the *same* chromosome. However, that view is probably a consequence of the fact that we have so far been able to study inactive chromosomes only during cell division. The spatial relationships of chromosomes with one another are hard to see when they are all squeezed together in the interphase nucleus. A region of one chromosome might very well lie cheek-by-jowl with another region of a different chromosome when they are together in the nucleus; yet such a relationship might be entirely unsuspected on the basis of the way they look in a dividing cell.

It is striking that the functional significance of banding patterns on chromosomes remains unknown. Yet the positions of these bands are very constant among members of a species or even of related species, showing that such patterning must be significant. It is entirely possible that one of the functions of the light bands (rich, as we have seen, in Z-DNA) is to maintain the proper spacing among functional genes so that they may interact spatially with one another in the interphase nucleus.

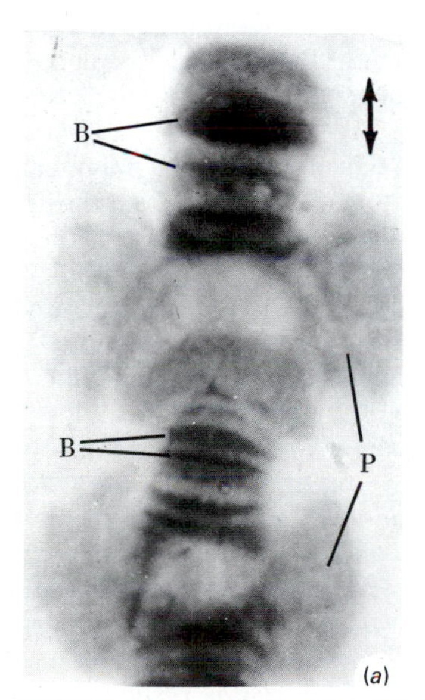
B
B
P
(a)

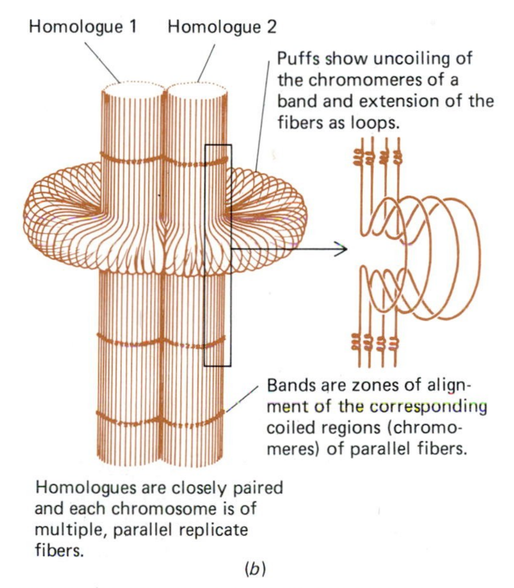
Homologue 1
Homologue 2
Puffs show uncoiling of the chromomeres of a band and extension of the fibers as loops.
Bands are zones of alignment of the corresponding coiled regions (chromomeres) of parallel fibers.
Homologues are closely paired and each chromosome is of multiple, parallel replicate fibers.
(b)

Salivary gland chromosomes. *(a)* A portion of a chromosome from the salivary gland of the fly Chironomus. The direction of the long axis of the chromosome is indicated by the double-headed arrow. The cross-banding (B) characteristic of these chromosomes is apparent as are two puffs (P) where the chromosome bands have been altered as indicated in part B of the diagram (×1900). *(b)* Diagram illustrating that homologous chromosomes are paired, gene for gene, and that each chromosome consists of many parallel longitudinally arranged fibers. (Courtesy of U. Clever.)

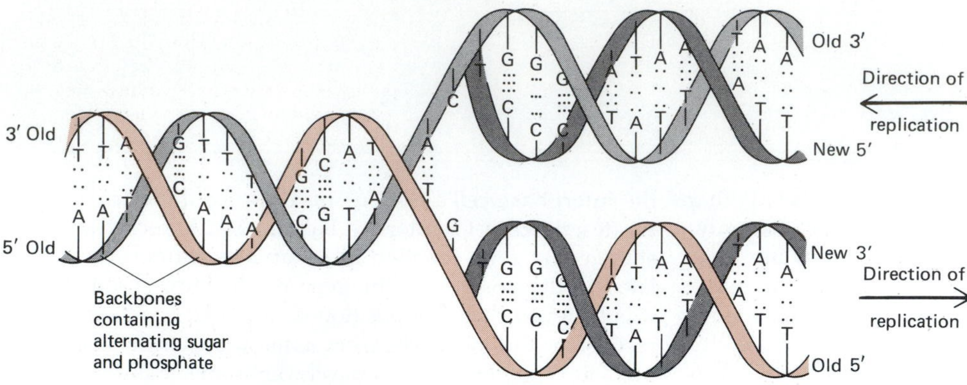

Figure 13–8 Diagrammatic scheme for the replication of duplex DNA based on the complementarity of AT and GC base pairs. Modeled on the original scheme presented by Watson and Crick.

DNA Replication

The most distinctive properties of the genetic material are that it carries information and undergoes replication. The Watson–Crick model explains how DNA molecules can carry out these two functions. When a DNA molecule undergoes replication, the two chains separate, and each one brings about the formation of a new chain that is complementary to it; thus two new chains are established (Fig. 13–8). The rungs of the ladder are "unzipped," and each half-ladder then finds the parts to replace the missing half. The result is two ladders, each identical with the original one.

The nucleotides in the new chain are assembled in a specific order, because each purine or pyrimidine in the original chain forms hydrogen bonds only with the complementary pyrimidine or purine nucleoside triphosphate from the surrounding medium and thus ultimately arranges them in a complementary order. Sugar-to-phosphate bonds formed by the reaction catalyzed by a specific enzyme system join the adjacent nucleotides, and a new polynucleotide chain results (Fig. 13–9). The new and the original chains then wind around each other, and two new DNA molecules are formed. Each chain, in other words, serves as a template or a mold against which a new partner chain is synthesized. The end result is two complete double-chain molecules, each identical with the original double-chain molecule and each containing one of the original chains. This method of information copying is known as **semiconservative** replication (see Focus on Replication: A Semiconservative Process).

The enzyme that catalyzes this process, **DNA polymerase,** must be one of the most important enzymes on the earth, for DNA synthesis could not take place without it. It appears actually to be a complex of several enzymes. It has a twofold function: Not only must it catalyze the polymerization (linking) of multiple nucleotides, but it must do this in a particular way. Ordinarily, an enzyme causes a thermodynamically possible reaction to proceed more rapidly than would occur in its absence, but does not otherwise influence the reaction. Since substrate specificity is usually very narrow, the nature of the products will almost always be the same at all times. The role of the DNA polymerase system in the polymerization of nucleotides is much more complex. At one time the product will have a freshly added terminal *guanine* nucleotide, at another time it will have a terminal *cytosine* nucleotide, and at still another time it will have a terminal *thymine* or *adenine* nucleotide. Thus, the DNA polymerase enzyme actually has *four* specific and mutually exclusive substrates, and it switches its preferences among them in accordance with whichever one of the template bases that it is currently encountering.

For example, if the DNA polymerase is resting on a preexisting strand of DNA that has, at that point, a projecting guanine base, the DNA polymerase will behave as a cytosine nucleotide polymerase and will add only a cytosine nucleotide to the forming strand at that point, and none other. It will then advance to the next base of the preexisting strand, which could be, say, thymine. At that point the DNA polymerase enzyme will behave as an adenine nucleotide polymerase, adding an adenine nucleotide to the previously added cytosine nucleotide. In this way the complex is directed by the preexisting base sequence of one strand to add each base on the lengthening new strand. The new strand will be complementary at all

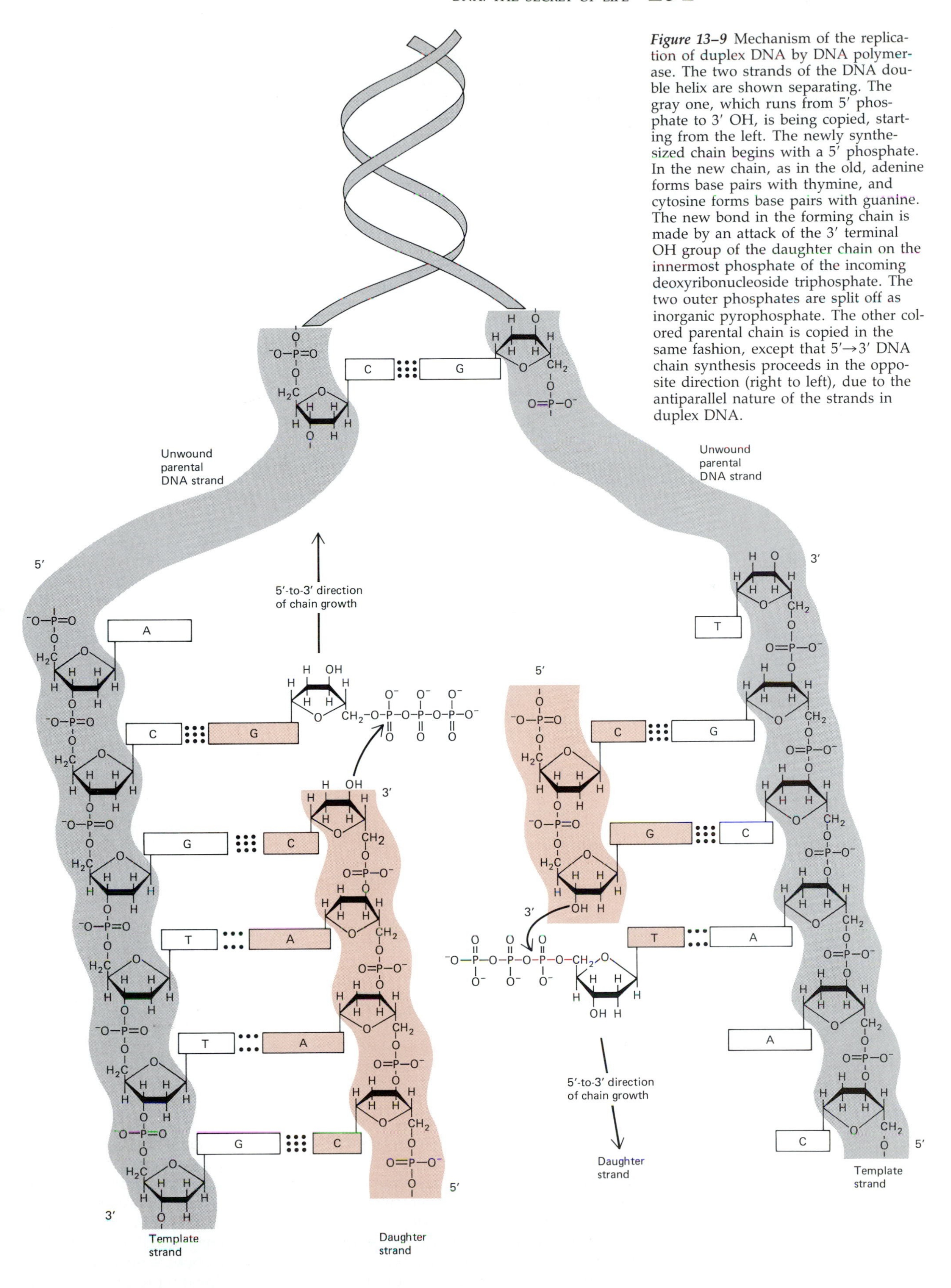

Figure 13–9 Mechanism of the replication of duplex DNA by DNA polymerase. The two strands of the DNA double helix are shown separating. The gray one, which runs from 5′ phosphate to 3′ OH, is being copied, starting from the left. The newly synthesized chain begins with a 5′ phosphate. In the new chain, as in the old, adenine forms base pairs with thymine, and cytosine forms base pairs with guanine. The new bond in the forming chain is made by an attack of the 3′ terminal OH group of the daughter chain on the innermost phosphate of the incoming deoxyribonucleoside triphosphate. The two outer phosphates are split off as inorganic pyrophosphate. The other colored parental chain is copied in the same fashion, except that 5′→3′ DNA chain synthesis proceeds in the opposite direction (right to left), due to the antiparallel nature of the strands in duplex DNA.

FOCUS ON

Replication: A Semiconservative Process

In semiconservative replication, the two original strands of DNA are retained or conserved in the product, one in each of the two daughter helices as shown. If replication were *conservative,* however, the double-stranded DNA would simply make a new double-stranded DNA helix.

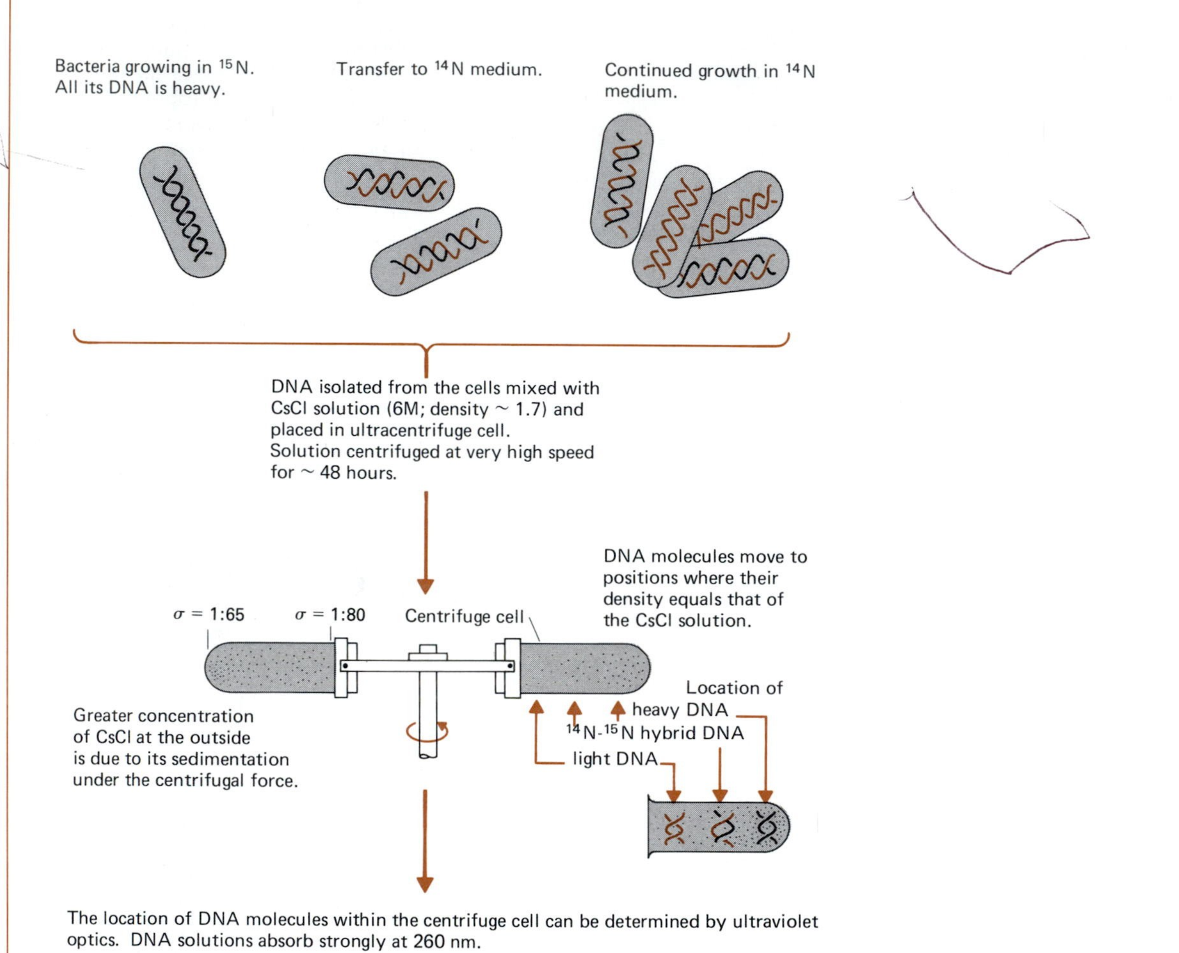
Bacteria growing in ^{15}N. All its DNA is heavy.
Transfer to ^{14}N medium.
Continued growth in ^{14}N medium.
DNA isolated from the cells mixed with CsCl solution (6M; density ~ 1.7) and placed in ultracentrifuge cell.
Solution centrifuged at very high speed for ~ 48 hours.
σ = 1:65
σ = 1:80
Centrifuge cell
DNA molecules move to positions where their density equals that of the CsCl solution.
Location of
heavy DNA
^{14}N-^{15}N hybrid DNA
light DNA
Greater concentration of CsCl at the outside is due to its sedimentation under the centrifugal force.
The location of DNA molecules within the centrifuge cell can be determined by ultraviolet optics. DNA solutions absorb strongly at 260 nm.
heavy DNA
Before transfer to ^{14}N.
^{14}N-^{15}N hybrid DNA
One cell generation after transfer to ^{14}N.
light DNA
^{14}N-^{15}N hybrid DNA
Two cell generations after transfer to ^{14}N.

Diagram of the experiment of Meselson and Stahl, which demonstrated that DNA is replicated by a semiconservative mechanism: The two original strands of DNA are retained in the newly formed daughter molecules, one strand in each daughter double helix.

points to the base sequence of its template. Although in general the "correct" complementary relationship might seem to be assured by the regularity of the DNA helix, it is really the other way around: The regularity of the DNA helix results from the classic Watson–Crick base-pairing relationship, and that in turn is assured by the DNA polymerase complex.

Consider all the things this amazing complex of enzymes must do! It must first recognize a specific base on a preexisting strand of DNA. It must then bring that nucleoside triphosphate to the 3′ end of the new, elongating strand of DNA; it

To distinguish between these possibilities it is necessary to be able to distinguish between new and old strands of DNA. This difficult feat was ingeniously accomplished in a classic experiment, by M. Meselson and F. Stahl, that employed heavy nitrogen, ^{15}N. Bacteria grown for several generations in a medium containing heavy nitrogen had DNA (and RNA and protein) that was labeled with ^{15}N. When a sample of the DNA was isolated and centrifuged in a gradient density tube, the DNA accumulated at the level in the tube that reflected its increased density caused by the presence of the heavy nitrogen atoms.

The bacteria were then transferred from the ^{15}N medium to a medium containing ordinary nitrogen, ^{14}N, and were allowed to divide once. When the DNA from this new generation of bacteria was isolated and centrifuged, all the DNA was lighter, with the density expected if it had just half as many ^{15}N atoms as the DNA of the parental generation. The result was as predicted on the basis of the Watson–Crick theory: replication was semiconservative, with one strand of the double-stranded DNA labeled with ^{15}N and the other containing only ^{14}N.

When these bacteria were allowed to divide a second time in the ^{14}N medium, each double helix of DNA in the progeny again received one parental strand and one new strand containing only ^{14}N. Some double-stranded DNA containing only ^{14}N was formed and, when centrifuged, proved to be light-density DNA. Strands containing ^{15}N made complementary strands containing ^{14}N, which after being centrifuged were sedimented; their density was characteristic of the half-^{15}N, half-^{14}N double-stranded state. Thus the original parental strands of DNA are not degraded during the replication process, but are conserved and passed to the next generation of cells. Each strand of the parental double helix is conserved in a *different* daughter cell; hence the process is termed *semi*conservative.

Since the two strands of replicating DNA must separate to replicate, they necessarily form a kind of Y-shaped structure temporarily while doing so. This is the **replication fork.** As shown in the figure, replication forks may occur in several places simultaneously on a DNA strand, but in prokaryotes they usually begin at one place and progressively separate the circle of DNA into two.

All of the prokaryote "chromosomal" DNA is replicated once the process has begun, so the bacterial "chromosome" is said to constitute a single unit of replication, or **replicon.** Accessory chromosomes called plasmids may or may not replicate at the same time, but since there is no physical connection between them and the "chromosome," the replication that begins in the "chromosome" does not simply continue into the plasmids, or vice versa. Thus the plasmids are independent replicons. Eukaryote DNA consists of many replicons within each chromosome. The significance of this finding is unknown, but it may well be connected with the more complex and sophisticated systems of genetic control, themselves as yet poorly understood, that eukaryotes possess (Chapter 15).

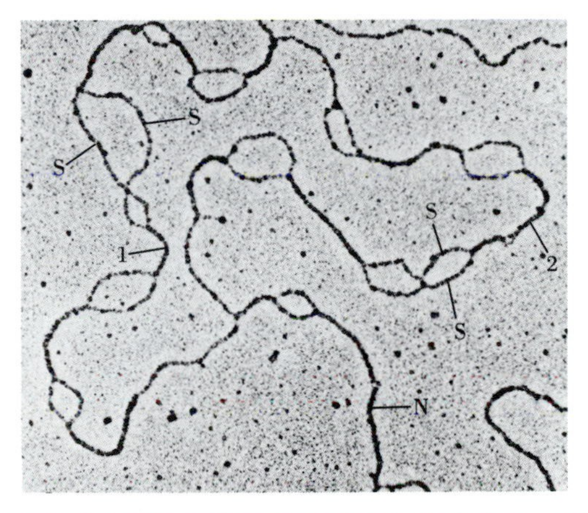

Because the eukaryotic genome is extremely large, many segments of its DNA must be replicated simultaneously for cell division to occur in a reasonable length of time. 1 and 2 are the two daughter double helices. N is a portion of the double helix not yet replicated. The double helix has been separated to show its constituent strands (S) where indicated to illustrate its double structure. (Courtesy of H. J. Kriegstein and D. S. Hogness.)

must bring about the necessary reaction to add this to the new strand; and it must release the reaction products. The enzyme next moves one base further along the templating preexisting strand so that it can repeat the process all over again. And it continues to do this at a rate of about 200 bases per second for hundreds or thousands of bases every time the cell divides, and occasionally at other times as well. Moreover, one form of the system is able to perform a certain amount of "proofreading," in which incorrect bases are removed from a growing strand of DNA and are replaced by the proper bases.

The synthesis of DNA occurs in the nuclei of cells of higher organisms only during the S phase of the cell cycle (Chapter 4), when chromosomes are in their extended form and are not readily visible. Some sort of biological signal must initiate DNA synthesis at this time and turn it off at other times. The enzyme (the DNA polymerase) and the substrates (the four deoxyribonucleoside triphosphates—dATP, dGTP, dCTP, and dTTP) are present in the cell all the time. The explanation currently believed to be most likely is that some sort of change in the DNA template initiates the synthesis of DNA at the appropriate time in the cell cycle and then turns it off.

Strong experimental evidence supports the concept that each eukaryotic chromosome is a single DNA molecule. For example, the DNA molecules of the fruit fly *Drosphila* are as long as 2.1 nm. But how can a cell replicate such an enormous molecule in the allotted time during the cell cycle? The answer is that replication does not simply begin at one end of the molecule and proceed to the other. Instead, replication begins at many sites along the chromosome (some 2 μm apart on the average in a rapidly dividing fertilized egg) and proceeds in *both* directions from the origin at about the same rate of 1 μm/min. Yet since replication always proceeds in a $5' \rightarrow 3'$ direction, some areas of replication will employ one of the strands as their template, but those proceeding in the opposite direction must employ the *other* strand as *their* template.

Cell division is not the only event that requires the replication of DNA. In the discussion of mutation (Chapter 14), we will see that a variety of accidents can change the base sequence of DNA, leading to random changes in the genetic material, and even to diseases such as sickle cell anemia and cancer. These genetic pathologies would be far more common than they are if DNA had no way of being repaired. In fact, the more complex forms of life such as ourselves might not exist were it not for DNA repair. We will examine the details of this process in the next two chapters.

DNA Information Density

When we consider the endless detail of anatomy—for example, the plethora of foramina, crevices, pores, fissures, and processes in a single bone—it is hard to see how the alleged 600 volumes of genetic information in the zygote's nucleus might be enough to specify all of the single features of an organism in complete detail. The resolution of this problem is a less formidable task than it might seem when we remember that all those anatomical details are the results of a developmental process. That process might be quite simple to set in motion. DNA is not so much like a draftsman who must draw every line as it is like a clerk who presses a button that sets some complex piece of machinery in motion. DNA also contains the plan for the machinery, but even so it is far simpler to make the machine and press the button than to specify every detail of the ultimate product.

The nucleus seems to have information storage space to spare, for not all the DNA contains genetic information. At least three classes of DNA can be distinguished: highly repetitive sequences (short sequences that may number more than 10^6); moderately repetitive sequences (longer sequences of nucleotides of which there are 5 to 100 copies in the nucleus); and unique sequences (sequences of which there are 5 or fewer copies in the entire genetic library of the individual (genome)). Whatever their function, the highly repetitive sequences do not appear to contain genes, but the moderately repetitive and unique sequences do. The genes for histones and the different kinds of RNA are present in the moderately repetitive sequences, while the genes for differentiated functions such as enzyme production are in the unique sequence class.

SUMMARY

I. In the Watson–Crick model of the DNA molecule, DNA consists of two very long helical chains of deoxyribonucleotides wound around a common central axis and extending in opposite directions. The deoxyribose phosphate backbone of each chain is on the outside of the double helix; the purines and pyrimidines are on the inside.

II. The two chains of the double helix are joined by hydrogen bonds between specific base pairs. Adenine (A) pairs with thymine (T), and guanine (G) pairs with cytosine (C), so that one strand of the double helix is the complement of the other.

III. The two strands of the double helix unwind and separate to permit the replication of DNA. Each strand acts as a template for the formation of a new complementary strand.
   A. The replication of DNA is semiconservative; each daughter molecule contains one strand from the parent molecule and one new molecule.
   B. Replication is a complex process catalyzed by several enzymes, primarily DNA polymerase. The substrates in the reaction are the four deoxyribonucleoside triphosphates.
   C. DNA polymerase functions only if the base on the nucleotide being added is complementary to the base in the template strand.
      1. DNA synthesis occurs in the cells of higher organisms principally during the S phase of the cell cycle.
      2. However, DNA is also synthesized to replace defective strands that are removed during the course of DNA repair.

POST-TEST

1. Each nucleotide chemically consists of a ______________, a ______________ ______________, and a ______________ ______________.

2. The synthesis of DNA requires as substrates ______________, ______________, ______________, and ______________.

3. DNA synthesis is catalyzed by the enzyme ______________.

4. The 0.34-nm periodicity of DNA represents the distance between adjacent ______________.

5. The 3.4-nm periodicity of DNA represents the distance between successive ______________ of the helix.

6. The clue that DNA is a double helix rather than a single helix came from its ______________.

7. The chief feature of the Watson–Crick model of DNA, which accounts for its ability to be replicated and to transfer genetic information, is the concept of specific ______________ ______________.

8. The two chains of the DNA double helix are ______________ and antiparallel.

9. In the DNA from all known organisms, the amount of ______________ equals the amount of ______________, and the amount of ______________ equals the amount of ______________.

10. Griffith's experiments with mice and pneumococci demonstrated that some sort of genetic message passed from ______________ ______________ bacteria to ______________ ______________ ones.

11. Avery showed that this genetic message was contained in ______________.

12. Studies by Franklin and Wilkins using ______________ ______________ provided the basic information needed by Watson and Crick to construct their model of DNA.

13. In the replication of DNA, the two chains separate, and each brings about the synthesis of a new chain that is ______________ to the old one.

14. The synthesis of DNA occurs in the nuclei of cells of higher organisms only during the ______________ ______________ of the cell cycle.

15. Each eukaryotic chromosome is a ______________ DNA molecule.

16. The process of replicating DNA begins at origin sites and proceeds in ______________ ______________ from the origin.

17. Viruses that infect bacteria are called ______________.

18. The process of DNA replication is said to be ______________ because the two original strands of DNA are retained in the products.

REVIEW QUESTIONS

1. Discuss the evidence that DNA is an integral part of the gene and transfers genetic information from one generation to the next.
2. What is the nature of the "transforming agent" studied by Avery and his colleagues? In what way may this phenomenon be of importance in our understanding of the chemical basis of inheritance?
3. What evidence regarding the nature of the genetic material has been obtained from experiments with bacteriophages?
4. Describe the chemical composition and structure of DNA.
5. How does the Watson–Crick model of DNA account for the observed properties of the molecule? How does this model explain the process of gene replication?
6. What are the two prime functions of DNA? How are these carried out by the DNA molecule?
7. Describe the mechanism by which the enormously long molecule of DNA in eukaryotic chromosomes is replicated rapidly.

14

The Bacterial Connection: Genetic Information and Protein Synthesis

OUTLINE

I. An overview of gene expression
II. Transcription: the synthesis of RNA
  A. The initiation of transcription
  B. The processing of RNA
III. Translation
  A. Activation of amino acids
  B. The role of transfer RNA
  C. The structure of transfer RNA
  D. The messenger RNA template
  E. The ribosomes
    1. Ribosome function
    2. Polyribosomes
  F. Protein synthesis
IV. Gene–enzyme relations
V. Changes in genes: mutations
VI. Some reflections on the genetic code
  A. Some experimental evidence for a triplet code
  B. Codon–amino acid specificity

LEARNING OBJECTIVES

After you have read this chapter you should be able to:

1. Outline the mechanisms by which a gene is transcribed to yield mRNA and the mechanisms by which mRNA is translated to yield a specific protein chain.
2. Define the term *reverse transcription,* and indicate the importance of this process in genetics.
3. Discuss the role of tRNA and of the enzymes that activate amino acids and transfer them to tRNA in the protein-synthesizing process.
4. Diagram the interactions of mRNA codons and tRNA anticodons in protein synthesis.
5. Describe the structure of a ribosome and the functions of its various parts in protein synthesis.
6. Distinguish the essential features of initiation, elongation, and termination in protein synthesis.
7. Give the significance of the relation between genes and the enzymes for which they code.
8. List and describe the various kinds of mutations.
9. Describe the characteristics of the genetic code, and explain why the code is said to be degenerate or redundant.
10. Discuss the implications of the universality of the genetic code.
11. Give the evidence that DNA and the protein chain for which it provides the code are colinear.

In the preceding chapter we saw that genetic information can be stored in DNA. We also saw how this genetic information is replicated so that it can be passed on to descendant cells and to the organisms of which those cells may be a part. This information is comparable to a description of how magnetic tapes are made and duplicated; for a full understanding of how a tape system produces sound, however, we must know how the tape player works also. Thus this chapter addresses the question of how the information in the DNA is encoded, decoded, and expressed. In Chapter 15 we will see how the total mechanism is controlled. By far the greatest part of the information in these two chapters has been gleaned from studies of gene function in bacteria. Prokaryotes are far simpler and easier to understand than are eukaryotes—never more so than in their molecular biology.

An Overview of Gene Expression

The collagen of the skeleton, the hemoglobin of the blood, and the actin and myosin that motorize muscles are all proteins. Even the nonprotein fats, sugars, and phospholipids are manufactured by intricately interacting enzymes which *are* proteins. Thus proteins are the strategic materials upon which all life functions of an organism converge—not only those of adult organisms but also those that govern the development of organisms. DNA governs the cell by governing the proteins. However, DNA has no known way, by itself, of influencing proteins. It cannot assemble amino acids into patterns, cannot establish peptide bonds among them, and cannot manufacture proteins any more than a magnetic tape can produce a symphony. Just as a tape requires the circuitry and mechanism of a player, DNA requires an intermediate readout mechanism. The several kinds of RNA constitute that mechanism.

Ribonucleic acid, RNA, is composed of nucleotides similar to those found in DNA. There are three basic kinds of RNA: **messenger RNA (mRNA), transfer RNA (tRNA),** and **ribosomal RNA (rRNA).** All are produced enzymatically by a process somewhat similar to DNA replication (Fig. 14–1). Thus each RNA molecule is complementary to the portion of the DNA that has supplied the information for its synthesis.

All three varieties of RNA play specific and different roles in protein production: rRNA and tRNA are parts of the machinery that synthesizes protein. The information that programs the machinery is carried by mRNA. Specific kinds of tRNA molecules are enzymatically attached to specific amino acids. Each species of

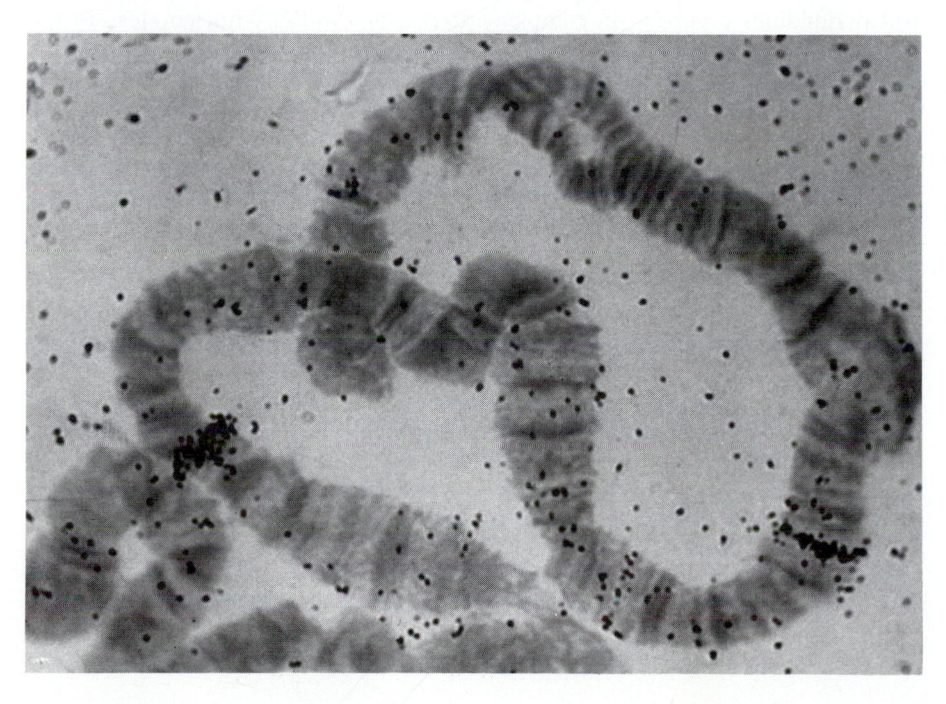

Figure 14–1 Production of RNA (transcription) along specific portions (genes) of DNA in a chromosome of the fruit fly *Drosophila melanogaster*. The dark particles are radioactively labeled tRNA. Note the two regions along the DNA where the particles are concentrated, indicating active tRNA synthesis. (Courtesy of T. A. Grigliatti.)

tRNA associates with only one species of amino acid. There are 20 different amino acids, but considerably more than 20 kinds of tRNA. Thus, although each kind of tRNA attaches to just one kind of amino acid, there are often several kinds of tRNA that can attach to one of the 20 amino acids.

The ribosome is composed of two different subunits, each of which contains both protein and rRNA. When the ribosome is inactive, the ribosomal subunits are separate. The active ribosome consists of both together; as they function, however, the subunits separate just enough to permit a strand of mRNA to pass between them. As it goes by, the strand of mRNA progressively binds to the smaller subunit. As random molecular motion brings tRNA–amino acid complexes close to the ribosome, these complexes are engulfed along with the moving strand of mRNA and appropriate complexes are selected by enzymes involved in the process. The ribosome forms peptide bonds between adjacent amino acids and releases the tRNA molecules that have carried them there. When the elongating protein is complete, the ribosome releases it and dissociates into its two constituent subunits once more.

None of this could occur in any organized or meaningful way without the genetic code that directs it. Yet the code is composed of only four characters, the four bases in DNA. How this language could govern the assembly of 20 amino acids into the vast variety of proteins that the cell contains is hard to imagine. But if we think of the DNA bases as serving as letters in a four-character alphabet, they are able to form 64 different three-letter words. This is more than enough to direct construction involving 20 amino acids. The meaningful unit of the genetic language is therefore the three-base "word," or **triplet.** The meaning of each of these triplets has been discovered by patient laboratory research.

There is an equivalent of the triplet at every level of gene function. A meaningful trio of DNA bases is known simply as a triplet, but the corresponding and complementary trio of mRNA bases is a **codon.** A group of three bases on a tRNA molecule that is complementary to the codon is known as the **anticodon.** The sequence of nucleotide triplets in DNA determines the sequence of codons in the mRNA that it produces. Various types of tRNA–amino acid complex are selected as determined by the specific mRNA-binding properties of the anticodon that each possesses. Since each species of tRNA carries only one kind of amino acid, the amino acid sequence in a protein is determined ultimately by the triplets of the original coding strand of DNA.

Transcription: The Synthesis of RNA

RNA differs from DNA in that it contains ribose instead of deoxyribose, and uracil instead of thymine (Fig. 14–2). Also, unlike DNA nucleotides, which are purines and pyrimidines occurring in complementary ratios, mRNA nucleotides are not present in complementary ratios or indeed in any other fixed proportions. This indicates that RNA generally is not a double helix like DNA but rather is single-stranded. The molecules of mRNA are unbranched and contain the four kinds of ribonucleotides, adenine (A), guanine (G), cytosine (C), and uracil (U), linked by 3′,5′ phosphodiester bonds. The synthesis of RNA is called **transcription,** for in this process information is transcribed from one kind of storage code (DNA nucleotides) into another (RNA nucleotides); the nucleotide triplet language in which it is conveyed, however, is not changed in the process.

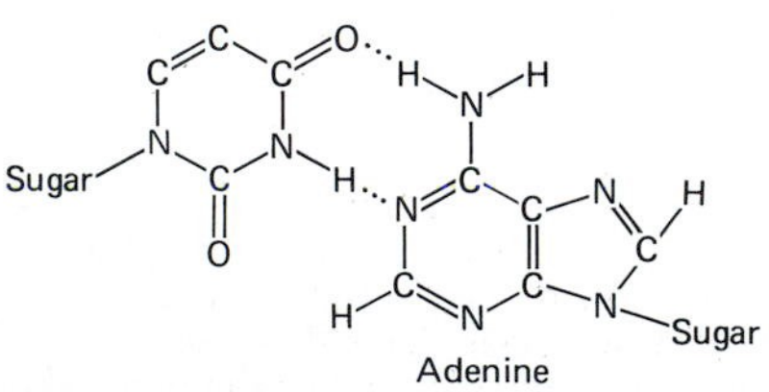

RNA is synthesized by **DNA-dependent RNA polymerases,** enzymes present in all cells. The best-understood of these occur in bacteria. These enzymes require DNA as the template and use as substrates the triphosphates of the four ribonucleosides commonly found in RNA: ATP, GTP, CTP, and UTP (Fig. 14–3). Bacterial polymerase is a very large complex enzyme composed of four subunits, each with specific functions in the synthesis of RNA. (The many kinds of tRNA and the several kinds of rRNA are also produced in the nucleus by DNA-dependent RNA-synthesizing systems; they are transcribed from complementary deoxynucleotide sequences in DNA. These genes render inadequate the definition of a gene as "the information necessary to produce a protein.") In contrast to bacterial cells, eukaryotic cells have three forms of RNA polymerase. These forms have evolved to divide the labor of transcription. Polymerase I transcribes rRNA genes, whereas polymerase II transcribes protein-coding genes. Polymerase III transcribes tRNA and other small nuclear RNA genes.

Uracil

Sugar

Adenine

Sugar

Figure 14–2 In RNA, uracil is incorporated in place of thymine. Like thymine, it binds with adenine by means of hydrogen bonds.

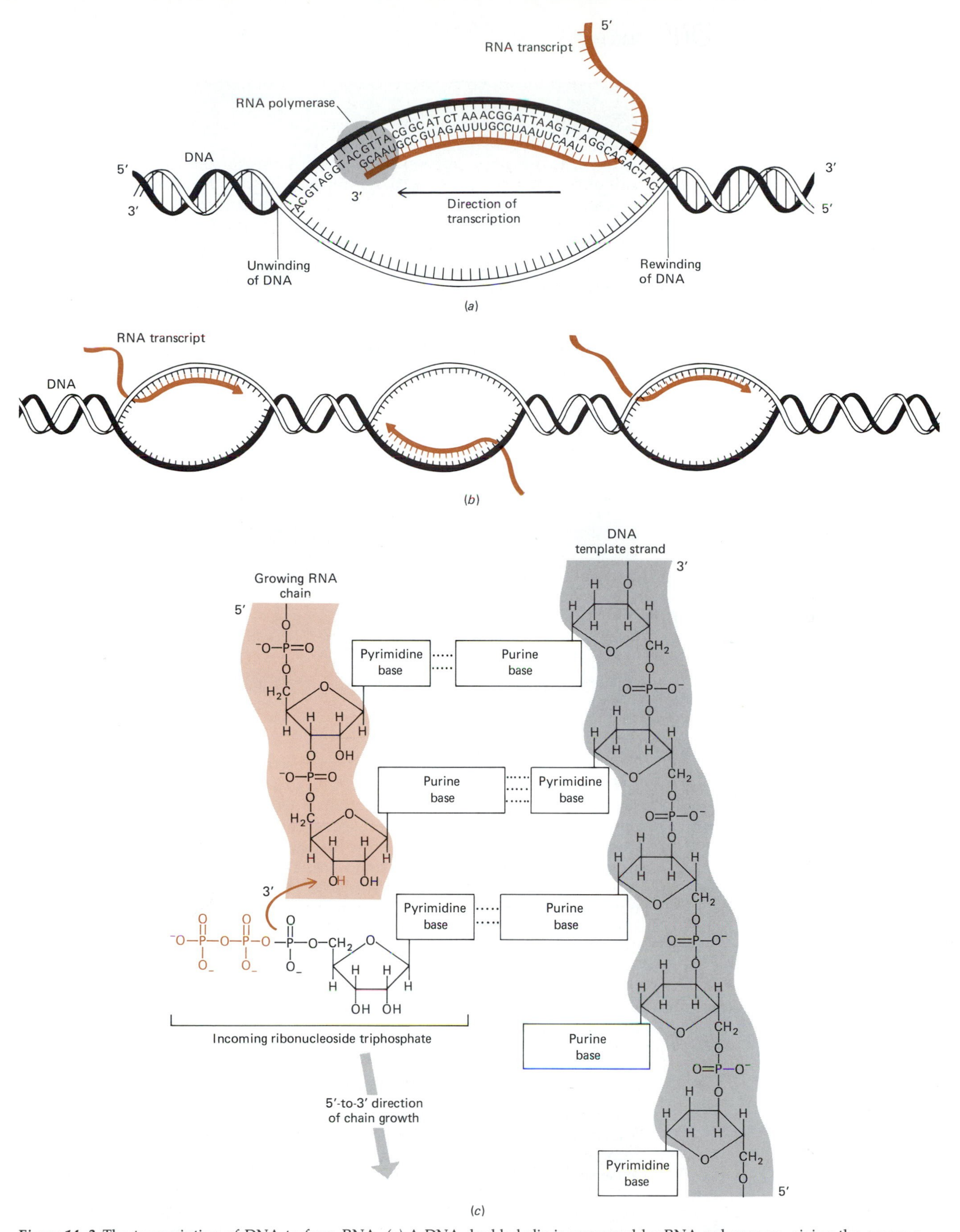

Figure 14–3 The transcription of DNA to form RNA. *(a)* A DNA double helix is unwound by RNA polymerase, giving the enzyme access to the nucleotide sequence. The DNA is then rewound behind the moving transcription complex. Initiation and termination sites, encoded in specific DNA sequences, determine where transcription starts and stops. RNA synthesis depends on base-pairing rules similar to those for DNA synthesis: adenine pairs with uracil, cytosine pairs with guanine. *(b)* Only one of the two strands is transcribed for a given gene, but the opposite strand may be transcribed for a neighboring gene. *(c)* A diagram of the reaction catalyzed by RNA polymerase. The exposed DNA strand on the right is being copied. Each incoming ribonucleoside triphosphate is selected for its ability to base-pair with the DNA template; a ribonucleoside monophosphate is then added to the 3′ OH end of the growing RNA chain.

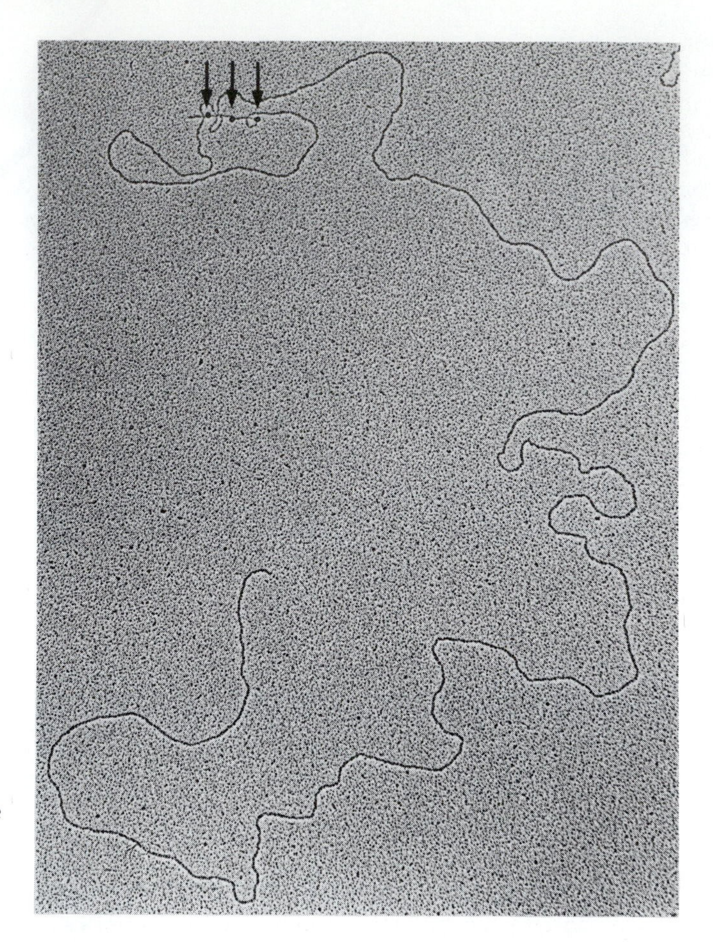

Figure 14–4 RNA polymerase binds to promoters along the chromosome of the *Escherichia coli* phage T7. Each arrow indicates an *E. coli* RNA polymerase molecule bound to a T7 promoter. (Courtesy of Dr. Th. Koller.)

Only one of the strands of DNA is transcribed.[1] If, for instance, a length of DNA contained the bases ATTGCCAGA (in that order), its complement would be TAACGGTCT—not the same thing at all. It would code for an entirely different protein containing a totally different sequence of amino acids, assuming that one could be translated from it at all. If the first strand could produce a useful protein, almost certainly its complement could not. This means that only *one* of the two strands—the coding strand—is transcribed into mRNA. The other strand is not transcribed and has no known function except in replication and repair (Chapter 15). The RNA polymerase tells the two strands apart by requiring specific "promoter" base sequences to be present before it will begin work. Discriminating between the two strands is clearly important, since in any double strand only one of the two strands serves as the coding strand at any particular point in its length. In a chromosome this varies, so that one strand may code for protein in one portion of its length while the other strand may code farther along.

THE INITIATION OF TRANSCRIPTION

The transcription process begins with the help of a subunit of RNA polymerase, the σ (sigma) factor, at specific sites termed **promoters** on the DNA (Fig. 14–4). The recognition signal at the promoter, which brings about the binding of RNA polymerase so that transcription can begin, consists of a special sequence of several base pairs. Like DNA, RNA is composed of nucleotides with distinguishable molecular ends. An mRNA chain is always synthesized by adding nucleotides to the 3′ end, so the process begins at the 5′ end and proceeds in the 5′→3′ direction. (The 5′ end of a new RNA chain always starts with a triphosphate—either pppG or pppA;

[1] These remarks apply to cellular life, both prokaryotic and eukaryotic. Viruses, however, must pack a large amount of genetic information into a very small nucleic acid memory space. In some of them, both strands evidently do produce RNA. In addition, some viruses have overlapping coded regions that carry information specifying several proteins at once. Future studies may disclose similar economies of information storage in cells as well.

therefore it has a triphosphate group at its 5′ terminus and a free OH group at its 3′ terminus. This difference permits the enzyme to distinguish the two directions.)

The termination of transcription, like its initiation, is under the precise control of specific sequences of nucleotides in the DNA. The DNA template contains nucleotide sequences that are interpreted by RNA polymerase as "stop" signals. Another specific protein, the **rho protein,** sometimes assists the polymerase in recognizing certain base pairs as the signal for terminating transcription. Information is generally transmitted from DNA to RNA. For an exception see Focus on Reverse Transcription.

THE PROCESSING OF RNA

Although bacterial mRNA is utilized immediately without further processing, in most cases the form in which RNA is first synthesized is not usually the form in which it reaches its ultimate destination. Most RNA undergoes a considerable amount of posttranscriptional modification. In eukaryotes all types of mRNA are initially longer than they appear in their final form. The segments that are to make up the final versions of these substances are split out of the initial strands, and the useless portions are discarded. The saved portions are united by ligase-type enzymes. We can readily appreciate how precise this process must be since single base mistakes in mRNA could result in the RNA equivalent of frameshift mutations (discussed later in this chapter), rendering the produced protein utterly useless. In the case of tRNA and mRNA, some of the bases are then enzymatically modified to form unusual purines or pyrimidines that do not form a part of the genetic code. (Their function appears to be that of increasing the specificity with which tRNA molecules are paired with mRNA codons, thus minimizing possible errors of translation.)

What is even harder to explain is that most eukaryotic genes contain very long base sequences that do not end up in mature RNA, and that some of these **interrupted genes** have their parts distributed among several such sets of sequences. The sequences that actually are retained in the mature RNA are called **exons,** and the discarded ones are called **introns** (Fig. 14–5). Consider these examples: The β-globin gene for one of the components of the hemoglobin molecule produces immature mRNA with two introns; the ovalbumin gene of egg white produces seven; and the gene specifying another egg white protein, called conalbumin, produces 16 introns. The apparently useless introns are often collectively much longer than the exons. The ovalbumin gene, for instance, contains about 7700 base pairs, despite the fact that the mature mRNA it produces is only 1859 bases in length. Why would a cell contain 5841 waste bases in one gene alone? Perhaps

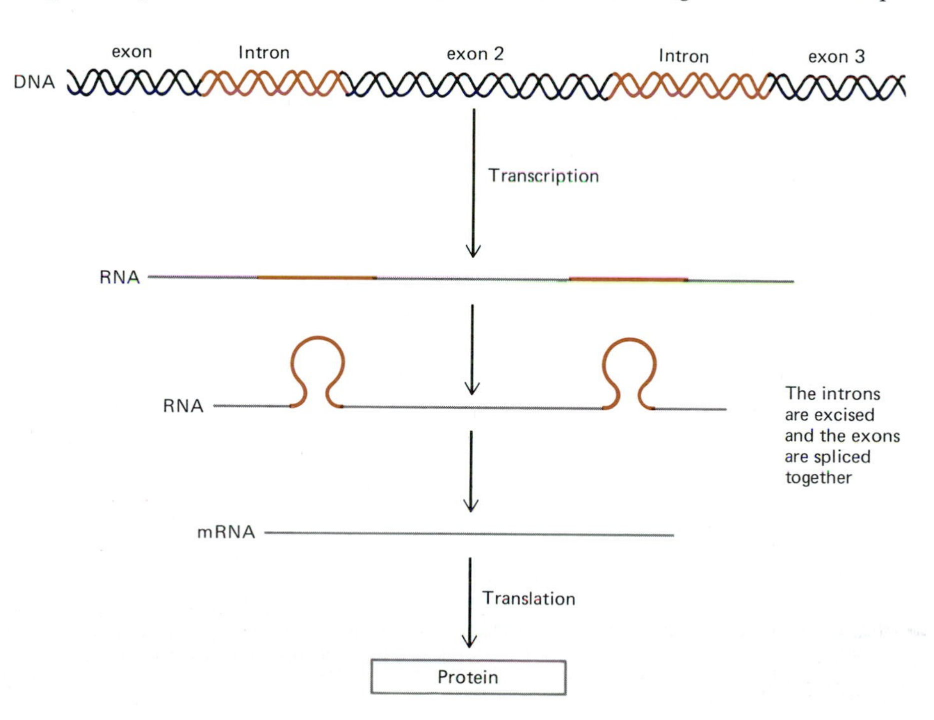

Figure 14–5 Introns are sequences that must be cut out of RNA before RNA can be translated into a protein. Are the sequences in the DNA that code for them "genetic junk," or do they have a function?

FOCUS ON

Reverse Transcription

Until recently, a **central dogma** of molecular biology stated that biological information always flows from DNA to RNA to protein. An important exception to this rule was discovered by Howard Temin in 1964, when he found that infection with certain RNA tumor viruses, such as the Rous sarcoma virus, is blocked by inhibitors of DNA synthesis and by inhibitors of DNA transcription. This suggested that DNA synthesis and transcription are required for the multiplication of RNA tumor viruses and that, in these organisms, information flows in the reverse direction—that is, from RNA to DNA. Temin proposed that a **DNA provirus** is formed as an intermediate in the replication of these RNA tumor viruses and in their cancer-producing effect. Temin's hypothesis required a new kind of enzyme—one that would synthesize DNA using RNA as a template. Just such an enzyme was discovered by Temin and by David Baltimore in 1970, discoveries for which they received the Nobel Prize in 1975. This RNA-directed DNA polymerase, also known as **reverse transcriptase,** was found to be present in all RNA tumor viruses.

After an infecting RNA virus enters the host cell, the viral reverse transcriptase forms a DNA strand complementary to the viral RNA. Subsequently a second DNA strand is synthesized. The double-stranded DNA provirus, is integrated into the host cell's genome. The provirus DNA is transcribed and the viral proteins are formed in the cytoplasm as this RNA is translated. Viral RNA molecules are formed and incorporated into mature virus particles with a protein coat. Because of their reversal of the usual direction of informational flow, such viruses have become known as **retroviruses.** (Some RNA viruses, however, replicate themselves directly without employing a DNA intermediate.)

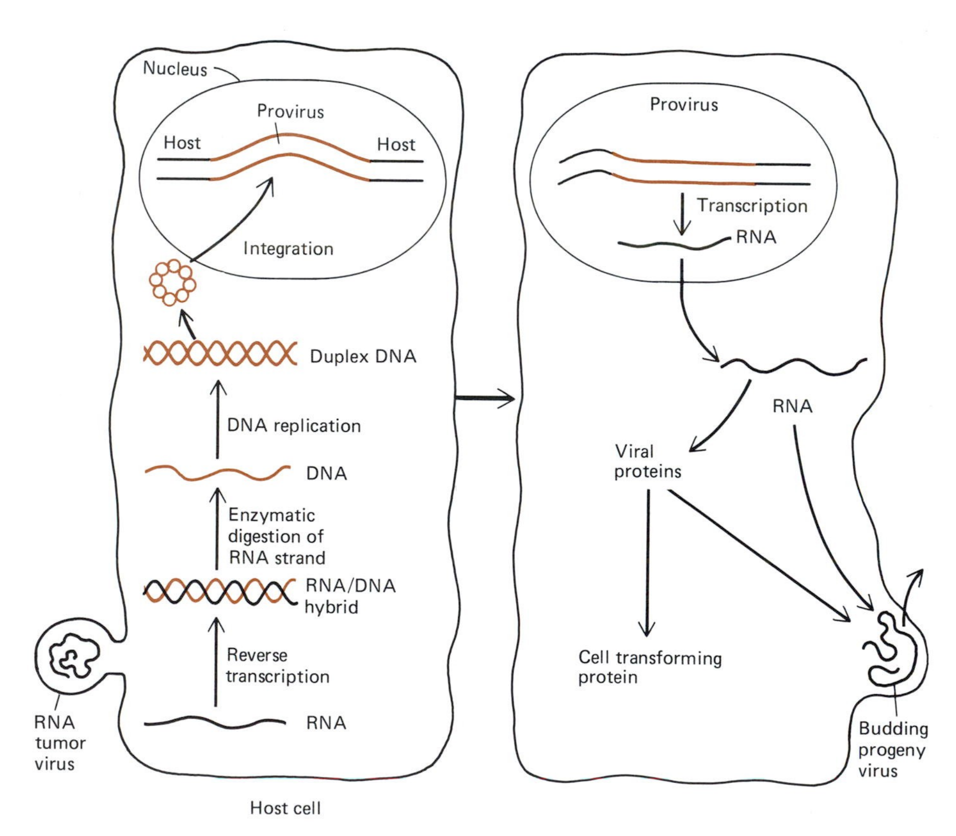

Replication and transcription events that occur during the life cycle of an RNA tumor virus. With the aid of the reverse transcriptase enzyme, the virus makes an intermediate DNA copy of its genome, which then transcribes viral RNA and the kinds of mRNA needed to make viral proteins.

intron–exon sequences play a role in controlling the flow of information from nucleus to cytoplasm and in controlling cellular differentiation. It has been suggested that introns may serve as the raw material from which many new functional genes are constructed, or from which those lost by mutation may perhaps be reconstructed. Moreover, by genetic recombination exons might be shuffled so as to produce novel proteins comprised of portions formerly incorporated in preexisting proteins.

Translation

As we saw in a previous section, the sequence of amino acids in a polypeptide chain is determined by the sequence of codons in mRNA. This is accomplished by matching the triplet anticodons in the tRNA molecules with the triplet mRNA codons. This process of protein synthesis on the ribosomes is termed **translation,** because the "language" of nucleic acids, the four base pairs, is translated into the different "language" of proteins, the 20 amino acids. The process of translation is complex, much more complicated than that involved in either the replication or the transcription of DNA. Translation involves the coordinated functioning of more than 100 kinds of macromolecules, including the ribosomes, soluble factors, mRNA, and the various tRNAs and their activating enzymes.

ACTIVATION OF AMINO ACIDS

Recall from Chapter 3 that amino acids are joined together to form proteins by linking the amino group of one to the carboxyl group of the adjacent one, forming a peptide bond. This reaction is endergonic and thus will not proceed spontaneously. For this reason, before the amino acids can be assembled into a peptide chain, each amino acid must be **activated** by an enzyme-mediated reaction with ATP (for more details, see Chapter 15). A separate specific enzyme, one for each kind of amino acid, activates each. The enzymes catalyze the reaction of the amino acid with ATP, forming an amino acid–AMP compound and releasing inorganic pyrophosphate (PP_i). The very same enzyme next catalyzes the transfer of that amino acid from the AMP with which it is presently associated to one of the specific tRNAs for that amino acid. Any amino acid not attached to tRNA cannot participate in protein synthesis. One end of each kind of tRNA molecule always contains the nucleotide sequence CCA, and it is here that the amino acid is attached to the 3′ hydroxyl group of the ribose of the terminal nucleotide.

THE ROLE OF TRANSFER RNA

Even before tRNA was discovered, Francis Crick had predicted on theoretical grounds that something like it must serve as an adaptor in protein synthesis. He argued that since there is no simple correspondence between the molecular structures of a polynucleotide chain and a polypeptide chain that would enable the nucleotide sequence to specify the amino acid sequence directly, the amino acids might be ordered in appropriate sequence by means of small RNA adaptor molecules. Attaching the amino acid to the tRNA activates its carboxyl group so that it can form a peptide bond; this attachment also ensures that it will be inserted at the correct place in the peptide chains. (It is not that only one species of amino acid will "fit" a particular kind of tRNA but rather that special enzymes insure a "correct" match.)

THE STRUCTURE OF TRANSFER RNA

Molecules of tRNA are considerably smaller than those of mRNA or rRNA. Each functions as a specific adaptor in protein synthesis, binding to and identifying one specific amino acid; that is, each of the 20 amino acids is enzymatically attached to a very specific kind of tRNA. Thus, a code specifying one species of tRNA also specifies one kind of amino acid. This property depends upon the fact that one portion of the nucleotide sequence in tRNA represents an anticodon, which as mentioned earlier is a nucleotide triplet complementary to a particular codon in mRNA. Since that codon specifies the anticodon complementary to it, ultimately it specifies the amino acid associated with the RNA that carries that amino acid. The anticodon and codon are bound to one another by hydrogen bonds between specific base pairs.

Transfer RNAs are polynucleotide chains of some 70 nucleotides. Each kind of tRNA has a unique, invariable sequence of nucleotides, and *all* of them have an identical sequence of nucleotides: CCA, at the 3′ end to which the amino acid is attached (Fig. 14–6). The chain is doubled back on itself to resemble a clover leaf, forming three or more loops of unpaired nucleotides. The folding is stabilized by hydrogen bonds between complementary bases in the intervening portions of the

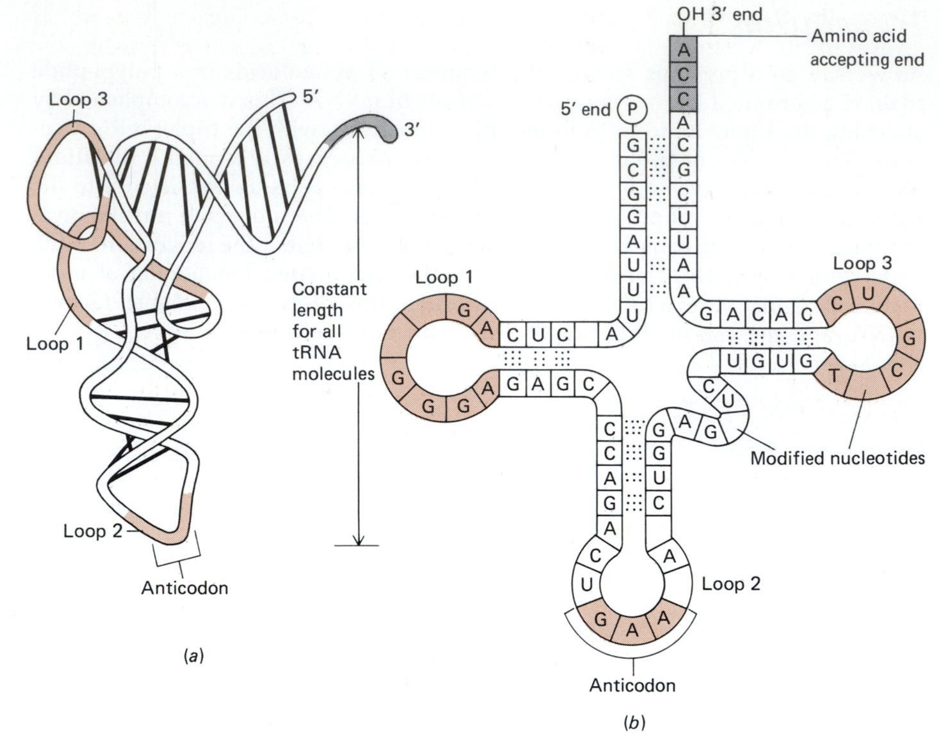

Figure 14–6 Two representations of the structure of a typical tRNA molecule; tRNA molecules are the compounds that "read" the genetic code. A diagram of the actual shape of a tRNA molecule is shown in *(a)*. Its three-dimensional shape is determined by intramolecular hydrogen bonds between base-paired regions, which are most clearly observed in the two-dimensional cloverleaf form depicted in *(b)*. One loop contains the triplet anticodon that forms specific base pairs with the mRNA codon. The amino acid is attached to the terminal ribose at the 3′ OH end, which has the nucleotide sequence CCA. Each tRNA also has guanylic acid, G, at the 5′ end (at P) and contains several modified nucleotides. The pattern of folding permits a constant distance between anticodon and amino acid in all tRNAs examined.

chain to form a short double helix (though the molecule as a whole is not double). The anticodon triplet is located at the midsection of a seven-membered middle loop.

THE MESSENGER RNA TEMPLATE

In the course of its processing in the nucleus, mRNA is given a long tail of polyadenylic acid (polyA) containing 150 to 200 molecules on its 3′ end. The significance of the polyA tail is unknown; it may play some role in transport through the nuclear membrane or in protecting the mRNA from destruction by ribonuclease. The polyA-rich RNA then passes out of the nucleus and becomes associated with the ribosomes. Eukaryote mRNAs also have a "cap" on the 5′ end, consisting of 7-methyl guanosine. The cap protects the mRNA from attack by certain enzymes and, more important, serves as a ribosome recognition site.

Prokaryote mRNA and eukaryote mRNA are quite different. For example, eukaryote mRNA contains the transcribed equivalent of only one gene. For a time, genes were often called "cistrons"; this obsolescent terminology persists in the name of single-gene mRNA, which is called **monocistronic.** Prokaryote mRNA, which usually contains the equivalent of several closely linked genes in a single strand, is called **polycistronic mRNA.**

THE RIBOSOMES

Ribosomes from the cells of different kinds of organisms may differ somewhat in their mass, in the size of their RNA, and in the ratio of RNA to protein, but there is a general similarity in their structures. Their subunit structure is apparent in electron micrographs, with the smaller subunit seeming to sit like a cap on the flat surface of the larger subunit. There are 21 proteins and 1 molecule of RNA in the smaller subunit of bacterial ribosomes, and 35 proteins and 2 RNA molecules in the larger subunit. With care, each subunit in turn can be separated into its constituent RNA and protein molecules. The 21 proteins and the RNA that compose the smaller subunit of bacterial ribosomes, under proper conditions, will undergo spontaneous reassembly and form a fully functional subunit. Similarly, the 35 proteins and the 2 RNA molecules of the larger subunit will also reassemble spontaneously to form a fully functional subunit. These findings demonstrate that all the information needed for the correct assembly of the ribosome from its parts is contained within the ribosome itself. (Traditionally, the larger ribosomal subunit is known as the 60S subunit, and the smaller as the 30S subunit.)

Ribosome Function

In the ribosomal synthesis of a polypeptide chain, the amino acid bound to its specific tRNA is transferred to the ribosomes. The ribosome accepts the tRNA–amino acid complex at the **A-site** with the aid of a special protein (elongation factor), the Ef • Tu factor. A ribosome also possesses a binding site (called the **P-site**) for the last, that is, the most recently added amino acid of the growing peptide chain. This permits it to add new amino acid molecules (which are bound to the A-site) to the end of the chain. Thus a ribosome accepts tRNA (attached at this time to the Ef • Tu protein), and attaches the amino acid it bears to the end of the peptide. The Ef • Tu protein and the tRNA are released, freeing the Ef • Tu protein to accept another amino acid–tRNA.

The role of the ribosome is, in sum, the proper orientation of the amino acid–tRNA precursor, the mRNA template, and the growing polypeptide chain, so that the genetic code on the mRNA can be read accurately. There are some 15,000 ribosomes in a rapidly growing cell of the bacterium *E. coli.* The ribosomes account for nearly one third of the total mass of the cell.

In eukaryotes, ribosomes are somewhat larger and are composed of two or four subunits; they also differ chemically from those of bacteria. Each ribosome contains several dozen kinds of proteins bound to several kinds of RNA. Protein-synthesizing particles somewhat similar to bacterial ribosomes are also found in the nucleus, in chloroplasts, and in mitochondria.

Polyribosomes

Protein synthesis has been studied intensively in preparations of rabbit reticulocytes, which were chosen because they synthesize primarily just one protein—hemoglobin. Alexander Rich and his coworkers showed that the ribosomes most active in protein synthesis are those that interact in clusters of five or more (Fig. 14–7). These clusters, termed **polyribosomes** (or sometimes **polysomes**), are held together by the strand of mRNA. Each ribosome in a polyribosome is actively engaged in protein synthesis directed by the mRNA strand. Like nucleic acids, peptide chains are elongated only at one end. Peptide chains are synthesized by the sequential addition of amino acids beginning at the NH_2-terminal end (that is, the end having a free amino group). Electron micrographs suggest that individual ribosomes become attached to one end of a polyribosome cluster and gradually move along the mRNA strand. At the same time the polypeptide chain attached to it increases in length by the sequential addition of amino acids.

PROTEIN SYNTHESIS

The incorporation of the first amino acid of a new polypeptide chain is called **initiation.** It requires a special (and complex) mechanism, since there is no preexisting polypeptide–tRNA complex to accept the amino acid at the combining site (Fig. 14–8). The NH_2-terminal amino acids of proteins are often substituted in their α-amino group with an acetyl or formyl group, particularly in prokaryotes. There is, in fact, a special bacterial enzyme that adds a formyl group to methionine *after* methionine has been linked to its tRNA. Such *N*-formyl methionine molecules are incorporated preferentially at the NH_2-terminal end of peptide chains, whereas methionines without the formyl group are incorporated in intermediate positions in the chain.

The initiation of the synthesis of a polypeptide requires the formation of an **initiation complex.** This complex forms when the smaller of the two ribosomal subunits, that is, the 30S subunit, binds to an initiation factor (IF-2), to a special amino acid, and to the strand of mRNA which is to be transcribed. The ribosome actually binds to the mRNA strand in response to a special signal codon (AUG) which codes for the special amino acid, methionine, which was just mentioned.

The addition of other amino acids to the forming peptide is known as **elongation.** This process begins with the enzymatic formation of peptide bonds, one after

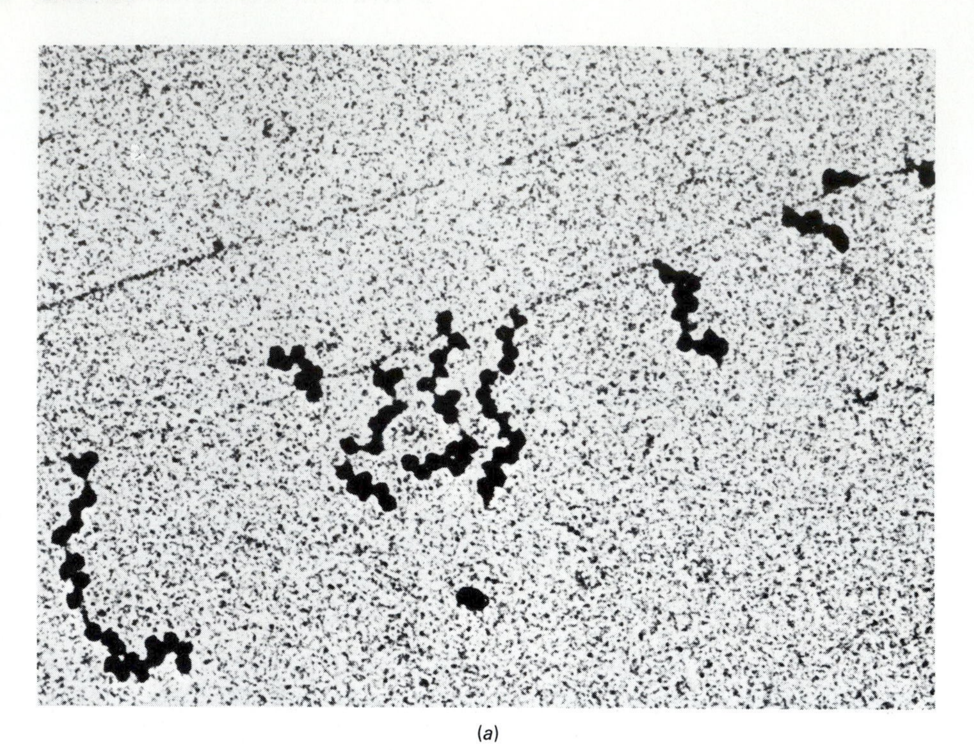

(a)

(b)

Figure 14–7 Transcription and translation in a bacterial cell. *(a)* Electron micrograph of two strands of DNA, one inactive and the other actively producing mRNA, which is attached to polyribosomes. *(b)* Diagrammatic representation of the processes of transcription and translation. (*(a)* Courtesy of O. L. Miller, Jr.; *(b)*, after Miller, Hamkalu, Thomas, Jr.)

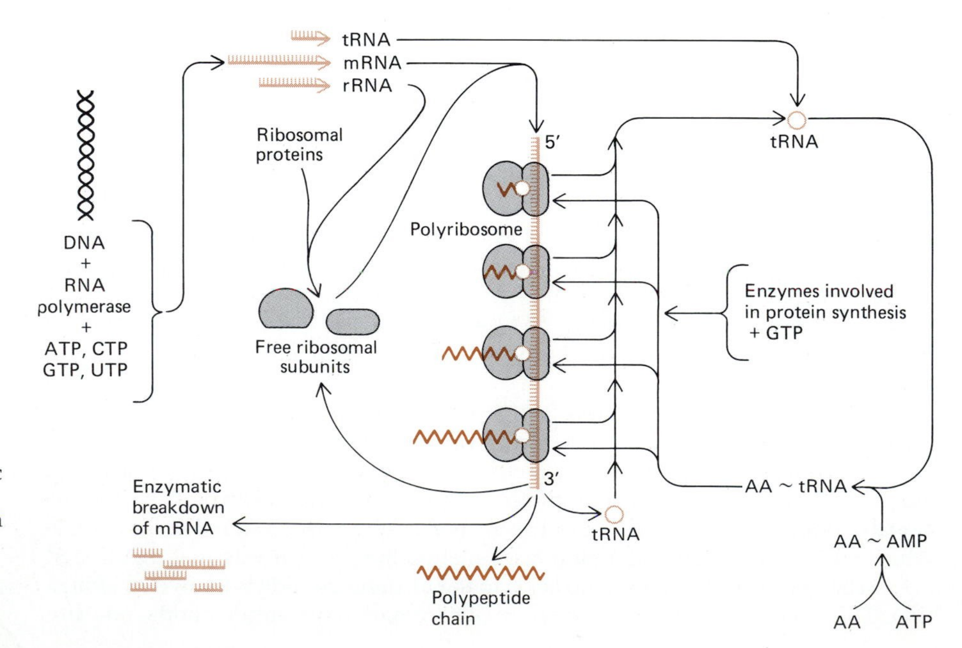

Figure 14–8 The sequence of reactions by which genetic information is translated into polypeptide products. Specific amino acids are activated, coupled to specific tRNAs (AA tRNA) and then, on the ribosome, transferred into defined sites in the growing polypeptide chain as directed by codons in a specific mRNA.

another; continues with the elimination of each uncharged tRNA (the tRNA that has transferred its amino acid); and ends with the transfer of the tRNA containing the peptide from the A-site within the ribosome to the P-site (Fig. 14–9). The A-site is left empty and ready to accept the next amino acid–tRNA complex.

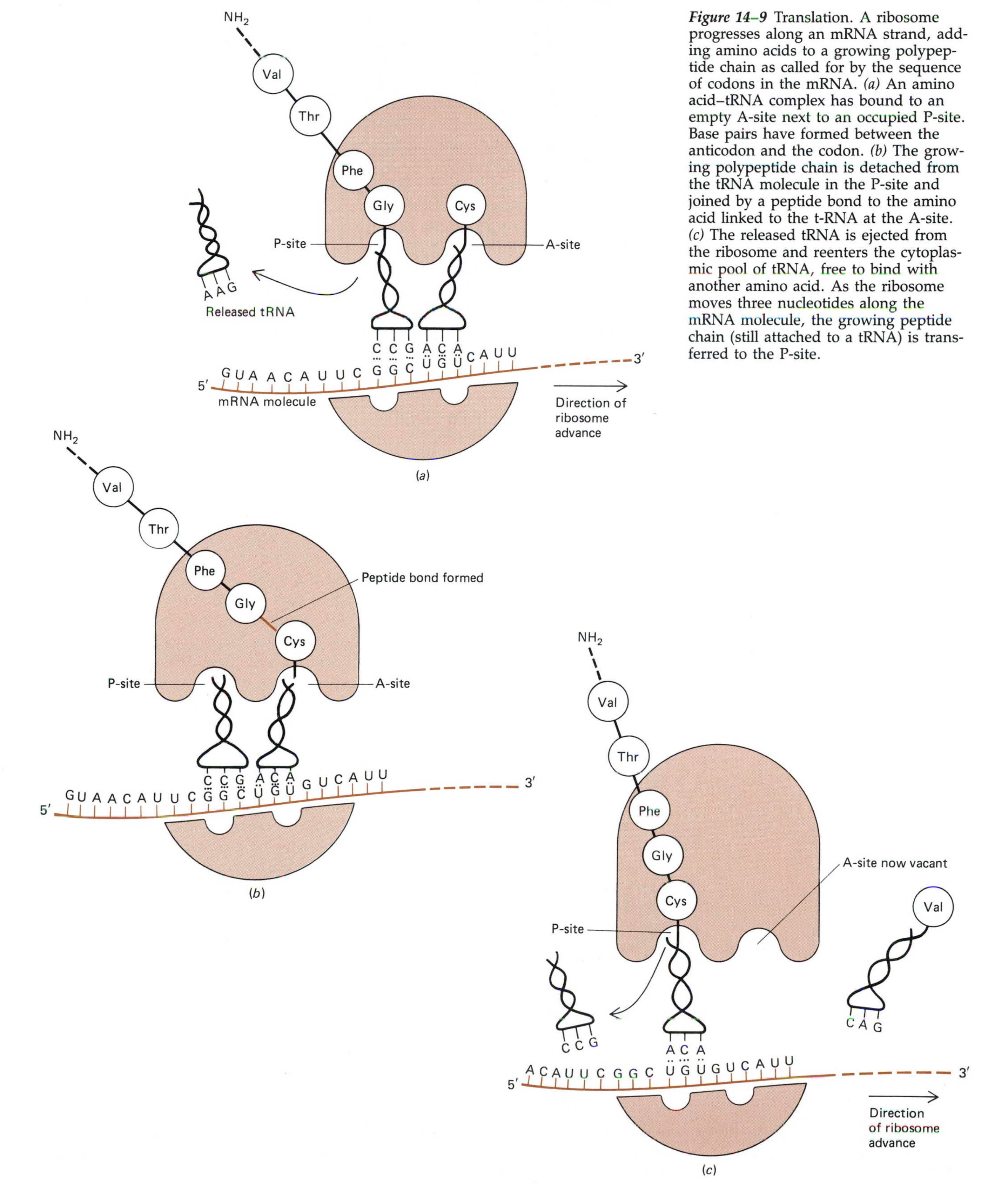

Figure 14–9 Translation. A ribosome progresses along an mRNA strand, adding amino acids to a growing polypeptide chain as called for by the sequence of codons in the mRNA. *(a)* An amino acid–tRNA complex has bound to an empty A-site next to an occupied P-site. Base pairs have formed between the anticodon and the codon. *(b)* The growing polypeptide chain is detached from the tRNA molecule in the P-site and joined by a peptide bond to the amino acid linked to the t-RNA at the A-site. *(c)* The released tRNA is ejected from the ribosome and reenters the cytoplasmic pool of tRNA, free to bind with another amino acid. As the ribosome moves three nucleotides along the mRNA molecule, the growing peptide chain (still attached to a tRNA) is transferred to the P-site.

Two different enzymes are required to carry out the transfer of the amino acids from tRNA to the peptide linkage in the peptide chain: One enzyme binds to the amino acid tRNA; the other is a transferase that forms the peptide bond. The reactions that require GTP are (1) the binding of the tRNA to the ribosome and (2) the translocation of the amino acid–tRNA complex from one site in the ribosome to the other, rather than the synthesis of the peptide bond. GTP is used, and the products are GDP and inorganic phosphate, P_i.

The synthesis of the peptide chain is **terminated** by "release factors" that recognize the terminator codons UAA, UGA, and UAG. This leads to the hydrolysis of the bond between the polypeptide and the transfer RNA. When the peptide chain is completed and released from the ribosome, the ribosome dissociates into its two subunits, and the process can begin again. The protein itself subsequently may undergo enzymatic modification and assumes a final configuration that is dictated by its amino acid sequence.

The several steps involved in protein synthesis can be specifically inhibited by certain toxins and antibiotics. Streptomycin, for example, inhibits the initiation process; tetracyclines inhibit the binding of amino acid tRNAs to the smaller subunit; and erythromycin binds to the larger subunit and inhibits the transfer of the growing peptide chain and its tRNA from one site within the ribosome to the next. Because prokaryotic ribosomes are more sensitive to these antibiotics than eukaryotic ribosomes are, bacterial protein synthesis is inhibited more than eukaryotic protein synthesis.

The ribosome is formed from its subunits each time a single peptide molecule is synthesized; then it must separate into its subunits before a new peptide can be manufactured. This repeated cyclic joining and separating of the ribosomal subunits is termed the **ribosomal cycle.**

Specific enzymes whose task is the destruction of mRNA exist in the cytoplasm. These enzymes ensure that outdated and inappropriate instructions will not interfere with present needs of the cell. If more copies of a particular kind of mRNA are needed, they can always be made. Although there is much variation, in the cells of higher animals the average half-life of mRNA is 6 to 24 hours. In bacteria it is only 2 minutes. Anyone who has had to file reams of outdated memoranda should appreciate the logic of this arrangement. In effect, the cell burns its memoranda as soon as their instructions have been carried out.

The overall process of initiation, elongation, and termination requires the coordinated action of more than 100 different macromolecules, including mRNA, all the specific tRNAs, activation enzymes, initiation factors, elongation factors, and termination factors, in addition to the ribosomes themselves, each made of subunits and many kinds of protein and RNA. Much remains to be learned about the biosynthesis of proteins. The rate of synthesis within an intact, living cell is 100 times faster than that in the best cell-free protein-synthesizing system.

Gene–Enzyme Relations

If a gene leads to the production of a specific enzyme or other protein by the method outlined earlier, we may next inquire how the presence or absence of a specific enzyme may affect the development of a specific trait. The expression of any structural or functional trait is the result of a number—perhaps a large number—of chemical reactions in series, with the product of each serving as the substrate for the next:

$$A \longrightarrow B \longrightarrow C \longrightarrow D$$

For example, the dark color of most mammalian skin or hair is due to the pigment melanin (D), produced from dihydroxyphenylalanine (C), derived in turn from tyrosine (B), derived from phenylalanine (A). The conversion of tyrosine ultimately to melanin is mediated by the enzyme tyrosinase. Albinism, characterized by the absence of melanin, results from the absence of tyrosinase (Fig. 14–10). The allele for albinism (*a*) does not code for production of tyrosinase, but the normal allele (*A*) does (Fig. 14–11).

The earliest attempts to connect the action of a specific gene with a specific enzymatic reaction were studies of the inheritance of flower colors in which specific flower pigments were extracted and analyzed. From studies of the inheritance of

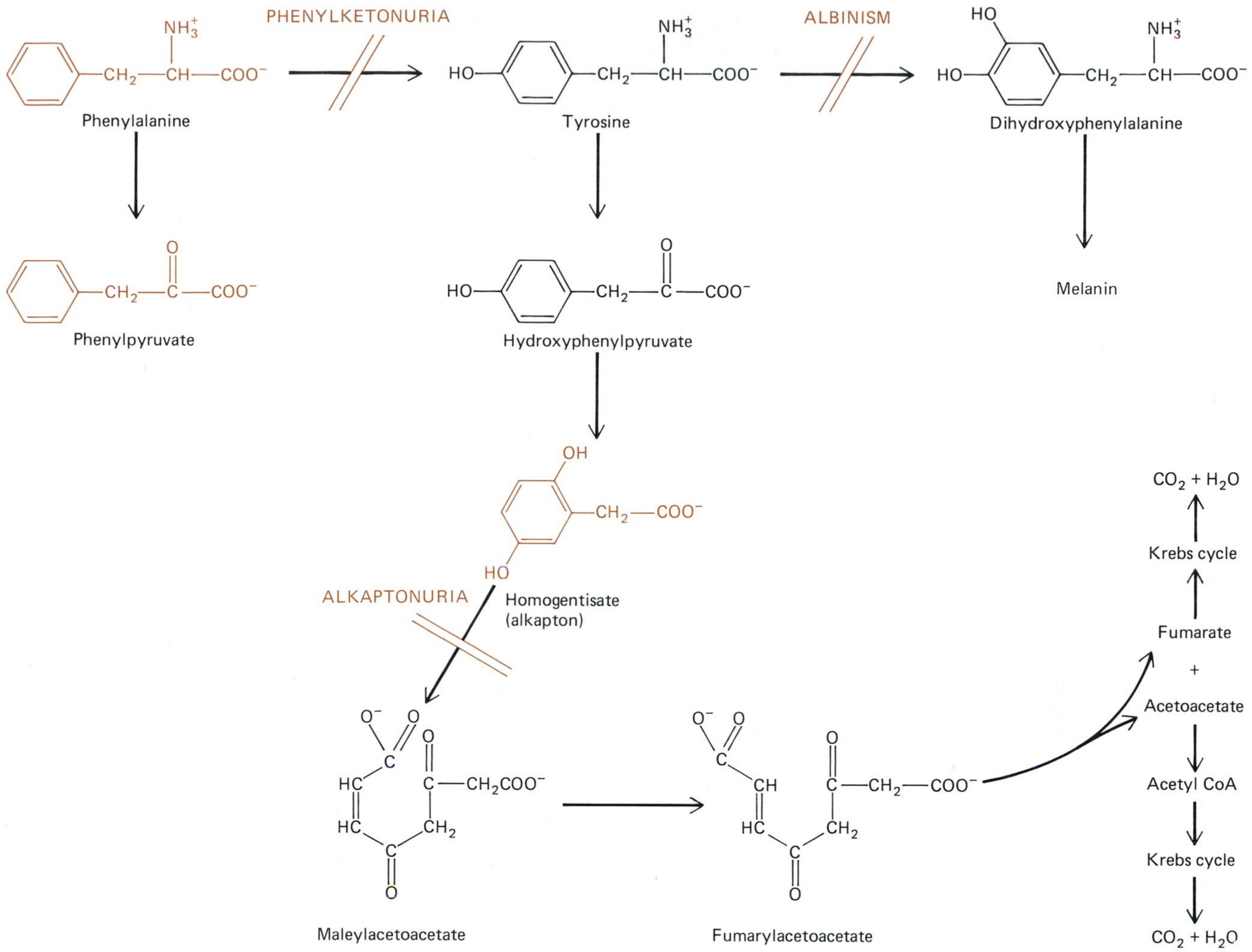

PHENYLKETONURIA

ALBINISM

Phenylalanine

Tyrosine

Dihydroxyphenylalanine

Phenylpyruvate

Hydroxyphenylpyruvate

Melanin

ALKAPTONURIA

Homogentisate (alkapton)

Maleylacetoacetate

Fumarylacetoacetate

Fumarate + Acetoacetate

Acetyl CoA

Krebs cycle

$CO_2 + H_2O$

Figure 14–10 Pathway by which phenylalanine and tyrosine are metabolized. Mutations that interfere with the activity of any of the enzymes that catalyze the individual steps shown may result in inborn errors of metabolism, such as phenylketonuria, albinism, and alkaptonuria. Certain of the compounds (shown in color) accumulate at abnormally high intracellular levels in such metabolic diseases.

coat color in mammals and of eye color in insects, researchers were also able to relate specific genes with specific enzymatic reactions in the synthesis of these pigments. A major advance in this field was made in 1949 when George Beadle and Edward Tatum, using the bread mold *Neurospora,* looked for mutations that interfere with reactions by which chemicals essential for its growth are produced.

The wild type *Neurospora* requires as nutrients only sugar, salts, inorganic nitrogen, and the vitamin biotin. A mixture of these makes up the so-called **minimal medium** for the growth of wild type *Neurospora. Neurospora* reproduces by means of haploid asexual spores, known as **conidia.** (The fact that they are haploid is important. It means that any effect produced by mutation cannot be masked by the presence of an allelic gene.) Exposure of the conidia to x rays or ultraviolet rays will produce mutations (which we will shortly consider; see Fig. 14–12).

In the study by Beadle and Tatum, after irradiation the mold was placed in a **complete medium,** an extract of yeast containing all the known biologically important amino acids, vitamins, purines, pyrimidines, and so on. Any nutritional mutant produced by the irradiation would be able to survive and reproduce when grown on this complete medium. The mold was next tested for its ability to grow on minimal medium. If the irradiated mold was unable to grow on minimal medium, these researchers concluded that the mutant was unable to produce some compound essential for growth. Substances were then added to the minimal medium in groups or singly until the mutant was able to grow. That identified the required substance. Genetic tests showed that the mutant strain produced by irradiation differed from the normal wild type by a single gene. Since chemical tests showed that the addition of a single chemical substance to the minimal medium enabled the mutant strain to grow normally, a one-to-one correspondence between the two seemed clear.

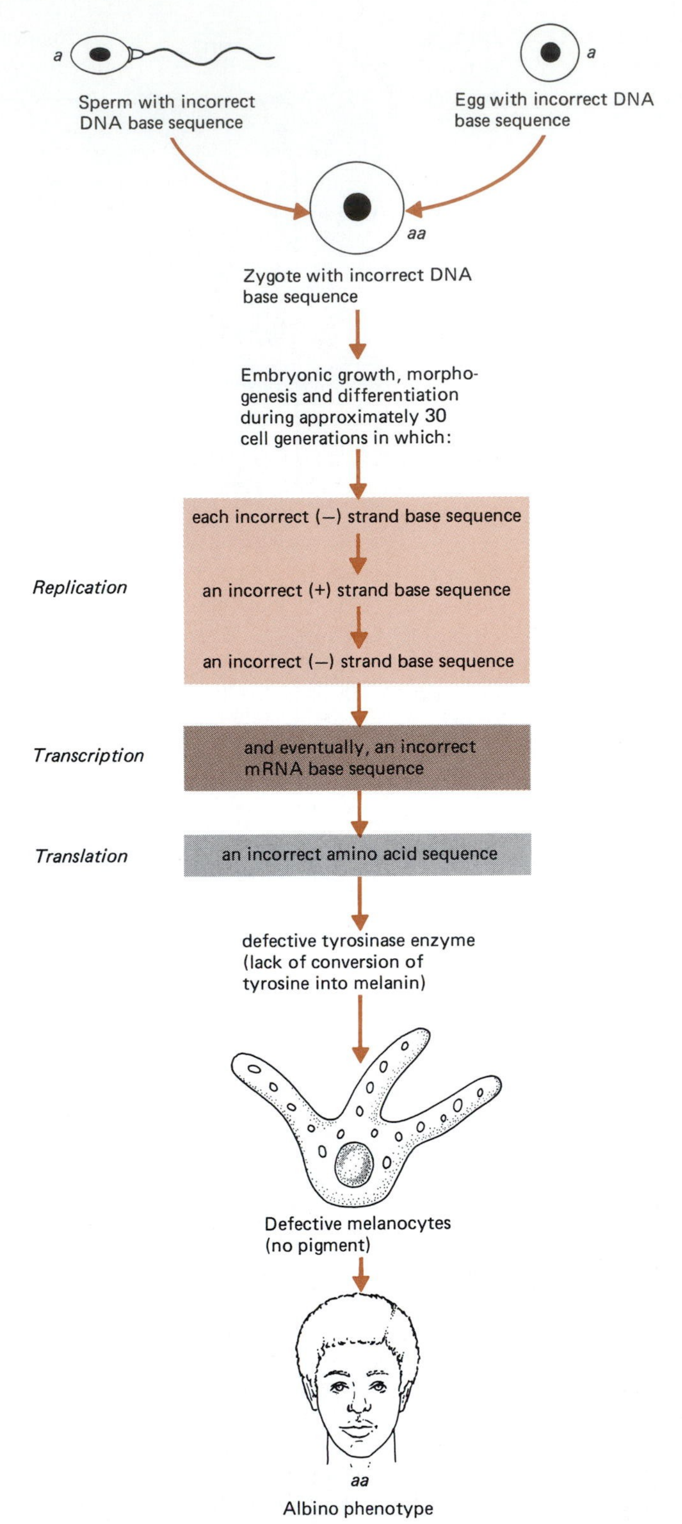

Figure 14–11 An error in DNA base sequence may produce a genetic disease—in this case, albinism. Other forms of albinism result from an inability of cells to absorb tyrosine in the first place.

Beadle and Tatum inferred that each normal gene produces a single enzyme that regulates a single step in the biosynthesis of that particular chemical. The mutant gene does not produce the enzyme, and therefore organisms with the mutant gene must be supplied with the product of the reaction that is impaired. In certain later studies it has been possible to extract the particular enzyme from the cells of normal *Neurospora,* but not from cells of the mutant strain. The biosynthesis of each compound involves a number of different steps, each mediated in turn by a separate gene-coded enzyme. Indeed, biologists estimate the minimal number of steps involved in the synthesis of a given substance from the number of different mutants that will interfere with its production.

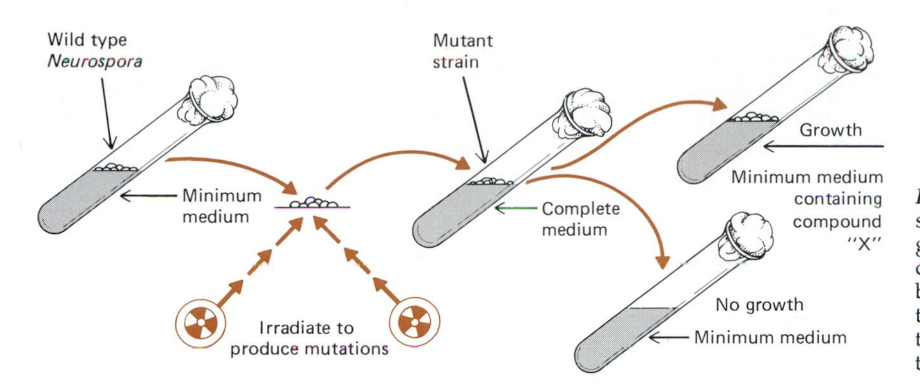

Figure 14–12 Production of mutant strains of *Neurospora* by irradiation with gamma rays. The mutant strains produced can grow on complete medium but not on minimal medium. However, they can grow on minimal medium if the medium is supplemented with X, the single missing nutrient.

Similar one-to-one relationships of gene, enzyme, and biochemical reaction in humans had been described by the English physician A. E. Garrod much earlier, in 1908. **Alkaptonuria** is an inherited condition in which a substance in the patient's urine turns black when exposed to air (see Table 12–3). Homogentisic acid, a normal intermediate in the metabolism of phenylalanine and tyrosine, is excreted in the urine of alkaptonurics. The tissues of normal persons have an enzyme that oxidizes homogentisic acid, so that it is ultimately converted to carbon dioxide and water. Persons with alkaptonuria lack this enzyme because they lack the gene that is necessary for its production; therefore, homogentisic acid accumulates in the tissues and blood and is excreted in the urine. Garrod coined the term "inborn errors of metabolism" to describe alkaptonuria and comparable conditions such as phenylketonuria and albinism. He formed an essentially accurate picture of the cause of such diseases even before the modern concepts of enzymes or genes were fully developed—a remarkable intellectual achievement.

Changes in Genes: Mutations

How do such genetic defects as alkaptonuria come into existence? Although genes are remarkably stable and are transmitted to succeeding generations with great fidelity, they do from time to time undergo changes called mutations. A **mutation** can be defined as any inherited change in the sequence of DNA. A mutation is not due to segregation or to the normal recombination of unchanged genetic material (Fig. 14–13). The definition of mutation includes the restriction that the change that

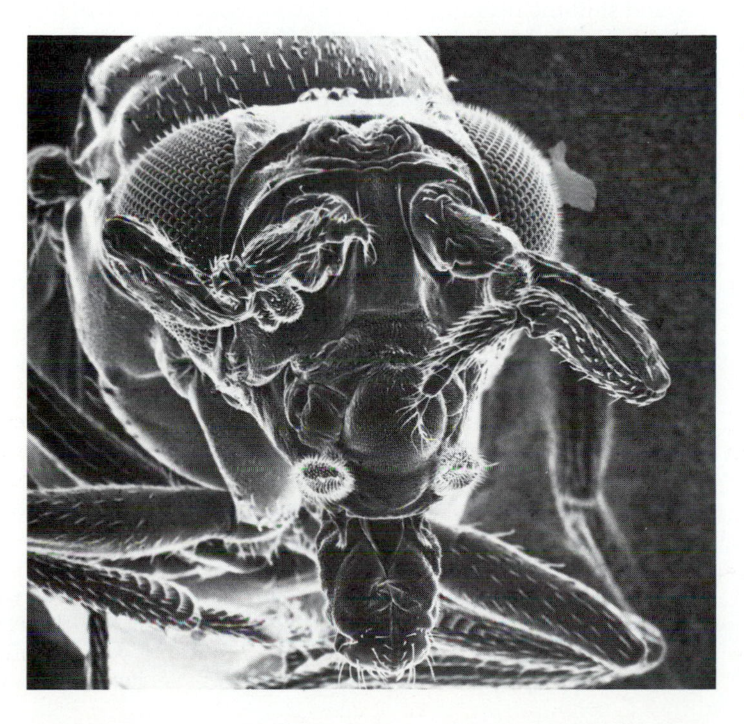

Figure 14–13 A mutant of *Drosophila* in which two legs emerge from between the large compound eyes. The antennae are normally located in this position. Normal legs can be seen at the bottom of the photograph. (Courtesy of Dr. Thomas Kaufman.)

has been introduced into the DNA molecule must be propagated subsequently for an indefinite number of generations. After a gene has mutated to a new form, the new form is stable and usually has no greater tendency than the original gene to mutate again. Mutations provide the diversity of genetic material that makes possible the study of the process of inheritance. Investigations of the mechanisms of the mutation process have provided important clues to the nature of the genetic material itself. As we will see later, mutations also are of fundamental importance in evolution.

Chromosomal mutations are sometimes accompanied by a visible change in the structure of the chromosome. A small segment of the chromosome may be missing (a **deletion**) or may be represented twice in the chromosome (a **duplication**). A segment of one chromosome may be transferred to a new position on a new chromosome (a **translocation**), or a segment may be turned end over end and attached to its usual chromosome (an **inversion**).[1]

Point mutations, or **base substitution** mutations, involve small changes in the molecular structure of the individual gene. Because these are not evident under the microscope, their existence must be inferred from their effects or by analysis of the sequence of bases in the DNA (or amino acids in the protein). These mutations involve some change in the sequence of nucleotides within a particular section of the DNA molecule, usually the substitution of one nucleotide for another in a given codon.

From our knowledge of the DNA molecule, we might predict that replacing one of the purine or pyrimidine nucleotides by an **analogue,** that is a structurally similar compound, but one that differs in certain properties, might result in mutation. In several experiments in which such analogues were incorporated into bacteriophage DNA, no mutations were evident. Because two or more codons may specify the same amino acid (see Table 14–1), a number of changes in base pairs could occur without changing the amino acid specified.

Gene mutations generally result from errors in base pairing during the replication process, such as replacing an AT base pair normally present with GC, CG, or TA pairs. The altered DNA may be transcribed to an altered mRNA; this may in turn be translated into a peptide chain with one amino acid different from the normal kind of peptide. If the amino acid substitution occurs at or near the active site of the enzyme, the altered protein may have markedly decreased or altered enzymatic effects. However, if the amino acid substitution occurs elsewhere in the enzyme molecule, it may have little or no effect on the properties of the enzyme and may be undetectable by current techniques. The true number of gene mutations may therefore be much greater than the number observed.

If a single nucleotide pair were *inserted into* or *deleted from* the DNA molecule, it would shift the reading of the genetic message, alter all of the codons lying to the right (that is, 3′-ward) of the substitution, and change completely the nature of the resulting peptide chain. Thus, if the normal sequence is CAGTTCATG (read CAG, TTC, ATG), the insertion of a G between the two Ts results in CAGTGTCATG (read CAG, TGT, CAT, G . . .). These **frameshift mutations,** as they are called, probably result some of the time in the *absence* of the involved enzyme, because a frameshift mutation involving the initiator codon would make it impossible for the affected mRNA to be translated or chance production of terminator codons could cause the protein to be abruptly terminated (Fig. 14–14).

Chemicals that cause mutations include nitrogen mustards, epoxides, nitrous acid, acridine dyes, and alkylating agents. These **mutagens** are chemicals that can react with specific nucleotide bases in the DNA and change their nature or which can disturb the DNA reading frame. Incorporation of such analogues into DNA may lead to mistakes in the pairing of nucleotides during subsequent replication processes. For example, when bromouracil is incorporated into DNA in place of thymine it will pair with guanine as well as with adenine, the normal pairing partner of thymine. This leads to the substitution of a GC pair of nucleotides at the point in the double helix previously occupied by an AT pair of nucleotides (Fig. 14–15).

[1] Why chromosomal rearrangements produce mutations is not clearly understood, but in deletions, genetic material is removed, resulting in absence or reduced quantities of whatever proteins the genes in the deleted areas may have specified. Translocations and inversions may result in excessive amounts of genetic material, or result in moving the genes involved away from their proper regulators so that they function at inappropriate times or manufacture too much or too little of their product.

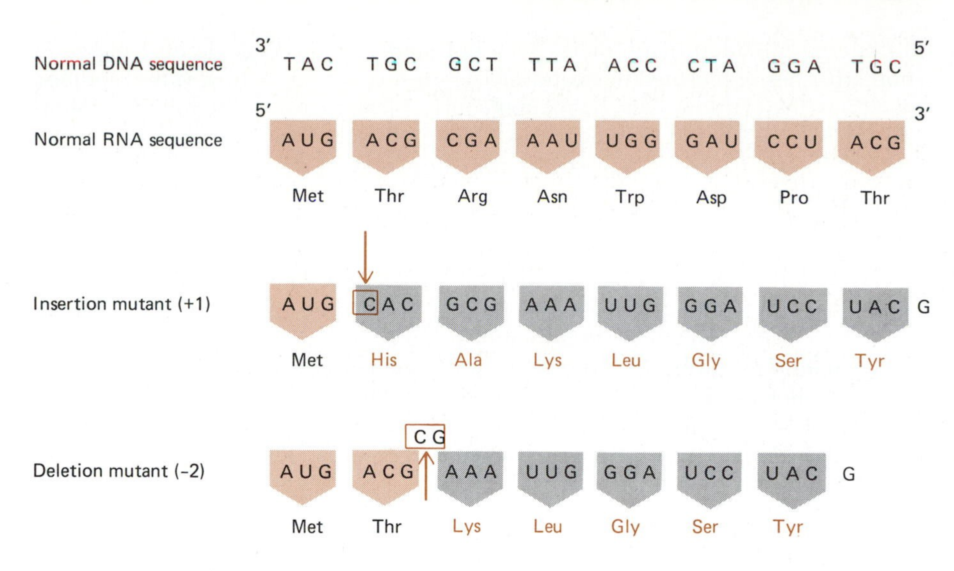

Figure 14–14 Insertions and deletions of just a few nucleotides can have a powerful frameshift effect on the sequence of the polypeptide produced. Nucleotides inserted into or deleted from the normal (wild type) sequence are indicated by arrows and boxes. The triplet reading frame is indicated by pentagons, with color representing normal codons and gray representing mutant codons. The amino acids corresponding to each codon are shown below the RNA sequence.

Gene mutations may also be induced by x rays, gamma rays, cosmic rays, ultraviolet rays, and other types of radiation. How radiation leads to changes in base pairs is not clear, but the radiant energy may react with water molecules, releasing short-lived, highly reactive substances that attack and react with specific bases. Mutations occur spontaneously at low but measurable rates that are characteristic of the species and of the gene. Some regions of DNA are much more prone

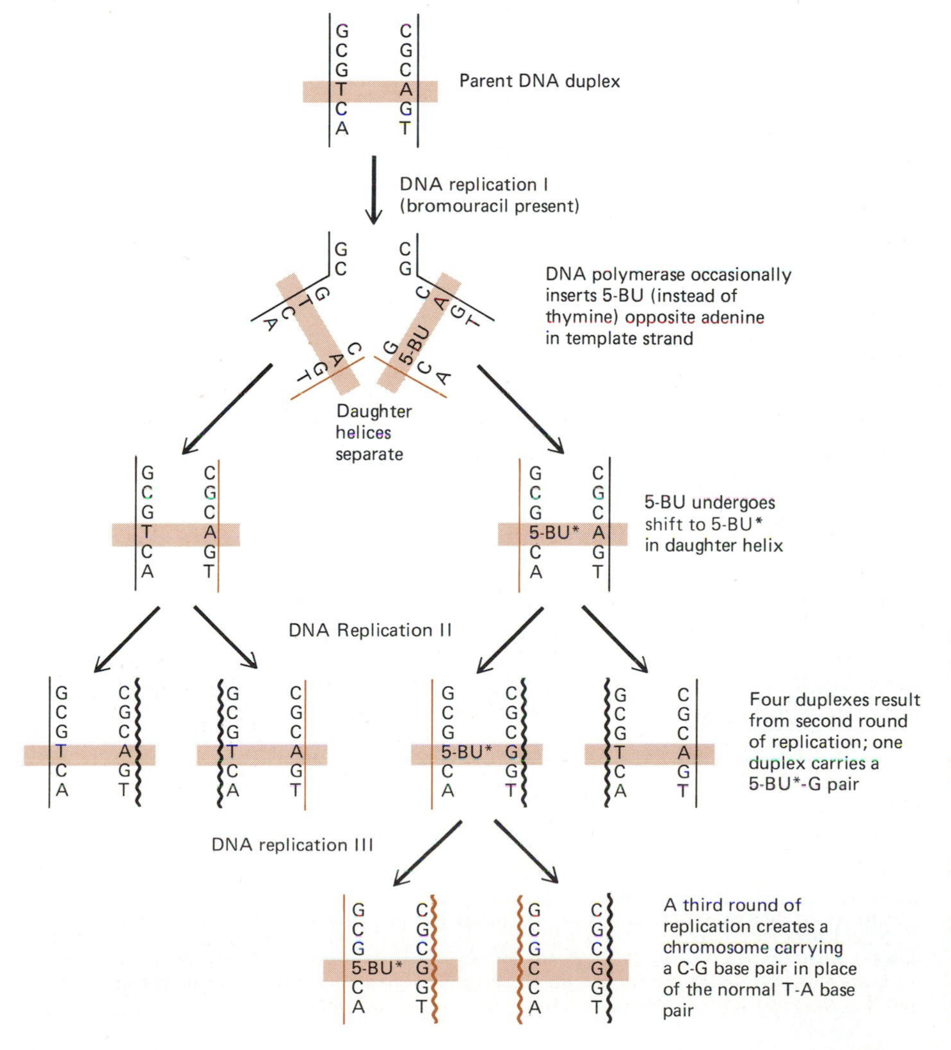

Figure 14–15 Scheme by which an analogue (a structurally similar compound but one that differs in certain properties) of a purine or pyrimidine might interfere with the replication process and cause a mutation. In this instance, the thymine analogue bromouracil (abbreviated 5-BU) is mistakenly substituted for a thymine. The 5-BU undergoes reversible shifts in spatial orientation and leads to the creation of TA→CG transition mutations. A single substitution of a GC pair for an AT pair would be sufficient to cause a mutation if it occurred in the triplet code at a point that changed the kind of amino acid specified.

to undergo mutation than others. Such **hot spots** are often single nucleotides or short stretches of repeated nucleotides. The most common of these is the modified base 5-methylcytosine which spontaneously deaminates to thymidine. Short stretches of repeated nucleotides cause DNA polymerase to "slip" causing errors. Certain mutations even increase the mutation rate—probably by rendering the DNA replication or repair enzyme systems less competent.

However, when all known causes of mutation such as natural background radiation or naturally occurring mutagenic substances are eliminated, a large number of unaccountable spontaneous mutations remains. The rates of spontaneous mutations of different human genes range from 10^{-3} to 10^{-5} mutations per gene per generation. Since humans have a total of some 25,000 genes, this means that the total mutation rate is on the order of one mutation per person per generation. Each of us, in other words, has some mutant gene that was not present in either of our parents. Most such mutations appear to be recessive or insignificant; they are not noticeably expressed. (See Focus on Genetic Repair and Proofreading.)

Some Reflections on the Genetic Code

The Watson–Crick model of the DNA molecule implied that genetic information is transmitted in the form of the specific sequence of its constituent nucleotides. In 1954, George Gamow, an imaginative physicist, was one of the first to suggest that the minimum coding relation between nucleotides and amino acids would be three nucleotides per amino acid. Four nucleotides taken two at a time provide for only 16 combinations ($4^2 = 16$), whereas four nucleotides taken three at a time provide for 64 combinations ($4^3 = 64$). At first glance, this would seem to provide many more code symbols than are needed, since there are only 20 different amino acids. It was believed at one time that some of these 64 combinations were simply "nonsense" codes that did not specify any amino acid. However, there is now strong evidence that all but three of the 64 combinations do, in fact, code for one or another amino acid. The genetic code is redundant, and as many as six different nucleotide triplets may specify the same amino acid. Such redundancy is known as **degeneracy;** it has the virtue of minimizing the effects of some mutations involving synonymous triplets.

SOME EXPERIMENTAL EVIDENCE FOR A TRIPLET CODE

We have seen that there is a direct sequential correspondence among the components in DNA, mRNA, and protein. Thus the three are **colinear.** From a mathematical analysis of the coding problem, Crick concluded early in 1961 that three consecutive nucleotides in a strand of mRNA provide the code that determines the position of a single amino acid in a polypeptide chain. Experimental evidence to support and extend this was quickly forthcoming from the laboratory of M. Nirenberg and H. Matthaei. Using purified enzyme systems, they studied the incorporation of specific labeled amino acids into protein under the direction of artificial mRNAs of known composition. Nirenberg and Matthaei prepared a synthetic polyuridylic acid (UUUU . . .). When this artificial mRNA was added to a system of purified enzymes for the synthesis of proteins, only phenylalanine—no other amino acid—was incorporated into the protein. The polypeptide that resulted contained only phenylalanine. The inference that UUU is the code for phenylalanine was inescapable. Similar experiments by Nirenberg and by Severo Ochoa showed that polyadenylic acid (AAAAA . . .) provided the code for the amino acid lysine, and that polycytidylic acid (CCCCC . . .) coded for the amino acid proline. Making mixed nucleotide polymers (such as polyAC) and using them as artificial messengers made it possible to assign many other nucleotide triplets to specific amino acids.

These experiments did not reveal the *order* of the nucleotides within the triplets, but this has been inferred from other kinds of experiments. It is possible to synthesize trinucleotides of known sequence. Using these, the coding assignment of all 64 possible triplets has been determined. For example, GUU, but not UGU or UUG, induces the binding of valine tRNA to ribosomes; UUG induces the binding of leucine tRNA. Since GUU and UUG code for different amino acids, it follows that the reading of the code in the mRNA strand makes sense only in one direction.

FOCUS ON

Genetic Repair and Proofreading

Random molecular motion is easily able to disrupt DNA, and so is base deamination resulting from several possible causes. If these and other such events were uncompensated, life would be rendered impossible because its informational basis could not be preserved from one cell generation to another. Observed mutations are comparatively rare not because the cell is protected from them, but because disruptions in DNA are swiftly and efficiently repaired once they do occur.

The enzyme DNA polymerase (Chapter 13) is responsible for the replication of DNA. Several forms of this enzyme are presently known—for example, in eukaryotes there are at least two cytoplasmic versions as well as a mitochondrial form. These three differ substantially from one another in the accuracy with which they replicate DNA base sequences, and in fact, none of them demonstrates the accuracy in vitro that can be observed in vivo. Presumably yet unknown additional mechanisms contribute additional accuracy to the replication process. Of the known several forms of DNA polymerase, some are known whose function is to repair damaged DNA. The simpler repair enzymes bind to DNA sequences that contain unnatural bases. For instance, deamination of cytosine produces the base uracil; the repair enzyme detects the uracil and automatically replaces it with a cytosine. Another such change involves the process of **depurination** (the removal of guanine or other purine): If the repair enzyme discovers a naked portion of the strand whose purine has been lost, it deletes that portion and substitutes whatever base is complementary to the one in the corresponding position of the other strand. Notice that the information redundancy of the two strands is what makes this repair possible.

Some mutations, such as those produced by carcinogenic chemicals, ultraviolet light, or nuclear radiation, produce large defects in one strand of DNA. A typical defect of this sort occurs when adjacent thymine bases on one strand of DNA unite with *one another* to produce a **dimer,** or double pyrimidine, as shown in the figure. Such a dimer cannot pair properly with bases on the other homologous strand, which has the effect of stopping replication or transcription at that point, often killing the cell. When such defects occur, the cell produces an **SOS signal** of uncertain origin that activates a repair complex consisting of perhaps 20 associated enzymes and other proteins, plus carbohydrate and RNA (its similarity to miniature ribosome suggests that the RNA perhaps serves some alignment function). These remove the many different chemically damaged bases—it takes a different kind of enzyme for each kind of damage to each kind of base—and replace them all with the proper sequence dictated by the (hopefully) undamaged complementary strand. For reasons unknown, visible light greatly enhances the effectiveness and speed of this repair process, which for that reason is called **photoreactivation.** Here again, notice the important role played by the complementary strand.

In bacteria the SOS signal may be received also by viruses that have been integrated into the bacterial genome. Evidently, it is on the whole to the virus' advantage to desert the damaged host cell without waiting to discover whether the repair process will be successful. The SOS signal derepresses a certain operon within the viral DNA; a series of enzymes are swiftly produced; and the viral DNA replicates explosively. The damaged host cell bursts, releasing a swarm of virus particles.

It could be that the part played by complementary DNA in the repair of mutations is just as important to living things as is its function in replication and cell division. We can gain some idea of that importance from the genetic disease **xeroderma pigmentosum,** in which there is a hereditary absence or malfunction of the enzymes of DNA repair. Sufferers from this disorder develop skin cancer with exposure to very modest amounts of ultraviolet light and often die from this cancer by early adulthood. To be sure, even in normal organisms the repair process is not perfect, and some errors are made not only here but in ordinary replication as well, at frequencies comparable to the error rate with which modern digital computers process data. While low, such a rate is still unacceptable over the kinds of time spans in which living things are thought to have existed on the face of the

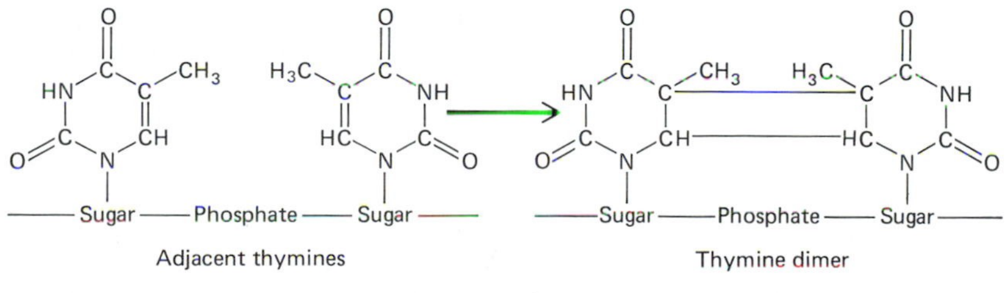

Adjacent thymines → Thymine dimer

(continued)

earth. However, there is a crude backup system that removes these remaining errors, not by acting on the nucleic acids, but by acting on the entire organism; this system is the process of natural selection.

RNA has no repair mechanism, so RNA viruses exhibit a mutation rate much in excess of DNA-based organisms; this is reflected in a very rapid rate of evolution. An RNA virus present in a person for most of a human lifetime actually has the opportunity to undergo considerable evolutionary change within the body of its host. Such a viral population may, within the span of months, elude any immunological defenses the host can mount just by rapid random mutation and natural selection.

Proofreading

We have just seen that damaged DNA is repairable, but it is also necessary for cells to avoid errors that might occur in the replication and other normal processing of genetic information. How this is accomplished is not well understood in eukaryotes. However, in bacteria the DNA polymerase I enzyme is able to detect an erroneous base replication. If such an error exists, instead of moving on to place the next base, the enzyme moves *backward* on the DNA strand, removes the incorrect base, and substitutes the correct one.

Errors also occur in transcription and translation, and proofreading mechanisms operate to minimize these. In translation the proper amino acid must attach to the appropriate molecular species of tRNA. The enzyme responsible for this must possess at least two active sites, one for the tRNA and one for the amino acid, as shown in the figure. To ensure accuracy, each active site must be absolutely and exclusively specific for its substrate. Unfortunately, in several cases the amino acid active site is not always sufficiently discriminating. Isoleucine tRNA synthetase, for example, sometimes accepts valine instead, and would join it to the tRNA that should bear isoleucine were there no additional discriminatory mechanisms to prevent this. Fortunately, there are. To use an example originating with Maurice Gueron,[1] let us assume that a tRNA synthetase enzyme erroneously accepts 1 valine molecule for every 99 molecules of isoleucine, thus producing an error rate of 1%. If the enzyme had two such sites, so that an amino acid was passed from one to the other before final attachment to tRNA, then their combined discrimination should be 1% X 1%, that is, .01%, so that only 1 out of a possible 10,000 errors should occur. The trouble is that there is no way of assuring the sequential action of these two active sites. An amino acid might very well bind to the second site right away so that little if any improvement would result.

The way this problem is solved is singularly ingenious. The first active site accepts the amino acid but does not combine it with tRNA. Instead the amino acid reacts with ATP, which changes it into an entirely new chemical. This requires the expenditure of ATP energy, but now the product (an amino acid adenylate) is accepted by a second active site that would not accept the amino acid itself. This second site discriminates between isoleucine adenylate and valine adenylate with 99% effectiveness, producing a final error rate, as we have seen, of only 1 in 10,000. Only after this second discriminatory step is the amino acid finally attached to the tRNA.

[1] Enhanced selectivity of enzymes by kinetic proofreading. *American Scientist* 66:202–208, 1978.

CODON–AMINO ACID SPECIFICITY

Careful examination of the coding relationships (Table 14–1) shows that there is a pattern to the degeneracy in the code. In 1965, Bernfield and Nirenberg found that the binding of phenylalanine tRNA is induced by either UUU or UUC in the mRNA. The binding of serine tRNA to mRNA is induced by either UCC or UCU, and proline tRNA is bound by either CCC or CCU. In all these cases the two alternative triplets are identical except for the substitution of one pyrimidine for the other (C or U) at the 3′ end of the triplet. In other instances, adenine and guanine can be interchanged in the 3′ position of the triplet. For example, the code for lysine may be either AAG or AAA. For a number of amino acids, the first two nucleotides of the codon are specific, but any of the four nucleotides may be present in the 3′ position. All of the four possible combinations will code for the same amino acid. In these, although the code may be read three nucleotides at a time, only the first two nucleotides appear to contain specific information. Only methionine and tryptophan have single triplet codes; all the other amino acids are specified by from two to as many as six different nucleotide triplets. Codons that specify the same amino acid are termed **synonyms.** As shown in Table 14–1, UCU and UCC are synonyms for serine, and CAU and CAC are synonyms for histidine.

That mRNA is read three nucleotides at a time was firmly established by experiments carried out by H.G. Khorana and his colleagues. A polyUC messenger containing the regularly alternating base sequence UCUCUCUC was synthesized

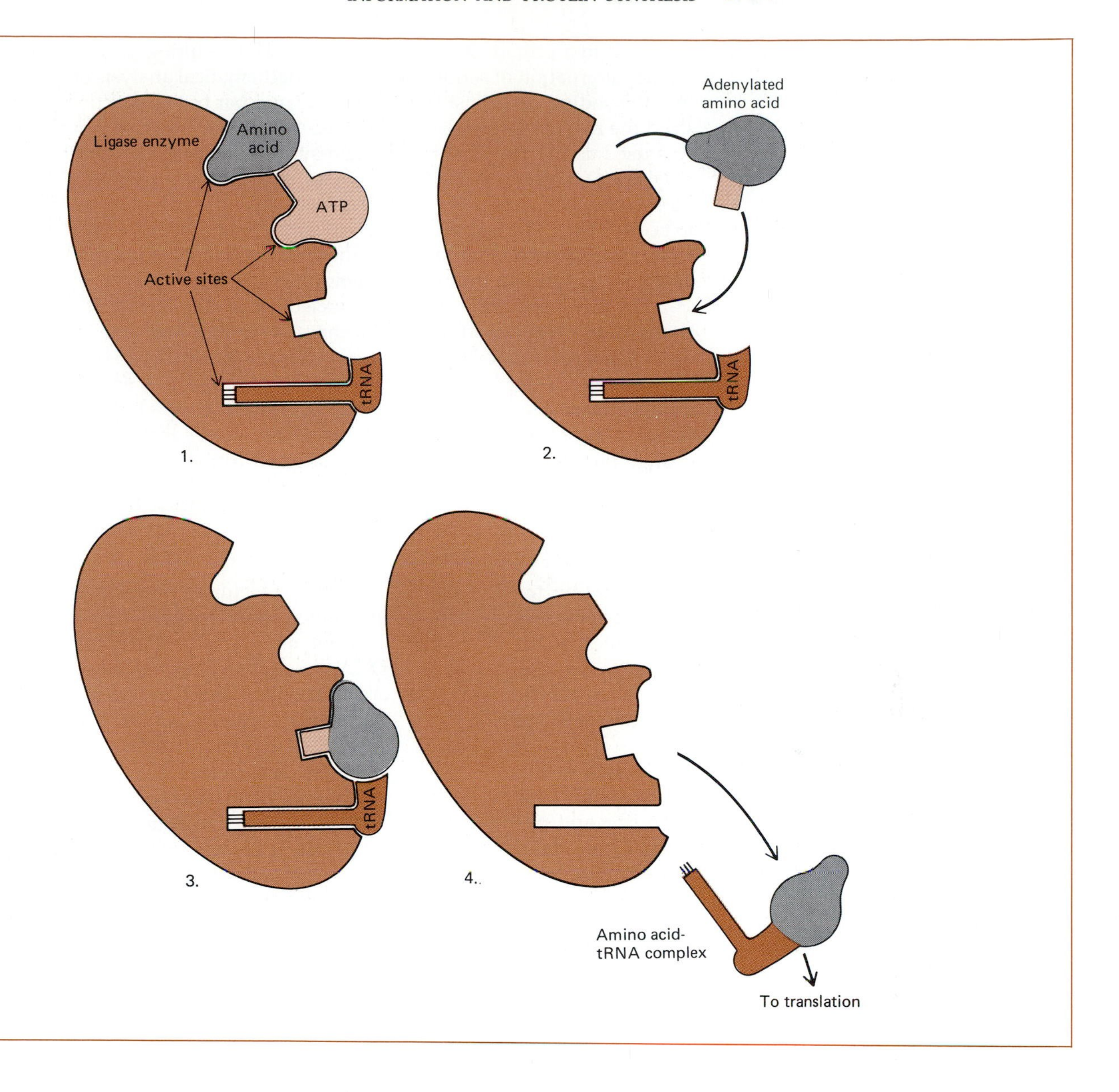

TABLE 14–1
The Genetic Code: The Sequence of Nucleotides in the Triplet Codons of mRNA That Specify a Given Amino Acid

First Position (5′ end)	*Second Position*	*Third Position (3′ end)* *U*	*C*	*A*	*G*
U	U	Phe	Phe	Leu	Leu
	C	Ser	Ser	Ser	Ser
	A	Tyr	Tyr	Terminator	Terminator
	G	Cys	Cys	Terminator	Trp
C	U	Leu	Leu	Leu	Leu
	C	Pro	Pro	Pro	Pro
	A	His	His	Glu–NH_2	Glu–NH_2
	G	Arg	Arg	Arg	Arg
A	U	Ileu	Ileu	Ileu	Met
	C	Thr	Thr	Thr	Thr
	A	Asp–NH_2	Asp–NH_2	Lys	Lys
	G	Ser	Ser	Arg	Arg
G	U	Val	Val	Val	Val
	C	Ala	Ala	Ala	Ala
	A	Asp	Asp	Glu	Glu
	G	Gly	Gly	Gly	Gly

and used in a protein-synthesizing system. The resulting polypeptide contained a regular alternation of serine and leucine. Mathematical analysis of this result shows that the coding unit must contain an odd number of bases. Khorana then synthesized the nucleotide sequence AAGAAGAAGAAG. When this nucleotide polymer was used as the template in a protein-synthesizing system, the result was either polylysine (AAG), polyglutamate (GAA), or polyarginine (AGA). The type of peptide synthesized depends on which nucleotide in the polynucleotide chain happens to be read first. The fact that a different amino acid is specified when the "frame" is shifted can be accounted for only if the chain is read in sequence, three nucleotides at a time, beginning from a fixed point.

In a few instances a specific nucleotide at the 5′ end of the codon can be changed and still provide a code for the same amino acid. For example, both UUG and CUG provide a code for leucine. Arginine is coded by six nucleotide triplets, two of which differ in containing A instead of C at the 5′ end. It appears that the middle nucleotide is the most informative one in the triplet, that the 5′ nucleotide is the next most informative, and that the 3′ nucleotide may not, in some instances, change the amino acid for which the affected triplet codes.

SUMMARY

I. The nucleotides of the DNA triplets determine the mRNA codons. These codons then determine the sequence of tRNA species that line up during translation, because they will associate only with complementary anticodons. The sequence of tRNAs will determine the sequence of amino acids in the finished peptide.

II. Genes are composed of DNA and are located in the chromosomes. Each gene contains information coded in the form of a specific sequence of purine and pyrimidine nucleotides, which is expressed by means of complex enzymatic systems of transcription and translation.
  - *A.* The genetic code is a triplet code; each codon is a group of three adjacent nucleotides that specify the location of a single amino acid in the polypeptide chain.
  - *B.* Genetic information flows from the DNA of the gene to mRNA and then to the specific sequence of amino acids in the peptide chains synthesized on the ribosomes; mRNA contains a sequence of ribonucleotides complementary to the sequence of deoxyribonucleotides in the gene.

III. To be incorporated into a peptide chain, amino acids are first activated by reacting with ATP and then transferred to a specific adaptor molecule, tRNA.
  - *A.* Each kind of tRNA has an anticodon, a sequence of three nucleotides complementary to the specific codon in mRNA.
  - *B.* The amino acid–tRNA complexes are arranged on the mRNA in an order dictated by the complementarity of the nucleotide triplets in the mRNA codon and the tRNA anticodon.
  - *C.* The information coded as a specific sequence of deoxyribonucleotides in DNA is transcribed as a specific sequence of ribonucleotides in mRNA and is ultimately translated into the specific order of the amino acids in the protein molecule.

IV. Messenger RNA is synthesized by a DNA-dependent RNA polymerase that uses DNA as the template and the triphosphates of the four ribonucleosides as the substrate.
  - *A.* The process yields as products an RNA chain and inorganic pyrophosphate (PP_i).
  - *B.* Transfer and ribosomal RNA are also produced by DNA-dependent RNA-synthesizing systems and are transcribed from complementary deoxyribonucleotide sequences in DNA.
  - *C.* After RNA has been synthesized, it may undergo reactions that cleave it or chemically modify it.
    1. The mRNA of eukaryotes may undergo extensive cleaving and splicing, with the removal of certain sequences called introns and the retention of others termed exons.
    2. A "cap" of 7-methyl guanosine is added at the 5′ end, and a polyadenylic acid tail of 150 to 200 adenylic acids is added at the 3′ end. They may serve to protect the RNA from attack by enzymes and to facilitate its transport across the nuclear membrane.

V. Protein synthesis occurs on the ribosomes in the cytoplasm of the cells.
  - *A.* Peptide chains are synthesized beginning at the amino-terminal end by the sequential addition of amino acids to the carboxyl end of the growing chain.
  - *B.* Each amino acid is bound to a specific tRNA, with the carboxyl group of the amino acid attached to the 3′ OH group of the ribose of the adenosine at the CCA (3′) end of the tRNA. The binding of each kind of amino acid to its respective tRNA is catalyzed by a specific enzyme.
  - *C.* Protein synthesis involves three separate processes: initiation, elongation, and termination.
    1. Initiation is marked by the formation of an initiation complex.
       - *a.* The first amino acid–tRNA complex, typically formyl methionyl tRNA, unites with mRNA, an initiation factor, and the smaller ribosomal subunit to form an initiation complex.
    2. Elongation begins with the binding of the second amino acid tRNA to the complex.
       - *a.* The elongation cycle continues with the en-

zymatic formation of a peptide bond, the release of the uncharged tRNA, and the movement of the tRNA containing the growing peptide chain from one site, the A-site, within the ribosome to another, the P-site leaving the A-site empty and ready to accept the next amino acid tRNA.

b. GTP is used in binding the tRNA to the ribosome and in moving the amino acid-tRNA complex from one site in the ribosome to the other.

3. The synthesis of the peptide chain is terminated by release factors that recognize the terminator codons on the mRNA.

VI. The ribosome is formed from its subunits each time a single peptide is synthesized. This permits the mRNA to become attached by complementary base pairing to the smaller ribosomal subunit at just the right point so that the message is read out correctly.

VII. Some genes code for enzymes that catalyze a single step in the biosynthesis of a particular chemical compound.

A. A mutant allele of the gene may produce an altered enzyme, and hence the biosynthesis of the chemical is prevented or hindered.

B. Conditions such as phenylketonuria, alkaptonuria, and albinism in humans result when a person is homozygous for the mutant allele instead of the normal one and lacks the enzyme coded for by the normal gene.

C. In these inborn errors of metabolism the lack of the enzyme interferes with the normal sequence of reactions and either a normal intermediate accumulates (as in alkaptonuria), or the normal product cannot be made (as in albinism).

VIII. Genes may undergo changes called mutations that result in changes in the gene products and thus the kind of protein synthesized.

A. Chromosomal mutations, involving a visible change in the chromosome, include deletions, duplications, inversions, or translocations of a segment of the chromosome.

B. Point mutations involve the substitutions of one nucleotide for another in a specific codon so that it codes for a different amino acid. Frameshift mutations result from the addition or deletion of bases to a triplet, and can drastically change or destroy the entire protein normally produced by the affected gene.

IX. The genetic code is said to be "degenerate," meaning that there is more than one code word for most amino acids; but the code is not ambiguous, for each codon specifies only one amino acid.

A. The genetic code is almost universal, but not quite. Thus a given codon specifies the same amino acid in the protein-synthesizing systems of almost all organisms.

B. The linear, unbranched polynucleotide chain of DNA is colinear with the linear unbranched polypeptide chain.

C. In the binding of an anticodon to the codon, the pairing of the first two bases is quite precise, but the factors recognizing the third base of the codon appear to be less stringent.

POST-TEST

1. The three major functional types of RNA are ______________, ______________, and ______________.

2. Information is transferred from DNA in the nucleus to the protein-synthesizing machinery of the ribosome by ______________.

3. The ribosome is composed of two subunits, each containing ______________ and ______________ ______________.

4. A sequence of three nucleotides in mRNA, the ______________, specifies the addition of a particular amino acid in the growing peptide chain.

5. The complementary sequence of nucleotides in tRNA, the ______________, specifies the amino acid to be added.

6. The synthesis of mRNA, tRNA, and rRNA on their DNA templates is termed ______________.

7. The synthesis of RNA by the enzyme ______________ ______________ ______________ requires a DNA template and, as substrates, ______________, ______________, ______________, and ______________.

8. The transcription process begins at specific ______________ sites on the DNA.

9. The initial RNA transcript undergoes further ______________ in becoming mRNA.

10. The genic DNA contains many more base pairs than those represented in the mRNA transcribed from it; the discarded sequences of bases are termed ______________.

11. The base sequences retained in mature mRNA are termed ______________.

12. The process of protein synthesis is named ______________ because the four-base-pair language of nucleic acids is converted into the 20-amino-acid language of the peptide chain.

13. The formation of an amino acid–RNA complex involves an initial ______________ of the amino acid by reaction with ______________ and then the transfer of the amino acid from AMP to tRNA.

14. The cap at the 5′ end of eukaryotic mRNA consists of ______________ ______________.

15. Protein synthesis occurs on the ribosomes, which brings the ______________ of the mRNA into proper register with the ______________ of the tRNA.

16. Clusters of ribosomes held together by a strand of mRNA are called ______________.

17. The process of protein synthesis can be divided into the stages of ______________, ______________, and ______________.

18. The energy to drive the reactions of protein synthesis is derived from ______________.

19. The several steps involved in protein synthesis can be inhibited by certain ______________ such as ______________, ______________, and ______________.

20. The repeated joining and separating of the ribosomal subunits in protein synthesis is termed the ______________ ______________.

21. The genetic lack of the enzyme tyrosinase results in ______________.

22. A mutation accompanied by a visible change in a chromosome is known as a ______________ ______________.

23. The insertion or deletion of a single base pair from a DNA molecule results in a ______________ ______________.

24. Gene mutations can be produced by ______________, or by exposure to certain ______________.

25. Codons that specify the same amino acid are termed ______________.

REVIEW QUESTIONS

1. Define the term mutation. What types of mutations can be distinguished? How may each be produced?
2. Discuss the problem of "coding" information in the gene.
3. Discuss the transfer of specific information from the gene to the site of protein synthesis in the cell.
4. Distinguish between mRNA, rRNA, and tRNA. What is the role of each in the synthesis of proteins?
5. Outline the steps involved in the initiation, elongation, and termination of a protein chain as it is synthesized on a ribosome.
6. Discuss the importance of specific base pairing between the triplets of the mRNA codon and the triplets of the tRNA anticodons.
7. Diagram the structure of a ribosome, and indicate the roles of each of its parts in peptide synthesis.
8. Describe the events of the ribosome cycle.
9. Outline the evidence that each normal gene produces a single enzyme that catalyzes a single biochemical reaction.

15 Gene Regulation

OUTLINE

I. Mechanisms of gene regulation
   A. The operon concept: control of protein synthesis
      1. Inducible enzymes
      2. Repressible enzymes
      3. Regulatory genes
   B. Posttranscriptional gene control
   C. Gene control in eukaryotes
II. Information transfer outside the nucleus
III. A genetic heresy

LEARNING OBJECTIVES

After you have read this chapter you should be able to:

1. Summarize current concepts of the control of gene function in prokaryotes and in eukaryotes, with emphasis on the operon model of prokaryote gene regulation.
2. Describe both enzyme induction and repression in prokaryotes.
3. Summarize the mechanism of end-product repression.
4. Briefly discuss the evident role of genetic induction and repression in eukaryote development.
5. Summarize the independent genetic mechanisms of mitochondria and of chloroplasts.

Only a small portion of a cell's DNA is likely to be undergoing transcription at any one time, so only a few of its genes are being expressed. Indeed, much of the total genetic information of a cell is more or less permanently suppressed, both in prokaryotes and in eukaryotes.

Mechanisms of Gene Regulation

The ancestry of any human cell—say, a muscle cell—could be traced through many cell generations back to a single, fertilized egg, the zygote. Other very different cells—such as epithelial cells—could also trace their ancestry back to the very same zygote. Except for sex cells, all the descendants of that primitive zygote were produced by mitosis, which, as we have seen, ensures that all daughter cells have exactly the same informational DNA. This would seem to indicate that all these very different cells share the same genetic information, and that in this regard they should behave as identical twins in all aspects of their structure and function.

Yet phenotypically they are obviously not identical twins at all. How can this be? The generally accepted view is that although the various differentiated cells of an organism probably have the same genetic information, different parts of that information are expressed in different tissues. Whatever is inappropriate—and that must be a vast fund—is repressed, not only in each cell but also in the descendants of that cell. In a differentiated cell that actively divides, such as an epithelial cell, the very suppression of much of its genetic information has become hereditary. As another example, the genetic information possessed by a male contains most if not all of the instructions for the production of a female; to be sure, these instructions are suppressed, but however deeply buried in the chromosomes, nevertheless they are there.

There is yet another aspect of this matter: Some cells express some of the information they contain only at certain times. For example, in insects the hormone ecdysone is secreted by the prothoracic glands in preparation for molting. This steroid hormone unlocks certain genes in the cells of numerous tissues (Fig. 15–1). Apparently, these genes function only when the cell is stimulated in just the proper way. How is gene suppression accomplished, and what makes it temporary or permanent? Those answers could trigger the most fundamental advances in every medical field from teratology (the study of malformations) to immunology.

THE OPERON CONCEPT: CONTROL OF PROTEIN SYNTHESIS

It appears from many kinds of experimental evidence that genes normally do not operate to produce the maximum number of enzymes or other proteins all the time. Each gene appears to be **repressed** (turned off) to a greater or lesser extent under normal conditions. Then, in response to some sort of environmental demand for that particular enzyme, the gene becomes **derepressed** (turned on). Derepression is the same as **induction**—it refers to the process of switching on transcription. When a single gene is fully derepressed it can cause the synthesis of fantastically large amounts of an enzyme; under these circumstances one kind of enzyme may comprise 5% to 8% or more of the total protein of the cell! The speed of biochemical reactions and the functional mapping of biochemical pathways depend, in part, on the quantities of specific enzymes present. If all enzymes were produced at a similar very high rate, metabolic chaos would ensue. Thus the phenomena of gene repression and derepression appear to be necessary to provide a means of increasing or decreasing the rate of synthesis of some particular enzyme.

For instance, the *Escherichia coli* bacteria living in the colon of an adult cow are not normally exposed to the milk sugar lactose, nor have their ancestors been for perhaps a hundred bacterial generations since their host was a calf. Yet if those *E. coli* bacteria were to end up on the foot of a fly, and that fly were to land in a bucket of milk, they would have lactose readily available as an energy source. This, in a way, poses a dilemma. Should the bacteria invest energy and materials to produce lactose-metabolizing enzymes all of their lives on the chance that they may fall into a bucket of milk? Yet if they do not produce those enzymes, they might one day starve in the midst of an abundant potential food resource.

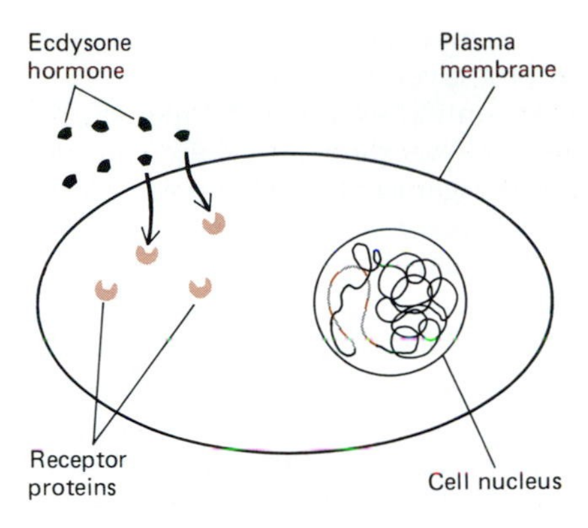

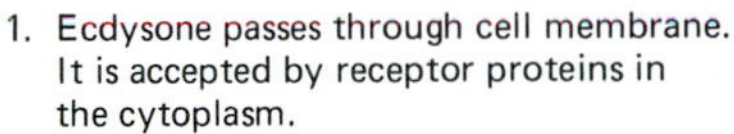

1. Ecdysone passes through cell membrane. It is accepted by receptor proteins in the cytoplasm.

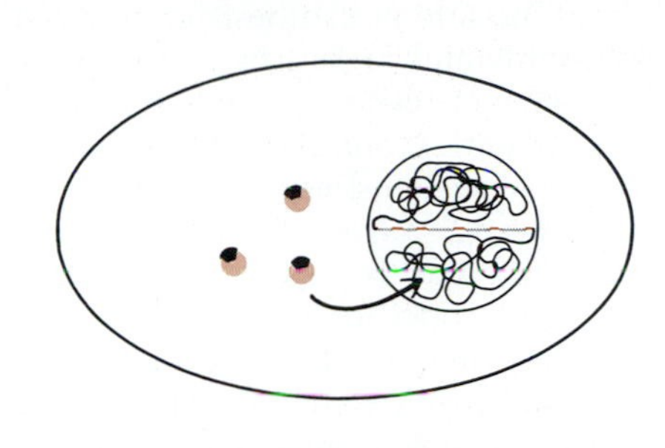

2. Receptor-ecdysone complex attaches to a segment of nuclear DNA

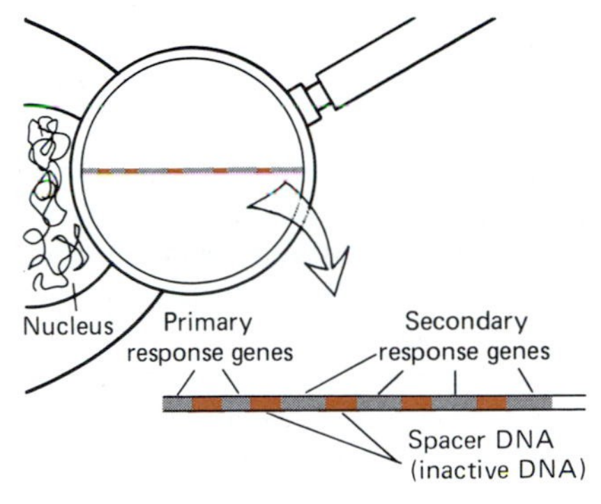

3. The nuclear DNA contains a portion that embodies specialized genes

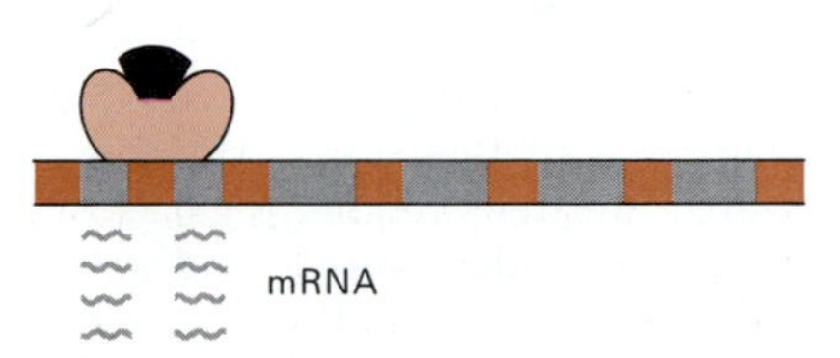

4. When the receptor-ecdysone complex attaches to the primary response genes transcription of these genes is activated. In due course the resulting mRNA is translated. It forms primary response proteins:

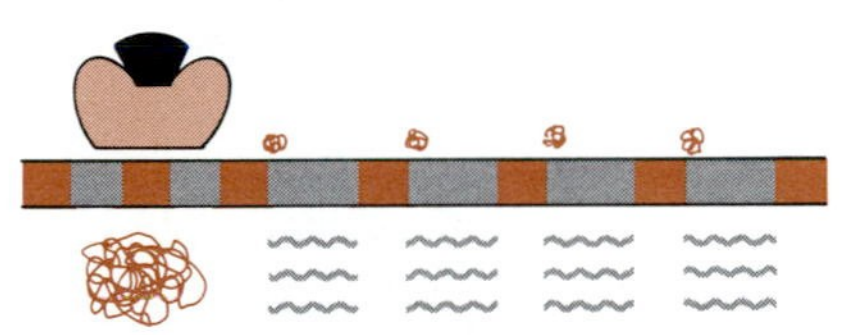

5. If a large amount of primary response protein is made, one of these newly synthesized proteins binds to the primary response genes and turns them off by displacing the receptor-ecdysone complex. (This is evidently a regulatory provision). Other primary response proteins cause the secondary response genes to be transcribed.

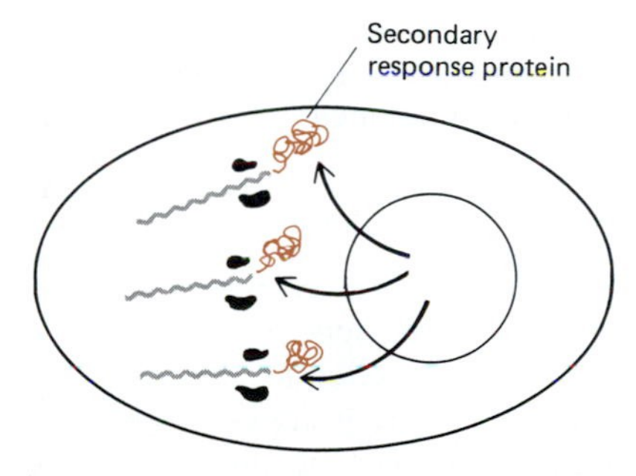

6. Secondary response proteins are translated from secondary response mRNA.

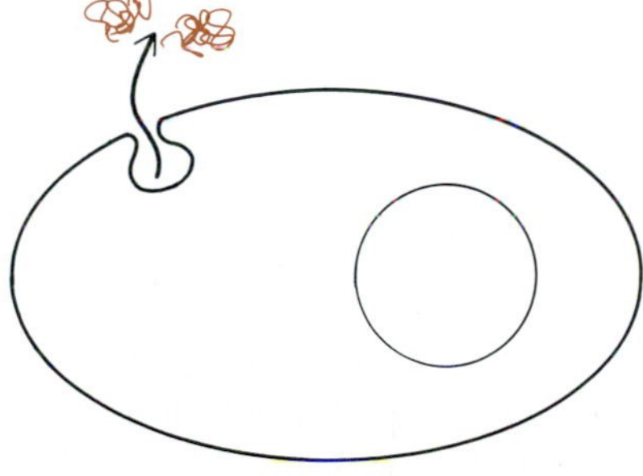

7. In some cases the secondary response proteins are exported from the cell.

Figure 15–1 How the steroid hormone ecdysone regulates genes in insect cells. The proteins produced in response to the ecdysone perform numerous functions in the body of the molting insect.

And what of the cow? As a calf it needed lactose-digesting intestinal enzymes, but as an adult that no longer drinks milk it produces little or no such enzyme. Both the bacterium and the cow need and have ways of regulating the enzymatic expression of their genetic information. Both in unicellular organisms and in the cells of multicellular organisms, there may be wide variations in the *number* of enzymes per cell and in the *amount* of a given enzyme per cell.

The rate of synthesis of a protein may be controlled in part by the genetic apparatus and in part by factors from the external environment. These controls have not yet been fully studied in a wide variety of organisms. Most of the information about the control of protein synthesis in cells has come from studies of simple microbial systems, especially those of the colon bacillus *E. coli.*

Even *E. coli* is extremely complex. From the length of the chromosome of this bacterium and the estimate that the average gene contains about 1500 nucleotide pairs (and codes for a polypeptide chain of 500 amino acids), it has been calculated that the genes of *E. coli* code for 2000 to 4000 different polypeptides. Estimates place the number of different enzymes required by an *E. coli* bacterium growing on glucose at about 800. Some of these must be present in large amounts, whereas others are required in only small quantities; but regardless of quantity, a specific plan of information storage and retrieval is required for the manufacture of each.

Inducible Enzymes

One of the most intensively studied enzyme systems in the *E. coli* bacterium is that concerned with the metabolism of the sugar lactose. *E. coli* cells growing on glucose contain very little of the enzyme **β-galactosidase,** which is necessary for the initial hydrolysis of this sugar. However, when grown on lactose as the sole carbon source, the cells require β-galactosidase to cleave the disaccharide lactose to the monosaccharides glucose and galactose. Under these conditions β-galactosidase makes up some 3% of the total protein of the cell. There are perhaps 3000 molecules of this enzyme in each cell of *E. coli,* which represents at least a thousandfold increase over the amount present in cells growing on glucose instead. Two other required enzymes, galactoside permease and galactoside transacetylase, respond with equally dramatic increases to the presence of lactose in the incubation medium. The permease is needed for an early stage in the degradation of galactose. Thus the presence of lactose stimulates the *entire* enzyme system necessary to metabolize it. A substance such as lactose that elicits the synthesis of new enzyme molecules is called an **inducer,** and enzymes that respond to inducers are termed **inducible enzymes** (Fig. 15–2).

Repressible Enzymes

Cells of *E. coli* can grow in a medium without any amino acids, because they contain the whole spectrum of enzymes required to synthesize all of the 20 known amino acids needed for the assembly of protein molecules. However, synthesizing an amino acid requires quite a bit of energy. It is to the bacterium's advantage to use them ready-made, if they are available. The introduction of one or more amino acids to the incubation medium greatly decreases the amount of those amino acids that the bacterium must produce, so as we might expect, this also decreases the amount of the biosynthetic enzymes required for the production of that amino acid. Biosynthetic enzymes that are reduced in amount by the presence of the end-product of such a biosynthetic sequence (e.g., the amino acid) are called **repressible enzymes,** and the small molecule (the amino acid) that brings about the repression is called a **corepressor.**

Regulatory Genes

Both enzyme induction and enzyme repression have now been explained by the work of several investigators. Differences in the amount of a particular protein found in bacteria are generally due to variations in the rate of synthesis of that protein rather than to changes in its stability or in the rate of its degradation. The rate of synthesis is controlled in turn by the amount of mRNA present per cell. How, though, could the rate of mRNA synthesis be regulated? The French researchers Jacob and Monod postulated that the amount of mRNA template that produces a given enzyme is controlled by a special kind of molecule, a protein called a **repressor** that blocks the synthesis of a specific kind of mRNA. They further suggested that these repressor proteins are coded for by special genes termed **regulatory genes.** Thus, regulatory genes control the expression of **structural genes,**

1. Normal situation, operon "off." Transcription and translation of the regulator gene produces a repressor protein that binds to the operator and blocks RNA polymerase from transcribing the operon. The RNA polymerase promoter site (not shown) overlaps the operator site.

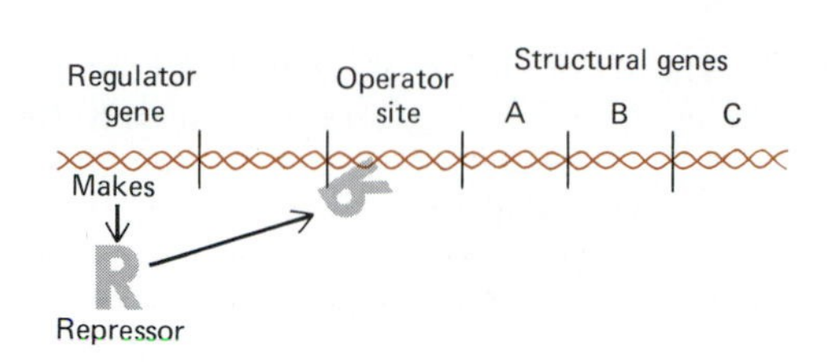

2. Inducer (from environment), which is usually a substrate of enzyme A, B, or C, inactivates repressor. Operator-promoter site now is free and transcription by RNA polymerase turns the operon "on."

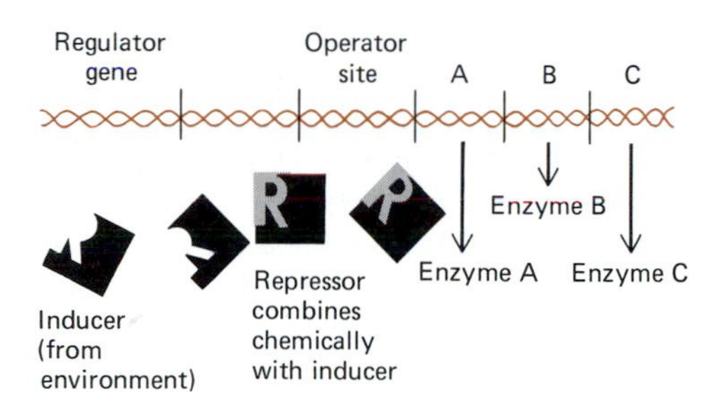

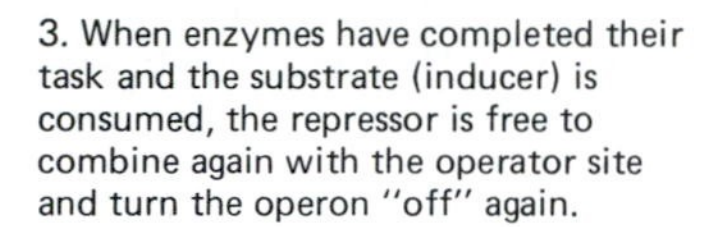
3. When enzymes have completed their task and the substrate (inducer) is consumed, the repressor is free to combine again with the operator site and turn the operon "off" again.

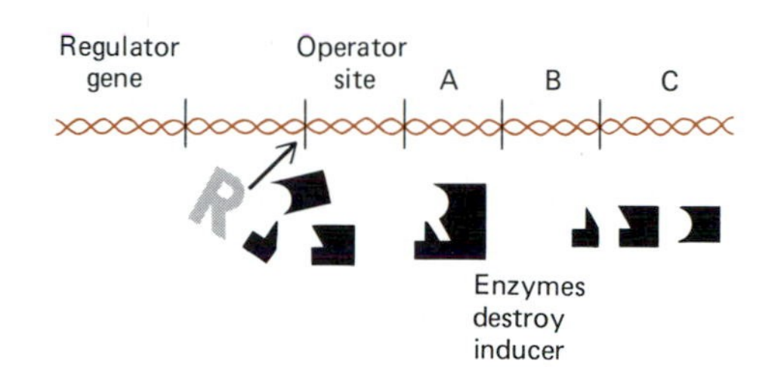

Figure 15–2 A typical mechanism for regulating prokaryotic gene expression. The operon is the basic unit of gene control. It consists of a group of structural genes, associated regulatory genes, and operator sites.

genes that code for any RNA or protein other than a regulator. This hypothesis has been abundantly confirmed. For example, Gilbert and his colleagues at Harvard reported the isolation and characterization of the repressor for β-galactosidase from *E. coli*, a medium-sized protein. From their data they calculated that there are about 10 molecules of β-galactosidase repressor per cell. This is not much—but then, bacteria are pretty small!

If regulatory genes made repressors all the time, it would follow that the synthesis of mRNA would always be inhibited. Since this is clearly not the case, it is necessary to postulate further that repressors exist in either active or inactive forms whose competence depends on whether the repressors are combined with inducers (such as lactose) or corepressors (such as amino acids). The attachment of an inducer inactivates the repressor by producing a conformational change in its structure. Thus the combination of lactose with the repressor neutralizes the repressor and permits the synthesis of the enzyme. The repressor itself binds to its specific inducer or corepressor by weak bonds such as hydrogen bonds located perhaps in an active site (Fig. 15–3).

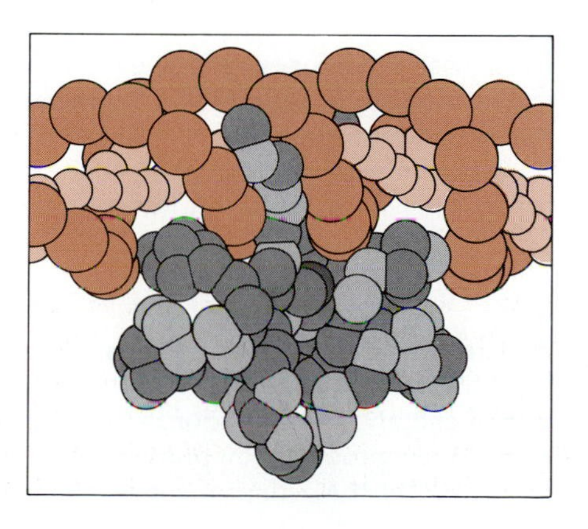

Figure 15–3 The relationship between a typical repressor protein and the DNA molecule. Note that the α-helices of the protein interpenetrate the major groove of the DNA double helix. Since mRNA transcriptase must attach to the downstream promoter site in this very major groove and overlaps the structural gene, the presence of the protein blocks transcription. (From Takeda, Y., Ohlendorf, D. H., Anderson, W. F., and Matthews, B. W.: DNA-binding proteins. *Science* 221:1020–1026, 1983. Copyright 1983 by the American Association for the Advancement of Science. Used by permission.)

The action of corepressors is the reverse of that of inducers. The attachment of the corepressor (e.g., an amino acid) to the repressor *activates* it. The resulting complexes then are able to bind to the DNA, decreasing the transcription of the DNA and the number of enzyme-coding mRNA molecules that are produced. Notice that it is the very same substance to be metabolized that combines with the repressor. Since the bacterium has access to this substance, it need not manufacture it. Thus all the enzymes needed to make it are suppressed when it is present ready-made.

In some instances (such as the case of lactose) the amounts of two or more enzymes may vary in a coordinate fashion, suggesting that they are all under the control of the same repressor system. It has been inferred that a single repressor may control the formation of the mRNA for these **coordinately repressed enzymes.** Frequently, but not always, genetic mapping shows that the genes for these coordinately repressed enzymes are closely linked on the chromosome. A single mRNA molecule may carry the message for the synthesis of two or more enzymes.

Originally it was believed that *all* the genes controlled by the same repressor must lie closely adjacent in a prokaryote chromosome, but in certain systems, widely separated genes appear to be coordinately repressed by the same repressor. However, in the bacteria in which these systems have been investigated, the "chromosome" is a single circle of DNA containing *all* the genes. Thus all bacterial genes are linked.

Just how does the repressor control the structural genes? Evidently it does so by means of **operator sites,** which are believed to lie adjacent to the set of structural genes. The operator site consists of a promoter, plus a base sequence to which the repressor can specifically bind. Active repressor molecules bind to the operator sites, thereby inhibiting the synthesis of mRNA by these structural genes (Fig. 15–3). The binding of the repressor blocks RNA polymerase because the operator overlaps the promoter. Regulatory genes, incidentally, are themselves controlled in much the same way that structural genes are. The ultimate control mechanism that masters the whole sequence is not well understood.

In the absence of active repressors, the structural genes are free to be transcribed, mRNA is formed, and enzymes are produced on the ribosomes. Mutations are known in which the operator is absent or incompetent. In these mutants, endless quantities of enzymes are produced regardless of the presence or absence of the inducer or corepressor substance.

The structural genes under the control of one repressor, together with an operator site, are collectively termed an **operon.** Thus according to the operon concept, protein synthesis is controlled by a coordinated unit consisting of a regulatory gene, an operator site, a promoter site for the binding of RNA polymerase, and a set of structural genes (Fig. 15–2). To sum it up (see Focus on Some Important Gene Regulation Definitions), in addition to the structural genes that provide the code for the synthesis of specific proteins, there must be regulatory genes that code for the synthesis of repressors. The repressors may be active or inactive, depending on whether they are bound to small molecules, the inducers, or corepressors. Active repressors bind to operator sites in the DNA and turn off the transcription of the adjacent structural genes.

Not all bacterial genes are subject to repressor control. There is clear evidence that the synthesis of many proteins in *E. coli* and in other organisms is not influenced by substances such as corepressors and inducers in the external environment.

The search for eukaryote equivalents of the operon has yielded some theoretical results, but thus far few of the numerous hypotheses that have been proposed have been confirmed (the case of ecdysone is a happy exception). One recent proposal is summed up in Figure 15–4. Even if it proves to be correct there are probably many other mechanisms in the complex eukaryotic cell.

POSTTRANSCRIPTIONAL GENE CONTROL

Posttranscriptional control mechanisms operate upon translation or upon the proteins themselves rather than upon DNA. Often, they involve some kind of regulation of enzyme function. For instance, enzymes can be inhibited by the accumulation of their products, by phosphorylation, or allosterically (by a change in shape) as a result of the action of some other triggering substance.

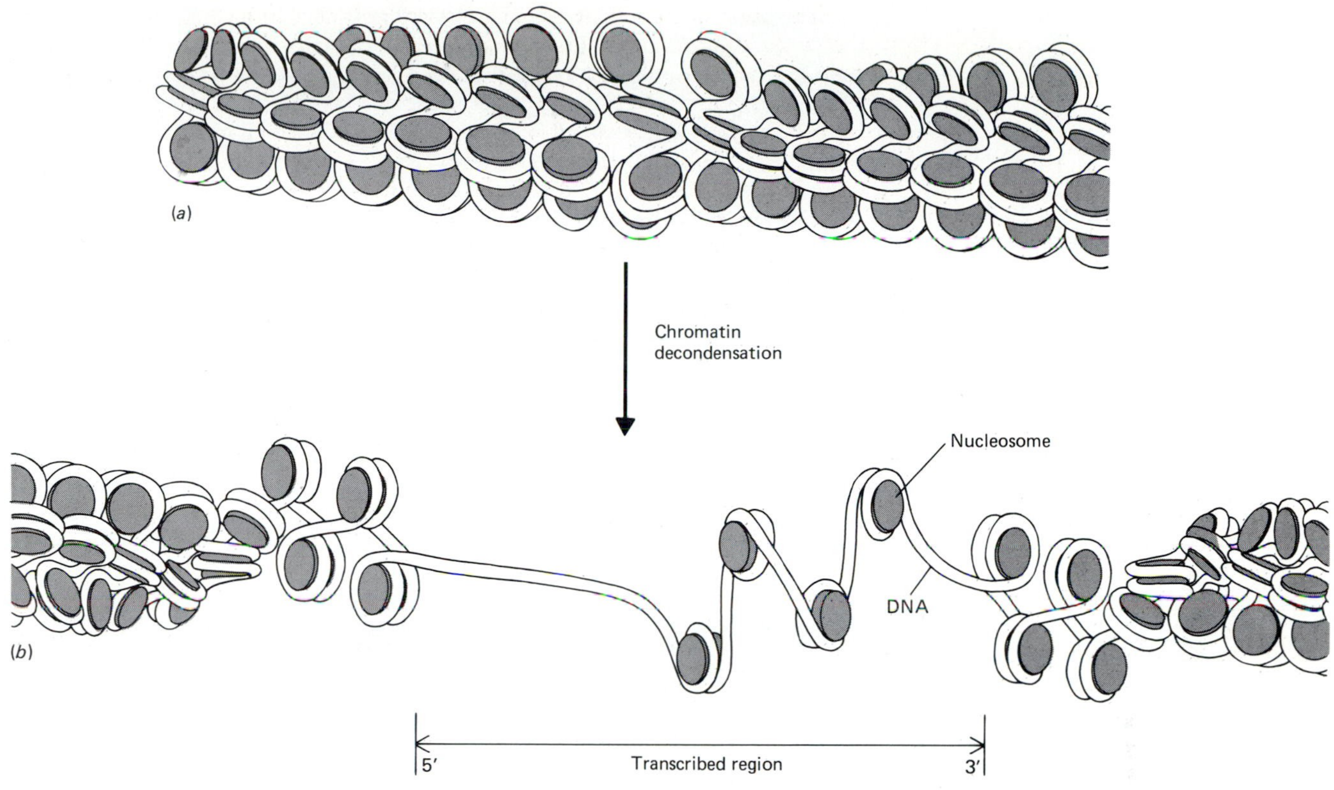

Figure 15–4 A possible mechanism of gene expression in eukaryotes. *(a)* An inactive section of eukaryotic DNA. The DNA is hightly condensed; it is organized in nucleosomes that, in turn, are tightly coiled into chromatin fibers. *(b)* The chromatin in the region of active genes is decondensed, sometimes in response to specific inducing signals, increasing the accessibility of the DNA to RNA polymerases required for transcription.

The most exciting discoveries in this area, however, have disclosed control mechanisms fully comparable to the operon that involve the regulation of *translation* from mRNA to protein. Although it is likely that a wide variety of cellular proteins (especially structural proteins) are regulated in this way, translational control of RNA expression was first demonstrated in the control of the production of the proteins that make up part of the ribosome itself (Fig. 15–5).

Even ribosomal proteins must be translated with ribosomal aid, although normally this does not constitute any kind of problem since each cell is derived from a preexisting cell from which it obtains an initial supply of ribosomes. Ribosomes, however, are expensive to produce in terms of the total chemical resources of the cell. It is advantageous to the cell to limit their production to the exact number needed at any time. The way this is done is to control the production of ribosomal RNA by means of the total number of ribosomes in the cell, and the amount of protein indirectly by the amount of rRNA available for the construction of ribosomes.

The *E. coli* ribosome consists of 3 molecules of RNA and 52 distinct protein particles. These must all be produced in equal quantities for greatest efficiency. Recent evidence[1] shows that a few key proteins repress the formation not only of themselves but of several other proteins as well. Even when present in great excess, however, they do not change the rate at which their mRNA is produced—suggesting that one protein of each group serves to inhibit the *translation* of all the members of the group. (In fact this has been shown to be the case.) The control protein blockades the strand of mRNA at a point called the **ribosome entry site,** which is the only place on the strand that a ribosome will initiate translation. The complements of several genes occur on the polycistronic strand, and if the ribosome entry site is blocked, none of their proteins will be produced since none of the genes can then be translated. That means that if there is an excessive amount of the control protein available, neither that protein itself nor other members of its polycistronic group will be manufactured until a shortage develops.

If unused rRNA is present, the control protein will combine with *it* and thus will not be able to block transcription of its mRNA strand (for all we know at

[1] For a more detailed discussion, see Masayasu Nomura, M.: The control of ribosome synthesis. *Scientific American,* January 1984, pp. 102–114.

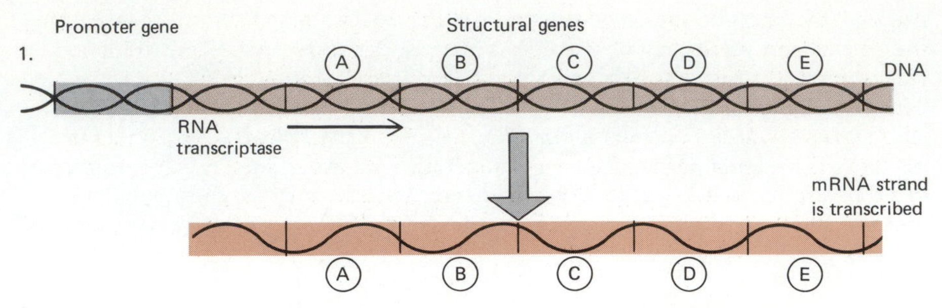

The structural genes for several ribosomal proteins are transcribed sequentially by RNA transcriptase, producing a strand of mRNA containing them all.

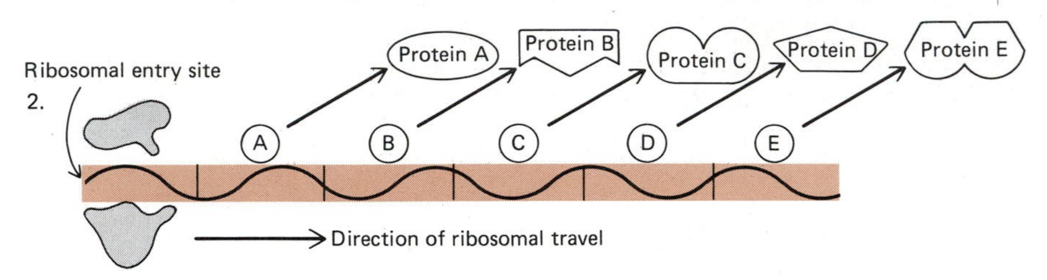

The mRNA strand is threaded into the ribosome at the ribosomal entry point. All the genes on the strand will be transcribed and *several* proteins will necessarily be produced in equal amounts (this is called polycistronic translation).

3. The proteins are assembled into a new ribosome. But if they cease being used, then

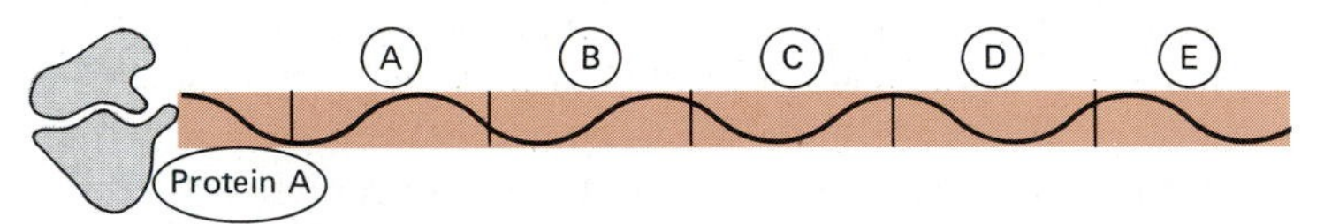

one of the proteins binds to the ribosomal entry site, preventing ribosomal entry. No further transcription can occur—*unless* protein A is used up before it can bind to the ribosomal entry site.

4. This ensures that only as much of each protein is made as can be used, and that they all are made in equal amounts.

Figure 15–5 The control of ribosomal protein manufacture.

FOCUS ON

Some Important Gene Regulation Definitions

Operon: in prokaryotes, a complete unit of genetic expression including a group of structural genes, associated regulatory genes, and operator sites.

Regulatory gene: section of DNA in an operon that codes for repressor substance whose function is to control rate of synthesis by associated structural genes. The repressor substance binds to the operator site.

Repressor substance: small protein that locks operator gene in "off" position; is capable of combining with inducer substance from environment, losing its effectiveness as a repressor.

Corepressor substance: in prokaryotes, an environmental trigger (such as intermediate metabolite) that represses transcription by binding to a repressor substance; combines with an inactive repressor substance in such a way as to activate it.

Operator site: in prokaryotes, a section of DNA whose function is to act as a switch, turning off structural genes when a repressor protein binds to it.

Inducer substance: in prokaryotes, environmental trigger (such as food substance that is attacked by the enzymes produced by the structural genes) that combines with the repressor substance, thus unlocking the structural gene or genes of the operon.

Structural gene: in prokaryotes, a section of DNA in an operon that codes for any RNA or protein product other than a regulator.

Promoter: a specific DNA sequence that serves to indicate where transcription should begin.

present, this may be the only function the control proteins have). Consequently, the production of ribosomal protein ultimately depends upon the production of rRNA, for when the cell runs out of rRNA the control proteins will have nothing to combine with and will stick themselves to the ribosome entry points of their mRNA (this leaves quantities of useless mRNA in the cytoplasm; presumably some mechanism exists to inhibit its continued production). It has been shown that the amount of rRNA produced depends on the number of ribosomes: When they are present in excess, some negative feedback mechanism prevents the further production of rRNA.

GENE CONTROL IN EUKARYOTES

Molecular geneticists used to say, "Whatever is true of *E. coli* must be true of elephants," reflecting the assumption that all life had a fundamental unity on the molecular level. By studying a simple system in a bacterium, researchers believed they could discover things much more readily than by studying a complicated eukaryote, and yet the results obtained could probably be extended to eukaryotes as well.

The odd thing, however, is that despite their admitted differences, what is true of *E. coli* or other bacteria *is* often true of elephants! Both use the same genetic code, for instance, which is not a trivial similarity. However, both do *not* employ the same system of gene control. How eukaryotes control gene expression is still unclear despite many years of research.

In many cases it is believed that cyclic adenosine monophosphate, or cAMP (see Fig. 3–15), is active in the release of selected regions of the chromosome from inhibition. In the thyroid gland, for instance, the peptide **thyroid-stimulating hormone** secreted by the anterior pituitary gland may stimulate certain receptor proteins in the thyroid gland's cell membranes. In response, these receptors stimulate the enzyme adenyl cyclase to catalyze the conversion of ATP into cAMP. The cAMP may then diffuse to the nucleus of each cell and there unlock the regions of its DNA that are concerned with thyroid hormone production. Of course this does not address the question of what it is that keeps it locked! One possibility is DNA methylation, in which methyl groups are attached to the bases in such a way as to prevent regional transcription. Actually, eukaryotic gene control is a very complex problem and it is now thought that several very different mechanisms may be employed.

The studies by Barbara McClintock (awarded a Nobel Prize in 1983 for her efforts) have sparked a now-widespread line of research that promises to cast light upon some of these difficult matters. In simple terms, McClintock discovered in the 1940s that the expression of certain genes governing kernel color in corn (maize) was expressed not merely in accordance with the usual dominance and recessiveness relationships but also in dependence upon *where they were located on the chromosome.* Since each kernel of corn is the product of a separate and single fertilization, all cells of a kernel share the same genotype. The variegation of color observable in various regions of each kernel could be explained only in terms of differential expression of the color genes in various cell lines in the kernel. McClintock suggested that chromosomes were sometimes broken and reformed in the course of cell division, and when this occurred certain genes that she called **controlling elements** would move about from place to place in the chromosome. When a controlling element happened to alight near a color gene it could turn the color gene's expression on or off. Eventually it was determined that the controlling elements also caused the breaks in the DNA of the chromosome that allowed them to jump about.

Despite the comparative simplicity of bacterial cells, and although such movable elements do occur in bacteria, they were not discovered in prokaryotes for another 20 years. Stanley Cohen, James Shapiro, and their coworkers found that certain bacterial DNA sequences, which they called **insertion sequences** (ISs), could "shoulder" into the middle of many bacterial genes; when this occurred the target gene ceased operation. Thus, an IS could function as a genetic switch in bacteria in much the way a controlling element could in maize. Eventually it became clear that genes conferring antibiotic resistance could jump about in the bacterial genome somewhat as the IS could. Such a gene might originate in the bacterial chromosome, jump into a plasmid (a kind of small accessory chromosome, discussed in Chapter 16), be transferred along with the plasmid to another bacterium, and then jump into the chromosome of its new owner. (Antibiotic resistance can even be transferred among bacteria of different species by similar mechanisms.) Such movable genetic elements are now called **transposons.**

McClintock found that some maize genes (**regulators**) are capable of governing the action of genes quite distant on the chromosome or perhaps even on other chromosomes. The target genes can apparently be of any variety provided that they contain **receptor** elements. The receptor elements are transposable. That means that a regulator can turn whole banks of genes on and off simultaneously—for example, in development. Membership in such a gene bank could also change in accordance with the needs of the organism. The facile changes in gene regulation thus made possible could produce more rapid evolutionary changes than classical Mendelian mechanisms would be able to accomplish.

Transposable elements not only make gene relocation and remote control easily possible, but also can produce rearrangements of coded sequences within a gene (see Focus on Four-winged Flies). One of the greatest objections to the clonal selection theory of antibody production (Chapter 36) has always been that there is no known way in which a particular region of a cell's genome could mutate as rapidly and consistently as would be required to produce lymphocytes that make antibodies to every imaginable antigen. Yet this is just what the clonal selection theory demands. Evidently the antibody-producing genes of lymphocytes are split genes in which the base sequences occur in little packets separated by intervening sequences of "nonsense" DNA. By transposing these in various combinations, the amino acid sequence of the immunoglobulin proteins they produce is rescrambled in every cell division so that the daughter cells make totally new antibodies, different both from one another and from their parent cell. When the lymphocyte becomes a plasma cell the ability to transpose these elements is lost so that the production of the current antibody becomes fixed for all subsequent generations of that cell line.

Information Transfer Outside the Nucleus

It has been known since early in this century that a few traits appear to be inherited not by the usual nuclear genes but by some mechanism restricted to the cytoplasm. Such traits are inherited exclusively from the maternal parent: The egg, but not the sperm, supplies cytoplasm to the zygote. This cytoplasmic heredity does involve DNA; thus, biological information may be transferred by DNA other than that in the nucleus. Both mitochondria and chloroplasts contain DNA (Fig. 15–6). For example, about 2% of the DNA of a liver cell is located in its mitochondria. Human mitochondrial DNA is very compact, containing no introns. Each mitochondrion has about the equivalent of the genetic information in a small virus or even less, hardly enough to manufacture or operate this organelle. Actually, most (95%) of the genes needed to construct and maintain mitochondria are nuclear. The amount of DNA in the mitochondria of beef heart cells has a total molecular mass of 3×10^7 amu. The DNA in mitochondria exists in several pieces, each a complete circle of molecular mass 1×10^7 amu. The replication of mitochondrial DNA is completely independent of the nuclear DNA.

Plastids such as chloroplasts contain even more DNA and RNA than do mitochondria, and, like mitochondria, have some capacity for independent growth and division and for the synthesis of specific proteins.

The protist *Euglena* normally has chloroplasts but can survive without them if supplied with appropriate nutrients. Euglenas that are deprived of their chloroplasts never develop new ones. However, if they are again supplied with chloroplasts these bodies will undergo division and appear in all daughter cells. If the *Euglena* is supplied with structurally different chloroplasts from a mutant strain, these unusual chloroplasts will undergo division and appear in the daughter cells, still retaining their odd characteristics. This is cogent evidence that the control of chloroplast structure is at least in part under the control of DNA in these plastids themselves.

But just how much of their needed information *is* supplied by the chloroplast's own DNA? Some idea of this could be gained by implanting chloroplasts into *animal* cells, which would presumably have no genetic information in their nuclei that would have any bearing on chloroplast structure or function. Margit Nass in 1969 showed that mouse cells grown in cell culture are able to engulf by phagocytosis chloroplasts prepared from spinach leaves. The chloroplasts survived

FOCUS ON

Four-winged Flies

According to the most widely accepted view, insects arose from millipede ancestors by a process of segment loss and fusion. Each primitive segment had a single pair of legs (unlike modern millipedes, in which segment pairs are fused so that each has four). Since two pairs of wings are considered to have developed subsequently from lateral folds or expansions of the thorax, surviving primitive insects (such as silverfish) have only six legs and no wings. In this view, most modern insects retain four wings (e.g., dragonflies, roaches, bees, and butterflies), but the true flies or Diptera have lost the second pair of wings ("Diptera" means "two-winged ones").

The second pair of wings is, however, not completely gone but persists in the form of a pair of balancing organs, or **halteres**, which help the fly to sense changes in flight direction. In 1923, C. Bridges reported the discovery of a strain of fruit flies (*Drosophila*) in which the halteres developed as a not entirely functional second pair of wings. This is the **bithorax trait**.

A series of **homeotic mutations** exists in each of which one segment is modified to resemble another one ahead of it or behind it. The genetic elements that govern segmental development in fruit flies appear to be chromosomally linked together in approximately the same order as the segments whose development they govern. A group of three such mutant genes produces the most perfect version of the four-winged bithorax trait. It is interesting that the mutant elements are for the most part not point mutations but insertions and deletions.

Indeed, bithorax-type traits have proved to result from the interposition of movable genetic elements such as those McClintock discovered in maize. It seems reasonable to think that the misplaced transposable elements interfere positionally with control systems governing developmental gene expression. In the case of the bithorax mutant, the genes that shape the halteres are evidently rendered inactive, whereas those that produce well-formed wings (possibly those that are normally active in the preceding segment) are permitted to influence a posterior segment over which they normally have no control.

Perhaps the ultimately most significant consequence of the existence of such genes may prove to be their evolutionary effect. The difference between two-winged and four-winged insects is a major one by any reasonable criterion. Transposable elements could produce such a change in one or a few generations, and in fact, they did! If four-winged flies had any significant advantage over the traditional two-winged variety, the four-winged insects could very swiftly become the norm. In such a case, evolution would not have occurred gradually, but in an abrupt fashion called **saltation** by some theorists (Chapter 45). Since "saltation" means "jumping," it would be appropriate if such mutations were produced by transposable, jumping genes.

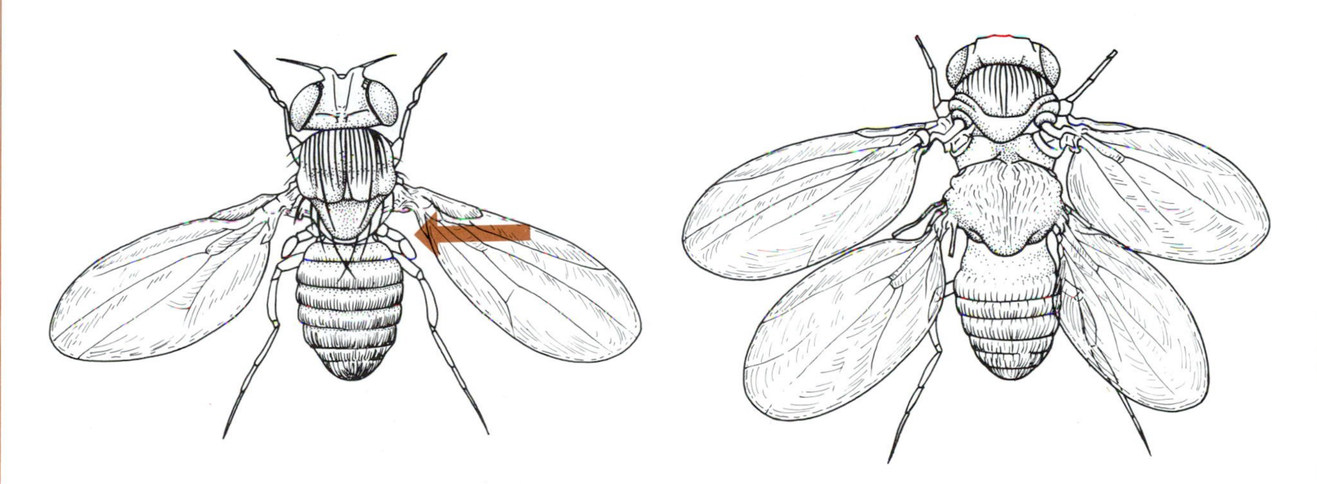

Normal (wild-type) fruit fly (left) compared with four-winged fly (right). The halteres (indicated by the arrow) of the normal fly are transformed into wings, from which they are thought to have evolved in the first place. Three separate bithorax mutations combine to produce a fruit fly with a second pair of wings. (After Edward Lewis, California Institute of Technology.)

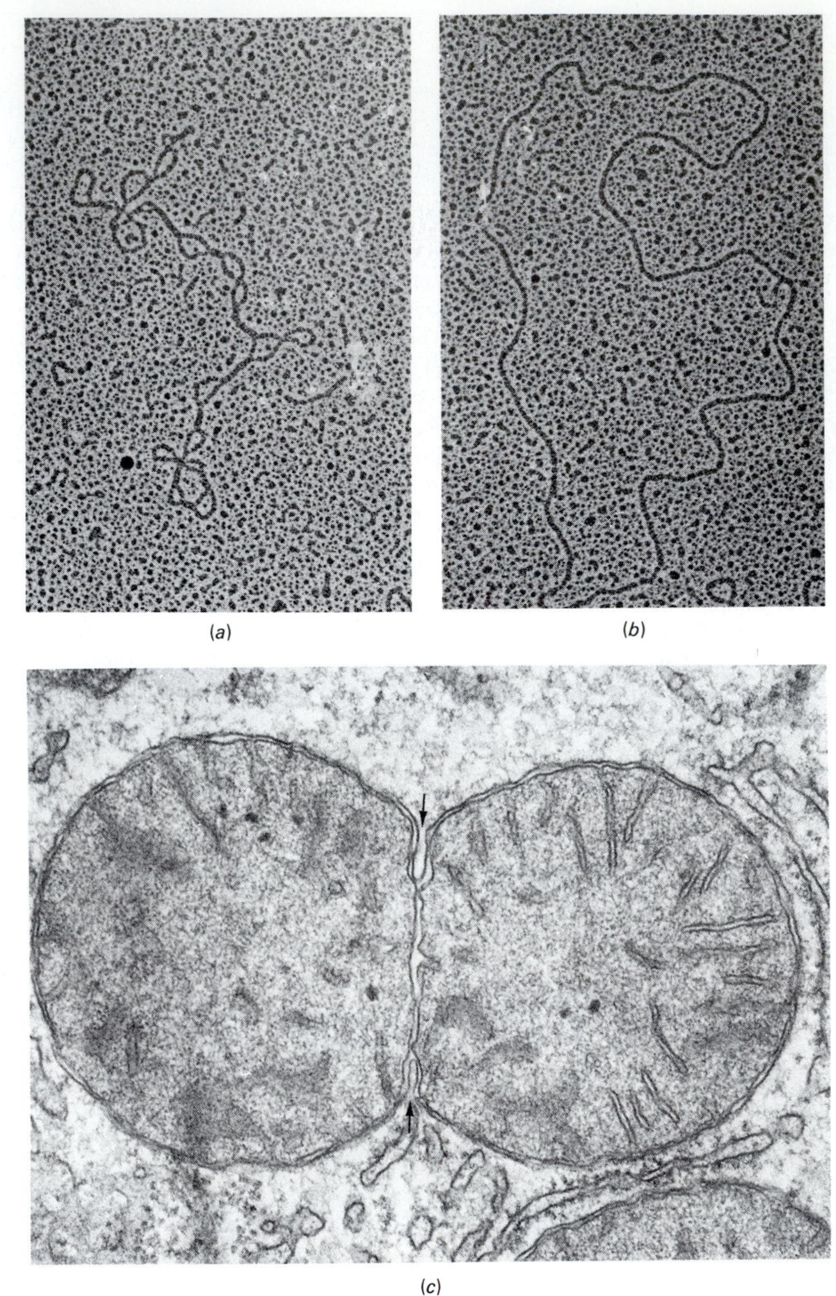

Figure 15–6 DNA extracted from rat liver mitochondria and visualized in the electron microscope. *(a)* A DNA molecule in the twisted circle configuration. *(b)* A DNA molecule in the open circle configuration. *(c)* New mitochondria arise from the growth and division of existing mitochondria. (Chloroplasts also duplicate themselves.) The dividing mitochondrion shown here is from a rat liver cell. *(a, b,* Courtesy of B. Stephens. *(c),* Daniel Friend, Photo Researchers, Inc.)

for at least six cell generations (six days) and retained their structural integrity (Fig. 15–7). At the appropriate time they divided and were assorted at random to the daughter cells.

These and other observations have led to the proposal of the endosymbiotic origin of subcellular organelles such as the mitochondria of eukaryotic cells. It postulates that early bacteria-like or alga-like organisms entered early eukaryotic organisms and took up permanent residence as mitochondria or chloroplasts. This hypothesis will be discussed further in Part VI.

Examination of the details of the coding relationships in the DNA of yeast and human mitochondria has led to the unexpected finding that some of these codons are different from the accepted "universal" code. For example, in yeast mitochondria the sequence AUA codes for methionine rather than for isoleucine and the sequence UGA codes for tryptophan instead of serving—as it does in the nucleus—as a terminator code. As far as is known, the chloroplast employs the "universal" code, and so does every other living thing.

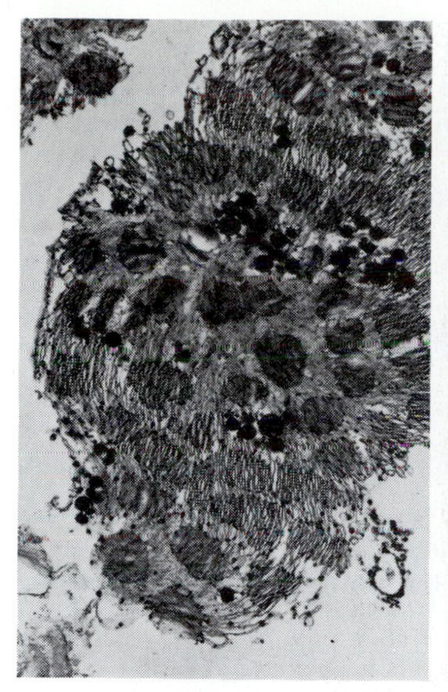
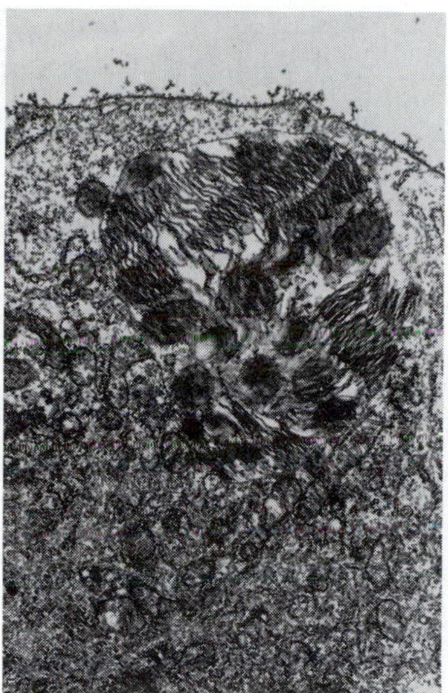
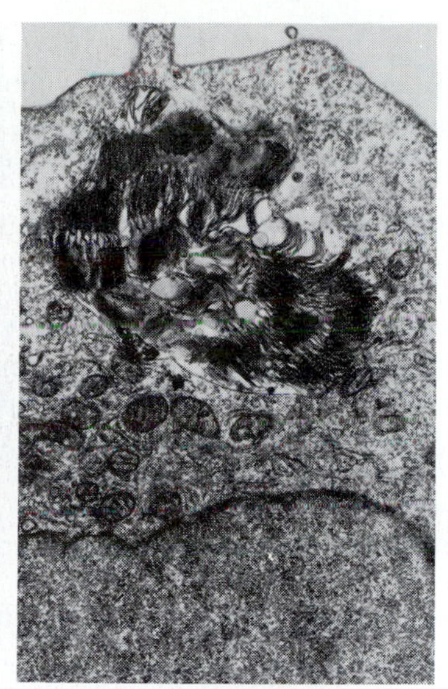

Figure 15–7 Electron micrographs of spinach chloroplasts that have undergone phagocytosis by mouse cells grown in culture (×12,000). The chloroplasts have retained their structural integrity. (Courtesy of Dr. Margit Nass.)

A Genetic Heresy

The central dogma of molecular genetics—that information flow is one way, from DNA to RNA—has an important exception in the case of RNA viruses, some of which, the retroviruses, are able to transcribe a DNA intermediate. An even more fundamental assumption of modern molecular genetics is now being challenged. Some recent evidence suggests that the infective agent of certain diseases (such as scrapie of sheep, kuru of New Guinea cannibals, or even Alzheimer's disease) thought to be virally caused may not be viral at all, but a tiny protein particle, called a **prion**, which *as yet* has not been found to contain any nucleic acid. Many researchers think that prions may cause disease by affecting genetic expression, but they do not actually transmit genetic information (i.e., they are not analogous to DNA as primary carriers of a genetic code).

Whether proof of the existence of such enigmatic agents will be possible cannot be predicted now. Yet there is nothing theoretically impossible about protein as a carrier of genetic information; it is certainly complex enough. Moreover, it is entirely likely that enzymes could be designed that would establish peptide bonds between amino acids without the need for templating by a nucleic acid.

SUMMARY

I. In prokaryotes, in addition to the structural genes that provide the code for the synthesis of specific proteins, there are regulatory genes that code for the synthesis of repressors.

A. Repressors bind to the operator sites in DNA and turn off the transcription of adjacent structural genes.

B. The repressor protein may be inactivated by the presence of a particular substrate. This provides a negative feedback mechanism for the control of the structural genes.

C. Structural genes under the control of a regulatory gene and operator sites are collectively termed an operon. All of the enzymes it produces are known as coordinately repressed enzymes.

II. Mechanisms analogous to enzyme induction can repress the formation of enzymes when their products are already present in the environment of the bacterium. A few examples of RNA equivalents of operons are also known.

III. The enzymes themselves may be controlled by post-transcriptional mechanisms, which produce conformational changes in the enzyme molecules.

IV. Little is as yet known of the control mechanisms of eukaryote genes, but some appear to be controlled positionally by insertion sequences (as in clonal antibody selection) and others can be shown to be switched on and off by promoter genes sensitive to such factors as steroid hormones and cyclic AMP.

V. Both mitochondria and chloroplasts contain DNA, which codes for some of the protein constituents of these organelles. The DNA in the mitochondria exists in several pieces, each a closed circle.

POST-TEST

1. A substance that elicits the synthesis of new molecules of a specific enzyme is termed a(n) ______________.

2. Enzymes that respond to such substances are called ______________ ______________.

3. Biosynthetic enzymes that are reduced in amount by the presence of end-product of the sequence of reactions in which they are involved are said to be ______________.

4. The small molecule, often the end-product of a biosynthetic sequence, that brings about a decreased production of the enzymes that have produced it is called a ______________.

5. The amount of mRNA template producing a given enzyme is regulated by a protein called a ______________, produced by a ______________ ______________.

6. A specific base sequence that is required for the initiation of transcription is called a ______________.

7. The unit consisting of a regulatory gene, an operator site, a promoter site for the binding of RNA polymerase, and a set of structural genes is termed an ______________.

8. DNA is present in small amounts in ______________ and in ______________, as well as in large amounts in the nucleus of the cell.

9. The genetic control of the structure of chloroplasts resides in the ______________ and in the ______________.

10. Movable genetic elements were first discovered in ______________ plants by ______________.

11. Movable genetic elements appear to be responsible for the bithorax trait in *Drosophila* in which the ______________ are transformed into a second pair of wings. It is thought that the transposable elements may block the proper ______________ of developmental genes.

REVIEW QUESTIONS

1. What is meant by a gene repressor? a derepressor?
2. Distinguish between structural genes, regulatory genes, and operator sites.
3. If mutations involving developmental genes can produce a four-winged fruit fly, do you think that mutations involving developmental genes could also produce a two-winged insect suddenly from a four-winged ancestor? Could such a mutation suddenly produce an elaborate organ, such as the haltere?
4. Describe the operon concept and the roles of regulatory genes, operator sites, and structural genes in the control of gene function.
5. Does the operon concept apply to eukaryotes? Summarize a suggested mechanism for gene regulation in eukaryotes.
6. What cytoplasmic organelles are now known to contain some of their own genetic information?

16

Genetic Frontiers

OUTLINE

I. The preparation of recombinant DNA molecules
   A. Restriction endonucleases
   B. Plasmids
II. Isolating a single eukaryotic gene
III. Cloning synthetic DNA
IV. Replacing defective genes
V. Other uses of recombinant DNA techniques
VI. Recombinant RNA
VII. Some worries

LEARNING OBJECTIVES

After you have read this chapter you should be able to:

1. Describe the primary techniques utilized in recombinant DNA experiments.
2. Summarize the problems involved in isolating, identifying, and cloning a single human gene.
3. List the several kinds of DNA recombination that occur in nature.
4. Discuss the kinds of reactions catalyzed by restriction endonucleases and their importance in recombinant DNA experiments.
5. Describe the nature of bacterial plasmids, and discuss their role in recombinant DNA experiments.
6. Outline the steps necessary to replace a defective gene.
7. Describe the special measures that have been employed to introduce genes experimentally into plant cells and animal cells.
8. Discuss the role of oncogenes in the production of cancer.
9. Summarize the ethical and other objections that have been raised against recombinant DNA studies, and give potential practical and research applications.

One of the most exciting aspects of molecular genetics at present is the development of methods for manipulating specific genes in the laboratory. It has been said that genetic engineering will be to the 21st century what nuclear physics is to the 20th century, and is likely to have at least as great an effect upon our lives and our world as such developments as thermonuclear fusion have had thus far.

The techniques of genetic engineering presently under development involve the formation of new and artificial combinations of DNA originating in quite different organisms. These **recombinant DNA** techniques permit investigators to identify one specific gene's DNA from among the totality (also called the library) of genes in a genome. The specific genic DNA can be isolated, transferred to a new organism, and incorporated into its genome so that it is replicated along with the host's DNA. If the gene can be inserted in such a fashion that its triplet reading frame is correct, and if the DNA transferred includes a regulatory region containing a promoter, it may be transcribed and translated, leading to the production of a new protein—one the host organism has not produced before.

One of the earliest such successes was achieved in the laboratory by isolating the gene coding for the production of insulin, first from the rat and then from the human, and inserting those genes into the DNA of a favorite genetic tool, the colon bacterium *Escherichia coli*. These bacteria can be grown in vast amounts in special fermentation tanks and will produce large amounts of insulin (Fig. 16–1). This can be separated from the bacterial proteins, purified, and used in the treatment of human diabetes. Although the industrial production of human and other specialized proteins has scarcely begun, it has an immense commercial and industrial potential apart from medical applications. Specially altered enzymes intended for use as industrial catalysts are only one of the possible (and profitable!) products.

In a way, genetic engineering is not entirely new, for some of its "techniques" have occurred naturally in the wild for uncounted years; in fact, some forms of cancer-producing viruses can best be accounted for in this way. The intelligently directed alteration of genomes is something new in the biosphere, however. Engineered bacteria serve our ends, not their own, and we will certainly not be content to restrict our engineering to bacteria.

The Preparation of Recombinant DNA Molecules

Recombinant DNA methodology became possible with the discovery of a class of nucleases that cleave the DNA at specific sequences—enzymes termed restriction endonucleases, discussed in the following section. In recombination experiments, investigators attempt to separate a single gene, or some well-defined portion of DNA, from all the other parts of the genome of a cell (Fig. 16–2). After it has been separated the gene is usually amplified, or **cloned,** to obtain enough of the material it produces for study of its properties. In addition, the amplified gene may be translated and transcribed, producing the protein for which it supplies the code.

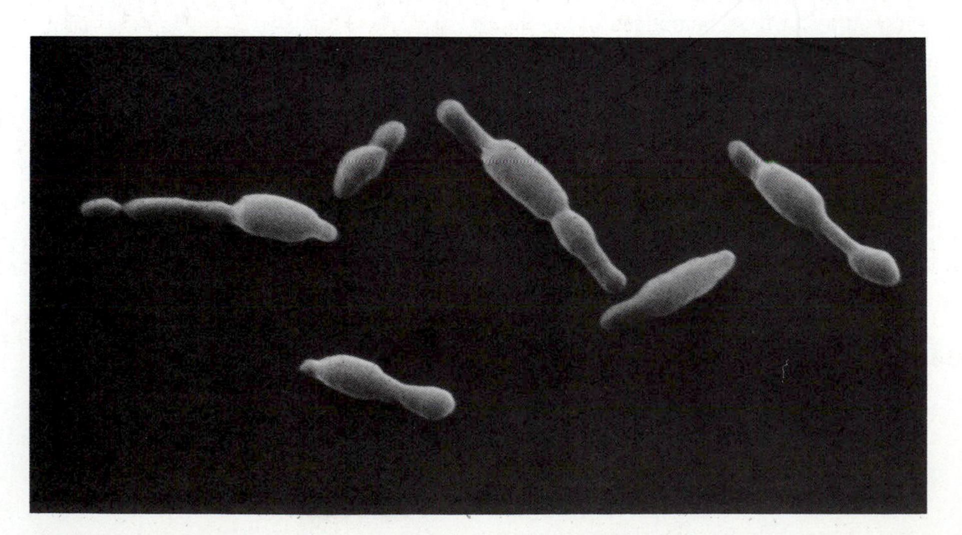

Figure 16–1 Scanning electron micrograph of *E. coli* cells "bulging" with human insulin. See also Figure 1–20. (Courtesy of Dr. Daniel C. Williams and the Lilly Microscope Laboratory.)

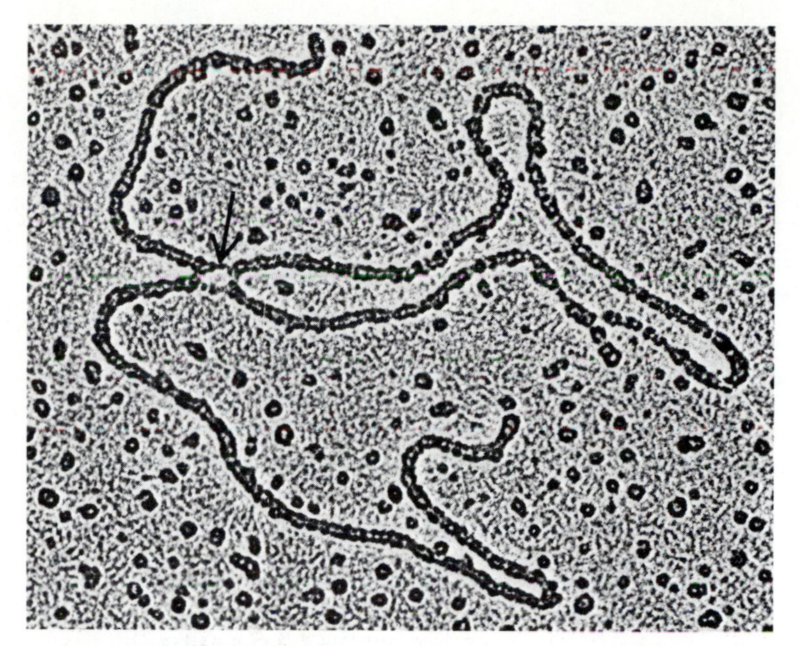

Figure 16–2 One method of preparing recombinant DNA. The DNA from a bacterial virus was treated in the test tube with *E. coli* proteins thought to mediate recombination. The arrow indicates the region in which the two double helices have undergone a rearrangement so that a segment from the strand from each molecule is exchanged. (From DasGupta, C., A. Wu, R. P. Cunningham, and C. M. Redding, *Cell* 25: 507, 1981. Courtesy of the authors and MIT Press.)

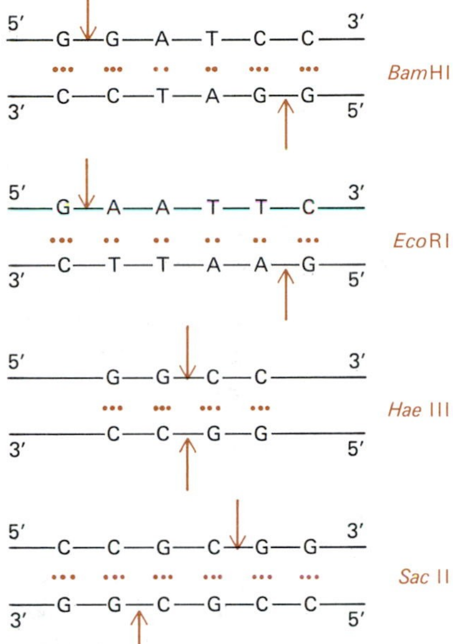

Figure 16–3 A few of the sequences in DNA that are recognized and cleaved by specific restriction endonucleases. Each endonuclease recognizes a specific sequence, generally four to six base pairs in length, and hydrolyzes a particular phosphodiester bond in each strand, as indicated by the colored arrows. Note the recognition sequence, read 5′→3′, is present in each strand. The labels refer to different kinds of restriction endonuclease obtained from different species of bacteria (EcoRI, for example, stands for **E. Coli Restriction I**).

RESTRICTION ENDONUCLEASES

Restriction endonucleases are bacterial enzymes that cleave DNA molecules in a very specific fashion. There are many different specific restriction endonucleases, each of which cleaves both strands of the double helix of DNA at the point or points containing a specific sequence of deoxynucleotides (Fig. 16–3). Restriction endonucleases are produced by a wide variety of bacteria, and their biological function is to protect the bacteria against invading bacteriophages by breaking the molecules of foreign DNA into nonfunctional bits. The cell's own native DNA is not split by the endonuclease because in it, the nucleotide sites that would be attacked by the endonuclease are protected by the addition of methyl groups to the nucleotide. Such sites are known as **recognition sites;** each consists of four to six base pairs. The nuclease hydrolyzes a specific bond in each strand in that recognition site. In the case of some restriction endonucleases, the resulting cleavage sites are characterized by symmetric sequences of base pairs that are palindromic.

A **palindrome** is a word or phrase that reads the same from right to left as it does from left to right—for example, "able was I ere I saw Elba," or "radar." The endonuclease splits the palindromic sequence at the same sites in the two strands, leaving short stubs of nucleotides that are antiparallel complements of each other (Fig. 16–4). This ensures their ability to recombine; that is, it produces "sticky" ends of the severed DNA strands. Therefore, even when two *different* DNA molecules are cut by the same restriction endonuclease, they necessarily yield segments with homologous terminal sequences. The close fitting of the two homologous segments permits them to recombine in a very specific fashion. The two homologous segments can then be spliced, or joined, by a different enzyme, **DNA ligase** (Fig. 16–5). Restriction endonucleases are indispensable tools for analyzing chromosome structure, for sequencing entire DNA molecules, for isolating specific genes, and for creating new DNA molecules.

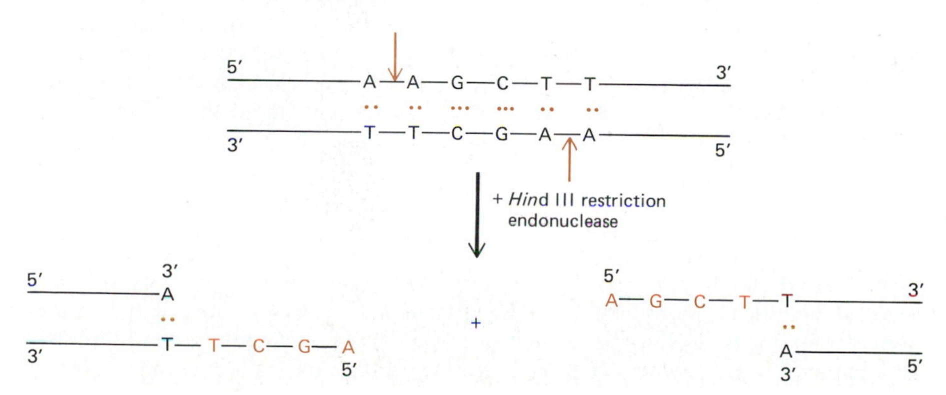

Figure 16–4 The cleavage of DNA by the restriction endonuclease Hind III. The enzyme cleaves bonds *within* the DNA double helix (thus it is referred to as an *endo*nuclease). The product DNA fragments have short single-stranded stubs (indicated in color) that are antiparallel complements of each other.

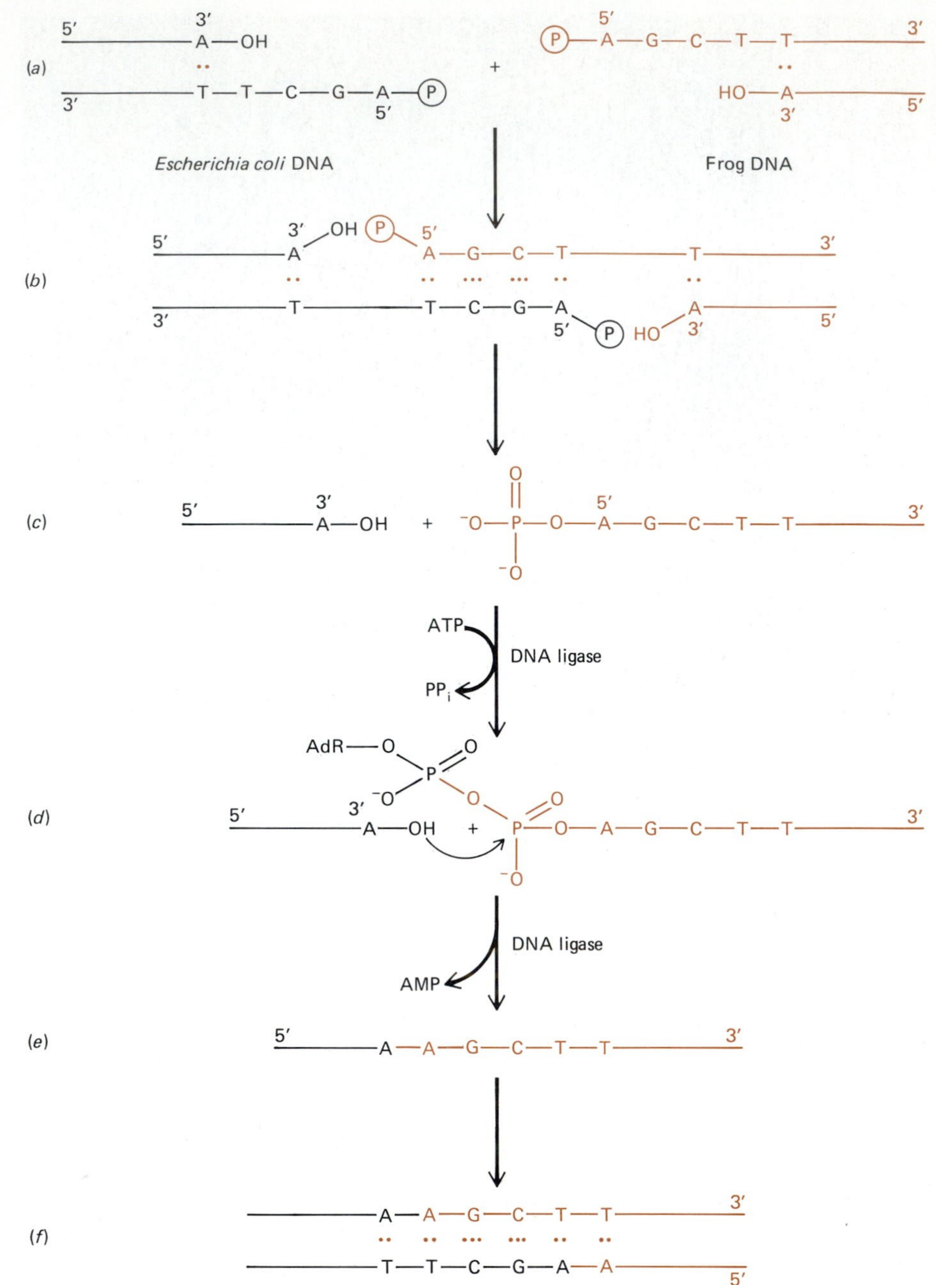

Figure 16–5 The mechanisms by which two double-stranded DNA fragments are covalently joined by the enzyme DNA ligase. *(a)* DNA samples isolated from *E. coli* bacteria and from frogs are mixed and cleaved with the restriction endonuclease Hind III to create DNA fragments with sticky ends. *(b)* Two different sticky ends anneal together via base pairing of complementary sequences. The enzyme DNA ligase catalyzes the formation of a phosphodiester bond between two adjacent nucleotides. The ligase can catalyze the repair of the break in DNA, provided that the DNA is in a properly base-paired double helix. Only the events in one strand are diagrammed. *(c)* Starting with ATP, the enzyme catalyzes the attachment of the AMP to the 5′ phosphate present at the strand interruption. *(d)* the AMP activates the 5′ phosphate group and permits it to form a bond with the free 3′ OH group on the other side of the nick. *(e)* The phosphodiester bond is formed, accompanied by the release of free AMP. *(f)* A similar reaction repairs the nick in the other DNA strand.

A piece of DNA produced by the action of one endonuclease can be specifically cleaved into smaller fragments by a second endonuclease, and into still smaller pieces by a third. Chromosomes can be mapped by using a series of restriction enzymes that split the large DNA molecules into small segments that can be sequenced. Small differences between related DNA molecules can be detected by separating their respective restriction fragments using gel electrophoresis. The electrophoretic mobility of a piece of DNA (that is, its movement in response to an electric current) is roughly inversely proportional to the number of base pairs present.

PLASMIDS

Many bacteria contain, in addition to the main chromosome, small, accessory, circular, double-helical chromosomes called **plasmids** (Fig. 16–6). These may be transferred from one bacterium to another during the process of conjugation (described in Chapter 19), the bacterial equivalent of sex. Plasmids usually contain 2,000 to

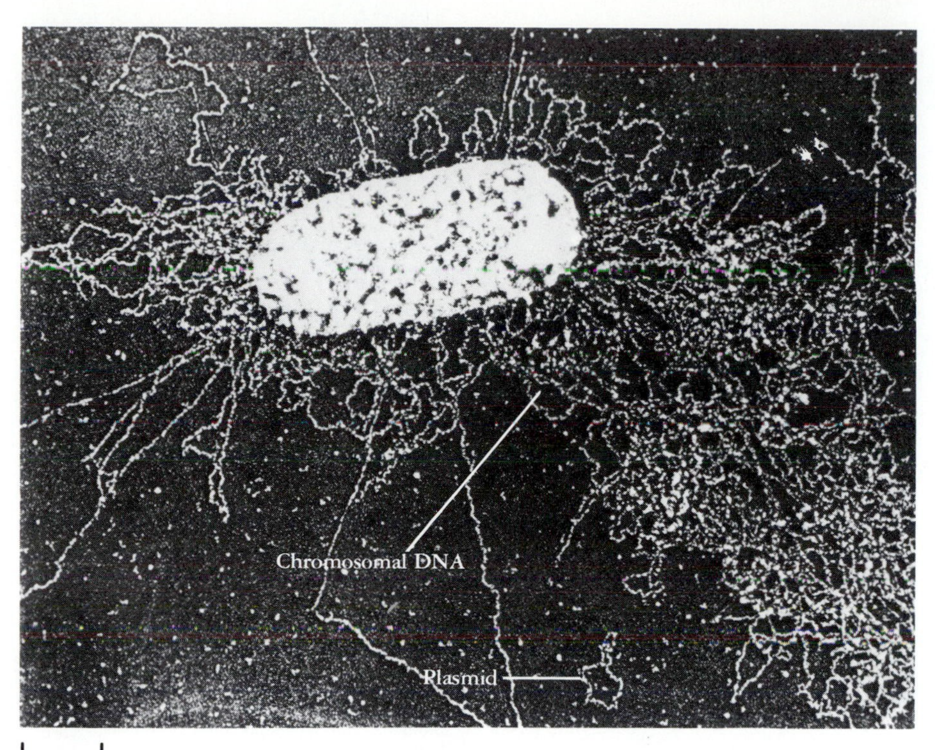

Figure 16–6 Chromosomal DNA and plasmid leaking out of a bacterial cell. Note the relative amounts of the two kinds of DNA. (Courtesy of J. Griffith.)

100,000 base pairs and typically carry genes for the inactivation of antibiotics, for the metabolism of certain natural products, or for the manufacture of toxins. Plasmids function in bacterial reproduction by producing genetic recombination and the transmission of genes from one bacterium to another. They serve as a very rapid method of gene transmission, and the genes typically carried in them are of the sort that would be needed to meet the demands of a rapidly changing environment. Plasmids may undergo replication independently of the host chromosome. An *E. coli* cell usually has about 20 copies of a given plasmid, along with 1 or 2 copies of the large chromosome.

The special properties of plasmids enable their use in recombinant DNA technology. First, plasmids are isolated from the bacterial host cell and treated with an endonuclease to open up the circular DNA, leaving complementary single-stranded ends. The DNA fragment to be inserted is formed by treating the DNA with the same endonuclease, so that it has single-stranded ends that are complementary to those of the cut plasmid. The DNA fragment and the cut plasmid are then joined by the enzyme DNA ligase (Fig. 16–7). Next, bacteria are exposed to the plasmid, and a small proportion of these take up the plasmid. The plasmid confers resistance to the antibiotic tetracycline; when the bacteria are exposed to a medium containing tetracycline, all of the ones without the plasmid are killed. In contrast, the ones with the plasmid survive and can be propagated. In this way any bit of DNA, even one synthesized chemically, can be added to a plasmid and inserted into a bacterial host.

Isolating a Single Eukaryotic Gene

Isolation of a single gene from the DNA of a eukaryotic organism such as a human involves a process something like the following: The entire genome is partially digested to generate random fragments having an average length of several thousand base pairs. The ends are treated to make them "sticky"—so that they have a single-stranded stub complementary to the ends of the plasmid. The ends are joined by ligases, and bacteria are then infected with the plasmid or phage containing the piece of DNA. The bacteria are grown so that they multiply many times and then are lysed (broken open) to release the DNA. The product is a cloned library of the DNA from the eukaryotic organism.

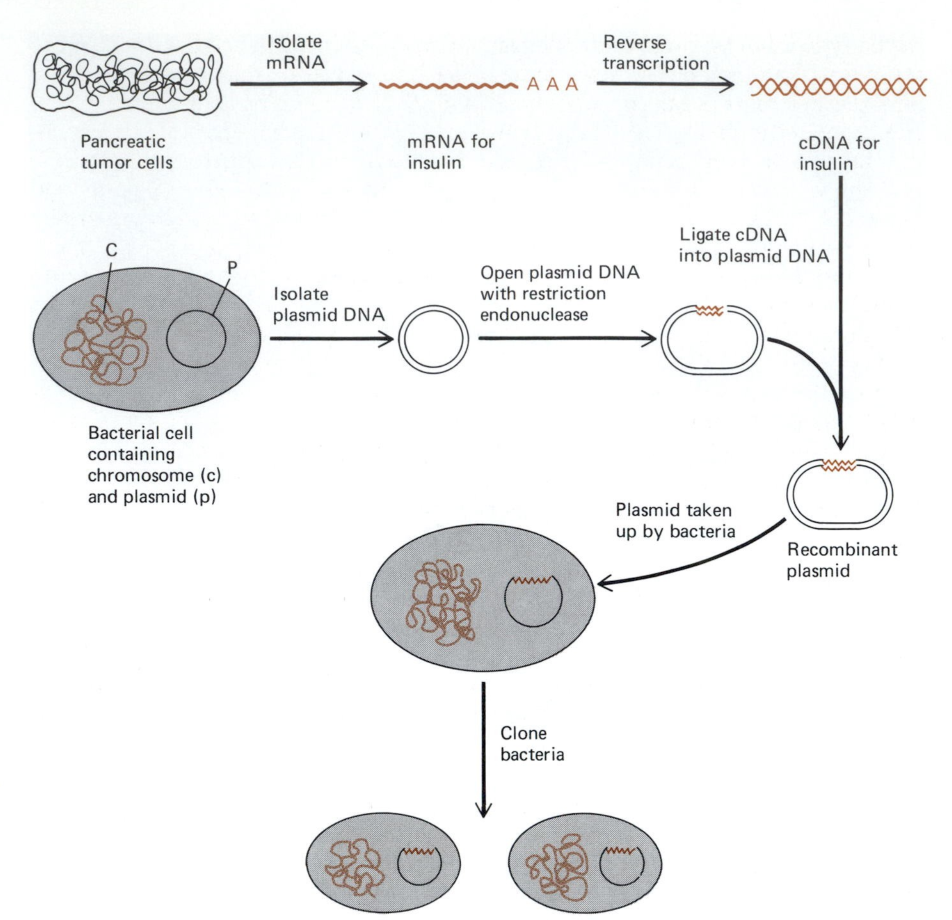

Figure 16–7 The process of enzymatically inserting a mammalian gene, here the gene for rat insulin, into a plasmid from the bacterium *E. coli,* replacing the plasmid in the bacterial cell, and cloning that cell to obtain many copies of the insulin gene. These cloned cells produce large amounts of insulin when they are incubated in growth medium.

In a more modern technique that does not require the use of plasmid vectors (carriers), the desired DNA is mixed with a calcium salt, which is then converted to calcium phosphate. The DNA is then **coprecipitated** with the insoluble calcium phosphate, which protects it from damage. The DNA itself is then induced to enter eukaryotic cells without a phage or plasmid vector.

Because the library contains many thousands of different genes, the next task is to screen the library to identify those particular plasmids or phages that may contain the desired gene. This is accomplished by **hybridizing** the DNA with a radioactively labeled RNA or a radioactively labeled copy DNA that is known to be complementary to the desired gene, and then visualizing by autoradiography the clones to which this **probe RNA** or **probe DNA** "sticks." These presumably have the gene. The particular clones identified in this way are then added to new *E. coli* hosts, and the gene is cloned along with the hosts so as to produce many copies.

Another way is to begin with an mRNA library: The mRNA is assayed by in vitro translation; the kind that produces the desired peptide is then copied into DNA using reverse transcriptase; the resulting DNA is known as **cDNA.** Then the RNA is deliberately degraded enzymatically, and the second cDNA strand is synthesized using DNA polymerase. Linking sequences are then added to the ends of the cDNA, and the cDNA is ligated into a plasmid, which is then introduced into a host cell.

Cloning Synthetic DNA

Advances in the methods for chemically synthesizing DNA with specific sequences of nucleotides have broadened the scope of recombinant DNA techniques even further. A gene with the desired sequence of nucleotides can be synthesized by chemical means, inserted into a plasmid or phage, taken up by *E. coli,* and expressed, producing the specific protein. Somatostatin, for example, is a peptide hormone secreted by the hypothalamus. It contains 14 amino acids and functions in the body to inhibit growth and in several other ways. The DNA coding for somatostatin was synthesized chemically and fused to the gene for β-galactosidase present on a plasmid. The β-galactosidase gene was attached to its normal operator gene and promoter region so that it was translated and transcribed normally. By attaching the somatostatin gene to the β-galactosidase gene, investigators avoided the problem of having to supply the somatostatin gene with its own operator and promoter. Bacteria containing this plasmid synthesized a hybrid protein containing somatostatin linked to β-galactosidase. The bond between the two was split chemically, and the free somatostatin obtained had all the properties of somatostatin synthesized normally in the mammalian hypothalamus. This experiment showed that cloning a chemically synthesized gene can lead to the production of a functional polypeptide. (Unfortunately the use of *E. coli* for commercial protein production has the drawback that *E. coli* will not normally secrete these abnormal products, so that they accumulate in the cell from which they must be removed by harsh mechanical disruption. It might be better to use *Bacillus subtilis* instead, since this bacterium naturally secretes protein into the culture fluid.)

More recent studies have demonstrated the feasibility of introducing sequences into *E. coli* from the bacterium *Pseudomonas* that are capable of producing the blue plant dye, indigo. This was done accidentally: Certain genes in the *Pseudomonas* plasmid were found to produce enzymes for the later parts of the metabolic pathway leading to the dye; other plasmids in the *E. coli* recipient were able to provide the earlier part of the chain of intermediates. The really significant aspect of this study may be that the indigo-producing gene was not derived from a plant source but was put together by combining certain *E. coli* genes already present with some that naturally occurred in a *Pseudomonas* plasmid. The equivalent of a eukaryote gene was thus constructed from prokaryote components! (It could be said that the experimenters "patched the bacterial blue genes.")

Eukaryotic genes can be introduced into mammalian cells as well as into bacterial cells. One way of doing this is to employ retroviruses such as the SV40 virus, into which a sequence such as that from the rabbit genome capable of producing the rabbit β-globin protein can be incorporated. The virus is then used as a vector to introduce the rabbit β-globin gene into monkey kidney cells maintained in culture; the infected cells will synthesize rabbit β-globin in substantial amounts, showing that the rabbit gene has been successfully introduced into the monkey cells.

Another, somewhat more up-to-date method is to use DNA directly to carry genetic information from one cell to another. Of several possible techniques, the most recent is microinjection of DNA into the recipient cell. In a recent study of this sort,[1] the human gene for growth hormone production was isolated from a library of human DNA, with the intention of transferring it to a fertilized mouse egg. In order to ensure high activity of this gene, its DNA was combined with a portion of mouse DNA ordinarily responsible for producing the metallothionein protein, a substance whose production is stimulated by the presence of heavy metals such as zinc. The metallothionein protein itself was not desired, but its promoter sequence, now associated with the gene for human growth hormone, would turn on the production of this hormone when exposed to small amounts of heavy metals. The metallothionein promoter could therefore be used as a kind of switch whereby the experimenters could turn human growth hormone production on and off at will.

The metallothionein–human growth hormone complex was then injected into mouse eggs. When these in due course became embryos, they were deliberately exposed to small amounts of zinc. In those in which the gene transplant had been successful there was enhanced growth due to the activity of the metallothionein gene's zinc-sensitive promoter. This promoter, which is linked to growth hormone genes in the experimental animal, caused those genes to be turned on in liver tissue, which is where metallothionein genes normally function. Thus the mouse's liver, rather than pituitary gland, produced very large quantities of growth hormone. One mouse originating from an egg that had received two copies of the human growth hormone gene grew to more than double the normal

[1] Palmiter, R. D., et al.: Metallothionein–human GH fusion genes stimulate growth of mice. *Science*, November 1983, pp. 809–814.

size (Fig. 16–8). As might be expected, such mice are also able to transmit their enhanced growth capability genetically to their offspring. What use might be made of giant mice is a puzzle, but we can imagine all sorts of practical potential applications of similar experiments in the breeding of domestic meat animals and in many other areas.

However, in addition to the more practical applications of recombinant DNA techniques in producing proteins of medical significance such as insulin, growth hormone, interferon, and so on, such techniques can provide abundant quantities of specific DNA segments for studies of the sequence of genes, studies of the structure of chromosomes, and studies of the roles of introns and exons[1] (see Focus on Localizing a Gene by Cell Fusion).

[1] The discussion of these research applications is beyond the scope of this book, but interested readers are urged to consult Watson, J. D., and Tooze, J.: *The DNA Story*, pp. 548–584. San Francisco, W. H. Freeman, 1981.

FOCUS ON

Localizing a Gene by Cell Fusion

Although computerization of medical records makes the detection of linkage among human genes more practical than we might suppose, the fact remains that humans cannot be experimentally bred and tend to have small families.

Cell fusion techniques can help in the detection of genes that transmit deleterious traits. These techniques depend upon the fact that some forms of virus cause cells to fuse with one another as a way of infecting new host cells. The Sendai virus, particularly noted for this property, can be inactivated so as to be harmless to the cells that it causes to fuse. The second fact upon which cell fusion depends is that when human and mouse cells are fused, for some reason with every cell generation there is a tendency for the compound cells that result to lose human chromosomes. The accompanying diagram explains how this can be used to isolate the human chromosome that contains a particular gene.

Notice that both classical and biochemical genetics are teamed to make this discovery. Methods similar to this were employed to identify the chromosome that contains the gene for Huntington's chorea, a serious, dominant hereditary disease that usually kills its victims in middle age.

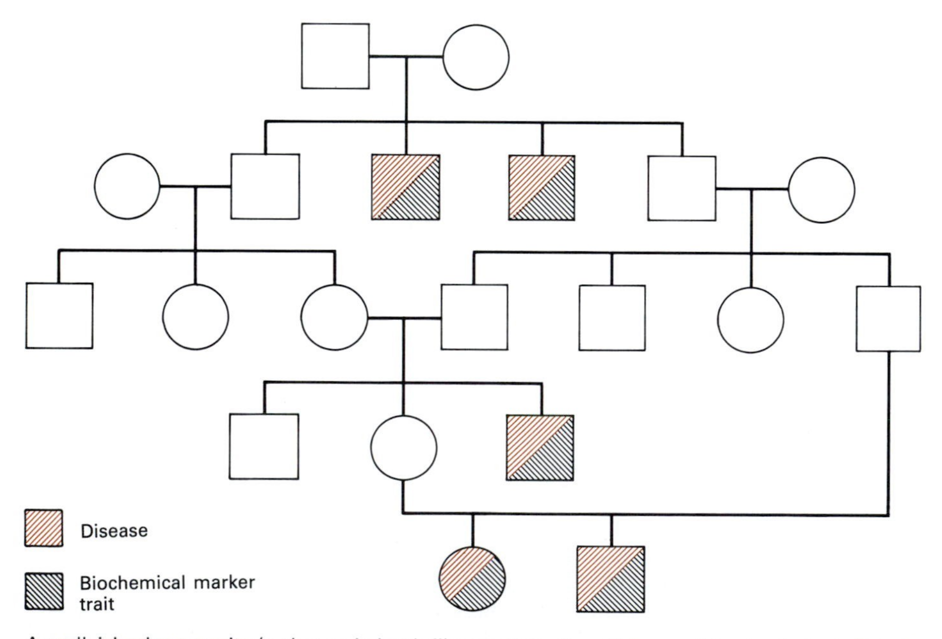

A small, inbred community (such as an isolated village or distinctive religious sect) is intensively studied. In one lineage it is observed that a genetic disease only occurs in people who also—incidentally—have another biochemical peculiarity or marker. Though the disease and the marker trait are unrelated in function, the fact that they are inherited together probably means that they occur on the same chromosome.

Figure 16–8 The giant mouse shown at left developed from an egg that received two copies of the human growth hormone gene. Exposure to the hormone resulted in growth of this experimental mouse to more than double the normal size. (Courtesy of R. L. Brinster.)

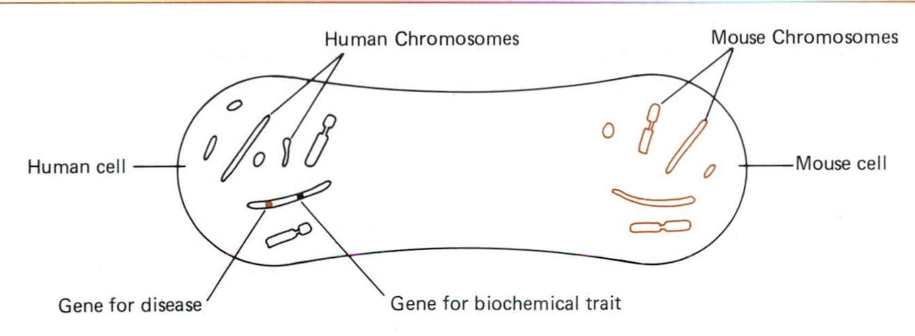

Cells from one of the disease victims from this lineage are fused with mouse cells.

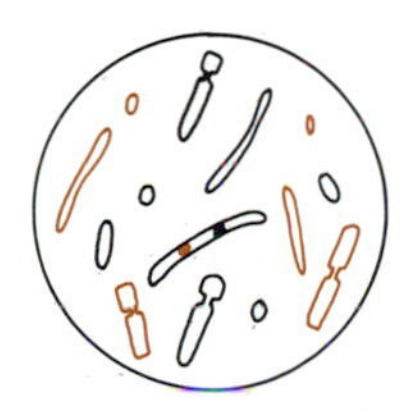

Hybrid cell contains both human and mouse chromosomes. However, as the hybrid cell reproduces, different human chromosomes are transmitted to different daughter cells so that sometimes certain human chromosomes fail to be transmitted. Mouse chromosomes, however, are always transmitted to daughter cells. After many cell generations, clones of mouse cells can be isolated, each of which has a single, different human chromosome remaining.

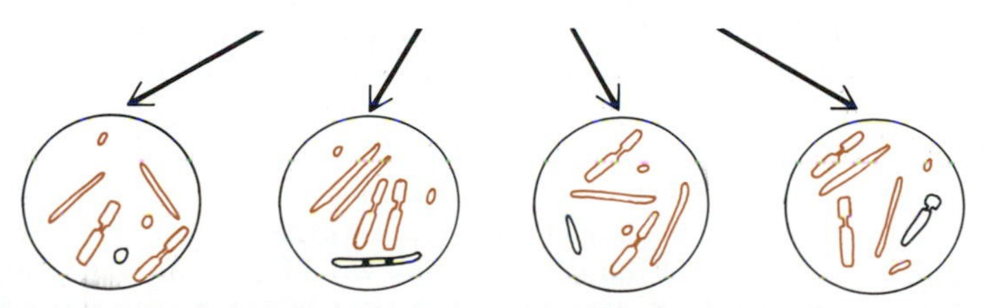

The clone that tests positively for the biochemical peculiarity must be the one which contains the chromosome bearing the gene for that peculiarity. That chromosome must also be the one that bears the gene for the disease.

Replacing Defective Genes

Recombinant DNA technology is also potentially capable of permanent repair of many genetic defects in human victims of inherited disease. It should be possible to select desired normal human genes from cells, clone them in bacteria to obtain large numbers of the gene, and then introduce them into the cells of patients lacking that particular gene. When this can be done it will provide a means of curing diseases caused by defective or deficient genes—diseases such as hemophilia, phenylketonuria, sickle cell anemia, glycogen storage disease, and others (see Focus on Molecular Genetics and Cancer). This requires, of course, much more than the means to produce large amounts of cloned DNA. Means and methods are being developed to

FOCUS ON

Molecular Genetics and Cancer

A cancer cell is an abnormal cell that lacks biological inhibitions. To it, the body is nothing more than a culture medium in which it can grow. It appears that cancer is a genetic disease, transmitted not so much between generations of hosts as between generations of cells.

In addition to lacking the normal contact inhibition that stops cell growth when its reasonable limits have been reached, cancer cells are immortal. By this is meant not that a cancer cell cannot die but that it cannot die of old age. Normal human fibroblasts, for example, can be maintained in tissue culture for perhaps 50 generations before they become enfeebled; yet the HeLa cancer cells isolated from a patient in the 1950s are today alive and well in tissue culture vessels the world over, still going strong after no telling how many cell generations. Given the proper conditions and care, there seems to be no reason why descendants of today's HeLa cells might not still be multiplying in the year 2100 or beyond.

Recombinant DNA techniques are currently being used in analyses of the process by which normal cells become transformed into cancer cells. These and other studies indicate that the process of carcinogenesis in humans and other mammals is a multistage process, one that involves several independent steps. Yet carcinogenesis does seem to be a unitary phenomenon in which a basic genetic change affects a cascade of processes to produce the multitude of specific properties that a cancer cell must have in order to become malignant. It begins to appear that cancer cells owe such traits to possession of at least one and probably several genes that are known as **oncogenes.**

In a typical study, a human line of cancer cells from a bladder tumor was shown to have undergone a mutation in a specific codon (a change from G to T) so that the glycine in position 12 of this gene's protein product is replaced by valine. This one change was shown to be critical to the conversion of the normal cell's normal gene into the cancer cell's oncogene, perhaps because the malignant gene somehow duplicated itself so that multiple copies came into existence, thus amplifying the effects of the original oncogene.

Oncogenes and oncogene-like DNA sequences are widespread among living things. It is thought that such genes, known as **proto-oncogenes,** function in normal cellular growth and embryogenesis. The fact that cancers of many kinds occur more frequently in sufferers from chromosomal abnormalities or are associated with specific known chromosomal defects suggests that proto-oncogene *position* may be important. Perhaps when dissociated from their normal regulatory regions such genes turn into actual oncogenes. Oncogenes that are of demonstrable animal origin are found in some viruses, especially retroviruses, which are able to transmit them to new cellular victims. Still other viruses carry oncogene sequences that do not seem to have originated in this way but which also effectively produce cancer. It seems likely, however, that several oncogenes, each requiring a separate induction, may usually be necessary to produce cancer. This could explain the observed multistep process of cancer induction required for the production of most experimentally developed cancers, and it could also explain the well-known fact that a tendency to cancer is often hereditary.

Imagine, for instance, that each one of a person's body cells contains an oncogene inherited from the parents. A viral infection may implant another, and later in life, mitotic malfunction may cause the original gene to be duplicated in one of the body's continually dividing cell lines. Carcinogenic food additives may produce yet another oncogene in one cell in that line, thereby completing its transformation into a cancer cell. Over the course of several years this ancestral cancer cell will refuse to obey normal contact inhibition and will grow without hindrance until an obvious tumor has formed and the cancer is diagnosed.

The concatenation of these and other random events would certainly take time. Perhaps that is why most cancers occur in greatest frequency in elderly people.

introduce the genic DNA into the appropriate cells of patients so that their cells are transformed. The cells could then take up and incorporate the new DNA into their genome so that it could be transcribed and translated under the control of its normal regulatory mechanisms.

Other Uses of Recombinant DNA Techniques

Recombinant DNA techniques may have many other effects on our lives. There are, for example, certain bacteria that can produce enzymes that will break down hydrocarbons. If the gene for this enzyme were isolated and cloned and then used to

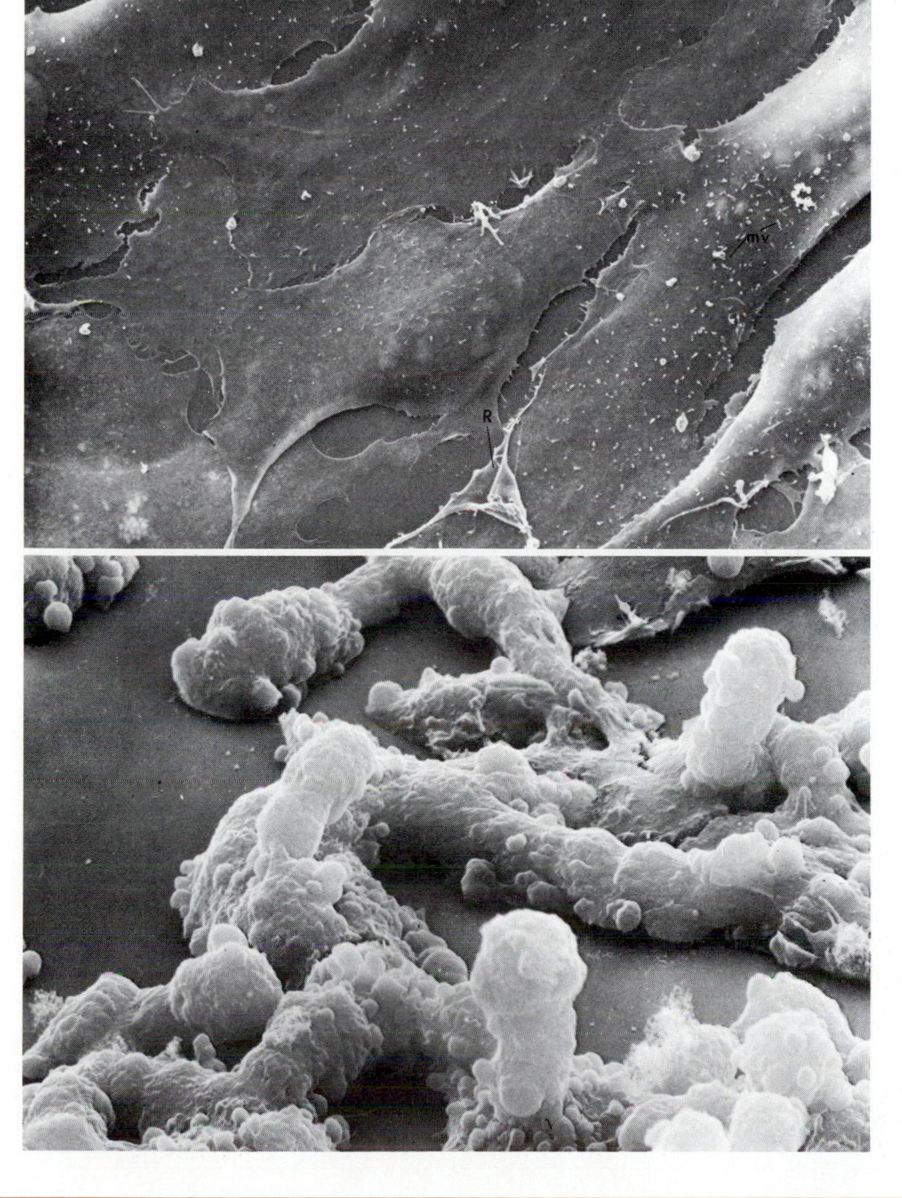

Scanning electron micrographs of cultured cells (approximately ×3000). *Above,* normal cells from a hamster embryo. A few microvilli *(mv)* and ruffles *(R)* are visible. *Below,* the same cell type after transformation by a cancer-causing virus. Note that the cells have blebs on their surfaces and aggregate to form several layers. (Courtesy of R. D. Goldman.)

produce large amounts of the enzyme, a powerful method for dealing with oil spills would become available. It is also conceivable that by introducing selected new genes into some of our present crop plants such as wheat or rice, their rate of production, or the areas of the world in which they can survive and grow, could be greatly extended (see Focus on Engineering Plants).

FOCUS ON

Engineering Plants

Plants have been selectively bred for centuries, if not millennia. The success of such efforts depends upon the presence of preexisting genes in the variety selected or in closely related wild or domesticated plants from which desirable traits may be introduced by crossbreeding. Unfortunately, the rarer agricultural plant varieties, especially primitive ones, are swiftly becoming extinct. This greatly reduces the size of the potential gene pool from which agricultural researchers may draw. Wild plants of all kinds are also threatened with extinction as every last acre of available agricultural land is brought into cultivation to feed the exploding human population. If genes could be introduced into plants from strains or species with which they do not ordinarily interbreed, or if totally synthetic genes could be introduced into them, the task facing agricultural researchers would be greatly eased.

Unfortunately, a suitable vector for the introduction of genes into plant cells has proved very difficult to find. Until recently, most genetic introductions have been performed by removing the plant cell wall to produce naked protoplasts (see the figure). If two such protoplasts are chemically induced to fuse, there is a chance that the resulting cell may give rise to mature, differentiated plants with some of the desirable genetic traits of the "parent" plants. Although natural reproductive barriers can sometimes be overcome by such methods (potatoes and tomatoes have been "crossed" in this way), protoplast fusion is essentially artificially facilitated sexual reproduction. Like the more traditional techniques of sexual propagation, it is essentially a hit-or-miss procedure whose results certainly cannot be predicted, let alone guaranteed.

A more recent technique employs the crown gall bacterium, *Agrobacterium tumefaciens,* which produces plant tumors by introducing a special plasmid into the cells of its host. The plasmid induces the abnormal growth by forcing the plant cells to produce abnormal quantities of growth hormone and also diverts the metabolism of the host cells to produce substances known as **opines,** simple derivatives of amino and keto acids. These opines are specifically preferred food substances for the bacterium.

The ability of the bacterium to introduce plasmids into a eukaryote host has suggested to researchers the possibility of incorporating desirable genes into that plasmid. The plasmid can be "disarmed" so that it does not also induce tumor formation; the cells into which such an inactivated plasmid is placed will be essentially normal except for the genes which the experimenter has intended to introduce. It has been shown that genes inserted in the plant genome in this fashion are transmitted sexually via seeds to the next generation, but they can also be propagated asexually if desired.

An additional but very interesting complication of plant genetic engineering is the substantial genome of the chloroplasts. Chloroplasts are pivotal in photosynthesis, and photosynthesis is the basis of plant productivity. It stands to reason that techniques must be developed for changing the portion of the chloroplast genetic information that resides within the chloroplast itself, although the chloroplast's nuclear genes can presumably be changed by "conventional" techniques.

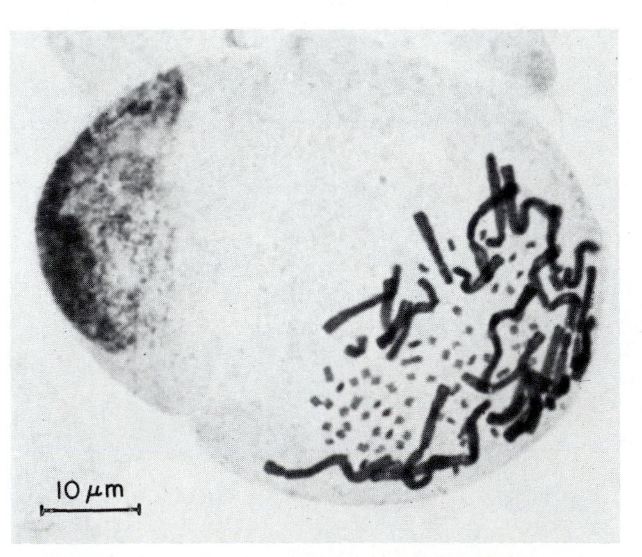

The hybrid cell undergoing mitosis was derived by protoplast fusion between soybean *(Glycine max)* and vetch *(Vicia hajastana)*. The large chromosomes are derived from vetch and the small ones from soybean. (From Constabel, F., et al.: *C. R. Acad. Sci. [D] (Paris)* 285:319–322, 1977.)

Recombinant RNA

It has proved possible to insert genes into RNA somewhat as is done with DNA, and to have the resultant recombinant RNA replicate itself directly without the need for a DNA intermediate. Why might this be desirable? DNA is usually thought necessary to produce RNA, but bear in mind that it is RNA, not DNA, that actually directs protein synthesis. RNA does not necessarily *need* DNA; some RNA viruses are capable of self-replication without the need for a DNA intermediate. Imagine that a strand of desired mRNA can be made by DNA in a minute. In 10 minutes, 10 strands will be manufactured. On the other hand, imagine a strand of mRNA that can replicate itself every minute. In 10 minutes, there will be, not 10, but 2^{10} strands of mRNA! The more strands of mRNA that are available, the more protein can be made. Moreover, the very simplicity of a self-replicating RNA system is attractive. Without the need for DNA it might prove more practical one day to carry out protein synthesis totally in vitro, under much better control than is possible in a complicated cellular medium.

Recombinant RNA technology is thus far based upon a small bacterial virus known impersonally as Q-beta. When infecting a host cell, Q-beta produces an enzyme, logically named Q-beta replicase, that immediately replicates the viral RNA. The two resulting particles are then replicated themselves, and so on, in an intracellular viral population explosion that proceeds exponentially to produce a cellful of virus.

In theory, a gene inserted in a strand of RNA should be rapidly replicated by the Q-beta replicase enzyme to produce a large number of copies, all of which could then be used to produce protein simultaneously. The problem has been, however, that Q-beta replicase is very specific; to avoid replicating the RNA of its bacterial host the enzyme is activated only by certain recognition codes built in to the Q-beta viral RNA. But this fact also suggests a potential solution: to link the desired sequence to be replicated with a fragment of viral RNA containing the required recognition code. This has been accomplished by preparing a DNA plasmid that incorporates the recognition sequences plus a desired gene. When this is transcribed, the RNA equivalent is automatically created. In the presence of Q-beta replicase, the RNA immediately makes multiple copies of itself.

Some Worries

While acknowledging the potential uses of recombinant DNA techniques as important and beneficial, many scientists regard the potential misuses as being at least equally potent. The development of an organism that produces undesirable ecological or other effects, not by design but by accident, is a real possibility. Totally new strains of bacteria or other organisms, with which the world of life has no previous experience, might be similarly difficult to control. For example, as described earlier, a strain of *E. coli* bacteria able to produce insulin is a boon to diabetes patients and to the pharmaceutical industry, which will soon be freed from the need to obtain animal pancreases from slaughterhouses. It is now possible to provide diabetics with human rather than animal insulin and should eventually be possible to develop special insulins tailored especially for drug use. But what if these insulin-producing *E. coli* bacteria were to escape and take up residence in our large intestines?

Safeguards against some of the potential hazards of genetic research are possible, and many have been implemented. For instance, at facilities in which genetically altered bacteria are handled, very stringent isolation measures reduce the chance of accidental escape, and the use of special strains of bacteria thought to be unable to live in the intestines of human beings or of animals reduces the chance that they will do harm even if they do escape.

The real problem, however, appears to lie in the permanence of the results of genetic manipulation. Most of what we human beings do is quite temporary, however severe the consequences of our actions may be in the short term. Real permanence involves the genetic information of organisms, for this is self-propagating in an active fashion that other environmental manipulations cannot match. Only the extinction of plants and animals has had permanent consequences for us to date, and of course that also involves genetic information, specifically its loss. Now, however, recombinant DNA technology has made possible the *creation* of genetic information and its incorporation into self-reproducing life forms, which, once created, might well persist for geological epochs.

Recent history has so far failed to bear out these genetic concerns. Experiments over the past decade have demonstrated that, at least until now, the experiments can be carried out in complete safety, and that some of the apprehensions voiced when recombinant DNA techniques were first introduced can perhaps be laid to rest for the present. It must also be acknowledged, however, that these very apprehensions led to the careful experimental design that experience has, thus far, shown to be safe.

SUMMARY

I. Recombinant DNA techniques enable investigators to isolate, identify, and manipulate—cleave, splice, and recombine—genes from the cells of organisms ranging from viruses, bacteria, and yeasts to green plants, animals, and humans.
- *A.* If the gene is inserted into a suitable host DNA it may be transcribed and translated, leading to the production of a protein not previously produced by the host organism.
- *B.* The DNA from different organisms undergoes, to some extent, a similar process of recombination in nature.
- *C.* Restriction endonucleases cleave double-stranded DNA at specific sites characterized by a specific, palindromic sequence of deoxynucleotides.
- *D.* Segments of DNA can be rejoined by the enzyme DNA ligase.
- *E.* Many bacteria contain small, accessory, circular, double-stranded helical "chromosomes" called plasmids.
  1. Plasmids may be transferred from one bacterium to another during conjugation.
  2. Plasmids undergo replication independently of the host chromosome.
  3. Plasmids typically carry genes for the inactivation of antibiotics.
- *F.* DNA synthesized in the laboratory can be inserted into a plasmid and undergoes transcription and translation under appropriate conditions.

II. Eukaryotic genes can be introduced into eukaryotic cells as well as into bacterial cells. It may be possible, with these techniques, to replace defective genes.

III. Recombinant DNA techniques may also be used to produce enzymes for cleaning oil spills, for enhancing production of crop plants, or for enabling them to grow in new areas of the world.

IV. Cancer may be caused by mutations or transpositions of certain growth-regulating genes called oncogenes.

POST-TEST

1. Techniques for manipulating specific genes in the laboratory are termed ________ ________ methods.

2. The class of bacterial nucleases that cleave DNA at specific sequences of deoxynucleotides are called ________ ________; they function in the intact cell as a defense against ________ invasion.

3. The specific sequences of nucleotides that are attacked by restriction endonucleases are called ________ ________.

4. A ________ is a word, phrase, or sequence of bases that reads the same from left to right as it does from right to left.

5. Small accessory double-helical chromosomes called ________ occur in many bacteria.

6. These typically contain genes for resistance to ________.

7. The DNA in plasmids undergoes ________ independently of that in the main "chromosome."

8. Single eukaryotic genes can be isolated by ________ them with radioactively labeled ________ that is complementary to the desired gene.

9. The gene can then be incorporated into a plasmid that has first been treated with a ________ ________ to break it and to produce "sticky" ends.

10. The plasmid is then rejoined with a ________ enzyme and incorporated into a bacterium; the bacterium is then ________.

11. The cloning of a synthetic gene for somatostatin was achieved by ________ it with the gene for ________, which was still associated with its normal operator gene and promoter region.

12. The disease ________ ________ has provided biologists with a vector capable of introducing recombinant genes into plants; this disease is produced by a ________ that injects a special ________ into its host cells.

13. One way of getting new genes into animal cells without the need for plasmid vectors is to adsorb the DNA to be transferred on a precipitate of ________ ________.

14. The principal advantage of recombinant RNA as opposed to DNA is that the viral RNA system employed is ________-________, so that much more of the desired ________ is produced in a short period of time.

REVIEW QUESTIONS

1. What is a reverse transcriptase, and what is its function? How might it be used for research in biochemical genetics?
2. Describe how genetic information from a eukaryote could be implanted in a bacterium. What is to be gained by this?
3. What is restriction endonuclease? Where is it likely to attack a strand of DNA? What practical application of this is employed in recombinant DNA research?
4. What has been the main difficulty thus far in introducing new genes into plant cells? How has this difficulty been overcome?
5. Cigarette smoke has apparently been shown to activate oncogenes. How are oncogenes affected by such mutagenic chemicals, by radiation, or by viral transduction to produce cancer?
6. In the experiment sequence in which functional human growth hormone genes were implanted in giant mice, why do you think it was necessary to include metallothionein genes?

Recommended Readings for Part III

Journal Articles about Human Genetics and the Principles of Heredity

Aledort, L. M. Current concepts in diagnosis and management of hemophilia. *Hospital Practice,* October 1982, 77–92. This sex-linked disease is no longer an automatic death warrant.

Bank, A., J. G. Mears, and F. Ramirez. Disorders of human hemoglobin. *Science* 207:486–493 (1 February 1983). There are more hemoglobinopathies than just sickle-cell anemia.

Brady, R. Inherited metabolic diseases of the nervous system. *Science* 193:733–739 (27 August 1976). A typical illustration of the enzymatic nature of much genetic disease.

Epstein, C., and M. Golbus. Prenatal diagnosis of genetic diseases. *American Scientist* 65:703–711 (November–December 1977). Though there has been much progress since this article was written, it gives the basics of amniocentesis in a clear and easily understood fashion.

Fedoroff, N. V. Transposable genetic elements in maize. *Scientific American,* June 1984. The jumping genes which Barbara McClintock discovered 40 years ago have been identified in bacteria, other plants, and animals. Several maize elements have now been characterized at the molecular level.

German, J., et al. Genetically determined sex-reversal in 46,XY humans, *Science* 202:53–56 (6 October 1978). A typical example of the complex workings of sex-determination and disorders of the sex chromosomes.

Gilat, T. Carbohydrate and milk intolerance. *Practical Gastroenterology* 4:35–40 (April 1980). A genetic "disorder" so common that most of the human race is affected.

Kolata, G. Fetal hemoglobin genes turned on in adults, *Science* 218:1295–1296 (24 December 1982). Accidental discovery of a drug that affects the suppression of fetal hemoglobin production which forms the basis of a potential method of treatment of sickle-cell anemia.

Langer, A. Practical genetics in office practice. *Hospital Medicine* 18:109–117 (August 1982). Excellent introduction to practical medical genetics. Especially useful summary of symptoms of Down syndrome.

Menozzi, P., A. Paizza, and L. Cavalli-Sforza. Synthetic maps of human gene frequencies in Europeans. *Science* 201:786–792 (1 September 1978). Inferring prehistoric population dynamics from present frequencies of human genes.

Tetrault, S. M. The student with sickle-cell anemia. *Today's Education,* April–May 1981, 54–57. A fine discussion of the practical problems faced by patients with sickle-cell anemia.

Woolf, C. M., and F. C. Dukepoo. Hopi Indians, inbreeding and albinism. *Science* 164:30–37 (4 April 1969). A typical example of the increased prevalence of a genetic trait in a small, genetically isolated population.

Books about Human Genetics and the General Principles of Heredity

Avers, C. *Genetics.* New York, D. Van Nostrand, 1980. A standard textbook of general genetics.

Edward, J. H. *Human Genetics.* New York, Methuen, 1978. A good text of general genetics with many examples and illustrations drawn from the genetics of human beings.

Goodenough, U. *Genetics,* 3d ed., Philadelphia, Saunders College Publishing, 1983. A clear presentation of the principles of genetics.

Jansen, J. D. *Child in a White Fog.* New York, Vantage Press, 1975. What it is like to have a child who is severely afflicted with the hereditary disease cystic fibrosis.

Mange, A. P., and E. J. Mange, *Genetics: Human Aspects.* Philadelphia, Saunders College Publishing, 1980.

Sutton, H. E. *An Introduction to Human Genetics,* 3d ed. Philadelphia, Saunders College Publishing, 1980. This and the preceding book are among the more interesting human genetics texts, readily understandable by the student.

Journal Articles on Nucleic Acids and their Function

Barnes, W. R., F. H. C. Crick, and J. D. White. Supercoiled DNA. *Scientific American,* July 1980, 118–133. The formation of a superhelix from double-helical DNA.

Bendiner, E. Avery: making sense of a 'stupid' nucleotide. *Hospital Practice,* October 1982, 195–219. A fascinating character study of an important participant in a little-appreciated chapter in the history of the double helix.

Brown, D. Gene expression in eukaryotes. *Science* 211:667–674 (13 February 1981). A summary of fairly recent thinking on the regulation of eukaryote gene expression—still much a mystery.

Chambon, P. Split genes. *Scientific American,* May 1981, 60–71. The role of introns and exons in the genetic message-coding for a protein.

Chilton, M. D. A vector for introducing new genes into plants. *Scientific American,* June 1983, 51–59. A eukaryote equivalent of transduction with possible application to genetic engineering.

Cohen, S. N., and J. A. Shapiro. Transposable genetic elements. *Scientific American,* February 1980, 40–49. Genes do not always stay put where they belong, and can even pass the prokaryote-eukaryote barrier. This poses interesting evolutionary questions and opens up new frontiers of genetic engineering.

Crick, F. Split genes and RNA splicing. *Science* 204:264–271 (20 April 1979). Why are DNA sequences that produce a single protein interrupted?

Dickerson, R. E., et al. The anatomy of A-, B-, and Z-DNA. *Science* 216:475–484 (30 April 1982). Newly characterized and unfamiliar forms of DNA that may function in gene control.

Dickerson, R. E. The DNA helix and how it is read. *Scientific American,* December 1983, 94–111. X-ray analysis of crystals of three types of DNA indicate that base-sequence information can be stored in the local structure of the helix.

Fedoroff, N. V. Transposable genetic elements in maize. *Scientific American,* June 1984. The jumping genes which Barbara McClintock discovered 40 years ago have been identified in bacteria, other plants, and animals. Several maize elements have now been characterized at the molecular level.

Gilbert, W., and L. Villa-Komaroff. Useful proteins from recombinant bacteria. *Scientific American,* April 1980, 74–96. Description of the techniques of cutting and splicing recombinant DNA by which bacteria are programmed to manufacture insulin, interferon, and other useful (to us) proteins.

Howard-Flanders, P. Inducible repair of DNA. *Scientific American,* November 1981, 72–100. Several systems of enzymes working together can repair damage to the genetic material.

Kornberg, R., and A. Klug. The nucleosome. *Scientific American* 244:52–64, 1981. The primary subunit of chromosome structure is helical DNA wound around a core of histone proteins.

Lake, J. A. The ribosome. *Scientific American,* August 1981, 84–97. Our emerging knowledge of the shape and much of the functioning of the ribosome.

Lewin, R. How mammalian RNA returns to its genome. *Science* 209:1052–1054 (4 March 1983). Routine violations of the old central dogma.

Maniatis, R., and Ptashne, M. A DNA operator-repressor system. *Scientific American,* January 1976, 64–78. A description of operators, promoters, and repressors and how they work in regulating the function of specific genes.

Marx, J. L. The case of the misplaced gene. *Science* 218:983–985 (1982). Loss of proximity between regulator and regulated genes may be involved in cancer initiation.

———. Newly made proteins zip through the cell. *Science* 207:164–167 (11 January 1980). Many proteins may find their way to special destinations in the cell by information encoded in their own structures.

———. A transposable element of maize emerges. *Science* 219:829–830 (18 February 1983). See paper by Cohen, cited above.

Miller, W. L. Recombinant DNA and the pediatrician. *The Journal of Pediatrics* 99(2):1–15 (1981). A much more useful article than the journal and title might lead one to believe. One of the best summaries of recombinant DNA principles presently in print.

Miller, W. L. Recombinant transposable genetic elements. *Scientific American,* February 1980, 40–49. Genes do not always stay put where they belong.

Motulsky, A. G. Impact of genetic manipulation on society and medicine. *Science* 219:135–140 (14 January 1983). A very respectable specimen of the recent discussions of the social and ethical issues raised by recombinant DNA techniques.

Nomura, M. The control of ribosome synthesis. *Scientific American* 250:102–114, 1984. A discussion of the structure and function of the ribosome, and how ribosome assembly is linked to the needs of the cell.

Novick, R. Plasmids. *Scientific American,* December 1980, 102–128. Small loops of circular DNA present in bacteria are very useful in the manipulation and cloning of DNA in genetic engineering.

Oppenheimer, J. H. Thyroid hormone action at the cellular level. *Science* 203:971–979 (9 March 1979). A classic problem—control of eukaryote gene function by hormones—seems near solution for the thyroid gland.

Prusiner, S. B. Novel proteinaceous infectious particles cause scrapie. *Science* 216:136–144 (9 April 1982). Can protein really carry genetic information? Or is the scrapie virus just too small to analyze properly?

Ptashne, M., A. D. Johnson, and C. O. Pabo. A genetic switch in a bacterial virus. *Scientific American,* November 1982, 128–140. A double pole, double throw operon.

Wang, J. C. Topoisomerism. *Scientific American,* July 1982, 94–108. The wonderfully precise enzymes that act on circular DNA molecules to link them together.

Weinberg, R. A. A molecular basis of cancer. *Scientific American* 249:126–144, 1983. A discussion of oncogenes and the molecular changes linked to them.

Books about DNA

Alberts, B., et al. *Molecular Biology of the Cell.* New York, Garland, 1983. An in-depth discussion of the molecular basis of genetics.

Facklam, H., and M. Facklam. *From Cell to Clone: The Story of Genetic Engineering.* New York, Harcourt Brace Jovanovich, 1979. An elementary discussion of genetic engineering intended for non-biologists.

Glover, D. M. *Genetic Engineering.* New York, Methuen, 1980. A paperback book with a simplified presentation of the principles of genetic engineering.

Holtzman, E., and A. B. Novikoff. *Cells and Organelles,* 3d ed. Philadelphia, Saunders College Publishing, 1984. Several chapters in this excellent paperback book focus on molecular genetics.

Judson, H. F. *The Eighth Day of Creation.* New York, Simon and Schuster, 1979. Aside from the famous (and controversial) *Double Helix,* this is perhaps the most readable and authoritative history of nucleic acid research to date.

Kornberg, A. *DNA Replication.* San Francisco, W. H. Freeman, 1980. A masterful discussion of DNA and its chemical properties by a noted scientist who is also an accomplished writer.

Lewin, B. *Genes.* New York, Wiley, 1983. An excellent treatment of molecular genetics.

Watson, J. D. *The Double Helix*. New York, Atheneum, 1968. Not everyone thinks this account to be a historically accurate one, and without doubt it is a highly colored and biased personal history. But there is nothing like it. One feels that, in essence if not in detail, this is how it really was to make a Nobel prize–winning discovery.

———, Tooze, J., and Kurtz, D. T. *Recombinant DNA: A Short Course*. San Francisco, W. H. Freeman, 1983. An introduction to the theory and current practice of genetic engineering.

———, and Tooze, J. *The DNA Story: A Documentary History of Gene Cloning*. San Francisco, W. H. Freeman, 1981. An invaluable historical record of major discoveries about the nucleic acids. Includes reprinted facsimiles of many historic papers and articles, as well as newspaper accounts and controversial materials.

———. *Molecular Biology of the Gene*. Menlo Park, Ca., W. A. Benjamin Inc., 1976 (and succeeding editions).

part IV

THE DIVERSITY OF LIFE

A cone-headed grasshopper, *Copiphora cornuta*, from South America. In their diversity of form and in their sheer numbers, insects challenge one's imagination. This grasshopper, though a vegetarian, has such powerful mouth parts that it can pierce human skin. (Chip Clark.)

17
The Classification of Living Things

OUTLINE

LEARNING OBJECTIVES

After you have read this chapter you should be able to:

1. Briefly summarize the development of the science of taxonomy.
2. Describe the general scheme of the Linnaean or binomial system of taxonomy.
3. Critically summarize the difficulties encountered in defining a species.
4. Apply the concept of shared derived characteristics to the classification of organisms.
5. List the domains and kingdoms of the modern five-kingdom scheme of biological classification, and give the traits of the organisms contained in each.

Much of the science of biology has, and will continue to be, associated with **taxonomy,** the science of naming and classifying organisms. Though there are literally millions of distinguishable kinds of organisms in existence, none comes with an already-attached label giving its name and evolutionary or ecological relationship with other organisms. It is up to human invention, therefore, to design some system that allows us to perceive order in the diversity of life.

Though designing such a system must come in part from our imagination, we will see that in actual use, a taxonomic system need not function arbitrarily; for among the individuals in any sampling of life forms, it is possible to discern numerous similarities and differences in appearances and lifestyles. It is assigning a relative weight to these characteristics that gives challenge and controversy to the work of the taxonomist.

The Development of Taxonomy

How would you use what you already know about living things if you wanted to start assigning them to categories? (Fortunately, you already are aware of the common names of many organisms.) Would you place insects, bats, and birds in one category (because they all have wings and fly), and squid, whales, fish, penguins, and Olympic backstroke champions in another because they all swim? Or would you classify organisms according to a culinary scheme, placing lobsters and tuna in the same part of the menu, perhaps, as "seafood"?

All of these schemes might be valid, depending on the purpose you might have in attempting to classify life. Similar methods have been used throughout history. Animals, for example, were classified by St. Augustine in the fourth century as useful, harmful, or superfluous—to human beings. Anthropologists have discovered that some cultures still employ a similar system of classification. Among organisms they find useful the Australian aborigines make sophisticated distinctions. Often these are much the same distinctions as biologists would make. However, among organisms they consider useless, aborigines draw only the most obvious distinctions.

In Renaissance times scholars began to abandon narrowly utilitarian schemes of classification and sought categories that might be inherent in the organisms themselves. These were originally thought to reflect natural groupings arrayed around an ideal type, an **archetype** in the mind of a creator, and were arranged roughly in an order that proceeded from the simple to the complex. Out of the many schemes of classification that were proposed, the system designed by Carolus Linnaeus in the mid-18th century, which we briefly encountered in Chapter 1, has survived, with some modification, to the present day.

Linnaeus and his colleagues had no conception of the vast number of living and extinct organisms that would be discovered since his time (Fig. 17–1). Additionally, the only characteristics available for use in classification were what we call morphological or physical traits, and for the most part, only those that were visible without the aid of a microscope. In frustration, Linnaeus gave up his attempts to classify the multitude of microorganisms that the newly invented microscope was just then disclosing.

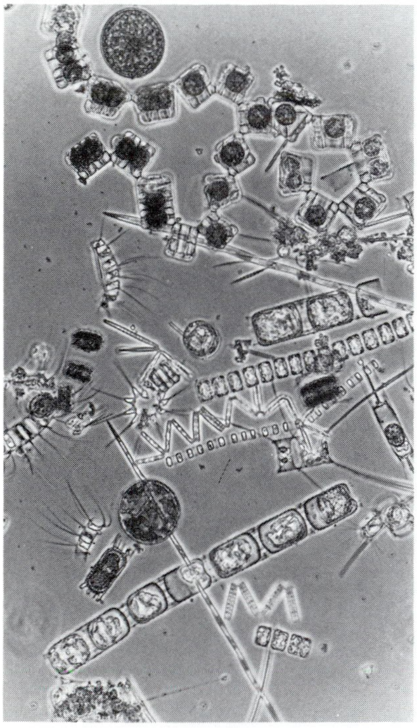

Figure 17–1 Photomicrograph of a concentrated sample of marine microorganisms (plankton) from the Rhode Island Sound (×225). Such variety, bewildering even to the modern graduate student, seemed to Linnaeus to be beyond hope of systematic classification. However, a modern planktonologist could identify and completely classify each one of the microorganisms (mostly diatoms) shown here. (Paul W. Johnson, Univ. of Rhode Island/BPS.)

The Linnaean System

The fundamental unit of classification that Linnaeus used for both plants and animals was the species. It is difficult to give a definition of this term that applies uniformly throughout the living world, but a **species** may be defined as a population of individuals, similar in their structural and functional characteristics, that in nature breed only with each other and share a close common ancestry.

Objectively defining a species *seems* easy. It is a "kind" of organism, and everyone thinks he or she knows what is meant by that. Yet consider the various breeds of dogs: We think of them as belonging to a single "kind"; yet how variable they are! (Would a future paleontologist excavating a pet cemetery think the Pekingese and the Great Dane buried nearby belong to the same species?) On the other hand, there are kinds of *Drosophila* fruit flies that only an expert could tell apart, but which apparently manage to distinguish one another without difficulty: different species steadfastly refuse to mate. The fruit fly, however, is probably the best judge of membership in its own "kind." For this reason, in their search for an objective criterion for separating species, biologists have emphasized the concept of **reproductive isolation,** the inability of members of one population of organisms to interbreed with members of another group. This definition is objective because it need only be observed; little interpretation is required. Reproductive isolation also receives particular emphasis because of its special importance in evolutionary theory (discussed in Chapter 45).

Unfortunately, although reproductive isolation is indeed the most objective criterion whereby species may be determined, it is not an absolute one because such "isolation" is often relative. For instance, the ordinary grass frog, *Rana pipiens* (found from one end of the East coast of the United States to the other), is considered a single species. Yet it has been shown that individuals from the extreme northern end of the range do not interbreed with individuals from the southern end of the range. Does this make them separate species? Maybe not, for the extreme northern and southern subpopulations *can* interbreed successfully with populations adjacent to them, and they in turn with the populations adjacent to *them,* and so on, until the middle of the range of the frog is reached. This means that in theory (and probably in practice as well), genetic traits can be exchanged between the non-interbreeding northern and southern populations via the genetically and geographically intermediate populations between them.

Another problem lies in defining the species that paleontologists discover in the fossil record. There are a few compellingly documented examples of one species changing to another in the course of time through a series of intermediate forms, each of which differs slightly from its ancestral and descendant forms. But where does the one species leave off and the next begin? And how can anyone be sure that the two, if they could somehow be brought together, would not interbreed? Defining them as separate species is a matter of fine biological judgement but lacks objective criteria of justification. To the definition of such **chronospecies,** separable from one another only by time, the concept of reproductive isolation is simply not relevant. Moreover, even living species must often be distinguished in much the same fashion that chronospecies are. How many biologists have ever actually tried to get two allegedly different species of jumping spiders to mate? If the conditions are not perfect, the likeliest result—even if they *are* members of the same reproductively isolated species—is mayhem rather than mating! Moreover, species are considered distinct if they are reproductively isolated *under natural conditions.* If that is the case, then the results of breeding experiments are almost automatically irrelevant because almost all such studies must necessarily be performed in some kind of captivity. Finally, the criterion of reproductive isolation cannot be applied to species that reproduce only asexually.

CATEGORIES OF CLASSIFICATION

Closely related species are grouped together in the next higher unit of classification, the **genus** (plural, **genera**). Recall from Chapter 1 that the scientific names of plants and animals consist of two words, the genus and the species, given in Latin. This system of naming organisms, called the **binomial** (two-name) **system,** was first used consistently by Linnaeus, though some of his predecessors employed something approaching it.

Just as several species may be grouped together to form a genus, so too in modern taxonomy a number of genera may be grouped together on the basis of structure, biochemical similarities, and other criteria to compose a **family.** Families may be grouped together into **orders,** orders into **classes,** and classes into **divisions**—for plants or fungi—or **phyla** (singular, **phylum**)—for animals or protists (Fig. 17–2). Housecats, for example, are members of the domain Eukaryota (domains will be discussed shortly), kingdom Animalia, phylum Chordata, subphylum Vertebrata, class Mammalia, subclass Eutheria, order Carnivora, family Felidae, genus *Felis,* species *catus.* For the classification of a different sort of organism, see Focus on Why You Are *Homo sapiens.*

Kingdom: Plantae, ANIMALIA, etc.

Phylum: Echinodermata, Arthropoda, Mollusca, Cnidaria, CHORDATA, etc.

Class: Aves, Amphibia, Reptilia, MAMMALIA, etc.

Order: Primates, Perissodactyla, Artiodactyla, Insectivora, CARNIVORA, etc.

Family: Canidae, Ursidae, Mustelidae, Viverridae, FELIDAE, etc.

Genus: *Panthera, FELIS,* etc.

Species: *concolor, CATUS,* etc.

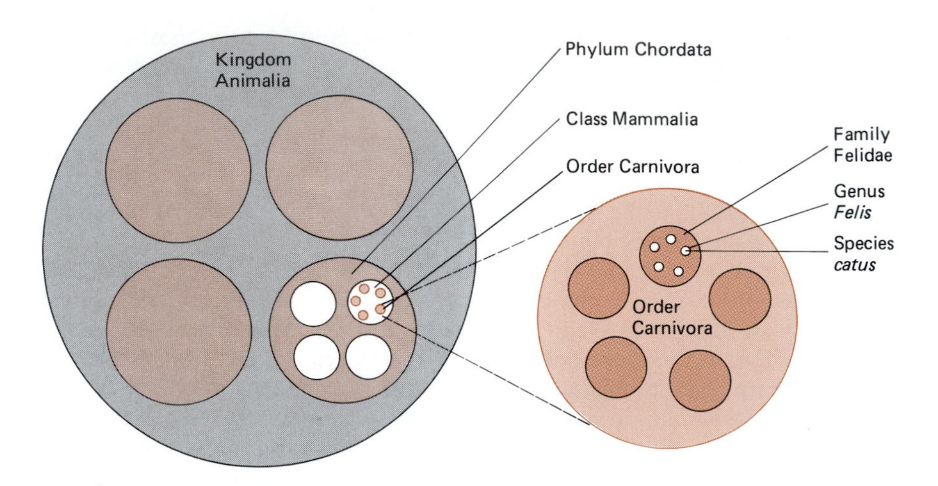

Figure 17–2 The principal categories used in classifying an organism.

SUBSPECIES

The species is the *fundamental* unit of classification but not the smallest in use. Geographically distinct populations within a species often display certain consistent characteristics that serve to distinguish them from other populations of the same species. If they do not form a reproductively isolated group, however, these are not truly separate species but are termed instead **subspecies** or **varieties.** Though subspecies do not ordinarily differ very much from one another, these differences may be sufficient to affect their behavior, biochemistry, or other characteristics important in biological research. This can be very confusing to scientists attempting to duplicate or extend one another's research findings using apparently the same, but actually different, subspecies as their experimental organisms. Deer mice, many species of oaks, and numerous other kinds of organisms occur in a number of such varieties. Although such subspecies are usually distinguishable from one another by experts, they may grade almost imperceptibly into one another at the borders of their geographical ranges where they are freely able to interbreed with one another. It is possible that some of these subspecies are in the process of becoming reproductively isolated and will, in the course of time, become clear-cut species (Fig. 17–3). Thus they can be utilized for studies of gene pools and of the speciation process.

SPLITTING AND LUMPING

Many plants and animals fall into easily recognizable, apparently natural groups, and their classification presents no obvious difficulty. Others, though, appear to lie on the borderline between two groups, having some characteristics in common with each. The number and inclusiveness of the principal groups vary according to the basis of the classification used and the judgment of the scientist making the classification. Some taxonomists like to group things in already existing units; these are called "lumpers." Others prefer to establish separate categories for forms that do not fall naturally into one of the existing classifications; they are called "splitters." Still others on the basis of new evidence deem it necessary to change the way organisms are classified by removing them from one category and putting them into another. Different taxonomists consider that there are from 10 to 35 animal phyla and from 4 to 12 plant divisions.

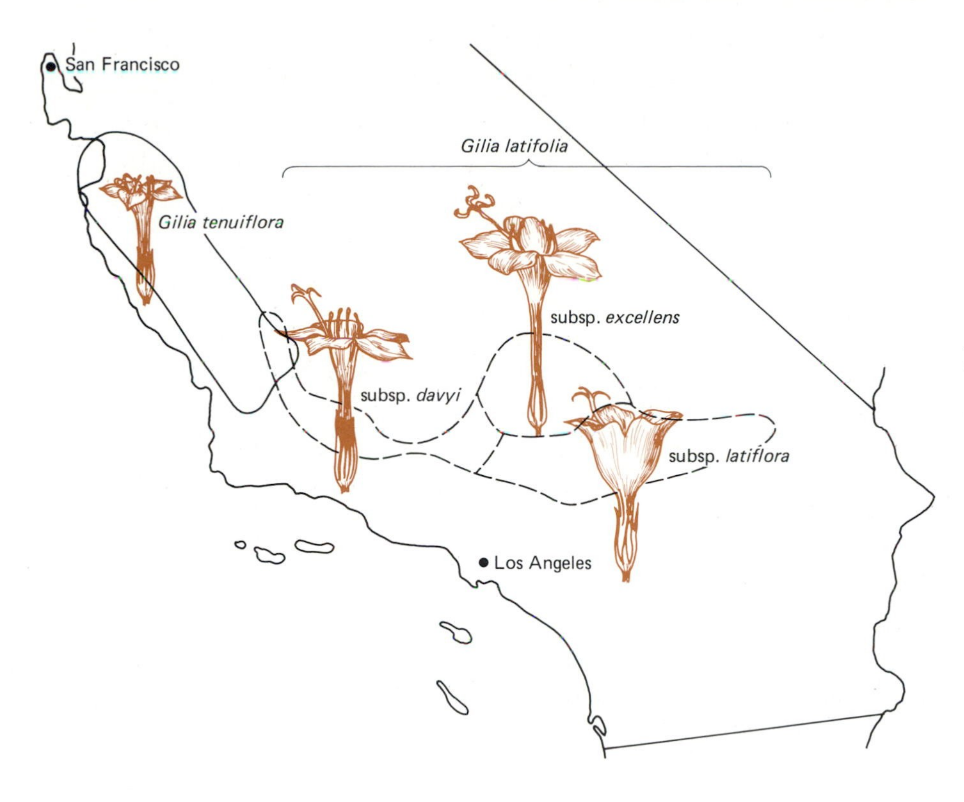

Figure 17–3 The ranges and distinguishing flower features of a subspecies of the California wildflower, *Gilia latifolia*, and of a closely related species, *G. tenuiflora*. From their similarities, it seems probable that *G. tenuiflora* was formerly a subspecies of *G. latifolia* but that, as shown by its geographical overlap with *G. latifolia* in one area without intergradation, it has become a distinct species. For simplicity, three other subspecies of *G. latifolia* with narrower distributions have been omitted. (The full name of a subspecies includes genus, species, and subspecies designations, e.g., *Gilia latifolia excellens*.) (From Grant, A., and Grant, V.: *El Aliso* 3: 203, 1956.)

Choosing Taxonomic Criteria

Deciding the appropriate weight for various traits in determining taxonomic categories is not always simple, even in apparently straightforward instances. What, for instance, are the most important invariable characteristics of a bird? We might list the egg-laying trait, beak, wings, absence of teeth, feathers, and warmbloodedness. Yet some mammals (the monotremes) also have beaks, lack teeth and lay eggs (see Fig. 26–32).

No mammal, however, has feathers. Is this trait *absolutely* diagnostic of birds? According to the conventional taxonomic wisdom, the presence of feathers could be used to decide what is and is not a bird. This applies only to modern birds, however. Some extinct reptiles may have been covered with feathers, while not being, in any meaningful sense, birds. Moreover, as Figure 17–4 shows, there are even some modern exceptions!

Usually, organisms are classified on the basis of a *combination* of characteristics rather than on the basis of one perhaps superficial trait such as the ability to live in water or to swim. The significance of these combinations is determined inductively, that is, by a kind of subconscious integration and interpretation of data. Such induction is a necessary first step in all science. The taxonomist proposes, for

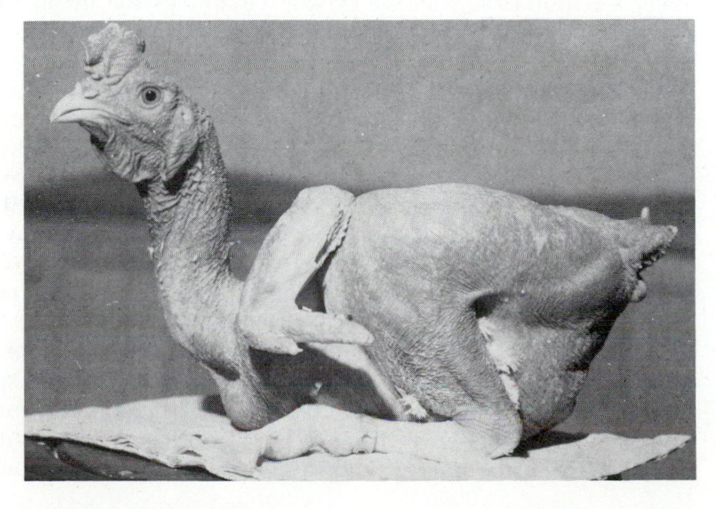

Figure 17–4 The possession of feathers is one of the key characteristics of birds, but as demonstrated by this naked hen it is possible for a bird to lack feathers entirely. Naked hens were bred in order to save the feed that would be used metabolically to produce feathers. The plan backfired when the uninsulated hens shivered so much that they used up more energy in muscular contraction than they would have used to produce feathers. (Courtesy of Dr. Max Rubin.)

FOCUS ON

***Why You Are* Homo sapiens**

1. At present there are five generally recognized kingdoms of organisms and two domains. Since human cells have discrete nuclei surrounded by nuclear membranes, you belong to the domain Eukaryota. Your cells lack chloroplasts and cell walls, and you are a multicellular heterotroph, with highly differentiated tissues and organ systems. That makes you a member of the kingdom Animalia.
2. What kind of an animal are you? You possess a spinal column composed of bony vertebrae that has largely replaced a cartilaginous rod you had as an embryo, the notochord. At that time you also had structures that had you been a fish would have developed into gill slits. You have a dorsal, nerve cord and brain, both of which still retain remnants of their embryonic cavities. These traits mark you out as a chordate and a vertebrate—that is, you belong to the phylum Chordata (because you either have or have had a notochord), and to the subphylum Vertebrata (because you have vertebrae that replaced the notochord).
3. Among the vertebrates there are several classes: cartilaginous fish, bony fish, jawless fish, amphibia, reptiles, mammals, and birds. You are homeothermic (warm-blooded) and so must be either a bird or a mammal. Lacking feathers and having teeth and (if you are female) the potential for nursing your young, you are a mammal. If you are male, do not be concerned; even if you cannot nurse, having hair is enough.
4. Within the mammals there are three subclasses: the Prototheria, the Metatheria, and the Eutheria. The Prototheria are confined either to zoos or to the Australian continent and its environs. In the present day they include the duck-billed platypus and the spiny anteater, both of which, in addition to other unusual traits, lay eggs. The Metatheria, most of which are also from Australia, usually carry their still-embryonic young around in a pouch and totally lack a placenta (an organ of exchange between mother and developing embryo). If you did not hatch from an egg or spend your infancy in a pouch, you can be confident of your status as a eutherian.
5. A number of orders exist within the Eutheria. The insectivores, for instance, include the moles and shrews, the Chiroptera are the bats, and the Carnivora include the dogs, cats and ferrets, among others. Your opposable thumbs, frontally directed eyes, flat fingernails, and several other characteristics identify you as a primate, along with monkeys, apes and tarsiers.
6. Primates include a number of families. You and the New World monkeys are obviously very different—they have prehensile tails, for instance, which you and all Old World monkeys and apes lack; indeed, you and the apes lack tails altogether. Your posture is upright, you have long legs and short arms, and not much body hair. You are blessed with your very own family that has no other modern occupants: the Hominidae.
7. Within the Hominidae anthropologists distinguish several species, all but one of which are known only as fossils. *Australopithecus* is one of these. If you are alive, you do not belong to any of those extinct genera, but to the genus *Homo.*
8. Again, the genus *Homo* has only one living species—*sapiens.* Since many taxonomists insist that the species name always includes the genus name, please think of yourself as *Homo sapiens.*

example, that birds should all have beaks, feathers, no teeth, and so on; this is really a hypothesis. Then he or she reexamines the living world and observes whether there are organisms that might reasonably be called birds that do not fit the definition of "birdness." If not, the definition is permitted to stand, at least until someone discovers too many exceptions to it. If so, the definition is abandoned or modified. Sometimes, however, it is necessary for the taxonomist to persuade the world that the apparent exception—a bat, for instance—resembles a bird only superficially and should not be considered to be one.

If taxonomy were entirely static it could not be considered a science, because science proceeds by the constant reevaluation of data, hypotheses, and theoretical constructs. As new data are discovered and old subjected to reinterpretation, the ideas of taxonomists are bound to change. Even in the 1980s, for example, organisms, the Lorcifera, Fig. 17–5, have been discovered whose combination of traits did not fit those of any established phylum. For that reason they required a new phylum of their own, initially just to accommodate a single species! To take another, perhaps more important example, as evolutionary conceptions about the origin of cellular life have changed, biologists have considered assigning the **Archaebacteria** to the rank of a kingdom (see Chapter 19). These bacteria do not visibly differ very much from others but are now thought by many to be much closer to the ancestry of the eukaryotes (organisms whose cells possess nuclei) than are the other bacteria. It is significant that this reassignment has been justified almost solely on the basis of their biochemical peculiarities.

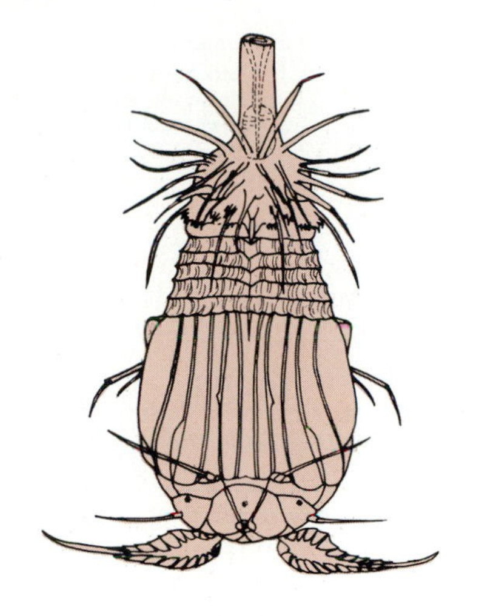

Improving Taxonomy

Sometimes the reinterpretation of taxonomy is merely a rethinking of facts long known but now viewed in a new way. In other cases, it depends upon the acquisition of totally new data or the construction of new theories. Perhaps the most obvious example of this is the Darwinian theory of evolution, which, by suggesting the ways in which the major groups of organisms might be related, also suggested seemingly natural ways of expressing this relationship through classification. This approach is not without its difficulties, as we will see in Part 6, but it does form the ostensible basis of all modern taxonomic thinking. Since its basis is also the basis of evolutionary thought, except for a few brief remarks, we must postpone detailed consideration until later.

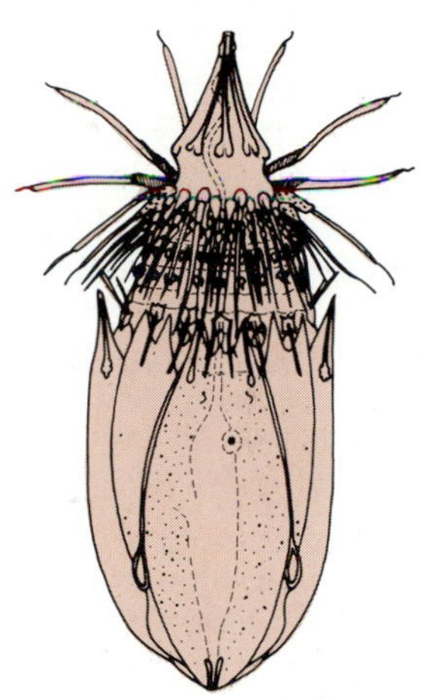

Figure 17–5 Only one new animal phylum—that for a group of microscopic marine animals—has been added to the animal kingdom since 1900. The recent discovery of the marine animal Lorcifera has resulted in the proposal that a second new phylum be added. Lorcifera larvae propel themselves with a pair of appendages attached to the body by a ball-and-socket joint. The tiny adults, about 230 μm long, lack appendages for swimming. Both larvae and adults have head spines and a flexible, retractable tubelike mouth. These animals live between grains of shell gravel in the ocean bottom. Are there other unique organisms yet to be discovered and classified?

Taxonomy and Ancestry

The basis for evolutionary taxonomy, formally designated **systematics,** is inferred relationships, so all organisms classified within a category should ideally share a common, or **monophyletic,** ancestry. Such inferences are based upon similarities and differences among organisms. Thus, closely similar organisms are held to be closely related, and less similar organisms are viewed as being more remotely related (Fig. 17–6). However, the choice of the similarities that are used in this approach is extremely important. The streamlined body form shared by fish and porpoises is viewed as less important than their dissimilarities; thus they are not classified together. As major "design" features, the ability to breathe air, nurse young, maintain a constant body temperature, and grow hair gives the porpoise more in common with mammals such as humans than with fish. Thus it is classified as a mammal and is viewed as having descended from a more typical mammalian ancestor.

But although the porpoise has more in common with us than it does with a fish, there are some characteristics—including a notochord (skeletal rod) and rudi-

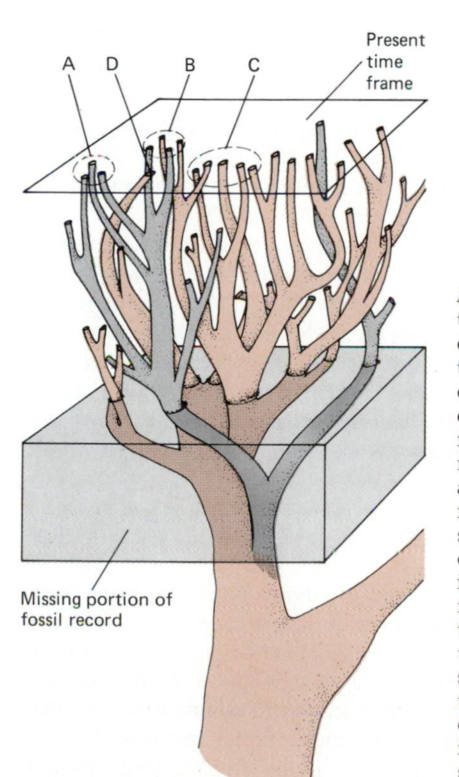

Figure 17–6 The difficulty with a system of taxonomy based upon evolution is that it can be only as good as our understanding of the actual course of evolution. Modern taxonomists, working solely with living forms, can compare them only on the basis of their modern similarities and differences. They might be inclined to create categories A, B, and C. Whether to classify D in A, B, or C might be highly controversial. In some cases similarities are actually not derived from close relationship but have resulted from coincidence. Paleontologists, investigating the fossil record, may confirm the findings of the taxonomists if all the organisms classified together taxonomically appear to share their similarities as far back in the fossil record as they can be traced. Alas, their very different origin could be concealed in a portion of the fossil record that geological events have destroyed.

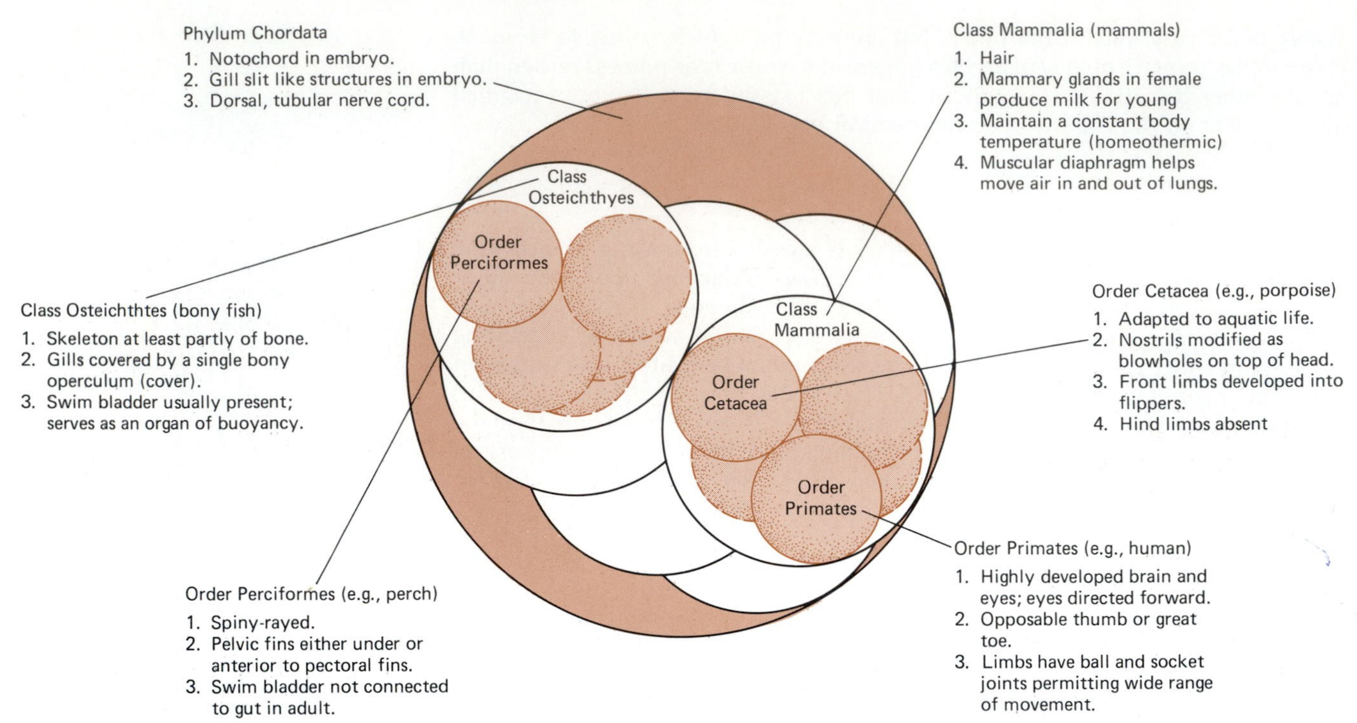
Phylum Chordata
1. Notochord in embryo.
2. Gill slit like structures in embryo.
3. Dorsal, tubular nerve cord.
Class Mammalia (mammals)
1. Hair
2. Mammary glands in female produce milk for young
3. Maintain a constant body temperature (homeothermic)
4. Muscular diaphragm helps move air in and out of lungs.
Class Osteichthyes
Order Perciformes
Class Osteichthtes (bony fish)
1. Skeleton at least partly of bone.
2. Gills covered by a single bony operculum (cover).
3. Swim bladder usually present; serves as an organ of buoyancy.
Class Mammalia
Order Cetacea
Order Primates
Order Cetacea (e.g., porpoise)
1. Adapted to aquatic life.
2. Nostrils modified as blowholes on top of head.
3. Front limbs developed into flippers.
4. Hind limbs absent
Order Perciformes (e.g., perch)
1. Spiny-rayed.
2. Pelvic fins either under or anterior to pectoral fins.
3. Swim bladder not connected to gut in adult.
Order Primates (e.g., human)
1. Highly developed brain and eyes; eyes directed forward.
2. Opposable thumb or great toe.
3. Limbs have ball and socket joints permitting wide range of movement.

Figure 17–7 Shared derived characteristics. Members of class Osteichthyes (bony fish) and class Mammalia (the mammals) share many more characteristics with one another and with the members of the other classes of phylum Chordata than they do with members of any other class of any other phylum. For example, a perch has more in common with a monkey than with a sea star or clam. Members of various orders of the same class share more characteristics than members of orders that belong to different classes. Thus, a porpoise shares more characteristics with a human being than with a perch. This indicates a more recent common ancestry for the porpoise and the human.

mentary gill slits in the embryo stage, a dorsal tubular nerve cord, and much else—shared by all three kinds of organisms. Such facts are used as a basis for classification, but additionally they can be interpreted as indicating a common ancestry—in this case, for ourselves, the porpoise, and the fish. This is, however, a more remote common ancestry than that between the porpoise and ourselves. Therefore, fish, humans, and porpoises are grouped together, but in the chordates, a much larger category than that containing porpoises and human beings (Chapter 26).

How can we make such decisions rationally? The order in which they are made is very important. First, we examine the characteristics that are common to the *largest* group of organisms and interpret them as indicating the remotest common ancestry. These are known as **shared derived characteristics.** The shared derived characteristics found in the next smaller group of organisms presumably indicate a somewhat more recent common ancestry for that group; those found in the next smallest group indicate an even more recent common ancestry for all members of *that* assemblage; and so on (see Focus on Approaches to Taxonomy). Organisms excluded from a group because they do not share its key traits are then grouped among themselves in accordance with the traits that they in turn have in common. In the end, we have, or hope that we have, a complete set of nested groupings that reflect the evolutionary relationships of all organisms (Fig. 17–7).

Today it is no longer necessary to restrict the traits we examine to physical structures. It is possible to consider a wide variety of similarities and differences in behavior, amino acid sequences within comparable proteins such as hemoglobin, immunological properties of proteins (see Focus on Taxonomy and Biochemistry), base sequences and hybridization coefficients of nucleic acids, computer-assisted statistical analyses of physical measurements of body parts, and far more. These techniques have greatly affected not only the way in which we classify particular organisms but have resulted in modifications of our basic system of taxonomy.

How Many Kingdoms?

Biologists since the time of Aristotle have divided the living world into two kingdoms, **Plantae** and **Animalia,** and this division is so deeply ingrained in our everyday modes of thought that at first there seems to be no alternative. Yet the German biologist Ernst Haeckel suggested more than a century ago that a third kingdom, the **Protista,** be established to include all the single-celled organisms that are intermediate in many respects between plants and animals. Some protists are very clearly plantlike and seem closely related to the true plants; other protists are ani-

FOCUS ON

Approaches to Taxonomy

Why are porpoises classified along with humans as mammals rather than as fish? Members of different schools of taxonomy might answer this question in different ways.

A taxonomist who follows the **phenetic** system, based on phenotypic similarities, would explain that even though porpoises are aquatic like fish, they share more characteristics with mammals. In the phenetic system, organisms are grouped according to the number of similarities they share, without trying to assign differential weight to any of them. Traits are assigned numbers, and designated as present (+) or absent (−) in the organisms of a particular category. This information is fed into a computer, which indicates which groups have the most traits in common.

A taxonomist who follows the **cladistic** system of classification might say that porpoises cannot be classified with fish because they evolved much later in time than fish. In fact, the whales appeared on earth millions of years after the fish. The cladistic approach to taxonomy emphasizes **phylogeny,** focusing upon how long ago one group branched off from another, but does not consider the significance or magnitude of specific adaptational differences between the descendants of a common ancestor. For example, birds (along with dinosaurs) are thought to be descendants of reptiles that shared a common ancestor with the modern crocodiles and alligators. Crocodiles and birds, then, might be considered sister groups, which some cladists might place in the same class. This decision might be made without considering that crocodiles appear to have much more in common adaptationally and ecologically with lizards, snakes, and turtles (all of which are descended from a different reptilian ancestor) than they do with birds.

If birds, alligators, and crocodiles are the sole surviving descendants of the same early group of reptiles, a case could be made for classifying them together in the same class. Modern lizards, snakes, and turtles would be excluded since they have a different ancestry.

More traditional taxonomists use a system of **phylogenetic** classification, and present evolutionary relationships in a phylogenetic tree (the type of diagram used in this book). A traditional taxonomist might explain that porpoises are mammals rather than fish because they share many characteristics with other mammals, and because these characteristics can be traced to a common ancestor. Organisms are classified in the same category according to their shared characteristics only if those traits are derived from a demonstrable common ancestor. This principle is applied in the light of the significance of the adaptations possessed by the various organisms under consideration. If, for instance, the egg-laying mammals could be shown to have a very different ancestry from the placental mammals, the phylogeneticist might erect a separate class to accommodate them. On the other hand, common ancestry, though necessary for inclusion in the same category, would not by itself be sufficient grounds for inclusion. A phylogeneticist would almost surely classify birds and crocodiles separately, for example, even though they share a common ancestor. In short, the phylogeneticist would be a splitter more than a lumper.

mal-like. Some have characteristics intermediate between those of animals and plants, and still others have characteristics that are distinctly different from those either of plants or of animals. Even the organisms included by different biologists in the kingdom Protista may differ. Most taxonomists include in the Protista only unicellular or colonial forms, but a minority include some fungi and multicellular algae as well.

In 1969 it was suggested by R. H. Whittaker that the **Fungi** be ranked as a separate kingdom. There is really no reason to classify fungi as plants; not one of them is photosynthetic, and most of them have cell walls made of chitin almost identical chemically with the chitin of arthropod exoskeletons and far different from the cellulose that occurs in plant cell walls. The life cycles of many of them are reminiscent of the algae, but many of them have an advanced tissue level of organization vastly different from that of any alga. If the fungi are indeed a monophyletic group (or even, perhaps, if they are not), they would seem to be entitled to kingdom status. That, however, was not the end of it. Whittaker also suggested that the

TABLE 17–1
The Five Kingdoms: Monera, Protista, Fungi, Plants, and Animals

Kingdom	*Characteristics*	*Ecological Role*
Monera	Prokaryotes (lack distinct nuclei and other membranous organelles); single-celled; microscopic	
Bacteria	Cell walls composed of peptidoglycan (a substance derived from amino acids and sugars); many secrete a capsule made of a polysaccharide material. In pathogenic bacteria capsule may protect against defenses of host. Cells may be spherical (cocci), rod-shaped (bacilli), or coiled (spirilla).	Decomposers; some chemosynthetic autotrophs; important in recycling nitrogen and other elements. A few are photosynthetic, usually employing hydrogen sulfide as hydrogen source. Some pathogenic (cause disease); some used in industrial processes.
Cyanobacteria	Specifically adapted for photosynthesis; use water as a hydrogen source. Chlorophyll and associated enzymes organized along layers of membranes in cytoplasm. Some can fix nitrogen.	Producers; blooms (population explosions) associated with water pollution
Protista	Eukaryotes; mainly cellular or colonial	
Protozoa	Microscopic; unicellular; depend upon diffusion to support many of their metabolic activities	Important part of zooplankton; near base of many food chains. Some are pathogenic (e.g., malaria is caused by a protozoan).
Eukaryotic algae	Sometimes hard to differentiate from the protozoa. Some have brown pigment in addition to chlorophyll.	Very important producers, especially in marine and fresh-water ecosystems
Fungi	Eukaryotes; plantlike but cannot carry on photosynthesis	Decomposers, probably to an even greater extent than bacteria. Some are pathogenic (e.g., athlete's foot is caused by a fungus).
Slime molds	Protozoan characteristics during part of life cycle; fungal traits during remainder	
True fungi (molds, yeasts, mildew, mushrooms, rust)	Body composed of threadlike hyphae; rarely, discrete cells. Hyphae may form tangled masses called mycelia, which infiltrate whatever the fungus is eating or inhabiting. Mycelium is often invisible, as in mushrooms.	Some used as food (yeast used in making bread and alcoholic beverages); some used to make industrial chemicals or antibiotics; responsible for much spoilage and crop loss
Plantae	Multicellular eukaryotes; adapted for photosynthesis; photosynthetic cells have chloroplasts. All plants have reproductive tissues or organs and pass through distinct developmental stages and alternations of generations. Cell walls of cellulose; cells often have large central vacuole. Indeterminate growth; often no fixed body size or exact shape	Almost entire biosphere depends upon plants in their role as primary producers; important source of oxygen in the atmosphere of the earth
Animalia	Multicellular eukaryotic heterotrophs, many of which exhibit advanced tissue differentiation and complex organ systems. Lack cell walls. Able as a rule to move about by muscle contraction; extremely and quickly responsive to stimuli, with specialized nervous tissue to coordinate responses; determinate growth	Almost the sole consuming organisms in the biosphere, some being specialized for herbivorous, carnivorous, and detrivorous (eating dead organisms or organic material such as dead leaves) life styles

bacteria and cyanobacteria, formerly considered to be plants or even fungi, deserved their own kingdom as well, the kingdom **Monera.** (For a summary of the five kingdoms now recognized by most biologists, see Table 17–1.)

The Domains

In classification, much depends upon what traits are viewed as most important. If "not-being-obviously-an-animal" is the most important thing, then bacteria are plants. If having a nucleus or not is the most important thing, then bacteria are not plants, whatever they may be. Yet is it enough merely to place the bacteria in a kingdom of their own?

Consider all the ways in which monerans—the bacteria and cyanobacteria—differ from other kinds of organisms: Not only do they lack a nucleus, but they also lack mitochondria, chloroplasts, lysosomes, centrioles, spindle fibers, microtubular flagellae, Golgi complex, nucleolus, and in fact virtually all the cytoplasmic organelles that most of the nucleated organisms *do have in common with one another!* In

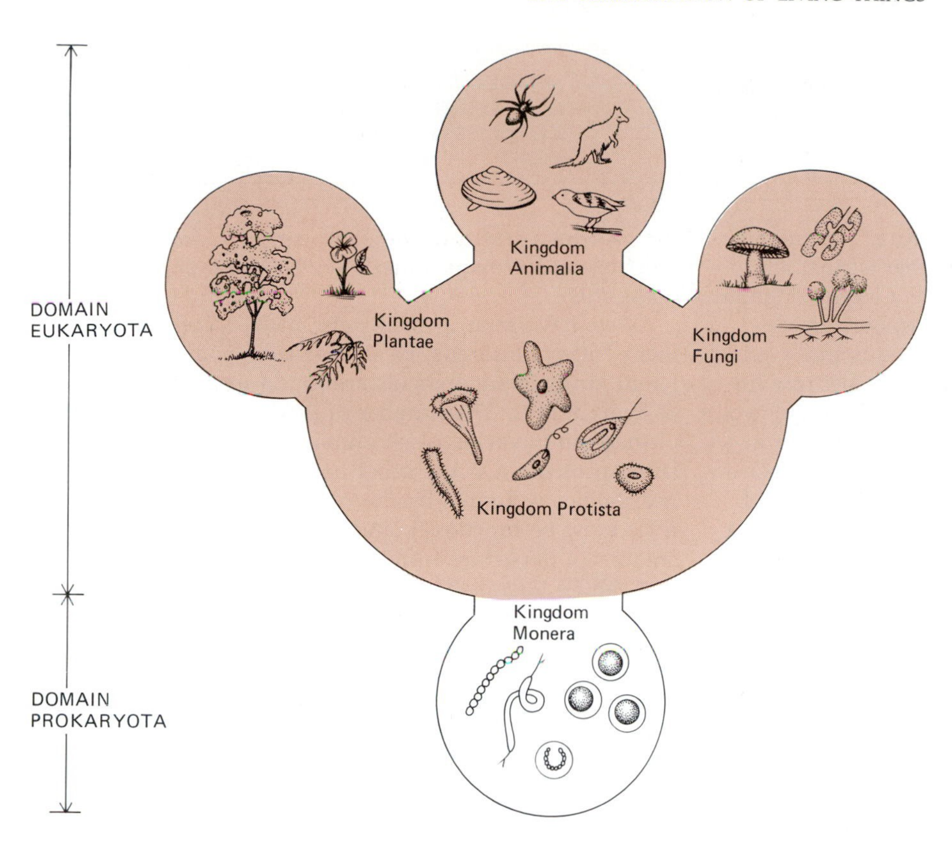

Figure 17–8 The five-kingdom system of classification.

accordance with the concept of shared derived characteristics, it would seem that the gulf between monerans and everything else alive is much broader than any gaps that may occur among the nucleated organisms. A bigger category than just the kingdom may be needed to express the magnitude of this difference, and the concept of the **domain** (synonym: **dominium**)[1] has been advanced to satisfy this lack. In this view, one domain would suffice to hold all the nucleated organisms, which do, after all, have a great deal in common with one another: the domain **Eukaryota,** named with the Greek words for "true" and "nucleus." The other domain is the **Prokaryota,** whose name means "before" and "nucleus" in Greek. Their names reflect the supposition that the Prokaryota came first, eventually giving rise to the Eukaryota[2] (Fig. 17–8).

We might wonder, with all this, what real progress has been made. The answer should be informed with a consideration of the place that frames of reference have in human thought. The odd name "intestinal flora" given to the microscopic inhabitants of the human gut is more than flowery language; it reflects the old idea that these microorganisms are really plants. Proposing a more appropriate classification for these organisms does more than generate scholarly argument; it helps us think more clearly about them and their significance. It is easier to see why antibiotics (or antibacterials) rather than herbicides are needed to treat bacterial infections when we remember the biochemical differences that exist between bacteria and plants (many antibiotics work by acting on the synthesis of components unique to the cell walls of bacteria). On a more fundamental level, promoting the protists to kingdom status helps us to avoid thinking of them as "simple," because their single-celled bodies can be actually more complex than the cells of some multicellular organisms.

Linnaeus probably intended to design a static system of classification, for he had no theory of evolution in mind when he set it up. Yet it is remarkable how flexible and adaptable to the influx of new biological knowledge and theory his system has proved to be. There are very few 18th-century inventions that survive today in a form that would still be recognizable by their originators.

[1] Taxonomic categories larger than kingdoms have been proposed by several scholars who have called them domains, superkingdoms, and dominiums. For example, R.T. Moore ("Proposal for the Recognition of Super Ranks," *Taxon* 23:650–652, 1974) suggested the use of the term dominium to comprise (1) the viruses, (2) the prokaryotes, and (3) the eukaryotes. C. R. Woese and G. E. Fox proposed that there should be a prokaryote domain consisting of the eubacteria and the methanogens.

[2] In some systems, the kingdom is retained as the highest category, so that Eukaryota and Prokaryota are each kingdoms, with the Eukaryota comprising four subkingdoms (plants, animals, protists, and fungi)!

FOCUS ON

Taxonomy and Biochemistry

The basic requirement for any "natural" taxonomic group is that it be monophyletic; that is, that all members of the group should have a *single* common ancestral species from which all members are descended. Sometimes, however, the traits that a group of organisms have in common may not be truly **homologous;** that is, they may not have the same evolutionary origin despite their superficial similarity in structure. Conclusions based on a limited number of traits—bone shapes, for instance—may be suspect. Unfortunately, bones may be all that remains of extinct organisms such as the dinosaurs. Among living forms, however, it is possible to take many other traits into consideration when proposing the evolutionary relationships of organisms. Comparisons of the proteins present in the organisms are among the most useful techniques that have been developed for doing this.

As we have seen, proteins are made up of chains of many, often hundreds, of amino acids. Some 20 different kinds of amino acids are found in the proteins of living things. The number of possible combinations of these amino acids is astronomical, especially for the larger proteins. On the other hand, the number of possibilities is limited by the consideration that proteins must be functional, and that sometimes the amount of departure from the basic plan of a protein that can be tolerated by the physiology of a living organism is limited. For example, to remain useful, there can be little permissible variation in the electron

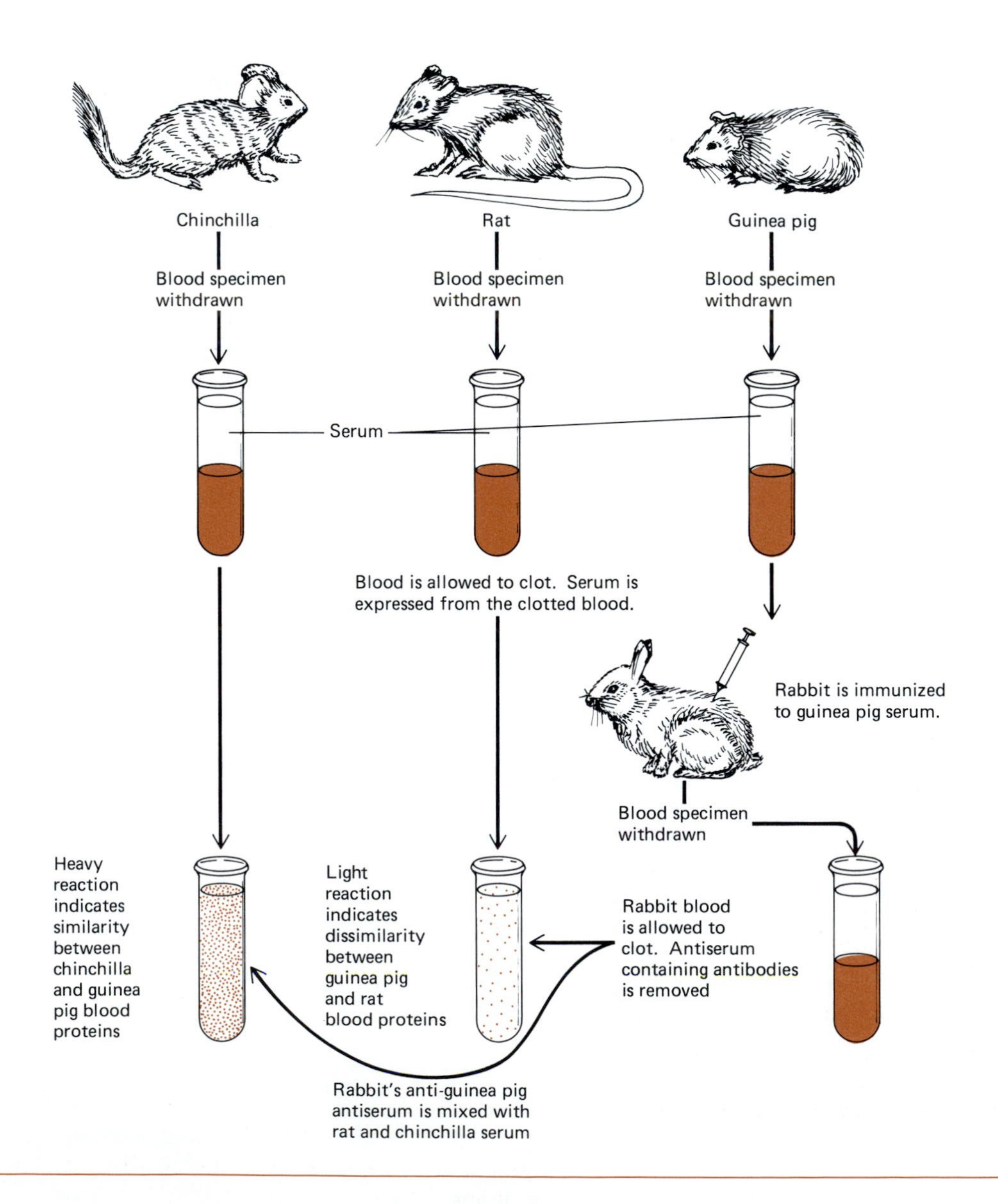
Chinchilla
Rat
Guinea pig
Blood specimen withdrawn
Blood specimen withdrawn
Blood specimen withdrawn
Serum
Blood is allowed to clot. Serum is expressed from the clotted blood.
Rabbit is immunized to guinea pig serum.
Blood specimen withdrawn
Heavy reaction indicates similarity between chinchilla and guinea pig blood proteins
Light reaction indicates dissimilarity between guinea pig and rat blood proteins
Rabbit blood is allowed to clot. Antiserum containing antibodies is removed
Rabbit's anti-guinea pig antiserum is mixed with rat and chinchilla serum

transport system protein cytochrome oxidase; somewhat more variation is seen in hemoglobin, however, and a great deal in albumin. Variation in the amino acids in any of these proteins can be produced only by genetic mutation.

Let us consider an example of variation resulting from genetic mutation: Human hemoglobin occurs in many variants, among which is one that produces the hereditary disease sickle cell anemia. It has been shown that if this abnormal hemoglobin comprises a *portion* of the hemoglobin that is present in the body's red blood cells, a partial immunity to malaria results. The trait is clearly advantageous in a malaria-prone environment, but carries the price that one fourth of children born to carriers of this gene will be homozygous for the gene and will have sickle cell anemia. Thus the gene that produces this hemoglobin variant is an advantage in those parts of Africa where malaria is endemic. Natural selection preserves it there because it is advantageous.

The sickling gene produces hemoglobin that differs from normal hemoglobin in only one amino acid. This sequence produces the disease or helps to kill malaria parasites, depending on whether normal hemoglobin is also present in the red blood cells. But there are also many places in the hemoglobin chain where amino acid substitutions are apparently not critical at all. Thus we are able to catalog a great deal of innocuous hemoglobin variation even *within* the human species.

It would seem that by comparing amino acid sequences of proteins we could gain some idea of degree of relatedness of two organisms by assuming that the greater the "agreement" in their amino acid sequences, the more closely they are related to one another. This can be done not only by sequencing the amino acid composition of the protein directly—an expensive and time-consuming endeavor—but by simpler comparisons using electrophoresis (discussed in Chapter 46) and serology.

Serological techniques involve the immunological comparison of proteins. Much of the original development of this technique was done shortly after the turn of the century by George Nuttall, who injected rabbits with the blood serum of other organisms. The rabbits developed antibodies to the alien serum, or to certain proteins in it, which acted as antigens (Chapter 36). When another specimen of the experimental serum was mixed with the blood serum of the injected rabbits, the antigen was bound to the antibody and a cloudy precipitate formed (see figure).

Antisera such as those from the immunized rabbit may **cross-react** with antigens that are similar to though not identical with the antigens used in eliciting the antibodies. If a rabbit had been immunized to guinea pig serum, for example, some (but less) precipitation would occur if its antiserum were then mixed with the serum of a chinchilla. This result indicates a similarity of the blood proteins, reflecting the fact that both the guinea pig and the chinchilla are rodents belonging to the same family. For a rodent whose exact taxonomic position was controversial, its degree of relationship to the guinea pig could be estimated by assessing the degree to which its serum cross-reacted with the anti–guinea pig serum. This can be done by using instruments that measure the density of the antibody–antigen precipitate or by using an agar-well technique, as shown in the figure. By similarly comparing the unknown animal immunologically with antigens from other rodents such as rats, hamsters and squirrels, its place within the order Rodentia could be determined.

Such techniques, however, cannot be used as ultimate arbiters of taxonomy. Serological and, to a lesser extent, electrophoretic techniques are often handicapped by the fact that they are applied to mixtures of unknown or poorly characterized proteins, so that which protein is cross-reacting, for example, may be unknown. Only the tedious and expensive amino acid sequencing techniques are really reliable as the basis for exact taxonomic determination, but even they do not necessarily settle all controversy. There is less biochemical variation between human beings and chimpanzees than that between many other closely related species (Chapter 46); yet we can easily appreciate the differences that do exist between these primates.

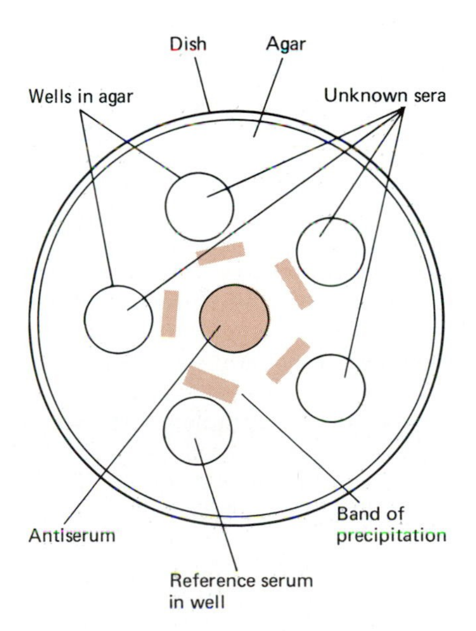

The agar-well technique for testing the degree of relationship of various organisms.

SUMMARY

I. Taxonomy is the science of naming and classifying organisms.

II. The modern system of scientific taxonomy is based on the binomial or Linnaean system.
  A. The fundamental unit of classification that Linnaeus used was the species.
  B. The categories of classification now used are as follows (using, as an example, the housecat): domain (Eukaryota), kingdom (Animalia), phylum (Chordata), subphylum (Vertebrata), class (Mammalia), order (Carnivora), family (Felidae), genus (*Felis*), species (*catus*). Subspecies may also in some cases be listed after species.

III. Shared derived characteristics indicate common ancestry.
  A. The human, porpoise, and fish are all classified in the same phylum, the chordates; the human and porpoise are also classified in the same class, which is a smaller category, indicating their closer relationship.
  B. More recent common ancestry is indicated by classification in smaller and smaller taxonomic groups.

IV. All organisms belong either to the domain Prokaryota or to the domain Eukaryota. Eukaryotes have clearcut nuclei equipped with nuclear membranes; prokaryotes do not and also lack most membranous cellular organelles.
  A. The four eukaryote kingdoms are Plantae, Animalia, Fungi, and Protista, all summarized in Table 17–1.
  B. The single prokaryote kingdom is the Monera, containing bacteria and cyanobacteria.

POST-TEST

1. The science of naming and classifying organisms is ______________.

2. A population of individuals, similar in their structural and functional characteristics, that in nature breed only with each other and share close common ancestry is a ______________.

3. The inability of members of one population to breed with members of another may be referred to as ______________ ______________.

4. The category of classification ranked just below the species is the ______________; the next higher is the ______________.

5. In the binomial system an organism is given both a ______________ and a ______________ name.

6. A taxonomist who prefers to establish separate categories for organisms that do not fall naturally into existing categories is referred to as a ______________.

7. Organisms classified within the same category should share a common, or ______________, ancestry.

8. Single-celled eukaryotic organisms are classified in kingdom ______________.

9. The Monera are included in domain ______________.

10. Multicellular, eukaryotic heterotrophs that lack cell walls belong to kingdom ______________.

A complete classification for a human being would be as follows:

11. Domain ______________
12. Kingdom ______________
13. Phylum ______________
14. Subphylum ______________
15. Class ______________
16. Subclass ______________
17. Order ______________
18. Family ______________
19. Genus ______________
20. Species ______________

REVIEW QUESTIONS

1. What advantage might there be in calling an organism by both its genus and species names? Why not employ the species name only?
2. Define "species" precisely, or as precisely as possible.
3. Give some reasons that the objective concept of the species does not always work out well in practice.
4. Upon what criteria were early systems of classification based?
5. Give, in outline form, the complete system of Linnaean classification.
6. What taxonomic problems was the five-kingdom scheme intended to solve? Has it created new problems?
7. How can shared derived characteristics be used to determine relationships among organisms?
8. In which kingdom would you classify each of the following?
   a. an oak tree
   b. an amoeba
   c. an *Escherichia coli* bacterium
9. Compare the phenetic, cladistic, and phylogenetic approaches to taxonomy.
10. In what way would it be helpful for a taxonomist to compare the amino acid sequences of proteins of various organisms?

18
Viruses

OUTLINE

I. Structure of a virus
II. Bacteriophages
III. Viral replication in lytic infections
IV. DNA integration in lysogenic infections
V. Another form of coexistence
VI. Animal viruses
VII. Viral diseases
VIII. Viruses and cancer
IX. Viroids

LEARNING OBJECTIVES

After you have read this chapter you should be able to:

1. Compare a virus with a living cell.
2. Describe the structure of a virus, or draw and label a virus.
3. Describe bacteriophages, and explain why they have been significant in the development of knowledge about viruses.
4. Trace the steps that take place in the process of viral infection.
5. Contrast a lytic infection with a lysogenic infection.
6. Give the basis of species and tissue specificity of viral infection.
7. Identify two viral infections of plants and six human viral infections.
8. Compare acute, chronic, latent, and slow virus infections.
9. Summarize what is currently known of the relationship between viruses and cancer.

Viruses lie on the threshold between life and nonlife. They are not cellular, cannot move about on their own, and cannot carry on metabolic activities independently. They can reproduce, but only within the complex environment of the living cells that they infect. In a sense, viruses come alive only when they infect a cell. For these reasons viruses cannot be classified in any of the five kingdoms of living things.

Should viruses be viewed as the simplest of all "living" things? Some taxonomists think that the ancestors of modern viruses were among the primitive, free-living heterotrophs that evolved in the primordial sea. These early "organisms" fed upon the organic nutrients that surrounded them. By the time these nutrients were depleted, some organisms had become autotrophs, while others had evolved the enzyme systems necessary to derive energy from feeding on the autotrophs. The early viruses developed neither of these life styles; instead, they adapted to a parasitic mode of life.

Other taxonomists suggest that viruses evolved from cellular ancestors, becoming highly specialized as parasites. During their evolution, they "lost" their cellular components—all but the nucleus. Still another hypothesis is that viruses are pieces of nucleic acid originally present in cellular organisms. Some viruses may trace their origin to animal cells, other to plant cells, and still others to bacterial cells. Their origin might explain the specificity with which they infect different types of organisms; perhaps they infect only those species closely related to the organisms from which they originated. Whatever their origin, we can agree that viruses are extremely successful at their particular mode of life—parasitism.

No system of virus classification has yet been agreed upon because so little is known about their evolutionary relationships. Although a system has been proposed for tentatively dividing them into families and genera, it is not universally accepted. At present viruses are usually grouped on the basis of four main criteria: (1) size; (2) shape; (3) presence or absence of an outer envelope; and (4) the type of nucleic acid—DNA or RNA—they contain, and whether it is single-stranded or double-stranded. They are also sometimes classified according to the types of diseases they cause or their mode of transmission.

Structure of a Virus

A **virus,** or **virion,** is a tiny particle consisting of a nucleic acid core surrounded by a protein coat called a **capsid.** Some viruses have an outer **envelope** containing lipids, carbohydrates, and traces of metals. All cellular forms of life contain both types of nucleic acid, but a virus contains *either* DNA or RNA, never both. Thus, there are DNA viruses (Fig. 18–1) and RNA viruses. Whichever type it has serves

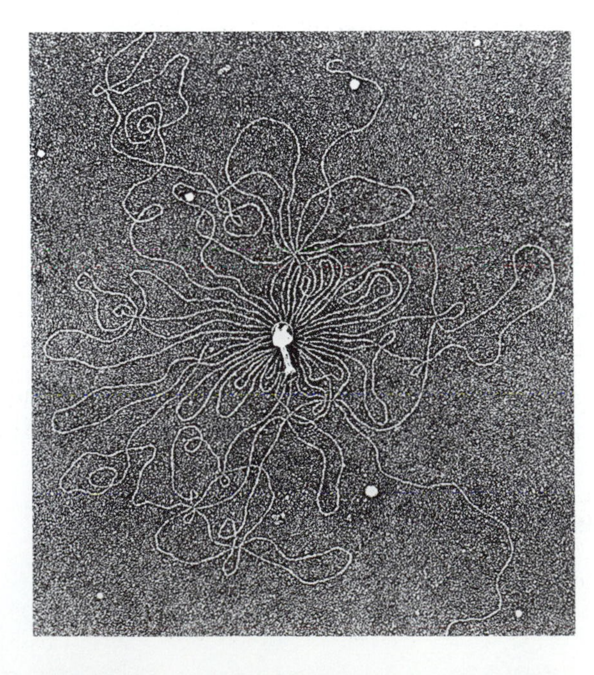

Figure 18–1 An osmotically shocked bacteriophage (a virus that attacks bacteria). Its single molecule of DNA has been released from the phage coat by breaking the coat protein. (From Kleinschmidt, A. K., Lang, D., Jacherts, D., and Zahn, R. K.: *Biochim. Biophys. Acta* 61:859, 1962.)

as its genetic material, or genome. The viral genome may consist of fewer than five genes or as many as several hundred. However, viruses never have tens of thousands of genes like the cells of more complex organisms.

Most viruses are much smaller than bacteria, and indeed some are scarcely larger than some very large single molecules of protein or nucleic acid. Even large viruses are only about 100 nm in diameter, so individual virus particles can be photographed only with an electron microscope. However, accumulations of viruses growing in the cytoplasm of an infected cell are visible with a light microscope.

The shape of a virus is determined by the organization of subunits that make up the capsid. Viruses are generally either helical or polyhedral in shape, or a complex combination of both (Fig. 18–2). Helical viruses, such as the tobacco mosaic virus, appear as long rods; their capsid is a hollow cylinder with a helical structure. The poliovirus is a polyhedral virus with 20 triangular faces and 12 corners. Unlike cells, viruses can be crystallized; then, when the inert crystals are put back into the appropriate host cells, they can undergo multiplication and produce the symptoms of disease.

Bacteriophages

Viruses that infect bacteria are called **bacteriophages** (bacteria eaters), or simply **phages.** There are many varieties of bacteriophages, and they are usually species-specific (or strain-specific), meaning that one type of phage generally only attacks one species (or strain) of bacteria. Because phages can be easily cultured within living bacteria in the laboratory, most of our knowledge of viruses has come from studying these bacterial viruses. It has now been shown that phages can serve as excellent models for the viruses that infect animal cells (see Focus on Culturing Viruses).

FOCUS ON
Culturing Viruses

Since viruses can multiply only when they have infected living cells, they cannot be cultured on a nonliving medium. One of the first, and still very useful, methods for culturing animal viruses was culturing them in developing chick embryos. A small piece of egg shell is removed about a week or two after fertilization of the egg, and the material containing the virus is injected through the opening. The virus can be injected into the embryo itself, or the virus can be injected onto one of the membranes surrounding the embryo—usually the yolk sac or chorioallantoic membrane (See figure). The opening in the shell can then be sealed with paraffin wax, and the egg incubated at 36°C. The virus multiplies within the living cells and can later be separated from the host cells by centrifugation. Cultivation of viruses in developing chick embryos has been used in production of virus for various vaccines including smallpox, influenza, and yellow fever vaccines. This method is also used for immunological and other research studies.

Currently the most widely used method for culturing viruses is tissue culture. Almost any type of animal cell can now be cultured in an appropriate culture medium in glass or plastic dishes. Appropriate viruses for the particular tissue obligingly infect these cells. Some vaccines are now prepared from viruses grown in tissue culture. This is advantageous to persons who may be allergic to eggs, and therefore to the vaccines prepared from viruses cultured in chick eggs. Viruses grown in tissue culture may induce characteristic changes in the tissue culture cells just as they do within the body, and so provide an important model for studying viral infection and its effects on cells.

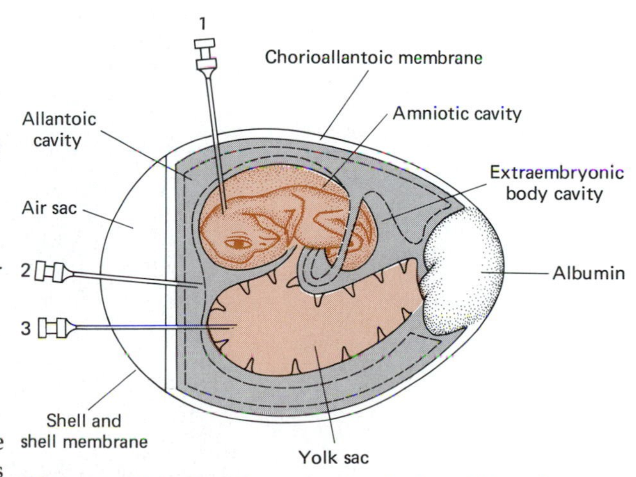

Diagrammatic section through a developing chick embryo from 10 to 12 days old, indicating how viruses can be inoculated into (*1*) the head of the embryo, (*2*) the allantoic cavity, or (*3*) the yolk sac.

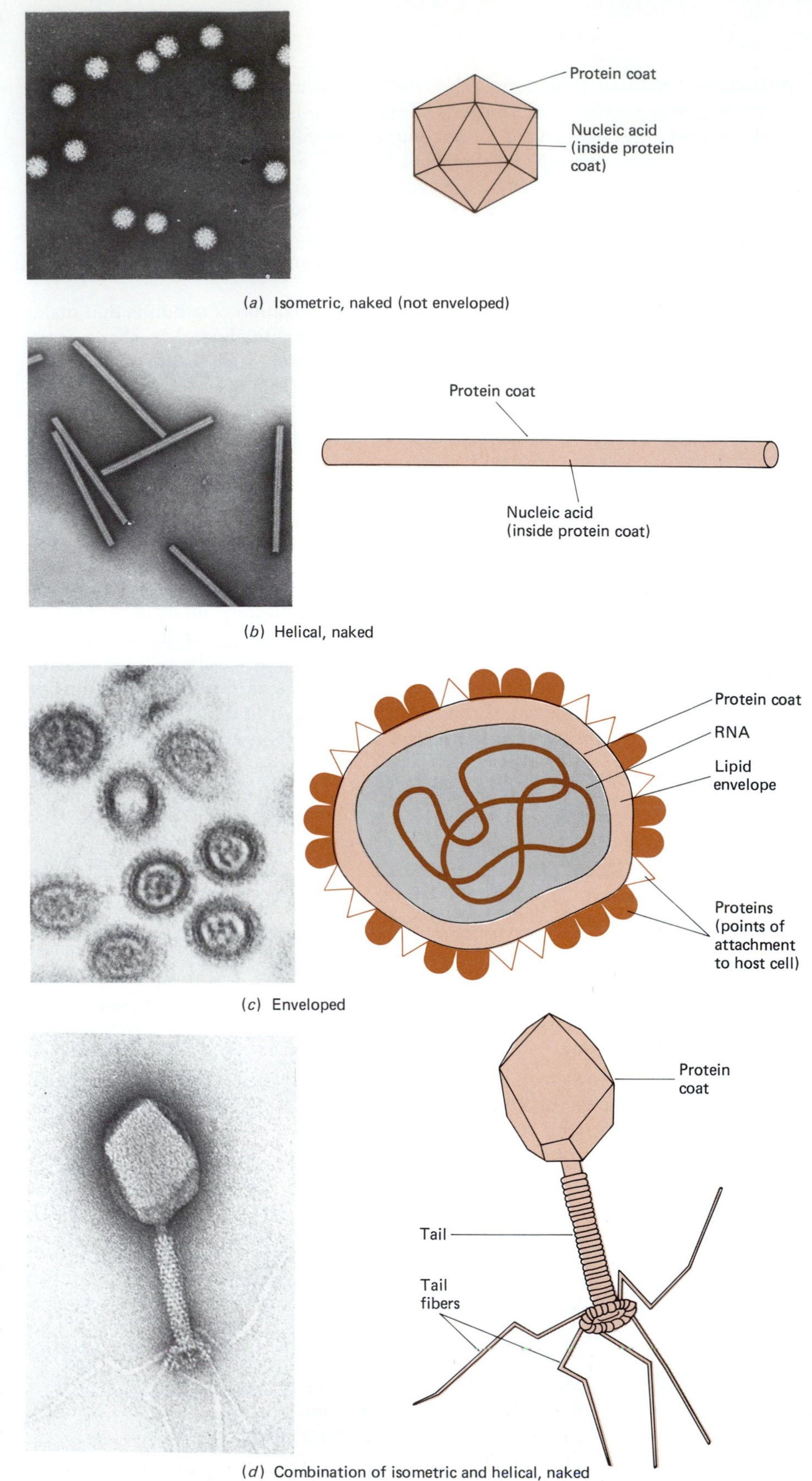

Figure 18–2 Viruses are generally either helical or polyhedral in shape, or a complex combination of both. (*a*) Bushy stunt virus, a polyhedral virus (approximately ×260,000). (*b*) Tobacco mosaic virus, a helical virus (approximately ×140,000). (*c*) Influenza virus, a polyhedral virus surrounded by an envelope (×200,000). (*d*) This bacteriophage is a complex combination of helical and polyhedral shapes (approximately × 275,000). ((*a*), (*b*), courtesy of R. Williams; (*c*), courtesy of E. Boatman; (*d*), courtesy of Dr. Lyle C. Dearden.)

Virulent, or **lytic,** bacteriophages cause lytic infections. After lytic viruses multiply they lyse (destroy) the host cell. **Temperate,** or **lysogenic,** viruses do not kill their host cells. Some temperate viruses integrate their nucleic acid into the DNA of the host cell and multiply whenever the host cell DNA replicates.

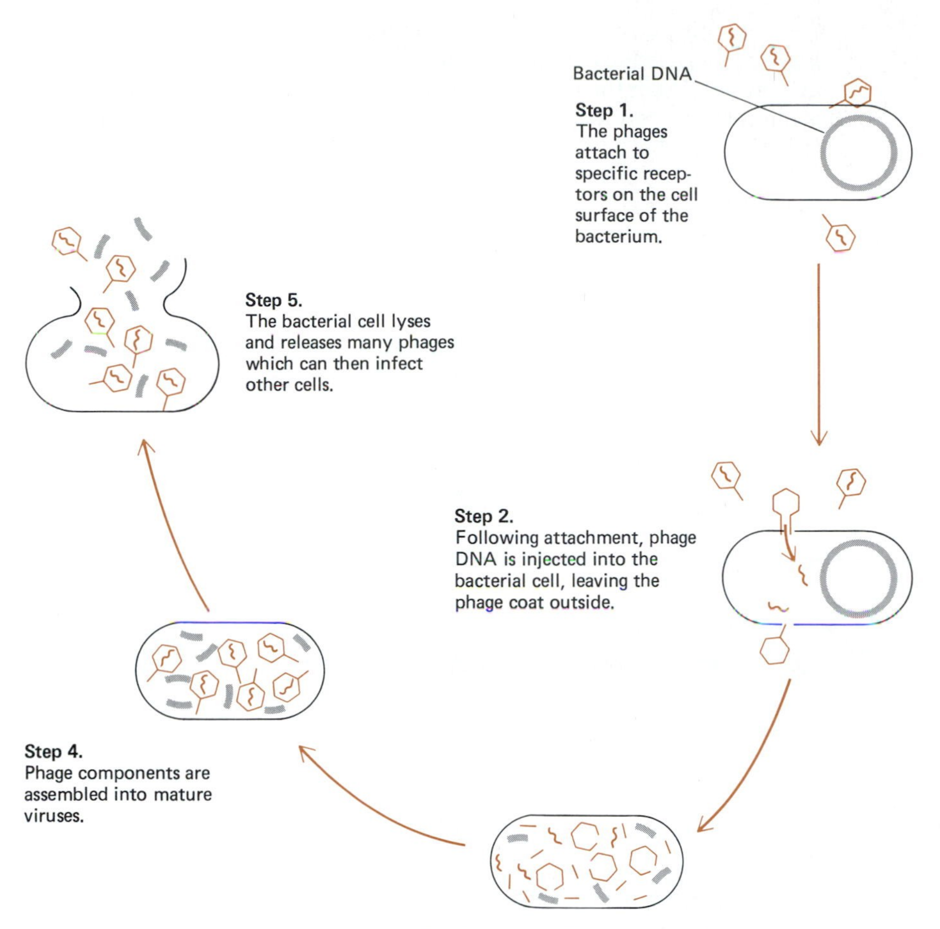

Bacterial DNA
Step 1.
The phages attach to specific receptors on the cell surface of the bacterium.
Step 2.
Following attachment, phage DNA is injected into the bacterial cell, leaving the phage coat outside.
Step 3. Bacterial DNA is degraded. Phage DNA is replicated. Phage components are synthesized.
Step 4.
Phage components are assembled into mature viruses.
Step 5.
The bacterial cell lyses and releases many phages which can then infect other cells.

Figure 18–3 The sequence of events in a lytic infection.

Viral Replication in Lytic Infections

Outside a living cell, a virus has no metabolic activity and cannot reproduce itself. When a virus infects a susceptible host cell, it uses the host cell's metabolic machinery to replicate its nucleic acid and produce its specific proteins.

There are several steps in the process of viral infection that are common to almost all bacteriophages (Fig. 18–3):

1. **Attachment** to the surface of the host cell. The virus attaches to specific receptors on the host cell surface (Fig. 18–4). Since each bacterial species, and sometimes each strain within a species, has different receptors, a virus will attach to only a specific species (or strain). A technique called bacterio-

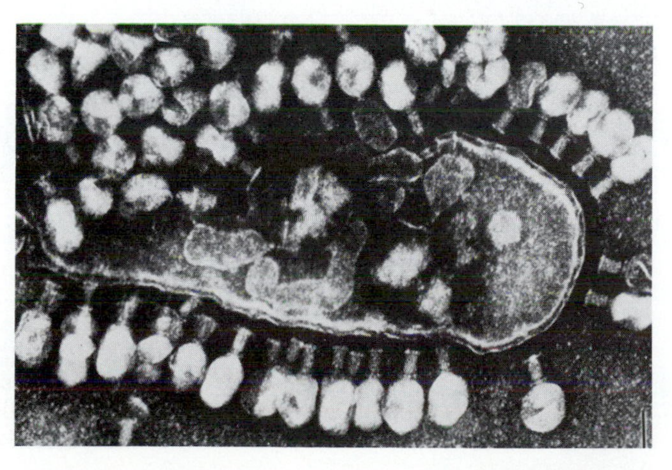

Figure 18–4 Phage infecting an *Escherichia coli* cell. Many phages are attached to the cell wall. The head, tail, and base plate of most virus particles are clearly visible. The tail core extending from the base plate to the cell wall is hollow and acts like a hypodermic needle in injecting viral DNA into the cell. The break in the bacterial cell wall is an artifact produced during preparation for viewing under the electron microscope. (Courtesy of Dr. Lee D. Simon, Institute for Cancer Research, Philadelphia.)

phage typing makes use of this discriminatory ability of viruses to distinguish among various strains of bacteria.

2. **Penetration.** After the virus has attached to the cell surface, its nucleic acid is injected through the cell membrane and into the cytoplasm of the host cell. The capsid of a phage remains on the outside. Most viruses that infect animal cells, in contrast, enter the host cell intact by phagocytosis.
3. **Replication.** Once inside, the virus takes over the metabolic machinery of the cell. The bacterial DNA is degraded so that the viral genes are free to dictate future biochemical operations. Using the host cell ribosomes, energy, and many of its enzymes, the virus replicates its own macromolecules. The viral genes contain all of the information necessary to produce new viruses.
4. **Assembly.** The newly synthesized viral components are assembled to produce complete new viruses.

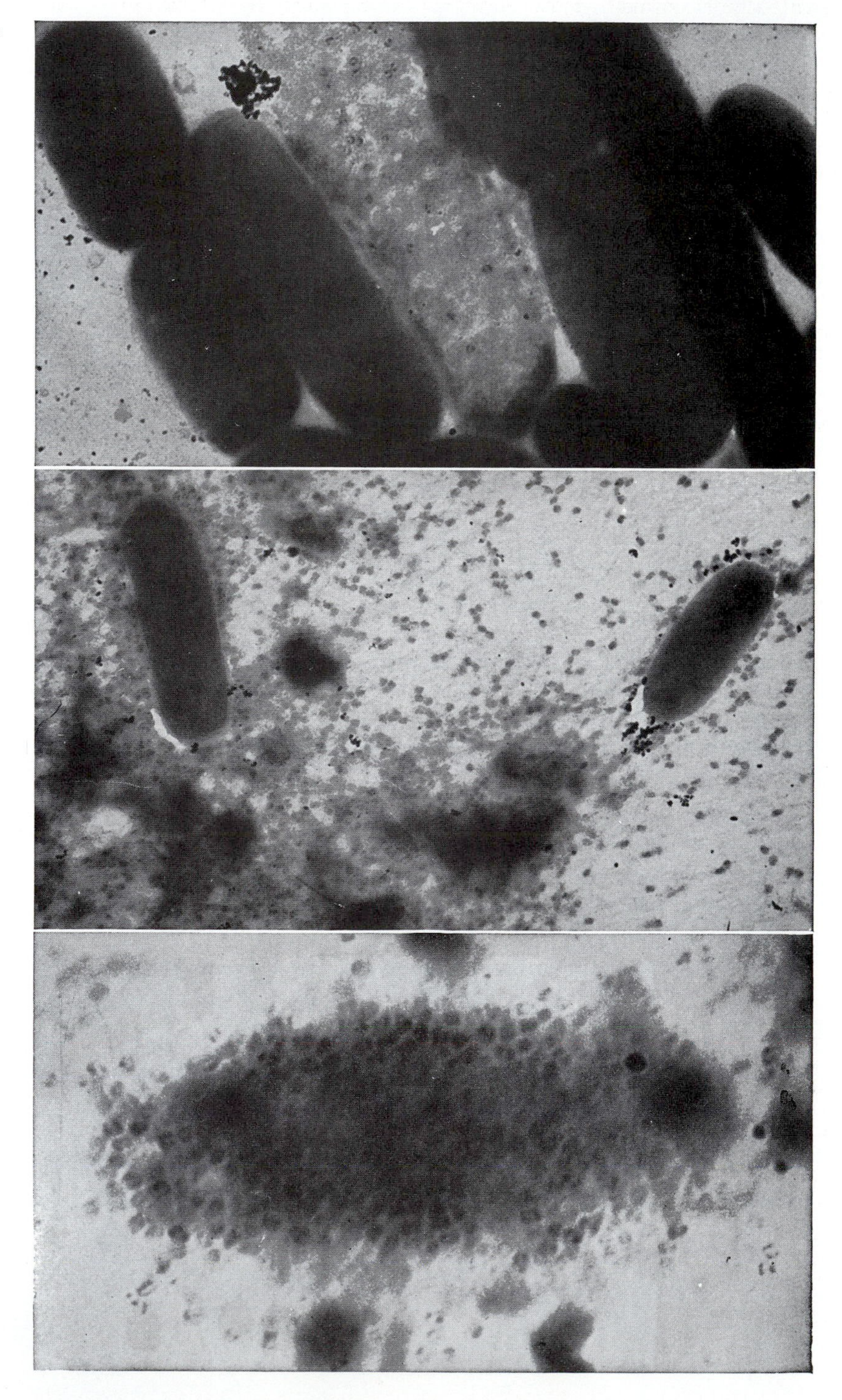

Figure 18–5 Electron micrographs of the destruction of *Escherichia coli* by bacteriophages. *Top,* Dark, sausage-shaped structures are normal bacteria; lighter one in middle has been attacked and destroyed by phage. Phage particles are evident within the cell. *Center,* Later stage with more phage particles visible and more bacteria destroyed. *Bottom,* Dense mass of phage particles occupying the space of the bacillus they have destroyed. (Courtesy of W. Burrows et al.)

5. **Release.** In a lytic infection, the virus produces a **lysozyme,** an enzyme that degrades the cell wall of the host cell. The host cell then lyses, releasing about 100 bacteriophages (Fig. 18–5). These can then infect other cells, and the process begins anew. Because infection results in lysis and death of the infected cell, viruses that cause lytic infections are referred to as virulent bacteriophages.

DNA Integration in Lysogenic Infections

Although virulent viruses always replicate and lyse their host cells, temperate viruses do not always lyse their host cells. They can integrate their DNA into the host DNA. When the bacterial DNA duplicates, the viral DNA, called a **prophage** when integrated, also replicates (Fig. 18–6). The viral genes that code for viral structural proteins may be repressed indefinitely. The bacterial (host) cell, on the other hand, may behave almost normally. Host cells carrying prophages are referred to as lysogenic.

In some cases bacterial cells containing temperate viruses may exhibit new properties. This is called **lysogenic conversion.** For example, the bacterium that causes diphtheria produces the toxin responsible for the disease symptoms only when inhabited by a specific phage. In the same way, a phage is responsible for producing the toxin associated with scarlet fever; only when the streptococci are lysogenic can they cause scarlet fever. *Clostridium botulinum* bacteria synthesize the toxin that causes botulism only when they are lysogenic for certain phages.

Certain external conditions can cause the phage nucleic acid to enter a lytic phase, releasing new phages. When a lysogenic cell does lyse, the phages released

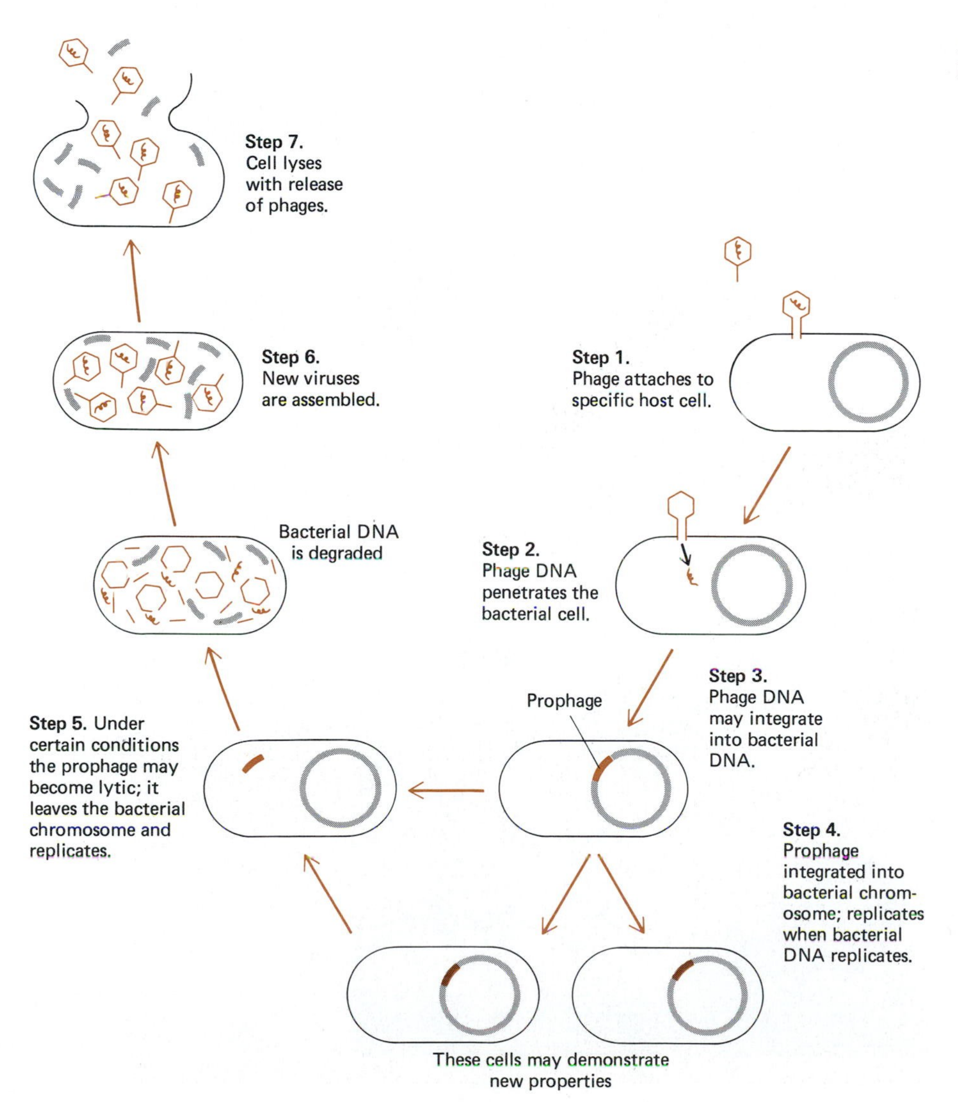
Step 7.
Cell lyses with release of phages.
Step 6.
New viruses are assembled.
Step 1.
Phage attaches to specific host cell.
Bacterial DNA is degraded
Step 2.
Phage DNA penetrates the bacterial cell.
Prophage
Step 3.
Phage DNA may integrate into bacterial DNA.
Step 5. Under certain conditions the prophage may become lytic; it leaves the bacterial chromosome and replicates.
Step 4.
Prophage integrated into bacterial chromosome; replicates when bacterial DNA replicates.
These cells may demonstrate new properties

Figure 18–6 The sequence of events in a lysogenic infection.

may contain some bacterial DNA in place of their own genetic material. When it infects a new bacterium, such a phage can introduce this DNA into the genome of the host. Known as **transduction,** this process permits genetic recombination in the new host cell (Fig. 18–7). This ability of some viruses to transfer DNA from one cell to another is taken advantage of in recombinant DNA experiments (Chapter 16).

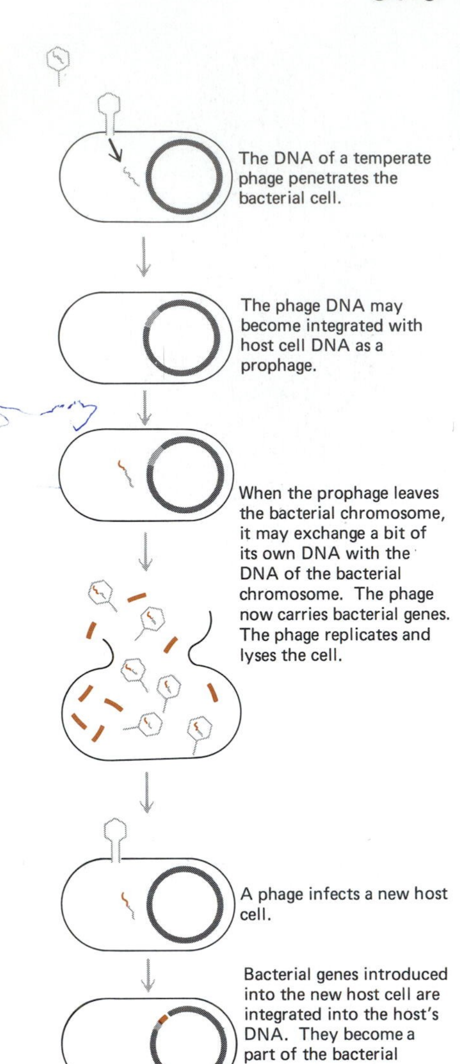

The DNA of a temperate phage penetrates the bacterial cell.
The phage DNA may become integrated with host cell DNA as a prophage.
When the prophage leaves the bacterial chromosome, it may exchange a bit of its own DNA with the DNA of the bacterial chromosome. The phage now carries bacterial genes. The phage replicates and lyses the cell.
A phage infects a new host cell.
Bacterial genes introduced into the new host cell are integrated into the host's DNA. They become a part of the bacterial chromosome and are replicated along with the rest of the bacterial DNA.

Figure 18–7 Transduction. A phage can transfer bacterial DNA from one bacterium to another.

Another Form of Coexistence

A few bacterial viruses (as well as some animal viruses) release new viruses slowly without destroying the host cell (Fig. 18–8). In such cases the host cell carries on its own metabolic activities, although some of its energy is used to produce new viruses. As they are assembled, mature viruses migrate to the cell membrane. They appear to bud off from the host cell in a process that may be the reverse of penetration.

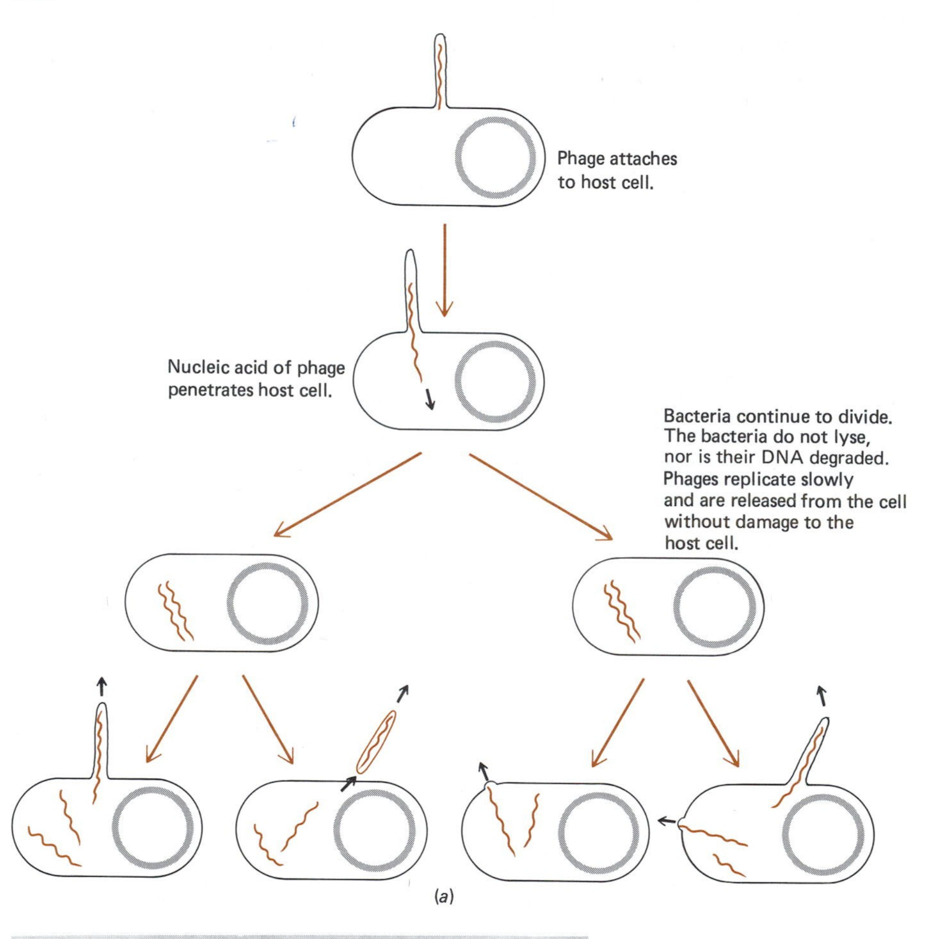

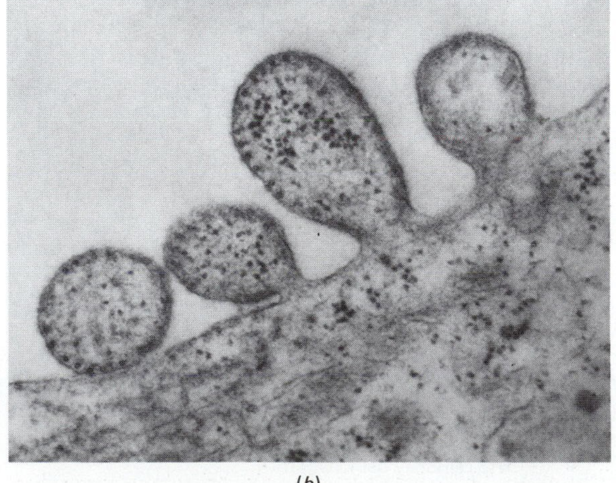

Phage attaches to host cell.

Bacteria continue to divide. The bacteria do not lyse, nor is their DNA degraded. Phages replicate slowly and are released from the cell without damage to the host cell.

Figure 18–8 Coexistence with viruses. (*a*) In some virus infections, assembly of new viruses and their release may occur slowly without destroying the host cell. A few types of phages coexist with their host cells in this way. (*b*) Virus particles budding from the surface of a human cell. Each new virus has incorporated some of the cell membrane of the host cell; however, the host is able to repair its membrane after the budding is complete. ((*b*), courtesy of J. Griffith.)

Animal Viruses

Unlike phages, animal viruses attach to specific receptor sites on the surface of a host cell. These receptors vary with each species and sometimes with each type of tissue. Accordingly, some viruses can infect only human cells, because only human cells have the necessary receptors. Some viruses, like the measles virus and poxviruses, can infect many types of tissues because the tissues all have the necessary receptors. However, only certain tissue cells—certain spinal cord cells, throat, and intestinal cells—have the receptors for the attachment of poliovirus.

Animal viruses enter the host cell by phagocytosis, and the envelope and capsid are removed within the host cell. As an animal virus leaves the host cell, it is often enclosed by a lipoprotein envelope composed of a bit of the host cell membrane.

Viral proteins synthesized within the host cell may damage the cell in a variety of ways. Such proteins may alter the permeability of the cell membrane, or may inhibit nucleic acid or protein synthesis. Viruses sometimes damage or kill their host cells by their sheer numbers. A poliovirus may produce 100,000 new viruses within a single host cell!

One way that cells react to viral infection is by the production of **interferons,** proteins that interfere with viral replication. Interferon is released from infected cells and helps to protect uninfected cells in the area. Interferon and other responses to virus infections are discussed in Chapter 36.

Figure 18–9 Tobacco plant infected with tobacco mosaic virus. The virus produces a yellow and green mottling (or mosaic pattern). Young leaves are especially vulnerable; however, the disease tends to reduce crop yields rather than killing the plants outright. (USDA Photo.)

Viral Diseases

In 1892, the Russian botanist Iwanowski found that tobacco mosaic disease (Fig. 18–9)—so called because the infected tobacco leaves have a spotted appearance—could be transmitted to healthy plants by daubing their leaves with the sap of diseased plants. The sap was infective even after it had been passed through filters fine enough to remove all bacteria. Other plant diseases now known to be caused by viruses include ring spot disease, "breaking" disease, and mosaic diseases of tomato, potato, lettuce, bean, cabbage, and sugarcane.

Animal diseases caused by viruses include hog cholera, foot-and-mouth disease, canine distemper, swine influenza, and Rous sarcoma in fowl. Humans are prone to a variety of viral diseases including smallpox, chickenpox, herpes simplex (one type of which is genital herpes), herpes zoster, mumps, rubella (German measles), rubeola (measles), rabies, warts, and influenza (see Focus on Animal Viruses). Indeed, it has been estimated that each of us suffers from two to six viral infections each year. Fortunately, most of these are relatively benign forms such as the common cold.

We are most familiar with **acute** viral infections, in which the disease is short-lived. However, some viruses cause **latent** infections, in which the viruses remain quietly in the body for years before becoming active. Sometimes, symptoms reappear periodically. After the initial infection, the herpes virus that causes cold sores can infect certain ganglia (groups of nerve cell bodies), where it may remain for years without causing symptoms. However, during times of stress or ill health, the virus may be activated and cause cold sores once again. Environmental factors such as exposure to the sun may also activate the virus. During the latent period the virus usually cannot be detected. It is during the active phase that the herpes virus can be transmitted through physical contact with the active sore.

In **chronic** viral infections, the virus can be shown to be present even though the carrier may not exhibit symptoms. Infected individuals can also transmit the disease to others even when no symptoms are present. **Slow** virus infections generally cause slow, progressive degeneration of the tissues involved. Such infections often lead to death. Multiple sclerosis is suspected of being a disease of this type.

Viruses and Cancer

Viruses are known to cause cancer in many kinds of animals. The nucleic acid of these cancer-causing viruses becomes integrated into the DNA of the host cells, which they transform into cancer cells. This viral nucleic acid probably codes for

FOCUS ON

*Animal Viruses**

Group	*Diseases Caused*	*Characteristics*
DNA Viruses		
Poxviruses	Smallpox, cowpox, and economically important diseases of domestic fowl	Large, complex, oval-shaped viruses that replicate in the cytoplasm of the host cell
Herpesviruses	Herpes simplex Type 1 (cold sores); herpes simplex Type 2 (genital herpes, a sexually transmitted disease); herpes zoster (chickenpox and shingles). The Epstein-Barr virus has been linked with infectious mononucleosis and Burkitt's lymphoma.	Medium to large, enveloped viruses; frequently cause latent infections; some cause tumors.
Adenoviruses	About 40 types known to infect human respiratory and intestinal tracts; common cause of sore throat, tonsillitis, and conjunctivitis; other varieties infect other animals.	Medium-sized viruses
Papovaviruses	Human warts and some degenerative brain diseases; cancer in animals other than humans	Small viruses
Parvoviruses	Infections in dogs, swine, arthropods, rodents	Very small viruses; some contain single-stranded DNA; some require a helper virus in order to multiply.
RNA Viruses		
Picornaviruses	About 70 types infect humans including polioviruses; enteroviruses infect intestine; rhinoviruses infect respiratory tract and are main cause of human colds; coxsackievirus and echovirus cause aseptic meningitis.	Diverse group of small viruses
Togaviruses	Rubella, yellow fever, equine encephalitis	Large, diverse group of medium-sized, enveloped viruses; many transmitted by arthropods
Myxoviruses	Influenza in humans and other animals	Medium-sized viruses that often exhibit projecting spikes
Paramyxoviruses	Rubeola, mumps, distemper in dogs	Resemble myxoviruses but somewhat larger
Reoviruses	Vomiting and diarrhea in children	Contain double-stranded RNA

*There are many ways to classify viruses. This classification is based on the one used in Nester, et al: *Microbiology*. Philadelphia, Saunders College Publishing, 1983.

enzymes that change important proteins in the host cell. Some viruses that cause cancer have one or a few genes, called **oncogenes,** responsible for their ability to transform host cells into cancer cells (Chapter 15). These generally give rise to tumors quickly, within days or weeks.

Many cancer viruses contain RNA, rather than DNA. These viruses cause the infected cell to produce a very unusual enzyme known as **reverse transcriptase,** which catalyzes the synthesis of a complementary DNA strand using the viral RNA as a template. The DNA then acts as the template for production of its complement, thus forming double-stranded viral DNA. This DNA can be used to produce copies of the viral RNA.

Viruses may eventually be shown to cause some forms of cancer in humans. There is some evidence that one type of herpesvirus, the **Epstein-Barr virus,** may cause a cancer called **Burkitt's lymphoma,** a cancer of the lymphatic system. The Epstein-Barr virus apparently infects almost all human beings. In some it causes infectious mononucleosis (popularly referred to as mono). In Central Africa this virus has been linked to Burkitt's lymphoma. The reason for this dramatic difference in virulence is not known but is suspected to result from its transmission by an insect, which injects the virus. It is suspected that viruses may also play a role in human leukemias, sarcomas, and certain breast cancers.

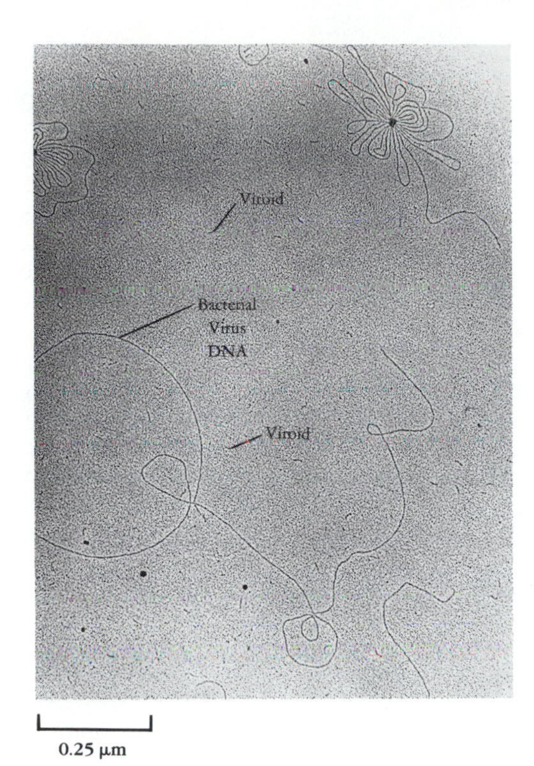

Figure 18–10 A viroid is a rodlike structure consisting of a single-stranded circular molecule of RNA. This electron micrograph compares the size of a viroid with molecules of bacteriophage DNA. (Courtesy of T. Koller and J. M. Sogo, Swiss Federal Institute of Technology, Zurich.)

Viroids

Viroids are even smaller and simpler than viruses. Each viroid consists of a very short strand of RNA without any sort of protective coat (Fig. 18–10). The amount of RNA present may be sufficient to code for a single, medium-sized protein. Viroids have been linked to several diseases of complex plants, including potato spindle-tuber disease and a disease that causes stunting of chrysanthemums.

Viroids had also been suspected of causing certain animal diseases including scrapie, a progressive neurological disease of sheep. Recently, however, another subviral pathogen referred to as a **prion** has been implicated. The prion (a coined name for a protein infectious particle)—even smaller than the viroid—appears to consist only of a hydrophobic protein. No nucleic acid has been demonstrated, but researchers continue to search for a small nucleic acid component.

SUMMARY

I. A virus is a tiny particle or virion consisting of a core of DNA or RNA surrounded by a capsid (protein coat).
  - *A.* Viruses are much smaller than bacteria.
  - *B.* Viruses are helical, polyhedral, or a combination of both shapes.

II. Bacteriophages are viruses that infect bacteria.
  - *A.* Some phages are virulent, or lytic.
  - *B.* Other phages are temperate, or lysogenic.

III. Viral infection includes the following processes: attachment to the host cell, penetration, replication, assembly, and release.
  - *A.* In a lytic infection, the virus produces a lysozyme, which causes the host cell to lyse, releasing the new viruses.
  - *B.* In lysogenic infections, temperate viruses integrate their DNA into the host DNA.
    1. Such nucleic acid integration may confer new properties on the host cell.
    2. Phages released from lysogenic cells may contain a portion of bacterial DNA, which can lead to transduction in a new host cell; this process has important application in recombinant DNA experiments.
    3. In some viral infections, the host cell is permitted to continue its metabolic activities, and viruses are slowly assembled and released with minimal damage to the host cell.

IV. Animal viruses generally enter the host cell by phagocytosis, and their capsids are removed within the host cell. Animal cells produce interferons in response to viral infection.

V. Many plant and animal diseases are caused by viruses. Human viral infections may be acute, chronic, latent, or slow.

VI. Viruses cause cancers in many types of animals, and there is some evidence that they cause certain human cancers. For example, the Epstein-Barr virus may cause Burkitt's lymphoma.

POST-TEST

1. The core of a virus consists of ______________ or ______________, but never both.
2. Bacteriophages are viruses that infect ______________.
3. The five main steps in bacteriophage infection are ______________, ______________, ______________, ______________, and ______________.
4. A lysozyme is an enzyme produced by the ______________, which degrades the ______________ ______________ of the host.
5. The part of the virus that actually enters the host cell is its ______________ ______________.
6. In ______________ ______________, bacterial cells containing temperate viruses exhibit new properties.
7. Virulent viruses cause ______________ infections.
8. Lysogenic viruses are also known as ______________ viruses.
9. Lysogenic phages can transfer nucleic acid from one virus to another, resulting in genetic recombination; this process is known as ______________.
10. ______________ are proteins produced by host cells that interfere with viral replication.
11. In a ______________ infection, viruses may live quietly in the body for years before becoming active.
12. Oncogenes are responsible for the ability of some viruses to transform ______________ cells into ______________ ______________.
13. ______________ are even smaller and simpler than viruses; each consists of a very short strand of RNA without any sort of protective ______________.
14. The portion of viral DNA integrated into the host DNA is referred to as a ______________.
15. The type of infection in which the host cell bursts open releasing many viruses is a ______________ infection.

REVIEW QUESTIONS

1. Why are viruses often looked upon as being on the threshold between nonlife and life? What characteristics does a virus share with a living cell? What characteristics of life are lacking in a virus?
2. Draw a diagram of a virus, and label its parts.
3. What is a bacteriophage? Why have phages been important in the development of knowledge about viruses?
4. List the steps in the process of viral infection, and briefly describe each step.
5. How do lysogenic infections differ from lytic infections?
6. Why is a virus limited in the number of species (or tissue types) that it can infect?
7. What is a latent viral infection? a slow infection?
8. What is the possible relationship between the Epstein-Barr virus and human cancer? Explain.
9. Define the following terms:
   a. capsid
   b. virion
   c. temperate virus
   d. lysozyme
   e. transduction

19
Kingdom Monera

OUTLINE

I. Cyanobacteria
- *A.* Structure of cyanobacteria
- *B.* Ecological importance

II. Bacteria
- *A.* Structure of bacteria
- *B.* Bacterial metabolism
- *C.* Reproduction
- *D.* Spore formation
- *E.* Some important groups of bacteria
  - *1.* Eubacteria
  - *2.* Myxobacteria (slime bacteria)
  - *3.* Spirochetes
  - *4.* Actinomycetes (moldlike bacteria)
  - *5.* Mycoplasmas (bacteria without cell walls)
  - *6.* Rickettsias
  - *7.* Chlamydias
- *F.* Bacteria in humans
- *G.* Bacteria and food
- *H.* Ecological importance of bacteria
- *I.* Bacterial diseases of plants

LEARNING OBJECTIVES

After you have read this chapter you should be able to:

1. Describe the distinguishing characteristics of members of kingdom Monera.
2. Compare the cyanobacteria with the bacteria.
3. Summarize the ecological importance of the cyanobacteria.
4. Describe the structure of a bacterial cell with emphasis on the cell wall and DNA.
5. Characterize bacteria as heterotrophs or autotrophs, and compare bacterial photosynthesis with plant photosynthesis.
6. Distinguish between facultative anaerobes and obligate anaerobes.
7. Describe the three mechanisms of genetic recombination (transformation, conjugation, and transduction) that take place in bacteria.
8. Describe the three main shapes of the eubacteria, and give examples of gram-positive and gram-negative eubacteria.
9. Characterize each of the following groups of bacteria: myxobacteria, spirochetes, actinomycetes, mycoplasmas, rickettsias, and chlamydias.
10. Identify human disease that may be caused by streptococci, staphylococci, *Clostridium, Haemophilus, Salmonella, Treponema pallidum*, mycobacteria, rickettsias, and chlamydias.
11. Describe how members of the normal bacterial community can sometimes cause disease, identify ways in which human pathogenic bacteria can be spread, and distinguish between exotoxins and endotoxins.
12. Give four examples of the use of bacteria in preparation of foods, and describe methods used to prevent bacterial spoilage of food.
13. Describe two important ecological roles of bacteria.
14. Describe two plant diseases caused by bacteria, and describe methods for controlling bacterial diseases in plants.

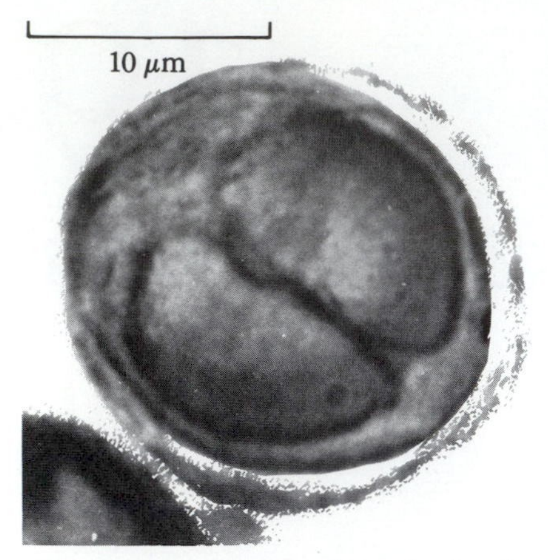

Figure 19–1 Fossil cyanobacteria, dated at approximately 850 million years. This four-celled colony of cyanobacteria surrounded by a sheath was discovered in a thin section of rock in Australia. (Courtesy of J. W. Schopf.)

For almost 3 billion years, the monerans may have been the only living organisms on our planet. The earth's oldest fossils, bacteria that resemble contemporary photosynthetic cyanobacteria, were found in rocks dated at 3.5 billion years old. In contrast, the oldest eukaryotic fossils are thought to be about 800 million years old (Fig. 19–1).

All of the prokaryotes are assigned to kingdom Monera. Most are unicellular organisms, but they often form colonies or filaments (ribbons) of independent cells. The cytoplasm contains ribosomes but lacks membrane-bound organelles typical of eukaryotic cells. Thus there are no mitochondria, endoplasmic reticulum, Golgi complex, or lysosomes (Fig. 19–2). The genetic material of a prokaryote is contained in a single circular DNA molecule that lies in the cytoplasm, not surrounded by a nuclear membrane. The DNA duplicates before the cell divides asexually by fission (splitting). Neither mitosis nor meiosis occurs in monerans.

Most prokaryotic cells possess a cell wall surrounding the cell membrane. The structure and composition of the cell wall is different from that found in eukaryotic cells. Some prokaryotes have flagella, but unlike the flagella of eukaryotes (which are composed of two central hollow microtubules surrounded by nine pairs of similar microtubules), these are solid structures made of long filaments of protein. None have cilia. Prokaryotic cells are much smaller than eukaryotic cells. Their cell volume is only about a thousandth of that of small eukaryotic cells, and their length only about one tenth.

Although the members of kingdom Monera are extremely diverse, they are generally classified in two large groups, or subkingdoms: the cyanobacteria and the bacteria.

Cyanobacteria

The **cyanobacteria** (formerly known as the blue-green algae) are found in ponds, lakes, swimming pools, and moist soil as well as on dead logs and the bark of trees. Some are also found in the oceans, and a few species even inhabit hot springs. Most are unicellular but many form large globular colonies or long filaments united by extracellular materials. Some species show a division of labor within members of a colony: In certain species, some cells are specialized for fixing nitrogen; other cells are specialized to reproduce; and still others may function to attach the colony to the substrate.

Most cyanobacteria are **photosynthetic autotrophs.** They contain chlorophyll *a*, which is also found in photosynthetic eukaryotes. They have several varieties of

Figure 19–2 Comparison of a eukaryotic cell with two prokaryotic cells. Note that the prokaryotic cells *(b)* and *(c)* lack a nuclear membrane and other membranous organelles characteristic of eukaryotic cells.

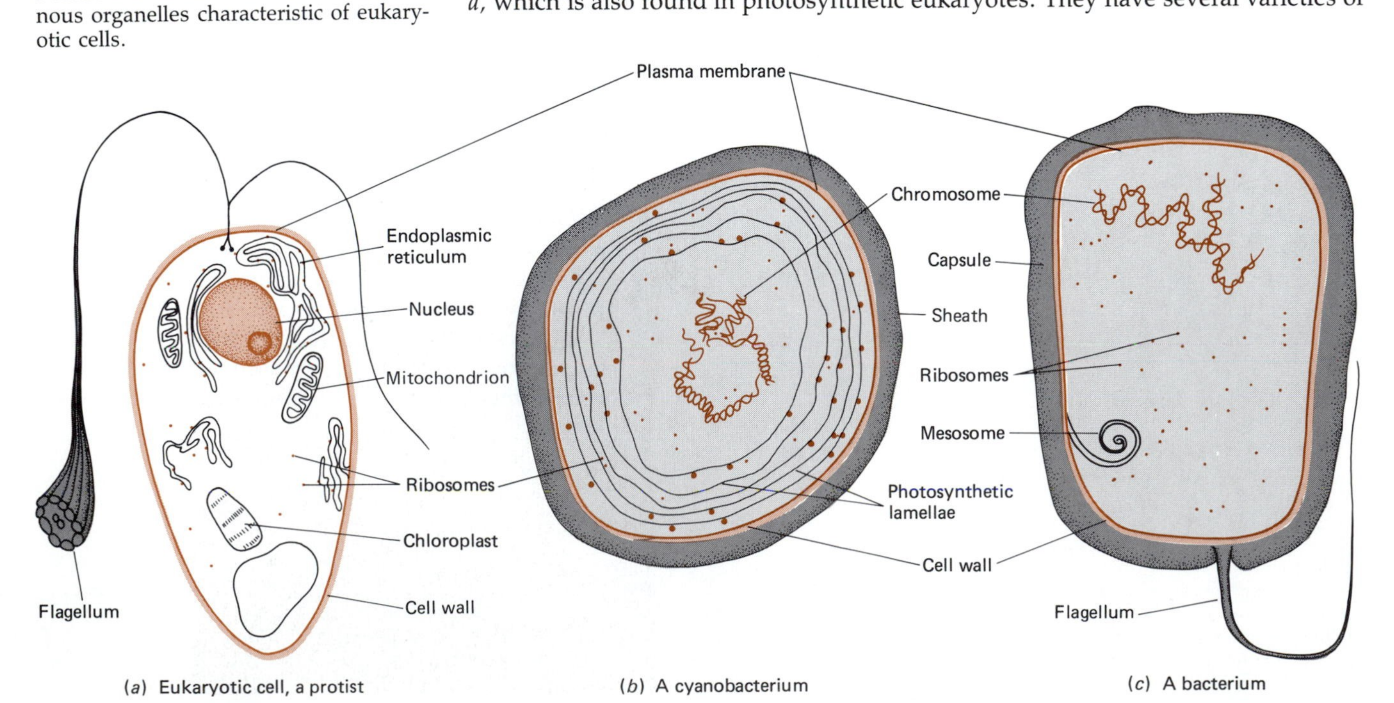

accessory pigments, including carotenoids, which are present in some eukaryotes and some bacteria. **Phycocyanin,** a blue pigment, is found only in cyanobacteria. Some have a red pigment, **phycoerythrin,** also present in red algae. Chlorophyll and the accessory pigments are not enclosed in plastids, as they are in algae and in plant cells, but are dispersed along membranes in the periphery of the cell or stacked in the cytoplasm.

As far as is known, cyanobacteria reproduce only asexually, usually by fission. Colonies may fragment, and then cells of each of the separated parts reproduce to form new colonies. In some species thick-walled spores form that are highly resistant to adverse conditions; they may remain dormant for many years until conditions are favorable for them to germinate and give rise to new colonies of cells.

STRUCTURE OF CYANOBACTERIA

Like other monerans, cyanobacteria lack a nuclear membrane and other membranous organelles such as mitochondria and chloroplasts. However, unlike the bacteria, cyanobacteria have internal membranes called **photosynthetic lamellae,** which contain chlorophyll and enzymes needed for photosynthesis (Fig. 19–3).

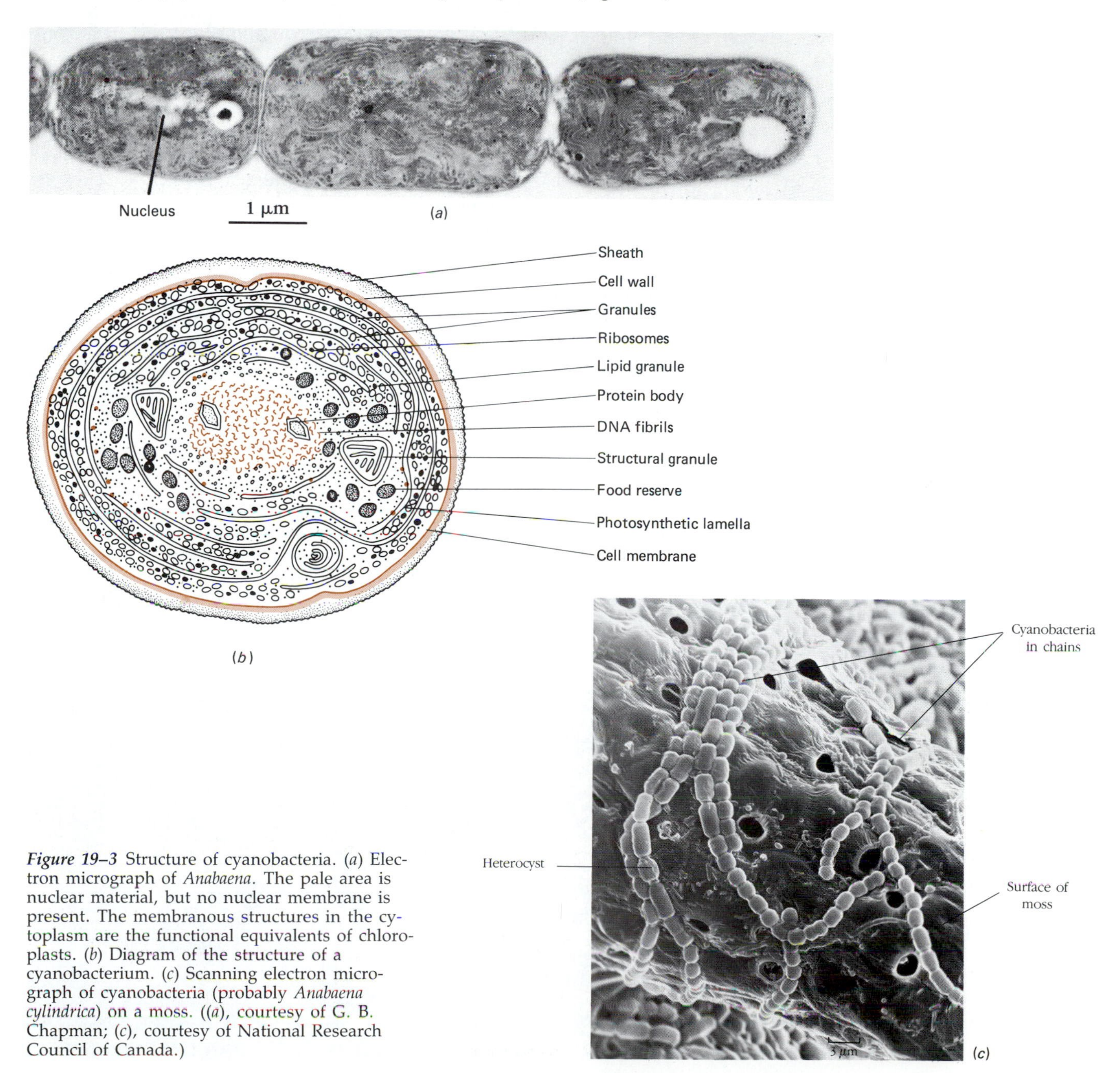

Figure 19–3 Structure of cyanobacteria. (*a*) Electron micrograph of *Anabaena.* The pale area is nuclear material, but no nuclear membrane is present. The membranous structures in the cytoplasm are the functional equivalents of chloroplasts. (*b*) Diagram of the structure of a cyanobacterium. (*c*) Scanning electron micrograph of cyanobacteria (probably *Anabaena cylindrica*) on a moss. ((*a*), courtesy of G. B. Chapman; (*c*), courtesy of National Research Council of Canada.)

The tough cell wall does not contain cellulose, but consists of other polysaccharides linked with polypeptides. Many of the cyanobacteria secrete a sticky gelatinous substance, which may form a sheath around the cell wall. The gelatinous material often contains pigments, and may also contain toxins that prevent fish and other organisms from feeding on them. Despite their name, only about half of the "blue-green algae" are actually blue-green. The coloration is modified by the photosynthetic pigments within the cell, producing brown, black, purple, yellow, blue, green, or even red individuals. The Red Sea acquired its name from red cyanobacteria, which sometimes occur there in such great numbers that they color the water red.

Cyanobacteria do not have flagella, but some of the filamentous species are capable of a curious back-and-forth oscillatory movement. Others have a slow, gliding motion.

ECOLOGICAL IMPORTANCE

As producers, cyanobacteria provide oxygen and organic material for other organisms. Many species can **fix nitrogen**—that is, incorporate atmospheric nitrogen into inorganic compounds that can be used by plants. This process enriches the soil. Owing to their presence in the rice paddies of Southeast Asia, rice can be grown on the same land for many years without adding nitrogen fertilizer. In some regions cyanobacteria are now being added to the soil to increase crop yield.

Cyanobacteria form symbiotic relationships with many organisms including protists, fungi, and some plants. Together with fungi they form lichens (discussed in Chapter 21). Those that form symbiotic relationships usually lack a cell wall and function as chloroplasts, producing food for their symbiotic partner.

Some cyanobacteria can tolerate extremes of salinity, temperature, and pH that kill algae and many other organisms. In fact, cyanobacteria are not only able to thrive in many polluted lakes and ponds but often become the dominant species under such conditions. Fish find them largely inedible, and cyanobacteria may reproduce so extensively that they form "blooms" in the water. The bloom may cause the water to become extremely turbid, limiting the penetration of sunlight. Many of the cyanobacteria die as a result of the crowding and shading. Their decomposition by heterotrophic bacteria consumes large amounts of oxygen from the water and may result in fish kills. Some cyanobacteria also produce toxic metabolic products that kill fish or any animal that takes in the water (Fig. 19–4).

Bacteria

Even the 19th-century pioneers in microbiology realized that bacteria are abundant in air, in liquids such as milk, and in and on the bodies of plants and animals, both living and dead (see Focus on Some Historical Highlights). In fact, there are relatively few places in the world that are devoid of bacteria, for they can be found in fresh and salt water, as far down as several meters deep in the soil, and even in the ice of glaciers.

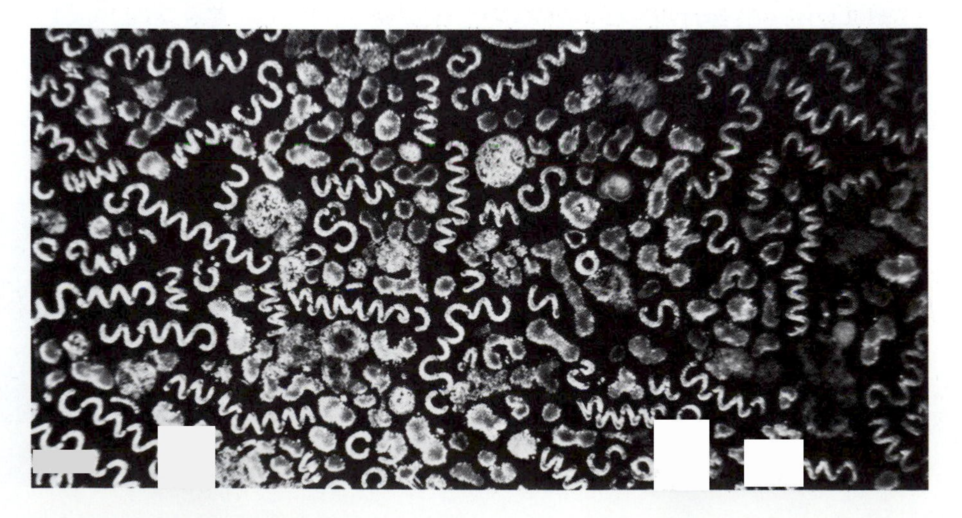

Figure 19–4 Living cyanobacteria (approximately ×60). *Anabaena spiroides*, a spiral cyanobacterium, and *Microcystis aeruginosa*, an irregularly shaped cyanobacterium. These species are often toxic. (Tom Adams.)

FOCUS ON

Some Historical Highlights

Many people associate the words disease and bacteria to such an extent that they cannot hear one without thinking of the other. Yet many diseases are not caused by bacteria, and most bacteria do not cause disease. Bacteria were probably first seen by Anton van Leeuwenhoek (1632–1723), who was a draper in Delft, Holland. With hand lenses he ground himself, he examined almost everything at hand—pond water, sea water, vinegar, pepper solutions (he wanted to find out what made peppercorns hot), feces, saliva, semen, and many other things; he described the objects he saw in letters to the Royal Society of London. In one letter written in 1683, his description of the size, the shape, and the characteristic motion of certain organisms he had observed leaves no doubt that they were bacteria.

The extensive research of Louis Pasteur in the 1870s and 1880s revealed the importance of bacteria as agents of disease and decay. This stimulated work by Robert Koch, Ferdinand Cohn, Joseph Lister, and others, and the science of bacteriology blossomed rapidly in the latter part of the 19th century. Pasteur's studies of the "diseases" of souring wine and beer showed that they were caused by microorganisms that entered the wine or beer from the air and brought about undesirable fermentations yielding products other than alcohol. By gently heating (a process now known as **pasteurization**) the grape juice or beer mash to kill the undesirable organisms and only then seeding the cooled juice with yeast, these "diseases" could be prevented.

Another of Pasteur's contributions to bacteriology was his unequivocal demonstration that bacteria cannot arise by spontaneous generation. After his study of wine, Pasteur was asked by the French government to investigate a disease of silkworms. When Pasteur found that this, too, was caused by microorganisms he reasoned that many animal and plant diseases might be caused by the invasion of "germs." During his investigation of anthrax, a disease of sheep and cattle, and of chicken cholera, he devised a method of treatment, that of immunization, which greatly reduced the death rate from these diseases.

Lord Lister, an English surgeon, was one of the first to understand the significance of Pasteur's discoveries and to apply the germ theory to surgical procedures. He initiated antiseptic techniques by dipping all his operating instruments into carbolic acid and by spraying the scene of the operation with that germicide. In this way he effected a marked decline in the number of fatalities following surgery.

In more recent decades, research with bacteria has been concerned with their physiological, nutritional, biochemical, and genetic mechanisms. Much of our present understanding of the molecular basis of life has been gained from studies of bacteria, especially of *Escherichia coli*.

Bacterial cells range in size from less than 1 to 10 μm in length and from 0.2 to 1 μm in width. It has been estimated that despite their small size, the total weight of all bacteria in the world exceeds that of all other organisms combined!

Among the secrets of the biological success of the bacteria are their small size, their impressive ability to reproduce, their rapid rate of mutation, and their ability to live almost anywhere. When conditions are especially adverse, bacteria of many species can form protected spores and remain in a state of suspended animation—sometimes for years—until environmental conditions are favorable. Bits of rock and ice drilled from depths of 430 m in Antarctica contained bacteria that had lain dormant at temperatures as low as −14°C for at least 10,000 years! When these bacteria were brought to more favorable temperatures they became activated and resumed normal metabolic activities.

STRUCTURE OF BACTERIA

Most bacterial species exist as single-celled forms, but some are found as colonies or as filaments of loosely joined cells. The cell membrane, the active barrier between the cell and the external environment, governs the passage of molecules moving into and out of the cell. Enzymes needed for the operation of the electron transport system (which in eukaryotic cells are found in mitochondria) are attached to the cell membrane.

The cell wall surrounding the cell membrane provides a strong, rigid framework that supports the cell, maintains its shape, and protects it from osmotic damage (Figs. 19–5 and 19–6). The great strength of the bacterial cell wall may be

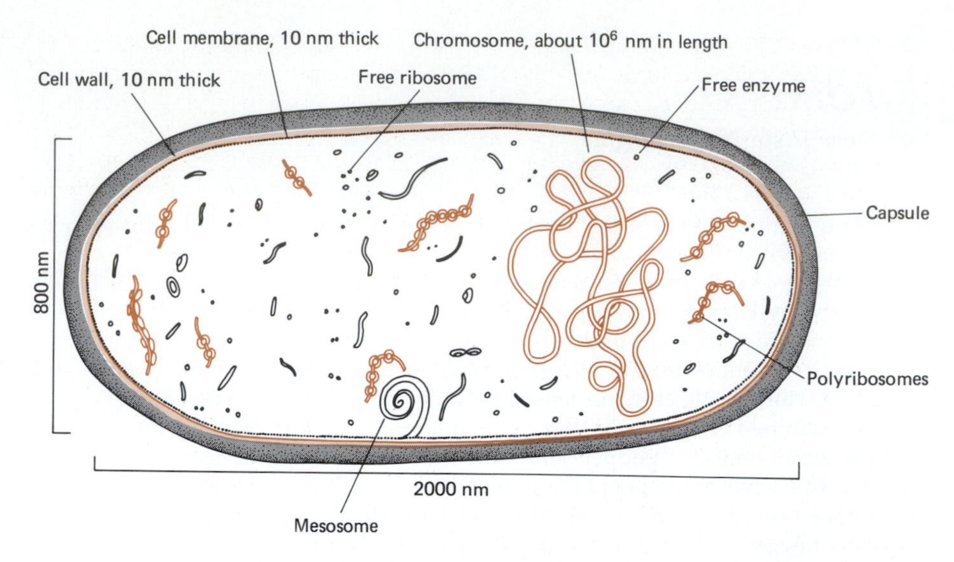

Figure 19–5 Structure of a typical bacterial cell.

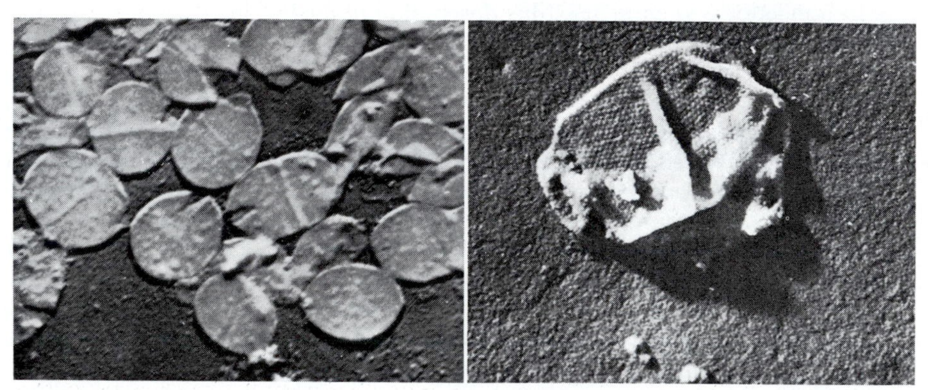

Figure 19–6 Cell walls of bacteria. *Left,* Walls of *Streptococcus fecalis* prepared by grinding the cells; splitting permitted the cell contents to escape (approximately ×12,000). *Right,* A portion of the cell wall of *Spirillum rubrum* showing the regular pattern of the spherical bodies with which this wall is coated (approximately ×42,000). (From Salton, M. R. J., and Williams, R. C.: *Biochemica et Biophysica Acta* 14:455, 1954.)

attributed to the properties of **peptidoglycan,** a unique macromolecule found only in prokaryotes. Peptidoglycan consists of two unusual types of sugar linked with short peptides. The peptides contain two amino acids found nowhere else but in bacterial cell walls. The sugars and peptides are linked to form a single, giant macromolecule that surrounds the entire cell membrane. (In an unusual group of bacteria, peptidoglycan is lacking in the cell walls; see Focus on The Archaebacteria.)

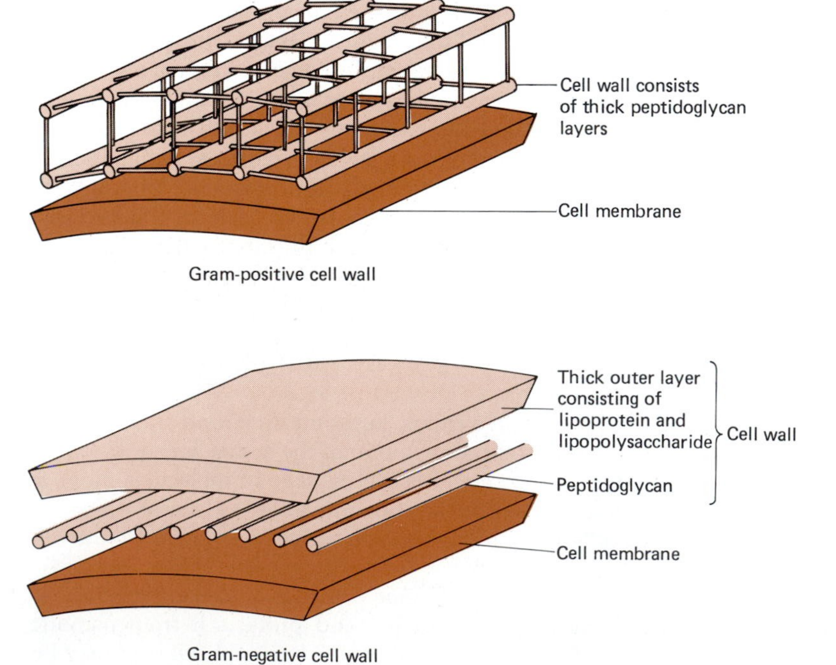

Figure 19–7 A simplified, schematic representation of the peptidoglycan layer in gram-positive and gram-negative bacterial cell walls. Note also the absence of a lipid layer in the gram-positive cell wall.

Almost 100 years ago the Danish physician Christian Gram developed a procedure that is still used to divide bacteria into two groups based on their staining properties. Bacteria that absorb and retain crystal violet stain during laboratory staining procedures are referred to as **gram-positive,** whereas those that do not retain the stain are **gram-negative.** The cell walls of gram-positive bacteria are very thick and consist primarily of peptidoglycan. Examples of gram-positive bacteria are *Streptococcus* and *Staphylococcus.* The envelope of a gram-negative bacterial cell consists of three layers: an inner cell membrane, a thin peptidoglycan layer, and a thick outer layer of lipoprotein and lipopolysaccharide, a lipid–polysaccharide complex (Fig. 19–7). *Escherichia coli* and *Salmonella* are among the gram-negative bacteria.

Differences in composition of the cell wall of gram-positive and gram-negative bacteria confer certain advantages on each kind of bacteria. The thick peptidoglycan layer of the gram-positive bacterium makes its cell stronger and less likely to break as a result of physical stress. However, the gram-positive cell wall is more susceptible to attack by lysozyme, a naturally occurring hydrolytic enzyme; the peptidoglycan layer of gram-negative bacteria is protected by the semipermeable outer membrane, which lysozyme does not penetrate.

The antibiotic penicillin interferes with peptidoglycan synthesis, resulting in a fragile cell wall that cannot effectively protect the cell (Fig. 19–8). Penicillin works most effectively against gram-positive bacteria. Nongrowing cells and cells that produce the enzyme penicillinase are not affected by penicillin.

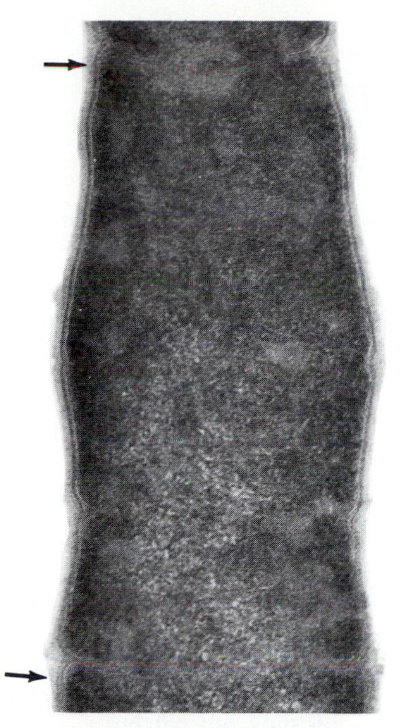

Figure 19–8 Bacterium treated with an antibiotic such as penicillin. The antibiotic interferes with the production of new cell walls. As a result, when the bacterium divides, the new cells are not able to separate and are unable to function. Arrows point to abnormal or abortive walls between daughter cells. (Courtesy of Victor Lorian, with permission of *The American Journal of Clinical Pathology.*)

A few species of bacteria produce a capsule or slime layer that surrounds the cell wall. Such a capsule may provide added protection for the cell. Some encapsulated disease-causing bacteria are able to resist phagocytosis by their host's white blood cells.

The dense cytoplasm of the bacterial cell contains ribosomes and storage granules containing glycogen, lipid, or phosphate compounds. Although the membranous organelles of eukaryotic cells are absent, in some bacterial cells the cell membrane is elaborately folded inwardly. Such complex extensions of the cell membrane are referred to as **mesosomes.** It is thought that the mesosome functions in a variety of metabolic processes and in reproduction. Respiratory enzymes may be associated with these membranes, and cellular respiration may take place there. During cell division the mesosome may be involved in the formation of a septum (wall) between the two new daughter cells.

The DNA is found mainly in a single long, circular molecule referred to as a chromosome (though histones and other proteins are not associated with it as they are in the eukaryotic chromosome; see Fig. 19–5). When stretched out to its full length the bacterial chromosome is about 100 times longer than the cell itself. A small amount of genetic information may be present as smaller DNA molecules, called **plasmids,** which replicate independently of the chromosome. Bacterial plasmids often bear genes involved in resistance to antibiotics.

Many types of bacteria have whiplike cellular outgrowths called **flagella** by which they propel themselves (Fig. 19–9). Some species have a single flagellum or a bundle of them at one end of the cell, while others have flagella distributed over the entire surface. Their arrangement is characteristic of the species and therefore use-

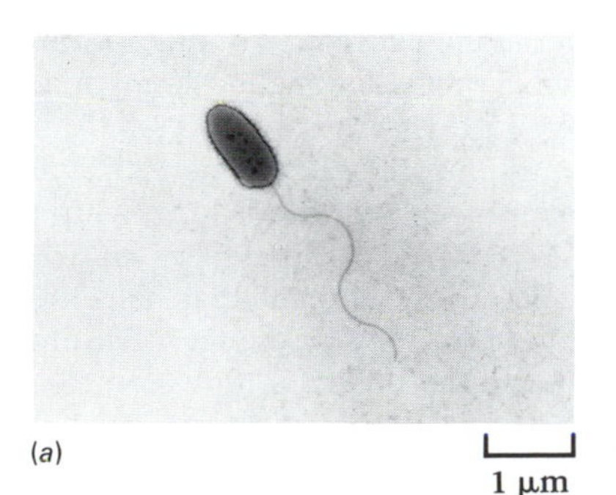

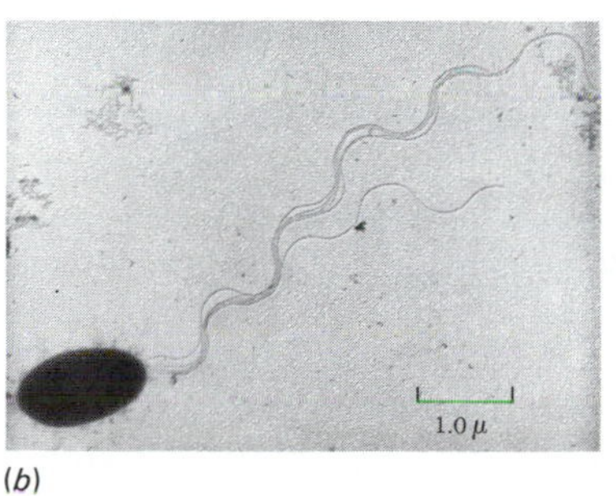

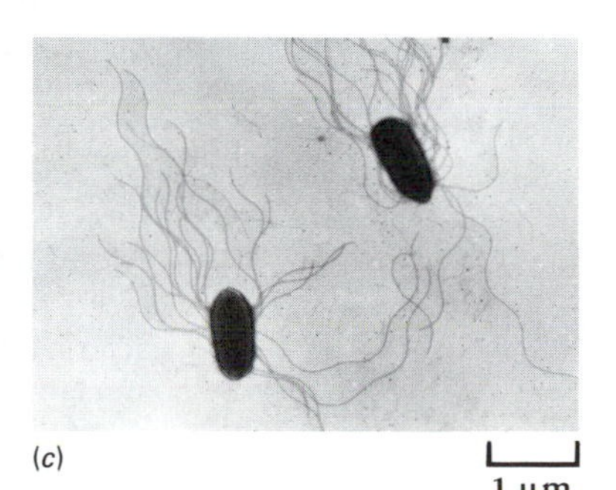

Figure 19–9 Bacterial flagella. (*a*) A single flagellum at the end of the bacterium *Pseudomonas aeruginosa* (approximately ×5000). (*b*) Some bacteria have a tuft of several flagella at one end of the cell. (*c*) In the bacterium *Proteus mirabilis,* flagella project from many surfaces of the cell (approximately ×3500). ((*a*), (*c*), courtesy of V. Chambers; (*b*), courtesy of E. S. Boatman.)

FOCUS ON

The Archaebacteria

All bacteria appear fundamentally similar when studied with the microscope. However, the archaebacteria are a group of organisms that are biochemically very different from the other bacteria. One of their most striking distinguishing features is the absence of peptidoglycan in their cell wall. There are other important differences in proteins and cell chemistry that set the archaebacteria apart from the other bacteria. Some of these biochemical differences may account for the fact that many archaebacteria inhabit extreme environments.

The archaebacteria include three groups:

1. **Extreme halophiles.** The halobacteria can live only in extremely salty environments such as salt ponds. They are sometimes found in brines used to cure fish, making their presence known by forming red patches.
2. **Methanogens.** These anaerobes produce methane from carbon dioxide and hydrogen. They inhabit sewage, swamps, and the digestive tracts of humans and other animals; in such habitats, organic material decomposes under extremely anaerobic conditions.
3. **Thermoacidophiles.** These organisms normally grow in hot, acid environments. Some are found in hot sulfur springs.

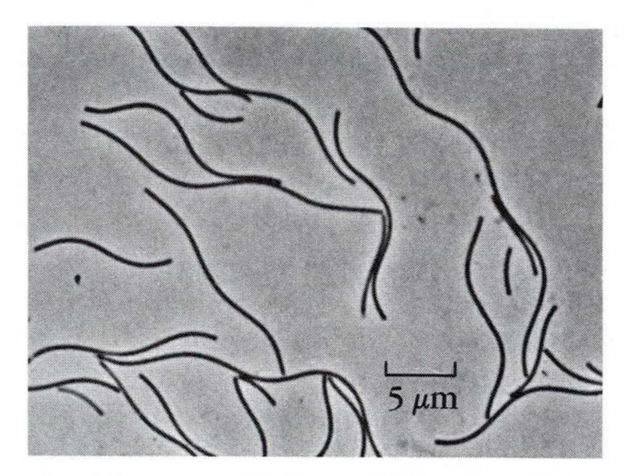

A methanogen, *Methanospirillum hungatei*. (Courtesy of J. G. Ferry.)

The archaebacteria are sometimes described as thermophilic bacteria; this term reflects their adaptations for environments with extreme temperatures.

Since life on earth possesses a certain degree of biochemical unity, the biochemical and metabolic differences between the archaebacteria and other bacteria suggest that these groups must have diverged from each other long ago—relatively early in the history of life on earth. This hypothesis is supported by the fact that many of the extreme conditions to which the archaebacteria are adapted resemble conditions thought to have existed on the primitive earth.

Because the archaebacteria are as biochemically unlike other prokaryotes as the prokaryotes in general are unlike the eukaryotes, some taxonomists have proposed that the archaebacteria be classified in a new domain, or as a second kingdom in the prokaryotic domain. However, other biologists fear that as biochemical analysis continues to challenge traditional taxonomic schemes, we may find ourselves splitting categories until our taxonomy is too fragmented to be useful. While the archaebacteria may be among the most newly discovered creatures in modern biology, they might be the organisms with the oldest history on earth.[1] They have thrived for billions of years, unaffected by existential questions about what they should be called.

[1] The name archaebacteria refers to the Archean era, a period from 3.9 to 2.6 billion years ago.

ful in identification. Bacterial flagella are distinctive in that they consist of a single fibril, whereas the flagella of eukaryotes are composed of 11 fibrils (two in the center surrounded by nine). At the base of a bacterial flagellum is a complex structure that produces a rotary motion, so that the cell is pushed along in the manner that a propeller pushes a ship through the water. In this way some bacteria can travel as much as 2000 times their own length in an hour. (Imagine a man 2 m tall swimming 4 km per hour in a viscous syrup simply by twirling a long whip!)

Many gram-negative bacteria have hundreds of hairlike appendages known as **pili** (Fig. 19–10). These structures are organelles of attachment that help the bacteria to adhere to certain surfaces, such as the cells they will infect. Pili are associated with determination of mating type and possibly with conjugation (the bacterial equivalent of sex) in certain types of bacteria.

BACTERIAL METABOLISM

A bacterial cell contains about 5000 different kinds of chemical compounds. What each of these does, how they interact, and how the bacterium synthesizes each of them from the nutrients it takes in are complex biochemical problems that have

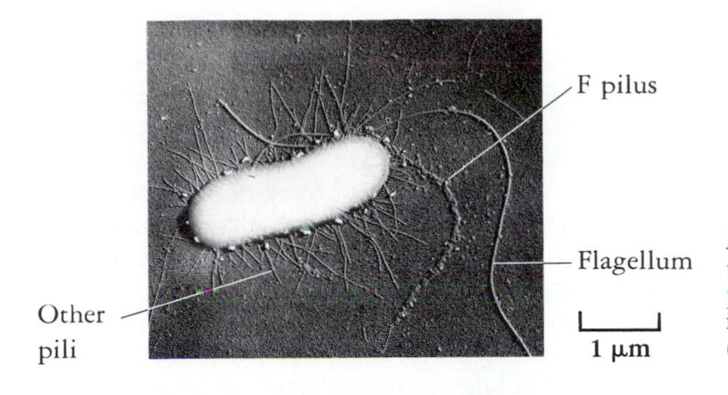

Figure 19–10 Pili of *Escherichia coli.* Note that there are many different kinds (approximately ×11,980). The long F pilus is concerned with the transfer of DNA. (Courtesy of C. Brinton, Jr.)

absorbed the efforts of researchers for many years. Much of the knowledge that has been gained from studying these mechanisms in bacterial cells has been successfully applied to cells of humans and other organisms, for there is surprising uniformity in basic biochemical processes.

Most bacteria are **heterotrophs,** obtaining preformed organic compounds from other organisms. And the majority of heterotrophic bacteria are free-living **saprobes,** organisms that get their nourishment from dead organic matter. Other species of heterotrophic bacteria live in **symbiosis** (living together; a close relationship between organisms of different species that is beneficial to at least one participant) with other organisms. These symbionts may be **commensals** that neither help nor harm their host (Fig. 19–11). A few are parasites that live at the expense of their host and cause diseases in plants and animals; disease-causing organisms are known as **pathogens.** Many symbiotic bacteria inhabit the human skin, digestive tract, and other areas of the body. They are part of the normal community of microorganisms that colonize our bodies. Most are commensals that normally do us no harm.

There are three groups of photosynthetic bacteria: the green sulfur bacteria, the purple sulfur bacteria, and the purple nonsulfur bacteria. Bacterial photosynthesis differs in two important ways from photosynthesis carried on by algae, plants, or cyanobacteria. First, their chlorophylls absorb light most strongly in the near-infrared portion of the light spectrum rather than in the visible light range. This enables them to carry on photosynthesis in red light that would appear very dim to human eyes. Second, bacterial photosynthesis does not produce oxygen because water is not used as a hydrogen donor; instead, the sulfur bacteria utilize sulfur compounds such as hydrogen sulfide (H_2S) as hydrogen donors.

Some bacteria are **chemosynthetic autotrophs,** or **chemoautotrophs.** Rather than carry on photosynthesis, they produce their own food from simple inorganic ingredients using energy obtained from oxidizing inorganic compounds. Chemosynthetic bacteria absorb carbon dioxide, water, and simple nitrogen compounds from their surroundings; using these they manufacture complex organic substances with energy they obtain from oxidation of ammonia to nitrites and nitrates, from

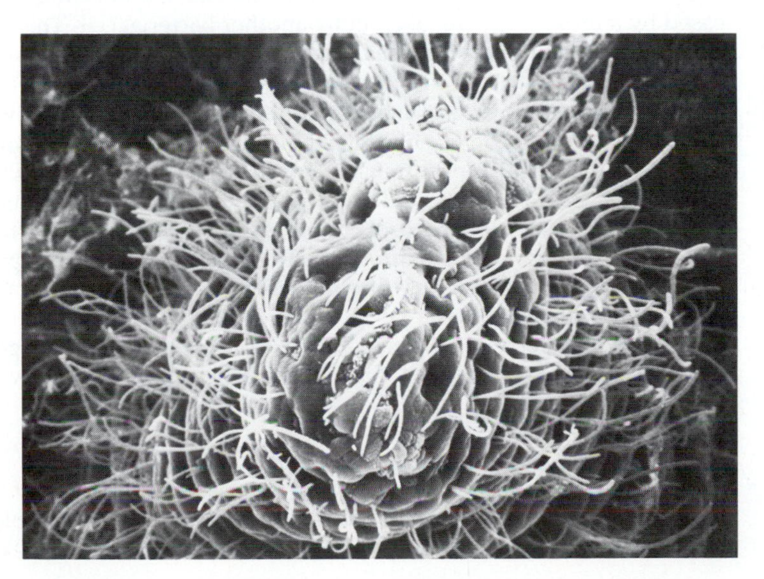

Figure 19–11 Numerous bacterial colonies attached to the intestinal lining of a dog (approximately ×550). (Courtesy of Hoskins, J. D., Henk, W. G., and Abdelbaki, Y. Z.: *American Journal of Veterinary Research,* Vol. 43, No. 10, 1982, pp. 1715–1720.)

the oxidation of sulfur or iron compounds, or from oxidation of gaseous hydrogen. Some of these bacteria play an important role in the nitrogen cycle (Chapter 30).

Whether they are heterotrophs or autotrophs, most bacterial cells, like animal and plant cells, are aerobic, requiring atmospheric oxygen for cellular respiration. Some bacteria are **facultative anaerobes;** they can use oxygen for cellular respiration if it is available, but can carry on metabolism anaerobically when necessary. Other bacteria are **obligate** (strict) **anaerobes,** which carry on energy-yielding metabolism only anaerobically. Some of these bacteria grow more slowly in the presence of oxygen; a few are actually killed by even low concentrations of oxygen gas.

In anaerobic metabolism, bacteria obtain their energy by the anaerobic degradation of carbohydrates or amino acids, and accumulate a variety of partially oxidized intermediates such as ethanol, glycerol, and lactic acid. The anaerobic metabolism of carbohydrates is termed **fermentation,** and the anaerobic metabolism of proteins and amino acids is called **putrefaction.** The foul smells associated with the decay of food, wastes, or corpses are due to nitrogen and sulfur-containing compounds formed during putrefaction.

REPRODUCTION

Bacteria generally reproduce asexually by **transverse binary fission,** in which the cell develops a transverse cell wall and then divides into two daughter cells. The transverse wall is formed by an ingrowth of both the cell membrane and cell wall. When the newly formed cell wall does not separate completely into two walls, a chain of bacteria may be formed.

Although the bacterial cell does not have a mitotic spindle, a copy of the single chromosome must be distributed to each new daughter cell. This appears to be facilitated by a connection between each chromosome and the cell membrane. The duplication of the chromosome and the division of the cell often get out of phase, so that a bacterial cell may have from one to four or even more identical chromosomes.

Bacterial cell division can occur with remarkable speed, and some species grown in an appropriately fortified and aerated culture medium (see Focus on Culturing Bacteria in the Laboratory) can divide every 20 minutes. At this rate, if nothing interfered, one bacterium would give rise to some 250,000 bacteria within 6 hours. This explains why the entrance of only a few pathogenic bacteria into a human being can result so quickly in the symptoms of disease. Fortunately, bacteria cannot reproduce at this rate for a very long time, because they are soon checked by lack of food or by the accumulation of waste products. Multiplication of bacteria that infect humans and other organisms is limited by the defense mechanisms of the host (Chapter 36).

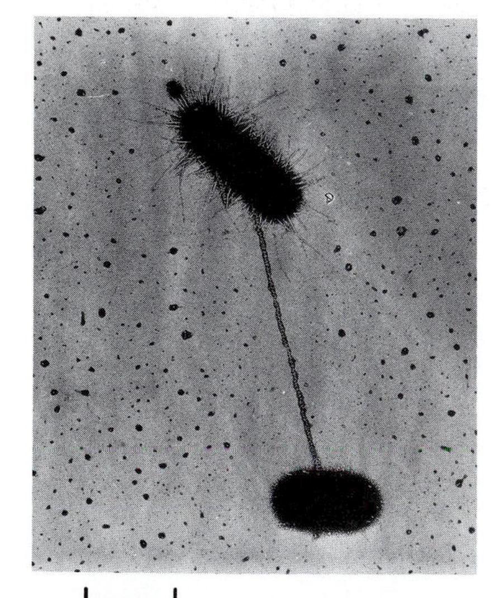

Figure 19–12 F pilus connecting *E. coli* bacteria during conjugation. DNA is transferred during this process. Bacterial viruses have attached to the pilus and are visible as tiny bumps. (Courtesy of C. Brinton, Jr., and J. Carnahan.)

Although complex sexual reproduction involving fusion of gametes does not occur in monerans, genetic material is sometimes exchanged between individuals. Such genetic recombination can take place by three different mechanisms: transformation, conjugation, and transduction. In **transformation,** fragments of DNA released by a broken cell are taken in by another bacterial cell. This mechanism was used experimentally to show that genes can be transferred from one bacterium to another and that DNA is the chemical basis of heredity.

In **conjugation,** two cells of different mating types (the equivalent of sexes?) come together and genetic material is transferred from one to another (Fig. 19–12). Conjugation has been most extensively studied in the bacterium *Escherichia coli,* of which there are F^+ strains and F^- strains. F^+ individuals contain a plasmid known as the F factor, which is capable of organizing special hollow pili. These pili serve as conjugation bridges that pass from the F^+ to the F^- cell. The F pili are long and narrow, and have an axial hole through which fragments of DNA might pass from one bacterium to the other.

In the third process of gene transfer, **transduction,** bacterial genes are carried from one bacterial cell into another within a bacteriophage, as we saw in Chapter 18.

SPORE FORMATION

When the environment of the bacterium becomes very unfavorable, for example, when it becomes very dry, many species can become dormant. The cell loses water,

FOCUS ON

Culturing Bacteria in the Laboratory

In nature, many species of bacteria are intermingled, forming mixed communities. In order to study one specific type, it is necessary to produce a pure culture consisting of a single species. In the laboratory bacteria can be cultured on a medium of agar (a solidifying agent extracted from seaweed) that is enriched with appropriate nutrients. The medium can be poured into a test tube or into a petri dish, a flat glass or plastic dish with a cover, as shown in the figure. When the agar medium is hot, it is liquid, but as it cools, it gels to form a solid. Bacteria cannot move very fast nor very far on a solid medium.

To obtain a pure culture of bacteria, a drop of saliva, feces, or some other material can be spread or streaked out on the surface of the agar medium using a wire loop or bent glass rod. This thins out the bacteria so that they are separated from one another. The petri dish containing the bacteria is then incubated at an appropriate temperature for several hours or days. Isolated bacteria multiply, each giving rise to a colony of daughter cells. When about 10 million to 100 million cells are present, the colony is visible to the naked eye as a small, raised, often circular area on the agar medium. The color, shape, and consistency of the colony varies with the species.

Some of the members of a colony can be carefully streaked on to a fresh dish of agar with a sterile wire loop. After incubation, this petri dish should contain only colonies formed by the desired species. In this way, any bacteria contaminating the original colony can be avoided. The culture is now pure; it contains only a single kind of bacteria.

The nutrient medium mixed with the agar can be a synthetic medium prepared by measuring specific amounts of sugars and various salts, or it can be a complex medium consisting of extracts of plant or animal tissues, milk, or other poorly defined materials. Bacteria of the genus *Hemophilus* and certain other bacteria grow best on chocolate agar, a medium enriched with blood.

(a)

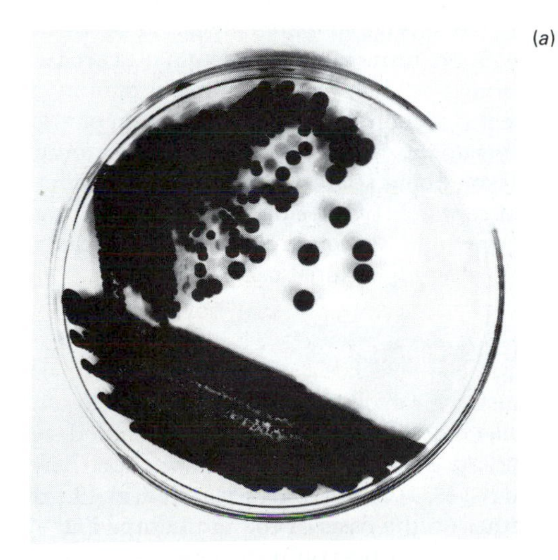

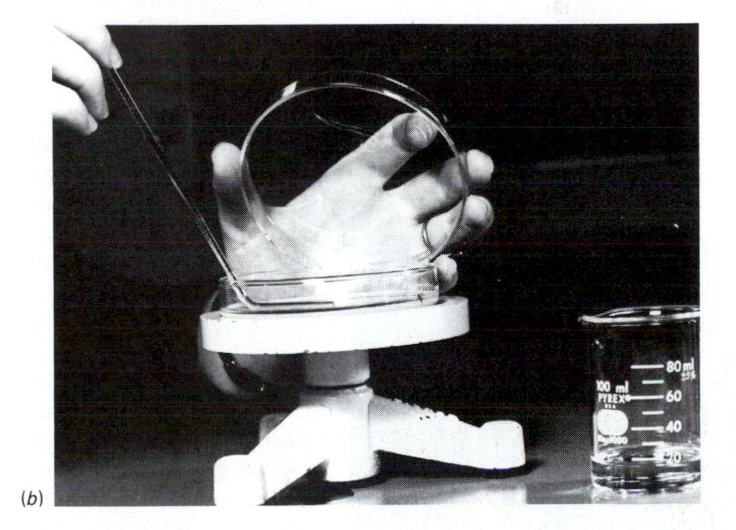

(b)

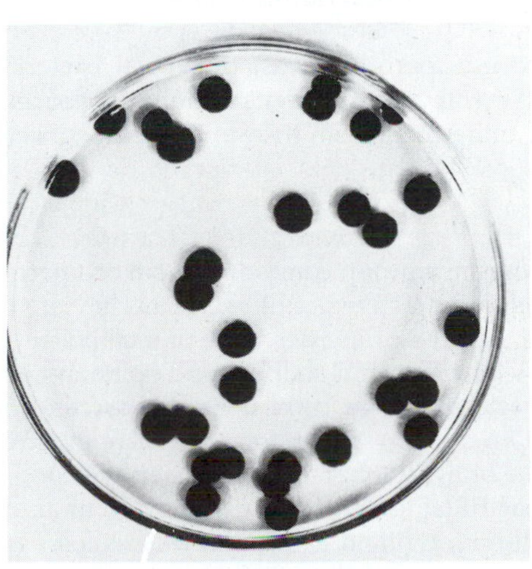

(c)

Techniques of bacterial culture. (*a*) Three-phase discontinuous streak plate; note isolated colonies (*S. marcescens* on nutrient agar). (*b*) Turntable for spread plate technique. Bent glass spreading rod is immersed in 95% alcohol and flame-sterilized. Inoculum is spread evenly over the surface of the agar. (*c*) Isolated colonies developed from spread plate (*S. marcescens* spread over nutrient agar). (Courtesy of Frobisher, M., Hinsdill, R. D., Crabtree, K. T., and Goodheart, C. R.)

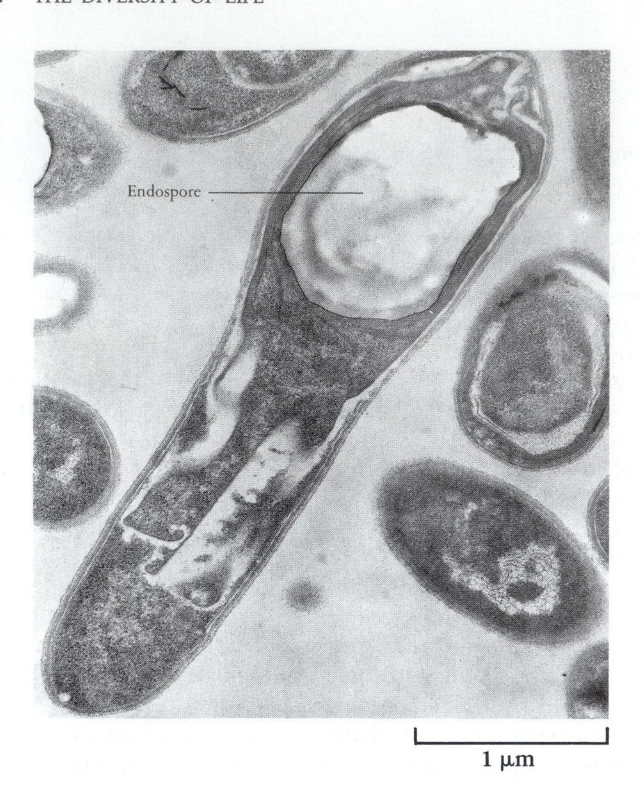

Figure 19–13 Endospore within a cell of *Clostridium*. (From Santo, L., Hohl, H., and Frank, H.: *Journal of Bacteriology* 99:824, 1969.)

shrinks a bit, and remains quiescent until water is again available. Other species form spores—dormant, resting cells—to survive in extremely dry, hot, or frozen environments or when food is scarce. Spore formation is not a kind of reproduction in bacteria, since only one spore is formed per cell. The total number of individuals does not increase as a result of spore production. During the formation of a spore, the outer coat of the spore develops inside the cell (Fig. 19–13). It forms around the DNA and a small amount of cytoplasm. Some spores are so strong that they can survive an hour or more of boiling, or centuries of freezing. When the environmental conditions are again suitable for growth the spore can absorb water, break out of its inner wall, and become an active growing bacterial cell again.

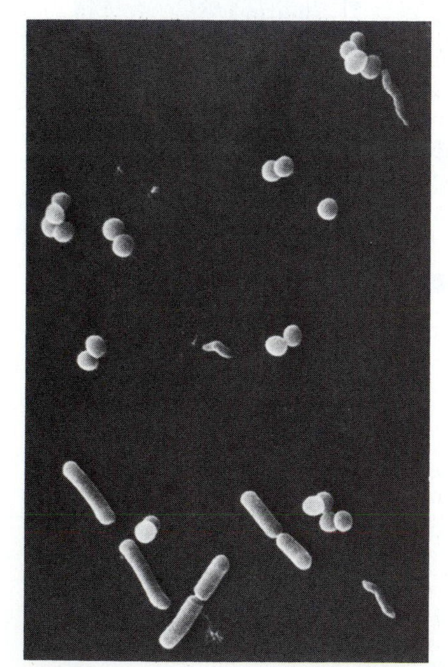

Figure 19–14 Three different types of bacteria, illustrating the coccus, bacillus, and spirillum shapes (approximately ×8300). The coccus bacteria are *Staphylococcus aureus*, a common cause of boils and wound infections. (Courtesy of Daniel C. Williams and Lilly Research Laboratories.)

SOME IMPORTANT GROUPS OF BACTERIA

Bacteria are difficult to classify because their evolutionary relationships to one another are not known. Taxonomic ranks higher than the genus have limited significance. According to the current standard work on bacterial classification (Bergey's Manual of Determinative Bacteriology, 1974) bacteria are classified in 19 groups called **parts.** They are grouped together on the basis of common shape and structure, biochemical properties, genetic characteristics, nutritional needs, habitat, and sensitivities to specific drugs.

In the following discussion, several parts (groups) of typical bacteria are lumped together as the eubacteria. We will also consider several other parts that each have unique characteristics separating them from the more typical eubacteria.

Eubacteria

The **eubacteria,** or true bacteria, are a diverse group consisting of several parts. All these bacteria conform to certain general characteristics of bacteria. They are often divided into parts on the basis of their staining properties, type of metabolism, and shape. There are three main shapes—spherical, rod-shaped, and spiral, as shown in Figure 19–14. Spherical bacteria, referred to as **cocci** (singular **coccus**), occur singly in some species, in groups of two in others, or in long chains (e.g., streptococci) or irregular clumps that look like bunches of grapes (e.g., staphylococci; Fig. 19–15). Rod-shaped bacteria, called **bacilli** (singular, **bacillus**), may occur as single rods or as long chains of rods. Diphtheria, typhoid fever, and tuberculosis are all caused by bacilli. Spiral bacteria are known as **spirilla** (singular, **spirillum**). Short, incomplete spiral-shaped bacteria are known as **comma bacteria** or **vibrios.**

Most eubacteria are not pathogenic. Rather, they are harmless saprobes ecologically important as decomposers. The lactic acid bacteria are gram-positive bacteria that produce lactic acid as the main end-product of their fermentation of sugars. Among these are the rod-shaped **lactobacilli,** which may be found in decomposing plant material, milk, yogurt, and other dairy products. They are commonly present in animals and are among the normal inhabitants of the human mouth and vagina.

Streptococci, also gram-positive bacteria, are found in the mouth as well as in the digestive tract. Among the harmful species of streptococci are those that cause "strep throat," scarlet fever, wound infections, and skin, ear, and other infections. **Staphylococci** are gram-positive bacteria that normally live in the nose and on the skin. Some species can cause disease when the immunity of the host is lowered. *Staphylococcus aureus* may cause boils and skin infections or may infect wounds. Certain strains (varieties) of *Staphylococcus aureus* cause food poisoning, and some are thought to cause toxic shock syndrome. The **clostridia** are a notorious group of anaerobic gram-positive eubacteria. One species causes tetanus; another causes gas gangrene; and *Clostridium botulinum,* an inhabitant of the soil, can cause botulism, an often fatal type of food poisoning.

Figure 19–15 Electron micrograph of a group of staphylococci, spherical bacteria that occur in bunches like grapes (photographed at approximately ×25,000; reduced ⅕ in printing). (Courtesy of Department of Physical Chemistry, Lilly Research Laboratories.)

The gram-negative eubacteria are a diverse group that exhibit differences in shape, structure, and metabolic processes. Among them are the **azotobacteria** which have the ability to fix nitrogen; others are chemoautotrophs, which as described earlier obtain energy from oxidizing inorganic compounds. Many important pathogens are gram-negative eubacteria: The gram-negative coccus **Neisseria gonorrhoeae** causes gonorrhea. **Hemophilus influenzae** is a gram-negative bacillus that can cause infections of the respiratory tract and ear as well as meningitis. The **enterobacteria** are a group of gram-negative rods that include free-living saprobes, plant pathogens, and a variety of symbionts that inhabit humans. *Escherichia coli,* a member of this group, inhabits the intestine of humans and other animals as part of the normal microbe population. Some strains of the enterobacterium *Salmonella* cause food poisoning.

Myxobacteria (Slime Bacteria)

The **myxobacteria** are unicellular short rods resembling the eubacterial bacilli, but they lack a rigid cell wall. They excrete slime, and when they are cultured in a petri dish, their growth is marked by a spreading layer of slime. These bacteria glide or creep along.

Most myxobacteria are saprobes that break down organic matter in the soil, manure, or rotting wood they call home. Some species can break down complex substrates such as cellulose and bacterial cell walls. A few of these prey on eubacteria.

In some species, reproduction is more complex than in other bacteria. Cells swarm together to form masses, which may develop into reproductive structures called **fruiting bodies** (Fig. 19–16). During this process, many bacterial cells are changed into **resting cells** (equivalent to spores) and are packaged within a protective wall, forming a cyst. When conditions are favorable, the cyst breaks open, and the resting cells become active.

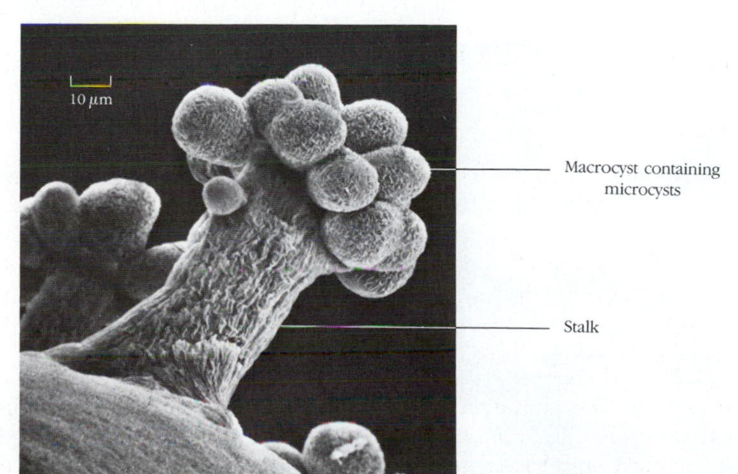

Figure 19–16 Myxobacteria aggregate at a certain time in their life cycle and form a fruiting body. Protective resting cells, called microcysts, are formed that are very resistant to heat and drying. The fruiting body shown here was formed by *Chondromyces crocatus.* (From Grilione, P. L., and Pangborn, J.: *Journal of Bacteriology* 124:1558, 1975.)

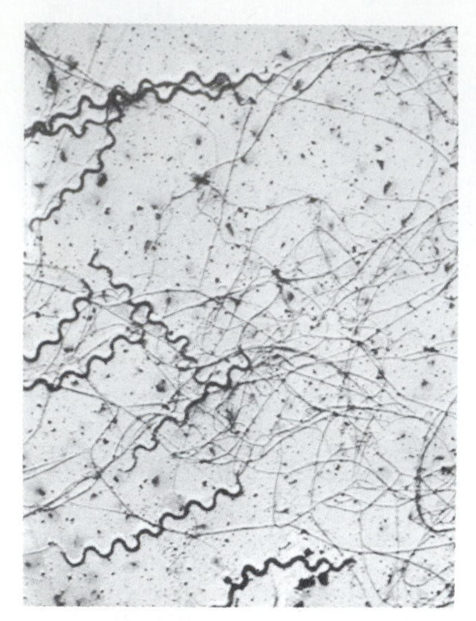

Figure 19–17 Electron micrograph of *T. pallidum*, the spirochete that causes syphilis (approximately ×6250). (From Thomas, M. et al.: *Applied Microbiology* 23:714, 1972.)

Spirochetes

Spirochetes are slender, spiral-shaped bacteria with flexible cell walls. They move by means of a unique structure called an **axial filament.** Some species inhabit freshwater and marine habitats, others form commensal associations, and a few are parasitic. The spirochete of greatest medical importance is ***Treponema pallidum*** (Fig. 19–17), the pathogen that causes syphilis (discussed in Chapter 43).

Actinomycetes (Moldlike Bacteria)

Actinomycetes resemble molds (which are fungi) in that their cells remain together to form branching filaments. Many produce moldlike spores, called **conidia.** However, they produce cell wall peptidoglycans, lack nuclear membranes, and have other bacterial characteristics that have convinced taxonomists that they are indeed bacteria.

The actinomycetes may be credited with much of the decomposition of organic materials in soil. In fact, most members of this group are saprobes, and some are anaerobic. Several species of the group collectively known as **Streptomyces** produce antibiotics; streptomycin, erythromycin, chloramphenicol, and the tetracyclines are among the antibiotic drugs derived from these bacteria.

Some actinomycetes cause disease. *Mycobacterium tuberculosis* is the bacillus that causes human tuberculosis. *Mycobacterium leprae* causes Hansen's disease (leprosy). Other actinomycetes can cause serious lung disease or generalized infections.

Mycoplasmas (Bacteria Without Cell Walls)

Also known as **PPLO** (for **P**leuro**p**neumonia-like **o**rganisms), **mycoplasmas** are tiny bacteria bounded by a pliable cell membrane, but lacking a typical bacterial cell wall. Some are so small that, like viruses, they pass through bacteriological filters. In fact, mycoplasmas are smaller than some viruses. Biologists have hypothesized that mycoplasmas may be the simplest form of life capable of independent growth and metabolism.

Mycoplasmas may be aerobic or anaerobic, depending on the species. Some live in soil, others in sewage. Still others are parasitic on plants or animals. Some species of *Mycoplasma* inhabit human mucous membranes but do not generally cause disease. One species causes a mild type of bacterial pneumonia in humans that until recently was thought to be caused by a virus (Fig. 19–18). This disease accounts for approximately 20% of all cases of pneumonia in some areas. Since the mycoplasmas lack a cell wall, they are resistant to penicillin and other antibiotics that act on the cell wall; however, they are sensitive to tetracycline and other antibiotics that inhibit protein synthesis.

Rickettsias

Rickettsias are small bacteria that lack the ability to carry on metabolism independently and so are obligate intracellular parasites (they are obligated to live as parasites in order to survive). Most parasitize certain arthropods such as fleas, lice,

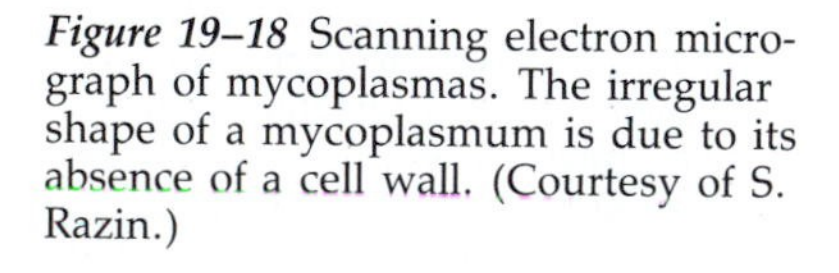

Figure 19–18 Scanning electron micrograph of mycoplasmas. The irregular shape of a mycoplasmum is due to its absence of a cell wall. (Courtesy of S. Razin.)

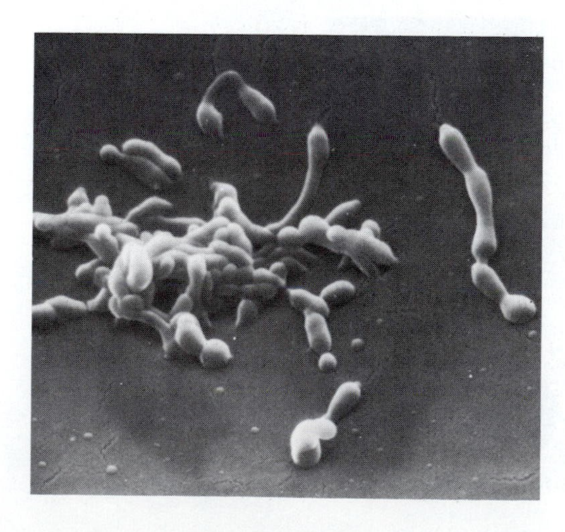

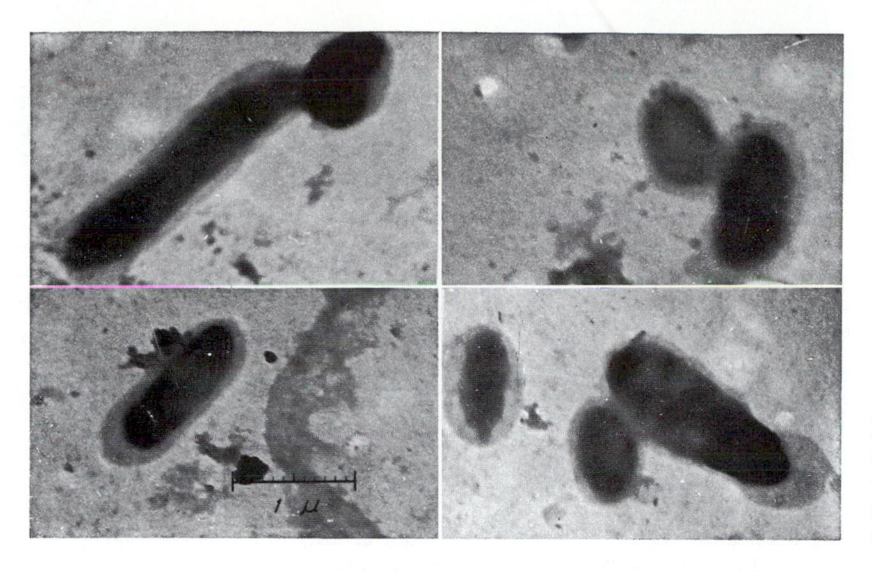

Figure 19–19 Electron micrographs of the rickettsia causing typhus. Note the variations in the size and shape of the particles and the less-dense capsule that surrounds each. (Courtesy of Lilly Research Laboratories.)

ticks, and mites without causing specific disease in them. Diseases caused by the few species known to be pathogenic to humans (and other animals) are transmitted by arthropod **vectors** (carriers), through bites or contact with their excretions; among these are typhus, Rocky Mountain spotted fever, and Q fever. In fact, Howard Ricketts, who discovered the rickettsia, died in Mexico in 1910 of typhus fever while studying the organisms that cause it.

Rickettsias are typically short gram-negative rods with rigid cell walls (Fig. 19–19). They are not motile and do not form spores. Reproduction takes place by binary fission.

Chlamydias

Chlamydias differ from rickettsias in that they are spherical rather than rod-shaped; moreover, they are not dependent upon arthropod vectors for transmission. Because they are obligate intracellular parasites, chlamydias were considered for many years to be large viruses. However, they are now classified as bacteria because they contain both DNA and RNA, possess ribosomes and synthesize their own proteins and nucleic acids, and are sensitive to a wide range of antibiotics. Although they do contain many enzymes and can carry on some metabolic processes, chlamydias are completely dependent on the host cell for ATP. For this reason they are sometimes referred to as **energy parasites.**

Studies indicate that chlamydias infect almost every species of bird and mammal. Perhaps 10% to 20% of the human population of the world is infected. Interestingly, though, individuals may be infected for many years without apparent harm. Sometimes chlamydias do cause acute infectious diseases. **Trachoma,** the leading cause of blindness in the world, is caused by a strain of *Chlamydia.* These bacteria also cause a contagious venereal disease (lymphogranuloma venereum) and psittacosis, a disease transmitted to humans by birds.

BACTERIA IN HUMANS

Many species of bacteria colonize our skin, digestive tract, upper respiratory tract, and other body openings. The number of bacteria that inhabit the lower digestive tract is so great that nearly half the dry weight of feces is bacteria. Some types of bacteria in our intestinal tract are our mutualistic partners; in exchange for food and shelter they provide us with vitamin K and some of the B vitamins. Bacteria of this normal microbial community (sometimes referred to as the normal flora) generally compete successfully against the pathogenic bacteria that occasionally happen upon the scene. By preventing the multiplication and spread of pathogens they help protect us against possible disease. Within a few days after birth an infant becomes "infected" with a variety of bacteria, and the normal microbial community is soon established.

A delicate balance between bacteria and host is normally maintained by the body's antibacterial defenses (discussed in Chapter 36). These prevent the bacteria from spreading to areas in which they could cause trouble. Among the normal community of microorganisms are some **opportunistic** species of bacteria ready to take advantage of the situation should the host's defenses falter. The stress of injury, illness, or poor diet may lower the host's normal resistance so that such opportunists are able to move into vulnerable territories and may then cause disease. Among the normally harmless bacteria that can cause disease under certain conditions are *Escherichia coli* and *Staphylococcus aureus.*

Opportunistic infections are becoming increasingly common as a result of certain clinical procedures such as purposely depressing the immune system after an organ transplant. Use of certain drugs, especially antibiotics, can result in the replacement of normal bacteria by varieties resistant to the drug. These bacteria can then multiply and cause secondary infections.

Pathogenic bacteria are often able to overwhelm the body's defense mechanisms and produce disease (see Focus on Bacteria That Infect Humans). They can be spread from one person to another in food, water, saliva, or nasal secretions carried as droplets in the air from a sneeze or cough. They can be spread by direct contact with an infected wound or lesion. Some bacterial diseases are transmitted by arthropod vectors, and a few are contracted from other animals.

How can such tiny organisms as bacteria cause symptoms of disease? Many details of how the infecting bacteria bring about their effects are not yet known, but there are two basic mechanisms involved: First, most pathogens damage the host cells and cause symptoms in the area of invasion; this is due to the rapid multiplication of the microorganisms themselves and to defense mechanisms (e.g., inflammation) triggered in the host in response to the presence of the invaders. The second mechanism involves the release of toxins by the bacteria.

Many pathogenic bacteria, especially gram-positive species, produce **exotoxins** and secrete them into the surrounding medium. Exotoxins are proteins; some of them are among the most powerful poisons known. Some pathogens do not even have to enter the host to cause disease. For example, *Clostridium botulinum* bacilli release their exotoxin in food, and ingestion of the toxin causes the symptoms of botulism. Other pathogens may infect only a localized area, but their toxins spread throughout the body. **Endotoxins** are part of the cell walls of gram-negative bacteria; they are lipid–polysaccharide complexes and are far less toxic than exotoxins. Endotoxins trigger immune responses of the host, thereby causing such symptoms as fever and decrease in blood pressure. (Immune responses are discussed in Chapter 36).

Some persons unknowingly carry highly dangerous bacteria and transmit them to other people. You may have heard of "Typhoid Mary," who was a carrier of the bacteria that cause typhoid fever. Though she herself did not exhibit symptoms of the disease she transmitted it to numerous people. The sexually transmitted disease gonorrhea is often spread by carriers who have the disease but do not exhibit symptoms.

BACTERIA AND FOOD

Humans have used to good advantage the abilities of specific strains of bacteria to produce chemicals such as ethanol and acetic acid (vinegar). Buttermilk and yogurt are produced by cultivating certain lactic acid bacteria in milk, thereby permitting fermentation. Bacteria are inoculated into cream to improve the flavor of butter. Numerous varieties of cheeses can be produced from the same batch of milk by the action of different types of bacteria.

Species of lactic acid bacteria (*Lactobacillus*) are used in the production of pickles and sauerkraut. Vinegar is prepared by permitting certain strains of bacteria to oxidize the ethyl alcohol in apple cider, wine, or other substances. Some kinds of sausages are flavored by specific types of bacteria.

On the negative side, bacteria are present in all food and can cause spoilage, as well as illness in those who partake. In fact, bacterial competition for our food supply has led to the development of many techniques for preventing contamination of food, inhibiting microbial growth and metabolism, and killing bacteria. From ancient times, people have struggled with the problem of food preservation. The early Egyptians and Romans knew how to salt, dry, and smoke food, and perishable foods were sometimes stored in the low temperatures of caves.

FOCUS ON

Bacteria That Infect Humans

Bacterium	*Characteristics*	*Importance*
Eubacteria		
Staphylococcus aureus	Cocci that often form clusters; gram-positive	Can live harmoniously as part of normal microbial community. Opportunistic; can cause boils. Also exotoxin is a major cause of food poisoning.
Streptococcus pyogenes	Cocci that form pairs and chains; gram-positive	Causes "strep throat," ear infections; scarlet fever. Induces rheumatic fever
Streptococcus pneumoniae	Cocci that form pairs or chains; gram-positive	Causes pneumonococcal pneumonia and meningitis
Clostridium tetani	Slender, gram-positive bacilli; strictly anaerobic; form spores	Causes tetanus (lockjaw); potent exotoxin affects nervous system.
Clostridium botulinum	Large gram-positive bacilli; anaerobic; form spores	A soil organism that causes botulism; potent exotoxin affects nervous system.
Neisseria gonorrhoeae	Gram-negative cocci that form pairs (diplococci); adhere to cells via pilli	Causes gonorrhea
Escherichia coli	Gram-negative bacilli; facultative anaerobes	Lives as part of normal intestinal microbial community; opportunistic strains among them can cause diarrhea, urinary tract infections, and meningitis.
Salmonella	Gram-negative bacilli	One species causes food poisoning (diarrhea, vomiting, fever); another species can cause typhoid fever; a third species causes infections of the blood.
Hemophilus influenza	Gram-negative small rods	Causes infections of upper respiratory tract and ear; can cause meningitis
Rickettsias		
Rickettsia rickettsii	Short rod-shaped; obligate intracellular parasite	Can cause Rocky Mountain spotted fever; transmitted by tick from dog or rodent
Spirochetes		
Treponema pallidum	Very slender, tightly coiled spirals; move via axial filament	Causes syphilis
Actinomycetes		
Mycobacterium tuberculosis	Slender, irregular rods	Causes tuberculosis of lungs and other tissues
Mycobacterium leprae	Slender, irregular rods	Causes Hansen's disease (leprosy)
Chlamydias		
Chlamydia trachomatis	Gram-negative cocci; obligate parasites	Causes trachoma (the leading cause of blindness); causes a sexually transmitted disease (lymphogranuloma venereum)

Modern techniques include killing bacteria by boiling, pasteurization, or irradiation and inhibiting bacterial growth by refrigeration, freezing, and dehydration. Chemicals such as benzoic acid are added to many foods to inhibit growth of microorganisms. Some of the time-tested methods are still used, but now we understand just how they work. For example, pickling inhibits bacterial growth by a combination of high salt content, which osmotically draws water out of the bacterial cells, and low pH, which discourages their growth. Preserving foods with sugar as in jellies and jams also results in plasmolysis of bacterial cells.

Despite all of the procedures available to preserve food, improper handling can result in bacterial contamination. One of the most common types of food poisoning is caused by *Staphylococcus aureus,* the same species that causes boils and other infections. The bacteria enter the food from the hands or sneezes of food handlers who may have acute infections or simply be carriers of the bacteria. Improperly stored mayonnaise, custards, processed meat, and milk provide excellent habitats for these bacteria. After only a few hours the bacteria produce sufficient toxin to cause food poisoning. Nausea, vomiting, and diarrhea are common symptoms that occur within a few hours after eating the contaminated food. The victim usually recovers within a day or two.

Botulism is a serious type of food poisoning characterized by paralysis, which is sometimes fatal. It is caused by an exotoxin released by the anaerobic bacterium *Clostridium botulinum*. Many outbreaks of botulism have been traced to improperly canned food. The spores of this bacterium are quite heat-resistant, so that cooking the food does not necessarily destroy them. (However, a few minutes of boiling does destroy the exotoxin.) Production of the toxin occurs only after lysogenic conversion of the bacterium (Chapter 18). The toxin is one of the most powerful poisons known; a few milligrams would be enough to kill several million people. More than a million people in the United States each year suffer from the inflammation of the digestive tract (gastroenteritis) that results from salmonella infection. Food products most often contaminated with *Salmonella* are eggs, egg products, and poultry.

ECOLOGICAL IMPORTANCE OF BACTERIA

Bacteria play an essential role as decomposers in our biosphere. Without the bacteria (and the fungi), all of the available carbon, nitrogen, phosphorus, and sulfur would eventually be tied up in the wastes and dead bodies of plants and animals. Life would soon cease because of the lack of raw materials for the synthesis of new cellular components.

Some bacteria can fix nitrogen from the air; they bring nitrogen gas into chemical combination by reducing it to ammonia. Other bacteria convert ammonia to nitrites and nitrates (Fig. 19–20). Some nitrogen-fixing bacteria are free-living. Others form symbiotic relationships with certain plants. Some grow as nodules on the roots of beans, peas, clover, and other leguminous plants; these bacteria are so

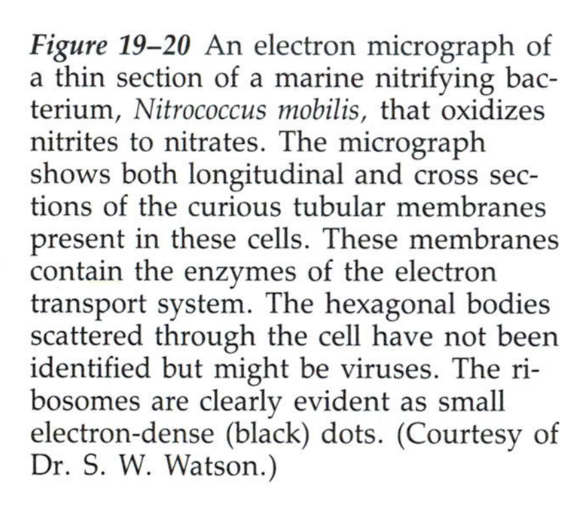

Figure 19–20 An electron micrograph of a thin section of a marine nitrifying bacterium, *Nitrococcus mobilis*, that oxidizes nitrites to nitrates. The micrograph shows both longitudinal and cross sections of the curious tubular membranes present in these cells. These membranes contain the enzymes of the electron transport system. The hexagonal bodies scattered through the cell have not been identified but might be viruses. The ribosomes are clearly evident as small electron-dense (black) dots. (Courtesy of Dr. S. W. Watson.)

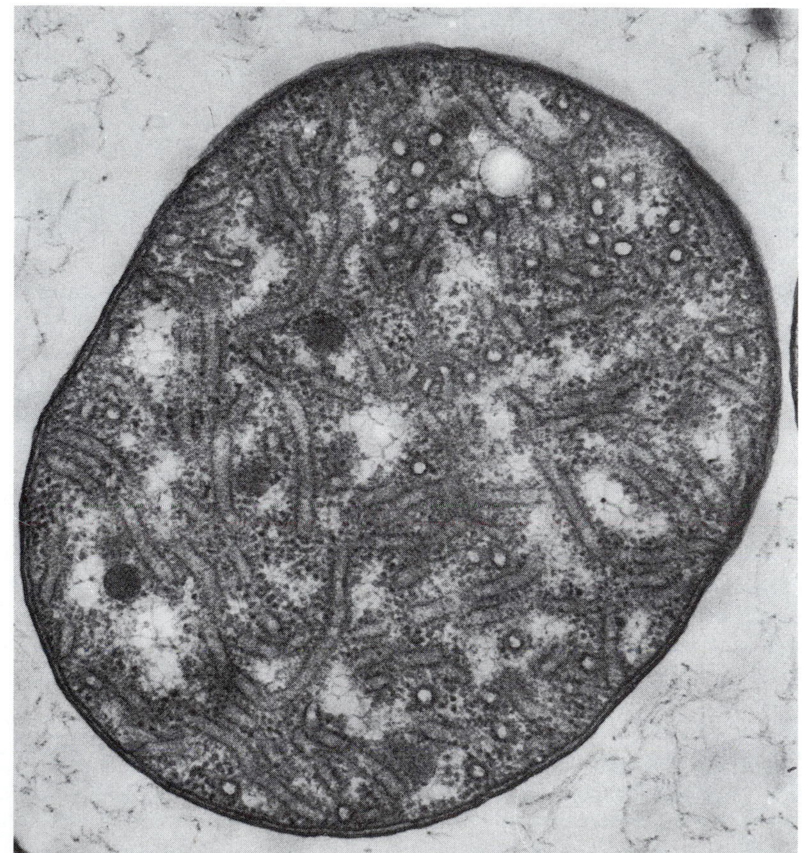

efficient at their task that some can fix as much as 100 or more kg of nitrogen per acre per year, producing nitrogen compounds required by plants.

As in soil, aquatic bacteria also recycle essential nutrients. Photosynthetic sulfur bacteria in the muddy bottoms of the water provide a valuable service by using the hydrogen sulfide produced by sulfide bacteria. This can be appreciated by considering the "rotten egg" odor produced by accumulated hydrogen sulfide in polluted ponds, lakes, or bays. When sewage or other organic material is dumped into a body of water, the sulfide bacteria multiply rapidly, but the photosynthetic sulfur bacteria grow more slowly and are unable to utilize the hydrogen sulfide quickly enough to prevent the odor.

Figure 19–21 A crown gall tumor growing on the base of the trunk of a young pecan tree. The growth of this tumor is induced by a plasmid carried by *Agrobacterium tumefaciens.* Agricultural stocks of such infected trees are usually burned to prevent the spread of the disease. (USDA Photo).

BACTERIAL DISEASES OF PLANTS

Although most plant diseases are caused by fungi rather than bacteria, about 200 bacterial plant diseases are known. Bacteria are responsible for the soft rot of a variety of vegetables, the black rot of cabbage, and the growth of crown galls on other plants. **Galls** are disorganized masses of plant cells that may form large swollen growths that cut off transport of water and nutrients within the plant and ultimately kill it (Fig. 19–21). **Crown gall,** sometimes referred to as plant cancer, can occur in sugar beets, fruit trees, and many other types of plants. Once the growth has begun to develop, killing the bacteria does not halt the progress of the disease. The explanation is that the bacteria transfer segments of DNA to the plant cells, transforming them into tumor cells. As we saw in Chapter 16, this ability has use in genetic engineering.

Cankers are another interesting type of bacterial disease in plants. The water-conducting tissue (xylem) of the plant is affected first, but the disease spreads into surrounding tissue, killing it and causing lesions. These lesions are surrounded by a gummy liquid produced by the plant around the infection. Other plants become infected when drops of this liquid are transported by insects. Bacteria enter the new plant through a wound or through a flower. A canker of apple and pear trees called fire blight is caused by bacteria that multiply so rapidly and kill the cells so quickly that the tree appears to be burned.

Plant diseases are controlled by trying to prevent the introduction of disease-causing bacteria into the area, by destroying the bacteria, and by keeping the plants healthy with a supply of water and proper nutrients so that they can resist infection. In addition, development of genetically resistant strains of plants greatly reduces the chances of bacterial infection, as well as the opportunity for the bacteria to spread.

SUMMARY

I. Kingdom Monera is made up of all of the prokaryotes—the cyanobacteria and the bacteria.

II. Most cyanobacteria are photosynthetic autotrophs, which occur mainly as colonial masses or filaments.
- *A.* Like other monerans, cyanobacteria lack a nuclear membrane; they have cell walls containing peptidoglycan.
- *B.* Cyanobacteria are ecologically important as producers, and some species can fix nitrogen. They often become the dominant microorganisms in polluted lakes and bays.

III. Bacteria are biologically successful organisms owing to their small size and their ability to mutate and reproduce quickly and to inhabit a wide range of ecological niches.
- *A.* The bacterial cell wall, composed mainly of peptidoglycan, supports the cell and protects it from chemical and osmotic damage.
  1. Gram-positive bacteria have a thick-layered cell wall of peptidoglycan.
  2. Gram-negative bacteria have a thick outer layer of lipid compounds surrounding the peptidoglycan layer in the cell wall.
- *B.* Typically, the bacterial cell lacks membranous organelles but may have a mesosome consisting of elaborate inward extensions of the plasma membrane.
- *C.* The DNA is a long, circular molecule found in the cytoplasm; plasmids may be present.

IV. Most bacteria are heterotrophic saprobes. Many form symbiotic relationships with other organisms. Some groups of bacteria are photosynthetic autotrophs; others are chemosynthetic autotrophs.

V. Some bacteria carry on respiration aerobically; others are facultative anaerobes, and still others are obligate anaerobes.

VI. Bacteria reproduce asexually by transverse binary fission. Genetic recombination can take place by transformation, conjugation, or transduction.

VII. Bacteria are classified in 19 groups called parts on the basis of their shape, structure, biochemical properties, genetic characteristics, nutritional needs, and habitat.

A. The eubacteria are highly diverse; the entire group consists of several parts. Among the eubacteria are gram-positive and gram-negative cocci and bacilli, and gram-negative spirilla. Most eubacteria are harmless saprobes, but a few are notorious for the human diseases they cause.

B. The myxobacteria (slime bacteria) are unicellular bacilli that resemble eubacteria but lack a rigid cell wall; they move by gliding.

C. Spirochetes are spiral-shaped bacteria with flexible cell walls; they move by means of an axial filament. *Treponema pallidum*, the pathogen that causes syphilis, is a spirochete.

D. Actinomycetes resemble molds; their cells form branching filaments, and many species produce conidia. Most species are saprobes that decompose organic material in soil. Streptomycin and other antibiotics are derived from them. Human tuberculosis and Hansen's disease are caused by actinomycetes.

E. Mycoplasmas are extremely small bacteria that lack a typical bacterial cell wall. One species causes a mild type of pneumonia.

F. Rickettsias are obligate intracellular parasites because they cannot carry on metabolism independently. Most parasitize certain arthropods; a few cause diseases transmitted to humans by arthropods.

G. Chlamydias are obligate intracellular parasites that differ from rickettsias in that they are spherical rather than rod-shaped and they do not depend upon arthropod vectors for transmission. Trachoma and psittacosis are among the diseases caused by chlamydias.

VIII. Ecologically, bacteria are important as decomposers; some species fix nitrogen.

IX. Bacteria are useful in the production of many foods including butter, many cheeses, vinegar, pickles, and some alcoholic beverages. On the other hand, bacteria can cause food spoilage and food poisoning.

X. Bacteria cause a variety of plant diseases including galls and cankers.

XI. Many types of bacteria are part of the normal microbial community on the human skin, in the digestive tract, and in other regions of the body. Some of these bacteria are opportunists that take advantage of their host's lowered resistance or of changes in the microbe community that permit them to multiply and cause diseases.

POST-TEST

1. All of the prokaryotes are assigned to kingdom ______________.
2. Prokaryotes lack a ______________ ______________ and other membranous ______________ characteristic of eukaryotic cells.
3. Ecologically, the cyanobacteria function as ______________.
4. Peptidoglycan is found in the bacterial ______________ ______________.
5. Bacteria that absorb and retain crystal violet stain are known as ______________ ______________ bacteria.
6. In some bacterial cells the cell membrane is elaborately folded inward to form a ______________.
7. ______________ are small molecules of DNA that replicate independently of the chromosome.
8. Hairlike organelles that help some gram-negative bacteria to attach to cells are called ______________.
9. The majority of heterotrophic bacteria are free-living ______________.
10. ______________ ______________ are bacteria that can carry on energy-yielding metabolism anaerobically when necessary.
11. In ______________, two cells of different mating types transfer genetic material from one to the other.
12. The three main shapes of bacterial cells are ______________, ______________, and ______________.
13. Spherical bacteria are referred to as ______________; rod-shaped bacteria, as ______________; and spiral bacteria, as ______________.
14. Species of ______________ (a group of anaerobic gram-positive bacteria) cause tetanus and botulism.
15. The pathogen that causes syphilis is a ______________.
16. Bacteria known as ______________ are sometimes referred to as energy parasites.
17. ______________ species of bacteria may become pathogenic only when the host's defenses are down.
18. ______________ ______________ is sometimes referred to as plant cancer.
19. Many gram-positive bacteria secrete ______________ into the surrounding medium. (Some of these are powerful toxins.)
20. Mycoplasmas are resistant to penicillin and some other antibiotics because they lack a ______________ ______________.

REVIEW QUESTIONS

1. Imagine that you discover a new microorganism. After careful study you determine that it should be classified in the kingdom Monera, with the cyanobacteria. What characteristics might lead you to each classification?
2. Contrast a typical cyanobacterium with a eubacterium.
3. How does bacterial photosynthesis differ from photosynthesis carried on by a plant?
4. How do the cyanobacteria sometimes contribute to fish kills?
5. Contrast the cell wall of a gram-positive bacterium with that of a gram-negative bacterium. Cite consequences of these differences.
6. Identify each of the following and tell why it is important:

a. peptidoglycan
b. mesosome
c. plasmid
d. pili
e. phycocyanin

7. Describe the process of bacterial conjugation, and compare it with transduction.
8. Give the distinguishing characteristics of each of the following parts:
   a. spirochetes
   b. actinomycetes
   c. rickettsias
   d. mycoplasmas
9. Identify three human diseases caused by members of the eubacteria, and identify at least one disease caused by each of the following:
   a. spirochetes
   b. chlamydias
   c. mycoplasmas
10. In what ways are bacteria important ecologically?
11. How are bacteria used in the production of certain foods? Give examples.
12. What types of bacteria cause food poisoning? How can food poisoning be prevented?
13. Why is crown gall sometimes referred to as plant cancer?
14. Under what circumstances do members of the body's normal microbial community cause disease?

20
Kingdom Protista

OUTLINE

LEARNING OBJECTIVES

After you have read this chapter you should be able to:

1. Describe the distinctive features of the protozoans with emphasis upon the specialized features of the protozoan cell.
2. Compare the phyla of Protozoa, giving at least one example of each phylum.
3. Describe the phyla of the algal protists, giving at least one example of the phylum and its typical features in each case.
4. Summarize proposed relationships among protists.

Kingdom Protista consists of unicellular or colonial eukaryotes, some of which are autotrophic and some heterotrophic. Protists are primarily aquatic, living in fresh or salt water; some inhabit the oceans, and others live in ponds, small puddles, or damp soils. The heterotrophs exhibit a great variety of lifestyles, some subsisting on bacteria, some on other protists (Fig. 20–1), and some living as parasites upon or within multicellular organisms.

A colonial protist can be distinguished from a multicellular animal or plant. The protist cells are similar to one another and none is specialized solely for feeding. Most of the individuals in a population of protists are produced by simple cell division of the parent, although sexual reproduction by the mating of two individuals does occur. Some protists are animal-like and are classified as protozoans. Those that carry on photosynthesis are referred to as algal or plantlike protists.

As we will see in this chapter, even the unicellular protists are not functionally simple organisms, for the protist body contains an array of specialized organelles capable of carrying out tasks assigned to the organs of complex multicellular organisms (Fig. 20–2).

Protozoa

The 25,000 or so species of animal-like protists were formerly classified as the phylum Protozoa within the animal kingdom. Today **Protozoa** is considered as a subkingdom of Protista and can be divided into four phyla that differ mainly in means of locomotion:

Mastigophora. Mastigophorans propel themselves by means of one or more long, whiplike flagella.

Sarcodina. Sarcodinians move (and also feed) by forming pseudopods.

Ciliata. Ciliates are characterized by the presence of many short hairlike cilia that beat in coordinated fashion and move the animal along.

Sporozoa. Sporozoans are parasites that lack locomotor structures and that reproduce by an unusual process of multiple fission.

The **contractile vacuole** is a prominent structure in freshwater protozoans that pumps excess water out of the cell. In freshwater protozoans the cytoplasm has a higher concentration of dissolved materials—salts, sugars, and organic acids—than in the surrounding water. Water tends to pass into the cytoplasm by osmosis.

Figure 20–1 A heterotrophic protist, *Didinium,* devouring another protist, *Paramecium* (also a heterotroph). (*a*) Scanning electron micrograph. (*b*) The same activity viewed under a light microscope. ((*a*), from H. Wessenberg and G. A. Antipa, J. Protozool., 1970, 17:250. Photo courtesy of G. A. Antipa; (*b*), E. R. Degginger.)

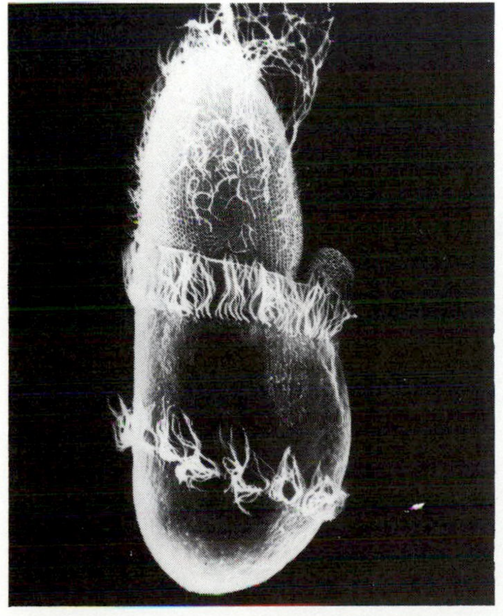

(*a*)

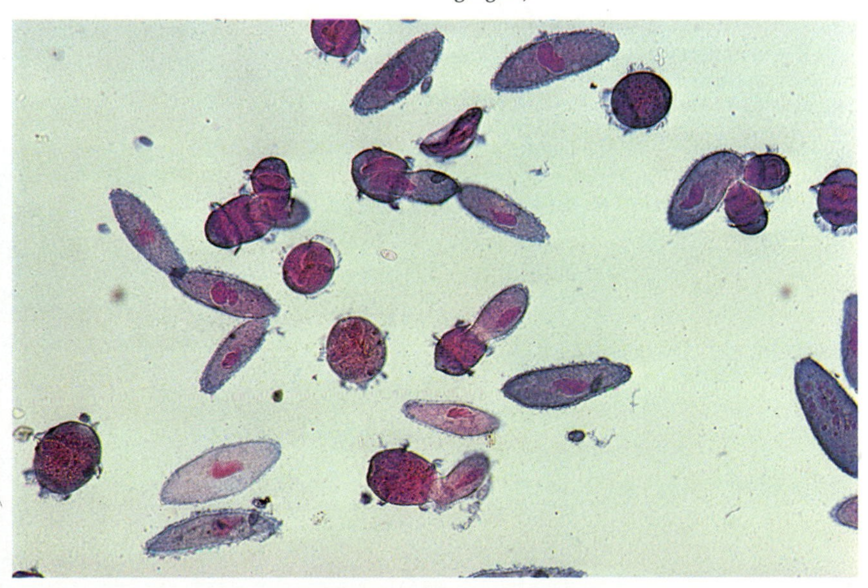

(*b*)

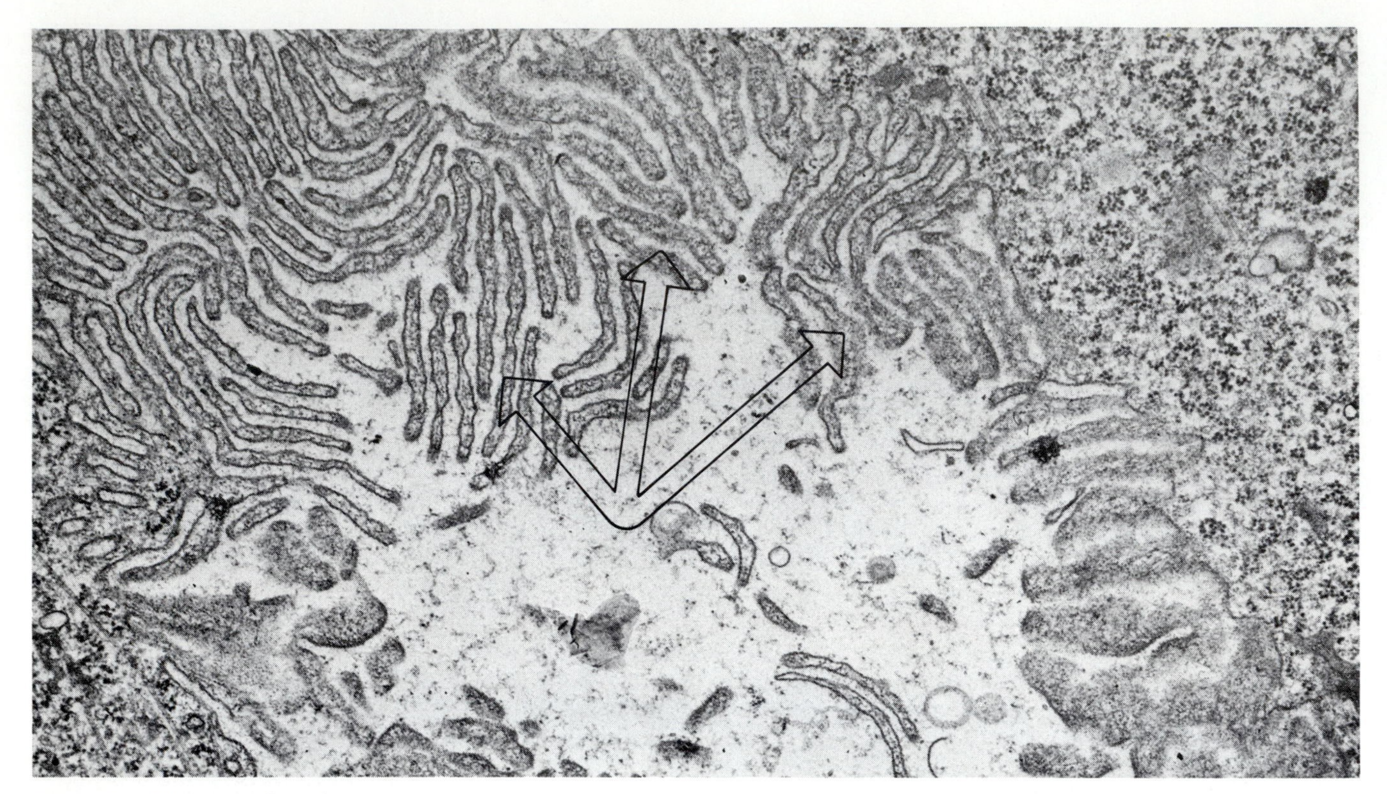

Figure 20–2 The organelles of some protists are astonishingly sophisticated. In the ciliate *Phacodinium metchnikoffi*, the walls of the digestive vacuole are thrown into a multitude of fingerlike projections that increase its absorptive area in much the same way that the villi and microvilli of the vertebrate small intestine increase its absorptive area. (From Didier, P., and Drogesco, J.: Organisation du cortex des vacuoles digestives de *Phacodinium metchnikoffi* (cilie heterotriche). *Transactions of the American Microscopical Society* 98:385–392, 1979.)

Without a pump to remove the excess water, the amoeba or other protozoan would swell and burst, just as human blood cells do when they are placed in distilled water. In contrast, most marine protozoans do not have or need a contractile vacuole, since the concentration of salts in the sea water is about the same as that in their cytoplasm.

PHYLUM MASTIGOPHORA

Phylum Mastigophora, the flagellates, includes more than half of all living species of Protozoa. Flagellates have spherical or elongate bodies, a single central nucleus, and one to many slender whiplike flagella at the anterior end that enable them to move (Fig. 20–3). Some flagellates are amoeboid and engulf food by forming pseudopods; others have a definite mouth, gullet, and other specialized organelles for

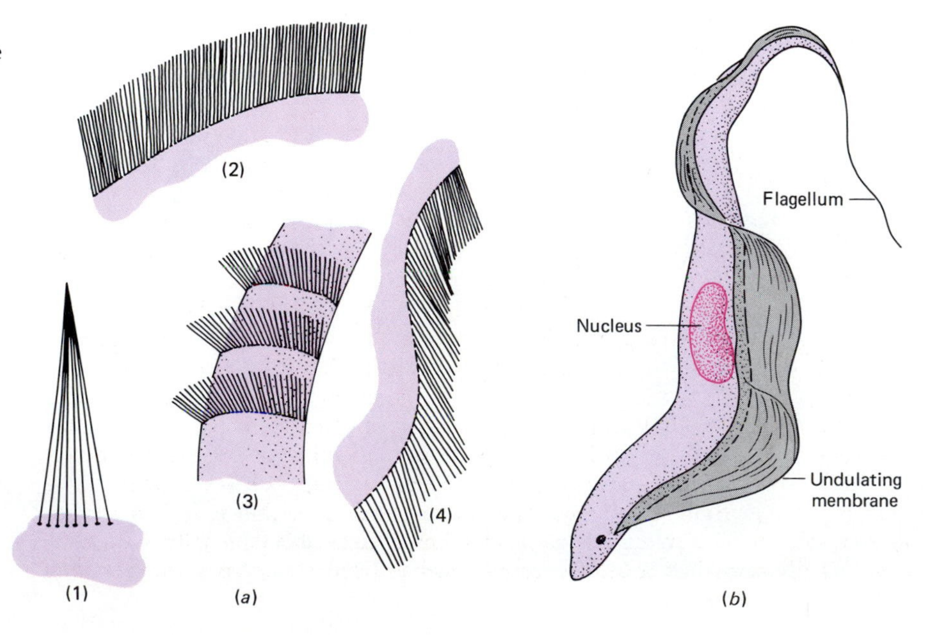

Figure 20–3 Two principal ways by which protozoans propel themselves are by means of cilia and flagella. (*a*) Various arrangements of cilia. (*b*) *Trypanosoma gambiense*, the cause of African sleeping sickness, propels itself by means of its single flagellum.

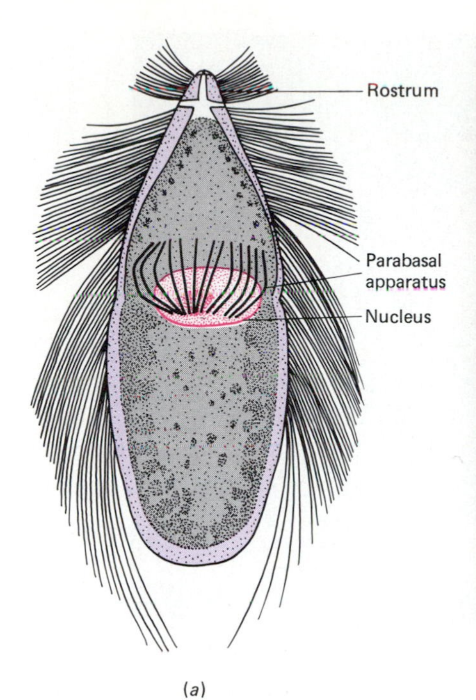

(a)

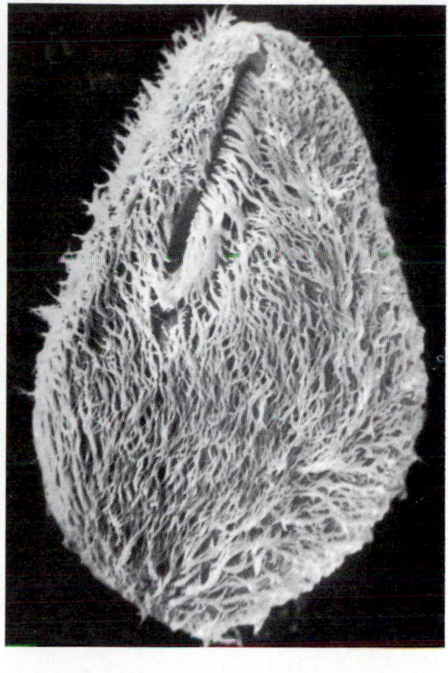

(b)

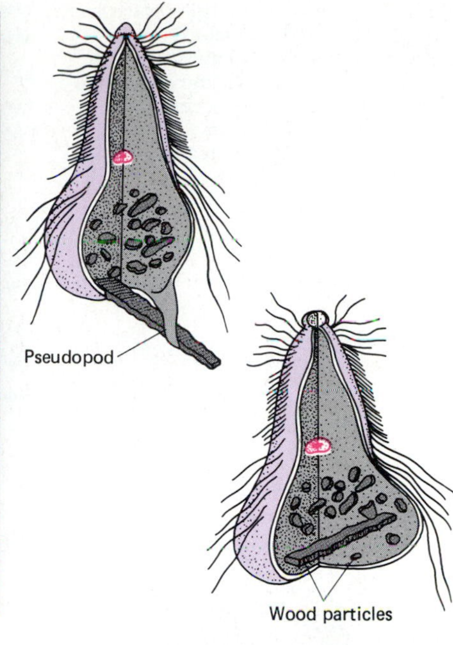

(c)

Figure 20–4 Representative mastigophorans (flagellates). (*a*) *Trichonympha agilis,* a symbiotic flagellate found in the gut of the termite. (*b*) A scanning electron micrograph of *Nyctotherus ovalis* from the hindgut of the cockroach *Blaberus discoidalis.* (*c*) *Trichonympha* engulfs wood particles by means of pseudopodia extended from the lower part of the animal.

processing food. Some species are completely autotrophic, others largely so, and still others heterotrophic. Among the last, some are saprobes that absorb their nutrients from dead or decomposing organic matter almost like fungi, while others are carnivores or detritus feeders much like animals. The mastigophorans with the largest number of flagella and the most specialized bodies are the ones living in the intestines of termites (Fig. 20–4).

Flagellates move rapidly, pulling themselves forward by lashing one or more flagella located at the anterior end. Each **flagellum** is a long, supple filament containing an axial fiber, composed of 11 tubules, a sheath of nine double tubules surrounding two single tubules in the center (see Fig. 4–18). In some flagellates, the flagellum forms, with the plasma membrane, a finlike **undulating membrane** (Fig. 20–3). A *Euglena* organism swimming with its single functional flagellum may be likened to a one-armed human trying to swim: At each stroke the flagellum bends toward the side bearing the pigment spot, not in a simple backward lash but obliquely toward the long axis of the organism. The body not only turns toward one side but also rotates a bit. Successive lashes of the flagellum thus move the organism forward in a spiral path with the pigment spot facing the outside of the spiral.

Because most flagellates are small, they are difficult to study, but a few, such as members of the genus ***Euglena*** (Fig. 20–5), are large and can be used to study representative features. The euglenids are unicellular organisms commonly found in fresh water but are also present in soil, brackish water, or even salt water.

As in other flagellates, the body of *Euglena* is covered with a delicate **pellicle** with spiral thickenings. Unlike the cellulose cell wall of plants and algae, the pellicle incorporates the cell's plasma membrane, which is outside a system of spirally arranged rigid strips made of protein. Beneath the pellicle is a layer of **contractile fibrils** that permit the organism to change its shape.

Scattered inside the cytoplasm are green chloroplasts that strike a somewhat jarring note in these otherwise animal-like organisms. Also present are transparent, colorless **paramylum bodies** containing stored polysaccharides. Euglenids have an eye spot that consists of a tiny light sensor and a patch of the red pigment astaxanthin, found elsewhere only in members of Crustacea such as lobsters and crabs. The shading of the light sensor by the pigment spot enables the organism to determine the direction of the source of light. The pigmented eyespot appears to be responsible for governing phototaxis in this organism, so that it swims toward moderate sources of illumination. Reproduction in euglenids is usually asexual by simple cell division, but sexual reproduction has been observed in at least one genus.

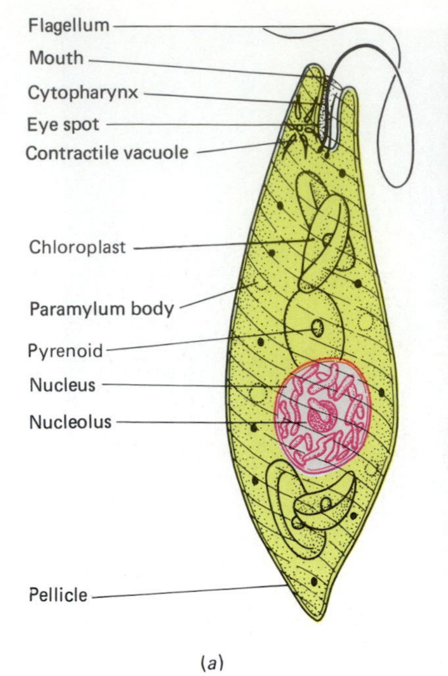

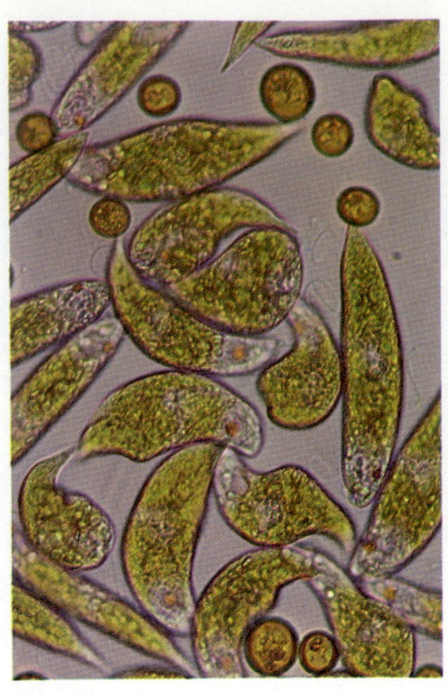

Figure 20–5 *Euglena*. (*a*) This protist has both plantlike and animal-like traits. It has at various times been classified in the plant kingdom (with the algae) and in the animal kingdom (when protozoans were considered to be animals). (*b*) Living euglenoids; note eyespots. ((*b*), Tom Adams.)

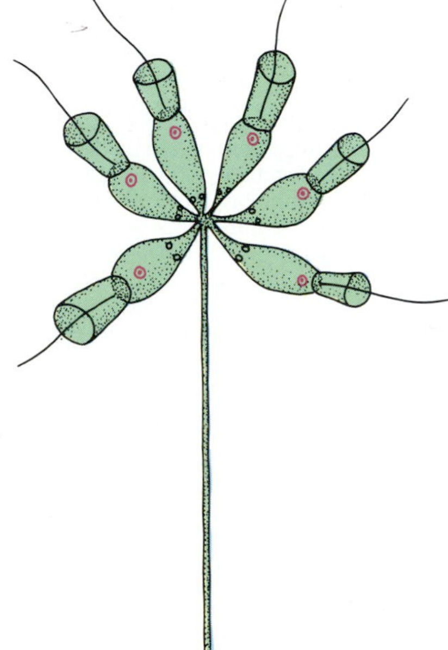

Figure 20–6 Colony of a typical choanoflagellate, *Codosiga botrytis*.

Although most euglenids contain chlorophyll, they are apparently not entirely autotrophic. They will not survive in a medium containing only inorganic salts but will flourish if small amounts of amino acids are added. When no light is available, *Euglena* is capable of functioning as a heterotroph. The euglenids, with their distinctive mixture of plant and animal characteristics, have often been cited as living examples of what early living things might have been like when the early autotrophs first evolved from heterotrophs before true plants and animals had evolved.

The more animal-like flagellates are small and rather uncommon. Of special interest, because of their resemblance to sponges, are the **choanoflagellates** (Fig. 20–6). These sedentary flagellates are attached to the substrate by a stalk, and their single flagellum is surrounded by a delicate collar of cytoplasm. As we will see in Chapter 24 the resemblance of certain cells of the sponges to these choanoflagellates suggests the choanoflagellates as possible sponge ancestors. There are some parasitic flagellates of medical importance, such as the trypanosomes responsible for African sleeping sickness. This disease is transmitted from one human to another by the bite of a tsetse fly which itself is parasitized by the trypanosomes (Fig. 20–7).

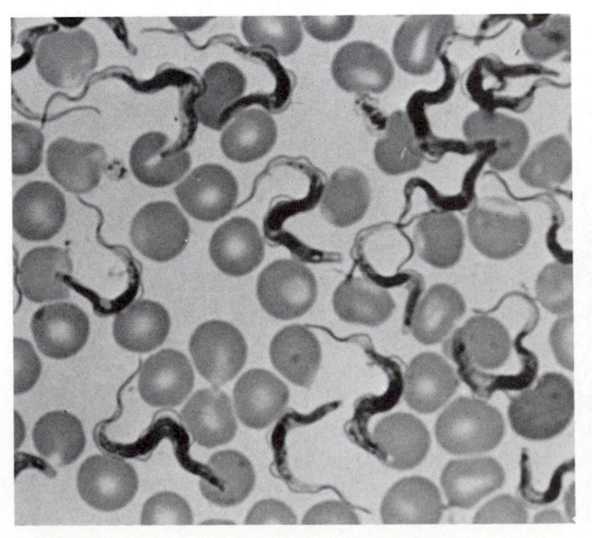

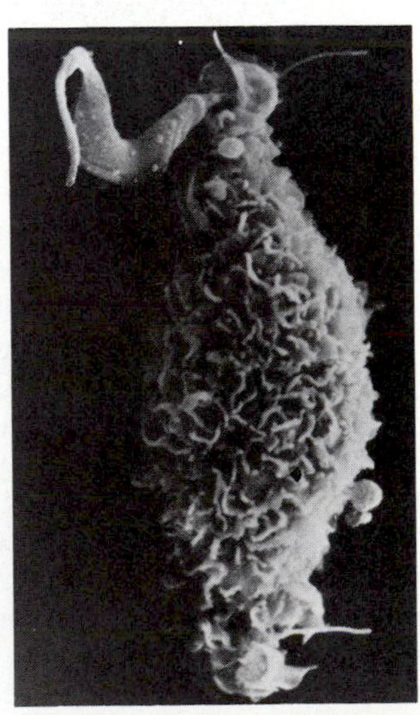

Figure 20–7 Parasitic flagellates. (*a*) Rat blood infested with a parasitic flagellate, *Trypanosoma brucei*, visible as dark wavy bodies among the red blood cells. Similar trypanosomes infest the central nervous system of human beings, causing sleeping sickness. (*b*) Scanning electron micrograph of a trypanosome, *Trypanosoma cruzi*, being engulfed by a large ruffled macrophage. ((*a*), from "Trypanosomiasis: an Approach to Chemotherapy by the Inhibition of Carbohydrate Catabolism," Clarkson, A. B., Jr., and Brohn, F. H.: *Science* 194:204–206, 1976. Copyright 1976 by the American Association for the Advancement of Science. (*b*), from Reed, S. G., Douglass, T. G., and Speer, C. A.: Surface interactions between macrophages and *Trypanosoma cruzi*. *Am. J. Trop. Med. Hyg*. 31:723–729, 1982. Copyright 1982 by the American Society of Tropical Medicine and Hygiene.)

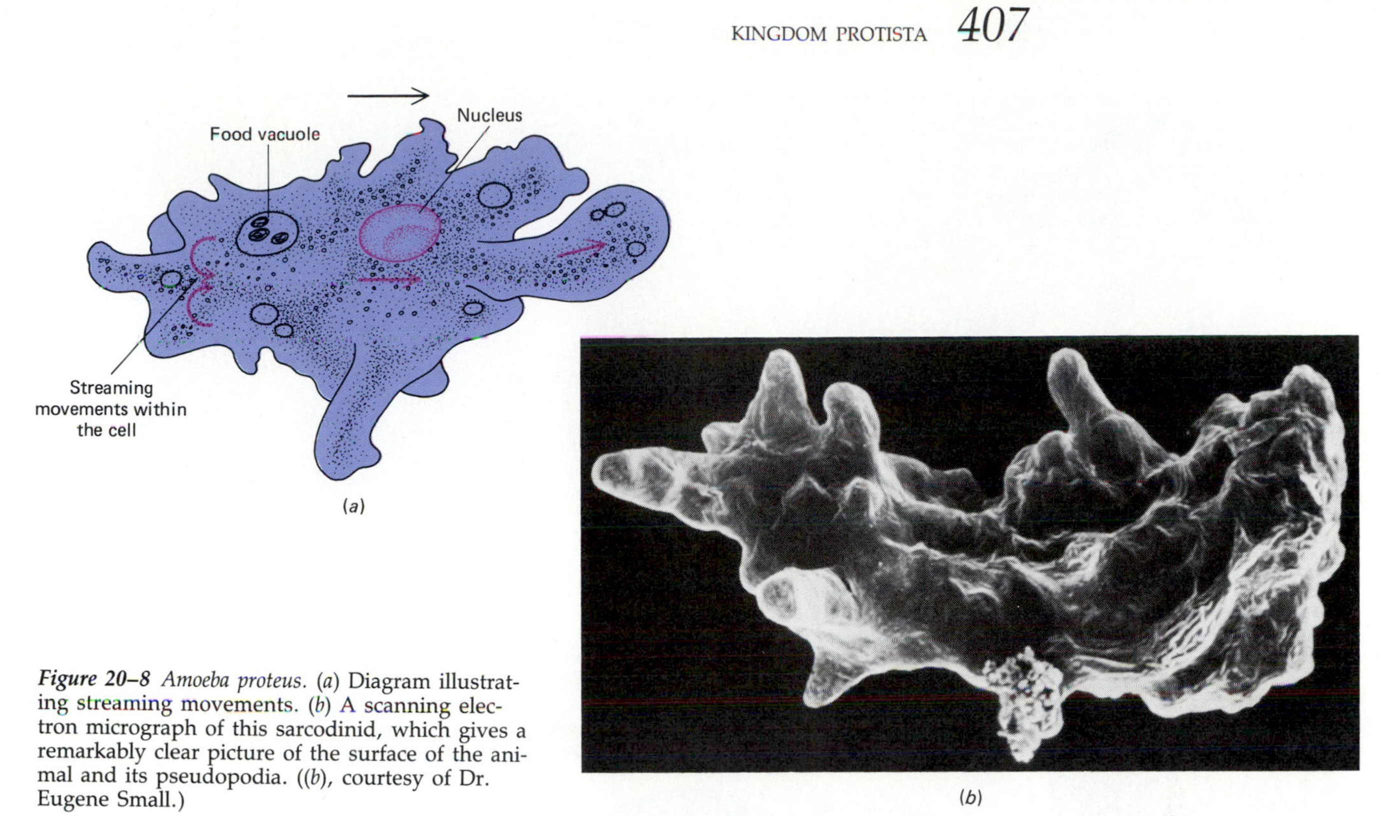

Figure 20–8 *Amoeba proteus.* (*a*) Diagram illustrating streaming movements. (*b*) A scanning electron micrograph of this sarcodinid, which gives a remarkably clear picture of the surface of the animal and its pseudopodia. ((*b*), courtesy of Dr. Eugene Small.)

PHYLUM SARCODINA

Members of the phylum Sarcodina, unlike other protozoans, have no definite body shape (Fig. 20–8). Their single cells change form as they move. The nucleus, contractile vacuole, and food vacuoles are shifted about within the cell as the animal moves.

An **amoeba**, a typical sarcodinian, moves by pushing out temporary cytoplasmic projections called **pseudopods** (false feet) from the surface of the body (Fig. 20–9). More cytoplasm flows into the pseudopods, enlarging them until all the cytoplasm has entered and the animal as a whole has moved. According to the view most widely accepted at present, contractile microfilaments at the "rear" and

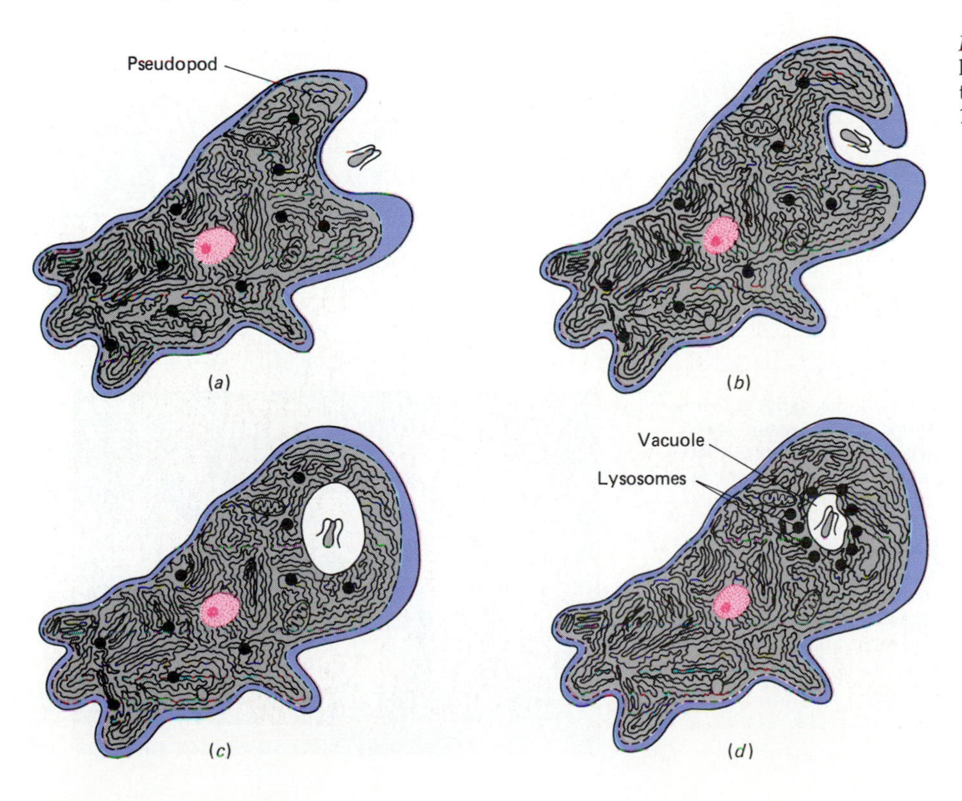

Figure 20–9 Ameboid movement. The lysosomes release digestive enzymes that disintegrate the prey. (See also Fig. 1–15(*a*).)

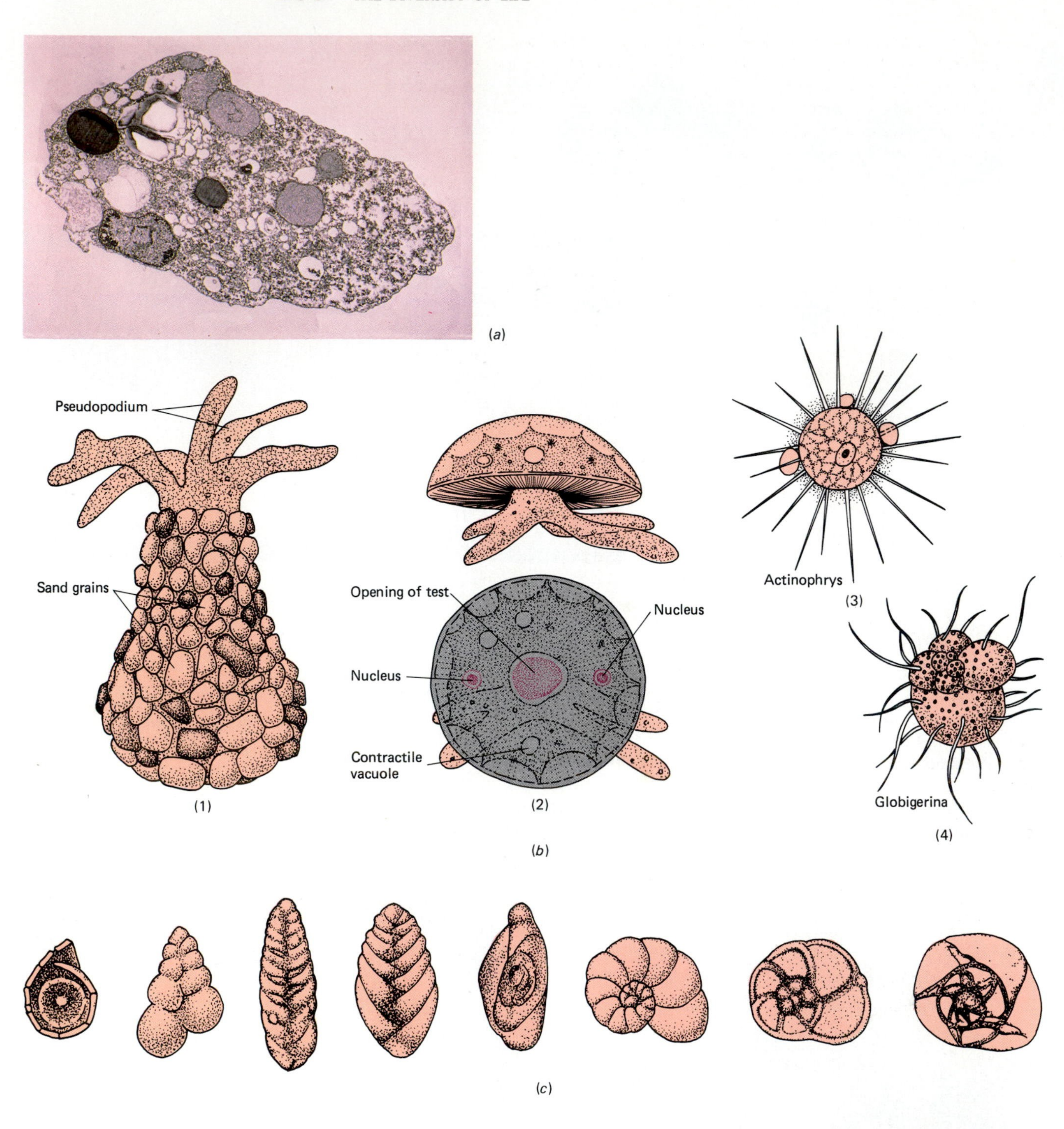

Figure 20–10 Some examples of the phylum Sarcodina. (*a*) A transmission electron micrograph of *Entamoeba histolytica.* This parasitic sarcodinian causes amoebic dysentery. The dark bodies are recently ingested red blood cells (×550); (*b*) *1. Difflugia* is a free-living amoeba that builds a protective layer around itself by cementing together grains of sand. *2. Arcella* has a turtle-like shell. *3. Actinophrys* is a member of the order Heliozoa that lives in fresh water. *4. Globigerina* is a marine form, a member of the order Foraminifera. (*c*) Different types of Foraminifera. (*d*) The shell of a foraminiferan, *Poneroplia perfusus.* ((*a*), Science Photo Library, Photo Researchers, Inc. Markell, Edward K., and Voge, Marietta. (*d*), Eric V. Grave, Photo Researchers, Inc.)

"sides" of the amoeba form a gel-like outer layer of cytoplasm that squeezes the more fluid inner cytoplasm; the flow of this fluid cytoplasm produces the pseudopods, which may be further shaped by other cytoskeletal components.

Pseudopods are used to capture food, two or more of them moving out to surround and engulf a bit of debris, another protist, or even a small animal. The food that has been engulfed is surrounded by a **food vacuole** and digested by enzymes added by lysosomes (Fig. 20–9). The digested materials are absorbed from the food vacuole, which gradually shrinks as it becomes empty. Any indigestible remnants are expelled from the body and left behind as the amoeba moves along.

The parasitic members of the phylum Sarcodina (Fig. 20–10) include the species causing amoebic dysentery in humans. Certain free-living species secrete shells around the body or cement sand grains together into a protective layer around their cells. The ocean contains untold trillions of amoeba-like protozoans, the **foraminiferans**, which secrete chalky, many-chambered shells with pores through which pseudopods can be extended. The dead foraminiferans sink to the bottom of the ocean, where their shells form a gray mud that is gradually transformed into chalk. Other amoeboid protozoans, the **radiolarians**, secrete elaborate and beautiful skeletons made of silica (Fig. 20–10); these skeletons become mud on the ocean floor and eventually are compressed and converted into siliceous rock.

PHYLUM CILIATA

Members of the phylum Ciliata, typified by ***Paramecium***, have a definite, permanent shape due to the presence of a sturdy flexible outer pellicle (Fig. 20–11). The surface of the cell is covered by several thousand fine cilia that extend through pores in the covering and move the animal along. The cilia beat with an oblique stroke so that the animal revolves as it swims; the coordination of the ciliary beating is so precise that the animal not only can go forward but can back up and turn around. A system of **neurofibrils** connects the rows of **basal bodies** at the inner end of each cilium; this feature may be involved in coordination of ciliary movements, though there is evidence that the plasma membrane is actually responsible. Near the surface of the cells of ciliates are many small **trichocysts**, organelles that can discharge filaments believed to aid the organism in trapping and holding its prey.

Ciliates differ from other protozoa in having at least two nuclei per cell, typically a **micronucleus** that functions in sexual reproduction, and a larger **macronu-**

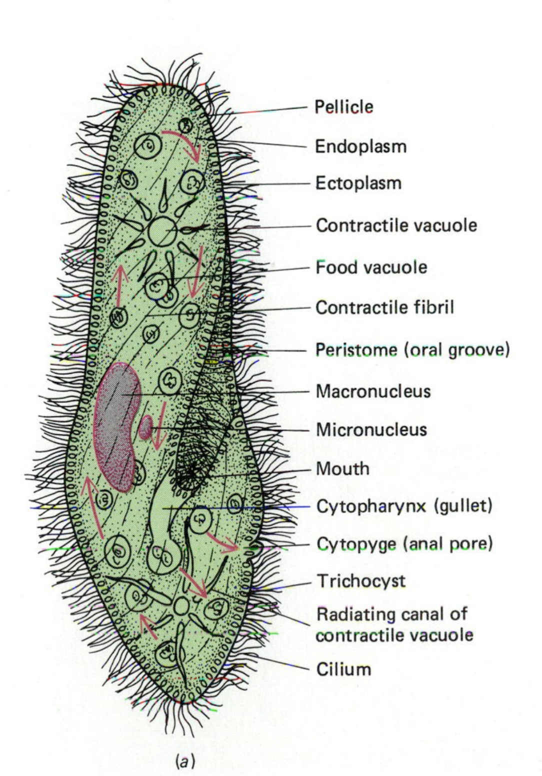

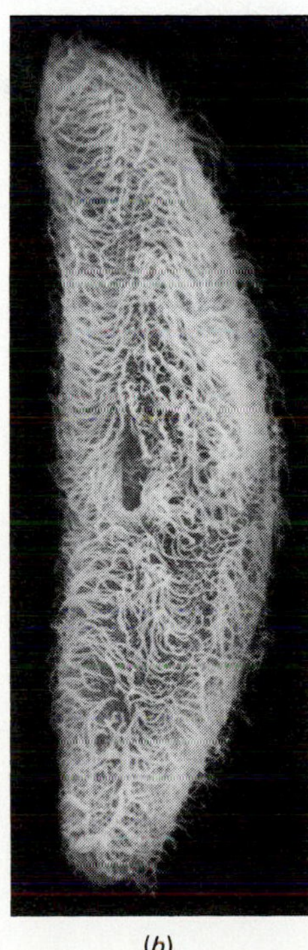

Figure 20–11 Paramecia. (*a*) A diagram of *Paramecium caudatum*. (*b*) A scanning electron micrograph of *Paramecium multimicronucleatum*. ((*b*), courtesy of Dr. Eugene Small.)

cleus that controls cell metabolism and growth. Both nuclei divide at each mitosis, but during sexual reproduction, the macronucleus disintegrates and the micronucleus gives rise to both nuclei of the offspring (Fig. 20–12). The macronucleus appears to be a compound structure formed by the amalgamation of many sets of chromosomes; it probably serves as a kind of genetic amplifier, containing as it does multiple copies of all the micronuclear genes. Ciliates are often quite large, highly differentiated cells (Figs. 20–13 and 20–14), and multiple copies of each gene may be required to control these cells. (They are comparable in complexity to some of

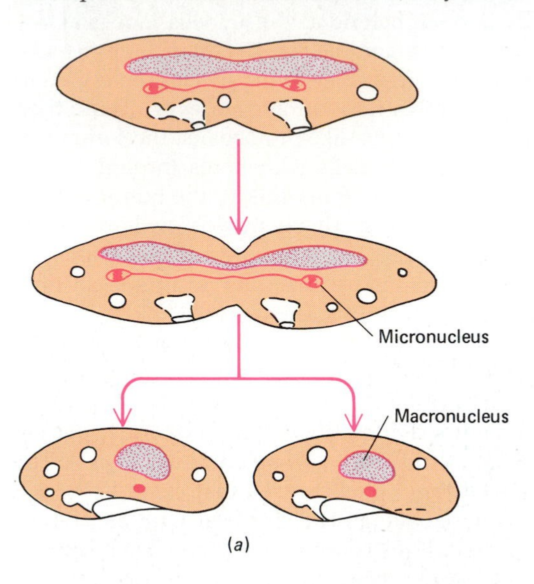

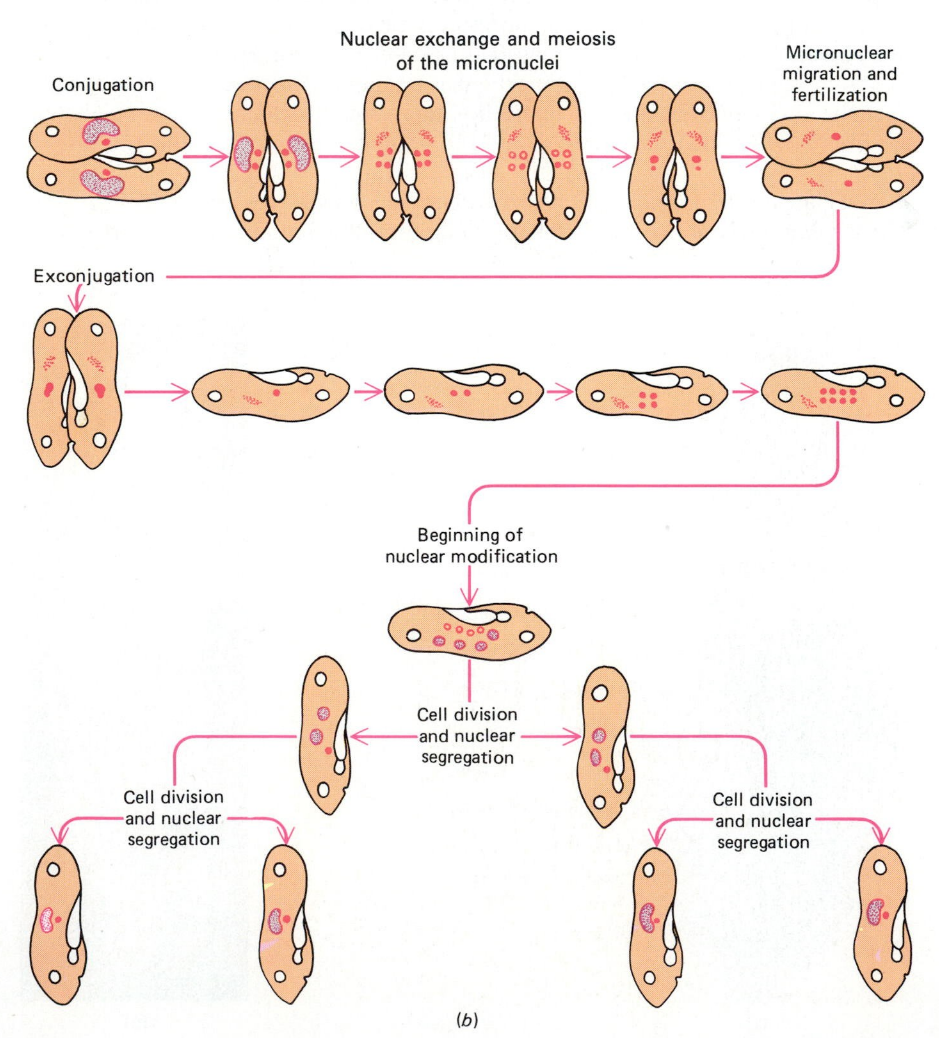

Figure 20–12 Different modes of reproduction in *Paramecium caudatum*. (*a*) Transverse binary fission, a form of asexual reproduction. The micronucleus divides by means of a mitotic spindle that is confined within the nuclear membrane; the macronucleus is simply pulled apart. (*b*) Conjugation. Follow the arrows: After the conjugants separate, only one of the cells is followed; however, a total of eight new individuals results ultimately from this one conjugation. ((*b*), courtesy of W. H. Johnson, L. E. Delanney, E. C. Williams, and T. A. Cole.)

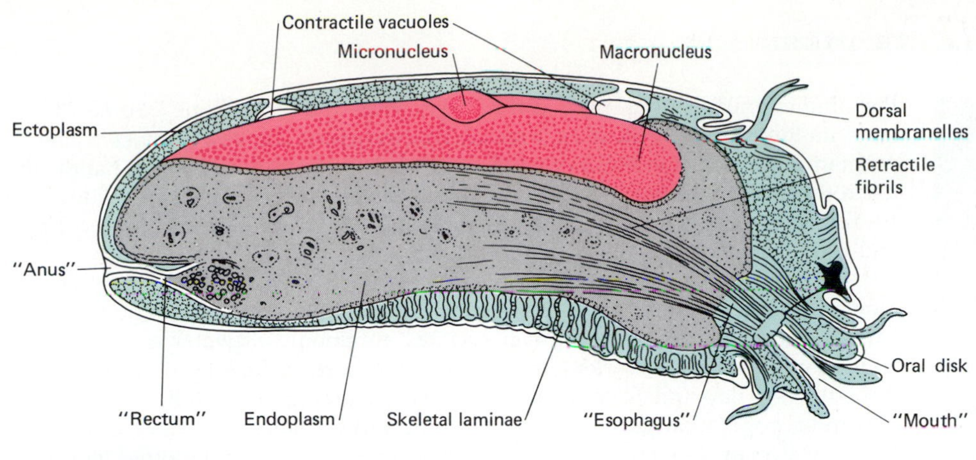

Figure 20–13 Even though composed of a single cell, *Epidininium ecaudatum* possesses a strikingly complex internal organization. *Epidininium* is a commensal in the rumen (stomach) of cattle.

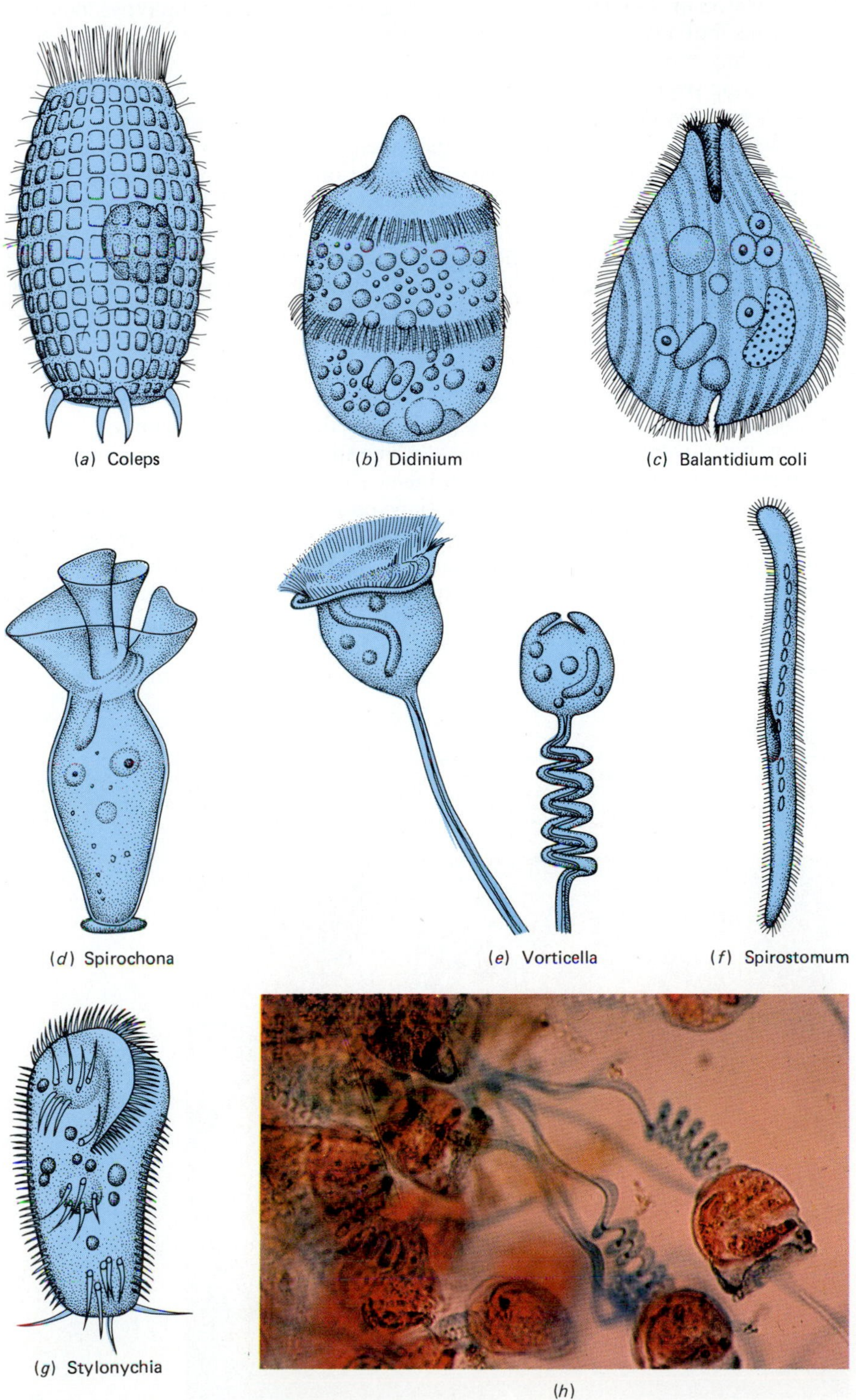

Figure 20–14 Ciliates. (*a*)–(*g*) Various species. (*h*) A colony of *Vorticella*, a stalked ciliate. (Carolina Biological Supply Company.)

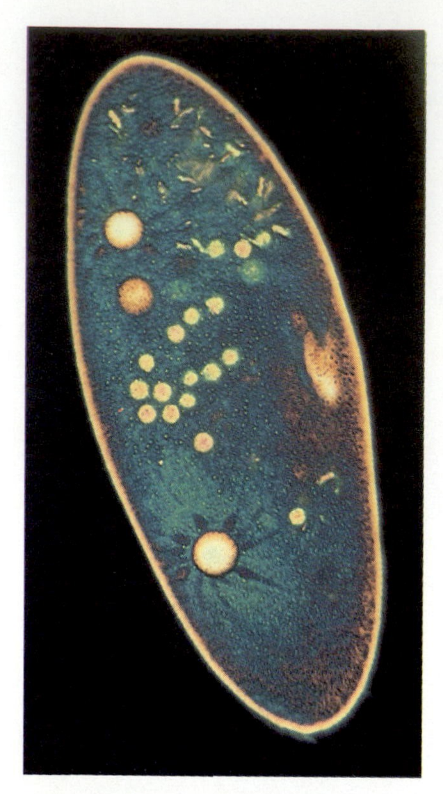

Figure 20–15 A micrograph of a stained *Paramecium caudatum*. Note the two large contractile vacuoles. (Eric V. Grave, Photo Researchers, Inc.)

the simpler animals, and a few can be seen without a microscope.) Some ciliates have multiple nuclei instead of a single macronucleus. Since the master genetic program is contained in the micronucleus, there is no adaptive reason to handle the macronucleus by the intricate process of mitosis when the organism divides. Accordingly, the macronucleus is simply pulled in half during asexual reproduction and roughly divided between the two daughter cells.

Paramecia have two contractile vacuoles (Fig. 20–15) that together can remove a volume of water equal to the total volume of the organism's body within half an hour. (In contrast, a human excretes an amount of water equal to body volume in about three weeks.) Well-fed paramecia reproduce by division two or three times a day and have often been used as subjects for studies of the laws governing population growth. Paramecia and other ciliates may have more than two, and as many as eight, "sexes" or mating types. All the mating types look alike to humans, but a *Paramecium* of one type will mate only with an individual of some other mating type.

During sexual reproduction, two individuals of different sexes conjugate and press their oral surfaces together (Fig. 20–16). Within each individual the macronucleus disintegrates and the micronucleus undergoes meiosis to form four daughter nuclei. Three of these four degenerate, leaving one viable haploid nucleus. This nucleus then divides once mitotically, and one of the two identical haploid nuclei remains within the cell. The other nucleus crosses through the oral region into the other organism and fuses with the haploid nucleus that is already there. Thus each conjugation yields two identical fertilizations, and the two new diploid nuclei are exact duplicates. This leads to two new cells, each an identical twin, but both of which are genetically different from the "parent" or, more properly, the **preconjugant** cells.

Not all ciliates swim about habitually. Some forms are stalked, for example, and others, such as *Stentor* (Fig. 20–17), while capable of some swimming, are more likely to remain attached to the substrate at one spot. Strong cilia set up currents in the surrounding fluid to bring food to them.

The suctorians, like other ciliates, have both a macronucleus and a micronucleus. Young individuals have cilia and swim about, but the adults are sedentary and, like some other ciliates, have stalks by which they are attached to the substrate. The body bears a group of delicate cytoplasmic tentacles (Fig. 20–18), some of which are pointed to pierce their prey and to suck out its cellular contents, and others that are tipped with rounded adhesive knobs to catch and hold their prey. The tentacles also secrete a toxic material that may paralyze the prey.

Figure 20–16 Apparent recognition behavior of two individuals of the ciliate protozoan species *Euplotes crassus* preparatory to conjugation. The organisms embrace one another with fused bundles of cilia known as cirri. (From Luporini, P., and Dallai, R.: Sexual interaction of *Euplotes crassus:* Differentiation of cellular surfaces in cell-to-cell union. *Developmental Biology* 77:167–177, 1980. Copyright 1980 by Academic Press, Inc.)

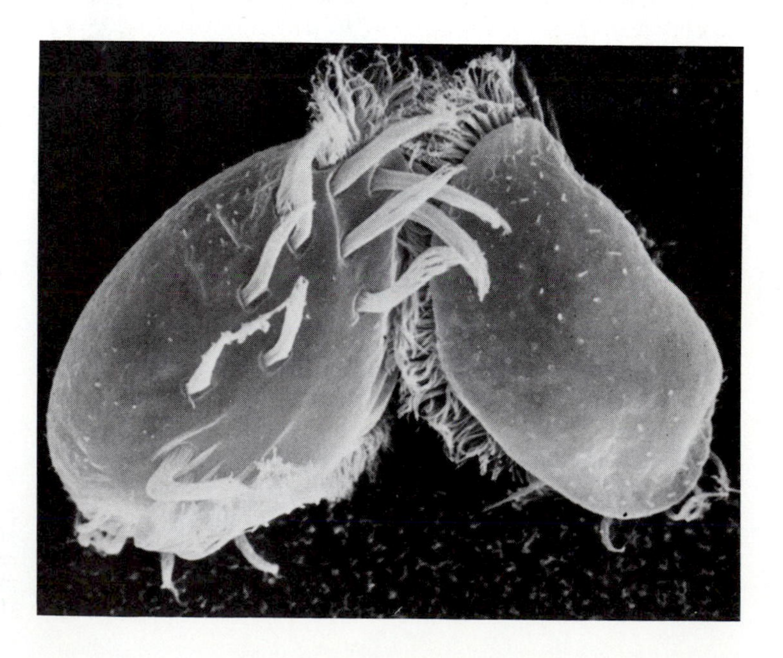

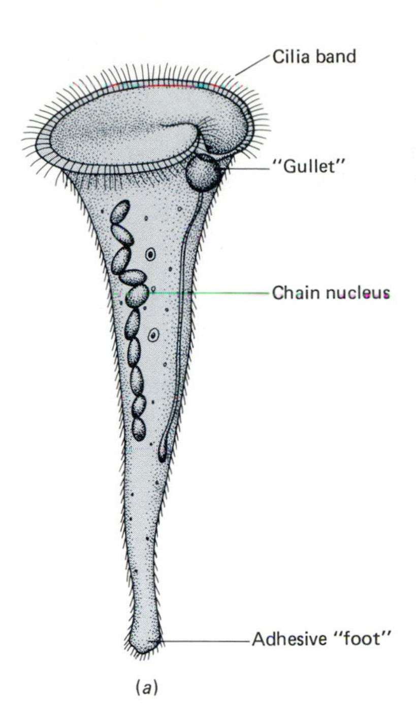

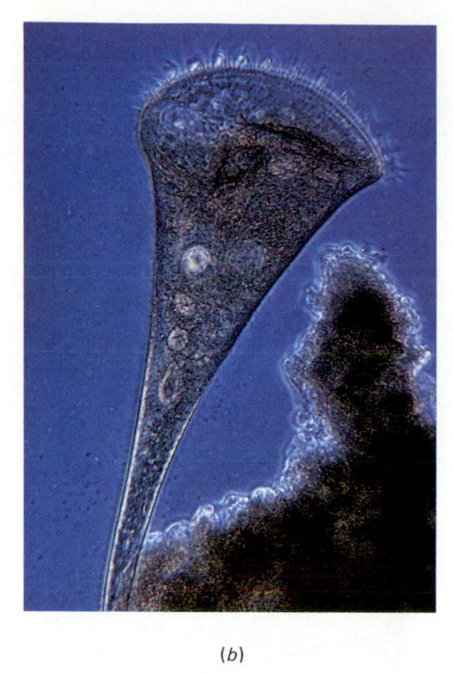

Figure 20–17 *Stentor*. (*a*) Internal anatomy. (*b*) Photomicrograph of living specimen. Note the numerous cilia that direct food particles into its "mouth." ((*b*), Eric V. Grave, Photo Researchers, Inc.)

Although adult suctorians lack cilia, they do possess basal bodies. During asexual reproduction, a bud forms on the suctorian, and the basal bodies multiply, become arranged in rows, and develop cilia. After the nucleus has divided, the bud separates from the parent and swims away. When it becomes attached to the substrate, the cilia disappear and tentacles develop.

PHYLUM SPOROZOA

The members of the phylum Sporozoa comprise a large group of parasitic protozoans, among which are the agents causing serious diseases such as coccidiosis in poultry and malaria in humans. Sporozoa have neither organelles for locomotion nor contractile vacuoles. Most sporozoans live as intracellular parasites in host cells during the growth phase of their life cycle.

The sporozoan causing malaria, *Plasmodium*, enters the human blood stream when a parasitized mosquito bites a human (Fig. 20–19). The plasmodia enter the red blood cells, and each divides into 4 to 36 new parasites, which are released when the red blood cell bursts. The released parasites, called **merozoites**, infect new red blood cells, and the process is repeated. The simultaneous bursting of millions of red cells causes a typical malarial chill followed by fever, as toxic substances are released and penetrate other organs of the body. If a second, unparasitized mosquito bites the infected person it will suck up some parasites along with its drink of blood; a complicated process of sexual reproduction then occurs within the mosquito's stomach, and new parasites develop, some of which migrate into the mosquito's salivary glands to infect the next person bitten. This process of sexual reproduction does not occur within the human. For this reason elimination of the mosquito hosts, though very difficult, would eradicate the disease. (Elimination of the human hosts would do the same, of course, but this suggestion has thus far not found much favor!)[1]

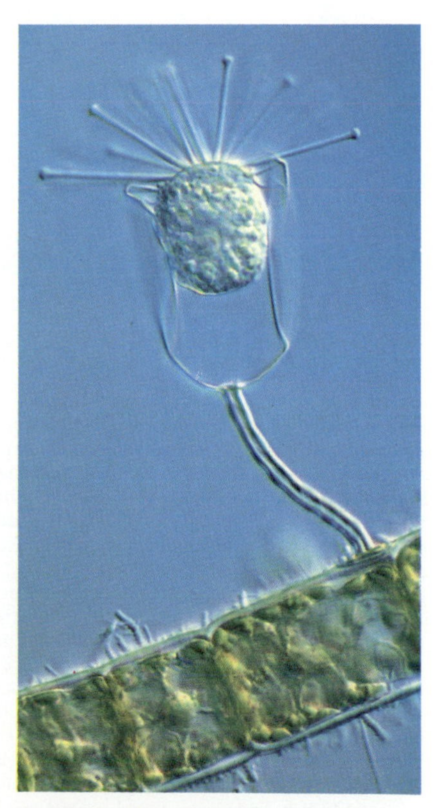

Figure 20–18 A suctorian, *Paracineta*, attached to a filamentous alga, *Spongomorpha*. Note the absence of cilia and the presence of tentacles. (Paul W. Johnson, Univ. of Rhode Island/BPS.)

The Algal Protists

The algal protists constitute some of the simple, unicellular algae of past systems of classification. However, the green algae display a remarkable range of organization, with some forms being unicellular and not too different from the flagellates

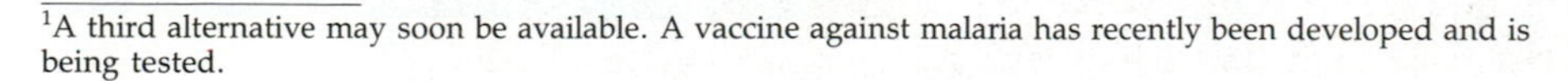

[1]A third alternative may soon be available. A vaccine against malaria has recently been developed and is being tested.

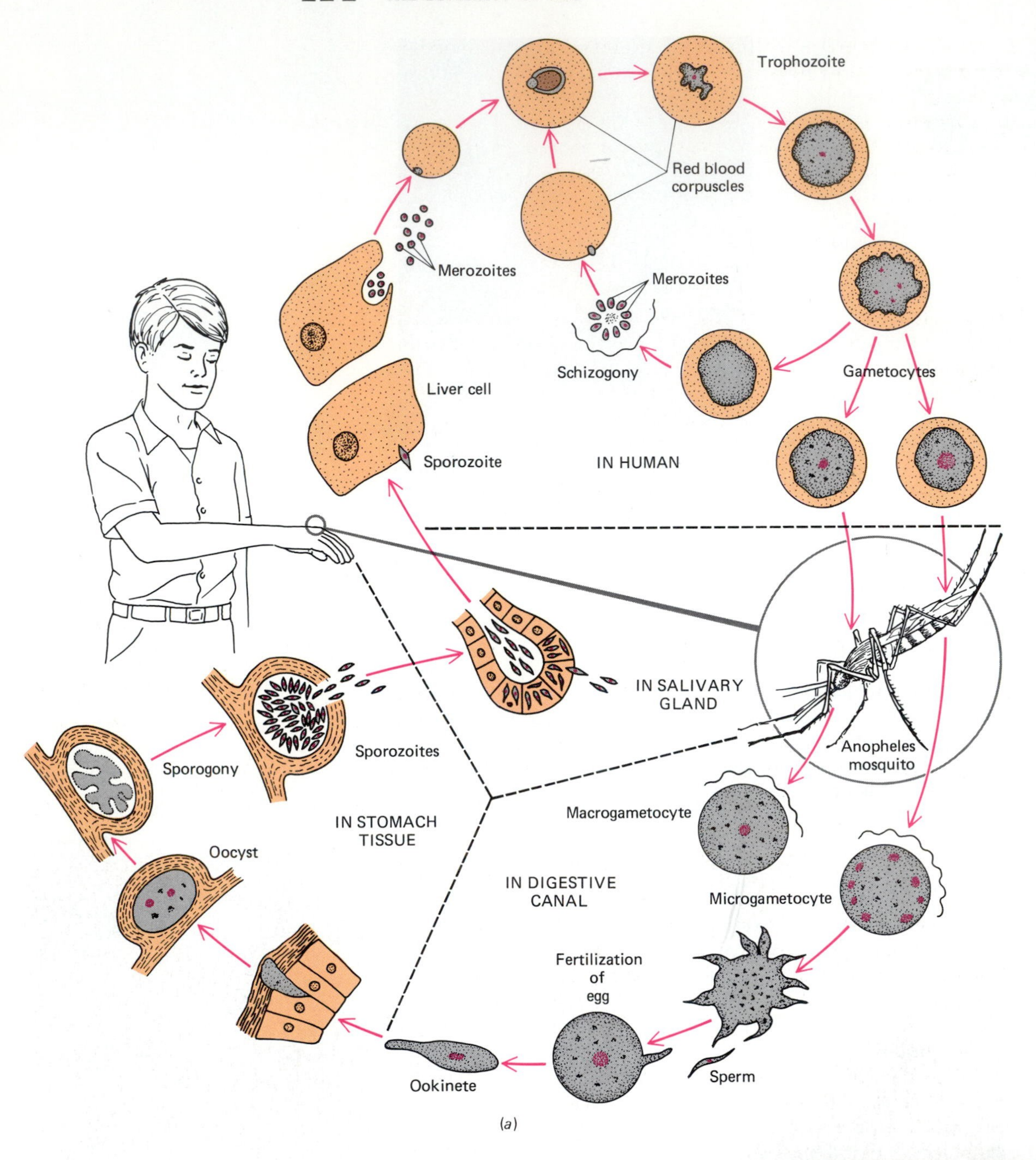

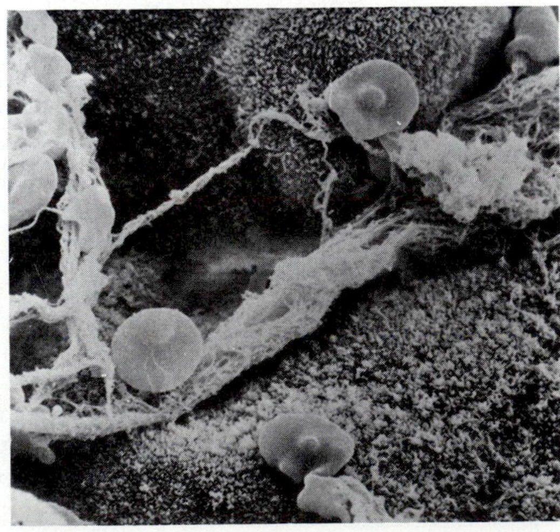

Figure 20–19 The sporozoan *Plasmodium* causes malaria in humans and other mammals. (*a*) Life cycle of *Plasmodium:* An infected mosquito bites a person and injects "spores" of *Plasmodium* into the bloodstream. These undergo sporulation and reproduce asexually within the red blood cells of the host. Periodically, the infected blood cells rupture, and the new crop of merozoites that are released then infects other red cells. The bursting of the red blood cells releases toxic substances that cause the periodic fever and chill. Some merozoites develop into gametocytes, which can infect another mosquito biting the infected person and so pass the disease on to other potential hosts. The gametocytes develop into eggs and sperm within the mosquito; these undergo sexual reproduction "in flight." The resulting zygote sporulates, producing "spores" that migrate to the mosquito's salivary glands and are ready to be injected when the mosquito bites the next person. (*b*) Scanning electron micrograph of red blood cells infected by malarial parasites. The blebs are caused by the presence of the sporozoan. (Courtesy of P. R. Walter, Y. Garin, and P. Blot. *The American Journal of Pathology* 109:330, 1982.)

and others approaching the true plants in the complexity and sophistication of their body plan and reproductive mechanisms. The five-kingdom system, by establishing the kingdom Protista, raised the problem of where to place the green algae. Most biologists have agreed to classify the green algae with the plants (Chapter 22), despite the fact that many of them are of unicellular organization. As we will see, the remaining two phyla of Protista, Chrysophyta and Pyrrophyta, have little in common. Pyrrophytes are somewhat mastigophoran-like, and chrysophytes are like nothing else in the world.

PHYLUM CHRYSOPHYTA

The phylum Chrysophyta includes a variety of diverse types, usually arranged in three classes. The first of these (Bacillariophyceae) contains the **diatoms**, which are microscopic and usually single-celled. Diatoms are found in fresh and salt water and are an important source of food for animals. The cell walls of diatoms are composed of silica and manganese rather than cellulose and are constructed in two overlapping halves that fit together like the two parts of a pillbox or Petri dish. These siliceous walls are ornamented with fine ridges, lines, and pores that are characteristic for each species (Fig. 20–20). Many of these markings are at the limit of resolution of the best light microscopes and are used as test objects to determine the quality of a lens.

Diatoms move with a slow, gliding movement apparently produced by the streaming of cytoplasm through **raphes** (grooves) on the surface of the cell wall. Diatoms store their food as the polysaccharide leucosin and as oil rather than as starch. It is thought that petroleum is derived from the oil of diatoms that lived in past geological ages. In any case, they are extremely important photosynthesizers in the present day. Probably three fourths of all the organic material synthesized in the world and a great part of the oxygen we breathe is produced by diatoms and other protists called dinoflagellates (belonging to the phylum Pyrrophyta). Diatoms are found in truly mind-boggling numbers in both freshwater and marine environments. They reproduce both asexually and sexually.

Figure 20–20 The structure of diatoms viewed through different methods of microscopy. (*a*), (*b*) Scanning electron micrographs of a diatom. Note the raphe (groove) and elaborate series of wall perforations. (*c*), (*d*) Diatoms viewed by a technique called interference optics microscopy. ((*a*) (*b*), from Hufford, T. L., and Collins, G. B.: Some morphological variations in the diatom *Cymbella cistula*. *Journal of Phycology* 8:192–195, 1972. (*c*) (*d*), Jonathan G. Izant.)

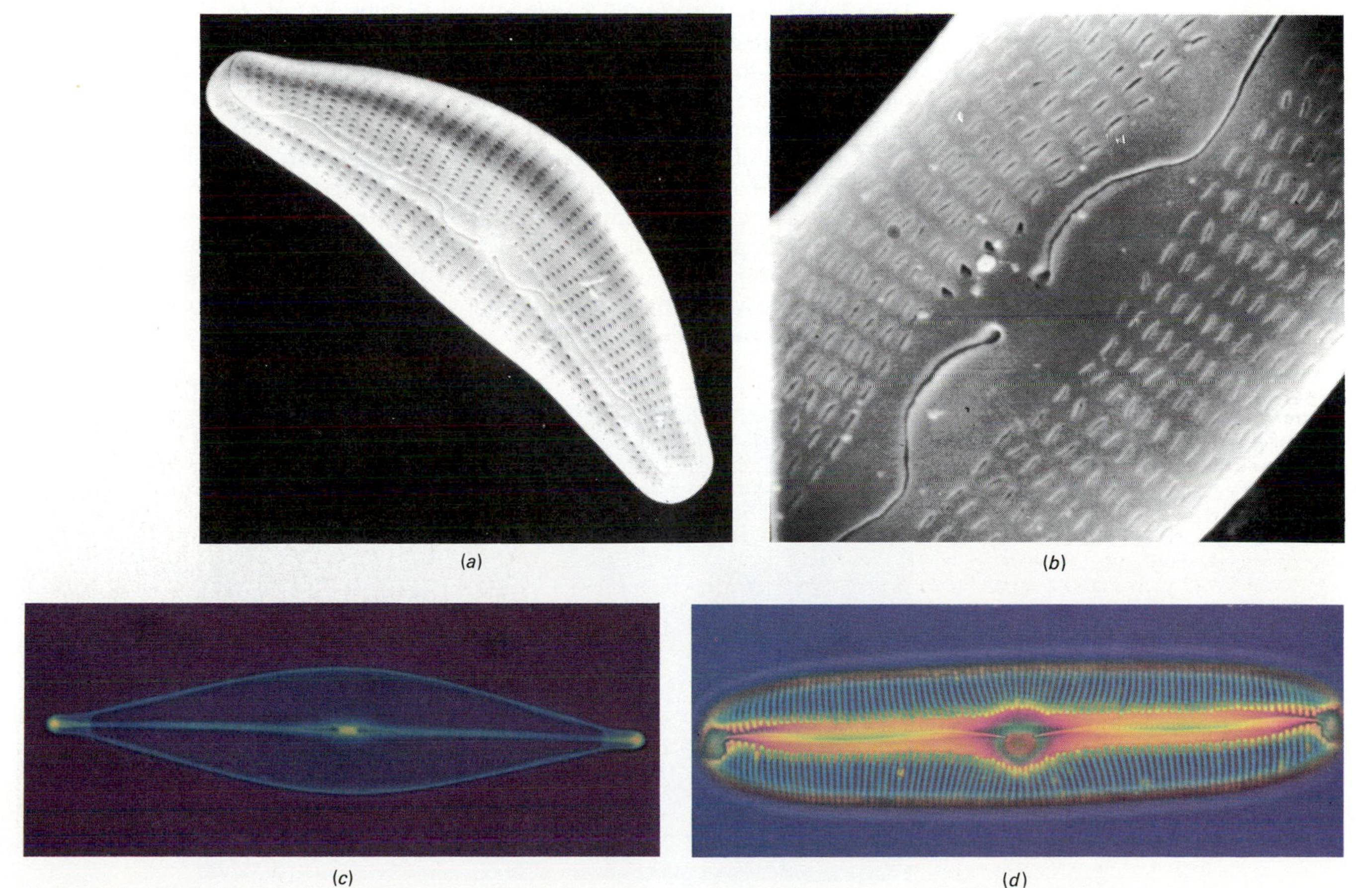

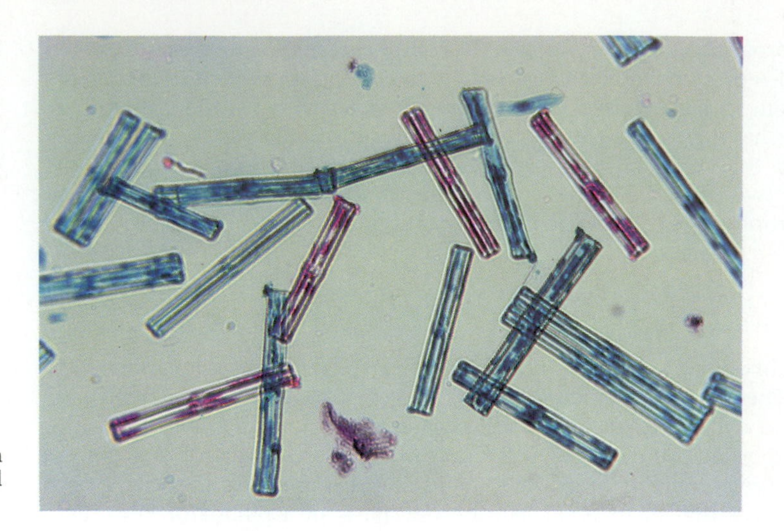

Figure 20–21 *Tabellaria,* a genus of yellow-green algae. These algae have been stained to show their rigid glasslike cell walls. (E. R. Degginger.)

The remains of the silica-containing cell walls have accumulated over millions of years as sediments in the oceans. Geological uplifts have brought layers of these deposits to the surface, allowing the diatomaceous earth to be mined for use in making such diverse products as insulating bricks, a filtering agent, and a fine abrasive used in silver polish and in some toothpastes. Some deposits of diatoms in California are more than 300 meters thick.

The other two classes of Chrysophyta are the yellow-green algae (Xanthophyceae) (Fig. 20–21), and the golden-brown algae (Chrysophyceae). Both of these have silica-impregnated, two-shelled cell walls and chloroplasts rich in carotenes and xanthophylls; these pigments give them their characteristic colors. Some members of each class lack cell walls and are amoeboid.

PHYLUM PYRROPHYTA

Dinoflagellates (Fig. 20–22) serve as a good introduction to the phylum Pyrrophyta. These are single-celled algae, most of which are surrounded by a shell made of thick, interlocking cellulose plates. They are all motile, having two flagella, one running in a longitudinal groove before projecting from one end, and the other running in a transverse groove that circles the cell (Fig. 20–23). The presence of a large amount of carotenoids give these cells a yellowish, reddish, or brownish color. Food reserves are stored as oils as well as polysaccharides. Most dinoflagellates are marine and are important photosynthesizers in the ocean.

Dense populations of marine dinoflagellates produce the phenomenon of **red tide** (Fig. 20–24), which can be disastrous to other marine life; these dense populations compete with other organisms for oxygen and produce toxic substances. These toxins tend to be concentrated by many particulate feeders such as mussels

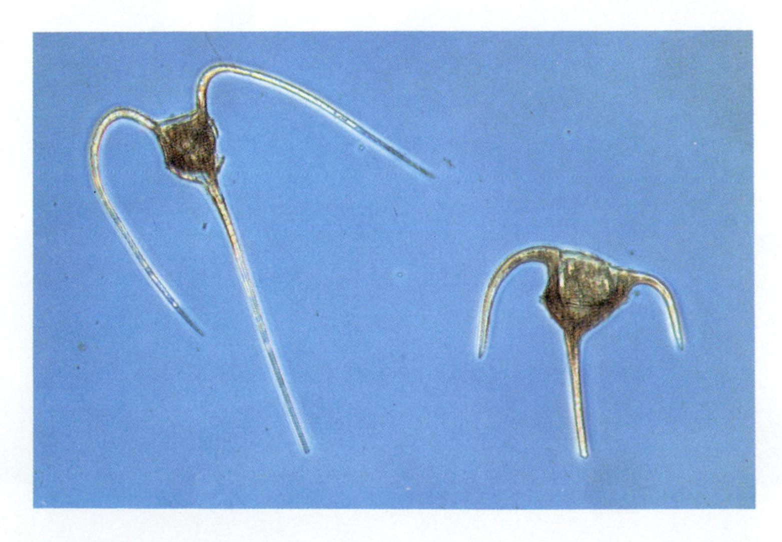

Figure 20–22 Two freshwater dinoflagellates of the genus *Ceratium.* (Paul W. Johnson, Univ. of Rhode Island/BPS.)

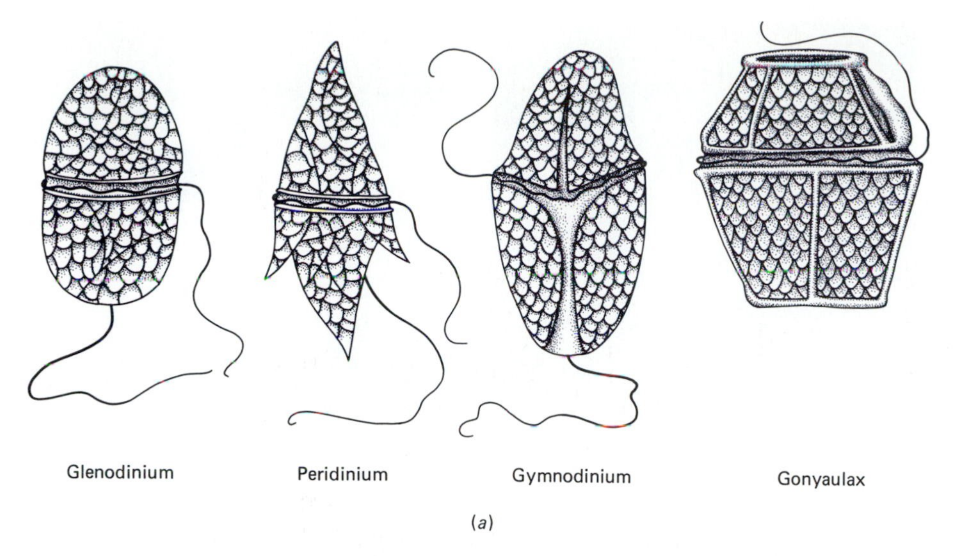

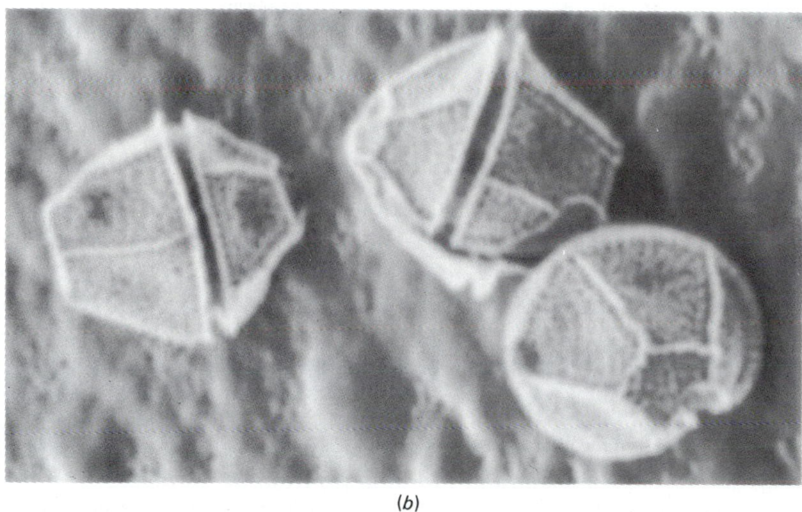

Figure 20–23 Some pyrrophytes. (*a*) Four species of dinoflagellates. Note the plates that encase the single-celled body and the characteristic two flagella, one of which is located in a transverse groove. (*b*) Scanning electron micrograph of *Gonyaulax*. (From Anikouchine, W., and Sterberg, R.: *The World Ocean: An Introduction to Oceanography*. Englewood Cliffs, NJ, Prentice-Hall, Inc., 1973.)

and can render them unfit for human consumption. Even Florida sea cows (manatees) have been killed as a result of eating tunicates (sea squirts) that had consumed the dinoflagellate *Gymnodinium brevis*. Severe outbreaks can even produce air pollution, resulting in respiratory complaints in people living nearby. The causes of red tide outbreaks are not fully understood. Although red tide may appear red in color in severe outbreaks because of a red pigment in *Gymnodinium*, this is not always the case; red tide algal blooms range from light tan through brown in color.

Figure 20–24 Red tide in Tampa Bay. The oddly patterned cloudiness in the water is produced by countless billions of organisms such as *Gymnodinium*. (Bill Morris, SELBYPIC.)

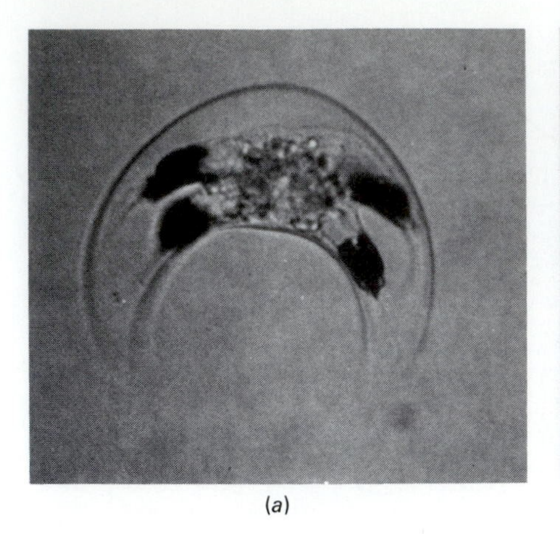
(a)

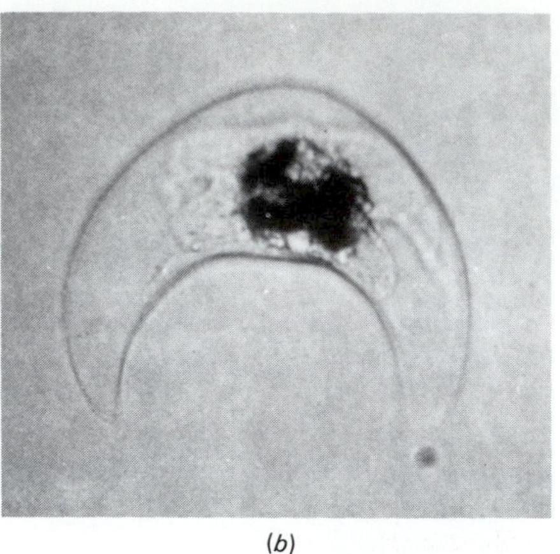
(b)

Figure 20–25 The chloroplasts (dark-colored objects) move about within the crescent-shaped dinoflagellate *Pyrocystis lunula*. (*a*) In the dark, the four chloroplasts draw apart. (*b*) When exposed to light, they move together in the center of the cell. The significance of this behavior is unknown. (Courtesy of Dr. Barry G. Cohen and Johns Hopkins University, Department of Biology.)

Many species of dinoflagellates have still other distinctive traits or effects. A great many species of dinoflagellates exhibit bioluminescence, emitting a faint glow visible even on a dark night. Many of the pyrrophytes have nuclei in which the individual chromosomes are visible always, rather than just during mitosis or meiosis. Their nuclear membrane remains intact during cell division. When the dinoflagellate *Pyrocystis lunula* is exposed to light its four chloroplasts move apart (Fig. 20–25).

Relationships of the Protists

The protist phyla are summarized in Table 20–1. The mastigophorans are generally considered the basic stock of organisms from which other protists and indeed the animals and plants (and perhaps the fungi) arose. The sarcodinians include several genera of ameboid organisms that have flagella. Several sarcodinid species also resemble typical flagellates when in open water, but lose their flagella and creep like amoebas when they contact a solid substrate. The existence of these intergrades suggests to some that the sarcodinians are polyphyletic—that is, that sarcodinians may have evolved at several different times from the flagellates.

The ciliates are a highly distinctive group, and if they arose from the flagellates, such evolution probably occurred only once. We can easily imagine the cilia as originating from multiple flagella, but the origin of the macronucleus is harder to

TABLE 20–1
The Protists

Phylum	*Description*	*Ecological Role*
Animal-like Protists–Protozoa		
Mastigophora	Flagellated forms; sometimes have ameboid stage	Some autotrophic, with chloroplast; all nutritional modes; some symbiotic forms very elaborate
Sarcodina	No definite body shape; some with elaborate shells (Formaminifera) and long, slender pseudopods	Consumers; some predatory and parasitic; usually aquatic, both freshwater and marine
Ciliata	Body covered with cilia; sometimes with protective trichocysts; free-swimming and stalked, usually with multiple nuclei	Mostly predatory, living on bacteria, yeast, and algae
Suctoria	Ciliated larvae; sometimes classed as ciliates but lack cilia as adults	Predators; lie in wait for prey which they pierce with tentacles
Sporozoa	No locomotory organs; all parasitic with complex life histories	Exclusively parasitic
Plantlike Protists–Algal Protists		
Chrysophyta	Golden-brown algae; cell wall composed of silica; no flagella	Siliceous shells of diatoms comprise extensive geological deposits.
Pyrrophyta	Flagellated forms having skeletons of cellulose plates.	Some species important in red tides

conceive, although it has been suggested that the macronucleus may have arisen from the fusion of multiple small nuclei via a chain nucleus, such as is found today in some ciliates like *Stentor*. Suctorians, on the other hand, virtually *are* ciliates and could possibly have developed from ciliates by losing cilia and other parts.

The sporozoans may well be a composite group. Some species are thought to show affinities with flagellates, while others more nearly resemble sarcodinians. The phenomenon of multiple fission is an adaptation to parasitism that sets these organisms apart from all others. There is little agreement at the present time about the origin of the sporozoans.

As for the origin of the algal protists, there is also little agreement—perhaps even less than in the case of the protozoans. If the euglenids are the rootstock from which the algal tree has also grown, a theory of algal evolution must account for the origin of cellulose and silica cell walls, for odd accessory photosynthetic pigments, and for aberrant kinds of flagella (not mentioned in this chapter but present in some of the algae).

SUMMARY

I. Most protists are unicellular, but some are colonial. The protists include the Protozoa and the algal protists.

II. The protozoans are animal-like; Protozoa contains the following phyla:

- *A.* Phylum Mastigophora: flagellated forms. Some are autotrophic, but a variety of nutritional modes is practiced within this group.
- *B.* Phylum Sarcodina: amoeba-like forms with no definite body shape except (in some groups) that imparted by a nonliving shell.
- *C.* Phylum Ciliata: The body is usually covered with cilia, although in some the cilia are restricted to definite body regions. Ciliates usually possess a micronucleus and at least one macronucleus.
- *D.* Phylum Suctoria: predatory ciliate-like forms without cilia as adults. They possess tentacles with which they remove the contents of their victims.
- *E.* Phylum Sporozoa: parasitic forms with complex life cycles; lack organs of locomotion.

III. The algal protists are the more plantlike of the two subkingdoms; they include two phyla, Chrysophyta and Pyrrophyta.

- *A.* Phylum Chrysophyta: the golden-brown algae. They possess siliceous shells. The diatoms are probably the ecologically most important chrysophytes.
- *B.* Phylum Pyrrophyta: flagellated forms having two functional flagella. Pyrrophytes have skeletons of cellulose plates. The dinoflagellates are ecologically very important.

IV. The probable consensus of evolutionary theorists places the mastigophorans at the base of the protist family tree. The mastigophorans are also thought to have given rise to the plants and animals, and perhaps also to the fungi.

POST-TEST

1. The kingdom Protista consists of ______________-celled or ______________ organisms that may be quite complex in their body organization.
2. In freshwater protozoans, the ______________ ______________ pumps excess water out of the cell.
3. The more animal-like protists are known as ______________, and the more plantlike of them as ______________ protists.

Match the characteristics in Column B with the phylum in Column A:

Column A	*Column B*
4. Phylum Sarcodina	a. locomotion by flagella; some with chloroplasts
5. Phylum Ciliata	b. locomotion by cilia
6. Phylum Mastigophora	c. locomotion by cilia in larval stage only
7. Phylum Suctoria	d. locomotion by pseudopods; some with elaborate skeletons or shells
8. Phylum Sporozoa	e. parasitic forms with involved life cycles usually involving two separate hosts

9. Dinoflagellates can be distinguished from ______________ because the dinoflagellates lack two overlapping silica "shells."
10. In the sporozoan malarial parasites, sexual reproduction occurs in the ______________ ______________ of the ______________ host.
11. Red tide is caused by ______________.

REVIEW QUESTIONS

1. What are the characteristics of a typical protist? How can a colonial protist be distinguished from a multicellular plant or animal?
2. Explain why *Euglena* is a taxonomically ambiguous creature.
3. How do the five phyla of Protozoa differ in their means of locomotion?
4. What practical importance do diatoms and dinoflagellates have for people?
5. Summarize reproduction in the paramecium.
6. Compare locomotion and cell coverings in Chrysophyta and Pyrrophyta.
7. Give arguments for or against the proposal that the ciliates evolved from the flagellates.
8. What is the main ecological role of each of the protist phyla?

21
Kingdom Fungi

OUTLINE

I. Body plan of a fungus: molds and yeasts
II. Metabolism and growth
III. Reproduction
IV. Classifying fungi
V. Division Eumycophyta
  A. Class Oomycetes (water molds)
  B. Class Zygomycetes
  C. Class Ascomycetes: the sac fungi
  D. Class Basidiomycetes: the club fungi
  E. Class Deuteromycetes (imperfect fungi)
VI. Division Myxomycophyta (slime molds)
VII. Symbiotic relationships of fungi
  A. The lichens
  B. Mycorrhizae: fungus-roots
VIII. Economic importance of fungi
  A. Fungi for food
  B. Fungi for drugs and useful chemicals
  C. Fungus diseases of plants
  D. Fungus diseases of animals

LEARNING OBJECTIVES

After you have read this chapter you should be able to:

1. Describe the distinguishing characteristics of the kingdom Fungi.
2. Contrast the structure of a yeast with that of a mold, and describe the body plan of a mold.
3. Trace the fate of a fungal spore that lands on an appropriate substrate such as an overripe peach, and describe conditions that permit fungal growth.
4. Give distinguishing characteristics for each of the five classes of division Eumycophyta (as given in the text), and give an example of each class.
5. Trace the black bread mold *Rhizopus nigricans* (a zygomycete) through the stages of its life cycle.
6. Trace a member of Ascomycetes (such as *Neurospora crassa*) through the stages of its life cycle.
7. Trace a member of Basidiomycetes (such as a mushroom) through the stages of its life cycle.
8. Contrast the plasmodial with the cellular slime molds, and describe the life cycle of a cellular slime mold.
9. Describe the ecological role of fungi as decomposers, and discuss the special ecological roles of lichens and mycorrhizae.
10. Summarize the economic significance of the fungi.
11. Identify several fungal diseases of plants and three human fungal diseases.

The tasty mushroom—delight of the gourmet—has much in common with the black mold that forms on stale bread and the mildew that collects on damp shower curtains. All of these life forms belong to the kingdom Fungi, a diverse group of more than 250,000 known species. Although they vary strikingly in size and shape, all of the fungi are eukaryotes. Their cells are encased in cell walls at least at some stage in the life cycle, a characteristic that accounted at least in part for their original classification in the plant kingdom. However, fungi lack chlorophyll, a very basic characteristic of plants. Fungi are heterotrophs that absorb their food through the cell wall and cell membrane. Some are **saprobes,** absorbing food from organic wastes or dead organisms, while others are parasites, absorbing food from the living bodies of their hosts (Fig. 21–1). Most fungi are nonmotile, although their reproductive cells may be motile. They reproduce by means of spores, which may be produced sexually or asexually.

Although the species that cause disease and destroy crops and stored goods, and even building materials, have given the fungi a bad name, these organisms make an important contribution to the ecological balance of our biosphere. Without their continuous decomposition of wastes and dead organisms, essential nutrients would soon become locked up in huge mounds of useless organic material, unavailable for recycling through new organisms. In addition, their many contributions to human society include yeasts for making bread, fungi that flavor cheese and ferment wine, the antibiotic penicillin, and of course, the mushroom, the morel, and the truffle.

Body Plan of a Fungus: Molds and Yeasts

The body structure of a fungus varies in complexity, ranging from the single-celled yeasts to the multicellular molds, a term used loosely to include the mildews, rusts and smuts, mushrooms and puffballs, and slime molds. The rigid cell wall encasing fungal cells is composed of cellulose, chitin, or other polysaccharides, or a combination of more than one carbohydrate. Cellulose, a polysaccharide made up of glucose subunits, is also found in the cell walls of plant cells. Chitin, which also is found in the external skeletons of insects and other arthropods, consists of subunits of a nitrogen-containing sugar, glucosamine. The other polysaccharides found in the fungal cell wall are glucose polymers that have the glucose subunits arranged in different ways than that in cellulose.

Yeast are unicellular fungi that reproduce asexually mainly by budding (Fig. 21–2), but also by fission, and sexually through spore formation, or fission. Each bud that separates from the mother yeast cell can grow into a new yeast. In general, yeast cells are larger than most bacteria. Some group together to form colonies. The yeasts are not classified as a single taxonomic group. Representatives are found in several of the classes of fungi discussed later in this chapter.

Figure 21–1 A parasitic fungus. After the ant ingests a spore of the fungus, the fungus begins to grow inside its body, absorbing nutrients from it. Eventually, the fungus affects the nervous system of the ant, causing it to climb to a leaf high in the tree. There the ant dies, and the fungus develops a specialized reproductive structure called the fruiting body. By altering the behavior of its host, the fungus ensures that the spores contained in the fruiting body are spread over a wide area. The effect of parasites on the behavior of host organisms is a rather new and interesting subject of biological research. (L. E. Gilbert, Univ. of Texas at Austin/BPS.)

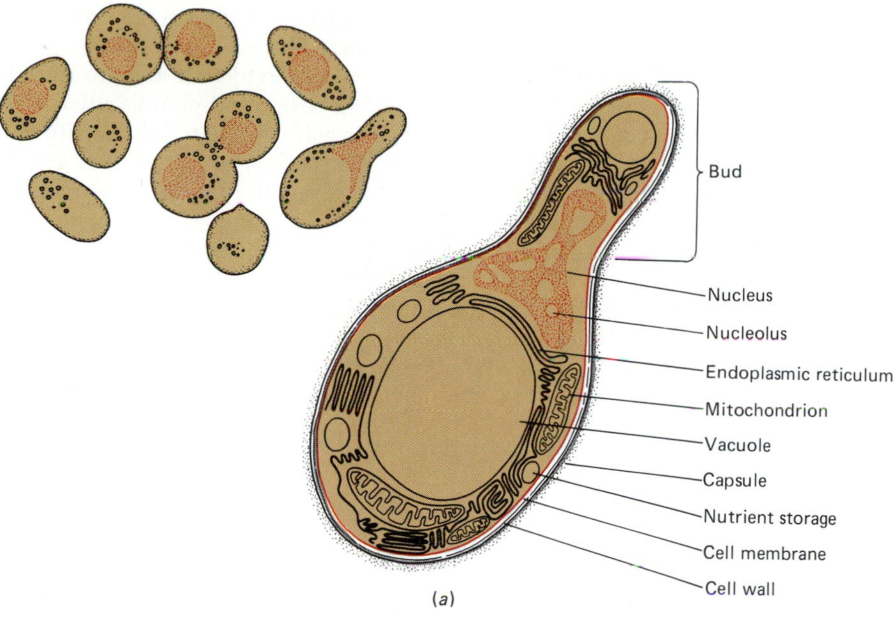

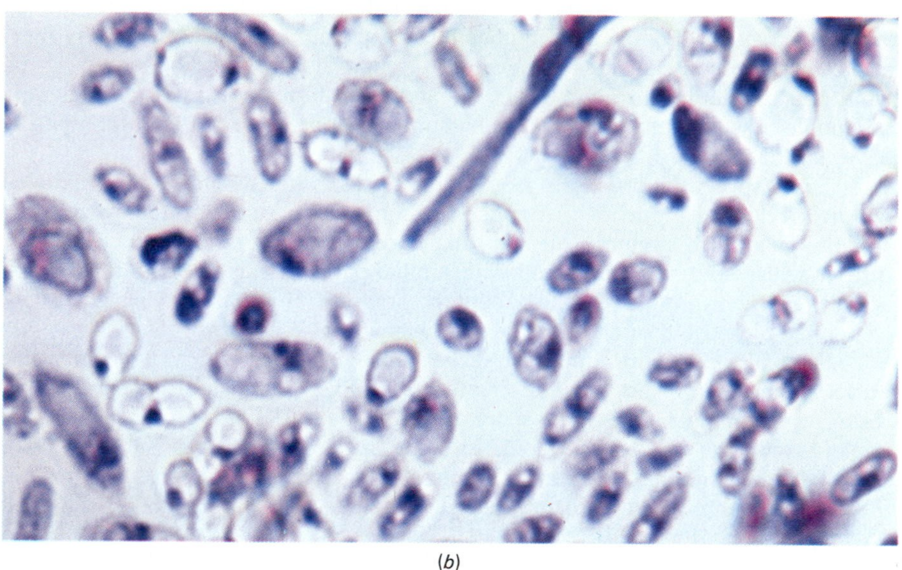

Figure 21–2 Yeasts are unicellular fungi that reproduce asexually mainly by budding. (*a*) Budding cells of the common bread yeast. (*b*) Photomicrograph of yeast cells (a common ascomycete). (Carolina Biological Supply Company.)

A mold consists of long, branched, threadlike strings of cells called **hyphae** (singular, hypha) that form a tangled mass or tissue-like aggregation known as a **mycelium** (Fig. 21–3). The cobweb-like mold sometimes seen on bread consists of the mycelia of mold colonies. What is not seen are the extensive mycelia that grow down into the substance of the bread. The color of the mold comes from the reproductive spores, which are produced in large numbers on the mycelia. Some hyphae are divided by walls, called **septa** (singular, septum), into individual cells containing one or more nuclei; others are **coenocytic,** undivided by septa, and are something like an elongated, multinucleated giant cell. Cytoplasm flows within the hypha, providing a kind of circulation. The whole fungus body is referred to as a **thallus.**

Many fungi, particularly those that cause disease in humans, are **dimorphic**—that is, they have two forms: They can change from the yeast form to the mold form in response to changes in temperature, nutrients, or other environmental factors.

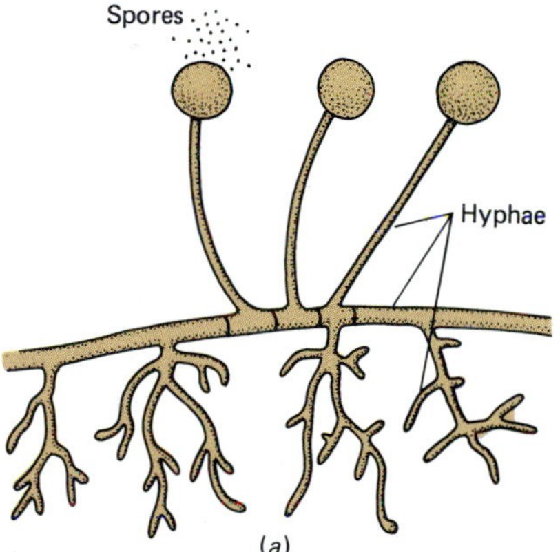

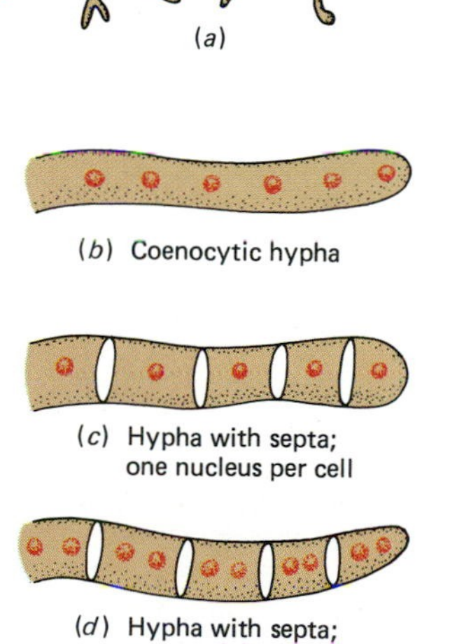

Figure 21–3 Molds. (*a*) Structure of a mold. (*b*) Coenocytic hypha. (*c*) Hypha divided into cells by septa; each cell has one nucleus. (*d*) Septate hypha in which each cell has two nuclei. In some classes the septa are perforated, permitting cytoplasm to stream from one cell to another.

Metabolism and Growth

Fungi grow best in dark, moist habitats, but they are found universally wherever organic material is available. Moisture is necessary for their growth, and they can obtain water from the atmosphere as well as from the medium upon which they

Figure 21–4 Fungi can thrive in a wide range of environmental conditions; even refrigerated foods, such as this peach, are not immune to fungal invasion.

live. When the environment becomes very dry, fungi survive by going into a resting stage or by producing spores which are resistant to desiccation. Although the optimum pH for most species is about 5.6, some fungi can tolerate and grow in pH ranging from 2 to 9. Fungi are not as sensitive to high osmotic pressures as bacteria. They can grow in concentrated salt solutions or sugar solutions such as jelly that discourage or prevent bacterial growth (Fig. 21–4). Fungi also thrive over a wide temperature range. Even refrigerated food is not immune to fungal invasion.

When a fungal spore comes into contact with an appropriate substrate, perhaps an overripe peach that has fallen to the ground, it germinates and begins to grow (Fig. 21–5). A threadlike hypha emerges from the tiny spore. Soon a tangled mat of hyphae infiltrates the peach, while other hyphae extend upward into the air. Cells of the hyphae secrete digestive enzymes into the peach, degrading its organic compounds to small molecules that the fungus can absorb. Fungi are very efficient at converting nutrients into new cell material. If adequate amounts of nutrients are available they are able to store them in the mycelium. Some excess nutrients may be excreted into the surrounding medium.

Reproduction

Fungal reproduction occurs in a variety of ways: asexually by fission, by budding, or by spore formation; or sexually by means that are characteristic for each group. Spores are usually produced on hyphae that project up into the air (aerial hyphae) above the food source. This arrangement permits the spores to be blown by the wind and distributed to new areas. The spores of terrestrial fungi are generally nonmotile cells dispersed by wind or by animals. Spores of aquatic fungi typically have flagella.

Some spores form directly from hyphal cells, whereas others are produced within specialized branches of hyphae. In some fungi such as mushrooms, the aerial hyphae form large complex **fruiting bodies,** specialized reproductive structures in which spores are produced. The familiar part of a mushroom or toadstool is actually a large fruiting body; the bulk of the organism is a nearly invisible network of hyphae buried out of sight in the rotting material it invades.

A sexual spore is produced by sexual reproduction in which two compatible nuclei come together and fuse to create a diploid zygote. Then meiosis occurs, restoring the haploid state to the spore nucleus. The nuclei of most fungal cells are haploid; only the zygote is diploid. In some species, sexual reproduction takes place between mycelia of different mating types referred to as plus and minus; these are like the two sexes of other organisms.

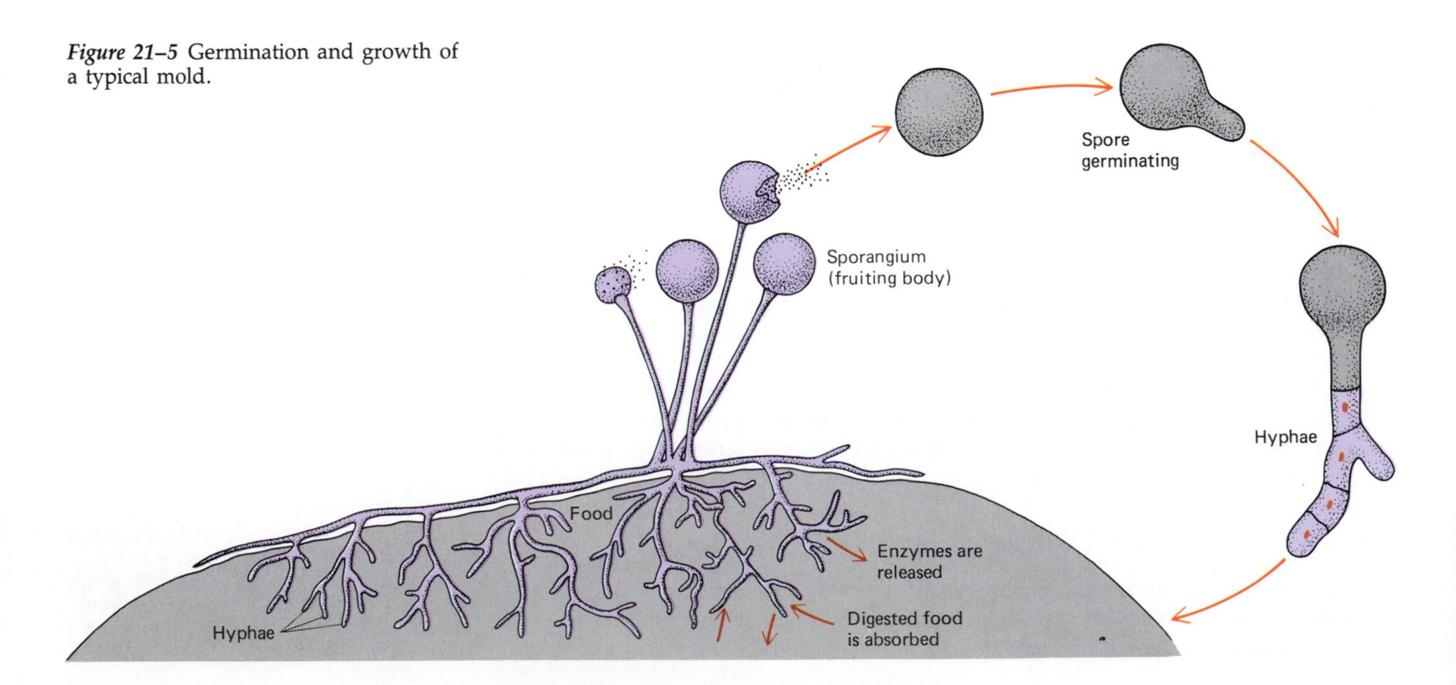

Figure 21–5 Germination and growth of a typical mold.

TABLE 21–1
Divisions and Major Classes of the Kingdom Fungi

Class	*Common Types and Examples*	*Type of Asexual Reproduction*	*Mechanism of Sexual Reproduction*
Division Eumycophyta	**True fungi**		
Oomycetes	Water molds, downy mildews, e.g., *Plasmopara* (causes downy mildew of grapes)	Zoospores with two flagella; motile	Fusion of male nucleus from antheridium with egg nucleus in oogonium to form thick-walled oospore
Zygomycetes	*Rhizopus nigricans* (black bread mold)	Nonmotile spores form in sporangium	Formation of zygospores from gametangia
Ascomycetes (sac fungi)	Yeasts, powdery mildews, molds, morels, truffles, *Neurospora crassa* (used in research)	Conidia pinch off from conidiophores.	Fusion of nuclei to form diploid nucleus; then meiosis to form haploid nuclei, which develop into ascospores
Basidiomycetes (club fungi)	Mushrooms, toadstools, bracket fungi, puffballs, rusts, smuts, e.g., *Amanita* (a genus that includes many poisonous mushrooms), *Agaricus campestris* (common, edible mushroom)	Uncommon in mushrooms	Fusion of cells at tips of hyphae to form a diploid nucleus; meiosis; development of basidiospores from the haploid nuclei
Deuteromycetes (imperfect fungi)	Molds, e.g., *Penicillium* (penicillin made from some species), *Candida albicans* (causes thrush and other infections of mucous membranes)	Conidia	Sexual stage not observed
Division Myxomycophyta	**Slime molds**		
Myxomycetes	Plasmodial slime molds	Mitosis and fission; spore formation following meiosis	Fusion of compatible haploid amoebas
Acrasiomycetes	Cellular slime molds	Mitosis and fission; spore formation	Sexual stage not known

Classifying Fungi

The classification of fungi is based mainly on the characteristics of the sexual spores and fruiting bodies. Not all authorities agree on how to classify these diverse organisms, but they are usually divided into two main divisions (equivalent to phyla in animal taxonomy): division Eumycophyta, which includes the true fungi, and division Myxomycophyta which comprises the slime molds. Table 21–1 summarizes the classification of the fungi.

The fungi were once thought to have evolved from the algae, but many biologists now think they can be traced back to the colorless flagellates. Certain features suggest that fungi are polyphyletic (i.e., derived from several groups)—that some evolved from algae, some from protozoans, and some from plants.

Division Eumycophyta

Members of the division **Eumycophyta,** or **true fungi,** almost always have cell walls and hyphae and reproduce by means of spores. They are divided into classes based on their characteristic sexual reproductive structures.

CLASS OOMYCETES (WATER MOLDS)

The fungi comprising the class **Oomycetes** are distinguished by their flagellated asexual spores, which can swim to new locations. Some oomycetes are unicellular; others consist of extensively branched mycelia with coenocytic hyphae. Their cell walls consist of cellulose. Commonly known as the water molds, many members of this class are aquatic (Fig. 21–6). Some are saprobes or parasites of fish, insect larvae, and seeds. Others feed on plant and animal debris. The fish mold, *Saprolegnia,* may parasitize fish and is a common pest in overcrowded aquaria. It appears as an infection in which cottony mycelia form over the fins and eventually over the entire bodies of aquarium fish.

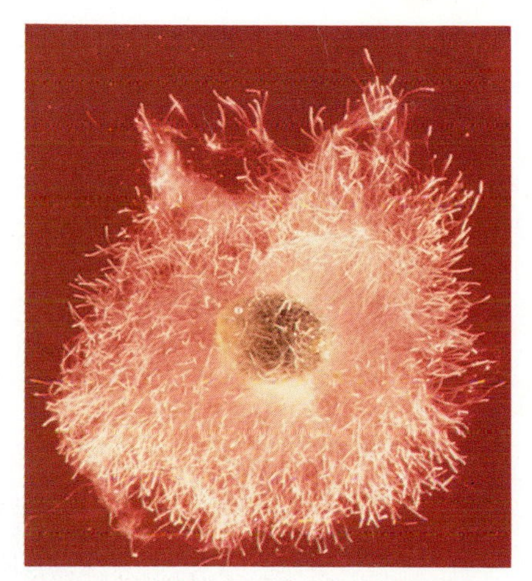

Figure 21–6 *Saprolegnia,* a member of Oomycetes, growing on hemp seed. (J. Robert Waaland, Univ. of Washington/BPS.)

Other oomycetes include a group of terrestrial pathogens that cause diseases such as the "damping off" of seedlings (*Pythium*) and the downy mildew of grapes (*Plasmopara*). The plant pathogen *Phytophthora infestans*, which causes potato blight, brought about the tragic potato famine that devastated Ireland in the 1840s. About a million people starved to death, and two million emigrated from Ireland as a consequence of the famine.

The name oomycete, meaning egg fungus, refers to the method of sexual reproduction characteristic of this class. Some of the hyphae develop enlarged tips called **oogonia.** Cytoplasm flows into these oogonia, and one to several egg cells form within each of them (Fig. 21–7). Other hyphae with slender, hooked tips are known as **antheridia;** these produce male gametes. Meiosis apparently occurs within the oogonia and antheridia. When an antheridial hypha contracts an oogonium, it forms a fertilization tube that grows into it and releases haploid nuclei. Each male nucleus fuses with an egg nucleus to form a diploid zygote.

The zygote develops into a thick-walled resistant **oospore,** which serves as a resting spore. When conditions are favorable, the oospore germinates and develops into a short hypha. A **sporangium,** a specialized structure in which spores are produced, forms, and flagellated **zoospores** develop (each bearing two flagella). When the spores are released, they swim about until they find a suitable location in

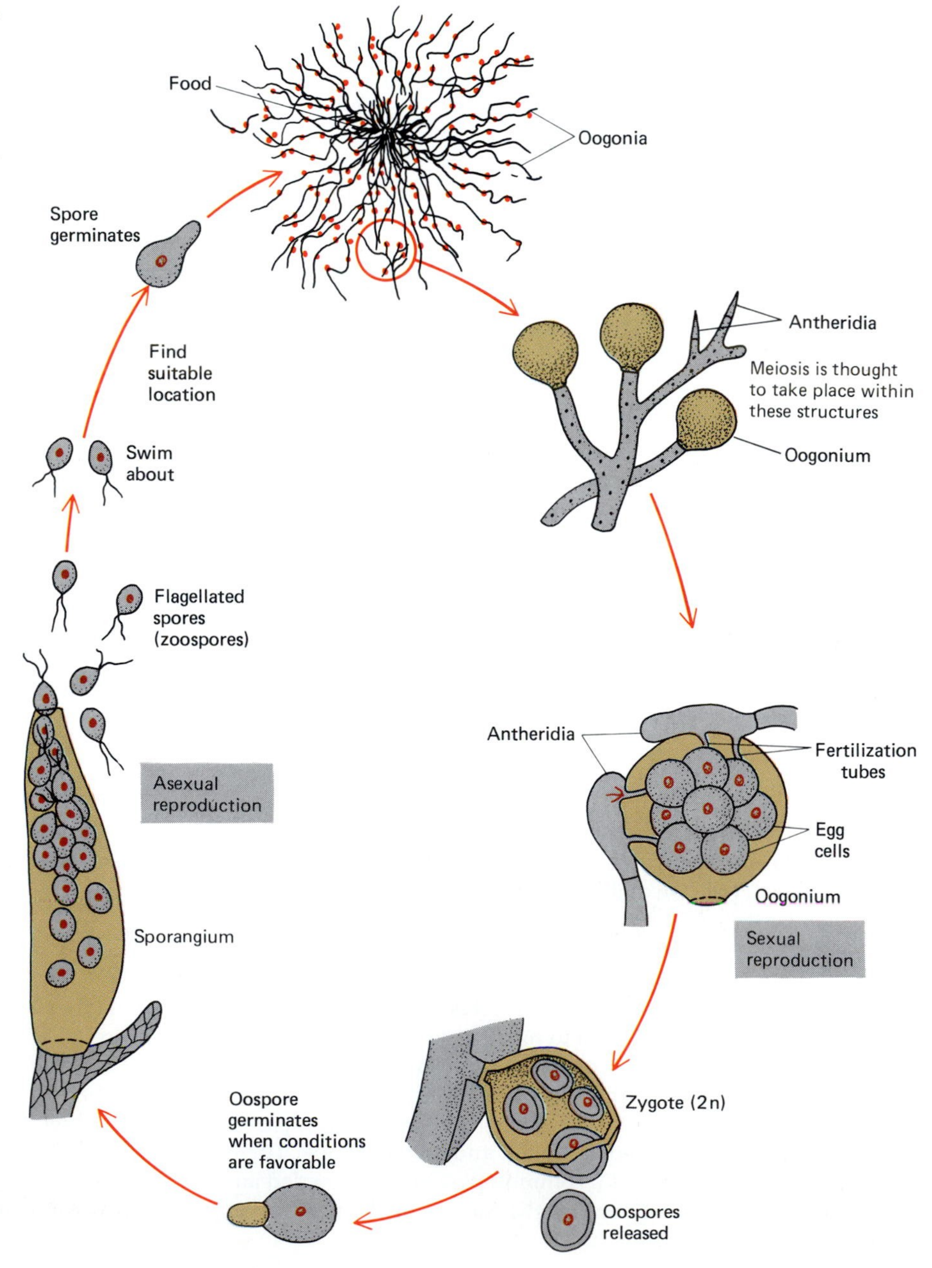

Figure 21–7 Life cycle of an oomycete, a water mold. During sexual reproduction, oogonia and antheridia form. Meiosis is thought to take place within these structures. Egg cells that form within the oogonia are fertilized by male nuclei produced within the antheridia. The diploid zygotes that form develop into oospores, which are resting spores. When conditions are favorable, an oospore germinates and then forms a short coenocytic hypha. Next a sporangium forms; within this structure, flagellated zoospores develop. Zoospores swim about until conditions are suitable for them to settle down and germinate. Each zoospore can give rise to a new mycelium.

which to settle down and germinate. With the formation of a new mycelium, the cycle begins anew.

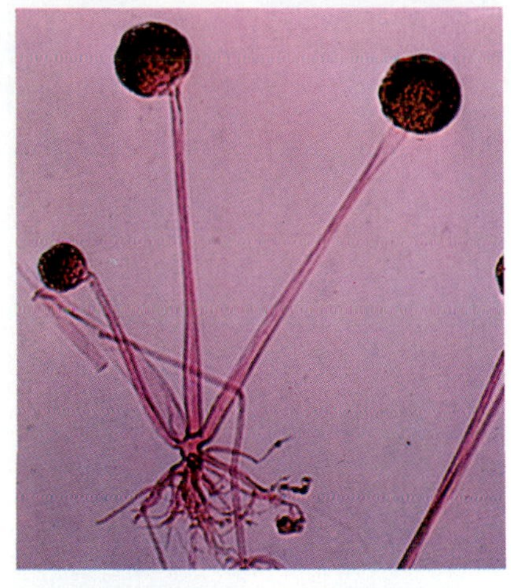

Figure 21–8 Bread mold *Rhizopus nigricans.* (Carolina Biological Supply Company.)

CLASS ZYGOMYCETES

The members of the class **Zygomycetes** are terrestrial fungi that produce sexual resting spores called **zygospores.** Their hyphae are coenocytic, and their cell walls consist mainly of chitin. Many zygomycetes live in the soil on decaying plant or animal matter. Some are parasites of plants and animals.

One common member of Zygomycetes is the black bread mold *Rhizopus nigricans* (Fig. 21–8). Bread becomes moldy when a mold spore falls upon it and then germinates and grows into a tangled mass of threads, the mycelium. Some of the hyphae, termed **rhizoids,** penetrate the bread and obtain nutrients; others, termed **stolons,** grow horizontally with amazing speed. Eventually certain hyphae grow upward and develop a **sporangium** (spore sac) at the tip. Clusters of black spherical spores develop within this sac and are released when the delicate sporangium ruptures (Fig. 21–9).

Sexual reproduction occurs when the hyphae of two different mating types come to lie side by side. The bread mold is **heterothallic** (having two mating types), and sexual reproduction can occur only between a member of a plus strain and one of a minus strain. This is a sort of physiological sex differentiation even though there is no morphological sex differentiation. However, it is not proper to refer to the two strains as male and female. When hyphae of opposite mating types meet, hormones are produced that cause the tips of the hyphae to come together and to form **gametangia,** structures that produce gametes. These structures become separated from the rest of the thallus by the formation of septa. Plus and minus nuclei then fuse, forming a young zygospore with several diploid nuclei.

The zygospore forms a thick black protective covering and may lie dormant for several months. Meiosis probably occurs at or just before germination of the zygospore. All the haploid nuclei but one then degenerate. An aerial hypha develops with a sporangium at the top. Haploid spores are produced, which eventually germinate to begin a new cycle. Only the zygote is diploid; all of the hyphae and the asexual spores are haploid.

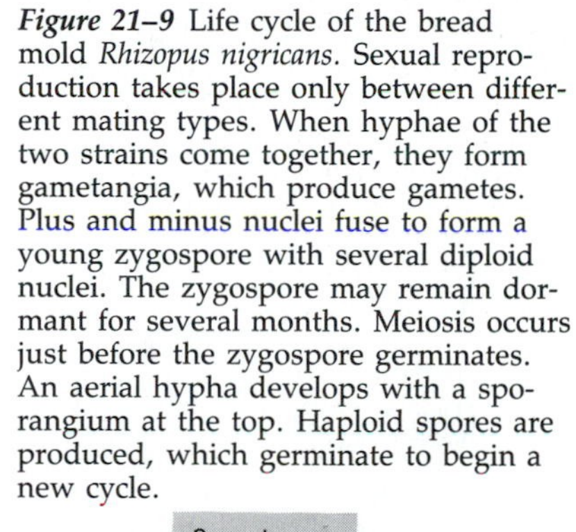

Figure 21–9 Life cycle of the bread mold *Rhizopus nigricans.* Sexual reproduction takes place only between different mating types. When hyphae of the two strains come together, they form gametangia, which produce gametes. Plus and minus nuclei fuse to form a young zygospore with several diploid nuclei. The zygospore may remain dormant for several months. Meiosis occurs just before the zygospore germinates. An aerial hypha develops with a sporangium at the top. Haploid spores are produced, which germinate to begin a new cycle.

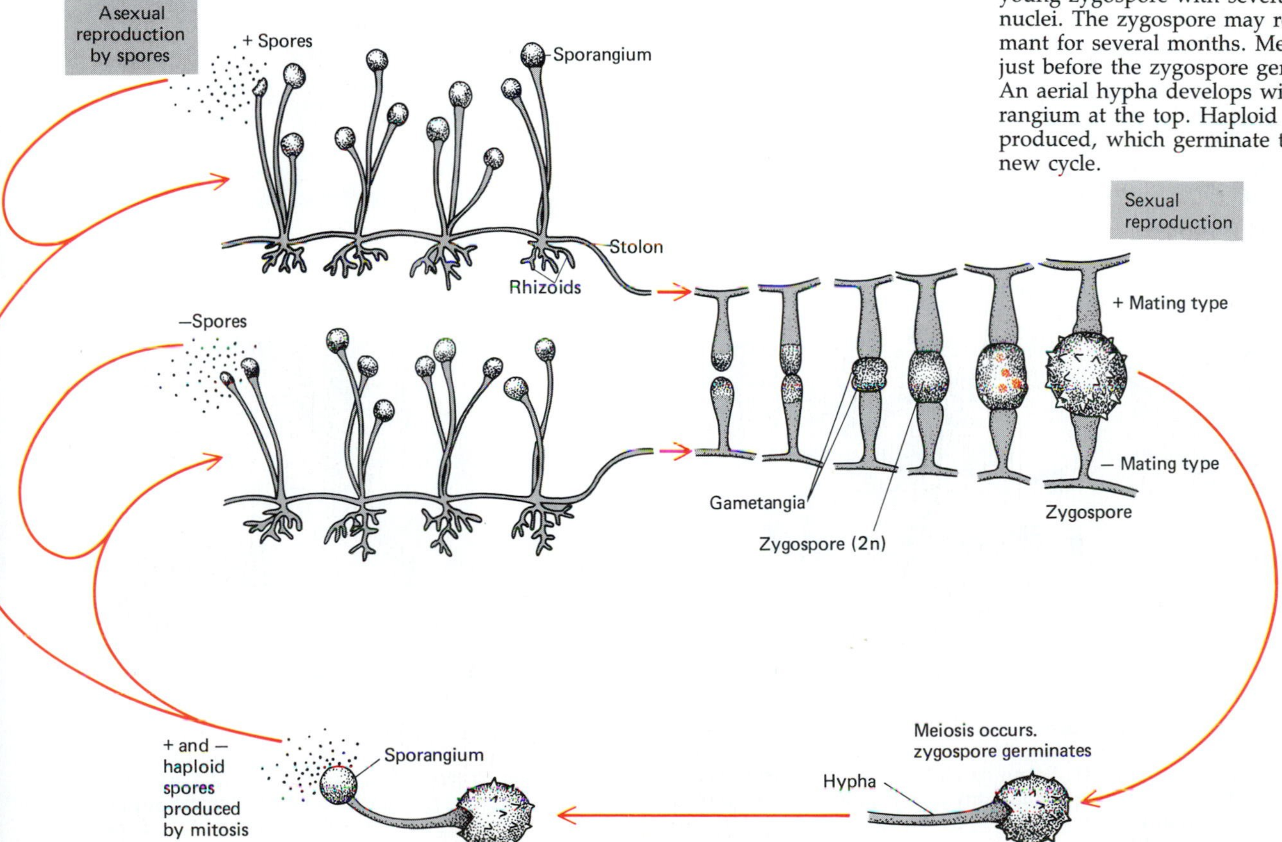

CLASS ASCOMYCETES: THE SAC FUNGI

The members of class **Ascomycetes,** the largest class of sexually reproducing fungi (about 30,000 described species), are called **sac fungi** because their spores are produced in little sacs called **asci** (singular, **ascus**). Their hyphae usually have septa, but these cross walls are perforated so that cytoplasm can move from one compartment to another.

The ascomycetes include some yeasts, the powdery mildews, most of the blue-green, red, and brown molds that cause food to spoil, and the edible morels and truffles (Fig. 21–10). The ascomycete *Neurospora crassa* is a saprobe that occurs on pies and cakes as a white fluff, which turns pink as it develops asexual pink spores. This fungus has been an important research tool in genetics and biochemistry. Some ascomycetes cause serious plant diseases including the Dutch elm disease (caused by *Ceratocystis ulmi,* a European fungus), chestnut blight (caused by *Endothia parasitica,* a fungus accidentally introduced from China), ergot disease of rye plants, and the powdery mildews that ruin fruits and ornamental plants.

The bodies of ascomycetes may be unicellular as in yeasts, multicellular and filamentous as in the powdery mildews, or thickened and fleshy, as in the truffle. There is much diversity in the class, both in the form of the vegetative cells and in the mode of asexual reproduction.

In most ascomycetes asexual reproduction involves production of spores called **conidia.** These spores are pinched off at the tips of certain specialized hyphae known as **conidiophores** (spore bearers) (Fig. 21–11). Sometimes called "summer spores," the conidia are a means of rapidly propagating new mycelia; they occur in

(*a*)

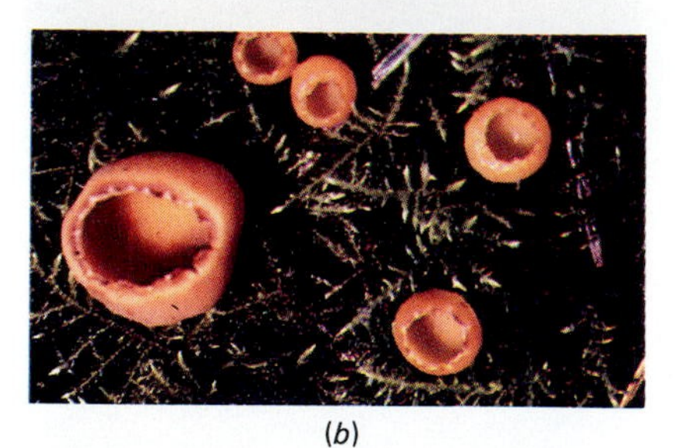

(*b*)

Figure 21–10 Common ascomycetes. (*a*) *Morchella deliciosa.* As you might guess from its Latin name, this is an edible fungus—more commonly called a morel. (*b*) *Peziza.* ((*a*), E. R. Degginger; (*b*) Carolina Biological Supply Company.)

(*a*) Ascocarps

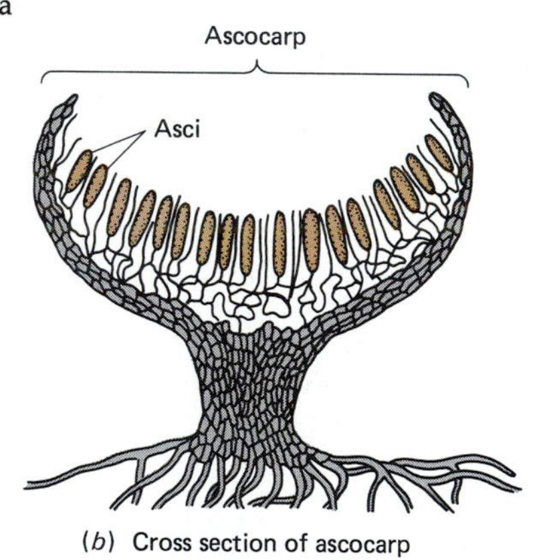

(*b*) Cross section of ascocarp

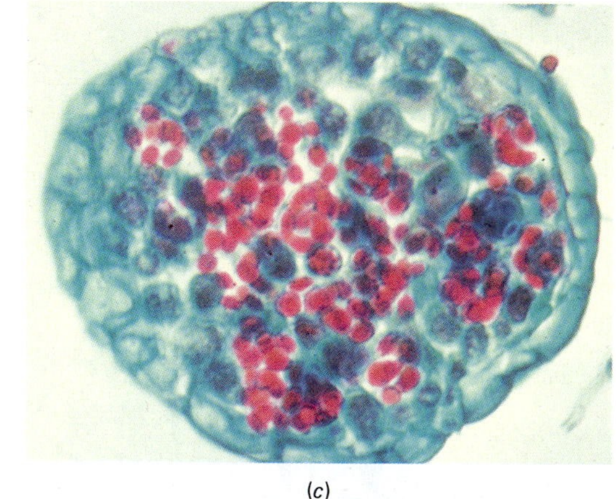

(*c*)

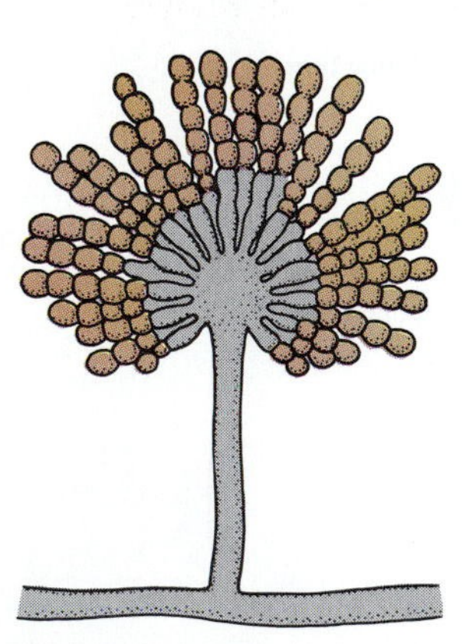

(*d*) Section through conidiophore

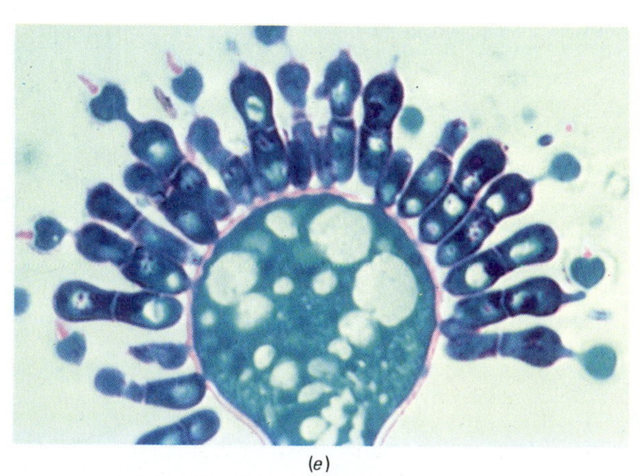

(*e*)

(*f*)

Figure 21–11 Ascomycete reproductive structures. (*a*) An ascocarp, a fruiting body containing asci. (*b*) Cross section through an ascocarp. (*c*) Photomicrograph of section through ascocarp with asci. (*d*) Section through conidiophore. (*e*) Photomicrograph through conidiophore showing conidia. (*f*) Scanning electron micrograph of conidia from *Penicillium purpurrescens.* Types of conidia are considered important in determining the identity and relationships of various species. (Bar = 1 μm.) ((*c*), (*e*), Carolina Biological Supply Company; (*f*), from Martinez, A. T., Calvo, M. A., and Ramirez, C.: Scanning electron microscopy of *Penicillium conidia. Antonie van Leeuwenhoek* 48, 245–255, 1982.)

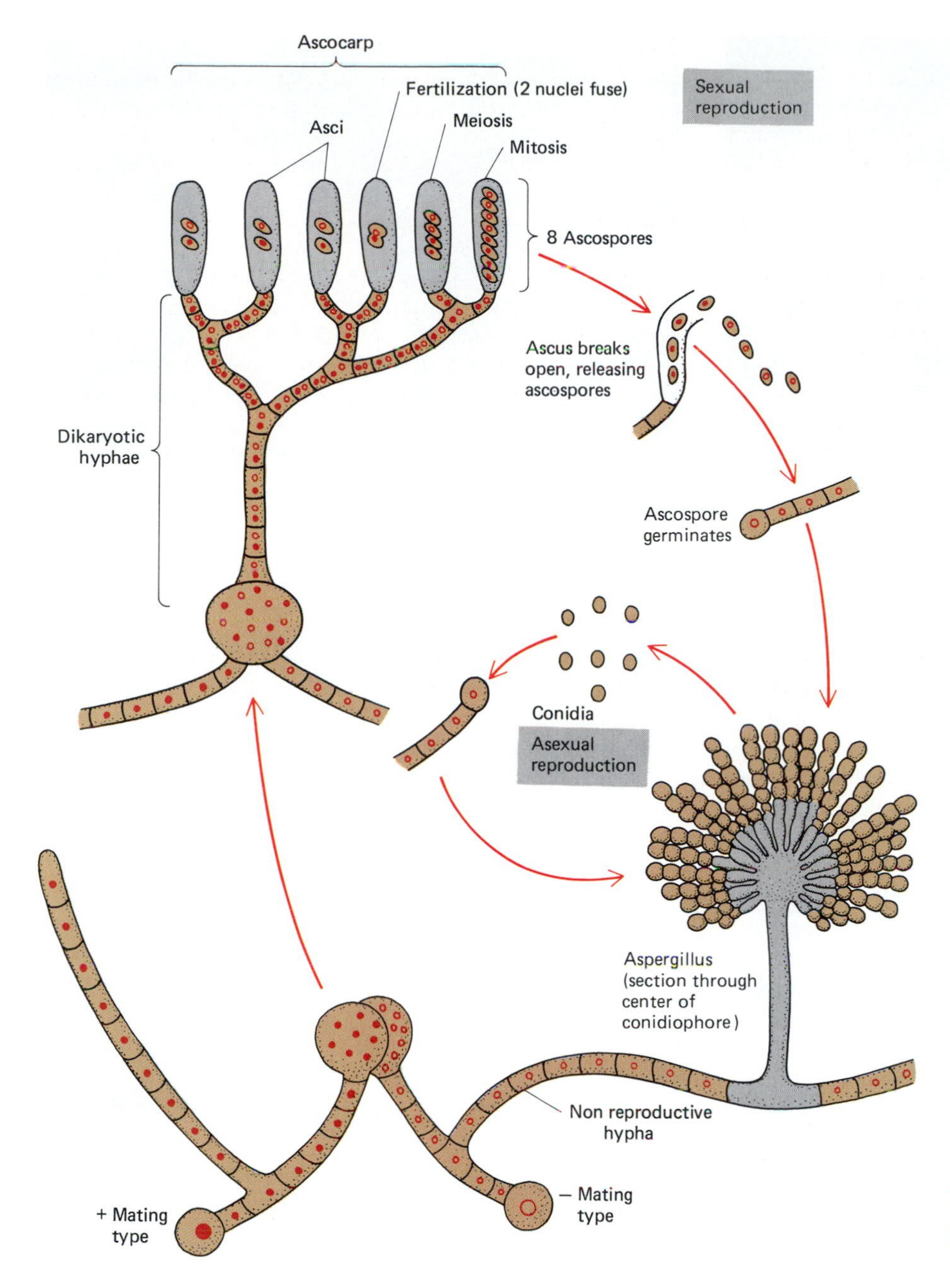

Figure 21–12 Life cycle of an ascomycete. Asexual reproduction involves production of spores called conidia by specialized hyphae called conidiophores. Sexual reproduction involves the formation of asci and ascospores.

various shapes, sizes, and colors in different species. The color of the conidia is what gives the characteristic black, blue, green, pink or other tint to many of these molds.

Some species of the class Ascomycetes are heterothallic; others are **homothallic** (similar mating types). Sexual reproduction takes place after two hyphae grow together and their cytoplasm mingles (Fig. 21–12). Within this fused structure the two nuclei come together, but they do not fuse. New hyphae develop from this fused structure; cells of these hyphae are **dikaryotic**—that is, each has two nuclei. These hyphae form a fruiting body, known as an **ascocarp,** characteristic of the species. This is where the asci develop. Within a cell that will form an ascus, the two nuclei fuse, forming a diploid nucleus. The nucleus then undergoes meiosis to form four haploid nuclei. This process is usually followed by one mitotic division of each of the four nuclei, resulting in formation of eight nuclei. The haploid nuclei surrounded by cytoplasm separate to form **ascospores,** so that there are eight haploid ascospores within the ascus. The ascospores are released when the tip of the ascus breaks open.

Figure 21–13 Some members of Basidiomycetes (*a*) *Gyrodon meruloides*. (*b*) *Tremella mesenterica*, also called witche's butter. (E. R. Degginger.)

CLASS BASIDIOMYCETES: THE CLUB FUNGI

The 25,000 or more species that make up the class **Basidiomycetes** include the most familiar of the fungi—mushrooms, toadstools, bracket fungi, and puff balls, as well as some important plant parasites, the rusts and smuts (Figs. 21–13 and 21–14). The basidiomycetes derive their name from the fact that they develop a **basidium,** a structure comparable in function to the ascus of ascomycetes. Each basidium is an enlarged, club-shaped, hyphal cell, at the tip of which develop four **basidiospores.** Note that basidiospores develop on the *outside* of the basidium, whereas ascospores develop *within* the ascus. The basidiospores are released, and when they come in contact with the proper environment, they develop into new mycelia. The vegetative (feeding) body of the fungus consists of hyphae divided into compartments by septa. As in Ascomycetes, the septa are perforated and allow cytoplasmic streaming.

Figure 21–14 Some types of basidiocarps. (*a*) *Fomes*, a bracket fungus. (*b*) *Calvatia*, a puffball. (*c*) *Geaster*, an earth star. (*d*) *Psilocybe*, a sacred mushroom. (*e*) *Amanita virosa*, the destroying angel, a poisonous species of gill mushroom. (*f*) *Agaricus campestris*, the nonpoisonous field mushroom. (*g*) *Clavaria*, a coral mushroom. (*h*) *Phallus*, a stinkhorn. (Courtesy of K. Norstog, and R. W. Long.)

The vegetative body of the cultivated mushroom, *Agaricus campestris*, consists of a mass of white, branching threadlike hyphae that occurs mostly below ground. Compact masses of hyphae, called buttons, develop along the mycelium (Fig. 21–15). The button grows into the structure we ordinarily call a mushroom, which consists of a stalk and cap. More formally, the mushroom is referred to as a fruiting body, or **basidiocarp.** The lower surface of the cap consists of many thin perpendicular plates called **gills,** extending radially from the stalk to the edge of the cap. The basidia develop on the surface of these gills (Fig. 21–16).

A typical basidiomycete life cycle includes three phases (Fig. 21–15). Each individual fungus produces millions of **basidiospores,** and each basidiospore has the potential, should it happen upon an appropriate environment, to give rise to a new **primary mycelium.** Hyphae of this mycelium consist of monokaryotic (having a single nucleus) cells. When in the course of its growth such a hypha encounters

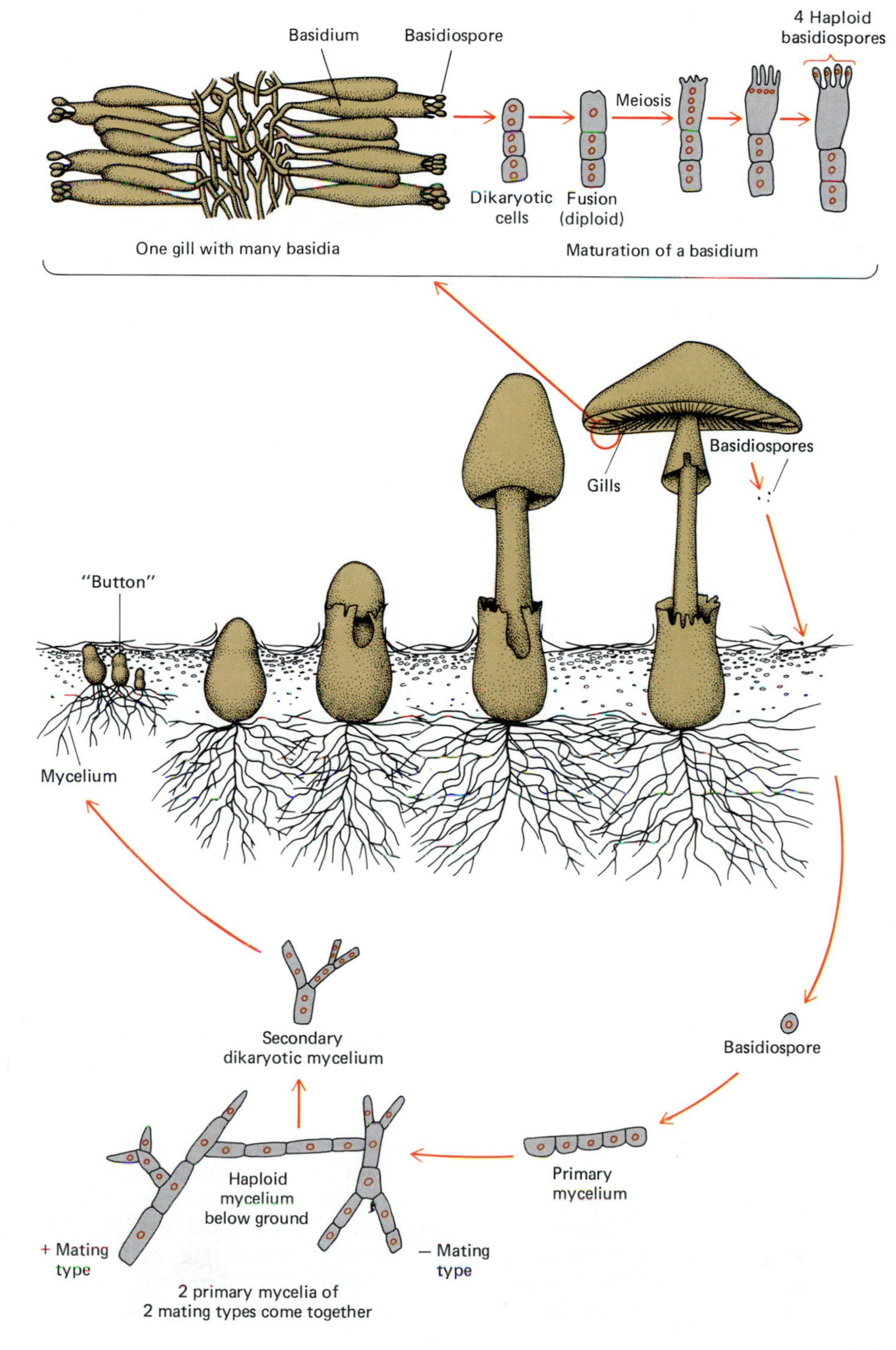

Figure 21–15 Life cycle of a basidiomycete. A mushroom develops from the mycelium, a mass of white, branching threads found underground. A compact button appears and grows into a fruiting body, or mushroom. On the undersurface of the fruiting body are gills, thin perpendicular plates extending radially from the stem. Basidia develop on the surface of these gills and produce basidiospores, which are shed. If these spores reach a suitable environment, they give rise to new mycelia.

Figure 21–16 A view of a basidiomycete, *Lepista nuda*, showing gills and basidium. (E. R. Degginger.)

another hypha of a different mating type, the two hyphae join to form a dikaryotic secondary mycelium in which each cell contains two nuclei, one of each mating type.

The hyphae of the mycelium grow extensively and eventually form basidiocarps. Each basidiocarp actually consists of intertwined hyphae matted together. On the gills of the mushroom the nuclei of the cells at the tips of the hyphae undergo nuclear fusion (fertilization), forming diploid nuclei. These are the only diploid nuclei that form in the life history of a basidiomycete. Meiosis then occurs, forming four haploid nuclei. These nuclei move to the outer edge of the basidium. The cell wall then forms finger-like extensions into which the nuclei and some cytoplasm move. The nuclei develop into basidiospores. Behind the nuclei the extensions come together, separating the basidiospores from the rest of the fungus by a delicate stalk. When the stalk breaks, the basidiospores are released.

CLASS DEUTEROMYCETES (IMPERFECT FUNGI)

About 25,000 species of fungi have been assigned to **class Deuteromycetes;** they are also called **imperfect fungi** because a sexual stage characteristic of the other eumycophytes has not been observed during their life cycle. (Should further study reveal a sexual stage, these species will be reassigned to a different class.) Most deuteromycetes reproduce only by means of conidia and so are closely related to the ascomycetes; a few appear to be related to the basidiomycetes.

The unique flavor of cheeses such as Roquefort and Camembert is produced by the action of members of the genus *Penicillium*. The mold *P. roquefortii* is found in caves near the French village of Roquefort; only cheeses produced in this area can be called Roquefort cheese. Another *Penicillium* species, of course, produces the antibiotic penicillin. *Aspergillus tamarii* and other imperfect fungi species are used in the Orient to produce soy sauce by fermenting soybeans.

Members of Deuteromycetes cause ringworm and athlete's foot. Another, *Candida albicans*, causes thrush and other infections of mucous membranes, and still others cause systemic fungus infections.

Division Myxomycophyta (Slime Molds)

The **slime molds**—division **Myxomycophyta**—are a group of curious organisms with little similarity to other molds except that they produce spores. In fact, they are sometimes classified as protists because for part of their life cycle they lack cell walls and they resemble amoebas somewhat. During part of their life cycle they ingest their food rather than absorb it.

The **plasmodial slime molds—class Myxomycetes**—are peculiar organisms that spend part of their life cycle as thin streaming masses, which creep along decaying leaves or wood (Fig. 21–17). They move by sending out pseudopods as do amoebas, and feed upon bacteria and small organic particles in their path. At this stage the organism, referred to as a **plasmodium,** is a multinucleate mass that may grow to about 30 grams. Spreading itself very thinly, the plasmodium may cover an area of a meter or more in diameter. Its shape varies as it grows and moves.

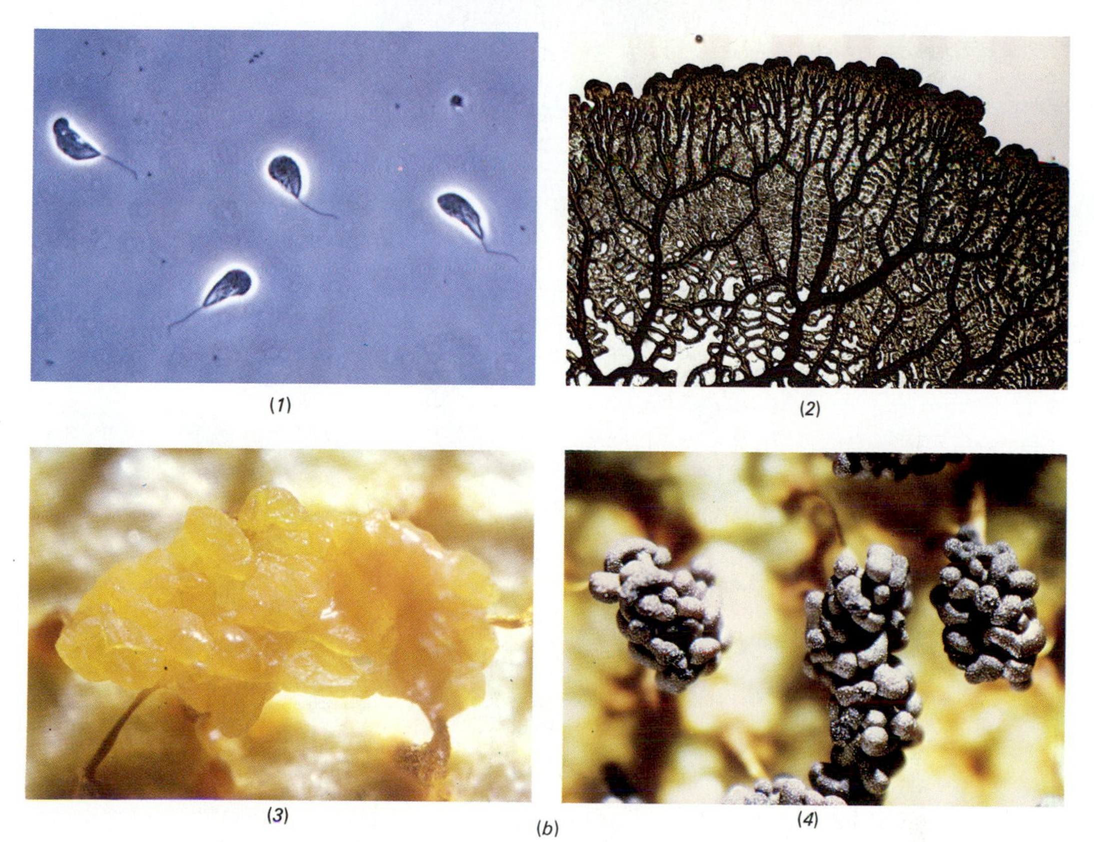

Figure 21–17 Plasmodial slime molds. (*a*) *Tubifera ferruginosa.* (*b*) Stages in the life cycle of *Physarum*. The amoeba-like cells (*1*) give rise to the plasmodium (*2*), a multinucleate mass, which forms sporangia (*3*). Mature sporangia (*4*) produce spores. ((*a*), E. R. Degginger; (*b*), Carolina Biological Supply Company.)

When the food supply dwindles or there is insufficient moisture, little mounds form on the plasmodium, each of which sends up a stalk with a sporangium at its tip. Meiosis takes place during spore formation, and the spores are haploid. The spores are extremely resistant to unfavorable environmental conditions, and years may pass before they germinate. Each spore produces one to four flagellated cells that eventually become amoeba-like and function as undifferentiated gametes. After fertilization the diploid zygote divides many times by mitosis to form a new plasmodium. Some 450 species of myxomycetes are known and may be distinguished by their sporangial and spore features as well as by the size, color, and texture of the plasmodium. In addition to their important role as decomposers, some plasmodial slime molds are parasites of plants: One species causes the disease known as clubroot in cabbages; another cause powdery scab of potato tubers.

The **cellular slime molds—class Acrasiomycetes**—are free-living and resemble amoebas through part of their life cycle (Fig. 21–18). The amoeba-like cells feed upon soil bacteria, and they grow and divide by mitosis repeatedly. After the food supply in their immediate vicinity has been depleted, the "amoebas" swarm to-

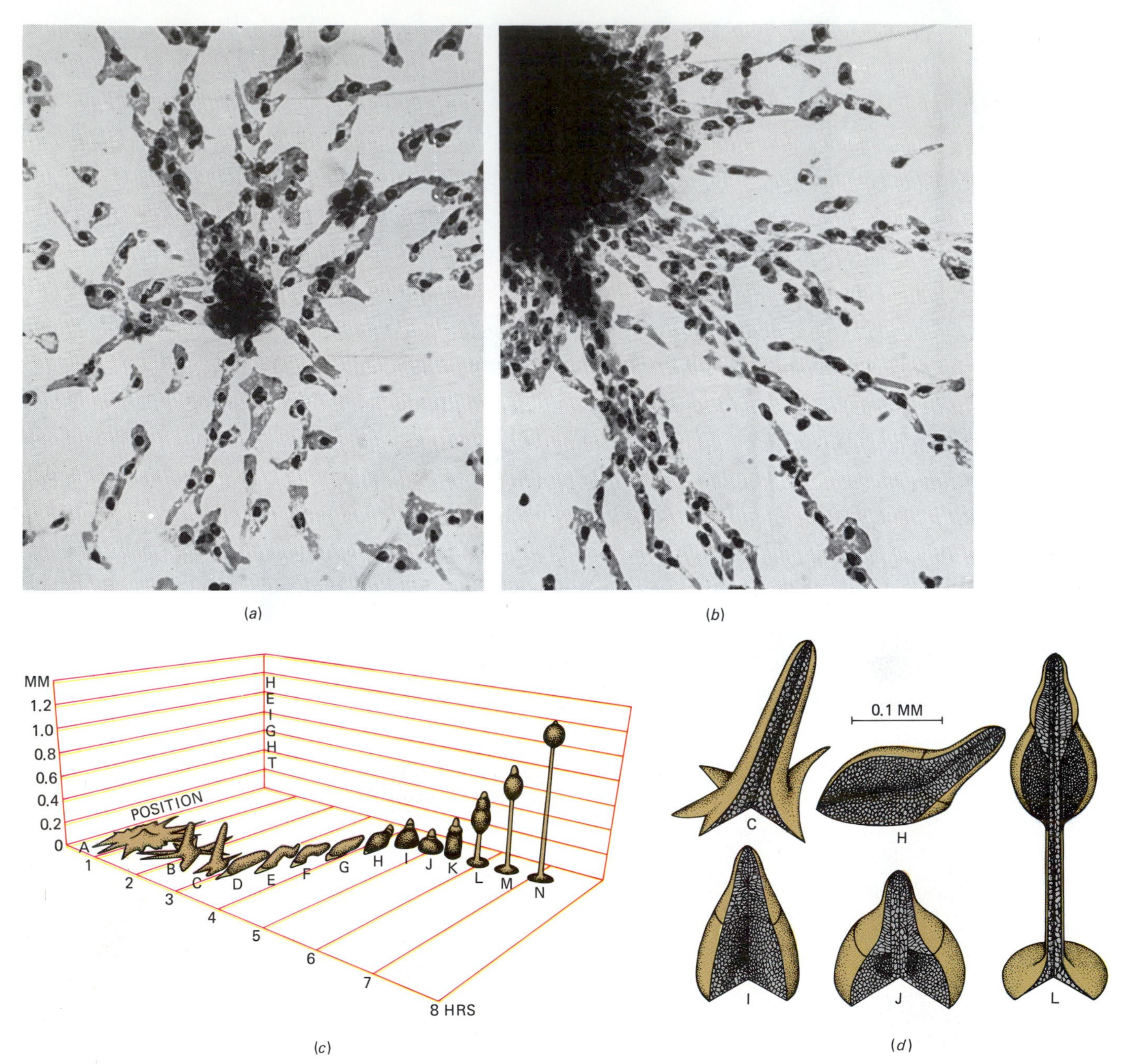

Figure 21–18 Photomicrographs showing the life cycle of a cellular slime mold. (*a*) Individual organisms are beginning to aggregate as a pseudoplasmodium. (*b*) A later stage in the aggregation process. *Center,* Further stages in the life cycle, some of which are shown in greater detail on the right. *A–C,* The pseudoplasmodium is formed by the aggregate of hundreds of individual amoeba-like cells. *D–H,* The pseudoplasmodium crawls along the substrate surface for some time. *I–N,* It grows a stalk that lifts the spore-producing portion off the surface. Finally, the spores are released. (Courtesy of Dr. John Bonner.)

gether, forming tiny heaps of cells. The signal for this aggregation process is the release of cyclic AMP (cyclic 3′,5′ adenosine monophosphate), a chemical that also serves as an intermediate in the action of several vertebrate hormones. The cells of each heap arrange themselves to form a slug-shaped organism called a **pseudoplasmodium,** composed of a few hundred to more than 100,000 cells, the number depending upon the population density of the "amoebas." This "slug" behaves like a multicellular animal, creeping about and responding in a coordinated manner to such stimuli as light and heat.

After a period of migration, which may involve search for a favorable habitat in which to continue its life cycle, the "slug" assumes a rounded shape and forms a fruiting body. Cells that had formed the anterior one third of the migrating slug now become stalk cells, while those from the posterior two thirds of the slug differentiate into spore cells. In this stage the slime mold consists of a thin stalk supporting a mass of hundreds of spores. These fruiting bodies look very much like tiny lollipops standing on end. Stalk cells die with each generation, but the spores are dispersed. When conditions are favorable, each spore cell liberates a new amoeba, and the life cycle begins anew. Sexual stages are not known.

Symbiotic Relationships of Fungi

Some fungi form symbiotic relationships with other organisms. Recall that symbiosis (living together) is an intimate living arrangement between members of two different species. When the relationship is beneficial to both species, it is referred to as **mutualism.** If beneficial to one species but without apparent effect on the second species, it is **commensalism.** When one species benefits and the other is harmed by the relationship, it is called parasitism.

THE LICHENS

Although a **lichen** looks like an individual plant, it is actually an intimate combination of a **phototroph** (an organism that carries on photosynthesis) and a fungus. The phototrophic component is usually either a green alga or cyanobacterium, and the fungus is most often an ascomycete (Fig. 21–19(a)). In some lichens from tropical regions the fungus partner is a basidiomycete. The phototrophs found in lichens are also found as free-living species in nature, but the fungal component of the lichen is generally found only as part of the lichen. In the laboratory the fungal and algal components can be separated and grown separately in appropriate culture media. The alga grows more rapidly when separated, but the fungus grows slowly and requires many complex carbohydrates. Generally, the fungus does not produce fruiting bodies when separated in this way. The phototroph and fungus can be reassembled as a lichen, but only if they are placed in a culture medium under conditions incapable of supporting either of them independently.

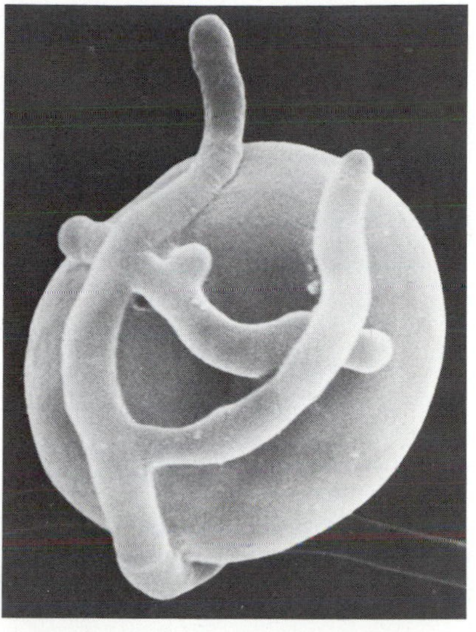

(*a*)

(*b*)

(*c*)

(*d*)

Figure 21–19 Lichens. (*a*) A fungal hypha encircling a single algal cell. (*b*) Lichen ascocarps. (*c*) A lichen from northern Michigan. (*d*) A multicolored lichen from Bylot Island, in northeastern Canada. ((*a*), from Ahmadjian, V., and Jacobs, J. B.: *Nature* 289:169–172, 1981. (*b*), Carolina Biological Supply Company. (*c*), courtesy of Leo Frandzel. (*d*), E. R. Degginger.)

What is the nature of this partnership? In the past it has been considered a definitive example of mutualism. The phototroph carries on photosynthesis, producing food for both members of the lichen. However, it is unclear how the phototroph benefits from the relationship. It has been suggested that the phototroph obtains water and minerals from the fungus as well as protection, mainly against desiccation. Some investigators have suggested that the lichen partnership is not really a case of mutualism but one of controlled parasitism of the phototroph by the fungus.

There are some 20,000 species of lichens (Fig. 21–19(b)–(d)). Resistant to extremes of temperature and moisture, lichens grow everywhere that life can be supported at all except in polluted, industrial cities. They exist farther north than any plants of the Arctic region and are equally at home in the steaming equatorial jungle. They grow on tree trunks, mountain peaks, and bare rock. In fact, they are often the first organisms to inhabit bare rocky areas and play an important role in the formation of soil. Lichens gradually etch the rocks to which they cling, facilitating disintegration of rocks by wind and rain.

The reindeer mosses of Arctic regions are lichens that serve as the main source of food for the reindeer and caribou of the region. Some lichens produce colored pigments. One of these, orchil, is used to dye woolens, and another, litmus, is widely used in chemistry laboratories as an acid–base indicator.

Lichens vary greatly in size. Some are almost invisible; others, like the reindeer mosses, may cover miles of land with a growth that is ankle deep. Growth proceeds slowly; the radius of a lichen may increase by less than a millimeter each year. Some mature lichens are thought to be thousands of years old.

Lichens absorb minerals mainly from the air and from rainwater, although some may be absorbed directly from their substrate. They have no means of excreting the elements they absorb, and perhaps for this reason they are very sensitive to toxic compounds. Lichen growth has been used as an air pollution indicator, especially of sulfur dioxide. Absorption of such toxic compounds results in damage to the chlorophyll. Return of lichens to an area indicates a reduction in air pollution.

When a lichen dries out, photosynthesis stops, and the lichen enters a state of suspended animation in which it can tolerate severely adverse conditions such as great extremes of temperature. Lichens reproduce mainly by asexual means, usually by fragmentation. Generally, bits of the thallus break off and land on a suitable substrate where they establish themselves as new lichens. Special dispersal units containing cells of both partners are released by some lichens. In others, the alga reproduces asexually by mitosis, while the fungus produces ascospores. The spores may be carried off by the wind and find an appropriate algal partner only by chance.

MYCORRHIZAE: FUNGUS-ROOTS

Mycorrhizae (fungus-roots) are fungi that form symbiotic relationships with the roots of higher plants (Fig. 21–20). Such relationships occur in more than 90% of all families of higher plants. The fungus may be an oomycete, a basidiomycete, or an ascomycete. The fungus benefits the plant by decomposing organic material in the soil, thus making the minerals available to the plant. The roots supply sugars, amino acids, and some other organic substances that may be used by the fungus.

The importance of mycorrhizae first became evident when horticulturalists observed that orchids do not grow unless they are colonized by an appropriate fungus. Similarly, it has been shown that many forest trees die from malnutrition when transplanted to nutrient-rich grassland soils that lack appropriate fungi. When forest soil that contains the appropriate fungi or their spores is added to the soil around these trees, they quickly assume a normal growth pattern.

Economic Importance of Fungi

The vital ecological role of fungi as decomposers has been mentioned previously. It is well to remember that without these organisms, life on earth eventually would become impossible. The same powerful digestive enzymes that enable fungi to decompose wastes and dead organisms also permit them to reduce wood, fiber, and food to their components with great efficiency. Various molds produce incalcu-

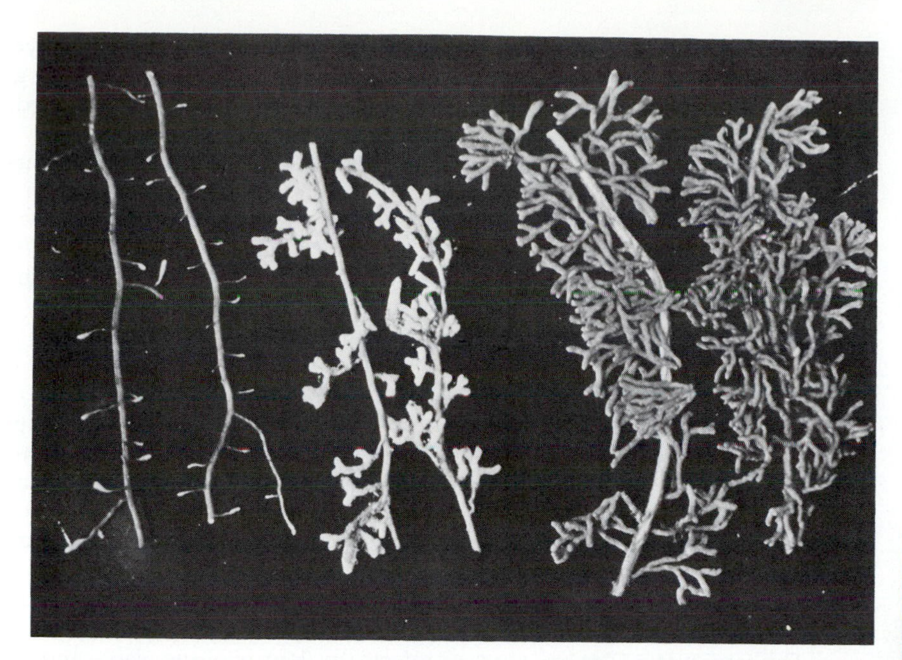

Figure 21–20 Mycorrhizae. (Dr. D. H. Marx, USDA Forest Service.)

lable damage to stored goods and building materials each year. Bracket fungi cause enormous losses by bringing about the decay of wood, both in living trees and in stored lumber. The amount of timber destroyed each year by these basidiomycetes approaches in value that destroyed by forest fires.

FUNGI FOR FOOD

Among the basidiomycetes, there are some 200 kinds of edible mushrooms and about 70 species of poisonous ones, sometimes called toadstools (Fig. 21–21). Edible mushrooms can be cultivated commercially: more than 60 thousand metric tons are produced each year in the United States alone. The morels, which are gathered and eaten like mushrooms, and truffles, which produce underground ascocarps, are ascomycetes. These delights of the gourmet are now being cultivated as mycorrhizae on the roots of tree seedlings.

Edible and poisonous mushrooms can look very much alike and may even belong to the same genus. There is no simple way to distinguish edible from poisonous mushrooms; they must be identified by an expert. The toxins in poisonous mushrooms may protect these fungi from predators. For humans, the most toxic substances in mushrooms are certain cyclopeptides (amatoxins and phallotoxins). One of these cyclopeptides strongly inhibits messenger RNA synthesis in animal cells. Some of the most poisonous mushrooms belong to the genus ***Amanita.*** Toxic species of this genus have been appropriately called such names as "destroying angel" *(Amanita virosa)* and "death angel" *(Amanita phalloides)*; ingestion of a single cap can kill a healthy adult.

Ingestion of certain species of mushrooms causes intoxication and hallucinations. The sacred mushrooms of the Aztecs, *Conocybe* and *Psilocybe,* are still used in religious ceremonies by Central American Indians and others for their hallucinogenic properties. The chemical ingredient psilocybin, chemically related to lysergic acid diethylamide (LSD), is responsible for the trancelike state and colorful visions experienced by those who eat these mushrooms.

The ability of yeasts to produce ethyl alcohol and carbon dioxide from glucose in the absence of oxygen is of great economic importance. The yeasts used in winemaking are usually the wild yeasts normally found on the skins of grapes. Some of the differences in flavors of the different kinds of wines are due to the kind of yeasts present in the grape-growing region. The yeasts used in brewing beer and in baking are cultivated yeasts carefully kept as strains to prevent contamination. Beer is made by fermenting grain, usually barley, flavored with hops (the dried conelike fruits of the female *Humulus lupulus* plant, a member of the mulberry family). During germination, the grain plant embryo degrades its starchy food supply to simple sugars, which are then fermented by the yeast.

(*a*)

(*b*)

Figure 21–21 Poisonous mushrooms. (*a*) *Amanita muscaria,* or fly amanita, is probably the best known of the poisonous mushrooms. Its colorful yellow cap is covered by white scales. This species is not as toxic as others in the same genus, and has been used in some cultures as a hallucinogen. (*b*) *Amanita phalloides,* the death angel. About 2 ounces (50 g) of this mushroom could kill a 68-kg man. ((*a*), (*b*), courtesy of Leo Frandzel.)

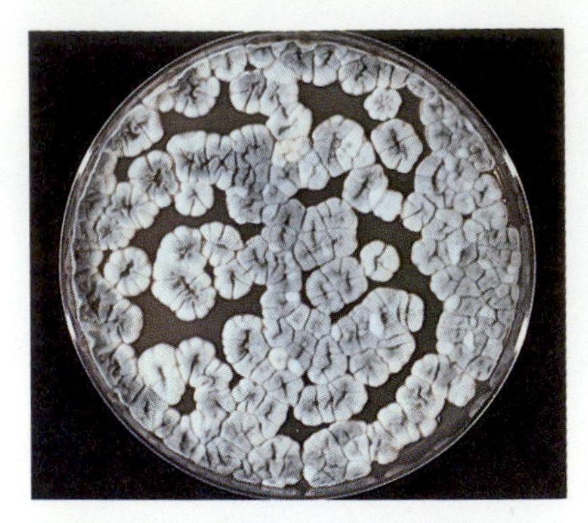

Figure 21–22 Colonies of *Penicillium chrysogenum* growing in a culture medium. Since Fleming's time, *P. chrysogenum* has become the fungus species most widely used to produce the antibiotic penicillin. (Courtesy of Fermenta Products, Inc.)

During the process of making bread, carbon dioxide produced by the yeast becomes trapped in the dough as bubbles that cause the dough to rise and give leavened bread its light texture. Both the carbon dioxide and alcohol produced by the yeast are driven off during baking.

As noted earlier, many cheeses owe their unique flavors to specific fungi used in their production, and authentic Chinese soy sauce is produced only with the help of an ascomycete that slowly ferments boiled soybeans. Soy sauce provides other foods with more than its special flavor; it also adds vital amino acids from both the soybeans and the fungi themselves to the low-protein rice diet. Fungi have been utilized in many cultures to improve the nutrient quality of the diet.

FUNGI FOR DRUGS AND USEFUL CHEMICALS

In 1928 Alexander Fleming noticed that one of his petri dishes containing staphylococci bacteria was contaminated by mold. The bacteria were not growing in the vicinity of the mold, leading Fleming to the conclusion that the mold was releasing some substance harmful to them. Within a decade of Fleming's discovery, penicillin produced by the deuteromycete *Penicillium notatum* was purified and used in treating bacterial infections (Fig. 21–22). Penicillin is still the most widely used and most effective antibiotic. Another fungus, *Penicillium griseofulvicum*, produces the antibiotic griseofulvin, which is used clinically to inhibit the growth of fungi.

The ascomycete *Claviceps purpurea* infects the flowers of rye plants and other cereals. It produces a structure called an ergot where a seed would normally form in the grain head (Fig. 21–23). When livestock eat this grain or when humans eat bread made from the infected rye they may be poisoned by the very toxic substances in the ergot. The toxic condition caused by eating grain infected with ergot often causes nervous spasms, convulsions, psychotic delusions, and even gangrene. This condition, called ergotism, was known as St. Anthony's fire during the Middle Ages, when it occurred often. In 994 A.D. an epidemic of St. Anthony's fire caused more than 40,000 deaths. In 1722, the cavalry of Czar Peter the Great was felled by ergotism on the eve of the battle for the conquest of Turkey. This was one of several recorded times that a fungus changed the course of history. Lysergic acid, one of the constituents of ergot, is an intermediate in the synthesis of lysergic acid diethylamide (LSD). Some of the compounds produced by ergot are now used clinically in small quantities as drugs to induce labor, to stop uterine bleeding, to treat high blood pressure, and to relieve one type of migraine headache.

Fungi can be used as biological control agents to prevent damage by many insect pests. They can replace some very toxic biocides that are environmentally damaging. Other fungi are used commercially to produce citric acid and other chemicals of high quality.

Figure 21–23 The ascomycete *Claviceps purpurea* infects the flowers of cereals. It produces a structure called an ergot where a seed would normally form in the grain head. (Carolina Biological Supply Company.)

FUNGUS DISEASES OF PLANTS

Fungi are responsible for many serious plant diseases, including epidemic diseases that spread rapidly and often result in complete crop failure, causing great economic loss and human suffering in some cases (Table 21–2). All plants are appar-

TABLE 21–2
Some Plant Diseases Caused by Fungi

Disease	*Symptoms*	*Causative Agent*
Chestnut blight	Rapid browning of leaves, branches, and twigs, as well as flowers; results in death	An ascomycete, *Endothia*
Rust of cereals	Many small, usually rusty-colored lesions on leaves and stems	A basidiomycete, *Puccinia*
Downy mildew of grapes	Loss of green color on leaves, stems, and fruit, or death of these structures; usually a layer of fungal mycelia covering these structures	An oomycete, *Plasmopara*
Wilt disease	Drooping of leaves or stems following loss of turgidity caused by disturbance in vascular system	A deuteromycete, *Fusarium*
Root rot of vegetables	Decay of part or all of the root system of the plant	An oomycete, *Aphanomyces*

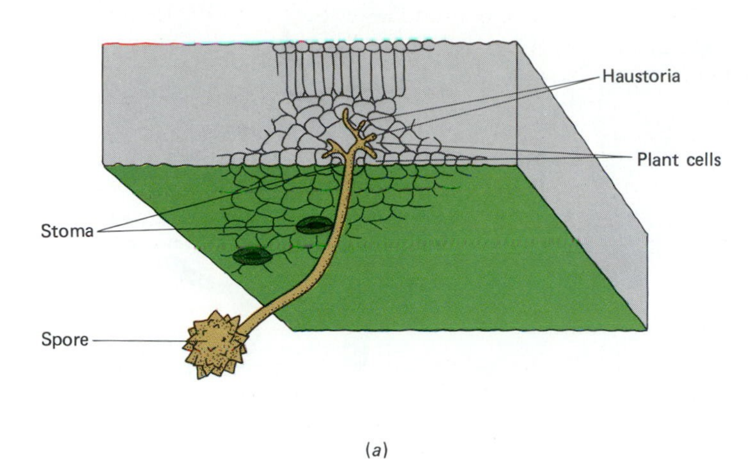

(a)

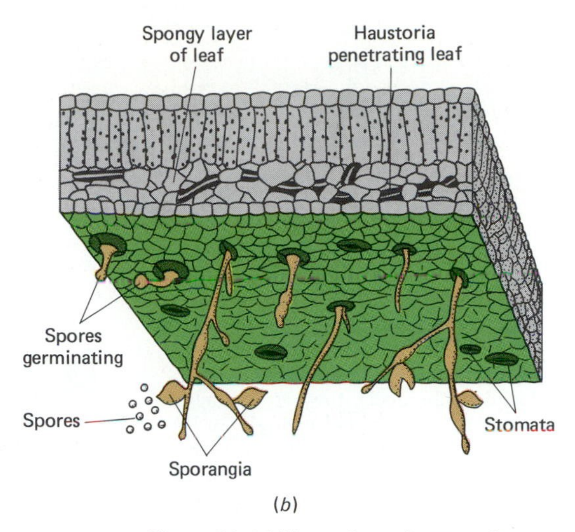

(b)

Figure 21–24 How plants become infected with fungi. (*a*) Germ tubes of hyphae enter through stomata or through wounds. Haustoria penetrate host cells and obtain nourishment from the cytoplasm. (*b*) The fungus may develop sporangia beneath the outer epidermis of the leaf.

ently susceptible to some fungal infection. Damage may be localized in certain tissues or structures of the plant, or the disease may be systemic, affecting the entire plant. Fungus infections may cause stunting of plant structures or of the entire plant; they may cause growths like warts; or they may kill the plant.

Generally, a plant becomes infected after germ tubes of hyphae enter through pores in the leaf or stem (stomata) or through wounds in the plant body. Fungal mycelia grow and may remain mainly between the plant cells or may penetrate the cells. Parasitic fungi often produce special hyphal branches called **haustoria,** which penetrate the host cells and obtain nourishment from the cytoplasm (Fig. 21–24). When the fungus is ready to reproduce, it may develop sporangia beneath the outer epidermis of the plant. Eventually, the epidermis ruptures, releasing the fungal spores into the air.

Some important plant diseases caused by ascomycetes are chestnut blight, Dutch elm disease, apple scab, and brown rot, which attacks cherries, peaches, plums, and apricots (Fig. 21–25). Grapes are attacked by a downy mildew introduced into France from the United States. This almost destroyed the French vineyards in wet years during the 1870s. An effective fungicide, called Bordeaux mixture, was accidentally discovered when a local vineyard owner sprayed his grapes with a disgusting-looking mixture of calcium hydroxide (lime) and copper sulfate to discourage people from stealing his grapes. In 1882 a University Professor named Pierre Millardet noticed that the sprayed vines did not have mildew. He mixed copper hydroxide and calcium hydroxide (lime), sprayed it on grape vines, and found that the fungal spores produced sufficient acid to dissolve the insoluble copper hydroxide. The copper kills the spores. This Bordeaux mixture, the first and still one of the best fungicides, also kills germinating spores of many other plant fungi.

Basidiomycetes include some 700 species of smuts and 6000 species of rusts that attack the various cereals—corn, wheat, oats, and other grains. In general,

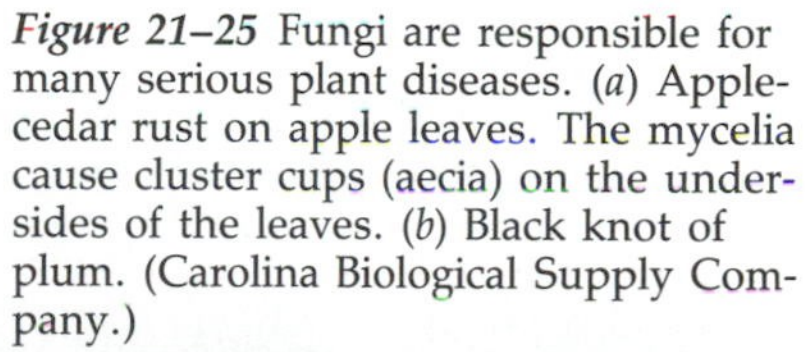
Figure 21–25 Fungi are responsible for many serious plant diseases. (*a*) Apple-cedar rust on apple leaves. The mycelia cause cluster cups (aecia) on the undersides of the leaves. (*b*) Black knot of plum. (Carolina Biological Supply Company.)

(a)

(b)

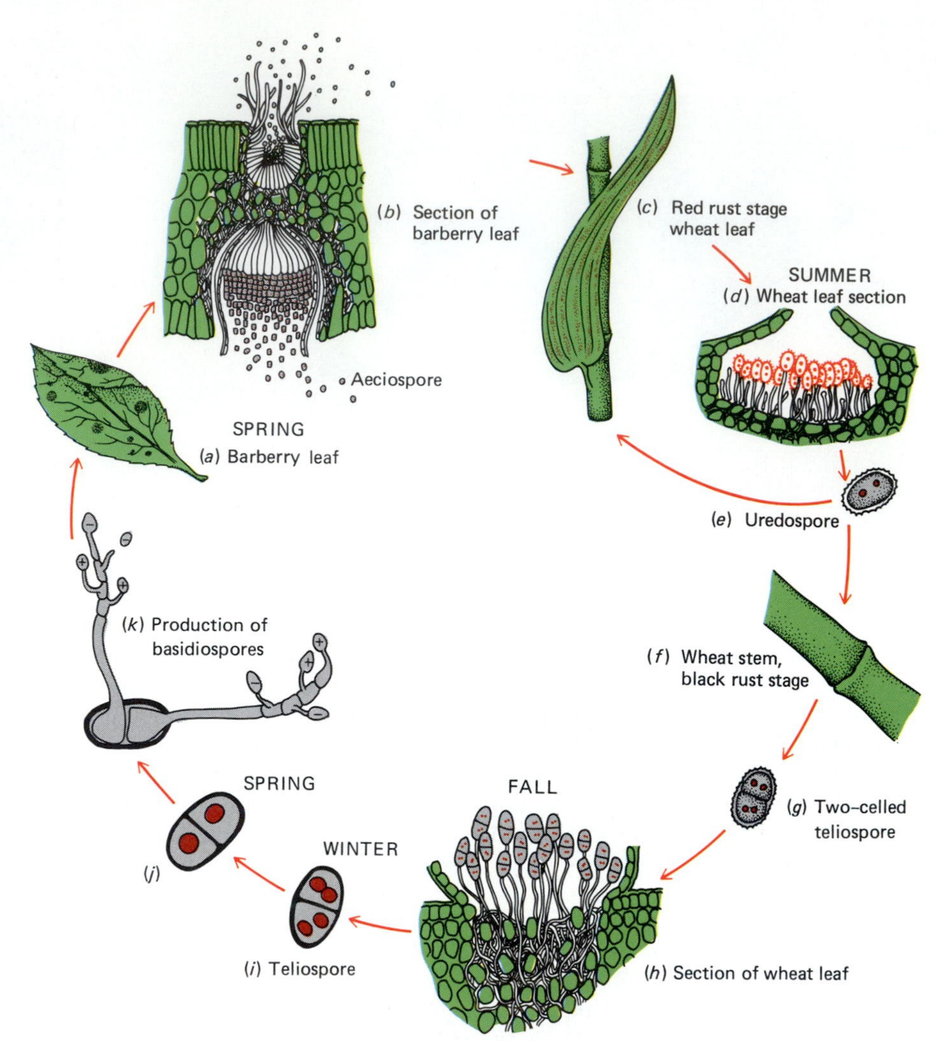

Figure 21–26 Life cycle of the wheat rust, *Puccinia graminis.* (*a*) In the spring basidiospores from infected wheat plants of the previous year infect leaves of the barberry plant, forming pycnia containing clusters of spermagonia on the upper surfaces and cluster cups, or aecia, on the lower surface of the leaf, as shown in (*b*). Aeciospores are produced in the aecia; they are binucleate, containing n + n (not a single 2n nucleus) chromosomes. (*c*) In early summer the aeciospores infect the leaves of young wheat plants. (*d*) They develop into clusters of red, single-celled uredospores, producing the red rust stage. (*e*) Uredospores are released and infect other wheat plants, producing more uredospores. In late summer, uredospores develop into dark brown, two-celled teliospores, as shown in (*g*), on the stems and leaf sheaths of wheat plants, forming the black rust stage, as shown in (*f*). (*h*) A section of the wheat stem shows the n + n teliospores. (*i*) The teliospores are thick-walled and remain dormant over the winter. (*j*) In the spring the n + n nuclei within each cell of the teliospore fuse to form a 2n nucleus. The teliospore, still attached to the wheat plant, germinates and undergoes meiosis. (*k*) Each teliospore cell produces four basidiospores. The haploid basidiospores then infect a barberry leaf to complete the cycle.

each species of smut is restricted to a single host species. Some of these parasites, such as the stem rust of wheat and the white pine blister rust, have complex life cycles that involve two or more different plants, during which several kinds of spores are produced. The white pine blister rust must infect a gooseberry or a red currant plant before it can infect another pine. The wheat rust must infect an American barberry plant[1] at one stage in its life cycle (Fig. 21–26). Since this has been known, the eradication of American barberry plants in wheat-growing regions has effectively reduced infection with wheat rust. However, the eradication must be complete, for a single barberry bush can support enough wheat rust organisms to infect hundreds of acres of wheat.

Even the complete eradication of barberry plants does not provide a final solution to the wheat rust problem, because there are different types of wheat rust spores. In the fall, the spores produced have thick walls, enabling them to survive a cold winter; these spores grow only if they land on a barberry. During most of the summer, thin-walled spores are produced that can infect other wheat plants directly, and infection spreads from plant to plant in a field in this way. If the winter is very mild, some of these thin-walled spores may survive and cause an infection the following year even in the absence of barberry plants.

FUNGUS DISEASES OF ANIMALS

Although the skin and mucous membranes of healthy animals present effective barriers to fungal penetration, some fungi cause disease in humans and other animals (Fig. 21–27). Some of these cause superficial infections in which the fungi

[1] The Japanese barberry plant commonly used today in hedges or as a decorative shrub, however, is resistant to infection by rust; only the American variety figures in the wheat rust life cycle.

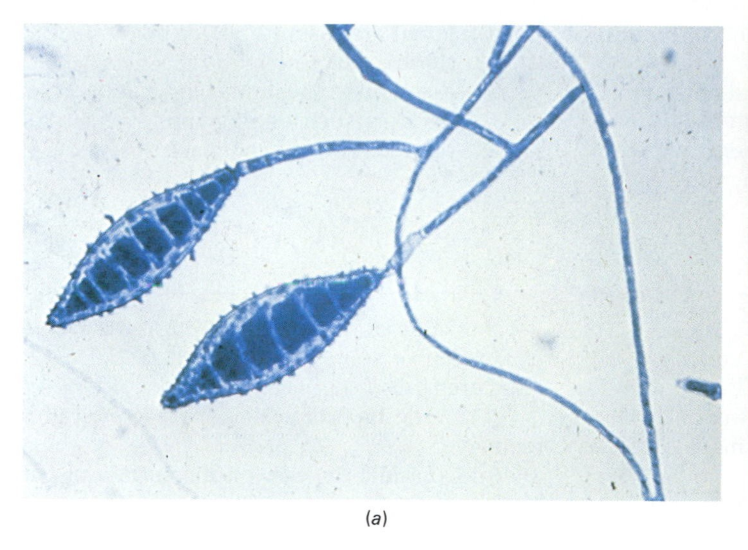

(a)

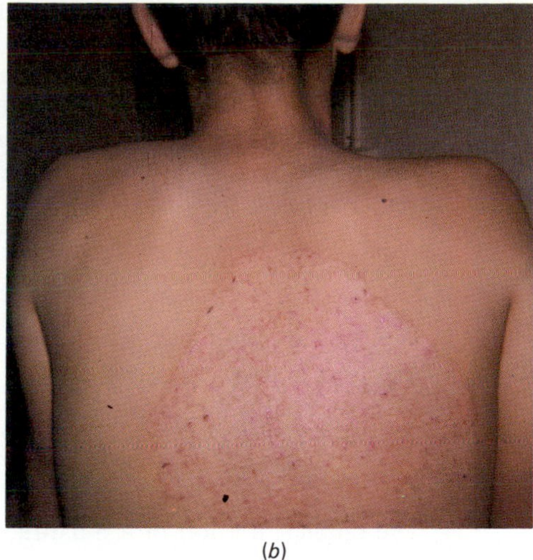

(b)

Figure 21–27 (*a*) A fungus (*Microsporum canis*) carried by pets that can cause fungus infections in humans, as shown in (*b*). (Courtesy of Wilfred D. Little.)

infect only the skin, hair, or nails. Others cause systemic infections, in which fungi infect deep tissues and internal organs and may spread through many regions of the body.

Ringworm and athlete's foot are examples of superficial fungus infections. Candidiasis is an infection of mucous membranes of the mouth or vagina and is among the most common fungal infections. Histoplasmosis is a serious human systemic fungus infection caused by a fungus that sporulates abundantly in soil containing bird droppings; a person who inhales the spores may then develop the infection. Most pathogenic fungi are opportunists that cause infections only when the body's immunity is lowered.

SUMMARY

I. Fungi are eukaryotes with cell walls: they lack chlorophyll and so are heterotrophic, absorbing their food through the cell wall and cell membrane. They reproduce by means of spores and function ecologically as decomposers.

II. A fungus may be unicellular, the yeast form, or multicellular, the mold form. A mold consists of long, branched hyphae, which form a mycelium. Some hyphae are coenocytic. The body of a fungus is called a thallus.

III. When a fungal spore comes into contact with an appropriate substrate, it germinates and begins to grow. Some hyphae infiltrate the substrate, digesting its organic compounds with enzymes. Spores are produced on aerial hyphae.

IV. Division Eumycophyta includes several classes of true fungi.

- *A.* The oomycetes are characterized by their flagellated spores. Sexual reproduction results in a zygote, which develops into a resting oospore. Members of this group include *Plasmopara,* which causes downy mildew of grapes, and *Phytophthora,* which causes potato blight.
- *B.* The zygomycetes produce sexual resting spores called zygospores. They have coenocytic hyphae and cell walls of chitin. The black bread mold *Rhizopus nigricans* is a common member of this group.
- *C.* The ascomycetes produce asexual spores called conidia at the tips of conidiophores. Sexual spores called ascospores are produced in asci. This group includes yeasts, morels, truffles, red and brown molds, and many others.
- *D.* The basidiomycetes produce basidiospores on the outside of a basidium. Basidia develop on the surface of gills in mushrooms. This class includes mushrooms, toadstools, rusts, and smuts.
- *E.* The deuteromycetes are the imperfect fungi, species for which a sexual stage has not been observed. Most reproduce by conidia. Members of this class include *Aspergillus tamarii,* used to produce soy sauce, and *Candida albicans,* which can cause human fungal infections.

V. Division Myxomycophyta includes the plasmodial and cellular slime molds, organisms with little similarity to other molds except that they produce spores.

VI. Some fungi establish symbiotic relationships with other organisms.

- *A.* A lichen is an intimate combination of a fungus and a phototroph in which the fungus benefits from the photosynthetic activity of the phototroph. Lichens play an important role in soil formation.
- *B.* Mycorrhizae are fungi that form symbiotic relationships with the roots of higher plants. The fungus decomposes organic material, making minerals available to the plant. The plant may secrete organic compounds needed by the fungus.

VII. Fungi are of both positive and negative economic importance.

- *A.* Mushrooms, morels, and truffles are used as food; yeasts produce ethyl alcohol and so are vital in production of wines and beer, and are also used to make bread; certain fungi are needed to produce cheeses and soy sauce.

B. Fungi are used to make penicillin and other antibiotics; ergot is used to produce certain drugs; fungi can be used as biological control agents; they can be used to make citric acid and many other industrial chemicals.

C. Fungi are responsible for many plant diseases including potato blight, wheat rust, Dutch elm disease, and chestnut blight; they cause human diseases such as ringworm, athlete's foot, histoplasmosis, and candidiasis.

POST-TEST

1. Ecologically, fungi serve as ________.
2. Yeasts reproduce asexually mainly by ________, but also by ________.
3. A mold consists of threadlike strings of cells called ________, which form a tangled mass called a ________.
4. The familiar portion of a mushroom is actually a large ________ ________.
5. Division Eumycophyta includes the ________ ________; Division Myxomycophta includes the ________ ________.
6. The thick-walled resistant spores formed by members of class Oomycetes (water molds) are known as ________.
7. The zoospores of oomycetes (water molds) are formed within a ________; they are characterized by the presence of two ________, which enable them to move about.
8. Rhizopus and other members of the zygomycetes form sexual resting spores called ________.
9. The term heterothallic means that there are two ________ ________.
10. In Ascomycetes (sac fungi), asexual reproduction involves formation of spores called ________, which are pinched off at the tips of ________.
11. Sexual reproduction in Ascomycetes involves production of spores known as ________ within structures called ________.
12. The type of sexual spore produced by a mushroom is a ________.
13. Basidia develop on the surface of perpendicular plates called ________.
14. The deuteromycetes are known as imperfect fungi because their ________ stage has not been observed.
15. The pseudoplasmodium of a cellular slime mold is formed by the aggregation of individual ________ ________.
16. A ________ consists of a phototroph and a fungus that are intimately related.
17. Mycorrhizae are fungi that form symbiotic relationships with the ________ of complex ________.
18. When the ascomycete *Claviceps purpurea* infects the flowers of cereals, it produces a structure called an ________ that is toxic.
19. Haustoria are hyphae produced by parasitic fungi that can ________.
20. Wheat rust must infect an American ________ ________ at one stage in its life cycle.

REVIEW QUESTIONS

1. How does a fungus differ from an alga? What are the distinguishing features of a fungus?
2. How does the body plan of a yeast differ from that of a mold? Describe the body structure of a mold.
3. What is the difference between a hypha and a mycelium? between an ascus and a basidium?
4. Describe the life cycle of a typical mushroom.
5. What is the ecological importance of fungi? of lichens? of mycorrhizae?
6. What measures can you suggest to prevent bread from becoming moldy?
7. Draw diagrams to illustrate the life cycle of the black bread mold *Rhizopus nigricans*.
8. In what ways are the cellular slime molds like true fungi? In what ways do they resemble animals?
9. Briefly describe three important fungal diseases of plants, and three fungal diseases of humans.
10. How could fungal disasters such as the Irish potato famine be prevented?
11. What conditions might permit opportunistic fungi to cause disease?
12. In what ways are the oomycetes different from the other groups of fungi?

22
The Simpler Plants

OUTLINE

I. Division Chlorophyta: green algae
- *A.* Lower green algae: unicellular and colonial forms
  - *1.* Motile colonial green algae
  - *2.* Filamentous green algae
  - *3.* Siphonous green algae
- *B.* Higher green algae: multicellular (plantlike) organization

II. Division Phaeophyta: brown algae

III. Division Rhodophyta: red algae

IV. Division Bryophyta: nonvascular terrestrial plants
- *A.* Musci: mosses
- *B.* Hepaticae: liverworts

LEARNING OBJECTIVES

After you have read this chapter you should be able to:

1. Summarize the principal features of the chlorophytes and their life cycles, with emphasis on gamete formation and the chromosome number characteristic of the stages.
2. Compare the plantlike green algae with the protist-like green algae.
3. Summarize the characteristics of the three divisions of plantlike algae.
4. Give the major traits of the bryophytes, and list the general adaptations of the bryophytes to the terrestrial environment.

What we commonly call *plants*—as opposed to the plantlike or algal protists—are truly multicellular photosynthetic organisms. In this chapter we will consider the more complex types of algae, which are sometimes classified as plants, and some more typical plants of still greater complexity. Later in this chapter Table 22–1 compares the characteristics of the members of the plant kingdom.

Distinctions between the complex algae and the simplest plants blur and are difficult to assess. Certain of the plantlike algae resemble elaborate protists more than plants and are classified as protists by some scientists. In fact, a single green alga cell is not greatly different from a cell taken from the spongy tissue of a maple leaf.

This raises a practical problem in classification. The plantlike algae include three divisions: Chlorophyta, Phaeophyta, and Rhodophyta. Each of these divisions contains some representatives that are protist-like. (Recall that a plant *division* corresponds roughly to an animal *phylum*.) For instance, some algae in division Chlorophyta are single-celled organisms that, but for their cellulose walls and single chloroplast, might be taken for euglenoid protists. However, since the bulk of each of these higher algal divisions *is* plantlike, it seem logical to classify all of their members as plants. (The alternative would be to consider part of each division protists and the rest plants—an even less satisfactory solution.) The taxonomic system employed in this book is only one of several in common use, and you should be aware that it is *not* perfect. There is no perfect taxonomic system, for each represents a need to impose an artificial human structure on nature, whose order is not ours.

Division Chlorophyta: Green Algae

The **green algae** comprise the division Chlorophyta. Some green algae are unicellular or colonial organisms resembling mastigophoran protists; others are plantlike, with structures comparable to stems, leaves, and other typical plant parts. For this reason, these complex algae may almost be thought of as straddling two kingdoms.

About 7000 species of green algae live in a wide variety of habitats ranging from salt to fresh water to damp soil. Most botanists believe that the complex plants probably evolved from algae similar to the present-day green algae, because the chlorophytes have the greatest number of characteristics in common with the complex plants. The chlorophyte cell has a distinct nucleus with a nuclear membrane; its pigments—chlorophylls *a* and *b*, carotene, and xanthophyll—are organized into chloroplasts; energy is stored in starch molecules (or in some cases as oil); and a cellulose cell wall is present.

The simplest green algae are unicellular and motile, but more complex members of the division have many-celled bodies in the shape of filaments or flat leaflike structures. Even in the more complex species, the cells of the algal body are almost all alike, and there is little differentiation of tissues. Many green algae have flagella, at least in some stage of their life cycle, but some species are nonmotile. Reproduction may be asexual (by cell division or by the formation of spores) or sexual (by the union of gametes).

A few species of green algae are terrestrial, living on the moist, shady sides of trees, rocks, and buildings. Some even inhabit the spaces among the grains of coarse sandstones in desert or polar regions. One species has become adapted to living on the surface of snow and ice; it has red pigment in addition to chlorophyll and may grow in patches thick enough to give the snow a red tinge. Perhaps even odder, the green alga *Desmococcus* is found in the fur of the three-toed sloth and nowhere else on earth; the greenish hue imported by this alga to the fur may help to camouflage the animal in its leafy habitat.

The many-celled green algae living in fresh water include the pond scums; these occasionally grow very thickly in ponds and streams. Among the multicellular marine green algae living near the low tide mark and in the upper 6 meters of water is the sea lettuce, *Ulva*, with a body some 30 cm long but only two cells thick. It resembles a crinkled sheet of green cellophane. The common, nonmotile, single-celled freshwater green algae known as desmids have symmetrical, curved, spiny,

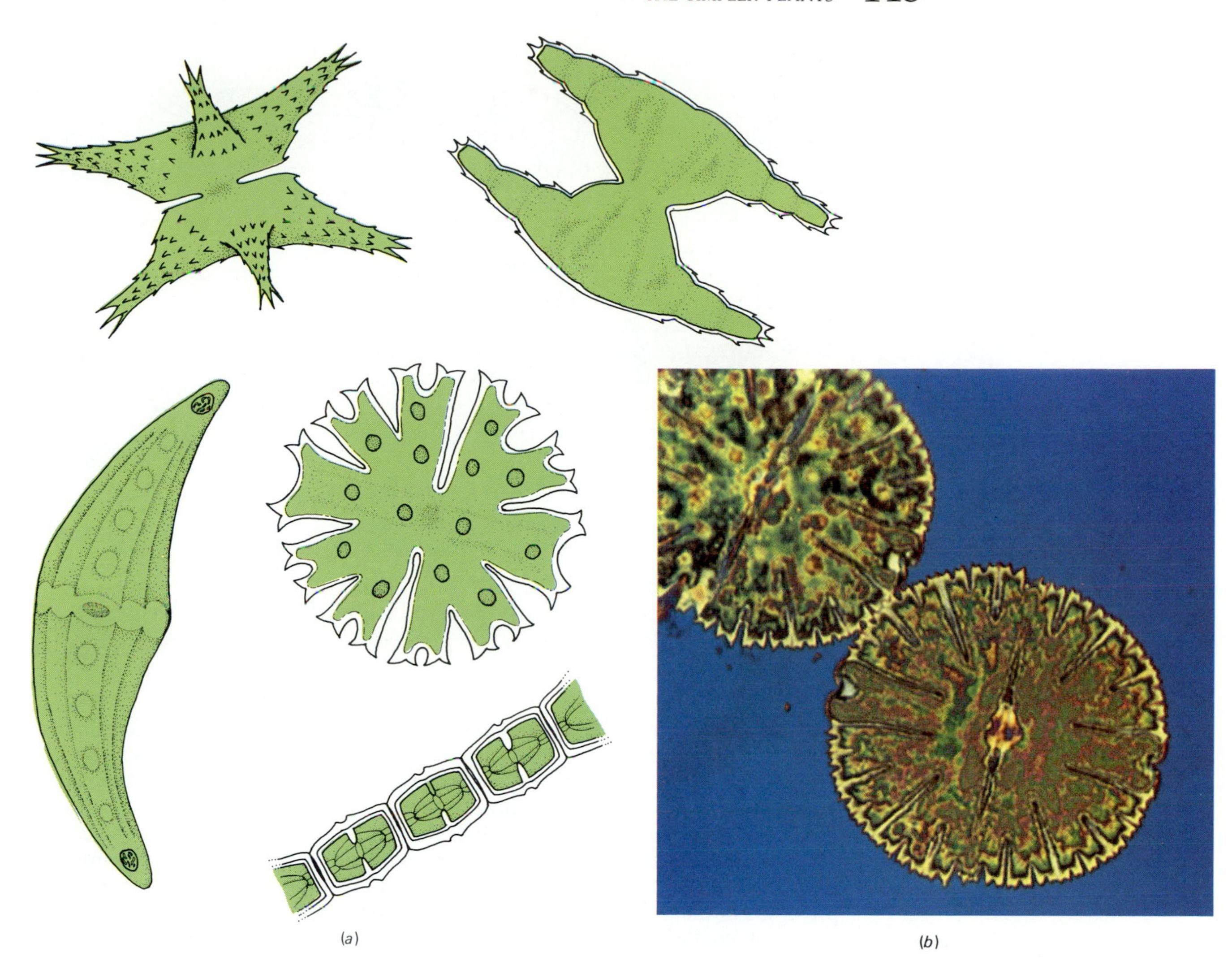

Figure 22–1 Desmids, unicellular green algae with some similarities to diatoms. (*a*) Several different species of desmids, highly magnified. Note the symmetry of the cells. (*b Micrasterias articulata.* ((*b*), courtesy of Jonathan G. Izant.)

or lacy bodies with a constriction in the middle of the cell; when seen under a microscope they look like snowflakes (Fig. 22–1).

The green algae exhibit an enormous variety of forms and reproductive processes. Asexual reproduction may occur by fragmentation of multicellular colonies. Many species of green algae form asexual spores termed **mitospores** (because they are produced by mitosis) or **zoospores** (if they are motile). Three variations in sexual reproduction among algae can be discerned: **Isogamy** refers to the union of morphologically identical, usually motile, gametes. The union of motile gametes, one of which is much larger than the other, is called **anisogamy,** and the union of a motile male gamete or sperm with a nonmotile egg is known as **oogamy.** (Isogamy is particularly characteristic of the simpler types of green algae.) In all green algae, gametes are formed in special cells called **gametangia.** After their release from the gametangia, the gametes, if they are motile, swim about for a short time, and then combine to form zygotes. The isogametes of some green algae are similar to the asexual zoospores, although generally smaller and physiologically weaker (see Fig. 22–2).

LOWER GREEN ALGAE: UNICELLULAR AND COLONIAL FORMS

The simpler chlorophytes are often referred to as the "lower" green algae, not because they should be considered inferior, but because they are not as highly differentiated or specialized as the other "higher" green algae. Three major groups of the lower green algae can be distinguished on the basis of their cellular organization: One group includes those forms with single motile cells or colonies in which all of the vegetative cells are motile. A second group includes those filamentous species in which the colony is composed of nonmotile cells. A third group contains algae with tubular multinucleate bodies.

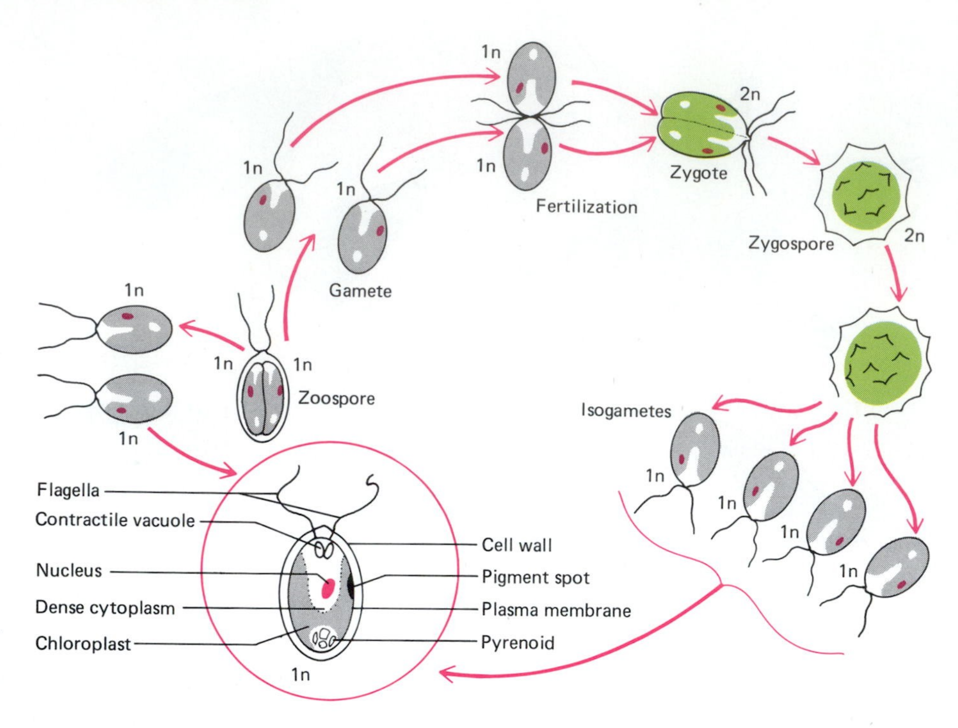

Figure 22–2 The life cycle of the green alga *Chlamydomonas. Left,* Asexual reproduction; *right,* stages in sexual reproduction. *Inset:* Enlarged view of a single individual showing body structures.

Motile Colonial Algae

The **motile colonial algae** are indeed all motile, but they may be either unicellular (e.g., *Chlamydomonas*) or colonial (e.g., *Volvox* or *Pandorina*).[1] One of the simplest green algae is the motile, freshwater *Chlamydomonas*, found in pools, lakes, and damp soil (Fig. 22–2). The vegetative cell bears two whiplash flagella at its anterior end and is protected by a heavy cellulose wall. (Contrast this with the somewhat similar protist *Euglena*, which has no cellulose cell wall and only one swimming flagellum.) Each cell has a single cup-shaped chloroplast containing a **pyrenoid** involved in the production and storage of starch. Other cellular structures are an eye spot, containing a red pigment, and two contractile vacuoles near the base of the flagella.

When nutrients are abundant and conditions of light and temperature are optimal in quiet standing pools, *Chlamydomonas* undergoes rapid asexual reproduction, discoloring the water and resulting in an algal "bloom" (Fig. 22–3). The cell divides to form two to eight zoospores within the cellulose wall; these are set free by the rupturing of the parental cell wall, and the zoospores swim away as independent individuals. When environmental conditions worsen, sexual reproduction occurs. Typically, the parent cell divides to form 8 to 32 gametes that resemble the zoospores and vegetative cells but are much smaller. Two of these gametes fuse beginning at the end bearing the flagella, forming a zygote. The fusion of the two isogametes results in a zygote with four flagella, two contributed by each gamete, but in time these are lost. The zygote becomes a dormant **zygospore;** it assumes a rounded shape and secretes a thick wall that enables it to survive long periods in an unfavorable environment. The isogametes are haploid cells; the zygote resulting from their fusion is diploid.

When favorable environmental conditions return, the zygospore becomes active and the zygospore nucleus divides by meiosis, forming four haploid zoospores. Zoospores formed directly by meiosis are called **meiospores.** The zoospores are motile and swim out of the zygospore when it breaks open, becoming independent vegetative cells; these cells reproduce asexually until environmental conditions lead to the generation of a new sexual cycle.

Chlamydomonas-like algae are sometimes colonial (Fig. 22–4), with a few or many cells living together in aggregates. However, in these colonies there is little specialization or division of labor among the cells. For instance, *Pandorina* forms a motile colony of 4 to 32 cells arranged as a hollow sphere in a jelly-like matrix. Each

Figure 22–3 Algal bloom. Green algae are growing just under the surface film of the water.

[1] Many classification schemes include *Chlamydomonas* and *Volvox* in the protists or protozoans (e.g., phylum Mastigophora, class Phytomastigophorea, order Volvocida; or phylum Protozoa, subphylum Sarcomastigophora, subclass Phytoflagellata, order Volvocida).

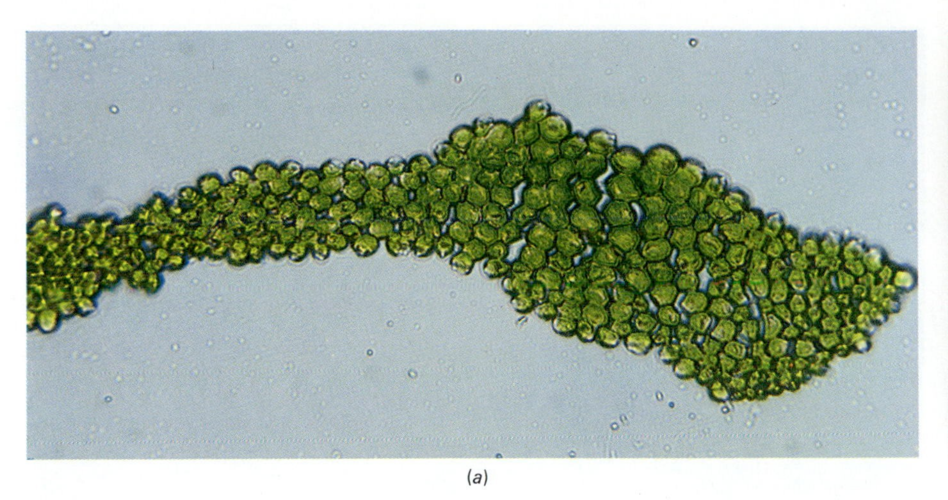

(a)

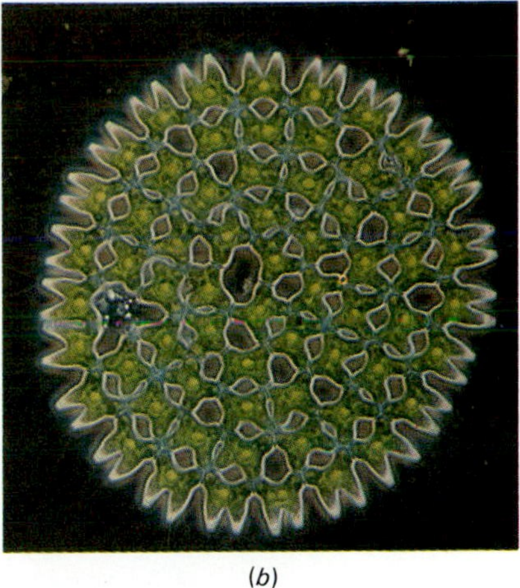

(b)

Figure 22–4 Colonial green algae. (*a*) Aggregation of *Chlamydomonas* cells. (*b*) *Pediastrum*, a motile colonial green alga. ((*a*), E. R. Degginger; (*b*), Walker of Stoke, Photo Researchers, Inc.)

cell resembles one cell of *Chlamydomonas*. Any of the cells may divide internally and produce from 4 to 32 zoospores. When these are released suddenly they escape as a unit and form a miniature new colony.

Volvox represents a further elaboration of this form of organization. A volvox colony is composed of a hollow ball of cells, each bearing two flagella and connected to its neighbors by fine cytoplasmic strands (Fig. 22–5). Each spherical colony may contain some 500 to 50,000 cells, most of which are alike and function only vegetatively. Small motile sperm cells that bear two flagella are produced only in special sperm-producing structures termed **antheridia.** A large single nonmotile

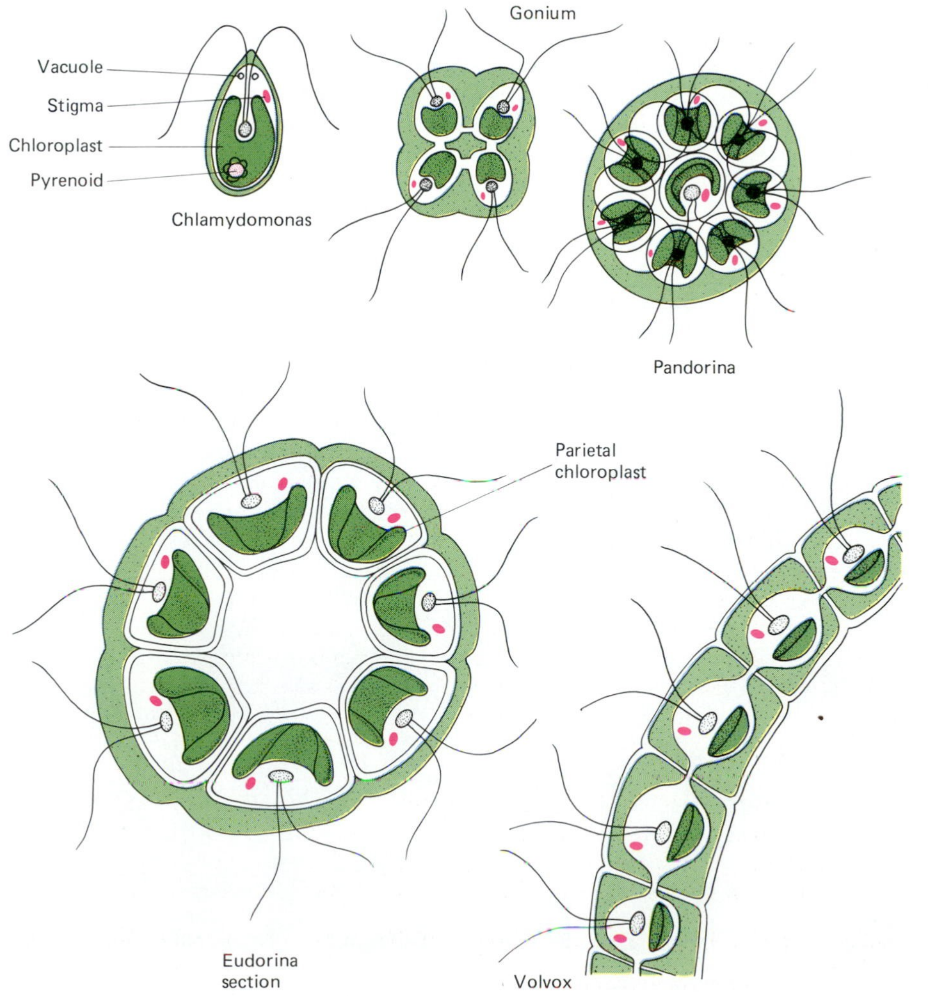

Figure 22–5 Some representative members of the volvox-like algae. Only a small section of the volvox colony is shown.

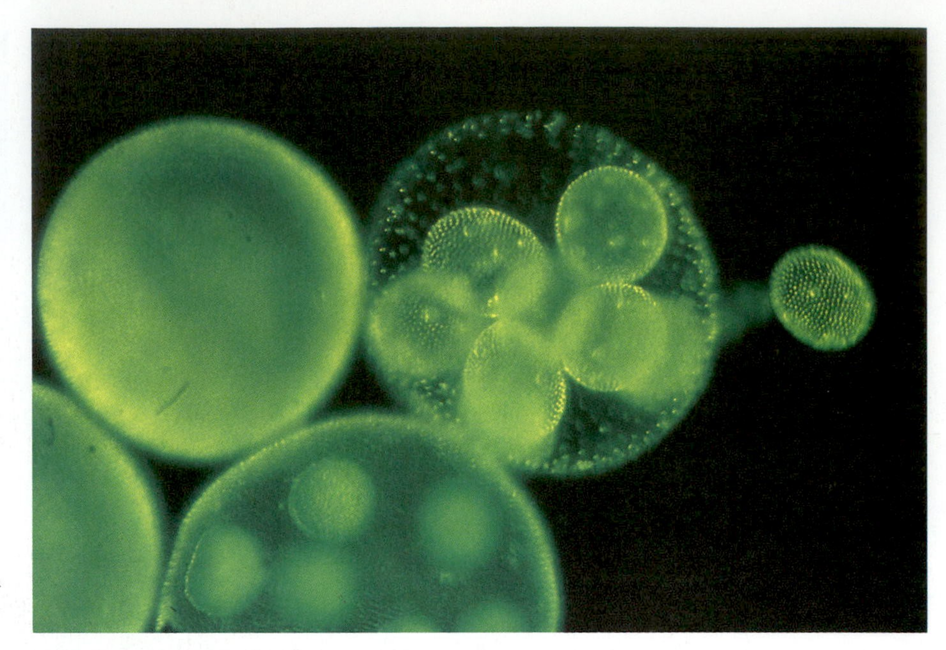

Figure 22–6 *Volvox* reproducing. A daughter colony can be observed leaving the mother colony through a rent in its wall. (James Bell, Photo Researchers, Inc.)

egg is produced within a special egg-producing **oogonium.** The motile sperm cells swim, if they can, to an egg; their union results in a diploid zygote, really a zygospore, which can resist harsh conditions such as winter or a dry spell. During zygospore germination, meiosis occurs, forming a haploid clone of colonial cells that develop over time into the mature colony—*all of whose cells are haploid.* This haploidy permits the production of gametes by mitosis.

Volvox may also reproduce asexually (Fig. 22–6). One cell in the colony may enlarge and then divide mitotically to form a mass of cells. These cells form a hollow sphere with their flagella directed inward. The sphere, which is complete except for a small opening, then turns itself inside out and floats freely within the parent colony. Several such daughter colonies may occur within a single mature mother colony of *Volvox* and become free-living when the mother colony breaks apart (Fig. 22–6).

Filamentous Algae

The **filamentous algae** include some nonmotile unicellular algae, more filamentous colonial forms, and some complex types, such as *Draparnaldia,* (Fig. 22–7), that

Figure 22–7 *Draparnaldia,* an unusual filamentous green alga. (*a*) Photomicrograph of section of filament. Note the frondlike lateral branches, each filament of which is composed of much smaller cells than those of the main filament. (*b*) Fruiting bodies. These develop from branches, representing a simple kind of cellular division of labor in these algae. Motile cells (zoospores), each provided with four flagella, are produced by these fruiting bodies. The zoospores swim away from the mother plant and eventually found new algal colonies. (Tom Adams.)

(*a*)

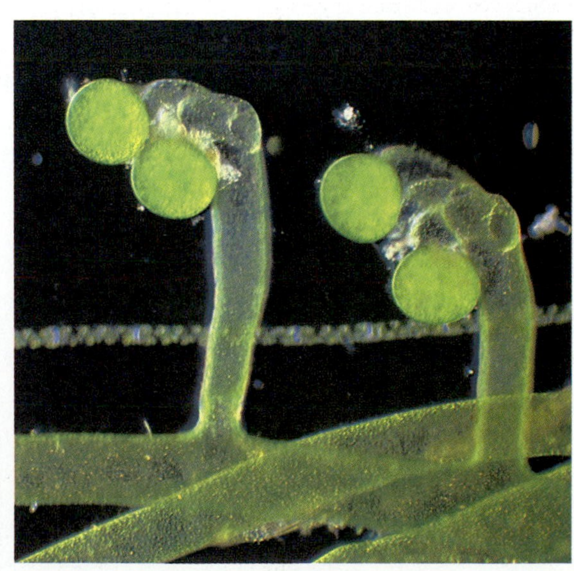

(*b*)

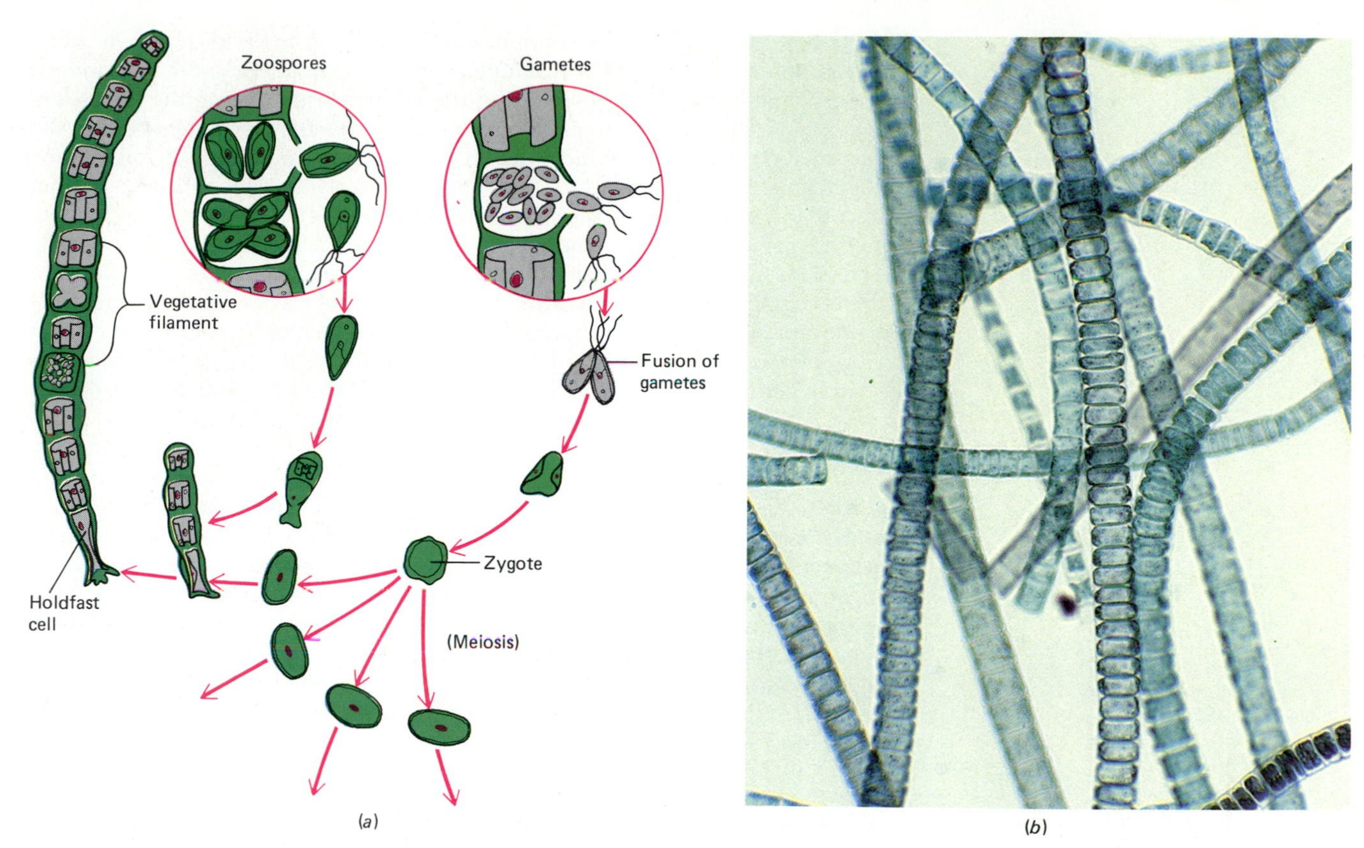

Figure 22–8 The filamentous green alga *Ulothrix*. (*a*) *Left*, a filament of *Ulothrix*. *Center*, An enlarged view of asexual reproduction by zoospores. *Right*, Sexual reproduction by the formation of gametes and the fusion of two gametes to form a zygote. (*b*) Filaments of *Ulothrix*. ((*b*), E. R. Degginger.)

approach a multicellular grade of organization. Some of the more complex types are really not filamentous, even though they are formally so classified. One of the simplest filamentous green algae is *Chlorococcus*, a nonmotile, unicellular alga that lives in soils and fresh water. It reproduces asexually by forming flagellated zoospores, which swim for a time and then lose their flagella and become nonmotile adults.

Often noticeable as greenish "frog scum" in still waters, or as hairlike masses attached to rocks in streambeds, the filamentous algae are ecologically very important members of freshwater communities. The freshwater green alga *Ulothrix* is an example of such filamentous forms. In this plant each haploid vegetative cell in the filamentous chain contains a single collar-shaped chloroplast and several pyrenoids that store starchlike polysaccharides. One cell may divide to form four to eight zoospores, each bearing four flagella, which are released and subsequently give rise to a new filament (Fig. 22–8). One of the cells of the filament may undergo several divisions to produce many small gametes resembling zoospores, but with two instead of four flagella. As in *Chlamydomonas*, two of the swimming gametes fuse, forming a zygote that initially has four flagella. Thus in *Ulothrix*, sexual reproduction is isogamous and occurs by the fusion of two morphologically identical cells. However, these cells are specialized gametes differing from the usual vegetative cells of the filament. The zygote eventually loses its flagella, secretes a thick cell wall, and becomes a zygospore that is capable of withstanding cold or drying. It later undergoes meiotic division and gives rise to four cells, which are liberated from the zygote wall and develop into new filaments. After swimming, the *Ulothrix* zoospore settles down on a suitable substrate and forms a **holdfast cell,** which anchors the plant and initiates growth of the new filament. Certain cells produce isogametes that are released into the surrounding water, where they fuse with other isogametes from another *Ulothrix* filament.

Another filamentous green alga is the pond scum, *Spirogyra*; its body consists of long filaments of haploid nonmotile cells arranged end to end. In the autumn, when reproduction usually occurs, two filaments come to lie side by side, and dome-shaped protuberances appear on the cells lying opposite (Fig. 22–9). These enlarge, fuse, and form a **conjugation tube** connecting the two cells. The contents of one cell become round and are then referred to as a gamete. This gamete oozes

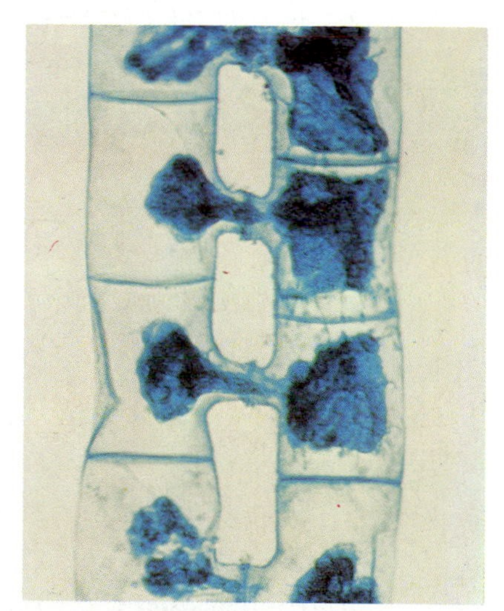

Figure 22–9 Sexual reproduction in *Spirogyra* occurs by fusion of nonflagellated gametes. Conjugation tubes form between cells of adjacent filaments of different mating types. Gametes of the donor filament migrate through the conjugation tubes and fuse with other gametes. The resulting zygotes form thick cell walls and become dormant. (Carolina Biological Supply Company.)

through the tube, and joins the second cell. The nuclei of the two cells then unite, and fertilization is complete. The resulting zygote develops a thick wall, becoming a zygospore, and is able to survive during the winter. In the spring the zygospore divides meiotically to form four haploid nuclei, three of which degenerate; the fourth remains, and after the thick wall breaks, it divides mitotically to form a new haploid filament. Sexual reproduction in *Spirogyra* involves unspecialized cells; that is, any cell in the filament can fuse with one from the neighboring filament, the two fusing cells being similar (isogamous). However, genetically identical *Spirogyra* cells will not mate with one another. In order to fuse, *Spirogyra* cells must belong to different mating types. These "sexes," however, are not physically distinguishable from one another. *Spirogyra* is characterized by its spiral chloroplasts; the fact that it employs no flagellated cells of any kind tends to remove it and other filamentous algae from consideration as possible ancestors of the higher plants.

Siphonous Algae

The **siphonous algae** are characterized by multinucleate cells. Each cell contains several or many nuclei rather than the single nucleus per cell characteristic of most other green algae. Multinucleate organisms, termed **coenocytes,** are found among the algae and fungi, but none of the higher green plants are coenocytic. Some of the siphonous or multinucleate marine green algae such as *Valonia* are easily seen with the unaided eye even though they are unicellular. *Acetabularia*, the "mermaid's wineglass," has been used in research on nuclear and cytoplasmic relationships, as was discussed in Chapter 4. It has one large nucleus near its base during most of its life cycle but becomes multinucleate later.

HIGHER GREEN ALGAE: MULTICELLULAR (PLANTLIKE) ORGANIZATION

The more complex chlorophytes are representatives of the higher algae, as are most species of Phaeophyta and Rhodophyta. The higher algae exhibit some degree of body differentiation and plantlike patterns of growth; they are certainly also multicellular.

The green alga *Ulva*, or sea lettuce, will serve both as an example of a complex chlorophyte and as an introduction to all the higher algae. *Ulva*, which is one of the simplest of the higher algae, is a flat, membranous marine plant composed of two layers of cells. It is sometimes classified with the filamentous algae. *Ulva* produces large numbers of isogametes, which, after release, fuse with those from other specimens to form zygotes (Fig. 22–10). The zygotes germinate to form diploid *Ulva* plants that are identical in appearance with the haploid *Ulva* plants, which produce isogametes. When the diploid *Ulva* plant matures, it produces haploid zoospores by meiosis. Since this generation produces spores, it is known as the **sporophyte generation.** These zoospores are released and settle upon a substrate, become attached, and grow into haploid *Ulva* plants, which subsequently produce isogametes. Since this generation produces gametes, it is known as the **gametophyte generation.** In *Ulva* there is an alternation of generations comparable to that in some of the complex plants, but because all the generations are **isomorphous** (that is, look alike), such alternation is termed **isomorphic.** As we will see, other plants also exhibit an alternation of gametophyte and sporophyte generations, but in almost all of them the generations are **heteromorphous**—that is, very different in form from one to another.

An even more plantlike group of green algae, the **stoneworts,** are freshwater forms with a worldwide distribution (Fig. 22–11). *Chara*, for example, is a comparatively large green alga with whorls of branches, both horizontal and vertical "stems," and specialized anchoring structures (rhizoids). Like the higher plants, *Chara* has an apical growth pattern—that is, an apical cell divides mitotically to form the cells further down the branch that subsequently develop into several kinds of structural and reproductive cells. Most stoneworts have calcified walls and complex parts specialized to produce gametes (gametangia). These multicellular algae resemble miniature trees with structures that superficially look like, and serve many of the functions of, roots, stems, leaves, and seeds, though they are not anatomically like their counterparts in the more typical plants. The "stem," for example, is only one cell in thickness.

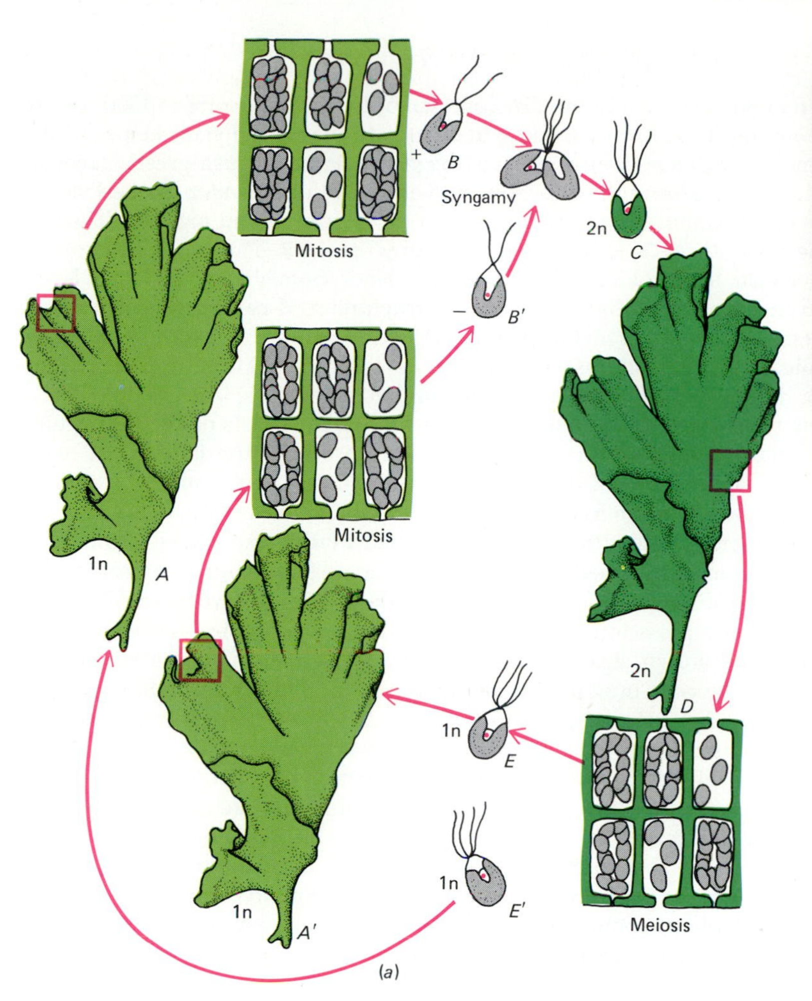

(b)

Figure 22–10 The green alga *Ulva.* (*a*) Life history of *Ulva,* shown diagrammatically. *A, A′,* Plus and minus gametophytes, respectively. *B, B′,* Plus and minus gametes. *C,* Zygote. *D,* Sporophyte plant. *E, E′,* Meiospores of the two mating types. (*b*) The thin, leaflike form of the blade has given *Ulva* its common name of "sea lettuce." ((*b*), J.N.A. Lott, McMaster University/BPS.)

The green algae have often been suggested as possible ancestors of the vascular plants because their cells and life cycles bear considerable resemblance to those of the complex plants. The remaining algal divisions are so different from both the green algae and the higher plants that most biologists feel there is no close evolutionary connection.

Figure 22–11 *Chara,* a stonewort or charophyte. Stoneworts are composed of long, single, coenocytic, internodal cells alternating with short nodal cells. The longer cells are up to several centimeters long. (J. N. A. Lott, McMaster University/BPS.)

Figure 22–12 *Laminaria,* a typical brown alga. This complex alga somewhat resembles a vascular plant in its differentiation of its body parts. (J. Robert Waaland, University of Washington/BPS.)

Division Phaeophyta: Brown Algae

The **brown algae,** comprising the division **Phaeophyta,** include about 1500 species of marine multicellular forms, ranging in size up to that of the giant kelps, whose bodies may be 100 meters long. Kelps are the prominent brownish-green seaweeds that usually cover rocks in the tidal zone and extend out into water 15 or so meters deep. The considerable amount of the golden-brown carotenoid pigment **fucoxanthin** in these plants tends to mask the chlorophyll present. The color of the plants thus ranges from light golden to dark brown or black. Some brown algae are large, highly advanced plants with complex body structures and parts that resemble the leaves, stems, and roots of higher plants (Fig. 22–12). In the algae, these parts are called **blade, stipe,** and **holdfast,** respectively, to indicate that they are not homologous with the corresponding structures of higher plants.

Brown algae are found in shallow waters along the coasts of all seas, but are larger and more numerous in cool waters. They are both the largest and most rugged of the algae. Some of the giant kelps are almost the aquatic equivalent of trees. Their stipes can reach 5 cm in diameter, and the whole organism can reach as much as 100 meters in length. They are attached by their holdfasts to the rocks beneath the surface and usually have gas-filled floats filled with a gas containing some carbon monoxide to buoy up the free ends toward the water surface. (The reason this usually poisonous gas is employed is not known.)

Among the brown algae, the plant body, or **thallus,** may be a simple filament, such as the soft brown tufts of *Ectocarpus,* commonly found on pilings; tough, ropelike, slimy strands, such as *Chorda,* the "Devil's shoelace"; or thick, flattened branching forms, such as *Fucus, Sargassum,* or *Nereocystis* (Fig. 22–13).

Most phaeophytes have a well-defined alternation of generations. *Ectocarpus,* like *Ulva,* consists of two kinds of plants that are similar in size and structure, but one produces gametes and the other produces spores. The diploid form produces haploid spores called zoospores that divide and grow into mature haploid plants. These produce haploid gametes that fuse to produce a diploid zygote; this develops into the diploid plant, completing the life cycle. Other brown algae have other kinds of life cycles, some exhibiting a great disparity in size between the sporophyte generation and the gametophyte generation.

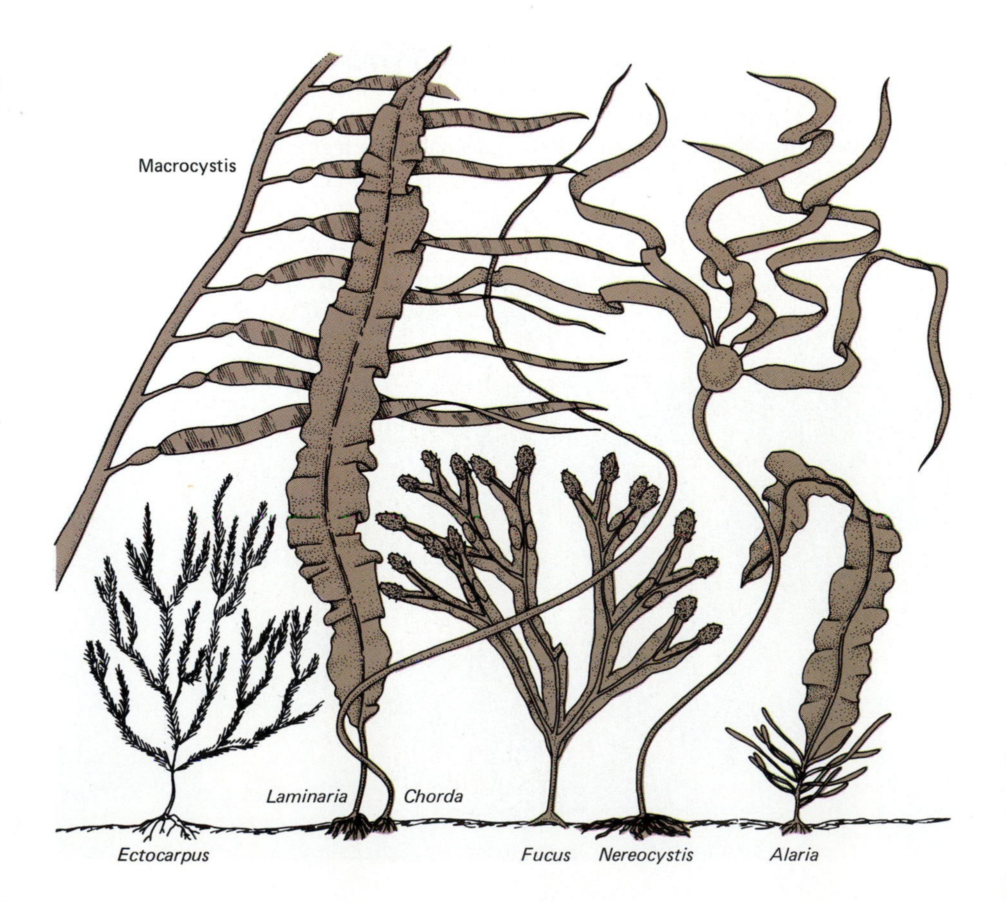

Figure 22–13 Some of the kinds of brown algae or kelps, all of which are multicellular marine plants. (The sketches are not drawn to the same scale.)

Brown algae furnish food and hiding places for many marine animals. In particular, the holdfasts of the largest forms are an important microhabitat for a variety of bottom-dwelling marine invertebrates. Some kelps are used as food in oriental countries. In addition to (and in some cases instead of) the more usual cellulose, kelps produce a number of unusual and commercially valuable polysaccharide carbohydrates. Kelps such as *Laminaria* (Fig. 22–12) are processed commercially to yield a colloidal carbohydrate known as **algin,** a pectin-like component of the cell wall. This has the ability to gel and thicken mixtures and is widely used in making ice cream, enabling ice cream manufacturers to use much less real cream and still have a smooth, apparently creamy product. Algin is also used in making candy, toothpaste, and cosmetics. Its value to the plant seems to lie in its ability to retain water and to minimize dehydration when these predominantly intertidal seaweeds are exposed to air by low tide.

Division Rhodophyta: Red Algae

Like the brown algae, the **red algae,** which make up the division **Rhodophyta,** are found almost entirely in the oceans. They are usually smaller and have more delicate bodies than the brown algae. A few species are unicellular, but most are multicellular filamentous forms or flattened sheets. Red algae have complex life cycles, with a marked alternation of sexual and asexual generations and specialized sex organs. They are unique among eukaryotes in having the red pigment **phycoerythrin** (Fig. 22–14) in addition to chlorophyll and phycocyanin and so are various shades of pink to purple. Recall that the cyanobacteria also have some phycoerythrin, as well as the bluish phycocyanin. In red algae, phycocyanin provides the bluish component of the more purple-colored species, but the reddish pigment phycoerythrin predominates.

Red algae can grow at greater depths than other algae and are found as deep as 100 meters beneath the surface. As sunlight penetrates water, first the red and

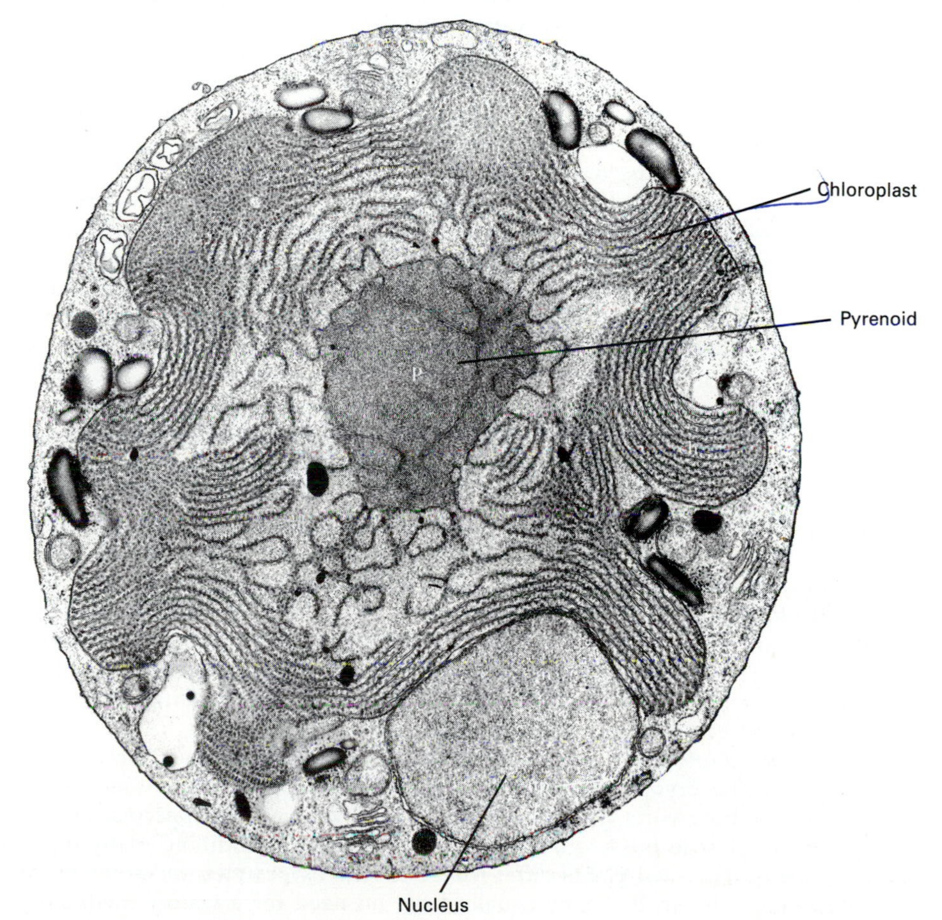

Figure 22–14 Electron micrograph of a section through a one-celled red alga, *Porphyridium cruentum* (×14,850). A single stellate chloroplast with a central pyrenoid food storage organelle occupies much of the cell volume. (From Gantt, E., and Conti, S. F. Reproduced from *The Journal of Cell Biology* by copyright permission of the Rockefeller University.)

then the orange, yellow, and green rays are filtered out, so that only the blue and violet rays remain. Phycoerythrin is more effective than chlorophyll in absorbing light energy of these short wavelengths, and red algae thus can live at greater depths than other plants that lack phycoerythrin (because of their absorption of available light, the red algae actually appear black in deeper water.) Although red algae occur as far up as the low tide line, they reach their greatest development in deeper tropical waters. Some 4000 species of rhodophytes are known.

Many rhodophytes have lacy, delicately branched bodies that are not as well adapted to survive in the intertidal zone as the tough, leathery brown algae, but they do well in the quieter deep waters. The **coralline red algae** accumulate calcium from the sea water and deposit it in their bodies as calcium carbonate. Coralline algae are abundant in tropical waters and are even more important in the formation of coral atolls than are coral animals.

Several kinds of red algae are used as food (Fig. 22–15). *Porphyra* is considered a great delicacy by the Japanese and is widely cultivated in submarine gardens (Fig. 22–16). Dulse (*Rhodymenia*) is boiled in milk and eaten by Scots. Agar, used in making culture media for bacteria, is extracted from the red algae *Gelidium* and *Gracilaria*; agar is also extensively used in baking and canning. *Carrageenin*, extracted from Irish moss (*Chondrus*), is used in brewing beer and in the preparation of chocolate milk to keep the chocolate from settling out.

The similarities in photosynthetic pigments have led to the speculation that red algae and cyanobacteria are closely related, a view that is buttressed further by ultramicroscopic similarities in the photosynthetic membranes of the two. However, the cyanobacteria are clearly prokaryotes, and the red algae are eukaryotes whose cells in some ways are more similar to the cyanobacteria than to other eukaryotes. This suggests to some theorists that the red and the green algae are the products of independent endosymbiotic unions (Chapter 47).

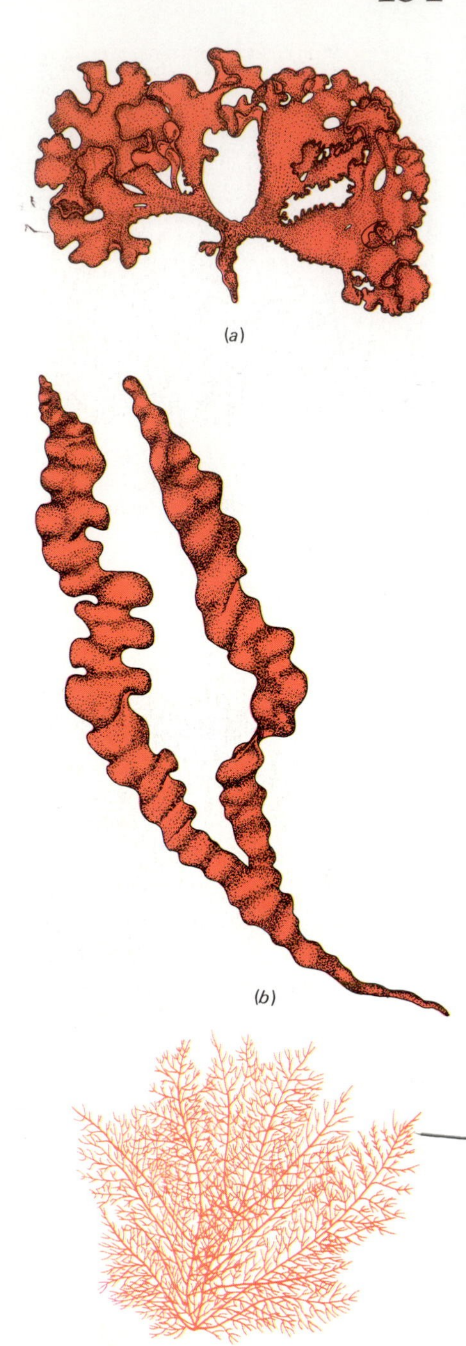

Figure 22–15 Some species of edible red algae. (*a*) *Chondrus crispus,* also called Irish moss, a red alga found in the North Atlantic and often used in the preparation of gels used in pudding mixes and similar products. (*b*) *Porphyra,* a red alga of the intertidal zone. A member of the subclass Florideophycidae, it is cultivated and harvested in Japan and used in the production of such foods as sushi, or nori. There are several species. (*c*) *Gelidium,* a red alga used in the production of agar. There are several other species of this genus that are useful to humans. (From K. Norstog and R. W. Long.)

Division Bryophyta: Nonvascular Terrestrial Plants

Aquatic plants can survive without many of the specialized structures found in terrestrial plants. The surrounding water keeps them supplied with nutrients, prevents the cells from drying out, buoys up and supports the plant body so that special supporting structures are unnecessary, and serves as a convenient medium for both the meeting of gametes in sexual reproduction and the dispersal of asexual spores. Indeed, the algae possess body structures adapted to expose a maximum of body surface for absorbing nutrients from the surrounding water.

The truly terrestrial plants must do without these advantages. They are nevertheless able to survive because they possess the following features: (1) the **cuticle,** a waxy, outer protective layer protecting the soft, watery tissues underneath; (2) **leaves** extending into the air to absorb light and carry on photosynthesis; (3) **roots** extending into the soil to provide anchorage and absorb water and salts; (4) **stems** supporting the leaves in the sunlight and connecting them with the roots to provide a two-way connection for the transfer of nutrients in the xylem and phloem; and (5) some means of reproduction, involving, for instance, flowers, pollen, and seeds, to minimize or eliminate the role of water in the union of male and female gametes and to enable the zygote to begin development while protected from desiccation.

The terrestrial habitat offers many advantages over the aquatic that make the genetic and metabolic cost of these adaptations well worthwhile. The intensity of sunlight is usually greater and more uniform on land, and suitable space for attachment is much more plentiful. It is striking that despite the buoyancy afforded by the aquatic habitat, the largest plants are terrestrial. A kelp the size of a redwood tree would have to begin life hundreds of meters below the surface in a region of eternal dusk. A redwood seedling, on the other hand, is likely to begin life bathed in all the light it needs, its growth bounded only by the structural strength of its components and the limits of its internal transport mechanisms.

The land plants are divided into two divisions: **Bryophyta** and **Tracheophyta** (Chapter 23). The bryophytes are the simplest of the land plants—the mosses, liverworts, and hornworts. They are distinguished by their lack of vascular tissues. The tracheophytes do possess these circulatory tissues; in addition, many reproduce by seeds. The seed plants differ from both the bryophytes and some of the other tracheophytes in that they usually have no need for a watery medium in order to reproduce. The land plants such as the ferns and the bryophytes that do not produce pollen partially evade the problem of reproduction without water by employing water only for gamete transport. However, all land plants, including the bryophytes, have a life cycle in which the zygote is retained within the female sex organ. There it obtains food and water from the surrounding parental tissues and is protected from drying while it develops into a multicellular **embryo.** For this reason the bryophytes and the tracheophytes, or higher plants, are classified together in the subkingdom **Embryonta.**

The division Bryophyta is made up of about 25,000 species of **mosses, liverworts,** and **hornworts** (which are not discussed in this book). The name "moss" is erroneously applied to a number of plants that are not bryophytes. The moss that grows on the bark of a tree may be an alga; reindeer moss is a lichen; and the Spanish moss hanging from trees in southern states is really a seed plant related to the pineapple. Even though mosses are widely distributed, they form an insignificant part of the vegetation. Some of the 15,000 species can live only in damp places; others can survive in a dormant state in dry, rocky places where enough moisture for growth is present only during a short part of the year. Liverworts, which are not as well protected against desiccation as mosses, are even more restricted. At least in temperate zones they are typically found in the deep shade of forests or on the shady side of a cliff, particularly where spring seepage occurs. Some are true aquatic plants (Fig 22–17).

Figure 22–16 *Porphyra* ("nori") beds ready to harvest at Sendai, Japan. (Courtesy of Prof. Isabella Abbott, University of Hawaii at Manoa.)

Figure 22–17 *Marchantia,* a thalloid or prostrate liverwort. The gametophyte is shown here. (L. Egede-Nissen/BPS.)

MUSCI: MOSSES

True **mosses** (class Musci) are all rather similar in structure, beginning as a filamentous green body, or **protonema,** on or in the soil. From the protonema grows an erect stemlike organ to which are attached a spiral whorl of "leaves" one cell thick. From the base of the stem extend many colorless, hairlike rootlets called **rhizoids** (these are not true roots). Mosses are never more than 15 to 20 cm high, owing to the inefficiency of the rhizoids as water absorbers and the absence (in most of them) of true vascular and supporting tissues.

The familiar, small green plants usually called mosses are the haploid gametophyte generation of the plant (Fig. 22–18). The gametophyte consists of a simple, central stem, bearing "leaves" arranged spirally and held in place in the ground by a number of rhizoids, which absorb water and salts from the soil. The "leaf" cells produce all the other compounds the plant needs for survival, so that each gametophyte is an independent organism. When the gametophyte has attained full growth, sex organs develop at the top of the "stem" in the middle of a circle of "leaves" and sterile hairs.

In some species the sexes are separate; in others, both male and female organs develop on the same plant. The male organs are sausage-shaped structures, **antheridia.** Each antheridium produces a large number of slender, spirally coiled sperm or **antherozoids,** each equipped with two flagella. After a rain or heavy dew the sperm are released and swim through the film of water covering the plant to a neighboring female sex organ, either on the same plant or on another one. The female organ, the **archegonium,** is shaped like a flask and has one large egg in its broad base. This organ releases a chemical substance that attracts sperm. Guided by this, the sperm swim down the neck of the archegonium and into its base, where one sperm fertilizes the egg. The resulting zygote is the beginning of the diploid sporophyte generation.

In contrast to the independent green gametophyte, the sporophyte is a leafless, single, spindle-shaped stalk called a **seta,** living in its later life as a parasite on the gametophyte and obtaining much of its nourishment by means of a **foot** that grows down into the gametophyte tissue. The sporophyte is capable of photosynthesis in young plants and also possesses certain simple conducting tissues and stomata, tiny pores in the epidermis (which closely resemble the stomata of vascular plants) used for gas exchange with the air. At the upper end of the sporophyte stalk a **sporangium,** or **capsule,** forms. Within the cylindrical cavity of this capsule, each of many diploid **spore mother cells** undergoes meiotic divisions to form four haploid spores. These are the beginning of the gametophyte generation (Fig. 22–19.)

Figure 22–18 Carpet of moss plants. It is possible that all of these gametophytes are a clone derived from a single protonema. (J. Robert Waaland, University of Washington/BPS.)

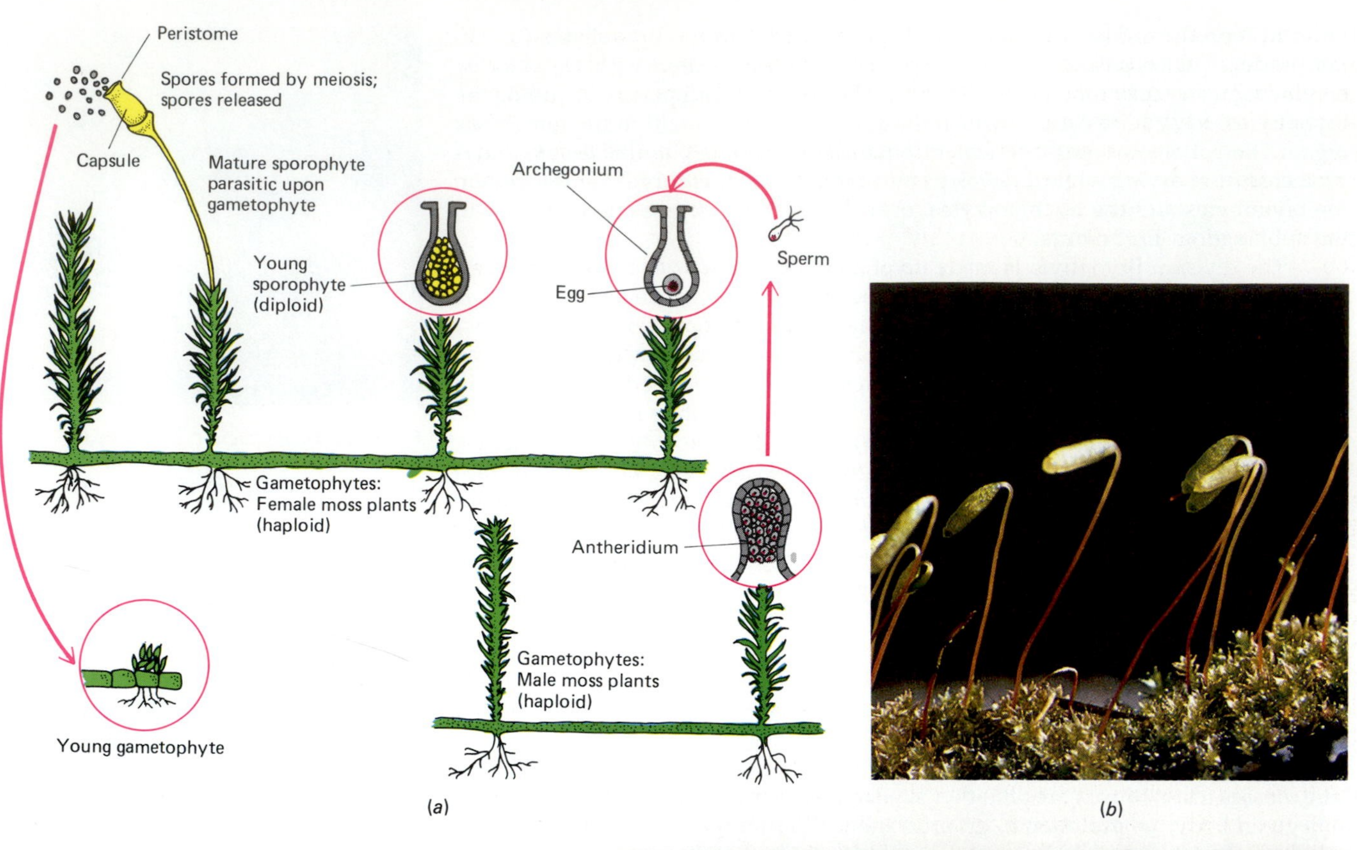

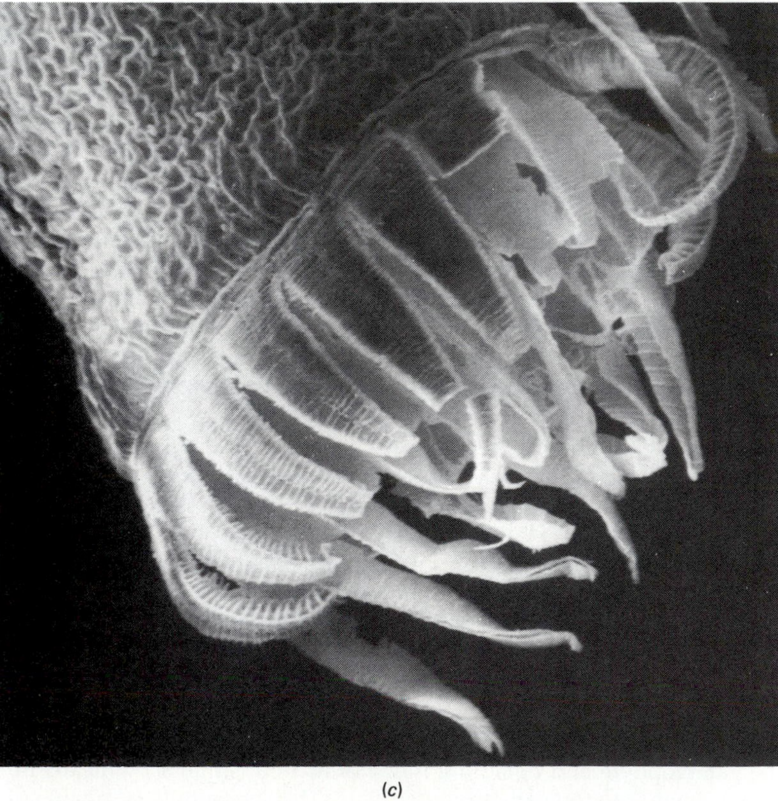

Figure 22–19 Mosses. (*a*) Life cycle of a moss. (*b*) Moss sporophytes, the elongated threadlike structures capped with spore cases. They are completely parasitic upon the gametophytes beneath them. (c) Scanning electron micrograph of a moss capsule showing the peristome teeth. ((*a*), from K. Norstog and R. W. Long. (*b*), J. Robert Waaland, University of Washington/BPS.) (*c*), courtesy of Dr. James A. McCleary.)

When the capsule matures, the upper end forms a lid, called a **calyptra,** that drops off. In some mosses the opening of the spore capsule is obstructed by one or two rings, called peristomes, of wedge-shaped teeth. In wet weather these bend inward and prevent the escape of the spores, but in dry weather they bend outward, permitting the liberation of spores when these are likely to be dispersed by the wind. When a spore drops in a suitable place, it germinates and develops into a protonema. The protonema then buds and produces several gametophytes, thereby completing the life cycle. Moreover, many moss plants can be produced asexually by a single protonema.

Some mosses can grow on bare places where few other plants can survive. Once mosses have become established and soil begins to accumulate, other plants can follow. An economically important plant is *Sphagnum*, which grows in boggy places. The remains of this plant accumulate under water and form **peat,** used as a fuel in many countries. Dried *sphagnum* moss can absorb and retain vast quantities of water and is used as a packing material for live plants.

The moss protonema somewhat resembles a filamentous alga. This fact, as we might predict, has led to the suggestion that the two are closely related.

HEPATICAE: LIVERWORTS

A second class of bryophytes, **Hepaticae,** or the liverworts, is simpler than the mosses. Some of the 9000 species of liverworts are flat, sometimes branched, ribbon-like structures that lie on the ground, attached to the soil by numerous rhizoids, and lacking a stem. Other species tend to grow upright and have a leaflike gametophyte. Still others, the leafy liverworts, have gametophytes that are differentiated into structures equivalent to stems, branches, and leaves but that lack vascular tissues. A few liverworts are strictly aquatic. Many of the leafy liverworts are epiphytes and form colonies that grow on the stems, branches, and leaves of trees in tropical rain forests. The leafy liverworts are believed to be the most primitive of the bryophytes.

The upper surface of the gametophyte is an epidermis (Fig. 22–20). It is one cell thick and punctured by many pores through which gas exchange occurs. These pores may be surrounded in some species by little "chimneys" whose function may be to limit the loss of water through the pores, but which cannot be as effective in this as the stomata of moss and tracheophytes are. The lower surface is an epidermis covered with many thin scales and from which grow many long, slender rhizoids. The recognizable plant is the gametophyte generation. As in mosses, the sporophyte grows as a parasite on the gametophyte (Fig. 22–21). The upper surface of some liverwort gametophytes may bear **gemma cups.** Small, flattened, ovoid gemmae are produced within these cups. The gemmae eventually separate from the parent plant and grow into new gametophytes—a process of asexual reproduction.

It is clear that the adaptations of the liverworts to the terrestrial environment are not as efficient as they perhaps could be, and this limits them to a very narrow terrestrial ecological niche indeed. The mosses are a lot tougher, often serving as the pioneer organisms that first colonize a very demanding habitat. But it is the vascular plants, the tracheophytes, that occupy the vast majority of terrestrial habitats, and they are the subject of the next chapter.

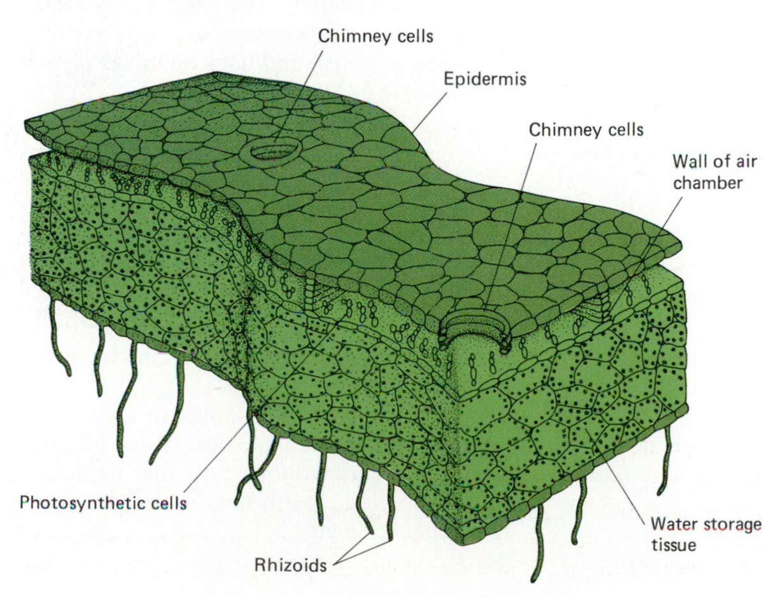

Figure 22–20 Anatomy of a liverwort. Though there are no true stomata, the chimneys of their pores give the liverwort thallus something like the structure of a leaf. The rhizoids absorb some water after the fashion of root hairs, there is a cuticle (not shown) and starch is stored in some of the deeper tissues of the thallus. All these are adaptations for terrestrial life.

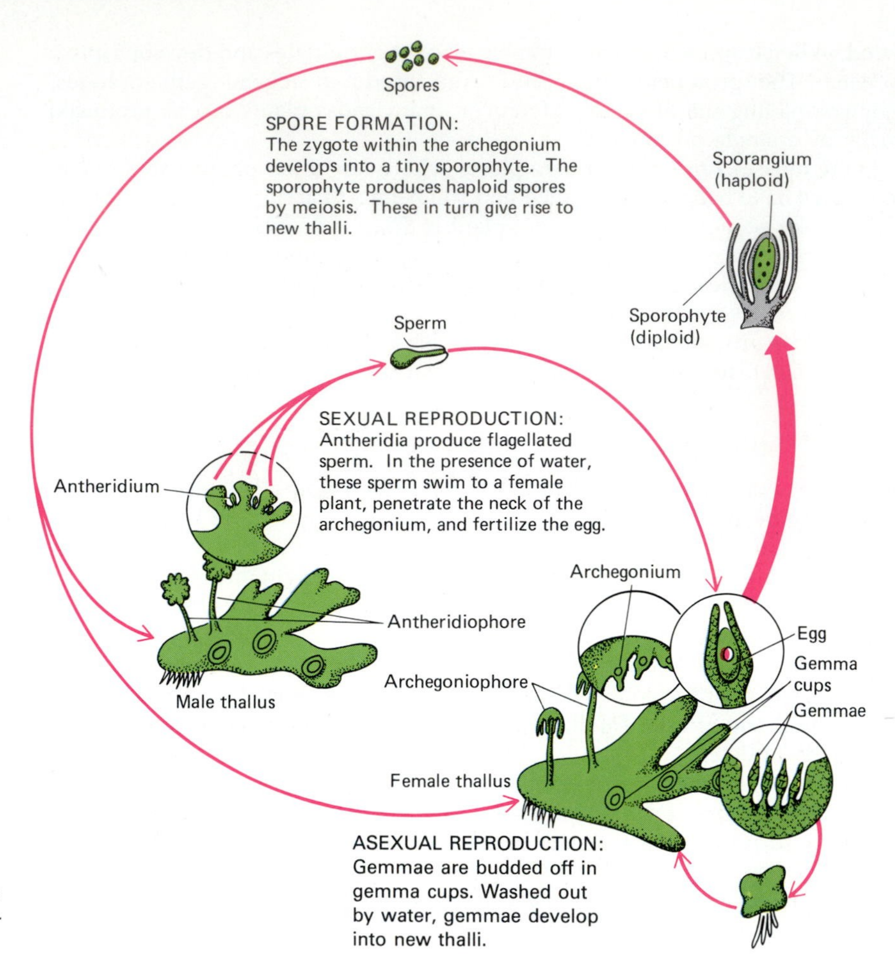
Spores
SPORE FORMATION:
The zygote within the archegonium develops into a tiny sporophyte. The sporophyte produces haploid spores by meiosis. These in turn give rise to new thalli.
Sporangium (haploid)
Sporophyte (diploid)
Sperm
SEXUAL REPRODUCTION:
Antheridia produce flagellated sperm. In the presence of water, these sperm swim to a female plant, penetrate the neck of the archegonium, and fertilize the egg.
Antheridium
Archegonium
Antheridiophore
Egg
Gemma cups
Gemmae
Archegoniophore
Male thallus
Female thallus
ASEXUAL REPRODUCTION:
Gemmae are budded off in gemma cups. Washed out by water, gemmae develop into new thalli.

Figure 22–21 Liverwort life cycle. Note the special method of asexual propagation.

SUMMARY

I. We cannot draw an absolute distinction between the simpler green algae of plantlike organization and the protist-grade algae. However, the plantlike green algae are truly multicellular, with differentiated body parts.
   A. The photosynthetic cells of green algae and their life cycles are similar to those of green plants.
   B. The green algae are dissimilar from the brown and red algae, and are almost certainly not closely related to them.

II. The plantlike algae consist of the divisions Chlorophyta, Phaeophyta, and Rhodophyta.
   A. In addition to chlorophyll *a*, the chlorophytes have chlorophyll *b*, carotene and xanthophyll as accessory pigments. They also have flagellated gametes.
   B. The phaeophytes possess the accessory pigments chlorophyll *c* and fucoxanthin (a carotenoid pigment) and also have flagellated gametes.
   C. The rhodophytes have as accessory pigments phycoerythrin, phycocyanin, and carotenoids and do not employ flagellated gametes.

III. Terrestrial plants are divided into the divisions Bryophyta (nonvascular plants) and Tracheophyta (vascular plants). The bryophytes include mosses, liverworts, and hornworts. All are small, employ flagellated gametes, and have an inconspicuous diploid sporophyte generation.

POST-TEST

1. The form of sexual reproduction that involves two cells that are morphologically indistinguishable from one another is known as ______________.

2. A large colonial alga consisting of a ball of hundreds of flagellated cells belongs to the genus ______________.

3. One way in which the flagellated green algae can be distinguished from the photosynthetic mastigophoran protists is that the algae have cell ______________ containing ______________.

4. The mature colony of *Volvox* is chromosomally ______________, so that when gametes are formed, meiosis is unnecessary.

5. In some filamentous green algae such as *Spirogyra*, sexual reproduction typically occurs by means of a conjugation tube through which a ______________ creeps to fertilize a similar cell in an adjacent filament.

6. The green alga *Ulva*, or sea lettuce, exhibits alternation of generations, but the haploid generation is ______________ with the diploid generation, an arrangement that is called ______________ alternation of generations.

TABLE 22–1
The Plant Kingdom

Classification	*Characteristics*	*Ecological Significance*
Algae (Lower Plants)	Adapted for photosynthesis	Producers
Chlorophytes	Unicellular, colonial, or multicellular; many with distinct alternation of generations (isomorphous)	Many are protist-like
Phaeophytes	Almost all multicellular, with pronounced alternation of generations; differentiated bodies and in a few cases vascular tissues	Important intertidal and shallow marine forms
Rhodophytes	Almost all multicellular with alternation of generations (heteromorphous); deeper-living, with aberrant accessory photosynthetic pigments	Sometimes thought to be of independent origin from other plants
Complex Plants	Multicellular; complex; adapted for photosynthesis; have reproductive tissues or organs; pass through distinct developmental stages; alternation of generations: sporophyte generation produces spores, which develop into gametophyte plants, which produce gametes	Producers
Bryophytes (mosses, liverworts, hornworts)	Small because they lack true stems, roots, or leaves, and any sort of circulatory system (must depend upon diffusion for distribution of materials); gametophyte dominant form; smaller sporophyte attached to gametophyte in mosses, is partially parasitic upon it	Often pioneer organisms in colonization of new habitats
Tracheophytes (vascular plants)	Have circulatory systems that enable them to attain large size; have true roots, stems, and leaves; multicellular embryos; sporophyte dominant form	Adapted for life on land
Ferns, etc.	Gametophyte develops as a tiny, independent plant; spores found in little clusters on underside of some sporophyte leaves	
Seed plants	Embryo encased within a seed equipped with nourishment (endosperm); may lie dormant for long periods until conditions are favorable; gametophyte reduced to a small structure on the sporophyte plant	
Gymnosperms	Leaves generally needle-like or scale-like; usually naked seeds borne in cones	Dominate entire northern forested area (much of Canada and Russia); produce much lumber used for construction
Angiosperms (flowering plants)	Possess sexual organs called flowers; seeds often surrounded by fleshy or hard envelope called a fruit (comprised mainly of the ovary wall); either annual (live only one season) or deciduous (lose their leaves seasonally); two types: monocots—seed with single small embryonic leaf and large endosperm, leaves with parallel veins; dicots—seed with two large embryonic leaves and little endosperm, leaves with netlike veins	Provide food for human beings and other consumers

7. The giant kelps belong to the division ________________ or brown algae. Some kelps produce a commercially valuable polysaccharide called ________________ used in the preparation of ice cream.

8. The red algae are sometimes suspected of having a different evolutionary origin from that of other plants, one reason for which is the fact that these algae employ the pigment ________________, which is also utilized by the modern ________________, but which does not occur in any other plants.

9. The nonvascular terrestrial plants discussed in this book consist of the ________________ and ________________. Together with the hornworts these make up the division ________________.

10. In moss, the ________________ generation possesses stemlike and leaflike structures, but only the ________________ generation has stomata-like structures.

11. A moss spore is ________________ (haploid/diploid).

12. Liverworts can propagate asexually via ________________, which are produced in special cuplike structures.

13. The ________________ of a moss plant quite closely resembles a ________________ alga.

REVIEW QUESTIONS

1. Describe the major features of the three groups of the lower green algae.
2. What reasons might there be to suppose that the red algae have originated independently of the green algae?
3. What features of the red algae set them apart from all other plants?
4. Compare the adaptations of the mosses and liverworts.
5. In your opinion, is *Volvox* best considered a protist or a plant? Explain.
6. Compare sexual reproduction in *Chlamydomonas*, *Ulothrix*, and *Spirogyra*.

23

Vascular Plants: Division Tracheophyta

OUTLINE

- *I.* Class Rhyniopsida: the rhinophytes
- *II.* Class Lycopsida: the club mosses
- *III.* Class Filicopsida: ferns and horsetails
  - *A.* Order Equisetophyta: the horsetails or scouring rushes
  - *B.* Order Filicophyta: the ferns
- *IV.* The seed plants
  - *A.* The gymnosperms
    - *1.* Class Coniferopsida: the conifers
    - *2.* Class Cycadopsida: the cycads
    - *3.* Class Lyginopteridopsida: the seed ferns
    - *4.* Class Gnetopsida and the gingko
  - *B.* Class Angiospermopsida: the angiosperms
    - *1.* Life Styles of the angiosperms
    - *2.* Monocots and dicots
- *V.* The origin and relationships of higher plants
  - *A.* The origin of the gymnosperms
  - *B.* The origin of the angiosperms

LEARNING OBJECTIVES

1. Summarize the nature of the physiological and reproductive adaptations of the tracheophytes to the terrestrial environment.
2. List and characterize the tracheophyte classes.
3. Describe four characteristics that distinguish angiosperms from gymnosperms.
4. Compare the features of the monocots and dicots.
5. Discuss the possible or proposed evolutionary origin of and relationships among the plants.

Vascular plants are the dominant plants on the earth's land surface today. Botanists have long debated whether the vascular plants are monophyletic or whether they had diverse ancestors. This controversy has led to a marked lack of consistency in classifying these plants. Some taxonomists classify all of the vascular plants together in division **Tracheophyta,** a system which we will follow here.[1] Except for the flowering plants, tracheids are the main water-conducting tubes of vascular plants. Even in the flowering plants, however, tracheids are present along with a more efficient system of vessels and fibers. The vascular plants are not necessarily dominant in the aquatic habitat, however, and when such producers as diatoms are taken into account, tracheophytes may not be responsible for the majority of photosynthesis that takes place on the face of the earth. It is clear, nevertheless, that without them complex terrestrial ecosystems and terrestrial life as we know it would be quite impossible. The vascular plants are surely the most complex plants that exist; their specialized structures, which permit survival in the land habitat where water is relatively scarce, include systems for conservation and transport of water and nutrients. As their name implies, their most distinctive and characteristic set of adaptations involves the tissues of their vascular or conductive system—phloem and xylem—which are discussed extensively in Chapter 27. Here it is enough to say that **xylem** is specialized to carry materials from the absorptive roots of the plant to the photosynthetic leaves, and **phloem** functions predominantly (but *not* exclusively) in the opposite direction. That is, it is specialized to carry traffic from the leaves to the roots, supplying the nonphotosynthetic parts of the plant with the products of photosynthesis. (Exceptions are discussed in Chapter 27.)

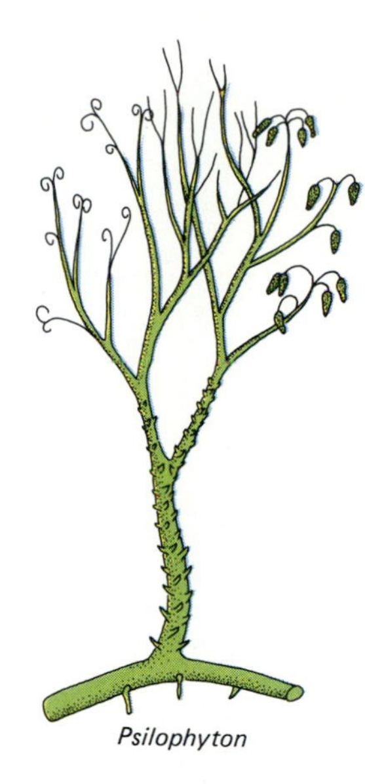

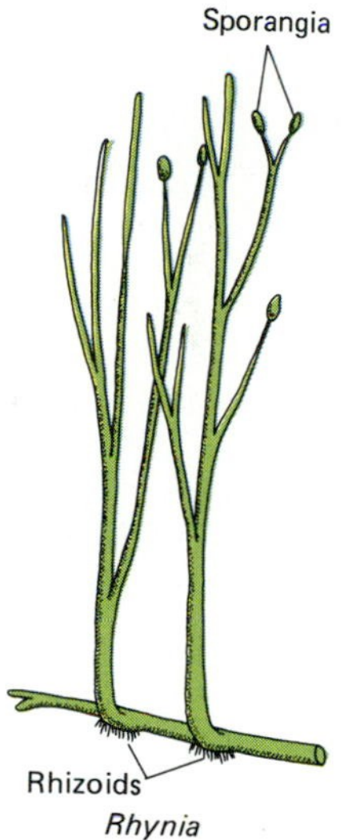

Figure 23–1 Three species of fossil vascular plants that lived during the Devonian period. *Rhynia* and *Psilophyton* are members of the class Rhynopsida; *Asteroxylon* is a member of the class Lycopsida.

Class Rhyniopsida: The Rhyniophytes

The simplest vascular plants were the members of the class **Rhyniopsida** (Fig. 23–1), which occur in Devonian geological deposits. The rhyniophytes are widely believed to be ancestors of the other vascular plants; they are now completely extinct. These plants had a creeping horizontal stem or **rhizome** from which grew branching, erect, presumably green stems attaining a height as great as 60 cm. The smaller branches were coiled at the tip and probably unrolled as they grew, as present-day ferns do. They had no roots and were either leafless or had small scalelike leaves. Fossil remains of these plants have been found in Scotland in such a good state of preservation that some details of the internal structures are visible. The center of the stem of the rhyniophytes was a cylinder of vascular tissue composed solely of xylem. Phloem, however, the other component of the vascular tissues of higher plants, cannot be distinguished in the fossil rhyniophytes found so far.

Class Lycopsida: The Club Mosses

The club mosses, quillworts, and their relatives comprise the class **Lycopsida.** Most lycopods are known only as fossils. Many extinct forms were tall and treelike, but today only five genera remain, all small forms usually less than 30 cm high. These inconspicuous plants consist of a creeping stem, from which grow true roots, and upright stems with thin, flat, spirally arranged true leaves called **microphylls** (Fig. 23–2).[2] Microphylls are believed to have evolved as superficial outgrowths of the stem, and they contain a single, unbranched, centrally located strand of vascular tissue (vein), a mesophyll of parenchyma cells with numerous chloroplasts, and an epidermis with stomata. (In contrast, leaves that occur on seed plants and ferns are called **megaphylls,** which are believed to have evolved by coalescence of several entire branches; megaphylls have several to many veins as well as a mesophyll of parenchyma cells with numerous chloroplasts, and an epidermis with stomata.)

The lycopod stem has a core of xylem tissue surrounded by a cylinder of phloem. In modern lycopods the stems cannot grow in diameter. Higher vascular plants possess a layer of permanently embryonic tissue, the **cambium,** (Chapter

[1] The system of classification of the higher plants used in this book is based on that of Ray, P. M., Steves, T. A., and Fultz, S. A.: *Botany*. Philadelphia, Saunders College Publishing, 1983.

[2] A "true" leaf is a leaf such as occurs on seed plants and ferns. It contains spongy photosynthetic tissue and the vascular tissues, phloem and xylem. The "leaves" of moss, for instance, lack phloem and xylem.

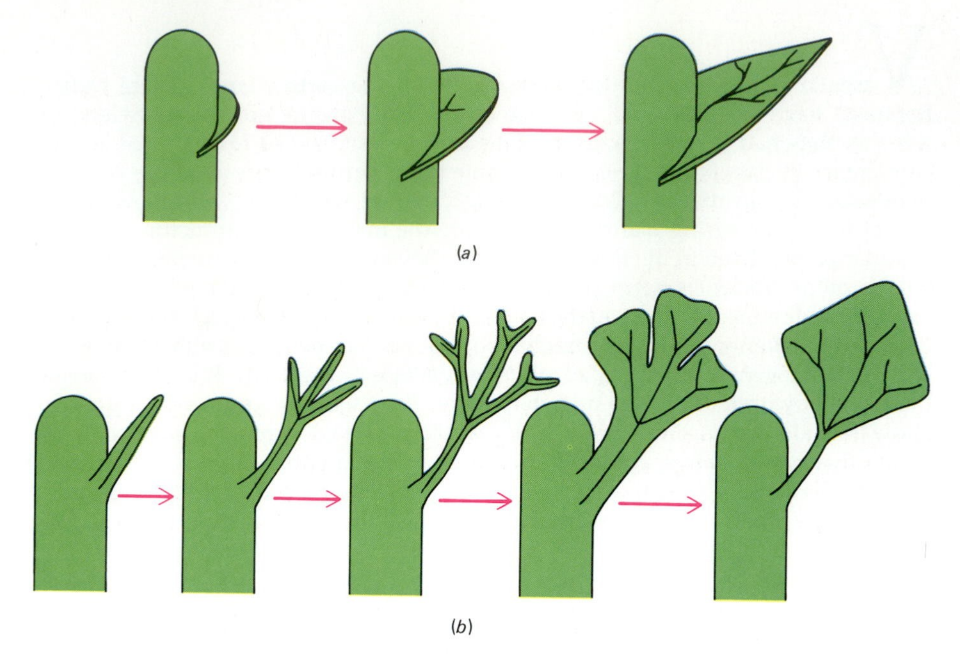

Figure 23–2 Microphyll (*a*) and megaphyll (*b*). Megaphylls are thought to develop from preexisting vascular branchlets, but microphylls (such as are still found on *Equisetum*) originate as flaps of stem tissue that develop vascular elements secondarily. The subject of microphyllous and megaphyllous origins of leaves is currently being restudied extensively.

27), which has this function, but the cambium is absent in present-day lycopods. However, some of the ancient club mosses found in Carboniferous deposits apparently did have a cambium and were treelike plants as much as a meter thick and 20 or more meters tall.

At the tip of the stem are specialized leaves, **sporophylls,** arranged somewhat in the shape of a pine cone. The spore-producing structures, **sporangia,** are born on the sporophylls. The sporophyte produces spores, which later germinate to form bisexual gametophytes. Sex organs develop on the gametophytes and produce eggs and biflagellate sperm that resemble the sperm of some of the green algae. After fertilization the developing embryo is, for a time, dependent upon the gametophyte for nutrition. In some lycopods, however, it is the gametophytes that are somewhat parasitic, as is also true in the seed plants.

When the haploid microspores are released from the sporophyte, they may drop near a megaspore. In the presence of moisture from dew or rain, the microspore wall splits, and the sperm within are free to swim to the megaspore and fertilize the haploid egg.

Lycopodium may be the most familiar example of a club moss (Fig. 23–3). One species, *L. complanatum,* grows horizontally on the ground but produces erect, bushy branches that are very attractive. This plant is endangered by the activities of collectors who gather them for holiday decorations.

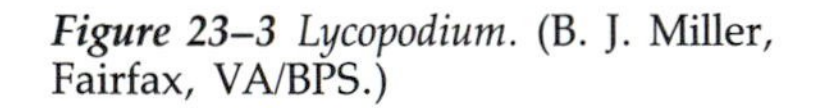

Figure 23–3 *Lycopodium.* (B. J. Miller, Fairfax, VA/BPS.)

Class Filicopsida: Ferns and Horsetails

Traditional taxonomy names the ferns and the horsetails as orders in the class **Filicopsida.** Whether the ferns and the horsetails ought truly to be classified together is debatable, however. Their reproductive patterns are somewhat similar, but their fossil histories appear to be quite different, and although both possess genuine leaves and stems, the details of these structures differ in the two orders.

ORDER EQUISETOPHYTA: THE HORSETAILS

Like Lycopsida, the order **Equisetophyta** includes many more fossil plants than living ones. This group of plants is believed to have originated during Devonian times and developed into a variety of species, some small and some that were gigantic, treelike plants as much as 30 meters tall and 50 cm in diameter (Fig. 23–4). The dead bodies of these treelike giant horsetails, together with those of certain other plants, are the major source of our present-day coal deposits.

The present-day equisetophytes—genus *Equisetum,* the "scouring rushes"—are widespread from the tropics to the Arctic on all continents except Australia. These plants, usually less than 40 cm tall, are found in both boggy and dry places. The name **horsetail** is appropriate because the multiple-branched, bushy structure of many species resembles a horse's tail. The popular name, "scouring rushes," reflects the fact that deposits of silica in the epidermis give the plants a harsh, abrasive quality; certain species were used to clean pots and pans for centuries before the invention of steel wool.

The sporophyte of *Equisetum* is made up of a horizontal, branching, underground rhizome from which grow slender, branching roots and jointed aerial stems. The stem contains many vascular bundles arranged in a circle around a

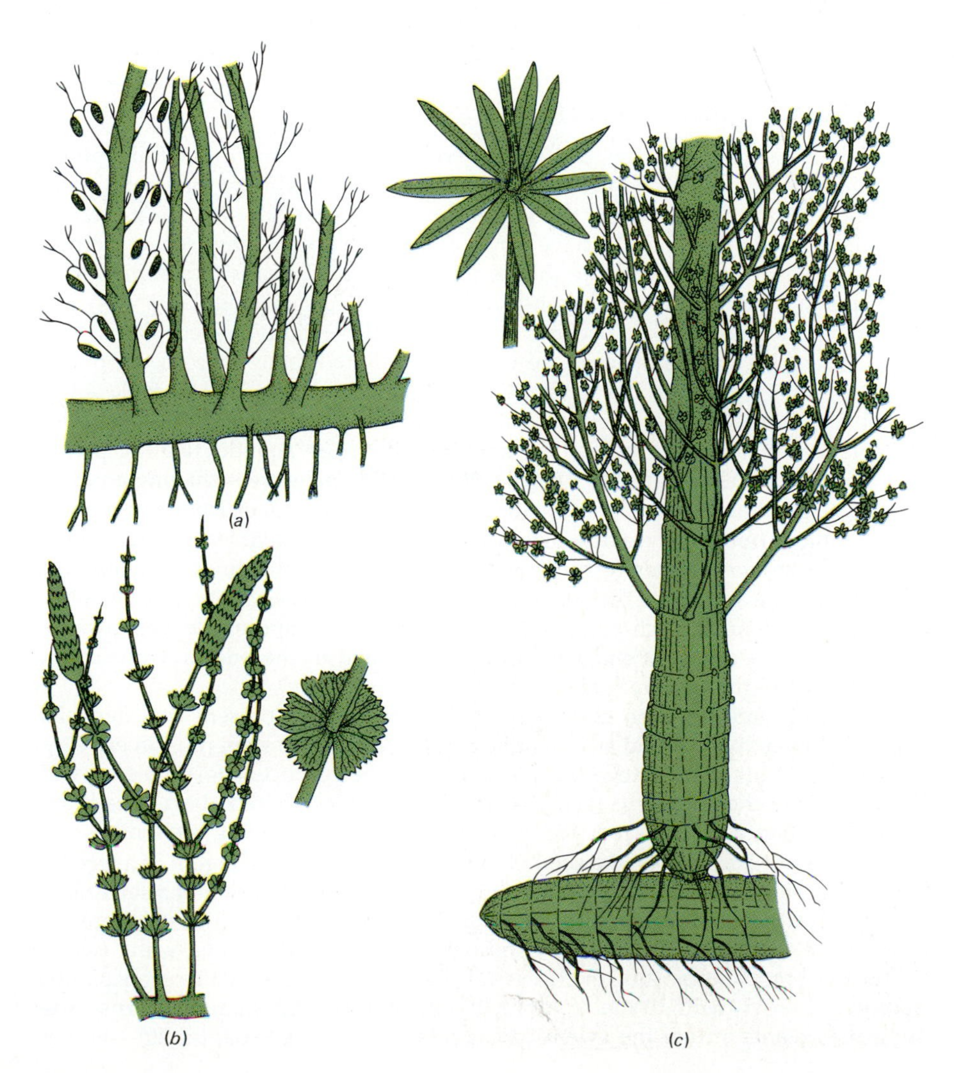

Figure 23–4 Equisetophytes of the Paleozoic era seen in diagrammatic form. (*a*) Reconstruction of *Hyenia,* a Middle Devonian species that had whorls of appendages but lacked cones. (*b*) *Sphenophyllum* and (*c*) *Calamites,* both upper Devonian and Carboniferous period plants. Calamites somewhat resembles the modern-day *Equisetum.* Many coal beds are made up largely of the remains of plants similar to these. Though they are reduced to comparative insignificance today, once they were forest giants. (From Smith.)

(a)

(b)

Figure 23–5 *Equisetum* species with different growth habits. (*a*) *Equisetum telematia*, the common horsetail, with unbranched, nongreen, nonphotosynthetic "fertile" shoots bearing cone-like strobili and separate highly branched green photosynthetic "sterile" shoots, both arising from an underground rhizome. (*b*) A dense growth of *Equisetum* on a bank. ((*a*), J. Robert Waaland, University of Washington/BPS; (*b*), Charles Seaborn.)

hollow center. The stems have conspicuous nodes that divide them into jointed sections. At each node there is a circlet or **whorl** of smaller secondary branches and a whorl of small, scalelike leaves (Fig. 23–5); in some species, these leaves are not capable of active photosynthesis, and this activity is carried out mainly by the stem. Some branches develop a conelike structure **(strobilus)** at their tip that contains numerous structures bearing spore sacs on their inner faces.

The spores released from these structures germinate into inconspicuous green gametophytes with egg- and sperm-producing organs. The zygote formed after fertilization develops into the sporophyte plant. At first this is parasitic on the gametophyte, but it quickly develops its own stem and roots.

ORDER FILICOPHYTA: THE FERNS

The characteristics that distinguish **ferns,** which comprise the order **Filicophyta,** from other lower, non-seed-bearing vascular plants include the structure of the leaf, the anatomy of the stem, the location of the sporangia, and the pattern of development. Ferns typically have large compound (branching) leaves termed **fronds** that develop by uncoiling (hence the name "fiddleheads" for some ferns); their sporangia are borne on the fronds in clusters called **sori** (singular, sorus), and their stems contain **pith,** a "packing" tissue that also occurs in seed plants.

Some 9000 species of ferns are widely distributed today in both the tropics and temperate regions. Those in the temperate regions thrive best in cool, damp, and shady places. Ferns are abundant in tropical rain forests, though some species are able to live in quite dry habitats, and still others are aquatic, floating plants. Some tropical ferns are tall and superficially resemble palm trees, having an erect, woody, unbranched stem with a cluster of compound leaves or fronds at the top. Some of these **tree ferns** reach a height of 16 meters and have leaves four meters long. The common temperate-zone ferns have horizontal rhizomes growing at or just beneath the surface of the soil, from which grow hairlike roots. The rhizomes are usually perennial; each year, new erect fronds or compound leaves grow on them. The leaves of ferns characteristically are coiled in the bud and unroll and expand to form the mature leaf (Fig 23–6).

Extinct forests of fern trees have contributed to our present coal deposits. Some of these fern trees had tall, slender trunks made of the stem plus an enveloping mass of roots matted together by hairs. Another group of fossil plants with fernlike leaves, the **seed ferns** (Fig. 23–16), are now known to have borne seeds and so are classified with the seed plants, described in the following section.

Ferns are remarkably like seed plants in many respects. The root of a fern has a root cap, meristematic elongation, and mature zones like the root of a seed plant. (These structures are described in detail in Chapter 27.) Its stem has a protective epidermis, supporting and vascular tissues, and leaves that have veins (whose pattern of branching is, however, very different from those of the seed plants), spongy photosynthetic tissue, a protective epidermis, and stomata. Ferns differ from seed plants in that the xylem contains only tracheids (Chapter 26)—no ves-

(*a*)

(*b*)

(*c*)

(*d*)

Figure 23–6 Representative ferns. (*a*) Branch of tree fern showing characteristic "fiddlehead" growth pattern of new leaves. (*b*) Cinnamon fern, showing brown reproductive spikes and green sterile leaves (the spikes bear the sporangia). (*c*) Staghorn fern, native to Australia although widely grown elsewhere. (*d*) Tree fern, *Cyathea.* These Southern Hemisphere plants are especially common in New Zealand. ((*a*) Carolina Biological Supply Company; (*b*) J. N. A. Lott, McMaster University/BPS; (*c*), J. Robert Waaland, University of Washington/BPS; (*d*), B. J. Miller, Fairfax, VA/BPS.)

sels—and the spores are produced in sporangia on the undersurfaces of certain leaves (Fig. 23–7). The sporophyte plant may live for several years and produce several yearly crops of spores.

The spores are released at the proper time, fall to the ground, and develop into flat, green, photosynthetic, heart-shaped gametophytes 5 to 6 mm in diameter.

(*a*)

(*b*)

Figure 23–7 In ferns, sporangia are borne on the fronds in clusters of sori. (*a*) Sorus without indusium. A multitude of sporangia can be seen. (*b*) Sori with indusium. The indusium, or shield, is the light membranous structure extending over the dark sori. ((*a*), J. N. A. Lott, McMaster University/BPS.)

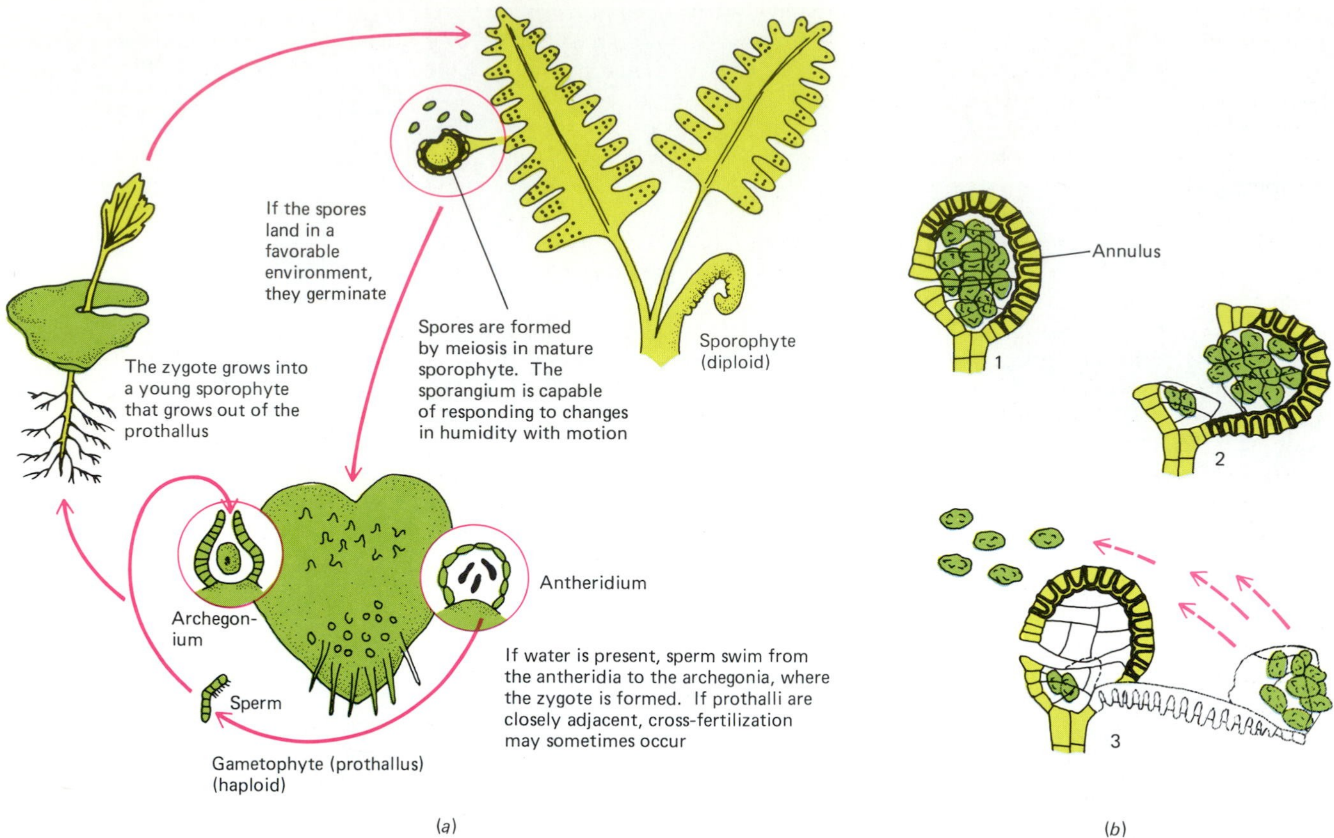
If the spores land in a favorable environment, they germinate
Spores are formed by meiosis in mature sporophyte. The sporangium is capable of responding to changes in humidity with motion
Sporophyte (diploid)
The zygote grows into a young sporophyte that grows out of the prothallus
Archegon-ium
Antheridium
Sperm
If water is present, sperm swim from the antheridia to the archegonia, where the zygote is formed. If prothalli are closely adjacent, cross-fertilization may sometimes occur
Gametophyte (prothallus) (haploid)
(a)
Annulus
1
2
3
(b)

Figure 23–8 (*a*) Life cycle of a fern. (*b*) Spore dispersal in ferns. As water evaporates from the cells of the stalklike annulus, the remaining content of water is subjected to tension from heavy cell walls. When this tension results in sufficient pressure reduction to cause boiling, the remaining water flashes into vapor in an instant. This allows the annulus to spring shut, scattering the spores. The action can be observed in some species of ferns in which the sporangia are enclosed in a case. When the case is broken open and exposed to the drying effects of the air, the sporophyte stalks begin to writhe like worms, and shortly the sporangia begin to go off like popcorn in a pan.

The gametophyte, called a **prothallus** (plural, prothalli), grows in moist, shady places, especially on decaying logs and on moist soil and rocks. A number of rhizoids grow into the soil from each gametophyte, anchoring it and absorbing water and salts. The male and female sex organs—antheridia and archegonia—develop on the undersurface of the gametophyte (Fig. 23–8).

Each **archegonium,** usually located near the notch of the heart-shaped prothallus, contains a single egg. The **antheridia,** located at the other end of the prothallus, develop a number of flagellated **antherozoids** or sperm; these are ovoid in shape and have many flagella on a spiral band at their anterior end. The antherozoids are released after a rain and, attracted by a chemical substance released by the archegonium, swim through the water on the undersurface of the gametophyte to reach the egg. The antheridia usually develop and discharge their antherozoids before the archegonia of that gametophyte plant have matured. The antherozoid of one plant usually fertilizes the egg of another plant, and a new sporophyte develops from the resulting zygote. The fertilized egg begins to develop into a sporophyte embryo within the archegonium. Initially the sporophyte develops as a parasite on the gametophyte, but it soon develops its own roots, stem, and leaves and becomes an independent sporophyte, completing the cycle.

The diploid fern sporophyte is well suited to terrestrial life. It has conducting and supporting tissues and, in contrast to the mosses, is nutritionally independent from the gametophyte. However, the gametophyte generation can survive only where there is plenty of moisture and shade, and the union of eggs and sperm in fertilization requires a watery medium. This is compensated for, however, by heavy reliance on asexual reproduction in many species of ferns. A kind of fern called bracken, for instance, rarely reproduces sexually but spreads by means of underground stolons and has become a serious pest in many parts of the world.

The Seed Plants

The remaining classes of the plant kingdom are often gathered into two informal groups: the gymnosperms and the angiosperms, which, as we will see, may not be closely related at all. Whatever their phylogeny, the seed plants differ from ferns in having no independent gametophyte generation. Their key characteristics are (1) the formation of seeds, structures enclosing the embryo during a resting stage, and (2) the union of male gametes with the egg by pollination, made possible by the growth of a pollen tube. Seeds provide for the wide and rapid dissemination of a species and are resistant to desiccation and to extreme temperatures. In the reproductive processes of lower plants, the sperm reaches the egg through an external supply of water—the ocean, a pond, a puddle, or perhaps simply a film of water from dew or rain. The pollen tube eliminates this need for water and provides a means for the direct union of male and female gametes. These two traits have undoubtedly been responsible in large measure for the wide distribution of seed plants as terrestrial organisms.

Seed plants characteristically produce two types of spores, **megaspores** and **microspores.** The megaspores develop into female gametophytes; the microspores develop into **pollen,** which gives rise to the male gametophytes. The female gametophyte is retained within the megaspore and gives rise to an egg, or ovule, that is fertilized there. The resulting zygote develops into an embryo with the rudiments of leaves, stems, and roots while still within its protective structure derived from the previous sporophyte. The **seed** thus consists of structures belonging to three distinct generations: the **embryo,** the new sporophyte; the "endosperm," nutritive tissue derived from the female gametophyte (in gymnosperms) or from a unique triple fertilization (in angiosperms); and the **seed coat,** derived from the old sporophyte.

There are more than 250,000 species of seed plants, adapted to survive in a variety of terrestrial environments and varying in size from the minute duckweed, a few millimeters in diameter, to giant redwood trees. They also are the plants of greatest value to humans as sources of food, shelter, drugs, and industrial products. The **gymnosperms** ("naked seeds") and the **angiosperms** ("enclosed seeds") differ in the relationship of the seeds to the structures producing them. The seeds of angiosperms are formed inside a **fruit,** and the seed covering is developed largely from the wall of the ovulary of the flower (Chapter 31), which encloses the ovules. Some angiosperms seem to violate this rule; for example, the seeds of wheat, corn, sunflowers, and maples do not appear to be enclosed. However, in these plants, the structure commonly called the "seed" actually is a fruit, enclosing the true seed. The seeds of gymnosperms are borne in various ways, usually on **cones,** but sometimes in association with fruitlike structures; yet they are never really enclosed as are angiosperm seeds. The presence of seeds has contributed immensely to the widespread distribution of the seed plants and offers the following advantages:

1. The stored food nourishes the embryo until it can lead an independent life.
2. The tough outer coat protects the embryo from heat, cold, drying, and parasites.
3. Seeds provide a means for the dispersal of a species.

THE GYMNOSPERMS

The diverse group of vascular plants often called gymnosperms includes the seed ferns (now extinct), the cycads, the conifers, and the gingko-like plants. Most species of gymnosperms are of little practical importance in human affairs, but the conifers are a notable exception.

Class Coniferopsida: The Conifers

Conifers, the familiar cone-bearing evergreens such as spruce and pine, comprise the class **Coniferopsida,** which includes the vast majority of the gymnosperms. They are important economically as the source of more than 75% of the wood used in construction and in the manufacture of paper, plastics, rayon, lacquer, photographic film, explosives, and many other materials. Some conifers produce resins used in the production of turpentine, tar, and oils. The seeds of a few conifers are used as food, and the berries of junipers produce aromatic oils used for flavoring alcoholic beverages such as gin.

Many of the conifers are trees, but some are shrubs; most are evergreen and have needle-shaped leaves. The conifers have a worldwide distribution and are of

(a)

(b)

(c)

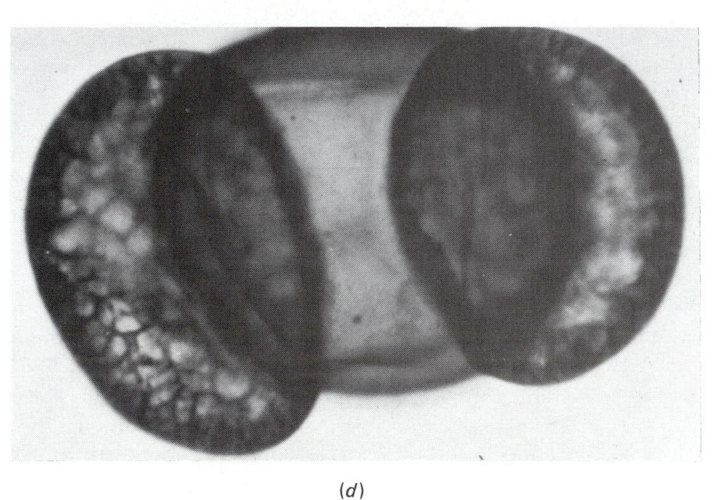

(d)

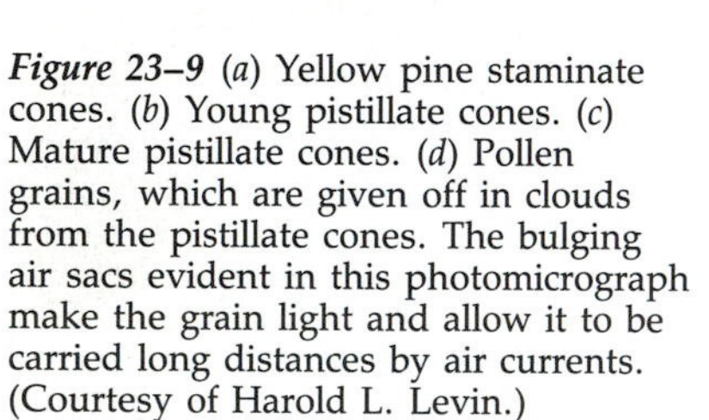

Figure 23–9 (*a*) Yellow pine staminate cones. (*b*) Young pistillate cones. (*c*) Mature pistillate cones. (*d*) Pollen grains, which are given off in clouds from the pistillate cones. The bulging air sacs evident in this photomicrograph make the grain light and allow it to be carried long distances by air currents. (Courtesy of Harold L. Levin.)

great importance both ecologically and economically. Conifers have no flowers but bear their seeds on the inner sides of scalelike leaves that are usually arranged spirally to form a cone.

Some conifers are deciduous, but in the evergreen species, the leaves may be retained for years. When they finally *are* shed, the leaves are not dropped all at one time. The conifers are almost all woody, perennial plants with a characteristic tree form. Their wood (xylem) is composed of tracheids with bordered pits (Chapter 27).

The pine tree, a typical conifer, bears two kinds of reproductive cones on the same tree (Fig. 23–9). Some of these, the male **staminate cones,** contain organs known as **microsporangia.** The microsporangia eventually produce the pollen. The remaining female **ovulate cones** contain organs known as the **megasporangia.** These eventually give rise to the female gametes of the tree and represent the locale of the early development of the zygote into an embryo.

In the staminate cones the microsporangia become filled with haploid **spore mother cells.** These are single, undivided cells but soon undergo mitosis to form two cells that degenerate and two that persist. The two surviving cells are known as the **generative cell,** which gives rise to the gametes, and the **tube cell.** These cells are contained in the pollen grain. The entire pollen grain then develops two hollow "wings" that help the breeze to waft it to the ovulate cone.

Meanwhile, within the ovulate cone, numerous megasporangia are maturing between the leaflets, more correctly termed **bracts,** of the cone. Each megasporangium contains a haploid mother cell that produces two kinds of cells: a large number of nutritive cells whose function is to form the "endosperm" tissue of the seed, storing food for the future use of the embryonic plant, and egg cells **(ovules),** usually two in number, only one of which will produce a surviving embryo. These haploid tissues represent the female gametophyte.

When the pollen grain reaches the megasporangium, it is drawn inside by the shrinkage of a drying mucilaginous **pollination fluid.** When the pollen grain comes into contact with the "skin" of the megasporangium that surrounds its haploid tissues, it proceeds to germinate, taking as long as a year to complete the process. First the pollen grain elongates, producing a tube which penetrates the megasporangium and grows toward an egg. The tube nucleus maintains a position at the head end of the pollination tube, producing RNA for the manufacture of digestive enzymes. These have the twofold function of clearing a path through the megasporangial tissue and of providing for the nutrition of the male gametophyte. The generative nucleus next divides into two sperm nuclei, and when the tube finally reaches an egg, the sperms are both discharged. However, only one sperm nucleus succeeds in fertilizing the egg. This process is summarized in Figure 23–10.

The fertilized seed develops in the ovulate cone until it is ready for release. Provided with wings and stored food, it is dispersed by wind and may remain dormant for a long time before favorable conditions release the growth inhibition from the embryo and a new sporophyte takes root. Not all coniferous gymno-

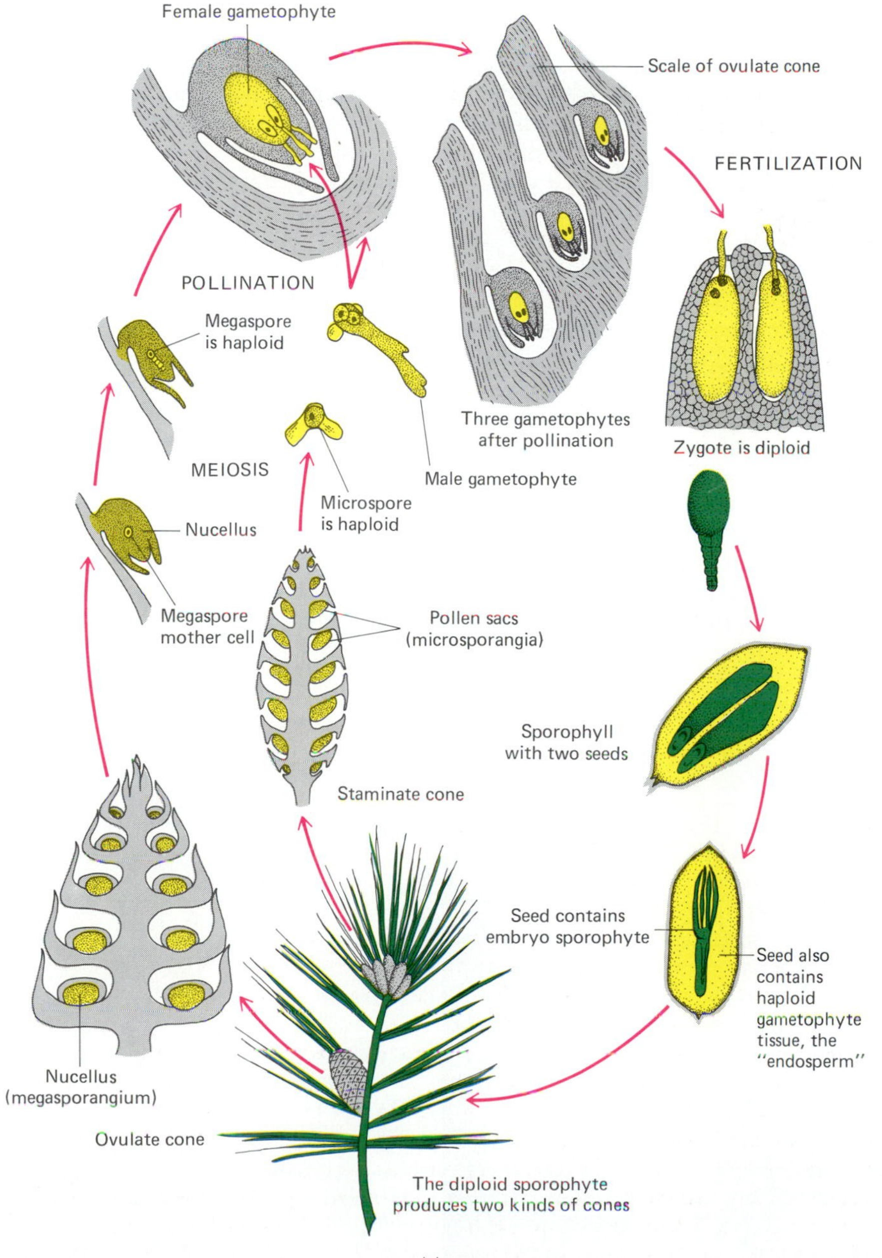

(*a*)

(*b*)

Figure 23–10 (*a*) Life cycle of a pine tree. (*b*) Bisected seed of bunya-bunya tree, a Southern Hemisphere gymnosperm. Note embryo in seed. The bunya-bunya matures for many years before bearing its cones. These are then dropped from the tree, unexpectedly bombarding everything beneath it—automobile roofs, for example.

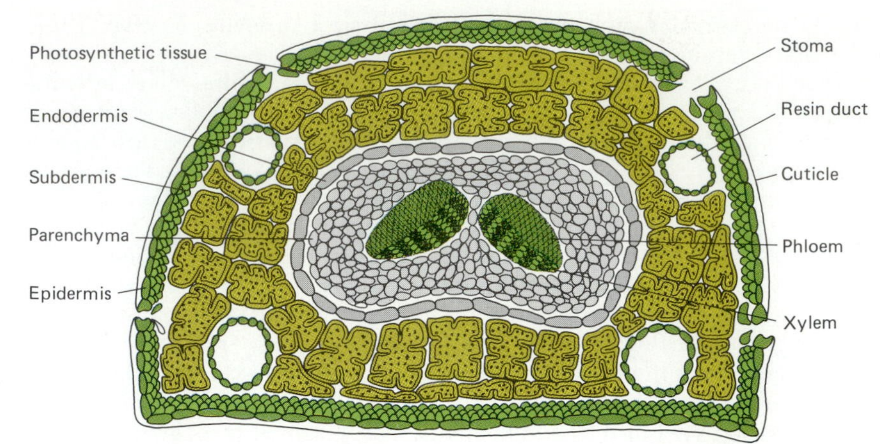

Figure 23–11 Cross-section of a pine needle. Very different from most angiosperm leaves, the subcylindrical shape of the pine needle minimizes its surface area and therefore also minimizes water loss. The resin ducts help to impregnate the needle with resin, which probably helps to discourage insect enemies.

Figure 23–12 Coniferous gymnosperms are the typical trees of northern latitudes. It is likely that their ability to retain their leaves and to carry out photosynthesis when the growing season is over for angiosperms gives them a critical advantage there and in some seasonally arid habitats.

sperms conform to this description: Some tropical or Southern Hemisphere species have immense cones that they drop as a whole; the cones then rot, and the seeds are dispersed by rainwater.

The needle-like leaves of the evergreen conifers are well adapted to withstand hot summers, cold winters, and the mechanical abrasions of storms. Since they are not all shed at one time, their retention may give evergreen conifers an advantage over angiosperms in many habitats that have a short growing season. Under the thick, heavily cutinized epidermal layer is a layer of thick-walled **sclerenchyma,** a supporting tissue. The stomata are set in deep pits that penetrate the sclerenchyma (Fig. 23–11). These probably serve to minimize evaporative water loss. Although conifers are found from tropical to subarctic regions, the most extensive coniferous forests are found in the Northern Hemisphere in Canada and Siberia, where they are the predominant trees of the taiga ecosystem (Fig. 23–12). However, the various species of conifers occur in a wide variety of habitats, ranging from very humid rain forests to dry semideserts. The northern coniferous forest region of the taiga is a rather arid environment with a low relative humidity during most of the year. Western North America has a most interesting group of conifers including the redwood, *Sequoia;* the California big tree, *Sequoiadendron,* the largest living organism; a variety of firs; and many species of pines, spruces, and junipers. Fossils of a dawn redwood were found in Japan and in other areas of the Northern Hemisphere, and botanists were greatly surprised when living dawn redwoods were found in remote valleys in central China in 1945. Seeds of the dawn redwood were collected and planted, and the trees can now be seen in a number of botanical gardens around the world and have recently become available to the public (Fig. 23–13).

Figure 23–13 Dawn redwood, *Metasequoia glyptostroboides,* with autumn coloration. Native to China and only recently discovered alive, this living fossil has been extensively propagated and now can be purchased for ornamental planting. Like the related bald cypress, this conifer is deciduous, losing its leaves in the winter. (Renee Purse, Photo Researchers, Inc.)

(a)

(b)

Figure 23–14 Reproductive structures in a cycad of the genus *Cycas.* (*a*) Female plant with ovulary-like megasporangia borne on modified cones. (*b*) Male plant with microsporangia-bearing cones. In many species these cones are immense and constitute a considerable disposal problem for the gardener. More distinctive is the fact that the sperm of these plants, borne in the pollen grain, is ciliated and swims actively to fertilize the female gametophyte.

Class Cycadopsida: The Cycads

Although far less economically important than conifers, the palmlike **cycads,** comprising the class **Cycadopsida,** are widely used as ornamental plants for landscaping in the tropics (Fig. 23–14). Some species provide food; the seeds of *Cycas,* for example, are gathered and eaten, and a starchy cycad flour is prepared from the fleshy underground stems of *Zamia.*

The cycads, found mainly in tropical and semitropical regions, have either short, tuberous underground stems or erect cylindrical stems above the ground. With their large compound divided leaves (Fig. 23–15), they resemble ferns or miniature palm trees, for which they are frequently mistaken.

The only cycad native to the United States is the sago "palm," *Zamia,* found in southern sections of Florida. *Zamia* is one of the smallest cycads, having a short fleshy underground stem, a long tap root, and leathery compound leaves, usually less than one meter long. In addition to the tap root, cycads have other roots that grow near the surface of the soil, closely branched and containing colonies of the nitrogen-fixing cyanobacteria, *Nostoc.* It seems likely that the nitrogen requirements of these plants are met at least in part by the cyanobacteria. Cycads grow very slowly but may live for an exceedingly long time. One very old cycad plant in western Australia is estimated to be about 5000 years old, making this one of the more ancient living organisms in the world.

The pine produces two kinds of cones on the same tree. In contrast, the cycad population consists of two kinds of trees: one producing only cones yielding pollen and the other producing only cones yielding seeds. As in conifers, these are not exactly "male" and "female" trees but are sporophytes that produce male and female gametophytes, respectively. The seed cones of cycads are larger than the pollen cones; in some species, they may be as much as a meter long and weigh 40 kg. Seed cones and the seeds themselves are usually quite colorful, ranging from yellow to deep red. With the remarkable exception of the fact that cycads have a motile swimming sperm inside their pollen grains, the life cycle of cycads resembles that of the conifers.

Figure 23–15 View of unfolding crown of new cycad leaves.

Class Lyginopteridopsida: Seed Ferns

Coal beds have yielded fossil remains of many large, leafy plants with fernlike leaves or fronds (Fig. 23–16). These were initially thought to be ferns, but more recently, seeds were found attached to some of these leaves, and the plants were named **seed ferns** and classified with the gymnosperms in the class **Lyginopteridopsida.** Some of the seed ferns were small trees; others were vinelike plants with

Figure 23–16 Seed ferns. (*a*) Reconstruction of *Medullosa noei,* which in life was about 5 meters tall. (*b*) A large seed fern seed, attached terminally to a part of a frond. (*c*) A microsporangial organ of *Medullosa,* composed of many longitudinally united microsporangia. (From T. Delavoryas.)

large fronds and slender stems. The seeds of the seed ferns, ranging in length from 4 mm to 11 cm, were attached to the leaves in several ways. Pollen-bearing organs were also attached to the leaves. Recently fossil pollen tubes in the ovule of a seed fern were described, which indicates that in them the process of fertilization resembled that in the modern cycads.

Figure 23–17 *Gingko biloba* tree showing "fruit." These evil-smelling (to us!) fruits clog sidewalks in urban environments where the pollution-resistant gingko trees are extensively planted. For this reason, staminate plants are preferred for ornamental use. (Courtesy of Lloyd Black.)

Class Gnetopsida and the Gingko

The maidenhair tree, *Gingko biloba* (Fig. 23–17), is the only living representative of a once numerous and widespread class with an extensive fossil record. Gingkos were cultivated in China and Japan as ornamental trees because of their distinctive fan-shaped leaves, which are shed in the fall.

These "living fossils" are tough and hardy trees under most conditions, and studies have shown that they are remarkably resistant to attack by insects and fungi and are also notably resistant to the polluted conditions that must be endured by urban trees. The wood, although brittle, has been used to make insect-proof cabinets, and the trees are widely used for decorative plantings along city streets and in parks. As with the cycads, there are two kinds of gingko trees: one producing only staminate cones and the other only ovulate cones. *Gingko* and the cycads are the only seed plants that produce swimming sperm rather than pollen tube nuclei. When the gingko ovule is fertilized and matures, the inner seed covering becomes hard, while the outer covering becomes a kind of soft, pulpy "fruit" with a rancid, sour odor. It has been suggested that gingkos may have declined in the wild because their seeds were dispersed by dinosaurs, now extinct themselves, that ate the fruit. If seeds are not effectively dispersed, of course, a plant's distribution will be reduced, and it may die out completely.

Class **Gnetopsida** is a group of plants containing a number of aberrant species, including *Welwitschia,* and the economically more important *Ephedra* of the

Figure 23–18 Welwitschia mirabilis, Southwest Africa (Namib desert). Note female cones. This odd gymnosperm absorbs condensation from the surfaces of its leaves. (F. J. Odendaal, Stanford University/BPS.)

American Southwest, from which the drug ephedrine is obtained. There is only one living species of *Welwitschia,* found in the deserts of coastal Southwest Africa. It has an exceedingly deep taproot by means of which it obtains subsurface water. The *Welwitschia* body consists of a short, broad, woody stem that is mostly underground except for a concave, disk-shaped crown that is covered with cork. The mature plant is rather like a large, woody turnip. Some specimens of *Welwitschia* are as much as 2000 years old and have crowns that are over 1 meter in diameter. During its entire lifespan each plant produces only a single pair of thick and leathery leaves, which grow on the periphery of the crown; these may become quite tattered as they sprawl over the surface of the soil. Although it may not rain in this region of Southwest Africa for periods of four or five years, there are frequent nightly fogs, which apparently supply these plants with much of the water they need by condensation on the leaf surfaces. The stomata on the leaves of *Welwitschia* open at night so as to take in this water from the fog and then remain closed during the daytime (Fig. 23–18).

CLASS ANGIOSPERMOPSIDA: THE ANGIOSPERMS

The true flowering plants are called angiosperms. In numbers of species, the angiosperms are the largest class in the plant kingdom and include more than a quarter million species of trees, shrubs, vines, and herbs adapted to almost every kind of habitat.

Life Styles of the Angiosperms

Some angiosperms live completely under water; others, in extremely arid conditions. The vast majority are autotrophic, but some, such as the Indian pipe, have little or no chlorophyll and are partially or wholly parasitic (Fig. 23–19). A few plants have devices for catching insects and other small animals and hence are carnivorous (Chapter 1), though not precisely **holozoic**—that is, the digestion involved is not the complete process seen in animals. Such plants evidently use the bodies of their victims only as sources of minerals and nitrogen (which would otherwise be scarce in the swampy environments where such plants typically occur); they do not incorporate organic substances from their "food" into their tissues, nor do they use these substances as a source of energy. The flowering plants provide us with much of our food, clothing, shelter, and drugs, and enrich the world with their beautiful colors and scented flowers.

Many angiosperms can complete an entire life cycle from the germination of the seed to the production of new seeds within a month; others require 20 to 30 years to reach sexual maturity. Some live for a single growing season; others, for centuries. The stems, leaves, and roots present a bewildering variety of forms, but all angiosperms develop flowers that have a fundamentally similar pattern.

Figure 23–19 Indian pipe, a flowering plant that is not a producer. It does not carry on photosynthesis but derives its energy from other organic matter, as do fungi. (Courtesy of Leo Frandzel.)

The following characteristics of angiosperms distinguish them from gymnosperms:

1. the abundance and prominence of the xylem vessels (most gymnosperms have only tracheids)
2. the formation of flowers and fruits
3. the presence of sepals, petals, or both in addition to the sporophylls
4. the formation of a pistil through which the pollen tube grows to reach the ovule and egg (in gymnosperms the pollen lands on the surface of the ovule and the pollen tube grows in directly)
5. a usually triploid rather than haploid endosperm that nourishes the embryo[1]

Monocots and Dicots

Flowering plants are assigned to one of two major subclasses: the **dicotyledons** and the **monocotyledons.** These are distinguished on the basis of their embryo structure, the form of their flowers, and the anatomy of their stem and leaf. The cereal plants such as wheat, corn, and rice—the basic agricultural crops of modern society—are monocotyledons; most vegetable, fruit, and nut crops are dicotyledons. Economically important angiospermous lumber trees are also dicotyledons. Other dicotyledons are medicinal plants such as *Digitalis,* which produces a heart stimulant, and *Rauwolfia,* which produces a compound used in the treatment of hypertension. Coffee, tea, and cocoa are also derived from dicotyledonous plants. The angiosperms include some 225,000 species of the dicotyledon subclass and 50,000 species of the monocotyledon subclass.

The two subclasses differ in the following respects:

1. In monocots the embryo has only one seed leaf or cotyledon; in dicots the embryo has two. In the cotyledon of dicots there are usually large deposits of starch and other foods that nourish the embryo and seedling until it is capable of making its own food by photosynthesis. The cotyledon of many monocots functions as a digestive and absorptive organ rather than as a storage organ.
2. Endosperm is typically present in the mature seed of the monocots, whereas endosperm is often absent from the mature seed of dicots.
3. In monocots, the leaves have parallel veins and usually smooth edges; in dicots, the veins of the leaves branch and rebranch (Fig. 23–20).
4. A cambium is usually absent in the monocot but is present in the dicot. Thus most monocots have little or no secondary growth. Woody monocots such as palms and yuccas, however, have a special thickening meristem.
5. In monocots, the flower parts (petals, sepals, stamens, and pistils) exist in threes or multiples of three; in dicots, the flower parts usually occur in fours or fives or multiples of these.
6. Monocots are generally herbaceous, with a few families (such as bamboos and palms) having woody parts.
7. In monocots, the vascular bundles of xylem and phloem are scattered throughout the stem. In dicots, the xylem and phloem occur either as a single solid mass extending up the center of the stem, or as a ring between the cortex and pith.
8. The roots of monocots are typically fibrous and multiple, whereas the root system of dicots usually consists of one or more primary taproots and secondary roots.

The many different families of monocots and dicots usually take their names from some conspicuous member. Representative monocot families are grasses, palms, lilies, orchids, and irises; some important dicot families are the buttercup, mustard, rose, maple, cactus, carnation, primrose, phlox, mint, pea, parsley, and aster. The rose family, for example, includes not only the familiar rose but also

[1]Conifers do not have a true endosperm. True endosperm (which develops from union of a sperm nucleus with two polar nuclei) is found only in angiosperms. In conifers, the "endosperm" is actually haploid female gametophyte tissue formed by mitotic divisions of one surviving megaspore. Most angiosperms have triploid endosperm, but some (e.g., members of the lily family) have *penta*ploid endosperm.

Figure 23–20 Some monocots and dicots. (*a*) Pasture rose, a dicot. (*b*) Maple flower, a wind-pollinated dicot. (*c*) Palm flower, a monocot. (*d*) Castor oil plant leaf. Note the typical dicot-branching veins. (*e*) Palm leaf. Note the parallel monocot venation (veining pattern). ((*a*), Courtesy of Leo Frandzel.)

trees—the apple, pear, plum, cherry, apricot, peach, almond, and hawthorn—and the strawberry, raspberry, and other shrubs.

The life cycle of angiosperms is similar to that of gymnosperms, although there are differences of detail in their reproduction. Flowers, which are the sexual reproductive organs of the angiosperms, are modified branches, and many of their parts are homologous with leaves. Although some species are **dioecious,** that is, consist of plants of two separate sexes, most are **monoecious,** having both male and female parts in each flower.

Reproduction in angiosperms has many similarities to that in gymnosperms. As in the gymnosperms, a pollen tube penetrates the female parts, and sperm move through it to fertilize the ovules. The ovules develop into seeds, and the ovulary (and sometimes other parts of the pistil) then develops into a fruit. (In gymnosperms, some species have the equivalent of a fruit, but usually this structure is absent.) The life cycle of angiosperms is diagrammed and discussed in greater detail in Chapter 31.

The Origin and Relationships of Higher Plants

In this book, the vascular plants are considered to be a major division of the plant kingdom, the tracheophytes. The tracheophytes are one of the two great divisions of the "higher" or terrestrial plants. They differ quite strikingly from the bryophytes or nonvascular plants, not only in vascularity but usually in their mode of reproduction as well. Among the vascular plants with mutally independent alternating generations, the sporophyte is by far the more conspicuous; indeed, among the gymnosperms and flowering plants, the sporophyte is the only stage that can be seen with the naked eye. Until the 19th century, in fact, the existence of a gametophyte generation in seed plants was unsuspected.

All the characteristics shared by the vascular plants were initially assumed to result from a common ancestry, but as we will see, such a common ancestry cannot be shown to be a fact at present. The realization of these facts cannot fail to influence the systematics of plant kingdom taxonomy. The bryophytes are now often viewed as an artificial assemblage of primitive plants united only on the basis of their nonvascularity; and some biologists consider the tracheophytes to be a possibly polyphyletic group whose diverse members have in common only a vascular "grade" of evolutionary advancement. The confused relationships among the various gymnosperm groups and between them and the angiosperms has produced an arbitrary system of classification that in some cases must abandon the very concepts upon which distinctions between families or classes are made.

How did the vascular and nonvascular plants originate? The nonvascular plants that seem most sophisticated in their general organization are the mosses, possessing as they do both stemlike and leaflike parts. They also resemble algae in some ways. The protonema of mosses is quite similar in appearance to many filamentous green algae, which suggests the conventional view that mosses evolved from green algae. To this let us add the facts that, as in the green algae, bryophytes store carbohydrates in the form of starch. Also as in the green algae, hornworts possess the food storage cells called pyrenoids. Some green algae also employ an alternation of haploid and diploid generations resembling that of higher plants and have a method of cell wall formation that higher plants also employ. Furthermore, the fine structure of some green algae and plant sperm flagella is closely similar. These resemblances are not trivial but are offset to some degree by the fact that most algae have unicellular sex "organs," whereas the multicellular archegonia and antheridia of bryophytes resemble nothing the algae possess. Of course it is always easy to imagine a hypothetical group of algae, now extinct, that gave rise to the bryophytes; however, fossil evidence for such a group is absent.

It is now widely believed that the vascular plants were directly derived from algal ancestors themselves, although this view was not always held. Originally it was thought that the vascular plants evolved from bryophyte ancestors, but the fossil evidence seems to indicate that the vascular plants are, if anything, older[1] than the bryophytes and so could not have been descendants of bryophytes. Moreover, as we have seen, the generational details of the life cycle of the two groups are very different. The ferns, for example, have a prominent sporophyte generation—just the opposite of the practice of the liverworts and mosses.

Figure 23–21 A restoration of *Archaeopteris.* (From Beck, C. B.: Reconstruction of *Archaeopteris,* and further consideration of its phylogenetic position. *American Journal of Botany* 49:373–382, 1962.)

THE ORIGIN OF THE GYMNOSPERMS

It is currently most widely believed that the conifers evolved separately from the seed ferns and the cycads, although all of these could have had a common ancestry. Seed ferns do not seem to be as good candidates for the ancestry of the rest of the gymnosperms as we might suppose, since trees that were clearly conifers have been found in the same geological strata as the seed ferns. The **Cordaites,** for example, a group of extinct, large, cone-bearing trees, had strap-shaped leaves and stems and were as much as 30 meters tall and a meter in diameter. These formed great forests during the Carboniferous period. The modern conifers are believed to have descended from *Cordaites* and related forms.

It is often suggested that the gymnosperms evolved from a group of Devonian plants known as **progymnosperms.** These had frondlike branches resembling those of early ferns, but in addition had the type of xylem generally associated with gymnosperms—one composed of tracheids with bordered pits. One of the progymnosperms, *Archaeopteris,* was a large tree with a trunk nearly 2 meters in diameter (Fig. 23–21). Its overall appearance may have been similar to that of present-day gymnosperms such as pines and cypresses, but its branches and leaves were far

[1] Plant paleontologists date the vascular plants as far back as the Silurian period, believed by geologists to have occurred some 410,000,000 years ago, whereas the earliest known bryophytes occur in Pennsylvanian deposits thought to have been laid down about 300,000,000 years ago.

different. In addition to tracheids with bordered pits, *Archaeopteris* had woody stems with extensive amounts of xylem, a characteristic typical of gymnosperms. However, the nature of the reproductive organs of the progymnosperms is not well understood, and it is by no means clear that they produced pollen or even seeds. Moreover, even if the conifers descended from them, it does not necessarily follow that the remaining gymnosperms all did so.

THE ORIGIN OF THE ANGIOSPERMS

Fossil remains of angiosperms have been found in rocks dated by geologists as belonging to the Cretaceous period during the Mesozoic era, and although it is usually thought that angiosperms evolved from some primitive gymnosperm, there is as yet no general agreement that the various groups of plants in older rocks that have been proposed as intermediates between gymnosperms and angiosperms are so in fact. Careful study of living plants has not been of much help. Observation of the flowers and fruits of the fossil and living angiosperms of the magnolia family discloses a somewhat conelike arrangement of parts, but the correspondence between these and the reproductive structures of gymnosperms is not close enough to establish confidently the homology of their parts. Among modern gymnosperms, the closest things to living representatives of a group possibly ancestral to the angiosperms are plants of the class Gnetopsida. They are unique among living gymnosperms because they have flower-like compound pollen cones, ovules with two integuments, and vessels in their wood. The latter are tubelike elements in the xylem composed of cylindrical xylem cells arranged end to end with sievelike or open-ended walls. Such vessels are unknown in other gymnosperms but are a characteristic component of the xylem of flowering plants.

The resemblances of Gnetopsida plants to flowering plants have suggested that they may represent an ancestral gymnosperm from which flowering plants evolved. However, most botanists now believe the similarities to flowering plants are coincidental or the result of parallel evolution.

SUMMARY

I. The vascular plants include a number of groups of greatly varying significance.

II. The most important vascular plants are the ferns and the seed plants.

- *A.* The ferns include two dramatically differing stages in their life cycle: a gametophyte and a sporophyte generation.
  1. The haploid gametophyte generation is small and inconspicuous. It produces flagellated sperm that give rise to the diploid sporophyte generation.
  2. The sporophyte is truly vascular, possessing phloem and xylem. The generational emphasis is exactly opposite to that of the bryophytes.
- *B.* The seed plants produce seeds containing living embryos, which are dispersed by a variety of mechanisms to grow independently of the parent plant. The existence of a gametophyte generation was unsuspected for many years, but it is now understood that the pollen grain and the macrogametophyte that produces the ovule constitute the extremely inconspicuous gametophyte generation.
- *C.* The seed plants include the gymnosperms and the angiosperms. The gymnosperms may well be a polyphyletic group.
  1. The gymnosperms usually lack anything resembling a fruit and have a haploid embryo-nourishing tissue ("endosperm").
  2. The angiosperms have fruit that develops principally from the ovary to enclose the seed; they also have a triploid endosperm produced by double fertilization. The angiosperms consist of two major subclasses: the monocotyledons and the dicotyledons.
     - *a.* The monocots have but a single embryonic or seed leaf. The endosperm is usually well developed, the leaf veins are parallel and unbranched, and the flowers are three-parted or have parts in multiples of three.
     - *b.* The dicots have two embryonic leaves, usually adapted to food storage and taking the place of the endosperm, which may be totally absent. Dicots have divergently branching leaf veins and flowers with four or five parts, or multiples of these.

POST-TEST

1. Members of the order ______________, whose common name is "horsetails," possess tiny leaves (that are apparently main stem derivatives) called ______________; these plants have skeletal tissue that is rich in ______________.

2. The ferns and their allies have a very prominent ______________ generation, in contrast to the mosses, in which this generation is semiparasitic.

3. In ferns, the gametophyte plants are known as ______________.

4. Fern gametophytes are ______________ (haploid/diploid).

5. The "endosperm" of gymnosperms differs from the true endosperm of flowering plants in that in gymnosperms it is chromosomally ______________ and is directly derived from the female gametophyte, whereas in angiosperms it is usually chromosomally ______________.

6. The ______________ are gymnosperms somewhat resembling palms that possess swimming sperm cells.

7. Pollen grains of conifers are carried by the wind to the ______________ cones of the same or other trees. There they germinate, producing a long ______________ ______________ in which the ______________ ______________ can travel near to an egg.

8. *Welwitschia,* an unusual African gymnosperm, absorbs most of the water it needs from condensation that forms on its ______________.

9. Generally, in angiosperms, the vascular tissue known as ______________ is better developed than in gymnosperms, and has more of the elements known as ______________ ______________ than is usually the case in gymnosperms.

10. The structure and development of flowers suggests that they are modified ______________.

11. Complete the following table:

	Leaf Vein Arrangement	*Number of Cotyledons*	*Endosperm (?)*	*Secondary Growth (?)*	*Number of Flower Parts*	*Vascular Bundles*
Monocots						
Dicots						

REVIEW QUESTIONS

1. Prepare a table comparing the major features of all the classes of tracheophytes.
2. How do pine trees reproduce?
3. What differences in means of fertilization exist between the bryophytes and the tracheophytes?
4. Give at least seven ways in which the dicotyledons differ from the monocotyledons, comparing endosperm, leaf and flower structure, xylem and phloem, and so forth.
5. In what ways are ferns like the seed plants? In what ways do they differ?
6. What similarities are there between the higher green algae and the bryophytes? between the higher green algae and the tracheophytes?
7. What evidence can be used to argue for an independent origin of the bryophytes and the tracheophytes?

24
The Animal Kingdom: Animals Without a Coelom

OUTLINE

I. What is an animal?
II. Their place in the environment
III. Animal relationships
  A. Acoelomates and coelomates
  B. Protostomes and deuterostomes
  C. Homologous and analogous structures
  D. Simple versus complex
IV. Phylum Porifera: the sponges
  A. Body plan of a sponge
  B. Life style of a sponge
V. Phylum Cnidaria
  A. Class Hydrozoa
  B. Class Scyphozoa (jellyfish)
  C. Class Anthozoa (corals)
VI. Phylum Ctenophora (comb jellies)
VII. Phylum Platyhelminthes (flatworms)
  A. Class Turbellaria
  B. Class Trematoda (flukes)
  C. Class Cestoda (tapeworms)
VIII. Phylum Nemertinea (proboscis worms)
IX. Phylum Nematoda (roundworms)
  A. *Ascaris*: a parasitic roundworm
  B. Other parasitic roundworms
X. Phylum Rotifera (wheel animals)

LEARNING OBJECTIVES

After you have read this chapter you should be able to:

1. List the characteristics common to most animals: using these characteristics, develop a brief definition of an animal.
2. Identify the ecological role of animals, and discuss their distribution; compare the advantages and disadvantages of life in the sea and of life in fresh water and on land.
3. Relate the animal phyla on the basis of symmetry, type of body cavity, and pattern of embryonic development, e.g., protostome versus deuterostome.
4. Identify the distinguishing characteristics of phyla Porifera, Cnidaria, Ctenophora, Platyhelminthes, Nemertinea, Nematoda, and Rotifera; compare the level of organization of each of these phyla.
5. Classify a given animal in the appropriate phylum (from among those listed in objective 4), and at the option of your instructor, identify the class to which it belongs.
6. Trace the life cycle of the following parasites: *Ascaris*, tapeworm, hookworm, and trichina worm. Identify several adaptations that these animals possess for their parasitic life style.
7. Explain the adaptive advantages of each of the following characteristics: bilateral symmetry, cephalization, a motile larva, digestive cavity with two openings, hermaphroditism.

More than a million species of animals have been described, and perhaps several million more remain to be identified. Most members of the kingdom Animalia are classified in about 30 different phyla. The animals most familiar to us—dogs, birds, fish, frogs, snakes—are **vertebrates,** animals with backbones. However, vertebrates account for only about 5% of the species of the animal kingdom. The majority of animals are the less familiar **invertebrates**—animals without backbones. The invertebrates include such diverse forms as sponges, jellyfish, worms, and insects.

What Is an Animal?

It is easy enough to distinguish between a horse and an oak tree, but there are so many diverse animal forms that exceptions can be found to almost any definition of an animal. Still there are some characteristics which describe at least most animals (Fig. 24–1):

1. All animals are multicellular.
2. Animals are eukaryotes.
3. The cells of an animal exhibit a division of labor. In all but the simplest animals, cells are organized to form tissues, and tissues are organized to form organs. In most animal phyla, specialized organ systems carry on specific functions.
4. Animals are exclusively heterotrophic; they ingest their food first and then digest it inside the body, usually within a digestive system.
5. Most animals are capable of locomotion at some time during their life cycle. However, there are some animals—for example the sponges—that are **sessile** (firmly attached to a substrate) as adults.
6. Most animals have well-developed sensory and nervous systems and respond to external stimuli with appropriate behavior.
7. Most animals reproduce sexually, with large, nonmotile eggs and small flagellated sperm. Sperm and egg unite to form a fertilized egg, or **zygote,** which goes through a series of embryonic stages before developing into a larva or immature form.

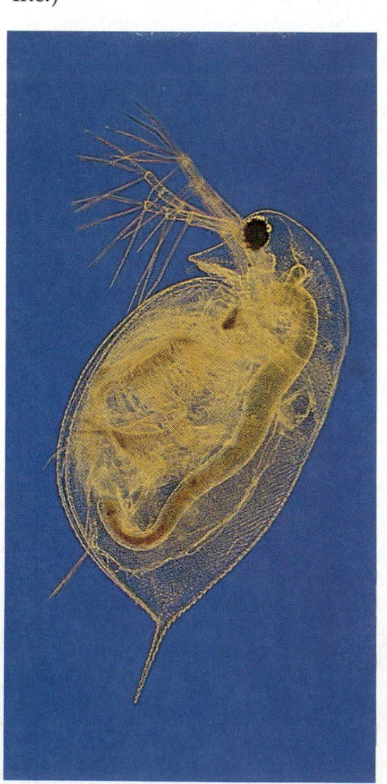
(a)

(b)

Figure 24–1 Despite their diversity, most members of the animal kingdom share several distinct traits. (*a*) The nearly transparent body of this freshwater crustacean, *Simocephalus vetulus,* or water flea, shows a complex of organ systems. (*b*) As heterotrophs, all animals must feed either on producers or on other animals that eat producers. Those that feed on other animals are often highly motile and complex in their behavior. Shown here is a fishing spider feeding on a small fish. The numerous black dots on the dorsal (back) surface of the spider are actually eyes. Though these eyes do not allow the spider to see especially sharp images, they do serve to make the animal an extremely efficient hunter. ((*a*), Herman Eisenbeiss, Photo Researchers, Inc.; (*b*), Robert Noonan, Photo Researchers, Inc.)

Their Place in the Environment

As consumers, all animals are ultimately dependent upon producers for their raw materials, energy, and oxygen. Animals are also dependent upon decomposers for recycling nutrients.

Animals are distributed in virtually every corner of the earth. Most early animals (and plants too) lived in the sea. Of the three environments—salt water, fresh water, and land—the sea is the most hospitable. Sea water is isotonic to the tissue fluids of most marine animals, so there is little problem in maintaining fluid and salt balance. The buoyancy of sea water supports its inhabitants, and the temperature is relatively constant owing to the large volume of water. **Plankton** (the organisms that are suspended in the water and float with the movement of the water) consists of tiny animals, plants, and protists that provide a ready source of food.

There are certain disadvantages to life in the sea. For example, the environment is continuously in motion. Although this motion brings nutrients to them and washes their wastes away, animals must be able to cope with the constant churning and currents that might sweep them away. Fish are strong enough swimmers so that they can direct their movements and location effectively. However, most invertebrates, owing to their body structure, are unable to swim strongly enough, and so have other adaptations. Some are sessile, attaching to some stable structure like a rock, so that they are not wafted about with the tides and currents. Others solve the problem by maintaining a small body size and becoming part of the plankton. Even though they are tossed about, their food supply continues to surround them, and so they are able to survive successfully. The brackish water of the estuarine environment, where fresh water meets the sea, presents its own special difficulties; the lower and changing salinities restrict the types of organisms that can thrive there.

Fresh water offers a much less constant environment than sea water. Oxygen content and temperature vary, and turbidity (due to sediment suspended in the water) and even water volume fluctuate. Fresh water is hypotonic to the tissue fluids of animals, and water tends to diffuse into the animal; therefore, freshwater animals must have some mechanism for pumping out excess water while retaining salts. This osmoregulation requires an expenditure of energy. Another disadvantage is that freshwater environments generally contain less food than marine environments. For these reasons far fewer animals make their homes in fresh water than in the sea.

Terrestrial life is even more difficult. Dehydration is a serious threat, because water is constantly lost by evaporation and is often difficult to replace. The many adaptations in terrestrial animals addressing this problem are discussed in Chapters 25 and 26. Only a few animal groups, most notably representatives of the arthropods (insects, spiders, and some related forms) and the higher vertebrates, have successfully made their homes on land.

Animal Relationships

Most biologists agree that the evolutionary origin of animals is obscure. Animals are thought to have arisen from the protists, probably from the flagellates. Although the relationships among the various phyla are a matter of conjecture, a few of the more widely held hypotheses are presented in this section.

The animal kingdom may be divided into two branches, or subkingdoms: **Parazoa,** which consists of the sponges, and **Eumetazoa,** which includes all of the other animals. This distinction is made because the sponges are so different from all other animals that most biologists think that they are not directly ancestral to any other animal phylum.

Members of Eumetazoa are often further classified on the basis of body symmetry. Two phyla, the cnidarians (jellyfish and relatives) and the ctenophores (comb jellies) are radially symmetrical and are included in the **Radiata.** In **radial symmetry,** similar structures are regularly arranged as spokes, or radii, from a central axis; any imaginary plane passing through the central axis divides the organism into two mirror images. All of the other animals are bilaterally symmetrical (at

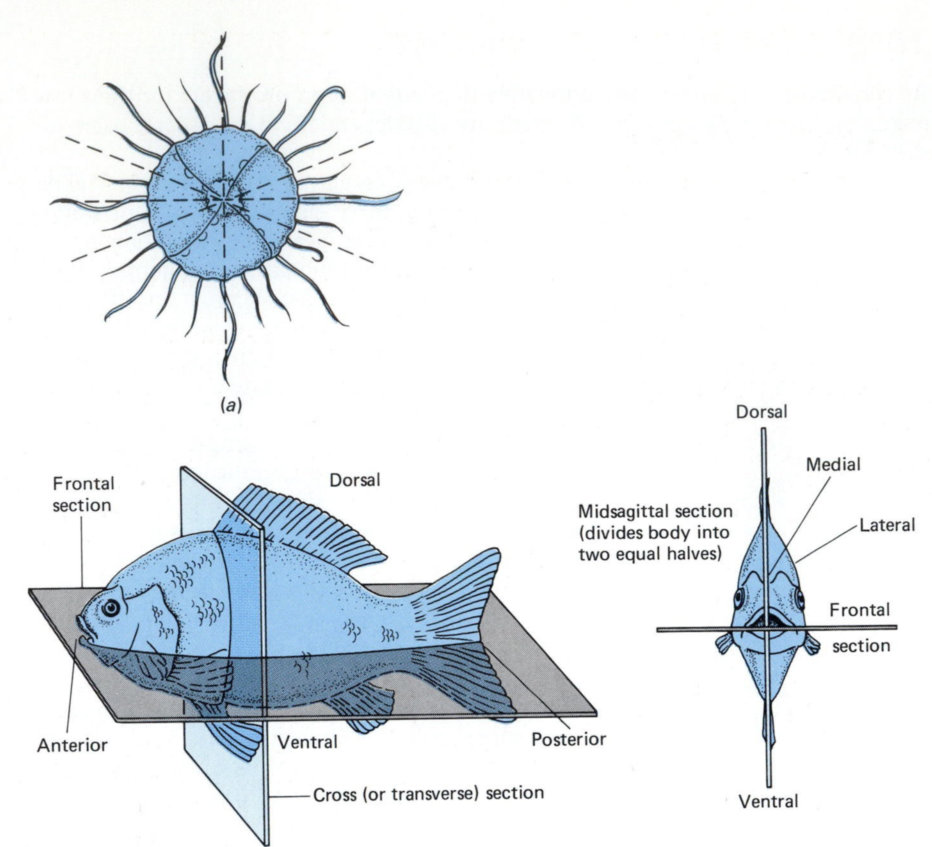

Figure 24–2 Types of body symmetry in animals. (*a*) In radial symmetry multiple planes can be drawn through the central axis; each divides the organism into two mirror images. (*b*) Most animals exhibit bilateral symmetry. A midsagittal cut (lengthwise vertical cut through the midline) divides the animal into right and left halves. The head end of the animal is generally its anterior end, and the opposite end is its posterior end. The back of the animal is its dorsal surface; the belly surface is ventral. The diagram also illustrates various ways in which the body can be sectioned (cut) in order to study its internal structure. Many cross sections and sagittal sections are used in illustrations throughout this book to show relationships among tissues and organs.

least in their larval stages) and belong to the **Bilateria.** In **bilateral symmetry,** the body is divided into roughly identical right and left halves when sliced down the midline. (See Fig. 24–2 and Focus on Body Plan and Symmetry.)

ACOELOMATES AND COELOMATES

A widely held system for relating the animal phyla to one another is based upon the type of body cavity. Some of the simpler groups of animals have no body cavity and are referred to as **acoelomates.** Others have a body cavity derived from the blastocoel, the cavity of the embryo; these animals are known as **pseudocoelomates.** Animals with a true coelom are **coelomates.** In order to understand what a true coelom is, we must digress briefly into the animal's embryonic origins. The structures of most animals develop from three embryonic tissue layers, called **germ layers.** The outer layer, called the **ectoderm,** gives rise to the outer covering of the body and to the nervous system. The inner layer, or **endoderm,** lines the digestive tract. **Mesoderm,** the middle layer, extends between the ectoderm and endoderm and gives rise to most of the other body structures, including the muscles, bones, and circulatory system.

In the simplest eumetazoans (cnidarians and platyhelminths) the body is essentially a double-walled sac surrounding a single digestive cavity with a single opening to the outside—the mouth. There is no body cavity, so these animals are acoelomates (Fig. 24–3). The more complex animals usually have a tube-within-a-tube body plan: The inner tube, the digestive tract, is lined with tissue derived from endoderm and is open at both ends—the mouth and the anus. The outer tube or body wall is covered with tissue derived from ectoderm. Between the two tubes is a second cavity, the body cavity. If the body cavity develops between the mesoderm and endoderm, it is called a **pseudocoelom.** If it forms within the mesoderm and is completely lined by mesoderm, the body cavity is a true **coelom.** The phylogenetic tree shown in Figure 24–4 indicates the relationships of the major phyla of animals based on the type of coelom they possess.

FOCUS ON

Body Plan and Symmetry

Most animals exhibit **bilateral symmetry,** a type of symmetry in which the body can be divided through only one plane (which goes through the midline of the body) to produce roughly equivalent right and left halves that are mirror images. Cnidarians (jellyfish, sea anemones, and their relatives) and adult echinoderms (sea stars, sea urchins, and their relatives) exhibit radial symmetry. In them, similar body parts are arranged around a central body axis. Radial symmetry is considered an adaptation for a sessile life style, for it enables the organism to receive stimuli equally from all directions in the environment.

Bilateral symmetry is considered an adaptation to motility. The front, or **anterior,** end of the animal generally has a head where sense organs are concentrated; this end receives most environmental stimuli. The **posterior,** or rear, end of the animal may be equipped with a tail for swimming, or may just follow along.

In order to locate body structures it is helpful to define some basic terms and directions (see Fig. 24–2). The back surface of an animal is its **dorsal** surface; the belly side is its **ventral** surface. (In animals that stand on two limbs, such as humans, the term **posterior** refers to the dorsal surface, and **anterior,** to the ventral surface.) A structure is said to be **medial** if it refers to the midline of the body and **lateral** if it is toward one side of the body. For example, in a human, the ear is lateral to the nose. The terms **cephalic** and **rostral** (and **superior,** in humans) refer to the head end of the body; the term **caudal** refers to structures closer to the tail. (In human anatomy, the term **inferior** is used to refer to structures located relatively lower in the body.)

A bilaterally symmetrical organism has three axes, each at right angles to the other two: an anterior–posterior axis extending from head to tail; a dorsoventral axis extending from back to belly; and a left–right axis extending from side to side. We can distinguish three planes (flat surfaces that divide the body into specific parts): The **midsagittal** plane (or section) divides the body into equal right and left halves; this plane passes from anterior to posterior and from dorsal to ventral. Any section or plane (cut) parallel to the midsagittal plane is parasagittal (or simply **sagittal**), and divides the body into unequal right and left parts. A **frontal** (or coronal) plane, or section, divides the body into anterior and posterior parts. A **transverse section,** or **cross section,** cuts at right angles to the body axis.

PROTOSTOMES AND DEUTEROSTOMES

A different family tree is shown in Figure 24–5. This important phylogenetic scheme is based on the pattern of embryonic development. In this scheme, the complex animals are divided into two groups: the protostomes and the deuterostomes. These groups reflect two main lines of evolution. Early during embryonic

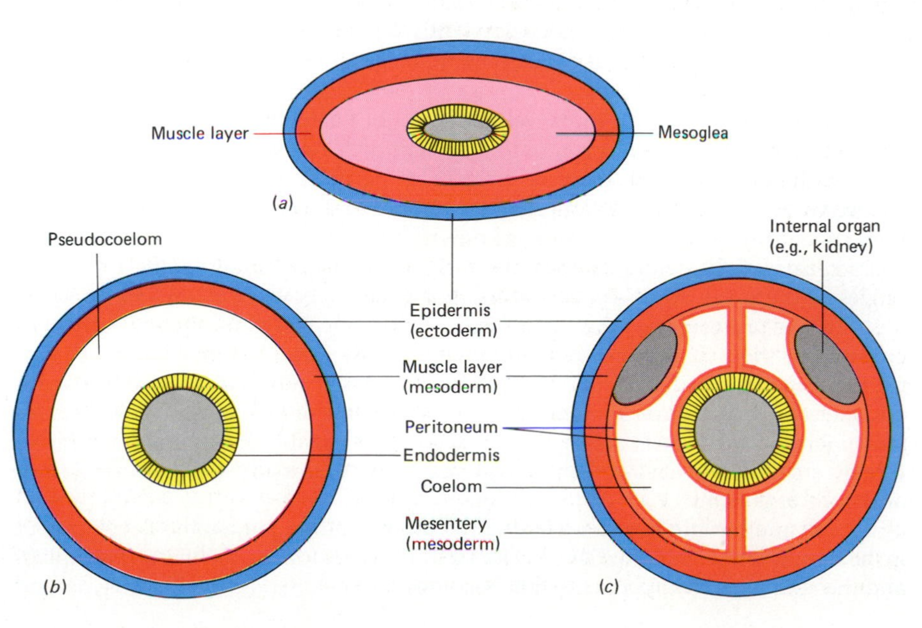

Figure 24–3 Three basic animal body plans are illustrated by these cross sections. (*a*) An acoelomate animal has no body cavity. (*b*) A pseudocoelomate animal has a body cavity that develops between the mesoderm and endoderm. The term body cavity refers to the space between the body wall and the internal organs. (*c*) In a coelomate animal the body cavity, called a coelom, is completely lined with tissue derived from mesoderm.

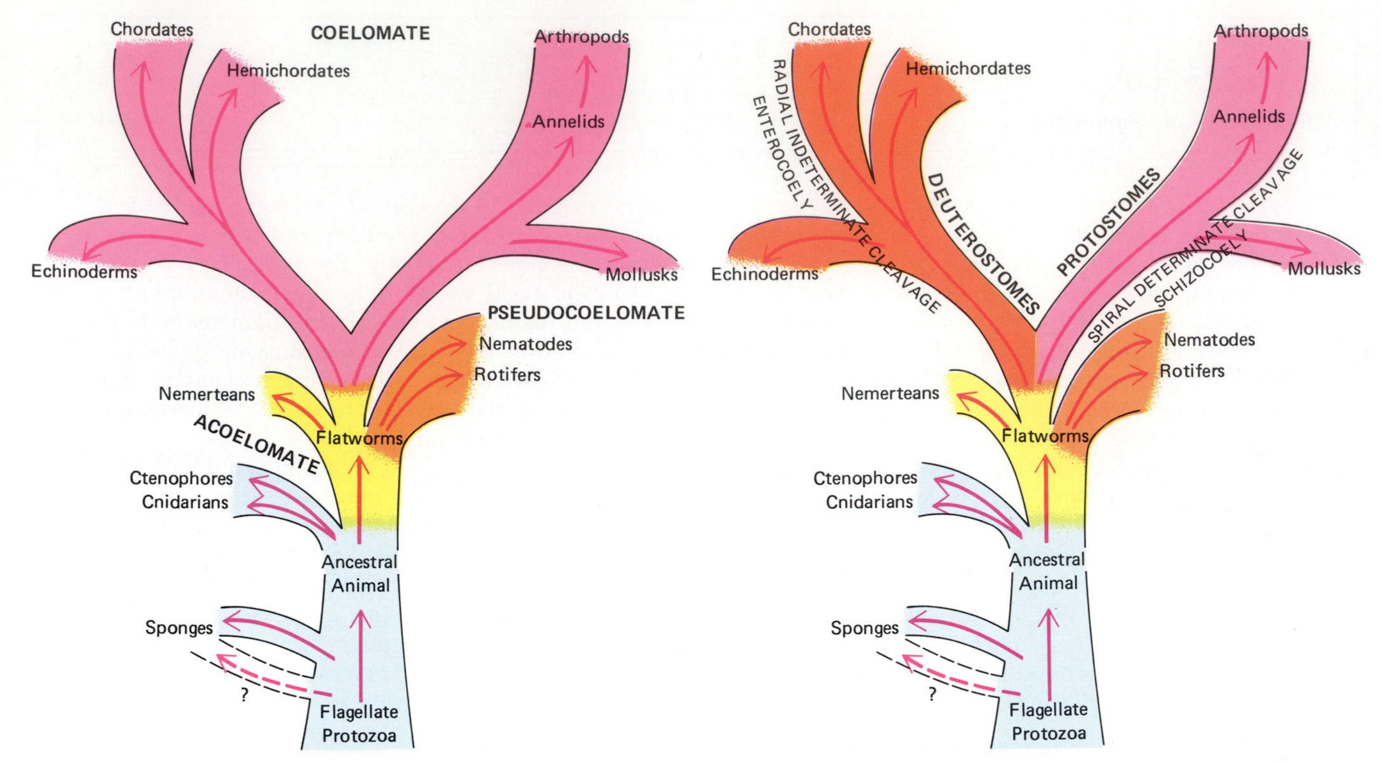

Figure 24–4 Proposed evolutionary relationships are illustrated by this phylogenetic tree indicating acoelomate, pseudocoelomate, and coelomate phyla. The flatworms and nemertenas have a solid body and so are referred to as acoelomate. Nematodes (roundworms) and rotifers do have a body cavity, but it develops from the embryonic cavity called a blastocoel and is located between the mesoderm and endoderm. All other bilateral animals have a true coelom, a body cavity lined by mesoderm. (After Barnes.)

Figure 24–5 A phylogenetic tree based on protostome–deuterostome characteristics. In the protostomes—the mollusks, annelids, and arthropods—the blastopore develops into the mouth, cleavage is generally spiral and determinate, and the coelom develops within the mesoderm when the mesoderm splits. In the deuterostomes—the echinoderms and chordates—the blastopore develops into the anus and the mouth develops from a second opening. Deuterostomes typically have radial, indeterminate cleavage, and the coelom develops from outpocketings of the gut. (After Barnes.)

development, a group of cells move inward to form an opening called the **blastopore.** In most of the mollusks, annelids, and arthropods, this opening develops into the mouth; these animals comprise the **protostomes** (meaning "first, the mouth"). Some taxonomists include the flatworms and pseudocoelomates as protostomes. In echinoderms (e.g., the sea star) and chordates (the phylum that includes the vertebrates), the blastopore develops into the anus; the opening that develops into the mouth forms later in development. These animals are the **deuterostomes** ("second, the mouth").

Another difference in the development of protostomes and deuterostomes is the pattern of **cleavage,** that is, the first several divisions of the embryo. In protostomes, the early cell divisions are oblique to the polar axis, resulting in a spiral arrangement of cells; any one cell is located between the two cells above or below it (Fig. 24–6). This pattern of division is known as **spiral cleavage.** In **radial cleavage,** characteristic of the deuterostomes, the early divisions are either parallel or at right angles to the polar axis; the cells are located directly above or below one another.

In the protostomes, the fate of each embryonic cell is fixed very early. For example, if the first four cells of an annelid embryo are separated, each cell will develop into only a fixed quarter of the larva; this is referred to as **determinate cleavage.** In deuterostomes, cleavage is **indeterminate:** If the first four cells of a sea star embryo, for instance, are separated, each cell is capable of forming a complete, though small, larva. Each cell of the early embryo has the potential to develop into an entire organism.

Still another difference between protostome and deuterostome development is the manner in which the coelom is formed. In protostomes, the mesoderm splits, and the split widens into a cavity that becomes the coelom (Fig. 24–7). This method

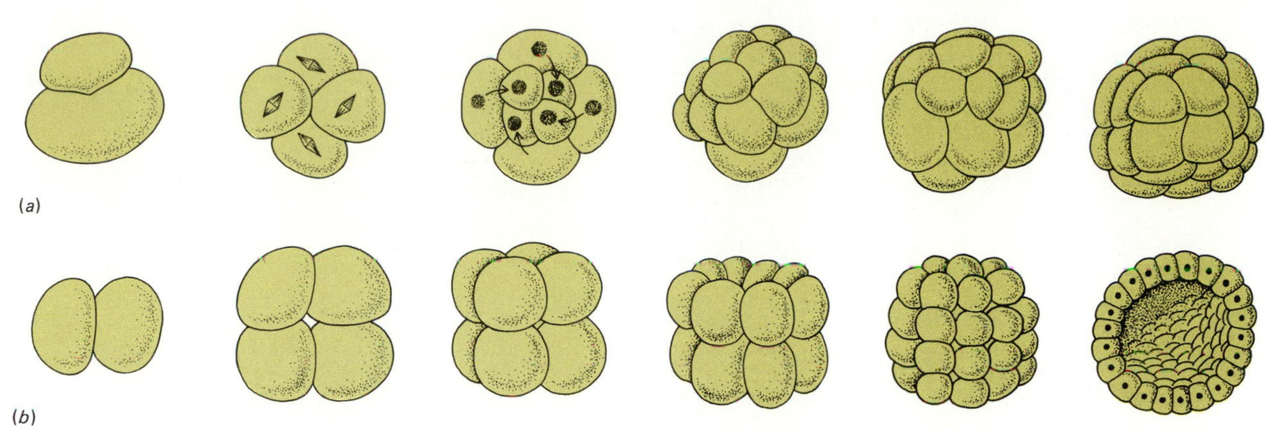

Figure 24–6 Types of cleavage in embryonic development. (*a*) Spiral cleavage is characteristic of protostomes. Note the spiral arrangement of the cells. (*b*) In the radial cleavage characteristic of deuterostomes the early divisions are either parallel or at right angles to the polar axis, so that the cells are stacked in layers.

of coelom formation is known as **schizocoely,** and for this reason the protostomes are sometimes called **schizocoelomates.** In deuterostomes, the mesoderm usually forms as outpocketings of the developing gut. These outpocketings eventually separate and form pouches; the cavity within these pouches becomes the coelom. This type of coelom formation is called **enterocoely,** and these animals are sometimes referred to as **enterocoelomates.**

HOMOLOGOUS AND ANALOGOUS STRUCTURES

In determining relationships among various animal groups, biologists have found it useful to study structures that these groups have in common. **Homologous struc-**

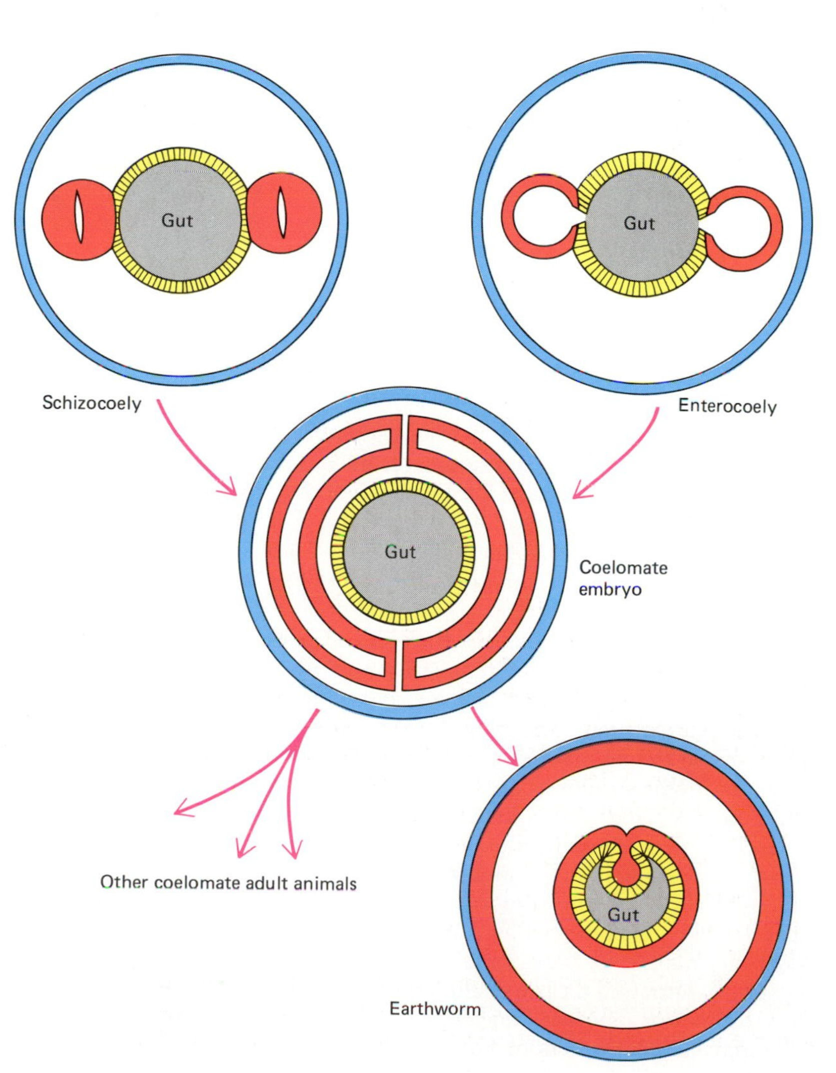

Figure 24–7 The coelom originates in the embryo as a block (or blocks) of mesoderm that split off of each side of the embryonic gut. In schizocoely the mesoderm splits; this split widens into a cavity that becomes the coelom. In enterocoely, the mesoderm outpockets from the gut, forming pouches. The cavity within these pouches becomes the coelom.

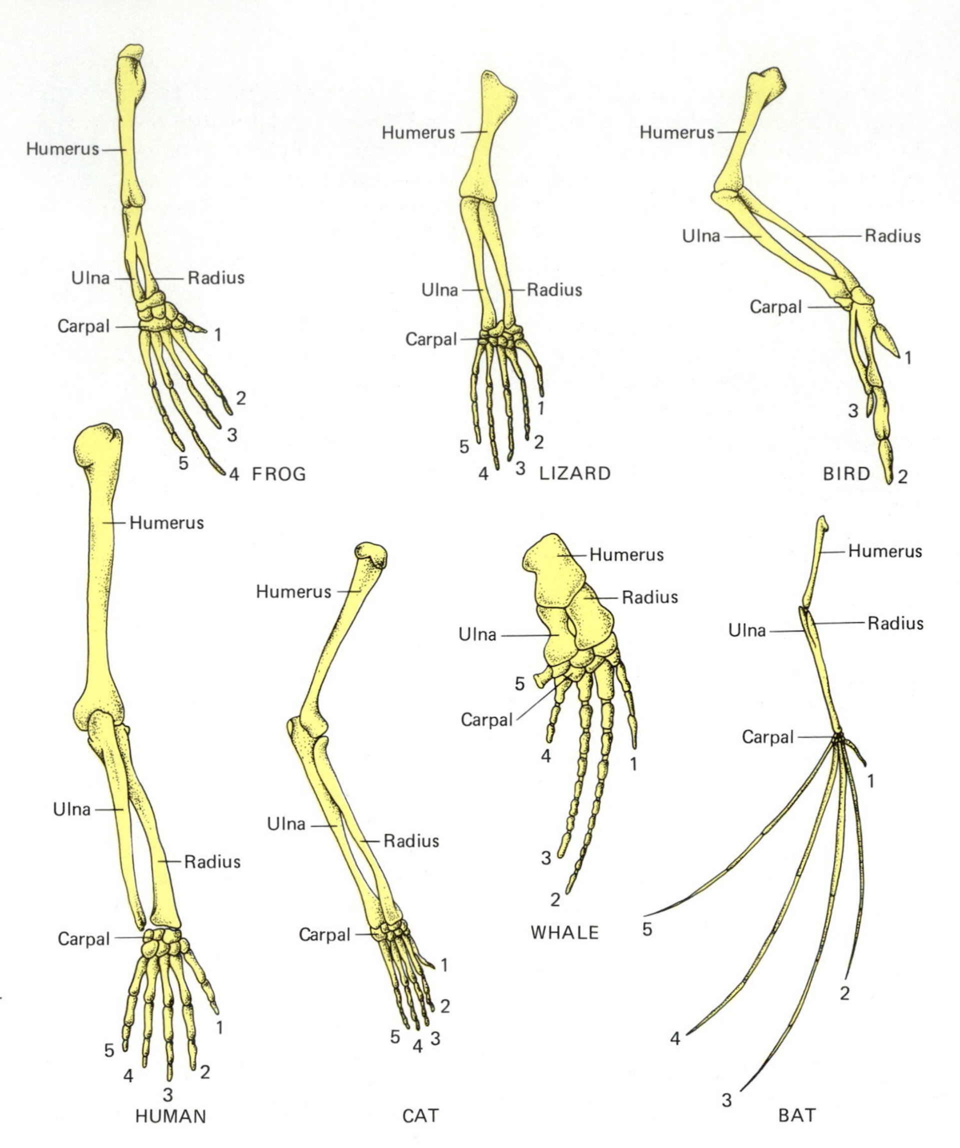

Figure 24–8 A classic example of homology: The bones of the forelimbs of a frog, lizard, bird, human, cat, whale, and bat, showing the arrangement of the homologous bones in these superficially different structures.

tures are those that develop from similar embryonic rudiments, and are similar in basic structural plan and development. Such structures reflect a common genetic endowment and are used to imply an evolutionary relationship. For example, the arm of a human, the wing of a bird, and the pectoral (front) fin of a whale are all homologous (Fig. 24–8). They have basically similar patterns of bones, muscles, nerves, and blood vessels, and similar embryonic origins, though rather different functions.

In contrast, **analogous structures** are superficially similar and serve a similar function, but have quite different basic structures and developmental patterns. The presence of analogous structures does not imply an evolutionary relationship in the animals bearing them. The wing of a bird and the wing of a butterfly are analogous. Both enable the animal to fly, but they have no developmental processes in common, and their structures are different. The wings of birds and the wings of bats, on the other hand, are homologous (both types of animals are vertebrates).

The terms homologous and analogous are also applied on the molecular level. The hemoglobins of different animals, the forms of cytochrome *c* present in different vertebrates, or the lactic dehydrogenases present in birds and mammals may be termed homologous proteins. The hemoglobins in different species, for example, may have very similar sequences of amino acids reflecting a common genetic pattern and a close evolutionary relationship. In contrast, hemoglobin and the respiratory pigment hemocyanin may be termed analogous molecules, since they have similar functions but quite different molecular structures.

SIMPLE VERSUS COMPLEX

In discussing and comparing the many different groups of animals, it is convenient to use such terms as *lower* and *higher*, *simple* and *complex*, and *primitive* and *advanced*. Such terms as higher, complex, or advanced do not imply that these animals are better or more nearly perfect than others. Rather, they are used in a comparative sense to describe their hypothesized evolutionary relationships. For example, the terms higher and lower usually refer to the level at which a particular group has diverged from a main line of evolution. It is customary, for instance, to refer to sponges and cnidarians as lower invertebrates because they are thought to have originated near the base of the phylogenetic tree of the animal kingdom. However, neither sponges nor cnidarians are primitive in all morphological or physiological characteristics. Each has become highly specialized in certain respects to its own particular life style. Furthermore, the terms lower and higher do not necessarily imply that the higher groups have evolved directly from or through the lower groups.

Phylum Porifera: The Sponges

About 5000 species of **sponges** have been identified and assigned to phylum **Porifera.** These simple, multicellular animals occupy aquatic, mainly marine, habitats. Living sponges may be drab, or bright green, orange, red, or purple (Fig. 24–9). They are usually slimy to the touch and may have an unpleasant odor. Sponges range in size from 1 to 200 cm in height and vary in shape from flat, encrusting growths to balls, cups, fans, or vases.

Sponges are usually thought to have evolved from certain flagellate protozoans, perhaps a hollow, free-swimming colonial flagellate. Sponge larvae resemble such flagellate colonies. In the sense that they apparently did not give rise to any other animal group, sponges seem to represent a dead end in evolution. Of course, sponges themselves continue to change as they are subjected to continual selective pressures from the environment.

(*a*)

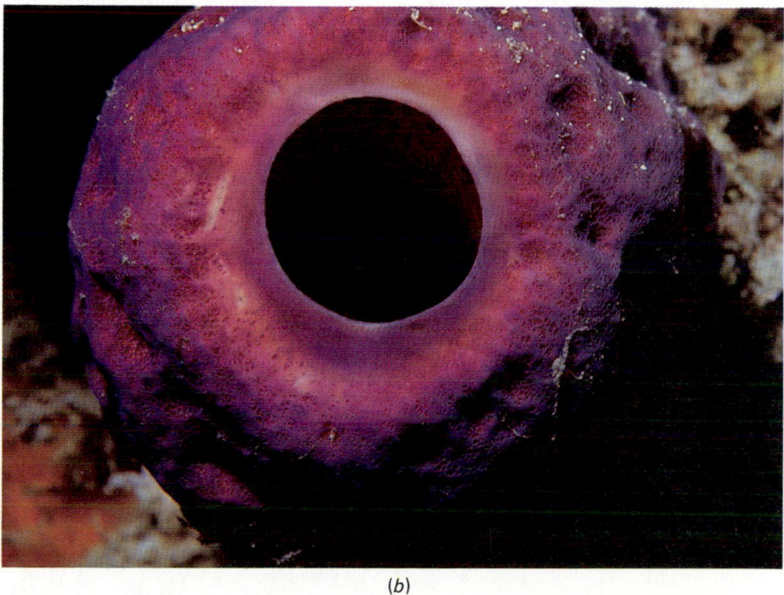

(*b*)

Figure 24–9 Sponges vary widely in size, shape, and color. (*a*) Tube sponge, a member of class *Demospongiae*. (*b*) View of the osculum of a large sponge collected in the Virgin Islands. ((*a*), Charles Seaborn; (*b*), Robin Lewis, Coastal Creations.)

Sponges are divided into three main classes on the basis of the type of skeleton they secrete: Members of the class **Calcispongiae** secrete a skeleton composed of small calcium carbonate spikes, or **spicules.** Members of class **Hexactinellida,** the glass sponges, have a skeleton of six-rayed siliceous spicules. Sponges that belong to the class **Demospongia** have a skeleton of **spongin** (a protein material) fibers or of siliceous spicules that do not have six rays. Bath sponges consist of the dried spongin skeleton of demospongians.

BODY PLAN OF A SPONGE

Porifera, meaning "to bear pores," aptly describes the sponge body, which resembles a sac perforated with tiny holes. In a simple sponge, water enters through these pores, passes into the central cavity, or **spongocoel,** and finally flows out through the sponge's open end, the **osculum.** Water is kept moving by the action of flagellated cells that line the spongocoel. Each of these **collar cells,** or **choanocytes,** is equipped with a tiny collar that surrounds the base of the flagellum. The collar is actually an extension of the cell membrane. The choanocytes of some complex sponges can pump a volume of water equal to the volume of the sponge each minute! In some types of sponges, the body wall is extensively folded, and there are complicated systems of canals (Fig. 24–10).

Most sponges are asymmetrical, but some exhibit radial symmetry. Although a sponge is multicellular, its cells are loosely associated and do not form definite tissues. There is a division of labor, with certain cells specialized to perform particular functions such as nutrition, support, or reproduction. The epidermal cells that make up the outer layer of the sponge are capable of contraction (Fig. 24–11). The choanocytes, which make up the inner layer, create the water current that brings

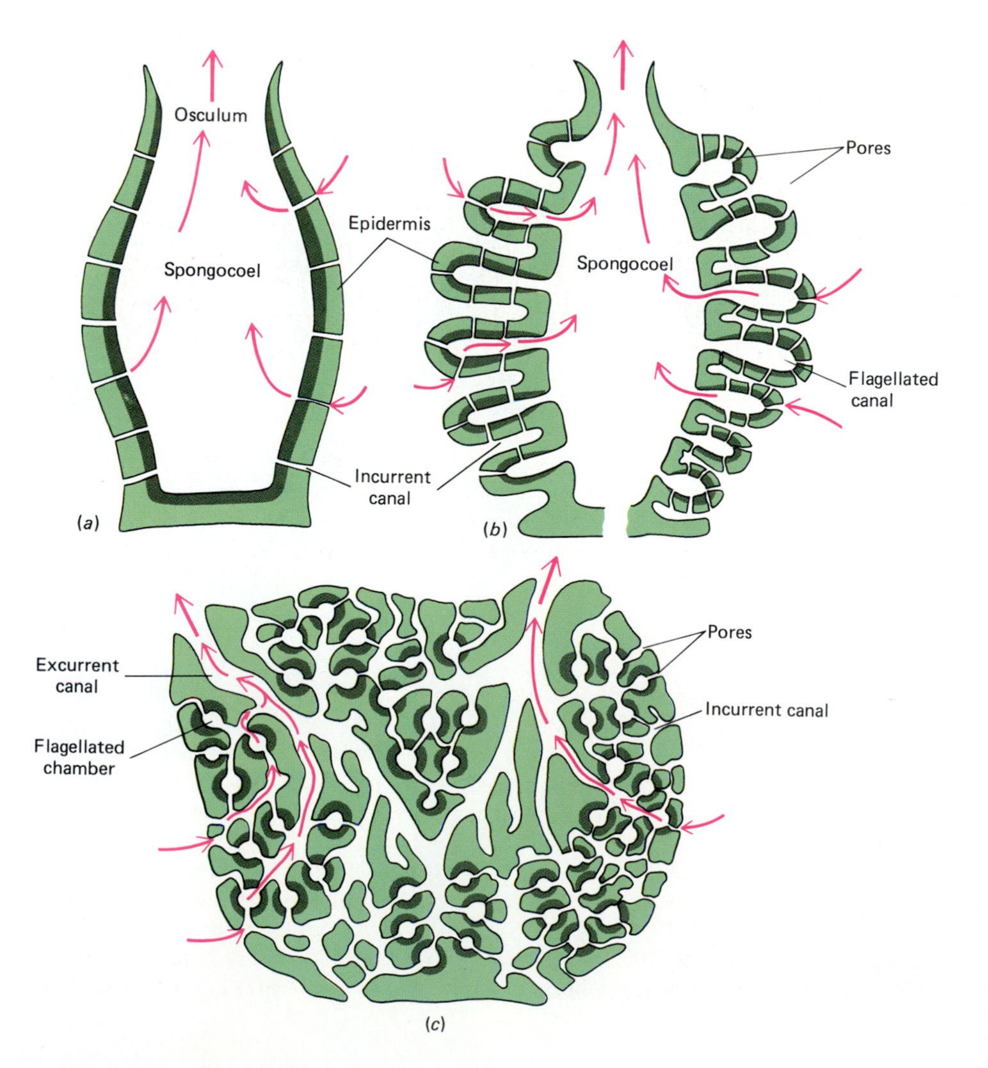

Figure 24–10 Three types of body plans in the sponge. (*a*) Simple unfolded or asconoid type of sponge. This type of sponge cannot grow very large because its surface area is limited and the rate of water flow is too slow. (*b*) In the syconoid type of sponge the body wall is folded, forming finger-like processes. This folding increases the surface area so that there are more choanocytes to circulate water. There is also a decrease in the size of the spongocoel so that there is less water that must be moved out through the osculum. (*c*) The most complex sponge body plan, the leuconoid type, is also the most efficient for circulating water, and most sponges have this intricately folded structure. Numerous channels replace the single, large spongocoel characteristic of the simpler types. (After Barnes.)

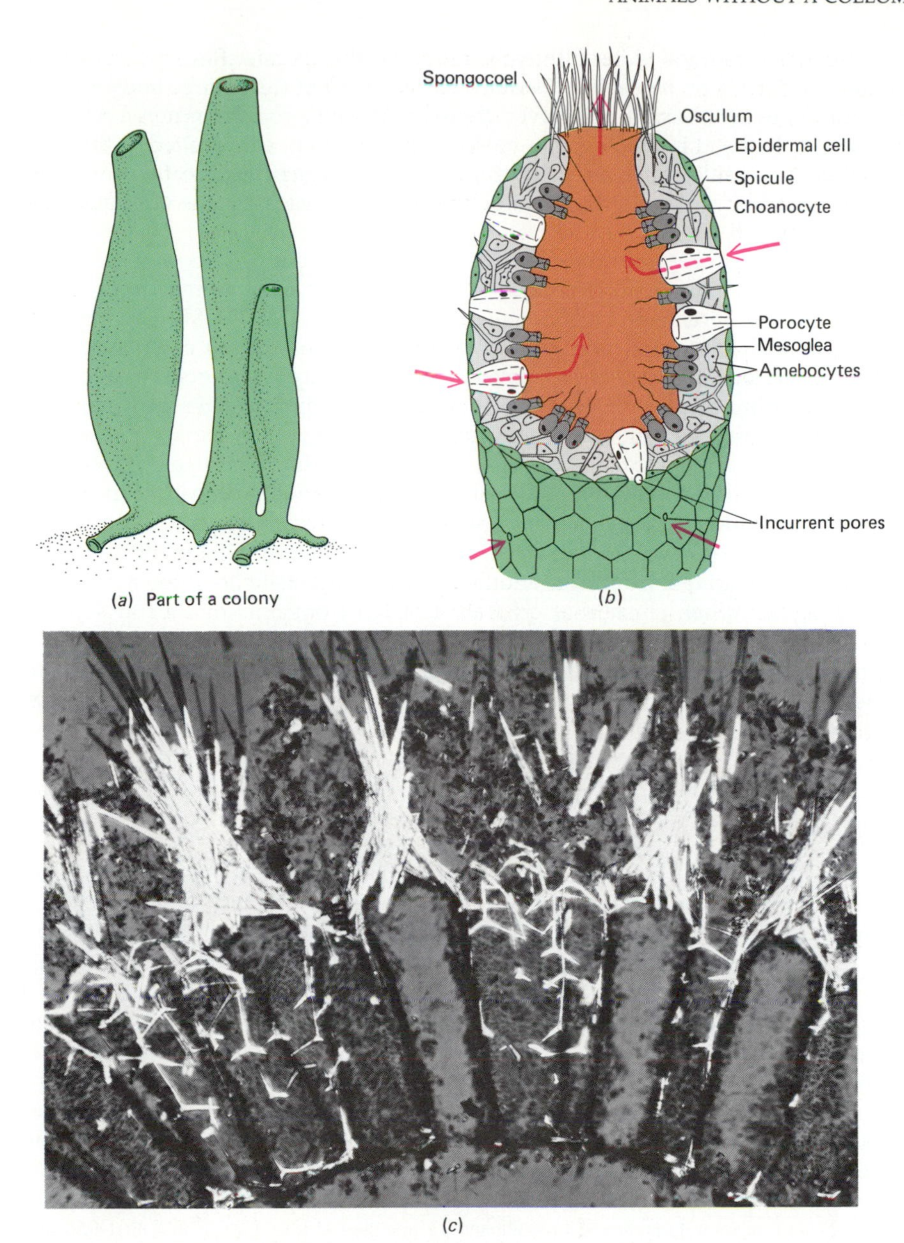

Figure 24–11 Sponge structure. (*a*) Part of a sponge colony. (*b*) A simple sponge cut open to expose its cellular organization. (*c*) A photomicrograph of a portion of a sponge (*Grantia*) showing spicules (the light structures). (Carolina Biological Supply Company.)

food and oxygen to the cells and carries away carbon dioxide and other wastes; they also trap food particles. Between the outer and inner cellular layers of the sponge body is a gelatin-like layer supported by skeletal spicules. Amoeba-like cells, aptly called **amoebocytes,** wander about in this layer; some of these amoebocytes secrete the spicules.

LIFE STYLE OF A SPONGE

Although larval sponges are flagellated and able to swim about, the adult sponge remains attached to some solid object on the sea bottom and is incapable of locomotion.

Since sponges cannot swim out in search of food, they are adapted for trapping and eating whatever food the sea water brings to them. Sponges are filter feeders. As water circulates through the body, food is trapped along the sticky collars of the choanocytes. Food particles pass down to the base of the collar, where they are taken by phagocytosis and either digested within the choanocyte or transferred to an amoebocyte for digestion. The amoebocytes help distribute food to other cells of the sponge. Direct absorption of nutrients from one cell to another may also help in distributing materials. Undigested food is simply eliminated into the water.

Oxygen from the water diffuses throughout the sponge. Respiration and excretion are carried on by each individual cell. Each cell of the sponge body is irritable and can react to stimuli. However, there are no sensory cells or nerve cells that would enable the animal to react as a whole. Behavior appears limited to the basic metabolic necessities such as procuring food and regulating the flow of water through the body. Pores and the osculum may be closed by the contraction of surrounding cells.

Sponges can reproduce asexually. A small fragment or bud may break free from the parent sponge and give rise to a new sponge, or may remain to form a colony with the parent sponge. Sponges also reproduce sexually. Most sponges are **hermaphroditic,** meaning that the same individual can produce both egg and sperm. Some of the amoebocytes develop into sperm cells, others into egg cells. Even hermaphroditic sponges can cross-fertilize with other sponges, however. Fertilization and early development take place within the jelly-like layer. Embryos eventually move out into the spongocoel and leave the parent along with the stream of excurrent water. After swimming about for a while as part of the plankton, the larva finds a solid object, attaches to it, and settles down to a sessile life.

Sponges possess a remarkable ability to regenerate. When the cells of a sponge are separated from one another, they reaggregate to form a complete sponge again. When clusters of cells are isolated from one another in separate containers of seawater, each cluster will reorganize and regenerate to form a new sponge. If the disaggregated cells of two different sponge species are mixed together, the cells will sort themselves out and reorganize separate sponges of the original species.

Phylum Cnidaria

Most of the 10,000 or so species of the phylum **Cnidaria**[1] are marine. They are grouped in three classes (Table 24–1): **Hydrozoa,** which includes the hydras, the hydroids such as *Obelia*, and the Portuguese man-of-war; **Scyphozoa,** which includes the jellyfish; and **Anthozoa,** which includes the sea anemones, true corals, and alcyonarians (sea fans, sea whips, and precious corals) (Figs. 24–12 and 24–13).

All of the cnidarians have stinging cells called **cnidocytes** from which they get their name (Cnidaria is from a Greek word meaning "sea nettles"). The cnidarian body is radially symmetrical and is organized as a hollow sac with the mouth and surrounding tentacles located at one end. The mouth leads into the digestive cavity called the **gastrovascular cavity.** The mouth is the only opening into the gastrovascular cavity and so must serve for both the ingestion of food and egestion of wastes.

[1]This phylum was formerly known as Coelenterata, a name derived from the fact that the body cavity serves as the digestive cavity: *coel* = hollow; *enteron* = gut.

TABLE 24–1
Classes of Phylum Cnidaria

Class and Representative Animals	*Characteristics*
Hydrozoa *Hydra* *Obelia* Portuguese man-of-war	Mainly marine, but some freshwater species; both polyp and medusa stage in many species (polyp form only in *Hydra*); formation of colonies by polyps in some cases
Scyphozoa Jellyfish (e.g., *Cyanea*)	Marine; inhabit mainly coastal waters; free-swimming jellyfish most prominent forms; polyp stage restricted to small larval stage
Anthozoa Sea anemones Corals Sea fans	Marine; solitary or colonial polyps; no medusa stage; gastrovascular cavity divided by partitions into chambers increasing area for digestion; sessile

Figure 24–12 Some common representatives of the three classes of phylum Cnidaria. Note the two basic types of body form. (See Fig. 23–14.)

Figure 24–13 Photographs of representative cnidarians. (*a*) A member of class Hydrozoa, *Obelia* forms a colony of polyps. (*b*) *Cyanea capillata,* a lion's mane jellyfish, a member of class Scyphozoa. The bell of this jellyfish can measure 2 meters; its tentacles trail over 10 meters. (*c*) This tree coral *Dendronephthya,* photographed in the South Pacific, is an anthozoan. (*d*) Polyps form the coral *Montastrea cavernosa* extended for feeding. (*e*) Opal bubble coral, *Plerogyra sinuosa,* from the South Pacific. ((*a*), Carolina Biological Supply Company; (*b*), (*d*), Charles Seaborn; (*c*), (*e*), Robin Lewis, Coastal Creations.)

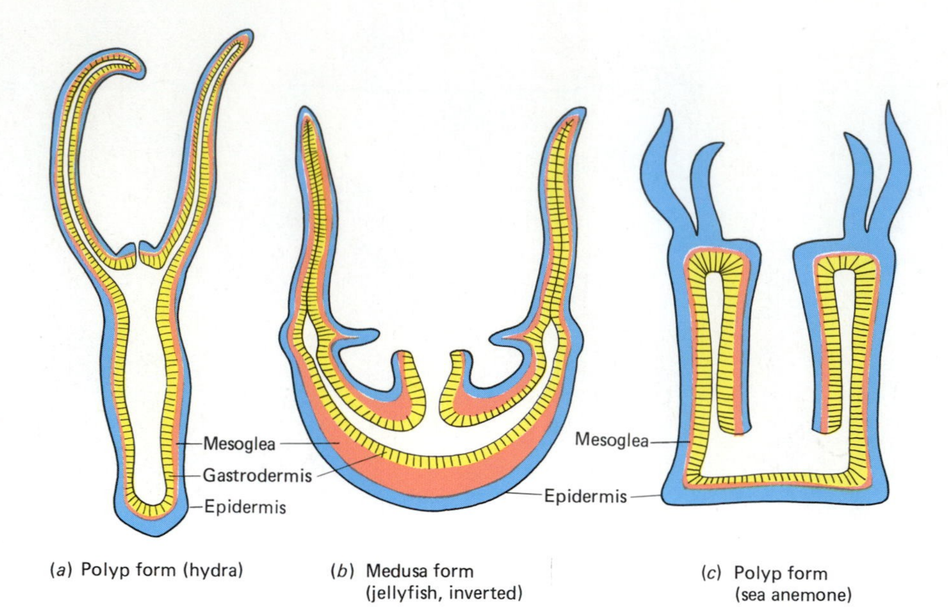

Figure 24–14 The polyp and medusa body forms characteristic of phylum Cnidaria are structurally similar. (*a*) The polyp form as seen in *Hydra.* (*b*) The medusa form is basically an upside-down polyp as shown by inverting a jellyfish. (*c*) In the anthozoan polyp the gastrovascular cavity is characteristically divided into chambers by vertical partitions.

Much more highly organized than the sponge, a cnidarian has two definite tissue layers. The outer **epidermis** and the inner **gastrodermis** are composed of several types of epidermal cells. These layers are separated by a gelatin-like **mesoglea.**

Cnidarians have two body shapes, the polyp and the medusa shapes (Fig. 24–14). The **polyp** form, represented by *Hydra*, resembles an upside-down, slightly elongated jellyfish. Some cnidarians have the polyp shape during their larval stage and later develop into the **medusa** (jellyfish) form. Though many cnidarians live a solitary existence, others group into colonies. Some colonies—for example, the Portuguese man-of-war—consist of both polyp and medusa forms.

CLASS HYDROZOA

As the most primitive class of the cnidarians, Hydrozoa is thought by some evolutionists to have given rise to both other classes. The cnidarian body plan is typified by a tiny animal, *Hydra*, found in freshwater ponds and appearing to the naked eye like a bit of frayed string (Fig. 24–15). This animal is named after the multiheaded monster of Greek mythology with the remarkable ability to grow two new heads for each head cut off. The cnidarian hydra also has an impressive ability to regenerate: When it is cut into several pieces, each piece may grow all the missing parts and become a whole animal.

The hydra's body, seldom more than 1 cm long, consists of two layers of cells enclosing a central gastrovascular cavity. The outer **epidermis** serves as a protective layer; the inner **gastrodermis** is primarily a digestive epithelium. The bases of cells in both layers are elongated into contractile muscle fibers; those of the epidermis run lengthwise, and those in the gastrodermis run circularly. By the contraction of one or the other, the hydra can shorten, lengthen, or bend its body. Throughout its life the animal lives attached to a rock, twig, or leaf by a pedal disk of cells at its base. At the outer end is the mouth, connecting the gastrovascular cavity with the outside and surrounded by a circlet of tentacles. Each tentacle may be as much as one and a half times as long as the body itself. The tentacles, composed of an outer epidermis and an inner gastrodermis, may be hollow or solid.

The cnidarians are unique in producing "thread capsules," or **nematocysts,** (Fig. 24–16) within cnidocytes (stinging cells) in the epidermis. The nematocysts, when appropriately stimulated, can release a coiled, hollow thread. Some types of nematocyst threads are sticky; others are long and coil around the prey; a third type is tipped with a barb or spine and can inject a protein toxin that paralyzes the prey. The nematocyst is shaped like a balloon, with a long tubular neck that develops tightly coiled and inside out within the cavity of the balloon.

Each cnidocyte has a small projecting trigger (cnidocil) on its outer surface that responds to touch and to chemicals dissolved in the water ("taste") and causes

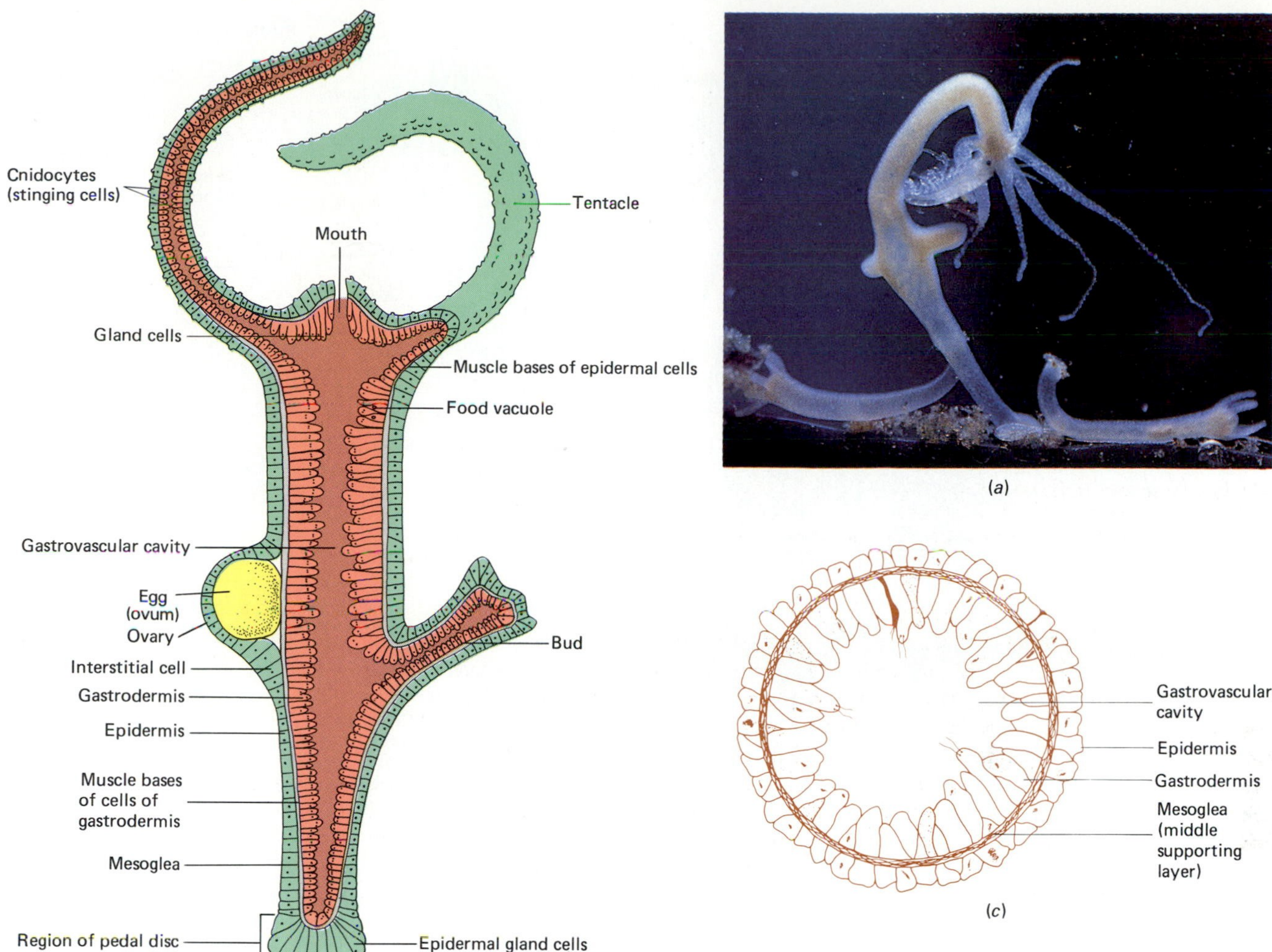

Figure 24–15 *Hydra* body structure. (*a*) A brown hydra, *Hydra oligactis*, capturing a small crustacean. Note the buds present on the hydra's body. One bud has already detached as a separate animal. (*b*) This *Hydra* is cut longitudinally to reveal its internal structure. Asexual reproduction by budding is represented on the right side of the figure; sexual reproduction is represented by the ovary on the left. Male hydras develop testes that produce sperm. (*c*) Cross section through the body of a *Hydra*. ((*a*), Tom Branch, Photo Researchers, Inc.)

the nematocyst to fire its thread. A nematocyst can be used only once; when it has been discharged, it is released from the cnidocyte and replaced by a new one, produced by a new cnidocyte. The tentacles encircle the prey and stuff it through the mouth into the gastrovascular cavity, where digestion begins. The partially digested fragments are taken up by pseudopods of the gastrodermis cells, and digestion is completed within food vacuoles in those cells.

Respiration and excretion occur by diffusion, for the body of a hydra is small enough that no cell is far from the surface. The motion of the body as it stretches and shortens circulates the contents of the gastrovascular cavity, and some of the gastrodermis cells have flagella whose beating aids in circulation. The hydra has no other circulatory device.

The first true nerve cells in the animal kingdom are found in the cnidarians. These animals have many nerve cells that form irregular **nerve nets** connecting the sensory cells in the body wall with muscle and gland cells. The coordination achieved thereby is of the simplest sort; there is no aggregation of nerve cells to form a brain or nerve cord, and an impulse set up in one part of the body passes in all directions more or less equally.

Hydras reproduce asexually by budding during periods when environmental conditions are optimal, but are stimulated to form sexual forms, males and females, in the fall, or when the pond water becomes stagnant. Females develop an ovary, which produces a single egg; males form a testis that produces sperm. After fertilization the egg becomes covered with a shell, leaves the parent, and remains within the protective shell throughout the winter.

Many of the marine cnidarians form colonies consisting of hundreds or thousands of individuals. A colony begins with a single individual that reproduces by

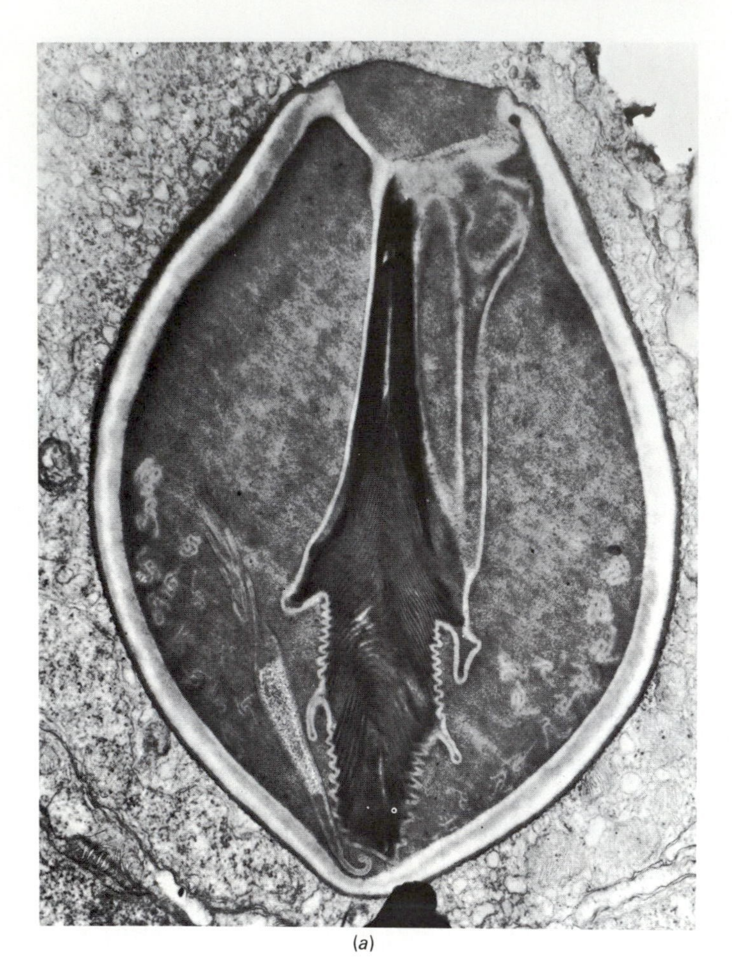

(a)

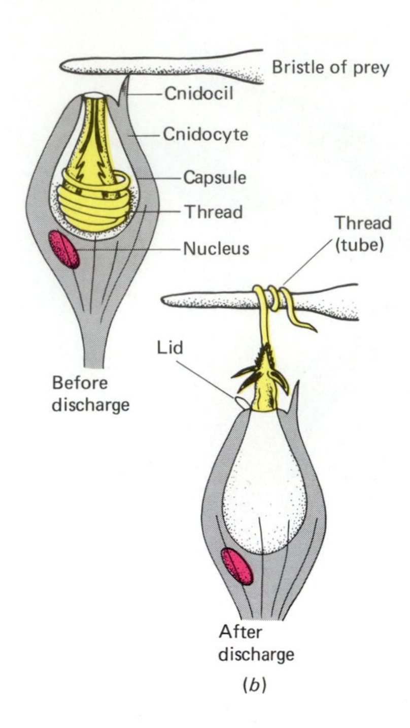

Figure 24–16 Nematocysts, the thread capsules within cnidarian cnidocytes. (*a*) Electron micrograph of an undischarged nematocyst of *Hydra* (sagittal section). (*b*) Discharge of a nematocyst. When an object comes in contact with the cnidocil, the nematocyst discharges, ejecting its thread, which may entangle the prey or secrete a toxic substance that immobilizes the prey. ((*a*), courtesy of G. B. Chapman, Cornell University Medical College. From Lenhoff, H. M., and Loomis, W. F. (eds.): *The Biology of* Hydra. Coral Gables, University of Miami Press, 1961.)

budding. However, instead of separating from the parent, the bud remains attached and continues to form additional buds. Several types of individuals may arise in the same colony, some specialized for feeding, some for reproduction, and others for defense. The existence of two or more different kinds of individuals within the same species is known as **polymorphism.**

The Portuguese man-of-war, *Physalia*, superficially resembles a jellyfish but is actually a colony of polyps and medusas. The individuals display remarkable polymorphism. A modified medusa acts as a float for the colony in the form of a gas-filled sac colored a vivid iridescent purple. The long tentacles of this animal, which may hang down for several meters below the float, are equipped with cnidocytes. These are capable of paralyzing a large fish and can wound a human swimmer severely.

Some of the marine cnidarians are remarkable for an alternation of sexual and asexual generations analogous to that in plants. This alternation of generations differs from that of plants in that both sexual and asexual forms are diploid. Only sperm and eggs are haploid. The cnidarian life cycle is illustrated by the colonial marine hydrozoan *Obelia* (Fig. 24–17). In this polyp colony, the asexual generation consists of two types of polyps: those specialized for feeding and those for reproduction. Free-swimming male and female medusas bud off from the reproductive polyps. These medusas eventually produce sperm and eggs, and fertilization takes place. The zygote develops into a ciliated swimming larva called a **planula.** The larva attaches to some solid object and begins to form a new generation of polyps by asexual reproduction.

CLASS SCYPHOZOA (JELLYFISH)

Among the jellyfish the medusa is the more prominent body form. It is like an upside-down hydra with a thick viscous mesoglea that gives firmness to the body. In scyphozoans, the polyp stage is restricted to a small larval stage. The largest

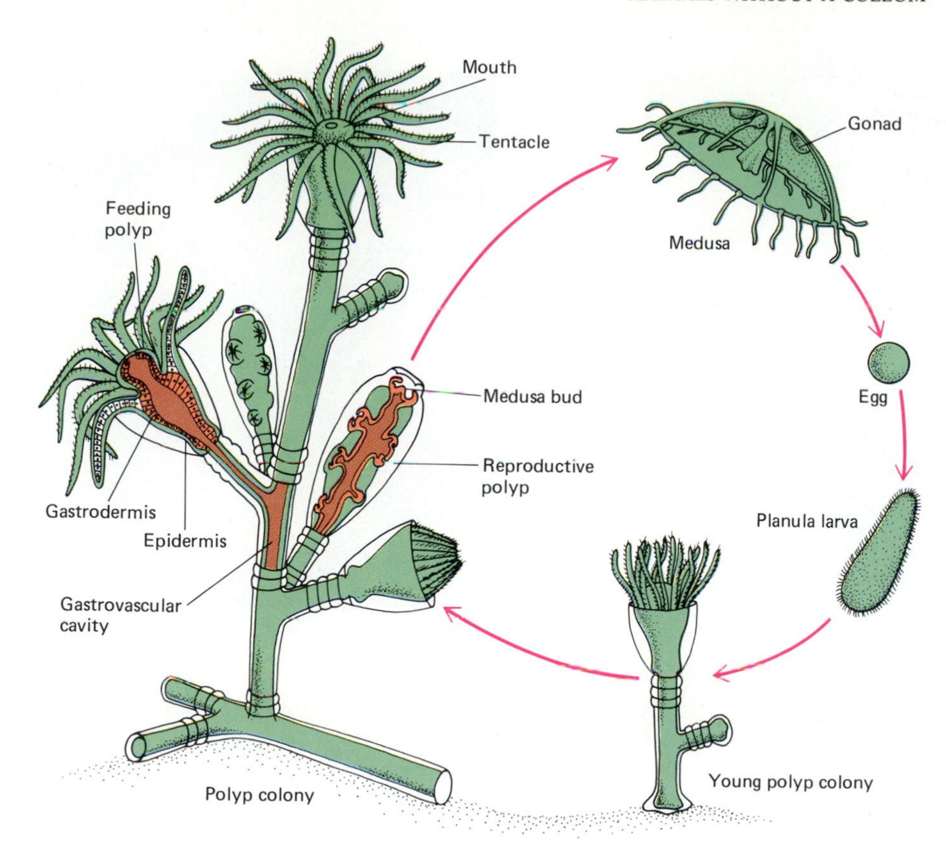

Figure 24–17 Life cycle of *Obelia,* a colonial marine hydrozoan. Note the specialization of individual members of the polyp colony. (After Barnes.)

jellyfish, *Cyanea,* may be more than 2 meters in diameter and have tentacles 30 meters long. These orange and blue monsters, among the largest of the invertebrate animals, are a real danger to swimmers in the North Atlantic Ocean.

CLASS ANTHOZOA

The sea anemones and corals have no free-swimming jellyfish stage, and the polyps may be either individual or colonial forms. These animals have a small ciliated larva, which may swim to a new location before attaching to develop into a polyp.

Anthozoans differ from hydrozoans in that the gastrovascular cavity is divided by a series of vertical partitions into a number of chambers, and the surface epidermis is turned in at the mouth to line a pharynx (Fig. 24–18). The partitions in the gastrovascular cavity increase the surface area for digestion, so that an anemone can digest an animal as large as a crab or fish. Although corals can capture prey, many tropical species depend mainly on photosynthetic dinoflagellates that live within their cells for nutrition.

In warm shallow seas, almost every square meter of the bottom is covered with coral or anemones, most of them brightly colored. The extravagant reefs and atolls of the South Pacific are the remains of billions of microscopic, cup-shaped calcareous skeletons, secreted during past ages by coral colonies and by coralline plants. Living colonies occur only in the uppermost regions of such reefs, adding their own calcareous skeletons to the mass.

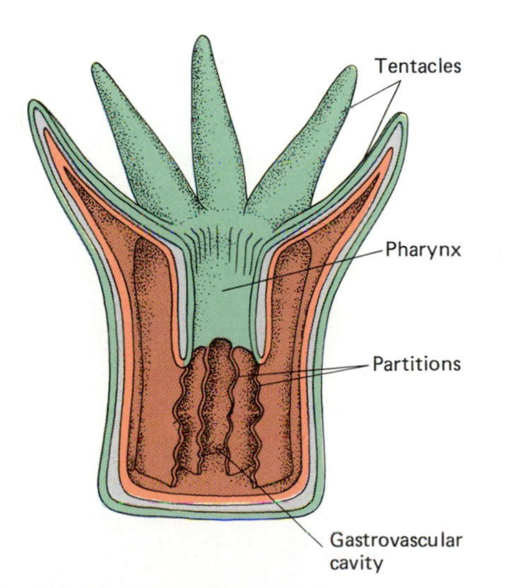

Figure 24–18 Structure of an anthozoan polyp. Longitudinal section. (After Barnes.)

Phylum Ctenophora (Comb Jellies)

The **ctenophores,** or comb jellies, are a phylum of about 50 marine species. They are fragile, luminescent animals that may be as small as a pea or larger than a tomato. They are biradially symmetrical, and their body plan is somewhat similar to that of a medusa. The body consists of two cell layers separated by a thick jelly-like mesoglea.

The outer surface of a ctenophore is covered with eight rows of cilia, resembling combs (Fig. 24–19). The coordinated beating of the cilia in these combs moves the animal through the water. At the upper pole of the body is a sense organ

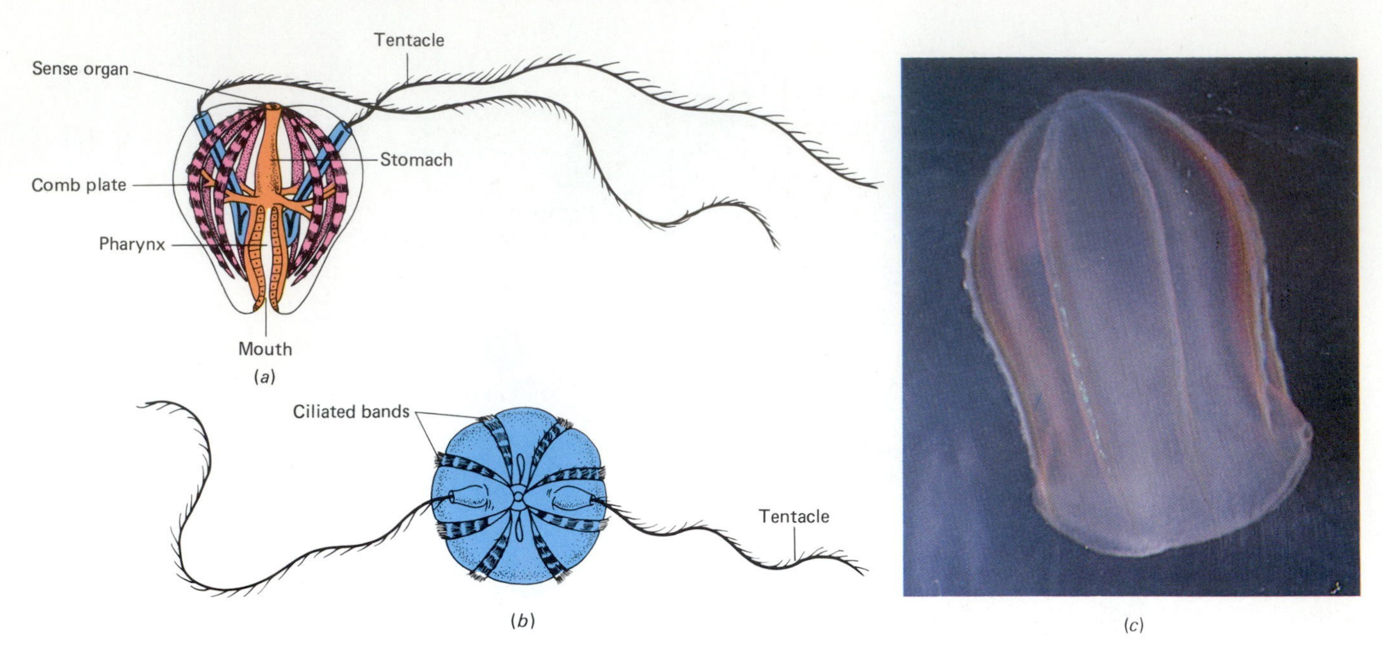

Figure 24–19 Ctenophore structure. (*a*) The ctenophore *Pleurobrachia* is shown in side view with the body cut open to expose internal structures. (*b*) Top view of the ctenophore shown in (*a*). (*c*) The ctenophore *Boreo*. ((*c*), Robin Lewis, Coastal Creations.)

containing a mass of limestone particles balanced on four tufts of cilia connected to sense cells. When the body turns, these particles bear more heavily on the lower cilia, stimulating the sense cells. This causes the cilia in certain of the combs to beat faster and bring the body back to its normal position. Nerve fibers extending from the sense organ to the cilia control the beating. If they are cut, the beating of the cilia below the incision is disorganized.

Ctenophores have only two tentacles, and most species lack the stinging cells characteristic of the cnidarians. However, their tentacles are equipped with adhesive glue cells called **colloblasts,** which trap their prey.

Phylum Platyhelminthes (Flatworms)

As their name implies, **flatworms,** which are members of the phylum **Platyhelminthes,** are flat, elongated, legless animals. They exhibit bilateral symmetry and are the simplest members of the Bilateria. Some zoologists classify them as acoelomate protostomes, while others just refer to them as acoelomates and reserve the term protostomes for the mollusks, annelids, and arthropods (animals with a complete digestive tract). The three classes of **Platyhelminthes** are **Turbellaria,** the free-living flatworms, including *Planaria* and its relatives; **Trematoda,** the flukes, which are either internal or external parasites; and **Cestoda,** the tapeworms, the adults of which are intestinal parasites of vertebrates (Table 24–2).

TABLE 24–2
Classes of Phylum Platyhelminthes

Class and Representative Animals	*Characteristics*
Turbellaria Planarians (e.g., *Dugesia*)	Free-living flatworms; mainly marine, some terrestrial forms living in mud; body covered by ciliated epidermis; usually carnivorous forms that prey on tiny invertebrates or on dead organisms
Trematoda Flukes (e.g., schistosomes)	All parasites with a wide range of vertebrate and invertebrate hosts; may require intermediate hosts; suckers for attachment to host
Cestoda Tapeworms	Parasites of vertebrates; complex life cycle with one or two intermediate hosts; larval host may be invertebrate; tapeworms have suckers and sometimes hooks on scolex for attachment to host; eggs produced within proglottids, which are shed; no digestive system

Some important characteristics of this phylum follow:

1. *Bilateral symmetry and cephalization.* Along with their bilateral symmetry, flatworms have a definite anterior end and a posterior end. This is a great advantage because an animal with a front end generally moves in a forward direction. With a concentration of sense organs in the part of the body that first meets the environment, an animal is able to detect an enemy quickly enough to escape; it is also more likely to see or smell prey quickly enough to capture it. A rudimentary head, the beginnings of **cephalization,** is evident in flatworms.
2. *Three definite tissue layers.* In addition to an outer epidermis, derived from ectoderm, and an inner endodermis, derived from endoderm, the flatworm has a middle tissue layer, derived from mesoderm, which comprises most of the body.
3. *Well-developed organs.* The flatworms are the simplest animals that have well-developed organs, functional structures made of two or more kinds of tissue. Among their organs is a muscular pharynx for taking in food, eyespots and other sensory organs in the head, a simple brain, and complex reproductive organs.
4. *A simple nervous system.* The simple brain consists of two masses of nervous tissue, called **ganglia,** in the head region. The ganglia are connected to two nerve cords that extend the length of the body. A series of nerves connect the cords like the rungs of a ladder; this type of system is sometimes called a ladder-type nervous system.
5. *Excretory structures* called **protonephridia** ending in specialized collecting cells, called flame cells.
6. A *gastrovascular cavity* in most species. It is often extensively branched and has only one opening, the mouth, usually located on the middle of the ventral surface.

CLASS TURBELLARIA

Members of the class Turbellaria are free-living, mainly marine, flatworms. **Planarians** are turbellarian flatworms found in ponds and quiet streams all over the world. The common American planarian *Dugesia* is about 15 mm long, with what appear to be crossed eyes and flapping ears called **auricles** (Fig. 24–20). These auricles actually serve as organs of smell.

Planarians are carnivorous, trapping small animals in a mucous secretion. The digestive system consists of a single opening (the mouth), a pharynx, and a branched intestine. A planarian can project its **pharynx** (the first portion of the digestive tube) outward through its mouth, using it to suck up small pieces of the prey. Extracellular digestion takes place in the intestine by enzymes secreted by gland cells. Digestion is completed after the nutrients have been absorbed into individual cells. Undigested food is eliminated through the mouth. The lengthy intestine (actually a highly branched gastrovascular cavity) helps to distribute food to all parts of the body, so that each cell is within range of diffusion. Flatworms can survive without food for months, gradually digesting their own tissues and growing smaller as time passes. Some flatworms confiscate intact nematocysts from the hydras they eat; they incorporate them into their own epidermis and use them for defense.

A planarian's flattened body ensures that gases can reach all the cells by diffusion. There are no specialized respiratory or circulatory structures. Although some excretion takes place by diffusion, an excretory system is present. It consists of two excretory tubes that extend the length of the body and give off branches called **protonephridia** throughout their length. Each of these tubules ends in a **flame cell,** a collecting cell equipped with cilia for channeling water containing wastes into the system of tubules. Planarians are capable of learning; memory is not localized within the brain, but appears to be retained throughout the nervous system.

Planarians can reproduce either asexually or sexually. In asexual reproduction an individual constricts in the middle and divides into two individuals. Each regenerates its missing parts. Sexually, these animals are hermaphroditic. During

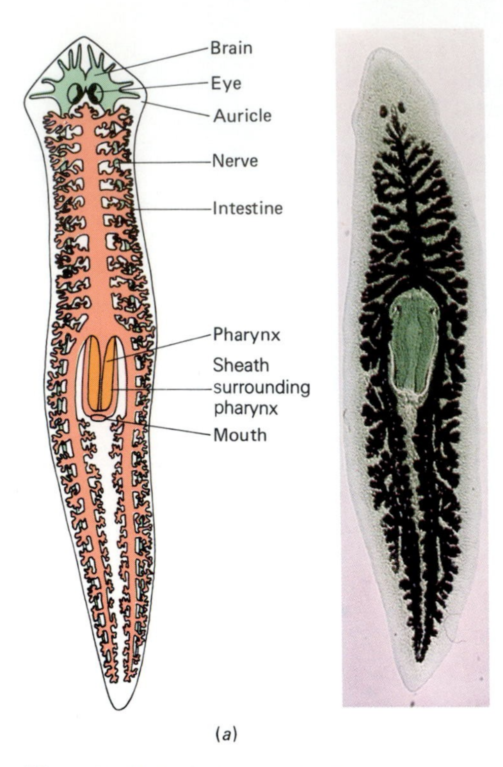

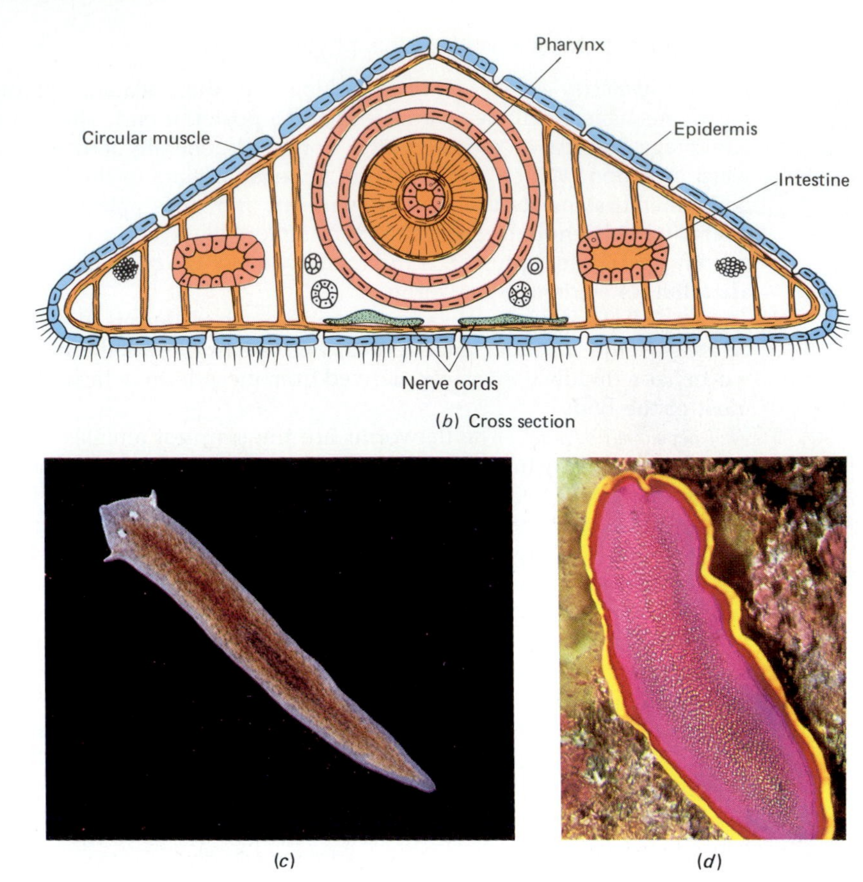

Figure 24–20 Turbellarians. (*a*) The common planarian, *Dugesia*. This stained specimen is shown with a line drawing. (*b*) Cross section through a body of a planarian. (*c*) A living *Dugesia*. (*d*) A marine turbellarian, *Pseudoceros fwerksii*, from Hawaii. ((*a*), photograph Carolina Biological Supply Company. (*c*), Tom Adams. (*d*), Charles Seaborn.)

the warm months of the year, each is equipped with a complete set of male and female organs. Two planarians come together in copulation and exchange sperm cells so that their eggs are cross-fertilized.

CLASS TREMATODA (FLUKES)

The **flukes,** which are members of the class **Trematoda,** are structurally like the free-living flatworms, but differ in having one or more suckers with which to cling to the host and in their lack of a ciliated epidermis (Fig. 24–21). The organs of digestion, excretion, and coordination are like those of the other flatworms, but the mouth is anterior rather than ventral. The reproductive organs are extremely complex.

The flukes parasitic in human beings are the blood flukes, widespread in China, Japan, and Egypt, and the liver flukes, common in China, Japan, and Korea. Blood flukes of the genus *Schistosoma* infect about 200 million people who live in tropical areas. Both blood flukes and liver flukes go through complicated life cycles, involving a number of different forms, alternation of sexual and asexual generations, and parasitism on one or more intermediate hosts such as snails and fishes (Fig. 24–22). When dams are built, marshy areas are created, which provide habitats for the aquatic snails that serve as intermediate hosts in the fluke life cycle.

CLASS CESTODA (TAPEWORMS)

As adults, members of the more than 1000 different species of the class Cestoda live as parasites in the intestines of probably every kind of vertebrate, including humans. Tapeworms are long, flat, ribbon-like animals strikingly specialized for their parasitic mode of life. Among their many adaptations are suckers and sometimes hooks on the head, or **scolex,** which enable the parasite to maintain its attachment to the host's intestine (Fig. 24–23). Their reproductive adaptations and abilities are extraordinary. The body of the tapeworm consists of a long chain of segments called **proglottids.** Each segment is an entire reproductive machine equipped with both male and female organs and containing as many as 100,000 eggs. Since an adult tapeworm may possess as many as 2000 segments, its reproductive potential

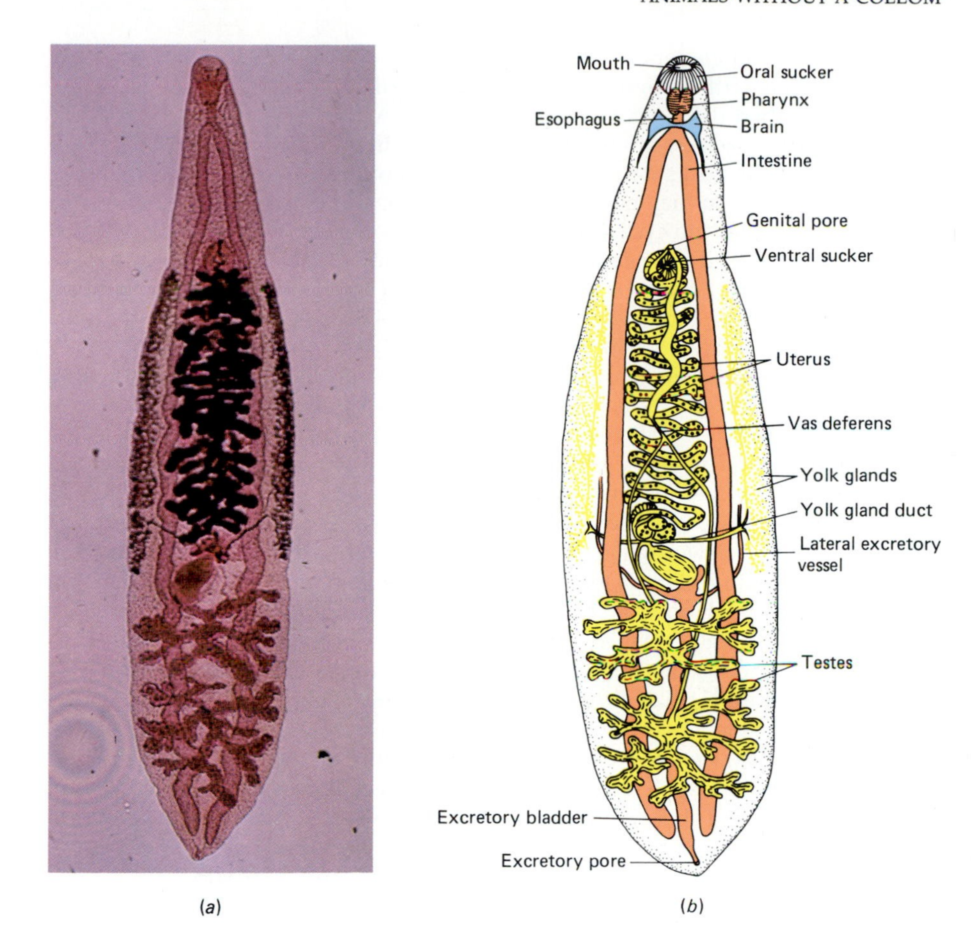

Figure 24–21 Flukes are adapted for a parasitic mode of life. (*a*) The Chinese liver fluke, *Clonorchis sinesis*. (*b*) Internal structure of a fluke. ((*a*), Carolina Biological Supply Company.)

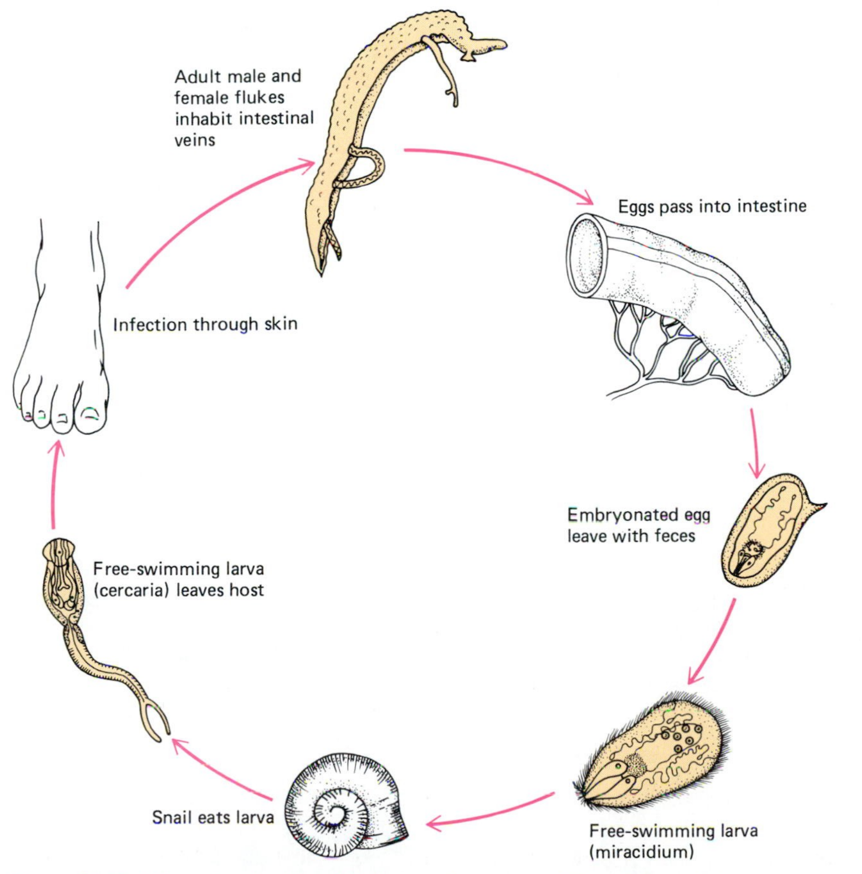

Figure 24–22 Life cycle of a blood fluke, a schistosome. The adult male has a long canal that holds the female during fertilization.

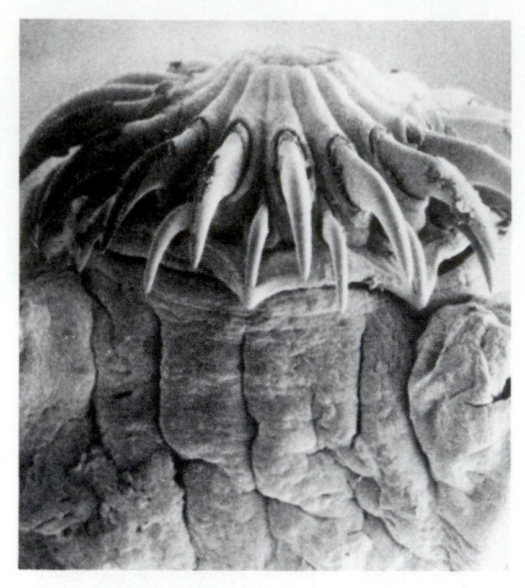

Figure 24–23 Some species of tapeworms are armed with powerful hooks that enable these parasites to maintain their attachment to the host. (Courtesy of Fred H. Whittaker and Bioscience.)

is staggering. A single tapeworm may produce as many as 600 million eggs in a year. Proglottids farthest from the tapeworm's head contain the ripest eggs; these segments are shed daily, leaving the host's body with the feces.

Tapeworms lack certain organs. They absorb their food directly through their body wall from the host's intestine, and have no mouth and no digestive system of their own. Neither do they possess any sense organs or a brain. Some tapeworms have rather complex life cycles, spending their larval stage within the body of an intermediate host and their adult life within the body of a different, final host. As an example let us consider the life cycle of the beef tapeworm, so named because human beings become infected when they eat poorly cooked beef containing the larvae (Fig. 24–24).

The microscopic tapeworm larva spends part of its life cycle encysted within the muscle tissue of beef. When a human being ingests infected meat, the digestive juices break down the cyst, releasing the larvae. The larva attaches itself to the intestinal lining and within a few weeks matures into an adult tapeworm, which may grow to a length of 50 feet. The parasite reproduces sexually within the human intestine and sheds proglottids filled with ripe eggs. Once established within a human host, the tapeworm makes itself very much at home and may remain there for the remainder of its life, as long as 10 years. A person infected with a tapeworm may suffer pain or discomfort, increased appetite, weight loss, and other symptoms, or may be totally unaware of its presence.

In order for the life cycle of the tapeworm to continue, its eggs must be ingested by an **intermediate host,** in this case, a cow. (This requisite explains why we are not completely overrun by tapeworms, and why the tapeworm must produce millions of eggs to ensure that at least a few will survive.) A cow eats grass or other food contaminated with human feces, and once ingested, the eggs hatch in the cow's intestine. The larvae make their way into muscle, where they encyst to await release by a **final host,** perhaps a human eating rare steak.

Two other tapeworms that infect humans are the pork tapeworm and the fish tapeworm. The pork tapeworm infects persons who eat poorly cooked, infected pork, and the fish tapeworm is contracted by ingesting raw or poorly cooked infected fish. Like most parasites, tapeworms tend to be species-specific; that is, each can infect only certain specific species. For example, the beef tapeworm can spend its adult life in no other host than a human being.

Figure 24–24 Life cycle of the beef tapeworm, a parasitic flatworm.

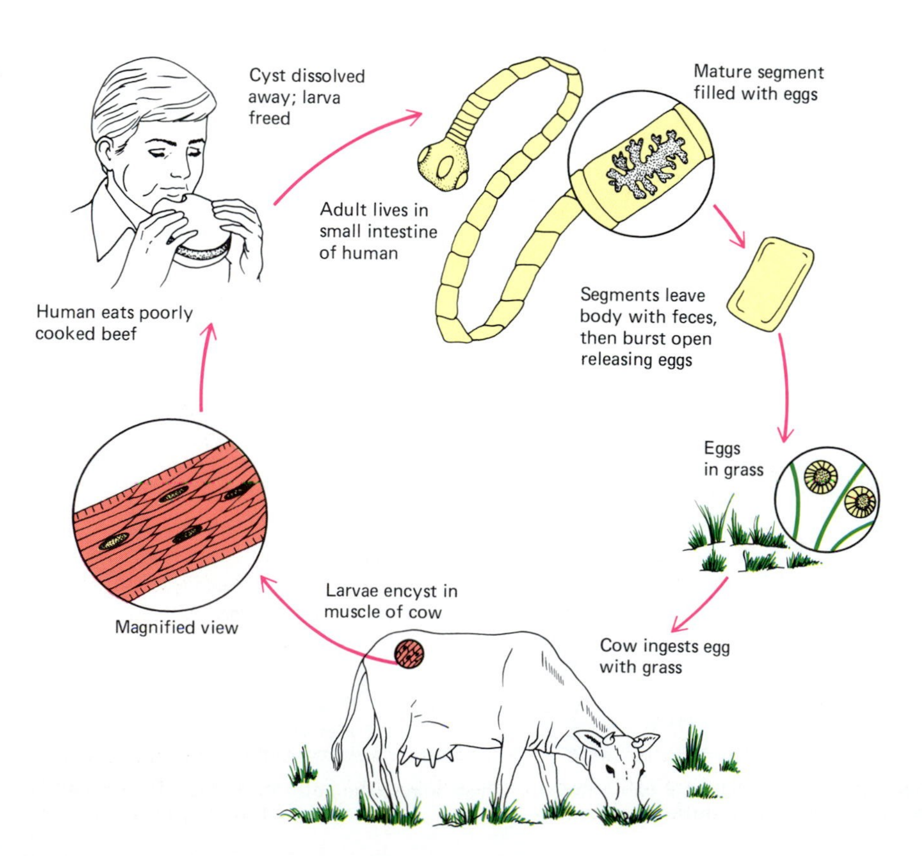
Cyst dissolved away; larva freed
Mature segment filled with eggs
Adult lives in small intestine of human
Segments leave body with feces, then burst open releasing eggs
Human eats poorly cooked beef
Eggs in grass
Magnified view
Larvae encyst in muscle of cow
Cow ingests egg with grass

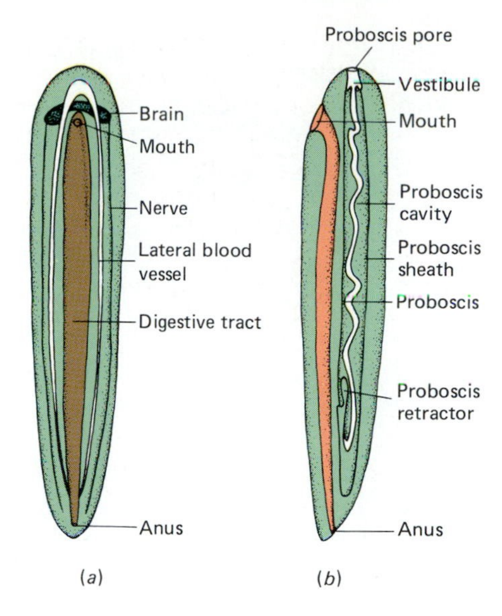

Figure 24–25 Structure of a typical proboscis worm or nemertean. (*a*) Dorsal view of the digestive, circulatory, and nervous systems. (*b*) Lateral view of the digestive tract and proboscis. Note the complete digestive tract that extends from mouth to anus, giving this animal a tube-within-a-tube body plan.

Phylum Nemertinea (Proboscis Worms)

The phylum **Nemertinea** is a relatively small group of animals (about 550 species) that is considered an evolutionary landmark because its members are the simplest animals to possess definite organ systems (Fig. 24–25). None of them is parasitic, and none is of economic importance. Almost all of them are marine, although a few inhabit fresh water or damp soil. They have long narrow bodies, either cylindrical or flattened, varying in length from 5 cm to 20 m. Some of them are a vivid orange, red, or green, with black or colored stripes. Their most remarkable organ—the **proboscis,** from which they get their common name, proboscis worms—is a long, hollow, muscular tube, which can be everted from the anterior end of the body for use in seizing food or in defense. The proboscis secretes mucus, helpful in catching and retaining the prey. In certain species, the proboscis is equipped with a hard point at the tip and poison-secreting glands at the base of this point. The outward movement of the proboscis is accomplished by the pressure of the surrounding muscular walls on the contained fluid; a separate muscle inside the proboscis retracts it.

The important advances displayed by the nemerteans are, first, a **tube-within-a-tube body plan.** The digestive tract is a complete tube, with a mouth at one end for taking in food and an anus at the other for eliminating undigested food. This is in contrast to the cnidarians and planarians, whose food enters and wastes leave by the same opening.

A second advance exhibited by the nemerteans is the separation of digestive and circulatory functions. These animals are the most primitive organisms to have a separate circulatory system. It is, of course, rudimentary, consisting simply of muscular tubes—the blood vessels—extending the length of the body and connected by transverse vessels. Some of these primitive forms have red blood cells filled with hemoglobin, the same red pigment that transports oxygen in human blood. Nemerteans have no heart, and the blood is circulated through the vessels by the movements of the body and the contractions of the muscular blood vessels. There are no capillaries. The nervous system is more highly developed than it is in the flatworm; there is a "brain" at the anterior end of the body, consisting of two groups of nerve cells (ganglia) connected by a ring of nerves extending around the sheath of the proboscis; two nerve cords extend posteriorly from the brain.

Phylum Nematoda (Roundworms)

Members of the phylum **Nematoda,** commonly called **roundworms,** are of great ecological importance because they are widely distributed in the soil, the sea, and in fresh water, and because they are so numerous. A spadeful of soil may contain

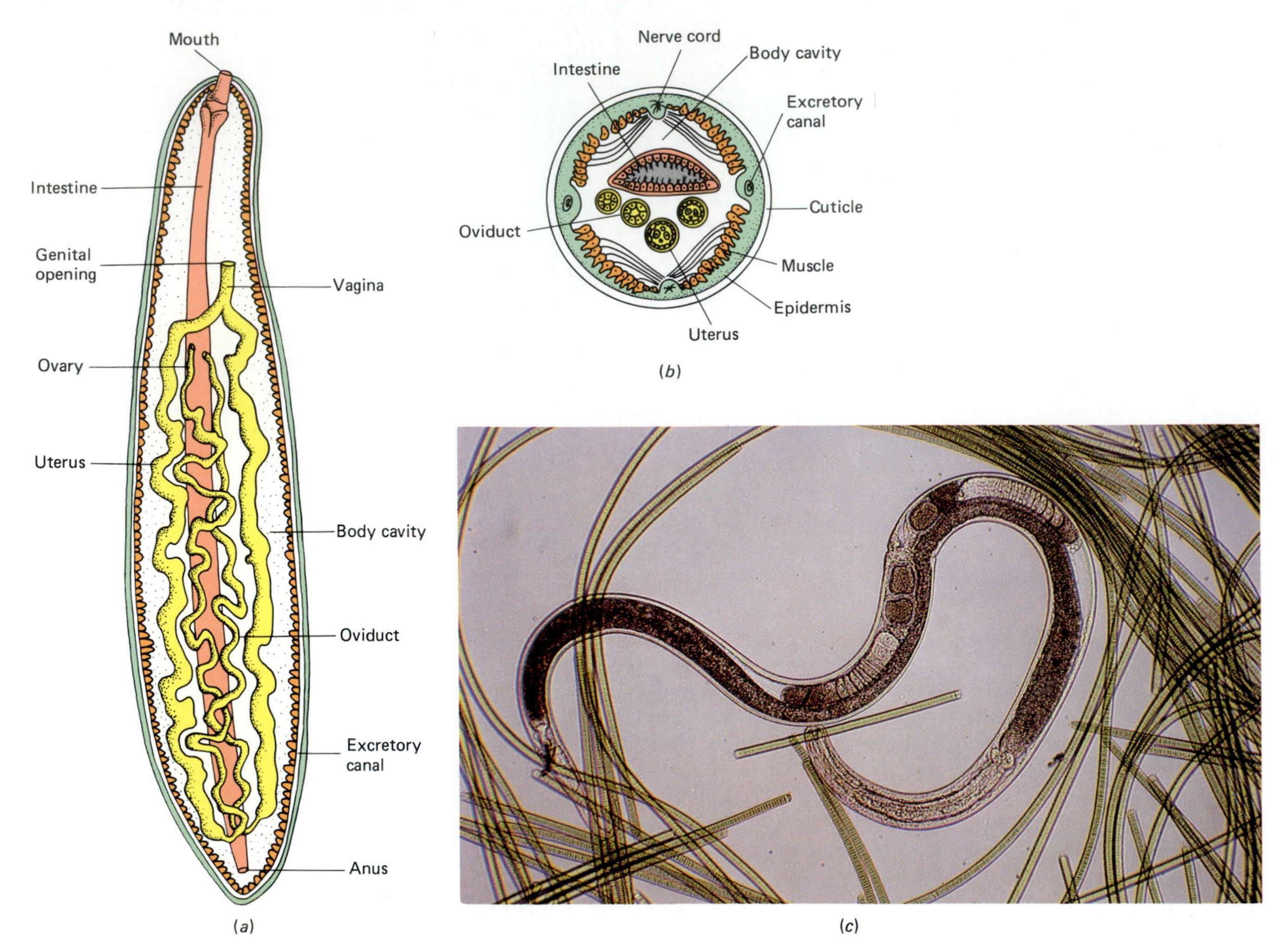

Figure 24–26 Nematodes. (*a*) Internal anatomy of a nematode such as *Ascaris* shown by longitudinal section (through the length of the body). Note the complete digestive tract that extends from mouth to anus. (*b*) Cross section through the body of a nematode such as *Ascaris*. (*c*) A free-living nematode among the cyanobacteria *Oscillatoria*, which it eats. ((*c*), Tom Adams.)

more than a million of these tiny white worms, which thrash around, coiling and uncoiling. Though many are free-living, others are important parasites in plants and animals. Among the human parasites belonging to phylum Nematoda are the hookworms, the intestinal roundworm *Ascaris*, pinworms, trichina worms, and filaria worms.

The elongate, cylindrical, threadlike nematode body is pointed at both ends and covered with a tough **cuticle** (Fig. 24–26). Nematodes are the most primitive animals to have a body cavity, but it is not a true coelom. Because the body cavity is not completely lined with mesoderm it is referred to as a pseudocoelom. Like the proboscis worms, the nematodes exhibit bilateral symmetry, a complete digestive tract, three definite tissue layers, and definite organ systems. However, they lack circulatory structures. The sexes are usually separate, and the male is smaller than the female. The characteristics of nematodes and other lower invertebrate phyla are summarized in Table 24–3.

ASCARIS: A PARASITIC ROUNDWORM

A common intestinal parasite of human beings, ***Ascaris*** is a whitish worm about 25 cm long. *Ascaris* spends its adult life in the human intestine, where it makes its living by sucking in partly digested food. Like the tapeworm, it must devote a great deal of effort to reproduction in order to ensure survival of its species. The sexes are separate, and copulation takes place within the host. A mature female may lay as many as 200,000 eggs a day.

Ascaris eggs leave the human body with the feces and, where sanitation is poor (that is, in most of the world), find their way onto the soil. In many parts of the world, human wastes are utilized as fertilizer—a practice that encourages the survival of *Ascaris* and of many other human parasites. People are infected when

TABLE 24–3
Comparison of Some Lower Invertebrate Phyla

Phylum and Representative Animals	*Level of Organization*	*Symmetry*	*Digestion*	*Circulation*	*Gas Exchange*	*Waste Disposal*	*Nervous System*	*Reproduction*	*Other Characteristics*
Porifera (pore bearers) Sponges	Multicellular but tissues loosely arranged	Radial or none	Intracellular	Diffusion	Diffusion	Diffusion	Irritability of cytoplasm	Asexual, by budding; sexual, most are hermaphroditic; larvae swim by cilia; adults incapable of locomotion	Filter feeders; skeleton of chalk, glass, or spongin (a protein material)
Cnidaria Hydra Jellyfish Coral	Tissues	Radial	Gastrovascular cavity with only one opening; intra- and extracellular digestion	Diffusion	Diffusion	Diffusion	Nerve net; no centralization of nerve tissue	Asexual by budding; sexual, sexes separate	Have cnidocytes (stinging cells) along their tentacles
Platyhelminthes (flatworms) Planarians Flukes Tapeworms	Organs	Bilateral; rudimentary head	Digestive tract with only one opening	Diffusion	Diffusion	Protonephridia; flame cells and ducts	Simple brain; two nerve cords; ladder type system; simple sense organs	Asexual, by fission; sexual, hermaphroditic, but cross-fertilization in some species	Three definite tissue layers; no body cavity; many parasitic
Nemertinea Proboscis worms	Organ systems	Bilateral	Complete digestive tract with mouth and anus	At least two pulsating longitudinal blood vessels; no heart; blood cells with hemoglobin	Diffusion	Two lateral excretory canals with flame cells	Simple brain; two nerve cords; cross nerves; simple sense organs	Asexual, by fragmentation; sexual, sexes separate	No body cavity; proboscis for defense and capturing prey
Nematoda (roundworms) Ascarids Hookworms Nematodes	Organ systems	Bilateral	Complete digestive tract with mouth and anus	Diffusion	Diffusion	Excretory canals	Simple brain; dorsal and ventral nerve cords; simple sense organs	Sexual, sexes separate	Have pseudocoelom (space between internal organs and body wall); many parasitic

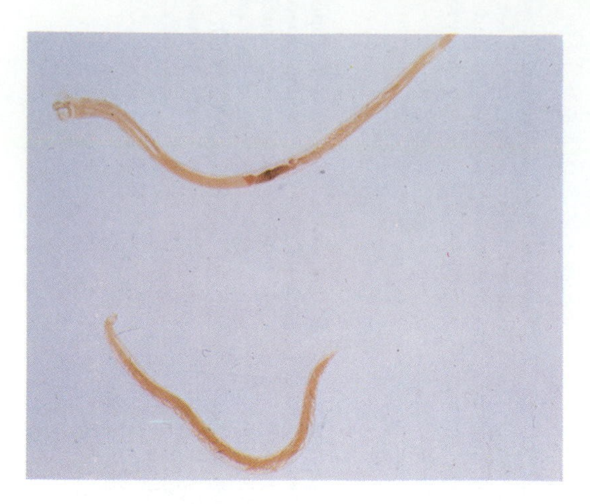

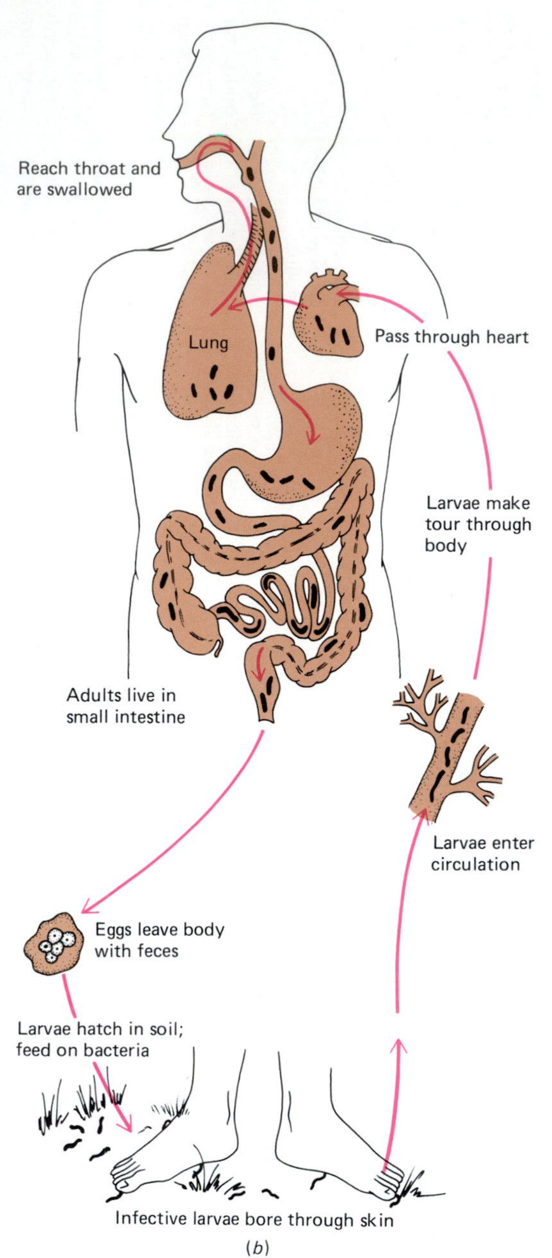

Figure 24–27 Hookworms are parasitic roundworms. (*a*) Male and female hookworm (*Necator americanus*). The female is the longer one. These hookworms are human parasites. (*b*) Life cycle of a hookworm. ((*a*), Carolina Biological Supply Company.)

they ingest *Ascaris* eggs. The eggs hatch in the intestine, and the larvae then journey through the body before settling down in the small intestine. During this migration the larvae can cause a great deal of damage to the lungs and other tissues.

OTHER PARASITIC ROUNDWORMS

The life cycle of a human **hookworm** (Fig. 24–27) is somewhat similar to that of *Ascaris*. Only one host is required. Adult worms, which are less than 1.5 cm (0.5 inch) long, live in the human intestine and lay eggs, which pass out of the body with the feces. Larvae hatch and feed upon bacteria in the soil. After a period of maturation they become infective. When a potential host walks barefoot on soil containing the microscopic larvae, or otherwise comes in contact with it, the larvae bore through the skin and enter the blood. They migrate through the body before finding their way to the intestine where they mature.

Human beings become infected with **trichina worms** by eating poorly cooked, infected pork. The trichina parasite is adapted to live inside many animals (pigs, rats, bears, and others), and the human being is an accidental host. Adult

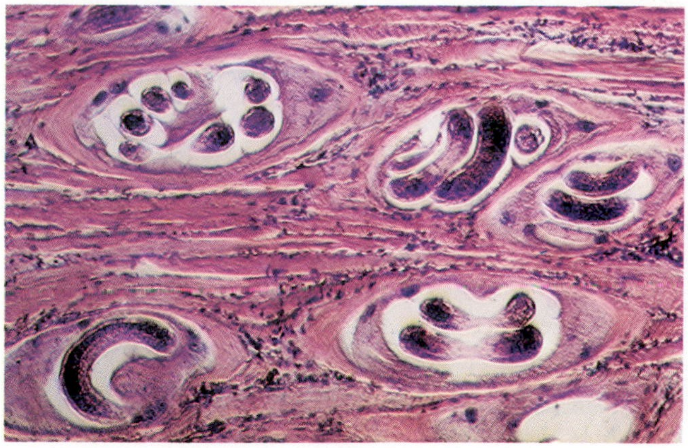

Figure 24–28 The roundworm *Trichinella spiralis* causes trichinosis. (*a*) Larvae of *Trichinella* encysted in skeletal muscle. (*b*) When poorly cooked pork infected with trichina worms is eaten, the larvae are released and grow rapidly to maturity in the intestine. After fertilization, the females produce tiny larvae, which burrow into the blood vessels and are carried to the muscles where they encyst. ((*a*), Carolina Biological Supply Company.)

(*a*)

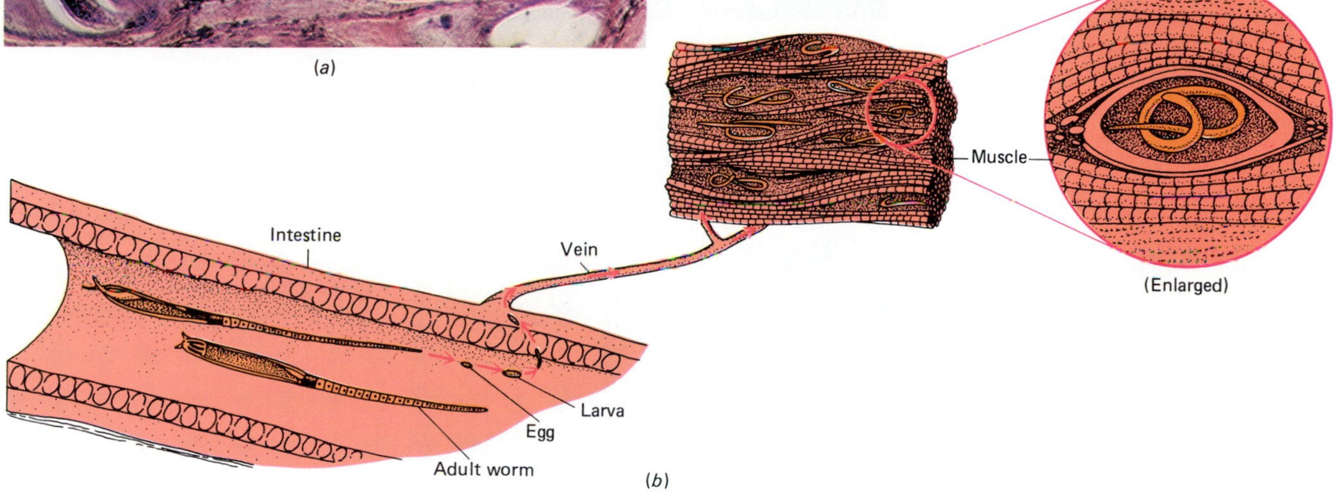

Intestine
Vein
Muscle
(Enlarged)
Larva
Egg
Adult worm

(*b*)

trichina worms live in the small intestine of the host. The females produce living larvae, which migrate through the body, making their way to skeletal muscle, where they encyst (Fig. 24–28). Continuance of the life cycle depends upon ingestion by another animal. Since human beings are not normally eaten, trichina larvae are not liberated from their cysts and eventually die. The cysts, however, become calcified and permanently remain in the muscles, causing stiffness and discomfort. Other symptoms are caused by the presence of the adults and by the migrating larvae. No cure has been found for this infestation.

Pinworms are the most common worms found in children. Adult worms, less than 1.3 cm (0.5 inch) long, live in the large intestine. Female pinworms often migrate to the anal region at night to deposit their eggs. Irritation and itching caused by this practice induce scratching, which serves to spread the tiny eggs. Eggs may be further distributed in the air and are in this way scattered throughout the house. The original host or other members of the household may be infected by ingesting the eggs. Eating with dirty hands facilitates the process. Mild infestations may go unnoticed, but those more serious may result in injury to the intestinal wall, discomfort, and irritation.

Phylum Rotifera (Wheel Animals)

Among the more obscure invertebrates are the "wheel animals" of the phylum **Rotifera.** These aquatic, microscopic worms, although no larger than many protozoans, have many-celled bodies. They have a complete digestive tract including a **mastax,** a muscular organ for grinding food; a pseudocoelom; an excretory system made up of flame cells and a bladder; a nervous system with a "brain" and sense organs; and a characteristic crown of cilia on the anterior end, which gives the appearance of a spinning wheel (Fig. 24–29).

Rotifers are "cell constant" animals: Each member of a given species is composed of exactly the same number of cells; indeed, each part of the body is made of

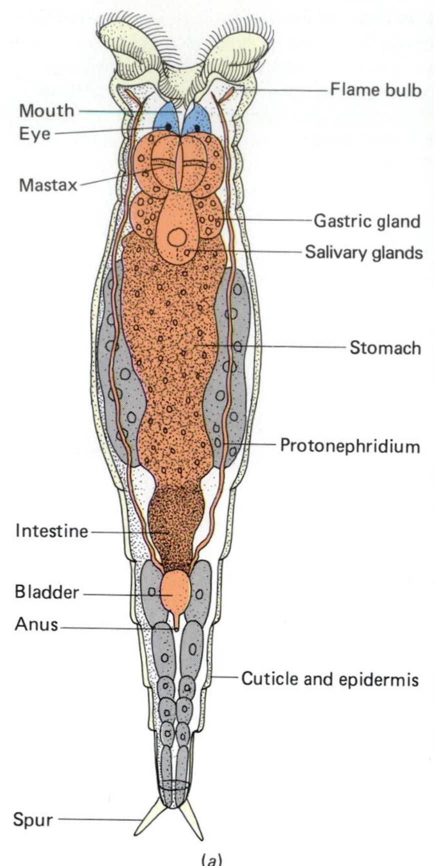

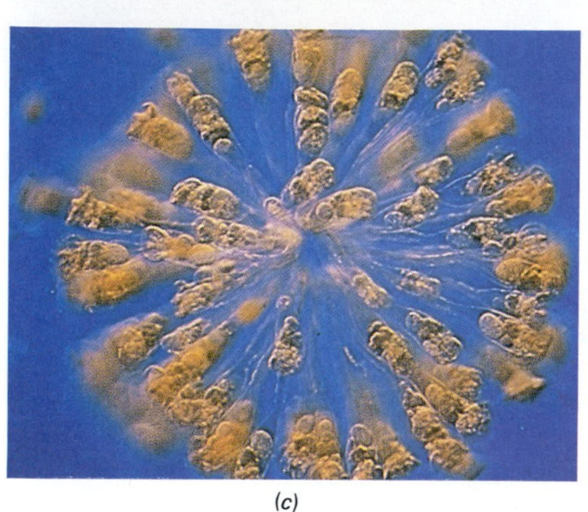

Figure 24–29 Wheel animals. (*a*) Structure of a rotifer. (*b*) A solitary rotifer, *Stephanboceros,* with cilia extended for feeding. (*c*) A colonial rotifer with an internal skeleton composed of silica. The lightweight skeleton helps keep the colony afloat as it travels with other plankton. ((*b*), Tom Adams; (*c*), James Bell, Photo Researchers, Inc.)

a precisely fixed number of cells arranged in a characteristic pattern. Cell division ceases with embryonic development, and mitosis cannot subsequently be induced; growth and repair are impossible. One of the challenging problems of biological research is the nature of the difference between such nondividing cells and the dividing cells of other animals. Do rotifers never develop cancer?

SUMMARY

I. Animals are eukaryotic, multicellular, heterotrophic organisms whose cells exhibit a division of labor. They generally are capable of locomotion at some time during their life cycle, can reproduce sexually, and can respond to external stimuli with appropriate behavior.

II. Animals are consumers that inhabit the sea, fresh water, and the land.

III. Animals may be classified as Parazoa or Eumetazoa; as Radiata or Bilateria; as acoelomates, pseudocoelomates, or coelomates; or as protostomes or deuterostomes.

IV. Phylum Porifera consists of the sponges.
- *A.* Sponges are divided into three main classes on the basis of the type of skeleton they secrete.
- *B.* The sponge body is a sac with tiny holes through which water enters, a central spongocoel, and an osculum.
- *C.* Metabolic processes in sponges depend upon diffusion of materials between the cells and the watery habitat that bathes them.

V. Phylum Cnidaria includes the hydras, jellyfish, and corals.
- *A.* Cnidarians are characterized by radial symmetry, cnidocytes, two definite tissue layers, and a nerve net.
- *B.* Many cnidarians have an alternation of generations in which sessile polyp forms give rise to free-swimming medusas.

VI. Phylum Ctenophora consists of the comb jellies, which are fragile, luminescent, biradially symmetrical marine animals.

VII. Phylum Platyhelminthes includes the planarians, the flukes, and the tapeworms.
- *A.* Flatworms are characterized by bilateral symmetry, cephalization, three definite tissue layers, well-developed organs, a simple brain and nervous system, and protonephridia.
- *B.* The flukes and tapeworms are parasites, adapted for their unusual life style.

VIII. Members of phylum Nemertinea (proboscis worms) have a tube-within-a-tube body plan, a complete digestive tract with mouth and anus, and a separate circulatory system.

IX. Phylum Nematoda, composed of the roundworms, includes species of great ecological importance and species that are parasitic in plants and animals.
- *A.* Nematodes are characterized by three definite tis-

sue layers, a pseudocoelom, bilateral symmetry, and a complete digestive tract.

B. Some nematodes parasitic in humans include *Ascaris,* hookworms, trichina worms, and pinworms.

X. Members of phylum Rotifera are aquatic, pseudocoelomate, microscopic worms that exhibit cell constancy.

POST-TEST

1. The majority of animals—those without backbones—are properly referred to as ______________.
2. Animals that remain stationary and attached to a substrate are described as ______________.
3. Acoelomate animals lack a ______________ ______________.
4. The germ layer that gives rise to the outer covering of the body and the nervous system is ______________.
5. A true coelom is completely lined with ______________.
6. In protostomes the blastopore develops into the ______________.
7. Radial symmetry is characteristic of ______________.
8. If the first four cells of a protostome embryo are separated, each cell will develop into ______________ of a larva.
9. Structures that develop from similar embryonic rudiments and are similar in basic plan and development are described as ______________.
10. Sponges have skeletal elements called ______________.
11. In a simple sponge water enters the body through many pores, passes into the central cavity called a ______________ and finally flows out through the sponge's open end, the ______________.
12. The two body forms found among cnidarians are the ______________ and the ______________.
13. The sea anemones and corals belong to phylum ______________ and class ______________.
14. The simplest animals to exhibit cephalization are the ______________.
15. The auricles of a planarian flatworm serve as organs of ______________.
16. Human beings may become infected with trichina worms by ______________.
17. Two important advances displayed by the nemerteans are ______________ and ______________.

Match the characteristics in Column B with the descriptions in Column A:

Column A	*Column B*
18. Body segments of a tapeworm	a. Cnidocytes
19. Cells that trap food particles in sponges	b. Proglottids
20. Cnidarian nervous system	c. Collar cells
21. Collecting cells in flatworm excretory system	d. Nerve net
22. Cnidarian stinging cells	e. Flame cells

Match the phyla in Column B with the animals or structures in Column A:

Column A	*Column B*
23. Comb jellies	a. Rotifera
24. Tapeworms	b. Platyhelminthes
25. Proboscis	c. Ctenophora
26. Wheel animals; have mastax	d. Nematoda
27. Hookworms, *Ascaris*	e. Nemertinea

REVIEW QUESTIONS

1. A fungus is a eukaryotic, multicellular heterotroph. Why isn't it classified as an animal? Why isn't a colonial protozoan an animal?
2. For centuries sponges were classified as plants. Justify their current classification as animals.
3. Why is the sea a more hospitable environment than the land or fresh water?
4. What advances do members of phylum Platyhelminthes exhibit over those animals that belong to phylum Cnidaria? In what ways are these animals alike?
5. What are the advantages of bilateral symmetry and cephalization?
6. What advances are exhibited by the nemerteans?
7. Identify the phyla (from among those studied in this chapter) that have the following characteristics:
   a. radial symmetry
   b. protonephridia
   c. acoelomate
   d. pseudocoelomate
   e. alternation of generations
   f. cnidocytes
   g. nerve net
   h. complete digestive tract
   i. circulatory system
8. Compare mesoglea with mesoderm. In which types of animals is mesoglea found?
9. Describe the alternation of generations exhibited by *Obelia.*
10. How do flatworms survive without specialized structures for gas exchange and internal transport of materials?
11. What special adaptations do tapeworms have to their parasitic mode of life?
12. Describe the life cycle of the following animals:
    a. a beef tapeworm
    b. *Ascaris*
    c. a hookworm

25

The Animal Kingdom: The Coelomate Protostomes

OUTLINE

I. Advantages of having a coelom
II. Adaptations for terrestrial living
III. Phylum Mollusca
- A. Relationship to the annelids
- B. Class Polyplacophora: chitons
- C. Class Gastropoda: snails and their relatives
- D. Class Bivalvia: clams, oysters, and their relatives
- E. Class Cephalopoda: squids, octopods, and their relatives

IV. Phylum Annelida
- A. Class Polychaeta
- B. Class Oligochaeta: the earthworms
- C. Class Hirudinea: the leeches

V. Phylum Onychophora
VI. Phylum Arthropoda
- A. The arthropod body plan
- B. The trilobites
- C. Subphylum Chelicerata
  1. Class Merostomata: horseshoe crabs
  2. Class Arachnida: spiders, scorpions, ticks, and mites
- D. Subphylum Crustacea: lobsters, crabs, shrimp, and their relatives
- E. Subphylum Uniramia
  1. Class Insecta
     - a. Secrets of insect success
     - b. Impact of insects on humans
  2. Classes Chilopoda and Diplopoda: centipedes and millipedes

LEARNING OBJECTIVES

After you have read this chapter you should be able to:

1. List several advantages of having a coelom.
2. Identify problems associated with terrestrial living, and describe adaptations that enable terrestrial animals to survive on the land.
3. Describe the distinguishing characteristics of mollusks, annelids, and echinoderms, and properly classify an animal that belongs to any of these phyla.
4. Describe the principal classes of mollusks, and give examples of animals that belong to each.
5. Describe and give examples of each of the three classes of annelids discussed.
6. Distinguish among the subphyla and classes of arthropods, and give examples of animals that belong to each class.
7. Discuss factors that have contributed to the great biological success of the insects.

What does an earthworm have in common with a clam or a butterfly? All of these animals have a coelom, and all are protostomes. The coelomate protostomes include the annelids, mollusks, and arthropods, as well as several smaller, related phyla. As explained in Chapter 24, in protostomes the mouth forms from the first opening in the embryonic gut. The true coelom is a space completely lined by mesoderm between the digestive tube and the outer body wall. Coelomate protostomes also have a complete digestive tract with separate mouth and anus, and most have well-developed circulatory, excretory, and nervous systems.

Advantages of Having a Coelom

Animals with a coelom, and to a lesser extent those with a pseudocoelom, have certain advantages over those that lack a body cavity. The coelom permits a clear separation between the muscles of the body wall and those in the wall of the digestive tract. This allows the digestive tube to move food along independently of body movements. Perhaps more significant, the coelom can be used as a hydrostatic skeleton, an enclosed compartment (or series of compartments) of fluid which can be manipulated by surrounding muscles. (Hydrostatic skeletons are discussed in Chapter 32.) In some animals fluid within the coelom helps transport materials such as food, oxygen, and wastes. Cells bathed by the coelomic fluid can exchange materials with it, obtaining food and oxygen for their use and excreting wastes into it. Some coelomates possess excretory structures that remove wastes directly from the coelomic fluid.

The coelom also serves as a space in which many organs develop and function. Perhaps the most notable example is the circulatory system. The heart and blood vessels develop within the coelom and can function freely there without being squeezed by other organs. The pumping action of the heart would not be possible without the surrounding space provided by the coelom. The gonads develop within the coelom; in animals with breeding seasons, these reproductive structures enlarge periodically as they fill with ripe gametes.

Adaptations for Terrestrial Living

Among the invertebrate coelomates, only certain arthropods are very successful terrestrial animals. It is true that the earthworm is a terrestrial animal, but most annelids are marine. There are a few land snails, but most mollusks also live in the sea. All echinoderms are marine. Among the arthropods, the crustaceans (crabs, lobsters, and their relatives) and the merostomes (horseshoe crabs) are largely marine forms. However, the other classes, which include the insects and spiders, are mainly terrestrial. Based on the fossil record, most evolutionists agree that the first air-breathing land animals were scorpion-like arthropods that came ashore in the Silurian period some 410,000,000 years ago. The first land vertebrates, the amphibians, did not appear until the latter part of the Devonian period some 60 million years later.

The chief problem of all terrestrial organisms is that of preventing desiccation in the absence of a surrounding watery medium. A body covering adapted to minimize fluid loss helps solve this problem in many land animals. Location of the respiratory surface deep within the animal also helps prevent desiccation. Thus, while gills are located externally, lungs and tracheal tubes (found in insects) are internal.

Supporting the body against the pull of gravity in the absence of the buoyant effect of water is another problem associated with life on land. Some animals, such as earthworms, do not have this problem because they live in the ground and because of their small body size. Larger burrowing animals and those living on the surface of the earth generally need some sort of supporting skeleton. The vertebrates have an **endoskeleton,** a supporting framework within the body. Arthropods and most mollusks have a tough **exoskeleton,** a skeleton outside of the body (Fig. 25–1). The exoskeleton serves at least four important functions: It provides stiffening, enabling the body to withstand the pull of gravity; it provides protection against desiccation; it serves as a coat of armor to protect the animal against predators; and it serves as a point of attachment for muscles.

Figure 25–1 The scorpion, like other arthropods, has a tough exoskeleton composed of chitin. It locates its prey within the soil by detecting vibrations. This scorpion is digging out a burrowing roach, which it will sting and consume. (Courtesy of Dr. Philip Brownell and *Science,* © 1977, The American Association for the Advancement of Science)

Reproduction on land poses still another problem. Aquatic forms generally shed their gametes in the water, and fertilization occurs there. The surrounding water serves as an effective shock absorber, protecting the delicate embryos as they develop. Some land animals, including most amphibians, return to the water for reproduction, and the larval forms—tadpoles—develop in the water. Earthworms, snails, insects, reptiles, birds, and mammals engage in internal fertilization. They transfer sperm from the body of the male directly into the body of the female by copulation. The sperm are surrounded by a watery medium or semen. The fertilized egg either is covered by some sort of tough, protective shell secreted around it by the female, or it develops within the body of the mother.

Phylum Mollusca

Mollusks are among the best known of the invertebrates—almost everyone has walked along the shore collecting their shells. Phylum **Mollusca,** with its more than 100,000 living species and 35,000 fossil species, is the second largest of all the animal phyla. It includes the clams, oysters, octopuses, snails, slugs, and the largest of all the invertebrates, the giant squid, which may achieve a weight of several tons (Fig. 25–2). Although most mollusks are marine, there are snails and clams that live in fresh water and many species of snails and slugs that inhabit the land. The major classes of mollusks are listed in Table 25–1.

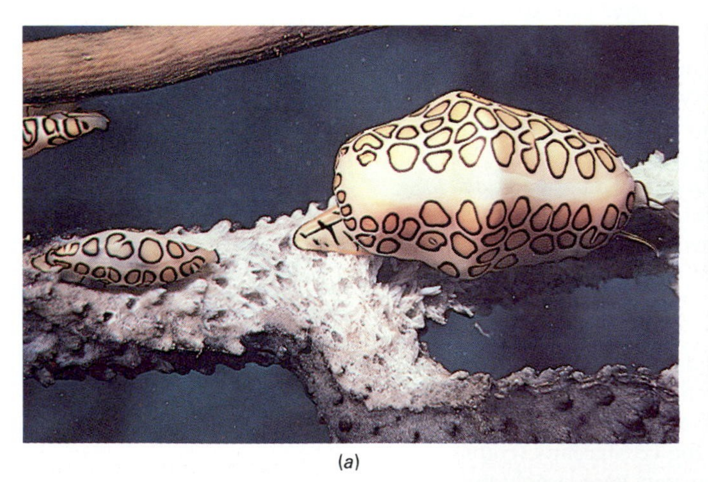

(*a*)

(*b*)

(*c*)

Figure 25–2 There are many beautiful forms of mollusks. (*a*) A flamingo tongue, *Cyphoma gibbosum*, photographed in the Virgin Islands. (*b*) A bay scallop, *Argopecten irradians*, photographed in a seagrass bed in Tampa Bay. (*c*) Lightning whelk, *Busycon contrarium*. ((*a*), (*b*), Robin Lewis, Coastal Creations; (*c*), James H. Carmichael, Jr., Coastal Creations.)

TABLE 25–1
Major Classes of Phylum Mollusca

Class and Representative Animals	*Characteristics*
Polyplacophora Chitons	Primitive marine animals with segmented shells; shell consists of 8 separate transverse plates; head reduced; broad foot used for locomotion
Gastropoda Snails Slugs Nudibranchs	Marine, freshwater, or terrestrial; body and shell is coiled; well-developed head with tentacles and eyes
Bivalvia Clams Oysters Mussels	Marine and freshwater; body laterally compressed; two shells hinged dorsally; hatched-shaped foot; filter feeders
Cephalopoda Squids Octopods	Marine; fast-swimming, predatory; foot divided into tentacles, usually bearing suckers; well-developed eyes

Although mollusks vary widely in outward appearance, most share certain basic characteristics (Fig. 25–3):

1. a soft body, usually covered by a dorsal shell
2. a broad, flat muscular **foot,** located ventrally, which can be used for locomotion
3. a **visceral mass,** located above the foot, that contains most of the organs
4. a **mantle,** a heavy fold of tissue covering the visceral mass and usually containing glands that secrete the shell. The mantle generally overhangs the visceral mass, forming a mantle cavity, which often contains gills.
5. a rasplike structure called the **radula,** which is a belt of teeth within the digestive system

All of the organ systems typical of complex animals are present in the mollusks. The digestive system is a tube, sometimes coiled, consisting of a mouth, buccal cavity, esophagus, stomach, intestine, and anus. The radula, located within the buccal cavity, can be projected out of the mouth and used to scrape particles of food from the surface of rocks or from the ocean floor. Sometimes the radula is used to drill a hole in another animal's shell or to break off pieces of a plant.

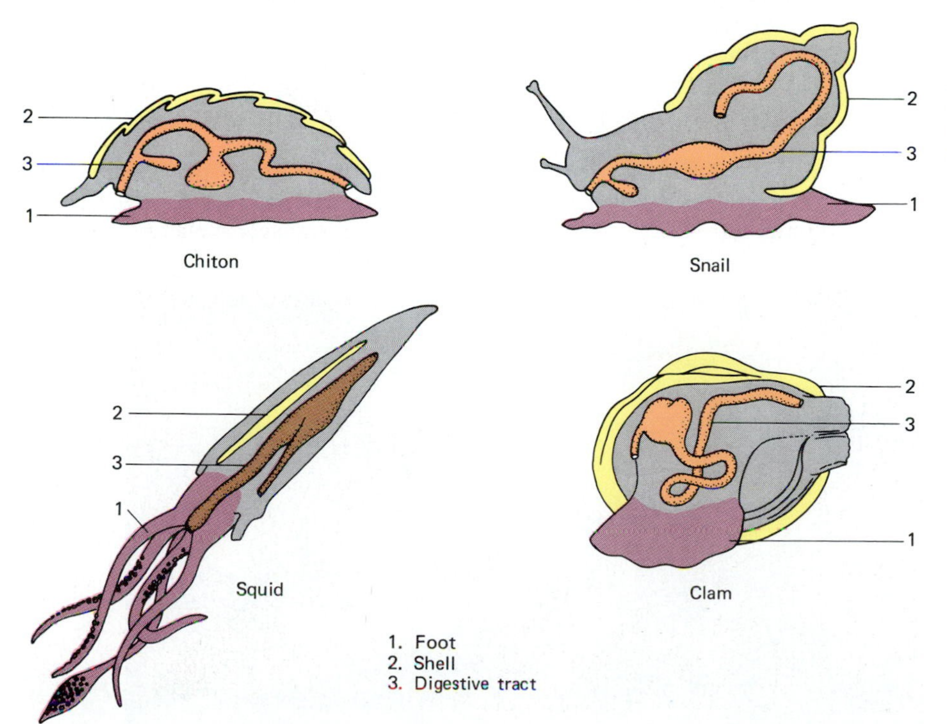

Figure 25–3 Variations in the basic molluskan body plan in chitons, snails, clams, and squids. Note how the foot (*1*), shell (*2*), and digestive tract (*3*) have changed their positions in the evolution of the several classes.

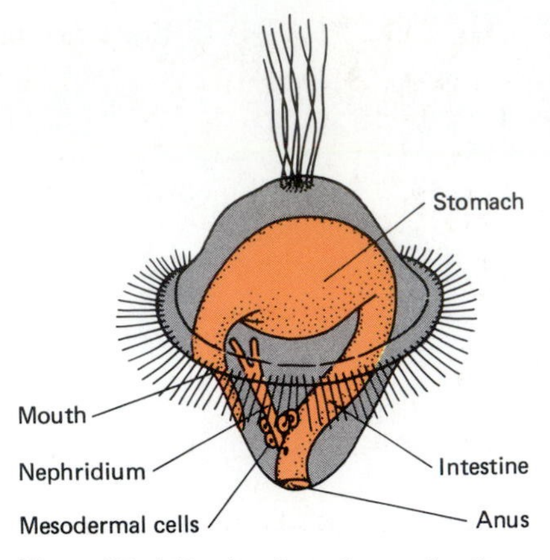

Figure 25–4 Trochophore larva, the first larval stage of a marine mollusk. This type of larva is also characteristic of annelids. Note the characteristic ring of ciliated cells just above the mouth.

The open circulatory system characteristic of most mollusks is well developed. The heart pumps blood into a single blood vessel, the **aorta,** which may branch into other vessels. Eventually, blood flows into a network of large sinuses, where the tissues are bathed directly; this network makes up the **hemocoel** (blood cavity). This system is referred to as an **open circulatory system,** in which the blood does not remain within a circuit of blood vessels but bathes the tissues directly. From the sinuses, blood drains into vessels that conduct it to the gills where oxygenation takes place. From the gills the blood returns to the heart. Thus, blood flow in a mollusk follows the pattern

Heart → aorta → smaller blood vessels → sinuses → gills → heart

Open circulatory systems are not very efficient. Blood pressure tends to be low, and tissues are not very efficiently oxygenated. However, because most mollusks are sluggish animals with low metabolic rates, this type of circulatory system is adequate. In the active cephalopods (a class of mollusks), the circulatory system is closed; the blood remains within a circuit of blood vessels.

The sexes are usually separate, with fertilization taking place in the surrounding water. Most marine mollusks pass through one or more larval stage. The first larval stage is typically a **trochophore larva,** a free-swimming, ciliated, top-shaped larva characteristic of mollusks and annelids (Fig. 25–4). In most of the mollusk classes, the trochophore larva develops into a **veliger larva,** which has a shell and foot. The veliger larva is unique to the phylum Mollusca.

RELATIONSHIP TO THE ANNELIDS

The striking similarities in the development of mollusks and annelids—the process of spiral cleavage and the appearance of a trochophore larva—had suggested that these two phyla were related in evolutionary origin and had a common coelomate ancestor. This view was supported by the discovery in 1952 of specimens of a primitive mollusk, *Neopilina,* in material dredged from a deep trench in the Pacific off Costa Rica. Since this discovery, specimens belonging to several related species have been collected from deep water in many parts of the world. *Neopilina* and its relatives have been assigned to class **Monoplacophora** (one plate) because they possess a single shell.

The most remarkable feature of the monoplacophorans is the segmental serial repetition of certain internal organs, a condition known as **metamerism.** They have five pairs of retractor muscles, six pairs of nephridia, and five pairs of gills. This has been interpreted by some zoologists as evidence of the segmental character of their ancestors. Moreover, they may be closely related to the annelids, which have a basically metameric body plan.

CLASS POLYPLACOPHORA: CHITONS

Chitons, which are members of the class **Polyplacophora** (many plates), are sluggish marine animals with flattened bodies (Fig. 25–5). Their most distinctive feature is a shell composed of eight separate, but overlapping, transverse plates. The head

Figure 25–5 Chitons are sluggish marine animals with shells composed of eight overlapping plates. These lined chitons, *Tonicella lineatus,* are from coastal waters off the Pacific Northwest. (Charles Seaborn.)

(*a*)

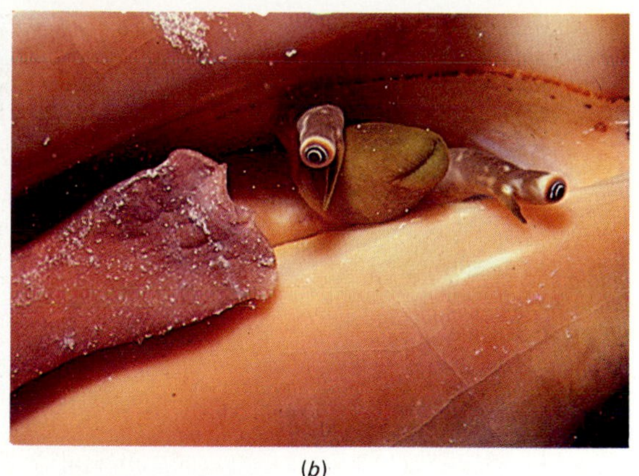

(*b*)

Figure 25–6 Representative gastropods. (*a*) The tulip snail (*Fasciolaria tulipa*). (*b*) A conch, *Strombus gigas*, showing foot, mouth, and eyes. ((*a*), James H. Carmichael, Jr., Coastal Creations; (*b*), Charles Seaborn.)

is reduced in this class, and there are no eyes or tentacles. Chitons inhabit rocky intertidal zones, feeding on algae and other small organisms, which they scrape off rocks and shells with the radula. The broad, flat foot not only functions in locomotion but also helps the animal adhere firmly to rocks. The mantle can also be pressed firmly against the substratum, and the chiton can lift the inner edge of the mantle to create a partial vacuum. The suction developed enables the animal to adhere powerfully to its perch.

CLASS GASTROPODA: SNAILS AND THEIR RELATIVES

The class **Gastropoda** is the largest and numerically the most successful group of mollusks (Fig. 25–6). In fact, this class, which includes the snails and their relatives, is the second largest class in the animal kingdom—second only to the insects. Gastropods inhabit a wide variety of habitats, including the seas, brackish water, fresh water, and many terrestrial areas, but they are most numerous and diverse in marine waters. Most land snails do not have gills; instead, the mantle is highly vascularized and functions as a lung. These snails are described as **pulmonate.**

We usually think of snails as having a single, spirally coiled shell, and many of them do. Yet many other gastropods, such as limpets and abalones, have shells like flattened dunce caps; others, such as the garden slugs and the marine snails known as **nudibranchs** have no shell at all (Fig. 25–2).

Gastropods characteristically have a well-developed head with tentacles. A pair of simple eyes are usually located on stalks that extend from the head. The broad flat foot is used for creeping. The most unique feature of this group is **torsion,** a twisting of the visceral mass. As the bilateral larva develops, the body twists permanently 180 degrees relative to the head. As a result, the digestive tract becomes somewhat U-shaped and the anus comes to lie above the head (Fig. 25–7). Subsequent growth is dorsal and usually in a spiral coil. The twist limits space in the body, and typically the gill, kidney, and gonad are absent on one side. The **viscera** (internal organs) of the slugs that lack shells also undergo torsion during development.

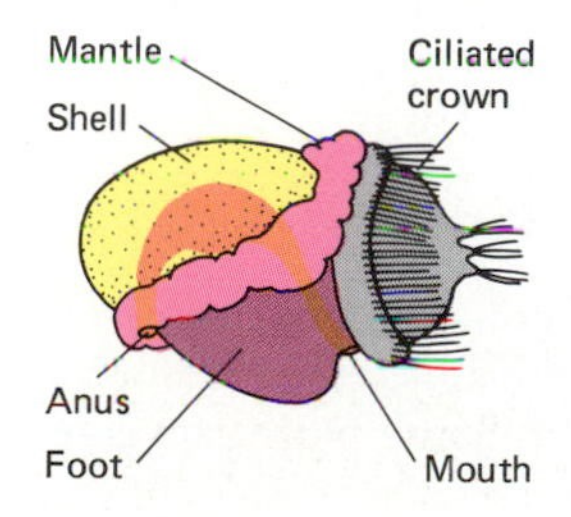

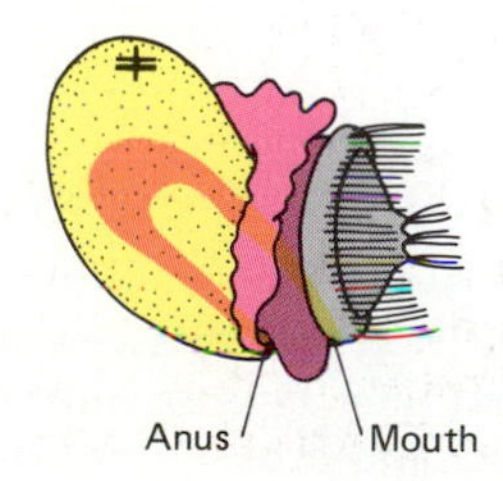

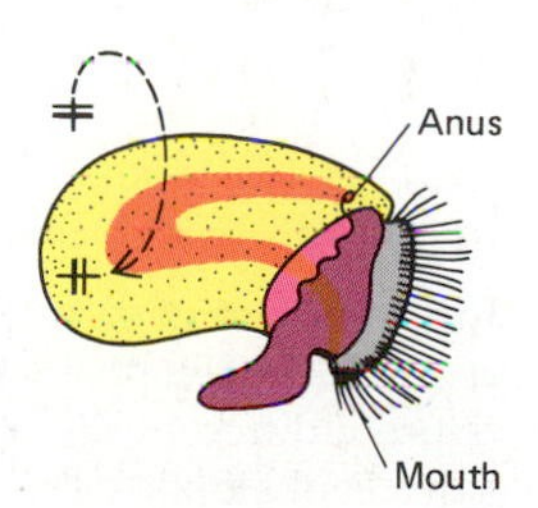

Figure 25–7 Embryonic torsion in the gastropod *Acmaea* (a limpet). (After Boutan, 1899.)

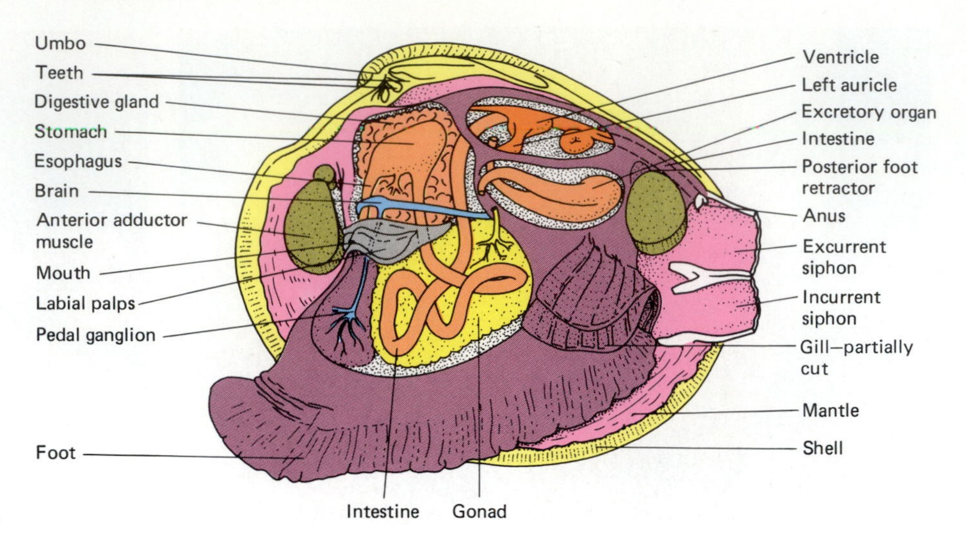

Figure 25–8 Internal anatomy of a clam.

CLASS BIVALVIA: CLAMS, OYSTERS, AND THEIR RELATIVES

The soft body of members of the class **Bivalvia** is laterally compressed and completely enclosed by two shells hinged dorsally and opening ventrally (Fig. 25–8). This arrangement allows the hatchet-shaped foot to protrude ventrally for locomotion. Apertures are also present for flow of water into and out of the mantle cavity. Extensions of the mantle, called siphons, permit bivalves to obtain water relatively free of sediment. There are both an **incurrent siphon** for water intake and an **excurrent siphon** for water output. Large, strong muscles attached to the shell enable the animal to close its shell.

The inner pearly layer of the bivalve shell is secreted in thin sheets by the epithelial cells of the mantle. Composed of calcium carbonate and known as mother-of-pearl, it is valued for making jewelry and buttons. Should a bit of foreign matter lodge between the shell and the epithelium, the epithelial cells are stimulated to secrete concentric layers of calcium carbonate around the intruding particle; this is how a pearl is formed.

Some bivalves, such as oysters, attach permanently to the substratum. Others, like clams, burrow slowly through rock or wood, seeking protected dwellings. (The shipworm, *Teredo,* which damages dock pilings and other marine installations, is just looking for a home.) Finally, some bivalves, such as scallops, swim with amazing speed by clapping their two shells together by the contraction of a large **adductor muscle** (the part of the scallop that is eaten by humans).

Clams and oysters are filter feeders. They obtain food by straining the sea water brought in over the gills by the siphon. The water is kept in motion by the beating of cilia on the surface of the gills. This stream of water carries food particles trapped in the mucus secreted by the gills to the mouth. An average oyster filters about 3 liters of sea water per hour. Because bivalves are filter feeders they have no need for a radula, and indeed they are the only group of mollusks that lack this structure.

Most bivalves have two distinct sexes. Gametes are usually discharged into the water, where fertilization takes place. In some marine and nearly all freshwater bivalves, sperm are shed into the water and fertilize the eggs within the mantle cavity of the female. In these species the female also broods her young within the mantle cavity. Development takes place among the gill filaments. In marine bivalves a trochophore larva typically develops, which then develops further into a veliger larva with shell and foot. Larvae of some freshwater species spend several weeks as parasites on the gills of fishes.

CLASS CEPHALOPODA: SQUIDS, OCTOPODS, AND THEIR RELATIVES

In contrast to most other mollusks, members of the class **Cephalopoda** (head-feet) are active, predatory animals. They are fast-swimming organisms, adapted for an entirely different life style than their filter-feeding relatives. Some biologists consider cephalopods the most advanced of the invertebrates.

Figure 25–9 *Octopus dofleini,* a cephalopod, from Puget Sound, Washington State. (Charles Seaborn.)

The octopus has no shell, and the shell of the squid is reduced to a small "pen" in the mantle. *Nautilus* has a flat, coiled shell consisting of many chambers built up year by year; each year the animal lives in the newest and largest chamber of the series. By secreting a gas resembling air into the other chambers, the *Nautilus* is able to float.

The cephalopod foot is divided into tentacles—ten in squids, eight in octopods. The tentacles, or arms, surround the central mouth of the large head (Fig. 25–9). Cephalopods have large well-developed eyes that form images. Although they develop differently, the eyes are structurally much like vertebrate eyes and function in much the same way.

The tentacles of squids and octopods are covered with suckers for seizing and holding prey. In addition to a radula, the mouth is equipped with two strong, horny beaks used to kill prey and tear it to bits. The mantle is thick and muscular and fitted with a funnel. By filling the mantle cavity with water and ejecting it through the funnel, the animal can attain rapid jet propulsion in the opposite direction.

Besides its speed, the cephalopod has developed two other important mechanisms that enable it to escape from its predators, which include the whales and moray eels. One is its ability to confuse the enemy by rapidly changing colors. By expanding and contracting pigment cells—**chromatophores**—in its skin, the cephalopod can display an impressive variety of mottled colors. Another defense mechanism is the **ink sac,** which produces a thick black liquid. This liquid is released in a dark cloud when the animal is alarmed; while its enemy pauses, temporarily blinded and confused, the cephalopod easily escapes. The ink has been shown to paralyze the chemical receptors of some predators.

The octopus feeds on crabs and other arthropods, catching and killing them with a poisonous secretion of its salivary glands. During the day, the octopus usually hides among the rocks; in the evening, it emerges to hunt for food. Its motion is incredibly fluid, giving little hint of the considerable strength in its eight arms.

Small octopods survive well in aquaria and have been studied extensively. They have a relatively high degree of intelligence and can make associations among stimuli. Their very adaptable behavior resembles more closely that of the vertebrates than the more stereotyped patterns of behavior seen in other invertebrates.

Phylum Annelida

Phylum **Annelida,** the segmented worms, includes the earthworms, leeches, and many marine and freshwater worms. The 10,000 or so species are divided into three main classes (Table 25–2 and Fig. 25–10). The term Annelida means ringed and refers to the series of rings, or segments, that make up the annelid body. Both the body wall and the internal organs are segmented. The segments are separated from one another by transverse partitions, the **septa.** The bilaterally symmetrical, tubular body may consist of about 100 segments. Some structures, such as the digestive

TABLE 25–2
Classes of Phylum Annelida

Class and Representative Animals	*Characteristics*
Polychaeta Sandworms Tubeworms	Mainly marine; each segment bears a pair of parapodia with many setae; well-developed head; separate sexes; trochophore larva
Oligochaeta Earthworm	Terrestrial and freshwater worms; few setae per segment; lack well-developed head; hermaphroditic
Hirudinea Leeches	Most are blood-sucking parasites that inhabit freshwater; lack appendages and setae; prominent muscular suckers

tract and certain nerves, extend the length of the body, passing through successive segments. Other structures are repeated in each segment (Fig. 25–11).

Segmentation is an advantage because not only is the coelom divided into segments, but each segment has its own muscles, enabling the animal to elongate one part of its body while shortening another part. The annelid's hydrostatic skeleton is discussed in Chapter 32. In the annelid the individual segments are almost all alike, but in many segmented animals—the arthropods and chordates—different segments and groups of segments are specialized to perform different functions. In some groups the specialization may be so pronounced that the basic segmentation of the body plan may be obscured.

Bristle-like structures called **setae** aid in locomotion Annelids have a well-developed coelom, a closed circulatory system, and a complete digestive tract extending from mouth to anus. Respiration takes place through the skin or by gills. Typically, a pair of excretory structures called **metanephridia** are found in each segment. The nervous system generally consists of a simple brain composed of a pair of ganglia and a double ventral nerve cord. A pair of ganglia and lateral nerves are repeated in each segment.

CLASS POLYCHAETA

The class **Polychaeta** includes marine worms, which swim freely in the sea, burrow in the mud near the shore, or live in tubes formed by cementing bits of shell and sand together with mucus and other secretions from the body wall. Each body

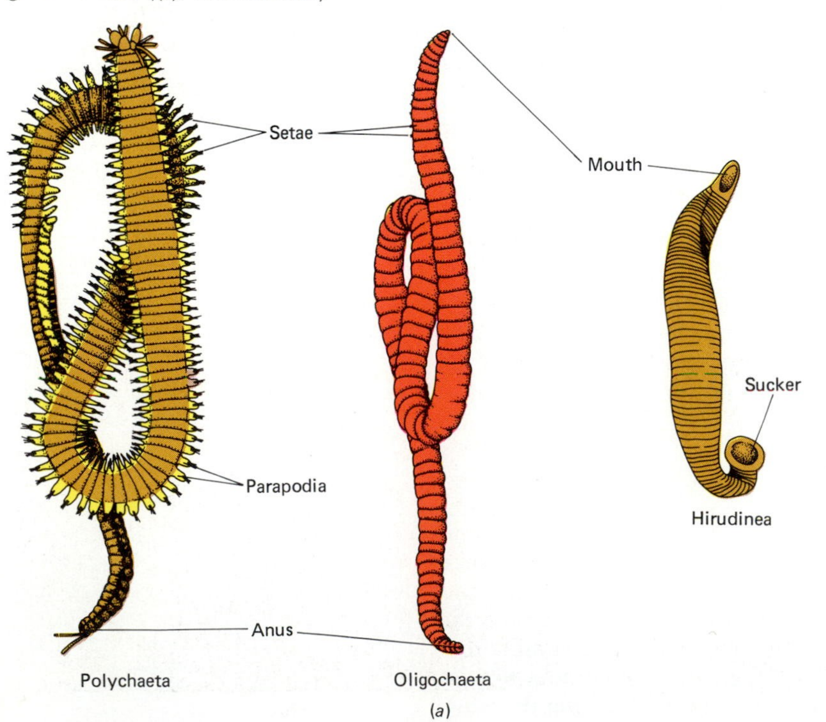

Figure 25–10 (*a*) Comparison of the classes of the phylum Annelida—Polychaeta, Oligochaeta, and Hirudinea. (*b*) Class Hirudinea is represented by this leech, *Helobdella stagnalis*. The dark area in its swollen body is recently ingested blood. ((*b*), Tom Adams.)

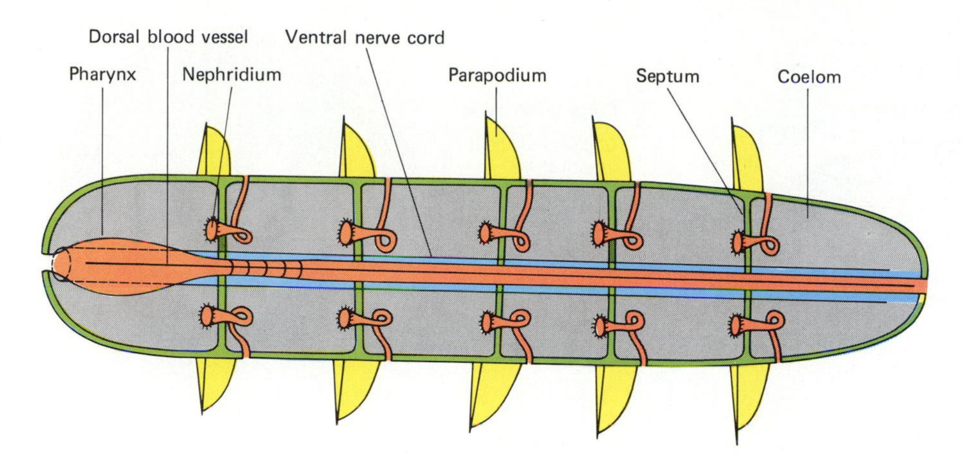

Figure 25–11 Metamerism in a generalized annelid. The body is segmented, and there is serial repetition of body parts (metamerism).

segment has a pair of paddle-shaped appendages called **parapodia** (singular, parapodium) that extend laterally and function in locomotion (Fig. 25–12). These fleshy structures bear many stiff setae (the name Polychaeta means many bristles). Most polychaetes have a well-developed head or **prostomium** bearing eyes and antennae. The prostomium may also be equipped with tentacles, bristles, and **palps** (feelers).

Many polychaetes have evolved behavioral patterns that ensure fertilization. By responding to certain rhythmic variations, or cycles, in the environment nearly all of the females and males of a given species release their gametes into the water at the same time. More than 90% of reef-dwelling *Palolo* worms of the South Pacific shed their eggs and sperm within a single two-hour period on one night of the year. In this animal the seasonal rhythm limits the reproductive period to November; the lunar rhythm, to a day during the last quarter of the moon when the tide is unusually low; and the diurnal rhythm, to a few hours just after complete darkness. The posterior half of the *Palolo* worm, loaded with gametes, actually breaks off from the rest, swims backward to the surface, and eventually bursts, releasing the eggs or sperm so that fertilization may occur. Local islanders eagerly await this annual event when they can gather up great numbers of the swarming polychaetes and broil them for dinner.

(*a*)

(*b*)

Figure 25–12 Polychaete annelids. (*a*) West Indian fireworm, *Hermodice carunculata*. (*b*) The Christmas tree worm (*Spiro branchus giganteus*) photographed in a Florida coral reef. ((*a*), Charles Seaborn; (*b*), James H. Carmichael, Coastal Creations.)

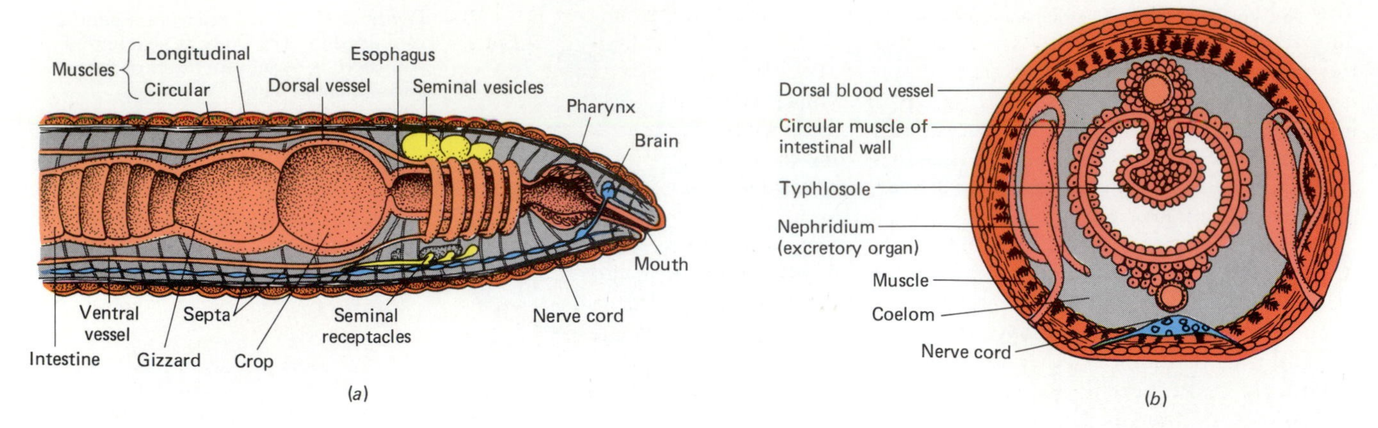

Figure 25–13 Internal structure of an earthworm (an oligochaete). (*a*) Diagrammatic longitudinal section of the anterior portion. (*b*) Cross section.

CLASS OLIGOCHAETA: THE EARTHWORMS

The 3000 or so species of the class **Oligochaeta** are found almost exclusively in fresh water and in moist terrestrial habitats. These worms lack parapodia, have few bristles per segment (the name Oligochaeta means few bristles), and lack a well-developed head. All oligochaetes are **hermaphroditic,** meaning that male and female reproductive systems are present in the same individual. Since the earthworm is among the most familiar of all invertebrates, let us examine this animal in more detail.

Lumbricus terrestris, the common earthworm, is about 20 cm long. Its body is divided into more than 100 segments separated externally by grooves and internally by septa. The mouth is located in the first segment, the anus in the last. The earthworm's body is protected from desiccation by a thin, transparent cuticle, secreted by the cells of the epidermis (Fig. 25–13). Mucus secreted by the glandular cells of the epidermis forms an additional protective layer over the body surface.

The body wall contains an outer layer of circular muscles and an inner layer of longitudinal muscles. The earthworm moves forward by contracting its circular muscles to elongate the body, grasping the ground or walls of the burrow with its setae, and then contracting its longitudinal muscles to draw the posterior end forward. Locomotion proceeds in waves. The muscles work against the hydrostatic skeleton provided by the coelomic fluid within the coelom of each segment. (For a more detailed discussion see Chapter 32.) Each segment except the first bears four pairs of setae supplied with tiny muscles that can move each seta in and out and change its angle. Thus in locomotion, the earthworm's body is extended, anchored by the setae, and then contracted.

An earthworm literally eats its way through the soil, ingesting its own weight in soil and decaying vegetation every 24 hours. During this process the soil is turned and aerated, and nitrogenous wastes from the earthworm enrich it. This is why earthworms are vital to the formation and maintenance of fertile soil. The earthworm's soil meal, containing nutritious decaying vegetation, is processed in the complex digestive system. Food is swallowed through the muscular **pharynx,** and passes through the **esophagus** to the **stomach.** The stomach consists of two parts: a thin-walled **crop** where food is stored and a thick-walled muscular **gizzard** where food is ground to bits. The rest of the digestive system is a long, straight **intestine,** where food is digested and absorbed. The surface area of the intestine is increased by a dorsal, longitudinal fold called the **typhlosole.** Wastes pass out of the intestine to the exterior through the **anus.**

The efficient, closed circulatory system consists of two main blood vessels that extend longitudinally. The dorsal blood vessel, just above the digestive tract, collects blood from numerous segmental vessels. It contracts, pumping blood anteriorly. In the region of the esophagus, five pairs of blood vessels propel blood from the dorsal to the ventral blood vessel. Located just below the digestive tract, the ventral blood vessel conveys blood both posteriorly and anteriorly. Small blood vessels branch from it and deliver blood to the various structures in each segment as well as to the body wall. Within these structures blood flows through very tiny blood vessels (capillaries) before returning the dorsal blood vessel.

Gas exchange takes place through the moist skin, and oxygen is usually transported by the respiratory pigment hemoglobin present in the blood plasma.

The excretory system consists of paired organs, the metanephridia, repeated in almost every segment of the body. Each metanephridium consists of a ciliated funnel (nephrostome) opening into the next anterior coelomic cavity and connected by a tube to the outside of the body (Fig. 25–13). Wastes are removed from the coelomic cavity partly by the beating of the cilia and partly by currents set up by the contraction of muscles in the body wall. The tube of the metanephridium is surrounded by a capillary network so that wastes are removed from the blood as well as from the coelomic fluid.

The metanephridia, open at both ends, are quite different from the protonephridia of the flatworms, which are blind tubules opening only to the exterior. In higher invertebrates, the adults typically have metanephridia, but larval forms usually have protonephridia. These protonephridia generally have a single long flagellum rather than a tuft of cilia. This developmental pattern of excretory structures is often used to support the concept that complex invertebrates (as well as vertebrates) evolved from forms similar to the lower invertebrates.

The nervous system consists of a pair of **cerebral ganglia** that serve as a brain, just above the pharynx in the third segment, and a **subpharyngeal ganglion,** just below the pharynx in the fourth segment. A ring of nerve fibers connects the ganglia. From the lower ganglion a double ventral nerve cord extends beneath the digestive tract to the posterior end of the body. In each segment along the nerve cord there is a pair of fused **segmental ganglia,** from which nerves extend laterally to the muscles and other structures of that segment. The segmental ganglia coordinate the contraction of the muscles of the body wall, so that the worm can creep along. The nerve cord contains a few giant axons that transmit nerve impulses more rapidly than ordinary fibers. When danger threatens, these stimulate the muscles to contract and draw the worm quickly back into its burrow.

The subpharyngeal ganglion is the main center controlling movement and vital reflexes. It exerts control over the other ganglia in the chain. When the subpharyngeal ganglion is destroyed, all movement stops. When the brain is removed, the subpharyngeal ganglion is able to continue to control movement, but the worm is no longer able to adjust its actions to conditions in the environment. Earthworm responses are limited to reflex (pre-programmed, stereotyped) actions. However, in the laboratory earthworms can be conditioned to perform simple acts such as contracting when exposed to bright light or vibration. Living a subterranean life as it does, the earthworm has no need for well-developed sense organs.

Like other oligochaetes, earthworms are hermaphroditic. During copulation two worms, heading in opposite directions, press their ventral surfaces together (Fig. 25–14). These surfaces become glued together by thick mucous secretions of the **clitellum,** a thickened ring of epidermis in segments 32 to 37. Sperm from each worm pass posteriorly to its clitellum and are stored in the **seminal receptacles** of the other worm. The worms then separate, and the clitellum secretes a membranous **cocoon** containing an albuminous fluid. As the cocoon is slipped over the worm's head, eggs are laid into it from the female pores and sperm are added as the cocoon passes over the seminal receptacles. When the cocoon is free, its openings constrict so that a spindle-shaped capsule is formed, and the eggs develop into tiny worms within the cocoon. This complex reproductive pattern is an adaptation to terrestrial life.

Figure 25–14 Two earthworms, genus *Lumbricus,* copulating. These animals are hermaphroditic, but cross-fertilize one another. (R. K. Burnard, Ohio State University/BPS.)

Figure 25–15 Medicinal leeches were used by physicians for bloodletting in the 17th and 18th centuries. Today, leeches are being used in some modern surgical procedures because they release a substance that prevents blood clotting in the immediate area where they are sucking. (Courtesy of World Health Organization.)

CLASS HIRUDINEA: THE LEECHES

Most **leeches,** which are members of the class **Hirudinea,** are blood-sucking parasites that inhabit fresh water. They differ from other annelids in having neither setae nor appendages. Prominent muscular suckers are present at both anterior and posterior ends for clinging to their prey. Most leeches attach themselves to a vertebrate host, bite through the skin of the host, and suck out a quantity of blood, which is stored in pouches in the digestive tract. An anticoagulant (hirudin), secreted by glands in its crop, ensures the leech a full meal of blood. Their meals may be infrequent, but they can store enough food from one meal to last a long time. The so-called medicinal leech, a freshwater worm about 10 cm long, was used by physicians for bloodletting in the 17th and 18th centuries (Fig. 25–15).

Phylum Onychophora

Only about 70 living species of **Onychophorans** are known, but this group of animals is considered important as a possible link between the annelids and the arthropods. In fact some zoologists classify onychophorans as annelids, and others classify them with the arthropods. These wormlike animals inhabit humid tropical areas such as rain forests.

Peripatus, a caterpillarlike creature, is the best known of the onychophorans (Fig. 25–16). About 5 to 8 cm long, *Peripatus* has a thin, soft cuticle covering its elongated muscular body. Its many pairs of unjointed legs bear claws. Like annelids, it is internally segmented, and many of its organs are duplicated serially. However, its jaws are derived from appendages as in arthropods, and like arthropods, it has an open circulatory system. Its coelom is reduced, for much of the body cavity is occupied by a hemocoel. Its respiratory system, which consists of air tubes (tracheal tubes), is also arthropod-like.

Some zoologists believe that terrestrial onychophorans gave rise to the insects, millipedes, and centipedes; others think that this group branched off the annelid–arthropod trunk after the annelids, but before the arthropods. In either case, the onycophorans are interesting creatures with a curious mixture of annelid and arthropod characteristics.

Figure 25–16 *Peripatus,* an onychophoran from moist forested regions of Costa Rica. This animal has features of both arthropods and annelids. Note the soft, sluglike body and the presence of a series of jointed legs. (D. R. Paulson, University of Washington/BPS.)

Phylum Arthropoda

The animals that make up phylum **Arthropoda** are, without doubt, the most biologically successful of all animals. There are more of them (about 800,000 described species); they live in a greater variety of habitats; and they can eat a greater variety of food than the members of any other phylum (Fig. 25–17). Among their most important characteristics are the following:

1. *Paired, jointed appendages,* from which they get their name (arthropod means jointed foot). These appendages function as swimming paddles, walking legs, mouth parts, or accessory reproductive organs for transferring sperm.
2. A hard, armorlike *exoskeleton* composed of chitin that covers the entire body and appendages. The exoskeleton provides protection against excessive loss of moisture as well as predators, and gives support to the underlying soft tissues. Distinct muscle bundles attach to the inner surface of the exoskeleton. These act upon a system of levers that permit the extension and flexion of parts at the joints. The exoskeleton has certain disadvantages, however. Body movement is somewhat restricted, and in order to grow, the arthropod must shed this outer shell periodically and grow another larger one, a process that leaves it temporarily vulnerable to predators.
3. A *segmented body,* like that of the annelid. In some arthropod classes, however, segments become fused together or lost during development.
4. An *open circulatory system* with a dorsal heart. A hemocoel (blood cavity) occupies most of the body cavity, and the coelom is small and made up chiefly of the cavities of the reproductive system.

Figure 25–17 The arthropods are considered the most highly successful animals. (*a*) Starry-eyed hermit crab photographed in the Virgin Islands. (*b*) Florida dragonfly. (*c*) A pair of horseshoe crabs, *Limulus polyphemus,* mating. (*d*) Hay mite. ((*a*), Robin Lewis, Coastal Creations; (*b*), James H. Carmichael, Coastal Creations; (*c*), Peter J. Bryant, UC – Irvine/BPS.)

(*a*)

(*b*)

(*c*)

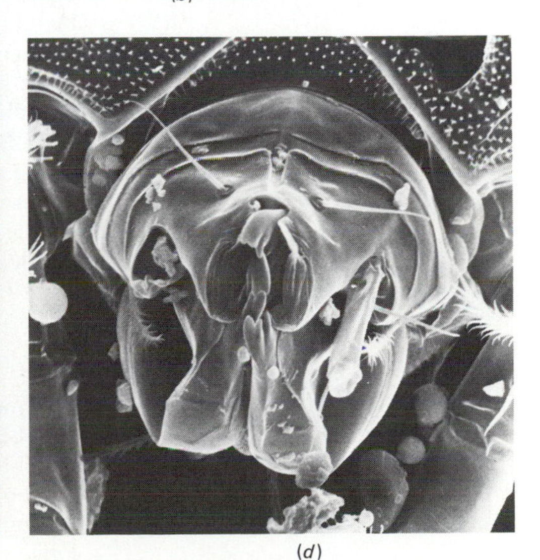

(*d*)

TABLE 25–3
Arthropod Subphyla and Classes

Subphylum and Classes	*Subphylum Characteristics*
Subphylum **Trilobitomorpha** Class Trilobita (fossil trilobites)	Extinct marine arthropods; covered by a hard, segmented shell; head, thorax, and abdomen; pair of biramous appendages on each body segment
Subphylum **Chelicerata** Class Merostomata (horseshoe crab) Class Arachnida (spiders, scorpions, ticks, mites)	First pair of appendages are the chelicerae used to manipulate food; body consists of cephalothorax and abdomen; no mandibles; no antennae
Subphylum **Crustacea** Class Crustacea (lobsters, crabs, shrimp, barnacles)	Mandibles; two pairs of antennae; biramous appendages
Subphylum **Uniramia** Class Insecta (grasshopper, roach) Class Chilopoda (centipedes) Class Diplopoda (millipedes)	Uniramous appendages; single pair of antennae; mandibles

There is much disagreement concerning arthropod classification. Although most zoologists believe that arthropods evolved either from an ancestor of the polychaetes or from the polychaetes themselves, some zoologists propose that there are actually several major groups of arthropods that evolved independently. Here we will use a scheme that divides the phylum into four subphyla: Trilobitomorpha, Chelicerata, Crustacea, and Uniramia (Table 25–3; also see Table 25–4).

THE ARTHROPOD BODY PLAN

The bodies of most arthropods are divided into three regions: the **head,** composed of exactly six segments; the **thorax;** and the **abdomen,** both of which consist of a variable number of segments. In contrast to most annelids, each arthropod has a fixed number of segments, which remains the same throughout life. The incredible range of variations in body plan and in the shape of the jointed appendages in the numerous species almost defies description.

The nervous system of the more primitive arthropods, like that of the annelids, consists of a ventral nerve cord connecting segmental ganglia. In the more complex arthropods, the successive ganglia usually fuse together. Arthropods have a variety of well-developed sense organs: complicated eyes, such as the compound eyes of insects; organs of hearing; **antennae** sensitive to touch and chemicals; and cells sensitive to touch on the surface of the body.

The open circulatory system includes a dorsal, tubular heart that pumps blood into a dorsal artery, and sometimes several other arteries. From the arteries blood flows into large sinuses, which collectively make up the hemocoel. Blood in the hemocoel bathes the tissues directly. No capillaries or veins are present. Eventually blood finds its way back into the heart through openings, referred to as **ostia,** in its walls.

Most of the aquatic arthropods have a system of gills for gas exchange. The land forms, in contrast, usually have a system of fine, branching air tubes called **tracheae** that conduct air to the internal organs. The digestive system typically is a simple tube similar to that of the earthworm. Both ends are lined with a waxy cuticle similar to the outer layer of the exoskeleton. Excretory structures vary somewhat from class to class.

THE TRILOBITES

The most primitive arthropods, the **trilobites,** which comprise the subphylum **Trilobitomorpha,** are extinct marine arthropods that were once abundant and widely distributed in Paleozoic seas. More than 4000 species have been described. Most of these lived on the sea bottom and dug into the mud or sand. They ranged in length from a millimeter to nearly a meter, but most were between 3 and 10 cm long.

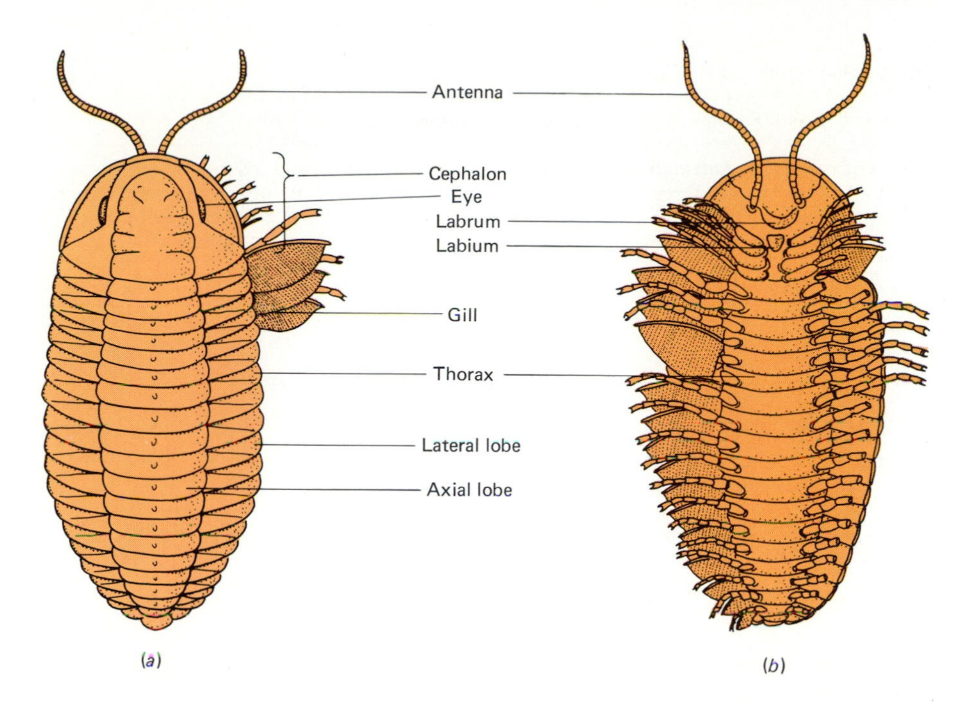

Figure 25–18 The trilobites are extinct marine arthropods that are considered the most primitive members of the phylum. (*a*) Dorsal view of a trilobite from the Ordovician period. (*b*) Ventral view of the same trilobite. (See also Fig. 46–13a.)

Covered by a hard, segmented shell, the trilobite body was a flattened oval divided into three parts: an anterior head of four fused segments bearing a pair of antennae and a pair of compound eyes; a thorax consisting of a variable number of segments; and a posterior abdomen composed of several fused segments (Fig. 25–18). At right angles to these divisions, two dorsal grooves extended the length of the animal, dividing the body into a median lobe and two lateral lobes. (The name trilobite derives from this division of the body into three longitudinal parts.) Each segment of the body had a pair of segmented **biramous** (two-branched) **appendages;** each appendage consisted of an inner walking leg and an outer branch bearing gills.

It is remarkable that fossil evidence has yielded information not only about the structure of the adult but also about the developmental stages of the trilobites. These animals went through three larval periods; during each one, the larvae underwent several molts. As molts occurred, additional segments were added to the body, and the body structure became more complex. The trilobites have a number of characteristics in common with the crustaceans, and others in common with the arachnids and horseshoe crabs. They may have been ancestors of both of these groups, but the exact evolutionary relationship is not clear.

SUBPHYLUM CHELICERATA

In members of subphylum **Chelicerata,** the body consists of a **cephalothorax** and an abdomen. Chelicerates have no antennae and no chewing mandibles, mouthparts that are characteristic of other arthropod subphyla (see Table 25–3). Instead, the first pair of appendages, which are located immediately anterior to the mouth, are the **chelicerae,** used to manipulate food and pass it to the mouth. The second pair of appendages, called **pedipalps,** are modified to perform different functions in various groups. Posterior to the pedipalps there are usually four pairs of legs.

Class Merostomata: Horseshoe Crabs

Almost all members of the class **Merostomata** are extinct. The only living merostomes, the **horseshoe crabs** in the subclass Xiphosura, have survived essentially unchanged for 350 million years or more. *Limulus polyphemus* is the species common in North America. As its common name describes, this animal is horseshoe-shaped. The long spikelike tail that extends posteriorly is used in locomotion, not for defense or offense. Horseshoe crabs feed on mollusks, worms, and other invertebrates that they find on the ocean floor.

TABLE 25–4
General Characteristics of the Principal Arthropod Classes

Class	Main Habitat	Body Divisions	Gas Exchange	Appendages			Development
				Antennae	Mouth parts	Legs	
Arachnida (about 60,000 species)	Mainly terrestrial	Cephalothorax and abdomen	Book lungs or tracheae	None	Chelicerae, pedipalps	4 pairs on cephalothorax	Direct, except mites and ticks
Crustacea (about 31,000 species)	Marine or fresh water; few on land	Cephalothorax and abdomen	Gills	2 pairs	Mandibles, 2 pairs of maxillipeds (for food handling)	1 pair per segment or less	Usually larval stages
Insecta (about 750,000 species)	Mainly terrestrial	Head, thorax, and abdomen	Tracheae	1 pair	Mandibles, maxillae	3 pairs on thorax; (+wings)	Usually larval stages; some with complete metamorphosis
Chilopoda (about 3000 species)	Terrestrial	Head with segmented body	Tracheae	1 pair	Mandibles, 2 pairs maxillae	1 pair per segment	Direct
Diplopoda (about 7500 species)	Terrestrial	Head, with segmented body	Tracheae	1 pair	Mandibles, maxillae	2 pairs (or 1) per segment	Direct

(*a*)

Class Arachnida: Spiders, Scorpions, Ticks, and Mites

The 60,000 or so species of class **Arachnida** include the spiders, scorpions, mites, ticks, and harvestmen or daddy long-legs (Fig. 25–19). The arachnid body consists of a cephalothorax (composed of fused head and thorax) and abdomen. Arachnids have six pairs of jointed appendages. In arachnids, the first pair, the chelicerae, are fanglike structures used to penetrate prey and suck out its body fluids; in some species, the chelicerae are used to inject poison into the prey. The second pair of appendages, the pedipalps, are used by spiders to hold and chew food and are modified as sense organs for tasting the food in some species. The other four pairs of appendages are used for walking. Most arachnids are carnivorous and prey upon insects and other small arthropods.

Gas exchange in arachnids takes place either by tracheal tubes or by book lungs, or by both. Each **book lung** consists of 15 to 20 plates, like pages of a book,

(*b*)

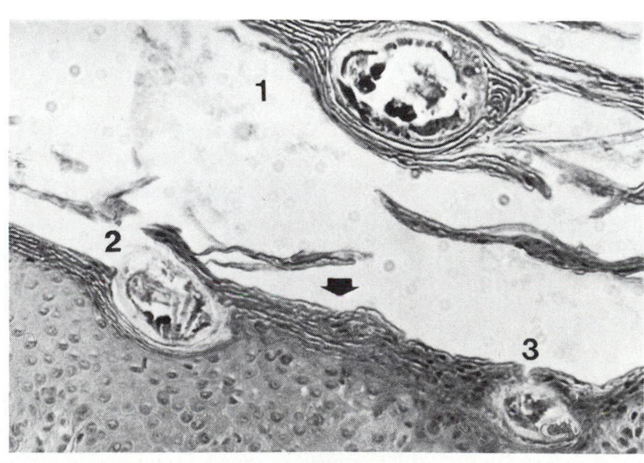

(*c*)

Figure 25–19 Class Arachnida includes the spiders, scorpions, mites, ticks, and harvestmen. (*a*) A silk spider. (*b*) A vinegarone, or whip scorpion. (*c*) The scabies mite is a parasite that burrows deep into the epidermis of the skin. ((*b*), Courtesy of Hubbard Scientific Company; (*c*) courtesy of D. Van Neste, *Archives of Dermatological Research*, 274, May 1981.)

Figure 25–20 A spider wrapping a webworm. (E. R. Degginger.)

that contain tiny blood vessels. Air enters the body through abdominal slits and circulates between the plates. As air passes over the blood vessels oxygen and carbon dioxide are exchanged. An arachnid may have as many as four pairs of book lungs.

Spiders have **silk glands** in their abdomen; these glands secrete an elastic protein that is liquid as it emerges from their spinnerets. It hardens as it is drawn out, and is used to construct webs for building nests, encasing eggs in a cocoon, and in some species, for trapping prey (Fig. 25–20). Many spiders lay down a silken dragline as they venture forth. This serves as a safety line and is also a means of communication between members of a species. From a dragline a spider can determine the sex and maturity level of the spider that spun it.

Although spiders do have poison glands for capturing their prey, only a few have poison that is toxic to humans. The most widely distributed poisonous spider in the United States is the black widow. Its poison is a neurotoxin that interferes with transmission of messages from nerves to muscles. Although painful, spider bites cause fewer than five fatalities per year in the United States; these usually occur in children.

Mites and ticks are among the greatest arthropod nuisances. They eat our crops, infect our livestock and pets, and inhabit our own bodies. Many live unnoticed, owing to their small size, but others cause disease. Certain mites cause mange in dogs and other domestic animals. Chiggers (red bugs), the larval form of red mites, attach themselves to the skin and secrete an irritating digestive fluid that may cause itchy red welts. Larger than mites, ticks are ectoparasites on dogs and other domestic animals. They can transmit diseases such as Rocky Mountain spotted fever, Texas cattle fever, and relapsing fever.

SUBPHYLUM CRUSTACEA: LOBSTERS, CRABS, SHRIMP, AND THEIR RELATIVES

Crustaceans—members of the subphylum **Crustacea**—are vital members of marine food chains, serving as primary consumers of algae and other producers, and serving as food for the many carnivores that inhabit the oceans. Countless billions of microscopic crustaceans swarm in the ocean and form the food (krill) of many fish and other marine forms such as whales.

Traditionally, crustaceans have been grouped with insects, centipedes, and millipedes in the subphylum Mandibulata. This was based on the presence in these animals of **mandibles** instead of chelicerae as the first pair of mouth parts. Unlike the clawlike chelicerae, mandibles are usually adapted for biting and chewing. The current trend in taxonomy is to separate the crustaceans because they are unique in several ways. For one thing, other than the trilobites they are the only group to have biramous appendages—that is, appendages that have two jointed branches at their ends. They are also the only group to have two pairs of antennae (Fig. 25–21).

(a) (b) (c) (d)

Figure 25–21 Crustaceans. (*a*) Banded coral shrimp (*Stenopus hispidus*) eating another shrimp (Carribean species). (*b*) Gooseneck barnacles, *Pollicipes polymerus*. (*c*) A single barnacle, *Balanus nubilis*, filtering water for food. (*d*) Crab. ((*a*), James Carmichael, Coastal Creations; (*b*), (*c*), (*d*), Charles Seaborn.)

The two pairs of antennae serve as sensory organs for touch and taste. The third pair of appendages, the mandibles, are located on each side of the ventral mouth and are used for biting and grinding food. Posterior to the mandibles are two pairs of appendages, the first and second **maxillae,** used for manipulating and holding food. Several other pairs of appendages are present. Usually five pairs are modified for walking. Others may be specialized for swimming, sperm transmission, carrying eggs and young, or sensation.

As the only class of arthropods that are primarily aquatic, crustaceans generally have gills for gas exchange. Two large **antennal** (green) **glands** located in the head remove metabolic wastes from the blood and body fluids and excrete them through ducts opening at the base of each antenna. The nervous system is somewhat similar to the annelid nervous system, but is proportionately larger, and ganglia are fused and large. Most adult members of the class have compound eyes (discussed in Chapter 41). Among the other sense organs present are **statocysts** for detecting gravity.

Crustaceans characteristically have separate sexes. During copulation the male uses specialized appendages to transfer sperm into the female. The fertilized eggs are usually brooded. The newly hatched animals pass by successive molts through a series of larval stages and finally reach the body form characteristic of the adult. The lobster, for example, molts seven times during the first summer; at each molt it gets larger and resembles the adult more. After it becomes a small adult, additional molts provide for growth. The process of molting is discussed in Chapter 32.

The barnacles are the only sessile crustaceans. They differ markedly in their external anatomy from other members of the class, and it was only in 1830, when the larval stages were investigated, that the relationship between the barnacles and other crustaceans was recognized. The barnacles are exclusively marine and secrete complex calcareous cups within which the animal lives. The larvae of barnacles are free-swimming forms that go through several molts and eventually become

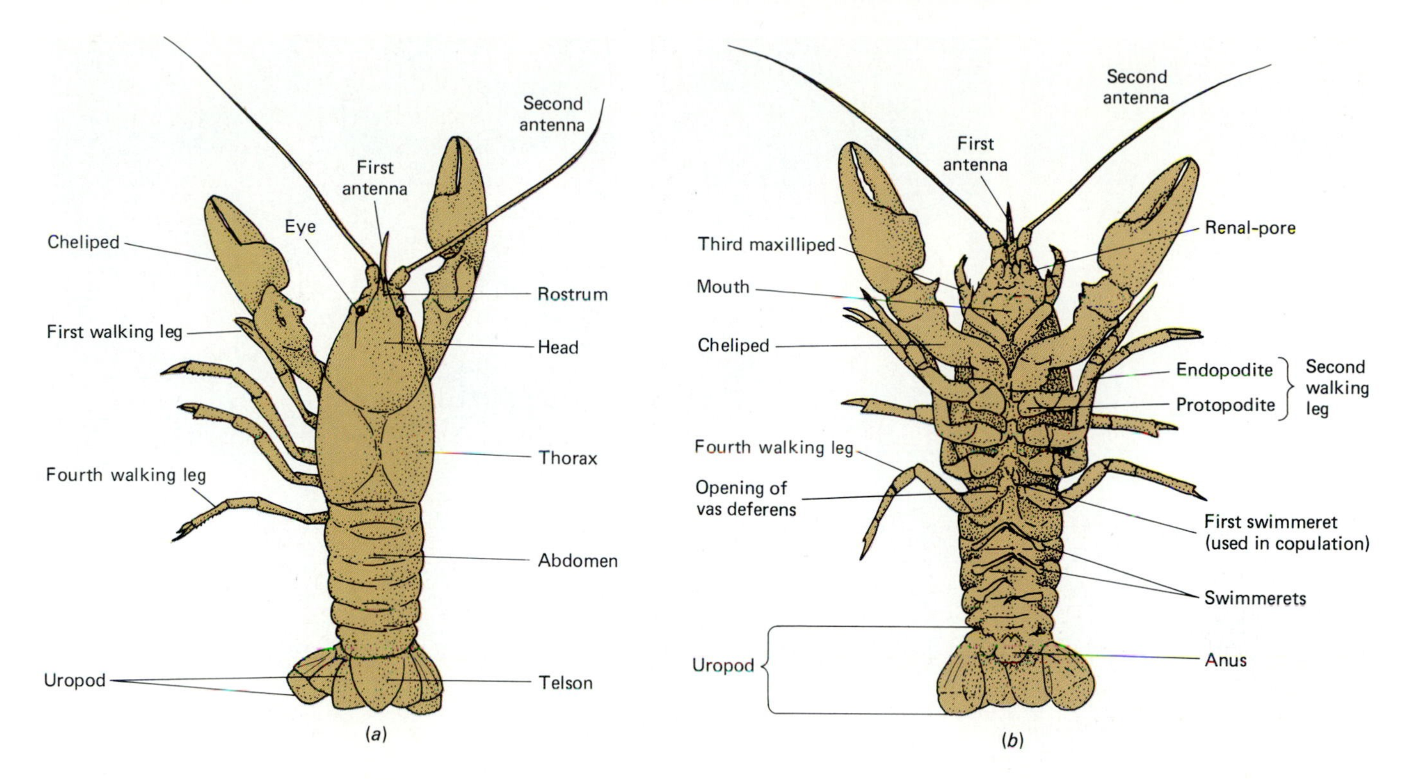

Figure 25–22 Anatomy of a crayfish. (*a*) Dorsal view; (*b*) ventral view.

sessile and develop into the adult form. Barnacles were described by the 19th century naturalist Louis Agassiz as "nothing more than a little shrimplike animal standing on its head in a limestone house and kicking food into its mouth."

The largest order of crustaceans, **Decapoda,** contains some 8500 species of lobsters, crayfish, crabs, and shrimp. Most decapods are marine but a few, such as the crayfish, certain shrimp, and a few crabs, live in fresh water. The crustaceans in general and the decapods in particular show in a striking way the specialization and differentiation of parts in the various regions of the animal. The segments of trilobites and perhaps of the earliest crustaceans bore appendages that were very similar. In the lobster, no 2 of the 19 pairs of appendages are identical, and the appendages in the different parts of the body differ markedly in form and function (Fig. 25–22).

The six segments of the lobster's head and the eight segments of the thorax are fused into a cephalothorax, covered on its top and sides by a shield, the **carapace.** The carapace is composed of chitin impregnated with calcium salts. The two pairs of antennae are the sites of chemoreceptors and tactile sense organs; the second pair of antennae are especially long. The mandibles are short and heavy, with opposing surfaces used in grinding and biting food. Behind the mandibles are two pairs of accessory feeding appendages, the first and second maxillae. The appendages of the first three segments of the thorax are the **maxillipeds,** which aid in chopping up food and passing it to the mouth. The fourth segment of the thorax has a pair of large **chelipeds,** or pinching claws. The last four thoracic segments have pairs of **walking legs.** The appendages of the first abdominal segment are part of the reproductive system and function in the male as sperm-transferring structures. On the following four abdominal segments are paired **swimmerets,** small paddle-like structures used by some decapods for swimming and by the females of all species for holding eggs. Each branch of the sixth abdominal appendages, which are called **uropods,** consists of a large flattened structure. Together with the flattened **telson,** the posterior end of the abdomen, they form a fan-shaped tail fin used for swimming backwards.

SUBPHYLUM UNIRAMIA

The insects, centipedes, and millipedes are grouped together in subphylum **Uniramia** because they all possess **uniramous,** or unbranched **appendages.** They also bear only a single pair of antennae rather than two pairs as in crustaceans. Many zoologists believe that this group evolved from the onychophorans.

(a)

(b)

(c)

Figure 25–23 Some insect adaptations include (*a*) Stick insect. (*b*) Thirteen-year *Cicada,* molting. This insect requires 13 years to mature. The nymphs live in the soil feeding on roots. (*c*) A maybug, *Melolontha melolontha,* preparing for a landing. ((*b*), Chris Simon; (*c*), Stephen Dalton, Photo Researchers, Inc.)

Class Insecta

With more than 750,000 described species, the class **Insecta** is the most successful group of animals on our planet in terms of diversity and number of species, as well as number of individuals (see Figs. 25–23 and 25–24). Insects are primarily terrestrial animals, but some species live in fresh water, a few are truly marine, and others inhabit the shore between the tides.

An insect may be described as an **articulated** (jointed), **tracheated** (having tracheal tubes for gas exchange) **hexapod** (having six feet). The insect body consists of three distinct parts—head, thorax, and abdomen (Fig. 25–25). Three pairs of legs emerge from the adult thorax, and usually two pairs of wings. One pair of antennae protrudes from the head, and the sense organs include both simple and compound eyes. A complex set of mouth parts is present; these may be adapted for piercing, chewing, sucking, or lapping. Excretion is accomplished by two to many slender **Malpighian tubules,** which receive metabolic wastes from the blood and after concentrating them, discharge them into the intestine.

The sexes are separate, and fertilization takes place internally. During development there are several molts. In some orders there are several developmental stages called nymphal stages and gradual metamorphosis (change in body form) to the adult form (see Focus on The Principal Orders of Insects). In others there is a complete metamorphosis with four distinct stages in the life cycle: egg, larva, pupa, and adult (Fig. 25–26).

Certain species of bees, ants, and termites exist as colonies or societies made up of several different types of individuals, each adapted for some particular function (Fig. 25–27). The members of some insect societies communicate with each other by "dances" and by chemicals called pheromones. Social insects and their communication are discussed in Chapter 49.

SECRETS OF INSECT SUCCESS. There are more species of insects than of all other classes of animals combined. What they lack in size, insects make up in sheer numbers. It has been calculated that if all the insects in the world could be weighed, they would weigh more than all of the remaining terrestrial animals. Because they have an extraordinary ability to adapt to changes in the environment, it has been predicted that these curious creatures may eventually inherit the earth.

What are the secrets of insect success? One important factor is their body plan, which can be modified and specialized in so many ways that insects have been able to adapt to an incredible number of life styles. They have filled almost every variety of ecological niche. One of their main secrets of success is their ability to fly. Unlike other invertebrates, which creep slowly along (or under) the ground,

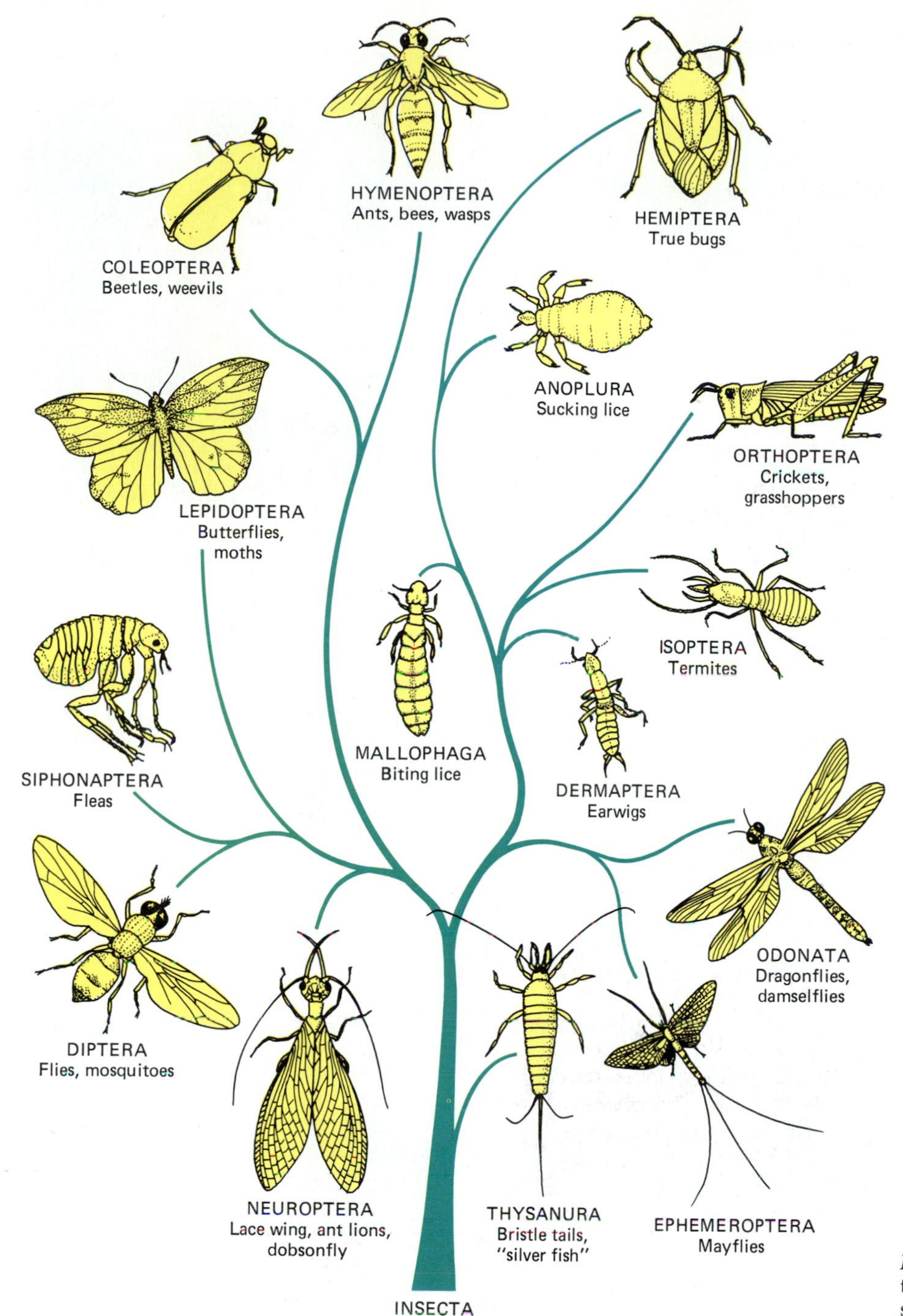

Figure 25–24 Representatives of some of the important orders of the class Insecta.

the insects fly rapidly through the air. Their wings and their small size facilitate their wide distribution.

The insect body is well protected by the tough exoskeleton, which also helps to prevent water loss by evaporation. Other protective mechanisms include mimicry, protective coloration, and aggressive behavior. Metamorphosis divides the insect life cycle into different stages, a strategy that has the advantage of placing larval forms into their own niches so that they do not have to compete with adults for food or habitats.

IMPACT OF INSECTS ON HUMANS. Not all insects compete with us for food or merely cause us to scratch, swell up, or recoil from their presence. Bees, wasps, mosquitos, and many other insects pollinate flowers of many crops and fruit trees. Some destroy harmful insects. For example, dragonflies eat mosquitoes; some organic farmers even purchase lady beetles, so adept are they at ridding plants of aphids and other insect pests. Insects are important members of many food chains. Many birds, mammals, amphibians, reptiles, and even some fish depend upon insects for dinner. Many beetles and the maggots of flies are detritus feeders that break down dead plants and animals and their wastes, permitting nutrients to be recycled.

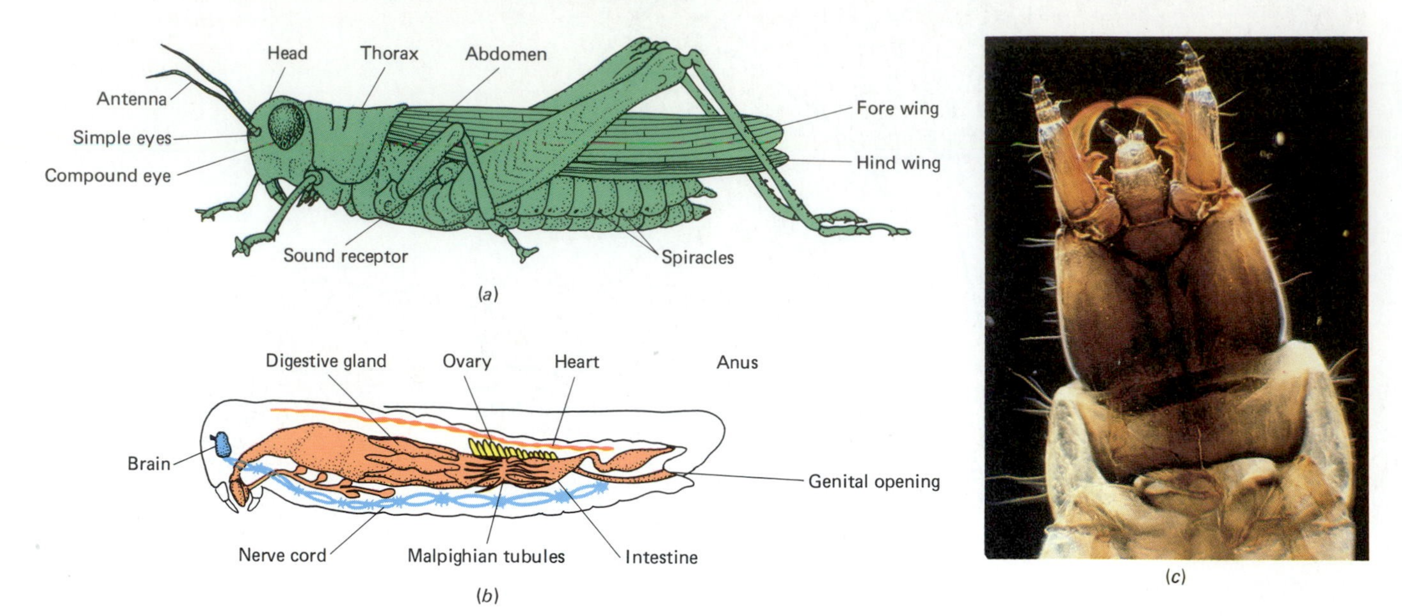

Figure 25–25 Insect body structure. (*a*) External anatomy of the grasshopper. Note the three pairs of articulated legs. (*b*) Internal anatomy of the grasshopper. (*c*) Head of the larva of a water scavenger beetle, an aquatic beetle. Can you identify the mandibles and antennae? ((*c*), Tom Adams.)

Many insect products are useful to us. Bees produce several thousand tons of honey used commercially each year, as well as a large amount of beeswax used in making candles, lubricants, chewing gum, and other products. Shellac is made from lac, a substance given off by certain scale insects that feed on the sap of trees. And the labor of silkworms provides us with millions of pounds of silk annually.

On the negative side, billions of dollars worth of crops are destroyed each year by insect pests. Whole buildings may be destroyed by termites, and clothing damaged by moths. Fire ants not only inflict painful stings but cause farmers serious economic loss because of their large mounds, which damage mowers and other farm equipment (Fig. 25–28). Such mounds also reduce grazing land because livestock quickly learn to avoid them.

Figure 25–26 The life cycle of a monarch butterfly. ((*a*), E. R. Degginger. (*b–e*), courtesy of Leo Frandzel.)

Figure 25–27 Model of a royal cell of the termite *Constrictotermes cavifrons*, from Guyana. The queen, with an enlarged abdomen, occupies the center of the chamber with her head toward the right. The king is at the lower left. Most of the individuals are workers. A few soldiers with "squirt gun" heads and reduced mandibles are at the left. (Courtesy of the Buffalo Society of Natural Sciences.)

Figure 25–28 Fire ant mounds such as the one shown here can interfere with farm equipment. Originally from South America, fire ants have now invaded much of the southeastern United States.

Blood-sucking flies, screwworms, lice, fleas, and other insects annoy and cause disease in both humans and domestic animals. Mosquitoes are vectors of malaria, yellow fever, and filiariasis. Body lice may carry the typhus rickettsia, and houseflies sometimes transmit typhoid fever and dysentery. Tsetse flies transmit African sleeping sickness, and fleas may be vectors of bubonic plague.

Classes Chilopoda and Diplopoda: Centipedes and Millipedes

Members of class **Chilopoda** are called **centipedes** (hundred-legged), and members of class **Diplopoda** are known as **millipedes** (thousand-legged). These animals are all terrestrial and are typically found beneath stones or wood in the soil in both temperate and tropical regions.

Centipedes and millipedes are similar in having a head and an elongated trunk with many segments, each bearing legs (Figs. 25–29 and 25–30). The centipedes have one pair of legs on each segment behind the head. Most centipedes do not have enough legs to merit their name—the most common number being 30 or so—although in a few species, the number of legs is 100 or more. The legs of centipedes are long, enabling them to run rapidly. Centipedes are carnivorous and feed upon other animals, mostly insects, but the larger centipedes have been known to eat snakes, mice, and frogs. The prey is captured and killed with poison claws located just behind the head on the first trunk segment. A pair of poison glands at the base of the claws empty into ducts that open at the tip of the pointed, fanglike claw.

The distinguishing feature of the millipedes is the presence of **diplosegments**—doubled trunk segments—resulting from the fusion of two original segments. Each double segment has two pairs of legs and contains two pairs of ganglia. The most anterior three or four segments have only a single pair of legs. The body of the millipede is cylindrical, whereas the body of the centipede tends to be flattened. Diplopods are not as agile as chilopods, and most species can crawl only slowly over the ground, though they can powerfully force their way through earth and rotting wood. Millipedes are generally herbivorous and feed on both living and decomposing vegetation. In both chilopods and diplopods, eyes may be completely lacking, or the animal may have simple eyes called **ocelli.** A few species of centipedes have eyes that are similar to the compound eyes of insects.

(*a*)

(*b*)

Figure 25–29 Chilopods and diplopods. (*a*) Centipede. (*b*) Millipede. (Carolina Biological Supply Company.)

Figure 25–30 Centipedes and millipedes have uniramous appendages. (*a*) Centipede, a member of the class Chilopoda. Centipedes have one pair of appendages per segment. (*b*) Millipede, a member of the class Diplopoda. Millipedes have two pairs of appendages per segment.

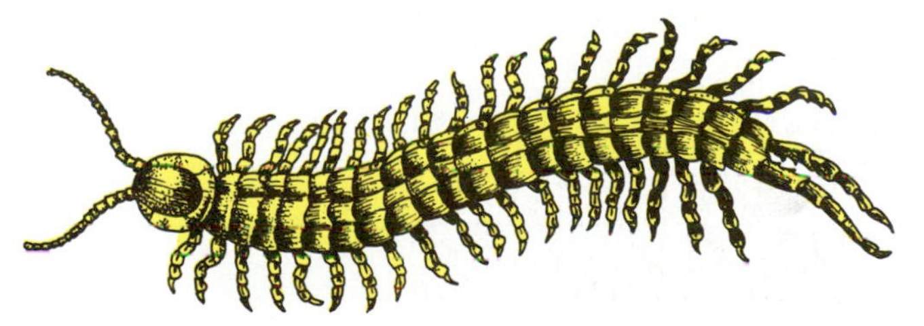

(*a*)

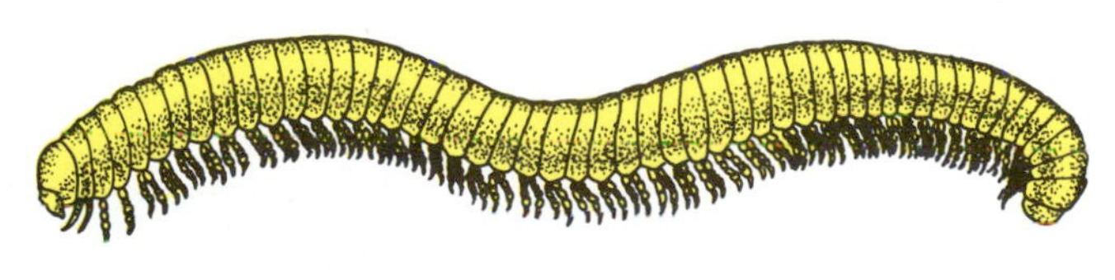

(*b*)

FOCUS ON

The Principal Orders of Insects

Order and Examples	*Mouth Parts*	*Wings*	*Other Characteristics*
*Ametabolous Insects**			
Thysanura Silverfish Bristletails	Chewing	None	Long antennae; 3 "tails" extend from posterior tip of abdomen; run fast; live in dead leaves and wood, or in houses, where they eat the starch in books and clothing
Collembola Springtails	Chewing	None	Abdominal structure for jumping; live in soil, dead leaves, rotting wood
Hemimetabolous Insects			
Odonata Dragonflies Damselflies	Chewing	2 pairs; long, narrow, membranous	Predators; large compound eyes; aquatic nymph
Ephemeroptera Mayflies	Chewing	2 pairs; membranous; forewings larger than hindwings	Small antennae; vestigial mouth parts in adult; 2 or 3 "tails" extending from tip of abdomen; nymph aquatic
Orthoptera Grasshoppers Crickets Roaches	Chewing	2 pairs or none; leathery forewings, membranous hindwings	Most herbivorous, some cause crop damage; praying mantis eats other insects.
Isoptera Termites	Chewing	2 pairs or none; wings shed by sexual forms after mating	Social insects, form large colonies; main diet wood; can be very destructive
Dermaptera Earwigs	Chewing	2 pairs; forewings very short; hindwings large, membranous	Forceps-like appendage on tip of abdomen; nocturnal
Anoplura Sucking lice	Piercing and sucking	None	Ectoparasistes of birds and mammals; head louse and crab louse are human parasites; vectors of typhus fever

*Insects may be divided into 3 groups based on their pattern of development. Ametabolous insects do not undergo metamorphosis (egg → immature form → adult). Hemimetabolous insects exhibit incomplete metamorphosis (egg → nymph (resembles adult in many ways but lacks functional wings and reproductive structures) → adult). Holometabolous insects undergo complete metamorphosis (egg → larva → pupa → adult). The wormlike larva, which is very different from the adult, hatches from the egg (Fig. 25–26). Typically, an insect spends most of its life as a larva. Eventually, the larva stops feeding, molts, and enters a pupal stage usually within a protective cocoon or underground burrow. The pupa does not feed and cannot defend itself. Its energy is spent remodeling its body form, so that when it emerges it is equipped with functional wings and reproductive organs.

Silverfish, *Lepisma saccharina*, adult and young. Silverfish are primitive ametablous insects. (Harry Rogers, Photo Researchers, Inc.)

Order and Examples	Mouth Parts	Wings	Other Characteristics
Hemiptera True bugs (chinch bugs, bedbugs)	Piercing and sucking	2 pairs; hindwings membranous	Mouth parts form beak; only order of insects properly called bugs
Homoptera Aphids Leaf hoppers Cicadas Scale insects	Piercing and sucking	Usually 2 pairs; membranous	Mouth parts form sucking beak; base of beak near thorax; some very destructive to plants
Holometabolous Insects			
Neuroptera Ant lions Dobson flies	Chewing	2 pairs; membranous	Larvae are predators.
Lepidoptera Moths Butterflies	Sucking	2 pairs, covered with overlapping scales	Larvae, called caterpillars, have chewing mouthparts; eat plants; adults that feed suck flower nectar.
Diptera (true flies) Houseflies Mosquitoes Gnats Fruitflies	Usually piercing and sucking	Forewings functional; hindwings small, knoblike halteres	Larvae are maggots or wigglers; may be damaging to domestic animals and food; adults often transmit disease.
Siphonaptera Fleas	Piercing and sucking	None	Legs adapted for jumping; lack compound eyes; ectoparasites on birds and mammals; vectors of bubonic plague and typhus
Coleoptera Beetles Weevils	Chewing	Usually 2 pairs; forewings modified as heavy, protective coverings	Largest order of insects (more than 300,000 species); majority herbivorous; some aquatic
Hymenoptera Ants Bees Wasps	Chewing but modified for lapping or sucking in some forms	2 pairs or none; transparent when present	Some are social insects; some sting.

An aphid, *Aphis nerii*, giving birth to a live young. Aphids are hemimetabolous insects. (Peter J. Bryant, UC–Irvine/BPS.)

An io caterpillar. This larva will undergo complete metamorphosis before becoming an adult (see Fig. 25–26). (Luci Giglio.)

SUMMARY

I. The coelomate protostomes include the mollusks, annelids, and arthropods, as well as some minor phyla.

II. The coelom provides space for many organs to develop and function, and permits the digestive tract to move independently of body movements; coelomic fluid helps transport materials and bathes cells that line the coelom.

III. In terrestrial animals, the body covering must prevent fluid loss, some sort of skeleton must be present to withstand the pull of gravity, and reproductive adaptations must be made such as internal fertilization, development within the mother's body, or shells that prevent desiccation of the developing embryo.

IV. Mollusks are soft-bodied animals usually covered by a shell; they possess a ventral foot for locomotion and a mantle that covers the visceral mass.
- *A.* Class Polyplacophora includes the sluggish marine chitons, which have segmented shells.
- *B.* Class Gastropoda, the largest and most successful group of mollusks, includes the snails, slugs, and whelks. In gastropods, the body is twisted, and the shell (when present) is coiled.
- *C.* Class Bivalvia includes the clams and oysters, animals enclosed by two shells, hinged dorsally.
- *D.* Class Cephalopoda includes the squids and octopods, which are active predatory animals; the foot is divided into tentacles that surround the mouth located in the large head.

V. Phylum Annelida, the segmented worms, includes many aquatic worms, earthworms, and leeches.
- *A.* Annelids have conspicuously long bodies that are segmented both internally and externally; their large compartmentalized coelom serves as a hydrostatic skeleton.
- *B.* Class Polychaeta consists of marine worms characterized by bristled parapodia, used for locomotion.
- *C.* Class Oligochaeta, which includes the earthworms, contains segmented worms characterized by a few setae per segment. The body is divided into more than 100 segments separated internally by septa.
- *D.* Class Hirudinea, which includes the leeches, is composed of animals that lack setae and appendages; they are equipped with suckers for sucking blood.

VI. Phylum Onychophora includes animals with both annelid and arthropod characteristics.

VII. Phylum Arthropoda is composed of animals with jointed appendages and an armor-like exoskeleton.
- *A.* The trilobites are extinct marine arthropods that were covered by a hard, segmented shell.
- *B.* Subphylum Chelicerata includes class Merostomata (the horseshoe crabs) and class Arachnida (spiders, mites, and their relatives).
  1. In the chelicerates the first pair of appendages are chelicerae, used to manipulate food. Chelicerates have no antennae and no mandibles.
  2. The arachnid body consists of a cephalothorax and abdomen; there are six pairs of jointed appendages, of which four pairs serve as legs.
- *C.* Subphylum Crustacea includes the lobsters, crabs, and barnacles. The body consists of cephalothorax and abdomen; often, five pairs of walking legs are present. Crustaceans have two pairs of antennae and mandibles for chewing.
- *D.* Subphylum Uniramia includes class Insecta, class Chilopoda, and class Diplopoda; members of this subphylum have unbranched appendages and a single pair of antennae.
  1. An insect is an articulated, tracheated hexapod; its body consists of head, thorax, and abdomen. Insects are the most ecologically successful group of animals.
  2. The centipedes have one pair of legs per body segment, whereas the millipedes have two pairs of legs per body segment.

POST-TEST

1. The ______________ of arthropods and mollusks provides protection and serves as a point of attachment for muscles.
2. The molluskan ______________ is a belt of teeth within the digestive system.
3. A hemocoel is characteristic of animals with an ______________ circulatory system.
4. The first larval stage of a mollusk is a free-swimming ______________ larva.
5. Segmental, serial repetition of structures within an animal is known as ______________.
6. In pulmonate snails, the highly vascularized mantle functions as a ______________.
7. Nudibranchs are mollusks that lack a ______________.
8. Setae are bristle-like structures that function in ______________.
9. In hermaphroditic animals, male and female reproductive systems are ______________.
10. The earthworm typhlosole functions to ______________.
11. The earthworm brain consists of a pair of ______________ ______________.
12. Onychophorans are considered a possible link between the annelids and the ______________.
13. Animals with an exoskeleton and paired, jointed appendages are ______________.
14. The mandibles of a crustacean are used for ______________.
15. Antennal (green) glands in the crustacean are ______________ organs.
16. The only sessile crustaceans are the ______________.
17. Chelipeds are large ______________ ______________.
18. Malpighian tubules are ______________ organs.

Select the appropriate animal in Column B for the description given in Column A:

Column A	*Column B*
19. An articulated, tracheated hexapod	a. Trilobite
20. An animal that uses book lungs	b. Insect
21. An animal with two pairs of legs per segment	c. Spider
22. A primitive arthropod with a hard, segmented shell	d. Millipede
	e. None of the above

Select the appropriate animal subphylum or class in Column B for the description given in Column A.

Column A	*Column B*
23. largest and most successful group of crustaceans	a. Polychaetes
24. an arthropod subphylum whose members have unbranched appendages	b. Chelicerates
25. arthropods with no antennae or mandibles	c. Gastropods
	d. Uniramia
	e. None of the above

REVIEW QUESTIONS

1. In what ways are mollusks and annelids alike? In what ways are they different?
2. Give two distinguishing characteristics for each:
   a. mollusks
   b. annelids
   c. arthropods
3. What are the advantages of each of the following?
   a. presence of a coelom
   b. the arthropod exoskeleton
   c. segmentation (metamerism)
4. Contrast the life style of a gastropod and a cephalopod. Identify adaptations possessed by each for its particular life style.
5. What is a trochophore larva? a veliger larva?
6. Describe some of the adaptations that have contributed to insect success.
7. Distinguish between insects and spiders.
8. What are the distinguishing features of each of the arthropod subphyla?
9. What are the distinguishing features of the principal arthropod classes?
10. Identify animals that belong to each class of arthropods.

26

The Animal Kingdom: The Deuterostomes

OUTLINE

I. Phylum Echinodermata
  A. Class Crinoidea: feather stars and sea lilies
  B. Class Asteroidea: sea stars
  C. Class Ophiuroidea: basket stars and brittle stars
  D. Class Echinoidea: sea urchins and sand dollars
  E. Class Holothuroidea: sea cucumbers
II. Phylum Chordata
  A. Subphylum Urochordata: tunicates
  B. Subphylum Cephalochordata: lancelets
  C. Subphylum Vertebrata
    1. Class Agnatha
    2. The earliest jawed fishes
    3. Class Chondrichthyes: sharks and rays
    4. Class Osteichthyes: bony fish
    5. Class Amphibia: frogs, toads, and salamanders
    6. Class Reptilia: turtles, lizards, snakes, alligators
      a. Modern reptiles
      b. Fossil reptiles
    7. Class Aves: birds
      a. Early birds
      b. Modern birds
    8. Class Mammalia
      a. Early mammals
      b. The monotremes
      c. Marsupials
      d. Placental mammals

LEARNING OBJECTIVES

After you have read this chapter you should be able to:

1. Discuss the relationship of the echinoderms and chordates, giving specific reasons for grouping them together.
2. Describe the distinguishing characteristics of the echinoderms.
3. Describe and give examples of each of the five main classes of echinoderms.
4. List the subphyla of phylum Chordata, and describe the characteristics that they have in common; describe the characteristics of subphylum Vertebrata.
5. List characteristics that would permit you to distinguish among the classes of vertebrates, and assign a given vertebrate to the correct class.
6. Trace the evolution of vertebrates according to current theory.
7. Identify adaptations that reptiles and other terrestrial vertebrates have made to life on land.
8. Contrast monotremes, marsupials, and placental mammals, and give examples of members that belong to each group.
9. Identify the major orders of placental mammals, and give an example of an animal that belongs to each.

It may seem strange to group the echinoderms—the sea stars and sand dollars—with the chordates, the phylum to which we belong. However, echinoderms and chordates are both deuterostomes, and they share similarities in their pattern of development. As discussed in Chapter 23, deuterostomes are characterized by radial and indeterminate cleavage, origin of the mesoderm as enterocoelous pouches, formation of the coelom from cavities within the mesodermal outpocketings, and formation of the mouth from a second opening that develops in the embryo. The characteristics of some of the higher animal phyla are summarized in Table 26–1.

Phylum Echinodermata

All of the members of phylum **Echinodermata** inhabit the sea. About 6000 living and 20,000 extinct species have been identified. The living species are divided into five principal classes (Fig. 26–1): Class **Crinoidea** includes the sea lilies and feather stars; class **Asteroidea,** the sea stars; class **Ophiuroidea,** the brittle stars; class **Echinoidea,** the sea urchins and sand dollars; and class **Holothuroidea,** the sea cucumbers.

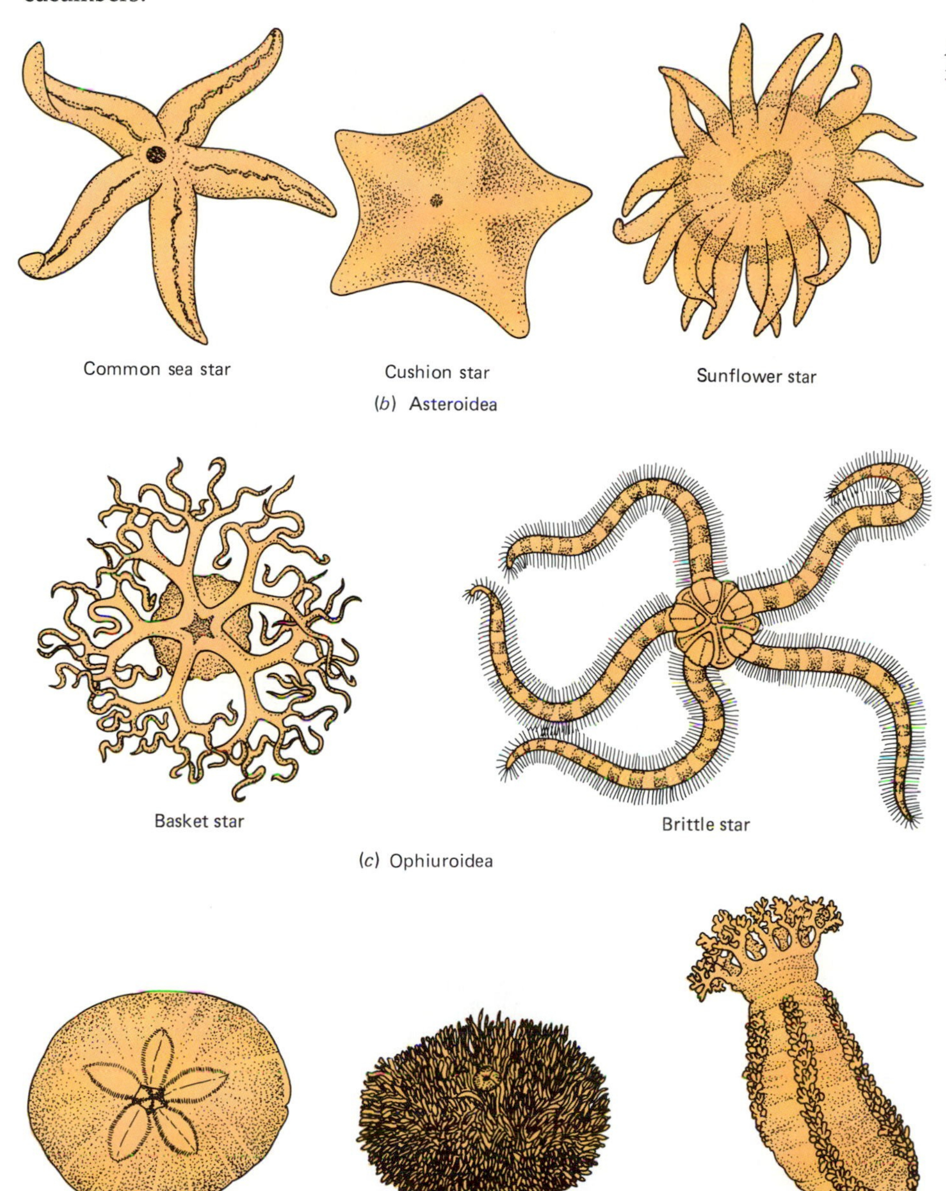

Sea lily

(a) Crinoidea

Figure 26–1 Some representatives of the five classes of phylum Echinodermata.

TABLE 26–1
*Comparison of Characteristics of Some Higher Animal Phyla**

Phylum	*Body Symmetry*	*Gas Exchange*	*Waste Disposal*	*Nervous System*	*Circulation*	*Reproduction*	*Other Characteristics*
Mollusca Clams Snails Squids	Bilateral	Gills and mantle	Kidneys	Three pairs of ganglia; simple sense organs	Open system	Sexual; sexes separate; fertilization in water	Soft-bodied; usually have shell and ventral foot for locomotion
Annelida (segmented worms) Earthworms Leeches Marine worms	Bilateral	Diffusion through moist skin; oxygen circulated by blood	Pair of metanephridia in each segment	Simple brain; ventral nerve cord; simple sense organs	Closed system	Sexual; hermaphroditic but cross-fertilize	Earthworms till soil.
Arthropoda (joint-footed animals) Crustaceans Insects Spiders	Bilateral	Tracheae in insects; gills in crustaceans; book lungs or tracheae in spider group	Malpighian tubules in insects; antennal (green) glands in crustaceans	Simple brain; ventral nerve cord; well-developed sense organs	Open system	Sexual; sexes separate	Hard exoskeleton; most diverse and numerous group of animals
Echinodermata (spiny-skinned animals) Sea stars Sea urchins Sand dollars	Embryo: bilateral; adult: modified radial	Skin gills	Diffusion	Nerve rings; no brain	Open system; reduced	Sexual; sexes almost always separate	Water vascular system; tube feet
Chordata Tunicates Lancelets Vertebrates	Bilateral	Gills or lungs	Kidneys and other organs	Dorsal nerve cord with brain at anterior end	Closed system; ventral heart	Sexual; sexes separate	(1) Notochord; (2) dorsal, tubular nerve cord; (3) pharyngeal gill slits

*Members of these phyla are at the organ system level of organization and have a complete digestive tract.

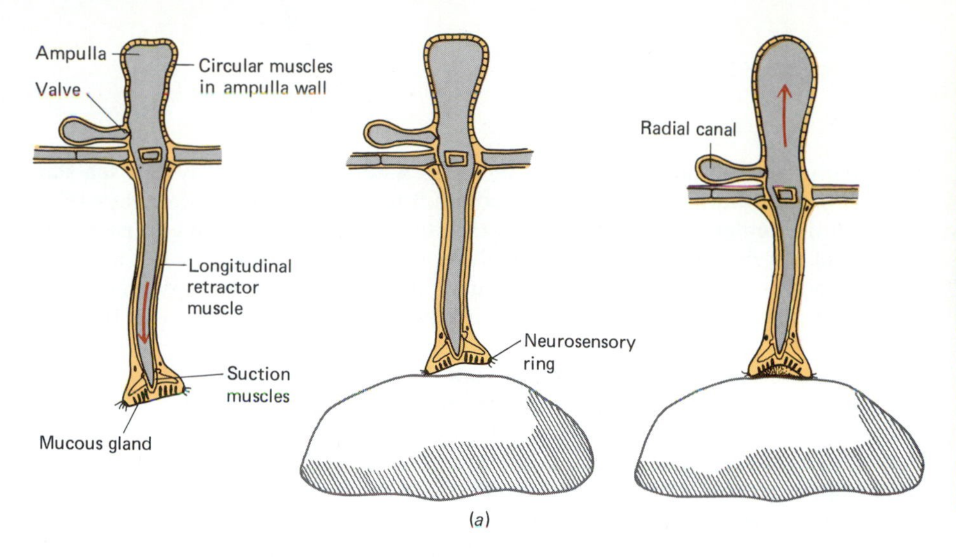

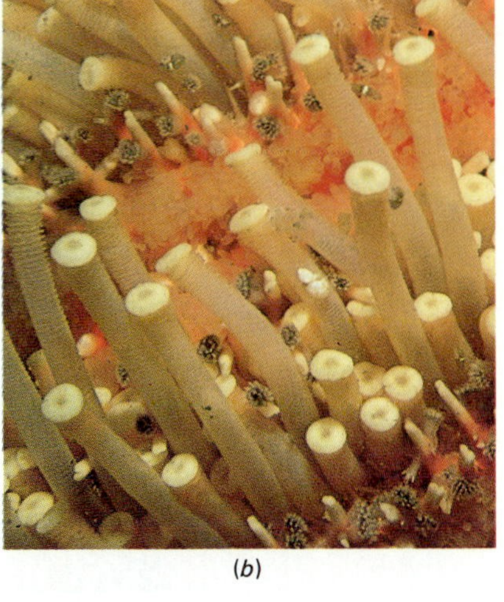

Figure 26–2 Tube feet in echinoderms. (*a*) Longitudinal section through the tube foot of a sea urchin. Sequence shows how the foot works to anchor the animal to a substrate: (*1*) When the valve to the radial canal is closed, there is a fixed volume of water in the tube foot. When the muscles in the ampulla contract, water is forced into the lower part of the foot, which elongates. (*2*) When the foot comes in contact with the substrate, the center of the sucker withdraws, creating a near-vacuum—or suction on the substrate. A secretion from the mucous glands aids in adhesion. (*3*) After adhesion of the sucker, the longitudinal muscles of the foot contract, shortening the foot and forcing fluid back into the ampulla. The valve to the radial canal remains closed as long as the hydrostatic pressure in the tube foot remains greater than the pressure within the canal. (*b*) Tube feet of a sea star. ((*b*), Charles Seaborn.)

The echinoderms are in many ways unique in the animal kingdom. Although their larvae have bilateral symmetry, the adults have **pentaradial symmetry.** This means that the body is arranged in five parts around a central disk where the mouth is located. A thin, ciliated epidermis covers the endoskeleton, which consists of small calcareous plates (composed of $CaCO_3$), typically bearing spines that project outward; the name Echinodermata, meaning spiny-skinned, reflects this trait.

A characteristic found only in echinoderms is the **water vascular system,** a network of canals through which sea water circulates. Branches of this system lead to numerous tiny **tube feet,** which extend when filled with fluid. The tube feet serve in locomotion and obtaining food, and in some forms, in gas exchange. The water vascular system serves as a hydrostatic skeleton for the tube feet. To extend a foot, a rounded muscular sac, or **ampulla,** at the upper end of the foot contracts, forcing water through a valve into the tube of the foot. At the bottom of the foot is a suction stucture that adheres to the substratum. The foot can be withdrawn by contraction of muscles in its walls, which forces water back into the ampulla (Fig. 26–2).

Echinoderms have a well-developed coelom, a complete digestive system, only a rudimentary circulatory system, and no specialized excretory structures. There are a variety of respiratory structures in the various classes, including the dermal gills in the sea stars and respiratory trees in sea cucumbers. The nervous system is simple, usually consisting of **nerve rings** about the mouth with radiating nerves.

CLASS CRINOIDEA: FEATHER STARS AND SEA LILIES

Class Crinoidea includes the feather stars and the sea lilies. The feather stars are free-swimming crinoids, whereas the sea lilies are sessile. Crinoids differ from other echinoderms in that the oral (mouth) surface is turned upward, and a number of arms extend upward (Fig. 26–3). The branched, feathery arms are used to trap food. The sessile sea lilies are attached to the ocean floor by a stalk. Although there are relatively few living species, a great many extinct crinoids are known.

Figure 26–3 A golden crinoid, a feather star, from the Pacific Ocean. (Robin Lewis, Coastal Creations.)

(a)

(b)

Figure 26–4 Sea stars. (*a*) *Linckia laevigate*. (*b*) *Choriaster granulatus*. (Charles Seaborn.)

CLASS ASTEROIDEA: SEA STARS

Sea stars, or starfish, are members of the class Asteroidea. The body of a sea star consists of a central disk from which radiate 5 to 20 or more **arms,** or **rays** (Fig. 26–4). In the center of the underside of the disk is the mouth. The endoskeleton consists of a series of calcareous plates that permit some movement in the arms. Around the base of the delicate skin gills used in gas exchange, are tiny pincer-like spines called **pedicellaria;** operated by muscles, these keep the surface of the animal free of debris (Fig. 26–5).

The undersurface of each arm is equipped with hundreds of pairs of tube feet. The cavities of the tube feet are all connected by radial canals in the arms; these in turn are connected by a circular canal in the central disk. The circular canal is connected by an axial stone canal to a button-shaped plate called the **madreporite** on the upper (aboral) surface of the central disk. As many as 250 tiny pores in the madreporite permit sea water to enter the water vascular system.

Most sea stars are carnivorous and feed upon crustaceans, mollusks, annelids, and even other echinoderms. Occasionally they catch and eat a small fish. To attack a clam or other shell fish, the sea star mounts it, assuming a humped position as it straddles the edge opposite the hinge. Then with its tube feet attached to the two shells, it begins to pull. The sea star uses many of its tube feet at a time, but can change and use new groups as active tube feet get tired. By applying a steady pull on both shells over a long period of time, the sea star succeeds in tiring the powerful muscles of the clam so that they are forced to relax, opening the shell.

To begin its meal, the sea star projects its stomach out through its mouth and into the soft body of its prey. Digestive enzymes are secreted into the clam so that it is partly digested while still in its own shell. The soft parts of the clam are digested to the consistency of a thick soup and pass into the body for further digestion by enzymes secreted from glands located in each arm. Its water vascular system does not enable the sea star to move rapidly, but since it usually preys upon slow-moving or stationary clams and oysters, the speed of attack is not as critical as in most other predators.

The blood circulatory system in sea stars is poorly developed and probably of little help in circulating materials. Instead, this function is assumed by the coelomic fluid, which fills the large coelom and bathes the internal tissues. Metabolic wastes pass to the outside by diffusion. The nervous system consists of a ring of nervous tissue encircling the mouth and a nerve cord extending from this into each arm; there is no aggregation of nerve cells that could be called a brain.

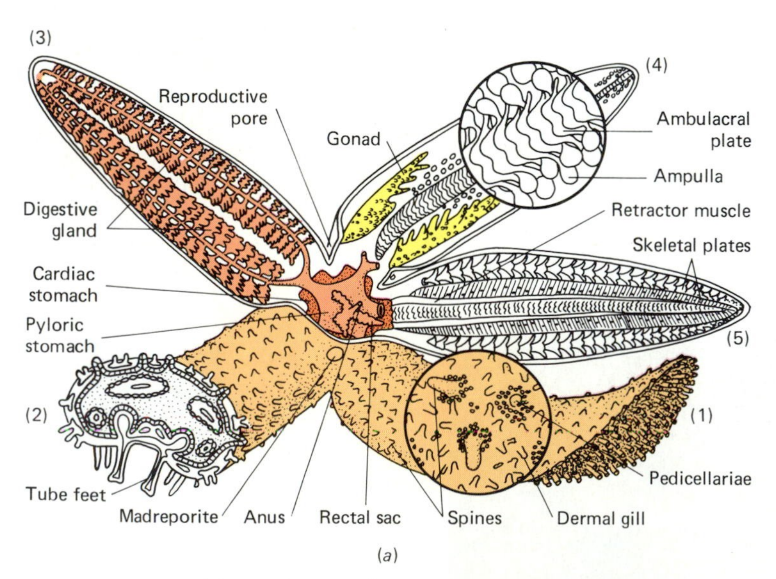

(a)

(b)

Figure 26–5 (*a*) The sea star *Asterias* viewed from above with the arms in various stages of dissection. (*1*) Upper surface with a magnified detail showing the features of the surface. The end is turned up to show the tube feet on the lower surface. (*2*) Arm shown in cross section. (*3*) Upper body wall of arm removed. (*4*) The upper body wall and digestive glands have been removed, and the ampullas and ambulacral plates are shown in magnified view. (*5*) All of the internal organs, except for the retractor muscles, have been removed to show the inner surface of the lower body wall. (*b*) Photomicrograph showing spines and pedicellaria on surface of an echinoderm.

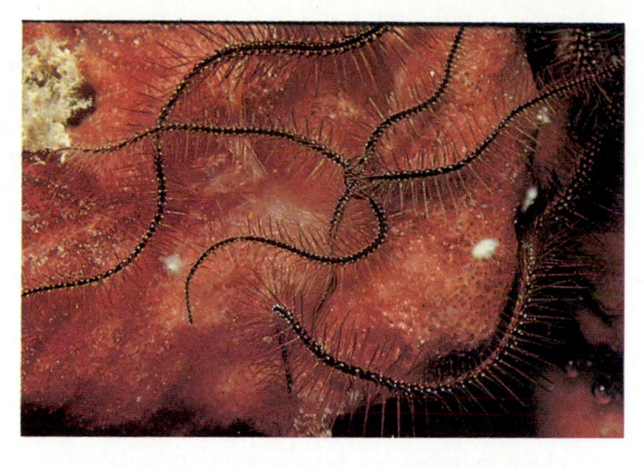

Figure 26–6 Brittle stars living on the surface of a sponge. (Charles Seaborn.)

CLASS OPHIUROIDEA: BASKET STARS AND BRITTLE STARS

Basket stars and brittle stars (serpent stars) are members of the class Ophiuroidea. They resemble asteroids in that their bodies also consist of a central disk with arms, but the arms are long and slender and more sharply set off from the central disk (Fig. 26–6). Ophiuroids can move more rapidly than asteroids, using their arms to perform rowing or even swimming movements. The tube feet are not used in locomotion and are thought to serve a sensory function, perhaps that of smell or taste. Tube feet are also used to collect and handle food.

CLASS ECHINOIDEA: SEA URCHINS AND SAND DOLLARS

The class Echinoidea includes the sea urchins and the sand dollars. Echinoids lack arms, and their skeletal plates are flattened and fused, forming a solid shell called a **test.** The sea urchin body is covered with spines (Fig. 26–7), which in some species can penetrate flesh and are difficult to remove. So threatening are these spines that swimmers on tropical beaches are often cautioned to put on their shoes before venturing off shore, where these animated pincushions lurk in abundance.

Sea urchins use their tube feet for moving and their movable spines for pushing themselves along. Most sea urchins graze, scraping the bottom with their teeth. They eat algae and tiny protists and plants as well as very small animals. Sand dollars have smaller and far fewer spines than sea urchins. Their flattened bodies are adapted for burrowing in the sand, and they feed on tiny organic particles in the sand.

Figure 26–7 A slate pencil urchin, *Heterocentrotus mammilatus,* photographed in a Hawaiian coral reef. (Charles Seaborn.)

CLASS HOLOTHUROIDEA: SEA CUCUMBERS

Sea cucumbers are members of the class Holothuroidea. Sea cucumbers are appropriately named, for many species are green and about the size of a small cucumber. The elongated body is a flexible, muscular sac (Fig. 26–8). The mouth is usually surrounded by a circle of tentacles that are actually modified tube feet. Another characteristic of these holothuroids is the reduction of the endoskeleton to microscopic plates. Like other echinoderms, sea cucumbers have a water vascular sys-

Figure 26–8 A sea cucumber, *Parastichopus californica,* from the waters of the Pacific Northwest. (Charles Seaborn.)

tem. Their blood circulatory system is more highly developed than that of other echinoderms; it functions to transport oxygen, and perhaps nutrients as well.

Sea cucumbers are sluggish animals that usually live on the bottom of the sea, sometimes burrowing in the mud. Some graze on the bottom with their tentacles; others stretch their branched tentacles out in the water and wait for dinner to float by. Algae and other tasty morsels are trapped in mucus along the tentacles; then, one at a time the tentacles are put into the mouth and the food particles removed and eaten.

An odd habit of some sea cucumbers is evisceration, in which the digestive tract, respiratory tree, and gonads are expelled from the body, usually when environmental conditions are unfavorable. When conditions improve the lost parts are regenerated. Even more curious, when some species are irritated or attacked, the rear end is directed toward the enemy, and red tubules are shot out of the anus! These unusual weapons are sticky (some release a toxic substance), and the offending animal may become hopelessly entangled.

Phylum Chordata

The phylum of animals to which humans belong, the phylum **Chordata,** is divided into three subphyla: subphylum **Urochordata,** which consists of marine animals called tunicates; subphylum **Cephalochordata,** which is comprised of the marine animals called lancelets; and subphylum **Vertebrata,** the animals with backbones.

Chordates are all coelomate animals with bilateral symmetry, three well-developed germ layers, a tube-within-a-tube body plan, and a segmented body. Most chordates have a tail that projects posterior to the anus, an endoskeleton, and a closed circulatory system with a ventral heart. In addition, there are three characteristics that distinguish them from all other groups (Fig. 26–9):

1. All chordates have a **notochord** during some time in their life cycle. The notochord is a dorsal longitudinal rod that is firm, yet flexible, and supports the body.
2. All chordates have a dorsal tubular **nerve cord.** The nerve cord differs from that of invertebrates not only in its position but in being single and hollow rather than double and solid.
3. All chordates have **pharyngeal gill slits** during some time in their life cycle. In the embryo a series of alternating **branchial arches** and **grooves** develop in the body wall in the pharyngeal region. **Pharyngeal pouches** extend laterally from the anterior portion of the digestive tract toward the grooves (see Chapter 44). In some chordates, these grooves perforate and become functional gill slits, but in terrestrial animals, they become modified to form entirely different structures more suitable for life on land. Biologists think that the earliest chordates were filter feeders, and that the arrangement of pharyngeal pouches and gill slits permitted them to take water in through the mouth, concentrate small particles of food in the gut, and let the water escape from the body through the gill slits.

No clear fossil record of the ancestors of the chordates exists, but they were probably small, soft-bodied animals. An impression of the lancelet-like animal, *Jamoytius,* has been found in rocks of the Silurian period (more than 400 million years old) in Scotland; some paleontologists have interpreted this as a primitive vertebrate. Current theories of the origin of chordates depend upon other types of evidence. The most widely held theory at present, based partly on their generally similar development of mesoderm and the coelom, and on the similarities of their larvae, is that echinoderms and chordates have a common evolutionary origin.

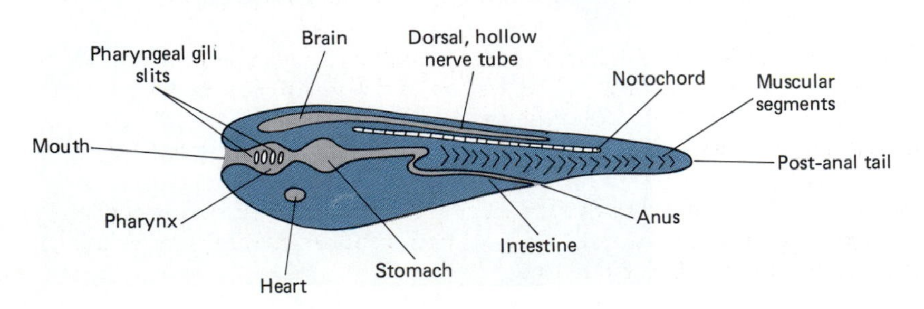

Figure 26–9 A generalized chordate illustrating the main chordate characteristics.

FOCUS ON

The Hemichordates

The **hemichordates** are a small group of sedentary, wormlike deuterostomes. These marine animals were once considered a subphylum of the chordates because they possess gill slits in the wall of the pharynx and a structure that was thought to be a notochord. However, it is now agreed that this structure is neither homologous with nor analogous to the chordate notochord, but simply an anterior projection of the digestive tract. The hemichordates are now assigned to their own phylum. Hemichordates have a ciliated larva that is very similar to the larva of some echinoderms.

The most common and best-known hemichordates are the acorn worms. Many species of these relatively large animals (up to 45 cm in length) burrow in mud along the shoreline or in shallow water. A distinguishing feature of the acorn worms is a short, conical proboscis that looks something like an acorn. In some species the proboscis is used for burrowing and movement within the burrow. Certain species construct mucus-lined burrows within the mud.

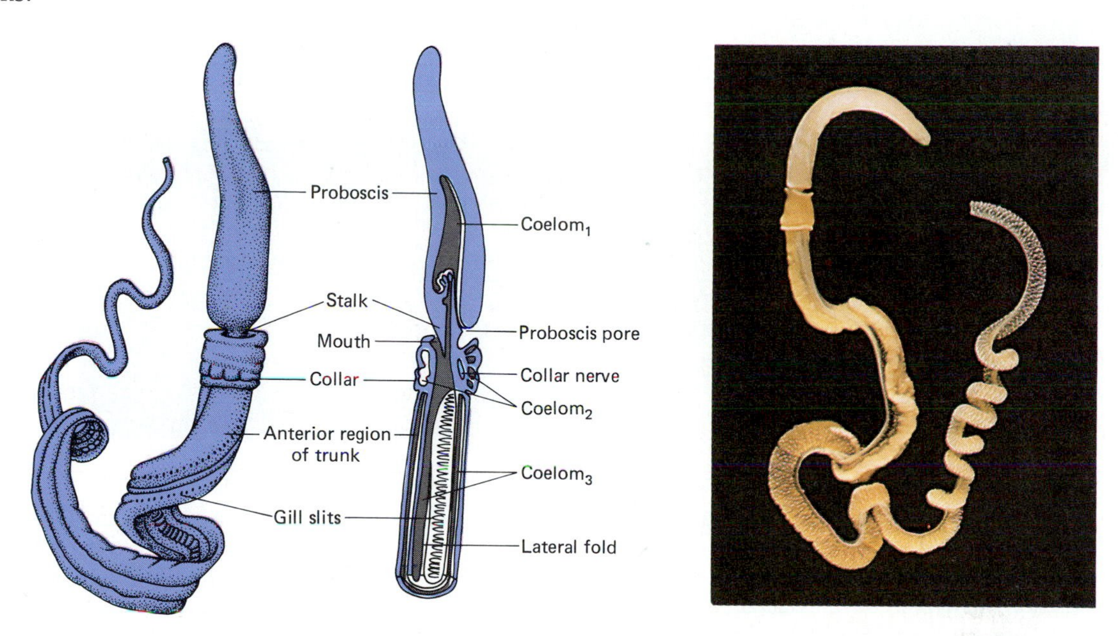

Saccoglossus kowalewskii, an acorn worm about 90 mm long. (C. R. Wyttenbach, University of Kansas/BPS.)

SUBPHYLUM UROCHORDATA: TUNICATES

The **tunicates,** which comprise the subphylum Urochordata, include the sea squirts and their relatives. Adult sea squirts are barrel-shaped, sessile, marine animals unlike other chordates; indeed, they are often mistaken for sponges or cnidarians (Fig. 26–10). Adult tunicates develop a tunic, quite thick in most species, that covers the entire animal. Curiously, the tunic is composed principally of cellulose. The tunic has two openings: the **incurrent siphon,** through which water and food enter, and the **excurrent siphon,** through which water, waste products, and gametes pass to the outside (Fig. 26–11).

Tunicates are filter feeders, removing plankton from the stream of water passing through the pharynx. Food particles are trapped in mucus secreted by cells of the **endostyle,** a groove that extends the length of the pharynx. Ciliated cells within the pharynx create a steady current of water that carries food particles down into the esophagus.

Some species of tunicates form large colonies in which members may share a common mouth and tunic. Colonial forms often reproduce asexually by budding. Sexual forms are usually hermaphroditic.

Figure 26–10 Clear tunicates, *Clavenina picta,* photographed in the Virgin Islands. (Robin Lewis, Coastal Creations.)

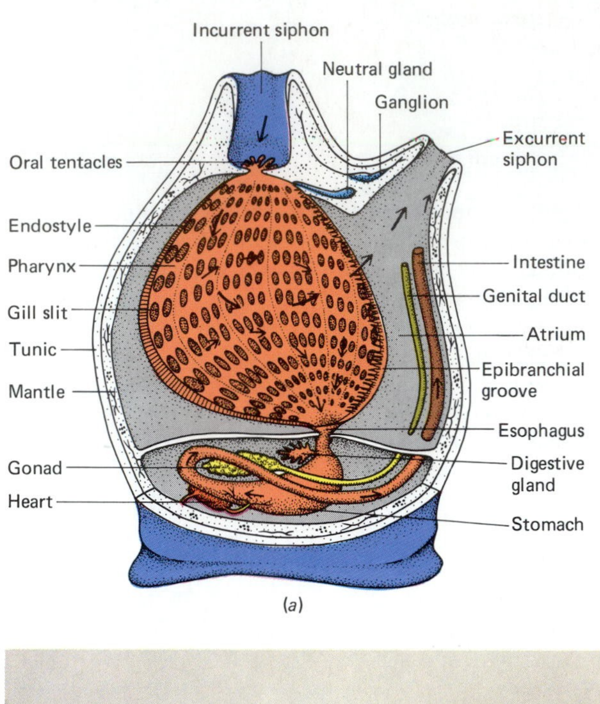

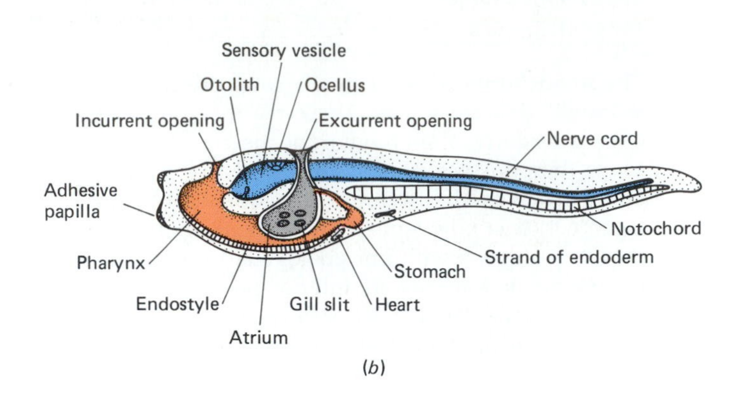

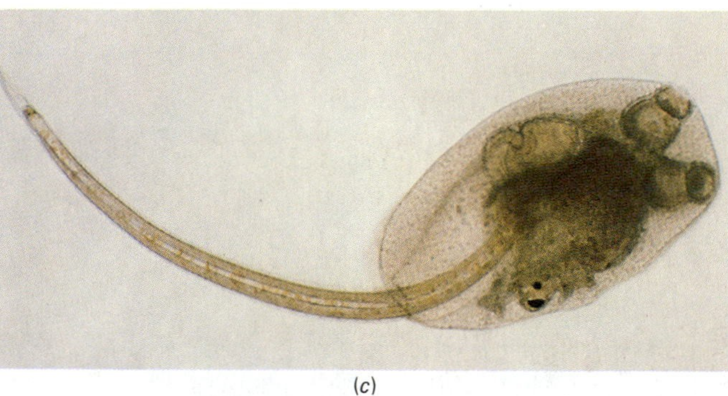

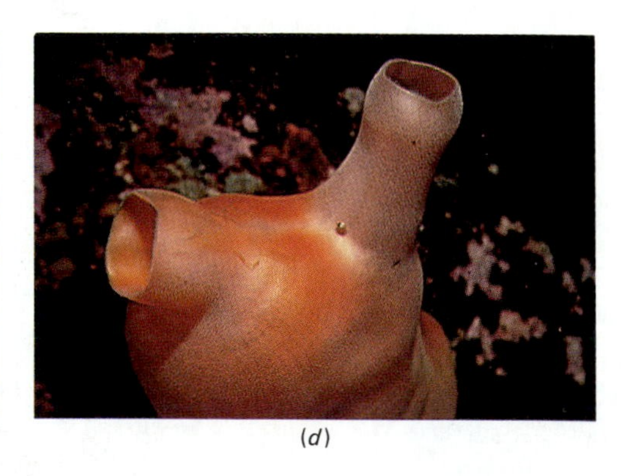

Figure 26–11 Structure of a tunicate. Lateral views of (*a*) an adult and (*b*) a larval tunicate, showing some of the internal structures. The large arrows in (*a*) represent the flow of water, and the small arrows represent the path of food. The stomach, intestine, and other visceral organs are embedded in the mantle. (*c*) Swimming tadpole stage of a tunicate, *Distaplia occidentalis*. (*d*) Incurrent (top) and excurrent (side) siphons of a sea peach tunicate, *Halocynthia auratium*. ((*c*), Richard A. Cloney, University of Washington; (*d*), Charles Seaborn.)

Larval tunicates are typically chordate, superficially resembling a frog tadpole. Their expanded bodies have a pharynx with gill slits, and the long muscular tail contains a notochord and dorsal nerve cord. Eventually the larva becomes attached to the sea bottom and loses its tail, notochord, and much of its nervous system. In the adult, only the gill slits suggest that the tunicate is a chordate.

SUBPHYLUM CEPHALOCHORDATA: LANCELETS

Subphylum Cephalochordata includes the **lancelets,** which are small, translucent, fishlike, segmented animals, 5 to 10 cm long and pointed at both ends. They are widely distributed in shallow seas, either swimming freely or burrowing in the sand near the low-tide line. All three chordate characteristics are highly developed in lancelets. The notochord extends from the tip of the head (hence the name Cephalochordata) to the tip of the tail; many pairs of gill slits are evident in the large pharyngeal region; and a hollow, dorsal nerve cord extends the entire length of the animal (Fig. 26–12).

The straight, rather simple digestive tract begins with the anterior **oral hood,** a vestibule surrounded by delicate tentacles bearing sensory cells. Under the hood and posterior to it, the mouth opening is surrounded by a membrane, or velum, equipped with a ring of sensory **cirri.** Like the tunicates, lancelets feed by drawing a current of water into the mouth by the beating of cilia, and straining out the microscopic plants, algae, and animals; food particles are trapped in the pharynx by mucus secreted by the endostyle and are then carried back to the intestine.

Water passes through the gill slits into the **atrium,** a chamber with a ventral opening, the **atriopore,** just anterior to the anus. Metabolic wastes are excreted by segmentally arranged, ciliated protonephridia that open into the atrium. In contrast to other invertebrates, in cephalochordates the blood flows anteriorly in the ventral vessel and posteriorly in the dorsal vessel.

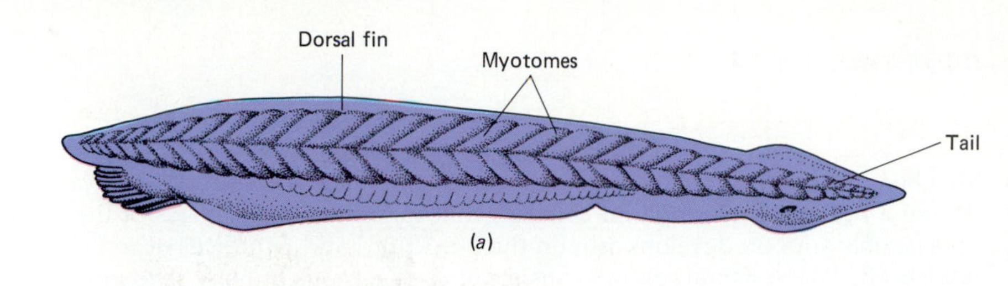

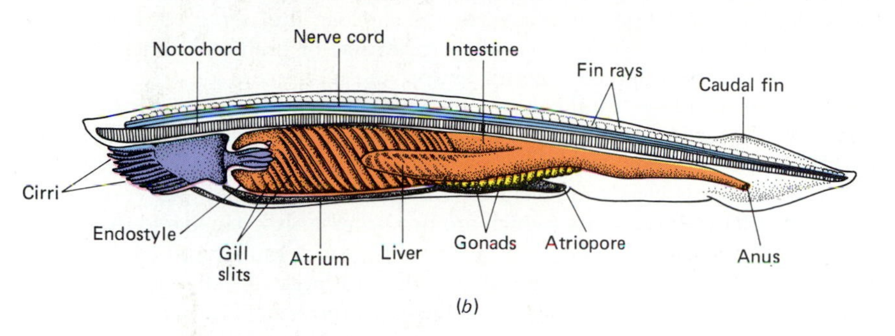

Figure 26–12 Amphioxus, a member of the subphylum Cephalochordata. (*a*) External view. (*b*) Longitudinal section.

Although superficially similar to fishes, lancelets are much more primitive, for they lack paired fins, jaws, sense organs, and a well-defined brain. It is generally believed that the cephalochordate *Branchiostoma* (commonly known as amphioxus) may be similar to the primitive ancestor from which the vertebrates evolved (Fig. 26–13). *Branchiostoma* admirably exhibits the basic chordate characteristics and for this reason is often the first animal studied in comparative anatomy courses.

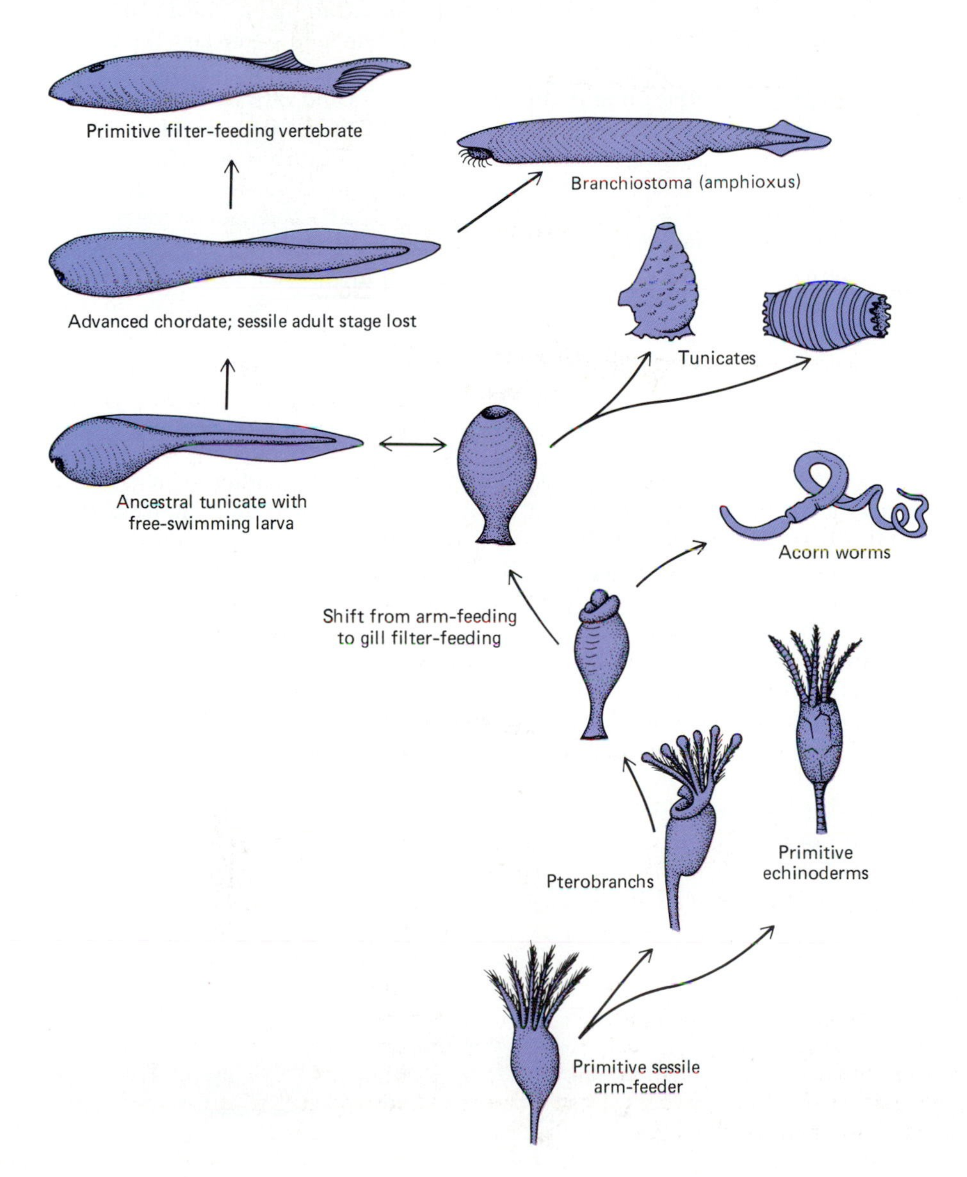

Figure 26–13 A diagram illustrating one hypothesis of the evolution of the chordates. (After Romer.)

SUBPHYLUM VERTEBRATA

Members of the subphylum Vertebrata are distinguished from other chordates in having a backbone, or **vertebral column,** that forms the skeletal axis of the body. This flexible support develops around the notochord and reinforces or replaces the notochord. The vertebral column consists of cartilaginous or bony segments called **vertebrae.** Dorsal projections of the vertebrae enclose the nerve cord along its length. Anterior to the vertebral column, a **cranium,** or braincase, encloses and protects the brain, the enlarged anterior end of the nerve cord.

The cranium and vertebral column are part of the endoskeleton, which—in contrast to the nonliving exoskeleton of invertebrates—is a living tissue that can grow with the animal. Another vertebrate characteristic is *pronounced* cephalization, concentration of nerve cells and sense organs in a definite head. All vertebrates share certain other characteristics (which are not necessarily exclusive to these animals): a closed circulatory system with a two-, three-, or four-chambered ventral heart; paired kidneys; a complete digestive tract and large digestive glands (liver and pancreas); muscles attached to the skeleton for movement; a brain that is regionally differentiated for specialized functions; 10 or 12 pairs of cranial nerves that emerge from the brain; an autonomic division of the nervous system that regulates involuntary functions of internal organs; well-developed organs of special sense (eyes; ears, which may serve as organs of equilibrium; organs of smell and taste); two pairs of appendages; and separate sexes.

The vertebrates are less diverse and much less numerous and abundant than the insects but rival them in their adaptation to a variety of modes of existence and excel them in the ability to receive stimuli and respond to them. The 43,000 or so living species are divided into superclass Pisces, the fish, and superclass Tetrapoda, the land vertebrates. Superclass **Pisces** includes class **Agnatha,** the jawless fish such as lamprey eels; class **Chondrichthyes,** the sharks and rays with cartilaginous skeletons; and class **Osteichthyes,** the bony fish. All fishes have highly vascular **gills** with a large surface for the transfer of oxygen and carbon dioxide.

The four-legged land vertebrates are placed together in the superclass **Tetrapoda,** which includes class **Amphibia,** frogs, toads, and salamanders; class **Reptilia,** lizards, snakes, turtles, and alligators; class **Aves,** birds; and class **Mammalia,** the mammals. Not all the tetrapods have four legs (e.g., the snakes), but they all evolved from four-legged ancestors. Not all the tetrapods now live on land (e.g., sea turtles, penguins, whales, seals); however, all the aquatic forms evolved from terrestrial ancestors.

Class Agnatha

The jawless fish comprise the class Agnatha (a = without; gnathos = jaw), which includes the extinct **ostracoderms** and the living lamprey eels and hagfishes (Fig. 26–14). Ostracoderms, the earliest known fossil chordates, have been found in rocks of the Ordovician, Silurian, and Devonian periods. These jawless fish were small, armored, bottom-dwelling freshwater filter feeders. The head was covered with thick bony plates, and the trunk and tail were covered with thick scales. Ostracoderms had median fins; some species had paired pectoral fins.

The living relatives of the ostracoderms are the lampreys and hagfishes. These animals have cylindrical bodies up to a meter long supported by a cartilaginous skeleton. Their smooth skin lacks scales, and they have neither jaws nor paired fins (Fig. 26–15). Most hagfish eat worms and other invertebrates, which they burrow for, or prey on dead and disabled fish.

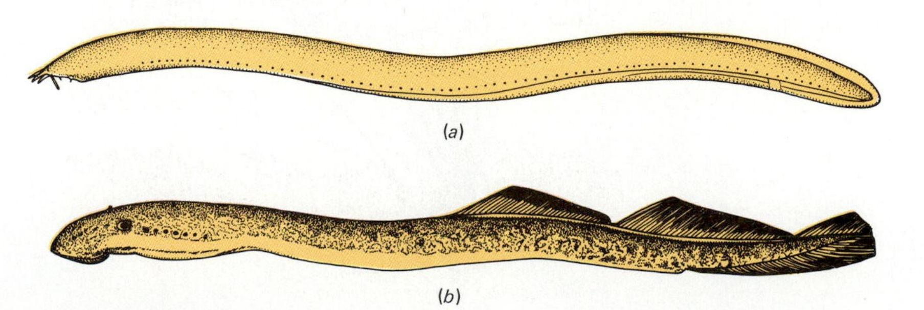

Figure 26–14 Two types of cyclostomes. (*a*) A hagfish. (*b*) A lamprey. Note the absence of jaws and paired fins.

Many species of adult lampreys (e.g., the sea lamprey *Petromyzon*) are parasites on other fish; they are the only parasitic vertebrates. Adult parasitic lampreys have a circular sucking disk around the mouth, which is located on the ventral side of the anterior end of the body. Using this disk to attach to a fish, the lamprey bores through the skin of its host with its horny teeth on the disk and tongue; it then injects an anticoagulant into its host and sucks its blood and soft tissues.

Adult lampreys leave the ocean or lakes and swim upstream to spawn. They build a nest, a shallow depression in the gravelly bed of the stream, into which eggs and sperm are shed; after spawning, they die. The fertilized eggs develop into larvae called **ammocoetes,** which resemble lancelets. They drift downstream to a pool and live as filter-feeders in burrows in the muddy bottom for up to seven years. Then they undergo a metamorphosis, become adult lampreys, and migrate back to the ocean or lake.

(*a*)

(*b*)

Figure 26–15 Lampreys. (*a*) A West Coast sea lamprey holds onto a rock with its mouth. (*b*) Suction-cup-like mouth of adult lamprey, *Estosphenus japonicus*. ((*a*), Carolina Biological Supply Company; (*b*), courtesy of Dr. Kiyoko Uehara, *Cell and Tissue Research*.)

The Earliest Jawed Fishes

Fossil evidence suggests that, during the Silurian and Devonian periods, some descendants of the ostracoderms evolved jaws and paired appendages and changed from filter-feeding bottom-dwellers to active predators. The earliest jawed fishes are placed in the subclass **Acanthodii.** Another group of primitive jawed fishes living in the Paleozoic era was the **placoderms,** small, armored, freshwater fish with a variable number (as many as seven) of paired fins (Fig. 26–16). The placoderms, all of which are now extinct, were probably the ancestors of both cartilaginous and bony fishes. One of the best-known placoderms is *Dinichthys*, a monster that attained a length of 9 meters; the head and anterior part of the trunk had a bony armor, but the remainder of the body was naked. The evolution of jaws from a portion of the gill arch skeleton enabled the placoderms and their descendants to become adapted to new modes of life. The success of the jawed vertebrates undoubtedly contributed to the extinction of the ostracoderms.

Class Chondrichthyes: Sharks and Rays

The ostracoderms and placoderms were primarily freshwater fishes; only a few ventured into the oceans. Members of class Chondrichthyes, the cartilaginous fishes, evolved as successful marine forms in the Devonian period. Most species have remained as ocean dwellers; only a few have secondarily returned to a freshwater habitat.

The chondrichthyes—sharks, rays, and skates (Fig. 26–17)—are characterized by a skeleton composed of cartilage, which may be strengthened by the deposit of calcium salts. A skeleton of cartilage represents the retention by the adult of the embryonic skeletal tissue; at the present time, such a skeleton is not regarded as a primitive adult condition because the adult ancestors had bony skeletons.

The dogfish shark is commonly used in biology classes to demonstrate the basic vertebrate characteristics in a simple, uncomplicated form. All the chrondrich-

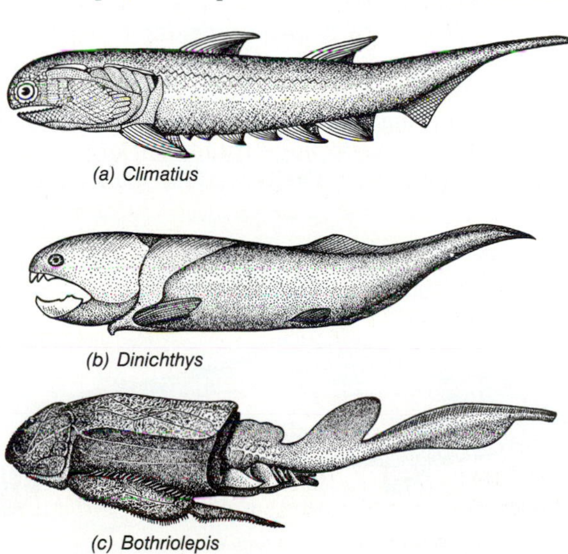

Figure 26–16 Acanthodians and placoderms from the Devonian period. (*a*) *Climatius*, a spiny-skinned shark with large fin spines and five pairs of accessory fins between the pectoral and pelvic pairs. (*b*) *Dinichthys*, a giant arthrodire that grew to a length of 9 meters. Its head and thorax were covered by bony armor, but the rest of the body and tail were naked. (*c*) *Bothriolepis*, a placoderm with a single pair of jointed flippers projecting from the body. (From A. S. Romer and T. S. Parsons.)

(a)

(b)

Figure 26–17 Some members of class Chondrichthyes. (*a*) Dorsal view of a skate, *Raja binoculara*. (*b*) Ratfish, *Hydrolagus colliei*. (Charles Seaborn.)

thyes have paired jaws and two pairs of fins. The skin contains **placoid scales** composed of an outer enamel layer and an inner dentine layer (Fig. 26–18). The lining of the mouth contains larger but essentially similar scales that serve as teeth. The teeth of the higher vertebrates are homologous with these shark scales. Shark teeth are embedded in the flesh and not attached to the jawbones; new teeth develop continuously in rows behind the functional teeth and migrate forward to replace any that are lost.

Cartilaginous fishes have five to seven pairs of gills. A current of water enters the mouth and passes over the gills and out the gill slits, constantly providing the fish with a fresh supply of dissolved oxygen. However, because it has no mechanism for propelling water over the gills, the shark must depend on its forward motion for gas exchange. In some species, the shark will suffocate if it stops moving.

Cartilaginous fishes tend to sink unless they are actively swimming, because their bodies are denser than water. (As we will see shortly, the bony fishes are able to compensate for this body density by means of a specialized structure called a swim bladder.) The large pectoral fins give a lift component to their forward motion; the sculling action of the tail provides additional lift.

The digestive tract of the shark consists of the mouth cavity, a long pharynx leading to a J-shaped stomach, a short, straight intestine, and a cloaca (Fig. 26–19). The **cloaca** receives digestive wastes as well as urine from the urinary system and, in the female, gametes from the male reproductive system. The cloaca, which is characteristic of many vertebrates, opens on the underside of the body; the opening is called a **vent.** The liver and pancreas discharge their digestive juices into the intestine. The intestine has a spiral fold called the **spiral valve** that slows the passage of food, allowing more time for digestion, and also increases the surface area for absorption.

The shark has a complex brain differentiated both structurally and functionally. The spinal cord is protected by vertebrae. Well-developed sense organs enable the shark to locate prey by smell and by vibrations in the water, as well as by sight. The **lateral line organ,** found in all fish, is a groove along each side of the body with many tiny openings to the outside. Sensory cells within the canals respond to movement of the water. On its head the shark has **electroreceptors;** these are found

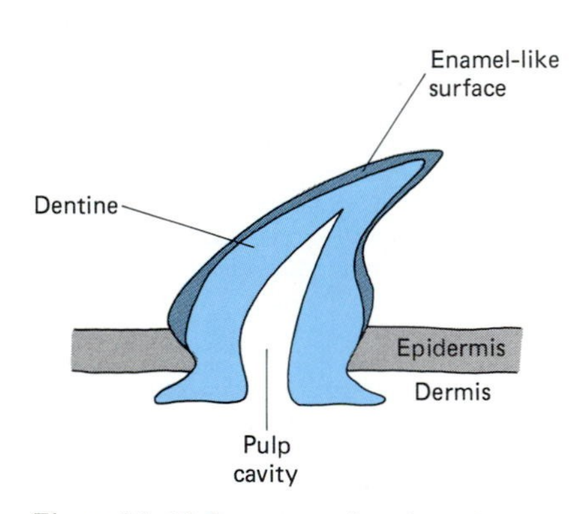

Figure 26–18 Structure of a placoid scale.

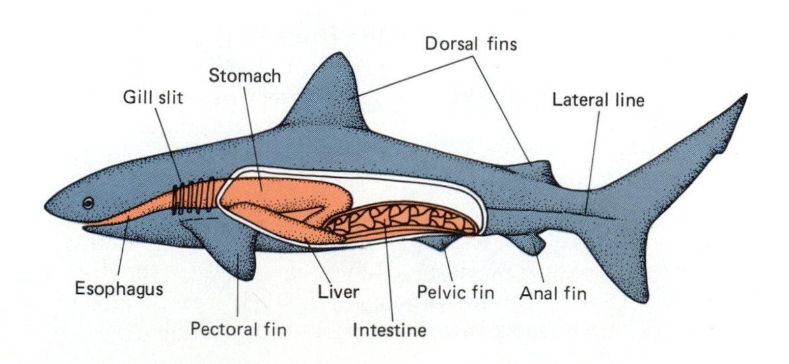

Figure 26–19 Internal anatomy of the shark.

in small chambers that connect by sensory canals to pores on the surface of the head. Using these receptors, a shark can sense electrical potentials generated by animals before it can detect them by sight or smell.

The sexes are separate, and fertilization is internal. In the mature male each pelvic fin has a slender, grooved section known as a **clasper,** which is used to open the female's vent and transfer sperm into her cloaca. The eggs are fertilized in the upper part of the female's oviducts. Part of the oviduct is modified as a shell gland, which secretes a protective coat around the egg. Skates and some species of sharks are **oviparous**—that is, they lay eggs. Many species of shark, however, are **ovoviviparous:** The eggs develop within a modified portion of the oviduct called the uterus, but most species depend on stored yolk for their nourishment. The young are born alive after they hatch from the eggs. A few species of sharks are truly **viviparous:** Not only do the embryos develop within the uterus, but they receive their nourishment from their mother's blood. An intimate relationship is established between the yolk sac surrounding each embryo and the blood vessels in the lining of the uterus.

Most sharks are streamlined predators that swim actively and catch other fish as well as crustaceans and some mollusks. The largest sharks, like the largest whales, dine on plankton. They gulp water through the mouth; as it passes through the pharynx and out the gill slits, food particles are trapped in a branchial sieve. The whale shark, which may reach a length of more than 12 meters, is the largest fish known.

Although there are only about 30 unprovoked shark attacks upon humans per year, most of us regard sharks as monstrous enemies. Sharks are attracted to blood so that a wounded swimmer or a skin diver towing speared fish is a target. Most sharks, however, do not go out of their way to attack humans.

Most rays and skates are sluggish, flattened creatures, living partly buried in the sand and feeding on mussels and clams. The undulations of its enormous pectoral fins propel the ray or skate along the bottom. The stingray has a whiplike tail with a barbed spine at the tip, which can inflict a painful wound. The electric ray has electric organs on either side of the head; these modified muscles can discharge enough electricity to stun fairly large fishes, as well as human swimmers.

If members of class Chondrichthyes occasionally inflict injury on humans, relations are strained in both directions. Sharks and rays are eaten by humans in many countries including the United States. Shark skin is tanned and used in making shoes and handbags, and shark liver oil is an important source of vitamin A.

Class Osteichthyes: Bony Fish

The class **Osteichthyes** includes some 20,000 species of bony fish, both freshwater and saltwater fishes of many shapes and colors (Fig. 26–20). Bony fish range in size from the Philippine goby, which is only about 10 mm (0.4 in) long, to the ocean sunfish (or mola), which may reach 900 kg (2000 lb). Paleontologists believe that bony fishes and chondrichthytes evolved from placoderm ancestors at about the same time (during the Devonian period), but independently; in any case, the bony fish did not evolve from the cartilaginous fish. Ancestors of the bony fish are thought to have been freshwater fish that had lungs. The bony fishes subsequently entered the oceans and became dominant there, too.

There were frequent seasonal droughts in the Devonian period during which ponds became stagnant or dried up completely. Fishes with some adaptation for breathing air, such as lungs, had a tremendous advantage for survival under those conditions. By the middle of the Devonian period, the bony fishes had diverged into two major groups: the **ray-finned fishes** (actinopterygians), which gave rise to most modern osteichthytes, and another group characterized by fleshy fins (the sarcopterygians), which includes the **lungfishes** and **lobe-finned fishes.** Both the lungfishes and lobe-finned fishes originally possessed lungs and had an armor of bony scales. Three genera of lungfishes have survived to the present day in the rivers of tropical Africa, Australia, and South America. The lobe-finned fishes, generally credited as being the ancestors of the land vertebrates, were almost extinct by the end of the Paleozoic era. However, a few specimens of marine lobe-finned fishes, **coelacanths,** (Fig. 26–21), have survived; some have been caught in the deep waters off the east coast of Africa near the Comoro Islands. Nearly 2 meters long, these giant living fossils have neither lungs nor functional swim bladders.

Figure 26–20 Fish and some of their adaptations. (*a*) Porcupine fish, *Diodon hysterix,* which has sharp barbs on its skin and can inflate itself with water, as shown in (*b*), making it look large and unappetizing to its predators. (*c*) Clownfish, *Amphiprion,* living in a sea anemone. The anemone will not sting a clownfish with a healthy mucous covering. The clownfish receives protection from predators, and it was once thought that the anemone benefited from the association by getting food brought to it from the clownfish. More often, however, the clownfish steals food captured in the tentacles of the anemone. In aquariums, anemones do better without resident clownfish. (*d*) This member of the scorpionfish family is adapted for bottom living and at first glance looks like debris. The feathery, wormlike tissues near its mouth, however, are able to lure small fishes. This slow-moving fish can be deadly to both prey and predators. The first several rays of its dorsal fin are pointed and hollow; when the fish is threatened, the spines stiffen and are filled with a poison similar in action to cobra venom. (Charles Seaborn.)

The ray-finned fishes are thought to have evolved slowly during the late Paleozoic era and then to have given rise in the Mesozoic era to the modern bony fishes, the **teleosts.** In them the lungs became modified as **swim bladders,** hydrostatic organs that may also store oxygen (Fig. 26–22). By secreting gases into the bladder or absorbing gases from it, the fish can change the density of its body and so hover at a given depth of water. Swim bladders are discussed further in Chapter 37.

In addition to having a swim bladder, most bony fish are characterized by a bony skeleton with many vertebrae (Fig. 26–23). Bits of the notochord may persist. The body is covered with overlapping bony dermal scales. Most species have both median and paired fins, with fin rays of cartilage or bone. A lateral protective flap of the body wall, the **operculum,** extends posteriorly from the head and covers the gills.

Figure 26–21 Ancestors of this coelacanth are thought to have given rise to the amphibians. The paired fins show the basic plan of a jointed series of bones that could evolve into the limbs of a terrestrial vertebrate. This coelacanth, *Latimeria chalumnae,* was photographed near Comoro Island, in deep waters off the coast of Southern Africa. (Peter Scoones, Seaphot.)

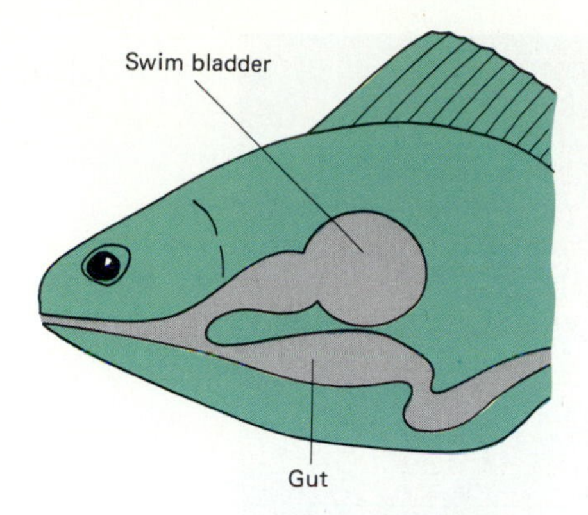

Figure 26–22 The swim bladder is a hydrostatic organ that enables the fish to change the density of its body and so to hover at a given depth.

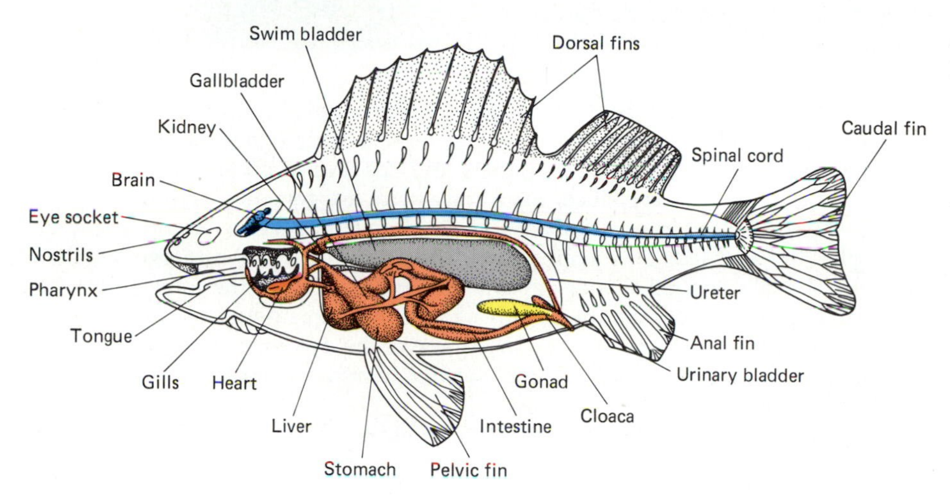

Figure 26–23 Internal anatomy of the perch, a bony fish.

Unlike the sharks, bony fish generally fertilize their eggs externally, and they are oviparous. Fish lay impressive numbers of eggs (see page 11). The ocean sunfish is said to lay over 300 million! Most of the eggs and young, of course, become food for other animals. Many species of fish build nests for their eggs and even watch over them.

Class Amphibia: Frogs, Toads, and Salamanders

The first successful tetrapods, or land vertebrates, were the **labyrinthodonts** (Fig. 26–24), clumsy, salamander-like animals with short necks and heavy muscular tails. These ancient members of class Amphibia closely resembled their ancestors, the lobe-finned fishes, but had evolved limbs strong enough to support the weight of the body on land. These earliest arms and legs were five-fingered, a pattern that has generally been kept by the higher vertebrates. There were many different kinds of ancient amphibians, all of which became extinct in the first part of the Mesozoic era. The labyrinthodonts ranged in size from small, salamander-sized animals up to ones as large as crocodiles. They gave rise to other primitive amphibians, to modern frogs and salamanders, and to the earliest reptiles, the **cotylosaurs,** or stem reptiles.

Modern amphibians are classified in three orders: Order **Urodela** includes the salamanders, mudpuppies, and newts—forms with long tails. Order **Anura** is

Figure 26–24 Primitive amphibians. (*a*) *Diplovertebron,* a primitive Paleozoic amphibian (labyrinthodont). (*b*) *Ophiacodon,* an early Permian pelycosaur. Although the pelycosaurs were primitive reptiles, they had characteristics indicating that they represented a first stage in the evolution of the mammals. (From A. S. Romer and T. S. Parsons.)

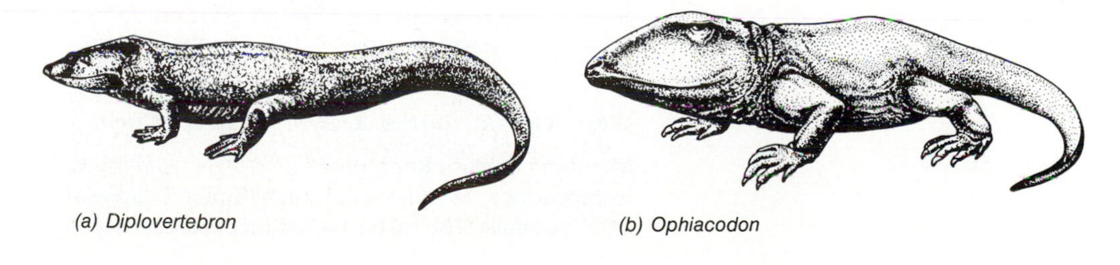

(a)

(b)

Figure 26–25 Modern amphibians. (*a*) Eastern spadefoot toad (*Scaphiopus holbrooki*). (*b*) The red salamander, *Pseudotriton ruber*, belongs to the family Plethodontidae. The plethodonts usually spend their entire lives in fairly moist environments. They have lost their lungs, and now rely entirely on their moist skin and the membranes that line their mouth and pharynx as organs for gas exchange. The red salamander is a species commonly found in the eastern United States. ((*a*), James H. Carmichael, Coastal Creations, 1983; (*b*), Carolina Biological Supply Company.)

made up of the tail-less frogs and toads, with legs adapted for hopping; and order **Apoda,** the wormlike, legless caecilians (Fig. 26–25). Although some adult amphibians are quite successful as land animals and can live in quite dry places, most return to the water to reproduce. Eggs and sperm are generally laid in the water.

The embryos of frogs and toads develop into larvae called tadpoles with tails and gills. Tadpoles feed on aquatic plants, and after a time they undergo metamorphosis. The gills and gill slits disappear, the tail is resorbed, the forelegs emerge, the digestive tract shortens, the mouth widens, a tongue develops, the tympanic membrane and eyelids appear, and the shape of the eye lens changes. Many biochemical changes occur to provide for the change from a completely aquatic life to an amphibious one.

As in insects, metamorphosis in amphibians is under hormonal control. Amphibian metamorphosis is regulated by thyroid hormones secreted by the thyroid gland. (The process can be prevented by removing the thyroid gland or the pituitary gland, which secretes a thyroid-stimulating hormone.) Amphibians undergo a single change from larva to adult, in contrast to the four or more molts involved in the development of arthropods to the adult form. Several species of salamanders, such as the mud puppy *Necturus,* do not undergo metamorphosis but grow to be very large "larvae" and reproduce in the larval state.

Adult amphibians do not depend solely on their primitive lungs for the exchange of respiratory gases. Their moist, glandular skin, which lacks scales and is plentifully supplied with blood vessels, also serves as a respiratory surface. Many mucous glands within the skin help to keep the body surface moist which is important in gas exchange and helps prevent desiccation. A number of amphibian species have skin glands that secrete poisonous substances to discourage predators.

The coloration of amphibians may conceal them in their habitat or may be very bright and striking. Many of the brightly colored species are poisonous, (see Fig. 51–16(a)); their distinctive colors warn predators that they are not encountering an ordinary amphibian. Some frogs have the ability to change color, from light to dark. The change in color is controlled by the melanocyte-stimulating hormone secreted by the intermediate lobe of the pituitary gland.

The amphibian heart has three chambers, two **atria** (receiving chambers) and one **ventricle** (a chamber that pumps blood into the arteries). Although there is some mixing of oxygen-rich and oxygen-poor blood, there is a double circuit of blood vessels. Blood passes through the **systemic circulation** to the various tissues and organs of the body. Then, after returning to the heart, it is directed through the **pulmonary circulation** to the lungs and skin to be recharged with oxygen. The oxygen-rich blood returns to the heart before being pumped out into the systemic circulation again. The comparative anatomy of the heart and circulation of various vertebrate classes is discussed in Chapter 35.

Class Reptilia: Turtles, Lizards, Snakes, Alligators

Members of class Reptilia are true terrestrial animals and need not return to water to reproduce, as most amphibians must. Many adaptations make the reptilian life style possible. The maternal oviduct secretes a protective leathery shell around the egg, which helps to prevent the developing embryo from drying out. Since a sperm cannot penetrate this shell, fertilization must occur within the body of the female before the shell is added. This internal fertilization requires additional adaptations—copulatory organs for transferring sperm from the body of the male into the female reproductive tract.

As the embryo develops within the protective shell, a membrane called the **amnion** forms and surrounds the embryo (see Fig. 44–11). The amnion holds **amniotic fluid,** providing the embryo with its own private pond. The amniotic fluid keeps the embryo moist and also serves as a shock absorber should the egg get bounced around. Like all terrestrial vertebrates, the reptile embryo has other membranes that protect and support its development (Chapter 44).

The reptile body is covered with hard, dry, horny scales, which protect the animal from desiccation and from predators. This dry, scaly skin cannot serve as an organ for gas exchange; however, reptile lungs are better-developed than the saclike lungs of amphibians. Divided into many chambers, the reptilian lung provides a greatly increased surface area for gas exchange. Most reptiles have a three-chambered heart, but there is an incomplete partition within the ventricle. (If this wall were complete, as it is in birds and mammals, the heart would have four separate chambers—two receiving chambers and two chambers that pump blood out into the arteries—and oxygenated and deoxygenated blood could be kept completely separate.) However, there is some separation of oxygen-rich and oxygen-poor blood, which facilitates oxygenation of body tissues. In the crocodile, the heart *is* four-chambered.

Another adaptation to life on land is a method of waste disposal that conserves water. Little water is removed from the blood with the wastes, and much of what is filtered out by the kidneys is reabsorbed in the kidney tubules and urinary bladder. In aquatic animals, nitrogenous wastes from protein and nucleic acid metabolism are excreted as ammonia. This waste product is quite toxic, and ammonia excretion requires large amounts of water. Reptiles (like many birds and terrestrial arthropods) convert ammonia to uric acid, which is much less toxic and can be excreted as relatively insoluble crystals.

Like fish and amphibians, reptiles lack metabolic mechanisms for regulating body temperature. They are **poikilotherms,** meaning that their body temperature more or less reflects the temperature of the surrounding environment. However, they do have behavioral adaptations that enable them to maintain a body temperature that is higher than that of their environment. You may have observed a lizard basking in the sun. Such behavior permits the lizard's body temperature to rise so that its metabolic rate increases and it can actively hunt for food. When the body temperature of a reptile is low, the metabolic rate is low and it is very sluggish. Because of this, reptiles are much more successful in warm than in cold climates.

Although a few species of turtles, tortoises, and lizards are herbivores, most reptiles are carnivores. Their paired limbs (usually with five toes) are well adapted for running and climbing, and their well-developed sense organs enable them to locate prey.

MODERN REPTILES. The reptiles living today are assigned to one of three orders: Order **Chelonia** includes the turtles, terrapins, and tortoises. Order **Squamata** is composed of the lizards, snakes, iguanas, and geckos, and Order **Crocodilia** includes the crocodiles, alligators, caimans, and gavials.

Members of order Chelonia are enclosed in a protective shell made up of bony plates overlaid by horny scales. The part of the shell that covers the back is the **carapace;** the ventral portion is called the **plastron.** Some species can withdraw their heads and legs completely into their shells. Turtles range in size from about 8 centimeters in length to the great marine species, which measure more than 2 m in length and may weigh 450 kg. The land species are usually referred to as tortoises, whereas the aquatic forms are called turtles (freshwater types are sometimes called terrapins).

Lizards and snakes are the most common of the modern reptiles (Fig. 26–26). These animals have rows of scales that overlap like shingles on a roof, forming a continuous armor that may be shed periodically. Lizards range in size from the gecko, which may weigh as little as 1 g, to the Komodo dragon of Indonesia, which may weigh 100 kg. Their body size and shape varies greatly. Some—for example, the glass snake, which is really a lizard—are legless. (The glass snake gets its name

(a)

(c)

(b)

Figure 26–26 Modern reptiles. (*a*) The desert collared lizard, *Crotaphytus collaris.* (*b*) Eastern hognose snakes, *Heterodon platyrhinos,* hatching. Many snakes, and some lizards, are actually live-bearing. (*c*) An Eastern diamondback rattlesnake, *Crotalus adamanteus.* Rattlesnakes bear their young alive. ((*a*), Charles Seaborn; (*b*), Animals, Animals, Zig Leszczynski; (*c*), James H. Carmichael, Coastal Creations.)

from its ability to detach from its tail when a predator grabs the tail. The long tail breaks—like glass—into many pieces that writhe around, distracting and confusing the predator while the tail-less lizard escapes.)

Snakes are characterized by a flexible, loosely jointed jaw structure that enables them to swallow animals larger than the diameter of their own jaws. Snakes lack legs, and their bodies are elongated. Their eyes, covered by a transparent cuticle, lack movable eyelids. They also lack an external ear opening, a tympanic membrane (eardrum), and a middle ear cavity.

The forked tongue of the snake, which often darts quickly from the mouth, is not poisonous. It is used as an accessory sensory organ for touch and smell; chemicals from the ground or air adhere to it, and then the tip is projected into Jacobson's organ, a sense organ located in the roof of the mouth that detects odors. Pit vipers and some boas also have a prominent sensory pit on each side of the head that enables them to detect heat. These sense organs enable them to locate and capture small nocturnal mammals.

Some snakes—for example, king snakes, pythons, and boa constrictors—capture their prey by rapidly wrapping themselves around the animal and squeezing it so that it cannot breathe. Others have fangs, which are hollow teeth connected with poison glands in the mouth. When the snake bites its prey, the poison is pumped through the fangs and into the prey's body. Some snake poisons cause the breakdown of red blood cells in the prey; others, such as that of the coral snake, are neurotoxins that interfere with nerve function. Poisonous snakes of the United States include the rattlesnakes, copperheads, water moccasins (also known as cottonmouths), and coral snakes. All except the coral snakes are pit vipers.

Three groups of crocodilians are (1) the crocodiles of Africa, Asia, and America, (2) the alligators of the southern United States and China (Fig. 26–27),

Figure 26–27 Alligator mississippiensis, the alligator native to the southeastern United States. (Courtesy of Dr. Sonja Lessne.)

and the caimans of Central America, and (3) the gavials of Southeast Asia. Most species live in swamps, in rivers, or along sea coasts, burrowing in the mud and feeding on various kinds of animals. Crocodiles are the largest living reptiles, some measuring more than 7 m in length.

FOSSIL REPTILES. The earliest reptiles are thought to have resembled amphibians. *Seymouria,* an extinct amphibian-like reptile, is thought to be the ancestral type. It looked something like a lizard, but its legs emerged from the sides of the body like those of salamanders.

The class Reptilia has many more extinct than living species. The **Mesozoic era,** which ended about 60 million years ago, is known as the age of reptiles; during that time reptiles were the dominant terrestrial animals. They had spread out into an impressive variety of ecological niches (Fig. 26–28): Some were able to fly, others became marine, and many filled terrestrial habitats. Some of the dinosaurs were among the largest land animals that have ever lived. *Brontosaurus* measured 23 m in length; *Diplodocus* was 27 m long. They may have weighed up to 32,000 kg (35 tons).

Figure 26–28 A evolutionary tree of the reptiles, drawn to emphasize the wide variations in life style found among fossil forms.

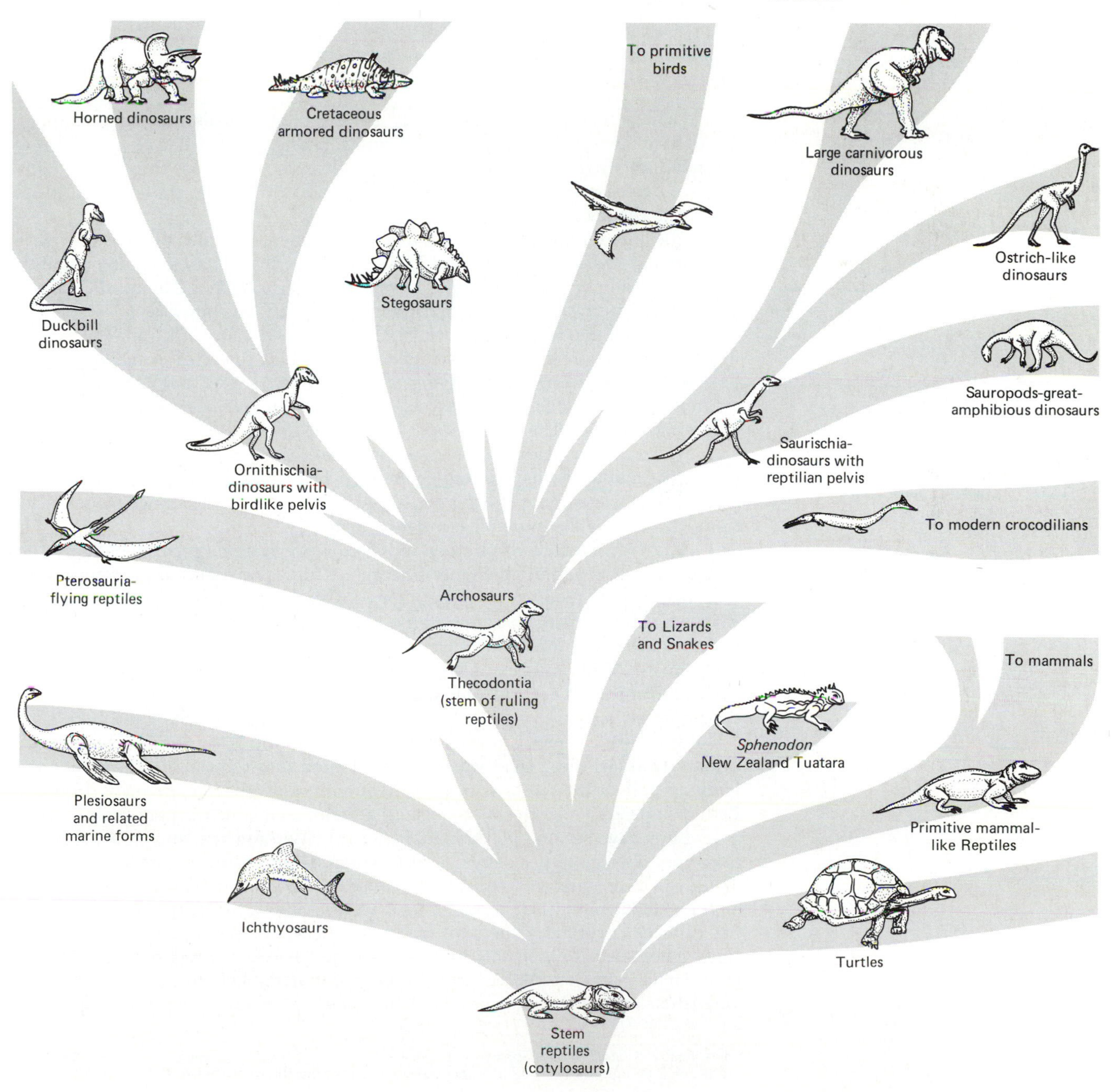

The reptiles were the dominant land animals for almost 200 million years. Then quite suddenly during the Cretaceous period, many of them, including all the dinosaurs, disappear from the fossil record. Many theories have been proposed to explain the sudden extinction of the large reptiles. Some biologists blame the mammals, which may have been successfully competing with dinosaurs and feasting on their eggs. This theory has been challenged by the fact that the large reptiles had successfully existed with the mammals for some time before their extinction. Other biologists blame changes in climate, for the environment was getting colder at the time the large reptiles disappeared. Low reproduction rate has also been suggested.

Another theory being debated suggests that 65 million years ago a shower of comets hit the earth. According to this theory, the impact raised a massive dust cloud that blocked out sunlight for many years, resulting in mass extinctions.[1] Gradually, as the dust settled, an iridium-rich layer of clay was deposited, marking the geological boundary between the Cretaceous and Tertiary periods. (The element iridium, extremely rare in the earth's crust, is comparatively abundant in extraterrestrial material. The presence of an iridium-rich layer is thus evidence for this theory.) Whatever the cause, by the end of the Mesozoic era, these large reptiles disappeared, leaving only three orders of reptiles. Long before their decline, however, the reptiles gave rise to both the birds and the mammals.

Class Aves: Birds

Birds, which comprise the class Aves, are the only animals that have feathers. Thought to have evolved from reptilian scales, feathers are flexible and very strong for their light weight. They protect the body, decrease water loss through the body surface, decrease the loss of body heat, and aid in flying by presenting a plane surface to the air.

The anterior limbs of birds are usually modified for flight; the posterior pair, for walking, swimming, or perching. Not all birds fly. Some, such as penguins, have small, flipper-like wings used in swimming (Fig. 26–29). Others, such as the ostrich and cassowary, have vestigial wings but well-developed legs.

In addition to feathers and wings, the birds have many other adaptations for flight. They have a compact, streamlined body, and the fusion of many bones gives the body the rigidity needed for flying. Their bones are strong, but very light; many are hollow, containing large air spaces. The jaw is light and instead of teeth, there is a light, horny beak. The very efficient lungs have thin-walled extensions called air sacs, which occupy spaces between the internal organs and within certain bones. Birds, like mammals, have a four-chambered heart and a double circuit of blood flow so that oxygen-poor blood is recharged with oxygen in the lungs before being pumped out into the systemic circulation again.

The very effective respiratory and circulatory systems provide enough oxygen to the cells to permit a high metabolic rate. This is necessary for the tremendous muscular activity required for flying. Some of the heat generated by metabolic activities is used to maintain a constant body temperature. This ability permits metabolic processes to proceed at constant rates and enables birds to remain active in cold climates. Birds and mammals are the only animals that can maintain a constant body temperature. They are sometimes called warm-blooded, but **homeothermic** is the preferred term.

Birds excrete nitrogenous wastes mainly as uric acid. Because they lack a urinary bladder, these solid wastes are delivered into the cloaca. They leave the body as part of the feces, which are frequently and unceremoniously dropped. Perhaps this is an adaptive mechanism that helps to maintain a light body weight.

Birds have become adapted to a variety of environments, and various species have very different types of beaks, feet, wings, tails, and behavioral patterns. Although all birds must eat frequently (because they do not store much fat but have a high metabolic rate), their choice of food varies widely among species. Many species eat seeds or fruits. Others eat worms, mollusks, or arthropods. Warblers and some other species eat mainly insects. Owls and hawks eat rodents, rabbits, and other small mammals. Vultures feed on dead animals. Pelicans, gulls, terns, and kingfishers catch fish. Some hawks catch snakes and lizards. Grouse and quail eat

[1]More than 75% of all species apparently became extinct at that time. This was one of the largest mass extinctions, though not the only one, that has occurred during the Earth's history.

(a)

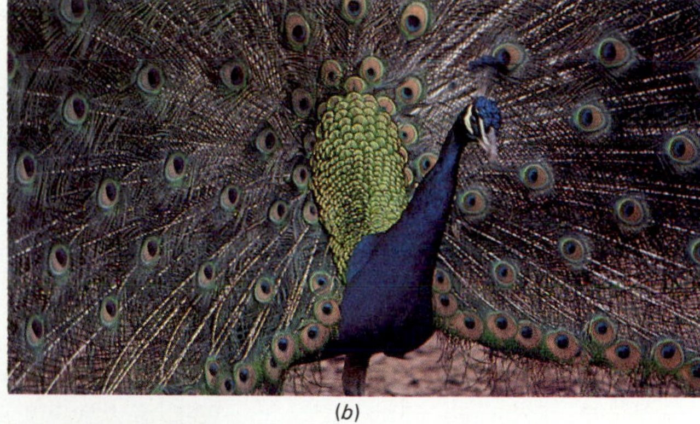

(b)

(c)

Figure 26–29 Modern birds. (*a*) Bald eagle, *Haliaeetus leucocephalus*. Once a greatly endangered species owing to the thinning of their eggshells because of the pesticide DDT, these animals are becoming more numerous in many sections of the United States. (*b*) Peacock, a male peafowl, *Pavo cristatus*. (*c*) A chinstrap penguin of the Antartic, keeping its nestlings warm. ((*a*), James H. Carmichael, Coastal Creations, 1983; (*b*), courtesy of Busch Gardens; (*c*), P. R. Ehrlich, Stanford University/BPS.)

leafy vegetation. Beaks are specifically adapted to the eating habits. An interesting feature of the bird digestive system is the **crop,** an expanded, saclike portion of the digestive tract below the esophagus, in which food is temporarily stored. The stomach is divided into a **proventriculus,** which secretes gastric juices, and a thick, muscular **gizzard** where food is ground. The bird swallows small bits of gravel, which act as "teeth" in the gizzard, mechanically breaking down food.

Birds have a well-developed nervous system with a brain that is proportionately larger than that of reptiles. Birds rely heavily on vision; their eyes are proportionately larger than those of other vertebrates. Hearing is also well developed.

In striking contrast to the silent reptiles, birds have developed the voice. Most birds have short simple calls that signal danger, or that influence feeding, flocking, or interaction between parent and young. Songs are usually more complex than calls and are performed mainly by males; they are related to reproduction, attracting and keeping a mate, claiming and defending territory. One of the most fascinating aspects of bird behavior is the annual migration that many species make. Some birds, such as the golden plover and Arctic tern, fly from Alaska to Patagonia, South America, and back each year, flying perhaps 25,000 miles en route. Others migrate only a few hundred miles south each winter; some, such as the bobwhite and great horned owl, do not migrate at all. A discussion of migration and navigation is included in Chapter 48.

EARLY BIRDS. At least two different groups of flying reptiles evolved, and one of these is thought to be the ancestor of the birds. Although the bones of birds are fragile and disintegrate quickly, a few fossils of early birds have been found. They looked very much like reptiles, even having teeth (which birds do not have), but their jaws were elongated into beaks, and they had feathers and wings.

The earliest known species of bird, ***Archaeopteryx,*** was about the size of a crow and had rather feeble wings, jawbones armed with reptilian-type teeth (so this early bird could get its worm?), and a long reptilian tail covered with feathers. Each of its wings was equipped with three digits bearing claws (Fig. 26–30). Several specimens of this species have been found in the Jurassic limestone of Bavaria, which was laid down about 150 million years ago (see Fig. 46–11).

Cretaceous rocks have yielded fossils of other early birds. *Hesperornis,* which lived in the United States, was a toothed aquatic diving bird with powerful hind legs and vestigial wings. *Ichthyornis* was a toothed flying bird about the size of a seagull. From the tertiary period (Eocene epoch) onward, the fossil record of birds

Figure 26–30 *Archaeopteryx,* the earliest known bird. This drawing represents the hypothesis that *Archaeopteryx* was at least a climbing animal that had some ability to use its wings and feathers for gliding. Other hypotheses suggest that *Archaeopteryx* was primarily a land animal, whose wings could be used to trap small insects, and whose feathers possibly served as insulation. (From a painting by Rudolph Freund, courtesy of the Carnegie Museum of Natural History.)

shows an absence of teeth and a progressive change in structure to the modern appearance.

MODERN BIRDS. Even modern birds possess some characteristics in common with the reptiles. For example, they have reptilian-type scales on their legs, and they lay eggs. (Although most male birds lack a copulatory organ, fertilization is internal. Copulation involves the approximation of male and female cloacas.)

About 9000 species of birds have been described; these have been classified in 27 orders. Birds inhabit a wide variety of habitats and can be found on all of the continents, most islands, and even the open sea. The largest living birds are the ostriches of Africa, which may be up to 2 m tall and weigh 136 kg, and the great condors of the Americas, with wingspreads of up to 3 m. The smallest known bird is Helena's hummingbird of Cuba, which is less than 6 cm long and weighs less than 4 g. Beautiful and striking colors are found among the birds. The color is due partly to pigments deposited during their growth and partly to reflection and refraction of light of certain wavelengths. Many birds, especially females, are protectively colored by their plumage. Brighter colors are often assumed by the male during the breeding season to help him attract a mate.

Class Mammalia

The distinguishing features of mammals are the presence of **hair, mammary glands,** which produce milk for the young, and the **differentiation of teeth** into incisors, canines, premolars, and molars. A muscular **diaphragm** helps to move air into and out of the lungs. Like the birds, but unlike other vertebrate groups, mammals maintain a constant body temperature. This process is supported by the covering of hair, which serves as insulation, by the four-chambered heart and separate pulmonary and systemic circulations, and by the presence of sweat glands. Red blood cells without nuclei are excellent oxygen transporters. The nervous system is more highly developed than in any other group. The cerebrum is especially large and complex, with an outer gray region called the **cerebral cortex.**

Fertilization is always internal, and except for the primitive monotremes that lay eggs, mammals are viviparous (bear their young alive). Most mammals develop a **placenta,** an organ of exchange between developing embryo and mother through which the embryo receives its nourishment and oxygen and rids its blood of wastes.

The limbs of mammals are variously adapted for walking, running, climbing, swimming, burrowing, or flying. In four-legged mammals the limbs are more directly under the body than in reptiles, which contributes to speed and agility. The organ systems of mammals are discussed in detail in Part VI.

EARLY MAMMALS. Mammals are thought to have evolved from a group of reptiles called **therapsids** (Fig. 26–31) probably during the Triassic period some 200 million years ago. The therapsids were doglike carnivores with differentiated teeth (a mammalian trait) and legs adapted for running. The fossil record indicates that the early mammals were small, about the size of a mouse or shrew.

How did the mammals manage to coexist with the reptiles during the 160 million or so years that the reptiles ruled the earth? Many adaptations permitted the mammals to compete for a place on the earth. The earliest mammals lived in trees and were nocturnal, searching for food (mainly insects and plant material and perhaps reptile eggs) at night while the reptiles were inactive. The life style is suggested by the large eye sockets in fossil species, which indicate that these animals had the large eyes characteristic of present-day nocturnal primates. By bearing their young alive, mammals avoided the hazards of having their eggs consumed by predators. And by nourishing the young and caring for them, the parents could offer both protection and an education, which probably focused upon how to avoid being eaten.

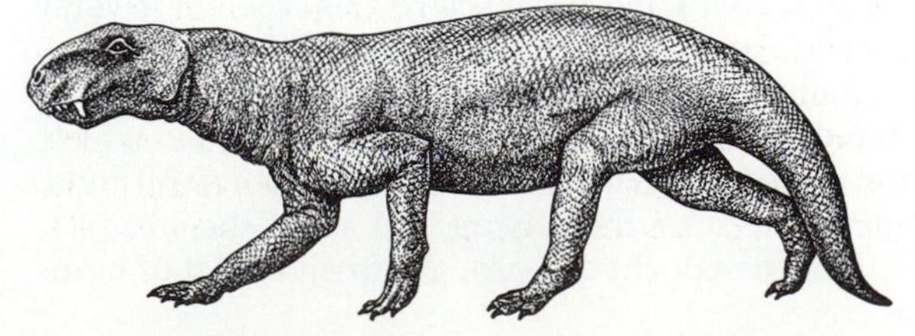

Figure 26–31 A mammal-like reptile, *Lycaenops,* from the late Permian period in South Africa. (From A. S. Romer and T. S. Parsons.)

Figure 26–32 The spiny anteater, *Tachyglossus*, a monotreme. (Photograph by Robert Anderson, reprinted with permission of Hubbard Scientific Company.)

(*a*)

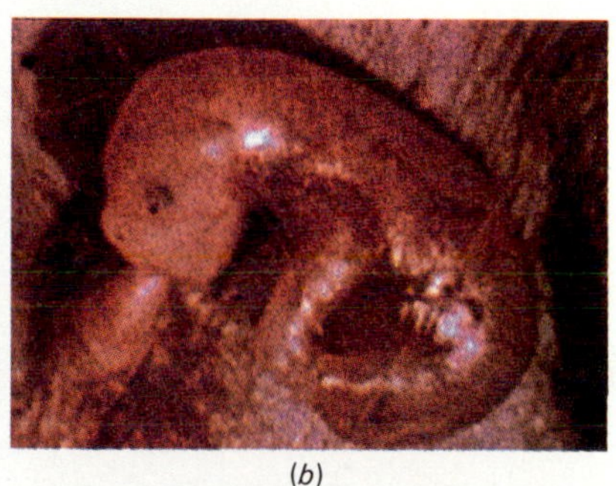

(*b*)

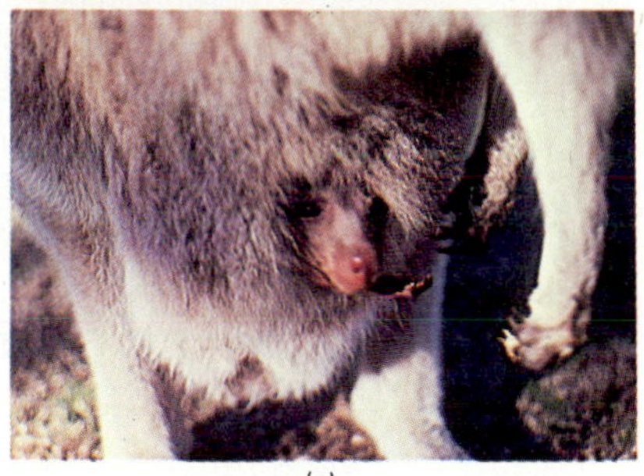

(*c*)

Figure 26–33 The kangaroo, a marsupial native to Australia. (*a*) Adult. (*b*) Young marsupials are born in a very immature state. (*c*) The young continue to develop in the safety of the marsupium. ((*a*), (*c*), Robin Lewis, Coastal Creations.) (*b*), photograph by Robert Anderson, reprinted with permission of Hubbard Scientific Company.)

As reptiles died out, the mammals began to move into their abandoned territories and ecological niches. The same was true for the birds. Numerous fossils of large birds exist that were adapted for life on the ground. Larger forms and numerous varieties evolved, and mammals became widely distributed and adapted to an impressive variety of ecological niches. Today mammals inhabit virtually every corner of the earth—on the land, in fresh and salt water, and in the air. They range in size from the tiny pigmy shrew, weighing about 25 g (less than an ounce) to the blue whale, which may weigh more than 90,000 kg and is thought to be the largest animal that ever lived.

By the end of the Cretaceous period there were three main groups of mammals. Today, these are classified in three subclasses: **Prototheria,** which includes the egg-laying mammals, also called **monotremes; Metatheria,** which includes the **marsupials,** or pouched mammals; and **Eutheria,** the **placental mammals.**

MONOTREMES. The monotremes are the only living order of subclass Prototheria. The two genera include the duck-billed platypus, *Ornithorhynchus,* and the spiny anteater or echidna, *Tachyglossus* (Fig. 26–32). Both are found in Australia and Tasmania; the spiny anteater is also found in New Guinea. The females lay eggs, which may be carried in a pouch on the abdomen or kept warm in a nest. When the young hatch they are nourished with milk from the mammary glands. As its name suggests, the spiny anteater feeds on ants, which it catches with its long sticky tongue. The duck-billed platypus lives in burrows along river banks. It has webbed feet and a flat, beaver-type tail, which aids in swimming. For food it catches freshwater invertebrates.

MARSUPIALS. Marsupials include pouched mammals such as kangaroos and opossums. Embryos begin their development in the mother's uterus, where they are nourished by yolk and from fluid in the uterus. After a few weeks, still in a very undeveloped stage, the young are born and crawl to the **marsupium** (pouch), where they complete their development. Each of the young attaches itself by its mouth to a mammary gland nipple and is nourished by its mother's milk (Fig. 26–33).

Figure 26–34 Comparison of marsupial and placental mammals. Placental and marsupial mammals have similar ways of life and occupy similar ecological niches. For every occupant of a given niche in one group there is a counterpart in the other group. This correspondence is not restricted to similarity of habit but also includes morphological features.

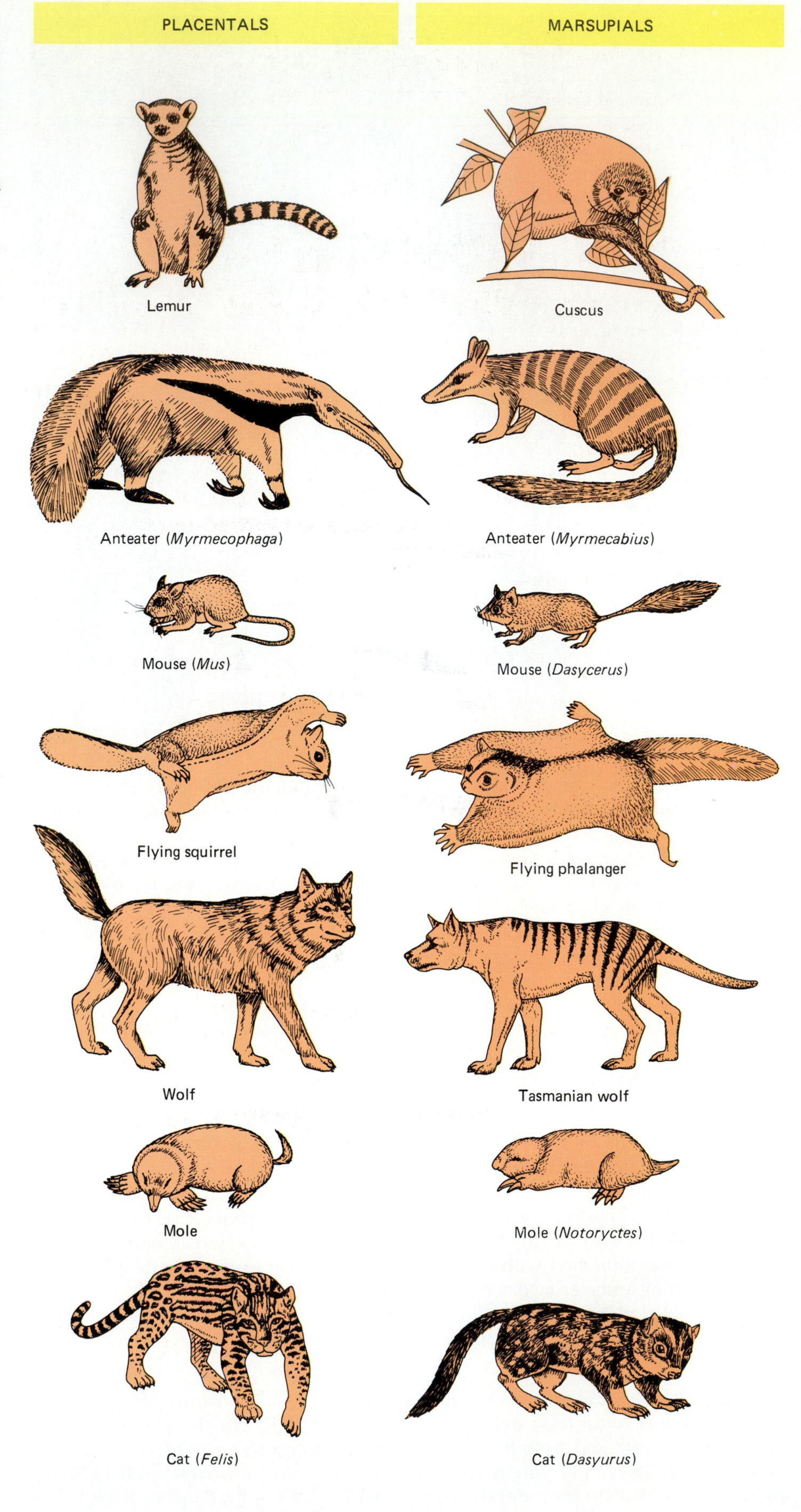

Like the monotremes, the marsupials are found mainly in Australia. Only the opossum is common in North America, although a few species, including a ratlike insectivorous marsupial found in the Andes, inhabit South America. At one time marsupials probably inhabited much of the world before they were supplanted by the placental mammals. Australia became geographically isolated from the rest of the world before placental mammals reached it, and there the marsupials remained the dominant mammals. Their evolution proceeded in many directions, fitting them for many different life styles, paralleling the evolution of placental mammals elsewhere. Thus in Australia and adjacent islands we find marsupials that correspond to our placental wolves, bears, rats, moles, flying squirrels, and even cats (Fig. 26–34).

PLACENTAL MAMMALS. Most familiar to us are the placental mammals, characterized by development of a placenta, the organ of exchange that enables the young to remain within the body of the mother until embryonic development is complete. The placenta forms from both embryonic membranes and the uterine wall. In it the blood vessels of the embryo come very close to the blood vessels of the mother so that materials can be exchanged by diffusion. (The two blood streams never normally mix.)

The young are born at a more mature stage than in the marsupials. Indeed, among some species the young can walk around and begin to interact with other members of the group within a few minutes of birth.

There are about 17 living orders of placental mammals. A brief summary of some of these is given in Focus on Some Orders of Living Placental Mammals. The probable family tree of the vertebrates is illustrated in Figure 26–35.

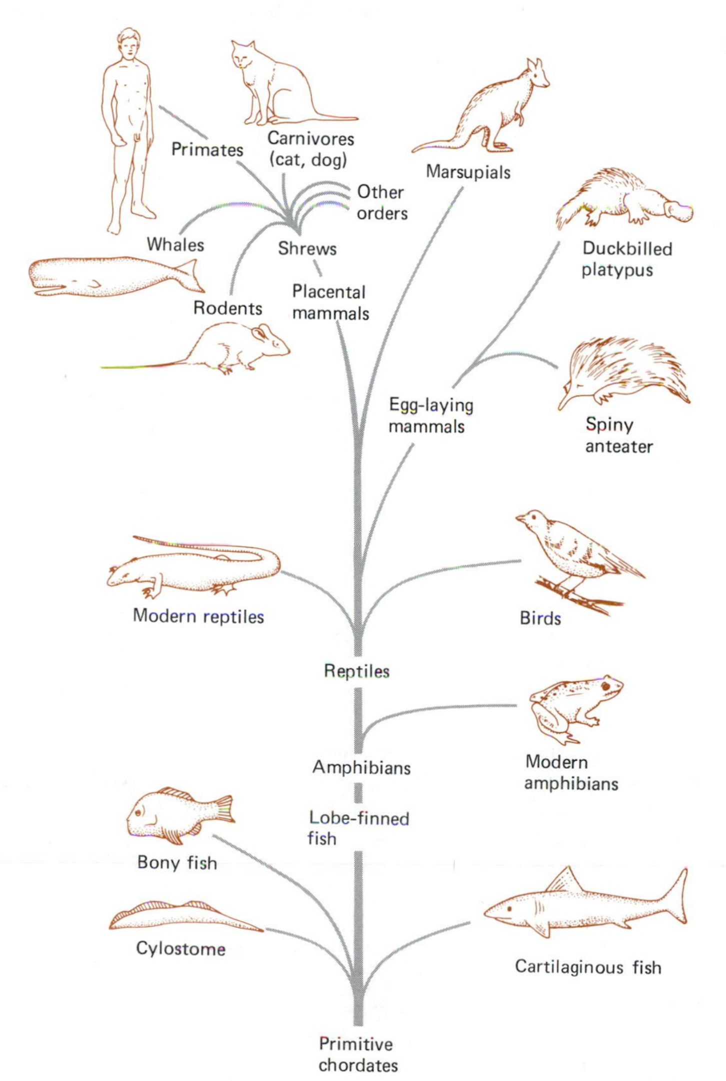

Figure 26–35 A vertebrate family tree.

FOCUS ON

Some Orders of Living Placental Mammals

Order Insectivora: moles, hedgehogs, and shrews. These are nocturnal insect-eating animals, considered to be the most primitive placental mammals and the ones closest to the ancestors of all the placentals. The shrew is the smallest living mammal; some weigh less than 5 g.

Order Chiroptera: bats. These mammals are adapted for flying; a fold of skin extends from the elongated fingers to the body and legs, forming a wing. Bats are guided in flight by a sort of biologic sonar: They emit high-frequency squeaks and are guided by the echoes from obstructions. These animals eat insects and fruit or suck the blood of other animals. Blood-feeding bats may transmit diseases such as yellow fever and paralytic rabies.

Order Carnivora: cats, dogs, wolves, foxes, bears, otters, mink, weasels, skunks, seals, walruses, and sea lions. Carnivores are flesh-eaters, with sharp, pointed canine teeth and shearing molars. In many species the canines are used to kill the prey. Carnivores have a keen sense of smell and exhibit complex social interactions. Its members are among the fastest, strongest, and smartest of animals. Limbs of the seals, walruses, and sea lions are modified as flippers for swimming. However, these animals are not completely adapted to life in the water; they come ashore to mate and bear their offspring.

Order Edentata: sloths, anteaters, armadillos. In these animals, the teeth are reduced to molars without enamel in the front part of the jaws, or no teeth are present. Sloths are sluggish animals that hang upside down from branches. They are often protectively colored by green algae that grow on their skin. Armadillos are protected by bony plates; they eat insects and small invertebrates.

Order Rodentia: squirrels, beavers, rats, mice, hamsters, porcupines, and guinea pigs. These are the gnawing mammals with chisel-like incisors that grow continually. As they gnaw, the teeth are worn down. The rodents are one of the most successful orders of mammals; about 3000 species have been described.

Order Lagomorpha: rabbits, hares, and pikas. Like the rodents, the lagomorphs have chisel-like incisors with enamel. Their long hind legs are adapted for jumping, and many have long ears.

Order Primates: lemurs, monkeys, apes, and humans. These mammals have highly developed brains and eyes, nails instead of claws, opposable great toes or thumbs, and eyes directed forward. Most species of primates are arboreal (tree-dwelling) and are thought to have evolved from the tree-dwelling insectivores. Primates may be divided into the prosimians, which include the lemurs, lorises, and tarsiers, and the anthropoids, which include monkeys, apes, and humans. (Primate evolution is discussed in Chapter 47.)

Order Perissodactyla: horses, zebras, tapirs, and rhinoceroses. These are herbivorous hoofed mammals with an odd number of digits per foot, usually one or three toes. (Hoofed mammals are often referred to as ungulates.) The teeth are adapted for chewing. These are usually large animals, with long legs.

Order Artiodactyla: cattle, sheep, pigs, deer, and giraffes. These herbivorous hoofed mammals have an even number of digits per foot. Most have two toes, but some have four. Many have antlers or horns on the head. Most are ruminants that chew a cud and have a series of stomachs in which bacteria that digest cellulose are incubated; this contributes greatly to their success as herbivores.

Order Proboscidea: elephants. These animals have a long, muscular trunk (proboscis) that is very flexible. Thick, loose skin is characteristic. The two upper incisors are elongated as tusks. Most are enormous with large heads and broad ears; the legs are like pillars. These are the largest land animals weighing as much as 7 tons. This order includes the extinct mastodons and wooly mammoths.

Order Sirenia: sea cows and manatees. These are herbivorous aquatic mammals with finlike forelimbs and no hind limbs. They are probably the basis for most tales about mermaids.

Order Cetacea: whales, dolphins, and porpoises. These mammals have become well adapted for their aquatic life style. They have fish-shaped bodies with broad, paddle-like forelimbs (flippers). Posterior limbs are absent. Many have a thick layer of fat called blubber covering the body. These very intelligent animals mate and bear their young in the water, and the young are suckled like those of other mammals. The blue whale is the largest living animal and probably the largest animal that has ever existed.

(a) (b) (c) (d) (e) (f)

Representative members of several orders of placental mammals. (*a*) *Trachops cirrhosus,* a frog-eating bat from Panama (family Chiroptera). After the rodents, bats are the largest order of mammals. (*b*) *Ursus maritimus,* a polar bear, photographed in the Kane Basin in the Arctic (family Carnivora). (*c*) *Erethizon dorsatum,* a porcupine (family Rodentia). Porcupine females bear only one offspring in a season. For mammals, young porcupines are unusually able to care for themselves; at the age of 2 days they are able to climb trees and find food. (*d*) *Leontopatheus rosalia,* the golden lion tamarin monkey. Only 150 of these primates survive in their native coastal rainforest in Brazil, which has been reduced to 2% of its original acreage. Thus, even though these animals have been able to breed well in captivity, their future as a species remains uncertain. (*e*) Two giraffes, *Giraffa camelopardalis* (family Artiodactyla), taking a cooling drink at an oasis in South Africa. Special vascular adaptations prevent the dangerous rise in cerebral blood pressure that would otherwise develop when the giraffe lowers its neck to drink. (*f*) A common dolphin, *Delthinus delthis,* photographed in the Sea of Cortez, Mexico. ((*a*), Merlin Tuttle, Photo Researchers, Inc.; (*b*), (*e*), (*f*), E. R. Degginger; (*c*), (*d*), Charles Seaborn.)

SUMMARY

I. The echinoderms and chordates are thought to be related because they are both deuterostomes and therefore share many developmental characteristics.

II. Phylum Echinodermata includes marine animals with spiny skins, a water vascular system, and tube feet; the larvae have bilateral body symmetry; most of the adults exhibit pentaradial symmetry.
   - *A.* Class Crinoidea includes the sea lilies and feather stars; in these animals the oral surface is turned upward; some are sessile.
   - *B.* Class Asteroidea is made up of the sea stars, animals with a central disk from which radiate five or more arms.
   - *C.* Class Ophiuroidea includes the brittle stars, which resemble asteroids but have longer, more slender arms, set off more sharply from the central disk.
   - *D.* Class Echinoidea includes the sea urchins and sand dollars, animals that lack arms; they have a solid shell, and their body is covered with spines.
   - *E.* Class Holothuroidea consists of sea cucumbers, animals with elongated flexible bodies; the mouth is surrounded by a circle of modified tube feet that serve as tentacles.

III. Phylum Chordata consists of three subphyla: Urochordata, Cephalochordata, and Vertebrata. At some time in its life cycle a chordate has a notochord, a dorsal tubular nerve cord, and pharyngeal gill slits.

IV. Subphylum Urochordata comprises the tunicates, which are sessile, filter-feeding marine animals that have tunics made of cellulose.

V. Subphylum Cephalochordata consists of the lancelets, small, segmented fishlike animals that exhibit all three chordate characteristics.

VI. Subphylum Vertebrata includes animals with a vertebral column that forms the chief skeletal axis of the body. Vertebrates also have a cranium that is part of the endoskeleton, pronounced cephalization, differentiated brain, muscles attached to the endoskeleton for movement, and two pairs of appendages.
   - *A.* Class Agnatha, the jawless fish, includes the lamprey eels and hagfishes.
   - *B.* Descendants of the ostracoderms (agnathans that are the earliest known fossil chordates) are thought to have evolved jaws and paired appendages and to have given rise to the modern jawed fishes.
   - *C.* Class Chondrichthyes, the cartilaginous fish, consists of the sharks, rays, and skates.
   - *D.* Class Osteichthyes, the bony fish, includes about 20,000 species of freshwater and saltwater fishes. The Osteichthyes and chondrichthyes are thought to have evolved from placoderm ancestors at about the same time. Most modern bony fish are ray-finned fishes with swim bladders.
   - *E.* Modern amphibians include the salamanders, frogs and toads, and wormlike caecilians.
     1. Most amphibians return to the water to reproduce; frog embryos develop into tadpoles, which undergo metamorphosis to become adults.
     2. Amphibians use their moist skin as well as lungs for gas exchange; they have a three-chambered heart with systemic and pulmonary circulations; and they have mucous glands in the skin.
   - *F.* Class Reptilia includes turtles, lizards, snakes, and alligators.
     1. Reptiles are true terrestrial animals.
     2. Fertilization is internal; most reptiles secrete a leathery protective shell around the egg; the embryo develops an amnion and other extraembryonic membranes, which protect it and keep it moist.
     3. A reptile has a dry skin with horny scales, lungs with many chambers, and a three-chambered heart (with some division of oxygen-rich and oxygen-poor blood) and excretes uric acid.
     4. Reptiles dominated the earth during the Mesozoic era; then during the Cretaceous period most of them, including all of the dinosaurs, became extinct.
   - *G.* Birds (class Aves) have many adaptations for flight including feathers, wings, and light hollow bones containing air spaces; birds have a four-chambered heart, very efficient lungs, a high metabolic rate, and a constant body temperature; they excrete solid wastes (uric acid).
     1. Birds have a well-developed nervous system and excellent vision and hearing.
     2. Birds have developed the voice, and communicate with simple calls and complex songs.
   - *H.* Mammals have hair, mammary glands, differentiated teeth, and maintain a constant body temperature. They have a highly developed nervous system and a muscular diaphragm.
     1. Monotremes, mammals that lay eggs, include the duck-billed platypus and the spiny anteater.
     2. Marsupials are pouched mammals such as kangaroos and opossums. The young are born in an immature stage and complete their development in the marsupium, where they are nourished with milk from the mammary glands.
     3. Placental mammals are characterized by an organ of exchange, the placenta, that develops between the embryo and the mother; this organ supplies oxygen and nutrients to the fetus and enables it to complete development within the uterus. There are about 17 living orders of placental mammals.

POST-TEST

1. Adult ______________ have pentaradial symmetry.

2. The ______________ ______________ system and ______________ feet are unique to echinoderms.

3. Echinoids (e.g., sea urchins) lack ______________, and their skeletal plates form a solid ______________.

4. The three distinguishing characteristics of a chordate are a ______________, a dorsal, ______________ ______________ ______________, and pharyngeal ______________ ______________.

5. _______________ are sessile, marine chordates often mistaken for sponges.

6. Vertebrates are distinguished from all other animals in having a _______________ _______________; anterior to this structure a _______________ encloses and protects the brain.

7. Fish belong to superclass _______________; amphibians, reptiles, birds, and mammals belong to superclass _______________.

8. _______________ scales are characteristic of sharks.

9. In sharks, the _______________ receives digestive wastes, urine, and gametes.

10. The shark's spiral valve slows the passage of _______________, permitting more time for _______________.

11. Modern fishes are thought to have descended from the _______________ fishes; the lobe-finned fishes are credited with being the ancestors of the _______________.

12. The operculum covers the _______________.

13. The labyrinthodonts are thought to have been the first successful _______________.

14. The amnion is an adaptation to _______________ life; it secretes a fluid that _______________.

15. The only homeothermic animals are the _______________ and the _______________.

16. Monotremes are mammals that _______________ _______________.

Match the answer in Column B with the description in Column A; there may be more than one answer for each question.

Column A	*Column B*
17. Have amnion	a. Bony fish
18. Have hair	b. Amphibians
19. Have four-chambered heart (two atria and two ventricles)	c. Reptiles
20. Have tube feet	d. Birds
21. Body covered with hard, dry, horny scales	e. Mammals
22. Bones contain air spaces; no teeth	f. None of the above
23. Have pharyngeal gill slits at some time in life cycle	
24. Agnathan with circular sucking disk	a. Cotylosaurs
25. Earliest known species of bird	b. Lamprey
26. A cephalochordate	c. Archaeopteryx
27. A lobe-finned fish	d. Coelacanth
28. Stem reptiles	e. Amphioxus

REVIEW QUESTIONS

1. Why are echinoderms thought to be more closely related to chordates than to other phyla?
2. What are the three principal distinguishing characteristics of a chordate? How are these evident in a tunicate larva? in an adult tunicate? in a lancelet? in a human?
3. What characteristics distinguish the vertebrates from the rest of the chordates?
4. How do lampreys and hagfishes differ from other fishes? Of what economic importance are agnathans?
5. What is the function of gills? In general terms, how do they work? Why do you suppose aquatic mammals do not possess them?
6. Compare the skins of sharks, frogs, snakes, and mammals.
7. Give the location and function of each of the following:
   a. swim bladder
   b. placenta
   c. operculum
   d. amnion
   e. marsupium
8. Give the phylum, subphylum, class, (and order if you can) for each of the following animals:
   a. a human being
   b. a turtle
   c. a lamprey eel
   d. *Branchiostoma* (amphioxus)
   e. dogfish shark
   f. whale
   g. frog
   h. pelican
   i. bat
9. Why are monotremes considered to be more primitive than other mammals? Some paleontologists consider them to be therapsid reptiles rather than mammals. Give arguments for and against this position.
10. Which vertebrate groups maintain a constant body temperature? How do they accomplish this? Why is this advantageous?
11. Which are more specialized animals, birds or mammals? Explain your answer.
12. According to current evolutionary theory, give the significance of each of the following:
    a. coelacanths
    b. placoderms
    c. labyrinthodonts
    d. *Seymouria*
    e. therapsids
    f. *Archaeopteryx*

Recommended Readings for Part IV

Articles for Taxonomy

Corliss, J. O. Consequences of creating new kingdoms of organisms. *Bioscience,* Vol. 33, May 1983. The objections of a holdout against the five-kingdom scheme of taxonomy.

Krogmann, D. W. Cyanobacteria (blue-green algae)—their evolution and relation to other photosynthetic organisms. *Bioscience,* Vol. 31, No. 2, February 1981. A good example of the application of modern taxonomic techniques.

Leedale, G. F. How many are the kingdoms of organisms? *Taxon,* Vol. 23, 1974. Discusses problems with assigning organisms to discrete kingdoms.

Mayr, E. Biological classification: toward a synthesis of opposing methodologies. *Science,* Vol. 214, October 1981. A discussion of phenetics, cladistics, and evolutionary classification suggesting that all three methods be utilized in taxonomy.

Moore, R. T. Proposal for the recognition of super ranks. *Taxon,* 23:650–652, 1974. A suggestion for using the term dominium to comprise the (1) viruses (2) prokaryota and (3) eukaryota.

Palleroni, N. J. The taxonomy of bacteria. *Bioscience,* Vol. 33, No. 6, June 1983. Modern approaches to microbial taxonomy are discussed following a discussion of the history of bacterial classification systems.

Whittaker, R. H. New concepts of kingdoms of organisms. *Science,* Vol. 163, 1969. A proposal for classifying living things according to a five-kingdom system.

Whittaker, R. H. On the broad classification of organisms. *Quarterly Review of Biology,* 34:210, 1959. The earliest rumble of what was to become the five-kingdom revolution.

Whittaker, R. H., and L. Margulis. Protist classification and the kingdoms of organisms. *Biosystems,* Vol. 10, 1978.

Woese, C. R., and G. E. Fox. Phylogenetic structure of the Prokaryote domain: the primary kingdoms. *Proceedings of the National Academy of Sciences,* 74 (11):5088–5090, 1977. Significant for its proposal of both the domain concept and the suggestion that the domain Prokaryota should contain two kingdoms, the eubacteria and the methanogens.

Books for Taxonomy

Dobzhansky, T., et al. *Evolution,* San Francisco, W. H. Freeman, 1977. An introduction to evolution that includes chapters on taxonomy.

Eldredge, N., and J. Cracraft. *Phylogenetic Patterns and the Evolutionary Process.* New York, Columbia University Press, 1980. A discussion of the interaction between taxonomy and evolutionary theory.

Gardner, E. J. *History of Biology*. Minneapolis, Burgess Publishing Company, 1965. Chapter 8 summarizes the work of Linnaeus and that of his lesser known contemporaries and predecessors in the establishment of our system of classification.

Books for Microbiology Chapters (Chapters 18, 19, 21)

Nester, E. W., N. N. Pearsall, J. B. Roberts, and C. E. Roberts. *The Microbial Perspective.* Philadelphia, Saunders College Publishing, 1982. A very readable and interesting presentation of microbiology including bacteria, viruses, fungi, and protozoa. The emphasis is on the role of microorganisms in human health and disease.

Nester, E. W., C. E. Roberts, M. E. Lidstrom, N. N. Pearsall, and M. T. Nester. *Microbiology,* 3d ed. Philadelphia, Saunders College Publishing, 1983. A microbiology textbook that emphasizes mechanisms by which microorganisms cause disease, and includes sections on environmental and industrial microbiology.

Pelczar, M. J., R. D. Reid, and E. C. S. Chan. *Microbiology,* 5th ed. New York, McGraw-Hill, 1983. An in-depth treatment of microbiology.

Ray, P., T. A. Steeves, and S. A. Fultz. *Botany.* Philadelphia, Saunders College Publishing, 1983. This comprehensive botany textbook includes chapters on prokaryotes and fungi, as well as on pathogens and plant diseases.

Articles for Microbiology Chapters (Chapters 18, 19, 21)

Ahmadjian, V. The nature of lichens. *Natural History,* March 1982. A beautifully illustrated summary of the algae-fungal partnership.

Anagnostakis, S. Biological control of chestnut blight. *Science,* Vol. 215, January 1982. An interesting account of the history of chestnut blight fungus in the United States and of virus-like agents that may be useful in controlling it.

Blakemore, R. P., and R. B. Frankel. Magnetic navigation in bacteria. *Scientific American,* December 1981. Certain aquatic bacteria have internal compasses that orient them in the earth's magnetic field.

Bonner, J. T. Chemical signals of social amoebae. *Scientific American,* April 1983. A discussion of the chemicals that cellular slime molds emit and respond to; these chemicals permit different species to coexist and yet maintain their own identity.

Burchard, R. P. Gliding motility of bacteria. *Bioscience,* Vol. 30, No. 3, March 1980. A study of the characteristics of gliding motility and possible mechanisms.

Butler, P. J. G., and A. Klug. The assembly of a virus. *Scientific American,* November 1978. An account of the assembly of the tobacco-mosaic virus.

Costerton, J. W., G. G. Geesey, and K. J. Cheng. How bacteria stick. *Scientific American,* January 1978.

Demain, A. L. Industrial microbiology. *Science,* Vol. 214, November 1981. A look at the applications of new genetics techniques to industrial microbiology.

Diener, T. O. Viroids: minimal biological systems. *Bioscience,* Vol. 32, No. 1, January 1982. A discussion of the smallest known agents of infectious disease.

Diener, T. O. The viroid—a subviral pathogen. *American Scientist*, Vol. 71, September-October, 1983. A discussion of the origin, structure, and pathogenesis of viroids, with speculation about prions.

Diener, T. O. Viroids. *Scientific American*, January 1981. A discussion of viroids and their possible role in causing diseases.

Fraser, D. W., and J. E. McDade. Legionellosis. *Scientific American*, October 1979. The story of Legionnaire's disease and the bacterium that causes it.

Henle, W., G. Henle, and E. T. Lennette. The Epstein-Barr virus. *Scientific American*, July 1979. An account of research investigating the Epstein-Barr virus, a common virus that causes infectious mononucleosis and has also been linked to human cancer.

Krogmann, D. W. Cyanobacteria (blue-green algae)—their evolution and relation to other photosynthetic organisms. *Bioscience*, Vol. 31, No. 2, February 1981. An interesting discussion of the cyanobacteria.

Marx, J. L. Cancer cell genes linked to viral *onc* genes. *Science*, Vol. 216, May 1982. A discussion of the hypothesis that cancers of both viral and nonviral origin may be caused by the inappropriate activation of similar cellular genes.

Miller, J. A. Slow viruses: the body's secret agents. *Science News*, Vol. 121, No. 20, May 1982. An interesting discussion of slow viruses.

Novick, R. P. Plasmids. *Scientific American*, December 1980. Are plasmids subcellular organisms poised on the threshold of life? The author of this interesting article presents evidence to support this hypothesis.

Strobel, G. A., and G. N. Lanier. Dutch elm disease. *Scientific American*, August 1981. An account of Dutch elm disease, a deadly fungal infection of elm trees, and of new biological techniques used to attack the fungus and the beetles that spread it.

Vidal, G. The oldest eukaryotic cells. *Scientific American*, February 1984. A review of the fossil record indicating that the eukaryotes originally evolved in the form of unicellular plankton some 1.4 billion years ago.

Wernick, R. From ewe's milk and a bit of mold: a fromage fit for a Charlemagne. *Smithsonian*, Vol. 13, No. 11, February 1983. The story of the production of Roquefort cheese.

Woese, C. R. Archaebacteria. *Scientific American*, June 1981. An approach to the evolutionary significance of the archaebacteria in which they are considered as a third major group along with the prokaryotes and the eukaryotes.

Books for Protist and Plant Chapters (Chapters 20, 22, 23)

Chapman, V. J., and D. J. Chapman. *The Algae*, 2d ed. London, Macmillan, 1973. A highly technical treatise of interest to botanists and algologists.

Good, R. *Features of Evolution in the Flowering Plants*. New York, Dover, 1974 (first published 1956). A technical but thought-provoking and iconoclastic discussion of theories of spermatophyte evolution.

Gray, A. *Gray's Manual of Botany*, 8th ed. New York, Van Nostrand Reinhold, 1950. A classic, nay *the* classic summary of the vascular plants.

Jones, S. B., and A. E. Luchsinger. *Plant Systematics*. New York, McGraw-Hill, 1979. A practical discussion of botanical classification of use in the field.

Klein, R. M. *The Green World*. New York, Harper and Row, 1979. An eminently readable brief introduction to botany.

Perry, F. *Flowers of the World*. New York, Crown, 1972.

Pickett-Heaps, J. *New Light on the Green Algae*. Burlington, N. C., Carolina Biological Supply Company, 1982. It is not easy to obtain detailed information on the microanatomy and life style of algae. Here is a convenient source.

Raven, P. H., R. F. Evert, and H. Curtis. *Biology of Plants*, 3d ed. New York, Worth Publishers, Inc., 1981. An excellent general botany textbook.

Ray, P., T. A. Steeves, and S. A. Fultz. *Botany*. Philadelphia, Saunders College Publishing, 1983. This comprehensive botany textbook includes a fine presentation of the evolution of plants. The very best available.

Thomas, B. *The Evolution of Plants and Flowers*. New York, St Martin's Press, 1981. A beautifully illustrated and imaginative summary of current thinking about plant origins and evolutionary development.

Tippo, O., and Stern, W. L. *Humanistic Botany*. New York, Norton, 1977. A fine and fascinating discussion of what plants mean to the human species.

Watson, E. V. *Mosses*. London, Oxford University Press, 1972. A brief Oxford Biology Reader that not only instructs but also debunks much of what "everyone knows" about mosses.

Articles for Protist and Plant Chapters

Christopher, T. A. The seeds of botany. *Natural History*, March 1981, pp. 50–56. The establishment of the first Western botanical garden in Padua, Italy.

Hunt, J. W. Algal biochemical tricks and classification. *The American Biology Teacher*, December 1978, pp. 528–532. A suggested method for teaching principles of classification using algae that also contains much useful information about the algae themselves.

Swaminathan, M. S. Rice. *Scientific American*, Vol. 250, No. 1, January 1984. Rice, along with wheat and maize, is a member of the grass family upon which the human species largely depends. Advances in plant genetics have greatly increased its yield per acre.

Books for Animal Chapters (Chapters 24–26)

Barnes, R. D. *Invertebrate Zoology*, 4th ed. Philadelphia, Saunders College Publishing, 1980. A comprehensive reference book for invertebrate zoology; the book is organized around the phyla and includes the protozoa.

Barth, R. H., and R. E. Broshears. *The Invertebrate World*. Philadelphia, Saunders College Publishing, 1982. A brief, but interesting parade through the phyla emphasizing the highlights of invertebrate zoology; the protozoans are included.

Daly, H. V., J. T. Doyen, and P. R. Ehrlich. *Introduction to Insect Biology and Diversity*. New York, McGraw-Hill, 1978. An introduction to the study of insects that emphasizes the major features of insects as living systems.

Fingerman, M. *Animal Diversity*, 3d ed. Philadelphia, Saunders College Publishing, 1981. Modern Biology Series. A very brief summary of both the invertebrates and the vertebrates, with emphasis on evolutionary relationships.

Gardiner, M. S. *The Biology of Invertebrates.* New York, McGraw-Hill, 1972. A functional, rather than a systematic, approach to invertebrate zoology. The book is organized around life processes rather than by phyla.

Orr, R. T. *Vertebrate Biology,* 5th ed. Philadelphia, Saunders College Publishing, 1982. This introduction to vertebrate biology focuses on biological processes.

Romer, A. S., and T. S. Parsons. *The Vertebrate Body,* 5th ed. Philadelphia, W. B. Saunders, 1977. An update of an important classic in biology, with emphasis on evolutionary relationships.

Vaughan, T. A. *Mammalogy,* 2nd ed. Philadelphia, Saunders College Publishing, 1978. An introduction to the mammals; a systematic approach.

Villee, C. A., W. F. Walker, and R. D. Barnes. *General Zoology,* 6th ed. Philadelphia, W. B. Saunders, 1984. An introductory zoology textbook with evolutionary emphasis.

Welty, J. C. *The Life of Birds,* 3rd ed. Philadelphia, Saunders College Publishing, 1982. An introduction to the biology of birds.

Articles for Animal Chapters

Adding flesh to bare therapsid bones. *Science News,* Vol. 119, No. 25, June 1981. A brief account of the therapsid reptiles thought to be our ancestors.

Alldredge, A. Appendicularians. *Scientific American,* July 1976. A description of a group of tunicates that build a house of mucus and filter food particles out of seawater.

Alldredge, A. L., and L. P. Madin. Pelagic tunicates: unique herbivores in the marine plankton. *Bioscience,* Vol. 32. No. 8, September 1982. An account of the unique adaptations of pelagic tunicates.

Bramble, D. M., and D. R. Carrier. Running and breathing in mammals. *Science,* Vol. 219, No. 4582, January 1983. A physiological study of the synchronization of locomotion and breathing in running mammals.

Burgess, J. W. Social spiders. *Scientific American,* March 1976. A few species of spiders interact socially and build large communal webs.

Camhi, J. M. The escape system of the cockroach. *Scientific American,* December 1980. A study of the mechanisms by which a roach rapidly escapes from predators.

Crews, D., and W. R. Garstka. The ecological physiology of a garter snake. *Scientific American,* Vol. 247, No. 5, November 1982. An account of the physiological and behavioral adaptations of the red-sided garter snake to its harsh environment.

Degabriele, R. The physiology of the koala. *Scientific American,* July 1980. A study of the survival adaptations of the koala, a marsupial that eats eucalyptus leaves, seldom drinks, and uses no shelter.

Goreau, T. F., N. I. Goreau, and T. J. Goreau. Corals and coral reefs. *Scientific American,* August 1979. An account of small coral polyps that live in symbiosis with photosynthetic algae and build large limestone reefs.

Foot, J. Squid swarm. *Natural History,* Vol. 91, No. 4, April 1982. An account of the schools of squid that gather off the California coast to spawn.

Heinrich, B. The regulation of temperature in the honeybee swarm. *Scientific American,* June 1981. A discussion of thermoregulation in a swarm of bees.

Miller, J. A. A brain for all seasons. *Science News,* Vol. 123, No. 17, April 23, 1983. A study of brain development during metamorphosis of a moth.

Miller, J. A. A skunk of a beetle. *Science News,* Vol. 115, May 19, 1979. An account of the defensive spray of a bombardier beetle.

Miller, J. A. Strike. *Science News,* Vol. 120, August 29, 1981. An account of rattlesnakes and vipers striking rodent prey and an exploration of the evolutionary implications of this behavior.

Mansour, T. E. Chemotherapy of parasitic worms: new biochemical strategies. *Science,* Vol. 205, August 1979. A review of the mechanisms by which various drugs act on parasitic worm infestations.

Merritt, R. W., and J. B. Wallace. Filter-feeding insects. *Scientific American,* April 1981. An account of insects that hatch underwater and gather food with nets and brushes, and a discussion of their ecological importance.

Moore, J. Parasites that change the behavior of their host. *Scientific American,* Vol. 250, No. 5, May 1984. Certain parasites, such as thorny-headed worms which infect pill bugs, make the host more vulnerable to predation by their next host.

Mossman, D. J., and W. A. S. Sarjeant. The footprints of extinct animals. *Scientific American,* Vol. 248, No. 1, January 1983. An account of vertebrate evolution with emphasis on information gained from animal tracks.

Myers, C. W., and J. W. Daly. Dart-poison frogs. *Scientific American,* Vol. 248, No. 2, February 1983. An account of poisonous frogs and the toxins they release.

Newman, E. A., and P. H. Hartline. The infrared "vision" of snakes. *Scientific American,* March 1982. Snakes of two families can detect and localize sources of infrared radiation.

Nijhout H. F. The color patterns of butterflies and moths. *Scientific American,* November 1981. A study of the development of the more than 100,000 different wing patterns of butterflies and moths.

Reid, R. G. B., and F. R. Bernard. Gutless bivalves. *Science,* Vol. 208, May 1980. A description of a new species of bivalve that lacks internal digestive organs.

Roper, C. F. E., and K. J. Boss. The giant squid. *Scientific American,* April 1982. A discussion of the anatomy and ecology of the giant squid.

Tangley, L. Tracing the roots of a gypsy. *Science News,* Vol. 122, No. 17, October 23, 1982. In a search for biological control mechanisms a team of scientists traveled to China, where the gypsy moth, a major threat to forests and ornamental trees, may have originated.

Ward, P., L. Greenwald, and O. E. Greenwald. The buoyancy of the chambered nautilus. *Scientific American,* October 1980. An account of the mechanisms by which the chambered nautilus divides its shell into compartments and removes water, thereby gaining mobility.

West, S. Moon history in a seashell. *Science News,* Vol. 114, No. 25, Dec. 16, 1978. A discussion of a theory linking the development of the chambered nautilus with lunar months.

Wicksten, M. K. Decorator crabs. *Scientific American,* February 1980. A discussion of species of spider crabs that camouflage themselves with materials that they attach to their exoskeletons.

Wursig, B. Dolphins. *Scientific American,* March 1979. An interesting description of dolphin behavior and learning ability.

part V

STRUCTURE AND LIFE PROCESSES IN PLANTS

Backlit maple leaves in Palo, Colorado. The veins of each leaf contain vascular elements that carry materials to and from the photosynthetic tissue of the leaf. The divergence of the veins indicates that maple trees are dicots. (C. W. May/BPS.)

27
The Plant Body

OUTLINE

I. The plant body: An overview
II. The leaf
- *A.* Epidermis
- *B.* Mesophyll
- *C.* Veins
- *D.* Water loss
- *E.* Guard cells and their control
  - *1.* The photosynthetic hypothesis
  - *2.* The role of abscisic acid

III. The stem
- *A.* The vascular tissues
  - *1.* Phloem
  - *2.* How phloem works
  - *3.* Xylem
  - *4.* How xylem works
  - *5.* Other xylem functions
- *B.* Monocot stems
- *C.* Dicot stems

IV. Roots
- *A.* Root structure and function
- *B.* Root respiration

LEARNING OBJECTIVES

After you have read this chapter you should be able to:

1. Describe the structure of a leaf, including all of its component tissues.
2. Give the functions of the epidermis, mesophyll, and veins.
3. Summarize the role of guard cells in the water economy of the plant, and describe how they function and are controlled.
4. Give the functions of plant stems.
5. Describe the arrangement of vascular tissues in the stems of monocots and dicots.
6. Summarize the functions of phloem and xylem.
7. Describe the structure of a typical root.
8. Briefly outline the process of absorption in the root, and trace the pathway materials take into the plant body after absorption has occurred.
9. Describe the typical respiratory adaptations of roots.

Plants are the antennae by which the energy of sunlight is captured in the ecosphere and then stored in food for later slow, catabolic release in the living cells of both plants and animals. To many of us, plants are little more than a green mass in the background of life. Yet more than 99% of our planet's living matter is composed of plants; this includes a tremendous variety of algae, vines, shrubs, trees, grasses, and herbs. Almost all of these diverse plant shapes are variations on a common theme, and almost all tracheophytes are composed of just three main parts: leaves, stems, and roots.[1]

The Plant Body: An Overview

The plant body can be divided into the underground root and the shoot, which consists of the stem and the leaves. **Leaves** are generally the main and often the only significant photosynthetic organs of the plant. They are usually thin and flat for maximum absorption of light energy and efficient internal diffusion of gases. Arranged on the plant so as to interfere minimally with one another's light supply, the leaves of plants form an intricate green mosaic, bathed in sunlight and atmospheric gases.

The **roots** perform two functions: They hold the plant firmly in the soil, and, even more important, they are the sole means the plant has to absorb water and to mine nutrients from the soil. The roots of all the plants in the world extract several thousand metric tons of minerals from the soil every minute.

The **stem** is the plant's midsection; it conducts commerce between the leaves and the roots. Water and minerals travel upward; carbohydrates travel downward (and sometimes upward, as we will see). Just as important, the stem supports both its own weight and that of the leaves and, by its growth, thrusts these above the leaves and stems of its competitors so that the plant may flourish.

Although these are the basic functions of the parts of the plant body, they are by no means the only ones. In some species, such as the African baobab tree (Fig. 27–1), the stems are modified for food storage; in others, such as the potato, portions of the stems lie underground, swollen with food. Leaves can be modified for protection as in the case of thorns, for support as in the tendrils of some vines, and for water storage as in the case of many succulent and salt marsh plants. In some curious species, like the Venus' fly-trap, leaves catch and digest animal prey. Leaves may even be missing altogether, leaving photosynthesis to the flattened stems (as in cacti), or even to the seedpods (as in one eccentric variety of garden peas).

Figure 27–1 The African baobab tree stores large volumes of water and starch in its stem tissues. (Courtesy of Dr. Pat Gill.)

[1] At various times, you might want to review the section and table on plant tissues in Chapter 6.

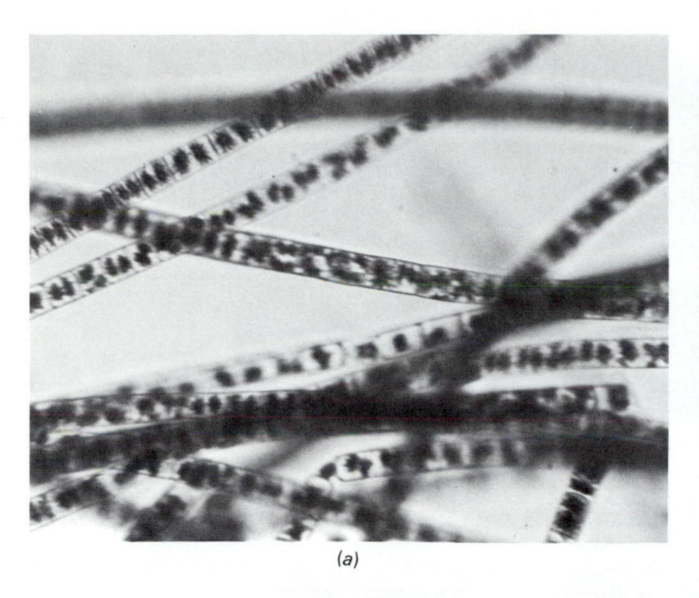

(a)

(b)

(c)

Figure 27–2 (*a*) The leaf of higher plants is specialized to perform the photosynthetic function that is performed by each cell of a simple green alga such as *Zygnema*. (*b*) Parts of a leaf. (*c*) Leaf mosaic. This morning glory vine is growing on a fence in such a way that leaves give one another minimal shade, promoting maximum absorption of solar energy.

The Leaf

Take a leaf from a tree or shrub, and observe it carefully in a good, strong light. The stemlike **petiole** attaches the leaf to the stem at a location called the **node.** The angle it forms with the stem is its **axil.** The petiole is continuous with the **midrib** of the leaf, which usually lies in a central valley that continues to the **tip.** The tip is often shaped so as to facilitate the drip of rainwater from the leaf **blade** (Fig. 27–2). The midrib gives rise to many tiny **veins** in the blade of the leaf, which branch in several fashions in dicots, or run parallel to one another in monocots. The two surfaces of the leaf may differ in appearance and somewhat in functions as well (Fig. 27–3).

EPIDERMIS

The **epidermis,** or outer covering of the leaf, has a smooth and glossy surface. Epidermal cells protect the wet tissues of the leaf from desiccation. Their major means of waterproofing is the layer of waxy, acellular **cuticle** that covers the entire leaf. In some tropical palms this layer of wax is thick enough to be of commercial value and is harvested for use in floor, car, and shoe polish.

Some leaves possess hairs that are outward extensions of epidermal cells. These hairs may have several functions. In some instances they help to retard evaporation of water from the leaf by interfering with the free flow of air over the leaf surface; in others they repel grazing animals. Some even sting. Still others reflect sunlight, helping the leaf to avoid overheating.

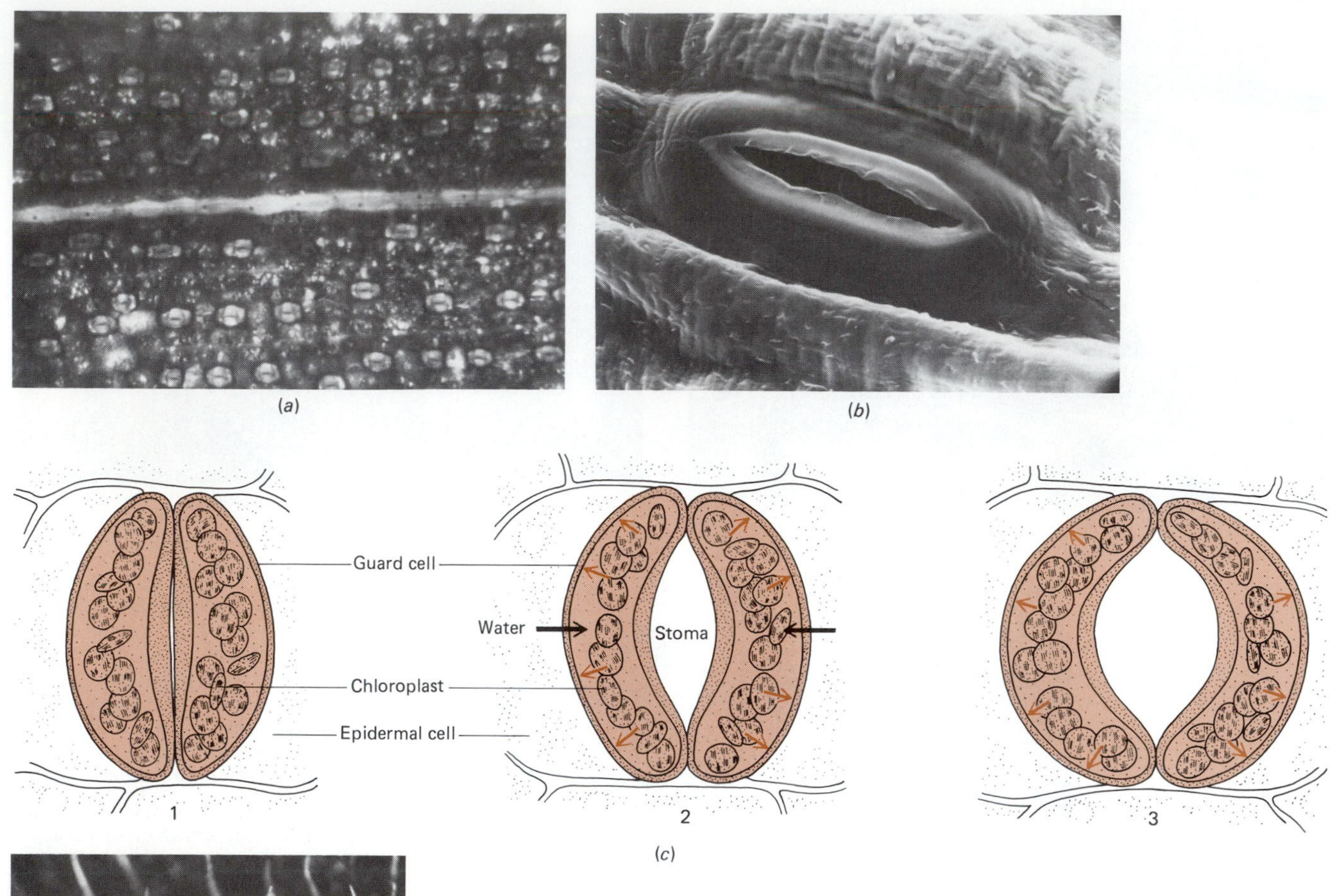

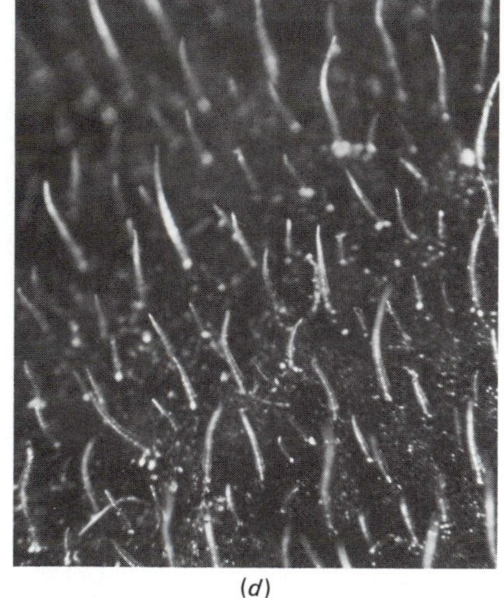

Figure 27–3 Leaf epidermis. (*a*) *Zebrina* leaf epidermis, seen from above (×100). Note guard cells. (*b*) Scanning electron micrograph of a single stoma in an onion leaf. Guard cells are located between hill-like epidermal folds which form a pit whose function may be to limit water loss. Photosynthetic parenchyma tissue cells are dimly visible through the stoma. (×2500). (*c*) Regulation of the size of the stoma: (*1*) Nearly closed condition. (*2*) When water enters the guard cells, turgor pressure increases and the guard cells buckle so as to increase the size of the stoma. (*3*) Stoma open. (*d*) Many leaves are equipped with hairs that sometimes function to limit the transpiration of water. ((*b*), E. Zeiger and N. Burstein, Stanford University/BPS.)

Microscopic openings called **stomata** are usually located on the leaf undersurface. Each stoma is like a tiny mouth surrounded by a pair of liplike **guard cells** (Fig. 27–4). All these epidermal structures, together with the much more numerous jigsaw puzzle–shaped flat cells that surround them, are the frontier between the semiaquatic inner world of the leaf and the comparatively dry terrestrial environment outside. Except for the guard cells of the stomata, the epidermal cells are not green and contain no chloroplasts.

MESOPHYLL

Sandwiched between upper and lower epidermis, the **mesophyll** is the photosynthetic tissue of the leaf. The cells of the mesophyll are not too different from those of many colonial algae, and like algae must be kept moist (if not wet) in order to live. The loosely packed cells of this tissue are green, a consequence of their abundant chlorophyll content. Extensive intercellular air spaces ensure the ready access of gases to each and every cell. As a result, oxygen for respiration or carbon dioxide for photosynthesis is made readily available by diffusion alone, as circumstances may require. Owing to the flattened shape of the leaf, no part of its fleshy mesophyll is very far from a stoma. Its transport system, profoundly different in design from that of almost any animal, does not carry oxygen to its tissues, so diffusion must suffice. The thinness of the leaf also ensures the penetration of light to all chlorophyll-containing cells in amounts adequate for photosynthesis (see Fig. 27–3).

In many leaves the mesophyll occurs in two distinguishable layers: the more tightly packed **palisade parenchyma,** whose arcades of elongated cells are usually located near the top surface of the leaf, and the somewhat looser **spongy parenchyma** below it. The significance of this arrangement, by no means universal among land plants, is unknown. However, the mesophyll cells, by their turgor, serve as a hydrostatic skeleton for the leaf, a function for which the architecture of the mesophyll may be necessary. It has been suggested that the long, thin palisade cells may function to conduct light efficiently into the center of the leaf. The spongy

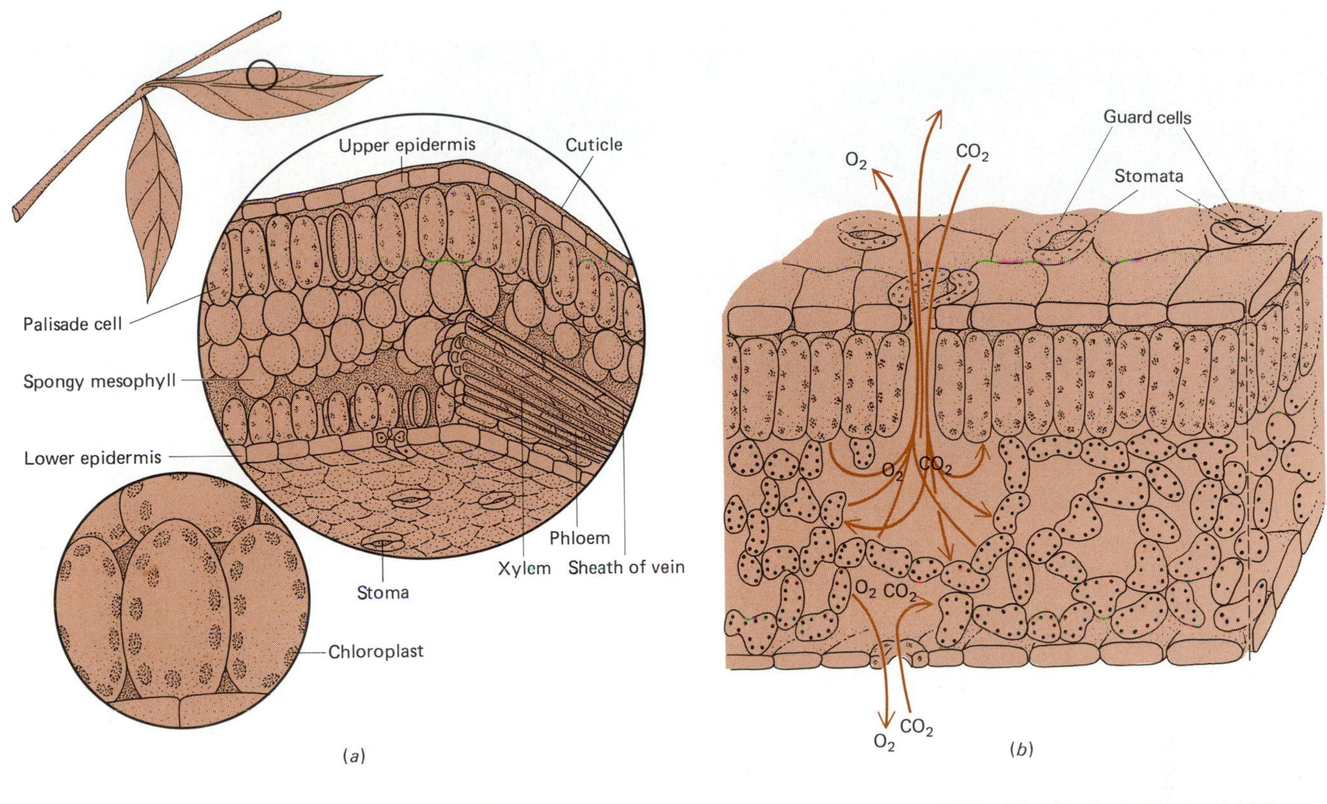

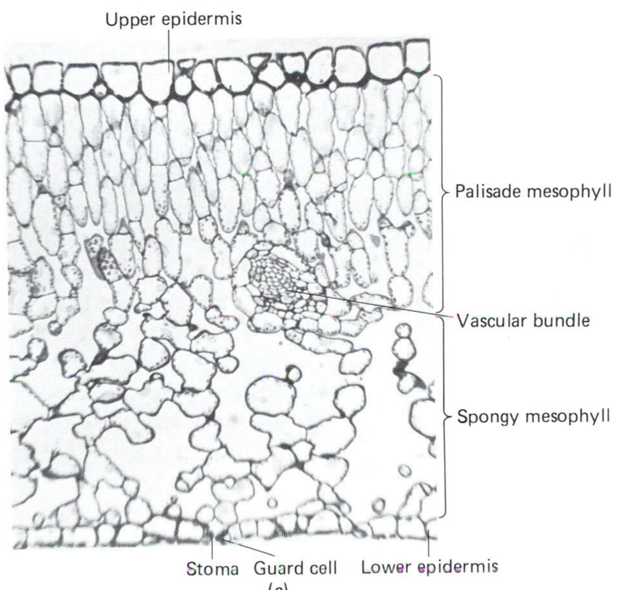

Figure 27–4 Leaf structures. (*a*) Diagram of the microscopic structure of a leaf. Part of a small vein is visible to the right. (*b*) An enlarged cross section of a leaf. The diffusion of carbon dioxide through the stomata to the interior of the leaf and the diffusion of oxygen from the photosynthetic cells through the stomata to the exterior are indicated by arrows. (*c*) Photomicrograph of a cross section of a privet leaf.

parenchyma, for its part, may scatter light throughout the leaf by multiple reflections and refractions so that only a small portion of the light energy escapes.

VEINS

The leaf veins serve two functions: to support the leaf as a skeleton, and to conduct materials to and from the metabolically active tissues (Fig. 27–5). Although skeletal elements may be lacking in the smaller veins, such structures are prominent in the midrib, whose upper portion may consist entirely of **sclerenchyma,** a tissue of mostly dead cells whose thick cell walls have become impregnated with hard waxy or gummy materials for strength. Even more important are the vascular tissues, **xylem** (which is usually located on the upper side of the vein) and **phloem.** These carry materials to and from the photosynthetic tissues of the leaf.

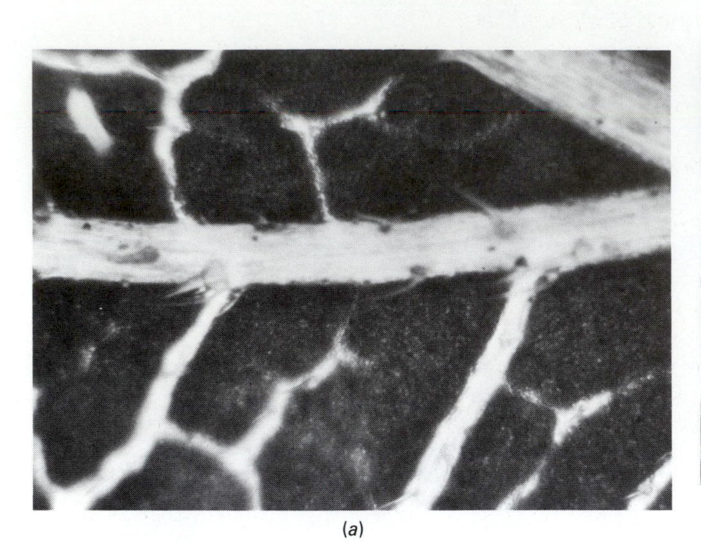

(a)

Phloem
Xylem
Parenchyma
Endodermis
Photosynthetic tissue
(b)

Figure 27–5 Leaf veins and vascular bundles. (*a*) Small veins in a dicot leaf. (*b*) Cross section of a vascular bundle in a pine needle.

WATER LOSS

The basic function of photosynthesis is served by almost every adaptation of the leaf, from its internal characteristics such as transparency, shape, and air spaces, to the way the leaf is positioned on the plant. Yet each of these photosynthetic adaptations serves also to promote the greatest hazard facing the plant: dehydration. Leaf structure in land plants represents a compromise between conflicting requirements: exposure to atmosphere and light versus protection against water loss. For instance, in the hottest part of the day the leaves of eucalyptus trees orient themselves parallel to the rays of the sun, limiting solar heating of their tissues and minimizing water loss by evaporation. (Unfortunately for us, this also has the effect of minimizing the shade they cast just when we would most enjoy it!)

A forest interior can be as much as 10° to 15°F cooler than the surrounding countryside, not just because of the shade, but as a result of the heat absorbed by the evaporation of hundreds of liters of water each day from large trees, each of which is estimated to have the cooling ability of five 10,000-watt air conditioners. This evaporative water loss, termed **transpiration,** occurs almost entirely through the leaves. We might think this prodigal use would signify an abundance of water; yet, except for a minority of swamp-dwelling plants or aquatic plants, water available to plants is in short supply for much or all of the year in most parts of the world.

GUARD CELLS AND THEIR CONTROL

Air comprises much of the volume of even the thinnest leaf. If a leaf is immersed in water and gradually warmed, tiny bubbles of gas can be seen to stream from the stomata as the trapped air expands. The total cellular surface exposed to this internal air may be more than 200 times greater than that of the leaf itself, but evaporation into it is relatively slow, since the internal air is at 100% relative humidity all of the time. Yet as the air escapes into the warmed water of our experiment, so water vapor escapes into the air from living plants by the same route.

The role of the stomata (see Fig. 27–4) is to permit the entry of carbon dioxide, without which growth and eventually life would be impossible for the plant. Yet when the stomata are wide open, they allow water vapor to leave the leaf at a rate only 50% less than would occur if the parenchyma cells were directly exposed to the air. Thus, the open stomata represent a glaring weakness in the water-conserving set of adaptations of the leaf.

Some simple plants—the liverworts—have leaflike structures equipped only with pores for gas entrance and exit, but these plants are restricted to wet or very damp habitats. The key stomatal adaptation that allows more complex plants to live even in the desert is the guard cell. These cells enable the plant to close the stomata when the risks of water loss exceed the benefits of continuing photosynthesis. Moreover, guard cells can close the stomata at night when photosynthesis does not normally occur, for oxygen, present in high concentration in the surrounding atmosphere, does diffuse into the leaf through the cuticle in amounts sufficient to support life.

How the guard cells are able to behave in this manner has been a puzzle for a long time. To understand the current explanation let us consider the microscopic anatomy of guard cells. Each one of them is a crescent-shaped semicylinder, and each pair is joined at the ends. Unlike other epidermal cells, they contain chloroplasts.

The cell wall of the guard cells is considerably thicker on the side nearer the stomatal opening. Additionally, fibers of cellulose in the cell wall are arranged in bundles, the **radial micelles,** that converge at the thickened, stomatal side of the cell wall. This arrangement of the cell wall fibers causes the concave surfaces of the two guard cells to pull apart when the volume and pressure of the cell's contents increase, thus opening the stoma. Conversely, when this pair of cells becomes flaccid, the cells relax, not uniformly, but in such a way that the stoma closes. What makes the volume and hence the pressure of the guard cells' contents change?

The Photosynthetic Hypothesis

One old and reasonable explanation of how guard cells operate to open and close the stomata is that water flows into and out of these cells in response to changes in their solute concentration—that is, by osmosis. Photosynthesis, by producing sugar, would therefore increase the tonicity of the guard cells, increase their turgidity, and lead to the opening of the stomata. This would cause them to be open during the day. At night, when the guard cell's own metabolism would have used up the sugar, the tonicity and turgidity of the cells would decrease. The stomata would then close, just as is, in fact, observed. This sweet hypothesis and its modifications are not credible, however, since in order to have this effect the solute concentration within the guard cells would have to change by about 50%. It is very doubtful that the guard cell chloroplasts could produce this much sugar by photosynthesis. However, more recent studies have disclosed that during opening of the stoma there is a major influx of potassium into the guard cells, which in many species that have been studied is more than enough to produce the necessary tonicity, which in turn leads to the observed turgidity.

In most species, stomata open in the light and are closed in the dark. (For an important exception, see Focus on Photosynthesis in Desert Plants: A Special Case.) Although mechanisms that control stomatal aperture are not yet fully understood, a widely accepted current explanation is that stomatal opening of most species involves two factors: carbon dioxide and ATP. When leaves are exposed to light, photosynthesis reduces CO_2 concentration in the leaf, and as a result also in guard cells. Guard cells lack significant levels of enzymes of the Calvin cycle, and do not ordinarily carry out the light-independent reactions of photosynthesis. However, guard cells *do* contain both photosystems, and guard cell chloroplast photophosphorylation can provide sufficient ATP to drive potassium intake.

Apparently, low guard cell CO_2 content and ATP energy are both needed to activate a K^+ uptake pump. This results in the active transport of K^+ into the guard cells (from surrounding epidermal cells), causing an increased turgor pressure and stomatal opening. In the dark, when CO_2 concentration in the guard cells becomes high owing to respiratory CO_2 production, the K^+ pump is inactivated, and guard cells passively lose most of their K^+, resulting in stomatal closure. Even in full sunlight, guard cell CO_2 content may increase, resulting in stomatal closure. For example, stomata typically begin closing gradually in the afternoon so that they are fully closed by evening. One explanation for this is that high midday temperatures result in a leaf respiratory rate that exceeds the leaf photosynthetic rate. This, in turn, causes an increased guard cell CO_2 concentration and gradual stomatal closure.

Under drought conditions, photosynthetic rates in most species fall to virtually zero, while respiratory rates are elevated. This increases guard cell CO_2 content and induces stomatal closure. However, increased CO_2 content in guard cells during water deficits does not fully account for stomatal closure.

FOCUS ON

Photosynthesis in Desert Plants: A Special Case

Having said that stomata open during the day and close at night, we must now note the exception of many desert plants—including the cacti—that reverse this process, opening their stomata at night and closing them during the day (see the figure). Such plants are known as **CAM plants,** for crassulacean acid metabolism. (Crassulaceae, for which this metabolic system is named, is an important family known for this adaptation, with many desert-living members.) In at least some CAM species that open their stomata in the dark and close them in the light, the synthesis of organic acids at night from CO_2 is apparently sufficient to result in opening of the stomata without need for K^+ uptake. This is clearly an advantage in water conservation, for the heat of the day is just when water will evaporate most from mesophyll or other green tissues. Yet it seems an impossibility as far as carbon dioxide fixation is concerned. How can the plant obtain the raw material of photosynthesis?

Carbon dioxide is as abundant at night as during the day. There is no reason it cannot enter the leaf during the hours of darkness, given open stomata. However, since light reactions are impossible at this time, there is also no way that the necessary ATP and protons can be generated for CO_2 fixation via the Calvin cycle. Some other fixation mechanism must be employed.

In the case of wet-leaved succulent desert plants and other CAM plants, the CO_2 is initially fixed by combining with phosphoenolpyruvate (PEP). The source of the PEP is ultimately starch, from which it is formed by glycolysis. When the PEP accepts CO_2, it is transformed into oxaloacetate. The oxaloacetate is next reduced to malate by NADH; this metabolite is stored in the central vacuoles of the mesophyll cells until daybreak. (Each of these steps is catalyzed by a specific enzyme.)

As dawn approaches and the stomata close, this process starts to operate in reverse, with the dehydrogenation of the malate to form oxaloacetate again, followed by decarboxylation, which yields PEP (which began the process) and CO_2. The PEP can now be made into starch once more, and the CO_2 is available for photosynthesis in the now sealed-off tissues of the parenchyma. It is simply refixed by conventional combination with ribulose bisphosphate, and from then on photosynthesis proceeds more conventionally.

It seems inefficient for CO_2 to be fixed in this roundabout fashion, at what must be a very substantial energy cost. But consider the alternative. Under desert conditions, conventional photosynthesis would be not merely inefficient, but fatal. Only under rare conditions of high humidity can the CAM plants employ conventional photosynthesis; but when they can, they do so. Most conventional plants are so ununadapted to life in the desert that they cannot live there at all. Much the same thing is true of C_4 plants. Because the C_4 process is *less* efficient in CO_2 fixation at low light intensities, C_3 plants actually have an advantage, as a rule, in northern well-watered locales; and only a few plant members—for example, St. Augustine grass, as illustrated—of such communities employ the C_4 process.

Figure 27–A Cactus, a typical CAM plant that grows in arid or excessively drained habitats.

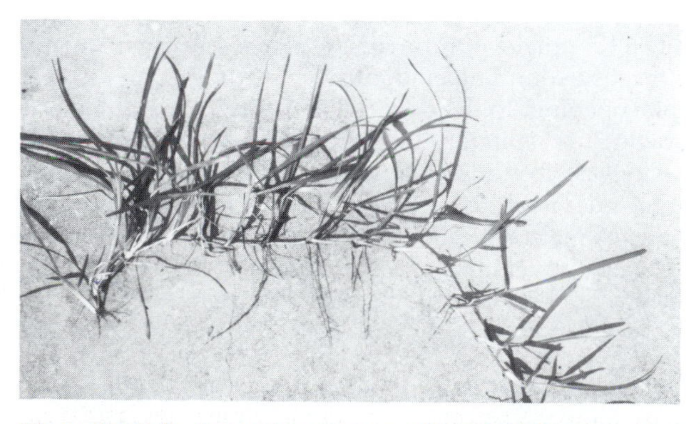

Figure 27–B St. Augustine grass, a subtropical C_4 monocot.

The Role of Abscisic Acid

Stomatal closure *due to water deficits* is apparently mediated by **abscisic acid (ABA)** as well (see Chapter 29). ABA appears to be present in an inactive form in mesophyll cells of well-watered plants. When water deficits in the photosynthetic tissues occur, ABA is converted to an active form and moves into the guard cells. Apparently the ABA causes stomatal closure by interfering with the K^+ pump. It is not clear whether elevated CO_2 concentration (which also closes stomata) and ABA interact or whether each acts independently. What is known is that, after periods of drought stress, ABA remains active in the guard cells for several days, which delays full stomatal opening during this period. Furthermore, a large amount of ABA produced over a long period of time helps to induce formation of an **abscission layer** of waterproof cells at the axil of the leaf. This eventually causes the leaf to drop off, thus protecting the plant against transpirational water loss under severe drought conditions. Even under less severe water stress conditions, ABA apparently plays a role in stomatal closure. As the normal day wears on, as a result of slight dehydration, the mesophyll of the leaf produces increasing amounts of ABA. In response, the stomata close by later afternoon.

The Stem

The stems of gymnosperms and woody dicots are of tremendous value as building materials. Even the monocot bamboo is pressed into service for scaffolding, fishing rods, and the old-fashioned slide rule. The strength and flexibility that give stems their desirable engineering properties are required by the plant to support its leaves above those of its competitors and to hold them up despite every challenge of wind and storm. The stems of even herbaceous plants are the strongest parts of those plants.

THE VASCULAR TISSUES

Plants, like animals, require a transport system, for the same general reasons (Fig. 27–6). The body of any large plant contains so much tissue that diffusion alone could not possibly carry fluids and the substances dissolved in them to all the cells that require those substances. There is some need for a transport system even in the larger algae, such as the giant kelp *Laminaria,* which has its leaflike parts just below the surface in sunlit upper waters and its rootlike holdfast anchored to rock in the darker depths 8 meters or more below. Sugars and other substances produced by the cells in the upper fronds may travel many meters to reach the holdfast, apparently by means of specialized conductive tissues. However, the kelp has no need to transport water or minerals to any part of its body. Surrounded as it is by seawater, the kelp's entire body obtains water and minerals directly from the sea. The transport system of a true land plant, however, not only carries the products of photosynthesis—**leaf sap** or **phloem sap**—in a generally downward direction but also conducts water and minerals—contained in the **root sap** or **xylem sap**—upward. Before reading further, please turn back to Table 6–5 and review the main plant tissue types.

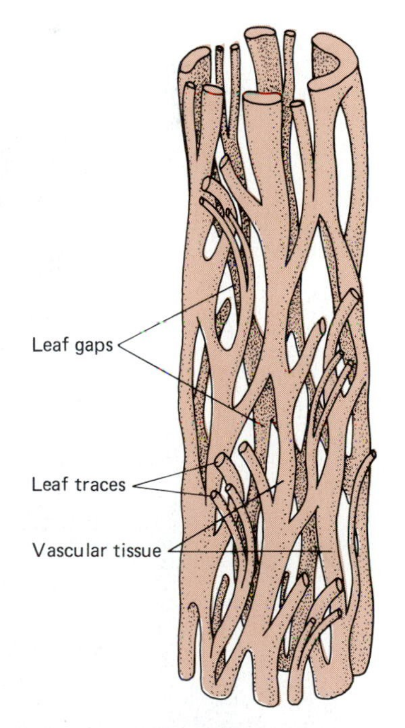

Figure 27–6 Three-dimensional view of a fern stem vascular system. Large leaf gaps result in a network that appears as a ring of bundles in cross section. (From Ray, Steeves and Fultz.)

Phloem

Phloem is a vascular tissue specialized to carry the products of photosynthesis from the leaves throughout the body of the plant. This process is known as **translocation.** The principal kind of phloem cell is the **sieve cell,** so called because of the perforated cell walls—the **sieve plates**—at each end of the cell that permit materials to pass from one such cell to another. The sieve cells of phloem are living, and although when mature most of them contain no nuclei, they do contain a layer of cytoplasm-like material just inside their cell walls. Sieve cells manage without nuclei because they are kept alive by the nucleated **companion cells** that always seem to occur adjacent to them. There are profuse cytoplasmic connections between the sieve and the adjacent companion cells. In addition, phloem tissue is strengthened by fiber cells that occur adjacent to the sieve and companion cells.

How Phloem Works

Phloem tissue is living and active. Each phloem cell contains a layer next to its cell wall of what is now often called **p-protein** or, more traditionally (if less elegantly), **slime.** The currently favored hypothesis of phloem function is the **pressure-flow mechanism,** which holds that materials pass through phloem by osmotically facilitated pumping.

Much of what is known about phloem transport of materials has been learned from aphids. Each aphid is equipped with a proboscis so fine that it can be inserted into an individual phloem cell. From this cell the aphid withdraws the food-laden and nourishing leaf sap, substantially free of slime or other contaminants. All the experimenter need do is wait until aphids are successfully feeding on an experimental plant and then snip off their proboscises. This will cause the leaf sap to ooze out of the severed proboscis by internal pressure. Chemical analysis of the phloem sap thus obtained reveals a nutrient content of up to 30% organic matter, containing mostly disaccharide sugars (especially sucrose), but also some amino acids and other substances. Radioactive tracer studies indicate that in intact phloem, the leaf sap can flow at rates up to 200 cm per hour. This is a high rate of travel, if we consider the microscopically fine diameter of the phloem cells.

Sugars are made rapidly in the daytime by the photosynthetic parenchyma tissue of the leaf. They are then pumped by some cell membrane mechanism of active transport into the sieve cells. This renders the phloem cells hypertonic to the surrounding leaf tissues. The leaf phloem then absorbs water osmotically from the surrounding parenchyma, producing high hydrostatic pressure in the phloem cells of the leaf.

The immensely long tube formed by the interconnected phloem cells connects with sugar-consuming cells somewhere else in the plant, usually in the root. These sugar-consuming tissues actively transport sucrose out of the phloem and then either metabolize the sugar or store it in the form of starch. (Conversion to starch has the effect of rendering the sugar osmotically inactive.) Thus, the sugar-utilizing end of the phloem will become hypotonic to its surroundings, and water will flow *out* of it by osmosis. At one end of the phloem pipeline, therefore, there is high hydrostatic pressure, and at the other, low pressure. There is usually no need for this pressure differential to oppose the force of gravity, because leaf sap flows mainly in a downward direction. Such a pressure difference therefore may easily produce the rapid cytoplasmic streaming and physical flow of leaf sap that we observe (Fig. 27–7).

It is important to note that phloem transport is not necessarily downhill, although it *is* "down" an osmotic gradient. In the spring, for instance, the sap rises in temperate-zone trees. What happens is that starch stored in roots and trunk is converted to sugar and transported to the parts of the tree that are now consuming it—the budding leaves. If a shallow hole is bored in the trunk of such a tree, the sweet sap will run out of the layer of phloem that is located just beneath the bark. The spring sap of sugar maple trees is collected in this way, boiled down, and sold as maple sugar or maple syrup. This sugar is largely sucrose, no different from the contents of an ordinary sack of table sugar. Differences in taste may be accounted for solely by the impurities maple sugar contains.

Not only food materials but also hormones may be distributed to various parts of the plant body by the flow of sap in the phloem. The direction of phloem transport is really "from" that of a source of material—usually sucrose—"to" a **sink** for that material, that is, a place where it is being consumed or converted to an osmotically inactive form such as starch.

Xylem

Xylem, in contrast to phloem, consists of hollow, dead cells laid end to end. There are several types of xylem cells, the most important being the fibers, vessels, and tracheids. The **fibers,** like sclerenchyma, are skeletal tissues and are vital for support in the stem, especially in plants composed of wood. Only the tracheids and vessels are conductive elements. The **vessels,** which with a few important exceptions are confined to flowering plants, are the simpler and presumably more efficient of the two types of water-conducting elements. They form from embryonic cells whose walls thicken and become impregnated with an impermeable and rigid substance known as **lignin.** Then the cells die, in the process losing their cyto-

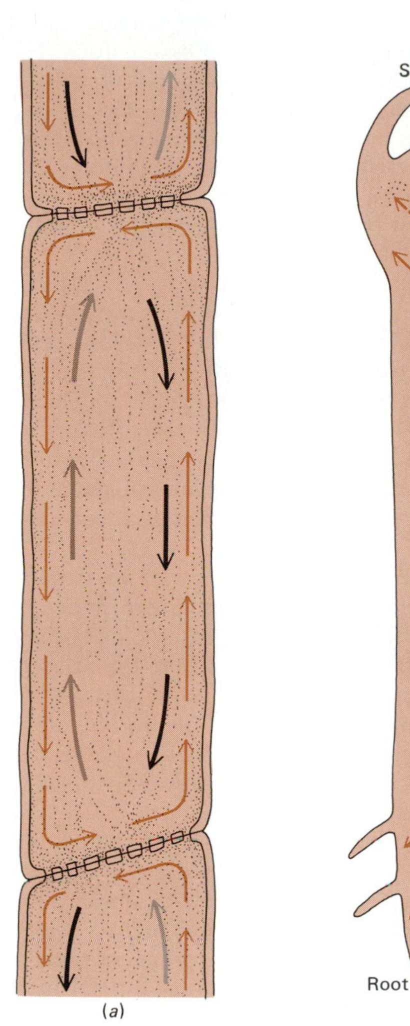

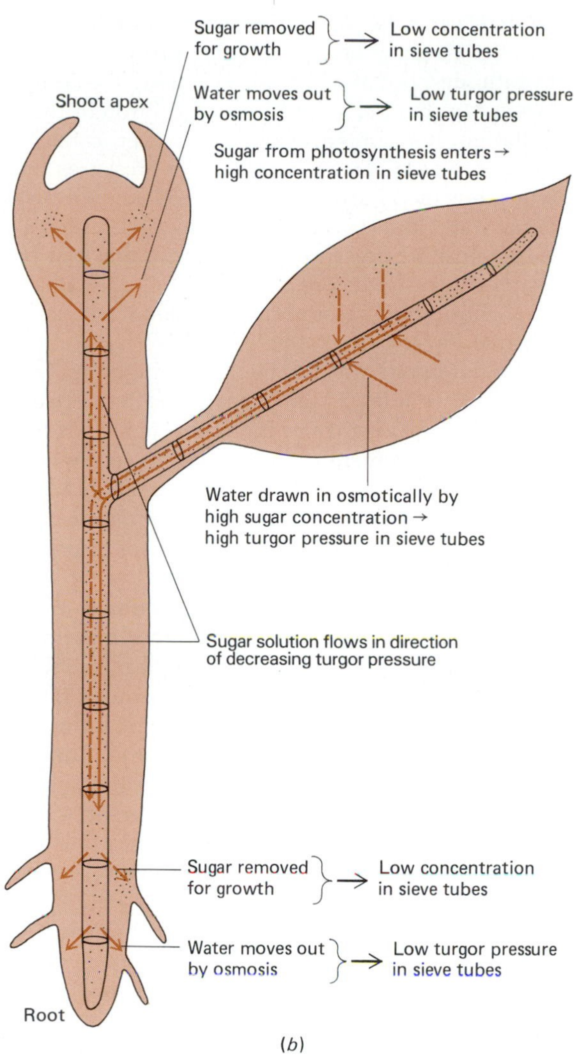

Figure 27–7 Phloem transport. (*a*) Cytoplasmic streaming in a phloem tube. The light and dark arrows illustrate that cytoplasmic streaming could account for the simultaneous transport of two different substances in opposite directions. (*b*) Diagram illustrating the pressure-flow theory of phloem transport in plants. (After Ray.)

plasm. The walls at the ends of these cells may persist after death in the form of perforated plates somewhat as in phloem, or as ringlike flanges surrounding the lumen of the vessel. Many vessels also have small pits in their walls, which we will consider later in the chapter.

The **tracheids** are smaller than the vessels and have acutely slanted end cell walls. Tracheids occur in all tracheophytes—in ferns, in flowering plants, and in gymnosperms. Tracheids are, in fact, the only kind of xylem tissue most gymnosperms possess.

How Xylem Works

The tallest trees approach 100 meters in height. A medium-sized apple tree transpires as much as 20 liters of water per day at the peak of the growing season, so a tree as huge as a redwood or giant blue gum transpires hundreds of gallons between dawn and sunset. Every milliliter of this water must be lifted to the leafy crown at a rate of as much as 75 cm per minute. The water is usually bound by capillary forces to the soil (Chapter 28) and must be collected from a large area around the tree base.

Much evidence indicates that root sap is not "pushed" up the tree. Mineral ions are absorbed by the root via active transport and move toward the center of the root. A special wax-impregnated area in the endodermis prevents the absorbed ions from flowing back into the soil. As a result of the accumulation of solutes in the root, an osmotic gradient is established in which water moves into the root to create a hydrostatic pressure called **root pressure.** If the soil contains ample water and an adequate oxygen supply (used in aerobic respiration that provides the ATP needed for active transport), root pressure can reach a value high enough to force water into the leaves of herbs and some shrubs. Plants possess modified stomata, called **hydathodes,** out of which the xylem sap forced upward by root pressure drips. The xylem sap (but *not* water) is exuded through the hydathodes, a process known as **guttation.** Root pressure alone, however, cannot account for the rise of xylem sap to the top of tall trees, because root pressures that develop can push sap upward only a few feet. Furthermore, it is well known that during the day, when transpirational water loss is occurring, xylem sap is under a *negative pressure* or tension rather than a positive pressure. (We will examine root pressure in more detail later in the chapter when roots are discussed.)

Since root pressure cannot account for movement of xylem sap to the top of a tall tree, is it, perhaps *pulled* up? We have already considered the properties of hydrogen bonds in Chapter 2. The highly polar structure of the water molecule (Fig. 2–8) ensures a significant attraction between the opposite partially charged portions of *adjacent* water molecules, which in turn produces a regular, somewhat ordered arrangement of these water molecules even in the liquid state. The hydrogen bonds also contribute to the force that holds water molecules together. Hydrogen bonds account for the fact that if a column of water is formed into a filament the diameter of a xylem cell, it has the tensile strength of steel—more than adequate to hang such a filament unbroken from the top of the tallest tree.

The columns of water rushing through the vessels and tracheids are of sufficiently small diameter to hold together indefinitely without forming a cavity. As water evaporates from the mesophyll of the leaf, the solutes in its cells become concentrated. Water then passes by osmosis into the mesophyll from the adjacent xylem located in the nearby vascular bundles of the leaf veins. As it leaves the xylem, each molecule of water exerts an infinitesimal pull upon the next in line in an unbroken chain stretching through the xylem of petiole and stem to the vast absorptive root surfaces among the interstices of soil particles below. Any break in this liquid thread, however small, would cause the entire mechanism to fail. Yet, there are fail-safe provisions against this very occurrence built into the microscopic anatomy of the xylem.

The cohesion system of xylem transport that vascular plants employ can function only if there is a continuous filament of water from one end of the xylem to the other. Though the absence of nuclei on which ice crystals can form does prevent freezing of the water in xylem down to about −30°C, in temperate and subarctic climates, that filament may be broken when sap in the xylem freezes. Such freezing expels any dissolved gas from the sap and produces gaps in the fluid column resulting from the long bubbles of gas thus released. How can xylem transport resume after thawing, when the cohesion of the water filament has been destroyed?

Let us look more closely at the anatomy of the xylem cell. Both tracheids and vessels are extensively supplied with tiny pits on their side walls. These pits connect *adjacent* cells with one another so that water can flow around an obstruction by traveling sideways and then upward. It is believed that the distinctive construction of the pits permits them to act as one-way valves, isolating the part of the water filament that has been disrupted by bubble formation and routing root sap around the area.

Other Xylem Functions

Xylem can also store food. In sugar maple trees, for example, transverse sheets of tissue, the **medullary rays,** unite the various layers of xylem, but also contain the starch that is converted to sugar in the spring.

MONOCOT STEMS

Stems of monocots and dicots are quite different from one another. The stem of a typical **herbaceous monocot,** such as a corn seedling (Fig. 27–8), is covered with an epidermis and cuticle not too different from that of the leaf. The stem may be provided with chloroplasts and even stomata. The bulk of its tissue is a pithy parenchyma with mainly skeletal function. In corn, for instance, vascular bundles are distributed rather randomly throughout this parenchyma, most of them in a vaguely defined layer around the periphery. In wheat and many other monocots, the center of the stem is actually hollow.

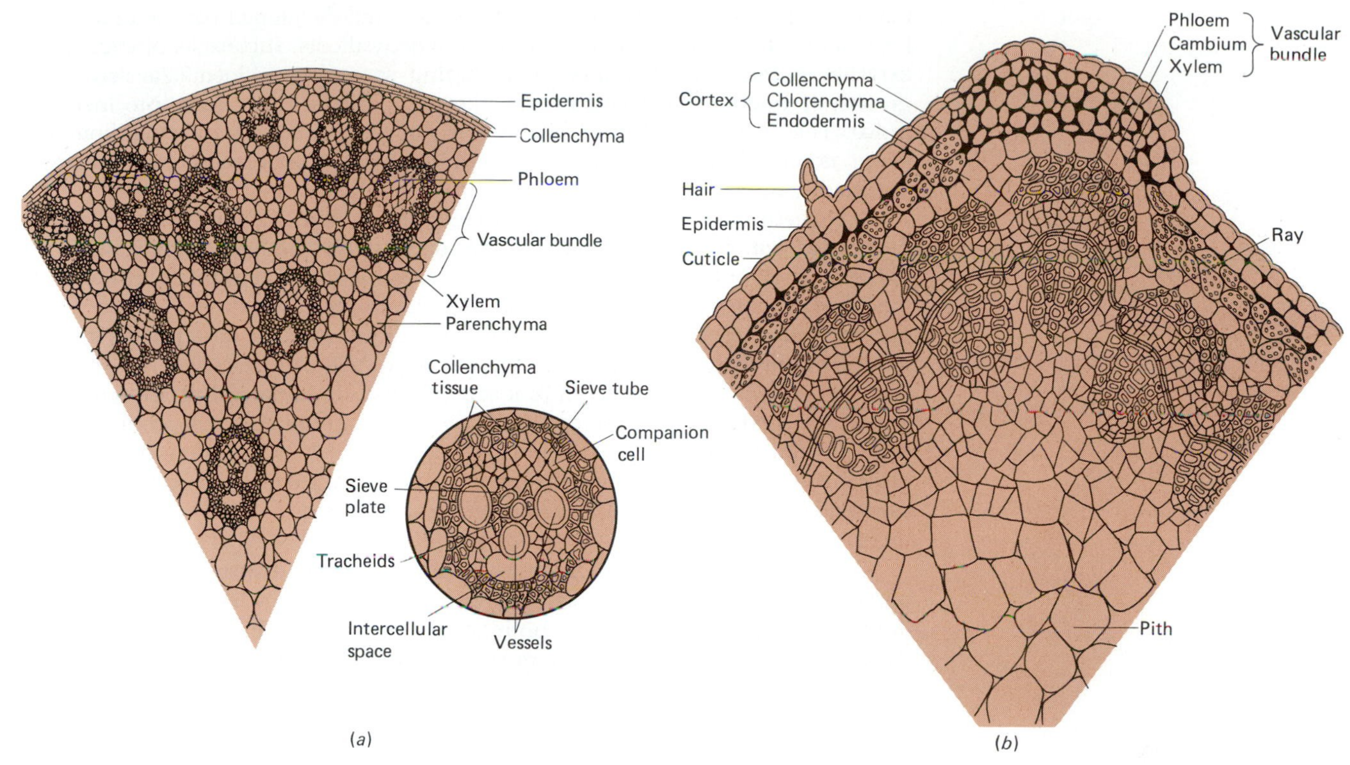

Figure 27–8 Stems in monocots differ from those in dicots. (*a*) A sector of a cross section of a stem from a monocot, corn. *Inset:* An enlarged view of a vascular bundle containing phloem (sieve tubes and companion cells) and xylem (tracheids and vessels). (*b*) A sector of a cross section of a stem from a herbaceous dicot, alfalfa.

In each vascular bundle, both phloem and xylem can be observed, with the xylem located on the interior aspect of the bundle and the phloem on the exterior aspect, that is, facing inward and outward, respectively. If it could be separated from the surrounding pithy parenchyma, each vascular bundle would appear to be a strand reaching from roots to leaves interwoven in complex fashion with other vascular bundles, but in cross section (as it is usually studied), each bundle appears as shown in Figure 27–8. Monocot stems grow by a variety of means, but do not usually have any means of secondary vascular growth; that is, once formed, the vascular bundle cannot enlarge further. In general, lateral growth of a monocot stem results from growth of the parenchyma, as can be seen by comparing the close-spaced vascular bundles in a young corn shoot with the widely spaced bundles of a mature stalk. This mode of growth results in a fairly constant stem diameter from the base to the top of the stem, such as can be observed in corn, bamboo, and palms, all of which are monocots.

DICOT STEMS

Many annual dicots have soft, green, non-woody stems. Such **herbaceous dicot** stems resemble those of monocots in exterior appearance, but their interiors are far different (see Fig. 27–8). Perennial dicots have a similar stem organization but also contain woody tissues that the short-lived herbaceous dicots do not usually develop.

In a typical herbaceous dicot, a skin of cork may cover the stem in place of the simple epidermis. **Cork** is composed of layers of dead cells impregnated with a waterproof substance, **suberin,** and with **tannins,** which are commercially extracted for use in leather production. In some ways cork is comparable to the outermost layers of mammalian skin, which are composed of dead cells impregnated with the waterproof protein keratin. When thick, the cork layer is called **bark.**

Figure 27–9 Lenticels on bark of young golden rain tree.

The cork layer is pierced by **lenticels** (Fig. 27–9), openings that permit gaseous exchange between the atmosphere and the respiring tissues of the stem. These lenticels permit gases to diffuse through even the thickest trunk, for in such a trunk only a layer a few millimeters thick of tissue beneath the bark is actually alive. Beneath the cork itself is a layer of **cork cambium,** whose function is to produce cells that differentiate, mature, and die, becoming converted into cork in the process. Cork cambium is an example of a **meristematic tissue,** a permanently embryonic tissue that can give rise to differentiated tissues throughout the life of the plant. Cork also contains dead phloem tissue. Well-developed bark is unusual in herbaceous dicots but is prominent in the woody dicots. It consists of everything external to the vascular cambium, including secondary phloem. As secondary growth of woody dicot stems occurs, the original cork layer thins into insignificance. The "cork" of commerce, which is not truly cork in the formal botanical sense, is cut from the very thick bark of a drought-resistant species of oak native to the Mediterranean region.

Beneath the cork cambium lies a layer of parenchyma known as the **cortex.** Embedded within this is a circular array of vascular bundles arranged with the phloem facing outward and xylem inward as in monocots, but totally confined to this layer, unlike those of monocots. In the center of the stem there is a core of **pith parenchyma.** Between the phloem and xylem of the vascular bundles is a layer of **vascular cambium.** This type of cambium is also a meristematic tissue, with a location that affords it access to both root sap and leaf sap to nourish its growth. As it grows, the cambium differentiates into a vascular cambium that will produce xylem centrally and phloem peripherally. This secondary growth produces considerable growth in stem diameter. However, herbaceous dicots are usually annuals, so their stems do not grow very thick as a rule, and true bark does not ordinarily develop in them.

Most woody dicots and gymnosperms are trees and shrubs. Since they are perennials, there is ample opportunity for their stems to grow in diameter. As they do, the cambium develops into a continuous sheet of cells. This becomes the vascular cambium, which produces a more-or-less continuous layer of phloem and xylem, so that the distinctions among the original vascular bundles become indistinct (Fig. 27–10). The vascular cambium itself becomes (and maintains itself as) a thin layer interior to the phloem. Interior to the vascular cambium is the xylem, which makes up more than 95% of the trunk of a typical tree. As successive seasons pass, ever more **secondary xylem** is laid down ring-fashion, while the phloem remains much the same thickness, because old phloem dies and is incorporated into the bark, so that the bulk of the stem comes to be made up of xylem. The oldest xylem, located near the center of the stem, becomes **heartwood,** the cells of which are so impregnated with resins and other strengthening substances that they are incapable of transporting root sap and become entirely skeletal in function. Heartwood may also serve as a repository for waste products of the tree's metabolism, which may also confer some degree of insect and fungus resistance upon the wood. The best cabinet and construction wood is heartwood.

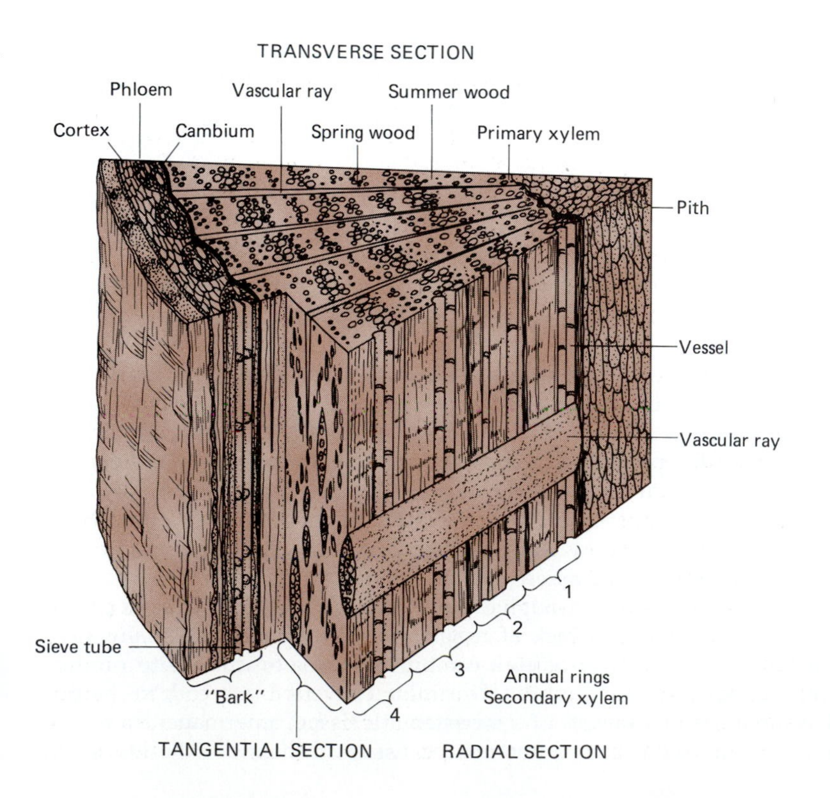

Figure 27–10 Diagram of a four-year-old woody stem, showing transverse, radial, and tangential sections and the annual rings of secondary xylem.

The younger xylem, located just interior to the vascular cambium, is **sapwood,** usually distinguishable by its much lighter color. If the heartwood is destroyed in an old tree by fungus or insect attack, the sapwood will continue to sustain the life of the tree until perhaps it is finally blown down and broken. In contrast, a tree cannot survive without sapwood nor, in the long run, without phloem. If the bark is removed in a complete circle around the trunk of a tree, the phloem is usually removed with it, and the roots starve for lack of leaf sap. The tree may survive for a season or so and may even produce a bumper crop of fruit (due to the accumulation of sugar in its crown), but it is doomed. Such an attack is called **girdling.** Farmers sometimes clear land in this fashion. Trees can also be killed by the girdling produced by the gnawing of beavers or of starving deer, and even by birds such as sapsuckers.

In temperate and desert climates, and also in some tropical, humid areas, there are often pronounced growing seasons. This results in differences in the amount of xylem produced by trees from season to season. In temperate-zone trees, there is often a big difference in color and texture between the less compact **spring wood** and the finer-pored summer wood. It is these differences that produce the familiar **growth rings** found not only in woody dicots but in gymnosperms such as pine trees. (Tropical trees such as jacaranda trees, not being subject to seasonal variations in growth, may lack growth rings entirely.)

Trees are often subjected to mechanical stress during their lifetimes. Other trees may press upon them, for example, and horizontally growing branches may be in danger of breaking from their own weight and the weight of the foliage they bear. In response to such stress, the cambium lays down wood more heavily on one side or the other in compensation. The formation of such **reaction wood** appears to be under hormonal control.

Stems represent a considerable investment of the products of photosynthesis, and a few plants possess adaptations that permit them to freeload, using the stems of other plants (see Focus on Freeloading Plants).

Roots

All substances the plant requires from its environment (except sunlight, oxygen, and carbon dioxide) must ordinarily be absorbed through the roots, and the anatomy of most roots is adapted to this function. Roots are by no means static structures, but grow in the soil, extending in directions where richer pockets of water and minerals may occur. Roots that have exhausted the possibilities of the surrounding soil simply die.

The total mass of a plant's roots is about equal to that of the branches, but can be very differently arranged. A small rye plant, for instance, has about 14 million primary and secondary roots, the total length of which is probably much in excess of 500 km, with a total surface area of about 275 square meters not even counting the surface of the root hairs. The taproot of an oak tree may extend 8 meters below the surface of the soil; the surface-feeding roots of a single prickly pear cactus extend no deeper than 15 cm but occupy a plot of ground 3 square meters in area.

ROOT STRUCTURE AND FUNCTION

Larger root branches much resemble the branches of the trunk of a tree and, like the trunk, are covered with cork provided with lenticels. The smallest roots are covered only with epidermis, many of whose cells are greatly modified to form long **root hairs** through which absorption takes place. If these root hairs are lost during transplanting, plants may be damaged or killed. Thus, when bare-root transplanting is used, virtually all foliage must also be removed, so as to greatly reduce the loss of water by transpiration until new root hairs can grow. The collective surface area of the root hairs is many times that of the plain epidermis.

The bulk of most roots consists of the **cortex,** composed mainly of parenchyma. In many roots food is stored in the form of starch grains deposited in the parenchyma cells (Fig. 27–11). The cortex also serves as a pathway for water movement, partially by the infiltration of water through the walls of its cells, but probably mostly by passage from one cell to another via intercellular plasmodesmata. The central core, or **stele,** of the root has no cytoplasmic connections with the cortical

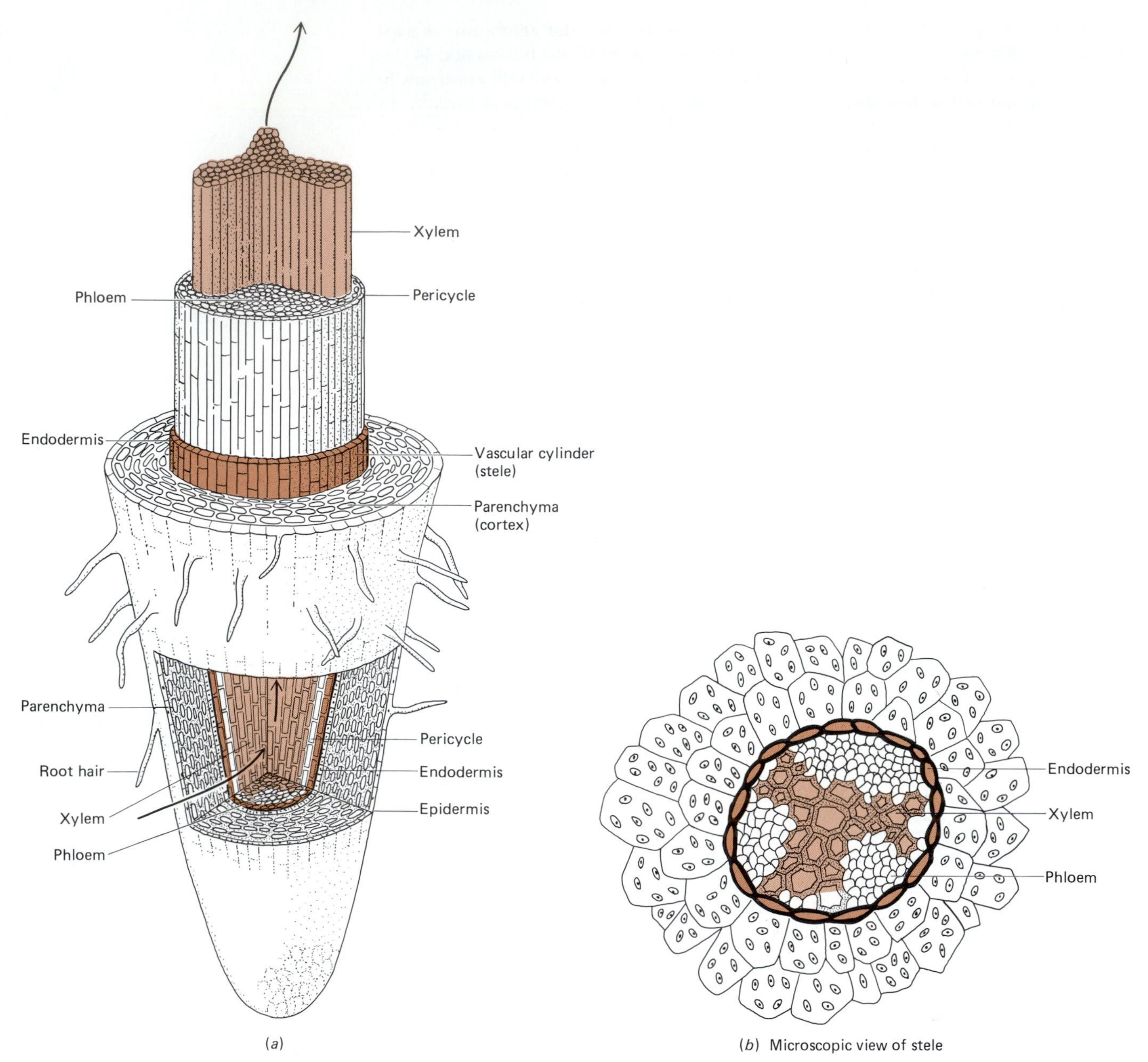

Figure 27–11 Root structures. (*a*) A typical young dicot root. Water, minerals, and certain organic substances enter the root hairs and cortical cells by diffusion and active transport. These then must diffuse through cortex, endodermis, and pericycle in order to reach the xylem. They then travel upward through the xylem. Food materials travel down through the phloem, and some food materials are stored in the form of starch in the cortex parenchyma cells. (*b*) Microscopic view of stele.

cells, however. Water and minerals are also prevented from passing between cells into the stele by means of the **Casparian strip,** a wax-impregnated area of the walls of the cells that surround the stele. These cells surround the stele in a single layer known as the **endodermis.** The Casparian strip of these cells prevents the inward (or outward) passage of fluid within their walls.

In addition, single-cell **root hairs,** located just behind the growing tip of very young roots, absorb water and minerals into their cytoplasm. These substances are then passed through cytoplasmic connections from cell to cell all the way to the center of the root, probably by-passing the Casparian strip of the cell walls when traveling into the root by this route.

If water with its dissolved mineral ions cannot pass directly around and between the cells of the endodermal barrier, it must pass *through* them, traveling first through their plasma membranes in order to do so. This permits the endodermal and probably also the stelar cells to regulate the chemical traffic into the plant. (Root absorption of nutrients is more fully discussed in Chapter 28.)

The importance of this arrangement in regulating the composition of a plant's body fluids is dramatically illustrated in mangrove trees. Several species of tropical shrubs and trees, collectively called **mangroves** (although they belong to several species), live in saline mud flats. Yet the sap in the xylem of these trees is far lower in salinity than the water surrounding their roots. The explanation for this is that during transpiration, so much pull is exerted on the water filaments in the xylem that water is forced through the membranes of the endodermis by atmospheric pressure. Yet these membranes are differentially permeable, permitting little sodium to pass. The result is that the salt is largely left behind, so that the plant functions as a solar-powered reverse-osmosis desalinization device. What salt does make its way to the leaves is excreted by tiny salt glands, often located in the axils.

It is more usual in typical terrestrial plants, however, for the roots to absorb minerals by way of active transport systems, causing water to follow by osmosis. This produces a root pressure that, as we have seen, forces water into the leaves of short plants; sometimes in amounts that threaten to drown the spongy parenchyma cells, or at least to fill the air spaces between them (Fig. 27–12).

Immediately inside the endodermis of the stele lies a layer of parenchyma tissue somewhat resembling cambium. This tissue is called the **pericycle.** The vascular tissues lie within this. In the roots of young dicots, xylem is usually arranged in a star-shaped configuration in the very center, with bundles of phloem occupy-

(a)

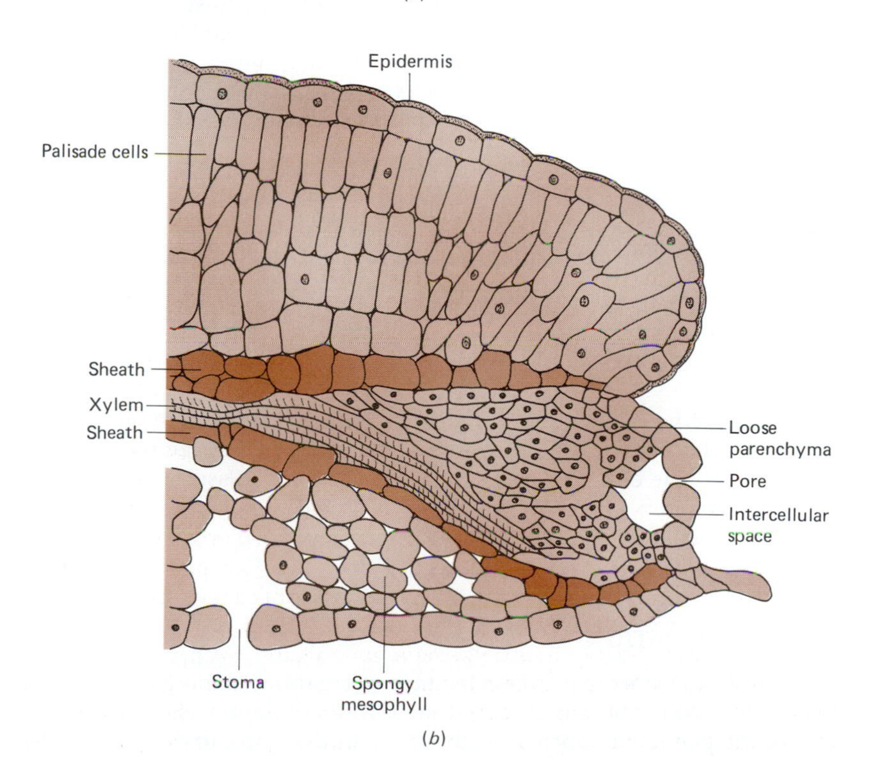

(b)

Figure 27–12 Too much water can pose a problem for plants. (*a*) Guttation in strawberry leaves. (*b*) A hydathode at the edge of a leaf of *Saxifraga lingulata.* Xylem sap, when under positive pressure, moves through intercellular spaces in the adjacent loose parenchyma tissue and is forced out through the hydathode's pore, a modified stoma. Water is prevented from moving into other leaf tissues by a compact cellular sheath lacking intercellular spaces. ((*a*), courtesy of J. Arthur Herrick, Kent State University, Kent, Ohio.) ((*b*), from Ray, Steeves and Fultz.)

FOCUS ON

Freeloading Plants

It is generally conceded that if possible it is better to get someone else to do your work for you. Quite a number of plants extend this strategy to having another plant provide some of their nutrients, or even some of their supporting tissues.

Most discussions of plant growth and development seem set in an ideal world where there is plenty of water, soil, and sunlight for the newly germinated seedling to use. In fact, it is often a world with intense competition for resources, preventing the success of all but a few seedlings. This is especially true of the floor of the tropical rain forest, which is a good bit like the lower depths of an aquatic habitat in that very little sunlight penetrates from the levels above. In the rainforest, the leaf mosaics of the crowns of the mature trees absorb or reflect to the sky most of the light; what is left has been filtered of almost all of the wavelengths most effective in photosynthesis.

If a newly germinated seedling survives at all, it is likely to persist as a sickly shadow of its proper self, and most such seedlings probably die eventually. Sometimes, however, chance brings down a nearby mature tree, letting in the sunlight long enough to allow a burst of growth that permits the seedling to thrust its crown of foliage into the sunlit heights.

The disadvantages of being a jungle seedling might be reduced if the new plant could minimize the amount of photosynthetic investment in its own supporting tissues. The major components of heartwood (and, for that matter, all the resins and hardeners found within it) are products of photosynthesis. As long as energetically expensive organic materials are tied up in supporting tissues, they cannot be used to produce buds, leaves, flowers, and fruit. Yet there is no way to do without a stem, is there?

Actually, there is. All that is needed is to utilize the stem of another, already established plant. Orchids and other epiphytes[1] like Spanish "moss," shown in the figure, are adapted to grow on the outside of a host plant's stem. There they can sometimes attain such masses as to break off entire branches with their weight, crashing disastrously to the forest floor. Usually, though, this does not happen, and the epiphytes prosper for years.

[1] Epiphytes are nonparasitic plants that grow on other plants, using the host for support.

Spanish "moss," an angiosperm epiphyte that does not usually harm its host tree.

The epiphyte's biggest problem is probably obtaining water and minerals, a problem that is solved by various adaptations. Some species simply absorb mineral matter from what little may be washed from the dust of the air by rainfall, but others may absorb some from the bark of their host. The roots of orchids themselves act as hosts for symbiotic microorganisms capable of at least fixing nitrogen from the atmosphere. Orchids also have a specialized spongy, water-absorbing tissue covering the roots, and most bromeliads form a cuplike rosette of leaves that can hold, in the larger species, gallons of water 50 meters above ground. Perhaps possessing the most specialized adaptation of all, the mistletoe absorbs at least some of its water and minerals from the host tree's

ing the spaces between the arms of the star. The phloem is often separated from the xylem by a layer of cambium, so that in perennial dicots, the roots as well as the stem are capable of secondary growth. The vascular tissues differentiate from a vascular cambium that originates by a fusion of the existing cambium with the pericycle to make a continuous layer. Branch roots also originate from parts of the pericycle.

ROOT RESPIRATION

The soil is a complex, particulate mixture of organic and inorganic components. A large proportion of the soil is composed of water or air. Roots respire like any other part of the plant and depend mostly upon diffusion to carry oxygen to their cells.

tissues, although it carries out its own photosynthesis. (For this reason, mistletoes are conspicuously green even in winter or other times of water shortage when most other deciduous plants lose their leaves, as can be seen in the accompanying photograph. Ancient peoples took this to be a reflection of magical powers, a vestige of which belief persists in modern holiday customs.)

Some plants are semi-epiphytes, remaining rooted in the forest floor, but climbing on host trees until they are able to place their crowns in lighted areas. Seedlings of the common greenhouse plant and ornamental *Monstera,* for instance, are negatively phototropic—that is, they are initially repelled by light. That causes them to grow into the shadow of a nearby host tree, where eventually they make contact with the trunk. At that point, other tropisms[2] (geotropisms and thigmotropism) dominate their growth so that they grasp the trunk of the host tree with thick tendrils and climb rapidly up it.

Strangler fig, growing on the trunk of a palm tree. The younger strangler fig stem is as yet too frail to support its own crown in the upper story of the forest, but the palm trunk serves this purpose temporarily.

Mistletoe, a rootless plant parasite or, perhaps, semi-epiphyte, still green on its leafless oak host in midwinter.

Perhaps most bizarre of all, the tropical strangler fig avoids investment in its own stem tissues by growing on the trunk of a host tree (as shown in the figure) until it reaches the sunlight, and then completely envelops and destroys the trunk of its hapless host. In time its own trunk becomes thick and strong enough to support its crown of foliage unaided by the host tree.

[2] A **tropism** is an automatic growth response to a stimulus, virtually the only behavior of which most plants are capable (Chapter 48). A geotropism is such a response to the force of gravity; thigmotropism involves response to touch; hydrotropism, to water.

This oxygen must be available to the roots in their immediate surroundings, or they will suffocate. If the water content of the soil is too great, this is exactly what will happen; perhaps it would be better to say that they will drown.

Some plants, however, are especially adapted to live in aquatic habitats or those with boggy soil. The water lily is an extreme example. This plant is equipped with floating leaves that possess stomata on their upper sides, as might be expected. Its tissues are also more spongy than usual, which aids the flotation of the leaves and also serves to carry air via the stems to the roots in the mud at the bottom. There are even parasitic insects that live on the submerged roots and stems, dependent on the plant not only for food but for air as well.

In certain plants such as some mangroves, air can reach the roots by special protruding organs called **pneumatophores** that reach above the water surface into

(a)

(b)

Figure 27–13 Pneumatophores, specialized organs that provide air to roots. (*a*) Cypress "knees," possibly respiratory structures that may function to provide oxygen for the metabolism of the roots, which are buried in anaerobic mud. (*b*) Pneumatophores of coastal mangrove trees. These serve as a habitat for barnacles, algae, and snails, in addition to their principal role in root respiration.

the air and can form quite a small forest around the base of the plant (Fig. 27–13). The protruding "knees" of cypress trees may have a similar function. The roots of some tropical trees also serve to prop up the plant; logically named **prop roots,** some of these may also serve as respiratory organs.

SUMMARY

I. The plant body may be divided into the root and the above-ground portion, the shoot. The shoot is in turn composed of stem and leaves.

II. The leaf possesses an upper and lower nonphotosynthetic epidermis overlain by a waxy cuticle. Guard cells and stomata permit gases and water vapor to pass in and out.

- *A.* The mesophyll is the photosynthetic tissue of the leaf. The spongy and palisade parenchymas are abundantly supplied with chloroplasts. Most of the volume of the mesophyll consists of air.
- *B.* The leaf veins contain phloem and xylem, which conduct materials to and from the parenchyma.
- *C.* To conserve water, the guard cells lose turgor pressure thus closing the stomata. This behavior appears to be partially under the control of the hormone abscisic acid.

III. The stem supports the leaves. In addition to pith, bark, and meristematic tissue, the stem is composed of the conducting tissues, phloem and xylem, whose arrangement varies between monocots and dicots.

- *A.* Phloem conducts sugars from areas of high concentration to areas of low concentration by a dynamic flow mechanism that is dependent upon an osmotic gradient. Since sugars are usually made in leaves and stored elsewhere in an osmotically inactive form, phloem flow is often downwards. Other materials such as hormones are also transported.
- *B.* Xylem carries water and minerals from where they are absorbed by the root to the above-ground portions of the plant body. It is transported by cohesion within the xylem cells in response to the force generated by the evaporation of water from the leaves.
- *C.* Especially in woody dicots, older layers of xylem become impregnated with resins and other materials to form heartwood, which supports the plant.
- *D.* The stems of dicots and monocots differ.
  - *1.* In monocot stems, vascular bundles are arranged throughout the pith of the stem.
  - *2.* In dicot stems, vascular bundles are confined to the periphery and may fuse with one another. In such a case the phloem is located just under the bark, and together with the cork cambium, helps to form the bark.

IV. The root serves to hold the plant in place and to absorb water and minerals from the ground. Root hairs actively absorb much of the needed materials, which then pass through the cortical tissues of the root to the central vascular cylinder. There they are pushed by root pressure and pulled by cohesive transpiration forces through the xylem up to the stem and leaves.

POST-TEST

1. The main functions of the plant root are ____________ and ____________ of water and nutrients.

2. The ____________ ____________ are the only photosynthetic cells located in the leaf epidermis.

3. The openings between the guard cells are

known as the ______; these close when ______ is in short supply and in response to certain other stimuli.

4. In many leaves, part of the photosynthetic layer is comprised by the long, cylindrical ______ ______ cells.

5. The turgor pressure within the guard cells changes in response to an influx of ______ into these cells. Water follows by osmosis.

6. Phloem tissue contains ______ ______, which are without nuclei, but which are served by adjacent ______ ______, which are nucleated. The former cells are lined with ______.

7. Movement of materials in phloem is from some high-sugar ______ to a low-sugar ______; this gradient, most of the time, produces a flow from the ______ to the lower portions of the plant, but this is not always true.

8. Xylem is non-living tissue; it generally transports ______ and dissolved ______ in response to a combination of transpiration and ______.

9. A minor function of xylem in some plants is ______ storage; it also performs a skeletal function when it turns into ______ wood, in which resins and other compounds are deposited.

10. Roots generally generate a positive ______ ______, which can result in the exudation of water, known as ______, from specialized stomata in the leaves.

REVIEW QUESTIONS

1. Compare the structure and cellular components of xylem and phloem.
2. What are the functions of roots?
3. Describe the processes by which a root absorbs water and salts from the surrounding soil and how these materials enter its vascular tissues.
4. Describe the three-dimensional structure of a woody stem. What are the functions of stems? Can stems and roots always be distinguished reliably from one another? If so, how?
5. Describe the functions of the following structures:
   a. cambium
   b. stomata
   c. heartwood
   d. lenticels
   e. abscission layer
   f. cuticle
   g. Casparian strip
6. Summarize the mechanisms by which guard cells regulate the size of the stomata.
7. What are the functions of leaves? What is the role of transpiration in the plant?
8. What special adaptations do desert plants employ in their photosynthesis, and why are these desirable from the plant's point of view?
9. What are the main differences in stem structure between monocots and dicots?

28
Growth and Development in Plants

OUTLINE

I. Seeds
- *A.* Seed structure
  - *1.* The monocot seed
  - *2.* The dicot seed
- *B.* Seed germination
  - *1.* Breaking dormancy
  - *2.* The early root and emergence

II. Meristems and buds
- *A.* Primary meristems
- *B.* The leaf bud
  - *1.* Leaf growth
  - *2.* Leaf fall
  - *3.* The ecology of abscission
- *C.* Secondary growth
  - *1.* In dicots
  - *2.* In monocots

LEARNING OBJECTIVES

After you have read this chapter you should be able to:

1. Give the main structural features of typical monocot and dicot seeds.
2. Describe the developmental events of dormancy, germination, and emergence.
3. Describe apical (primary) meristems and their derivatives.
4. Describe or diagram a typical leaf bud.
5. Summarize the processes of leaf growth, senescence, and abscission.
6. Summarize the main adaptive features of leaf loss.

Although a plant begins life as a seed, the seed has a history of development all its own. The fundamental body plan of a plant is laid down even before a seed leaves the parent plant. If an ordinary dried bean from the supermarket is soaked in water and then opened (Fig. 28–1), the future root, stem, and leaves can easily be seen already formed in the embryo within. What cannot be seen are the hundreds of branches of root and stem, and all the leaves and flowers that are potentially there and would eventually result from genetically programmed growth and differentiation.

Seeds

The seed plants are, you will recall, classified into two major groups, the gymnosperms and the angiosperms—the gymnosperms usually having nothing that could be called fruit, and the angiosperms always exhibiting this trait. The angiosperms are, in turn, divided into the monocots and the dicots, mostly on the basis of their seed structure.

SEED STRUCTURE

All angiosperm seeds have a similar fundamental plan: On the exterior is found a seed coat, and inside that are the endosperm food storage tissue and the embryo. Where they differ most strikingly is in the size (or even the presence) of the endosperm, for in some plants the task of food storage is accorded to special organs—the **cotyledons**—of the embryo itself.

The Monocot Seed

In a typical monocot such as a grass, the seed (Fig. 28–1) is surrounded with a protective **seed coat.** The **aleurone layer,** which lies under the seed coat, functions

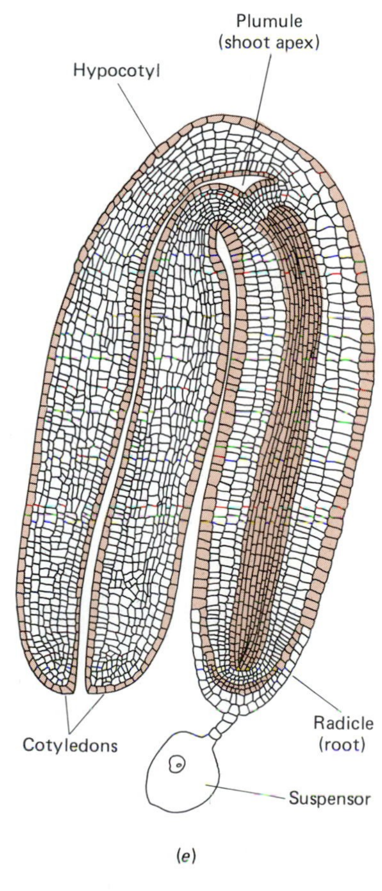

Figure 28–1 Examples of seed structure. (*a*) Castor bean, a dicot that possesses endosperm. (*b*) Corn, a monocot with endosperm. (*c*), (*d*) Common bean (*Phaseolus vulgaris*), a dicot with no endosperm, its food reserves stored instead in the embryo's thick fleshy cotyledons. An internal side view with one of the cotyledons removed is shown in (*c*); an external view, edge on, in (*d*), showing by dotted lines the location of cotyledons and embryo axis within the seed. (*e*) Embryo of shepherd's purse, a herbaceous dicot. (From Ray, Steeves, and Fultz.)

as a kind of digestive organ in seed germination. Most of the seed consists of starchy **endosperm,** a food storage tissue that forms separately from the embryo while the seed is still attached to the parent plant. In some monocots the embryo proper is demarked from the endosperm by a tough scutellum, and often by another layer known as the **coleorhiza,** which protects the future root, or **radicle,** as it grows out of the seed into the surrounding soil. All of these structures will be discarded once they have served their function, leaving only the radicle, the **mesocotyl** (future stem), and the **shoot apex,** which by then will have developed into the mature versions of these structures.

Monocot means "single leaf," which refers to its only seed structure that appears to be homologous with a leaf—the single cotyledon. The monocot cotyledon absorbs digested food from the endosperm. In grass embryos a very large cotyledon develops, which is known as the **scutellum.**

The Dicot Seed

Dicots, as their name implies, have *two* cotyledons (Fig. 28–1). Sometimes, as in the castor bean plant, these do function as actual leaves. More typically, as in beans, peas, and maple seeds, the cotyledons serve as food storage organs in place of the endosperm. In a bean, for instance, the seed coat surrounds two vast cotyledons, which are so engorged with food that they dwarf the comparatively tiny embryo that lies between them.

Other dicot seed structures include the radicle and the **plumule,** which corresponds to the shoot apex in monocots. In most dicot seeds, however, there is nothing comparable to the aleurone or coleorhiza, although some do have a functional endosperm.

The embryo and its cotyledon(s) are the result of one of two distinct fertilizations: In one of these, the endosperm (if any) is produced. The other fertilization process, which gives rise to the embryo, yields a zygote, which then undergoes cleavage. The cell divisions that the zygote undergoes following fertilization first produce a basal cell and a terminal cell. From the basal cell develops a filament of cells called a **suspensor.** The terminal cell divides, forming a rounded mass of cells; from this mass grow the two cotyledons and a central **axis.** In dicots, the part of the axis below the point of attachment of the cotyledons is called the **hypocotyl** and the part above it, the **epicotyl.** The embryo is in this state of development when the seed becomes dormant.

SEED GERMINATION

The embryonic plant is kept in a state of suspended development called **dormancy,** which ends when germination takes place. The length of time germination can be delayed varies greatly. As the seed develops there is a progressive loss of water from its tissues. Absorption of water from surrounding soil to replace this is usually the main requirement for germination. For many seeds, however, there are other requirements. Certain seeds germinate under the influence of light; others require extensive soaking, scarring of the seed coat, or even charring by fire. These seemingly perverse requirements make ecological sense. For instance, a seed that will sprout only after exposure to fire will do so at a time when competition by mature plants has been reduced by fire. A desert annual that sprouts after a brief rain will shrivel and die, but one that will sprout only when enough rain has fallen to leach a germination inhibitor out of its seed coat will probably find enough water in the ground to support growth throughout its brief life cycle.

Some seeds will die if they do not germinate rapidly—for example, sugar maple seeds, which live less than a week, or peach seeds, which die after a month or so of dormancy. At the other extreme, Indian lotus seeds more than a thousand years old have been induced to germinate by careful abrasion of their seed coats followed by a good soaking. Whenever the requirements for germination are provided, the seed initiates a series of events that breaks dormancy, speeding up the rate of its life processes hundreds of times and leading to the development of a mature organism.

Breaking Dormancy

As a first step in breaking dormancy, the embryo emits a hormone called **gibberellin,** which diffuses through the seed. In monocots, the hormone then triggers the production of digestive enzymes by the aleurone (Fig. 28–2); in dicots, the digestive

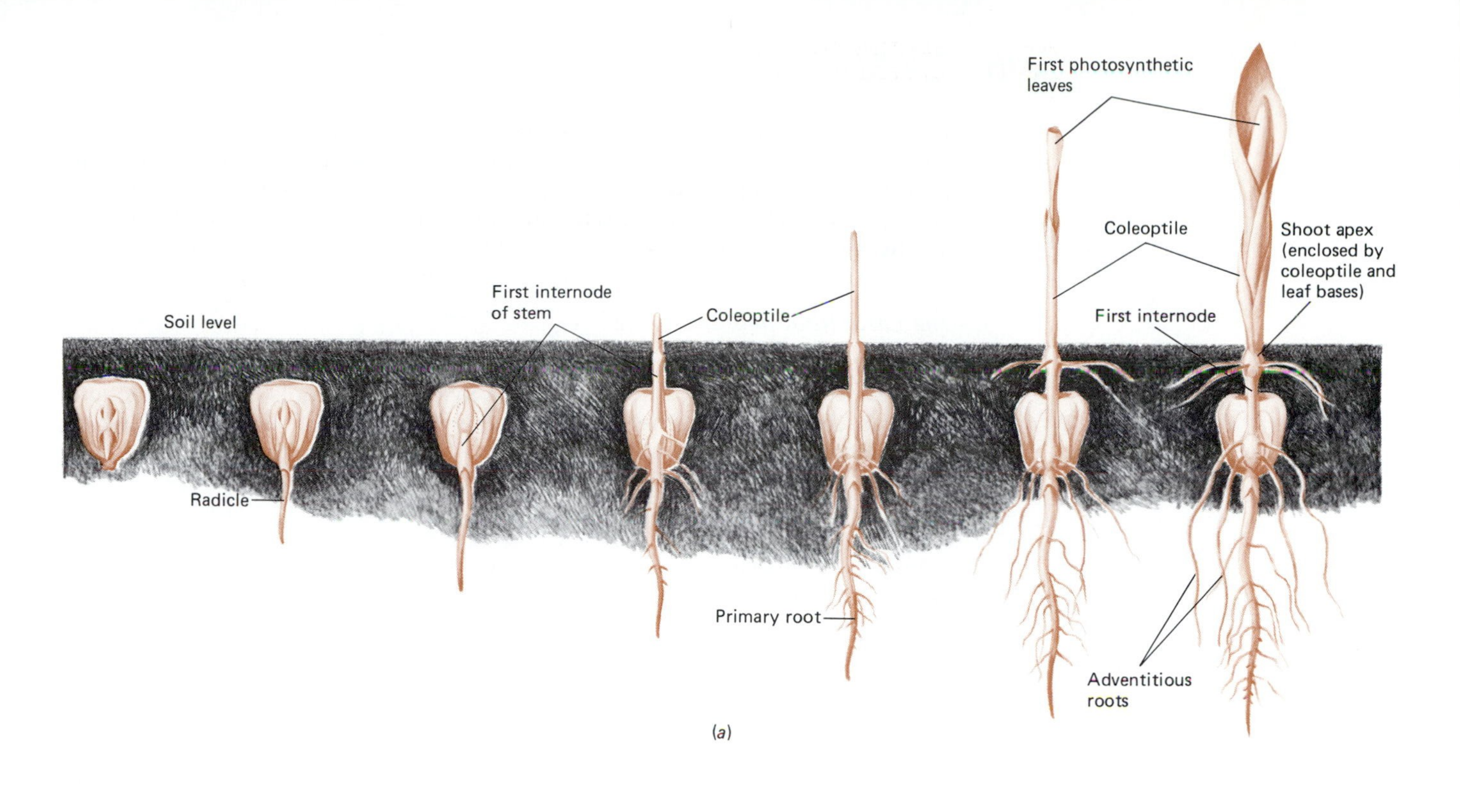

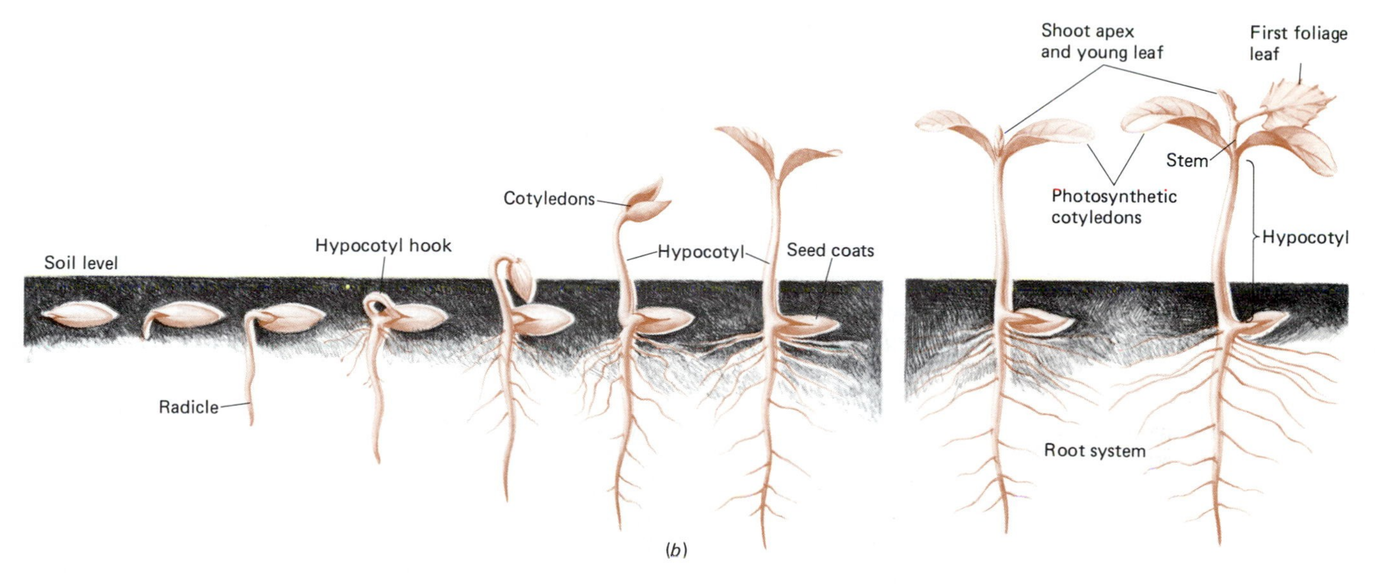

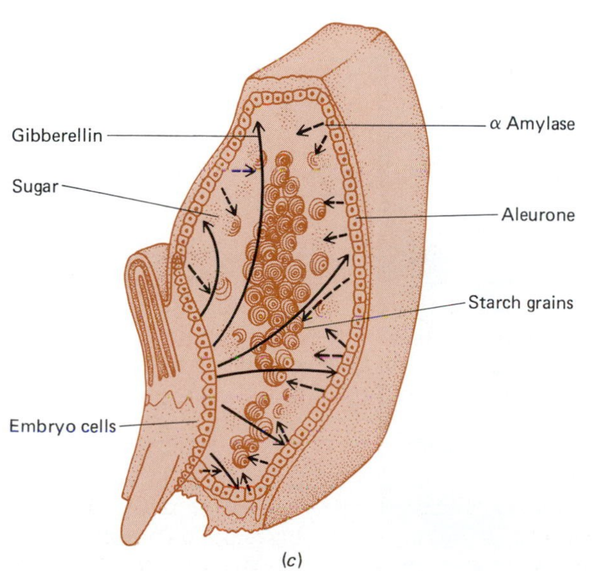

Figure 28–2 Seed germination. (*a*) Hypogeous germination in corn. In this form of germination, the seed remains underneath the surface of the soil. (*b*) Epigeous germination in squash (*Cucurbita*). In this form of germination the seed is forced out of the earth exposing the cotyledons. Note that at the base of the hypocotyl the squash seedling develops a spur (probably a modified root) that holds the seed coats, helping the hypocotyl hook to pull the cotyledons and shoot tip out of the seed coats. The seed coats are left below ground. Most epigeously germinating seeds do not have this feature. (*c*) The action of gibberellin in the germination of barley grains (highly diagrammatic). Following water imbibition, cells of the embryo produce and secrete gibberellin. The gibberellin in turn activates the synthesis of starch-digesting enzymes (amylases) in the living cells of the aleurone. The amylases are secreted into the starchy endosperm, where the hydrolysis of starch to sugar occurs. ((*a*), (*b*), (*c*), from K. Norstog and R. W. Long.)

enzymes are produced by the cotyledons. These enzymes then proceed to break down the stored food in the endosperm or cotyledons. For example, amylase breaks down starch to form maltose, which is then cleaved by maltase to yield glucose; and other enzymes attack stored proteins, fats, and oils, mobilizing them for the seedling's use.

The Early Root and Emergence

If a bean is placed on a piece of moist blotting paper, the initial swelling caused by imbibed and osmotically absorbed water is followed by wrinkling and cracking of the seed coat. The white root tip then emerges and elongates greatly, forming elaborate convolutions and developing a fuzzy appearance from the multitude of tiny root hairs that form on it (Fig. 28–3).

Microscopic examination of the root tip will disclose a cap consisting partly of dead or moribund cells at its apex, whose function is largely to protect the tender tissues from abrasion by soil particles, and in addition to serve as a source of growth hormones. The actual living tissue, however, is undergoing vigorous mitosis. For that reason this part of the root tip is known as the **zone of division.** It is an **apical meristem,** a growing tip of embryonic, differentiating tissue. Despite this division, very little growth takes place at the apex of the tip, because the cells there have not yet begun to enlarge to any great extent. By the time they have begun to do so, they are left behind in a **zone of elongation,** which is responsible for most of the lengthwise growth of the root tip. Since new cells are constantly added to this zone by mitosis, however, growth continues indefinitely. At the same time, the older portions of the zone of elongation cease to grow and become incorporated into the **zone of maturation,** where tissue differentiation now begins.

It is in the zone of maturation that the root hairs make their appearance. The root hairs absorb much of the water and minerals that enter the root. They are long, cylindrical extensions of some of the outermost cells of the former zone of elongation. In still older portions of the root, the outermost cells lose their root hairs and may die, becoming converted into corklike epidermis.

Root growth does more than push the root down deeper into the soil. It is also responsible (along with the other seedling parts) for pushing the rest of the embryo and the cotyledons out of the soil and into the sunlight. In dicots, at the point of attachment to the cotyledons the hypocotyl develops a sharp hairpin curve—the **hook** (see Fig. 28–2). The convex side of this hook bears the brunt of

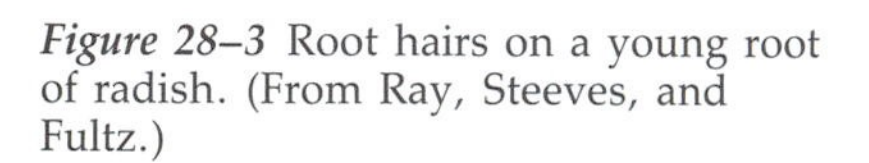
Figure 28–3 Root hairs on a young root of radish. (From Ray, Steeves, and Fultz.)

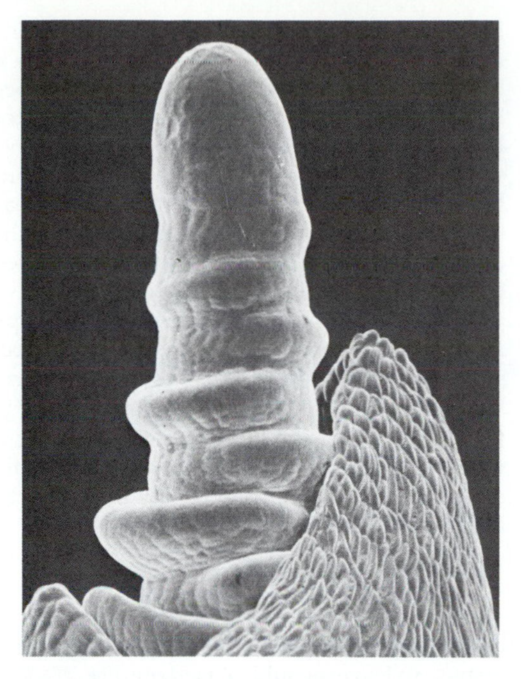

Figure 28–4 Scanning electron micrograph of the shoot apical meristem of Arawa wheat (×200). The tip of the apex consists of apical initial cells, which divide to form the shoot; the leaves are initiated lower down on the side of the apex. The leaves first appear as a ridge, but with successive cell divisions and expansions, the size of the ridge increases, and the ridge takes on the more familiar shape of a leaf. The spiral arrangement of the future leaves is already evident in that of the primordia. (From Troughton, J., and Donddson, L. A.: *Probing Plant Structure.* New York, McGraw-Hill, Book Co., 1972.)

breaking a pathway in the soil, dragging the cotyledons and the rest of the new plant behind it.[1]

As soon as the hypocotyl and cotyledons become exposed to sunlight, the red light wavelengths excite the photoreceptor pigment **phytochrome,** which occurs in the cells of the seedling. (Phytochrome also has many other roles in the control of the life processes of the plant.) The phytochrome triggers a series of only partly understood processes that result in a straightening of the hook, the unfolding of the cotyledons to expose the leaves of the seedling, the inhibition of further emergence from the soil, and the production of chlorophyll. By now the food stored in the cotyledons is exhausted, or nearly so, and although the cotyledons may turn green and live for a while, their function is fulfilled and they soon drop off. The proliferating leaves of the seedling are now in business, and photosynthesis rather than digestion of stored foods fuels the metabolism of the young plant.

Meristems and Buds

The root tip apex will persist in its rapidly dividing state throughout the life of the plant, as will those of the branch roots that will eventually develop. A similar, permanently embryonic bit of tissue occurs at the apex of the above-ground stem, or shoot (Fig. 28–4), and also at the tips of the terminal buds of all the above-ground branches that will eventually develop. *All* of the plant's above-ground body is ultimately derived from these apically located **primary meristems.** Each small branch has almost a segmental layout, with buds clustering together at one level, the **node,** which is separated from the next adjacent node by a budless length of stem, the **internode.**

PRIMARY MERISTEMS

For the most part, the primary meristems give rise to differentiated tissues such as phloem and xylem. However, some tissues derived from the primary meristem remain undifferentiated and form the cambium layers of the stem and the similar pericycle layer of the root. These eventually form differentiated tissues themselves, but in such a way as to produce lateral growth and increase in girth—for example,

[1]The details of seedling emergence differ in the grasslike monocots. For example, they push a sheaf of tightly rolled leaves into the sunlight. The leaves are protected from abrasion by a sheath, the **coleoptile** (of which more will be said later). The coleoptile and its enclosed leaves grow principally from the base, rather than the tip as in dicots. Once the coleoptile has emerged, it is ruptured by the growth of the leaves. The leaves then expand into the air and commence photosynthesis.

by giving rise to new vascular tissue in the stem. Portions of this secondary meristematic tissue, however, can also give rise to new apical meristems in the lateral buds, which can produce branches if they become active. Typically, these lateral buds form in the axils of the leaves, but lateral buds are not restricted to those locations in all species. Similar branching occurs among the roots, particularly when older roots die, but not by the formation of buds. Branch roots are produced by cell divisions originating in the pericycle and subsequent breakthrough of the branch root as it penetrates all the tissues of the root that are external to the xylem. (The pericycle also can give rise to the cork cambium of the root, and to portions of the vascular cambium.)

Lateral stem branching tends to be suppressed as long as the topmost bud is intact, a phenomenon called **apical dominance.** (The tendency is not absolute, and varies with the growth habits of various species of trees and shrubs.) This is why bushy branching of shrubs is encouraged by pruning their tops. By removing the uppermost buds, the inhibition of branching is reduced. The internal control system that produces lateral bud inhibition is unknown, but is thought to involve the secretion of a hormone, probably an auxin, by the uppermost bud, as discussed later in the next chapter.

THE LEAF BUD

In the very early spring, if we were to examine a twig of a deciduous tree such as a maple, no leaves would be evident (Fig. 28–5). The previous season's leaves would have long since dropped off, each leaving behind it a **leaf scar.** Close inspection would disclose the axillary buds just above each leaf scar. Most of them have not yet developed and never will. Often, though, on the opposite side of the twig from the leaf scar, a still smaller branch that *did* develop from an axillary bud can be found. At the end of this branch there will probably be three buds—a flower bud, which perhaps may actually be in bloom, and two leaf buds. Each leaf bud is covered with a mosaic of overlapping scales—tiny modified leaves that serve only to protect the dormant meristem beneath them. In the summer these are absent, the bud being surrounded by green **leaf primordia,** which will grow into functional photosynthetic organs. In the very center of the bud is the terminal meristem, hardly bigger than a pinpoint. As the days grow longer and warmer, the entire bud swells with the growth of the meristem within and the tightly rolled bundle of leaves it is producing.

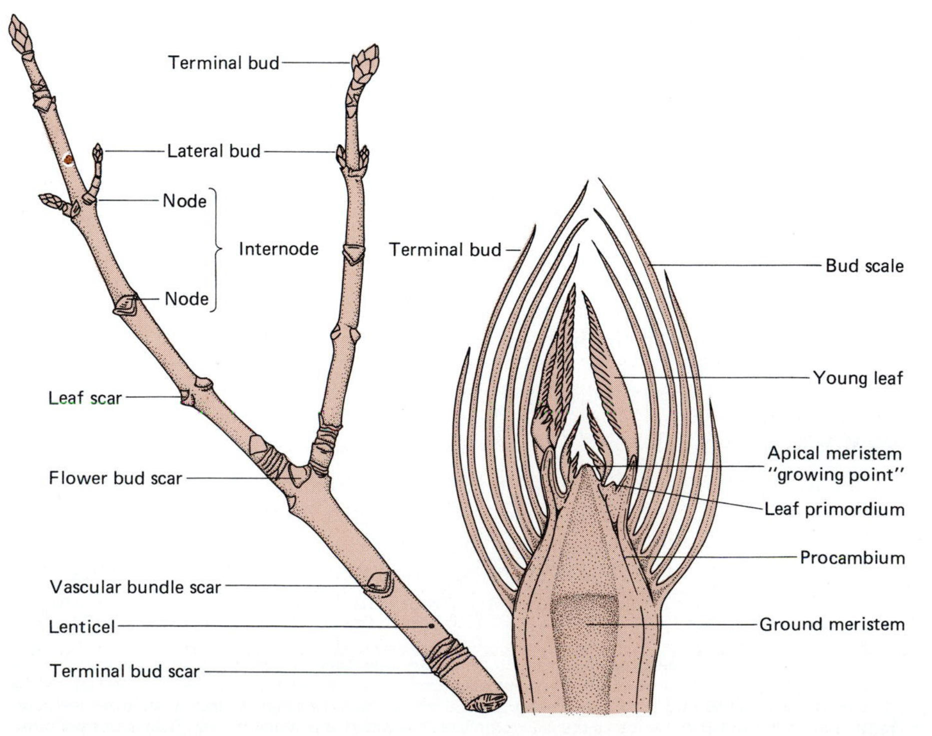

Figure 28–5 Leaf bud structure. (*a*) A twig from a horse chestnut tree, showing buds and scars. (*b*) An enlarged longitudinal section through a terminal bud from a hickory tree.

Leaf Growth

Each leaf originates on the side of the meristem, growing upward as it enlarges and begins to differentiate. Once its growth is well under way, another group of cells next to it and somewhat above it begins to grow outward and upward. Meanwhile the tip of the meristem itself is still growing, so that now there is room on it for the beginning of yet another region of growth next to and somewhat above the last one. This pattern of growth produces a spiral arrangement of new leaves around the branch. As each leaf differentiates into blade and petiole, a bud forms in the axil that may, a year later, give rise to yet another branch. Thus the secondary branch arrangement reflects the original alternate or spiral leaf arrangement. (In some plants, leaf primordia form on opposite sides of the stem. The description here really only applies fully to those with alternate-side leaf formation.)

Leaf Fall

Leaves age and die in all plants, and in some cases the plant senesces as a whole. Even the oldest plants, such as redwoods, experience the aging process. As xylem ages, for example, it becomes clogged with resins and turns to heartwood. As cells produced by the cork cambium age, they die and become converted to cork. In fact, even in the oldest redwoods, the tissue that is alive at any one time is only (at the most) a few years old.

The part of a perennial plant that ages most obviously—and sometimes spectacularly—is the leaf. Even in evergreen trees leaves constantly fall from the plant, as the heaps of needles at the base of a pine tree attest. In deciduous trees, however, all leaves age and die at about the same time in what appears to be basically a water conservation measure.

The Ecology of Abscission

The deciduous habit can often be observed in places like the tropical northeast areas of Brazil, where there is a great deal of rainfall during part of the year, but a pronounced dry season during another part of the year. The jacaranda tree is a typical member of this Brazilian forest community and has also been introduced into tropical and subtropical habitats all around the world. The jacaranda loses its leaves during the dry season both in its native Brazil and in Florida, where it is also now common. But in both places the dry season is not notably cold. It is the *dryness* of the season—the fact that it is a time of water scarcity—that gives adaptive significance to the leaf loss. Without leaves the tree cannot lose water by transpiration, so its loss of leaves helps the tree to conserve water.[1]

Similarly, in cool, temperate climates, deciduous trees (such as maples, sweet gum, and many species of oaks) lose their leaves during winter. The many feet of snow that accumulate in a northern woodland make it difficult to think of winter as a dry season in such a habitat, but it is indeed dry physiologically and ecologically. The frozen water in the snow and ice cannot be absorbed by plants, and even when the water is still liquid, it can be absorbed only slowly, if at all, when temperatures are near the freezing point. In the plants that *do* retain their leaves in winter, such as conifers, the leaves are often small and needle-like, or are covered with waxy resin (like magnolia leaves), or have other adaptations that minimize water loss.

Oak trees are good examples of plants with varying adaptations to water shortage. The post oak tree, for example, has broad, large leaves that are relatively wasteful of water, so that it cannot compete well in really dry habitats. In August most of the leaves are elderly, and few new ones are still being produced. Photosynthesis has greatly declined, and chemical analysis will disclose a reduction in protein and often sugar content of the leaves as vital substances are withdrawn by the main body of the plant to be conserved for future use. When the leaf seems almost dead, it may begin to change color. Color change is accelerated dramatically if the days are sunny and the nights cool, a combination that is rare in most of the world, but which produces spectacular autumnal coloring in much of North America, Japan, and some regions of China. The appearance of these colors is partly due to the loss of green chlorophyll, which unmasks the carotene and anthocyanin

[1] In some tropical habitats where there is seasonal flooding, the roots of some trees are unable to respire at the height of the wet season due to soil saturation and resultant anoxic conditions. These trees cannot then absorb water, and they lose leaves during *flooding*.

(a)

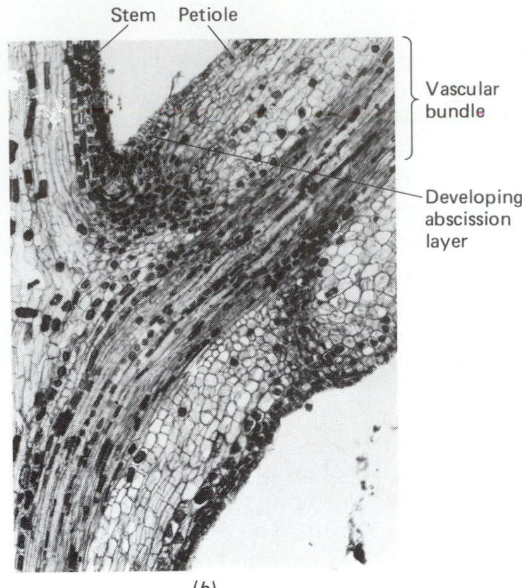

(b)

(c)

Figure 28–6 (*a*) Leaf scars resulting from the loss of the giant leaves of *Monstera deliciosa*, a tropical plant. (*b*) Abscission layer of a dicot leaf. (*c*) The regular geometric arrangement of bud primordia on a stem is reflected in the branching of this East Indian rosewood. ((*b*), Carolina Biological Supply Company.)

pigments that were present all along. Some of the color, however, results from the production of additional quantities of these yellow and red pigments.

At about this time an **abscission layer** (Fig. 28–6) is beginning to form at the base of the petiole. It develops from cells that become impregnated with lignin and therefore impermeable to water as they mature. **Abscission** is the process by which plants shed one of their parts, and an abscission layer is an adaptation that specifically permits loss of leaves. The abscission layer also represents a point of weakness, so that in due time, the yellowed leaves are swirled away by the winds of winter. If for any reason (such as an unseasonable frost) the leaves are killed without having had time to form an abscission layer, they can hang on for months in their brown unsightliness.

What triggers this process is not known, although plant hormones are surely involved. For example, the hormone abscisic acid (ABA), which is produced by the parenchyma of a stressed leaf, will stimulate abscission in laboratory plants. Whether this relationship is enough to explain abscission in nature is not at all certain. Another class of hormones, the **cytokinins** (Chapter 29), may be involved, but the major influence is probably exerted by the growth hormone **auxin.** Auxin production declines in the senescent leaf; this apparently is the major influence that produces abscission. Under laboratory conditions, cytokinin delays abscission and ethylene accelerates it. The extent to which these hormones operate in nature is not known.

SECONDARY GROWTH

As discussed in the preceding chapter, the trunk and branches of large woody plants grow not only in length but also in diameter. The growth in length is usually confined to the apical meristems of stem, branches, and roots, so that an inscription on a beech tree's bark remains the same height above the ground for the life of the tree instead of, as perhaps the inscriber might hope, being carried up into the crown of the tree!

In Dicots

If root and shoot (stem) meristems give rise to all plant tissues, a process of differentiation must occur behind the growing tips of the stem and the root. In stems, the earliest trace of differentiation occurs just behind the apical zone of active mitosis. In the center of the apical bud is a cylinder that develops into the vascular tissues as it matures. It is appropriately known as the **provascular cylinder.** On the exterior, potential epidermis called the **protoderm** develops. Between the two lies a layer called **ground meristem** or **ground tissue,** which becomes the cortex and the pith.

Initially, the differentiating vascular bundles are widely separated by parenchyma. The parenchyma remaining in the center becomes pith. The cambium will also give rise to a continuous layer of meristematic cells, a vascular cambium. This vascular cambium will now produce secondary xylem toward the center of the stem and secondary phloem toward the exterior. The original xylem, or **primary xylem,** remains in the center of the stem. By the time the stem is a few years old the secondary xylem will make up the bulk of the stem and will be surrounded by the vascular cambium and a thinner, peripheral layer of secondary phloem.

As the stem increases in diameter owing to increased secondary growth, the epidermis ruptures. This triggers some of the cells of the cortex beneath it to form the cork cambium, which produces the cork cells of the bark. At each node a bit of apical meristem is left behind to give rise to axillary buds and flower buds (Fig. 28–7).

The root has a different anatomy and thus develops somewhat differently than does the stem. Primary phloem and xylem differentiate from the provascular cylinder to form the typical star-shaped pattern of root vascular tissues, the primitive phloem lying between the arms of the primary xylem. In roots a vascular cambium may also develop to produce secondary growth, much as in the stem. The cambium that occurs between the arcs of xylem and phloem will fuse with the pericycle found in association with the tips of the xylem arcs and thus form a continuous layer, which becomes the vascular cambium. As might be expected, secondary xylem will be produced toward the inside of the vascular cambium, and secondary phloem will be laid down external to the vascular cambium. A root bark analogous to the stem bark may also be produced. Tissue differentiation in roots and stems is summarized in Table 28–1.

What controls this differentiation? If a bit of pith from a tobacco plant is placed in tissue culture, it can be made to develop roots or stems, depending upon the relative concentrations of auxin and cytokinin in the medium. If the auxin concentration is relatively high, roots tend to develop, but not stems; if the cytokinin concentration is relatively high, stems develop, but not roots. Thus the stimulus that triggers the main direction of differentiation is somewhat subtle—the proportions of two hormones. What determines these proportions in the first place is still a mystery.

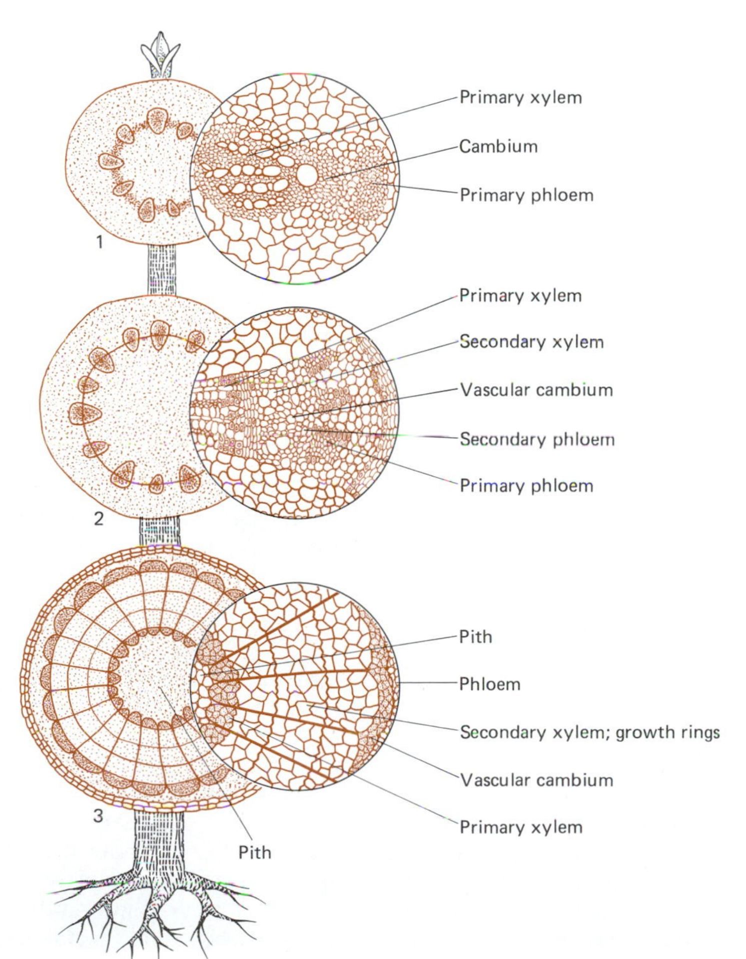

Figure 28–7 Maturation of vascular bundles in a woody dicot. (*1*) Near the growing bud the elements are immature, and vascular bundles are widely separated. (*2*) Farther down the stem the bundles are more mature. Vascular cambium gives rise to secondary tissues. Note that primary xylem is confined to the center of the stem. (*3*) Vascular bundles have coalesced in the three-year-old portion of the stem to form a thick central layer of secondary xylem and a thinner peripheral layer of phloem. (Cork cambium not shown.)

TABLE 28–1
Tissue Differentiation in Stems and Roots

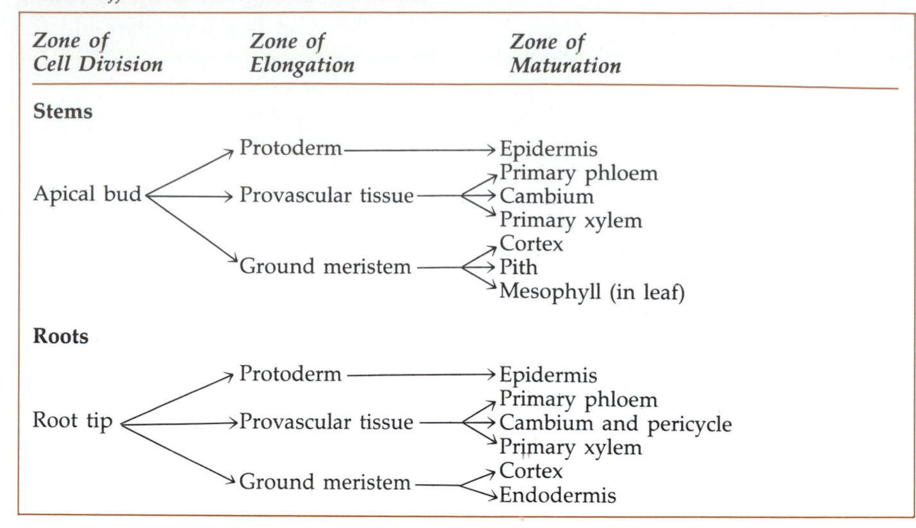

Zone of Cell Division	*Zone of Elongation*	*Zone of Maturation*
Stems		
Apical bud	Protoderm	Epidermis
	Provascular tissue	Primary phloem
		Cambium
		Primary xylem
	Ground meristem	Cortex
		Pith
		Mesophyll (in leaf)
Roots		
Root tip	Protoderm	Epidermis
	Provascular tissue	Primary phloem
		Cambium and pericycle
		Primary xylem
	Ground meristem	Cortex
		Endodermis

Meristematic tissue of some monocots is atypical. Grasses, for instance, possess intercalary meristems internodally, as has already been mentioned. In them, therefore, the processes of differentiation we have been considering do not occur at the apex of the plant (if they did, mowing the lawn would be all too effective!), but at the base of each blade.

In Monocots

Secondary growth usually does not occur in monocots. The trunk of a treelike monocot such as bamboo or a palm tree remains roughly the same diameter from base to crown; in fact, the diameter of a palm trunk can be somewhat smaller at the base than midway between it and the crown. Yet the trunk of a palm seedling is obviously not nearly as great in diameter as is that of the mature tree. What happens is a bit different than what occurs in dicots. Immediately behind the apical meristem is a **primary thickening meristem** that is as great in diameter as the trunk. It is derived from the apical meristem, which continuously enlarges to produce it. The primary thickening meristem then lays down vascular and other differentiated tissues to form the stem of the palm tree.

As we will see in the following chapter, plants can be said to respond to environmental changes with simple forms of behavior, which is for the most part based on differential growth. We will also consider the ways in which plants regulate their growth by means of hormones and cues from their external environment.

A coconut germinating on a Pacific coast beach in Costa Rica. (L. E. Gilbert, Univ. of Texas at Austin/BPS.)

SUMMARY

I. Seed plants propagate sexually by means of embryonic plants supplied with food for independent development.
   A. Seed structure differs in monocots and dicots.
      1. The monocot seed consists of, in addition to the embryo, such specialized structures as the seed coat and aleurone. Its greatly developed endosperm is specialized food storage tissue.
      2. The dicot seed typically possesses no endosperm but stores food in a pair of specialized embryonic leaves, the cotyledons.
   B. Seeds germinate by breaking dormancy. During development seeds become progressively dehydrated. Soaking reverses this, resulting in germination. Additional requirements for germination may include light, *prolonged* soaking, scarification, and charring. The hormone gibberellin mediates germination.
   C. The root is the first seedling structure to emerge. It develops a sharp hook (in dicots) that enables it to force its way through the soil, dragging the rest of the seed out into the sunlight. At the same time, root hairs develop on the end of the root and elongation begins. Most of this elongation occurs in a zone just behind the zone of division.

II. The earliest areas of cell division in the seedling are the root and apical meristems.
   A. At the tip of the seedling, and later at the tip of each branch, is a small meristem surrounded by leaf primordia, the leaf bud.
   B. New branches grow from the meristem.
      1. Leaf growth takes place mainly in the blade as the bud unfolds. Actively growing buds may absorb cytokinins from the root and definitely do produce auxins.
      2. When leaves are fully mature they senesce, usually producing an abscission layer at the base of the petiole, which facilitates leaf fall and seals off the leaf scar.
      3. Deciduous trees lose all their leaves at about the same time. This is an adaptation to seasonal water shortage during dry or cold seasons.

POST-TEST

1. In the grass seed, the cotyledon is a food-absorbing structure known as a ________.

2. The portion of the seedling below the cotyledons is known as the ________. One of the structures included in it is the future root, or ________.

3. The seeds of desert annuals may contain a ________ ________, which is washed out by rains that are heavy enough to ensure sufficient moisture to support a complete life cycle of the plant.

4. The root hairs form in the zone of ________ of the growing root tip.

5. When the "hook" of a dicot seedling is exposed to light during germination, it reacts by ________; the receptor substance that initiates this reaction is a pigment known as ________.

6. The growing nodes of embryonic tissue found at the shoot and root tips of a plant are its ________ ________.

7. Though the means whereby the apical bud of the growing plant suppresses the growth of all others is unknown, it is thought principally to involve the production of the plant hormone class known as ________.

8. The arrangement of secondary branches on a trunk usually reflects the arrangement of the ________ on the parent branch during the preceding season.

9. The basic ecological and physiological function of seasonal leaf loss appears to be that of ________ conservation.

10. Secondary growth in monocots is minimal; in some of the treelike monocots such as palms there is a ________ ________ meristem located just behind the apical meristem.

REVIEW QUESTIONS

1. Compare the monocot seed with the dicot seed.
2. What is the aleurone layer? How does it function?
3. Describe the events of germination in sequence.
4. What might be the adaptive significance of a germination requirement that a seed be frozen in order to break dormancy?
5. Describe each of the following: *a.* zone of elongation in a root tip, *b.* radicle, and *c.* abscission.
6. Describe the way in which the primary meristem produces apical bud growth.
7. What is a secondary meristem? What part do secondary meristems play in dicot growth? Do they have any role in monocot growth?

29
Plant Hormones and Tropisms

OUTLINE

I. Rapid plant responses
II. Tropisms (growth responses)
III. Plant hormones
  A. Auxins
    1. Apical dominance
    2. Other auxin actions
  B. Gibberellins
  C. Ethylene
  D. Cytokinins
  E. Abscisic acid
IV. Photoperiodism: florigens and phytochromes
  A. Florigen
  B. Phytochrome

LEARNING OBJECTIVES

After you have read this chapter you should be able to:

1. Describe both behavioral and growth responses in plants, using the proper terminology.
2. Summarize the actions of the major plant hormones: auxins, gibberellins, ethylene, cytokinins, and abscisic acid.
3. Outline the relationship between auxin action and plant tropisms.
4. Summarize plant photoperiodism with respect to both phytochromes and the control of flowering.

Plants are capable of much more "behavior" than we might at first imagine. Though incapable of muscular movement, all plants do grow, and some can perform specialized rapid movements. The swamp-growing Venus's fly-trap (see Fig. 1–7), for instance, lacks a ready source of nitrogen and obtains what it needs from the bodies of insects it entraps by rapidly folding its leaves around its victims. The more common sensitive plant (Fig. 29–1) also folds its leaves whenever they are irritated. Other plants, such as the prayer plant (*Maranta*), fold their leaves at night, in what is sometimes called a **sleep movement** or nyctinasty (Fig. 29–2); such movements function to minimize water and heat loss from the leaf.

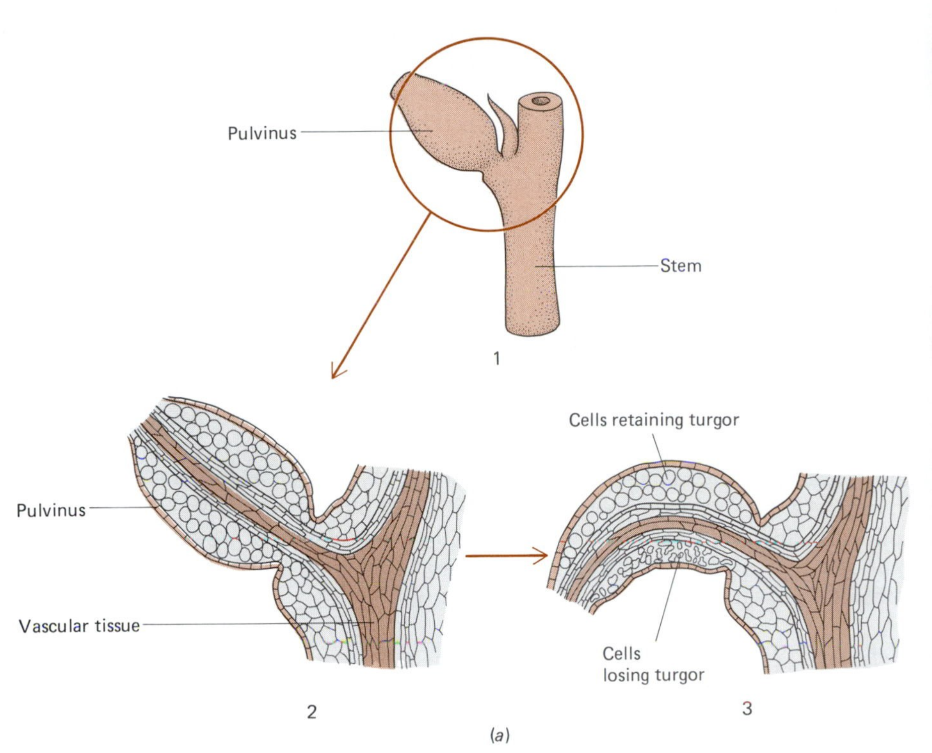

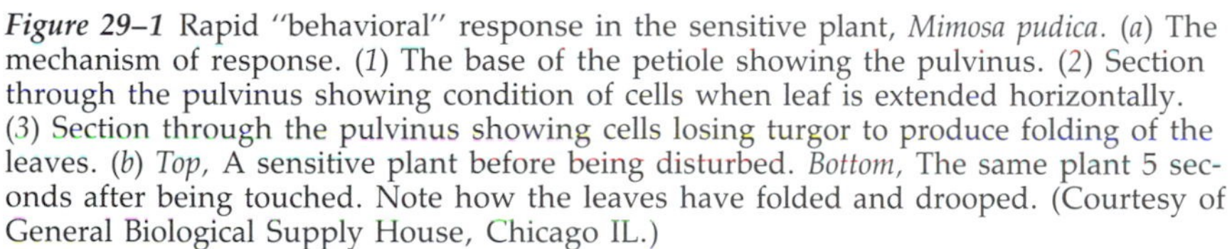

Figure 29–1 Rapid "behavioral" response in the sensitive plant, *Mimosa pudica*. (*a*) The mechanism of response. (*1*) The base of the petiole showing the pulvinus. (*2*) Section through the pulvinus showing condition of cells when leaf is extended horizontally. (*3*) Section through the pulvinus showing cells losing turgor to produce folding of the leaves. (*b*) *Top,* A sensitive plant before being disturbed. *Bottom,* The same plant 5 seconds after being touched. Note how the leaves have folded and drooped. (Courtesy of General Biological Supply House, Chicago IL.)

Figure 29–2 Sleep movements of (*Maranta*) leaves. (*a*) The plant in daylight. (*b*) The same plant at night.

Rapid Plant Responses

The more rapid plant responses are generally produced not by differentiated growth but by turgor changes in specialized cells. The combination of tough cellulose cell wall and large central vacuole in a typical plant cell results in turgor, because the vacuolar fluid presses against the walls of the cell in somewhat the same way that the inflated inner tube of a tire presses against its casing. In both cases the result is a structure capable of supporting weight. If the tire tube is deflated, however, the tire will collapse and the vehicle will sag to the ground on that side; similarly, loss of turgor in plant cells will cause changes that produce movement of the leaves or stems.

At the base of the petiole of the sensitive-plant leaf is a specialized collection of turgid cells, the **pulvinus.** The cells of this structure are able to lose turgor very rapidly, collapsing and permitting the leaf to be pulled out of its normal position by the elasticity of the petiolar tissues or by gravity. A similar loss of turgor pressure permits the closure of the Venus's fly-trap. The collapse of these specialized cells is triggered by an electrical charge resembling the action potential of animal nerve and muscle cells, which is passed from one cell to another until it reaches the pulvinar cells. If the plant is small enough, the entire plant may respond to the injury of a single leaf.

Tropisms (Growth Responses)

If observed in speeded-up motion pictures, the growth responses or **tropisms** of plants are striking. A geranium in the window stretches after the sun, roots seem to burrow through the soil, and tendrils whip around objects. Many Arctic and Alpine flowers have petals shaped like a solar collector, allowing them to focus sunlight on the ovaries and thus to keep the developing seeds warm and to attract pollinating insects. As the sun travels across the sky it is followed by the flower, whose stem, following the sun, slowly twists about like the shaft of a radar antenna. All of these motions are produced by differential growth of the plant parts, and this growth is controlled by hormones.

Plant Hormones

Hormones were originally discovered in animals, where most are produced by glands specialized for the purpose, then transported through the body by the circulatory system. Exceptions to this generalization exist, however, even in animals. In plants, hormones appear to be generated not by specialized glands of internal secretion but by tissues that are functioning primarily in some other way. Another difference is that more often the target tissues of plant hormones are the very ones that are producing them. Here we will consider five known classes of plant hormones.

AUXINS

Auxins are the longest known and probably best understood of the plant hormones. Between 1910 and 1930 a golden age of auxin research existed in Europe. Long before, none other than Charles Darwin and his son Francis had undertaken a series of investigations into the growth response of young canary grass seedlings. Their observations (Fig. 29–3) provided a data base for later work by such pre–World War I investigators as the Danish Boysen-Jensen and the Dutch Frits Went. The Darwins caused seedlings of a monocot, canary grass, to sprout. Recall that the seeds of such grasses as oats, canary grass, wheat, and others do not send up a stem initially. Instead, they produce a tightly rolled bundle of leaf blades ensheathed in a protective coleoptile. As in the adult plant, these early leaves elongate at the base rather than the tip.

It seemed logical to assume that the growing part of the plant would determine the direction of growth, since plants tend to grow toward the light, and grass seedlings share this tendency. Imagine Darwin's surprise when he and his son discovered that even if the base of the coleoptile was shielded from the light, the shoot grew toward the source of light anyway. Only when the Darwins covered the shoot *tip* was the tropism abolished. These results suggested to Went and Boysen-Jensen that the tip produced something—perhaps an electrical impulse or a chemi-

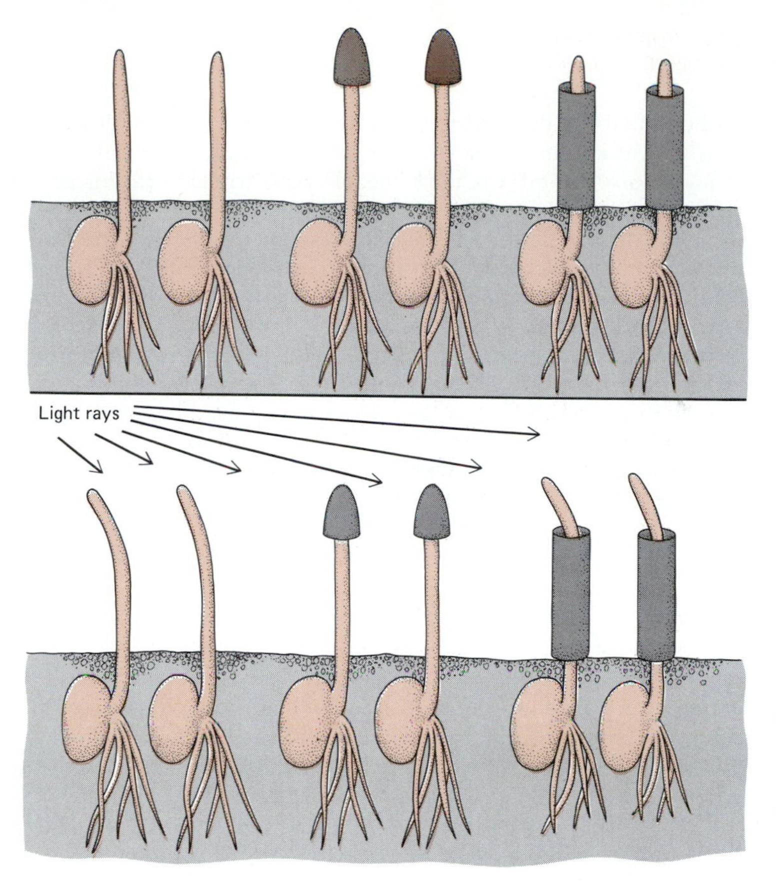

Figure 29–3 The Darwins' experiment with canary grass seedlings. *Upper row,* Some plants were uncovered, some were covered only at the tip, and others were covered everywhere but at the tip. After exposure to light coming from one direction (*lower row*), the uncovered plants and the plants with uncovered tips (*right*) grew in a bent fashion. The plants with covered tips (*center*) grew straight up. Darwin and his son concluded that: the tip of the seedling is sensitive to light and gives off some "influence" that moves down the stem and causes the bending.

cal substance—that governed growth, and that it produced this factor differentially so as to produce differential growth.

As a first step in investigating this question, the experimenters removed the coleoptile tip, as shown in Figure 29–4. Growth stopped. When the tip then was replaced, growth resumed. When a very thin flake of the mineral mica was interposed between tip and stalk, growth stopped once more. It appeared that the hy-

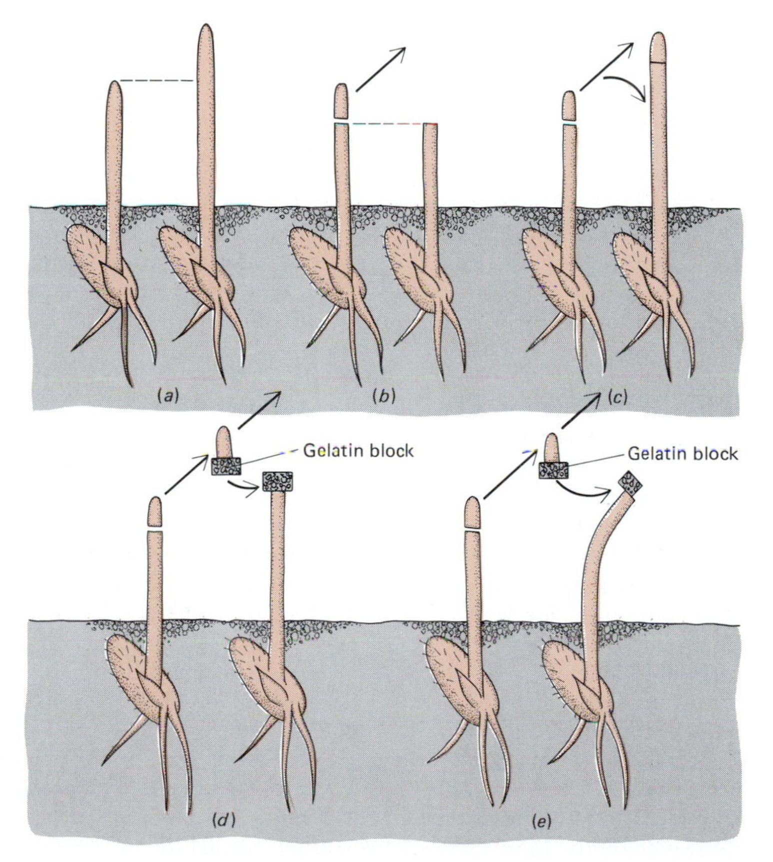

Figure 29–4 A series of experiments that demonstrate the existence and mode of action of plant growth hormone in oat coleoptiles. In each pair of drawings, the figure on the left indicates the experiment performed and the figure on the right the growth after a period of time. (*a*) Control: No operation performed, normal growth. (*b*) If the tip of the coleoptile is cut off and removed, no growth occurs. (*c*) If the tip is cut off and then replaced, normal growth ensues. (*d*) If the tip is cut off and placed on a block of gelatin for a time, and then the gelatin block, but not the coleoptile tip, is placed on the seedling, growth occurs. (*e*) If the tip is placed on a gelatin block for a time and then the gelatin block is placed asymmetrically on the seedling, curved growth results.

pothesis that the tip produced a factor required for growth was confirmed. However, it was still not clear whether the factor was chemical, electrical, or of some other nature, for mica is impermeable both to electrical forces and to any likely chemical substances.

The investigators next tried interposing a substance that would likely be permeable both to electrical forces and to chemical substances; they chose agar, a complex polysaccharide obtained from certain seaweeds. When mixed with water, agar forms a gel that is mostly water. By interposing a block of this jelly-like agar between the tip and shaft of the coleoptile, the investigators provided, in effect, a water bridge between the two plant parts. Any water-soluble chemical could easily diffuse through it. When an agar block was placed between the tip and the shaft, growth did resume. Finally, when a tip was left in contact with an agar block for some time, the block itself was able to stimulate growth even without the tip. This observation was consistent with the hypothesis of a water-soluble control substance that, during the time that elapsed, might have become concentrated in the block. The observation was obviously not, however, consistent with a hypothesis based on some electrical force. The postulated substance was the first known plant hormone. Went named it **auxin** after a Greek word (*auxein*) meaning "to grow."

This series of studies, a beautiful example of the scientific method at work, was only the beginning. It was soon discovered that if an auxin-impregnated block of agar was placed off-center on a decapitated coleoptile tip, cell elongation on that side was stimulated and the coleoptile bent toward the *opposite* side. This suggested that some tropisms might result from auxin accumulation on the side of a stem or other structure so that it would bend or grow toward the opposite side.

Such a hypothesis was hard to demonstrate, considering the small quantities of auxin that we now know to be involved. The growing shoot of the pineapple plant, for instance, contains about 6 micrograms of indoleacetic acid (the most common form of the hormone) per kilogram of plant material. As J. P. Nitsch remarked, this quantity is comparable to the weight of one needle in a 22-ton haystack! Eventually it was shown that radioactively labeled auxin did migrate to the shaded side of an illuminated stem, but no ordinary method of chemical analysis available to Went in the 1930s could have detected that.

It is now believed that indoleacetic acid may in fact be the only naturally occurring form of auxin. It is synthesized by actively growing plant tissue from the amino acid tryptophan and is actively transported toward the base of the plant by the cells of the stem at rates ranging from 0.5 to 1.5 cm per hour. Ordinarily indoleacetic acid stimulates the production of hydrogen ions by individual plant cells, and the transportation of these hydrogen ions into the cell walls. The resulting lowered pH optimizes the action of enzymes that attack the crosslinks between adjacent chains of cellulose polymers in the cell wall. The loosened cell walls are then stretched by the turgor pressure of the cell. Due to the orientation of the cellulose fibers in the cell wall, this mechanism acts mainly along the longitudinal axis of the cell so that the result is elongation rather than simple enlargement of the cells (Fig. 29–5). As the cells elongate, so does the stem. Most tropisms of the above-ground portions of plants can be explained in this fashion (Fig. 29–6).

Apical Dominance

Although auxin stimulates cell elongation, it does not necessarily stimulate mitosis. In fact, it tends to inhibit meristematic tissue from growing. Yet meristematic tissue produces auxin, so that in locations like buds the auxin content is very high—almost at inhibitory concentrations. As mentioned earlier, in order to make a tree or shrub grow in spreading fashion, the topmost buds should be removed. Part of the reason is that the apical bud produces large amounts of auxin, which diffuse throughout the plant body. An excess of auxin is actually growth-inhibiting. Since the growth of the lateral buds is on the verge of being inhibited by the buds' own auxin production, that auxin plus what originates from the apical bud is sufficient to prevent bud growth, or at least to slow it. Different degrees of apical bud dominance account for the many differences in tree shape (Fig. 29–7).

It is likely that auxin production and inhibition do not fully account for apical bud dominance, however. It is strongly suspected that two other factors may play a part. First, actively growing plant tissue tends to draw nutrients out of other parts of the plant body. If the metabolic rate of the apical bud were high enough, it might

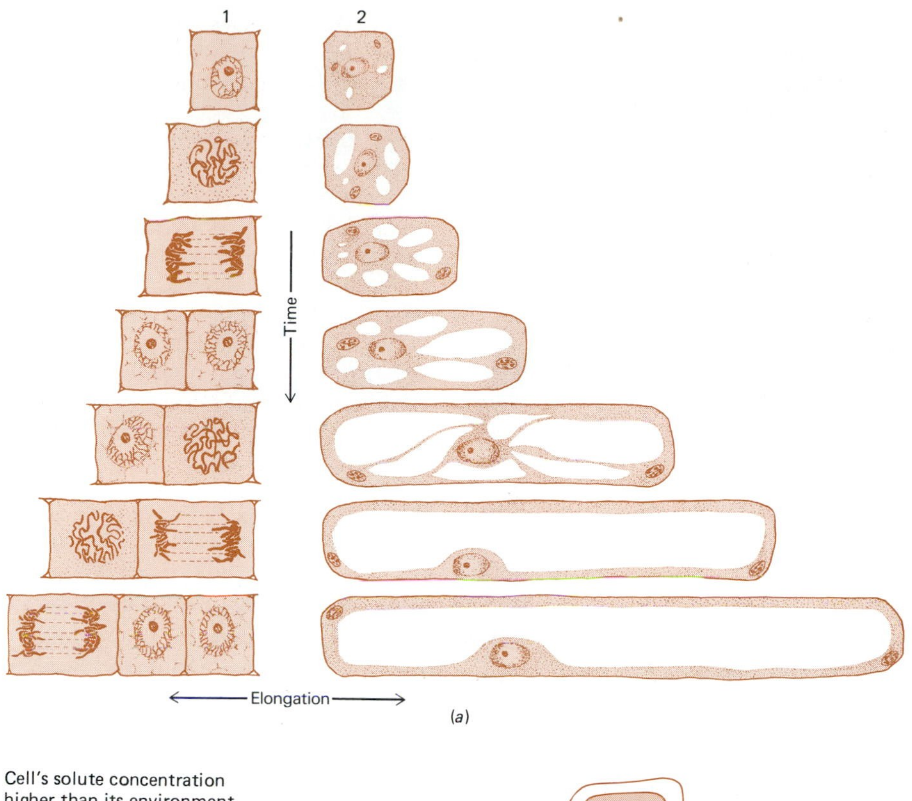

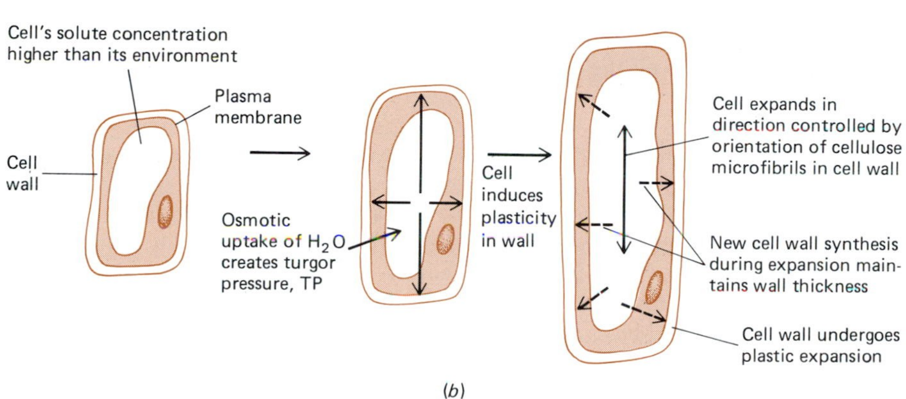

Figure 29–5 Growth, especially in plants, does not always involve cell division. Cell enlargement is responsible for much differential plant growth. (*a*) Diagram illustrating growth of plant cells with (*1*) and without (*2*) cell division. Lightly colored material is cytoplasm; clear areas are vacuoles. Cells in sequence *1* are meristematic cells; cells in sequence *2* are elongating cells from a cell enlargement zone. (*b*) Physical principles of plant cell enlargement. During actual growth, the processes shown in (*a*) and (*b*) occur together, with those in (*a*) continuously driving the plastic expansion of the primary wall involved in (*b*). This process is under the control of auxins, as explained in the text. (Courtesy of Ray, Steeves and Fultz.)

be able to produce some degree of lateral bud inhibition just by starving the lateral buds. Second, and much more controversial, a mitosis-promoting hormone, cytokinin, is produced in the roots and is transported preferentially to rapidly growing tissues such as the apical bud. Perhaps the lateral buds cannot compete successfully with the apical bud for the limited cytokinin supply.

How does the apical bud get into its dominant position in the first place? Theoretically, the apical bud of a thousand-year-old redwood tree is the same apical meristem with which it was equipped about 1000 AD, when it germinated and became a seedling. Still, anything from insect damage to lightning could have eradi-

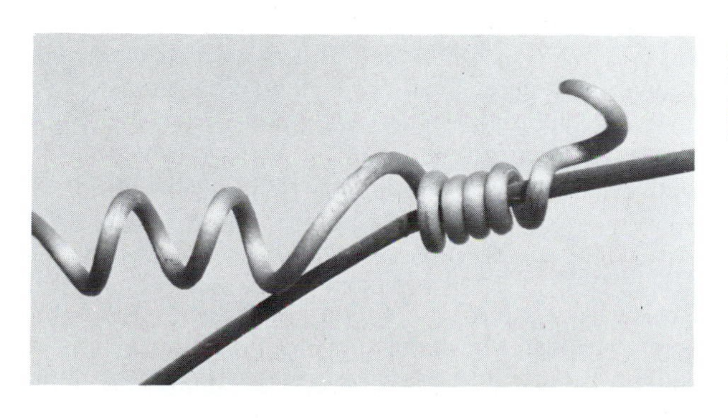

Figure 29–6 A tendril of *Smilax* wrapped around the stem of another plant. (Jack Dermid, Photo Researchers, Inc.)

(a)

(b)

Figure 29–7 Differences in plant growth habits result largely from differences in apical bud dominance. (*a*) Camphor tree. This form of growth habit results from little apical bud dominance. (*b*) Ornamental cedar, whose cylindrical form results from marked apical bud dominance.

cated that original apical bud at any time, and probably did. This would have released all the lateral buds from inhibition. The one that began to grow most vigorously would have produced enough auxin and absorbed enough food and cytokinin to suppress all the others, so that it would have become the new apical bud.

At least in angiosperms, the new apical bud is generally the highest or among the highest of the lateral buds, which seems to suggest that position is another factor in determining dominance. Light exposure also seems to be involved, since individual plants of the same species may branch in widely different ways when growing under different degrees of illumination. Moreover, especially in trees with a spreading habit of growth, lateral branches do elongate slowly, although they do not ordinarily grow upward. Apical bud dominance may influence the geotropic (gravity response) abilities of the lateral buds even more than it inhibits their growth.

Why the apical bud does not inhibit itself with all the auxin it produces is partly explained by the fact that, at least in some plants, if lateral buds are allowed to start to grow vigorously, auxin will not inhibit them. In other words, in these plants meristematic tissue is inhibited by auxin only from *starting* to grow, not from growth as such.

Other Auxin Actions

Auxins stimulate the growth of cambium so as to promote the formation of reaction wood in response to stress, as mentioned in Chapter 27. Artificial auxins can be used as herbicides, the weed killer 2,4-D being a familiar example. It is sometimes said that herbicidal auxins increase the metabolic rate of plants so that they use up their reserves of stored food and, in effect, starve. Although auxins do increase metabolic rate, the mode of action of these herbicides is not well understood. In any case, 2,4-D is absorbed well by broad-leafed dicots and kills them effectively, but monocots, especially grasses, tend to be resistant to its action. Thus, a dose of 2,4-D that will kill broad-leaved weeds is nearly harmless to the grass of the lawn where those weeds are growing. Auxins also have a variety of other agricultural and horticultural uses, for instance, encouraging cuttings to root.

GIBBERELLINS

To speak of a foolish seedling might seem unscientific, though most of us who garden probably have called plants much worse things than that. Nevertheless, rice suffers from a fungus disease that is known in Japan as the "foolish seedling disease." Infected seedlings grow abnormally tall but are weakened and agriculturally worthless. The Japanese biologist Kurosawa discovered that the fungus produced a chemical substance that was responsible for these symptoms. The fungus was named *Gibberella fujikori,* so it seemed logical to name its secretion **gibberellin,** or **gibberellic acid.**

It soon became evident, however, that gibberellic acid is not a mere toxin but a naturally occurring plant hormone existing in several forms, mostly produced by the chloroplasts of the leaves. Some dwarf varieties of common plants, when treated with gibberellic acid, grow to normal height. This suggests that such plants have a genetically determined deficiency of this growth hormone. Along with auxin, gibberellin stimulates the elongation of stem cells, and there is evidence that gibberellins often stimulate the production of auxin in the first place. Gibberellins also work along with auxin and cytokinin in fruit development. Unlike auxins, however, gibberellins greatly increase leaf growth, and we have already noted their role in seed germination. Gibberellins also stimulate bolting[1] (Fig. 29–8), and even flowering in some plants when they are applied artificially. Because gibberellin-induced flowering is abnormal in some respects, it is unclear to what extent gibberellins influence flowering normally, even in those plants that are sensitive to their action. Our understanding of this class of hormones is complicated by the existence of more than a dozen of them, all with somewhat differing actions. The total picture of the role of gibberellins in the regulation of plant growth is still unclear at this time.

Figure 29–8 Gibberellic acid (*GA*) causes bolting of cabbage. *Left,* Untreated. *Right,* Sprayed several times with 0.1% GA. (From Wittwer, S. H., and Bukovac, M. J.: in *Economic Botany.*)

ETHYLENE

That one rotten apple will spoil the barrel is more than just an old saying. It is literally true. The reason is not merely that decay spreads from one rotten apple to a sound, adjacent one, but that ripening and eventual overripening of fruit can be induced by a substance given off by a ripe one. The fruits need not even be of the same species.

Fruits initially develop in response to the production of auxin by the developing seed, or even by the pollen that fertilizes the flower. In fact, the seeds are not necessary if auxin can be provided in some other way, such as by anointing the flower with an auxin-containing cream. Some fruits develop spontaneously without any special auxin stimulus, as in the case of navel oranges or bananas. Since such **parthenocarpic** fruits are incapable of reproduction, the plants must be propagated asexually. Bananas may be propagated by cuttings or by dividing their tubers, and all the navel oranges in the world are derived from an original plant that appeared by mutation in a Brazilian grove in the 19th century. That plant has been propagated by grafting to the roots of other varieties of citrus ever since.

The substance that ultimately causes ripening is gaseous, almost the only known example of a gaseous hormone. It is **ethylene,** a simple, unsaturated hydrocarbon. Since ethylene is readily available from chemical companies, artificial ripening of such fruits as oranges and tomatoes is easy, provided that they are sufficiently mature. Thus, fruit can be picked in a green state, shipped without fear of spoilage, and ripened as needed near the point of consumption. Ripening can also be delayed, if desired, by vigorous ventilation of stored fruit, or by storing it in an atmosphere of carbon dioxide, which seems to antagonize the action of ethylene.

It is interesting that fruit ripening is one of the few examples of positive feedback in biology. Ethylene stimulates ripening; as a consequence of ripening, more ethylene is produced, which further accelerates ripening; and so on. The process proceeds rapidly and irreversibly to completion.

CYTOKININS

Accidentally discovered in bottles of partially deteriorated nucleic acids, the cytokinins occur as normal plant hormones. Even before they were specifically identified, it was known for years that coconut milk contains a growth-promoting substance that, when added to the formulas for plant tissue culture media, would

[1] Bolting is a phenomenon exhibited by many temperate-zone biennials. Biennials develop only vegetatively the first season, and do not flower. After a period of winter dormancy, flower stalks elongate (this is called bolting) during the second summer. Beets, cabbage, carrots, and celery are all biennials that exhibit bolting.

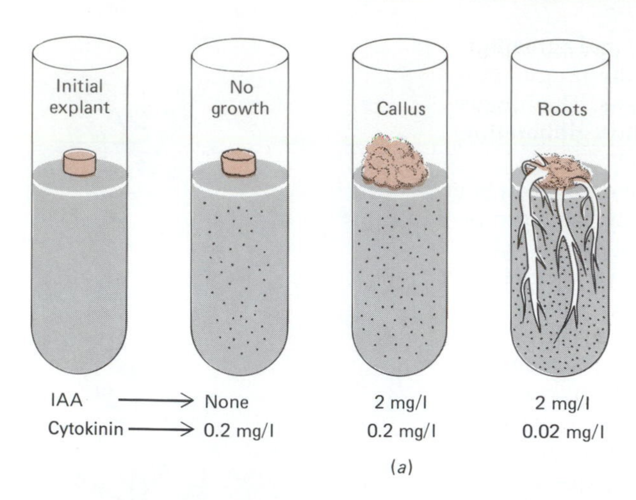

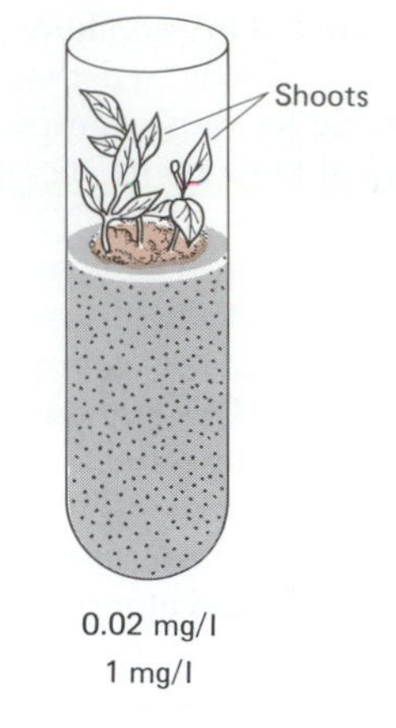

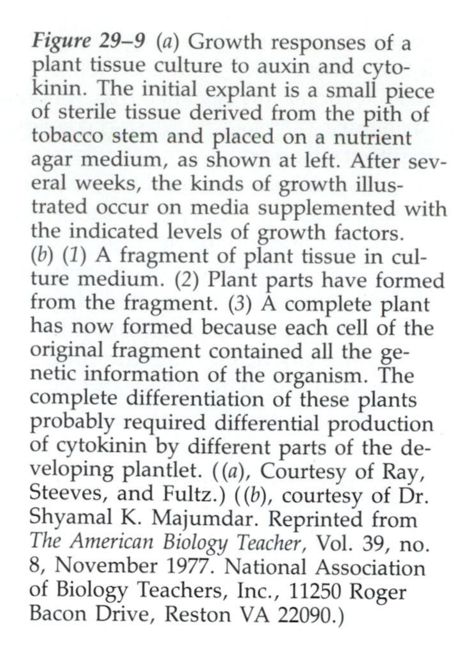

(*a*)

1

2

3

(*b*)

Figure 29–9 (*a*) Growth responses of a plant tissue culture to auxin and cytokinin. The initial explant is a small piece of sterile tissue derived from the pith of tobacco stem and placed on a nutrient agar medium, as shown at left. After several weeks, the kinds of growth illustrated occur on media supplemented with the indicated levels of growth factors. (*b*) (*1*) A fragment of plant tissue in culture medium. (2) Plant parts have formed from the fragment. (3) A complete plant has now formed because each cell of the original fragment contained all the genetic information of the organism. The complete differentiation of these plants probably required differential production of cytokinin by different parts of the developing plantlet. ((*a*), Courtesy of Ray, Steeves, and Fultz.) ((*b*), courtesy of Dr. Shyamal K. Majumdar. Reprinted from *The American Biology Teacher*, Vol. 39, no. 8, November 1977. National Association of Biology Teachers, Inc., 11250 Roger Bacon Drive, Reston VA 22090.)

assure growth of the plant cells. This mysterious substance was later determined to be a cytokinin.

Chemical relatives of the nucleic acid component adenine, the cytokinins promote mitosis in both intact plants and tissue cultures. Cytokinins are also active in plant wound healing and in causing plant parts to differentiate. In addition, they work together with auxins in fruit development, and apparently promote the growth of all kinds of meristematic tissue.

In tissue culture, if auxin is supplied in a relatively high concentration in proportion to cytokinin, an undifferentiated mass of plant tissue, called a **callus,** will tend to produce roots (Fig. 29–9). On the other hand, if the proportion of cytokinin is high in comparison to auxin, buds and leaves are more likely to form. Cytokinins also slow the processes of plant aging (especially in annuals) and, if applied to leaves, interfere with their senescence and abscission. As mentioned previously, the cytokinins seem to be produced mainly in the roots.

ABSCISIC ACID

We have already examined the role of abscisic acid (ABA) in a plant's seasonal loss of leaves (Chapter 28), as well as its role in the control of transpiration via the opening and closing of the stomata (Chapter 27). ABA also induces dormancy and controls root growth. Generally, ABA functions as an inhibitory hormone that antagonizes the auxins, gibberellins, and cytokinins, all of which promote growth in one way or another. ABA also affects differentiation: When produced by senile leaves, it induces the nearby leaf primordia to form protective bud scales instead of actual leaves.

Many articles in the older botanical literature argue that root behavior—specifically, the negative phototropism characteristic of root growth—is under the control of auxins. The argument begins with the true statement that in high concentrations, auxin inhibits rather than stimulates plant growth, and then goes on to state that the auxin concentration in plant roots is so high already that any added auxin would inhibit growth. The explanation proposed for the negative phototropism is that auxin accumulates on the dark side of the root, just as it accumulates on the dark side of the stem. Because, however, of the high auxin concentrations already existing in the root, growth is inhibited on that side so that the root grows *toward*, rather than away from, the dark.

This explanation may actually be correct in the particular case of the negative *photo*tropism of roots, but it does not seem to explain the positive geotropism of roots. It is easy to demonstrate that roots tend to grow downward, even in the absence of differences in light or moisture, and it has been shown that ABA accumulates on the downward-facing sides of roots, inhibiting their growth on that side so that the growth that does occur tends to be on the upper sides, directing the roots in a downward direction. The mechanism that ultimately directs this root growth is not well understood. However, it has been observed that relatively heavy starch grains (amyloplasts) found in the cells of many roots accumulate on the bottom of each root cell. These granules may help the cells to tell up from down. How the differential intracellular distribution of the starch granules produces the differential accumulation of ABA in the root as a whole is still an unanswered question.[1]

Photoperiodism: Florigens and Phytochromes

Flowering and the formation of fruit are also developmental processes that are regulated by hormones. The initiation of these processes is similar to that in leaf development. Like the leaf, the flower begins life as a bud. As it unfolds it spreads out a multitude of leaflike parts. Eventually it will age and die, but because the flower functions as a reproductive organ, and if fertilization takes place, new plants may grow from its seeds. Unlike the leaf bud, the flower bud is not able to give rise to a stem. Once seeds and fruit have formed, the story of the flower is ended. However, specialized meristems near the flowers may give rise to new flower buds, which usually remain dormant until the next season. (The parts of the flower and their functions are discussed in the following chapter.)

Different kinds of plants flower at different seasons of the year. The time of flowering in some plants can be related to day length (the number of daylight hours per day), the **photoperiod.** The variation in response to the photoperiod is termed **photoperiodism.**

The role of the ratio of daylight to darkness in determining the time of flowering of plants was unknown until 1920, when Garner and Allard, of the United States Department of Agriculture, demonstrated that the time of flowering of tobacco plants could be altered by changing the photoperiod. The discovery began with the abrupt appearance of a mutant tobacco plant that grew to an unheard-of size. It was named the Maryland Mammoth, and there were high hopes of increasing the yield of tobacco by making this variety widely available to farmers. Unfortunately, the plant did not flower, and without flowers it was not possible to obtain seed or to perform experimental crosses with other varieties.

Experimenters decided to try asexual propagation and moved the plant into a greenhouse. True to its freakish nature, the Maryland Mammoth finally did flower—in the middle of the winter. The explanation, surprisingly, did not involve temperature, humidity, or soil differences. It depended upon day length. Maryland Mammoth could never have survived in a natural state because it required a day so short for flowering that it would have been killed by frost long before days of the right length came to pass.

Some species of plants (e.g., asters, cosmos, chrysanthemums, dahlias, poinsettias, and potatoes) are **short-day plants** and will not produce flowers if the pho-

[1] For a discussion of recent research, see "How roots perceive and respond to gravity" by Randy Moore, *The American Biology Teacher* 46: 257–265, May 1984.

Figure 29–10 Light exposure and flowering. All petunias received eight hours of daylight each day. Plant on left with small buds got in addition eight hours of fluorescent light, which contains flower-suppressing red light but no infrared. Flowering plant in center was given an extra eight hours of incandescent light, containing both red and flower-stimulating infrared light. Only the plant that received additional infrared light was able to flower. (Courtesy of U.S. Dept. of Agriculture.)

toperiod exceeds a certain critical length, that is, if there are more than a certain number of hours of daylight per day. Such plants normally flower in the early spring, late summer, or fall. Others, termed **long-day plants** (e.g., beets, clover, coreopsis, corn, delphinium, and gladiolus), require a photoperiod that *exceeds* a certain critical length for flowering to take place. The length of the critical photoperiod varies among species, ranging from 9 to 16 hours, and is not necessarily "short" in a short-day plant. Indeed, some short-day plants have a longer critical photoperiod than that of some long-day plants.

The important difference is in the nature of the physiological message that the photoperiod sends. The photoperiod must be *less* than the critical period in short-day plants to induce flowering but *greater* than the critical period in long-day plants. Short-day plants are actually "long-night" plants, for the controlling factor is the length of the period of uninterrupted darkness.

Long-night plants will flower only if the dark period is longer than 24 hours minus its critical photoperiod. For example, a short-day species with a 10-hour critical photoperiod will flower only if the dark period exceeds 14 hours (24 hours − 10 hours = 14 hours). They can be made to flower earlier than usual by decreasing their daily exposure to light or by covering them, and they can be kept from flowering by giving them artificial illumination (Fig. 29–10). A long-day ("short-night") plant can be prevented from flowering if it is covered for part of each day, thereby reducing its daily exposure to light to less than its critical photoperiod.

Carnations, cotton, dandelions, sunflowers, and tomatoes are examples of **day-neutral** plants. The flowering of these plants is relatively unaffected by the amount of daylight per day. The time of flowering even in short-day and long-day plants is not controlled *solely* by the photoperiod. Temperature, moisture, soil nutrients, and the amount of crowding may also play a role.

FLORIGEN

How the length of darkness can affect the time of flowering is not known in detail, but the results of some experiments suggest that a flower-producing hormone named **florigen** is involved. In a typical experiment, using cocklebur (a short-day plant), one plant is grown exposed to 12 hours of light per day until it produces flowers (Fig. 29–11). It is then grafted to another plant that was grown exposed to 18 hours of light per day (and thus was inhibited from producing flowers). The two parts, though grafted, are separated by a light-tight partition, with the first part continuing to receive 12 hours and the second 18 hours of sunlight. The short-day part of the plant continues to produce flowers, and in time the long-day part of the plant also begins to produce flowers, usually starting at the point nearest the graft.

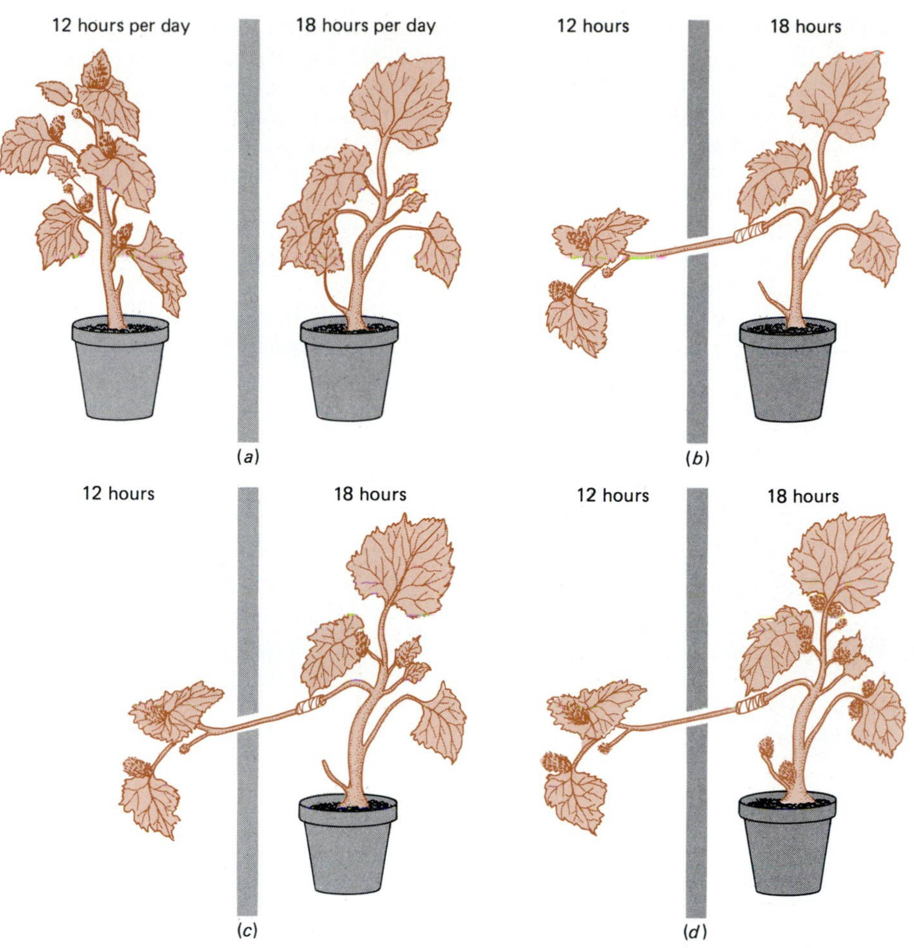

Figure 29–11 An experiment to demonstrate the existence of a flower-inducing hormone. (*a*) Two cocklebur plants are grown in pots separated by a light-tight partition, exposed to 12 and 18 hours, respectively, of light per day. The 12-hour plant has flowered; the 18-hour plant has not. (*b*) The 12-hour plant is cut off, inserted through a light-tight hole in the partition, and grafted to the 18-hour plant. The two parts continue to receive 12 and 18 hours of light respectively. The 18-hour plant gradually develops flowers, first on the twigs nearest the graft, as shown in (*c*), and eventually on all twigs, as seen in (*d*). If no graft had been made, the 18-hour plant would not have developed flowers.

This is taken as evidence for a diffusible, flower-inducing hormone produced in the leaves and transported in the phloem to the buds. Nothing is known of the chemical composition of this hormone, nor of how it might act to induce flowering. In fact, florigen is a *postulated* hormone; we cannot really be certain it exists. The action attributed to florigen may result from complex quantitative interactions of other hormones.

PHYTOCHROME

Although the sense receptors involved in many plant responses are not known, flowering is a happy exception. It is initiated by the absorption of light by a special protein pigment, **phytochrome.** Phytochrome is not equally sensitive to all wavelengths of light, and some of the light to which it is most sensitive is not visible to us, although it is abundantly present in sunlight.

Red light inhibits flowering in short-day plants but induces flowering in long-day plants. Infrared light induces flowering in short-day plants and inhibits flowering in long-day plants. The reason is that each is absorbed by a different form of the phytochrome pigment. This pigment exists in two forms. One, P_{660}, is sensitive to red light (660 nm); the other, P_{735}, is sensitive to the shortest wavelengths of infrared light, termed far red (735 nm), which are barely visible to the human eye. P_{660} appears to be the quiescent form in which the plant stores the potentially active compound, and P_{735} is the active material. Red light (or white light that contains red wavelengths) converts inactive P_{660} to active P_{735}, and infrared light converts P_{735} back to P_{660}. That is,

$$P_{660} \underset{\text{infrared light}}{\overset{\text{red light}}{\rightleftarrows}} P_{735}$$

These photochemical reactions are not the only ones that occur, for P_{735} reverts to P_{660} in the dark. This might provide the plant with a means of detecting whether it is light or dark. The rate at which P_{735} is converted to P_{660} could provide the plant with a "clock" for measuring the duration of darkness. Each plant has an internal biological clock of some kind—perhaps based on a series of relatively temperature-independent chemical reactions that ultimately produce florigen. *These reactions are evidently set in motion in long-day plants by phytochrome P_{735}.* If the day is long enough, sufficient florigen accumulates to induce flowering. If not, P_{735} becomes P_{660}, and florigen does *not* accumulate in sufficient amounts to induce flowering. Short-day plants would reverse this process.

The role of phytochrome in the photoperiodic control of plant development is virtually universal (Fig. 29–12). It functions, for instance, in the control of leaf

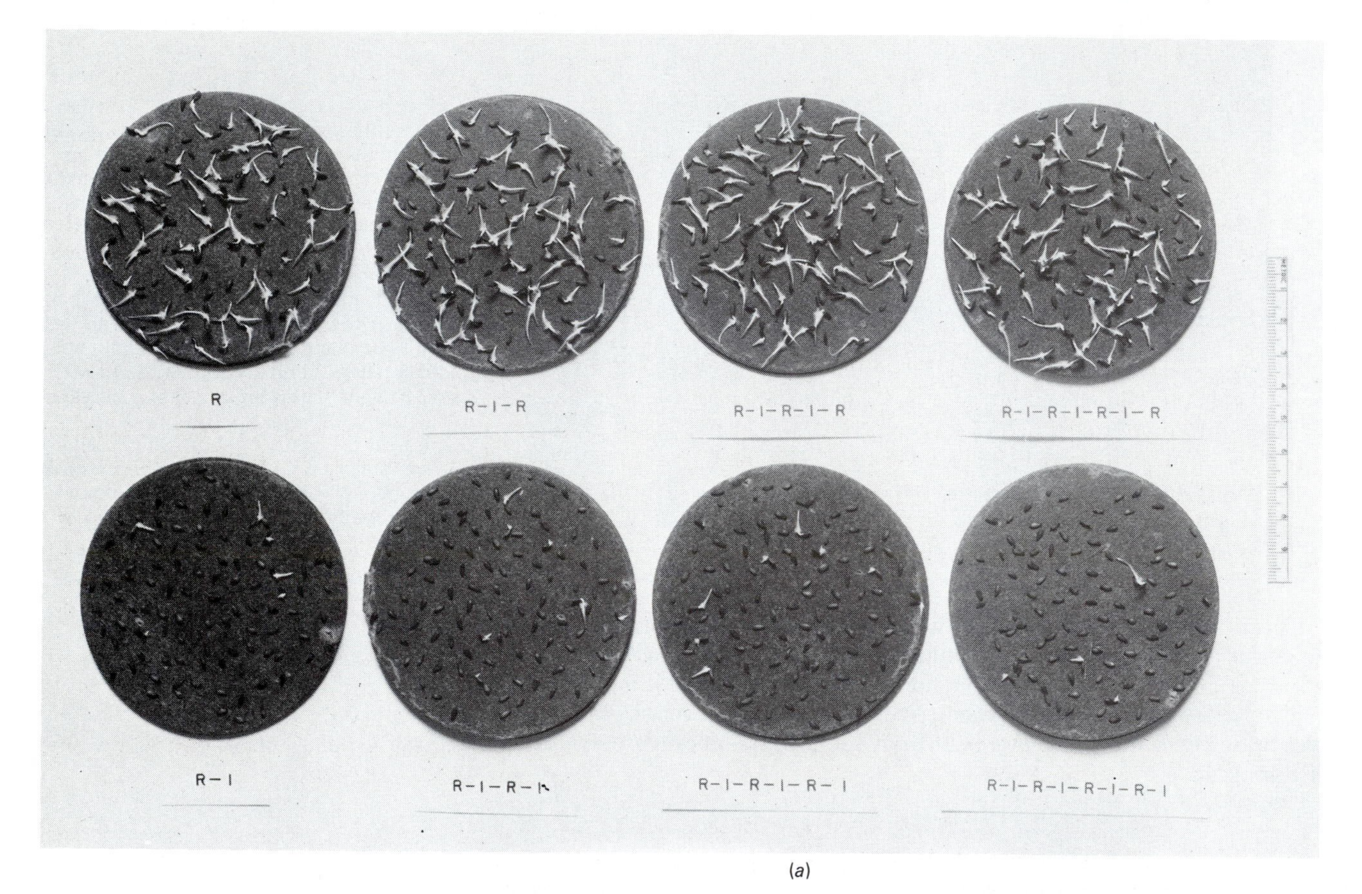

(a)

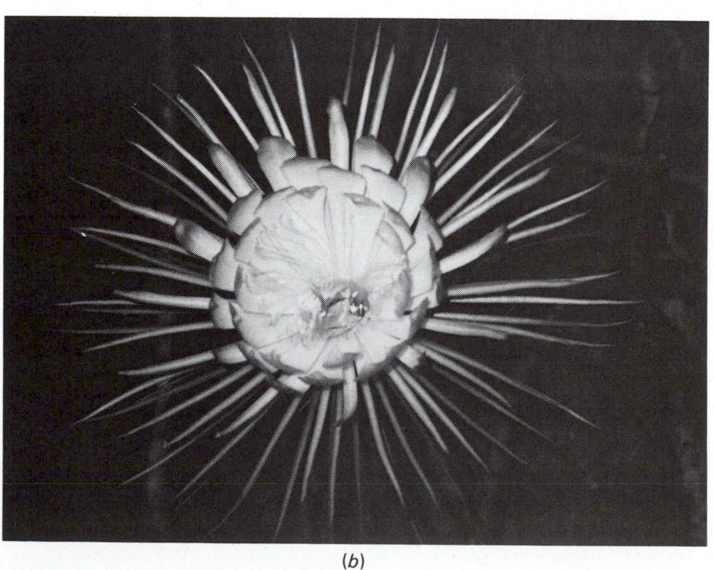

(b)

Figure 29–12 (*a*) The control of lettuce seed germination by red (*R*) and far-red (*1*) light. Seeds are moistened and then exposed to red light (for 1 min each exposure) and far-red light (for 4-min exposure) in the sequences indicated. If the last exposure is to red, most of the seeds germinate; if to far-red, they remain dormant. (*b*) Night-blooming *Cereus* flower. This bat-pollinated flower must open during the same week-long period of blooming for all the other *Cereus* plants in the neighborhood, and it must open at night. The flower will close early the next day. ((*a*), from Borthwick, H. A., et al.: *Proceedings of the National Academy of Science* USA 38:662–666, 1952.)

abscission, leaf sleep movements, pigment formation in ripening fruit, and in seed germination.

A number of plant functions other than flowering may also be affected by the daily photoperiod. The formation of tubers by Irish potato plants, for example, is accelerated when the daily exposure to light is shortened. Since the growth of the tuber (the part we know as the potato) involves the deposition of starch, the photoperiod must in some way stimulate the transfer to carbohydrates from the leaves to the tuber, where starch synthesis and deposition occur.

SUMMARY

I. Plants exhibit behavior, both by rapid responses to stimuli and by growth responses known as tropisms.
   A. Rapid responses such as leaf folding result from the abrupt loss of turgor in a specialized collection of cells, the pulvinus. These are stimulated by the propagation of electric impulses resembling action potentials.
   B. Tropisms result from the action of auxin produced by the apical meristem or other bud tissues and transported to one side of an elongating stem. The auxin stimulates cell elongation by lowering pH within the cell wall. This leads to selective weakening of cell walls, permitting cell expansion along the longitudinal axis.
   C. There are several important plant hormones.
      1. Auxin stimulates fruit growth and stem elongation. It inhibits lateral bud development by causing, in large part, apical dominance. It may inhibit root growth; in moderate concentrations it initiates root growth.
      2. Gibberellins are an important stem-growth stimulant. They mediate germination in seeds and aid in fruit development.
      3. Ethylene is given off by ripening fruits and promotes ripening.
      4. Cytokinins are involved in wound healing and probably in tissue differentiation. They reverse senescence in leaves and may govern direction of nutrient flow in plants. Cytokinins are involved in fruit development and, because they stimulate mitosis, in bud growth as well. They are produced principally in the roots.
      5. Abscisic acid is a stress-engendered hormone that sometimes promotes leaf senescence and abscission. It also mediates stomatal closure.

II. Plants exhibit several responses to day length. This ability is known as photoperiodism.
   A. Day length can determine seed germination, time of flowering, and setting of the biological clock governing sleep movements. Flowering photoperiodism is governed by a photoreceptive pigment, phytochrome.
   B. The flower bud resembles a leaf bud, but is not meristematic. The stimulus that provokes flower development is not known, but photoperiodism and perhaps florigen, a postulated hormone, are involved.

POST-TEST

1. The ________________ at the base of the leaves of sensitive plants loses ________________ ________________ abruptly, causing the leaves to collapse.

2. In the phototropic responses of the above-ground parts of plants, auxins accumulate on the ________________ side of the stem.

3. Auxin ________________ the pH of elongating cells, which optimizes the attack of enzymes on the ________________ ________________ of the cell walls.

4. Fruit ripening is accelerated by the gaseous plant hormone ________________.

5. The gravity sensors of roots that help them to grow in a downward direction in positive geotropism are probably ________________, tiny granules of ________________ within the root cells.

6. The photoreceptor pigment that controls flowering in long- and short-day plants is ________________, which is sensitive to ________________ light.

7. Actually, this pigment exists in two forms, one of which is sensitive to ________________ light, and the other, to ________________ light; generally, their actions are mutually antagonistic.

8. ________________ stimulate stem growth and mediate germination in seeds.

9. The weed killer 2,4-D is an artificial ________________ that kills ________________ leafed weeds, but is nearly harmless to grass.

REVIEW QUESTIONS

1. How might it be to the advantage of a sensitive plant to fold its leaves when stimulated? Similarly, what function might be served by sleep movements? (Hint: Consider water conservation and heat loss.)

2. What is a tropism? How does auxin contribute to phototropism?

3. Compare the actions of auxin and gibberellin.

4. What is meant by photoperiodism? How would you determine experimentally whether a plant is a long-day, short-day, or day-neutral plant?

5. What is the evidence for the existence of florigens?

6. What is the difference between the two kinds of phytochrome? Under what circumstances is each apt to be produced?

30
Plant Nutrition

OUTLINE

I. Limiting factors in plant nutrition
II. Plant nutrient requirements
- *A.* The macronutrients
- *B.* The micronutrients
- *C.* Fertilizers

III. Origin of soil nutrients
- *A.* History of a typical soil
- *B.* Ecochemical cycles
  - *1.* The nitrogen cycle
  - *2.* The phosphate cycle

IV. How plants obtain nutrients from the soil
- *A.* The soil solution
- *B.* Exchangeable nutrients
- *C.* Mineralization of organic matter
  - *1.* Detritus feeders
  - *2.* Decomposing microorganisms
  - *3.* Mycorrhizae

V. Soils
- *A.* Sand
- *B.* Clay
- *C.* Silt
- *D.* Humus
- *E.* Loam

LEARNING OBJECTIVES

After you have read this chapter you should be able to:

1. Summarize the known plant macronutrient requirements.
2. Give the physiological role (to the extent that it is understood) of each macronutrient in the life processes of plants.
3. Repeat Objectives 1 and 2 for the known plant micronutrients.
4. Summarize the concept of limiting factors for plant growth.
5. Summarize the role of the soil in the ecochemical cycles of plant nutrients.
6. Outline the major features of the phosphorus and nitrogen cycles.
7. Discuss the roles of the soil solution and exchangeable nutrients in plant nutrition.
8. Summarize the role of decomposers, detritus feeders, and mycorrhizae in the liberation of mineral nutrients from organic matter in the soil.
9. List the main soil components, and give the characteristics and ecological significance of each.

The basic photosynthetic reaction combines carbon dioxide with water in two stages to yield glucose and oxygen. Glucose contains the elements carbon, oxygen, and hydrogen—the only elements occurring in complex sugars and in most fats. Thus the carbohydrates and fats occurring in plants can be made from water and carbon dioxide alone. However, plants do also require various other inorganic nutrients. Biosynthesis of compounds such as amino acids, phospholipids, nucleic acids, and ATP requires the elements phosphorus, nitrogen, and sulfur.

Certain other compounds essential to plant metabolism may require still other special ingredients. Chlorophyll, for instance, contains magnesium, and phytochrome contains iron. Furthermore, the enzymes necessary to synthesize some of these materials may require certain metallic ions in order to function. One stage of chlorophyll synthesis, to take another example, requires iron even though chlorophyll itself contains none. The adaptive reason for many of these requirements is, however, not known. All plants have an inflexible requirement for boron, for instance, but the metabolic role of this element is unknown.

Often the very nutrients a plant requires in modest amounts can poison it when present in excess. Overuse of fertilizer, for instance, can make the soil hypertonic to the plant, resulting in osmotic stress that injures or kills the plant. Plants, like animals, have *optimum* ranges in which nutrients are present in concentrations that best promote their health and growth.

Limiting Factors in Plant Nutrition

A basic concept of plant nutrition often overlooked just because it *is* simple, is that plants (as well as all other living things) need a sufficient supply of everything they require and they need it at all times. A logical consequence of this principle is that, as long as it is not poisonous, it does not matter how much excess of one essential nutrient is present if there is an insufficiency of another. Whatever factor is in the shortest supply will limit growth. No matter how much sunlight a plant may have, for instance, insufficient water will kill or dwarf it, a fact well known to oriental bonsai gardeners and used by them deliberately to produce dwarfed trees. Natural bonsais also can often be seen in very hostile environments such as mountaintops, where they are stunted because of severe weather conditions or by lack of soil and nutrients. Healthy plant communities can occur only when *all* mineral nutrients are provided in sufficient amounts.

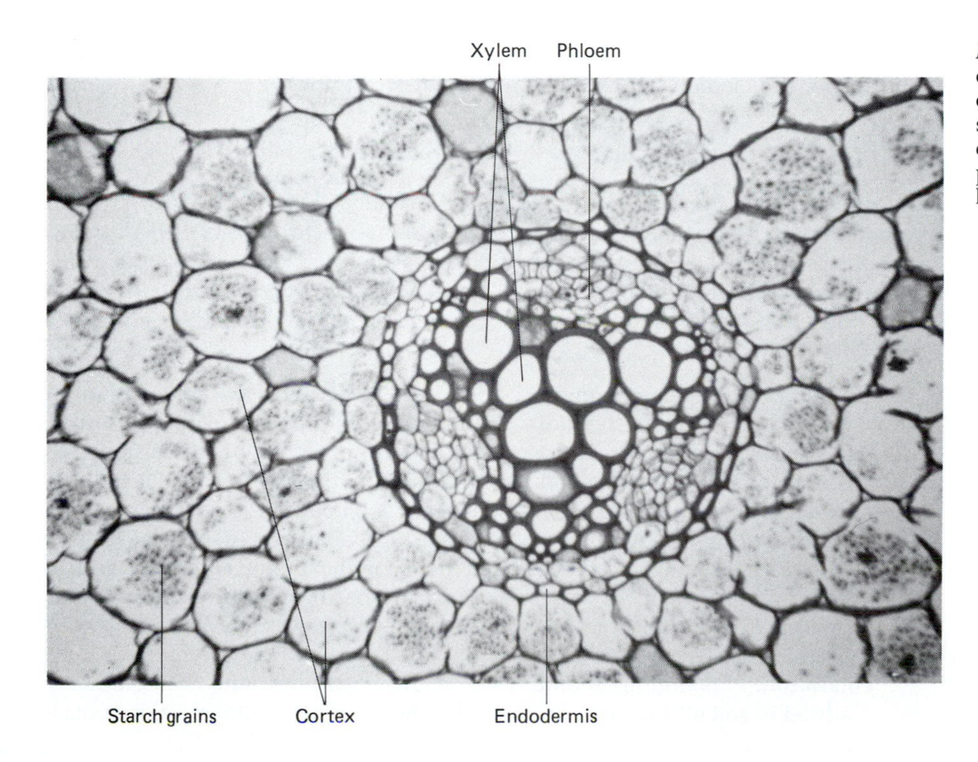

Figure 30–1 Photomicrograph of a vascular cylinder, or stele, in a root. The endodermis with its Casparian strip serves as a barrier to the free diffusion of ions into the vascular system of the plant and also serves to retain high hydrostatic pressure within.

Figure 30–2 Tobacco plants illustrating the effects of deficiencies of specific elements. The plant in the center (*Ck.*) received all essential elements; the others were supplied with all essential elements except the one indicated on the label. All plants are the same age and variety. Note the marked growth inhibition resulting from lack of N and K, the green but distorted leaves of the plant lacking Ca, the general chlorosis (except along the veins) of the −Mg plant, and the chlorosis in older leaves only when P was lacking. The other leaves of the −P plant are dark blue-green. The leaves of the −S plant show different degrees of chlorosis, but more commonly the younger leaves are the most chlorotic. Note also the necrosis in the −N, −P, and −K plants. (Courtesy of W. R. Robbins, Rutgers University.)

Plant Nutrient Requirements

Determining the needs of a species of plant is simple in theory. Most plants can be raised **hydroponically,** meaning without soil but in an aerated solution of mineral nutrients. The roots are allowed to hang into a container of solution, which is renewed periodically to make up for any loss of nutrients. It is possible to hydroponically produce plants that are indistinguishable from those grown in soil. This method can also be used for experimental purposes. Plants can be raised hydroponically in a culture medium of known composition that is able to support their growth throughout their life cycle. Then one of the components of the medium is eliminated and the plant is observed for ill effects, if any (Fig. 30–2). If the lack of some element in its nutrition results in the inability of the plant to complete its life cycle, the element is termed an **essential element.**

THE MACRONUTRIENTS

Macronutrients are those nutrients needed by virtually all plants in relatively large amounts. Historically, they were the first requirements to be established, using hydroponics combined with methods of chemical purification and analysis available in the earliest days of plant physiological research. The following list of required nutrients should be considered as of more than passing interest, for our lives depend upon them.

Nitrogen. Nitrogen was one of the first nutrients to be discovered. Although it can be provided in the form of nitrate or ammonia, it is usually absorbed in the form of nitrate. Nitrogen in the form of organic material must be converted into ammonia and then often into nitrate by bacterial action before it is absorbed by the plant. Nitrogen is needed for proper leaf growth and development. A deficiency may produce yellowing of older leaves or a general lightening of all the green parts of the plant, combined with a stunting of growth. An excess produces hypertrophy of foliage and suppresses fruit production. Nitrogen is mainly important as a component of proteins and nucleic acids.

Potassium. Potassium, another early known plant nutrient, is traditionally added to soil in the form of wood ash. The main intracellular cation, potassium, is probably most important for maintaining the membrane potential of plant cells, and perhaps their turgidity as well (especially in the guard cells of the stomata). Potassium is essential also for the proper functioning of many enzymes and biosynthetic pathways. Deficiency produces general symptoms of poor health, which can include localized **chlorosis** (low chlorophyll content), or mottling, of leaves, with small spots of dead tissue at the tips and between the veins of lower leaves. These lower leaves do not usually dry up and shrivel, as they do in nitrogen deficiency.

Phosphorus. Deficient in a great many soils, phosphorus is essential for the production of such vital compounds as the nucleic acids and ATP. It is needed for flowering, fruiting, and root development. Although phosphorus occurs naturally in some sedimentary rock formations in the form of calcium phosphate, it is not readily absorbed in this form. If the calcium phosphate is treated with sulfuric acid, however, several more soluble phosphorus compounds are formed; it is this "superphosphate" that is usually used for fertilizer. Deficiency results in small dark green leaves over the entire plant and the abnormal presence of red and purple colors in the leaves and stalks.

Sulfur. Sulfur is an essential component of protein because of its occurrence in the amino acids cysteine and methionine. Deficiency produces chlorosis in new leaves and buds, usually without spotting, and poor root development. Sulfur cannot be absorbed in elemental form but must be present as sulfate, SO_4^{2-}.

Calcium. Calcium deficiency results in abnormal growth and cell division, since calcium is an important component of the middle lamella of cell walls (along with pectin). Typically the terminal bud dies following a period in which small leaves with dried-up tips are produced. Calcium has a multitude of cellular functions in the plant body.

Magnesium. Magnesium is required for the action of many enzymes and is needed also in the synthesis of chlorophyll, which contains it. Deficiency, therefore, produces mottled chlorosis. Some plants (e.g., Florida Key lime trees) have especially high magnesium requirements.

THE MICRONUTRIENTS

Certain substances, termed **micronutrients,** are as vital to the plant as macronutrients but are required only in extremely small amounts. The need for such micronutrients was not suspected in early studies to determine plant nutrient requirements because their presence as impurities in water, or in the chemicals used to make up the hydroponic solution, was enough to sustain plant growth. The following elements in trace amounts are now known to have a significant role in maintaining plant health.

Iron. Iron is needed in several of the electron transport substances of the cell (ferredoxin, cytochromes), and in some other materials (e.g., phytochrome). It is also required for chlorophyll synthesis. Although it is usually present in the soil in sufficient amounts, deficiency in iron absorption can occur in soils with high or low pH. If there is a deficiency, a characteristic form of chlorosis—**interveinal chlorosis,** characterized by yellowing of the leaf along the veins, that is confined to the youngest leaves—is the result.

Boron. The function of boron is unknown. Deficiency results in abnormally dark foliage, growth abnormalities, and malformations, such as the heart rot of beets and the stem cracks of celery. Root tip elongation also slows.

Zinc. Zinc is required for the production of the amino acid tryptophan. Since auxins are derived from tryptophan, zinc is indirectly required for the production of auxins as well; it is also required as a cofactor for some of the DNA polymerase enzymes. Deficiency produces small leaves and stunted stems owing to short internodes (Fig. 30–3). In excess, zinc is poisonous to plants.

Manganese. Manganese is required as a cofactor for enzymes in oxidative metabolism (particularly the citric acid cycle), and in photosynthetic oxygen production. Its deficiency produces a mottled, characteristic form of chlorotic leaf yellowing.

Figure 30–3 Zinc deficiency in potatoes (*Solanum tuberosum*) grown in the state of Washington. The plants in the background received zinc and phosphorus fertilizers; the dying plants (*arrow*) in the foreground received phosphorus but no zinc. (USDA Misc. Publication 923.)

Chlorine. Probably required for ionic balance and maintenance of cellular membrane potentials, chlorine (in the form of chloride) is apparently also needed for oxygen production in photosynthesis. Its deficiency results in very small leaves and slow growth. Leaves become wilted, chlorotic, or even necrotic and may eventually become bronze-colored.

Molybdenum. Molybdenum is needed as part of the denitrifying and nitrogen-fixing enzymes of microorganisms. Nevertheless, molybdenum is not needed by plants or their communities if an artificial source of fixed nitrogen is provided. It is far less expensive to provide the molybdenum than the nitrogen fertilizer. Molybdenum is also needed by the nitrate reductase enzyme present in most plant roots. Plants must utilize this enzyme if they are to employ *nitrate* as a nitrogen source. However, plants that absorb ammonia as a nitrogen source do not need molybdenum.

Molybdenum is absent from much of the soil in parts of Australia, which greatly limits agricultural productivity because it limits the nitrogen content of the soil indirectly. The deficiency can be repaired by applying a mere ounce (20 to 30 g) of molybdenum fertilizer per acre every five years. If the affected parts of Australia had never been farmed, no one might have suspected that their low productivity was related to molybdenum deficiency.

Copper. Copper is a component of some enzymes and cytochromes. Its deficiency results in a lowered rate of protein synthesis and sometimes in chlorosis. Young leaves may be dark green and twisted, with dead spots.

FERTILIZERS

A **fertilizer** is a plant nutrient or mixture of plant nutrients applied to the plant (usually by mixing with the soil) to compensate for a potential deficiency of that nutrient. Since plants can absorb only inorganic materials, it is necessary to supply the fertilizer in inorganic form or, if slow release is desired, incorporated into some organic material from which it is liberated by decomposition. The addition of organic matter to soil also benefits humus content, as we will see.

Most commonly, commercial fertilizers consist of a mixture of potassium, phosphate, and nitrate salts to which trace nutrients are sometimes added. The proportions of the major nutrients may vary considerably in accordance with intended use. For example, large amounts of nitrate tend to encourage foliage growth at the expense of fruit production. A fertilizer for use on citrus or tomatoes, therefore, should be low in nitrogen, as would a fertilizer (for different reasons) intended for soybeans.

The use of ammonia for fertilizer deserves special mention. It is possible to inject ammonia gas into the ground, from which it is directly absorbed by plant roots. Since ammonia is directly usable in the production of amino acids, the plants need not absorb nitrate, which requires enzymatic reduction to ammonia within the plant before it can be used in amino acid synthesis. The **nitrate reductase** enzyme that plants employ for this purpose disappears from the roots when ammonia is added to the soil; the enzyme is no longer needed. Yet nitrate reductase reappears when ammonia fertilization is discontinued. This is an example of enzyme induction that has been observed among eukaryotes (Chapter 15).

Although it can incidentally serve as a source of calcium and magnesium, **lime** (calcium hydroxide) is usually not used as a fertilizer. Lime is generally added to soils to increase their pH. If soil pH becomes low as a result of marked production of organic acids by humus decomposition, for instance, or from excessive or acid rain (see Chapter 52), some mineral nutrients are liberated in large amounts by ion exchange and may be leached out of the soil. Moreover, such toxic substances as aluminum or manganese may be mobilized from clay soils by acidic ion exchange, and lowered pH may reduce the biological availability of phosphorus.

Origin of Soil Nutrients

In theory, soil nutrients originate from the parent rocky material, which gave rise to the soil by weathering. This is such a slow process, however, that the mineral needs of mature plant communities cannot be sustained solely by continued weathering of the soil. However, the very earliest plant communities to colonize a habitat may indeed have obtained their mineral nutrients from the rocky materials upon which they grew.

HISTORY OF A TYPICAL SOIL

Imagine a bare rock surface on a moderately tall mountain in a warm climate. Very few organisms could do more than briefly rest on its surface, for it would provide almost none of the conditions needed for plant or any other life. One form of life could exist and even thrive there, however: lichens (Fig. 30–4). They are important members of many pioneer communities.

As has already been discussed in Chapter 21, a lichen is not really a plant. In fact, taxonomists classify lichens by their fungal partners alone. A lichen is a compound organism consisting of one of several species of fungi in association with one or another species of algae or cyanobacteria. Biologists once considered this relationship to be one of mutualism, in which both partners benefited equally. It is now usually viewed as a case of parasitism, in which the fungus extracts the products of photosynthesis from the alga, giving little or nothing in return. Lichens are amazingly hardy, tolerating extremes of dehydration that would kill any true plant. This obviously suits them to the harsh, bare-rock habitat. What is just as important, lichens have an incredible ability to digest or solubilize that rock and to concentrate nutrients from it and from the surrounding air, probably via rainfall.

Figure 30–4 Lichens growing on bare rock surfaces in the mountains of Georgia. Ideally adapted to pioneer life, lichens accelerate the process of soil formation.

Lichens are not immortal, however, and when they die, the minerals they have accumulated in their bodies are released by decay processes and taken up by living lichens, in addition to whatever the new generation of lichens can liberate afresh from the rock. Thus the lichen community gradually accumulates an ever-increasing capital of minerals. These minerals are stored in the bodies of living lichens and the animals that are now living on them, and, to a lesser extent, in the organic detritus that is beginning to accumulate under and around them. By now fine particles of rock have weathered away from the surface, but have been trapped by the lichen community and cannot wash away as they did before the lichens were established on that spot. Eventually the mineral particles and the organic detritus form a thin layer of true soil capable of holding some moisture for a brief period of time. This soil contains a substantial part of the accumulating mineral capital of the developing community of organisms.

As the soil thickness increases and more minerals accumulate in it, it becomes suited to the requirements of some higher plants—mosses. They now take hold among the lichens and grow vigorously. In fact, these mosses are better adapted to the less rigorous conditions the lichens have created than are the lichens themselves. Gradually they crowd the lichens out, going through many cycles of life and decay themselves in the process. Minerals continue to weather out of the parent rock, now buried, after perhaps a thousand years, beneath several centimeters of soil, and continue to be added over the course of years to the accumulating capital of the now radically changed community. Few or no minerals leave the community, for as soon as they are released from the bodies of dead organisms by decay bacteria and fungi, the minerals are either immediately reincorporated into young organisms or are held in the soil temporarily in adsorbed form.

ECOCHEMICAL CYCLES

By the time a mature or near-climax plant community[1] has developed it has accumulated a large store of mineral nutrients. In fact, the community has developed *because* it has this accumulated capital of nutrients. If the community had to depend only on what it could extract afresh from the rock, it would be restricted to lichens. As the nutrient storehouse increases, the potential range for plant variation also increases. Each new seedling obtains the minerals it requires from the soil as the young plant grows, with little or none of those coming immediately from the parent rock, now several meters below the surface of the soil. Almost all of them have been recycled by the action of decomposers from the bodies of dead organisms and from waste products.

Soils develop in different ways, but in all cases—together with the living organisms they support—they act as a kind of bank into which and from which deposits and withdrawals are constantly made. This constant cash flow of nutrients is best illustrated by the two best-understood **ecochemical cycles,** also known as **biogeochemical cycles**—the nitrogen cycle and the phosphorus cycle.

The Nitrogen Cycle

We and other consumers ultimately depend upon photosynthetic autotrophs for chemically combined or **fixed nitrogen**—in the form of proteins and nucleic acids. Plants depend upon both chemosynthetic autotrophs and decomposers for recycling the nitrogen upon which their survival depends (Fig. 30–5).

At first glance it would appear that there is no danger of nitrogen starvation. After all, the surrounding ocean of atmosphere contains about 80% nitrogen. But the element nitrogen occurs combined in stable diatomic molecules (N_2) and zealously resists any tendency to unite with any other common substance. Biologically, molecular nitrogen is a nearly inert gas. The overall reaction that breaks up diatomic nitrogen and combines the nitrogen with such elements as oxygen or hydrogen is strongly endergonic; that is, it requires much energy. Photosynthesis might seem an excellent source of the required energy, and so it is, but not at first hand. The oxygen liberated by photosynthesis destroys the enzyme **nitrogenase.** This enzyme chemically welds nitrogen into chemical combination in biological nitrogen fixation.

[1] A climax community (as we will see in Chapter 50) is a stable association of organisms that, if left undisturbed, undergoes no significant change.

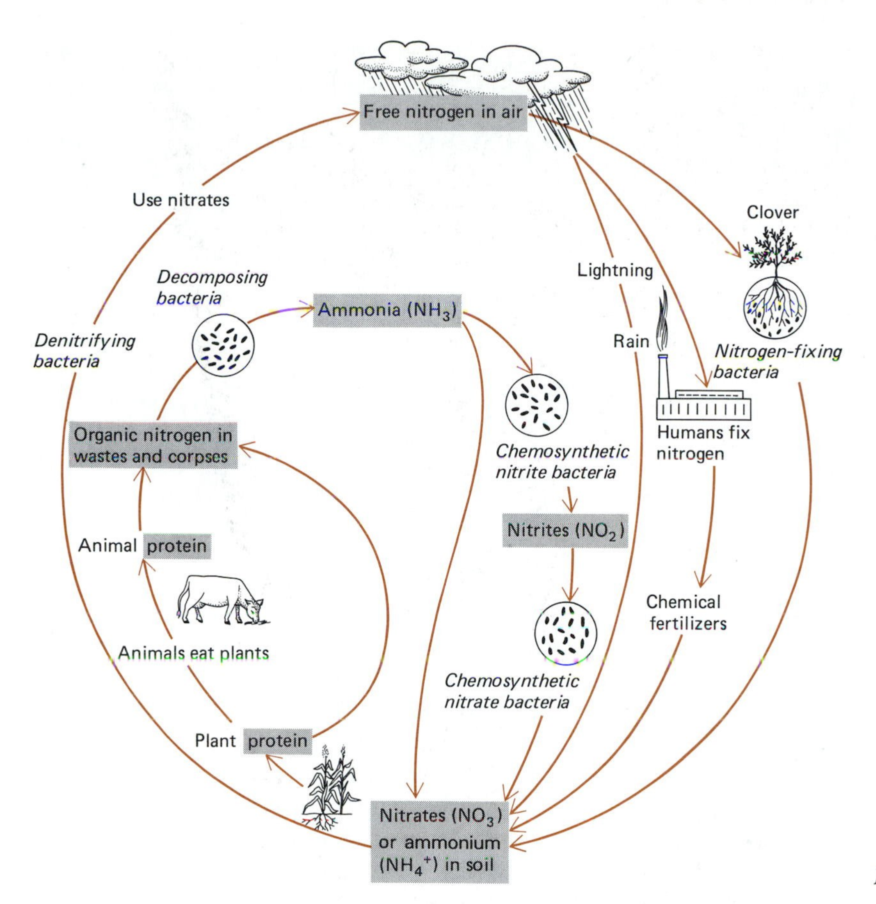

Figure 30–5 The nitrogen cycle.

Most plants, as has been seen, require nitrogen in the form of nitrate, NO_3^-, although many plants absorb enough of the minute quantities of ammonia present in the atmosphere to comprise a significant portion of their nitrogen budget. Plants require nitrate in order to manufacture amino acids. Not only can they make proteins from amino acids, but they also can make nucleic acids from some of them (aspartate and glycine). Without a source of nitrate this is impossible. It is easy to see that the existence of all life on earth demands that plants have access to nitrate, for without it there could be no proteins, no DNA and RNA, and thus no continuance of present-day life forms.

Nitrate is produced by chemosynthetic bacteria known as **nitrifying** bacteria, which oxidize nitrite, NO_2^-, to nitrate, NO_3^-. The oxidation furnishes them with the energy by which they live. Nitrite, in its turn, is produced from ammonia by other bacteria that oxidize ammonia as their source of energy. Ammonia itself is produced by still other bacteria that reduce amino acids to ammonia, water, and carbon dioxide; amino acids are the products of decay of plant and animal proteins such as occur in corpses and wastes. Furthermore, a large amount of an animal's protein intake is biologically oxidized in its cells. This involves the removal of the amino group, which is then usually converted either to urea or to uric acid and excreted in the urine. Certain bacteria convert urea or uric acid to ammonia.

Denitrifying bacteria produce free nitrogen gas from fixed nitrogen as the ultimate result of their work, reversing the action of the nitrifying and nitrogen-fixing bacteria. Denitrifying bacteria are anaerobic. Apparently they live chiefly in the deepest reaches of the soil near the water table. They employ nitrate or nitrite as electron acceptors in place of oxygen. The action of the denitrifying bacteria would produce a net deficit of chemically combined nitrogen in the biosphere were it not for the action of the nitrogen-fixing bacteria, which bring nitrogen into chemical combination, and therefore stand between the world of life and ecological disaster by nitrogen starvation.

Most biological nitrogen fixation is carried out by nitrogen-fixing microorganisms, some of which live beneath layers of oxygen-excluding slime on the roots of a number of higher plants. But the most important of these microorganisms live in special swellings, or **nodules,** on the roots of leguminous plants such as beans or

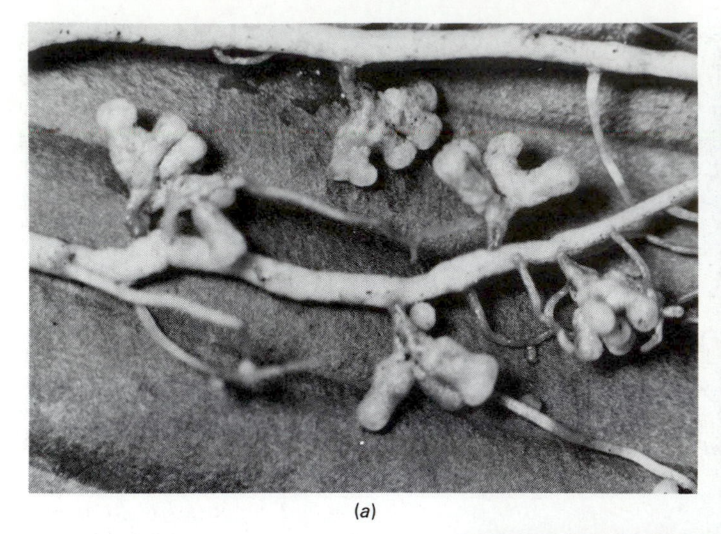

(a)

(b)

Figure 30–6 Legumes serve as hosts for symbiotic nitrogen-fixing bacteria. (*a*) Nitrogen-fixing bacteria nodules on clover roots. (*b*) Nodulated and control soybean plants. The plant on the right was grown in sterilized soil from seed inoculated with specific nodule-forming bacteria; that on the left was grown from uninoculated seed. (*c*) A look into a plant cell inside a soybean nodule. Note that the cell is packed with *Rhizobium,* symbiotic bacteria capable of nitrogen fixation. ((*a*), Carolina Biological Supply Company; (*b*), U.S. Dept. of Agr. Farmers Bull. 2003; (*c*), courtesy of Dr. Winston Brill.)

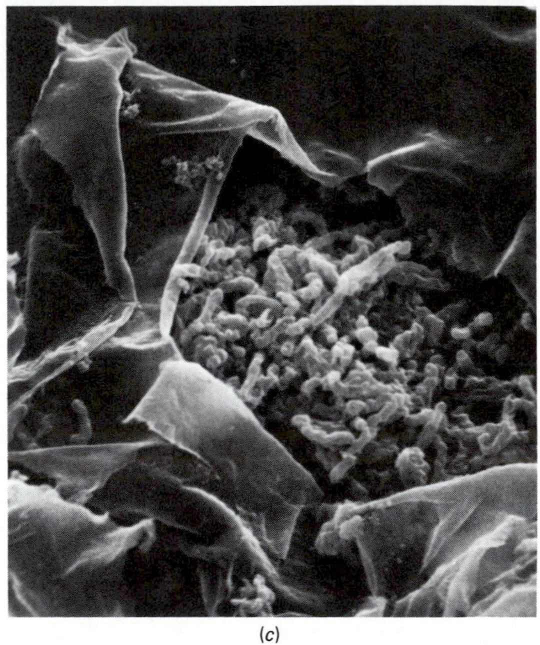

(c)

peas (Fig. 30–6) and certain other woody plants in several families (the actinorhizal plants). It is estimated that in terrestrial environments these mutualistic bacteria can fix nitrogen 100 times faster than other, less vigorous nitrogen-fixing organisms, employing the energy released from carbohydrate provided by the photosynthesis of their host plant.

The reduction of nitrogen gas to ammonia by nitrogenase is a remarkable accomplishment of biological industry, which manages without the tremendous heat, pressure, and energy consumption of the commercial processes developed by technology. Even so, it takes the consumption of 12 grams of glucose or the equivalent to fix a single gram of nitrogen biologically.

In addition to its energy requirements, nitrogenase demands an almost anaerobic environment, which root nodules do provide. Oxygen is transported to the actively respiring tissues within by **leghemoglobin,** an oxygen carrier that is quite similar to animal hemoglobin. However, the chemical constants of the leghemoglobin are so balanced that little oxygen is available to the nitrogenase enzyme itself. Legumes often are planted to introduce ammonia (and ultimately nitrate) into nitrogen-poor soils. (They cannot, however, make up other deficiencies, such as those of phosphorus or potassium.)

A large part of the nitrogen fixation that goes on in aquatic habitats is done by cyanobacteria. In filamentous cyanobacteria occasional heterocyst cells occur, which explicitly function to fix nitrogen and which are evidently not themselves photosynthetic. (If they were, the oxygen they generated would inactivate their

(a)

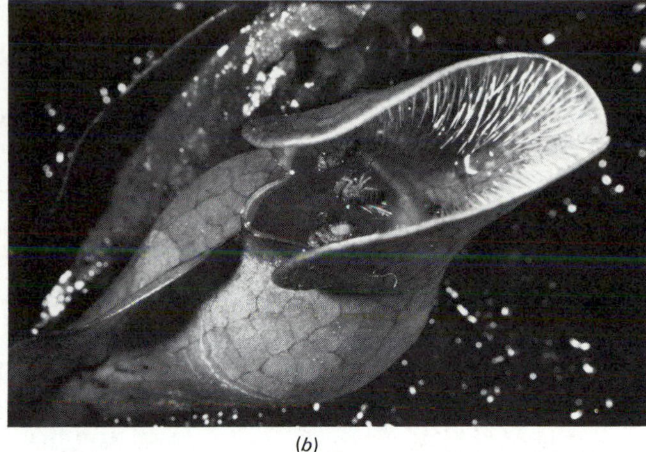

(b)

Figure 30–7 A novel way of obtaining nitrogen. (*a*) A bladderwort (*Ultricalaria minor*) trapping a mosquito larva (×6). Other carnivorous plants include the tropical pitcher plant (*Nepenthe*), shown in (*b*), from whose modified leaves insects cannot escape and must therefore drown in the contained water, and Venus's fly-trap, shown in Figure 1–7. ((*a*), Kim Taylor, Bruce Coleman Inc.; (*b*), E. R. Degginger.)

nitrogenase.) Some water ferns have cavities in which such cyanobacteria live, somewhat as nitrogen-fixing bacteria live in the root nodules of leguminous plants on land. Cyanobacteria also fix nitrogen in symbiotic association with cycads and some other terrestrial plants.

In summary, the four main groups of microorganisms involved in the nitrogen cycle perform, respectively, the following functions: (1) they produce ammonia from decaying proteins, urea, or uric acid; (2) they oxidize ammonia—ultimately to nitrates; (3) they defix nitrogen by liberating it from nitrate; and (4) they fix molecular nitrogen directly and turn it into ammonia. (This they do by employing the energy from carbohydrate and the enzyme nitrogenase; the carbohydrate is obtained from a host plant.) Ecologists believe that virtually all nitrogen now in the earth's atmosphere has been fixed and liberated many times by these organisms. In light of this, it is remarkable that at any one time there are probably only a few kilograms of the enzyme nitrogenase on the entire planet. Those few kilograms have sustained the biosphere for millennia.

Many of the agricultural problems of the world are due to nitrogen-deficient soils. Such deficiencies can largely be remedied either by planting legumes and plowing them into the soil, by applying human or animal wastes, or by applying nitrogen in the form of ammonia or nitrate that has been manufactured industrially (Fig. 30–5). The production of nitrogen fertilizer consumes energy, just as does biological nitrogen fixation. But unlike the latter process, industrial nitrogen fixation depends not upon the solar power of photosynthesis but upon dwindling supplies of gas and petroleum, and is therefore the less desirable of the two remedies. The solar energy that has always powered the nitrogen cycle is virtually inexhaustible.

The Phosphate Cycle

As water runs over rocks, it gradually wears away the surface and carries off a variety of minerals, either in solution or in suspension. Phosphate minerals are among the most important of those that are lost from the land in this way. Because the most common phosphorus minerals are quite insoluble, there is no quick way they can be returned to the land. Instead, they tend to be deposited permanently on the sea floor, although geological processes of uplift may some day expose these sediments on new land surfaces from which they can once again be eroded.

Phosphorus in the soil of plant communities is taken in by plant roots in the form of inorganic phosphates. Once in the plant body, it is converted into a variety of organic phosphate intermediates in the metabolism of carbohydrates, fats, and other compounds. Some of these are incorporated into nucleic acids and related compounds. Animals obtain most of their phosphate as inorganic or organic compounds in the food they eat, although in some localities the drinking water may contain substantial amounts of inorganic phosphate.

Very little phosphorus is probably lost from natural communities, but all too few communities of organisms today can reasonably be considered in a "natural state." If there were some way that phosphorus could be returned to communities via the atmosphere (as with nitrogen), its loss would not be of critical importance. As it is, on a human time scale, phosphorus, once lost, is lost for good, for it ends up and remains in the sea. Moreover, when phosphorus loss is accelerated from soil by such practices as the clear-cutting of timber, or as a result of erosion from agricultural or residential land use, the phosphorus can produce ecological damage of the aquatic communities it passes through by contributing to eutrophication (Chapter 52). Eutrophication involves an abnormal overgrowth of plants or algae in aquatic habitats, which can lead to habitat damage.

Phosphorus enters aquatic food chains mostly by being absorbed by algae and plants, which are in turn consumed by large organisms and by microorganisms and other plankton. These, then, are eaten by various fin and shell fish, which may also feed upon one another. A small portion of the fish and invertebrates are eaten by sea birds, which may defecate on their own roosting places. **Guano,** the manure of sea birds, contains large amounts of phosphorus and nitrate. The phosphate contained in guano thus finds its way back to the land, and some of it may be able to enter terrestrial food chains in this way, although the amounts involved are obviously trivial.

The bulk of phosphorus transport in the food chain is in the opposite direction. Corn grown in Iowa may be used to fatten cattle in an Illinois feedlot. Beef from these cattle may be consumed by people living far away—for instance, in New York City. Part of the phosphorus absorbed from the Iowa soil by the roots of the corn plants ends up in feedlot wastes, which probably eventually wash into the Mississippi River. More ends up flushed down toilets in places like Manhattan. To replace this steady loss, Iowa farmers add phosphate fertilizer to their fields. That fertilizer is made (more than likely) in Florida from deposits of phosphate rock that will probably be exhausted somewhere around the year 2000, at present rates of use. We have thus made phosphorus into a nonrenewable resource.

How Plants Obtain Minerals from the Soil

Minerals in the soil would do plants little good if they could not be absorbed. Roots absorb water and minerals by means of root hairs, mycorrhiza, and (to some extent) the root's epidermis. Minerals available to plants occur in three forms in the soil: (1) dissolved in the soil solution, (2) as part of the inorganic matter of the soil, and (3) incorporated into the organic matter of humus.

THE SOIL SOLUTION

Among the particles of soil are many microscopic spaces, which may contain either air or liquid (Fig. 30–8). When water is present it fills many of these spaces, but some water may drain by gravity flow to join the ground water, often out of reach of plant roots. What remains is bound by forces of capillary attraction. Since this water dissolves many mineral ions and holds them in solution, it is called the **soil solution.** When soil contains as much water as it can hold by capillary action, it is said to be at **field capacity.**

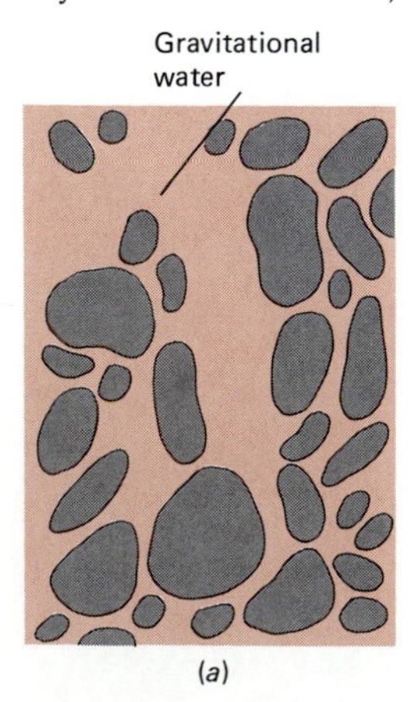

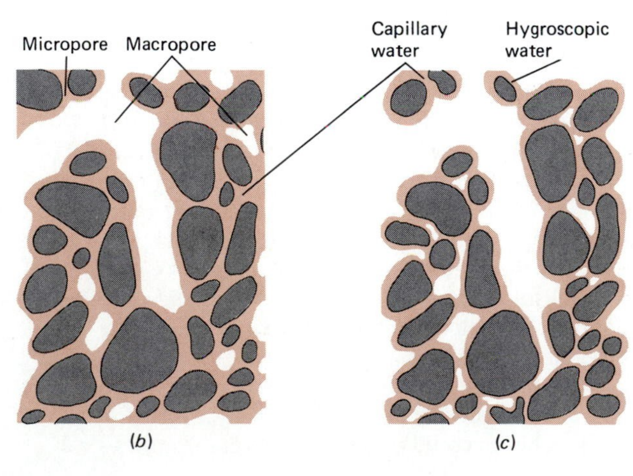

Figure 30–8 A "structured" soil (containing crumbs) under three conditions of hydration. (*a*) Water-saturated, all pores filled with water. (*b*) Field capacity, macropores air-filled and micropores water-filled. (*c*) Permanent wilting point, pores air-filled, hygroscopic water films around soil particles, small amounts of capillary water at points of particle contact. (Courtesy of Ray, Steeves, and Fultz.)

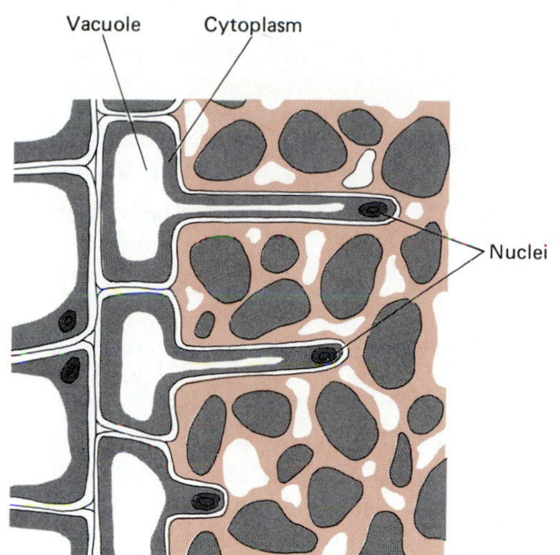

Figure 30–9 Root epidermal cells, showing stages in root hair development and interpenetration of soil particles. (Courtesy of Ray, Steeves, and Fultz.)

The soil solution or its components can enter the root in three ways: by active absorption through the root hairs, by imbibition through the epidermis, and, in some cases, by mycorrhizae (which we will consider presently).

In the case of active absorption by the root hairs, materials entering through the root hairs (Fig. 30–9) usually pass from cell to cell until they reach the endodermis. This is known as the **symplast** route of absorption. The cells of the endodermis have no cytoplasmic connections with those of the parenchyma. Thus, to pass on to the vascular tissues, the materials must be absorbed again at that point across the cell membranes of the endodermis.

In the case of imbibition through the epidermis, the soil solution is imbibed by the epidermal wall of the root (this is possible over large areas of its surface). It then passes through the cortical cell walls *without* ever being absorbed by the cells until it reaches the endodermis. This is known as the **apoplast** route of absorption. There it is able to go no further by the cell wall route because it is stopped by the Casparian strip, which was discussed in Chapter 27 (Fig. 30–10). Any of its compo-

Figure 30–10 Absorption of nutrients from the soil. (*a*) Diagram illustrating the symplast theory of ion uptake by roots. Ions diffusing into the cell walls of epidermal or cortical cells can be actively taken up into the cytoplasm through plasma membranes. These ions can then pass without crossing any membranes until the ions reach the endodermis, at which point they must cross the cell membranes of the endodermis in order to be released into the xylem. Plasmodesmata are represented much larger relative to cell size, and fewer in number, than in actuality. (*b*) Diagrammatic cross section of a root, showing alternative pathways for uptake of water and nutrients. For simplicity the cortex is shown as being only a few cells thick, although in typical roots it is much thicker. The thickness of cell walls of living cells is exaggerated here. Note how the Casparian thickenings in cell walls of the endodermis block off pathway A all the way around the vascular core of the root. (Courtesy of Ray, Steeves, and Fultz.)

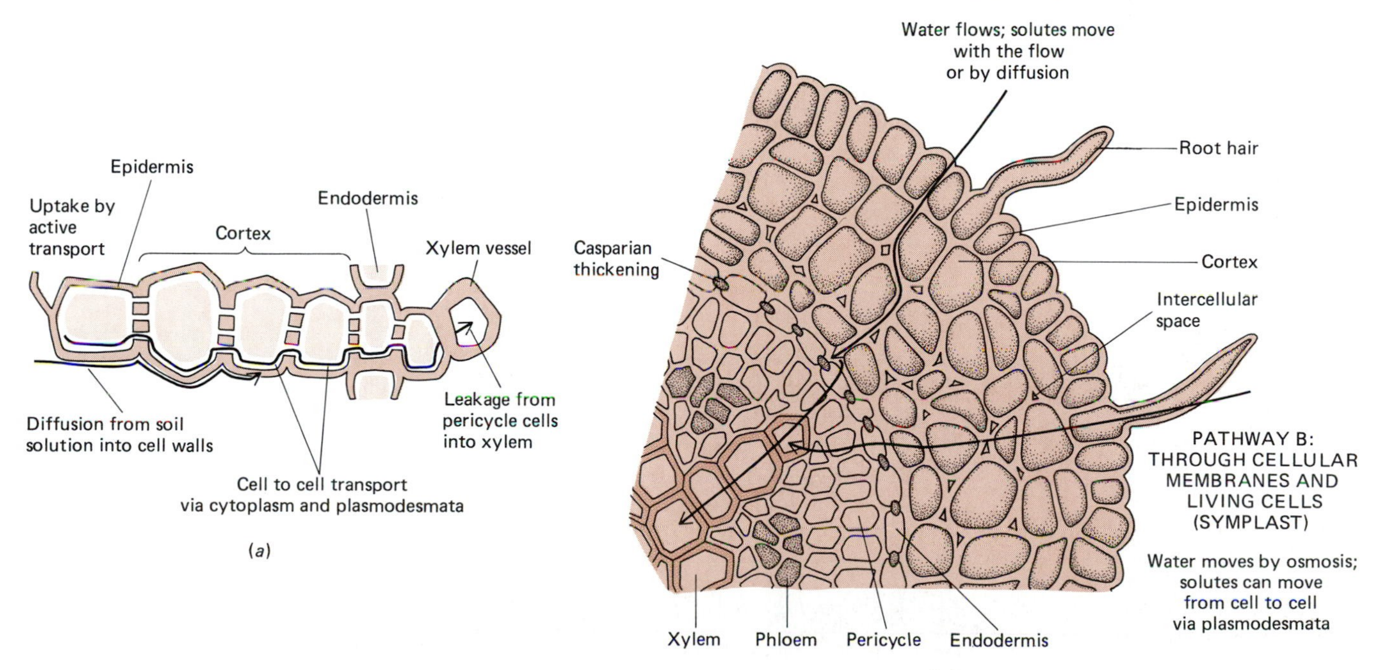

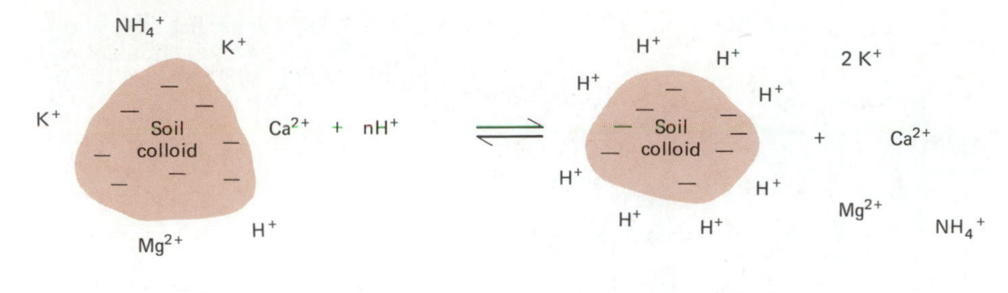

Figure 30–11 Cation exchange with soil particles. Arrows show that cation binding is reversible. (Courtesy of Ray, Steeves, and Fultz.)

nents that are absorbed must now be taken up by the cells of the endodermis. Therefore, any material entering the vascular tissues of the root must first be absorbed across the membranes of at least one set of cells, even if it is not absorbed by the root hairs. Active transport mechanisms in these membranes are capable of concentrating some materials many times as compared to their original concentration in the soil solution.

EXCHANGEABLE NUTRIENTS

Some minerals are so tightly bound to soil particles that they do not readily enter the soil solution. Such minerals are usually unavailable to plants, but at least in the case of cations they often can be dislodged by exchange with hydrogen ions. Root hairs produce carbon dioxide, which forms carbonic acid; this then dissociates, forming hydrogen ions and bicarbonate ions. An additional quantity of hydrogen ions may be actively secreted by the root hairs. The result is the liberation of such ions as potassium and magnesium and their extraction from the soil in much greater quantities than simple water percolation (without living plants) could ever accomplish (Fig. 30–11).

MINERALIZATION OF ORGANIC MATTER

Soil is not an inanimate substance but a complex community in its own right, consisting of a bewildering array of inhabitants. Except for the roots of plants, however, all of these creatures are heterotrophs. A few of them eat or are parasitic upon the roots of plants, but most of them are either decomposer organisms (usually fungi) or detritus feeders that consume organic matter arising from dead or discarded plant parts.

The organic matter of the soil contains mineral elements that must be liberated from it and converted to inorganic ions before they can be utilized by plants. Some of these ions were incorporated into the organic matter when that was first synthesized by the organisms of which it originally formed a part. In addition, however, organic matter tends to absorb and hold minerals from such sources as inorganic fertilizer, detaining them in the soil and preventing them from being leached out.

Detritus Feeders

An earthworm will serve here as an example of a **detritus feeder.** It actually eats its way through the earth, digesting almost all the organic matter the soil may contain. Earthworms even possess enzymes capable of digesting cellulose. When this partially digested soil is eventually defecated by the worm, the minerals that were originally incorporated into such organic materials as DNA are now in the form of simple inorganic compounds like phosphate or ammonia, readily available for absorption by plants.

Decomposing Microorganisms

The process of mineral liberation can take place even without worms. Dead wood is invaded first by sugar-metabolizing fungi that consume such readily available sim-

Figure 30–12 Fruiting body of *Amanita,* a typical mycorrhizal fungus associated with oak trees. This mushroom is extremely poisonous to human beings.

ple carbohydrates as glucose or maltose. Then, when these carbohydrates are exhausted, cellulose-digesting fungi complete the digestion of the wood, often aided by termites and bacteria. The fungus itself may be eaten by a variety of fungivores like millipedes, whose waste products consequently contain minerals liberated from the fungus. Thus, the complex food chains within the soil constantly break down organic matter, liberating its minerals for recycling into plant and animal bodies once more.

Mycorrhizae

Naturalists have long appreciated that some kinds of fungi appear only in association with certain species of trees. The extremely poisonous death angel mushroom, *Amanita,* for example, is often found in oak woodlands (Fig. 30–12). These mushrooms are derived from mycelial masses that infest the cortical region of the oak roots but are not parasitic upon them. The growth of oak trees is enhanced by the presence of the fungus rather than retarded by it, as we would expect from a parasitic relationship. Such infested roots have few or no root hairs, which suggested to investigators that the fungus might be absorbing minerals from the soil much as root hairs otherwise would. Investigation with radioactive isotopes of phosphorus and other important mineral nutrients disclosed that this was indeed the case. Many materials are far better absorbed by fungal hyphae than by ordinary plant root hairs. The fungus, for its part, benefits from the food materials provided to it by the tree, so the association is a classic example of mutualism.

Almost all plants that have been investigated have been shown to be mycorrhizal. Many of them are obligately mycorrhizal; they will not survive past the seedling stage without their fungal symbionts (Fig. 30–13). Mycorrhizae are particularly important in tropical ecosystems. The soil in most such communities is deficient in organic matter and usually in available minerals as well. Because of the fungal partner, mycorrhizal plants are able to decompose cellulose and other organic material that roots by themselves could not access. Therefore, because of their fungal partners, the roots of living plants are able to invade the bodies of freshly dead organisms by proxy and to extract the minerals immediately, before they can be leached away by heavy tropical rainfalls. So effective are these tropical mycorrhizae that the runoff from a rain forest often contains no more minerals than did the rain that fell upon it.

Figure 30–13 Cultivated citrus species show a marked dependency on an endomycorrhizal association for adequate growth. Shown here are rough lemon seedlings after six months, with and without mycorrhizae, and with weekly applications of 0, 1/2, or full-strength nutrient solution without phosphate. (Dr. J. Menge, University of California at Riverside.)

Soils

The number and distribution of plants are influenced by several factors, including **climatic factors** and **edaphic factors** (soil factors). The kinds of plants, together with climatic and edaphic factors, influence the number and distribution of the various kinds of animals, and this, in turn, may influence plants.

The soil on the surface of the earth varies widely in composition, with weathering, erosion, and sedimentation all producing marked geochemical differences in the earth's surface. Although it is true that the distribution of organisms is greatly affected by the kind of soil present, it is also true that the organisms present make a major contribution to the kind of soil. Organisms are the sources of the great deposits of fossil fuels, such as peat, coal, and oil.

In addition to the water and air they may contain, soils are made up of varying proportions of four components: sand, clay, silt, and humus. Each of these components influences the properties of the soil. Some soils consist almost solely of one of these components, and most are mixtures of at least two of them. Most soils remain where they were formed from the parent rock, but some have been carried from their place of origin to another location by the wind (sand dunes or loess), by water (alluvial deposits at the deltas of rivers), or by glaciers. Most of the soils of Canada and the northeastern United States were deposited by the action of glaciers.

After its formation, a soil undergoes development, controlled both directly and indirectly by the climate. The temperature and rainfall determine the rate at which materials in solution and suspension are transported by percolating water out of the soil (and sometimes into it). Climate determines the kind of vegetation present, and this in turn determines what kind of organic materials will be available to be incorporated into the soil. In cold, humid regions where precipitation is greater than evaporation, the vegetation produces an acid humus, giving an ash-gray soil termed **podzol.** In tropical regions with high temperatures and heavy rainfall, there is little acidity from the decay of tropical vegetation; this results in red

lateritic soil with a high iron content. In regions with low rainfall unevenly distributed over the year and high rates of evaporation, the soil tends to be calcified, rich in calcium carbonate.

Soil characteristics determine the kinds of plants that can grow in the region, and soils supply anchorage for plants. Soils also supply water, mineral nutrients, and air for the roots. Some of the important characteristics of a soil are its texture (for instance, whether it is rocky, sandy, or hard-baked clay), its organic content, the amount of soil water present, the amount of air entrapped in the soil, and its acidity and salinity.

As soils develop they tend to become stratified and form more-or-less well defined layers known as **horizons.** The uppermost layer is one from which nutrients have been removed by water percolating through it. Below this is a layer of accumulated materials derived from the layer above. The bottom layer is composed of unweathered parent material.

SAND

Sand is usually derived from the weathering of siliceous rocks and is made up of silica, SiO_2. Some special sands, such as those found in coral reefs and islands, are composed of other minerals such as calcium carbonate. Sand particles are relatively coarse (2 to 0.05 mm in diameter); the resulting porosity permits large amounts of oxygen to reach the roots of plants, but the soil can hold little water by capillary action. Water that is held in this fashion is measured as **field capacity;** by weight the field capacity of sand is about 6%. Sand has almost no ability to capture exchangeable minerals. Sandy soils, therefore, tend to be dry and infertile.

CLAY

Clay is an aluminum-containing mineral derived from the weathering of many common rocks such as granite. Clay is able to hold large amounts of water (field capacity about 40%), but not all of that water is available to plants because much of it may be so tightly bound to the clay particles that the osmotic and hydrostatic forces generated by roots are unable to extract it. Clay absorbs and releases exchangeable minerals readily, but it is difficult for roots to penetrate clay. Since clay is composed of extremely fine particles (less than 2 μm in diameter), once compacted, clay contains little or no air space. Roots embedded in clay thus tend to suffocate, particularly in wet weather. In dry weather they may be unable to grow into new, moister areas.

SILT

Silt is composed of particles intermediate in size (2 to 50 μm in diameter) between those of clay and sand. It is derived from several kinds of parent minerals and is deposited in soil by wind or water action, as, for example, in a river delta. Silt resembles clay in its soil properties, but is less prone to compaction.

HUMUS

Humus is the organic matter of soil. It is composed of partly decomposed plant and animal remains. Eventually all humus becomes completely decomposed, but in a natural community it is constantly being replaced as plants die and shed their parts. While it lasts, humus preserves air spaces in clay soils and retains water in sandy soils. Humus also delays the leaching out of mineral nutrients by percolating rainwater. If, because of overgrazing or destructive agricultural practices, organic matter is not returned to the soil at the same rate that it decays, the soil eventually will lose its humus content and may become unsuited to agriculture or livestock grazing.

LOAM

Loam is a mixture of two or more types of inorganic soil components. The desirable properties of each component may be combined in loam. In loam composed of sand and clay, for example, the resulting soil contains ample air because of the sand and

Figure 30–14 Serious soil erosion resulting from a flood. Notice how the rocks have protected pedestals of soil beneath them. Soils that have lost their humus and root structure are especially vulnerable to all forms of erosion. (Courtesy of U.S. Forest Service.)

retains water and minerals well because of the clay. Most loam soils also contain considerable humus—between 5% and 20% or even more. Humus tends to prevent soil compaction while retaining minerals and water well, so that a soil that is mostly sand and humus, or clay and humus, would probably serve as well as a growth medium for plants.

Humans tend to look upon soils primarily from the standpoint of their agricultural potential. There are many soils with poor agricultural potential that nevertheless support thriving plant communities. Those of the evergreen taiga of Canada or Siberia afford us one example, and those of the tropical rain forest provide us another. The rich agricultural loams such as those of the North American Midwest or the Ukraine have proved so suitable for agriculture that their original plant communities have been virtually expunged. It remains to be seen whether these soils will be protected from erosion (Fig. 30–14) and mineral depletion in the absence of their original communities.

Each raindrop (*a*) makes its contribution to the erosion of unprotected soil. Notice the solid particles in the splash shown in (*b*). (Naval Research Laboratory Photograph.)

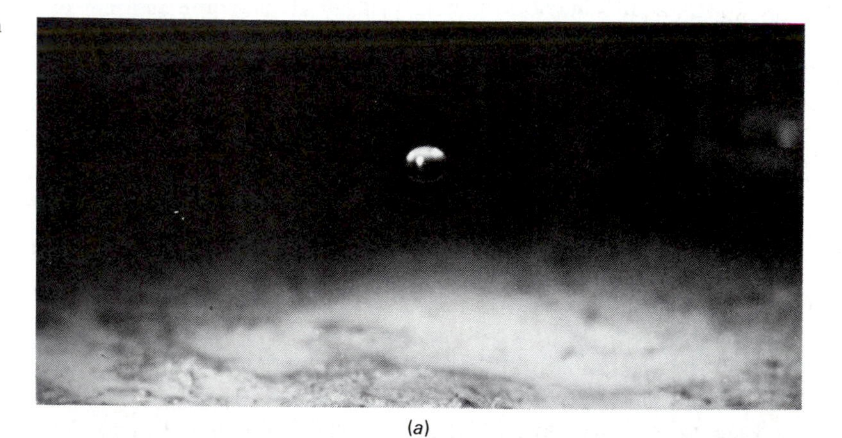

(*a*)

(*b*)

SUMMARY

I. Plants have no requirements for organic food. They do require carbon, hydrogen, and oxygen for photosynthesis.

II. Plants also require certain mineral nutrients for incorporation into organic compounds and for synthesis of organic compounds.

III. Macronutrients are those nutrients required in large amounts; micronutrients are needed in very small amounts.

- *A.* Macronutrients include nitrogen, potassium, phosphorus, sulfur, calcium, and magnesium.
- *B.* Micronutrients include iron, boron, zinc, manganese, chloride, molybdenum, and copper.

IV. Plant growth may be limited by whatever essential nutrient is in shortest supply.

V. Most nutrients absorbed by a plant originate ultimately in the weathering of rocks, but have been recycled many times since then. The intermediate source of most plant nutrients is decomposing organic matter in the soil.

- *A.* Nitrogen serves as one example. Most of the nitrogen absorbed by plants has been liberated from decomposing organic matter, but denitrifying bacteria return some of it to the atmosphere. This produces a net deficit in the nitrogen capital of the community.
- *B.* Nitrogen-fixing bacteria, especially those associated with legume root nodules, make up this deficit.
- *C.* Phosphorus, unlike nitrogen, cannot be returned to most communities (except by artificial fertilizer) after it has been lost by erosion or other processes, because it forms insoluble sedimentary products that accumulate on ocean bottoms.

VI. Plants obtain nutrients from the soil solution and also from clay particles on which they are adsorbed by ion exchange. In addition, nutrients are liberated from organic material by detritus feeders, decomposer microorganisms, and mycorrhizae.

VII. Soils are composed of various mixtures of clay, sand, silt, and humus. Sandy soils are well oxygenated but hold little water or exchangeable nutrients. Silt and clay soils are poorly oxygenated but retain water and nutrients well. Loam soils are mixtures of these components and often combine their desirable characteristics. Almost any soil can be improved by the addition of humus, which must, however, be continuously renewed.

POST-TEST

1. A limiting factor is whatever requirement for an organism's life is in the ______________ supply.

2. The metal ______________ is an essential ingredient in the metalloenzyme nitrogen reductase; this enzyme is necessary for biological ______________ ______________ to take place.

3. Phosphorus is often a limiting factor both in agriculture and in natural ecology, largely because there are no ______________ compounds of phosphorus that occur naturally in the biosphere, so that phosphorus is poorly recycled.

4. ______________ is the commonest intracellular metallic ion.

5. A lichen is a compound organism composed of a unicellular ______________ or ______________ plus ______________; lichens are of special importance in ______________ communities.

6. As the years pass, lichens live and die resulting in the accumulation of accessible ______________ within the community and its immediate environment.

7. As a rule, nitrogen is first liberated from decaying proteins in the form of ______________. Some plants absorb it in this form without change, but it is more usual for them to absorb it in partially oxidized form, that is, as ______________.

8. Most nitrogen fixation from the atmospheric supply of nitrogen is probably accomplished by bacteria, and in aquatic (or even some terrestrial) instances by ______________.

9. When absorbed materials and water are passed from cell to cell to reach the vascular tissues of the root, this is referred to as the ______________ route of absorption.

10. The ______________ strip prevents the absorption of materials into the stele by the apoplast route.

11. Decomposers and detritus feeders help to liberate ______________ from the partially decayed organic matter of the soil's ______________.

12. Fungal microorganisms called ______________ aid the roots of many if not most plants to absorb minerals from the surrounding soil.

13. The principal shortcoming of sandy soils is their very low ______________ capacity.

14. Clay, on the other hand, contains very little ______________, and when dry, it is difficult or impossible for roots to penetrate.

REVIEW QUESTIONS

1. What are the symptoms of deficiencies of each of the plant macronutrients?
2. Why is it desirable to have soil analyses performed? Why not simply infer the nature of a soil nutrient deficiency by observing garden plants for symptoms?
3. Describe an experiment designed to determine whether rubidium is a required plant micronutrient.
4. In a typical soil, is the phosphorus that the plants are presently absorbing newly liberated from the underlying rock, or has it mostly been recycled from organic sources? Explain.
5. Describe or diagram the nitrogen and phosphorus cycles. What are their most significant differences?

6. What is the difference between the soil solution and exchangeable nutrients?
7. What are mycorrhizae and how do they function in plant nutrition?
8. Describe the properties of an "ideal" soil, and give its constituents.
9. If plants have no requirements for organic food, what role does organic matter in the soil play in plant nutrition?
10. A "law of the minimum" was presented in this chapter. Can you formulate a similar "law of the maximum" regarding plant nutrients?

31
Reproduction in Seed Plants

OUTLINE

I. Asexual propagation in nature
II. Artificial asexual propagation
III. Sexual reproduction in seed plants
   A. Sexual reproduction in gymnosperms
   B. Sexual reproduction in flowering plants
      1. Flower structure
      2. Fertilization
         a. Strategies of pollination
         b. Self-pollination
      3. Fruits

LEARNING OBJECTIVES

After you have read this chapter you should be able to:

1. Discuss the adaptive advantages and disadvantages of both sexual and asexual reproduction.
2. Describe two natural and two artificial methods of asexual reproduction in plants.
3. Describe and compare sexual reproduction in gymnosperms and in angiosperms.
4. Describe a typical flower, and identify its parts, using appropriate scientific terminology.
5. Describe the production of pollen and ovules in angiosperms, and summarize the process of fertilization as it occurs in them.
6. Give three methods of pollination, and contrast wind-vectored with animal-vectored methods of pollination as adaptations.
7. Describe the main varieties of fruits, and summarize the processes by which they are formed.

Sexual and asexual reproduction are not mutually exclusive options. Reproductive styles can be mixed to ensure an opportunity for recombination of genes (the advantage of sexual reproduction) plus reliable propagation of the best combinations (an advantage of asexual reproduction). Thus many plants reproduce asexually as a rule, and sexually only as an exception.

Asexual Propagation in Nature

A tree covering several acres may seem like a science fiction fantasy, but the East Indian banyan tree, which can grow to occupy such an expanse, is a fact of nature (Fig. 31–1). This tropical member of the fig family extends adventitious[1] roots from its horizontal branches. When these reach the ground they develop into accessory trunks and eventually enlarge to become comparable to the original trunk of the tree. They, in turn, produce branches and more roots, which eventually become trunks, and so on. Thus over hundreds or thousands of years such a tree can become its own forest. But it is all one individual tree, and its every part is derived from the apical meristem of one original seedling. All that is needed to make this one tree into many is a lumberjack with an ax, a bolt of lightning, high winds, or anything that would produce breaks in the horizontal branches that originally bound it into one mass. For the banyan, asexual propagation is just an uninhibited form of growth.

A somewhat similar mode of growth can be seen in strawberries, which send runners, or **stolons**—horizontal stems—across the ground. These take root at every other node and give rise there to new plants. The aquatic water hyacinth (commonly considered a pest in tropical and subtropical freshwater habitats) produces underwater runners, or **rhizomes,** which grow into plants that can cover acres of water surface in months, shutting out sunlight and oxygen from the communities below (Fig. 31–2). Crabgrass and St. Augustine grass also spread by means of underground rhizomes. The roots of trees and shrubs are sometimes capable of similar behavior, growing horizontally under the surface of the ground and sending up shoots at some distance from the "parent," so that at a casual glance the young plants look like seedlings of the parent plant (Fig. 31–3).

In all these cases and many more, offspring are merely extensions of the parent body. In other instances, however, parent plants produce specialized

[1] An adventitious root forms from a part of a plant that is not itself a root.

(a)

(b)

Figure 31–1 (*a*) *Ficus indicus,* the banyan tree, a tropical fig that spreads extensively by asexual reproduction based on adventitious roots. (*b*) The banyan, like most asexually propagating plants, also reproduces sexually. These small figs contain numerous seeds.

(a)

(b)

Figure 31–2 (*a*) Water hyacinths. (*b*) The Asiatic kudzu vine. Both the kudzu and the water hyacinth are pests that spread very rapidly by asexual propagation. In each case all the plants visible may be members of a single clone. The kudzu vine does flower, though it rarely if ever actually forms seeds.

propagules—bits of their bodies specialized for breakaway and dispersal. Among some species of the genus *Kalanchoe* (often grown as exotic houseplants), young plants form on the borders of attached or abscissed leaves (Fig. 31–4). Such leaves or plantlets can drop from the parent plant and be carried (probably by water, as a rule) far from their point of origin. Other species produce miniature plants at the ends of branchlike leaves and drop them in the area surrounding the parent plant. Perhaps the ultimate in propagule production occurs in the common dandelion. The showy yellow flowers always produce seed asexually (never sexually, as far as is known), so that each dandelion seed is genetically identical with its parent plant and, borne by the winds, carries that genotype to far places (Fig. 31–5).

Figure 31–3 Root propagation, a common form of asexual reproduction in plants.

Figure 31–4 Asexual propagation in *Kalanchoe*. Tiny plantlets develop along or at the ends of the branches. When they are old enough, these propagules drop off the "mother" plant.

Artificial Asexual Propagation

Tissue culture represents an ultimate in artificial asexual propagation of plants. In this technique, a bit of mature plant tissue such as pith is placed in an appropriate culture medium along with suitable quantities of plant hormones. When the technique is successful, both stems and roots differentiate, and a new but genetically identical individual results. A propagation technique of such sophistication is not needed to produce ordinary sweet potatoes, for example, but is frequently used when growers desire to preserve the genetic composition of hybrid plants, which cannot be propagated asexually in any other way. In orchid culture, to take another example, growers find that hybrid specimens (Fig. 31–6) are the most attractive. These are obtained from crosses of what in nature are separate species. The genetic reassortment and recombination that sexual reproduction necessarily produces usually result in the loss of the distinctive characteristics of these flowers, so they must be propagated asexually. For a variety of reasons, the only practical way to do this is by tissue culture. Since prize specimens of orchids and other plants can be worth thousands of dollars, there is great motivation to do so!

Another, simpler means of asexual propagation involves the **rooting of stems** or cuttings from a plant; this is a means employed by many plants in their natural state. Sweet potatoes, for example, can be allowed to sprout and the vines that

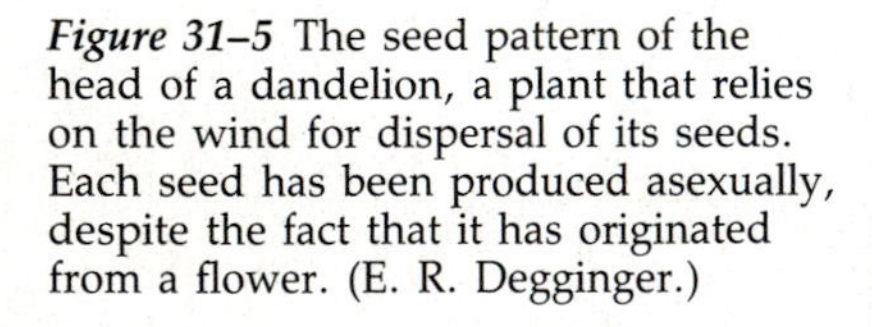

Figure 31–5 The seed pattern of the head of a dandelion, a plant that relies on the wind for dispersal of its seeds. Each seed has been produced asexually, despite the fact that it has originated from a flower. (E. R. Degginger.)

Figure 31–6 Hybrid (*Laeliocattleya*) orchid. To retain its desirable combination of horticultural characteristics, this plant must be propagated asexually. (E. R. Degginger.)

result can be cut off and planted separately; each will put out roots and, in perhaps six months, will have produced a new mass of sweet potato vegetables. Similarly, it is easy to root house plants such as *Coleus* or some kinds of ivy by leaving a sprig in water to which a little auxin has been added or, often, simply by leaving it in plain water or moist soil.

Grafting is a method of propagation that rarely if ever occurs in nature but is widely practiced in agriculture and horticulture. To produce a graft, a sprig or bud, of the desired species, called a **scion,** is attached to a host plant, the **stock;** the stock is sometimes nothing more than a short stump attached to the root. It is important that the cambial layers of scion and stock be in contact so that their vascular tissues can fuse; the graft can then live as a parasite until it is able to make a contribution by means of photosynthesis to the nutrition of the host. Eventually the scion may become the dominant portion of the resulting plant, and the graft may be almost undetectable. The flowers, foliage, and fruit are all of the scion; only the unseen root is genetically different. Almost every rosebush or citrus tree that is commercially produced for sale has been grafted to a wild-type rootstock in this way. If the scion should be killed, however, adventitious buds on the stock may be released from apical dominance and may give rise to shoots bearing vegetative parts characteristic of itself. Many a newcomer to the Citrus Belt has been confused by the apparent change from a prized orange tree to a lumpy lemon tree following a heavy frost and severe pruning of the damaged parts.

Sexual Reproduction in Seed Plants

The seed plants, both gymnosperms and angiosperms, have a life cycle that is less vulnerable to unfavorable conditions than that of such simpler terrestrial plants as ferns and mosses. Two key adaptations are crucial to this life cycle: seed propagation and pollen production. No film of water is required to transport a sperm to the female gametophyte, because the sperm are, in effect, airborne, genetically enclosed in the package of pollen grain that is either carried by the wind or transported by insects, usually flying insects. Thus water, at least in large quantities, is not needed for reproduction in the most fully adapted terrestrial plants. (There are some aquatic seed plants with planktonic pollen.) The seed can be thought of as a live young plant shipped away from the parent by some dispersal method and provided with protection by seed coat (and sometimes by fruit). The seed is also provided with an abundant supply of stored food, so that when the seed germinates, the seedling has a far better start in life than any spore could. Since the life requirements of the seedling are essentially the same as those of the adult plant, no special environments are needed for an alternative generation, as they are in the case of ferns, for instance. In fact, the seed itself may well be able to survive periods of unfavorable environmental conditions (such as winter) that would kill the mature plants.

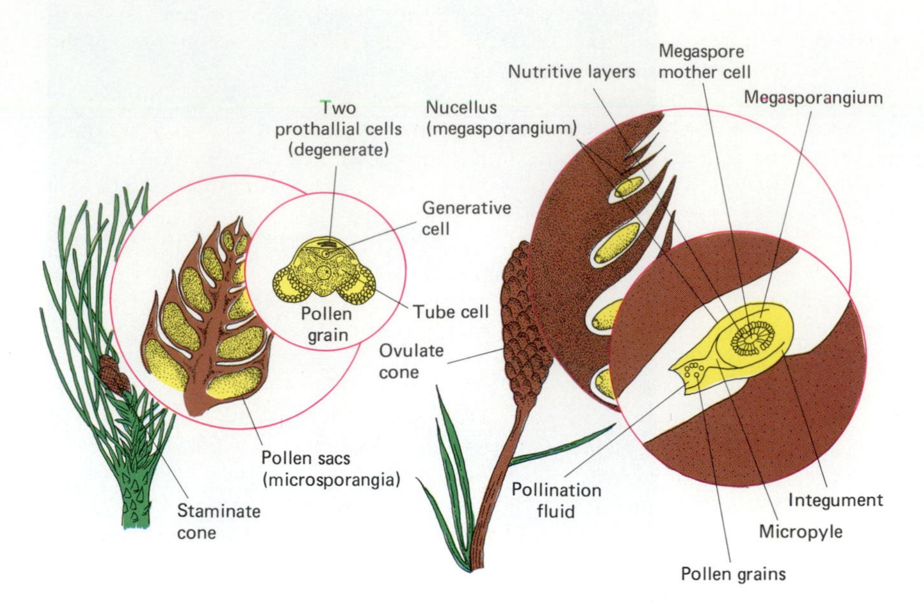

Figure 31–7 Reproduction in a pine tree.

SEXUAL REPRODUCTION IN GYMNOSPERMS

A pine tree, perhaps the most familiar kind of gymnosperm, is both male and female; that is, each reproductive organ is *either* male *or* female. In conifers, the reproductive organs are called **cones.** Thus each tree has both male, or **staminate,** cones and female, or **ovulate,** cones; the male cones produce pollen, and the female cones produce eggs and provide for the development of the seeds that result.

In gymnosperms, the terms staminate and ovulate are used to conform to the terminology applied to the male and female parts, respectively, of true flowers, which will be discussed presently. In pines, staminate cones are inconspicuous wormlike objects often less than 3 cm long (Fig. 23–9), but the ovulate cones can be as much as 45 cm long in some species. The female cone is composed of many scales attached in spiral fashion to a central stemlike structure. Two **ovules** are attached to the margins of each scale near its base, where they are protected by the shingle-like arrangement of the other overlapping scales. When ready to be fertilized, the mature cone separates its scales slightly so that pollen grains can enter between them to reach the ovules. The ovules mature to form seeds. Each ovule is a megasporangium completely enclosed by one or two coverings.

Within each megasporangium is a diploid **megaspore mother cell.** This divides by meiosis to form four haploid **megaspores.** Only one of these becomes functional; it grows into a multicellular **megagametophyte.** On each megagametophyte are two or three female sex organs (**archegonia**), each containing a single, large egg (Fig 31–7).

On the underside of each scale of the staminate cone are two **microsporangia** or pollen sacs. Within each microsporangium are many **microspore mother cells,** each of which divides by meiosis to form four **microspores.** While still within the microsporangium, each microspore divides mitotically to form a four-celled microgametophyte, or **pollen grain.** These grains are released and carried by the wind. When a pollen grain reaches an ovulate cone it enters the ovule through an opening, the **micropyle,** and comes in contact with the megasporangium. In some species the pollen grains are trapped by a sticky secretion that accumulates around the micropyles of the ovules.

A year or more may elapse before one cell of the pollen grain elongates into a pollen tube and grows through the megasporangium to the megagametophyte. Thus in gymnosperms fertilization can take place a year after pollination! Another cell of the pollen grain divides to form two male **gamete nuclei,** not motile sperm as in lower plants. (Electron microscopy has disclosed that the gamete nuclei are true cells in their own right, enclosed in plasma membranes that demarcate them from the surrounding cytoplasm.) When the end of the pollen tube reaches the neck of the archegonium and bursts open, the two male nuclei are discharged near the egg. One fuses with the egg nucleus to form the diploid zygote; the other disintegrates. This event constitutes the actual fertilization, or properly, **syngamy** (union of gametes). After syngamy, the zygote divides and differentiates to produce a sporophyte embryo, surrounded by the tissues of the megagametophyte and those of the parent sporophyte. This entire structure is the **seed.**

The tissues of the megagametophyte provide nutrition for the developing embryo and are called **endosperm.** However, they are haploid cells and are quite different fom the endosperm cells of angiosperms, though both serve the same nutritive function for their respective embryos. After a short period of growth and differentiation, resulting in the development of several leaflike cotyledons, the epicotyl, and the hypocotyl, the embryo becomes quiescent, remaining so even after the seed is shed and drops to the ground. Then, when conditions are favorable, it germinates and develops into a mature sporophyte, the pine tree. The **epicotyl** develops into the stem, and the primary root develops from the tip—the **radicle**—of the **hypocotyl.**

You may have wondered why some of these structures are called gametophytes. Because the diploid seed plant itself resembles the sporophyte stage in ferns, it is often called the **sporophyte generation** of the seed plant, despite the fact that seed plants do not possess typical spores resembling those of ferns and mosses. By this thinking the early pollen grain would correspond roughly to the spore. This is reflected in its name—microspore. The germinated pollen grain, containing as it does the generative nuclei, would be a gametophyte, thought to be the last vestige of the ancient days when the ancestors of seed plants might have had an independent gametophyte generation. It is interesting that gingkos and cycads (which are also gymnosperms) have ciliated motile sperm that actually swim down the pollen tube. The early ovule, then, would also be a spore (which is reflected in its name, megasporangium). However, no fossil evidence has been found for the existence of such independent gametophytes in the ancestry of the seed plants.

SEXUAL REPRODUCTION IN FLOWERING PLANTS

A traditional symbol of romance, the flower is appropriately a sex organ (Fig. 31–8). Some flowering plants, or angiosperms, like holly trees and some squash, are **dioecious,** meaning they have either male or female flowers, but not both, on any one individual plant. Others are **monoecious,** with flowers on the same plant that contain only male parts and flowers that contain only female parts. More commonly, angiosperms have **perfect flowers,** which contain both male and female parts in the same flower, although these may function at different times, or be adapted in some way to prevent self-fertilization.

Flower Structure

The flower of an angiosperm consists of a modified stem and leaves, but in place of the usual foliage it bears **whorls** (concentric circles) of structures specialized for reproduction. Internodal elongation does not usually take place in flowers; thus, several sets of nodal structures are crowded together to form this compound sexual

Figure 31–8 Imperfect (*a*) and perfect (*b*) flowers. Part (*a*) shows the personal parts of a staminate (left) and a pistillate (right) squash flower. Part (*b*) is a leopard lily flower. ((*a*), J. Robert Waaland, University of Washington/BPS; (*b*), R. Humbert/BPS.)

(*a*)

(*b*)

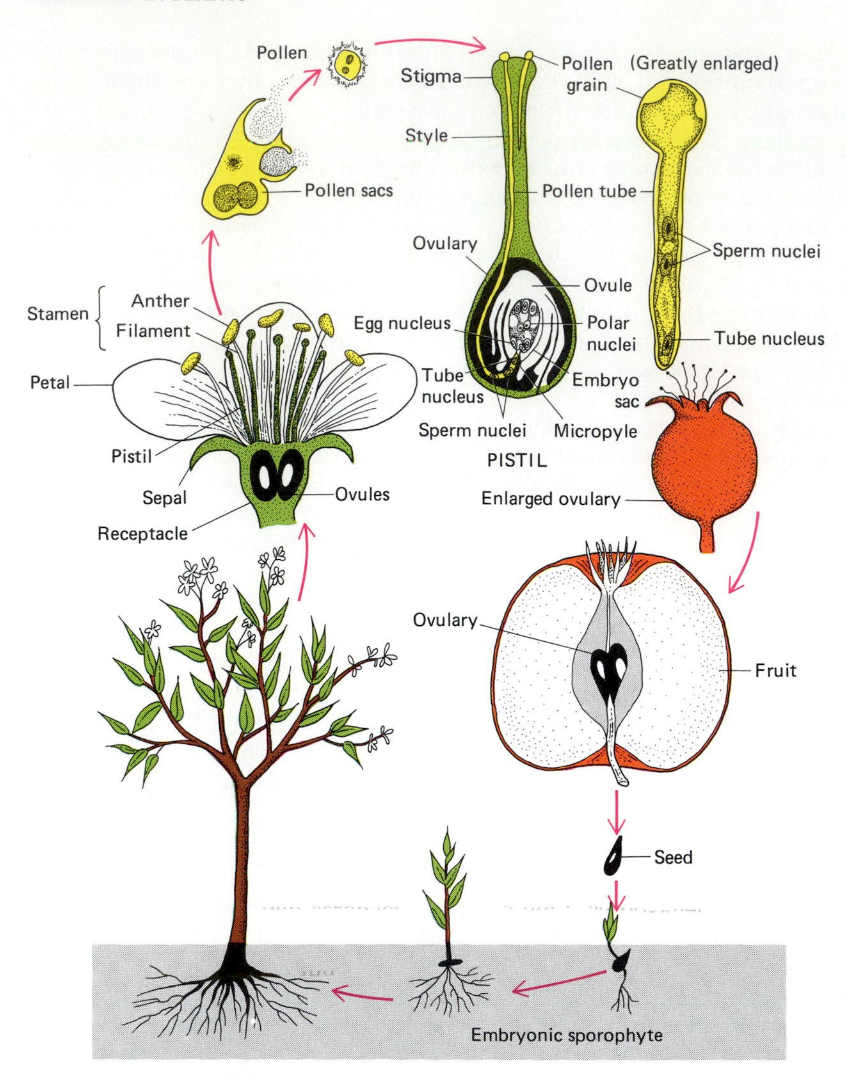

Figure 31–9 The life cycle of an angiosperm.

organ. A typical flower consists of four whorls of parts (Fig. 31–9) attached to the **receptacle,** the expanded end of the flower stem. The outermost parts, usually green and resembling ordinary leaves, are called **sepals.** Within the circle of sepals are the **petals,** typically brilliantly colored to attract insects or birds and ensure pollination. The petals collectively are called the **corolla,** and the sepals, the **calyx.** The corolla and calyx together comprise the **perianth.** Just inside the circle of petals are the **stamens,** the male parts of the flower. Each stamen consists of a slender filament with an **anther** at the tip. The anther is a group of **pollen sacs** (microsporangia). Each pollen sac contains a group of microspore mother cells—**pollen mother cells.** Each diploid pollen mother cell divides meiotically and produces four haploid microspores. A further mitotic division converts each microspore into a microgametophyte, or **pollen grain.**

At maturity each pollen sac typically splits along a seam of weakness and turns inside out so as to expose the pollen grains that have developed within (Fig. 31–10). The shape of pollen grains is highly distinctive and often elaborately sculptured. This sometimes permits the identification of the flora of bygone eras even when the plants themselves have not survived. The remarkably enduring pollen grains can often be seen even in partially metamorphosed sedimentary rocks (Fig. 31–11).

In the very center of the flower is either a ring of **pistils** or a single fused one. Each pistil has a swollen hollow, basal part, the **ovulary** (or ovary), a long slender portion above this, the **style,** and at the top a flattened part, the **stigma.** The ovulary is often composed of several fused **carpels,** each of which contains a cavity that holds one to several **ovules,** the equivalent of ova. In many flowers the stigma secretes a moist, sticky substance to trap and hold the pollen grains that reach it. Different species of angiosperms show great variation in the number, position, and

(a)

(b)

(c)

Figure 31–10 Some "male" and "female" flower parts. The hibiscus (*a*) bears the stigma (*c*) at the end of a stamen tube, which bears stamens and anthers (*b*) proximally. Each yellow anther carries pollen grains, just visible at this magnification, which stick to the fuzzy surface of the stigma of another flower. ((*a*), E. R. Degginger.)

shape of the various parts. Recall that a flower that has both stamens and pistils is called a perfect flower; one that lacks one or the other, however, is called an **imperfect flower.** Flowers with stamens but no pistils are **staminate flowers;** those with pistils but no stamens are called **pistillate flowers.** In dioecious species, such as willows, poplars, and date palms, the two kinds of plants can be distinguished by these flower types: The male plant bears only staminate flowers; the female plant bears only pistillate flowers. Monoecious plants bear imperfect flowers; that is, the plant has male and female flowers, which often open at different times.

As in the gymnosperms, the angiosperm ovule is a megasporangium completely enclosed by one or two integuments. Each ovule typically contains one megaspore mother cell. Each diploid megaspore mother cell divides by meiosis to form four haploid megaspores. One of the megaspores develops into the megagametophyte; the other three disintegrate. There is some variation in different species in the details of megagametophyte development, but typically the megaspore enlarges greatly and its nucleus divides by mitosis. The two daughter nuclei migrate to opposite ends of the cell; then each divides, and the daughter nuclei divide

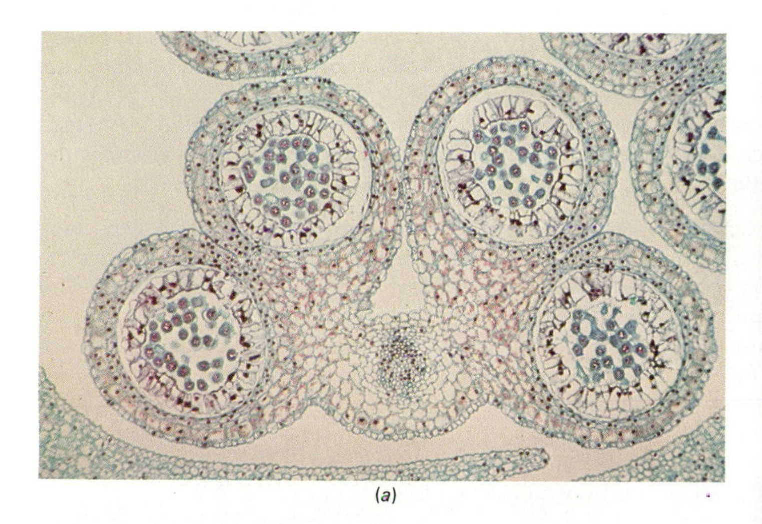
(a)

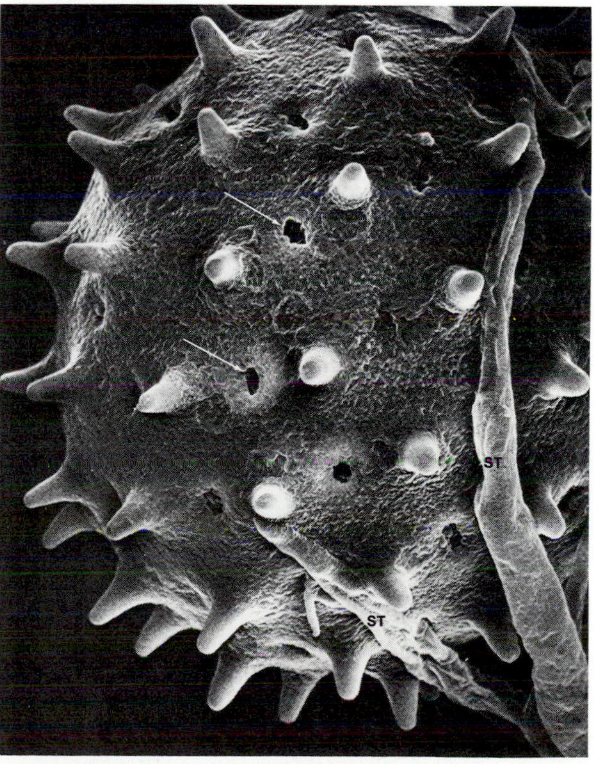

(b)

Figure 31–11 (*a*) Pollen grains form inside the saclike anther, shown here in cross section. The developing pollen can be seen within the round cavities. The wall later everts, turning the sac inside out and exposing the pollen grains. (*b*) Scanning electron microscope of a *Hibiscus* pollen grain. The outer covering of a pollen grain is called the exine, which in the hibiscus has a number of thin-walled apertures (arrows). When the pollen grain germinates, the pollen tube emerges through one of the apertures. The pollen grain shown here has become trapped by some of the tiny "hairs" of the stigma (st). ((*a*), J. Robert Waaland, Univ. of Washington/BPS; (*b*), John A. Lott, McMaster Univ./BPS.)

again. The resulting megagametophyte is initially a very simple structure called an **embryo sac** with three cells at each end and a large binucleate cell in the middle. These two nuclei are called the **polar nuclei.** One of the three cells at one end of the megagametophyte becomes the egg; the other two and the three cells at the other end all disintegrate. Consult Figure 31–9 for the specialized terminology that is applied to these cells.

Fertilization

The haploid microspore develops into the pollen grain or young microgametophyte while still within the pollen sac. The nucleus of the microspore divides into a larger **tube nucleus** and a smaller **generative nucleus.** Most pollen grains are released while in this state and are carried by the wind, insects, or birds to the stigma of the same or a neighboring flower. This is **pollination.** The pollen grain then germinates, with the pollen tube growing out of the pollen grain and down the style to the ovule. The tip of the pollen tube produces enzymes that dissolve the cells of the style, thus making room for the pollen tube to grow. The tube nucleus remains in the tip of the pollen tube as it grows. The generative nucleus migrates into the pollen tube and divides to form two cells, known as the **sperm** or **germinative nuclei.** The mature male gametophyte consists of the pollen grain and tube, the tube nucleus, and the two sperm cells.

When the tip of the pollen tube penetrates the megagametophyte through the micropyle, it bursts, and the two sperm cells are discharged into the megagametophyte. In fertilization one of the sperm cells migrates to the egg nucleus and fuses with it. The resulting diploid cell is the **zygote,** the beginning of the new sporophyte phase, or generation. The other sperm nucleus migrates to the two polar nuclei and, in a remarkable and unusual process, all three fuse to form a 3n **endosperm nucleus.** In some species the two polar nuclei have fused to form a single one before the sperm nucleus arrives. This phenomenon of **double fertilization,** which results in a diploid zygote and a triploid endosperm, is peculiar to, and characteristic of, the flowering plants.[1] The triploid state serves as a developmental switch that causes the zygote to develop into a source of food for the other embryo, rather than into a young plant in its own right.

After syngamy the zygote undergoes a number of divisions and forms a multicellular embryo. The endosperm nucleus also undergoes a number of divisions and forms a mass of endosperm cells, gorged with nutrients. The endosperm fills the space around the embryo and provides it with nourishment. The sepals, petals, stamens, stigma, and style usually wither and fall off after fertilization. The ovule with its contained embryo becomes the seed; its walls become thick and form the **integuments** (outer coverings) of the seed, which are usually very thin but which can become thick, tough, or even fruitlike in some species. (The fruit itself is derived not from the seed but principally from the ovulary, as we will see.)

STRATEGIES OF POLLINATION. We have seen that gymnosperms (and even some angiosperms such as the grasses and oak and maple trees) are pollinated by the wind. The flowers of wind-pollinated angiosperms are often extremely small and inconspicuous and do not resemble gymnosperm cones (Fig. 31–12). All wind-pollinated seed plants produce pollen in large quantities; it is a matter of chance whether any of the pollen will reach the pistil of another plant of the same species. If a pollinating conifer is shaken, yellow clouds of pollen are often dislodged from the staminate cones; pollinating oaks and maples are responsible for a large part of the springtime misery of hayfever and asthma sufferers. (A few aquatic angiosperms disperse their pollen by water currents.)

The vast majority of angiosperms employ animal-vectored pollination rather than the less efficient wind pollination. The most common form of animal-vectored pollination relies upon insects, and among them bees are the most used. Animal-vectored pollination does have some drawbacks. It requires much less investment of the plant's resources in pollen production, but does require the production of elaborate petals, scent apparatus, and above all, nectar. There is also the possibility that pollinating insects may eat pollen (or other parts of the flower) rather than disseminate it, as is the case with the yucca moth larva, and to some extent, with the honeybee.

[1] As pointed out in Chapter 23, some angiosperms have pentaploid endosperm.

Figure 31–12 A willow catkin, a wind-pollinated flower. (Leo Frandzel.)

The common jack-in-the-pulpit plant of many deciduous woodlands serves as an example of an insect-vectored pollinator whose adaptations are relatively simple (Fig. 31–13). The imperfect flowers of this and other members of the genus *Arum* are examples of **spadices** (singular, spadix), in which a single, central stalk, or **rachis,** bears tiny flowers and is wrapped in a leaflike, hooded structure, called a **bract,** marked with brown or purple stripes. Jack-in-the-pulpit plants are male when young, bearing only staminate flowers on the spadix. When older and larger they change sex and bear pistillate flowers exclusively. In a typical woodland, plants of a variety of ages provide both pollen and ovules for reproduction.

The flowers emit a fragrance that attracts tiny fungus gnats, relatives of the better-known fruit flies used in genetic research. The gnats enter the bracts and are unable either to fly or crawl out because of the shape and slickness of the bract's inner surface. If it is a male flower, the gnats are eventually able to escape from an opening at the base of the bract, but not without first coming into contact with the stamens of the flowers. If it is a female plant, however, there is no escape, and their doom is sealed. In their death throes, however, the gnats manage to spatter the pistillate flowers with pollen, and fertilization then occurs.

A highly sophisticated type of insect-vectored pollination is used by several species of tropical orchids (Fig. 31–14). The very name "orchid" means testis, so perhaps it should come as no surprise that these flowers are sexually exploitative sex organs. A portion of the flower is shaped and colored so that it resembles the abdomen of a certain kind of female wasp. When the orchids have begun to produce pollen there are no available female wasps of these species, for they have not yet emerged from pupation. There are, however, many sex-deprived male wasps buzzing about that as yet lack appropriate sex objects. In the absence of an appropriate sex object they attempt to copulate with the seductively shaped flowers, depositing their sperm there to no avail. The flower deposits its pollen on *them,* however, and the insects carry it to the next orchid with which they also attempt to mate. By the time the female wasps emerge the orchids have set seed and the male wasps lose interest in the flowers, favoring the female wasps instead.

Figure 31–13 Jack-in-the-pulpit flowers. The "sex" of these plants changes with age. (E. R. Degginger.)

Figure 31–14 These orchids are adapted to resemble female insects.

Figure 31–15 Flowers employ numerous devices that attract bees and help them to learn the locations of nectar (and pollen). (*a*) Bougainvillea. The blossoms are tiny and inconspicuous, but the brightly colored leaves—bracts—serve to attract insects. (*b*) Black-eyed susans guide bees to nectaries and anthers by patterns of contrasting pigment. (*c*) Many flowers have such bee guides visible only to the bee, which is able to detect ultraviolet light.

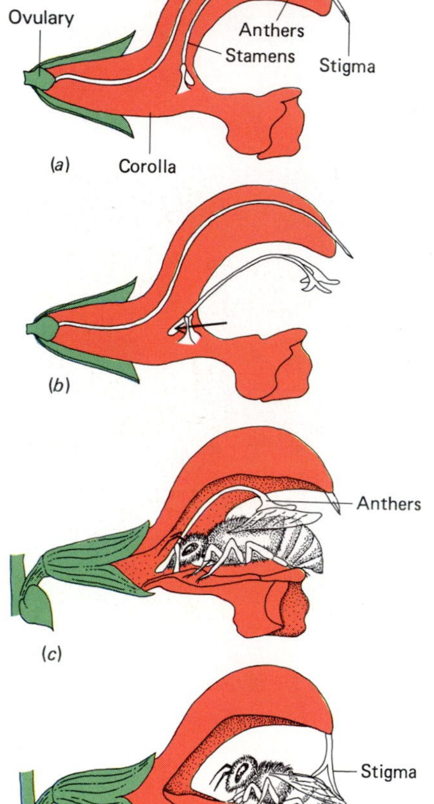

Figure 31–16 Flowers of sage (*Salvia*), showing specific structural arrangements for transferring pollen via its pollinators (bees). (*a*) Longisection of flower, showing stamens attached to the tubular corolla (fused petals) by a hinge bearing a basal projection. (*b*) When the basal projection is pushed (*arrow*), the stamens tip downward. (*c*) A bee entering the flower to reach the nectar at its base pushes against the projection, causing stamens to swing down and deposit pollen on bee's abdomen. (*d*) When a bee alights on a flower whose stigma is exposed and receptive to pollen, the pollen-dusted part of the bee's abdomen contacts the stigma and pollinates it. (Courtesy of Peter Ray.)

The vast variety of bee-pollinated flowers take advantage of the highly specialized adaptations of the honeybee. Among these are a furry body (to which pollen grains readily stick) and even pollen "baskets" made of hairs on the legs. (The bees take pollen as well as nectar back to the hive to feed their larvae. The pollen is the chief source of protein in the diet of most bees.) Bees are behaviorally complex. By means of dances they are able to communicate the location of pollen-producing flowers to other members of the hive, so that if one bee finds a field of a particular flower open on a given day, it can communicate this fact to many others, who will then go there themselves. Moreover, once a bee has begun to find nectar in a particular species of flower, it remembers the color and scent, returning to that species and none other as long as its nectar flow continues. A particular field of flowers might well have the greater part of several hives of bees working on it at once, carrying pollen grains from one flower to another without going to any other species. It is hard to imagine a more efficient scheme of pollination.

Although there are many variations on this common theme, bee-pollinated flowers usually exhibit a characteristic combination of adaptations. First, they generally are colored in a range to which insect vision is sensitive, that is, from yellow to ultraviolet (Fig. 31–15). Bees cannot see the color red. To them, it probably appears black. Second, they generally possess a sweet, perfume-like scent. Third, the flowers have some way of positively reinforcing the bees, almost always with a sweet, sugary nectar, which the bees later convert to honey. This **nectar** is secreted by specialized **nectaries,** also found in parts of some plants outside the flower. Finally, flowers often possess adaptations, some quite bizarre, that force the insects to crawl over stamens and pistil to reach the nectar or, in some cases, in order to escape (Fig. 31–16).

(a)

(b)

(c)

(d)

Figure 31–17 Various animal pollinators. (*a*) A honey bee, *Apis mellifera*. (*b*) A moth, *Haemorraghia thysbe*. (*c*) A greater short nosed fruit bat, *Cynopterus sphinx*, pollinating a banana plant. The pollen grains on the bat's fur will be carried to the next plant, where cross pollination will occur. Bats are important in the pollination of many fruit plants. (*d*) A ruby-throated hummingbird pollinating a flower from a trumpet vine. ((*a*), Stephen Dalton, Photo Researchers, Inc.; (*b*), R. Humbert/BPS; (*c*), Merlin Tuttle, Photo Researchers, Inc.; (*d*), Steve Maslowski, Photo Researchers, Inc.)

Some flowers are pollinated by beetles; such flowers are often spicy in odor. Others, pollinated by flies, smell like carrion. Still others are pollinated by moths and open at night to accommodate them (Fig. 31–17). Some are even pollinated by vertebrates. Hummingbirds, for instance, are major pollinators of red-colored flowers with long throats. These birds insert their bills into flowers to sip their nectar. Since the sense of smell is poorly developed in birds, such flowers do not need to

Figure 31–18 Fly-pollinated flowers of skunk cabbage, which smells as you might expect. To a fly, these very foul-smelling flowers may suggest a festering wound. (Alan R. Bleeker, The Kings College.)

have a strong scent. Some of the more spectacular night-flowering plants, such as the night-blooming *Cereus,* are bat-pollinated. There are even mouse- and possum-pollinated flowers. The leading pollinators of the animal kingdom, however, remain the bees.

Ordinarily, plants with perfect flowers employ a variety of means that prevent self-pollination. Though some fruit species are capable of self-pollination, most are not. This makes it necessary for fruit growers to plant different varieties of a fruit species, since a grove containing genetically identical trees could not be pollinated. For example, all the members of any one variety of apple tree are actually members of a clone that was asexually propagated by grafting from a single ancestor that possessed an agriculturally desirable combination of properties. The Anna apple, for instance, is one of the few varieties that will yield fruit in subtropical climates. An entire orchard of Anna apples would be undesirable, however, because a genetically determined self-incompatibility prevents successful self-fertilization, so that all the genetically identical members of this clonal variety cannot fertilize one another. By planting several trees of the less desirable Ein Sheymer variety, which have been derived from a different scion, fertilization of the Anna trees can be assured.

SELF-POLLINATION. Despite all the advantages of biparental reproduction, some plants employ self-fertilization instead. The reasons for this are obscure, but it is true of so common a plant as the garden pea, and Gregor Mendel took advantage of it in his early genetic experiments. By using peas whose male flower parts had been removed, he could be sure that the male parent of his choice was the one to fertilize the emasculated pea flower. Although bees do visit some pea flowers, there are many varieties that attract no pollinating insects. In them, the flowers are tiny as a rule and very inconspicuous.

Some plants employ both self- and cross-pollination, with self-pollination serving as a kind of last resort when cross-pollination has not taken place during the flowering period. Often in these flowers, differential growth eventually brings the pistil and stamens of the same flower into contact with one another. This does not occur, however, if cross-fertilization has already produced developing seeds in the ovulary.

Fruits

The ovulary, the basal part of the pistil containing the ovules, enlarges and forms the **fruit.** The fruit thus contains as many seeds as there were ovules in the ovulary. In the strict botanical sense of the word, a fruit is a mature ovulary containing seeds—the matured ovules. Although we usually think of only such sweet, pulpy things as grapes, berries, apples, peaches, and cherries as fruits, bean and pea pods, corn kernels, tomatoes, cucumbers, and watermelons are also fruits, as are nuts, burrs, and the winged "seeds" of maple trees. A **true fruit** is one developed solely from the ovulary. If the fruit develops from sepals, petals, or the receptacle as well as from the ovulary, it is known as an **accessory fruit.** The apple fruit consists mostly of an enlarged, fleshy receptacle; only the core is derived from the ovulary.

The three types of true and accessory fruits are simple fruits, aggregate fruits, and multiple fruits (Fig. 31–19). **Simple fruits** (e.g., cherries, dates, palms) mature from a flower with a single pistil; **aggregate fruits** (raspberries and blackberries) mature from a flower with several pistils; and **multiple fruits** (pineapples) are derived from a cluster of flowers, all of which contribute structures that unite to form a single fruit. Fruits are also classified as **dry fruits** if the mature fruit is composed of rather hard dry tissues or as **fleshy fruits** if the mature fruit is soft and pulpy. Dry fruits represent adaptations for dispersal by the wind or for attachment to animal bodies by hooks. Birds, mammals, and other animals eat fleshy fruits and their enclosed seeds. The seeds pass through the animal's digestive tract and are dropped with the feces in a new place. Thus fleshy fruits also represent an adaptation for the dispersal of the species.

In one type of dry fruit, termed a **nut,** the wall of the ovulary develops into a hard shell that surrounds the seed. The edible part of a chestnut is the seed within the fruit coat or shell. A Brazil nut is really a seed; there are about 20 such seeds borne within a single fruit. An almond is not a "nut" at all but the seed or "stone" of a fleshy fruit related to the peach. The outer woody covering of a peach stone or

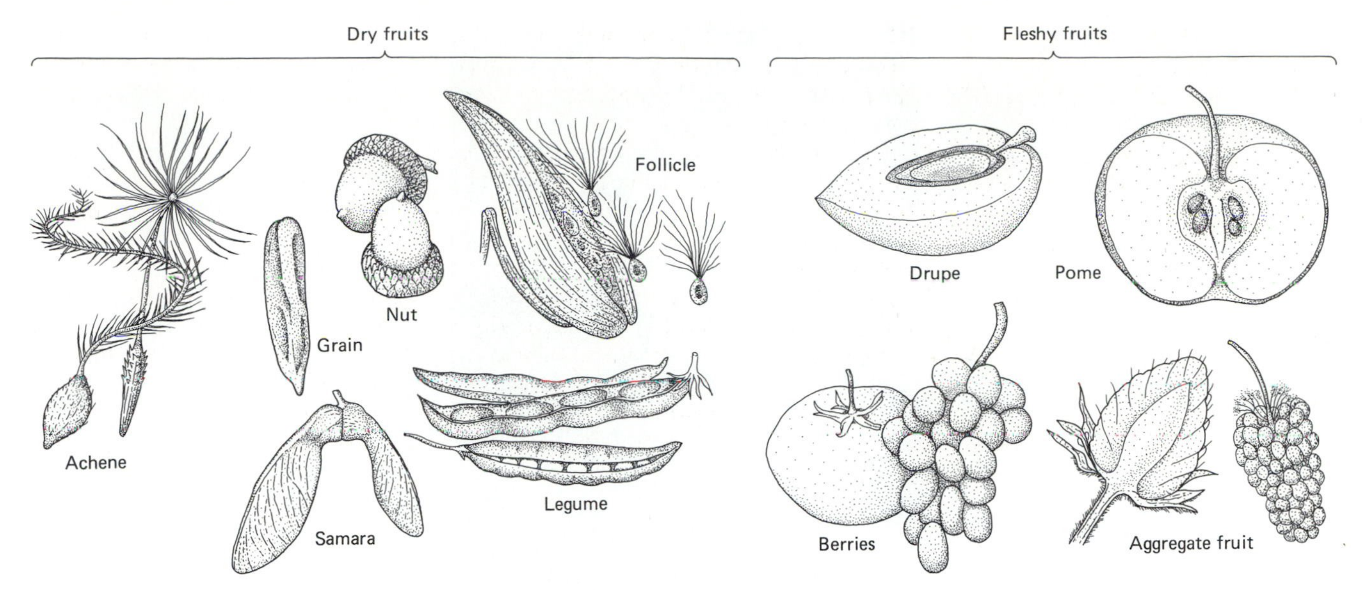

Figure 31–19 Principal types of fruits. The samaras and the plume-bearing dandelion achenes are examples of wind-dispersed fruits; wind-dispersed seeds are illustrated by the plume-bearing seeds escaping from the milkweed follicle. Raspberries and strawberries are derived from similar types of flowers, but the pistils of raspberries become fleshy drupes and the pistils of strawberries become dry achenes, the yellow seedlike spots scattered over the surface of the fruit. (Courtesy of Ray, Steeves and Fultz.)

almond is, incidentally, part of the fruit. Only the meat within is the seed. The seeds of apricots, nectarines, and many other members of the rose family are poisonous and should not be eaten, for they contain cyanide.

Grapes, tomatoes, bananas, oranges, watermelons, although superficially different, are all examples of fleshy fruits (Fig. 31–20). In these the entire wall of the

Figure 31–20 Flowering and fruit production in a citrus tree.

Figure 31–21 A few angiosperms release their seed only after it has germinated. This mangrove plant will soon release the spear-shaped seedling that, if it becomes stuck in the muddy estuarine bottom, will sprout leaves and grow into a young tree.

ovulary becomes pulpy. Such fruits are technically called **berries.** Peaches, plums, cherries, and apricots, in contrast, are stone fruits, or **drupes.** In a drupe the outer part of the ovulary wall forms a skin, the middle part becomes fleshy and juicy, and the inner part forms a hard pit or stone around the seed. There are, then, many kinds of fruit, differing in the number of seeds present, in the part of the flower from which they are derived, as well as in color, shape, water and sugar content, and consistency.

SUMMARY

I. Asexual reproduction does not change the genetic makeup of an offspring from that of its parent. Thus it can be used to preserve favorable combinations of genes and to propagate them rapidly.
  A. In some plants, asexual reproduction is little different from growth; other plants employ specialized propagules to disseminate the genotype of the parent.
  B. Sexual and asexual reproduction are not mutually exclusive options. Many plants reproduce by both modes. In this way they are able to form new advantageous combinations of genes which they are also able to propagate unchanged.

II. Some of the forms of artificial asexual propagation include tissue culture, the cutting and rooting of stems, and grafting.

III. Seed propagation is the form of sexual reproduction employed by many plants. It frees them from dependence on special weather conditions or the presence of water as requirements for fertilization, and provides the young plant with both a means of dispersal and stored food with which to start a new life.

IV. Fertilization differs in gymnosperms and angiosperms.
  A. Gymnosperms reproduce by means of staminate cones and ovulate cones. The staminate cones produce pollen, which is carried by the wind to the ovulate cones. There it may enter an ovule through a micropyle opening. When a pollen grain contacts the megasporangium it elongates into a pollen tube. One of the nuclei fertilizes the egg nucleus. The megagametophyte then gives rise to a haploid endosperm. The fertilized egg nucleus produces an embryo consisting of cotyledons (first leaves), epicotyl (stem), and hypocotyl (usually the future root).
  B. Angiosperms possess flowers, most of which are adapted for animal-vectored pollination. A flower is a highly specialized branch consisting of the male and female parts plus a series of modified leaves, the calyx and corolla. Pollen forms in anthers and is then transferred to the sticky stigma of the female pistil. There it produces a pollen tube containing a tube nucleus plus two sperm, or germinative nuclei. One germinative nucleus fertilizes the egg nucleus, to form a diploid embryo. The other fuses with the two polar nuclei to form a triploid endosperm.
  C. Pollination and later the development of the embryo stimulate the production of fruit, which develops from the ovulary and its associated parts to form a structure that is often specialized for the dispersal of its contained seeds.

V. Wind-pollinated flowers are small and inconspicuous, usually lacking well-developed parts (such as calyx and corolla) that are not directly concerned with pollen and seed formation. Animal-vectored flowers are usually showy and conspicuous, often provided with scent glands and nectaries to attract their usual pollinators.

POST-TEST

1. The main advantage of asexual reproduction, in addition to its speed, is that it permits the propagation of advantageous ______________ makeups since the offspring are genetically ______________ with the parent.

2. Sexual reproduction, on the other hand, permits ______________ ______________ of traits to arise from the contributions of multiple different ancestors.

3. Certain plants produce body parts specialized for breakaway and dispersal that are known as ______________.

4. In grafting, the plant that is to be asexually propagated is known as the ______________; this is attached to the ______________, which may be little more than a root. It is important that the two have their ______________ layers in contact with each other.

5. The production of ______________ by the seed plants largely circumvents the requirement found in ferns, mosses, and the like for ______________ as a requirement for fertilization.

6. In pines, the ______________ cones produce pollen.

7. If all goes well, the pine pollen grain lands in an ______________ cone; there it develops a ______________ ______________, which conveys two gamete nuclei to the neck of the ______________. One survives to fertilize the ovule, producing the sporophyte ______________.

8. In a flower, the ______________ are usually the most brilliantly colored parts, although other flower parts may sometimes be brightly colored in addition to or instead of them.

9. Perfect flowers must contain both ______________ and ______________ organs.

10. The tendency of honeybees to remain faithful to a particular species of plant whose flowers are currently in bloom is important to ensure proper and efficient ______________.

11. A cherry is an example of a ______________ fruit.

12. The hard coat of the stone of a peach is part of the ______________.

REVIEW QUESTIONS

1. From a biological standpoint, what are the advantages and disadvantages of sexual reproduction in plants?
2. What is a propagule?
3. Describe plant grafting. When is it likely to be employed?
4. What is the difference between monoecious and dioecious plants?
5. Compare fertilization and endosperm formation in gymnosperms and angiosperms.
6. Obtain a flower of any normal kind (not a "double" horticultural variety) and describe it as completely as you can.
7. Contrast the adaptations of a typical bee-pollinated flower with those of a typical bird-pollinated flower.
8. Give a common example of a drupe and a berry.
9. What is the difference between a perfect and an imperfect flower? between an imperfect flower and an incomplete flower?
10. Why is it inappropriate to speak of gymnosperm cones as "flowers" since they contain staminate or pistillate parts and produce pollen just as flowers do?

Recommended Readings for Part V

Books

Galston, A. W., P. J. Davies, and R. L. Satter. *The Life of the Green Plant,* 3d ed. Englewood Cliffs, N.J., Prentice-Hall, Inc., 1980. A description of plant function written on a basic level.

Raven, P. H., R. F. Evert, and H. Curtis. *Biology of Plants,* 3d ed. New York, Worth Publishers, Inc., 1981. A botany textbook that covers all of the basics of plant structure, function, evolution, and ecology.

Ray, P., T. A. Steeves, and S. A. Fultz. *Botany.* Philadelphia, Saunders College Publishing, 1983. For those who would like to delve more deeply into structure and life processes of plants, this well-written, comprehensive textbook is highly recommended.

Articles

Echlin, P. Pollen. *Scientific American,* April 1968. How pollen develops and functions.

Heslop-Harrison, Y. Carnivorous plants. *Scientific American,* February 1978. The adaptations of carnivorous plants permit them to survive in habitats where few other plants can live.

Janzen, D. H., and P. S. Martin. Neotropical anachronisms: the fruits the gomphotheres ate. *Science,* Vol. 215, January 1982. A discussion of seed dispersal by fruit-eating animals now extinct. Contains the amusing suggestion that dinosaurs may have dispersed gingko seeds.

Rick, C. M. The tomato. *Scientific American,* August 1978. How we have adapted a flowering—and fruiting—plant to our needs.

Shepard, J. F. The regeneration of potato plants from leaf-cell protoplasts. *Scientific American,* Vol. 246, No. 5, May 1982. A new approach to cloning, or asexual propagation, of plants yields variants that promise future crop improvements.

Taylor, T. N. Reproductive biology in early seed plants. *Bioscience,* Vol. 32, January 1982. Speculations regarding the origin of the seed plants.

part VI

STRUCTURE AND LIFE PROCESSES IN ANIMALS

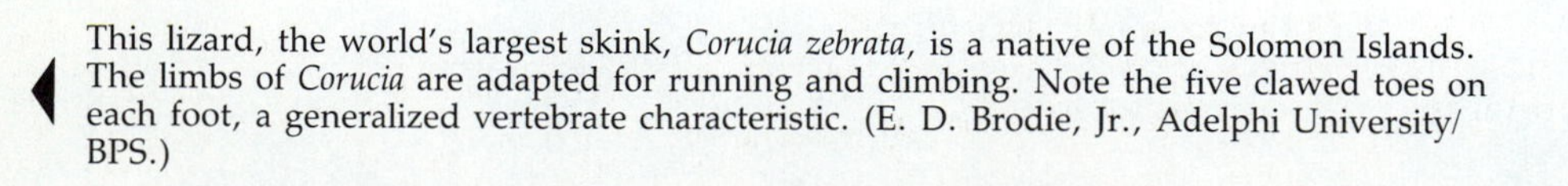

This lizard, the world's largest skink, *Corucia zebrata,* is a native of the Solomon Islands. The limbs of *Corucia* are adapted for running and climbing. Note the five clawed toes on each foot, a generalized vertebrate characteristic. (E. D. Brodie, Jr., Adelphi University/ BPS.)

32

Skin, Bones, and Muscles: Protection, Support, and Locomotion

OUTLINE

I. Skeletons and skin
- *A.* Hydrostatic skeletons
- *B.* External skeletons
- *C.* The vertebrate skin
  - *1.* The dermis
  - *2.* The epidermis
- *D.* Internal skeletons
- *E.* The skeleton itself
  - *1.* Parts of the skeleton
  - *2.* The bones of vertebrates

II. Muscles
- *A.* Muscle actions
- *B.* Posture and movement
- *C.* Muscle physiology
  - *1.* The all-or-none effect
  - *2.* The simple twitch
  - *3.* Tetanus
  - *4.* Tonus
- *D.* The biochemistry of muscular contraction
  - *1.* How muscle obtains its energy
  - *2.* Fatigue
  - *3.* What makes muscles move?
  - *4.* Triggering contraction
- *E.* Smooth and cardiac muscle
- *F.* A muscular zoo

LEARNING OBJECTIVES

After you have read this chapter you should be able to:

1. Summarize the role of the hydrostatic skeleton in the transmission of muscular forces.
2. Describe the arthropod exoskeleton and its structure.
3. Describe the structure of the skin of a typical vertebrate, and give its principal derivatives.
4. List the major divisions of the vertebrate skeleton.
5. Describe the basic structure of a skeletal muscle cell and the fundamental characteristics of a simple twitch and of tetanus.
6. Describe the cross-bridge theory of muscular contraction.
7. Summarize the process of muscular contraction, including the role of the muscle, cell membrane, T-system, sarcoplasmic reticulum, and contractile proteins.
8. Describe the distinctive characteristics of insect flight muscle.

Some run, some jump, some fly. Others remain rooted to one spot, sweeping their surroundings with tentacles. Many contain internal circulating fluids, pumped by hearts and contained within hollow vessels that maintain their pressure by gentle squeezing. Digestive systems push food along with peristaltic writhings. All of these activities of multicellular animals are powered by a specialized tissue that, however varied its effects, has but *one* action: it can contract. This tissue is called **muscle.**

Skeletons and Skin

A muscle must have something to act upon, something by means of which its contractions can be transmitted to leg, body, or wing. In some of the simplest animals this is no more than the glutinous jelly-like substance of the body itself, or perhaps a fluid-filled body cavity. More complex animals, however, require a true skeleton to receive, transmit, and transform the simple movement of their muscular tissues. In some animals this skeleton is internal—plates or shafts of calcium-impregnated tissue that develop in the mesoderm. In others the skeleton is not a living tissue at all but a deposit atop the epidermis—a shell, or **exoskeleton.**

The exoskeleton is able not only to transmit muscular forces, but also to protect the delicate structures beneath. Yet its rigidity limits the growth of the organism so that in some animals it must be shed periodically and replaced. In other organisms it strictly limits the size and shape of the body.

In this chapter we will consider muscle, skeleton, and outer covering together because they are closely interrelated in function and significance, and because they work together in motions such as hopping, crawling, running, and flying.

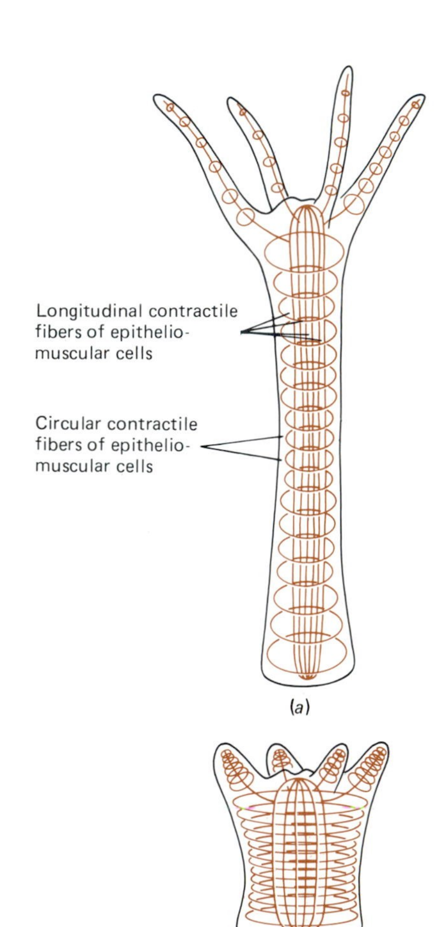

Figure 32–1 Muscular action in *Hydra.* The longitudinally arranged fibers are antagonistic to the circular fibers. (*a*) Contraction of the circular muscles elongate the body. (*b*) Contraction of the longitudinal muscles shortens the body.

HYDROSTATIC SKELETONS

Consider the classic *Hydra,* mechanically little more than a simple bag of fluid (Fig. 32–1). The body of a hydra is mostly empty space—space that extends to the very tips of the animal's tentacles. This space is the gastrovascular cavity. The living tissue of a hydra is thin and consists of just two layers of cells, an outer **epidermis,** and an inner **endodermis.** All of the animal's activities—digestion, defense, transmission of impulses, or movement—are carried out by these two layers of cells.

Examination of the individual cells of the endodermis and epidermis that make up the **hydrostatic skeleton** (or hydraulic skeleton) of a hydra discloses that all of them are capable of contraction. These cells are often referred to as epitheliomuscular cells. The cells in the endodermis shorten preferentially in a direction at right angles to the main body axis of the hydra. Since they are arranged in a circular pattern, this has the effect of reducing the circumference of the animal and thus its diameter as well. The epidermal cells are so arranged that when they contract the entire body of the hydra shortens. Imagine an elongated balloon full of water. Pulling on the balloon would lengthen it, but squeezing it would accomplish the same thing. Conversely, it would shorten if the ends were pushed. The circular and longitudinal epitheliomuscular cells of *Hydra* work in much this fashion. When the epidermal, longitudinal layer contracts, the hydra shortens, and because of the fluid present in its gastrovascular cavity, force is transmitted so that it thickens as well. When the endodermal circular layer contracts, the hydra becomes thinner, but its fluid contents force it to lengthen as well. In other words, the two layers are arranged in antagonistic fashion: What each can do, the other can undo. This principle holds not only for *Hydra* but for virtually all animals. A muscle can only contract, that is, can only exert a pull; so its action depends on its relation to the skeleton. The primary function of the skeleton is to transmit muscular force, although in more complex organisms it may have additional functions.

In most organisms with hydrostatic skeletons, only rather crude mass movements of the body or its appendages are possible. Delicate movements are difficult because in a fluid, force tends to be transmitted equally in all directions and hence throughout the entire fluid-filled body of the animal. A hydra cannot easily thicken one part of its body while thinning another (Fig. 32-2).

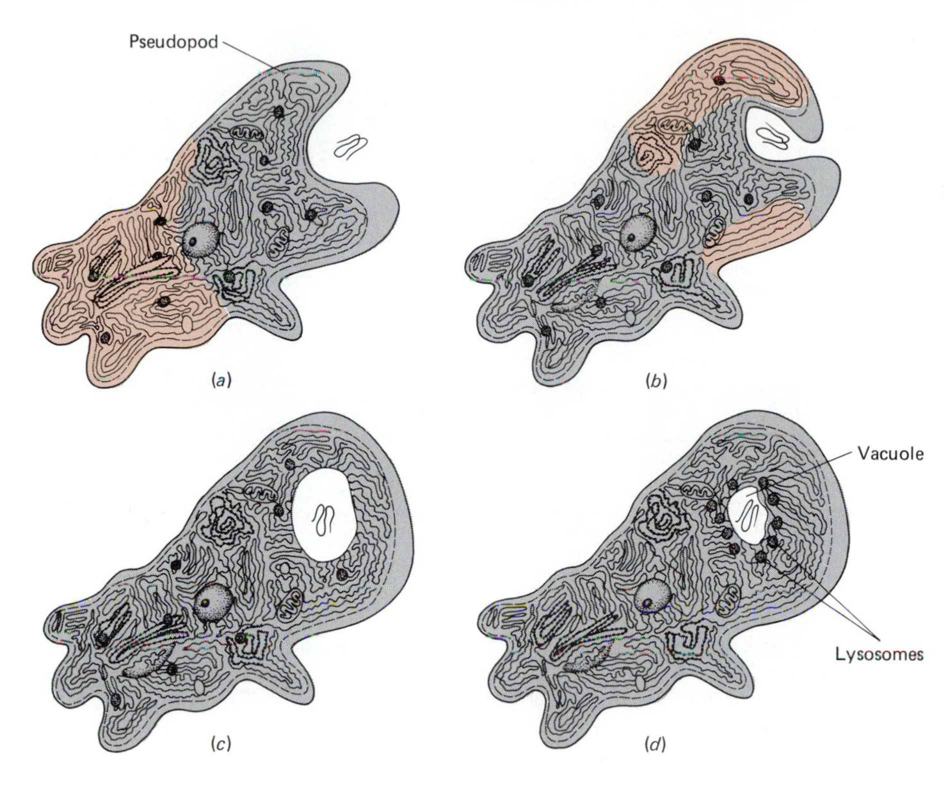

Figure 32–2 Amoeboid movement. The amoeba flows along by thrusting forth extensions of cytoplasm called pseudopods. One swells forth, then another; finally most of the cytoplasm has flowed into the pseudopod, and the amoeba has moved forward. These inconstant, shifting appendages are also used to capture food. One theory of amoeboid movement holds that contractile microfilaments at the rear—wherever that may currently happen to be—squeeze the cytoplasm so that pseudopods are thrust forward at the front. If this is true, the semifluid cytoplasm acts as a kind of hydrostatic skeleton, transmitting the force generated at the rear to the active appendages at the front. The lysosomes release digestive enzymes that disintegrate the prey.

An annelid worm is capable of body movements far more versatile than those possible in a hydra, despite their common lack of a hard skeleton. The anterior end of an earthworm may protrude from its burrow on damp evenings to reach bits of decayed vegetation on the surface for food. Its posterior end may then protrude in order to defecate worm castings. Obviously this practice is not without its hazards, for if the hungry worm waits too long, an equally hungry early bird is likely to come upon it with a peck and a jerk. If the bird is quick enough, the story ends right there; if not, the giant axons of the worm's ventral, solid nerve cord swiftly transmit impulses to its muscles, inhibiting the circular muscles and stimulating the longitudinal ones. The longitudinal muscles contract, pulling the body of the worm toward the safety of its burrow. If despite this evasive move the bird still retains a firm hold, the worm holds on too, with its swollen, contracted posterior now fitting the burrow like a cork in a bottleneck. If the bird releases its hold and the worm escapes, it will rapidly crawl down the burrow to safety. But how?

If the worm is progressing anterior end first, it must protrude a thinned portion of its body into the burrow ahead. Then, while remaining anchored posteriorly by a thickened portion of itself, the worm must cause the anterior end to swell. Having gripped the burrow ahead, the worm releases its posterior grip and, by contraction of the longitudinal muscle, drags the whole body toward the anchored anterior end. It repeats this process again and again (Fig. 32–3).

All this is made possible by the transverse partitions, or **septa,** that extensively subdivide the body cavity of the worm. These septa isolate portions of the body cavity and its contained fluid, permitting the hydrostatic skeletons of each segment to be independent of one another in large measure. Thus the contraction of the circular muscle in the elongating anterior end need not interfere with the action of the longitudinal muscle in the segments of the as yet anchored posterior end.

For animals that do more than drag themselves along on their bellies, the hydrostatic skeleton is insufficient. Yet some examples of it occur to a small extent in the higher invertebrates and even in the vertebrates. Among the mollusks, for example, the feet of bivalves are extended and anchored by a hydrostatic blood pressure mechanism not too different from that used by the earthworm. The many tube feet of echinoderms such as the sea star and sea urchin (Fig. 25–2) are protruded by a different version of the hydrostatic skeleton, and in male mammals, the penis becomes erect and stiff because of the turgidity of blood under pressure in its cavernous spaces.

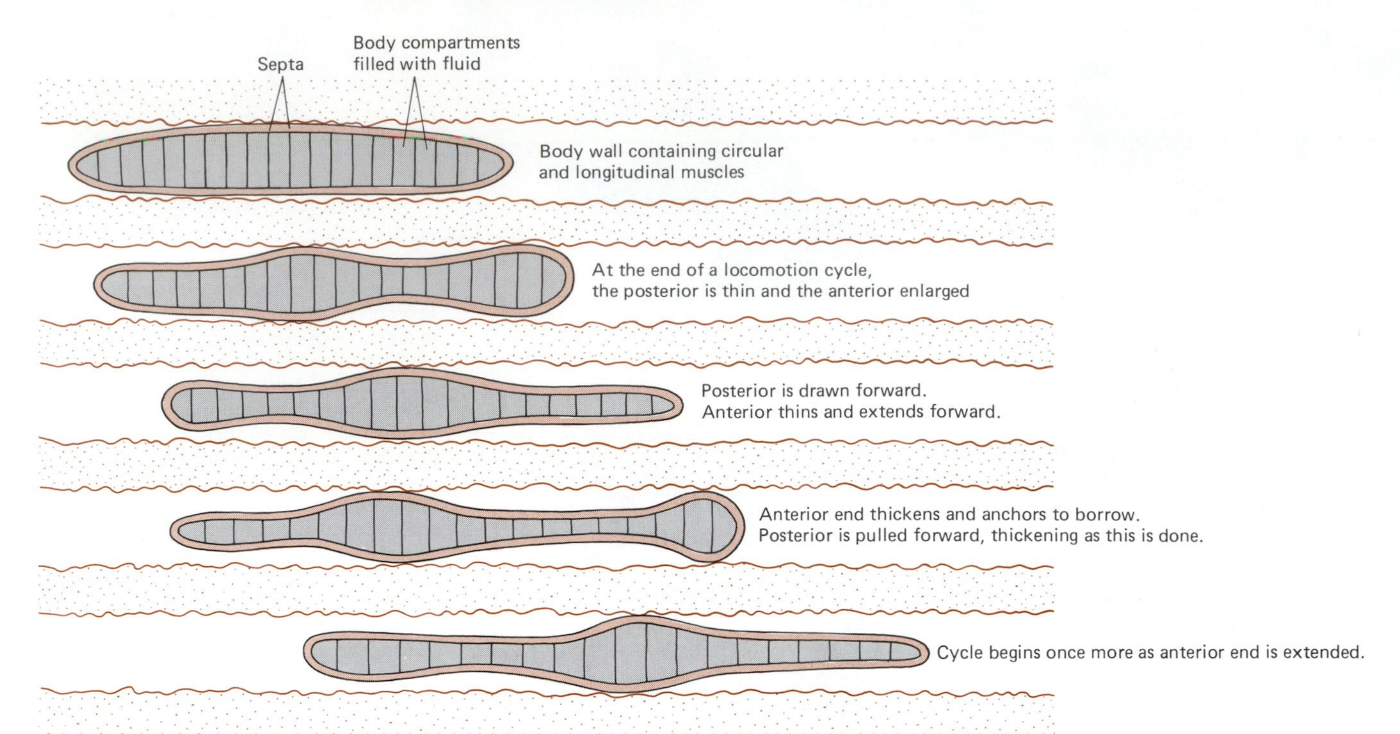
Septa
Body compartments filled with fluid
Body wall containing circular and longitudinal muscles
At the end of a locomotion cycle, the posterior is thin and the anterior enlarged
Posterior is drawn forward. Anterior thins and extends forward.
Anterior end thickens and anchors to borrow. Posterior is pulled forward, thickening as this is done.
Cycle begins once more as anterior end is extended.

Figure 32–3 Annelid locomotion. Can you infer the segments in which longitudinal or circular muscle is active in each stage? The worm is aided in anchoring itself by bristle-like setae (not shown).

EXTERNAL SKELETONS

Although there are others, the two major groups of animals with shells are the mollusks and the arthropods. In both the mollusks and the arthropods the shell is a non-living product of the cells of the epidermis, but it differs substantially in function between the two phyla. In the mollusks the exoskeleton typically provides protection. The muscles attached to the skeletion are used to aid in that function. Thus the clam has a pair of muscles whose major function is to hold the two valves of the shell tightly shut against the onslaughts of sea stars and chowdermakers. The contraction of similar muscles in the scallop enable it to clap its shells together rapidly and swim. The muscle attachments to the shells of gastropods such as snails serve to retract the body into the shell in emergencies.

The arthropod exoskeleton has an excellent ability to transmit muscular forces and gives form to the body. It can give full body protection while still allowing locomotion and flexibility, and it can be made into all manner of specialized tools, weapons, and other adaptations suited to a vast variety of life styles. Whereas in mollusks the shell serves primarily as an emergency retreat, with the bulk of the body nakedly and succulently exposed at other times, in arthropods the exoskeleton covers every bit of the body. It even extends inward as far as the stomach on one end and for a considerable distance inward past the anus on the other. The arthropod exoskeleton is a continuous one-piece sheath, but it varies greatly in thickness and flexibility, with large, thick inflexible plates separated from one another by thin, flexible joints arranged segmentally.

The arthropod exoskeleton, however, has a profound disadvantage: The rigid exoskeleton prevents growth. To overcome this disadvantage it is necessary for arthropods to **molt,** that is, to cast off their old integument from time to time to accommodate new growth. In order to understand molting we must first examine the tissue structure of the arthropod skeleton in some detail.

The living **epidermis** of an arthropod (Fig. 32–4) consists of a thin single layer of more-or-less cuboidal cells interspersed with a variety of glandular and sensory cells. Directly above this living layer, and in contact with it, is the non-living **cuticle.** The cuticle consists of protein and chitin. The deepest layer, known as the **endocuticle** is fairly soft. In the outer portion of the cuticle, the **exocuticle,** cross-linkages are formed among the chains of chitin by the action of tanning agents containing phenols, which are secreted by epidermal glands.

Atop the exocuticle lies the **epicuticle,** which is composed of proteins, waxes, and sometimes oils. The epicuticle retards the evaporation of water. It is particularly likely to be present and well-developed in terrestrial arthropods such as in-

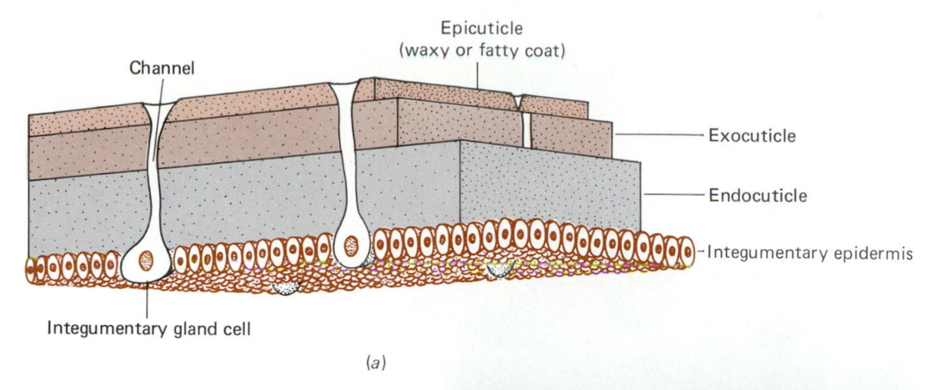

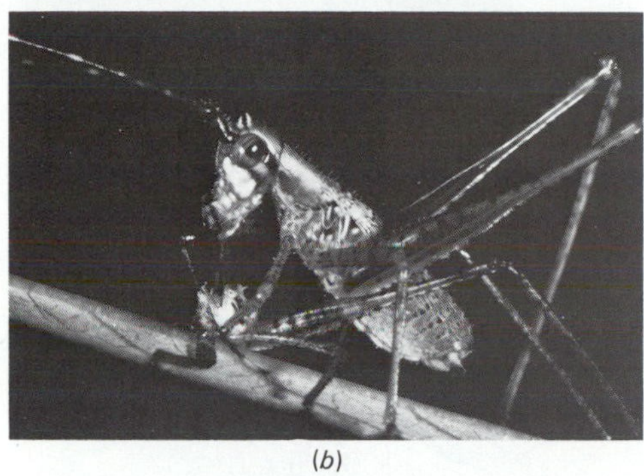

Figure 32–4 The arthropod exoskeleton. (*a*) Generalized arthropod exoskeleton. (*b*) A fork-tailed katydid, *Scudderia mexicana*, eating and thereby recycling some of the chitin and protein from its recently shed cuticle. ((*b*) P. J. Bryant, UC-Irvine/BPS.)

sects and millipedes. If the greasy epicuticle is experimentally removed from a cockroach—little more than wiping is required—the insect dies in a few hours from dehydration. In most aquatic and a few terrestrial arthropods the exocuticle is impregnated with calcium salts and forms a shell almost as hard as a mollusk's. In the joint membranes, these hard layers are reduced or absent for the sake of flexibility.

The intervals between moltings are known as **instars.** Somewhere near the end of an instar, an endocrine gland produces a hormone, called molting hormone or **ecdysone,** which initiates the molting, that is, the **ecdysis.** In response to ecdysone the epidermis secretes both a new cuticle and enzymes that attack the old cuticle at its base. It is not clear why the new cuticle is not digested also, but when the old one has been sufficiently loosened and dissolved, it splits open along predetermined seams and the animal slowly wriggles and pulls every appendage and all the other details of its complicated anatomy out of the old shell. Even the internal linings of mouth and anus must be detached. At last the exhausted animal lies still and recuperates. Eventually it drinks or swallows air to stretch its new suit of hardening chitin in preparation for future tissue growth.

Ecdysis obviously produces a very weak animal, unable to defend itself and almost unable to move, with a soft integument open to attack by any predator. Accordingly, this process is usually carried out in some very sheltered location. During the ensuing hours or days, epidermal glands secrete tanning agents, which harden the exocuticle; in some arthropods, channels in the cuticle carry other glandular secretions to the surface, which becomes the new epicuticle.

THE VERTEBRATE SKIN

You may not be used to thinking of skin as a skeleton, and of course technically it is not, but in many fish and in some reptiles, the skin is developed into a set of scales formidable enough to be considered armor. Even in human beings, the skin has considerable strength mostly attributable to the dermal layer, as we will see shortly.

An examination of our own body surface discloses a variety of structures in addition to the skin itself, including finger and toe nails, hair of various types, and the specialized coverings of the eyeball, mouth, and teeth. The skin proper consists of layers something like that of the arthropod cuticle, but the layers are not rigid and are all cellular (Fig. 32–5). The outer layers of our skin, however, are no more alive than the outer layers of a lobster's shell.

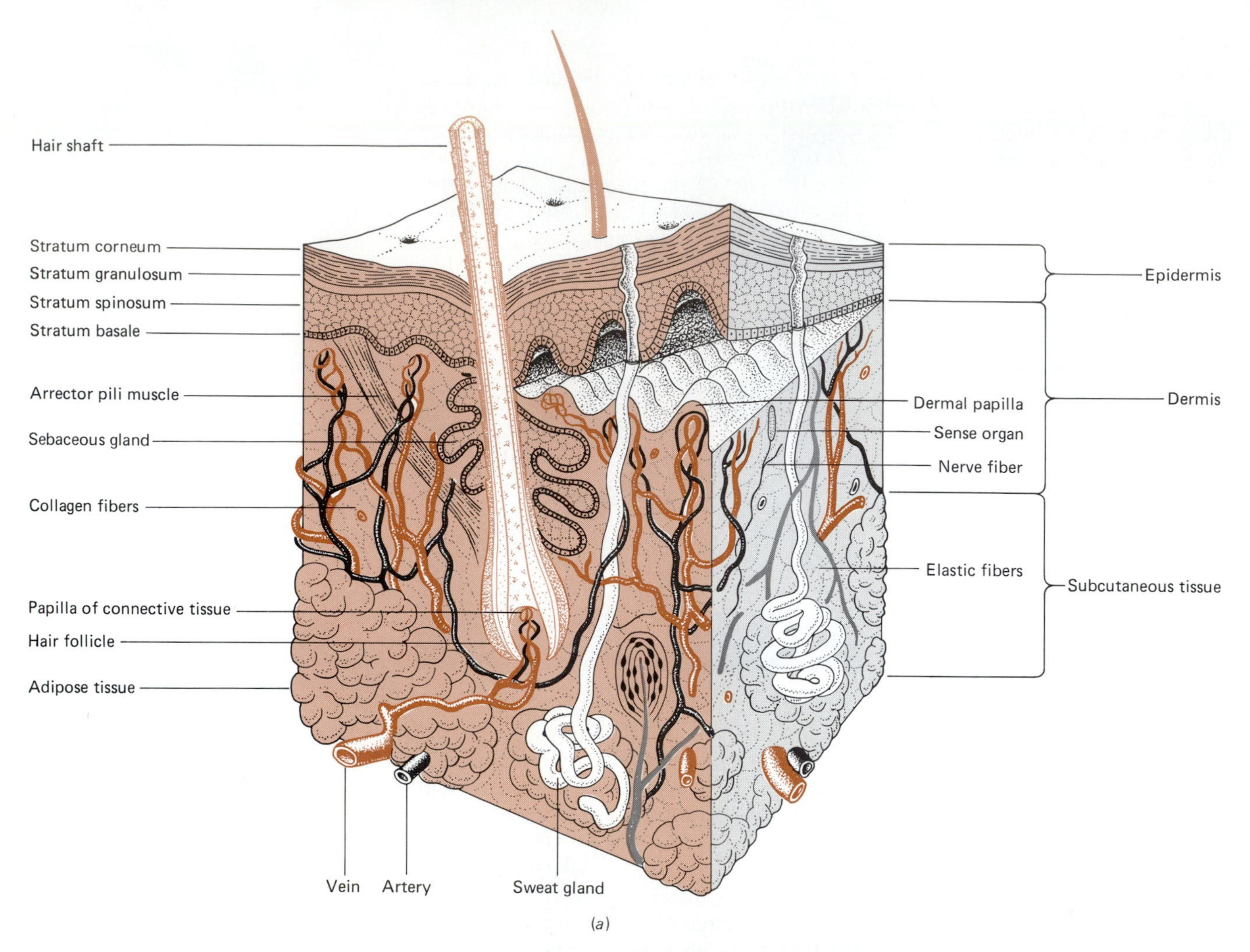

Figure 32–5 (*a*) Microscopic structure of skin. (*b*) Arctic mammals, such as this harbor seal, have thick layers of subcutaneous fat that serve as insulation. ((*b*) E. Simmons, The Image Bank.)

The Dermis

Mammalian skin rests on a layer of subcutaneous fat that insulates the body from outside temperature extremes. Localized thickening of the subcutaneous tissue produces some of the distinctive body contours that enable us at a glance to distinguish between the sexes in humans and many other mammals. Above this we find the foundational layer of the skin proper, the **dermis.** The dermis consists principally of a tough, fibrous connective tissue composed largely of the protein collagen, necessary for holding together the integument. The major part of each sweat gland is embedded in the dermis, and the hair follicles (which we will presently consider) reach down into it. The dermis also contains blood vessels which nourish the skin and sense organs concerned with what we vaguely call "touch." Above the dermis lies the real frontier—the epidermis—the interface between our more vulnerable tissues and the hostile universe.

The Epidermis

The epidermis consists of several strata, the lowest of which is the **stratum basale** and the outermost of which is the **stratum corneum.** The cells in the stratum basale divide continually and rapidly. The stratum basale cells are continuously forced upward by the pressure of yet other cells being produced below them. The epidermal cells mature as they move upward in the skin. There are no capillaries in the epidermis, and so as they move upward the cells are progressively deprived of nourishment and hence become less active metabolically. The cells manufacture the skin protein, **keratin,** a tough, coiled protein. The coils of keratin are themselves coiled into what is called a **supercoiled structure.** This arrangement produces considerable mechanical strength combined with flexibility, for the coils are capable of stretching or bending much like springs. This is very important in derivatives of skin such as hair, horn, and nails. But perhaps the most important characteristic property of keratin is its insolubility. It is so insoluble that it actually serves as an excellent body surface sealant. What we call fully mature epidermal cells are actually dead cells—as dead and as waterproof (due to their keratin and lipid content) as shingles on a roof. Like shingles, the cells of stratum corneum, the outermost layer, continually wear off, but unlike shingles, they are continually replaced (Fig. 32–6).

The rate of cell replacement in the epidermis is exactly regulated and responds to rates of wear, so that a heavily abraded area may even form thick calluses that are nothing more than enormously hypertrophied layers of dead epidermis. Not only the epidermis but also the dermis varies in thickness and toughness in different parts of the body, being thickest on the soles of the feet and on the hands, but of almost gossamer thinness on the head of the penis.

Since the skin is usually not rigid and is replaced continuously as it wears away, most vertebrates do not undergo ecdysis, but there are exceptions, most notably the snakes. These reptiles from time to time shed the entire stratum corneum all at once, exposing a polished bright new body surface beneath it. Even the transparent covering of the eyes is shed at these times.

The skin in other vertebrates may show considerable variations. The skin of birds possesses feathers, which develop in a manner comparable to the hairs of mammals. Among the poikilothermic vertebrates we find epidermal scales (as in reptiles), naked skin covered with mucus (as in many amphibians and fish), and skin with bony or toothlike scales, which are partly of dermal origin (as in other fish). Some skin, such as that in certain tropical frogs, is provided with poison glands so potent that many arrows can be tipped with the poison of a single tiny frog. Skin and its derivatives are often brilliantly colored as warning coloration in connection with courtship rituals, territorial displays, and various kinds of communication.

The skin also functions as a thermostatically controlled radiator, regulating the elimination of heat from the body. Heat is constantly produced by the metabolic processes of the cells and distributed by the blood stream. A certain amount of heat must be lost all the time to maintain a constant temperature within the body. Some heat leaves the body in expired breath and some in feces and urine, but approximately 90% of the total is lost through the skin. When the external temperature is low, arterioles in the skin are constricted, thereby decreasing the flow of blood through the skin and decreasing the rate of heat loss.

In a warm environment the reverse occurs: The arterioles dilate and the increased flow of blood, which causes the skin to appear flushed, results in increased heat loss. In very warm environments this mechanism cannot eliminate enough heat from the body, and the sweat glands of the skin are stimulated to increase the amount of perspiration secreted. The evaporation of sweat from the surface of the skin lowers the body temperature by removing from the body the heat necessary to convert liquid sweat into water vapor (540 kilocalories are required to convert a liter of water to water vapor). About 2.5 million sweat glands secrete this water, which can amount to several liters per day in hot, dry climates during heavy exercise.

The skin contains several types of sense receptors responsible for our ability to feel pressure, temperature, and pain. The skin of mammals also contains sebaceous and mammary glands. In humans sebaceous glands are especially numerous on the face and scalp (Fig. 32–7). The oil secreted keeps the hair moist and pliable and prevents the skin from drying and cracking. The mammary glands also are derivatives of the skin, specialized for the secretion of milk.

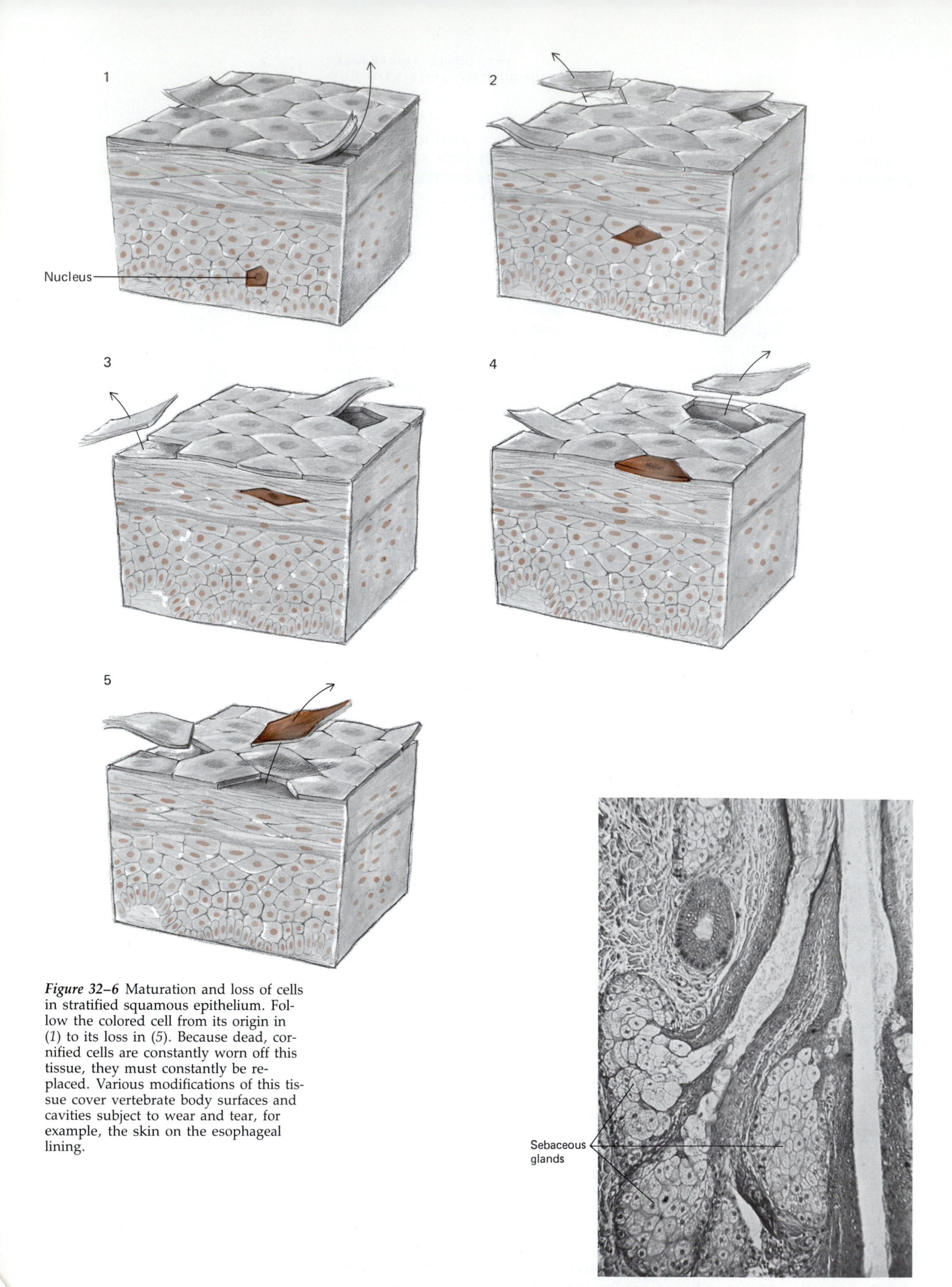

Figure 32–6 Maturation and loss of cells in stratified squamous epithelium. Follow the colored cell from its origin in (*1*) to its loss in (*5*). Because dead, cornified cells are constantly worn off this tissue, they must constantly be replaced. Various modifications of this tissue cover vertebrate body surfaces and cavities subject to wear and tear, for example, the skin on the esophageal lining.

Figure 32–7 Hair follicle showing sebaceous glands (×200).

INTERNAL SKELETONS

Some mollusks—a few cephalopods, such as the squid—have internal skeletons of sorts, but the only phyla in which these are extensively developed are the echinoderms and the chordates. The echinoderms have spicules and plates of calcium salts (calcite) embedded in the mesodermal tissues of the body wall. These serve mainly for support and protection in those animals. But it is the vertebrates that employ the internal skeleton for its full range of potential—for support and, to a lesser extent, for protection, but primarily for the transmission of forces.

THE SKELETON ITSELF

The most obvious function of an internal skeleton, like an exoskeleton, is to give support and "shape" to the body. For an animal to rise off the ground and move around, some hard, durable substance is needed to maintain the soft tissues against the pull of gravity and act as a firm base for the attachment of muscles. The vertebrate skeleton protects the underlying organs such as the brain and lungs from injury. The marrow tissue, found within the cavity of the bones, usually performs the special task of manufacturing all red blood cells and some kinds of white ones.

Vertebrates characteristically have an **endoskeleton.** The skeleton of sharks and rays is made of cartilage, but in the bony fishes and other vertebrates most of the cartilage is transformed to bone in the course of development. The human skeleton (Fig. 32–8) consists of approximately 206 bones. The exact number varies at different periods of life as some of the bones, at first distinct, gradually become fused. Most of the bones are hollow and contain bone marrow cells.

Parts of the Skeleton

The vertebrate skeleton may be divided into the **axial skeleton** (the bones and cartilages in the middle or axis of the body) and the **appendicular skeleton** (the bones and cartilages of the fins or limbs). The axial skeleton includes the **skull, vertebral column** (spine or backbone), **ribs,** and **sternum** (breastbone).

The skull consists of the cranial and the facial bones. Some of the cranial bones participate in the formation of the face, but what really distinguishes them is that they surround the brain. The sides and roof of the skull consist of membrane bones, ones that originate embryonically in connective tissue membranes and do not replace cartilage. Most other bones, including the long bones of the appendicular skeleton, originate in the embryo in the form of cartilage models that are eventually replaced by bone.

The Bones of Vertebrates

Many other vertebrates possess bones that humans lack, such as the skeleton of the gill arches of fish. Careful studies of the embryos of humans and other mammals have shown, however, that a number of elements of the skull originate embryonically in the same way as do the gill arches of fishes. The tiny middle ear bones—malleus, incus, and stapes—are examples of such elements. Similarly, the cranium of fish contains many bones that are not found in the human cranium but which originate in the same way during the course of early development. Indeed, several cranial bones that are single in the adult human result from the fusion of two or more bones that were originally separate in the embryo or even in the newborn (Fig. 32–9).

The human spine is made of 33 separate **vertebrae,** which differ in size and shape in different regions of the spine. A typical vertebra consists of a basal portion, the **centrum,** and a dorsal ring of bone, the **neural arch,** which surrounds and protects the delicate spinal cord. Different vertebrae have different projections for the attachment of ribs and muscles and for articulating (joining) with neighboring vertebrae. The first vertebra, the **atlas** (named for the mythical Greek who held the world on his shoulders), has rounded depressions on its upper surface into which fit two projections from the base of the skull. Since there are no ribs in the human neck, the cervical vertebrae lack the little facets whereby ribs are attached to the vertebrae of the thorax. The several sacral vertebrae of the pelvic region are probably the most specialized of all and are fused into what is functionally a single bone.

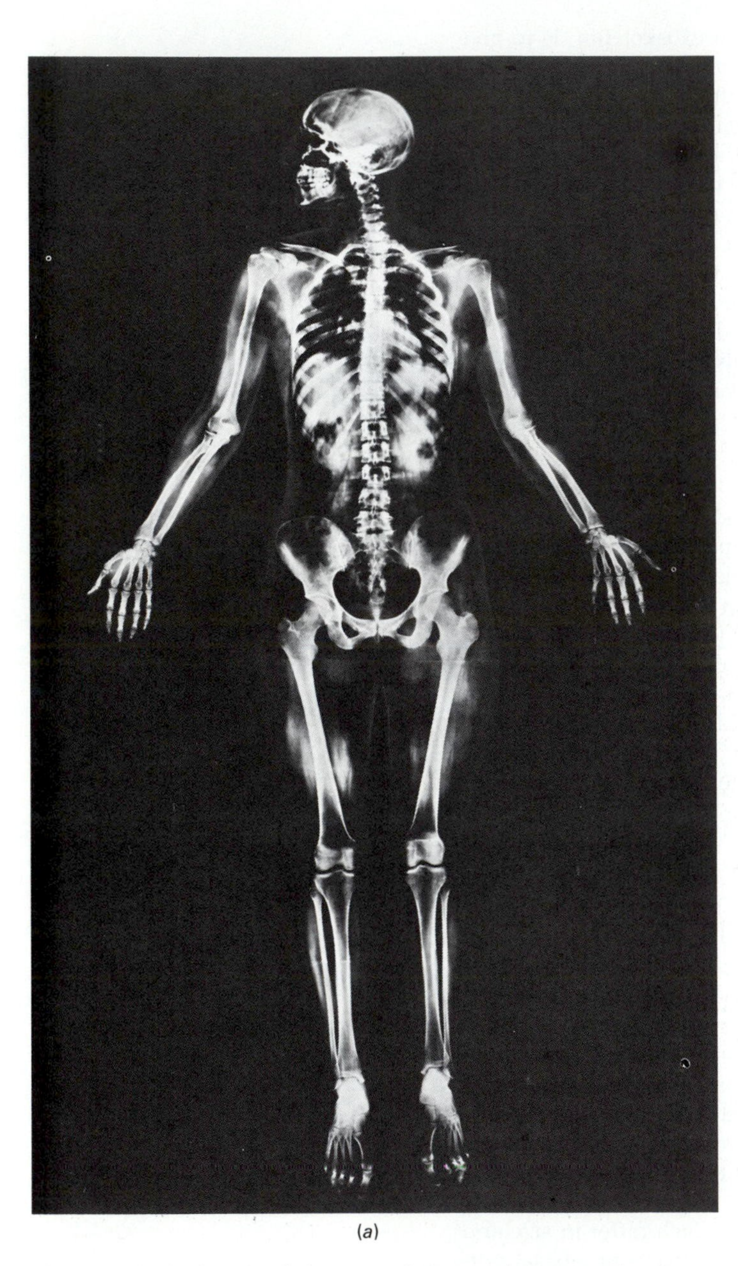

(a)

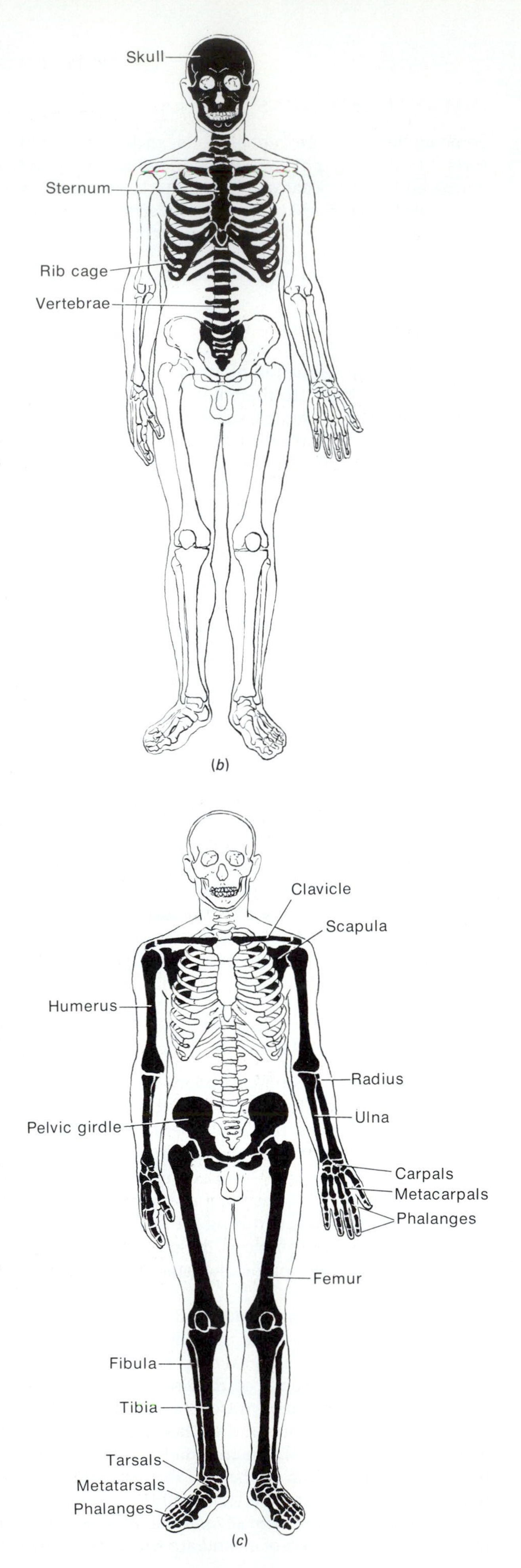

Figure 32–8 The human skeleton. (*a*) Compounded series of radiographs of the whole body of a young adult female. (*b*) Diagram showing the bones of the axial skeleton. (*c*) Diagram showing the bones of the appendicular skeleton. ((*a*), published with the permission of ILFORD Limited, England.)

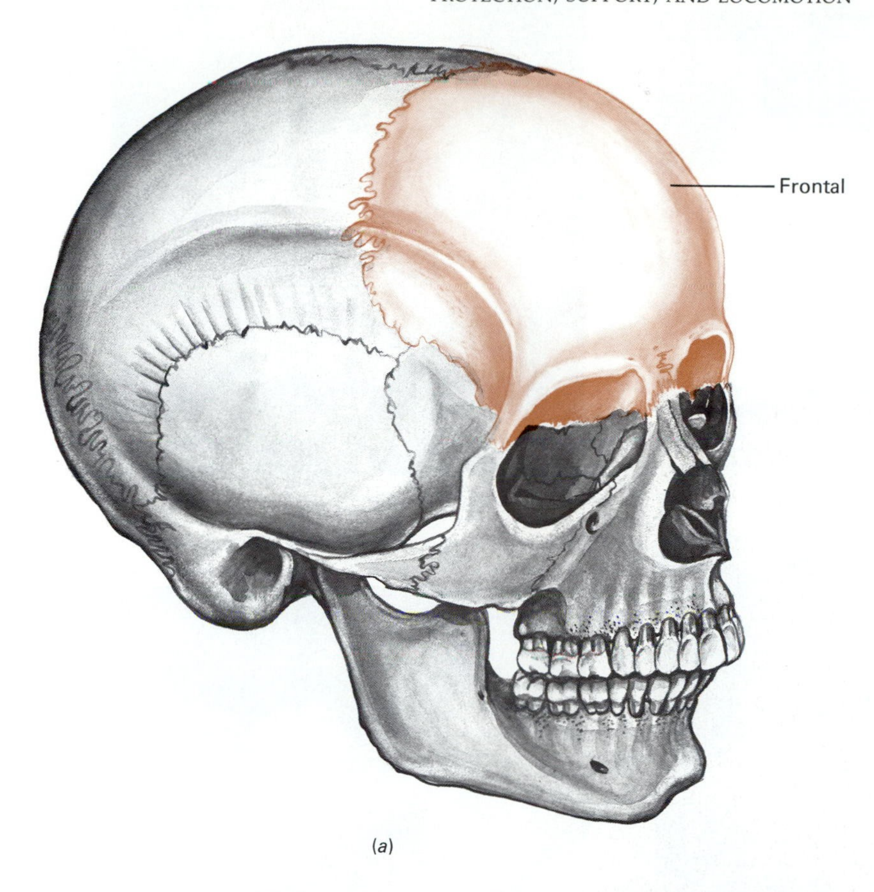

(a)

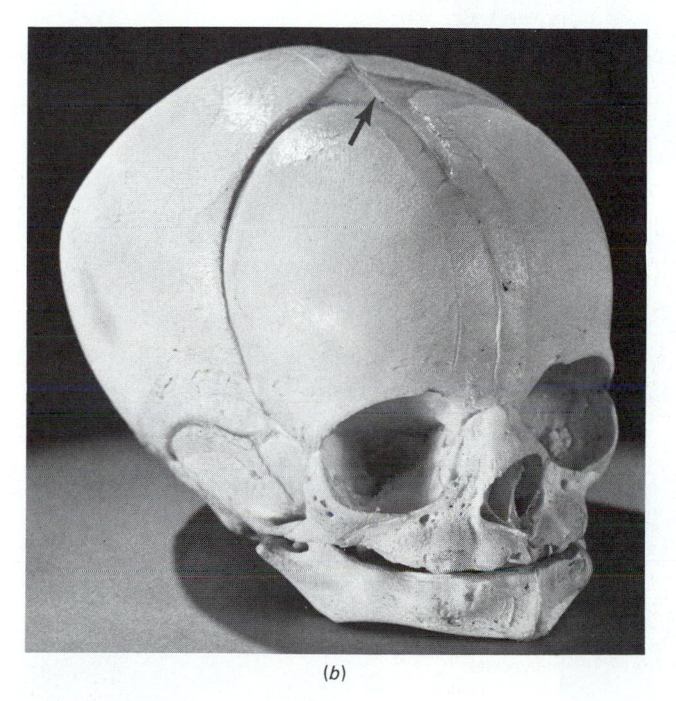

(b)

Figure 32–9 The human skull. (*a*) A mature skull. (*b*) A fetal skull. Note the differences in proportions between the two. In the fetal specimen the frontal bones have not yet fused to produce the single adult frontal bone. The same is true of the lower jaw. The skull of an infant also has a **fontanel** (*arrow*), or gap, between the skull bones at the superior surface of the skull, which closes in adult life. ((*b*), Carolina Biological Supply Company.)

The rib basket, composed of a series of flat bones, supports the chest wall and keeps it from collapsing as the diaphragm contracts. Each pair of ribs is attached dorsally to a separate vertebra. Of the 12 pairs of ribs in the human, the first seven are attached ventrally to the sternum (breastbone), the next three are attached indirectly by cartilages, and the last two, called "floating ribs," have no attachments to the sternum.

The bones of the appendages (the arms and legs) and the **girdles** that attach them to the rest of the body make up the appendicular skeleton. The **pelvic girdle**

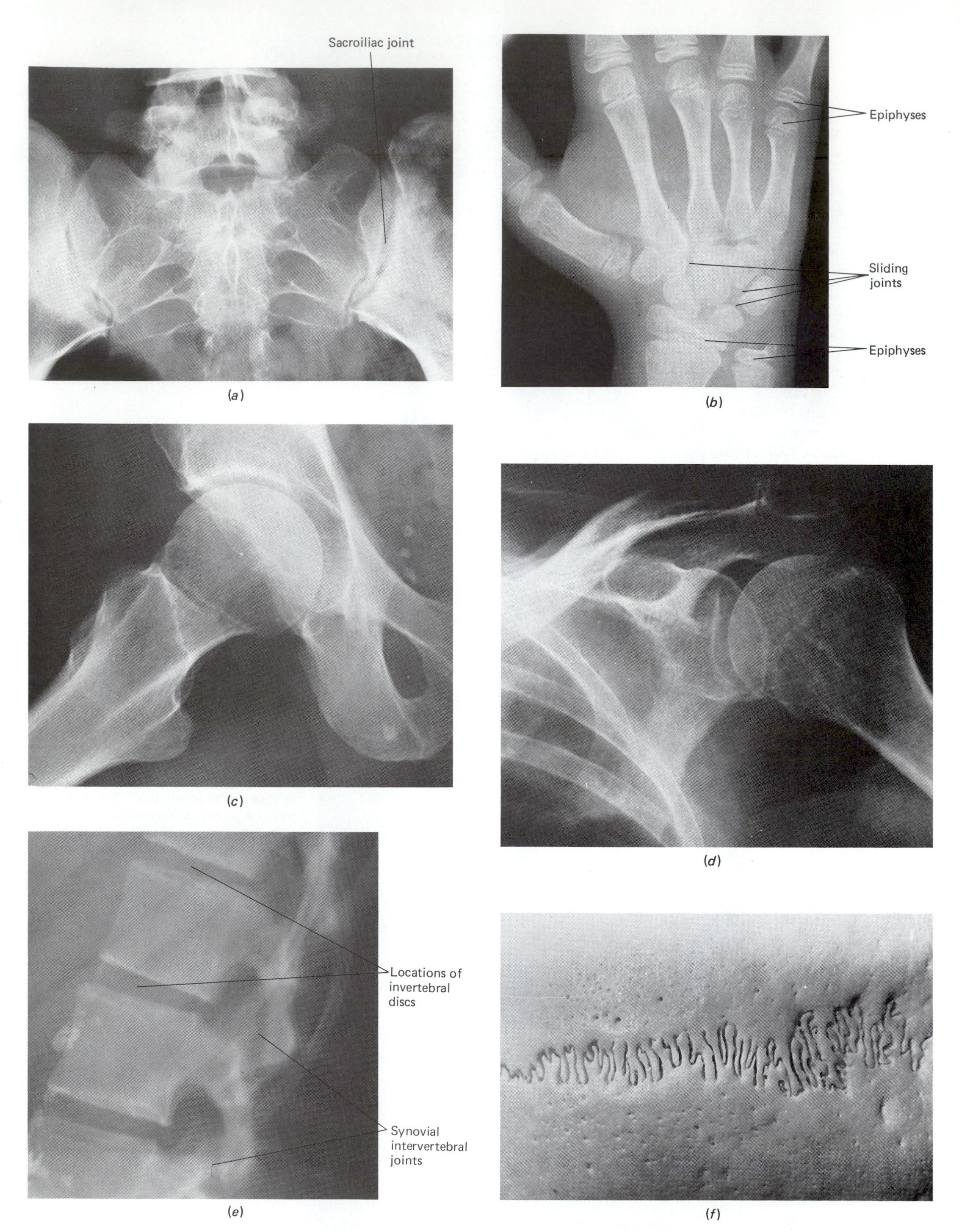

Figure 32–10 Representative joints of the human skeleton. (*a*) Sacroiliac joint, a very tight and almost inflexible joint. (*b*) Sliding joints between the wrist bones. (*c*) Ball-and-socket hip joint. (*d*) Ball-and-socket shoulder joint, one of the most freely movable joints and also the loosest joint in the body. The shoulder joint is held together partly by the steady contraction of the surrounding muscles. (*e*) Intervertebral joints. The synovial joints are plane joints, that is, their articulating surfaces are flat and slide past one another. (*f*) A skull suture, a highly immovable joint. Note the elaborate jigsaw puzzle–like interdigitations of this joint.

consists of a pair of large bones, each composed of three fused hipbones, and the **pectoral girdle** consists of the two collarbones, or **clavicles,** and the two shoulderblades, or **scapulas.** The pelvic girdle is securely fused to the vertebral column, whereas the pectoral girdle is loosely and flexibly attached to it by muscles.

The appendages of human beings are comparatively generalized, each terminating in five **digits**—the fingers and toes, whereas the more specialized appendages of other animals may be characterized by four digits (as in the pig), three (as in the rhinoceros), two (as in the camel), or one (as in the horse).

Great apes and humans do, however, have a highly specialized feature—the opposable thumb. (In addition, great apes have an opposable big toe; and in humans, though it is not opposable, the big toe is similar enough in structure to the thumb that it can be used as a surgical substitute.) The opposable thumb can readily be wrapped around objects such as a tree limb in climbing, but it is especially useful in grasping and manipulating objects. It can be opposed to each finger singly, or to all of them collectively. The muscles that move the thumb are almost as powerful as those of all the other fingers put together.

In humans, the anterior appendages (which contain our thumbs) are not used for locomotion as in other mammals, including the great apes. Our hands have been emancipated. The combination of our opposable thumbs and our upright posture enables us to use our hands to shape and build; to effect changes on our environment to a greater extent than any other organism on Earth.

Muscles

In all the multicellular animals, the principal effectors that provide for movements in response to stimuli are muscles composed of specialized contractile cells. Most of our body muscles consist of individual muscles attached to bones. This kind of muscle is called **skeletal muscle.** It is also known as **voluntary muscle** because, to a large extent, the muscle can be controlled at will (Fig. 32–11). In the human body there are nearly 700 named skeletal muscles. Characteristics of various types of muscle are summarized in Table 6–3.

A typical skeletal muscle of a vertebrate is an elongated mass of tissue composed of millions of individual **muscle fibers** bound together by connective tissue fibers. The entire structure is surrounded by a tough, smooth sheath of connective tissue. The two ends of the muscle in vertebrates are typically attached to two different bones, and the contraction of the muscle draws one bone toward the other, with the bones acting as levers and the joint between the two acting as the fulcrum of the lever system. A few muscles pass from a bone to the dermis of the skin or, as in some of the muscles of facial expression, from one part of the dermis to another. Many of the muscles that do act upon bones are attached to them indirectly by means of **tendons.** (You can observe the tendons of some of the arm muscles that move the fingers by clenching your fist and watching the tendons tense in your wrist.) The end of the muscle that remains relatively fixed when the muscle contracts is called the **origin;** the end that moves is called the **insertion;** and the thick part between the two is called the **belly.** The two origins of the biceps, for example, are on the shoulder; its insertion is on the bone in the forearm called the radius. When the biceps contracts, the shoulder remains fixed, but the radius is pulled toward the shoulder so that the elbow is bent (Fig. 32–12).

MUSCLE ACTIONS

Muscles typically contract in groups rather than singly. Without special training you cannot, for instance, contract the biceps alone; you can only bend the elbow, which involves the contraction of a number of other muscles in addition to the biceps. Muscles can exert a pull but not a push; hence muscles are typically arranged in **antagonistic pairs:** One pulls a bone in one direction, and the other pulls it in reverse. The biceps muscle bends or flexes the arm and is termed a **flexor.** Its antagonist, the triceps muscle, straightens or extends the arm and is termed an **extensor.** Similar sets of opposing flexors and extensors are found at the wrist, knee, ankle, and other joints. When a flexor contracts, the opposing extensor, if contracted, must relax to permit the bone to move. This requires proper coordina-

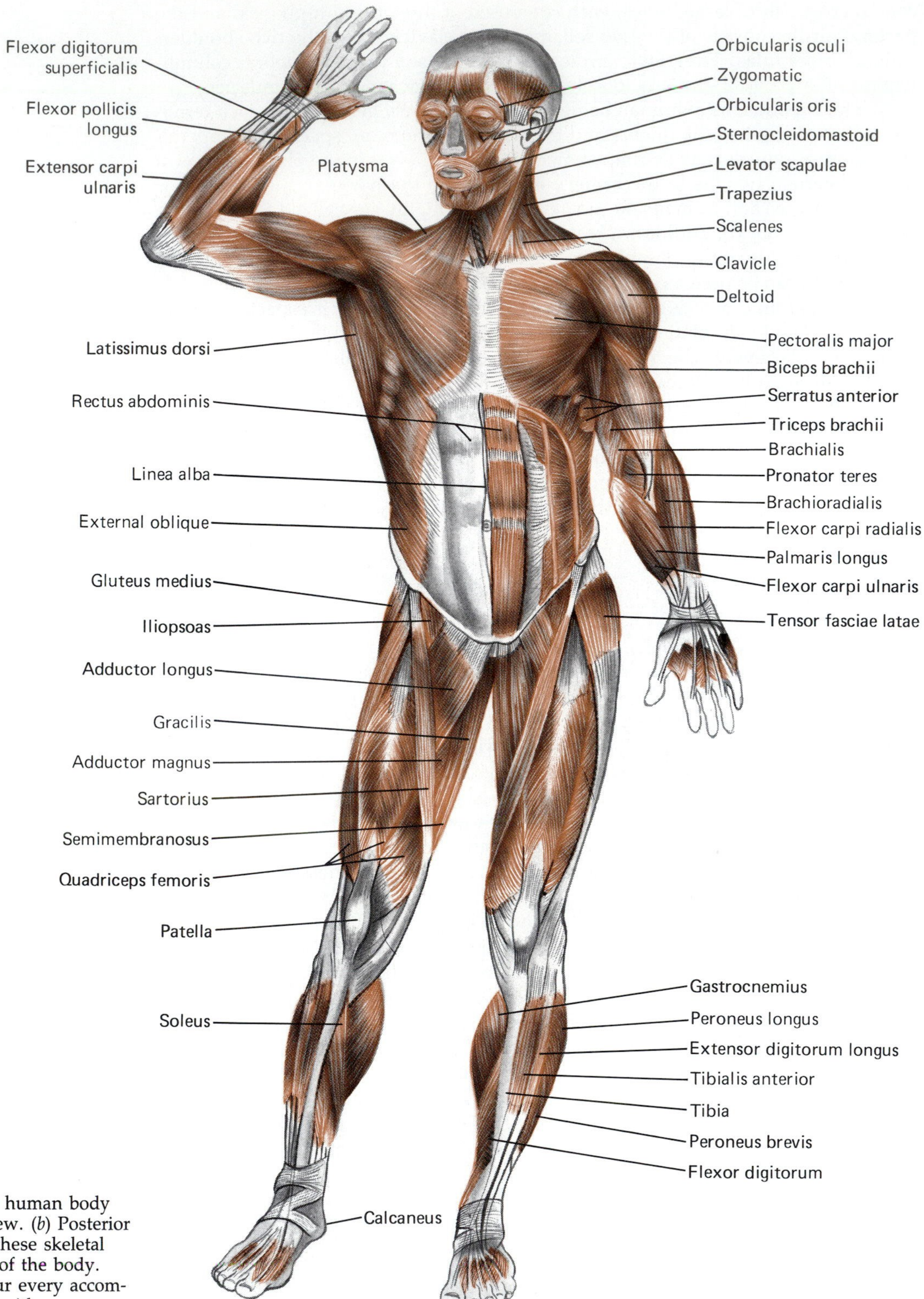

Figure 32–11 Superficial human body muscles. (*a*) Anterior view. (*b*) Posterior view. Several layers of these skeletal muscles comprise most of the body. Directly or indirectly, our every accomplishment is performed with our muscles.

tion of the nerve impulses going to the two sets of muscles. Other antagonistic pairs of muscles are **adductors** and **abductors,** which move parts of the body toward or away from the central axis of the body; **levators** and **depressors,** which raise and lower parts of the body; **pronators,** which rotate body parts downward and backward; **supinators,** which rotate them upward and forward; and **sphincters** and **dilators,** which decrease and enlarge, respectively, the size of an opening.

Of course there is far more to muscle action than simple antagonism. The specific action of each muscle, since it can only contract, depends entirely upon the

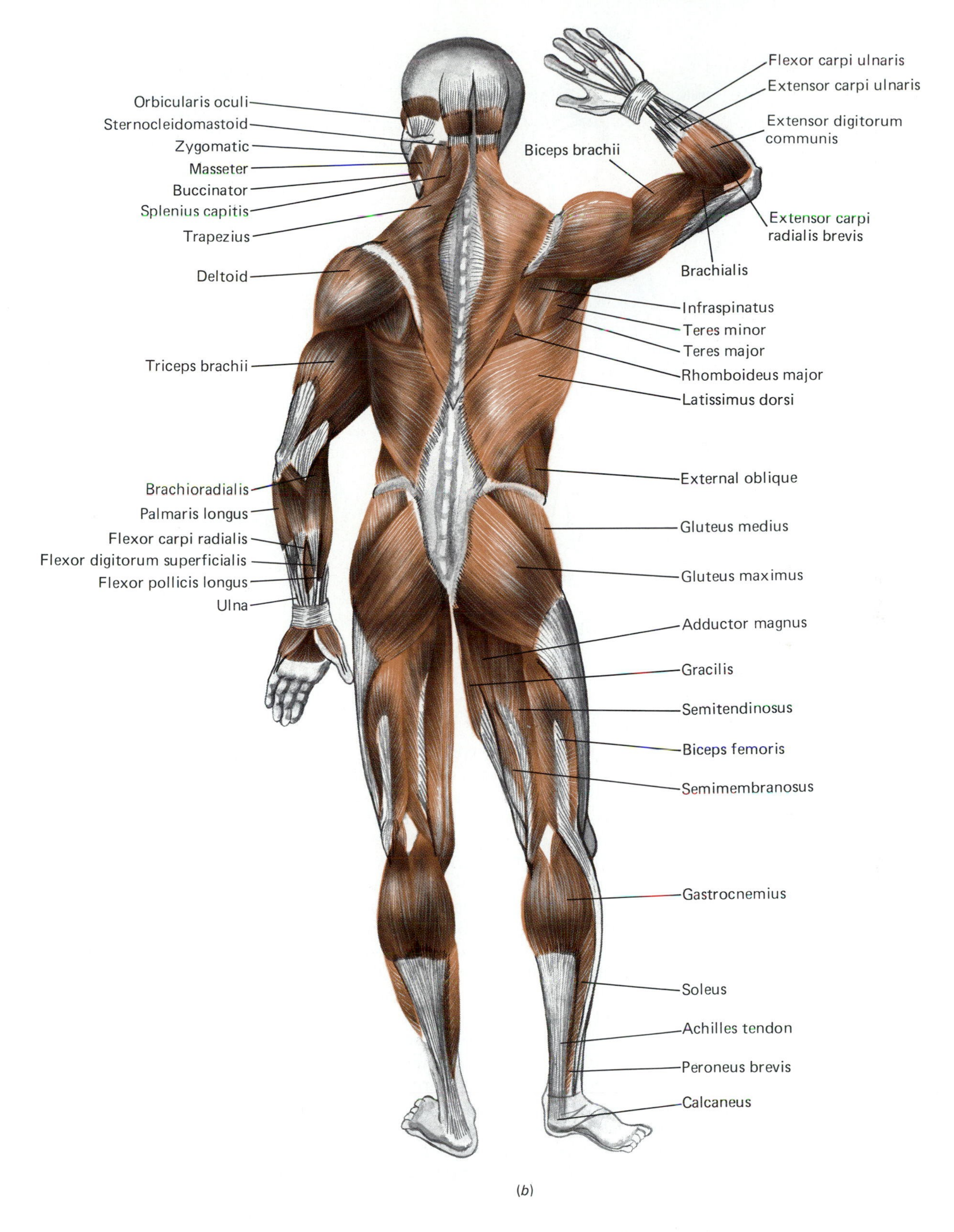

(*b*)

bone it is attached to and *how it is attached*. Even simple actions such as flexion and extension require the cooperation of many muscles. Although each muscle produces a single type of motion, many muscles contract and still others relax to produce a movement such as swinging an arm. Taking a single step, for instance, requires the participation of almost every muscle below the shoulders, perhaps as many as 200 of them. It is not surprising that learning to walk is a long and tedious process. Some of the superficial muscles (those toward the body surface) of the human body are shown in Figure 32–11.

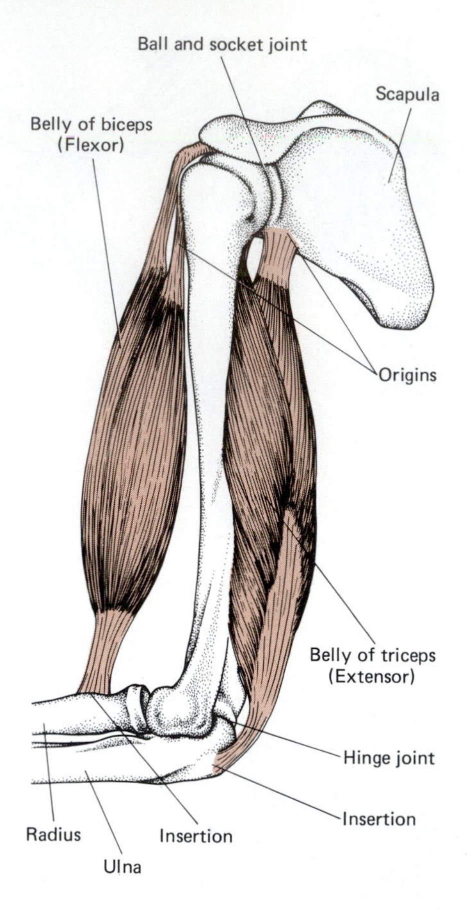

Figure 32–12 Muscles and bones of the upper arm, showing origin, insertion, and belly of a muscle and the antagonistic arrangement of the biceps and triceps.

POSTURE AND MOVEMENT

Even when a muscle is not contracting to effect movement, it is not completely relaxed. As long as a person is conscious, all his muscles are contracted slightly (a condition known as tonus). Posture is maintained by the partial contraction of the muscles of the back and neck and of the flexors and extensors of the legs. When a person stands, both the flexors and extensors of the thigh must contract simultaneously so that the body sways neither forward nor backward on the legs. The simultaneous contraction of the flexors and extensors of the shank locks the knee in place and holds the leg rigid to support the body.

Some of the larger muscles of the human body are remarkably strong—for example, the gastrocnemius muscle in the calf of the leg. The gastrocnemius has its origin at the knee and its insertion via the tendon of Achilles on the heel bone. Because the distance from the toes to the ankle joint is at least six times that from the ankle joint to the heel, the gastrocnemius is working against an adverse lever ratio of 6 : 1. Thus when a woman weighing 60 kg stands on one leg and rises on her toes, the one gastrocnemius muscle must exert a force of 360 kg. When one ballet dancer holds another in his arms and rises on the toes of one leg, his gastrocnemius is exerting a force of nearly 1000 kg.

Normal nervous coordination prevents muscles from contracting maximally, but when nervous control is impaired as in an epileptic seizure, a muscle may contract forcefully enough to rip tendons and break bones. You may have read of tremendous feats of strength performed under great emotional stress; in such circumstances, normal nervous control, as well as the corresponding normal limits of function, have been overridden.

MUSCLE PHYSIOLOGY

The functional unit of vertebrate muscles, the **motor unit,** consists of a single motor neuron and the group of muscle cells innervated by its axon, all of which will contract when an impulse travels down the motor neuron to the **motor end plate** (Fig. 32–13). It is estimated that the human body has some 250,000,000 muscle cells, but only some 420,000 motor neurons in spinal nerves. Obviously some motor

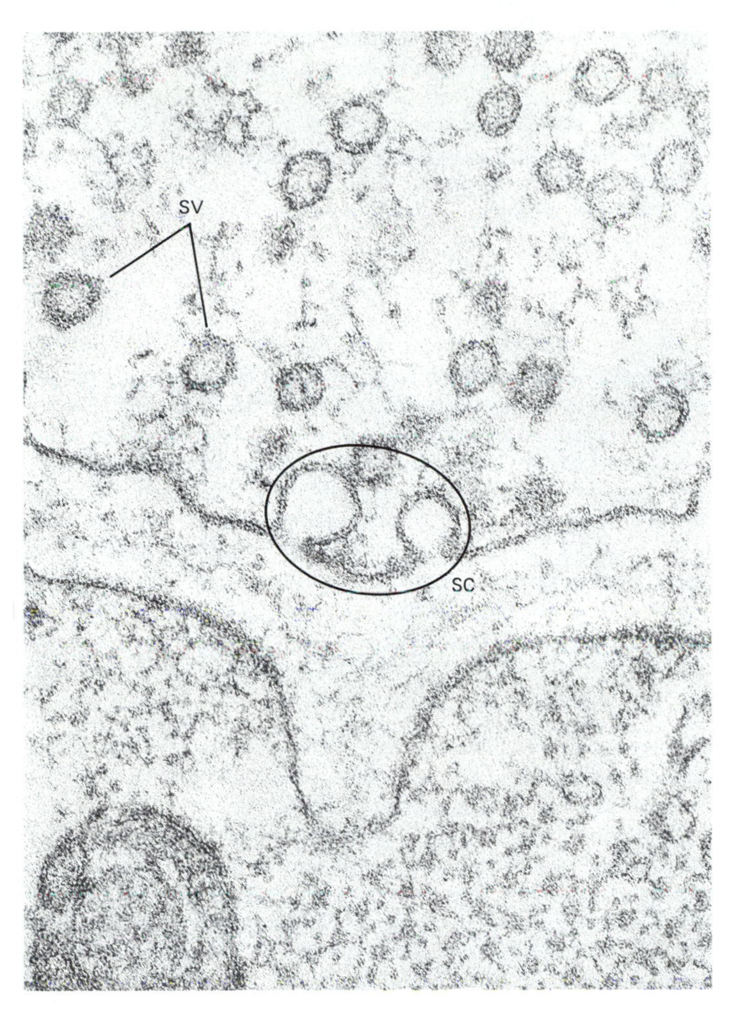

Figure 32–13 Electron micrograph of a synaptic knob and cleft (approximately ×125,000). In the circled area, two synaptic vesicles are merging with the membrane of the synaptic knob and discharging neurotransmitter into the cleft. The membrane of a skeletal muscle cell synapsing with this knob is shown below. *SV*, synaptic vesicles; *SC*, synaptic cleft; a neuron is shown above the cleft, a muscle cell below. (Courtesy of Dr. John Heuser.)

neurons must innervate more than one muscle fiber (Fig. 32–14). The degree of fine control of a muscle depends upon the number of motor units it possesses, because in effect, each motor unit behaves as an independent muscle. The exact degree of contraction of an entire muscle, therefore, depends upon the number of motor units that have been recruited. The more of these there are, the greater the potential variation in strength of contraction. Thus the greater the delicacy of control re-

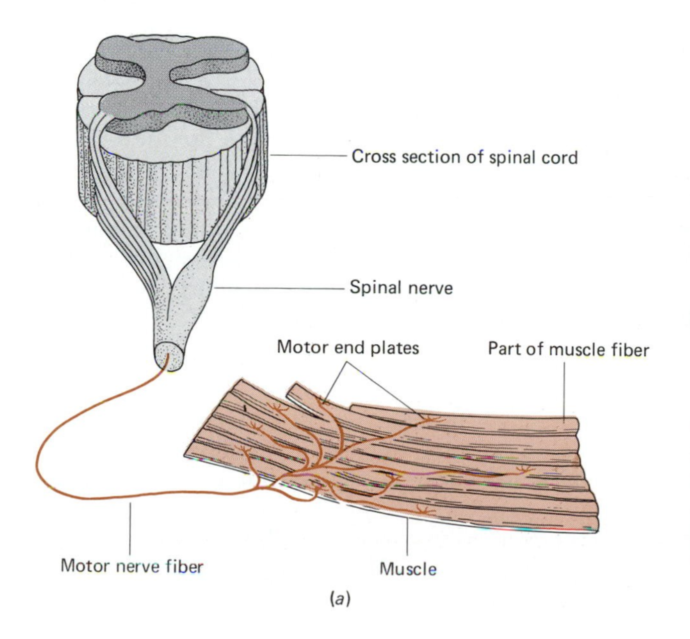

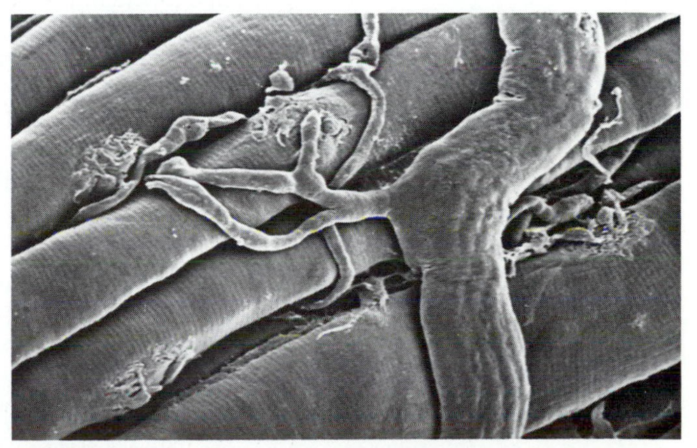

Figure 32–14 Motor units. (*a*) A motor unit typically includes many more muscle fibers than appear here, averaging about 150 muscle fibers each, but some units have less than a dozen fibers, while others have several hundred. (*b*) Scanning electron micrograph of some of the cells in a motor unit (×900). Note how the large neuron branches send subdivisions to each cell in the motor unit. (From Desaki, J.: "Vascular autonomic plexuses and skeletal neuromuscular junctions: A scanning electron microscopic study." *Biomedical Research* (Supplement), 139–143, 1981.)

quired in a muscle, the more motor units it must have—that is, the fewer muscle fibers there must be in each motor unit. Look up from this page and fix your gaze on an object near the wall. The positions of your eyeballs had to be very exactly adjusted to make it possible for you to see each of these views with clarity. This is possible because the muscles of the eyeball have a very high number of motor units for such tiny muscles. In fact, they have so many that there are as few as three to six fibers per motor unit in them, whereas the leg muscles have as many as 650 fibers per unit.

The All-or-None Effect

If a single motor unit is isolated and stimulated with brief electric shocks of increasing intensity, beginning with stimuli too weak to cause contraction, there will be no response until a certain intensity is reached at which point the response is maximal. This phenomenon is termed the **all-or-none effect.** In contrast, an entire muscle, composed of many individual motor units, can respond in a graded fashion depending on the number of motor units contracting at any given time. Although an entire muscle cannot usually contract maximally, a single motor unit can contract only maximally or not at all. The strength of contraction of an entire muscle composed of thousands of motor units depends upon the number of its constituent motor units that are contracting and upon whether the motor units are contracting simultaneously or alternately.

The Simple Twitch

Muscles continue to live and retain their ability to contract for some time after they have been removed from the body. Their contraction may be studied in detail by mechanical or electronic means in the laboratory.

When a muscle is given a single stimulus, a single electric shock, it responds with a single, quick contraction, a **simple twitch,** lasting about 0.1 second in a frog's muscle and about 0.05 second in a human muscle. A graphic record of a simple twitch reveals that it consists of three separate phases: (1) the **latent period,** lasting less than 0.005 second, an interval between the application of the stimulus and the beginning of the visible shortening of the muscle; (2) the **contraction period,** about 0.04 second in duration, during which the muscle shortens and does work; and (3) the **relaxation period,** the longest of the three, lasting 0.05 second, during which the muscle returns to its original length.

Muscle fibers also have a **refractory period,** a very short period of time immediately after one stimulus during which they will not respond to a second stimulus. However, the refractory period in skeletal muscle is so short (about 0.002 second) that muscle can respond to a second stimulus while still contracting in response to the first. The superposition of the second contraction on the first results in a greater shortening of the muscle fiber than in a simple twitch. This effect is known as **summation.**

The first event after the stimulation of a muscle is the initiation and propagation of an electrical change in its cell membranes known as an **action potential,** followed by changes in the structure of the contractile proteins (actin and myosin). After a twitch, the muscle consumes oxygen and gives off carbon dioxide and heat at a rate greater than during rest, marking a recovery period in which the muscle is restored to its original state. This recovery period lasts for several seconds, and if a muscle is stimulated repeatedly so that successive contractions occur before the muscle has recovered from the previous ones, the muscle becomes fatigued and the twitches grow feebler and finally stop. If the fatigued muscle is allowed to rest for a time it regains its ability to contract.

Tetanus

Muscles normally contract not in single twitches but in sustained contractions evoked by volleys of nerve impulses reaching them in rapid succession. Such a sustained contraction is called **tetanus;**[1] during a tetanic contraction the stimuli

[1] Although dictionaries do not always draw a distinction, this term should not be confused with tetany, the muscular spasms occurring in deficiencies of the parathyroid hormone, or with the disease tetanus ("lock-jaw"), characterized by abnormal muscular contractions and caused by a toxin produced by the tetanus bacillus.

occur so rapidly (several hundred per second) that relaxation cannot occur between the contractions of successive twitches. In most tetanic contractions the individual motor units are stimulated sequentially rather than simultaneously, so that although individual groups of muscle fibers contract and relax, the muscle as a whole remains partially contracted.

Tonus

The term **tonus,** or muscle tone, refers to the state of sustained partial contraction present in all normal skeletal muscles as long as the nerves to the muscle are intact. Tonus is a mild state of tetanus, present at all times and involving only a small number of the fibers of a muscle at any moment. The individual fibers contract in turn, working in relays, so that each fiber has an opportunity to recover completely while other fibers are contracting before it is called upon to contract again. A muscle under slight tension can react more rapidly and contract more strongly than one that is completely relaxed.

THE BIOCHEMISTRY OF MUSCULAR CONTRACTIONS

Muscles are able to use between 20% and 40% of the chemical energy of glucose in the mechanical work of contraction. The remainder is converted into heat but is not wholly wasted since it is used to help maintain the body temperature. If you refrain from contracting your muscles, the heat produced elsewhere in your body is insufficient to keep you warm in a cold place. In these circumstances your muscles contract involuntarily (you "shiver"), and heat is thereby produced to restore and maintain normal body temperature.

The problem of how a muscle can exert a pull has been attacked enthusiastically by physiologists and biochemists for many years, and great gains have been made in our understanding of the process. However, some of the chemical and physical events involved in muscle contraction are still a matter of conjecture, particularly in the involuntarily controlled muscles.

About 80% of muscle is water; the remainder is mostly protein. Small amounts of fat and glycogen, as well as two phosphorus-containing substances, **phosphocreatine** and adenosine triphosphate (ATP), are also present. The actual contractile part of a muscle fiber is composed of protein chains, which shorten by a sliding together of their parts. Two proteins, **myosin** and **actin,** have been extracted from muscle, neither of which is capable of contracting alone. When they are combined to form a thread of actomyosin, and potassium, calcium, and adenosine triphosphate are added, the thread undergoes contraction.

This demonstration of contraction in a test tube, made by the American biochemist Albert Szent-Gyorgyi, was one of the most exciting discoveries made in biochemistry, for it showed how the chemical energy of ATP could be converted into the mechanical energy of muscle contraction. But of course this was just a start. It left unanswered the questions of just how the actin and myosin interacted in the living cell, how the ATP transferred its energy to them, what molecular changes might take place, how the energy was obtained, and how the entire process was triggered. Most of these details are now known, as we will see.

How Muscle Obtains Its Energy

Skeletal muscle contains considerable glycogen. Recall that glycogen is a polymer of glucose, one of the basic fuels of the body. When a muscle cell is allowed to contract in an atmosphere containing no oxygen, glycogen disappears and lactate appears in a chemically equivalent amount. The obvious inference is that the glycogen is converted to lactate and that the energy difference between the two compounds is what powers the contraction of the muscle cell. However, if oxygen is present lactate does not accumulate; instead, carbon dioxide is produced. Investigation has disclosed that, as we might expect, lactate is produced by glycolysis in the absence of oxygen. When oxygen is added this lactate disappears. One fifth of it is oxidized to carbon dioxide and water; the energy liberated is used to drive a series of endergonic reactions that convert the remainder of the lactate to glycogen once again.

Even though the muscle must function anaerobically at times, lactate is easily reconverted to glucose. If oxygen is consistently scarce, the muscle tolerates the accumulation of lactate for a time. The resting muscle cell relies principally upon beta oxidation of fatty acids to power its maintenance metabolism.

During heavy exercise, the oxygen available to the muscle cells is not sufficient to support the aerobic respiration necessary to provide the large amounts of ATP needed for muscle contraction. At this time glycolysis provides a significant amount of the needed ATP. Even though the ATP yield per molecule of glucose is far less, glycolysis is a much more rapid source of energy than any aerobic pathway in muscle. However, after strenuous exertion, a runner might have to pant and puff for some time after a 100-yard dash to supply enough oxygen to dispose of all the excess blood lactate that the exertion has produced. The sensitivity of the brain's respiratory center to reduced blood pH produced by both lactate and carbon dioxide will ensure the necessary rapid and deep respiration. In all, it is very much as if vigorous muscular activity incurs an **oxygen debt,** which must subsequently be repaid.

An odd fact discovered some years ago shows that muscle is metabolically distinctive in another way: Researchers treated muscle with the chemical iodoacetate, which is a specific poison affecting the reactions of glycolysis in all cells. Despite this, they discovered that the poisoned muscle was still capable of some contraction, though nowhere near as much as normal muscle would be. That indicated that though glycolysis was the ultimate source of energy for muscular contraction, there must be some way in which energy from glycolysis was temporarily stored before it was transferred to the ATP that was the immediate power source of contraction. Further work resulted in the discovery of phosphocreatine, a kind of chemical storage battery that normally contains enough energy for some 60 to 70 contractions. Glycolysis, in effect, charges the phosphocreatine battery; the ATP consumption of contraction discharges it.

In sum, we now know that muscular contraction involves the following chemical reactions:

1. ATP $\rightleftharpoons$ Inorganic phosphate + ADP + Energy (used in actual contraction)
2. Phosphocreatine + ADP $\rightleftharpoons$ Creatine + ATP
3. Glycogen $\rightleftharpoons$ Intermediates $\rightleftharpoons$ Lactate + Energy
4. Part of lactate + $O_2 \longrightarrow CO_2 + H_2O$ + Energy ($\sim$P, used in resynthesis of ATP and phosphocreatine)

Fatigue

A muscle that has contracted many times, exhausted its stores of organic phosphate and glycogen, and accumulated lactate is unable to contract further and is said to be **fatigued.** Fatigue is induced in part by this accumulation of lactate, although fatigue is felt before the muscle reaches the exhausted condition. The reason is that the junction between a muscle cell and its nerve fiber—the **neuromuscular junction**—is fatigued before the muscle itself.

What Makes Muscles Move?

A striking feature of skeletal muscle cells is its stripes, or **striations,** alternating light and dark bands that can be seen under both light and electron microscopes (Fig. 32–15). These define the contractile units of the muscle cell. Each complete set of bands is known as a **sarcomere;** it is all the sarcomeres contracting together that produces the contraction of the muscle cell as a whole.

Electron micrographs show that muscle cells are composed of even smaller cylindrical fibers, the **myofibrils,** which in turn are made of longitudinal filaments, the **myofilaments.** These are of two kinds: **thick filaments** (10 nm thick and 1.5 nm long) and **thin filaments** (5 nm thick and 2 nm long). The *thick filaments consist of* **myosin** *and the thin filaments consist of* **actin.** The actin and myosin filaments are arranged in such a fashion that, seen in cross section, each myosin filament is surrounded by six actin filaments. The filaments form an interlocking series of hexagonal prisms all across the muscle cell.

The banding pattern of skeletal muscle is described by a series of standard names (Figs. 32–15 and 32–16). Each sarcomere consists of an **A band** bounded on

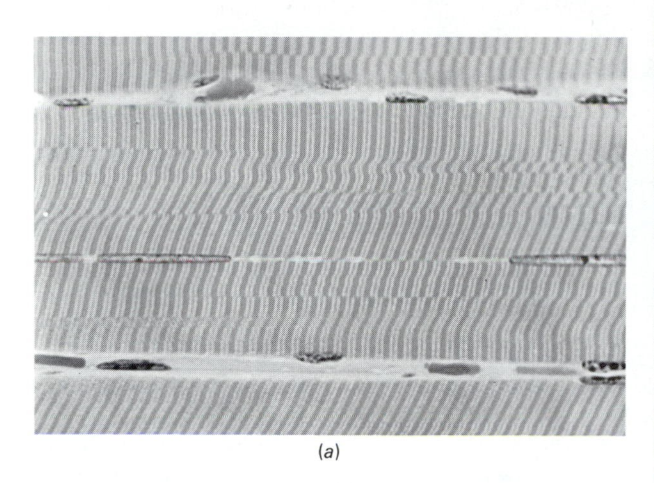

(a)

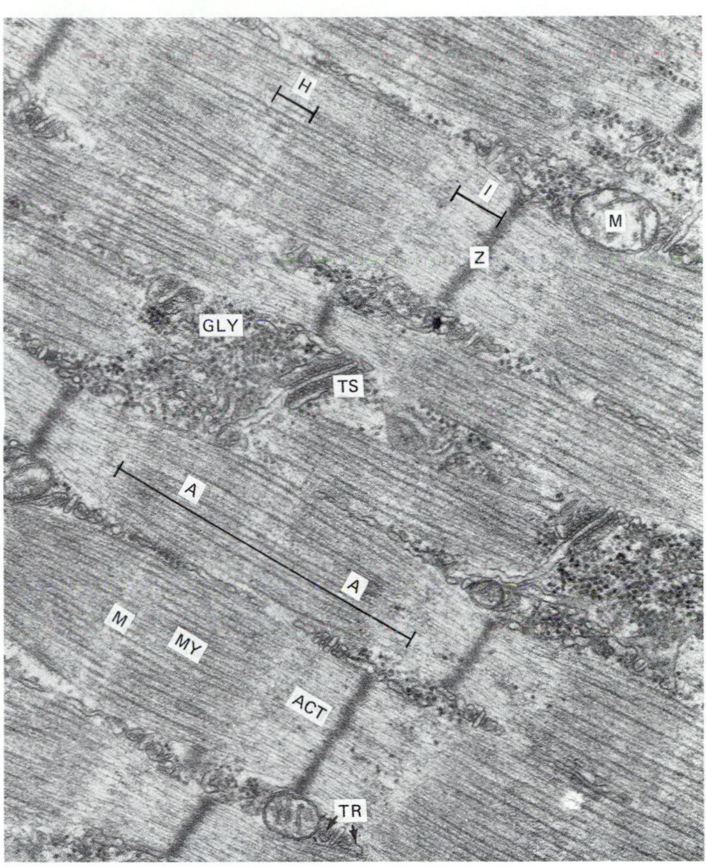

(b)

Figure 32–15 Human skeletal muscle. (*a*) Light photomicrograph showing striations (approximately ×200). (*b*) Electron micrograph (approximately ×30,000); note that striations persist at this much higher magnification. Black line indicates 1 μm. *GLY,* glycogen; *MY,* myosin filaments; *ACT,* actin filaments; *M,* mitochondria; *TS,* transverse tubule or T-system; *A, H, I,* and *Z,* the zones and bands in the muscle tissue. ((*a*), Carolina Biological Supply Co.; (*b*), courtesy of Dr. Lyle C. Dearden.)

each side by an **I band** and separated from the adjacent sarcomere by a **Z line,** a thin, dense line through the center of the I band. The central portion of the A band is somewhat less dense and is called the **H zone.** Electron micrographs reveal that the thick primary myosin filaments are found only in the A band and that the I band contains only thin secondary actin filaments. The actin filaments, however, are not limited to the I band but extend for some distance into the A band, interdigitating with the myosin filaments and overlapping them. Thus, at either end of the A band are both myosin and actin filaments interdigitating, but the central part of the A band (the H zone) contains only myosin filaments. The actin filaments are smooth, but the myosin filaments have minute spines every 6 to 7 nm along their length that project toward the adjacent filament. These spines look like bridges connecting the two sets of filaments and are called **cross-bridges.**

Since there are six actin filaments arranged hexagonally around the myosin filament, the sets of cross-bridges are repeated six times around the circumference of each myosin filament. Actin filaments do not have projections; each filament is composed of two long actin molecules wrapped helically about each other, with a complete turn every 70 nm. Every 40 nm along the actin filament there are reactive sites that interact with the ends of the myosin cross-bridges to provide the force to draw the actin filaments along the myosin filaments.

As these matters first began to be understood by muscle physiologists, the question began to emerge of *how* the actin and myosin filaments were able to pull themselves past one another. It was now obvious that the early experiment by Szent-Gyorgyi, though exciting, did not cast as much light on the matter as was originally supposed. Actin and myosin do not form a stable chemical compound with one another in the living cell, but they *are* associated chemically. It now seems almost certain that this association is formed by the cross-bridges (Figs. 32–17 and 32–18). These are now known to be extensions of the myosin molecules that attach themselves to some kind of active sites on the actin filaments. The cross-bridges also function as ATPase enzymes, and the breakdown of ATP which they catalyze powers muscular contraction. In brief, it appears that each cross-bridge attaches to an active site. Then it flexes, somewhat as a leg might. This action moves the two

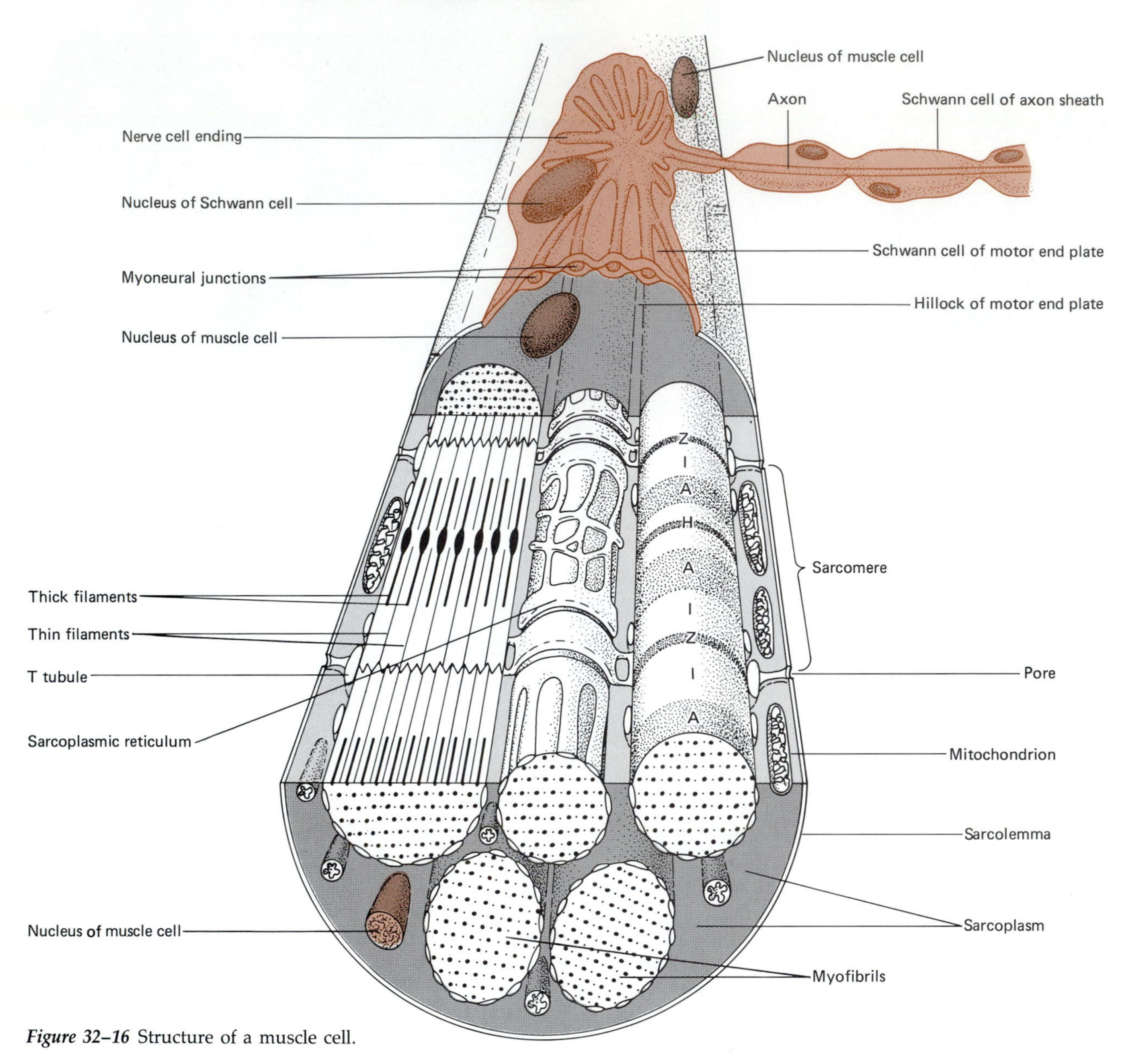

Figure 32–16 Structure of a muscle cell.

filaments past one another. To get an idea of the movements involved, imagine a centipede walking on a pencil. As all its legs take one step, it and the pencil move past one another. Then each leg would have to release its hold temporarily and reach for the next step. In this manner, presumably, each cross-bridge physically flexes, releases its active site, and reaches for the next one that has just been vacated by the cross-bridge ahead. The orientation of the molecules of the thin filament determines the direction of stepping.

Since ATP is required not only to move the cross-bridges but to release their grip on the thin filament, in death, when ATP production ceases, the cross-bridges become permanently fixed to the thin filaments. The muscular rigidity this produces is known as **rigor mortis,** a condition of stiffness, not true contraction.

Triggering Contraction

Not only is ATP necessary for the release of the connection between a cross-bridge and an active site, it also induces a change in the angles of weak chemical bonds in the myosin molecule—called a **conformational change**—that produces the necessary mechanical flexing. It is the repeated flexing of these trillions of submicroscopic legs that allows us to type, speak, or roller-skate.

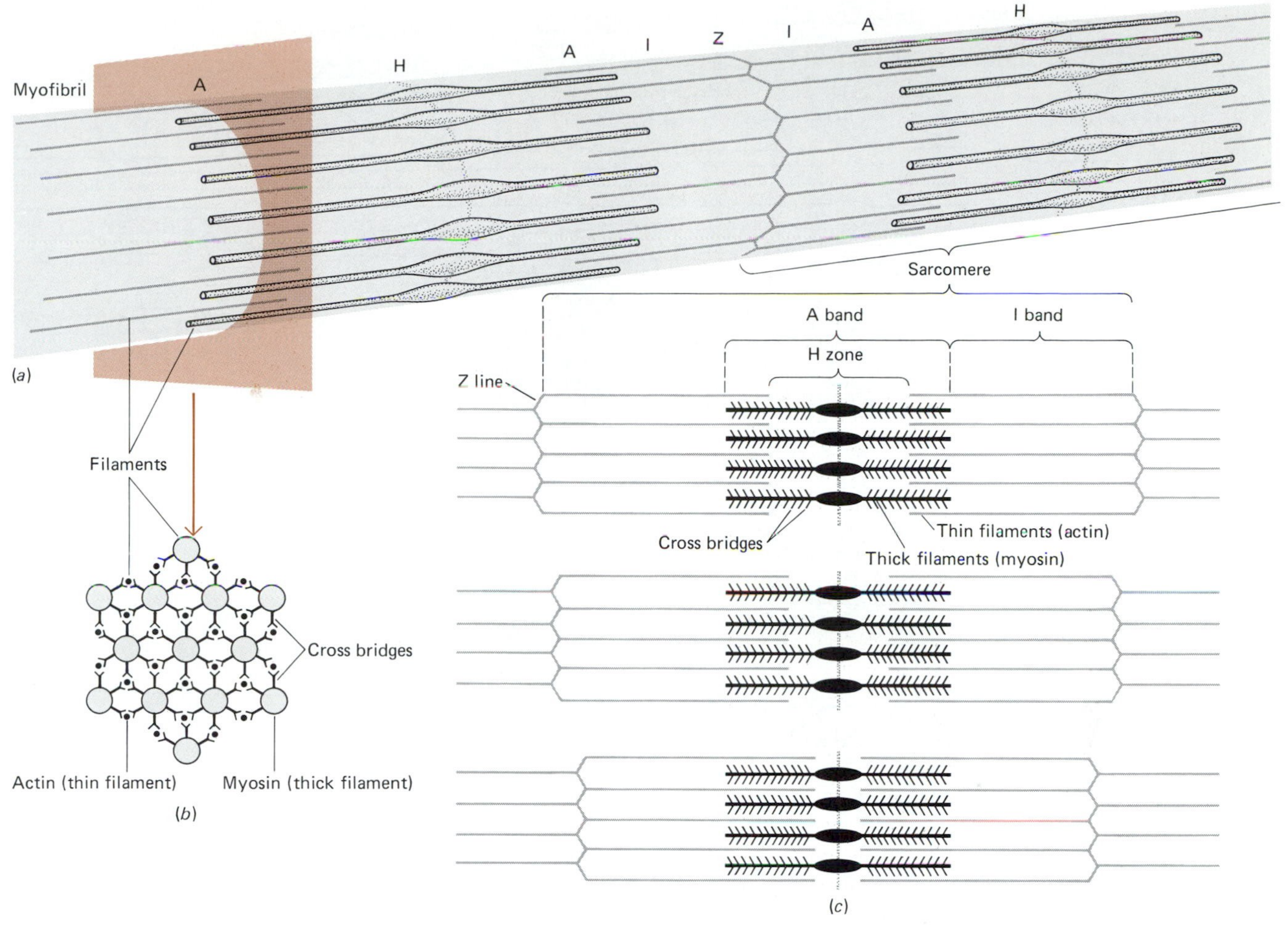

Figure 32–17 Functional arrangement of actin and myosin in a muscle cell. (*a*) A myofibril stripped of the accompanying membranes. The Z lines mark the ends of the sarcomeres. (*b*) Cross section of myofibril shown in (*a*). (*c*) Filaments slide past each other during contraction. Note that the thick filaments run the width of the A band; thin filaments run through the I band and overlap the thick filaments within the A band. The H band, which lies in the center of the A, is an area of thick filaments only. *Top,* The myofibril is relaxed. *Middle,* The filaments have slid toward each other, increasing the amount of overlap and shortening the sarcomere. *Bottom,* Maximum contraction has occurred; the sarcomere has shortened considerably.

But what triggers the contraction of the muscle cell? Apparently that question would better be put negatively: What prevents the actin and myosin filaments from interacting when the muscle is not contracting? The answer involves a third kind of protein complex, the **tropomyosin system.** Molecules of tropomyosin occupy the active sites and interfere with the attachment of the cross-bridges to the actin filaments. Muscle contracts when calcium is released into the cytoplasm of the muscle cell, because calcium induces the tropomyosin to leave the active sites so that the cross-bridges can attach to them (Fig. 32–18). Normally, calcium occurs in the muscle cell only within the vesicles of a system of membranes called the **sarcoplasmic reticulum** (SR). This is considered to be the muscle cell equivalent of the smooth ER. The SR is associated with a system of transverse tubules, the T-tubule system or **T-system.** The T-system tubules are lined with a unit membrane that is continuous with the muscle cell membrane or **sarcolemma.** From the outside, the entrances to the T-system appear to be so many pores leading into tunnels, with the sarcolemma seeming to continue into the tubules without a break. The sarcolemma is depolarized by the action of **acetylcholine,** the neurotransmitter released by the axon of a motor neuron (see Focus on Depolarization). That same electrochemical impulse spreads into the cellular interior via the membranous linings of the T-system tubules. In some manner not yet discovered, this causes the release of calcium from the SR, and contraction ensues. At the end of contraction calcium is actively pumped back into the SR by an ATP-powered mechanism similar to that of the sodium pump.

SMOOTH AND CARDIAC MUSCLE

The basic mechanism of the contraction of cardiac and smooth muscles is probably very similar to the sliding filament mechanism that operates in skeletal muscle. Although neither is under conscious control, both cardiac and smooth muscles

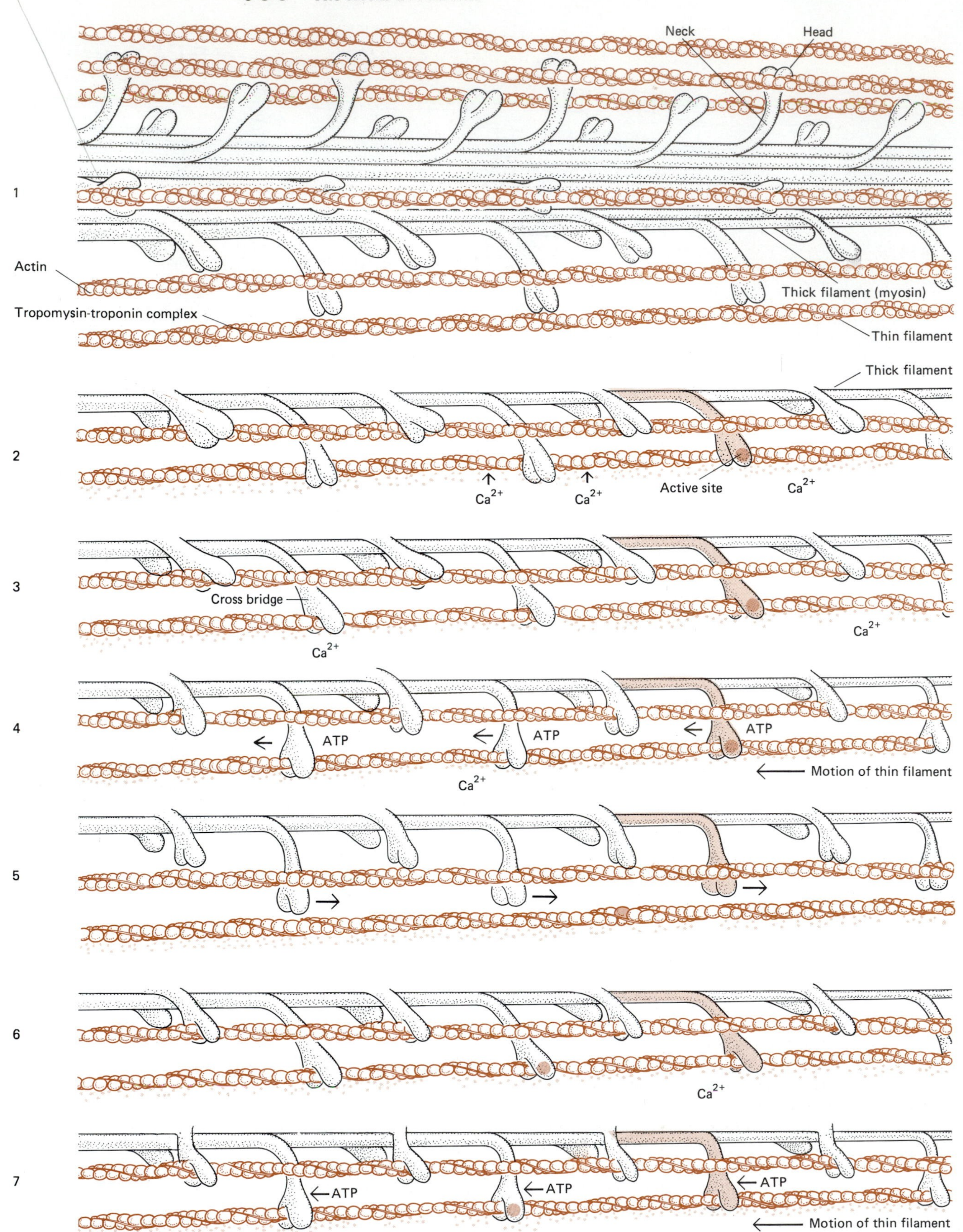

Figure 32–18 Proposed mechanism for movement of the thin and thick filaments past each other by cross-bridges in muscle contraction.

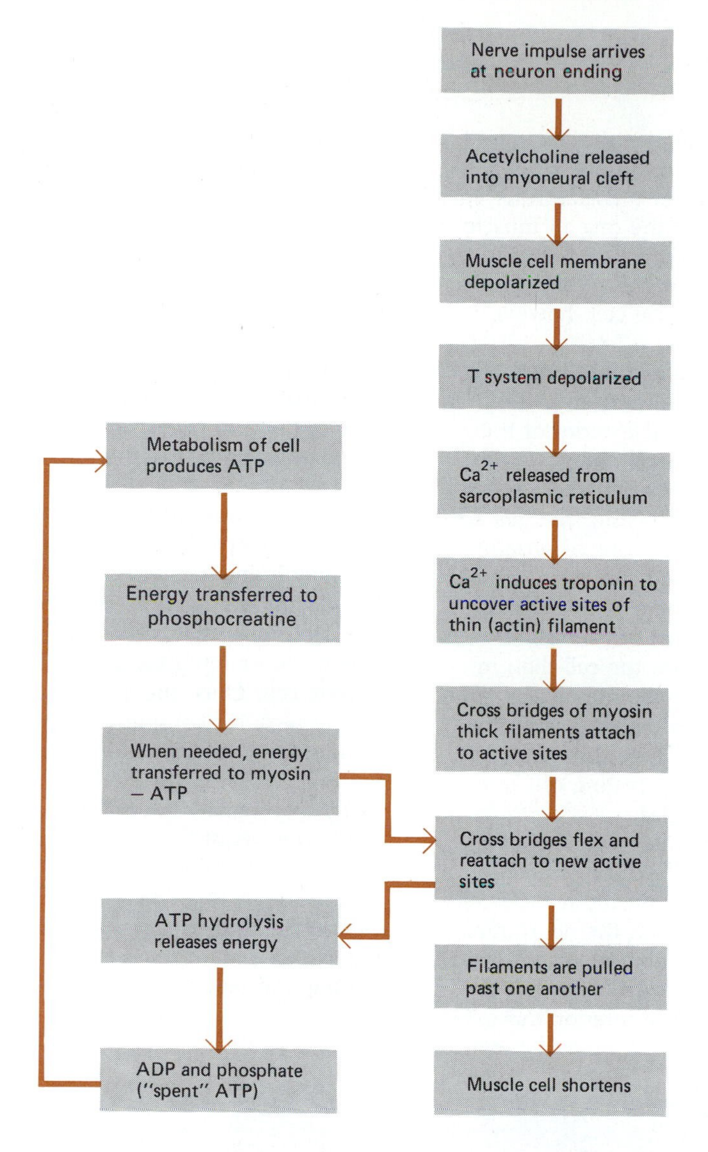

Figure 32–19 Summary of the major events of muscle contraction.

contain actin and myosin, and the contraction process involves the hydrolysis of ATP and an interaction between actin and myosin initiated by calcium ions. However, the arrangement of actin and myosin filaments is nowhere near as precise in smooth muscle as in skeletal, so smooth muscle exhibits no cross-striations.

Smooth muscle exhibits wide variations in tonus; it may remain almost relaxed or highly contracted for long periods of time. Despite the fact that typically, smooth muscle in human beings is not under direct conscious control, it can certainly be influenced consciously, as seen in the deliberate inhibition of the normal emptying mechanisms of bladder and bowel that every baby learns upon graduation from the nursery. More specialized training can also permit a person to exercise considerable control over such smooth muscle functions as maintenance of blood pressure (Fig. 32–20). Smooth muscle also performs such diverse tasks as the peristaltic transportation of food through the digestive tract, the erection of our hair when we bristle in rage (or get "goose bumps" of fear), and the contraction of the uterus in childbirth.

Smooth muscle is far less extensively innervated than skeletal. Smooth muscle also contracts in response to such hormones as epinephrine, and also in response to stretching, as occurs in the urinary bladder. Smooth muscle cells apparently transfer impulses that stimulate contraction among themselves by numerous gap junctions.

Each beat of the heart represents a simple twitch. **Cardiac muscle** (which is unique to the vertebrates) has a long refractory period in which it will not contract. Consequently it is unable to contract tetanically, since one twitch cannot follow another quickly enough to maintain a contracted state.

FOCUS ON

Depolarization

Depolarization is the term for the electrical changes that transmit information along a cell membrane. Depolarization occurs in all types of muscle, in nerve cells, and perhaps in all tissue. In the case of muscle cells it transmits the signal for contraction from the myoneural junction (the junction of nerve fiber and muscle cell) to all parts of the muscle cell, however long that cell may be.

Although depolarization will be discussed in more detail in Chapter 39, it may be summarized as follows: When resting, the interior and exterior of the cell membrane bear different electrical charges, with the exterior being positively charged and the interior being negatively charged. The contrasting charges are accounted for by differing concentrations of certain ions, themselves electrically charged, on the two sides of the membrane. For example, there is considerably more potassium inside the cell than outside it, and there is much more sodium outside the cell than inside it. These ionic differences are maintained by special proteins in the sarcolemma that actively pump ions into or out of the cell. The charge difference between the interior and exterior of a resting cell sarcolemma is called the **resting potential.**

When the motor nerve cell secretes a neurotransmitter, acetylcholine, this substance diffuses across the tiny space between the nerve and the muscle cells known as the **myoneural cleft.** When the neurotransmitter reaches the muscle cell membrane (sarcolemma) there is a change in membrane permeability that produces a charge reversal between the interior and exterior of the membrane, so that temporarily the outside becomes *negatively* charged and the inside becomes *positively* charged. This charge disturbance is known as **depolarization.** If the muscle cell is small, its entire cell membrane may become depolarized at one time, but it is more usual for a wave of depolarization to sweep over the cell, followed immediately by restoration of the resting potential. Thus, at any one time only a moving zone of the cell membrane is actually depolarized. The depolarization of the muscle cell is not confined to the sarcolemma but passes inward to the membrane of the T-tubules and then causes the cell to contract.

Depolarization, at least in skeletal muscle cells, is begun by **acetylcholine receptors.** These receptors are cell membrane–protein complexes on the surface of the plasma membrane. They combine with acetylcholine molecules that diffuse across the minute gap between the membranes of the nerve ending and the muscle cell. Once the acetylcholine has been bound by the receptor complex, the protein composing it undergoes a conformational change, causing a channel to open within it. Sodium and potassium, normally maintained in very different concentrations on the two sides of the sarcolemma by the **sodium pump** (and possibly by a separate potassium pump), now rush through this channel by diffusion. This massive electrolyte shift triggers the initial depolarization of the sarcolemma, which immediately becomes self-sustaining and sweeps down the length of the muscle cell.

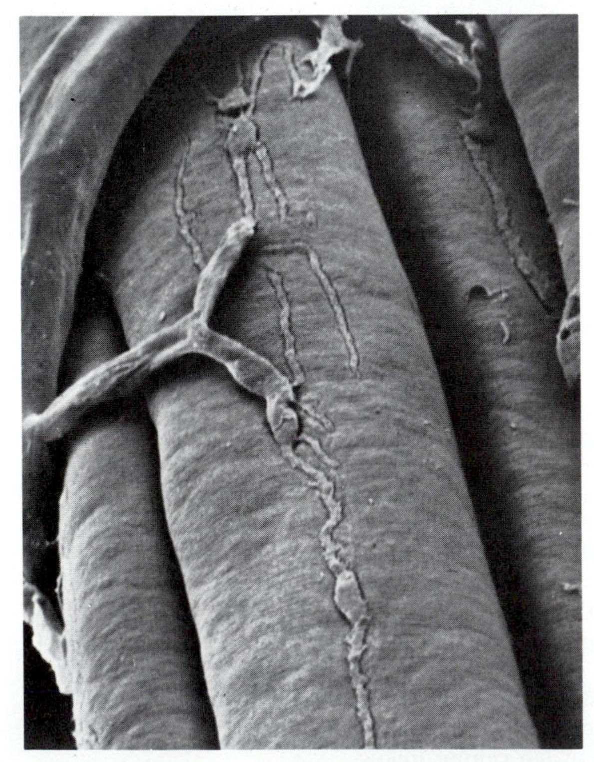
(*a*)

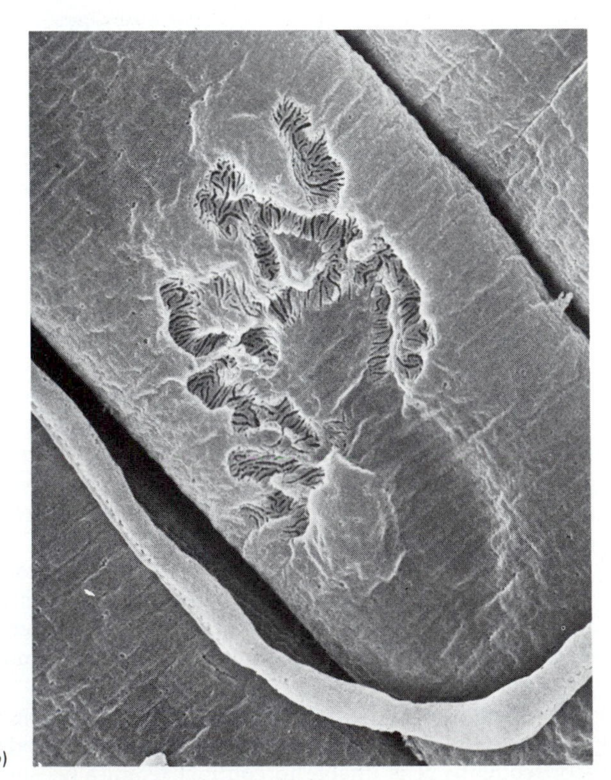
(*b*)

In these remarkable scanning electron micrographs,* a number of **neuromuscular junctions** are visible in (*a*); one of these has been stripped away in (*b*) to disclose the features of the muscle side of the neuromuscular junction. Notice that the sarcolemma of this region is elaborately sculptured to form a multitude of tiny ridges and valleys. The acetylcholine receptors are located at the crests of the ridges (×2000 and ×5700, respectively).

*Taken by Yasuo Uehara, Junzo Desaki, and Takashi Fukiwara of the Ehime University School of Medicine in Japan. (From Desaki, J., and Uehara, Y.: "The overall morphology of neuromuscular junctions as revealed by scanning electron microscopy." *Journal of Neurocytology* 10:101–110, 1981.)

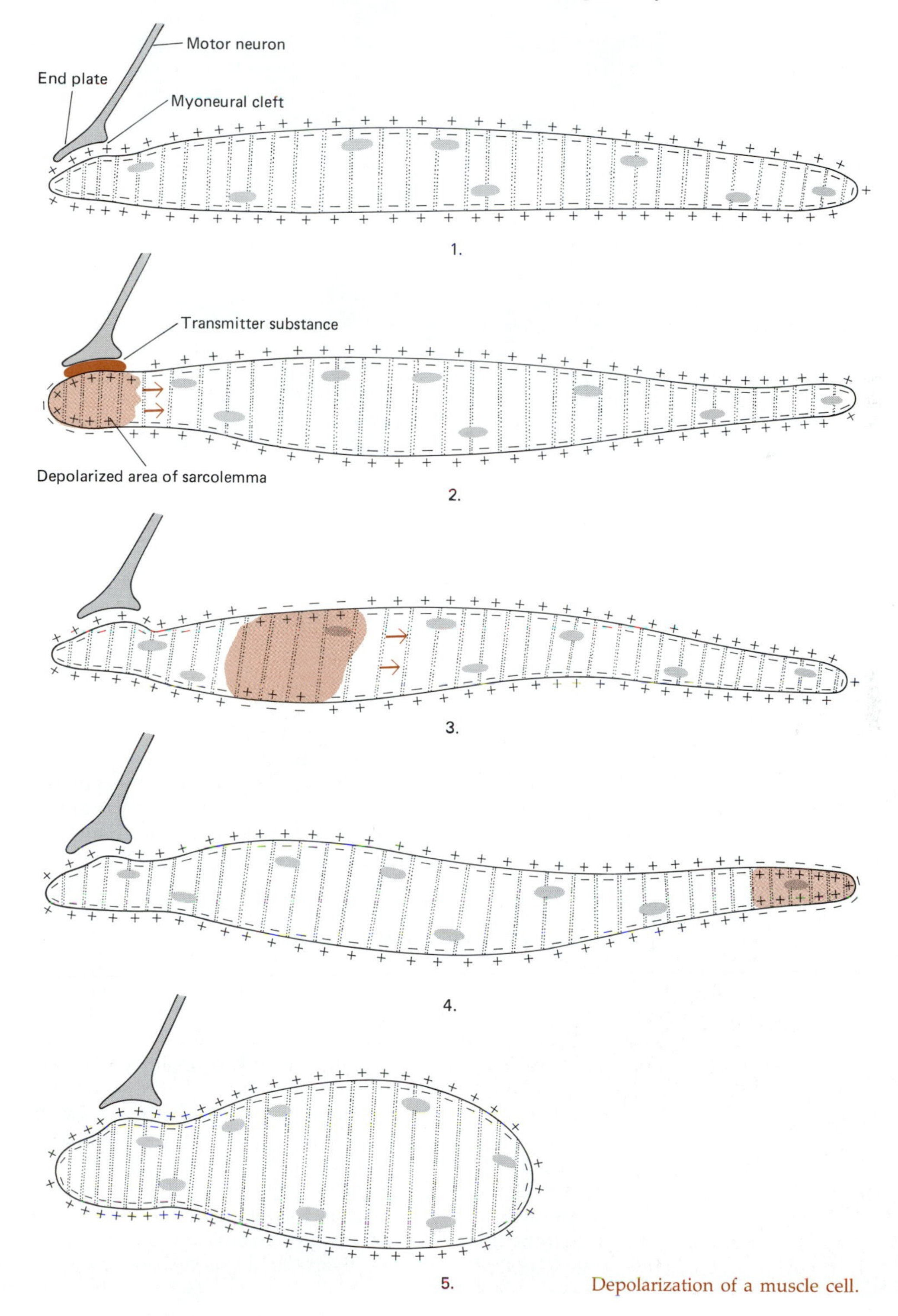

Depolarization of a muscle cell.

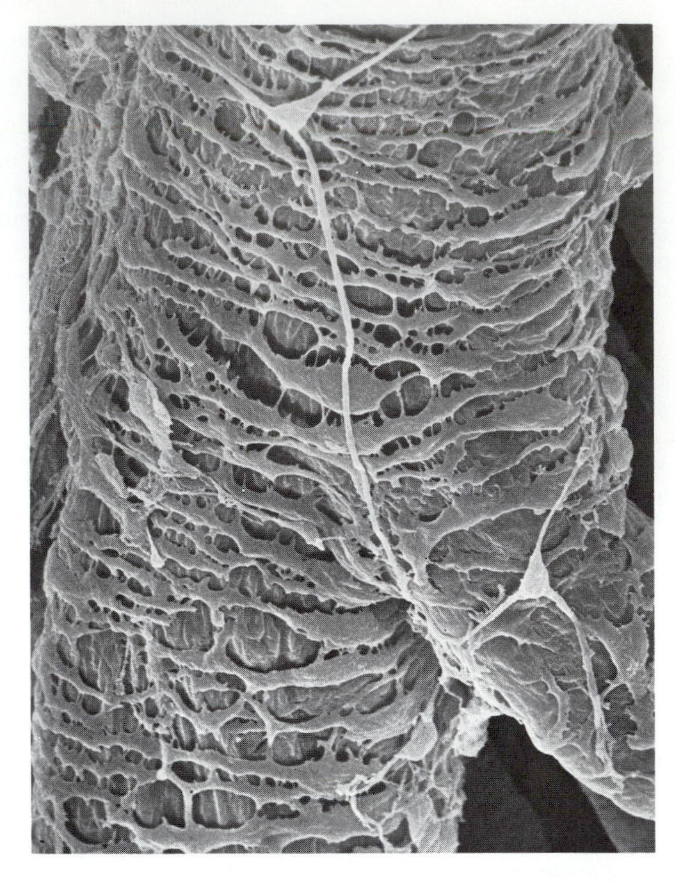

Figure 32–20 Scanning electron micrograph of the autonomic innervation of the smooth muscle cells of a small vein (×900). The thin white fibers are autonomic nerve axons; the triangular bodies where these branch are Schwann cells. The neurons are arranged to stimulate large numbers of smooth muscle cells (arranged in rings around the vein) at the same time. (From Desaki, J.: "Vascular autonomic plexuses and skeletal neuromuscular junctions: A scanning electron microscopic study." *Biomedical Research* (Supplement), 139–143 1981).

A unique feature of cardiac muscle is its inherent rhythmicity; it contracts at a rate of about 72 times per minute even when denervated and removed from the body. Cardiac muscle discharges its membrane potential each time this has built up to a certain level. After each impulse has passed, the membrane becomes repolarized but then suddenly becomes permeable again, initiating the transmission of the next action potential.

A MUSCULAR ZOO

The humblest of animals has the ability to move, and usually to change shape. It has been shown that even amoebas and similar protists have actin (and probably myosin) strands in their cytoplasm that squeeze their viscid body substance in such a way as to produce the pseudopods that these simple organisms use for locomotion. Similar strands have been discovered in motile cells of higher organisms. For example, they occur in undifferentiated embryonic cells that must crawl into position during **morphogenesis** (the production of the basic form of an organism.) Even in the adult body, white blood cells creep about throughout tissues by amoeboid motion, using actin and myosin filaments.

The simple cnidarian *Hydra* has only two layers of cells, both consisting of epithelial tissue. Yet this tissue is muscular and nervous as well. Such multipurpose tissue sometimes occurs in higher organisms as well, for example, in the ducts of most sweat glands. Muscle does occur as a differentiated tissue, however, in flatworms and other wormlike animals, where it is organized into definite layers of antagonistic or specialized function.

With the conspicuous exception of arthropods, invertebrates are characterized by the presence of much more smooth muscle than striated muscle. Smooth muscle contracts very slowly compared with striated muscle, but tends to contract to a greater extent and so is very suitable for substantial changes in body shape—which is essentially what the hydrostatic skeleton of the simpler invertebrates accomplishes. Smooth muscle also is highly suitable for long-maintained contraction. This property is exploited by the bivalve mollusks such as clams.[1] Anyone who has

[1] It is striated muscle, however, that enables the clam to close its shell rapidly.

shucked a clam knows how hard it is to get the halves of the shell apart. But this neat little package of meat can be opened easily by inserting a knife between the shells and cutting the great muscles that hold the two halves together.

The only invertebrate phylum that possesses striated muscle to any great extent is the arthropods. In some cases arthropod striated muscle may be more highly differentiated even than that of vertebrates. This is seen best in the **flight muscles** of insects.

In most flying insects the flight muscles are attached not directly to the wings but to the flexible portions of the exoskeleton that articulate with the wings. In such insects, each contraction of the muscles produces a dimpling of the exoskeleton in association with a downstroke, and, depending on the exact arrangement of the muscles, sometimes on the upstroke as well. When the dimple springs back into its resting position, the muscles attached to it are stretched. The stretching initiates another contraction immediately, and the cycle is repeated. The deformation of the cuticle is transmitted as a force to the wings and they beat—so fast that we may perceive the sound as a musical note. In the common blowfly, for instance, the wings may beat at a frequency of 120 cycles *per second*. Yet in that same blowfly, the neurons that innervate those furiously contracting flight muscles will be delivering impulses to them at the astonishingly low frequency of 3 per second. It seems very likely that the mechanical properties of the musculoskeletal arrangement are what provides the stimuli for contraction, by stretching the muscle fibers at the resonant frequency of the system. But the nerve impulses are needed to maintain it.

The metabolic rate of insect flight muscle in action is very high. Accordingly it has more mitochondria than any known variety of muscle, and is elaborately infiltrated with tiny air-filled tracheae that carry oxygen directly to each cell (Fig. 32–21). Many insects have special adaptations to rid the body of the excess heat produced by the flight muscles. The rapidly flying hawk moth, for example, has what amounts to a radiator in its abdomen, a great blood vessel that carries heat from the thorax where it is generated and emits it into the cool of the night.

Flight muscles must be kept at operating temperature if they are to function. You have probably noticed the constant twitching of the wings of such insects as wasps even when they are crawling instead of flying. Probably this behavior is

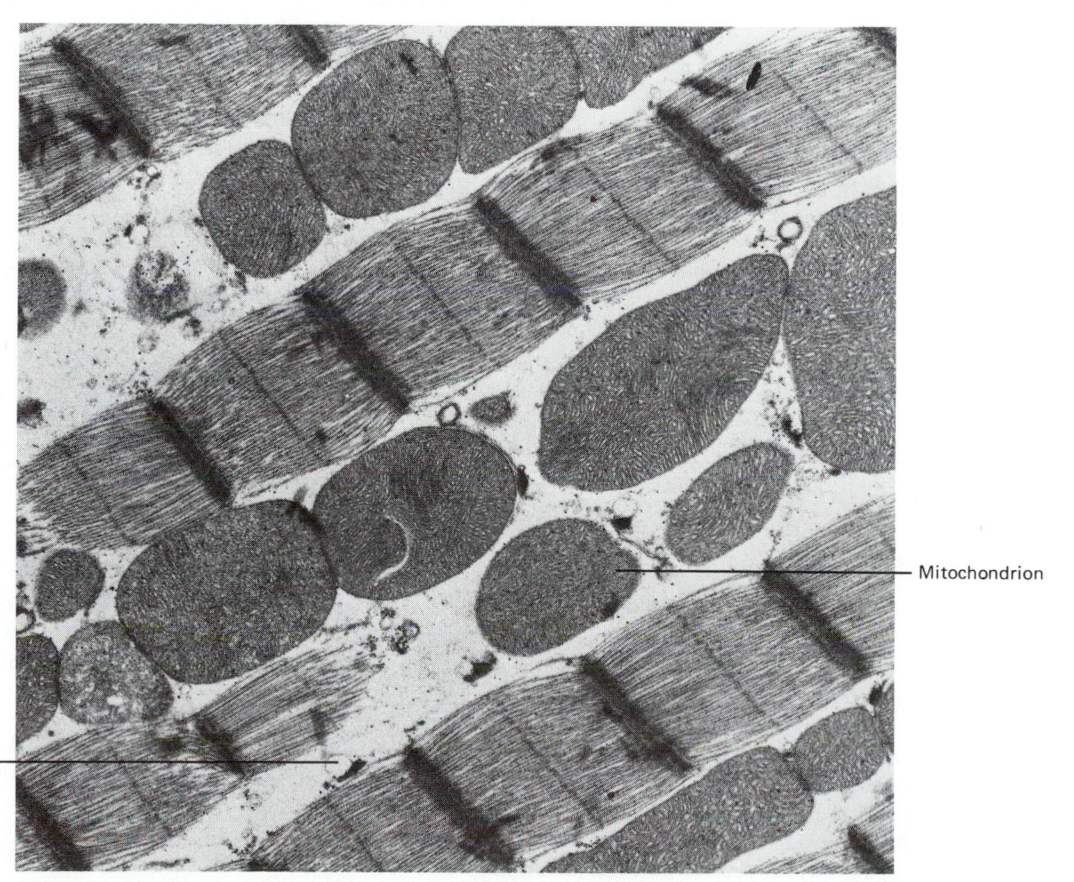

Figure 32–21 The ultimate muscle. Insect flight muscle, such as that of the bumblebee shown here, may be the most powerful found in any living thing. Oxygen is brought directly to the muscle by the tracheal tubes, which convey air into the muscle cell itself. Note the prominent striations and the tremendously convoluted internal membranes of the numerous mitochondria. (From Heinrich, B.: *Bumblebee Economics*. Boston, Harvard University Press, 1978. Electron micrograph by Mary Ashton. Courtesy of Dr. Heinrich, used by permission.)

necessary to keep the temperature of the flight muscles high enough for instant combat readiness. You may also have observed that the bodies of many moths are quite furry; this "fur," more properly called **pilus,** serves the same function as fur in a mammal—to conserve body heat: When the moth awakens and prepares for flight it shivers its flight muscles at a low frequency to warm them up, constricting its abdominal blood vessel to keep the heat in its thorax. Gradually the frequency of the shivering increases until at a critical moment, when enough heat has been generated, it is able to spread its wings and hum off into the darkness.

SUMMARY

I. The main functions of animal skeletons are to transmit muscular force and to support and protect the body. Most animals possess an exoskeleton, although soft-bodied organisms have a hydrostatic skeleton. Vertebrates and echinoderms have an endoskeleton.
  A. The arthropod exoskeleton is composed mainly of chitin. It is molted periodically and a new exoskeleton is secreted by the cells of the epidermis.
  B. The vertebrate skeleton is usually composed of bone or cartilage. It is divided into axial and appendicular portions.

II. In vertebrates the skin consists of a tough connective tissue dermis covered by a multilayered epithelial epidermis and the derivatives of that epidermis such as hair and feathers. (Fishes have scales of dermal origin.)

III. Muscle, the motor tissue of the body, has the ability to contract. It moves body parts by pulling on them. Complex body motions result from the way in which muscles are attached to bones, and usually from their multiple cooperative action.
  A. The smallest functional unit of muscle is the muscle cell. In skeletal muscle this cell is striated, reflecting the interdigitations of its thin (actin) and thick (myosin) filaments. These interact by means of their cross-bridges, which exert a pull, resulting in muscle contraction.
  B. Calcium acts to uncover the active sites of the thin filaments that permit cross-bridge action. The cross-bridges are powered by ATP. Energy is stored in phosphocreatine between contractions. Depolarization of the muscle cell membrane initiates contraction by spreading inward via the membranes of the T-tubule system. This leads to the release of calcium and then to muscular contraction.
  C. There are three types of muscle: skeletal, smooth, and cardiac. The last two are largely involuntary in their action, and though cardiac muscle resembles skeletal muscle functionally, little is known about the mechanism of smooth muscle contraction.

POST-TEST

1. A ________ skeleton has the function of transmitting muscular force, and does not function substantially to produce body form or to bestow protection.

2. The cells of the stratum ________ of the vertebrate skin are dead and almost waterproof.

3. Since an exoskeleton tends to limit size, arthropods must ________ from time to time in order to grow.

4. The foot bones are considered to be part of the ________ skeleton.

5. Almost all voluntary muscles have at least two points of attachment to the skeleton: the less movable ________, and the more movable ________.

6. A ________ muscle decreases the size of a body opening or duct.

7. A particular muscle cell either contracts ________ or not at all.

8. Unscramble this list of the events of muscle contraction into the correct sequence:
   a. calcium release
   b. T-system depolarization
   c. acetylcholine release
   d. nerve impulse
   e. tropomyosin uncovers the active sites of the thin filaments.
   f. cross-bridges flex
   g. cross-bridges release active sites

9. Phosphocreatine serves as an ________ ________ device in the muscle cell.

10. ________ is a chemical storage battery in muscle cells that transfers energy to ATP.

REVIEW QUESTIONS

1. What are the functions of the human skeleton?
2. What is a hydrostatic skeleton? Which of the general skeletal functions does it perform?
3. What are the disadvantages of an exoskeleton?
4. Contrast smooth and skeletal muscle.
5. Outline the sequence of events that causes a muscle cell to contract, beginning with the stimulation of its nerve and including cross-bridge action.
6. Should the dermis of the skin be considered part of the skeleton? If so, is it an endoskeleton or an exoskeleton?
7. Many plant cells maintain their shape by a combination of high internal pressure and support from the cellulose cell wall. What kind of skeleton might such a cell be said to possess?
8. Refer back to Chapter 6. How does cartilage differ from bone? What might be the advantages of having cartilage rather than bone as a skeletal material in the nose, external ear, and air ducts of the lung?

33
Processing Food

OUTLINE

I. Modes of nutrition
  A. Adaptations of herbivores
  B. Adaptations of carnivores
  C. Omnivores
  D. Symbionts
II. Invertebrate digestive systems
III. Vertebrate digestive systems
  A. Wall of the digestive tract
  B. Inside the mouth
    1. The teeth
    2. The salivary glands
  C. Through the pharynx and esophagus
  D. In the stomach
  E. Inside the small intestine
  F. The pancreas
  G. The liver
  H. Enzymatic digestion
    1. Carbohydrate digestion
    2. Protein digestion
    3. Lipid digestion
  I. Control of digestive juice secretion
  J. Absorption
  K. Through the large intestine

LEARNING OBJECTIVES

After you have read this chapter you should be able to:

1. Describe, in general terms, the following steps in processing food: ingestion, digestion, absorption, and elimination.
2. Compare the food habits of herbivores, carnivores, and omnivores, and compare adaptations that they possess for their particular mode of nutrition.
3. Compare the nutritional life styles of parasites, commensals, and mutualistic partners.
4. Compare the types of digestive systems found in cnidarians and flatworms with the digestive system of an earthworm.
5. Trace a bite of food through each structure of the human digestive tract, and describe the changes that take place en route.
6. Describe the four layers of the wall of the vertebrate digestive tract.
7. Describe the types of teeth found in mammals, and give their functions.
8. Draw and label a tooth, and give the function of each structure.
9. Describe the accessory digestive glands of terrestrial vertebrates, and how each promotes digestion.
10. Describe the anatomic features of the small intestine that increase its surface area, and discuss their advantages.
11. Trace the step-by-step digestion of a carbohydrate, a lipid, and a protein.
12. Draw a diagram of an intestinal villus, and label its parts.
13. Describe the absorption of glucose, amino acids, and fat.

Nutrients are the substances present in food that are needed by organisms as building blocks for growth and repair of the body, and as an energy source for running the machinery of the body. Obtaining nutrients is so important that many organisms have undergone adaptations in response to their nutritional needs and the means by which they obtain and process food. An organism's body plan and its life style are adapted to its particular mode of nutrition.

All animals are **heterotrophs,** organisms that must obtain their energy and nourishment from the organic molecules manufactured by other organisms. Because heterotrophs eat the macromolecules made by other organisms, they must break down these molecules and refashion them for their own needs. For example, humans cannot incorporate the proteins in steak directly into muscle proteins. The body must break down the steak proteins into their component amino acids and deliver these to the muscle cells, and then the cells must arrange these components into human muscle proteins, or into other structural proteins and enzymes.

After foods are selected and obtained they are **ingested,** that is, taken into the body. In a human, ingestion includes taking the food into the mouth and swallowing it. Once the food is inside the organism it must be **digested** or broken down before its nutrients can be utilized by the animal. Typically, large pieces of food are first mechanically broken down into smaller ones. Then, macromolecules are hydrolyzed by digestive enzymes into small molecules that can be absorbed and utilized.

In animals that are equipped with digestive tracts, **absorption** involves the passage of nutrients through the cells lining the digestive tube and into the blood or other body fluids. In sponges and other organisms with intracellular digestion, absorption involves merely the passage of nutrients from the phagosome (the vesicle enclosing the food) into the cytoplasm. Nutrients are then distributed to all parts of the organism and utilized for metabolic activities within each cell. Undigested and unabsorbed food is eliminated from the body, a process referred to as **egestion** in simpler forms and as **elimination** or **defecation** in more complex animals.

Modes of Nutrition

Animals are consumers, ultimately dependent upon plants for their food, energy, and oxygen. Some animals are **herbivores,** or **primary consumers,** which eat plant materials. Herbivores may be consumed by flesh-eating **carnivores,** which also may consume one another. Omnivores eat both plants and animals. Some animals are **symbionts,** organisms that form intimate nutritional relationships with members of other species.

ADAPTATIONS OF HERBIVORES

Herbivores eat only algae or plants. They may restrict their diet to a particular plant part—the leaves, for example, or perhaps the roots, seeds, or nectar. A herbivore may eat only one or a few species of plants. Many aquatic herbivores are filter feeders. As large volumes of water flow by, their filtering system traps tiny plants or protists in a mucous secretion. Cilia then sweep the mucus containing the captured food to the mouth.

Terrestrial plants contain a great deal of supporting material, including cell walls made of cellulose. Animals cannot digest cellulose, so it is somewhat of a problem for them to obtain nutrients from the plant material they eat. The adaptations utilized to exploit plant food sources are strikingly varied (Fig. 33–1). Some herbivorous insects have piercing and sucking mouthparts so that they can pierce through the tough cell walls and suck the sap or nectar within the plant cells. Other herbivores simply eat great quantities of food. Grasshoppers, locusts, elephants, and cows, for example, all spend a major part of their lives eating. Most of what they eat is not efficiently digested but moves out of the body as waste, almost unchanged. However, by eating large enough quantities, sufficient material is digested and absorbed to provide the nourishment necessary to sustain their life processes.

Many herbivores are equipped with jaws and teeth or toothlike structures for ingesting food. The teeth of herbivorous mammals include wide molars for grinding plant food, which often have enamel specially adapted to heavy wear. Herbi-

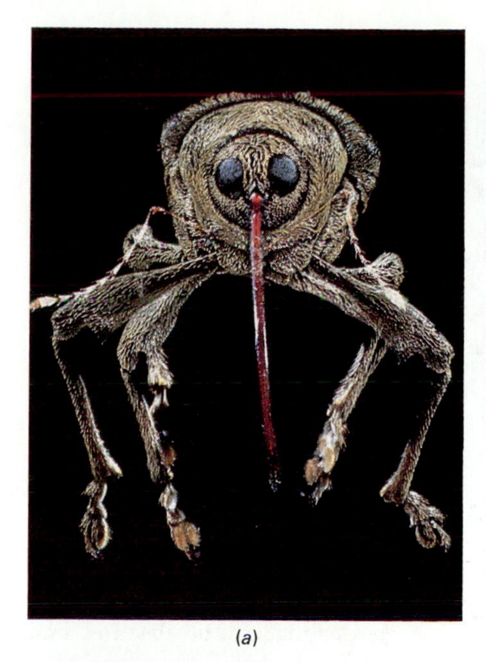

(a)

(b)

Figure 33–1 Adaptations of herbivores. (*a*) An acorn weevil. The long "snout" of this beetle is used both for feeding and to make a hole in the acorn through which an egg is deposited. When hatched, the larva feeds upon the contents of the acorn seed. All members of this insect superfamily (Curculionoidea) have elongated rostrums, with the movable mouthparts located on its very tip. (*b*) The parrot has a powerful beak for cracking nuts and seeds. Unlike most birds, the parrot can use one foot to manipulate food and feed itself. (*c*) The rhinoceros can use its horn to overturn small trees and bushes; it then eats the leaves. Members of some species use their lips to break off grass. (*d*) Camels can live on seeds, dried leaves, and desert plants. They can eat sharp cactus thorns without injury because their mouth lining is very tough. ((*a*), Darwin Dale, Photo Researchers, Inc.; (*b*), (*c*), and (*d*), courtesy of Busch Gardens.)

(c)

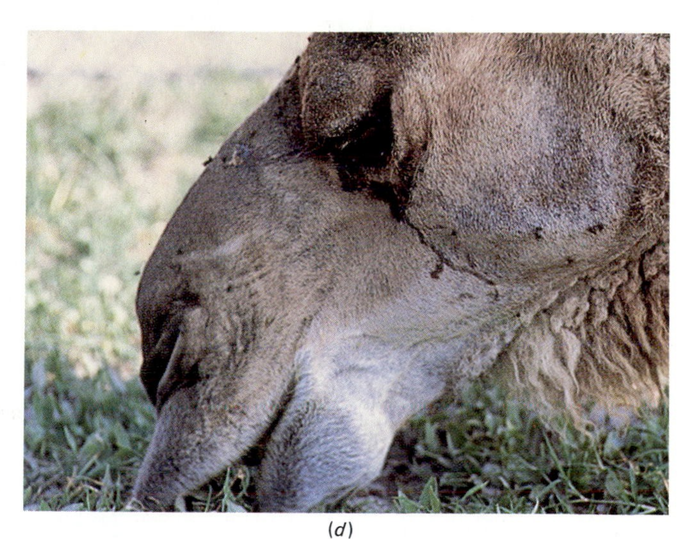

(d)

vores also have longer and more elaborate digestive tracts than carnivores, so that food can be retained for digestion for a long time.

A common and very interesting adaptation of herbivores is a symbiotic relationship with microorganisms in their digestive tracts that can digest cellulose for them. Such symbionts occur in termites, for instance, and in cows and horses. Vertebrate herbivores generally have a specialized section of the digestive tract called the **cecum** in which live the bacteria capable of digesting cellulose. The stomach of ruminant mammals (e.g., cows, sheep, and deer) is divided into chambers (Fig. 33–2). When a cow eats grass, very little chewing occurs before swallowing. The food descends into the esophagus and then into the **rumen** and **reticulum,** where it is mixed and churned. Microorganisms that inhabit these chambers secrete enzymes that can break down cellulose into simple carbohydrates. Many of these carbohydrates are then used by the bacteria themselves. The bacteria, for their part, produce acetic acid and other organic acids during their metabolism, some of which are absorbed by the cow's rumen.

Stomach muscles of the cow push the bolus of partly digested food (called a "cud") back up into the cow's mouth, where it is chewed again and then re-swallowed. Food and fluid then move from the rumen into the **omasum,** where water is absorbed. Finally, the partly digested food and the microorganisms mixed with it enter the **abomasum,** or true stomach. Like the stomach of other vertebrates, the abomasum produces digestive enzymes. Both the food and the symbiotic bacteria

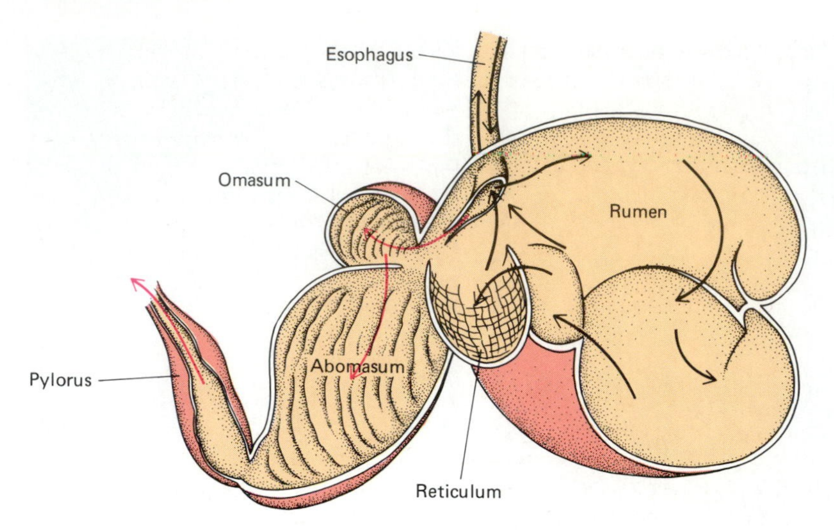

Figure 33–2 The ruminant stomach is divided into chambers. Food passes from the esophagus to the rumen and reticulum where microorganisms digest cellulose. The partly digested food is pushed back into the animal's mouth where it is re-chewed. After re-swallowing, the partly digested food moves into the omasum and finally into the abomasum, or true stomach.

are digested in the abomasum so that many of the nutrients absorbed by the bacteria are recovered by the cow. From the abomasum food moves into the intestine, where digestion continues.

ADAPTATIONS OF CARNIVORES

Carnivores eat meat, and some eat large amounts of bones as well. As predators their first problem is to find and capture their prey. Even simple invertebrate carnivores have adaptations for this purpose. For example, *Hydra* and its relatives have long tentacles equipped with stinging cells called cnidocytes. Should an organism brush by a cnidocyte, a long threadlike lasso springs forward, entangling and perhaps paralyzing the prey. The tentacles then deliver the meal into the waiting mouth. Dinner is captured without the hydra's ever having to move from its perch.

(*a*)

(*b*)

(*c*)

Figure 33–3 Adaptations of carnivores. (*a*) The long-nose butterfly fish, *Forcipiger longirostris,* has a mouth adapted for picking small worms and crustaceans from tight spots in coral reefs. (*b*) With lightning speed the Burmese python strikes at its prey and then suffocates it before consuming it whole. (*c*) With a wide field of vision and fast reflexes, the California mantis is a very able carnivore. ((*a*), Charles Seaborn; (*b*), courtesy of Mical Solomon and Trudi Segal; (*c*), P. J. Bryant, UC-Irvine/BPS.)

Vertebrate carnivores have many interesting adaptations for catching prey (Fig. 33–3). The fast-moving tongue of the frog captures many a fly, while the long, quick legs, sharp teeth, and claws of the lion enable it to catch and kill gazelles. All carnivorous mammals have well-developed canine teeth for stabbing during combat; their molars are modified for shredding meat into small chunks that can be swallowed easily. The digestive juices of the stomach break down proteins, and because meat is more easily digested than plant food, carnivores' digestive tracts are shorter.

OMNIVORES

Omnivores, such as bears and humans, include both plant material and meat in their diet. They obtain food by a wide variety of mechanisms. Most aquatic filter-feeding organisms ingest both tiny plants and animals. Earthworms take in large amounts of soil containing both animal and plant material; they utilize the organic material for food and egest the rest. Omnivores are generally equipped to distinguish among a wide range of smells and tastes, which enable them to select various foods.

SYMBIONTS

A symbiont is an organism that lives in intimate association with a member of another species. One or both of the organisms usually derive nutritional benefit from the association. Three types of symbionts are parasites, commensals, and mutualistic partners.

A **parasite** lives in or on the body of a living plant or animal, the **host** species, and obtains its nourishment from the host (Fig. 33–4). Since the parasite's environment is its host, an effective parasite does not kill its host. For example, a tapeworm might live within the intestine of its host for many years. **Ectoparasites,** such as fleas and ticks, live on the host's body; **endoparasites,** such as tapeworms and hookworms, live inside the host. Whether the parasite nourishes itself from food ingested by its host or by sucking the host's blood, it is strictly a freeloader.

A **commensal** is an organism that derives benefit from its host without either doing harm to or benefiting the host. Commensalism is especially common in the ocean. Practically every worm burrow and hermit crab shell contain some uninvited guests that take advantage of the shelter and abundant food supplied by the host.

Mutualistic partners are two species of organisms that live together for their mutual benefit. They may be unable to survive separately. For example, the cellulose-digesting bacteria that inhabit the digestive tracts of herbivores help provide nourishment for their hosts in exchange for a place to live. A classic example of mutualism is the flagellate protozoan that lives in the intestine of the termite. The termite eats wood but, like most animals, does not produce the enzymes necessary to digest it. Its mutualistic partner, the flagellate, cannot chew wood and cannot survive outside the termite's gut, but it does produce the enzymes necessary to digest cellulose. Termites cannot survive without these intestinal inhabitants, and newly hatched termites lick the anus of another termite to obtain a supply of flagellates. This requires these insects to live together in social groups. Symbionts are discussed further in Chapter 50.

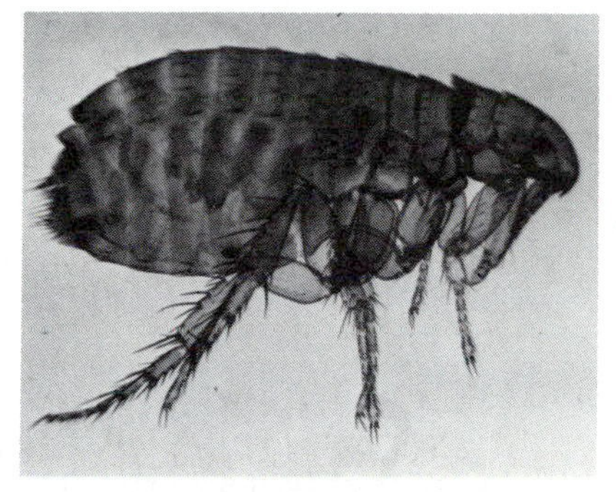

(*a*)

(*b*)

Figure 33–4 Parasitic symbionts. (*a*) The flea is a well-known ectoparasite. Its body shape is adapted for slipping through fur, and its long hind legs are built for jumping. Note the claws, which help the flea hold onto hairs on the host's body. (*b*) A most interesting form of predation is practiced by the brachonid wasp. The adult wasp lays her eggs under the skin of an insect such as this saddleback caterpillar of the flannel moth. When the eggs hatch, the larvae feed on the body fluids of the caterpillar until they are ready to spin cocoons. This caterpillar is almost covered by cocoons and has little chance of developing into an adult moth. Note that wasps are beginning to emerge from their cocoons. ((*b*), Luci Giglio.)

Invertebrate Digestive Systems

Sponges, the simplest animals, are filter feeders. They obtain food by filtering microscopic plants and animals out of sea water or pond water. Individual cells (both choanocytes and amoebocytes) phagocytize the food particles. Digestion is intracellular within a food vacuole (phagosome), and nutrients can be transferred to other cells. Wastes are egested into the water that continuously circulates through the sponge body.

Cnidarians, such as *Hydra* and jellyfish, capture small aquatic animals with the help of their cnidocytes and tentacles. The prey is pushed into the mouth by the tentacles. The mouth opens into a large gastrovascular cavity lined by cells that secrete proteolytic enzymes (Fig. 33–5). The gastrodermal cells are equipped with flagella, which mix the food. During digestion within the gastrovascular cavity, proteins are enzymatically cleaved to polypeptides. Digestion continues intracellu-

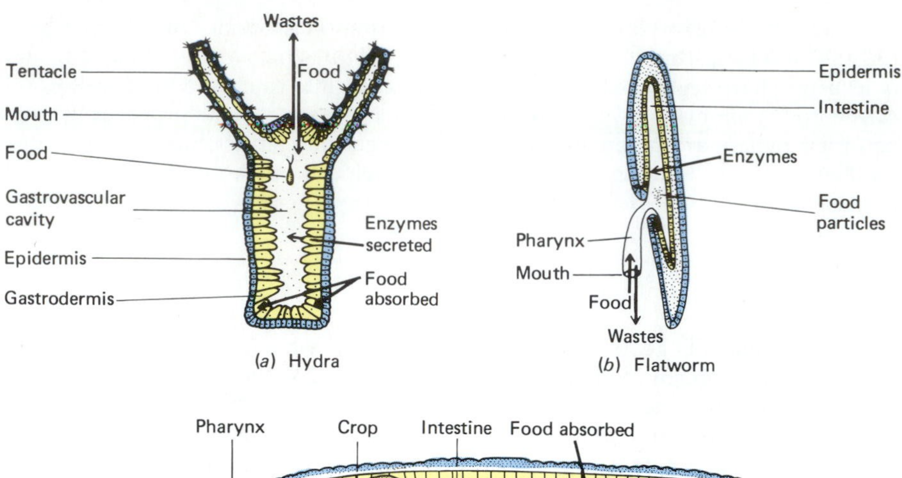

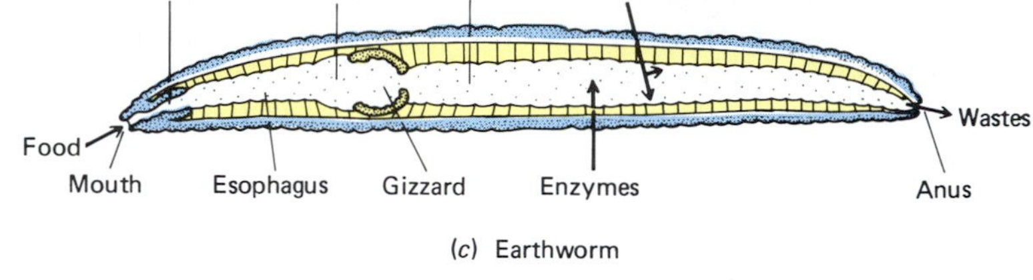

Figure 33–5 The structural basis of the process of digestion in several types of invertebrates. (*a*) *Hydra* and (*b*) flatworms have a digestive tract with a single opening that must serve as both mouth and anus. (*c*) The earthworm, like most complex animals, has a complete digestive tract extending from mouth at one end of the body to anus at the other.

larly within food vacuoles, and the products diffuse to the epidermal cells. Undigested food is ejected through the mouth by contraction of the body.

Free-living flatworms generally feed on small invertebrates such as crustaceans, rotifers, and annelid worms. Planarians begin to digest their prey even before ingesting it, by extending the pharynx through the mouth and secreting digestive enzymes onto the prey. The food, when ingested, is pumped into the branched intestine by waves of muscular contraction (peristalsis). Extracellular digestion proceeds as intestinal cells secrete digestive enzymes. Partly digested food fragments are then phagocytized by cells of the intestinal lining, and digestion is completed intracellularly within food vacuoles. The highly branched intestine facilitates the distribution of digested food, but there is no separate circulatory system. Like cnidarians, flatworms have no anal opening, so undigested wastes are eliminated through the mouth.

In most other invertebrates, and in all vertebrates, the digestive tract is a complete tube with two openings. Food enters through the mouth and undigested food is eliminated through the anus. Peristaltic movements push the food in one direction, so that more food can be taken in while previously eaten food is being digested and absorbed farther down the tract. Various parts of the tube are specialized to perform specific functions. For example, the digestive tract of the earthworm includes a mouth, a muscular pharynx that secretes a mucous material to lubricate food particles, an esophagus, a thin-walled crop where food is stored, a thick muscular gizzard where food is ground against small stones, and a long, straight intestine in which extracellular digestion occurs. The intestine terminates in an anus through which food wastes are eliminated. Some invertebrates—certain worms, insects, mollusks, crustaceans, and sea urchins—have hard, toothed mouthparts that can tear off and chew bits of food.

Vertebrate Digestive Systems

The vertebrate digestive tract is a complete tube extending from mouth to anus (Fig. 33–6). In simple vertebrates the digestive tract, or **gut,** may be a rather simple tube. In more complex vertebrates, however, various regions of the gut have specialized

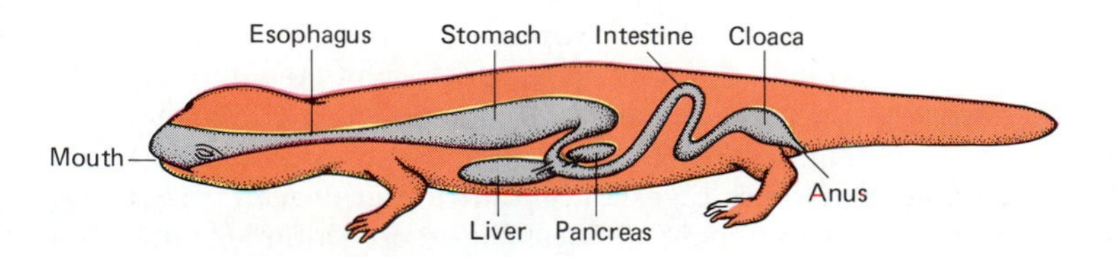

Figure 33–6 Like all vertebrates, the salamander has a complete digestive tract. The liver and pancreas are digestive glands that secrete digestive juices into the digestive tract.

structures and functions. Food succesively passes through these parts of the digestive tract: mouth, pharynx (throat), esophagus, stomach, small intestine, large intestine, and anus. All vertebrates have accessory glands that secrete digestive juices into the digestive tract. These include the liver and the pancreas and, in terrestrial vertebrates, the salivary glands (Fig. 33–7).

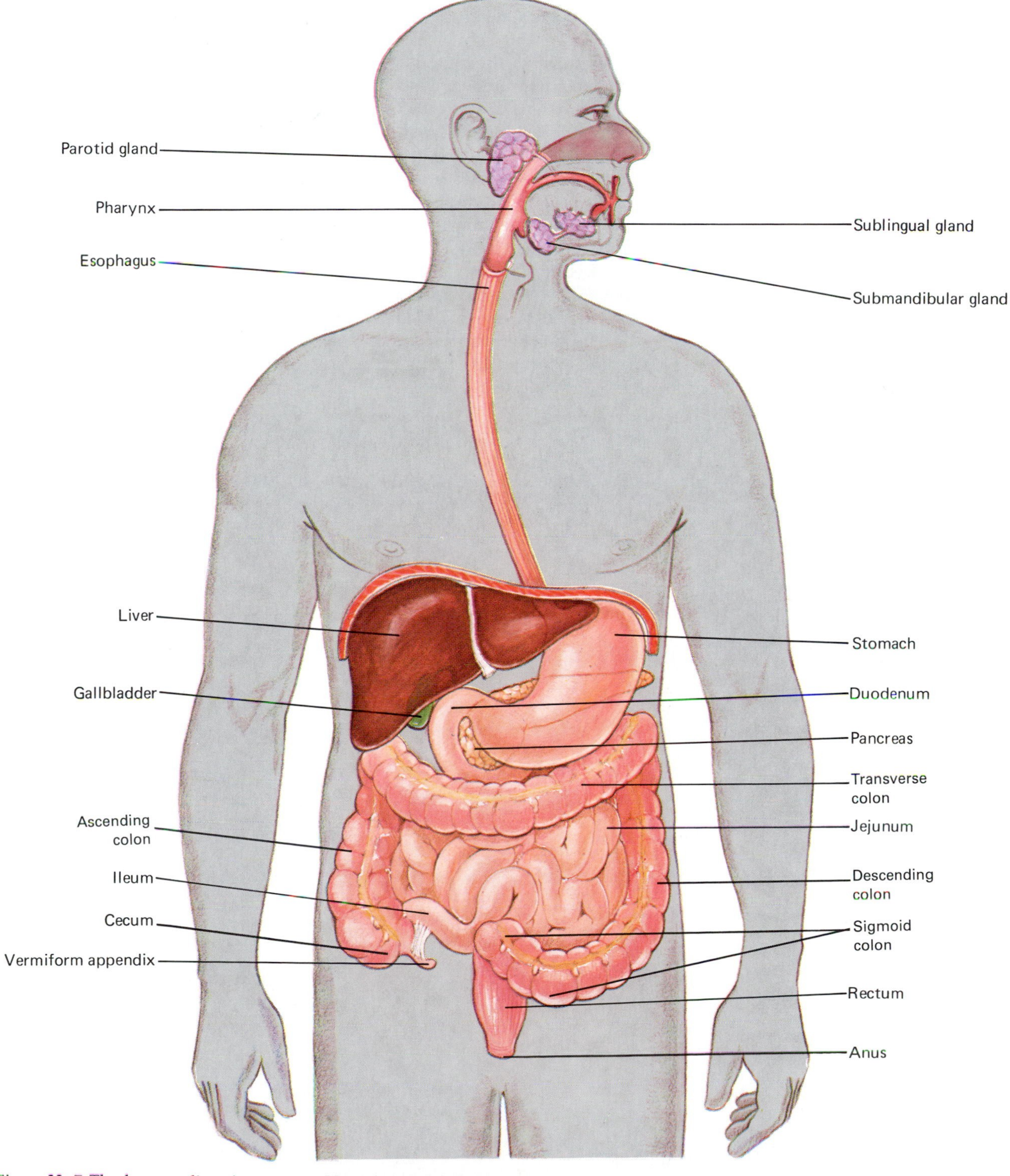

Figure 33–7 The human digestive system. Note the complete digestive tract, a long, coiled tube extending from mouth to anus. Locate the three types of accessory glands.

WALL OF THE DIGESTIVE TRACT

The wall of the mammalian digestive tract, from the esophagus to the rectum, has a similar structure and is composed of the same four layers. From the **lumen** (inner space) outward, they are the mucosa, submucosa, muscularis, and adventitia (Fig.33–8). The **mucosa** lines the digestive tract. It consists of epithelial tissue resting upon a layer of connective tissue. **Goblet cells** in the epithelial tissue secrete mucus, which protects and lubricates the inner surface of the digestive tract. The multicellular glands of the digestive tract are formed as inpocketings of the mucosa. In the stomach and intestine the mucosa is greatly folded to increase the secreting and absorbing surface of the digestive tube.

The **submucosa,** made up of connective tissue, binds the mucosa to the muscle layer beneath. The submucosa is rich in blood vessels, lymph vessels, and nerves. Along most of the digestive tract, the **muscularis** consists of two layers of smooth muscle, an inner one, which has muscle fibers arranged circularly, and an outer layer with fibers arranged longitudinally. Localized contractions of these muscles help to mechanically break down food and to mix it with digestive juices. Rhythmic waves of contraction of these muscles push food along through the digestive tract, the process of **peristalsis** (see Fig. 33–13).

The **adventitia** is the outer connective tissue coat of the digestive tract. Below the level of the diaphragm it is covered by a layer of squamous epithelium and is called the **visceral peritoneum.** By various folds it is connected to the **parietal peritoneum,** a sheet of connective tissues that lines the walls of the abdominal and pelvic cavities. Between the visceral and parietal peritoneums there is a potential space, the **peritoneal cavity.** Inflammation of the peritoneum, called **peritonitis,** can be very serious because infection can spread along the peritoneum to most of the abdominal organs.

Figure 33–8 Cross section through the wall of the small intestine illustrating the mucosa, submucosa, muscularis, and visceral peritoneum.

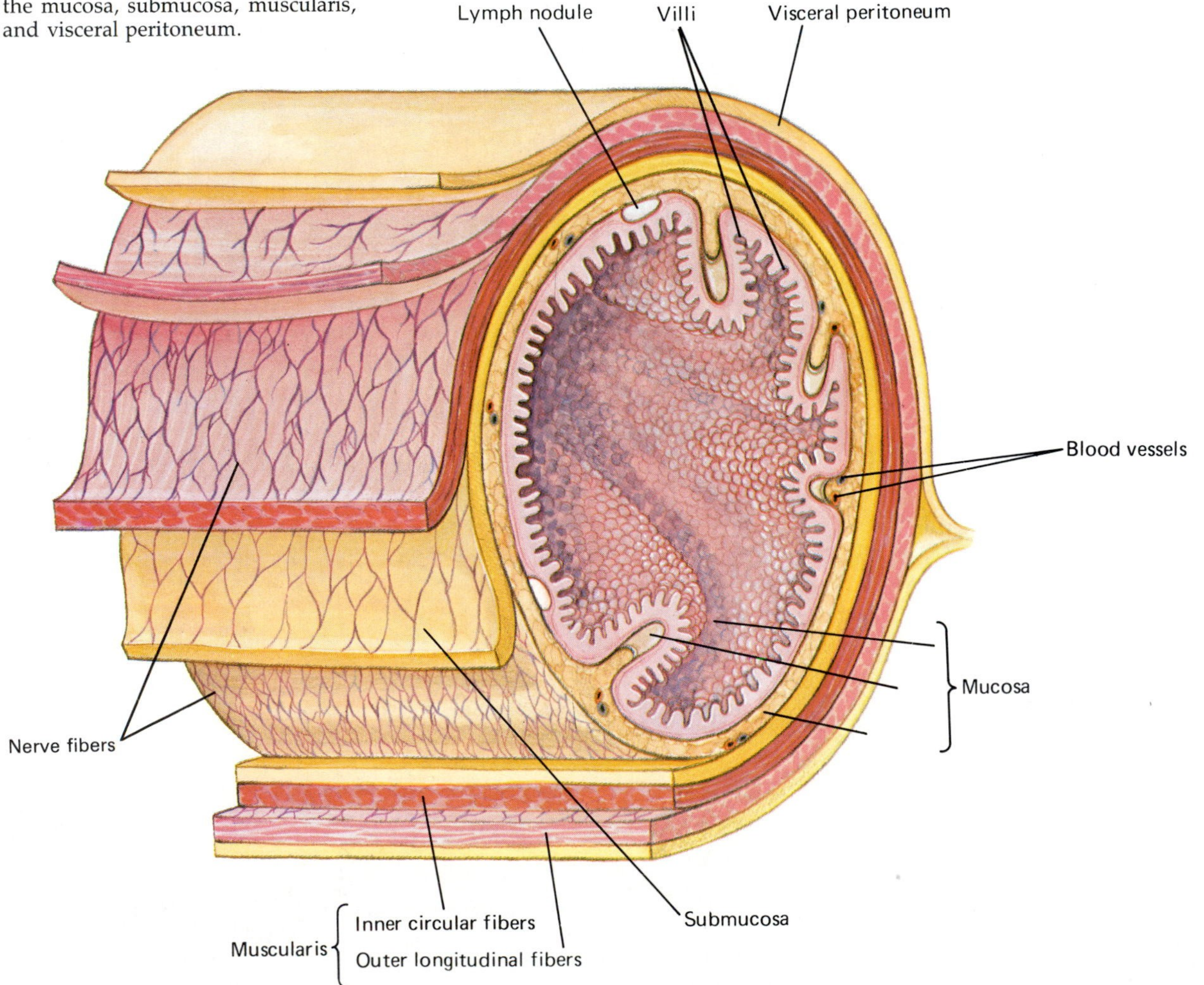

INSIDE THE MOUTH

Food is ingested into the mouth, and both mechanical and chemical digestion begin there. In humans the mouth, and especially the tongue and teeth, also aids in speech.

The fleshy, sensitive lips that surround the opening of the mouth help guide food into the mouth. The mouth cavity is supported by jaws and is bounded on the sides by teeth, gums, and cheeks, and on the bottom by the tongue. Its roof, the **palate,** is a shelf that separates the mouth cavity from the nasal cavity. The anterior bony portion is the **hard palate,** and the posterior fleshy part, the **soft palate.**

A muscular, mobile **tongue** is characteristic of most terrestrial vertebrates. In many amphibians and reptiles, and in some birds, the tongue is used to help capture food. You may have seen the long tongue of a frog dart outward with lightning speed to catch an insect on its sticky tip. Such catapulting tongues are attached anteriorly and are free in the back. The mammalian tongue functions mainly to manipulate food, pushing it between the teeth to be chewed, and then shaping it into a mass, called a **bolus,** which is then swallowed.

In mammals the **taste buds** are concentrated in the **papillae,** tiny projections on the tongue's surface. Sensory cells in the taste buds respond to different chemicals, enabling humans to distinguish four primary tastes: sweet, sour, salty, and bitter.

The Teeth

The **teeth** are used to bite, tear, crush, and grind food. Unlike the simple, pointed and conical teeth of fish, amphibians, and reptiles, the teeth of mammals vary in size and shape and are specialized to perform specific functions (Fig. 33–9). The eight front (two in each quadrant of the jaw) chisel-shaped **incisors** are used for biting, and are especially large in gnawing animals such as mice, rats, squirrels, and beavers. In herbivores the lower incisors are well-developed for cutting off grass.

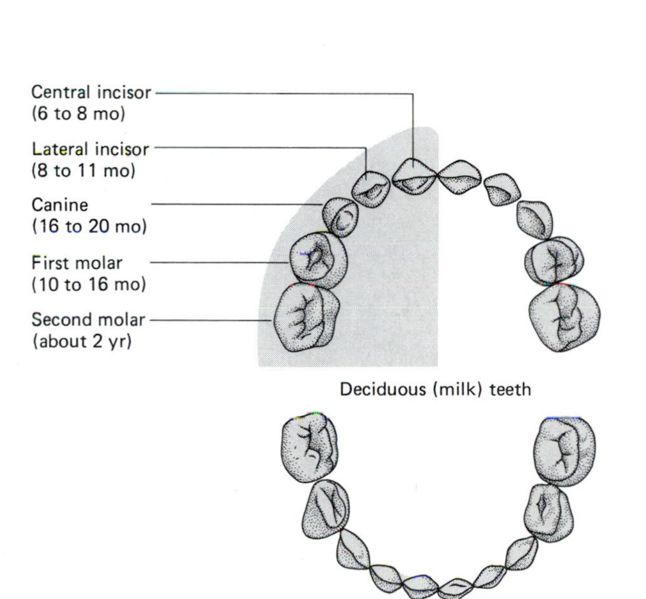

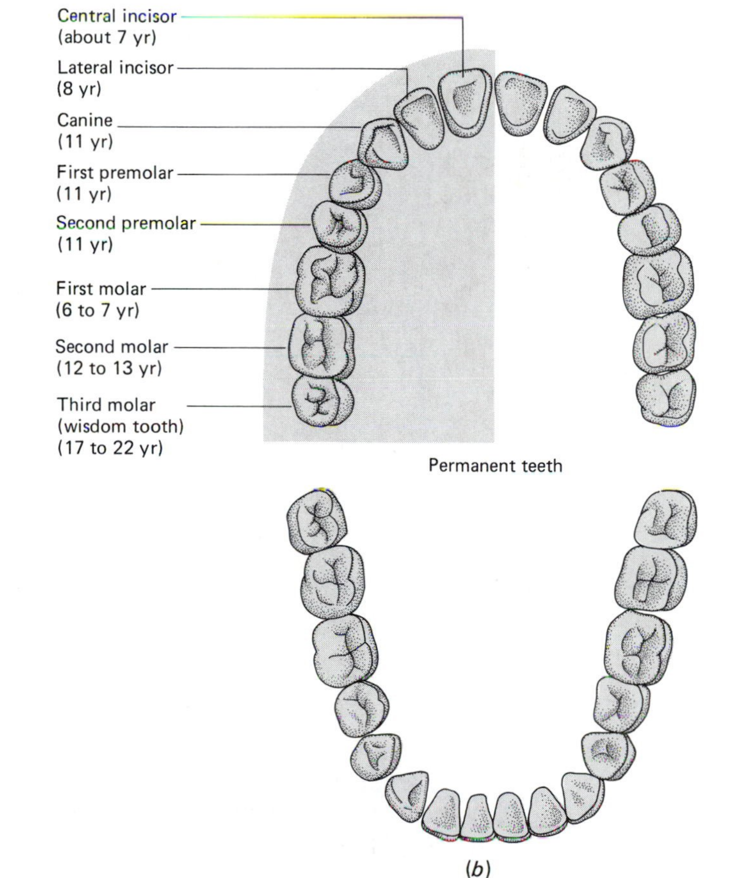

Figure 33–9 Human deciduous and permanent teeth. (*a*) Deciduous teeth. Approximate time of eruption is shown in parentheses. (*b*) Permanent teeth. One quadrant of the jaw has been shaded.

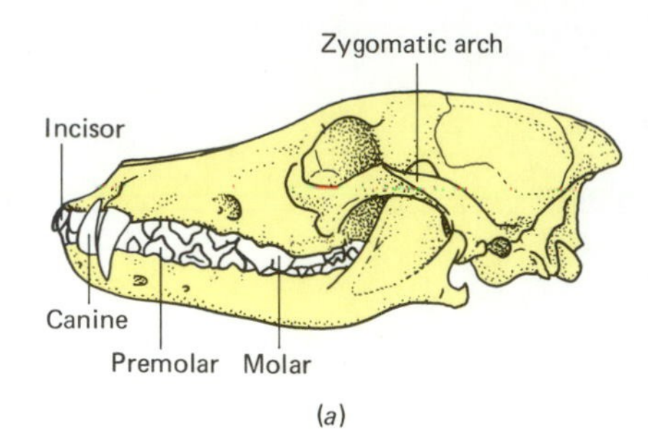

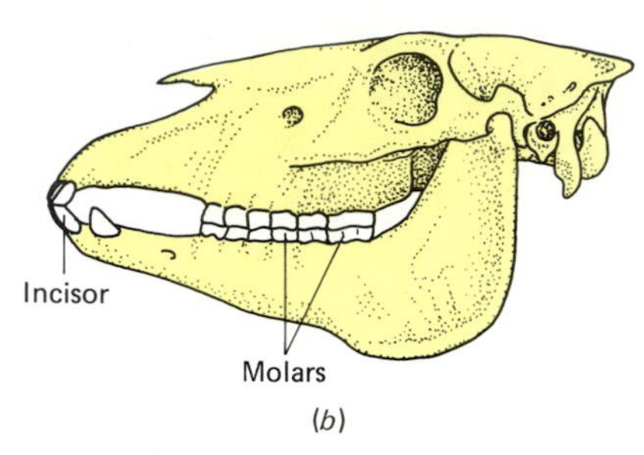

Figure 33–10 Comparison of the teeth of a carnivore and herbivore. (*a*) Skull of coyote. (*b*) Skull of domestic horse.

Upper incisors do not develop in ruminants. These animals crop grass by pulling it with their tongue and upper lip across the cutting edge of the lower incisors.

The four long, pointed **canines** (one in each quadrant of the jaw just lateral to the incisors) are used for stabbing and tearing food. Carnivores such as wolves, dogs, and lions have prominent canine teeth, sometimes referred to as fangs (Fig. 33–10). (In fact, these teeth are called canines because they are so large in dogs.) Male baboons and some other animals use canines in dominance displays and in defense.

Humans have eight **premolars** and 12 **molars,** arranged with two premolars and three molars on each side of the upper and lower jaw (see Fig. 33–9). These teeth have flattened surfaces for crushing and grinding food. Herbivores have large, flat molars for grinding plant material. In humans the third molars, called **wisdom teeth,** frequently fail to erupt or, if they do emerge from the gum, are often crooked and useless. This is usually interpreted as reflecting a trend in evolution of modern humans toward a shortening of the jaws with a resulting crowding of the teeth, which leaves inadequate space for the last molar.

Although shaped somewhat differently among the species, all mammalian teeth have the same basic structure. The part of the tooth projecting above the gum is the **crown** (Fig. 33–11). One or more **roots** are hidden beneath the gumline, and the somewhat constricted junction between the crown and root is the **neck.** Each root is embedded in a socket of the jawbone.

In the crown and upper neck region a hard, outer **enamel** covers the tooth. Enamel is the hardest substance in the body. Beneath the enamel is a thick layer of **dentin,** which makes up most of the tooth and resembles bone in its composition and hardness. In the neck and root region a calcified connective tissue called **cementum** covers the dentin and attaches it, via connective tissue, to its socket.

Beneath the dentin is a **pulp cavity** filled with **pulp,** a soft connective tissue containing blood vessels, lymph vessels, and nerves. Narrow extensions of the

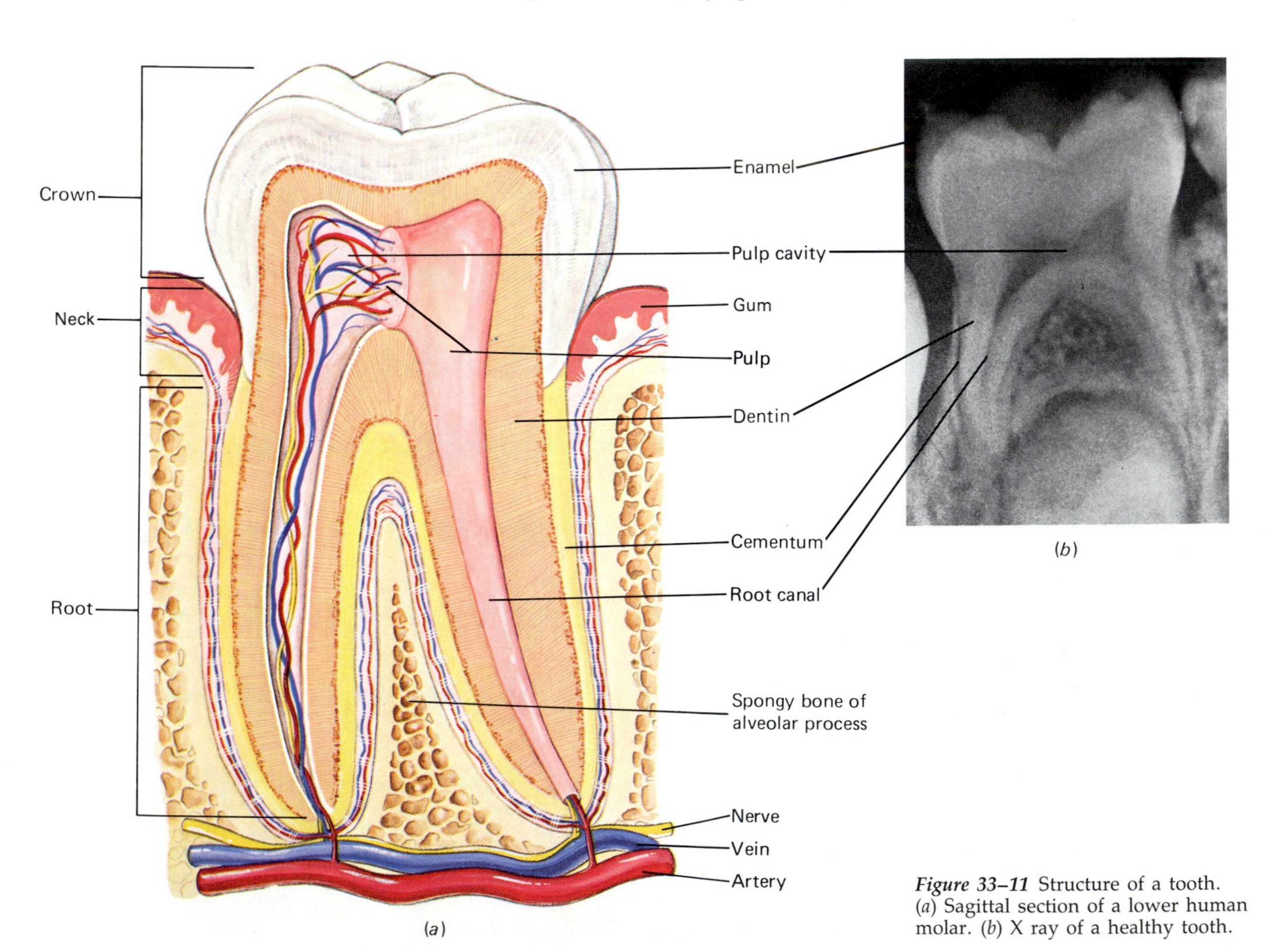

Figure 33–11 Structure of a tooth. (*a*) Sagittal section of a lower human molar. (*b*) X ray of a healthy tooth.

pulp cavity, called **root canals,** pass through the roots of the tooth. Blood vessels, lymph vessels, and nerves reach the pulp cavity by way of the root canals.

When allowed to accumulate, bacteria that inhabit the mouth form dental plaque on the surfaces of the teeth, particularly near the gum line. Carbohydrates (especially sucrose) deposited on the teeth are fermented by the bacteria of the plaque. Organic acids produced during fermentation demineralize the outer layers of the teeth and cause **dental caries,** or cavities.

The Salivary Glands

Aquatic animals have plenty of water available to moisten food and aid in swallowing. Terrestrial vertebrates possess **salivary glands,** which produce saliva to perform these functions. In mammals there are three main pairs of salivary glands: the parotid, submandibular, and sublingual glands. The **parotid glands,** the largest salivary glands, are located in the tissue below and in front of the ears (see Fig. 33–7). These glands often become infected and swell when a person has the mumps. The **submandibular glands** lie below the jaw and the **sublingual glands** lie under the tongue. Secretions from these glands are delivered into the mouth cavity through tiny ducts.

Saliva consists of a thin, watery component containing the digestive enzyme **salivary amylase** and a mucous component that lubricates the passage of the bolus during swallowing. Saliva also contains salts and substances that kill bacteria. Salivary amylase begins the digestion of carbohydrates by hydrolyzing starch to maltose. Saliva is normally slightly acidic with a pH of about 6.7, and amylase works best at this pH. After it reaches the stomach, the bolus is penetrated by the very acidic gastric juice, and the amylase is then inactivated.

Humans secrete about one liter of saliva daily. Secretion is regulated by control centers in the brain that send messages to the glands by way of nerves. Feeling food in the mouth or tasting it stimulates these control centers. Even smelling, seeing, or thinking about food may stimulate an increase in saliva secretion. Food such as sour pickles and lemons are the strongest stimuli. Very little saliva is secreted during sleep, and should the body become dehydrated, salivation slows or stops altogether. The dry feeling in the mouth that results is one of the stimuli that indicates thirst, motivating us to ingest fluids and thus restore homeostasis.

THROUGH THE PHARYNX AND ESOPHAGUS

During swallowing, food passes from the mouth cavity into the pharynx (throat region), the muscular cavity where the digestive and respiratory systems cross. The **esophagus** is a muscular tube extending from the pharynx to the stomach. It passes between the lungs behind the heart, and penetrates the diaphragm.

The movement of food from the mouth to the stomach is aided by a series of reflex actions. The first part of swallowing is under voluntary control. The tongue is raised against the roof of the mouth, and the bolus of food between the tongue and palate is pushed into the pharynx by a wavelike movement of the tongue (Fig. 33–12). When swallowing begins, breathing is stopped momentarily by a reflex mechanism that prevents food from passing into the respiratory passageways.

Several openings in the pharynx close by reflex action before food reaches the pharynx. This ensures that the food will pass only into the esophagus. The hard bump in the ventral midline of the neck, perhaps known to you as the Adam's apple, is the **larynx.** Contraction of muscles raises the larynx so that its opening (the glottis) is sealed off against a flap of tissue called the **epiglottis.** This action prevents food from entering the respiratory passageway. You can observe the raising of the larynx by watching someone swallow—it bobs upward with each swallow.

The reflex movements that propel the bolus through the pharynx and into the esophagus are so fast that it takes only about a second for food to move from the mouth to the esophagus. As the bolus enters the esophagus a peristaltic wave pushes it downward toward the stomach (Fig. 33–13). This journey takes only about 10 seconds. Gravity is not necessary to pull food through the esophagus. Astronauts at zero gravity are able to swallow and even if you are standing on your head, food will reach the stomach!

The opening from the esophagus to the stomach is controlled by a portion of circular muscle that acts as a sphincter. This sphincter is generally closed, prevent-

Figure 33–12 Position of the tongue and epiglottis during (*a*) breathing and (*b*), (*c*) swallowing. In (*b*), note how a bolus is pushed from the mouth into the pharynx by the tongue to initiate swallowing.

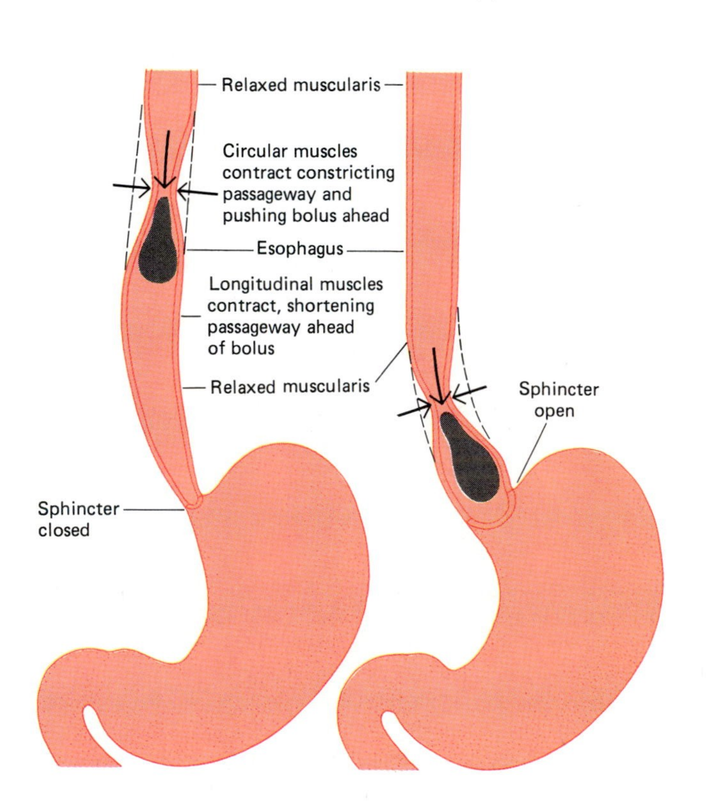

Figure 33–13 Peristalsis. Food is moved along through the digestive tract by waves of muscular contraction referred to as peristalsis.

ing the highly acidic gastric juice from splashing up into the esophagus as it does during heartburn when the sphincter fails to close. When a peristaltic wave reaches the lower portion of the esophagus, the ring of muscle relaxes, permitting the bolus to enter the stomach.

IN THE STOMACH

The **stomach** is a thick-walled, muscular sac on the left side of the body just beneath the lower ribs. The muscular layers of the stomach wall are very thick and include a diagonal layer of fibers in addition to the circular and longitudinal fibers found elsewhere in the wall of the digestive tract (Fig. 33–14). When empty, the stomach is collapsed and shaped somewhat like a hotdog. As a meal is eaten the stomach stretches, assuming the shape of a football; its capacity is about one liter. The lining of the empty stomach appears wrinkled because of prominent folds of the mucosa called **rugae.** As the stomach fills with food, the rugae gradually flatten out.

The stomach is lined with simple columnar epithelium that secretes large amounts of mucus. Millions of microscopic **gastric glands** extend deep down into

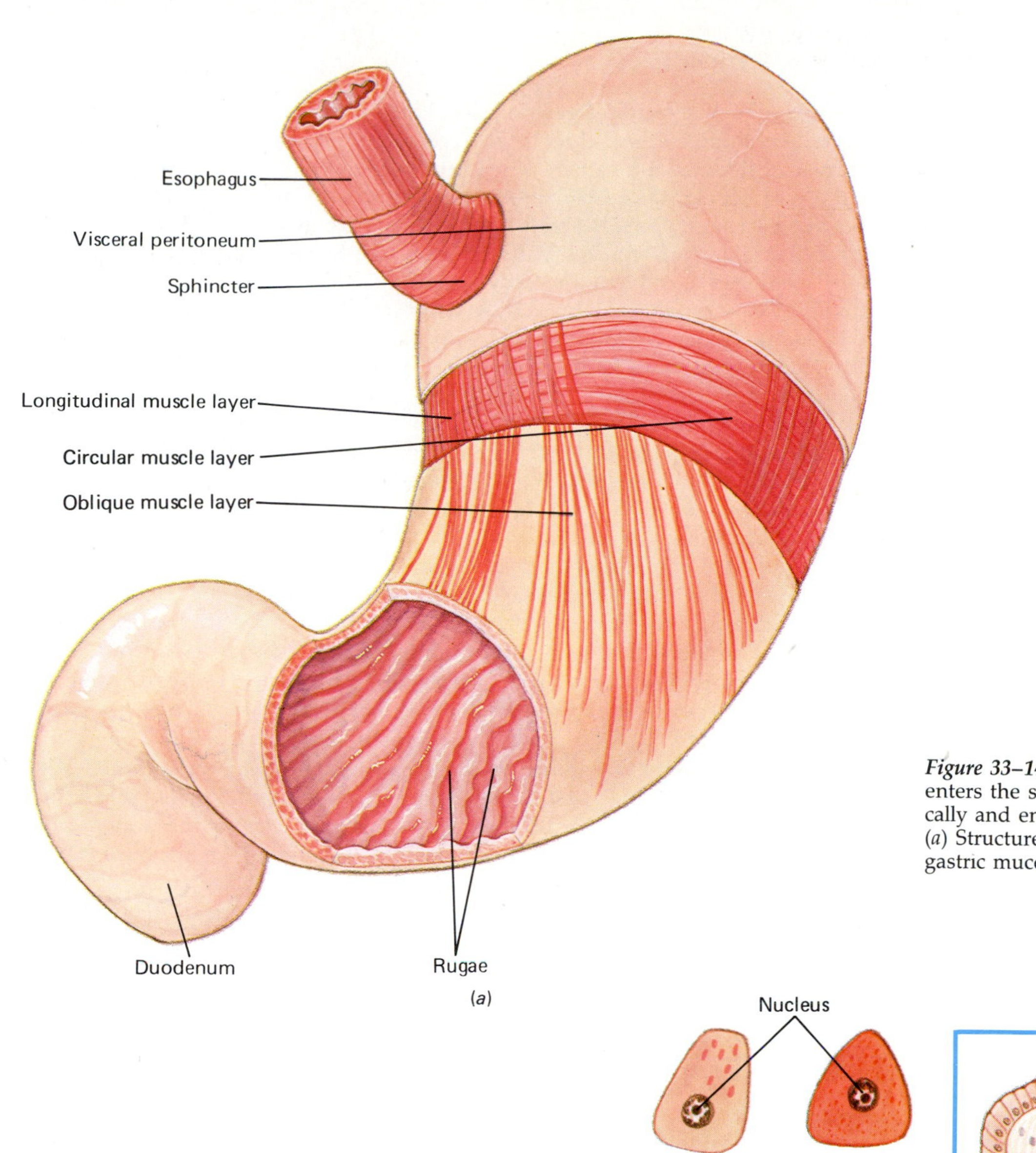

Figure 33–14 From the esophagus, food enters the stomach where it is mechanically and enzymatically digested. (*a*) Structure of the stomach. (*b*) The gastric mucosa and gastric glands.

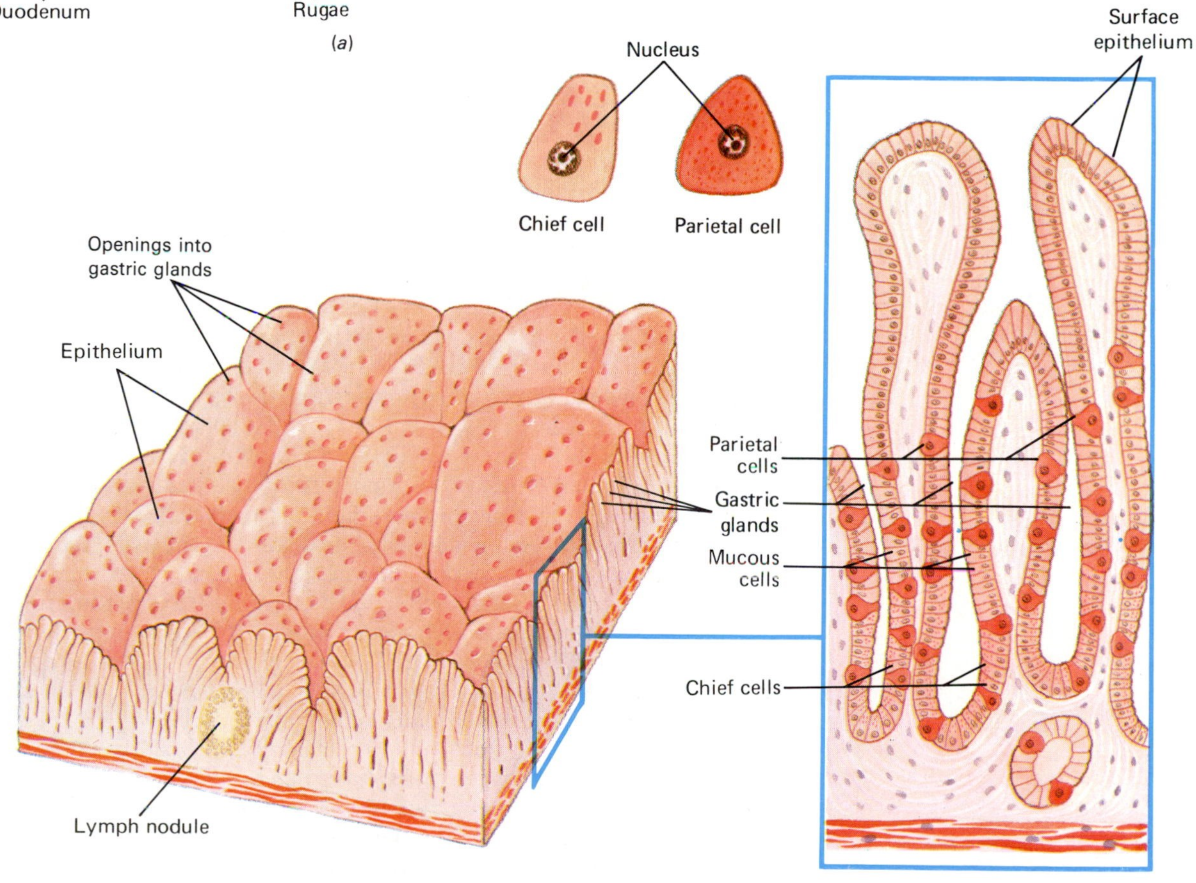

the mucosa. They secrete the **gastric juice. Parietal cells** of the gastric glands produce hydrochloric acid, and **chief cells** produce a protein, **pepsinogen,** the inactive precursor of the enzyme pepsin. The gastric juice is highly acidic, with a pH of about 0.8. However, the gastric juice mixes with mucus and food so that the final pH of the gastric contents is about 2. This is sufficiently acidic to kill most of the bacteria that enter the stomach with the food (see Focus on Peptic Ulcers).

Pepsinogen is converted to active **pepsin** by the removal of a portion of the molecule. This reaction is catalyzed by hydrochloric acid (HCl) and by pepsin itself. The HCl also provides an optimum pH for pepsin action. Pepsin, the major enzyme of the gastric juice, initiates the digestion of proteins. One of its most important actions is to specifically digest the protein collagen found in the connective tissue of meat. As collagen is broken down by pepsin, the protein within the muscle cells becomes more accessible to the digestive juices and can be digested.

The activities of the stomach are regulated by both the nervous and endocrine systems. When food is contemplated, smelled, viewed, or tasted, the brain sends messages stimulating the gastric glands. By the time food arrives in the stomach, gastric juices have already been released. Then, when food presses against receptors in the stomach wall, the gastric glands are further stimulated. Stretching the stomach with food also stimulates the stomach mucosa to release a hormone called **gastrin.** This hormone is absorbed into the blood and transported to the gastric glands, where it stimulates release of gastric juice. The presence of partly digested proteins, caffeine, or moderate amounts of alcohol in the stomach also stimulates secretion of gastrin.

After a meal, food may remain in the stomach for more than four hours. As it is churned, mashed, and digested by gastric juice, the food is converted into a soupy mixture called **chyme.** Peristaltic waves slowly push the chyme along toward the exit of the stomach. Very few substances are absorbed through the stomach wall. Only water, salts, and lipid-soluble substances such as alcohol are absorbed from the stomach. The exit of the stomach is generally kept closed by contraction of a ring of muscle, the **pyloric sphincter.** When digestion in the stomach is completed, the pyloric sphincter relaxes so that chyme can be pushed, a few milliliters at a time, into the small intestine.

INSIDE THE SMALL INTESTINE

The length of the small intestine is correlated with the type of diet. Herbivores have a very long small intestine, while carnivores have a shorter one, and omnivores have one of intermediate length. The frog larva (tadpole) is herbivorous and has a long small intestine, but the carnivorous adult frog has a much shorter one.

The human small intestine is a coiled tube about 2.6 m long by 4 cm in diameter. Curved like the letter C, the first 21 cm or so of the small intestine is the **duodenum** (see Fig. 33–17). As the small intestine turns downward it is called the **jejunum.** The jejunum extends for about 0.9 m before becoming the **ileum.** The duodenum is held in place by connective tissue ligaments that attach to the liver, stomach, and dorsal body wall. The rest of the small intestine (and most of the large intestine) is loosely anchored to the dorsal body wall by a thin transparent membrane called the **mesentery.** The inner lining of the small intestine is not smooth like the inside of a water pipe. Instead, the inner surface is intricately folded in three ways: First, the mucous membrane is thrown into visible **circular folds.** Then, the mucous membrane is pushed up into millions of microscopic fingerlike projections called **villi** (Figs. 33–15 and 33–16). Finally the intestinal surface is further increased by thousands of **microvilli,** which are folds in the cell membrane of the exposed borders of the epithelial cells (Fig. 33–17). The microvilli give the epithelial lining a fuzzy appearance, termed a **brush border,** when viewed with an electron microscope. Together, the circular folds, villi, and microvilli increase the surface area of the small intestine so extensively that if the lining could be completely unfolded and spread out, its surface would approximate the size of a tennis court!

Most of the enzymatic digestion of food takes place in the duodenum. Bile from the liver and pancreatic juice from the pancreas are secreted into the duodenum. As we will see, these secretions play important roles in digestion. In addition, millions of tiny **intestinal glands** in the intestinal mucosa secrete intestinal juice, which serves as a medium for digestion and absorption of nutrients. The intestinal epithelial cells produce several enzymes that catalyze the final steps in digestion.

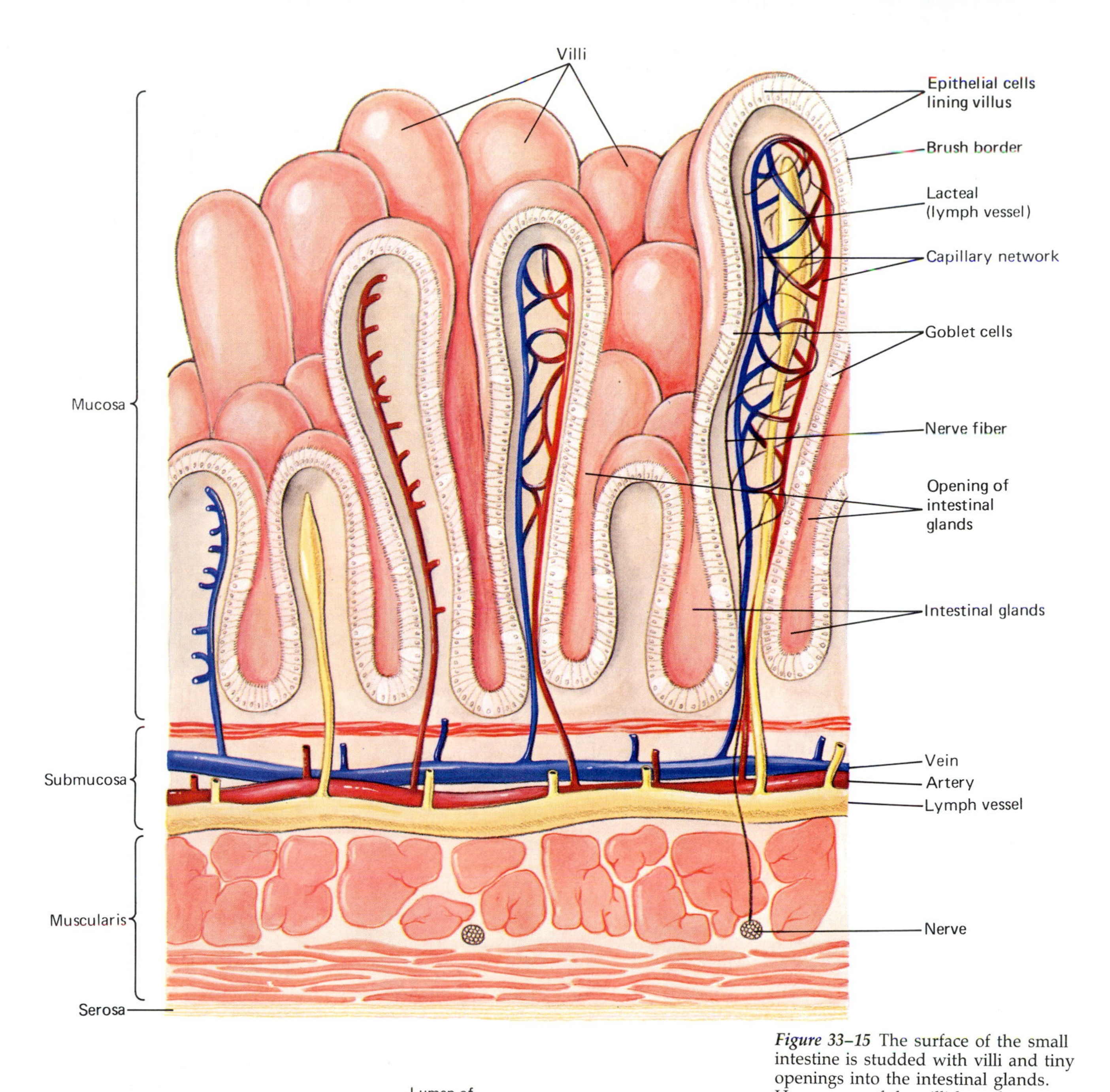

Figure 33–15 The surface of the small intestine is studded with villi and tiny openings into the intestinal glands. Here, some of the villi have been opened to show the blood and lymph vessels within.

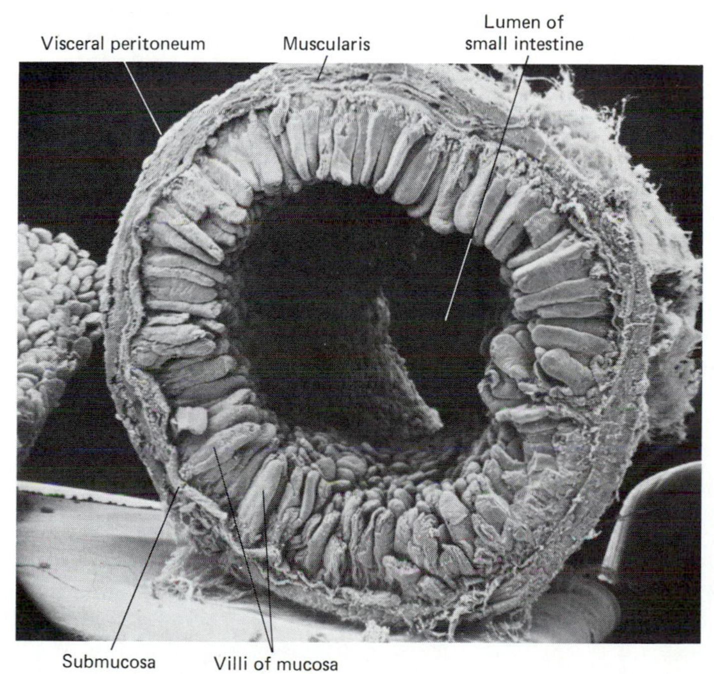

Figure 33–16 Scanning electron micrograph of a cross section of the small intestine (approximately ×30). (From Kessel, R. G., and Kardon, R. H.: *Tissues and Organs: A Text–Atlas of Scanning Electron Microscopy*. San Francisco, W. H. Freeman and Co., 1979.)

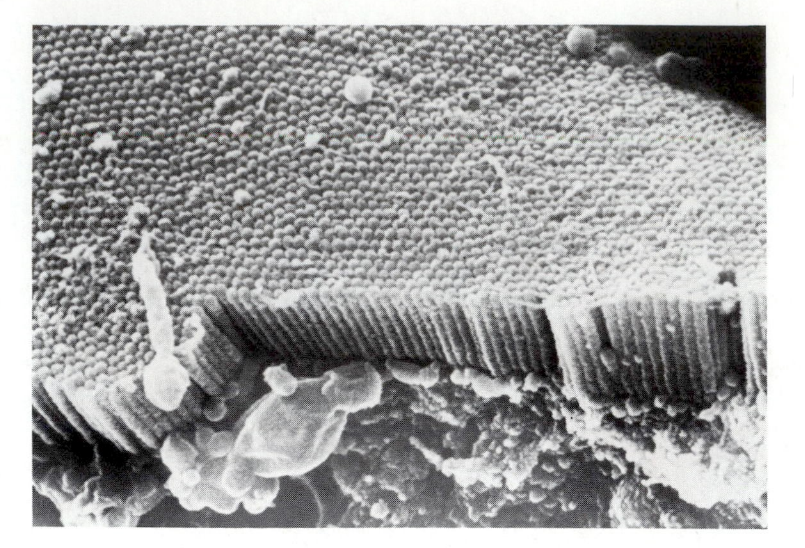

Figure 33–17 Scanning electron micrograph of the surface of an epithelial cell from the lining of the small intestine, showing microvilli (approximately ×14,000). The epithelium has been cut vertically so that the microvilli can be viewed from the side as well as from above. (Courtesy of J. D. Hoskins, W. G. Henk, and Y. Z. Abdelbaki, from the *American Journal of Veterinary Research,* Vol. 43, No. 10.)

As elsewhere in the digestive tract, contractions in the small intestine produce both mixing movements and peristaltic waves. Several hours are required for chyme to be propelled through the length of the small intestine and into the large intestine. Movement and digestion in the small intestine are regulated by neural messages and by hormones. When acidic chyme from the stomach comes in contact with the mucosa of the duodenum, a hormone called **secretin** is released from the duodenal mucosa. Secretin stimulates both the pancreas and the liver to release some of their secretions. The presence of fatty acids or partly digested proteins in the duodenum also stimulates the duodenal mucosa to release a hormone known as **cholecystokinin,** or **CCK.** This hormone stimulates the pancreas and gallbladder (see Table 33–5) and is thought to affect the appetite control centers in the brain.

THE PANCREAS

The pancreas and liver are large accessory digestive glands that develop in the embryo as outgrowths of the digestive tract. The **pancreas** is an elongated gland that lies in the abdominal cavity between the stomach and duodenum (Fig. 33–18). The cells that secrete the pancreatic enzymes are arranged in units called **acini,**

FOCUS ON

Peptic Ulcers

One of the wonders of physiology is that gastric juice does not normally digest the stomach wall itself. Several protective mechanisms prevent this from happening. Cells of the gastric mucosa secrete an alkaline mucus that coats the stomach wall and also neutralizes the acidity of the gastric juice along the lining. In addition, the epithelial cells of the lining fit tightly together, preventing gastric juice from leaking between them and onto the tissue beneath. Should some of the epithelial cells be damaged, they are quickly replaced. In fact, the lifespan of an epithelial cell in the gastric mucosa is only about three days. About a half million of these cells are shed and replaced every minute.

Still, these mechanisms sometimes malfunction or prove inadequate, and a small bit of the stomach lining is digested, leaving an open sore or **peptic ulcer.** Substances such as alcohol and aspirin reduce the resistance of the stomach mucosa to digestion by gastric juice. Peptic ulcers occur more often in the duodenum than in the stomach. They also sometimes occur in the lower part of the esophagus.

Peptic ulcers may bleed, leading to anemia. If the ulcer extends into the muscularis, large blood vessels may also be damaged, resulting in hemorrhage. A **perforated ulcer** is one that extends all the way through the wall of the stomach or other affected organ. The opening created may allow bacteria and food to pass through to the peritoneum, leading to peritonitis and shock. Perforation is the main cause of death from ulcers.

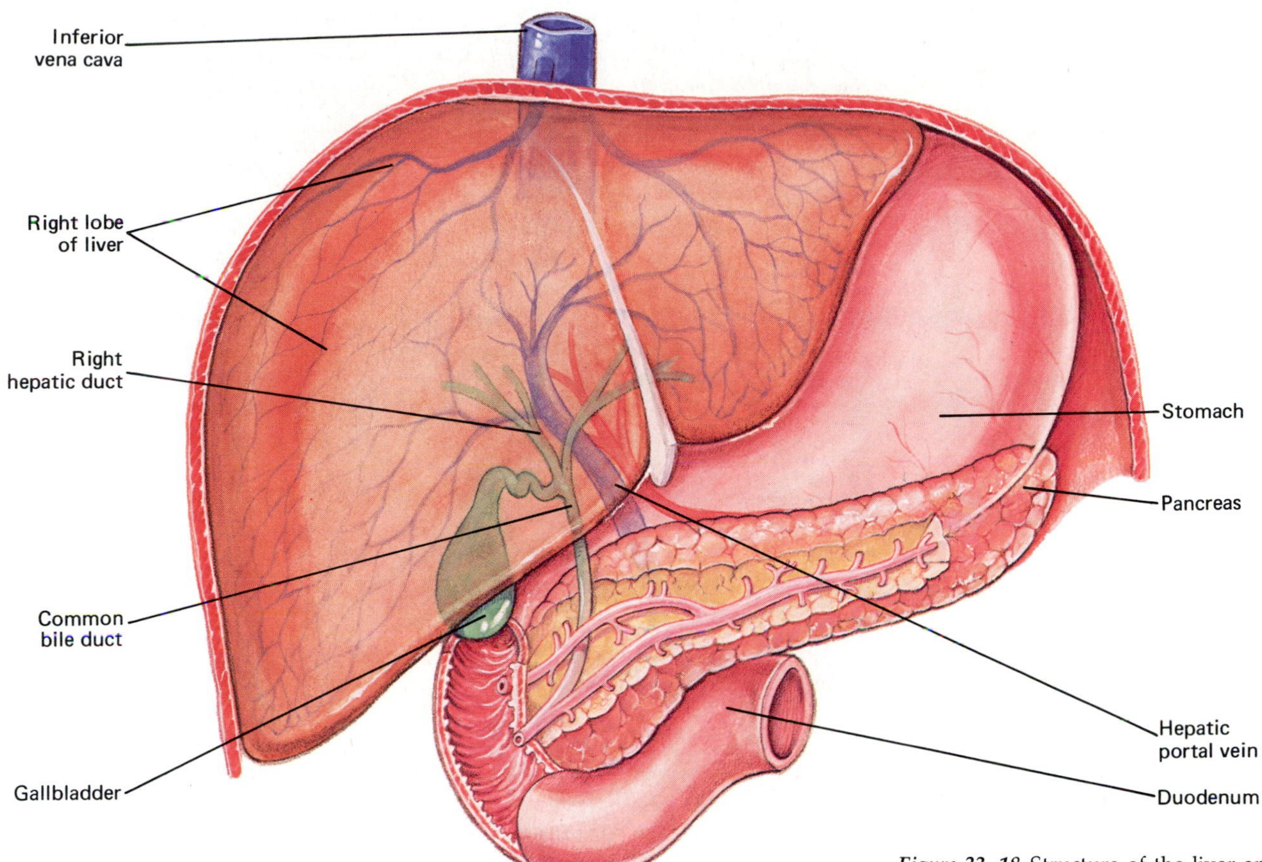

Figure 33–18 Structure of the liver and pancreas. Note the gallbladder and ducts.

which look like clusters of grapes. The ducts leading from the acini not only conduct the enzymes, but also secrete a sodium bicarbonate solution, which makes the pancreatic juice somewhat alkaline. The pancreatic enzymes include (1) the proteolytic enzymes **trypsin, chymotrypsin,** and **carboxypeptidase;** (2) **pancreatic lipase,** which hydrolyzes neutral fats; (3) **pancreatic amylase,** which degrades almost all carbohydrates except cellulose to disaccharides; (4) an esterase that splits cholesterol esters; and (5) ribonuclease and deoxyribonuclease, which split RNA and DNA into free nucleotides (Table 33–1).

All of the proteolytic enzymes are secreted as inactive precursors. Trypsin is activated in the duodenum when it comes in contact with the enzyme **enterokinase** secreted by the intestinal mucosa. Enterokinase splits off a portion of the precursor molecule trypsinogen to yield the active enzyme trypsin and an inactive molecular fragment. The activated trypsin then activates the other proteases. To further protect itself from digestion by the proteases it secretes, the pancreas produces an internal trypsin inhibitor, which inactivates any trypsin that might accidentally become activated in the pancreas. If the pancreas is damaged (as in alcoholism), or if the duct is blocked, large amounts of pancreatic enzymes may accumulate. The trypsin inhibitor system then may be overwhelmed and the proteases may digest the tissues of the pancreas. This can result in **acute pancreatitis,** which is often fatal.

The pancreas is an endocrine, as well as an exocrine, gland. Its endocrine component, the **islets of Langerhans,** secretes the hormones insulin and glucagon, which regulate the concentration of glucose in the blood.

THE LIVER

The **liver** is one of the largest and functionally most complex organs in the body. Each liver cell can carry on hundreds of metabolic activities. The liver (1) secretes **bile,** important in fat digestion; (2) removes nutrients from the blood; (3) converts glucose to glycogen for storage and converts glycogen to glucose as needed; (4) stores iron and certain vitamins; (5) converts amino acids to keto acids and urea;

TABLE 33–1
Enzymes Important in Digestion

Enzyme	*Source*	*Optimum pH*	*Substrate*	*Product*
Salivary amylase	Saliva	Neutral	α-Glycosidic bonds of starch and glycogen	Maltose
Pepsin	Stomach	Acid	Peptide bonds within chain and adjacent to tyrosine or phenylalanine	Peptides
Rennin	Stomach	Acid	Peptide bonds in casein	Coagulated casein
Trypsin	Pancreas	Alkaline	Peptide bonds within chain adjacent to lysine or arginine	Peptides
Chymotrypsin	Pancreas	Alkaline	Peptide bonds within chain adjacent to tyrosine, phenylalanine, or tryptophan	Peptides
Lipase	Pancreas	Alkaline	Ester bonds of fats	Glycerol, fatty acids, mono- and diacylglycerols
Amylase	Pancreas	Alkaline	α-Glycosidic bonds of starch and glycogen	Maltose
Ribonuclease	Pancreas	Alkaline	Phosphate esters of RNA	Nucleotides
Deoxyribonuclease	Pancreas	Alkaline	Phosphate esters of DNA	Nucleotides
Carboxypeptidase	Pancreas	Alkaline	Peptide bond adjacent to free carboxyl end	Free amino acids
Aminopeptidase	Intestinal glands	Alkaline	Peptide bonds adjacent to free amino end	Free amino acids
Enterokinase	Intestinal glands	Alkaline	Trypsinogen	Trypsin
Maltase	Intestinal glands	Alkaline	Maltose	Glucose
Sucrase	Intestinal glands	Alkaline	Sucrose	Glucose and fructose
Lactase	Intestinal glands	Alkaline	Lactose	Glucose and galactose

(6) manufactures many proteins found in the blood; (7) detoxifies many drugs and poisons that enter the body; (8) phagocytizes bacteria and worn-out red blood cells; and (9) performs countless functions in the metabolism of amino acids, fats, and carbohydrates.

Liver cells continuously secrete small amounts of bile, which pass by a system of ducts into the **common bile duct.** The common bile duct empties into the duodenum, but the exit from the duct is usually closed by a sphincter muscle. When the sphincter is constricted, bile is shunted into the pear-shaped **gallbladder** for storage. When fat enters the duodenum it stimulates release of the hormone CCK from the intestinal mucosa. CCK stimulates the gallbladder to contract and relaxes the sphincter so that bile is released into the duodenum.

Bile consists of water, bile salts, bile pigments, cholesterol, salts, and lecithin (a phospholipid). **Bile salts** are made by the liver from cholesterol. They act as detergents, emulsifying (mechanically breaking down into droplets) the fats in the intestine. When large fat globules are mechanically dispersed into many small globules, their surface area is increased, and lipase can come into contact with individual fat molecules and cleave off fatty acids. (When the bile duct is obstructed and bile salts are absent from the intestine, both the digestion and absorption of fats are impaired, causing much of the fat eaten to be wastefully eliminated in the feces.) Bile salts are conserved by the body. They are reabsorbed in the lower part of the intestine and transported back to the liver by the blood to be secreted once again.

Cholesterol is synthesized in the liver; its concentration in the bile reflects the amount of lipid in the diet. Cholesterol is rather insoluble in water but combines with bile salts and lecithin to form soluble molecular aggregates called **micelles.** Under certain abnormal conditions cholesterol precipitates and produces hard little pellets called **gallstones** (Fig. 33–19). Persons on high-fat diets over a period of years tend to develop gallstones more readily than do those on low-fat diets.

The color of bile results from the presence of bile pigments (green, yellow, orange, or red in different animal species). Bile pigments are formed from the heme portion of hemoglobin by enzymatic processes in the liver. In the intestine, bile pigments are metabolized further by bacterial enzymes and become a brownish color. These brown pigments are responsible for the color of feces. Sometimes the excretion of bile pigment is prevented by some obstruction of the bile duct such as gallstones. The pigments then accumulate in the blood and tissues, imparting a yellowish tinge to the skin, a condition known as **jaundice.** Absence of the pigments from the intestinal contents leaves the feces clay-colored.

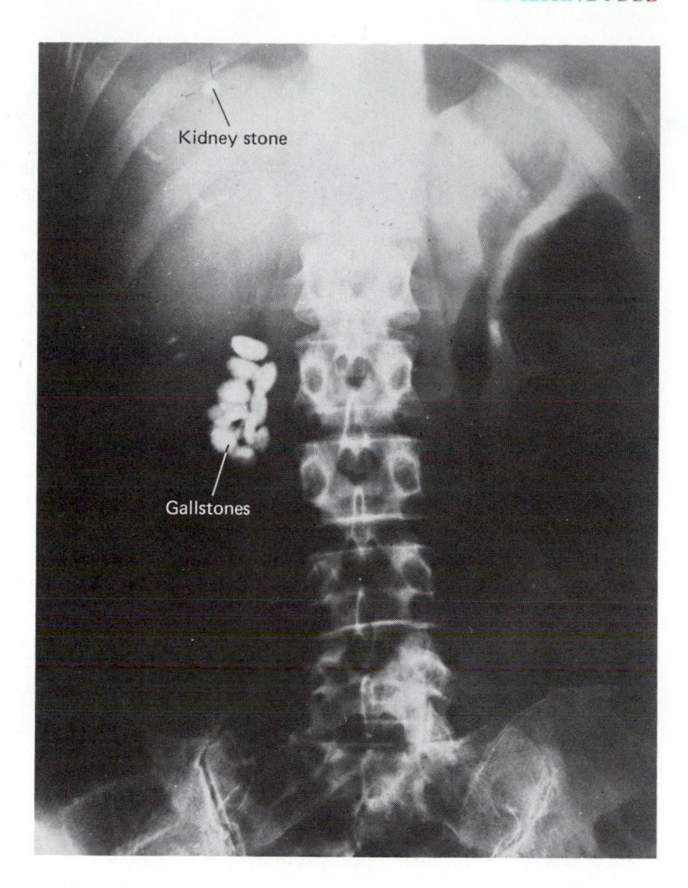

Figure 33–19 X-ray film showing calicified gallstones in gallbladder. Also note kidney stone.

ENZYMATIC DIGESTION

As chyme is moved through the digestive tract by peristaltic action, mixing contractions, and motion of the villi, enzymes come into contact with the nutrients and digest them.

Carbohydrate Digestion

Polysaccharides such as starch and glycogen are an important part of the food ingested by humans and most other animals. The glucose units of these large molecules are joined by glycosidic bonds linking carbon 4 (or 6) of one glucose molecule with carbon 1 of the adjacent glucose molecule. These bonds are hydrolyzed by **amylases,** enzymes that digest polysaccharides to the disaccharide maltose (Table 33–2). Amylases cannot split the bond between the two glucose units of maltose. Although the amylases of the digestive tract can split the α-glycosidic bonds present in starch and glycogen, they cannot split the β-glycosidic bonds present in cellulose. In most vertebrates amylase is secreted only by the pancreas. However, in humans and certain other mammals, amylase is also secreted by the salivary glands.

Enzymes produced by the cells lining the small intestine break down disaccharides to monosaccharides. These enzymes are found in the brush border of the epithelial cells and are thought to catalyze hydrolysis while the disaccharides are being absorbed through the epithelium. **Maltase,** for example, splits maltose into two glucose molecules.

Protein Digestion

Several kinds of proteolytic enzymes are secreted into the digestive tract (Table 33–3). Each is specific for peptide bonds in a specific location in a polypeptide chain. Three main groups are exopeptidases, endopeptidases, and dipeptidases. **Exopeptidases** split the peptide bond joining the terminal amino acids to the peptide chain. For example, carboxypeptidase cleaves the peptide bond joining the amino acid with the free terminal carboxyl group to the peptide chain. Aminopeptidase splits off the amino acid with a free terminal amino group.

TABLE 33–2
Summary of Carbohydrate Digestion

Location	*Source of Enzyme*	*Digestive Process**
Mouth	Salivary glands	Polysaccharides (e.g., starch) —salivary amylase→ Maltose + Dextrin
Stomach		Action continues until salivary amylase is inactivated by acidic pH
Small intestine Lumen	Pancreas	Undigested polysaccharides and dextrins —pancreatic amylase→ Maltose
Brush border	Intestine	Disaccharides hydrolyzed to monosaccharides as follows: Maltose (malt sugar) —maltase→ Glucose + Glucose Sucrose (table sugar) —sucrase→ Glucose + Fructose Lactose (milk sugar) —lactase→ Glucose + Galactose

* ⬭ = monosaccharide (complete structures are shown in Chapter 3).

Endopeptidases cleave only peptide bonds *within* a peptide chain. Pepsin, trypsin, and chymotrypsin are endopeptidases. They differ in their requirements for specific amino acids adjacent to the bond to be split (Fig. 33–20). These endopeptidases split peptide chains into smaller fragments, which are then cleaved further by exopeptidases. The combined action of the endopeptidases and exopeptidases results in splitting the protein molecules to dipeptides. **Dipeptidases** in the

TABLE 33–3
Summary of Protein Digestion

Location	*Source of Enzyme*	*Digestive Process**
Stomach	Stomach (gastric glands)	Protein —pepsin→ Polypeptides
Small intestine Lumen	Pancreas	Polypeptides (A—A—A—A—A / A—A—A—A—A) —trypsin, chymotrypsin→ Tripeptides + Dipeptides (A—A—A) Dipeptides (A—A) —carboxypeptidase→ Free amino acids (A A A)
Brush borders (and within cytoplasm of epithelial cells)	Small intestine	Tripeptides + Dipeptides (A—A—A A—A) —peptidases→ Free amino acids (A A A A A A)

*A = amino acid units or, when standing alone, a free amino acid.

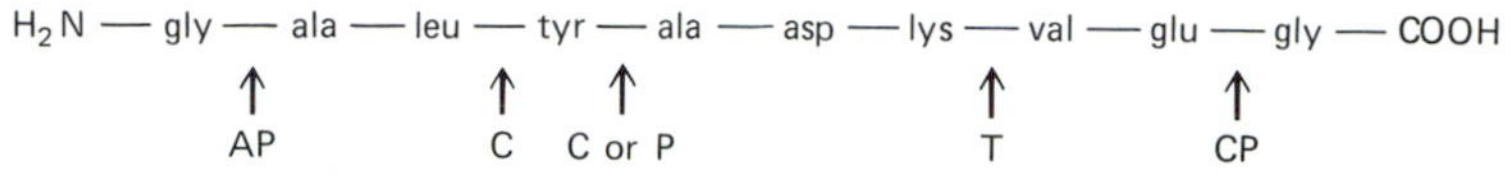

H$_2$N — gly — ala — leu — tyr — ala — asp — lys — val — glu — gly — COOH

(Arrows: AP at gly–ala; C at leu–tyr; C or P at tyr–ala; T at lys–val; CP at glu–gly)

Figure 33–20 Formula of a peptide indicating points of attack of pepsin (*P*), trypsin (*T*), chymotrypsin (*C*), aminopeptidase (*AP*), and carboxypeptidase (*CP*).

brush borders of the duodenum then split these to free amino acids. The free amino acids, and some dipeptides and tripeptides, are absorbed through the epithelial cells lining the villi and enter the blood.

Lipid Digestion

Lipids are usually ingested as large masses of triacylglycerols. They are digested largely within the duodenum by pancreatic lipase (Table 33–4). Like other proteins, lipase is water-soluble, but its substrates are not. Thus, the enzyme can attack only those molecules of fat at the surface of a mass of fat. The bile salts are detergents that reduce the surface tension of fats, breaking the large masses of fat into smaller droplets. This greatly increases the surface area of fat exposed to the action of lipase and so increases the rate of lipid digestion.

Conditions in the intestine are usually not optimal for the complete hydrolysis of lipids to glycerol and fatty acids. The products of lipid digestion therefore include monoacylglycerols and diacylglycerols, as well as glycerol and fatty acids. Undigested triacylglycerols remain as well, but some of these can be absorbed without digestion.

CONTROL OF DIGESTIVE JUICE SECRETION

Most digestive enzymes are produced only when food is present in the digestive tract. The quantity of each enzyme secreted reflects the amount needed for digestion of the food materials present. The salivary glands are controlled entirely by the nervous system, but secretion of other digestive juices is regulated by both nervous and endocrine mechanisms. For example, gastric juice is secreted in response both to neural messages and to the hormone gastrin. Both mechanisms are triggered by the presence of food in the stomach. Table 33–5 summarizes the actions of the principal hormones of the digestive system. Several other substances are suspected of being digestive system hormones, but just what they do and how they do it are not yet clear.

ABSORPTION

After the digestive enzymes have split the large molecules of protein, polysaccharides, lipids, and nucleic acids into their constituent subunits, the products are absorbed through the wall of the intestine. These ingested materials, however,

TABLE 33–4
Summary of Lipid Digestion

Location	*Source of Enzyme or Digestive Substance*	*Digestive Process**
Small intestine	Liver	Glob of fat —bile salts→ Emulsified fat (individual triacylglycerols) 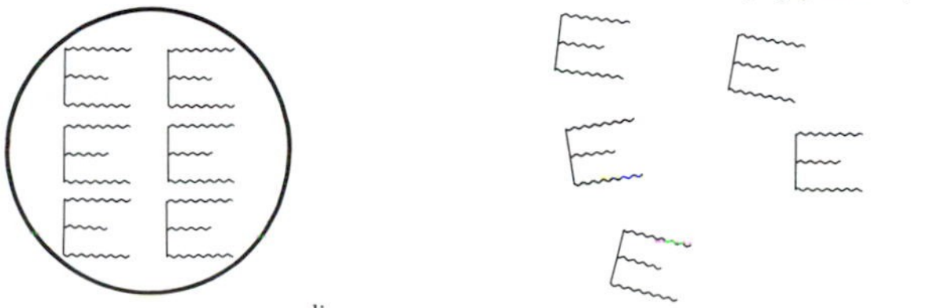
	Pancreas	Triacylglycerol —lipase→ Fatty acids + Glycerol

* [symbol] = triacylglycerol; E = glycerol; [wavy line] = fatty acid.

TABLE 33–5
Hormones for the Digestive Tract

Hormone	*Source*	*Target Tissue*	*Actions*	*Factors That Stimulate Release*
Gastrin	Stomach (mucosa)	Stomach (gastric glands)	Stimulates gastric glands to secrete pepsinogen	Distension of the stomach by food; certain substances such as partially digested proteins and caffeine
Secretin	Duodenum (mucosa)	Pancreas	Stimulates release of alkaline component of pancreatic juice	Acidic chyme acting on mucosa of duodenum
		Liver	Increase rate of bile secretion	
Cholecystokinin (CCK)	Duodenum (mucosa)	Pancreas	Stimulates release of digestive enzymes	Presence of fatty acids and partially digested proteins in duodenum
		Gallbladder	Stimulates contraction and emptying	
Gastric inhibitory peptide	Duodenum (mucosa)	Stomach	Decreases stomach motor activity, thus slowing emptying	Presence of fat or carbohydrate in duodenum

account for only a small part of the total amount of fluid absorbed daily (about 1.5 liters of a total of approximately 9 liters). The rest consists of mucus and digestive juices released by the digestive system itself.

Most substances are absorbed through the villi in the wall of the small intestine. Figure 33–15 illustrates the structure of a villus. Each of these tiny structures consists of a single layer of epithelial cells covering a network of blood capillaries. A central lymph vessel called a **lacteal** is also present within each villus.

Absorption occurs in part by simple diffusion, in part by facilitated diffusion, and in part by active transport. Glucose and other monosaccharides and the amino acids are absorbed by active transport. Absorption of these nutrients is linked to the absorption of sodium. It is thought that each protein carrier in the cell membrane that transports these nutrients has two receptor sites, one for the amino acid or monosaccharide, and the other for sodium. Both receptor sites must be filled for the carrier molecule to be activated.

After nutrients such as amino acids are actively transported into the epithelial cells lining the villi, they accumulate within the cells and then diffuse into the blood of the intestinal capillaries. Amino acids and glucose are transported to the liver by the **hepatic portal vein.** In the liver this vein gives rise to a vast network of sinusoids (tiny blood vessels similar to capillaries), which allow the nutrient-rich blood to course slowly through the liver tissue. This gives the liver cells opportunity to remove nutrients and certain toxic substances from the circulation.

The products of lipid digestion are absorbed by a different process and a different route (Fig. 33–21). Recall that fatty acids and monoacylglycerols form soluble complexes with bile salts called micelles. This greatly facilitates absorption, because the micelles transport the fatty substances to the brush borders. When the micelles come into contact with the epithelial cells of the villi, the monoacylglycerols and fatty acids (both soluble in the lipid of the cell membrane) diffuse into the cell, leaving the rest of the micelle behind to combine with new fatty acids and monoacylglycerols. In the epithelial cells, free fatty acids and glycerol are assembled once again into triacylglycerol by the endoplasmic reticulum. These triacylglycerols then are packaged into globules with absorbed cholesterol and phospholipids and covered with a thin coat of protein. Such protein-covered fat globules are called **chylomicrons.** They pass out of the epithelial cell and into the lacteal of the villus. The chylomicrons are transported by the lymph and eventually are emptied with the lymph into the blood. About 90% of absorbed fat enters the blood circulation in this indirect way. The rest, mainly short-chain fatty acids such as those in butter, are absorbed directly into the blood. After a meal rich in fats the great number of chylomicrons in the blood may give the plasma a turbid, milky appearance for a few hours.

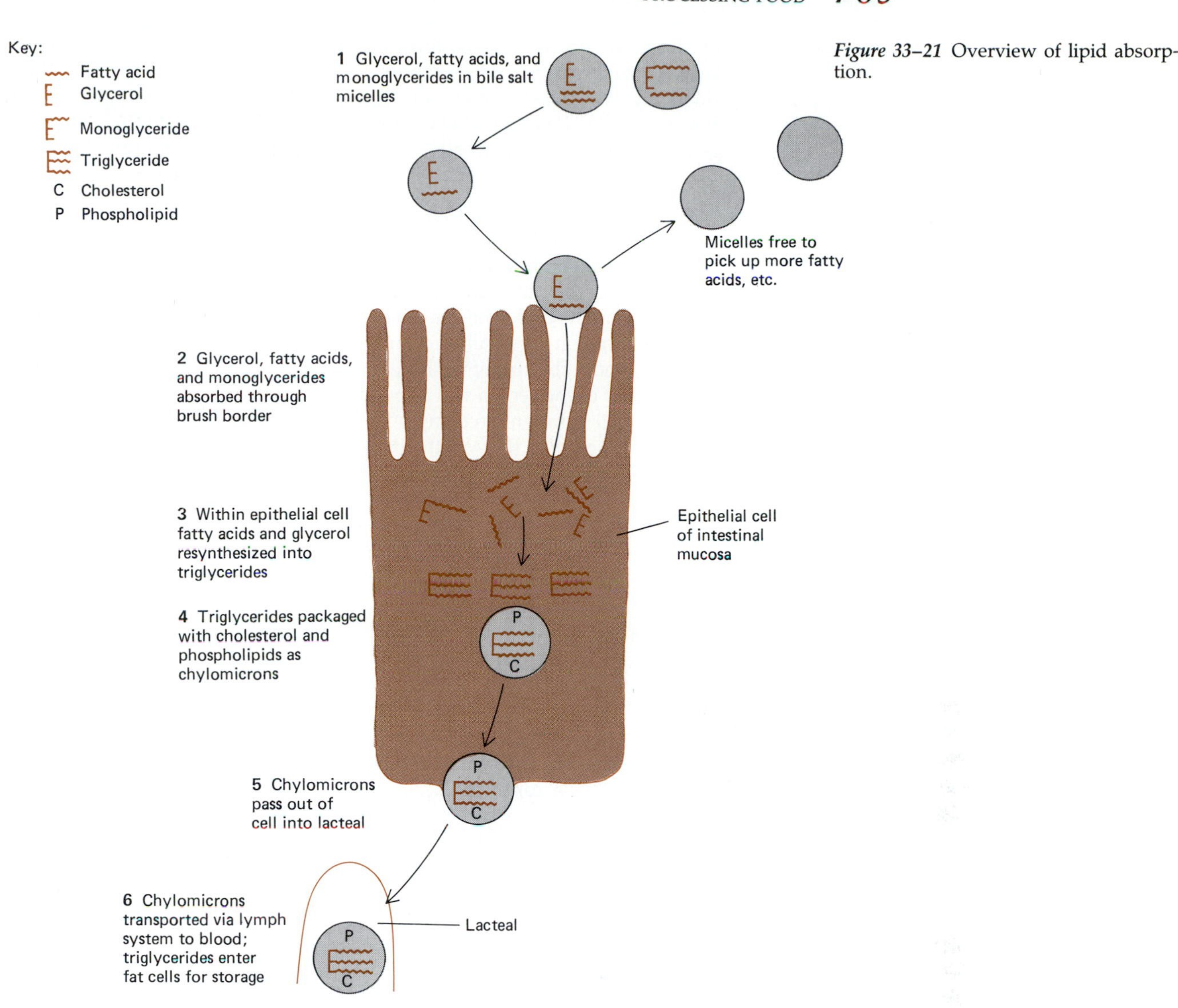
Key:
Fatty acid
Glycerol
Monoglyceride
Triglyceride
C Cholesterol
P Phospholipid
1 Glycerol, fatty acids, and monoglycerides in bile salt micelles
Micelles free to pick up more fatty acids, etc.
2 Glycerol, fatty acids, and monoglycerides absorbed through brush border
3 Within epithelial cell fatty acids and glycerol resynthesized into triglycerides
Epithelial cell of intestinal mucosa
4 Triglycerides packaged with cholesterol and phospholipids as chylomicrons
5 Chylomicrons pass out of cell into lacteal
6 Chylomicrons transported via lymph system to blood; triglycerides enter fat cells for storage
Lacteal

Figure 33–21 Overview of lipid absorption.

Most of the nutrients in the chyme are absorbed by the time the chyme reaches the end of the small intestine. What is left of the chyme (mainly waste) passes through a sphincter, the **ileocecal valve,** and into the large intestine.

THROUGH THE LARGE INTESTINE

Approximately nine hours elapse from the time food is ingested until its remnants reach the large intestine. From one to three days or even longer may be required for the slow journey through the large intestine. The large intestine functions in the following ways:

1. It absorbs sodium and water from the chyme. Sodium is absorbed by active transport, and water follows by osmosis. The chyme is slowly solidified into the consistency of normal feces.
2. It incubates bacteria. The movements of the large intestine are sluggish, giving bacteria time to grow and reproduce there. Some kinds of intestinal bacteria are mutualistic partners with their human hosts. They produce certain vitamins (vitamin K, thiamine, riboflavin, and vitamin B_{12}) in exchange for a place to live and the remnants of their host's last meal. The presence of harmless bacteria in the intestine inhibits the growth of pathogenic varieties. Should the normal ecology of the large intestine be disturbed, however, as sometimes happens when a person takes certain antibiotics, harmful bacteria may multiply and cause disease.
3. It eliminates wastes. Undigested and unabsorbed food, as well as cells that are sloughed off from the intestinal mucosa, are eliminated from the body

by the large intestine in the form of feces. A distinction should be made between elimination and excretion. **Elimination** is the process of getting rid of digestive wastes, materials that never left the digestive tract and so never participated in metabolic activities. **Excretion** refers to the process of getting rid of metabolic wastes, and is mainly the function of the kidneys. The large intestine does, however, excrete bile pigments.

The large intestine is shorter in length but larger in diameter than the small intestine. Its regions include the cecum, ascending colon, transverse colon, descending colon, sigmoid colon, rectum, and anus, the opening for elimination of feces (see Fig. 33–7). The small intestine empties into the side of the ascending colon about 7 cm from its end. This leaves a blind pouch, the **cecum,** which hangs down below the junction of small and large intestines. The **appendix,** a worm-shaped blind tube about the size of the little finger, hangs down from the end of the cecum. The presence of the cecum and appendix is something of a mystery, as they have no known function in human beings. They were probably larger in our remote ancestors and did function in them in the digestion of plant foods. Herbivores such as rabbits and guinea pigs have a large, functional cecum, containing bacteria that digest cellulose.

Both mixing and peristaltic movements occur in the large intestine, but both are ordinarily slower and more sluggish than those in the small intestine. Periodically, usually after eating, more vigorous peristaltic movements force the contents along. When a mass of fecal material reaches the weak sphincter at the entrance to the rectum, it relaxes, allowing feces to enter the rectum. Distention of the rectum stimulates nerves in its walls and brings about the impulse to **defecate,** that is, expel feces. This results in relaxation of the internal anal sphincter, which is composed of smooth (involuntary) muscle. However, the external anal sphincter, which is composed of skeletal muscle, remains contracted until voluntarily relaxed. Thus defecation is a reflex action that can be voluntarily inhibited by keeping the external sphincter muscle contracted.

The feces of a healthy human contains about 75% water by weight. The solid portion consists of about 30% bacteria, both live and dead. The rest is cellulose and other undigested and unabsorbed remnants of food, dead cells, salt, and bile pigments.

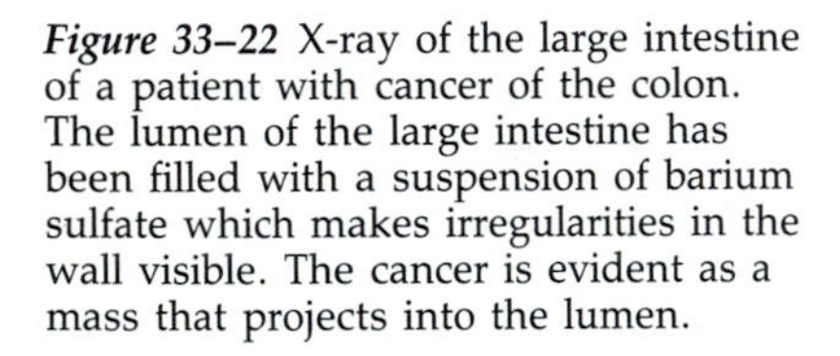

Figure 33–22 X-ray of the large intestine of a patient with cancer of the colon. The lumen of the large intestine has been filled with a suspension of barium sulfate which makes irregularities in the wall visible. The cancer is evident as a mass that projects into the lumen.

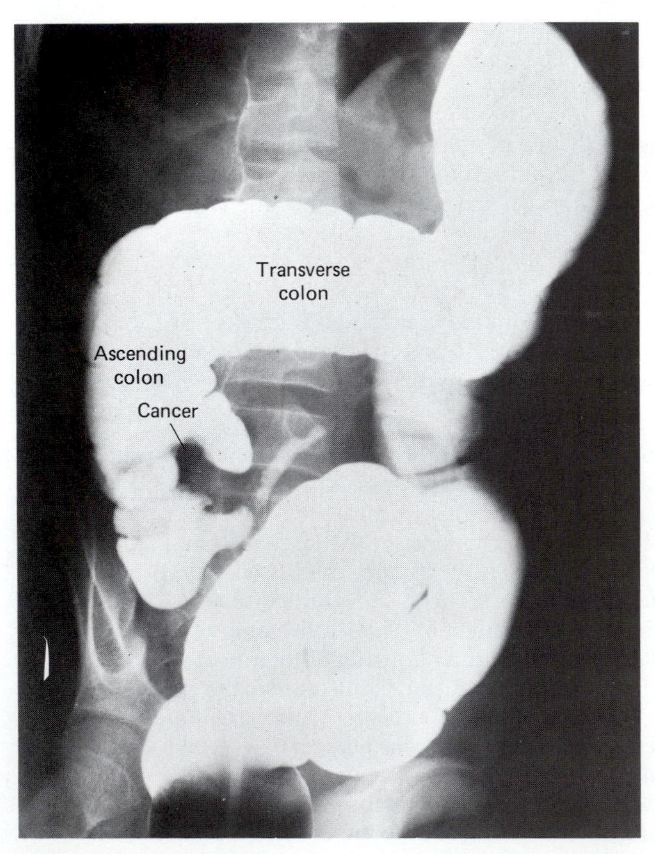

If the lining of the colon is irritated, as in certain infections, motility of the large intestine may be increased while absorption is decreased. The intestinal contents pass rapidly through the colon and only a small amount of water is absorbed from them. This condition, known as **diarrhea,** results in frequent defecation and watery feces. Prolonged diarrhea can result in loss of water and of needed electrolytes such as sodium and potassium. Dehydration, especially in infants, may be very serious, or even fatal. The opposite condition, **constipation,** results when the contents pass through the colon too slowly, so that too much water is removed from them. The feces then become excessively hard and dry. Constipation may be caused by a diet containing too little bulk.

Cancer of the colon is one of the most common types of cancer in the United States and many other industrialized countries (Fig. 33–22). This type of cancer is thought to be related to diet, since the disease is more common in people whose diets are very low in fiber. It has been suggested that a low-fiber diet results in less frequent defecation, allowing prolonged contact between the mucosa of the colon and carcinogenic substances in foods.

SUMMARY

I. Animals are heterotrophs and they are consumers.
   A. Herbivores, the primary consumers, eat algae or plants.
   B. Carnivores consume meat and are usually predators.
   C. Omnivores, such as bears and humans, eat both meat and plant material.
   D. Symbionts are members of two different species that live in intimate association with one another. Three types of symbionts are parasites, commensals, and mutualistic partners.

II. Digestion of food is accomplished in a variety of ways in invertebrates.
   A. Sponges have no digestive system; digestion is carried on intracellularly.
   B. Cnidarians and flatworms have digestive systems with only one opening, which serves as both mouth and anus.
   C. In more complex invertebrates the digestive tract is a complete tube with an opening at each end.

III. The vertebrate digestive system is a complete tube extending from mouth to anus.
   A. The wall of the digestive tract from the lumen outward consists of mucosa, submucosa, muscularis, and adventitia.
   B. Ingestion takes place through the mouth, and both mechanical breakdown and chemical digestion begin there.
      1. Each tooth consists mainly of dentin covered by enamel in the crown region and by cementum in the root.
      2. Three pairs of salivary glands produce saliva in terrestrial vertebrates. Saliva moistens food, aids in swallowing, and contains salivary amylase, which initiates carbohydrate digestion.
   C. During swallowing, food passes from the mouth cavity through the pharynx and into the esophagus. Peristaltic action moves it through the esophagus and into the stomach.
   D. In the stomach food is mechanically broken down. There, the enzyme pepsin in the gastric juice initiates protein digestion. The food is reduced to a soupy mixture called chyme.
   E. Most enzymatic digestion of food takes place in the duodenum, which receives secretions from both the liver and pancreas.
      1. The liver secretes bile, which emulsifies fat.
      2. The pancreatic juice contains proteolytic enzymes, lipase, pancreatic amylase, and other enzymes.
   F. Polysaccharides are digested to maltose by salivary and pancreatic amylases. Maltase in the brush border of the intestine splits maltose into glucose, the main product of carbohydrate digestion.
   G. Proteins are split by pepsin in the stomach and by proteolytic enzymes in the pancreatic juice. The dipeptides produced are then split by dipeptidases in the brush borders of the duodenum. Free amino acids are the end-products of protein digestion.
   H. Lipids are emulsified by bile salts and then hydrolyzed by lipase in the pancreatic juice.
   I. Secretion of digestive enzymes is regulated by nerves and hormones.
   J. Most digested nutrients are absorbed through the villi of the small intestine. Monosaccharides and amino acids enter the blood; glycerol, fatty acids, and monoacylglycerols enter the lymph.
   K. The large intestine absorbs sodium and water from the intestinal contents, incubates bacteria, and eliminates wastes.

POST-TEST

1. Once food has been ingested it must be ______________ and then ______________ through the lining of the digestive tube.

2. Vertebrate herbivores generally have a specialized section of the digestive tract called the ______________ in which live ______________.

3. When a ruminant animal chews its cud it is actually ________________.

4. Animals that eat mainly meat are ________________; those that include both meat and plant material in their diets are ________________.

5. A flea is an ________________ parasite; a tapeworm is an ________________ parasite.

6. The inner lining of the digestive tract, also called the ________________, contains ________________ cells that secrete mucus.

7. The rhythmic waves of contraction that move food through the digestive tract are referred to as ________________.

8. ________________ are teeth used for biting; they are especially large in gnawing animals.

9. The parotid glands are the largest ________________ ________________.

10. Food passing through the pharynx next enters the ________________; food leaving the stomach next enters the ________________.

11. The prominent folds in the stomach lining are called ________________.

12. The gastric juice is secreted by ________________ ________________ in the mucosa of the ________________.

13. The common bile duct conducts bile into the ________________.

Select the most appropriate answer from Column B for each entry in Column A. The same answer may be used more than once, and each entry may have more than one answer.

Column A	*Column B*
14. Hormone that stimulates release of gastric juice	a. Pepsin
15. Initiates digestion of proteins	b. Bile
16. Splits maltose	c. Dipeptidases
17. Secreted by pancreas	d. Gastrin
18. Secreted by epithelial cells lining small intestine	e. Amylase
19. Emulsifies fats	f. Maltase

20. Chylomicrons are protein-covered globules of ________________.

REVIEW QUESTIONS

1. Contrast adaptations in herbivores and carnivores that help each group to be successful in their particular nutritional life style.
2. Give examples of parasites, commensals, and mutualistic partners, and describe how each relates to its host.
3. Compare food processing in a sponge, a hydra, a flatworm, and an earthworm.
4. Trace a bit of lettuce through the human digestive system, listing in sequence each structure through which it would pass.
5. Trace the ingredients of a hamburger sandwich through the human digestive system, indicating all changes that occur en route. (Hint: The hamburger contains both protein and fat; the bun contains starch, a polysaccharide.)
6. What prevents the stomach from being digested by gastric juice? What happens when these protective mechanisms fail?
7. How does the absorption of fat differ from the absorption of glucose?
8. Why is it advantageous that the inner lining of the digestive tract is not smooth like the inside of a water pipe? What structures increase its surface area?
9. Give four functions of the liver.
10. How are the movements and secretion of the digestive system regulated? Give specific examples.

34
Nutrition and Metabolism

OUTLINE

- *I.* Metabolic work
  - *A.* Metabolic rate
  - *B.* Energy requirement and body weight
- *II.* Carbohydrate metabolism
  - *A.* Carbohydrate utilization by the cells
  - *B.* Glucose storage
- *III.* Lipid metabolism
  - *A.* Lipid storage
  - *B.* Using fat as fuel
  - *C.* Other uses of lipid
- *IV.* Protein metabolism
  - *A.* Vegetarian diets
  - *B.* Use of amino acids by the cells
  - *C.* Amino acid catabolism
- *V.* Minerals
- *VI.* Vitamins
- *VII.* Water

LEARNING OBJECTIVES

After you have read this chapter you should be able to:

1. Compare anabolism and catabolism, giving examples of each.
2. Define basal metabolic rate.
3. Define total metabolic rate, and differentiate it from basal metabolic rate.
4. Write the basic energy equation for maintaining body weight, and give the consequences that result when it is imbalanced in either direction.
5. List the principal types of carbohydrate ingested, and describe carbohydrate utilization by the cells.
6. Describe the role of the liver in regulating the level of glucose in the blood, and define the terms glycogenesis, glycogenolysis, and gluconeogenesis.
7. Trace the fate of a fat from the time of absorption until it is deposited for storage in a fat cell.
8. Trace the fate of a fatty acid from the time it is released from a fat cell until it is oxidized by a body cell such as a muscle cell.
9. Describe the fate and functions of fat, cholesterol, and phospholipids in body cells.
10. Identify types of food rich in saturated fats and cholesterol and types rich in unsaturated fats; critically summarize hypotheses that relate dietary intake of lipids to health.
11. Explain why essential amino acids must be included in the diet, and give examples of foods containing complete and incomplete proteins.
12. List three reasons why it is more difficult to obtain balanced amino acid intake from a vegetarian diet, and describe a balanced vegetarian diet.
13. Describe the effects of protein deficiency on health.
14. Describe cellular utilization of amino acids, and trace the fate of amino acids.
15. Identify minerals required by the body (as listed in Table 34–2), and give their functions in the body.
16. List the actions of the vitamins, and describe the effects of specific vitamin deficiencies.
17. List the functions of water in the body.

In the last chapter we focused upon the digestive system and how it processes food. In this chapter we will study how nutrients are utilized by the cells of the body. Nutrients, you may recall, are the chemical substances in food that can be utilized as fuel for the release of energy, as building blocks for growth, repair, and maintenance, and as substrates for metabolic processes. Nutrients for these needs are obtained by eating a balanced diet containing carbohydrates, lipids, proteins, minerals, vitamins, and water. Although specific needs may vary among species, all animals have certain common nutritional requirements.

The term **metabolism** describes all the chemical and energy transformations that take place in the organism. The building, or synthetic, phases of metabolism are called **anabolism.** Through **anabolic reactions** nutrients are assembled to form macromolecules. Some of these are incorporated into cell organelles; others serve as enzymes to catalyze specific chemical reactions. Anabolic activities require energy, which is derived from using some of the nutrients as fuel. As carbohydrates, fatty acids or amino acids are oxidized; energy stored within their chemical bonds is released, and some is conserved in the energy-rich phosphate bonds of ATP. As energy is needed, the phosphate ester bonds of ATP are cleaved and the energy becomes available to drive anabolic reactions and other forms of cellular work. The process by which larger molecules are split to smaller ones is known as **catabolism,** and its individual reactions are known as **catabolic reactions.**

Metabolic Work

The greater the metabolic activity of an organism, the more oxygen is consumed. Thus the amount of oxygen utilized by an aerobic organism is a measure of its metabolic work.

METABOLIC RATE

An organism's **total metabolic rate** is the sum of its **basal metabolic rate (BMR)** and the energy used to carry on all of its daily activities. BMR reflects the amount of energy an organism must expend just to survive. When oxygen is in short supply many organisms reduce their activity to (or close to) the BMR. They do this by withdrawing into their shells or tubes, or by retreating to a place where they can remain quiet until the environmental oxygen supply again becomes sufficient to support normal activity. Some adventurous persons have had themselves buried alive in locked boxes. Often such persons are capable of astonishing endurance. They must remain completely inactive and in a very relaxed state, approximating basal conditions, so that oxygen consumption remains at a minimal level.

The BMR for a young man is about 1600 kilocalories (kcal) per day, and for a woman, about 5% less. Total metabolic rate varies markedly with a person's life style. Someone with a sedentary life style may use about 2000 kcal per day, whereas a construction worker engaged in heavy physical labor may expend 6000 kcal or more each day. Table 34–1 gives the approximate kcal expended per hour for various activities.

ENERGY REQUIREMENT AND BODY WEIGHT

An average-sized male student who spends his days studying and does not engage in physical exercise uses about 2000 kcal daily. If he ingests 2000 kcal in food each day, his body will be in a state of energy balance, with energy (Calorie) input equaling energy output. When energy input equals energy output body weight remains constant.

Middle-aged people tend to gain weight because physical activity, but not appetite, decreases with age. When energy input is greater than energy output, the excess nutrients are stored primarily as fat and the person gains weight. An excess of only 10 kcal per day over the energy requirement leads to an increase in weight of about 1 kg (2.2 lb) in the course of a year.

When energy output is greater than energy intake, the body must draw upon its stored materials. Fats are mobilized and fatty acids are oxidized to yield energy. Body weight then decreases.

TABLE 34–1
Energy Expenditure per Hour of a 70-kg Man During Different Types of Activity

Form of Activity	*Kilocalories per Hour*
Sleeping	65
Awake, lying still	77
Sitting at rest	100
Standing relaxed	105
Dressing and undressing	118
Typewriting rapidly	140
"Light" exercise	170
Walking slowly (4.2 km per hour)	200
Carpentry, metal-working, industrial painting	240
"Active" exercise	290
"Severe" exercise	450
Sawing wood	480
Swimming	500
Running (8.5 km per hour)	600–800
Walking very fast (8.5 km per hour)	650
Walking up stairs	1100

From Guyton, A. C.: *Textbook of Medical Physiology*, 6th ed. Philadelphia, W. B. Saunders Co., 1981. Data compiled by Professor M. S. Rose.

When no food is eaten for a period of time the body mobilizes its stored materials in a definite sequence. The first to be used are the carbohydrates, stored as glycogen in the liver and muscles (Fig. 34–1). Glycogen stores are depleted within a few hours of fasting. The cells then turn to fat and protein for fuel. An average adult male has about 9 kg of stored fat. When this is utilized, it will supply enough energy for 5 to 7 weeks of life. During this time, proteins are also mobilized from muscle cells, split to their component amino acids, and converted to keto acids, which can be metabolized in the citric acid cycle. After several weeks of

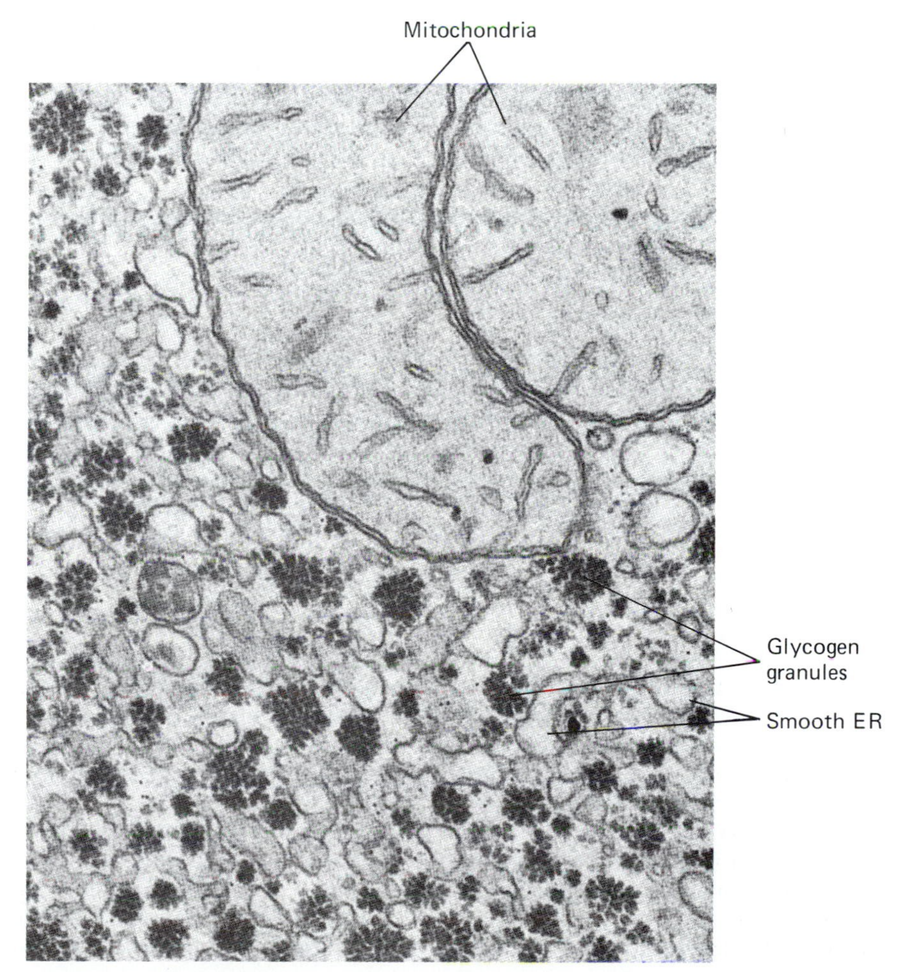

Figure 34–1 Electron micrograph showing glycogen granules stored in the cytoplasm of a liver cell. Note the extensive smooth endoplasmic reticulum. (From D. W. Fawcett.)

fasting, so much protein is catabolized from skeletal muscle, heart muscle, and other internal organs that the body is unable to function. When body proteins have been depleted to about half their normal level, death occurs, usually from failure of the heart muscle.

Carbohydrate Metabolism

Sugars and starches are the principal sources of energy in the ordinary human diet, but they are not essential nutrients because if necessary, energy can be obtained instead from a mixture of proteins and fats. Foods rich in carbohydrates (such as bread, cereals, potatoes, and other root vegetables) are usually less expensive than those rich in fats and protein. Consequently, economic factors often play a role in determining the percentage of carbohydrates in the diet. Carbohydrates, mainly starch and sugar, account for about 45% of the kilocalories ingested daily in the average American diet.

Starch, you may recall, is the form in which plant cells store glucose. Much carbohydrate is also ingested in the form of cellulose, the material that makes up the cell walls of plant cells. Humans do not have the enzymes necessary to split the β-1,4-glycosidic bonds in cellulose, so they are unable to digest this carbohydrate. Cellulose does, however, provide bulk needed for proper function of the large intestine and formation of feces. Along with certain other indigestible carbohydrates found in plant tissues, cellulose is a principal component of dietary **fiber.** Fruit, vegetable pulp and skin, stalks, leaves, and the outer coverings of nuts, grains, seeds, and legumes are all rich in cellulose.

In affluent societies the disaccharide sucrose (cane or beet sugar) accounts for about 25% of the carbohydrate intake, with the percentage even higher in children. Sucrose is the refined sugar put in coffee and used to make desserts. Other important dietary carbohydrates are lactose, the sugar in milk, and fructose, the sugar in fruit.

During digestion, starch and disaccharides are cleaved to the monosaccharides glucose, fructose, and galactose. These are then absorbed and transported to the liver via the hepatic portal vein. In the liver all of the monosaccharides are converted to glucose.

One of the important jobs of the liver is to help maintain a steady blood sugar level. The normal fasting concentration of blood glucose is about 90 mg/100 ml (also expressed 90 mg%). After a meal the concentration may increase briefly to about 140 mg/100 ml. The liver removes glucose from the blood after a meal and stores it, and then returns glucose to the blood when needed.

CARBOHYDRATE UTILIZATION BY THE CELLS

As blood circulates through the body, it transports glucose to all of the cells. Glucose diffuses into the interstitial fluid and is taken up by cells almost continuously. Once inside individual cells, glucose is used primarily as fuel for cellular respiration (Chapter 9).

Glucose is also used by cells for synthesizing other molecules. Some glucose is used to make ribose and deoxyribose, which are components of nucleic acids. Portions of the carbon skeleton of glucose are utilized in the synthesis of amino acids and other substances.

GLUCOSE STORAGE

As blood courses through the liver, liver cells remove glucose. In the liver cells, molecules of glucose are linked to produce glycogen, a highly branched polysaccharide of high molecular weight. Glycogen is a stable molecule that usually precipitates out of solution, forming granules (Fig. 34–1). The process of glycogen synthesis is called **glycogenesis** (Fig. 34–2).

Liver cells are able to store large amounts of glycogen, up to 8% of their weight. Muscle cells can store about 1% of their weight, and other types of cells store smaller amounts. Brain cells have only scanty stores of glycogen and only a limited ability to use fatty acids or amino acids as sources of energy; for these reasons, the brain is the first organ to suffer when the concentration of glucose in

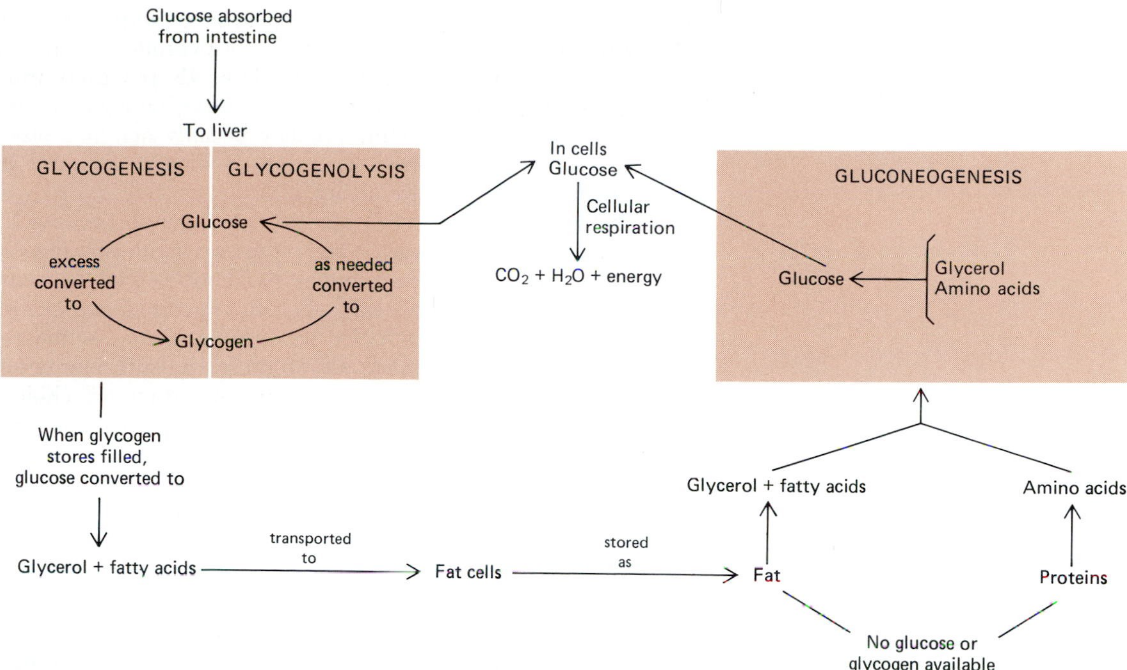

Figure 34–2 The fate of glucose in the body.

the blood falls below homeostatic levels. The brain then is not adequately supplied with fuel and cannot carry on the metabolic processes that yield energy for its normal functioning. Symptoms of a low glucose concentration **(hypoglycemia)** resemble those of oxygen deficiency: The victim may appear mentally confused, become drowsy and pass out, suffer convulsions, or even die.

When the blood glucose level begins to fall between meals, glycogen is reconverted to glucose, a process known as **glycogenolysis** (breaking down glycogen). The glucose then is released into the blood. The reactions of glycogenolysis are not the reverse of those used to produce glycogen. Each glucose molecule is split off from the glycogen by a phosphorylation process; that is, a phosphate is added to the glucose. This requires an enzyme called **phosphorylase.**

The liver can store enough glycogen to maintain blood glucose levels for several hours. After that, liver cells begin to convert amino acids and the glycerol portions of fat molecules into glucose. This process is called **gluconeogenesis** (production of new glucose).

The liver's ability to maintain the concentration of glucose in the blood is regulated by several hormones, principally insulin and glucagon, which are both secreted by the islets of Langerhans in the pancreas. Their actions will be discussed in Chapter 42.

When carbohydrate-rich food is eaten, the liver cells may become fully loaded with glycogen. The remaining glucose may then be converted to fatty acids and glycerol by liver and fat cells. These compounds are then assembled into triacylglycerols and stored in adipose tissue for use as fuel at some later time. It has long been known that eating large amounts of starches or sugar is fattening. The starch of corn or wheat eaten by cattle and pigs is converted into the fat of butter and bacon. With the use of radioactive or stable isotopes, it can be shown that a particular carbon or hydrogen atom that enters the body as a carbohydrate can be recovered as fat in adipose tissue or liver. The metabolic pathways by which carbohydrates are converted to fats are outlined in Figure 34–2.

Lipid Metabolism

The average adult American eats about 65 kg of lipids (mainly fat) each year, receiving about 40% of the total kilocalories from this source. Among poor people this percentage falls to less than 10%, because most foods rich in lipid—meats, eggs, and dairy products—are relatively expensive. Most lipids in the diet are ingested as triacylglycerols. Foods of animal origin tend to be rich in both saturated fats and cholesterol, whereas foods of plant origin contain mainly unsaturated fats and no cholesterol. Commonly used polyunsaturated vegetable oils are safflower oil (with about 78% polyunsaturated fatty acids), soybean (with about 62% polyunsaturated fatty acids), and corn oil (with about 58%). Coconut oil is an exception in that it is mainly saturated. Butter contains only about 4% polyunsaturated fatty acids.

The average American diet includes about 700 mg cholesterol daily, far more than the recommended 300 mg. Foods especially rich in cholesterol are egg yolk, liver, and kidney. Cholesterol is present in smaller amounts in the fat of meat, butter, milk, cheese, and ice cream.

Dietary deficiencies of lipids are not common because lipids are so widespread in foods, and because the body is able to synthesize most of them from other organic compounds. However, the human body is unable to synthesize enough of three polyunsaturated fatty acids—linoleic, linolenic, and arachidonic acids—and these must be obtained in the diet. Because these are essential to the diet, they are termed the **essential fatty acids.** Given these and sufficient nonlipid nutrients, the body can make all of the lipid compounds (including fats, cholesterol, phospholipids, and prostaglandins) that it needs.

LIPID STORAGE

Recall from the last chapter that lipids are absorbed into the lymph circulation and are transported as tiny droplets called chylomicrons. A **chylomicron** consists of triacylglycerols, cholesterol, phospholipids, and a small amount of protein present as a covering around the lipid droplet. From the lymph circulation the chylomicrons enter the blood. As they pass through the capillaries of the adipose tissue and sinusoids of the liver, most of the chylomicrons are removed.

Cholesterol is stored primarily in the liver cells. Phospholipids are degraded to yield fatty acids, which are reassembled into triacylglycerols within the liver and fat cells. These triacylglycerols, along with those from the chylomicrons, are stored in the fat and liver cells. The fat cells of adipose tissue can store large amounts of triacylglycerols, up to 95% of their volume (see Fig. 6–5). For this reason adipose tissue is often referred to as the fat depot of the body.

The cells of adipose tissue are very active metabolically. Triacylglycerols are constantly being synthesized and degraded. This occurs so rapidly that the triacylglycerol content in the fat cells is entirely changed within a three-week period. Thus, the fat molecules stored in your fat cells are not the same ones that were there last month. Sad to say, the total amount of fat is probably much the same! (See Focus on Obesity.)

USING FAT AS FUEL

The main function of adipose tissue is to store triacylglycerols until they are utilized as fuel by the cells. Fats are used as the body's principal storage form of energy because they are a very concentrated source of energy and because they are not water-soluble. The metabolism of a gram of fat to carbon dioxide and water yields up to 9 kcal of energy, more than twice as much as the metabolism of a gram of protein or carbohydrate. Most cells can use fatty acids as a source of energy. Between meals, in fact, most cells shift their metabolism so that fatty acids are oxidized in preference to glucose. This shift spares glucose for the brain cells which are less able to utilize fatty acids as fuel.

Normally, the fasting level of free fatty acids in the blood is about 15 mg/100 ml. As cells remove fatty acids from the blood and the concentration decreases, fats are mobilized from the fat cells. Fat cells degrade triacylglycerol and release free fatty acids into the blood. These fatty acids combine with a certain type of albumin—a plasma protein—and are transported in this way to the liver.

Fatty acids are metabolized by **β-oxidation** (beta-oxidation; Chapter 9), which takes place mainly in the liver cells. Some of the acetyl coenzyme A formed is oxidized in the citric acid cycle, by the liver cells, but some is transported to other cells (Fig. 34–3). Before these molecules can leave the liver cells they must be converted to a form that can diffuse freely out of the liver cells and be transported in the blood. Two acetyl coenzyme A molecules are combined and the coenzyme A portions of the molecule split off to form one of three types of **ketone bodies** (acetoacetic acid, β-hydroxybutyric acid, and acetone). Ketone bodies are taken up from the blood by heart and skeletal muscle cells, are converted back to acetyl coenzyme A, and then may be used as fuel.

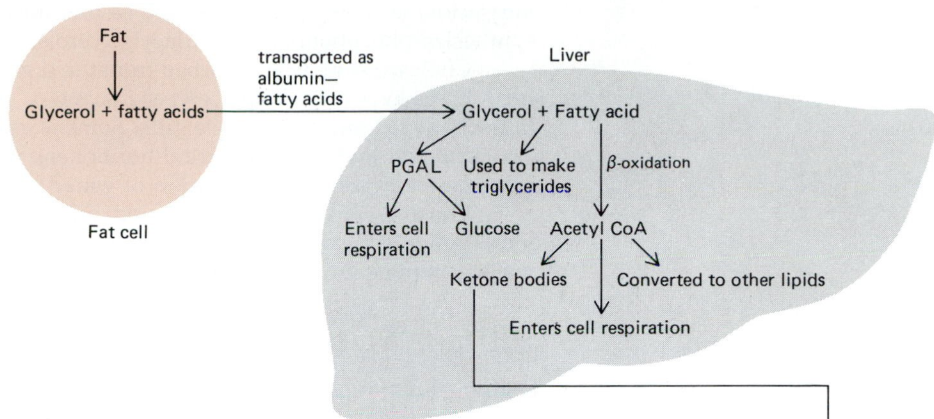

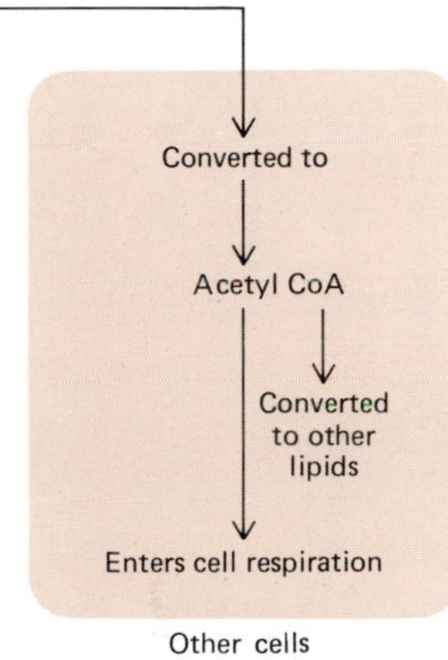

Figure 34–3 Overview of lipid metabolism.

Normally the concentration of ketone bodies in the blood is only about 3 mg/100 ml. When a person has untreated diabetes mellitus, is starving, or is on a high-fat diet, however, the rate of lipid metabolism is greatly increased. Ketone bodies are then produced in much greater quantity and their concentration in the blood may increase, producing a condition known as **ketosis,** in which the normal acid–base balance of the body is disturbed. The accumulated acids (acetoacetic acid and β-hydroxybutyric acid) render the blood and tissues acidic. Without appropriate treatment this can lead to coma and even death.

OTHER USES OF LIPID

Triacylglycerols are important components of adipose tissue and as such contribute to its functions. In the subcutaneous layer, adipose tissue provides insulation, which helps maintain body temperature. Adipose tissue also helps to protect the body against mechanical injury.

Fats are also important in promoting the absorption and transport of certain vitamins that are fat-soluble. Cells use acetyl coenzyme A (which is produced from

FOCUS ON

Obesity

At present obesity is one of the most important nutritional problems in the United States. A person who has an excessive accumulation of body fat, and who is more than 20% overweight, is obese. About 20% of persons in the United States fit into this category. Obesity predisposes to a number of diseases, including cardiovascular disease and diabetes, and thus also decreases life expectancy.

The number of fat cells in the adult body appears to be determined mainly by the amount of fat that was stored initially during infancy and childhood. Overfed infants may develop up to three times more fat cells than those fed a diet more balanced in calories. Such a person may also develop a higher setting of the brain center that controls food intake. When an overweight person goes on a diet, fat is mobilized and the fat cells may shrink, but they do not disappear; should the person begin to overeat once again, the waiting fat cells simply fill up like tiny balloons with new fat. Persons who become overweight during middle age or old age probably do not develop new fat cells. Those they have just become larger as they store more fat.

Most overweight people overeat owing to a combination of poor eating habits and psychological factors. For many, eating is a way of releasing tension.

The only cure for obesity is to adjust food intake to meet energy needs. To lose weight, energy intake must be less than energy output. The body will then draw on its fat stores for the needed kilocalories, and body weight will decrease. This is best done by a combination of increased exercise and decreased total caloric intake.

Despite all of the claims by proponents of fad reducing diets, most nutritionists agree that the best reducing diet contains no special kinds or proportions of food items but is a well-balanced diet that provides a normal proportion of fats, carbohydrates, and proteins. In other words, the dieter should eat everything, but in small quantities.

fatty acids) to synthesize cholesterol, steroid hormones, and the fatty acids used to synthesize phospholipids and triacylglycerols.

Cholesterol may be absorbed from the digestive tract or produced in the cells (mainly the liver cells) from acetyl coenzyme A. As much as 80% of the cholesterol in the body is used to make bile salts. Some is used to make steroid hormones (the sex hormones and adrenal cortical hormones). Cholesterol and other lipids in the skin also help prevent evaporation of water from the skin.

Phospholipids are manufactured primarily in the liver cells. They play an important role as structural components of cell membranes. Phospholipids are also major components of the myelin sheaths of nerve cells.

Protein Metabolism

High-quality protein is the most expensive and least available of all nutrients. For this reason the amount of protein consumed is often an index of a country's as well as a person's economic status. Twenty-two amino acids have been identified as constituents of most proteins. The body is able to synthesize most of these by transferring an amino group to the corresponding keto acid. Of these 22 amino acids, 8 (10 in children) cannot be synthesized by the body cells. These, which must be provided in the diet, are termed **essential amino acids.** (Note, however, that although they *are* essential in the diet, they are no more essential metabolically than the other amino acids.)

Not all proteins contain the same varieties or quantities of amino acids, and many proteins are deficient in some of the essential ones. **Complete proteins** contain all of the essential amino acids in sufficient quantity for protein synthesis and growth in the young. Such proteins include casein (the main protein in milk), along with the main protein in egg, and proteins found in meat, fish, and poultry. **Partially incomplete proteins,** such as those found in legumes (peanuts, soybeans), grains, and nuts, provide sufficient amino acids for maintaining protein metabolism, but not growth. **Totally incomplete proteins,** such as those found in gelatin and corn, lack essential amino acids and by themselves support neither normal protein metabolism nor growth. Most plant proteins are deficient in one or more essential amino acids, usually lysine, methionine, threonine, or tryptophan.

The recommended daily amount of protein in the diet is about 56 gm for a 70-kg man (about 0.8 gm of protein per kg of body weight per day). In the United States and other developed countries most persons eat more protein than they need. The average American eats about 54 kg of meat each year. This is about 150 grams of meat per day. Since the protein content of meat is about 25%, daily meat consumption includes about 38 grams of protein. Additional protein intake comes from vegetables, nuts, grains, fish, and dairy products. In contrast those residing in the less economically developed countries consume an average of only about 1 kg of meat per year.

Millions of human beings throughout the world, especially in the less developed countries, suffer from symptoms of **protein deficiency.** Without an adequate supply of amino acids in the diet, children cannot synthesize proteins and so cannot develop properly, and both physical and mental development may be retarded. Without adequate dietary proteins the body cannot produce antibodies against disease organisms, so resistance to disease is lowered. Common childhood diseases such as measles, chickenpox, and whooping cough are often fatal in malnourished children. A characteristic symptom of protein deficiency is **edema** (swelling caused by abnormal amounts of fluid in the tissues). Severely malnourished children may have swollen bellies (Fig. 34–4). With inadequate protein intake the liver cannot manufacture the needed plasma proteins that normally help to maintain a balance between the fluids in the blood and tissues. And without sufficient plasma proteins, fluid accumulates in the tissues.

VEGETARIAN DIETS

Most of the world's population depends almost entirely upon plants, especially cereal grains—usually rice, wheat, or corn—as the staple food. None of these foods contains adequate amounts of essential amino acids. Besides being deficient in some of the essential amino acids, plant foods contain a lower percentage of protein than do animal foods. Meat contains about 25% protein, whereas even the new

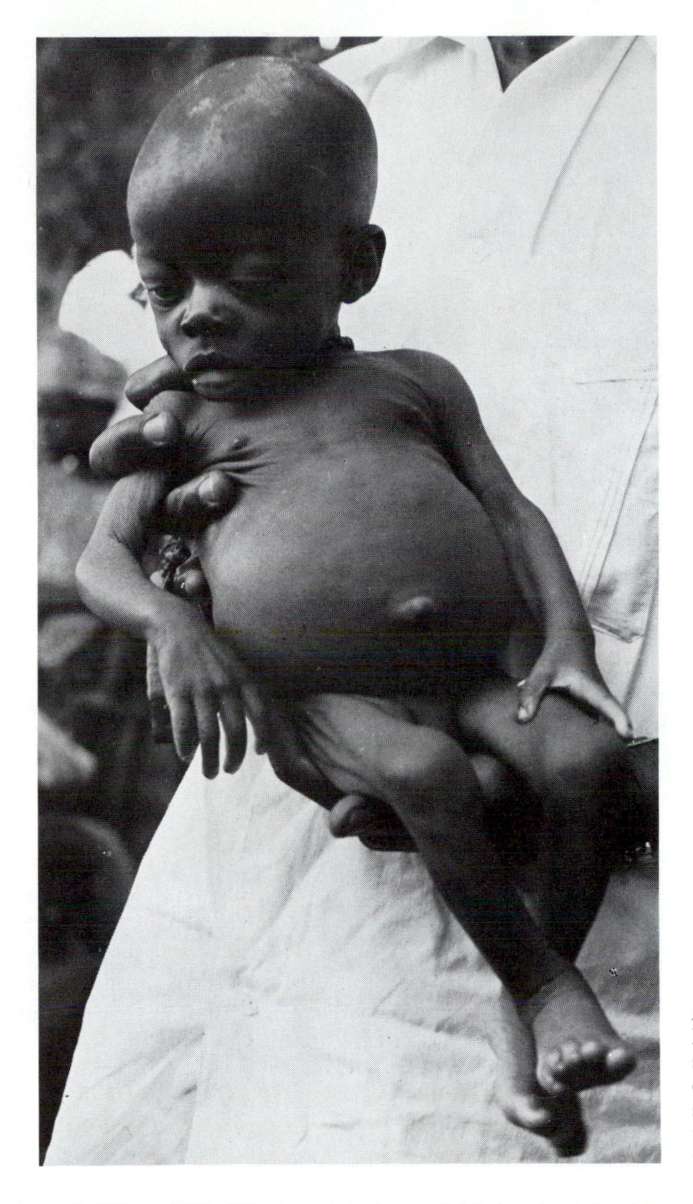

Figure 34–4 Child suffering from kwashiorkor, a disease caused by severe protein deficiency. Note the characteristic swollen belly, which results from fluid imbalance. (United Nations, Food and Agricultural Organization photo.)

high-yield grains contain only 5% to 13%. What protein is available in plant food is also less digestible than that in animal foods. Because much of the protein is encased within the indigestible cellulose cell walls, a great deal of it passes right through the digestive tract.

Despite these potential nutritional problems, more and more people are turning to vegetarian diets. Meats are becoming increasingly expensive because they are ecologically expensive to produce. About 21 kg of protein in grain, for example, is required to produce just 1 kg of beef protein. If the human population of our planet continues to expand at a much greater rate than food production, more grain will be diverted for human food and less for animal feed. The price of meat will continue to soar and may become unaffordable for many of us.

Can a vegetarian diet be nutritionally balanced? With an awareness of the special nutritional problems associated with a vegetarian diet (especially in growing children), they can be overcome. The most important rule is to select foods that complement each other. This requires knowledge of which amino acids are deficient in each kind of food. Since the body cannot store amino acids, all of the essential amino acids must be ingested at the same meal. For example, if rice is eaten for dinner, and beans for lunch the next day, the body will not have all of the essential amino acids that are needed at the same time to manufacture proteins. If beans and rice are eaten together, however, all of the needed amino acids will be provided, because what one food lacks, the other provides. Similarly, if dairy products are not excluded from the vegetarian diet, then macaroni can be paired with cheese, or cereal with milk, and all the essential amino acids will be obtained.

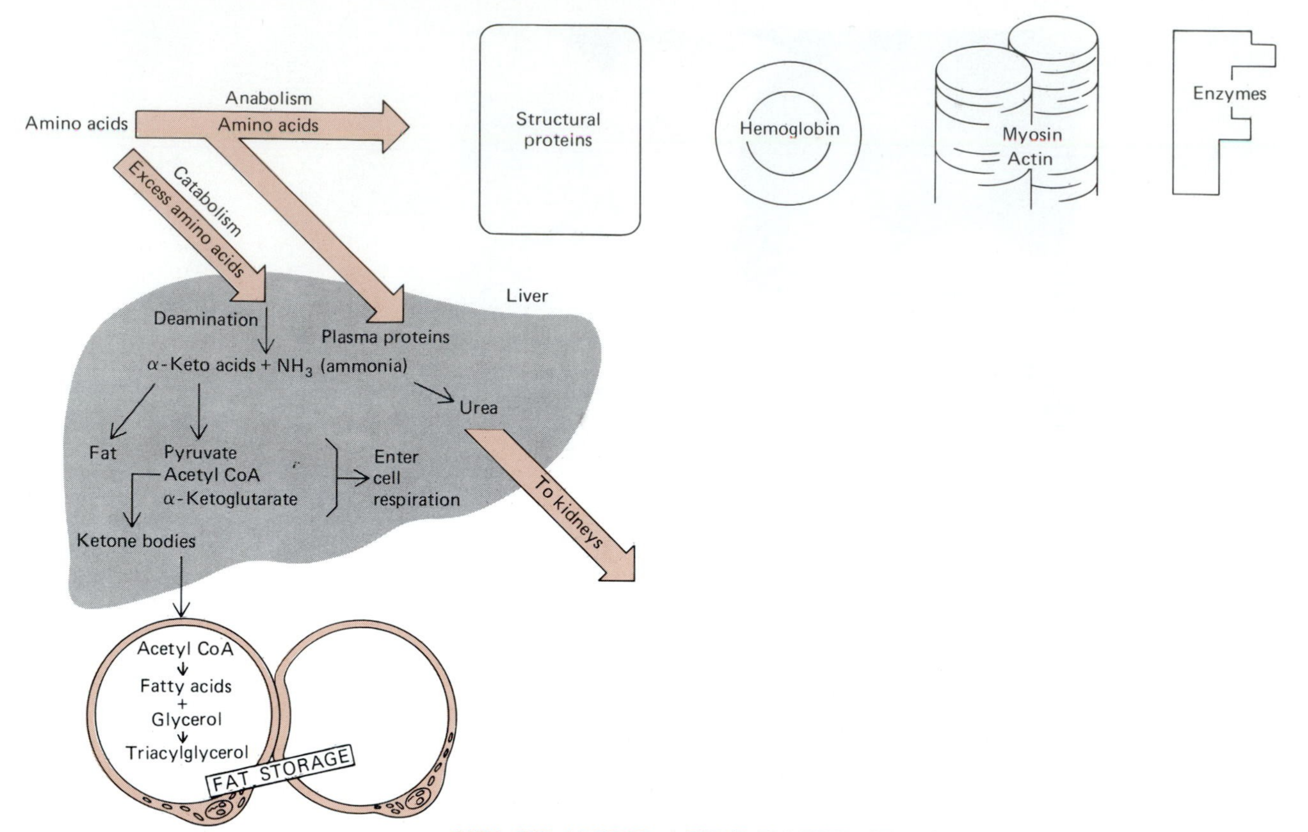

Figure 34–5 Overview of protein metabolism.

USE OF AMINO ACIDS BY THE CELLS

Amino acids are utilized primarily in the anabolic processes of the body, for example, in the synthesis of proteins and nucleic acids (Fig. 34–5). Amino acids are used to manufacture all of the thousands of enzymes needed by each cell. They are also used to make plasma proteins, including the immunoglobins needed to defend the body against disease organisms. They are the essential ingredients of hemoglobin, and of the muscle proteins actin and myosin. Amino acids are also components of several hormones and of some body secretions such as mucus and milk. Amino acids can also be used as a source of energy. Like glucose, they supply about 4 kcal per gram.

AMINO ACID CATABOLISM

Although small pools of amino acids are present in the cytoplasm of all cells, there is no mechanism for storing large quantities of these nutrients. Amino acids are removed from the blood by liver cells and deaminated; that is, the amino group is split off from the carbon skeleton.

$$\underset{\text{Amino acid}}{R_1-\overset{\displaystyle NH_2}{\overset{|}{\underset{\underset{\displaystyle H}{|}}{C}}}-COOH} + H_2O + NAD^+ \longrightarrow \underset{\text{Keto acid}}{R_1-\overset{\displaystyle O}{\overset{\|}{C}}-COOH} + \underset{\text{Ammonia}}{NH_3} + NADH + H^+$$

As they are split from the amino acid, the amino groups are converted to ammonia (NH_3). However, ammonia is toxic, and is rapidly converted to urea, which diffuses into the blood and is excreted by the kidneys. The carbon chain of the amino acid remaining after deamination is a keto acid. Some keto acids can be converted to glucose and either used as fuel or stored as glycogen. Other keto acids are converted into acetyl coenzyme A and used as fuel, or converted into fatty acids and stored as fat.

Even if you drastically reduce your carbohydrate and lipid intake in favor of proteins, you may still gain weight. An excess of *any* organic nutrient can be converted to fat. Figure 34–6 summarizes the metabolic relationships among proteins, lipids, and carbohydrates.

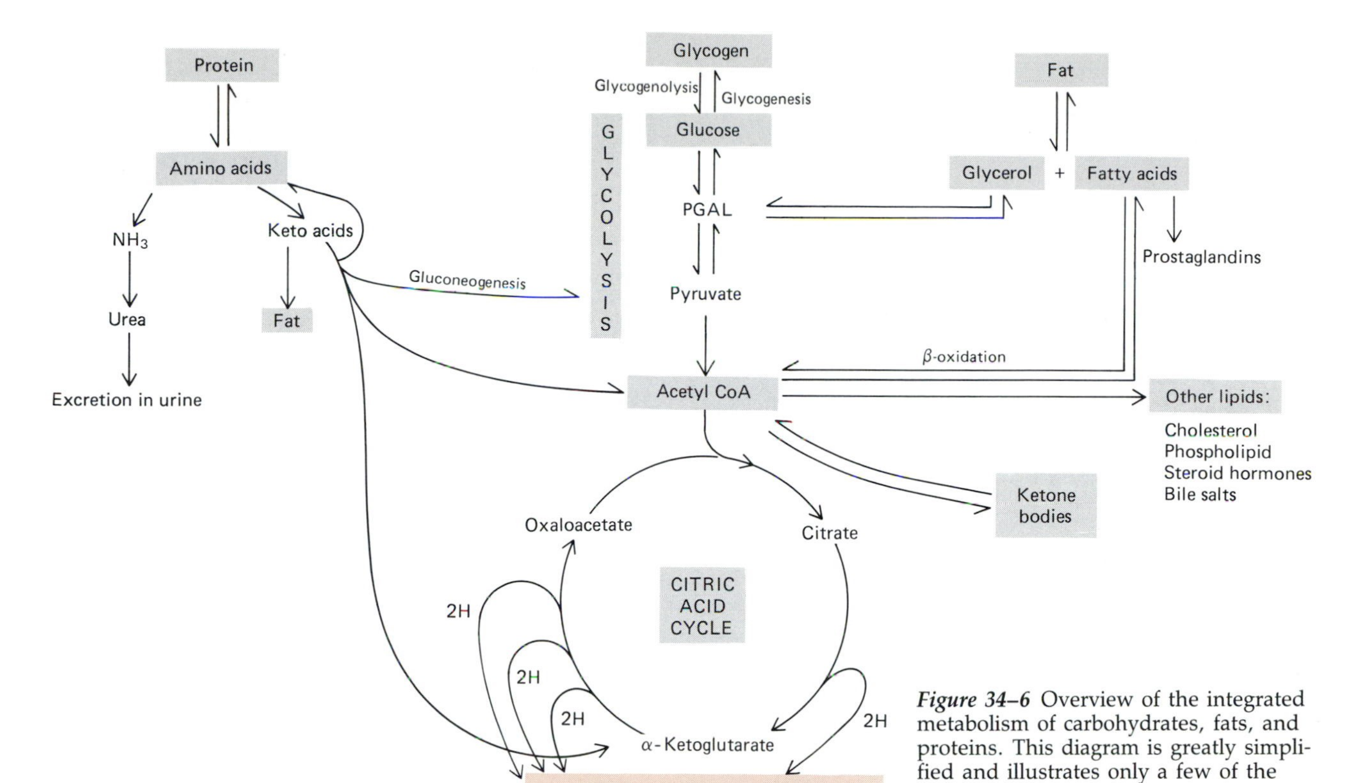

Figure 34–6 Overview of the integrated metabolism of carbohydrates, fats, and proteins. This diagram is greatly simplified and illustrates only a few of the principal pathways.

Minerals

Minerals are inorganic nutrients ingested as salts dissolved in water or food, or as part of organic compounds. Twenty-one minerals are known to be essential to human nutrition. Some are important in maintaining fluid balance; others help to regulate metabolism by functioning as a component of enzymes. Certain minerals are essential for normal nerve and muscle function, for blood clotting, and as components of some hormones and other organic compounds. Other minerals are structural components of tissues such as bone and teeth.

Calcium, phosphorus, sulfur, potassium, sodium, chlorine, and magnesium are needed in relatively large amounts (more than 100 mg per day). Other minerals, such as iron and zinc, are needed in smaller amounts (less than 20 mg per day). Still others, required in minute quantities, are known as **trace elements.** Table 34–2 lists some important minerals and gives their functions in the body. Some of these minerals are discussed in more detail in chapters dealing with the functions that they help perform.

Iron, calcium, and iodine are the minerals most likely to be deficient in the diet. Because iron is an essential ingredient of hemoglobin, dietary deficiency of this mineral causes anemia. The blood of anemic individuals cannot transport sufficient oxygen, and metabolism is slowed. One of calcium's many functions is to act as a structural component of bone and teeth; it is also essential for transmission of neural impulses and for muscle contraction. Iodine is an essential component of the thyroid hormones; too little iodine may result in goiter, a condition in which the thyroid gland enlarges, sometimes to monstrous size.

Vitamins

The discovery of vitamins and the analysis of their properties and functions in metabolism have been among the most notable achievements in science since the turn of this century. **Vitamins** are a group of unrelated organic compounds, required in the diet in very small amounts, and essential for normal metabolism and good health. Some vitamins help to regulate metabolism by serving as part of coenzymes. Niacin, for example (one of the B complex vitamins), is a component of the coenzyme NAD, which is essential to cellular respiration and many other biological processes.

TABLE 34–2
Some Important Minerals and Their Functions

Mineral	*Functions*	*Comments*
Calcium	Component of bone and teeth; essential for normal blood clotting; needed for normal muscle and nerve function	Good sources: milk and other dairy products, green leafy vegetables. Bones serve as calcium reservoir.
Phosphorus	As calcium phosphate, an important structural component of bone; essential in energy transfer and storage (component of ATP) and in many other metabolic processes; component of DNA and RNA	Performs more functions than any other mineral; absorption impaired by excessive intake of antacids
Sulfur	As component of many proteins (e.g., insulin), essential for normal metabolic activity	Sources: high-protein foods such as meat, fish, legumes, nuts
Potassium	Principal positive ion within cells; influences muscle contraction and nerve excitability	Occurs in many foods
Sodium	Principal positive ion (cation) in interstitial fluid; important in fluid balance; essential for conduction of nerve impulses	Occurs naturally in foods; sodium chloride (table salt) added as seasoning; too much ingested in average American diet; in excessive amounts, may lead to high blood pressure
Chlorine	Principal negative ion (anion) of interstitial fluid; important in fluid balance and in acid–base balance	Occurs naturally in foods; ingested as sodium chloride
Copper	Component of enzyme needed for melanin synthesis; component of many other enzymes; essential for hemoglobin synthesis	Sources: liver, eggs, fish, whole wheat flour, beans
Iodine	Component of thyroid hormones (hormones that stimulate metabolic rate)	Source: seafoods, iodized salt, vegetables grown in iodine-rich soils. Deficiency results in goiter (abnormal enlargement of thyroid gland).
Cobalt	As component of vitamin B_{12}, essential for red blood cell production	Best sources are meat and dairy products. Strict vegetarians may become deficient in this mineral.
Manganese	Necessary to activate arginase, an enzyme essential for urea formation; activates many other enzymes	Poorly absorbed from intestine; found in whole-grain cereals, egg yolks, green vegetables
Magnesium	Appropriate balance between magnesium and calcium ions needed for normal muscle and nerve function; component of many coenzymes	Occurs in many foods
Iron	Component of hemoglobin, myoglobin, important respiratory enzymes (cytochromes), and other enzymes essential to oxygen transport and cellular respiration	Mineral most likely to be deficient in diet. Good sources: meat (especially liver), nuts, egg yolk, legumes. Deficiency results in anemia.
Fluorine	Component of bones and teeth; makes teeth resistant to decay	In areas where it does not occur naturally, fluorine may be added to municipal water supplies (fluoridation). Excess causes tooth mottling.
Zinc	Component of at least 70 enzymes, including carbonic anhydrase; components of some peptidases, and thus important in protein digestion; may be important in wound healing	Occurs in many foods

About 20 vitamins are known to be (or thought to be) important in human nutrition. They are classified into two groups: the fat-soluble vitamins and the water-soluble vitamins. The **fat-soluble vitamins,** found in association with lipids, are vitamins A, D, E, and K. The **water-soluble vitamins** include vitamin C and those belonging to the B complex.

Vitamin requirements of different animals are not identical. For example, humans require vitamin C in the diet, but most animals can synthesize vitamin C from glucose and so do not require a dietary source. (Plants synthesize all of the vitamins they need.) Essential roles for vitamins D and K have not been demonstrated in invertebrates. Thus a vitamin essential for one animal is not necessarily required for another, either because the animal can synthesize it or does not require it for metabolism. Most animals and plants require many of the same vitamins for their metabolic activities.

Vitamin deficiency results in predictable metabolic disorders and clinical symptoms (Figs. 34–7 through 34–12). For example, vitamin A deficiency results in night blindness, vitamin D deficiency in rickets (Fig. 34–12), and vitamin K deficiency in prolonged clotting time. Table 34–3 lists the vitamins, their actions, and the effects of deficiency.

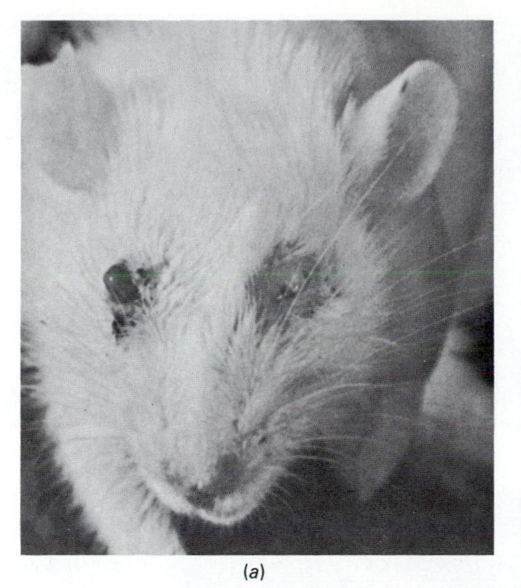
(a)

(b)

Figure 34–7 Effects of vitamin A deficiency. (*a*) This rat is suffering from a typical eye disorder produced by lack of vitamin A. (*b*) The eyes have been restored to normal by the feeding of 3 IU (about 0.001 mg) of vitamin A daily. (Courtesy of E. R. Squibb and Sons.)

(a)

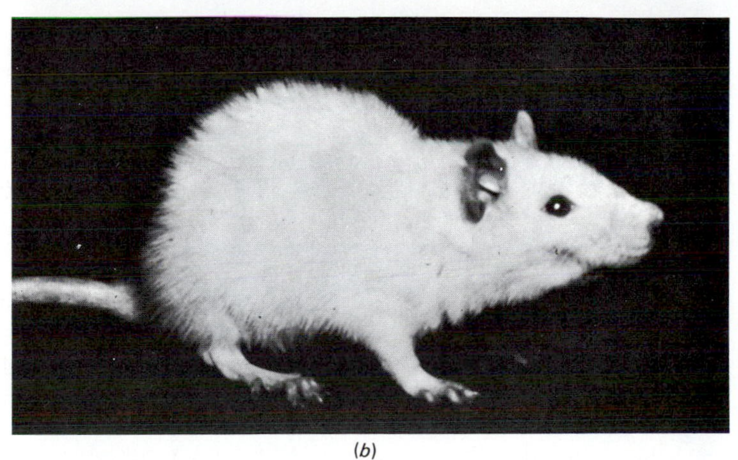
(b)

Figure 34–8 Effects of thiamine deficiency. (*a*) Polyneuritis (beriberi) in a rat raised on a diet deficient in thiamine. Note that the back is arched and the hind legs are stretched and far apart. Such animals have a peculiar halting gait and are particularly awkward in turning, readily losing their balance. When rotated, they have great difficulty in regaining equilibrium, probably because of degeneration of the nerves to the semicircular canals in the inner ears. (*b*) The same rat eight hours after receiving an adequate dose of thiamine; the back and hind legs are normal, and the animal readily regains equilibrium when spun. (Courtesy of the Upjohn Company.)

Vitamin production is a multimillion-dollar industry. Yet there is disagreement on whether healthy individuals should take vitamin supplements. Some authorities think that people who eat a balanced diet do not require vitamin pills. Others argue that most of us do not eat enough fresh fruits and vegetables and should take vitamin supplements. There is also great controversy regarding the

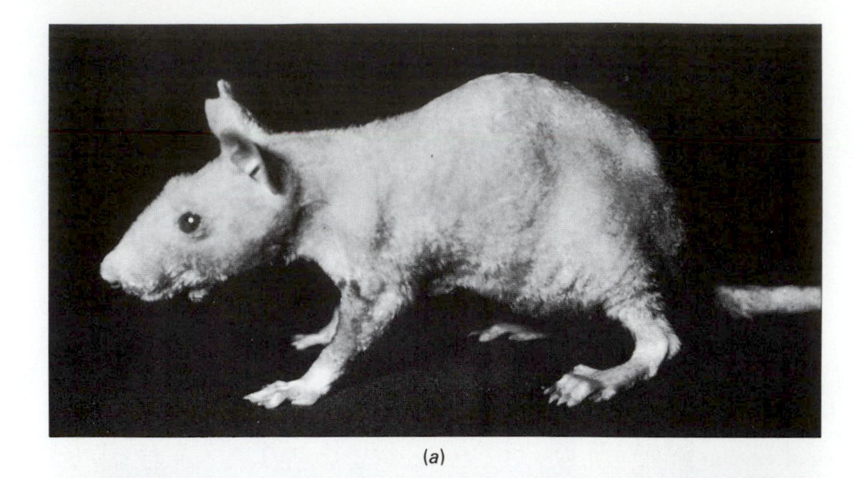

(a)

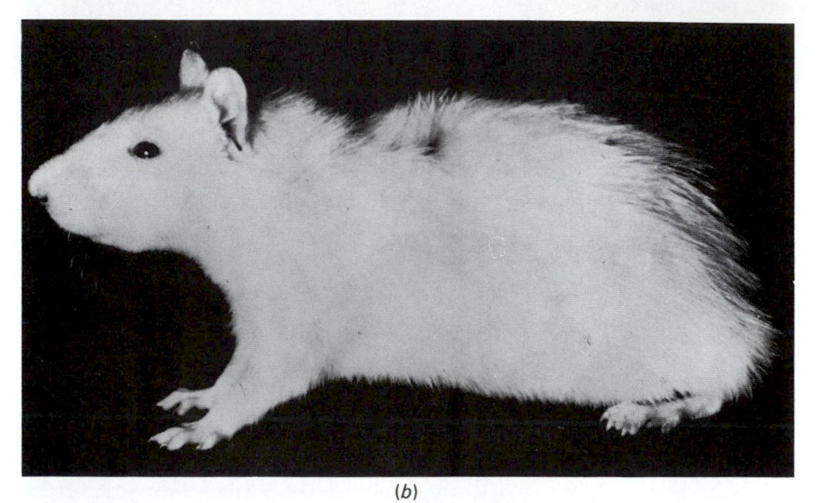

(b)

Figure 34–9 Effects of riboflavin deficiency. (*a*) Riboflavin-deficient rat with stunted growth, general inflammation of the skin (note the open sore on the left front leg), scanty hair, and inflammation of the eyes. (*b*) The same rat after two months of treatment with riboflavin; no signs of the deficiency are visible; growth has been resumed, and the lesions of the skin and eyes are healed. (Courtesy of the Upjohn Company.)

(a)

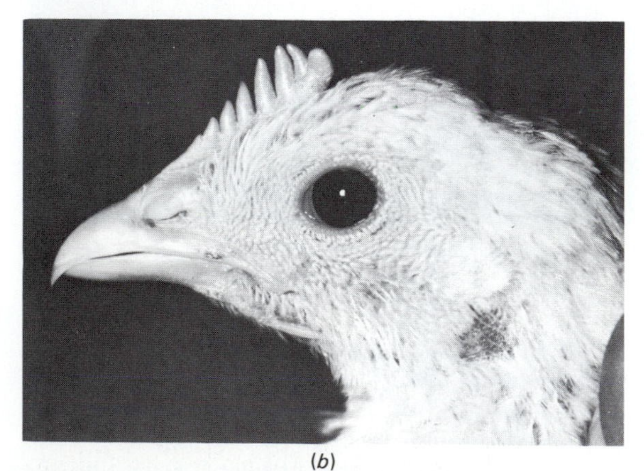

(b)

Figure 34–10 Effects of pantothenic acid deficiency. (*a*) Chick after being fed a diet deficient in pantothenic acid. The eyelids, corners of the mouth, and adjacent skin are inflamed. The growth of feathers is retarded, and the feathers are rough. (*b*) The same chick after three weeks on a diet with pantothenic acid; the lesions are completely cured. (Courtesy of the Upjohn Company.)

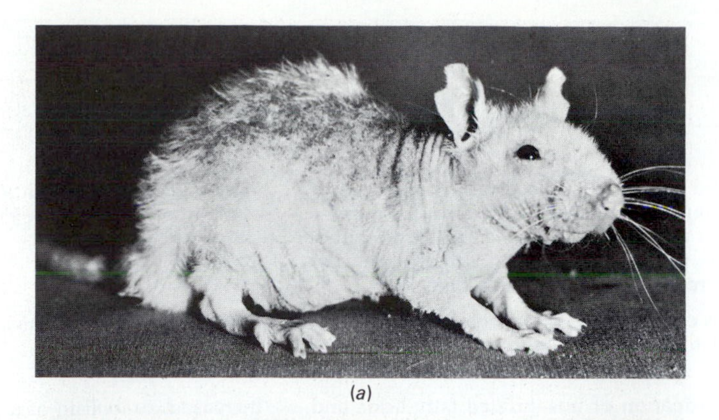

(a)

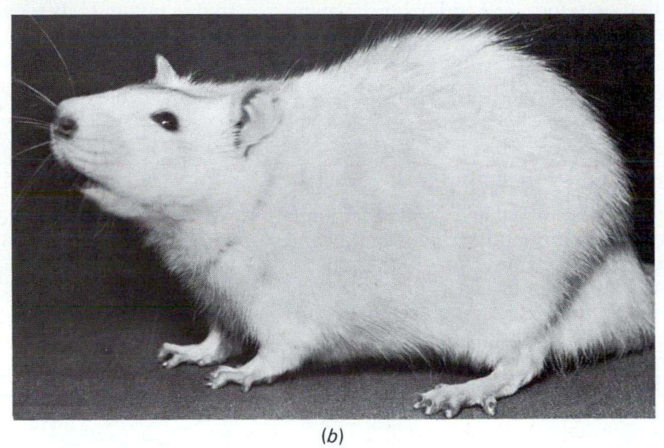

(b)

Figure 34–11 Effects of biotin deficiency. (*a*) Rat after being fed a diet deficient in biotin, to which raw egg white was added. Growth has been retarded, and there is generalized inflammation of the skin. (*b*) The same rat after three months on a diet containing adequate amounts of biotin; growth is normal and the skin lesions are completely healed. (Courtesy of the Upjohn Company.)

effects of megadoses (massive doses) of individual vitamins such as vitamin C (often taken to prevent or minimize colds) or vitamin E (frequently taken as a protection against vascular disease and other processes associated with aging). To date, research studies have generated conflicting results.

Too much vitamin A or D can be harmful. Moderate overdoses of the water-soluble vitamins are excreted in the urine, but the fat-soluble vitamins are not easily excreted and may accumulate in the body tissues. Overdoses of vitamin A may result in skin ailments, retarded growth, enlargement of the liver and spleen, and painful swelling of the long bones. An excess of vitamin D can cause weight loss, mineral loss from the bones, and calcification of soft tissues, including the heart and blood vessels. Very high doses of vitamin D ingested by pregnant women have been linked to a form of mental retardation in the offspring.

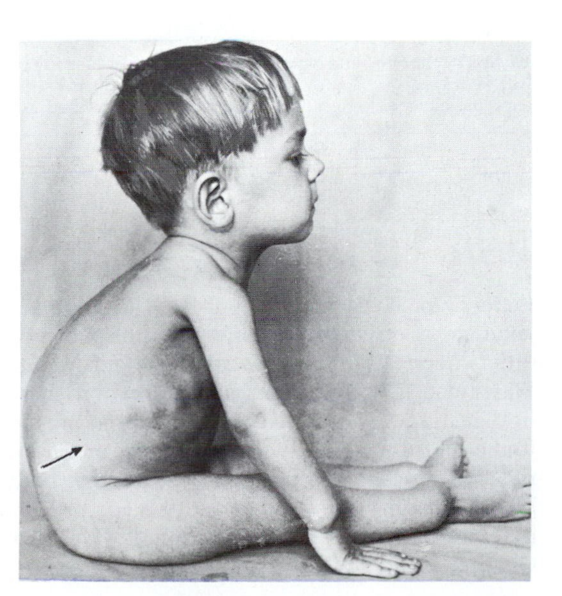

Figure 34–12 A child with rickets. A deficiency of vitamin D decreases the body's ability to absorb and use calcium and phosphorus and produces soft, malformed bones. These are most clearly evident in the ribs (*arrow*) and in wrists and ankles. Note the bowed legs. (Courtesy of Dr. Niilo Hallman.)

TABLE 34–3
The Vitamins

*Vitamins and U.S. RDA**	*Actions*	*Effect of Deficiency*
Fat-soluble		
Vitamin A, Retinol 5000 IU†	Converted to retinal, a necessary component of retinal pigments, essential for normal vision; essential for normal growth and integrity of epithelial tissue; promotes normal growth of bones and teeth by regulating activity of bone cells	Failure of growth; night blindness; atrophy of epithelium; epithelium subject to infection; scaly skin
Vitamin D, Calciferol 400 IU	Promotes calcium absorption from digestive tract; essential to normal growth and maintenance of bone	Bone deformities; rickets in children; osteomalacia in adults
Vitamin E, Tocopherols 30 IU	Inhibits oxidation of unsaturated fatty acids and vitamin A that help form cell and organelle membranes; precise biochemical role not known	Increased catabolism of unsaturated fatty acids, so that not enough are available for maintenance of cell membranes and other membranous organelles; prevents normal growth
Vitamin K probably about 1 mg	Essential for blood clotting	Prolonged blood clotting time
Water-soluble		
Vitamin C (ascorbic acid) 60 mg	Needed for synthesis of collagen and other intercellular substances; formation of bone matrix and tooth dentin, intercellular cement; needed for metabolism of several amino acids; may help body withstand injury from burns and bacterial toxins	Scurvy (wounds heal very slowly and scars become weak and split open; capillaries become fragile; bone does not grow or heal properly)
B-complex Vitamins		
B_1, Thiamine 1.5 mg	Derivative acts as coenzyme in many enzyme systems; important in carbohydrate and amino acid metabolism	Beriberi (weakened heart muscle, enlarged right side of heart, nervous system and digestive tract disorders)
B_2, Riboflavin 1.7 mg	Used to make coenzymes (e.g., FAD) essential in cellular respiration	Dermatitis, inflammation and cracking at corners of mouth; mental depression
Niacin 20 mg	Component of important coenzymes (NAD and NADP) essential to cellular respiration	Pellagra (dermatitis, diarrhea, mental symptoms, muscular weakness, fatigue)
B_6, Pyridoxine 2 mg	Derivative is coenzyme in many reactions in amino acid metabolism	Dermatitis, digestive tract disturbances, convulsions
Pantothenic acid 10 mg	Constituent of coenzyme A (important in cellular metabolism)	Deficiency extremely rare
Folic acid 0.4 mg	Coenzyme needed for reactions involved in nucleic acid synthesis and for maturation of red blood cells	A type of anemia
Biotin 0.3 mg	Coenzyme needed for carbon dioxide fixation	
Vitamin B_{12} 6 mg	Coenzyme important in nucleic acid metabolism	Pernicious anemia

*RDA is the recommended dietary allowance, established by the Food and Nutrition Board of the National Research Council, to maintain good nutrition for healthy persons.
†International Unit: the amount that produces a specific biological effect and is internationally accepted as a measure of the activity of the substance.

Water

Water, which makes up about two thirds of the human body, and about 98% of a jellyfish, is an essential component of every cell. It is the medium in which the other chemicals of the body are dissolved and in which all chemical reactions occur. Water also serves as an active participant in many chemical reactions. For example, in digestion a water molecule is required for each sugar, amino acid, or fatty acid unit split from a carbohydrate, protein, or fat. Water is also used to transport materials within cells and from one place in the body to another. It is the fluid part of blood, lymph, urine, and sweat. It helps distribute and regulate body heat and, as perspiration, cools the body surface. (The properties of water are discussed in Chapter 2.)

Sources	*Comments*
Liver, fish-liver oils, egg; yellow and green vegetables	Can be formed from provitamin carotene (a yellow or red pigment); sometimes called anti-infection vitamin because it helps maintain epithelial membranes; excessive amounts harmful
Liver, fish-liver oils, egg yolk, fortified milk, butter, margarine	Two types: D_2, a synthetic form; D_3, formed by action of ultraviolet rays from sun upon a cholesterol compound in the skin; excessive amounts harmful
Oils made from cereals, seeds, liver, eggs, fish	
Normally supplied by intestinal bacteria; green leafy vegetables	Antibiotics may kill bacteria; then supplements needed in surgical patients
Citrus fruits, strawberries, tomatoes	Possible role in preventing common cold or in the development of acquired immunity(?); harmful in very excessive dose
Liver, yeast, cereals, meat, green leafy vegetables	Deficiency common in alcoholics
Liver, cheese, milk, eggs, green leafy vegetables	
Liver, meat, fish, cereals, legumes, whole-grain and enriched breads	
Liver, meat, cereals, legumes	
Widespread in foods	
Produced by intestinal bacteria; liver, cereals, dark green, leafy vegetables	
Produced by intestinal bacteria; liver, chocolate, egg yolk	
Liver, meat, fish	Contains cobalt; intrinsic factor secreted by gastric mucosa needed for absorption

Although the exact amount varies widely with individual activities and climate, an average of about 2.4 liters of water is lost from the body daily. This loss must be replaced promptly. Humans can live for several weeks without food, but only a few days without water. Much of the daily requirement for water can be satisfied by eating foods, because all foods contain some water. Certain fruits and vegetables contain as much as 95% water.

Water is one of the main factors that limits the distribution of animal populations. Most animals have mechanisms for controlling the water content of their bodies, but different animals have different tolerances for both water loss and water gain. Some small invertebrates can withstand long periods of dehydration by forming waterproof cysts around themselves. When water in the environment is again available they emerge none the worse for the experience. Certain desert vertebrates

An "89" butterfly, *Diaethria clymena*, extending its proboscis to take in moisture beside a stream. This butterfly is so named because of the unusual pattern of its wings. An "88" variety of this butterfly also exists. (P. R. Ehrlich, Stanford University/BPS.)

live indefinitely without drinking water by obtaining it from the foods they eat and from the oxidation of their food. Problems of water balance in aquatic animals are discussed in Chapter 38.

SUMMARY

I. Metabolism is the sum of all chemical and energy transformations that take place in the organism.
   A. Anabolism is the synthetic phase of metabolism in which large molecules are synthesized from smaller ones.
   B. Catabolism is the phase of metabolism in which larger molecules are split into smaller ones.

II. Basal metabolic rate (BMR) reflects the amount of energy an organism must expend just to survive. Total metabolic rate is the sum of BMR and the energy needed to carry on daily activities.

III. When energy input equals energy output, body weight remains constant.
   A. When energy input exceeds output, the excess is stored in fat, and body weight increases.
   B. When energy input is less than output, the body draws on its fuel reserves (fat), and body weight decreases.

IV. Glucose, the end-product of carbohydrate digestion, is utilized by the cells mainly as fuel for cellular respiration.

V. The liver maintains a relatively constant concentration of glucose in the blood, with a fasting level equal to 90 mg/100 ml.
   A. When the concentration of glucose in the blood exceeds steady-state conditions, the liver removes glucose from the blood and converts it to glycogen for storage (the process of glycogenesis).
   B. As glucose is removed from the blood by the cells, its concentration falls below the steady state. Liver cells then convert glycogen back to glucose (the process of glycogenolysis) and return glucose to the blood.
   C. When glycogen stores are depleted, liver cells convert amino acids and glycerol to glucose (the process of gluconeogenesis).

VI. Fat is stored in adipose tissue. When the concentration of glucose in the blood falls below the steady state, fat is mobilized and can be used as an energy source.
   A. Fatty acids are degraded via β-oxidation to molecules of acetyl coenzyme A.
   B. Acetyl coenzyme A can be used as fuel by the liver cells or converted to ketone bodies for transport to other cells.
   C. Acetyl coenzyme A can be used to synthesize steroid hormones, cholesterol, or the fatty acids of phospholipids and triacylglycerols.

VII. Amino acids are utilized mainly in anabolism.
   A. Amino acids are used to make structural proteins, enzymes, functional proteins such as hemoglobin, and nucleic acids.
   B. Amino acids are deaminated in the liver, and the remaining keto acids can be used as an energy source, or converted to glucose or fatty acids for storage as glycogen or fat.

VIII. Minerals required by the body include iron (a component of hemoglobin), iodine (a component of thyroid hormones), calcium and phosphorus (components of bones and teeth), and sodium and chlorine (needed for maintaining fluid balance).

IX. Many vitamins serve as part of coenzymes. Their actions, sources, and the effects of various vitamin deficiencies are listed in Table 34–3.

X. Water is a vital component of all organisms. Homeostasis depends upon fluid balance.

POST-TEST

1. The building or synthetic phases of metabolism are termed ______________; the breaking-down aspects of metabolism are called ______________.

2. The rate at which the body uses energy under resting conditions is the ______________ ______________ rate.

3. The total metabolic rate is the sum of the BMR and the ______.

4. When energy input equals energy output, body weight ______; when energy input exceeds energy output, body weight ______.

5. The principal sources of energy in the human diet are ______ and ______.

6. Glucose is utilized by the cells mainly as ______.

7. Glycogenesis is the process of ______ ______.

8. During glycogenolysis, glycogen is ______ ______.

9. Liver cells convert certain amino acids and other nutrients to glucose during the process of ______.

10. Cholesterol is stored mainly in the ______ cells; triacylglycerols are stored mainly in ______ cells.

11. Ketone bodies are produced during the metabolism of ______; they are taken up by the cells, are converted back to ______ ______, and then may be used as fuel.

12. In ketosis the pH of the blood and tissues becomes ______.

13. Essential amino acids are amino acids that must be ______.

14. Complete proteins contain ______.

15. Black beans and rice are considered a nutritious meal because together these foods contain all of the ______ ______ ______.

16. During deamination of an amino acid, the ______ ______ is split off the ______ skeleton.

17. The fat-soluble vitamins are vitamins __, __, __, and __.

Select the most appropriate substance from Column B for the description in Column A.

Column A	*Column B*
18. Component of thyroid hormones	a. Vitamin B_{12}
19. Component of hemoglobin	b. Iodine
20. Component of retinal pigments	c. Vitamin K
21. Deficiency results in pernicious anemia	d. Iron
22. Essential for blood clotting	e. Vitamin A

REVIEW QUESTIONS

1. What types of nutrients are utilized as fuel for cellular respiration? What other types of nutrients are required for a balanced diet?
2. Give two examples of catabolic reactions and two of anabolic reactions.
3. What is the difference between basal metabolic rate and total metabolic rate?
4. What happens when energy input exceeds energy output? when energy input equals energy output?
5. How does the liver help to maintain a constant concentration of glucose in the blood?
6. When and why is gluconeogenesis important?
7. What are the products of β-oxidation? What is the fate of these products?
8. What function do ketone bodies serve? What is ketosis?
9. What is the fate of excess amino acids? Explain.
10. Why are each of the following nutrients needed by the body?
    a. essential amino acids
    b. calcium
    c. iodine
    d. iron
    e. vitamin A
    f. vitamin D
11. What happens when each of the following is deficient in the diet?
    a. essential amino acids
    b. vitamin D
    c. vitamin K
    d. cholesterol
12. What types of foods contain complete proteins? unsaturated fatty acids?
13. What would be the approximate caloric content and the specific types of nutrients needed for an active young man? Describe how the diets of each of the following persons should differ from this:
    a. a 10-year-old boy
    b. a 65-year-old man
    c. a pregnant woman

35
Internal Transport

OUTLINE

I. Invertebrates with no circulatory systems
II. Invertebrates with open circulatory systems
III. Invertebrates with closed circulatory systems
IV. The vertebrate circulatory system
- *A.* Blood
  - *1.* Plasma
  - *2.* Erythrocytes
  - *3.* Leukocytes
  - *4.* Platelets and blood clotting
- *B.* The blood vessels
  - *1.* Arteries
  - *2.* Capillaries
  - *3.* Veins
- *C.* The heart
  - *1.* Comparison of vertebrate hearts
    - *a.* Fish
    - *b.* Amphibians
    - *c.* Reptiles
    - *d.* Birds and mammals
  - *2.* Structure of the human heart
  - *3.* How the heart works
    - *a.* Heart sounds
    - *b.* Electrocardiograms
    - *c.* Cardiac output
      - *(1)* Regulation of stroke volume
      - *(2)* Regulation of heart rate
- *D.* Pulse and blood pressure
- *E.* The pattern of circulation
  - *1.* Pulmonary circulation
  - *2.* Systemic circulation
    - *a.* Coronary circulation
    - *b.* Circulation to the brain
    - *c.* Hepatic portal system

V. The lymphatic system
- *A.* Design of the lymphatic system
- *B.* Role in fluid homeostasis

LEARNING OBJECTIVES

After you have read this chapter you should be able to:

1. List seven functions of a circulatory system.
2. Compare invertebrates that can function with no specialized circulatory system with those that require a circulatory system.
3. Compare open and closed circulatory systems and give examples of invertebrates that have an open system.
4. List and briefly describe the principal components of human blood, giving the function of each component.
5. Describe the life cycle of a red blood cell.
6. Describe and identify in illustrations five types of leukocytes, and give the function of each type.
7. Summarize the events involved in blood clotting.
8. Compare the structures and functions of arteries, arterioles, capillaries, sinusoids, and veins.
9. Compare the hearts of a fish, amphibian, reptile, bird, and mammal.
10. Describe the external and internal structure of the human heart.
11. Describe the heart's conduction system and cardiac muscle.
12. Briefly describe the events of the cardiac cycle, and relate normal heart sounds to the events of this cycle.
13. Define cardiac output, and describe how it is regulated or affected.
14. Explain the physiological basis for arterial pulse and for blood pressure.
15. Trace a drop of blood from one organ or body part to another through each part of the heart, and through the pulmonary and systemic circulations.
16. Describe atherosclerosis, giving its complications and risk factors.
17. List the functions of the lymphatic system, and describe how this system operates to maintain fluid balance.

All living cells require a continuous supply of nutrients and oxygen, and continuous removal of waste products. Very small organisms living in a watery environment can accomplish this by simple diffusion and do not need specialized circulatory structures. In larger organisms, diffusion cannot supply enough raw materials to all of the cells, and other mechanisms are required to transport materials to and from cells not in direct contact with the water in which the animal lives. In complex animals, specialized structures are present to accomplish internal transport. These structures make up the **circulatory system.**

A circulatory system typically consists of the following:

1. **blood,** a fluid connective tissue consisting of cells and cell fragments dispersed in fluid
2. a pumping device, usually called a **heart**
3. a system of blood vessels or spaces through which the blood circulates

In the **closed circulatory system** present in many animals, blood passes through a continuous network of blood vessels. In the arthropods and most mollusks, the circulatory system is an **open circulatory system,** so named because the heart pumps blood into vessels that have open ends. Blood spills out of them, filling the body cavity, and bathing the cells. Blood finds its way back into the circulatory system through openings within the heart (or through open-ended blood vessels that lead to the heart). Movement of blood through an open system is not as rapid or as efficient as through a closed system.

A circulatory system may perform the following functions:

1. Transport nutrients from the digestive system and from storage depots to each of the cells of the body
2. Transport oxygen from gills or lungs to the cells of the body, and carbon dioxide from the cells to the lungs or gills
3. Transport wastes from each cell to the excretory organs
4. Transport hormones from endocrine glands to target tissues
5. Help maintain fluid balance
6. Defend the organism against invading microorganisms
7. Help regulate body temperature in homeothermic ("warm-blooded") animals

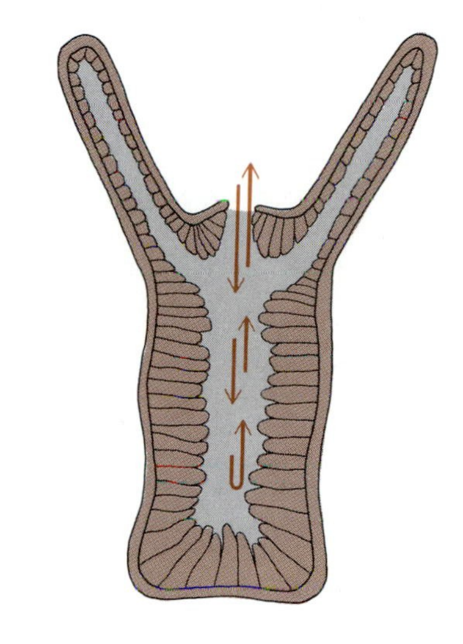

(*a*) Hydra

(*b*) Planarian flatworm

Figure 35–1 Some invertebrates with no circulatory system. (*a*) In *Hydra* and other cnidarians, the gastrovascular cavity serves a circulatory function, permitting nutrients to come in contact with the body cells. (*b*) In planarian flatworms, the branched intestine conducts food to all of the regions of the body.

Invertebrates with No Circulatory System

No circulatory systems are present in sponges, cnidarians, ctenophores, flatworms, and nematodes. In aquatic forms, the tissues are bathed in water laden with oxygen and nutrients. Wastes simply diffuse into the water and are washed away. Because the bodies of these invertebrates are only a few cells thick, diffusion is an effective mechanism for distributing materials to and from their cells.

In cnidarians, the central gastrovascular cavity serves as both a digestive organ and a circulatory organ (Fig. 35–1). The animal's tentacles capture prey and stuff it through the mouth and into the cavity, where digestion occurs. The digested nutrients then pass into the cells lining the cavity, and through them to cells of the outer layer. Movement of the animal's body, as it stretches and contracts, stirs up the contents of the central cavity and aids circulation.

In the flatworm planaria, the branched intestine brings nutrients to all regions of the body. Nutrients diffuse into the tissue fluid of the mesenchyme, and then to cells of the mesenchyme and the outer layer of the body. As in cnidarians, circulation is aided by contractions of the muscles of the body wall, which agitate the fluid in the intestine and the tissue fluid. Oxygen distribution and waste removal from individual cells depend mainly upon diffusion.

Fluid in the pseudocoelom of nematodes and other pseudocoelomate animals helps to circulate materials. Nutrients, oxygen, and wastes dissolve in this fluid and diffuse through it to and from the individual cells of the body. Body movements of the animal result in movement of the fluid itself, which distributes these materials to all parts of the body.

Invertebrates with Open Circulatory Systems

In mollusks (except for the cephalopods) and in arthropods, the circulatory system is open. In many mollusks the heart is surrounded by a pericardial cavity. The heart typically consists of three chambers, two **atria,** which receive blood from the gills, and a **ventricle,** which pumps this oxygen-rich blood to the tissues. Blood vessels leading from the heart open into large spaces called sinuses, enabling the blood to bathe the body cells. Such blood-filled spaces comprise a **hemocoel** (blood cavity). From the hemocoel blood passes into vessels that lead to the gills; there blood is recharged with oxygen and passes into blood vessels that return it to the heart. In the clam, the ventricle pumps blood both forward and backward (Fig. 35–2).

Some mollusks, as well as arthropods, have blood that contains a pigment, **hemocyanin,** which contains copper. Hemocyanin transports oxygen and imparts a bluish color to the blood of these animals (the original bluebloods!).

Figure 35–2 Mollusks and arthropods have an open circulatory system. (*a*) Circulatory system of the clam. (*b*) Circulatory system of the crayfish. Lateral view and cross section.

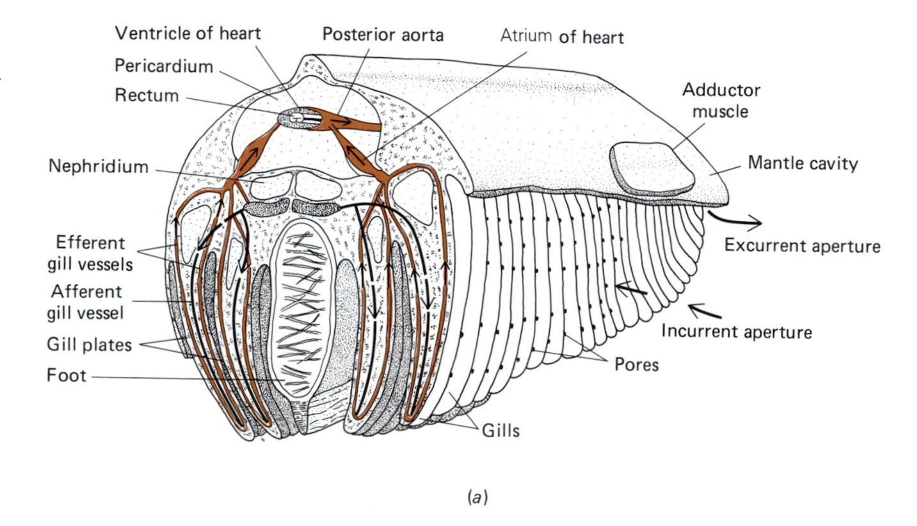

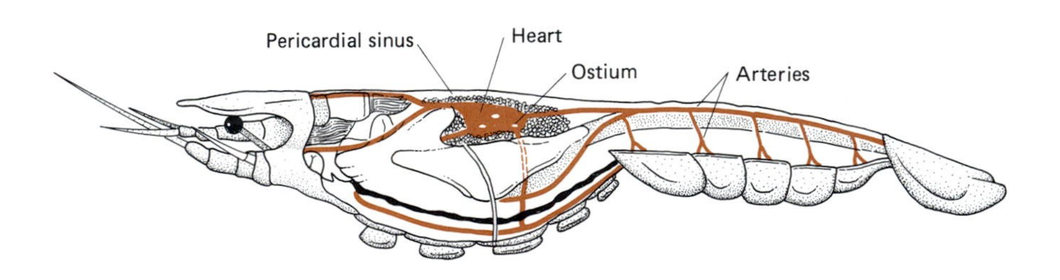

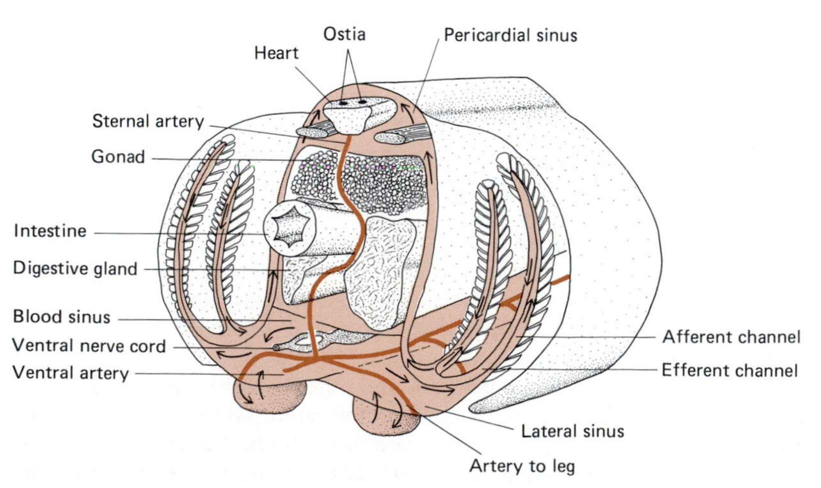

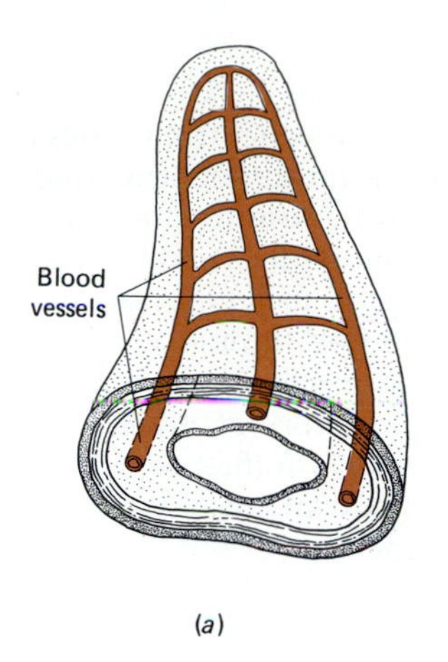

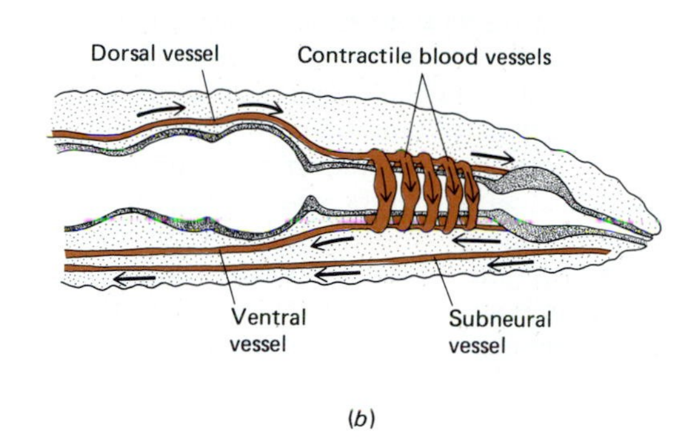

Figure 35–3 Examples of a closed circulatory system. (*a*) A rudimentary closed circulatory system is present in nemertines. (*b*) Earthworms have a complex closed circulatory system with five pairs of contractile blood vessels that deliver blood from the dorsal vessel to the ventral vessel.

In insects and other arthropods, a tubular heart pumps blood into blood vessels (arteries) that eventually deliver blood to the sinuses that comprise the hemocoel. Blood then circulates through the hemocoel, eventually finding its way back to the pericardial cavity surrounding the heart. It enters the heart through tiny openings called **ostia,** which are equipped with valves to prevent backflow. Some insects have accessory "hearts" that help pump blood through the extremities. Circulation of the blood is faster during muscular movement. Thus, when an animal is active and most in need of nutrients for fuel, its own movement ensures effective circulation. In insects, oxygen is transported directly by the tracheae of the respiratory system, rather than by the circulatory system.

Invertebrates with Closed Circulatory Systems

A rudimentary closed circulatory system is found in the proboscis worms (phylum Nemertinea; Fig. 35–3). This system consists of a complete network of blood vessels including at least two vessels that extend the length of the body. These communicate anteriorly and posteriorly by means of connecting blood sinuses. Branches from the main vessels extend into the tissues. No heart is present; instead, blood flow depends upon movements of the animal and upon contractions in the walls of the large blood vessels. Blood may move in either direction. Blood cells containing colored pigments are found in this group.

Earthworms and other annelids have a complex closed circulatory system. Two main blood vessels extend lengthwise in the body. The ventral vessel conducts blood posteriorly, while the dorsal vessel conducts blood anteriorly. Dorsal and ventral vessels are connected by lateral vessels in every segment. Branches of the lateral vessels deliver blood to the skin, where it is oxygenated, and to the various tissues and organs. In the anterior part of the worm are five pairs of contractile blood vessels (sometimes referred to as hearts) that connect dorsal and ventral vessels (Fig. 35–3). Contractions of these paired vessels and of the dorsal vessel, as well as contraction of the muscles of the body wall, circulate the blood. Earthworms have hemoglobin, the same red pigment that transports oxygen in vertebrate blood; however, their hemoglobin is not within red blood cells but is dissolved in the blood plasma. The circulatory system of the earthworm apparently does little to remove wastes from the body cells. This function is delegated instead to the coelomic fluid, which transports wastes to the nephridia.

Although other mollusks have an open circulatory system, the fast-moving cephalopods (squid, octopus) require a more efficient means of internal transport. They have a closed system made even more effective by the presence of "hearts" at the base of the gills, which speed the passage of blood through the gills.

The circulatory system of the sea cucumbers (holothuroids) is the most highly developed system of any of the echinoderms. Its vessels parallel the tubes of the water vascular system, and it appears to transport both nutrients and oxygen.

Invertebrate chordates have a closed circulatory system, usually with a ventral heart. An exception is the cephalochordate amphioxus, which has no heart and must depend upon certain pulsating blood vessels to move its blood.

The Vertebrate Circulatory System

The circulatory systems of all vertebrates are fundamentally the same, from fishes, frogs, and reptiles, to birds and human beings. All have a muscular heart that pumps blood into a closed system of blood vessels (Fig. 35–4). **Arteries** are blood vessels that branch and rebranch, carrying blood away from the heart and to the various organs of the body. The tiniest branches deliver blood to the **capillaries,** vessels with walls so thin that nutrients and oxygen can diffuse through them to the individual cells of the body. Wastes from the cells diffuse into the blood and are carried to the excretory organs. After passing through a network of capillaries, blood flows into **veins,** the vessels that conduct the blood back to the heart. This basic plan is similar in all vertebrates, so that the human circulatory system can be studied by dissecting an animal such as a shark or a frog.

BLOOD

In humans the total circulating blood volume is about 8% of the body weight—5.6 liters (6 quarts) in a 70-kg (154-lb) person. This is about the amount of oil in the crank case of most cars! Although blood appears to be a homogeneous crimson fluid as it pours from a wound, it is composed of a pale yellowish fluid, called **plasma,** in which red blood cells, white blood cells, and blood platelets are suspended. About 55% of the blood is plasma; the remaining 45% is made up of blood cells and platelets. The loss of water in profuse sweating may reduce the plasma volume to 50% of the blood, and drinking large quantities of fluid may increase it to 60%. Because cells and platelets are heavier than plasma, the two may be sepa-

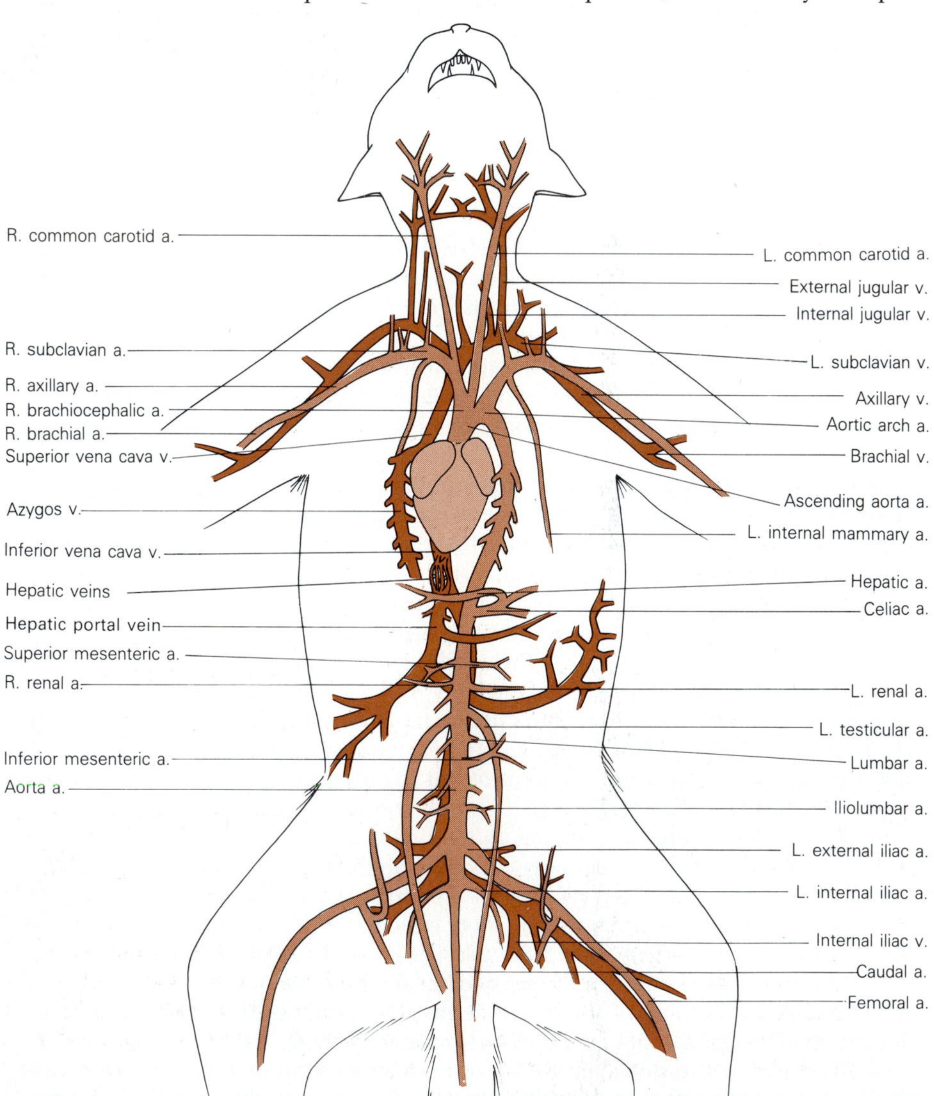

Figure 35–4 The circulatory system of the cat is a typical vertebrate system in which a ventral heart pumps blood into a system of arteries, which branch into smaller and smaller vessels until blood flows through the thin-walled capillaries, where materials are exchanged between blood and body cells. Blood returns to the heart through a system of veins. Arteries are shown in red, veins in blue.

rated from plasma by the process of centrifugation. Plasma does not separate from blood cells in the body because the blood is constantly mixed as it circulates in the blood vessels.

Plasma

Plasma is composed of water (about 92%), proteins (about 7%), salts, and a variety of materials being transported, such as dissolved gases, nutrients, wastes, and hormones. Plasma is in dynamic equilibrium with the interstitial fluid bathing the cells and the intracellular fluid within the cells. As blood passes through the capillaries, substances constantly move into and out of the plasma. Changes in its composition initiate responses on the part of one or more organs of the body to restore its normal steady state.

Plasma contains several kinds of proteins, each with specific properties and functions: fibrinogen; alpha, beta, and gamma globulins; albumin; and lipoproteins. **Fibrinogen** is one of the proteins involved in the clotting process. When the proteins involved in blood clotting have been removed from the plasma, the remaining liquid is called **serum.** The **gamma globulin** fraction contains many types of antibodies that provide immunity to diseases, such as measles and infectious hepatitis. Purified human gamma globulin is sometimes used to treat certain diseases, or to reduce the possibility of contracting a disease. **Albumins** and **globulins** help to regulate fluid balance.

Plasma proteins are too large to pass readily through the walls of blood vessels, and thus can exert an osmotic pressure, which is important in maintaining an appropriate blood volume. These proteins therefore play a necessary role in regulating the distribution of fluid between plasma and tissue fluid. Plasma proteins (along with the hemoglobin in the red blood cells) are also important acid–base buffers, helping to keep the pH of the blood within a narrow range—at its normal, slightly alkaline pH of 7.4.

Erythrocytes

Erythrocytes, also called **red blood cells** (RBCs), are highly specialized for transporting oxygen. In all vertebrates except mammals erythrocytes have nuclei. During development of a RBC in a mammal, the nucleus is pushed out of the cell. Each erythrocyte is a flexible, biconcave disk, 7 to 8 μm in diameter and 1 to 2 μm thick (Fig. 35–5). An internal elastic framework maintains the disk shape and permits the cell to bend and twist as it passes through blood vessels even smaller than its own

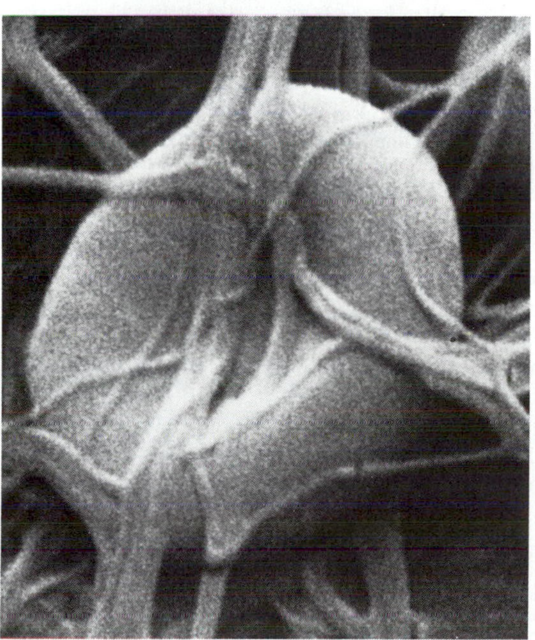

(a)

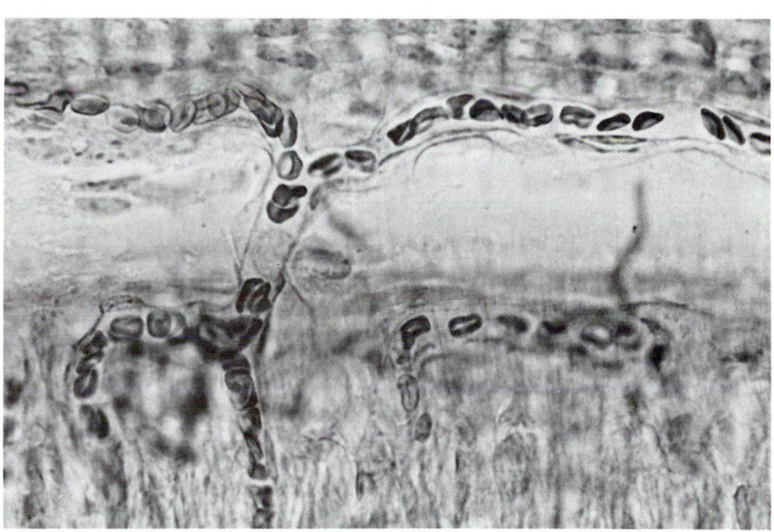

(b)

Figure 35–5 Red blood cells. (*a*) A scanning electron micrograph of a red blood cell. (*b*) Red blood cells pass through capillaries in single file. (*a*) from E. Bernstein: *Science* 173, 1971. Copyright 1971, the American Association for the Advancement of Science.

diameter. In a human, about 30 trillion erythrocytes circulate in the blood—approximately 5.4 million per μl (microliter) in an adult male, 5 million per μl in an adult female.

RBCs are produced within the red bone marrow of certain bones (the vertebrae, ribs, breast bone, skull bones, and long bones). As an RBC develops it produces great quantities of hemoglobin, the oxygen-transporting pigment that gives vertebrate blood its red color. (Oxygen transport is discussed in Chapter 37.) The lifespan of a human erythrocyte is about 120 days. As blood circulates through the liver and spleen, phagocytic cells remove worn-out erythrocytes from the circulation. These erythrocytes are then disassembled and some of their components are recycled. In the human body, 2.4 million RBCs are destroyed every second, so an equal number must be produced in the bone marrow to replace them.

Anemia is a deficiency in hemoglobin (often accompanied by a decrease in the number of RBCs). When the amount of hemoglobin is insufficient, the amount of oxygen transported is inadequate to supply the body's needs. An anemic person may complain of feeling weak and may become easily fatigued. Anemia is caused by loss of blood, decreased production of hemoglobin (as in iron-deficiency anemia), or rapid erythrocyte destruction (in the hemolytic anemias such as sickle cell anemia).

An increase in the number of circulating RBCs is called **polycythemia.** In **polycythemia vera,** a serious disease of unknown cause, the number of circulating erythrocytes may double. The blood becomes viscous and tends to plug the small blood vessels.

Leukocytes

The **leukocytes,** or **white blood cells** (WBCs), are specialized to defend the body against invading bacteria and other threatening intruders. Leukocytes are amoeba-like cells, capable of independent movement. They can move against the current of the bloodstream, and some types routinely slip through the walls of blood vessels and enter the tissues.

Human blood contains five kinds of leukocytes, which may be classified as either granular or agranular (Figs. 35–6 and 35–7). The **granular** leukocytes are manufactured in the red bone marrow and have large distinctive granules in their cytoplasm.

The three varieties of granular leukocytes are the neutrophils, eosinophils, and basophils. **Neutrophils,** the principal phagocytic cells in the blood, are especially adept at seeking out and ingesting bacteria. They also phagocytize the remains of dead tissue cells, a clean-up task that must be performed after injury or infection. **Eosinophils** have large granules that stain bright red with eosin, an acidic dye. These cells increase in number during allergic reactions and during parasitic infections such as tapeworm infections. **Basophils** exhibit deep blue granules when stained with basic dyes. Like eosinophils, these cells are thought to play a role in allergic reactions. Basophils contain large amounts of the chemical histamine, which they release in injured tissues and in allergic responses. Because they contain the anticlotting chemical **heparin,** basophils may play a role in preventing blood from clotting inappropriately within the blood vessels.

Agranular leukocytes lack large distinctive granules. In these cells the nucleus is rounded or kidney-shaped. Lymphocytes and monocytes are agranular leukocytes. Some **lymphocytes** are specialized to produce antibodies, while others attack foreign invaders such as bacteria or viruses directly. Just how they manage these feats is discussed in the next chapter.

Monocytes are the largest WBCs, reaching 20 μm in diameter. They are manufactured in the bone marrow. After spending about 24 hours in the blood, a monocyte leaves the circulation. Development is completed within the tissues, where the monocyte greatly enlarges and becomes a **macrophage,** a giant scavenger cell. All of the tissue macrophages develop in this way. Macrophages voraciously engulf bacteria, dead cells, and any debris littering the tissues.

In human blood there are normally about 7000 WBCs per μl of blood (only one for every 700 red blood cells). During bacterial infections the number may rise sharply so that a **white blood cell count** is a useful diagnostic tool. The proportion of each kind of WBC is determined by a **differential** WBC count. The normal distribution of leukocytes is indicated in Table 35–1.

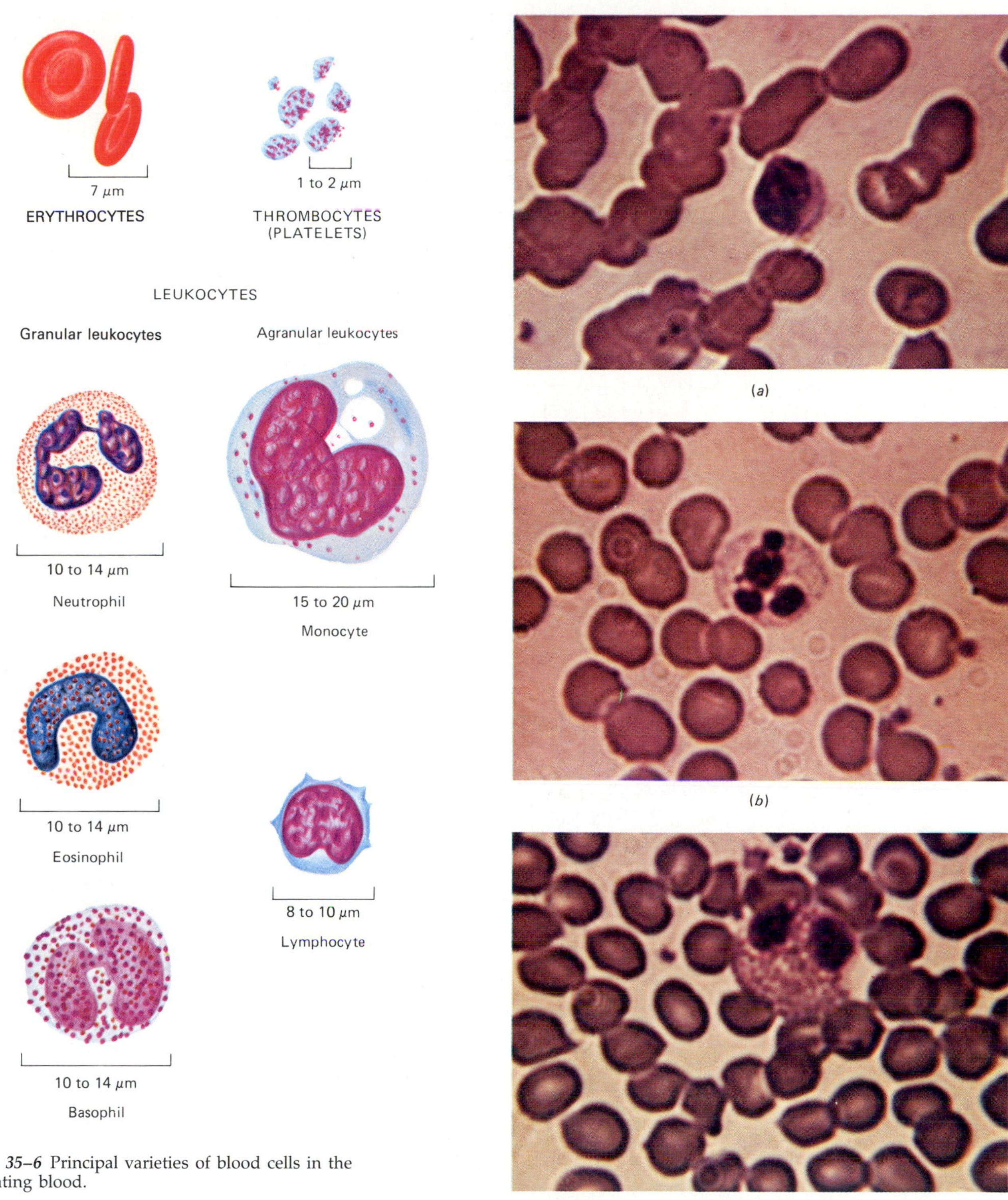

Figure 35–6 Principal varieties of blood cells in the circulating blood.

Figure 35–7 Photomicrographs of circulating blood cells. (*a*) A lymphocyte surrounded by red blood cells. (*b*) A neutrophil. Note the lobed nucleus. (*c*) An eosinophil.

Leukemia is a form of cancer in which any one of the kinds of white cells multiply rapidly within the bone marrow. Many of these cells do not mature, and their large numbers crowd out developing red blood cells and platelets, leading to anemia and impaired clotting. A common cause of death from leukemia is internal hemorrhaging, especially in the brain. Another frequent cause of death is infection because, although there may be a dramatic rise in the white cell count, the cells are immature and abnormal, and unable to defend the body against disease organisms. Although no cure for leukemia has been discovered, radiation treatment and therapy with antimitotic drugs can induce partial or complete remissions lasting as long as 15 years in some patients.

TABLE 35–1
Cellular Components of Blood

Component	*Normal Range*	*Function*	*Pathology*
Red blood cells	Male: 4.2–5.4 million/μl Female: 3.6–5.0 million/μl	Oxygen transport: carbon dioxide transport	Too few: anemia Too many: polycythemia
Platelets	150,000–400,000/μl	Essential for clotting	Clotting malfunctions; bleeding; easy bruising
White blood cells (total)	5000–10,000/μl		
Neutrophils	About 60% of WBCs	Phagocytosis	Too many: may be due to bacterial infection, inflammation, leukemia (myelogenous)
Eosinophils	1–3% of WBCs	Some role in allergic response	Too many: may result from allergic reaction, parasitic infection
Basophils	1% of WBCs	May play a role in prevention of clotting in body	
Lymphocytes	25–35% of WBCs	Produce antibodies; destroy foreign cells	Atypical lymphocytes present in infectious mononucleosis; too many may be due to leukemia, certain viral infections.
Monocytes	6% of WBCs	Differentiate to form macrophages	May increase in monocytic leukemia, tuberculosis, fungal infections

Platelets and Blood Clotting

In most vertebrates other than mammals, the blood contains small, oval cells called **thrombocytes,** which have nuclei. In mammals thrombocytes are tiny, spherical, or disk-shaped bits of cytoplasm that lack a nucleus. They are usually referred to as blood **platelets.** About 300,000 platelets per μl are present in human blood.

Platelets are formed from bits of cytoplasm that are pinched off from very large cells (megakaryocytes) in the bone marrow. Thus, a platelet is not a whole cell but a fragment of cytoplasm enclosed by a cell membrane.

Platelets play an important role in **hemostasis** (the control of bleeding). When a blood vessel is cut, it constricts, preventing loss of blood. Platelets stick to the rough, cut edges of the vessel, physically patching the break in the wall. As platelets begin to gather they release ADP, which attracts other platelets. Within about five minutes after injury a complete platelet patch, or temporary clot, has formed.

At the same time that the temporary clot is formed, a stronger, more permanent clot begins to develop. More than 30 different chemical substances interact in this very complex process. The series of reactions that leads to clotting is triggered when one of the clotting factors in the blood is activated by contact with the injured tissue. In **hemophiliacs** (persons with "bleeder's disease") one of the clotting factors is absent due to an inherited genetic mutation.

In a very simplified way the clotting process can be summarized as follows:

$$\underset{\text{a plasma protein}}{\text{Prothrombin}} \xrightarrow[\text{compounds released from platelets}]{\text{several clotting factors, } Ca^{2+},} \underset{\text{active form of prothrombin}}{\text{Thrombin}}$$

$$\text{Fibrinogen} \xrightarrow{\text{thrombin}} \text{Fibrin}$$

Prothrombin is a globulin manufactured in the liver and requiring vitamin K for its production. **Thrombin,** the activated form of prothrombin, is an enzyme that catalyzes the conversion of the soluble plasma protein fibrinogen to an insoluble protein, **fibrin.** Once formed, fibrin polymerizes, producing long threads that stick to the damaged surface of the blood vessel and form the webbing of the clot. These threads trap blood cells and platelets, which help to strengthen the clot.

THE BLOOD VESSELS

The circulatory system of a vertebrate includes three main types of blood vessels: arteries, veins, and capillaries (Fig. 35–8). A blood vessel wall, like the wall of the

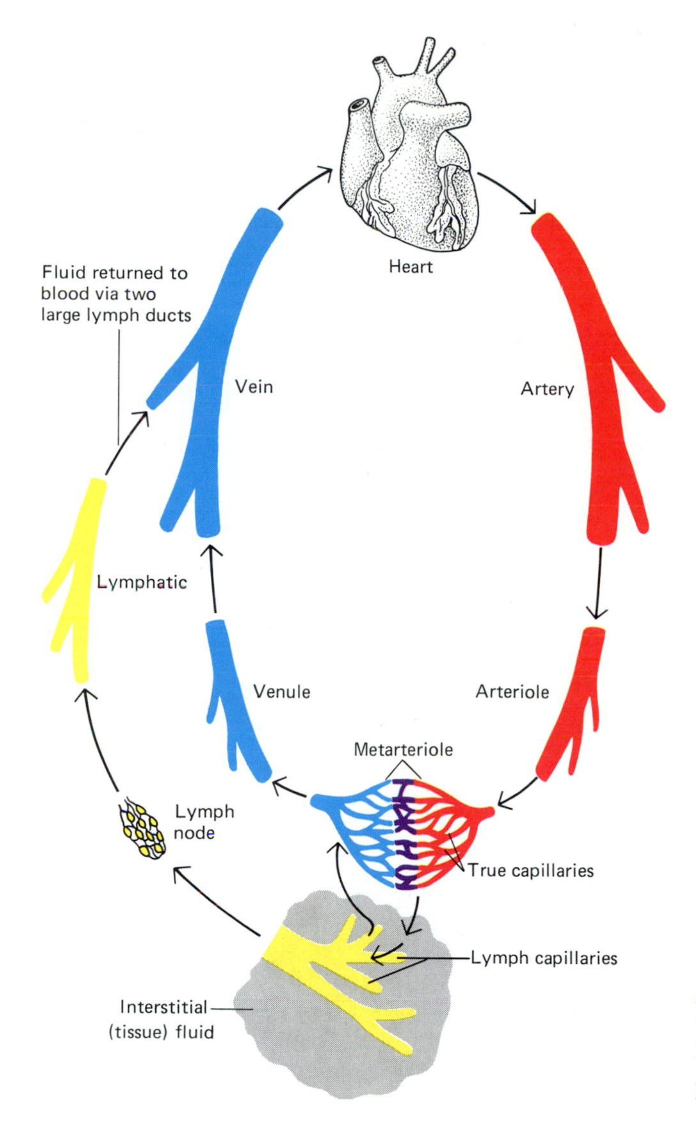

Figure 35–8 Types of blood vessels and their relationship to one another. Lymphatic vessels return interstitial fluid to the blood by way of ducts that lead into large veins in the shoulder region.

heart, has three layers (Fig. 35–9). The innermost layer (**tunica intima**), which lines the blood vessel, consists mainly of **endothelium,** a tissue that resembles squamous epithelium.

Arteries

In addition to the endothelial lining, the inner layer of most arteries contains a strong **internal elastic membrane,** which gives additional strength to the walls. The largest arteries have *very* elastic walls that stretch as the arteries fill with blood delivered to them with each heartbeat. These elastic arteries branch into smaller distributing arteries that deliver blood to specific organs. Within an organ or tissue, a distributing artery branches to form very small arteries called **arterioles,** which are important in determining the amount of blood distributed to a tissue and in maintaining blood pressure.

Smooth muscle in the wall of an arteriole can contract or relax, changing the diameter of the vessel and the volume of blood that can pass through it. Arteriolar contraction produces **vasoconstriction;** relaxation causes **vasodilatation.** Such changes in blood flow are under control of the nervous system; these changes help to maintain appropriate blood pressure and to determine the amount of blood passing to a particular organ (Table 35–2).

Capillaries

Arterioles deliver blood into **capillaries,** microscopic vessels with walls just one cell thick. Only the walls of the capillaries are thin enough to permit the exchange of

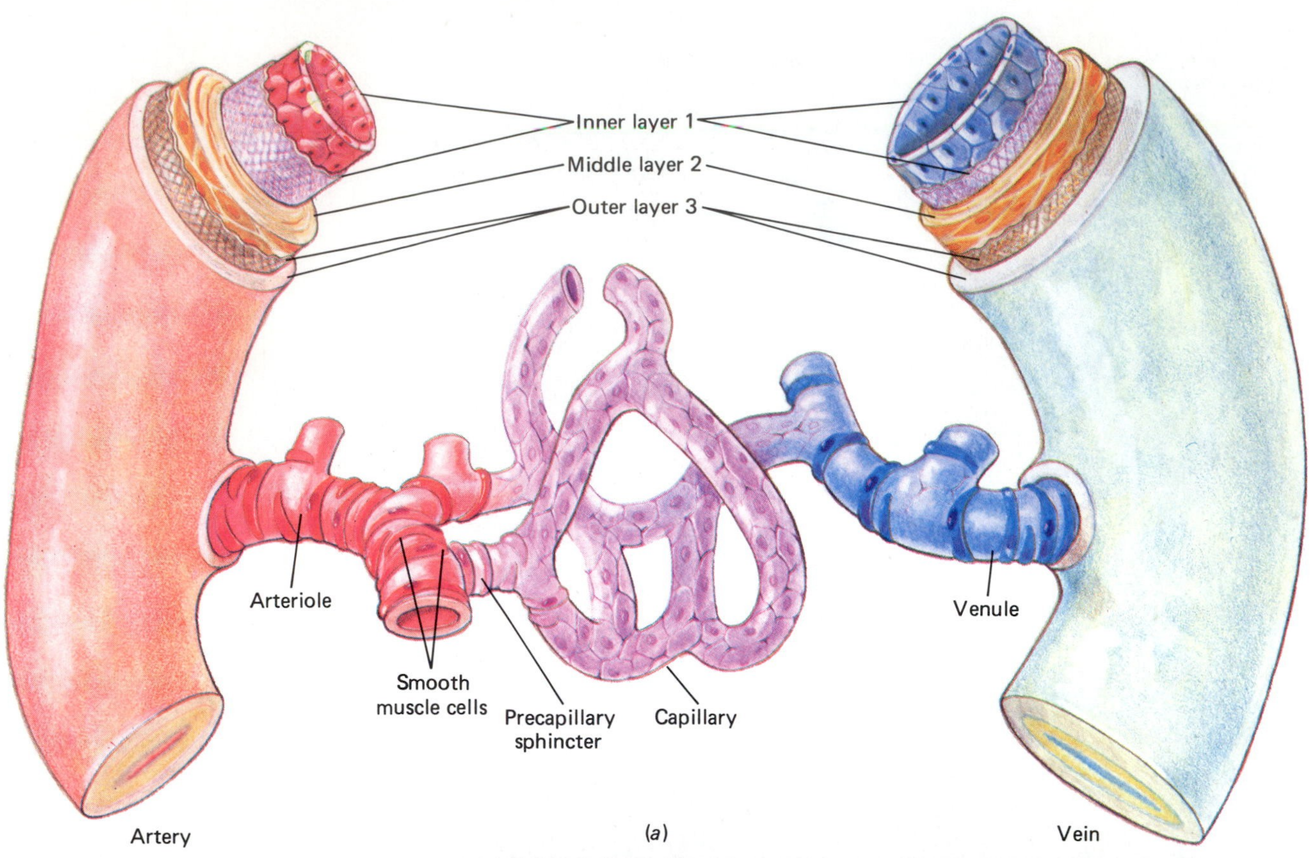

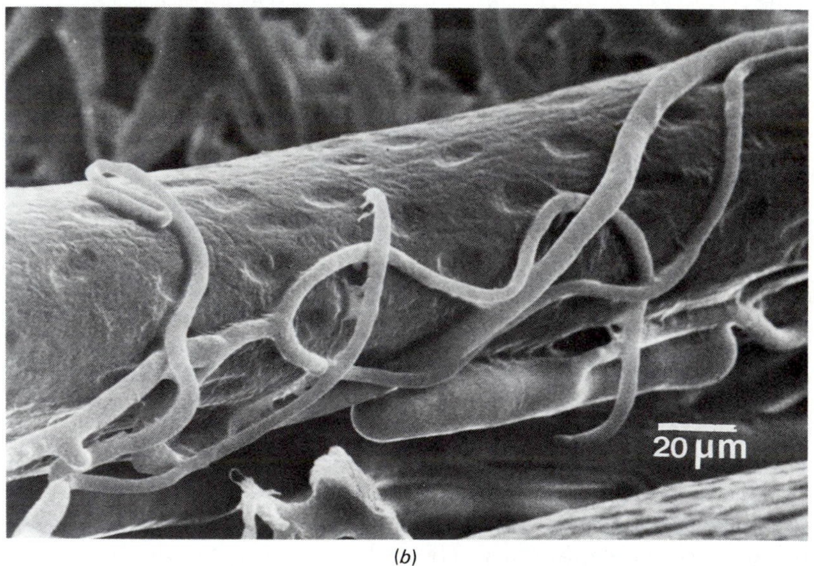

(b)

Figure 35–9 Blood vessel structure. (*a*) Comparison of the wall of an artery, vein, and capillary. (*b*) Scanning electron micrograph of a branch of the hepatic artery with small capillaries. Notice the elliptical depressions in the artery wall. This preparation was made by injecting the blood vessels with a special plastic, followed by treatment with a corrosive that removed all surrounding tissues. The nuclei of the flattened cells that lined the artery in life made the oval depressions. (Courtesy of Johan G. Hanstede and *The Anatomical Record*.)

nutrients, gases, and wastes between blood and tissues. Capillary walls consist of a single layer of cells, the endothelium, continuous with the endothelial lining of the artery and vein on either side. Each capillary is only about 1 mm (0.04 in.) long; yet there are so many of these tiny vessels that almost every cell in the body is within two or three cells of a capillary—close enough for oxygen and nutrients to diffuse from the blood to every cell. The number of capillaries in the body is almost beyond calculation. In tissues with a high metabolic rate they are very close together. One investigator places the number of capillaries in muscle at about 240,000 per square centimeter. Less active tissues are not so well supplied; fatty tissue has few capillaries, and the lens of the eye has none at all.

The amount of blood that the capillaries in the body could hold is so great that you would need about 40% more blood to completely fill them all. Actually, at any moment only about 5% of your blood can be found within capillaries. Thus at any

TABLE 35–2
Blood Flow to Regions of the Human Body Under Basal Conditions and During Strenuous Exercise

	*Basal Conditions**		*Exercise*	
	ml/min	*% of Total*	*ml/min*	*% of Total*
Brain	700	14	750	4.2
Heart	200	4	750	4.2
Bronchi	100	2	200	1.1
Kidneys	1100	22	600	3.3
Liver	1350	27	600	3.3
Via portal vein	1050	21		
Via hepatic artery	300	6		
Skeletal muscles	750	15	12,500	70.3
Bone	250	5	250	1.4
Skin	300	6	1,900	10.7
Thyroid gland	50	1	50	0.3
Adrenal glands	25	0.5	25	0.2
Other tissue	175	3.5	175	1.0
	5000	100	17,800	100

*Data from Guyton, A. C.: *Function of the Human Body*, 4th ed. Philadelphia, W. B. Saunders Co., 1974. Based on data complied by Dr. L. A. Sapirstein.

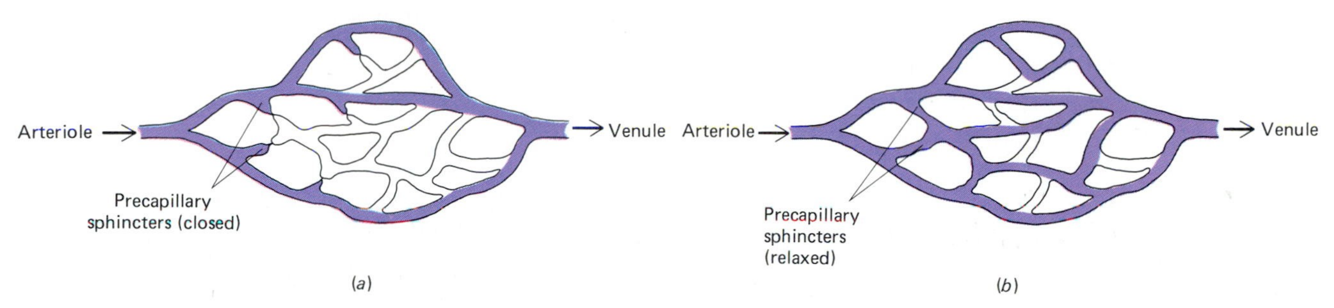

Figure 35–10 Changes in blood flow through a capillary bed as the tissue becomes active. (*a*) When the tissue is inactive, only the metarterioles are open. (*b*) When the tissue becomes active, the decreased oxygen tension in the tissue brings about a relaxation of the precapillary sphincters, and the capillaries become open. This increases the blood supply and the delivery of oxygen to the active tissue.

time most of these tiny vessels are not filled with blood. During periods of intense activity of a particular organ, most of its capillary networks fill with blood (Fig. 35–10).

Capillaries are somewhat "leaky" because some of their endothelial cells have tiny pores, and because others overlap slightly. As blood passes through a capillary some of the plasma passes through its walls and out into the tissues. This fluid, which bathes the tissues, is called **tissue fluid,** or **interstitial fluid.** It may be laden with nutrients and oxygen, which pass out of the blood by diffusion (Fig. 35–11).

The small vessels that directly link arterioles with **venules** (small veins) are **metarterioles.** The so-called **true capillaries** branch off from the metarterioles then rejoin them (Fig. 35–8). True capillaries also interconnect with one another. Wherever a capillary branches from a metarteriole, a smooth muscle cell called a **precapillary sphincter** is present. These sphincters can open or close to regulate passage of blood. Precapillary sphincters open and close continuously, directing blood first

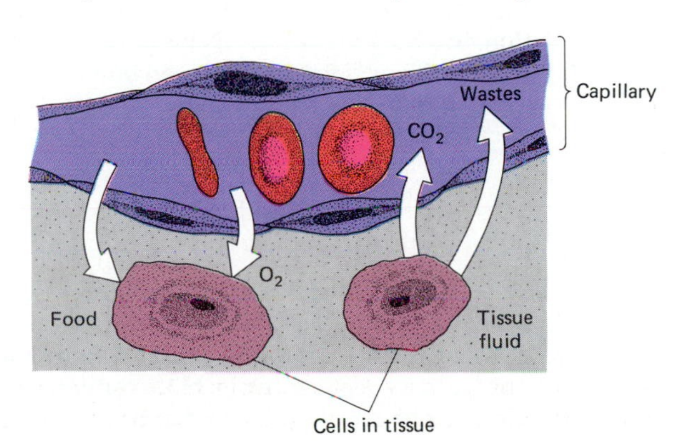

Figure 35–11 Nutrients, oxygen, and other materials diffuse out of the blood and through the tissue fluid that bathes the cells. Carbon dioxide and other waste products diffuse out of the cells and enter the blood through the capillary wall.

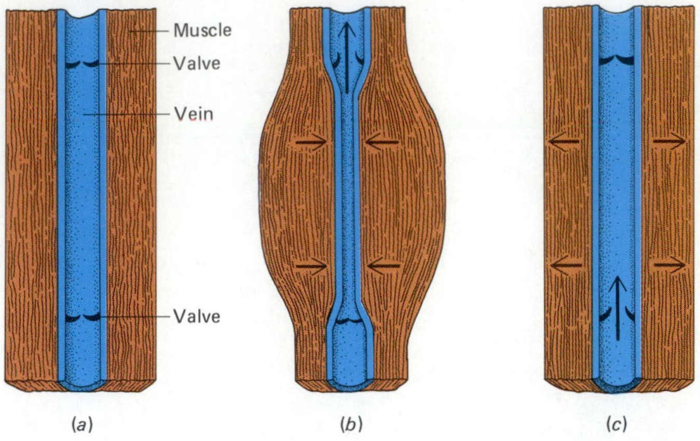

Figure 35–12 The action of skeletal muscles in moving blood through the veins. (*a*) Resting condition. (*b*) Muscles contract and bulge, compressing veins and forcing blood toward the heart. The lower valve prevents backflow. (*c*) Muscles relax, and the vein expands and fills with blood from below. The upper valve prevents backflow.

to one and then to another section of tissue. These sphincters also (along with the smooth muscle in the walls of arteries and arterioles) regulate the blood supply to each organ and its subdivisions. Such mechanisms ensure that blood flow meets the changing needs of the body as a whole, as well as the metabolic needs of the tissue being serviced. For example, during exercise, the increased metabolic rate of muscle cells demands a greater blood supply. Arterioles serving the muscle dilate, permitting a 10-fold increase in the amount of blood delivered to these cells.

Veins

In mammals about 50% of the blood may be found within the veins at any given moment. Vein walls are much thinner and less elastic than those of arteries. By the time blood flows through capillaries and into veins, its pressure is quite low. What, then, keeps the blood flowing through the veins? Breathing and other forms of muscular activity contribute to venous blood flow. When muscles contract, the veins within them are compressed; this helps to push the blood along (Fig. 35–12). Most veins larger than 2 mm (0.08 in.) in diameter that conduct blood against the force of gravity are equipped with **valves** that prevent backflow of blood. Such valves usually consist of two **cusps** formed by inward extensions of the wall of the vein.

When a person stands still for a long period of time, blood tends to accumulate in the veins of the legs. Excessive pooling of blood may stretch the veins so that the cusps of their valves no longer meet. This is especially likely to occur in persons whose occupations require them to stand for long hours each day. **Varicose veins** are likely to result from stretched veins, particularly in overweight persons or in those who have inherited weak vein walls. A varicose vein appears dilated, twisted, and elongated. **Hemorrhoids** are varicose veins in the anal region. They develop when venous pressure in that area is elevated abnormally, as in chronic constipation (due to straining) or during pregnancy, when the increased size of the uterus results in added pressure on the veins of the legs and the lower abdomen.

THE HEART

In vertebrates, the heart consists of one or two chambers called **atria,** which receive blood returning from the tissues, and one or two chambers called **ventricles,** which pump blood into the arteries. Additional chambers are present in some animals.

Comparison of Vertebrate Hearts

Surveying the vertebrates, we find that the structure of the heart and circulatory system reflects evolutionary development. Fish, the earliest vertebrates, have a simple heart; amphibians, reptiles, and birds and mammals have hearts of increasingly complex structure (Fig. 35–13).

FISH. Because it has only one atrium and one ventricle, the fish heart is usually described as a two-chambered heart. Actually, two accessory chambers are present. A thin-walled **sinus venosus** receives blood returning from the tissues and pumps it into the atrium. The atrium then contracts, sending blood into the ventricle. The

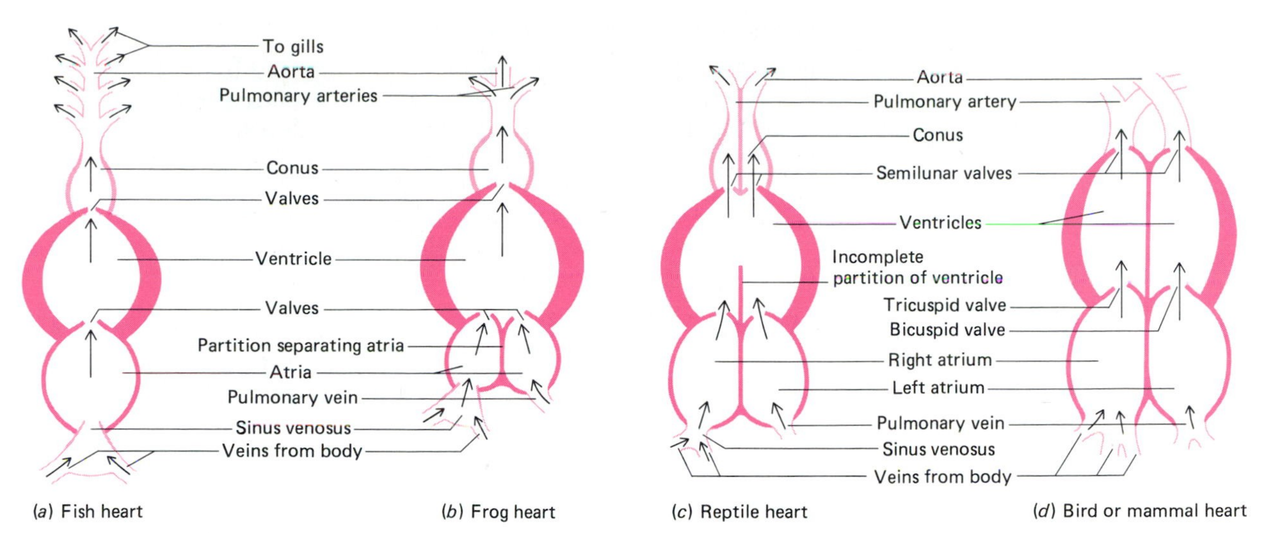

Figure 35–13 The evolution of the vertebrate heart. (*a*) In the fish heart there is one atrium and one ventricle. (*b*) The amphibian heart consists of two atria and one ventricle. (*c*) The reptilian heart has two atria and two ventricles, but the wall separating the ventricles is incomplete so that blood from the right and left chambers mix to some extent. (*d*) Birds and mammals have two atria and two ventricles, and oxygenated blood is kept completely separate from oxygen-poor blood.

ventricle in turn pumps the blood into an elastic **conus arteriosus,** which does not contract. These four compartments are separated by valves that prevent blood from flowing backward. From the conus, blood flows into a large artery, the ventral aorta, which branches to distribute blood to the gills. Because blood must pass through the capillaries of the gills before flowing to the other tissues of the body, blood pressure is low through most of the system. This low-pressure circulatory system permits only a low rate of metabolism in the fish and helps to explain the inability of most fish to maintain a body temperature much higher than that of the surrounding water.

AMPHIBIANS. The three-chambered amphibian heart consists of two atria and a ventricle. A thin-walled sinus venosus collects blood returning from the veins and pumps it into the right atrium. Blood returning from the lungs passes directly into the left atrium. Both atria pump blood into the single ventricle. In the frog heart, oxygenated and deoxygenated blood are kept somewhat separate. Deoxygenated blood is pumped out of the ventricle first and passes into the tubular conus arteriosus, which has a spiral fold that helps to keep the blood separate. Much of the deoxygenated blood is directed to the lungs and skin, where it can be charged with oxygen. Oxygenated blood is sent into arteries, which conduct it to the various tissues of the body.

REPTILES. In reptiles the heart consists of two atria and two ventricles. In all reptiles except the crocodiles, however, the wall between the ventricles is incomplete, so that some mixing of oxygenated and deoxygenated blood does occur. Mixing is minimized by the timing of contractions of the left and right side of the heart and by pressure differences.

BIRDS AND MAMMALS. The hearts of birds and mammals have completely separate right and left sides. The wall between the ventricles is complete, preventing the mixture of oxygenated blood in the left side with deoxygenated blood in the right side. The conus has split and become the base of the aorta and pulmonary artery. No sinus venosus is present as a separate chamber (although a vestige remains as the sinoatrial node, or pacemaker, described later in the chapter).

Complete separation of right and left hearts makes it necessary for blood to pass through the heart twice each time it makes a tour of the body. As a result, blood in the aorta of birds and mammals contains more oxygen than that in the aorta of the lower vetebrates. Hence the tissues of the body receive more oxygen, a higher metabolic rate can be maintained, and the homeothermic (warm-blooded) condition is possible. Birds and mammals can maintain a constant, high body temperature even in cold surroundings.

The pattern of blood flow in birds and mammals may be summarized as follows:

veins → right atrium → right ventricle →
one of the pulmonary arteries → capillaries in the lung →
one of the pulmonary veins → left atrium → left ventricle → aorta

Structure of the Human Heart

In an average lifetime of 70 years, the human heart beats about 2.5 billion times, pumping about 180 million liters of blood. Yet this remarkable organ is only about the size of a fist and weighs only about 400 g (less than a pound). The heart is a hollow, muscular organ located in the chest cavity directly under the breastbone (Fig. 35–14). Enclosing it is a tough connective tissue sac, the **pericardium.** The inner surface of the pericardium and the outer surface of the heart are covered by a smooth layer of epithelium-type cells. Between these two surfaces is a small **pericardial cavity** filled with fluid, which reduces friction to a minimum as the heart beats.

The right atrium and ventricle are separated from the left atrium and ventricle by a wall known as the **septum** (Fig. 35–15). Between the atria the wall is known as the **interatrial septum;** between the ventricles it is the **interventricular septum.** On the interatrial septum a shallow depression, the **fossa ovalis,** marks the place where an opening, the **foramen ovale,** was located in the fetal heart. In the fetus, the foramen ovale permits the blood to move directly from right to left atrium so that very little blood passes to the as yet nonfunctional lungs. At the upper surface of each atrium lies a small, muscular pouch called an **auricle.**

To prevent blood from flowing backward, the heart is equipped with valves that close automatically. The valve between the right atrium and ventricle is called the right **atrioventricular (AV) valve** (also known as the tricuspid valve). The left AV valve is referred to as the **mitral valve.** The AV valves are held in place by stout cords, or "heart-strings," the **chordae tendineae.** These cords attach the valves to the **papillary muscles** that project from the walls of the ventricles. When blood returning from the tissue fills the atria, blood pressure upon the AV valves forces them to open into the ventricles. Blood then fills the ventricles. As the ventricles contract, blood is forced back against the AV valves, pushing them closed. However, contraction of the papillary muscles and tensing of the chordae tendineae prevent them from opening backward into the atria. These valves are like swinging doors that can open in only one direction.

Semilunar valves (named for their flaps, which are shaped like half-moons) guard the exits from the heart. The semilunar valve between the left ventricle and the aorta is known as the **aortic valve,** and the one between the right ventricle and

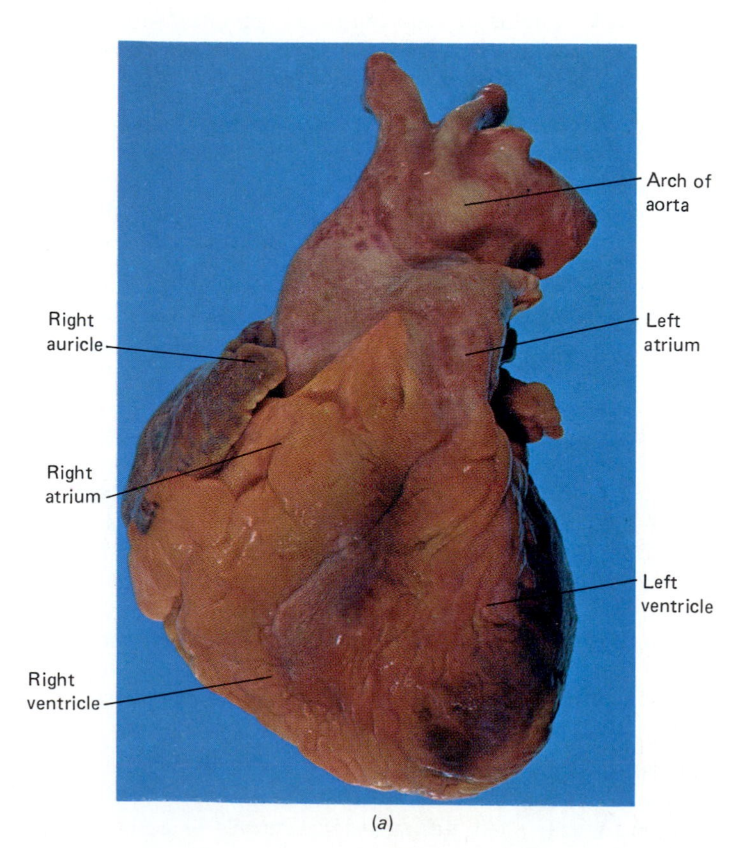

Figure 35–14 The human heart. (*a*) Photograph of anterior external view. (*b*) Anterior view. Note the coronary blood vessels that bring blood to and from the heart muscle itself. (*c*) Posterior view. ((*a*), courtesy of Phil Horne, Stanford University School of Medicine.)

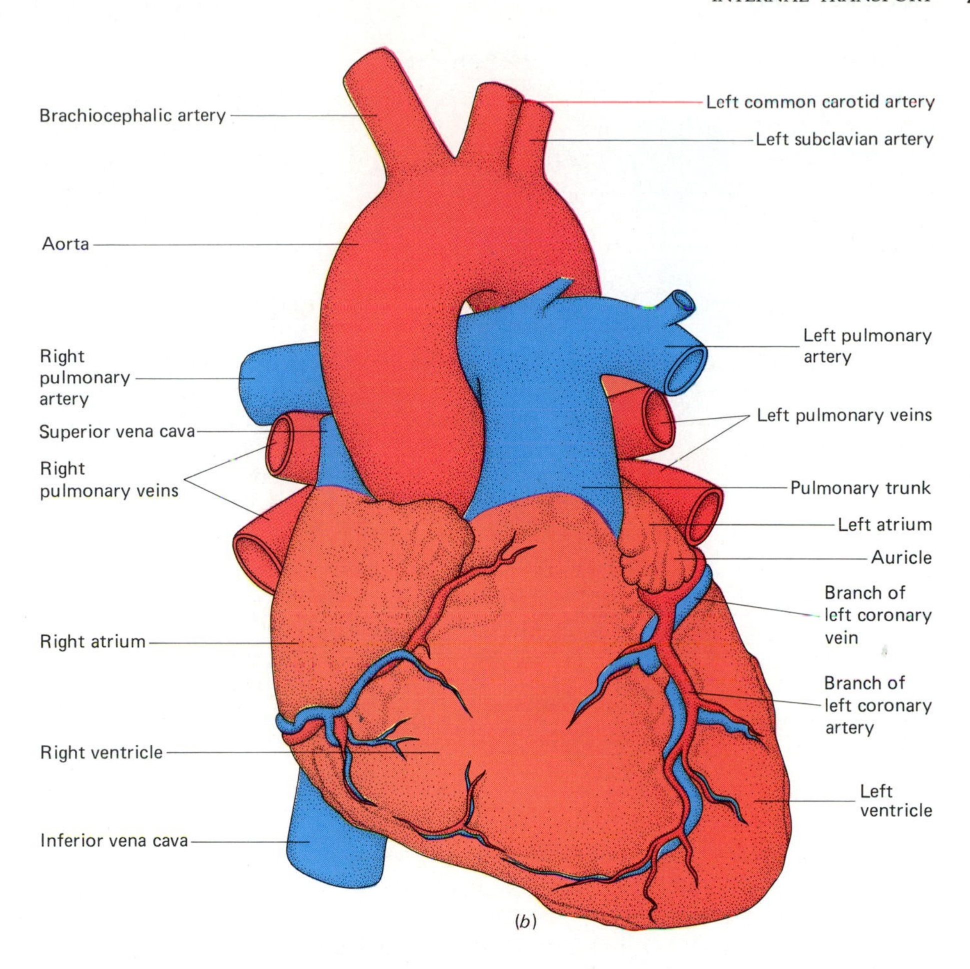
Brachiocephalic artery
Left common carotid artery
Left subclavian artery
Aorta
Left pulmonary artery
Right pulmonary artery
Superior vena cava
Left pulmonary veins
Right pulmonary veins
Pulmonary trunk
Left atrium
Auricle
Branch of left coronary vein
Right atrium
Branch of left coronary artery
Right ventricle
Left ventricle
Inferior vena cava

(*b*)

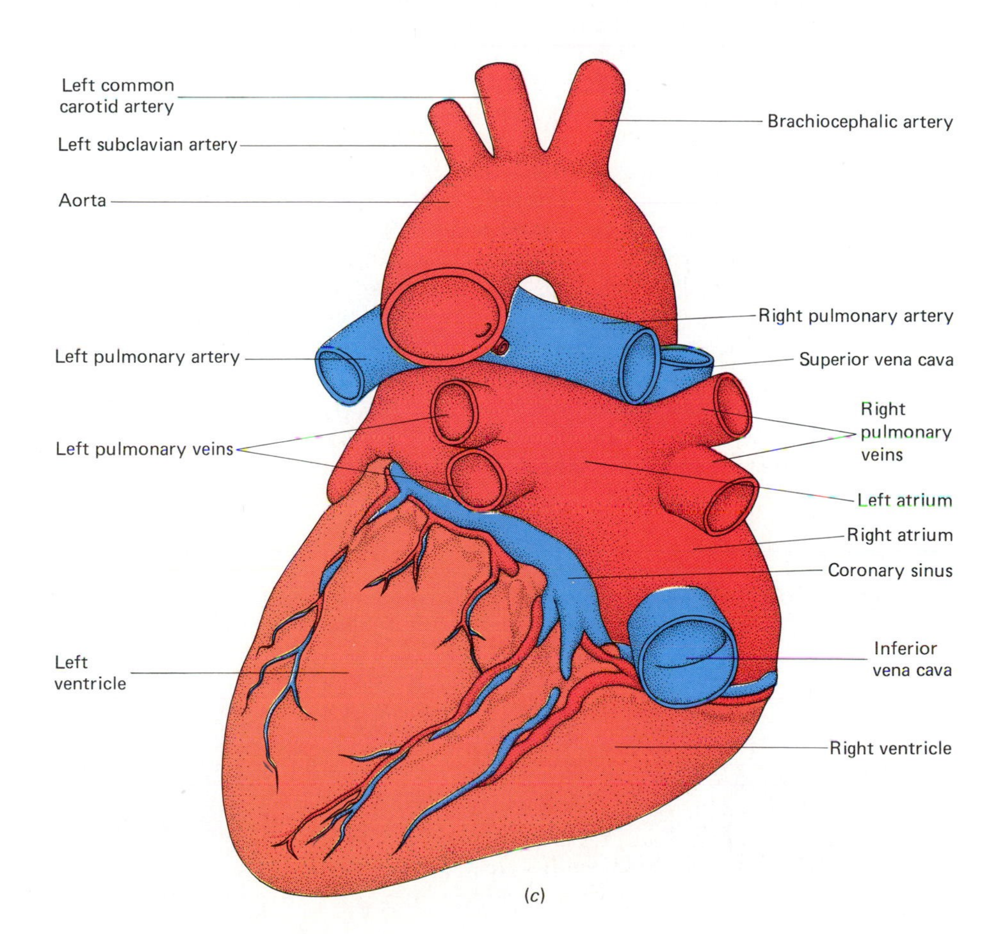
Left common carotid artery
Brachiocephalic artery
Left subclavian artery
Aorta
Right pulmonary artery
Left pulmonary artery
Superior vena cava
Right pulmonary veins
Left pulmonary veins
Left atrium
Right atrium
Coronary sinus
Left ventricle
Inferior vena cava
Right ventricle

(*c*)

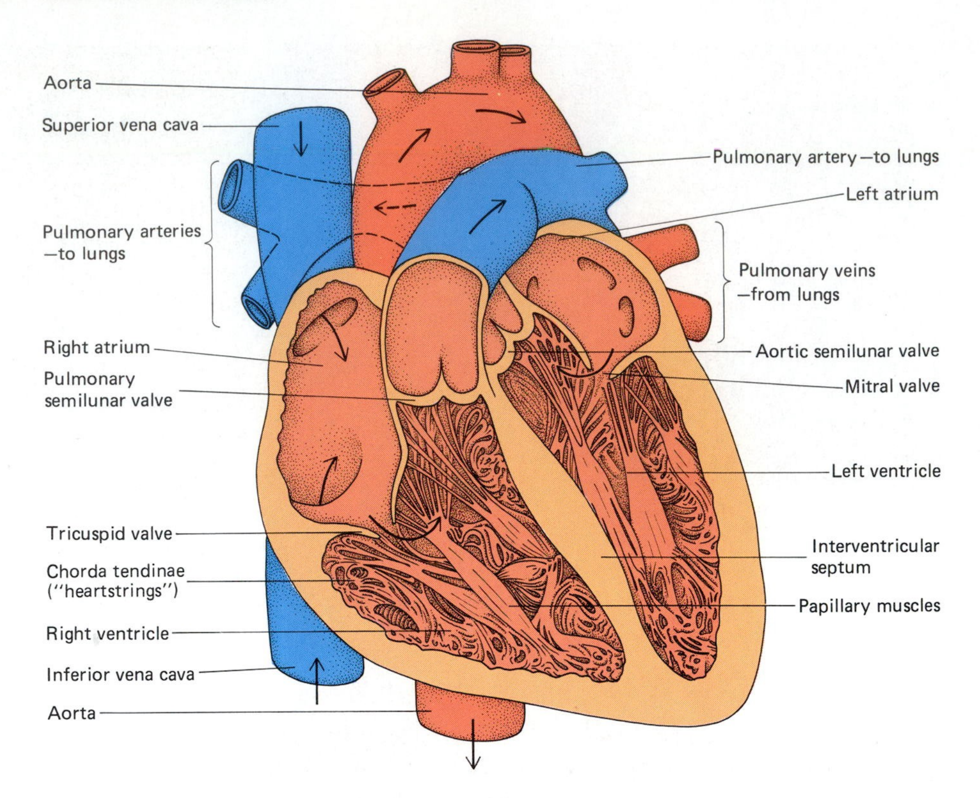

Figure 35–15 Section through the human heart showing chambers, valves, and connecting blood vessels.

the pulmonary artery as the **pulmonary valve.** When blood passes out of the ventricle, the flaps of the semilunar valve are pushed aside and offer no resistance to blood flow. But when the ventricles are relaxing and filling with blood from the atria, the blood pressure in the arteries is higher than that in the ventricles. Blood then fills the pouches of the valves, stretching them across the artery so that blood cannot flow back into the ventricle (Fig. 35–16).

Valve deformities are sometimes present at birth or may occur as a result of certain diseases such as rheumatic fever or syphilis. As a consequence of inflammation and scarring, valves may be thickened so that the passageway for blood is narrowed. Sometimes the valve tissues are eroded so that the flaps cannot close tightly, causing blood to leak backward and reducing the efficiency of the heartbeat. Diseased valves can now be surgically replaced with artificial valves.

The wall of the heart is composed mainly of cardiac muscle attached to a framework of collagen fibers. At their ends cardiac muscle cells are joined by dense bands called **intercalated disks** (Fig. 35–17). Each disk is a type of gap junction (see Chapter 6) in which two cells overlap slightly. This type of junction is of great physiological importance because it offers very little resistance to the passage of an action potential. Ions move easily through the gap junctions, allowing the entire atrial (or ventricular) muscle mass to contract as one giant cell. Because it acts as a single unit, cardiac muscle is sometimes referred to as a *functional syncytium.* The atrial syncytium is separated from the ventricular syncytium by fibrous connective tissue.

How the Heart Works

Horror films not infrequently feature a scene in which a heart cut out from the body of its owner continues to beat. Scriptwriters of these tales actually have some fac-

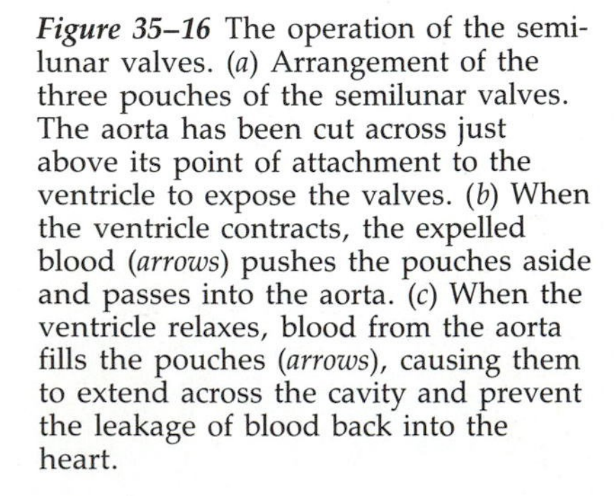

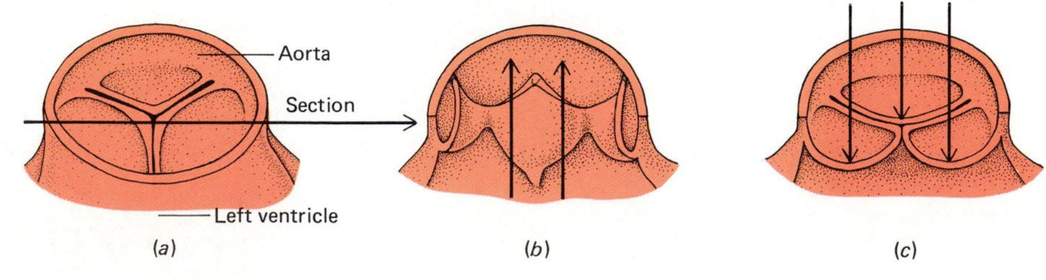

Figure 35–16 The operation of the semilunar valves. (*a*) Arrangement of the three pouches of the semilunar valves. The aorta has been cut across just above its point of attachment to the ventricle to expose the valves. (*b*) When the ventricle contracts, the expelled blood (*arrows*) pushes the pouches aside and passes into the aorta. (*c*) When the ventricle relaxes, blood from the aorta fills the pouches (*arrows*), causing them to extend across the cavity and prevent the leakage of blood back into the heart.

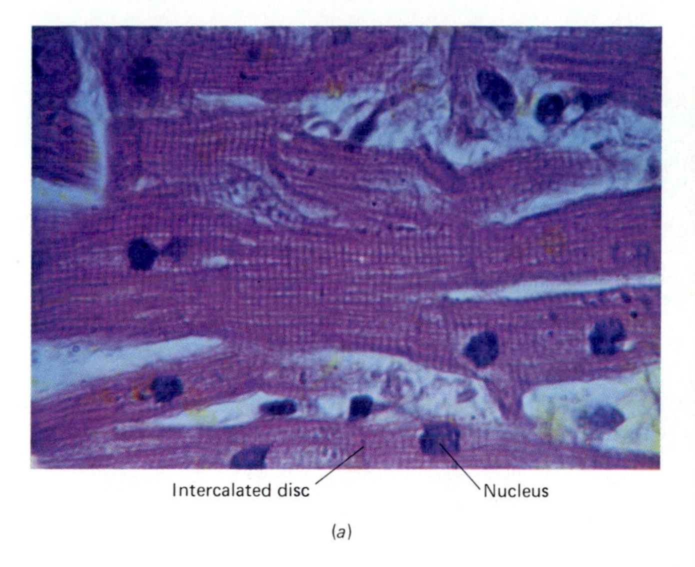

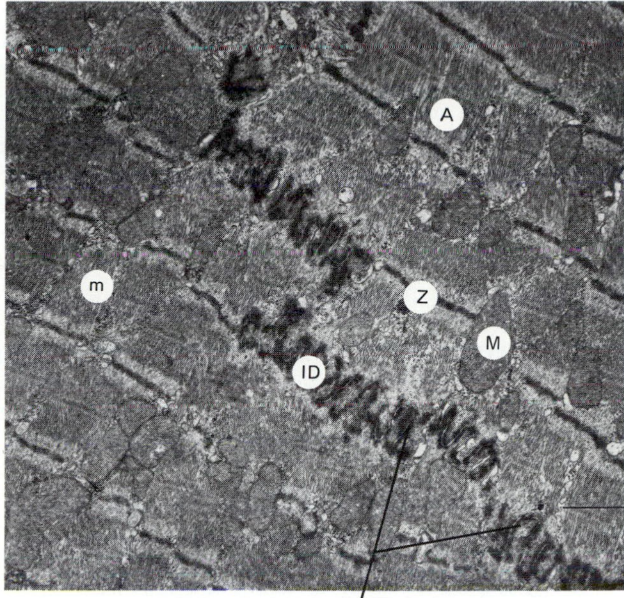

Figure 35–17 Cardiac muscle. (*a*) Cardiac muscle as seen with the light microscope (approximately × 400). (*b*) An electron micrograph of cardiac muscle; *M*, mitochondrion. ((*b*), courtesy of Dr. Lyle C. Dearden.)

tual basis for their gruesome fantasies, for when carefully removed from the body, the heart does continue to beat for many hours if kept in a nutritive, oxygenated fluid. This is possible because the contractions of cardiac muscle begin within the muscle itself and can occur independently of any nerve supply.

To ensure that the heart beats in a regular and effective rhythm, there is a specialized conduction system. Each beat is initiated by the **pacemaker,** called the **sinoatrial (SA) node** (Fig. 35–18). This is a small mass of specialized cardiac muscle in the posterior wall of the right atrium near the opening of the superior vena cava. Ends of the SA node fibers fuse with surrounding ordinary atrial muscle fibers so that the action potential spreads through the atria, producing atrial contraction.

Figure 35–18 The conduction system of the heart.

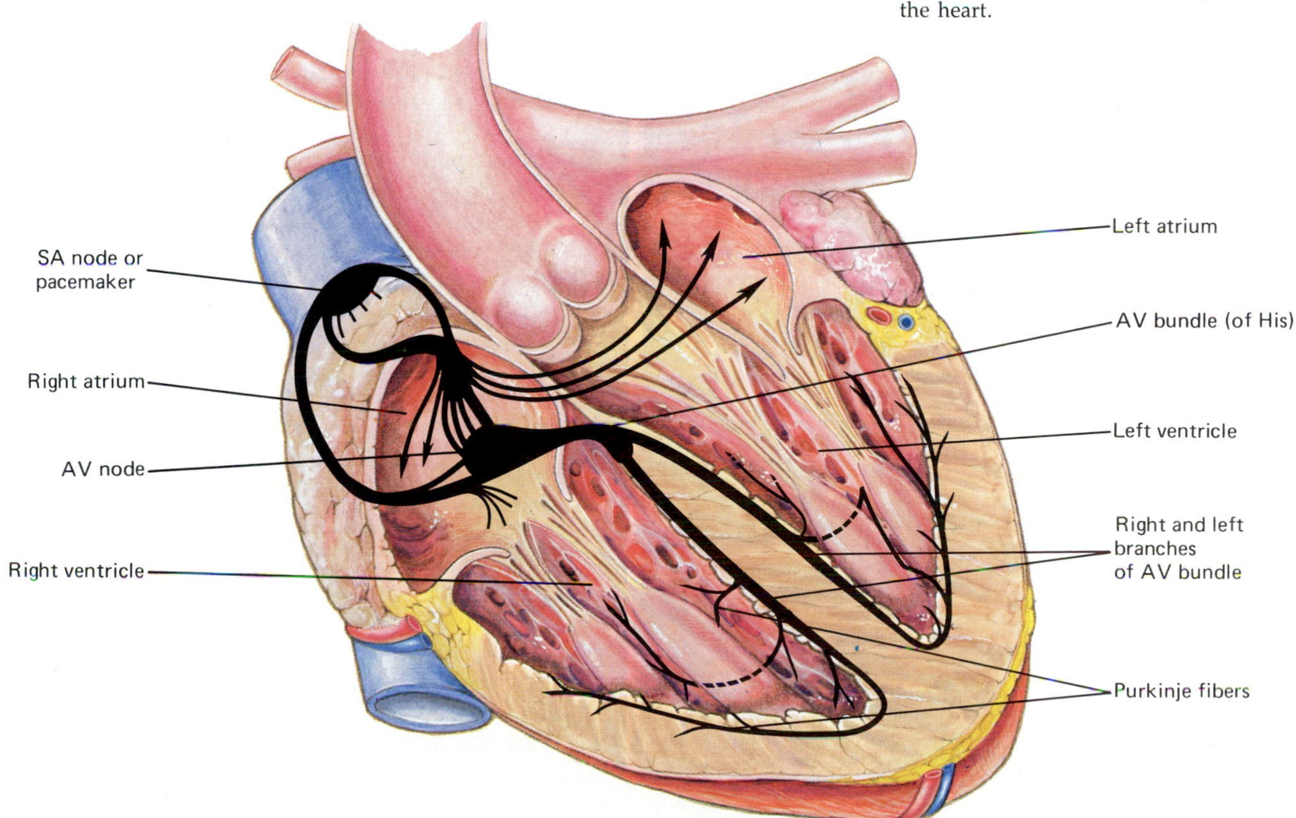

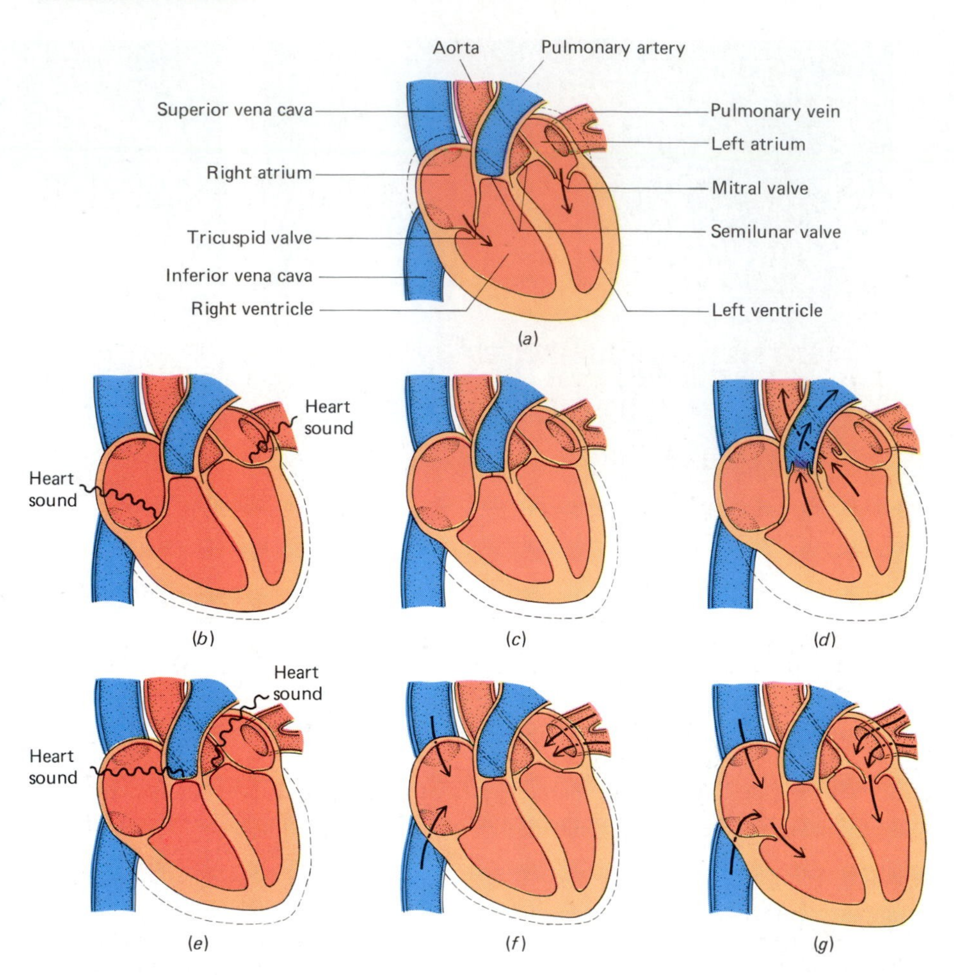

Figure 35–19 The cardiac cycle. Arrows indicate the direction of blood flow; dotted lines indicate the change in size as contraction occurs. (*a*) Atrial systole: Atria contract, and blood is pushed through the open tricuspid (right AV) and mitral valves into the ventricles. The semilunar valves are closed. (*b*) Beginning of ventricular systole: Ventricles begin to contract, and pressure within the ventricles increases and closes the tricuspid and mitral valves, causing the first heart sound. (*c*) Period of rising pressure. (*d*) The semilunar valves open when the pressure within the ventricles exceeds that in the arteries, and blood spurts into the aorta and pulmonary artery. (*e*) Beginning of ventricular diastole: When the pressure in the relaxing ventricles drops below that in the arteries, the semilunar valves snap shut, causing the second heart sound. (*f*) Period of falling pressure: Blood flows from the veins into the relaxed atria. (*g*) The tricuspid and mitral valves open when the pressure in the ventricles falls below that in the atria; blood then flows into the ventricles.

One group of atrial muscle fibers conducts the action potential directly to the **atrioventricular (AV) node** located in the right atrium along the lower part of the septum. Here transmission is delayed briefly, permitting the atria to complete their contraction before the ventricles begin to contract. From the AV node the action potential spreads into specialized muscle fibers called **Purkinje fibers.** These large fibers make up the **atrioventricular (AV) bundle.** The AV bundle then divides, sending branches into each ventricle. When an impulse reaches the ends of the Purkinje fibers it spreads through the ordinary cardiac muscle fibers of the ventricles.

Each minute the heart beats about 70 times. One complete heart beat takes about 0.8 second and is referred to as a **cardiac cycle.** That portion of the cycle in which contraction occurs is known as **systole;** the period of relaxation is **diastole.** Figure 35–19 shows the sequence of events that occur during one cardiac cycle.

HEART SOUNDS. When you listen to the heartbeat with a stethoscope you can hear two main heart sounds, lub-dup, which repeat rhythmically. The first heart sound, **lub,** is low-pitched, not very loud, and of fairly long duration. It is caused mainly by the closing of the AV valves and marks the beginning of ventricular systole. The lub sound is quickly followed by the higher-pitched, louder, sharper, and shorter **dup** sound. Heard almost as a quick snap, the dup marks the closing of the semilunar valves and the beginning of ventricular diastole.

The quality of these sounds tells a discerning physician much about the state of the valves. When the semilunar valves are injured, a soft hissing noise ("lub-shhh") is heard in place of the normal sound. This is known as a **heart murmur** and may be caused by any injury that has affected the valves so that they do not close tightly, permitting blood to flow backward into the ventricles during diastole.

ELECTROCARDIOGRAMS. As each wave of contraction spreads through the heart, electrical currents spread into the tissues surrounding the heart and onto the body surface. By placing electrodes on the body surface on opposite sides of the heart, the electrical activity can be amplified and recorded by either an oscilloscope or an

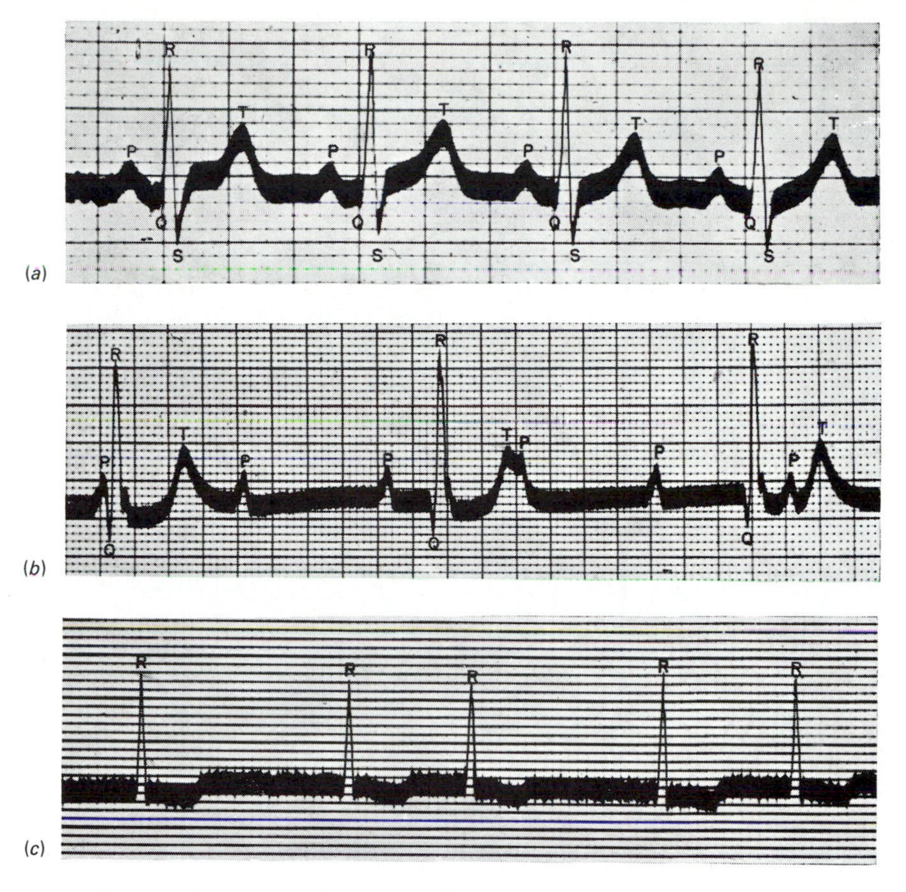

Figure 35–20 Electrocardiograms. (*a*) Tracing from a normal heart. The P wave corresponds to the contraction of the atria, the QRS complex to the contraction of the ventricle, and the T wave to the relaxation of the ventricle. (*b*) Tracing from a patient with a complete block of the atrioventricular node, so that the atria and ventricles beat independently, each at their own rate. Note that the P waves appear at regular intervals and QRS and T waves appear at regular but longer intervals, but that there is no relation between the P and QRS waves. (*c*) Tracing from a patient with atrial fibrillation. The individual muscle fibers of the atrium twitch rapidly and independently. There is no regular atrial contraction and no P wave. The ventricles beat independently and irregularly, causing the QRS wave to appear at irregular intervals. (Courtesy of Dr. Lewis Dexter and the Peter Bent Brigham Hospital, Boston, Mass.)

electrocardiograph. The written record produced is called an **electrocardiogram,** or ECG (Fig. 35–20). Malfunctioning of the heart causes abnormal action currents, which in turn produce an abnormal ECG.

CARDIAC OUTPUT. The volume of blood pumped by one ventricle during one beat is called the **stroke volume.** By multiplying the stroke volume times the number of times the ventricle beats in one minute, the cardiac output can be computed. In other words, the **cardiac output** is the volume of blood pumped by one ventricle in one minute. For example, in a resting adult the heart may beat about 72 times per minute and pump about 70 ml of blood with each contraction.

$$\text{Cardiac output} = \text{Stroke volume} \times \underset{\text{(number of ventricular contractions per minute)}}{\text{Heart rate}}$$

$$= \frac{70 \text{ ml}}{\cancel{\text{stroke}}} \times \frac{72 \ \cancel{\text{stroke}}}{\text{minutes}}$$

$$= 5040 \text{ ml/min (or a little more than 5 liters/min)}$$

The cardiac output varies dramatically with the changing needs of the body. During stress or heavy exercise, the normal heart can increase its cardiac output four- to fivefold, so that up to 25 liters of blood can be pumped per minute. Cardiac output varies with changes in either stroke volume or heart rate.

Regulation of Stroke Volume. Stroke volume depends mainly upon venous return, the amount of blood delivered to the heart by the veins. According to **Starling's law of the heart,** the greater the amount of blood delivered to the heart by the veins, the more blood the heart pumps (within physiological limits). When extra amounts of blood fill the heart chambers, the cardiac muscle fibers are stretched to a greater extent and contract with greater force, pumping a larger volume of blood into the arteries. This increase in stroke volume increases the cardiac output (Fig. 35–21).

The release of **norepinephrine** by sympathetic nerves also increases the force of contraction of the cardiac muscle fibers. **Epinephrine** released by the adrenal glands during stress has a similar effect on the heart muscle. When the force of

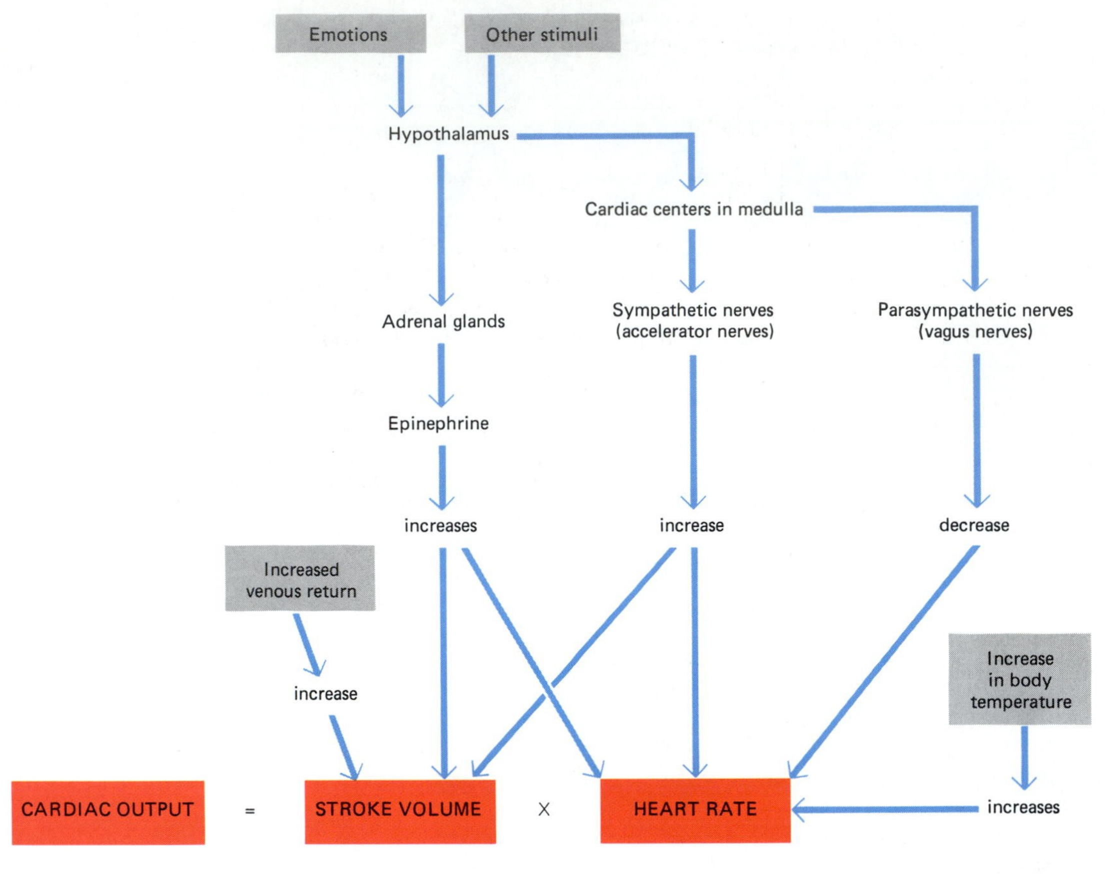

Figure 35–21 Some factors that influence cardiac output.

contraction increases, the stroke volume increases, and this in turn increases cardiac output.

Regulation of Heart Rate. Although the heart is capable of beating independently of its control systems, its rate is, in fact, carefully regulated by the nervous system. Sensory receptors in the walls of certain blood vessels and heart chambers are sensitive to changes in blood pressure. When stimulated they send messages to **cardiac centers** in the medulla of the brain. These cardiac centers maintain control over two sets of autonomic nerves that pass to the SA node. Sympathetic nerves release norepinephrine, which speeds the heart rate and increases the strength of contraction. Parasympathetic nerves release **acetylcholine,** which slows the heart and decreases the force of each contraction.

Hormones also influence heart rate. During stress, the adrenal glands release epinephrine and norepinephrine, which speed the heart. An elevated body temperature can greatly increase heart rate; during fever, the heart may beat more than 100 times per minute. As you might expect, heart rate decreases when body temperature is lowered. This is why a patient's temperature may be deliberately lowered during heart surgery.

PULSE AND BLOOD PRESSURE

You have probably felt your pulse by placing a finger over the radial artery in your wrist or over the carotid artery in the neck. Arterial **pulse** is the alternate expansion and recoil of an artery. Each time the left ventricle pumps blood into the aorta, the elastic wall of the aorta expands to accommodate the blood. This expansion moves down the aorta and the arteries that branch from it in a wave (Fig. 35–22). When the wave passes, the elastic arterial wall snaps back to its normal size. Every time the heart contracts, a pulse wave begins, so the number of pulsations you count in an artery per minute indicates the number of heartbeats.

Blood pressure is the force exerted by the blood against the inner walls of the blood vessels. It is determined by the blood flow and the resistance to that flow. **Blood flow** depends directly upon the pumping action of the heart. When cardiac output increases, blood flow increases, causing a rise in blood pressure. When

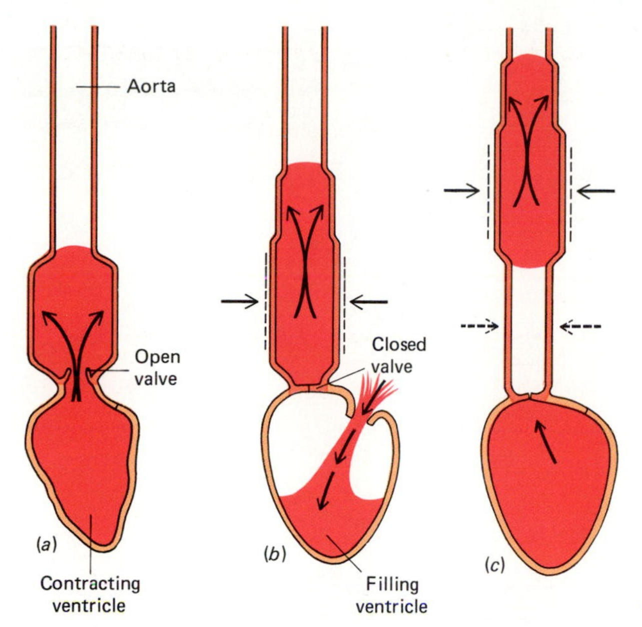

Figure 35–22 The movement of blood from the ventricle through the elastic arteries. For simplicity, only one ventricle and artery are shown, and the amount of stretching of the arterial wall is exaggerated. (*a*) As the ventricle contracts, blood is forced through the semilunar valves, and the adjacent wall of the aorta is stretched. (*b*) As the ventricle relaxes and begins to fill for the next stroke, the semilunar valve closes and the expanded part of the aorta contracts, causing the adjacent part of the aorta to expand as it is filled with blood. (*c*) The pulse wave of expansion and contraction is transmitted to the next adjoining section of the aorta.

cardiac output decreases, blood flow decreases, causing a fall in blood pressure. The volume of blood flowing through the system also affects blood pressure. If blood volume is reduced by hemorrhage or by chronic bleeding, the blood pressure drops. On the other hand, an increase in blood volume results in an increase in blood pressure. For example, a high dietary intake of salt causes retention of water. This may result in an increase of blood volume and lead to higher blood pressure.

Blood flow is impeded by resistance; when the resistance to flow increases, blood pressure rises. **Peripheral resistance** is the resistance to blood flow caused by viscosity and by friction between the blood and the wall of the blood vessel. In the blood of a healthy person, viscosity remains fairly constant and is only a minor factor influencing changes in blood pressure. More important is the friction between the blood and the wall of the blood vessel. The length and diameter of a blood vessel determine the surface area of the vessel in contact with the blood. The length of a blood vessel does not change, but the diameter, especially of an arteriole, does. A small change in the diameter of a blood vessel causes a big change in blood pressure. The resistance is inversely proportional to the fourth power of the vessel radius. Thus if the radius is doubled, the resistance would be dramatically reduced to one sixteenth of its former value, and the flow would increase 16-fold.

Blood pressure in arteries rises during systole and falls during diastole (Fig. 35–23). Normal blood pressure for a young adult is about 120/80 millimeters of mercury, abbreviated mm Hg, as measured by the sphygmomanometer (Fig. 35–24). Systolic pressure is indicated by the numerator, diastolic by the denominator.

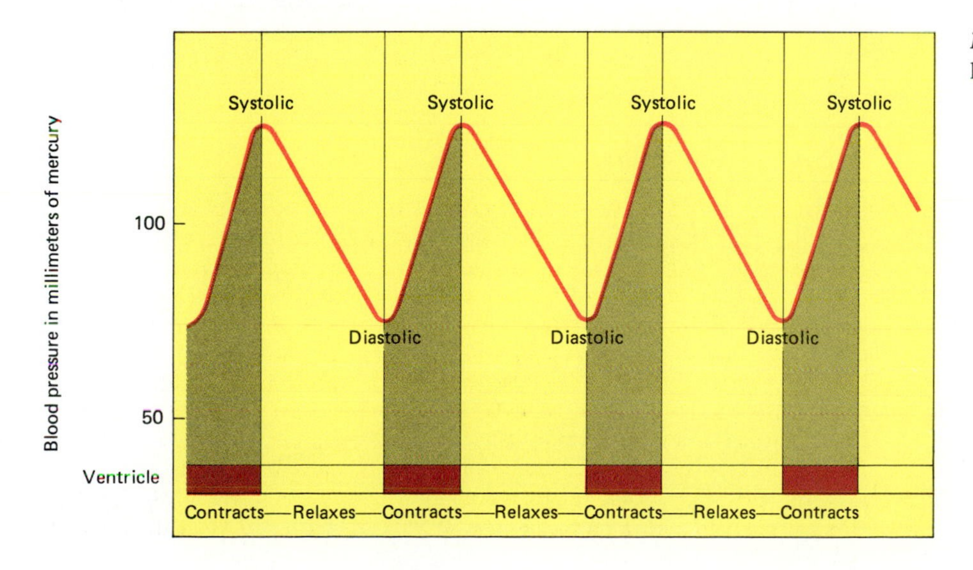

Figure 35–23 The changes in blood pressure in an artery as the heart beats.

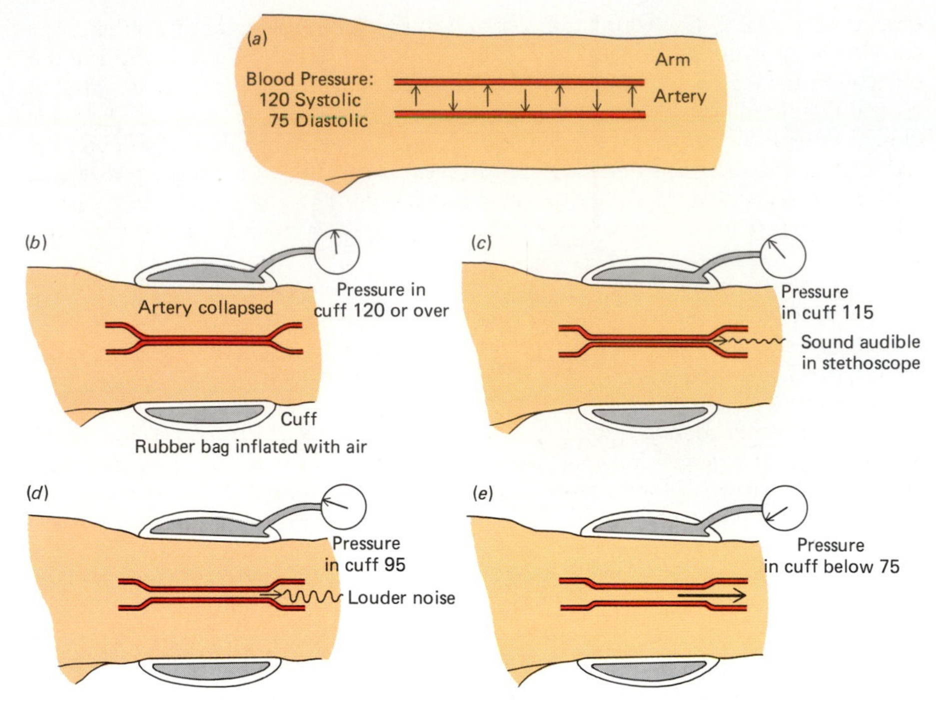

Figure 35–24 The measurement of blood pressure by the sphygmomanometer. (*a*) Blood pushes against the arterial walls. (*b*) When the pressure in the rubber cuff is 120 mm Hg, the artery is collapsed and no blood passes. (*c*) With the pressure in the cuff just below 120 mm Hg, the artery is collapsed except for a small period during systole, when a small amount of blood squirts through, producing a sound audible in the stethoscope. (*d*) With the pressure in the cuff at 95 mm Hg, the artery is open for a longer time during systole, a greater amount of blood passes through, and a louder noise is produced. (*e*) When the pressure in the cuff drops below 75 mm Hg, the artery is open continuously, blood passes through continuously, and no noise is heard.

When the diastolic pressure consistently measures more than 95 mm Hg, the patient may be suffering from high blood pressure, or **hypertension.** In hypertension, there is usually increased vascular resistance, especially in the arterioles and small arteries. Work load of the heart is increased because it must pump against this increased resistance. As a result, the left ventricle increases in size and may begin to deteriorate in function. Heredity, obesity, and possibly high dietary salt intake are thought to be important factors in the development of hypertension. Recent research suggests that inadequate intake of calcium, potassium, vitamin A, and vitamin C are associated with hypertension.

As you might guess, blood pressure is greatest in the large arteries and decreases as blood flows through the smaller arteries and capillaries (Fig. 35–25). By the time blood enters the veins its pressure is very low, even approaching zero. As described earlier, flow of blood through veins depends upon several factors, including muscular movement, which compresses veins, and valves, which prevent backflow.

When a person stands perfectly still for a long time, as when a soldier stands at attention, blood tends to pool in the veins. This is so because when fully dis-

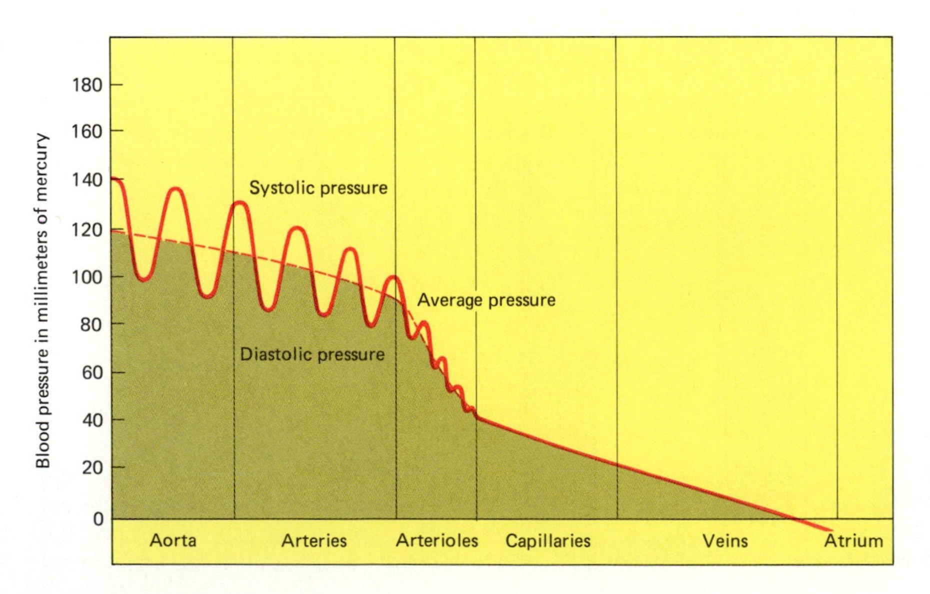

Figure 35–25 Blood pressure in different types of blood vessels of the body. The systolic and diastolic variations in arterial blood pressures are shown. Note that the venous pressure drops below zero (below atmospheric pressure) near the heart.

tended with blood, the veins can accept no more blood from the capillaries. Pressure in the capillaries increases, and large amounts of plasma are forced out of the circulation through the thin capillary walls. Within just a few minutes, as much as 20% of the blood volume can be lost from the circulation in this way—with drastic effect. Arterial blood pressure falls dramatically, so that blood flow to the brain is reduced. Sometimes the resulting lack of oxygen in the brain causes fainting, a protective response aimed at increasing blood supply to the brain. Lifting a person who has fainted to an upright position can result in circulatory shock and even death.

Each time you get up from a horizontal position, changes occur in your blood pressure. Several complex mechanisms interact to maintain normal blood pressure so that you do not faint when you get out of bed each morning or change position during the day. When blood pressure falls, sympathetic nerves to the blood vessels stimulate vasoconstriction so that pressure rises again.

The **baroreceptors** present in the walls of certain arteries and in the heart wall are sensitive to changes in blood pressure. When an increase in blood pressure stretches the baroreceptors, messages are sent to the **vasomotor center** in the medulla. This center stimulates parasympathetic nerves that slow the heart, lowering blood pressure. Also, the vasomotor center inhibits sympathetic nerves that constrict arterioles, thereby lowering blood pressure. These neural reflexes act continuously to maintain blood pressure.

Hormones are also involved in regulating blood pressure. The **angiotensins** are a group of hormones that act as powerful vasoconstrictors. They are formed from a plasma protein (angiotensinogen) when the kidneys release the hormone **renin.** Release of renin is stimulated by low blood pressure within the kidneys. The kidneys also act *indirectly* to maintain blood pressure by influencing blood volume. This is accomplished by varying the rate at which the kidneys excrete salt and water, which in turn is regulated by hormones.

THE PATTERN OF CIRCULATION

One of the main jobs of the circulation is to bring oxygen to all of the cells of the body. In humans, as in other mammals and in birds, blood is charged with oxygen in the lungs. Then it is returned to the heart to be pumped out into the arteries that deliver it to the other tissues and organs of the body. There is a double circuit of blood vessels—(1) the **pulmonary circulation,** which connects the heart and lungs, and (2) the **systemic circulation,** which connects the heart with all of the tissues of the body. This general pattern of circulation may be traced in Figure 35–26.

Pulmonary Circulation

Blood from the tissues returns to the right atrium of the heart partly depleted of its oxygen supply. This oxygen-poor blood, loaded with carbon dioxide, is pumped by the right ventricle into the pulmonary circulation. As it emerges from the heart, the large **pulmonary trunk** branches to form the two **pulmonary arteries,** one going to each lung. These are the only arteries in the body that carry oxygen-poor blood. In the lungs the pulmonary arteries branch into smaller and smaller vessels, which finally give rise to extensive networks of **pulmonary capillaries** bringing blood to all of the air sacs of the lung. As blood circulates through the pulmonary capillaries, carbon dioxide diffuses out of the blood and into the air sacs. Oxygen from the air sacs diffuses into the blood so that by the time blood enters the **pulmonary veins** leading back to the left atrium of the heart, it is charged with oxygen. Pulmonary veins are the only veins in the body that carry blood rich in oxygen.

In summary, blood flows through the pulmonary circulation in the following sequence:

right atrium → right ventricle → pulmonary artery →
pulmonary capillaries (in lung) → pulmonary vein → left atrium

Systemic Circulation

Blood entering the systemic circulation is pumped by the left ventricle into the **aorta,** the largest artery of the body. Arteries that branch off from the aorta conduct blood to all of the regions of the body. Some of the principal branches include the **coronary arteries** to the heart wall itself, the **carotid arteries** to the brain, the **subcla-**

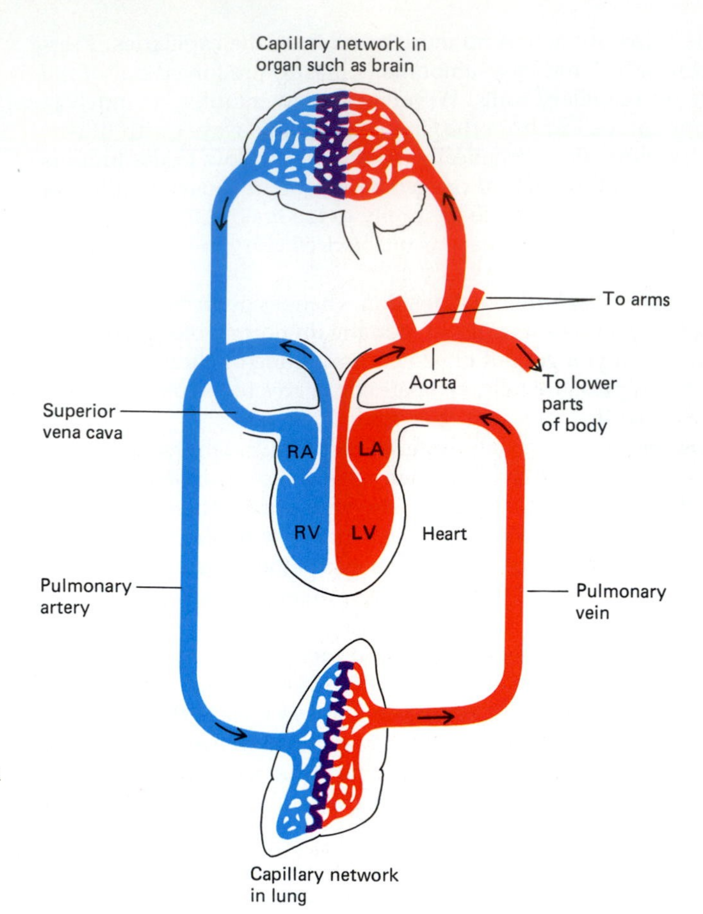

Figure 35–26 Highly simplified diagram showing the pattern of circulation through the systemic and pulmonary circuits. Red represents oxygenated blood; blue represents deoxygenated blood.

vian arteries to the shoulder region, the **mesenteric artery** to the intestine, the **renal arteries** to the kidneys, and the **iliac arteries** to the legs (Fig. 35–27). Each of these arteries gives rise to smaller branches, which in turn give rise to smaller and smaller vessels somewhat like branches of a tree that divide until they form tiny twigs. Eventually blood flows into the capillary networks within each tissue or organ.

Blood returning from the capillary networks within the brain passes through the **jugular veins.** Blood from the shoulders and arms drains into the **subclavian veins.** These veins and others returning blood from the upper portion of the body merge to form the **superior vena cava,** a very large vein that empties blood into the right atrium. **Renal veins** from the kidneys, **iliac veins** from the lower limbs, **hepatic veins** from the liver, and other veins from the lower portion of the body return blood to the **inferior vena cava,** which delivers blood to the right atrium.

As an example of blood circulation through the systemic system, let us trace a drop of blood from the heart to the right leg and back to the heart:

> left atrium → left ventricle → aorta → right common iliac artery →
> smaller arteries in leg → capillaries in right leg →
> small vein in leg → right common iliac vein → inferior vena cava →
> right atrium → right ventricle → into pulmonary circulation

CORONARY CIRCULATION. The heart muscle is not nourished by the blood within its chambers, because its walls are too thick for nutrients and oxygen to diffuse through them to all of the cells. Instead, the cardiac muscle is supplied by coronary arteries branching from the aorta at the point where that vessel leaves the heart. These arteries branch, giving rise to a network of blood vessels within the wall of the heart. Nutrients and gases are exchanged through the coronary capillaries. Blood from these capillaries flows into coronary veins, which join to form a large vein, the **coronary sinus.** The coronary sinus empties directly into the right atrium; it does not join either of the venae cava.

When one of the coronary arteries is blocked, the cells in the area of the heart muscle served by that artery are deprived of oxygen and nutrients, and die. The affected muscle stops contracting; if sufficient cardiac muscle is affected, the heart

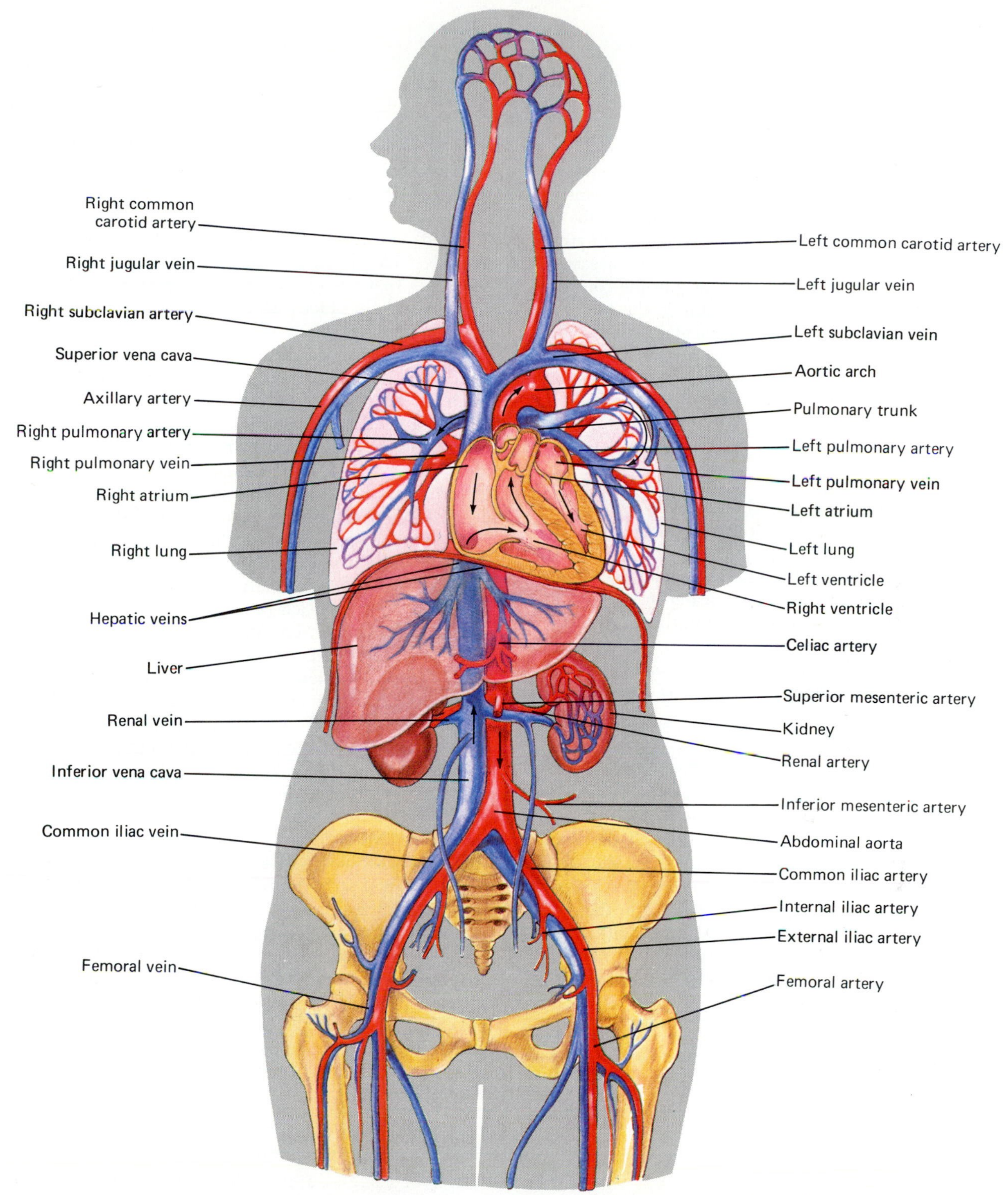

Figure 35–27 Circulation of blood through some of the principal arteries and veins of the human body. Blood vessels carrying oxygenated blood are red; those carrying deoxygenated blood are blue.

may stop beating entirely. This is a common cause of "heart attack" and is often the result of atherosclerosis (see Focus on Cardiovascular Disease).

CIRCULATION TO THE BRAIN. Four arteries—two internal carotid arteries and two vertebral arteries (branches of the subclavian arteries)—deliver blood to the brain. At the base of the brain, branches of these arteries form an arterial circuit called the **circle of Willis.** In the event that one of the arteries serving the brain becomes blocked or injured in some way, this arterial circuit helps ensure blood delivery to the brain cells via other vessels. Blood from the brain returns to the superior vena cava by way of the internal jugular veins at either side of the neck.

FOCUS ON

Cardiovascular Disease

Cardiovascular disease is the number one cause of death in the United States and in most other industrial societies. Most often death results from some complication of **atherosclerosis*** (hardening of the arteries as a result of lipid deposition). Although atherosclerosis can affect almost any artery, the disease most often develops in the aorta and in the coronary and cerebral arteries. When it occurs in the cerebral arteries it can lead to a **cerebrovascular accident (CVA),** commonly referred to as a stroke.

Although there is apparently no single cause of atherosclerosis, several major risk factors have been identified:

1. Elevated levels of cholesterol in the blood, often associated with diets rich in total calories, total fats, saturated fats, and cholesterol.
2. Hypertension. The higher the blood pressure, the greater the risk.
3. Cigarette smoking. The risk of developing atherosclerosis is two to six times greater in smokers than nonsmokers and is directly proportional to the number of cigarettes smoked daily.
4. Diabetes mellitus, an endocrine disorder in which glucose is not metabolized normally.

The risk of developing atherosclerosis also increases with age. Estrogen hormones are thought to offer some protection in women until after menopause, when the concentration of these hormones decreases. Other suggested risk factors that are currently being studied are obesity, hereditary predisposition, lack of exercise, stress and behavior patterns, and dietary factors such as excessive intake of salt or refined sugar.

In atherosclerosis, lipids are deposited in the smooth muscle cells of the arterial wall. Cells in the arterial wall proliferate and the inner lining thickens. More lipid, especially cholesterol from low-density lipoproteins, accumulates in the wall. Eventually calcium is deposited there, contributing to the slow formation of hard plaque. As the plaque develops, arteries lose their ability to stretch when they fill with blood, and they become progressively occluded (blocked), as shown in the figure. As the artery narrows, less blood can pass through to reach the tissues served by that vessel and the tissue may become **ischemic** (lacking in blood). Under these conditions the tissue is deprived of an adequate oxygen supply.

When a coronary artery becomes narrowed, **ischemic heart disease** can occur. Sufficient oxygen may reach the heart tissue during normal activity, but the increased need for oxygen during exercise or emotional stress results in the pain known as **angina pectoris.** Persons with this condition often carry nitroglycerin pills with them for use during an attack. This drug dilates veins so that venous return is reduced. Cardiac output is lowered so that the heart is not working so hard and requires less oxygen. Nitroglycerin also dilates the coronary arteries slightly, allowing more blood to reach the heart muscle.

Myocardial infarction (MI) is the very serious, often fatal, form of ischemic heart disease that often results from a sudden decrease in coronary blood supply. The portion of cardiac muscle deprived of oxygen dies within a few minutes and is then referred to as an **infarct.** The term myocardial infarction is used as a synonym for heart attack. MI is the leading cause of death and disability in the United States. Just what triggers the sudden decrease in blood supply that causes MI is a matter of some debate. It is thought that in some cases an episode of ischemia triggers a fatal arrhythmia such as **ventricular fibrillation,** a condition in which the ventricles contract very rapidly without actually pumping blood. In other cases, a **thrombus** (clot) may form in a diseased coronary artery. Because the arterial wall is roughened, platelets may adhere to it and initiate clotting. If the thrombus blocks a sizable branch of a coronary artery, blood flow to a portion of heart muscle is impeded or completely halted. This condition is referred to as a

*Atherosclerosis is the most common form of arteriosclerosis, any disorder in which arteries lose their elasticity.

HEPATIC PORTAL SYSTEM. Blood almost always travels from artery to capillary to vein. An exception to this sequence is made by the **hepatic portal system,** which delivers blood rich in nutrients to the liver. Blood is conducted to the small intestine by the superior mesenteric artery. Then, as it flows through capillaries within the wall of the intestine, blood picks up glucose, amino acids, and other nutrients. This blood passes into the mesenteric vein and then into the **hepatic portal vein.** Instead of going directly back to the heart (as most veins would), the hepatic portal vein delivers nutrients to the liver. Within the liver, this vein gives rise to an extensive network of tiny blood sinuses. As blood courses through the hepatic sinuses, liver cells remove nutrients and store them. Eventually liver sinuses merge to form hepatic veins, which deliver blood to the inferior vena cava. Although laden with food

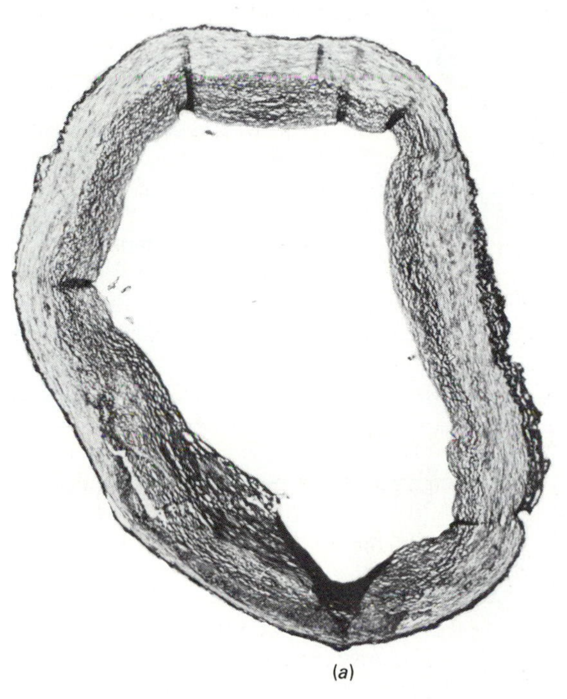

(a)

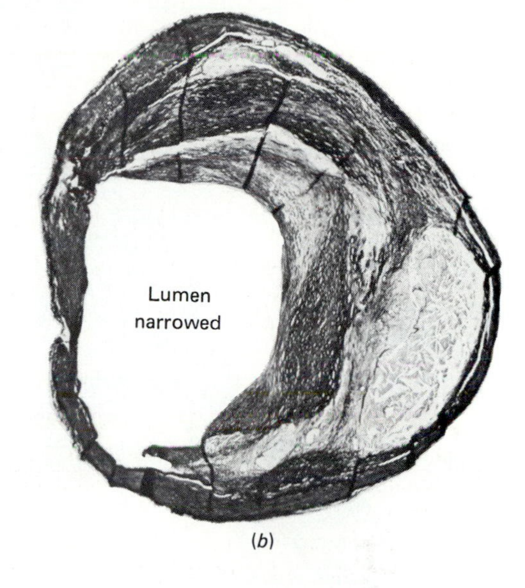

(b)

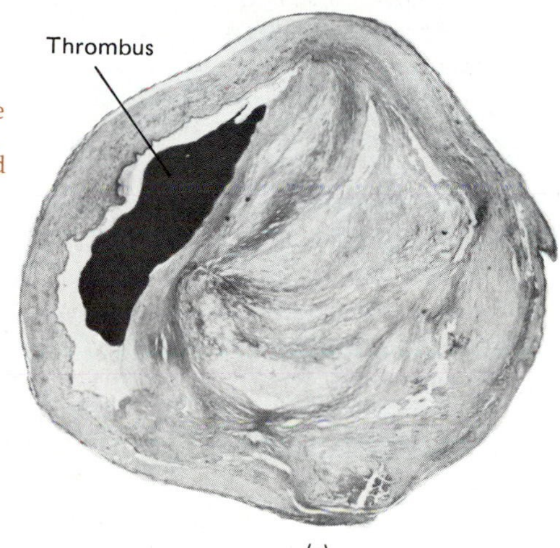

(c)

Progression of atherosclerosis. Cross sections through three arteries showing changes that take place in atherosclerosis. (*a*) Early stage of atherosclerosis. Inner lining has thickened slightly at lower left. (*b*) Pronounced changes have taken place in this artery. Note the marked thickening of the wall. (*c*) This artery is almost completely blocked with atherosclerotic placque. (Courtesy of American Heart Association.)

coronary occlusion. If the coronary occlusion prevents blood flow to a large region of cardiac muscle, the heart may stop beating—that is, **cardiac arrest** may occur—and death can follow within moments. If only a small region of the heart is affected, however, the heart may continue to function. Cells in the region deprived of oxygen die and are replaced by scar tissue.

materials, the hepatic portal vein contains blood that has already given up some of its oxygen to the cells of the intestinal wall. Oxygenated blood is supplied to the liver by the hepatic artery.

The Lymphatic System

In addition to the blood circulatory system, vertebrates have a **lymphatic system,** a subsystem of the circulatory system (Fig. 35–28). The lymphatic system has three important functions: (1) to collect and return interstitial fluid to the blood, (2) to defend the body against disease organisms by way of immune mechanisms, and (3) to absorb lipids from the digestive tract. In this section we will focus upon the

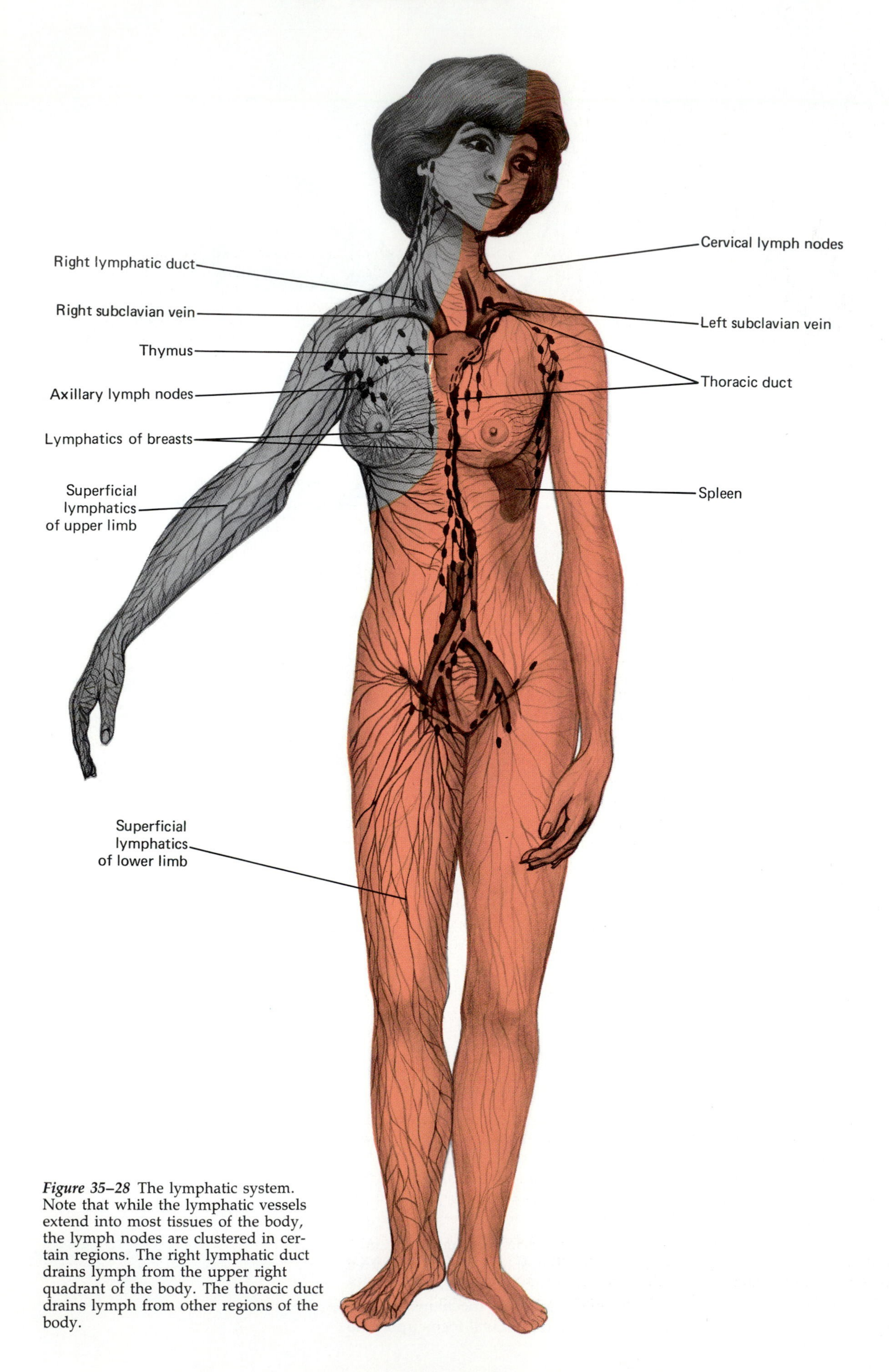

Figure 35–28 The lymphatic system. Note that while the lymphatic vessels extend into most tissues of the body, the lymph nodes are clustered in certain regions. The right lymphatic duct drains lymph from the upper right quadrant of the body. The thoracic duct drains lymph from other regions of the body.

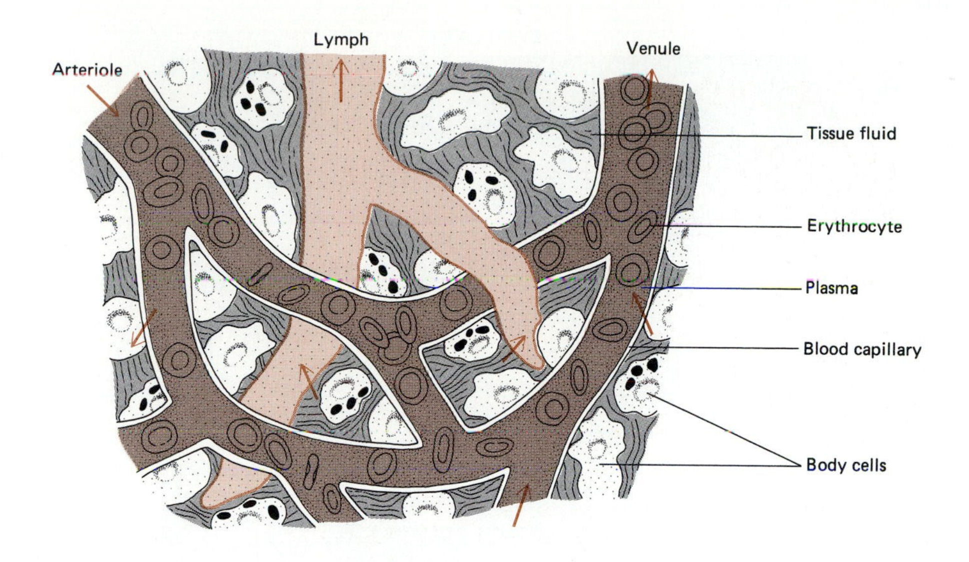

Figure 35–29 The relation of lymph capillaries to blood capillaries and tissue cells. Note that blood capillaries are connected to vessels at both ends, whereas lymph capillaries, outlined in color, are "dead-end streets" and contain no erythrocytes. The arrows indicate direction of flow.

first function. Immunity is discussed in Chapter 36 and lipid absorption is discussed in Chapter 33.

DESIGN OF THE LYMPHATIC SYSTEM

The lymphatic system consists of (1) an extensive network of lymphatic vessels that conduct **lymph,** the clear, watery fluid formed from interstitial fluid, and (2) **lymph tissue,** a type of connective tissue that has large numbers of lymphocytes. Lymph tissue is organized into small masses of tissue called **lymph nodes** and **lymph nodules.** Tonsils and adenoids are examples of lymph tissue. The **thymus** and **spleen,** which consist mainly of lymph tissue, are also part of the lymphatic system.

Tiny "dead-end" capillaries of the lymphatic system extend into almost all of the tissues of the body (Fig. 35–29). Lymph capillaries converge to form larger **lymphatics** (lymph veins). There are no lymph arteries.

Interstitial fluid enters lymph capillaries and then is conveyed into lymphatics. At certain locations the lymphatics empty into lymph nodes, where lymph is filtered. Bacteria and other harmful matter are removed from the lymph. The lymph then flows into lymphatics leaving the lymph node. Lymphatics from all over the body conduct lymph toward the shoulder region. These vessels join the circulatory system at the base of the subclavian veins by way of ducts—the **thoracic duct** on the left side, and the **right lymphatic duct** on the right.

Some nonmammalian vertebrates such as the frog have lymph "hearts," which pulsate and squeeze lymph along. However, in mammals the walls of the lymph vessels themselves pulsate, pushing lymph along. Valves within the lymph vessels prevent the lymph from flowing backward. Lymph flow is increased when anything compresses body tissue, for this has the effect of squeezing the lymph vessels. When muscles contract or when arteries pulsate, for example, pressure on the lymph vessels enhances the flow of lymph. The rate at which lymph flows is slow and variable, but the total lymph flow is about 100 ml per hour—very much slower than the 5 liters per minute of blood flowing in the vascular system.

ROLE IN FLUID HOMEOSTASIS

Recall that when blood enters a capillary network it is under rather high pressure, so that some plasma is forced out of the capillaries and into the tissues. Once it leaves the blood vessels, this fluid is called interstitial fluid, or tissue fluid. It is somewhat similar to plasma but contains no red blood cells or platelets, and only a few white blood cells. Its protein content is about one fourth of that found in plasma. This is because proteins are too large to pass easily through capillary walls. Smaller molecules dissolved in the plasma do pass out with the fluid leaving the blood vessels. Thus, interstitial fluid contains glucose, amino acids, other nutrients, and oxygen, as well as a variety of salts. This nourishing fluid bathes all the cells of the body.

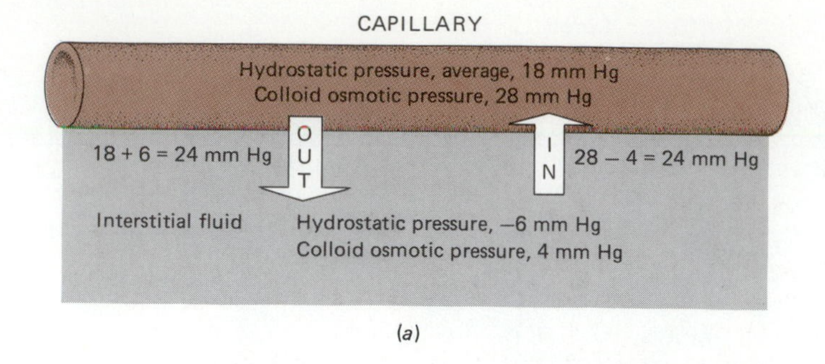

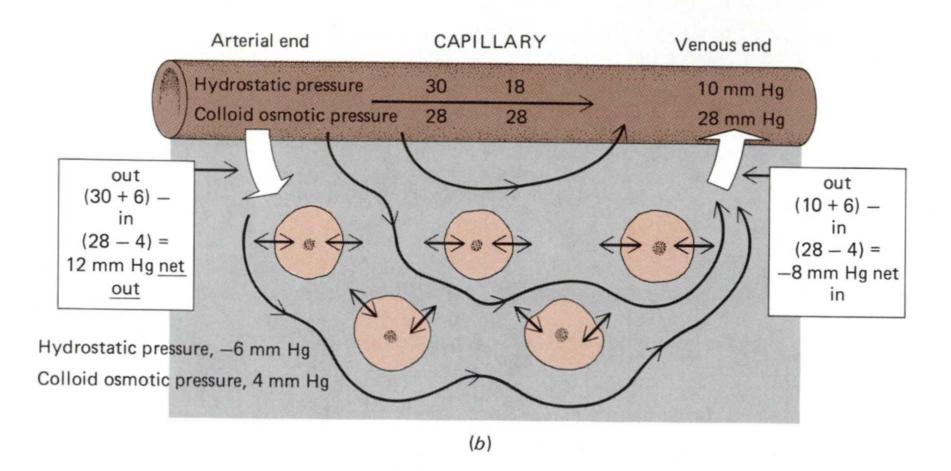

Figure 35–30 Materials are exchanged between blood and tissue fluid through the thin walls of the capillaries. (*a*) Hydrostatic and osmotic pressures are responsible for the exchange of materials between blood and tissue fluid. (*b*) The path by which water flows in and out of capillaries to the tissue fluid.

The main force pushing plasma out of the blood is hydrostatic pressure of the blood, that is, the pressure exerted by the blood on the capillary wall caused by the beating of the heart (Fig. 35–30). At the venous ends of the capillaries the hydrostatic pressure is much lower. Here the principal force is the osmotic pressure of the blood, which acts to draw fluid back into the capillary. However, osmotic pressure is not entirely effective for two reasons: First, not as much fluid is returned to the circulation as escapes; second, protein does not return effectively into the venous capillaries and instead tends to accumulate in the interstitial fluid. These potential problems are so serious that fluid balance in the body would be significantly disturbed within a few hours and death would occur within about 24 hours if it were not for the lymphatic system.

The lymphatic system preserves fluid balance by collecting about 10% of the interstitial fluid and the protein that accumulates in it. Once it enters the lymph capillaries, the interstitial fluid is called lymph.

The walls of the lymph capillaries are composed of endothelial cells that overlap slightly. When interstitial fluid accumulates, it presses against these cells, pushing them inward like tiny swinging doors that can only swing in one direction. As fluid accumulates within the lymph capillary, the cell doors are pushed closed.

Obstruction of the lymphatic vessels can lead to **edema,** a swelling due to excessive accumulation of interstitial fluid. Lymphatic vessels can be blocked as a result of injury, inflammation, surgery, or parasitic infection. For example, when a breast is removed (mastectomy) because of cancer, lymph nodes in the underarm region are also often removed in an effort to prevent the spread of cancer cells. The patient's arm may swell greatly because of the disrupted lymph circulation. Fortunately, new lymph vessels develop within a few months and the swelling slowly subsides.

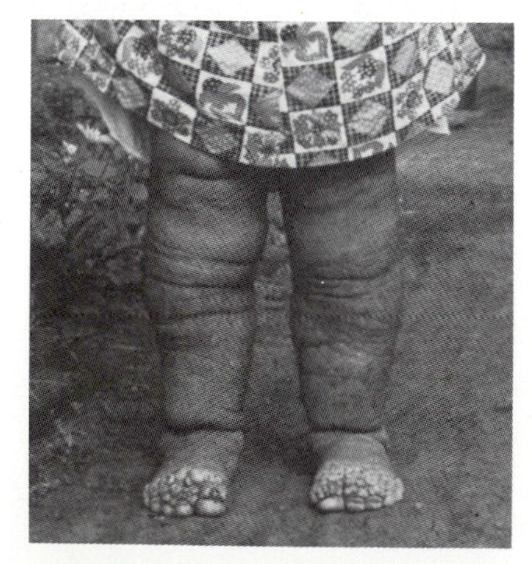

Figure 35–31 Lymphatic drainage is blocked in the limbs of this individual because of the parasitic infection known as filariasis. The condition characterized by such swollen limbs is elephantiasis. (From E. K. Markell and M. Voge.)

Filariasis, a parasitic infection caused by a larval nematode that is transmitted to humans by mosquitoes, also disrupts lymph flow. The adult worms live in the lymph veins, blocking lymph flow. Interstitial fluid then accumulates, causing tremendous swelling. The term **elephantiasis** has been used to describe the swollen legs sometimes caused by this disease, because they resemble the huge limbs of an elephant (Fig. 35–31).

SUMMARY

I. The circulatory system in most animals transports nutrients, oxygen and carbon dioxide, wastes, and hormones; it helps in maintaining fluid balance, regulating body temperature in warm-blooded animals, and defending the organism against invading microorganisms.

II. Sponges, cnidarians, ctenophores, flatworms, and nematodes have no specialized circulatory structures.

III. Arthropods and most mollusks have an open circulatory system in which blood flows into a hemocoel.

IV. Other invertebrates and all vertebrates have a closed circulatory system in which blood flows through a continuous network of blood vessels.

V. In the vertebrate circulatory system a muscular heart pumps blood into a system of arteries, capillaries, and veins.

VI. Human blood consists of red blood cells, white blood cells, and platelets suspended in plasma.
- *A.* Plasma proteins include fibrinogen, globulins, albumin, and lipoproteins.
- *B.* Red blood cells transport oxygen and carbon dioxide.
- *C.* White blood cells defend the body against disease-causing organisms.
- *D.* Platelets function in hemostasis.

VII. Arteries carry blood away from the heart; veins return blood to the heart.
- *A.* Arterioles constrict and dilate to regulate blood pressure and the distribution of blood to the tissues.
- *B.* Capillaries are the exchange vessels with thin walls through which materials pass back and forth between the blood and tissues.
- *C.* Veins are equipped with valves that prevent backflow of blood.

VIII. In vertebrates the heart consists of one or two atria, which receive blood, and one or two ventricles, which pump blood into the arteries.
- *A.* The four-chambered hearts of birds and mammals have completely separate right and left sides, which separate oxygenated blood from deoxygenated blood.
- *B.* The heart is enclosed by a pericardium and is equipped with valves that prevent backflow of blood.
- *C.* The sinoatrial node initiates each heartbeat. A specialized conduction system ensures that the heart will beat in a coordinated manner.
- *D.* Cardiac output equals stroke volume times heart rate.
  - *1.* Stroke volume depends upon venous return and upon neural messages and hormones.
  - *2.* Heart rate is regulated mainly by the nervous system but is influenced by hormones, body temperature, and certain other factors.

IX. Pulse is the alternate expansion and recoil of an artery; blood pressure is the force exerted by the blood against the inner walls of the blood vessel.
- *A.* Blood pressure is greatest in the arteries and decreases as blood flows through the capillaries.
- *B.* Baroreceptors sensitive to changes in blood pressure send messages to the vasomotor center in the medulla. When informed of an increase in blood pressure, the vasomotor center stimulates parasympathetic nerves that slow heart rate, thereby reducing blood pressure.
- *C.* Angiotensins raise blood pressure.

X. The pulmonary circulation connects heart and lungs; the systemic circulation connects the heart with all of the tissues.
- *A.* In the systemic circulation the left ventricle pumps blood into the aorta, which branches into arteries leading to all of the body organs; after flowing through capillary networks within various organs, blood flows into veins that conduct it to the right atrium.
- *B.* The coronary circulation supplies the heart muscle with blood.
- *C.* The hepatic portal system circulates blood rich in nutrients through the liver; it consists of the hepatic portal vein and a network of tiny blood sinuses in the liver from which blood flows into hepatic veins.

XI. Atherosclerosis leads to ischemic heart disease in which the heart muscle does not receive sufficient blood. Myocardial infarction is a very serious form of ischemic heart disease.

XII. The lymphatic system collects interstitial fluid and returns it to the blood. It plays an important role in homeostasis of fluids.

POST-TEST

Select the most appropriate answer or answers from Column B for the description in Column A.

Column A	*Column B*
1. No circulatory system	a. Insect
2. Open circulatory system	b. Bird
3. Closed circulatory system	c. Flatworm
4. Heart with two atria and two ventricles	d. Earthworm
	e. Cnidarian
	f. Clam

5. Hemocyanin is a pigment that transports ______________.

6. When the proteins involved in blood clotting are removed from plasma, the remaining fluid is called ______________.

7. The ______________ ______________ fraction of plasma proteins contains many types of antibodies.

8. A deficiency in hemoglobin is referred to as ______________.

Select the most appropriate term in Column B to fit the description in Column A.

Column A	*Column B*
9. Transport oxygen	a. Platelets
10. Principal phagocytic cells in blood	b. Red blood cells
11. Release histamine	c. Macrophages
12. Develop from monocytes	d. Basophils
13. Initiate clotting	e. Neutrophils

14. Prothrombin requires vitamin _____ for its production.

15. In the presence of thrombin, fibrinogen is converted to the insoluble protein ______________.

16. ______________ are blood vessels important in maintaining appropriate blood pressure.

17. The vessels through which nutrients and other materials are exchanged are ______________.

18. In birds and mammals blood leaving the right ventricle enters one of the ______________ ______________.

19. The ______________ valves guard the exits of the heart.

20. The portion of the cardiac cycle during which the heart contracts is referred to as ______________; the relaxation phase is ______________.

21. The cardiac output is the ______________.

22. According to Starling's law of the heart, the greater the amount of blood delivered to the heart by the veins, ______________________________.

23. The force exerted by the blood against the inner walls of the blood vessels is known as ______________ ______________.

24. Blood pressure is determined by ______________ ______________ and by ______________.

25. The largest artery in the human body is the ______________.

26. The carotid arteries deliver blood to the ______________.

27. The renal veins deliver blood from the ______________; the hepatic veins deliver blood from the ______________.

28. The term myocardial infarction (MI) is used as a synonym for ______________ ______________.

29. Baroreceptors are sensitive to changes in ______________ ______________.

30. The angiotensins are a group of hormones that are powerful ______________.

31. Label the diagram below.

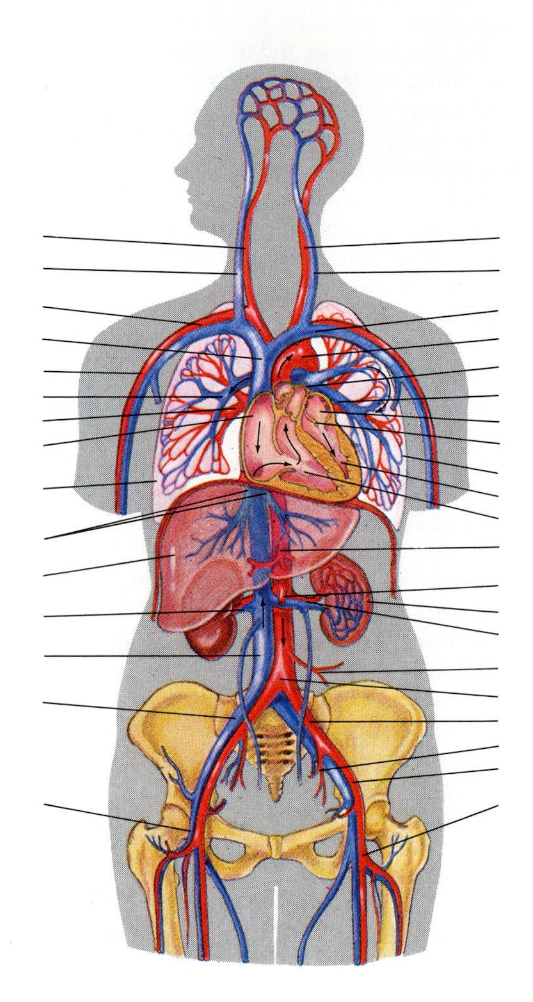

REVIEW QUESTIONS

1. List five functions of the circulatory system.
2. Compare the manner in which nutrients and oxygen are transported to the body cells in a hydra, a planarian, an earthworm, a clam, and a frog.
3. Contrast an open with a closed circulatory system.
4. List the functions of the main groups of plasma proteins.
5. Contrast the structure and functions of red and white blood cells.
6. What is anemia? What are the causes of this condition?
7. Summarize the process by which blood clots.
8. Compare the functions of arteries, capillaries, and veins. Why are arterioles important in maintaining homeostasis?
9. Compare the heart of a fish, frog, reptile, and bird.
10. Define cardiac output, and describe factors which influence it.
11. What is the relationship between blood pressure and peripheral resistance? What mechanisms regulate blood pressure?
12. Trace the path of a red blood cell from the inferior vena cava to the aorta, and from the jugular vein to the kidney.
13. What is the function of the hepatic portal system? How does its sequence of blood vessels differ from that in most other circulatory routes?
14. List four risk factors associated with the development of atherosclerosis. How is atherosclerosis linked to ischemic heart disease? Describe myocardial infarction.
15. How does the lymph system help maintain fluid balance? What is the relationship between plasma, interstitial fluid, and lymph?

36
Internal Defense

OUTLINE

I. Self and nonself
II. Internal defense mechanisms in invertebrates
III. Internal defense mechanisms in vertebrates
- *A.* Nonspecific defense mechanisms
  - *1.* Inflammation
  - *2.* Phagocytosis
- *B.* Specific defense mechanisms
  - *1.* T and B lymphocytes
  - *2.* Cell-mediated immunity
  - *3.* Antibody-mediated immunity
    - *a.* Antibodies and their structure
    - *b.* Classes of antibodies
    - *c.* How antibodies work
    - *d.* Antibody diversity
  - *4.* Function of the thymus
- *C.* Primary and secondary responses
- *D.* Active and passive immunity
- *E.* How the body defends itself against cancer
- *F.* Rejection of transplanted tissue
- *G.* Immunologically privileged sites
- *H.* Immunological tolerance
- *I.* Hypersensitivity
  - *1.* Allergic reactions
  - *2.* Autoimmune disease

LEARNING OBJECTIVES

After you have read this chapter you should be able to:

1. Compare in general terms the types of internal defense mechanisms in invertebrates and vertebrates.
2. Distinguish between specific and nonspecific defense mechanisms.
3. Describe the physiological changes and clinical symptoms associated with inflammation, and summarize the role of inflammation in the defense of the body.
4. Describe the process of phagocytosis.
5. Contrast T and B lymphocytes with respect to life cycle and function.
6. Describe the mechanisms of cell-mediated immunity, including development of memory cells.
7. Define the terms *antigen* and *antibody,* and describe how antigens stimulate immune responses.
8. Give the basic structure of an antibody, and list the five classes of antibodies and their biological roles.
9. Describe the mechanisms of antibody-mediated immunity, including the effects of antigen–antibody complexes upon pathogens; include a discussion of the complement system.
10. Describe the functions of the thymus in immune mechanisms.
11. Recount the clonal selection theory, and explain why organisms do not normally develop antibodies to their own tissues.
12. Contrast active and passive immunity, giving examples of each.
13. Contrast a secondary with a primary immune response.
14. Explain the theory of immunosurveillance, and describe how the body destroys cancer cells.
15. Describe the immunological basis of graft rejection, and explain how the effects of graft rejection can be minimized.
16. Describe situations in which foreign tissue may be accepted by the body.
17. Explain the immunological basis of allergy, and briefly describe the events that occur during a hayfever response and during systemic anaphylaxis.
18. Describe the immunological basis of autoimmune diseases, give two examples, and list possible causes.

All animals have defense mechanisms that provide protection against disease-causing organisms, or **pathogens.** Defense mechanisms can be nonspecific or highly specific. **Nonspecific defense mechanisms** deter a variety of pathogens, by preventing their entrance into the body or by destroying them quickly in the event they do penetrate the body's outer covering (Fig. 36–1). Phagocytosis of invading bacteria is an example. **Specific defense mechanisms** are collectively referred to as **immune responses.** The term immune is derived from a Latin word meaning *safe.* Immune responses are tailor-made to the particular type of pathogen that infects the body and are highly effective. The production of antibodies (highly specific proteins that help destroy pathogens) is one of the body's most important specific defense mechanisms. **Immunology,** the study of these specific defense mechanisms, is one of the most exciting fields of medical research today.

Self and Nonself

Immune responses depend upon the ability of an organism to distinguish between itself and foreign matter. Such recognition is possible because organisms are biochemically unique. Many cell types have surface macromolecules (proteins or large carbohydrates) that are slightly different from the surface macromolecules on the cells of other species or even other organisms of the same species. An organism "knows" its own macromolecules and "recognizes" those of other organisms as foreign.

A single bacterium may have from 10 to more than 1000 distinct macromolecules on its surface. When a bacterium invades another organism, these macromolecules stimulate the organism to launch an immune response. A substance capable of stimulating an immune response is called an **antigen.**

Internal Defense Mechanisms in Invertebrates

All invertebrate species that have been studied demonstrate the ability to distinguish between self and nonself. However, most invertebrates are able to make only nonspecific immune responses, such as phagocytosis and the inflammatory response.

Figure 36–1 Summary of the body's nonspecific and specific defense mechanisms.

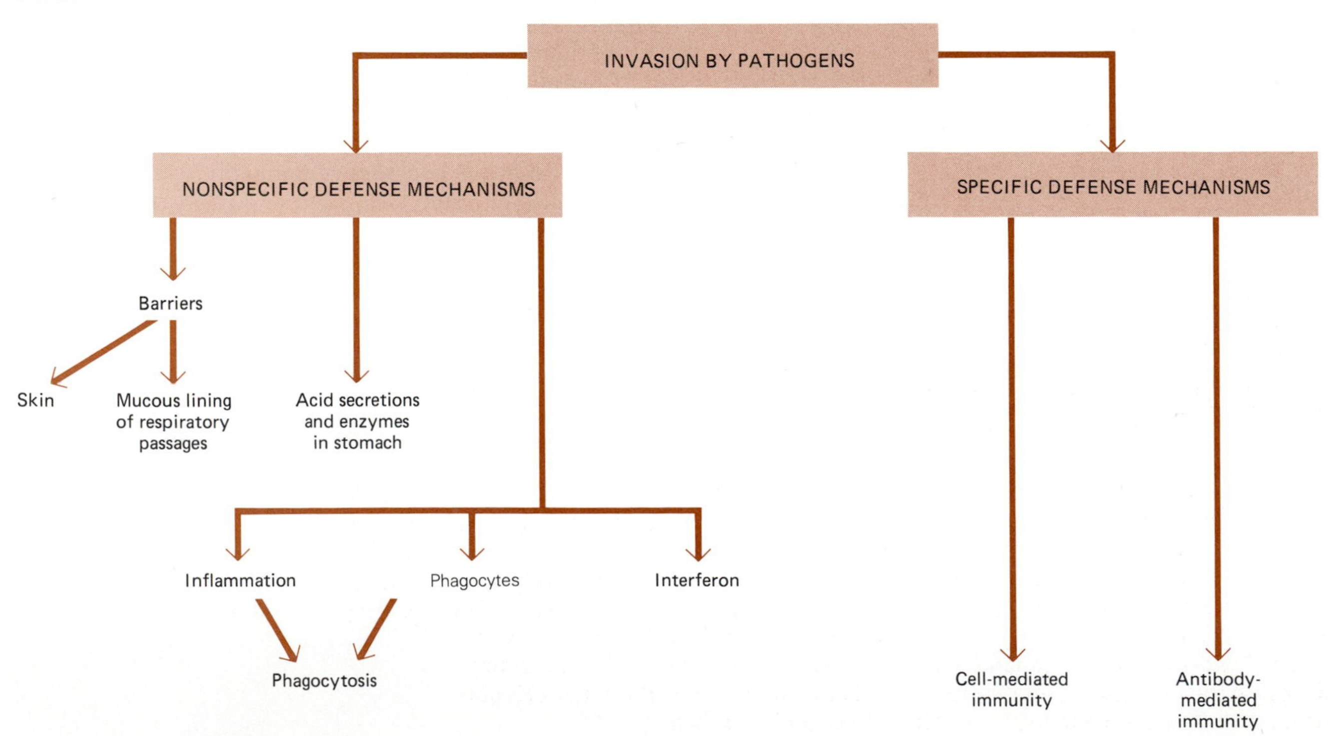

Sponge cells possess specific glycoproteins on their surfaces that enable them to distinguish between self and nonself. When cells of two different species of sponges are mixed together, they reaggregate according to species. Cnidarians also possess this ability and can reject grafted tissue and cause the death of foreign tissue.

In all coelomate invertebrates, **coelomocytes** (wandering amoeboid phagocytes) very effectively phagocytize bacteria and other foreign matter. Any particle too large to be phagocytized is walled off or encapsulated by the coelomocytes. Coelomate invertebrates also have nonspecific substances in the hemolymph that kill bacteria, inactivate cilia in some pathogens, and cause the agglutination (clumping) of some foreign cells. In mollusks these hemolymph substances enhance phagocytosis by the coelomocytes.

Echinoderms and tunicates are the simplest animals known to possess differentiated white blood cells that perform immune functions. Tunicates also have nodules of lymphatic tissue. Among the invertebrates, only certain annelids (e.g., earthworms) and cnidarians (e.g., corals) are thought to possess specific immune mechanisms and immunological memories. In them and in some echinoderms and simple chordates, the body appears to remember antigens for a short period of time and can respond to them more effectively in a second encounter.

Internal Defense Mechanisms in Vertebrates

All vertebrates can launch both nonspecific and specific immune responses. Vertebrates possess many of the basic mechanisms present in invertebrates but have in addition more sophisticated defense mechanisms made possible by the development of a specialized lymphatic system. In the discussion that follows we will focus on the human immune system, with references to those of other vertebrates.

NONSPECIFIC DEFENSE MECHANISMS

The outer covering of an animal, the first line of defense against pathogens, is often more than just a mechanical barrier. The human skin, for example, is populated by millions of harmless microorganisms that appear to inhibit the multiplication of potentially harmful organisms that happen to land on it. In addition, sweat and sebum contain chemicals that destroy certain kinds of bacteria.

Microorganisms that enter with food are usually destroyed by the acid secretions and enzymes of the stomach. Pathogens that enter the body with inhaled air may be filtered out by hairs in the nose or trapped in the sticky mucous lining of the respiratory passageways. They are then destroyed by phagocytes.

Should pathogens invade the tissues, other nonspecific defense mechanisms are activated. Certain kinds of cells, for instance, when infected by viruses or other intracellular parasites (some types of bacteria, fungi, and protozoa), respond by secreting proteins called **interferons.** This group of proteins stimulates other cells to produce antiviral proteins, which prevent the cell from manufacturing macromolecules required by the virus. The virus particles produced in cells exposed to interferon are not very effective at infecting cells. Drug companies have invested millions of dollars trying to develop an inexpensive, effective method of producing human interferon. Success has been achieved using recombinant DNA techniques. Research has established that interferon is useful in treating some viral infections. Recent studies suggest that interferon might be helpful in treating certain forms of cancer. Two nonspecific defense mechanisms that we will now consider in some detail are inflammation and phagocytosis.

Inflammation

When pathogens invade tissues, they trigger an **inflammatory response** (Fig. 36–2). Blood vessels in the affected area dilate, increasing blood flow to the infected region. The increased blood flow makes the skin look red and feel warm. Capillaries in the inflamed area become more permeable, allowing more fluid to leave the circulation and enter the tissues. As the volume of interstitial fluid increases, **edema** (swelling) occurs. The edema (and also certain substances released by the injured cells) cause the pain that is characteristic of inflammation. Thus, the clinical characteristics of inflammation are *redness, heat, edema, and pain.*

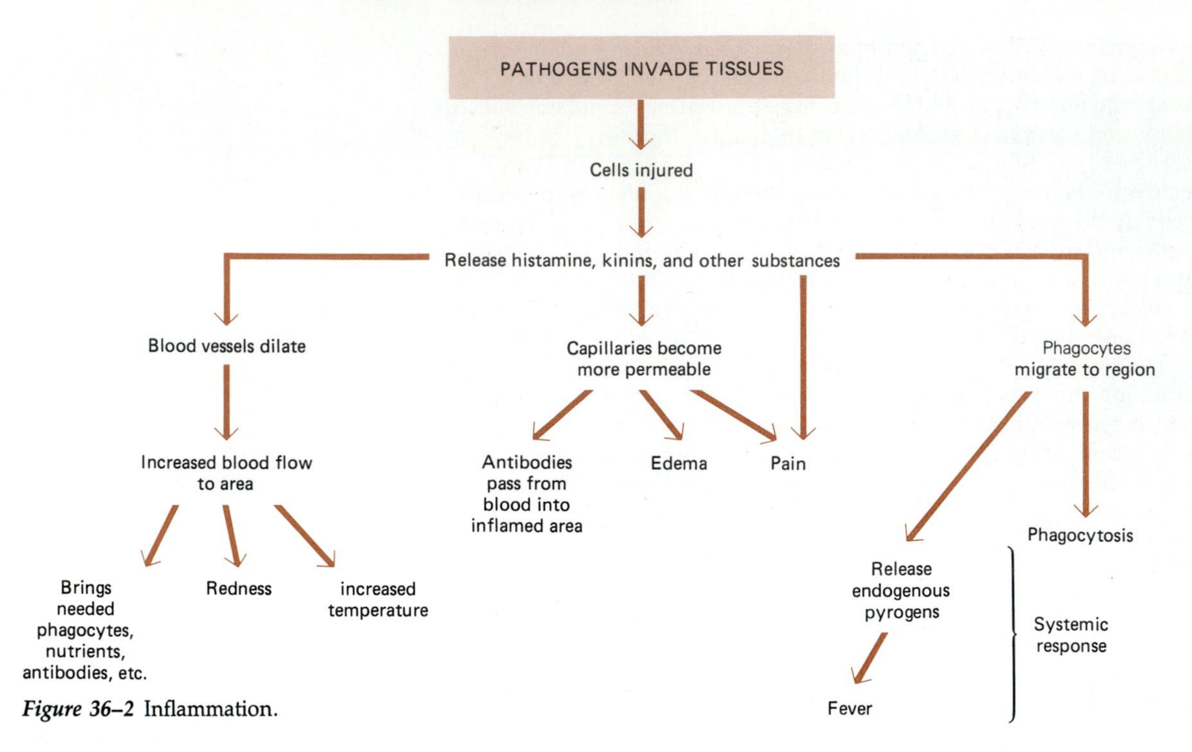

Figure 36–2 Inflammation.

The increased blood flow that occurs during inflammation brings great numbers of phagocytic cells (first, neutrophils and, later, monocytes; see Chapter 35) to the infected area. The increased permeability of the blood vessels allows needed gamma globulins, which serve as antibodies, to leave the circulation and enter the tissues. As fluid leaves the circulation, it also brings with it needed oxygen and nutrients.

Although inflammation is often a local response, sometimes the entire body is involved. **Fever** is a common clinical symptom of widespread inflammatory response. Proteins called **endogenous pyrogens** are released by neutrophils and macrophages and somehow reset the body's thermostat in the hypothalamus. Substances known as prostaglandins are also involved in this resetting process. Fever interferes with bacterial iron uptake and places the invaders at a metabolic disadvantage.

Phagocytosis

One of the main functions of inflammation appears to be increased phagocytosis. A phagocyte ingests a bacterium or other invading organism by flowing (amoeboid fashion) around it and engulfing it. As it ingests the organism, the cell wraps it within membrane pinched off from the cell membrane. The vesicle containing the bacterium is called a **phagosome.** One or more lysosomes adhere to the phagosome membrane and fuse with it. Under a microscope the lysosomes look like granules, but when they fuse with the phagosome the granules seem to disappear. For this reason, the process is referred to as **degranulation.** The lysosome releases potent digestive enzymes onto the captured bacterium, and the phagosome membrane releases hydrogen peroxide onto the invader. These substances destroy the bacterium and break down its macromolecules to small, harmless compounds that can be released or even utilized by the phagocyte.

After a neutrophil phagocytizes 20 or so bacteria it becomes inactivated (perhaps by leaking lysosomal enzymes) and dies. A macrophage can phagocytize about 100 bacteria during its lifespan. Can bacteria counteract the body's attack? Certain bacteria are able to release enzymes that destroy the membranes of the lysosomes. The powerful lysosomal enzymes then spill out into the cytoplasm and may destroy the phagocyte. Other bacteria, such as those that cause tuberculosis, possess cell walls or capsules that resist the action of lysosomal enzymes.

Some macrophages wander through the tissue phagocytizing foreign matter and, when appropriate, release antiviral agents (Fig. 36–3). Others stay in one place

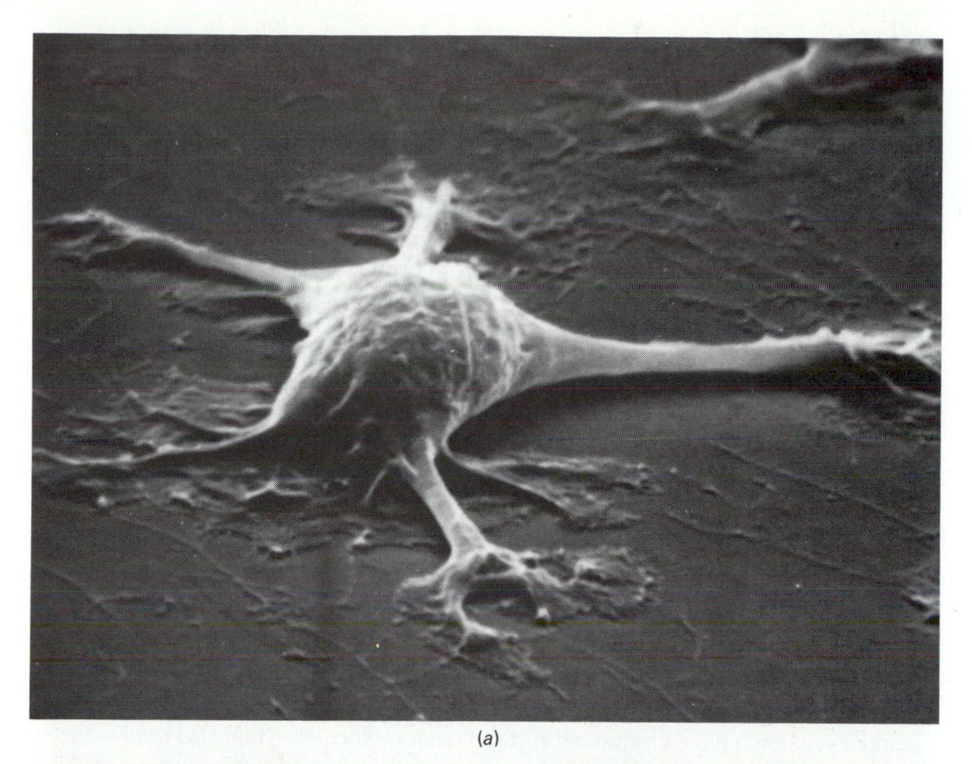

(a)

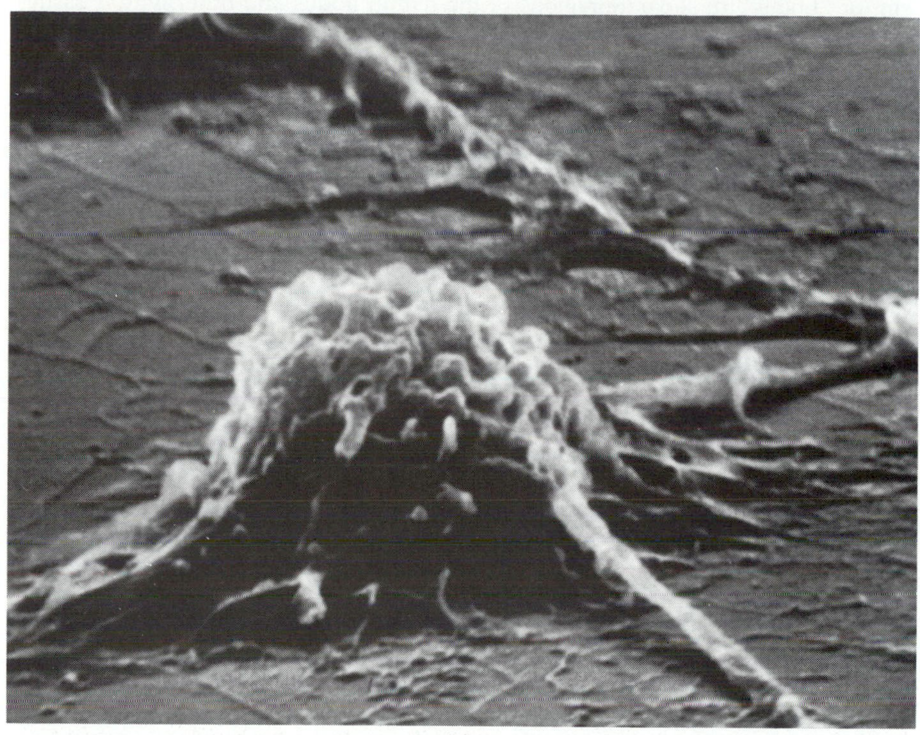

(b)

Figure 36–3 Scanning electron micrographs of macrophages recovered from the peritoneal cavity of mice. The upper photograph (*a*, ×14,400) is of an unstimulated macrophage. The macrophage in the lower picture (*b*, ×16,200) was taken from a mouse that had received an injection of mineral oil in the peritoneal cavity a few days earlier. The mineral oil acts as an irritant, producing "angry" macrophages with greatly increased metabolic rates, folding of the plasma membrane and increased ability to ingest bacteria. The irregular surface and many lateral microprojections of the "angry" macrophage are evident. The macrophages were grown in coverslips in tissue culture medium for a brief period before photography. (From Albrecht et al.: *Experimental Cell Research* 70: 230–232, 1971.)

and destroy bacteria that pass by. For example, air sacs in the lungs contain large numbers of tissue macrophages that destroy foreign matter entering with inhaled air.

SPECIFIC DEFENSE MECHANISMS

Nonspecific defense mechanisms destroy pathogens and prevent the spread of infection while the specific defense mechanisms are being mobilized. Several days are required to activate specific immune responses, but once in gear, these mechanisms are extremely effective. There are two main types of specific immunity: **cell-mediated immunity,** in which lymphocytes attack the invading pathogen directly, and

antibody-mediated immunity, in which lymphocytes produce specific antibodies designed to destroy the pathogen.

T and B Lymphocytes

The main warriors in specific immune responses are the trillion or so lymphocytes stationed strategically in the lymphatic tissue throughout the body. In amphibians and more complex vertebrates there are two types of lymphocytes, T lymphocytes and B lymphocytes. Both types are thought to originate either in the bone marrow or in the embryonic liver (Fig. 36–4). On their way to the lymph tissues, the **T lymphocytes,** or **T cells,** stop off in the thymus gland for processing. The T in T lymphocytes refers to thymus-derived; somehow the thymus gland influences the differentiation of lymphocytes so that they become capable of immunological response. T lymphocytes are responsible for cellular immunity.

B lymphocytes, or **B cells,** are responsible for antibody-mediated immunity (Fig. 36–5). In birds they are processed in the **bursa of Fabricius,** a lymphatic organ located near the cloaca. The B in B lymphocytes refers to bursa-derived, because these cells are processed in the bursa; other vertebrates do not have a bursa, however, and an equivalent organ has not yet been identified. In mammals, B lymphocytes may be processed as they form in the bone marrow, or they may be processed in the fetal liver or spleen.

Although T and B lymphocytes have different functions and life histories, when viewed with a light microscope they are similar in appearance. More sophisticated techniques such as fluorescence microscopy, however, have shown that the B and T cells each have unique cell surface macromolecules. They also tend to locate in (or "home" to) separate regions of the lymph nodes and other lymph tissues.

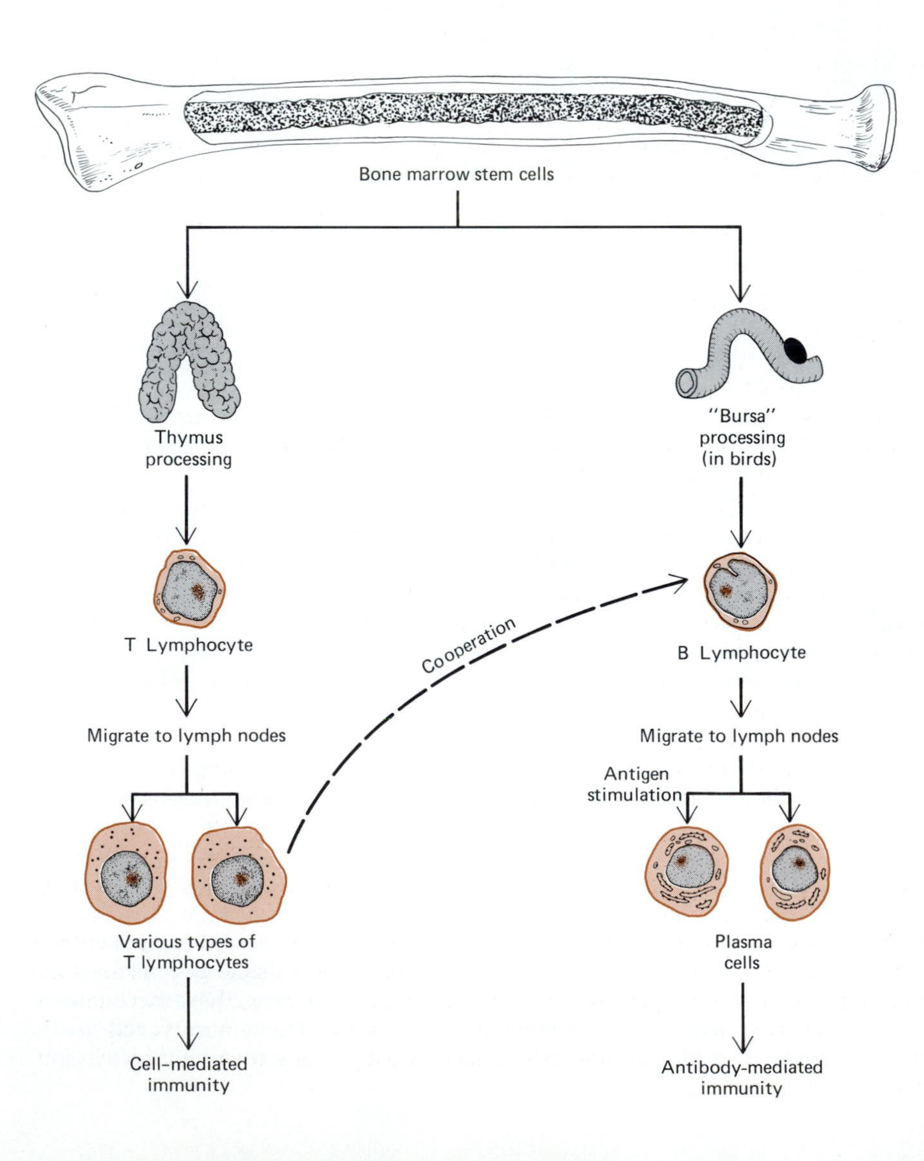

Figure 36–4 Origin and functions of T and B lymphocytes.

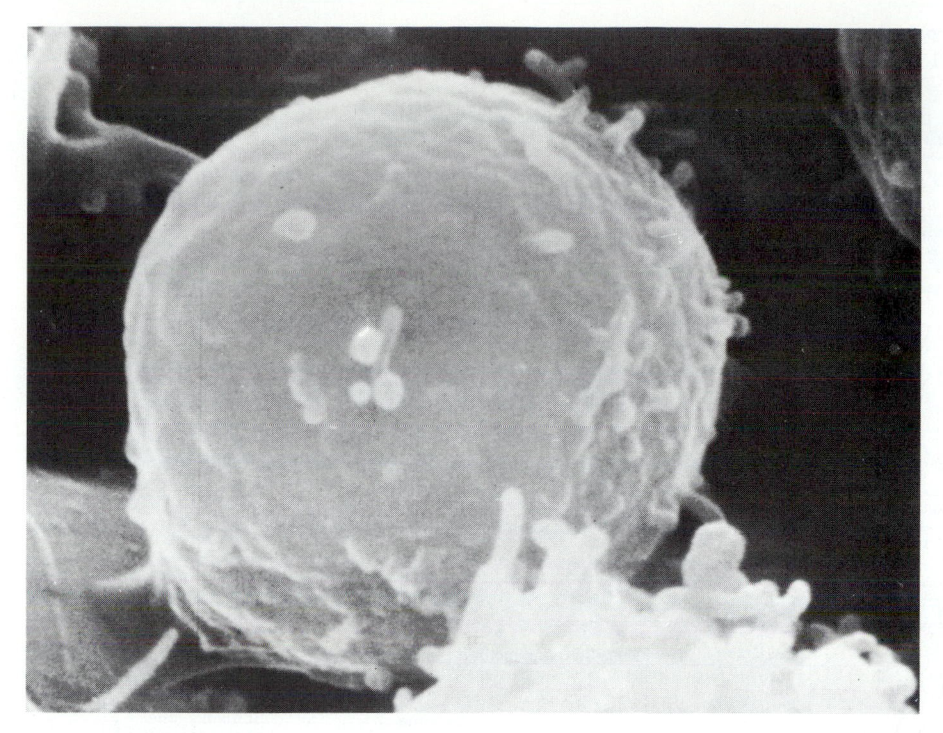

Figure 36–5 Scanning electron micrograph of a typical circulating lymphocyte. This type of rough cell surface is frequently associated with lymphocytes. (From Polliack, A., Lampen, N., Clarkson, B. D., DeHarven, E., Bentwich, Z., Siegal, F. P., and Kunkel, H. G.: Identification of human B and T lymphocytes by scanning electron microscopy. *Journal of Experimental Medicine* 138: 607, 1973.)

Cell-mediated Immunity

Cell-mediated immunity is the responsibility of the T cells and macrophages. These cells destroy viruses or foreign cells that enter the body. Most lymphocytes are usually in an inactive state. These are called "small" lymphocytes. There are thousands of different varieties of "small" lymphocytes, each capable of responding to a specific type of antigen. When an antigen invades the body, macrophages engulf it and bring the antigen to the lymphocytes. The variety of lymphocyte able to react to that antigen—that is, the **competent lymphocyte**—becomes activated, or **sensitized.** Once stimulated, lymphocytes increase in size. Then they divide by mitosis, each giving rise to a sizable clone of cells identical with itself (Fig. 36–6). These T lymphocytes then differentiate to become killer T lymphocytes, helper T lymphocytes, suppressor T lymphocytes, or memory cells. These specialized cells are referred to as T cell subsets. Members of this cellular infantry then leave the lymph nodes and make their way to the infected area.

Killer T lymphocytes combine with the antigen on the surface of the invading cell and then release powerful enzymes directly into the attacked cell. These enzymes destroy the foreign or malignant cell by disrupting its cell membrane. **Helper T lymphocytes,** the most numerous type of T cell, secrete substances known as lymphokines that enhance the response of killer T cells, suppressor T cells, and B cells. One kind of lymphokine inhibits macrophages from leaving the infection site and stimulates them to be more effective at phagocytosis so that they destroy more invading pathogens. Such stimulated macrophages are sometimes called "angry macrophages." **Suppressor T lymphocytes** inhibit immune defenses several weeks after an infection activates them.

T lymphocytes are especially effective in attacking viruses, fungi, and the types of bacteria that live within host cells. How do the T cells know which cells to attack? Once a pathogen invades a body cell, the host cell's macromolecules may be altered. The immune system then regards that cell as foreign, and T lymphocytes destroy it. Killer T cells also destroy cancer cells, and, unfortunately, the cells of transplanted organs.

When T lymphocytes are activated and give rise to a clone, not all of the sensitized cells produced leave the lymph tissue. Some remain behind as **memory cells.** Such cells, or their progeny, live on for many years. Should the invading pathogen ever attack again, the memory cells launch a far more rapid response than was possible during the first invasion. In this type of response (called a secondary immune response), the pathogens are usually destroyed before they have time to establish themselves and cause symptoms of the disease. This is why we usually do not suffer from the same disease several times. Most persons get measles or chicken pox, for example, only once.

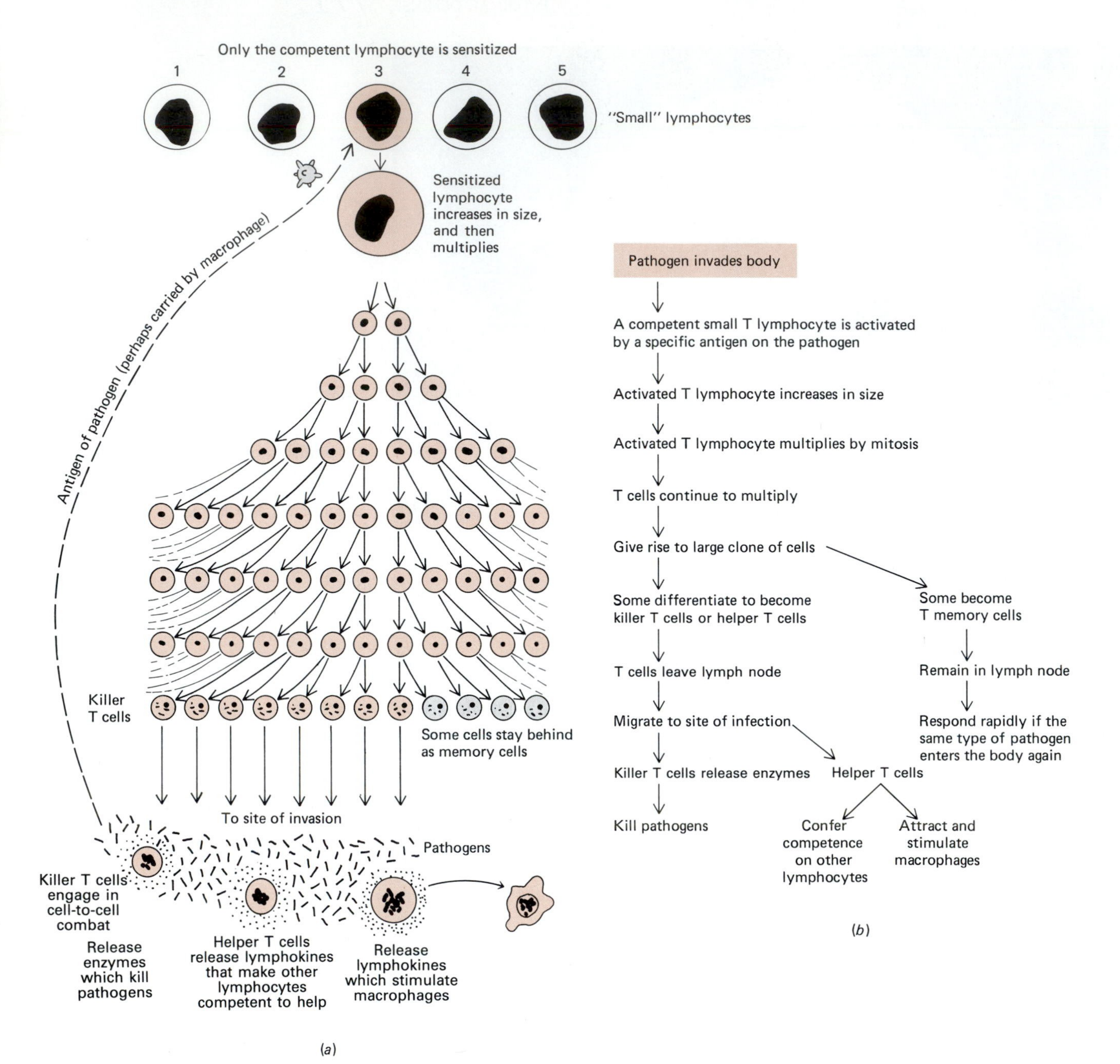

Only the competent lymphocyte is sensitized
1
2
3
4
5
"Small" lymphocytes
Sensitized lymphocyte increases in size, and then multiplies
Antigen of pathogen (perhaps carried by macrophage)
Killer T cells
Some cells stay behind as memory cells
To site of invasion
Pathogens
Killer T cells engage in cell-to-cell combat
Release enzymes which kill pathogens
Helper T cells release lymphokines that make other lymphocytes competent to help
Release lymphokines which stimulate macrophages
(a)
Pathogen invades body
A competent small T lymphocyte is activated by a specific antigen on the pathogen
Activated T lymphocyte increases in size
Activated T lymphocyte multiplies by mitosis
T cells continue to multiply
Give rise to large clone of cells
Some differentiate to become killer T cells or helper T cells
Some become T memory cells
T cells leave lymph node
Remain in lymph node
Migrate to site of infection
Respond rapidly if the same type of pathogen enters the body again
Killer T cells release enzymes
Helper T cells
Kill pathogens
Confer competence on other lymphocytes
Attract and stimulate macrophages
(b)

Figure 36–6 Cell-mediated immunity. When activated by an antigen, a T lymphocyte gives rise to a large clone of cells. Many of these differentiate to become killer T cells that migrate to the site of infection and attempt to destroy the invading pathogens.

However, if this is true, how can a person get flu or a cold more than once? Unfortunately, there are many varieties of the common cold and of flu, each caused by a slightly different virus with slightly different antigens. To make matters worse, viruses mutate often (a survival mechanism for them), which may change their surface antigens. Even a slight change may prevent recognition by memory cells. Each "different" antigen is treated by the body as a new immunological challenge.

Victims of acquired immune deficiency syndrome (AIDS) have a deficiency of T cells, and an abnormally high ratio of suppressor T cells to helper T cells. As a result, the ability to resist infection is severely depressed, and AIDS victims die of diseases such as pneumonia or cancer. A variant of a virus that triggers a rare form of leukemia is suspected of causing AIDS; it is usually transmitted by sexual contact (particularly male homosexual contact) or blood transfusions.

Antibody-mediated Immunity

The B lymphocytes are responsible for antibody-mediated immunity, also called **humoral immunity.** Just as with the T lymphocytes, there are thousands of varieties of B lymphocytes, each capable of responding to a specific type of antigen.

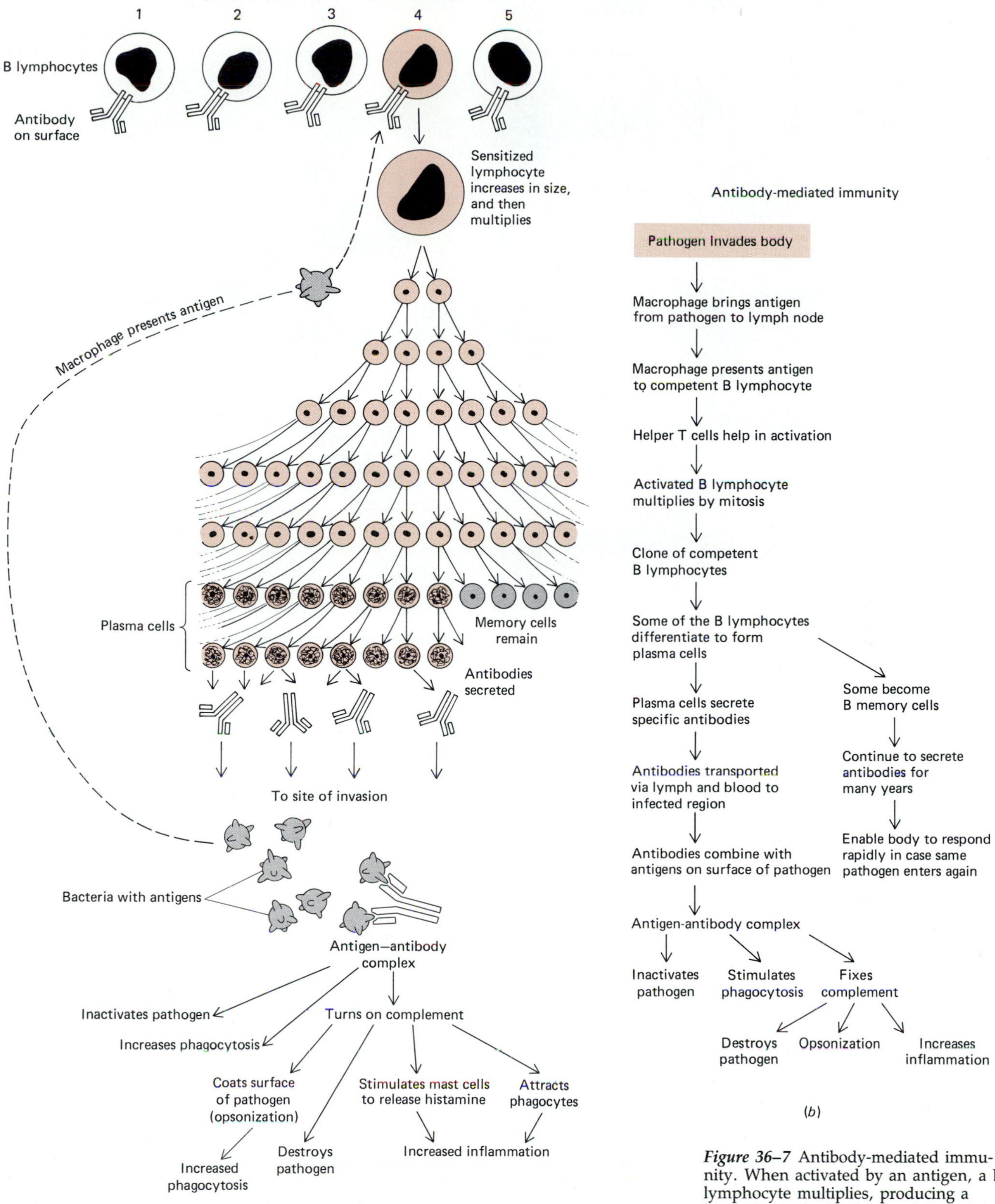
Only the competent lymphocyte is sensitized
1
2
3
4
5
B lymphocytes
Antibody on surface
Sensitized lymphocyte increases in size, and then multiplies
Macrophage presents antigen
Plasma cells
Memory cells remain
Antibodies secreted
To site of invasion
Bacteria with antigens
Antigen–antibody complex
Inactivates pathogen
Increases phagocytosis
Turns on complement
Coats surface of pathogen (opsonization)
Stimulates mast cells to release histamine
Attracts phagocytes
Increased phagocytosis
Destroys pathogen
Increased inflammation
(a)
Antibody-mediated immunity
Pathogen invades body
Macrophage brings antigen from pathogen to lymph node
Macrophage presents antigen to competent B lymphocyte
Helper T cells help in activation
Activated B lymphocyte multiplies by mitosis
Clone of competent B lymphocytes
Some of the B lymphocytes differentiate to form plasma cells
Some become B memory cells
Plasma cells secrete specific antibodies
Continue to secrete antibodies for many years
Antibodies transported via lymph and blood to infected region
Enable body to respond rapidly in case same pathogen enters again
Antibodies combine with antigens on surface of pathogen
Antigen-antibody complex
Inactivates pathogen
Stimulates phagocytosis
Fixes complement
Destroys pathogen
Opsonization
Increases inflammation
(b)

Figure 36–7 Antibody-mediated immunity. When activated by an antigen, a B lymphocyte multiplies, producing a large clone of cells. Many of these differentiate and become plasma cells, which secrete antibodies. The plasma cells remain in the lymph tissues, but the antibodies are transported to the site of infection by the blood. Antigen–antibody complexes form that directly inactivate some pathogens and also turn on the complement system. Some of the B lymphocytes become memory cells that continue to secrete small amounts of antibody for years after the infection is over.

Activated B lymphocytes divide to produce large clones of immunologically identical lymphocytes (Fig. 36–7). Most of these cells increase in size and differentiate into **plasma cells,** which may be thought of as mature B lymphocytes. Plasma cells have an extensive, highly developed rough endoplasmic reticulum for the synthesis of proteins (Fig. 36–8), because they are the cells that produce antibodies. Plasma cells do not leave the lymph nodes, as do T cells; only the antibodies they secrete leave the lymph tissues and make their way via the lymph and blood to the infected area.

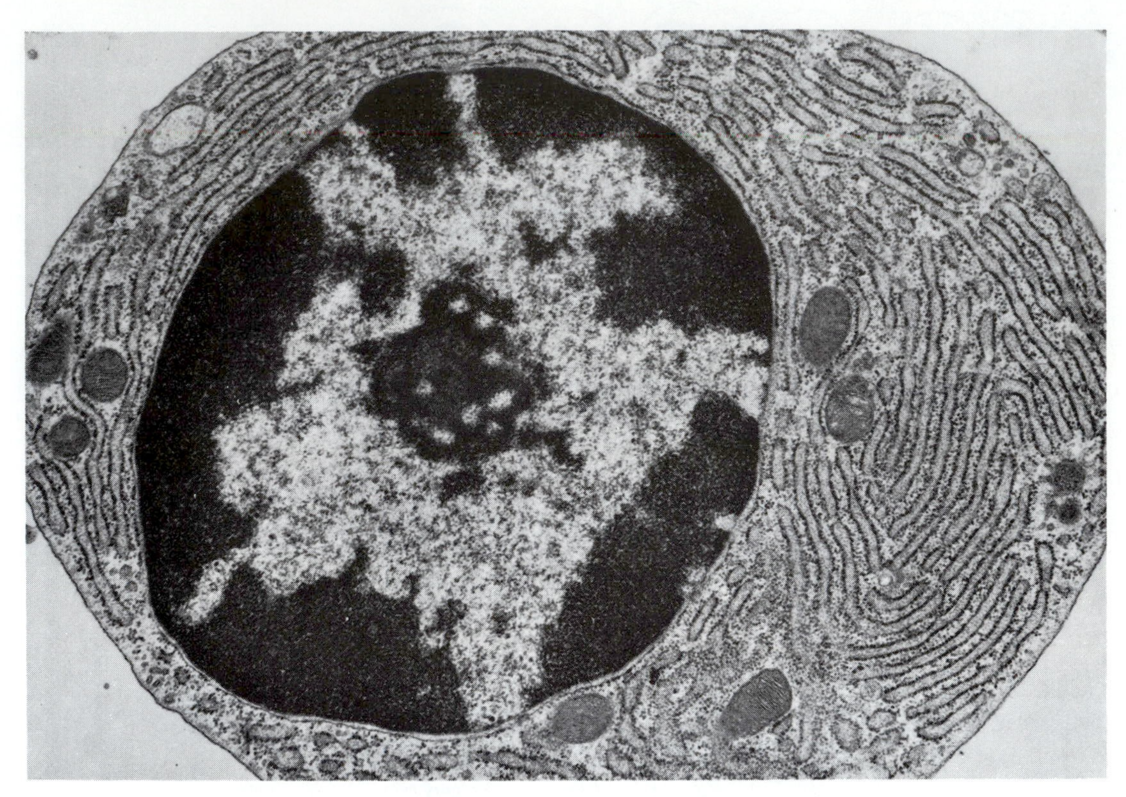

Figure 36–8 An electron micrograph of a guinea pig plasma cell showing the greatly developed endoplasmic reticulum. (From W. Bloom and D. Fawcett.)

Some activated B lymphocytes do not differentiate into plasma cells, but instead become memory cells similar in function to T lymphocyte memory cells. B memory cells continue to produce small amounts of antibody long after an infection has been overcome. This antibody, part of the gamma globulin fraction of the plasma, becomes part of the body's arsenal of chemical weapons. Should the same pathogen enter the body again, this circulating antibody is immediately available to destroy it. At the same time memory cells would quickly divide to produce new clones of the appropriate plasma cells.

ANTIBODIES AND THEIR STRUCTURE. Antibodies are highly specific proteins called **immunoglobulins** that may be produced in response to specific antigens. A typical immunoglobulin consists of four polypeptide chains: two identical heavy chains and two identical light chains (Fig. 36–9). Each light chain is made up of 214 amino acids, and each heavy chain of more than 400. The polypeptide chains are held together and their configurations are stabilized by disulfide (—S—S—) linkages and by noncovalent bonds. Each chain has a constant region and a variable region.

The **constant region,** or **C region,** may be thought of as the standard part of a key, the handle that is held. At its variable regions the antibody folds three-dimensionally, assuming a shape that enables it to recognize and combine with a specific antigen. The **variable region,** or **V region,** is the part of the key that is unique for a specific antigen (the lock). When they meet, antigen and antibody fit together *somewhat* like a lock and key and must fit in just the right way for the antibody to be effective. However, the fit is not as precise as with an enzyme and its substrate. A given antigen can bind with different strengths, or affinities, with different antibodies. In the course of an immune response better, stronger (higher-affinity) antibodies are generated. A typical antibody is a Y-shaped molecule that contains two binding sites (Fig. 36–9), enabling the antibody to combine with two antigen molecules. This permits formation of **antigen–antibody complexes.**

How does an antibody "recognize" a particular antigen? In a protein antigen there are specific sequences of amino acids that constitute an **antigenic determinant.** These amino acids give part of the antigen molecule a specific configuration that can be "recognized" by an antibody or cell receptor. However, the mechanism is even more complicated. Usually, an antigen has 5 to 10 antigenic determinants

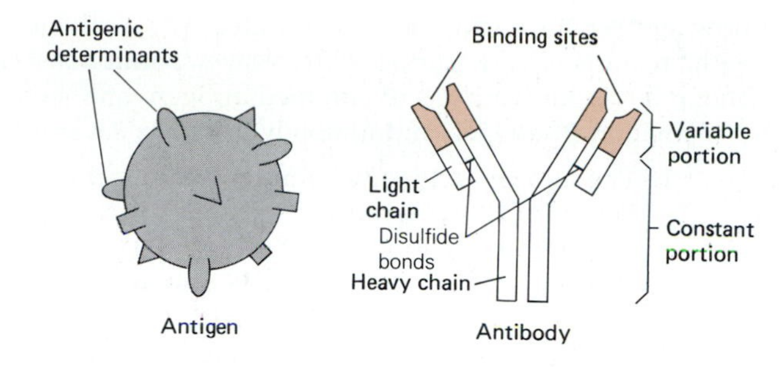

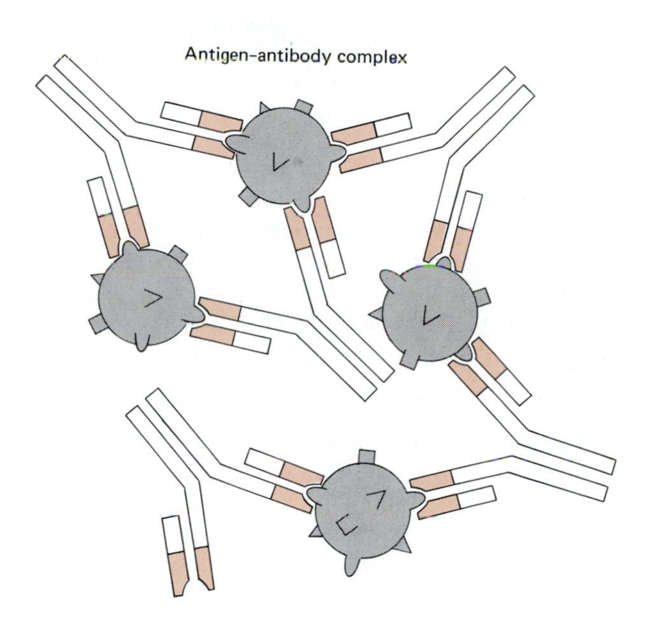

Figure 36–9 Antigen, antibody, and antigen–antibody complex. Note that the antibody molecule is composed of two light chains and two heavy chains, which are joined together by disulfide bonds. The constant (C) and variable (V) regions of the chains are indicated.

on its surface. Some have 200 or even more. These antigenic determinants may differ from one another, so that several different kinds of antibodies can combine with a single complex antigen.

Some substances found in dust and certain drugs are too small to be antigenic; yet they do stimulate immune responses. These substances, called **haptens,** become antigenic by combining with a larger molecule, usually a protein.

CLASSES OF ANTIBODIES. Immunoglobulins are grouped in five classes according to their structure. At its constant end, the heavy chains of antibody have an amino acid sequence characteristic of the particular antibody class. Using the abbreviation Ig for immunoglobulin, the classes are designated IgG, IgM, IgA, IgD, and IgE. In the simpler vertebrates, only IgM is present. In the amphibians, IgM and IgG are characteristic. These classes and also IgA are found in mammals, and all five classes are produced in humans. The classes of antibodies have different functions. Perhaps as more complex animals evolved, it was advantageous to have a variety of antibody classes with specialized functions.

In humans about 75% of the antibodies in the blood belong to the **IgG** group; these are part of the gamma globulin fraction of the plasma. IgG contributes to immunity against many blood-borne pathogens, including bacteria, viruses, and some fungi. IgG, along with the IgM antibodies, are responsible for complement fixation, as we will see in the following section. **IgM** immunoglobulins are effective against bacteria and also function as the antibodies characterizing blood type.

IgA immunoglobulins are the principal antibodies found in body secretions, such as mucous secretions of the nose, respiratory passageways, digestive tract, and tears, saliva, and vaginal secretions. IgA antibodies are thought to be important in protecting the body from infections by inhaled or ingested pathogens. Although the function of the **IgD** type of immunoglobulin is not known, it, along with IgM, is found on the surface of B lymphocytes. **IgE** immunoglobulins, the mediators of allergic responses, are discussed in a later section.

HOW ANTIBODIES WORK. Antibodies identify a pathogen as foreign by combining with an antigen on its surface. Often several antibodies combine with several such antigens, creating a mass of clumped antigen–antibody complex (Fig. 36–9). The combination of antigen and antibody activates several defense mechanisms:

1. The antigen–antibody complex may inactivate the pathogen or its toxin. For example, when an antibody attaches to the surface of a virus, the virus loses its ability to attach to a host cell.
2. The antigen–antibody complex stimulates phagocytosis of the pathogen by macrophages and neutrophils.
3. Antibodies of the IgG and IgM groups work mainly through the **complement system.** This system consists of about 11 proteins present in plasma and other body fluids. Normally, complement proteins are inactive, but an antigen–antibody complex stimulates a series of reactions that activate the system. The antibody is said to "fix" complement. Proteins of the complement system then work to destroy pathogens. Some complement proteins digest portions of the pathogen cell. Others coat the pathogens, a process called **opsonization.** This seems to make the pathogens "tastier" so that the macrophages and neutrophils rush to phagocytize them. Complement proteins also increase the extent of inflammation.

   Complement proteins are not specific. They act against any antigen, provided they are activated by antigen–antibody complex. Antibodies identify the pathogen very specifically; then complement proteins complement their action by destroying the pathogens.

ANTIBODY DIVERSITY. One of the most puzzling problems in immunology has been accounting for the tremendous diversity and numbers of antibodies. The immune system has the potential to produce millions of different antibodies, each programmed to respond to a different antigenic determinant. Two principal theories explaining this diversity have been debated. The **germ line theory** proposed that we are born with a different gene for the production of every possible antibody variable region. One criticism of this theory has been that a mammal has only about a million genes, and only a small number of these could be devoted to coding antibodies. Thus, we do not have sufficient numbers of genes to account for the millions of different antibodies.

The **somatic mutation theory** held that we inherit only a few types of genes for making antibody variable regions and that mutations occur within the DNA of the lymphocytes, producing thousands of slightly different lymphocytes in terms of variable regions. Each somatically generated V (variable) region gene could combine with a C (constant) region gene to form the immunoglobulin molecule. Thus, each variety of lymphocyte would be programmed to produce certain specific antibodies. These mutations might occur in the thymus or bursa equivalent.

Recent research indicates that the correct explanation is probably a combination of these theories. The ability to make many different antibodies is inherited, but this diversity is probably increased by mutation, as well as by recombination. According to current theory, there is only one gene that codes for the C region portion of each type of antibody. However, one of several hundred genes can code for the V region of the light chain, and one of several hundred can specify the V region of the heavy chain. There are also four or five other genes that code for joining segments that lie between the C and V regions of each chain.

The formation of an active antibody involves the recombination of one of many possible V region sequences with joining segments. These are then combined with a C region segment. This process occurs in the production of both the light and heavy chains. Then the two chains associate to form the completed antibody. During the course of development there is also a very high mutation rate in the V region DNA. This, of course, contributes to the diversity in antibody structure. However, even without these mutations, it has been estimated that from about 300 separate genes, about 18 billion possible antibodies could be produced by the recombination of the various segments of protein that make up the antibody.

Function of the Thymus

Present in all vertebrates, the **thymus gland** is thought to have at least two functions. First, in some unknown way, the thymus confers immunological competence upon T lymphocytes. Within the thymus these cells develop the ability to differentiate into cells that can respond to specific antigens. This "instruction" within the thymus is thought to take place just before birth and during the first few months of postnatal life. When the thymus is removed from an experimental animal before this processing takes place, the animal is not able to develop cellular immunity. If the thymus is removed after that time, cellular immunity is not seriously impaired.

The second function of the thymus is that of an endocrine gland. It secretes several hormones, including one known as **thymosin.** Although not much is known about these hormones, thymosin is thought to affect T cells *after* they leave the thymus, stimulating them to complete their differentiation and to become immunologically active. Thymosin has been used clinically in patients who have poorly developed thymus glands. It is also being tested as a modifier of biological response in patients with certain types of cancer; by stimulating cellular immunity in such patients, it may help to prevent the spread of the disease.

PRIMARY AND SECONDARY RESPONSES

The injection of an antigen into a immunocompetent animal causes specific antibodies to appear in the blood plasma in 3 to 14 days. The first exposure to an antigen stimulates a **primary response.** After injection of the antigen there is a brief **latent period,** during which the antigen is recognized and appropriate lymphocytes begin to form clones. Then there is a **logarithmic phase,** during which the antibody concentration rises logarithmically for several days until it reaches a peak (Fig. 36–10). Finally, there is a **decline phase,** during which the antibody concentration decreases to a very low level.

A second injection of the same antigen, even years later, evokes a **secondary response** (Fig. 36–10). Owing to the presence of memory cells, this is generally a much more rapid response with a shorter latent period and a rapid production of antibodies and T lymphocytes. More antibodies are produced than in a primary response, and the decline phase is slower. In a secondary response the predominant antibody is IgG (rather than IgM). The affinity, or strength of fit, of the antibody also increases following secondary exposure. The amount of antigen necessary to evoke a secondary response is much less than that needed for a primary response. Booster shots of vaccine are given in order to elicit a secondary response, reinforcing the immunological memory of the disease-producing antigens.

ACTIVE AND PASSIVE IMMUNITY

We have been considering **active immunity,** immunity developed following exposure to antigens. After you have had measles as a young child, for example, you develop memory cells and immunity that keep you from contracting measles again. Active immunity can be *naturally* or *artificially induced* (Table 36–1). If someone with measles sneezes near you and you contract the disease, you develop the immunity naturally. However, such immunity can also be artificially induced by **immunization,** that is, by injection of a vaccine. In this case, the body launches an immune response against the antigens contained in the measles vaccine and develops memory cells so that future encounters with the same pathogen will be dealt with swiftly.

Effective vaccines can be prepared in a number of ways. A virus may be attenuated (weakened) by successive passage through cells of nonhuman hosts. In the process, mutations occur that adapt the pathogen to the nonhuman host, so

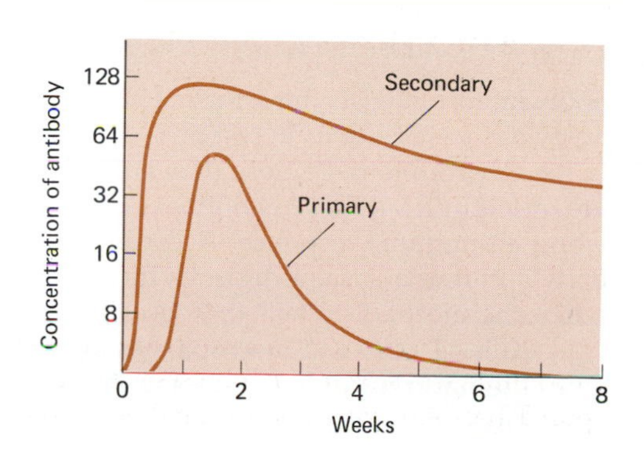

Figure 36–10 Primary and secondary responses of antibody formation to successive doses of antigens.

TABLE 36–1
Active and Passive Immunity

Type of Immunity	*When Developed*	*Development of Memory Cells*	*Duration of Immunity*
Active			
Naturally induced	Pathogens enter the body through natural encounter (e.g., person with measles sneezes on you)	Yes	For many years
Artificially induced	After immunization	Yes	For many years
Passive			
Naturally induced	After transfer of antibodies from mother to developing baby	No	Only a few months
Artificially induced	After injection with gamma globulin	No	Only a few months

that it can no longer cause disease in humans. This is how polio vaccine, smallpox vaccine, and measles vaccine are produced. Whooping cough and typhoid fever vaccines are made from killed pathogens that still have the necessary antigens to stimulate an immune response. Tetanus and botulism vaccines are made from toxins secreted by the respective pathogens. The toxin is altered so that it can no longer destroy tissues, but its antigens are still intact. When any of these vaccines is introduced into the body, the immune system actively develops clones, produces antibodies, and develops memory cells.

In **passive immunity,** an individual is given antibodies actively produced by another organism. The serum or gamma globulin containing these antibodies can be obtained from humans or animals. Animal sera are less desirable because nonhuman proteins can act as antigens, stimulating an immune response that may result in a clinical illness termed serum sickness.

Passive immunity is borrowed immunity, and its effects are not lasting. It is used to boost the body's defense temporarily against a particular disease. For example, during the Vietnam War, in areas where hepatitis was widespread, soldiers were injected with gamma globulin containing antibodies to the hepatitis pathogen. Such injections of gamma globulin offer protection for only a few months. Because the body has not actively launched an immune response, it has no memory cells and cannot produce antibodies to the pathogen. Once the injected antibodies wear out, the immunity disappears.

Pregnant women confer natural passive immunity upon their developing babies by manufacturing antibodies for them. These maternal antibodies, of the IgG class, pass through the placenta (the organ of exchange between mother and developing child) and provide the fetus and newborn infant with a defense system until its own immune system matures. Babies who are breast-fed continue to receive immunoglobulins, particularly IgA, in their milk. These immunoglobulins provide considerable immunity to the pathogens responsible for gastrointestinal infection, and perhaps to other pathogens as well.

HOW THE BODY DEFENDS ITSELF AGAINST CANCER

Some immunologists think that a few normal cells are transformed into cancer cells every day in each of us in response to viruses, hormones, radiation, or carcinogens in the environment. Because they are abnormal cells, some of their surface proteins are different from those of normal body cells. Such proteins act as antigens, stimulating an immune response. According to the **theory of immunosurveillance,** the body's immune system destroys these abnormal cells whenever they arise. Only when the mechanism fails do these abnormal cells divide rapidly, causing cancer.

Killer T cells and macrophages attack cancer cells (Fig. 36–11). Another cell type, the **natural killer cell,** is now thought to be important in destroying cancer cells. These natural killer cells are capable of killing tumor cells or virally infected

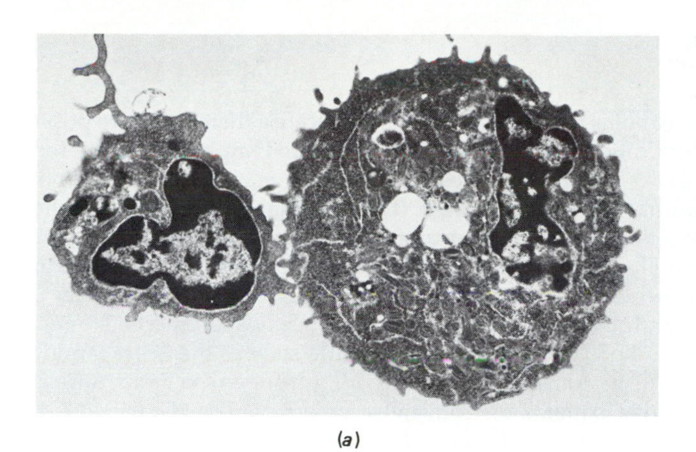

(a)

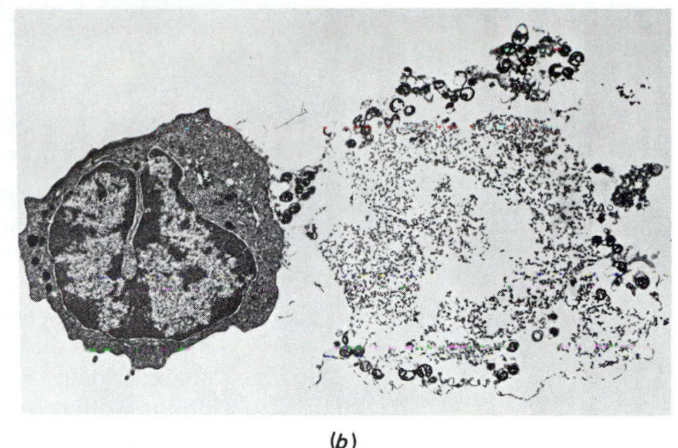

(b)

Figure 36–11 Killer T cell activity. (*a*) An electron micrograph of a killer T cell (the smaller of the two cells) approaching a cancer cell. (*b*) After about two hours of contact, a killer T cell has destroyed a cancer cell. (Courtesy of Prof. Daniel Zagury.)

cells upon first exposure to the foreign antigen. Patients with advanced cancer are thought to have less natural killer cell activity than do normal persons.

What prevents killer T cells, macrophages, and natural killer cells from effectively destroying cancer cells in some persons? The immune system cells may fail to recognize the cancer cells as foreign, or they may recognize them but be unable to destroy them. Sometimes the presence of cancer cells stimulates B cells to produce IgG antibodies that combine with antigens on the surfaces of the cancer cells. These **blocking antibodies** may block the T cells so that they are unable to adhere to the surface of the cancer cells and destroy them. For some unknown reason, the blocking antibodies are not able to turn on the complement system which would destroy the cancer cells.

An exciting new approach in cancer research involves the production of **monoclonal antibodies.** In this procedure, mice are injected with antigens from human cancer cells. After the mice have produced antibodies to the cancer cells, their spleens are removed and cells containing the antibodies are extracted from this tissue. These cells are fused with cancer cells from other mice. Because of the apparently unlimited ability of cancer cells to divide, these fused hybrid cells will continue to divide indefinitely. Researchers select hybrid cells that are manufacturing the specific antibodies needed and then clone them in a separate cell culture. Cells of this clone produce large amounts of the specific antibodies needed—hence the name monoclonal antibodies. Such antibodies can be injected into the very same cancer patients whose cancer cells were used to stimulate their production, and are highly specific for destroying the cancer cells. (Monoclonal antibodies specific for a single antigenic determinant can now be produced.) In trial studies such antibodies are being tagged with toxic drugs that are then delivered specifically to the cancer cells.

REJECTION OF TRANSPLANTED TISSUE

Skin can be successfully transplanted from one part of the body to another. However, when skin is taken from one individual and grafted onto the body of another, the skin graft is rejected and it sloughs off. Why?

Each of us has several groups of antigens called **histocompatibility antigens.** In humans the most important group is the **HLA** (human leukocyte antigen) **group,** determined by five different linked genes. These genes are all polymorphic—that is, there are multiple alleles at each locus. Tissues from the same individual or from identical twins have the same histocompatibility alleles and thus the same histocompatibility antigens. Such tissues are compatible.

Tissue transplanted from one location to another in the same individual is called an **autograft.** Owing to the polymorphism of the histocompatibility genes, it is difficult to find identical matches among strangers. If a tissue or organ is taken from a donor and transplanted to the body of an unrelated host, several of the HLA antigens are likely to be different. Such a graft made between members of the same species but of different genetic makeup is called a **homograft.** The host's immune system regards the graft as foreign and launches an effective immune response called **graft rejection.** T lymphocytes attack the transplanted tissue and can destroy it within a few days (Fig. 36–12).

Before transplants are performed, tissues from the patient and from potential donors must be typed and matched as well as possible. However, not all of the HLA antigens can be readily typed by serological means. (Cell typing is somewhat

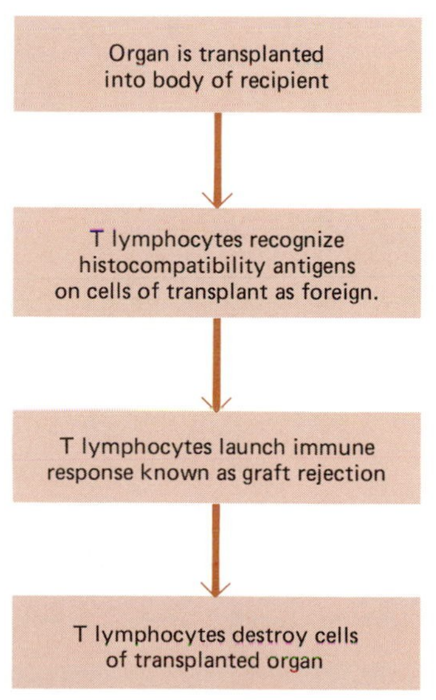
Organ is transplanted into body of recipient
T lymphocytes recognize histocompatibility antigens on cells of transplant as foreign.
T lymphocytes launch immune response known as graft rejection
T lymphocytes destroy cells of transplanted organ

Figure 36–12 Graft rejection.

similar to blood typing but is more complex, in part because there are more possible alleles.) In fact, identification of a single locus, by in vitro lymphocyte proliferation test, takes about 5 days. Therefore, the results of a tissue match may not be known until after the organ has been transplanted. The information is still useful, however, because it gives the physician an idea of how serious the graft rejection may be and how to treat it. If all five of the HLA group of antigens are matched, the graft has about a 95% chance of surviving the first year. Unfortunately, not many persons are lucky enough to have an identical twin to supply spare parts, so perfect matches are difficult to find. Furthermore, some parts such as the heart cannot be spared. Most organs to be transplanted, therefore, are removed from unrelated donors, often from dying patients or from those who have just died.

To try to prevent graft rejection in less compatible matches, drugs and x rays are used to kill T lymphocytes. These methods do not kill T lymphocytes selectively, however, so all types of lymphocytes are indiscriminately destroyed. Unfortunately, lymphocyte destruction suppresses not only graft rejection but other immune responses as well, so that many transplant patients succumb to pneumonia or other infections. In immunosuppressed patients there is also an increased incidence of certain types of tumor growths. A new antibiotic called cyclosporin A appears to suppress T cells that have been activated by antigens on the graft, but has little effect on B cells. Thus, the graft is not rejected and the patient can still resist infection.

IMMUNOLOGICALLY PRIVILEGED SITES

There are a few immunologically privileged locations in the body in which foreign tissues will be accepted by a host. Cornea transplants, for example, are highly successful because the cornea has almost no blood or lymphatic vessels associated with it and so is out of reach of most lymphocytes. Furthermore, antigens in the corneal graft probably would not find their way into the circulatory system, and so would not stimulate an immune response. The uterus is another immunologically privileged site. There the human fetus is able to develop its own biochemical identity in safety.

IMMUNOLOGICAL TOLERANCE

As we have seen, an organism is normally able to distinguish between self and nonself. It is thought that lymphocytes capable of producing antibodies against self are manufactured, but are somehow inactivated or suppressed, perhaps during embryonic development. In this way individuals develop **self-tolerance.**

Immunological tolerance to foreign tissue can be induced experimentally. Mice of a specific genetic strain, say strain A, are genetically homogeneous, almost like identical twins, and so are tolerant to tissue transplanted among their group. However, a mouse from strain A differs genetically from a mouse of another strain, say strain B, and normally rejects tissue transplanted from a strain B donor.

If embryonic cells from a mouse of one genetic strain, strain A, are infused into a *newborn* mouse of another genetic strain, strain B, the strain B mouse will develop **immunological tolerance.** It will develop normally, but when adult will not reject tissue transplants from a strain A mouse. By exposing the strain B mouse to the antigens from strain A before it develops immunological competence to produce a specific immune response to them, the development of that ability is delayed. If the antigens are present continuously, the ability to respond to them will be postponed indefinitely. However, this same animal will respond quite normally to the presence of other antigens, will form antibodies to them, and will reject grafts from other strains of mice.

Immunological tolerance can be induced in this manner only in fetal or neonatal animals. If the thymus gland of a newborn animal is removed, the animal's lymph nodes remain small and the animal is deficient in cellular immunity. An animal treated in this way will accept tissue grafts from other animals that differ from it genetically.

HYPERSENSITIVITY

The immune system normally functions to defend the body against pathogens and to preserve homeostasis, but sometimes the system malfunctions. **Hypersensitivity** is a state of altered immune response that is harmful to the body. Two familiar types of hypersensitivity are allergic reactions and autoimmune disease.

Allergic Reactions

About 15% of the population are plagued by a major allergic disorder such as allergic asthma or hayfever. There appears to be an inherited tendency to these disorders. Allergic persons have a tendency to manufacture antibodies against mild antigens, called **allergens,** that do not stimulate a response in nonallergic individuals. In many kinds of allergic reactions, distinctive IgE immunoglobulins called **reagins** are produced.

Let us examine a common allergic reaction—a hayfever response to ragweed pollen (Fig. 36–13). When an allergic person inhales the microscopic pollen, allergens stimulate the release of reagin from sensitized plasma cells in the nasal passages. The reagin attaches to receptors on the membranes of mast cells, large connective tissue cells filled with distinctive granules. Each mast cell has thousands of receptors to which the reagin molecules may attach. Each reagin molecule attaches to a mast cell receptor by its C region end, leaving the V region end of the reagin free to combine with the ragweed pollen allergen.

When the allergen combines with IgE antibody, the mast cell secretes histamine, serotonin, and other chemicals that cause inflammation. These substances produce dilation of blood vessels and increased capillary permeability, leading to edema and redness. Such physiological responses cause the victims' nasal passages to become swollen and irritated. Their noses run, they sneeze, their eyes water, and they feel generally uncomfortable.

Certain foods or drugs act as allergens in some persons, causing the swollen red welts known as **hives.** Here the allergen–reagin reaction takes place in the skin, with the histamine released by mast cells causing the hives. In **allergic asthma,** an allergen–reagin response occurs in the bronchioles of the lungs. Mast cells release SRS-A (slow-reacting substance of anaphylaxis), which causes smooth muscle to constrict. The airways in the lungs can constrict for several hours, making breathing difficult.

Systemic anaphylaxis is a dangerous kind of allergic reaction that can occur when a person develops an allergy to a specific drug such as penicillin, or to compounds present in the venom injected by a stinging insect. Within minutes after the substance enters the body, a widespread allergic reaction takes place. Large amounts of histamine are released into the circulation, causing extreme vasodilatation and permeability. So much plasma may be lost from the blood that circulatory shock and death can occur within a few minutes.

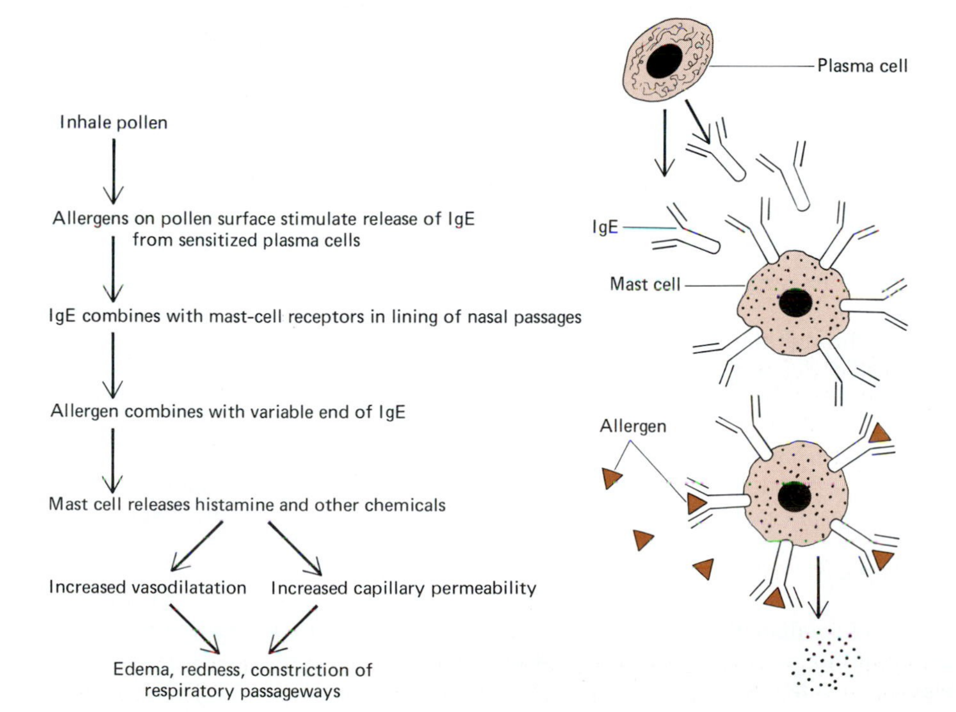
Inhale pollen
Allergens on pollen surface stimulate release of IgE from sensitized plasma cells
IgE combines with mast-cell receptors in lining of nasal passages
Allergen combines with variable end of IgE
Mast cell releases histamine and other chemicals
Increased vasodilatation
Increased capillary permeability
Edema, redness, constriction of respiratory passageways
Plasma cell
IgE
Mast cell
Allergen

Figure 36–13 A common type of allergic response.

The symptoms of allergic reactions are often treated with **antihistamines,** drugs that block the effects of histamines. These drugs compete for the same receptor sites on cells targeted by histamine. When the antihistamine combines with the receptor, it prevents the histamine from combining, and thus prevents its harmful effects. Antihistamines are useful clinically in relieving the symptoms of hives and hayfever. They are not completely effective, however, because mast cells release substances other than histamines that cause allergic symptoms.

In serious allergic disorders patients are sometimes given **desensitization therapy.** Very small amounts of the very antigen to which they are allergic are either injected or administered in the form of drops daily over a period of months or years. This stimulates production of IgG antibodies against the antigen. When the patient encounters the allergen, the IgG immunoglobulins combine with the allergen, blocking its receptors so that the IgE cannot combine with it. In this way a less harmful immune response is substituted for the allergic reaction.

Autoimmune Disease

Sometimes regulatory mechanisms malfunction and the body reacts immunologically against its own tissues, causing an **autoimmune disease.** Some of the diseases that result from such failures in self-tolerance are rheumatoid arthritis, multiple sclerosis, myasthenia gravis, systemic lupus erythematosus (SLE), and perhaps infectious mononucleosis.

Myasthenia gravis is an autoimmune disease in which function at the neuromuscular junctions is impaired. Affected persons experience muscle weakness and are easily fatigued. Sometimes respiratory muscles are affected to a life-threatening extent. Most myasthenia gravis patients have a circulating antibody that combines with receptors of acetylcholine in the motor end plates. This antibody can damage or even destroy the receptors.

What causes the production of abnormal antibodies in myasthenia gravis and in other autoimmune diseases? No one really knows. Some investigators have suggested genetic predisposition; others speculate that prior damage to the tissue is involved. Recent studies suggest that a viral infection in the involved tissue previously stimulated the body to manufacture antibodies against the infected cells. Then, after the virus has been destroyed, the body continues to manufacture harmful antibodies capable of attacking the body cells—even though they are no longer infected.

SUMMARY

I. Immune responses depend upon the ability of an organism to distinguish between self and nonself.

II. Most invertebrates are capable only of nonspecific responses such as phagocytosis and the inflammatory response.

III. All vertebrates can launch both nonspecific and specific responses.

- *A.* Nonspecific defense mechanisms that prevent entrance of pathogens include the skin, acid secretions in the stomach, and the mucous lining of the respiratory passageways.
- *B.* Should pathogens succeed in breaking through the first line of defense, other nonspecific defense mechanisms are activated to destroy the invading pathogens.
    - *1.* When pathogens invade tissues, they trigger an inflammatory response, which brings needed phagocytic cells and antibodies to the infected area.
    - *2.* Neutrophils and macrophages phagocytize and destroy bacteria.
- *C.* Specific immune responses include cell-mediated immunity and antibody-mediated immunity.
    - *1.* Both T and B lymphocytes respond to antigens.
    - *2.* In cell-mediated immunity, specific T lymphocytes are activated by the presence of specific antigens; these activated T lymphocytes multiply, giving rise to a clone of cells.
        - *a.* Some T cells differentiate to become killer T cells; some of these then migrate to the site of infection and destroy pathogens.
        - *b.* Some sensitized T cells remain in the lymph nodes as memory cells; others become helper T cells or supressor T cells.
    - *3.* In antibody-mediated immunity, specific B lymphocytes are activated by the presence of specific antigens; these activated B lymphocytes multiply, giving rise to clones of cells which differentiate to become plasma cells, which secrete specific antibodies.
        - *a.* Antibodies are highly specific proteins called immunoglobulins produced in response to specific antigens; they are grouped in five classes according to their structure.
        - *b.* Antibodies diffuse into the lymph and are transported to the site of infection by lymph and blood.

c. Antibody combines with a specific antigen to form an antigen–antibody complex, which may inactivate the pathogen, stimulate phagocytosis, or activate the complement system. The complement system increases the inflammatory response and phagocytosis; some complement proteins digest portions of the pathogen cell.

D. Second exposure to an antigen may evoke a secondary immune response, which is more rapid than the primary response.

E. Active immunity develops as a result of exposure to antigens; it may occur naturally or may be artificially induced by immunization. Passive immunity develops when an individual is injected with antibodies produced from another person or animal, and is temporary.

F. According to the theory of immunosurveillance, the immune system destroys abnormal cells whenever they arise; diseases such as cancer develop when this immune mechanism fails to operate effectively.

G. Transplanted tissues possess antigens that stimulate graft rejection, an immune response launched mainly by T cells that destroys the transplant.

H. Immunological tolerance to foreign tissues can be induced experimentally under certain conditions.

I. Hypersensitivity is a state of altered immune response that is harmful to the body.
   1. In an allergic response, an allergen can stimulate production of IgE antibody, which combines with the receptors on mast cells; the mast cells then release histamine and other substances, causing inflammation and other symptoms of allergy.
   2. In autoimmune diseases the body reacts immunologically against its own tissues.

POST-TEST

1. An antigen is a substance capable of ______________ ______________.
2. Specific proteins produced in response to specific antigens are called ______________.
3. When infected by viruses, some cells respond by producing proteins called ______________.
4. The clinical characteristics of inflammation are ______________, ______________, ______________, and ______________.
5. T lymphocytes are thought to originate in the ______________ ______________, get processed in the ______________, and then proliferate in the ______________ tissues.
6. Lymphokines are released by ______________ ______________ ______________.
7. When the body is invaded by the same pathogen a second time, the immune response can be launched more rapidly owing to the presence of ______________ cells.
8. The cells that produce antibodies are ______________ ______________.
9. An antigenic determinant gives the antigen molecule a specific configuration that can be "recognized" by an ______________.
10. The ______________confers immunological competence upon T lymphocytes.
11. The complement system is activated when an ______________ complex is formed.
12. In opsonization, complement proteins ______________.
13. Although artificially induced , immunization is a form of ______________ immunity.
14. An individual injected with antibodies produced by another organism is receiving ______________ immunity.
15. In humans the most important histocompatibility antigens are the ______________ group.
16. An autograft consists of tissue transplanted from ______________.
17. In graft rejection the host launches an effective ______________ ______________ against ______________ tissue.
18. Cornea transplants are highly successful because the cornea is an immunologically ______________ site.
19. An ______________ is a mild antigen that does not stimulate a response in an individual who is not ______________.
20. In a typical allergic reaction mast cells secrete ______________ and other compounds that cause ______________.

REVIEW QUESTIONS

1. How does the body distinguish between self and nonself? Are invertebrates capable of making this distinction?
2. Contrast specific and nonspecific defense mechanisms. Which type confronts invading pathogens immediately?
3. How does inflammation help to restore homeostasis?
4. Give two specific ways in which cell-mediated and antibody-mediated immune responses are similar, and three ways in which they are different.
5. Describe three ways in which antibodies work to destroy pathogens.
6. John is immunized against measles. Jack contracts measles from a playmate in nursery school before his mother gets around to having him immunized. Compare the immune responses in each child. Five years later, John and Jack are playing together when Judy, who is coming down with measles, sneezes on both of them. Compare the response in Jack and John.
7. Why is passive immunity temporary?
8. What is immunological tolerance?
9. What is graft rejection? What is the immunological basis for it?
10. List the immunological events that take place in a common type of allergic reaction such as hayfever.
11. Explain the theory of immunosurveillance. What happens when immunosurveillance fails?
12. What is an autoimmune disease? Give two examples.

37

Gas Exchange

OUTLINE

I. Adaptations for gas exchange
  A. The body surface
  B. Tracheal tubes
  C. Gills
  D. Lungs
II. Air versus water
III. Respiratory pigments
IV. The human respiratory system
  A. The mechanics of breathing
  B. The quantity of air respired
  C. Exchange of gases in the lungs
  D. Oxygen transport
  E. Carbon dioxide transport
  F. Regulation of respiration
  G. Hyperventilation
  H. High flying and low diving
  I. Effects of smoking and air pollution

LEARNING OBJECTIVES

After you have read this chapter you should be able to:

1. Describe the following adaptations for gas exchange: the body surface of some animals such as annelids; tracheal tubes; gills; and lungs.
2. Compare gas exchange in various types of animals, e.g., flatworms, earthworms, insects, fishes, and mammals.
3. Compare air and water as sources of oxygen.
4. State the function of respiratory pigments.
5. Trace a breath of air through the human respiratory system from external nares to air sacs.
6. Summarize the sequence of events that takes place in breathing.
7. Define the various volume measures of respiratory function (e.g., tidal volume and minute respiratory volume), and define vital capacity.
8. Describe how oxygen and carbon dioxide are exchanged in the lungs and in the tissues.
9. Explain the role of hemoglobin in oxygen transport, and identify factors that determine and influence the oxygen–hemoglobin dissociation curve.
10. Outline the mechanisms by which carbon dioxide is transported in the blood.
11. Draw a diagram to illustrate how respiration is regulated.
12. Describe the physiological effects of each of the following: hyperventilation; sudden decompression at 12,000 meters; surfacing too quickly from a deep-sea dive; and breathing compressed air for a long period of time.
13. Describe the following effects of breathing polluted air: bronchial constriction; chronic bronchitis; emphysema; and lung cancer.

Most cells die quickly if deprived of oxygen. Oxygen is required as the final electron acceptor in the process of biological oxidation by which cells obtain energy for life processes. Mammalian brain cells are especially sensitive and are damaged beyond repair if their supply is cut off for only a few minutes. To provide this supply of oxygen, gases must be exchanged continuously between the organism and its environment.

Oxygen is taken from the environment and delivered to the cells of the organism, while the carbon dioxide generated during cellular respiration must be excreted into the environment. This exchange of gases between organism and environment is known as **respiration.** To be more precise, respiration may be divided into two processes: **Organismic respiration** is the process of getting oxygen into the body and to the cells, and of delivering carbon dioxide from the cells into the environment. **Cellular respiration** is the complex series of reactions by which the cell breaks down fuel molecules, releasing carbon dioxide and energy, and generally requiring oxygen.

Gases move into and out of cells by diffusion. If the air or water supplying the oxygen to the cells can be continuously renewed, more oxygen will be available. For this reason animals carry on **ventilation;** that is, they actively move their air or water supply over their cells or over the surfaces of specialized respiratory structures such as gills or lungs. Breathing is one form of ventilation; movement of water over the respiratory surface by ciliary action is another.

Adaptations for Gas Exchange

The exchange of gases is a fairly simple process in small, aquatic organisms such as sponges, hydras, and flatworms. Dissolved oxygen from the surrounding water diffuses into the cells, and carbon dioxide diffuses out of the cells and into the water. No specialized respiratory structures are needed.

Oxygen diffuses thousands of times more rapidly through air than through tissues. In an organism less than 1 mm thick, diffusion alone provides a satisfactory method of gas exchange. However, in larger, more complex forms, the cells deep within the body cannot efficiently exchange gases directly with the environment, since oxygen cannot diffuse rapidly enough through muscle, bone, and other tissues to reach all of the cells. Specialized respiratory structures are required (Fig. 37–1).

Specialized respiratory structures must have thin walls so that diffusion can easily occur. They must be kept moist so that oxygen and carbon dioxide can be dissolved in water. In addition, respiratory structures are generally richly supplied with blood vessels to ensure transport of respiratory gases. Four principal types of respiratory structures used by animals are the body surface, tracheal tubes, gills, and lungs (Fig. 37–1; Table 37–1).

THE BODY SURFACE

Gas exchange occurs through the entire body surface in many animals including nudibranch mollusks, most annelids, small arthropods, and a few vertebrates. All of these animals are small, with a high ratio of surface to volume. Most animals that exchange gas through the body surface also have a relatively low metabolic rate, so that only small quantities of oxygen are needed.

How does an animal such as the earthworm exchange gases through its body surface? Movements of the worm and air currents ventilate the air so that fresh oxygen-rich air is brought in contact with the body surface. Gland cells in the epidermis secrete mucus, which keeps the body surface moist. Oxygen from tiny air pockets in the loose soil that the earthworm inhabits dissolves in the mucus and then diffuses through the body wall. The oxygen diffuses into blood circulating in a network of capillaries just beneath the outer cell layer. Oxygen is transported by the blood to all of the cells of the earthworm body. As blood circulates past these cells, oxygen diffuses out of the blood and into them. Carbon dioxide from the body cells diffuses into the blood and is transported to the body surface, from which it diffuses out into the environment.

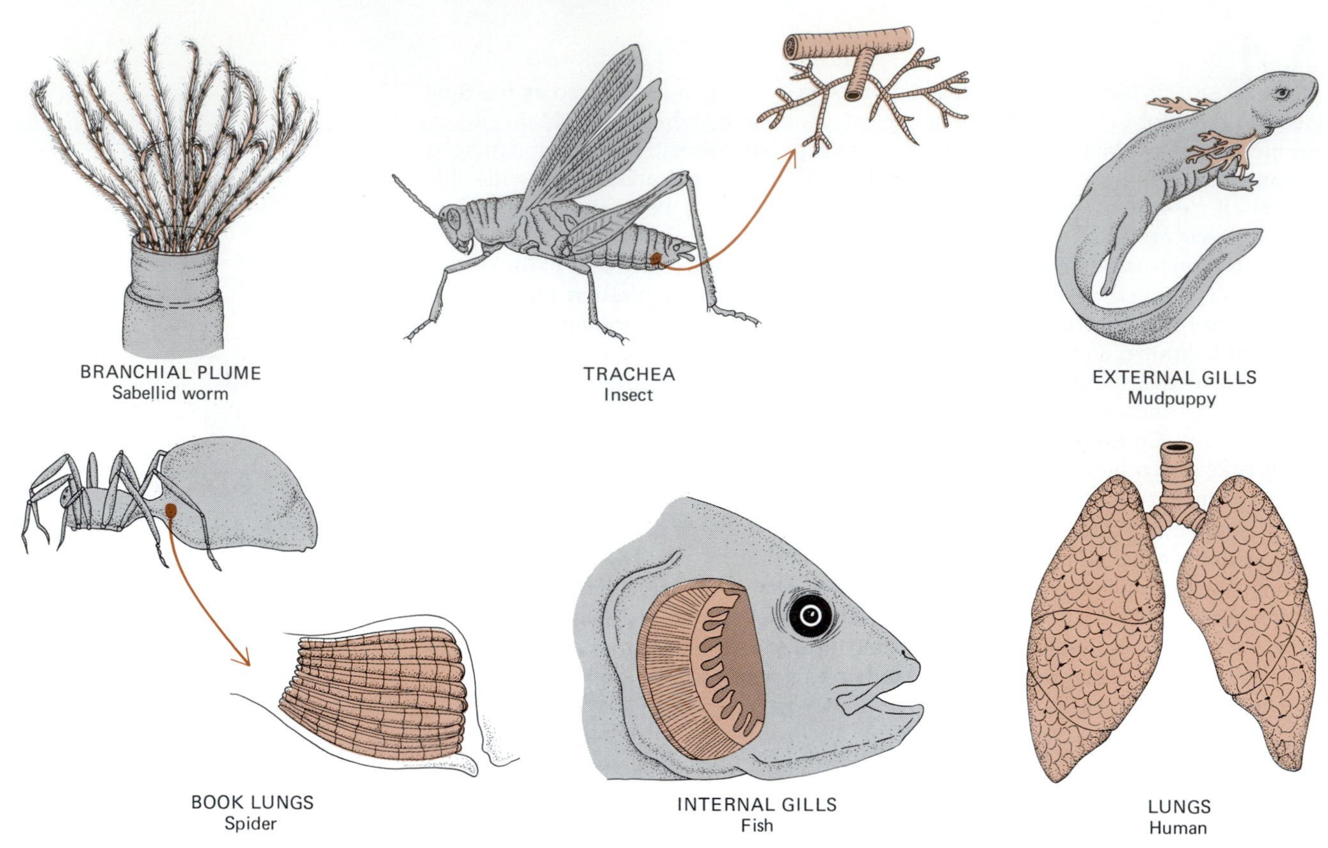

Figure 37–1 Types of respiratory structures found in animals.

When it rains, water fills the air pockets in the soil. Water holds less oxygen than air, so the oxygen available may not be sufficient at such times. This is why earthworms often come to the surface of the soil after it rains—to come in contact with the air above the ground.

TABLE 37–1
Principal Types of Respiratory Structures

Structure	*Animals that Utilize Such Structures*	*Description*	*Comment*
Body surface	Nudibranch mollusks, most annelids, small arthropods, some vertebrates	In terrestrial forms the body surface is kept moist by mucus secretion. Blood vessels or coelomic fluid is present just below the surface to receive oxygen and transport it to other body regions.	These animals have high surface-to-volume ratio, low metabolic rate. In some animals (e.g., frogs), other respiratory structures are present.
Tracheal tubes	Insects, some mites, some spiders, diplopods, chilopods	Air enters tracheal tubes through spiracles, passes through the branching tracheal tubes to all parts of the body, terminating in fluid-filled tracheoles. Gas exchange occurs by diffusion between tracheoles and body cells.	In large or active animals ventilation occurs by movements of the body or of tracheal tubes.
Gills	Found mainly in aquatic animals, some annelids, mollusks, crustaceans, fish, and amphibians	Moist, thin structures that grow out from the body surface. In gills of bony fish, each gill consists of many filaments containing blood vessels.	A great deal of energy must be expended in ventilation of the gills.
Lungs	Arachnids, some mollusks, most vertebrates	Respiratory structures that develop as ingrowths of the body surface or from the wall of a body cavity. In vertebrates a series of air passageways may terminate in thin-walled air sacs within the lungs.	Most modern fish have swim bladders instead of lungs.

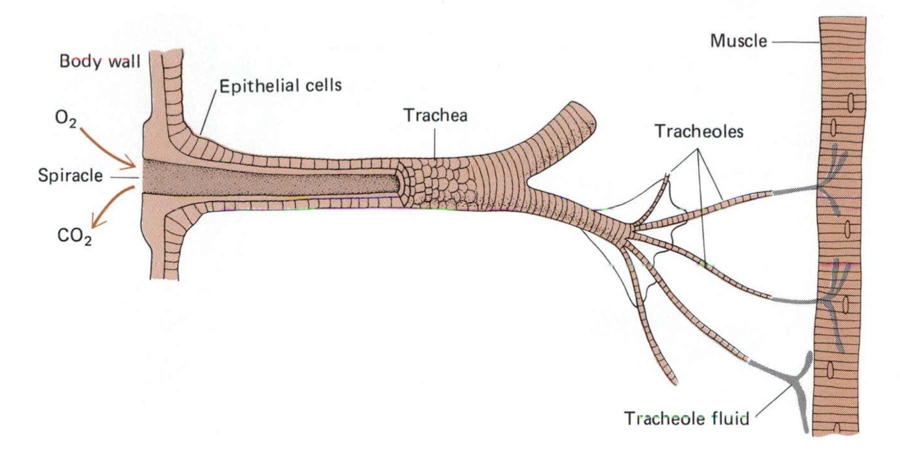

Figure 37–2 Tracheal tubes are characteristic of insects and some other invertebrates.

TRACHEAL TUBES

In insects and some other arthropods (e.g., chilopods, diplopods, some mites, and some spiders) the respiratory system consists of a network of **tracheal tubes** (Fig. 37–2). Air enters the tracheal tubes through a series of tiny openings called **spiracles** along the body surface. The maximum number of spiracles in an insect is 20—two thoracic pairs and eight abdominal pairs—but the number and position vary in different species. In large or active insects air moves into and out of the spiracles by movements of the body or by rhythmic movements of the tracheal tubes. For example, the grasshopper draws air into its body through the first four pairs of spiracles when the abdomen expands. Then the abdomen contracts, forcing air out through the last six pairs of spiracles.

Once inside the body, the air passes through the branching tracheal tubes, which extend to all parts of the animal. The tracheal tubes terminate in microscopic, fluid-filled tracheoles, and gases are exchanged between these tracheoles and the body cells. Although hemolymph fills the body cavity in these organisms, it does not function in the transport of gases. The tracheal system itself conducts air deep within the insect body, near enough to each cell so that gas exchange can take place directly between the respiratory system and the body cells. In some insects the branching tracheal tubes end in elastic air sacs, which expand and contract, aiding gas exchange.

GILLS

Gills are respiratory structures found mainly in aquatic animals, because gills are supported in water, but tend to collapse in air. They are moist, thin structures that extend out from the body surface. In many animals the outer surface of the gills is exposed to water, while the inner side is in close contact with networks of blood vessels (Fig. 37–3).

Sea stars and sea urchins have simple **dermal gills** that project from the body wall. The beating of the cilia of the epidermal cells ventilates the gills by moving a stream of water over them. Gases are exchanged through the gills between the water and the coelomic fluid inside the body. Various types of gills are found in some annelids, mollusks, crustaceans, fish, and amphibians. Molluskan gills are folded, providing a large surface for respiration. In bivalve mollusks and in simple chordates, the gills are adapted for trapping and sorting food. Rhythmic beating of cilia draws water over the gill area, and food is filtered out of the water as gases are exchanged. In mollusks, gas exchange also takes place through the mantle.

In chordates, the gills are usually internal. A series of slits perforates the pharynx, and the gills are located along the edges of these gill slits. In bony fish, the fragile gills are protected by an external bony plate, the **operculum.** Movements of the operculum help to pump water in through the mouth. The water flows over the gills and then leaves through the gill slits.

Each gill in the bony fish consists of many **filaments,** which provide an extensive surface for gas exchange. A capillary network delivers blood to the gill membranes, facilitating diffusion of oxygen and carbon dioxide between blood and water. The direction of blood flow increases the efficiency of this system. Blood

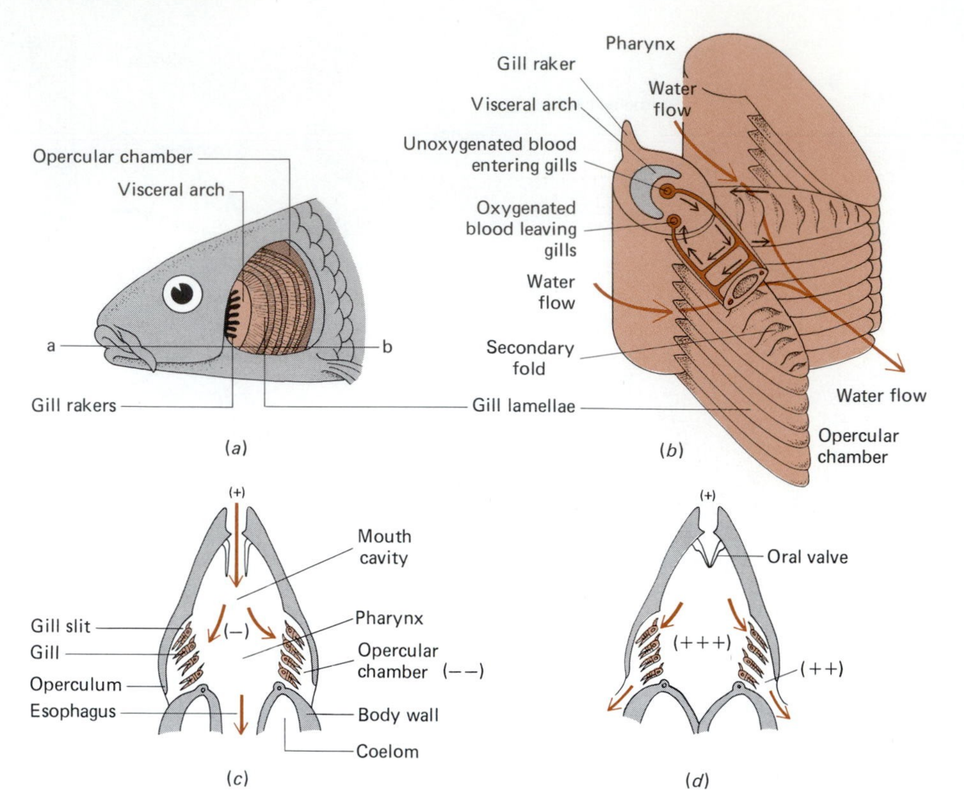

Figure 37–3 In bony fish, each gill consists of many filaments that provide an extensive surface for gas exchange. Blood flows in a direction opposite to that of water providing a countercurrent exchange system. (*a*) The operculum has been cut away to show the gills in the operculum chamber. (*b*) An enlargement of a portion of one gill. (*c*), (*d*) Frontal section through the mouth and pharynx in the plane of line *a–b* in (*a*). Inspiration occurs in (*c*) and expiration in (*d*). Relative water pressures in the various parts of the system are shown by + and −.

flows in a direction opposite to that of the water. This arrangement is referred to as a **countercurrent exchange system.** Water entering the gills is oxygen-rich. Blood, on the other hand, is oxygen-poor when it enters the gill circulation. Blood entering the gills picks up some oxygen from the water passing over the gills, but it can hold still more oxygen. As it leaves the gills, the blood comes in contact with oxygen-rich water entering the gills, and diffusion increases the concentration of oxygen in the blood. As the water passes on over the gill surface it contains less oxygen. However, the water is coming in contact with blood containing less and less oxygen, so oxygen continues to diffuse into the blood. This effective mechanism ensures that a great deal of the oxygen in the water will diffuse into the blood.

LUNGS

Lungs are broadly defined as respiratory structures that develop as ingrowths of the body surface or from the wall of a body cavity such as the pharynx (Fig. 37–4). Arachnids and some small mollusks (particularly terrestrial snails and slugs, but also some others) have lungs that depend almost entirely on diffusion for gas exchange. The book lungs of spiders are enclosed in an inpocketing of the abdominal wall. These lungs consist of a series of parallel, thin plates filled with blood. The plates are separated by air spaces that are connected to the outside environment through a spiracle.

Larger mollusks and vertebrates with lungs have some means of forcefully moving air across the lung surface, that is, of ventilating the lung. This helps to ensure adequate oxygenation in active animals.

As far as can be ascertained from the evidence of fossils, crossopterygian fish, thought to be the ancestors of the amphibians, had lungs somewhat similar to those of modern lungfish. Three genera of lungfish are known today. They live in the headwaters of the Nile, in the Amazon, and in certain Australian rivers. Streams inhabited by the lungfish dry up during seasonal droughts; during these dry periods lungfish remain in the mud of the streambed, exchanging gas by means of their lungs. These fish are also equipped with gills, which they use when swimming. The African lungfish uses both its gills and lungs all year around, and cannot survive if deprived of air.

Remains of crossopterygian fish occur extensively in the fossil record. Those found in Devonian strata are believed to be similar to the ancestors of amphibians.

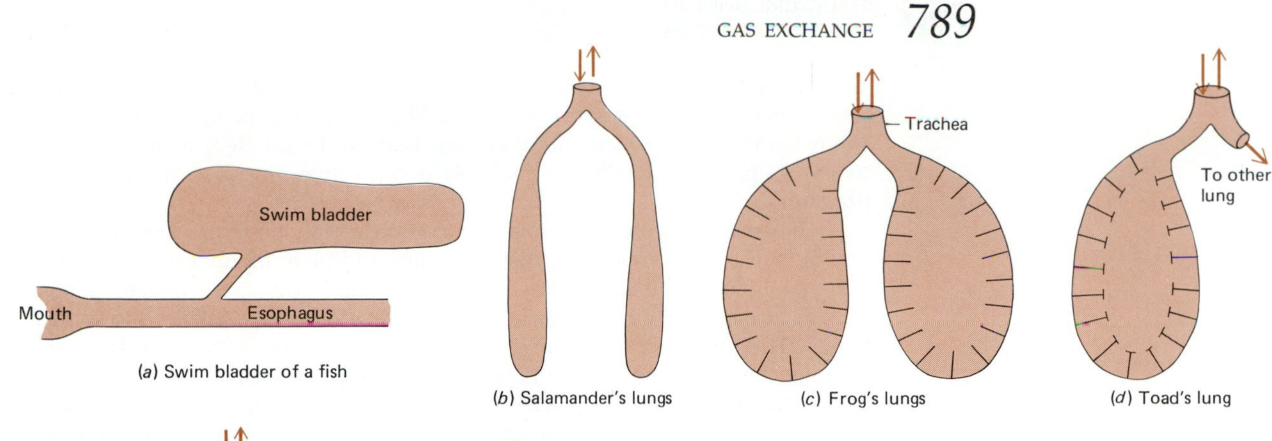

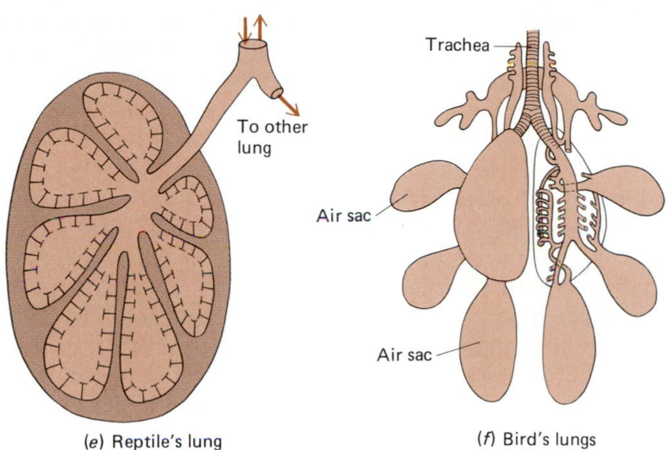

Figure 37–4 The structure of the lung varies in the different vertebrate classes. Note the increased surface area in the lungs of reptiles and birds. Mammalian lungs (pp. 791–795) have a very large surface area.

Numerous amphibian fossils also occur in adjacent ancient strata. The geological evidence suggests that periodic droughts occurred in Devonian times. Some paleontologists hypothesize that lungs may have evolved originally as an adaptation to these conditions, and that lungs or lunglike structures may have been present in all early bony fish.

Most modern fish have no lungs, but almost all of them do possess homologous **swim bladders.** These are hydrostatic organs that may also store oxygen. By regulating the amount of air in the swim bladder, the fish can control its own density and so become more or less buoyant. The fish can rise or sink, or maintain a particular level in the water without muscular effort. The swim bladder is usually single, although it may be paired. Its size and shape vary greatly in different species. Like the vertebrate lung, the swim bladder develops as an outpocketing of the digestive tract. The connection between the two is often lost in adulthood.

How does the fish control the amount of air in the swim bladder? Cells at the anterior end of the swim bladder have the ability (found nowhere else in the animal kingdom) to secrete oxygen from the blood into the bladder; other cells remove oxygen from the swim bladder and return it to the blood. By "pumping" oxygen in or out of the swim bladder, the fish can vary its buoyancy.

A swim bladder that is connected to the pharynx can serve as an accessory respiratory organ (see Fig. 37–4). When the oxygen level in the pond or lake is low, the fish may come to the surface and gulp air into its swim bladder. From there, oxygen passes into the blood circulating through the swim bladder wall.

The lungs of the simplest modern amphibians, the mud puppies, are two long, simple sacs, covered on the outside by capillaries. Frogs and toads have ridges containing connective tissue on the inside of the lung, which increase the respiratory surface somewhat. Since frogs have no diaphragm or rib muscles, their method of breathing is quite different from that of humans; it depends upon the action of valves and muscles in the nostrils and throat—so frogs gulp the air they breathe. Some amphibians have no lungs at all. Among plethodontid (lungless) salamanders (see Fig. 26–25), for example, all gas exchange takes place in the pharynx, or across the thin, wet skin.

The lungs of most reptiles are rather simple, with only some folding of the wall to increase the surface for gas exchange. Because gas exchange is not very effective, these reptiles are unable to sustain long periods of activity. In some lizards and in turtles and crocodiles, the lungs are more complex, having many subdivisions that give them a spongy texture. The lungs of some lizards such as African chameleons have supplementary air sacs that can be inflated, enabling the animal to swell up. Such swelling is probably a protective device to frighten would-be predators.

Birds are very active animals with high metabolic rates. They require large amounts of oxygen to sustain their activities and have highly effective respiratory systems. In them the lungs have developed several extensions (usually nine) called **air sacs,** which reach into all parts of the body, and even penetrate into some of the bones. The respiratory system is arranged so that there is a one-way flow of air through the lungs, and the air is renewed during each inspiration. Instead of alveoli, the lungs have tiny, thin-walled ducts, the **parabronchi,** which are open at both ends. Gas exchange takes place across the walls of these ducts. The direction of blood flow in the lungs is opposite that of air flow through the parabronchi. This countercurrent flow increases the amount of oxygen that enters the blood.

When the bird is at rest, a forward and upward movement of the ribs and a forward and downward movement of the sternum expand the trunk volume, drawing air into the body. When the bird is flying, the chest wall must be held rigid to form an anchor for the flight muscles. However, air sacs lying between certain flight muscles are squeezed and relaxed on each stroke of the wing. These act as bellows to move air into and out of the lungs. The faster the bird flies, the more rapid is the circulation of air through the lungs.

The lungs of mammals are very complex and have an enormous surface area. Let us trace a breath of air through the respiratory system of a mammal. Air enters through the nose or mouth and passes through the pharynx (throat region), and into the **larynx** (voice box). Air then passes through the **trachea** (windpipe) and into either the right or the left **bronchus.** The trachea and bronchi are supported by C shaped rings of cartilage that prevent the tubes from collapsing as air is drawn in. The bronchus conveys the air into the lung and then branches and gives rise to thousands of tiny **bronchioles.** The smallest bronchioles end in clusters of microscopic **alveoli.** Gas exchange takes place through the thin walls of the alveoli. Each alveolus is surrounded by a network of capillaries, which brings blood into close contact with the air inside the alveolus. Oxygen diffuses from the alveolus into the blood, while carbon dioxide diffuses from the blood into the alveolus.

Mammals have a muscular **diaphragm** that forms the floor of the thoracic cavity. When the diaphragm contracts (together with certain rib muscles, the intercostal muscles), the volume of the chest cavity increases. As a result, air rushes into the elastic lungs, filling the alveoli with air. When the diaphragm and rib muscles relax, air rushes back out of the lungs. (Exhalation can be assisted by various chest muscles in running, vocalization, or other activities.)

Air Versus Water

We have considered a variety of adaptations for gas exchange. Although there are exceptions, some respiratory structures, like tracheal tubes and lungs, seem best adapted for exchanging gases with air, and others, like gills, function best in water. Gas exchange with air is more effective than gas exchange with water because air contains far more oxygen than water, and oxygen diffuses much more rapidly through air than through water. Because water has a much greater density and viscosity (resistance to flow) than air, an animal must expend more energy to ventilate its respiratory surface with water than air. A fish uses up to 20% of its total energy expenditure to perform the muscular work needed to ventilate its gills. An air breather expends much less energy, only 1% or 2% of the total, to ventilate the lungs.

Air is not salty, so air breathers do not have the problem of ions' diffusing into their body fluids with their oxygen. Nor do they have the fluid balance problems of freshwater fish or amphibians. Aquatic animals also find it difficult to maintain a body temperature that is higher than that of their watery habitat. (Only a few of them do and most of those [such as whales] actually breathe air.) Thus, breathing air has many advantages: It conserves energy, makes it easier to maintain homeostasis with respect to ion composition, and gives the animal a better opportunity to maintain a body temperature higher than that of the surroundings.

All the same, breathing air is not without disadvantages. Its chief problem is water loss. Respiratory surfaces must be kept moist because molecules must be in solution in order to cross cell membranes. Loss of water from the respiratory surfaces would present a serious threat to fluid balance. This is perhaps why lungs are located deep within the body rather than exposed like gills. Air must pass through a long sequence of passageways before reaching the respiratory surface, and expired air must pass through these passageways before leaving the body. The lungs are thus partially protected from the drying effects of air.

Respiratory Pigments

Animals also have **respiratory pigments** that combine reversibly with oxygen and greatly increase the capacity of blood to transport oxygen. For example, the hemoglobin in human blood increases its capacity to transport oxygen by about 75 times. Oxygen enters the pulmonary capillaries and combines with hemoglobin in the red blood cells. Then, as blood circulates through tissues where the concentration of oxygen is low, hemoglobin releases oxygen. This oxygen diffuses out of the blood and into the tissue cells.

Hemoglobin is the respiratory pigment characteristic of vertebrates. It is also present in many invertebrate species from several phyla, including annelids, nematodes, mollusks, and arthropods. In some of these animals the hemoglobin is dispersed in the plasma rather than confined to blood cells.

Hemoglobin is actually a general name for a group of related compounds, all of which consist of an iron–porphyrin, or heme, group bound to a protein, known as a globin (Fig. 3–12(*d*)). The protein portion of the molecule varies in size, amino acid composition, and physical properties in various species. Three other types of respiratory pigments are hemocyanins, chlorocruorins, and hemerythrins. **Hemocyanins** are blue, copper-containing proteins found in many species of mollusks and arthropods. These do not have a heme (porphyrin) group. When oxygen is combined with the copper, the compound appears blue. Without oxygen, it is colorless. Hemocyanins are dispersed in the blood rather than confined within cells.

The Human Respiratory System

The respiratory system in humans and other air-breathing vertebrates includes the lungs and the system of tubes through which air reaches them (Fig. 37–5). Air enters the body through the **external nares,** or nostrils, which open into the **nasal cavities.** The paired nasal cavities are separated by a wall of cartilage, the **nasal septum.** Three projections called **conchae** extend from the lateral wall of each nasal cavity, increasing the surface area over which air travels as it moves through the nasal cavities. The **olfactory epithelium,** the organ of smell, is located in the roof of the nasal cavity.

The nasal cavities are lined with ciliated epithelium complete with mucus cells. These cells produce more than 400 ml of mucus a day, providing a thin layer of mucus that traps dirt particles in the inhaled air. The beating of cilia moves a continuous band of mucus along toward the throat, from which it is swallowed together with saliva. In this way, dirt particles are delivered to the digestive system, which disposes of them, and the delicate lower part of the respiratory system is protected from contact with such particles.

The nasal cavities are connected to the **sinuses,** small cavities in the bones of the skull. Mucus produced by the epithelial lining of the sinuses drains into the nose. When the capillaries in the sinuses or nasal cavities dilate during infection or allergic response, fluid accumulates in these tissues, which then swell. This, along with secretion of extra amounts of mucus, produces the "stopped-up" feeling that accompanies colds or hayfever.

Air passes by way of the internal nares to the pharynx. Whether breathing begins through the nose or the mouth, air finds its way into the **pharynx.** Nose breathing is more desirable, however, because as air passes through the nose it is filtered, moistened, and brought to body temperature.

An opening in the floor of the pharynx leads into the **larynx,** which is evident externally as the "Adam's apple." In most mammals the larynx contains **vocal cords,** folds of epithelium that vibrate and produce sounds as air passes over them. Muscles adjust the tension of the cords to produce sounds of varying pitch.

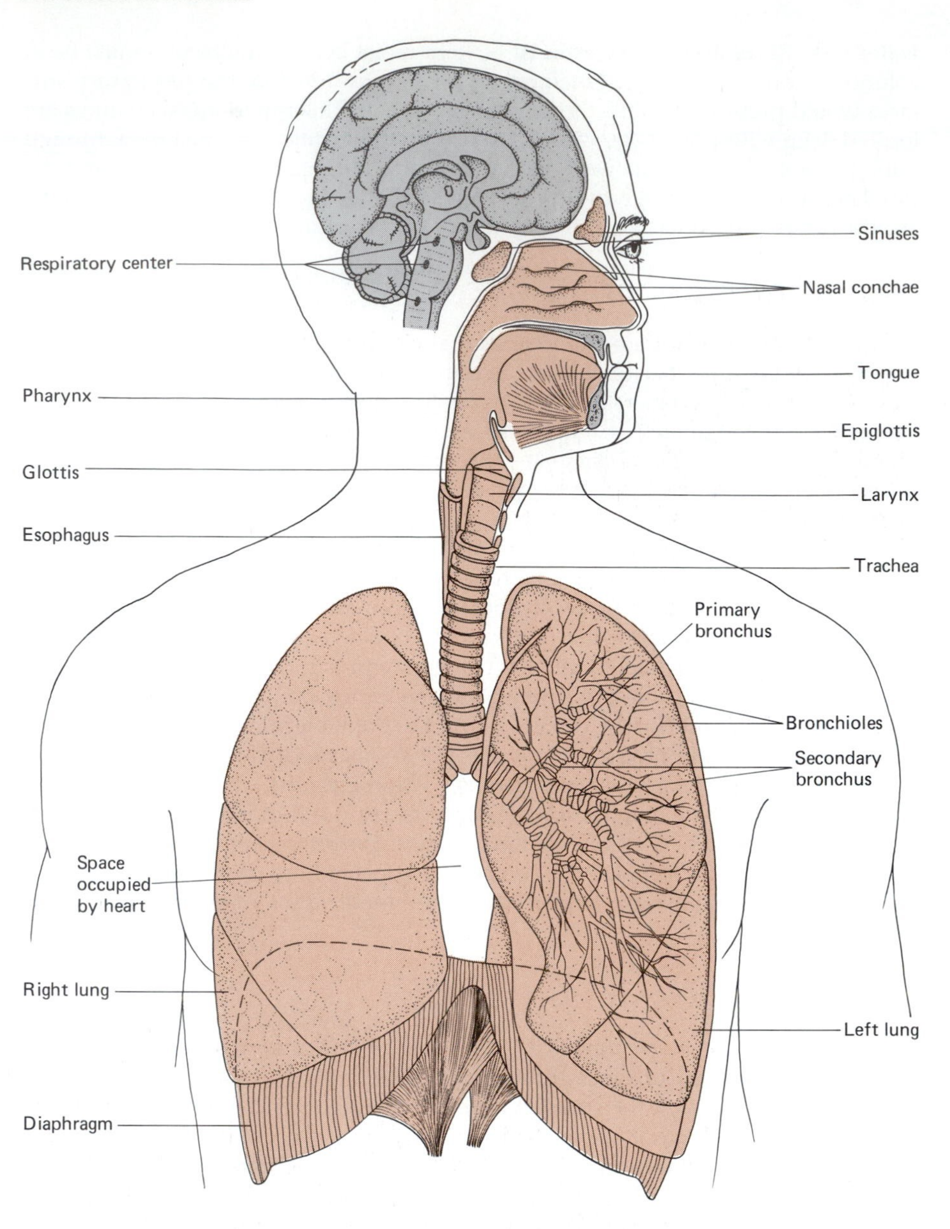

Figure 37–5 The human respiratory system.

The **epiglottis** is a flap of tissue that automatically seals off the larynx during swallowing so that neither food nor water can enter the airway. Occasionally this automatic mechanism fails, and food goes down into the larynx. When any foreign matter comes in contact with the larynx, a cough reflex is initiated. Coughing expels the foreign matter from the respiratory system. If coughing does not push out the food or material that has entered, the larynx may be blocked, which results in choking (see Focus on Choking).

From the larynx air passes into the **trachea** (the windpipe). Like the larynx, the trachea is kept from collapsing by rings of cartilage in its wall. These are necessary because air pressure in the trachea is less than atmospheric pressure during inspiration. At the level of the first rib the trachea branches into two cartilaginous **bronchi,** one extending to each lung. Inside the lung, each bronchus branches, giving rise to smaller bronchi, which, after several generations of branching, give rise ultimately to the tiny **bronchioles.** These branch repeatedly into smaller and smaller passageways, leading eventually into clusters of **alveoli.**

The very thin walls of the alveoli (only one cell thick) permit gases to diffuse easily (Fig. 37–6). A thin film of detergent-like lipid protein known as **surfactant** coats the alveoli and facilitates their dilation by lowering the surface tension. Each alveolus is enveloped in a network of capillaries. Thus only two membranes separate the air in the alveolus from the blood: the epithelium of the alveolar wall and the endothelium of the capillary.

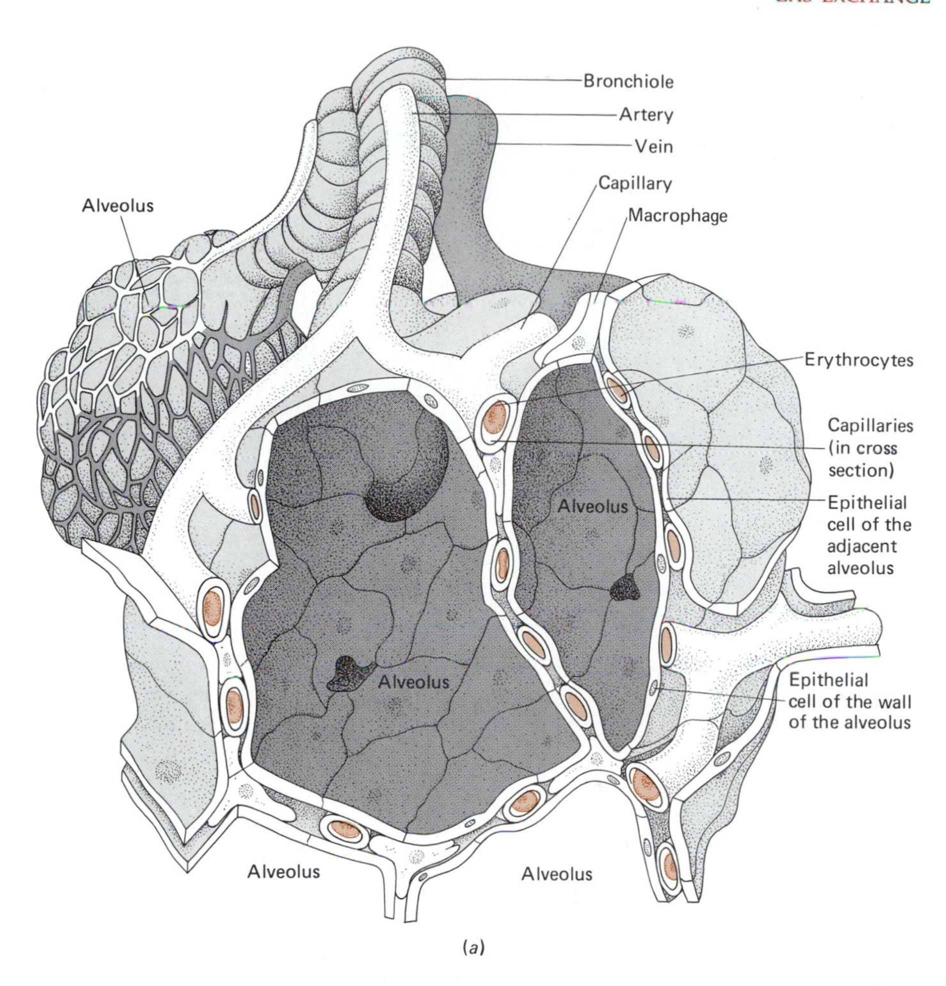

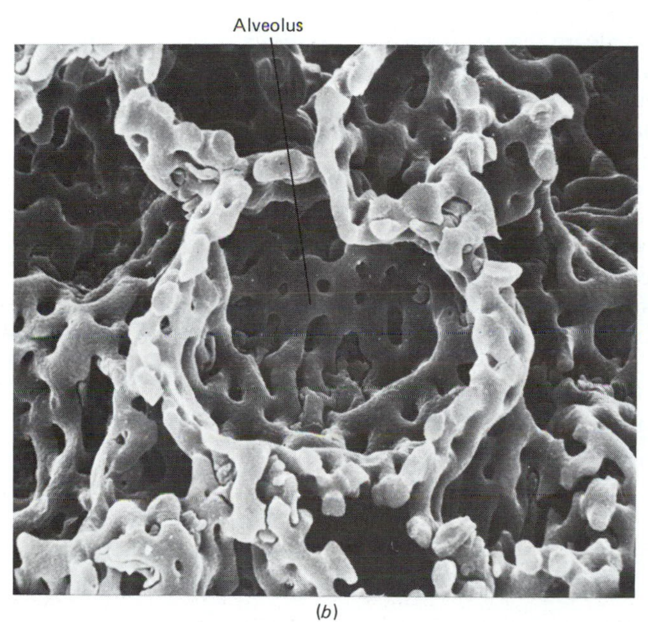

Figure 37–6 Structure of the alveolus. (*a*) Note that the alveolar wall consists of extremely thin squamous epithelium. Each alveolus is enmeshed by a network of capillaries, allowing free exchange of gases between the alveolus and the blood. (*b*) Scanning electron micrograph showing the capillary network surrounding a portion of one alveolus, seen in the center (approximately ×850). The vascular system has been injected to provide a clear, three-dimensional picture. Such casts are prepared by injecting a resin into the pulmonary artery. The resin fills all branches of the vessel; after it sets, the cellular and fibrous elements of the lung, including the blood vessels are digested, leaving only casts of the circulatory system. (From Kessel, R. G., and Kardon, R. H.: *Tissues and Organs, A Text-Atlas of Scanning Electron Microscopy*. San Francisco, W. H. Freeman Co., 1979.)

Both the trachea and bronchi are lined by ciliated epithelium, which contains mucus cells (Fig. 37–7). Particles of dust and bacteria are trapped in the film of mucus covering these cells. The beating of cilia moves a stream of dirty mucus back up to the pharynx, from which it is swallowed. This mechanism keeps foreign matter out of the lungs, serving as a cilia-propelled mucus escalator.

Neither mucus nor ciliated cells are present in the smallest bronchioles or in the air sacs. Foreign particles such as those from tobacco smoke that find their way

FOCUS ON

Choking

Choking kills an estimated 8,000 to 10,000 people per year in the United States. Many of these have long suffered from some degree of paralysis or other malfunction of the muscles involved in swallowing, often without consciously realizing it. Swallowing is a very complex process in which the mouth, pharynx, esophagus, and vocal cords must be coordinated with great precision. Functional muscular disorders of this mechanism can originate in a variety of ways—as birth defects, for example, or from brain tumors or vascular accidents involving the swallowing center of the brainstem.

Choking is more likely to occur in restaurants, where social interactions and unfamiliar surroundings are likely to distract a person's attention from swallowing and where alcohol is more likely to be taken with the meal. A large number of choking victims have a substantial blood alcohol content upon autopsy, which suggests the possibility that in them a marginally effective swallowing reflex has been further and fatally compromised by the effects of alcohol on the brain.

Anyone who begins to choke and gasp during a meal should be asked if he or she can speak. If not, as indicated by shaking the head or other gestures, the person is probably suffering a laryngeal obstruction rather than a coronary heart attack. If you are present during such an episode, be aware that you can take certain steps that can save the person's life: As a first step, deliver a strong blow to the victim's back with the open hand. If this fails, stand behind the victim, bring your arms around the person's waist, and clasp your hands just above the beltline. Your thumbs should be facing inward against the victim's body. Then squeeze abruptly and strongly in an upward direction. In most instances the residual air in the lungs will pop the obstruction out like a cork from a bottle. This is called the abdominal thrust, or **Heimlich maneuver.**

If the Heimlich maneuver must be performed in the prone position, place the victim face up. Kneel astride the victim's hips and, with one of your hands on top of the other, place the heel of your bottom hand on the abdomen slightly above the navel but below the rib cage. Press into the victim's abdomen with a quick upward thrust. This may be repeated if necessary. If there is no response within 15 or 20 seconds, it may be necessary to start cardiopulmonary resuscitation, or CPR. (See Focus on Cardiopulmonary Resuscitation [CPR].)

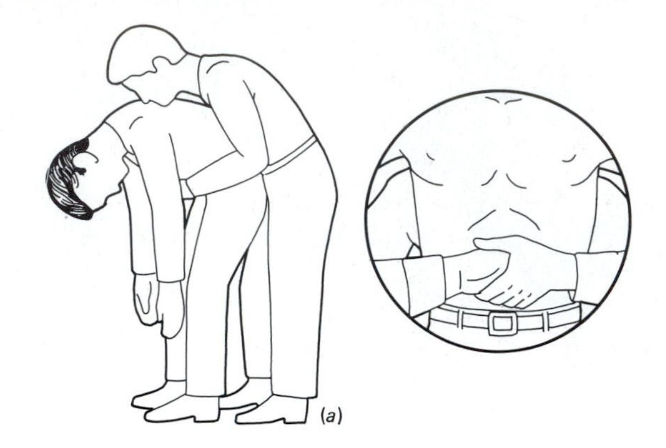
(a)

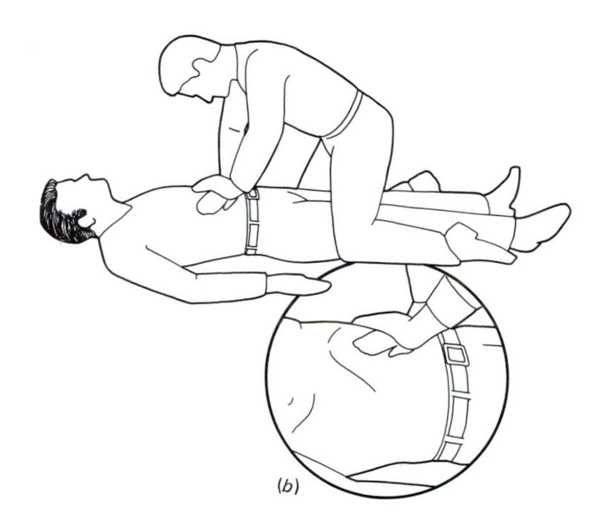
(b)

into the air sacs may remain there indefinitely, or may be phagocytized by macrophages. Such macrophages may accumulate in the lymph nodes of the lungs, permanently blackening them.

The lungs are large, paired spongy organs that occupy the thoracic cavity. Each lung, as well as the cavity in which the lung rests, is covered by a thin sheet of smooth epithelium called the **pleura.** The potential space between the pleura covering the lungs and the pleura lining the chest cavity is the **pleural cavity.** A film of fluid over the pleura keeps the membrane moist and enables the lung to move in the chest cavity during breathing with little friction. Inflammation of the pleura,

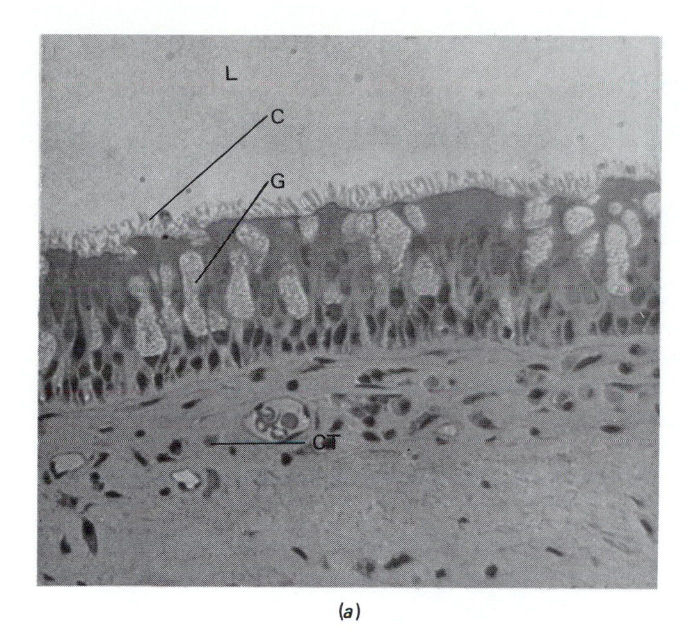

(a)

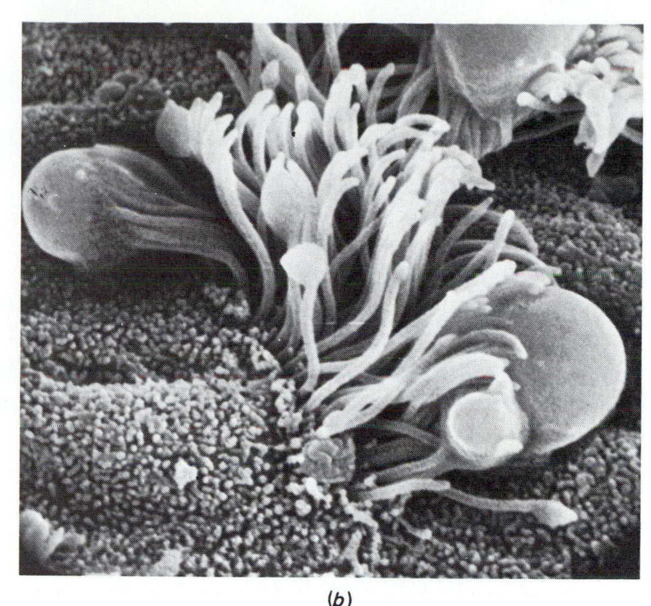

(b)

Figure 37–7 The airway is lined with ciliated epithelium. (*a*) Pseudostratified columnar epithelium of a bronchus (approximately ×4000). *L*, lumen of bronchus; *C*, cilia; *G*, goblet cell; *CT*, connective tissue. (*b*) Ciliated epithelial cells from a hamster trachea, viewed from above by a scanning electron microscope that magnifies them about 7000 times. The round bodies are mucus droplets. (From Port, C., and Corvin, I.: cover photo. *Science*, 177 (No. 4054), September 22, 1972. © 1972 by The American Association for the Advancement of Science.)

called **pleurisy,** results in the secretion of fluid into the pleural cavity, causing considerable pain during breathing.

The thoracic cavity is closed and has no communication with the outside atmosphere or with any other body cavity. It is bounded on the top and sides by the chest wall, which contains the ribs, and on the bottom by the strong, dome-shaped diaphragm.

Inside, each lung consists of bronchioles, alveoli, and great networks of capillaries, all supported by connective tissue rich in elastic fibers. Lymph tissue and nerves are also present. The surface area available for gas exchange in the lung is tremendous—more than 50 times the area of the skin, or an area that approximates the size of a tennis court.

THE MECHANICS OF BREATHING

Breathing is the mechanical process of taking air into the lungs—**inspiration**—and letting it out again—**expiration.** Oxygen continuously moves from the air in the alveoli into the blood, while carbon dioxide constantly moves from the blood into the alveoli. If oxygen is to be continuously available, the air in the alveoli must be continuously replaced with fresh air. In the resting adult the normal breathing cycle of inspiration and expiration is repeated about 12 times each minute.

In humans and other mammals the ribs, chest muscles, and diaphragm are easily movable, and the volume of the chest cavity can be increased or decreased at will. During inspiration, the rib muscles contract, drawing the front ends of the ribs upward and outward, an action made possible by the hingelike connection of the ribs with the vertebrae (Fig. 37–8). The diaphragm contracts, becoming less convex in shape, which also has the effect of enlarging the chest cavity. As the chest expands, the film of fluid on the pleura pulls the membranous walls of the lungs outward, along with the chest walls. This increases the space within each lung. The air molecules in the lung now have more space in which to move about, and the pressure of the air in the lung decreases to 2 or 3 mm Hg below atmospheric pressure. Air from outside the body then rushes in through the air passageways and fills the lungs until the two pressures are equal once again.

Expiration occurs when the diaphragm and chest muscles relax. When the rib muscles relax, the ribs return to their original position. The simultaneous relaxation of the diaphragm permits the abdominal organs to push it back up to its initial convex shape. The volume of the chest cavity decreases, and the pressure in the lungs consequently increases (to 2 or 3 mm Hg above atmospheric pressure). The distended elastic air sacs return to their normal size, expelling the air that was inhaled and bringing the pressure back to atmospheric levels. Thus, in inspiration, the millions of tiny air sacs fill with air like so many balloons, while during expiration, the air rushes out of the alveoli, partially deflating the balloons.

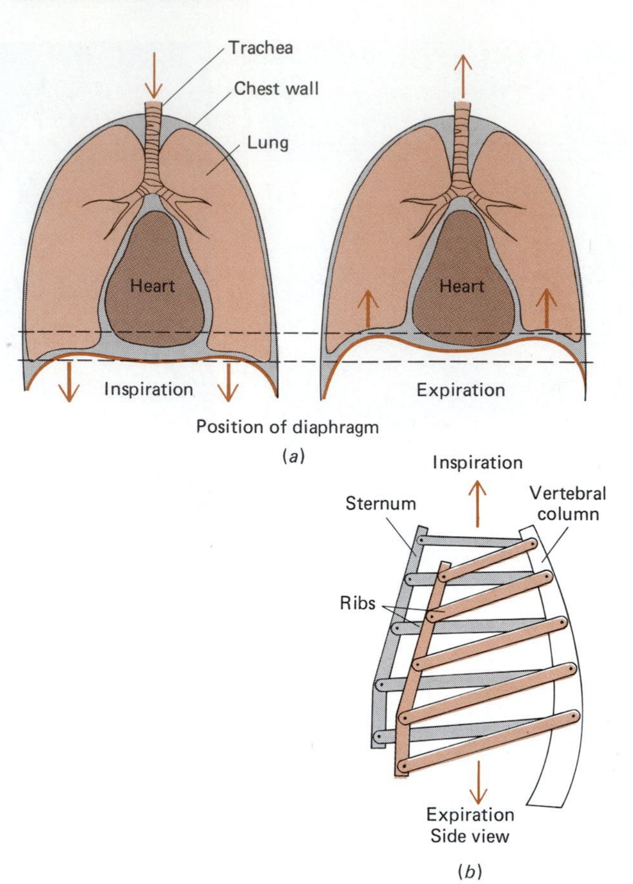

Figure 37–8 The mechanics of breathing. (*a*) Changes in the position of the diaphragm in expiration and inspiration, which result in changes in the volume of the chest cavity. (*b*) Changes in the position of the rib cage in expiration and inspiration. The elevation of the front ends of the ribs by the chest muscles causes an increase in the front-to-back dimension of the chest, and a corresponding increase in the volume of the chest cavity. When the volume of the chest cavity is increased, a corresponding amount of air is taken into the lungs.

THE QUANTITY OF AIR RESPIRED

The amount of air moved into and out of the lungs with each normal breath is called the **tidal volume.** The normal tidal volume of a young adult male is about 500 ml. The **vital capacity** is the maximum amount of air a person can exhale after filling the lungs to the maximum extent.

Vital capacity is greater than tidal volume. This means that the lungs are not completely emptied of stale air and filled with fresh air with each breath. For this reason, alveolar air contains less oxygen and more carbon dioxide than does atmospheric air (Table 37–2). Expired air has had less than one fourth of its oxygen removed and can be breathed over again—a good thing for those in need of mouth-to-mouth resuscitation!

EXCHANGE OF GASES IN THE LUNGS

After the lungs are ventilated, the oxygen in the alveoli must pass into the pulmonary capillaries, and carbon dioxide from the blood in the pulmonary capillaries must pass in the reverse direction. Since there is normally a greater concentration of oxygen in the lung alveoli than in the blood within the pulmonary capillaries, oxygen diffuses from the alveoli into the capillaries (Fig. 37–9; Table 37–2). On the other hand, the concentration of carbon dioxide is greater in the pulmonary capil-

TABLE 37–2
Composition of Inhaled Air Compared with That of Exhaled Air

	% Oxygen (O_2)	*% Carbon Dioxide (CO_2)*	*%Nitrogen (N_2)*
Inhaled air*	20.9	0.04	79
Exhaled air†	14.0	5.6	79

*The same as atmospheric air.
†Sometimes referred to as alveolar air.
Note: As indicated, the body uses up about one third of the inhaled oxygen. The amount of CO_2 increases more than 100-fold because it is produced during cellular respiration.

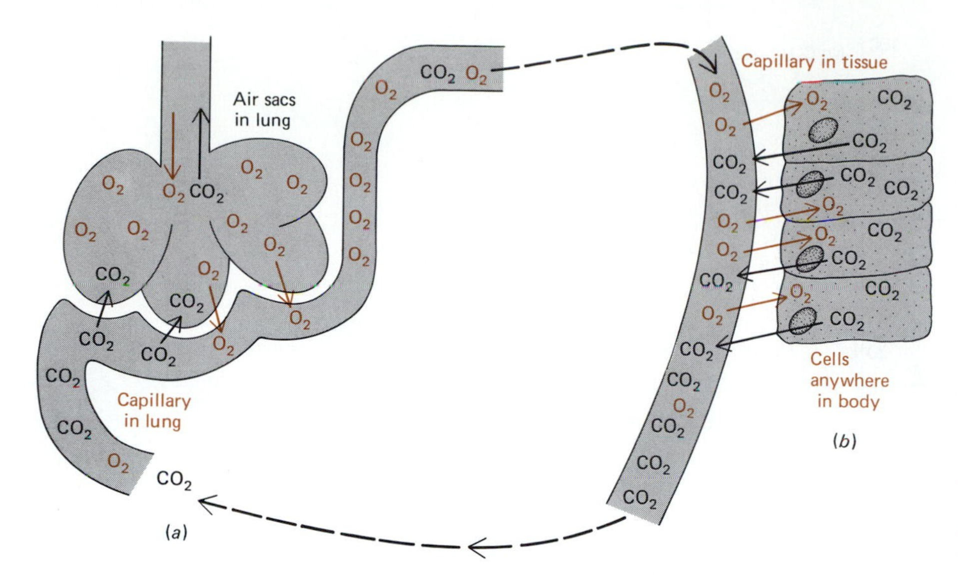

Figure 37–9 Regulation of oxygen and carbon dioxide concentration in the body. (*a*) In the lung. The concentration of oxygen is greater in the alveoli than in the pulmonary capillaries, so oxygen moves from the alveolus into the blood. Carbon dioxide is more concentrated in the blood than in the air sacs, so it moves out of the capillaries and into the air sacs. (*b*) In the tissues. Here, oxygen is more concentrated in the blood than in the cells, so it moves out of the capillary into the cell. Carbon dioxide is more concentrated in the cells, and so leaves the cells and moves into the blood.

laries than in the alveoli, so the carbon dioxide diffuses in the opposite direction—from blood to alveolar air.

The factor that determines the direction and rate of diffusion is the pressure or tension of the particular gas. According to Dalton's law of partial pressures, in a mixture of gases the total pressure of the mixture is the sum of the pressures of the individual gases. Each gas exerts, independently of the others, the same pressure it would exert if it were present alone. The pressure of earth's atmosphere, that is, the *barometric pressure,* at sea level is able to support a column of mercury (Hg) 760 mm high. Because the atmosphere is made up of about 21% oxygen, oxygen's share of that pressure is $0.21 \times 760 = 160$ mm Hg. Thus, 160 mm Hg is the partial pressure of oxygen, abbreviated P_{O_2}.

Blood passes through the lung capillaries too rapidly to become completely equilibrated with the alveolar air. The partial pressure of oxygen in arterial blood is about 100 mm Hg. The P_{O_2} in the tissues ranges from 0 to 40 mm Hg, so that oxygen diffuses out of the capillaries and into the tissues. Not all of the oxygen leaves the blood, however. The blood passes through the tissue capillaries too rapidly for equilibrium to be reached; thus, the partial pressure of oxygen in venous blood returning to the lungs is about 40 mm Hg.

The continuous metabolism of glucose and other substrates in the cells results in the continuous production of carbon dioxide and utilization of oxygen. Consequently, the concentration of oxygen in the cells is lower than that in the capillaries entering the tissues, and the concentration of carbon dioxide is higher in the cells than in the capillaries. Thus, as blood circulates through capillaries of a tissue such as brain or muscle, oxygen moves by diffusion from the capillaries to the cells, and carbon dioxide moves from the cells into the blood.

Throughout the system, from lungs to blood to tissues, oxygen moves from a region of higher concentration to one of lower concentration: Oxygen moves from the air to the blood, and then to the tissue fluid, and is finally used in the cells. Carbon dioxide also moves from a region of greater to one of lesser concentration. It moves from the cells where it is produced through the tissue fluid and blood to the lungs and then out of the body.

OXYGEN TRANSPORT

At rest, the cells of the human body utilize about 250 ml of oxygen per minute, or about 300 liters every 24 hours. With exercise or work this rate may increase as much as 10- or 15-fold. If oxygen were simply dissolved in plasma, blood would have to circulate through the body at a rate of 180 liters per minute to supply enough oxygen to the cells at rest. This is because, as we have seen, oxygen is not very soluble in blood plasma. Actually, the blood of a human at rest circulates at about 5 liters per minute and supplies all of the oxygen the cells need. Why are only 5 liters per minute rather than 180 liters per minute required?

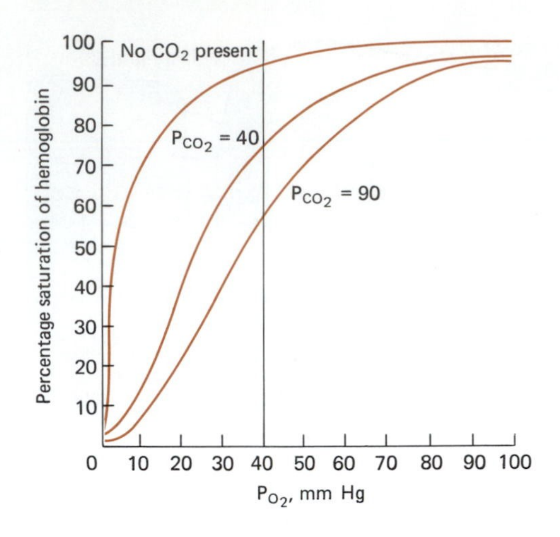

Figure 37–10 Oxygen dissociation curves. These curves show that as oxygen concentration increases, there is a progressive increase in the amount of hemoglobin that is combined with oxygen. The curves also show how carbon dioxide affects the dissociation of oxyhemoglobin. Look at the vertical axis, labeled *percentage saturation.* If the blood contains 20% O_2 by volume, which is one fourth of the amount it could contain, it is said to be 25% saturated. The left curve shows what happens if no carbon dixoide is present. The middle curve shows the situation when the P_{CO_2} is 40, which is typical of arterial blood. The right curve indicates a P_{CO_2} of 90, which is quite unhealthy. Now find the location on the horizontal axis where the partial pressure of oxygen is 40, and follow the line up through the curves. Notice how the saturation of hemoglobin with oxygen differs among the three curves even through the partial pressure of oxygen is the same.

The answer, of course, is hemoglobin, the respiratory pigment in red blood cells. Hemoglobin transports about 97% of the oxygen. Only about 3% is dissolved in the plasma. Plasma in equilibrium with alveolar air can take up in solution only 0.25 ml of oxygen per 100 ml, but the properties of hemoglobin enable whole blood to carry some 20 ml of oxygen per 100 ml. The protein portion of hemoglobin is composed of four peptide chains, typically two α and β chains, to which are attached four **heme** (porphyrin) **rings.** An iron atom is bound in the center of each heme ring.

Hemoglobin has the remarkable property of forming a loose chemical union with oxygen. An oxygen molecule may attach to each of four iron atoms. In the lung (or gill), oxygen diffuses into the erythrocyte and combines with hemoglobin (Hb) to form oxyhemoglobin (HbO_2).

$$Hb + O_2 \rightleftharpoons HbO_2$$

Hemoglobin would, of course, be of little value to the body if it could only take up oxygen. It must also *release* the oxygen where needed. The reaction goes to the right in the lungs, forming oxyhemoglobin, and to the left in the tissues, releasing oxygen. Oxyhemoglobin is a bright scarlet color, giving arterial blood its bright red color; reduced hemoglobin is purple, giving venous blood a darker hue.

The ability of oxygen to combine with hemoglobin and to be released from oxyhemoglobin is influenced by several factors, including the concentration of oxygen, the concentration of carbon dioxide, the pH, and the temperature. The **oxygen–hemoglobin dissociation curves** shown in Figure 37–10 illustrate the fact that as oxygen concentration increases, there is a progressive increase in the amount of hemoglobin that is combined with oxygen. This is known as the **percent saturation** of the hemoglobin. The percent saturation is highest in the pulmonary capillaries where the concentration of oxygen is greatest. In the capillaries of the tissues where there is less oxygen, the oxyhemoglobin dissociates, releasing oxygen. There, the percent saturation of hemoglobin is correspondingly less.

The extent to which oxyhemoglobin dissociates is determined mainly by oxygen concentration, but is also influenced by carbon dioxide. Carbon dioxide reacts with water in the plasma to form carbonic acid, H_2CO_3. An increase in the carbon dioxide concentration increases the acidity and lowers the pH of the blood. Oxyhemoglobin dissociates more readily in a more acidic environment. Lactate released from active muscles also lowers the pH of the blood and has a similar effect on the oxygen–hemoglobin dissociation curve. Displacement of the oxyhemoglobin dissociation curve by a change in partial pressure of carbon dioxide or in pH is known as the **Bohr effect.**

Some carbon dioxide is transported by the hemoglobin molecule. Although it attaches to the hemoglobin molecule in a different way and at a different site than oxygen, the attachment of a carbon dioxide molecule causes the release of an oxygen molecule from the hemoglobin. Thus carbon dioxide concentration affects the oxygen–hemoglobin dissociation curve in two ways. This results in an extremely efficient transport system. In the capillaries of the lungs (or gills in fishes), carbon dioxide concentration is relatively low and oxygen concentration is high, so oxygen saturates a very high percentage of hemoglobin. In the capillaries of the tissues, carbon dioxide concentration is high and oxygen concentration is low, so oxygen is released from the hemoglobin.

CARBON DIOXIDE TRANSPORT

When carbon dioxide enters the blood, a small percentage of it becomes dissolved in the plasma. Most of it enters the red blood cells, where an enzyme called **carbonic anhydrase** catalyzes the following reaction:

$$CO_2 + H_2O \xrightarrow{\text{carbonic anhydrase}} H_2CO_3 \longrightarrow H^+ + HCO_3^-$$

This reaction takes place slowly in the plasma, but the carbonic anhydrase in the red blood cells accelerates the rate of the reaction by about 5000 times. (It also accelerates the reverse reaction in the lungs by the same factor.)

Most of the hydrogen ions released from the carbonic acid combine with hemoglobin, which is a strong buffer. Many of the bicarbonate ions diffuse into the plasma. Chloride ions diffuse into the red blood cells to replace the bicarbonate ions, a process known as the **chloride shift.**

Some of the carbon dioxide that enters the red blood cell combines with hemoglobin. The bond between the hemoglobin and carbon dioxide is very weak; the reaction, therefore, is readily reversible. By far, most—about 70%—of the carbon dioxide is transported as the bicarbonate ion. About 23% is transported in combination with hemoglobin, and about 7% is dissolved in the plasma.

Any condition (such as pneumonia) that interferes with the removal of carbon dioxide by the lungs leads to an increased concentration of carbon dioxide in the form of carbonic acid and bicarbonate ions in the blood. This condition is called **respiratory acidosis.** Although the pH of the blood is not actually acidic in this state, it is more acidic than normal.

REGULATION OF RESPIRATION

In the medulla and pons of the brain there are special groups of cells referred to collectively as the **respiratory center** (Fig. 37–11). Some of these neurons govern mainly inspiration, some mainly expiration. A group of neurons in the pons makes up the **pneumotaxic center,** which helps control the rate of respiration. Since inspiration is an active muscular process and expiration usually a passive one, the respiratory center basically works by producing inspiration and then by inhibiting inspiration. Expiration is produced by default. Each time the inspiratory neurons produce an inspiration they also send impulses to the pneumotaxic center (Fig. 37–11). Neurons from the pneumotaxic center send messages to the expiratory neurons in the medulla. The expiratory neurons inhibit the inspiratory neurons, allowing expiration to occur.

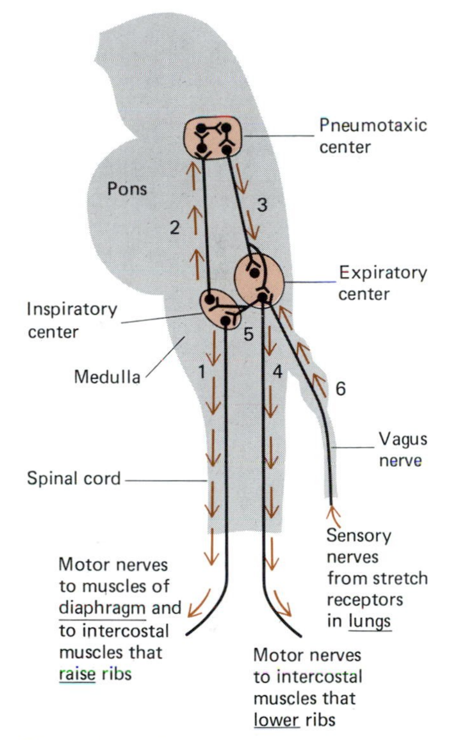

Figure 37–11 Regulation of breathing. Nerve impulses (*1*) from the inspiratory center in the medulla stimulate contraction of the diaphragm and of the intercostal muscles, raising the ribs. Other impulses (*2*) pass to the pneumotaxic center in the pons, around its neuronal circuits, and eventually to the expiratory center (*3*) in the medulla. The expiratory center is excited and sends impulses (*4*) to intercostal muscles that lower ribs. Other impulses (*5*) pass to inspiratory center to inhibit it momentarily. When impulses from the pneumotaxic center die out, inspiration begins again and the cycle is repeated. In addition, sensory nerve endings in the lungs stimulated by stretching during inspiration send impulses via the vagus nerves (*6*) that stimulate the expiratory center and inhibit the inspiratory center. This reflex from the stretch receptors in the lungs provides a safety mechanism that helps prevent overinflation of the lungs.

Carbon dioxide concentration is the most important factor controlling respiration. During exercise, when muscle cells produce large quantities of carbon dioxide, the carbon dioxide concentration in the blood rises. This stimulates the respiratory center, increasing both the depth and rate of breathing. As carbon dioxide concentration increases, more hydrogen ions are produced (from carbonic acid). An increase in hydrogen ions also affects the respiratory center, increasing ventilation. As carbon dioxide is removed by the lungs, the hydrogen ion concentration in the blood and other body fluids decreases, and homeostasis is restored. Then, since the respiratory center is no longer stimulated, the rate and depth of breathing return to normal.

In addition to the direct effects of carbon dioxide and hydrogen ions on the respiratory center, there is another mechanism for regulating respiration: Special nerve endings called **chemoreceptors** located within the walls of the aorta and carotid arteries are sensitive to the concentration of oxygen in the blood. When the partial pressure of oxygen falls below normal, these chemoreceptors are stimulated. They send messages to the respiratory center via afferent nerves. Messages from these chemoreceptors then stimulate respiratory activity. It is interesting that oxygen concentration does not affect the respiratory center directly but influences it by way of the aortic and carotid chemoreceptors. In healthy persons living at sea level, oxygen concentration rarely plays an important part in controlling respiration.

Although breathing is an involuntary process, the action of the respiratory center can be consciously influenced for a short time by either stimulating or inhibiting it. For example, you can inhibit respiration by holding your breath. You cannot hold your breath indefinitely, though, because eventually you feel a strong impulse to breathe. Even if you were able to ignore this, you would soon pass out and breathing would ensue.

Individuals who have stopped breathing because of drowning, smoke inhalation, electric shock, or cardiac arrest can be sustained by mouth-to-mouth resuscitation until their own breathing reflexes can be initiated again. Cardiopulmonary resuscitation (CPR) is a method for aiding victims who have suffered respiratory and or cardiac arrest. For a summary of the procedure, see Focus on Cardiopulmonary Resuscitation (CPR).

HYPERVENTILATION

Underwater swimmers and some Asiatic pearl divers voluntarily **hyperventilate** before going under water. By taking a series of deep inhalations and exhalations, the carbon dioxide content of the alveolar air and of the blood can be markedly reduced. As a result, it takes longer before the impulse to breathe becomes irresistible.

When hyperventilation is carried on too long, the person may feel dizzy and sometimes becomes unconscious. This is because a certain concentration of carbon dioxide is needed in the blood to maintain normal blood pressure. (This is accomplished by way of the vasoconstrictor center in the brain, which maintains the muscle tone of blood vessel walls.) Furthermore, if divers hold their breath too long, the low concentration of oxygen may result in unconsciousness and drowning.

HIGH FLYING AND LOW DIVING

The barometric pressure decreases at progressively higher altitudes. Since the concentration of oxygen in the air remains at 21%, the partial pressure of oxygen decreases along with the barometric pressure. At an altitude of 6000 m (19,500 ft), the barometric pressure is about 350 mm Hg, the partial pressure of oxygen is about 75 mm Hg, and the hemoglobin in arterial blood is about 70% saturated with oxygen. At 10,000 m, the barometric pressure is about 225 mm Hg, the partial pressure of oxygen is 50 mm Hg, and arterial oxygen saturation is only 20%. Thus, getting sufficient oxygen from the air becomes an ever-increasing problem at higher altitudes.

When a person moves to a high altitude, the body adjusts over a period of time by producing a greater number of red blood cells. In a person breathing pure oxygen at 10,000 m, the oxygen would have a partial pressure of 225 mm Hg, and the hemoglobin would be almost fully saturated with oxygen. Above 13,000 m, however, the barometric pressure is so low that even breathing pure oxygen would not permit complete oxygen saturation of arterial hemoglobin.

A person can remain conscious only until the arterial oxygen saturation falls to 40% to 50%. This level is reached at about 7000 m when the person is breathing air, or 14,500 m when pure oxygen is used. All high-flying jets have cabins that are airtight and pressurized to an altitude of about 2000 m.

Hypoxia, a deficiency of oxygen, results in drowsiness, mental fatigue, headache, and sometimes euphoria. The ability to think and to make judgments is impaired, and there is a loss in ability to perform tasks requiring coordination. If a jet were flying at about 11,700 m and underwent sudden decompression, the pilot would lose consciousness in about 30 seconds and become comatose in about 1 minute.

In addition to the problems of hypoxia, a rapid decrease in barometric pressure can cause **decompression sickness** (the bends). Whenever the barometric pressure drops below the total pressure of all gases dissolved in the blood and other body fluids, the dissolved gases tend to come out of solution into a gaseous state and form bubbles. A familiar example of such bubbling occurs each time you uncap a bottle of soda, thus reducing the pressure in the bottle. The carbon dioxide is released from solution and bubbles out into the air. In the body, it is nitrogen that causes the problem. Nitrogen comes out of solution and forms bubbles that may block capillaries, interfering with blood flow. Other tissues may also be damaged. The clinical effects of decompression sickness are pain, dizziness, paralysis, unconsciousness, and even death.

FOCUS ON

Cardiopulmonary Resuscitation (CPR)

Cardiopulmonary resuscitation, or **CPR,** is a method for aiding victims of accidents or heart attacks who have suffered cardiac arrest and respiratory arrest. It should not be used if the victim has a pulse or is able to breathe. It must be started immediately, because irreversible brain damage may occur within about 3 minutes of respiratory arrest. Here are its ABCs:

Airway Clear airway by extending victim's neck. This is sometimes sufficient to permit breathing to begin again.
Breathing Use mouth-to-mouth resuscitation.
Circulation Attempt to restore circulation by using external cardiac compression.

The procedure for CPR may be summarized as follows.

I. Establish unresponsiveness of victim.
II. Procedure for mouth-to-mouth resuscitation:
1. Place victim on his or her back on firm surface.
2. Clear throat and mouth and tilt head back so that chin points outward. Make sure that the tongue is not blocking airway. Pull tongue forward if necessary.
3. Pinch nostrils shut and forcefully exhale into victim's mouth. Be careful, especially in children, not to overinflate the lungs.
4. Remove your mouth and listen for air rushing out of the lungs.
5. Repeat about 12 times per minute. Do not interrupt for more than 5 seconds.

III. Procedure for external cardiac compression:
1. Place heel of hand on lower third of breastbone. Keep your fingertips lifted off the chest. (In infants, two fingers should be used for cardiac compression; in children, use only the heel of the hand.)
2. Place heel of the other hand at a right angle to and on top of the first hand.
3. Apply firm pressure downward so that the breastbone moves about 4 to 5 cm (1.6 to 2 in.) toward the spine. Downward pressure must be about 5.4 to 9 kg (12 to 20 lb) with adults (less with children). Excessive pressure can fracture the sternum or ribs, resulting in punctured lungs or a lacerated liver. This rhythmic pressure can often keep blood moving through the heart and great vessels of the thoracic cavity in sufficient quantities to sustain life.
4. Relax hands between compressions to allow chest to expand.
5. Repeat at the rate of at least 60 compressions per minute. (For infants or young children, 80 to 100 compressions per minute are appropriate.) If there is only one rescuer, 15 compressions should be applied, then two breaths, in a ratio of 15 : 2. If there are two rescuers, the ratio should be 5 : 1.

Decompression sickness is even more common in deep-sea diving than in high-altitude flying. As a diver descends, the surrounding pressure increases tremendously—1 atmosphere for each 10 m. To prevent the collapse of the lungs, a diver must be supplied with air under pressure, thereby exposing the lungs to very high alveolar gas pressures (Fig. 37–12).

Figure 37–12 A diver's tanks containing compressed air. (Charles Seaborn.)

At sea level an adult human has about 1 liter of nitrogen dissolved in the body, with about half of that in the fat and half in the body fluids. After a diver's body has been saturated with nitrogen at a depth of 100 m, the body fluids contain about 10 liters of nitrogen. To prevent this nitrogen from rapidly bubbling out of solution and causing decompression sickness, the diver must be brought to the surface gradually, with stops at certain levels on the way up. This allows the nitrogen to be expelled slowly through the lungs.

Some mammals can spend rather long periods of time in the ocean depths without coming up for air; see Focus on Adaptations of Diving Mammals.

FOCUS ON

Adaptations of Diving Mammals

The Weddell seal can swim under the ice for more than an hour without coming up for air. The bottle-nosed whale can remain in the ocean depths for as long as two hours. Porpoises, whales, seals, beavers, mink, and several other air-breathing mammals have adaptations that permit them to dive for food or disappear below the water surface for several minutes to elude their enemies.

Diving mammals have about twice the volume of blood, relative to their body weight, as nondivers. Many have high concentrations of myoglobin, an oxygen-binding pigment similar to hemoglobin found in muscles. They do not, however, have larger lungs than those in nondiving mammals.

When a mammal dives, a group of physiological mechanisms known collectively as the **diving reflex** are activated. Breathing stops. Bradycardia (slowing of the heart rate) occurs. The heart rate may decrease to one tenth of the normal rate, reducing the body's consumption of oxygen and energy. Blood is redistributed, with the lion's share going to the brain and heart, the organs that can least withstand anoxia. Skin, muscles, digestive organs, and other internal organs can survive with less oxygen, and so receive less blood while an animal is submerged.

Diving mammals do not take in extra air before a dive. In fact, seals exhale before they dive, and the lungs of whales are compressed during diving. These adaptations are thought to reduce the chance of decompression sickness, because with less air in the lungs there is less nitrogen in the blood to dissolve during the dive.

The diving reflex is present to some extent in human beings, where it may act as a protective mechanism during birth when an infant may be deprived of oxygen for several minutes. Many cases of near-drownings have been documented in which the victim had been submerged for several minutes (as long as 45 minutes) in very cold water before being rescued and resuscitated. In many of these survivors there was no apparent brain damage. The shock of the icy water slows the heart rate, increases blood pressure, and shunts the blood to the internal organs of the body that most need oxygen (blood flow in the arms and legs decreases). Metabolic rate decreases so that less oxygen is required.

Mother and baby bottle-nose porpoise (*Tursiops truncatus*). (Courtesy of the Miami Seaquarium.)

Oxygen poisoning can also result from inhaling compressed air. One of its worst effects is convulsions, which can be lethal to divers deep beneath the surface. Just why oxygen is toxic is not understood, but several possibilities have been suggested. One hypothesis is that high concentrations of oxygen may inactivate some of the essential respiratory enzymes. There is also evidence that high concentrations of oxygen decrease blood flow through the brain by 25% to 50%. Still another suggestion is that too much oxygen in the cell may cause the production of oxidizing free radicals. (A free radical is a particle with an unpaired electron and is generally very reactive.) Free radicals may oxidize certain essential cell components, thereby damaging the cell's metabolic machinery.

EFFECTS OF SMOKING AND AIR POLLUTION

A human being breathes about 20,000 times each day, inhaling about 35 pounds of air—six times the amount of food and drink consumed. Many of us breathe dirty urban air laden with particulates, carbon monoxide, sulfur oxides, and other harmful substances (Fig. 37–13). How do these air pollutants affect the respiratory system?

Mucus cells in the epithelial lining of the respiratory passageways react to the irritation of inhaled pollutants by secreting increased amounts of mucus. The ciliated cells, damaged by the pollutants, are unable to effectively clear the mucus and trapped particles from the airways. The body attempts to clear the airways by coughing. (Smoker's cough is a well-known example.) The coughing causes increased irritation, and the respiratory system becomes susceptible to viral, bacterial, and fungal infections. Particles of dirt also accumulate in the lungs over a period of years (Fig. 37–14).

One of the body's fastest responses to inhaling dirty air is **bronchial constriction.** The bronchial tubes narrow, providing increased opportunity for the inhaled particles to be trapped along the sticky mucous lining of the airway. Unfortunately, the narrowed bronchial tubes allow less air to pass through to the lungs so that oxygen available to the body cells is decreased. Chain smokers and those who live in polluted industrial areas experience almost continuous bronchial constriction.

Chronic bronchitis and emphysema are **chronic obstructive pulmonary diseases** that have been linked to smoking and air pollution. More than 75% of patients with chronic bronchitis have a history of heavy cigarette smoking. Advanced emphysema is virtually unknown in those who have never smoked regularly.

Figure 37–13 Air pollution contributes to respiratory disorders.

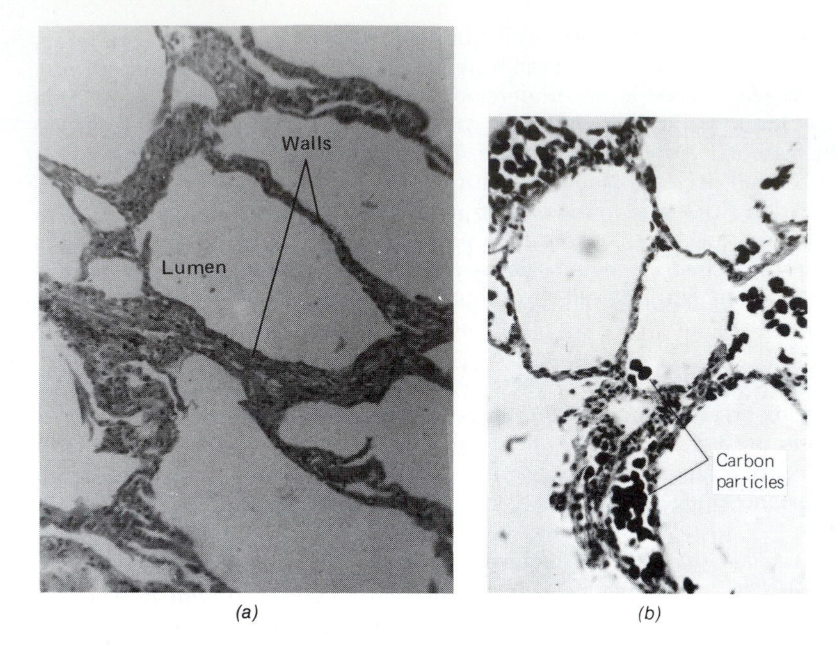

Figure 37–14 Lung tissue. (*a*) Normal lung tissue. (*b*) Lung tissue with accumulated carbon particles. This results from breathing smoky, polluted air.

In **chronic bronchitis,** the bronchioles secrete too much mucus and are constricted and inflamed. Respiratory cilia cannot clear the passages of the mucus and particles that partly clog them. Affected persons are short of breath and often cough up mucus. They are susceptible to frequent episodes of **acute bronchitis,** a temporary inflammation usually caused by viral or bacterial infection. Persons with chronic bronchitis often develop emphysema.

In **emphysema,** obstruction of the bronchioles results in increased airway resistance and increased work in breathing, especially during expiration. The alveolar walls break down so that several air sacs become joined to form larger and less elastic alveoli (Fig. 37–15). The number of capillaries in the lungs is reduced. Because the air sacs are not as elastic as they should be, air is not expelled effectively, and stale air accumulates in the air sacs. This decreases the amount of oxygen available to combine with hemoglobin in the blood. To compensate, the right ventricle of the heart pumps harder and becomes enlarged. Emphysema patients often die of heart failure.

Cigarette smoking is also the main cause of **lung cancer.** About 80% of deaths from lung cancer would not occur if people did not smoke cigarettes. (See Focus on Facts About Smoking.) In fact, cigarette smoking is the major single cause of all cancer deaths in the United States. More than 10 of the compounds in the tar of tobacco smoke have been shown to cause cancer in animals. These carcinogenic compounds irritate the cells lining the respiratory passages and alter their metabolic balance. Normal cells are transformed into abnormal cancerous ones, which multiply rapidly and invade surrounding tissue.

Figure 37–15 Whole-mount lung sections. (*a*) From normal lung. (*b*) From a patient with advanced emphysema. (Courtesy of Dr. Oscar Auerbach and *The New England Journal of Medicine.*)

(*a*)

(*b*)

FOCUS ON

Facts About Smoking

- The life of a 30-year-old who smokes 15 cigarettes a day is shortened by more than 5 years.
- If you smoke more than one pack per day, you are about 20 times more likely to develop lung cancer than a nonsmoker. According to the American Cancer Society, cigarette smoking causes more than 75% of all lung cancer deaths.
- If you smoke, you are more likely to develop atherosclerosis, and you double your chances of dying from cardiovascular disease.
- If you smoke, you are 20 times more likely to develop chronic bronchitis and emphysema than a nonsmoker.
- If you smoke, you are seven times more likely to develop peptic ulcers (especially malignant ulcers) than a nonsmoker.
- If you smoke, you have about 5% less oxygen circulating in your blood (because of carbon monoxide–hemoglobin) than a nonsmoker.
- If you smoke when you are pregnant, your baby will weigh about 6 ounces less at birth, and there is double the risk of miscarriage, stillbirth, and infant death than if you did not smoke.
- Workers who smoke one or more packs of cigarettes per day are absent from their jobs because of illness 33% more often than nonsmokers.
- Risks increase with the number of cigarettes smoked, inhaling, smoking down to a short stub, and use of nonfilter or high-tar, high-nicotine cigarettes. Cigar and pipe smokers have lower risks than cigarette smokers because they do not inhale as much. Cigarette smokers who switch to cigars and continue to inhale actually increase their risks.
- Nonsmokers confined in living rooms, offices, automobiles, or other places with smokers are adversely affected by the smoke. For example, when parents of infants smoke, the infant has double the risk of contracting pneumonia or bronchitis in its first year of life.
- When smokers quit smoking, their risk of dying from chronic pulmonary disease, cardiovascular disease, or cancer decreases. (Precise changes in risk figures depend upon the number of years the person smoked, the number of cigarettes smoked per day, the age of starting to smoke, and the number of years since quitting.)
- If everyone in the United States stopped smoking, more than 300,000 lives would be saved each year.

SUMMARY

I. In organisms less than 1 mm thick, gas exchange can occur effectively by diffusion, but in larger more complex forms, specialized respiratory structures are required.
  - *A.* In nudibranch mollusks, most annelids, small arthropods, and some vertebrates, gas exchange occurs through the entire body surface.
  - *B.* In insects and some other arthropods, the respiratory system consists of a network of tracheal tubes.
  - *C.* Gills are respiratory structures characteristic of aquatic animals.
    - *1.* In echinoderms, simple dermal gills project from the body wall.
    - *2.* In chordates, gills are usually internal, located along the edges of the gill slits.
    - *3.* In bony fish, a countercurrent exchange system promotes diffusion of oxygen into the blood and diffusion of carbon dioxide out of the blood and into the water.
  - *D.* Large mollusks and terrestrial vertebrates have lungs with some means of ventilating them.
  - *E.* Most modern fish do not have lungs but possess homologous swim bladders.

II. Gas exchange with air is more efficient than gas exchange with water because air contains more oxygen than water and because oxygen diffuses more rapidly through air than through water.

III. Respiratory pigments greatly increase the capacity of blood to transport oxygen.

IV. The human respiratory system includes the lungs and a system of tubes through which air reaches them. A breath of air passes through the nose, pharynx, larynx, trachea, bronchus, bronchioles, and alveoli.
  - *A.* During breathing, the diaphragm and rib muscles contract, expanding the chest cavity. The membranous walls of the lungs move outward along with the chest walls, decreasing the pressure within the lungs. Air from outside the body rushes in through the air passageways and fills the lungs until the pressure once more equals atmospheric pressure.
  - *B.* Tidal volume is the amount of air moved into and out of the lungs with each normal breath. Vital ca-

pacity is the maximum volume that can be exhaled after the lungs are filled to the maximum extent.

C. Oxygen and carbon dioxide are exchanged between alveoli and blood by diffusion.

D. About 97% of the oxygen in the blood is transported as oxyhemoglobin.
   1. As oxygen concentration increases there is a progressive increase in the amount of hemoglobin that combines with oxygen.
   2. Owing to the Bohr effect, oxyhemoglobin dissociates more readily as carbon dioxide concentration increases.

E. About 70% of the carbon dioxide in the blood is transported as bicarbonate ions.

F. The respiratory center in the medulla stimulates inspiration and then inhibits inspiration.
   1. The respiratory center is activated by increased carbon dioxide concentration.
   2. The respiratory system is also stimulated by an increase in hydrogen ions and, under some circumstances, by signals from chemoreceptors sensitive to low oxygen concentration.

G. Hyperventilation reduces the concentration of carbon dioxide in the alveolar air and in the blood.

H. As altitude increases, barometric pressure decreases and less oxygen enters the blood. This situation can lead to hypoxia. A rapid decrease in barometric pressure can cause decompression sickness.

I. Inhaling polluted air results in bronchial constriction, increased mucus secretion, damaged ciliated cells, and coughing. It eventually can lead to chronic bronchitis, emphysema, or lung cancer.

POST-TEST

1. Specialized respiratory structures must have ________ walls so that ________ easily occurs; they must be ________ so that gases can be dissolved; and they are typically richly supplied with ________ ________ to ensure transport of gases.

2. In insects, air enters a network of ________ tubes through openings called ________.

3. The operculum is a bony plate that protects the ________ in ________ ________.

4. Respiratory structures that develop as ingrowths of the body surface are called ________.

5. A fish can rise or sink, or maintain a particular level in the water without muscular effort, by regulating the amount of air in its ________ ________.

6. In birds, the lungs have several extensions referred to as ________ ________.

7. In the mammalian respiratory system, inhaled air passing through the larynx next enters the ________, and then passes into a ________.

8. In the mammalian respiratory system, gas exchange takes place through the thin walls of the ________.

9. In mammals, the floor of the thoracic cavity is formed by the ________.

10. The chief problem with breathing air is ________ ________.

Select the most appropriate answer in Column B for each description in Column A.

Column A	*Column B*
11. Seals off larynx during swallowing	a. Sinuses
12. Cavities in bones of skull	b. Larynx
13. Initiates cough reflex	c. Alveoli
14. Covers lung	d. Pleura
15. Coated with surfactant	e. Epiglottis

16. The maximum amount of air a person can exhale after filling the lungs to the maximum extent is the ________ ________.

17. The extent to which oxyhemoglobin dissociates is determined mainly by the ________ concentration.

18. An increase in carbon dioxide concentration lowers the blood pH and results in greater dissociation of ________; this is known as the ________ ________.

19. Most carbon dioxide is transported in the blood as ________ .

20. Hypoxia is a deficiency in ________.

21. Bronchial constriction is one of the body's most rapid responses to ________.

22. In ________, the alveolar walls break down so that several air sacs join to form larger, less elastic alveoli.

23. The main cause of lung cancer is ________ ________.

24. Label the following diagram. (Refer to Fig. 37–5 as necessary.)

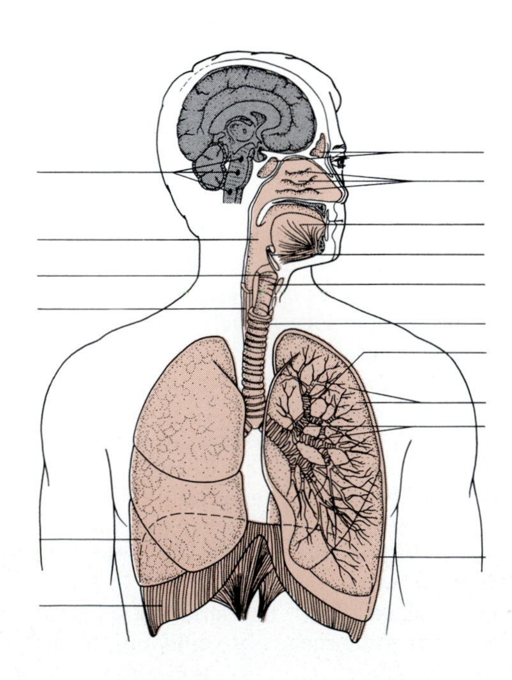

REVIEW QUESTIONS

1. Why are specialized respiratory structures necessary in a tadpole but not in a flatworm?
2. Compare ventilation in a sea star with ventilation in a human.
3. Compare gas exchange in the following animals.
   a. earthworm
   b. grasshopper
   c. fish
   d. frog
4. Why are lungs more suited for an air-breathing vertebrate and gills more effective in a fish? Why are lungs internal?
5. What respiration problem is solved by respiratory pigments? How?
6. Why does alveolar air differ in composition from atmospheric air?
7. What physiological mechanisms bring about an increase in rate and depth of breathing during exercise? Why is such an increase necessary?
8. What is the advantage of having lungs with millions of alveoli rather than lungs consisting of simple sacs, as in the mud puppies?
9. In what way might it be an advantage for a fish to have lungs as well as gills? What function do the "lungs" of most modern fish serve?
10. How does the countercurrent exchange system increase the efficiency of gas exchange between gills and blood?
11. What mechanisms does the human respiratory system have for getting rid of inhaled dirt? What happens when so much dirty air is inhaled that these mechanisms cannot function effectively?
12. What factors affect the dissociation of oxyhemoglobin?
13. What happens to a deep-sea diver who surfaces too quickly?

38

Fluid Balance and Disposal of Metabolic Wastes

OUTLINE

I. Functions of the excretory system
II. Waste products
III. Waste disposal and osmoregulation in invertebrates
  A. Nephridial organs
  B. Antennal glands
  C. Malpighian tubules
IV. Osmoregulation and waste disposal in vertebrates
  A. The vertebrate kidney
  B. Osmoregulation
V. Fluid balance and excretion in humans
  A. The human kidney and its ducts
  B. The nephron
  C. Urine formation
    1. Filtration
    2. Reabsorption
    3. Secretion
    4. Countercurrent exchange
  D. Composition of urine
  E. Fluid homeostasis
    1. Regulation of urine volume
    2. Regulation of sodium reabsorption

LEARNING OBJECTIVES

After you have read this chapter you should be able to:

1. Identify the principal types of metabolic wastes (including three types of nitrogenous wastes), and describe how each is produced and how each is excreted.
2. Compare the excretory organs of a flatworm, a crustacean, an insect, and a vertebrate, explaining briefly how each functions.
3. Compare the adaptations of marine bony fish and cartilaginous fish that enable them to carry on osmoregulation.
4. Compare the osmoregulation problem of a freshwater fish with that of a marine fish, and describe how the freshwater fish solves the problem.
5. Describe the adaptations that enable each of the following animals to solve their osmoregulatory problems: amphibians; marine birds and reptiles; and a dehydrated human being.
6. Label on a diagram and give the functions of each organ of the human urinary system.
7. Label on a diagram the principal parts of the nephron, and give the function of each structure. Include the following structures: Bowman's capsule, glomerulus, proximal convoluted tubule, loop of Henle, distal convoluted tubule, collecting duct, afferent arteriole, and efferent arteriole.
8. Describe the process of urine formation, comparing filtration and reabsorption.
9. Trace a drop of filtrate from Bowman's capsule to its release from the body as urine.
10. Describe countercurrent exchange, and explain the importance of this mechanism.
11. Describe the effects of each of the following on the volume and composition of urine: ADH; low fluid intake; alcohol intake; and eating a bag of potato chips.
12. Summarize the functions of the kidneys in maintaining homeostasis.

Water shapes the basic nature of life and its distribution on earth. Most organisms consist mainly of water, and it is the medium in which metabolic reactions take place. The simplest forms of life are small organisms that live in the sea. They obtain their food and oxygen directly from the sea water that surrounds them, and they release waste products into it. Larger, more complex animals and most terrestrial animals must have their own internal sea—the blood and interstitial fluid—to bathe their cells and to transport and dissolve nutrients, gases, and waste products.

Terrestrial animals have a continuous stream of water flowing through them. Water is taken in with food and drink and is formed in metabolic reactions. Some of this water becomes part of the blood, which transports materials throughout the organism. Then, interstitial fluid is formed from the blood plasma to bathe all of the cells of the body. Excess water evaporates from the body surface or is excreted by specialized structures.

The water content of the body, as well as the concentration and distribution of ions, must be precisely regulated, a process called **osmoregulation.** Freshwater protozoans have contractile vacuoles that periodically pump out the excess water that is osmotically drawn into the cell. In many organisms, the same structures that rid the body of excess water and ions often are also adapted to rid the body of metabolic wastes. These organs make up the **excretory system.**

As cells carry on metabolic activities, waste materials are generated. If permitted to accumulate, such waste products would reach toxic concentrations and threaten homeostasis. Therefore, waste must be continuously removed from the body. **Excretion** is the process of removing metabolic wastes from the body. The term excretion is sometimes confused with the term **elimination** (Fig. 38–1). Undigested and unabsorbed food materials are *eliminated* from the body in the feces. Such substances never actually participated in the organism's chemical metabolism or entered body cells, but merely passed through the digestive system.

Functions of the Excretory System

Typically, an excretory system helps maintain homeostasis in three ways: (1) It excretes metabolic wastes; (2) it carries on osmoregulation (regulating both the fluid and salt content of the body); and (3) it regulates the concentrations of most of the constituents of body fluids. To carry out these functions, an excretory organ collects fluid, generally from the blood or interstitial fluid. It then adjusts the composition of this fluid by *reabsorbing* substances the body needs. Finally, the adjusted excretory product (urine, for example) must be expelled from the body.

Waste Products

The principal waste products generated by metabolic activities in most animals are water, carbon dioxide, and nitrogenous wastes. Carbon dioxide is mainly excreted by gills, lungs, or other respiratory surfaces. Excretory organs such as kidneys excrete water and nitrogenous wastes.

Nitrogenous wastes include ammonia, uric acid, and urea. The first step in the catabolism of amino acids is **deamination,** a process in which the amino group is removed and converted to **ammonia** (Fig. 38–2). Aquatic invertebrates and freshwater vertebrates excrete nitrogen mainly as ammonia. The water that surrounds these animals dilutes the ammonia and quickly washes it away. Ammonia is highly toxic and cannot be permitted to accumulate in the body or in the organism's surroundings. In many organisms it is quickly converted to some less toxic nitrogenous waste product such as uric acid or urea.

Uric acid is produced from ammonia and also when nucleotides from nucleic acids are metabolized. Uric acid is poorly soluble and precipitates from a supersaturated solution, forming crystals. Thus, uric acid can be excreted as a paste with little fluid loss. Among the animals that excrete their nitrogenous wastes mainly as uric acid are insects, birds, pulmonate snails, and certain reptiles.

Urea is the principal nitrogenous waste product of amphibians and mammals. It is produced mainly in the liver. The sequence of reactions by which the urea molecule is synthesized from ammonia and carbon dioxide is known as the **urea cycle** (Fig. 38–3). These reactions require specific enzymes and the input of

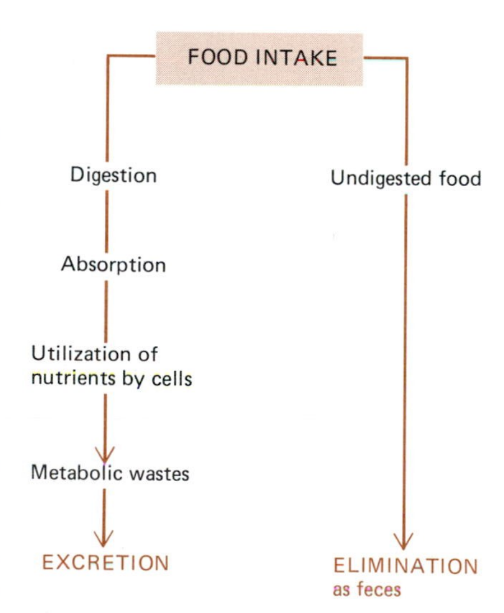

Figure 38–1 Excretion contrasted with elimination.

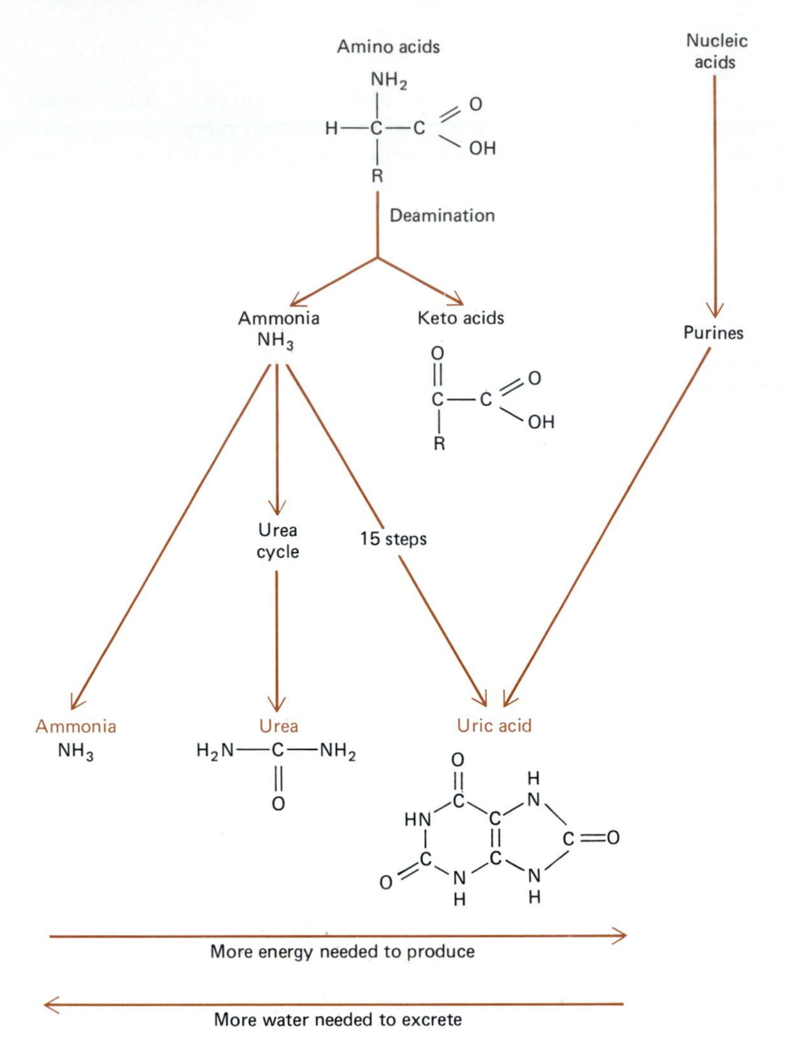

Figure 38–2 Nitrogenous wastes are formed by deamination of amino acids and by metabolism of nucleic acids. Ammonia is the first metabolic product of deamination. In many animals, ammonia is converted to urea via the urea cycle (see Fig. 38–3). Other animals convert ammonia to uric acid. Energy is required to convert ammonia to urea and uric acid, but less water is required to excrete these wastes.

energy by the cells. However, urea is far less toxic than ammonia and so can accumulate in higher concentrations without causing tissue damage. Thus, urea can be excreted in more concentrated form.

Waste Disposal and Osmoregulation in Invertebrates

Sponges and cnidarians have no specialized excretory structures. Their wastes pass by diffusion from the intracellular fluid to the external environment. For this reason certain environments such as coral reefs are especially prone to damage when changes in water current or stagnation occur. The vast majority of sponges and cnidarians are marine organisms living in an isotonic environment. They have no special problems of excess water intake. There are a few freshwater sponges and a few freshwater cnidarians such as *Hydra* that live in a medium that is very hypotonic to their intracellular fluid. How they prevent inflow of excess water or how they pump it out remains a mystery.

NEPHRIDIAL ORGANS

Nephridial organs are a common type of excretory organ found among invertebrates. They consist of simple or branching tubes that open to the outside of the body through nephridial pores. Flatworms and nemertines are the simplest animals with specialized excretory organs. Although metabolic wastes do pass by diffusion through the body surface, these animals also have nephridial organs that consist of tubules with enlarged blind ends containing cilia. These organs are known as **protonephridia** and have **flame cells.** A branching system of excretory ducts connects the protonephridia with the outside (Fig. 38–4). The flame cells lie in the fluid that bathes the body cells, and wastes diffuse into the flame cells and from there into the

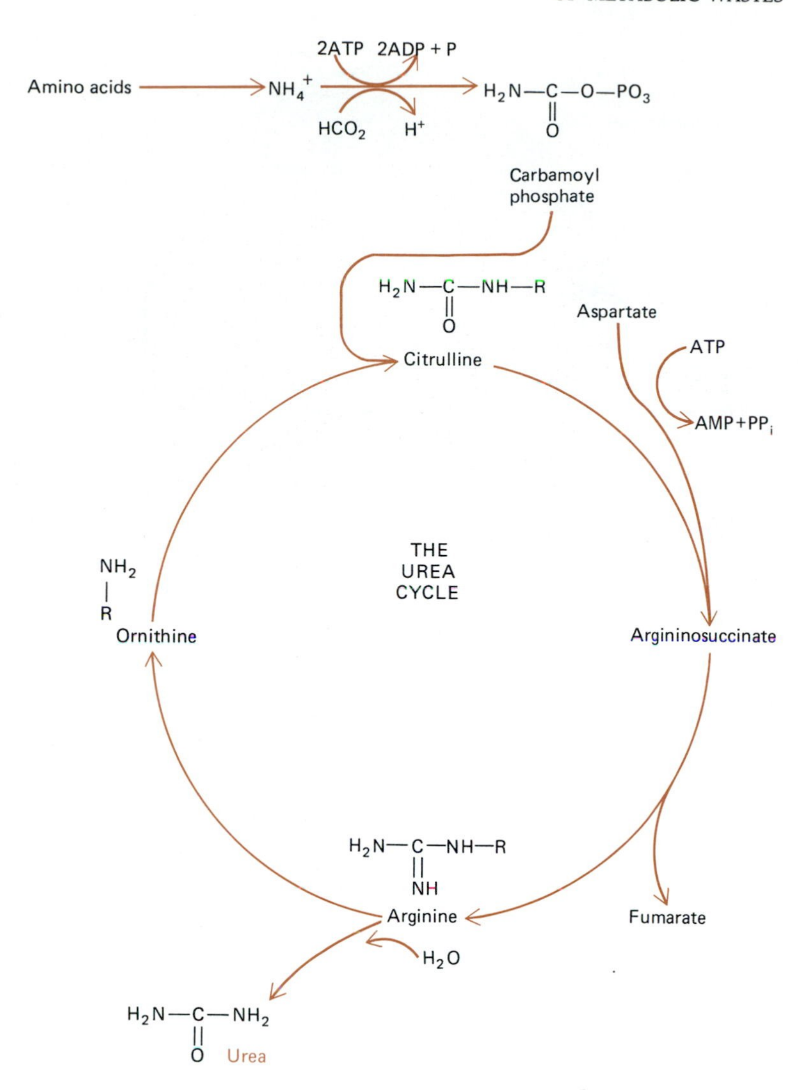

Figure 38–3 The urea cycle is a sequence of enzymatic reactions by which the amino groups from amino acids are converted to urea. Carbon dioxide and ammonia are combined to form carbamoyl phosphate, which then combines with ornithine to form citrulline. Aspartate (the ionic form of the amino acid aspartic acid) then reacts with the citrulline to form the intermediate argininosuccinate. This intermediate compound is cleaved to yield arginine and fumarate. Thus, the amino group of aspartate is transferred to arginine. The arginine is hydrolyzed by the enzyme arginase to yield urea and ornithine, which can be used in the next cycle. The energy to drive the cycle and synthesize urea is provided by the two ATPs used in the synthesis of carbamoyl phosphate and the ATP used in the synthesis of argininosuccinate.

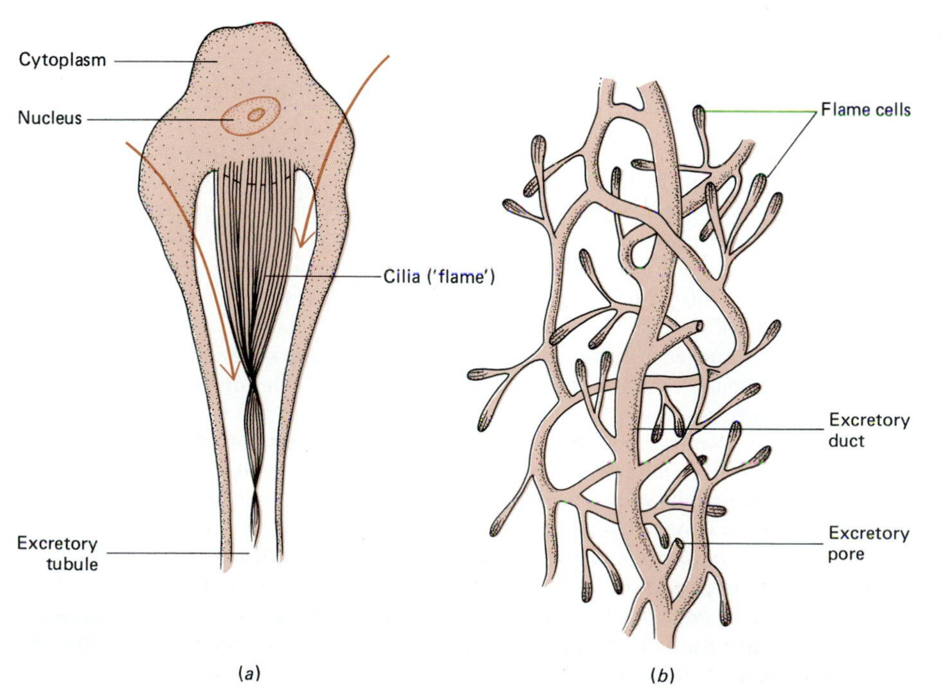

Figure 38–4 The excretory organs of the flatworm. (*a*) A single flame cell. (*b*) Flame cells are connected by excretory ducts to excretory pores that open to the outside of the body.

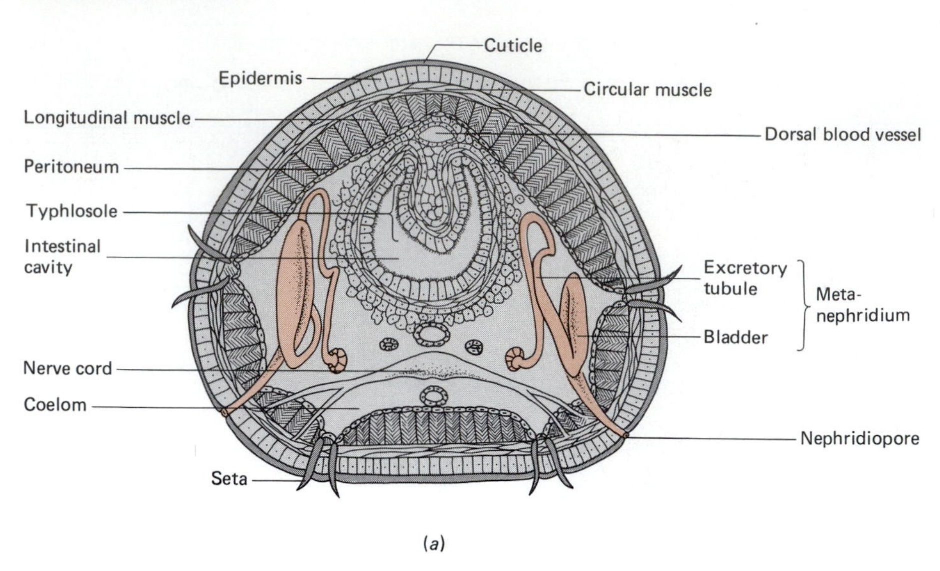

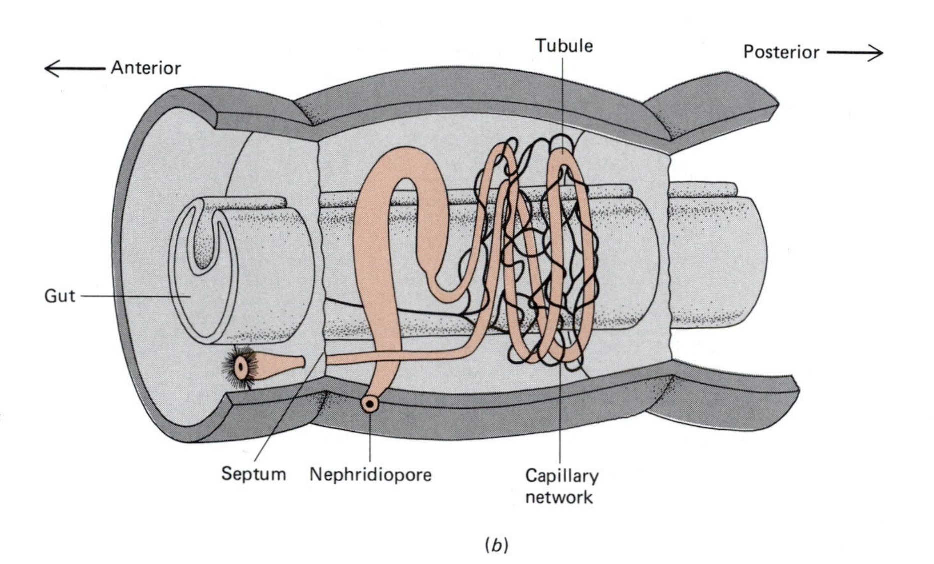

Figure 38–5 The excretory structures of the earthworm are a series of paired metanephridia. Each consists of a ciliated funnel opening into the coelom, a coiled tubule, and a pore opening to the outside.

excretory ducts. The beating of the cilia (which suggests a flickering flame when seen under the microscope) presumably moves fluid in the ducts out toward the excretory pores.

The chief role of the flame cells is to regulate the water content. In fact, the number of protonephridia in a planarian is adjusted to the salinity of the environment. Planaria grown in slightly salty water develop fewer flame cells, but the number quickly increases if the concentration of salt in the environment is reduced.

Earthworms have nephridial organs called **metanephridia** in each segment of their bodies. In contrast to the flame cells of flatworms, the metanephridium is a tubule open at both ends. The inner end opens into the coelom as a ciliated funnel (Fig. 38–5). Around each tubule is a network of capillaries, which permits the removal of wastes from the blood. Fluid from the coelom passes into the metanephridium. As this fluid passes through the long, looped tubule, its composition is adjusted. Water and substances such as glucose and salts are reabsorbed into the coelom and ultimately returned to the blood. Wastes are concentrated and pass out of the body as part of the very dilute, copious urine. The earthworm excretes urine at a rate of about 60% of its total body weight each day.

Molluskan nephridial organs are sometimes called "kidneys." Coelomic fluid enters the metanephridium through the ciliated nephrostome. As it passes along the metanephridium, some substances are selectively reabsorbed and others are secreted into the fluid. In aquatic forms, the resulting urine is expelled into the mantle cavity.

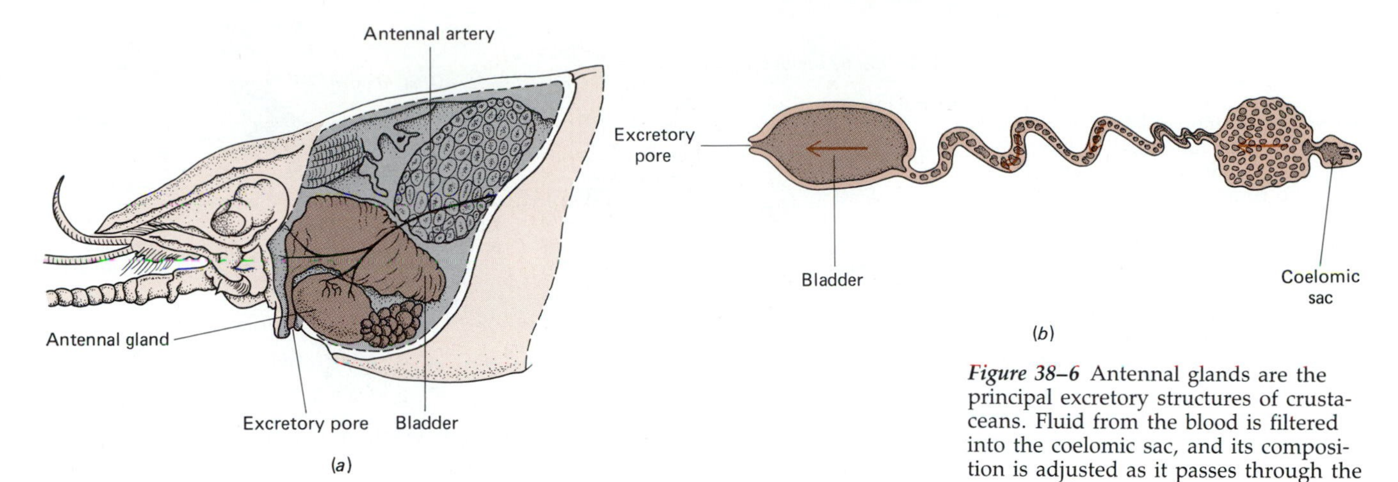

Figure 38–6 Antennal glands are the principal excretory structures of crustaceans. Fluid from the blood is filtered into the coelomic sac, and its composition is adjusted as it passes through the excretory ducts. The exit duct may be enlarged to form a bladder.

ANTENNAL GLANDS

Antennal glands, also called **green glands,** are the principal excretory organs of crustaceans. A pair of these structures is located in the head, typically at the base of the antennae. Each gland consists of a coelomic sac, a greenish glandular chamber with folded walls, an excretory tubule, and an exit duct that is enlarged to form a bladder in some species (Fig. 38–6). Fluid from the blood is filtered into the coelomic sac, and its composition is adjusted as it passes through the excretory organ. Needed materials are reabsorbed into the blood. Wastes can also be actively secreted from the blood into the filtrate within the excretory organ.

MALPIGHIAN TUBULES

The excretory system of insects consists of organs called **Malpighian tubules** (Fig. 38–7). There may be from two to several hundred tubules, depending upon the species. Malpighian tubules have blind ends that lie in the hemocoel, bathed in blood. Their cells transfer wastes by diffusion or active transport from the blood to the cavity of the tubule. Each tubule has a muscular wall, and its slow writhing movements assist the passage of wastes down its lumen to the gut cavity. The Malpighian tubules empty into the intestine between the midgut and the hindgut. Water is reabsorbed into the hemocoel both from the tubule and from the digestive tract. The major waste product, uric acid, precipitates as the water is reabsorbed, and is excreted as fairly dry pellets. This adaptation helps to conserve the insect's body fluids.

Osmoregulation and Metabolic Waste Disposal in Vertebrates

In most vertebrates not only the urinary system but also the skin, lungs or gills, and digestive system function to some extent in metabolic waste disposal and fluid balance. In addition, some vertebrates have special salt glands, which excrete salt.

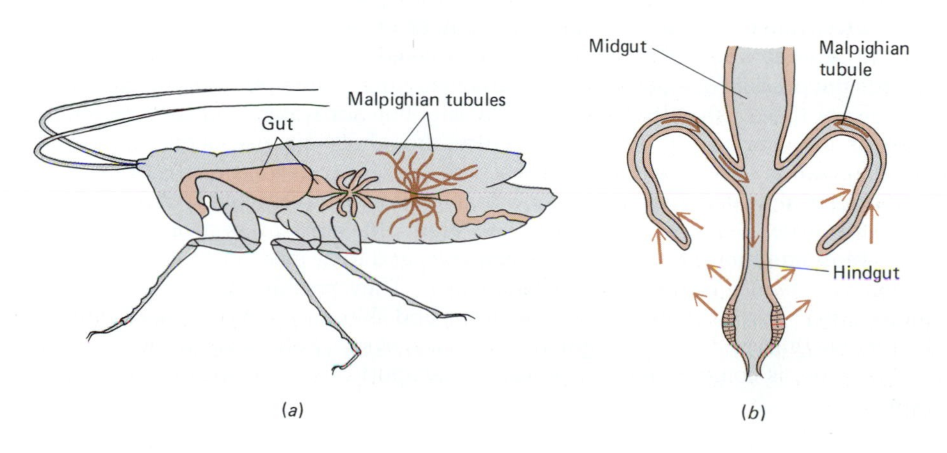

Figure 38–7 The slender Malpighian tubules of insects have blind ends that lie in the hemocoel. Their cells transfer wastes from the blood to the cavity of the tubule. Uric acid, the major waste product, is discharged into the gut.

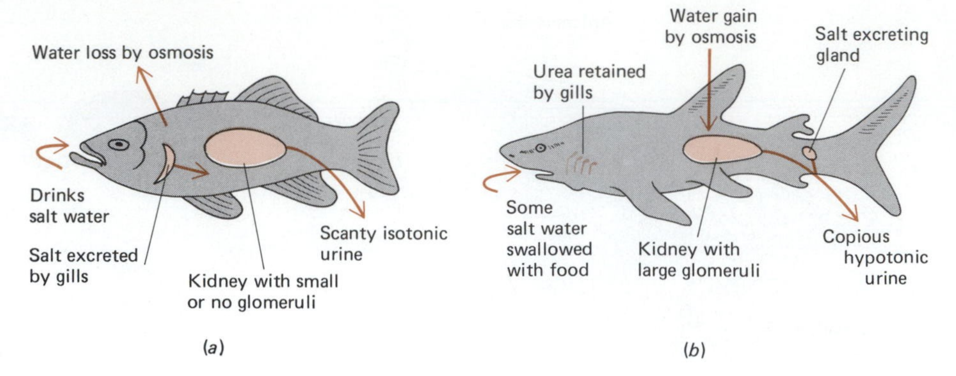

Figure 38–8 Osmoregulation in marine fish and sharks. (*a*) A marine fish in a hypertonic medium loses water by osmosis. To compensate it drinks salt water, excretes the salt, and produces very little urine. (*b*) The shark accumulates urea in high enough concentration to become hypertonic to the surrounding medium. As a result some water enters the body by osmosis. A large quantity of hypotonic urine is excreted.

By removing water, salts, and potentially toxic waste materials, these organs help to maintain homeostasis.

THE VERTEBRATE KIDNEY

The main excretory organs in vertebrates are **kidneys,** which excrete nitrogenous wastes, water, a variety of salts, and some other substances. Together with the urinary bladder and associated ducts they form the **urinary system.**

The vertebrate kidney is composed of functional units called **nephrons** which are described in a later section. Most vertebrate kidneys function by filtration and reabsorption. Blood is filtered, and the initial filtrate that enters the nephron contains all of the substances present in the blood except large compounds such as proteins. (Blood cells and platelets are, of course, too large to be filtered out of the blood.) As the filtrate passes through the coiled tubules of the nephron, needed materials such as glucose, amino acids, salts, and water are selectively reabsorbed into the blood. Thus the composition of the filtrate is slowly adjusted, and the urine that is finally excreted consists of metabolic waste products and materials such as salts.

OSMOREGULATION

Aquatic vertebrates have a continuous problem of osmoregulation. The marine bony fishes have blood and body fluids that are hypotonic to the sea water. They tend to lose water osmotically and, like Coleridge's "Ancient Mariner," are in danger of becoming dehydrated even though surrounded by water. Many bony fish compensate by drinking sea water constantly. They retain the water and excrete salt by the action of specialized cells in their gills. Only a small volume of urine is produced, and the kidneys of such fish are generally very small (Fig. 38–8).

Marine chondrichthyes (sharks and rays) solve their osmotic problem in a different way (Fig. 38–8). They accumulate urea and retain it in a concentration high enough to render their blood and interstitial fluid slightly hypertonic to sea water. Some water enters the body osmotically through the gills. Their tissues are adapted to function at concentrations of urea that would be toxic to most other animals. Such animals are able to excrete a hypotonic urine despite the high salinity of their environment. Excess sodium is excreted both by the kidneys and by a special rectal gland in the posterior region of the intestine.

Freshwater fish are in continuous danger of drowning due to the osmotic inward flow of water. Although they are covered with a mucous secretion that retards the passage of water into the body, water does enter through the gills. The kidneys of these fish selectively reabsorb salts but not water, and they excrete a copious, dilute urine (Fig. 38–9). Water entry, though, is only part of the problem of osmoregulation in freshwater fish. These animals also tend to lose salts to the surrounding fresh water. To compensate, special cells in the gills actively transport salt from the water into the body. Salts also enter the body as part of their food.

Most amphibians are at least semiaquatic, and their mechanisms of osmoregulation are similar to those of freshwater fish. They produce a large amount of dilute urine. A frog can lose through its urine and skin an amount of water equivalent to one third of its body weight in a day. Loss of salt, both through the skin and in the urine, is compensated for by active transport of salt by special cells in the skin.

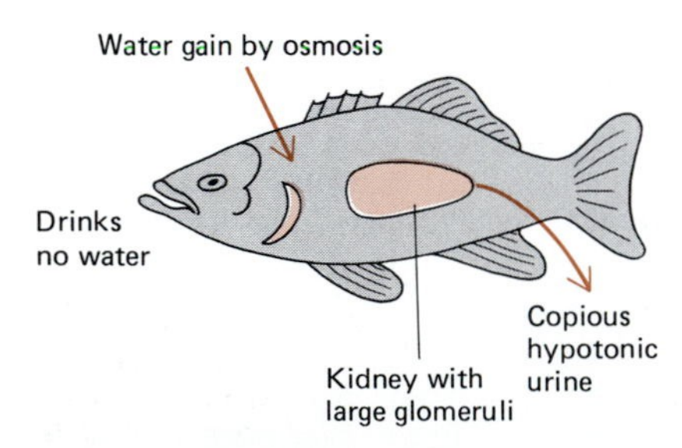

Figure 38–9 Freshwater fish live in a hypotonic medium, and water continuously enters the body by osmosis. They excrete a large quantity of dilute urine.

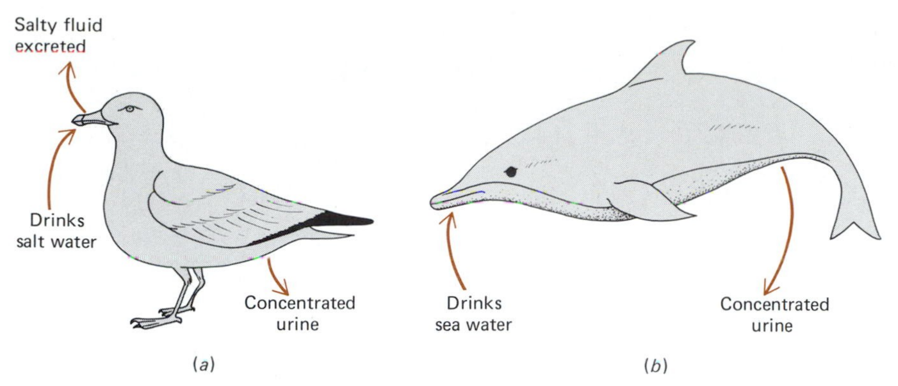

Figure 38–10 Osmoregulation in marine birds and mammals. (*a*) Marine birds have salt glands in the head that can excrete the excess salt that enters the body with the sea water they drink. They excrete a concentrated urine. (*b*) Marine mammals drink sea water along with their food. Their kidneys produce a very salty urine.

Certain reptiles and marine birds have salt glands in the head that can excrete the salt that enters the body with the sea water they drink (Fig. 38–10). Salt glands are usually inactive and only begin to function in response to osmotic stress. Thus, when sea water or salty food is ingested, the salt glands excrete a fluid laden with sodium and chloride.

Whales, dolphins, and other marine mammals ingest sea water along with their food. Their kidneys produce a very concentrated urine, much more salty than sea water (Fig. 38–10). This is an important physiological adaptation, especially for marine carnivores. The high-protein diet of these animals results in production of large amounts of urea, which must be excreted in the urine.

Fluid Balance and Excretion in Humans

In human beings, as in many other vertebrates, the kidneys, skin, lungs, and digestive system all function to some extent in waste disposal (Fig. 38–11). The lungs excrete water and carbon dioxide. The liver excretes bile pigments (the breakdown products of hemoglobin), which pass into the intestine and then pass out of the body with the feces. Though primarily concerned with the regulation of body temperature, the sweat glands also excrete 5% to 10% of all metabolic wastes. Sweat contains the same substances (salts, urea, and water) as those in urine but is much more dilute, having only about an eighth as much solid matter. The volume of

Figure 38–11 (*a*) In terrestrial mammals, the kidney conserves water by reabsorbing it. (*b*) Disposal of metabolic wastes in humans and other terrestrial mammals. In order to conserve water, a small amount of hypertonic urine is usually produced. Nitrogenous wastes are produced by the liver and transported to the kidneys. All cells produce carbon dioxide and some water during cellular respiration. Kidneys, lungs, skin, and digestive system all participate in disposal of metabolic wastes.

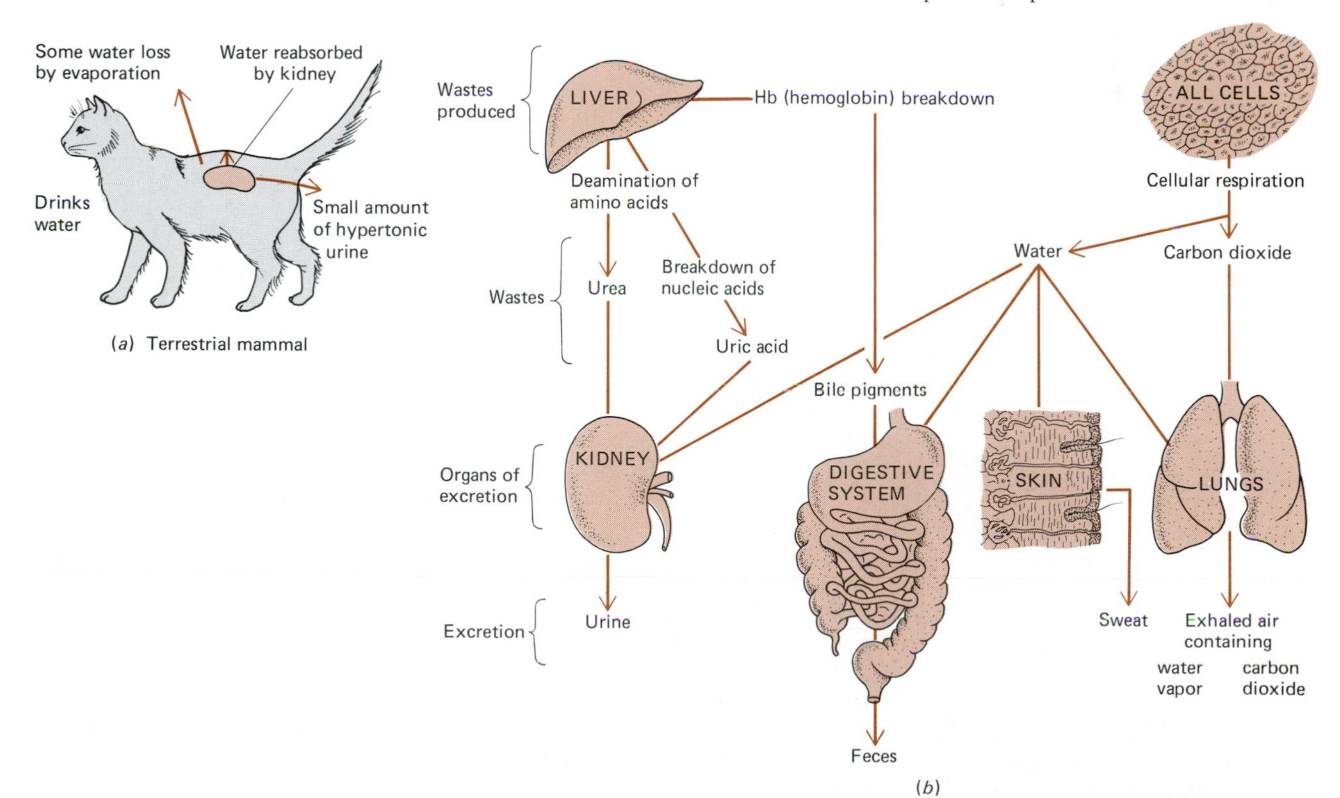

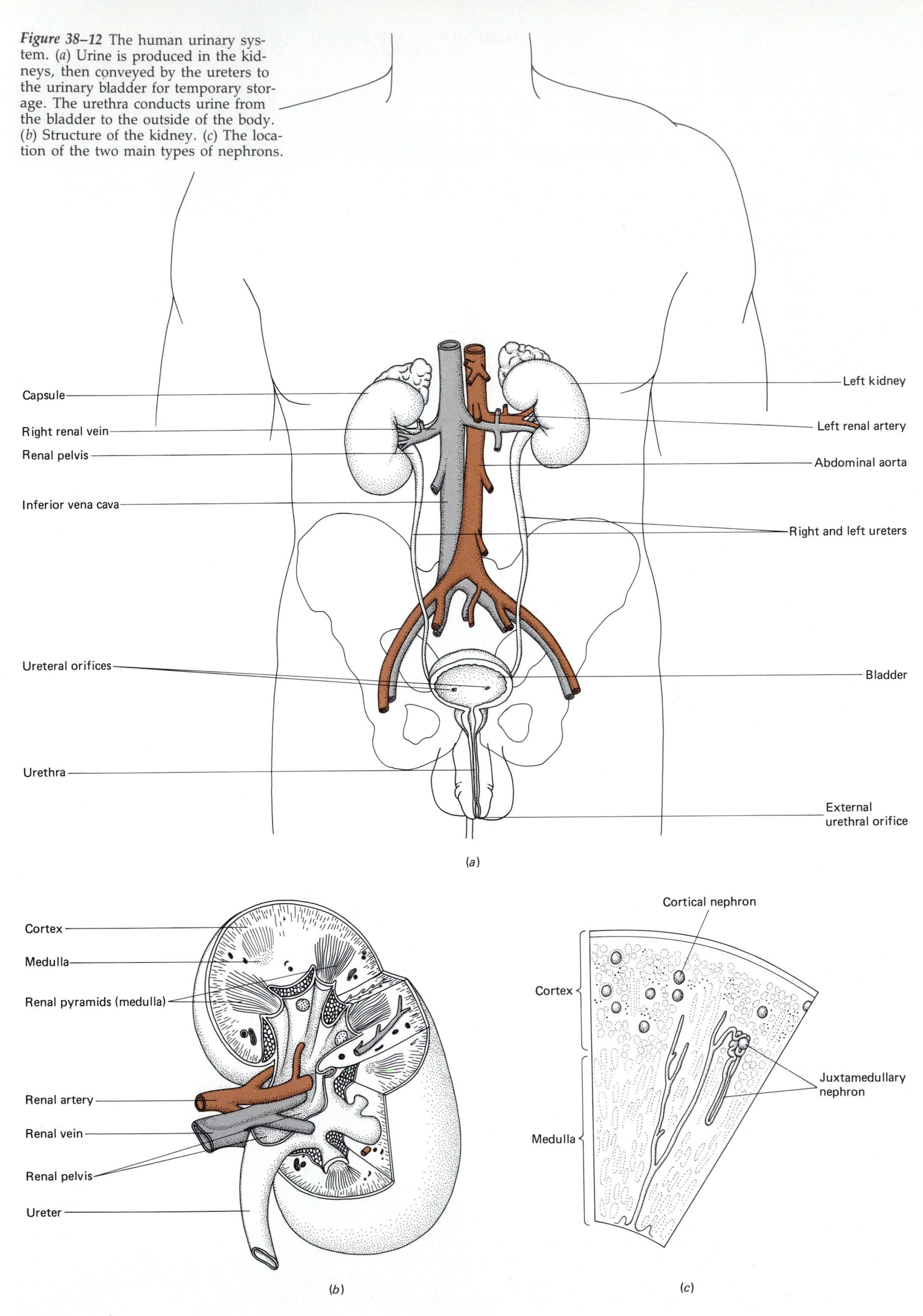

Figure 38–12 The human urinary system. (*a*) Urine is produced in the kidneys, then conveyed by the ureters to the urinary bladder for temporary storage. The urethra conducts urine from the bladder to the outside of the body. (*b*) Structure of the kidney. (*c*) The location of the two main types of nephrons.

perspiration varies, ranging from about 500 ml on a cool day to as much as 2 or 3 liters on a hot one. While doing hard work at high temperatures, a human may excrete from 3 to 4 liters of sweat in an hour!

THE HUMAN KIDNEY AND ITS DUCTS

The human kidneys are paired, bean-shaped structures about 10 cm long, each about the overall size of a fist. They are located just below the diaphragm in the abdominal cavity (Fig. 38–12). The outer portion of the kidney is called the **cortex;** the inner portion is the **medulla.** On the medial concave side of each kidney is a funnel-shaped chamber called the **renal pelvis.**

The kidneys produce urine from the water, salts, wastes, and other materials that they filter from the blood. By adjusting the amount of water, salts, and other materials excreted, the kidneys help to maintain the internal chemical balance of the body. Indeed, they are indispensable homeostatic organs.

The urine, excreted by the kidney in a continuous trickle, collects in the renal pelvis. From the pelvis, urine passes through one of the paired **ureters,** moved along by peristaltic contractions. Each ureter, about 25 cm long, connects its kidney with the **urinary bladder,** a hollow, muscular organ located in the lower, anterior portion of the pelvic cavity. The smooth muscle and special transitional epithelium of the urinary bladder enable it to stretch to accommodate up to 800 ml (about a pint and a half) of urine. As the volume of urine in the bladder increases, the distension of its muscular walls stimulates nerve endings to send impulses to the brain, producing the sensation of fullness. Impulses may be sent back to the bladder, causing **micturition** (urination), the expulsion of urine from the bladder.

The **urethra** is a duct leading from the urinary bladder to the outside of the body. In the female, the urethra is short and transports only urine. In the male, however, the urethra is lengthy and passes through the penis. Semen as well as urine is transported through the male urethra (although at different times). The length of the male urethra discourages bacterial invasion of the bladder, so that such infections are less common in males than in females.

Bladder control depends upon the learned ability to facilitate or inhibit the reflex action that causes micturition. Humans can learn to empty the bladder voluntarily at a convenient time even before it is full. Similarly, micturition can be inhibited for some time even after the bladder is full. Such voluntary control cannot be exerted by an immature nervous system, and most babies are unable to develop complete urinary control until about 2 years of age.

THE NEPHRON

Each kidney contains more than a million functional units called nephrons that regulate the composition of the blood and excrete wastes. A nephron consists of two main parts: a **renal corpuscle** and a **renal tubule** (Fig. 38–13). The renal corpuscle is made up of a double-walled, hollow cup of cells, called **Bowman's capsule,** and a spherical tuft of capillaries, the **glomerulus,** which projects into the capsule. The renal tubule consists of three main regions: the **proximal** (near) **convoluted tubule, loop of Henle,** and the **distal** (far) **convoluted tubule.** Each distal convoluted tubule delivers its content into a **collecting duct.**

The inner wall of Bowman's capsule consists of specialized epithelial cells called **podocytes.** The podocytes possess elongated foot processes, which cover the surfaces of most of the glomerular capillaries (Fig. 38–13). Spaces between the "toes" of these foot processes are called **slit pores.**

URINE FORMATION

Urine is produced by a combination of three processes: filtration, reabsorption, and tubular secretion.

Filtration

Filtration occurs at the junction between the glomerular capillaries and the wall of Bowman's capsule (Fig. 38–14). Blood is delivered to the kidneys by the renal arteries. Branches of the renal arteries give rise to afferent arterioles. (An afferent arteriole conducts blood into a structure; an efferent arteriole conducts blood out of a structure.) In the kidney, an **afferent arteriole** delivers blood into the capillaries of

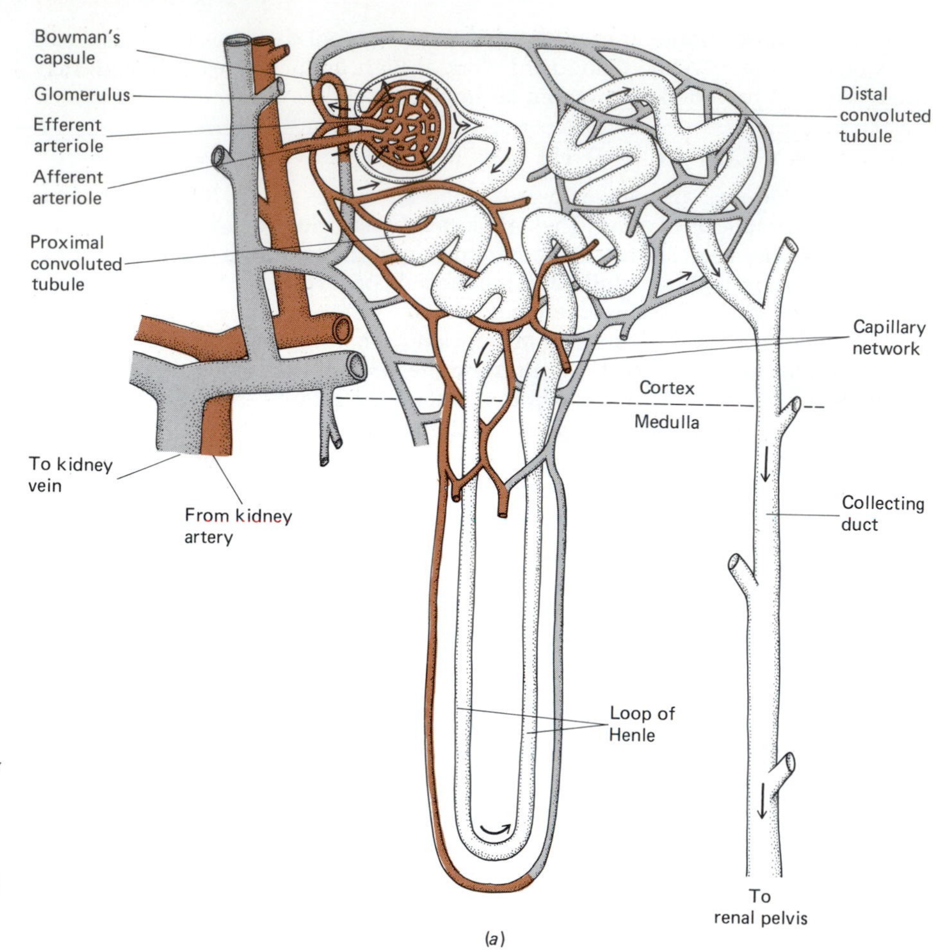

Figure 38–13 Each kidney is composed of more than a million microscopic nephrons. (*a*) Highly diagrammatic view of the basic structure of a nephron. (*b*) Detailed view of a renal corpuscle. Note that the distal convoluted tubule is actually adjacent to the afferent arteriole. (*c*) Photomicrograph of a renal corpuscle.

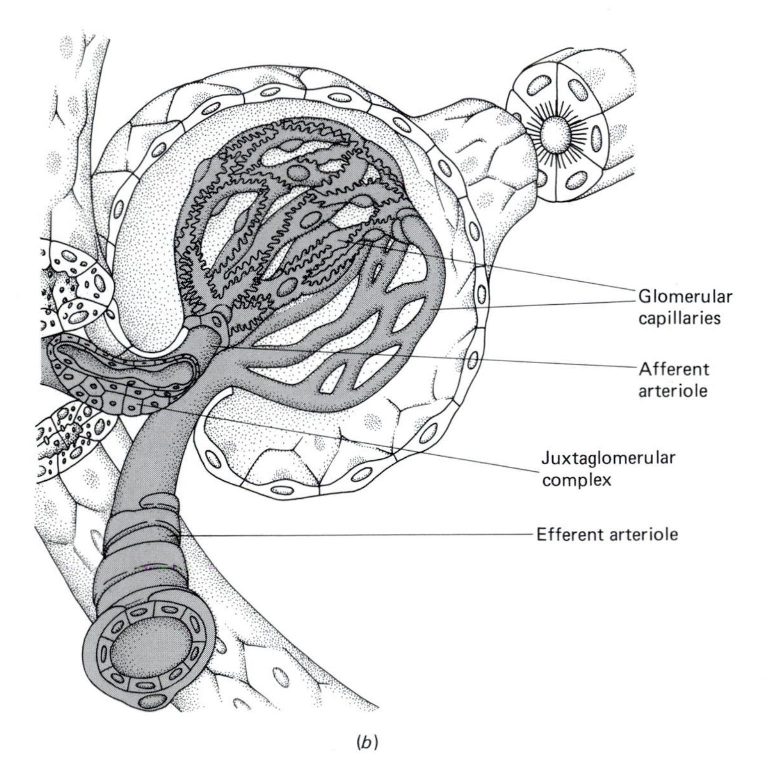

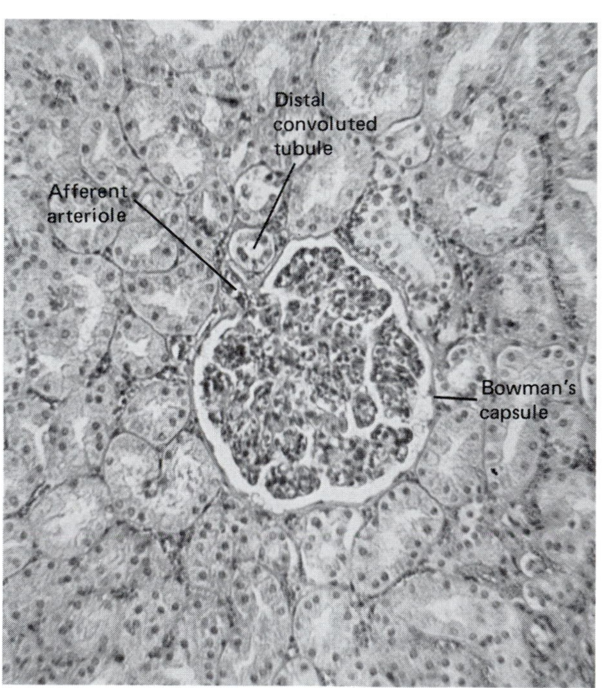

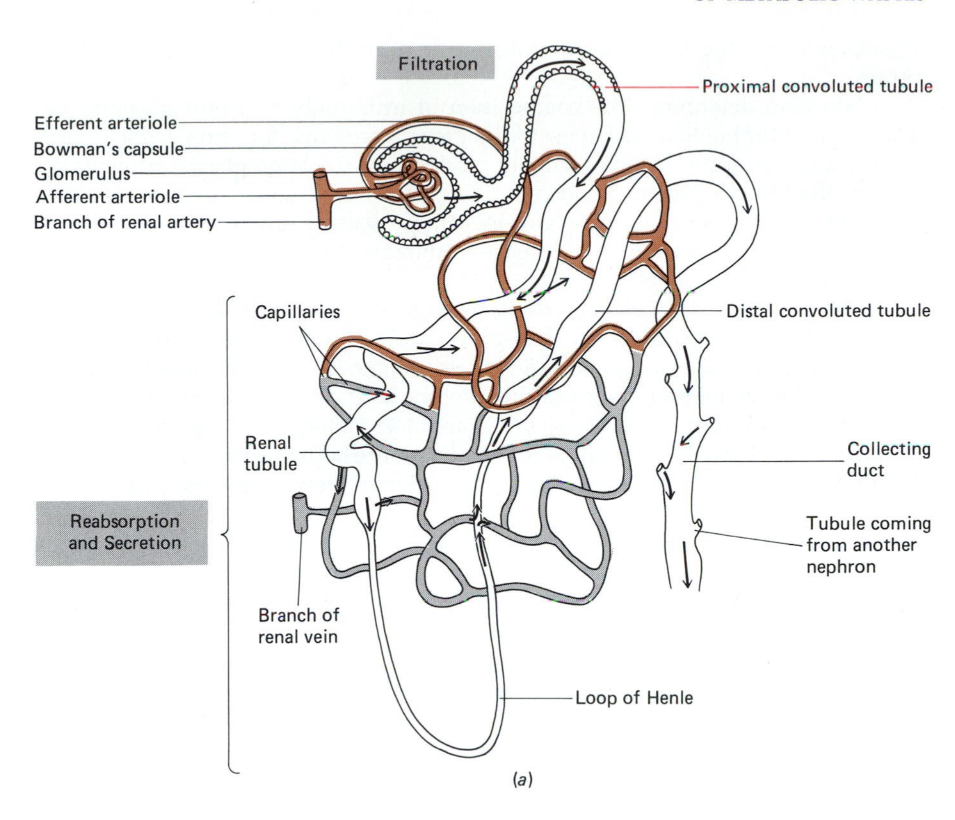

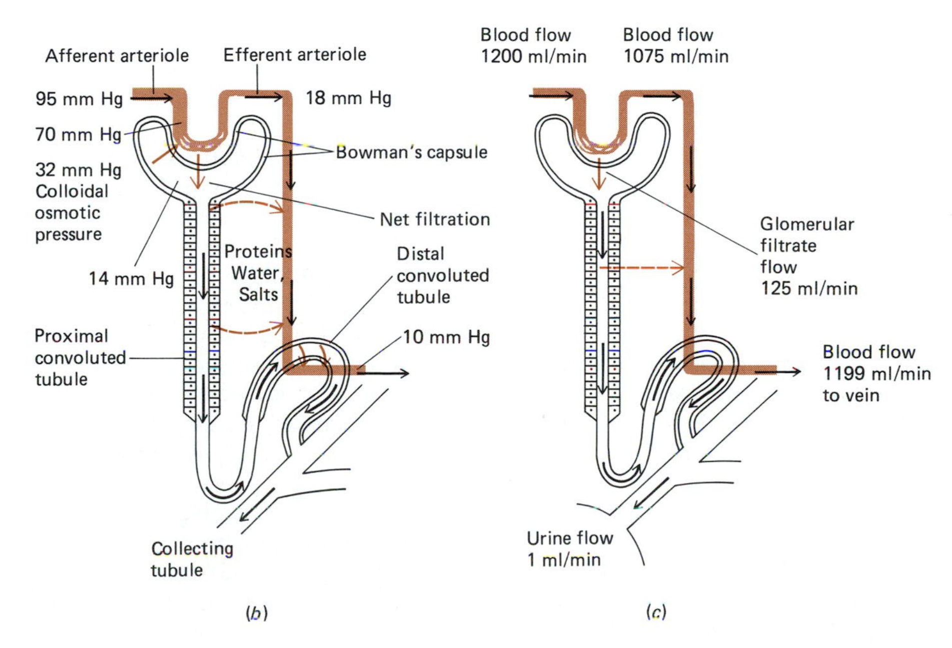

Figure 38–14 Nephron function. (*a*) Diagrammatic view of a nephron showing where filtration, reabsorption, and secretion take place. The dashed, color arrows indicate reabsorption of water, glucose, salts, and other substances in the capillaries. (*b*) Diagram of a nephron showing the pressure gradients that move fluids from blood to glomerular filtrate (filtration pressure). Substances are reabsorbed, largely in the proximal convoluted tubules, by processes of active transport. (*c*) The total fluid movement in all of the nephrons.

the glomerulus. The blood then drains from the glomerulus by a much narrower **efferent arteriole.** Constriction of the efferent arteriole produces high hydrostatic pressure in the glomerulus. This hydrostatic pressure forces fluid to leave the capillaries of the glomerulus. The fluid passes through the slit pores and enters the urinary pipeline. Once it enters Bowman's capsule, this fluid is referred to as the **glomerular filtrate.**

The amount of fluid that enters Bowman's capsule depends on the **effective filtration pressure,** the combination of mechanical and osmotic forces that determines filtration. The main force *favoring* filtration is the hydrostatic pressure of the blood in the glomerulus. Most of the *resistance* to this push results from (1) the resistance of the capillary wall and the wall of Bowman's capsule to the passage of material, (2) the hydrostatic pressure of the fluid already in the lumen of Bowman's capsule, and (3) the difference in osmotic pressure between the blood and the filtrate.

Filtration determines the composition of urine only to a minor extent, because, except for holding back most of the plasma proteins, filtration is not a selective process. Except for a small amount of albumin, the large plasma proteins remain in the blood with the blood cells and platelets. However, smaller materials dissolved in the plasma—such as glucose, amino acids, sodium, potassium, chloride, bicarbonate, other salts, and urea—are filtered out of the blood and become part of the filtrate.

The total volume of blood passing through the kidneys is about 1200 ml per minute, or about one fourth of the entire cardiac output. The plasma passing through the glomerulus loses about 20% of its volume to the glomerular filtrate; the rest leaves the glomerulus through the efferent arteriole. The normal glomerular filtration rate amounts to about 180 liters (about 45 gallons) each 24 hours. This is four-and-a-half times the amount of fluid in the entire body! Common sense tells us that urine could not be excreted at that rate. Within a few moments, dehydration would become a life-threatening problem.

Reabsorption

The threat to homeostasis posed by the vast amounts of fluid filtered by the kidneys is avoided by **reabsorption.** About 99% of the filtrate is reabsorbed into the blood through the renal tubules, leaving only about 1.5 liters to be excreted as urine. Reabsorption permits precise regulation of blood chemistry by the kidneys. Needed substances such as glucose and amino acids are returned to the blood, while wastes and excess salts and other materials remain in the filtrate and are excreted in the urine. Each day the tubules reabsorb more than 178 liters of water, 1200 g (2.5 lb) of salt, and about 250 g (0.5 lb) of glucose. Most of this is, of course, reabsorbed many times over.

After leaving the glomerulus the efferent arteriole delivers blood to a second capillary network in the nephron. This capillary network surrounds the renal tubule and receives materials returned to the blood by the tubule. Blood from these capillaries then flows into small veins, which eventually merge to form the large renal vein draining each kidney.

The simple epithelial cells lining the renal tubule are well adapted for reabsorbing materials. They have abundant microvilli, which increase the surface area for reabsorption (and give the inner lining a "brush border" appearance; see Fig. 38–15). These cells also contain large numbers of mitochondria needed to provide the energy for running the cellular pumps that actively transport materials.

About 65% of the filtrate is reabsorbed as it passes through the proximal convoluted tubule. Glucose, amino acids, vitamins, and other substances of nutritional value are reabsorbed there, as are many ions, including sodium, chloride, bicarbonate, and potassium. Some of these ions are actively transported; others follow by diffusion. Reabsorption continues as the filtrate passes through the loop of Henle and the distal convoluted tubule. Then the filtrate is further concentrated as it passes through the collecting duct leading to the renal pelvis.

Normally, substances that are useful to the body such as glucose or amino acids are reabsorbed from the renal tubules. If the concentration of a particular substance in the blood is high, however, the tubules may not be able to reabsorb all of it. The maximum rate at which a substance can be reabsorbed is called its **tubular transport maximum,** or **Tm.** For example, the Tm for glucose averages 320 mg/min for an adult human being. Normally, the tubular load of glucose is only about 125 mg/min, so almost all of it is reabsorbed. However, if there is an excess above the Tm, that excess will not be reabsorbed but will instead pass into the urine.

Each substance that has a Tm also has a **renal threshold** concentration in the plasma. When a substance exceeds its renal threshold, the portion not reabsorbed is excreted in the urine. In a person with diabetes mellitus, the concentration of glucose in the blood exceeds its threshold level (about 150 mg glucose per 100 ml of blood), so glucose is excreted in the urine. Its presence in the urine is evidence of this disorder.

Secretion

Some substances, especially potassium, hydrogen, and ammonium ions, are **secreted** from the blood into the filtrate. Some drugs, such as penicillin, are also

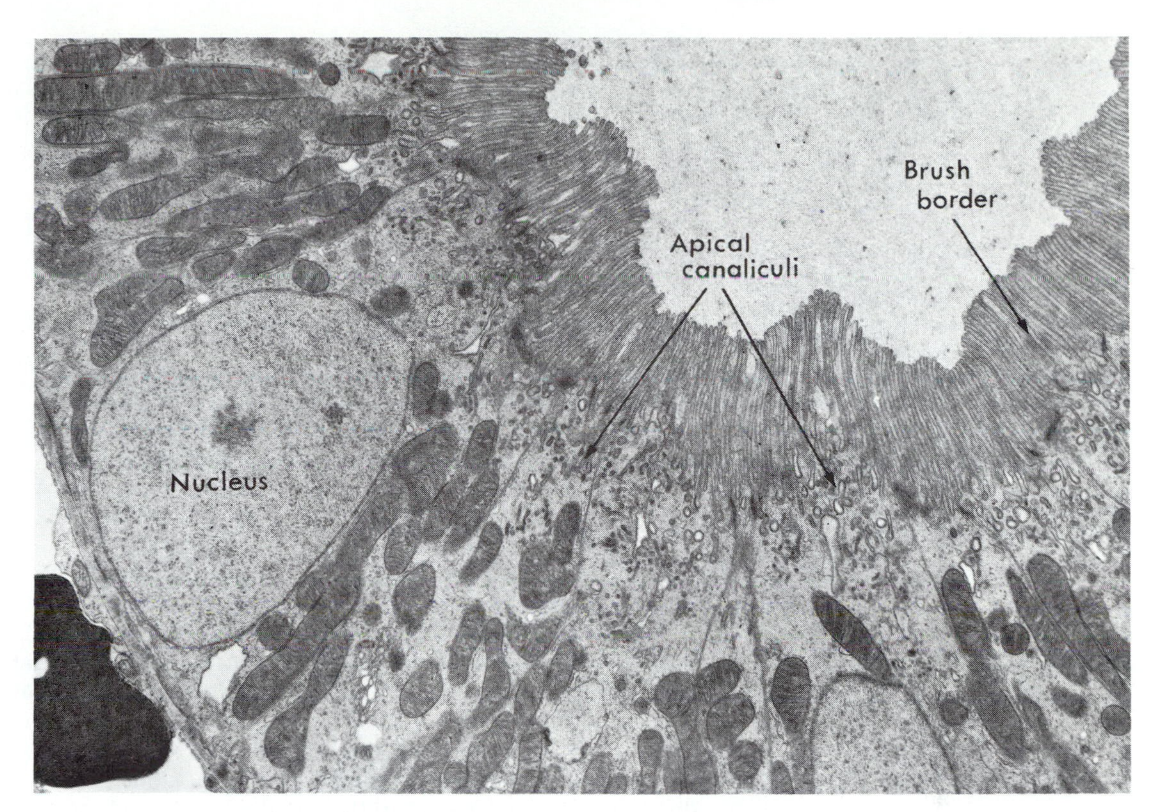

Figure 38–15 An electron micrograph of the cells of the proximal tubule of the kidney showing the brush border of the lumen (approximately ×6000). (From W. Bloom and D. W. Fawcett.)

removed from the blood by secretion. Secretion occurs mainly in the region of the distal convoluted tubule.

Secretion of hydrogen ions is an important homeostatic mechanism for regulating the pH of the blood. When the blood becomes too acidic, more hydrogen ions are secreted into the filtrate. The secretion of potassium is also very important. Excess potassium ions in the blood stimulate the adrenal cortex to secrete aldosterone. This hormone speeds the sodium pumps, which are also thought to function as potassium pumps. However, potassium is pumped into the tubule, whereas sodium is reabsorbed. This mechanism helps prevent the accumulation of excess potassium in the blood, which can cause cardiac arrhythmias and even cardiac arrest.

Countercurrent Exchange

There are two types of nephrons: the cortical nephrons and the more internal juxtamedullary nephrons. In the **cortical nephrons,** sodium is transported from the filtrate to the interstitial fluid, making that fluid slightly hypertonic to the filtrate. As a result, water flows out of the tubule into the interstitial fluid by osmosis. Water, salts, and other materials in the interstitial fluid then enter the capillary network surrounding the renal tubule and leave the kidney through the blood. The cortical nephrons do not produce urine that is significantly hypertonic to blood since they reabsorb both salt *and* water. They do, however, concentrate wastes in the fluid that remains, since wastes are reabsorbed, only at a low rate.

The **juxtamedullary nephrons** have a very long loop of Henle that extends deep into the medulla. In them the loop of Henle is specialized to produce a highly hypertonic interstitial fluid near its apex. The juxtamedullary loop of Henle consists of a descending loop, into which filtrate flows, and an ascending loop, through which the remaining filtrate passes on its way to the distal convoluted tubule. The ascending loop is impermeable to water but highly permeable to sodium. When the sodium concentration becomes greater outside the renal tubule than inside, water moves out of the filtrate in the collecting duct by osmosis.

Much of the sodium that leaves the ascending loop passes into the descending loop (Fig. 38–16). This causes a great increase in sodium concentration near the apex. However, since sodium ions are removed in the ascending loop, by the time the filtrate flows through the distal convoluted tubule, it is isotonic to blood or even hypotonic to it. This mechanism of action is called **countercurrent exchange.**

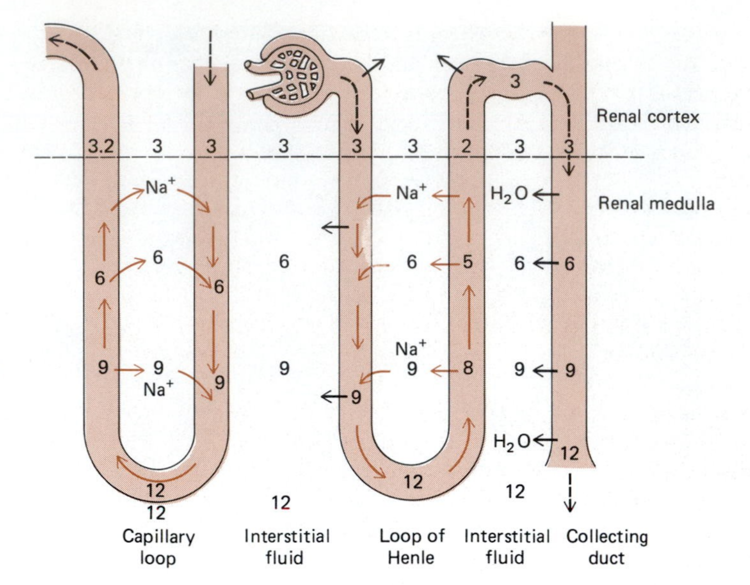

Figure 38–16 Countercurrent exchange: The general direction of sodium and fluid movement. The direction of movement of the fluid is indicated by dashed black arrows. Sodium movement is indicated by colored arrows, and water movement by solid black arrows. The numerals refer to the relative concentrations of osmotically active solutes. When two zeros are added, they refer to the concentration of solutes in milliosmoles per liter.

Countercurrent exchange establishes a very hypertonic interstitial fluid near the renal pelvis that draws water osmotically from the filtrate in the collecting ducts. The collecting ducts are so routed as to pass through this zone on their way to the renal pelvis. This loss of water from the duct's contents concentrates the urine to such an extent that it can be hypertonic to blood. *A hypertonic urine has a low concentration of water, and so conserves water.* The urine becomes most hypertonic in thirst (in fact, thirst is a signal that the fluid content of the blood is low). In thirst, the permeability of the collecting duct walls is very high.

The water that diffuses from the filtrate into the interstitial fluid is removed by blood vessels known as the **vasa recta** and carried off in the venous drainage of the kidney. The vasa recta are long looped extensions of the efferent arterioles of the juxtamedullary nephrons. They extend deep into the medulla, only to return hairpin fashion to the cortical venous drainage of the kidney.

COMPOSITION OF URINE

By the time the filtrate reaches the renal pelvis, its composition has been precisely adjusted. Useful materials have been returned to the blood by reabsorption. Wastes and unneeded materials that became part of the filtrate by filtration or secretion have been retained by the renal tubules. The adjusted filtrate, called **urine,** is composed of about 96% water, 2.5% nitrogenous wastes (mainly urea), 1.5% salts, and traces of other substances.

The composition of the urine yields many clues to body function or malfunction. **Urinalysis,** the physical, chemical, and microscopic examination of urine, is a very important diagnostic tool and is used to monitor many disorders such as diabetes mellitus.

FLUID HOMEOSTASIS

If salt and water intake were always the same, and if an organism were always exposed to the same conditions (such as heat and humidity) that influence water loss, the kidneys would not have to vary their output. However, both the ingestion of water and its loss vary greatly from time to time, so that in order to maintain homeostasis, the kidney output must be appropriately regulated.

Regulation of Urine Volume

As we have seen, most water reabsorption takes place in the proximal convoluted tubule, and the rate of this absorption does not vary. Similarly, the glomerular filtration rate is usually relatively constant. What does vary is the rate of reabsorption of the collecting ducts. The permeability of the collecting ducts to water is regulated by **antidiuretic hormone (ADH),** a hormone synthesized in the hypothalamus and released by the posterior lobe of the pituitary gland. ADH increases the permeability of the collecting ducts so that more water is reabsorbed.

When fluid intake is low, the body begins to become dehydrated (Fig.38–17). The concentration of salts dissolved in the blood becomes greater, causing an increase in osmotic pressure of the blood. Specialized receptors (osmoreceptors) in the brain and in large blood vessels are sensitive to such change. The posterior lobe of the pituitary responds by releasing increased amounts of ADH. As a result, the walls of the collecting ducts become more permeable and more water is reabsorbed. In this way, more water is conserved for the body, blood volume increases, and conditions are restored to normal. Thus, the more ADH is secreted, the less water is lost from the body. ADH promotes a small, concentrated urine volume.

On the other hand, intake of large amounts of fluid results in dilution of the blood and a fall in osmotic pressure. Release of ADH decreases, lessening the amount of water reabsorbed from the collecting ducts. A large amount of dilute urine is then produced.

ADH release is increased during sleep, and is reduced by some diuretic agents, such as alcoholic beverages, that increase urine volume. (Caffeine is also a diuretic but works by a different mechanism.) In the disorder **diabetes insipidius,** the body cannot make sufficient ADH, or the kidneys lack receptors for ADH. Large volumes of urine are produced, and the sufferer must drink immense quantities of water to compensate for this fluid loss. Diabetes insipidus can now be treated with injections of ADH or with an ADH nasal spray. (Diabetes insipidus should not be confused with diabetes mellitus, a disorder of carbohydrate metabolism; see Chapter 42. Diabetes mellitus also increases urine flow, and with some thought you can explain why. The clue is that the increased glucose content of the urine increases its tonicity.)

Regulation of Sodium Reabsorption

Fluid balance is linked closely with sodium balance, because sodium is the most abundant extracellular ion, accounting for about 90% of all positive ions outside the cells. Recall that when salt concentration increases, water is drawn to the region

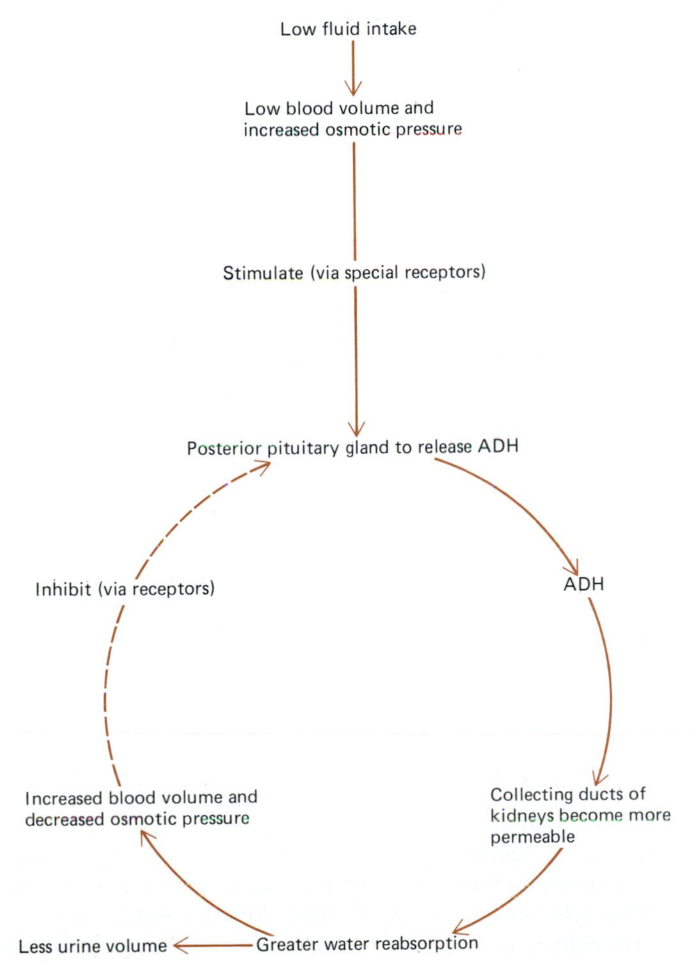

Figure 38–17 Regulation of urine volume. When fluid intake is low, blood volume decreases and osmotic pressure increases. The posterior lobe of the pituitary gland releases the hormone ADH, which stimulates increased water reabsorption.

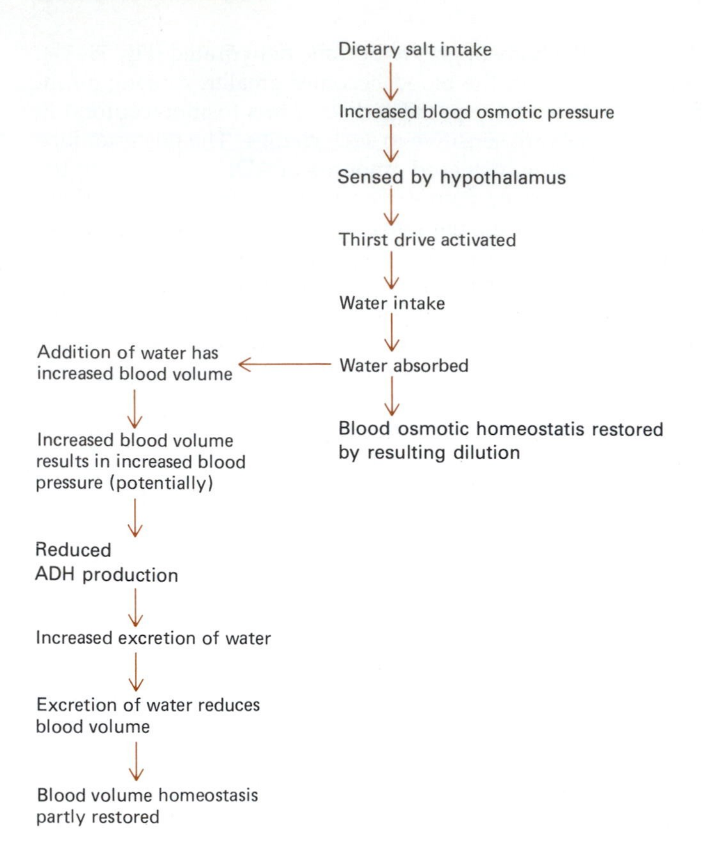
Dietary salt intake
Increased blood osmotic pressure
Sensed by hypothalamus
Thirst drive activated
Water intake
Water absorbed
Blood osmotic homeostatis restored by resulting dilution
Addition of water has increased blood volume
Increased blood volume results in increased blood pressure (potentially)
Reduced ADH production
Increased excretion of water
Excretion of water reduces blood volume
Blood volume homeostasis partly restored

Figure 38–18 Regulation of sodium reabsorption.

osmotically. An increase in dietary salt intake thus increases the osmotic pressure of the blood. Special receptors (the thirst center) in the hypothalamus sense this increase in osmotic pressure and cause the sensation of thirst.

When the thirst drive is activated, an animal drinks water. This water intake restores the osmotic balance of the blood. However, the addition of water increases the blood volume (Fig. 38–18), and now the proper blood volume must be restored. This is accomplished by a decrease in ADH production, which results in increased water excretion.

SUMMARY

I. Three functions of the excretory system are excretion of metabolic wastes, osmoregulation, and regulation of concentrations of the constituents of body fluids.

II. Principal metabolic wastes in most animals are water, carbon dioxide, and nitrogenous wastes such as ammonia, urea, and uric acid.

III. Various mechanisms for waste disposal and osmoregulation are found in invertebrates.
   A. Nephridial organs consist of tubes that open to the outside of the body through nephridial pores. The protonephridia of flatworms and the metanephridia of earthworms are nephridial organs.
   B. Antennal glands, found in crustaceans, consist of a coelomic sac, a greenish glandular chamber, an excretory tubule, and an exit duct. Fluid filtered into the coelomic sac is adjusted as its passes through the excretory organ.
   C. The Malpighian tubules of insects have blind ends that lie in the hemocoel. Wastes are transferred from the blood to the tubule by diffusion and active transport. Water is reabsorbed into the hemocoel, and uric acid is excreted into the intestine.

IV. The vertebrate kidney functions in excretion and osmoregulation and is vital in maintaining homeostasis.

V. Aquatic vertebrates have a continuous problem of osmoregulation.
   A. Marine bony fish lose water osmotically. They compensate by drinking sea water and excreting salt through their gills; only a small volume of urine is produced.
   B. Marine cartilaginous fish retain large amounts of urea, which enables them to take in water osmotically through the gills. This water can be used to excrete a hypotonic urine.
   C. Freshwater fish take in water osmotically. They excrete a large volume of dilute urine.
   D. Other aquatic vertebrates have specific adaptations for dealing with osmoregulation. Marine birds and reptiles, for example, have salt glands that excrete excess salt.

VI. The urinary system is the principal excretory system in human beings and other vertebrates.
   A. The kidneys produce urine, which passes through the ureters to the urinary bladder for storage. Dur-

FOCUS ON
Kidney Disease

Kidney disease ranks fourth among major diseases in the United States. Kidney function can be impaired by infections, by poisoning caused by substances such as mercury or carbon tetrachloride, by lesions, tumors, kidney stone formation, shock, or by many circulatory diseases.

The most serious consequence of kidney disease is **renal failure.** In this condition the nephrons do not function effectively and the kidneys are unable to maintain homeostasis of the blood. Edema develops owing to retention of water, and the concentration of hydrogen ions increases, causing acidosis. Nitrogenous wastes accumulate in the blood and tissues, giving rise to **uremia.** If untreated, the acidosis and uremia can cause coma and eventually death. Chronic kidney failure can be treated by kidney dialysis or by kidney transplant.

Dialysis

In extracorporeal **dialysis,** a plastic tube is surgically inserted into both an artery and a vein in the patient's arm or leg. These tubes can then be connected to a circuit of plastic tubing from a dialysis machine. The patient's blood flows through the tubing, which is immersed in a solution containing most of the normal blood plasma constituents in their normal proportions. The walls of the plastic tubing constitute a semipermeable membrane, and wastes from the patient's blood dialyze through minute pores in the tubing and into the surrounding solution. As the blood flows repeatedly through the tubing in the machine, dialysis continues, eventually adjusting most of the values of the patient's blood chemistry to normal ranges. Although much improved by recent engineering advances, machine dialysis is very expensive, clumsy, and inconvenient and may produce serious side effects such as osteoporosis.

A different dialysis technique, continuous ambulatory peritoneal dialysis (CAPD), makes use of the fact that the peritoneum (the lining of the abdominal cavity) is a differentially permeable membrane. A plastic bag containing dialysis fluid is attached to the patient's abdominal cavity, and the fluid is allowed to run into the abdominal cavity. After several hours, the fluid is withdrawn into the bag and discarded, and fresh fluid is placed in the abdominal cavity. This type of dialysis is much more convenient but poses the threat of peritonitis, should bacteria enter the body cavity with the dialysis fluid.

Kidney Transplant

Long-term use of dialysis, of course, is not as desirable for the patient as would be a functioning kidney. With a successful kidney transplant a patient can live a more normal life with far less long-term expense. At present more than two thirds of kidney transplants are successful for several years, although physicians must routinely treat the problems of graft rejection (discussed in Chapter 36).

ing micturition the urine passes through the urethra to the outside of the body.

B. The functional units of the kidneys are the nephrons. Each nephron consists of a renal corpuscle and a renal tubule.
   1. The renal corpuscle consists of a glomerulus surrounded by Bowman's capsule.
   2. The renal tubule consists of the proximal convoluted tubule, the loop of Henle, and the distal convoluted tubule.

C. Urine formation is accomplished by filtration of plasma, reabsorption of needed materials, and secretion of a few substances into the renal tubule.
   1. Plasma is filtered out of the glomerular capillaries and into Bowman's capsule, but macromolecules such as protein are retained in the blood. Filtration is not a selective process. Thus, needed materials such as glucose and amino acids become part of the filtrate, as well as wastes and excess substances.
   2. About 99% of the filtrate is reabsorbed from the renal tubules into the blood. This is a very selective process that returns needed materials to the blood and adjusts the composition of the filtrate.
   3. Potassium ions, hydrogen ions, ammonia, and certain drugs are secreted from the blood into the renal tubule.
   4. Countercurrent exchange establishes a very hypertonic interstitial fluid surrounding the nephron. This draws water osmotically from the filtrate in the collecting ducts and permits concentration of urine in the collecting ducts so that urine hypertonic to the blood can be produced.

D. The adjusted filtrate, called urine, consists of water, nitrogenous wastes, salts, and a variety of other substances.

E. Urine volume is regulated by the hormone ADH, which is released in appropriate amounts by the posterior lobe of the pituitary. ADH increases the permeability of the collecting ducts so that more water is reabsorbed.

POST-TEST

1. The process of removing metabolic wastes from the body is ______________.
2. The principal nitrogenous waste product of insects and birds is ______________ ______________.
3. The principal nitrogenous waste product of amphibians and mammals is ______________.
4. Flatworms have excretory structures called ______________, which are characterized by ______________ cells.
5. Earthworms have ______________ in each of their body segments.
6. The principal excretory organs of crustaceans are ______________ ______________.
7. The excretory system of insects consists of organs called ______________ ______________.
8. The vertebrate kidney consists of functional units called ______________.

For each group, select the most appropriate answer from Column B for each description in Column A.

Column A	*Column B*
9. Drink water, constantly; small kidneys produce small urine volume	a. Freshwater fish
10. Accumulate urea; excrete hypotonic urine; excrete sodium	b. Marine bony fish
11. Kidneys produce concentrated urine, saltier than surrounding water	c. Marine mammals
12. Excrete large volume of dilute urine; transport salt into body	d. Marine chondrichthyes (sharks)
13. Have salt glands in head that excrete excess salt	e. Marine birds

Column A	*Column B*
14. Outer portion of human kidney	a. Cortex
15. Delivers urine to bladder	b. Medulla
16. Funnel-shaped chamber that receives urine from collecting ducts	c. Ureter
	d. Urethra
	e. Renal pelvis

Column A	*Column B*
17. Site of filtration	a. Collecting duct
18. Site of most reabsorption	b. Proximal convoluted tubule
19. Site of most renal secretion	c. Distal convoluted tubule
	d. Bowman's capsule
	e. Ureter

20. The glomerulus consists of a tuft of ______________ that project into ______________ ______________.
21. Blood is delivered to the glomerulus by an ______________ ______________ and leaves the glomerulus in an ______________ ______________.
22. Once the fluid that has left the glomerular capillaries enters Bowman's capsule, it is known as the ______________ ______________.
23. The tubular transport maximum is the maximum rate at which a substance can be ______________.
24. When a substance exceeds its renal threshold concentration, the portion not reabsorbed is ______________.
25. A ______________ urine is low in water, and so conserves water.
26. Antidiuretic hormone (ADH) increases the permeability of the ______________ ______________ so that more water is ______________ and the volume of urine is ______________ (*increased/decreased*).
27. A person with diabetes insipidus does not produce sufficient ______________ and as a result ______________.

REVIEW QUESTIONS

1. Compare mechanisms of metabolic waste disposal in the following:
   a. earthworms
   b. flatworms
   c. crayfish
   d. insects
   e. vertebrates
2. Define osmoregulation. What type of osmoregulatory problem is faced by marine fish? How do they solve it?
3. How do freshwater fish solve their osmoregulatory problems? How is this accomplished by marine birds?
4. In the human urinary system, name the structure associated with each of the following:
   a. filtration
   b. absorption
   c. urea formation
   d. temporary storage of urine
   e. conduction of urine out of the body
5. What substances are present in normal human urine?
6. Compare the following nitrogenous wastes:
   a. ammonia
   b. uric acid
   c. urea
   Explain how each is formed, and describe the advantages and disadvantages of each as an excretory product.
7. Compare filtration, reabsorption, and secretion.
8. Trace a molecule of urea from its formation in the liver to its excretion in the urine.
9. Why is glucose normally not present in urine? Why is it present in the urine of an individual with diabetes mellitus?
10. How is urine volume regulated? Explain.
11. In what ways are the kidneys vital homeostatic organs?

39
Neural Control: Neurons

OUTLINE

I. Information flow through the nervous system
II. Organization of the vertebrate nervous system
III. Structure of the neuron
  A. Dendrites
  B. The axon
IV. Nerves and ganglia
V. Transmission of a neural impulse
  A. The resting potential
  B. Local changes in potential
  C. The action potential
  D. Saltatory conduction
  E. Substances that affect excitability
  F. Synaptic transmission
    1. Excitatory and inhibitory signals
    2. Neurotransmitters
  G. Direction and speed of conduction
VI. Integration
VII. Organization of neural circuits
VIII. Reflex action

LEARNING OBJECTIVES

After you have read this chapter you should be able to:

1. Trace the flow of information through the nervous system.
2. Describe the principal divisions of the vertebrate nervous system.
3. Compare the functions of neurons and neuroglia.
4. Draw a diagram of a neuron, giving the name and function of each structure.
5. Describe the structure of a nerve and of a ganglion.
6. Explain how the resting potential of a neuron is maintained.
7. Contrast local changes in potential with an action potential.
8. Describe the effects of calcium imbalance, local anesthetics, and such agents as DDT on neuron excitability.
9. Describe synaptic transmission, and explain how it regulates the direction of neural transmission.
10. List the neurotransmitters described in the chapter, and give an example of where each is secreted.
11. Draw diagrams to illustrate convergence, divergence, facilitation, and reverberating circuits.
12. Define and describe the process of neural integration.
13. Draw a reflex pathway consisting of three neurons, label each structure, and indicate the direction of information flow; relate reflex action to the processes of reception, transmission, integration, and actual response.

Vibrations from approaching footsteps provoke an earthworm to retreat quickly into its burrow. When a crayfish is hungry, it seeks out and devours a wiggling worm. A child learns to look both ways before crossing a busy street, and a college student learns the principles of biology or calculus. All of these activities, and countless others, are made possible by the nervous system. In complex animals the endocrine system works with the nervous system in regulating many behaviors and physiological processes. The endocrine system generally provides a relatively slow and long-lasting regulation, whereas the nervous system permits a very rapid response.

Changes within the body or in the outside world that can be detected by an organism are termed **stimuli.** The ability of an organism to survive and to maintain homeostasis depends largely upon how effectively it can respond to stimuli in its internal or external environment.

Information Flow Through the Nervous System

The nervous system is bombarded with thousands of stimuli each day. It receives information, transmits messages, sorts and interprets incoming data, and then sends the proper messages so that the responses will be coordinated and homeostatic. Even the simplest response to a stimulus generally requires a sequence of information flow through the nervous system that includes reception, transmission of impulses to the brain or spinal cord, integration, transmission of impulses from the brain or spinal cord, and response by an effector—usually a muscle or gland.

Imagine that you are very hungry and an obliging friend sets a delicious steak dinner before you. You cannot lift the first forkful to your mouth until **reception, transmission, integration,** and **response** by an effector have taken place (Fig. 39–1). First, you must detect the food—the stimulus. At least two types of **receptors** (your eyes and your olfactory epithelium) receive the information. Second, these messages must be sent to your brain, informing you that they have received a stimulus. **Afferent (sensory) neurons** transmit this information in the form of neural impulses *from* the sense organs *to* the brain.

When a decision to eat the food has been made, **efferent (motor) neurons** transmit the message from the brain to the appropriate effector cells—in this case, certain muscle fibers in your arm and hand. The last process in this sequence is the actual contraction of the muscle fibers necessary to carry out the response. Now, finally, you spear the food with the fork and lift it into your mouth.

Organization of the Vertebrate Nervous System

In vertebrates the **central nervous system (CNS)** consists of a complex brain that is continuous with a single, dorsal, tubular nerve cord (spinal cord). The central nervous system integrates all incoming information and determines appropriate responses. Within the CNS, afferent neurons **synapse,** that is, make functional connections with **association neurons,** also called **interneurons.** Each association neuron may synapse with many other association neurons. At these synapses, incoming neural messages are sorted out and interpreted.

The **peripheral nervous system (PNS)** consists of the sensory receptors and the nerves lying outside the brain and spinal cord. Twelve pairs of cranial nerves (ten in fish and amphibians) and several pairs of spinal nerves (31 pairs in humans) link the CNS with various parts of the body. Afferent neurons in these nerves continually inform the CNS about stimuli that impinge upon the body. Efferent neurons then transmit the impulses from the CNS to appropriate effector cells—muscles and glands—which make the adjustments necessary to preserve homeostasis.

For convenience, the PNS can be subdivided into somatic and autonomic systems. The **somatic system** consists of the receptors and nerves concerned with stimuli in the outside world. The **autonomic system** consists of the receptors and nerves operating to regulate the internal environment. In the autonomic system there are two types of efferent nerves: sympathetic and parasympathetic. These divisions of the vertebrate nervous system are discussed in Chapter 40.

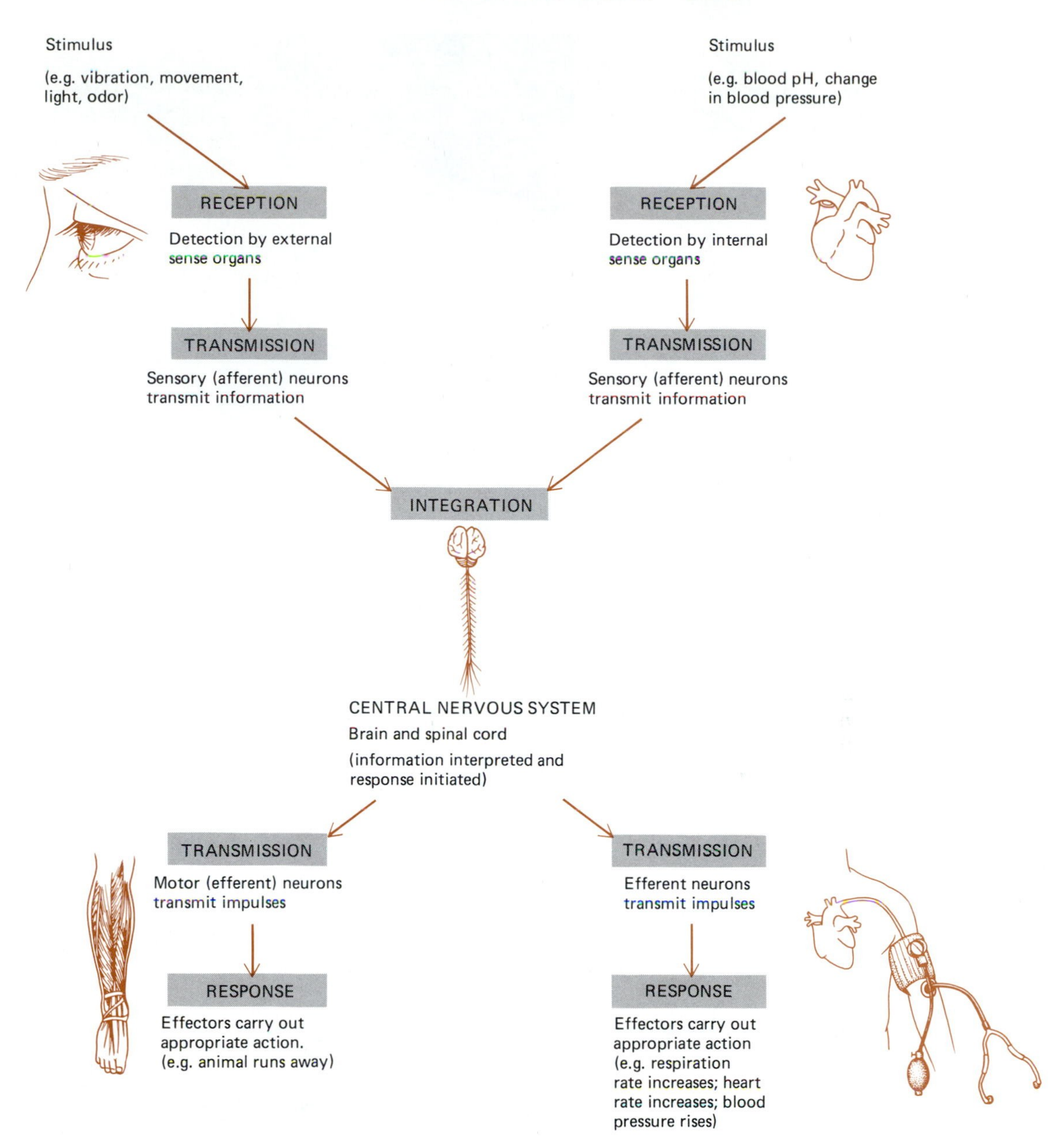
Stimulus
(e.g. vibration, movement, light, odor)
Stimulus
(e.g. blood pH, change in blood pressure)
RECEPTION
Detection by external sense organs
RECEPTION
Detection by internal sense organs
TRANSMISSION
Sensory (afferent) neurons transmit information
TRANSMISSION
Sensory (afferent) neurons transmit information
INTEGRATION
CENTRAL NERVOUS SYSTEM
Brain and spinal cord
(information interpreted and response initiated)
TRANSMISSION
Motor (efferent) neurons transmit impulses
TRANSMISSION
Efferent neurons transmit impulses
RESPONSE
Effectors carry out appropriate action.
(e.g. animal runs away)
RESPONSE
Effectors carry out appropriate action
(e.g. respiration rate increases; heart rate increases; blood pressure rises)

Figure 39–1 Flow of information through the nervous system.

Structure of the Neuron

The **neuron,** or nerve cell, is the structural and functional unit of the nervous system. The specialized supporting cells of nervous tissue are called **neuroglia.** A typical **multipolar neuron** consists of a **cell body,** which has bushy cytoplasmic extensions called **dendrites,** and a single, long cytoplasmic extension, the **axon** (Fig. 39–2). Other types of neurons are illustrated in Figure 39–3.

The cell body contains the nucleus and many of the other organelles as well as the bulk of the cytoplasm. It is concerned with the metabolic maintenance and growth of the neuron. The cell body contains **Nissl substance,** which consists of deeply staining regions of cytoplasm rich in rough endoplasmic reticulum and free ribosomes. This is the site of protein synthesis. Microtubules and microfilaments are distributed throughout the cytoplasm. The microtubules help to maintain the shape of the neuron, especially of the dendrites and axon, and play a role in transporting materials through the axon. The short microfilaments, composed of actin, are also thought to function in transporting materials.

DENDRITES

Dendrites are highly branched extensions of the cytoplasm that project from the cell body. Dendrites and the surface of the cell body are specialized to receive stimuli.

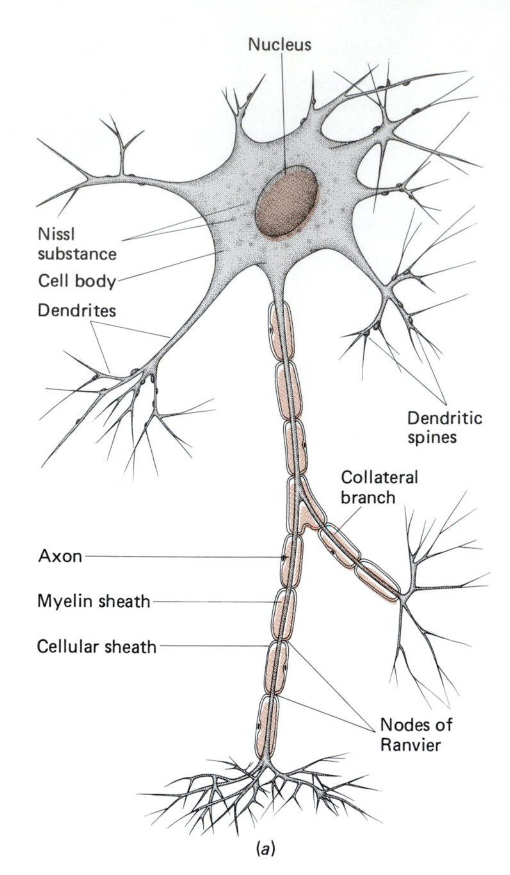

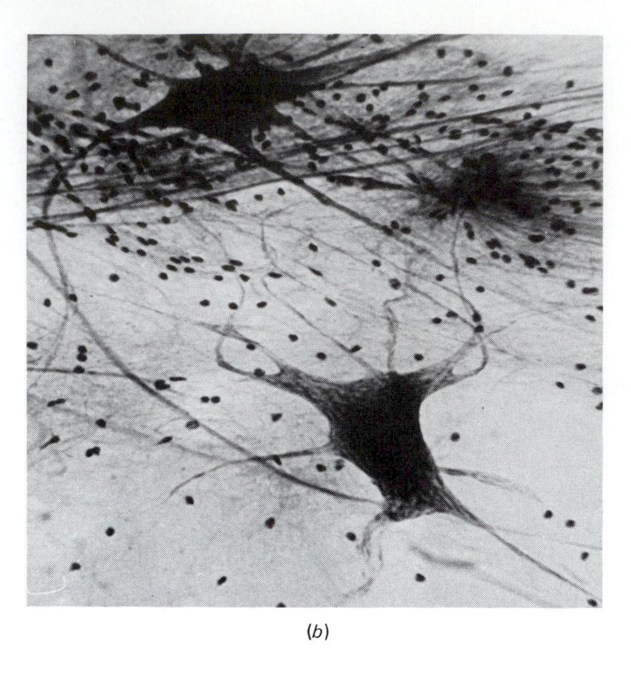

Figure 39–2 A multipolar neuron consists of a cell body, numerous dendrites, and a single, long axon. (*a*) Structure of a multipolar neuron. The axon of this neuron is myelinated, and therefore the myelin sheath is shown as well as the cellular sheath. (*b*) Photomicrograph of nerve tissue showing two multipolar neurons.

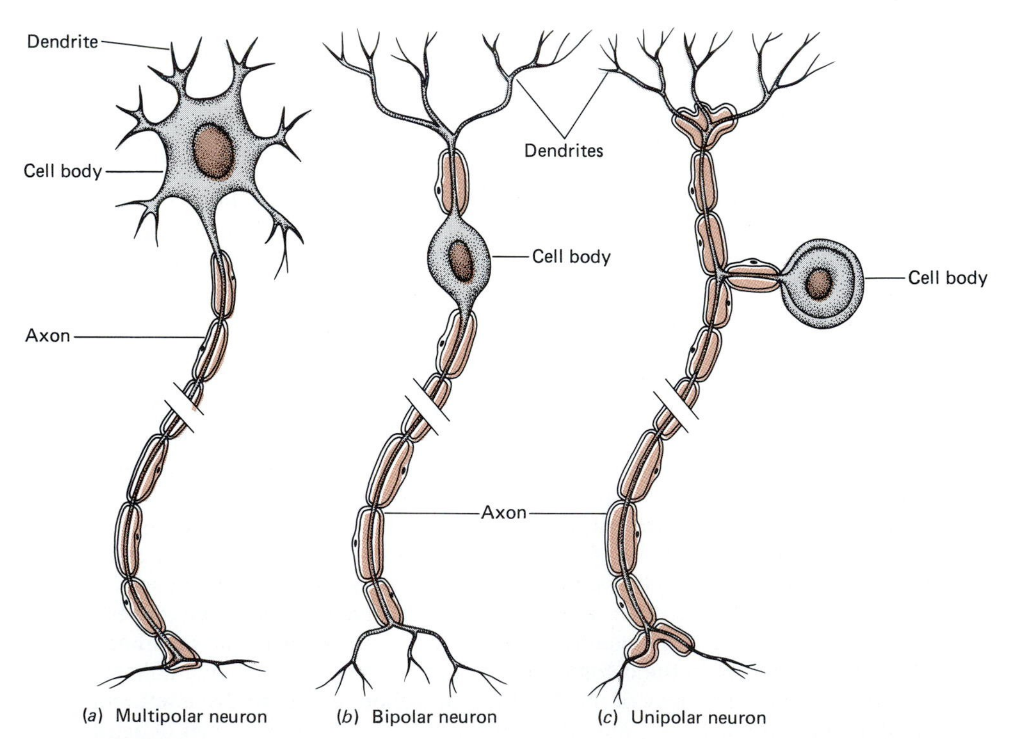

Figure 39–3 Types of neurons classified according to the number of extensions. (*a*) A multipolar neuron. This type of neuron has many short dendrites and one long axon. Motor neurons are of this type. (*b*) A bipolar neuron. A bipolar neuron has one axon and one dendrite. This type of neuron is found in the retina of the eye, in the olfactory nerve, and in the nerves coming from the inner ear. (*c*) A unipolar neuron has a short process that divides into two long processes. The distal process may be called either a dendrite or an axon. Since it functions like a dendrite, it is called a dendrite here. Sensory neurons are unipolar.

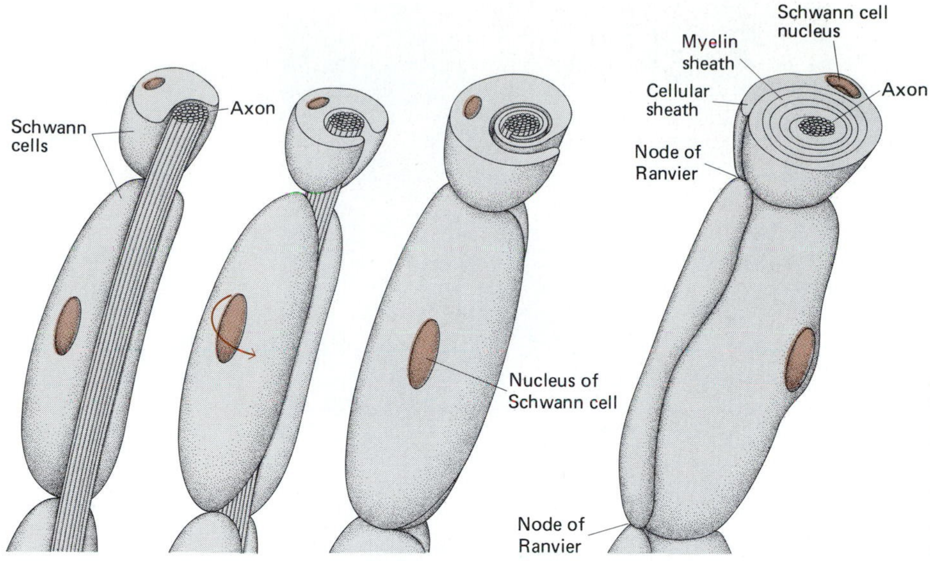

Figure 39–4 Formation of the myelin sheath around the axon of a peripheral neuron. A Schwann cell wraps its cell membrane around the axon many times to form the insulating myelin sheath. The rest of the Schwann cell remains outside the myelin sheath, forming the cellular sheath.

Their surfaces are dotted with thousands of tiny **dendritic spines,** which are the sites of specialized junctions with other neurons.

THE AXON

The axon arises from a thickened area of the cell body, the **axon hillock.** It transmits neural impulses from the cell body to another neuron or effector cell. Although microscopic in diameter, the axon may be very long. A sensory cell located in the toe of a giraffe may be barely 0.1 mm in diameter but spans a distance of several meters before ending in the spinal cord. Perhaps because of its impressive length, the axon is sometimes referred to as a **nerve fiber.**

At its distal end the axon branches extensively. At the ends of these branches are tiny enlargements called **synaptic knobs.** These knobs release a substance (a neurotransmitter) that transmits messages from one neuron to another. Branches called **collaterals** may arise along the length of the axon, permitting extensive interconnections among neurons.

Axons of peripheral neurons are enveloped by a **cellular sheath,** or **neurilemma,** composed of supporting cells called **Schwann cells.** The cellular sheath is important in regeneration of injured nerves (see Focus on Regeneration of an Injured Neuron). In forming the cellular sheath, Schwann cells line up along the axon and wrap themselves around it. On some axons the Schwann cells produce an insulating inner sheath called the **myelin sheath.** This covering forms when the Schwann cell membrane becomes wrapped around the axon several times (Figs. 39–4 and 39–5). Myelin is a white, lipid-rich substance that makes up the cell membrane of the Schwann cell. This fatty material is an excellent electrical insulator and speeds the conduction of nerve impulses (see under Saltatory Conduction, later in chapter). Between adjacent Schwann cells there are gaps called **nodes of Ranvier.** At these points, which are from 50 to 1500 μm apart, the axon is not insulated with myelin. Almost all axons more than 2 μm in diameter are myelinated, that is, possess myelin sheaths. Those with a smaller diameter are generally unmyelinated.

Axons within the CNS have no neurilemma. Their myelin sheaths are formed by another type of neuroglial cells (oligondendrocytes) rather than by Schwann cells. Certain areas of the brain and spinal cord are composed principally of myelinated axons. The myelin imparts a whitish color to these areas, so that they are referred to as white matter.

When myelin is destroyed, nerve function is impaired. **Multiple sclerosis** is a neurological disease in which patches of myelin deteriorate at irregular intervals along neurons in the CNS. The myelin is replaced by a type of scar tissue, and the affected neurons are not able to conduct impulses. This leads to impaired neural function, including loss of coordination, tremor, and paralysis of affected body parts.

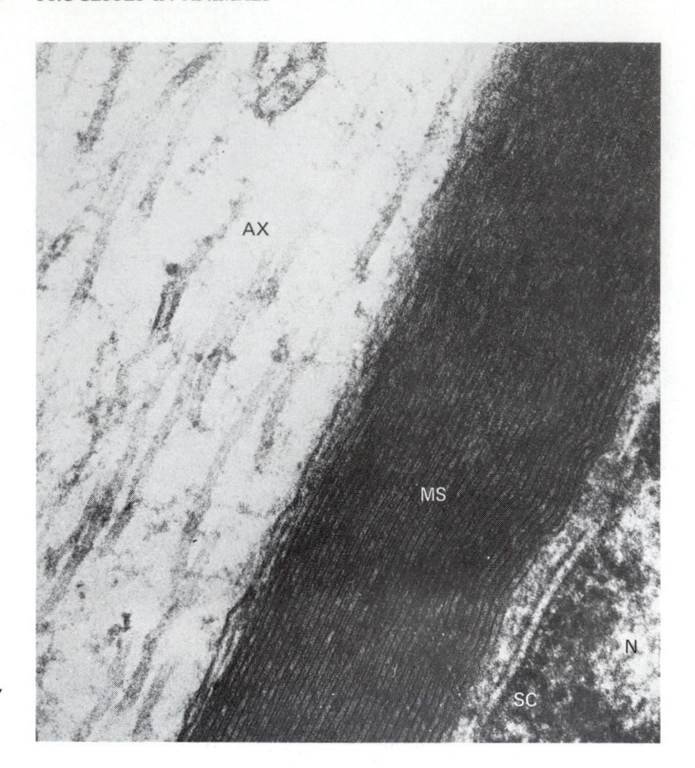

Figure 39–5 Electron micrograph of a section through a single myelinated axon. *AX*, axon; *MS*, myelin sheath; *SC*, Schwann cell; *N*, nucleus of Schwann cell. (Courtesy of Dr. Lyle C. Dearden.)

Nerves and Ganglia

The nerves observed in gross anatomical dissection consist of hundreds or even thousands of axons wrapped in connective tissue (Fig. 39–6). A nerve can be compared to a telephone cable. The individual axons correspond to the wires that run through the cable, and the sheaths and connective tissue coverings correspond to the insulation. Within the CNS, bundles of axons are referred to as **tracts** or **pathways,** rather than nerves. The cell bodies of neurons are usually grouped together in masses called **ganglia.**

Transmission of a Neural Impulse

Once a receptor has been stimulated, the neural message must be transmitted to the CNS and then back to appropriate effector cells. Information must be conducted through a sequence of neurons. Transmission of a nerve impulse down the length of a neuron is an electrochemical process that depends upon changes in ion distribution. Transmission from one neuron to another across a synapse is generally a chemical phenomenon involving the secretion of neurotransmitter by the axon and the action of chemoreceptors in the dendrite.

THE RESTING POTENTIAL

In a **resting neuron**—one not transmitting an impulse—the inner surface of the plasma membrane is negatively charged compared with the interstitial fluid surrounding it (Fig. 39–7). The resting neuron is said to be **electrically polarized,** that is, oppositely charged along the inside of the membrane compared with the interstitial fluid outside. When electric charges are separated in this way, they have the potential of doing work should they be permitted to come together. The difference in electric potential on the two sides of the membrane may be expressed in millivolts (mV). (A millivolt is a thousandth of a volt and is a unit for measuring electrical potential.)

The **resting potential** of a neuron is about 70 mV. By convention this is expressed as −70 mV because the inner surface of the plasma membrane is negatively charged relative to the interstitial fluid. The resting membrane can be measured by placing one electrode, insulated except at the tip, inside the cell and a second electrode on the outside surface. The two electrodes are connected with a suitable recording instrument such as a galvanometer, which measures current by electro-

FOCUS ON

Regeneration of an Injured Neuron

When an axon is separated from its cell body by a cut, it soon degenerates. A hollow tube of Schwann cells remains, but myelin eventually disappears. As long as the cell body of the neuron has not been injured, however, it is capable of regenerating a new axon. Sprouting begins within a few days following cutting (see accompanying figure). The growing axon enters the old sheath tube and proceeds along it to its final destination. Axons can grow in the absence of sheaths if some conduit is provided for them. They can, for example, be made to grow within sections of blood vessels or extremely fine plastic tubes. The length of time required for regeneration depends on how far the nerve has to grow and may require as long as 2 years. When cuts occur within the spinal cord or brain, regeneration, if if occurs at all, is very feeble. It is thought that growth of new sprouts in the CNS is prevented by scar tissue formed by neuroglial cells at the site of injury.

It is remarkable that (if not blocked by scar tissue or other barrier) each regenerating axon of a cut peripheral nerve finds its way back to its former point of termination, whether this be a specific connection in the central nervous system or a specific muscle or sense organ in the periphery. If, during the early stages of development of an amphibian, an extra limb bud is transplanted next to the normally developing limb, both will grow to maturity. The extra limb then moves synchronously with the normal one. Anatomical examination reveals that the nerve that innervates the normal limb sends out branches to the extra one. Clearly, the extra limb exerts some stimulating influence on the growing nerves to produce more branches, and some directive influence as well.

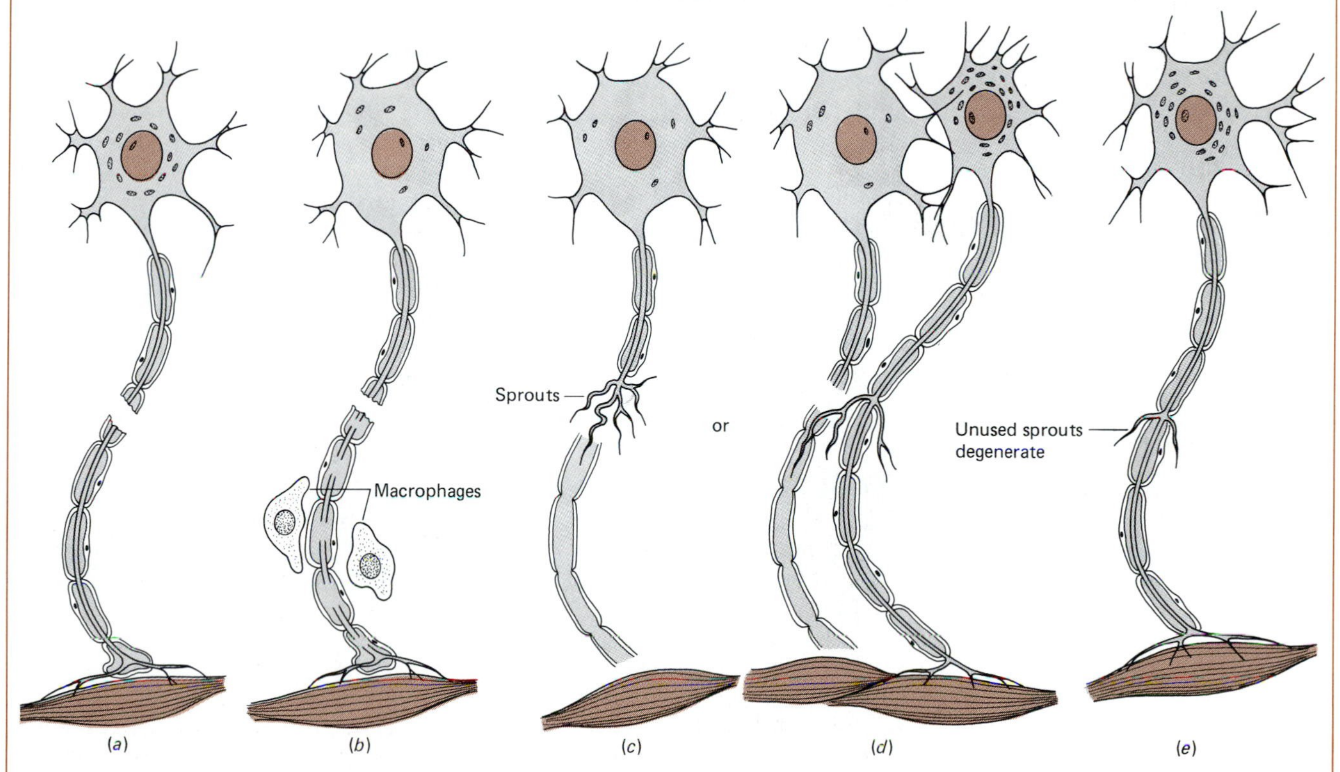

Regeneration of an injured neuron. (*a*) A neuron is severed. (*b*) The part of the axon that has been separated from its nucleus degenerates. Its myelin sheath also degenerates, and macrophages phagocytize the debris. The cell body enlarges and the Nissl substance breaks down, a sign of increased protein synthesis. (*c*) The tip of the severed axon begins to sprout, and one or more sprouts may find their way into the empty cellular sheath, which has remained intact. The sprout grows slowly and becomes myelinated. (*d*) Sometimes an adjacent undamaged neuron may send a collateral sprout into the cellular sheath of the damaged neuron. (*e*) Eventually the neuron may regenerate completely, so that function is fully restored. Unused sprouts degenerate.

magnetic action. If both electrodes are placed on the outside surface of the neuron, no potential difference between them is registered; all points on the outside are at equal potential.

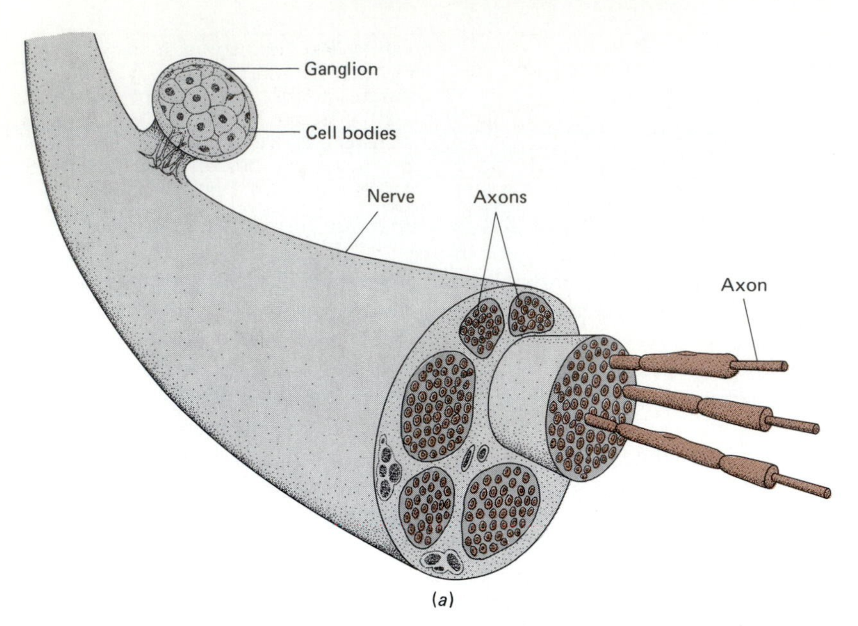

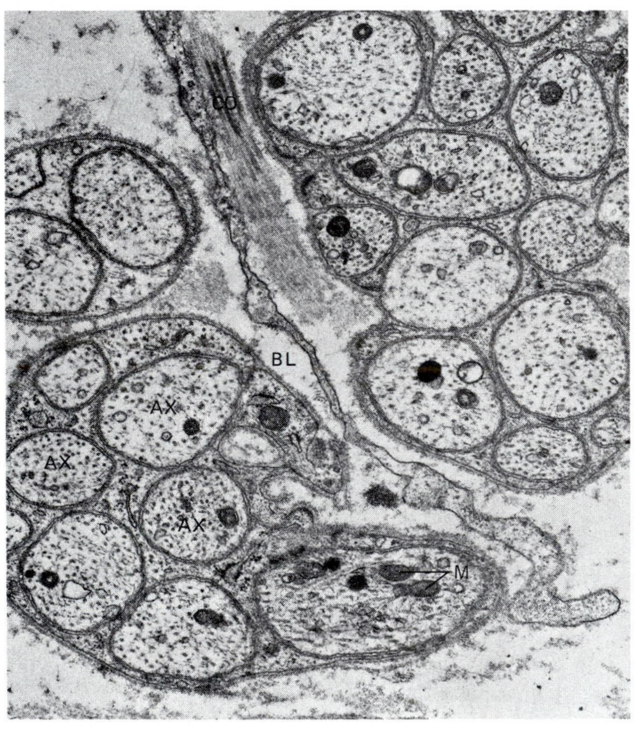

Figure 39–6 A nerve consists of bundles of axons held together by connective tissue. The cell bodies belonging to these axons are grouped together in a ganglion. (*a*) Structure of a nerve and a ganglion. (*b*) Electron micrograph showing cross section through a portion of the sciatic nerve (approximately ×30,000). The axons shown here are unmyelinated. *AX*, axon; *M*, mitochondria; *CO*, collagen fibers; *BL*, basal lamina. (Courtesy of Dr. Lyle C. Dearden.)

What is responsible for the resting potential? It results from the presence of a slight excess of *negative* ions *inside* the plasma membrane and a slight excess of *positive* ions *outside* the plasma membrane. This imbalance in ion distribution is brought about by several factors. The plasma membrane of the neuron has very efficient sodium pumps that actively transport sodium out of the cell against a concentration and an electrochemical gradient. The sodium pump requires energy and uses ATP derived from metabolic processes within the nerve cell. The same pumps may also actively transport potassium ions into the cell. About three sodium ions are pumped out of the neuron for every two potassium ions that are pumped in. Thus, more positive ions are pumped out than in.

The neuron membrane is much more permeable to potassium than to sodium. For this reason, sodium cannot easily diffuse back into the resting neuron, but potassium ions are able to diffuse out. Potassium ions leak out through the membrane along a concentration gradient until the positive charge outside the membrane reaches a level that repels the outflow of additional positively charged potassium ions. A steady state is reached when the potassium outflow equals the inward flow of sodium ions. At this point the resting potential is about −70 mV.

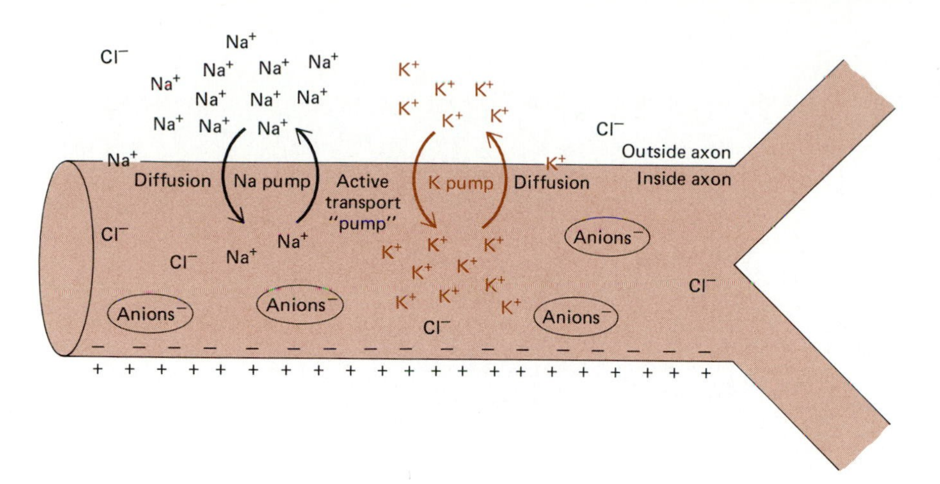

Figure 39–7 Segment of an axon of a resting (nonconducting) neuron. Sodium is actively pumped out of the cell, and potassium is pumped in. Sodium is unable to diffuse back to any extent, but potassium does diffuse out along its concentration gradient. Because of the unequal distribution of ions, the inside of the axon is negatively charged compared with the outer tissue fluid. The presence of negatively charged proteins and other large anions in the cell contributes to this polarity.

Contributing to the overall ionic situation are large numbers of negatively charged proteins and organic phosphates within the neuron that are too big to diffuse out. The plasma membrane is permeable to negatively charged chloride ions, but because of the positively charged ions that accumulate outside the membrane, chloride ions are attracted to the outside and tend to accumulate there.

The resting potential is due mainly to the outward diffusion of potassium ions along their concentration gradient. The conditions for this diffusion, however, are set by the action of the sodium–potassium pumps. The active transport of ions by these pumps is a form of cellular work, and so requires energy.

LOCAL CHANGES IN POTENTIAL

An electrical, chemical, or mechanical stimulus may alter the resting potential by increasing the permeability of the plasma membrane to sodium, or by making the neuron more negative relative to the interstitial fluid. Such local responses in the membrane potential are called **postsynaptic potentials.** When they occur in receptor cells, they are called *receptor potentials.*

Neurophysiologists think that the neuronal membrane contains specific sodium gates. Apparently, these gates lead into channels through the membrane that are formed by proteins. **Excitatory stimuli** open sodium gates, permitting sodium ions to rush into the cell. This passage of positive sodium ions into the cell depolarizes the membrane, that is, causes the membrane potential to become less negative for a brief moment. **Inhibitory stimuli** may hyperpolarize the membrane, that is, increase the resting potential. This occurs because of an increase in permeability of the membrane to potassium, which permits potassium ions to flow out of the neuron. With additional positively charged potassium ions outside the membrane, the neuron becomes more negative relative to the outside than when it was at rest.

Local changes in potential can cause a flow of electric current. The greater the change in potential, the greater the flow of current. Such a local current flow can function as a signal only over a very short distance, because it fades out within a few millimeters of its point of origin. As we will see, however, postsynaptic potentials can be added together, resulting in action potentials.

THE ACTION POTENTIAL

The membrane of a neuron can depolarize as much as 15 mV (which changes the resting potential to about −55 mV) without initiating an impulse. When the extent of depolarization reaches about −55 mV, a critical point called the **threshold level** or firing level is reached. At this point the resulting depolarization is self-propagating; that is, it spreads down the axon as a wave of depolarization without fading. This wave of depolarization is called a **nerve impulse** or **action potential** (Fig. 39–8).

When threshold level is reached, an almost explosive action occurs as the action potential is produced. The neuron membrane quickly reaches zero potential and even overshoots to about +35 mV so that there is a momentary reversal in

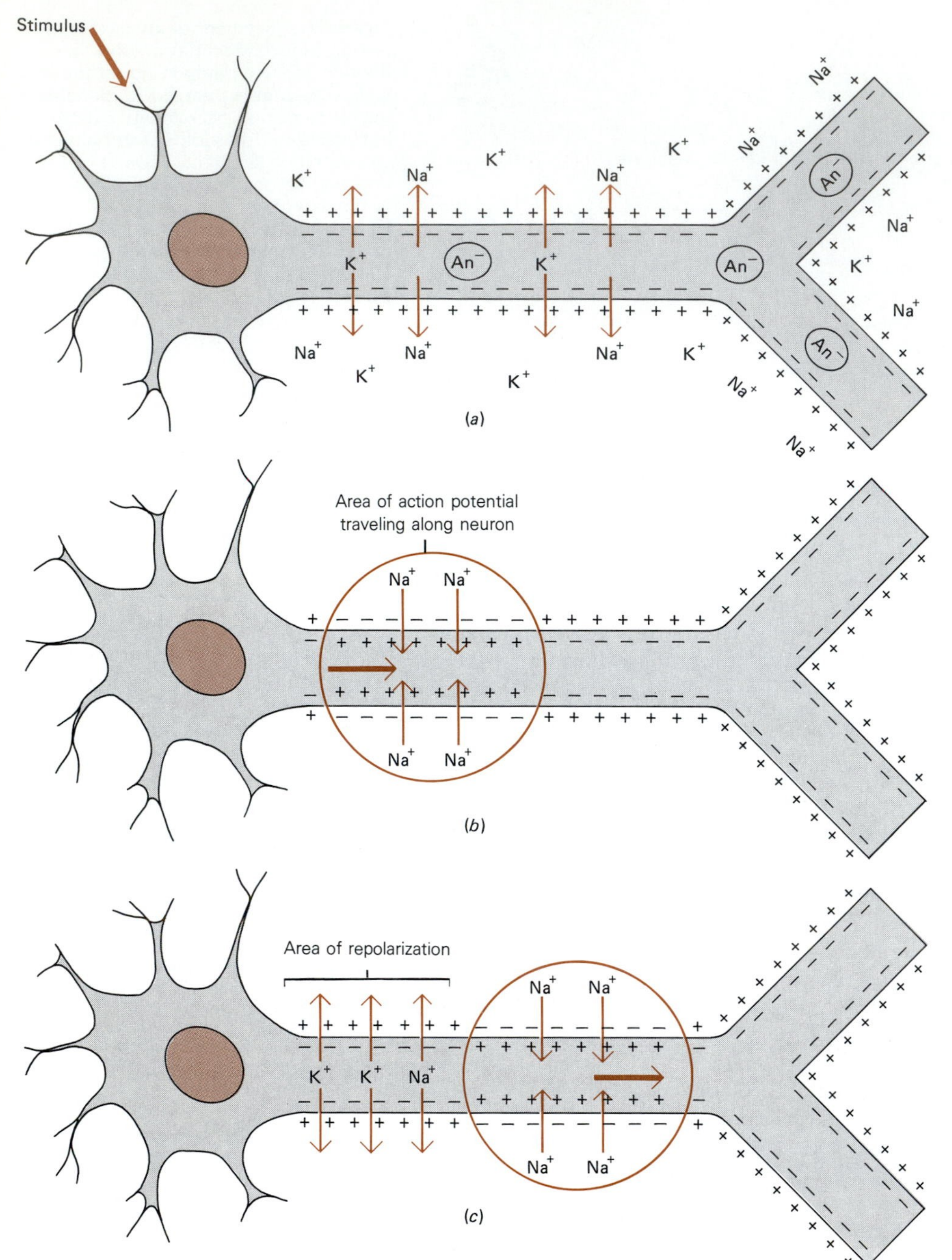

Figure 39–8 Transmission of an impulse along an axon. (*a*) The dendrites (or cell body) of a neuron are stimulated sufficiently to depolarize the membrane to firing level. The axon shown is still in the resting state, and has a resting potential. (*b*), (*c*) An impulse is transmitted as a wave of depolarization that travels down the axon. At the region of depolarization, sodium ions diffuse into the cell. As the impulse passes along from one region to another, polarity is quickly reestablished. Potassium ions flow outward until the resting potential is restored. Sodium is slowly pumped back out of the axon so that resting conditions are reestablished.

polarity. The sharp rise and fall of the action potential is referred to as a **spike.** Figure 39–9 illustrates an action potential.

The action potential is an electric current of sufficient strength to induce collapse of the resting potential in the adjacent area of the membrane. The impulse moves along the axon at a constant velocity and amplitude for each type of neuron. The neuron is said to obey an **all-or-none law** since there are no variations in intensity of the action potential: Either the neuron fires completely, or it does not fire at all.

As the wave of depolarization moves along the axon, the normal polarized state is quickly reestablished behind it. By the time the action potential moves a few millimeters along the axon, the membrane over which it has just passed begins to repolarize. The sodium gates close, so that the membrane becomes impermeable to sodium. Potassium gates in the membrane then open, permitting potassium to leave. The accumulation of potassium ions outside the membrane results in repolarization.

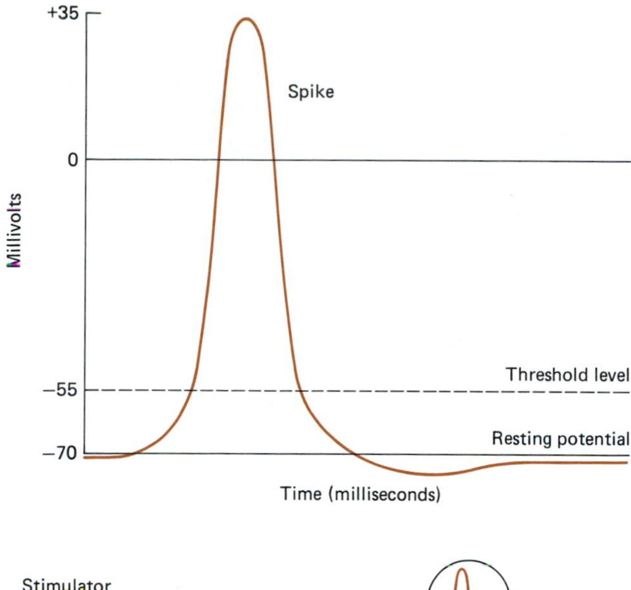

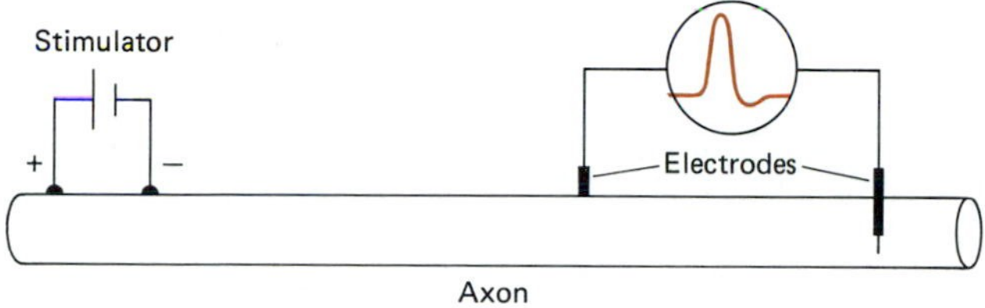

Figure 39–9 An action potential recorded with one electrode inside the cell and one just outside the plasma membrane. When the axon depolarizes to about −55 mV, an action potential is generated.

Repolarization occurs very rapidly, but redistribution of sodium and potassium ions back to the resting condition requires more time. Resting conditions are reestablished when the sodium pump actively transports excess sodium out of the neuron.

During the millisecond or so during which it is depolarized, the axon membrane is in an **absolute refractory period,** when it cannot transmit an action potential no matter how great a stimulus is applied. Then, for two or three additional milliseconds, while the resting conditions are being reestablished, the axon is said to be in a **relative refractory period.** During this time another potential can be generated if the stimulus is stronger than the normal threshold stimulus. Even with the limits imposed by their refractory periods, neurons can transmit several hundred impulses per second!

The electrical and chemical processes involved in the transmission of a nerve impulse are similar in many ways to those involved in muscle contraction. Compared with a contracting muscle, however, a transmitting nerve expends little energy; the heat produced by one gram of nerve stimulated for one minute is equivalent to the energy liberated by the oxidation of 10^{-6} grams of glycogen. This means that if a nerve contained only 1% glycogen to serve as fuel, it could be stimulated continuously for a week or more without exhausting the supply. Nerve fibers are practically incapable of being fatigued as long as an adequate supply of oxygen is available. Whatever "mental fatigue" may be, it is not due to the exhaustion of the energy supply of nerve fibers!

SALTATORY CONDUCTION

The smooth, progressive impulse transmission just described is characteristic of unmyelinated neurons. In myelinated neurons the myelin insulates the axon except at the nodes of Ranvier, where the membrane makes direct contact with the interstitial fluid. Depolarization skips along the axon from one node of Ranvier to the next (Fig. 39–10). The ion activity at the node depolarizes the next node along the axon. Known as **saltatory conduction,** this type of impulse transmission is more rapid than the continuous type.

Saltatory conduction requires less energy than continuous conduction. Only the nodes depolarize, so fewer sodium and potassium ions are displaced, and the

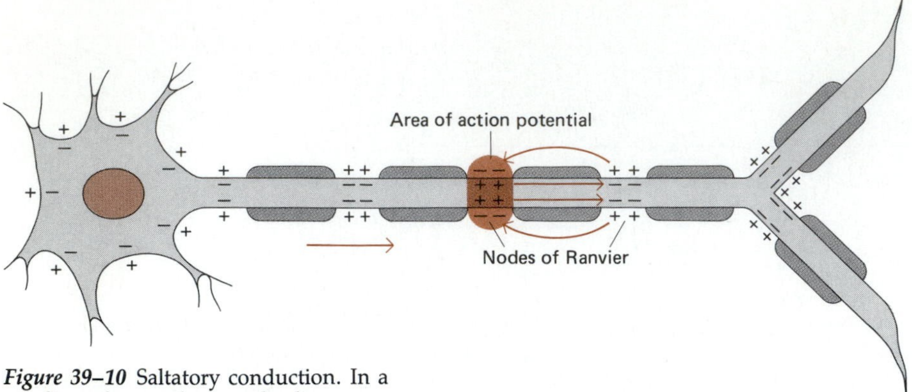

Figure 39–10 Saltatory conduction. In a myelinated axon the impulse leaps along from one node of Ranvier to the next.

cell does not have to work as hard to reestablish resting conditions each time an impulse is transmitted.

SUBSTANCES THAT AFFECT EXCITABILITY

Substances that increase the permeability of the membrane to sodium make the neuron more excitable than normal. Other substances make the neuron less excitable. Calcium balance is essential to normal neural function. When insufficient calcium ions are present, the sodium gates apparently do not close completely between action potentials. Sodium ions then leak into the cell, lowering the resting potential and bringing the neuron closer to firing. The neuron fires more easily—sometimes even spontaneously. As a result, the muscle innervated by the neuron may go into spasm, a condition known as low-calcium tetany. When calcium ions are too numerous, neurons are less excitable and more difficult to fire.

Local anesthetics such as procaine and cocaine are thought to decrease the permeability of the neuron to sodium. Excitability may be so reduced that the neuron cannot propagate an action potential through the anesthesized region. DDT and other chlorinated hydrocarbon biocides interfere with the action of the sodium pump. When nerves are poisoned by these chemicals, they are unable to transmit impulses. Fortunately insects are poisoned by much lower concentrations of DDT than those harmful to humans.

SYNAPTIC TRANSMISSION

A **synapse,** you will recall, is a junction between two neurons. The neurons are generally separated by a tiny gap from 0.002 to 0.02 μm wide (less than a millionth of an inch) called the **synaptic cleft.** A neuron that ends at a specific synapse is referred to as a **presynaptic neuron,** whereas a neuron that begins at a synapse is a **postsynaptic neuron.** A neuron may be postsynaptic with respect to one synapse and presynaptic with respect to another.

Some presynaptic and postsynaptic neurons come very close together (within 2 nm of one another) and form low-resistance gap junctions. Such junctions permit an impulse to be electrically transmitted directly from one cell to another. Electrical synapses of this sort are found in cnidarian nerve nets and in parts of the nervous systems of earthworms, crayfish, and fish.

At most synapses, however, a gap of more than 20 nm separates the two plasma membranes, and the impulse is transmitted by special substances called **neurotransmitters.** When an impulse reaches the synaptic knobs at the end of a presynaptic axon, it stimulates the release of neurotransmitter into the synaptic cleft. This chemical messenger rapidly diffuses across the narrow synaptic cleft and affects the permeability of the membrane of the postsynaptic neuron.

The synaptic knobs continuously synthesize neurotransmitter and store it in little vesicles. Mitochondria in the synaptic knobs provide the ATP required for this synthesis. The enzymes needed are produced in the cell body and move down the axon to the synaptic knobs. Each time an action potential reaches a synaptic knob, calcium ions pass into the cell. This apparently induces several hundred vesicles to fuse with the membrane and to release their contents into the synaptic cleft (Figs. 39–11 and 39–12). After diffusing across the synaptic cleft, the neurotransmitter

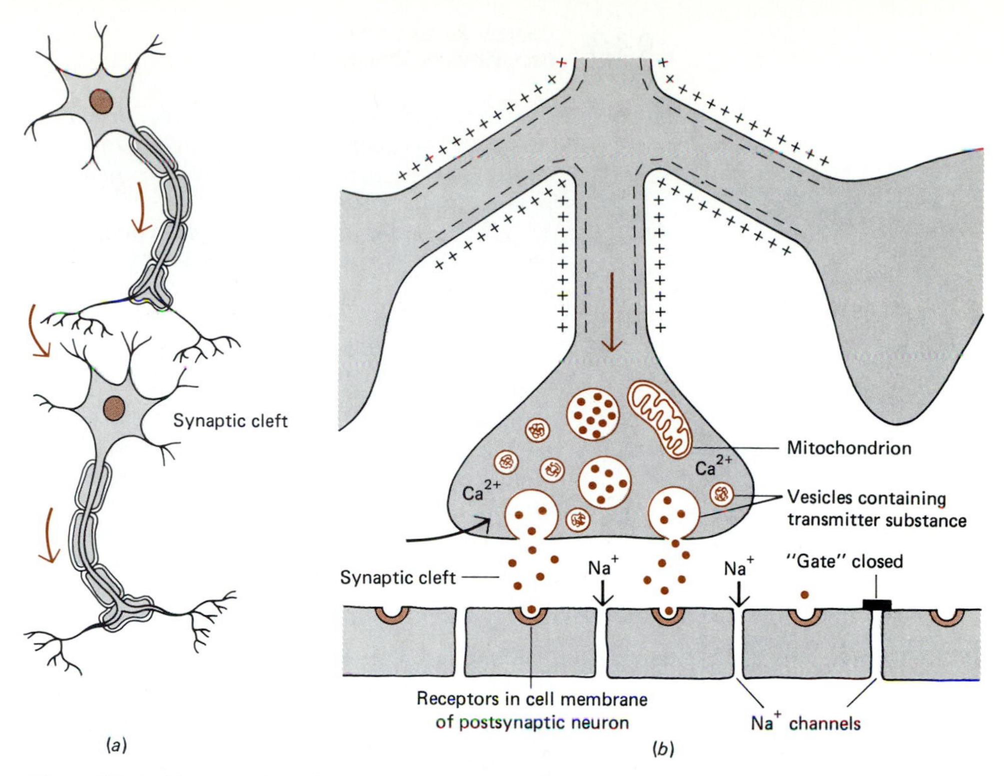

Figure 39–11 Transmission of an impulse between neurons, or from a neuron to an effector. (*a*) In most synapses, the wave of depolarization is unable to jump across the synaptic cleft between the two neurons. (*b*) The problem is solved by the release of neurotransmitter from vesicles within the synaptic knobs of the axon. The neurotransmitter diffuses across the synaptic cleft and may trigger an impulse in the postsynaptic neuron. It is thought that when the neurotransmitter combines with receptors in the membrane of the postsynaptic neuron, sodium gates open, permitting sodium to rush into the axon.

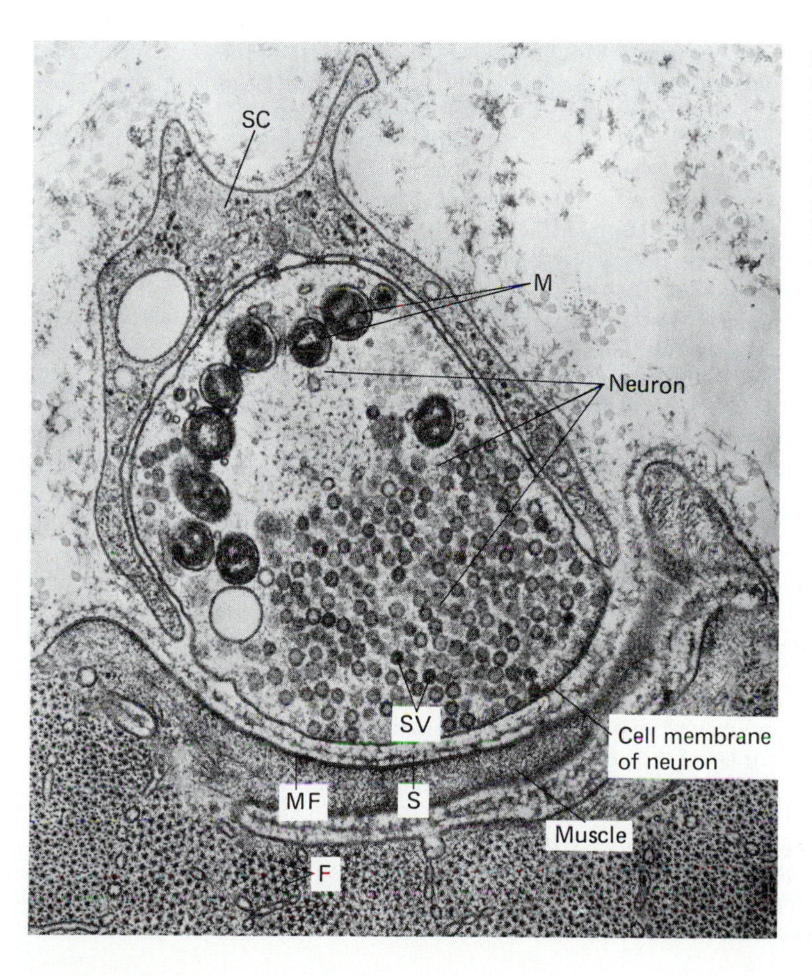

Figure 39–12 Electron micrograph of a synaptic knob filled with synaptic vesicles (approximately ×20,000). This is a motor neuron synapsing with a muscle fiber. *SC*, Schwann cell; *M*, mitochondria; *SV*, synaptic vesicles; *S*, synaptic cleft; *M*, membrane of muscle fiber; *F*, filaments of muscle. (Courtesy of Dr. John Heuser.)

combines with specific receptors on the dendrites or cell bodies of postsynaptic neurons. This opens gates in the membrane, and the resulting redistribution of ions affects the electrical potential of the membrane, either depolarizing or hyperpolarizing it.

In order for repolarization to occur quickly, any excess neurotransmitter must be removed. It is either inactivated by enzymes or reabsorbed into the synaptic vesicles.

Excitatory and Inhibitory Signals

If the effect of a neurotransmitter is to partially depolarize the membrane, the change in potential is called an **excitatory postsynaptic potential,** or **EPSP.** On the other hand, if the effect of the neurotransmitter is to hyperpolarize the postsynaptic membrane, the change in potential is called an **inhibitory postsynaptic potential,** or **IPSP.**

One EPSP by itself is usually too weak to trigger an action potential in the postsynaptic neuron. Its effect is subliminal, that is, below threshold level. However, EPSPs may be added together, a process known as **summation. Temporal summation** occurs when repeated stimuli cause new EPSPs to develop before previous ones have faded. By adding several EPSPs the neuron may be brought to threshold level. In **spatial summation,** several synaptic knobs release transmitter substance at the same time, so that the postsynaptic neuron is stimulated at several points simultaneously. The added effects of this stimulation can also bring the postsynaptic neuron to the critical firing level.

Neurotransmitters

About 30 different substances are now known (or suspected) to be neurotransmitters. Many types of neurons secrete two or even three different types of neurotransmitters. Furthermore, a postsynaptic neuron can have receptors for more than one type of neurotransmitter. Some of its receptors may be excitatory and some inhibitory.

The two neurotransmitters that have been investigated most extensively are acetylcholine and norepinephrine. **Acetylcholine** triggers muscle contraction. It is released not only from motor neurons that innervate skeletal muscle but also by some neurons in the autonomic system and by some neurons in the brain. Cells that release acetylcholine are referred to as **cholinergic neurons.** Acetylcholine has an excitatory effect on skeletal muscle but an inhibitory effect on cardiac muscle. Whether a neurotransmitter excites or inhibits is apparently a property of the postsynaptic receptors with which it combines.

After acetylcholine is released into a synaptic cleft and combines with receptors on the postsynaptic neuron, the remaining molecules must be removed to prevent repeated stimulation of the muscle. The enzyme **cholinesterase** catalyzes the breakdown of acetylcholine into choline and acetate.

Nerve gases and organophosphate biocides inactivate cholinesterase. As a result, the amount of acetylcholine in the synaptic cleft increases with each successive nerve impulse. This causes repetitive stimulation of the muscle fiber and may lead to life-threatening muscle spasms. Should the muscles of the larynx go into spasm, for example, a person may die of asphyxiation.

Norepinephrine is released by sympathetic neurons as well as by many neurons in the brain and spinal cord. Neurons that release norepinephrine are called **adrenergic neurons.** Norepinephrine and the neurotransmitters epinephrine and dopamine belong to a class of compound known as **catecholamines.** After their release from synaptic knobs, catecholamines are removed mainly by re-uptake into the vesicles in the synaptic knobs. Some are inactivated by enzymes such as monoamine oxidase (MAO). Catecholamines affect mood, and many drugs that modify mood do so by altering the levels of these substances in the brain. Other neurotransmitters are listed in Table 39–1.

DIRECTION AND SPEED OF CONDUCTION

In the laboratory it can be demonstrated that an impulse can move in both directions within a single axon. However, in the body an impulse generally stops when it reaches the dendrites, because there is no neurotransmitter present there to con-

TABLE 39–1
Some Neurotransmitters

Substance	*Origin*	*Comments*
Acetylcholine	Myoneural (muscle–nerve) junctions; preganglionic autonomic endings;* postganglionic parasympathetic nerve endings; parts of brain	Inactivated by cholinesterase
Norepinephrine	Postganglionic sympathetic endings; reticular activating system; areas of cerebral cortex, cerebellum, and spinal cord	Reabsorbed by vesicles in synaptic axons knob; inactivated by MAO (monoamine oxidase); norepinephrine level in the brain affects mood
Dopamine	Limbic system; cerebral cortex; basal ganglia; hypothalamus	Thought to affect motor function; may be involved in pathogenesis of schizophrenia;† amount reduced in Parkinson's disease
Serotonin (5-HT, 5-hydroxytryptamine)	Limbic system; hypothalamus; cerebellum; spinal cord	May play a role in sleep; LSD antagonizes serotonin; thought to be inhibitory
Epinephrine	Hypothalamus; thalamus; spinal cord	Identical with the hormone released by the adrenal glands
GABA (γ-Amino-butyric acid)	Spinal cord; cerebral cortex; Purkinje cells in cerebellum	Thought to act as inhibitor in brain and spinal cord
Glycine	Released by neurons mediating inhibition in spinal cord	Acts as an inhibitor
Endorphins and enkephalins	Many parts of CNS	Group of compounds that affect pain perception and other aspects of behavior

*These and other structures listed in this table are discussed in Chapter 40.
†Studies suggest that the brains of schizophrenics have more dopamine receptors than those of nonschizophrenics.

duct it across a synapse. This limitation imposed by the location of the neurotransmitters makes neural transmission unidirectional. Neural pathways thus function as one-way streets, with the usual direction of transmission from the axon of the presynaptic neuron across the synapse to the dendrite or cell body of the postsynaptic neuron.

Compared with the speed of an electric current or the speed of light, the rate of nerve impulse travel is rather slow. Still, some neurons can transmit impulses at a rate of more than 120 meters per second. The rate of conduction of impulses increases as the diameter of the axon increases. Such an increase in diameter decreases the internal electrical resistance along the length of the axon, permitting sodium ions to spread rapidly when they enter the axon. In some invertebrates, giant axons that can transmit impulses very rapidly are employed to conduct danger signals. In vertebrates, the presence of the myelin sheath[1] permits rapid saltatory conduction of impulses along myelinated neurons. The largest, most heavily myelinated neurons conduct impulses most rapidly.

When considering speed of conduction through a sequence of neurons, the number of synapses must be taken into account. Each time an impulse is conducted from one neuron to another, there is a slight synaptic delay (about 0.5 msec). Synaptic delay is due to the time required for the release of transmitter substance, its diffusion, its binding to postsynaptic membrane receptors, and the generation of the action potential in the postsynaptic neuron.

Integration

Neural integration is the process of adding and subtracting incoming signals and determining an appropriate response. Each neuron synapses with hundreds of other neurons. It is the job of the dendrites and cell body of every neuron to inte-

[1] By separating the nodes, where depolarization is possible, the myelin sheath forces an electric current to flow between a depolarized node and the next as yet undepolarized node. This current immediately produces depolarization of that node. Since current flow is faster than depolarization, the fewer and farther apart the nodes are, the more quickly the nerve impulse is conducted.

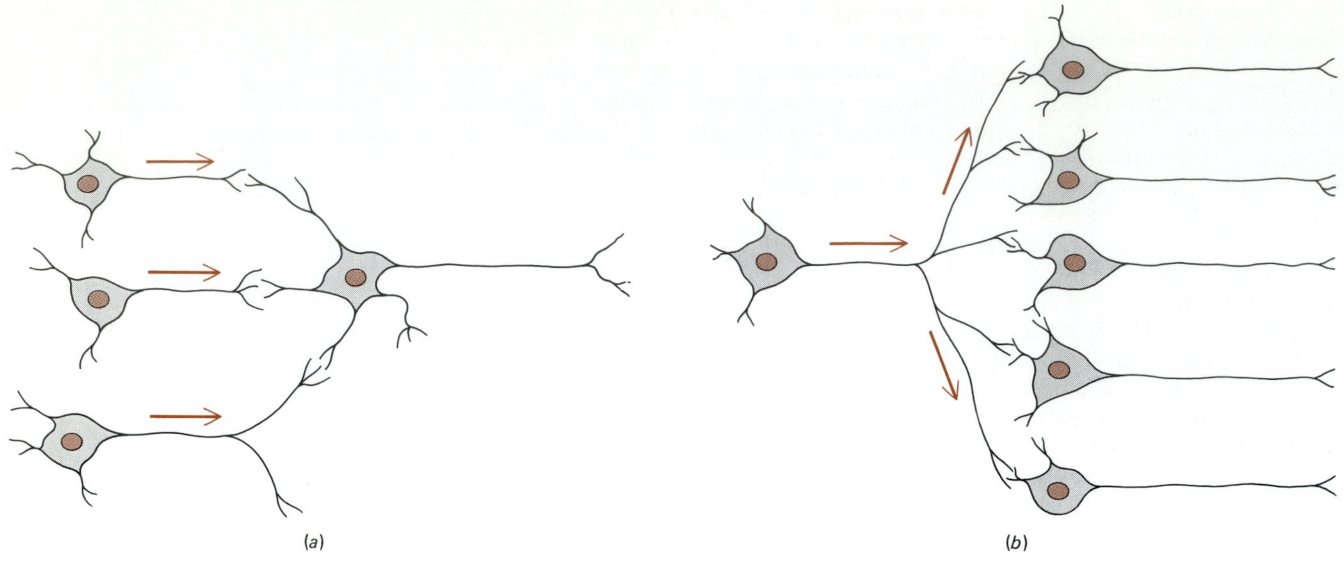

Figure 39–13 Organization of neural circuits. (*a*) Convergence of neural input. Several presynaptic neurons synapse with one postsynaptic neuron. This organization in a neural circuit permits one neuron to receive signals from many sources. (*b*) Divergence of neural output. A single presynaptic neuron synapses with several postsynaptic neurons. This organization allows one neuron to communicate with many others.

grate the numerous messages that they are continually receiving. Hundreds of EPSPs and IPSPs may be tabulated before an impulse is actually transmitted. When sufficient excitatory neurotransmitter predominates, the neuron is brought to the threshold level and an action potential is generated. Such an arrangement permits the neuron and the entire nervous system far greater flexibility and a wider range of response than would be possible if every EPSP generated an action potential.

Every neuron acts as an integrator, sorting through the thousands of bits of information continually bombarding it. Since more than 90% of the neurons in the body are located in the brain and spinal cord most neural integration takes place there.

Organization of Neural Circuits

Neurons are organized into specific pathways, or **circuits.** Within a neural circuit many presynaptic neurons may converge upon a single postsynaptic neuron. In **convergence,** the postsynaptic neuron is controlled by signals from two or more presynaptic neurons (Fig. 39–13). An association neuron in the spinal cord, for instance, may receive information from sensory neurons entering the cord, from neurons originating at other levels of the spinal cord, and even from neurons bringing information from the brain. Information from all of these converging neurons must be integrated before an action potential is generated in the association neuron and an appropriate motor neuron is stimulated.

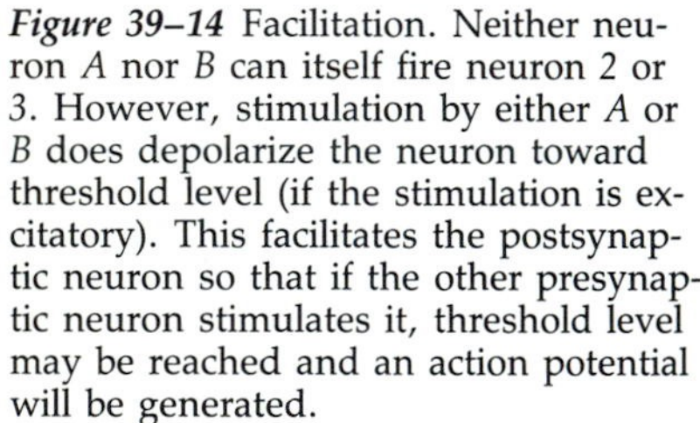

Figure 39–14 Facilitation. Neither neuron *A* nor *B* can itself fire neuron 2 or 3. However, stimulation by either *A* or *B* does depolarize the neuron toward threshold level (if the stimulation is excitatory). This facilitates the postsynaptic neuron so that if the other presynaptic neuron stimulates it, threshold level may be reached and an action potential will be generated.

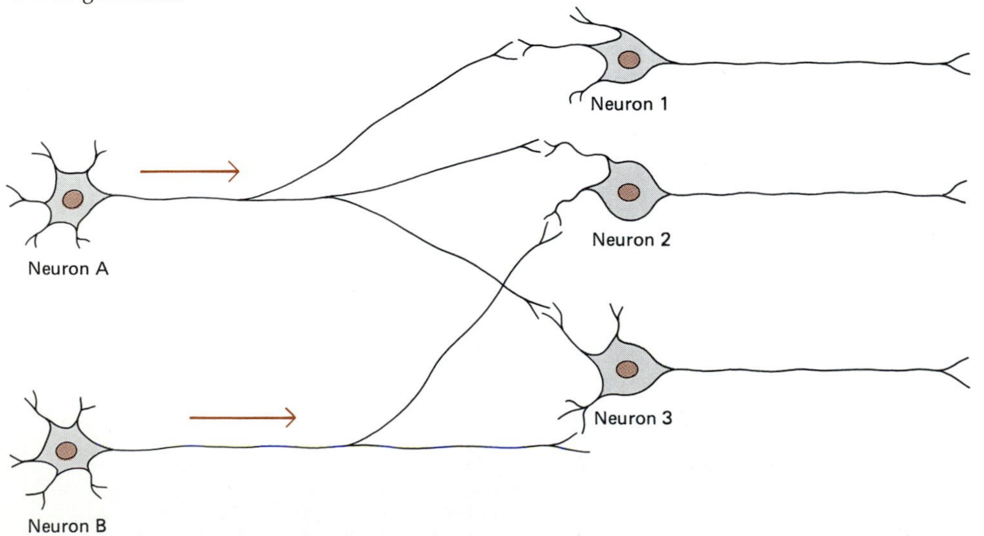

In **divergence,** a single presynaptic neuron stimulates many postsynaptic neurons (Fig. 39–13). Each presynaptic neuron may synapse with up to 25,000 or more different postsynaptic neurons. In **facilitation,** the neuron is brought close to threshold level by EPSPs from various presynaptic neurons but is not yet at the threshold level. The neuron can be easily excited by a new EPSP. Figure 39–14 illustrates facilitation: Neither neuron *A* nor neuron *B* can by itself fire neuron *2* or *3*. However, when either neuron *A* or *B* is fired, neurons *2* and *3* are facilitated. Then when the other presynaptic neuron fires, the postsynaptic neuron receives sufficient neurotransmitter to generate an action potential.

A very important type of neural circuit is the **reverberating circuit.** This is a neural pathway arranged so that a neuron collateral synapses with an association neuron (Fig. 39–15). The association neuron synapses with a neuron in the sequence that can send new impulses again through the circuit. New impulses can be generated again and again until the synapses involved become fatigued (owing to depletion of neurotransmitter), or until they are stopped by some sort of inhibition. Reverberating circuits are thought to be important in rhythmic breathing, in maintaining alertness, and perhaps in short-term memory.

Reflex Action

A **reflex action** is a stereotyped, automatic response to a given stimulus that depends only on the anatomic relationships of the neurons involved. Reflexes are functional units of the nervous system, and many physiological mechanisms depend upon reflex actions. A reflex typically involves part of the body rather than the whole. Constriction of the pupil in response to bright light is an example. Breathing, heart rate, salivation, and regulation of blood pressure and temperature are other examples of reflex actions. A change in body temperature, for instance, acts as a stimulus, causing the temperature-regulating center in the brain to mobilize homeostatic mechanisms that bring body temperature back to normal. Many responses to external stimuli, such as withdrawing from painful stimuli, are also reflex actions.

The simple knee jerk, or patellar reflex, is an example of a very simple type of reflex requiring a chain of only two sets of neurons. Because only one group of synapses is involved, this type of reflex is called a **monosynaptic reflex.** Yet even this simple reflex action requires the sequence of information flow through the nervous system discussed earlier—reception, transmission, integration, and response by an effector. In the knee jerk the receptors are muscle spindles that respond to stretch stimuli when the tendon is tapped suddenly. Afferent neurons transmit the impulses to the spinal cord, where integration takes place at the synapses between afferent and efferent neurons. An efferent neuron then transmits the impulse to the effector cells. The muscle fibers contract, resulting in a sudden straightening of the leg.

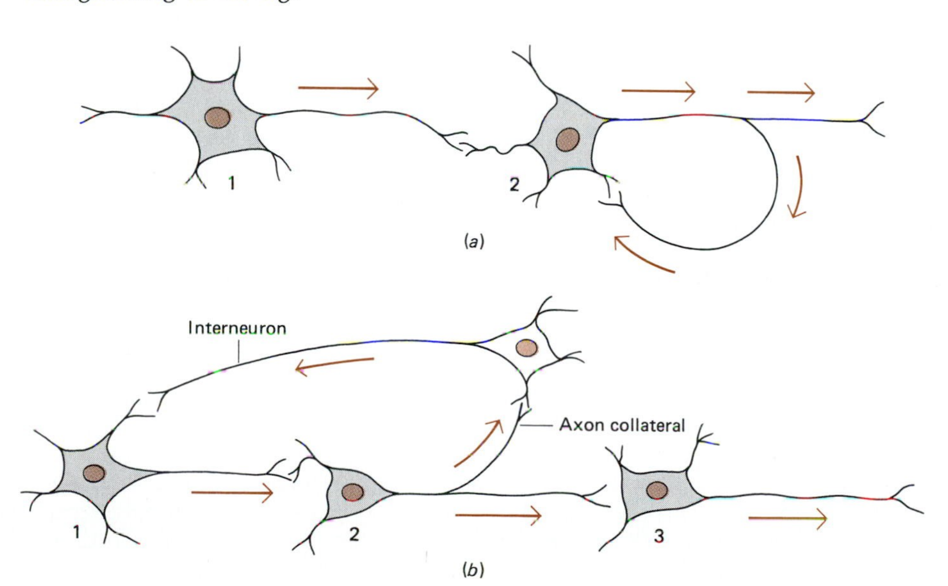

Figure 39–15 Reverberating circuits. (*a*) A simple reverberating circuit in which an axon collateral of the second neuron turns back and synapses with one of its own dendrites so that the neuron continues to stimulate itself. (*b*) In this neural circuit an axon collateral of the second neuron synapses with an interneuron. The interneuron synapses with the first neuron in the sequence. New impulses are triggered again and again in the first neuron, causing reverberation. See text for explanation of factors that can bring a halt to this reverberation.

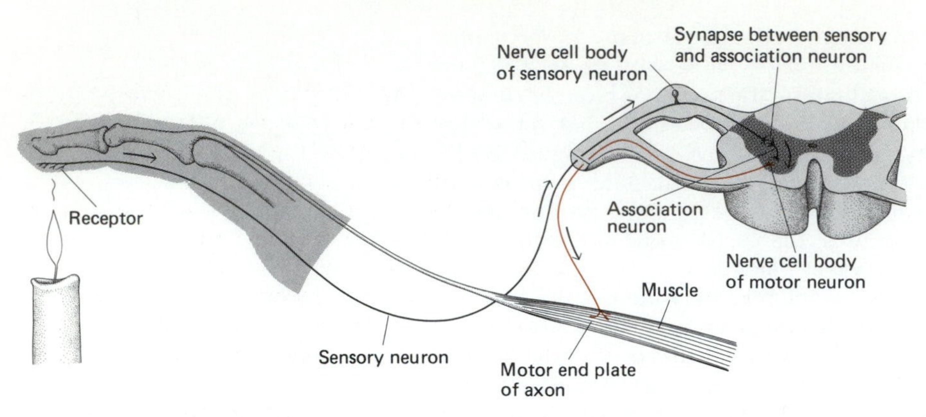

Figure 39–16 A withdrawal reflex is polysynaptic. The one shown here involves a chain of three neurons. A sensory neuron transmits the message from the receptor to the CNS, where it synapses with an association neuron. Then an appropriate motor neuron (shown in color) transmits an impulse to the muscles that move the hand away from the flame (the response).

Withdrawal reflexes are **polysynaptic,** requiring participation of three sets of neurons (Fig. 39–16). For example, an accidental burn on your finger would cause you to jerk your hand away from the painful stimulus even before you became aware of the pain. The pain receptors (dendrites of sensory neurons) in your fingers send messages through afferent neurons to the spinal cord. There each neuron synapses with an association neuron. Integration takes place and impulses are sent via appropriate efferent neurons to muscles in the arm and hand, instructing them to contract, jerking the hand away from the harmful stimulus. At the same time that the association neuron sends a message to the motor neuron, it may also dispatch a message up the neurons in the spinal cord to the brain. Very quickly you become conscious of your plight and can make the decision to hold your hand under cold water. This is not part of the reflex action, however.

That the brain is not essential to many reflex actions can be demonstrated by an experiment often performed in college physiology laboratories. The brain of a frog is destroyed, creating a "spinal" animal. Then a piece of acid-soaked paper is applied to the animal's back. No matter how many times the piece of paper is placed on the skin, one leg will invariably move upward and flick it away. This experiment also demonstrates that reflex actions are stereotyped and automatic. A frog with a brain might make the response two or three times, but eventually would try a different response—perhaps hop away.

Some reflex actions—for example, the pupil reflex—do involve the brain, but only the so-called lower parts, which are functionally similar to the spinal cord and have nothing to do with conscious thought. Some reflex actions can be consciously inhibited or facilitated. An example is the reflex that voids the urinary bladder when it fills with urine. In babies, urination is a reflex action; when the bladder fills with urine and is stretched to a critical point, a sphincter muscle relaxes and urine passes out of the body. In early childhood we learn to facilitate the reflex consciously, stimulating it before bladder pressure reaches a critical level. We also learn to inhibit the reflex consciously if the bladder becomes full at an inconvenient time or place.

Much of the behavior in a relatively simple animal such as a sea star can be explained in terms of reflex actions. The sea star can extend and retract its tube feet and make postural movements associated with ambulation. The extension and retraction of the tube feet are unoriented reflex responses. The apparent coordination of all the tube feet, retracting in unison when a wave washes over the animal, is simply the sum of individual responses to a common stimulus. Numerous aspects of crustacean behavior are primarily reflex responses—the withdrawal of the eye stalk, the opening and closing of the claws, and movements associated with escape, defense, feeding, and copulation.

Although we might, in theory, construct an animal capable of many responses simply out of reflexes, reflexes in complex animals account for only a small part of total behavior. Reflex actions and behavior are discussed further in Chapter 48.

SUMMARY

I. Information flow through the nervous system begins with reception. Information is then transmitted to the CNS via afferent neurons. Integration takes place within the CNS, and appropriate nerve impulses are then transmitted by efferent neurons to the effectors that carry out the response.

II. The vertebrate nervous system is divided into central nervous system (CNS) and peripheral nervous system (PNS).
  A. The CNS consists of brain and spinal cord.
  B. The PNS consists of sensory receptors and nerves; it may be divided into somatic and autonomic systems.

III. The neuron is the structural and functional unit of the nervous system.
  A. A typical multipolar neuron consists of a cell body from which project many branched dendrites and a single, long axon.
  B. The axon is surrounded by a neurilemma. Many axons are also enveloped in a myelin sheath.

IV. A nerve consists of hundreds of axons wrapped in connective tissue; a ganglion is a mass of cell bodies.

V. Transmission of a neural impulse is an electrochemical mechanism.
  A. A neuron that is not transmitting an impulse has a resting potential.
    1. The inner surface of the plasma membrane is negatively charged as compared to the outside.
    2. Sodium–potassium pumps continuously transport sodium ions out of the neuron, and transport potassium ions in.
    3. Potassium ions are able to leak out more readily than sodium ions are able to leak in.
  B. Excitatory stimuli are thought to open sodium gates in the plasma membrane. This permits sodium to enter the cell and depolarize the membrane. Inhibitory stimuli hyperpolarize the membrane.
  C. When the extent of depolarization reaches threshold level, an action potential may be generated.
    1. The action potential is a wave of depolarization that spreads along the axon.
    2. The action potential obeys an all-or-none law.
    3. As the action potential moves down the axon, repolarization occurs very quickly behind it.
  D. Saltatory conduction takes place in myelinated neurons. In this type of transmission, depolarization skips along the axon from one node of Ranvier to the next.
  E. Excitability of a neuron can be affected by calcium concentration and by certain substances such as local anesthetics and biocides.
  F. Synaptic transmission generally depends upon release of a neurotransmitter from vesicles in the synaptic knobs of the presynaptic neuron.
    1. The neurotransmitter diffuses across the synaptic cleft and combines with specific receptors on the postsynaptic neuron.
    2. This may cause an excitatory postsynaptic potential (EPSP) or an inhibitory postsynaptic potential (IPSP).
    3. Temporal or spatial summation may bring the postsynaptic neuron to the threshold level.
  G. The largest, most heavily myelinated neurons conduct impulses most rapidly.

VI. Neural integration is the process of adding and subtracting EPSPs and IPSPs and determining an appropriate response.

VII. Complex neural pathways are possible because of such neuronal associations as convergence, divergence, and facilitation.

VIII. A simple reflex action includes reception of a stimulus, transmission of impulses to the CNS via an afferent neuron, integration within the CNS, transmission of impulses via a motor neuron to an effector, and response by the effector.

POST-TEST

1. Afferent neurons transmit information from the ________ ________ to the ________ ________ ________.

2. Functional connections between neurons are called ________.

3. Changes in the environment that can be detected by an organism are termed ________.

4. The peripheral nervous system consists of sensory ________ and ________.

5. A nerve cell is properly referred to as a ________; the supporting cells of nervous tissue are called ________.

6. Dendrites are specialized to ________ ________; the axon functions to ________ from the ________ ________ to a ________.

7. The cellular sheath is important in ________________.

8. In some peripheral neurons, Schwann cells produce both a ________ ________ and a ________ ________.

9. A ganglion consists of a mass of ________ ________.

10. The ________ ________ of a neuron is due mainly to the outward diffusion of potassium ions along their concentration gradient.

11. ________ stimuli are thought to open sodium gates, thereby permitting sodium to rush into the cell.

12. The passage of sodium ions into the neuron ________ the cell membrane.

13. A wave of depolarization that travels down the axon is called a nerve impulse or ________ ________.

14. Because there is no variation in the intensity of an action potential the neuron is said to obey an

_______________________________ law.

15. In saltatory conduction, depolarization skips along the axon from _______________.

16. When insufficient calcium ions are present, the _______ _______ is lowered and the neuron fires _______ (*more/less*) easily.

17. When an impulse reaches the synaptic knobs, it stimulates the release of _______.

18. The neurotransmitter that stimulates muscle contraction is _______.

19. Adrenergic neurons release the neurotransmitter _______.

20. In _______, a single presynaptic neuron stimulates many postsynaptic neurons.

21. A stereotyped, automatic response to a given stimulus that depends only on the anatomic relationships of the neurons involved is called a _______ _______.

22. In a typical withdrawal reflex, pain receptors send messages through _______ neurons to the _______ _______, where _______ takes place.

23. Label the following diagram. (Refer to Fig. 39–2(*a*) as necessary.)

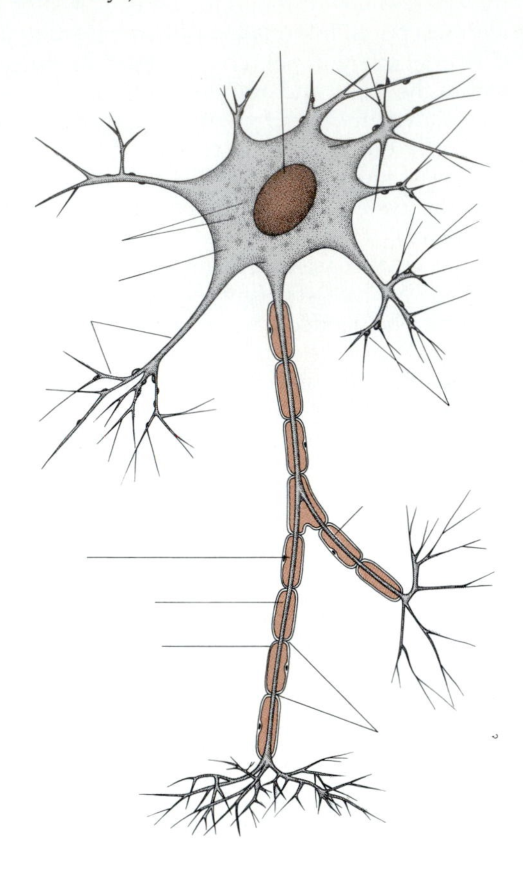

REVIEW QUESTIONS

1. Distinguish between a neuron and a nerve.
2. Imagine that a very unfriendly-looking monster suddenly appears before you. What processes must take place within your nervous system before you can make your escape?
3. Contrast the functions of an afferent and efferent neuron.
4. Describe the functions of the following:
   a. myelin
   b. a ganglion
   c. neuroglia
   d. dendrites
   e. axon
5. What is meant by the resting potential of a neuron? How do sodium-potassium pumps contribute to the resting potential?
6. What is an action potential? What is responsible for it?
7. What is the all-or-none law?
8. Contrast saltatory conduction with conduction in an unmyelinated neuron.
9. How is neural function affected by the presence of too much calcium? too little calcium?
10. Describe the functions of the following substances:
    a. acetylcholine
    b. cholinesterase
    c. norepinephrine
11. Contrast convergence and divergence.
12. What is summation? Describe facilitation.
13. Draw a diagram illustrating a withdrawal reflex, and label each neuron involved. Indicate the direction of neural transmission.

40

Neural Control: Nervous Systems

OUTLINE

I. Invertebrate nervous systems
- *A.* Nerve nets and radial systems
- *B.* Bilateral nervous systems

II. The vertebrate brain
- *A.* The hindbrain
- *B.* The midbrain
- *C.* The forebrain

III. The human central nervous system
- *A.* The spinal cord
- *B.* The brain
  - *1.* The reticular activating system
  - *2.* The limbic system
  - *3.* Brain waves
  - *4.* Sleep
  - *5.* Learning and memory
  - *6.* Effects of environmental experience on the brain

IV. The peripheral nervous system
- *A.* The cranial nerves
- *B.* The spinal nerves
- *C.* The autonomic system

V. Effects of drugs on the nervous system

LEARNING OBJECTIVES

After you have read this chapter you should be able to:

1. Compare nerve nets with bilateral nervous systems.
2. Describe five specific advances characteristic of bilateral nervous systems, and compare them with systems found in radially symmetric animals.
3. Compare the nervous system of a planarian with that of an arthropod, a mollusk, and a vertebrate.
4. Trace the development of the principal parts of the vertebrate brain (e.g., cerebrum, cerebellum, medulla) from the forebrain, midbrain, and hindbrain.
5. Discuss the differences in relative size and function of the principal parts of the brain in fish, amphibians, reptiles, birds, and mammals.
6. Describe the structures that protect the human brain and spinal cord.
7. Describe the spinal cord, and list its principal functions.
8. Describe the three functional areas of the human cerebral cortex and give their functions.
9. Give the functions of the reticular activating system and the limbic system.
10. Identify three main types of brain wave patterns, and relate them to specific types of activity.
11. Contrast REM sleep with non-REM sleep.
12. Review current theories of learning and memory as presented in this chapter.
13. Cite experimental evidence linking environmental stimuli with demonstrable changes in the brain and with learning and motor abilities.
14. Identify the structures and functions of the peripheral nervous system. (At the option of your instructor, list the cranial nerves and give their function.)
15. Describe the functions of the autonomic system and contrast the functions of the sympathetic and parasympathetic systems, giving examples of the effects of these systems on specific organs.
16. Describe the actions and effects of various drugs on mood states.

Even though some organisms do not have a brain, or even nerves, they still must make at least some response to stimuli in order to survive. In a one-celled organism such as an amoeba, the entire cellular surface may be sensitive to stimuli such as light, heat, touch, and certain chemicals. Cytoplasm itself is **irritable,** that is, sensitive to changes in the environment. Thus, information may be conducted throughout the cytoplasm of even a single-celled organism enabling it to respond appropriately, either by moving toward a stimulus or by withdrawing from it. Some protozoa do have specialized receptors, and in ciliates there is a system for coordinating ciliary action. However, in such organisms the range and types of responses are very limited and stereotyped. For varied and sophisticated responses to be possible, an organism must have a complex nervous system.

Invertebrate Nervous Systems

There is no nervous system in the sponge. Whatever responses it makes are at the cellular level. Among other invertebrates there are two main types of nervous systems: nerve nets and bilateral nervous systems.

NERVE NETS AND RADIAL SYSTEMS

The simplest organized nervous system is the **nerve net** found in *Hydra* and other cnidarians (Fig. 40–1). In a nerve net the nerve cells are scattered throughout the body. No central control organ and no definite pathways are present. Sensory cells, specialized to receive stimuli, transmit information to ganglion cells (interneurons), which are the main cells of the nerve net. From the ganglion cells, information is passed on somewhat haphazardly to neurosecretory cells, which apparently send chemical messages to effector cells such as the cnidocytes (stinging cells). Impulses are transmitted in all directions, becoming less intense as they spread from the region of initial stimulation. If the stimulus is strong, the number of neurons of the net receiving the message will be more than if the stimulus is weak.

Since it produces responses that involve the body as a whole, or large parts of it at the same time, such a diffuse pattern of transmission is permissible in a radially symmetric animal with sluggish, slow means of locomotion. Responses in cnidarians are limited to discharge of nematocysts and contractions that permit movement. In some cnidarians several nerve nets are present, some mediating quick responses and some mediating slow responses. These nerve nets may work together to integrate information.

The somewhat more complex nervous system of the echinoderm consists of a circumoral (around the mouth) nerve ring from which a large radial nerve extends into each arm. These nerves coordinate movement of the animal. In asteroids (sea stars), a nerve net mediates the responses of the dermal gills to tactile stimulation.

BILATERAL NERVOUS SYSTEMS

In bilaterally symmetric animals the nervous system is usually more complex than in radially symmetric animals. A bilateral form of symmetry usually reflects a more active way of life, with the need to respond quickly to the environment in a sophisticated manner. The following trends can be identified:

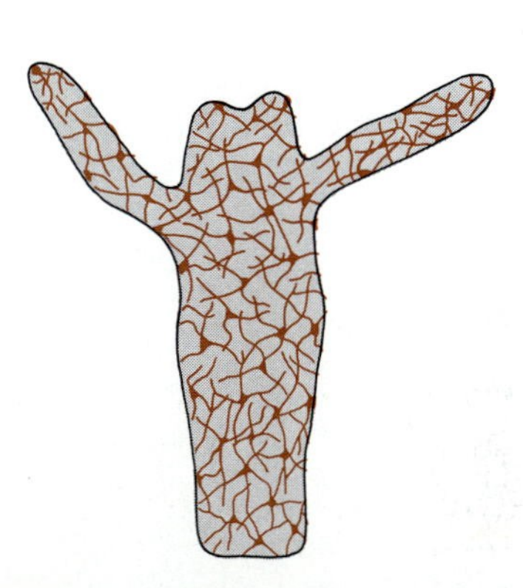

Figure 40–1 The nerve net of *Hydra* and other cnidarians is the simplest organized nervous system. No central control organ and no definite neural pathways are present.

1. Increased number of nerve cells.
2. Concentration of nerve cells to form thick cords or masses of tissue, which become nerves, nerve cords, ganglia, and brain.
3. Specialization of function. For example, transmission of nerve impulses in one direction requires both **afferent** nerves, which conduct impulses toward a central nervous system, and **efferent** nerves, which transmit impulses away from the central nervous system and to the effector cells. Various parts of the central nervous system are usually specialized also to perform specific functions, so that distinct structural and functional regions can be identified.
4. Increased number of association neurons and complexity of synaptic contacts. This permits much greater integration of incoming messages, provides a greater range of responses, and allows far more precision in responses.

5. Cephalization, or formation of a head. A bilaterally symmetric animal generally moves in a forward direction, so concentration of sense organs at the front end of the body is important for detection of an enemy quickly enough to escape or for seeing or smelling food in time for its capture. Response can be more rapid if these sense organs are linked by short pathways to decision-making nerve cells nearby. Therefore, nerve cells are also usually concentrated in the head region, comprising a definite brain.

In planarian flatworms, there are concentrations of nerve cells in the head region referred to as **cerebral ganglia** (Fig. 40–2). These serve as a primitive "brain" and exert some measure of control over the rest of the nervous system. Two ventral longitudinal nerve cords extend from the ganglia to the posterior end of the body. Transverse nerves connect the brain with the eyespots and anterior end of the body. This arrangement is referred to as a **ladder-type** nervous system.

In annelids and arthropods, there is also typically a pair of ventrally located longitudinal nerve cords (Figs. 40–3 and 40–4). The cell bodies of the nerve cells are massed into pairs of ganglia located in *each* body segment. Afferent and efferent neurons are located in lateral nerves that link the ganglia with muscles and other body structures.

When the earthworm brain is removed, the animal can move almost as well as before, but when it bumps into an obstacle it persists in futile efforts to move forward instead of turning aside. The brain is therefore necessary for adaptive movements; it enables the earthworm to respond appropriately to environmental change.

In complex arthropods, especially in insects, some of the abdominal ganglia move anteriorly in the course of embryonic development and fuse with the thoracic ganglia. In some arthropods the cerebral ganglia are somewhat brainlike in that specific functional regions have been identified in them (Fig. 40–4).

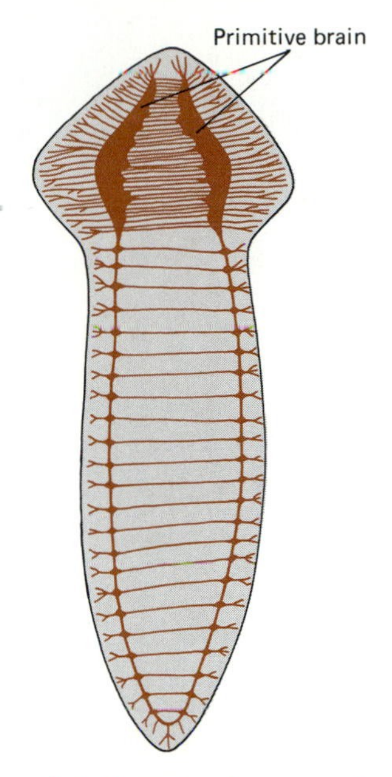

Figure 40–2 Planarian flatworms have a ladder-type nervous system. Cerebral ganglia in the head region serve as a simple brain and, to some extent, control the rest of the nervous system.

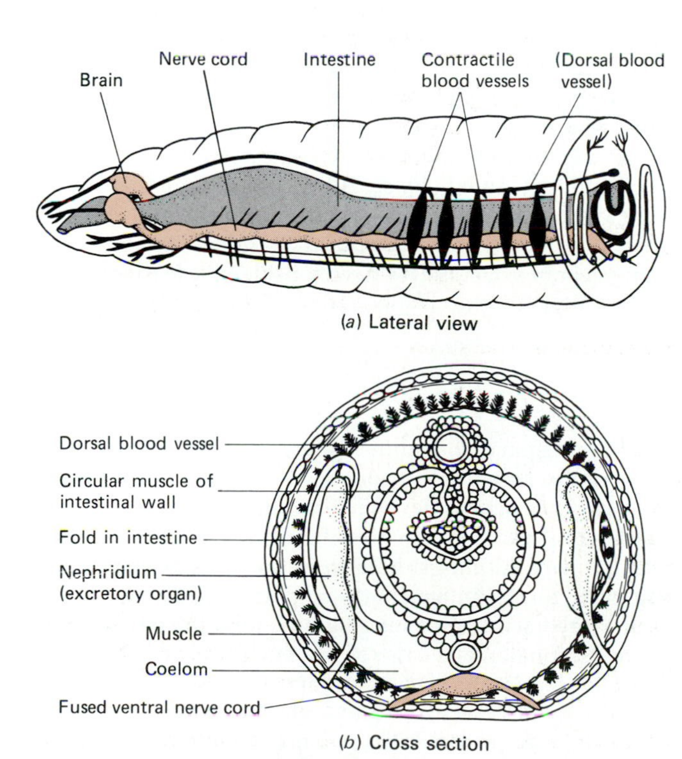

Figure 40–3 The nervous system of the earthworm is typical of those found in other annelids. The cell bodies of the neurons are located in ganglia found in each body segment. They are connected by the paired ventral longitudinal nerve cords.

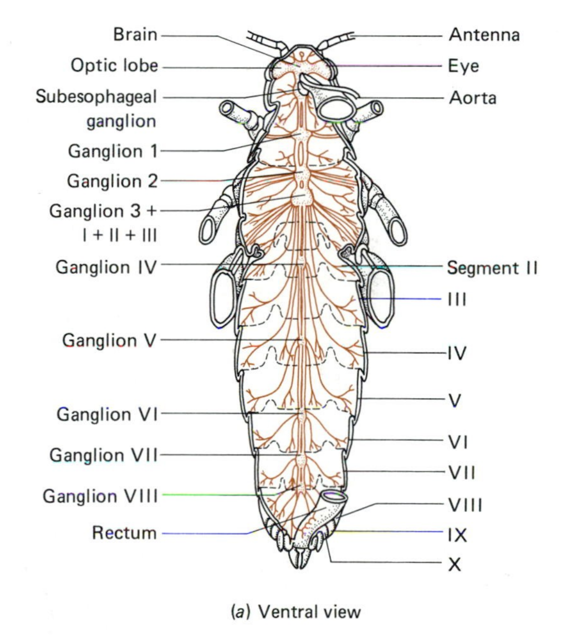

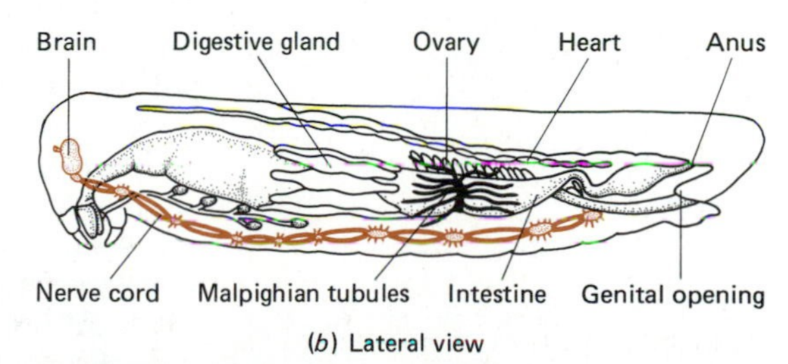

Figure 40–4 In the insect nervous system the cerebral ganglia serve as a simple brain. Two ventral nerve cords are present.

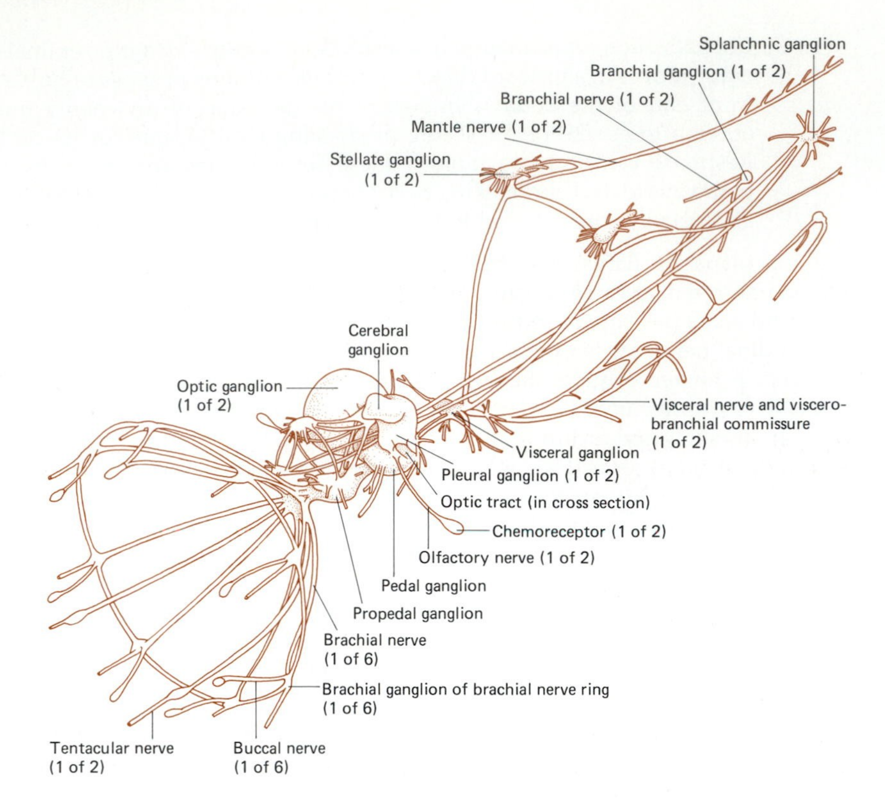

Figure 40–5 The cephalopod nervous system. Several ganglia including the cerebral, optic, and pedal ganglia make up the "brain." These structures contain millions of nerve cell bodies.

In mollusks there are typically at least three pairs of ganglia: **cerebral ganglia,** found dorsal to the esophagus (which serve as a coordinating center for complex reflexes and which also have a motor function); **visceral ganglia,** located among the organs (which control shell opening and closing); and **pedal ganglia,** located in the foot (which control the movement of the foot). The visceral and pedal ganglia are connected to the cerebral ganglia by nerve cords.

In cephalopods, such as the octopus, there is a tendency toward concentration of the nerve cells in a central region (Fig. 40–5). All the ganglia are massed in the **circumesophageal ring,** which contains about 168 million nerve cell bodies. With this complex brain, it is no wonder that the octopus is considered to be among the most intelligent of the invertebrates. Octopuses have considerable learning abilities and can be taught quite complex tasks.

The Vertebrate Brain

In the early embryo the brain and spinal cord differentiate from a single tube of tissue, the **neural tube.** Anteriorly, the tube expands and differentiates into the structures of the brain. Posteriorly, the tube develops into the spinal cord. Brain and spinal cord remain continuous and their cavities communicate. As the brain begins to differentiate, three primary swellings develop in the anterior end of the neural tube. These give rise to the **forebrain, midbrain,** and **hindbrain** (Fig. 40–6). As indicated in Table 40–1, the forebrain further subdivides to form the **telencephalon** and **diencephalon.** The telencephalon gives rise to the cerebrum, and the diencephalon to the thalamus and hypothalamus. The hindbrain subdivides to form the **metencephalon,** which gives rise to the cerebellum and pons, and the **myelencephalon,** which gives rise to the medulla. The medulla, pons, and midbrain make up the **brain stem,** the elongated portion of the brain that looks like a stalk holding up the cerebrum.

At the most posterior part of the brain, the **medulla** is continuous with the spinal cord. Its cavity, the **fourth ventricle,** communicates with the **central canal** of the spinal cord. The fourth ventricle communicates with the **third ventricle** (located within the diencephalon) by means of a channel, the cerebral aqueduct, that runs

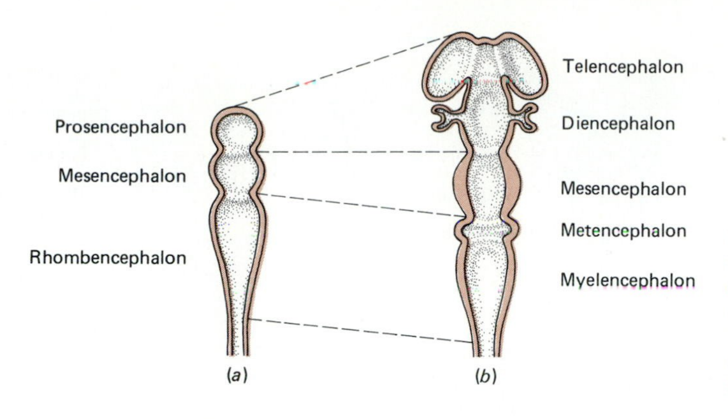

Figure 40–6 Early in the development of the vertebrate embryo, the anterior end of the neural tube differentiates into the forebrain, midbrain, and hindbrain. These primary divisions subdivide and then give rise to specific structures of the adult brain (see Table 40–1).

through the midbrain. The third ventricle in turn is connected with the **lateral** (first and second) **ventricles** within the cerebrum by way of the **interventricular foramen.**

As illustrated in Figure 40–7, all vertebrates from fish to mammals have the same basic brain structure. Certain parts of the brain are specialized to perform specific functions, and some regions, such as the cerebellum and cerebrum, are vastly more complex in the higher vertebrates.

THE HINDBRAIN

The walls of the medulla are thick and made up largely of nerve tracts that connect the spinal cord with various parts of the brain. In complex vertebrates the medulla contains discrete nuclei that serve as **vital centers** regulating respiration, heart beat, and blood pressure. Other reflex centers in the medulla regulate such activities as swallowing, coughing, and vomiting.

The size and shape of the **cerebellum** vary greatly among the vertebrate classes. Development of the cerebellum in different animals is correlated roughly with the extent and complexity of muscular activity. In some fish, birds, and mammals the cerebellum is highly developed, whereas it tends to be small in cyclostomes, amphibians, and reptiles. The cerebellum coordinates muscle activity and is responsible for muscle tone, posture, and equilibrium. Injury or removal of the cerebellum results not in paralysis but in impairment of muscle coordination. A bird without a cerebellum is unable to fly, and its wings thrash about jerkily. When the human cerebellum is injured by a blow or by disease, muscular movements are uncoordinated. Any activity requiring delicate coordination, such as threading a needle, is very difficult, if not impossible.

In mammals, a large mass of fibers known as the **pons** connects various parts of the brain. The pons forms a bulge on the anterior surface of the brain stem. The pons contains nuclei that relay impulses from the cerebrum to the cerebellum.

TABLE 40–1
Differentiation of CNS Structures

Early Embryonic Divisions	*Subdivisions*	*Derivatives in Adult*	*Cavity*
Brain			
Forebrain (prosencephalon)	Telencephalon	Cerebrum	Lateral ventricles (first and second ventricles)
	Diencephalon	Thalamus, hypothalamus, epiphysis (pineal body)	Third ventricle
Midbrain (mesencephalon)	Midbrain	Optic lobes in fish and amphibians; superior and inferior colliculi	Cerebral aqueduct
Hindbrain (rhombencephalon)	Metencephalon Myelencephalon	Cerebellum, pons Medulla oblongata	Fourth ventricle
Spinal cord		Spinal cord	Central canal

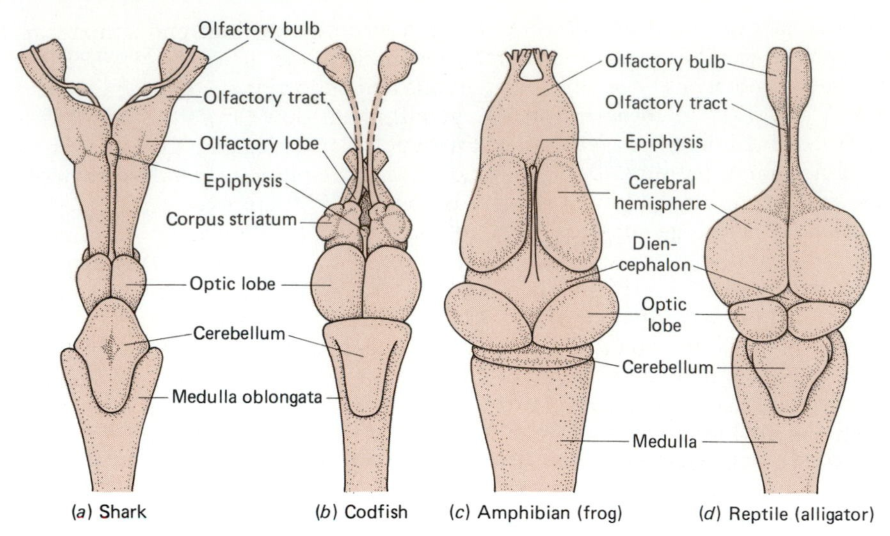

Figure 40–7 Comparison of the brains of members of six vertebrate classes indicates basic similarities and evolutionary trends. Note that different parts of the brain may be specialized in the various groups. For example, the large olfactory lobes in the shark brain (*a*) are essential to this predator's highly developed sense of smell. During the course of evolution, the cerebrum and cerebellum have become larger and more complex. In the mammal (*f*) the cerebrum is the most prominent part of the brain; the cerebral cortex, the thin outer layer of the cerebrum, is highly convoluted (folded), which greatly increases its surface area.

THE MIDBRAIN

In fish and amphibians the midbrain is the most prominent part of the brain. In these animals the midbrain is the main association area. It receives incoming sensory information, integrates it, and sends decisions to appropriate motor nerves. The dorsal portion of the midbrain is differentiated to some degree in these lower vertebrates. For example, the **optic lobes,** specialized for visual interpretations, are part of the midbrain. In reptiles, birds, and mammals, many of the functions of the optic lobes are assumed by the cerebrum. In mammals, the midbrain consists of the **superior colliculi,** which are centers for visual reflexes such as pupil constriction, and the **inferior colliculi,** which are centers for certain auditory reflexes. The mammalian midbrain also contains the **red nucleus,** a center that integrates information regarding muscle tone and posture.

THE FOREBRAIN

The forebrain consists of the diencephalon and telencephalon. The diencephalon contains the thalamus and hypothalamus (Fig. 40–7). In all vertebrate classes the **thalamus** is a relay center for motor and sensory messages. In mammals all sensory messages (except those from the olfactory receptors) are delivered to the thalamus before being relayed to the sensory areas of the cerebrum.

Below the thalamus, the **hypothalamus** forms the floor of the third ventricle. The hypothalamus is the principal integration center for the regulation of the viscera (internal organs). It provides input to centers in the medulla and spinal cord that regulate activities such as heart rate, respiration, and digestive system function. The hypothalamus also links the nervous and endocrine systems. In fact, the pituitary gland (an important endocrine gland) hangs down from the hypothalamus. **Releasing hormones** produced by the hypothalamus regulate the secretion of several hormones produced by the anterior lobe of the pituitary gland. In reptiles, birds, and mammals, body temperature is controlled by the hypothalamus. The hypothalamus also regulates appetite and water balance, and is involved in emotional and sexual responses.

The telencephalon differentiates to form the cerebrum, and, in most vertebrate groups, the **olfactory bulbs.** The olfactory bulbs are concerned with the chemical sense of smell, the dominant sense in most aquatic and terrestrial vertebrates. In fact, much of brain development in vertebrates appears to be focused upon the integration of olfactory information. In fish and amphibians, the cerebrum is almost entirely devoted to the integration of such incoming sensory information.

Birds are an exception among the vertebrates in that their sense of smell is generally poorly developed. In them, however, a part of the cerebrum called the **corpus striatum** is greatly developed. This structure is thought to control the innate, stereotyped, yet complex action patterns characteristic of birds. Just above the corpus striatum is a region thought to govern learning in birds.

In most vertebrates the cerebrum is divided into right and left hemispheres. Most of the cerebrum is made of **white matter** consisting mainly of axons connecting various parts of the brain. In mammals and most reptiles there is a layer of **gray matter** called the **cerebral cortex** that makes up the outer portion of cerebral tissue. Certain reptiles possess a different type of cortex, not found in lower vertebrates, known as the **neopallium;** it serves as an association area, a region that links sensory and motor functions and is responsible for higher functions such as learning. The neopallium is much more extensive in mammals, making up the bulk of the cerebrum,[1] which becomes the most prominent part of the brain. In mammalian embryonic development, in fact, the cerebrum expands and grows backwards, covering many of the other brain structures.

In mammals the cerebrum is responsible for many of the functions that are performed by other parts of the brain in lower vertebrates. In particular, it has many complex association functions lacking in reptiles, amphibians, and fish. In small or simple mammals, the cerebral cortex may be smooth. However, in large, complex mammals, the surface area is greatly expanded by numerous folds called **convolutions,** or **gyri.** The furrows between them are called **sulci** when shallow and **fissures** when deep.

The Human Central Nervous System

As in other vertebrates, the human central nervous system (CNS) consists of the brain and spinal cord (Fig. 40–8). These soft, fragile organs are carefully protected. Both are encased in bone and wrapped in three layers of connective tissue collectively termed the **meninges.** The three meningeal layers are the tough, outer **dura mater,** the middle **arachnoid,** and the thin, vascular **pia mater** that adheres closely to the tissue of the brain and spinal cord (Fig. 40–9). **Meningitis** is a disease in which these coverings become infected and inflamed.

Between the arachnoid and the pia mater is the **subarachnoid space,** which contains **cerebrospinal fluid.** This shock-absorbing fluid cushions the brain and spinal cord and prevents them from bouncing against the bones of the vertebrae or skull with every movement. Cerebrospinal fluid also circulates through the ventricles of the brain. It is produced by special networks of capillaries called the **choroid plexuses** that project from the pia mater into the ventricles. After circulating through the CNS, cerebrospinal fluid is reabsorbed into the blood.

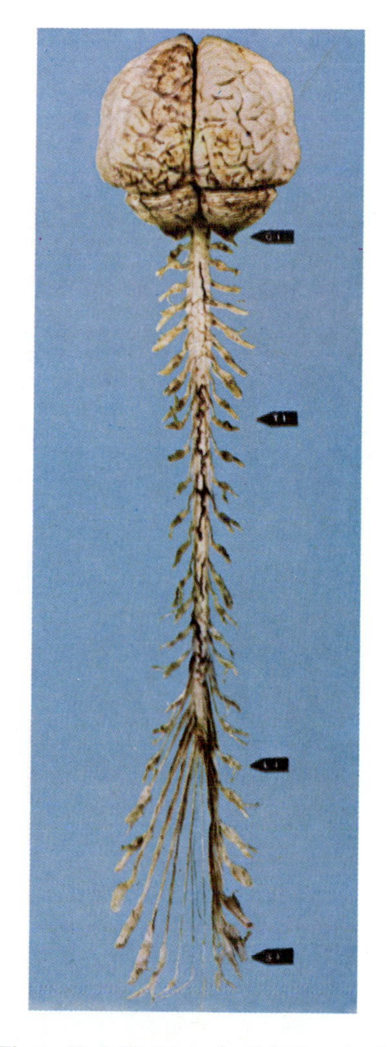

Figure 40–8 Photograph of human brain and spinal cord. The roots of the spinal nerves are still attached. Note the group of nerves that extend caudally from the lower region of the cord. Because they resemble a horse's tail they are referred to as the cauda equina. These nerves have been left undisturbed on the right but have been fanned out on the left. (Dissection by Dr. M. C. E. Hutchinson, Department of Anatomy, Guy's Hospital Medical School, London, England. From Williams and Warwick (eds.): *Gray's Anatomy.*)

THE SPINAL CORD

The tubular **spinal cord** extends from the base of the brain to the level of the second lumbar vertebra. It has two main functions: (1) to transmit impulses to and from the brain and (2) to control many reflex activities. A cross section through the spinal

[1] In humans about 90% of the cerebral cortex is neopallium and consists of six distinct cell layers.

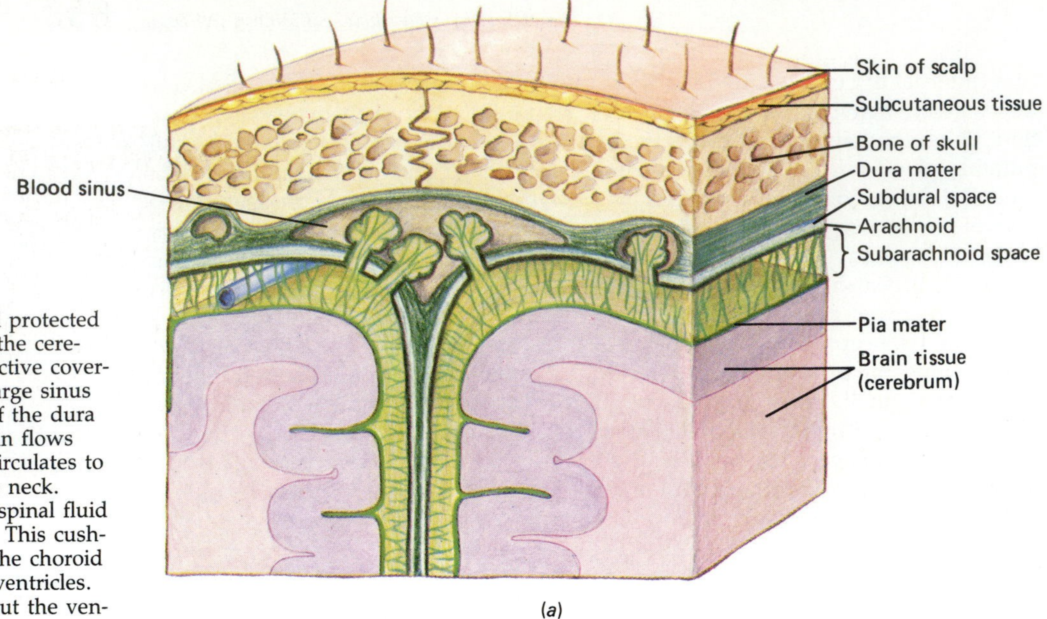

Figure 40–9 The brain is well protected by several coverings and by the cerebrospinal fluid. (*a*) The protective coverings of the brain. Note the large sinus shown between two layers of the dura mater. Blood leaving the brain flows into such sinuses and then circulates to the large jugular veins in the neck. (*b*) Circulation of the cerebrospinal fluid in the brain and spinal cord. This cushioning fluid is produced by the choroid plexuses in the walls of the ventricles. The fluid circulates throughout the ventricles and subarachnoid space. It is continuously produced and continuously reabsorbed into the blood of the dural sinuses.

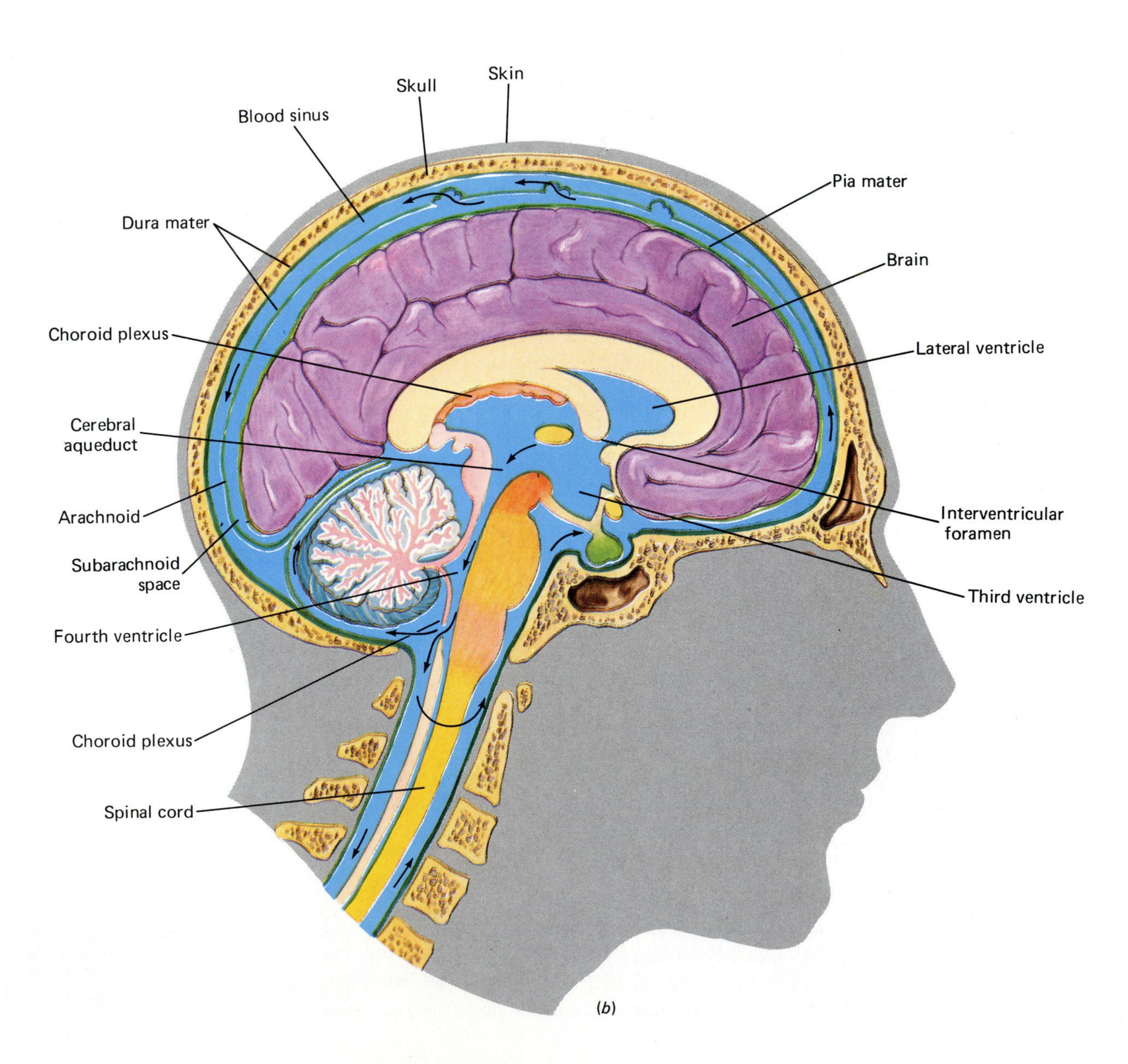

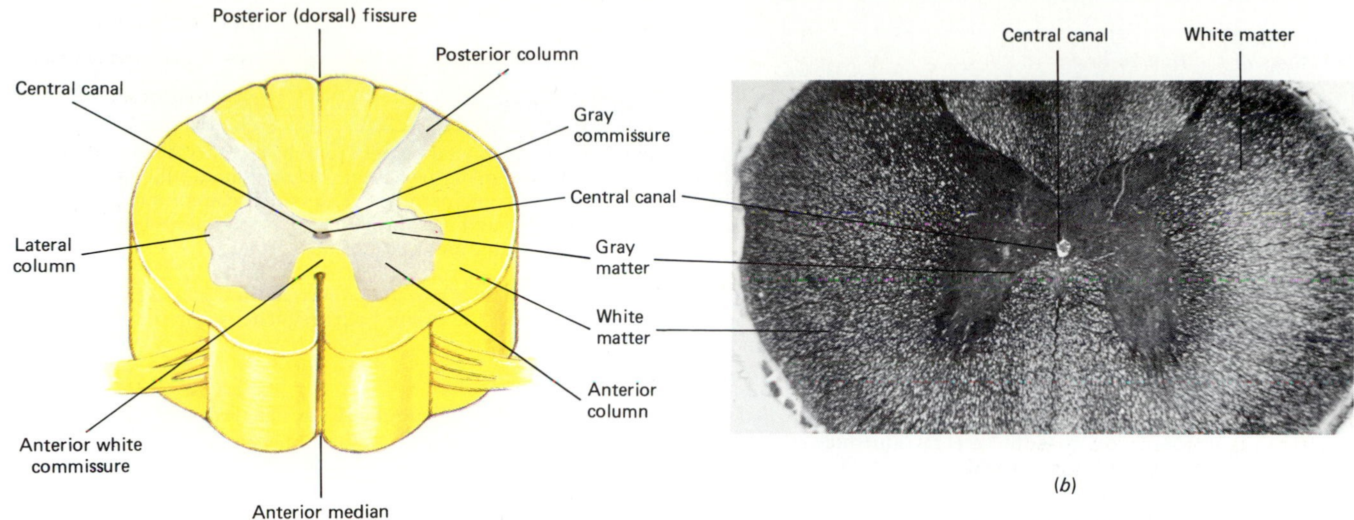

Figure 40–10 The spinal cord consists of gray matter and white matter. (*a*) Cross section through the spinal cord. (*b*) Photomicrograph of a cross section through the spinal cord (approximately ×25).

cord reveals a small **central canal** surrounded by an area of gray matter shaped somewhat like the letter H (Fig. 40–10). Outside the gray matter, the spinal cord consists of white matter. The gray matter is composed of large masses of cell bodies, dendrites, and unmyelinated axons, as well as glial cells and blood vessels. The gray matter is subdivided into sections called **columns.** The white matter consists of myelinated axons arranged in bundles called **tracts,** or **pathways.** Long **ascending tracts** conduct impulses up the cord to the brain. For example, the **spinothalamic tracts,** located in the anterior and lateral columns of the white matter, conduct pain and temperature information from sensory neurons in the skin. The **pyramidal tracts** are descending tracts that convey impulses from the cerebrum to spinal motor nerves at various levels in the cord.

THE BRAIN

The structure and functions of the main parts of the human brain are summarized in Table 40–2. The human brain is illustrated in Figures 40–11 and 40–12. In humans, as in other mammals, the cerebral cortex is functionally divided into three areas: (1) the **sensory areas,** which receive incoming sensory information; (2) the **motor areas,** which control voluntary movement; and (3) the **association areas,** which link the sensory and motor areas and are responsible for thought, learning, language, memory, judgment, and personality.

Experimental evidence has established that there is a considerable amount of localization of function in the cortex. By surgically removing particular regions of the cortex from experimental animals, it has been possible to localize many functions exactly. Functions have also been localized by observing the paralysis or loss of sensation in a patient with a brain injury or tumor and then examining the brain after death to determine the location of the injury. During operations on the brain, surgeons have electrically stimulated small regions and observed which muscles contracted. Since brain surgery can be carried on under local anesthesia, a patient can be asked what sensations are felt when a particular region is stimulated. Curiously, the brain itself has no nerve endings for pain, so that stimulation of the cortex is not painful. Brain activity can be studied by measuring and recording the electrical potentials or "brain waves" given off by various parts of the brain when active.

By combining the data obtained in several ways, investigators have been able to map the human brain, locating the areas responsible for different functions. The posterior **occipital lobes** contain the visual centers. Stimulation of these areas, even by a blow on the back of the head, causes the sensation of light; their removal causes blindness. The centers for hearing are located in the lateral **temporal lobes** of the brain above the ear; stimulation by a blow causes a sensation of noise. Although removal of both auditory areas causes deafness, removal of one does not cause

TABLE 40–2
Divisions of the Human Brain

Division	*Description*	*Functions*
Medulla	Most inferior portion of the brain stem; continuous with spinal cord. Its white matter consists of nerve tracts passing between spinal cord and various parts of the brain; its gray matter consists of nuclei. Its cavity is the fourth ventricle.	Contains vital centers (within its reticular formation) that regulate heartbeat, respiration and blood pressure; contains reflex centers that control swallowing, coughing, sneezing, and vomiting; relays messages to other parts of the brain
Pons	Consists mainly of nerve tracts passing between the medulla and other parts of the brain; forms a bulge on anterior surface of brain stem; contains a respiratory center	Serves as a link to connect and integrate various parts of the brain; helps regulate respiration
Midbrain	Just superior to the pons; contains red nucleus; cavity is the cerebral aqueduct; posteriorly consists of superior and inferior colliculi	Superior colliculi mediate visual reflexes; inferior colliculi mediate auditory reflexes. Red nucleus integrates information regarding muscle tone and posture.
Diencephalon		
Thalamus	Located on each side of the third ventricle; consists of two masses of gray matter partly covered by white matter; contains many important nuclei	Main relay center conducting information between spinal cord and cerebrum. Incoming messages are sorted and partially interpreted within thalamic nuclei before being relayed to appropriate centers in the cerebrum.
Hypothalamus	Forms ventral floor of third ventricle; contains many nuclei. Optic chiasms mark crossing of the optic nerves. The pituitary stalk connects pituitary gland to hypothalamus.	Contains centers for control of body temperature, appetite, and fluid balance; secretes releasing hormones that regulate pituitary gland; helps control autonomic functions; involved in some emotional and sexual responses
Cerebellum	Second largest part of the brain; consists of two lateral cerebellar hemispheres; superior to the fourth ventricle	Responsible for smooth, coordinated movement; maintains posture and muscle tone; helps maintain equilibrium
Cerebrum	Largest, most prominent part of brain. Longitudinal fissure divides cerebrum into right and left hemispheres, each containing a lateral ventricle and each divided into six lobes: frontal, parietal, occipital, temporal, limbic, and insular.	Center of intellect, memory, language, and consciousness; receives and interprets sensory information from all of the organs; controls motor functions
Cerebral cortex	Convoluted, outer layer of gray matter, functionally divided into three areas:	
	Motor areas	Control voluntary movement and certain types of involuntary movement
	Sensory areas	Receive incoming sensory information from eyes, ears, touch and pressure receptors, and other sense organs. Sensory association areas interpret sensory information.
	Association areas	Responsible for thought, learning, language, judgment, and personality; store memories; connect sensory and motor areas
White matter	Consists of association fibers that interconnect neurons within the same hemisphere; fibers that interconnect the two hemispheres (e.g., corpus callosum), and fibers that are part of ascending and descending tracts. Basal ganglia are located within the white matter.	Link various areas of the brain

deafness in one ear but rather produces a decrease in the auditory acuity of both ears.

A fissure called the **central sulcus** crosses the top of each hemisphere from medial to lateral edge. This partially separates the **primary motor areas** in the **frontal lobes** controlling the skeletal muscles from the **parietal lobes** just behind the furrow. The parietal lobes are responsible for the sensations of heat, cold, touch, and pressure that result from stimulation of sense organs in the skin. In both motor and sensory areas there is a further specialization along the furrow from the top of the brain to the side. Neurons at the top of the cortex control the muscles of the

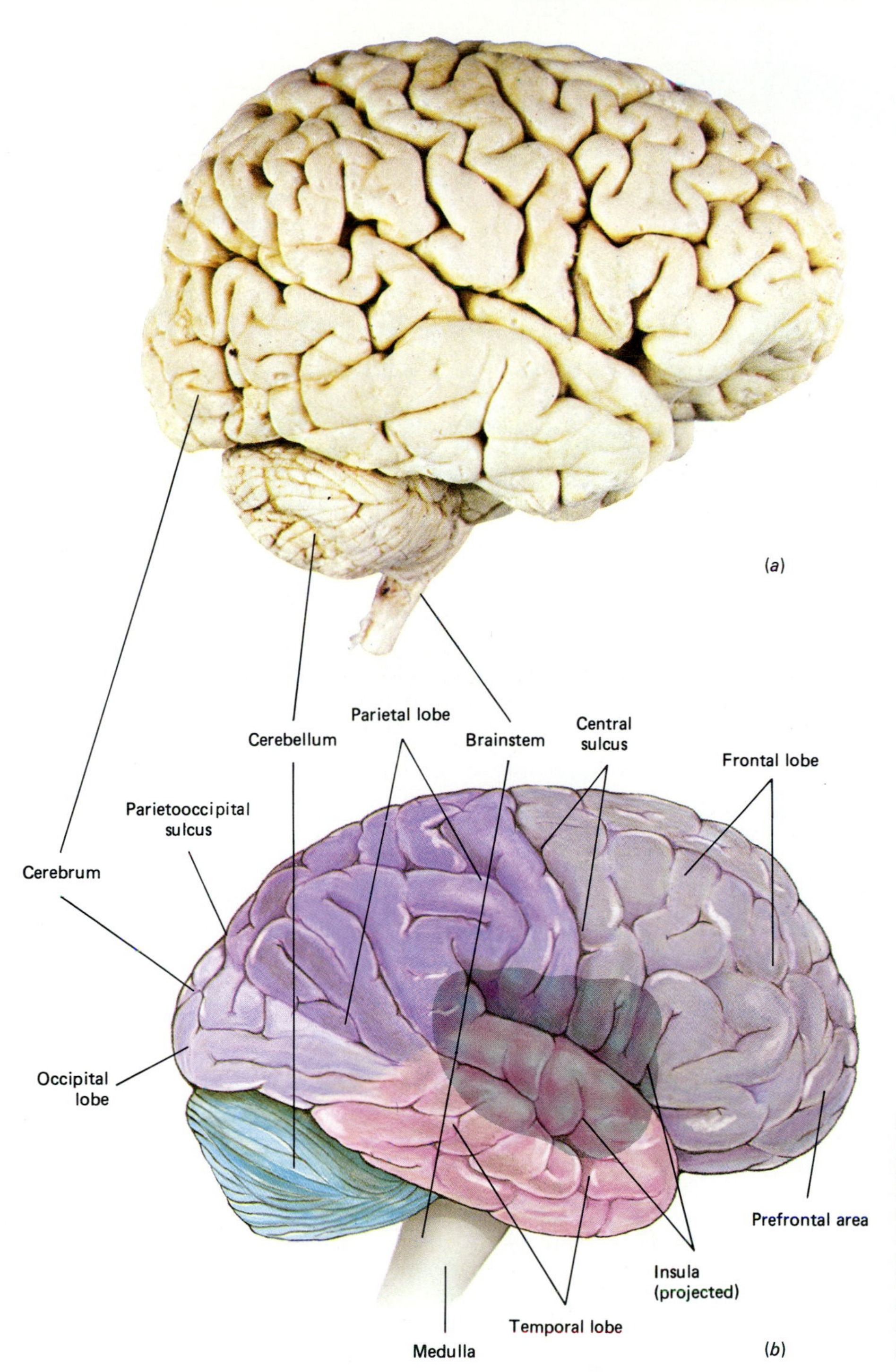

Figure 40–11 Structure of the human brain. (*a*) Photograph of the human brain, lateral view. Note that the cerebrum covers the diencephalon and part of the brain stem. (*b*) Lateral view of the human brain showing the lobes of the cerebrum. Part of the brain has been made transparent so that the underlying insular lobe can be located. ((*a*), from Williams and Warwick (eds.): *Gray's Anatomy*.)

feet; the neurons next in line control those of the shank, thigh, abdomen, and so on; and the neurons farthest around to the side control the muscles of the face.

The size of the motor area in the brain for any given part of the body is proportional not to the amount of muscle but to the elaborateness and intricacies of movement involved. Predictably, there are large areas for the control of the hands and face (Fig. 40–13). There is a similar relationship between the parts of the sensory area and the region of the skin from which it receives impulses. In connections between the body and the brain, not only is there a crossing of the fibers, so that one side of the brain controls the opposite side of the body, but a further "reversal" makes the uppermost part of the cortex control the lower extremities of the body.

When all the areas of known function are plotted, they cover almost all of the rat's cortex, a large part of the dog's, a moderate amount of the monkey's, but only a small part of the total surface of the human cortex (Fig. 40–14). The remaining

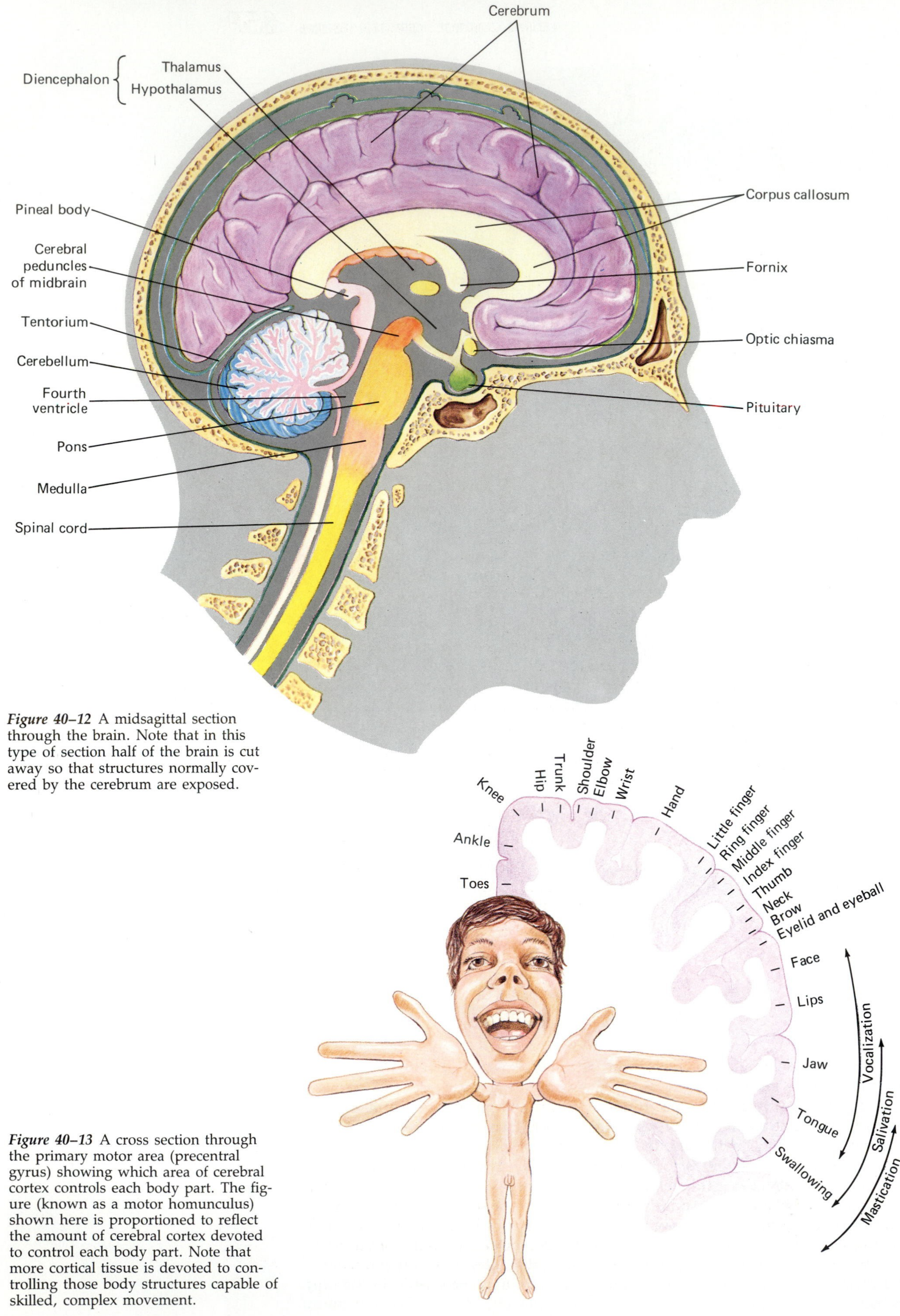

Figure 40–12 A midsagittal section through the brain. Note that in this type of section half of the brain is cut away so that structures normally covered by the cerebrum are exposed.

Figure 40–13 A cross section through the primary motor area (precentral gyrus) showing which area of cerebral cortex controls each body part. The figure (known as a motor homunculus) shown here is proportioned to reflect the amount of cerebral cortex devoted to control each body part. Note that more cortical tissue is devoted to controlling those body structures capable of skilled, complex movement.

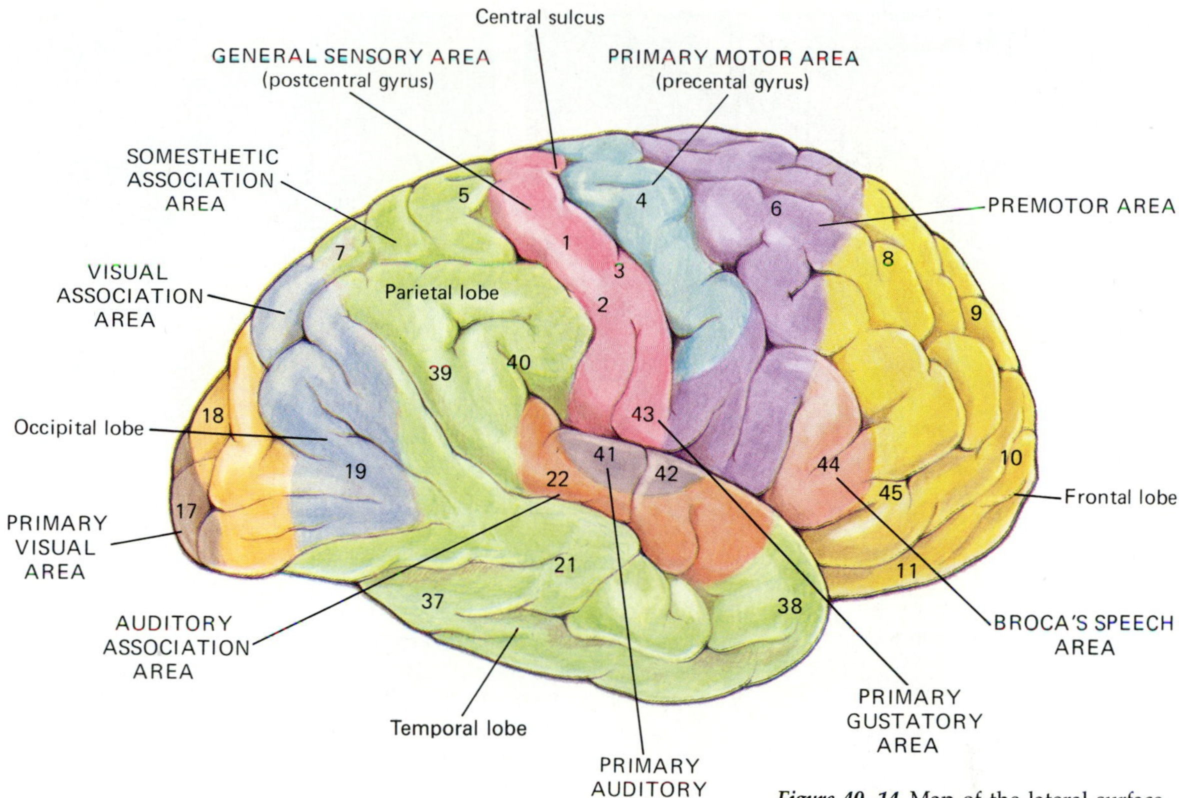

Figure 40–14 Map of the lateral surface of the cerebral cortex showing some of the functional areas. Areas *4*, *6*, and *8* are motor areas; areas *1*, *2*, *3*, *17*, *41*, *42*, and *43* are primary sensory areas; and areas *9*, *10*, *11*, *18*, *19*, *22*, *38*, *39*, and *40* are association areas.

cortical areas are the regions responsible for the higher intellectual faculties of memory, reasoning, learning, imagination and personality; these are the association areas. In some way, the association regions integrate all the diverse impulses constantly reaching the brain into a meaningful unit, so that the proper response is made. When disease or accident destroys the functioning of one or more association areas, **aphasia** may result, a condition in which the ability to recognize certain kinds of symbols is lost. The names of objects may be forgotten, for example, although their functions are remembered and understood.

The Reticular Activating System

The **reticular activating system (RAS)** is a complex neural pathway within the brain stem and thalamus. It receives messages from neurons in the spinal cord and from many other parts of the nervous system and communicates with the cerebral cortex by complex neural circuits. The RAS is responsible for maintaining wakefulness. When the RAS is very active and sending many messages into the cerebrum, a state of mental and physical alertness is noted. When RAS activity slows, however, sleepiness results. If the RAS is severely damaged, a deep, permanent coma may result.

The Limbic System

The **limbic system,** another action system of the brain, consists of certain structures of the cerebrum and diencephalon. This system affects the emotional aspects of behavior, sexual behavior, biological rhythms, autonomic responses, and motivation, including feelings of pleasure and punishment. Stimulation of certain areas of the limbic system in an experimental animal results in increased general activity, and may cause fighting behavior or extreme rage.

When an electrode is implanted in the so-called reward center of the limbic system, a rat will press a lever that stimulates this area as many as 15,000 times per hour (Fig. 40–15). Stimulation of this area is apparently so rewarding that an animal will forego food and drink and may continue to press the lever until it drops from exhaustion. When an electrode is implanted in the punishment center of the limbic system, an experimental animal quickly learns to press a lever to *avoid* stimulation. The reward and punishment centers are thought to be important in influencing motivation and behavior.

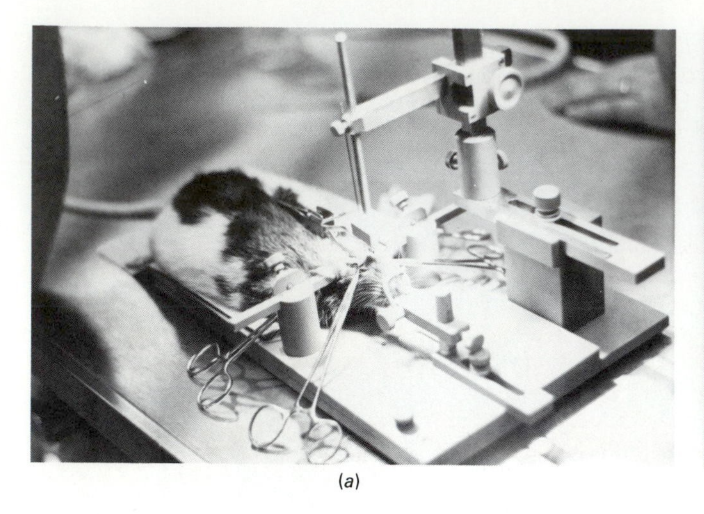
(a)

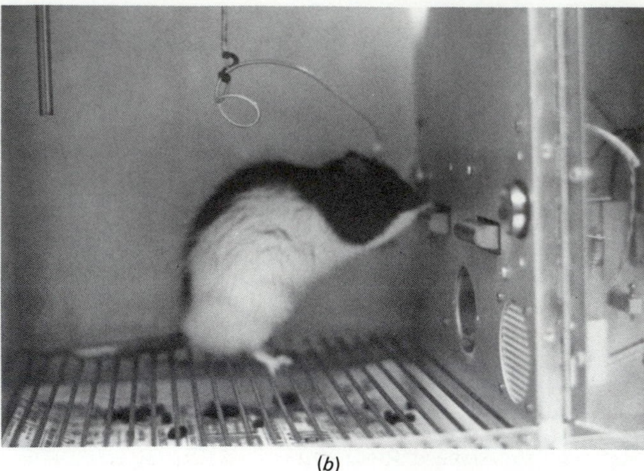
(b)

Figure 40–15 Electrodes can be implanted in the pleasure center of a rat's brain as shown in (*a*), so that when the rat depresses a lever, a stimulating electric current is delivered directly to the pleasure center, as seen in (*b*).

Brain Waves

Metabolism is invariably accompanied by electrical changes, and the electrical activity of the brain can be recorded by a device known as an **electroencephalograph.** To obtain a recording, called an **electroencephalogram** or **EEG,** electrodes are taped to different parts of the scalp and the activity of the underlying parts of the cortex is measured. The electroencephalograph shows that the brain is continuously active. As seen on the EEG, the most regular manifestations of activity, called **alpha waves,** come mainly from the visual areas in the occipital lobes when the person being tested is resting quietly with eyes closed. These waves occur rhythmically at the rate of 9 or 10 per second and have a potential of about 45 mV (Fig. 40–16).

When the eyes are opened, alpha waves disappear and are replaced by more rapid, irregular waves. That the latter are produced by objects seen can be demonstrated by presenting to the eyes some regular stimulus, such as a light blinking at regular intervals; brain waves with a similar rhythm will appear. As you are reading this biology text your brain should be emitting **beta waves,** which have a fast-frequency rhythm most characteristic of heightened mental activity such as information processing. During sleep, brain waves become slower and larger as the person falls into deeper unconsciousness; these slow, large waves associated with normal sleep are called **delta waves.** The dreams of a sleeping person are mirrored in flurries of irregular waves.

Certain brain diseases alter the character of the waves. Epileptics, for exam-

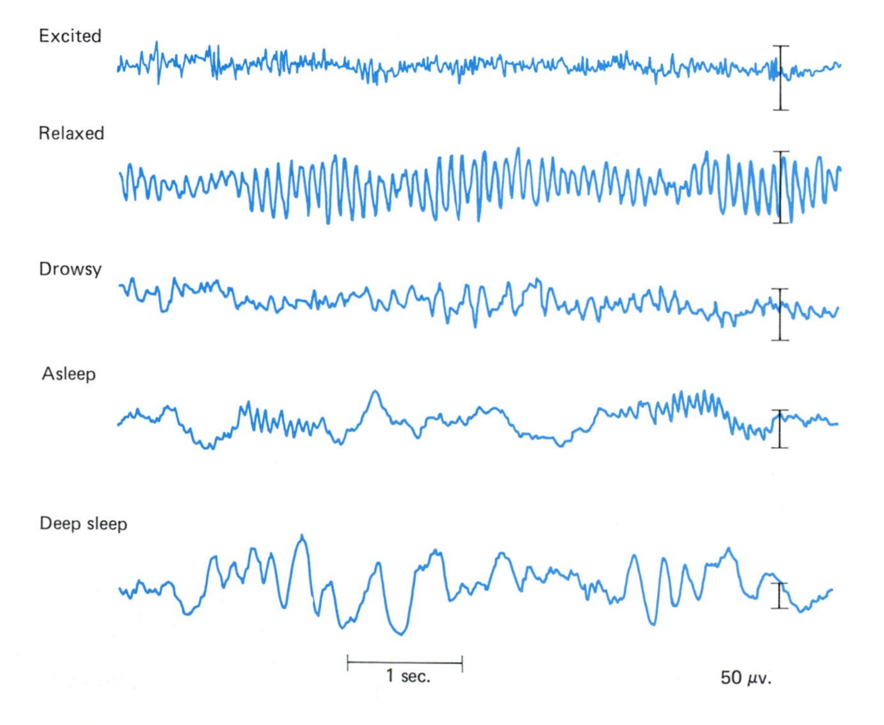

Figure 40–16 Electroencephalograms made while the subject was excited, relaxed, and in various stages of sleep. Recordings made during excitement show brain waves that are rapid and of small amplitude, whereas in sleep the waves are much slower and of greater amplitude. The regular waves characteristic of the relaxed state are called alpha waves. (From Jasper: In Penfield and Erickson: *Epilepsy and Cerebral Localization)*

ple, exhibit a distinctive, readily recognizable wave pattern, and even people who have never had an epileptic attack, but might under certain conditions, show similar abnormalities. The location of brain tumors or the sites of brain damage caused by a blow to the head, for instance, can sometimes be determined by noting the part of the brain showing abnormal waves.

Sleep

Sleep is a state of unconsciousness during which there is decreased electrical activity of the cerebral cortex, and from which a person can be aroused by external stimuli. When signals from the RAS slow down so that the cerebral cortex is deprived of activating input, a person may lapse into sleep. For this reason, we find it easy to go to sleep, even when we are not particularly tired, if there is nothing interesting to occupy the mind. But although we tend to be wakeful in the presence of attention-holding stimuli, there is a limit beyond which sleep is inevitable.

There are thought to be sleep centers in the brain stem. When stimulated, their neurons release the neurotransmitter **serotonin** (5-hydroxytryptamine, or 5HT). Serotonin is thought to inhibit signals passing through the RAS, thus inducing sleep.

Two main stages of sleep are recognized: non-REM and REM. The letters REM are an acronym for rapid eye movements. During **non-REM** sleep, sometimes called normal sleep, metabolic rate decreases, breathing slows, and blood pressure decreases. Delta waves, thought to be generated spontaneously by the cerebral cortex when it is not driven by impulses from other parts of the brain, are characteristic of non-REM sleep.

Every 90 minutes or so, a sleeping person enters the **REM** stage for a time. During this stage, which accounts for about one fourth of total sleep time, the eyes move rapidly about beneath the closed but fluttering lids. Brain waves change to a desynchronized pattern of beta waves. Sleep researchers claim that everyone dreams during REM sleep. Dreams may result from release of norepinephrine within the RAS, which generates stimulating impulses that are fed into the cerebral cortex.

Why sleep is necessary is not understood. Apparently only higher vertebrates with fairly well-developed cerebral cortices sleep. When a person stays awake for unusually long periods, fatigue and irritability result, and even routine tasks cannot be performed well. Perhaps certain waste products accumulate within the nervous system, and sleep gives the nervous system opportunity to dispose of them. When deprived of sleep for several days, a person becomes disoriented and may eventually exhibit psychotic symptoms.

Not only is normal non-REM sleep required, but REM sleep is apparently also essential. In sleep deprivation experiments performed with human volunteers, lack of REM sleep makes subjects anxious and irritable. After such experiments, when the subjects are permitted to sleep normally again, they go through a period when they spend more time than usual in the REM stage. Many types of drugs alter sleep patterns and affect the amount of REM sleep. For example, sleeping pills may increase the total sleeping time, but decrease the time in REM sleep. When a person stops taking such a drug, several weeks may be required before normal sleep patterns are reestablished.

Learning and Memory

Learning is a relatively long-lasting adaptive change in behavior resulting from experience. It is a modification of behavior that cannot be accounted for by sensory adaptation, central excitatory states, biological rhythms, motivational states, or maturation. Laboratory experiments have shown that members of every animal phylum can learn, and field observations indicate that learning is important for a wide variety of natural situations. Even some unicellular organisms that completely lack a nervous system are capable of some simple types of learning.

Learning involves the storage of information in the nervous system and its retrieval on demand. Many recent studies have attempted to discover the physical and chemical basis of memory and learning. Somewhere in the nervous system there must be stored a more-or-less permanent record of what has been learned that can be recalled on future occasions. This record has been termed the memory trace, or **engram.**

Electrical stimulation of the cerebral cortex of a patient undergoing brain surgery can cause vivid recollection of long-forgotten events. From this it had been inferred that the items of memory, the engrams, were filed away in specific parts of the brain. Some years ago, however, Karl Lashley investigated the retention of maze learning in rats by removing portions of the cortex after the rats had learned to solve various problems. Lashley's results indicated that the extent of the memory removed by the operation was a function of how much of the cortex was removed and not of what specific part of the cortex was removed. Lashley concluded that engrams were not sorted out at specific cortical sites but were in some way present throughout its substance. He speculated that memory might be a system of impulses in reverberating circuits (Chapter 39).

Just how the brain stores information and retrieves the memory on command is still the subject of speculation. According to current theory there are several levels of memory. **Short-term memory** involves recalling information for only a few moments. When you look up a telephone number in a directory, for example, you generally remember it just long enough to dial it. Should the line be busy and you turn your attention to another task before returning to try again, you would probably have to look up the number again. Short-term memory may depend upon circuits that continue to reverberate for several minutes until they fatigue, or until new signals that interfere with the old are received.

When a decision is made to store information in **long-term memory,** the brain apparently rehearses the material and then stores it in association with similar memories. Some investigators think that changes take place in the presynaptic knobs or postsynaptic neurons, which permanently facilitate the transmission of impulses within a newly formed neural circuit. Perhaps specific neurons become more sensitive to neurotransmitter. According to this theory, each time a memory is stored, a new neural pathway is facilitated. Such a facilitated circuit is the engram. Several other theories have been proposed involving glial cells, RNA, or protein as memory molecules, or rhythms of firing.

Several minutes are required for a memory to become consolidated in the long-term memory bank. Should a person suffer a brain concussion (or undergo electroshock therapy), memory of what happened immediately before the blow (or delivery of the shock) may be lost. The limbic system is important in processing stored information. When the hippocampus (part of the limbic system located on the lower inner margin of each cerebral hemisphere) is removed, a person can recall information stored in the past but loses the ability to convert new short-term memories to long-term memories. No new information can be stored.

Retrieval of information stored in the long-term memory is of considerable interest—especially to students. Some investigators think that once information is deposited in the long-term memory bank, it remains within the brain permanently. When you seem to forget something, the problem may be that you have actually forgotten the search routine that would permit you to retrieve that item for conscious use.

Effects of Environmental Experience on the Brain

Environmental experience can cause physical as well as chemical changes in brain structure. In one series of studies, a group of rats was provided with an enriched environment, while another group was placed in a deprived environment. Rats in the enriched environment were provided with toys, rat and human interaction, and opportunity to learn. Those in the deprived environment, while given sufficient food, water, and shelter, were deprived of intellectual stimulation and social interaction.

After several weeks rats from each group were killed and their brains were studied. Those exposed to enriched environments exhibited large cell bodies, greater numbers of glial cells, increased numbers of synaptic contacts, and biochemical changes. Some investigators reported that the cerebral cortex actually became thicker and heavier. Animals reared in a complex environment may also be able to process and remember information more quickly than those reared in a deprived environment.

Early environmental stimulation can also promote development of motor areas in the brain. For example, rats encouraged to exercise show increased cerebellar development. Such studies linking the development of the brain with environmental experience support the concept that early stimulation is important for the neural, motor, and intellectual development of children, as well.

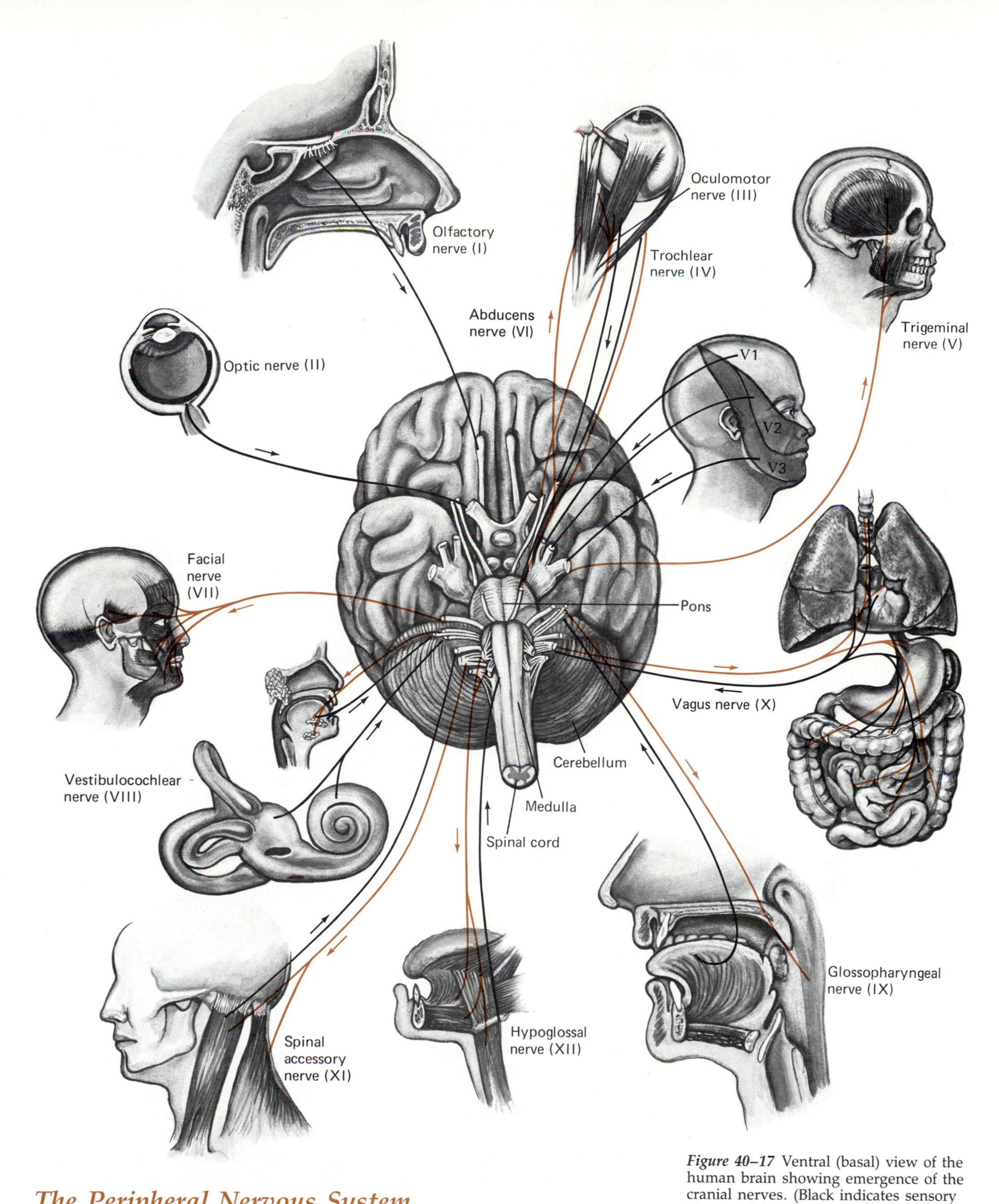

Figure 40–17 Ventral (basal) view of the human brain showing emergence of the cranial nerves. (Black indicates sensory fibers; color indicates motor fibers.)

The Peripheral Nervous System

The **peripheral nervous system (PNS)** consists of the sensory receptors, the nerves that link them with the CNS, and the nerves that link the CNS with the effectors.

THE CRANIAL NERVES

Twelve pairs of **cranial nerves** originate in various parts of the brain and innervate the sense organs, muscles, and glands of the head, as well as many of the internal organs (Fig. 40–17). The same 12 pairs, innervating homologous structures, are

TABLE 40–3
The Cranial Nerves of Mammals

Number	*Name*	*Origin of Sensory Fibers*	*Effector Innervated by Motor Fibers*
I	Olfactory	Olfactory epithelium of nose (smell)	None
II	Optic	Retina of eye (vision)	None
III	Oculomotor	Proprioceptors* of eyeball muscles (muscle sense)	Muscles that move eyeball (with IV and VI); muscles that change shape of lens; muscles that constrict pupil
IV	Trochlear	Proprioceptors of eyeball muscles	Other muscles that move eyeball
V	Trigeminal	Teeth and skin of face	Some of muscles used in chewing
VI	Abducens	Proprioceptors of eyeball muscles	Other muscles that move eyeball
VII	Facial	Taste buds of anterior part of tongue	Muscles used for facial expression; submaxillary and sublingual salivary glands
VIII	Auditory (vestibulocochlear)	Cochlea (hearing) and semicircular canals (senses of movement, balance, and rotation)	None
IX	Glossopharyngeal	Taste buds of posterior third of tongue, lining of pharynx	Parotid salivary gland; muscles of pharynx used in swallowing
X	Vagus	Nerve endings in many of the internal organs: lungs, stomach, aorta, larynx	Parasympathetic fibers to heart, stomach, small intestine, larynx, esophagus, and other organs
XI	Spinal accessory	Muscles of shoulder	Muscles of shoulder
XII	Hypoglossal	Muscles of tongue	Muscles of tongue

*Proprioceptors are receptors located in muscles, tendons, or joints that provide information about body position and movement.

found in all reptiles, birds, and mammals. Fish and amphibia have only the first 10 pairs. Some cranial nerves consist only of sensory neurons (nerves I, II, and VIII), some are composed mainly of motor neurons (nerves III, IV, VI, XI, and XII), and the others are mixed nerves, containing both sensory and motor neurons. The names of the cranial nerves and the structures they innervate are given in Table 40–3. For example, cranial nerve X, the vagus nerve (which forms part of the autonomic system), innervates the internal organs of the chest and upper abdomen.

THE SPINAL NERVES

In human beings, 31 symmetric pairs of **spinal nerves** emerge from the spinal cord (Fig. 40–18). All the spinal nerves are mixed nerves, containing both motor and sensory neurons. Each nerve innervates the receptors and effectors of one region of the body. Each spinal nerve has two **roots,** points of attachment with the cord. All of the sensory neurons enter the cord through the **dorsal root;** all motor fibers leave the cord through the **ventral root** (Fig. 40–19). Just before the dorsal root joins the spinal cord, it is marked by a swelling called the **spinal ganglion,** or **dorsal root**

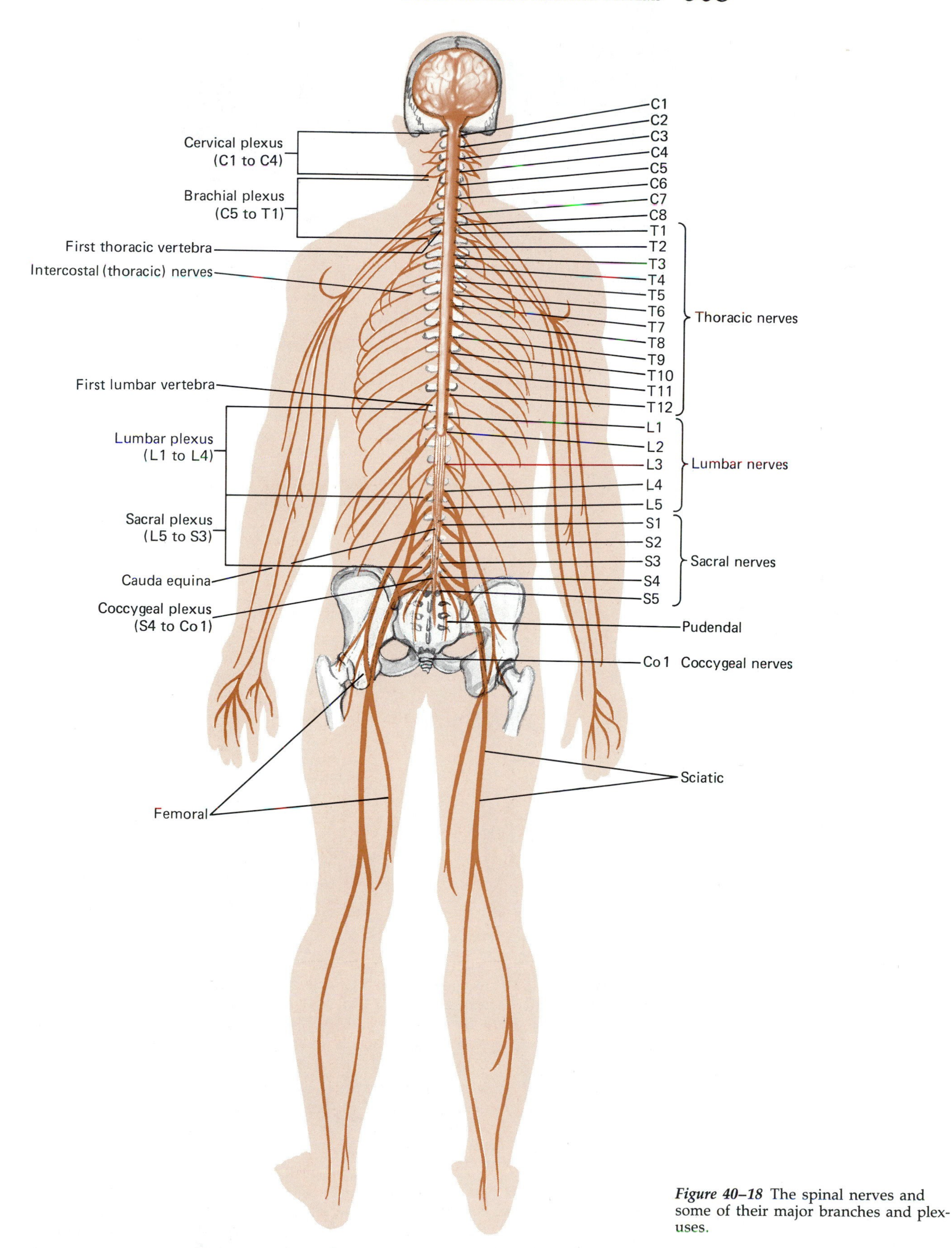

Figure 40–18 The spinal nerves and some of their major branches and plexuses.

ganglion, which consists of the cell bodies of the sensory neurons. Cell bodies of the motor neurons are located within the gray matter of the cord. Dorsal and ventral roots unite, forming the spinal nerve.

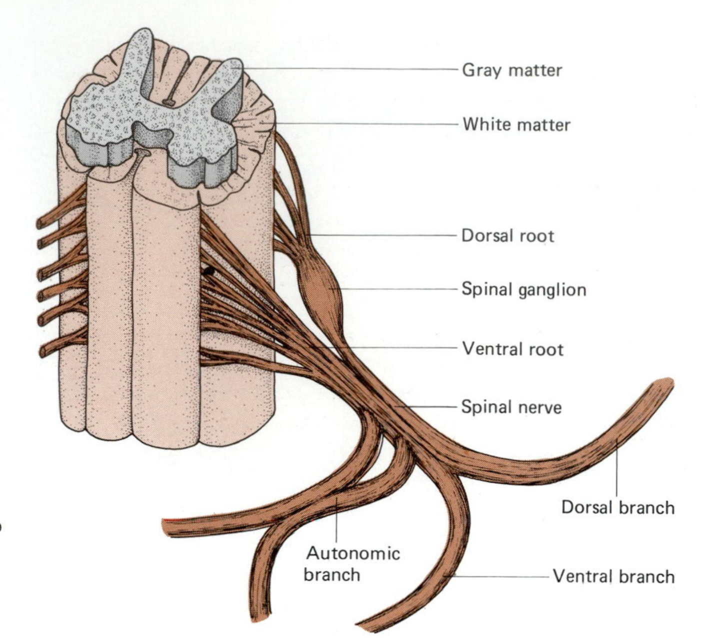

Figure 40–19 Dorsal and ventral roots emerge from the spinal cord and join to form a spinal nerve. The spinal nerve divides into several branches. (Black indicates sensory fibers; color indicates motor fibers.)

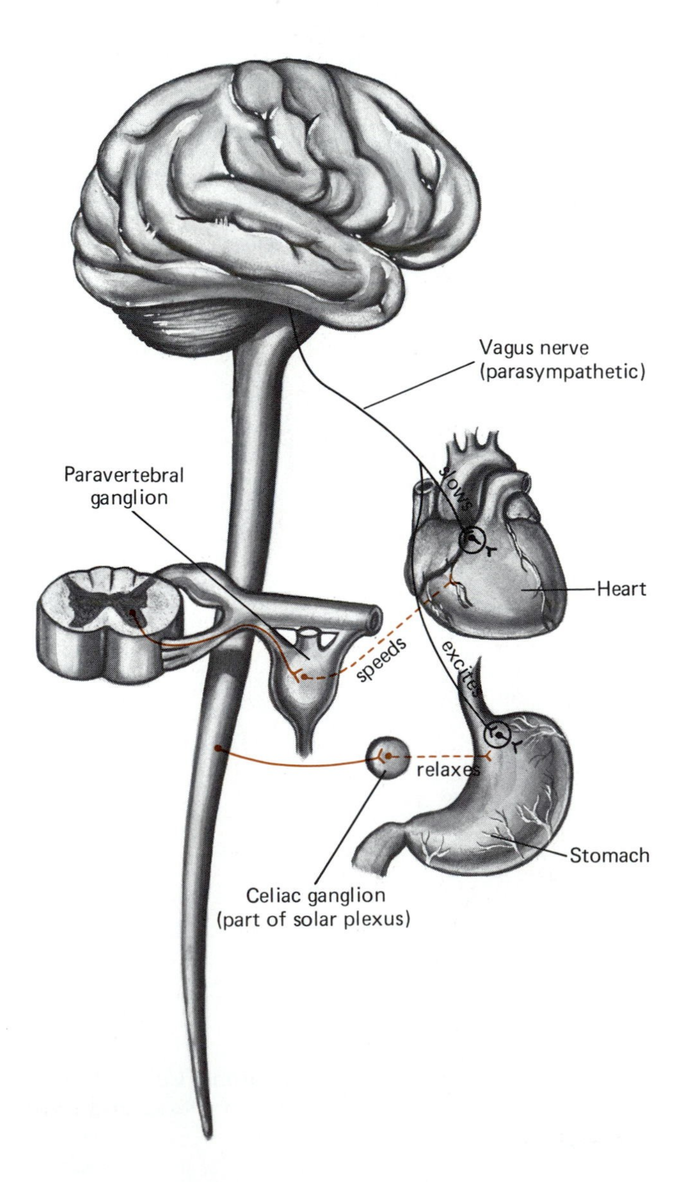

Figure 40–20 Dual innervation of the heart and stomach by sympathetic and parasympathetic nerves. (Sympathetic nerves are shown in color; postganglionic fibers are shown as dotted lines.)

If the dorsal root is severed, the part of the body innervated by that nerve suffers complete loss of sensation without paralysis of muscles. If the ventral root is cut, there is complete paralysis of muscles innervated by that nerve, but the senses of touch, pressure, temperature, pain, and kinesthesis (muscle sense) are not impaired.

Beyond the junction of the dorsal and ventral roots, each spinal nerve divides into branches. The dorsal branch serves the skin and muscles of the back. The ventral branch serves the skin and muscles of the sides and ventral part of the body. The autonomic branch innervates the viscera. The ventral branches of several spinal nerves form tangled networks called **plexuses** (Fig. 40–18). Within a plexus, the fibers of a spinal nerve may separate and then regroup with fibers that originated in other nerves. Thus, nerves emerging from a plexus consist of neurons that originated in several different spinal nerves. Among the principal plexuses are the cervical plexus, the brachial plexus, the lumbar plexus, and the sacral plexus.

THE AUTONOMIC SYSTEM

The **autonomic system** helps to maintain a steady state within the internal environment of the body. For example, it helps to maintain a constant body temperature and to regulate the rate of the heartbeat. Receptors within the viscera relay information through afferent nerves to the CNS, and the impulses are transmitted along efferent neurons to the appropriate muscles and glands.

The efferent portion of the autonomic system is subdivided into sympathetic and parasympathetic systems. Many organs are innervated by both (Figs. 40–20 and 40–21). In general, the **sympathetic system** mobilizes energy and enables the body to respond to stress (Table 40–4). Its nerves speed the heart's rate and force of

Figure 40–21 Sympathetic and parasympathetic nervous systems. For clarity, peripheral and visceral nerves of the sympathetic system are shown on separate sides of the cord. Complex as it appears, this diagram has been greatly simplified. (Colored lines represent sympathetic nerves, black lines represent parasympathetic nerves, and dotted lines represent postganglionic nerves.) See Table 40–5 for specific action of the nerves.

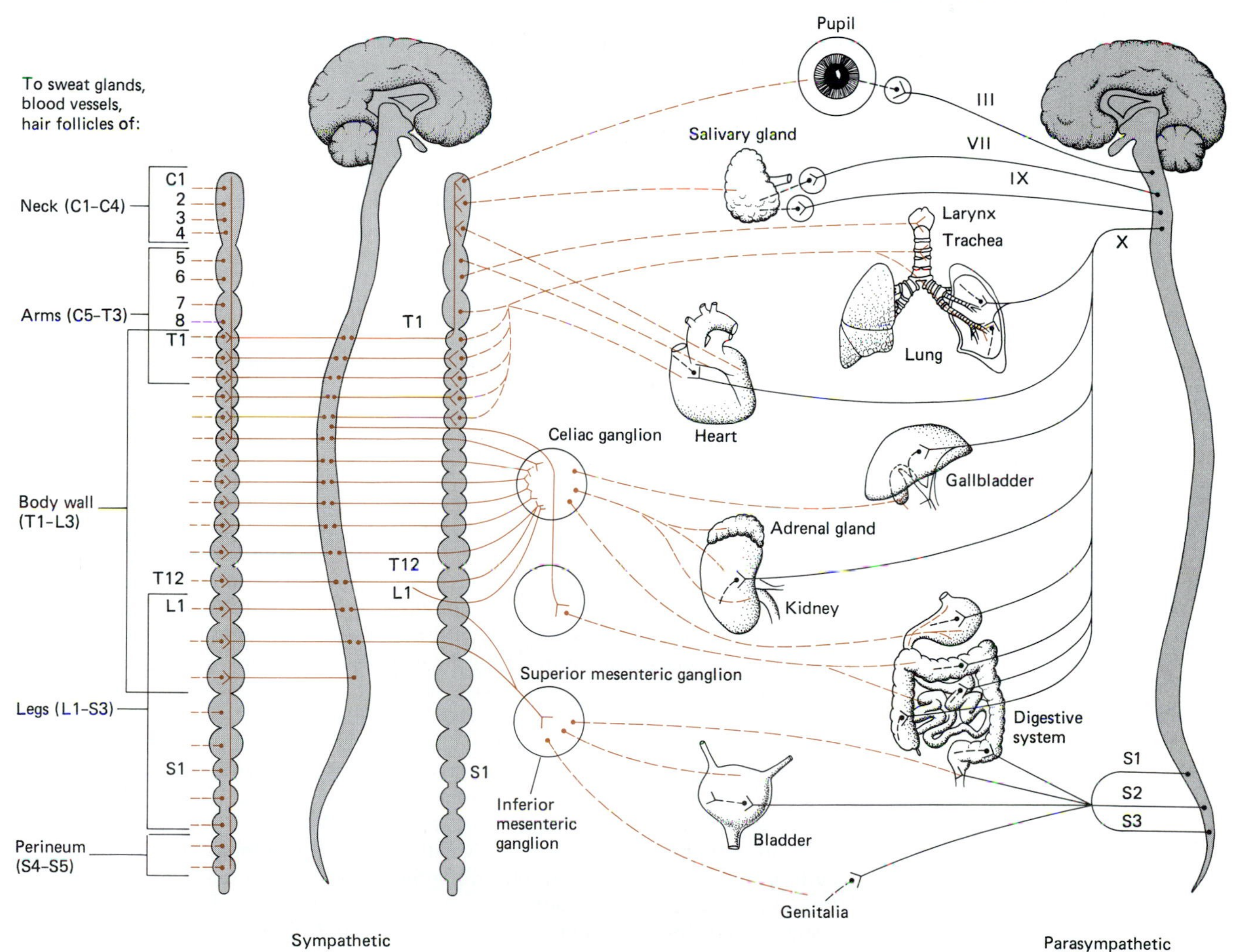

TABLE 40–4
Comparison of Sympathetic with Parasympathetic System

Characteristic	*Sympathetic System*	*Parasympathetic System*
General effect	Prepares body to cope with stressful situations	Restores body to resting state after stressful situation; actively maintains normal configuration of body functions.
Extent of effect	Widespread throughout body	Localized
Transmitter substance released at synapse with effector	Norepinephrine (usually)	Acetylcholine
Duration of effect	Lasting	Brief
Outflow from CNS	Thoracolumbar levels of spinal cord	Craniosacral levels (from brain and spinal cord)
Location of ganglia	Chain and collateral ganglia	Terminal ganglia
Number of postganglionic fibers with which each preganglionic fiber synapses	Many	Few

contraction, increase blood pressure, increase blood sugar concentration, and reroute blood circulation when required so that skeletal and cardiac muscles receive the added amounts of blood needed to support their maximum effort.

The **parasympathetic system** is most active in ordinary, restful situations. After a stressful episode, it decreases the heart rate, decreases blood pressure, and stimulates the digestive system to process food. The parasympathetic system is dominant during relaxation or calm, quiet activities. The sympathetic and parasympathetic systems work together to orchestrate the numerous complex activities continuously taking place within the body (Table 40–5).

Instead of utilizing a single efferent neuron, as in the somatic system, the autonomic system uses a relay of two neurons between the CNS and the effector. The first neuron, called the **preganglionic neuron,** has a cell body and dendrites within the CNS. Its axon, part of a peripheral nerve, ends by synapsing with a **postganglionic neuron.** The dendrites and cell body of the postganglionic neuron are located within a ganglion outside the CNS. Its axon terminates near or on the effector. The sympathetic ganglia are paired, and there is a chain of them on each side of the spinal cord from the neck to the abdomen, the **paravertebral sympathetic ganglion chain.** Some sympathetic preganglionic neurons do not end in these ganglia, but instead pass on to ganglia located in the abdomen close to the aorta and its major branches. These ganglia are known as **collateral ganglia.** Parasympathetic preganglionic neurons synapse with postganglionic neurons in **terminal ganglia** located near or within the walls of the organs they innervate.

The sympathetic and parasympathetic systems also differ in the neurotransmitters they release at the synapse with the effector. Sympathetic postganglionic neurons release norepinephrine (although preganglionic neurons secrete acetylcholine). Both preganglionic and postganglionic parasympathetic neurons secrete acetylcholine. Table 40–4 compares the sympathetic and parasympathetic systems.

The autonomic system got its name from the original belief that it was independent of the CNS, that is, autonomous. Physiologists have shown that this is not so, and that the hypothalamus and many other parts of the CNS help to regulate the autonomic system. Although the autonomic system usually functions automatically, its activities can be consciously influenced. **Biofeedback** provides visual or auditory evidence to a person concerning the status of an autonomic body function; for example, a tone may be sounded when blood pressure reaches a desirable level. Using such techniques, subjects have learned to control certain autonomic activities

TABLE 40–5
*Comparison of Sympathetic and Parasympathetic Actions on Selected Effectors**

Effector	*Sympathetic Action*	*Parasympathetic Action*
Heart	Increases rate and strength of contraction	Decreases rate; no direct effect on strength of contraction
Bronchial tubes	Dilates	Constricts
Iris of eye	Dilates pupil	Constricts pupil
Sex organs	Constricts blood vessels; ejaculation	Dilates blood vessels; erection
Blood vessels	Generally constricts	No innervation for many
Sweat glands	Stimulates	No innervation
Intestine	Inhibits motility	Stimulates motility and secretion
Liver	Stimulates glycogenolysis (conversion of glycogen to glucose)	No effect
Adipose tissue	Stimulates free fatty acid release from fat cells	No effect
Adrenal medulla	Stimulates secretion of epinephrine and norepinephrine	No effect
Salivary glands	Stimulates thick, viscous secretion	Stimulates profuse, watery secretion

*Refer to Figure 40–21 as you study this table. Note that many other examples could be added to this list.

such as brain wave pattern, heart rate, blood pressure, and blood sugar level. Even certain abnormal heart rhythms can be consciously modified.

Effects of Drugs on the Nervous System

Among the most widely used drugs in the United States today are tranquilizers, sedatives, alcohol, stimulants, and antidepressants—all drugs that affect mood. Many of these drugs act by altering the levels of neurotransmitters within the brain. For example, amphetamines increase the amount of norepinephrine within the RAS, thus stimulating the CNS. Table 40–6 describes the effects of several types of commonly used and abused drugs.

Many mood drugs, including alcohol, are taken in order to induce a feeling of **euphoria,** or well-being. Habitual or prolonged use of almost any mood drug may result in **psychological dependence,** in which the user becomes emotionally dependent upon the drug. When deprived of it, the user may become irritable and feel unable to carry out normal activities.

Some drugs induce **tolerance** when they are taken continuously for several weeks. This means that increasingly large amounts are required in order to obtain the desired effect. Tolerance is thought to occur when the liver cells are stimulated to produce larger quantities of the enzymes that metabolize and inactivate the drug. Use of some drugs, such as heroin, alcohol, or barbiturates, may also result in **physical addiction,** in which physiological changes take place in body cells, making the user dependent upon the drug. When the drug is withheld, the addict suffers physical illness and characteristic **withdrawal symptoms.**

Physical addiction can also occur because certain drugs, such as morphine, have components similar to substances that body cells normally manufacture on their own. The continued use of such a drug, followed by sudden withdrawal, causes potentially dangerous physiological effects because the body's natural production of these substances has been depressed. It may be some time before homeostasis is reestablished.

TABLE 40–6
Effects of Some Commonly Used Drugs

Name of Drug	*Effect on Mood*	*Actions on Body*	*Dangers Associated with Abuse*
Barbiturates, e.g., Nembutal, Seconal	Sedative-hypnotic;* "downers"	Inhibit impulse conduction in RAS: depress CNS, skeletal muscle, and heart; depress respiration; lower blood pressure; cause decrease in REM sleep	Tolerance, physical dependence, death from overdose, especially in combination with alcohol
Methaqualone, e.g., Quāālude, Sopor	Hypnotic	Depresses CNS; depresses certain polysynaptic spinal reflexes	Tolerance, physical dependence, convulsions, death
Meprobamate, e.g., Equanil, Miltown (minor tranquilizers)	Antianxiety drug;[†] induces calmness	Causes decrease in REM sleep; relaxes skeletal muscle; depresses CNS	Tolerance, physical dependence; coma and death from overdose
Diazepam, e.g., Valium; chlordiazepoxide, e.g., Librium (mild tranquilizers)	Reduce anxiety	May reduce rate of impulse firing in limbic system; relax skeletal muscle	Minor EEG abnormalities with chronic use; physical dependence with very large doses
Phenothiazines, e.g., chlorpromazine (major tranquilizers)	Antipsychotic; highly effective in controlling symptoms of psychotic patients	Affect levels of catecholamines in brain (block dopamine receptors, inhibit uptake of NE,** dopamine, and serotonin); depress neurons in RAS and basal ganglia	Prolonged intake may result in parkinsonian symptoms
Antidepressants, e.g., Elavil	Elevate mood; relieve depression	Block uptake of NE so that more is available to stimulate nervous system	Central and peripheral neurological disturbances; uncoordination; interference with normal cardiovascular function
Alcohol	Euphoria; relaxation; release of inhibitions	Depresses CNS; impairs vision, coordination, judgment; lengthens reaction time	Physical dependence; damage to pancreas; liver cirrhosis; possible brain damage
Narcotic analgesics, e.g., morphine, heroin	Euphoria; reduction of pain	Depress CNS; depress reflexes; constrict pupils; impair coordination; block action of pain-transmitting neurons	Tolerance; physical dependence; convulsions; death from overdose

*Sedatives reduce anxiety; hypnotics induce sleep.
[†]Antianxiety drugs reduce anxiety but are less likely to cause drowsiness than the more potent sedative-hypnotics.
**NE = norepinephrine; MAO = monoamine oxidase.

SUMMARY

I. Among invertebrates, nerve nets and radial nervous systems are typical of radially symmetric animals, and bilateral nervous systems are characteristic of bilaterally symmetric animals.
  A. A nerve net consists of nerve cells scattered throughout the body; no CNS is present. Response of these animals to stimuli is generally slow and imprecise.
  B. Echinoderms typically have a nerve ring and nerves that extend into various parts of the body.
  C. In a bilateral nervous system there is a concentration of nerve cells to form nerves, nerve cords, ganglia, and (in complex forms) a brain. There is also an increase in numbers of neurons, especially of the association neurons. This permits greater precision and a wider range of responses.

II. In the vertebrate embryo the brain and spinal cord arise from the neural tube. The anterior end of the tube differentiates into forebrain, midbrain, and hindbrain.
  A. The hindbrain subdivides into the metencephalon and myelencephalon.
    1. The myelencephalon develops into the medulla, which contains the vital centers and other reflex centers.
    2. The metencephalon gives rise to the cerebellum and pons.
      a. The cerebellum is responsible for muscle tone, posture, and equilibrium.
      b. The pons connects various parts of the brain.
  B. The midbrain is the largest part of the brain in fish and amphibians. It is their main association area, linking sensory input and motor output. In reptiles, birds, and mammals the midbrain has a lesser function and is used as a center for certain visual and auditory reflexes. It also contains the red nucleus that integrates information about muscle tone and posture.
  C. The forebrain differentiates to form the diencephalon and telencephalon.

Name of Drug	*Effect on Mood*	*Actions on Body*	*Dangers Associated with Abuse*
Cocaine	Euphoria; excitation followed by depression	CNS stimulation followed by depression; autonomic stimulation; dilates pupils; local anesthesia	Mental impairment; convulsions; hallucinations; unconsciousness; death from overdose
Amphetamines, e.g., Dexedrine	Euphoria; stimulant; hyperactivity; "uppers," "pep pills"	Since chemical structure is almost identical with that of NE, compete with NE for receptor sites; block reuptake of NE into neurons; inhibit MAO;** enhance flow of impulses in RAS; increase heart rate; raise blood pressure; dilate pupils	Tolerance; possible physical dependence; hallucinations; death from overdose
Caffeine	Increases mental alertness; decreases fatigue and drowsiness	Acts upon cerebral cortex; relaxes smooth muscle; stimulates cardiac and skeletal muscle; increases urine volume (diuretic effect)	Very large doses stimulate centers in the medulla (may slow the heart); toxic doses may cause convulsions.
Nicotine	Psychological effect of lessening tension	Stimulates sympathetic nervous system; combines with receptors in the postsynaptic neurons of autonomic system; effect similar to that of acetylcholine, but large amounts result in blocking transmission; stimulates synthesis of lipid in arterial wall	Tolerance; physical dependence; stimulates development of atherosclerosis
LSD (lysergic acid diethylamide)	Overexcitation; sensory distortions; hallucinations	Alters levels of transmitters in brain (may inhibit serotonin and increase NE); potent CNS stimulator; dilates pupils (may appear unequal in size); increases heart rate; raises blood pressure	Irrational behavior
Marijuana	Euphoria	Impairs coordination; impairs depth perception and alters sense of timing; inflames eyes, peripheral vasodilation; exact mode of action unknown	In large doses, sensory distortions, hallucinations; evidence of lowered sperm counts and testosterone (male hormone) levels

1. The diencephalon develops into thalamus and hypothalamus.
   a. The thalamus is a relay center for motor and sensory information.
   b. The hypothalamus controls autonomic functions, links nervous and endocrine systems, controls temperature, appetite, and fluid balance, and is involved in some emotional and sexual responses.
2. The telencephalon develops into the cerebrum and olfactory bulbs.
   a. In fish and amphibians the cerebrum functions mainly to integrate incoming sensory information.
   b. In birds the corpus striatum controls stereotyped but complex behavior patterns. Another part of the cerebrum is thought to govern learning.
   c. The simplest animals possessing a neopallium are certain reptiles. The neopallium is present in birds and mammals, and accounts for about 90% of the cerebral cortex in the human brain. In mammals the cerebrum has complex association functions.

III. The human brain and spinal cord are protected by bone and by three meninges, and are cushioned by cerebrospinal fluid.
   A. The spinal cord consists of ascending tracts, which transmit information to the brain, and descending tracts, which transmit information from the brain. Its gray matter consists of many nuclei that serve as reflex centers.
   B. The human cerebral cortex consists of motor areas, which control voluntary movement; sensory areas, which receive incoming sensory information; and association areas, which link sensory and motor areas and also are responsible for learning, language, thought, and judgment.
      1. The reticular activating system is responsible for maintaining consciousness.
      2. The limbic system affects the emotional aspects of behavior, motivation, sexual behavior, autonomic responses, and biological rhythms.
      3. Alpha wave patterns are characteristic of relaxed

states, beta wave patterns of heightened mental activity, and delta waves of non-REM sleep.
4. Metabolic rate slows during non-REM sleep. REM sleep is characterized by dreaming.
5. Short-term memory may depend upon reverberating circuits in the brain. Mechanisms of long-term memory are not understood.
6. Environmental experience can cause physical and chemical changes in the brain.

IV. The peripheral nervous system consists of sensory receptors and nerves, including the cranial and spinal nerves and their branches.

V. The autonomic system regulates the internal activities of the body.
A. The sympathetic system enables the body to respond to stressful situations.
B. The parasympathetic system influences organs to conserve and restore energy.

VI. Many drugs alter mood by increasing or decreasing the concentrations of specific neurotransmitters within the brain.

POST-TEST

1. The simplest organized nervous system is the ______________ ______________ found in *Hydra* and other cnidarians.

2. ______________ nerves transmit impulses toward a central nervous system whereas ______________ nerves transmit impulses away from the CNS.

3. The cerebral ganglia in a flatworm serve as a simple ______________.

4. In addition to cerebral ganglia, mollusks typically have ______________ ganglia located among the organs and ______________ ganglia located in the foot.

5. In vertebrates the embryonic neural tube expands anteriorly to form the ______________ and develops posteriorly into the ______________ ______________.

6. The medulla, pons, and midbrain make up the ______________ ______________.

7. The fourth ventricle, located within the ______________, communicates with the ______________ ______________ of the spinal cord.

For each group, select the most appropriate answer from Column B for the description given in Column A.

Column A	*Column B*
8. Most prominent part of amphibian brain	a. Cerebellum
9. Links nervous and endocrine systems	b. Cerebrum
10. Most prominent part of mammalian brain	c. Midbrain
11. Coordinates muscle activity	d. Medulla
12. Contains vital centers	e. Hypothalamus

13. Controls innate, complex action patterns in birds	a. Neopallium
14. Convey voluntary motor impulses from cerebrum down spinal cord	b. Corpus striatum
15. Shallow furrows between gyri	c. Meninges
16. Makes up most of human cerebral cortex	d. Pyramidal tracts
17. Layers of connective tissue that protect CNS	e. Sulci

18. Contains primary motor areas	a. Limbic system
19. Action system concerned with emotional behavior and with reward and punishment	b. RAS
20. Contains the visual centers	c. Frontal lobe
21. Maintains wakefulness	d. Temporal lobe
	e. None of the above

22. As you answer these questions your brain should be emitting ______________ waves.

23. Dreaming takes place during ______________ sleep.

24. Cranial nerve X, the ______________ nerve, innervates the ______________.

25. The sympathetic nervous system mobilizes ______________ and helps the body respond to ______________.

26. Sensory nerves enter the spinal cord through the ______________ root.

27. Some drugs induce tolerance, which means that ______________.

28. Label the following diagram. (Refer to Fig. 40–12 as necessary.)

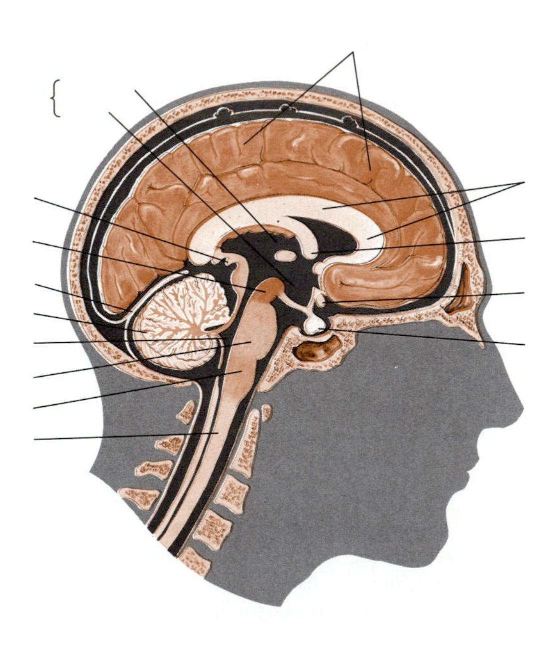

REVIEW QUESTIONS

1. Compare the nervous system of a hydra with that of a planarian flatworm.
2. Compare the flatworm nervous system with that of a vertebrate.
3. Describe characteristics of the bilateral nervous system, and compare it with the radial system.
4. The forebrain, midbrain, and hindbrain give rise to which parts of the brain?
5. What are the functions of each of the following structures in the human brain: medulla, midbrain, cerebellum, thalamus, hypothalamus, and cerebrum.
6. Compare the fish midbrain and cerebrum with that of the mammal.
7. Describe the protective coverings of the human CNS, and give the function of the cerebrospinal fluid.
8. Cite experimental evidence supporting the view that environmental experience can alter the brain.
9. Contrast the structure and function of the sympathetic system with that of the parasympathetic system.
10. Describe how each of the following drugs affects the CNS:
    a. alcohol
    b. phenothiazines
    c. barbiturates
    d. amphetamines

41
Sense Organs

OUTLINE

I. What is a sense organ?
II. How sense organs are classified
III. How sense organs work
IV. Sensory coding and sensation
V. Mechanoreceptors
- *A.* Tactile receptors
- *B.* Gravity receptors: statocysts
- *C.* Lateral line organs
- *D.* Proprioceptors
- *E.* Equilibrium
  - *1.* Halteres in flies
  - *2.* The labyrinth of the vertebrate ear
- *F.* Auditory receptors

VI. Chemoreceptors: taste and smell
- *A.* The sense of taste in insects
- *B.* The human sense of taste
- *C.* The sense of smell

VII. Thermoreceptors
VIII. Photoreceptors
- *A.* The human eye
  - *1.* The chemistry of vision
  - *2.* Color vision
  - *3.* Binocular vision and depth perception
  - *4.* Defects in vision
- *B.* The compound eye

LEARNING OBJECTIVES

After you have read this chapter you should be able to:

1. Distinguish among exteroceptors, proprioceptors, and interoceptors, and explain the importance of each group.
2. Name the five types of receptors that are classified according to the types of energy to which they respond. Give examples of specific sense organs of each type.
3. Describe how a sense organ functions, including definitions of energy transduction, receptor potential, and adaptation in your answer.
4. Describe how the following mechanoreceptors work: tactile receptors, statocysts, lateral line organs, and proprioceptors.
5. Compare the function of the saccule and utricle with that of the semicircular canals in maintaining equilibrium.
6. Trace the path taken by sound waves through the structures of the ear, and explain how the organ of Corti is able to function as an auditory receptor.
7. Describe the receptors of taste and smell.
8. Describe the ways thermoreceptors are advantageous in various types of animals.
9. Label the structures of the human eye on a diagram, and give the functions of each of the accessory structures.
10. Name the two types of photoreceptors in the human retina, and compare their functions, taking into account the role of rhodopsin.
11. Compare the vertebrate eye with the compound eye of an insect.

Sense organs link organisms with the outside world and enable them to receive information about their environment. The kinds of sense organs an animal has determine just how it perceives the world. We humans live in a world of rich colors, multishapes, and varied sounds. But we cannot hear the high-pitched whistles audible to dogs and cats, or the ultrasonic echoes by which bats navigate (Fig. 41–1). Nor do we ordinarily recognize our friends by their distinctive odors. And although vision is our dominant and most refined sense, we are blind to the ultraviolet hues that light up the world for insects.

What Is a Sense Organ?

A **sense organ** is a specialized structure consisting of one or more **receptor cells** and, sometimes, **accessory cells.** For example, the receptor cells of the human eye are the rod and cone cells located in the retina. The accessory structures include the cornea, lens, iris, and ciliary muscles. Accessory structures enhance the versatility of the sense organ, but in some instances may limit its performance. The lens, for example, enhances the ability of the eye to see by adjusting the focus of the light rays. On the other hand, the lens filters out ultraviolet light before it reaches the retina, so we cannot see it. The cells in the retina can respond to light of this wavelength, however.

Receptor cells may be either neuron endings or specialized cells that are in close contact with neurons. Human taste buds are modified epithelial cells connected to one or more neurons.

How Sense Organs Are Classified

Sense organs can be classified in more than one way. One type of classification focuses on the location of the stimuli affecting the sense organ—the exteroceptors, proprioceptors, and interoceptors. Sense organs that reveal the outside world to

Figure 41–1 Bats navigate by ultrasonic echoes inaudible to the human ear. (Courtesy of Frederic Webster.)

the organism are known as **exteroceptors.** They enable an animal to search for food, find and attract a mate, find shelter, detect enemies, recognize friends, explore the world, and even learn. Exteroceptors are obviously of great importance to the survival of the individual and of the species. **Proprioceptors** are sense organs within muscles, tendons, and joints that enable the animal to perceive the position of the arms, legs, head, and other body parts, along with the orientation of the body as a whole. With the help of our proprioceptors we humans can get dressed or eat in the dark.

Interoceptors are sense organs *within* body organs that detect changes in pH, osmotic pressure, body temperature, and the chemical composition of the blood. We are usually not conscious of messages sent to the CNS by these receptors as they play a continuous role in maintaining homeostasis. We do become aware of their activity when they enable us to perceive such diverse internal conditions as thirst, hunger, nausea, pain, and orgasm. Interoceptors are described not in this chapter but in conjunction with discussions of blood pressure, temperature regulation, respiration, and other specific body functions.

Another way that sense organs can be classified is according to the type of energy to which they respond. **Mechanoreceptors** respond to mechanical energy—touch, pressure, gravity, stretching, or movement. **Chemoreceptors** respond to certain chemical stimuli, while **photoreceptors** detect light energy. **Thermoreceptors** respond to heat or cold. Some fish have well-developed **electroreceptors,** which detect electrical energy.

Traditionally, humans are said to have five senses: touch, smell, taste, sight, and hearing. These are all made possible by sense organs that are classified as exteroceptors. Today, balance is also recognized as a sense, and touch is viewed as a compound sense that involves detection of pressure, pain, and temperature. In this chapter we will also consider some proprioceptors that enable us to sense muscle tension and joint position.

How Sense Organs Work

All receptor cells absorb energy, **transduce** (convert) that energy into electrical energy, and produce a receptor potential (Fig. 41–2). In its capacity as a detector or sensor, a receptor receives a small amount of energy from the environment. Each kind of receptor is especially sensitive to one particular form of energy. Rods and cones in the retina absorb the energy of photons. Temperature receptors respond to radiant energy transferred by radiation, conduction, or convection. Electricity is detected by the energy of electrons. Taste buds and olfactory cells detect the change in energy accompanying the binding of specific molecules to their chemical receptors.

Receptor cells are remarkably sensitive to appropriate stimuli. The rods and cones of the eye, for example, are stimulated by an extremely faint beam of light, whereas only a very strong light can stimulate the optic nerve directly. The negligible amount of vinegar that can be tasted, or the amount of vanilla that can be smelled, would have no effect if applied directly to a nerve fiber.

The various kinds of environmental energy act as triggers, causing the receptor cells to perform biological work. These relationships are best exemplified by a very simple sense organ, the tactile hair of an insect. This hair plus its associated cells constitutes a complete sense organ. But only the bipolar neuron at the base of the hair is a receptor cell. The dendrite of the neuron is attached to the base of the hair near the socket, and the neuron's axon passes directly to the CNS without synapsing.

In its unstimulated state, the neuron maintains a steady resting potential; that is, there is a potential difference between the inside and outside of the neuron. This potential difference exists because the ionic compositions of the fluids on each side of the selectively permeable cell membrane are different. The difference is maintained by sodium-potassium pumps and by metabolic work performed by the cell. When the hair is touched (a mechanical stimulus), its shaft moves in the socket and mechanically deforms the dendrite. This mechanical energy increases the permeability of the neuron membrane to ions, with the result that the potential difference between the two sides of the membrane decreases, disappears, or increases. If it increases, the cell is hyperpolarized. If it decreases or disappears, the cell is said to be **depolarized.**

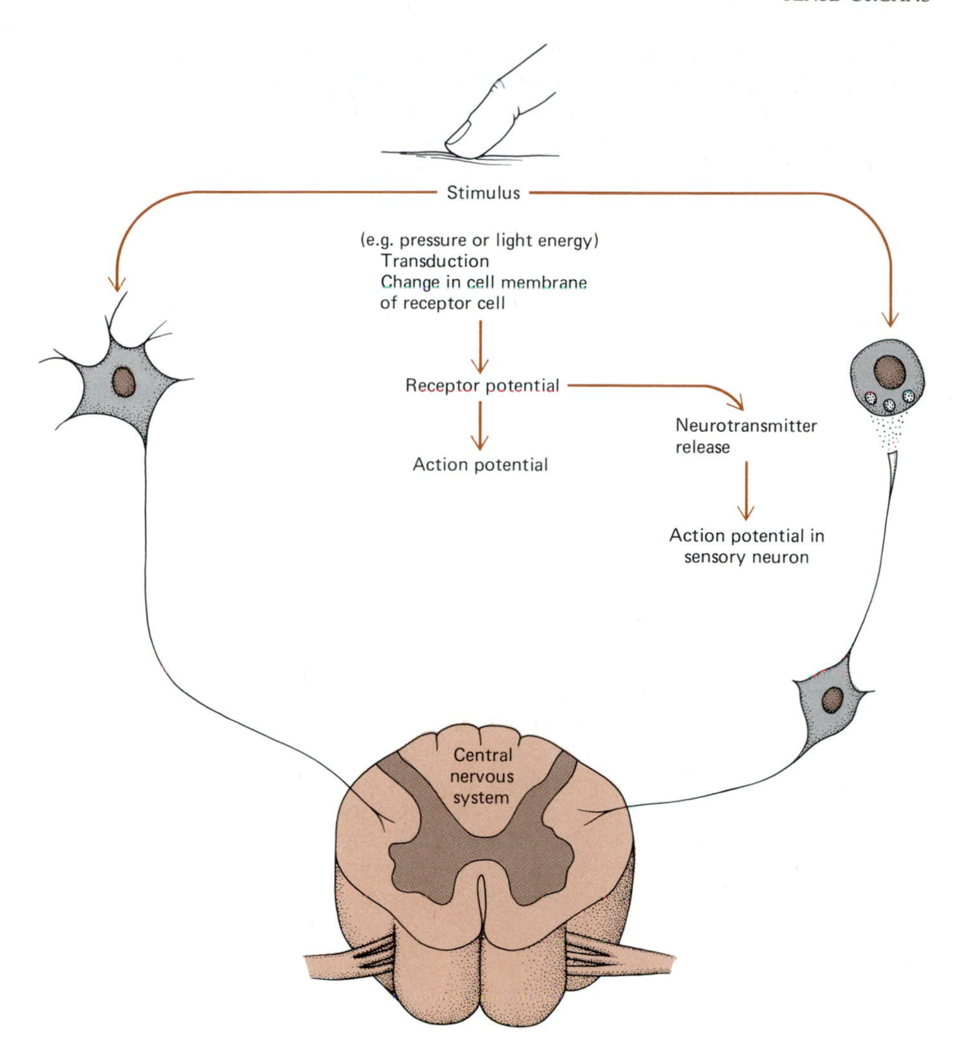

Figure 41–2 How a sense organ works. All receptor cells absorb energy, transduce that energy into electrical energy, and produce a receptor potential. Receptor cells may be either neuron endings themselves or specialized cells in close contact with neuron endings. Both types are shown in the diagram.

The state of depolarization caused by a stimulus is called the **receptor potential.** It spreads relatively slowly down the dendrite, decaying exponentially as it goes. When a special area of the cell near the axon—the axon hillock—becomes depolarized, the threshold level may be reached and an **action potential** is generated. The action potential then travels along the axon to the central nervous system. The receptor thus performs all the essential functions of a sense organ: It detects an event in the environment (a force acting on the hair) by absorbing energy; it converts the energy of the stimulus into electrical energy; and it produces a receptor potential, which may result in an action potential that transmits the information to the CNS. With minor variations, this is how all receptors operate.

The amplitude and duration of the receptor potential are related to the strength and duration of the stimulus. A strong stimulus causes a greater depolarization of the receptor membrane than does a weak one. The action potentials are repetitive, and the frequency at which they are generated is related to the magnitude of the receptor potential. The strength of a stimulus is reflected in the frequency of the action potentials. According to the all-or-none law, the amplitude of each action potential bears no relation to the stimulus; it is characteristic of the particular neuron under the usual recording conditions. In contrast, the receptor potential is a graded response.

Once a stimulus has triggered a receptor to generate action potentials, the stimulus has no further control over them. The situation is analagous to lighting a fuse. The heat of the match is the stimulus. When the end of the fuse reaches the combustion point, the fuse begins to burn, and utilizing its own energy, it ignites adjacent parts of itself. In this way the "message" travels the length of the fuse independently of the temperature of the match flame.

Many receptors do not continue to respond at their initial rate, even if the stimulus remains unabated in intensity. With time, the frequency of action potentials in the sensory neuron decreases. This may occur because the sensory neuron

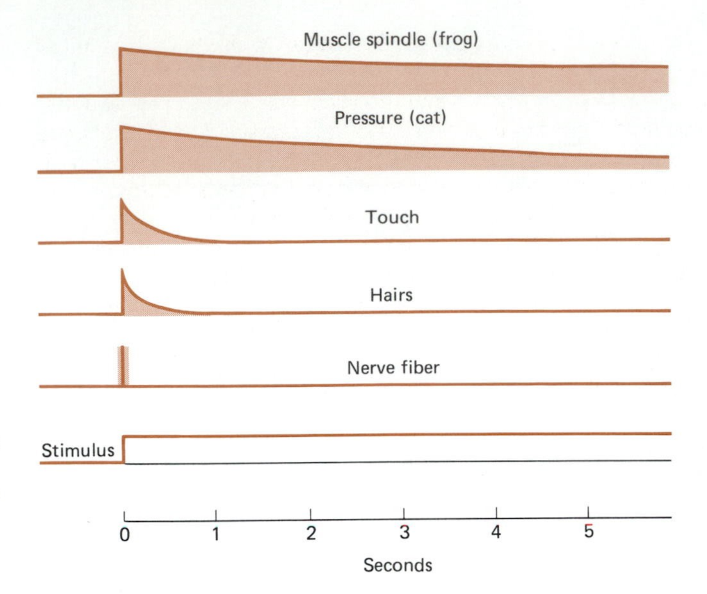

Figure 41–3 The relationship between the stimulus and the different rates of adaptation for different receptors and a nerve fiber. The heights of the curves indicate the rates of discharge of action potentials. (From Adrian, E. D.: *The Basis of Sensation*. London, Chatto and Windus Ltd., 1949.)

becomes less responsive to stimulation, or because the receptor produces a smaller receptor potential, or for both reasons. This diminishing response to a continued, constant stimulus is called **adaptation.**

Some receptors, such as those for pain or cold, adapt so slowly that they continue to trigger action potentials as long as the stimulus persists. Other receptors adapt rapidly, permitting an animal to ignore persistent unpleasant or unimportant stimuli (Fig. 41–3). For example, when you first pull on a pair of tight jeans your pressure receptors let you know that you are being squished and you may feel uncomfortable. Soon, though, these receptors adapt, and you hardly notice the sensation of the tight fit. In the same way we quickly adapt to odors that at first smell seem to assault our senses.

Sensory Coding and Sensation

The stimulation of any sense organ initiates what might be considered a coded message, composed of action potentials transmitted by the nerve fibers and decoded in the brain. Impulses from the sense organ may differ in any of several ways: (1) the total number of fibers transmitting, (2) the specific fibers carrying action potentials, (3) the total number of action potentials passing over a given fiber, (4) the frequency of the action potentials passing over a given fiber, or (5) the time relations between action potentials in specific fibers. These are the possibilities in the "code" sent along the nerve fiber; how the sense organ initiates different codes and how the brain analyzes and interprets them to produce various sensations are not yet understood.

All action potentials are qualitatively the same. Light of the wavelength 400 nanometers (blue), sugar molecules (sweet), and sound waves of 440 hertz (A above middle C) all cause action potentials to be sent to the brain via the appropriate nerves; these action potentials are identical. How can the organism assess its environment accurately? The qualitative differentiation of stimuli must depend either upon the sense organ itself, upon the brain, or upon both. In fact, it depends upon both. Primarily, our ability to discriminate red from green, hot from cold, or red from cold is due to the fact that particular sense organs and their individual sensitive cells are connected to specific cells in particular parts of the brain.

The frequency of the repetitive action potential codes the intensity of the stimulus. Since each receptor normally responds to only one category of stimuli (i.e., light, sound, taste, and so forth), a message arriving in the central nervous system along this nerve is interpreted as meaning that a particular stimulus occurred. Interpretation of the message and, in the case of humans, of the quality of sensation depends upon which central association neurons receive the message. Sensation, when it occurs, occurs in the brain. Rods and cones do not see; only the combination of rods, cones, and centers in the brain see. Furthermore, many sensory messages never give rise to sensations. For example, chemoreceptors in the carotid sinus and the hypothalamus sense internal changes in the body but never stir our consciousness.

Since only those nerve impulses that reach the brain can result in sensations, any blocking of the impulse along the nerve fibers by an anesthetic has the same effect as removing the original stimulus entirely. The sense organs, of course, will continue to initiate impulses that can be detected by the proper electrical apparatus, but the anesthetic prevents them from reaching their destination.

Spatial localization of stimuli impinging on the body, especially mechanical and pain stimuli, also depends upon the destination of specific nerves in the brain. The importance of the brain in localization and in making sensations possible is emphasized by the phenomenon of referred pain. A well-known example of referred pain is that often experienced by persons suffering from heart pains; such persons may complain of pain in the shoulder, upper chest, or left arm. Actually, the stimuli originate in the heart, but the nerve impulses terminate in the same part of the brain as do impulses genuinely originating in the shoulder, chest, or arm.

Cross-fiber patterning, another method of coding information, is probably the one used in olfactory organs. An olfactory organ does not contain a specific receptor for each of the thousands of individual odors that can be recognized. Instead, there is evidence that a limited number of categories of receptors exists. Several receptors are thought to react to each odor. The brain probably makes a statistical analysis of the pattern of responding receptors and from this pattern infers the odor.

The **temporal pattern** of action potentials generated in a single neuron may serve as a code for different stimuli. The single taste receptors of flies, for example, generate action potentials at an even, regular frequency when the stimulus is salt, but generate irregular frequencies when the stimulus is acid.

In invertebrates it is usual for the axon of a sensory neuron to extend all the way to the CNS without synapsing. In these circumstances the message generated at the periphery arrives unaltered. However, in the compound eye of the arthropod, as in many vertebrate sense organs, many interneurons are interposed between the receptor and the CNS. The vertebrate retina or olfactory bulb has an exceptionally complicated neural circuitry. As a consequence of all these synaptic connections, the original message is altered and may lose or gain some of its information.

Mechanoreceptors

Mechanoreceptors respond to touch, pressure, gravity, stretch, or movement. Some of these sense organs are concerned with enabling an organism to maintain its primary body attitude with respect to gravity (for us, head up and feet down; for a dog, dorsal side up and ventral side down; for a tree sloth, ventral side up and dorsal side down).

Mechanoreceptors are also concerned with maintaining postural relations (i.e., the position of one part of the body with respect to another). This information is essential for all forms of locomotion and for all coordinated and skilled movements, from spinning a cocoon to completing a reverse one-and-a-half dive with twist. In addition, mechanoreceptors provide information about the shape, texture, weight, and topographical relations of objects in the external environment. Finally, mechanoreceptors affect the operation of some internal organs. They supply, for example, information about the presence of food in the stomach, feces in the rectum, urine in the bladder, or a fetus in the uterus.

TACTILE RECEPTORS

The simplest mechanoreceptors are free nerve endings in the skin that are directly stimulated by contact with any object on the body surface. Somewhat more complex are the tactile receptors that lie at the base of a hair or bristle (Fig. 41–4). They are stimulated indirectly when the hair is bent or displaced. A receptor potential then develops, and a few action potentials may be generated. Because this type of receptor responds only when the hair is moving, it is known as a **phasic** receptor. Even though the hair may be maintained in a displaced position, the receptor is not stimulated unless there is motion. Such tactile hairs are found in many inverte-

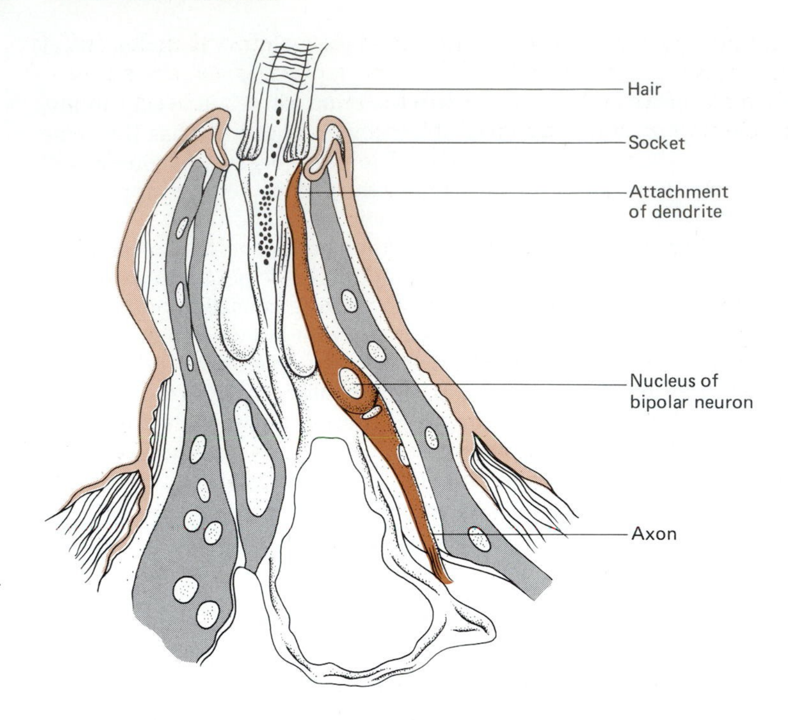

Figure 41–4 A tactile hair from a caterpillar, showing the attachment of the dendrite of the bipolar neuron (the mechanoreceptor) at the point where the shaft of the hair enters the socket.

brates as well as vertebrates and are involved in orientation to gravity, in postural orientation, and in the reception of vibrations in air and water, as well as in contacts with other objects.

The remarkable tactile sensitivity of human skin, especially on the fingertips and lips, is due to a large and diverse number of sense organs (Fig. 41–5). By making a careful point-by-point survey of a small area of skin, using a stiff bristle to test for touch, a hot or cold metal stylus to test for temperature, and a needle to test for pain, it has been found that receptors for each of these sensations are located at different spots. By comparing the distribution of the different types of sense organs and the types of sensations produced, it has been found that the free nerve endings are responsible for pain perception, that a variety of tiny sense organs (e.g., Meissner's corpuscles, Ruffini's end organs, and Merkel's disks) are responsible for touch, that Krause's corpuscles may be responsible for sensations of cold and warmth, and that pacinian corpuscles mediate the sensation of deep pressure.

The pacinian corpuscle has been particularly well studied. The bare nerve ending is surrounded by connective tissue layers (lamellae) interspersed with fluid. Compression causes displacement of the lamellae, which provides the deformation

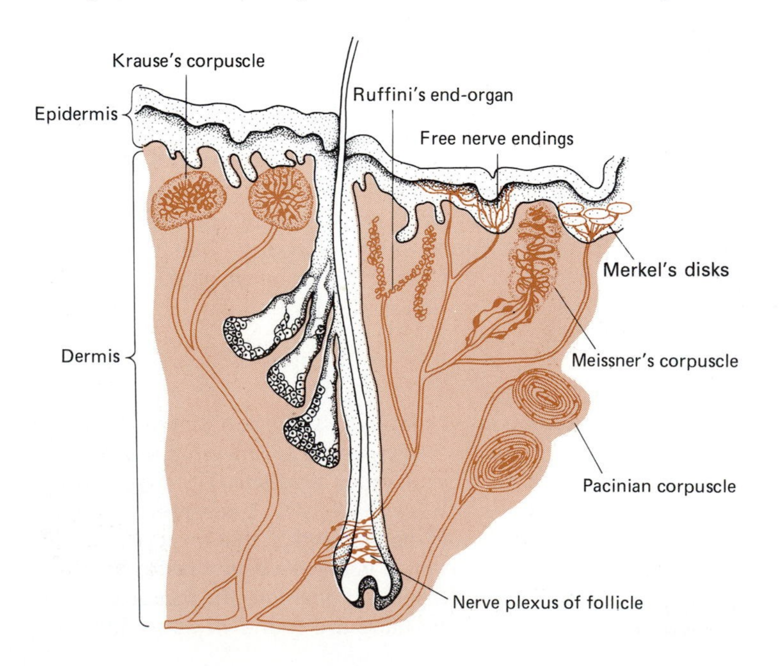

Figure 41–5 Diagrammatic section through the human skin showing the types of sense organs present. The free nerve endings respond to pain; Krause's corpuscles are thought to respond to hot and cold stimuli; tactile hairs, Merkel's disks, Ruffini's end organs, and Meissner's corpuscles respond to touch; pacinian corpuscles respond to deep pressure.

stimulating the axon. Even though the displacement is maintained under steady compression, the receptor potential rapidly falls to zero and action potentials cease—an excellent example of adaptation. The pacinian corpuscle is a phasic receptor responding to velocity (rapid movement of the tissue).

GRAVITY RECEPTORS: STATOCYSTS

All organisms are oriented in a characteristic way with respect to gravity. When displaced from this normal position, they quickly adjust the body to reassume it. To accomplish this, receptors must continually send information regarding the position and movements of the body to the CNS.

Many invertebrates have specialized sense organs called **statocysts** that serve as gravity receptors. A statocyst is basically an infolding of the epidermis lined with receptor cells that have hairs (Fig. 41–6). The cavity contains a **statolith** (sometimes more than one), which is a tiny granule of loose sand grains or calcium carbonate. The particles are held together by an adhesive material secreted by cells of the statocyst. Normally the particles are pulled downward by gravity and stimulate the hair cells. When the position of the statolith changes, the hairs of the receptor cells are bent. This mechanical displacement results in receptor potentials and action potentials that inform the CNS of the change in position. By "knowing" which hair cells are firing, the animal knows where down is, and so can correct any abnormal orientation.

In a classic experiment on crayfish, the function of the statocyst was demonstrated by substituting iron filings for sand grains in the statocyst. The force of gravity was overcome by holding magnets above the animals. The iron filings were

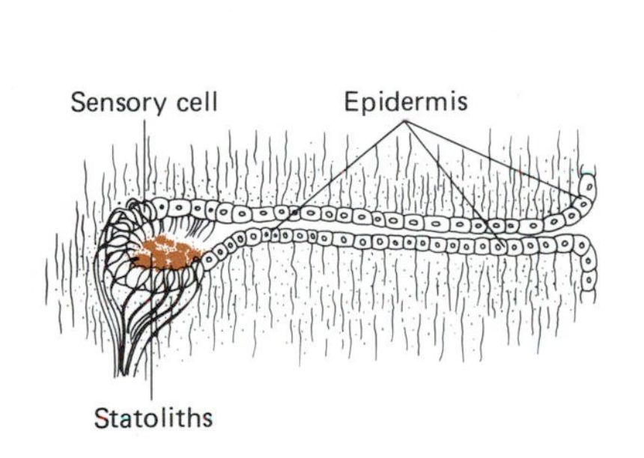

(*a*) A type of statocyst found in mollusks

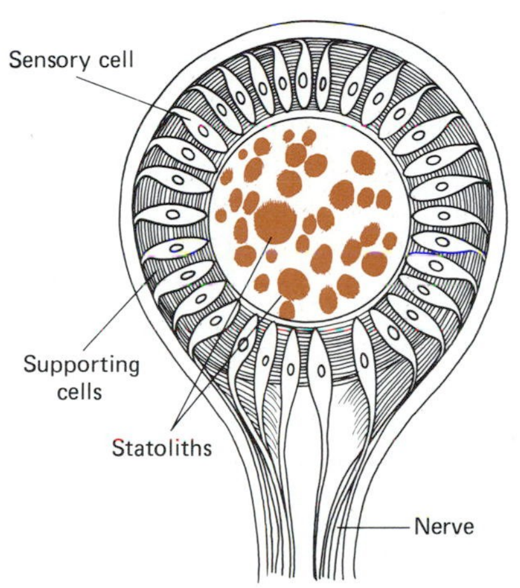

(*b*) A statocyst from an annelid polychaete

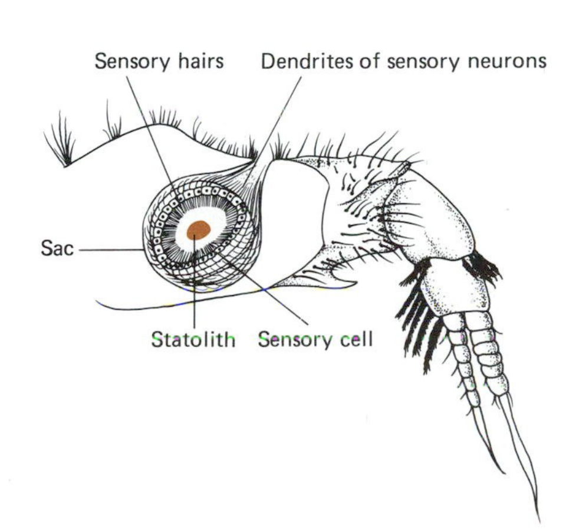

(*c*) Statocyst in the antennule of a decapod crustacean

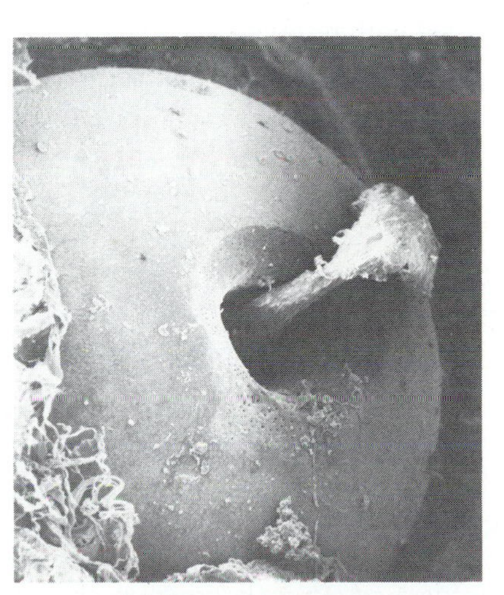

(*d*) Sea urchin statocyst

Figure 41–6 Many invertebrates have statocysts, which serve as sensors of gravitational force. ((*d*), Courtesy of J. Derrenbacher.)

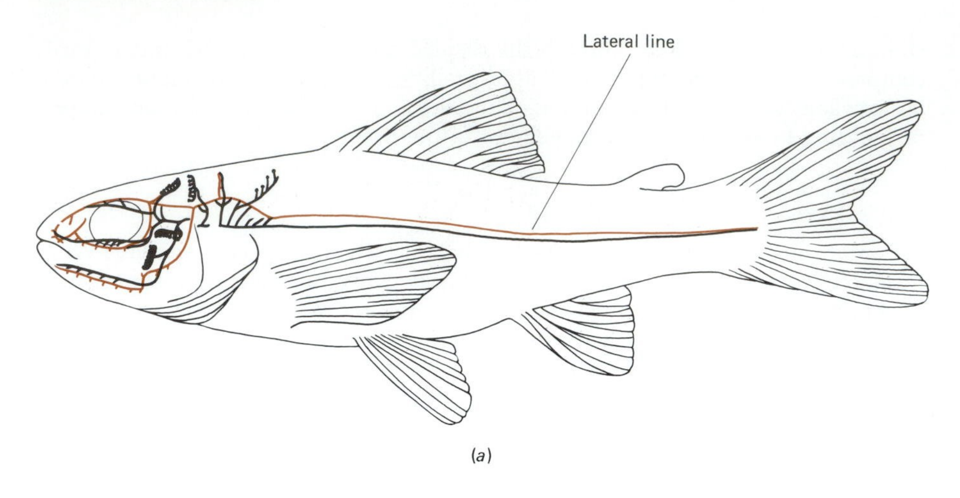

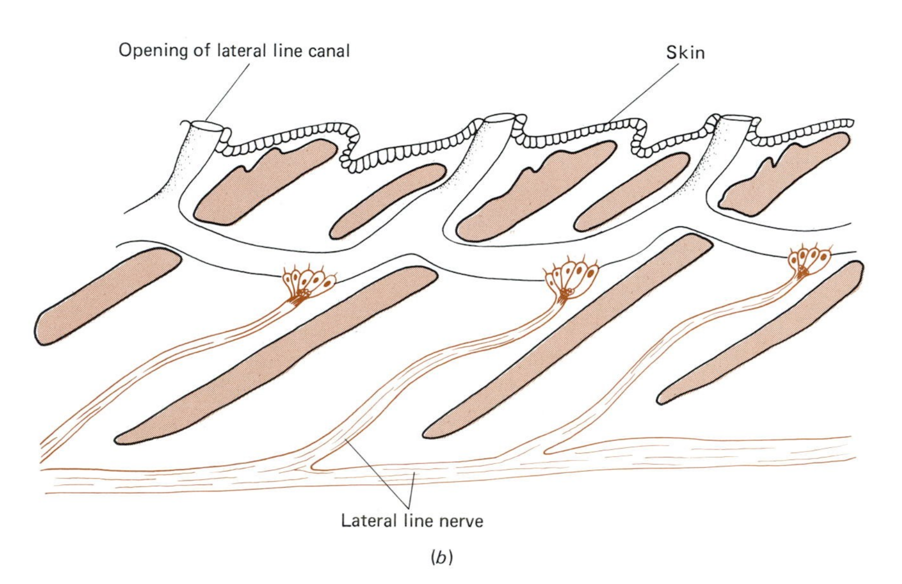

Figure 41–7 The receptor cells of the lateral line organ respond to waves, currents, or disturbances in the water, informing the fish of obstacles or moving objects.

attracted upward toward the magnets, and the crayfish began to swim upside down in response to the new information provided by their gravity receptors.

LATERAL LINE ORGANS

Lateral line organs are found in fishes, and in aquatic and larval amphibians. Typically this sense organ consists of a long canal running the length of the body and continuing into the head on both sides of the animal (Fig. 41–7). The canals are lined with receptor cells that have hairs. Above the hairs and enclosing their tips is a mass of gelatinous material, called a **cupula,** secreted by the receptor cells.

The receptor cells are thought to respond to waves, currents, or disturbances in the water. The water moves the cupula and causes the hairs to bend. This results in messages being dispatched to the CNS. The lateral line organ is thought to supplement vision by informing the fish of obstacles in its way or of moving objects such as prey or enemies.

PROPRIOCEPTORS

Proprioceptors are sense organs that respond continuously to tension and movement in muscles and joints. Vertebrates have three main types: **muscle spindles,** which detect muscle movement (Fig. 41–8); **Golgi tendon organs,** which determine stretch in the tendons that attach muscle to bone; and **joint receptors,** which detect movement in ligaments. These are **tonic** (static) sense organs. In contrast to that in phasic receptors, the receptor potential is maintained (though not at constant magnitude) as long as the stimulus is present and action potentials continue to be generated. Thus, information about the position of the organ concerned is continuously supplied.

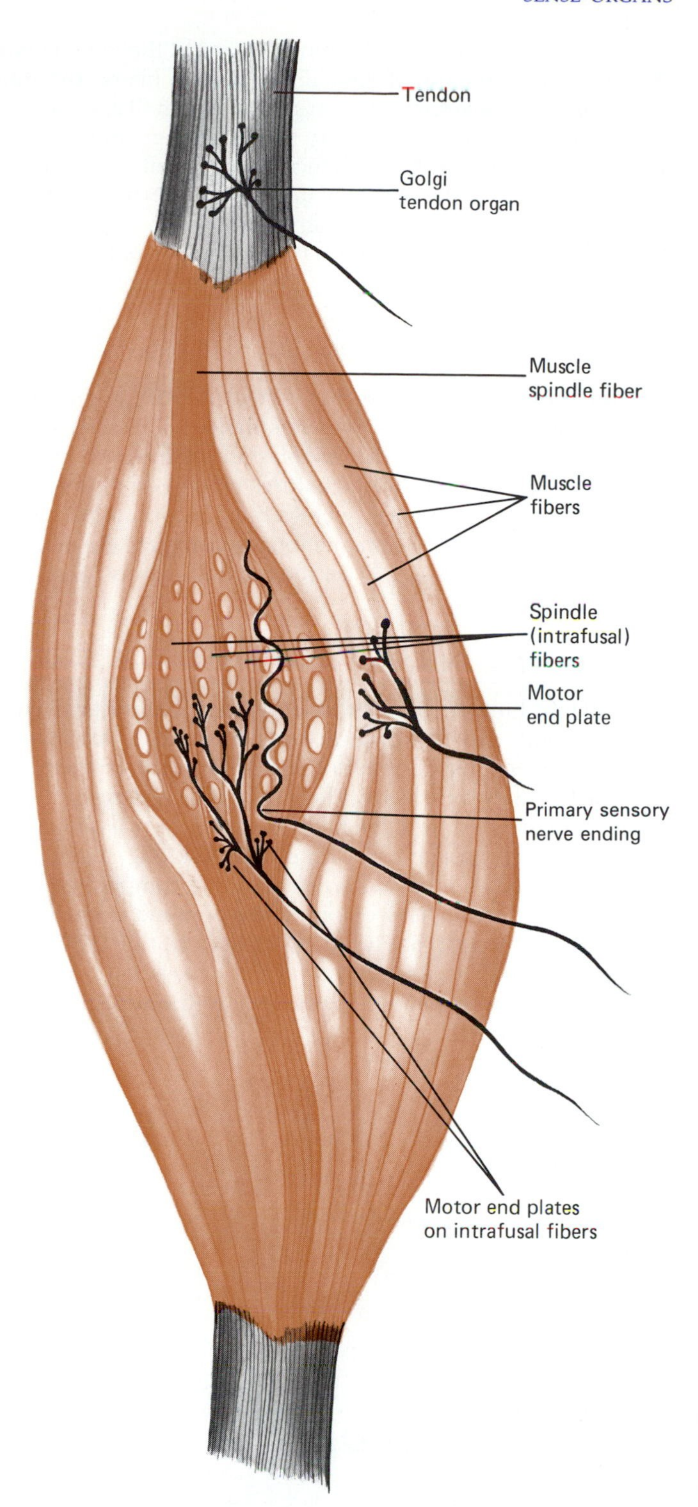

Figure 41–8 A muscle spindle and Golgi tendon organ. Muscle spindles detect muscle movement; Golgi tendon organs determine stretch in tendons.

By means of these sense organs we can, even with our eyes closed, perform manual acts such as dressing or tying knots. Impulses from the proprioceptors are also extremely important in ensuring the harmonious contraction of different muscles involved in a single movement; without such receptors, complicated skillful acts would be impossible. Impulses from these organs are also important in maintaining balance. Proprioceptors, which were discovered only a little more than 100 years ago, are probably more numerous and more continuously active than any of the other sense organs, although we are less aware of them than most of the others. We obtain some idea of what life without proprioceptors would be like when a leg or arm "goes to sleep"—a feeling of numbness, which results in part from the lack of proprioception.

The mammalian muscle spindle is one of the more versatile stretch receptors. It consists of a bundle of specialized muscle fibers, the intrafusal fibers. In the center of the muscle spindle is a region in which filaments are absent. This region is encircled by two types of sensory nerve endings: primary endings and secondary endings. Both types respond statically; they continue to transmit signals for a prolonged period of time, and they do so in proportion to the degree of stretch. The primary endings also exhibit a strong dynamic response. They respond very actively to a rapid rate of change in length, but only while the length is actually increasing.

EQUILIBRIUM

In addition to monitoring events within its body and outside, an organism must have a way of sensing its own orientation. The state of balance or adjustment between opposing forces that enables an organism to maintain this orientation is known as **equilibrium.** Most creatures employ the force of gravity to provide this information, but other cues, such as the direction of the prevailing light, may also be used. Two examples of orientational sensing are the more typical human mechanism, which is based largely on the force of gravity, and a more unusual method employed by flies.

Halteres in Flies

Long before humans invented the gyroscope, flies evolved a balancing organ to stabilize flight. Any flying machine must maintain stability if it is to be controllable in the air. Flies must be able to control lift and stabilize in all three planes of rotation; that is, they must correct for pitch, roll, and yaw. They accomplish this with information derived from the **halteres,** a pair of marvelously modified hind wings. Each is a heavy mass of tissue on a thin stalk (Fig. 41–9) and resembles an Indian club. The base is folded and articulated in a complicated fashion and is equipped with about 418 mechanoreceptors. These respond to strains produced in the cuticle by gyroscopic torque produced by the beating of the halteres. These oscillating masses generate forces at the base of the stalk as the whole fly rotates. They probably do not act as stabilizing gyroscopes of the sort placed in ships to offset their

(*a*)

(*b*)

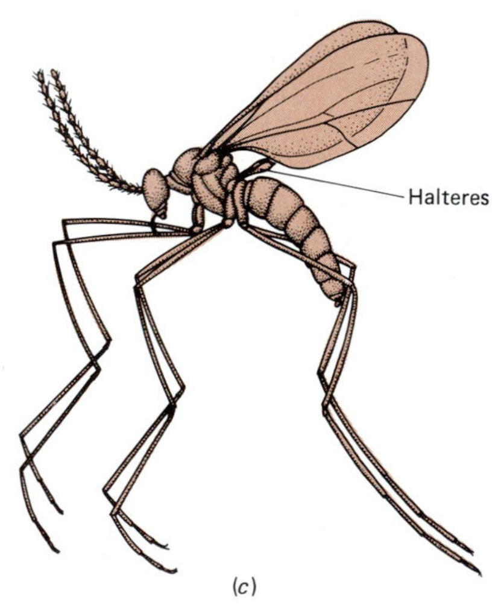

(*c*)

Figure 41–9 In Dipterans, the hind wings are reduced to halteres, small stalks expanded at the distal end. (*a*) Ventral and (*b*) dorsal views of a model of the left haltere of the blowfly *Lucilla sericata*. Rows of mechanoreceptors are visible at the base (×130). (*c*) The halteres of a gnat. ((*a*), (*b*), courtesy of J.W.S. Pringle.)

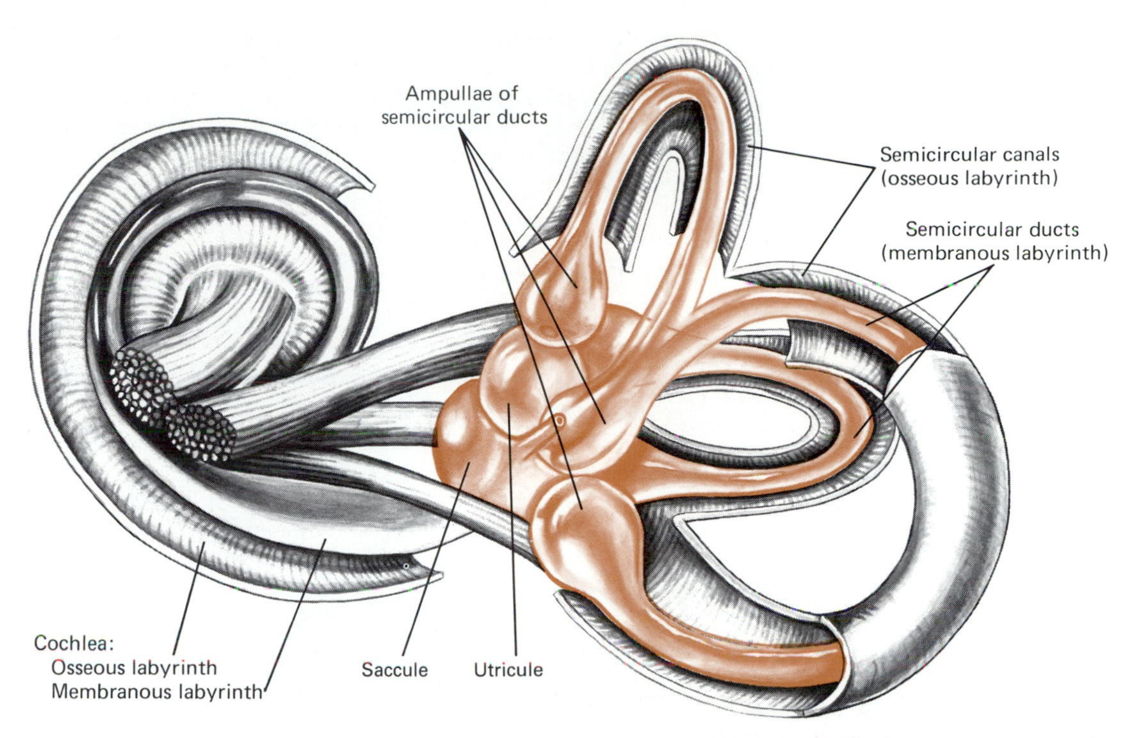

Figure 41–10 The human inner ear with the membranous labyrinth exposed. Because this is a posterior view the utricle and saccule can be seen. Note that the membranous labyrinth is shown in color only within the semicircular canals and is not colored in the cochlea.

movement. Their action is indirect in that their mechanoreceptors signal the CNS to make the necessary corrections in flight.

The Labyrinth of the Vertebrate Ear

When we think of the ear, we think of hearing. However, fishes do not use their ears for hearing. In fact, in all vertebrates the basic function of the ear is to help maintain equilibrium. Typically the ear also contains gravity receptors. Although many vertebrates do not have outer or middle ears, all of them have inner ears.

The inner ear consists of a complicated group of interconnected canals and sacs, often referred to as the **labyrinth.** In jawed vertebrates the labyrinth consists of two saclike chambers, the **saccule** and **utricle,** and three **semicircular canals,** as well as the **cochlea.**

Collectively, the saccule, utricle, and semicircular canals are referred to as the **vestibular apparatus** (Fig. 41–10). Destruction of the vestibular apparatus leads to a considerable loss of the sense of equilibrium. A pigeon in which these organs have been destroyed is unable to fly, but in time can relearn how to maintain equilibrium using visual stimuli. Equilibrium in the human depends not only upon stimuli from the organs in the inner ear but also upon the sense of vision, stimuli from the proprioceptors, and stimuli from cells sensitive to pressure in the soles of the feet.

The saccule and utricle house gravity detectors in the form of small calcium carbonate ear stones called **otoliths** (Fig. 41–11). The sensory cells of these structures are similar to those of the lateral line organ. They consist of groups of hair cells surrounded at their tips by a gelatinous cupula. The receptor cells in the saccule and utricle lie in different planes. Normally, the pull of gravity causes the otoliths to press against particular hair cells, stimulating them to initiate impulses sent to the brain by way of sensory nerve fibers at their bases. When the head is tilted, or in linear acceleration (change in speed when the body is moving in a straight line), the otoliths press upon the hairs of other cells and stimulate them. This enables the animal to perceive the direction of gravity and of linear acceleration or deceleration when the head is in any position.

Information about turning movements is furnished by the three semicircular canals. Each of these is connected with the utricle, and lies in a plane at right angles to the other two. Each canal is a hollow ring filled with fluid called **endolymph.** At one of the openings of each canal into the utricle is a small, bulblike enlargement, the **ampulla.** Within each ampulla is a clump of hair cells called a **crista,** similar to those in the utricle and saccule, but lacking otoliths. These receptor cells are stimulated by movements of the endolymph in the canals (Fig. 41–12).

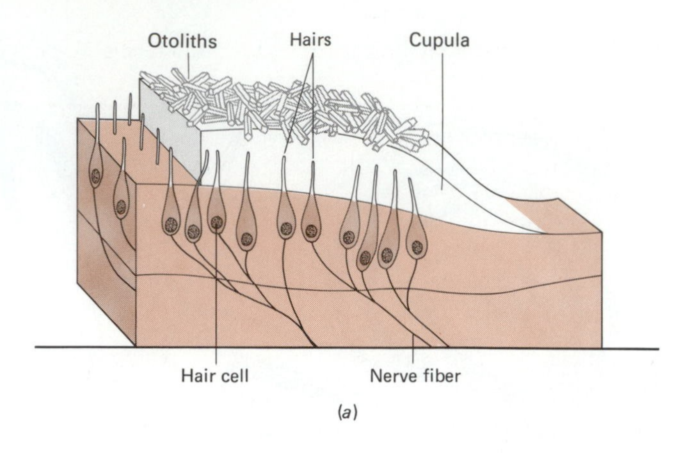

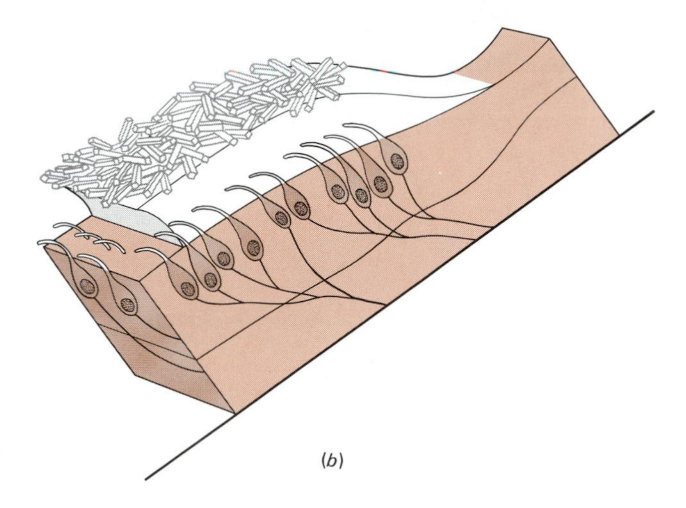

Figure 41–11 The saccule and utricle. Compare the positions of the otoliths and hairs in (*a*) with those in (*b*). Changes in head position cause the force of gravity to distort the cupula, which in turn distorts the hairs of the hair cells; the hair cells respond by sending impulses down the vestibular nerve (part of the auditory nerve) to the brain.

When the head is turned, there is a lag in the movement of the fluid within the canals, so that the hair cells move in relation to the fluid and are stimulated by its flow. This stimulation produces not only the consciousness of rotation but also certain reflex movements in response to it—movements of the eyes and head in a direction opposite to the original rotation. Since the three canals are located in three

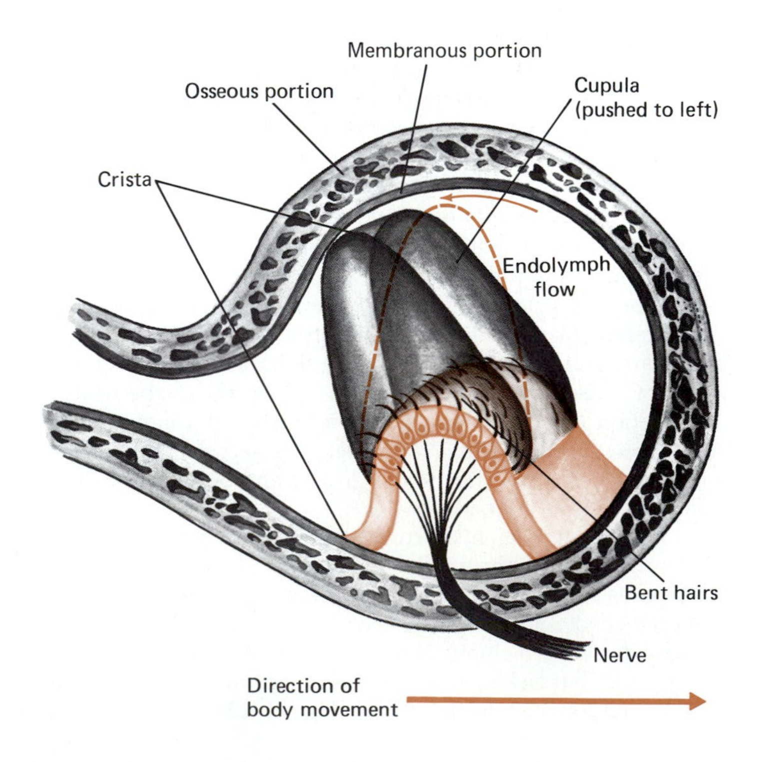

Figure 41–12 How movement of the endolymph within the semicircular ducts distorts the cupula inside the ampulla. The hair cells of the cupula then are bent, reporting any change to the brain via the vestibular nerve.

different planes, a movement of the head in any direction will stimulate the movement of the fluid in at least one of the canals.

We humans are used to movements in the horizontal plane, which stimulate certain semicircular canals, but we are unused to vertical movements parallel to the long axis of the body. Movements such as the motion of an elevator, or of a ship pitching in a rough sea, stimulate the semicircular canals in an unusual way, and may cause sea sickness or motion sickness, with their resulting nausea or vomiting. When a person so affected lies down, the movement stimulates the semicircular canals in a different way, and nausea is less likely to occur.

AUDITORY RECEPTORS

Some fish hear by means of receptors in the utricle, but hearing does not seem to be important to them. Hearing is an important sense in tetrapods, however. Both birds and mammals have a highly developed sense of hearing based in the cochlea, a structure in the inner ear that contains mechanoreceptor hair cells that detect pressure waves.

Part of the labyrinth of the inner ear, the cochlea is a spiral tube coiled two-and-a-half turns and resembling a snail's shell (Fig. 41–13). If the cochlea were uncoiled, as in Figure 41–14, it would be seen to consist of three canals separated from each other by thin membranes and coming almost to a point at the apex. Two of these canals, or ducts, the **vestibular duct** and the **tympanic duct,** are connected with one another at the apex of the cochlea and are filled with a fluid known as **perilymph.** The middle canal, the **cochlear duct,** is filled with endolymph and contains the actual auditory receptor, the **organ of Corti.** Each organ of Corti contains about 24,000 hair cells arranged in five rows extending the entire length of the coiled cochlea. Each cell is equipped with hairlike projections extending into the cochlear duct. These cells rest upon the **basilar membrane,** which separates the cochlea from the tympanic duct. Overhanging the hair cells is another membrane, the **tectorial membrane,** attached along one edge to the membrane on which the hair cells rest, with its other edge free. The hair cells initiate impulses in the fibers of the cochlear (auditory) nerve.

In terrestrial vertebrates, accessory structures in the outer and middle ear change sound waves in air to pressure waves in the cochlear fluid. In the human

Figure 41–13 The anatomy of the human ear.

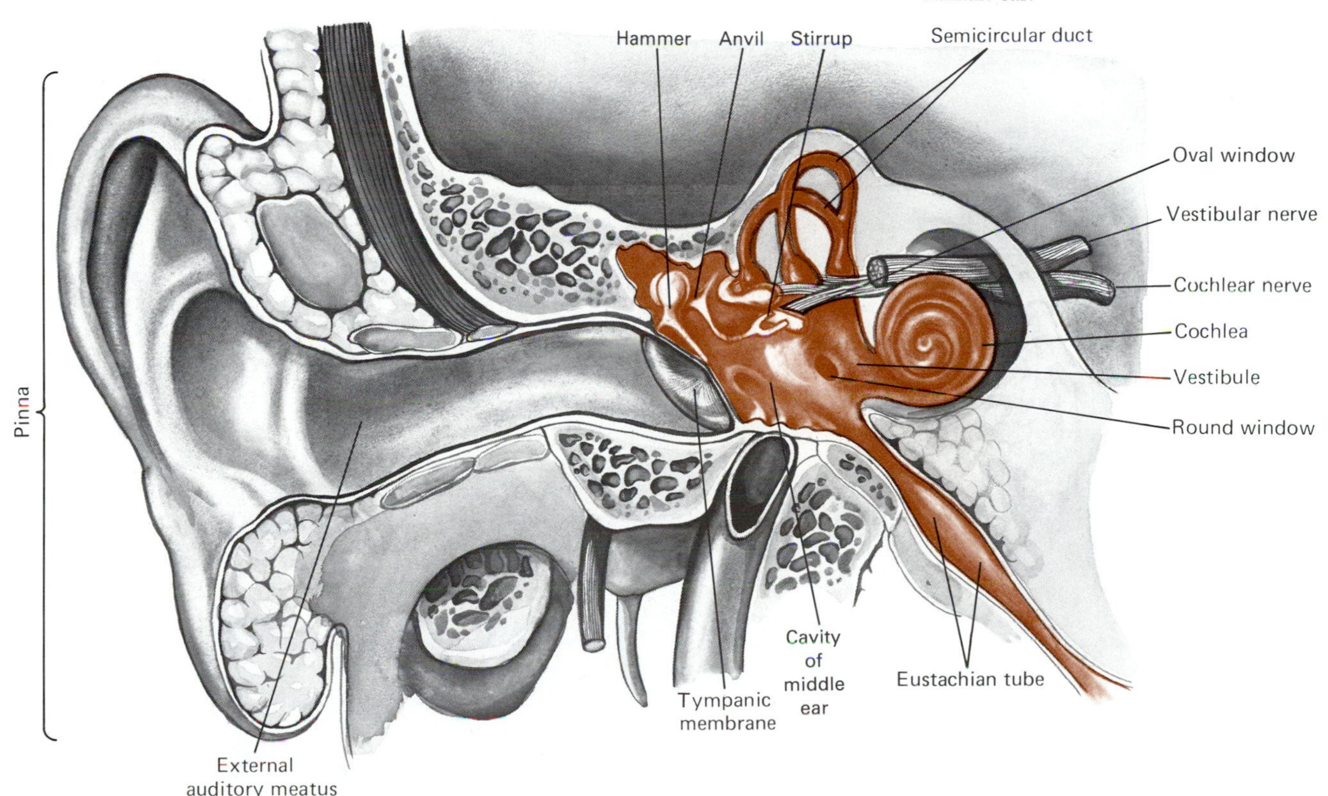

Figure 41–14 The cochlea is the part of the inner ear concerned with hearing. (*a*) Cross section through the cochlea to show the organ of Corti resting on the basilar membrane and covered by the tectorial membrane. (*b*) Scanning electron micrograph of guinea pig organ of Corti showing inner hair cells, *IHC*, and three rows of outer hair cells, *OHC 1–3* (magnification ×1,790). *N*, nucleus; *S*, stereocilia; *TM*, tectorial membrane; *SM*, scala media (cochlear duct); *ST*, scala tympani (tympanic duct); *black arrows*, basilar membrane; *HC*, *DC*, *PC*, various supporting cells. (Courtesy of Dr. L. G. Duckert, University of Washington.) (*c*) Diagram of the cochlea uncoiled and drawn out in a straight line. (*d*) The organ of Corti. Vibrations transmitted by the hammer, anvil, and stirrup set the fluid in the vestibular canal in motion; these vibrations are transmitted to the basilar membrane and the organ of Corti. The hair cells of the organ of Corti are the receptor cells for hearing and are innervated by the cochlear nerve, a branch of the auditory nerve.

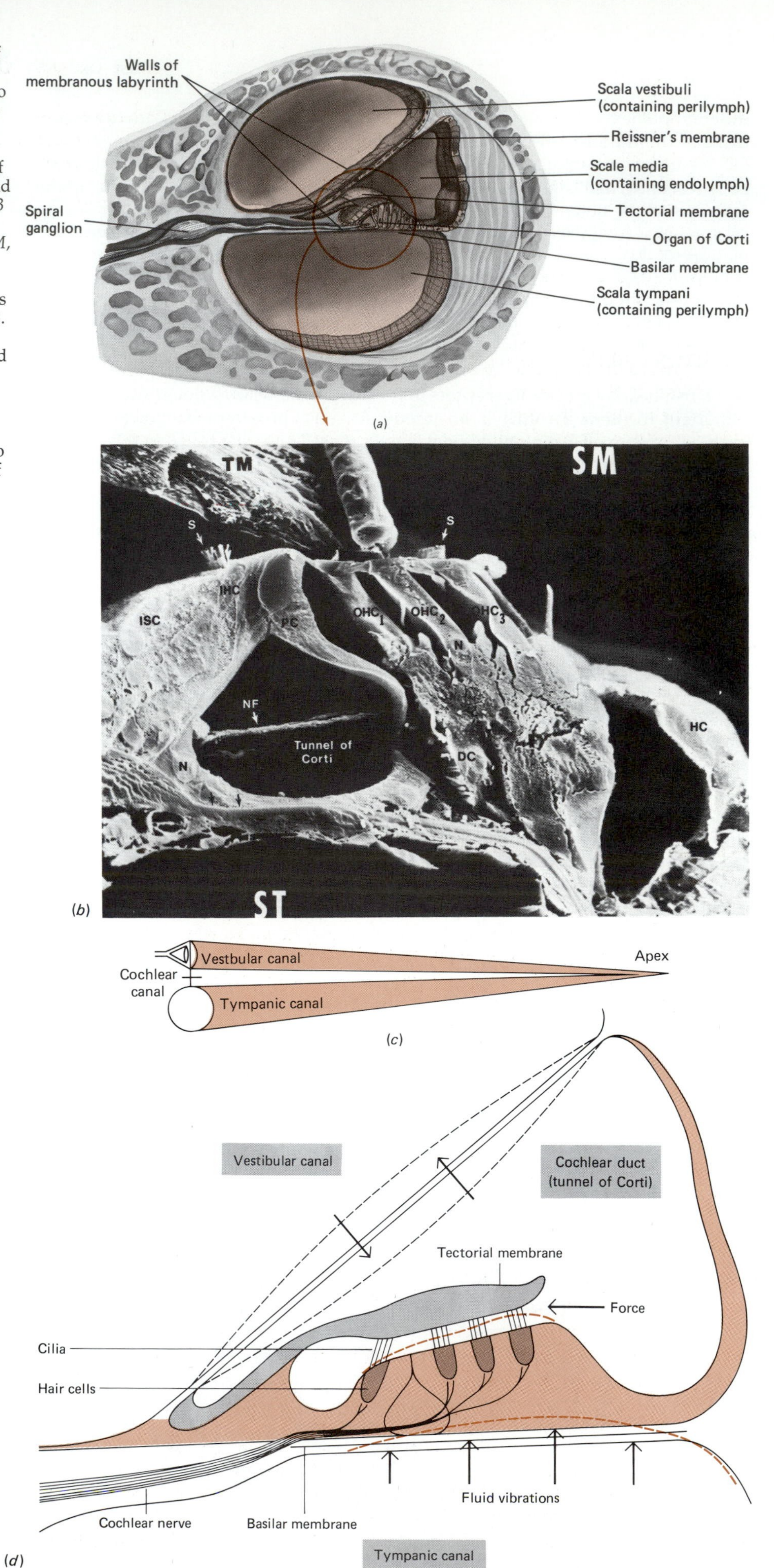

ear, for example, sound waves pass through the **external auditory meatus** (the canal of the outer ear) and set the **eardrum** (the membrane separating outer ear and middle ear) vibrating. These vibrations are transmitted across the middle ear by three tiny bones, the **hammer, anvil,** and **stirrup** (so called because of their shapes). The hammer is in contact with the eardrum, and the stirrup is in contact with the membrane at the opening of the inner ear called the **oval window.** The vibrations pass through the oval window to the fluid in the vestibular canal.

Since liquids cannot be compressed, the oval window could not cause movement of the fluid in the vestibular duct unless there were an escape valve for the pressure. This is provided by the **round window** at the end of the tympanic duct. The pressure wave presses upon the membranes separating the three ducts, is transmitted to the tympanic canal, and causes a bulging of the round window. The movements of the basilar membrane produced by these pulsations are believed to rub the hair cells of the organs of Corti against the overlying tectorial membrane, thus stimulating them and initiating nerve impulses in the dendrites of the cochlear nerve lying at the base of each hair cell.

Since sounds differ in pitch, intensity, and quality, any theory of hearing must account for the ability to differentiate between these characteristics of sound. Microscopic examination of the organ of Corti reveals that the fibers of the basilar membrane along the coiled cochlea are of different lengths, being longer at the apex and shorter at the base of the coil, thus resembling the strings of a harp or piano. Sounds of a given frequency (and pitch) set up resonance waves in the fluid in the cochlea that cause a particular section of the basilar membrane to vibrate. The vibration stimulates the particular group of hair cells in that section. In this way the brain infers the pitch of a sound by taking note of the particular hair cells that are stimulated. Loud sounds cause resonance waves of greater amplitude and lead to a more intense stimulation of the hair cells and to the initiation of a greater number of impulses per second passing over the auditory nerve to the brain. The nerve impulses produced by particular sounds have the same frequency as those sounds; thus the brain may recognize particular pitches by the frequency of the nerve impulses reaching it, as well as by the identity of the nerve fibers conducting the impulses.

Variations in the quality of sound, such as are evident when an oboe, a cornet, and a violin play the same note, depend upon the number and kinds of **overtones,** or **harmonics,** present. These stimulate different hair cells in addition to the main stimulation common to all three. Thus, differences in quality are recognized by the *pattern* of the hair cells stimulated. Careful histological work has shown that the nerve fibers from each particular part of the cochlea are connected to particular parts of the auditory area of the brain, so that certain brain cells are responsible for the perception of sensations of high tones, others for low tones.

The human ear is equipped to register sound frequencies between about 20 and 20,000 cycles per second, although there are great individual differences. Some animals—dogs, for example—can hear sounds of much higher frequencies. The human ear is more sensitive to sounds between 1000 and 2000 cycles per second than to higher or lower ones. Within this range, the ear is extremely sensitive. In fact, when the energy of audible sound waves is compared with the energy of visible light waves, the ear is 10 times more sensitive than the eye.

The normal human ear is extremely efficient. Any further increase in sensitivity would probably be useless, since it would pick up the random movement of air molecules, which would result in a constant hiss or buzzing. Similarly, if the eye were more sensitive, a steady light would appear to flicker because the eye would be sensitive to the individual photons (light particles) impinging upon it.

There is little fatigue connected with hearing. Even though the ear is constantly assailed by noises, it retains its acuity and fatigue disappears after a few minutes. When one ear is stimulated for some time by a loud noise, the other ear also shows fatigue (i.e., loses acuity), indicating, not unexpectedly, that some of the fatigue is in the brain rather than in the ear itself.

Deafness may be caused by injuries or malformations of either the sound-transmitting mechanisms of the outer, middle, or inner ears, or of the sound-perceiving mechanism of the latter. The external ear may become obstructed by wax secreted by the glands in its wall; the middle-ear bones may become fused after an infection; or, more rarely, the inner ear or auditory nerve may be injured by a local inflammation or by the fever accompanying some disease.

When the ear is subjected to intense sound, the organ of Corti is injured. This was demonstrated by an experiment in which guinea pigs were exposed to continuous pure tones for a period of several weeks. When their cochleas were examined microscopically after death, it was found that the guinea pigs subjected to high-pitched tones suffered injury only in the lower part of the cochlea, while those subjected to low-pitched tones suffered injury only in the upper part of the cochlea. Members of rock bands, boilermakers, and other workers subjected to loud, high-pitched noises over a period of years frequently become deaf to high tones because the cells near the base of the organ of Corti become injured.

Chemoreceptors: Taste and Smell

Throughout the animal kingdom many feeding, social, sexual, and reproductive activities are initiated, regulated, or influenced in some way by specific chemical cues in the environment. Insects, for example, use many chemicals in communication, in defense from predators, and in recognizing specific foods. Many vertebrates employ chemical secretions to mark territory, to attract their sexual partners, and to defend themselves. Chemoreception is also used to help carnivores track prey and to help the intended prey elude the carnivores.

The sensitivity of chemoreceptors varies greatly. Some, such as those in the skin of a frog, may be gross and nonspecific. A frog is thus unable to differentiate between certain stimuli and will scratch its back when dilute acid or concentrated solutions of inorganic salts are applied to the skin. Free nerve endings are the chemoreceptors involved. This common chemical sense is widely distributed among aquatic animals. Among mammals it is restricted to moist areas of the body. Recall how your eye smarts and waters in the presence of ammonia fumes or a peeled onion, and how a broken blister stings if touched by a nonphysiological solution.

Two highly sensitive chemoreceptive systems are the senses of **taste** and **smell** (olfaction). These are easily distinguishable in humans and other terrestrial organisms. However, in aquatic organisms, especially members of lower phyla, it becomes increasingly difficult to decide what is taste and what is olfaction.

THE SENSE OF TASTE IN INSECTS

One of the most thoroughly studied organs of taste is the taste hair of the fly (Fig. 41–15). The terminal segments of the legs and the mouthparts of flies, moths, butterflies, and a number of other insects are equipped with very sensitive hairs. In the fly, each one of these contains four taste receptors and a tactile receptor. All are primary neurons. One taste receptor is more-or-less specific to sugars, one to water, and two to salts. If water is placed on one hair of a thirsty fly, action potentials generated by the water cell pass directly to the CNS and cause the fly to respond by extending its retractable proboscis and drinking. Similarly, sugar on a particular hair stimulates the sugar receptor and causes feeding. Salt causes the fly to reject the solution.

THE HUMAN SENSE OF TASTE

The organs of taste in humans are budlike structures known as **taste buds,** which are located predominantly on the tongue and soft palate. They are found mainly in tiny elevations, or papillae, on the surface of the tongue. There is a rapid turnover of taste bud cells; every 10 to 30 hours the cells are completely replaced.

A taste bud is an oval body that consists of an epithelial capsule containing several taste receptors. Each taste receptor is an epithelial cell with a border of microvilli at its free surface (Figs. 41–16 and 41–17). A hairlike projection extends to the external surface of the taste bud through an opening called the taste pore. The connections between the taste receptors and the nerve cells that innervate them are complicated, with each taste receptor being innervated by more than one neuron. Furthermore, some neurons are connected with one taste cell, while others are connected with many. This complexity of connections renders interpretation of taste-sensory physiology difficult.

Traditionally there are four basic tastes: sweet, sour, salty, and bitter. Although it is true that the greatest sensitivity to each of these tastes is restricted to a

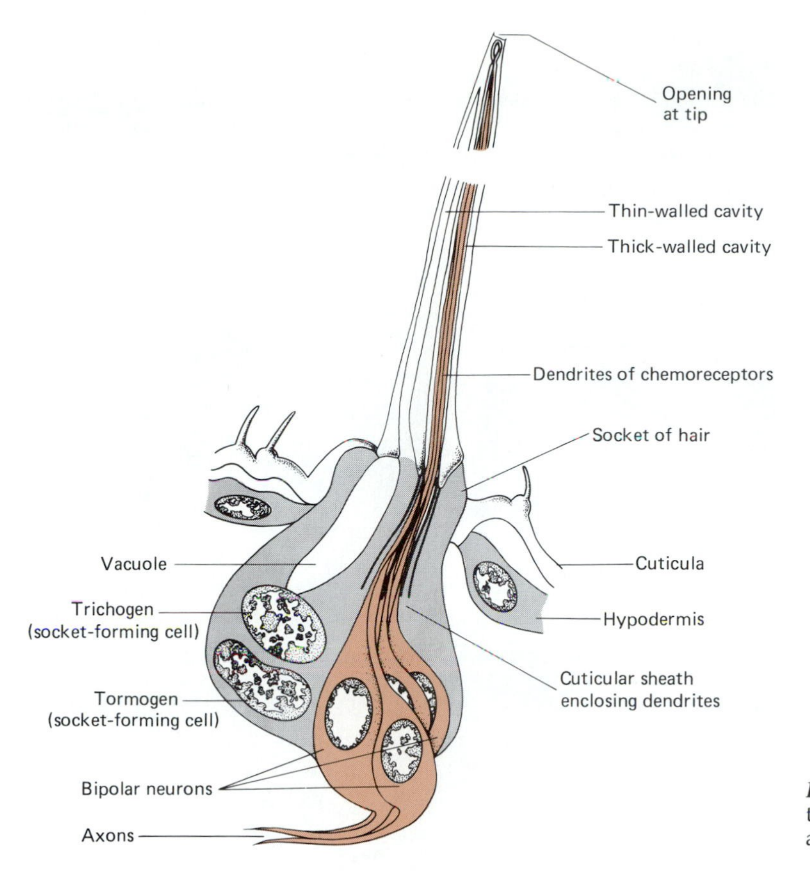

Figure 41–15 A chemoreceptive hair of the blowfly. Four of the five neurons are shown.

given area of the human tongue (Fig. 41–16), not all papillae are restricted in their sensitivity to a single category of taste. Some indeed are responsive specifically to salty, bitter, or sweet taste, but the majority respond to two or more categories of taste. Nor is a single taste bud restricted in its sensitivity to a single type of chemi-

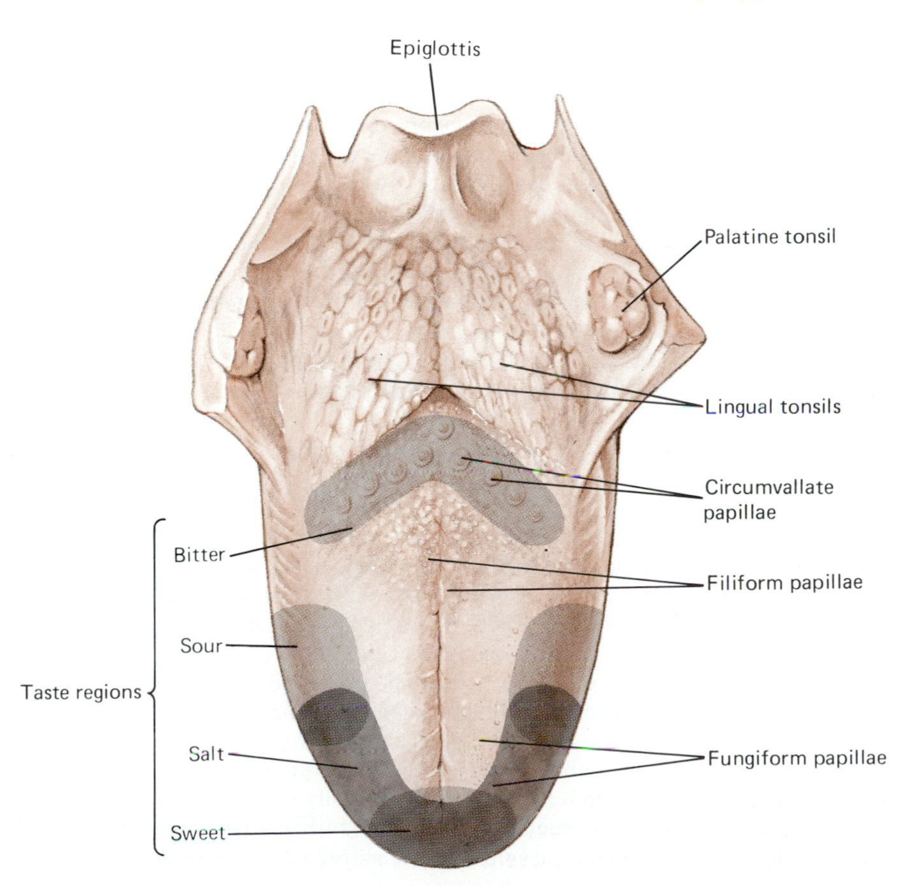

Figure 41–16 The distribution on the surface of the tongue of taste buds sensitive to sweet, bitter, sour, and salt.

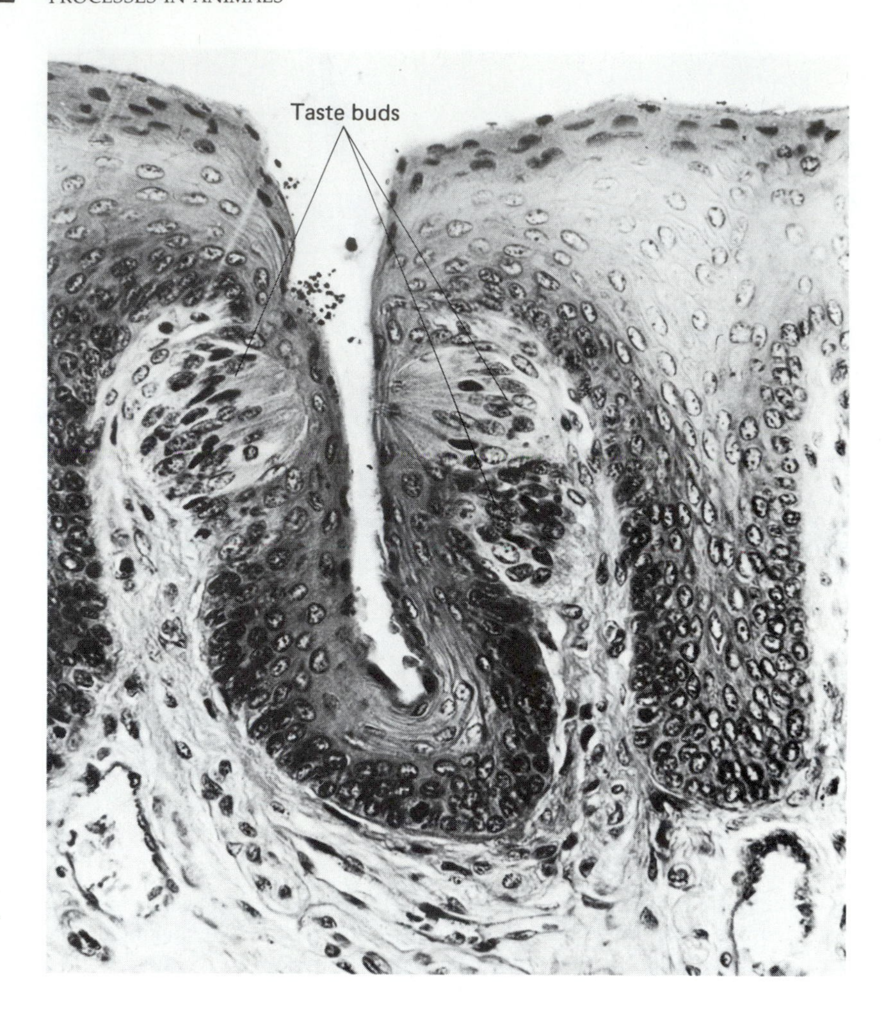

Figure 41–17 Electron micrograph of papillae of the tongue containing taste buds. The taste buds are the oval bodies (approximately ×400). (Biophoto Associates, Photo Researchers, Inc.)

cal. A single taste receptor may respond to more than one category of taste. Thus the detection and processing of information in the taste organs of the tongue are very complex. Taste discrimination probably depends on a code that consists of cross-fiber patterning; that is, each receptor responds to more than one kind of chemical, but no two respond exactly alike, so that the total pattern of messages going to the brain is different for different solutions.

Flavor does not depend on the perception of taste alone. It is actually compounded of taste, smell, texture, and temperature. Smell affects flavor because odors pass from the mouth to the nasal chamber via the internal nares. No doubt you have observed that when you have a cold, food seems to have little "taste." Actually, the taste buds are not affected, but the blockage of nasal passages severely reduces the participation of olfactory reception in the composite sensation of flavor.

THE SENSE OF SMELL

In terrestrial vertebrates, olfaction occurs in the nasal epithelium. In humans the **olfactory epithelium** is located in the roof of the nasal cavity (Fig. 41–18). This epithelium contains specialized olfactory cells with axons that extend upward as the fibers of the olfactory nerves. These fibers penetrate the exceedingly thin cribriform plate of the ethmoid bone on the cranial floor through many sievelike pores. The end of each olfactory cell on the epithelial surface bears several olfactory hairs that are believed to react to odors (chemicals) in the air.

Unlike the taste buds, which are sensitive to only a few chemical sensations, the olfactory epithelium is thought to react to as many as 50. Mixtures of these primary smell sensations produce the broad spectrum of odors that we are capable of perceiving. The olfactory organs respond to remarkably small amounts of a substance. For example, ionone, the synthetic substitute for the odor of violets, can be detected by most people when it is present in a concentration of only one part in more than 30 billion parts of air.

Despite its sensitivity smell is perhaps the sense that adapts most quickly. The olfactory receptors adapt about 50% in the first second or so after stimulation, so that even offensively odorous air may seem odorless after only a few minutes. Part of this adaptation is thought to take place in the CNS.

Thermoreceptors

Heat is another form of radiant energy to which all living things respond. Although not much is known about specific thermoreceptors in invertebrates, many invertebrates are sensitive to gradations in temperature. In very small animals, the CNS itself may respond to temperature change. In others, free nerve endings in the integument may be responsible for detection of heat. Mosquitoes and other blood-sucking insects and ticks must use thermoreception in their search for a warm-blooded host. Some have temperature receptors on their antennae sensitive to changes of less than 0.5°C.

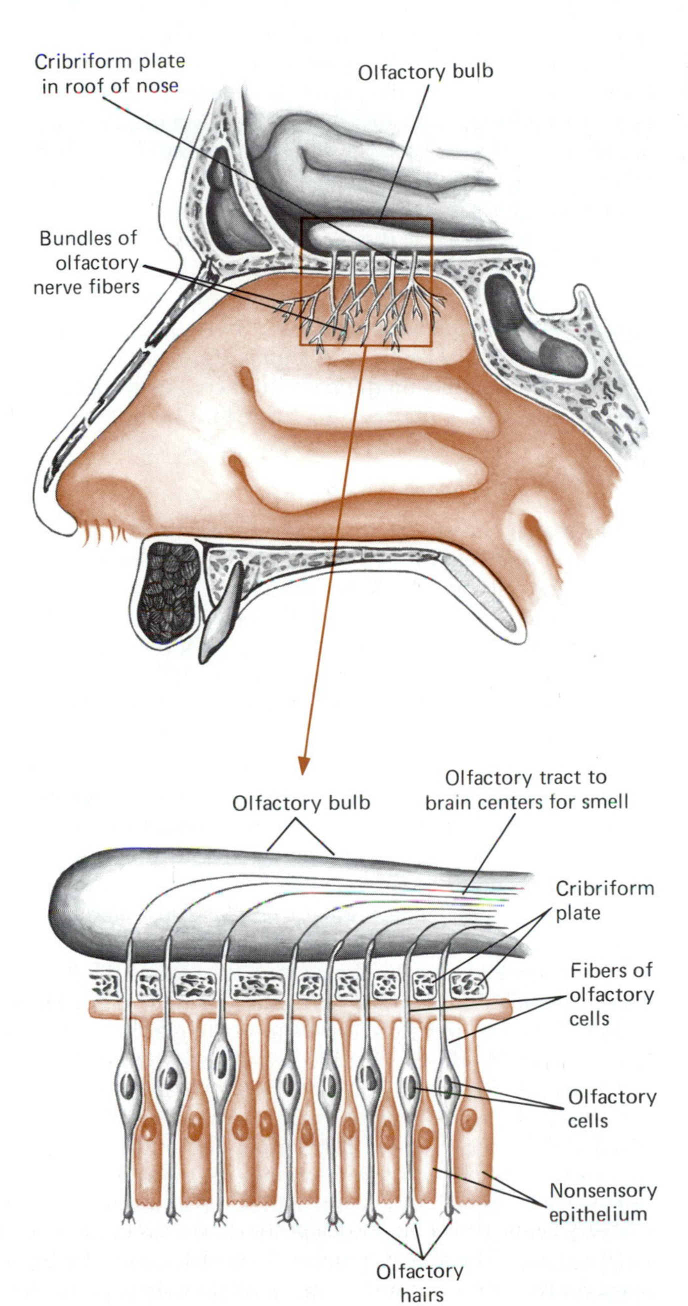

Figure 41–18 Location and structure of the olfactory epithelium. Note that the receptor cells are located in the epithelium itself.

Figure 41–19 A golden eye-lash viper showing a pit organ, a sensory structure located between each eye and nostril. The pit organ can detect the heat from a warmblooded animal up to a distance of one to two meters. (Tom McHugh, Steinhart Aquarium, Photo Researchers, Inc.)

Certain snakes (pit vipers and boas) use thermoreceptors to locate warmblooded prey. Pits in the heads of these snakes are thermoreceptors that can detect heat generated by a small animal as far as a half meter away (Fig. 41–19). The snake orients its head so that the pits on both sides of its head detect the same amount of heat. This position indicates that the snake is facing its prey directly, and presumably improves the snake's "aim" when striking. Each pit organ consists of a cavity 1 to 5 mm in diameter in the head of the snake. In rattlesnakes, a very thin membrane is suspended across the cavity. About 7000 heat sensitive axons (of the trigeminal nerve) are located along the membrane surface. When warmed, each of these axons can send action potentials to the snake's brain.

In mammals, free nerve endings and specialized receptors in the skin and tongue detect temperature changes. Thermoreceptors in the hypothalamus of the brain detect internal changes in temperature and receive and integrate information from thermoreceptors on the body surface. The hypothalamus then initiates homeostatic mechanisms that ensure a constant body temperature.

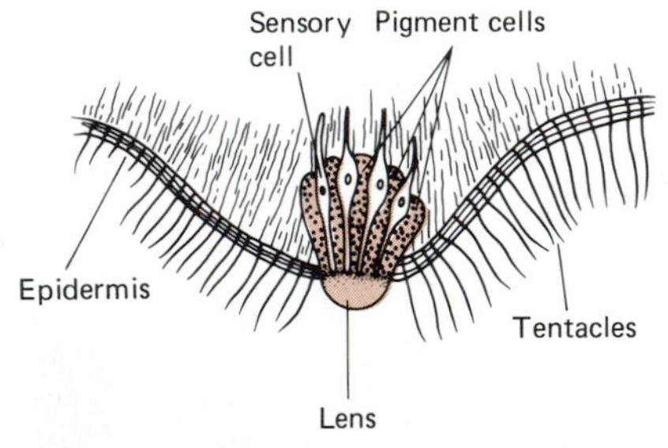

(*a*) Simple invertebrate eye (Ocellus from cell of jellyfish)

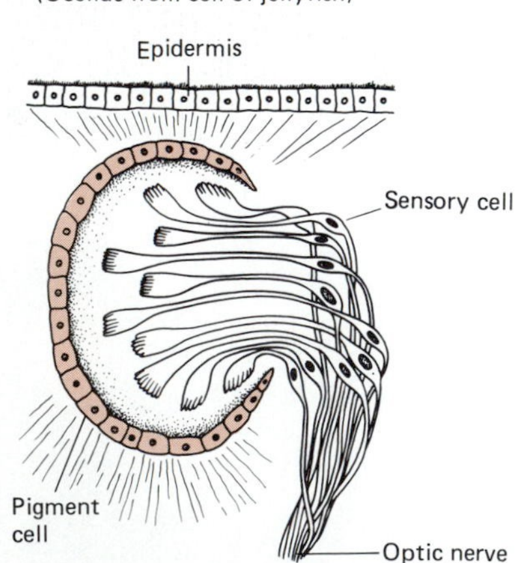

(*b*) Ocellus of planarian worm

Figure 41–20 Simple invertebrate eyes. (*a*) Ocellus from bell of jellyfish. (*b*) Ocellus of planarian worm.

Photoreceptors

Just as matter consists of atoms, light consists of units called **photons.** The energy content of one photon is defined as one **quantum.** In photoreception, quanta of light energy striking the light-sensitive cells trigger the receptor cell to transmit a nerve impulse. Light energy is absorbed by certain pigments. **Rhodopsins** are the photosensitive pigments found in the eyes of cephalopod mollusks, arthropods, and vertebrates.

Cells are sensitive to radiant energy, including light, but this general sensitivity (found even among protozoa) is not considered photoreception. Some protozoa have eyespots, which are more sensitive to light than the rest of the cell surface, but on the evolutionary scale, the first true light-sensitive organs are found in certain cnidarians and in flatworms. Their photoreceptor organs, comparable to simple eyes, are called **ocelli.** In planarian flatworms they are bowl-shaped structures containing black pigment (Fig. 41–20). At the bottom of the pigment are clusters of light-sensitive cells. The pigment shades these light-sensitive cells from all light except that coming from above and slightly to the front. This arrangement enables the planarian to detect the direction of the source of light. These photoreceptors can also distinguish light intensity. Planarians have other light-sensitive cells over the body surface and therefore can continue to react to light even after their ocelli have been destroyed. However, their responses become random and slow.

Animals with eyespots or very simple eyes can detect light, but they cannot see objects. Effective image formation, called **vision,** requires a more complex eye, usually with a lens that can focus an image on the light-sensitive cells. A necessary first step in the evolution from photoreceptor to true eye, therefore, was the development of a lens to concentrate light on a group of photoreceptors. As better lens systems evolved, the photoreceptors were able to form images, and an eye in the strict sense of the word evolved. Two fundamentally different types of complex eyes are the camera eye of some cephalopods (squids and octopuses) and verte-

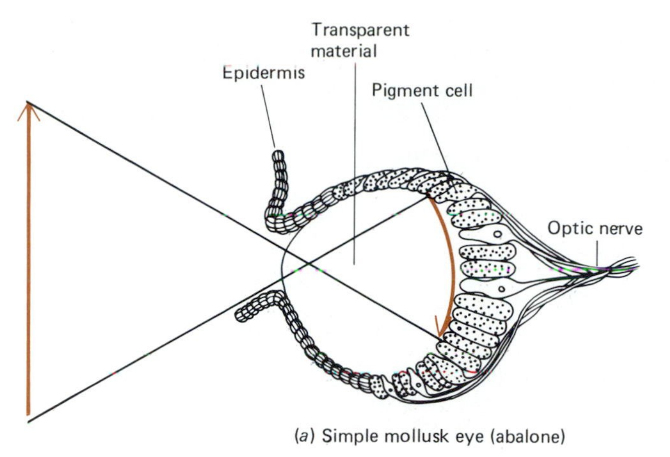

(a) Simple mollusk eye (abalone)

Figure 41–21 Comparison of different types of eyes. (*a*) Some mollusks have a simple eye that lacks a lens, and works like a pinhole camera. (*b*), (*c*) Some cephalopods have eyes that work like vertebrate eyes. An adjustable lens is present. (*d*) A unit from the compound eye of an insect or crustacean. This type of eye registers changes in light and shade so that the insect can detect movement. The compound eyes of most of these organisms do not actually form images. (*e*) The eyes of a darning needle (an insect). ((*c*), Charles Seaborn; (*e*), Luci Giglio.)

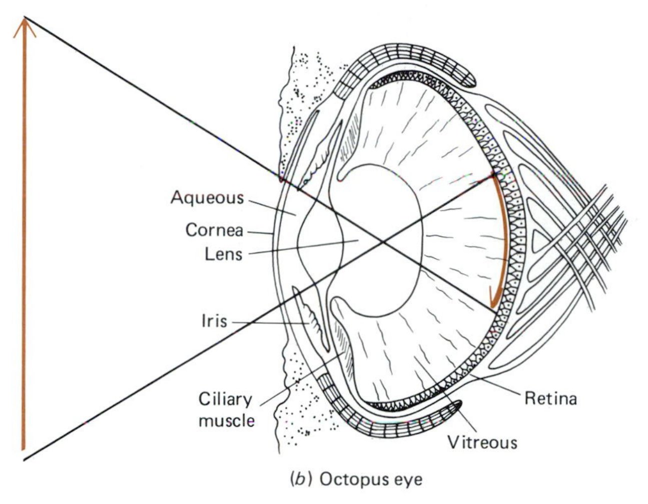

(b) Octopus eye

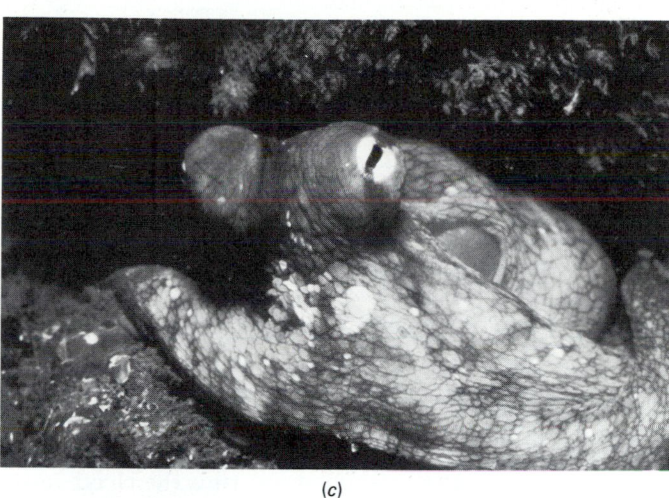
(c)

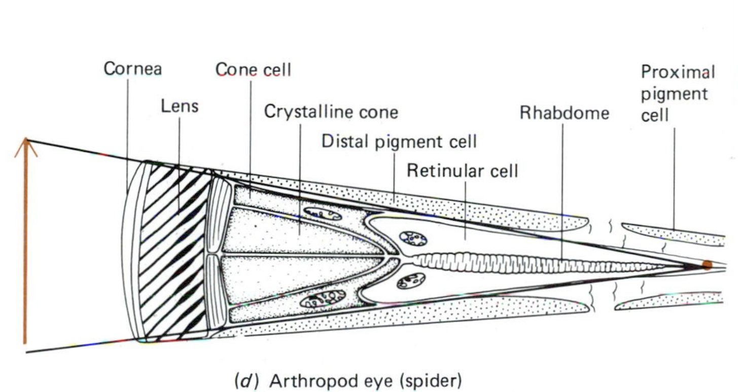

(d) Arthropod eye (spider)

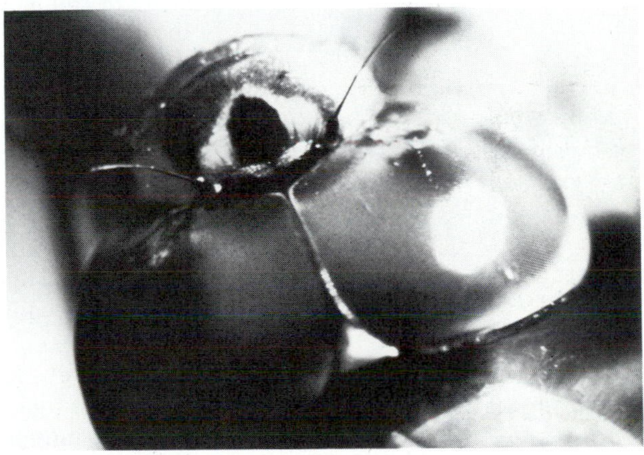
(e)

brates, and the compound eye of the arthropods (Fig. 41–21). (The vertebrate eye and the cephalopod eyes are analogous structures; that is, they evolved independently of one another and are functionally but not structurally similar.)

THE HUMAN EYE

The eye of a squid or octopus is rather like a simple box camera equipped with "slow" (requiring much light for development) black-and-white film, whereas the human eye is like a 35-millimeter camera loaded with extremely sensitive color film.

The analogy between the human eye and a camera is an especially apt one. The eye (Fig. 41–22) has a **lens** that can be focused for different distances; a diaphragm, the **iris,** which regulates the size of the light opening (the pupil); and a light-sensitive **retina** located at the rear of the eye, corresponding to the film of the

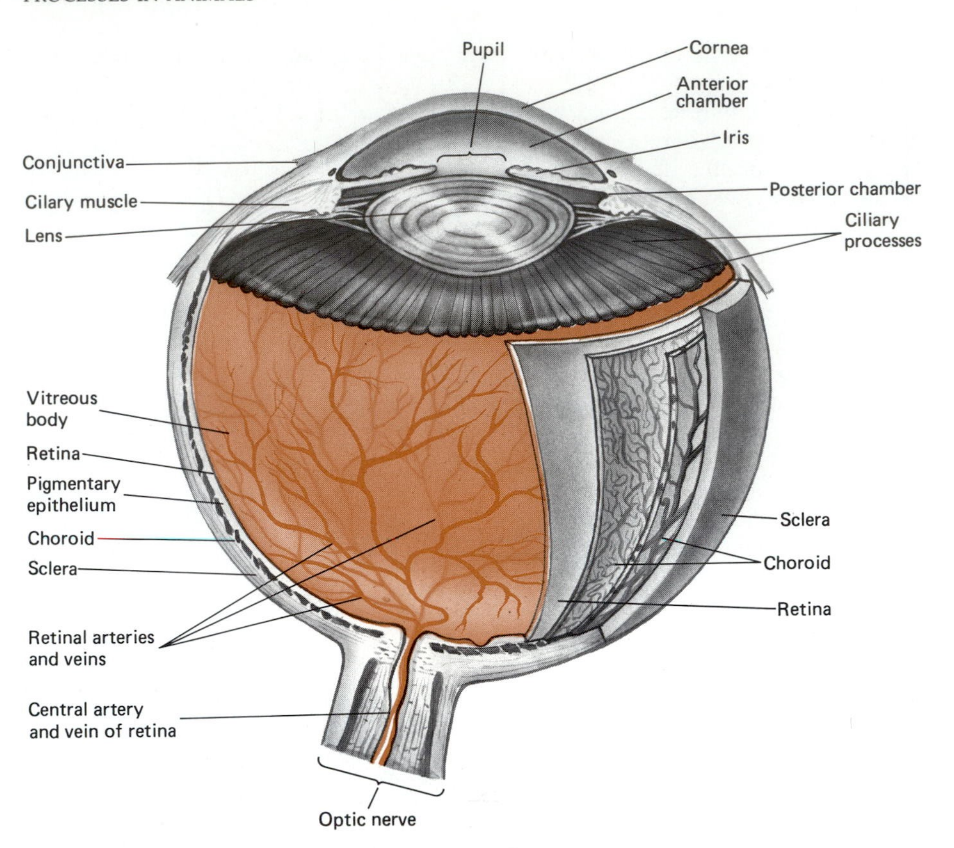

Figure 41–22 Structure of the human eye.

camera. Next to the retina is a sheet of cells filled with black pigment that absorbs extra light and prevents internally reflected light from blurring the image (cameras are also painted black on the inside). This sheet, called the **choroid coat,** also contains the blood vessels that nourish the retina.

The outer coat of the eyeball, called the **sclera,** is a tough, opaque, curved sheet of connective tissue that protects the inner structures and helps to maintain the rigidity of the eyeball. On the front surface of the eye this sheet becomes the thinner, transparent **cornea,** through which light enters.

The lens is a transparent, elastic ball located just behind the iris. It bends the light rays coming in and brings them to a focus on the retina. The lens is aided by the curved surface of the cornea and by the refractive properties of the liquids inside the eyeball. The cavity between the cornea and the lens is filled with a watery substance, the **aqueous fluid.** The larger chamber between the lens and the retina is filled with a more viscous fluid, the **vitreous body.** Both fluids are important in maintaining the shape of the eyeball. The aqueous fluid is secreted by the **ciliary body,** a doughnut-shaped structure that attaches the ligament holding the lens to the eyeball.

The eye has the power of **accommodation,** meaning it can change focus for near or far vision by changing the curvature of the lens. This is made possible by the stretching and relaxing of the lens by the **ciliary muscle,** a part of the ciliary body that is attached to the lens by tiny fibers called zonules. Because of the pressure of the fluids within, the eyeball is under tension transmitted by the ciliary muscle to the lens. Relaxation of the ciliary muscle fibers places tension on the zonules and flattens the lens; this focuses the eye for far vision, the condition of the eye at rest. When contracted, tension on the zonules lessens, the elastic lens assumes a more spherical shape for near vision.

As people grow older, the lens enlarges and becomes less elastic and thereby less able to accommodate for near vision. When this occurs, spectacles with one portion ground for distance vision and one portion ground for near vision (bifocals) may be worn to accomplish what the eye can no longer do.

The amount of light entering the eye is regulated by the **iris,** a ring of smooth muscle that appears as blue, green, or brown, depending on the amount and nature of pigment present. The iris is composed of two mutually antagonistic sets of muscle fibers; one that is arranged circularly and contracts to decrease the size of the pupil, and one that is arranged radially and contracts to increase the size of the pupil. The response of these muscles to changes in light intensity is not instantaneous, but requires from 10 to 30 seconds. Thus when a person steps from a light to a dark area, some time is needed for the eyes to adapt to the dark, and when a person steps from a dark room to a brightly lighted area, the eyes are dazzled until the size of the pupil is decreased. The retina of the eye (soon to be discussed) is also able to adapt to changes in light intensity.

Each eye has six muscles stretching from the surface of the eyeball to various points in the bony socket. These enable the eye as a whole to move and be oriented in a given direction. These muscles are innervated by cranial nerves in such a way that the eyes normally move together and focus on the same area.

The light-sensitive part of the human eye is the retina, a hemisphere made up of an abundance of receptor cells called, according to their shape, **rods** and **cones.** There are about 125 million rods and 6.5 million cones. In addition, the retina contains many sensory and connector neurons and their axons. Curiously, the sensitive cells are at the *back* of the retina; to reach them, light must pass through several layers of connecting neurons (Fig. 41–23).

At a point in the back of the eye, the individual axons of the sensory neurons unite to form the **optic nerve,** which then passes out of the eyeball. Here there are no rods and cones. This area is called the "blind spot" since images falling on it cannot be perceived. Its existence can be demonstrated by closing the left eye and focusing the right eye on the X in Figure 41–24. Starting with the page about 13 cm from the eye, move it away until the circle disappears. At that position the image of the circle is falling on the blind spot and so is not perceived.

In the center of the retina, directly in line with the center of the cornea and lens, is the region of keenest vision, a small depressed area called the **fovea.** Here are concentrated the light-sensitive cones, responsible for bright-light vision, for the perception of fine detail, and for color vision.

The other light-sensitive cells, the rods, are more numerous in the periphery of the retina, away from the fovea. These function in twilight or dim light and are

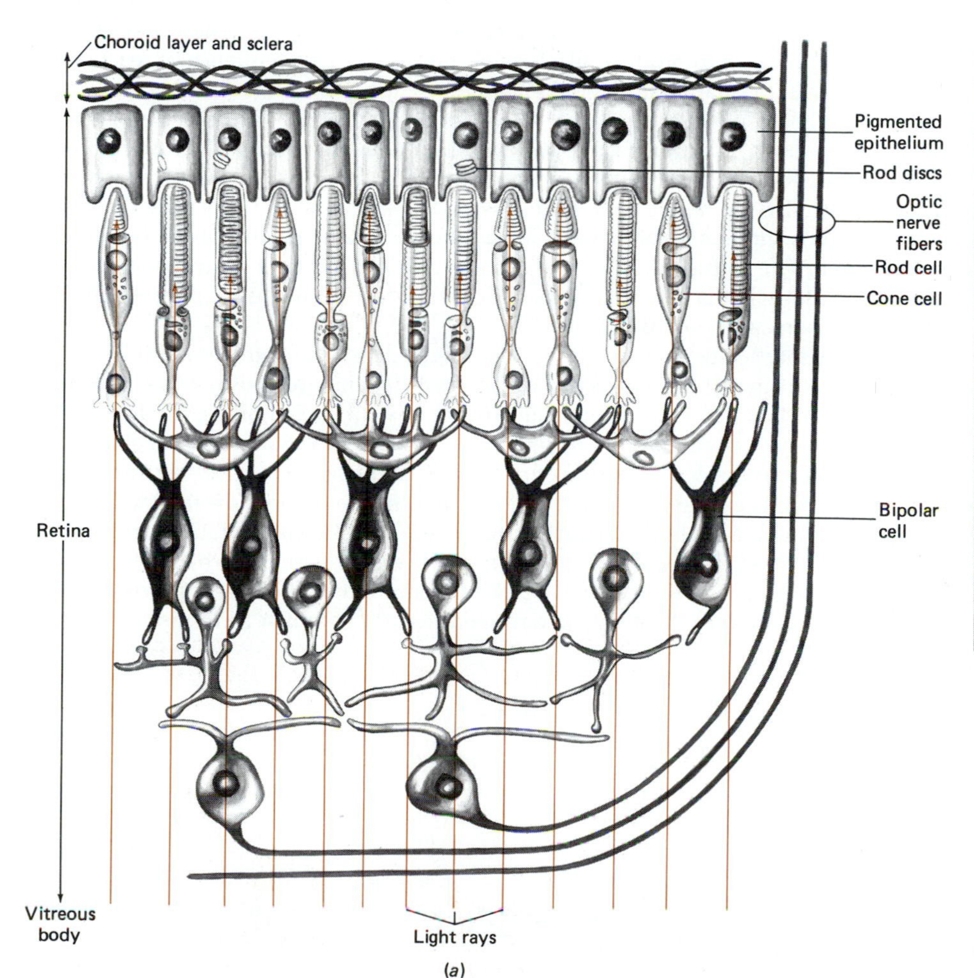

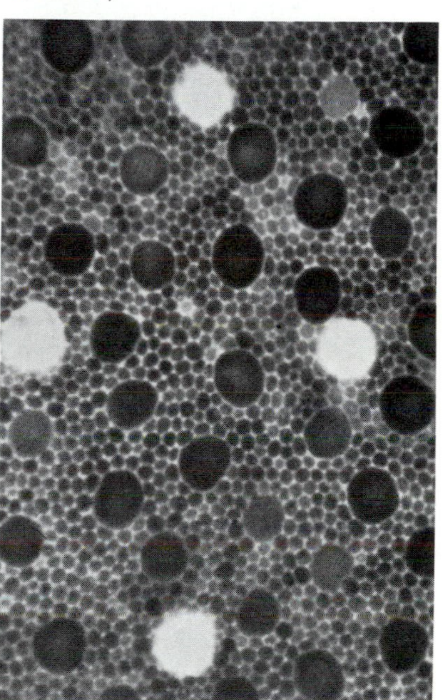

Figure 41–23 The retina. (*a*) Neuronal connections in the retina. The elaborate interconnections among the various layers of cells allow them to interact and to influence one another in a number of ways. (*b*) High-power view of the retinal surface, showing the geometric, spatial arrangement of the cone cells. Blue-sensitive cones appear as bright spots in this photomicrograph. Cones sensitive to others colors of light appear as dark, round holes. Rods appear as a dark cobblestone mesh interspersed among the various cones. The brain infers the color of light by observing the pattern of cone response in the part of the retina on which the light is falling. If the bright cones were responding more than the others, the light would appear blue. (Courtesy of Francisco M. de Monasterio, S. J. Schein, and E. P. McCrane.)

Figure 41–24 Demonstration of the blind spot on the retina. See text for details.

insensitive to colors. We are not ordinarily aware that only those objects more or less directly in front of our eyes can be perceived in color, but this can be demonstrated by a simple experiment. Close one eye and focus the other on some point straight ahead. As a colored object is gradually brought into view from the side, you will be aware of its presence and of its size and shape before you are aware of its color. Only when the object is brought closer to the direct line of vision, so that its image falls on a part of the retina containing cones, can its color be determined. The rods are actually more sensitive in dim light than are the cones. Since the rods are located not in the center but in the periphery of the retina, it is a curious fact that you can see an object better in dim light if you look at it not directly (when its image would fall on the cones in the center of the retina) but slightly to one side of it, so that its image falls on the rods in the periphery of the retina.

The Chemistry of Vision

The substance primarily responsible for the ability to see is **rhodopsin** (also called visual purple), found only in the rod cells, together with a series of very closely related pigments found in the cone cells. Rhodopsin consists of a large protein called opsin, chemically joined with a carotenoid called **retinal,** which is made from vitamin A. Two isomers of retinal exist: the 11-*cis* form and the all-*trans* form.

When light strikes rhodopsin, it transforms **11-*cis* retinal** to **all-*trans*-retinal.** This change in shape causes the breakdown of rhodopsin into its components, opsin and retinal. The breakdown somehow triggers a depolarization of the rod cell that contains the rhodopsin; this in turn produces an impulse that is transmitted to other neurons in the retina, and, if not inhibited, ultimately to the brain. The all-*trans* retinal is converted back to the 11-*cis* form by an enzyme. Then the retinal combines with opsin to produce rhodopsin once again. This sequence of reactions is known as the visual cycle (Fig. 41–25).

It has been shown that a single quantum of light can be absorbed by a single molecule of rhodopsin and lead to the excitation of a single rod. When the eye is exposed to a flash of light lasting only a millionth of a second, it sees an image of light that persists for nearly a tenth of a second. This is the length of time that the retina remains stimulated following a flash. This persistence of images in the retina enables your eye to fuse the successive flickering images on a motion picture or screen, so that what is actually a rapid succession of still pictures is perceived as moving persons and objects.

The ability to see an exceedingly faint light depends on the amount of rhodopsin present in the retinal rods. This in turn depends on the relative rates of synthesis and breakdown of rhodopsin. In bright light, much of the rhodopsin is broken down to free retinal and opsin. The synthesis of rhodopsin is a relatively slow process, and the concentration of rhodopsin in the retina is never very great so long as the eye is exposed to bright light. When the eye is suitably shielded from light, the breakdown of rhodopsin is prevented and its concentration gradually builds up until essentially all of the opsin has been converted to rhodopsin. The sensitivity of the eye to light, a function of the amount of rhodopsin present, can increase 1 million–fold if the eye is dark-adapted for as much as an hour.

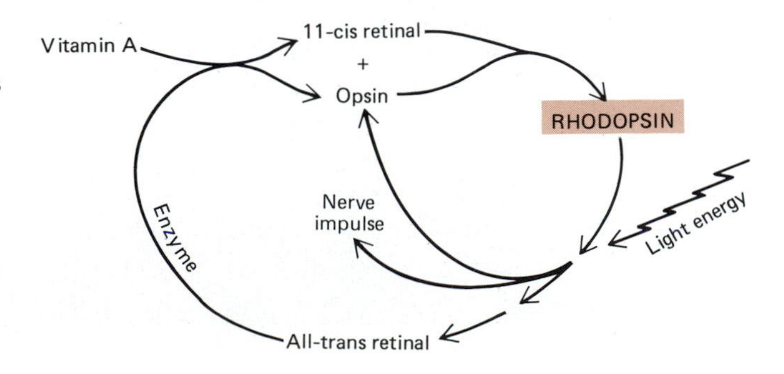

Figure 41–25 The visual cycle. When light strikes rhodopsin, it breaks down depolarizing the rod cell which contains it. This produces an impulse. See text for further explanation.

Color Vision

The chemistry of the cones and of color vision is less well understood. There are three different types of cones and three different cone pigments. One cone pigment, **iodopsin,** is composed of retinal and an opsin different from that found in the rods. The cones are considerably less sensitive to light than the rods and cannot provide vision in dim light. The prime function of the cones is to perceive colors. The evidence from certain psychological tests is consistent with the hypothesis that there are three different types of cones, which respond respectively to blue, green, and red light. This has been substantiated recently by the extraction of three kinds of color receptors—red, green, and blue—from human and other primate retinas. Each type can respond to light with a considerable range of wavelengths. The green cones, for example, can respond to light of any wavelength from 450 to 675 nanometers (i.e., blue, green, yellow, orange, and red light), but they respond to green light more strongly than to any of the others. Intermediate colors other than blue, green, and red are perceived by the simultaneous stimulation of two or more types of cones. The "red" cones are actually more sensitive to yellow light than to red, but they respond to red before any of the others do, and therefore behave as red receptors. By a comparison of the rate at which various receptors respond, the brain is able to detect light colors of intermediate frequency. Color blindness results when one or more of the three types of cones is absent. This is an inherited sex-linked condition (see Chapter 11).

Binocular Vision and Depth Perception

The position of the eyes in the head of humans and certain other higher vertebrates permits both of them to be focused on the same object (Fig. 41–26). This **binocular vision** is an important factor in judging distance and depth. To focus on a near object, the eyes must converge (turn inward so that the animal becomes slightly cross-eyed). The proprioceptors in the eye muscles causing this convergence are

(*a*)

(*b*)

(*c*)

Figure 41–26 The location of the eyes varies in different vertebrates, resulting in differences in vision. (*a*) The eyes of the addax are positioned laterally, enabling the animal to see on both sides; even while grazing, it can spot a predator approaching from behind. (*b*) The orbits (bony cavities that contain the eyeballs) of the hippopotamus are elevated, enabling the animal to see even when most of its head is under water. (*c*) Like many other nocturnal animals, the night monkey has large eyes. Its eyes are positioned at the front of the head, and it has binocular vision, permitting it to judge distances. (Courtesy of Busch Gardens.)

stimulated by this contraction to send impulses to the brain. Thus part of judgment of distance and depth depends upon impulses originating when the sensory fibers in those muscles are stimulated. In addition, the eyes, being some distance apart (a little over 5 cm in humans), see things from slightly different angles and therefore get slightly different views of a close object. The images of a given object that the two eyes perceive differ most for a near object and least for a distant object. By comparing the differences, the brain is able to infer distances. Depth perception is also made possible by the differential size of near and far objects on the retina, by perspective, by overlap and shadow, by distance over the horizon, and by the increasing dimness of distant objects.

Defects in Vision

The most common defects of the human eye are **nearsightedness** (myopia), **farsightedness** (hypermetropia), and **astigmatism.** In the normal eye, as shown in Figure 41–27(*a*), the shape of the eyeball is such that the retina is the proper distance behind the lens for the light rays to converge in the fovea. In a nearsighted eye, illustrated in Figure 41–27(*b*), the eyeball is too long and the retina is too far from the lens. The light rays converge at a point in front of the retina, and are again diverging when they reach it, resulting in a blurred image. In a farsighted eye, the eyeball is too short and the retina too close to the lens (part (*c*) of the figure). Light rays strike the retina before they have converged, again resulting in a blurred image. Concave lenses correct for the nearsighted condition by bringing the light rays to a focus at a point farther back, and convex lenses correct for the farsighted condition by causing the light rays to converge farther forward.

In astigmatism the cornea is curved unequally in different planes, so that the light rays in one plane are focused at a different point from those in another plane, as shown in Figure 41–27(*d*). To correct for astigmatism, lenses must be ground unequally to compensate for the unequal curvature of the cornea.

THE COMPOUND EYE

Compound eyes are found in crustaceans and insects. Not only do these eyes look different from vertebrate eyes, but they also see differently (Fig. 41–28). The surface of a compound eye appears faceted. Each **facet** is the convex cornea of one of its

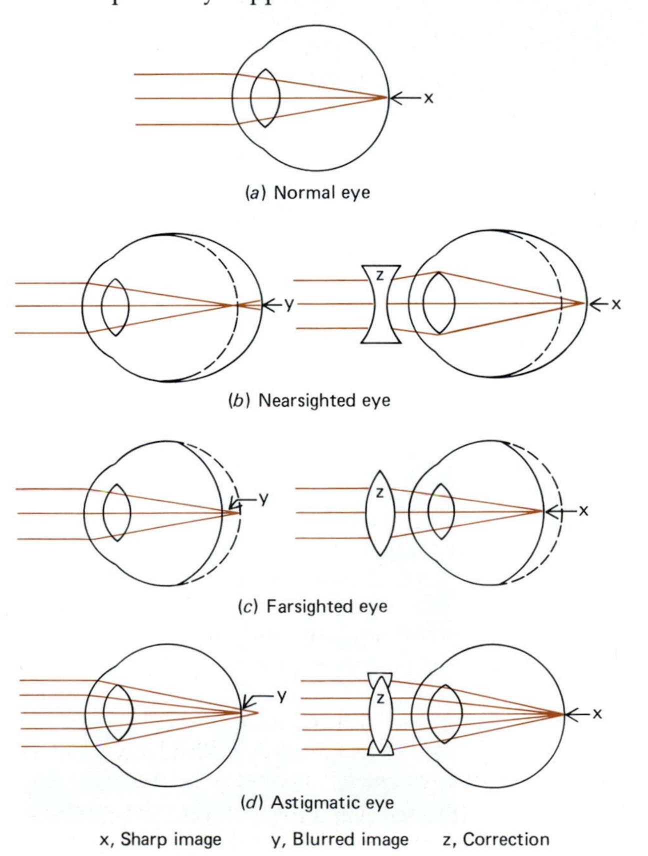

Figure 41–27 Common abnormalities of the eye that result in defects in vision. (*a*) Normal eye, in which parallel rays coming from a point in space are focused as a point on the retina. (*b*) Nearsighted eye, in which the eyeball is elongated so that parallel light rays are brought to a focus in front of the retina (on dotted line, which represents the position of the retina in a normal eye) and so form a blurred image on the retina. This is corrected by placing a concave lens in front of the eye, which diverges the light rays, making it possible for the eye to focus these rays on the retina. (*c*) Farsighted eye, in which the eyeball is shortened and light rays are focused behind the retina. A convex lens converges the light rays so that the eye focuses them onto the retina. (*d*) Astigmatic eye, in which light rays passing through one part of the eye are focused on the retina, while light rays passing through another area of the lens are not focused on the retina. This is a result of the unequal curvature of the lens or cornea. A cylindrical lens corrects this by bending light rays going through only certain parts of the eye.

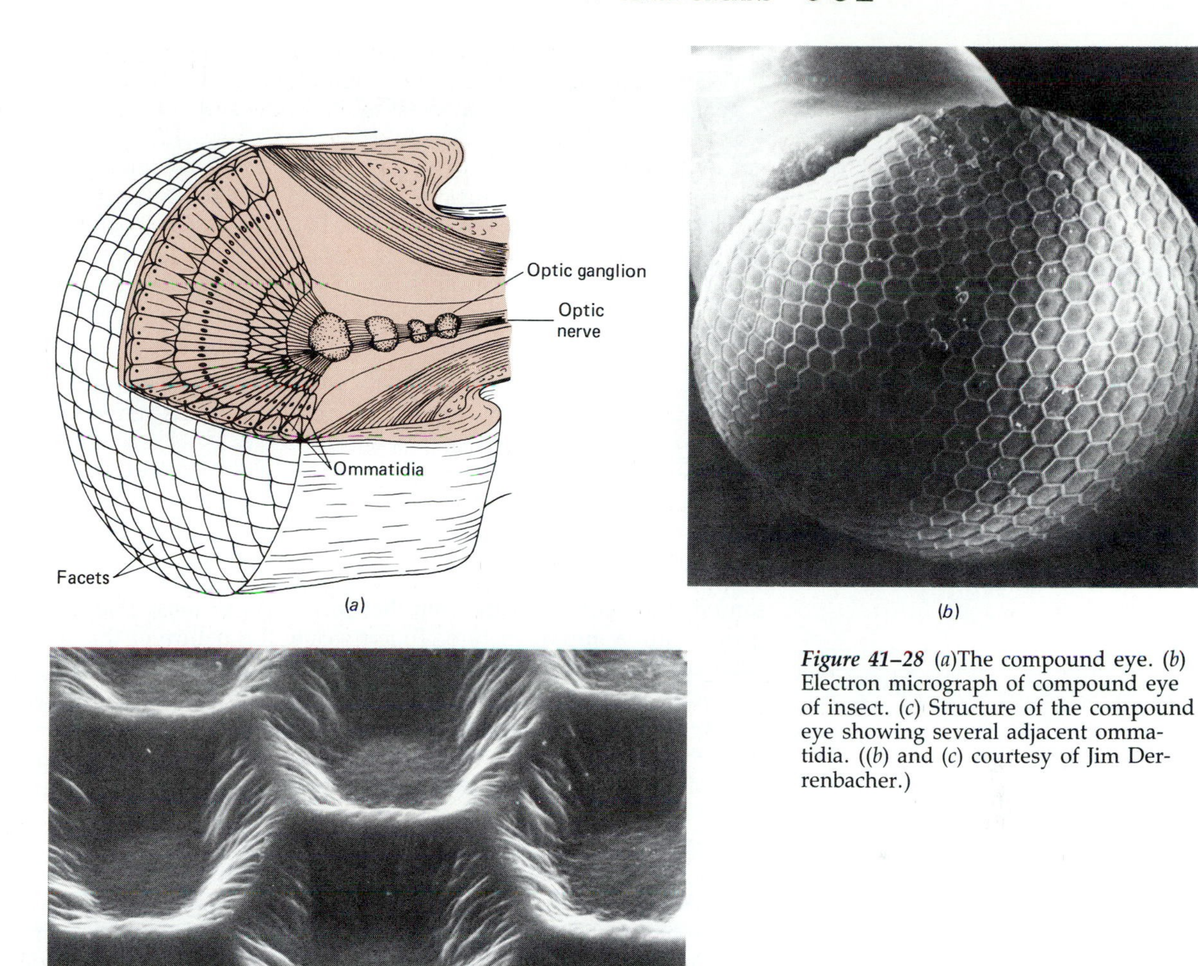

Figure 41–28 (*a*)The compound eye. (*b*) Electron micrograph of compound eye of insect. (*c*) Structure of the compound eye showing several adjacent ommatidia. ((*b*) and (*c*) courtesy of Jim Derrenbacher.)

visual units called an **ommatidium.** The number of ommatidia varies in different species. For example, there may be only 20, as in the eye of certain crustaceans, or as many as 28,000, as in the eye of a dragonfly.

Each ommatidium consists of a cornea, a lens, and a light-sensitive **rhabdome,** the central core of the ommatidium. The rhabdome is surrounded by retinular cells, which transmit the sensory stimulus (Fig. 41–29). A sheath of pigmented cells envelops each ommatidium. Arthropod eyes usually adapt to different intensities of light. In nocturnal and crepuscular (dark- or dusk-loving) insects and many crustaceans, pigment is capable of migrating proximally and distally. When the pigment is in the proximal position, each ommatidium is shielded from its neighbor and only light entering directly along its axis can stimulate the receptors. When the pigment is in the distal position, light striking at an angle may pass through several ommatidia and stimulate many retinal units. As a result, in dim light, sensitivity of the eye is increased, and in bright light, the eye is protected from excessive stimulation. Pigment migration is under neural control in insects and under hormonal control in crustaceans. In some species it follows a daily rhythm.

Compound eyes do not perceive form well. Although the lens system of each ommatidium is adequate to focus a small inverted image on the retinular cells, there is no evidence that such images are actually perceived as images by the organism. However, all the ommatidia together do produce a composite image. Each ommatidium, in gathering a point of light from a narrow sector of the visual field, is in fact

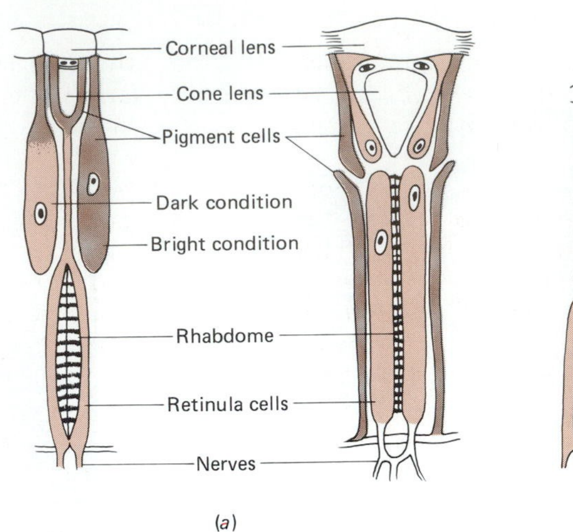

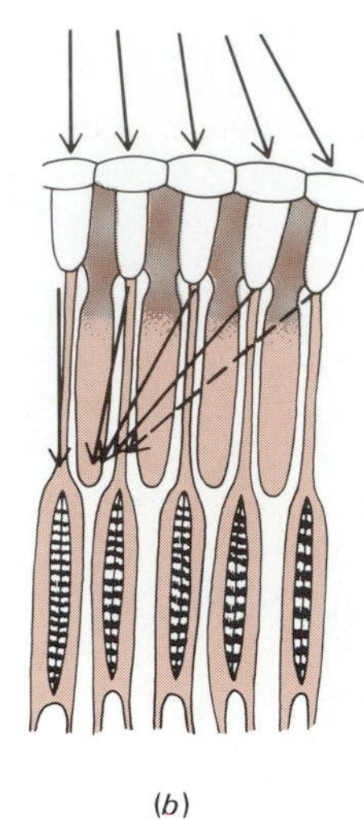

Figure 41–29 (*a*) Insect ommatidia, showing a diurnal type (*left*) and a nocturnal type (*right*). In the diurnal type, the pigment is shown in two positions: adapted for very dark conditions on the left side, and for relatively bright conditions on the right. (*b*) Nocturnal type of eye adapted for dark conditions, showing how light can be concentrated upon one rhabdome from several lenses. If the pigment moved downward, light from peripheral lenses would be screened out.

sampling a mean intensity from that sector. All of these points of light taken together form a **mosaic** picture. To appreciate the nature of this mosaic picture we need only look at a newspaper photograph through a magnifying glass; it is a mosaic of many dots of different intensities. The clearness and definition of the picture will depend upon how many dots there are per unit area—the more dots, the better the picture. So it is with the compound eye. The image as perceived by the animal is probably much better in quality than might be suspected from the structure of the compound eye. The nervous system of an insect is apparently capable of image processing similar to that employed to improve the quality of photographs sent to the earth by robot spacecraft.

Although the compound eye can form only coarse images, it compensates for this by being able to follow **flicker** to higher frequencies. Flies are able to detect flickers up to about 265 per second. In contrast, the human eye can detect flickers of only 45 to 53 per second; because flickering lights fuse above these values, we see motion pictures as smooth movement and the ordinary 60-cycle light in the room as steady. To an insect, both motion pictures and room lighting must flicker horribly. However, because the insect has such a high critical flicker fusion rate, any movement of prey or enemy is immediately detected by one of the eye units. Hence the compound eye is peculiarly well-suited to the arthropod's way of life.

Compound eyes are superior to our eyes in two other respects. They are sensitive to different wavelengths of light ranging from the red into the ultraviolet (UV), and they are able to analyze the **plane of polarization** of light. Accordingly, an insect can see UV light well, and its world of color is much different from ours. Since different flowers deflect UV light to different degrees, two flowers that appear identically colored to us may appear strikingly different to insects. How the world appears to an insect with UV vision can be appreciated by viewing the landscape through a television camera with an UV light–transmitting lens (Fig. 41–30). A sky that appears equally blue to us in all quadrants reveals quite different patterns to an insect, because the plane of polarization of the light is not the same in all parts of the sky, and the insect's eye can detect the difference. Honey bees and some other arthropods employ this ability as a navigational aid.

Figure 41–30 (*a*) UV video-viewing of marsh marigolds, *Caltha palustris*, in the field. (*b*) The marsh marigold appears to be uniformly yellow to the human eye. (*c*) When viewed with UV film the flowers have darker areas that represent light-absorbing centers. (From Eisner, T.: *Science* 28: 1172, 1969.)

(*a*)

(*b*)

(*c*)

SUMMARY

I. A sense organ consists of one or more receptor cells, and sometimes, accessory cells. Receptor cells may be neuron endings or specialized cells in close contact with neurons.

II. Exteroceptors are sense organs that receive information from the outside world. Proprioceptors are sense organs within muscles, tendons, and joints, which enable the animal to perceive orientation of the body and position of its parts. Interoceptors are sense organs within body organs.

III. Sense organs also may be classified on the basis of the type of energy to which they respond. Thus there are mechanoreceptors, chemoreceptors, photoreceptors, thermoreceptors, and electroreceptors.

IV. Receptor cells absorb energy, transduce that energy into electrical energy, and produce receptor potentials.

V. Adaptation of a receptor to a continuous stimulus results in diminished perception. For this reason, adaptation to an unpleasant odor or noise occurs after a few moments.

VI. Mechanoreceptors respond to touch, pressure, gravity, stretch, or movement.

- *A.* The tactile receptors in the skin are mechanoreceptors that respond to mechanical displacement of hairs or of the receptor cells themselves.
- *B.* Statocysts are gravity receptors found in many invertebrates.
  1. When the position of the statolith within the statocyst changes, hairs of receptor cells are bent.
  2. Messages sent to the CNS inform an animal which hairs have been stimulated; from this the animal can determine where "down" is, and can correct for any abnormal orientation.
- *C.* Lateral line organs supplement vision in fish and some amphibians by informing the animal of moving objects or objects in its path.
- *D.* Muscle spindles, Golgi tendon organs, and joint receptors are proprioceptors that respond continuously to tension and movement in the muscles and joints.
- *E.* The saccule and utricle of the vertebrate ear contain otoliths that change position when the head is tilted or when the body is moving forward. The hair cells stimulated by the otoliths send impulses to the brain, enabling the animal to perceive the direction of gravity.
- *F.* The semicircular canals of the vertebrate ear inform the brain about turning movements. Their cristae are stimulated by movements of the endolymph.
- *G.* The organ of Corti within the cochlea is the auditory receptor in birds and mammals.
  1. Sound waves pass through the external auditory meatus, cause the eardrum to vibrate, and are transmitted through the middle ear by the hammer, anvil, and stirrup.
  2. Vibrations pass through the oval window to fluid within the vestibular duct. Pressure waves press upon the membranes separating the three ducts of the cochlea.
  3. Movements of the basilar membrane rub the hair cells of the organ of Corti against the overlying tectorial membrane, thus stimulating them.
  4. Nerve impulses are initiated in the dendrites of the auditory neurons lying at the base of each hair cell.

VII. Chemoreceptors include receptors for taste and smell.

- *A.* Taste receptors are specialized epithelial cells located in taste buds.
- *B.* The olfactory epithelium contains specialized olfactory cells with axons that extend upward as fibers of the olfactory nerves.

VIII. Thermoreceptors are important in warm-blooded animals in providing cues about body temperature. In some invertebrates they are used to locate a warm-blooded host.

IX. Photoreceptors in very simple eyes detect light, but these eyes do not form images effectively. Effective image formation and interpretation is called vision.

- *A.* In the human eye, light enters through the cornea, is focused by the lens, and sensed as an image by the retina. The iris regulates the amount of light that can enter.
- *B.* When light strikes rhodopsin in the rod cells, a chemical change in retinal occurs that breaks down the rhodopsin, triggering depolarization of the rod cell.
- *C.* The rods form images in black and white, whereas the cones function in color vision.
- *D.* The compound eye found in insects and crustaceans consists of ommatidia, which collectively form a mosaic image.

POST-TEST

1. A sense organ consists of one or more ________ cells and, sometimes, ________ cells.

2. Exteroceptors are sense organs that ________; proprioceptors enable an animal to perceive ________, along with the ________ of the body as a whole.

3. ________ detect light energy; ________ respond to touch, gravity, or movement.

4. Receptor cells absorb ________, transduce this energy into ________ energy, and produce a ________ ________.

5. The diminishing response of a receptor to a continued, constant stimulus is called ________.

6. Statocysts serve as ________ receptors; their action depends upon mechanical displacement of receptor cell hairs by change in position of a ________.

7. The lateral line organ of fishes is thought to supplement ________; its canals are lined with recep-

tor cells that have ______________; the receptor cells secrete a mass of gelatinous material called a ______________.

8. Three main types of vertebrate proprioceptors are ______________ ______________, which detect muscle movement; ______________ ______________ organs, which determine stretch in tendons; and ______________ receptors, which detect movement in ligaments.

9. Phasic receptors respond only to ______________; tonic sense organs respond as long as the ______________ is present.

10. ______________ are balancing organs that stabilize flight in flies.

11. The basic function of the vertebrate ear is to help maintain ______________.

12. The inner ear consists of interconnected canals and sacs called the ______________; in jawed vertebrates this structure consists of two saclike chambers, the ______________ and ______________, and three ______________ canals, as well as the cochlea.

13. The rocks in your head (within the saccule and utricle) called ______________ are actually ______________ detectors.

14. Each semicircular canal is filled with fluid called ______________; at one of the openings of each canal into the utricle is a small enlargement, the ______________.

15. The cochlea, located in the ______________ ear, contains mechanoreceptor hair cells that detect ______________ waves.

16. The actual auditory receptor is the organ of ______________; it is located within the ______________ canal.

17. The senses of taste and smell depend upon ______________.

18. The photosensitive pigments in the eyes of vertebrates and arthropods are ______________.

Select the most appropriate answer in Column B for each description in Column A.

Column A	*Column B*
19. Light-sensitive part of human eye	a. Ommatidium
20. Regulates size of pupil	b. Cones
21. Perceive color	c. Retina
22. Region of keenest vision	d. Iris
23. Visual unit in compound eye	e. Fovea

24. Label the diagram below. (Refer to Fig. 41–22 as necessary.)

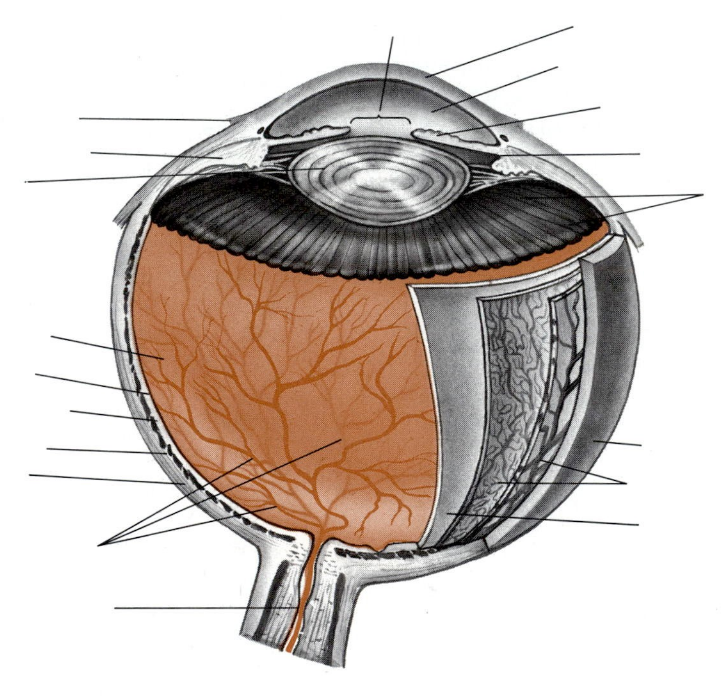

REVIEW QUESTIONS

1. What are mechanoreceptors? Give three examples.
2. What is a proprioceptor? What is its function in the mammalian body?
3. How do anesthetics reduce or eliminate the sensation of pain?
4. Draw a diagram of the human eye, labeling all parts. How are rods and cones distributed in the retina?
5. Discuss the mechanism by which photoreceptors are stimulated by light. What is the function of rhodopsin? How is it regenerated?
6. Describe the anatomical abnormality that produces each of the following visual defects:
   a. myopia
   b. hypermetropia
   c. astigmatism
7. How does the human eye adjust to near and far vision and to bright and dim lights?
8. Draw a diagram of the human ear, labeling all parts.
9. Discuss the mechanism by which the sensory cells of the ear are stimulated by sound waves.
10. What are otoliths and what is their role in maintaining equilibrium?
11. Contrast the function of the insect's compound eye with that of the vertebrate eye.

42

Animal Hormones: Endocrine Regulation

OUTLINE

I. The chemical nature of hormones
II. Mechanisms of hormone action
  A. Activation of genes
  B. Action through a second messenger
  C. Prostaglandins
III. Regulation of hormone secretion
IV. Invertebrate hormones
  A. Endocrine regulation of reproductive development in cephalopods
  B. Color change in crustaceans
  C. Hormonal control of insect development
V. Vertebrate hormones
  A. Endocrine disorders
  B. The hypothalamus and pituitary gland
  C. Growth and development
    1. Growth hormone
      a. Regulation of growth hormone secretion
      b. Abnormal growth
    2. Thyroid hormones
      a. Regulation of secretion
      b. Thyroid hormones and growth disorder
  D. Regulation of blood sugar level: insulin and glucagon
    1. Regulation of insulin and glucagon secretion
    2. Diabetes mellitus
    3. Hypoglycemia
  E. Response to stress
    1. Hormones of the adrenal medulla
    2. The effects of cortisol

LEARNING OBJECTIVES

After you have read this chapter you should be able to:

1. Define the terms *hormone* and *endocrine gland,* and identify sources of hormones other than endocrine glands.
2. Characterize hormones chemically, giving examples of hormones derived from amino acids and fatty acids, and examples of peptide, protein, and steroid hormones.
3. Describe mechanisms of hormone action, including activation of genes and the role of cyclic AMP.
4. Summarize the regulation of hormone secretion by negative feedback mechanisms, and draw diagrams illustrating the regulation of secretion of each of the hormones discussed in this chapter.
5. Give the general sources and actions of invertebrate hormones.
6. Describe the interaction of hormones that control development in insects.
7. Locate the principal vertebrate endocrine glands, list the hormones secreted by each, and summarize their actions.
8. Explain why the hypothalamus is considered the link between nervous and endocrine systems, and describe the mechanisms by which the hypothalamus influences the anterior and posterior lobes of the pituitary gland.
9. Describe the actions of growth hormone and thyroid hormones in promoting growth, and explain the consequences of hyposecretion and hypersecretion of these hormones.
10. Compare the actions of insulin and glucagon in regulating the concentration of glucose in the blood; describe the physiological problems associated with diabetes mellitus and hypoglycemia.
11. Describe the roles of the adrenal medulla and the adrenal cortex in enabling the body to deal with stress.

The endocrine system works closely with the nervous system to maintain the steady state of the body. The long-term adjustments of metabolism, growth, and reproduction are generally under endocrine control. Hormones play a major role in regulating the concentrations of glucose, sodium, potassium, calcium, and water in the blood and interstitial fluid. The endocrine system is also essential in enabling the body to cope with stress.

An **endocrine gland** is a ductless gland that produces and secretes one or more specific hormones. Endocrine glands are to be distinguished from exocrine glands, such as sweat glands or salivary glands, which release their secretions into ducts for transport to some body surface or cavity.

Traditionally, hormones have been defined in animals as chemical messengers produced by endocrine glands and transported by the blood to other tissues, where they stimulate a change in some metabolic activity. The tissues influenced by a specific hormone are referred to as its **target tissues.** In recent years **endocrinology,** the branch of biology dealing with endocrine activity, has broadened its scope to include chemical messengers not secreted by discrete endocrine glands. Some hormones, for example, are produced by neurons. (See Focus on Identification of Endocrine Tissues and Hormones.) Hormones can be classified in the following groups:

1. Hormones secreted by endocrine glands.
2. Hormones secreted by neurons (Fig. 42–1). These are referred to as **neurohormones,** and the cells that secrete them are **neurosecretory cells.** Antidiuretic hormone (ADH) and oxytocin, produced by specialized neurons in the hypothalamus, are examples of vertebrate neurohormones. Many neurohormones are known among the invertebrates. Some endocrinologists consider neurotransmitters to be hormones because they qualify as chemical messengers.
3. Assorted chemical messengers released by individual cells or tissues. Almost all cells produce substances by which they communicate with other cells. Various growth factors and prostaglandins are examples. Histamine, released during inflammation and allergic responses, is another example. Some of the hormones in this group simply diffuse through the interstitial fluid and into their target tissues.

In order to include chemical messengers produced by neurons and other cells as hormones, we must broaden the traditional definition of a hormone. A more contemporary definition follows: A **hormone** is a chemical messenger produced by

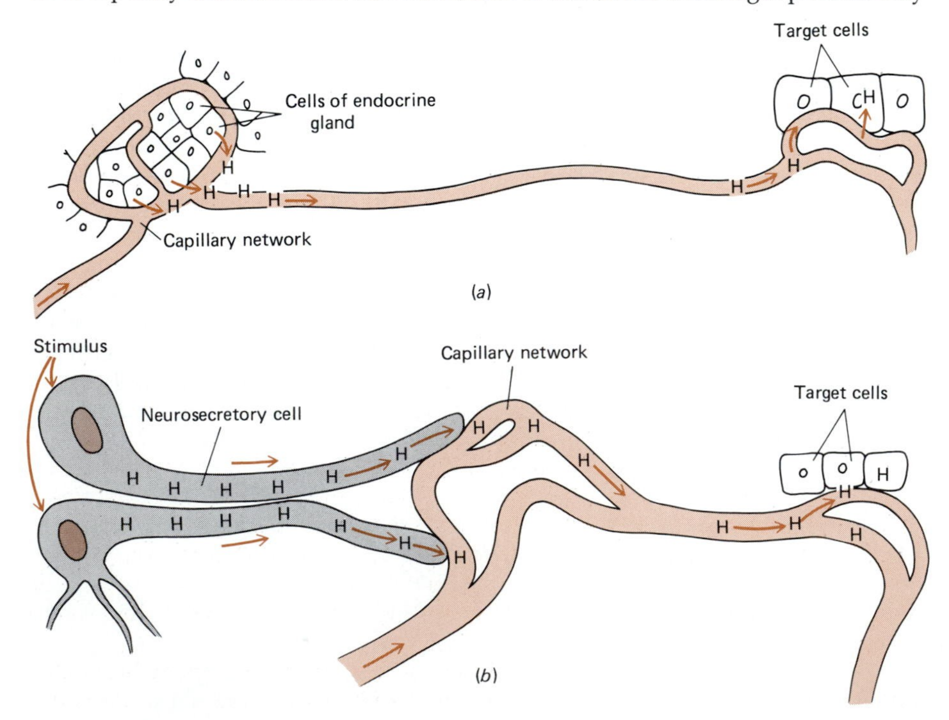

Figure 42–1 Hormones may be secreted by the cells of endocrine glands, as in (*a*), or by neurosecretory cells, as in (*b*). Hormones are generally transported by the blood and are taken up by target cells.

FOCUS ON

Identification of Endocrine Tissues and Hormones

How are endocrine tissues identified? How can an investigator prove that any particular hormone exists? Traditionally, there are three main experimental procedures used to make such determinations:

1. The suspected endocrine tissue is removed. If it is an endocrine tissue, removal should produce deficiency symptoms in the experimental animals.
2. The tissue is then replaced by transplanting similar tissue from another animal, often to a different location within the body cavity. The changes induced by removing the tissue should be reversed by replacing it.
3. The suspected hormone is extracted from the endocrine tissue of one animal and injected into an experimental animal from which the suspected endocrine tissue has been removed. Deficiency symptoms should be relieved by replacing the suspected hormone.

Most of the known hormones and endocrine glands were established by these criteria. Today these techniques are supplemented by investigation with the electron microscope and by sophisticated biochemical procedures.

In 1849 the German physiologist Berthold performed the first clear-cut and successful experiments in endocrinology. He removed the testes from cockerels (young roosters) and observed that the comb became atrophic. He then transplanted a testis into some of the birds and observed that the combs then grew back to normal size. He postulated that these male sex glands secrete a substance into the bloodstream that is essential for the normal male secondary sex characteristics.

one type of cell that has a specific regulatory effect on the activity of another type of cell. As a group, hormones are very important in maintaining steady states within the body, and they are effective at very low concentrations.

Pheromones are chemicals produced by an animal for communication with other animals of the same species. Since pheromones are generally produced by exocrine glands and do not regulate metabolic activities within the animal that produces them, they are not hormones in the classical sense, and they will not be considered here. Their role in regulating behavior is discussed in Chapter 49.

The Chemical Nature of Hormones

Epinephrine and norepinephrine released by the adrenal glands of vertebrates (as well as by certain neurons) are amines derived from the amino acid tyrosine (Fig. 42–2). The thyroid hormones are also derived from tyrosine. Oxytocin and antidiuretic hormone, produced by neurosecretory cells in the hypothalamus, are short peptides composed of nine amino acids. Seven of the amino acids are identical in the two hormones; yet the substances have quite different actions. One of the pigment-concentrating hormones found in crustaceans is an octapeptide.

Insulin, glucagon, adrenocorticotropic hormone (ACTH), and calcitonin are somewhat longer peptides with about 30 amino acids in the chain. Insulin, secreted by the β cells of the islets of Langerhans in the pancreas, consists of two peptide chains joined by disulfide bonds. The pioneering work of Fred Sanger and his colleagues at the University of Cambridge determined the sequence of the 21 amino acids in one chain and the 30 amino acids in the other. Insulin was the first protein whose amino acid sequence was determined. These studies earned Sanger a Nobel prize, for they not only clarified the structure of insulin but also established methods by which the amino acid sequence of other proteins could be determined. Insulin is synthesized in the β cells of the pancreas as a single peptide composed of 84 amino acids. This compound, called **proinsulin,** undergoes folding; three disulfide bonds are formed; and a connecting peptide, containing 33 amino acids, is removed from the center of the original chain by enzymatic cleavage. This leaves two peptide chains joined by disulfide bridges (Fig. 42–3).

Proinsulin is an example of a **prohormone.** Several other peptide hormones are synthesized as prohormones and undergo partial cleavage to yield smaller pep-

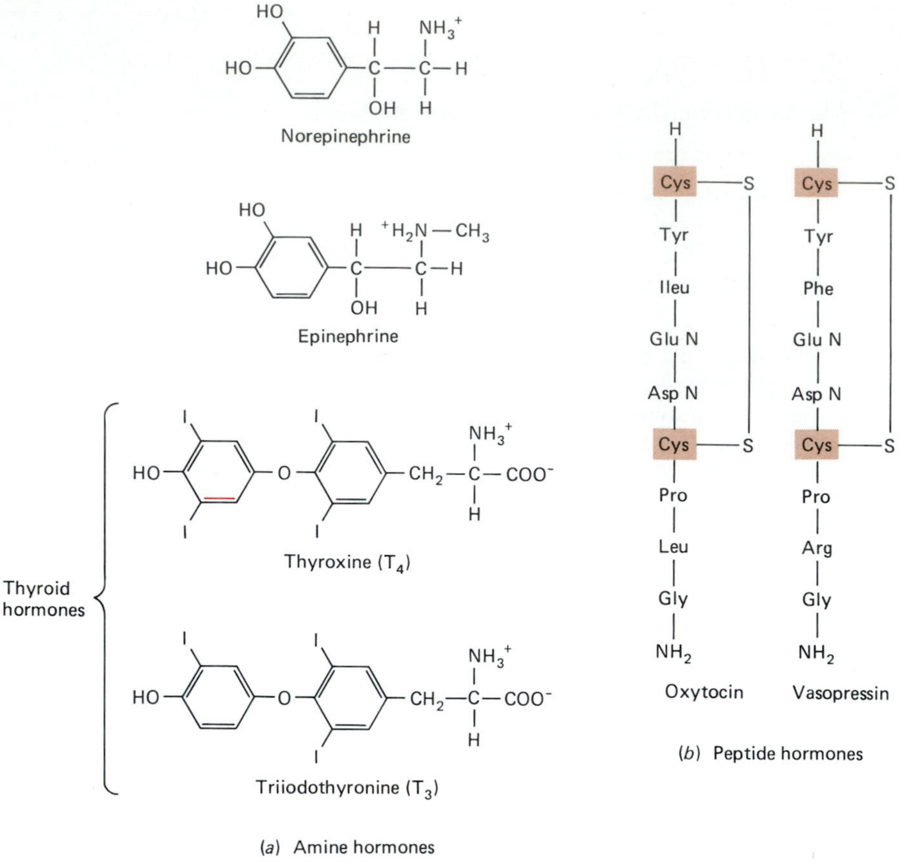

Norepinephrine

Epinephrine

Thyroid hormones

Thyroxine (T_4)

Triiodothyronine (T_3)

(*a*) Amine hormones

Oxytocin: H – Cys – Tyr – Ileu – Glu N – Asp N – Cys – Pro – Leu – Gly – NH_2

Vasopressin: H – Cys – Tyr – Phe – Glu N – Asp N – Cys – Pro – Arg – Gly – NH_2

(*b*) Peptide hormones

Figure 42–2 (*a*) Some hormones derived from amino acids. Note the presence of iodine in the thyroid hormones. (*b*) Peptide hormones produced in the hypothalamus and secreted by the posterior pituitary. Note that they differ by only two amino acids.

tides with full hormonal activity. Growth hormone, thyroid-stimulating hormone, and the gonadotropic hormones, all secreted by the anterior lobe of the pituitary gland, are proteins with molecular weights of 25,000 or more.

Prostaglandins and the juvenile hormones of insects are examples of hormones derived from fatty acids (Fig. 42–4). The molting hormone (also called ecdysone) of insects is a steroid with 27 carbons arranged in four rings, and a tail as in the cholesterol molecule. In vertebrates, the adrenal cortex, testis, ovary, and placenta secrete steroids synthesized from cholesterol. Progesterone, one of the female sex hormones, has 21 carbons; it is a precursor of the hormones produced by the adrenal cortex (Fig. 42–5).

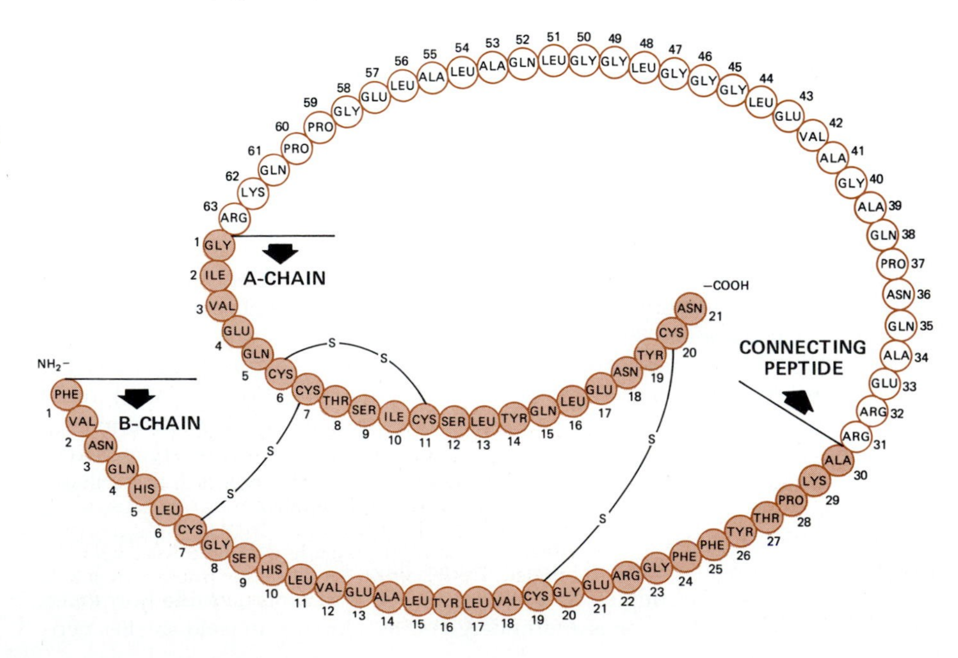

Figure 42–3 The molecular structure of proinsulin, a single polypeptide chain, and its conversion to insulin, composed of two peptide chains. This is achieved by the formation of three disulfide bonds and the removal of a section of the peptide chain between peptide A and peptide B. (Courtesy of R. W. McGilvery.)

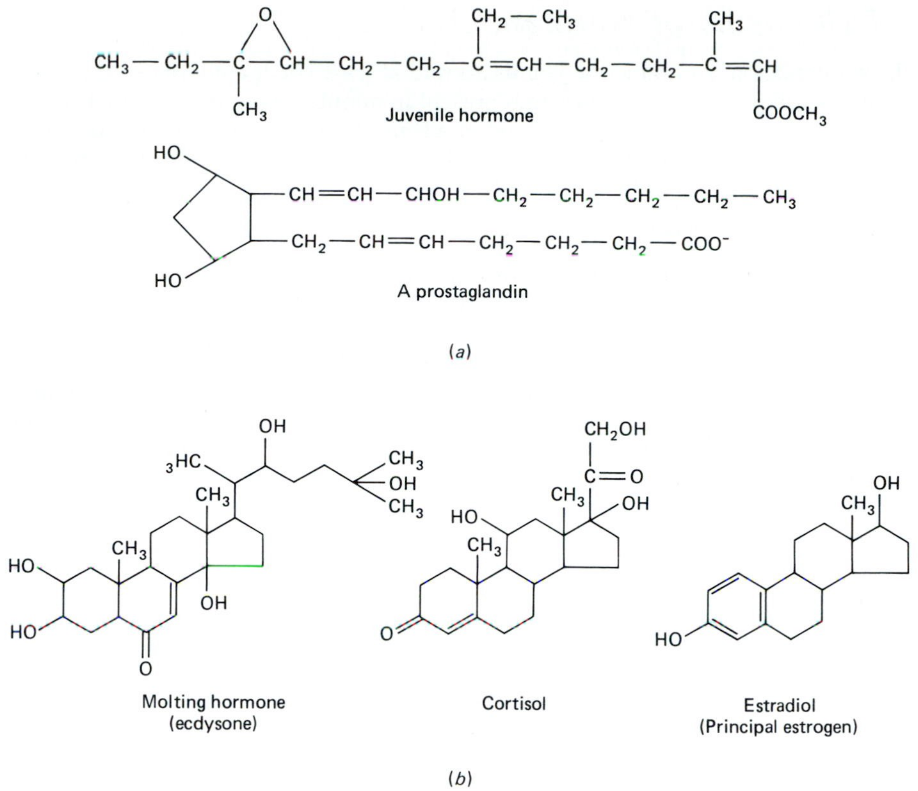

Juvenile hormone

A prostaglandin

(a)

Molting hormone (ecdysone)

Cortisol

Estradiol (Principal estrogen)

(b)

Figure 42–4 Some lipid hormones. (*a*) Juvenile hormone and prostaglandins are derived from fatty acids. (*b*) Steroid hormones.

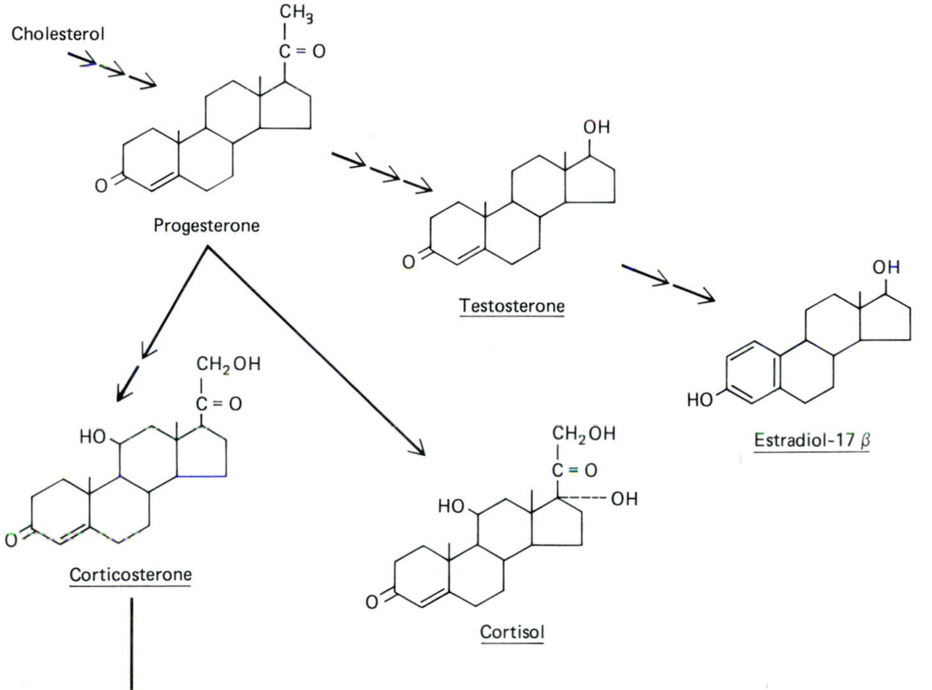

Cholesterol

Progesterone

Testosterone

Estradiol-17 β

Corticosterone

Cortisol

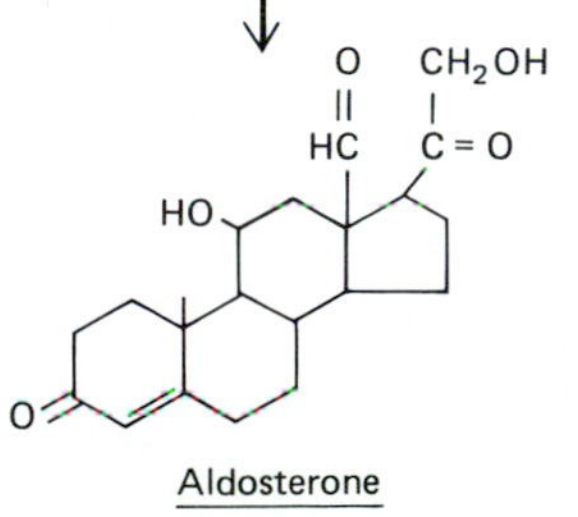

Aldosterone

Figure 42–5 The sequence of reactions by which testosterone (the principal male sex hormone), estrogens (female sex hormones), glucocorticoids, and mineralocorticoids are synthesized from progesterone (another female sex hormone). Progesterone is synthesized from cholesterol. Note that the methyl (CH_3) groups attached to each line extending upward have been omitted. Double arrows indicate intermediate steps that have been omitted from diagram. Many of these steps are reversible.

Mechanisms of Hormone Action

In vertebrates most endocrine glands secrete at least small amounts of their hormones continuously. Thus, although present in minute amounts, 30 or 40 different hormones are circulating in the blood at all times. Only small amounts of a hormone are free in the plasma—most of the hormone present is transported bound to plasma proteins. The following steady state exists:

$$\text{Hormone—Plasma protein} \rightleftharpoons \text{Free hormone} + \text{Plasma protein}$$

Free hormone molecules are continuously removed from the circulation by target tissues. They are also removed by the liver, which inactivates some hormones, and by the kidneys, which excrete them. The rate of secretion of a specific hormone can be estimated by analyzing its rate of excretion in the urine.

Hormone molecules diffuse from the blood into the interstitial fluid and may pass through many tissues before they are taken up by their target tissue. How does the target tissue recognize its hormone? Specialized receptor proteins in the target tissue cells specifically bind the hormone. The receptor protein is compared to a lock, and the hormones to different keys. Only the hormone that fits the lock—the specific receptor—can influence the metabolic machinery of the cell.

Most cells have receptors for more than one type of hormone, as several different hormones may be involved in regulating their metabolic activities. Some hormones act synergistically on their target tissues; that is, the presence of one enhances the effect of another.

ACTIVATION OF GENES

Steroid hormones are relatively small, lipid-soluble molecules that easily pass through the cell membrane of a target cell into the cytoplasm (Fig. 42–6). Specific soluble protein receptors in the cytoplasm (or perhaps in the nucleus) combine with the hormone, forming a hormone–receptor complex. The hormone–receptor complex moves into the nucleus and combines with yet another receptor, a protein associated with the DNA. This combination activates certain genes and leads to synthesis of the messenger RNAs coding for specific proteins. Thyroid hormones are also small, hydrophobic (though not steroid) molecules that pass through the plasma membrane and bind to receptors within the nucleus.

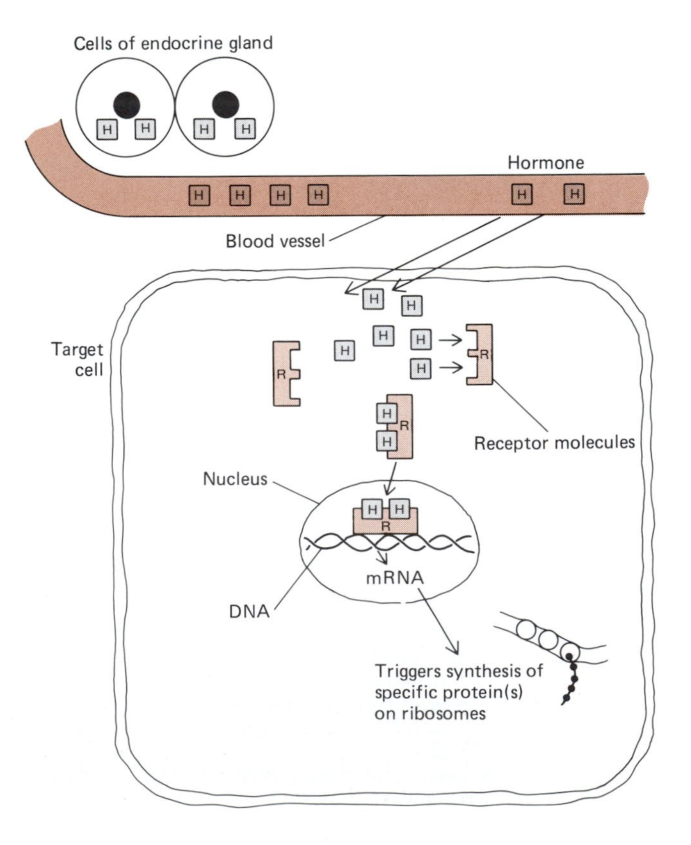

Figure 42–6 Activation of genes by steroid hormones. Steroid hormones are small lipid soluble molecules that pass freely through the cell membrane. Some steroid hormones combine with receptors within target cells. The steroid hormone–receptor complex moves into the nucleus and combines with a protein associated with the DNA. This activates specific genes, leading to the synthesis of mRNA coding for specific proteins. The proteins cause the response recognized as the hormone's action.

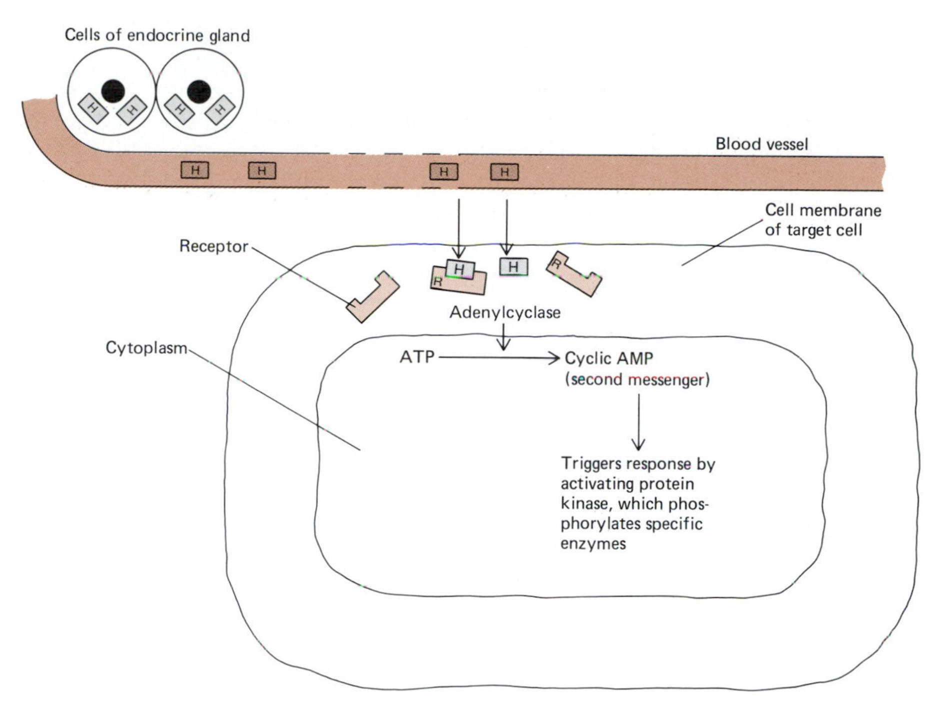

Figure 42–7 Second-messenger mechanism of hormone action. Many hormones (especially large protein hormones) combine with receptors in the cell membrane of target cells. The hormone–receptor complex stimulates an enzyme, adenyl cyclase, which catalyzes the conversion of ATP to cyclic AMP, a second messenger. Cyclic AMP then activates protein kinase, which phosphorylates specific enzymes.

ACTION THROUGH A SECOND MESSENGER

Many protein hormones combine with receptors in the cell membrane of the target cell. The hormonal message is then relayed to the appropriate site within the cell by a second messenger. In the 1960s Earl Sutherland identified **cyclic AMP** (adenosine monophosphate) as a hormone intermediary. He suggested that when certain hormones combine with a receptor, a membrane-bound enzyme called adenyl cyclase is activated. Adenyl cyclase catalyzes the conversion of ATP to cyclic AMP (Figs. 42–7 and 42–8). Cyclic AMP then acts as a second messenger and stimulates specific cyclic AMP–dependent protein kinases. These enzymes add phosphate groups to proteins and either activate or inhibit their enzymatic activity.

PROSTAGLANDINS

Prostaglandins are derivatives of polyunsaturated fatty acids, each with a five-carbon ring in its structure. Prostaglandins have many hormone-type effects and may help to regulate the action of other hormones by stimulating or inhibiting the formation of cyclic AMP.

Prostaglandins are synthesized and released by many different tissues, including the seminal vesicle, the lungs, liver, and digestive tract. They are classified on the basis of their chemical structure. Among the major classes are PGE and PGF. A subscript number is used to identify the number of double bonds in the side chains of the molecule; thus, PGE_1 has one double bond, whereas PGF_2 has two double bonds. Prostaglandins are rapidly inactivated, and their half-life is only a few minutes at most.

Various prostaglandins dilate the bronchial passageways, inhibit gastric secretion, increase intestinal motility, stimulate contraction of the uterus, raise or lower blood pressure, regulate metabolism, and cause inflammation. Prostaglandins are released when blood clots, and are involved in the process of inflammation. Those synthesized in the temperature-regulating center of the hypothalamus cause fever. In fact, the effect of aspirin in reducing fever and reducing inflammation depends upon its ability to inhibit prostaglandin synthesis.

Because they are involved in the regulation of so many metabolic processes, prostaglandins may prove effective as drugs to control many different disorders. At present, prostaglandins are used clinically to initiate labor and to induce abortion, and their use as a birth control drug is being investigated.

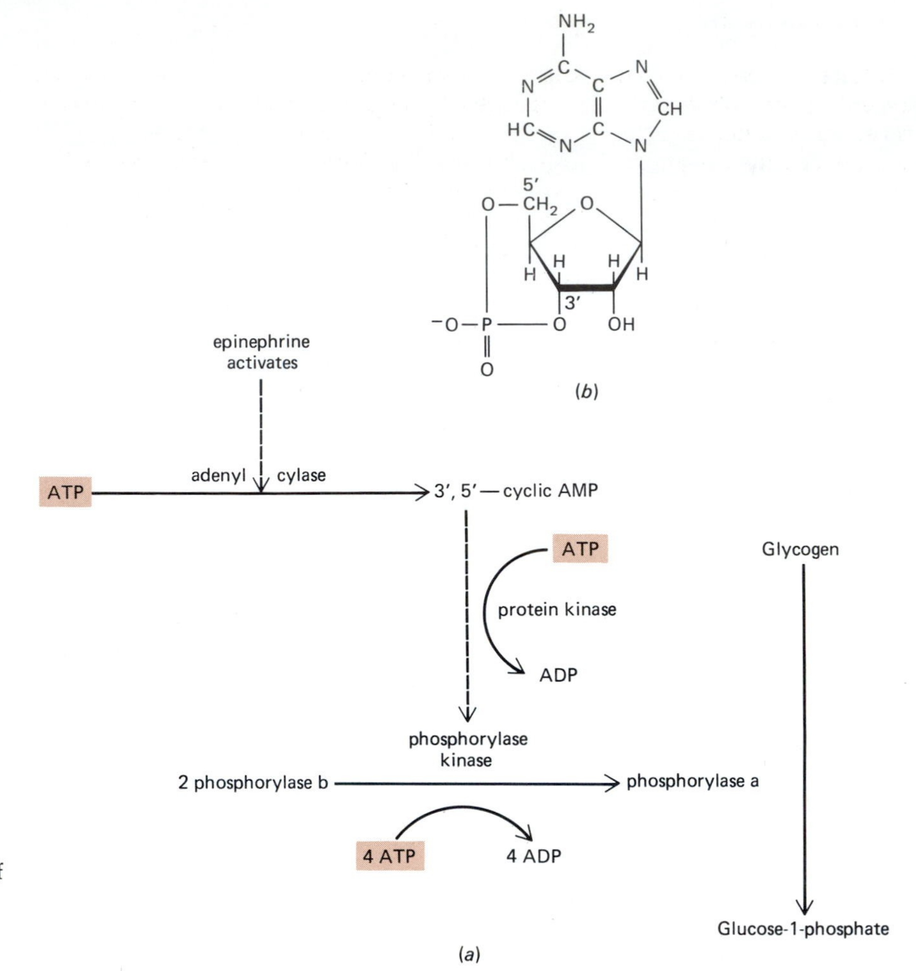

Figure 42–8 (*a*) The sequence of enzymatic events by which epinephrine or glucagon stimulates adenyl cyclase, bringing about the synthesis of cyclic AMP. This in turn activates a protein kinase that phosphorylates the enzyme phosphorylase kinase. Phosphorylase kinase then activates phosphorylase. This finally brings about the cleavage of glycogen and the secretion of glucose. (*b*) Structure of cyclic AMP (adenosine 3′,5′-monophosphate).

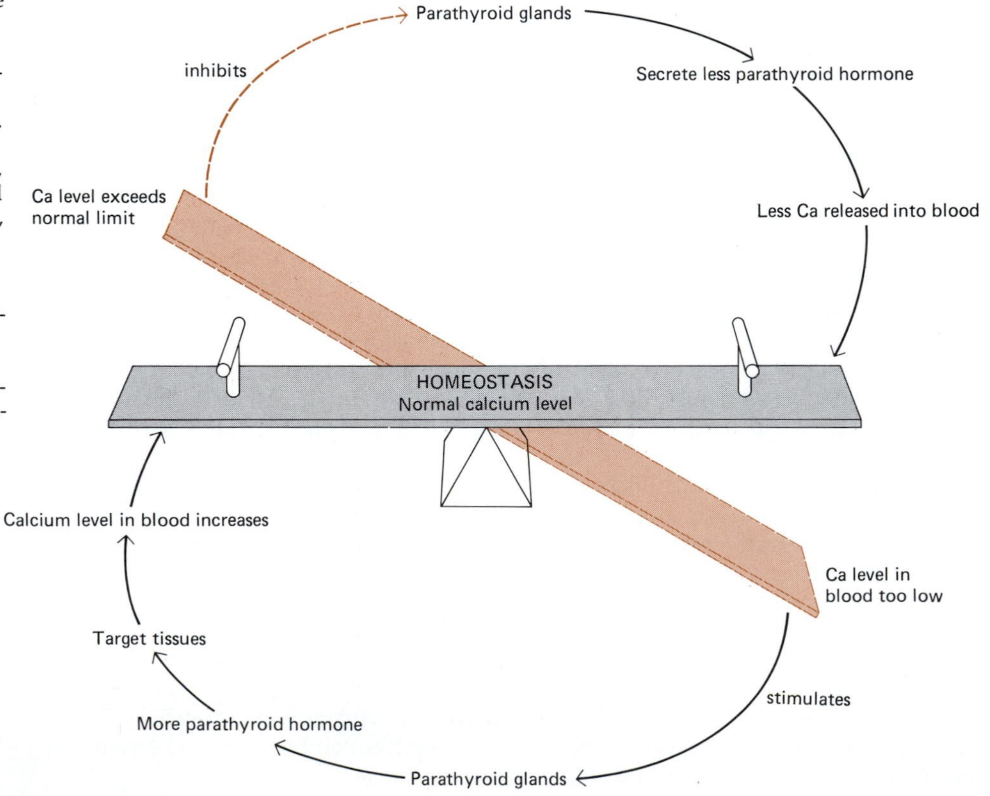

Figure 42–9 Regulation of hormone secretion by negative feedback. When the calcium level in the blood falls below normal, the parathyroid glands are stimulated to secrete more parathyroid hormone. This hormone acts to increase the calcium concentration in the blood, thus restoring homeostasis. Should the calcium content exceed normal, the parathyroid glands are inhibited and slow their secretion of hormone. This diagram has been greatly simplified; calcitonin, a hormone secreted by the thyroid gland, works antagonistically to parathyroid hormone and is important in lowering blood calcium concentration.

Regulation of Hormone Secretion

How does an endocrine gland "know" how much hormone to secrete at any given moment? Hormone secretion is regulated by negative feedback control mechanisms. Information regarding the hormone level or its effect is fed back to the gland, which then responds homeostatically. The parathyroid glands, located in the neck of tetrapod vertebrates, secrete parathyroid hormone, which helps regulate the calcium level in the blood. Even a slight decrease in calcium concentration is sensed by the parathyroid glands, which respond by increasing their rate of secreting parathyroid hormone (Fig. 42–9). This hormone stimulates release of calcium from the bones and increases reabsorption of calcium by the kidney tubules, increasing the concentration of calcium in the blood.

When the calcium concentration rises above normal, the parathyroid glands respond by decreasing their secretion of parathyroid hormone. Both responses are negative feedback mechanisms, because in both cases the effects are opposite (negative) to the stimulus. Negative feedback is the basis of hormonal regulation. Two or three different hormones may interact in the regulatory process (see Fig. 42–21).

Invertebrate Hormones

Most invertebrate hormones are secreted by neurons rather than by endocrine glands. These neurohormones regulate such processes as regeneration in *Hydra,* flatworms, and annelids; molting and metamorphosis in insects; color changes in crustaceans; and gamete production, reproductive behavior, and metabolism in other groups.

ENDOCRINE REGULATION OF REPRODUCTIVE DEVELOPMENT IN CEPHALOPODS

The cephalopods have **optic glands** on the optic stalks on each side of the brain (Fig. 42–10). These glands secrete a gonadotropic hormone that stimulates the ovary and testis to enlarge. The optic gland is regulated by an inhibitory nerve from the brain. The inhibitory centers in the brain appear to be regulated by changes in photoperiod (day and night length). Light stimuli received by the eyes activate the inhibitory nerve centers in the brain, which in turn inhibit the optic glands. When the optic nerves are cut (so that the cephalopod is blinded), the optic glands enlarge and secrete gonadotropin, causing the ovaries to develop precociously.

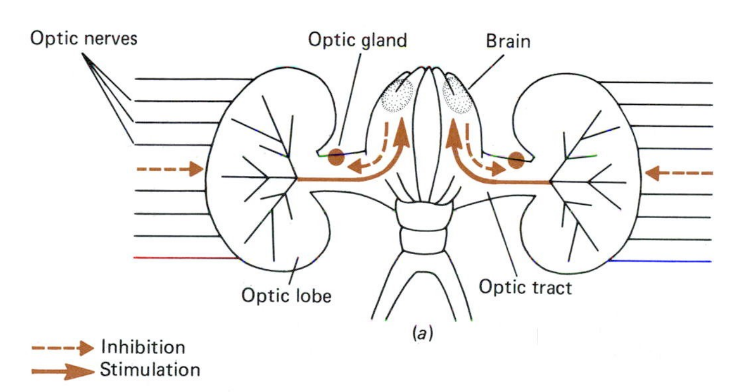

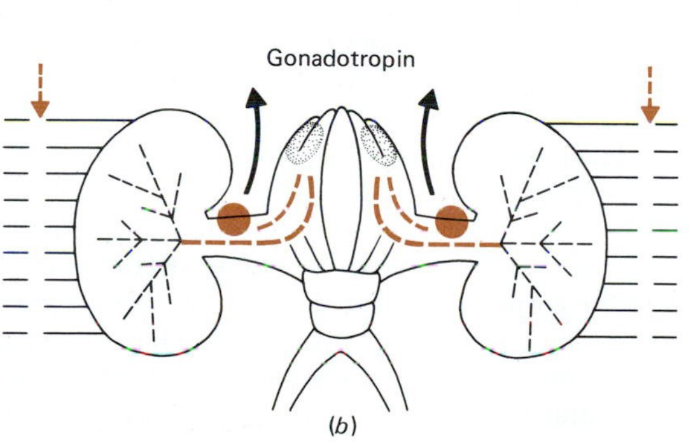

Figure 42–10 Neuroendocrine control of gonad maturation in the octopus. (*a*) In an immature octopus, light acts on the brain, which inhibits the optic glands. (*b*) When the optic nerves are cut, the inhibitory nerve centers in the brain are no longer stimulated. The optic glands are not inhibited, and they secrete gonadotropin, which stimulates growth of the gonads.

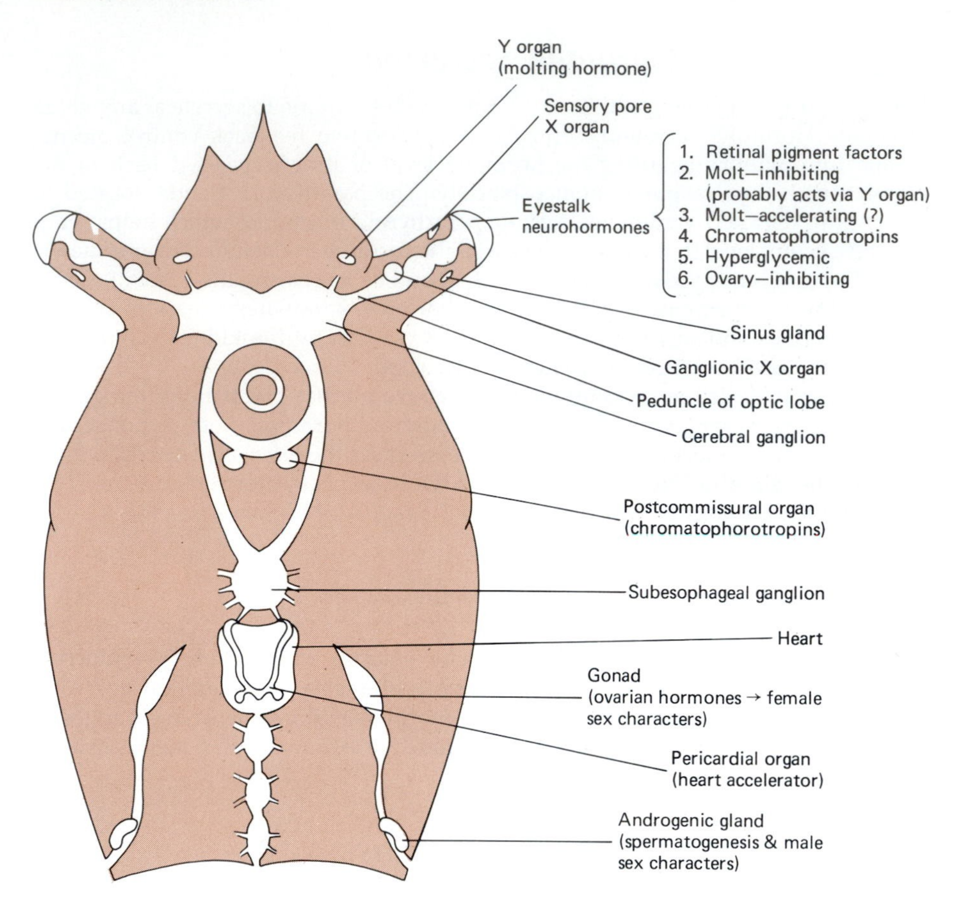

Figure 42–11 Locations of the most extensively studied endocrine glands of a crustacean.

COLOR CHANGE IN CRUSTACEANS

Crustaceans possess true (non-neural) endocrine glands, as well as masses of neurosecretory cells. Their hormonal regulation is complex and affects many activities, including molting, migration of retinal pigment, reproduction, heart rate, and metabolism (Fig. 42–11). One of the most interesting—and novel—activities regulated by hormones is color change.

Pigment cells of crustaceans are located in the integument beneath the exoskeleton. Pigment may be black, yellow, red, white, or even blue. Color changes are produced by changes in the distribution of pigment granules within the cells. When the pigment is concentrated near the center of a cell, its color is only minimally visible; but when it is dispersed throughout the cell, the color shows to advantage. These pigments provide protective coloration. By appropriate condensation or dispersal of specific types of pigments a crustacean can approximate the color of its background. Distribution of pigment, that is, color display, is regulated by hormones produced by neurosecretory cells.

HORMONAL CONTROL OF INSECT DEVELOPMENT

Like crustaceans, insects have endocrine glands as well as neurosecretory cells. The various hormones interact with one another to regulate growth and development, including molting and morphogenesis. Hormones also help regulate metabolism and reproduction.

Hormonal control of development in insects is complex and varies among the many species. Generally some environmental factor (e.g., temperature change) affects neurosecretory cells in the brain. Once activated, these cells produce a hormone referred to as **brain hormone (BH),** which their axons secrete into organs called the **corpora cardiaca** (Fig. 42–12). BH is stored in the corpora cardiaca. When released, it stimulates the **prothoracic glands,** endocrine glands in the prothorax, to produce **molting hormone,** also called **ecdysone.** Molting hormone stimulates growth and molting.

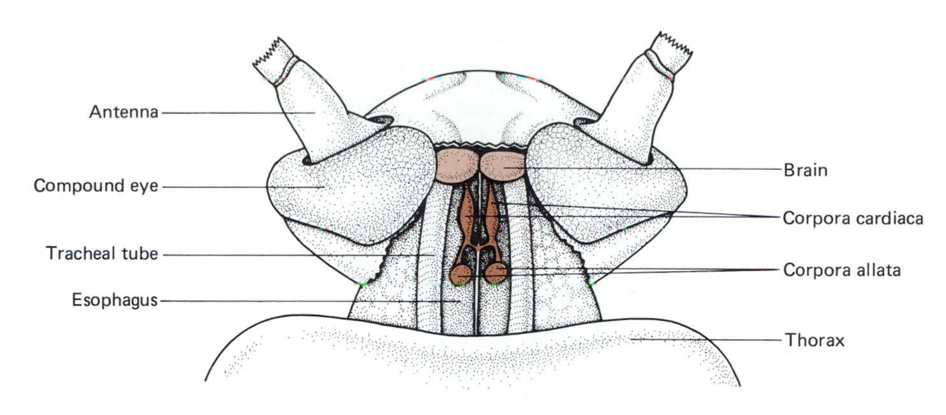

Figure 42–12 Dissection of an insect head showing the paired corpora allata and corpora cardiaca, as well as the brain.

In the immature insect, endocrine glands called **corpora allata** secrete **juvenile hormone.** This hormone suppresses metamorphosis at each larval molt so that the insect retains its immature state (Fig. 42–13). After the molt the insect is still in a larval stage. When the concentration of juvenile hormone decreases, metamorphosis occurs and the insect is transformed into a pupa (Chapter 26). In the absence of the juvenile hormone, the pupa molts to become an adult. The secretory activity of the corpora allata is regulated by the nervous system, and the amount of juvenile hormone decreases with successive molts.

Vertebrate Hormones

In vertebrates, hormones regulate such diverse activities as growth, metabolic rate, utilization of nutrients by cells, and reproduction. They are largely responsible for regulating fluid balance and blood homeostasis, and they help the body to cope with stress. The principal human endocrine glands are illustrated in Figure 42–14. Most vertebrates possess similar endocrine glands. Table 42–1 gives the physiological actions and sources of some of the major vertebrate hormones.

ENDOCRINE DISORDERS

When a disorder or disease process affects an endocrine gland, the rate of secretion may become abnormal. If **hyposecretion** (reduced output) occurs, target cells are deprived of needed stimulation. If **hypersecretion** (abnormal increase in output)

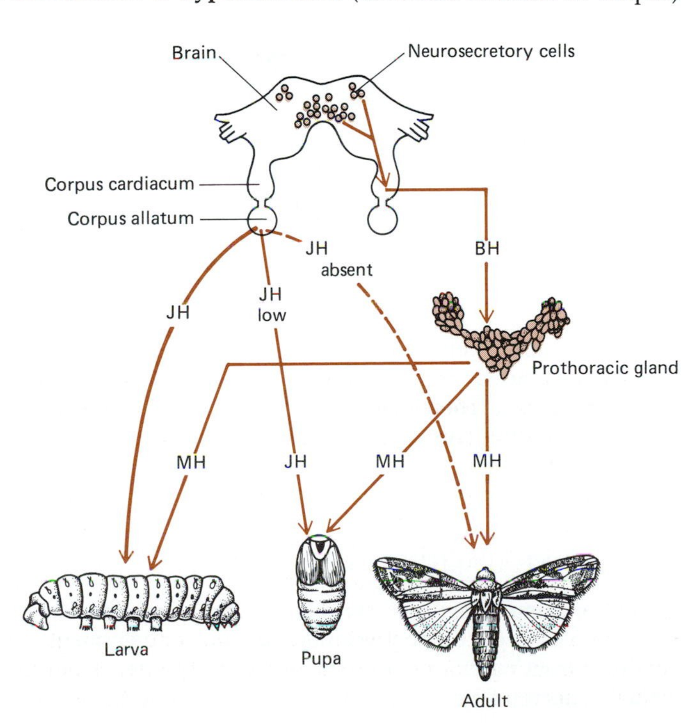

Figure 42–13 The neural and endocrine control of growth and molting in a moth. Neurosecretory cells in the brain secrete a hormone, BH, which stimulates the prothoracic glands to secrete molting hormone (*MH*). In the immature insect, the corpora allata secrete juvenile hormone (*JH*), which suppresses metamorphosis at each larval molt. Metamorphosis to the adult form occurs when molting hormone acts in the absence of juvenile hormone.

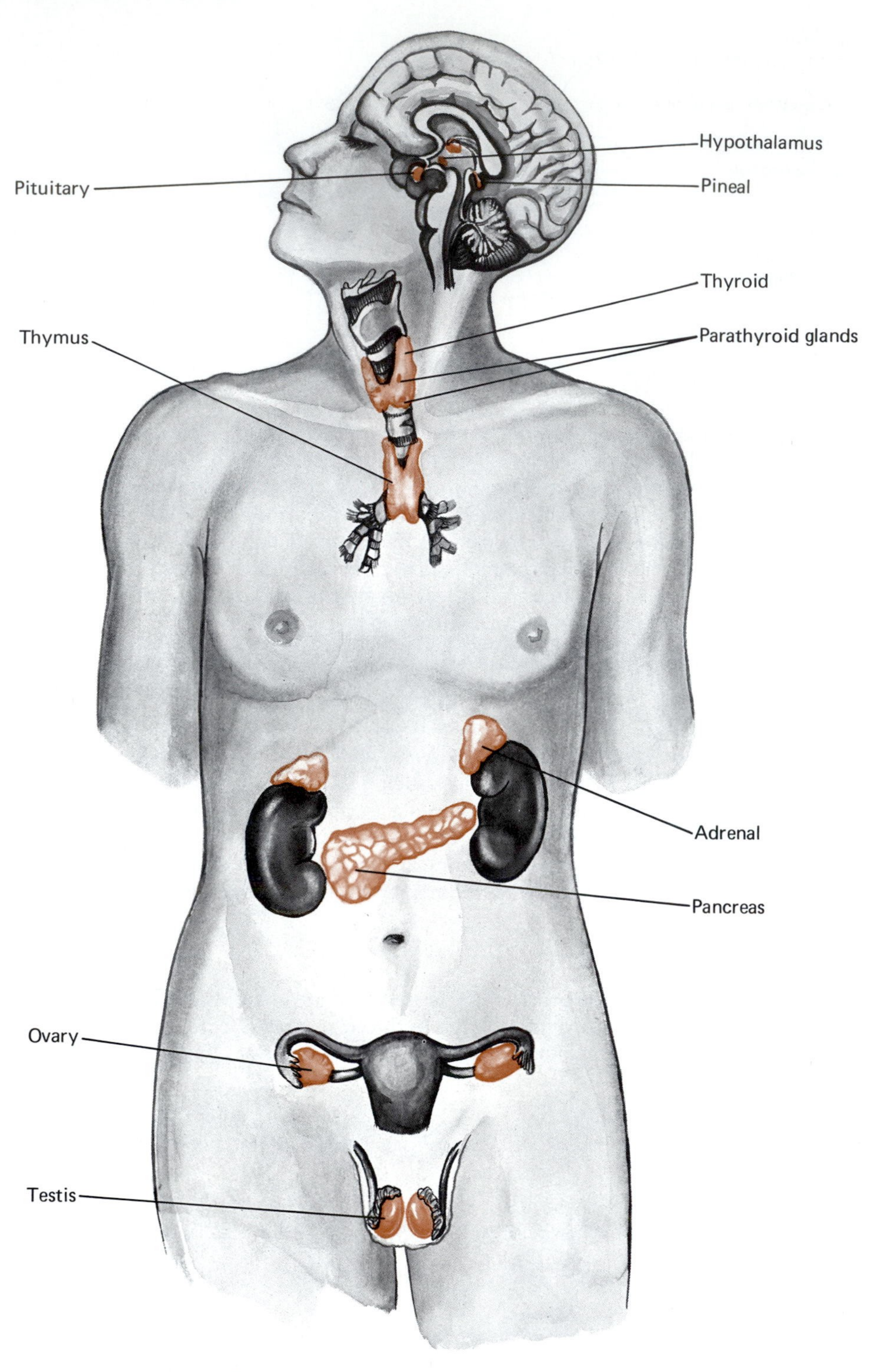

Figure 42–14 Location of the principal endocrine glands in the human male and female.

occurs, the target cells may be overstimulated. In some endocrine disorders, an appropriate amount of hormone is secreted, but target cells may not be able to take it up and utilize it. There may be too few protein receptors, or the receptors may not function properly. Any of these abnormalities leads to predictable metabolic malfunctions and clinical symptoms (see Focus on Consequences of Endocrine Malfunction).

THE HYPOTHALAMUS AND PITUITARY GLAND

Much hormonal activity is controlled directly or indirectly by the **hypothalamus,** which serves as the link between the nervous and endocrine systems. After receiving input from other areas of the brain and from hormones in the blood, the hypothalamus secretes hormones that regulate the release of hormones from the pitui-

TABLE 42–1
*Some Endocrine Glands and Their Hormones**

Endocrine Gland and Hormone	*Target Tissue*	*Principal Actions*
Hypothalamus		
Releasing and inhibiting hormones	Anterior lobe of pituitary gland	Stimulates or inhibits secretion of specific hormones
Hypothalamus (production) Posterior lobe of pituitary (storage and release)		
Oxytocin	Uterus	Stimulates contraction
	Mammary glands	Stimulates ejection of milk into ducts
Antidiuretic hormone	Kidneys (collecting ducts)	Stimulates reabsorption of water; conserves water
Anterior lobe of pituitary		
Growth hormone	General	Stimulates growth by promoting protein synthesis
Prolactin	Mammary glands	Stimulates milk production
Thyroid-stimulating hormone	Thyroid gland	Stimulates secretion of thyroid hormones; stimulates increase in size of thyroid gland
Adrenocorticotropic hormone	Adrenal cortex	Secretion of adrenal cortical hormones
Gonadotropic hormones (follicle-stimulating hormone, luteinizing hormone)	Gonads	Stimulate gonad function and growth
Thyroid gland		
Thyroxine (T_4) and triiodothyronine (T_3)	General	Stimulate metabolic rate; essential to normal growth and development
Calcitonin	Bone	Lowers blood calcium level by inhibiting bone breakdown by osteoclasts
Parathyroid glands		
Parathyroid hormone	Bone, kidneys, digestive tract	Increases blood calcium level by stimulating bone breakdown; stimulates calcium reabsorption by kidneys; activates vitamin D
Islets of Langerhans of pancreas		
Insulin	General	Lowers glucose concentration in the blood by facilitating glucose uptake and utilization by cells; stimulates glycogenesis; stimulates fat storage and protein synthesis
Glucagon	Liver, adipose tissue	Raises glucose concentration in the blood by stimulating glycogenolysis and gluconeogenesis; mobilizes fat
Adrenal medulla		
Epinephrine and norepinephrine	Muscle, cardiac muscle, blood vessels, liver, adipose tissue	Help body cope with stress; increase heart rate, blood pressure, metabolic rate; re-route blood; mobilize fat; raise blood sugar level
Adrenal cortex		
Mineralocorticoids (aldosterone)	Kidney tubules	Maintain sodium and phosphate balance
Glucocorticoids (cortisol)	General	Help body adapt to long-term stress; raise blood glucose level; mobilize fat
Pineal gland		
Melatonin	Gonads, pigment cells, other tissues(?)	Influences reproductive processes in hamsters and other animals; pigmentation in some vertebrates; may control biorhythms in some animals; may help control onset of puberty in humans
Ovary†		
Estrogens	General; uterus	Develop and maintain sex characteristics in female; stimulate growth of uterine lining
Progesterone	Uterus; breast	Stimulates development of uterine lining
Testis‡		
Testosterone	General; reproductive structures	Develops and maintains sex characteristics of males; promotes spermatogenesis; responsible for adolescent growth spurt

*The ovaries and testes and their hormones are discussed in Chapter 43. The digestive hormones are described in Chapter 33.
†For more detailed description see Table 43–2.
‡For more detailed description see Table 43–1.

FOCUS ON

Consequences of Endocrine Malfunction

Hormone	*Hyposecretion*	*Hypersecretion*
Growth hormone	Pituitary dwarf	Gigantism if malfunction occurs in childhood; acromegaly if in adult
Thyroid hormones	Cretinism (in children) Myxedema, a condition of pronounced adult hypothyroidism (basal metabolic rate reduced by about 40%; patient feels tired all of the time and may be mentally slow) Goiter, enlargement of the thyroid gland (see figure)	Graves' disease; goiter
Parathyroid hormone	Spontaneous discharge of nerves; spasms; tetany; death	Weak, brittle bones; kidney stones
Insulin	Diabetes mellitus	Hypoglycemia
Adrenal cortical hormones	Addison's disease (body cannot synthesize sufficient glucose by gluconeogenesis; inability to cope with stress; sodium loss in urine may lead to shock)	Cushing's disease (edema gives face a full-moon appearance; fat is deposited about trunk; rise in blood glucose level; immune response depressed)

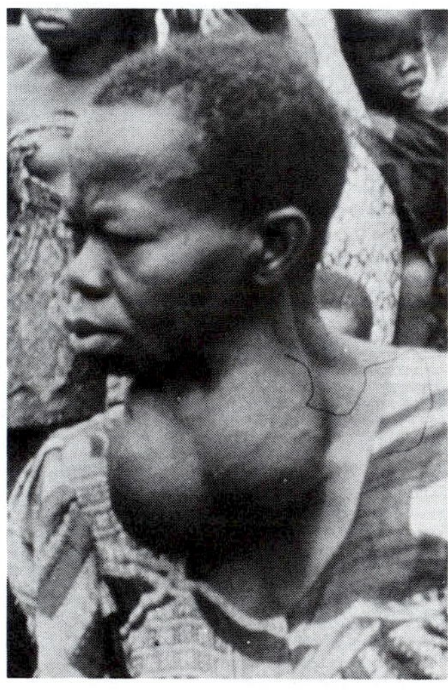

Goiter resulting from iodine deficiency. (Courtesy of United Nations Food and Agricultural Organization.)

tary gland. The hypothalamus produces two hormones, **ADH** and **oxytocin,** which pass down the axons of neurons into the posterior lobe of the pituitary gland (Fig. 42–15). These neurosecretions are stored within the tips of the axons in the posterior lobe until the neurons are stimulated. They are then released from the posterior lobe and enter the circulation.

The hypothalamus regulates the pituitary by secreting several **releasing** and **inhibiting hormones,** each more-or-less specific for one type of cell in the pituitary. These peptide neurohormones are released by the hypothalamus, enter capillaries, and pass through special portal veins that connect the hypothalamus with the anterior lobe of the pituitary. (A portal vein does not deliver blood to a larger vein directly, but connects two sets of capillaries.) Within the anterior lobe of the pituitary, the portal veins divide into a second set of capillaries from which the hypothalamic hormone can diffuse into the tissue of the anterior lobe of the pituitary (Fig. 42–16). The releasing and inhibiting hormones from the hypothalamus regulate secretion of the hormones produced by the anterior lobe of the pituitary gland.

Only about the size of a pea, the **pituitary gland** is a remarkable organ that secretes at least nine hormones. These hormones regulate the activities of several

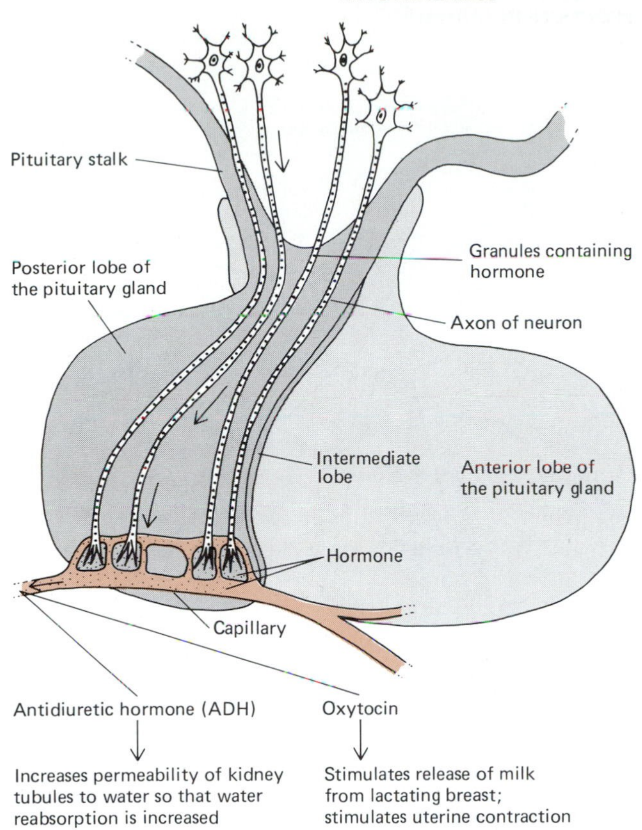

Figure 42–15 The hormones secreted by the posterior lobe of the pituitary are actually manufactured in cells of the hypothalamus. The axons of these neurons extend down into the posterior lobe of the pituitary. The hormones are packaged in granules that flow through these axons and are stored in their ends. The hormone is secreted into the interstitial fluid as needed and then transported by the blood.

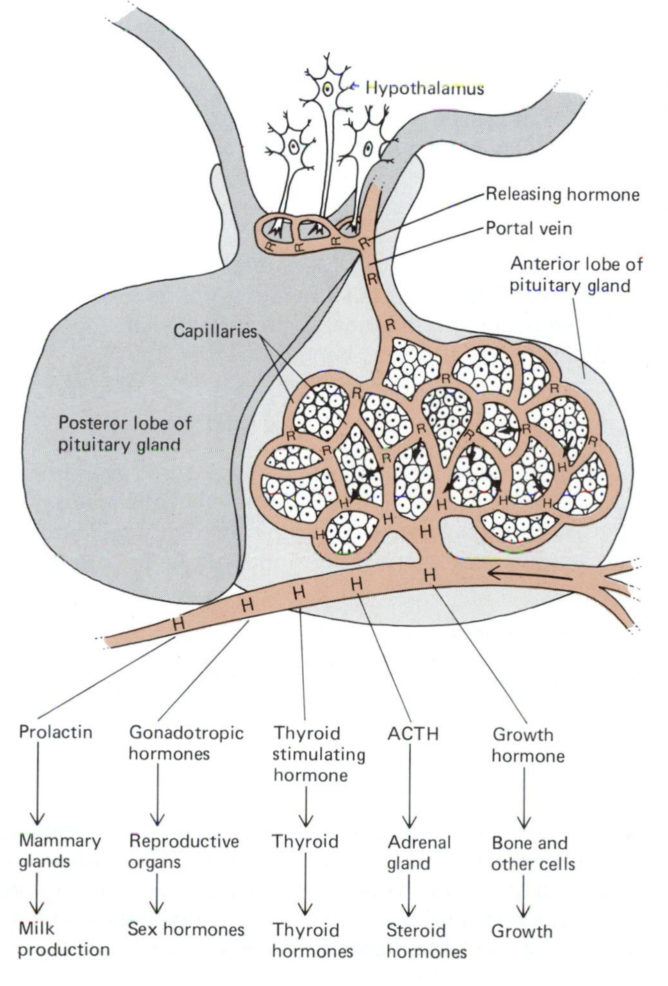

Figure 42–16 The hypothalamus secretes several specific releasing and inhibiting hormones, which reach the anterior lobe of the pituitary gland by way of portal veins. Each releasing hormone stimulates the synthesis of a particular hormone by the cells of the anterior lobe. (*R*, releasing hormone; *H*, hormone.)

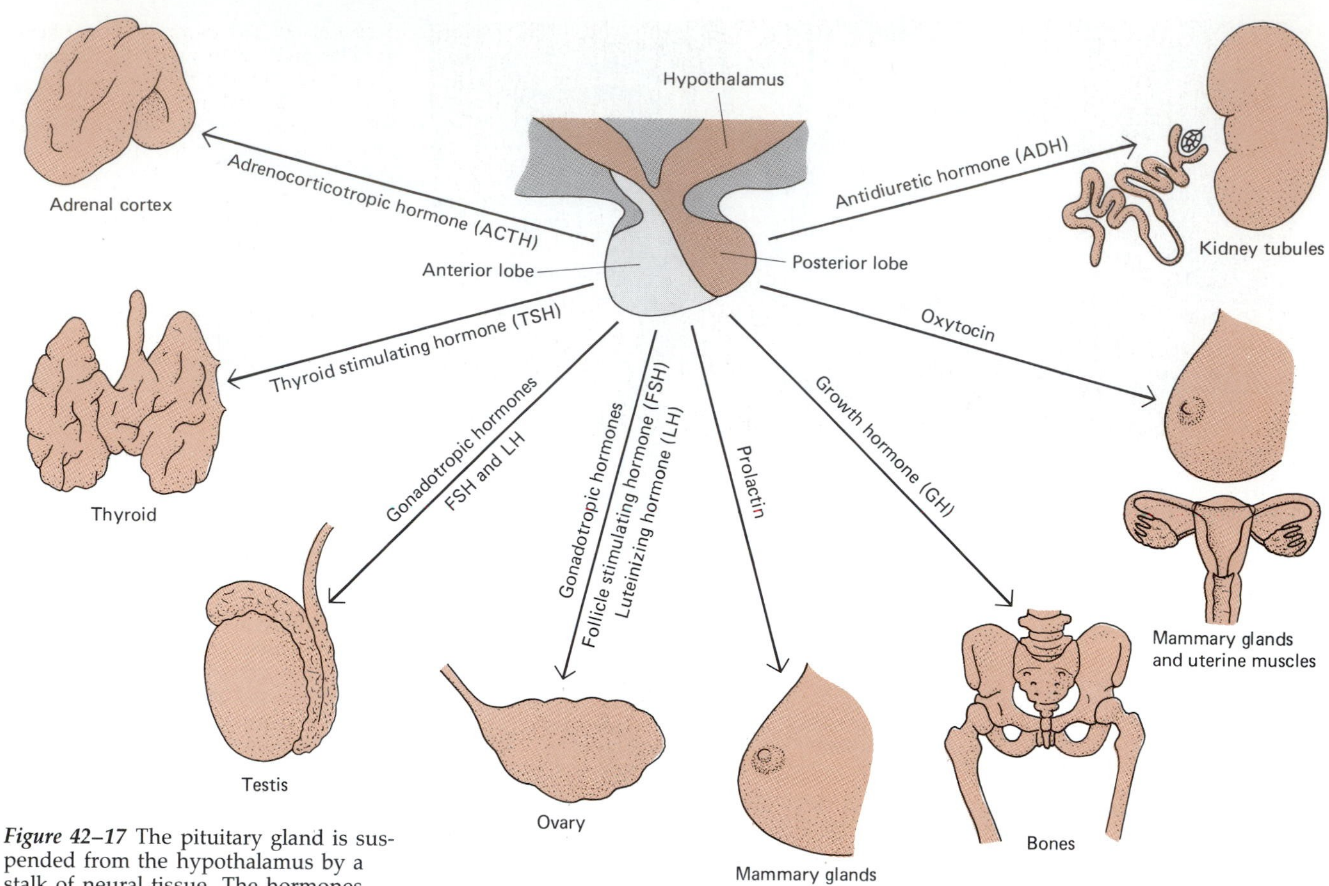

Figure 42–17 The pituitary gland is suspended from the hypothalamus by a stalk of neural tissue. The hormones secreted by the anterior and posterior lobes of the pituitary gland and the target tissues they act upon are shown.

other endocrine glands and influence a wide range of physiological processes (Fig. 42–17). The **posterior lobe** of the pituitary gland releases the hormones ADH and oxytocin produced by the hypothalamus. The **anterior lobe** produces growth hormone, prolactin, and several **tropic hormones,** hormones that regulate other endocrine glands. The anterior lobe also secretes a peptide called **β-lipotropin,** the precursor of endorphins and enkephalins (substances that function in pain perception).

In some vertebrates, the pituitary contains an intermediate lobe that secretes two forms of **melanocyte stimulating hormone (MSH).** These hormones regulate skin pigmentation. In humans, the intermediate lobe is poorly developed, and the function of MSH, if any, is not known.

GROWTH AND DEVELOPMENT

Growth and development are influenced by several hormones. Among the most important are growth hormone, secreted by the anterior lobe of the pituitary gland; thyroid hormones, secreted by the thyroid gland; and sex hormones. The role of the sex hormones, especially important at puberty, is discussed in Chapter 43.

Growth Hormone

Growth hormone (GH) stimulates body growth mainly by increasing uptake of amino acids by the cells and by stimulating protein synthesis. The effects of GH on the growth of the skeleton is indirect. GH stimulates the liver (and perhaps certain other tissues) to produce peptides called **somatomedins,** which actually mediate the growth response.

REGULATION OF GROWTH HORMONE SECRETION. Secretion of GH is regulated by a GH-releasing hormone and a GH-inhibiting hormone secreted by the hypothalamus. A high level of GH in the blood signals the hypothalamus to secrete the inhibiting hormone; this leads to a decrease in the secretion of GH by the pituitary. A low level of GH in the blood stimulates the hypothalamus to secrete its releasing

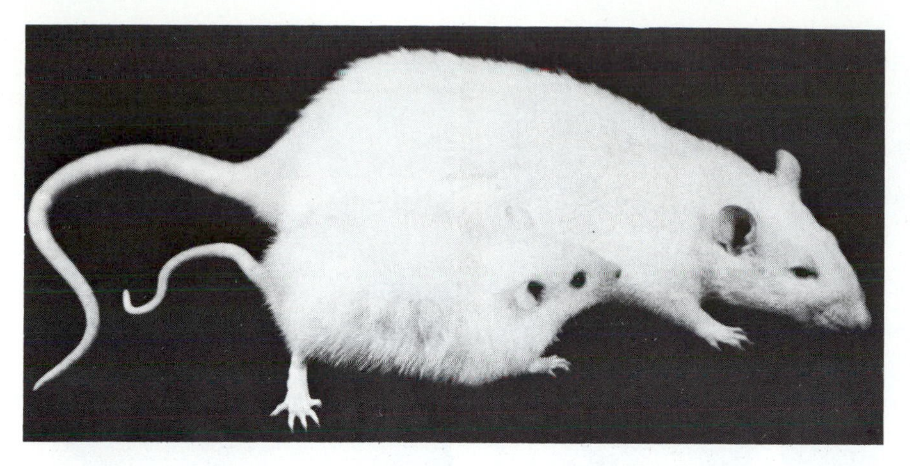

Figure 42–18 The two rats shown here are littermate brothers. The pituitary gland was removed from the smaller one (experimental) at 28 days of age. Each weighed 72 g at that time. When this photograph was taken the animals were 10 months of age; the experimental animal weighed 81 g, and the control animal, 465 g. The experimental animal never developed an adult coat, and the testes did not descend. The juvenile hair was lost rapidly and bald areas frequently appeared on the back, particularly near the base of the tail. When the pituitary gland is removed from adult rats, the coarse adult hair is gradually replaced by the soft fluffy hair characteristic of juveniles. (From C. D. Turner and J. T. Bagnara.)

hormone, stimulating the pituitary to secrete more GH. Many other factors influence GH secretion: It is increased by a low concentration of glucose or by a high amino acid concentration in the blood, and by stress, and is inhibited by high blood sugar levels.

You may recall your parents' telling you that you must get plenty of sleep and exercise in order to grow properly. These age-old notions have been supported by research. Secretion of growth hormone does increase during exercise, probably because rapid metabolism by muscle cells lowers the blood sugar level. Secretion of GH markedly increases during non-REM sleep.

Emotional support is also necessary for proper growth. Growth is often retarded in children who are deprived of cuddling, playing, and other forms of nurture, even when their physical needs (food and shelter) are amply met. Some emotionally deprived children exhibit abnormal sleep patterns, which may be the basis for decreased secretion of GH. In extreme cases, emotional deprivation can produce dwarfism.

ABNORMAL GROWTH. Secretion of too little or too much GH results in growth abnormalities (Fig. 42–18). Most circus midgets are **pituitary dwarfs,** individuals whose pituitary gland did not produce enough GH during childhood. Though miniature, a pituitary dwarf has normal intelligence and is usually well-proportioned. If this condition is diagnosed when the growth centers in the long bones are still open, it can be treated clinically by injection of GH, which can now be synthesized commercially by recombinant DNA techniques.

Not all abnormally slow growth is the result of hyposecretion of GH. The hypothalamus may not secrete releasing hormones appropriately, or somatomedin may not be released from the liver. In the African pygmy, the usual amounts of GH and somatomedin are secreted, but body tissues are not as sensitive to these hormones as the tissues of most other people. Growth is limited because the target tissues lack effective receptors for GH.

When the anterior pituitary secretes excessive amounts of GH during childhood, the condition known as **gigantism** results, producing abnormally tall stature in the affected person (Fig. 42–19). If hypersecretion of growth hormone begins during adulthood, the affected person cannot grow taller, but connective tissue proliferates, and bones—especially those in the hands, feet, and face—increase in diameter. The first sign of the disorder may be the need for a wider shoe size, or fingers so thickened that a customarily worn ring no longer fits. This condition is known as **acromegaly** (Fig. 42–20).

Figure 42–19 The world's tallest woman, Sandy Allen, standing next to her 12-year-old brother. Ms. Allen's growth (she is 2.22 meters, or 7 ft, 7¼ inches tall) is due to an excess of growth hormone. (Bettina Cerone, Photo Researchers, Inc.)

Thyroid Hormones

The **thyroid hormones** are secreted by the **thyroid gland,** located in the neck, anterior to the trachea and just below the larynx (see Fig. 42–14). The two hormones generally referred to as the thyroid hormones are **triiodothyronine (T_3),** which has three iodine atoms, and **thyroxine (T_4),** which has four iodine atoms (see Fig. 42–2). Both are synthesized from the amino acid tyrosine and from iodine.

Thyroid hormones are essential for normal growth and development and stimulate the rate of metabolism in most body tissues. These hormones are also

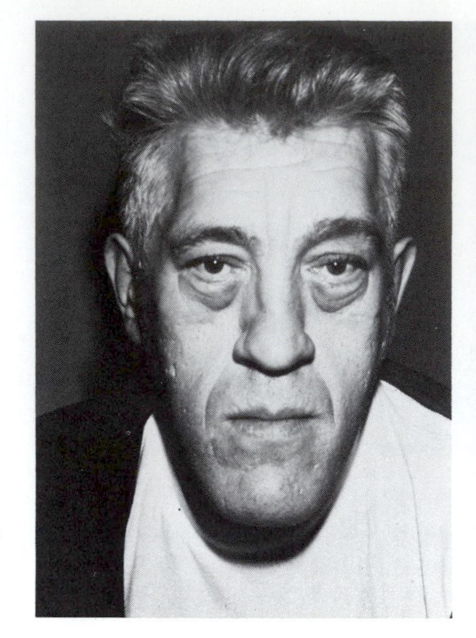

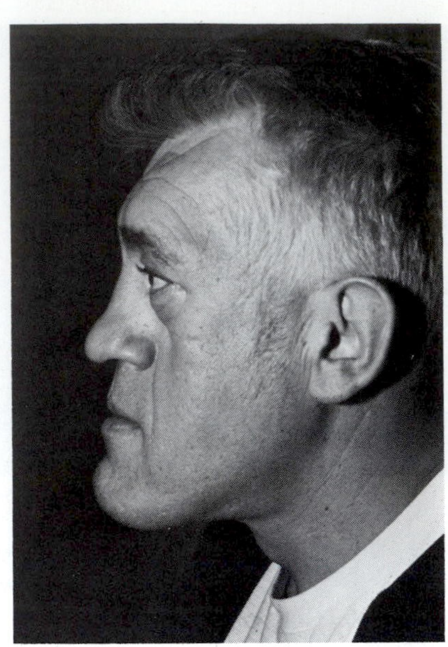

Figure 42–20 A case of acromegaly. Instead of producing growth in height, the hypersecretion of GH in the adult causes a thickening of connective tissue. Note the enlarged nose, jaw, and cheekbones. (Courtesy of Dr. Gordon Williams.)

necessary for cellular differentiation. The development of tadpoles into adult frogs cannot take place without thyroxine; the hormone appears to regulate selectively the synthesis of specific proteins. Thyroid hormones stimulate growth by promoting protein synthesis and enhancing the effect of growth hormone.

REGULATION OF SECRETION. The regulation of thyroid hormone secretion depends mainly upon a feedback system between the anterior lobe of the pituitary and the thyroid gland (Fig. 42–21). The pituitary gland secretes **thyroid-stimulating hormone (TSH),** which acts by way of cyclic AMP to promote synthesis and secretion of thyroid hormones, as well as to increase the size of the thyroid gland itself. When the level of thyroid hormones in the blood rises above normal, the cells in the anterior pituitary are inhibited and the release of TSH is decreased.

Too much thyroid hormone in the blood may also affect the hypothalamus, inhibiting secretion of TSH-releasing hormone. However, the hypothalamus is thought to exert its regulatory effects mainly in certain stressful situations, such as exposure to extreme weather conditions.

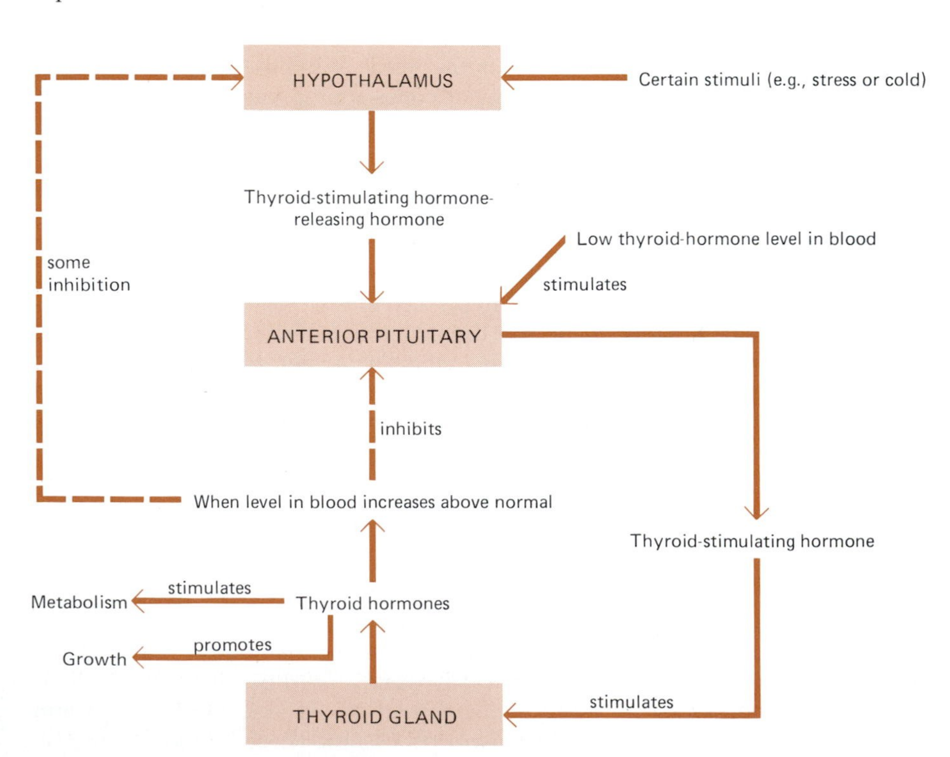

Figure 42–21 Regulation of thyroid hormone secretion. Colored arrows indicate inhibition.

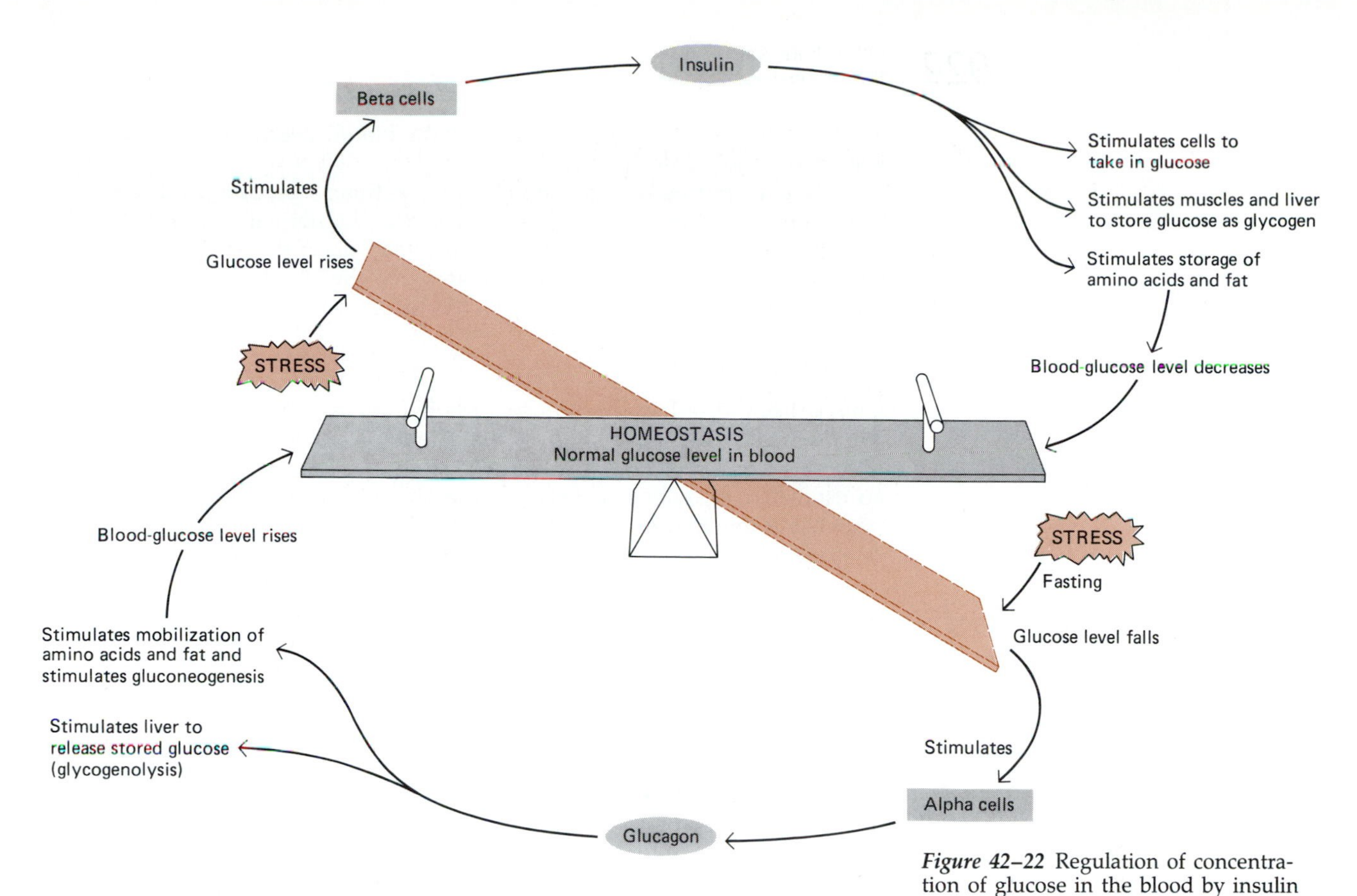

Figure 42–22 Regulation of concentration of glucose in the blood by insulin and glucagon.

THYROID HORMONES AND GROWTH DISORDER. Extreme hypothyroidism during infancy and childhood results in low metabolic rate and retarded mental and physical development, a condition known as **cretinism.** A cretin dwarf is very different from a pituitary dwarf. When diagnosed and treated early enough, cretinism can be prevented by clinical administration of synthetic thyroid hormone.

REGULATION OF BLOOD SUGAR LEVEL: INSULIN AND GLUCAGON

More than a million small clusters of cells known as the **islets of Langerhans** are scattered throughout the pancreas. About 70% of the islet cells are **beta cells,** which produce the hormone **insulin. Alpha cells** secrete the hormone **glucagon.** Both insulin and glucagon are proteins.

Insulin exerts widespread influence on metabolism, but its principal effect is to facilitate the uptake of glucose into most cells, especially muscle and fat cells. By stimulating cells to take up glucose from the blood, insulin lowers the concentration of glucose in the blood (Fig. 42–22). Liver cells are very permeable to glucose and do not require insulin to take it up from the blood. However, insulin increases the amount of glucokinase, an enzyme that phosphorylates glucose so that it cannot diffuse out of the liver cells. In this way insulin acts to trap glucose inside liver cells. The hormone also stimulates glycogenesis (glycogen formation) and storage of glycogen in the liver. Insulin stimulates protein synthesis and promotes fat storage.

Glucagon acts antagonistically to insulin. Its principal action is to increase the concentration of glucose in the blood. It does this by stimulating glycogenolysis (conversion of glycogen to glucose) in the liver, and by stimulating gluconeogenesis (production of glucose from other nutrients). Glucagon mobilizes fatty acids and amino acids as well as glucose.

The insulin–glucagon system is a powerful, fast-acting mechanism for keeping the concentration of glucose in the blood within a narrow range. Brain cells are especially dependent upon a continuous supply of glucose because they are normally unable to utilize other nutrients as fuel.

Regulation of Insulin and Glucagon Secretion

Secretion of insulin and glucagon is controlled by the concentration of blood sugar. After a meal, when the blood glucose concentration rises as a result of intestinal absorption, the beta cells are stimulated to increase insulin secretion (Fig. 42–22).

Then, as the cells remove glucose from the blood, decreasing its concentration, insulin secretion decreases.

When a person has not eaten for several hours, the concentration of glucose in the blood begins to fall. When it falls from its normal fasting level of about 90 mg/100 ml to about 70 mg/100 ml, the alpha cells of the islets secrete glucagon. Glucose is mobilized from the liver stores, and the blood sugar concentration returns to normal.

Diabetes Mellitus

One of the most serious disorders of carbohydrate metabolism is **diabetes mellitus.** Although many of the symptoms of diabetes can be controlled, the long-term complications of this disorder reduce life expectancy by as much as one third. There are an estimated 10 million diabetics in the United States alone, and almost 40,000 persons die annually as a result of this disorder, making it the third most common cause of death.

More than 90% of all cases of diabetes are of the maturity-onset type of diabetes, which develops gradually, usually in overweight persons over 40 years of age. In many cases of maturity-onset diabetes, sufficient insulin is secreted by the islets of Langerhans. The problem is that the target cells are not able to take up the insulin and use it. One current hypothesis is that as food intake increases in obesity, excessive amounts of insulin are secreted, and through a negative feedback mechanism, the number of insulin receptors on the target cells is decreased. With fewer receptors, the cells are not effective in taking up the insulin from the blood and using it.

Juvenile-onset diabetes usually develops before age 20. This disorder is marked by a dramatic decrease in the number of beta cells in the pancreas, resulting in insulin deficiency. Daily insulin injections are necessary to relieve the carbohydrate imbalance that results. At present juvenile diabetes is thought to be induced by a combination of genetic and environmental factors.

Similar metabolic disturbances occur in both forms of diabetes mellitus: (1) decreased utilization of glucose, (2) increased fat mobilization, and (3) increased protein utilization. Let us examine each of these effects.

Decreased Utilization of Glucose. In diabetics, cells dependent upon insulin can take in only about 25% of the glucose they require for fuel. Glucose accumulates in the blood, causing **hyperglycemia** (abnormally high blood glucose level). The blood sugar load is further increased by the liver, which cannot effectively trap glucose or store glycogen without insulin. Instead of the normal fasting level of about 90 mg/100 ml, the diabetic's blood may contain from 300 to more than 1000 mg/100 ml of glucose.

The concentration of glucose in the blood is so high in the untreated diabetic that it exceeds the renal threshold, and glucose is excreted in the urine. A simple test for detecting glucose in the urine is used as a screening test for diabetes and for evaluating control of glucose metabolism in known diabetics.

Increased Fat Mobilization. Despite the large quantities of glucose in the blood, the cells cannot take it up and must turn to other sources of fuel. However, the absence of insulin promotes mobilization of fat stores, so that the concentration of fatty acids in the blood rises, providing nutrients for cellular respiration. Unfortunately, however, the blood lipid content may reach five times the normal value, leading to atherosclerosis. The increased fat metabolism also increases formation of ketone bodies (acetone and other breakdown products of fatty acid metabolism). Ketone bodies accumulate in the blood, producing a condition known as ketosis. The pH can become too low, causing acidosis, and if this is sufficiently marked can result in coma and death. The presence of ketone bodies in the urine provides another useful clinical indication of diabetes.

When ketones and glucose are excreted in the urine, they take water with them because of increased osmotic pressure. Urine volume increases, and the body becomes dehydrated. Constant thirst is a clinical symptom of diabetes.

Increased Protein Mobilization. Lack of insulin also causes protein wasting. Normally, proteins are continuously synthesized and broken down, but without insulin to stimulate protein synthesis, the balance is disturbed and there is a shift in the direction of protein catabolism. Amino acids are deaminated in the liver, and their carbon chains are converted to glucose, further compounding the excess glucose problem. The untreated diabetic becomes thin and emaciated, despite (usually) a hearty appetite.

Hypoglycemia

Hypoglycemia (low blood sugar) is sometimes diagnosed in people who later develop diabetes. This condition may be an overreaction by the islets to glucose challenge. Too much insulin is secreted in response to carbohydrate absorption. About 3 hours after a meal, the concentration of glucose in the blood falls below normal, making the hypoglycemic person feel very drowsy. If this response is severe enough, the person may become uncoordinated or even unconscious.

RESPONSE TO STRESS

Stressful stimuli threaten homeostasis and must be dealt with quickly and effectively. A stressful stimulus can be anything from a broken bone or disease to anxiety caused by taking a test for which one is not fully prepared. The brain responds to stress by signaling the adrenal medulla and the adrenal cortex.

Hormones of the Adrenal Medulla

The paired **adrenal glands** are small masses of tissue that lie in contact with the upper ends of the kidneys (see Fig. 42–14). Each gland consists of a central pink portion, the **adrenal medulla,** and a larger, yellow outer section, the **adrenal cortex.** The adrenal medulla is often referred to as the emergency gland of the body. It secretes **epinephrine** (adrenaline) and **norepinephrine** (noradrenaline). Recall that norepinephrine is also secreted as a neurotransmitter by many neurons. The hormone secreted by the adrenal medulla has much longer-lasting effects (about ten times longer) because it is removed from the blood more slowly than it is removed from synapses. Still, the nervous system and the adrenal glands reinforce one another's action during stress or excitement (such as participation in a competitive athletic event).

During sudden stress the brain sends messages through sympathetic nerves to the adrenal medulla. The medulla initiates an **alarm reaction.** This response enables a person to think more quickly, fight harder, or run much faster than usual. Metabolic rate increases as much as 100%. The adrenal medullary hormones stimulate the heart to beat faster, raise blood pressure, and lower thresholds in the RAS of the brain (Fig. 42–23). These hormones also enlarge airways, permitting more effective breathing. This property is medically useful; epinephrine and related drugs are used clinically to relieve nasal congestion and asthma.

Hormones of the adrenal medulla cause blood to be re-routed to those organs essential for emergency action. Blood vessels to the skin, digestive organs, and kidneys are constricted, while those that deliver blood to the brain, muscles, and heart are dilated. Constriction of blood vessels in the skin has the added advantage of decreasing blood loss in case of hemorrhage. It also explains the sudden paling that comes with fear or rage. Epinephrine and norepinephrine also raise the concentration of glucose in the blood by stimulating glycogenolysis in the liver. They raise fatty acid concentrations in the blood by mobilizing fat stores from adipose tissue. These actions provide additional fuel molecules for the rapidly metabolizing muscle cells.

The Effects of Cortisol

Cortisol (also known as hydrocortisone) is one of the principal steroid hormones produced by the adrenal cortex. It belongs to a group of hormones known as **glucocorticoids.** Cortisol reinforces the actions of epinephrine and norepinephrine by making nutrients available for gluconeogenesis; it increases the transport of amino acids into liver cells, and increases the liver enzymes necessary to convert amino acid skeletons to glucose. These actions contribute to the formation of large amounts of glycogen in the liver and help to ensure adequate fuel supplies for the cells, especially when the body is under stress.

Cortisol also reduces inflammation by inhibiting the synthesis of prostaglandins and decreasing the permeability of capillary walls, thereby reducing swel-

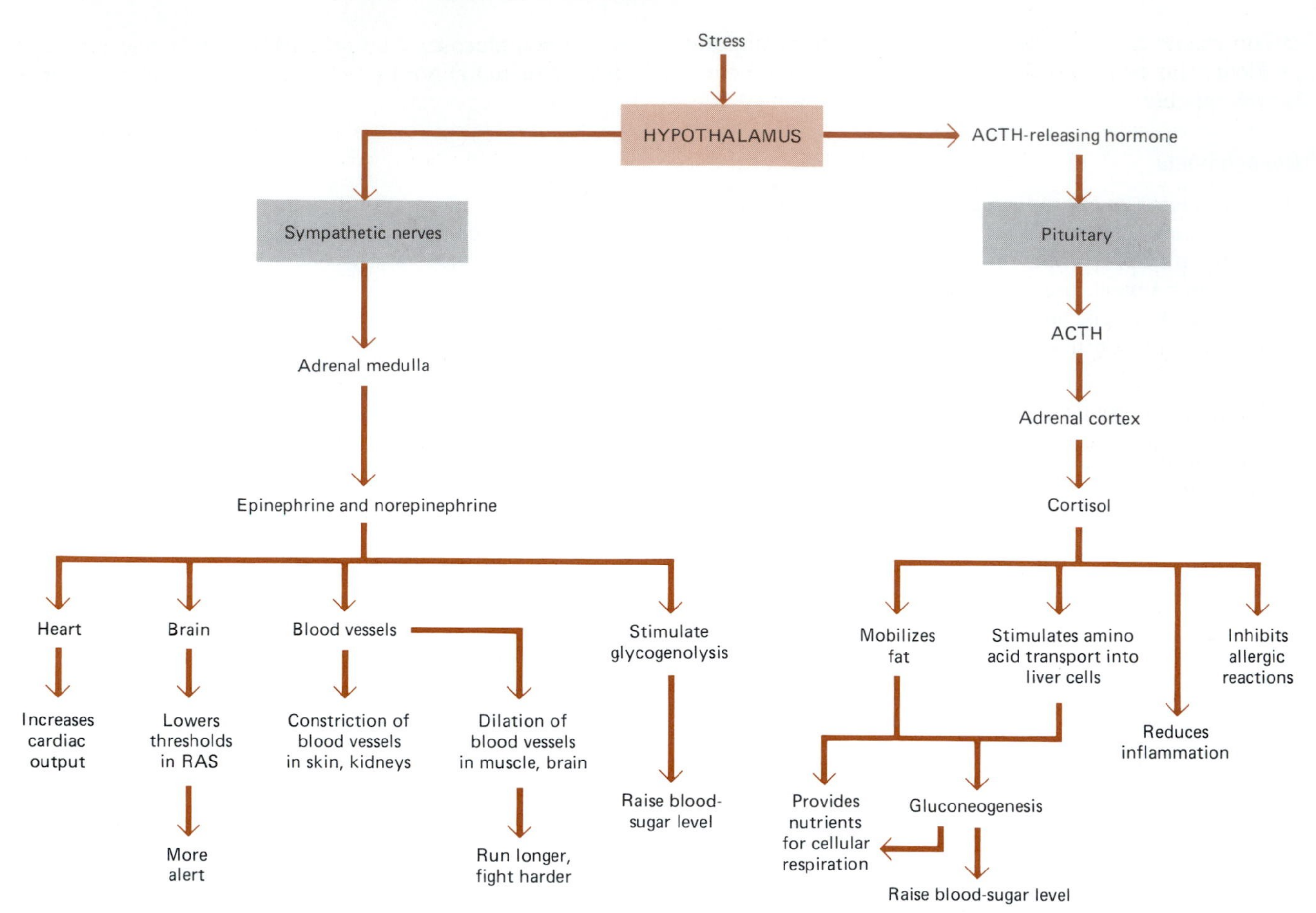

Figure 42–23 Some effects of stress.

ling. Cortisol is thought to help stabilize lysosome membranes so that their potent enzymes do not leak out and damage tissues. Because of these actions, cortisol and related glucocorticoids are used clinically to reduce inflammation in allergic reactions, arthritis, and certain types of cancer and with caution in infections.

Any type of stress stimulates the hypothalamus to secrete adrenocorticotropic hormone–releasing hormone, abbreviated ACTH-releasing hormone (Fig. 42–23). This hormone stimulates certain cells in the anterior lobe of the pituitary to secrete **adrenocorticotropic hormone,** or **ACTH.** In turn, this tropic hormone stimulates the adrenal cortex to grow and to increase its output of cortisol. So potent is the effect of ACTH that it can result in as much as a 20-fold increase in cortisol secretion within minutes. Cortisol has direct negative feedback effects on both the hypothalamus and the pituitary.

During prolonged stress, such as might be caused by a chronic disease or an unhappy marriage, the blood levels of adrenal hormones, especially of cortisol, remain elevated. Blood pressure may remain abnormally high, and metabolism in general remains geared to help the body cope with stress. Such chronic stress is thought to be harmful because of the side effects of long-term elevated levels of adrenal hormones. Chronic high blood pressure may contribute to heart disease, and increased levels of fat in the blood may promote atherosclerosis. High levels of cortisol over a long period of time can interfere with normal immune responses, increasing susceptibility to infection. Such effects are seen when large doses of steroids are administered clinically to human patients, and when experimental animals are injected with large amounts of cortisol. Among the diseases that have been linked to excessive amounts of cortisol (or the closely related steroids used clinically) are ulcers, hypertension, atherosclerosis, and possibly diabetes mellitus.

SUMMARY

I. Hormones are chemical messengers that help to regulate homeostasis; many hormones are produced by endocrine glands and transported to their target tissues by the blood.

II. Hormones may be steroids, proteins, or derivatives of fatty acids or amino acids.

III. Special receptors in the cells of target tissues bind specific hormones.

A. Steroid hormones pass through the plasma membrane of a target cell and combine with specific receptors in the cytoplasm. These hormone-receptor complexes are thought to activate specific genes.

B. Many protein hormones combine with receptors in the cell membrane of the target cell and act by way of a second messenger such as cyclic AMP.

C. Prostaglandins may help regulate hormone action by regulating cyclic AMP formation.

IV. Hormone secretion is regulated by negative feedback control mechanisms.

V. Many invertebrate hormones are secreted by neurons rather than by endocrine glands. They help to regulate regeneration, molting, metamorphosis, reproduction, and metabolism.

A. The optic glands of mollusks are endocrine organs that stimulate gonadal development.

B. Distribution of pigment in crustaceans is regulated by hormones secreted by neurosecretory cells.

C. Hormones control development in insects.

1. When stimulated by some environmental factor, neurosecretory cells in the insect brain secrete a hormone known as BH.
2. BH is stored in the corpora cardiaca; when released, it stimulates the prothoracic glands to produce molting hormone (ecdysone), which stimulates growth and molting.
3. In the immature insect the corpora allata secrete juvenile hormone, which suppresses metamorphosis at each larval molt.
4. The amount of juvenile hormone decreases with successive molts; when its concentration becomes low, the larva develops into a pupa; in its absence the pupa changes into an adult.

VI. In vertebrates the hypothalamus is the link between the nervous and endocrine systems.

A. The hypothalamus regulates the anterior lobe of the pituitary gland by producing several releasing and inhibiting hormones that regulate secretion of pituitary hormones.

B. The hypothalamus produces ADH and oxytocin, the hormones released by the posterior lobe of the pituitary.

C. Among the hormones that influence growth and development are growth hormone (GH), secreted by the anterior pituitary, and thyroid hormones, secreted by the thyroid gland.

1. Both GH and the thyroid hormones promote protein synthesis.
2. Hypersecretion of GH during childhood results in gigantism; hyposecretion may result in a pituitary dwarf. Hyposecretion of thyroid hormones during infancy and childhood may result in cretinism.

D. Glucose concentration in the blood is regulated mainly by insulin and glucagon.

1. Insulin lowers the concentration of glucose in the blood by stimulating uptake of glucose by the cells, and by promoting glycogenesis.
2. Glucagon raises the blood sugar level by stimulating glycogenolysis and gluconeogenesis.
3. In diabetes mellitus, cells are unable to utilize glucose properly and turn to fat and protein for fuel.

E. The adrenal medulla and the adrenal cortex secrete hormones that help the body cope with stress.

1. The adrenal medulla secretes epinephrine and norepinephrine, which increase heart rate, metabolic rate, and strength of muscle contraction, and re-route the blood to the organs that require more blood in time of stress.
2. The adrenal cortex releases cortisol, which promotes gluconeogenesis in the liver, thereby raising the glucose concentration; this hormone provides fuel needed for the increased metabolic activities stimulated by the adrenal medullary hormones.

POST-TEST

1. The tissues influenced by a specific hormone are referred to as its ______________ ______________.
2. Neurosecretory cells are neurons that secrete ______________.
3. A hormone may be defined as a ______________ ______________________________.
4. Many protein hormones combine with receptors in the ______________ ______________ of the target cell; the hormonal message is relayed by a second messenger such as ______________ ______________.
5. Hormone secretion is regulated by ______________ ______________ control mechanisms.
6. The insect hormone known as BH stimulates the prothoracic glands to secrete ______________ hormone.
7. Releasing and inhibiting hormones that regulate pituitary gland activity are secreted by the ______________.
8. Somatomedins mediate ______________ responses.

Select the most appropriate answer in Column B for each description in Column A; the same answer may be used more than once.

Column A	*Column B*
9. Stimulates protein synthesis	a. Norepinephrine
10. A steroid hormone	b. Growth hormone
11. Increases heart rate; raises blood pressure	c. Cortisol
12. Lowers blood glucose level	d. Glucagon
13. Helps body deal with long-term stress	e. None of the above
14. Hyposecretion may cause cretinism	
15. Used clinically to reduce inflammation	

16. ACTH stimulates the adrenal ______________ to increase secretion of ______________.

17. Acromegaly results from ______________ secretion of ______________.

18. The ______________ hormones stimulate the rate of metabolism.

19. In diabetes mellitus, there are a decreased utilization of ______________ and an increased utilization of fat and ______________.

20. When the concentration of thyroid hormones in the blood rises above normal levels, the ______________ is inhibited and slows its release of ______________.

REVIEW QUESTIONS

1. Why is it necessary to modify the traditional definition of a hormone?
2. How are most hormones transported? How do their target tissues recognize them?
3. Draw a diagram illustrating how a steroid hormone can activate a gene.
4. What is the role of cyclic AMP in hormone action?
5. Why are prostaglandins of interest to both endocrinologists and drug companies?
6. Draw a diagram illustrating how negative feedback regulates the secretion of the following hormones:
   a. parathyroid hormone
   b. cortisol
   c. thyroid hormones
7. Draw a diagram illustrating the hormonal control of development in insects.
8. Locate each of the following endocrine glands in the human body:
   a. pituitary
   b. thyroid
   c. parathyroids
   d. islets of Langerhans
   e. adrenal glands
9. List the hormones secreted by each of the following glands, and give their actions:
   a. the posterior lobe of the pituitary
   b. the anterior lobe of the pituitary
   c. the thyroid
   d. the islets of Langerhans
   e. the adrenal medulla
   f. the adrenal cortex
10. Explain how the hypothalamus influences endocrine activity.
11. How do GH and thyroid hormones stimulate growth?
12. Describe the abnormalities of growth associated with hyposecretion or hypersecretion of GH and thyroid hormones.
13. Describe the role of insulin and glucagon in regulating the concentration of glucose in the blood.
14. What physiological disturbances are associated with diabetes mellitus?
15. How do the adrenal glands help the body to cope with stress?

43
Reproduction

OUTLINE

I. Asexual reproduction
II. Sexual reproduction
III. Reproductive systems
IV. Some reproductive variations
  A. Metagenesis
  B. Parthenogenesis
  C. Hermaphroditism
V. Human reproduction
  A. The male reproductive system
  B. The female reproductive system
  C. Reproductive hormones in the male
  D. Hormonal control of the menstrual cycle
  E. Physiology of sexual response
  F. Fertilization
  G. Sterility
VI. Birth control
  A. Oral contraceptives: the pill
  B. The intrauterine device (IUD)
  C. Other common methods
  D. Sterilization
  E. Abortion

LEARNING OBJECTIVES

After you have read this chapter you should be able to:

1. Compare asexual and sexual reproduction, and give two examples of asexual reproduction.
2. Give examples of metagenesis, parthenogenesis, and hermaphroditism.
3. Trace the passage of sperm cells through the male reproductive system from their origin in the seminiferous tubules until they leave the body in the semen.
4. Label the structures of the male reproductive system on a diagram, and give the functions of each.
5. Label the structures of the female reproductive system on a diagram, and give the functions of each.
6. Trace the development of an ovum and its passage through the female reproductive system until it is fertilized.
7. Describe the actions of testosterone and of the gonadotropic hormones in the male.
8. Describe the hormonal regulation of the menstrual cycle, and identify the time of important events of the cycle such as ovulation and menstruation.
9. Describe the physiological changes that occur during sexual response in the male and female.
10. List the functions of fertilization, and describe the process.
11. Compare the methods of birth control in Table 43–3 with respect to mode of action, effectiveness, advantages and disadvantages.

If there is any one feature of a living system that is most unique, it is its ability to reproduce and perpetuate the species. The survival of each species requires that its individual members multiply, producing new individuals to replace those that die.

At the molecular level, reproduction is a function of the unique capacity of the nucleic acids to replicate themselves. At the level of the whole organism, reproduction ranges from the simple fission of bacteria or protists to the incredibly complex structural, functional, and behavioral processes of reproduction in higher animals. Reproduction in numerous organisms has been discussed throughout this book. In this chapter we will summarize some major features of animal reproductive processes and then will focus upon human reproduction.

Asexual Reproduction

In **asexual reproduction** a single parent splits, buds, or fragments to give rise to two or more offspring that have hereditary traits identical with those of the parent. Sponges and cnidarians can reproduce by **budding,** in which a small part of the parent's body separates from the rest and develops into a new individual (Fig. 43–1). It may split away from the parent and establish an independent existence or it may remain attached and become a more-or-less independent member of a colony.

Salamanders, lizards, sea stars, and crabs can grow a new tail, leg, arm, or certain other organs if the original one is lost. In some cases, this ability to regenerate a part becomes a method of reproduction. The body of the parent may break into several pieces; each piece then regenerates the missing parts and develops into a whole animal. Such reproduction by **fragmentation** is common among flatworms. Similarly, oyster farmers learned long ago that when they tried to kill sea stars by chopping them in half and throwing the pieces back into the sea, the number of sea stars preying on the oyster bed doubled! A sea star can, in fact, regenerate an entire new individual from a single arm.

Sexual Reproduction

Most animals reproduce most of the time by sexual means. Sexual reproduction involves two parents. Each contributes a specialized **gamete** (an egg or sperm); these fuse to form the fertilized egg, or **zygote.** The egg is typically large and nonmotile, with a store of nutrients to support the development of the embryo. The sperm is usually small and motile, adapted to swim actively by beating its long, whiplike tail.

Sexual reproduction has the biological advantage of promoting genetic variety among the members of a species, because the offspring is the product of genes contributed by both parents, rather than a genetic copy of a single individual. By making possible the genetic recombination of the inherited traits of two parents, sexual reproduction gives rise to offspring that may be better able to survive than either parent. Advantageous genes can spread rapidly through a population that reproduces sexually. This permits the rapid and effective spread of adaptations that enable a species to survive in an ever-changing environment.

Fertilization, the fusion of sperm and egg, may take place inside the body—**internal fertilization**—or outside—**external fertilization** (Fig. 43–2). Most aquatic animals practice external fertilization. Mating partners usually release eggs and sperm into the water simultaneously. Many gametes are lost—to predators, for

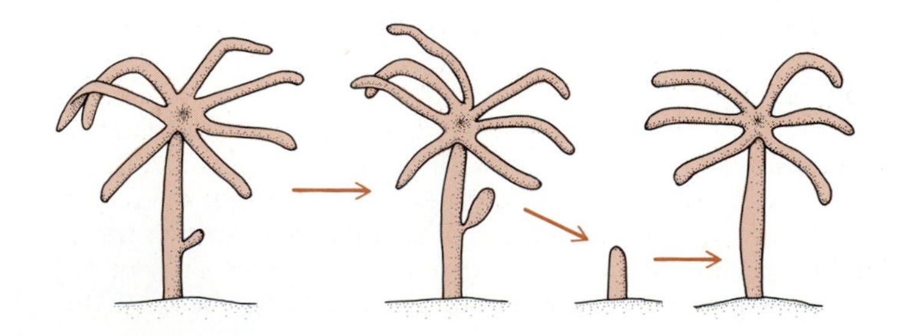

Figure 43–1 *Hydra* and many other organisms reproduce asexually by budding. A part of the body grows outward and then separates and develops into a new individual. The portion of the parent body that buds is not specialized exclusively for performing a reproductive function.

(a)

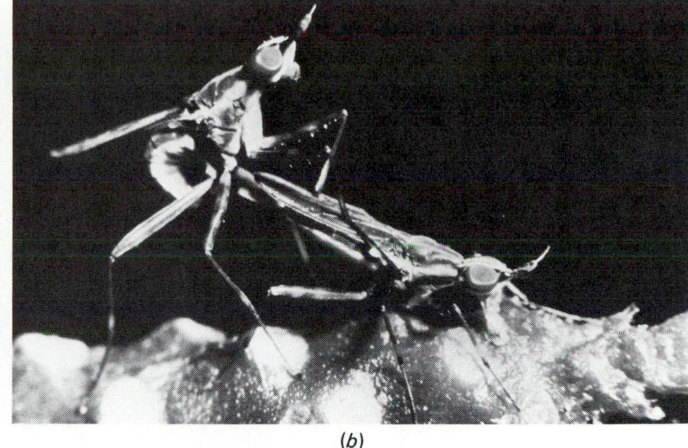
(b)

(c)

Figure 43–2 External fertilization (*a*) and internal fertilization (*b*). (*a*) Spawning pickerel frogs. Most amphibians must return to water for mating. The female lays a mass of eggs, while the male mounts her and simultaneously deposits his sperm in the water. (*b*) A male depositing sperm in the genital orifice of a female fly while she disinterestedly chews on a rotten twig. (*c*) A butterfly, *Heliconius ismenius*, ovipositing (depositing eggs) on a passionflower plant. ((*a*), E. R. Degginger; (*b*), (*c*), L. E. Gilbert, University of Texas at Austin/BPS.)

example—but so many are released that a sufficient number of sperm and egg cells meet and unite to perpetuate the species.

In internal fertilization, matters are left less to chance. The male generally delivers sperm cells directly into the body of the female. Her moist tissues provide the watery medium required for movement of sperm. Most terrestrial animals, as well as a few fish and some other aquatic animals, practice internal fertilization.

Reproductive Systems

Although the details of the reproductive process vary tremendously from one organism to another, some generalizations about animal reproductive systems may be helpful in understanding the variations (Fig. 43–3). The basic components of the male reproductive system are the male gonad, or **testis,** in which sperm are produced, and the **sperm duct** for the transport of sperm to the exterior of the body. Parts of the sperm duct, or areas adjacent to it, may be modified for specific functions. A part of the duct may be given over for sperm storage; such a part is often called a seminal vesicle. (The human organ termed the seminal vesicle, however, does *not* store sperm.) There also may be glandular areas for the production of semen, which serves as a vehicle for the sperm and may also activate, nourish, and protect them. The terminal part of the sperm duct may open onto or into a copulatory organ, a **penis,** which in internal fertilization provides for the transfer of sperm to the female.

The basic parts of the female system are the female gonad, or **ovary,** and the **oviduct,** a tube for the transport of eggs. Like the sperm duct, the oviduct may be modified in any of several ways in different species. It may have a glandular portion for the secretion of an egg shell, a case, or a cocoon. A section of the oviduct, a **seminal receptacle,** may be modified for the storage of sperm following their transfer from the male. Another portion of the oviduct may be modified as a **uterus** for egg storage or for the development of the fertilized egg within the body of the

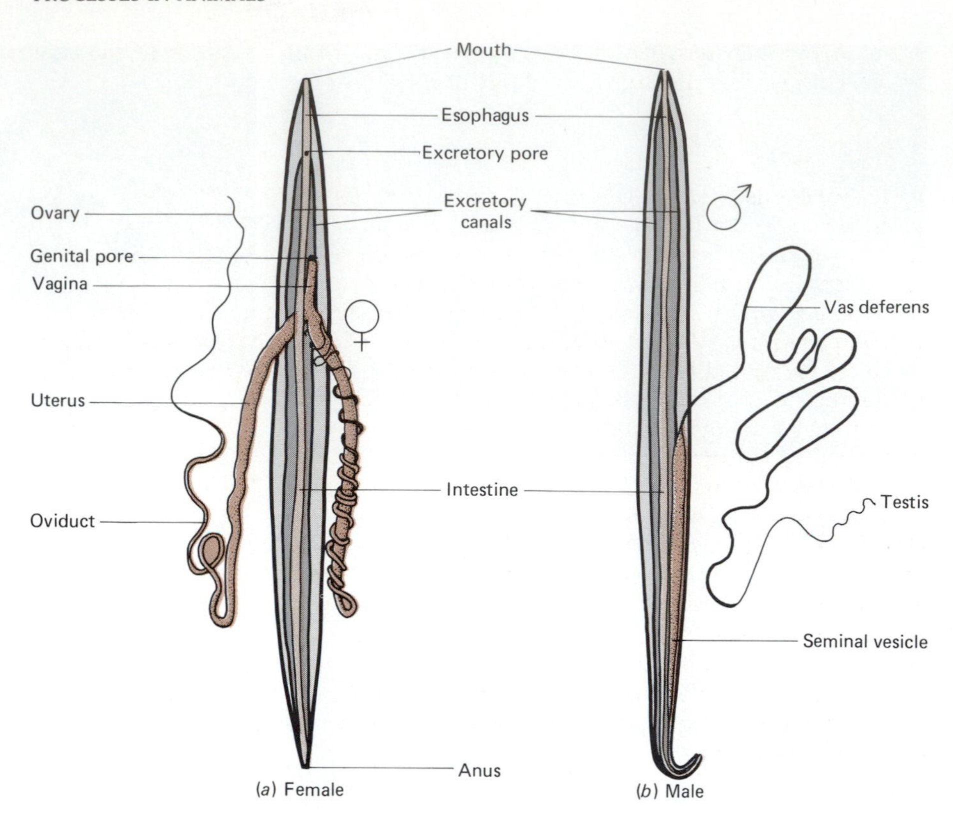

Figure 43–3 Male and female reproductive systems in *Ascaris*.

female. The terminal portion of the oviduct may be adapted as a **vagina** for receiving the male copulatory organ.

The male and female systems of most animals do not possess all these modifications, and these structures cannot be arranged in any kind of progressive sequence of evolutionary changes. Some otherwise simple animals have extremely complex reproductive systems. This is especially likely to be true of parasites. The presence or absence of some specific adaptation of the reproductive system is correlated with the circumstances of reproduction—whether the animal lives in the sea, in fresh water, or on land; whether fertilization is external or internal; whether the eggs are liberated singly into the water to develop or are deposited onto the ocean floor or river bottom within envelopes; and whether or not the developing individual passes through a larval stage.

Most species of fishes utilize a simple form of reproduction in which the female sheds into the water enormous numbers of eggs, which may or may not be fertilized by sperm from some passing male. Thereafter, both male and female fishes ignore the eggs. In such fishes, the uterus is missing and the male does not have a copulatory organ. In many species, however, the male undertakes much responsibility for the care of the eggs and young. In the sea catfish, for instance, the fertilized eggs are deposited in a clump that the male picks up and carries in his mouth during most of their incubation (presumably fasting and yielding not to temptation!).

Some Reproductive Variations

Animals show tremendous diversity in their methods of reproduction. Even members of the same class may differ markedly in their reproductive processes.

METAGENESIS

Metagenesis (transformation development) refers to an alternation of asexual and sexual generations. For example, in the hydrozoan *Obelia* a polyp generation gives rise by budding to a generation of medusas (Fig. 24–17). The motile medusas produce gametes and reproduce sexually, giving rise to a new generation of polyps.

Thus, there is an alternation of generations—polyp, medusa, polyp, medusa, and so on. Both generations consist of diploid organisms.

PARTHENOGENESIS

Parthenogenesis (virgin development) is a form of reproduction in which an unfertilized egg develops into an adult animal. Parthenogenesis is common among rotifers, some gastropod mollusks, some crustaceans, insects, especially honeybees and wasps, and some reptiles. Some species of arthropods (and even some vertebrates—lizards) consist entirely of females that reproduce in this way. More commonly, parthenogenesis occurs for several generations, after which males develop, produce sperm, and mate with the females to fertilize their eggs. In some species, parthenogenesis is advantageous in maintaining the social order; in others, it appears to be an adaptation for survival in times of stress or when there is a serious decrease in population.

A special form of parthenogenesis occurs in honeybees. The queen honeybee is inseminated by a male during the "nuptial flight." The sperm she receives are stored in a little pouch connected with her genital tract but closed off by a muscular valve. As the queen lays eggs, she can either open this valve, permitting the sperm to escape and fertilize the eggs, or keep the valve closed, so that the eggs develop without fertilization. Generally, fertilization occurs in the fall, and the fertilized eggs are quiescent during the winter. The fertilized eggs become females (queens and workers); the unfertilized eggs become males (drones). Some species of wasps give birth alternately to a parthenogenetic generation and a generation developing from fertilized eggs.

HERMAPHRODITISM

Many invertebrate species are **hermaphroditic,** meaning that a single individual produces both eggs and sperm. In some hermaphrodites, ovaries and testes are both present, so that eggs and sperm may be produced at the same time. A few, such as the parasitic tapeworms, are capable of self-fertilization. Although this form of reproduction is still classified as sexual, (since both eggs and sperms are involved), it is an exception to the important generalization that sexual reproduction involves two different individuals. Hermaphroditism of this type in tapeworms is obviously an adaptation suited to the tapeworm's mode of life, in which only a single tapeworm may infect a host, but must nonetheless be able to reproduce in order for the species to survive.

Most hermaphrodites do not reproduce by self-fertilization. Rather, as in earthworms, two animals copulate, and each inseminates the other. In some hermaphroditic species, self-fertilization is prevented by the development of testes and ovaries at different times. In the clam *Mercenaria mercenaria,* 98% of the population are males when they first reach maturity; later in their lives they produce eggs and become functional females. A few continue to produce only sperm and remain functional males. In the American oyster *Crassostrea virginica,* sperm are mainly produced when the animal first matures. The next year the animal produces eggs, and so on, in a regular annual alternation of sexes.

Certain fish exhibit a somewhat similar reproductive pattern in which sex is related to dominance. In one species of wrasse, the dominant fish is always male, and he lords his position over a harem of females. If he is removed or dies, one of the remaining female fish reverses sex and becomes the new patriarch. In another species of wrasse, the opposite situation exists: The dominant fish is always female, and the males change sex to achieve dominance if she is removed.

Human Reproduction

In human beings, as in other mammals, reproductive processes include formation of gametes, cyclic changes in the female body in preparation for pregnancy, sexual intercourse, fertilization, pregnancy, and lactation (producing milk for nourishment of the young). These events are precisely regulated and coordinated by the interaction of hormones secreted by the anterior lobe of the pituitary gland and by the gonads.

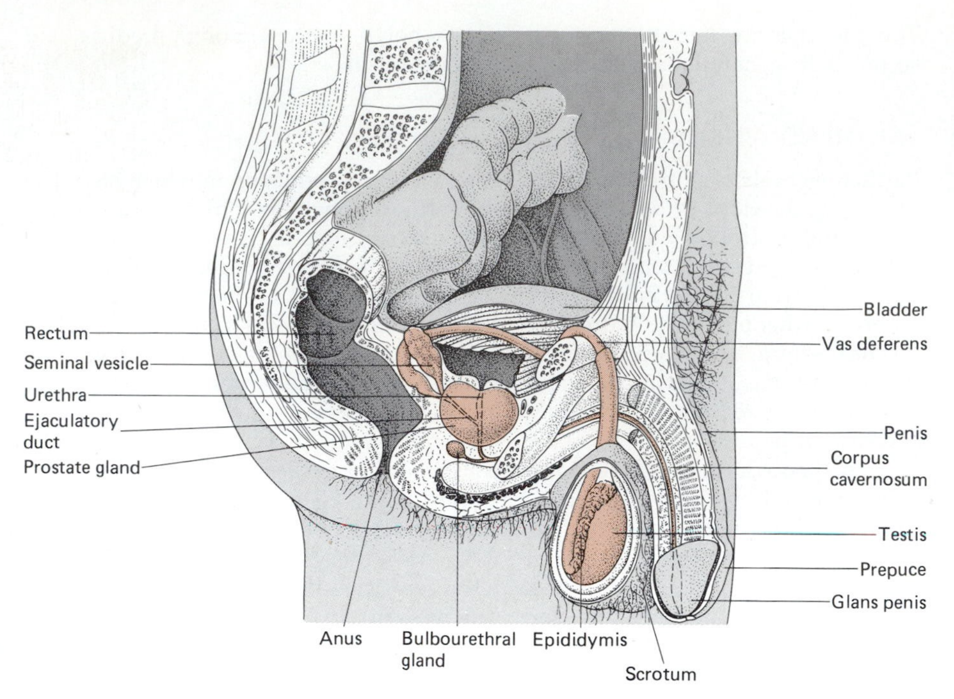

Figure 43–4 Anatomy of the human male reproductive system. The scrotum, penis, and pelvic region are shown in sagittal section to illustrate their internal structures.

THE MALE REPRODUCTIVE SYSTEM

In the adult male, millions of sperm are produced daily by the paired male gonads, the testes (Figs. 43–4 and 43–5). Each **testis** is a small oval organ about 5 cm long, packed with about 1000 coiled **seminiferous tubules.** Sperm are produced within these tubules from stem cells called **spermatogonia** found next to the wall of the tubule (Fig. 43–6). During their development sperm cells are intimately associated with large supporting cells called **Sertoli cells,** which nourish them. (Review spermatogenesis discussed in Chapter 10.) Between the tubules lie the **interstitial cells,** which produce the male sex hormone **testosterone.**

The testes develop within the abdominal cavity of the male embryo, but they descend about 2 months before birth into the **scrotum,** a skin-covered sac suspended from the groin. The scrotum is an outpocketing of the pelvic cavity and is connected to it by the **inguinal canals.** As they descend, the testes pull their blood vessels, nerves, and conducting tubes after them. These structures, encased by the cremaster muscle and by layers of connective tissue, constitute the **spermatic cord.**

The inguinal region remains a weak place in the abdominal wall. Straining the abdominal muscles by lifting heavy objects may result in tearing the inguinal tissue; a loop of intestine can then bulge into the scrotum through such a tear. This is called an **inguinal hernia.**

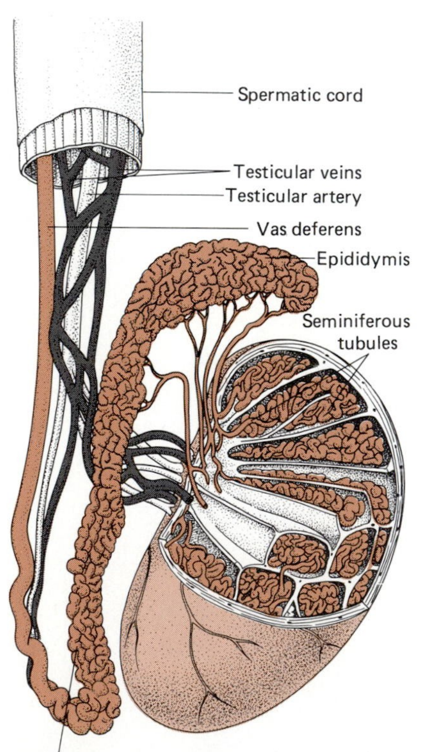

Figure 43–5 Structure of the testis, epididymis, and spermatic cord. The testis is shown in sagittal section to illustrate the arrangement of the seminiferous tubules.

The descent of the testes into the scrotum is necessary for sperm production. Sperm cells are not able to develop at body temperature, and the scrotum serves as a cooling unit, maintaining them at about 2°C below body temperature. In rare cases where the testes fail to descend, the seminiferous tubules eventually degenerate and the male becomes **sterile,** that is, unable to father offspring. The abdominal testis remains able to secrete male sex hormones, however.

Sperm cells leave the seminiferous tubules of the testes through fine tubules (vasa efferentia) that pass from the testis and empty into a larger tube, the epididymis. The **epididymis** of each testis is a complexly coiled tube (as much as 6 m long in humans) in which sperm complete their maturation and are stored.

From each epididymis a sperm duct, the **vas deferens,** passes from the scrotum through the inguinal canal and into the pelvic cavity. Each vas deferens empties into a short **ejaculatory duct,** which passes through the prostate gland and then opens into the urethra. The single **urethra,** which at different times conducts either urine or semen, passes through the penis to the outside of the body.

About 3.5 ml of **semen** is ejaculated during sexual climax. Semen consists of about 400 million sperm cells suspended in the secretions of the accessory glands. These glands include the paired **seminal vesicles,** which empty into the vasa deferentia, and the single **prostate gland,** which releases its alkaline secretion into the

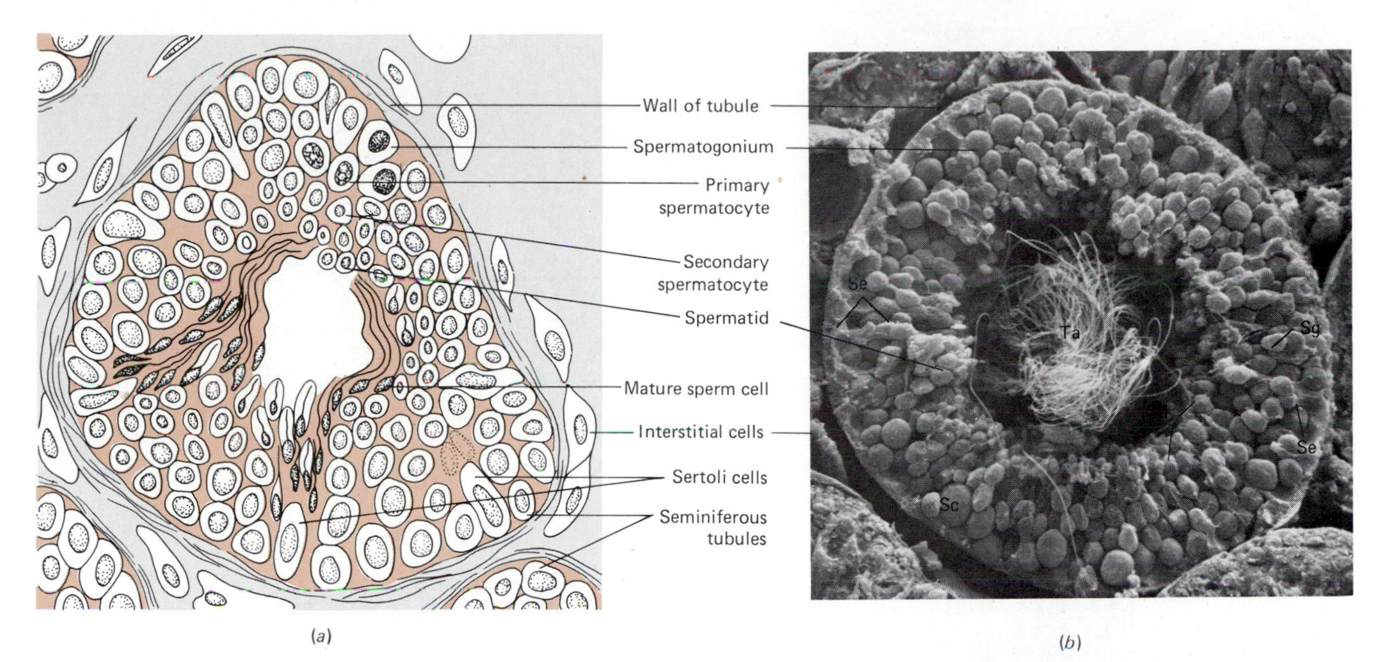

Figure 43–6 Structure of a seminiferous tubule showing developing sperm cells in various stages of spermatogenesis. (*a*) Identify the sequence of sperm cell differentiation. Note the Sertoli cells and the interstitial cells. (*b*) A scanning electron micrograph of a transverse section through a seminiferous tubule (×580). *Se*, Sertoli cell; *Sc*, primary spermatocyte; *Sg*, spermatogonium. (Richard G. Kessel and Randy H. Kardon: *Tissues and Organs: A Text-Atlas of Scanning Electron Microscopy*. San Francisco, W. H. Freeman and Co., 1979.)

urethra. During sexual arousal the paired **bulbourethral glands** release a few drops of alkaline fluid, which can neutralize the acidity of the urethra and aid in lubrication.

The **penis** is an erectile copulatory organ adapted to deliver sperm into the female reproductive tract. It consists of a long **shaft** that enlarges to form an expanded tip, the **glans.** Part of the loose-fitting skin of the penis folds down and covers the proximal portion of the glans, forming a cuff called the **prepuce,** or **foreskin.** In the operation termed **circumcision** (commonly performed on male babies for either hygienic or religious reasons), the foreskin is removed.

Under the skin, the penis consists of three parallel columns of **erectile tissue,** sometimes called the **cavernous bodies,** (corpora cavernosa) (Fig. 43–7). One of these columns surrounds the portion of the urethra that passes through the penis. The erectile tissue consists of large blood vessels called venous sinusoids. When the male is sexually stimulated, nerve impulses cause the arteries of the penis to dilate. Blood rushes into the vessels of the erectile tissue. As the erectile tissue fills with blood it swells, compressing veins that conduct blood away from the penis and slowing the outflow of blood through them. Thus, more blood enters the penis than can leave, causing the erectile tissue to become further engorged with blood. The penis thus becomes erect, that is, longer, larger in circumference, and firm. Though the human penis does not contain any bone, penis bones do occur in some other mammals, such as bats.

THE FEMALE REPRODUCTIVE SYSTEM

Like the male gonads, the female gonads, or **ovaries,** produce both gametes and sex hormones. About the size and shape of large almonds, the ovaries are located close to the lateral walls of the pelvic cavity (Figs. 43–8 and 43–9). The ovaries are held in position by several connective tissue ligaments. Each ovary is covered with a single layer of epithelium. Internally the ovary consists mainly of connective tissue called **stroma,** through which are scattered **ova** (eggs) in various stages of maturation (see Fig. 43–12). (You may want to reread the discussion of oogenesis in Chapter 10.)

Each month an ovum is released from one of the ovaries into the pelvic cavity, a process called **ovulation.** Almost immediately the ovum passes into the funnel-shaped opening of the **uterine tube** (also called fallopian tube or oviduct). Peristaltic contractions of the muscular wall of the uterine tube and action of the cilia in its lining help to move the ovum along toward the uterus. Fertilization takes place within the uterine tube. If fertilization does not occur, the ovum degenerates there.

The uterine tubes empty into the upper corners of the pear-shaped uterus (see Fig. 43–9). About the size of a fist, the **uterus** occupies a central position in the pelvic cavity. This organ has thick walls of smooth muscle and a mucous lining, the

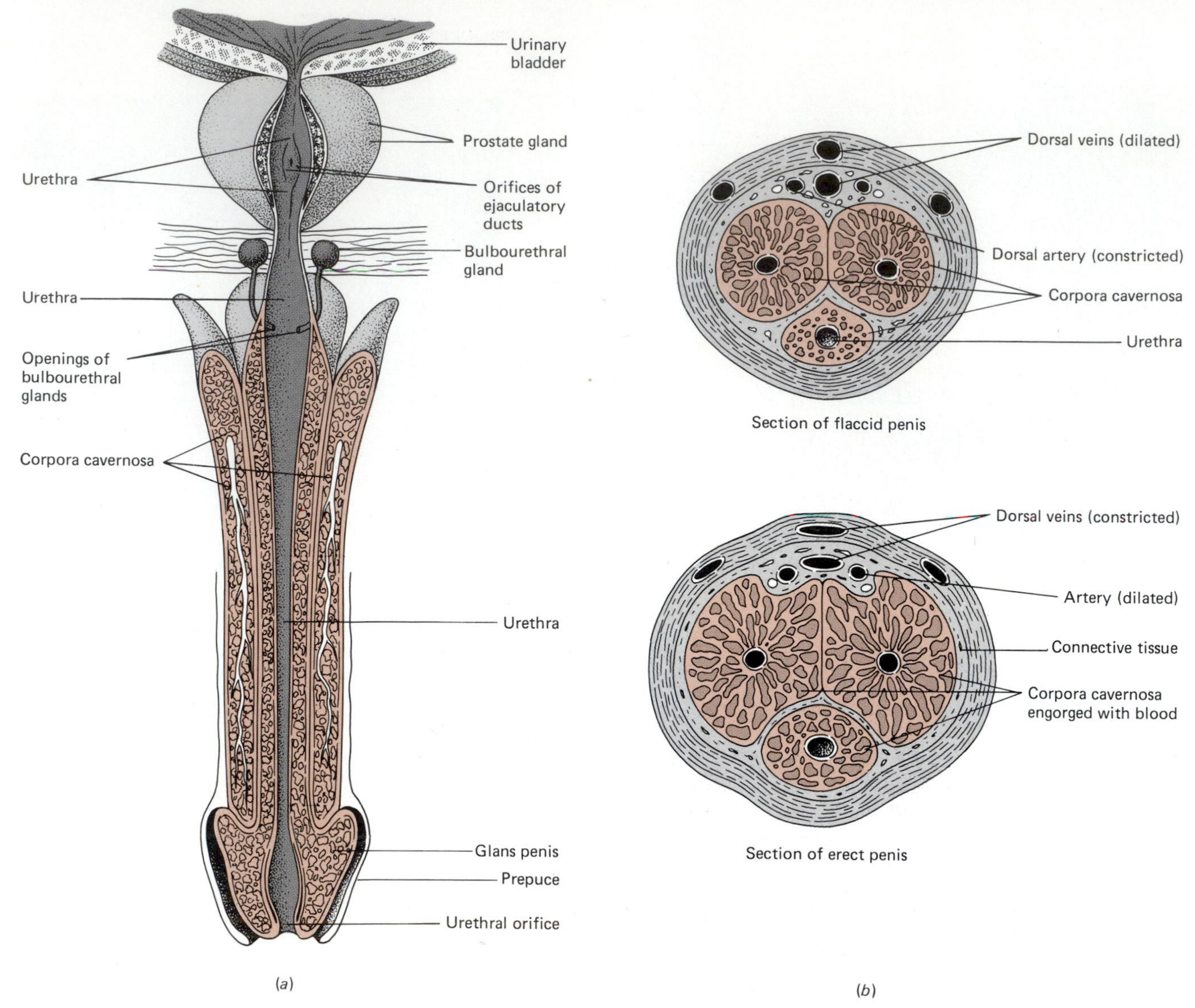

Figure 43–7 Internal structure of the penis. (*a*) Longitudinal section through the prostate gland and penis. (*b*) Cross section through flaccid and erect penis. Note that the erectile tissues of the corpora cavernosa (cavernous bodies) are engorged with blood in the erect penis.

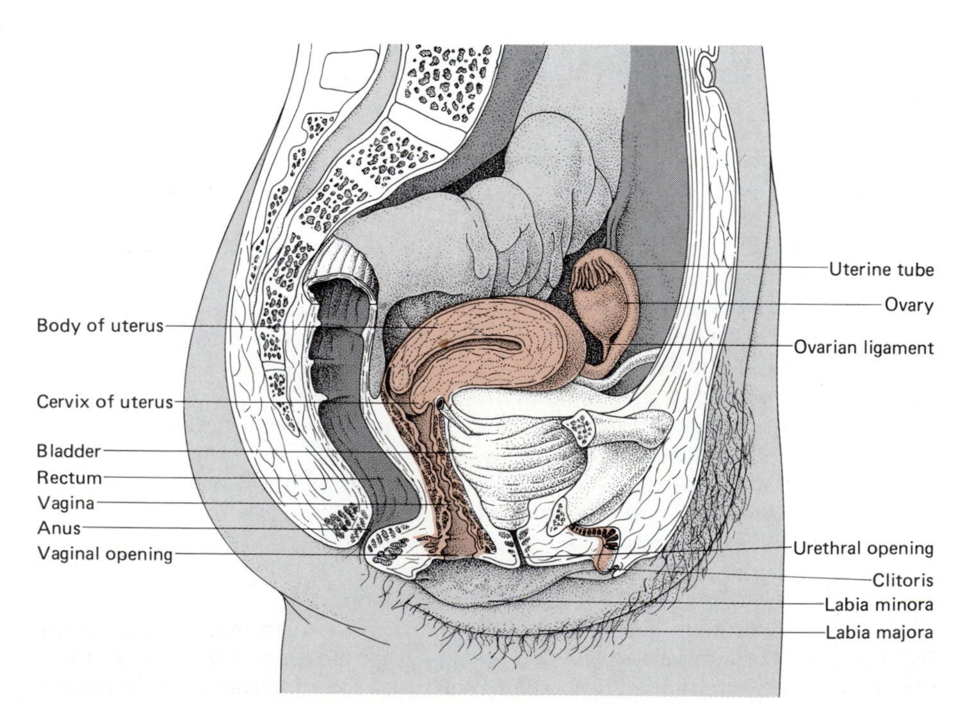

Figure 43–8 Midsagittal section of female pelvis showing reproductive organs. Note the position of the uterus relative to the vagina.

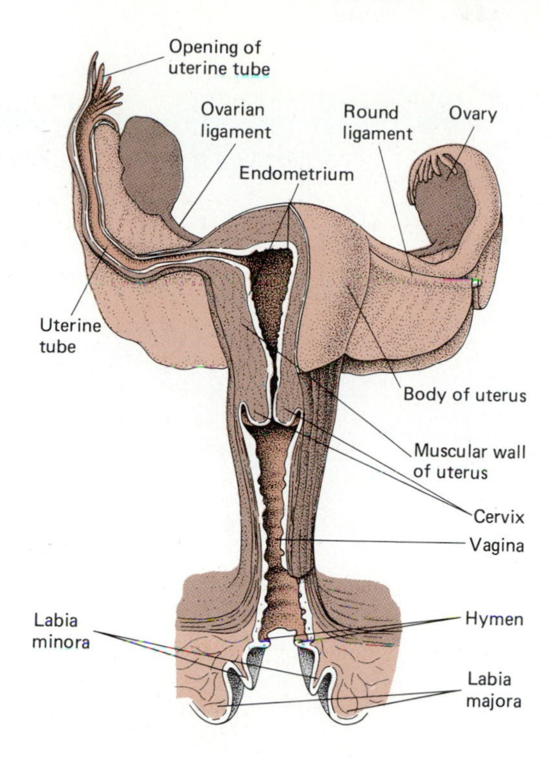

Figure 43–9 Anterior view of female reproductive system. Some organs have been cut open to expose the internal structure.

endometrium, which thickens each month in preparation for possible pregnancy. If an ovum is fertilized, the tiny embryo finds its way into the uterus and is implanted in the endometrium. Here it grows and develops, sustained by nutrients and oxygen delivered by surrounding maternal blood vessels. If fertilization does not occur during the monthly cycle, the endometrium sloughs off and is discharged. This process is called **menstruation.**

The lower portion of the uterus, called the **cervix,** projects slightly into the vagina. The cervix is a common site of cancer in women. Detection is usually possible by the routine Papanicolaou test (Pap smear) in which a few cells are scraped from the cervix during a regular gynecological examination and studied microscopically. About 50% of cervical cancer is now detected in this way at very early stages of malignancy, when the patient can still be cured.

The **vagina** is a single, elastic, muscular tube that extends from the uterus to the exterior of the body. The vagina serves as a receptacle for sperm during sexual intercourse and as part of the birth canal when development of the fetus is complete.

The external female sex organs, collectively known as the **vulva,** include liplike folds, the **labia minora,** which surround the opening to the vagina and the opening to the urethra (Fig. 43–10). External to the delicate labia minora are the heavier, hairy **labia majora.** Anteriorly the labia minora merge to form the prepuce of the **clitoris,** a very small erectile structure comparable to the male glans penis. Like the penis, the clitoris contains erectile tissue that becomes engorged with blood during sexual excitement. Rich in nerve endings, the clitoris serves as a center of sexual sensation in the female.

The **hymen** is a thin ring of tissue that may partially block the entrance to the vagina. It is often ruptured during a woman's first sexual intercourse. However, the hymen may be destroyed by strenuous exercise during childhood, or by the use of tampons inserted into the vagina to absorb the menstrual flow.

The **mons pubis** is the mound of fatty tissue just above the clitoris at the junction of the thighs and torso. At puberty it becomes covered by coarse pubic hair. Although the **breasts** are also female reproductive organs, they are discussed in the section on lactation in Chapter 44.

REPRODUCTIVE HORMONES IN THE MALE

At the age of about 10 years, the hypothalamus begins to mature in its function of regulating sex hormones. It secretes releasing hormones, which stimulate the anterior pituitary to secrete the gonadotropic hormones **follicle-stimulating hormone**

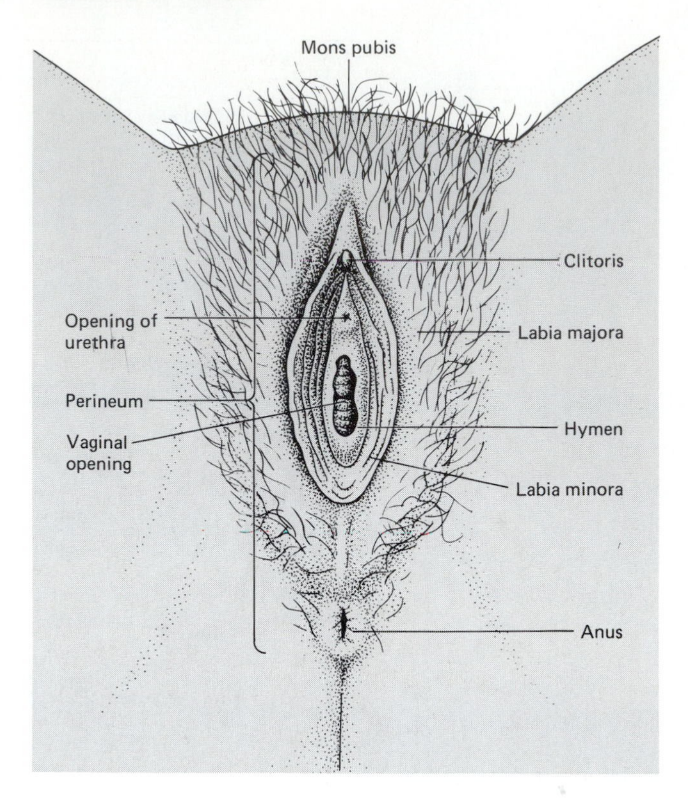

Figure 43–10 The vulva, the external genital structures of the female.

(FSH) and **luteinizing hormone (LH).** FSH stimulates development of the seminiferous tubules and promotes spermatogenesis (Table 43–1). LH stimulates the interstitial cells to secrete the male hormone testosterone.

Testosterone is responsible for the adolescent growth spurt, which occurs at about age 13 years. This hormone stimulates growth of the male reproductive organs and is also responsible for the secondary sexual characteristics that develop at puberty. The beard begins to grow, and pubic and axillary hair appears. Vocal cords increase in length and thickness, causing the voice to deepen, and muscle development is stimulated. If the testes are removed—a procedure known as **castration**—before puberty, the male becomes a **eunuch,** a person whose secondary sexual characteristics do not develop because he has been deprived of the male sex hormone that is secreted by his testes.

TABLE 43–1
Principal Reproductive Hormones and Their Actions in Males

Endocrine Gland and Hormones	*Principal Site of Action*	*Principal Actions*
Anterior pituitary		
Follicle-stimulating hormone (FSH)	Testes	Stimulates development of seminiferous tubules; may stimulate spermatogenesis
Luteinizing hormone (LH) (also called interstitial cell–stimulating hormone (ICSH)	Testes	Stimulates interstitial cells to secrete testosterone
Testes		
Testosterone	General	Before birth: stimulates development of primary sex organs and descent of testes into scrotum At puberty: responsible for growth spurt; stimulates development of reproductive structures; stimulates development of secondary sex characteristics (male body build, growth of beard, deep voice, etc.) In adult: responsible for maintaining secondary sex characteristics; may stimulate spermatogenesis

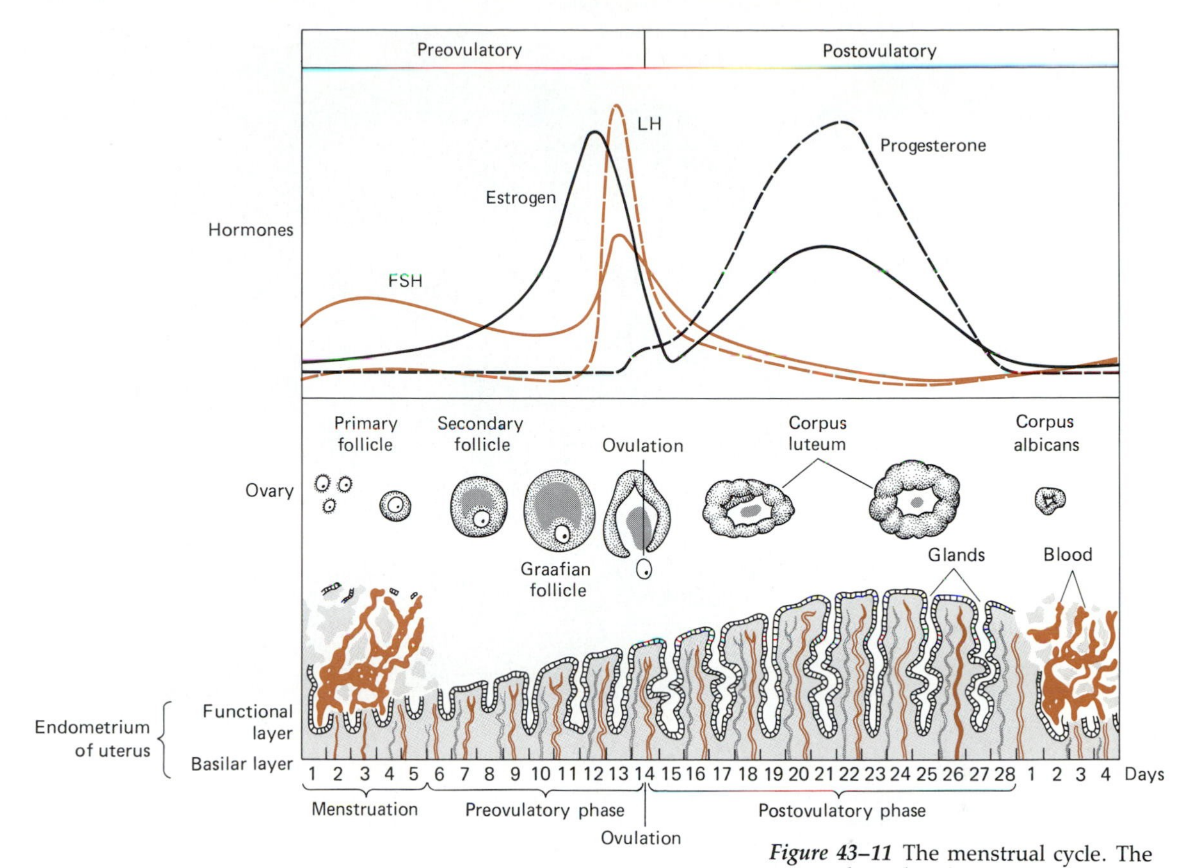

Figure 43–11 The menstrual cycle. The events that take place within the pituitary, ovary, and uterus are precisely synchronized. When fertilization does not occur, the cycle repeats itself about every 28 days. Compare this illustration with Figure 43–13.

HORMONAL CONTROL OF THE MENSTRUAL CYCLE

As a female approaches puberty the anterior pituitary gland secretes the gonadotropic hormones FSH and LH, which signal the ovaries to become active. Interaction of FSH and LH with estrogen and progesterone from the ovaries regulates the **menstrual cycle,** the monthly chain of events that prepares the body for possible pregnancy. The menstrual cycle runs its course every month from puberty until **menopause,** the end of a women's reproductive (though not sexually active) life.

Although there is wide variation, a typical menstrual cycle is 28 days long (Fig. 43–11). The first day of the cycle is marked by menstruation, the monthly discharge through the vagina of blood and tissue from the endometrium. Ovulation occurs on about the 14th day of the cycle.

During the **menstrual phase** of the cycle, which lasts about 5 days, the pituitary gland releases FSH. This hormone stimulates a few follicles to develop in the ovary. As an oocyte develops, it becomes separated from its surrounding follicle cells by a thick membrane, the **zona pellucida.** The follicle cells themselves proliferate so that the follicle grows in size (Fig. 43–12). As the follicle develops, follicle cells secrete fluid, which collects in a space (antrum) created between them. Connective tissue from the stroma surrounds the follicle cells; this tissue contains cells that secrete the steroid sex hormones called **estrogens**[1] during the preovulatory phase of the menstrual cycle.

As a follicle matures, it moves closer to the surface of the ovary, eventually resembling a fluid-filled blister on the ovarian surface. Such mature follicles are called **Graafian follicles.** Normally only one follicle fully matures each month. Several others may develop for about a week, then deteriorate. These remain in the ovary as atretic (degenerate) follicles.

Estrogen stimulates growth of the endometrium, which begins to thicken and develop new blood vessels and glands. The rise in estrogen levels also stimulates the anterior lobe of the pituitary to secrete LH. The LH together with the FSH stimulates ovulation. LH then stimulates the portion of the follicle that remains in the ovary after the ovum has been ejected to develop into a corpus luteum.

[1] Estrogens are a group of closely related 18-carbon steroid hormones. The principal estrogen is β-estradiol.

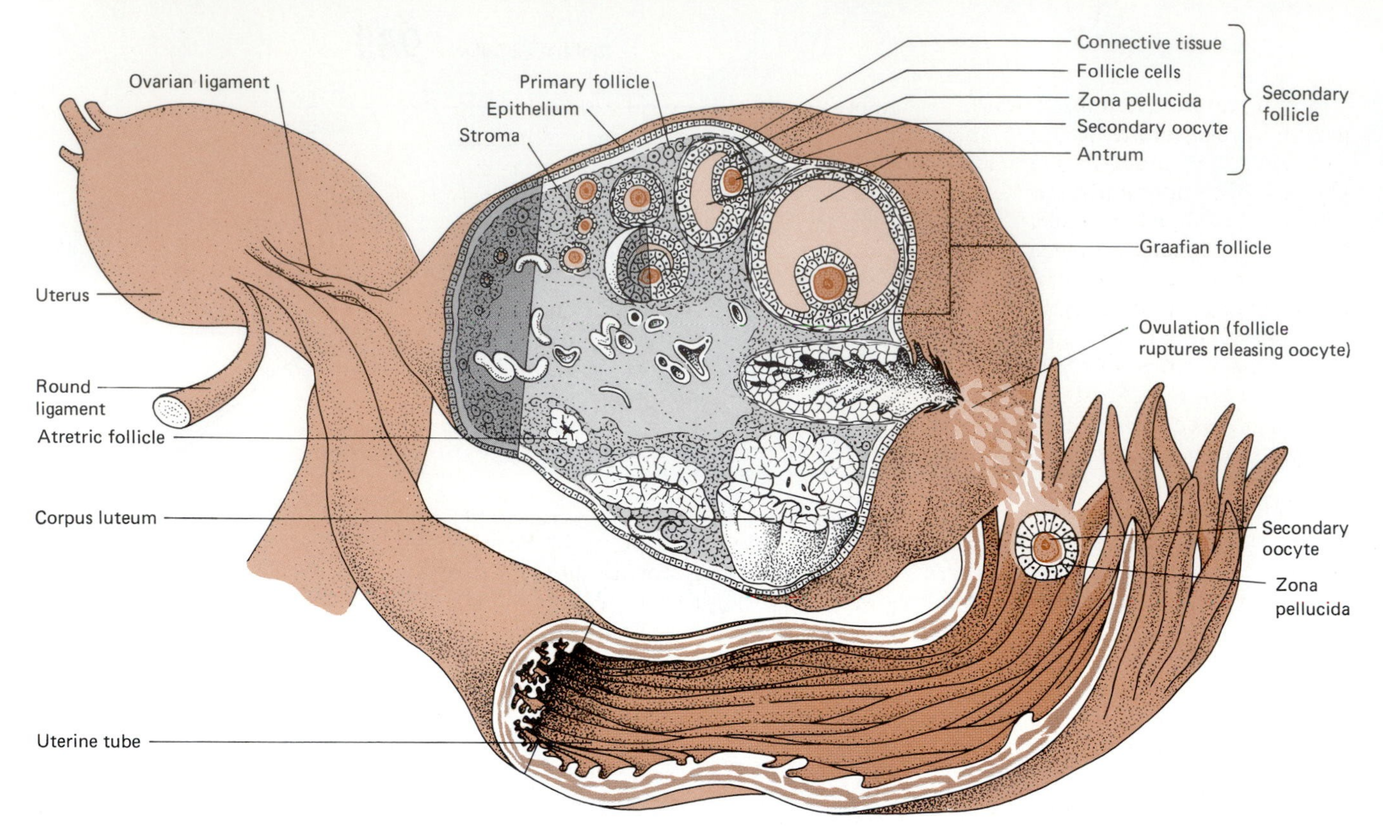

Figure 43–12 Microscopic structure of the ovary. Follicles in various stages of development are scattered throughout the ovary.

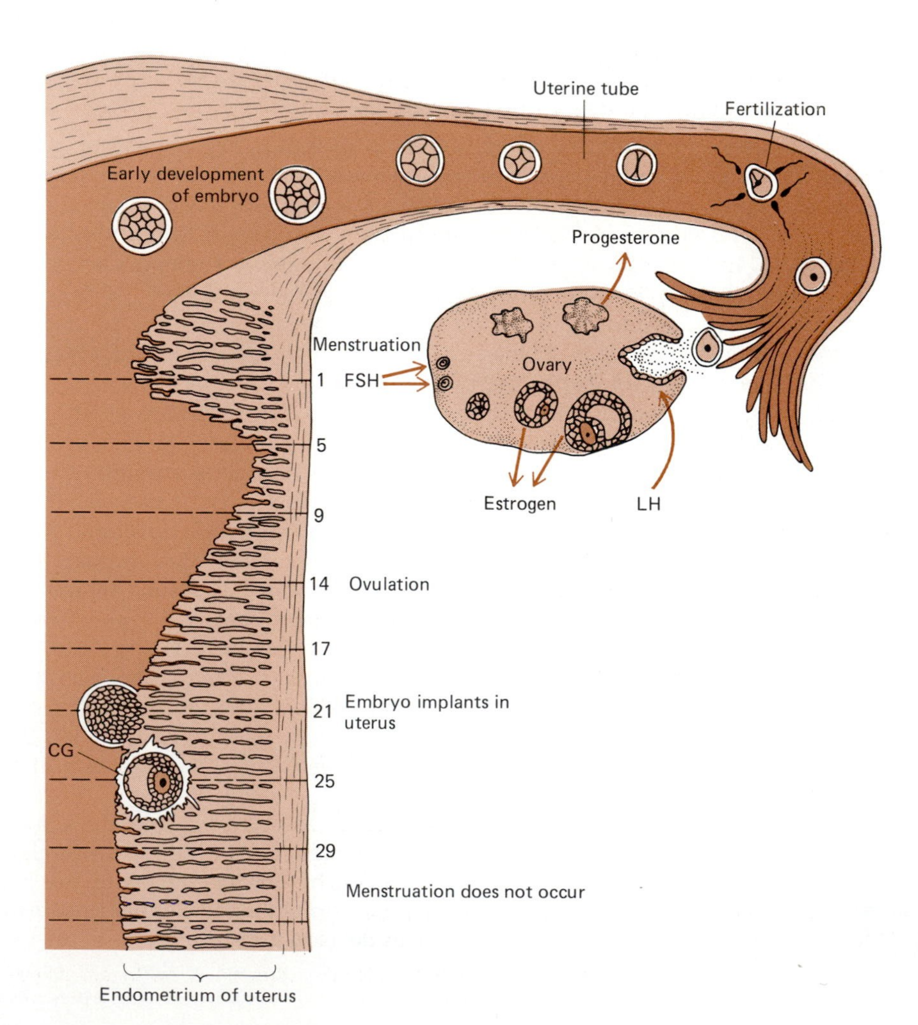

Figure 43–13 The menstrual cycle is interrupted when pregnancy occurs. The corpus luteum does not degenerate, and menstruation does not take place. The wall of the uterus remains thickened so that the embryo can develop within it.

The **corpus luteum** is a temporary endocrine gland that produces both estrogen and **progesterone** during the **postovulatory phase.** These hormones stimulate the uterus to continue its preparation for pregnancy. Progesterone stimulates tiny glands in the endometrium to secrete a fluid rich in nutrients. If the ovum is fertilized, this nutritive fluid nourishes the early embryo when it arrives in the uterus on about its fourth day of development (Fig. 43–13). On about the seventh day after fertilization the embryo begins to implant in the thick endometrium. Membranes that develop around the embryo secrete a hormone called **human chorionic gonadotropin (HCG),** which signals the mother's corpus luteum to continue to function.

If the ovum is not fertilized, the corpus luteum begins to degenerate and the concentrations of progesterone and estrogens in the blood fall dramatically. Small arteries in the endometrium constrict, and the tissue they supply becomes ischemic (oxygen-deprived). As cells die and damaged arteries rupture and bleed, menstruation begins once again.

Table 43–2 lists the actions of the pituitary and ovarian reproductive hormones. Note that like testosterone in the male, estrogens are responsible for the growth of the sex organs at puberty, for body growth, and for the development of secondary sexual characteristics. In the female these include breast development, broadening of the pelvis, and the characteristic development and distribution of muscle and fat responsible for the shape of the female body.

Some testosterone is produced in the female, mainly in the adrenal cortex. This hormone is largely responsible for the development of pubic and axillary hair in the female (as well as in the male).

PHYSIOLOGY OF SEXUAL RESPONSE

Human reproduction is accomplished sexually by the union of ovum and sperm. During copulation, also called **coitus** or sexual intercourse in humans, the male deposits semen into the upper end of the vagina. The complex structures of the male and female reproductive systems and the complex physiological, endocrine, and psychological phenomena associated with sexual activity can be looked upon as adaptations that promote the successful union of sperm and ovum and the subsequent development and nurturing of the resulting embryo.

Sexual stimulation results in two basic physiological responses: vasocongestion and increased muscle tension (myotonia). **Vasocongestion** occurs as blood flow

TABLE 43–2
Principal Reproductive Hormones and Their Actions in Females

Endocrine Gland and Hormone	*Principal Target Tissue*	*Principal Actions*
Anterior pituitary		
Follicle-stimulating hormone (FSH)	Ovary	Stimulates development of follicles; with LH, stimulates secretion of estrogen and ovulation
Luteinizing hormone (LH)	Ovary	Stimulates ovulation and development of corpus luteum
Prolactin	Breast	Stimulates milk production (after breast has been prepared by estrogens and progesterone)
Ovary		
Estrogens	General	Growth of body and sex organs at puberty; development of secondary sex characteristics (breast development, broadening of pelvis, distribution of fat and muscle)
	Reproductive structures	Maturation; monthly preparation of the endometrium for pregnancy; makes cervical mucus thinner and more alkaline
Progesterone	Uterus	Completes preparation of and maintains endometrium for pregnancy
	Breasts	Stimulates development of mammary glands

is increased to the genital structures and to certain other tissues such as the breasts and skin. During vasocongestion, erectile tissues within the penis and clitoris, as well as in other areas of the body, become engorged with blood.

Sexual response can be divided into four phases: **excitement, plateau, orgasm,** and **resolution.** Sexual excitement can result from psychological or physical stimulation, or usually both. Such stimulation, often referred to as sexual foreplay, promotes vasocongestion and increased muscle tension. Before the penis can enter the vagina and function in coitus it must be erect; accordingly, penile erection is the first male response to sexual excitement. In the female, vaginal lubrication is the first response to effective sexual stimulation; the lubricating fluid is produced in the vaginal wall owing to vasocongestion (glands are not present in the vagina). During the excitement phase, the vagina lengthens and expands in preparation for receiving the penis; the clitoris and breasts become vasocongested, and the nipples become erect.

If erotic stimulation continues, sexual excitement heightens to the plateau phase. Vasocongestion and muscle tension increase markedly. In the female the inner two thirds of the vagina continues to expand and lengthen. The walls of the outer one third of the vagina become greatly vasocongested so that the vaginal entrance becomes somewhat constricted. In this vasoconstricted state the outer one third of the vagina is referred to as the orgasmic platform. In the male the penis increases in circumference. In both sexes, blood pressure increases and heart rate and breathing are accelerated.

Coitus may begin during the excitement or plateau phase. During coitus the penis is moved back and forth in the vagina by movements known as pelvic thrusts. Physical and psychological sensations resulting from this friction (and, from the entire intimate experience between the partners) lead to orgasm, the climax of sexual excitement.

Although lasting only a few seconds, orgasm is the phase of maximum sexual tension and its release. In the male, orgasm is marked by **ejaculation** of semen. Contractions of the vas deferens propel sperm into the ejaculatory ducts. The accessory glands contract, adding their secretions; then, contractions of the ejaculatory ducts, urethra, and certain muscles of the pelvic floor eject the semen from the penis. Muscular contractions continue at about 0.8-second intervals for several seconds. After the first few contractions, their intensity decreases, and they become less regular and less frequent.

In the female, orgasm is marked by rhythmic contractions of the pelvic muscles and the orgasmic platform, starting at approximately 0.8-second intervals and recurring 5 to 12 times. After the first 3 to 6 contractions their intensity decreases, and the time interval between them increases. No fluid ejaculation accompanies orgasm in the female. Stimulation of the clitoris is thought to be important in heightening the sexual excitement that leads to orgasm. In both sexes, heart rate and respiration more than double, and blood pressure rises markedly both just before and during orgasm.

Orgasm is followed by the resolution phase, during which muscle relaxation and detumescence (the subsiding of swelling) restore the body to its normal state. In most males, there is a refractory period during which physiological response to sexual stimulation does not occur. Duration of the refractory period varies in different individuals and also in different situations. Most women are able to repeat the sexual cycle immediately, reaching orgasm again if appropriately stimulated.

Erection of the penis is necessary for effective coitus. Chronic inability to sustain an erection is termed **erectile dysfunction** (formerly referred to as impotence); in men so affected, the disorder is often associated with psychological problems.

FERTILIZATION

Fertilization is the union of sperm and ovum to produce a zygote. Fertilization serves three functions: (1) The diploid number of chromosomes is restored as the sperm contributes its haploid set of chromosomes to the haploid ovum. (2) The sex of the offspring is determined by the sperm cell. (3) Fertilization provides the stimulation necessary to initiate development.

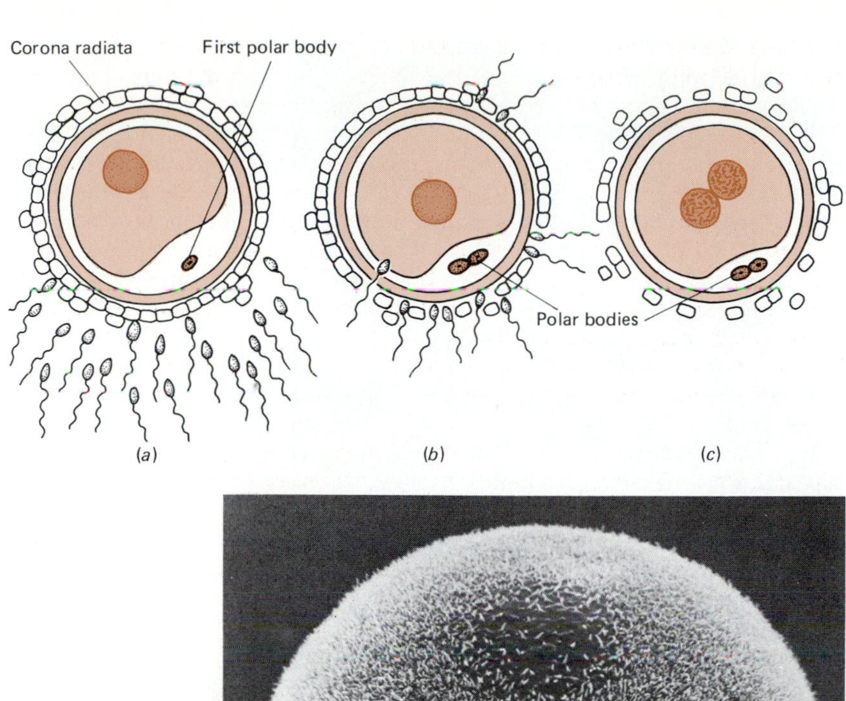

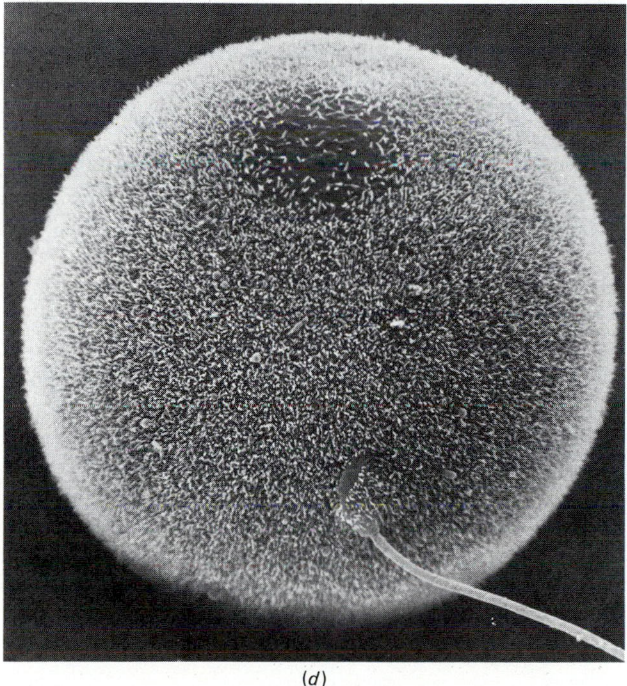

Figure 43–14 Fertilization. (*a*) Each sperm is thought to release a small amount of enzyme that helps to disperse the follicle cells surrounding the ovum. (*b*) After a sperm cell enters the ovum, the ovum completes its second meiotic division, producing an ovum and a polar body. (*c*) Pronuclei of sperm and ovum combine, producing a zygote with the diploid number of chromosomes. (*d*) A scanning electron micrograph of sperm cell fertilizing a hamster ovum. (David M. Phillips, The Population Council.)

When conditions in the vagina and cervix are favorable, sperm begin to arrive at the site of fertilization in the upper uterine tube within 5 minutes after ejaculation. Contractions of the uterus and uterine tubes (perhaps stimulated by prostaglandins in the semen) are thought to help transport the sperm. The sperms' own motility is probably important in approaching and fertilizing the ovum.

If only one sperm is needed to fertilize an ovum, why are millions involved in each act of coitus? For one thing, sperm movement is undirected, so that many lose their way. Others die as a result of unfavorable pH or phagocytosis by leukocytes and macrophages in the female tract. Only a few thousand succeed in traversing the correct uterine tube and reaching the vicinity of the ovum. Additionally, large numbers of sperm may be necessary to penetrate the covering of follicle cells, the **corona radiata,** that surrounds the ovum. Each sperm is thought to release small amounts of enzymes from its acrosome that help break down the cement-like substance holding the follicle cells together.

Once one sperm has entered the ovum, there is an electrical change in the surface layer of the ovum that prevents the entrance of other sperm. As the fertilizing sperm enters the ovum, it usually loses its tail (Fig. 43–14). Sperm entry stimulates the ovum to complete its second meiotic division. The head of the sperm then swells to form the **male pronucleus,** and the nucleus of the ovum becomes the **female pronucleus.** The haploid pronuclei fuse, forming the diploid nucleus of the zygote (Fig. 43–15). The process of fertilization and the subsequent establishment of pregnancy together are referred to as **conception.**

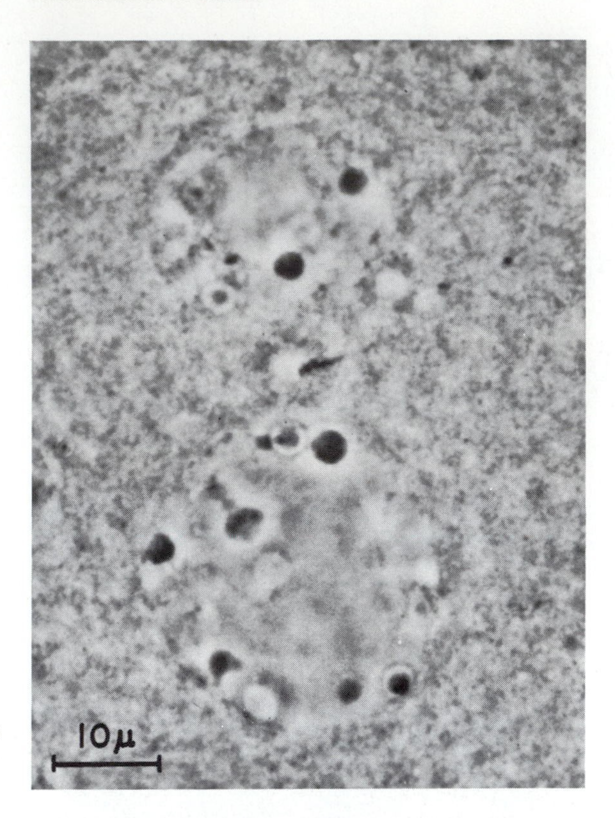

Figure 43–15 Photomicrograph of living human ovum recovered from the uterine tube, showing the male and female pronuclei before their fusion. The sperm tail is evident in the center of the photomicrograph. (Courtesy of Dr. Robert Noyes.)

After ejaculation into the female reproductive tract, sperm remain alive and retain their ability to fertilize an ovum for about 24 hours. The ovum itself remains fertile for only about 24 hours after ovulation. Thus, there are only about three days during each menstrual cycle (days 12 to 15 in a very regular 28-day cycle) when sexual intercourse is likely to result in fertilization.

In view of the many factors working against fertilization it may seem remarkable that it ever occurs! Yet the frequency of coitus and the large number of sperm deposited at each ejaculation enable the human species not only to maintain itself but to increase its numbers at an alarming rate.

STERILITY

Men with fewer than 20 million sperm per milliliter of semen are usually considered to be sterile. When a couple's attempts to produce a child are unsuccessful, a sperm count and analysis may be performed in a clinical laboratory. Sometimes semen is found to contain large numbers of abnormal sperm, or occasionally none at all. In about one fourth of the cases of mumps in adult males, the testes become inflamed; in some of these cases the spermatogonia die, resulting in permanent sterility. Low sperm counts also have been linked to cigarette smoking, and studies show that men who smoke tobacco are also more likely than nonsmokers to produce abnormal sperm. Exposure to chemicals such as DDT and PCBs may also result in low sperm count and sterility.

One cause of female sterility is scarring of the uterine tubes. Inflammation of the uterine tubes, frequently caused by gonorrhea (see Focus on Sexually Transmitted Diseases), may result in scarring, which blocks the tubes so that the ovum can no longer pass to the uterus. Sometimes partial constriction of the uterine tube results in tubal pregnancy, in which the embryo begins to develop in the wall of the uterine tube because it cannot progress to the uterus. Uterine tubes are not adapted to bear the burden of a developing embryo; thus, the uterine tube and the embryo it contains must be surgically removed before it ruptures and endangers the life of the mother.

Women with blocked uterine tubes usually can produce ova and can incubate an embryo. However, they need clinical assistance in getting the ovum from the ovary to the uterus. The ovum can be removed from the ovary, fertilized with the husband's sperm in laboratory glassware (in vitro fertilization), and then placed in the woman's uterus, where it may develop normally (see Focus on Novel Origins).

FOCUS ON

Sexually Transmitted Diseases

The most common **sexually transmitted diseases,** or STD (also called **venereal diseases**), are gonorrhea, syphilis, genital herpes, and pelvic inflammatory disease.

Gonorrhea

More than 2 million patients with **gonorrhea** are treated each year in the United States, making it the most prevalent communicable disease except for the common cold. One reason for its frequency is that there are many strains of the causative organism. Although a person may develop immunity to one strain after infection, reinfection may occur following exposure to another strain.

Gonorrhea is caused by a bacterium, *Neisseria gonorrhoeae.* Infection most often results from direct sexual contact, but it may rarely be transmitted indirectly by contact with towels or toilet seats bearing the infective bacteria.

From 2 to 31 days after exposure, the bacteria produce exotoxin (antigens), which may cause redness and swelling at the site of infection. In the male the infection usually begins in the urethra. Symptoms include frequent urination, accompanied by a severe burning sensation, and a profuse discharge from the penis. Infection may spread to the prostate, seminal vesicles, and epididymides. Abscesses may develop in the epididymides, causing extensive destruction that may lead to sterility.

In about 60% of infected women no symptoms occur, so that treatment is sought only after the woman learns that a sexual partner is infected. The urethra and cervix provide favorable environments for development of the bacteria. If untreated, the infection may spread to the uterine tubes. This type of infection tends to heal, leaving dense scar tissue, which may effectively close the walls of the uterine tubes and result in sterility. In both sexes, untreated gonorrhea may spread to other parts of the body, including heart valves, meninges, and joints. Joint involvement causes a type of arthritis.

Diagnostic criteria include clinical symptoms and positive identification of *N. gonorrhoeae* bacteria in cultures of body secretions. Penicillin is the drug of choice, but tetracyclines and other antibiotics have been used effectively. Strains of gonorrhea bacteria that are resistant to penicillin have been known since about 1976, and the tetracycline spectinomycin has been used as a back-up drug since that time. Recently, however, a strain of gonorrhea resistant to both of these drugs has been reported.

Syphilis

Although not as common as gonorrhea, **syphilis** is a more serious disease, causing death in 5% to 10% of those infected. It is caused by a slender, corkscrew-shaped spirochete, *Treponema pallidum,* and is contracted almost exclusively by direct sexual contact. *T. pallidum* enters the body through a tiny break in the skin or mucous membrane; within 24 hours it invades the lymph or circulatory system and spreads throughout the body.

The course of the disease may be divided into four stages: primary, secondary, latent, and advanced. The primary stage begins about 3 weeks after infection (although the time may vary considerably) with the appearance of a **primary chancre,** a small, painless ulcer at the site of infection, usually the penis in the male or the vulva, vagina, or cervix in the female (see figure). The primary chancre is highly infectious, and sexual contact at this time will almost certainly transmit the disease to an uninfected partner. In many cases the chancre goes unnoticed and spontaneously heals within about a month.

About 2 months after initial infection the secondary stage occurs, sometimes marked by an influenza-like syndrome accompanied by a mild rash. Sometimes annular (circular) scaling lesions develop. Mucous membranes of the mouth and genital structures may exhibit lesions containing large populations of spirochetes. The secondary stage is the most highly contagious because the many lesions literally teem with spirochetes.

If untreated, the disease enters a latent (hidden) stage, which may last for 20 years or longer. Although there may be no obvious symptoms at this time, the spirochetes invade various organs of the body.

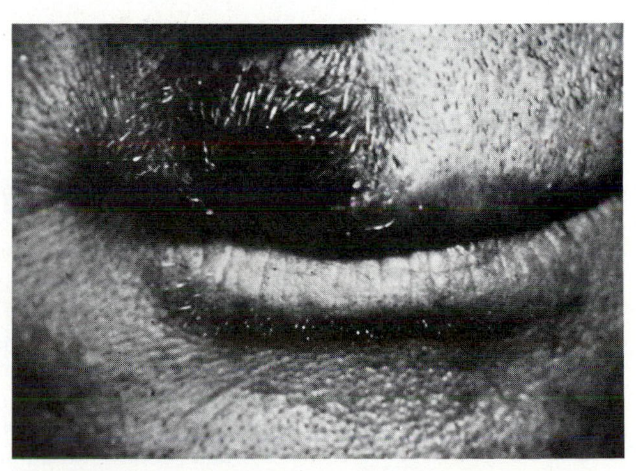

Clinical symptoms of syphilis. Primary syphilitic chancre. This is usually the first symptom. (U.S. Department of Health, Education and Welfare, Center for Disease Control, Atlanta, Georgia)

(*Continued*)

Continued

About a third of untreated patients recover spontaneously. Another third retain the disease in the latent form. The remaining third develop lesions, usually in the cardiovascular system, and less frequently in the CNS or elsewhere. Lesions known as **gummas** are characteristic of advanced syphilis; they occur mainly in skin, liver, bone, and spleen. Involvement of the cardiovascular system is the principal cause of death from syphilis.

In the early primary stage, syphilis can be diagnosed by darkfield examination, which involves identifying the spirochete (using microscopic techniques) in serum collected from a lesion. Within 2 weeks after the appearance of the primary chancre, blood tests may be utilized for diagnoses. One of the most common tests in current use is the VDRL (Venereal Disease Research Laboratory) serological test, which is based on the presence of an antibody-like substance, **reagin,** that appears in the blood in response to the infection.

As with gonorrhea, penicillin is the drug of choice. When treated in its early stages, syphilis can be completely cured. Late syphilis is more difficult to cure, and lesions may have already caused permanent damage to various organs.

Genital Herpes

Genital herpes is an increasingly common disease caused by herpes simplex virus type 2. (This is not to be confused with the herpes simplex type 1 virus, which causes ordinary cold sores.) Symptoms usually appear within a week after infection. First infections are often very mild, however, and may even go unnoticed. When they occur, symptoms take the form of tiny, painful blisters on the genital structures. The blisters may break down into ulcers. Fever, enlarged lymph nodes, and other influenza-like symptoms may appear. The ulcers usually heal within a few days or weeks.

In some patients the disease leaves permanently; in others it recurs periodically (although usually with decreasing severity) for many years. It is thought that the virus may be harbored in the dorsal root ganglia, which receive sensory fibers from the reproductive structures.

There is no cure for genital herpes, although some drugs, such as acyclovir, lessen the duration and severity of initial infections, and oral forms of the drug show promise in preventing recurrence. Pain medications are often prescribed, and sulfa drugs are used to treat secondary bacterial infections that may develop. Because a link has been shown between genital herpes and cervical cancer, women with this disease should have annual Pap smears.

Trichomoniasis

Trichomoniasis is a common infection of the genital tract caused by the flagellate protozoan *Trichomonas vaginalis.* It can easily be transmitted by dirty toilet seats and towels as well as sexual contact. This disease affects about 20% of women during their reproductive years. Symptoms include vaginal itch and vaginal discharge in women. Males may not have symptoms.

Pelvic Inflammatory Disease

Pelvic inflammatory disease (PID) is an infection of the cervix, uterus, uterine tubes, or ovaries. It results most commonly from infection transmitted by intercourse but also from abortion procedures performed under unsanitary conditions. Patients with IUDs are especially susceptible. PID, now the most common sexually transmitted disease in the United States, is caused by chlamydial infections. PID may now be the most common cause of sterility in women, since it leaves women sterile in more than 15% of cases.

Birth Control

Populations of various animal species are limited under natural conditions by a variety of mechanisms, even apart from such things as disease and food shortage. Under crowded conditions, parental neglect, feeding only the strong offspring, and cannibalism are three ways in which populations of organisms may be controlled. In some species spontaneous abortion, genetic deterioration, or death by stress may occur.

Among human populations, some use of abortion as a means of birth control has been found among every population studied. Even infanticide has been practiced, especially in primitive societies. For centuries, however, human beings have searched for more desirable methods of birth control. Highly effective contraceptive (literally, "against conception") methods are now available (Fig. 43–16), but all have side effects, inconveniences, or other disadvantages (see Table 43–3). The ideal contraceptive has not yet been developed.

FOCUS ON

Novel Origins

About 10,000 children born each year are products of **artificial insemination.** Usually this procedure is sought when the male partner of a couple desiring a child is sterile or carries a genetic defect. Although the sperm donor remains anonymous to the couple involved, his genetic qualifications are screened by physicians.

In vitro fertilization is a technique by which an ovum is removed from a woman's ovary, fertilized in a test tube, and then reimplanted in her uterus. Such a procedure may be attempted if a woman's fallopian tubes are blocked, or if they have been surgically removed. With the help of this technique a healthy baby was born in England in 1978 to a couple who had tried unsuccessfully for several years to have a child. Since that time, many other children have been conceived within laboratory glassware.

Another novel procedure is **host mothering.** In this procedure, a tiny embryo is removed from its natural mother and implanted into a female substitute. The foster mother can support the developing embryo either until birth or temporarily until it is implanted again into the original mother or into another host. This technique has already proved useful to animal breeders. For example, embryos from prize sheep can be temporarily implanted into rabbits for easy shipping by air, and then reimplanted into a foster mother sheep, perhaps of inferior quality. Host mothering also has the advantage of allowing an animal of superior quality to produce more offspring than would be naturally possible. In one recent series of experiments mouse embryos were frozen for up to 8 days and then successfully transplanted into host mothers. Host mothering may someday be popular with women who can produce embryos but are unable to carry them to term.

Someday society may have to deal with **cloning** (not yet a reality in the case of humans). In this process the nucleus would be removed from an ovum and replaced with the nucleus of a cell from a person who wished to produce a human copy of himself. Theoretically, any cell nucleus could be used, even a white blood cell nucleus. The fertilized ovum would then be placed into a human uterus for incubation; the resulting baby would be an identical, though younger, twin to the individual whose nucleus was used.

ORAL CONTRACEPTIVES: THE PILL

More than 80 million women worldwide use **oral contraceptives** (more than 8 million in the United States alone). The most common preparations consist of a combination of progestin (synthetic progesterone) and synthetic estrogen. (Natural hormones are destroyed by the liver almost immediately, but the synthetic ones resist destruction.) Starting on the fifth day of the menstrual cycle, a woman takes one pill each day for about three weeks. She then stops taking the pill, and menstrua-

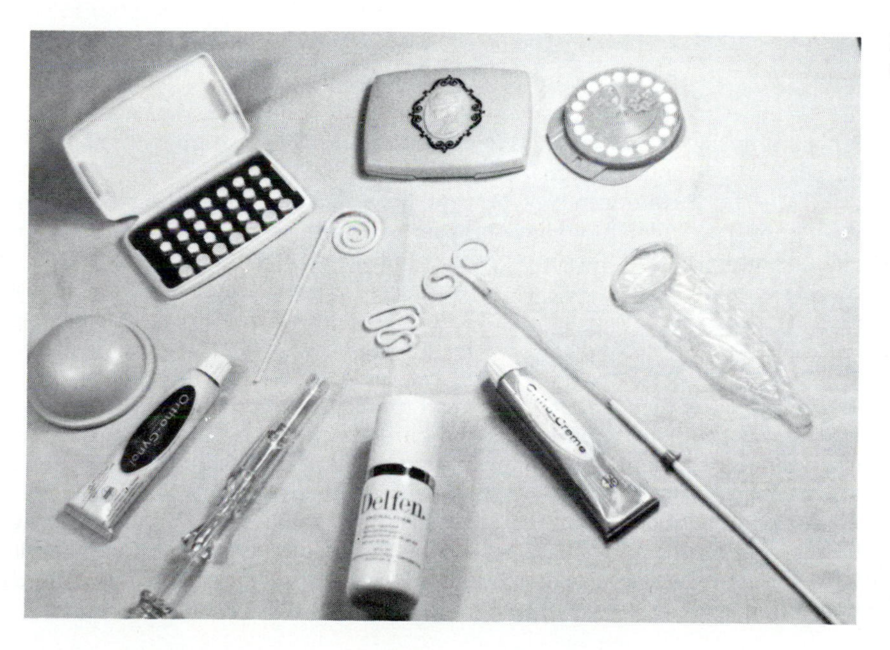

Figure 43–16 Some commonly used contraceptive devices.

TABLE 43–3
Birth Control Methods

Method	*Failure Rate**	*Mode of Action*	*Advantages*	*Disadvantages*
Oral contraceptives	0.3; 5	Prevents ovulation; may also affect endometrium and cervical mucus and prevent implantation	Highly effective; sexual freedom; regular menstrual cycle	Minor discomfort in some women; possible thromboembolism; hypertension, heart disease in some users
Intrauterine device (IUD)	1; 5	Not known; probably stimulates inflammatory response	Provides continuous protection; highly effective	Cramps; increased menstrual flow; spontaneous expulsion
Spermicides (foams, jellies, creams)	3; 22	Chemically kill sperm	No side effects (?)	Not reliable. Best used in conjunction with diaphragm
Contraceptive diaphragm (with jelly)**	3; 13	Diaphragm mechanically blocks entrance to cervix; jelly is spermicidal.	No side effects	Must be prescribed (and fitted) by physician; must be inserted prior to coitus and left in place for several hours after intercourse
Condom	2.6; 10	Mechanical; prevents sperm from entering vagina	No side effects; some protection against STD	Interruption of foreplay to fit; slightly decreased sensation for male; could break
Rhythm†	35, but varies greatly	Abstinence during fertile period	No side effects (?)	Not very reliable
Douche	40	Flush semen from vagina	No side effects	Not reliable; sperm beyond reach of douche within seconds
Withdrawal (coitus interruptus)	20?	Male withdraws penis from vagina prior to ejaculation.	No side effects	Not reliable; contrary to powerful drives present when an orgasm is approached. Sperm present in fluid secreted before ejaculation may be sufficient for conception.
Sterilization				
Tubal ligation	0.04	Prevents ovum from leaving uterine tube	Most reliable method	Usually not reversible
Vasectomy	0.15	Prevents sperm from leaving scrotum	Most reliable method	Usually not reversible
Chance (no contraception)	About 90			

*The lower figure is the failure rate of the method; the higher figure is the rate of method failure plus failure of the user to utilize the method correctly. Based on number of failures per 100 women who use the method per year in the United States.
**Failure rate is lower when used together with spermicidal foam.
†There are several variations of the rhythm method. For those who use the calendar method alone, the failure rate is about 35. However, by taking the body temperature daily and keeping careful records (temperature rises after ovulation), the failure rate can be reduced. Also, by keeping a daily record of the type of vaginal secretion, changes in cervical mucus can be noted and used to determine time of ovulation. This type of rhythm contraception is also slightly more effective. When women use the temperature or mucus method and have intercourse *only* more than 48 hours *after* ovulation, the failure rate can be reduced to about 7.

tion begins about three days later. Ovulation does not occur if she resumes the pills on the fifth day of her cycle. When taken correctly, these pills are about 99.9% effective in preventing pregnancy.

Most oral contraceptives prevent pregnancy by preventing ovulation. By maintaining postovulatory levels of ovarian hormones in the blood, the pituitary gland is inhibited and does not produce the surge of FSH and LH that stimulates ovulation. Some advantages of oral contraceptives are summarized in Table 43–3.

THE INTRAUTERINE DEVICE (IUD)

The **intrauterine device (IUD)** is a small plastic loop or coil that must be inserted into the lumen of the uterus by a medical professional (Fig. 43–17). Once in place, it can be left in the uterus indefinitely or until the woman wishes to conceive. Newer types of IUDs are about 99% effective. The mode of action of the IUD is not well understood, but it is thought that the IUD sets up a minor local inflammation in the uterus, attracting macrophages, which destroy the tiny embryo before it implants. IUDs that contain copper may interfere with both migration of sperm and implantation of the embryo.

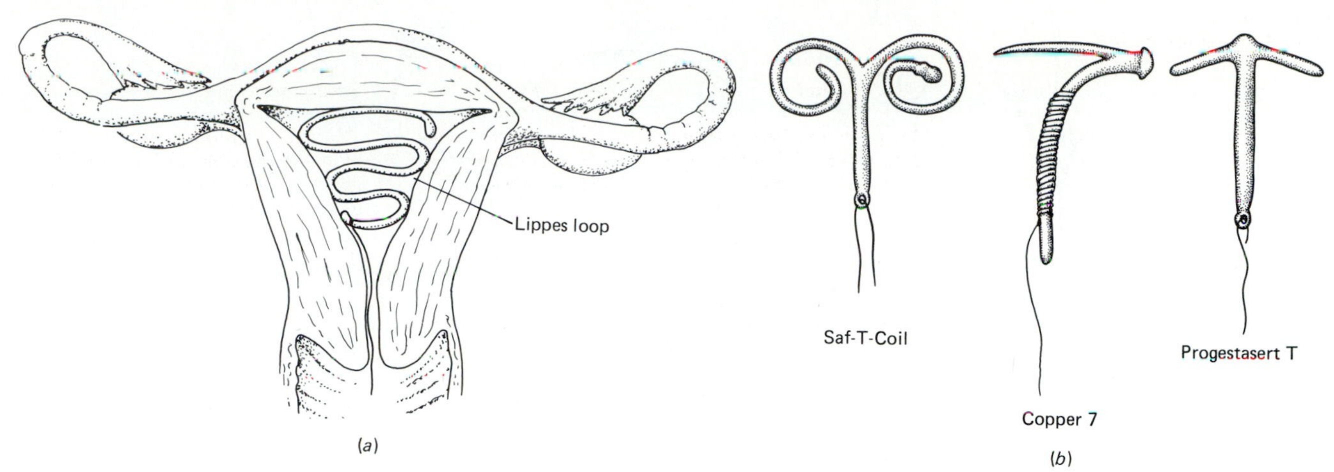

Figure 43–17 (*a*) An IUD (intrauterine device) in place within the uterus. The type shown is a Lippes loop. (*b*) Other styles of IUD in current use include the Saf-T-Coil, the Copper 7, and the Progestasert T.

OTHER COMMON METHODS

Other common contraceptive methods are listed and compared in Table 43–3. For instance, the contraceptive diaphragm mechanically blocks the passage of sperm from the vagina into the cervix. It is usually covered with spermicidal jelly or cream and inserted just prior to coitus (Fig. 43–18). The condom is also a mechanical method of birth control. The most commonly used male contraceptive device, the condom holds the semen so that sperm cannot enter the female tract. When properly used, both of these methods are highly effective.

STERILIZATION

Short of total abstinence, the only foolproof method of birth control is **sterilization,** which renders an individual permanently incapable of parenting offspring. About 75% of sterilization procedures are currently performed on men.

The most common method of male sterilization is **vasectomy** (Fig. 43–19) in which each vas deferens is severed and tied off. Vasectomies can be performed in a physician's office using a local anesthetic. A small incision is made on each side of the scrotum. Each vas deferens is then separated from the other structures of the spermatic cord and is cut; its ends are tied or clipped so that they cannot grow back together.

Testosterone secretion is not affected by vasectomy, so this procedure in no way affects masculinity. Sperm continue to be produced, though at a much slower rate. Instead of being ejaculated, however, they eventually die and are destroyed by macrophages in the testis or epididymis. Because sperm account for very little of the semen volume, there is no noticeable change in the amount of semen ejaculated. In cases where a reversal of the procedure was requested, surgeons have been able to reverse the sterilization successfully in about 30% of the cases at-

Figure 43–18 Insertion of a contraceptive diaphragm.

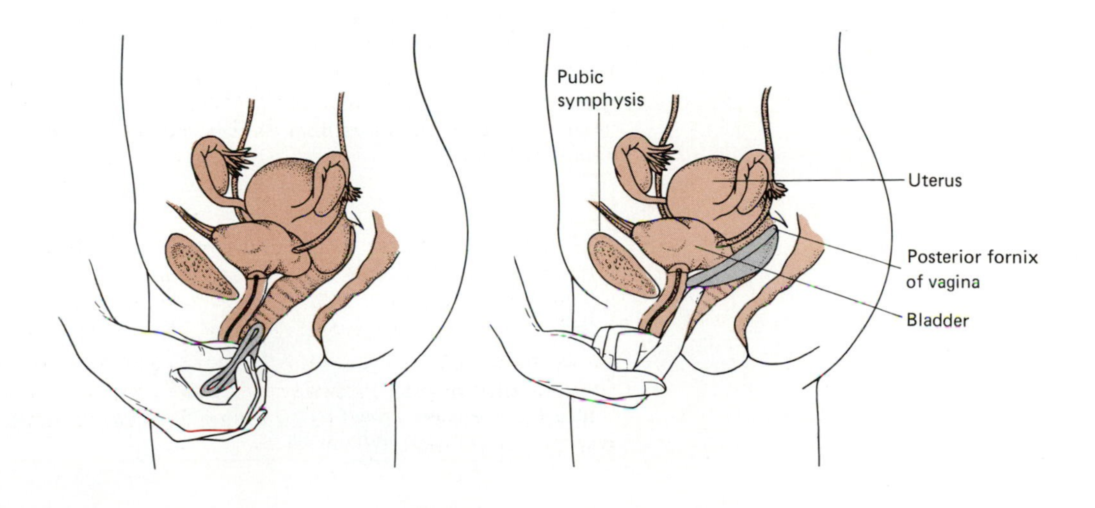

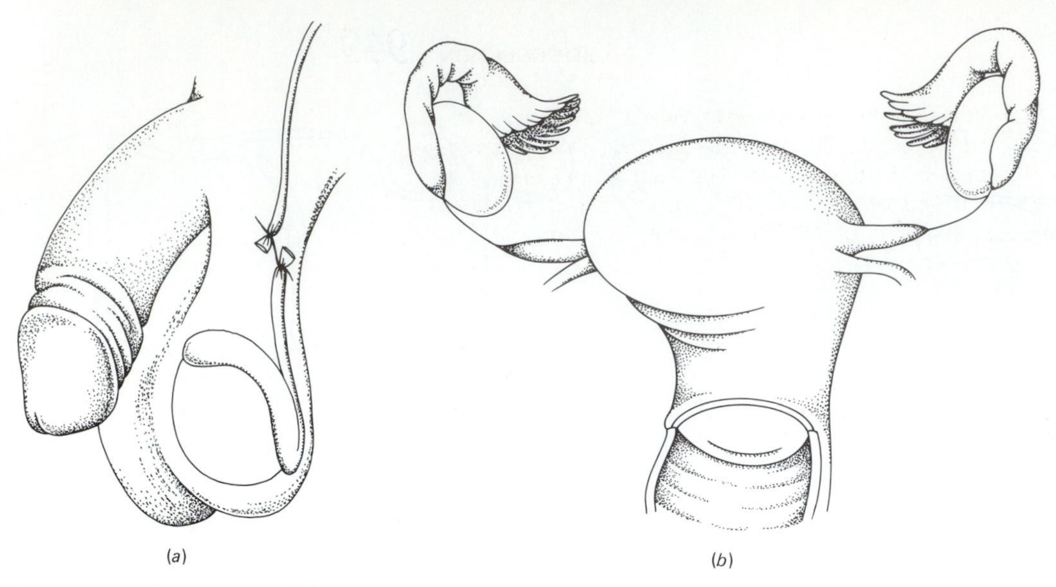

Figure 43–19 Sterilization. (*a*) In vasectomy, the vas deferens (sperm duct) on each side is cut and tied. (*b*) In tubal ligation, each uterine tube is cut and tied so that ovum and sperm can no longer meet.

tempted. The low success rate may be because some sterilized men eventually develop antibodies against their own sperm, making the sperm nonviable.

Female sterilization generally involves **tubal ligation,** cutting and tying the uterine tubes (Fig. 43–19). As in the male, hormone balance and sexual performance are not affected.

ABORTION

Abortion is the termination of pregnancy resulting in the death of the embryo or fetus. An estimated 40 million abortions are deliberately performed each year worldwide (more than a million in the United States). Three kinds of abortions may be distinguished: spontaneous abortions, therapeutic abortions, and those undertaken as a means of birth control. **Spontaneous abortions** (popularly called miscarriages) occur without intervention and often are nature's way of destroying a defective embryo. **Therapeutic abortions** are performed in order to maintain the health of the mother, or when there is reason to suspect that the embryo is grossly abnormal. The third type of abortion, the type performed as a means of birth control, is the most controversial.

Most first-trimester abortions (those performed during the first three months of pregnancy) are done using a suction method. After the cervix is dilated, a suction aspirator is inserted in the uterus, and the embryo and other products of conception are evacuated. In another method, dilation and curettage ("D & C"), the endometrium is scraped out. (D & Cs may also be performed to remove polyps from the uterus and for diagnosis and treatment of a number of other uterine disorders.)

Later in pregnancy abortions are performed using saline injections of saline and prostaglandins. The amniotic fluid surrounding the embryo is removed with a needle and replaced with a solution containing salt and prostaglandins. The fetus dies within 1 to 2 hours; labor begins several hours later.

TABLE 43–4
Deaths in the United States from Pregnancy and Childbirth and from Various Birth Control Methods

	Death Rate per 100,000 Women
Pregnancy and childbirth	20
Oral contraceptive	4
IUD	0.5
Legal abortions	
First trimester	3
After first trimester	26
Illegal abortion performed by medically untrained individuals	about 100

When abortion is performed during the first trimester by skilled medical personnel, the mortality rate is about 3 per 100,000. After the first trimester this rate rises to 26 per 100,000 (Table 43–4). The death rate from illegal abortions performed by medically untrained individuals is about 100 per 100,000. In contrast to these figures, the death rate from pregnancy and childbirth is about 20 per 100,000.

SUMMARY

I. In asexual reproduction, a single parent endows its offspring with a set of genes identical with its own. In sexual reproduction, each of two parents contributes a gamete containing half of the offspring's genetic endowment.

II. Unusual variations of reproduction include metagenesis, parthenogenesis, and hermaphroditism.

III. Human reproductive processes include gametogenesis, cyclic changes in the female body in preparation for pregnancy, sexual intercourse, fertilization, pregnancy, and lactation.

- *A.* The male reproductive system includes the testes, which produce sperm and testosterone; a series of conducting tubes; accessory glands; and the penis.
  1. The testes, housed in the scrotum, contain the seminiferous tubules, where the sperm are produced, and the interstitial cells, which secrete testosterone.
  2. Sperm complete their maturation and are stored in the epididymis and may also be stored in the vas deferens.
  3. During ejaculation sperm pass from the vas deferens to the ejaculatory duct, and then into the urethra, which passes through the penis.
  4. Semen contains about 400 million sperm suspended in the secretions of the seminal vesicles and prostate gland.
  5. The penis consists of three columns of erectile tissue; when this tissue becomes engorged with blood, the penis becomes erect.
- *B.* The female reproductive system includes the ovaries, which produce ova and the hormones estrogens and progesterone, the uterine tubes, the vagina, the vulva, and the breasts.
  1. After ovulation the ovum enters the uterine tube, where it may be fertilized.
  2. The uterus serves as an incubator for the developing embryo.
  3. The vagina is the lower part of the birth canal; it receives the penis during coitus.
  4. The clitoris is the center of sexual sensation in the female.
- *C.* The gonadotropic hormones FSH and LH stimulate sperm production and testosterone secretion. Testosterone is responsible for establishing and maintaining primary and secondary sex characteristics in the male.
- *D.* The first day of menstrual bleeding marks the first day of the menstrual cycle. Ovulation occurs at about day 14 in a typical 28-day menstrual cycle. Events of the menstrual cycle are coordinated by the gonadotropic and ovarian hormones.
  1. FSH stimulates follicle development; FSH and LH together stimulate ovulation; LH promotes development of the corpus luteum.
  2. The developing follicles release estrogen, which stimulates development of the endometrium and is responsible for the secondary female sex characteristics.
  3. The corpus luteum secretes progesterone, which stimulates final preparation of the uterus for possible pregnancy.
- *E.* Vasocongestion and increased muscle tension are two basic physiological responses to sexual stimulation. The cycle of sexual response includes the excitement stage, plateau stage, orgasm, and resolution.
- *F.* Fertilization restores the diploid chromosome number, determines the sex of the offspring, and triggers development.
- *G.* Effective methods of birth control include oral contraceptives, intrauterine devices, condoms, contraceptive diaphragms, and sterilization.

POST-TEST

1. The type of reproduction in which an animal divides into several pieces and then each piece develops into an entire new animal is called ______________.

2. In metagenesis there is an alternation of ______________________________.

3. Parthenogenesis is a type of reproduction in which an unfertilized egg ______________________________.

4. An individual that can produce both eggs and sperm is described as ______________.

5. A sex cell (either egg or sperm) is properly called a ______________; a fertilized egg is a ______________.

6. An adult who is unable to parent offspring is said to be ______________.

For each group, select the most appropriate answer from Column B for each description in Column A.

Column A	Column B
7. Sperm produced here	a. Seminiferous tubules
8. Produce testosterone	b. Prostate gland
9. Columns of erectile tissue	c. Interstitial cells
10. Secretes alkaline fluid into urethra	d. Cavernous bodies
11. Sac that holds the testes	e. None of the above

Column A	*Column B*
12. Produces gametes	a. Uterine tube
13. Thickens each month in preparation for pregnancy	b. Ovary
14. Lower portion of uterus	c. Cervix
15. Fertilization takes place here	d. Endometrium
	e. None of the above
16. Produces FSH	a. Hymen
17. Produces progesterone	b. Corpus luteum
18. Center of sexual sensation	c. Anterior lobe of pituitary
19. Extends from uterus to exterior of body	d. Clitoris
	e. None of the above
20. Responsible for secondary sexual characteristics in female	a. Testosterone
21. Responsible for secondary sexual characteristics in male	b. Estrogen
22. Produced by pituitary	c. LH
23. Stimulates glands in endometrium to develop	d. Progesterone
	e. None of the above

Column A	*Column B*
24. Prevents ovulation	a. IUD
25. Prevents sperm from entering vagina	b. Contraceptive diaphragm
26. Procedure in which vas deferens is severed	c. Oral contraceptive
27. Blocks passage of sperm from vagina into uterus	d. Condom
	e. None of the above

28. A ______________ abortion is performed in order to maintain the mother's health, or when the embryo is thought to be grossly abnormal.

29. Tubal ligation is a common method of ______________ ______________.

30. The menstrual cycle runs its course every month from puberty until ______________.

31. Label the following diagrams. (Refer to Figures 43-4 and 43-9 in the text as necessary.)

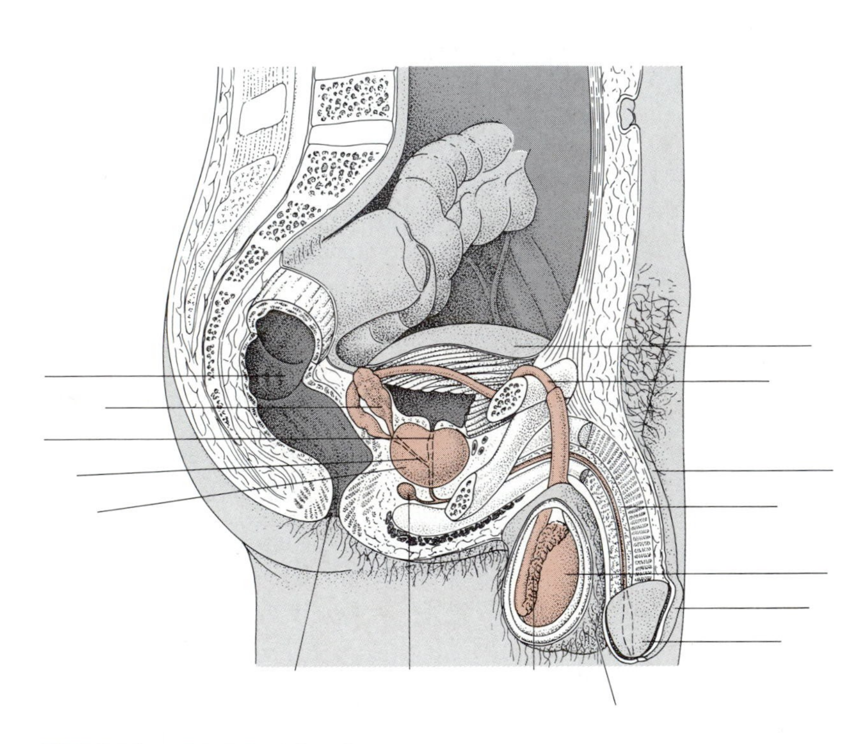

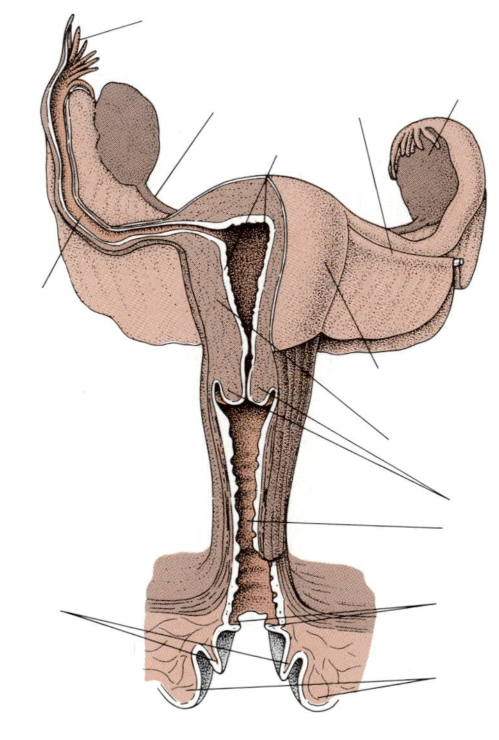

REVIEW QUESTIONS

1. Compare asexual with sexual reproduction, and give specific examples of asexual reproduction.
2. Give an example of metagenesis and one of parthenogenesis.
3. Can you think of any biological advantages of parthenogenesis? of hermaphroditism? Explain.
4. Compare the functions of the ovaries and testes.
5. Trace the passage of sperm from a seminiferous tubule through the male reproductive system until it leaves the male body during ejaculation. Assuming that ejaculation takes place within the vagina, trace the journey of the sperm until it meets the ovum.
6. What are the actions of testosterone? of estrogens? of progesterone?
7. What is the function of the corpus luteum? Which hormone stimulates its development?
8. What are the actions of FSH and LH in the female?
9. Why are so many sperm produced in the male and so few ova produced in the female?
10. Which methods of birth control are most effective? least effective?
11. Draw a diagram of the principal events of the menstrual cycle, including ovulation and menstruation. Indicate on which days of the cycle sexual intercourse would most likely result in pregnancy.
12. Distinguish between the following terms: erectile dysfunction, sterility, and castration.
13. Give the physiological basis for penile erection.

44
Development

OUTLINE

LEARNING OBJECTIVES

After you have read this chapter you should be able to:

1. Relate the preformation theory and the theory of epigenesis to current concepts of development.
2. Describe the roles of growth, morphogenesis, and cellular differentiation in the development of an organism.
3. Describe the principal characteristics of each of the early stages of development: zygote, cleavage, blastocyst, gastrula, nervous system development, and development during the first month of human embryonic life.
4. Describe implantation and the development of the extraembryonic membranes, giving the functions of each membrane (including the placenta).
5. Describe the general course of the later development of the human being from one month after conception until the time of birth.
6. Identify the principal events of each stage of labor.
7. Contrast postnatal with prenatal life, describing several adaptations that the neonate must make in order to live independently.
8. Identify the hormones involved in lactation, and discuss the advantages of breastfeeding an infant.
9. Discuss genetic and nongenetic factors that interact to regulate development, relating experiments discussed in this chapter.
10. List specific steps that a pregnant woman can take to promote the well-being of her developing child, and describe how the embryo can be affected by nutrients, drugs, cigarette smoking, pathogens, and ionizing radiation.
11. Give the stages of the human life cycle.
12. Identify anatomical and physiological changes that occur with aging, and discuss current theories of aging.

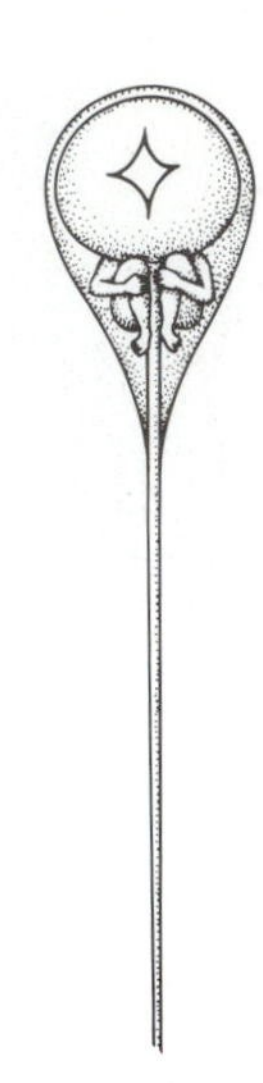

Figure 44–1 The preformed "little man" within the sperm as visualized by 17th-century scientists. (After Hartseeker's drawing from "Essay de Dioptrique," Paris, 1694.)

Many scientists of the 17th century thought that the egg cell contained a completely formed, though miniature, human being. They believed that all the parts were already present, so that the embryo had only to grow in size. This concept is known as the **preformation theory.**

By the end of the 17th century two competing groups of preformationists emerged. One group, the ovists, thought that the preformed organism resided within the egg; the opposing group, the spermists, were certain that the "little man" was housed in the sperm. Using their crude microscopes some investigators even imagined that they could see a completely formed tiny human being within the head of the sperm (Fig. 44–1).

Some scientists carried the theory to an extreme form, arguing that every woman contained within her body a miniature of every individual who would ever descend from her. Her children, grandchildren, great grandchildren, and so on were thought be be preformed, each within the reproductive cells of the other. Some investigators of that time even computed mathematically how many generations could fit, one within the other's gametes. They concluded that when all these generations had lived and died, the human species would end. Jan Swammerdam, a renowned preformationist, felt that this concept explained original sin. He wrote, "In nature there is no generation, but only propagation, the growth of parts. Thus, original sin is explained, for all men were contained in the organs of Adam and Eve. When their stock of eggs is finished the human race will cease to be."

The idea of preformation was not restricted in its application to human beings alone. All plant and animal species were included. For almost 200 years this theory was seriously debated by scientists and philosophers.

An opposing view, the **theory of epigenesis,** gained experimental support as better techniques for investigation were developed. This theory held that the embryo develops from an undifferentiated zygote and that the structures of the body emerge in an orderly sequence, developing their characteristic forms only as they emerge.

Today we know that development is largely epigenetic. No microscopic human waits preformed in either gamete. Development proceeds from one cell to billions, from a formless mass of cells to an intricate, highly specialized and organized organism. However, a spark of truth can be found in the preformationist view. Although the "little man" itself is not to be found within the zygote, its blueprint *is* there, precisely encoded in the form of chemical specifications within the DNA of the genes.

Early Development

How does a microscopic, unspecialized zygote give rise to blood, bones, brain, and all the other complex structures of an organism? As we will see, development is a balanced combination of three processes: growth, morphogenesis, and cellular differentiation.

Growth of the embryo includes both cellular growth and mitosis. An orderly pattern of growth provides the cellular building blocks of the organism. But growth alone would produce only a formless heap of cells. Cells must arrange themselves into specific structures and appropriate body forms. The precise and complicated cellular movements designed to bring about the form of a multicellular organism with its intricate pattern of tissues and organs constitute **morphogenesis.**

Not only must cells be arranged into specific structures, but they must also be able to perform varied and specialized functions. Cells must be specialized as well as organized. In order to function in different fashions, body structures must be composed of different components. During early development, cells begin to differentiate from one another, specializing biochemically and structurally to perform specific tasks. More than 200 distinct types of cells can be found in the human body. The process by which cells become specialized is known as **cellular differentiation.**

As you read the following section on development, bear in mind that growth, morphogenesis, and cellular differentiation are intimately interrelated. Although this chapter centers on human development, the pattern of early development is basically similar for all vertebrates.

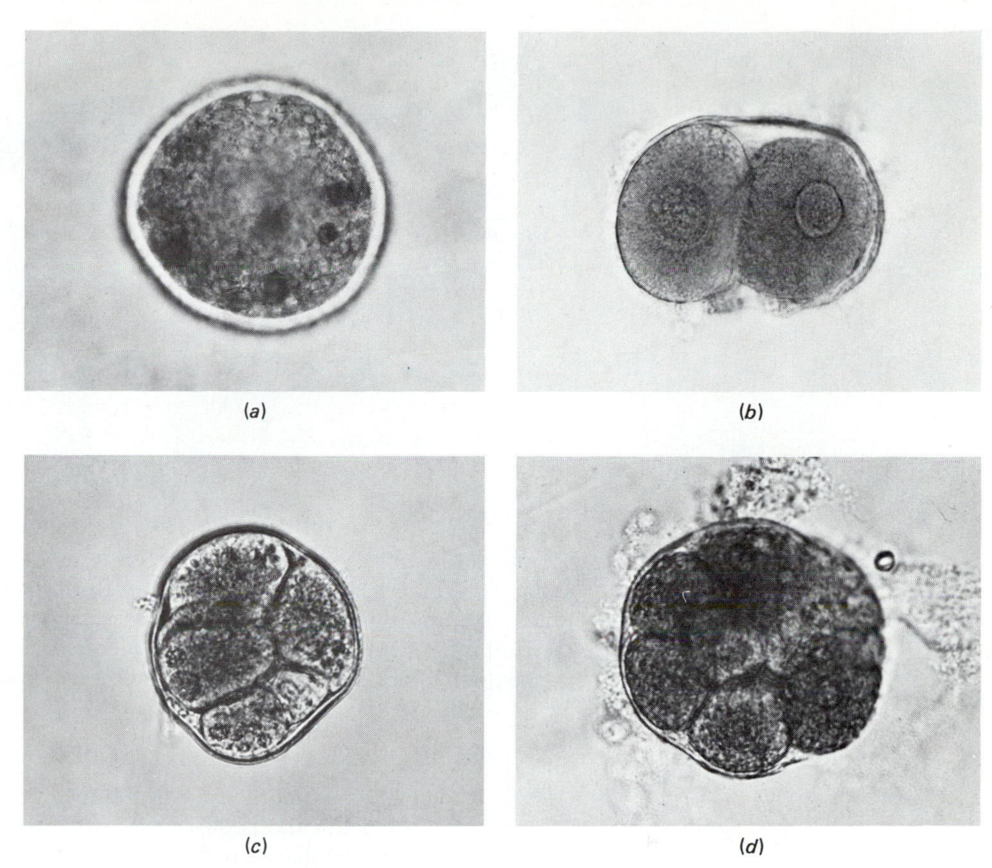

Figure 44–2 Early human development (approximately ×250). (*a*) Human zygote. This single cell contains the genetic instructions for producing a complete human being. (*b*) Two-cell stage. (*c*) Eight-cell stage. (*d*) Cleavage continues, giving rise to a cluster of cells called the morula. (Courtesy of Carnegie Institution of Washington.)

THE ZYGOTE

Although it appears to be a relatively simple cell, the **zygote,** or fertilized egg, has the potential to give rise to all the diverse cell types of the complete individual (Fig. 44–2). Since the sperm cell is quite tiny in comparison with the egg, the bulk of the zygote cytoplasm comes from the ovum. However, it contains the chromosomes contributed by both sperm and egg.

In most vertebrate zygotes, the cytoplasm contains **yolk,** which serves as food for the developing embryo. The amount and distribution of yolk vary among different animal groups. Whereas bird eggs contain great quantities of yolk, those of mammals have only small amounts. Yolk is absent from the human zygote. In organisms that possess large amounts of yolk, the pole of the cell that contains the largest amount of this metabolically inert food substance is known as the **vegetal pole;** the opposite, more metabolically active pole is the **animal pole.** The animal pole gives rise to the cells that become the actual embryo. Animal and vegetal poles can be distinguished even in yolkless eggs.

CLEAVAGE: FROM ONE CELL TO MANY

Shortly after fertilization the zygote undergoes a series of rapid mitoses, collectively referred to as **cleavage.** By about 24 hours after fertilization, the human zygote has completed the first mitotic division and reached the two-cell stage. Each of the cells of the two-cell-stage **embryo** undergoes mitosis, bringing the number of cells to four. Repeated divisions continue to increase the number of cells making up the embryo. At about the 16-cell stage the embryo consists of a tiny cluster of cells called the **morula** (Fig. 44–2). As cleavage takes place, the embryo is pushed along the uterine tube by ciliary action and muscular contraction. By the time the embryo reaches the uterus, on about the fifth day of development, it is in the morula stage (Fig. 44–3).

During cleavage each interphase period is so brief that the cells do not have time to grow. For this reason the mass of cells produced is not larger than the original zygote. The principal effect of early cleavage, then, is to partition the zygote into many small cells. These serve as basic building units that determine the early form of the organism. Their small size allows the cells to move about with relative

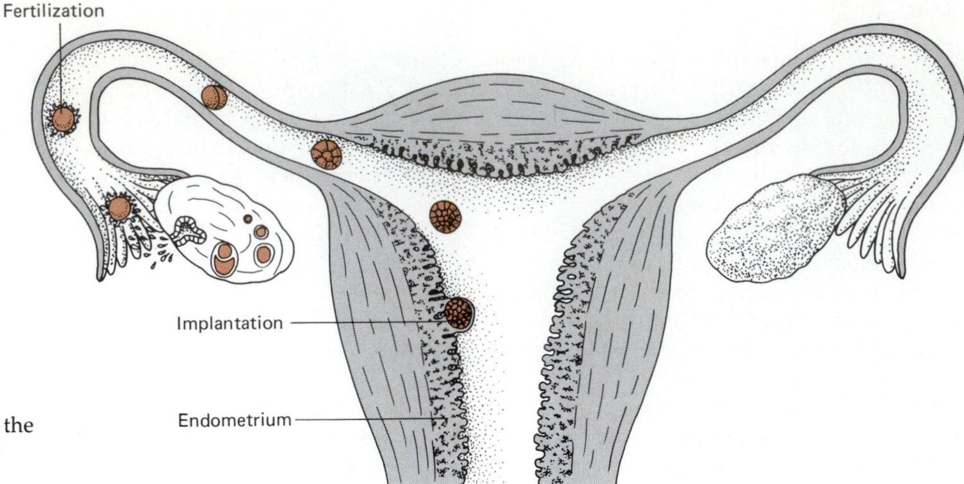

Figure 44–3 Cleavage takes place as the embryo is moved along through the uterine tube to the uterus.

ease, arranging themselves into the patterns necessary for further development. Each cell moves by amoeboid motion, probably guided by the proteins of its own cell coat and those of other cell surfaces. Specific properties conferred upon cell membranes by these surface proteins are important in helping cells to "recognize" one another and therefore in determining which ones will adhere to form tissues.

THE BLASTOCYST

When the embryo enters the uterus, the membrane that has surrounded the embryo, the **zona pellucida,** is dissolved. Now the embryo is bathed in a nutritive fluid secreted by the glands of the uterus. Nourished in this manner, the embryo continues its development for two or three days, floating free in the uterine cavity. During this period its cells arrange themselves into the form of a hollow ball called a **blastocyst** (the comparable stage is called a blastula in many other animal groups) (Fig. 44–4). The outer layer of cells, the **trophoblast,** eventually forms protective and nutritive membranes (the chorion and placenta) that surround the embryo. A little cluster of cells, the **inner cell mass,** projects into the cavity of the blastocyst. These cells give rise to the embryo itself.

Occasionally, the inner cell mass subdivides to form two independent groups of cells, and each develops into a complete organism. Since each cell has an identical set of genes, the individuals formed are exactly alike—**identical twins.** Very rarely, the two inner cell masses are not completely separated and give rise to **conjoined** (Siamese) **twins. Fraternal twins** develop when a woman ovulates two eggs and each is fertilized by a separate sperm. Each zygote has its own distinctive genetic endowment, so the individuals produced are not identical. Triplets (and other multiple births) may similarly be either identical or fraternal. In the United States twins are born once in about 88 births, triplets once in 88^2 (or 7744), and quadruplets once in 88^3 (or 681,472).

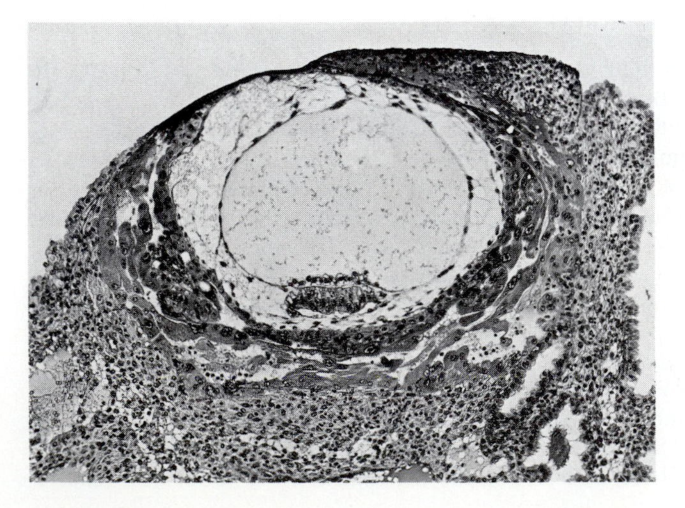

Figure 44–4 Implanted blastocyst at 12 days after fertilization. (Courtesy of Carnegie Institution of Washington.)

IMPLANTATION

On about the seventh day of development the embryo begins to **implant** in the endometrium (lining) of the uterus. The trophoblast cells in contact with the uterine lining secrete enzymes that erode an area just large enough to accommodate the tiny embryo. Slowly the embryo works its way down into the underlying connective and vascular tissues. The opening through which the blastocyst enters the lining of the uterus is closed, first by a blood clot, and eventually by the overgrowth of regenerated epithelial cells. All further development of the embryo takes place *within* the endometrium of the uterus.

During implantation, enzymes destroy some tiny maternal capillaries in the wall of the uterus. Blood from these capillaries comes in direct contact with the trophoblast of the embryo, temporarily providing a rich source of nutrition.

Implantation is completed by the ninth day of development. This would correspond to about the twenty-third day of a woman's menstrual cycle, so that despite all the developmental activities taking place in her uterus, a woman would probably not even know that she was pregnant at this time.

FORMATION OF GERM LAYERS

The cells of the inner cell mass of the blastocyst arrange themselves to form a two-layered disk (Fig. 44–5). The cells of the lower level then merge to line an inner cavity, the **primitive gut,** or **archenteron,** which will eventually develop into the digestive tract and certain other structures. These cells make up the **endoderm,** while the cells that remain to cover the embryo and become its outermost layer form the **ectoderm.** A third layer of cells, the **mesoderm,** proliferates between the ectoderm and endoderm. Ectoderm, mesoderm, and endoderm are known as the three **germ layers,** or embryonic tissue layers. Each gives rise to specific structures in all vertebrate embryos (Table 44–1).

The process by which the blastocyst becomes a three-layered embryo, called a **gastrula,** is termed **gastrulation.** Cleavage and formation of the blastocyst is thought to depend mainly upon the organization of the cytoplasm in the ovum. Among protostomes, for example, cleavage proceeds according to a spiral pattern

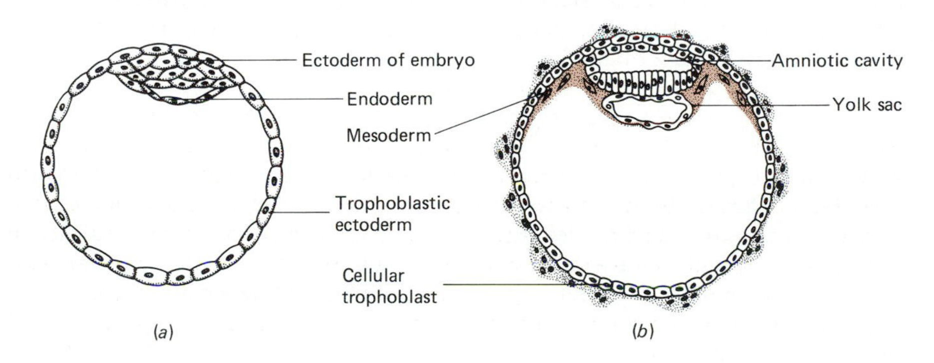

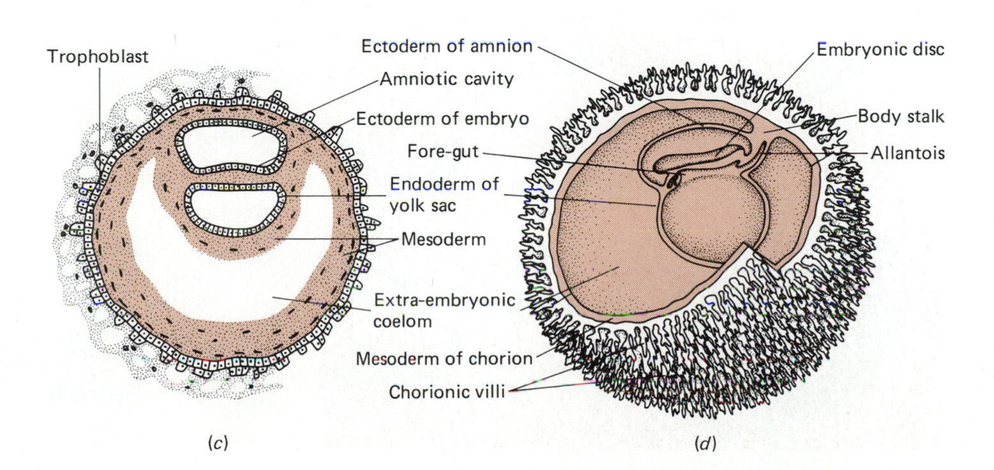

Figure 44–5 Human embryos ten (*a*) to twenty (*d*) days showing the formation of the amnion and yolk sac and the origin of the embryo.

TABLE 44–1
Fate of the Germ Layers

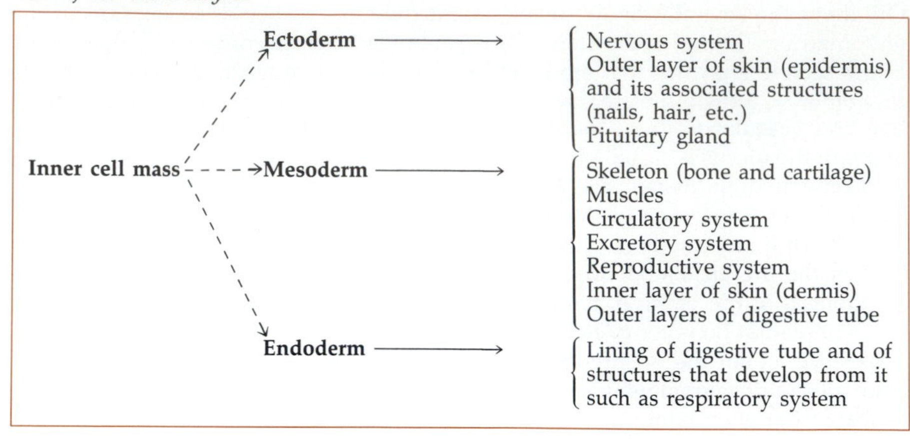

	Germ layer	Structures formed
Inner cell mass	**Ectoderm** →	Nervous system Outer layer of skin (epidermis) and its associated structures (nails, hair, etc.) Pituitary gland
	Mesoderm →	Skeleton (bone and cartilage) Muscles Circulatory system Excretory system Reproductive system Inner layer of skin (dermis) Outer layers of digestive tube
	Endoderm →	Lining of digestive tube and of structures that develop from it such as respiratory system

(see Focus on Early Development: A Comparative Study) which is completely absent in echinoderms and vertebrates. The complex morphogenetic movements of cells and their ability to form the germ layers depend upon instructions from the genes of the embryo itself. The spiral cleavage characteristic of annelids and other protostomes is illustrated in Focus on Early Development: A Comparative Study. In the human embryo, the first visible sign of differentiation occurs during gastrulation when one group of cells of the inner cell mass becomes flattened and the remaining cells become columnar. The flattened cells develop into the endoderm, whereas the columnar cells become ectoderm.

The amount of yolk present in an egg plays an important role in determining the pattern of early development. (See Focus on Early Development: A Comparative Study.) The egg of *Amphioxus,* for example, has a small amount of evenly distributed yolk, and cleavage divides the embryo into cells of fairly equal size. In contrast, the large amount of yolk in the frog egg slows cleavage, so that the blastula consists of many small cells at the end of the embryo derived from the animal pole of the egg, and a few large ones at the opposite end, derived from the vegetal pole. In eggs with a very large amount of yolk, such as hen's eggs, cleavage occurs *only* in the small disk of cytoplasm at the animal pole.

DEVELOPMENT OF THE NERVOUS SYSTEM

The brain and spinal cord are among the first organs to develop in the early embryo (Fig. 44–6). During the second week of human development a cylindrical rod of cells grows forward along the length of the embryo, forming a structure called the **notochord.** The notochord serves as a flexible skeletal axis in all chordate embryos. In vertebrate embryos it is eventually replaced by the vertebral column. The notochord **induces** (stimulates) the overlying ectoderm to thicken, forming the **neural plate.**

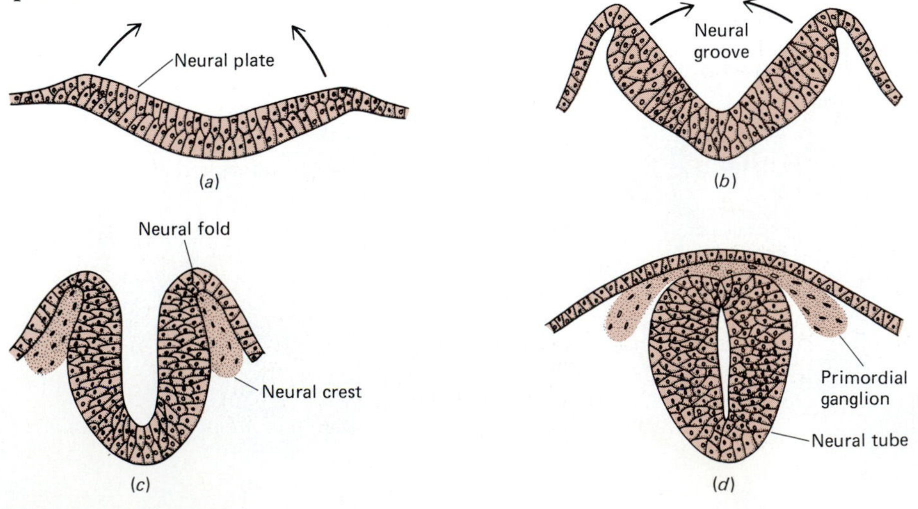

Figure 44–6 Cross sections of the ectoderm of human embryos at successively later stages, illustrating the early development of the nervous system. The neural crest cells form the dorsal root ganglion and the sympathetic nerve ganglia.

FOCUS ON

Early Development: A Comparative Study

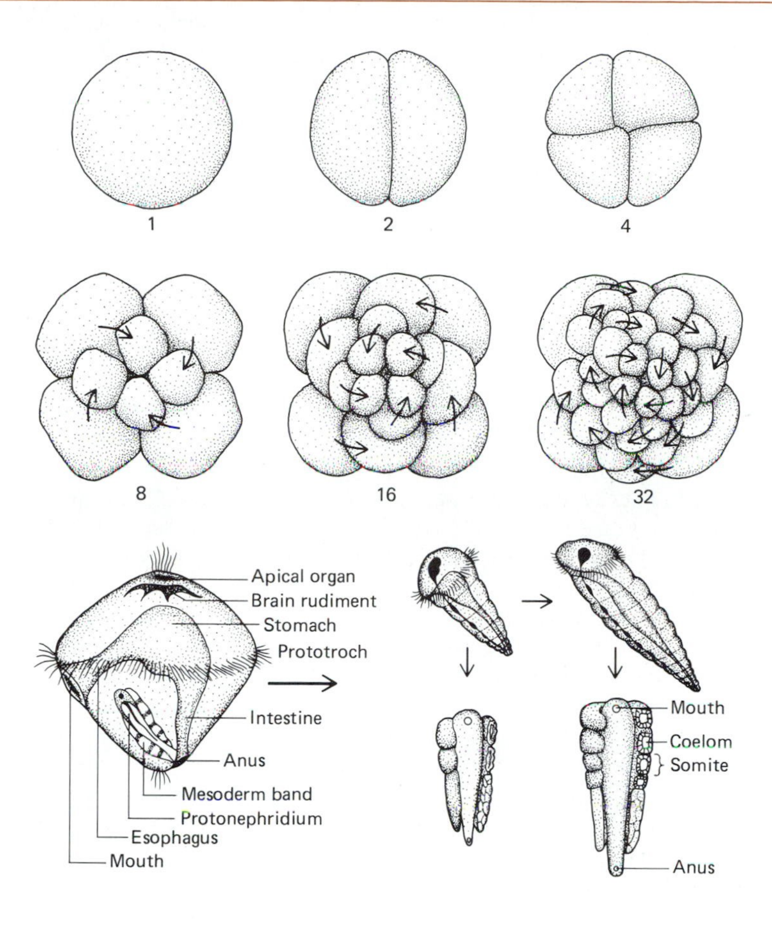

Development in annelids. *Upper half,* The successive cleavage divisions occur in a spiral pattern as indicated. The numbers indicate the number of cells in the developing organism. *Lower left,* A typical trochophore. The upper half of the trochophore develops into the extreme anterior end of the adult worm; all the rest of the adult body develops from the lower half. A series of cavities appears within each mesodermal somite (blocklike mass of tissue) *(lower right),* which coalesce to form the coelom.

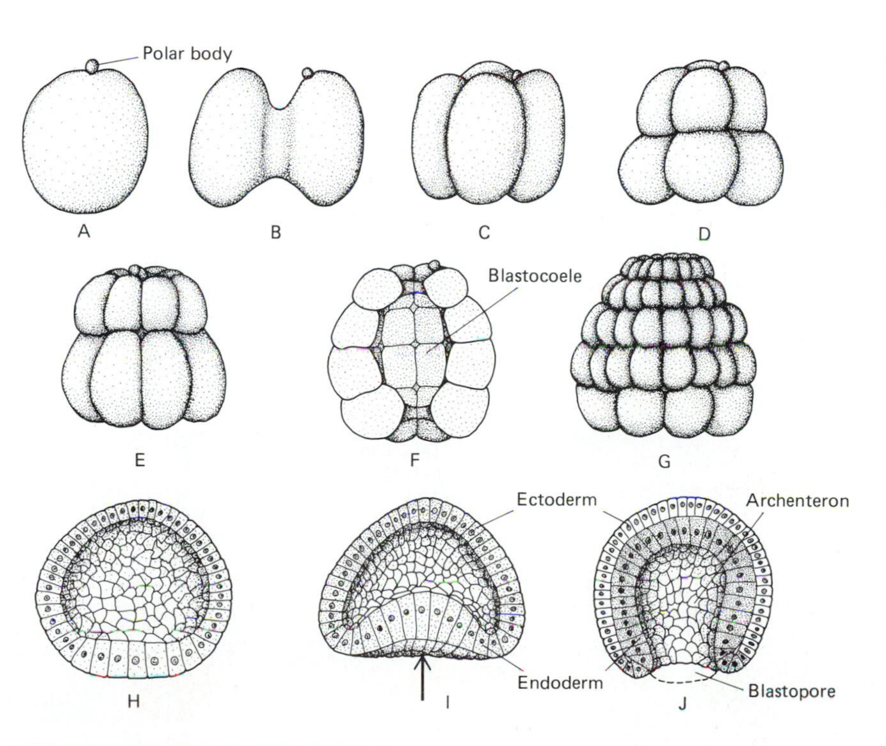

Isolecithal cleavage and gastrulation in *Amphioxus* viewed from the side. *A,* Mature egg with polar body. *B–E,* Two-, four-, eight-, and 16-cell stages. *F,* Embryo at 32-cell stage cut open to show the blastocoele. *G,* Blastula. *H,* Blastula cut open. *I,* Early gastrula showing beginning of invagination at vegetal pole *(arrow)*. *J,* Late gastrula. Invagination is completed and blastopore has formed.

(Continued on page 960)

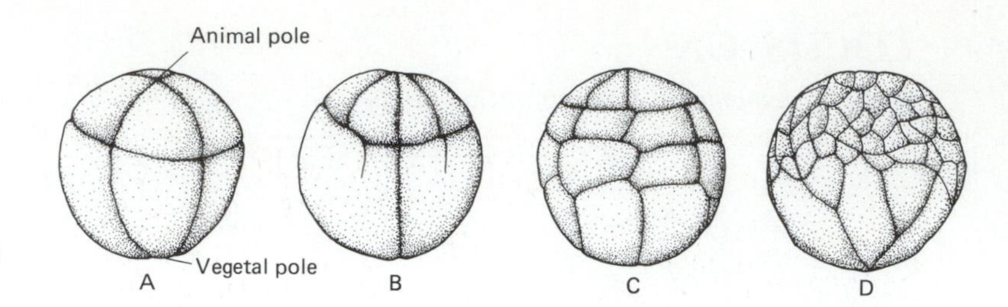

Successive stages in telolecithal cleavage in the frog, viewed from the side.

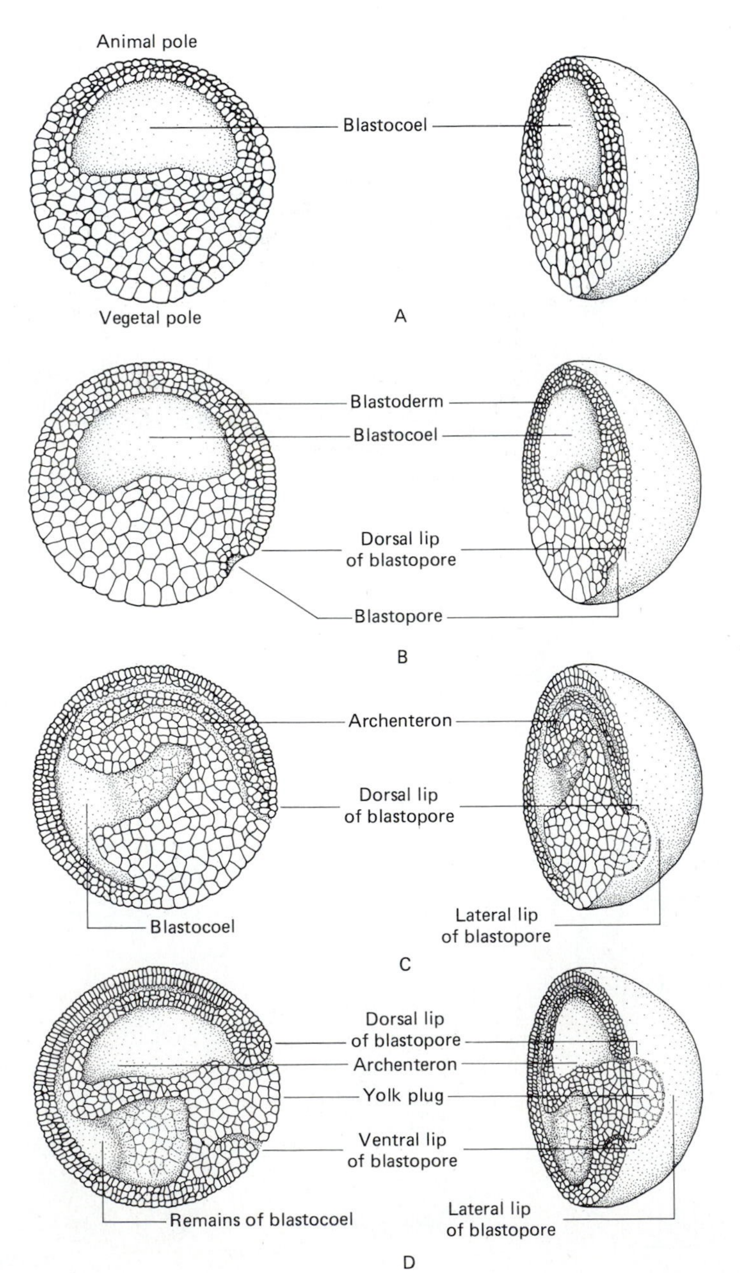

Stages in the development of a frog embryo. *A,* Late blastula. *B,* Early gastrula. *C,* Middle gastrula. *D,* Late gastrula.

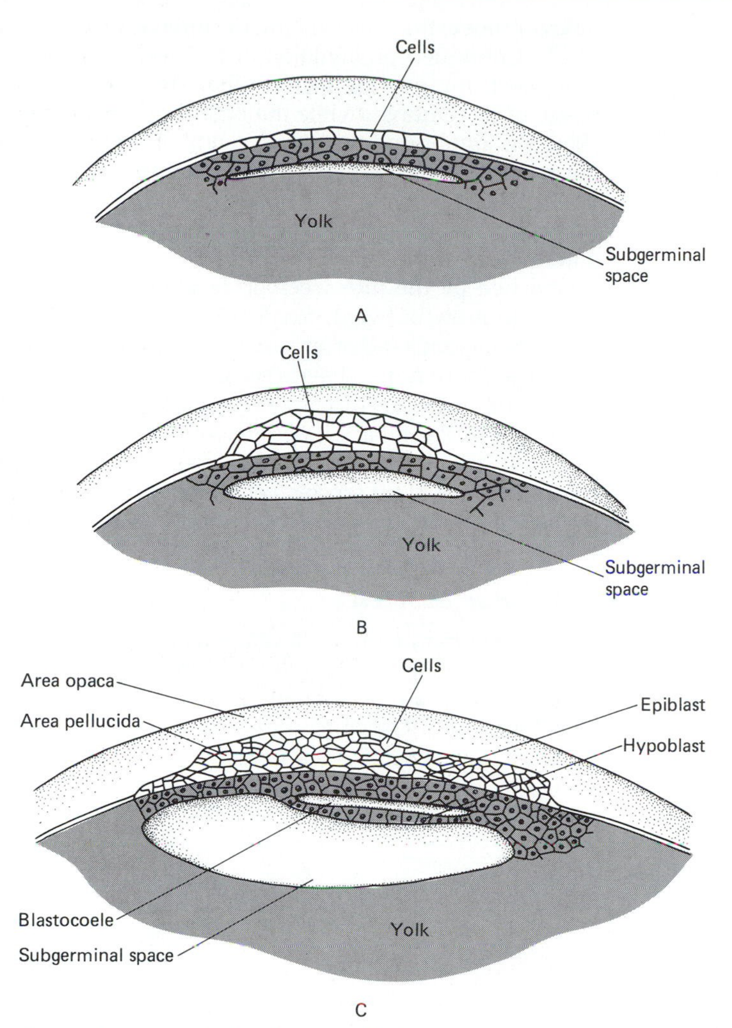

Successive stages in the cleavage of a hen's egg. *A,* Cleavage is restricted to a small disk of cytoplasm on the upper surface of the egg yolk called the blastodermic disk. *B,* A subgerminal space appears beneath the blastodermic disk, separating it from the unsegmented yolk. *C,* The blastodermic disk cleaves into an upper epiblast and a lower hypoblast separated by the blastocoele.

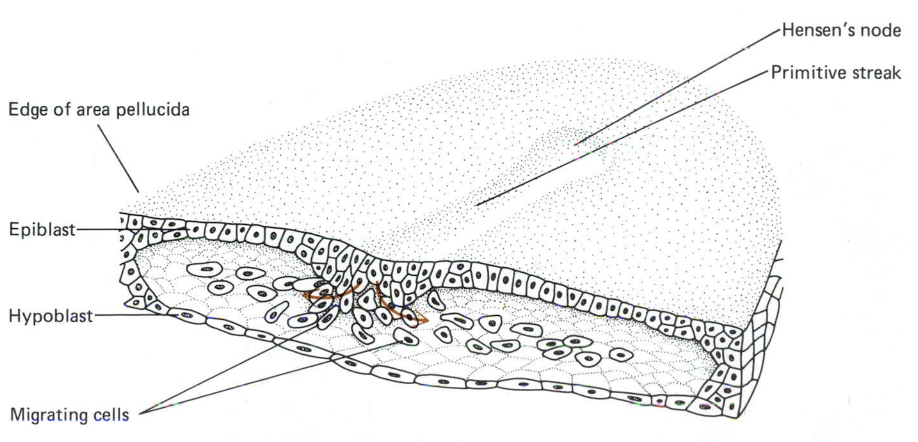

Gastrulation in the bird. The anterior half of the area pellucida of a chick embryo is cut transversely to show the migration of mesodermal and endodermal cells from the primitive streak.

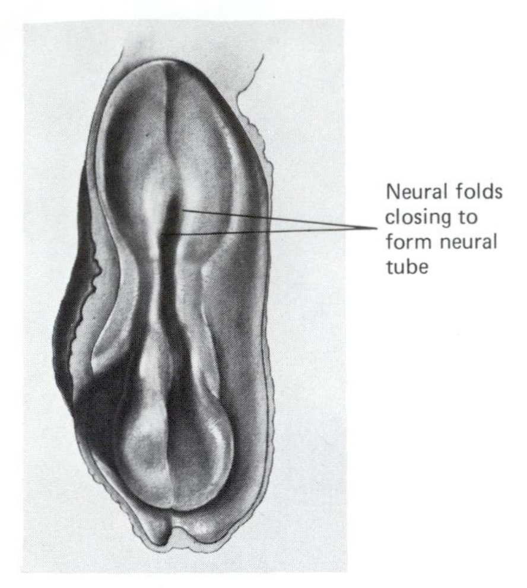

Figure 44–7 A 20-day-old human embryo. Note the developing nervous system. (Courtesy of Carnegie Institution of Washington.)

Central cells of the neural plate move downward, forming a depression, the **neural groove;** the cells flanking the groove on each side form the **neural folds** (Fig. 44–7). Continued proliferation of their cells brings the folds closer together until they meet, forming the **neural tube.** The neural tube sinks slowly down into the tissue of the embryo, while the outer ectoderm grows, forming the outer layer of skin. The anterior portion of the neural tube grows and differentiates into the brain; the remainder of the tube develops into the spinal cord.

The anterior part of the neural tube is much larger than the posterior part and continues to grow so rapidly that the head region comes to bend down at the anterior end of the embryonic disk. The forebrain, midbrain, and hindbrain are all established by the fifth week of development. A week or so later, the forebrain begins to grow outward, forming the rudiments of the cerebral hemispheres.

The various motor nerves grow out of the developing brain and spinal cord, but the sensory nerves have a separate origin. When the neural folds fuse, forming the neural tube, bits of nervous tissue known as the **neural crest** are left over on each side of the tube. These migrate downward from their original position and form the dorsal root ganglia of the spinal nerves and the postganglionic sympathetic neurons. From sensory cells in the dorsal root ganglia, dendrites grow out, to the sense organs, and axons grow in, to the spinal cord.

THE FIRST MONTH

During the first month of development, the S-shaped tubular heart develops and begins to beat—about 60 times each minute (Fig. 44–8). In the pharyngeal region, **branchial arches** have formed. These can be seen in Figure 44–9 and in Focus on Development of Organ Systems. The **branchial grooves** lie between these arches.

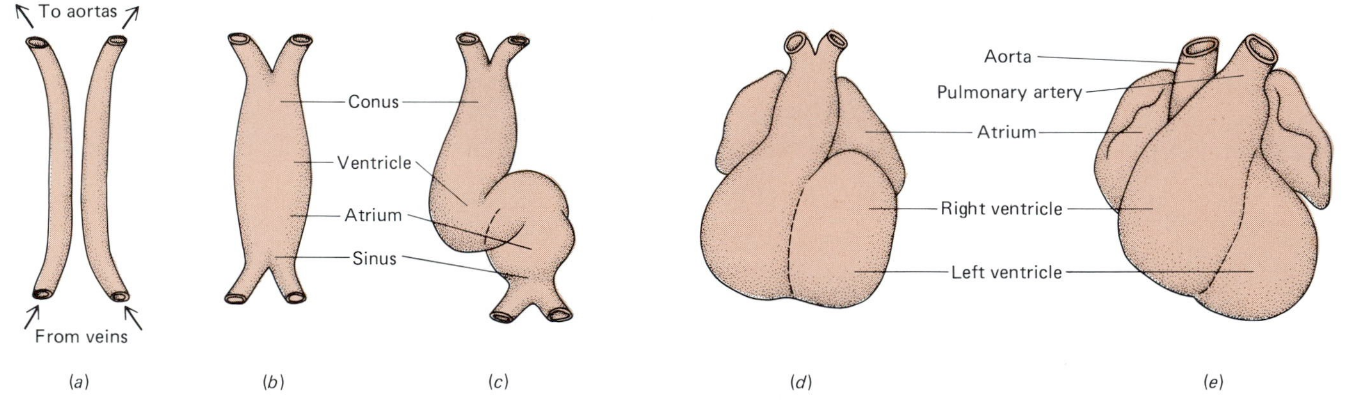

Figure 44–8 Ventral views of successive stages in the development of the heart. The heart forms from the fusion of two blood vessels. At first it is upside down, with the receiving end at the bottom. As it develops, the heart twists and turns in position, and the chambers divide to form the right-side-up four-chambered heart. The heart begins to beat spontaneously without nervous stimulation.

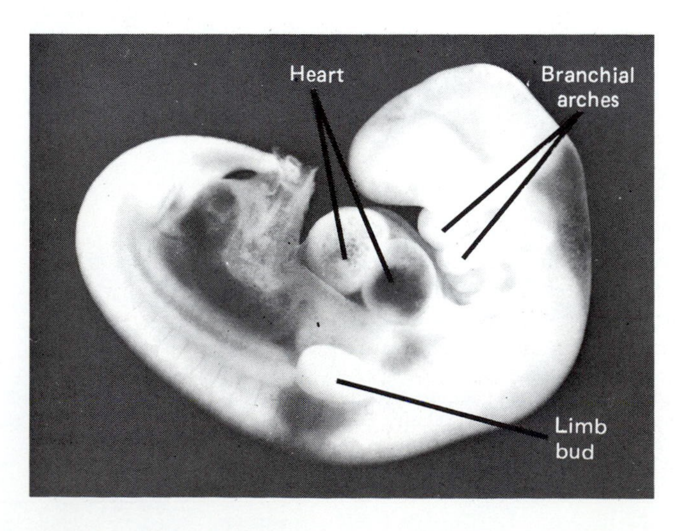

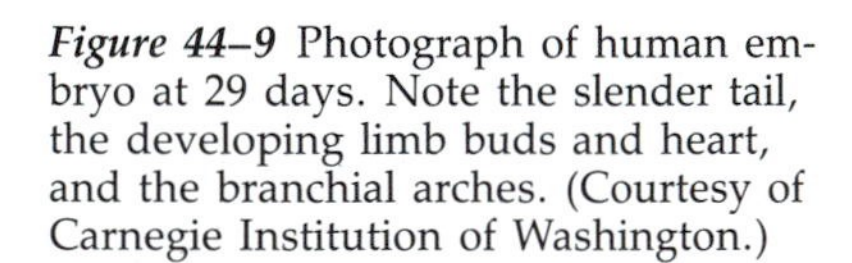

Figure 44–9 Photograph of human embryo at 29 days. Note the slender tail, the developing limb buds and heart, and the branchial arches. (Courtesy of Carnegie Institution of Washington.)

FOCUS ON

Development of Organ Systems

Many organ systems begin to take shape during the fifth week of development. Limb buds are evident and branchial arches have formed (Fig. 44–9). These arches contain the rudiments of skeletal, neural, and vascular elements of the face, jaws, and neck. Branchial grooves between the arches extend inward toward the pharyngeal pouches pushing outward (Fig. A). In aquatic animals with gills these cavities meet to form gill slits. In terrestrial forms each branchial groove remains separated from the corresponding pharyngeal pouch by a thin membrane of tissue. Structures more appropriate for life on land develop. For example, the first branchial groove becomes the external ear canal, and the first pharyngeal pouch develops into the middle ear cavity and eustachian tube.

Esophagus, stomach, and intestine are already evident, and the liver and pancreas have begun to develop (Fig. B). The already functioning heart continues to differentiate, and blood cells and blood vessels develop. The respiratory system develops as an outgrowth of the floor of the pharynx. The trachea and paired lung buds become evident during the fifth week. Nervous, endocrine, and other organ systems continue to develop.

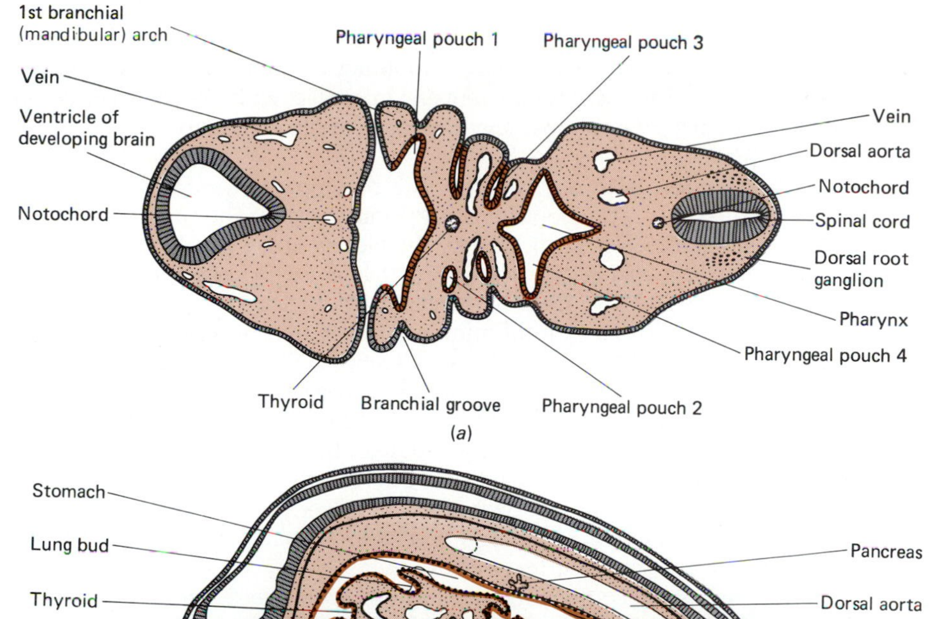

Cross section through a human embryo during the fifth week of development. Note that because the embryo is flexed, both the brain and spinal cord are present in the cross section. Note the branchial grooves and pharyngeal pouches.

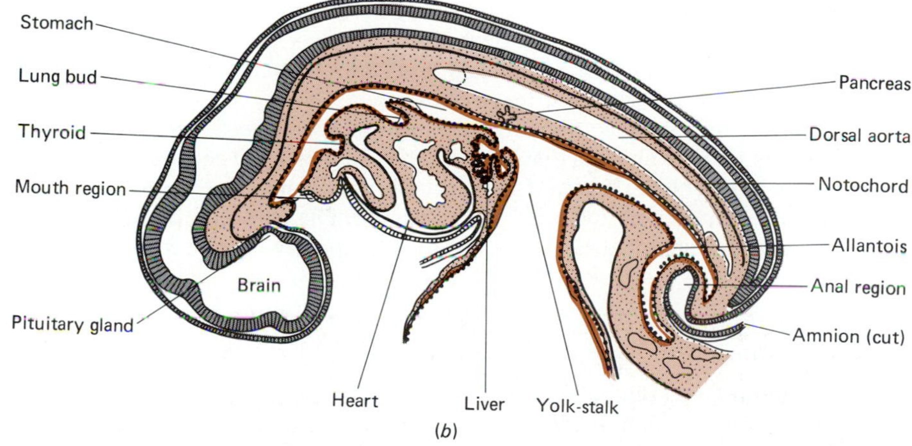

Sagittal section through human embryo showing some of the structures developing during the fifth week. Note that the liver, pancreas, and respiratory systems develop as outpocketing from the digestive tract.

Interior to each branchial groove lies a small cavity called a **pharyngeal pouch.** In fish and some amphibians, gills develop along the branchial arches. In humans and other terrestrial vertebrates, these structures are modified, forming structures more appropriate to life on land; for example, the tissue separating the first branchial groove and interior pharyngeal pouch becomes the eardrum, the groove itself becomes the external auditory meatus, and the pharyngeal pouch becomes the eustachian tube. In the floor of the pharynx at the level of the fourth pharyngeal pouches a tube of cells grows downward, forming the primordial trachea, which gives rise to the lung buds.

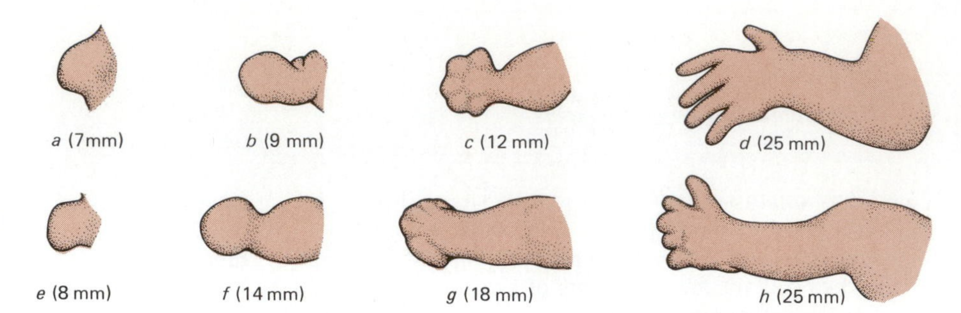

Figure 44–10 Stages in the development of the human arm *(upper row)* and leg *(lower row)* between the fifth and eighth weeks.

During the first month, the limb buds begin to differentiate, and will give rise to arms and legs (Fig. 44–10). The digestive system gives rise to outgrowths that will develop into the liver, gallbladder, and pancreas.

EXTRAEMBRYONIC MEMBRANES AND PLACENTA

All terrestrial vertebrates have four **extraembryonic membranes:** the chorion, the allantois, the yolk sac, and the amnion (Fig. 44–11). These membranes are not part of the embryo proper, and are discarded at birth. During development they protect the embryo and help in obtaining food and oxygen and eliminating wastes.

The **chorion** develops from an outfolding of the body wall. In the eggs of reptiles and birds, this membrane remains in contact with the inner surface of the shell. In mammals, it develops from the trophoblast and lies next to the cells of the uterine wall.

The **allantois** is an outgrowth of the developing digestive tract. In reptiles and birds, it serves as a depot for nitrogenous wastes. The products of nitrogen metabolism are excreted as uric acid by the kidney of the developing embryo. The poorly soluble uric acid is deposited as crystals in the cavity of the allantois and is discarded along with the allantois when the young hatch out of the egg shells.

The allantois fuses with the chorion to form the **chorioallantoic membrane,** which is rich in blood vessels. In the chick embryo, blood in these vessels provides oxygen and receives carbon dioxide and other wastes from the embryo. Gases are exchanged through the shell. In the human, the allantois is small and nonfunctional except that its blood vessels contribute to the formation of umbilical vessels. When the chick hatches or the child is born, most of the allantois, like the other extraembryonic membranes, is discarded. However, the base of the allantois, the part originally connected to the digestive tract, remains within the body and is converted into part of the urinary bladder.

Like the allantois, the **yolk sac** forms as an outpocketing of the developing digestive tract. In animals that produce yolk, the yolk sac encloses the yolk, slowly digests it, and makes it available to the embryo. Even though there is no yolk in the human ovum, a yolk sac forms and is discernible between the second and sixth weeks of development. Its walls serve as temporary centers for the formation of blood cells.

The Amnion

The **amnion** is an outfolding of the body wall of the embryo that grows around the embryo, meeting and fusing above it. Eventually it encloses the entire embryo (Fig. 44–11). The space between the embryo and the amnion, known as the amniotic cavity, becomes filled with clear, watery **amniotic fluid** secreted by the membrane. Embryos of terrestrial vertebrates develop within this pool of fluid. The amniotic fluid prevents desiccation of the embryo and acts as a protective cushion that absorbs shocks and prevents the amniotic membrane from sticking to the embryo. It also permits the embryo a certain freedom of motion.

During the birth process in mammals, pressure of the contracting uterus is transmitted via the amniotic fluid and helps dilate the cervix. Normally the amnion ruptures shortly before birth, releasing about a liter of amniotic fluid. Sometimes it fails to burst and the child is born with its head still enveloped by the amnion. The amnion is then popularly known as a "caul" and is the source of many odd superstitions.

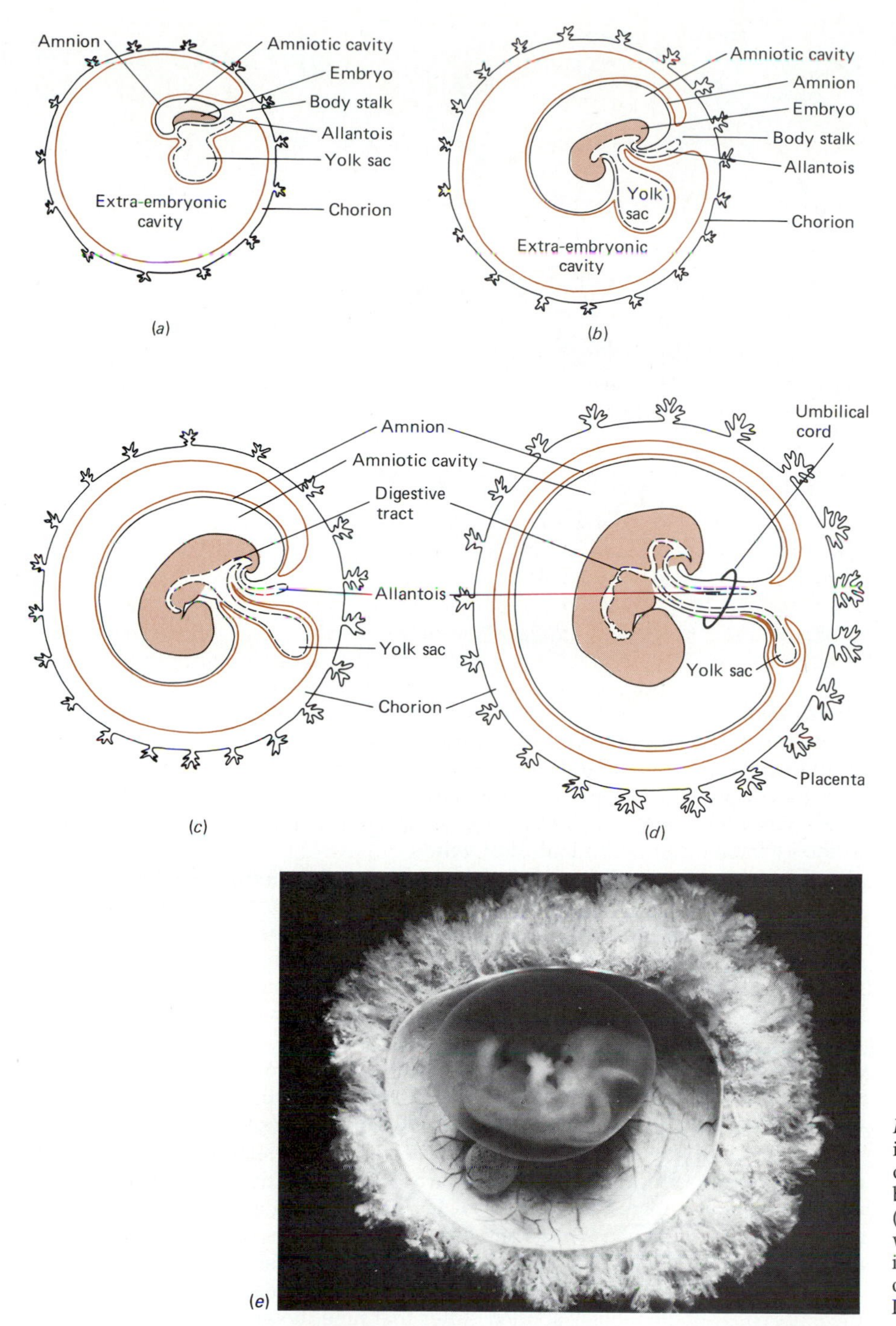

Figure 44–11 (*a*)–(*d*) Successive stages in the development of the extraembryonic membranes, umbilical cord, and body form in the human embryo. (*e*) Photograph of embryo in seventh week of development, surrounded by its intact amnion and other extraembryonic membranes. (Courtesy of Carnegie Institution of Washington.)

Placenta

In humans and other placental mammals, the **placenta** is the organ of exchange between mother and fetus, providing nutrients and oxygen for the fetus, and removing wastes from the developing organism for excretion by the mother (Fig. 44–12). In addition, the placenta serves as an endocrine organ.

The placenta develops from both the embryonic chorion and the maternal uterine tissue. After implantation, the chorion continues to grow rapidly, invading the endometrium and forming villi, which become vascularized (infiltrated with blood vessels) as the embryonic circulation develops. Enzymes released by the invading embryonic cells destroy some of the endometrial tissue, including many small blood vessels. Small amounts of blood oozing from these damaged vessels form pools about the chorionic villi.

As the human embryo grows, the region on the ventral side from which the folds of the amnion, yolk sac, and allantois grew becomes relatively smaller, and

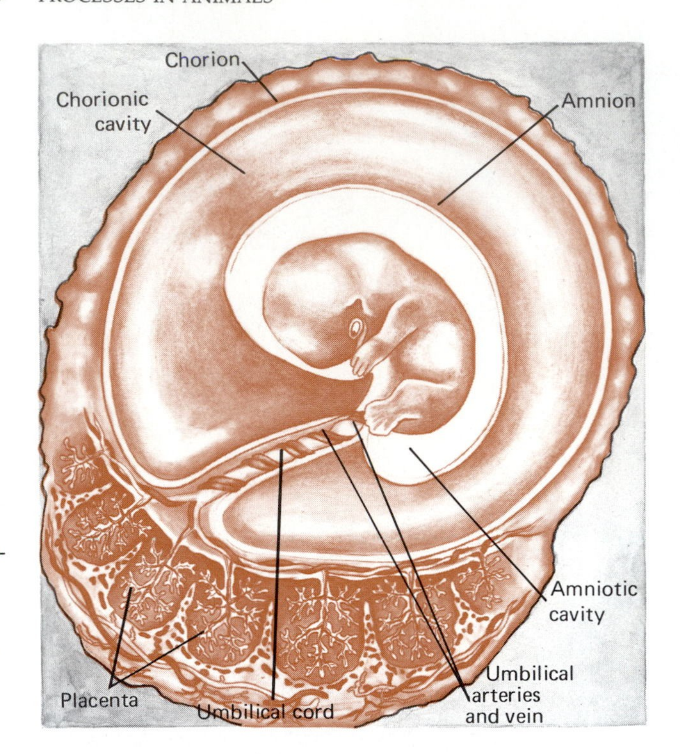

Figure 44–12 The placenta is the organ of exchange between mother and developing child. In the placenta the blood of the mother comes close enough to the blood of the child so that materials can effectively diffuse between the two circulatory systems. However, the maternal and fetal blood itself does not mix.

the edges of the amniotic folds come together to form a tube that encloses the other membranes. This tube, the **umbilical cord,** connects the embryo with the placenta. In addition to the yolk sac and allantois, the umbilical cord contains the two umbilical arteries and the umbilical vein. The **umbilical arteries** connect the embryo with a vast network of capillaries developing within the villi; blood from the villi returns to the embryo through the **umbilical vein.**

The placenta actually consists of the portion of the chorion that develops villi, together with the uterine tissue between the villi, which contains maternal capillaries and small pools of maternal blood. The blood of the fetus in the capillaries of the chorionic villi comes in close contact with the mother's blood in the tissues between the villi. However, they are always separated by a membrane through which substances may diffuse or be actively transported. *Maternal and fetal blood do not normally mix in the placenta or any other place.*

Several hormones are produced by the placenta. From the time the embryo first begins to implant itself, its trophoblastic cells release **human chorionic gonadotropin (HCG),** which signals the corpus luteum that pregnancy has begun. In response, the corpus luteum increases in size and releases large amounts of progesterone and estrogen, which in turn stimulate continued development of endometrium and placenta. Without HCG, the corpus luteum would degenerate and the embryo would be aborted and flushed out with the menstrual flow. In such a case the woman would probably not even know that she had been pregnant. When the corpus luteum is removed before about the eleventh week of pregnancy, the embryo is spontaneously aborted. After that time, however, the placenta itself produces enough progesterone and estrogens to maintain pregnancy.

Later Development

All of the organs continue to develop during the second month. A thin tail becomes prominent during the fifth week (see Fig. 44–9) but fails to keep pace with the rapid growth of the rest of the body and so becomes inconspicuous by the end of the second month. Muscles develop, and the embryo is capable of some movement (Table 44–2). The brain begins to send impulses to regulate the functions of some organs, and a few simple reflexes are evident. After the first two months of development the embryo is referred to as a **fetus** (Fig. 44–13).

By the end of the **first trimester** (first three months of development), the fetus is recognizably human. The external genital structures have differentiated, indicating the sex of the fetus. Ears and eyes approach their final positions. Some of the bones become distinct, and the notochord degenerates. The fetus performs breath-

TABLE 44–2
Some Important Developmental Events

Time from Fertilization	*Event*
24 hours	Embryo reaches two-cell stage
3 days	Morula reaches uterus
7 days	Blastocyst begins to implant
2.5 weeks	Notochord and neural plate are formed; tissue that will give rise to heart is differentiating; blood cells are forming in yolk sac and chorion.
3.5 weeks	Neural tube forming; primordial eye and ear visible; pharyngeal pouches forming; liver bud differentiating; respiratory system and thyroid gland just beginning to develop; heart tubes fuse, bend, and begin to beat; blood vessels are laid down.
4 weeks	Limb buds appear; three primary vesicles of brain formed.
2 months	Muscles differentiating; embryo capable of movement. Gonad distinguishable as testis or ovary. Bones begin to ossify. Cerebral cortex differentiating. Principal blood vessels assume final positions.
3 months	Sex can be determined by external inspection. Notochord degenerates. Lymph glands develop.
4 months	Face begins to look human. Lobes of cerebrum differentiate. Eye, ear, and nose look more "normal."
Third trimester	Lanugo appears, then later is shed; neurons become myelinated; tremendous growth of body
266 days (from conception)	Birth

ing movements, pumping amniotic fluid into and out of its lungs, and even carries on sucking movements. By the end of the third month, the fetus is almost 56 mm (about 2.2 in.) long and weighs about 14 g (0.5 oz).

During the **second trimester** (Fig. 44–14), the fetal heart (now beating about 150 times per minute) can be heard with a stethoscope. The fetus moves freely through the amniotic cavity; during the fifth month, the mother usually becomes aware of fetal movements ("quickening").

Figure 44–13 The human embryo at 10 weeks of development. (*a*) Note position in the uterine wall. (*b*) Photograph at 10 weeks. (Courtesy of Carnegie Institution of Washington.)

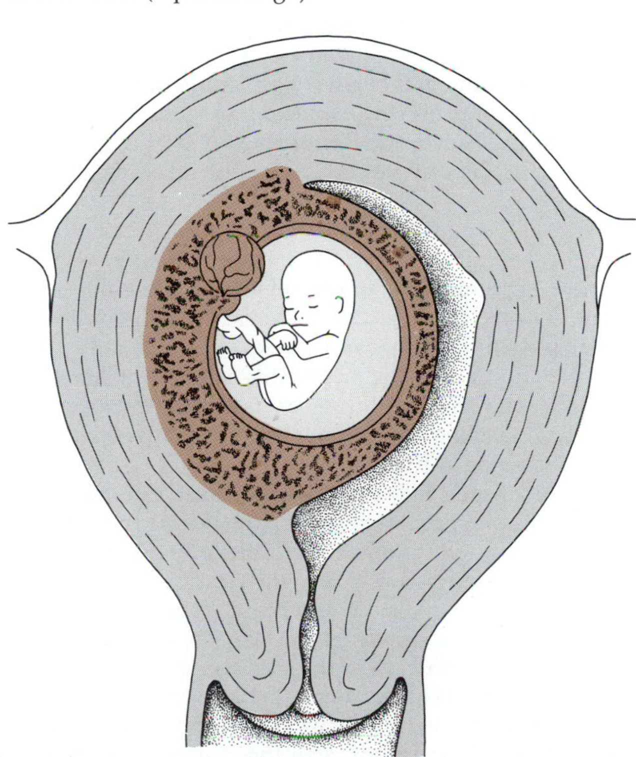

(a)

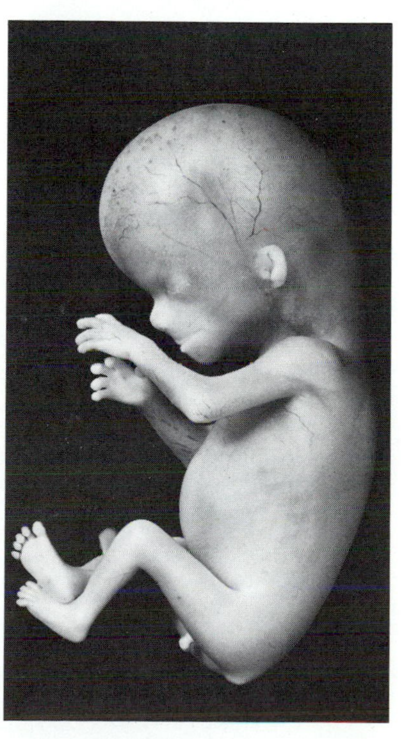

(b)

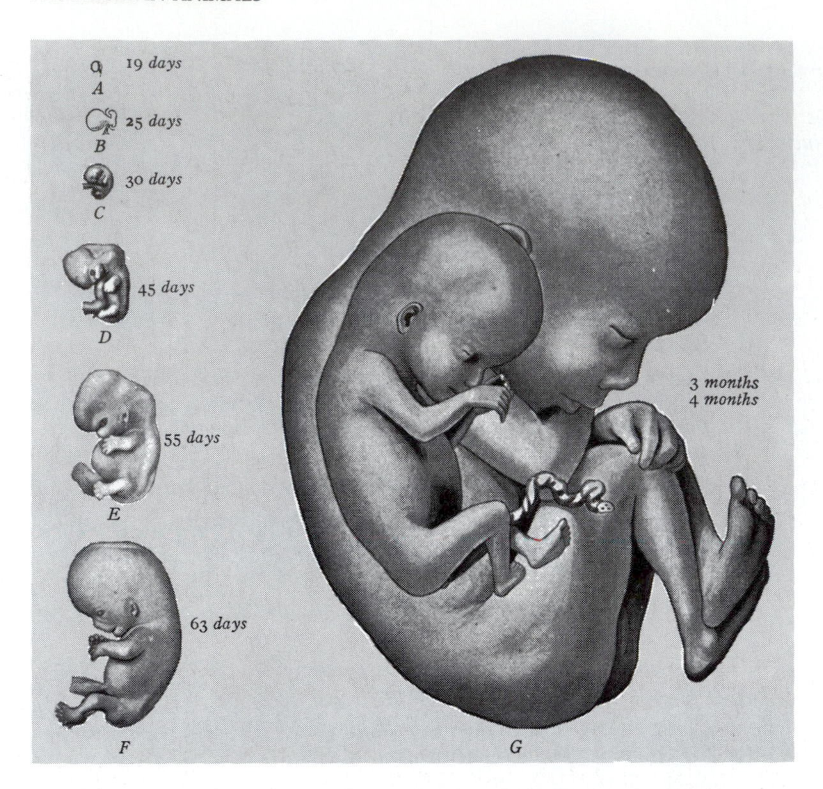

Figure 44–14 A graded series of human embryos. Note the characteristic position of the arms and legs in the 4-month fetus. (From L. B. Arey.)

During the **final trimester,** the fetus grows rapidly, and final differentiation of tissues and organs occurs. At the beginning of the sixth month the skin has a wrinkled appearance, perhaps because it is growing faster than the underlying connective tissue. If born prematurely at this age the fetus attempts to breathe and is able to move and cry, but almost always dies because its brain is not sufficiently developed to sustain vital functions such as rhythmic breathing and the regulation of body temperature.

During the seventh month the cerebrum grows rapidly and develops convolutions. The grasp and sucking reflexes are evident, and the fetus may suck its thumb. Most of the body is covered by a downy hair called **lanugo,** which is usually shed before birth. Occasionally the lanugo is not shed until a few days after birth.

During the last months of prenatal life, a protective creamlike substance, the **vernix,** covers the skin, and hair begins to grow on the scalp. At birth the average full-term baby weighs about 3000 g (7 lb) and measures about 52 cm (20 in.) in total length (or about 35 cm from crown to rump).

The Birth Process

The human **gestation period,** the duration of pregnancy, is normally 280 days, from the time of the last menstrual period to the birth of the baby (or 266 days from conception). Babies born as early as 28 weeks or as late as 45 weeks after the last menstrual period may survive. The factors that actually initiate the process of birth, or **parturition,** after the period of gestation is complete are not well understood. Childbirth begins with a long series of involuntary contractions of the uterus, experienced as the contractions of **labor.**

Labor may be divided into three stages. During the **first stage,** which typically lasts about 12 hours, the contractions of the uterus move the fetus down toward the cervix, causing it to dilate. The cervix also becomes effaced, losing its normal shape and flattening so that the fetal head can pass through. During the first stage of labor the amnion usually ruptures, releasing the amniotic fluid, which flows out through the vagina. In the **second stage,** which normally lasts between 20 minutes and an hour, the fetus passes through the cervix and vagina and is born, or "delivered" (Fig. 44–15). With each uterine contraction the woman holds her breath and bears down so that the fetus is expelled from the uterus by the combined forces of uterine contractions plus the contractions of the muscles of the abdominal wall.

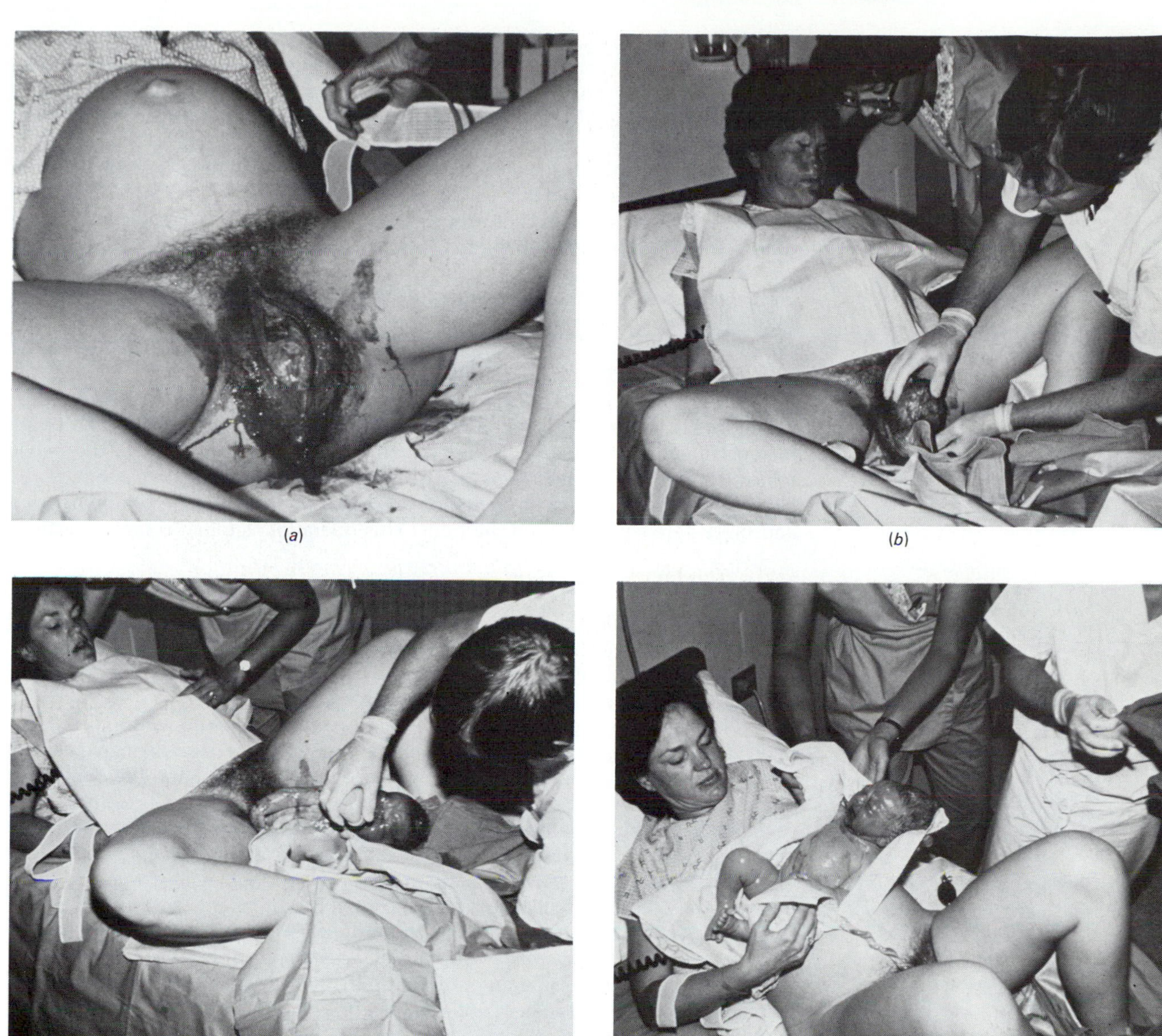

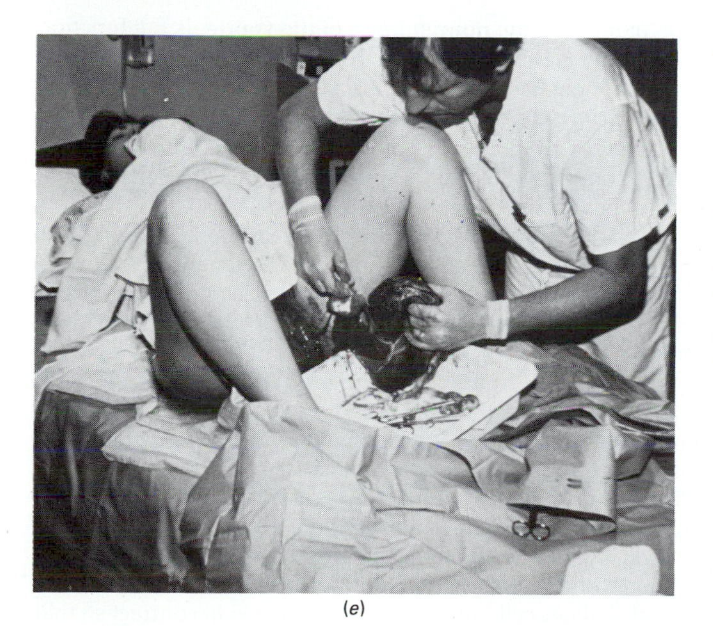

Figure 44–15 Birth of a baby. In about 95% of all human births the baby descends through the cervix and vagina in the head-down position. (*a*) The baby's head appears. (*b*) The mother bears down hard with her abdominal muscles helping to push the baby out. When the head fully appears, the physician or midwife can gently grasp it and guide the baby's emergence into the outside world. (*c*) Once the head has emerged, the rest of the body usually follows readily. The physician gently aspirates the mouth and pharynx to clear the upper airway of any amniotic fluid, mucus, or blood. At this time the neonate usually takes its first breath. (*d*) The baby, still attached to the placenta by its umbilical cord, is presented to its mother. (*e*) During the third stage of labor the placenta is delivered. (Courtesy of Dan Atchison.)

After the baby is born, the contractions of the uterus squeeze much of the fetal blood from the placenta back into the infant. After the pulsations in the umbilical cord cease, the cord is tied and cut, severing the child from the mother. The stump of the cord gradually shrivels until nothing remains but the depressed scar, the **navel.**

During the **third stage** of labor, which lasts 10 or 15 minutes after the birth of the child, the placenta and the fetal membranes are loosened from the lining of the uterus by another series of contractions and expelled. At this stage they are called collectively the **afterbirth.** In humans and certain other mammals in which the placenta forms a very tight connection with the uterine lining, the expulsion of the placenta is accompanied by some loss of blood. In other mammals in which the connection between the fetal membranes and uterine wall is not close, the placenta can pull away from the uterine wall without causing bleeding. Following birth, the size of the uterus decreases, and its lining is rapidly restored.

During labor an obstetrician may administer drugs such as oxytocin or prostaglandins to increase the contractions of the uterus, or may assist with special forceps or other techniques. In some women, the aperture between the pelvic bones through which the vagina passes is too small to permit the passage of the baby, so the child must be delivered by an operation in which the abdominal wall and uterus are cut open from the front. This operation is called **cesarean section.**

Adaptations to Birth

Great changes take place within a short time after a baby is born. During prenatal life, the fetus received both food and oxygen from the placenta. Now the newborn's own digestive and respiratory systems must function. Correlated with these changes are several major changes in the circulatory system.

Normally the **neonate** (newborn infant) begins to breathe within a few seconds of birth, and cries within a half a minute. If anesthetics have been given to the mother, however, the fetus may also have been anesthetized, and breathing and other activities may be depressed. Some infants may not begin breathing until several minutes have passed. This is one of the reasons for the current trend toward natural childbirth, in which as little medication as possible is used.

It is believed that the first breath of the neonate is initiated by the accumulation of carbon dioxide in the blood after the umbilical cord is cut. This stimulates the respiratory centers in the medulla. The resulting expansion of the lungs enlarges its blood vessels (which previously were partially collapsed), and blood from the right ventricle flows in increasing amounts through the pulmonary vessels instead of through the arterial duct that connected the pulmonary artery and aorta during fetal life.

Lactation

Lactation, the production of milk for the nourishment of the young, is the function of the breasts, which contain the **mammary glands.** The breasts overlie the pectoral muscles and are attached to them by connective tissue. Fibrous bands of tissue called **ligaments of Cooper** firmly connect the breasts to the skin.

Each breast is composed of 15 to 20 lobes of glandular tissue, further subdivided into lobules made of connective tissue in which gland cells are embedded. The secretory cells are arranged in little grapelike clusters called **alveoli** (Fig. 44–16). Ducts from each cluster unite to form a single duct from each lobe, so that there are 15 to 20 tiny openings on the surface of each nipple. The amount of adipose tissue around the lobe of the glandular tissue determines the size of the breasts and accounts for their soft consistency. The size of the breasts does not affect their capacity to produce milk.

During pregnancy, high concentrations of estrogens and progesterone produced by the corpus luteum and by the placenta stimulate the glands and ducts to develop, resulting in increased breast size. For the first couple of days after childbirth, the mammary glands produce a fluid called **colostrum,** which contains protein and lactose but little fat. After birth, **prolactin,** a hormone secreted by the anterior lobe of the pituitary, stimulates milk production, and usually by the third day after delivery milk itself is produced. When the infant suckles at the breast, a reflex action in the mother results in release of prolactin and oxytocin from the pituitary gland. Oxytocin stimulates cells surrounding the alveoli to contract so that

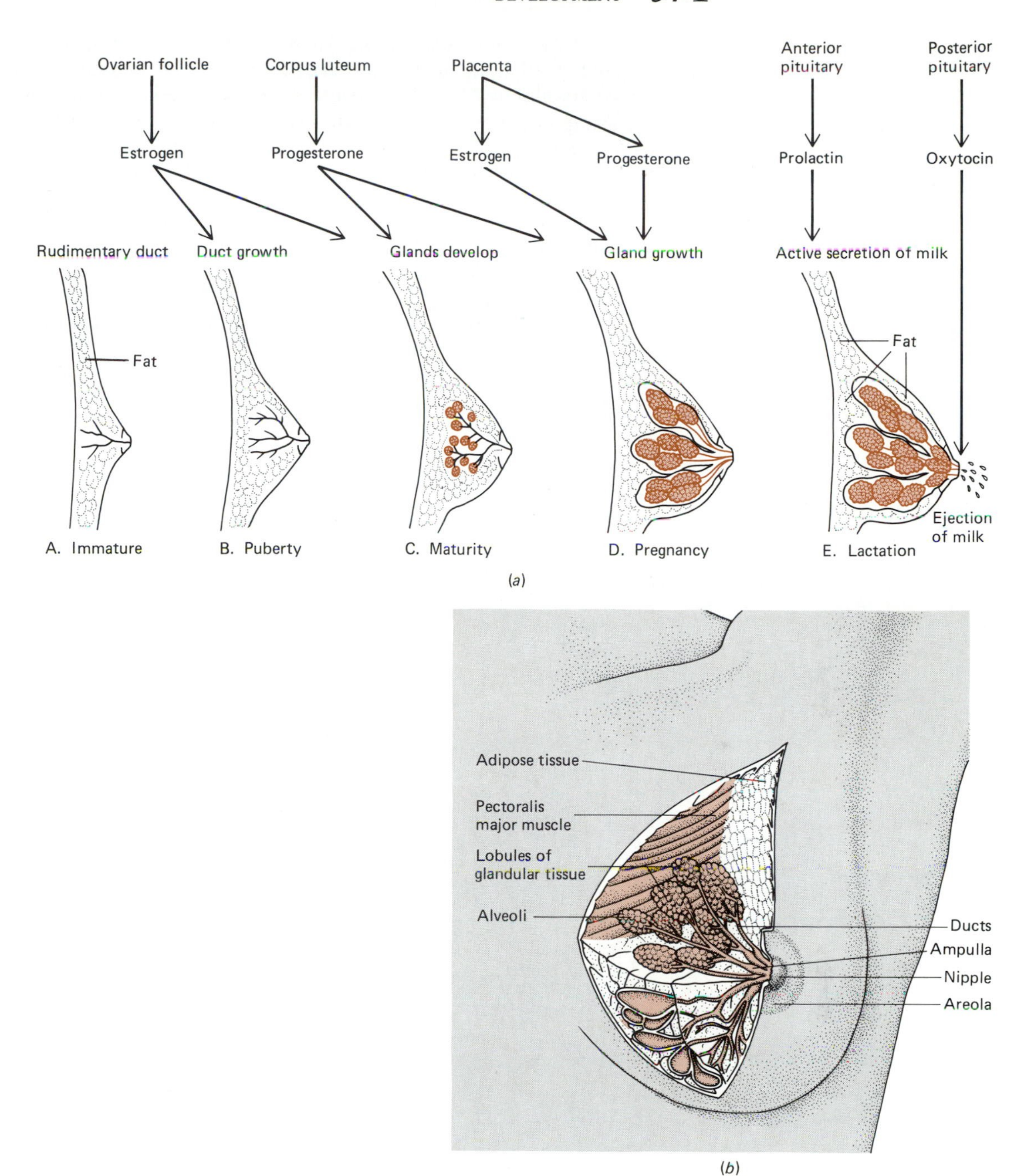

Figure 44–16 The development of the female breast is under hormonal control.

the alveoli are compressed. This forces milk from the alveoli into the ducts where it can be sucked by the baby.

Breastfeeding a baby offers many advantages besides promoting a close bond between mother and child. Breast milk is tailored to the nutritional needs of the human infant, whereas cow's milk is more likely to produce allergies to dairy products. Furthermore, breastfed babies receive antibodies in the colostrum and mother's milk that are thought to play a protective role, resulting in a lower incidence of infantile diarrhea and even of respiratory infection during the second six months of life. Breastfeeding also promotes recovery of the uterus, because the oxytocin released in conjunction with breastfeeding stimulates the uterus to contract to nonpregnant size.

What Regulates Developmental Processes?

One of the important unsolved problems of modern biology is the nature of the mechanisms that regulate developmental processes so that each organ appears at the proper time and in the proper spatial relations to the other organs. How can a single cell, the fertilized egg, give rise to the many different types of cells that make up the adult organism—cells that differ so widely in their structure, functions, and chemical properties? The advances in biochemical genetics in the past 25 years have permitted new intellectual and experimental approaches to this problem.

CYTOPLASMIC FACTORS

Perhaps the initial influence upon differentiation is the distribution of materials in the cytoplasm of the zygote itself. If the zygote cytoplasm is not homogeneous, then during cell division the cytoplasmic substances portioned out to each new cell might also vary. The two new cells would differ from one another with regard to their cytoplasmic content; such differences could influence the course of development (Fig. 44–17).

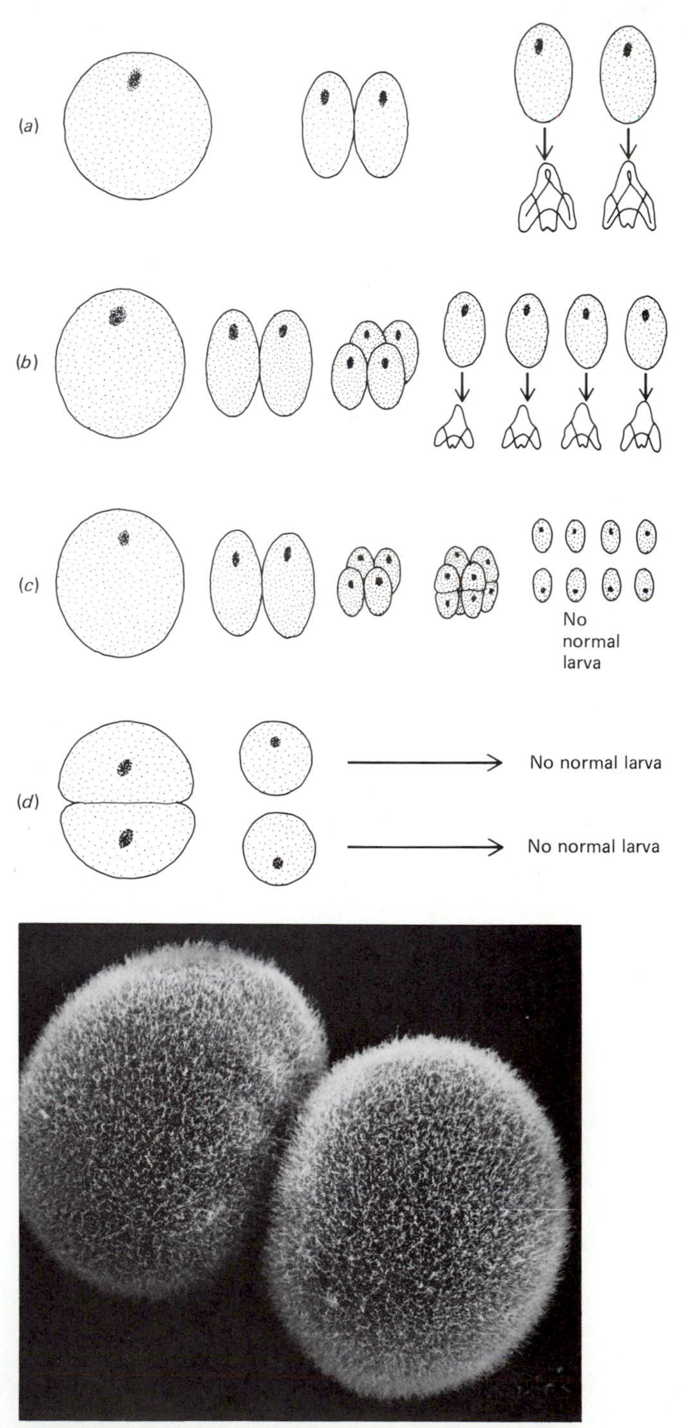

Figure 44–17 Experiments that illustrate the influence of cytoplasmic factors upon development. (*a*) When the first two cells of a sea urchin embryo are separated from one another, each cell is able to give rise to a normal larva. (*b*) If each of the first four cells of the sea urchin embryo are separated from one another, four normal larvae develop. (*c*) When the cells of the eight-cell stage are separated from one another, not even one is able to develop into a normal larva. (*d*) If an experimenter causes the plane of cleavage of the first division to extend horizontally, instead of from pole to pole of the zygote, neither cell of the two-cell stage is able to develop into a normal larva. Apparently, certain cytoplasmic substances necessary for normal development are heterogeneously distributed in the zygote so that neither half has all of the needed ingredients. (*e*) Scanning electron micrograph of the first cleavage division of the egg of a sea urchin ($\times$1300). ((*e*), Courtesy of Dr. Everett Anderson.)

Most of the cytoplasm in the zygote is contributed by the egg. It contains most of the RNA needed for synthesizing the proteins necessary for cleavage. When the zygote is injected with actinomycin D, a drug that inhibits RNA synthesis, cleavage is normal, although the embryo ultimately dies. When the zygote is treated with cycloheximide, a drug that inhibits protein synthesis, cleavage does not proceed normally. Thus, protein synthesis is necessary for normal cleavage, but the messenger RNA necessary for this synthesis is already present in the ovum before fertilization.

ROLE OF THE GENES

The genes contained within the nucleus of the zygote are an elaborate set of instructions for making a specific organism. A human zygote always develops into a human baby, never into a calf or puppy, because information in its genes is specific for making a human organism. Furthermore, one human may be tall with blue eyes, another short with brown eyes. These, as well as thousands of other characteristics, are determined by the specific genes present in the zygote.

Cellular differentiation might be explained if genetic material were parcelled out differently at cell division and the daughter cells received different kinds of genetic information. Although there are a very few clear instances of differential nuclear division in animals such as *Ascaris* and *Sciara,* this does not appear to be a general mechanism of differentiation. By and large, genes are neither lost nor gained during developmental processes. The generalization that the mitotic process ensures the exact distribution of genes to each cell of the organism is a valid one.

When the zygote divides by mitosis, forming two cells, a complete set of genes is distributed to each new cell. As cleavage and later development proceed, more and more cells are formed by mitosis, and every one contains the identical number of genes and genetic information present in the zygote. Experiments with nuclear transplants show that even the nucleus from a differentiated cell taken from an advanced stage of embryonic development can, when placed in an enucleated egg, lead to the development of a normal embryo. Clearly, a full set of genetic information is retained in such a cell (Fig. 44–18). Now if all the cells of the organism have an identical set of genes with the same instructions, how can they become different from one another?

Genes control the activities of the cell by providing the instructions for making specific types of proteins. Apparently, the differences in kinds of enzymes and other proteins found in different cells of the same organism arise by differences in the activity of the same set of genes in different cells. In a muscle cell, for example, the genes for synthesizing actin and myosin are active, while in a nerve cell, genes specifying muscle proteins do not operate. Instead, genes with instructions for making proteins specific to nerve cells are active. Some striking evidence regarding the differential activity of genes comes from cytological studies of chromosome puffs in insect tissues (see Chapter 13).

INFLUENCE OF OTHER CELLS: ORGANIZERS

The position of a cell in relation to its cellular neighbors may be critical in determining its fate. One cell, or a group of cells, releases a substance that diffuses out among neighboring cells. This chemical is most concentrated in the vicinity of the

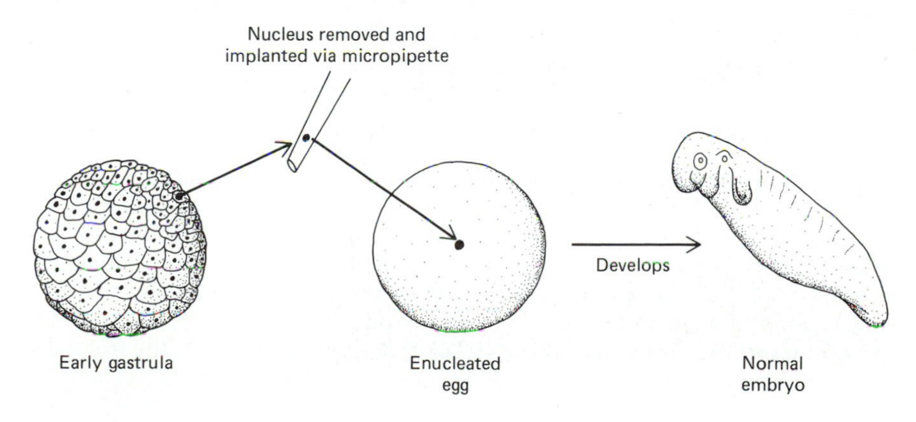

Figure 44–18 When a nucleus from the cell of an early gastrula of a frog is transplanted to an enucleated egg, a normal larva develops.

cell that produced it and less concentrated farther away. Such a concentration gradient could provide information to the cells regarding their position in relation to the group, and this information could trigger the biochemical events that would determine differentiation.

A chemical that determines the future differentiation of nearby cell groups is called an **organizer.** Let us look at a specific example, the development of the nervous system. In the late 1920s, the biologist Hans Spemann performed a series of experiments that helped to explain why ectoderm cells in the specific region of the neural plate develop into nerve tissue, while other ectoderm cells give rise to the outer layer of the skin.

Using early frog gastrulas, Spemann transplanted a tiny bit of ectoderm from the region that would normally develop into neural tube to the belly region of the embryo. The transplanted pre-neural tube ectoderm no longer formed neural tissue; instead, it joined forces with its new cellular neighbors to form outer skin. Likewise, when belly ectoderm was transplanted to the pre-neural tube region, its fate was determined by its new surroundings—it contributed to the formation of neural tube.

When these same experiments were performed on older gastrulas a few hours later in development, the results were entirely different. Then, when a piece of pre-neural tube tissue was transplanted into the belly region, it differentiated into neural tissue, even though this is out of place in the midst of the belly skin. And, as might be expected, transplanted belly ectoderm cells proceeded to form skin cells even in the midst of the developing neural tube.

Apparently something happened between the early and late gastrula stages that accounted for the difference in results. Spemann solved this mystery by showing that the notochord cells lying just beneath the preneural tube ectoderm are responsible for the change. The notochord cells induce the ectoderm cells to form neural tube. By producing and secreting a chemical organizer, the notochord cells communicate with the ectoderm cells, stimulating them to differentiate into neural tissue. In the early gastrula the pre-neural tube cells have not yet been subjected to the organizer, so their fate has not yet been determined. Once these cells have been influenced by the organizer during late gastrulation, their fate is decided and cannot easily be altered. When a piece of notochord tissue is transplanted beneath the belly ectoderm of an early gastrula, a second neural tube forms in this region. These and thousands of similar experiments have provided evidence that development depends upon chemical communication among cells and the resulting differential activity of their genes.

ENVIRONMENTAL FACTORS

Embryonic tissues growing in vivo have differential sensitivities to changes in nutrients, to the presence of inhibitors and antimetabolites, and to various environmental agents. Any of these factors, applied during the appropriate critical period in development, may change the course of development and differentiation. (Environmental factors are discussed further later in the chapter.)

Such factors as temperature, humidity, light, gravity, pressure, and chemicals in the surrounding environment are known to affect the embryo. It is logical, then, to suspect that if some cells in the embryo are subjected to slightly different environmental conditions than other cells, they might be affected in different ways. For example, the cells on the outside of the gastrula experience different conditions than those lining the inside of the embryo. Such differences may influence the internal workings of the cells so that the genes are affected.

INTERACTION OF NONGENETIC WITH GENETIC FACTORS

Although nongenetic influences are critical in determining differentiation, the genes exercise ultimate control. Consider a specific experimental example. When ectoderm from the mouth region of an early frog embryo is transplanted to the mouth region of a salamander, the surrounding salamander tissue induces the transplanted cells to develop into mouth structures. However, the mouth that forms is not the characteristic mouth of a salamander. Instead it bears the horny rows of teeth and horny jaws of a frog. Why? As you might guess, the frog cells are not *competent* to form structures specific to salamanders.

We have seen that variations in cytoplasmic substances, exposure to organizers diffusing from cellular neighbors, and differences in environmental conditions can initiate differences among cells. Once two cells become different from each other by virtue of these influences, their internal chemical composition is affected. Sooner or later, the genes in the nucleus are also affected; certain genes are activated, while others are repressed, resulting in the production of different types of proteins. Cellular differentiation, then, is an expression of genetic activity, and genetic activity in turn is influenced by a variety of factors within and without the cell. Although all cells of an organism possess the same set of genes, differential gene activity results in variations in chemistry, behavior, and structure among cells. In this way an embryo comes to be composed of more than 200 types of cells, each exquisitely designed to perform its specific functions.

Environmental Influences upon the Embryo

Although it is obvious that a baby's growth and development are influenced by the food he eats, the air he breathes, the disease organisms that infect him, and the chemicals or drugs to which he is exposed, it is less obvious that *prenatal* development is also affected by these environmental influences. Life before birth is, in fact, even more sensitive to environmental changes than it is in the fully formed baby.

About 5% of all babies born alive, or 175,000 babies per year, arrive with a defect of clinical significance. Such birth defects account for about 15% of deaths among newborns. Birth defects may be caused by genetic or environmental factors or a combination of the two. Genetic factors were discussed in Part 3. In this section we will examine the environmental conditions that affect the well-being of the embryo.

Substances or conditions that intrude upon the developing embryo from the outside environment may cause significant damage at one period in development, yet appear to be harmless during a later stage. Timing is important. Each developing structure has a critical period during which it is most susceptible to unfavorable conditions. Generally this critical period occurs early in the development of the structure, when interference with cell movements or divisions may prevent formation of normal shape or size, resulting in permanent malformation. Since most structures form during the first three months of embryonic life, the embryo is most susceptible to environmental conditions during this early period. During a substantial portion of this time the mother may not even realize that she is pregnant and may therefore take no special precautions to minimize potentially dangerous influences.

Anything that circulates in the maternal blood—nutrients, drugs, or even gases—may find its way into the blood of the fetus. Some drugs, as well as other types of agents are **teratogens;** they can interfere with normal development (Fig. 44–19). Table 44–3 describes some of the environmental influences upon development.

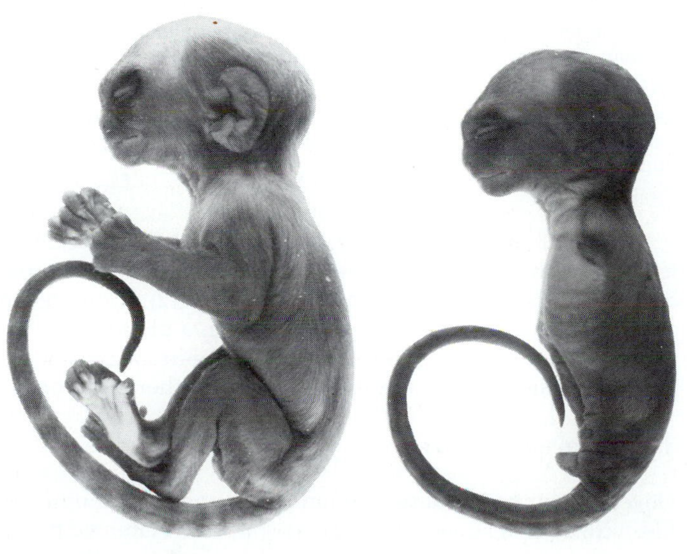

Figure 44–19 Thalidomide administered to the marmoset *(Callithrix jacchus)* produces a pattern of developmental defects similar to those found in humans. (*a*) Control marmoset fetus obtained from an untreated mother on day 125 of gestation. (*b*) Fetus (same age as control) of marmoset treated with 25 mg/kg thalidomide from days 38 to 52 of gestation. The drug suppresses limb formation, perhaps by interfering with the function of cholinergic nerves. (Courtesy of Dr. W. G. McBride and P. H. Vardy, Foundation 41; from *Development, Growth and Differentiation* 25(4):361–373, 1983.)

TABLE 44–3
Environmental Influences on the Embryo

Factor	*Example and Effect*	*Comment*
Nutrition	Severe protein malnutrition doubles the risk of birth defects. Fewer brain cells are produced, and learning ability may be permanently affected.	Growth rate is mainly determined by rate of net protein synthesis by embryo's cells.
Excessive amounts of vitamins	Vitamin D is essential, but excessive amounts may result in a form of mental retardation. Too much vitamin A and K may also be harmful.	Vitamin supplements are normally prescribed for pregnant women, but some women mistakenly reason that if one vitamin pill is beneficial, four or five might be even better.
Drugs	Many drugs affect the development of the fetus. Even aspirin has been shown to inhibit growth of human fetal cells (especially kidney cells) cultured in the laboratory. It may also inhibit prostaglandins, which are concentrated in growing tissue.	Common prescription drugs are generally taken in amounts based on the body weight of the mother, which may be hundreds or thousands of times too much for the tiny embryo.
Alcohol	When a woman drinks heavily during pregnancy, the baby may be born with fetal alcohol syndrome, that is, deformed and mentally and physically retarded. Low birth weight and structural abnormalities have been associated with as little as two drinks a day. It is thought that some cases of hyperactivity and learning disabilities may be caused by alcohol intake of the pregnant mother.	Fetal alcohol syndrome is thought to be one of the leading causes of mental retardation in the United States.
Heroin	Maternal heroin use is associated with a high mortality rate and high prematurity rate.	Infants that survive are born addicted to the drug and must be treated for weeks or months.
Methadone	Methadone use by mother results in addiction also.	
Thalidomide	Thalidomide, marketed as a mild sedative, was responsible for more than 7000 grossly deformed babies born in the late 1950s in 20 different countries. Principal defect was **phocomelia,** a condition in which babies are born with extremely short limbs, often with no fingers or toes.	Interferes with cellular metabolism; most hazardous when taken during the fourth to sixth weeks, when limbs are developing.
Cigarette smoking	Cigarette smoking reduces the amount of oxygen available to the fetus because both maternal and fetal hemoglobin are combined with carbon monoxide. May slow growth and can cause subtle forms of damage. (In extreme form carbon monoxide poisoning causes gross defects such as hydrocephaly.)	Mothers who smoke deliver babies with lower than average birth weights. They also have a higher incidence of spontaneous abortions, stillbirths, and neonatal deaths. Studies also indicate a possible link between maternal smoking and slower intellectual development in the offspring.
Pathogens	Rubella (German measles) virus crosses the placenta and infects the embryo; interferes with normal metabolism and cell movements. Causes a syndrome that involves blinding cataracts, deafness, heart malformations, and mental retardation. The risk is greatest (about 50%) when a pregnant woman contracts rubella during the first month of pregnancy; it declines with each succeeding month.	A rubella epidemic in the United States in 1963–65 caused about 20,000 fetal deaths and 30,000 infants born with gross defects.
	Syphilis is transmitted to the fetus in about 40% of infected women. Fetus may die or be born with defects and congenital syphilis.	Pregnant women are routinely tested for syphilis during prenatal examinations.
Ionizing radiation	When mother is subjected to x rays or other forms of radiation during pregnancy, infant has higher risk of birth defects and leukemia.	Radiation was one of the earliest teratogens to be recognized.

Recent advances in medicine have enabled physicians to diagnose some defects while the embryo is in the uterus. In some cases treatment is possible before birth. **Amniocentesis,** discussed in Chapter 12, is one technique used to detect certain defects. Figure 44–20 is a **sonogram,** a photograph taken of the embryo by using ultrasound. Such previews are helpful in diagnosing defects and also in determining the position of the fetus, and whether a multiple birth is pending.

The Human Life Cycle

Development begins at conception and continues through the stages of the human life cycle until death (see Table 44–4). We have examined briefly the development of the embryo and fetus, the birth process, and the adjustments it requires of the

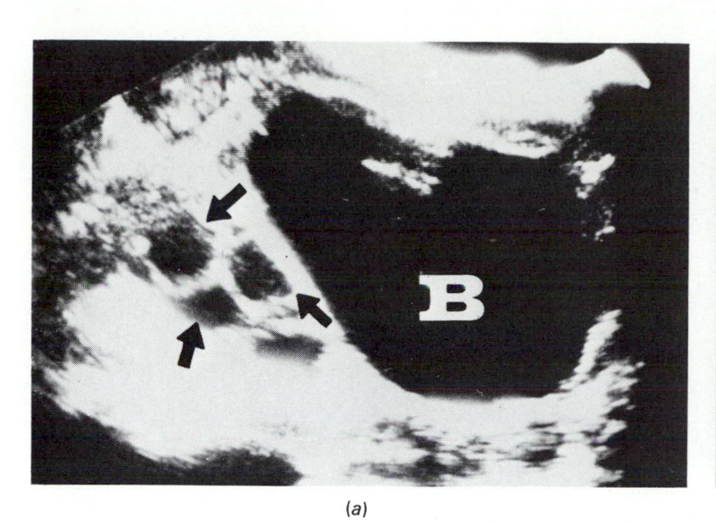

(a)

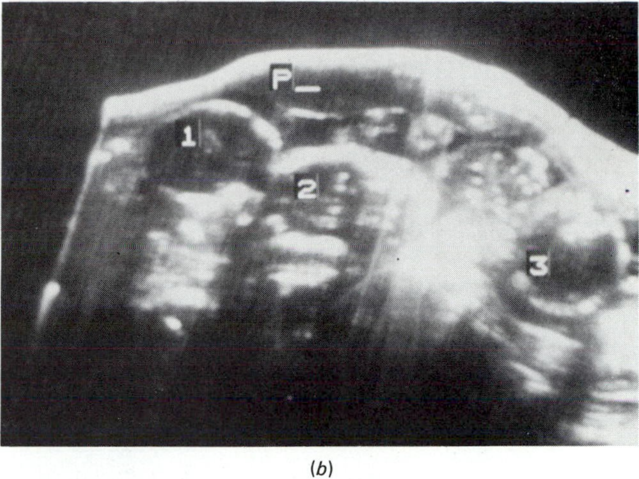

(b)

Figure 44–20 Ultrasonic techniques can be used to monitor follicle maturation and ovulation, as well as to give the physician information about the fetus. (*a*) Sonogram taken with ultrasound techniques, showing three follicles of equal maturation in the left ovary of a human. (*b*) Triplets in the same patient at 16 weeks of pregnancy. *P*, placenta. Such previews are valuable to the physician in diagnosing defects and predicting multiple births. (Courtesy of Biserka Funduk-Kurjak and Asim Kurjak, from *Acta Obstetrics and Gynecology Scand* 61:1982.)

neonate. The **neonatal period** is usually considered to extend from birth to the end of the first month of extrauterine life. **Infancy** follows the neonatal period and lasts until the rapidly developing infant can assume an erect posture (i.e., can walk), usually between 10 and 14 months of age. Some regard infancy as extending to the end of the first 2 years. **Childhood,** also a period of rapid growth and development, continues from infancy until adolescence.

Adolescence is the time of development between puberty and adulthood. During adolescence a young person experiences the physical and physiological changes that result in physical and reproductive maturity (Figs. 44–21 and 44–22). This is also a time of profound psychological development, as young people make adjustments that help prepare them to assume the responsibility of adulthood.

TABLE 44–4
Stages in the Human Life Cycle

Stage	*Time Period*	*Characteristics*
Embryo	Conception to end of eighth week of prenatal development	Development proceeds from single-celled zygote to an embryo that is about 30 mm long, weighs 1 g, and has the rudiments of all its organs.
Fetus	Beginning of ninth week of prenatal development to birth	Period of rapid growth, morphogenesis, and cellular differentiation, changing the tiny parasite to a physiologically independent organism.
Neonate	Birth to 4 weeks of age	Neonate must make vital physiological adjustments to independent life—i.e., now must process its own food, excrete its wastes, obtain oxygen, and make circulatory changes appropriate to its new mode of life.
Infant	End of fourth week to 2 years of age. (Sometimes ability to walk is considered the end of infancy.)	Rapid growth; deciduous teeth begin to erupt; nervous system develops (myelinization), making coordinated activities possible; language skills begin to develop.
Child	Age 2 years to puberty	Rapid growth; eruption of deciduous teeth, which are slowly shed and replaced by permanent teeth; development of muscular coordination and of language skills and other intellectual abilities
Adolescent	Puberty (approximately age 11–14 years) to adult	Growth spurt; development of primary and secondary sexual characteristics; development of motor skills; development of intellectual abilities; psychological changes as adolescent prepares to assume adult responsibilities
Young adult	End of adolescence (approximately age 20 years) to about age 40 years	Peak of physical development reached; individual assumes adult responsibilities that may include marriage, fulfilling reproductive potential, and establishing a career. After age 30, physiological changes associated with aging begin.
Middle age	Age 40 years to about age 65 years	Physiological aging continues through this period, leading to menopause in women and physical changes associated with aging in both sexes (e.g., graying hair, decline in athletic abilities, skin wrinkling). This is a period of adjustment for many as they begin to face their own mortality.
Old age	Age 65 to death	Period of senescence (old age). Physiological aging continues; homeostatic return to steady state is more difficult when the body is challenged by stress. Death often results from failure of the cardiovascular or immune system.

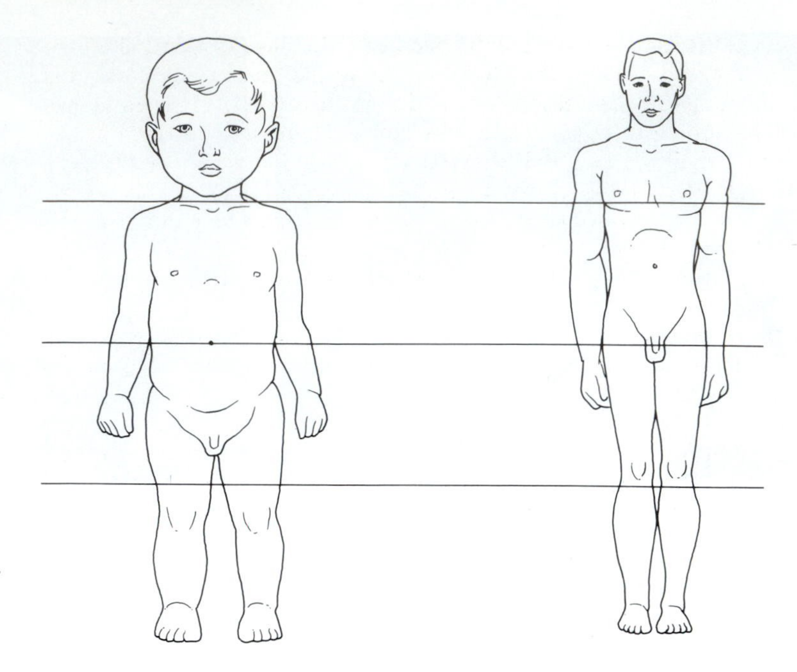

Figure 44–21 Relative sizes of the head, arms, and legs in a young child and in an adult.

Young adulthood extends from adolescence until about age 40. Middle age is usually considered to be the period between ages 40 and 65. Old age begins after age 65.

The Aging Process

Since development in its broadest sense includes any biological change with time, it also includes those changes that result in the decreased functional capacities of the mature organism, the changes commonly called **aging.** The declining capacities of the various systems in the human body, though most apparent in the elderly, may begin much earlier in life, during childhood, or even during prenatal life. The newborn female has only 400,000 oocytes remaining of the 4 million she had three months earlier in fetal life!

The aging process is far from uniform in various parts of the body. Various systems of the body may begin their decline at quite different times. A 75-year-old man, for example, has lost 64% of his taste buds, 44% of the renal glomeruli, and

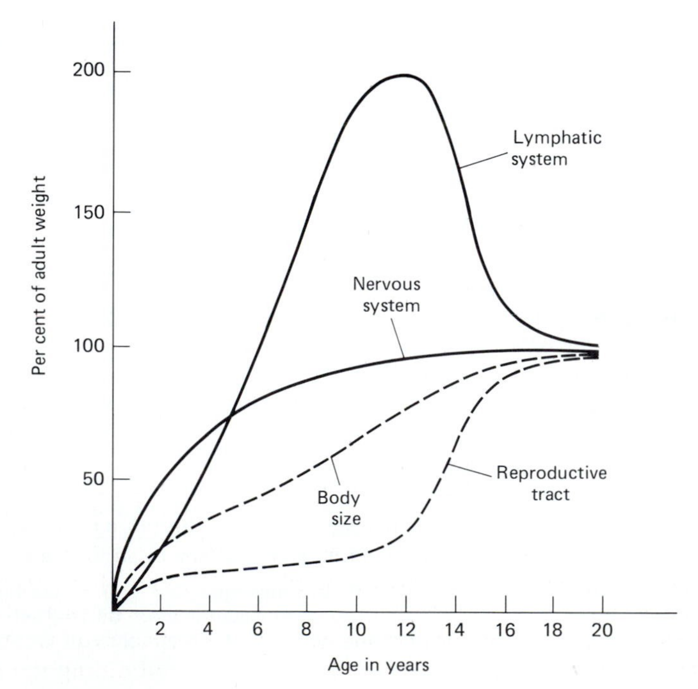

Figure 44–22 Relative rates of growth of several different organ systems during human development.

37% of the axons in his spinal nerves that he had at age 30. His nerve impulses are propagated at a rate 10% slower, the blood supply to his brain is 20% less, his glomerular filtration rate has decreased 31%, and the vital capacity of his lungs has declined 44%. The aging process is also marked by a progressive decrease in the body's homeostatic ability to respond to stress.

Although relatively little is known about the aging process itself, this is now an active field of scientific investigation. Although marked improvements in medicine and public health have led to survival of a larger fraction of the total human population to an advanced age, there has been no concomitant increase in the maximum life expectancy for either men or women.

A remarkable model of the aging process in humans is provided by a rare, inherited type of abnormal development called **progeria.** In this condition, babies develop more or less normally until they are about 1 year old; they then begin to undergo changes that are considered typical of aging, including loss of hair and arrest of the growth process. The physical appearance becomes that of a wizened old man or woman. The collagen in the connective tissues of the skin becomes highly cross-linked, as seen in old age. Affected children usually die at age 10 to 15 of coronary artery disease secondary to extensive atherosclerosis.

Cells that differentiate and stop dividing appear to be more subject to the changes of aging than are those that continue to divide throughout life. Nerve and muscle cells, which lose the capacity for cell division at an earlier age, show a decline in their respective functional capacities at an earlier age than do tissues such as liver and spleen, which retain the capacity to undergo cell division.

Several theories have been advanced regarding the nature of the aging process—that it involves hormonal changes; that it involves the development of autoimmune responses (immune responses against certain components of the organism's own body that result in destruction of those components by antibodies); that it involves the accumulation of specific waste products within the cell (the "clinker" theory); that it involves changes in the molecular structure of macromolecules such as collagen (an increased cross-linkage between the helical chains); that there is a decrease in the elastic properties of connective tissues owing to an accumulation of calcium, which results in stiffening of the joints and hardening of the arteries; that it involves the peroxidation of certain lipids by free radicals; or that cells are destroyed by hydrolases released by the breaking of lysosomes.

Other current theories suggest that aging involves the accumulation of somatic mutations caused by continued exposure to cosmic radiation and x-radiation, mutations that decrease the ability of the cell to carry out its normal functions at the normal rate. In all likelihood, aging is both part of and due to the same kinds of developmental processes that bring about the increasing functional capacities of the various systems of the body during earlier development. The processes may be part of the program of timed development built into the genome. Like other developmental processes, aging may be accelerated by certain environmental influences and may occur at different rates in different individuals because of inherited differences. For example, there is some experimental evidence that aging, at least in rats, can be delayed by dietary means, specifically, by caloric restriction. Thin rats, by and large, live longer than fat rats! For now, however, genetic predisposition may be the best guarantee of a long life.

SUMMARY

I. Development proceeds as a balanced combination of growth, morphogenesis, and cellular differentiation.
  - *A.* Cleavage occurs as the human embryo passes along the uterine tube to the uterus.
  - *B.* In the uterus, the embryo develops into a blastocyst and implants itself in the endometrium.
  - *C.* During gastrulation, the ectoderm, mesoderm, and endoderm form; each of these embryonic tissues gives rise to specific adult structures.
  - *D.* The nervous system is one of the first systems to develop. The brain and spinal cord develop from the neural tube.
  - *E.* The extraembryonic membranes protect the embryo and give rise to the placenta.
    - *1.* The amnion is a fluid-filled sac that surrounds the embryo and keeps it moist; it also acts as a shock absorber.
    - *2.* The embryonic chorion and maternal tissue give rise to the placenta, the organ of exchange between mother and developing child.

II. Development proceeds as an orderly sequence of predictable events. After the first two months of development, the embryo is referred to as a fetus.

III. Parturition takes place after about 280 days from the time of the mother's last menstrual period. During the first stage of labor the cervix becomes dilated and effaced; during the second stage the baby is delivered; and during the third stage the placenta is delivered.

IV. Lactation is the function of the breasts. Breastfeeding a baby offers several advantages over bottle feeding.
V. Development is regulated by the interaction of genes with cytoplasmic factors, organizers, and environmental factors such as pressure, temperature, and certain substances.
VI. By controlling environmental factors such as nutrition, vitamin and drug intake, cigarette smoking, and exposure to disease-causing organisms, a pregnant woman can help ensure the well-being of her unborn child.
VII. The human life cycle can be divided into the following stages: embryo, fetus, neonate, infant, child, adolescent, young adult, middle age, and old age.
VIII. The aging process is marked by a progressive decrease in the body's homeostatic abilities to respond to stress.

POST-TEST

1. Movement of cells to form a tube such as the neural tube is an example of ______________; specialization of cells to form neurons or some other cell types is called ______________ ______________.
2. The rapid series of mitoses that converts the zygote to a morula is referred to as ______________.
3. The cluster of cells that projects into the cavity of the blastocyst is the ______________ ______________ ______________; it gives rise to the ______________.
4. The process by which the blastula becomes a three-layered embryo is called ______________.
5. The tissue layer that gives rise to the nervous system is the ______________; the germ layer that gives rise to the lining of the digestive tract is the ______________.
6. The notochord induces the overlying ectoderm to form the ______________ ______________.
7. The ______________ ______________ prevents the embryo from drying out and acts as a shock absorber.
8. The ______________ is the organ of exchange between mother and fetus.
9. The duration of pregnancy is known as the ______________ period.
10. The term neonate refers to the ______________ ______________.
11. Lactation is the ______________.
12. An organizer is a chemical that ______________.
13. When ectoderm from the mouth region of an early frog embryo is transplanted to the mouth region of a salamander, the transplanted cells develop into mouth structures characteristic of a ______________ (*frog/salamander*).
14. Teratogens are agents that can interfere with normal ______________.
15. Changes that result in decreased functional capacities of the organism are characteristic of the process of ______________.

REVIEW QUESTIONS

1. Contrast the preformation theory with the theory of epigenesis, and relate these theories to current concepts of development.
2. Trace the development of an embryo from zygote to gastrula; draw and label diagrams to illustrate your description.
3. Trace some developmental process such as the formation of the neural tube, and explain how growth, morphogenesis, and cellular differentiation are an integral part of the process.
5. Give examples of adult structures that develop from each of the germ layers.
6. Describe implantation.
7. What happens during each stage of labor?
8. What kinds of adaptations must the neonate make immediately after birth?
9. What are some advantages of breastfeeding an infant?
10. What are some of the nongenetic factors that influence development?
11. What steps can the pregnant woman take to help ensure the safety and well-being of her developing child?
12. Trace development through the stages of the human life cycle.
13. Describe some of the changes that take place during the aging process.

Recommended Readings for Part VI

Books for Structure and Life Processes Chapters

Barnes, R. D. *Invertebrate Zoology*, 4th ed. Philadelphia, Saunders College Publishing, 1980. This comprehensive textbook discusses life processes of each invertebrate phylum.

Eckert, R. *Animal Physiology*. San Francisco, W. H. Freeman, 1978. An in-depth textbook of animal physiology.

Ganong, W. F. *Review of Medical Physiology*, 10th ed. Los Altos, CA, Lange, 1981. A textbook of human physiology.

Gardiner, M. S. *The Biology of Invertebrates*. New York, McGraw-Hill, 1972. An excellent textbook organized around life processes of invertebrates.

Guyton, A. C. *Textbook of Medical Physiology*, 6th ed. Philadelphia, W. B. Saunders, 1981. A well-written standard reference work in human physiology.

Ralph, C. L. *Introductory Animal Physiology*. New York, McGraw-Hill, 1978. An interesting account of animal physiology.

Romer, A. S., and T. S. Parsons. *The Vertebrate Body*, 5th ed. Philadelphia, W. B. Saunders, 1977. A well-respected, classic textbook that takes a comparative approach to life processes in vertebrates.

Solomon, E. P., and P. W. Davis. *Human Anatomy and Physiology*. Philadelphia, Saunders College Publishing, 1983. A very readable presentation of human anatomy and physiology.

Welty, J. C. *The Life of Birds*, 3rd ed. Philadelphia, Saunders College Publishing, 1982. A very readable ornithology textbook that includes several chapters on the life processes of birds.

Articles for Life Processes in Animals: Chapters 32–38

Bodde, T. Coping in space: The body's answer to zero gravity. *Bioscience*, Vol. 32, No. 4, April, 1982. An interesting discussion of the physiological changes experienced by astronauts.

Buisseret, P. D. Allergy. *Scientific American*, August 1982. A discussion of the cellular and biochemical changes that occur during an allergic response.

Dantzler, W. H. Renal adaptations of desert vertebrates. *Bioscience*, Vol. 32, No. 2, Feb., 1982. A description of the mechanisms found in desert vertebrates for maintaining fluid balance.

Garmon, L. A bubble is born. *Science News*, Vol 118, September 20, 1980. A report on bubble formation in decompression sickness and its applications to diving medicine.

Kolata, G. Cell biology yields clues to lung cancer. *Science*, Vol. 218, 1 Oct., 1982. By studying lung cancer cells in culture, researchers are gaining insights into how the cells multiply and how they can be prevented from multiplying.

Kooyman, G. L. The Weddell seal. *Scientific American*, An account of the diving behavior of the Weddell seal.

Lerner, R. A. Synthetic vaccines. *Scientific American*, Vol. 48, No. 2, Feb., 1983. A report on experiments on the preparation of synthetic vaccines.

Levy, R. I., and J. Moskowitz. Cardiovascular research: Decades of progress, a decade of promise. *Science*, Vol. 217, 9 July, 1982. A review of current techniques for dealing with cardiovascular disease, and a discussion of the decrease in mortality due to these diseases.

Marx, J. L. "Human T-Cell Leukemia Virus Linked to AIDS." *Science*, Vol. 220, 20 May, 1983. Does the virus cause AIDS?

Marx, J. L., "Monoclonal Antibodies in Cancer." *Science*, Vol. 216, No. 4543, 16 April, 1982. An interesting account of the clinical use of monoclonal antibodies.

West, S. "The Deepest Dive." *Science News*, Vol. 119, March 21, 1981. A report of experiments in simulated dives.

Articles for Structure Animal Life Processes: Regulation and Sense Organs. Chapters 39–42

Baker, M. A. A brain-cooling system in mammals. *Scientific American*. May 1979. A discussion of the cooling mechanism present in the brains of some mammals.

Bloom, F. E. Neuropeptides. *Scientific American*, October 1981. An account of the discovery and actions of neuropeptides that help regulate body activities, in some cases acting as both neurotransmitters and hormones.

Constantine-Paton, M., and Margaret Law. The development of maps and stripes in the brain. *Scientific American*. December 1982. Vol. 247, No. 6. An account of studies of the visual system and its organization.

Crews, D. The hormonal control of behavior in a lizard. *Scientific American*. August 1979. An interesting account of how the brain and sex glands interact in regulating the sexual behavior of male and female anole lizards.

Dolin, B. J. Alcoholism and its treatment: 1983. *Hospital Medicine*, January 1983. A brief overview of the state of the art.

Epps, G. The brain biologist and the mud leech. *Science 82*, Vol. 3, No. 1, Jan.-Feb., 1982. An account of research on neuron interaction using the leech as an experimental animal.

Heinrich, B. and G. A. Bartholomew. Temperature control in flying moths. *Scientific American*, June 1972. A description of the mechanisms of temperature regulation in moths and of the relationship of temperature to flight.

Hudspeth, A. J. The hair cells of the inner ear. *Scientific American,* January 1983. A description of the mechanism by which hair cells in the inner ear respond to convey information about acoustic tones and acceleration.

Kandel, E. R., and J. H. Schwartz. Molecular biology of learning: modulation of transmitter release. *Science,* Vol. 218, 29 October 1982. A review that focuses on the biochemistry of learning in a marine mollusk.

Keynes, R. D. Ion channels in the nerve-cell membrane. *Scientific American,* March 1979. A discussion of the generation of a nerve impulse by the flow of sodium and potassium ions through channels in the neuron membrane.

Levi-Montalcini and P. Calissano. The nerve-growth factor. *Scientific American,* June 1979. A description of some important characteristics and effects of nerve-growth factor.

Levine, J. S., and E. F. MacNichol, Jr. Color vision in fishes. *Scientific American,* February 1982. A discussion of retinal pigments in fishes and their relationship to the evolution of the eye.

Llinas, R. R. Calcium in synaptic transmission. *Scientific American,* Vol. 247, No. 4, October 1982. The role of calcium is studied in the giant synapse of a squid.

Miller, J. A. Cells of Babel. *Science News,* Vol. 122, December 18 and 25, 1982. A report on the release of several different neurotransmitters by single neurons in the brain.

Miller, J. A. Colorful views of vision. *Science News,* Vol. 120, Oct. 3, 1981. An account of a technique for mapping components of the visual system.

Miller, J. A. Wiretap on the nervous system. *Science News,* Vol. 123, No. 9, February 26, 1983. A report on a technique for analyzing neural function in conscious humans.

Morell, P., and W. T. Norton. Myelin. *Scientific American,* May 1980. A description of myelin and its functions.

Morrison, A. R. A window on the sleeping brain. *Scientific American,* Vol. 248, No. 4, April 1983. A study of REM sleep and accompanying paralysis.

Patterson, P., D. Potter, and E. Furshpan. The chemical differentiation of nerve cells. *Scientific American,* July 1978. A discussion of the differentiation of adrenergic and cholinergic nerve cells.

Pfaff, D. W., and B. S. McEwen. Actions of estrogens and progestins on nerve cells. *Science,* Vol. 219, 18 Feb., 1983. An account of experiments showing that estrogens and progestins alter electrical and chemical features of neurons, especially in the hypothalamus.

Pollie, R., The educated nervous system. *Science News,* Vol. 123, No. 5, January 29, 1983. A summary of research on invertebrates that supports the hypothesis that learning results from changes in nerve transmission.

Popper, A. N., and S. Coombs. Auditory mechanisms in teleost fishes. *American Scientist,* Vol. 68, July–August 1980. A discussion of the variations in both hearing capabilities and auditory structures among species of bony fish.

Schwartz, J. H. The transport of substances in nerve cells. *Scientific American,* April 1980. A discussion of the movement of substances long distances between the cell body and the neuron endings.

Shepherd, G. M. Microcircuits in the nervous system. *Scientific American,* February 1978. An account of synapses involving dendrite to dendrite associations.

Stent, G. S., and D. A. Weisblat. The development of a simple nervous system. *Scientific American,* January 1982. Research on the development of the nervous system in leech embryos gives clues to its functioning in the adult.

Wurtman, R. J. Nutrients that modify brain function. *Scientific American,* April 1982. An exciting discussion of the effect of nutrients such as choline, tryptophan, and tyrosine on brain function.

Journal articles for Chapters 43 and 44

Beaconsfield, P., G. Birdwood, and R. Beaconsfield. The placenta. *Scientific American,* August 1980. A summary of the development of the placenta and of its functions.

Cates, W. Legal abortion: The public health record. *Science,* Vol. 215, No. 4540, 26 March, 1982. An interesting analysis of the effects of the increasing availability and utilization of legal abortion in the United States during the 1970s.

Davidson, E. H., B. R. Hough-Evans, and R. J. Britten. Molecular biology of the sea urchin embryo. *Science,* Vol. 217, 2 July, 1982. Research on the development of the sea urchin suggests roles of RNA in early development.

De Robertis, E. M., and J. B. Gurdon. Gene transplantation and the analysis of development. *Scientific American,* December 1979. The amphibian oocyte serves as a living test tube for studying the biochemistry of gene regulation in development.

Garcia-Bellido, A., P. A. Lawrence, and G. Morata. Compartments in animal development. *Scientific American,* July 1979.

Hayflick, L. The cell biology of human aging. *Scientific American,* January 1980. A very interesting theory of the aging of cells and of the relationship of aging cells to human life span.

Herbert, W. Premenstrual changes. *Science News,* Vol. 122, No. 24, December 11, 1982. An interesting discussion of premenstrual syndrome.

Johnson, E. M., and B. E. G. Gabel. Application of the hydra assay for rapid detection of developmental hazards. *Journal of the American College of Toxicology,* Vol. 1, No. 3, 1982.

Johnson, E. M. Screening for teratogenic hazards. *Annual Review of Pharmacological Toxicology,* 1981, Vol. 21.

Jones, M. D., Jr, et al. Oxygen delivery to the brain before and after birth. *Science,* Vol. 216, 16 April, 1982.

Langone, J. The quest for the male pill. *Discover,* October 1982.

Marx, J. L. Electric currents may guide development. *Science,* Vol. 211, 13 March, 1981.

Small, S. A. Birth control from amulets to the pill. *MD,* March 1983.

part VII

THE BIOLOGY OF POPULATIONS: EVOLUTION, BEHAVIOR, AND ECOLOGY

"Blood lust" among fireflies. The firefly on the left is a female of the genus *Photuris;* on the right is a male of the genus *Photinus.* This embrace will be his last, for she is gnawing through his neck, having lured him from afar by imitating the light signals of a female of his own species. Courtship rituals have evolved among many animals as a mechanism of reproductive isolation; that is, they help ensure that mating takes place only among members of the same species. Additionally, as we will see in Chapter 49, courtship rituals usually limit aggression between individuals—at least until copulation has been successfully completed. In this case, however, we are seeing in operation a form of mimicry in which an organism of one species has somehow evolved a means of copying the behavior of another species, and in so doing has secured an easy supply of food. (From James E. Lloyd, SCIENCE 187:452, Copyright 1975 by the American Association for the Advancement of Science.)

45
The Genetic Mechanisms of Evolution

OUTLINE

I. Evolution and genetics
  A. Changes in gene frequency
    1. Mutation
    2. Natural selection
    3. Differential immigration or emigration
    4. Genetic drift
    5. Founders and bottlenecks
  B. Evolution: the failure to maintain genetic equilibrium
    1. What disturbs genetic equilibrium
    2. Examples of selection
      a. Protective coloration and industrial melanism
      b. Mimicry
II. Speciation
  A. Isolating mechanisms
  B. The origin of species by hybridization
  C. Stabilizing selection
III. Historical development of the concept of evolution
  A. Jean Baptiste de Lamarck
  B. Charles Darwin and Alfred Russel Wallace
  C. The Darwin-Wallace theory of natural selection

LEARNING OBJECTIVES

After you have read this chapter you should be able to:

1. State the Hardy-Weinberg law, and discuss its significance and consequences in terms of population genetics and evolution.
2. Define evolution in genetic terms, discuss the relationship between the Hardy-Weinberg law (discussed in Chapters 11 and 12) and evolution, and describe the dependence of evolution upon genetics. In particular, describe the role of mutation, natural selection and related mechanisms, genetic drift and related mechanisms, and stabilizing selection.
3. Summarize the "modern" concept of evolution, discussing its mechanisms (mutation, genetic recombination, genetic drift, natural selection, reproductive isolation) and the adaptive results, particularly speciation.
4. Discuss the history of evolutionary thought listing the most important contributors and summarizing their ideas.

In previous chapters we had a glimpse of the immense variety of living organisms that inhabit every nook and cranny of the sea and land. The concept of evolution has been implicitly and explicitly involved in many of the subjects discussed previously and is therefore not a new topic at this point. The general meaning of the word evolution is simply gradual change. In biology, **evolution** refers to **organic evolution,** the theory that living things, and the populations of which they form a part, change gradually over the course of their history, and that all living things are, to some extent, genetically related to one another.

Modern evolutionary theory is a synthesis of the natural history[1] approach used by Charles Darwin and modern studies of inheritance. This theory, termed **Neo-Darwinism,** is supported by the knowledge we now have of how traits are passed from one generation to another and by a weighty amount of "circumstantial" evidence of continuity and change presented in the fossil record, and through descriptive and comparative studies in subjects such as anatomy, physiology, and (most recently) biochemistry. This chapter first surveys evolutionary theory as it is now, and then traces its historical development.

Evolution and Genetics

Darwin, as we will see, viewed evolution as a process of descent with modification. These modifications, according to Darwin, are selected for or against by the environment of the organism; they are not the product of any predetermined plan and have no specific direction or goal. The modern understanding of evolution as accepted by most biologists explains evolution by the slow accumulation of small genetic changes. It can be considered to be a logical extension of Darwin's original insight of natural selection (discussed later in the chapter). But it owes as much to another 19th-century worker, Gregor Mendel, who, as you may recall, was the first to discover the genetic principles that have led to the study of inheritance—not only in individuals but in populations as well. Most biologists use the term **microevolution** to refer to changes in gene frequencies that occur within populations, so that ordinarily microevolution refers to changes that take place within a species. **Macroevolution** generally refers to evolution above the species level that might give rise to new genera or higher-level categories of organisms.

CHANGES IN GENE FREQUENCY

Mendelian genetics replaced the older "blending theory" of inheritance. The blending theory held that offspring represented a mixture of, or compromise between, their parents, not only in their traits but *also in their heredity;* but it did not explain how genetic changes could spread throughout a population of organisms or exert any long-term influence whatever. If such blending occurred, any new trait would be swamped in the sea of older traits that surrounded the organism possessing the novelty, for in its immediate offspring the trait would be weakened by blending with the more conventional ones. The offspring possessing those weakened traits would then interbreed, with the probable result that *their* offspring would have the new trait in further weakened form, and so on, until it would (for practical purposes) be lost altogether. Mendel's laws, however, implied the indefinite survival of any genetic unit as long as it had indeed been passed on to future generations by reproduction, and the Hardy-Weinberg law, as we saw in Chapters 11 and 12, gave this implication an explicit formulation. Yet the Hardy-Weinberg principle (see Focus on The Hardy-Weinberg Law: A Review) seems to raise another problem for evolutionary theory: If genes survive indefinitely, will not the gene pools of populations remain static? How, then, can genetic change take place?

The Hardy-Weinberg principle holds true, however, only in the absence of perturbing forces of mutation, natural selection, genetic drift, or differential migration of organisms possessing particular genotypes. *Without* those phenomena (which are actually always present), genetic frequencies in a large freely interbreeding population will not change from generation to generation. Thus the Hardy-Weinberg law really establishes a *base line* from which evolutionary departures must

[1] Natural history is a nontechnical study of the physical world and the life it contains.

take place, and states the rules that govern such departures. The resolution of the evolutionary paradox that the Hardy-Weinberg law seems to raise depends upon the examination of the special ideal conditions under which the Hardy-Weinberg Law can be expected to hold true.

Mutation

Quite clearly, genetic frequencies *would* change if one kind of gene changed into another form of the same allele by mutation. This, in fact, does happen in all known populations. Nevertheless, gene frequencies will remain stable even in the presence of mutation if **genetic equilibrium** is established. By genetic equilibrium is meant a dynamic state of genetic stability in which a factor potentially capable of changing a genetic frequency is counterbalanced by one or more factors with the opposite effect. Thus, genetic equilibrium can occur either in the absence of mutations or when the rates of forward and reverse mutations are the same.

By way of illustration, let us consider mutation in a gene pair A and a. Either there is no change of allele A into allele a, or else the rate at which allele A mutates to allele a—**forward mutation**—is equal to the rate at which allele a mutates to allele A—**reverse mutation.** However, because the rates of forward and reverse mutations are rarely equal, there is usually a tendency, termed **mutation pressure,** for one of the alleles to increase in frequency and for the other to decrease in frequency. On the other hand, mutation pressure may be countered by some other factor such as natural selection (discussed in the following section). Nevertheless, even though mutations occur constantly, they occur at random. Mutations are seldom the major factor in producing changes in gene frequencies in a population. They do increase genetic variability and *ultimately* provide the raw material of evo-

FOCUS ON

The Hardy-Weinberg Law: A Review

Imagine a population of organisms that is panmictic for two allelic genes: A, whose frequency is p, and a, whose frequency is q. Their combined frequency is obviously 1, so we may write

$$p + q = 1 \tag{1}$$

so

$$p = 1 - q$$
$$q = 1 - p$$

Now a gamete bearing gene A may combine with another one to form an AA zygote. Similarly, $A \times a \rightarrow Aa$, $a \times A \rightarrow aA$, and $a \times a \rightarrow aa$. The frequency of each combination is the product of the frequencies of its component genes, as follows:

	$p(A)$	$q(a)$
$p(A)$	$p^2(AA)$	$pq(aA)$
$q(a)$	$pq(Aa)$	$q^2(aa)$

Adding these up, the totality of all genotypes is:

$$p^2 + 2pq + q^2 = 1 \tag{2}$$

where

p^2 = frequency of AA
$2pq$ = frequency of Aa
q^2 = frequency of aa

Now in the *next* generation, the present Aa organisms will be able to produce *either* A or a gametes, that is, half the gametes will be A and half a from this source. The AA organisms will, of course, only yield A gametes; the aa only a gametes. Thus the total frequency of all A and a gametes may be expressed by the following equations, where p' stands for the new frequency of A and q' for the new frequency of a:

$$p' = p^2 + \tfrac{1}{2}(2pq) \tag{3}$$
$$= p^2 + pq$$
$$= p^2 + p(1 - p)^*$$
$$= p^2 + p - p^2 = p$$
$$q' = q^2 + \tfrac{1}{2}(2pq) \tag{4}$$
$$= q^2 + pq$$
$$= q^2 + q(1 - q)^*$$
$$= q^2 + q - q^2 = q$$

Therefore, if left undisturbed, gene frequencies do not change from generation to generation in a panmictic population, regardless of dominance or recessivity.

*See equation (1).

lution, but mutations alone are unlikely to determine the nature or direction of evolutionary change, although they are essential for any evolution to occur. (In almost all cases, however, sufficient genetic variation exists within a population at any one time to fuel evolutionary mechanisms for a long time to come.)

A mutation establishes an alternative allele at a given locus and thus makes possible an alternative phenotype. Evolutionary changes are possible only when there *are* alternative phenotypes. The process of selection, however, operates not gene by gene, but rather individual by individual, and on the basis of the phenotypic expression of the individual's entire genetic system. The forces of natural selection operate on the entire individual and not on single traits. In the end, the survival and reproduction of many individuals determine the frequency of genes in the population they comprise.

When a mutation first appears, only one or a very few organisms in the population will bear the mutant gene. How rapidly the gene will spread depends largely (but not solely) on whether it is advantageous, neutral, or harmful in its effects. The change in the gene pool so that the mutant gene appears with greater and greater frequency in the population is a gradual process that may occur over many generations. In a large population, the success or lack of success of some new mutant gene will depend mostly on its **selective advantage.** By this is meant its ability to confer on its possessor the capacity to leave a larger number of surviving individuals in the next generation than do the possessors of alternative alleles.

At one time, mutation was thought to be a more important evolutionary force than is now believed. The term mutation was coined by the Dutch botanist Hugo de Vries, one of the rediscoverers of Mendel's laws in the early part of this century. De Vries believed that he had found the major cause of evolution. He carried out genetic experiments with the evening primrose and other plants that grew wild in Holland. When he transplanted these into his garden and crossed them, some of the resulting plants were unusual and differed markedly from the original wild plant. These unusual forms bred true in subsequent generations. De Vries conceived of mutations as just such sudden major changes in the character of an organism. Darwin had described such sudden changes much earlier, but believed that they occurred too rarely to be of importance in evolution.

According to our modern understanding of genetics, **mutation** has come to refer to a sudden, random, discontinuous change in a gene (or chromosome). The vast number of genetic experiments with plants and animals carried on since De Vries' studies have shown that mutations do occur constantly and that the changes in the phenotype produced by such mutations may rarely be of adaptive value and contribute to the survival of the organism. Instead, such mutations are more likely to have a deleterious or even disastrous effect. A random change in a computer chip or even a computer program is hardly likely to improve it. Mutations are random changes in nucleic acids and usually represent not merely an alteration but an actual *loss* of genetic information. When we consider the complexity of living things, the wonder is not that mutations are usually harmful, but that any mutation is ever advantageous.

Hundreds of different mutations have been observed in the plants and animals most widely used in genetic experiments—corn and fruit flies. Among the mutations that have been observed in the fruit fly (Fig. 45–1) are unusual body colors, such as brown, gray, and black instead of the normal yellow; eye colors ranging from the normal red, to brown, purple and white; wings that are curled, crumpled, shortened, or completely absent; oddly shaped legs and bristles; and such remarkable changes as the development of a pair of legs on the forehead in place of the antennae (also see Focus on Genetic Mapping in Chapter 11.) The six-toed cats of Cape Cod (normal cats have five toes) and the short-legged breed of Ancon sheet (normal sheep have longer legs) are examples of mutations among domestic animals.

Mutations that change the nutritional requirements of bacteria have also been extensively studied. These can have life-or-death consequences for their possessors. Mutations that increase bacterial resistance to antibiotics (e.g., penicillin-resistant strains of staphylococci) greatly increase the bacteria's ability to survive. Coping with these antibiotic-resistant forms has been a severe problem in many hospitals.

Some mutations produce barely distinguishable changes in the structure or function of the organism in which they occur. Other mutations produce a major

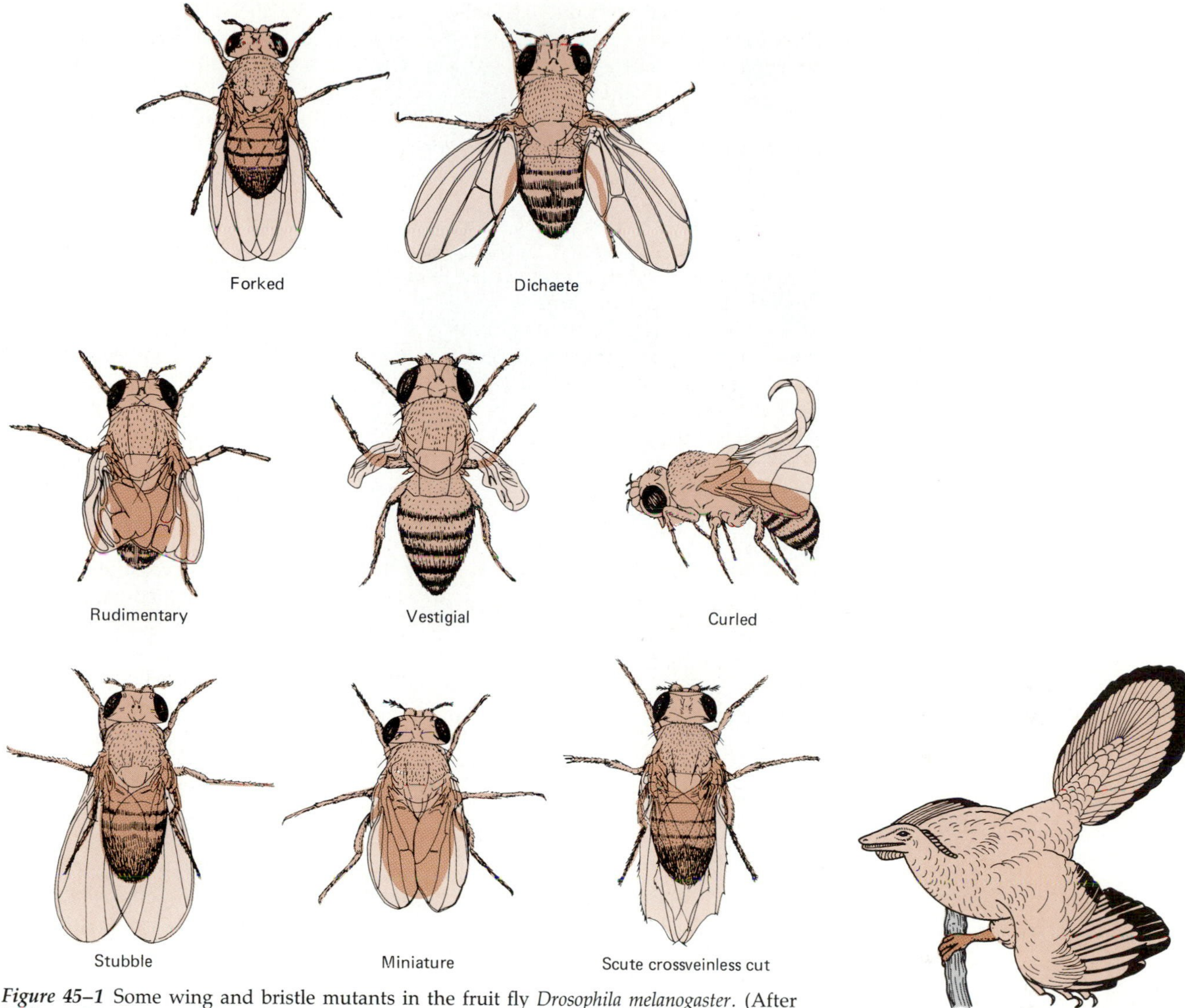

Figure 45–1 Some wing and bristle mutants in the fruit fly *Drosophila melanogaster*. (After Wallace, Sturtevant, and Beadle.)

change early in development and lead to multiple marked changes in the resulting body form or function. Usually when such a major change occurs in the control of some early stage of development, the result is a nonviable monster that dies almost immediately. A few such major changes may give rise to forms that are able, by virtue of their mutation, to occupy some new environment. These constitute what Richard Goldschmidt of the University of California called "hopeful monsters."

Goldschmidt suggested, for example, that the ancestral type of bird, *Archaeopteryx*, evolved into the modern bird by a mutation that in a single step changed the shape of its tail. *Archaeopteryx* (see Fig. 45–2) had a long, reptile-like tail that was covered with feathers. If a single mutation caused a shortening of the tail, then a "hopeful monster" with the fan-shaped arrangement of feathers characteristic of modern birds might have resulted. This fan-shaped tail, better suited for flying than the former long, reptile-like tail, would give its possessors a selective advantage during evolution. Observation of more modern species has confirmed that such changes *can* occur rarely as the result of a single mutation. The Manx cat, for example, owes its stubby tail to a mutation that caused the shortening and fusing of most of the tail vertebrae. We will return to Goldschmidt and his hopeful monsters in Chapter 47. Let us note for the time being that his views are not widely shared.

Some mutations—**chromosomal mutations**—are accompanied by a visible change in the structure of the chromosome or in the total number of chromosomes per cell. A single chromosome may be added to or deleted from the usual diploid set, or the entire set of chromosomes may be doubled or tripled, yielding organisms

Figure 45–2 A comparison of the structure of the tail of the extinct birdlike reptile *Archaeopteryx* (*top*) with that of a modern bird (*bottom*). The odd thing about *Archaeopteryx* is that, though it may well have been unable to actually fly, it had feathers that were not just partially but superbly adapted to flight.

Figure 45–3 Polyploidism in Easter lilies. The polyploid Easter lily *(left)* is larger and more robust than the diploid plant *(lower right)*. (U.S. Department of Agriculture.)

called **polyploids.** Polyploid plants are usually larger and more robust than their diploid parents (Fig. 45–3).

In plants, chromosomal duplication does far less harm, as a rule, than it does in animals (recall the example of Down syndrome). Some cultivated varieties of tomatoes, corn, wheat, and other plants owe their vigor and large fruit size to the fact that they are polyploids. Animal species that have apparently originated by polyploidy do, however, exist.

Natural Selection

A far stronger influence that tends to disturb the genetic equilibrium of populations is **natural selection.** Darwin saw the demands of the environment as exercising a selective effect upon the organisms present in that environment. As an example, recall the case of sickle cell anemia from Chapter 12. In this disorder an abnormal form of hemoglobin occurs in the red blood cells of persons homozygous for the trait. In the absence of treatment the condition is almost invariably fatal. Yet a heterozygote suffers no more than a mild, harmless anemia. In East Africa, where the trait probably originated, the heterozygote enjoys some protection against the malaria parasite (Fig. 45–4).

Most, perhaps all, genes have many different effects on the phenotype; they are said to be **pleiotropic.** Some of the effects of a given gene may be advantageous for survival and thus exert what is known as **positive selection pressure.** Others

Figure 45–4 Map showing the distribution of sickle cell anemia *(bars)* compared with the distribution of falciparum malaria *(shaded region)*. This correlation strongly suggests that the resistance of heterozygous individuals to malaria has served to balance the deleterious effects of sickle cell anemia.

may be disadvantageous, exerting what is known as **negative selection pressure.** Whether the frequency of a given allele increases or decreases will depend on whether the *sum* of the positive selection pressures due to its advantageous effects is greater or less than the sum of the negative selection pressures due to its harmful effects. As another consequence, it is possible for a neutral or even deleterious trait to persist or increase in a population as a side effect of strongly positive selection in favor of some other trait with which the neutral or deleterious one is associated genetically. We will presently return to the subject of selection, since it appears to be by far the most important of the forces favoring or retarding the change of genetic frequencies in populations.

Differential Immigration or Emigration

Theoretically, selective or **differential immigration** or **emigration** can change genetic frequencies, since it represents the arrival or departure of organisms possessing one genotype at a greater rate than others in the population. This changes the proportional representation of certain genes in the gene pool. Odd as it may seem, genes, too, can flow differentially by themselves, if the organisms that bear them mate outside their normal population limits. This is easily possible for migratory animals.

Genetic Drift

Genetic frequencies can remain constant only if a population is fairly large. In reproduction within small populations, chance alone may play a considerable role in determining the composition of the succeeding generation. In this way, even traits that are adaptively neutral can change their frequency in a population very markedly. Within a small interbreeding population, heterozygous gene pairs tend to become homozygous for one allele or the other by chance rather than by selection, so that one or the other alternative allele is lost. This may lead to the accumulation of certain disadvantageous characteristics and the subsequent elimination of the entire group possessing those characteristics.[1]

How can this happen? There are many possibilities. For instance, if a population consists of only a few individuals, predators could destroy the only representatives of a particular genotype and miss the others purely at random. Such an event would be most unlikely in a large population. Then too (and probably more important), it is a matter of chance whether a particular genotype will be represented in those gametes that do manage to unite in fertilization—a chance that depends upon the random distribution of maternal and paternal chromosomes in meiosis. The production of changes in gene frequency by random events is known as **genetic drift.**

The role that genetic drift actually plays in the evolution of organisms in nature has been a subject for debate among biologists, but there seems little doubt that it does play at least a minor role. (Some think that it may be actually very important, as we will see in Chapter 47.) Certainly many animal and plant populations in nature are divided into subgroups small enough to be affected by the chance events underlying genetic drift. Genetic drift may explain the common observation that closely related species in different parts of the world frequently differ in curious, even bizarre, ways that appear to have no particular adaptive value.

Founders and Bottlenecks

If a new habitat is accidentally colonized by a small number of organisms, the only genes that will be represented among their descendants will be those few that they chanced to possess. Thus, isolated populations may have very different gene frequencies from those characteristic of the species elsewhere, and these differences may very well be random rather than adaptive. The disproportionate effect exerted upon a population by a limited number of ancestors is termed the **founder effect.** Like genetic drift, it can produce great changes in gene frequency even in the absence of natural selection (Fig. 45–5).

[1] This possibility is of great concern in the case of small populations of endangered species such as might be found in zoos or nature preserves.

This effect is most apparent on islands and other areas of geographical isolation and helps to account for the differences evident in island populations as compared to their mainland relatives. When a species is expanding continuously, the populations at the edge of the range, invading new areas, are likely to be small and differ genetically from the main body of the population (in other words, their genetic diversity is more limited). In all of these situations, when the breeding population is small, chance rather than selection may play a large role in determining the evolution of a particular group.

In some species, very few individuals survive some critical stage in their life cycle. For example, among houseflies in northern areas, only a few representatives survive the winter months; these give rise to most of the summer population. In principle, this is similar to the founder effect. Only a few individuals (which are perhaps not truly representative of the genetics of the population from which they came) will give rise to the entire future population. They will by chance exert a disproportionate influence over its prospective genetic frequencies. Since we can think of this phenomenon as a periodic squeezing out of some of the genes in a gene pool in random fashion, it is termed the **bottleneck effect.** Like genetic drift and the founder effect, it can change gene frequencies and limit genetic diversity. Although natural selection (Fig. 45–6) is not a part of mechanisms such as the bottleneck effect or genetic drift, it does operate on the populations produced as a result of such mechanisms.

EVOLUTION: THE FAILURE TO MAINTAIN GENETIC EQUILIBRIUM

In evolution, an organism is not necessarily eliminated because of a single bad trait, nor does it get a gold star for a single good adaptive trait. Natural selection operates not upon the phenotypes of single genes, but upon the total phenotypic effect of the entire array of genes present. One group of organisms may survive despite some clearly disadvantageous trait. Another group may be eliminated despite certain traits that are highly advantageous for survival. The organisms that ultimately survive and serve as parents of the next generation are ones whose *total* spectrum of qualities renders them a little better able to survive and reproduce their kind. That means that many genetic traits may succeed by a kind of piggy-back effect that results only from their association (especially by pleiotropism) with other traits that are highly adaptive for the organism.

What Disturbs Genetic Equilibrium

In summary, the Hardy-Weinberg principle prevails only under certain conditions:

1. Absence of mutation
2. Absence of selection
3. Presence of **panmixis** (random mating)
4. Absence of selective immigration or emigration
5. Large population

Changes in gene frequencies can be produced only by departures from basic Hardy-Weinberg conditions, but under natural conditions such departures nearly always do take place.

Mutations are always occurring, as is natural selection. Panmixis seldom prevails (after all, many organisms are able to choose their mates, and that choice is hardly random in those capable of elaborate courtship behavior [Chapter 49]), and many collections of organisms in nature are divided into small, more-or-less isolated gene pools separated by geographical or topographical barriers. Thus, in most instances, microevolutionary changes in genetic frequencies are inevitable in natural populations of organisms.

A population with a gene pool that is constant from one generation to the next is in genetic equilibrium because the frequency of each allele in the population remains unchanged in successive generations. In contrast, a population undergoing evolution is one in which the gene pool is changing from generation to generation.

The gene pool of a population may be changed by mutations, by the introduction of genes from some outside population into this population, by one of the genetic drift–like mechanisms, or by natural selection. **Recombinations** brought

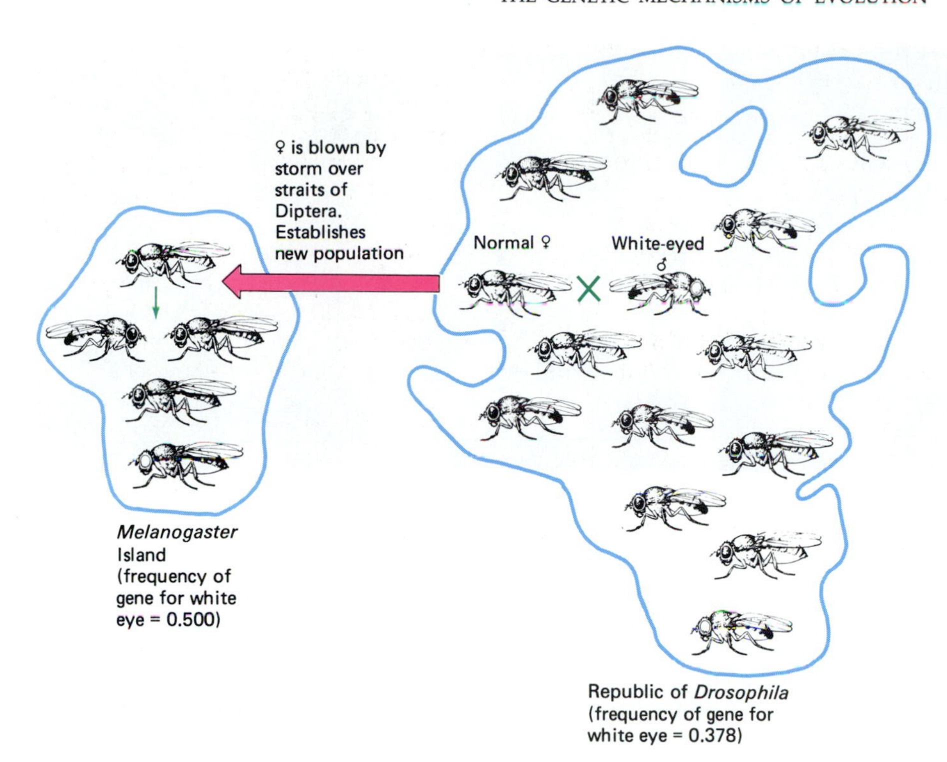

Figure 45–5 Founder effect. In this example, the genetic frequencies of a population have been determined by the genotypes that happened to be possessed by its founders but were not characteristic in frequency of the population as a whole.

about by crossing over or by the assortment of chromosomes in meiosis may also lead to new *combinations* of genes. The new phenotypes resulting from such recombinations may have some specific advantage or disadvantage for survival or for superior reproduction (which is not quite the same thing). Then the genes they possess would also have some advantage or disadvantage—an increased or lessened probability of being transmitted to the next generation. Ultimately, this would be reflected in a change in the gene pool.

Examples of Selection

In recent years a number of examples of natural selection that have come to light in the laboratory or in the field have afforded an incomparable opportunity to study this evolutionary mechanism in action. Among these can be listed the development of DDT resistance in insect populations exposed to this pesticide, industrial melanism in moths (described shortly), and the development of resistance to drugs among disease bacteria.

Figure 45–6 Bottleneck effect. Since only a small population of flies survives the winter, its genotypes, not necessarily resulting from natural selection, determine the genetic frequencies of the entire succeeding summer population.

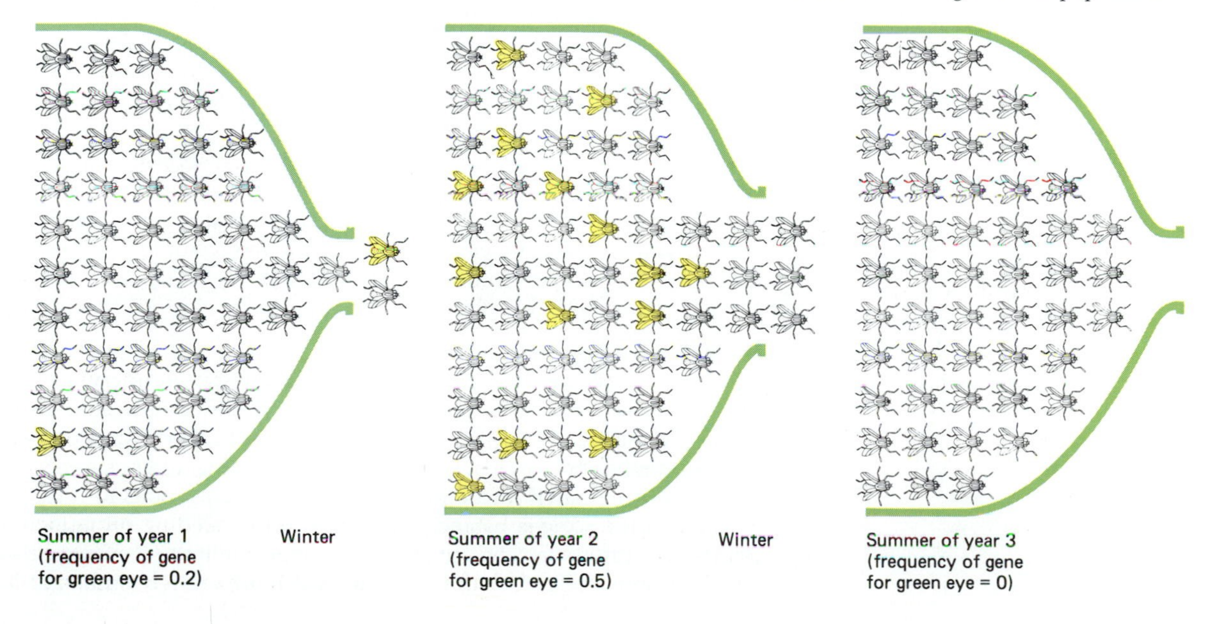

(a)

(b)

Figure 45–7 Camouflage conferred by protective coloration. (*a*) A leaf-mimicking katydid from Peru. The coloring of this insect matches the mottling on the leaf it rests on. (*b*) As with many ground-nesting birds, in their natural surroundings the chicks of nighthawks are almost invisible both to us and to predators. ((*a*), James L. Castner.)

PROTECTIVE COLORATION AND INDUSTRIAL MELANISM. A subtle example of natural selection is the development of protective coloration. **Protective coloration** exists when an organism blends in with its surroundings in such a way as to make it hard to see. This has the effect of protecting it from its predators or, in those that *are* predators, of keeping their victims from noticing them until it is too late.

Examples of protective form and coloration abound in nature (Fig. 45–7). Some katydids resemble leaves so closely that you would never guess that they are animals—until they start to walk. The chicks of ground-nesting birds usually have feathers that blend in with the surrounding weeds and earth so that they simply cannot be discerned from a distance. And there are species of praying mantis that resemble flowers—pity the poor bee that comes to visit! The suggestion is obvious that such protective coloration has been preserved and accentuated by means of natural selection. How might these adaptations originate?

In the past there were many moths whose colors blended in with the light-colored lichens on the trunks of trees. However, air pollution kills lichens so readily that the numbers of lichens are often considered an index of air quality. As discussed in Chapter 1, prior to about 1850 in England dark-colored specimens of the moth *Biston betularia* were very unusual, and they were extremely rare in museum collections made before that date. Nowadays, however, it is hard to find any of the formerly predominant light-colored specimens, at least in industrialized portions of that country. **Industrial melanism** is the name used for the development of such dark coloration in populations exposed to air pollution. The dark color is not dirt, it is a genetically based increase in the dark pigment melanin in their bodies (see Fig. 1–9).

When air pollution kills light-colored lichens on tree trunks, light-colored moths stand out against the dark tree bark. Under these circumstances a resemblance to light-colored lichens is no longer an advantage. Birds eat the light-colored moths, and those that are darker are more likely to survive under the changed conditions. This is a well-documented case of the development of protective coloration occurring just since the Industrial Revolution began to produce substantial air pollution in the 19th century. Is natural selection responsible? Light-colored moths released experimentally in polluted districts are known to be eaten more readily by birds than dark ones, and the reverse is true in nonpolluted districts. Many moths in the vicinity of large cities now even have a dark-colored larval stage (the caterpillar). Since air pollution control was instituted in the 1950s in certain British districts, the proportion of light-colored moths is again increasing, however.

MIMICRY. An organism's close imitation of the appearance of a model to which it is unrelated is called **mimicry.** One example is **Batesian mimicry,** or the resemblance of a harmless or palatable species to one that is dangerous, obnoxious, poisonous, or revolting (Fig. 45–8). There are many examples. For instance, a harmless moth may resemble a bee so closely that even a biologist would hesitate to pick it up. The well-known monarch, or milkweed, butterfly is poisonous to birds and mammals because of the toxic substances it absorbs while feeding on poisonous milkweed plants as a caterpillar. The monarch has many imitators among other butterflies, which closely resemble it in color, but which are entirely edible, at least by birds!

(a)

Figure 45–8 Batesian mimicry. (*a*) At the right is Jordan's salamander, *Plethodon jordani*, a distasteful species. To its left is *Desmognathus imitator*, a palatable species. (*b*) (left) Few would want to get close enough to this insect to discover that it is actually a moth (note the wide, frilly antennae). (right) A genuinely noxious insect, the golden paper wasp. ((*a*), E. D. Brodi, Jr., Adelphi University/BPS; (*b*), (left) L. E. Gilbert, University of Texas at Austin/BPS, (right) P. J. Bryant, UC-Irvine/BPS.)

(b)

Natural selection has maintained a resemblance that gives its possessor almost as much protection as the model, because as soon as predators learn to associate the distinctive markings of the model with its undesirable characteristics, they tend to avoid all similarly marked animals.

Another form of mimicry, **Müllerian mimicry,** involves several noxious species that share a similar warning coloration. This type of mimicry operates to the advantage of all the species in that a predator, having had a distasteful experience with one individual, will avoid preying on any similarly colored organism (Fig. 45–9).

Figure 45–9 Müllerian mimicry. These various lepidopterans are all unpalatable. Their similar coloration makes it easier for potential predators to learn their system of warning coloration than if each species had its own distinctive pattern. (L. E. Gilbert, University of Texas at Austin/BPS.)

FOCUS ON

Symbiosis

The most delicate and precise adaptations organisms must have are probably adaptations to one another. This is evident even in such ecological relationships as exist between predator and prey, but the kind of mutual relationship called symbiosis is probably the most exacting of them all.

Symbiosis, which means "living together," is a term that encompasses a spectrum of relationships. At the one extreme are the **parasitic** relationships, where one member of the association benefits at the expense and to the harm of the other. At the other extreme we find mutualism in which *both* parties benefit, equally as far as one can judge. Somewhere in the middle lies **commensalism** in which the benefits of the relationship are one-sided, yet there is little harm to the other party (Fig. A).

How would you describe the relationship of the pilot fish to the shark (Fig. B)? These fish cling to the body of the shark host, feeding on its scraps that it probably would not eat anyway. They might do a little harm to the shark in slowing it down, or causing it to expend more energy in swimming, yet it is possible that they aid it in subtle ways as well, perhaps by sensing potential enemies.

B. A shark with several pilotfish attached. (Charles Seaborn.)

A. This Caribbean shrimp (genus *Pereclemines*) might receive some leftover food from the host anemone, but, primarily, it is given protection from predators. The anemone probably neither benefits nor is harmed by the relationship. (Charles Seaborn.)

Algae live inside the body cells of *Hydra viridissima.* The hydra subsists on the organic products of the alga's photosynthesis. Is it a parasite? Perhaps the nitrogenous wastes of the hydra's metabolism make an important contribution to the nutrition of the algae (Fig. C).

Cleaning symbiosis is an especially interesting case of mutual adaptation. A tiny, brilliantly colored wrasse, which could be swallowed at a gulp by the much larger fish (such as a grouper) it services, swims about them with impunity (Fig. D). The host may even open its mouth so that the little cleaner may pick decaying food from its teeth, or devour parasites deep in the pharynx. The bright color of the wrasse serves as a negative releaser (see Chapter 48) which inhibits predation from its fish hosts. Indeed so strong is the host's inhibition that the relationship serves as the basis of a kind of mimicry in which a brightly colored imitator of the wrasse swims up to the host fish. When the host presents itself for grooming, the imposter steals, perhaps, into its mouth. Quickly grasping a mouthful of succulent

Speciation

Clearly it is not enough to speak of mutations and the change of gene frequencies as accounting for all the major differences among organisms. By themselves, these processes tell us little about the mode of formation of new species, and still less about the production of higher taxonomic categories.

Changes of any kind in organisms occur over many generations during which individuals are born and die, but the population continues. Thus, even though natural selection operates on individuals, the unit in evolution is not the individual but rather a *population* of individuals. A population of similar individuals living

C. A budding *Hydra* with symbiotic algae. (Walker England, Photo Researchers, Inc.)

E. A scanning electron micrograph (falsely colored to show contrasts) of a wingless fly on the back of a bee. (Dr. Brad Adams, Photo Researchers, Inc.)

flesh, the mimic darts off before the startled host can react.

Perhaps the most interesting relationships, at least from an evolutionary point of view, are the frankly parasitic ones. Often parasites lack body parts that their nonparasitic relatives do possess. An example is the wingless fly (*Braula*) shown (Fig. E) riding on the back of a bee host. What use are wings to the fly? The bee does all the flying; why carry excess baggage?

D. A cleaner fish with a grouper. (Charles Seaborn.)

Yet such adaptations do severely limit many parasites in their choice of a host. Bird lice, for example, are best transmitted from generation to generation of their hosts by nest contact, but this tends to minimize the chance that they will be transmitted to other species of bird. Through this and many other adaptations, parasites become not only host-dependent but host-specific as well. If hosts evolve a resistance to parasites, then the parasites may well have to coevolve, thus keeping up with the new "design features" of their hosts. In fact, evolutionary taxonomists have used the more easily interpreted features of the related species of parasites to give them clues as to the possible evolution of their hosts. If two species of parasites seem closely related, the reasoning goes, so must be their hosts.

within a geographically defined area and interbreeding is termed a **deme,** or **genetic population.** The territorial limits of any given deme may be vague and difficult to define, and the number of individuals in the deme may fluctuate widely from time to time. The next larger unit of population in nature is the species, which is composed of all demes among which gene exchange normally occurs.

A **species** may be defined, as we have seen, as a reproductively isolated group of organisms incapable of genetic exchange with other groups.[1] It follows

[1] Strictly speaking, this definition best describes *biological* species only. Organisms that look different from one another are often classified taxonomically as belonging to separate species even if they are potentially capable of interbreeding.

from this definition that the key to the development of a species—**speciation**—is the development of its reproductive isolation so that it cannot interbreed with other species. It is thought that there are two main ways in which such a situation may arise. One is through the chance effects of long physical isolation and independent development. For example, populations may become isolated on islands or mountain tops or in streams belonging to separate watersheds. This is called **allopatric speciation** or **geographic speciation.** In allopatric speciation, reproductive isolation often occurs only *after* organisms have begun to exploit a new habitat. Recent studies indicate that most species of organisms are composed of populations that are more-or-less isolated from one another and among which there is actually little or no gene exchange. Allopatric speciation could take place easily among such populations, producing several species from one.

For example, early in the 15th century a litter of domesticated rabbits was released on Porto Santo, a small island near Madeira. There were no other rabbits and no carnivorous enemies on the island, so the rabbits multiplied with amazing speed. By the 19th century they were strikingly different from the ancestral European stock: They were only half as large as their European relatives, with a different color pattern, and a more nocturnal way of life. More important, they could not produce offspring when bred with members of the European species. Within 400 years, then, a new species of rabbit had developed.

Geographic isolation of organisms need not be extreme to be effective. In a mountainous region the individual mountain ranges afford effective barriers between the valleys. Valleys only a short distance apart but separated by ridges always covered with snow typically have species of plants and animals that are peculiar to those valleys. Thus there are usually more different species in a given area of a mountainous region than on open plains. For example, in the mountains of the western United States there are 23 species and subspecies of rabbits, whereas in the much larger plains area of the Midwest and the East there are only 8 species of rabbits.

Geographic isolation is usually not permanent, and two previously isolated groups may come into contact again and resume interbreeding unless genetic isolation or interspecific sterility has arisen in the meantime, as happened in the case of the Porto Santo rabbits. Genetic isolation results from mutations that occur independently of mutations for structural or functional features. Genetic isolation may develop only after a long period of geographic isolation has produced striking differences between two groups of organisms, or it may originate within a single otherwise homogeneous group of organisms. As generations pass and different mutations accumulate in each group by chance and by selection, the two groups will probably become increasingly different. It is not hard to imagine these differences accumulating to the extent that substantial differences in courtship behavior, breeding season, or even the mechanical features of copulation could prevent successful mating.

The other principal means of speciation is **sympatric speciation,** in which two populations occupy the same territory. Sometimes such sympatric species can be detected only by virtue of the fact that they will not interbreed. How might this come about?

Imagine two somewhat differing potential ecological niches, that is, life styles that are available in the same location. A versatile organism is able to exploit them *both* satisfactorily. However, certain mutants are capable of exploiting one of them—let us call it niche A—better than the normal members of the population. It is obviously to their advantage if no genes from the parent population find their way into the gene pool of this new group, which we will now call population A. Meanwhile, a population B is developing, specialized for a life style suited to niche B. It is to their mutual advantage that they not interbreed, and natural selection can be expected to favor those genotypes that render interbreeding impossible, so that adaptations that prevent or minimize it will develop.

Less complex mechanisms of sympatric speciation also occur. Among plants, for example, hybridization, either by itself or in combination with polyploidy, can produce instant speciation in the very same area even, at least at first, without significant niche differentiation such as that just described. In fact, speciation by polyploidy can also occur among animals capable of asexual reproduction or reproduction by parthenogenesis. We will examine this topic in greater depth shortly.

Yet another variety of speciation is **stasipatric speciation,** which occurs when species form a gradient or cline such that geographically extreme populations in this gradient cannot interbreed with one another directly. However, since they are able to breed with adjacent populations they can exchange genes with one another, however slowly. The best-known example of this phenomenon (mentioned in Chapter 17) occurs among grass frogs, *Rana pipiens,* as studied by John Moore and others. This common species extends from the northeastern United States all the way down the coast to the extreme South. When introduced in the laboratory, frogs from extreme Southern populations cannot interbreed successfully with those of extreme Northern populations, probably in large part because of variations in the timing of embryonic events suited to life in differing climatic zones. Yet those frogs have no problem in interbreeding with the populations next to their normal location, and these in turn can interbreed freely with populations next to *them.* Where can we draw the line to separate these incipient species from one another? Moreover, this is no rare phenomenon; other examples are known in fruit flies, salamanders, and many other organisms (see also Fig. 17–3).

Two groups of organisms living in the same geographic area may be **ecologically isolated** if they occupy different habitats to which each is narrowly adapted. Marine animals living in the intertidal zone are effectively isolated from other organisms living only a few feet away below the low-tide mark. Ecological isolation may also result from the simple fact that two groups of organisms breed at different times of the year.

The ultimate in isolation is achieved when one species does evolve into another—because individuals in ancestral and descendant populations are unable to breed because they are not alive at the same time. Chronologically differentiated species—that is, **chronospecies**—do, however, pose a practical and philosophical problem for the paleontologist. Both Neanderthal and modern human beings are now considered part of the same species, *Homo sapiens,* but if they could somehow be accorded the opportunity, would they interbreed successfully?

ISOLATING MECHANISMS

In the process of speciation, organisms employ various mechanisms to achieve reproductive isolation from one another. These are known as **isolating mechanisms.**

There are well-established behavior patterns of courtship and mating in many species that lead to the acceptance or refusal of one individual by another in mating. Such behavior patterns constitute one of the forces that may direct differential reproduction through **nonrandom mating.** For example, in many fishes or birds, some brightly colored part on the male (Fig. 45–10) serves as a necessary stimulus to the female before copulation can be begun. Mutations that lead to the formation of bigger, brighter spots tend to make those males more attractive to females and may confer selective advantages on their possessors. Conversely, mutations that lead to the formation of smaller, duller spots would have a negative selection pres-

Figure 45–10 Red-winged blackbirds. (*a*) The male is easily identified by its red and white markings on its wings. (*b*) A female red-winged blackbird. ((*a*), Russ Kinne, Photo Researchers, Inc.; (*b*), Gregory K. Scott, Photo Researchers, Inc.)

(*a*)

(*b*)

(a)

(b)

(c)

Figure 45–11 Courtship and mating in the fruit fly, *Drosophila.* (*a*) The male follows the female and vibrates his right wing. (*b*) If all goes well, the male licks the genitalia of the female while continuing to vibrate. (*c*) Copulation, which can last as long as 17 minutes, follows. (Courtesy of H. C. Bennett-Clark.)

sure. This kind of evolutionary force results from a failure of panmixis. It is termed **sexual selection,** but it is simply one kind of natural selection—one factor that may result in differential reproduction. It does not necessarily adapt the organism more perfectly to its ecologic niche but may, however, have significance in species maintenance. Sexual selection is no doubt important in the maintenance of the sign stimuli and courtship rituals that precede mating in many species, so the specificity of courtship and sign stimuli related to mating certainly should serve well as an isolating mechanism.

Isolating mechanisms that interfere with mating are known as **prezygotic isolating mechanisms,** because they act before the zygote can be formed and in fact prevent its formation. Fruit flies, for instance, exhibit a definite courting behavior that is species-specific. Unless both partners behave in just the proper instinctive fashion, they will not mate successfully. Part of the behavior is a love song—a series of buzzes of just the right pitch and rhythm performed by the male. If he does not have the right beat, the encounter ends right there. These differences in love song keep some *Drosophila* species apart. Even geneticists often cannot tell all the species apart in any other way. Similar, though more elaborate, mechanisms isolate various species of crickets, birds, and many other animals that have instinctive, highly stereotyped patterns of courtship behavior. Such courtship rituals must be performed with precision in order for mating to take place (Fig. 45–11).

Probably the most obvious prezygotic isolating mechanisms are physical barriers to mating—more-or-less gross inappropriateness of structure, so that reproductive parts simply will not fit each other and even an *attempt* at mating rarely takes place.

Perhaps the most obvious **postzygotic isolating mechanism** is embryonic lethality, in which the zygote or embryo does not develop properly and is aborted. Another, more subtle mechanism is hybrid sterility. In this case two species can mate and the offspring may be healthy or even unusually vigorous; but if the offspring is sterile, then in the long run genetic mixture between the species does not truly take place. A common example of a sterile hybrid is the mule, the vigorous but sterile offspring of a horse and a donkey. The most effective isolating mechanisms, however, operate early in courtship, for they save the energy that otherwise would have to be futilely expended in mating, pregnancy, and the rearing of sterile young.

THE ORIGIN OF SPECIES BY HYBRIDIZATION

Although, as has been emphasized, members of different species are usually not interfertile, it is best to think of separate species as reproductively isolated—that is, incapable of gene flow between their populations. Thus, it does not really violate the concept of species to observe that occasionally—perhaps as the result of extraordinary circumstances—members of two different but closely related species may interbreed to produce yet a third species by **hybridization.**

Often new combinations of genes are more favorable than old. It has long been recognized that when two genetically uniform inbred strains of organisms are crossed, the offspring of the union display **hybrid vigor;** that is, they are usually more vigorous than the parents (see Fig. 11–21). By hybridization, the best characters of each of the original species may be combined into a single descendant, thereby creating a new type better able to survive than either of its parents. (If the new form combined the worst characteristics of both parents it would obviously be at a serious disadvantage and would be unlikely to survive.)

When different species with different chromosome numbers are crossed (as in the production of mules from a horse–donkey cross), the offspring are usually sterile. The unlike chromosomes cannot pair in meiosis, and the resulting eggs and sperm do not receive the proper assortment of chromosomes—one of each kind. However, in plant hybridization, in which the number of chromosomes of such **interspecific hybrids** is doubled, meiosis can take place in a normal fashion, and normal fertile ovules and pollen will be produced. Thereafter, the hybrid species will breed true, and indeed will not produce fertile offspring when bred with either of the parental species. Moreover, the well-known tendency for plants to reproduce vegetatively by asexual means makes hybridization a more practical way of forming new species in them than it is in most animals. Although it is doubtful that the common dandelion originated in this way, loss of the ability to reproduce sexually would be of no significance to the ancestors of this weed, because the current model produces all of its seeds by a modified kind of parthenogenesis. Many related species of higher plants have chromosome numbers that are multiples of some basic number. The various species of wheat include ones with 14, 28, and 42 chromosomes; there are species of roses with 14, 28, 42, and 56 chromosomes, and species of violets with every multiple of 6 from 12 to 54.

That such natural series arise by hybridization and doubling of the chromosomes is supported by laboratory experiments yielding similar series. One of the more famous of these experimental crosses was made by Karpechenko, who crossed the radish with a cabbage, hoping perhaps to get a plant with a cabbage top and a radish root. Radishes and cabbages belong to different genera, but both have 18 chromosomes. The resulting hybrid also had 18 chromosomes, 9 from its radish parent and 9 from its cabbage parent. Since the radish and cabbage chromosomes were unlike, they could not pair during meiosis, and the hybrid was almost completely sterile. By chance, however, a few of the eggs and pollen formed contained all 18 chromosomes, and a mating between two of these resulted in a plant with 36 chromosomes. This plant was fertile, because during meiosis, the pairs of radish chromosomes underwent synapsis, as did the pairs of cabbage chromosomes. The hybrid exhibited some of the characteristics of each parent and bred true for these characteristics. Unfortunately it had a radish-like top and a cabbage-like root! The significant conclusion, however, is that since it could not be crossed readily with either of its parent species, it was, in effect, a new species produced by hybridization, followed by the doubling of the number of chromosomes.

A similar occurrence in nature has been documented in marsh grasses. One new species, *Spartina townsendii,* first appeared more than 100 years ago in the harbor of Southampton, England, in company with two others, *Spartina maritima* and *Spartina alterniflora.* The new species, *S. townsendii,* was sexually sterile but well able to reproduce asexually. In fact, it was much more vigorous than either of the parents and was soon widespread. It was especially valuable in collecting and holding soil and was transplanted to the Holland dykes and to other parts of the world. Because in many characteristics it was intermediate between the two species with which it was first found, it was believed to have originated as a hybrid. When examination of the chromosome numbers became possible, the hypothesis was confirmed: *S. townsendii* was found to have 62 chromosomes, some of which were derived from *S. maritima,* which was 56, and others from *S. alterniflora,* which has 62. There is also, incidentally, another sexually fertile species, *S. anglia,* which has 122 chromosomes. There seems no doubt that this new species arose by hybridization and doubling of the chromosomes. The following diagram illustrates the crosses:

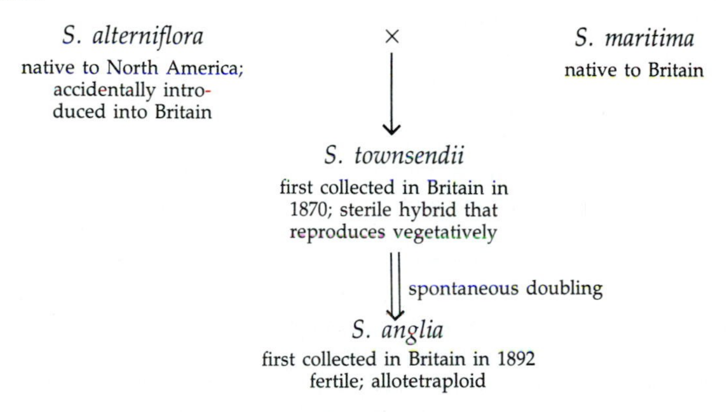

STABILIZING SELECTION

Though Darwin is credited with originating the concept of evolution by natural selection, he did have a number of less well-known predecessors, one of whom, Edward Blyth, saw natural selection as a force that would tend to preserve the originally created organisms from evolutionary change. This view of natural selection as a negative force was soon eclipsed, however, by views of Darwin, who held it to be a positive force—in fact, the very driving force of evolution. Eventually, however, the negative aspects of natural selection were restudied and the concept was revived.

Toward the end of the 19th century an ornithologist named Hermon Bumpus collected a group of dead sparrows, killed in an exceptionally severe snowstorm. He compared these dead birds in nine different ways with those that survived the storm, using such characteristics as wingspread and body weight as the bases for his comparison.

Bumpus discovered that the dead birds tended to be abnormal in many of their characteristics; that is, they represented the extreme ends of the normal range of variation in a sparrow population. Bumpus concluded that there is a more-or-less standard body build suitable for a bird with the life style of a sparrow. Though extreme deviants from this standard may get along well when conditions are not rigorous, extreme stresses—such as the blizzard—periodically arise that tend to weed out the unsuitable phenotypes, together with, presumably, the genotypes that produce them.

Although Bumpus' data have been subjected to several reanalyses (the observation cannot be duplicated until a comparable blizzard bird-kill recurs), his basic thesis seems to stand. The phenomenon is known not as Bumpus' law (thank goodness) but as **stabilizing selection.** It is, in a sense, anti-evolutionary, for it tends to maintain a standard phenotype in a population of organisms. In other words, stabilizing selection is genetically homeostatic. The set point around which the homeostasis operates is determined, however, by the environment. More recent observations show, for instance, that sparrows living in cold climates have an average body size larger than that of sparrows living in warm climates. The range of variation in both groups, however, overlaps extensively.

As long as the environment of a species does not undergo longterm changes, natural selection will tend to stabilize the genetic composition of populations. Should the environment change, however, or should the organism find itself able to expand its range into a new kind of environment or life style, called an **ecological niche,** then and only then will natural selection produce evolutionary changes. Therefore, species are preserved as well as produced by natural selection. However, if the conditions of natural selection change in a consistent fashion over a long period of time, a population of organisms may be altered according to what appears to be a consistent trend. This is called **directional selection.** Under other circumstances, as has been discussed, **disruptive selection** may produce divergence and differentiation of one original population into two adaptively distinguishable groups.

Historical Development of the Concept of Evolution

Our understanding of the concept of evolution can emerge logically from our understanding of genetics. This is not, however, how the concept of evolution arose historically.

The idea that present forms of life arose from earlier, simpler ones was not new when Darwin's *The Origin of Species* was published in 1859. Long before Darwin, odd fragments resembling bones, teeth, and shells had been discovered buried in the ground. Some of these corresponded to parts of familiar, living animals, but others were strangely unlike any known form. Many of the objects found in rocks high in the mountains resembled parts of marine animals. In the 15th century, Leonardo da Vinci interpreted these curious finds as being the remains of animals that had existed in previous ages but that had become extinct. Gradually others accepted his explanation. In Europe at that time the Biblical book of Genesis was almost universally accepted as historically accurate, and for most Europeans, what we today would call creationism was simply a fact. Since this book contains the account of a world-wide catastrophic flood, the evidence of former life suggested the theory of **catastrophism,** the idea that catastrophes such as fires and floods have periodically destroyed all or most living things and necessitated the repopulation of the world by successive acts of creation.

Three Englishmen in the 18th and early 19th centuries laid the foundations of modern geology. In 1785 James Hutton developed the concept, termed **uniformitarianism,** that the geological forces at work in the past were the same as those of the present. After a careful study of the erosion of valleys by rivers and of the formation of sedimentary deposits at the mouths of rivers, he concluded that the processes of erosion, sedimentation, disruption, and uplift must have required vast periods of time to take place at anything like their present rate of occurrence. Hutton held that these processes, operating over long enough periods of time, could account for the formation of fossil-bearing rock strata. In 1802 John Playfair's *Illustrations of the Huttonian Theory of the Earth* was published, in which he gave further explanation and examples of the idea of uniformitarianism in geological processes.

Sir Charles Lyell, one of the most influential geologists of his time, did much in his *Principles of Geology* (1832) to establish the principle of uniformity, convincing many previously skeptical scientists of his time that the earth is much older than a few thousand years, old enough for the process of organic evolution to have occurred. He was a personal friend of Darwin and had great influence on his thinking, and through his work he paved the way for, and made possible, the ideas presented in *The Origin of Species.*

JEAN BAPTISTE DE LAMARCK

The earliest theory of organic evolution to be logically developed was that of Jean Baptiste de Lamarck, the great French zoologist whose *Philosophie Zoologique* was published in 1809. Lamarck, like most biologists of his time, believed that organisms are guided through their lives by an innate and mysterious life force that enables them to adapt and to overcome adverse environmental forces. This view today would be called **teleology;** it implies the existence of a plan and foresight in nature, so that evolution has a goal in each species and proceeds through a series of primitive stages until finally that goal is reached. As Lamarck would have seen it, the horseshoe crab, which seems to have persisted unchanged in its present form since the days of the dinosaurs, reached its goal long ago.

Lamarck believed that, once made, these adaptations, called acquired characteristics, are transmitted from generation to generation. In developing this idea, Lamarck went on to state that new organs arise in response to demands of the environment and that their size is proportional to their "use or disuse." Such changes in size were believed to be inherited by succeeding generations. Lamarck explained the evolution of the giraffe's long neck, for instance, by suggesting that an antelope-like ancestor took to browsing on the leaves of trees, instead of on grass, and in reaching up, stretched and elongated its neck. Its offspring then supposedly inherited the longer neck.

The Lamarckian theory of the inheritance of acquired characteristics is an attractive one, and appeals to many people intuitively. It would explain the complete adaptation of many plants and animals to the environment, but it is unacceptable because overwhelming genetic evidence indicates that *acquired characteristics* (that is, characteristics modified by the environment) *cannot be inherited* (see Focus on Natural Selection in Bacteria). Many experiments have been performed in attempts to demonstrate the inheritance of acquired traits, but all have ended in failure. From what we now know about the mechanism of heredity, it is obvious that acquired traits cannot be inherited, because such characteristics are in the *body* cells only, whereas an inherited trait is transmitted by the gametes—the eggs and sperm. No way is known whereby the adaptive responses of the organism to the stresses of life could be recorded in the nucleic acids of any cells, including the gametes. (Newer techniques of genetic engineering may change this, however, permitting human goals to influence future evolution through the deliberate acquisition of truly inheritable characteristics, as discussed to some extent in Chapter 16.)

Figure 45–12 Charles Darwin (1809–1882). Darwin's studies in natural history had made him well known even before the publication of *The Origin of Species* in 1859. Darwin's role as the leader of an intellectual revolution was accepted reluctantly. He confronted his critics only in his letters and books. Most of his adult years were spent in Down, England, where he continued to do investigations in botany, ecology, and animal behavior until the time of his death. This photograph was taken in 1881. (Mary Evans Picture Library/Photo Researchers, Inc.)

CHARLES DARWIN AND ALFRED RUSSEL WALLACE

Darwin's contribution to the body of scientific knowledge was twofold: He presented a mass of detailed evidence and cogent argument to prove that organic evolution has occurred, and he devised a theory—that of natural selection—to explain how it operates.

The son and grandson of successful physicians, Darwin initially attempted to attend medical school and carry on the family tradition. However, he had no evident aptitude for or interest in the subject and dropped out of the medical curriculum. His father feared that he was destined for a life of parasitic idleness, for he was at best an indifferent and poorly motivated student. Eventually he agreed to be trained for the ministry and did obtain a bachelor's degree from Cambridge University, although he was never actually ordained. During the course of his university stay he vigorously pursued natural history as a hobby, under the guidance of professors and fellow students who had similar interests. (Biology and geology as aca-

FOCUS ON

Natural Selection in Bacteria

Does evolution proceed by the direct and directive effect of the environment upon genes? If so, the genes should change in all bacteria subjected to an environmental stress in such a way as to adapt them to that stress. To test this prediction, a series of experiments of a type first developed by Joshua and Esther Lederberg may be performed.

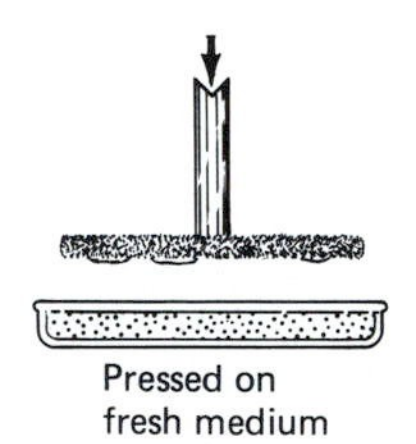

Velvet surface (sterilized)

Tool is pressed on surface of a culture plate of bacteria

1. A dish is filled with culture medium, and bacteria are grown upon it. Among the carpet of bacteria there may be a few clones derived from individuals that are already suited by mutation of certain genes to resist a given antibiotic. The others, the vast majority, are not.

Pressed on fresh medium

2. Since these mutants, if they exist, cannot be detected by inspection, it is necessary to discover them by means of their antibiotic resistance. A sterilized velvet carrier is placed in contact with the original plate, and bacteria of all kinds adhere to it.

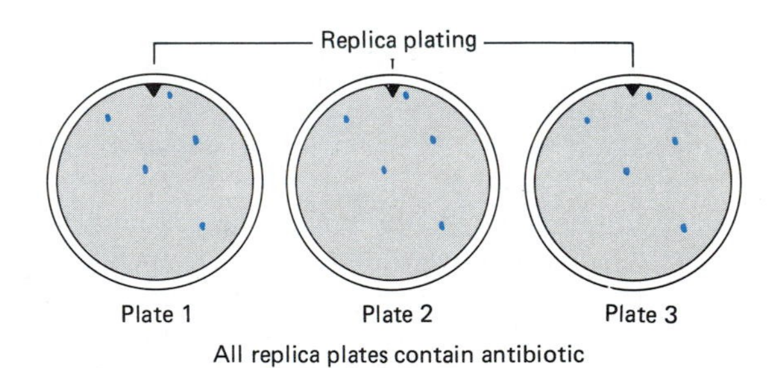

3. When the carrier is placed on a series of fresh plates *containing antibiotic,* only representatives of the original antibiotic-resistant clones will survive. These are shown to have been present in the original culture plate, rather than having developed in the antibiotic plates, by the fact that the resistant colonies always occur in the same position in all the replica plates. This can reflect only their *original* distribution in the *antibiotic-free* original plate.

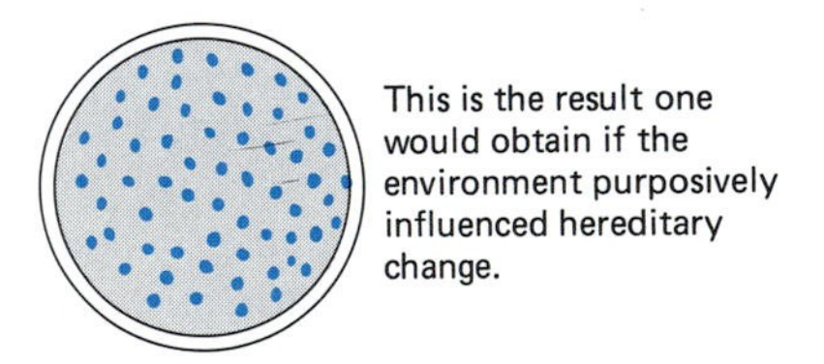

This is the result one would obtain if the environment purposively influenced hereditary change.

4. If the antibiotic *stimulated* bacteria to become resistant to it, a random and probably much denser pattern of colonies would be found on all replica plates. Since this result is not observed, Lamarckian evolution would seem not to occur in these bacteria.

demic disciplines were in their infancy and really did not exist as professions in their own right.) By the time of his graduation, Darwin had learned a great deal about geology and biology by this informal route.

While at Cambridge, Darwin became acquainted with Professor Henslow, the naturalist. Through his help, Darwin, just out of college and only 22 years old, was appointed to the unpaid position of naturalist on the *HMS Beagle,* a ship that was to make a five-year cruise around the world (Fig. 45–13) to gather data for oceanographic charts for the British navy. Darwin studied the animals, plants, and geological formations of the east and west coasts of South America, making extensive collections and notes. The *Beagle* then went to the Galapagos Islands, west of Ecuador, where Darwin was fascinated by the diversity of the giant tortoises and the finches that lived on each of the islands. Each island had its own distinctive species of tortoise; yet overall the tortoises were quite similar. The finches, too, were simi-

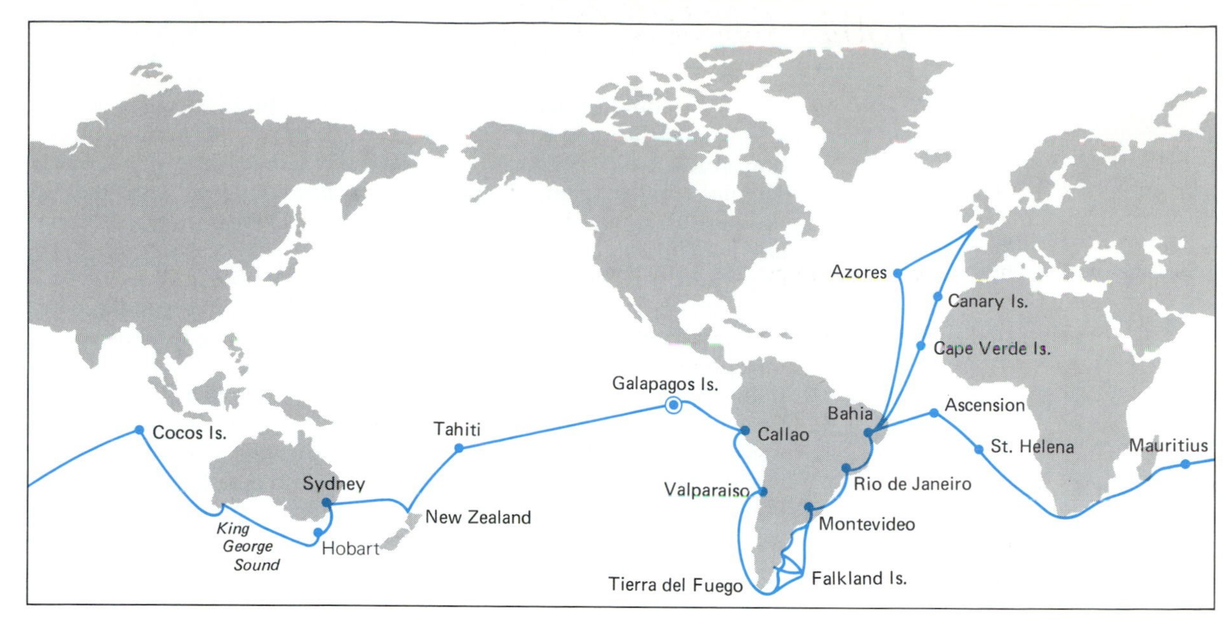

Figure 45–13 Voyage of the *Beagle*.

lar to one another but were anatomically or behaviorally adapted to a variety of life styles (Fig. 45–14). As Darwin mused over these observations, he came to reject the theory of special creation of each of these species and to seek an alternative explanation for his observations.

(a)

(b)

(c)

Figure 45–14 Three species of Darwin's famous Galapagos Island finches. As you can see, these are drab, unremarkable appearing birds which are evidently all derived from a common ancestry. Despite this common origin, the twelve known species are variously specialized for a variety of life styles, or ecological niches, that are elsewhere filled by birds of different families. The continental equivalents of the various Galapagos finches are presumably better adapted to these life styles than most or all of the finches are, but as chance had it, were never afforded the opportunity to colonize the Galapagos Islands. The likely derivation of such different birds from a common ancestry helped to suggest to Darwin that species were not invariable (unchanging) and that they originated by natural selection. (*a*) Cactus finch, *Geospiza scandens*. (*b*) A large ground finch, *Geospiza magnirostra*. This bird has an extremely heavy nutcracker-type bill adapted for eating heavy-walled seeds. (*c*) Woodpecker finch, *Camarhynchus pallidus*. This remarkable bird has insectivorous habits similar to those of woodpeckers but lacks the complex beak and tongue adaptations that permit woodpeckers to reach their prey. The adaptations of the woodpecker finch to this life style are almost entirely behavioral. In one of the few known instances of animal tool use, this bird digs insects out of bark and crevices using cactus spines, twigs or even dead leaves as implements. ((*a*), Jeanne White, Photo Researchers, Inc.; (*b*), (*c*), Miguel Castro, Photo Researchers, Inc.)

According to his journal, the idea of natural selection occurred to Darwin shortly after his return to England in 1836, but some historians think it occurred to him much earlier. According to Darwin himself, however, it was upon reading Thomas Malthus' pamphlet *Essay on the Principles of Population*[1] that he realized that the very large number of offspring most organisms produce cannot possibly all survive. This suggested to him a struggle for existence, with only the best-endowed organisms surviving to give rise to other generations. In Darwin's words, "Favourable variations would tend to be preserved and unfavourable ones destroyed."

He spent the next 20 years or so accumulating the vast body of facts and interpretations that eventually became *The Origin of Species*. During this time he published other material and became well-known as a naturalist. In 1858, before Darwin himself published his evolutionary proposals, he received a manuscript by post from Alfred Russel Wallace, a young naturalist who was studying the distribution of plants and animals in the East Indies and the Malay Peninsula. In this paper Wallace set forth his concept of natural selection, which he had reached independently—stimulated, as Darwin had been, by Malthus' book on population growth and pressure and the struggle for existence. By mutual agreement, Darwin and Wallace had a joint paper on their theory presented at the meeting of the Linnean Society in London in 1858, and Darwin's monumental work, *The Origin of Species,* was published the following year.

The *Origin* is only an abstract of Darwin's vastly longer archives, which he never did publish in complete form. In the *Origin,* Darwin sought to establish two things: first, that evolution had indeed occurred (by no means universally accepted in the scholarly world of that day), and second, that natural selection produced it. Since up until that time no one had proposed a truly credible mechanism for evolution, his proposal tended to overcome much of the skepticism toward evolution that had formerly existed. Wallace later said that since Darwin had accumulated so much more evidence in favor of natural selection than he had, and had been working on the hypothesis for so much longer, a fact corroborated by correspondence between Darwin and Lyell, Darwin should be credited as the originator of the concept.

THE DARWIN-WALLACE THEORY OF NATURAL SELECTION

The explanation advanced by Darwin and Wallace of the way evolution occurs may be summarized as follows:

1. Variation is characteristic of every group of animals and plants. Darwin and Wallace assumed that variation was one of the innate properties of living things. Inherited and noninherited variations can now be distinguished. We now realize that only inherited variations, produced by mutations, are important in evolution.
2. More of each kind of organism are produced than can possibly obtain food, survive, and reproduce. Since the number of members of each species remains fairly constant under natural conditions, it must be assumed that most of the offspring in each generation perish.
3. Since a larger number of individuals are born than can survive, there is a struggle for survival, a competition for food and space. This may be an active kill-or-be-killed contest, or one less immediately apparent but no less real, such as the struggle of plants or animals to survive drought, cold, and other unfavorable environmental conditions.
4. Those organisms with variations that better equip them to survive in a given environment will be favored over other organisms that are less well-adapted. The ideas of the struggle for survival and of "survival of the fittest" (a term borrowed from Herbert Spencer) are the core of Darwin's and Wallace's theory of natural selection.
5. The surviving individuals will give rise to the next generation; in this way the "successful" variations are transmitted to the next generation, and the next, and so on.

[1]Malthus noted that human population tended to increase geometrically (that is, logarithmically), as does compound interest. Yet agricultural production at best tended to increase arithmetically, which supposedly would lead to widespread misery and starvation. Malthus viewed this disparity with profound pessimism, seeing it only as a mechanism of hopeless social decay. But Darwin, by concentrating on those who *survive,* converted Malthus' negative view into a positive propulsive force for evolution.

This process will tend to provide successive generations of organisms with better adaptations to their environment. Indeed, as the environment changes, further adaptations will follow. The operation of natural selection over many years can lead ultimately to the development of descendants that are quite different from their ancestors—different enough to be recognized as a separate kind of animal or plant. Certain members of the population with one group of variations may become adapted to environmental changes in one way, while other members with a different set of variations may become adapted in a different way. Thus, two or more different kinds of organisms may arise from a single ancestral group. Darwin and Wallace recognized that animals and plants may also exhibit variations that are neither a help nor a hindrance to them in their survival in a given environment. These variations will not be affected directly by natural selection, and the transmission of such neutral variations to succeeding generations will be governed by chance.

SUMMARY

I. The term evolution can be used and understood in more than one sense. One of these meanings refers to the change in gene frequencies within populations and is more precisely termed microevolution. In a broader sense, evolution is used to refer to the concept that all life is commonly descended from one or a few ancestors and is therefore genetically related; this is macroevolution.

II. The modern concept of evolution continues to be based upon the idea of natural selection, but emphasizes changes in gene frequencies within populations.

A. Such changes will not take place, according to the Hardy-Weinberg law, if mating is panmictic (at random), if there is no selection against a genotype, if there is no selective immigration or emigration, and if the population is relatively large.

B. Since these conditions are rarely met, changes in gene frequency usually do take place in most populations of organisms over a period of time.

III. An intrafertile group of organisms that cannot exchange genes with other groups is termed a species. The origin of species, that is, speciation, involves the development of this genetic isolation. In addition, speciation usually involves the genetic divergence of population, resulting in the division of a single ancestral population into populations specialized for different ecological niches, or for life in different geographical areas, or both.

IV. Stabilizing selection leads to stable genetic frequencies within a population and preserves a standard type of organism if the environmental conditions remain the same. If environmental conditions change systematically, however, genetic frequencies will tend to change also.

V. Recognizable evolutionary theories in the modern sense were first propounded in the 18th and 19th centuries; the principal theory that has survived to the present day was proposed by Darwin and Wallace.

A. Darwin and Wallace did not originate the concept of evolution, but proposed a plausible mechanism whereby it might take place.

B. Their view included four main points: overproduction of offspring, variation among those offspring, inheritability of that variation, and natural selection for the variants.

POST-TEST

1. Mendelian genetics provide a theoretical basis for evolution since traits are not ________ when passed on to offspring, even though they may not always be expressed.

2. The process of ________ provides the genetic variability that is the raw material of evolution.

3. It is considered likely that most diploid organisms possess sufficient alternative genes in their gene pool so that mutation is probably a less important force in evolution than is ________ or other mechanisms of genetic change.

4. If a heterozygote is adaptively superior to either homozygote, natural selection tends to preserve ________ genes in the population. An example of this is the trait for the disease ________ ________.

5. In addition to selective mechanisms, random genetic events, such as ________, the ________ effect, or the ________ effect, can make a major impact upon gene frequencies, at least if populations are ________.

6. ________ is an example of natural selection that has occurred as an indirect consequence of air pollution.

7. ________ mimicry exists in the case of a harmless organism whose color or other characteristics resemble those of a noxious model in such a way as to mislead potential predators.

8. The Porto Santo or Madeira rabbits have become incapable of ________ with the ancestral European rabbit population as a result of long-standing geographical isolation.

9. The commonest kind of isolating mechanism between species is probably incompatibility of courtship behaviors; this would be an example of a ________ isolating mechanism.

10. Species do not always originate by divergence; hybridization seems to be a common form of species pro-

duction among ______________, and among animals that reproduce asexually or by ______________.

11. For the most part, ______________ selection tends to minimize genetic change provided that the environmental factors responsible for the selection do not change themselves.

12. The concept of ______________, that is, that the geological past was influenced by much the same kinds of events that we observe today, was originated by James Hutton.

13. Jean Baptiste de Lamarck believed in evolution by the inheritance of adaptive ______________ characteristics.

14. As a young naturalist, Darwin sailed on what we today would call an oceanographic expedition on the ship *HMS* ______________.

15. Darwin's early thinking on the matter of natural selection was much influenced by an essay by the economist ______________, in which widespread ______________ was predicted.

16. Give the four main points of the Darwin-Wallace theory of natural selection: ______________ ______________ ______________ ______________.

REVIEW QUESTIONS

1. What assumptions underlie the theory of uniformitarianism?
2. Explain briefly the concept of organic evolution.
3. In what ways does Lamarck's theory of adaptation not agree with present evidence?
4. What contributions did Darwin make to the theory of evolution?
5. Describe briefly the Darwin-Wallace theory of natural selection. What is meant by "survival of the fittest"? Do you think this adequately expresses the theory?
6. Why is it that only inherited changes are important in the evolutionary process?
7. After a mutation has occurred in a population, what events must take place if the mutant trait is to become established in the population?
8. What is meant by the term genetic drift? balanced polymorphism? gene pool?
9. Discuss the current theory of the steps involved in the establishment of a new species of plant or animal. Do you think most new species arise by the accumulation of small mutations or by a few mutations with large phenotypic effects? Give reasons for your answer.
10. Why do nearly all of the mutations occurring at the present time (and presumably the past) have a detrimental effect on the organisms in which they occur?
11. Contrast hybridization with other ways in which new species may be produced.
12. What contributions to the principles of evolution were made by Alfred Russel Wallace, Thomas Malthus, and Jean Baptiste de Lamarck?
13. Discuss the role of isolation in the origin of species.
14. Summarize the role of stabilizing selection in evolution.
15. What bearing do dominance and recessiveness have upon change in gene frequencies? List the factors that can be expected to change the frequencies of genes within a population of panmictic organisms.

46
Evolutionary Evidence

OUTLINE

I. Interpreting evolutionary evidence
II. The nature of the evidence
  A. Microevolution
  B. Evidence from morphology
    1. Homology
    2. Vestigial organs
  C. Biochemistry
  D. Evidence from embryology
  E. Biogeography and distribution
  F. Structural parallelism
  G. Adaptive radiation
  H. Fossil record

LEARNING OBJECTIVES

After you have read this chapter you should be able to:

1. Summarize the nature of the special application of the scientific method that is required in the study of evolution.
2. Summarize the evidence that can be used to support the occurrences of evolution, with particular reference to arguments that can be advanced on the basis of the following categories: microevolution; comparative morphology (including vestigial organs); comparative biochemistry; comparative embryology; biogeography and organismic distribution; and the fossil record.
3. Define the following and give at least two examples of each: homology, analogy, and structural parallelism.

In the preceding chapter we considered the mechanisms by which small genetic changes are made in populations. Let us now turn to the concept that much more extensive genetic changes have occurred in organisms since the time that life originated on the earth. The **general theory of organic evolution** states that all organisms have gradually developed from a common, simple ancestral type. The theory is based on data from such areas as morphology, biochemistry, the fossil record, and observed microevolutionary processes.

It is sometimes disputed whether evolution in this general sense has occurred at all, or if it has, whether it has occurred by the mechanisms presented in Chapter 45. We will consider these two controversial issues separately: In this chapter, we will examine the nature of the evidence bearing upon the question of whether the general theory of evolution is valid. In the following chapter, we will deal with the question of whether natural selection and population genetics could have produced general evolution, or could have produced all of it.

Interpreting Evolutionary Evidence

You may wonder how scientific investigation can be directed to evolutionary theory—as opposed to, say, the particulars involved in the reactions of photosynthesis. Obviously, the study of evolution requires the indirect application of the scientific method to *past events*—events that cannot be reproduced at will to test a particular hypothesis. In contrast, whenever a scientist wishes to test a hypothesis about photosynthesis, he or she can collect the necessary chemicals and apparatus, assemble them properly, and run the experiment. If the results are as predicted, the hypothesis is taken to be confirmed, at least tentatively. If the scientist can predict other results and design experiments to produce them, and if these experiments do indeed yield the predicted results, the confirmation is strengthened. Notice the stress that is laid on *prediction* in the scientific method. But how can we "predict" past events?

Actually, there *are* ways of "predicting" the past, irrational as this may sound, provided that the word prediction is defined in a special way. In science, a **prediction** is merely a logical consequence of a hypothesis; it is something that is true if a hypothesis is true. Unfortunately, it can also be true in some cases by coincidence; that is, an expected outcome may occur, but it may not be due to the causes that are ascribed to it. The possibility of coincidence, therefore, introduces an element of uncertainty into the scientific investigation. However, in a typical investigational experiment, often carried out by several researchers working independently, it is usually possible to test a *number* of predictions. If some of them fail to occur as predicted, serious doubt is cast on the hypothesis, but if all *do* come to pass, it is unlikely that coincidence could be responsible. Granted, this is not the same as saying the hypothesis is certain to be true, but under those circumstances, the hypothesis is *likely* to be valid.

Suppose we propose a hypothesis about the cause of a past event. Perhaps this proposed cause would have made other events occur as well. Since these events are foretold by the hypothesis, they are predictions by our definition, even if they occurred in the past; sometimes such past-time predictions are given the technical name of **retrodictions.**[1] If these events have left unambiguous evidence of their occurrence that we can discover today, then our hypothesis is confirmed. This is the method used by detectives and laboratory technicians—both in real-life police work and on television shows depicting such investigations—in reconstructing a crime.

But as the television detective show may also illustrate, interpretive mistakes are very possible, and this is the case in science as well.[2] The main reasons are that (1) evidence is not always unambiguous; (2) the variety of evidence that thorough confirmation may require can be unavailable, either because it simply cannot be found or because it has been destroyed; and (3) innocent error or even deliberate

[1] Retro means "backward" or "reversed"; inferring such predictions requires reasoning backward from present effects to presumed past causes.

[2] For an excellent example of the interpretation of such evidence, see the amusing article by Paul Hoffman (who has taught physics at Harvard University), Asteroid on trial, *Science Digest*, June 1982, pp. 58–63.

fraud may pass unchallenged, since the opportunity for independent experimental confirmation of results is often lacking, especially where fossil material is concerned.

The Nature of the Evidence

What *is* the evidence for evolution? It falls into several categories, gleaned from a number of sciences, among which are microevolution, morphology, biochemistry, paleontology, geographic distribution, structural parallelism, and embryology. Such evidences are used not only as arguments to establish the occurrence of evolution but also as clues to the way in which it might have occurred.

MICROEVOLUTION

Many cases of naturally occurring microevolution have been observed. For instance, bacterial populations have become resistant to antibiotics, flies and mosquitos have become resistant to pesticides, and industrial melanism has occurred in moths (Fig. 46–1). The selection and breeding of domesticated animals and cultivated plants for the past several thousand years provide us with other examples. All the varieties of present-day dogs are descended from one or a few related species of wild dog or wolf; yet they vary tremendously in many characteristics. Compare, for example, the size of the Chihuahua and the St. Bernard or Great Dane; the head shape of the bulldog and collie; the body proportions of the cocker spaniel, dachshund, and Russian wolfhound. If these varieties were found in the wild, they would probably be assigned to different species, and perhaps even to different genera. If they were known only as fossils this would undoubtedly be the case. But since all are known to come from common ancestors, and since all are interfertile, they are regarded as varieties of a single species.

The domestic rabbit and the house mouse are subspecies that have evolved during recent years. Reproductively isolated species have even been produced experimentally. For instance, Dobzhansky and his collaborators produced almost complete reproductive isolation between strains of *Drosophila* by breeding them in large numbers in the same population cages but selectively destroying the hybrids in each generation. If microevolution has been witnessed in a relatively short span of years, it is quite plausible that over millions of years, many microevolutionary changes have accumulated to produce macroevolution. Furthermore, where can we draw the line between micro- and macroevolution? If a species can evolve, why not a whole genus, order, class, or phylum, given enough time?

Figure 46–1 Examples of microevolution. (*a*) Arsenic-resistant *Andropogon* grass, growing in mine wastes intolerable to most specimens of this species and of other species of plants. (*b*) Antibiotic resistance in bacteria. An even coating of bacteria covers the surface of this culture dish except where an antibiotic to which the bacteria are sensitive prevents this growth. Varieties of bacteria that have developed resistance to certain antibiotics are common today, especially in hospital environments. ((*a*), courtesy of Dr. Sue Ellen B. Rocovich.)

(*a*)

(*b*)

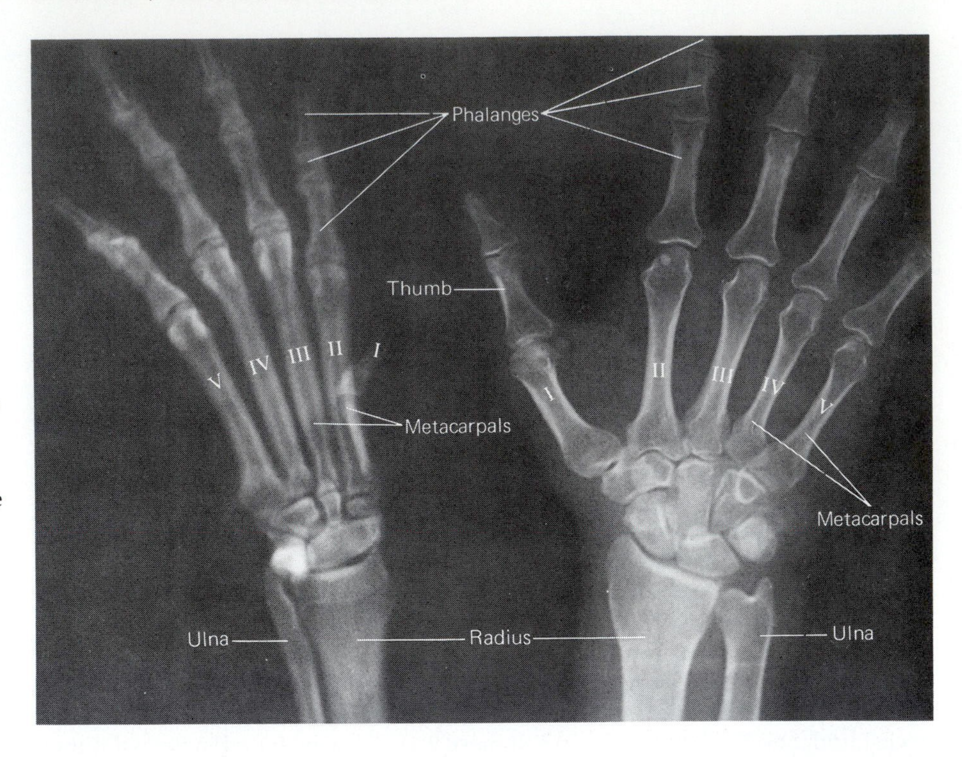

Figure 46–2 Skeletal adaptations to life styles. The forelimb of a dog *(left)* is employed for locomotion; the human hand, for manipulation. The dog walks upon its toes, the phalanges; its long metacarpal bones serve as part of the leg. Human metacarpals are much shorter in proportion, and have much more mobile joints at the junctions of the fingers, an adaptation for grasping. Note that digit I of the dog is very small, for it plays little part in walking. That of the human being, the thumb, is large and is opposed to all the other fingers, an adaptation for grasping. Note, however, that though they differ in shape, proportions, and function, the same bones are found in the hands of the two organisms. These x-ray photographs have been made the same size for ease of comparison, although that showing the forelimb of the dog was originally much smaller. (Courtesy of Dr. J. Koser, Dr. R. Bessmer, and Mrs. Jean Robertson.) (For another example, see Fig. 23–8.)

EVIDENCE FROM MORPHOLOGY

By comparing the structures of groups of animals and plants we are able to classify them into taxonomic groups. Within each such group, organ systems have a fundamentally similar pattern that is varied to some extent among its members. The skeletal, circulatory, and excretory systems of vertebrates provide particularly clear illustrations of this.

Homology

Only similarities based on homologous organs are valid in attributing evolutionary relationships. Recall that **homologous organs** are basically similar in their structure, in their relationships to adjacent structures, usually in their embryonic development, and in their nerve and blood supply. A seal's front flipper, a bat's wing, a cat's paw, a horse's front leg, and the human hand and arm (Fig. 46–2; see also Fig. 24–8), though superficially dissimilar and adapted for quite different functions, are homologous organs. Each consists of almost the same number of bones, muscles, nerves, and blood vessels, arranged in the same pattern, and with very similar modes of development. They differ principally in minor details of proportion and shape. The very existence of such homologous organs is a strong argument for a common evolutionary origin.

Vestigial Organs

Most plants and animals contain organs or parts of organs that are seemingly useless and degenerate, often undersized or lacking some essential part, as compared to homologous structures in related organisms. In the human body there are more than 100 such organs that have from time to time been viewed as vestigial, including the appendix, the coccyx (fused tail vertebrae), the wisdom teeth, the body hair, and muscles that move the ears and nose. Whales and pythons have vestigial hind leg bones embedded in the flesh of the abdomen; wingless birds have vestigial wing bones; many blind, burrowing or cave-dwelling animals may have vestigial eyes; and so on.

Vestigial organs are usually explained as being remnants of ones that were functional in some ancestral animal. Because of a change in the environment or mode of life of the species, the organ became unnecessary for survival and, gradually, nonfunctional. Ironically, vestigial organs have been held to justify Lamarck's concept of the role of "use and disuse" in evolution. But mutations are constantly occurring that decrease the size and function of various organs. If the organs *are* necessary for survival, the organisms undergoing such mutations will be eliminated. If the organs are not necessary for survival, they may become reduced in size or nonfunctional and eventually will be eliminated if they prove to be a handicap to survival. The Lamarckians were answered, after a fashion, by the proposal that a vestigial organ is one that is now *in the process* of being eliminated by natural selection.

BIOCHEMISTRY

Evolutionary relations can be estimated by studies of similarities and differences in *molecular* structure, as well as by studies of gross structure. All organisms have the same fundamental biochemical mechanisms: All employ DNA, and most use the citric acid cycle, cytochromes, and so forth. It seems inconceivable that the biochemistry of living things would be so similar if all life did not develop from a single common ancestral group. Furthermore, the amino acid sequence of proteins is very similar in organisms thought to be genetically related. For example, the sequence of the 300 amino acids in hemoglobin appears to be *identical* in humans and in chimpanzees. In the seemingly less closely related gorilla, two of the amino acids in the sequence are different. Monkey hemoglobin differs in a sequence of 12 amino acids. Since DNA codes for protein synthesis, protein similarity is a strong indicator of genetic similarity. Further biochemical evidence is accumulating that usually supports the evolutionary relationships that have been previously proposed.

The degree of similarity between the plasma proteins of various animals may also be tested by the antigen–antibody technique (see Chapter 17). Thousands of tests involving different animals have revealed a basic conformity among the blood proteins of all the mammals, the degree of similarity being indicated by how much the antigen and antibody solutions can be diluted and still result in visible precipitation. The closest "blood relations" of humans, as determined in this way are, in descending order, the great apes, the Old World monkeys, the New World prehensile-tailed monkeys, and the tarsioids (Fig. 46–3). This order is about the same as that suggested by these animals' physical similarities to human beings on the basis of anatomical studies. The biochemical relationships of a variety of animals and plants tested in this way correlate well with the evolutionary relationships determined by other means.

Figure 46–3 Arranged in the order of their blood protein similarity to human beings, the chimpanzee (*a*) is most similar, and the lemur (*e*) the least. Pigmy chimpanzee is shown in (*a*). (*b*) Baboons are Old World (Eastern Hemisphere) monkeys. (*c*) Spider monkey, a New World monkey with a strong prehensile tail, used in swinging from tree to tree. (*d*) The tailless potto, an Old World primate. The large, forward-directed eyes adapted for binocular night vision. (*e*) The ring-tailed lemur, one of the most primitive of living primates. ((*a*), courtesy Jim Leggett, Busch Gardens; (*b, d*) courtesy Busch Gardens; (*c*), Leonard Lee Rue III, The Image Bank; (*e*), Russ Kinne, Photo Researchers, Inc.)

	Human	Monkey	Pig, Bovine, sheep	Horse	Dog	Rabbit	Kangaroo	Chicken, turkey	Duck	Rattlesnake	Turtle	Tuna fish	Moth	Neurospora	Candida	Yeast
Human	0															
Monkey	1	0														
Pig, bovine, sheep	10	9	0													
Horse	12	11	3	0												
Dog	11	10	3	6	0											
Rabbit	9	8	4	6	5	0										
Kangaroo	10	11	6	7	7	6	0									
Chicken, turkey	13	12	9	11	10	8	12	0								
Duck	11	10	8	10	8	6	10	3	0							
Rattlesnake	14	15	20	22	21	18	21	19	17	0						
Turtle	15	14	9	11	9	9	11	8	7	22	0					
Tuna fish	21	21	17	19	18	17	18	17	17	26	18	0				
Moth	31	30	27	29	25	26	28	28	27	31	28	32	0			
Neurospora	48	47	46	46	46	46	49	47	46	47	49	48	47	0		
Candida	51	51	50	51	49	50	51	51	51	51	53	48	47	42	0	
Yeast	45	45	45	46	45	45	46	46	46	47	49	47	47	41	27	0

Figure 46–4 A diagram illustrating the differences in amino acid sequences in cytochrome *c* obtained from different species of animals, plants, and microorganisms. The numbers refer to the number of different amino acids in the cytochrome *c* from the species compared. (From Dayoff, M. O., and Eck, R. V.: *Atlas of Protein Sequence and Structure.* Silver Spring, MD, National Biomedical Research Foundation, 1968.)

Investigations of the sequence of amino acids in the α and β chains of hemoglobins from different species have revealed great similarities and specific differences, the pattern of which demonstrates the order in which the underlying changes in nucleotide base pairs is presumed to have occurred in evolution. The evolutionary relationships inferred from these studies agree for the most part with those based on anatomical studies, but this agreement is by no means perfect, and in some cases there is a substantial difference. Analyses of the sequence of amino acids in the protein portion of the cytochrome enzymes are taken as further corroborating evidence of evolutionary relationships (Fig. 46–4).

By cytological methods somewhat related to biochemical studies, the number and the detailed band arrangement of the chromosomes of related species can be compared. Such studies have provided useful evidence concerning the evolutionary history of fruit flies, jimson weeds, primroses, and many other plants and animals. Comparative studies of animal proteins may also be carried out by the technique of gel electrophoresis, illustrated in Figure 46–5.

EVIDENCE FROM EMBRYOLOGY

The importance of embryological evidence for evolution was stressed by Darwin and brought into even greater prominence by the German biologist Ernst Haeckel in 1866. Haeckel developed the theory that embryos, in the course of development, repeat the evolutionary history of their ancestors in some abbreviated form, a phenomenon known as **recapitulation.** Thus, in human embryology there is a single cell (protozoan) stage, a fishlike stage with gill pouches, a late stage covered with hair (lanugo), and so on. This idea, succinctly stated as "ontogeny recapitulates phylogeny," embodies the so-called **biogenetic law;** it stimulated research in embryology and focused attention on the general resemblance between embryonic development and the evolutionary process. The original formulation of the biogenetic law has, however, been shown to be false, which is not surprising. How could a human embryo possibly compare with a fully adapted and developed fish? It is

(a)

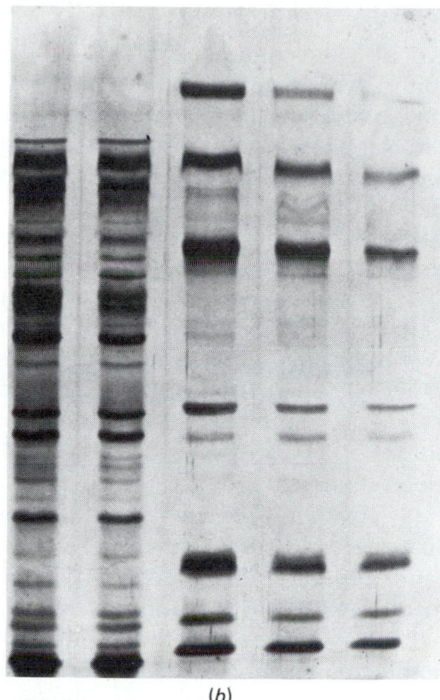

(b)

Figure 46–5 Comparison of proteins by gel electrophoresis. (*a*) In this apparatus, a strong electric potential is applied to a tube of gel containing a protein. The protein migrates in the gel in response to the electric potential, but its speed of response depends upon its chemical makeup. (*b*) When exposed to an appropriate reagent, the proteins can be observed and their patterns compared. Similar proteins are assumed to indicate a close evolutionary relationship between the organisms being compared. (NASCO.)

now said that the embryos of the higher animals resemble the *embryos* of lower forms, not the adults, as Haeckel had believed. It is true that the early stages of all vertebrate embryos are remarkably similar (Fig. 46–6), and it is not easy to differentiate a human embryo from the embryo of a pig, chick, frog, or fish during the early stages of development.

The usual present position is that in recapitulating its evolutionary history in a few days, weeks, or months, the embryo eliminates some steps and alters and distorts others. Also, although early mammalian embryos have many characteristics in common with those of fish, amphibians, and reptiles, they have in addition other structures that their ancestors never had which enable them to survive and develop within the mother's uterus, rather than within an eggshell.

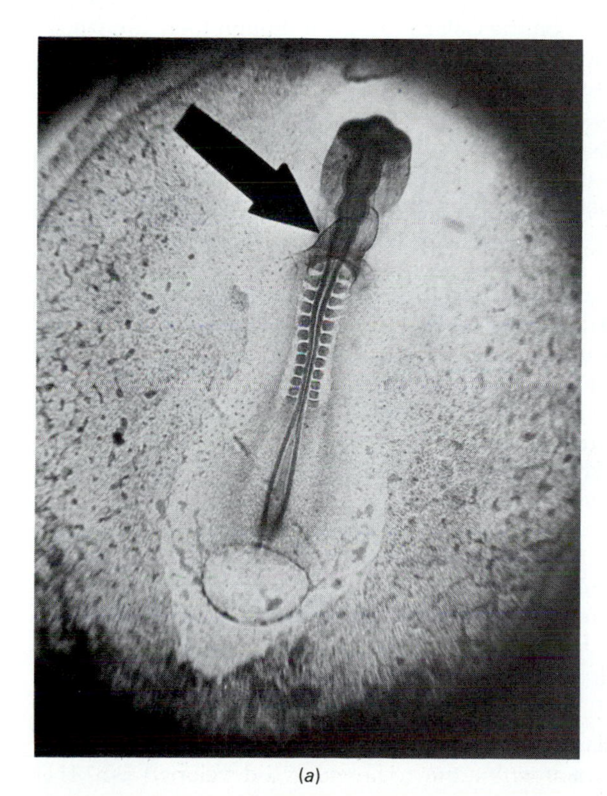

(a)

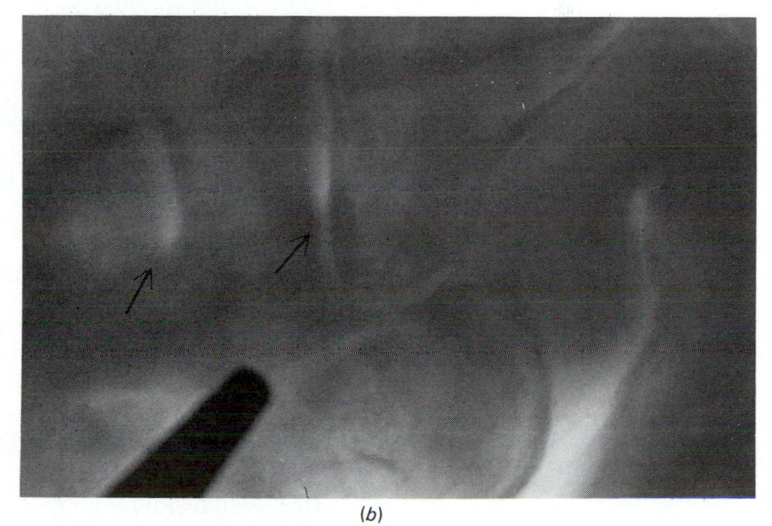

(b)

Figure 46–6 Some embryonic structures that are taken to reflect functional ancestral organs or states. (*a*) The arrow points to the single ventricle of the heart of an embryonic chick. The adult chick has two ventricles, but the ancestral state of the heart is taken to be reflected in the embryo. Note also the segmental arrangement of the developing musculoskeletal system. (*b*) Branchial grooves in embryonic chick.

BIOGEOGRAPHY AND DISTRIBUTION

Not all plants and animals are found in all parts of the world. They are not even found everywhere that they could survive, as we would expect if climate and topography were the only factors determining distribution. Central Africa, for example, has elephants, gorillas, chimpanzees, lions, and antelopes, while Brazil, with a similar climate and other environmental conditions, has none of these, but does have prehensile-tailed monkeys, sloths, and tapirs. The present distribution of organisms seems understandable only on the basis of evolution, which explains it as follows:

The **range** of a given species—that is, the portion of the earth over which it is found—may be only a few square miles or, as with humans, almost the entire world. In general, closely related species do not have identical ranges, nor are their ranges far apart. They are usually adjacent, but separated by a barrier of some sort, such as a mountain or a desert. This generalization, formulated by David Starr Jordan and known as **Jordan's rule,** follows from the role of isolation in the formation of species.

As we might expect, then, regions such as Australia and New Zealand, which have been separated from the rest of the world for a long time, have a flora and fauna peculiar to these areas. Australia has a population of monotremes and marsupials (Chapter 26) found nowhere else. During the Mesozoic era,[1] Australia was isolated from the rest of the world. Its primitive mammals, therefore, never had any competition from the generally better-adapted placental mammals, which are thought to have competitively eliminated the monotremes and most of the marsupials everywhere else they may have existed. By the founder effect (Chapter 45), the original Austrialian mammals gave rise to a variety of forms that were able to take advantage of the different habitats available (Fig. 46–7; see also Fig. 26–34).

The kinds of animals and plants found on oceanic islands resemble, in general, those of the nearest mainland; yet they include some species found nowhere else. Darwin studied the flora and fauna of the Cape Verde Islands, some 400 miles west of Dakar, Africa, and of the Galapagos Islands, a comparable distance west of Ecuador. On each archipelago the plants and terrestrial animals were indigenous, but those of the Cape Verdes resembled African species and those of the Galapagos resembled South American ones.

Darwin concluded that organisms from the neighboring continent migrated or were carried to the island and subsequently evolved into new species. The animals and plants found on oceanic islands are only those that could survive the trip there. There are no frogs or toads on the Galapagos, even though there are woodland spots ideally suited for such creatures, because neither the animals nor their eggs can survive exposure to sea water. There are no terrestrial mammals either, although there are many bats, as well as land and sea birds. The occurrence of these particular forms—closely related to, yet not identical with, those of the Ecuador coast—suggests strongly that after the first animals and plants arrived on the islands, mutations took place that changed the species slightly; these changes were retained because of isolation, natural selection, and genetic drift.

Alligators are found only in the rivers of the southeastern United States and in the Yangtze River of China, and sassafras, tulip trees, and magnolias grow only in the eastern United States, Japan, and eastern China. Geologists infer from the various fossil layers that early in the Cenozoic era, the northern hemisphere was low-lying and much flatter than it is now, and the North American continent was connected with eastern Asia by a land bridge at the Bering Strait and possibly with Greenland. The climate of this region was much warmer than at present, as judged by the fossil evidence, which shows that alligators, magnolia trees, and sassafras were distributed over the entire region. Later in the Cenozoic, as the Rockies increased in height, the western part of North America became colder and dry, causing the plants adapted to a warm, humid climate to become extinct.

Then, with the Pleistocene glaciations, the ice sheets moving from the North met the desert and mountain regions in western North America, eliminating any surviving temperate-zone plants. In Europe the polar glaciations nearly met the glaciers spreading from the Alps, so that many of the temperate-zone plants there became extinct. In southeastern United States and eastern China there were regions untouched by the glaciation in which the magnolia trees and alligators survived.

[1] For a guide to geological time, see Table 47–1.

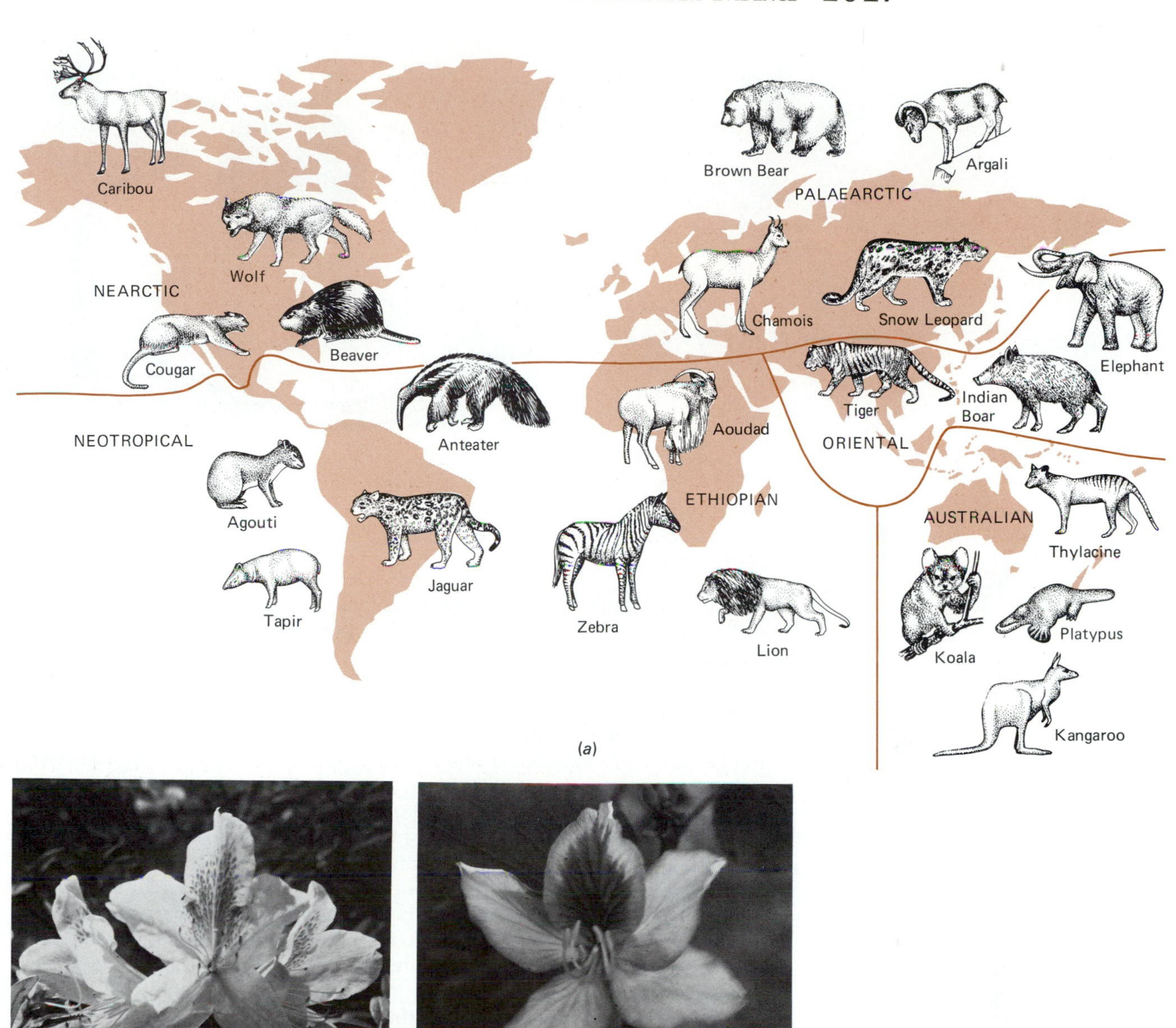

Figure 46–7 (*a*) The biogeographical realms of the world, with some of their characteristic animals. (*b*) Possible evolutionary parallelism among plants. The azalea *(left)* and sea grape *(right)* flowers have closely similar patterns of pigment that aid bees in finding the nectaries and in the transfer of pollen; yet the two plants are not at all closely related, and this adaptation cannot be taken as indicating a common ancestry.

Because the alligators and magnolia trees of the two regions have been separated for several million years, they have followed separate evolutionary pathways and are slightly different, although they are still closely related species of the same genera.

The study of the distribution of plants and animals constitute the science of **biogeography.** One of its basic tenets is that each species of animal and plant (and for that matter, each genus, family, order, and perhaps class and phylum) originated only once. The particular place where this occurred is known as the species' **center of origin.** The center of origin is not a single point but the range of the population when the new species was formed. From its center of origin, each species spreads out until halted by a barrier of some kind—physical, such as an ocean or mountain; environment, such as an unfavorable climate; or ecological, such as the absence of food or the presence of organisms that prey upon it or compete with it for food or shelter.

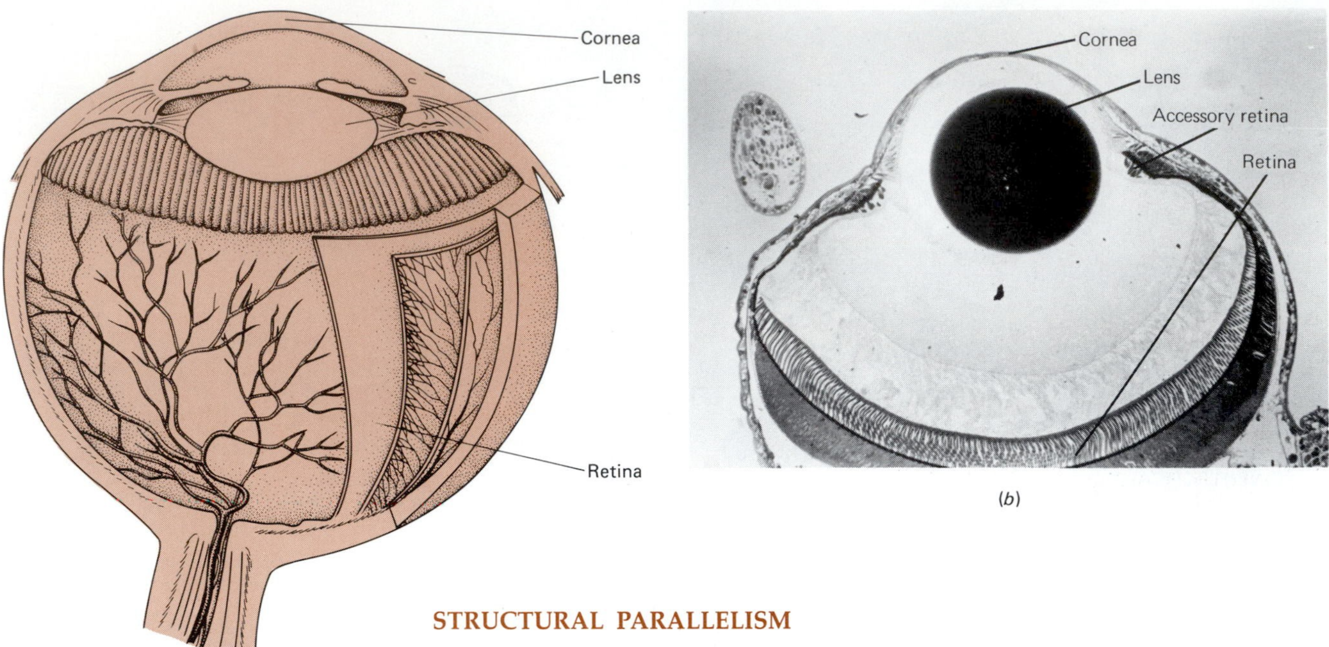

Figure 46–8 Structural parallelism. (*a*) The human eye, partly dissected to show internal structure. (*b*) The eye of an annelid worm. Compare with (*a*). (*c*) This cathedral plant, a member of the crown-of-thorns family, although not a cactus, nevertheless resembles a cactus. Note the reduced leaves, which perhaps may be considered vestigial organs. ((*b*), courtesy of Dr. George Wald and *Science*. From Wald, G.: Vision in annelid worms. *Science*, July 24, 1970, pp. 1434–1439. Copyright 1970 by the American Association for the Advancement of Science.)

STRUCTURAL PARALLELISM

Unrelated or distantly related organisms sometimes possess very similar adaptations. This is known as **structural parallelism.** Structural parallelism might at first be taken as evidence that such organisms are, in fact, closely related, except for the fact that their organs, when closely examined, may turn out to have vastly different embryonic origins. See Figure 46–8 for a case in point.

Close study of the annelid eye discloses that when compared with the vertebrate eye, there are substantial differences of detail. The lens is not adjustable, there is no equivalent of an iris, there are accessory retinas that have no vertebrate equivalent, and the cellular arrangement of the retina is far different. Functionally, though, they are much the same. That is, they are **analogous** but not homologous.

How can this extraordinary similarity be taken as evidence for evolution? It is argued that since the embryonic origins of the two structures are different, they must have had different ancestral origins. Yet the fact that they are mechanically similar shows that natural selection in similar life styles tends to produce the same or similar adaptations. But since unrelated animals are genetically different, the forces of natural selection must operate on different sets of genes, so that even though the end results may be similar, the genetic basis of the two sets of similar adaptations is different. This is reflected in the differences of detail that we have seen, and also differences of embryonic origin. A nonevolutionary view of the origins of parallel adaptations would have difficulty accounting for these differences. A marsupial with a life style like that of a wolf, for example, would possess adaptations similar to those of a wolf, so natural selection would produce a wolflike body in both cases. Since different evolutionary pathways lead to much the same result in such instances, the development of similar adaptations in unrelated lineages is called **convergent evolution.**

ADAPTIVE RADIATION

Every potential habitat or ecological niche represents a source of resources that could be available to an organism. If the habitat or niche could be exploited, it would afford an advantage to any organism (and to its genes) that could become adapted to its demands. Because of the constant competition for food and living space, each group of organisms does tend to spread out and occupy as many different habitats and ecological niches[1] as possible. Within each habitat, those genotypes that produced superior phenotypic adaptations would propagate themselves better than those less well-suited to the demands of the environment or the life styles appropriate to it. Since the habitats and the ecological niches differ, the ap-

[1] As will be discussed in much greater detail in Chapter 50, an ecological niche is a potential life style. Organisms are adapted to their life styles as well as to their habitat. Human beings and cockroaches are both well-adapted to live inside houses (but not, of course, in the same way!). Thus, they share the same habitat but not the same ecological niche, and their adaptations are quite different.

Figure 46–9 Adaptive radiation. All the various mammals shown are deduced, on the basis of comparative anatomy and to a somewhat lesser extent, the fossil record, to have evolved from the common shrewlike ancestor depicted in the center. Each of the organisms is specifically adapted to a different ecologic niche.

propriate adaptations would differ as well, resulting in a variety of physical and behavioral specializations. This process of evolution from a single ancestral species to a variety of forms that occupy somewhat different habitats and ecological niches is termed **adaptive radiation.**

One of the classic examples of adaptive radiation is the evolution of placental mammals (Fig. 46–9). Apparently the earliest known mammal was an insect-eating, five-toed, short-legged creature that walked with the soles of its feet flat on the ground. Today we see a great variety of mammalian types. These include dogs and deer adapted for a terrestrial life in which running rapidly is important for survival; squirrels and primates adapted for life in the trees; bats equipped for flying; beavers and seals, which maintain an amphibious existence; the completely aquatic whales, porpoises, and sea cows; and the burrowing animals—moles, gophers, and shrews. In each of these, the number and shape of the teeth, the length and number of leg bones, the number and attachment sites of muscles, the thickness and color of fur, the length and shape of the tail, and so on are specifically adapted to the animal's life style and environment.

Adaptive radiation that gives rise to several different types of descendants, adapted in different ways to different environments, is a result of **divergent evolution.** It is the opposite, in a way, of the convergent evolution that results in structural parallelism.

Let's return to the example of the Galapagos Islands. Darwin took them to be a microcosm of adaptive radiation, represented by the variety of ground finches present there today. Some of these birds live on the ground and feed on seeds, others feed mainly on cactus, and still others have taken to living in trees and eating insects. These variations in feeding have been accompanied by changes in the size and structure of the beak. This suggested to Darwin that the essence of adaptive radiation is the evolution from a single ancestral form to a variety of different forms, each of which is adapted and specialized in some unique way to survive in a particular habitat.

FOSSIL RECORD

Exploration of fossil beds in sedimentary rocks often reveals that the fossils were deposited in striking sequence. In the deepest and oldest strata are found the most primitive fossils. Buried in successive layers from the bottom to the top is a progression of fossils from the simplest to the most complex. Since the bottom sediments were presumably laid down first, **stratigraphy** has been accepted since before Darwin's time as a way of dating the relative age of rocks and the fossils contained in them.

Careful studies have indicated that, under the most typical present-day circumstances, approximately a foot of sediment is deposited every 5000 years. By measuring the depth at which fossils are buried in a deposit, scientists are able to arrive at an estimate of their age, though perhaps not a very accurate estimate. Among the many problems that have been cited, erosion will serve as one example. If deposited sediments are partially eroded, and then others deposited on what is left, the age of the total formation could be grossly underestimated, perhaps by a factor of as much as tenfold, on the average. Valid or not, such objections do point up the need for an absolute method of dating sedimentary rocks. Such methods have been developed, but in the opinion of many geologists, the task remains an extraordinarily difficult one, which demands the use of many sophisticated scientific tools.

A more recent and more accurate method for determining the age of fossils is **radioactive dating.** Radioactive elements decay into stable products at specific, constant rates. This decay rate is not altered by such factors as temperature or pressure (as far as is known), and appears to proceed at the same rate over time. One of the most useful systems is the uranium–lead method. When igneous rocks are first formed by solidifying from molten lava or magma, some contain uranium-238. The uranium continuously decays at a constant rate to form lead-206. The half-life of uranium-238 is 4.5 billion years, which means that half the atoms in a particular sample will be converted to lead during that time period. By measuring the ratio of uranium-238 to lead present in a rock sample, scientists can approximate the age of such a rock, at least in many instances.

The potassium–argon method is based upon the decay of radioactive potassium-40 into argon and calcium. Potassium-40 has a half-life of 1.3 billion years. For archeological artifacts or fossils less than 30,000 years old, a radiocarbon method is utilized. The half-life of carbon-14 is 5568 years. Since the amount of radioactive carbon generated in the earth's atmosphere by the sun is variable, radioactive carbon dating has not been as easy to use as was initially thought, and has required revision in the light of such objective methods of dating as tree-ring counts. Unfortunately, it is much harder to check the accuracy of most other dating methods.

Advances in isotopic techniques have made possible some astonishing conclusions in the field of geology. For example, the proportion of the various oxygen isotopes in the calcium carbonate secreted by living organisms depends upon temperature. Consequently, by analyzing the oxygen isotopes in the calcium carbonate of fossil shells, it is possible to estimate the temperature of the ancient seas in which those animals lived.

The earliest-dated fossils are ones resembling cyanobacteria. They are believed to be about 3 billion years old (Fig. 46–10). An extensive sequential record of fossils has been identified from an estimated 600 million years ago to recent times. However, whole sequences of fossil organisms that can be traced throughout geological time, including the evidence for evolutionary change, are rare.

The science of **paleontology** deals with the finding, cataloging, and interpretation of the abundant and diverse evidence of life in former times. The term **fossil** (Latin *fossilum,* something dug up) refers not only to the bones, shells, teeth, and other hard parts of a plant or animal body that have been preserved, but also to any impression or trace left by a previously existing organism (Fig. 46–11). In view of the large number of fossils of plants and animals that have been found, it is sober-

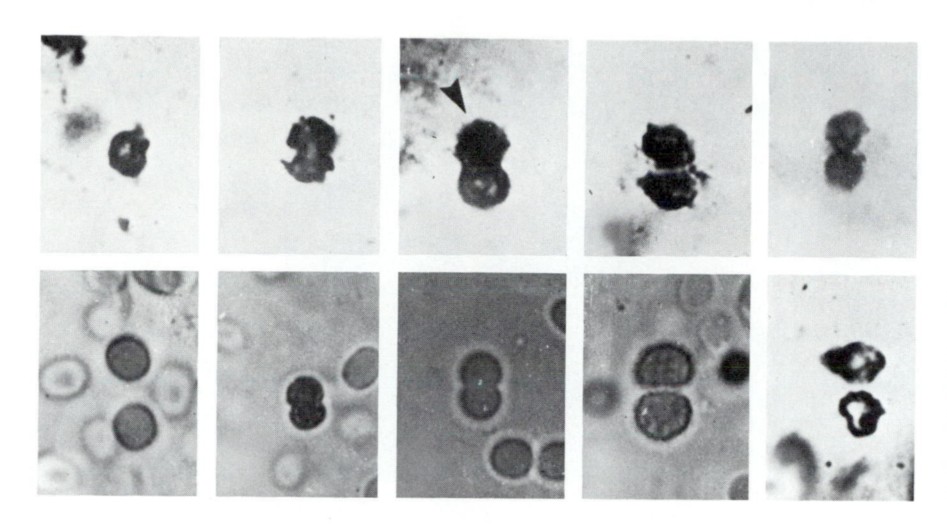

Figure 46–10 Precambrian cyanobacteria microfossils. (Courtesy Dr. E. S. Barghoorn and *Science*. From Knoll, A. H., and Barghoorn, E. S., Archaean microfossils showing cell division from the Swaziland System of South Africa, *Science*, Oct 28, 1977, pp. 396–398. Copyright 1977 by the American Association for the Advancement of Science.)

ing to realize that only a small fraction of all the organisms that ever lived have been preserved as fossils, and that only a small fraction of these fossils have been dug up and studied to date.

Footprints or trails made in soft mud, which subsequently hardened, are a common type of fossil—according to some estimates, the most common type of vertebrate fossil (Fig. 46–12). From such remains scientists can infer something of the structure and body proportions of the animals that made them, and even their behavior. From the evidence of dinosaur footprints, for example, it seems that some dinosaurs traveled in herds in which young were included and probably protected.

Most of the vertebrate fossils are skeletal parts, from which it is possible to deduce the animal's posture and style of walking. From the bone scars, indicating muscle attachments, paleontologists can deduce the general position and size of the muscles, and from this the contours of the body. On the bases of such considerations, reconstructions are made of the animal as it is imagined to have looked in life (see Fig. 46–9). Such qualities as the texture and color of the fur or scales can, of course, only be guessed at.

In one interesting and striking type of fossil, the original hard parts and, rarely, soft tissues have been replaced by minerals—a process known as **petrifaction.** The minerals that replace the tissues may be iron pyrites, silica, calcium carbonate, or other substances. The petrified muscles from a shark dated at more than

Figure 46–11 One of the more famous examples of a fossil, the remains of *Archaeopteryx*, a tailed, toothed birdlike dinosaur from the Jurassic period. (Compare this with the reconstructions shown in Figs. 26–30 and 45–2.)

Figure 46–12 Dinosaur footprints occurring in sedimentary rocks in Texas. Human feet included for size comparison.

300 million years old were so well preserved by this process that not only individual muscle fibers but their cross-striations could be observed in thin sections under the microscope. (But such a fine state of preservation is a great rarity.) The Petrified Forest in Arizona is a famous example of the process of petrifaction.

Molds and casts are superficially similar to petrified fossils but are produced differently. **Molds** were formed by the hardening of the material surrounding the buried organism, followed by the decay and removal of the organism by seepage of the groundwater. Sometimes the molds were subsequently filled with minerals, which in turn hardened to form **casts**—replicas of the original structures (Fig. 46–13).

Occasionally, paleontologists are fortunate to find organisms frozen in the soil or ice of the far North, usually in Siberia and Alaska. The remains of woolly mammoths more than 25,000 years old have been found so well preserved that their flesh has been fed to dogs. Other forms—plants, insects, and spiders—have been preserved in amber, a fossil resin from pine trees. Originally the resin was a sap soft enough to engulf a fragile insect, for instance, and penetrate every part; then it gradually hardened, preserving the animal intact (Fig. 46–14).

The formation and preservation of a fossil require that some structure be buried. This may take place at the bottom of a body of water or on land by the accumulation of windblown sand, soil, or volcanic ash. The people and animals in Pompeii were preserved almost perfectly by the volcanic ash from the eruption of Vesuvius. Sometimes animals were trapped and entombed in a bog, quicksand, or an asphalt pit. The famous Rancho La Brea tar pits in Los Angeles have provided superb fossils of Pleistocene animals (Fig. 46–15).

In the following chapter we will look at some of the fossil evidence and will summarize the current consensus of just how evolution is thought to have produced our present world of life. We will also trace the history of the stages of life's development.

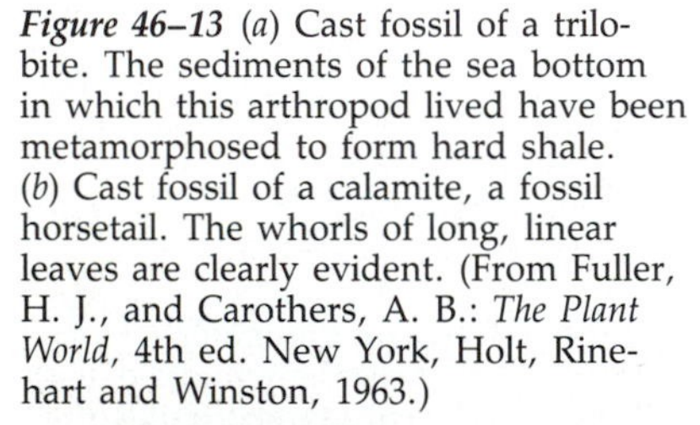

Figure 46–13 (*a*) Cast fossil of a trilobite. The sediments of the sea bottom in which this arthropod lived have been metamorphosed to form hard shale. (*b*) Cast fossil of a calamite, a fossil horsetail. The whorls of long, linear leaves are clearly evident. (From Fuller, H. J., and Carothers, A. B.: *The Plant World*, 4th ed. New York, Holt, Rinehart and Winston, 1963.)

(*a*)

(*b*)

Figure 46–14 Two termites embedded in amber. These insects, dating from the Middle Tertiary (perhaps 38 million years ago) have been preserved almost perfectly. (From Buchsbaum, R.: *Animals Without Backbones,* rev. ed. Chicago, University of Chicago Press, 1948. Photograph by P. S. Tice.)

Figure 46–15 Re-creation of a scene at the Rancho La Brea tar pits (now a part of Los Angeles, California) in the Pleistocene epoch. Fossilized skeletal remains of all of these animals were actually found together in this locality, allowing us to form some picture of the paleo-ecology of the area. In the left foreground are two saber-toothed tigers, and in the right foreground, three large ground sloths; the giant vultures, now extinct, had a wingspread of 3 meters. In the background are mastodons and dire wolves. (Copyright American Museum of Natural History, New York. From a painting by Charles R. Knight.)

SUMMARY

I. Evolution can, for the most part, only be inferred on the basis of indirect evidence that, because of its historical nature, is often susceptible to a variety of interpretations.

II. Much of the evidence for evolution falls into comparative and historical categories; other evidence is based on biogeographical studies, for example, and modern experiences with microevolution.

A. Evidence from comparative anatomy and other comparative studies: It is possible to arrange organisms into groups whose anatomy, physiology, biochemistry, and behavior are similar. Usually these groups coincide. With certain reservations, this convergence of similarity is taken as evidence for genetic relationship.

B. Evidence from the fossil record: The fossil record represents the actual course that the history of earthly life has taken, and demonstrates that most extinct and living forms are different; yet it is usually possible to identify putative ancestors for living forms in the fossil record and, often, to suggest relationships among fossil organisms.

C. Evidence from biogeography: Geographically separate areas tend to be inhabited by organisms that are ecologically similar but taxonomically distinct. This is understood as indicating that each set of organisms represents a separate adaptive radiation.

D. Evidence from vestigial structures: Organs that have no apparent use in an organism, but which did have adaptive value in the presumed ancestor, are hard to interpret in any way other than indicating genetic relationship.

E. Evidence from embryology: Arguments based on vestigial structures are similar to those based on embryonic structures that do not persist to form functional organs in the adult. The embryonic structures are similar to structures that were functional in the adults of the presumed ancestral organism.

F. Evidence from current examples of microevolution: Laboratory studies of microevolution and historical examples of speciation give evidence that microevolution does occur. It seems reasonable to suppose that, given enough time in which to work, microevolutionary changes will accumulate sufficiently to produce major differences between ancestral and descendant organisms.

POST-TEST

1. The proposal that organisms that appear to possess body parts that are the "same" in embryonic origin, and which contain the same basic components, are probably related is an argument based on the concept of __________.

2. An organ that appears to have slight or no function in an organism, but which is similar to a fully functional equivalent in the organism's presumed ancestor or relatives, is known as a __________ organ.

3. The amino acid sequence of a number of important proteins, for example, hemoglobin, is identical in human beings and in __________.

4. According to the so-called biogenetic law popularized by Haeckel, structures in the embryos of modern forms are similar to corresponding __________ structures in their ancestors.

5. The existence of unrelated ecological equivalents in different mutually isolated geographical areas is an example of an argument from __________, and is presumed to reflect past __________ of the forms concerned.

6. The eye of vertebrates and that of the squid are remarkably similar but not __________; this is an example of __________ __________.

7. When a group of organisms differentiates into species that fill a multitude of ecological niches, the phenomenon is referred to as __________ __________.

8. The methods of dating geological formations and their contained fossils that are considered most reliable are those involving __________ __________ of unstable __________.

REVIEW QUESTIONS

1. List the various kinds of paleontological evidence used to establish the general theory of evolution.
2. What are some of the factors that interfere with our obtaining a complete and unbiased picture of life in the past from a study of the fossil record?
3. Explain how an estimate of the age of a rock is made on the basis of the radioactive elements present.
4. What is the theory of recapitulation? How has it been modified and what is its significance in its present form?
5. Why are marsupials widespread in Australia and almost nonexistent elsewhere?
6. What could be advanced as an explanation for the observation that the animals and plants of England and Japan are very similar, despite the fact that they lie on nearly the opposite sides of the world?
7. Name some organs found in the human body that are taken to be vestigial, and suggest the functional organs of which they might be the remains. (Spare your instructor the male nipple as an example, unless, of course, you are prepared to argue that *Australopithecus* nursed *his* young.)
8. Define the terms *range* and *center of origin.*
9. What methods are used to determine evolutionary relationship from the nature of serum proteins? from the nature of tissue enzymes?
10. Discuss critically the proposition that the hierarchical scheme of animal and plant classification is evidence for organic evolution. Is this reasoning circular?
11. Define homology with precision, if possible. Must homologous organs be produced by the same genes in the various organisms that have them? Can you suggest a means, based upon molecular genetics, by which it might objectively be determined whether they *are* indeed the same?

47

The Fossil Record and the Controversies

OUTLINE

I. The geological time table
II. The origin and history of life
  A. How did life originate?
  B. The origin of cells
  C. The origin of eukaryotes
  D. Precambrian life
    1. The Archeozoic era
    2. The Proterozoic era
  E. The Paleozoic era
    1. The Cambrian period
    2. The Ordovician period
    3. The Silurian period
    4. The Devonian period
    5. The Carboniferous period
    6. The Permian period
  F. The Mesozoic era
  G. The Cenozoic era
III. Fossil humans
IV. Human evolution
V. Gaps in the fossil record: punctuated equilibrium

LEARNING OBJECTIVES

After you have read this chapter you should be able to:

1. Summarize the geological time table, with emphasis on the names of its divisions and the time span believed to be occupied by each (as summarized in Table 47–1).
2. Summarize the chemical hypotheses for the origin of life that are discussed in this chapter, and give their experimental basis.
3. Summarize the endosymbiotic theory of the origin of eukaryotes, and list the evidence used to support it.
4. Give the main varieties of fossil organisms occurring in the principal geological formations, and summarize the events in the history of life that are inferred from these.
5. Summarize the classification and characteristics of the primates.
6. Summarize the fossil evidence bearing upon human evolution, identify the principal varieties of hominids (employing the terminology given in this chapter), and discuss the current consensus regarding their significance.
7. Summarize the theory of punctuated equilibrium, contrast it to conventional neo-Darwinian evolutionary theory, and assess its possible significance to evolutionary thought.

The past history of life may be read in the rocks. The fossil record of life is immense. Some of it is quite clear, consisting of such things as a complete dinosaur skeleton, and some less so—like mysterious worm tracks or an isolated human jaw. But it is all hard to interpret. Just as it was in Darwin's time, the study of evolution is in a state of intellectual ferment, with new ideas and research results surfacing almost every week.

In this chapter we try to avoid dogmatism. What is presented hopefully represents the current consensus, and in some instances, more than one view—subject, of course, to change as research reveals new evidence.

The Geological Time Table

As discussed in Chapter 4, the cell theory holds that all cells arise only from previously existing cells. Also known as the principle of **biogenesis,** this principle explains that all organisms arise from living parents. This concept is firmly established in biology today. Yet according to evolutionary theory, life did ultimately originate from non-living molecules, that is, by **abiogenesis.** This apparent paradox is explained by the assertion that conditions on earth were far different billions of years ago when life first began to evolve. Then, when living things came into being, they changed the conditions of their environment so that abiogenesis was no longer probable, at least on most parts of the earth's surface.

The Origin and History of Life

The sediments of the earth's crust consist of five major systems of rock strata, each subdivided into minor strata, lying one on top of the other. These sheets of rock were formed by the accumulation of mud or sand at the bottom of oceans, seas, and lakes, and each contains certain characteristic fossils that serve to identify deposits made at approximately the same time in different parts of the world. Geological time has been divided into **eras,** which are subdivided into **periods,** which in turn are composed of **epochs** (Table 47–1), according to the succession of these rock strata one on the other.

Between the major eras, and serving to distinguish them, there were widespread geological disturbances, called **revolutions,** which raised or lowered vast regions of the earth's surface and created or eliminated shallow inland seas. These revolutions altered the distribution of sea and land organisms and may have caused the abrupt and simultaneous extinction of many life forms. The era known as the Paleozoic ended with the revolution that raised the Appalachian Mountains and, it is thought, killed all but 3% of the then-existing species of life by changing the patterns of air flow and thus the climate throughout the planet. Similarly, the Rocky Mountain Revolution, which raised the Andes, the Alps, and the Himalayas, as well as the Rockies, could have caused the annihilation of most of the reptiles of the Mesozoic era. (Recently there has been a revival of proposals that catastrophic mechanisms may have been responsible. Supernova radiation and asteroidal bombardment are two that have been proposed.)

The raising and lowering of portions of the earth's crust result from the slow movements of the enormous **tektonic plates** that compose the crust and float on the underlying molten core. These movements are continuous and build up tensions in the crust that, when finally released, result in earthquakes. Although their role has only been generally accepted in the last 25 years, the movements of these plates have produced tremendous changes in the earth's geography, including the migration and rising or sinking of the land masses, termed **continental drift** (Fig. 47–1).

HOW DID LIFE ORIGINATE?

Some theorists have suggested that abiogenesis did not necessarily figure in the origin of life on earth. Perhaps some kind of spore or germ may have been carried through space from some other planet to this one. This extraterrestrial explanation seems unsatisfactory, not only because it begs the question of the ultimate source of the spores, but also because it seems most unlikely that any sort of living thing could survive the extreme cold and intense irradiation encountered in interplane-

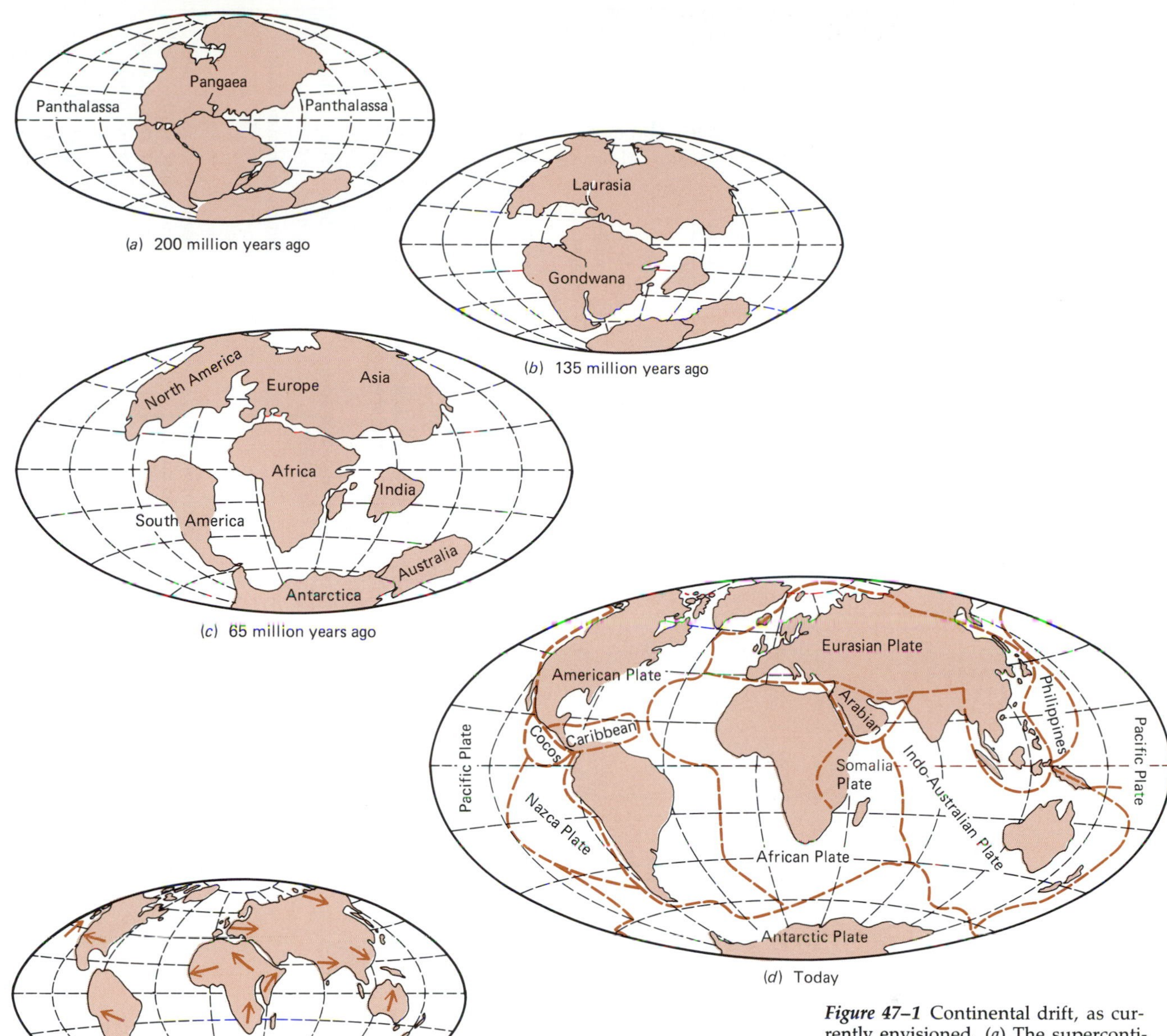

Figure 47–1 Continental drift, as currently envisioned. (*a*) The supercontinent Pangaea of the Triassic period, about 200 million years BP (Before Present). (*b*) Breakup of Pangaea into Laurasia (Northern Hemisphere) and Gondwana (Southern Hemisphere) 135 million years BP in the Cretaceous period. (*c*) Further separation of land masses, which occurred in the Tertiary period, 65 million years BP. Note that Europe and North America are still joined and that India is a separate land mass. (*d*) The continents today. (*e*) Projected positions of the continents in 50 million years. (From K. Norstog and R. W. Long.)

tary travel. *Ambiguous* evidence for life in other parts of the cosmos came from the discovery in 1961 of what could be interpreted as fossils of microscopic organisms, somewhat like algae, in meteorites, but this of course is no proof that *living* organisms could be transported though space.[1]

The concept that the first living things evolved from non-living things, and suggestions as to what the sequence of events may have been, were put forward by J. B. S. Haldane, R. Beutner, and particularly the Russian biochemist A. I. Oparin in his book *The Origin of Life* (1938). According to their view, the earth originated some 5 billion years ago, either as a part broken off from the sun or by the gradual condensation of interstellar dust. Since the earth was probably very hot and molten when it was first formed (or shortly became so as a result of radioactive heating), conditions consistent with life might have appeared only perhaps 3 billion years ago. Twenty-two different amino acids were isolated recently from Precambrian rocks from South Africa that are dated at about 3.1 billion years of age. There is evidence (though strongly disputed[2]) that at that time the earth's atmosphere con-

[1] This theory is discussed in many books on the origin of life, for example, *Lifecloud: The Origin of Life in The Universe,* by F. Hoyle and N.C. Wickrama-Singhe (Harper & Row, N.Y., 1978).

[2] For a summary of arguments against the conventional view of an oxygen-free atmosphere of the early earth, see *Life Itself,* by Francis Crick, New York, Simon and Schuster, 1981.

TABLE 47–1
Some Important Biological Events in Geological Time*

Era	*Period*	*Epoch*	*Time from Beginning of Period to Present (millions of years)*	*Duration (millions of years)*	*Geological Conditions*	*Plants and Microorganisms*	*Animals*
Cenozoic (Age of Mammals)	Quaternary	Recent	(Last 10,000 years)		End of last Ice Age; warmer climate	Decline of woody plants; rise of herbaceous plants	Age of *Homo sapiens*
		Pleistocene	1.9	1.9	Four Ice Ages; glaciers in Northern Hemisphere; uplift of Sierras	Extinction of many species	Extinction of many large mammals
	Tertiary	Pliocene	6	4	Uplift and mountain-building; volcanoes; climate much cooler	Development of grasslands; decline of forests; flowering plants	Large carnivores; many grazing mammals; first known human-like primates
		Miocene	25	19	Climate drier, cooler; mountain formation		Many forms of mammals evolve
		Oligocene	38	13	Rise of Alps and Himalayas; most land low; volcanic activity in Rockies	Spread of forests; flowering plants, rise of monocotyledons	Apes evolve; all present mammal families are represented
		Eocene	54	16	Climate warmer	Gymnosperms and angiosperms dominant	Beginning of Age of Mammals; modern birds
		Paleocene	65	11	Climate mild to cool; continental seas disappear		Evolution of primate mammals
Mesozoic (Age of Reptiles)	Cretaceous		135	70	Two major land masses begin to separate; formation of Rockies; other continents low; large inland seas and swamps	Rise of angiosperms; gymnosperms decline	Dinosaurs reach peak, then become extinct; toothed birds become extinct; first modern birds; primitive mammals
	Jurassic		181	46	Climate mild; continents low; inland seas; formation of mountains	Ferns and gymnosperms common	Large, specialized dinosaurs; first toothed birds; insectivorous marsupials

Era	Period	Millions of years since start	Duration (millions of years)	Physical conditions	Plants	Animals
	Triassic	230	49	Many mountains form; widespread deserts	Gymnosperms and ferns dominate	First dinosaurs; egg-laying mammals
Paleozoic (Age of Ancient Life)	Permian	280	50	Glaciers; formation of Appalachians; continents rise	Conifers evolve	Modern insects appear; mammal-like reptiles; extinction of many Paleozoic invertebrates
	Pennsylvanian	320	40	Lands low; great coal swamps	Forests of ferns and gymnosperms	First reptiles; spread of ancient amphibians; many insect forms
	Mississippian	345	25	Climate warm and humid; later cooler	Club mosses and horsetails dominant; gymnosperms	Ancient sharks abundant; many echinoderms
	Devonian	405	60	Glaciers; inland seas	Terrestrial plants established; first forests; gymnosperms appear	Age of Fish; amphibians appear; wingless insects and millipedes appear
	Silurian	425	20	Continents mainly flat; flooding	Vascular plants appear; algae dominant	Fish evolve; marine arachnids dominant; first insects
	Ordovician	500	75	Sea covers continents; climate warm	Marine algae dominant; terrestrial plants first appeared	Invertebrates dominant; first fish appeared
	Cambrian	600	100	Climate mild; lands low; oldest rocks with abundant fossils	Algae dominant	Age of marine invertebrates; most modern phyla represented
Precambrian (Archeozoic and Proterozoic Eras)		3800	2930	Planet cooled; glaciers; formation of earth's crust; mountains form	Primitive algae and fungi, marine protozoans	Toward end, marine invertebrates
Evidence of first bacterial cells		3.5 billion years ago				
Origin of the earth		4.6 billion years ago				
Origin of the universe		15–20 billion years ago				

*You may want to study this table starting from the bottom and working your way up through time.

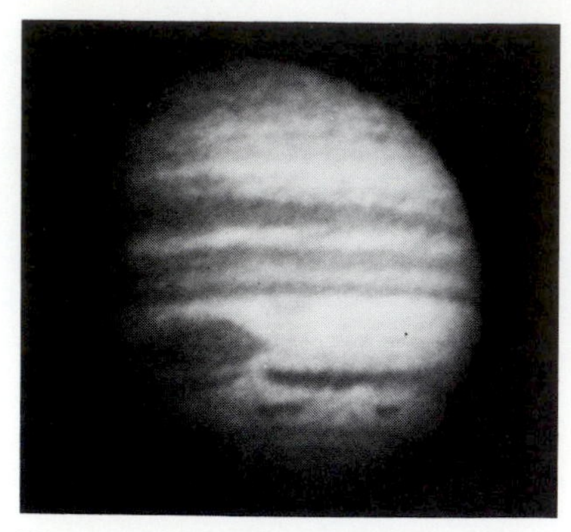

Figure 47–2 The planet Jupiter. The reducing, oxygen-free atmosphere of the lifeless planets suggests to theorists, along with geochemical evidence, that the early atmosphere of the earth may also have been without oxygen. It has been suggested that organisms similar to those of the early earth may some day be found on the outer planets or their satellites. (NASA.)

tained essentially no free oxygen—all the oxygen atoms were combined as water or as oxides. The primitive atmosphere, therefore, would have been strongly reducing, composed of methane, ammonia, and water originating by "out-gassing" from the earth's interior (Fig. 47–2).

Oparin and his school assumed that originally the carbon atoms in the earth's crust were present mainly as metallic carbides. These would react with water to form acetylene, which can polymerize to form compounds with long chains of carbon atoms. High-energy radiation, such as cosmic rays, can catalyze the synthesis of organic compounds. This was shown by Melvin Calvin's experiments in which solutions of carbon dioxide and water were irradiated in a cyclotron, and formic, oxalic, and succinic acids, which contain one, two, and four carbons respectively, were obtained. (These compounds, you may recall, are intermediates in certain metabolic pathways of living organisms.)

Irradiating solutions of inorganic compounds with ultraviolet light, or passing electric charges through the solutions to simulate lightning, also produces organic compounds. Stanley Miller and Harold Urey in 1953 exposed a mixture of water vapor, methane, ammonia, and hydrogen gases—the compounds they thought made up the earth's early atmosphere—to electric discharges for a week and demonstrated the production of single organic compounds including a mixture of simple D- *and* L-amino acids such as alanine (*only* L-amino acids now occur in the proteins of living things).[1] Based on computer simulations, it is now thought that ammonia and methane would have been rapidly broken down by ultraviolet radiation. Revised theory suggests that the early atmosphere consisted of water vapor, carbon dioxide, carbon monoxide, nitrogen, and some free hydrogen. Using this combination of gases, the Miller-Urey experiment has been repeated. Even greater amounts and combinations of organic compounds have formed, including nucleotide bases of DNA and RNA. Amino acids and other compounds could be produced on earth even at the present time by lightning discharges or ultraviolet radiation. However, any organic compound produced in this way might undergo spontaneous oxidation, or it would be phagocytized by protists or degraded by the molds or bacteria that now abound on earth. Under the presumed original sterile, anoxic conditions, these compounds could have persisted.

The details of the chemical reactions that could give rise without the intervention of living things to carbohydrates, fats and amino acids have been worked out by Oparin and extended by Calvin and others. They believe that most, if not all, of the reactions by which the more complex organic substances were formed probably occurred in the sea, in which the inorganic precursors and the organic products of the reaction were dissolved and mixed. The sea became a sort of dilute broth in which these molecules collided, reacted, and aggregated to form molecules of increasing size and complexity. As more has been learned of the role of hydrogen bonds and other weak intramolecular forces in the pairing of specific nucleotide bases, and the effectiveness of these processes in the transfer of biological information, it has become clear that similar forces could have operated early in evolution before living organisms first appeared.

Oparin suggested that the forces of intermolecular attraction and the tendency for certain molecules to form liquid crystals might provide a means by which large, complex specific molecules can be formed spontaneously. He thought that a kind of natural selection could have operated in the evolution of these complex molecules before anything recognizable as life was present. As Oparin's school developed their theory, they evolved the view that when molecules came together to form colloidal aggregates such as membrane-bound protein spheroids, these aggregates began to compete with one another for raw materials. Some of the aggregates that had some particularly favorable internal arrangement would acquire new molecules more rapidly than others and would eventually become the dominant types.

Once some protein molecules had been formed, and had achieved the ability to catalyze reactions, the rate of formation of additional molecules would be greatly speeded up. When combined with nucleic acids (especially RNA), these complex protein molecules should eventually acquire the ability to catalyze the synthesis of molecules like themselves. These hypothetical **autocatalytic** particles made of nucleic acids and proteins could have been *something like* a modern virus or perhaps a

[1] D- and L-amino acids are enantiomers, isomers that are nonsuperimposable mirror images of each other (like right and left hands).

plasmid. Of course, present-day viruses can reproduce only within the cells of other organisms, so the similarity could not be a close one. Another obvious reservation is that there is a great difference between a molecular group of assorted chemicals and even the simplest living cell.

THE ORIGIN OF CELLS

In order to produce cells, a major step required in the evolution of these prebiotic aggregates would necessarily have been the development of a protein–lipid *membrane* surrounding the aggregate that would permit the accumulation of some molecules and the exclusion of others. Another major evolutionary step that would be required is the development of the genetic code. No feature of a living cell could be maintained for more than one generation, if that long, without an informational basis. Thus, any credible theory on the origin of life must suggest ways whereby not only the nucleic acids but the information content of the nucleic acids, as well as the readout mechanisms by which this information is translated into cellular structures, would have originated.

The first living organisms, having arisen in a sea of organic molecules and in contact with an atmosphere lacking oxygen, presumably obtained energy by the fermentation of certain of these organic substances. The first organisms, therefore, were almost certainly heterotrophs. This initial population could survive only as long as the supply of organic molecules that had previously existed in the sea lasted. However, long before that supply was exhausted, some of the heterotrophs might have evolved into autotrophs, which were able to make their own organic molecules by chemosynthesis or photosynthesis. One of the by-products of photosynthesis is gaseous oxygen. All the oxygen in the atmosphere is now produced by photosynthesis, and according to this view, always has been produced by photosynthesis.[1]

An explanation of how an autotroph may have evolved from one of these primitive, fermenting heterotrophs was proposed by N. H. Horowitz in 1945. Horowitz postulated that an organism would acquire, by successive gene mutations, the enzymes needed to synthesize complex substances from simple substances, but these enzymes would be acquired in the *reverse* order of the sequence in which they are ultimately used in normal metabolism. For example, let us suppose that our first primitive heterotroph required an organic compound, Z, for its growth. This substance, Z, and a vast variety of other organic compounds, Y, X, W, V, U, and so forth, were present in the organic sea broth[2] that was the environment of this heterotroph. They had been synthesized previously by the action of non-living factors of the environment. The heterotroph would be able to survive as long as the supply of compound Z lasted. If a mutation occurred for a new enzyme enabling the heterotroph to synthesize Z from substance Y, the strain of heterotroph with this mutation would be able to survive when the supply of substance Z was exhausted. A second mutation that established an enzyme catalyzing a reaction by which substance Y could be made from substance X would again have survival value when the supply of Y was exhausted. Similar mutations, setting up enzymes enabling the organism to use successively simpler substances, W, V, U . . . , and eventually some inorganic substance, A, would result in an organism able to make substance Z, which it needs for growth, out of substance A. When, by other series of mutations, the organism could synthesize all its requirements from simple inorganic compounds, as the algae and plants can, it would have become an autotroph. Once the first simple autotrophs had evolved, the way would have been clear for the further evolution of the enormous variety of protists, bacteria, fungi, plants, and animals that now inhabit the earth.

The most widely accepted view of the origin of life, then, can be summarized as follows:

1. Organic substances were formed from inorganic substances by the action of physical factors in the environment.

---

[1] Some theorists believe that photolysis of water vapor by ultraviolet light and other inorganic sources of oxygen produced an early atmosphere with much more oxygen in it than that allowed for by the Oparin-Haldane theory we have been considering.

[2] The organic content of canned chicken broth is about equal to what is postulated for the primitive ocean, so that this is sometimes called the "chicken soup" hypothesis!

2. They interacted to form more complex substances, and finally enzymes and self-reproducing systems (genes).
3. These free genes diversified and united to form primitive heterotrophs.
4. Lipid–protein membranes evolved to separate these prebiotic aggregates from the surrounding environment.
5. Autotrophs then evolved from the primitive hereotrophs.

This theory has the virtue of being plausible; many of its components have in fact been subjected to experimental verification, but only in the way of proving that certain events *could* happen—not that they actually *did*. What is described here may not be the only possibility in the origin of life. To ascribe more certainty to it would be an act not of science but of faith. It seems unlikely that we will ever have certain scientific knowledge of how life originated.

THE ORIGIN OF EUKARYOTES

It is logical to see the ancestors of modern organisms as being very simple. Among modern organisms, the very simplest forms of cellular life are prokaryotes. That is one reason why evolutionists think that the earliest cells were prokaryotes. Recall from Chapter 19 that prokaryotic cells lack nuclear membranes as well as other membranous organelles such as mitochondria, endoplasmic reticulum, chloroplasts, and the Golgi complex. Of the two ways proposed for the evolution of eukaryotes, the older view hypothesized that membranous organelles arose by multiple inpocketing and infolding of the cell membrane in an ancestral prokaryote.

A second hypothesis, the **endosymbiotic theory,** suggests that mitochondria, chloroplasts, and perhaps even centrioles and flagella may have originated by the union of prokaryotes. Thus, mitochondria are seen as former bacteria, and chloroplasts as former cyanobacteria (Fig. 47–3). The theory further stipulates that each of these partners brought to the union something that the others lacked. For example, mitochondria provided the ability to employ oxidative metabolism, which was lacking in the original host cell, and writhing spiral bacteria provided the ability to swim, eventually becoming flagellae.

The principal evidence in favor of the endosymbiotic theory is that mitochondria and chloroplasts do possess *some* (but *not* all) of their own genetic apparatus distinct from that of the cell's nucleus. Thus they have their own DNA and their own ribosomes. In chloroplasts there is as much DNA as that in the average virus particle. Moreover, the DNA and ribosomes of these organelles are rather similar to those found in prokaryotes, although not as similar as might be predicted. However, neither are they very similar to those of the nucleus and cytoplasm.[1]

Two momentous events occurred in the evolutionary history of life on earth: first, the origin of cells an estimated 3.5 billion years ago, and second, the origin of complex multicellular animals and plants about 700 million years ago. According to evolutionary theory, about 70% of the history of life (and half of the duration of the planet earth) unfolded between these two events. During or at the end of this vast span of time, many crucial mechanisms and organisms evolved, including eukaryotic cells, photosynthesis, mitosis, meiosis, and cellular respiration. During these billions of years a multitude of one-celled organisms and other life processes also evolved. It is thought that the majority of these evolutionary experiments in life forms were discarded in favor of the familiar varieties that survived, so that the species we now know are isolated "islands of life in a sea of death." The rest of the story—the widely accepted course of evolution of eukaryotes—has already been told briefly in Part IV of this book. But a summary would be useful at this point also.

PRECAMBRIAN LIFE

The richest deposits of fossils date from the beginnings of the "explosion of life" that occurred during the Cambrian period, some 500 to 600 million years ago. However, there is evidence that life existed long before the Cambrian period. Signs of **Precambrian** life date from the Archeozoic era, which began about 3.6 billion years ago.

[1] One cannot make too much of these nucleic acid similarities and differences, however, because recent studies have disclosed a considerable genetic interchange between mitochondria and nucleus. In other words, organelle genes simply do not stay put and may have had more than one origin.

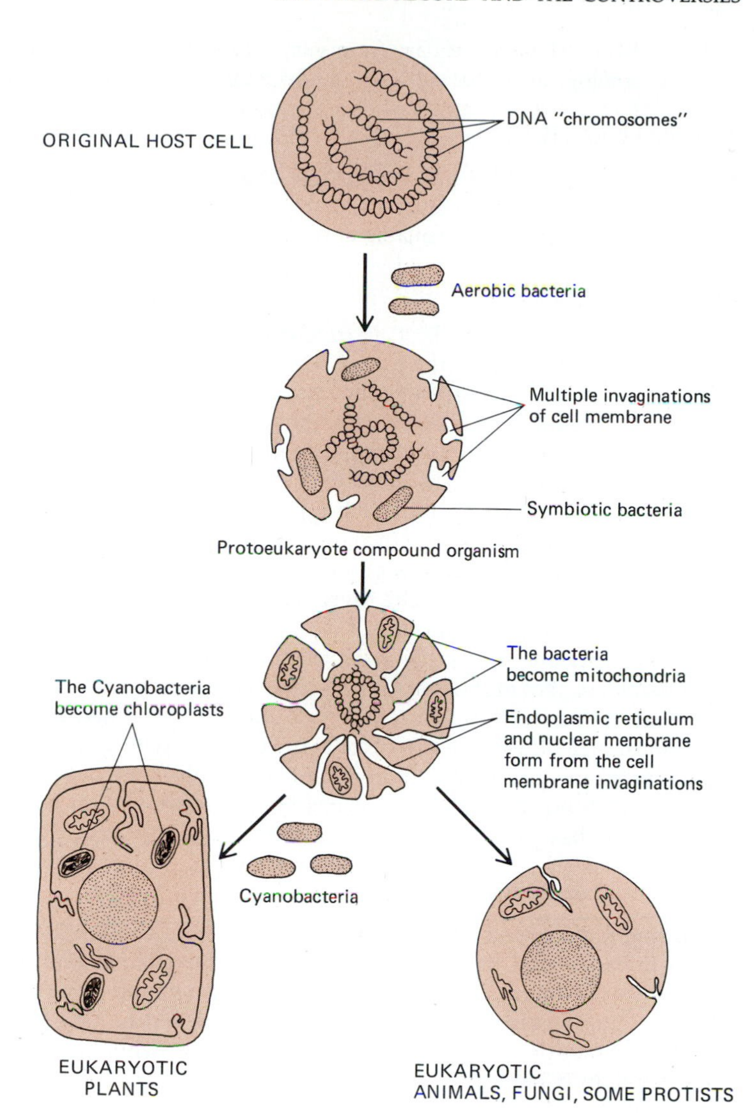

Figure 47–3 The endosymbiotic theory of the origin of the eukaryotes.

The Archeozoic Era

The oldest era, the **Archeozoic era,** begins not with the origin of the earth but with the formation of the earth's crust, when rocks and mountains were already in existence and the processes of erosion and sedimentation had begun. Because the rocks of the Archeozoic era are very deeply buried in most parts of the world, they are considered to be the most ancient. However, Archeozoic rocks are exposed at the bottom of the Grand Canyon and along the shores of Lake Superior.

The 2-billion-year-long Archeozoic era was characterized by catastrophic and widespread volcanic activity, together with giant upheavals that climaxed in the raising of mountains. The heat, pressure, and churning associated with these movements probably destroyed most of whatever fossils may have been formed, but some evidence of life still remains. This evidence consists of traces of graphite or pure carbon, which are possibly the transformed remains of primitive life forms. These remains are especially abundant in what were the oceans and seas of that era. Fossils of what appear to be cyanobacteria have been recovered from several Archeozoic formations.

The Proterozoic Era

The second era, the **Proterozoic era,** thought to be about a billion years in length, was characterized by the deposition of large quantities of sediment, reflecting massive erosion and perhaps the result of at least one great period of glaciation. The fossils found in the later Proterozoic rocks show clear-cut examples of some major groups of plants and animals. One source of rich deposits of Precambrian fossils has been South Australia. The forms of life found there include jellyfish, corals, segmented worms, and two animals with no resemblance to any other known fossil or living form. These fossils are from very late in Precambrian time; except for an arbitrary geological boundary, they might well be considered early Cambrian.

THE PALEOZOIC ERA

Between the strata of the late Proterozoic and the lowest layers of the third major era, the **Paleozoic era** is a considerable gap, possibly caused by a geological revolution or by an increase in available atmospheric oxygen.

The Cambrian Period

The lowest subdivision of the Paleozoic era, the **Cambrian period,** is represented by rocks rich in fossils. All the present-day animal phyla, except the chordates, are represented, at least in marine sediments. There were arachnid-like forms, some of whose descendants (such as the horseshoe crab) exist almost unchanged today. The sea floor was covered with simple sponges, corals, crinoid echinoderms growing on stalks, snails, bivalves, primitive cephalopods, brachiopods, and trilobites.

Except for chordates, the major types of animal body plans were established so early in the history of the eukaryotes that very little further change of a basic nature is seen since. In fact, a number of groups occurring in these rocks soon disappeared, which reduced the diversity of animal body plans to an even lower level—a level that persists today.

This does not necessarily mean that no other patterns of animal organization are possible, or that mutations for new patterns did not occur. It probably indicates only that by the early Cambrian period, animal forms had reached a degree of adaptation that allowed them to exploit the earth's existing environments and cope with future changes in the environment with only limited modifications of their body plans. Since the Paleozoic, it has been said, evolution has concerned itself only with endless relatively minor variations on a few basic themes.

The Ordovician Period

According to geologists, during the Cambrian period the continents gradually had begun to be covered with water, and in the **Ordovician period** this submergence reached its maximum, so that much of what is now land was covered by shallow seas. The total marine biomass of the time is believed to have been much greater than that of today because these seas, being shallow, would have been easily penetrated by light (Chapter 50). Thus, plant life would have grown over vast areas of sea bottom.

Inhabiting the seas were giant cephalopods—squid or nautilus-like animals with straight shells 5 to 7 m long and 30 cm in diameter. The first traces of the early vertebrates, the jawless, bony-armored **ostracoderms,** are found in Ordovician rocks (Fig. 47–4). They lived in fresh water, and the armor of these finless, bottom-dwelling animals might have served as a defense against **eurypterids,** giant (up to 3 m long), carnivorous water scorpions.

The Silurian Period

Two life forms of great biological significance appeared in the **Silurian period:** the land plants and the air-breathing animals. The first-known land plants resembled ferns rather than mosses (which, as you may recall, is an argument *against* the mosses as vascular plant ancestors), and ferns were the dominant plants of the Devonian and Mississippian periods that followed. The only air-breathing land animals that have been discovered in Silurian rocks were arachnids, resembling to some extent modern scorpions (and there is some debate as to whether these might not have been shallow water dwellers that did not actually breathe air). The continental areas, which had been low during the Cambrian and Ordovician, apparently rose—especially in what is now Scotland and northeastern North America—and the climate became much colder.

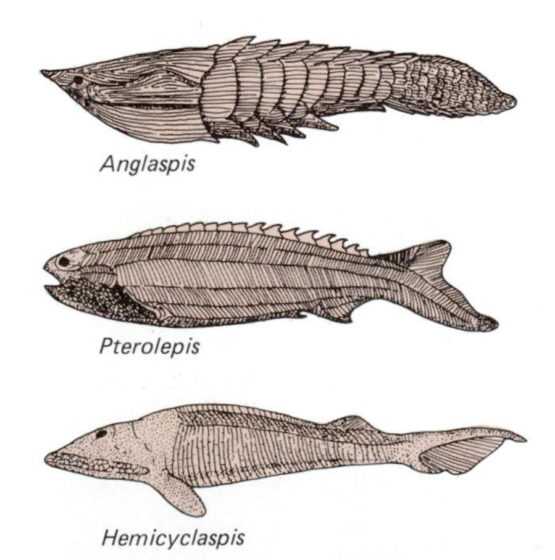

Figure 47–4 Three fossil ostracoderms, which were primitive jawless, limbless fishes.

The Devonian Period

A great variety of fishes appear in **Devonian period** formations; in fact, the Devonian is frequently called the "Age of Fishes." Unlike the ostracoderms, the Devonian fishes typically had jaws, an adaptation that enables a vertebrate to chew and, of course, to bite, as ostracoderms could not.

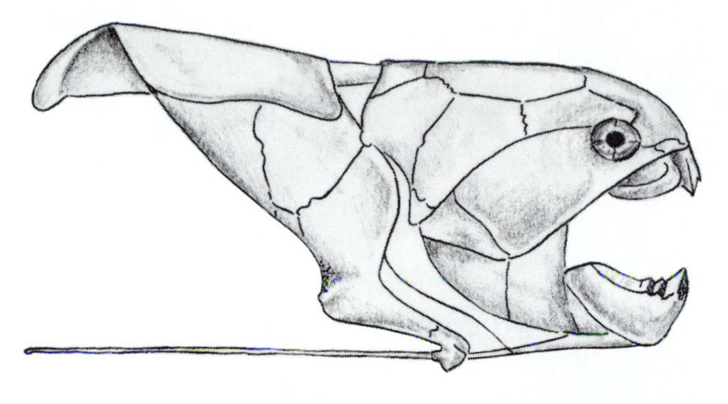

Figure 47–5 *Dunkelosteus,* a predatory placoderm from the later Devonian period.

Arising during this time were the **placoderms,** armored freshwater fishes with a variable number (as many as seven) of paired fins (Fig. 47–5). However, also appearing in Devonian deposits are sharks (Chondrichthyes) and the three main types of bony fish (Osteichthyes): lungfishes, lobe-finned fishes, and the ray-finned fishes. A few lungfishes have survived to the present. The ray-finned fishes later gave rise to the major modern orders of fishes. The lobe-finned fishes, some of which are considered ancestral to the land vertebrates, were thought to have become extinct by the end of the Mesozoic; then, in 1939, the first living coelacanth was discovered off the coast of Africa (Fig. 47–6).

Upper Devonian sediments contain fossil remains of **labyrinthodonts,** clumsy salamander-like, ancient amphibians that were often very large, with short necks and heavy muscular tails (Fig. 47–7). These creatures, whose skulls were encased in bony armor, were quite similar in many respects to the lobe-fins. They differed in that they had limbs strong enough to support the weight of the animal on land, although these limbs were attached to the side of the body (see Fig. 47–9) and could not have been much used for terrestrial locomotion.

The Devonian was the first period characterized by forests. Ferns, club mosses, horsetails, and the seed ferns all flourished along with labyrinthodonts (Fig. 47–8). Wingless insects and millipedes are believed to have originated in the late Devonian from unknown ancestors.

The Carboniferous Period

The **Mississippian** and **Pennsylvanian periods** are frequently grouped together as the **Carboniferous** period, so named for the great swamp forests whose remains persist today as major coal deposits of the world. The land during this time was covered with low swamps filled with horsetails, ferns, seed ferns, and large-leaved evergreens. The first reptiles, the **cotylosaurs** or **stem reptiles,** appeared in the Pennsylvanian period, flourished in the final Paleozoic period (the Permian) and became extinct early in the Mesozoic era. (Recently it has been suggested, however,

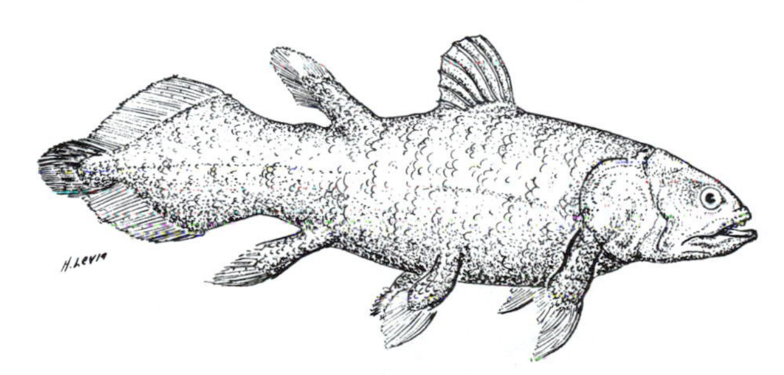

Figure 47–6 *Latimeria,* a modern coelacanth. (A living coelacanth is shown in Fig. 26–21.)

Figure 47–7 *Eryops,* a labyrinthodont.

Figure 47–8 A restoration of a Middle Devonian forest in the eastern United States. (A) An early lycopod. (B) An early horsetail. (C) Early tree fern. (Courtesy of the Field Museum of Natural History; painting by Charles R. Knight.)

Figure 47–9 *Seymouria,* one of the stem reptiles. Its heavier skeleton represents a transitional form between the amphibians and reptiles.

that the early reptiles, though they resembled amphibians, may have had a separate origin from other coelacanth ancestors.) *Seymouria* (Fig. 47–9), a cotylosaur, is an example of a typical carboniferous reptile. Its short, stubby legs extended laterally from the body as do a salamander's or an alligator's, instead of being closer together and extending directly down to form pillar-like supports for the body, as they do in most mammals of today.

Two important groups of winged insects occur for the first time in the Carboniferous: cockroaches (Fig. 47–10), the largest of which reached a length of 10 cm (much as the largest species of today), and dragonflies, some of which had a wingspread of 75 cm and are thought to have resembled birds in their general lifestyle. (There were also dragonflies of modern size and even smaller.)

The Permian Period

The final period of the Paleozoic, the **Permian period,** was characterized by great changes in climate and topography. The level of the continents rose all over the

Figure 47–10 A primitive Pennsylvanian-period cockroach.

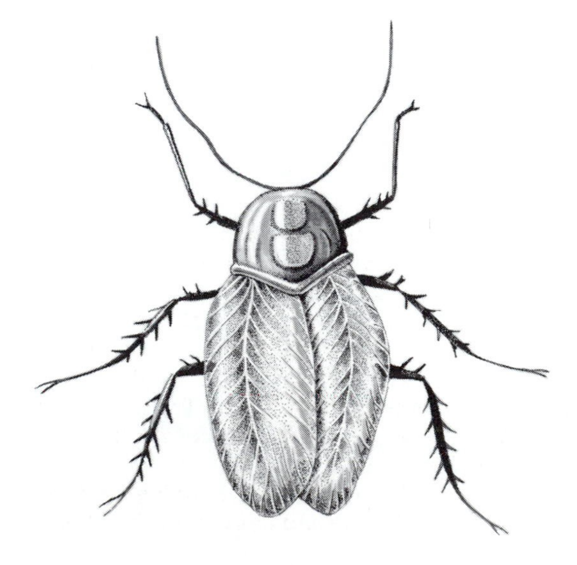

Figure 47–11 *Cynognathus,* a Permian therapsid that might have had many mammalian characteristics.

world, so that the shallow seas that covered the region from Nebraska to Texas at the beginning of the period drained off or were isolated, evaporating to leave the land a salt desert. At the end of the Permian a general folding of the earth's crust, called the **Appalachian Revolution,** raised the great mountain chain from Nova Scotia to Alabama. These mountains originally were higher than the present Rockies and, in the absence of the erosion that wore them down to their current height, would have produced a far different continental climate than we have in the present-day United States.

Other ranges were brought into existence in Europe at this time. A great glaciation, spreading from the Antarctic, covered most of the Southern Hemisphere, extending almost to the equator in Brazil and Africa. Many Paleozoic forms of life may have been unable to adapt to the climatic or topographical changes and became extinct. Even many marine forms became extinct, perhaps owing to the cooling of the water and the decrease in the amount of space available caused by the diminishing of the shallow seas.

During the late Carboniferous and early Permian, a cotylosaur-like group of reptiles appeared, believed to be in (or close to) the direct line giving rise to the mammals. These were the **pelycosaurs,** carnivorous reptiles that were more slender and lizard-like than the stem reptiles. In the latter part of the Permian there also evolved, probably from the pelycosaurs, another group of reptiles with more mammalian characteristics—the **therapsids.** One of these, *Cynognathus* (Fig. 47–11) the "dog-jawed" reptile, was a slender, lightly built animal about 150 cm long, with a skull intermediate between that of a reptile and a mammal. Its teeth, instead of being conical and all alike, as reptilian teeth are, were differentiated into incisors, canines, and molars. In the absence of information about the animal's soft parts, whether it had scales or hair, whether or not it was homeothermic, and whether it suckled its young, it is called a reptile. The therapsids were widespread in the late Permian, but were crowded out in the early Mesozoic by the great variety of other reptiles.

THE MESOZOIC ERA

The **Mesozoic era** is dated as beginning about 230 million years ago and lasting some 167 million years. It is divided into *Triassic, Jurassic,* and *Cretaceous* periods. The outstanding feature of the Mesozoic era was the origin, differentiation, and final extinction of a great variety of reptiles. For this reason the Mesozoic is commonly called the "Age of Reptiles."

There were six major evolutionary lines of reptiles whose remains are present in Mesozoic formations. The most primitive line includes the ancient cotylosaurs and the turtles. Turtles are first seen in Permian strata. They have the most complicated armor of any land animal, consisting of scales derived from the epidermis fused to the underlying ribs and breastbone. With this protection, both marine and land forms have survived with few structural changes since before the time of the dinosaurs. Their legs extend laterally, making locomotion difficult and slow, and their skulls are unpierced behind the eye sockets, a feature essentially the same as in the ancient cotylosaurs.

Most of the snakes and lizards found in Mesozoic formations are also similar to their present-day descendants. A notable exception to this are the **mosasaurs,** marine lizards of the **Cretaceous period** (Fig. 47–12), which attained a length of 13 m and had a long tail useful in swimming.

The main group of Mesozoic reptiles was the **archosaurs,** or "ruling reptiles." The only living members of this group are the alligators and crocodiles. The earliest

Figure 47–12 Though quite large, *Tylosaurus* belongs not to the archosaurs but to the group ancestral to modern lizards.

of the archosaurs occur in the same formations as those of the mammal-like reptiles. The archosaurs were then evidently about 1 m long and were adapted to two-legged locomotion.

Some archosaurs continued to use two-legged locomotion, but others became adapted to walking on all fours. These descendants include the **phytosaurs**—aquatic, alligatorlike reptiles common during the Triassic period; the **crocodiles,** which occur during the Jurassic period and which today have replaced the phytosaurs as aquatic forms; and the **pterosaurs,** or flying reptiles. The pterosaurs included animals the size of a robin, as well as the largest animal ever to fly, a pterosaur with a wingspread of 15.5 m, discovered in 1975 in western Texas. Although many pterosaurs evidently had wing muscles too weak for active flight, presumably they could soar like the modern albatross or buzzard. Most pterosaurs, however, apparently could fly as well as modern birds. Certain pterosaur fossils show evidence of fur, an indication that some species might have been homeothermic.

Of all the reptilian branches, the most famous are the **dinosaurs,** or "terrible reptiles." These were divided into two main types: one with a birdlike pelvis, the other with a more typically reptilian pelvis.

The **saurischians,** which were characterized by a reptilian pelvis, were fast, two-legged forms ranging from the size of a dog to the ultimate representative of this group, the gigantic carnivore of the Cretaceous, *Tyrannosaurus* (Fig. 47–13). Other saurischians were adapted to a plant diet and had a four-legged gait. Some of these were among the largest animals that have ever lived: *Brontosaurus,* with a length of 21 m; *Diplodocus,* with a length of 29 m; and *Brachiosaurus,* until recently thought to be the biggest of them all, with an estimated weight of in excess of 50 tons. Some of these animals appear to have been amphibious, but most were probably fully terrestrial.

The other group of dinosaurs, the **ornithischians** (with a birdlike pelvis), were entirely herbivorous. Although some of them walked upright, the majority had a four-legged gait. Some had no front teeth and may have possessed a stout, horny, birdlike beak. In some forms this was broad and ducklike (hence the name "duck-billed" dinosaurs). Webbed feet were characteristic of this type; other species had great armor plates, possibly as protection against the carnivorous saurischians. *Ankylosaurus* (Fig. 47–14), dubbed "the reptilian tank," had a broad, flat

Figure 47–13 *Tyrannosaurus,* the largest of the flesh-eating dinosaurs. *Tyrannosaurus* reached a length of 15 meters and a height of 6 meters. Its head was as much as 2 meters long (that's right, the *head*) and was equipped with many sharp teeth, whose edges were serrated like the blades of steak knives. The front legs are small and are often considered vestigial, but it has been suggested that *Tyrannosaurus* may have used them to steady itself when arising from a reclining position. The long tail was probably held straight and well off the ground—not as shown here. It served as a counterweight to the immense head.

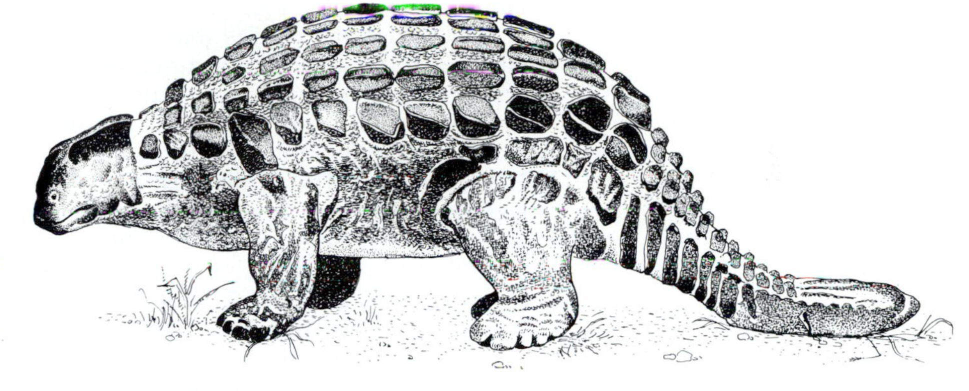

Figure 47–14 *Ankylosaurus,* a heavily armoured ornithischian. Though one would not guess from external appearance of this creature, the ornithischians were so named for their birdlike pelvis.

body covered with armor plate and large, laterally projecting spines. As suggested earlier, at least some of these archosaurs may have been homeothermic (although this suggestion is, at present, the subject of vigorous controversy).

Two other groups of Mesozoic reptiles, separate from each other and from the dinosaurs, were the marine **plesiosaurs** and **ichthyosaurs** (Fig. 47–15). The extremely long neck of the plesiosaurs took up over half of their total length of 15 m. The porpoise-like ichthyosaurs (fish reptiles) had a body form superficially like that of a fish or a whale, with a short neck, large dorsal fin, and shark-type tail. They swam by wiggling their tails, and used their feet only for steering. The ichthyosaur young were apparently born alive, after having hatched from eggs within the mother, because the adults were too specialized to come out on land, and a reptilian egg will drown in water. The presence of skeletons of the young within the body cavity of adult fossils has strengthened this hypothesis.

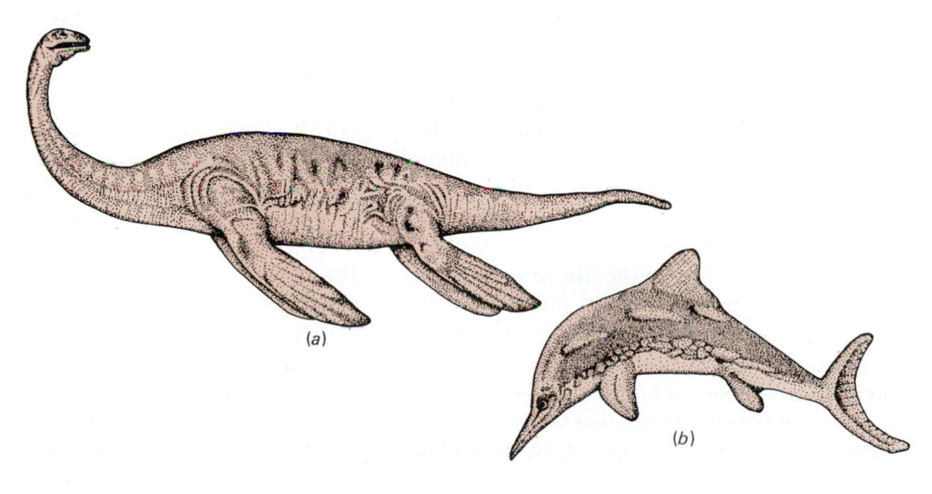

Figure 47–15 Two types of marine reptiles. (*a*) *Plesiosaurus.* (*b*) *Ichthyosaurus.* Notice the similarity of *Plesiosaurus* to animals such as modern seals, and the similarity of *Ichthyosaurus* to modern porpoises. During their long reign as the dominant animals on earth, the reptiles had radiated into almost every conceivable environment. Note the huge eyes of the ichthyosaurs, the largest of any known vertebrate. They evidently did not possess the "sonar" that modern, similar whales and porpoises use to "see" underwater. There is evidence that ichthyosaurs were live-bearing. It does seem unlikely that they could have crawled out on land to lay eggs. (Copyright, Chicago Natural History Museum; from the painting by Charles R. Knight.)

As discussed in Chapter 26, at the end of the Cretaceous a great many reptiles abruptly became extinct. The **Rocky Mountain Revolution** brought changes in climate that the great reptiles might not have been able to cope with; other explanations have also been proposed. The problem is that the dinosaurs were, as far as we can tell, always well-adapted to their environment and lifestyle. In fact, even as the types of vegetation on the earth's surface began to change, the dinosaurs were still increasing in both their diversity and numbers. Additionally, the mammals that have replaced them today did not originate after the extinction of the dinosaurs but coexisted unobtrusively with them throughout their history. If the mammals put the dinosaurs out of business in a classic Darwinian competition, why didn't they do so much earlier? Even more mysterious, why did the dinosaurs die out so suddenly? In the past, the demise of the dinosaurs was said to have been gradual and to have taken tens of thousands of years, if not a million or more. Modern stratigraphic techniques, however, seem to indicate that their extinction was truly sudden; some feel that a time span of a year or less cannot be ruled out, although, of course, it could have been far longer. This has led to a modern revival of catastrophism, at least where the dinosaurs are concerned. Most of the proposed catastrophic explanations are extraterrestrial: collision of the earth with particles of a comet or asteroid, dramatic reduction or other change in solar radiation owing to an epidemic of sunspots, or lethal radiation from a nearby (in cosmic terms) supernova.

Although the reptiles were the dominant animals of the Mesozoic, many other important organisms occur in the same formations: Most of the modern orders of insects appeared early during that era; snails and bivalves increased in number and kind; sea urchins reached their peak; mammals first appear in the Triassic; and teleost (modern) fishes and birds first appear in Jurassic formations. During the early Triassic the most abundant plants were seed ferns, cycads, and conifers; but by the Cretaceous, many others resembling present-day species had appeared—sycamores, magnolias, palms, maples, and oaks.

Excellent bird fossils, some even showing the outlines of feathers, have been preserved from the Jurassic. *Archaeopteryx* is the classic example. As we saw in previous chapters, this animal was about the size of a crow, had rather feeble "wings," jawbones armed with teeth, and a long reptilian tail covered with feathers. Judging from the available skeletal attachments for flight muscles, *Archaeopteryx* probably could not fly, although the feathers themselves *were* similar to those of modern flying (not flightless) birds. Increasingly, *Archaeopteryx* is seen as representative of a rather rare group of ground-dwelling reptiles, one branch of which gave rise to the birds, but not as a bird itself. True birds do occur in the Cretaceous rocks, some of them apparently even older than *Archaeopteryx*.

THE CENOZOIC ERA

With equal justice, the **Cenozoic era** could be called the "Age of Mammals," the "Age of Birds," the "Age of Insects," or the "Age of Flowering Plants," for it is marked by the appearance of all these forms in great variety and numbers of species. It extends from the Rocky Mountain Revolution, dated at some 65 million years ago, to the present, and is subdivided into two periods: the earlier **Tertiary period,** encompassing some 62 million years, and the **Quaternary period,** which covers the last million or million and a half years.

The Tertiary is subdivided into five epochs, named, from earliest to latest, **Paleocene, Eocene, Oligocene, Miocene,** and **Pliocene.** The Quaternary period is subdivided into the **Pleistocene epoch** and **Recent epoch.** The Rockies, formed at the beginning of the Tertiary, were considerably eroded by the time of the Oligocene, giving the North American continent a gently rolling topography. In the Miocene another series of uplifts raised the Sierra Nevadas and a new set of Rockies, and resulted in the formation of the western deserts.

The uplift begun in the Miocene continued in the Pliocene and, coupled with the ice ages of the Pleistocene, may have killed many of the contemporary mammals and other forms. The final elevation of the Colorado Plateau, which also caused the cutting of Grand Canyon, occurred almost entirely in the short Pleistocene and Recent epochs.

A variety of mammal-like reptiles occur in Permian formations, but the earliest fossils of true mammals were deposited late in the Triassic. By the Jurassic there were four orders of mammals in evidence, all about the size of a rat or small dog and resembling modern insectivores. The earliest mammals are traditionally held to have been **monotremes,** egg-laying mammals at present confined to Australia and its environs. It could be, however, that the main mammalian stock was viviparous from the start, with the monotremes having a separate origin. Monotremes are hardly represented in the fossil record at all; thus we know little of their history.

During the Tertiary the appearance of grasses, which served as food, and dense forests, which afforded protection from predatory dinosaurs, may have been important factors in leading to changes in the mammalian body pattern. Concomitant with a tendency toward increased size, the mammals all displayed quite consistent tendencies toward an increase in the relative size of the brain and toward certain changes in the teeth and feet.

The first-known carnivores, the **creodonts,** occur in Paleocene and Eocene formations (Fig. 47–16). They were replaced in the Eocene and Oligocene by more modern forms ancestral to the present-day carnivores, such as cats, dogs, bears, and weasels, as well as the web-footed marine carnivores—the seals and walruses. One of the most famous fossil carnivores, the saber-toothed tiger, became extinct only recently in the Pleistocene. These animals had tremendously elongated, knife-like upper canine teeth and a lower jaw that could be swung down and out of the way, allowing the teeth to be used as sabers for stabbing the prey. (The interpretation of the feeding habits of the saber-toothed tigers has been undergoing some recent reevaluation.)

The larger herbivorous mammals, most of which have hooves, are sometimes referred to as the **ungulates.** They do not form a single, natural group, but consist of several independent lines. Although both horses and cows have hooves, they are no more closely related than either one is to a tiger. The molar teeth of ungulates are flattened and enlarged to facilitate the chewing of leaves and grass. Their legs are elongated and adapted for the rapid movement necessary to escape predators. The earliest ungulates, the **condylarths,** appear in Paleocene formations. They had long bodies and tails, flat, grinding molars, and short legs ending in five toes, *each* of which bore a hoof. Corresponding archaic carnivores, or creodonts, are believed to have given rise not only to the modern carnivores, but to the whales as well. The archaic ungulates called **uintatheres** were contemporaneous with the condylarths and creodonts. During the Paleocene and Eocene some of these were as large as elephants, and some had three large horns projecting from the top of the head.

The Pleistocene epoch of the Quaternary period was marked by four periods of glaciation, between which the sheets of ice temporarily retreated. At their greatest extent these ice sheets covered nearly 4 million square miles of North America, extending south as far as the Ohio and Missouri rivers. It is estimated that in the past, when the Mississippi drained lakes as far west as Duluth and as far east as Buffalo, its volume was more than 60 times as great as at present. During the Pleistocene glaciations enough water was removed from the sea and locked in the

Figure 47–16 Early mammals. (*a*) *Oxyaena,* a small, weasel-like creodont. (*b*) Re-creation of an archaic meat-eating mammal, a creodont from the Eocene period, eating a tiny horse, *Eohippus,* that is thought to have been ancestral to modern horses. There are no modern creodonts. ((*b*), Copyright, American Museum of Natural History, New York.)

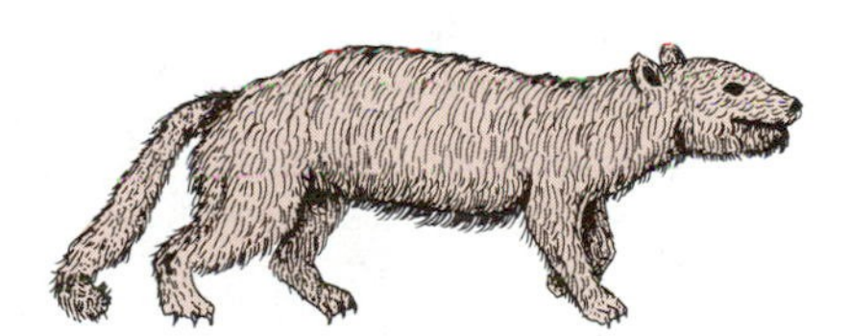

(*a*)

(*b*)

Figure 47–17 *Megatherium*, a ground sloth that was nearly the size of a modern elephant. (Courtesy and copyright, Chicago Museum of Natural History; painting by Charles R. Knight.)

ice to lower the sea level by 65 to 100 meters. This created land connections, highways for the dispersal of many land forms, such as that between Siberia and Alaska at the Bering Strait and between England and the continent of Europe.

The plants and animals of the Pleistocene were similar to those alive today. For this reason, it is sometimes difficult to distinguish between Pleistocene and recent deposits. A considerable number of New World mammals, including the saber-toothed tiger, the mammoth (Fig. 47–17), and the giant ground sloth, became extinct during the Pleistocene, very possibly as a result of early human hunting. The Pleistocene was marked by the extinction of many species of plants, especially woody ones, and the appearance of numerous herbaceous forms.

Fossil Humans

Repeatedly we have studied the body systems, organs, and tissues of humans as examples of what other animals also possess. But there are differences as well as similarities. The anatomical differences between the great apes and ourselves are rather small differences, however, confined mainly in the proportion of parts. *Gray's Anatomy* would suffice quite well as a guide to the body of a chimpanzee. Some of the characteristics that distinguish the human are (1) a brain that is two-and-one-half to three times larger than the gorilla's (more, when difference in body size is taken into account); (2) a nose with a prominent bridge and peculiar elongated tip; (3) lips that roll outward, revealing the mucous membranes, and an upper lip with a median furrow; (4) a jutting chin; (5) a great toe that is not opposable but is in line with the others; (6) a foot adapted for bearing weight by being arched both lengthwise and crosswise; (7) relatively hairless skin; (8) canine teeth that project very little, if at all, beyond the line of the other teeth and a dental arcade that is V-shaped rather than U-shaped, as in the apes; (9) an erect posture; (10) legs that are longer than arms. Additionally, humans differ from contemporary apes in being well-adapted to a bipedal gait and in being able to use hands not for locomotion, but for the manufacture and manipulation of tools (see Focus on A Distinctive Human Adaptation: The Grips of the Hand).

Antigen–antibody tests of similarities in serum proteins show that, of all the apes and monkeys, gorillas and chimpanzees have serum proteins most nearly like those of the human. The amino acid sequence in the chimpanzee's hemoglobin is identical with that of the human; that of the gorilla and rhesus monkey differ from the human's in 2 and 15 amino acids, respectively. With regard to biochemistry, or to structure or proportion of parts, the difference between human, gorilla, and

FOCUS ON

A Distinctive Human Adaptation: The Grips of the Hand

Prehension Patterns	*Description*
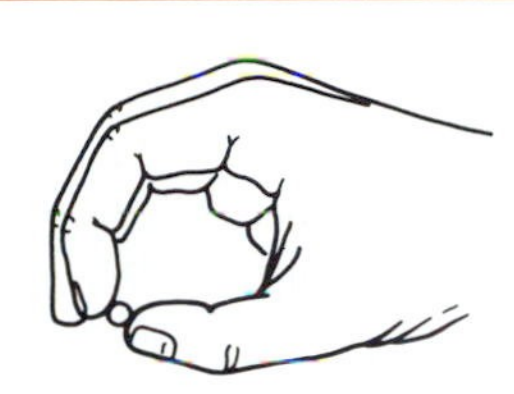	**Fingertip Prehension.** This grip is used for picking up tiny items. It involves primarily the thumb and index finger. It is not a very stable type of pinch but requires the most coordination.
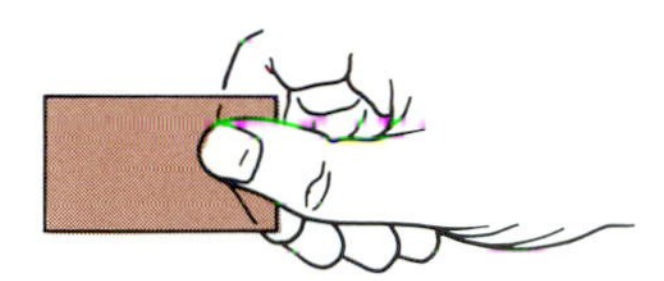	**Lateral Prehension.** In this grip the thumb is *adducted* and *flexed* against the lateral surface of the index finger. This grip is used to turn a key, hold a plate or a hand of cards, or wind a watch. A larger surface area is involved; thus the grip is more stable than fingertip prehension but produces less of an opening between digits, so smaller objects can be held with lateral prehension. The grip usually involves a stronger pinch, with somewhat larger muscles coming into play.
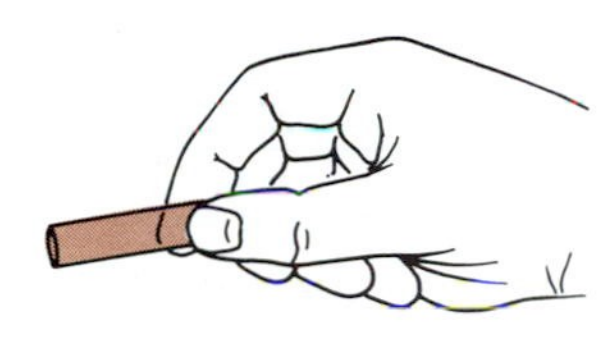	**Palmar Prehension.** This grip is the most versatile pinch, the one used most often for picking up and holding objects. The pads of the index and middle fingers and thumb are used for pinching. This stable pinch can begin with a very wide opening, making it possible to pick up large objects.
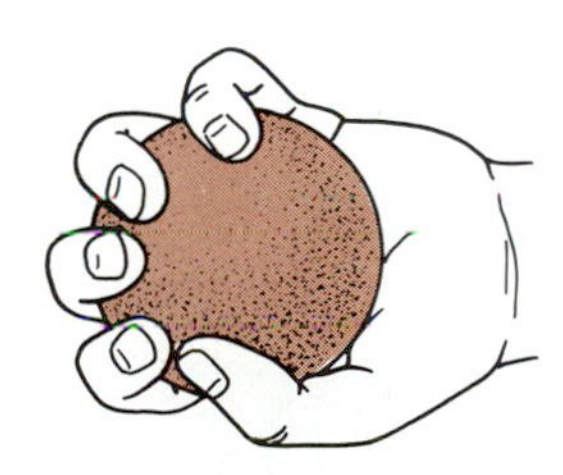	**Spherical Grasp.** This grip is used when holding anything round.
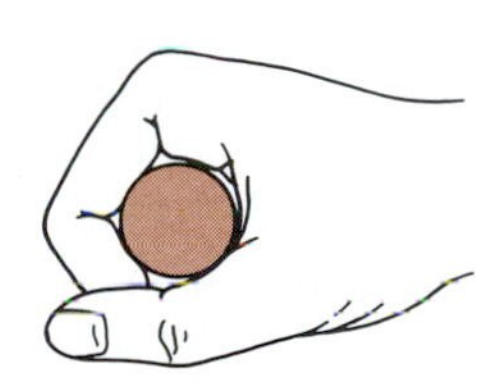	**Cylindrical Grasp.** Crudest type of prehension. It is a stable grasp because the width of the palm gives stability. It requires wrist extension with flexion of the fingers and thumb and is used when grasping a cup, telephone, or stair rail.
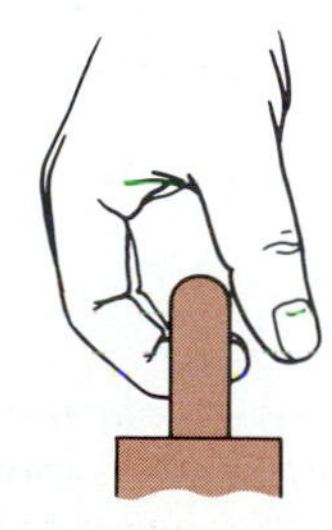	**Hook Grasp.** This grip does not require the use of the thumb but does require that the fingers be able to flex completely. It is used when carrying a suitcase or a bucket of water, or when hanging onto a cliff.

chimpanzee is less than the difference between any of these and monkeys. These similarities have led evolutionists since before Darwin's time to propose that of all the lower animals, the primates and especially the apes are the most closely related to humans and that an ape or ape-like primate was the common ancestor of all the members of the superfamily Hominoidea.

Human Evolution

In 1855 the British anatomist Richard Owen proposed an argument that, he believed, would show that humans and the great apes are *not* related. With the exception of the orangutan, all apes possess a heavy ridge of bone above the eyes, the **supraorbital torus.** This occurs rarely in humans, and then only feebly. There is no obvious reason why such a ridge would be a disadvantage to humans, he reasoned. If apes were indeed the ancestors of humans, the torus should still be present in either living or dead varieties of humanity.

Unfortunately for Owen's choice of argument, quarry workers discovered the bones of a human being in a cave in the Neander valley of Germany. The word for valley is *thal* in German, so this ancient gentleman became known to the world as **Neanderthal man.** Among other distinctive features, Neanderthal man had a very heavy supraorbital torus. He also suffered from skeletal deformities, perhaps of arthritic origin, that would in life have produced a stooped, shambling, and somewhat ape-type gait. Although the discovery of numerous other Neanderthal skeletons has shown that these people stood as erect as we, the supraorbital torus has stood the test of time as a distinctive Neanderthal trait, along with a virtual absence of a chin and an extremely heavy skeleton generally. The size of Neanderthal man's brain was fully equal to ours, and probably ran a little larger on the average. However, the forehead was much lower than that of ours. Nevertheless, properly manicured and attired, Neanderthal man would probably not have stood out in a shopping mall, although he probably would attract attention (not necessarily admiring) on a beach.

Neanderthal man was but the first in a series of discoveries of fossil humans and others not clearly human that are now collectively known as **hominids.**

The next putative human ancestor to be discovered was what we now know as **Homo erectus** (Fig. 47–18). That was, however, not its original name. (Unless the sex of a fossil hominid is relevant to the discussion, from here on it is referred to with a neuter pronoun.) Even before any actual bones had come to light, the German evolutionist Haeckel had proposed a scientific name for the as yet–unknown

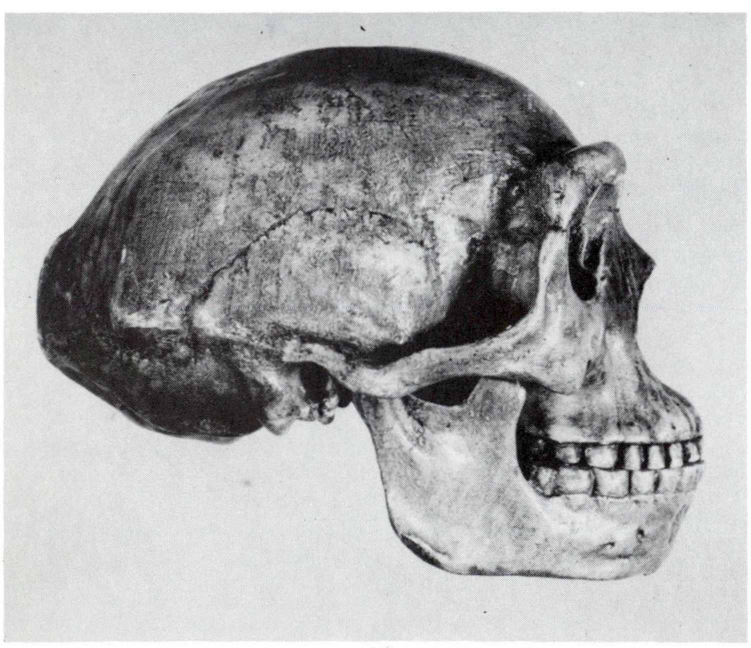

(*a*)

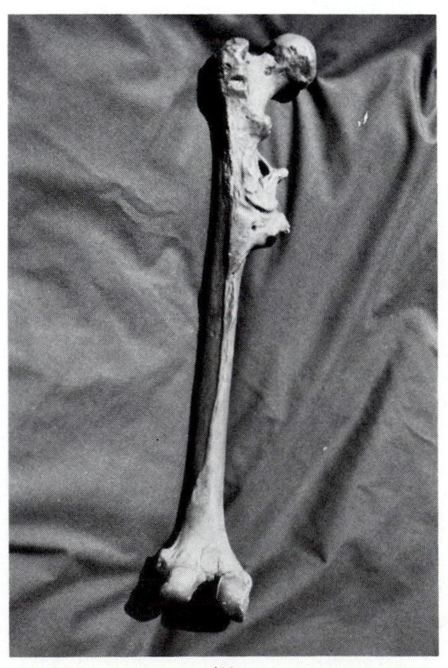

(*b*)

Figure 47–18 *Homo erectus.* (*a*) A replica of the skull. Note the massive bony ridges over the eyes, the receding forehead, the protruding jaw, and absence of a chin. (*b*) A view of the femur of *Homo erectus* (*Pithecanthropus*), discovered by Eugene Dubois. The linea aspera is well-developed, indicating the presence of a highly developed gluteal musculature, which indicates in turn that this hominid stood with a fully erect stance. Note the advanced bony tumor on the bone, possible evidence that sick *Homo erectus* hominids were aided by well ones. ((*a*), photograph of Wenner-Ren Foundation replica by David G. Gantt.)

hominid linking humanity and the apes. In Greek, ape is rendered *pithecos*, and human being, *anthropos*. An ape-man, then, would be *Pithecanthropus*. Haeckel, nevertheless, was not destined to discover the bones of his ape-man. This privilege was reserved by history to Eugène Dubois.

Despite his French name, Dubois was Dutch. Although he trained as a physician, Dubois' ambition was to discover *Pithecanthropus*. He reasoned that since, in the then-majority view, the East Indian ape known as the orangutan (Fig. 47–19) resembles humanity more closely than do the African apes, it was likely that *Pithecanthropus* would be found in the East Indies where orangutans occur. Dubois joined the Dutch army and obtained an assignment to the East Indies (then a Dutch possession) as a military surgeon. Once there, he persuaded both the army and the colonial government to support his research, and in due course unearthed the remains of what he was pleased to call *Pithecanthropus erectus*. There wasn't much left of it. *Pithecanthropus* consisted of a skullcap, a femur, and two teeth. The femur, except for what appeared to be a developing bone tumor, was very similar to a modern femur. The skull, as far as anyone could tell, would have had a cranial capacity of 855 cc—very low indeed—pathologically so—for an adult human. Yet the brain was far too large for an ape, at least any known ape.

Figure 47–19 *Orangutan*. Dubois thought these asiatic apes to be more human-like than any others, which impelled him to seek *Pithecanthropus* in Indonesia. (The Image Bank.)

Since Dubois' time, many fossils have been unearthed that seem to belong to the same general group as *Pithecanthropus*. Like *Pithecanthropus*, these occur in the East Indies and also in China (where they were known for a while as *Sinanthropus pekinensis*), Africa, and even Europe, represented by a single German jaw that, unfortunately, also received its own scientific name (which you are spared here).

It is instructive to note the features of *Pithecanthropus*[1] as depicted in a statue that Dubois had made as a reconstruction, in light of what is now known from far more complete fossil remains than those available to him. This somewhat battered sculpture shows *Pithecanthropus* with arms long enough to reach below his knees (be assured that the statue, at least, is definitely male), and opposable big toes like a chimpanzee. The thumb also resembles that of a chimpanzee, being located much nearer the wrist than it is in humans and apparently imagined by Dubois as not fully opposable to the other fingers. The face is protruding and chinless, the supraorbital torus pronounced, the expression dim-witted but benign. There is no beard, and the hair is combed. This mosaic of mismatched traits is what Dubois *expected* from such a "missing link." But the skeletons of *Homo erectus* (as *Pithecanthropus* is now known) discovered since are much the same as those of modern humans—at least below the neck. As Dubois expected, however, the brain cavity is small and the jaw chinless. On the whole, *Homo erectus* was smaller than *Homo sapiens* but, like the Neanderthal, quite powerfully built. The intermediate traits of *Homo erectus* are almost entirely confined to the head, particularly the cranial cavity.

We may quickly pass over Piltdown man, a fraudulent concoction of the 1920s composed of an old *Homo sapiens* skull and the jaw of an orangutan, except to note that it is odd that this humbug received the acceptance that it did, for it numbered among its supporters some very eminent scientists. Perhaps the explanation is that, as in Dubois' reconstruction of *Pithecanthropus*, their preconceptions of the traits that an early form of humanity would possess were extremely powerful and shaped their perception of reality.

The next major discovery made in the investigation of humanity's possible origins was *Australopithecus*, a hominid whose investigation is the most actively pursued area of human paleontology even today.

Darwin and his cohorts had proposed that the origin of the human race was probably on the African continent, a suggestion that did not sit well with Dubois, but which did find favor with the Australian physician Raymond Dart. After service in World War I, Dart was able to secure a position as professor of anatomy in the then new and struggling University of Witwatersrand near Johannesburg, South Africa. The presence of a rich store of mammalian fossils was brought to his attention in 1924. Although the early fossil discoveries at the site, the Taung quarry, were in the tradition of Neanderthal man, quite a different creature was eventually found, a child with a dentition comparable to that of a modern small child of perhaps 6 years of age. In due course, the skull was named *Australopithecus africanus*. Since there were no other specimens from which a more complete estimation of its skeletal anatomy could be gleaned, *Australopithecus* was initially of unclear signifi-

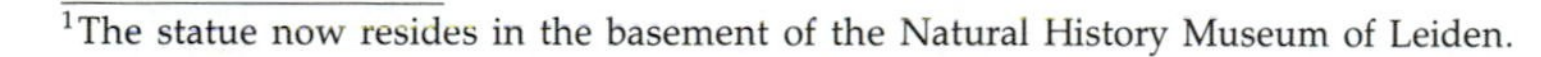

[1]The statue now resides in the basement of the Natural History Museum of Leiden.

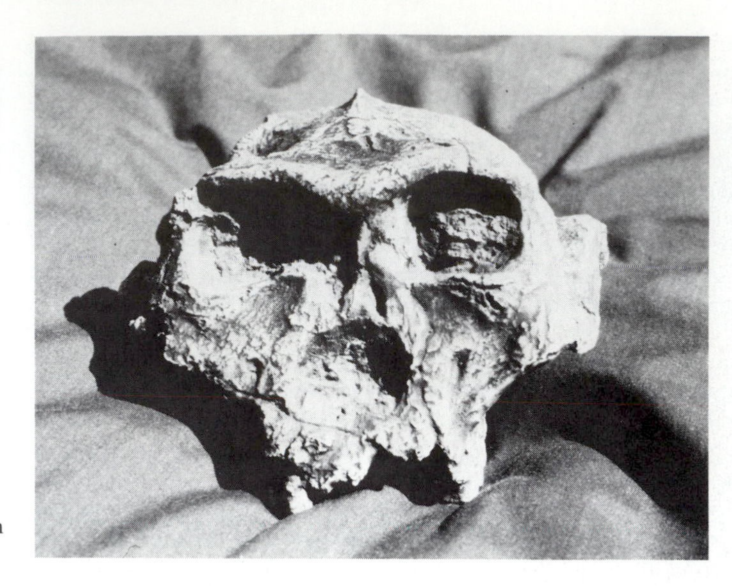

Figure 47–20 *Australopithecus robustus,* one of the first australopithecines to be discovered. This is a specialized form not considered ancestral to other known hominids.

cance and highly controversial. It was to be years before a better understanding of this hominid could be had; indeed, there is a great deal about it that is unclear to this day.

The next chapter in the hominid story was written by Robert Broom, a South African physician and paleontologist who was much impressed by Dart's *Australopithecus* baby. Broom's attention was drawn to another quarry, located near the village of Sterkfontein, especially two sites known as Kromdraai and Swartkrans. Here he found more *Australopithecus* material, including a fair amount of the postcranial skeleton such as the pelvis and the upper half of one femur. The femur alone was enough to establish that *Australopithecus* could walk upright. Further work led to the conclusion that two, not one, species of *Australopithecus* were represented in Broom's collection. The second one is today known as *Australopithecus robustus* (Fig. 47–20). It has a generally heavier skeleton, an immense jaw, and dental evidence of a very coarse diet. It is considered to be a later comer than *A. africanus,* although the two species apparently coexisted for a considerable time. *A. robustus* is universally agreed as having died without evolutionary issue, a specialized animal more intelligent, perhaps, than modern chimpanzees, but of little more biological significance. *A. africanus,* whose brain size range also overlaps that of modern apes, is thought by at least a large number of anthropologists to be directly ancestral to the genus *Homo,* including *H. erectus* and *H. sapiens.*

Louis Leakey broke with that tradition. The child of missionary parents stationed near Nairobi, Kenya, he was educated in England and received a degree in anthropology from Cambridge University. Having conceived an interest in human paleontology in boyhood, Leakey was able to organize expeditions to Kenya and Tanzania initially in the 1920s. He was rewarded in a locale now famous throughout the world—the Olduvai Gorge in Tanzania. Expeditions by Leakey family members lasting up to the present have unearthed a wealth of fossil hominid material, including, but not limited to, *Australopithecus.* Some of these were *A. africanus,* some were *A. robustus* (or, as Leakey called it, *Zinjanthropus*), and some were *Homo.* Of the *Homo* specimens, some had a brain size intermediate between that of *A. africanus* and *Homo erectus;* others were clearly *Homo erectus.* Leakey named the small-brained specimen *Homo habilis.* He felt that it was distinct from *Australopithecus* and in this way established, or attempted to establish, a human lineage that did not include *Australopithecus* at all. Had *Homo habilis* been discovered by another group of paleontologists, it probably would have been placed in the *Australopithecus* genus. In fact, something very like this was about to occur.

After a series of expeditions to the Afar triangle region of Ethiopia, Tim White and Donald Johansen announced in 1978 a new species of *Australopithecus* that, because of its age and to some degree its anatomy, they consider ancestral not only to other species of *Australopithecus* but to *Homo habilis* as well. The Leakey family agrees that this new hominid was probably ancestral to *Homo habilis,* but does not see it as an australopithecine. They prefer to believe that the *Homo* and *Australopithecus* lineages have been distinct since the Miocene, perhaps. Johansen and White

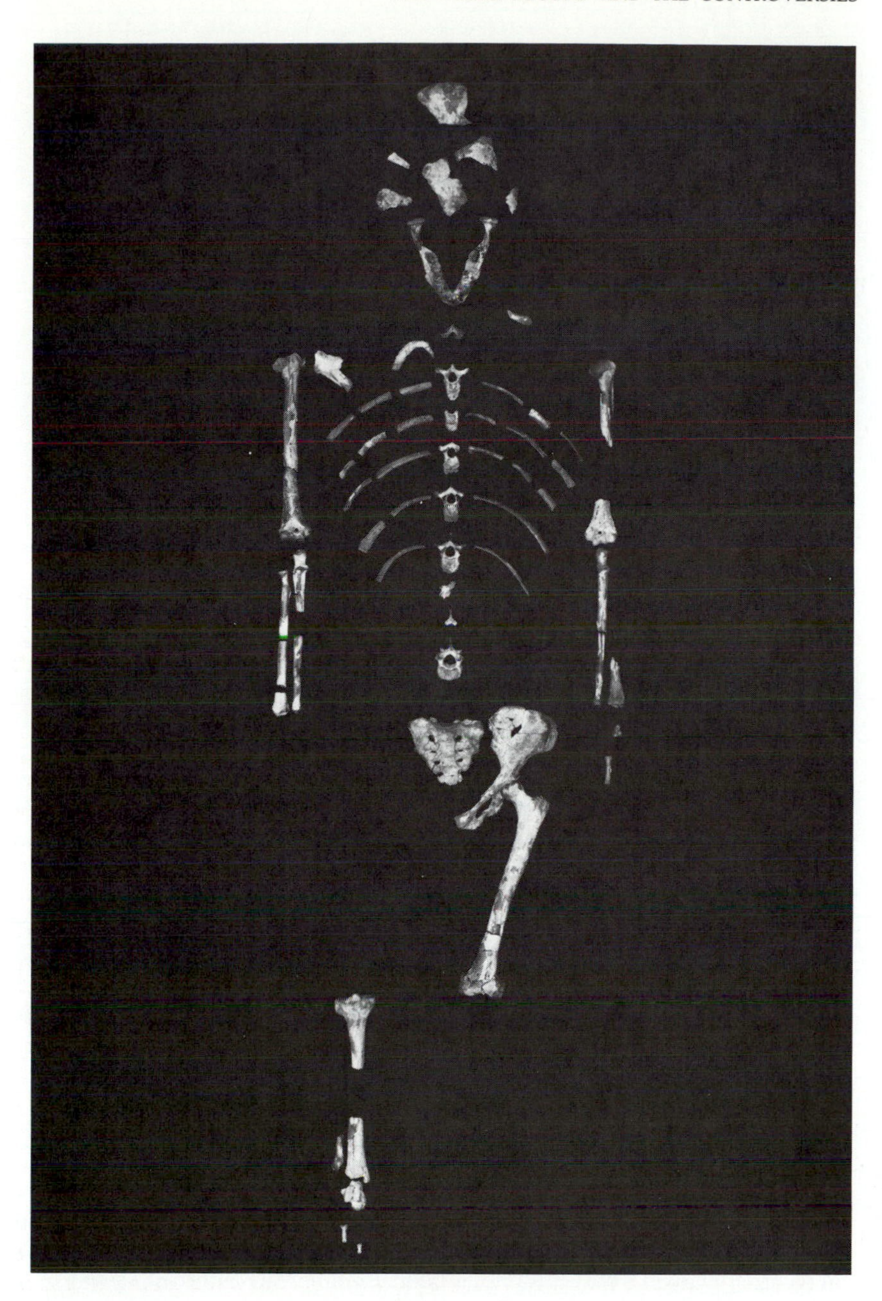

Figure 47–21 The skeletal remains of Lucy, a hominid approximately 3.5 million years old. (Photograph courtesy of the Cleveland Museum of Natural History, with permission.)

have had the privilege as discoverers of naming their fossil, however, and have thus registered their opinion.[1] They called her *Australopithecus afarensis* or, more familiarly, just Lucy[2] (Fig. 47–21).

The majority of archeologists, anthropologists, and paleontologists knowledgeable in the field of hominid fossils would probably support a summary view of human evolution something like the following (Fig. 47–22):

Australopithecus afarensis was ancestral to all later hominids. *A. afarensis* itself was derived from some apelike ancestor such as *Dryopithecus* (not described here), which also gave rise to the modern African and possibly Asiatic apes. This view is bolstered by the fact that the similarity between the proteins of chimpanzees and human beings is very great. *A. afarensis* in turn gave rise to the australopithecines, on one hand, and to the *Homo* lineage, on the other. The australopithecines coexisted with *Homo* for as long as 2 million years, eventually becoming extinct. In their early history, there were at least two species of *Australopithecus: A. africanus* and *A. robustus*. It is possible that these hominids made a few simple tools and were eliminated by *Homo*, particularly *H. erectus*, who was known to have used a variety of

[1]The first to publish a scientific description of an organism is entitled to name it.

[2]Legend has it the name came from a popular Beatles song, "Lucy in the Sky with Diamonds."

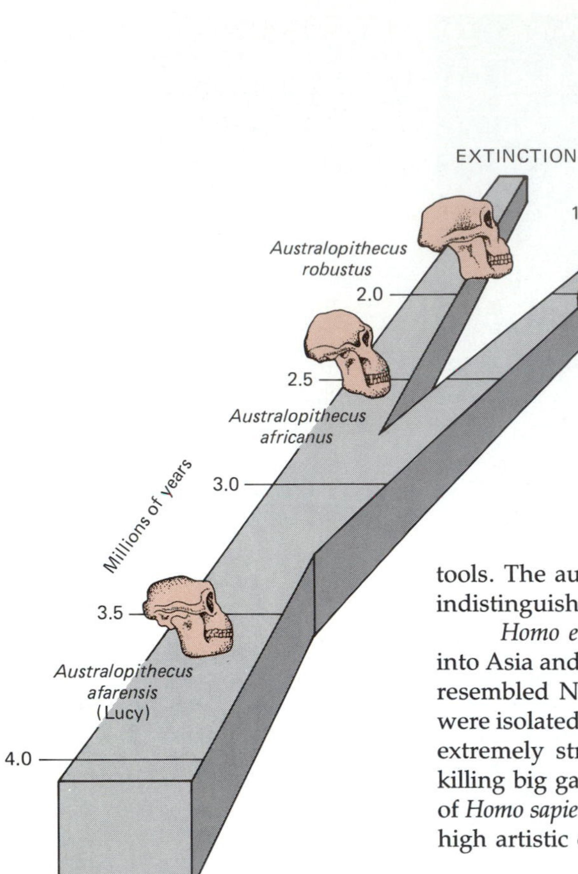

Figure 47–22 A hypothetical hominid family tree. Note the overlapping time periods of various species. Whether one hominid group forced the extinction of the other, or whether certain species were simply less well adapted to their environment, is unknown. (After Johanson, D. C., and White, T. D.: *Science*, 203:321–330, 1979.)

tools. The australopithecines and early *Homo* were fully bipedal, with feet almost indistinguishable from those of modern human beings.

Homo erectus was the first hominid to spread beyond the African continent into Asia and Europe. It gave rise to *Homo sapiens*. Possibly the earliest *Homo sapiens* resembled Neanderthal people, but the most extreme Neanderthals (Fig. 47–23) were isolated in glacial cul-de-sacs in Europe, where they became specialized for an extremely strenuous life style, perhaps involving close-encounter techniques of killing big game. The Neanderthals were eventually replaced by modern versions of *Homo sapiens* such as the famous Cro-Magnon people, who have left paintings of high artistic quality in caves in France and Spain.

Gaps in the Fossil Record: Punctuated Equilibrium

A falsification of a particular aspect of a general theory does not necessarily invalidate the theory, but is more likely to lead to its modification or improvement. Science proceeds by just this kind of internal readjustment. It was just such a critical reevaluation that led to the general *acceptance* of evolution when Darwin partly rejected Lamarckian explanations as the main driving force of evolution.

Writing in one of the later editions of *Origin of Species*, Darwin noted, "Geology assuredly does not reveal any such finely graded organic chain; and this, perhaps, is the most obvious and serious objection which can be urged against the theory. The explanation lies as I believe, in the extreme imperfection of the geological record." Darwin's observation has been abundantly confirmed up to this day by

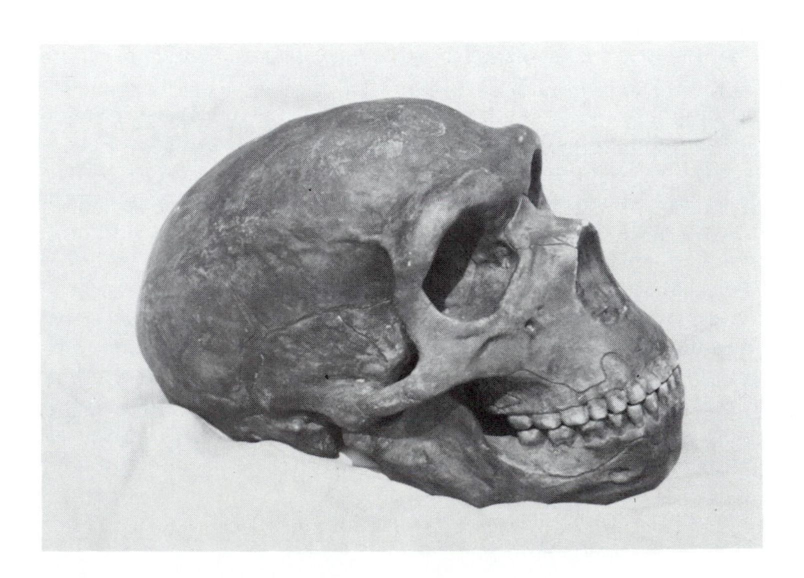

Figure 47–23 Neanderthal skull. Note the eyebrow ridges, almost as heavy as those of *Homo erectus*. The brain size, however, was if anything greater than that of modern varieties of *Homo sapiens*.

(a)

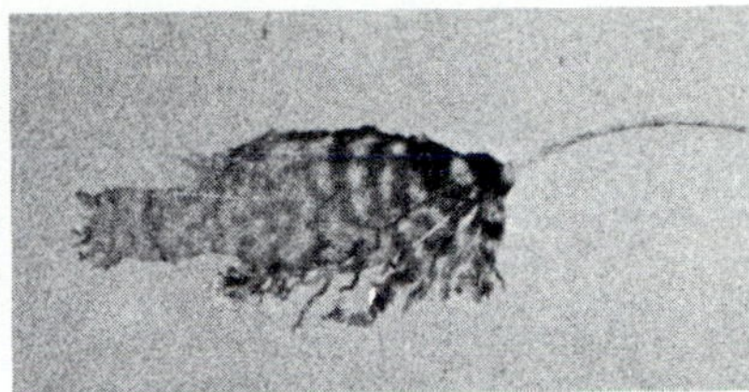

(b)

Figure 47–24 (*a*) *Antennarius*, an anglerfish. Using the fish-shaped filamentous proboscis, shown in (*b*), this sluggish predator lures its victims into reach. Instead of catching the "fish," the victim is caught itself and swiftly eaten. Other species of anglerfish are not this perfectly adapted to their way of life (or so it seems to us; the species shown here is by no means as common as some other kinds). However, the proboscis can be imagined to have reached this perfected form by a series of individually small genetic modifications, each of which slightly increased its attractiveness to prey species. But see also Figure 47–25. (From Pietsch, T. W., and Grobecker, D. B.: The compleat angler, aggressive mimicry in an antennarid anglerfish. *Science* 201:369–370, 1980. Copyright 1980 by the American Association for the Advancement of Science.)

a multitude of studies. Almost a hundred years later, in 1953, the paleontologist George Gaylord Simpson noted that "nearly all categories above the level of families appear in the record suddenly and are not led up to by known, gradual, completely continuous transitional sequences."

Certainly a convenient example is the modern version of *Homo sapiens*. We appear suddenly in the fossil record, or so it seems to many paleontologists. Some dispute this, pointing to apparently transitional forms in the Middle Eastern sites of Skhul and Tabun, but these fossils can also be interpreted as hybrids between modern-type *Homo sapiens (Homo sapiens sapiens)* and the Neanderthal variety. Thus there is no unambiguous record of the emergence of modern humankind—though, obviously, we are here! The science fiction writer James P. Hogan even based a series of novels on the idea that the evolution of *Homo sapiens sapiens* took place on another planet! This alternative, if such it be, is surely not popular among biologists. However, what has been seriously proposed is that, when major transitions occur, they occur suddenly and not as a response to natural selection. Aside from humans, where, for example, are the "prototurtles"? the "quasi-bats"? All the old bats are fully adapted flying mammals; they are in no way transitional forms.

There is another problem in accounting for major evolutionary changes. One is that major adaptations would be of use to an organism only when more-or-less fully developed, though not necessarily fully refined (Figs. 47–24 and 47–25). The lensed eye of vertebrates, cephalopod mollusks, and some polychaete annelid worms are examples. Eye lens and retina must coexist to be of any use at all.[1] It is difficult to reconstruct a series of stages in which *each is adaptive*, whereby such an eye might have come into existence, or, following a proposal of such stages, to show that this is indeed the way in which it actually happened.

There have been several evolutionary attempts at a solution of these gap problems. One of the first to take them seriously was the paleontologist George Gaylord Simpson. Simpson noted that most of the time, fossil species changed rather slowly and in relatively minor ways; this he labeled the **bradytelic** mode of evolution (brady is a Greek word meaning "slow"). Abrupt changes unrepresented by intermediate fossil forms took place rapidly, he thought; these represented the **tachytelic** mode (tachy means "fast"). But what might be responsible for the tachytelic mode? Moreover, if tachytelism is characteristic of major evolutionary change (not of just minor rearrangements of traits), might it not be a *necessary* feature of such changes?

A truly *major* change, when only partially complete, might well have the effect of *un*fitting an organism for the ecological niche of its ancestors *without* fitting that organism for the niche later to be exploited by its descendants. If major changes are indeed as adaptively ambiguous as they appear to us, then they would

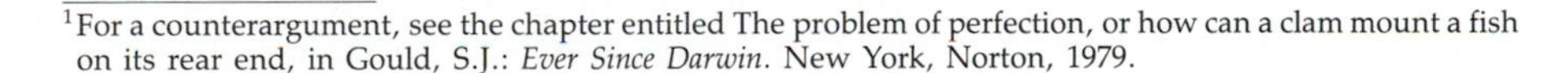

[1] For a counterargument, see the chapter entitled The problem of perfection, or how can a clam mount a fish on its rear end, in Gould, S.J.: *Ever Since Darwin*. New York, Norton, 1979.

(a)

(b)

(c)

Figure 47–25 A bombardier beetle in action. (*a*) An ant approaches the captive beetle and (*b*) attacks it. (*c*) The beetle mixes a peroxide and a catalyst in a special chamber, and sprays the ant in the face with the boiling, noxious fluid. Since neither the peroxide nor the catalyst is useful by itself, we might well ask from what beginnings the beetle's present mechanism could have evolved. Can you suggest what their original, quite different function might perhaps have been? (Courtesy of T. Eisner and D. Aneshansley and *Science*. From Aneshansley, D. J.: Biochemistry at 100°C: Explosive secretory discharge of bombardier beetles (*Brachinus*). *Science,* July 4, 1969, pp. 61–63. Copyright 1969 by the American Association for the Advancement of Science.)

surely have to take place rapidly or, like the half-formed monsters out of Ovid's *Metamorphoses,* they would quickly perish without issue. Yet here, as we have observed, we are. Were our ancestors monsters?

According to Richard Goldschmidt (whose concept of "hopeful monsters" we met in Chapter 45), they might have been. In his 1940 book *The Material Basis of Evolution,* Goldschmidt proposed that extreme mutations, **macromutations,** might take place from time to time. These would produce nothing so trivial as a changed number of bristles on a fruitfly's posterior, or an altered enzyme with a new temperature optimum. Goldschmidt pointed out that just one or a few changes in genes governing early development could produce proportionately large results. Were he alive today (he died in 1958) Goldschmidt might point to recent studies showing that humans and chimpanzees have more than 90% of their protein amino acid sequences in common (probably 99%, in fact), and therefore, since genes make proteins, the two species must have most genes in common as well. The differences between the two must rest on a few key genes—**regulatory genes.** Such genes presumably would regulate such things as the relative rate of growth of brain parts, limb bones, and the forehead. In principle, then, genetic engineering might make something virtually human out of a chimpanzee with rather little effort.

Goldschmidt thought that something like this may have taken place naturally; he called such a major evolutionary leap **saltation.** Goldschmidt's work has not been taken seriously until recently when some of his views have been discussed with favor. No serious evolutionist today would claim, as Goldschmidt thought, that the first bird hatched from a reptilian egg. Nevertheless, a view somewhat similar to Goldschmidt's saltation theory has gained some credence of late; it is called **punctuated equilibrium.** According to the main promoters of this view, the major events of evolution and many of the minor ones as well occurred in isolated, out-of-the-way habitats such as islands or glacially isolated terrain. These changes result not so much from natural selection as from genetic drift, the founder effect, and similar mechanisms fostered by intense inbreeding in isolated populations.

Natural selection, according to these views, would come into play mainly when the isolating barriers were broken down, and it would operate rapidly. At that point the evolutionary novelties would be loosed on the world. If their bizarre new combinations of traits rendered them adaptively inferior to their more normally endowed cousins in the outside world, nothing more would be heard of them. If they were *superior,* however, they would swiftly take over any available ecological niche. Thus, punctuated equilibrium *predicts* the very gaps in the fossil record that have continued to bother scientists since Darwin's time, for if the major changes took place with unusual rapidity in small isolated areas, then only a very few if any transitional forms would be represented in the fossil record.

Other theorists blend aspects of gradualism—classical Darwinism—with catastrophism. A sudden change in environment, whether of terrestrial or extraterrestrial origin, could decimate the populations of otherwise successful species, while affording surviving organisms an opportunity to radiate into newly vacated niches. Such is the scenario whereby the mammals might have replaced the reptiles as the dominant land animals; and it is a scenario that might have occurred during other periods of time for which evolutionary change is rapid and the evidence of transitional forms is poor.

The main difficulty with punctuated equilibrium as a scientific hypothesis is that proposal of an observational test whereby it might be shown to be false—if indeed it is false—is very difficult. We could have confidence in it only if it survived the application of such a test. Since punctuated equilibrium, if it occurs, is a very singular event not subject to duplication or widespread observation, it is almost as difficult to assimilate into conventional, gradualistic evolutionary theory as is the theory that organisms have arisen through successive, individual acts of creation. As yet, however, it is too early to see how well punctuated equilibrium will survive the attacks being mounted against it.[1]

SUMMARY

I. The geological time table is summarized in Table 47–1.

II. The discrediting of spontaneous generation is a singularly important chapter in the history of biology, and nothing in biology is so surely established as the proposition that only life begets life, at least under present-day conditions.

III. According to current theories, life may nevertheless have originated on the earth from naturally occurring macromolecules by a kind of spontaneous generation. Eukaryotic cells may have originated from a symbiotic union of several prokaryotes.

IV. Focusing on fossil hominids, the current consensus on their evolution is that an ape-like ancestor gave rise to both the hominid and ape lineages. The earliest hominids resembled *A. afarensis,* which gave rise both to the australopithecines and to the genus *Homo.* The earliest known member of this genus was *H. habilis,* which gave rise to *H. erectus. H. erectus* then gave rise to both the Neanderthal and modern versions of *H. sapiens.*

V. The theory of punctuated equilibrium holds that major evolutionary changes occurred rapidly in isolated habitats and involved many nonselectional mechanisms such as genetic drift. Natural selection did determine, however, whether these novel organisms would subsequently prevail.

POST-TEST

1. Conventional scientific thinking regards the atmosphere of the earliest earth as containing no ________________ gas.

2. The primitive oceans might have contained a dilute solution of complex ________________ compounds, including such materials as ________________ and ________________ ________________.

3. The earliest organisms are now considered to have been ________________, and would have employed ________________ metabolism.

4. Enzymes employed today in cellular life processes would have been acquired in the ________________ order of the sequence with which they are now employed in metabolism.

5. According to the ________________ hypothesis of the origin of the eukaryotes, mitochondria originated as moderately conventional ________________, and ________________ as cyanobacteria.

6. ________________ are the only fossils obtainable from most Precambrian deposits.

7. Most of the basic animal body plans occur first in ________________ formations.

8. The land plants and animals appear initially in ________________ deposits.

9. ________________ amphibians occur in Devonian rocks; these are thought to have descended from ________________ bony fishes.

10. Together, the ________________ and ________________ periods comprise the ________________ period, a time of formation of present-day coal beds.

11. Remains of ________________-like reptiles occur in late Carboniferous and early Permian formations.

12. Among the catastrophic explanations that have been proposed for the demise of the ________________, one currently popular idea involves the collision of the earth with an ________________ from outer space.

13. The ________________ era is conventionally called the Age of Mammals, but it also includes a vast variety of species of ________________, ________________, and ________________ ________________.

[1]For a summary, see Rensberger, B.: The evolution of evolution. *Mosaic,* September/October 1981, pp. 14–22.

FOCUS ON

Creationism and Evolution

Darwin never actually proved that life on earth had evolved from simple molecules to complex cellular and multicellular organisms. He merely presented a theory whereby the great complexity and diversity of life on earth could be understood in light of reason and some observations that could be made in the world around us. An explanation regarding evolution had been sought for some time. The fossil record left indisputable evidence that the planet's life forms had not always been as they appeared in the middle of the 19th century. Darwin postulated that if, in a thousand years, artificial selection by humans could derive from a wolflike animal such creatures as toy poodles and Saint Bernards, then there was likely a mechanism in nature that, operating over millions (we now know billions) of years, could produce from some ancestral being such organisms as *Homo sapiens, Paramecia,* and maple trees. This great inductive leap, as we have discussed, provided two of the major organizing theories in modern biology: (1) the theory of natural selection; and (2) the theory that, through the mechanisms of natural selection, all life forms evolved from a common ancestor.

Since the time of Darwin our record of fossil life has become more complete. With the advent of molecular genetics, we now know how seemingly minor mutations can cause large changes in morphology and biochemistry. Yet there continues to be debate over the specific mechanisms and the speed by which natural selection operates. Ironically, as our knowledge of biology grows, and as we continue to pose questions about life in an evolutionary framework, we are left with the facts that no one has observed macroevolution in progress, and no one has proven such theories as gradualism or punctuated equilibrium true or false.

To view the world as operating according to strict physical laws is to have a materialistic outlook. The modern theory of evolution, which focuses on chemical changes in genes that then get expressed in organisms subject to natural selection by the physical environment, is very much a materialistic explanation of the history of life on earth. Yet, as individuals, scientists are by no means materialists in their personal outlook. In fact, many are, like Aristotle, *theistic evolutionists.* Implicit in their study of the world is a belief that it displays an order brought into being by some superior force. However, in their actual practice of science, they are concerned only with what is observable, what is measurable, and what can be tested. Because of this, science and what is called philosophical materialism are often considered one and the same. This notion of the scientist as being a materialist might not be accurate; however, it is true that from a purely scientific point of view, the overall Darwinian account of evolution remains the most appealing. This is because it is simple in concept and testable in at least some of its implications (such as we saw in the experiment on natural selection and microevolution in bacteria; see Chapter 45).

You will recall from our discussion of retrodictions, predictions, etc. that when looking at past events, there is no way to completely rule out the influence of coincidence, chance, or some factor that—even in the present—we cannot see in operation. It is not difficult to understand, therefore, that there are interpretations of the history of life on earth that are not wholly materialistic. To this day, some people state that the imprecision of the fossil record (see figure), the gaps in the record, and so forth leave open or even demand the postulation that some creator, some supernatural, or some extra-terrestrial force has had a hand at various times in causing the extinction and creation of life forms. Some of these individuals are in fact scientists, and some are referred to as "scientific creationists."

14. The apparently earliest human beings belonged to the genus ______________.

15. The Neanderthal people are now considered to be members of the species ______________ ______________.

16. According to the theory of ______________ ______________ the initial evolution of organisms that depart in major ways from their forebears takes place very rapidly and often in out-of-the-way places; this theory, according to its proponents, accounts for the presence of apparent ______________ in the fossil record.

REVIEW QUESTIONS

1. In what geological formations do vertebrates first appear? What did they look like and where did they live?
2. Compare and contrast the ostracoderms and the placoderms as to structure and mode of life.
3. What significant life forms appeared during the Silurian period? What is the significance of the name Carboniferous?
4. What was the Appalachian Revolution? What geological formations are associated with it? What were its effects?

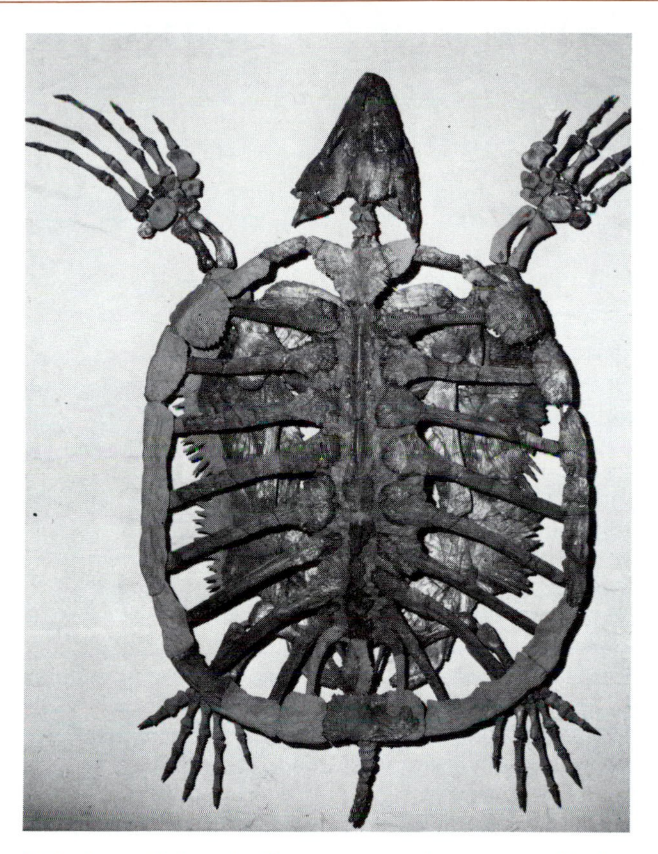

Little changed since the Cretaceous, turtles appear suddenly (without transitional forms) in the fossil record. Since there are few clues to their ancestry, this leaves any interpretation of their origin open to controversy. This marine turtle, *Protostega gigas,* was a contemporary of even larger seagoing lizards and other aquatic reptiles of the age of dinosaurs. (E. R. Degginger.)

Much of the criticisms that scientific creationists have of modern theories of evolution are based on negativism; that is, our reconstruction of the history of life on earth is not quite airtight. In addition to citing problems already acknowledged by evolutionary theorists, creationists maintain that even if the current methods of dating fossils are accurate (which they doubt), 3.5 or 3.8 billion years is too little time to account for the diversity of life that we see today. As critics, the creationists remind us that the theory of evolution is indeed a theory, subject and in need of much scrutiny and reevaluation. As scientists, however, the creationists offer us an alternative to evolution that amounts to an unfalsifiable hypothesis. For in the study of macro- and microevolution the evidence for or against the operation of a superior being is simply not observable, and testing for the operation of a superior being is beyond the means of scientific inquiry. Ultimately, therefore, the creationists leave scientists with no possibility but to continue to study the history of life as being controlled by materialistic mechanisms. This is not the same thing as saying that the Darwinian account of evolution is "true," but that it does continue to be the only theory that has held up to the types of investigations that lie within the domain of science.

In the next few chapters, we will study the interrelationships of the vast numbers of organisms that inhabit our planet. In Chapter 52, we will examine the impact of human beings on the world ecosystem. Whether viewed through a materialist perspective, a theological-spiritual perspective, or any sort of mix of the two, the diversity of life is stunning. As we mentioned in Chapter 1 regarding art, science, and religion, we need not stand on one another's toes in order to be amazed by it, or lose any sense of the urgency with which we must act to preserve it.

5. What is meant by the term "stem reptile"? What relation do turtles evidently bear to these animals?
6. What factors may have contributed to the extinction of the dinosaurs?
7. If mammals originated during a time when dinosaurs also existed, why didn't the more "advanced" mammals become dominant?
8. Briefly trace the evolution of the various phyla from the beginning of life to the present. Choosing one of these, discuss whether its proposed relationships appear reasonable to you.
9. How does *Australopithecus robustus* differ from *Homo habilis?*
10. How does punctuated equilibrium differ from the more gradualistic evolutionary concepts presented in Chapter 45? Does it make use of any of them? What is the evidence in favor of it? Can you think of evidence against it?

48

The Behavior of Organisms

OUTLINE

I. What is behavior?
II. Behavior as adaptation
III. Simple behavior
  A. Tropisms
  B. Taxes
IV. Biological rhythms and clocks
  A. Lunar cycles
  B. Circadian rhythms
  C. What controls the biological clock?
V. Sign stimuli
VI. The genetic basis of behavior
VII. Physiological readiness
VIII. Learning
IX. Imprinting
X. Migration

LEARNING OBJECTIVES

After you have read this chapter you should be able to:

1. Support the concept of behavior as adaptive, homeostatic, and flexible.
2. Define tropisms and taxes, and give examples of each.
3. Cite examples of biological rhythms, and suggest some of the mechanisms known or thought to be responsible for them.
4. Using appropriate examples, summarize the role of sign stimuli (releasers) in the expression of simple and complex programmed behavior.
5. Summarize the contributions of heredity, environment, and maturation to behavior.
6. Compare learning ability with innate (genetically determined) behavior as adaptational systems, and give at least one example of their interaction.
7. Discuss the adaptive significance of imprinting.
8. Postulate biological advantages for migration.

Suppose that your instructor were to arm you with a hypodermic syringe full of poison and demand that you find a particular type of insect (which you have never seen and which can fight back), that you inject the ganglia of its nervous system (about which you have been taught nothing) with just enough poison to paralyze your victim, but *not* enough poison to kill it! You would be hard put to accomplish these tasks, but a solitary wasp no longer than the first joint of your thumb does it all with elegance and surgical precision, and without instruction.

The bee-killer wasp *Philanthus* captures bees, stings them, and places the paralyzed insects in burrows excavated in the sand (Fig. 48–1). *Philanthus* then lays an egg on her victims, which are devoured alive by the larva that hatches from that egg. From time to time *Philanthus* returns to her hidden nest to reprovision it until the larva becomes a hibernating pupa in the fall. The next generation of her offspring will repeat this behavior, never having seen it done, but able to do it to perfection.

When *Philanthus* covers a nest with sand, she takes precise bearings on the location of the burrow before flying off again to hunt. There is no way in which knowledge of the location of the burrow could be genetically pre-programmed in the wasp. How to dig it, how to cover it, how to kill the bees—these behaviors appear to be genetically programmed. But since a burrow must be dug wherever a suitable location occurs, the location of the burrow *must* be learned after it is dug. That this is so was determined by the Dutch investigator Nikko Tinbergen.

Tinbergen surrounded the wasp's burrow with a circle of pine cones, on which the wasp took her bearings. Before she returned with another moribund bee, Tinbergen moved the circle of pine cones. The wasp could not find her burrow—the cones no longer surrounded it. Only when the experimenter restored the cones to their original location could *Philanthus* find her burrow successfully!

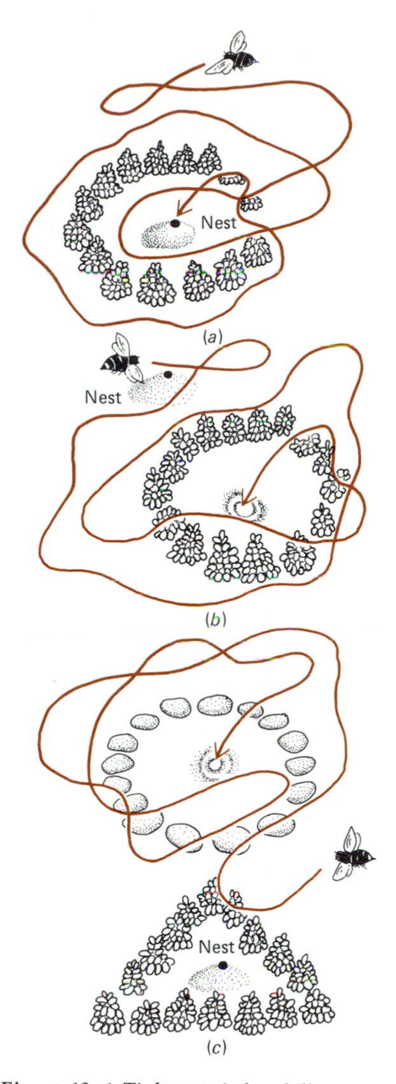

Figure 48–1 Tinbergen's bee-killer wasp experiment. When the ring of pine cones is moved from the position shown in (*a*) to that shown in (*b*), the wasp behaves as if her nest were still located at the center. She has therefore learned its position in relation to the cones. That it is the arrangement of the cones rather than the cones themselves that the wasp reacts to is shown by the substitution of a ring of stones for cones in (*c*). The learning ability of *Philanthus* is very limited but adequate for situations that normally arise in a state of nature. (After Tinbergen, N.: *The Study of Instinct*. Clarendon Press, Oxford, 1951.)

What Is Behavior?

Behavior refers to the responses of an organism to signals from its environment. Notice how efficiently the wasp carried out a complex, though largely genetically programmed, sequence of behaviors. Very little of her behavior had to be learned—that is, gained from experience. In contrast, the very existence of pre-programmed behavior is hard to demonstrate in humans. We owe the complexity of our behavior to a *generalized* ability to learn. *Philanthus'* intelligence, however, is as narrowly specialized as her stinger.

Much of what organisms do can be analyzed in terms of specific behavior patterns that occur in response to stimuli (changes) in the environment. A dog may wag its tail, a bird may sing, or a butterfly may release a volatile sex attractant. Movements of the entire body are involved in some behaviors such as the stalking movements of a cougar, the courtship ritual of a stickleback fish, or the crawling of a snail. Other behavior patterns may involve an organism's not moving, such as a possum playing dead when it spots an enemy, or a young bird remaining motionless on the ground in the presence of a predator. Behavior is just as diverse as biological structure and is just as characteristic of a given species as structure and biochemistry.

In studying behavior, objectivity is needed when looking for the physical and chemical factors that really do produce the actions we observe. Many higher vertebrates have brains similar to those of humans, with centers known in humans to be associated with specific emotions and conscious experiences. These animals may experience states analogous to rage, fear, sexual drive, and the like. However, a balanced, truly scientific approach to behavior must try to read as little human emotion and feeling into that behavior as possible, while keeping in mind that the behaviors of humans and of other animals may have much in common. The persistent problem, to which we will return again and again, is just how much that may be.

Behavior as Adaptation

The study of behavior in natural environments from the point of view of adaptation is called **ethology.** In recent years biologists have come to view behavior as an adaptation, part of the total adaptational package of any organism and as necessary

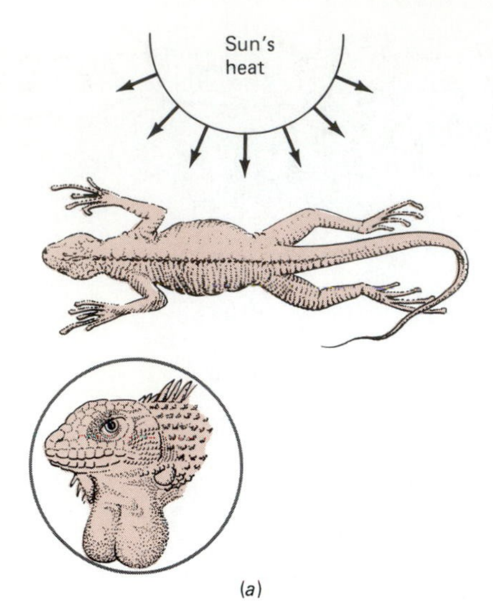

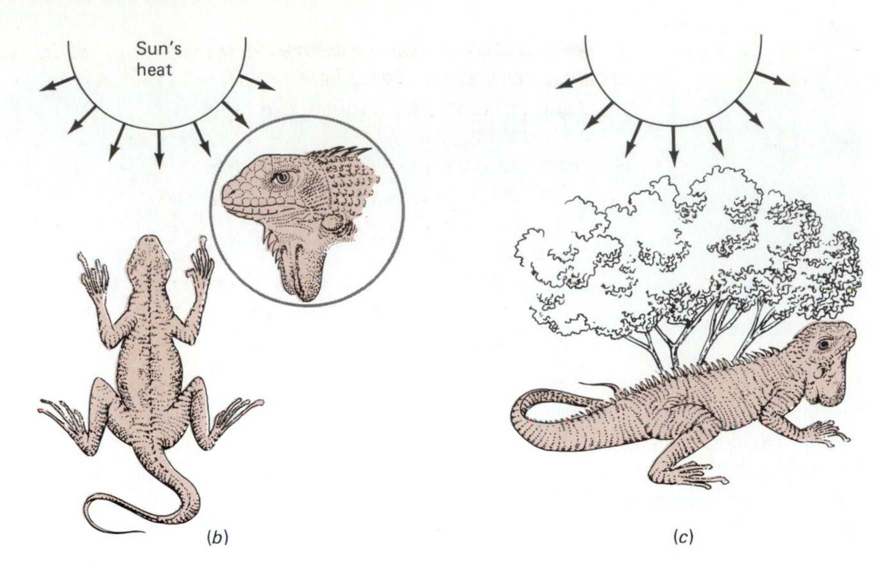

Figure 48–2 Behavioral thermoregulation in a lizard. (*a*) The cold lizard lies at right angles to the sunlight, puffing up its body to increase the surface area available for heat absorption. (*b*) When too warm, the lizard will orient itself parallel to the sun's rays and deflate the body. (*c*) Eventually, it seeks shade. (Illustration concept courtesy of Pam Godfrey.)

for its survival as its heartbeat. The behavior may help the organism obtain food or water, acquire and maintain territory in which to live, protect itself, or reproduce. Certain behavioral responses may even lead to the death of the individual, but will be propagated anyway if, despite this, they increase the chance of survival of the genes that determine them through the survival of the offspring.

Behavior tends to be homeostatic, as well as adaptive. The body of a homeothermic organism has a collection of behavioral as well as physiological responses that help to keep body temperature constant. A bird, when cold, may fluff up its feathers to increase their insulating value. A human may shiver to generate more heat, or perspire when too hot. A dog may pant, cooling the blood in the blood vessels of the respiratory tract.

Many poikilotherms (''cold-blooded'' animals) can regulate their body temperature solely by behavioral adaptations. Lizards, for example, may warm their bodies by basking in the sunlight. To absorb the maximum amount of heat the lizard places its body at right angles to the sun's rays, puffs itself up and spreads out all body membranes (Fig. 48–2). If the lizard becomes too warm, it may then orient the body parallel to the rays of sunlight, decreasing the area exposed directly to sunlight. It may also retract its body membranes and shrink its body as much as possible. If that proves insufficient, the lizard will seek shade, spreading out all body membranes and the body itself to the maximum to radiate excess heat. This behavioral mechanism of thermoregulation is surprisingly effective. It has been shown that lizards infected with dangerous bacteria can maintain their temperatures several degrees above normal. In effect, these ''cold-blooded'' creatures run a fever. The fever and behavior responsible for it can be abolished with aspirin!

Simple Behavior

Even bacteria ''make decisions'': whether to move toward a concentration of food (Fig. 48–3) or away from a toxic substance; toward a region with a certain temperature, or away from it if it is too high or too low. But no bacterium has a nervous system, specialized sense organs, or muscles. How does it sense stimuli and make an appropriate response?

Evidence is accumulating that some proteins such as those responsible for transporting food materials into a bacterium through its cell membrane also function as receptors capable of detecting food and other substances. Some bacteria are sensitive to other stimuli, such as the earth's magnetic field (Fig. 48–3), or light.

Bacteria respond to many stimuli by moving toward or away from them. When the flagella rotate counterclockwise, they rotate together and the bacterium travels in a fairly straight line. Clockwise rotation pulls the bundle of flagella apart and results in the bacterium's dancing in place. Such microscopic dances are known as **twiddles** (Fig. 48–4). Resumption of the counterclockwise movement sends the bacterium in a straight line again, but not necessarily in the original direction. The bacterium responds to gradients of a stimulating substance in the surrounding water. In the presence of a stimulus to which it responds positively, the bacterium employs less twiddling (random motion), so that on the whole it tends to approach the source of the stimulus. Negative stimuli cause the bacterium to reverse this, so that it tends to move away from the source of the noxious material or situation.

Such simple behavior appears to be basically a matter of physics and chemistry, little more complicated, perhaps, than the guidance systems of a military missile. But the physics and chemistry are so organized as to adapt the bacterium actively to the changes that are constantly occurring in its environment. Comparable mechanisms probably govern the phagocytosis of food by amoebas or of bacteria by white blood cells. And even multicellular organisms display simple behavior, some of which—but not all—may be simply explained.

TROPISMS

Plants have neither muscles nor a nervous system, so how can they be said to behave? Yet plants certainly do grow, and time-lapse motion pictures demonstrate stimulus-oriented growth. Growth responses toward or away from a stimulus are known as **tropisms.** As discussed in Chapter 29, plant tropisms depend upon hormones; phototropisms occur in response to light, geotropisms in response to gravity, and thigmotropisms in response to a solid object.

Plant behavior is not limited to tropisms. We have already seen in previous chapters that a fair number of plants employ effectors—such as traplike leaves with which they can capture insect prey! Gordon H. Orians and David F. Rhoades of the University of Washington have gathered evidence that both willows and alders are capable of a unique form of behavior modification: Within each species, individual trees communicate—evidently by airborne chemical cues—the fact that they are being attacked by insects. Trees a considerable distance away from the specimens under attack respond by developing the same kinds of insect-resistant chemicals in their edible parts as those being developed by the trees that are actually being chewed upon. Not only does this appear to be behavior, but it borders upon the kind of pheromone-mediated social behavior that is exhibited by many animals, humans included.

TAXES

Many animals far more complex organizationally than bacteria or plants respond to stimuli in much the same way as do these simple organisms with simple orienta-

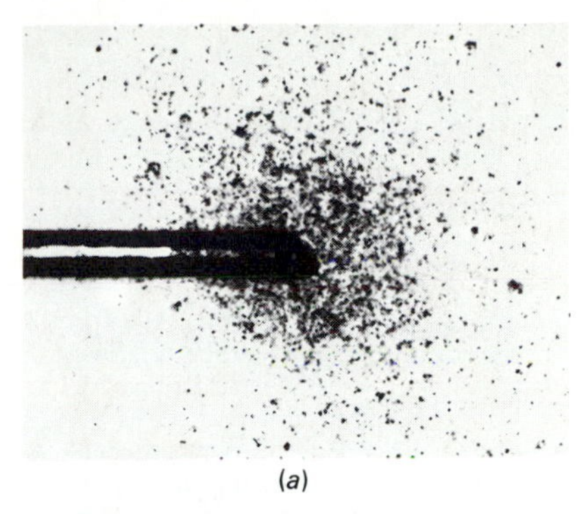
(*a*)

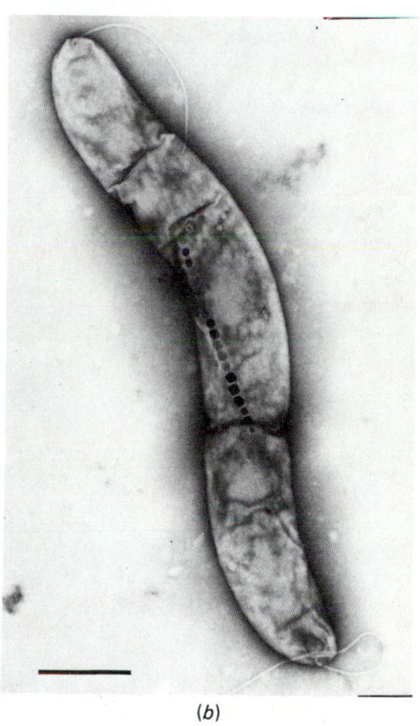
(*b*)

Figure 48–3 Chemotaxis in bacteria. (*a*) In this highly magnified view, bacteria cluster about the opening of a capillary tube containing a chemical that is attractive to them. (*b*) This bacterium contains magnetic granules (line of dots) whose attraction by the earth's magnetic field tells the bacterium which direction is "down" (×62,000). Similar granules occur in honeybees and birds, probably providing them with a built-in magnetic compass (see Fig. 48–5). ((*a*), courtesy of Dr. Julius Adler and *Science.* Copyright 1972 by the American Association for the Advancement of Science. (*b*), courtesy of Denise Maratea.)

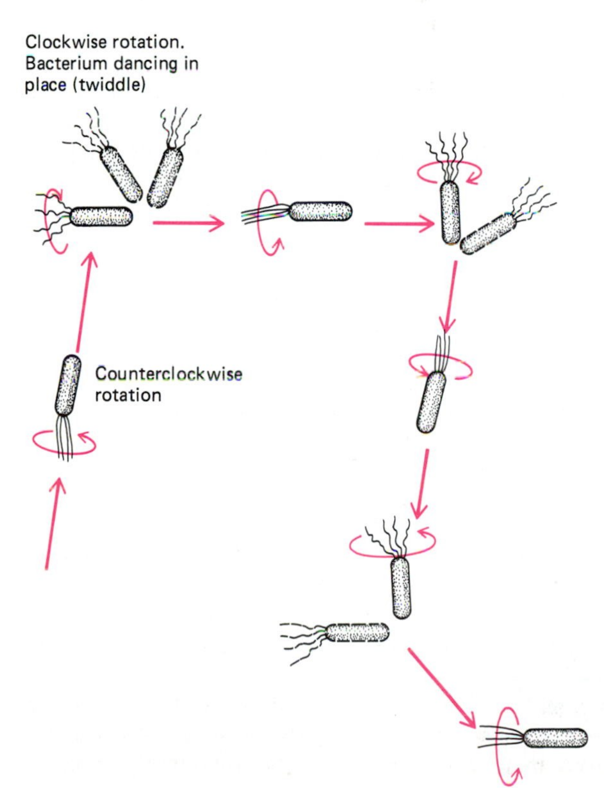

Figure 48–4 Bacteria swim in straight lines when their flagella rotate counterclockwise, but twiddle briefly when the rotation is reversed, for flagellar rotation cannot be coordinated when it is clockwise.

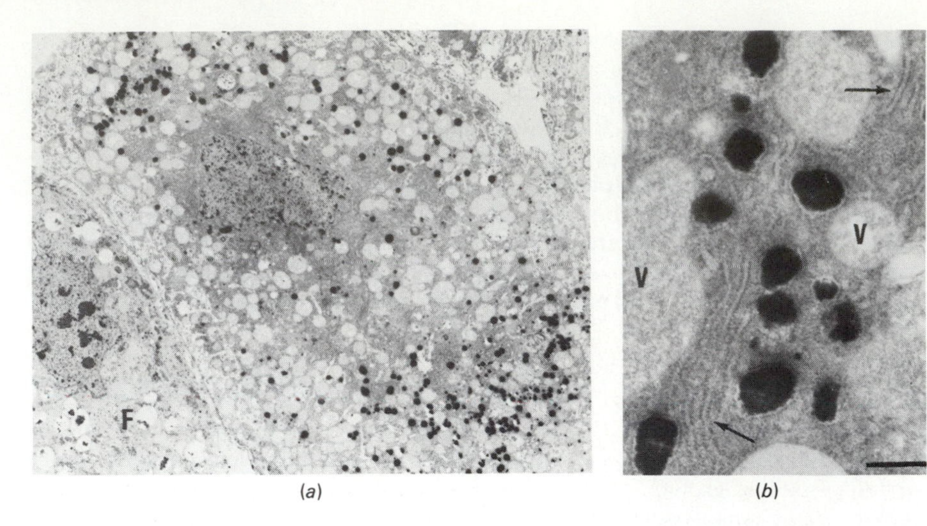

Figure 48–5 Iron-containing cells from the abdomen of a honeybee (*a*). The dark spots are particles of iron oxide. Although magnetic navigation in bees has not yet been demonstrated, evidence suggests that iron oxide might be a component of the compass system used for navigation by other eukaryotes, such as homing pigeons. (*b*) Enlarged view showing iron oxide particles within a single cell. (From Kuterbach, D. A., Walcott, B., Reeder, R. J., and Frankel, R. B.: Iron-containing cells in the honeybee (*Apis mellifera*). *Science* 218:695, 1982.)

tional behaviors known as **taxes.** A taxis generally involves the reception of a stimulus and a movement toward or away from that stimulus (Fig. 48–5). Thus, a positive geotaxis is a movement downward in response to gravity, while a negative phototaxis is a movement away from a light source. Positive chemotaxis has been demonstrated in flatworms. These worms will congregate on a piece of raw meat left overnight in a stream; they need no other stimulus than the chemical cues given off by the meat. If a flatworm is placed in the apparatus shown in Figure 48–6 and meat extract is placed in one of the bottles, the flatworm will swim into the arm of the trough that receives water from the bottle containing meat extract. By adjusting the lighting and the rate of flow from both bottles so that all other possible stimuli are equal on both sides of the apparatus, it is possible to show that the animal is responding to the meat extract alone. Even such complex animals as insects and mammals have a large collection of simple orientational behaviors in their repertoires.

Biological Rhythms and Clocks

We have already considered photoperiodism and the phytochrome pigments in plants, but even in plants, some behavior (such as "sleep" movement in which the leaves fold to conserve water and heat at night) lacks obvious triggers such as physiological imbalance or external cues, and follows a regular daily or longer cycle. In human beings also, physiological processes seem to follow an intrinsic rhythm. Human body temperature, for example, follows a typical daily curve.

It is to an organism's advantage that its metabolic processes and behavior be synchronized with the cyclic changes in the external environment. Cyclic control mechanisms change these processes at repetitive intervals, giving rise to daily rhythms, monthly cycles, or annual rhythms. The activities of many marine animals that live along the shore are linked to the cycle of the tides. Fiddler crabs on the eastern coast of the United States emerge from their burrows to feed at each low tide (twice every 24 h). Certain intertidal snails, on the other hand, release their eggs only at the very high spring tides, which occur twice each month. Many animals migrate to and from breeding grounds twice a year, and many have annual cycles of reproduction or hibernation.

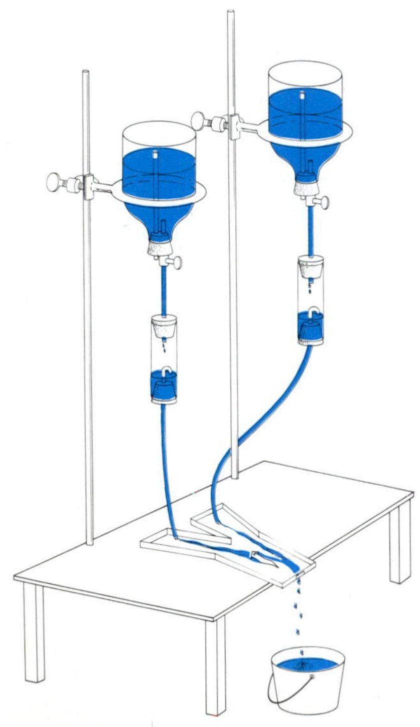

Figure 48–6 A simple maze for studying chemotaxis in a flatworm. The worm will turn left or right depending on the location of a source of an attractive chemical.

LUNAR CYCLES

Some biological rhythms of animals, and perhaps plants, reflect the lunar cycle. The most striking ones are those in marine organisms that are tuned to the changes in the tides due to the phases of the moon. In many externally fertilized invertebrates, it is necessary to synchronize the breeding of all members of a local population to ensure the meeting of their gametes. The swarming of the *Palolo* worm at a particular time of year is governed by a combination of tidal, lunar, and annual rhythms. The Atlantic fireworm swarms in the surface waters surrounding Bermuda for 55 minutes after sunset on days of the full moon during the three summer months. The grunion, a small pelagic fish that lives off the Pacific coast of the United States, swarms from April through June on those three or four nights when the spring tide occurs. At precisely the high point of the tide, the fish squirm onto the beach, deposit eggs and sperm in the sand, and return to the sea in the next wave. By the time the next tide reaches that portion of the beach 15 days later, the young fish have hatched and are ready to enter the sea.

CIRCADIAN RHYTHMS

Periods of activity and sleep, feeding and drinking, body temperature, and many other processes have a cycle approximately 24 hours long. Hence they are called **circadian rhythms** (From Latin *circa,* approximately, and *dies,* day). Some animals are **diurnal,** exhibiting their greatest activity during the day, whereas others are **nocturnal,** being most active during the hours of darkness. Still, others are **crepuscular,** having their greatest activity during the twilight hours, at dawn, or both. If an animal's food is most plentiful in the early morning, for example, its cycle of activity must be regulated so that the organism becomes active shortly before dawn, even though dawn changes slightly from day to day. As the adage says, "The early bird catches the worm."

Certain insects exhibit diurnal variations in pigmentation and continue to show these when placed in continuous darkness. Some spiders show a diurnal rhythm of neurohormone secretion; preparations of nervous tissue isolated from the animal and maintained *in vitro* continue to show this cyclic production of neurohormones. Many of our own physiological processes exhibit circadian rhythms. Our body temperature is highest at 4:00 or 5:00 PM and lowest (as much as 1°C lower) at 4:00 or 5:00 AM. The secretions of a number of hormones, the heart rate and blood pressure, and the rate of excretion of sodium and potassium all show circadian peaks and valleys. Even parasites within the bodies of their hosts have persistent rhythms. The nematode that causes elephantiasis in humans remains in the circulation of the deep tissues during the day and migrates to the peripheral circulation at night. This behavior is synchronized with that of the intermediate host, a nocturnal mosquito that transmits the parasite from one human host to the next.

WHAT CONTROLS THE BIOLOGICAL CLOCK?

Current evidence seems to indicate that there is no single mechanism that operates the biological clock in most organisms. Instead, the interaction of a number of biochemical processes may produce the timed accumulation of certain substances to critical levels. These unknown substances may be responsible for governing behavioral and physiological rhythms. The pineal gland is thought to play a role in the timing systems of rats, birds, and some other vertebrates. Regions of the hypothalamus have been shown to be a part of the biological clock in mammals.

In many organisms the biological clock appears to have an entirely genetic basis. Normal fruit flies, *Drosophila,* have a clock that has a free running period of 24.2 hours. The running period is the clock's repetitive cycle when the animals are isolated from environmental cycles and kept under constant conditions. Mutant fruit flies have been discovered with free running periods of 19 and 28 hours. Each mutation has been traced to the same locus on the X chromosome.

Some investigators hold that biological rhythms are **endogenous,** that is, regulated internally by a biological clock capable of detecting the passage of time. According to this theory, no regular environmental stimulus is needed to keep the clock running. Snails and some other marine organisms whose activities vary with the tide continue to show the cyclic variations in activity when removed to an aquarium and protected from changes in light, temperature, and other factors. This persistence of rhythmic changes in activity, coordinated with the cyclic changes in the environment from which the animal was removed, is strong evidence for the endogenous explanation of biological clocks.

Other investigators argue that biological rhythms are **exogenous,** that is, controlled by environmental stimuli detected by the organisms. Studies suggest that biological clocks often do interact to some extent with *both* external and internal stimuli, and often can be reset by environmental cues. If animals that breed in the spring are transported from the Northern to the Southern Hemisphere, their cycle eventually shifts over to coincide with the occurrence of spring in their new home.

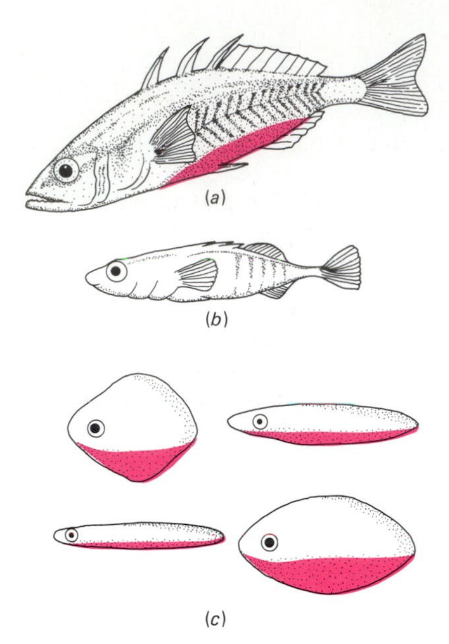

Figure 48–7 A sign stimulus is a particular feature that triggers an innate response. A male stickleback fish, shown in (*a*), will not attack a realistic model of another male stickleback, as seen in (*b*), but will attack another model, however unrealistic, that has a red-colored "belly," depicted in (*c*). Therefore, it is the specific red stimulus, rather than a recognition based on a combination of features, that triggers the aggressive behavior.

Sign Stimuli

Certain segments of the retina of the frog's eye are specialized to react to small, dark, convex objects. The principal item in the diet of frogs is flies, and frogs respond more readily to small, dark, circular moving objects near them than to larger, more diffuse or distant objects. Such **sign stimuli,** also called **releasers,** often serve as triggers for fixed action patterns of behavior (Fig. 48–7). When quick action is essential, as in escaping from a predator, a danger sign is more useful than a detailed description of the danger. Alarm signs, whether they are sights or sounds, are usually simple and contrast sharply to the environment. Small birds typically show an immediate flight reaction to animals with large eyes (understandably, considering that their major predators, owls and hawks, have large eyes). Small birds flee any staring eye, and many moths, the prey of birds, have evolved eye-shaped spots on their hind wings. These "eyes" are concealed by the forewings when the moth is at rest, but spring into view when the moth is disturbed and extends its forewings (Fig. 48–8).

By systematically varying the characters of a stimulus, the ethologist can discover the specific sign. In the spring, male stickleback fish establish territories from which they drive other males. By presenting males with various models, investigators found that very simple models with red undersides were more effective in releasing defense behavior than more realistic models lacking the red underside. Apparently, the red belly of the male is the important sign stimulus responsible for releasing territorial defense behavior in sticklebacks.

The Genetic Basis of Behavior

An insect such as a bee can maintain an elaborate society because the instructions for that society are genetically inherited and pre-programmed. A bee is capable of only the most limited learning—that required by the immediate demands of its environment. The complexity of some genetically pre-programmed behavior is wondrous, but no more so than the genetically predetermined complexity of the anatomy and physiology of any organism.

Although the distinction is not always clear, ethologists recognize two sorts of behavior, instinctive and learned. **Instinctive,** or **innate, behavior** is genetically based; it may be composed of pre-programmed motor patterns. In contrast, **learned behaviors** develop as a result of experience. Some instinctive behavior appears to be functional from the moment that the neural circuitry is in place, and does not seem to be modified by environmental factors. The first web spun by an orb-weaving spider, for example, is complete in all detail and repeatedly built in the same manner throughout the life of the spider.

In many species, instinctive behavior is capable of modification as a result of interaction with the environment. Herring gull chicks peck the beaks of the parents, which then regurgitate partially digested food for the young. The chick is attracted by a red spot on the beak and by the beak's shape and downward movement (sign stimuli). This "begging behavior" is sufficiently functional to get the chick its first meal, but much energy is wasted in pecking. Some pecks are off target and fail to reach the parent's beak. However, the begging behavior becomes more efficient over time, and in experiments with models of the parent's beak, the chicks become increasingly selective of the shape necessary to evoke the begging response. Thus, the initial functional instinct is perfected by environmental interaction.

Behavior is essentially a property of the coordinating mechanisms of the body (that is, of the nervous and endocrine systems). The capacity for behavior is therefore subject to whatever genetic characteristics govern the development and *range of function* of these systems. We may think of behaviors as on a kind of continuous scale ranging from the most rigidly pre-programmed, genetically inherited types through those that are somewhat modifiable, to those that, though containing a genetic component, are extensively developed through experience.

Figure 48–8 A releaser stimulus. (*a*) The moth, when threatened by a predatory bird, abruptly exposes two "eyespots" on its lower wings (*b*). These resemble the eyes of an owl or other carnivore, and apparently trigger an innate avoidance response in the insectivorous bird. Notice that the entire owl need not be present to trigger this behavior—just the specific eyespot stimulus.

Physiological Readiness

Before any pattern of behavior can be exhibited, an organism must be physiologically ready to produce it. Breeding behavior does not ordinarily occur among birds

FOCUS ON

The Genetic Basis of Hygienic Behavior in Bees

Honeybees of the Van Scoy strain are very susceptible to a serious epidemic disease known as American foul-brood, which kills immature bees as they develop in brood cells of the honeycomb. Bees of the Brown strain are less susceptible because worker bees of this strain remove dead larvae from their waxen cells and discard them. This hygienic behavior has two components: (1) the removal of the wax cap of the cell, and (2) the removal of the dead larva. The Van Scoy bees leave the dead larvae to rot. Appropriately, this contrasting behavior is called "unhygienic." The existence of these two variety-specific traits implied that hygienic behavior in honeybees is under genetic control.

In 1964, W. C. Rothenbuhler investigated the genetic basis of the behavior. Rothenbuhler crossed hygienic and unhygienic bees. The F_1 generation was unhygienic, so the hygienic trait was evidently recessive (see figure). Back-crossing these hybrid unhygienic bees with hygienic ones, he obtained four behavioral phenotypes in approximately equal proportions:

1. Worker bees that would neither uncap the cells of dead larvae nor remove the corpses even if the experimenter opened the cells for them.
2. Workers that both uncapped the cells and removed larvae.
3. Workers that opened caps of cells but left larvae untouched.
4. Workers that did not uncap cells but that would remove larvae if they were uncapped by the experimenter.

These results can be explained easily by the hypothesis that the ability to uncap is controlled by a single pair of allelic genes, and that the ability to remove is controlled by an independently assorted pair. Notice that since the total behavior is sequential, when the first part—the uncapping—is complete, it serves as a releaser to trigger the second component—removal. But the appropriate neural pathways can develop in the nervous systems of the insects only if functional copies of the genes are present and unsuppressed by abnormal alleles.

Note that Rothenbuhler did not demonstrate that the total uncapping behavior was completely specified by a single gene, nor that all of the removal behavior was encoded in another single gene. Rather, he showed that there was at least one key gene necessary for the presence of each of these behaviors. Removing that gene was like removing a necessary cog in a machine. Even though the machine becomes nonfunctional, most of it is still there.

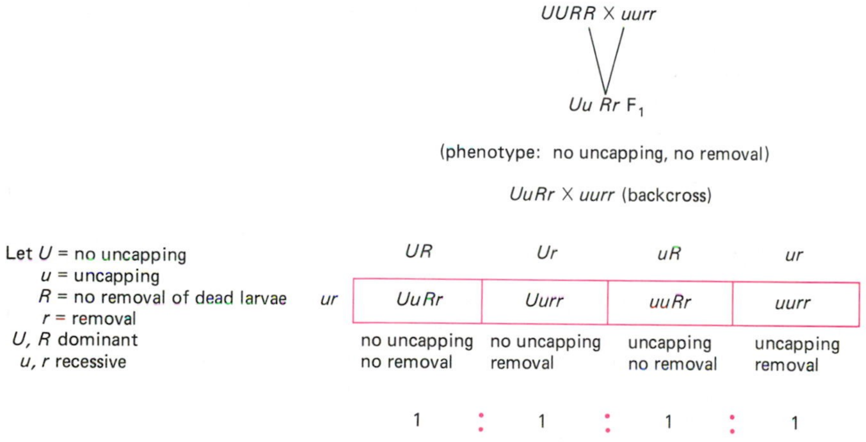

UURR × *uurr*

Uu Rr F_1

(phenotype: no uncapping, no removal)

UuRr × *uurr* (backcross)

Let *U* = no uncapping
u = uncapping
R = no removal of dead larvae
r = removal
U, R dominant
u, r recessive

	UR	*Ur*	*uR*	*ur*
ur	*UuRr*	*Uurr*	*uuRr*	*uurr*
	no uncapping no removal	no uncapping removal	uncapping no removal	uncapping removal
	1	1	1	1

(ratio 1 : 1 : 1 : 1)

Inheritance of hygienic behavior in honeybees. The genes *u* and *r* represent the hygienic traits (see key); *U* and *R* represent unhygienic traits.

or most mammals unless steroid sex hormones are present in their blood at certain concentrations. A human baby cannot walk unless its reflex and muscular development permits it. And yet these states of physiological readiness are produced not only by genetic predisposition but also by a continuous interaction with the environment. The level of sex hormones in a bird's blood may be determined by seasonal variations in day length. The baby's muscles develop in response to exercise; without the trial and error involved in learning how to walk, walking would be retarded.

Perhaps the best example of such interaction between readiness and environment is afforded by the white-crowned sparrow, which exhibits considerable regional variation in its song.[1] This bird, even if kept in isolation, eventually will sing

[1] Different species of birds vary greatly in the extent to which their songs are learned or innate. Among cowbirds, for example, which rely on birds of other species to raise their young, the song is *entirely* innate. Some song birds, on the other hand, are able to learn hundreds of songs in a single breeding season.

a very poorly developed but recognizable white-crowned-sparrow song. However, if it is allowed to grow up under the care of its parents for the first three months of life, when it matures it will sing in the local "dialect" characteristic of its parents or foster parents. If such learning does not take place in those three months, it never will, and if the sparrow consorts with birds of other species, it will *not* learn *their* songs.

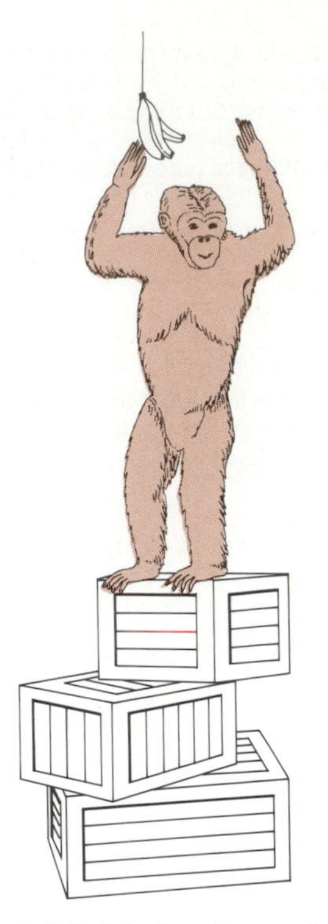

Figure 48–9 Insight learning, and in this case, simple tool use. Confronted with the problem of reaching food hanging from the ceiling, the chimpanzee stacks boxes until it can climb up and reach the food. Many other examples of apparent insight are known from the behavior of these animals.

Learning

Learned behavior can be defined as behavior that is modified as a result of interaction with the environment. The simplest form of learning is **habituation,** learning to ignore repeated stimuli that are not followed by either benefit or obvious cost. Learning capabilities reflect the specialized mode of life that an animal leads. The same rat that has difficulty learning the artificial task of pushing a lever to get an immediate reward learns in a *single* trial to avoid a food that has made it ill as long as 6 hours after the food was eaten. Those who poison rats to get rid of them can readily appreciate the adaptive value of this learning talent to the rat! Such quick aversive learning forms the basis of warning (aposematic) coloration and behavior, found in many poisonous insects and in brilliantly colored but distasteful bird eggs. Once made ill by such foods, the predators avoid them thereafter.

The most complex learning is **insight learning,** the ability to remember past experiences, which may involve different stimuli, and to adapt these recalled events to solve a new problem. Insight learning is most easily demonstrated in primates (Fig. 48–9). An experimental animal can be placed in a blind alley that it must *circumvent* in order to reach a reward (food). The difficulty that must be overcome by the animal seems to be that it must move *away* from the reward in order to get *to* it. If the experimental animal is a dog (a carnivore), at first it typically flings itself at the barrier nearest the food. Eventually, by trial and error, the frustrated dog may find its way around the barrier and reach the reward. A baboon placed in the same kind of situation, however, is likely to "see" the solution immediately. It is the immediacy of this kind of learning that has produced its name—insight.

Learning is widely believed to depend upon changes in the readiness of individual neurons to form circuit relationships with one another and to transmit impulses in preferential directions. In order to learn, an organism's neurons must have a large number of potential interactions with one another; hence, there must be a large number of synapses, not just the few required by pre-programmed behavior. Since innate behavior is really a consequence of the biophysical properties of individual neurons and of the topology of their interconnections, the more narrowly stereotyped a system of behavior is, the more obvious is its genetic control (Fig. 48–10). Yet without the necessary preexistence of the proper neural circuitry that can be interconnected in various ways, even learned behavior would be impossible. Moreover, the kind of learned behavior that the organism typically and most easily develops also depends upon the layout of that circuitry.

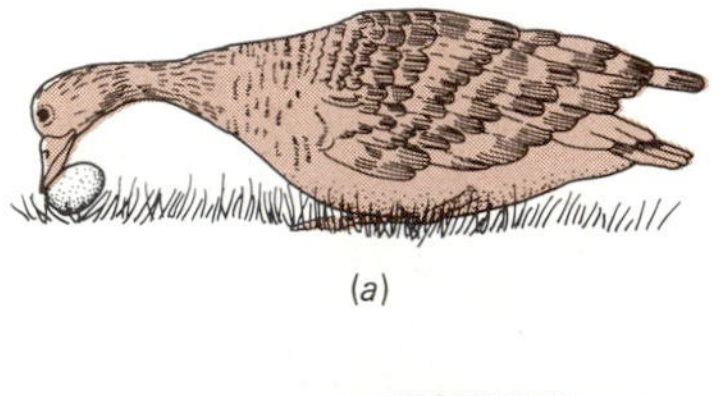

(a)

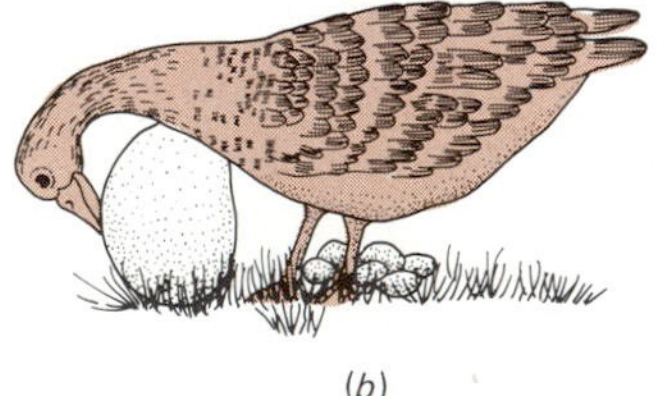

(b)

Figure 48–10 An example of highly stereotyped behavior. (*a*) A greylag goose returning an egg to its nest. The utility of such behavior is obvious; adaptatively, therefore, it seems that the advantage goes to birds who have such behavior pre-programmed, rather than to those birds who would have to learn by trial and error and possibly lose a few eggs in the process. In fact, the behavior needed to retrieve an egg from outside a nest *is* pre-programmed in many ground-nesting birds. (*b*) The stereotypic nature of the behavior of the goose is emphasized by the fact that it will retrieve an egg (or egg-like object) that is not its own. (After Lorenz and Tinbergen.)

Imprinting

Anyone who has watched a mother duck with a swarm of ducklings must have wondered how she can keep track of such a horde of almost identical little creatures, tumbling about in the weeds and grass, let alone tell them from those belonging to another hen[1] (Fig. 48–11). Although she is capable of recognizing her offspring to an extent, basically it is the ducklings that have the responsibility of keeping track of her, which, since there is only one hen per brood, is a far simpler chore; this arrangement is ecologically more efficient in that development of a far more complex maternal behavior—involving the equivalent of counting, perhaps—would be required if responsibility for keeping the ducklings safely together fell on the mother alone. Thus, each duckling stays close to her, like a nail attracted to a magnet. Even if it gets into trouble, it is the duckling that usually must take the initiative. The duckling emits distress cries, and in response the mother will rescue it if she can. Clearly, the survival of the duckling requires an extremely rapid establishment of the behavioral bond between it and its parent. In fact, such a behavioral

[1] All mother birds, not just barnyard chickens, are referred to as hens. Likewise, their young are called chicks.

(a)

(b)

Figure 48–11 The formation of parent–offspring bonds. (*a*) Through imprinting, some young animals follow the first moving object they encounter. Usually the object is their mother, although it is possible experimentally for such infants to imprint upon unnatural objects. Note that in this instance the ducklings are with one, not both, of the two hens. Even though the two hens would probably be indistinguishable to us, the ducklings know which one is their mother. (*b*) In large populations, such as this colony of seals at Cape Cross, Namibia, an offspring that is separated from its parent will probably remain lost and will starve to death. The bond, therefore, between offspring and parent must be quickly established and must be maintained by a complex system of behavioral cues. ((*b*), Mitchel L. Osborne, The Image Bank.)

bond is usual between parent and offspring, especially if the newborn offspring are **precocial,** that is, able to follow the parent about. This form of learning, in which a young animal forms a strong attachment to an individual, usually one of the parents, within a few hours of birth (or hatching), is known as **imprinting.** An early investigator of imprinting, ethologist Konrad Lorenz, discovered that a newly hatched bird will imprint upon a human, or even an inanimate object, if its parent is not present.

If imprinting does not take place during a critical period, it will probably never take place at all no matter what opportunities are later afforded. In the chick this critical period is associated with a temporary increase in the blood concentrations of adrenal steroid hormones in the hatchling. Among many kinds of birds, especially ducks and geese, imprinting is established even before hatching. The older embryos in these species are able to exchange calls with their nestmates and parents right through the porous eggshell. When they hatch, at least one parent is normally on hand, emitting the characteristic vocalizations that the hatchlings have already learned. During a brief critical period after hatching, the chicks learn to associate these vocalizations with the appearance of the parent.

Imprinting establishes the bond between mother and offspring among many mammals, as well as among birds. Among many species the mother establishes a bond with her offspring while the offspring is imprinting upon her. The mother in some species of hoofed mammals will initially accept her offspring only during a few hours after birth. If they are kept apart past that time, the young are thereafter rejected by the mother. Normally this behavior results in the mother's ability to distinguish her own offspring from those of others, evidently by olfactory cues. Thus, by her behavior, the mother insures the propagation of her own genes but not those of an unrelated animal.

Migration

The dramatic seasonal migrations of birds have long excited human curiosity, but it is only quite recently that even its rudiments were properly understood. Until Renaissance times, for instance, the winter disappearance of many migratory birds from Europe was thought to be due to their hibernation in such unlikely places as the muddy bottoms of ponds. Widespread human travel and communication over long distances has led to our modern view that even without any obvious immediate motivation (such as hunger), many animals regularly travel long distances to breed or just to spend certain seasons of the year. Some migrations involve astonishing feats of endurance and navigation. Ruby-throated hummingbirds cross the vast reaches of the Gulf of Mexico twice each year, and the sooty tern travels across the entire South Atlantic from Africa to reach its tiny island breeding grounds south of Florida.

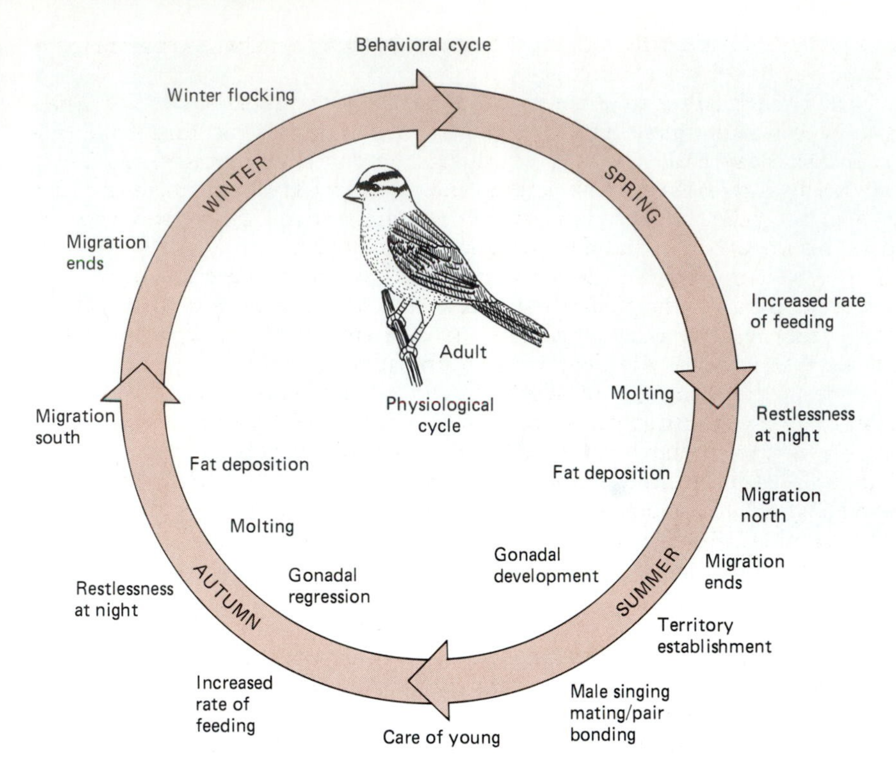

Figure 48–12 Seasonal changes in the physiology and behavior of the white-crowned sparrow. Note the increased rate of feeding and the restlessness (Zugunruhe) that precedes each period of migration. (From Alcock, J.: *Animal Behavior: An Evolutionary Approach,* 2nd ed. Sunderland MA, Sinauer Associates, 1979.)

Though there is no reason to think that migration is purposeful, it can seem carefully planned. Birds may feed heavily during weeks before those food reserves will be needed, and often fly south even *before* the weather turns cold or food becomes in short supply (Figs. 48–12 and 48–13). Salmon, normally marine fish, swim into fresh water toward the end of their life cycles. Monarch butterflies fly southward, and the *next generation* flies north in the spring. The propensity for this behavior must be inherited, and maintained by natural selection. Migration, in other words, must be a specific adaptation in the life styles of many organisms. Yet the adaptive significance of migratory behavior is not always clear. It is obviously to

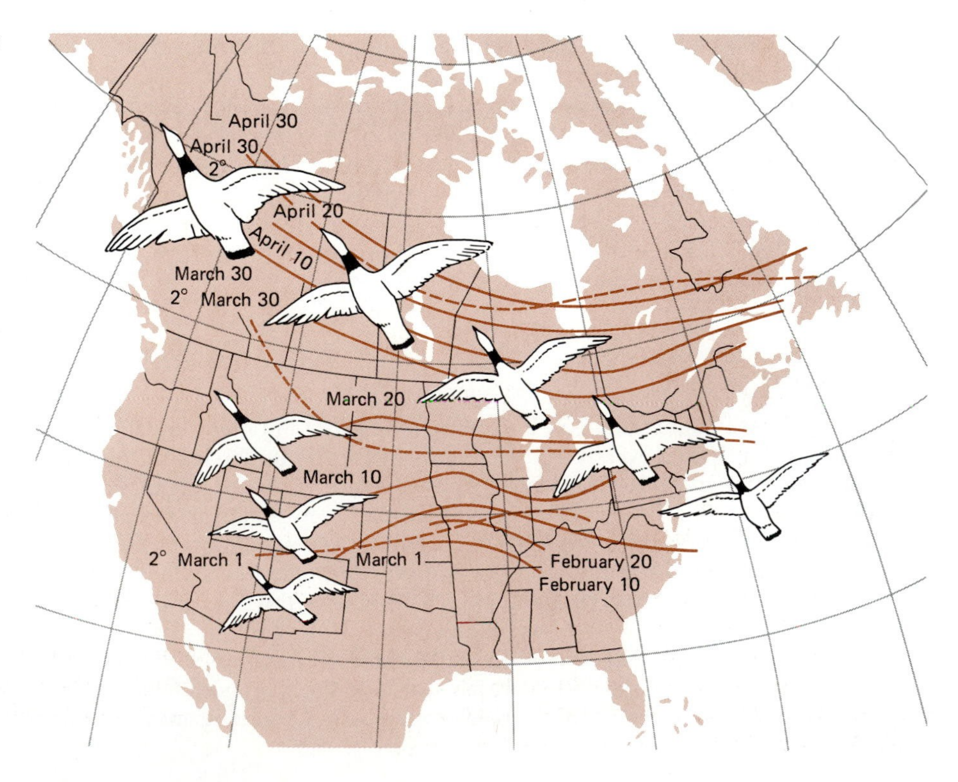

Figure 48–13 The migration north of the Canada goose keeps pace with the arrival of spring in different parts of the North American Continent. The geese are shown following lines that connect different points on the map according to mean temperature of 2°C (35°F). (Modified after Lincoln.)

the advantage of birds to fly south in the winter, but why certain eels migrate to the Sargasso Sea to spawn is a mystery.

The behavioral trigger that sets off migratory behavior varies. Some animals migrate upon maturation. In others, explicit environmental cues trigger the process. In migratory birds, for example, changes in day length are sensed directly through skin and feathers by the pineal gland. These trigger the characteristic restless behavior called **Zugunruhe**—migratory restlessness. The bird evinces increased readiness to fly, and flies for longer periods of time.

The *direction* of travel is also important, and in some cases requires a precision comparable to that of the guidance system of a ballistic missile and this raises the general problem of animal navigation. The cues employed by many animals to negotiate their migratory journeys are not understood.

One of the organisms in which migration *is* fairly well understood is the whitethroat, a small European warbler. Working in the 1950s, Franz and Eleonore Sauer hand-reared a number of these birds (presumably to rule out any possibility of culture-like transmission of information from parent to offspring). When (and only when) the birds could see the star patterns of the night sky, they attempted to fly in the normal direction of migration for this species, a direction that they had no opportunity to learn. When the birds were brought into a planetarium, and the night sky of a different locale was simulated on the planetarium dome, they attempted to fly in a direction that would have been appropriate to take them to their normal wintering grounds from that locality. The conclusion seemed inescapable: Though the direction of migration was unlearned, the birds were able to find it by means of celestial navigation.

Other, similar studies with buntings disclosed that these birds are not guided by *all* the stars, but only by some of the brighter Northern stars. Why Northern? These are easily identifiable from the fact that of all stars, these change position the least as the night passes.

The "need" to migrate and the direction of migration in these birds is unlearned. The star patterns that make navigation possible *are* learned. But how to learn them is innate. The entire mechanism is so constructed, though, that the learned behavior is dependent upon the unlearned, so that under normal circumstances all buntings learn the same thing and behave, for the most part, identically during migration. This intricate interaction of learned and unlearned behavior does have one common theme: that of adaptation. If any component were to fail, the birds would not reach their destination.

SUMMARY

I. Behavior consists of the responses of an organism to signals from its environment.

II. Ethology is the scientific study of behavior under natural conditions from the point of view of adaptation.
  A. Behavior tends to be adaptive.
  B. Behavior tends to be homeostatic.

III. Tropisms and taxes are examples of simple behavior.
  A. Plant behavior consists largely of tropisms, growth responses toward or away from a stimulus.
  B. Taxes are simple orientational behaviors such as moving toward or away from a bright light.

IV. It is adaptive for an organism's metabolic processes and behavior to be synchronized with the cyclic changes in the environment.
  A. Some biological rhythms reflect the lunar cycle, or changes in tides due to phases of the moon.
  B. In many species physiological processes and activity follow circadian rhythms.
  C. No single biological clock has been found; biological rhythms are thought to be regulated by both endogenous and exogenous factors.

V. Released behavior is unlearned and is triggered by a specific unlearned sign stimulus, or releaser. Releasers and the responses they evoke are sometimes arranged in complex chains. Increasingly, behavior originally thought to be entirely innate has been shown to be modified by environmental interaction; that is, the final behavior is affected by learning.

VI. Instinctual behavior is genetic; learned behaviors develop as a result of experience. The actual development of behavior is generally a product of a complex interaction between heredity and environment. Virtually all behavior is modifiable to some extent, and possesses some genetic component or predisposition.

VII. Before any pattern of behavior can be exhibited, an organism must be physiologically ready to produce it.

VIII. Learning can be defined as behavior that is modified as a result of interaction with the environment.

IX. In imprinting, a bond is established between the offspring and mother during a critical period of development.

X. In some birds the need to migrate and the direction of migration appears to be genetically programmed, but how to navigate may be learned.

POST-TEST

1. Behavior may be defined as responses of an organism to ______________.

2. ______________ is the study of behavior in natural environments from the point of view of adaptation.

3. A positive growth response of a plant to gravity is a ______________.

4. The movement of flatworms toward a piece of raw meat is an example of ______________ ______________.

5. A biological rhythm with approximately a 24-hour cycle is a ______________ rhythm.

6. Animals that are most active at dawn or twilight are described as ______________.

7. Another name for a sign stimulus is a ______________.

8. ______________ behavior is mainly genetic; ______________ behavior develops as a result of experience.

9. Precocial offspring are able to ______________.

10. The term Zugunruhe refers to ______________ ______________.

REVIEW QUESTIONS

1. In what ways are the behaviors of *Philanthus,* the bee-killer wasp, adaptive?
2. What is the difference between a tropism and a taxis? Could plants be capable of taxes, or animals of tropisms?
3. What is imprinting and why is it considered a form of learning? What is its significance as an adaptation?
4. How does the response of a gull chick to feeding develop? Which components of the parent–chick interaction are learned and which innate? How do the innate components ensure uniformity of the learned responses?
5. What is Zugunruhe? Describe Zugunruhe and the determination of migratory direction in buntings.
6. Sensitivity to the earth's magnetic field has been observed in birds, bacteria, and possibly bees. What possible reason would bacteria have to be sensitive to this stimulus?
7. Why is it adaptive for some species to be diurnal while others are nocturnal or crepuscular?
8. How does physiological readiness affect instinctual behavior? learned behavior?
9. When Konrad Lorenz kept a greylag goose isolated from other geese for the first week of its life, the goose persisted in following humans about in preference to other geese. How could this behavior be explained?

49
Social Behavior

OUTLINE

LEARNING OBJECTIVES

After you have read this chapter you should be able to:

1. Distinguish between social and other aggregative behavior.
2. Given a description of an animal society, identify the cooperative result of the mutual action of the participating organisms, the suppression of aggression, and the modes of communication employed by the interacting animals.
3. Present the concept of a dominance hierarchy, giving at least one example, and speculate upon its possible general adaptive significance and social function.
4. Define territorial behavior, distinguish between home range and territory, be able to recognize territorial behavior when given an example, and give three theories relating to the adaptive significance of territoriality.
5. Discuss the adaptive value of courtship behavior, and describe a pair bond.
6. Compare the society of a social insect with human society.
7. Define *kin selection,* and summarize its proposed role in the maintenance of insect and other animal societies.
8. Summarize the principal emphases of sociobiology.

In a pioneering work in 1938,[1] W. C. Allee showed that many animals are far more resistant to noxious environments in groups than alone (Fig. 49–1). Schools of fish are less vulnerable to predators than single fish because large numbers tend to confuse their predators. Flocks of birds may be able to find food better than single individuals. Insects are able to construct elaborate nests and raise young by mass production methods when they cooperate. Each member of a pack of wolves or a pride of lions has far greater success in hunting than the individual animal would when hunting alone. Animals that are hunted may be better able to detect or discourage predators when in a group where some individuals are always on watch. It seems clear that social behavior is often a biological advantage.

Figure 49–1 The advantages in numbers. (*a*) This large school of colorful grunts makes it difficult for a predator to focus on and chase any one individual at a time. (*b*) This group of elephants is clustered in what is called a "circle of protection." Large elephants, which few predators would dare attack, stand on the perimeter of the circle, while the young elephants are protected inside. If you look closely, you can see an infant standing at the feet of one of its protectors. (*c*) In many herds of mammals, one or several animals watch for predators while others carry out other activities, such as grazing or drinking. These giraffes were photographed at a waterhole in Namibia. ((*a*), Charles Seaborn; (*b*), (*c*), E. R. Degginger.)

(*a*)

What Is Social Behavior?

The mere presence of more than one individual does not mean that the behavior observed is social. Many factors of the physical environment bring animals together in **aggregations,** but whatever interaction they experience is circumstantial. A light shining in the dark is a stimulus that causes large numbers of moths to aggregate around it. The high humidity under a log may attract aggregations of wood lice, and a water hole in the desert is a common focus for large number of birds and mammals. None of these groups is, strictly speaking, social.

Ethologists generally define **social behavior** as adaptive, conspecific (among members of the same species) interactions (Fig. 49–2). A **society** is a group of individuals belonging to the same species that cooperate in an adaptive manner. That is, they engage in social behavior. A hive of bees, a flock of birds, a pack of wolves, and a school of fish are examples of societies. In highly organized societies, there is cooperation and a division of labor among animals of different sexes, age groups, or castes. A complex system of communication reinforces the organization of the society. The members of a society tend to remain together and to resist attempts by outsiders to enter the group.

Like most behavior, social behavior is often homeostatic, both for the individual organism concerned and for the population as a whole. Even on a cold day, the temperature of the interior of a beehive is likely to be in the range of 33° to 38°C, because the bees huddle together, sharing their meager individual body heats in such a way that the society as a whole is virtually homeothermic. On a hot day, on the other hand, water is brought into the hive by foraging bees. The evaporation of this water, aided by a stream of air fanned through the hive by guard bees, keeps the temperature below that of melting beeswax. This temperature control tends to maintain homeostasis both in the hive and in the individual bees themselves.

[1] Allee, W. C. *The Social Life of Animals.* New York, W. W. Norton, 1938.

(*b*)

(*c*)

(a)

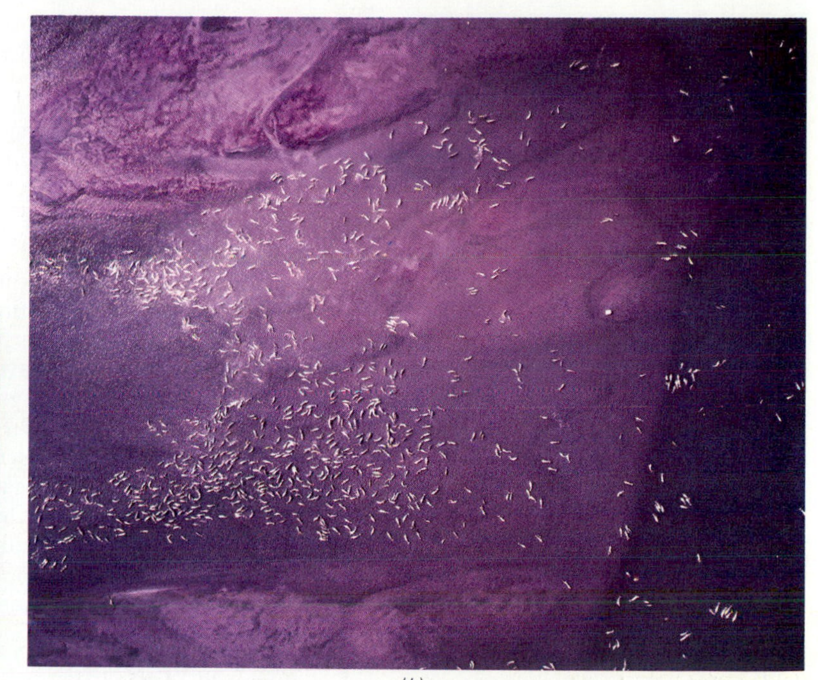

(b)

Figure 49–2 The distinction between aggregative behavior and social behavior is not always clear-cut; however, social behavior implies communication among individuals and often a sharing or division of labor. (*a*) The aposematic (warning) coloration of these unpalatable butterfly larvae, *Eumaeus minyas,* is more effective—that is, visible—when the organisms are present in large numbers. However, what probably brought these numerous individuals together was the deposition of the eggs from which they all developed on a host cycad plant upon which the larvae could continue to feed until they undergo metamorphosis. They continue to stay together not because of a reinforcing system of communication but because of their common attraction to a physical stimulus. (*b*) White whales of the Canadian Northwest Territories are thought to seek out the warmer waters of river mouths to reduce the thermal shock of birth to their newborn young. Which aspects of this behavior would you consider social and which would you consider aggregative? (The unusual color of this photograph results from the use of a special heat-sensitive film used in aerial photography). ((*a*), L. E. Gilbert, University of Texas at Austin/BPS; (*b*), courtesy of J. D. Heyland, Quebec Wildlife Service.)

Communication

The ability to communicate is an essential ingredient of social behavior, for only by exchanging mutually recognizable signals can one animal influence the behavior of another. Communication occurs when an animal performs an act that changes the behavior of another organism. Communication may facilitate finding food, as in the elaborate dances of the bees. It may hold a group together, warn of danger, indicate social status, solicit or indicate willingness to provide care, identify members of the same species, or indicate sexual state, such as maturity and receptivity (Fig. 49–3).

MODES OF COMMUNICATION

Animal communication differs significantly from most human communication in that it is not symbolic. As you read, information is conveyed to your mind by words; yet the words themselves are not the information—they only stand for it. The relationship between the word "cat" and the animal itself is a learned one; a person who could read only Japanese would not recognize it. This is not to say that *signals* in some sense are not employed by animals. In a way, all releasers are signals. However, releasers are not necessarily learned, whereas true symbols are.

Although in humans some body communication is culturally determined learned behavior, a large part of it (such as smiling) is truly universal and appears to be physiologically determined as well as innately understood. The pupil of the human eye dilates in certain emotional situations such as sexual interest or excitement. Without realizing it, people do respond to such subtly transmitted cues. In

Figure 49–3 A male hylid frog (tree frog) of Costa Rica calling to locate a mate. (L. E. Gilbert, University of Texas at Austin/BPS.)

Figure 49–4 A pronghorn antelope, *Antilocapra americana,* marking territory with scent from facial glands. (Harry Engels, Bruce Coleman Inc.)

experiments, a photograph of a woman's face with the pupils retouched to appear greatly dilated was far more attractive to men than one in which they had been shown as pinpoints. Such dilation of the pupils seems to serve as a communication sign of the releaser (or possibly, the conditioned reflex) type in human beings, and one that may bear little relationship to symbolic language.

Signals are often transmitted involuntarily as an accompaniment of the physiological state of the organism. Information about an animal's emotional or mental state may be transmitted even if no other members of the species are nearby. For example, a bird automatically gives an alarm call when it sights a predator. Certainly there are times when we humans would rather not communicate our true feelings—yet there may be instances in which we do not really have any choice, as with pupil dilation. And who has not blushed at a time when he or she would have given almost anything not to have done so?

Do animals ever employ symbols, or are animal signals totally restricted to the equivalent of gasps of alarm? The matter is controversial. Many ethologists feel that even dogs respond to spoken language not as we do, but by deducing the commanded behavior from an astute reading of human facial expression, voice intonation, and bodily attitude. On the other hand, chimpanzees have been taught to speak a very few words meaningfully and, to a limited extent, to use sign language appropriately. Whether apes employ symbolic language in nature is unlikely, although the potential seems to exist.

The singing of birds is an obvious example of auditory communication, serving to announce the presence of a territorial male. Some animals communicate by scent rather than sound. Antelopes rub the secretions of facial glands on conspicuous objects in their vicinity (Fig. 49–4). Dogs mark territory by frequent urination. Certain fish, the gymnotids, use electric pulses for navigation and communication (Fig. 49–5), including territorial threat, in a fashion similar to bird vocalization. As the sociobiologist Edward O. Wilson has said, "The fish, in effect, sing electrical songs." Who would have guessed it?

PHEROMONES

Pheromones are chemical signals that convey information among members of a species. They provide a simple, widespread means of communication. Secretion of pheromones is the only communication mechanism available to unicellular organisms and to many simple invertebrates because other communication channels require rather complex sending and receiving mechanisms. Pheromone communication has been discovered in nearly all organisms studied including plants.

An advantage to pheromone communication is that little energy must be expended to synthesize the simple but distinctive organic compounds involved.

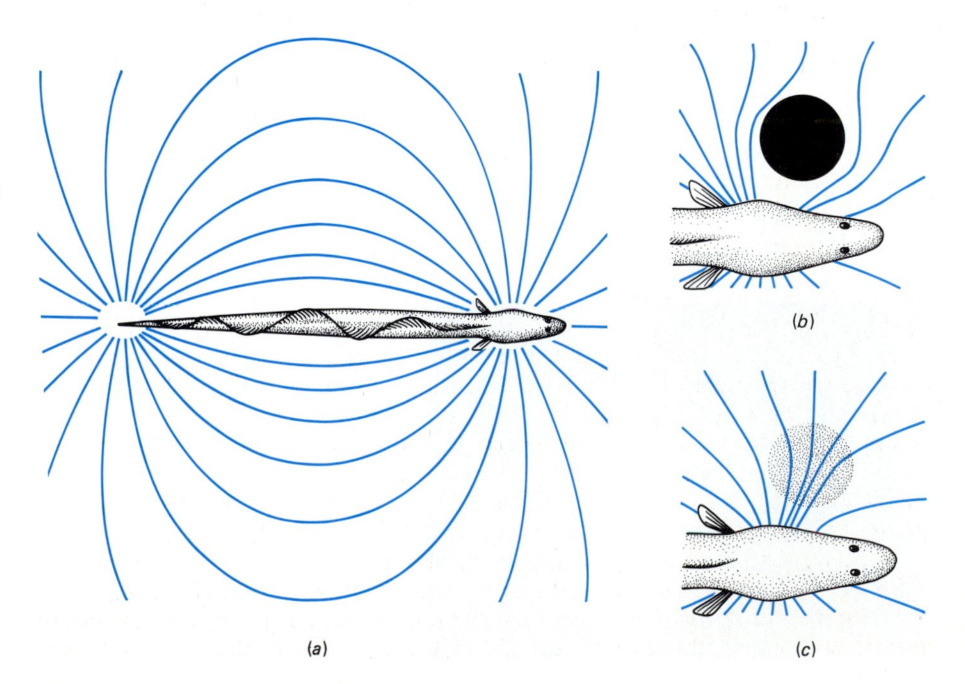

Figure 49–5 Certain fish navigate and communicate through electrical signals. In *Gymnarchas,* an electric field is radiated from its head to its tail, as shown in (*a*). Objects that conduct less (*b*) or more electricity (*c*) than the surrounding water distort the electric field. Special sense organs along the sides of the fish help it use this information to navigate and to locate inanimate objects, rivals, and potential mates.

Conspecific individuals have receptors attuned to the molecular configuration of the pheromone, and it is ignored by other species. Pheromones are effective in the dark, they can pass around obstacles, and they last for several hours or longer.

Major disadvantages of pheromone communication are the slow transmission and limited information content of such substances. Some animals compensate for the latter disadvantage by secreting different pheromones with different meanings. Black-tailed deer release pheromones in their urine and feces, and from glands located on different parts of the feet and head. Pheromones may be deposited as scent markers on the ground or be carried by water or wind. Wind-blown ones are molecules small enough to be volatile, yet large enough to be distinctive.

Many types of signals can be conveyed by pheromones. Most act as releasers, eliciting a very specific, immediate but transitory type of behavior. Others act as primers, triggering hormonal activities that may result in slow but long-lasting responses. Some may act in both ways. In social insects, some releaser pheromones supplement visual and tactile clues in leading members of the colony back to a food source, as in the trail pheromones of ants. Other pheromones enable members of a colony to recognize each other, or warn the colony of foreign intruders. When termites from another nest enter a colony, some members of the resident colony release an alarm pheromone that attracts members of the soldier caste. The soldiers then fend off the invaders.

Pheromones are important in attracting the opposite sex and in sex recognition in many species. Many female insects produce pheromones that attract conspecific males. An example is bombykol, secreted at night by female silkworm moths. Some sex attractant pheromones have been used commercially to help control such pests as gypsy moths by luring the males to traps baited with synthetic analogues of the female pheromone. Male butterflies who mate in the daytime are attracted to any flapping object that more or less resembles a female. After approaching the potential mate, the male engages in "hair-pencilling" behavior: Delicate pencil-like processes covered with pheromones are dangled in front of the prospective mate. If the flapping object is indeed a receptive female of the appropriate species, she will settle down and permit copulation.

Some aspects of the sexual cycle of vertebrates are affected by pheromones. The odor of a male mouse introduced among a group of females causes their estrous cycles to become synchronized. In some species of mice, the odor of a strange male—a sign of high population density—causes a newly impregnated female to abort.

Dominance Hierarchies

In the spring, a nest may be founded cooperatively by a number of female paper wasps that have survived their winter hibernation (Fig. 49–6). During the early course of construction a series of squabbles among the females takes place, in which the combatants bite one another's bodies or legs, and (rarely) sting. In the end, one of the young potential queens gets the upper hand over all the rest, and thereafter she is hardly ever challenged. The paramount queen spends more and more time on the nest, and less and less time out foraging for herself. She takes the food she needs from the others as they return, and if they do not like it, they can leave; some do.

Figure 49–6 A nest of hornets. Like paper wasps, these insects maintain their society for the most part by dominance relationships. (Courtesy Leo Frandzel.)

The paramount queen then begins to take an interest in raising a family—*her* family. Since she is almost always at hand, she is able to prevent other wasps from laying eggs in the brood cells by rushing at them, jaws agape. At the same time, she cannot be stopped from laying all the eggs she wants, since she has already demonstrated that she cannot be successfully challenged. Furthermore, if any of the other females *do* manage to lay, she eats their eggs. *Her* eggs are undisturbed! Those subordinates that stay experience regression of the ovaries and eventually become sterile workers. The female young that have been raised in the nest grow up as sterile workers from the outset, although they retain the potential of becoming reproductively competent in the event that the paramount queen dies.

A careful analysis of this aggressive behavior discloses that the paramount queen can bite any other wasp without serious fear of retaliation. However, within the nest is another wasp that, while usually not a queen, can nevertheless bite any wasp she chooses (other than the queen) without fear of retaliation. Thus, although

Figure 49–7 A "yawn" threat display from a dominant male baboon. (Courtesy of Busch Gardens.)

the queen can bite any wasp in the nest, the other wasps are not equal in their relationships with one another. All the members of the nest can be arranged in a definite **dominance hierarchy,** determined by the status that regulates aggressive behavior within the society:

Queen ⟶ Wasp A ⟶ Wasp B ⟶ . . . Wasp E ⟶ Wasp F

SUPPRESSION OF AGGRESSION

Once a dominance hierarchy is established, little or no time is wasted in fighting. Subordinate wasps, upon challenge, generally exhibit submissive poses that inhibit the aggressive behavior of the queen toward them. Consequently, few or no colony members are lost through wounds sustained in fighting (Fig. 49–7).

Animals that do not exhibit a dominance hierarchy under ordinary conditions may develop such a system when placed in stressful situations such as captivity.

PHYSIOLOGICAL DETERMINANTS

In some animals, dominance is a simple function of aggressiveness, which is itself often influenced directly by sex hormones. Among chickens the cock is the most dominant. If a hen receives testosterone injections, her rank in the dominance hierarchy shifts upward. If she is spayed, the reverse takes place. Recent tests on rhesus monkeys have shown that in dominant males, testosterone levels are much higher than in defeated males. Thus, not only can estrogen reduce dominance and testosterone increase dominance, but dominance may increase testosterone. It is not always easy to unscramble cause and effect!

In many species, males and females have separate dominance systems, and in many monogamous animals, especially flocking birds, the female takes on the dominance status of her mate by virtue of their relationship. However, this is not always the case. Like many fishes, Labrid coral reef fishes are capable of sex reversal. What is odd is that the most dominant individual is always male, and the remaining fishes within his territory are always female. If the male dies or is removed, the most dominant female will become the new male. Should anything now happen to "him," the next ranking female will become the new sultan of the harem. Still other fishes, incidentally, reverse this behavior completely. In them the most dominant fish becomes a female, demonstrating that animal behavior studies must be interpreted with caution, especially when the intent is to make inferences about human behavior.

Dominance hierarchies may be observed in the relationships of one species to another. Perhaps you have seen this at a song bird winter feeder. It has been carefully studied among great blue herons, for example, which are dominant over most other species of herons. Such inherited status minimizes conflict between the various species and therefore might operate to their mutual advantage.

Territoriality

Organisms do not usually crowd together. Virtually all animals, and even some plants, maintain a minimum personal distance from their neighbors, as you may have observed in the even spacing among the members of a flock of birds resting on

Figure 49–8 The coral reef environment allows many secluded areas in which a territorial animal can establish a home range. Among the most territorial of coral reef animals are the moray eels. Eels such as the one pictured here will attack any animal (including a human diver) that might enter too close to its shelter. (Charles Seaborn.)

a telephone line. Often, especially in those of more solitary habits, there is a geographical area that they seldom or never leave. Such an area is called a **home range.** Since the animal has opportunity to become familiar with everything in that range, it has an advantage over both its predators and its prey in negotiating cover and finding food. Some, but not all, animals defend a portion of the home range against other individuals of the same species (**conspecifics**), and even against individuals of other, ecologically similar species. Such a defended area is called a **territory.** The tendency to defend such a territory is known as **territoriality** (Fig. 49–8).

Territoriality is easily studied in birds. Typically, the male chooses a territory at the beginning of the breeding season. This behavior results from high concentrations of sex hormones in the blood. The males of adjacent territories fight until territorial boundaries become fixed. Generally, the dominance of a cock varies directly with his nearness to the center of his territory. Thus, close to "home" he is a lion. When invading some other bird's territory, he is likely to be a lamb. The interplay of dominance values among territorial cocks eventually produces a neutral line at which neither is dominant. That line is the territorial boundary. Bird songs announce the existence of a territory and often serve as a substitute for violence. Furthermore, they announce to eligible females that a propertied male resides in the territory. Typically, male birds take up a conspicuous station, sing, and sometimes display striking patterns of coloration to their neighbors and rivals.

Territoriality among animals may be adaptive in that it tends to reduce conflict, to control population growth, and to ensure the most efficient utilization of environmental resources by encouraging dispersion and spacing organisms more or less evenly throughout a habitat. Certainly, not all forms of territoriality can be shown to be directly sexual in origin. Usually territorial behavior is related to the specific life style of the organism that displays it, and to whatever aspect of its ecology is most critical to its reproductive success. For instance, sea birds may range over hundreds of square miles of open water, but exhibit territorial behavior that is restricted to nesting sites on a rock or island, their resource that is in the shortest supply and for which competition is keenest.

Sexual Behavior and Reproduction

The minimum social contact—and, for many species of animals (for example, certain spiders), the only social contact—is the sex act. For many species that do not otherwise exhibit much social behavior, fertilization and perhaps the rearing of young are almost the only forms of sociality. Let us consider the sex act as a basic example of social behavior, for the elements to which it can be reduced are also the least common denominators of most social behavior.

Like other social relationships, the sex act is adaptive in that it promotes the welfare of the species. It requires (1) cooperation, (2) the temporary suppression of aggressive behavior, and (3) a system of communication. Among some jumping spiders, for example, mating is preceded by a ritual courtship on the part of the male, the effect of which is to produce temporary paralysis in the female. While she is thus enthralled, the male inseminates her. Should she recover before he makes good his escape, he becomes the main course at his own wedding feast. Perhaps in most species, the male makes only a genetic contribution to his offspring. In this instance, whether he appreciates the opportunity or not, the male is able to make

(a)

(b)

Figure 49–9 Courtship displays. (*a*) A male waved albatross from the Galapagos Islands. (*b*) Male and female magnificent frigate birds, from the Galapagos Islands. The brilliant red plumage of the male is an important feature in attracting females and thus plays a part in sexual selection. (*c*) Courtship signals by male fiddler crabs are specific to each species. This sequence of the motion of the large right claw is characteristic of the species *Uca lactea*. ((*a*), D. R. Paulson, University of Washington/BPS; (*b*), E. R. Degginger; (*c*), after Crane, J.: *Zoologica* 42:69–82, 1957.)

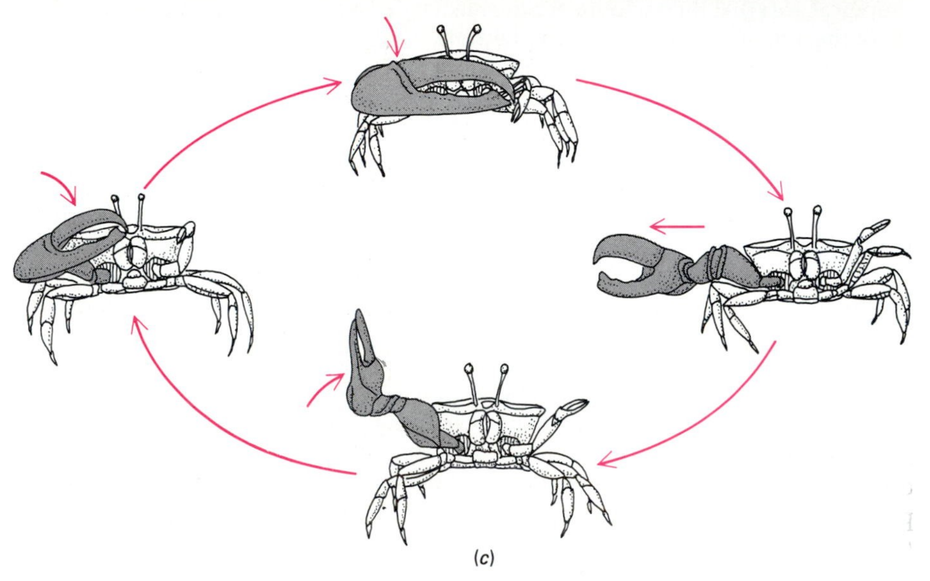

(c)

the ultimate in material contributions to the eggs the female will presently produce. She would otherwise have to bear the metabolic burden of their production all by herself. If, as a result of the male's contribution, more young are successfully produced, he has helped to assure the survival of his genes.

Since an individual that reproduces perpetuates its genes, it is not surprising that natural selection has favored mechanisms, including behavior, that promote successful reproduction. In order to fertilize as many females as possible, the males of many species compete intensely with one another. Sexual competition among males of the same species often has contributed to the evolution of large size, brilliant breeding colors, ornaments, antlers, and other features that give a male an advantage in establishing dominance among his peers and attracting females (Fig. 49–9). It has also led to strategies by which less successful males may occasionally be able to mate. A low-status male in some fish species may mimic the behavior of a female, gain access to the dominant male's nesting territory, and spawn newly laid eggs of the resident female.

Sexual selection has also led to strategies whereby a successful male protects an inseminated female from copulation with other males. After copulation a male damselfly continues to grasp and fly with the female until she has deposited her eggs (Fig. 49–10). A successful drone honeybee discharges much of his genital apparatus into the queen's genital passages (a remarkable chastity belt), thereby blocking them temporarily against insemination by another male.

Since the female usually chooses the mate, epigamic (Greek *epi*, upon, and *gamos*, marriage) selection has affected those male attributes that enhance a male's attractiveness. Selection has also favored female characteristics that enable the female to ascertain that the quality of the male is worthy of her time and energy investment and that the male is indeed a member of the same species. Success of a

Figure 49–10 A male damselfly holding his mate with abdominal claspers while she deposits her fertilized eggs. (Courtesy of John Alcock, Arizona State University.)

Figure 49–11 A female baboon, genus *Chacma,* showing sexual readiness. The swelling of the posterior is partly due to vasodilation, produced in turn by the increased female sex hormones associated with estrus. (Clem Haagner, Bruce Coleman Inc.)

male in dominance encounters with other males is an important indicator to the female of male fitness. Some females accept the first male that comes courting, but other females test the males by provoking encounters. Female baboons and chimpanzees in estrus have enlarged, brilliantly colored genital swellings that attract all males and incite competition among them (Fig. 49–11). A female frog grasped by a small young male during the breeding season may swim off in search of a larger male before shedding eggs.

The victorious male courts the female. A primary function of courtship is to ensure that the male is a member of the same species, but it also provides the female further opportunity to assess the quality of the male. Courtship may also be necessary as a sign signal to trigger nest building or ovulation. Courtship rituals may be long and elaborate. The first display of the male releases a counterbehavior of a conspecific female. This, in turn, releases additional male behavior, and so on until the pair are psychologically and physiologically ready for copulation. Certain male spiders make an offering of food to the female during courtship. This inhibits any aggressive tendencies that the female may have on being approached and also provides the female with some of the food needed for egg production. The male praying mantis often offers his own body, and the female munches on his head while the rest of his body copulates with her (Fig. 49–12).

Figure 49–12 A female mantis eating and copulating with a male. Not all male mantids necessarily "lose their heads" during the courtship ritual; however, the removal of the head eliminates certain inhibitory signals from the brain and causes the male to copulate more vigorously. (E. S. Ross, California Academy of Natural Sciences.)

PAIR BONDS

A **pair bond** is a stable relationship between animals of the opposite sex that ensures cooperative behavior in mating and the rearing of the young (Fig. 49–13). In some species, a newly arrived female is initially treated as a rival male. Then, through the use of instinctive appeasement postures and gestures by both male and female, the initial hostility is dissipated and mating takes place. Such sexual appeasement behavior may be very elaborate and gives rise to mating dances in some birds.

The releaser mechanisms involved in the establishment and maintenance of the pair bond are often remarkably detailed. A male flicker possesses a black mustache-like marking under the beak. This is lacking in the female. If a "happily married" female flicker is captured, and such a mustache painted on her, her mate

Figure 49–13 A pair of nesting albatrosses. In many species, pair bonds are maintained by grooming or other displays of affection. (E. R. Degginger.)

will vigorously attack her as if she were a rival male. He will accept her again if it is removed. Such cues enable courtship rituals to function as behavioral genetic isolating mechanisms among species.

CARE OF THE YOUNG

Care of the young is an additional component of successful reproduction in many species, but it too requires a parental investment (Fig. 49–14). The benefit of parental care is the increased likelihood of the survival of the offspring, but the cost is a reduction in the number of offspring that can be produced. Because of her time spent carrying the developing embryo, the female has more to lose than the male if the young conceived do not develop. Thus, females are more likely to care for eggs and young than males, and usually the females invest more in parental care.

A high investment in parental care is usually less advantageous to a male, for time spent in parenting is time lost from inseminating other females. With a given female, which of her mating partners fathered the offspring may not be certain; raising another male's offspring is to the male's genetic disadvantage. Perhaps for this reason, when a male lion is victorious over a rival, he not only takes over the rival's former harem but also kills any cubs that may have been fathered by his predecessor. Under some circumstances, however, it may be to the species' advantage for the male to help rear the young: Receptive females may be few and far between; only a few young of certain paternity may be produced; food and other resources may be scarce and require more effort than a single pair of parents can provide; and all young may need more protection against severe predation than a single parental pair can give them.

Play

Play is an important aspect of the development of behavior in many species, especially young mammals. It serves as a means of practicing adult patterns of behavior (Fig. 49–15), or perfecting means of escape, prey killing, and even sexual conduct. In true play, the behavior may not be actually consummated. Thus, a kitten pounces upon a dead leaf but of course does not kill it, even though it administers a typical carnivore neck bite. When interacting with a littermate, the same kitten may practice the disemboweling stroke with its hind claws, but the littermate is not intentionally injured in the process.

(*a*)

(*b*)

Figure 49–14 Examples of parental investment. (*a*) Cougars and black bears are normally mortal enemies and ordinarily actively avoid each other. This confrontation was initiated when the cougar intruded in the area where a female bear was raising her cubs. (*b*) A young albatross begging food from a parent. Such an investment of time and energy on the part of the parent does not benefit the parent directly. However, it does help ensure the transmission of the parent's genes into succeeding generations. Such "selflessness" of individuals and "selfishness" of genes is discussed again later in the chapter. ((*a*), E. R. Degginger; (*b*), Paul Ehrlich, Stanford University/BPS.)

Figure 49–15 Young lions playing in southern Africa. Play is behavior that is not consummated, and which often serves as a means of practice behavior that will be used in earnest in later life, possibly in hunting, fighting for territory, or competition for mates. (Susan McCartney, Photo Researchers, Inc.)

Displacement Activity

When an animal's interests are in conflict, it may behave in a manner completely irrelevant to the stimuli. For example, when a male confronts an opponent and cannot decide whether to fight or flee, he may at first perform an entirely irrelevant act. When fear and aggression are balanced, gorillas will nibble on a leaf, birds may peck at the ground, and people may scratch their heads. Irrelevant behavior stemming from a conflict situation is known as **displacement activity** (Fig. 49–16). Displacement activity forms a significant part of many courtship rituals, which are often rather tense situations even among humans, and may involve the opposition of many needs among animals.

Displacement activity should not be confused with **redirected behavior.** True displacement activity does not resemble any of the behaviors that would otherwise result from any of the conflicting needs, so that in displacement, behavior is transformed. In redirection, behavior remains the same but finds another, often inappropriate object. For example, if an animal cannot "decide" whether or not to attack a threatening conspecific, it might redirect its aggression against a less formidable opponent.

Elaborate Societies

Some animal societies exhibit elaborate and complex patterns of social interactions. In such societies there is considerable division of labor not directly connected with the care of the young.

INSECT SOCIETIES

Although many insects cooperate socially (tent caterpillars, for example, spin a communal nest), the most elaborate insect societies are found among the ants, wasps, and termites (see Fig. 25–27).

Insect societies are held together by an elaborate system of releasers, and in consequence tend to be quite rigid. Many social insects secrete pheromones that

Figure 49–16 Displacement activity. When two forms of behavior (such as fighting or fleeing) are equally appropriate, an organism may engage in some completely irrelevant behavior, as with this gray timberwolf chewing on a branch. (Robert J. Ashworth, Photo Researchers, Inc.)

Figure 49–17 A queen honeybee on comb. Note the numerous workers that surround her. They constantly lick secretions from the queen bee. These secretions are transmitted throughout the hive and act to suppress the activity of the workers' ovaries. (R. K. Burnard, Ohio State University/BPS.)

accomplish such tasks as suppressing the ovaries of worker honeybees or alerting an ant hill to the presence of an enemy. (An "alarm substance" is given off from a special abdominal gland by excited worker ants.) Virtually all the intercommunication within insect societies is comparable to the hormonal communication within a single organism. The social coordination achieved approaches the behavior of the cells of a loosely organized individual organism of about the level of organization of a sponge.

The social organization of honeybees has been studied more extensively than that of any social insect. A honeybee society generally consists of a single adult **queen,** up to 80,000 **worker bees** (all female), and at certain times, a few males called **drones** that fertilize newly developed queens (Fig. 49–17). The queen deposits as many as 1000 fertilized eggs per day in the wax cells of a comb (up to 1 million in her lifetime). Each fertilized egg develops into a larva that is fed by worker bees. After about 6 days, the workers seal the cell off with wax, and the larva develops into a pupa. About 12 days later an adult worker bee emerges from the wax cell.

Division of labor in the bee society is mostly determined by age. The youngest hive bees serve as nurse bees. They have special glands on the head that secrete "royal jelly" essential for the early nutrition of all larval bees, and which, in larger quantities, produces queens. When larvae are about 3 days old, the workers begin feeding them a mixture of honey and pollen (called beebread). After about a week as nurse bees, workers begin to produce wax and build and maintain the wax cells. Older workers are foragers, bringing home the nectar needed to make honey (a half-teaspoonful per bee lifetime) and pollen (the larva's protein source). Most worker bees die at the ripe old age of 42—days, that is. The queen may live as long as 5 years.

Behavioral cues tip off the bees when there is a labor shortage in any category. Thus if there are too many larvae for the nurse bees, foragers will pitch in for the duration of the emergency, redeveloping royal jelly glands, if need be.

The composition of a bee society is controlled by an anti-queen pheromone secreted by the queen. It acts as a releaser, inhibiting the workers from raising a new queen; it also has a primer effect: It inhibits the development of the ovaries in the workers. If the queen dies, or if the colony becomes so large that the inhibiting effect of the pheromone is dissipated, the workers begin to feed some developing larvae only royal jelly, the special food that promotes their development into new queens. If more than one queen hatches at the same time, they fight until one stings the other to death.

The most sophisticated known mode of communication among bees is a stereotyped series of body movements known as a **dance.** If a worker runs across a rich source of nectar, it can communicate this fact to the other bees within the hive by dancing on a vertical comb surface. If the food supply is nearby, the bee performs a **round dance** (Fig. 49–18), which generally excites the other bees and causes them to fly about in all directions till they have found the source of nectar. But if the source is distant, the bee performs a **waggle dance.** This "step" has a figure-eight configuration. As the bee treads the long axis of the figure-eight, she emits a series of distinctive sounds and waggles her abdomen from side to side. The angle between the straight long axis of the figure-eight and the force of gravity is the same as the angle between the sun's rays and the shortest flight direction to the food source (Fig. 49–19). The bee is also able to communicate other information during the

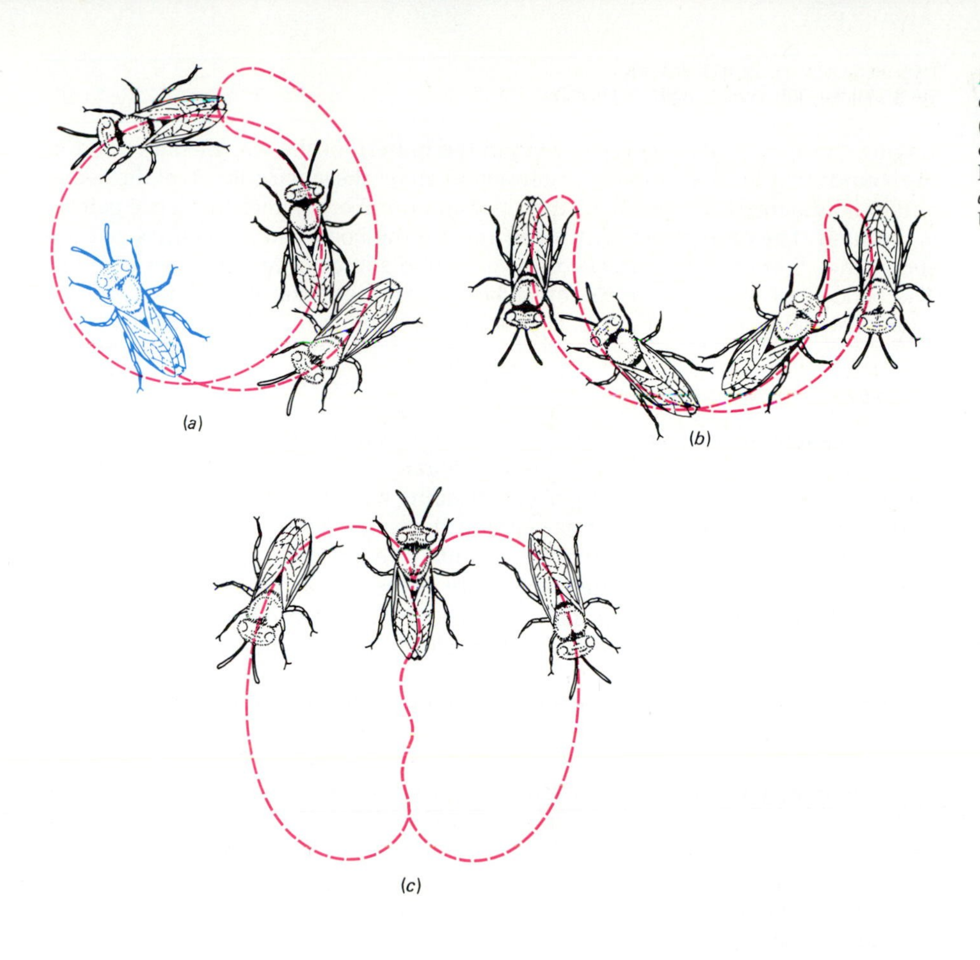

Figure 49–18 Three kinds of communication dances performed by honeybees. (*a*) The round dance. (*b*) The sickle dance. (*c*) The waggle dance. (From von Frisch, K., *The Dance Language and Orientation of Bees.* Cambridge, Harvard University Press, 1967.)

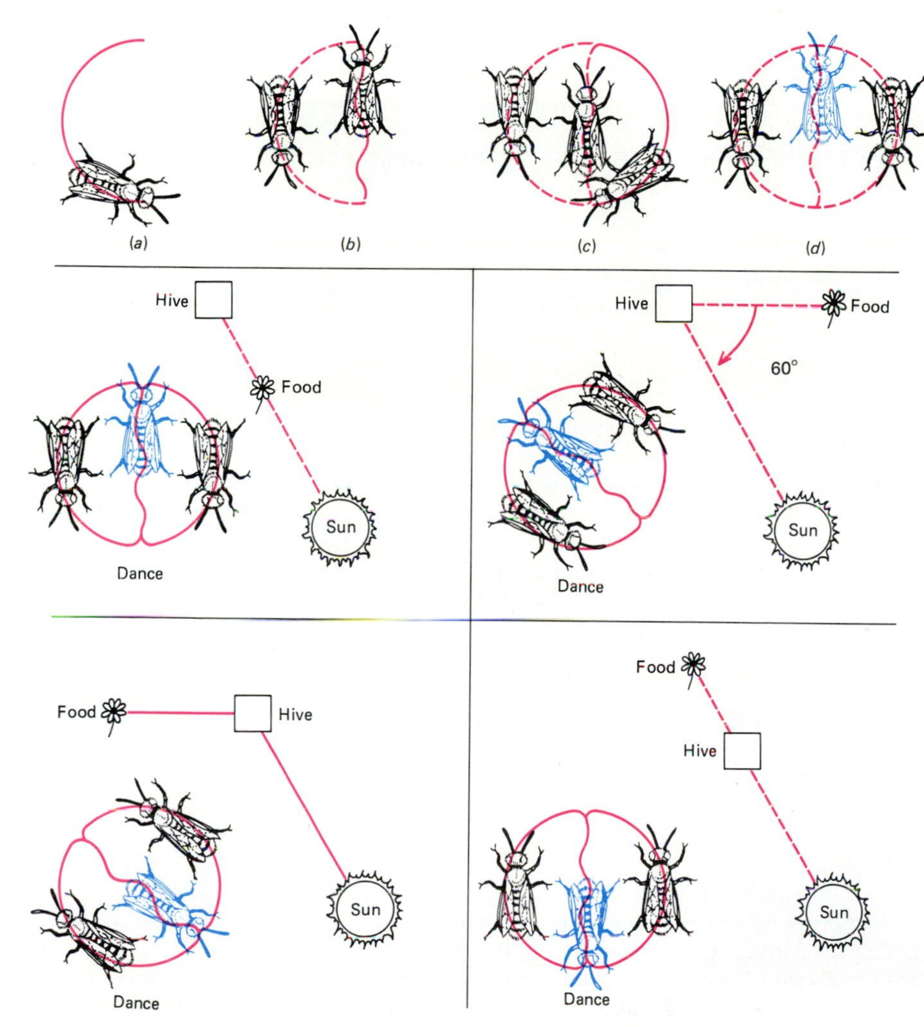

Figure 49–19 Indication of direction by the waggle dance.

waggle dance. The distance is encoded in the pattern of the waggle dance and in the sounds that the bee produces; information about the sugar content of the nectar and its abundance is passed on by the vigor and number of times the dance pattern is repeated. The odor of the plant's perfume on the body and fur of the scout bee serves the others as confirmation of the indicated source once they have found it. The behavioral repertoire of the honeybee is extensive. Yet it is almost entirely innate, perhaps specified by the sequence of bases in the DNA of the cell nuclei.

VERTEBRATE SOCIETIES

Among vertebrate societies we find a far greater range and plasticity of potential behavior than among insect societies. Vertebrate societies are far less rigid and much more adaptable to changing needs. In some ways they are also less complex. Vertebrate societies usually contain nothing comparable to the physically and behaviorally specialized castes of termites or ants. What is more, except for humans, individual members of vertebrate societies are not as specialized in their tasks as are the social insects. A beehive may be considered collectively a kind of superorganism, but it is hard to interpret a wolf pack or a herd of red deer in that fashion (Fig. 49–20).

Whereas the elaborate societies of social insects result from genetic programming of their behavior, human society results from **culture,** behavior that is symbolically transmitted from generation to generation. We humans do have a genetically determined capacity for culture. However, the medium for cultural transmission is not the information contained in DNA but the information contained in language. The principal distinguishing characteristics of human societies are (1) reliance upon culture, (2) symbolic language, and (3) use of tools.

Some analogies to culture do exist in some animal societies. For example, among Japanese macaques, patterns of eating candy and washing sweet potatoes have been maintained in experimental populations by nongenetic, "cultural" transmission.

(a)

(b)

(c)

Figure 49–20 Social interactions among primates. (*a*) Nit-picking (the removal of small parasites) is not only a hygienic social action but a mechanism of reinforcing bonds among closely related infants and adults. The animals shown here are Java monkeys. (*b*) Most primate young have a long period of dependency on their parents. During this time numerous social and even "cultural" skills are learned from the parents. This lar gibbon is shown with its five-month-old baby. (*c*) Baby baboons ride on the mother's back during early infancy and come to inherit some of her social status. ((*a*), E. R. Degginger; (*b*), Margarette Mead, The Image Bank; (*c*), courtesy of Busch Gardens.)

Figure 49–21 Prairie dog. The rodents live in large colonies in which a few act as sentries. Though it places the life of the sentry in grave danger to expose itself outside its burrow, the sentry acts to protect its siblings and by so doing helps ensure that the genes they all have in common will be perpetuated in the population. (Courtesy of Tina Waisman.)

Kin Selection

Altruistic behavior, in which one individual appears to behave in such a way as to benefit others rather than itself, is frequently observed in the more complex social groups (Fig. 49–21). A particularly clear case of altruistic behavior has been observed by biologists Watts and Stokes in the mating of wild turkeys. Several differing groups of males, in each of which there is a dominance hierarchy, gather in a special mating territory and go through their displays of tail spreading, wing dragging, and gobbling in front of females who come to the area to copulate. One group attains dominance over other groups as a result of cooperation among the males within the group. The dominant member of the dominant group is the one to copulate most frequently with the females; seemingly, the males who helped establish the dominant group but have low status within it gain nothing. Close analysis, however, has shown that members of a group are brothers from the same brood. Since they share many genes with the successful male, they are indirectly perpetuating many of their genes. In this case, altruism is closely related to **kin selection,** vicarious gene propagation among closely related individuals.

Kin selection may account for the evolution of the complex societies of social insects in which some individuals specialize for reproduction and other close relatives do the chores of the colony. In the bee society, the workers are sterile females, and the queen functions vicariously as their reproductive organ. If the queen successfully produces offspring, at least a large portion of the genes shared by the queen and worker will have been passed on to the next generation, even though the worker herself has not reproduced.

This is particularly true of bees, wasps, and ants. Since male bees are haploid, each male passes on his *entire* complement of genes to his offspring without the intervention of meiosis, and all sperm are genetically identical. Thus each female bee has three fourths of her genes in common with her sisters, instead of the more usual half. This is especially true since the queen bee mates infrequently during her lifetime, being able to store sperm for long periods of time between copulations. Every bee alive in a hive at the same time may well have the same father. Expressed more exactly, any given worker bee will probably have half of her genes derived mitotically from the same father and therefore identical with half of the genes possessed by any other worker in the hive—plus half of the *remaining* half in common with any other worker in the hive that had, as it probably did, the same mother. That gives all workers a **coefficient of relatedness** of 75%, that is, 0.75.

Among organisms in which both sexes are diploid, such as humans, there is only a 0.50 coefficient of relatedness between parent and child. To be sure, it is not important how closely related bee workers may be to one another, since they never have offspring in the normal course of events. However, sooner or later some of those larvae being reared so assiduously by the worker bees will become drones, and another will become a new queen. These sexual forms *will* pass on those genes to the next generation. Apparently it is more efficient for the worker to propagate her genes by proxy and the mass production of larvae than it would be for her to lay her own eggs and go through the complete process of raising them to maturity herself.

Kin selection is not limited to social insects. Among some birds (Florida jays and others), nonreproducing individuals aid in the rearing of the young. Nests tended by these additional helpers as well as parents produce more young than nests with the same number of eggs overseen only by parents. By helping to care for their siblings' children, therefore, these individuals have a better chance of ensuring that at least some of their genes (the genes shared with their siblings) will be maintained in future populations. (Helpers also stand a good chance of later "inheriting" the parents' territory.)

Sociobiology

Sociobiology is the school of ethology that focuses upon the evolution of social behavior through natural selection. It is a synthesis of population genetics, evolution, and ethology. Like many biologists of the past (such as Darwin), Edward O. Wilson and other sociobiologists emphasize the animal roots of human behavior, but in accordance with modern thinking, they have attempted to place their discipline firmly in the area of population genetics, with particular emphasis on the effect of kin selection on patterns of inheritance. Many of the concepts discussed in this chapter, such as altruism and paternal investment in care of the young, are based on contributions made by sociobiologists.

For the sociobiologist, the organism and its adaptations—including its behavior—are ways its genes have of making more copies of themselves. The cells and tissues of the body carry out functions that ensure the functioning of the reproductive system. The reproductive system's job is the transmission of genetic information through time to succeeding generations.

As might be guessed, there are unique pitfalls in attempting to reconstruct the evolution of behavior, because behavior rarely leaves an explicit fossil record. Additionally, by applying human social terms to behavior in animals that may be only superficially similar, we create the perhaps entirely false impression that it is the *same* behavior. It is an easy step from that to the assumption that the causes and utility of these behaviors are the same as those of corresponding human behavior. Consider, for instance, the question of whether humans are territorial. We do tend to preserve space between us as individuals, to defend our homes, and as groups to defend larger, political areas. However, do these behaviors have the same genetic and adaptive value in humans as that in animals? And is human territoriality homologous with that of other animals, or is it merely analogous?

Also, problems of objectivity can exist. Any *a priori* assumptions we may have about our own territoriality can have the effect of our looking at the behavior of animals as a mirror of ourselves. Among closely related species of primates, social organization and the degree of territoriality and aggressive behavior vary widely. Which of these species should we choose as models for studying human behavior?

Most of the controversy that has been triggered by sociobiology seems related to its possible or alleged ethical implications. Sociobiology is often taken as denying that human behavior is flexible enough to permit substantial improvements in the quality of our social lives. Yet sociobiologists do not disagree with their critics that human behavior is flexible. The debate therefore seems to rest on the *degree* to which human behavior is genetic and the *extent* to which it can be modified.

As sociobiologists acknowledge, people through culture possess the ability to change their way of life far more profoundly in a few years than a hive of bees or a troop of baboons could accomplish in hundreds of generations of genetic evolution. This very ability is indeed genetically determined, and that is a very great gift. But how we use it and what we accomplish with it is not a gift, but a responsibility upon which our own well-being and the well-being of other species of the biosphere depend.

SUMMARY

I. Social behavior is adaptive conspecific interaction. A society is a group of individuals of the same species that cooperate in an adaptive manner.
   A. In a society there are a means of communication, cooperation, division of labor, and a tendency to stay together.
   B. Animals form societies because it is adaptive for them to do so.
   C. In aggregations, interaction is incidental; organisms are drawn together coincidentally by a common environmental stimulus.

II. Animal communication involves the transmission of signals but does not utilize (so far as is known) symbolic language in the human sense.
   A. Animals often transmit signals involuntarily as a result of their physiological state.

B. Pheromones are chemical signals that convey information between members of a species.

III. Dominance hierarchies result in the suppression of aggressive behavior.

A. Dominance relationships may be physiologically or socially determined.

B. One species may establish dominance over another.

IV. Organisms often inhabit a home range, from which they seldom or never depart. This range, or some portion of it, may be defended against members of the same (or occasionally different) species.

A. Defended areas are called territories, and the defensive behavior is territoriality.

B. Often territorial defense is carried out by display behavior rather than actual fighting.

V. Courtship behavior ensures that the male is a member of the same species, and permits the female to assess the quality of the male.

A. A pair bond is a stable relationship between a male and female that ensures cooperative behavior in mating and rearing the young.

B. Parental care increases the probability that the offspring will survive. A high investment in parenting is less advantageous to the male than the female.

VI. Play gives the young animal a chance to practice adult patterns of behavior.

VII. Insect societies depend upon releasers and therefore tend to be rigid, with the role of the individual narrowly defined.

A. In a bee society there is a marked division of labor.

B. Bees communicate by means of body movements known as dances.

VIII. Vertebrate societies are far less rigid than insect ones. Though innate behavior is important in them, in general the role of the individual is learned.

A. Human society is by far the most complex of all vertebrate societies. It is almost uniquely characterized by the possession of language, culture, and tool use.

B. Some analogies to culture exist in some animal societies.

IX. In altruistic behavior one individual appears to behave in such a way as to benefit others rather than itself.

A. Altruism may be closely related to kin selection.

B. Kin selection may account for the evolution of complex societies of social insects in which only a few members reproduce.

X. Sociobiology is a school of ethology that focuses on the evolution of behavior through natural selection.

POST-TEST

1. Adaptive conspecific interactions are known as ____________ ____________.

2. A ____________ is a group of individuals belonging to the same species that cooperate in an adaptive manner.

3. A swarm of flies in a cow pasture is an example of an ____________.

4. An important difference between human and animal communication is that animal communication is not generally ____________.

5. ____________ are chemical signals that convey information between members of a species.

6. An arrangement of members of a population by status is called a ____________ ____________.

7. The geographical area that members of a population seldom leave is the ____________ ____________.

8. Territoriality tends to reduce ____________ and control ____________ growth.

9. A ____________ ____________ is a stable relationship between animals of the opposite sex which ensures cooperative behavior in mating and rearing the young.

10. In a bee hive the youngest bees serve as ____________ bees; they secrete ____________ ____________.

11. The extensive behavioral repertoire of the bee is almost entirely ____________ (*genetic/learned*).

12. Human society differs from other animal societies in that it depends mainly on the transmission of ____________.

13. In ____________ behavior one individual appears to act to benefit others rather than itself.

14. ____________ selection favors the indirect perpetuation of an animal's genes by a relative.

15. According to sociobiology, an organism and its adaptations are ways that its genes have of ____________.

REVIEW QUESTIONS

1. What distinguishes an organized society from an aggregation? Cite an example of an organized society, and describe characteristics that qualify the society as organized.
2. How many similarities between the transmission of information by symbolic language and by heredity can you think of? how many differences?
3. Contrast the "language" of bees with human language.
4. How does an organism learn its place in a dominance hierarchy? What determines this place? What are the advantages of a dominance hierarchy?
5. What is territoriality? What functions does it seem to serve?
6. What is sociobiology?
7. What is kin selection? How is kin selection used to explain the evolution of altruistic behavior?
8. How is play behavior adaptive? Give specific examples.
9. How are pair bonds adaptive? Give specific examples.
10. What are some advantages of courtship rituals?

50 Principles of Ecology

OUTLINE

- *I.* Ecology, communities, and ecosystems
- *II.* The ecology of nutrition
  - *A.* Kinds of producers
  - *B.* Kinds of consumers
  - *C.* Kinds of decomposers
- *III.* The cyclic use of matter and the flow of energy
  - *A.* The carbon cycle
  - *B.* The water cycle
  - *C.* Other elements
  - *D.* Energy flow
    - *1.* Solar radiation
    - *2.* Energy flow in ecosystems
- *IV.* Food chains and pyramids
- *V.* Limits and limiting factors
  - *A.* Temperature
  - *B.* Light
  - *C.* Water
  - *D.* Other environmental factors
- *VI.* Habitat and ecological niche
- *VII.* Interspecific interactions
  - *A.* Competition and cooperation
  - *B.* Negative interactions
- *VIII.* The ecology of stress
- *IX.* Community succession
  - *A.* Some examples of succession
  - *B.* Why successions succeed

LEARNING OBJECTIVES

After you have read this chapter you should be able to:

1. Define, distinguish among, and give examples of *communities* and *ecosystems.*
2. List the main varieties of nutrition employed by living things; define, and give examples of each.
3. List, define, and give examples of nutritional styles of consumer organisms.
4. Summarize the carbon and water cycles.
5. Summarize the energy relationships of the biosphere (with particular attention to the laws of thermodynamics), and give their principal ecological applications.
6. Summarize the climatic effects of the spatial relationships between the earth and the sun.
7. Summarize the concept of the food chain, give an example, and describe or diagram the typical reduction in biomass expected with increasing trophic level.
8. Define *ecological niche,* relate it to the concept of adaptation, and give several examples of niche equivalents.
9. Put the "law of the minimum" into your own words; give at least one example of its operation, and relate it to the concepts of ecological niche, population limitation, and community distribution.
10. Describe the main categories of interactions between species and within species of organisms, giving the ecological significance of each.
11. Describe the general effect of stress on communities, and give an example.
12. Summarize the concept of ecological succession; give a definition plus an example of primary succession and secondary succession.

For many people, ecology has a flavor somewhat between a political movement and a religion. Of course it is neither; it is a *science*. This is not to say that commitments to its implications and applications may not be emotional, but the same is true for any important body of knowledge—nuclear physics, for example. If ecology were a political philosophy, then its principles could be legitimately reversed by political change. Since ecology is a science, however, debates and disagreements may take place about value judgments connected with its implications and applications, but we cannot intelligently speak of "believing" or "not believing" in ecology itself—unless we are willing to speak of nuclear physics or electronics in the same way.

The objective of these last, ecologically oriented chapters is to present ecology as a science. Perhaps that will result in a stronger and more positive commitment to the values that may be derived from it than rhetoric alone ever could accomplish.

Ecology, Communities, and Ecosystems

Ecology (from Greek *oikos,* house or place to live, and *logos,* study) is literally the study of organisms "at home" in their native environment. The term was proposed by the German biologist Ernst Haeckel in 1869, but many of the concepts of ecology antedated the term by a century or more. Ecology is concerned with the biology of groups of organisms and their relations to the environment. The study of ecology also has many subdivisions. The term **autecology** refers to studies of individual organisms, or of populations of a single species and their relations to their environment. A large amount of the material that has already been covered in this book is actually the autecology of one organism or another. The contrasting term, **synecology,** refers to studies of interactions among groups of organisms associated in communities (Fig. 50–1).

(*a*)

Figure 50–1 Ecology is the study of relationships; however, it is approached in two main ways: One, autecology, focuses on the individual adaptations of organisms to the demands of their environment or life style. (*a*) The "living stone," an African desert plant, reduces water loss by the reduction of its leaves and most of its stem. (*b*) The water lily is adapted to disposing of excess water through stomata on the topsides of its leaves and through the spongy, air-conducting tissue in its stems. The second approach to ecology, synecology, studies whole communities, such as the coral reef community shown in (*c*). ((*c*), Robin Lewis, Coastal Creations.)

(*b*)

(*c*)

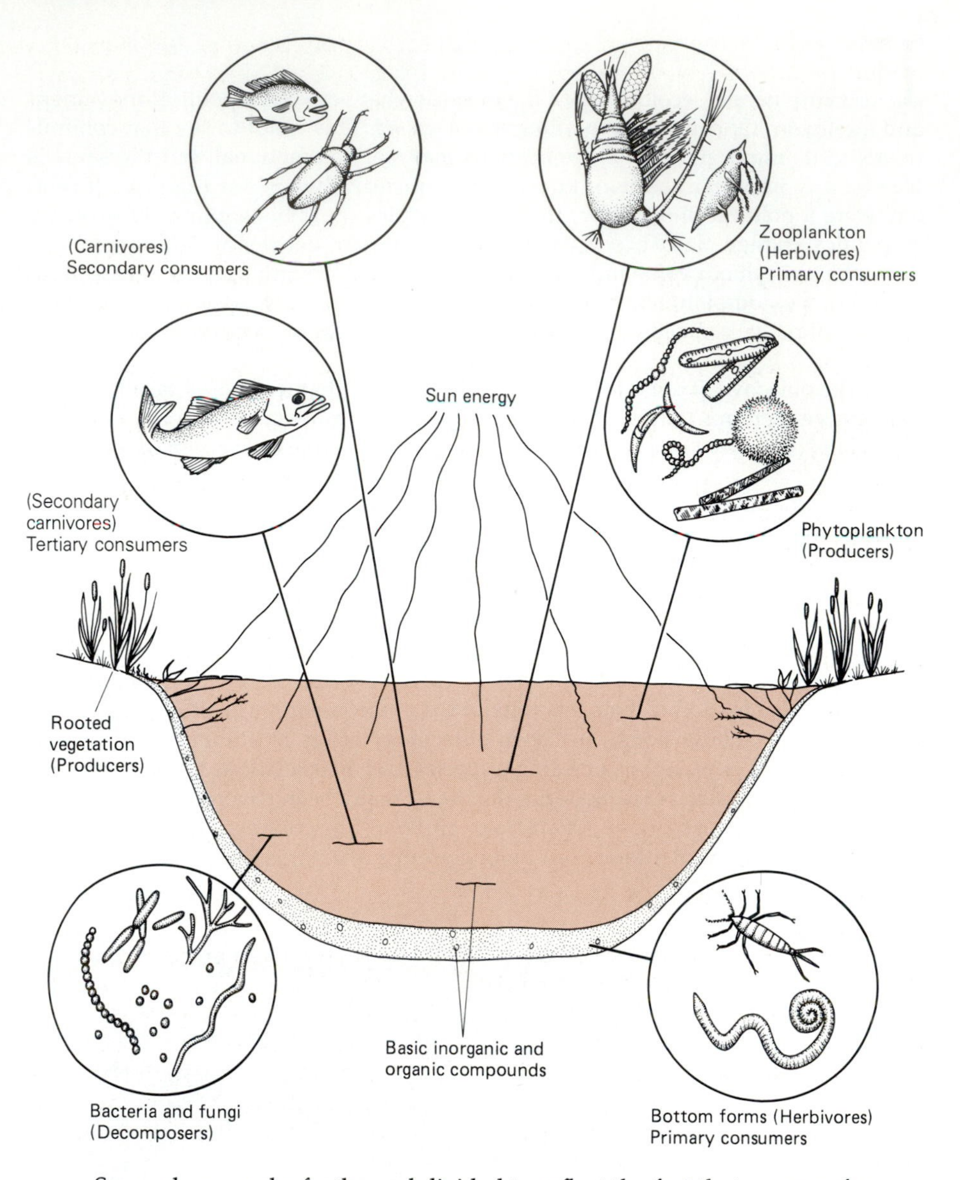

Figure 50–2 A small freshwater pond as an example of an ecosystem. The component parts—producer, consumer, and decomposer, or reducer, organisms—plus the non-living parts are indicated.

Synecology can be further subdivided to reflect the fact that groups of organisms may be associated in three main levels of organization—populations, communities, and ecosystems. In ecological usage, a **population** is a group of individuals of any one kind of organism, a group of individuals of a single species. A community in the ecological sense, a **biotic community,** includes all the populations occupying a given defined habitat. An **ecosystem** consists of the community and the physical, non-living environment, between which energy and materials are exchanged.

In an ecosystem (Fig. 50–2), the living and non-living parts interact as a stable system in which the exchange of materials between living and non-living parts follows a closed, more-or-less cyclic path. An ecosystem may be as large as an ocean or a forest, or it may be as small as an aquarium jar containing tropical fish, plants, bacteria, fungi and snails.

In addition to these ecological definitions, groups of organisms may make up **societies,** if their members interact in such a way as to modify one another's behavior. To illustrate, several colonies of termites (societies) may make up the population of termites in a decaying log. The termites plus the decay organisms—fungi and bacteria—plus the beetles, millipedes, and all the other populations of organisms within the decaying log make up a community. The log community is not self-sufficient, however, for when the wood has been consumed, all the members of the community will have to disperse or die. It is, however, a member of the forest ecosystem, which, if undisturbed, will renew itself indefinitely from the air, water, and sunlight of its surroundings. Synecology concerns itself with all this—populations, societies, communities, ecosystems, and more besides. But it is also true that as biologists learn more about what each species is and does, distinctions among the various ecological groupings tend to blur as it becomes ever more apparent that each organism or group of organisms is part of a system of interdependent and interacting parts that form these larger units.

The Ecology of Nutrition

The interaction of autotrophs (producers) and heterotrophs (consumers and decomposers) is a virtually *universal* feature of all known ecosystems, whether they are located on land, in fresh water, or in the ocean. Frequently the interacting autotrophic and heterotrophic components are, however, partially separated in space. The greatest amount of autotrophic metabolism usually occurs in a **green belt** stratum, in which light energy is available. Below this lies a **brown belt,** in which the most intense heterotrophic metabolism takes place. In the brown belt, organic matter tends to accumulate both in soils and in sediments. The two functions may also be separated in time, for there may be a considerable delay in the heterotrophic utilization of the products of autotrophic organisms. In a forest, for example, the products of photosynthesis tend to accumulate in the forms of leaves, wood, and the food stored in seeds and roots. Much time may elapse before these materials become litter and soil and available to the heterotrophic system. There can also be a considerable spatial separation as well. Consider, for instance, an underground cave. Since detritus from the forest above falls into the streams that enter the cave, and supports all the underground food chains there (perhaps hundreds of feet below the soil surface), the cave is in effect a part of the forest ecosystem.

The two major biomass[1] circuits in almost any ecosystem are the **grazing circuit,** in which animals eat living plants or parts of plants, and the contrasting **organic detritus circuit,** in which dead materials accumulate and are decomposed by bacteria and fungi. The living and non-living parts of ecosystems are operationally very tightly interwoven and difficult to separate. Both inorganic compounds and organic compounds not only are found within and without living organisms, but also are in a constant state of flux between living and non-living conditions. A few substances, such as ATP, are found uniquely inside living cells. In contrast, the partially decomposed organic substances of humus are never found inside living cells, yet are a major and characteristic component of all ecosystems. DNA and chlorophyll may occur both inside and outside organisms, but are nonfunctional when outside the cell. Ecologists can measure the amount of ATP, humus, and chlorophyll in a given area or volume to provide an index of the biomass, the decomposition, and the production in that ecosystem.

KINDS OF PRODUCERS

To survive, **autotrophs** require only water, carbon dioxide, inorganic salts, a source of energy, and perhaps some minerals. It is their source of energy that distinguishes the kinds of autotrophs. Green plants, algae, and certain bacteria are **photosynthetic autotrophs,** deriving the energy needed for biosynthetic processes from sunlight. A few bacteria are **chemosynthetic autotrophs,** which obtain energy by oxidizing certain inorganic substances such as ammonia or hydrogen sulfide. These bacteria have special enzyme systems that catalyze the oxidation of these substances and couple the oxidation with the generation of energy-rich phosphates. For example, nitrite bacteria (*Nitrosomonas*) oxidize ammonia to nitrites; nitrate bacteria (*Nitrobacter*) oxidize nitrites to nitrates; iron bacteria oxidize ferrous to ferric iron; and still other bacteria oxidize hydrogen sulfide to sulfates. The energy derived from these oxidations is utilized to synthesize all the organic materials necessary to maintain life and growth. The nitrite and nitrate bacteria are also important in the cyclic use of nitrogen, for together they convert ammonia to nitrate, a form more readily used by many plants.

For the most part, chemosynthetic autotrophs are ecologically far less important than the photosynthetic autotrophs, but some, such as the nitrate bacteria, are of vital significance to the biosphere. Chemosynthetic autotrophic bacteria also form the basis of the deep-sea marine ecosystems located near hot springs where

[1] Biomass refers to the total weight of the living organisms in a particular habitat, or in a particular ecological category in that habitat.

no light can penetrate (see Focus on Life Without the Sun). The energy source for all the organisms in these communities is not, therefore, sunlight at all. Although recent evidence suggests that they are more widespread than previously believed, it seems that deep-sea oasis communities make only a negligible contribution to the total energy budget of the biosphere.

KINDS OF CONSUMERS

In contrast to autotrophs, **heterotrophs** are organisms that are unable to synthesize their own foodstuffs from inorganic materials. Heterotrophs *must* live at the ex-

FOCUS ON

Life Without the Sun

In the mid 1970s an oceanographic expedition was dispatched to study the deep cleft in the ocean floor known as the Galapagos Rift, off the coast of Ecuador. The expedition discovered a series of extremely hot springs on the floor of the abyss in which seawater apparently had penetrated to be heated by the hot rocks below. During its sojourn within the earth, the water had also been charged with mineral compounds, including hydrogen sulfide, H_2S. Certain bacteria that are chemosynthetic autotrophs extract the energy they need to build carbon dioxide into carbohydrates not from light by photosynthesis, but from the oxidation of inorganic compounds. Some of these bacteria can cause H_2S to react with oxygen to produce water and sulfur or even sulfate. These reactions are exergonic and serve to provide the energy to operate a version of the light-independent reactions of photosynthesis.

At the tremendous depths of the Galapagos Rift there is no light available for photosynthesis, and therefore little plankton except that which rains down dead from the lighted surface layers. But the hot springs support a rich and bizarre life, in great contrast to the surrounding lightless desert of the abyssal floor. Many of the species occurring in these oases of life were new to science; some could not even be assigned with confidence, at that time, to known phyla. Some had astonishing adaptations; one species of clam, for example, is unique among invertebrates in possessing red blood cells that contain hemoglobin. What did these species live on?

The basis of the food chains in these aquatic oases was the chemically autotrophic bacteria, which are themselves remarkable in being able to survive and multiply in the extremely high temperatures of the hot springs. They function as primary producers. As an astonishing exception, the energy that activated their living communities originated not in the thermonuclear reactions of the sun but in the heat released by slow radioactive decay in the depths of the earth.

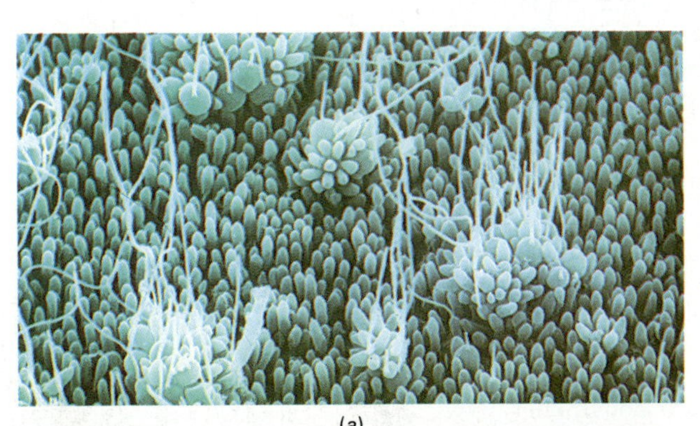

(*a*)

(*b*)

The inhabitants of the Galapagos Rift. (*a*) Scanning electron micrograph of chemoautotrophic bacteria (×5200). Such bacteria serve at the base of the food chain in hydrothermal vent systems. Although they probably cannot reproduce at such temperatures (at least as evidenced by laboratory studies), some hydrothermal vent bacteria can live at temperatures exceeding 300°C in high pressure conditions; that is, at pressures high enough so that water does not boil. (*b*) Chemoautotrophic bacteria living in the tissues of these "beard worms" extract hydrogen sulfide and carbon dioxide from water to manufacture organic compounds. Beard worms are thought to be related to the annelids, but are assigned to the phylum Pogonophora. Though some species of pogonophorans reach a length of 1 to 1.5 meters (their bodies are supported by an outer tube containing chitin), they completely lack a digestive system. They depend to a large extent on the organic compounds provided by the endosymbiotic bacteria, along with materials filtered from the surrounding water and digested extracellularly. Also visible in the photograph are some filter-feeding mollusks. ((*a*), Carl D. Wirson, Woods Hole Oceanographic Institution; (*b*), J. Fredrick Grassle, Woods Hole Oceanographic Institution.)

pense of autotrophs or upon decaying organic matter. All animals, all fungi, most bacteria, and some protists are heterotrophs.

As we saw in Chapter 33, there are several types of heterotrophic nutrition. When food is obtained as solid particles that must be eaten, digested, and absorbed (as in most animals), the process is termed **holozoic nutrition.** Holozoic organisms must constantly find, catch, and eat other organisms. To do this, animals possess a variety of sensory, nervous, and muscular structures to find and catch food, as well as a digestive system to convert food into molecules small enough to be absorbed.

Holozoic organisms can be subdivided into three main categories. **Herbivores** are animals that eat plants and algae and obtain their energy-rich compounds from the contents of the plant and algal cells. Other animals are **carnivores,** eating animals (that eat plants), or **omnivores,** eating either plant or animal material. All heterotrophic organisms obtain their energy-rich nutrients ultimately from autotrophic organisms that trapped the radiant energy of sunlight to synthesize those compounds.

Detritus feeders, usually known as **scavengers,** eat dead and decomposing organic matter. They have some of the characteristics of decomposers even though they are holozoic and (as we will see) true decomposers are not. Many detritus feeders, like earthworms, prepare material for attack by decomposers. Others, like fly maggots, eat the decomposers themselves.

Yet another type of heterotrophic nutrition, found among both plants and animals, is parasitism. A **parasite** lives in or on the living body of a plant or animal, called the **host,** and obtains its nourishment from it *without* necessarily killing the host (although a parasite may weaken its host and thus hasten its death). Almost every living organism is the host for one or more kinds of parasites.

Parasites may obtain their nutrients by ingesting and digesting solid particles or by absorbing organic molecules through their cell walls from the body fluids or tissues of the host. Some parasites cause little or no harm to the host. Others are known to produce diseases, destroying the host's cells, or producing toxic substances that interfere with the host's metabolic processes. The **pathogenic** (disease-producing) **parasites** of humans and other animals include viruses, bacteria, fungi, protozoa, and an assortment of worms. Most plant diseases are caused by parasitic fungi; a few are due to viruses, worms, or insects.

Nearly every major group of organisms has some parasitic representatives. There are parasitic mollusks, for instance, and even a parasitic vertebrate (the lamprey). Viruses, bacteria, fungi, flatworms, and nematode worms, however, are among the organisms that seem to have a particular flair for a parasitic mode of life.

Parasitism takes many forms. **Occasional parasites** (like fleas, which spend much of their time living rather inactively in the host's nest or carpets) feed on their hosts only when their metabolic needs are high, such as when they are young or are about to reproduce. A few plants, such as mistletoe, are partially parasitic as well as autotrophic, for although they have chlorophyll and make their own organic food, their roots grow into the stems of other plants, and they absorb some of their nutrients (water and minerals) from their host. A few organisms are **opportunistic parasites;** for example, many types of leeches suck blood when it can be obtained, but normally live a predatory life.

Finally, a number of forms of social parasitism exist, such as **brood parasitism,** exemplified by the European cuckoo and North American cowbird (Fig. 50–3). These birds lay their eggs in the nest of a host. The parasitic chick hatches first and either kills or competitively starves the host chicks. It then is raised by the host parents. Insect social parasites infest colonies of social insects, often feeding on the hosts or their larvae, yet are immune to attacks.[1]

Figure 50–3 A young cuckoo (*left*) being fed by a hedge sparrow. After the cuckoo has reached nearly full size, the smaller sparrow, reacting to the sign stimulus of a gaping mouth, continues to act as the cuckoo's "natural" parent. (Stephen Dalton, Photo Researchers, Inc.)

KINDS OF DECOMPOSERS

Yeasts, molds, and most bacteria can neither make their nutrients by autotrophic processes nor ingest solid food. They must absorb their required organic nutrients directly through the cell membrane. This type of heterotrophic nutrition is known as **saprobic** nutrition. Saprobes can grow only in places where there are decomposing bodies of animals or plants, or masses of plant and animal by-products. Such

[1]Though old, *Ants: Their Structure, Development and Behavior* by W. M. Wheeler (New York, Columbia University Press, 1910) is still widely available and contains many fascinating examples. See also *Sociobiology* by Edward O. Wilson (Cambridge, Harvard University Press, 1975).

Figure 50–4 The biosphere. No significant loss or gain of matter occurs on the earth as a whole. All materials utilized by living things must be continually recycled within this vast, closed system. (NASA.)

places are the humus of the uppermost layers of the forest floor, and the bottom "ooze" of some aquatic habitats.

The Cyclic Use of Matter and the Flow of Energy

The first law of thermodynamics states that matter is neither created nor destroyed. Obviously, the carbon and nitrogen atoms that comprise the bodies of living things must have been used over and over again in the formation of new generations of plants and animals. The biosphere neither receives any great amount of matter from other parts of the universe, nor loses significant amounts of matter to outer space. The atoms of each element—carbon, hydrogen, oxygen, nitrogen, and the rest—are taken from the planetary environment, made a part of some cellular component, and finally (perhaps by a quite circuitous route involving several other organisms) are returned to the non-living environment to be used again (Fig. 50–4).

THE CARBON CYCLE

In the atmosphere over each acre of the earth's surface, there are about six *tons* of carbon in the form of carbon dioxide. Yet in a single year an acre of luxuriant plant growth such as sugar cane can extract as much as 20 tons of carbon from the atmosphere and incorporate it into the plant body.

If there were no way to renew the supply, producers would eventually, perhaps in a few centuries, use up the entire atmospheric supply of carbon. However, carbon dioxide is continuously returned to the air by the decarboxylations that occur in cellular respiration. Plants carry on respiration continuously; in addition, plant tissues are eaten by animals that, by respiration, return more carbon dioxide to the air. But plant and animal respiration alone would not return enough carbon dioxide to the air to balance what is withdrawn by photosynthesis. Vast amounts of carbon would eventually accumulate in the compounds making up the dead bodies of plants and animals and their wastes. Thus the carbon cycle is balanced by the decay bacteria and fungi, which cleave the carbon compounds of dead plants and animals and convert the carbon to carbon dioxide (see Fig. 1–13).

If carbon dioxide–removing processes did not exist, volcanic action over the millennia would also add very large amounts of carbon dioxide to the atmosphere, greatly disturbing the heat balance of the earth. Aside from photosynthesis, probably the most important carbon dioxide–removing process is the production of calcium carbonate skeletal structures by aquatic organisms, which then eventually become sedimentary rocks. Also, however, when the bodies of plants are compressed under water for long periods of time they are not decayed by bacteria, but undergo a series of chemical changes to form **peat,** later brown coal or **lignite,** and finally **coal.** The bodies of certain marine organisms undergo somewhat similar changes to form **petroleum.** Although these processes remove some carbon from the cycle, geological changes or human mining and drilling may eventually bring the coal and oil to the surface to be burned to carbon dioxide and restored to the cycle. It is, in fact, a matter of some concern that the artificially increased supply of carbon dioxide brought about by burning fuels, combined with that released by uncompensated decay processes in recently cleared land, may exceed the ability of plants to absorb it, leading to increased global temperatures and unforeseeable ultimate consequences (see Chapters 1 and 52).

THE WATER CYCLE

Of prime importance to living things as a solvent and source of hydrogen for photosynthesis, water is fortunately in abundant supply. But if it were not for the **hydrologic cycle,** water would remain in its great reservoir, the ocean. The sun's heat vaporizes water and forms clouds. These, moved by winds, may pass over land, where they are cooled enough to precipitate the water as rain or snow. Some of this water soaks into the ground, and some runs off the surface into streams that go directly back to the sea. The ground water is returned to the surface by springs, by pumps, and by the activities of producers. Water inevitably ends up back in the sea, but it may become incorporated into the bodies of several different organisms, one after another, en route. The energy to run the cycle—that is, the heat needed to evaporate water—comes from sunlight.

OTHER ELEMENTS

Elements such as phosphorus, nitrogen, and iron must also be recycled, as they too are essential to life. The phosphorus and nitrogen cycles have already been discussed in Chapter 30 (which you should review at this point). However, although the elements required for life can be continuously recycled, the energy that operates the life processes of living things cannot.

ENERGY FLOW

The cycles of all types of matter are **closed,** meaning that the atoms are used over and over again. To keep the cycles going does not require new matter, but it does require the continued input of energy, for energy does not cycle but flows in a single direction. Although energy is neither created nor destroyed, recall (Chapter 7) that whenever energy is transformed from one kind to another, or performs any work, there is an increase in entropy, and a decrease in the energy available to do further work; this is the second law of thermodynamics.

Only a small fraction of the light energy reaching the earth is trapped. Considerable areas of the earth have no plants, and even where they are present, photosynthetic plants can utilize only about 3% of the incoming solar energy. This radiant energy is converted into the potential energy of the chemical bonds of the organic substances made by the plant.

When an animal eats a plant (or when bacteria decompose it), these organic substances are oxidized. The energy liberated is equal to the amount of energy used in synthesizing the substance, but some of the energy is lost as heat. If this animal in turn is eaten by another one, a further decrease in useful energy occurs as the second animal oxidizes the organic substances of the first and liberates energy to synthesize its own cellular constituents.

Eventually, all the energy originally trapped by plants in photosynthesis is converted to heat and dissipated to outer space, and all the carbon of the organic compounds ends up as carbon dioxide. Solar energy, however, not only powers the world of life, but also determines the climate, which is also of great biological importance.

Solar Radiation

A most outstanding feature of the earth is the nonuniformity of its physical conditions, which range from Arctic tundra to tropical rain forests. Even the oceans are surprisingly patchy, nonuniform places. Although the earth derives nearly all its energy from the sun, even the sun's energy is not uniformly distributed over the face of the globe. The solar radiation reaching the surface of the earth varies—with (1) the length of the path that the sun's rays take through the atmosphere; (2) the angle at which those rays strike the earth; (3) the distance of the earth from the sun (which changes seasonally because of the elliptical orbit of the earth around the sun); (4) the amount of water vapor, dust, and pollutants in the atmosphere; and (5) the total amount of daylight hours per day—the **photoperiod.** At higher latitudes the angle of incidence of the sun's rays is less than at middle latitudes, and the energy is spread more thinly. The rays must also pass through a thicker layer of atmosphere (Fig. 50–5), and consequently the polar regions receive less radiant energy in the course of a year than do equatorial regions.

It is estimated that as much as 40% of the heat in the atmosphere comes from the condensation of water vapor derived from the evaporation of water from the surface of the ocean. The moisture-laden air rises, moves to higher latitudes where it is cooled, and gives up its moisture as clouds or rain. The heat is then absorbed by the wet atmosphere. The atmosphere, heated from below and radiating heat back to the surface of the earth, serves as a heat trap, as does a greenhouse, in which the glass acts as the clouds and water vapor.

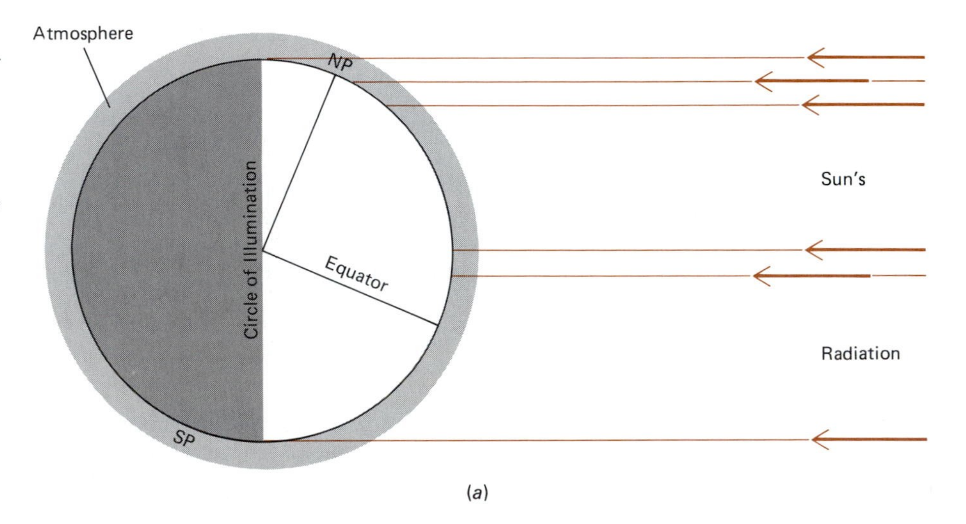

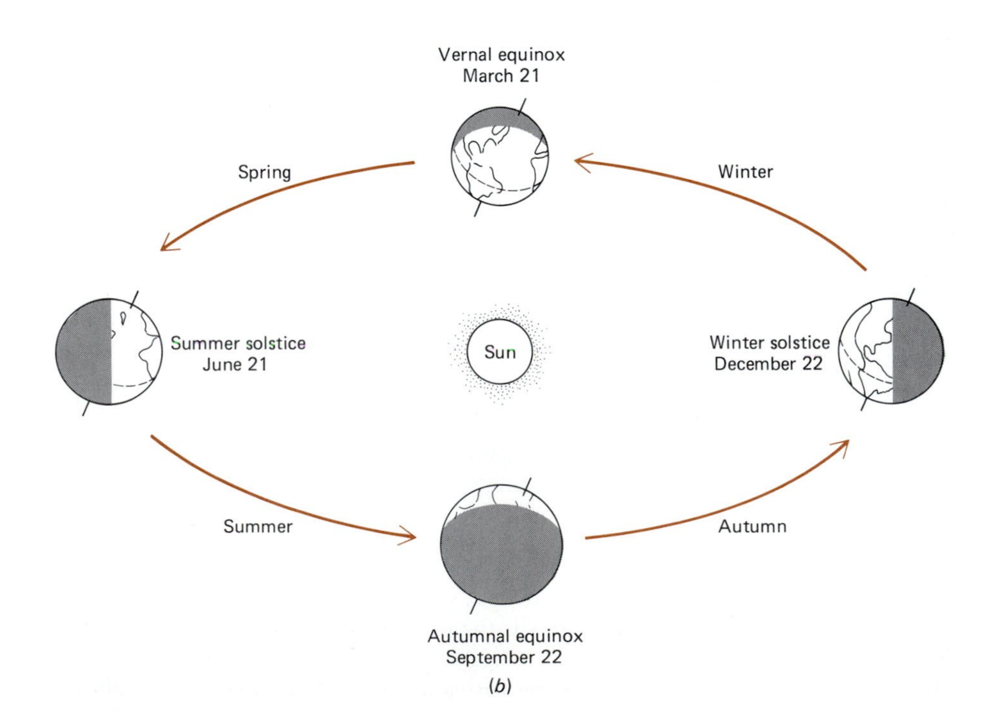

Figure 50–5 Photoperiods. (*a*) Circle of illumination, areas of daylight and darkness, angles of sun's rays at different latitudes, together with differences in areas affected and thickness of atmosphere penetrated at the time of the summer solstice. (*b*) The sunlit portions of the Northern Hemisphere are seen to range from greater than one half in summer to less than one half in winter. The proportion of any latitude that is sunlit is also the proportion of the 24-hour day between sunrise and sunset.

Energy Flow in Ecosystems

Carbon, nitrogen, and various other cycles in nature operate to conserve the limited amount of usable matter on the earth. In contrast, the amount of energy available is very great and is constantly being renewed in the form of sunlight. By measuring the amount of energy absorbed and given off by each kind of organism, ecologists can determine the total energy budget and also the functional relationships of the organisms living together in a community. From this it can be calculated how much life can ideally be supported in a given area, and how many individuals of each species the area can support.

The transfer of energy through a biological community begins when the energy of sunlight is fixed in a plant or algal cell by photosynthesis. It is estimated that only 8% of the energy of the sun reaching the planet strikes plants, and that only 3% is utilized in photosynthesis. The rest of the energy leaves the planet as part of the earth's general heat loss.

The rate at which energy is used by producers over a certain period of time is referred to as **gross productivity.** Part of the energy used in photosynthesis is used by the plant itself to drive the many processes required for its maintenance. The amount that is stored and reflected in growth and seed production is called the **net productivity.** This measurement is obtained by subtracting the rate of energy utilization (cellular respiration) from gross productivity.

The net production of a field of sugar cane in Hawaii in a recent year was 190 Kcal per m^2 per day, and the average exposure to solar energy was about 4000 Kcal/m^2 per day. From this we can calculate that the net efficiency of the sugar cane was about 4.8%. Such values can be achieved only by certain crops, and then only under optimal conditions. On an average annual basis, even the plants in tropical forests have a collective efficiency of about 2%.

The stored energy accumulates as living material. Part of this energy is recycled each season by the death and decomposition of the organisms. The part that remains in living tissue is called the **standing crop biomass.** This, of course, varies with the season and the type of ecosystem. The most productive ecosystems from an energy standpoint are coral reefs and estuaries (coastal bodies of water that connect with the ocean). The least productive are deserts and the open ocean (Fig. 50–6).

The various kinds of consumers all depend upon the energy production of plants and algae. As has been emphasized, not all the net production available to herbivores is assimilated. For example, a grasshopper assimilates only about 30% of its food. Most of this goes into the maintenance of the organism and is lost as heat in the process of respiration. A small residue is stored in the form of new tissue and new individuals. It is only the energy stored in the herbivore body that is available to the next **trophic level** (a level in a food chain), the carnivores.

The various kinds of organisms in nature are balanced with their environment, but in many cases this balance is a precarious one. Each organism is in a dynamic state, and its constituents are constantly being degraded and rebuilt. Thus each organism can be considered as a sort of transformer, through which energy and matter flow. Members of the human species have upset nature in an alarming number of instances, and in many cases, once undone, the balance is difficult (if not impossible) to reestablish. The reasons for this will become more apparent when we examine the concept of community succession later in the chapter, but for now, let us consider an example.

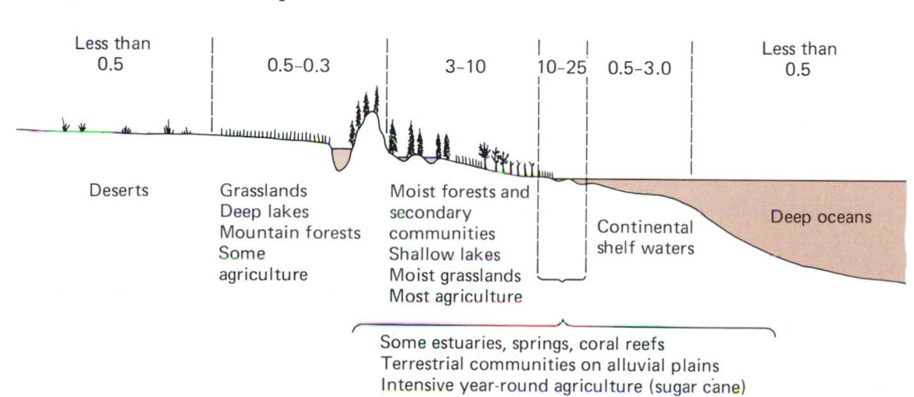

Figure 50–6 The world distribution of gross production, in grams of dry matter per square meter per day, as indicated by the daily rates of gross production in major ecosystems. (After Odum.)

In tropical Africa and South America the original inhabitants for many generations have cleared portions of the rain forest for fields to grow crops. Without fertilization the cleared area may produce a crop for only a few years. When it becomes unproductive, the area is abandoned, and more forest is felled to provide new crop land. If this is practiced with restraint, little harm is done. (Often social taboo systems and other customs operate to limit the amount of time that a plot of land may be farmed in primitive agricultural societies.) If, however, the same land is reused too frequently, or for too long at a time, sooner or later a point is reached where the abandoned area can probably never again be covered with a mature rain forest. The thin tropical soils have very meager supplies of mineral nutrients, and most of them are quickly leached out of the soil by heavy rains as soon as the land is cleared. Most of the minerals usable by the plants are already part of living plants! The mature rain forest is in precarious balance with the soil and can maintain itself only as long as the balance is not disturbed. Once the balance has been upset, the forest is irretrievably lost. The worst situations of all occur when intensive agriculture is practiced in the tropics. All too often a tropical desert is the quick result.

Many tragic examples could be given of human blunders in upsetting critical ecological balances. The organisms living in nature are parts of complex interacting communities of many species and are not single, isolated species. The well-being of these organisms is determined not simply by limitations and peculiarities of temperature, soil, pH, salinity, and other factors of the abiotic environment, but also by their relationships with other organisms living in that region. It is true that organisms make the community; but it is equally true that the community makes the organisms.

Food Chains and Pyramids

The number of organisms of each species—or, more precisely, their total biomass—is determined by the rate of flow of energy through the biological part of the ecosystem that includes them. (Recall that the total mass of the organisms living in a particular area is referred to as the **biomass.**)

The transfer of energy from its initial reception in plants and algae, through a series of organisms in which each eats the preceding and is eaten by the following, is known as a **food chain.** The number of steps in the series is limited to perhaps four or five, at the most, partly because of the great decrease in available energy at each step. The percentage of food energy consumed that is converted to new cellular material, and thus is available as food energy for the next animal in the food chain, is known as the **percentage efficiency of energy transfer.**

The flow of energy in ecosystems, from sunlight through photosynthesis in autotrophic producers, through the tissues of herbivorous primary consumers, to the tissues of carnivorous secondary consumers, determines the number and biomass of organisms at each level in the ecosystem. The flow of energy is greatly reduced at each successive level of nutrition because of the energy utilization by the organisms and heat losses at each transformation of energy; this largely accounts for the decrease in biomass in each level. In addition, no predator is completely efficient at capturing its prey; some energy is lost in the hunt.

Some animals eat only one kind of food and therefore are members of a single food chain. Other animals eat many different kinds of food and not only are members of different food chains but also may occupy different positions in different food chains. An animal may be a primary consumer in one chain, eating plants, but a secondary or tertiary consumer in other chains, eating herbivorous animals or other carnivores.

Humans are at the end of a number of food chains. For example, a human might eat a fish such as a black bass, which in turn ate little fishes, which ate small invertebrates, which ate algae. The ultimate size of the human population (or the population of any animal) is limited by the length of the food chain, the percentage efficiency of energy transfer at each step in the chain, and ultimately by the amount of light energy falling on the earth.

Since humans can do nothing about increasing the amount of incident light energy and very little about the percentage efficiency of energy transfer, they can increase their food energy only by shortening their food chain—that is, by eating the primary producers, plants, rather than animals. In overcrowded countries, peo-

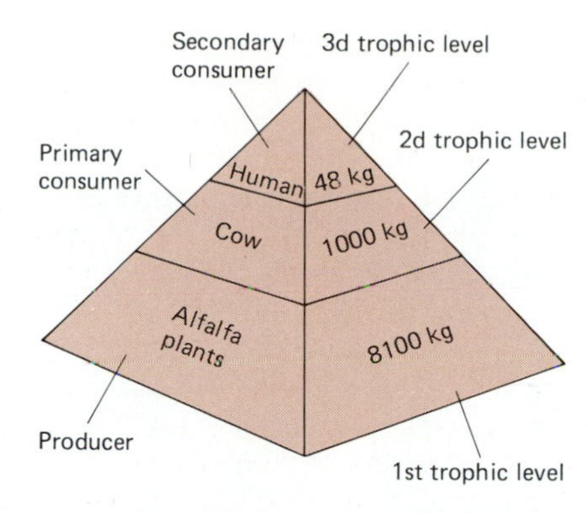

Figure 50–7 A simple food pyramid.

ple tend to be vegetarians because then the food chain is shortest and a given area of land can in this way support the greatest number of people. Steak is a luxury in both ecological and economic terms, but hamburger is just as much an ecological luxury as steak is. Religious and legal measures prohibit eating either one in some overpopulated nations.

A food chain may be visualized as a **food pyramid,** with each step in the pyramid much smaller than the one on which it feeds (Figs. 50–7 and 50–8). Ecologist H. T. Odum has calculated that 8100 kg of alfalfa plants is required to provide the food for 1000 kg of calves, which provide enough food to keep one 12-year-old 48-kg boy alive for one year. (Although boys eat many things other than veal, and calves other things besides alfalfa, these numbers illustrate the principle of a food chain.) Since the predators are individually larger as a rule than their prey, the pyramid of numbers of individuals in each step of the chain is even more striking than the pyramid of the mass of individuals in successive steps; one boy requires 4.5 calves, which require 20 million alfalfa plants.

There are certain exceptions to the progressive decline in biomass with upper positions in the food chain. Sometimes the primary producers reproduce very rapidly but are consumed almost as rapidly as they reproduce. This very rapid turnover rate results in a small **standing crop** at any one time. The longer-lived consumers thus can have, at least temporarily, a larger biomass at any one time than the producers.

In addition to predator food chains, such as the human–black bass–minnow–crustacean one, there are parasite food chains. For example, mammals and birds are parasitized by fleas; in the fleas live protozoa, which are in turn hosts for bacteria.

A third type of food chain is one in which living organisms or their wastes are converted into dead organic matter, **detritus,** before being eaten by animals such as

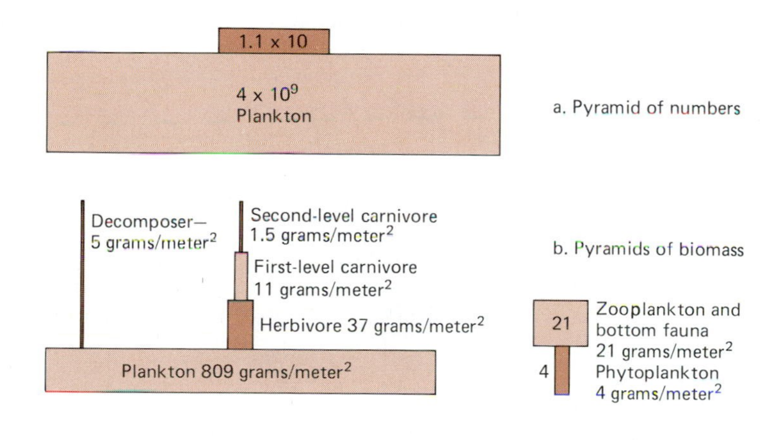

Figure 50–8 Ecological pyramids. (*a*) A pyramid based on numbers of organisms on each trophic level. (*b*) Pyramids based on the biomass of organisms consumed. (*c*) A pyramid based on energy transfer from each level of the pyramid. (From Pequegnat, Odum, Harvel and Teal.)

(a)

(b)

Figure 50–9 The beginning of a detritus food chain. Coastal mangrove trees (*a*) shed their leaves which decay (*b*) under water. The decaying leaves and the decay organisms themselves are eaten by tiny crustaceans which in turn are eaten by fish.

millipedes and earthworms on land, by marine worms and mollusks, or by bacteria and fungi (Fig. 50–9).

In a community of organisms in the shallow sea, about 30% of the total energy flows via detritus chains, but in a forest community, with a large biomass of plants and a relatively small biomass of animals, a much larger proportion of energy flow may be via detritus pathways. In an intertidal salt marsh, where most of the animals—shellfish, snails, and crabs—are detritus eaters, 90% or more of the energy flow is via detritus chains.

In any natural community there are many species with the ability to eat a number of different kinds of food. Thus, food chains tend to intersect to form a kind of network or **food web** (Fig. 50–10). The redundancy this represents produces a large margin of safety, so that most natural communities may not be greatly disturbed if one or another component organism dies out. There is obviously a limit, however, beyond which such homeostatic mechanisms of a community will not protect it.

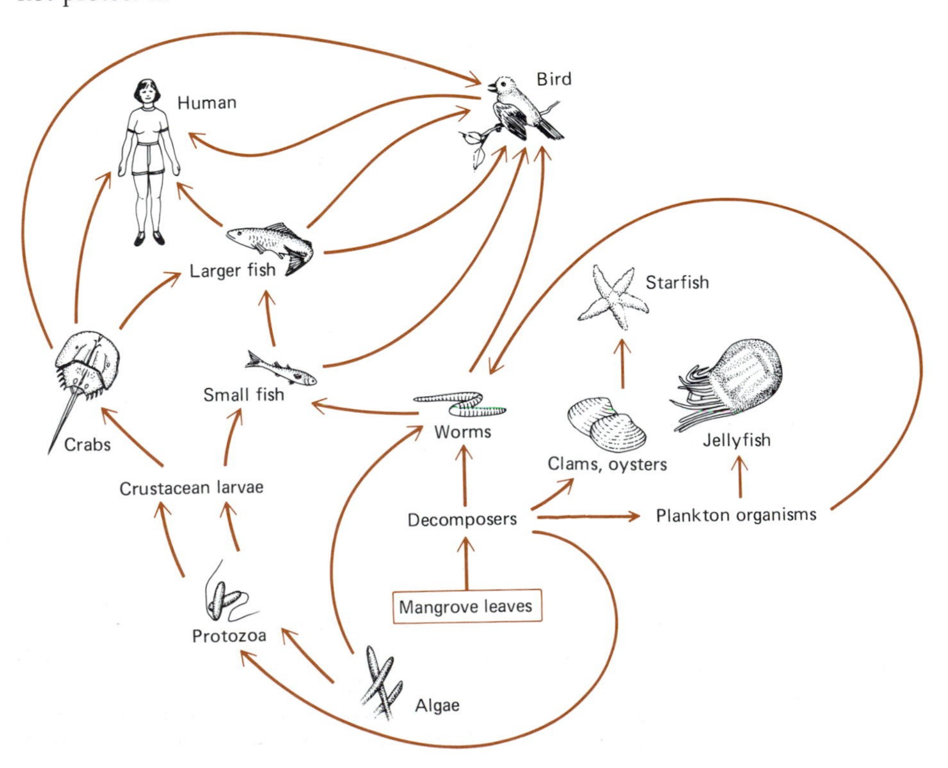

Figure 50–10 A marine food web. Arrows point from each organism to its consumer. Note that even this web is simplified. Humans, for example, often eat algae, thereby acting as primary consumers.

Limits and Limiting Factors

Probably no species of plant or animal is found everywhere in the world; some parts of the earth are too hot, too cold, too wet, too dry, or too something else for the organism to survive there. Even if the environment does not kill the adult directly, it can effectively keep the species from becoming established if it prevents reproduction or kills off the egg, embryo, or some other stage in the life cycle.

Most species of organisms are not even found in all the regions of the world where they could survive. The existence of barriers prevents their further dispersal and enables us to distinguish the major biogeographic realms, characterized by certain assemblages of plants and animals.

Strange as it may now seem to us, biologists only became aware in the early 19th century that each species requires certain materials for growth and reproduction and will be restricted in growth or population if the environment does not provide a certain minimal amount of each one of these materials. The pioneer agricultural chemist Justus Liebig stated in 1840 what is now known as the **law of the minimum,** which holds that the rate of growth of each organism is limited by whatever essential nutrient is present in a minimal amount, that is, is in the shortest supply (Fig. 50–11). Liebig, who studied the factors affecting the growth of plants, found that the yield of crops was often limited not by a nutrient required in large amounts, such as water or carbon dioxide, but by something needed only in trace amounts, such as boron or manganese.

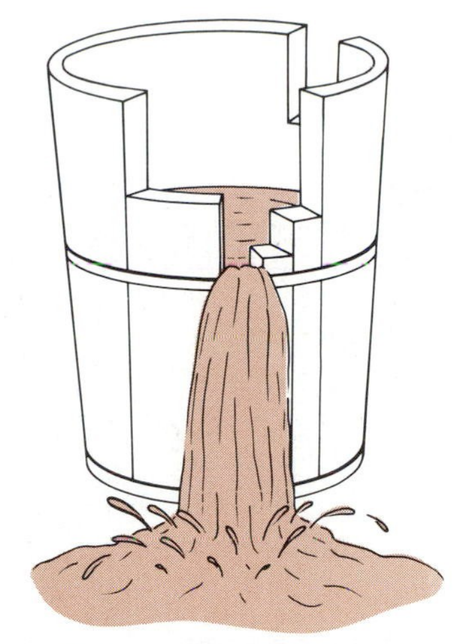

Figure 50–11 Liebig's concept of the law of the minimum. Regardless of how tall the rest of the staves may be, the shortest stave in the bucket determines the water level. So too, the requirement that is in the shortest supply, the limiting factor, governs the life of an organism, or a population of organisms. (Illustration redrawn after Liebig's original version.)

We now realize that, though Liebig was fundamentally correct, there may in addition be interactions between factors such that a very high concentration of one nutrient or some other essential may alter the rate of utilization of another (the rate-limiting one) and hence alter the effective minimal amount required. Certain plants, for example, require less zinc when growing in the shade than when growing in the sunlight. The biologist V. E. Shelford pointed out also in 1913 that *too much* of a certain factor would act as a limiting factor just as well as too little of it, so that the distribution of each species is really determined by its **range of tolerance** to variations in each environmental factor.

Limiting factors are sometimes effective only during part of the life cycle of an organism. For instance, seedlings and larvae are usually more sensitive than adult plants and animals. Adult blue crabs can survive in water with a low salt content and can migrate for some distance upriver from the sea, but their larvae cannot, and so the species could not become permanently established there if a population were to become permanently landlocked.

In effect, therefore, limiting factors work over the entire life cycle of an organism. Some creatures go through an ecological bottleneck at one stage or another of their life histories. The eggs of birds represent one such vulnerable stage. Although the ring-necked pheasant has been introduced into the southern United States a number of times and the adults survive well, the developing eggs are apparently killed by the high daily temperatures and are unable to complete development.

Some organisms have very narrow ranges of tolerance to environmental factors; others can survive within much broader limits (Fig. 50–12). Any given organism may have narrow limits of tolerance for one factor and wide limits for another. Ecologists use the prefixes **steno-** and **eury-** to refer to organisms with narrow and wide, respectively, ranges of tolerance to a given factor. A stenothermic organism, for example, is one that will tolerate only narrow variations in temperature. The housefly is a eurythermic organism, for it can tolerate temperatures ranging from 5°C to 45°C. The adaptation to cold of the Antarctic fish *Trematomus bernacchi* is remarkable. It is extremely stenothermic and will tolerate temperatures only between −2°C and +2°C. At 1.9° this fish is immobile from heat prostration!

Figure 50–12 A small quantity of soil and water trapped in the crevices of this rock has enabled this small plant to grow and flower in a dry, subalpine habitat. (Tina Waisman.)

TEMPERATURE

Temperature is an important limiting factor, as the relative sparseness of life in desert and arctic regions demonstrates. Most of the animals that do live in the desert are behaviorally adapted to the rigors of the environment by living in burrows during the day and coming out to forage only at night. Many animals escape the bitter cold of the northern winter not by migrating south but by burrowing beneath the snow. Measurements made in Alaska show that when the surface temperature is −55°C, the temperature 60 cm under the snow, at the surface of the soil, is −7°C. On the other hand, certain bacteria are able to survive in hot springs that are almost at the boiling point of water. Still other bacteria that are associated with deep-sea hot springs (where because of the pressure even extremely hot water will not boil) can reproduce at a temperature of 250°C; how their proteins are spared from denaturation will be extremely interesting to know.

LIGHT

The amount of light is an important factor in determining the distribution and behavior of both plants and animals. The complete absence of light in caves or deep-water habitats in most cases restricts their inhabitants to heterotrophs, which must subsist on leftovers carried into their communities from the outside. Light is, of course, the ultimate source of energy for life on this planet; yet prolonged exposure of cells to light of high intensity or short wavelength may be fatal. Both plants and animals have mechanisms and responses to protect them against too much or too little light.

The photoperiod has a marked influence on the time of flowering of plants, the time of migration of birds, the time of spawning of fish, and the seasonal color changes observed in certain birds and mammals. Knowledge of photoperiod phenomena (Chapter 29) has proved to be of considerable economic importance. Since photoperiod varies not only seasonally but also geographically, it can prove impossible to grow certain crops in some localities not because the climate is unsuitable, but because of an inappropriate photoperiod. Modification of photoperiod requirements is an important part of plant selective breeding.

WATER

Water is a physiological necessity for all living things and is also a limiting factor, primarily for land organisms. The total amount of rainfall, its seasonal distribution, the humidity, and the supply of ground water (water beneath the earth's surface) are some of the most important factors limiting distribution of plants and animals. Some lakes and streams, particularly in the western and southwestern United States, periodically become dry or almost dry, and the fish and other aquatic animals are killed. During periods of low water, the water temperature may rise enough to kill off the aquatic forms. Many protozoa survive the drying of the puddles in which they normally live by forming thick-walled protective cysts. As mentioned before, some animals are adapted to desert conditions by digging and living in burrows in which the temperature is lower and the humidity is higher than at the surface. Measurements have shown that the temperature in the burrow of a kangaroo rat 60 cm underground may be only 16°C when the surface temperature is over 38°C. The desert plants must necessarily stay on the surface, but they have structures that minimize water loss and resist high temperatures.

An excess of water is fatal to certain animals. Earthworms may be driven from their burrows by heavy rainfall. Oxygen is only sparingly soluble in water, and the earthworm cannot get enough oxygen when immersed. Knowledge of the limits of water tolerance can be used by people to regulate insect pests. For example, wire worms, pests attacking West Coast crops, were found to have rather narrow limits of tolerance to water and to be most sensitive as larvae and pupae. They can be destroyed by exceeding the maximum limit of tolerance—by flooding irrigated fields—or by planting alfalfa or wheat, which dry out the soil below the limit of tolerance of the larvae.

OTHER ENVIRONMENTAL FACTORS

Atmospheric gases are usually not limiting for land organisms, except for forms living deep in the soil, on mountain heights, or within the bodies of other animals. In aquatic environments, the amount of dissolved oxygen present may vary considerably and is a common limiting factor. For instance, the biologically available oxygen content in stagnant ponds or in streams fouled by sewage and industrial wastes may become so low that it is incompatible with many forms of life.

The amount of carbon dioxide in the air is remarkably constant, but the amount dissolved in water varies widely. An excess of carbon dioxide may be a limiting factor for fish and insect larvae. The hydrogen ion concentration, pH, of water is related to the carbon dioxide concentration, and it too may be an important limiting factor in aquatic environments. Even water currents are limiting factors in certain aquatic plants and animals. There are marked differences, for example, between the life forms of a still pond and those of a swiftly flowing creek.

The trace elements necessary for plant and animal life may also be limiting factors. Deficiencies of cobalt and copper produce severe deficiency diseases in plants and grazing animals—making certain regions of Australia unsuitable for raising cattle or sheep, for instance. Other trace elements that may be limiting factors are manganese, zinc, iron, sulfur, selenium, and boron.

The type of soil, the amount of topsoil and its pH, porosity, slope, water-retaining properties, and so on are limiting factors for many plants. The ability of many animals to survive in a given region depends on the presence of certain plants to provide shelter, cover, and food. Grasses, shrubs, and trees each provide shelter for certain kinds of land animals, and seaweeds and freshwater aquatic plants play a similar role for aquatic animals. Even fire is a factor of ecological importance. The continued existence of the fine forests of long-leaf pines in the southeastern states is due to their superior resistance to fire. In the absence of occasional small ground fires, these pines are gradually replaced by small hardwoods, much less valuable as timber and much more readily killed by fire.

Still another limiting factor is air pressure. By its effect on the ability of oxygen to enter the body fluids, air pressure can restrict many terrestrial animals to specific altitude ranges, functioning particularly to limit life forms at high altitudes to specially adapted organisms in sparse populations. Pressure is even more important in determining the depths at which marine organisms live. Certain deep-sea bacteria, for instance, can grow only at a pressure in the vicinity of 500 atmospheres. Similar pressure-loving, or **barophilic,** bacteria live in some deep geological formations, particularly in association with some petroleum and other mineral deposits.

Some limiting factors are **density-dependent** and tend to regulate population density of an organism for that reason. For example, when certain animal species become overcrowded, they tend to suffer from stress diseases engendered from excessive interaction with one another, or from diseases that are more readily transmitted due to overcrowding. You may recognize this as an example of negative feedback, and it may almost be considered a homeostatic mechanism that operates on the population level. **Density-independent** limiting factors, such as, for instance, an occasional blizzard that kills organisms without regard to their previous population density, do not *regulate* population although they of course do affect it, usually temporarily. Density-independent mechanisms are usually more important in determining the species composition of a community, its geographical distribution, and its succession (as discussed in the following sections).

In summary, whether or not an animal or plant can become established in a given region is the result of a complex interplay of physical factors, such as temperature, light, water, winds, and salts, and biotic factors, such as the plants and other animals in that region that may serve as food, compete for food or space, or act as predators or parasites.

Habitat and Ecological Niche

What is a limiting factor for one organism may not affect a second one at all. These differences in requirements result from differences in the adaptations of the organisms, adaptations not only to the differing physical and biological environments of these organisms but also to their different life styles.

It is useful to distinguish between *where* an organism lives and what it *does* as part of its ecosystem. The **habitat** of an organism is the place where it lives, a physical area, some specific part of the earth's surface, air, soil, or water. It may be as large as the ocean or a prairie, or as small and restricted as the underside of a rotten log or the intestine of a termite, but it is always a tangible, physically demarcated region. Many different kinds of organisms may live in a particular habitat.

In contrast, the **ecological niche** is the status or role of an organism within the community or ecosystem. It depends on the organism's structural adaptations, physiological responses, and behavior. It may be helpful to think of the habitat as an organism's address (where it lives), and of the ecological niche as its profession (what it does biologically). The ecological niche is not a physically demarcated space but an abstraction that includes all the physical, chemical, physiological, and biotic factors that an organism requires. To describe an organism's ecological niche, we must know what it eats, its range of movement, and its effects on other organisms and on the non-living parts of its surroundings. Of course, ecological niche does *determine* habitat; we would not expect to find a penguin in the middle of an Amazon rain forest.

Communities are by no means completely uniform habitats. Most of them contain a variety of **microhabitats,** each of which contains its characteristic assemblage of organisms. The water trapped in the leaves of a pitcher plant, for instance, contains a characteristic assemblage of protozoans and even mosquito larvae that somehow manage to escape digestion.

All biotic communities show marked **vertical stratification,** determined in large part by vertical differences in physical factors such as temperature, light, and oxygen. The operation of such physical factors in determining vertical stratification in lakes and in the ocean is quite evident. In a forest there is a vertical stratification of plant life, from mosses and herbs on the ground to shrubs, low trees, and tall trees. Each stratum has a distinctive animal population. Even such highly motile animals as birds are more or less restricted to certain layers. Some species of birds are found only in shrubs, others only in the tops of tall trees. There are daily and seasonal changes in the populations found in each stratum, and some animals are found first in one and then in another layer as they pass through their life histories. These strata are interrelated in diverse ways, and most ecologists consider them to be subdivisions of one large community rather than separate communities. Vertical stratification, by increasing the number of ecological niches in a given surface area, reduces competition between species and enables more species to coexist in a given area.

An ecological niche includes the functional role of an organism as a member of the community—that is, its trophic position and its position with respect to the gradients of temperature, moisture, pH, and other conditions of the environment. The ecological niche of an organism also depends on what it does—how it transforms energy, how it behaves in response to and modifies its physical and biotic environment, and how it is acted upon by other species. To describe the complete ecological niche of any species thus requires detailed knowledge of that organism and its environment. This is very difficult to obtain. However, it is often relatively easy to discover the *differences* between ecological niches. Therefore, the concept of ecological niche is most useful when describing adaptational differences between species.

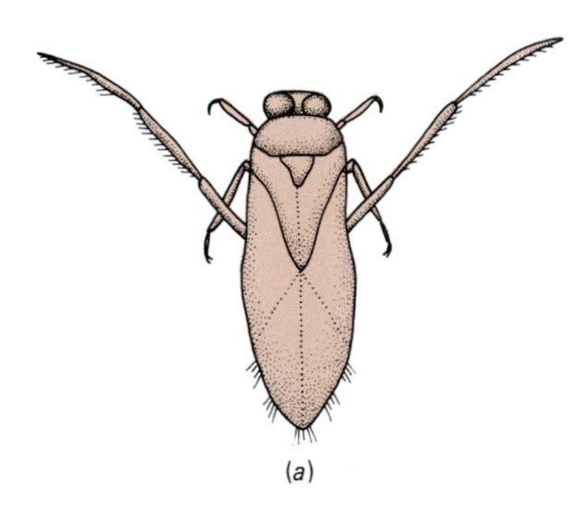
(a)

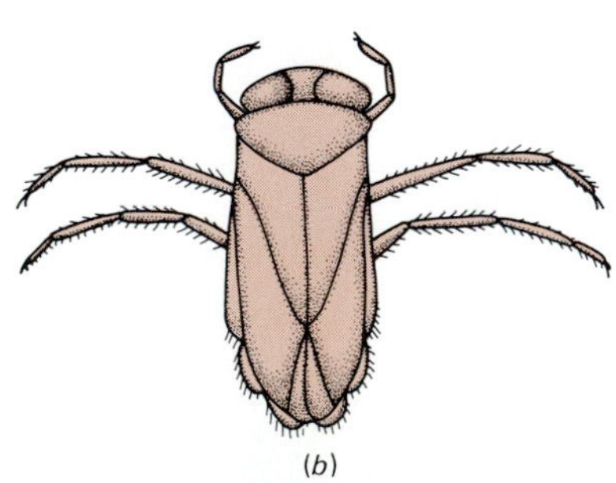
(b)

Figure 50–13 *Notonecta,* the backswimmer, shown in (*a*), and *Corixa,* the water boatman, in (*b*), are two aquatic bugs occupying the same habitat—the shallow, vegetation-choked edges of ponds and lakes—but having different ecological niches.

Two different species of aquatic insects (Fig. 50–13) may live in the same habitat, such as the waters of a small, shallow, vegetation-choked pond, but occupy quite different ecological niches. The backswimmer, *Notonecta,* is a predator that swims about catching and eating other animals. The water boatman, *Corixa,* looks very much like the backswimmer but plays a very different role in the community since it feeds largely on decaying vegetation.

Two species of organisms that occupy the same or similar ecological niches in different geographical locations are termed **ecological equivalents.** The array of species present in a given type of community in different biogeographic regions may differ greatly. However, similar ecosystems tend to develop wherever there are similar physical habitats, with the equivalent functional niches being occupied by whatever biological groups happen to be present in the region. Thus, a savanna biome tends to develop wherever the climate permits the development of extensive grasslands, but the species of grass may be quite different in different parts of the world. On each of the four continents there at one time were grasslands with large grazing herbivores present. These herbivores were all ecological equivalents. However, in North America the grazing herbivores were bison and pronghorn antelope; in Eurasia, the saiga antelope and wild horses; in Africa, other species of antelope and zebra; and in Australia, the large kangaroos. In all four regions these native herbivores have been replaced to a greater or lesser extent by human-domesticated sheep and cattle.

One important generalization of ecology is that no two species may occupy the same ecological niche in the same geographical location. This principle is known as the **competitive exclusion principle,** or **Gause's law.** (Gause was an early twentieth century biologist.) According to this principle, one of the species will eventually be forced to extinction by the other because one of them is bound to possess a superior group of adaptations to meet the demands of their common life style. (Competitive exclusion is discussed more fully in the following section.) A single species may occupy somewhat different niches in different regions, depending on such things as the available food supply and the number and kinds of competitors. Some organisms, such as animals with distinctly different stages in their life history, occupy different niches in succession. The frog tadpole is a primary consumer, feeding on plants, but an adult frog is a secondary consumer, feeding on insects and other animals. In contrast, young river turtles are secondary consumers, eating snails, worms, and insects, whereas the adult turtles are primary consumers and eat green plants such as tape grass.

Such an arrangement prevents direct competition between young and adults, a competition that would be bound to work to the disadvantage of the young, were it to occur. The occupation of different niches allows young and adults to fully specialize in the exploitation of two entirely different kinds of food resources. Gause's law appears, however, to have many exceptions. It has come to be viewed with some skepticism in recent years and may require extensive modification and reinterpretation in the future.

Interspecific Interactions

The interaction of members of the same species has been discussed in detail in Chapters 48 and 49. The members of two *different* species of animals or plants may interact with each other in any of several different ways, both positive and negative. Many of these interactions are clearly cooperative or competitive in their nature.

COMPETITION AND COOPERATION

If each species is adversely affected by the other in its search for food, space, or some other need, the interaction is one of **competition.** Two species may compete for the same space, food, or light, or in avoiding predators or disease. They are, in a sense, competing for the same ecological niche. Competition may cause one species to die off or to change its ecological niche—to move away or to utilize a different source of food. Careful ecological studies usually confirm that there is only one species in an ecological niche: complete competitors cannot coexist (the "4-c" rule, or, as we saw earlier, Gause's law). One of the clearest examples of ecological competition was provided by classic experiments with populations of the protozoan *Paramecium* carried out by Gause. When either of two closely related species, *Paramecium caudatum* or *Paramecium aurelia,* was cultured separately on a fixed amount of food (bacteria), it multiplied and finally reached a constant level. But when both species were placed in the same culture with a limited amount of food, only *P. aurelia* was left at the end of 16 days. The *P. aurelia* had not attacked the other species nor secreted any harmful substance. It simply had a slightly greater growth rate and thus had been more successful in competing for the limited food supply.

Studies in field ecology also often confirm Gause's law. The cormorant and the shag are two fish-eating, cliff-nesting sea birds that have long coexisted and survived despite the fact that they seemed to occupy the same ecological niche. However, the cormorant feeds on bottom-dwelling fish and shrimp, whereas the shag hunts eels and typical fish in the upper levels of the sea. Further study showed that these birds typically choose slightly different nesting sites on the cliffs. Thus, they are not really complete competitors but have developed considerable **niche differentiation.**

The relationship in which two species habitually live together, one deriving benefit from the association and the other unharmed by it, is known as **commensalism.** Commensalism is especially common in the ocean. Practically every worm burrow and each shellfish contain some uninvited guests—the commensals—that take advantage of the shelter and possibly of the abundant food provided by the host organism without doing it good or harm. Some flatworms live attached to the gills of the horseshoe crab and get their food from the scraps of the crab's meals. They receive shelter and transportation from the host but apparently do not injure

Figure 50–14 A hermit crab carrying several small anemones on its shell. Other examples of such interspecies interactions are given in Focus on Symbiosis, in Chapter 45. (Charles Seaborn.)

it. One of the more startling examples of commensalism is that of a small fish that lives in the posterior end of the digestive tract of the sea cucumber, entering and leaving it at will. These fish are quickly eaten by other fish if removed from their sheltering host.

If both species gain from an association but are able to survive without it, the association is sometimes termed **protocooperation.** Several kinds of crabs put cnidarians of one sort or another on top of their shells, presumably as camouflage (Fig. 50–14). The cnidarians benefit from the association by obtaining particles of food when the crab captures and eats an animal. Neither crab nor cnidarian is absolutely dependent on the other.

When both species gain from an association and are unable to survive separately, the association is called **mutualism.** A striking example of a mutualistic association is that of termites and their intestinal flagellates. Termites are famous for their ability to eat wood; yet they have no enzymes to digest it. In their intestines, however, live certain flagellate protozoa that do have the enzymes to digest the cellulose of wood to sugars. Although the flagellates use some of this sugar for their own metabolism, there is enough left over for the termite. Termites cannot survive without their intestinal inhabitants; newly hatched termites instinctively lick the anus of another termite to obtain a supply of flagellates. Since a termite loses all its flagellates along with most of its gut lining at each molt, termites must live in colonies so that a newly molted individual will be able to get a new supply of flagellates from its neighbor. The flagellates also benefit by this arrangement: They are supplied with plenty of food and the proper environment. They can, in fact, survive only in the intestines of termites.

NEGATIVE INTERACTIONS

In certain types of interspecific associations, one of the species is harmed by the other. If one is harmed but the second is unaffected, the relationship is termed **amensalism.** Organisms that produce antibiotics and the species inhibited by antibiotics are examples of amensalism. The mold *Penicillium* produces **penicillin,** a substance that inhibits the growth of a variety of bacteria. The mold presumably benefits by having a greater food supply when the competing bacteria have been removed. Humans, of course, take advantage of this and culture *Penicillium* and other antibiotic-producing molds in huge quantities to obtain bacteria-inhibiting substances to combat bacterial infections. The use of these bacteria-inhibiting agents has had the unexpected effect of increasing the incidence of fungus-induced diseases in humans. Fungi are normally kept in check by the presence of bacteria; when the bacteria are killed off by antibiotics, pathogenic fungi have a golden opportunity to multiply in the host.

Much the same sort of thing is practiced on a macroscopic scale by plants, especially desert plants. It is striking how evenly spaced desert plants are from one another, which is highly adaptive from their point of view. Each plant needs a very extensive root system to capture the sparse rainfall. If the plants were closely spaced, their roots would interfere with one another so much that none would receive enough water for proper growth. The adaptive spacing is achieved by **allelopathy,** in which the established plants emit substances from their roots or fallen leaves that inhibit the germination of competing plants nearby, often including their own seedlings.

We would be quite wrong if we assumed that the host–parasite or predator–prey relationship was invariably harmful to the host or prey *as a species.* This is usually true when such relationships are first set up, but in time, the forces of natural selection tend to decrease the detrimental effects. If the detrimental effects continued, the parasite would eventually kill off all the hosts, and unless it found a new species to parasitize, it would die itself.

Studies of hundreds of different examples of parasite–host and predator–prey interrelations show that in general, where the associations are of long standing, the long-term effect on the host or prey is not very detrimental and may even be beneficial. Rarely does the activity of a predator have a very substantial impact on the population structure of its prey; other limiting factors are usually more important. Predators do typically remove the old, the sick and the infirm from the prey population and therefore help to maintain a population and genetic structure that is ultimately beneficial to the prey species. The size of the predator population in the wild often varies with the size of the population preyed upon. The swings in the size of the predator population tend to lag a bit behind those of the prey.

On the other hand, newly arrived predators or parasites are usually quite damaging. The plant parasites and insect pests that are most troublesome to us and our crops are usually those that have recently been carried into some new area and thus have a new group of organisms to attack.

The Ecology of Stress

A regularly mowed lawn is a much more highly productive community than might be supposed, with a greater species diversity than is at first evident. Close study would reveal that among the plants, only two growth habits can be found: low and recumbent, as with dandelions or plantain, or tall but able to grow basally rather than apically, as in the grass. Some have both habits, as crabgrass does. It is natural to wonder why only these two growth habits are found. Every time a lawn is mowed, tall-growing plants are chopped off, and with repeated mowing, they are killed. If they possess a basal meristem, as grass does, they can avoid this fate, and if they are low-growing, they elude the mower blade entirely. Rather few species fit these categories, however, and a plant census that compared a regularly mowed lawn to a neglected meadow or to a woodland would disclose a far greater species diversity in those latter, less-stressed environments. Stress, then, of whatever kind, permits only those organisms that are preadapted to it to survive. Different kinds of stress will produce different kinds of degraded communities, with different species compositions, but stress usually simplifies a community and reduces its species diversity. The species that remain may do very well for themselves indeed, achieving high biomass and numbers of individuals.

Other examples of stress include aquatic pollution, chronic exposure to ionizing radiation, continual exposure to pesticides or herbicides, or, in the hospital environment, continuing exposure of bacterial populations to antibiotics. In all of these cases the result is the same in principle: Few species survive the stress, and the community is simplified by it (Fig. 50–15). Moreover, because of that simplification, the community becomes much less resistant to further stress of a different type. In other words, it becomes potentially unstable. If the stress is removed, the community will undergo succession (as we will see shortly) and may eventually be restored to its original state.

Community Succession

Any given area usually has an orderly sequence of communities that change together with the physical conditions and lead finally to a stable, mature community, or **climax community,** which supposedly undergoes no further substantial change if left undisturbed. The entire sequence of communities characteristic of a given region is termed a **sere,** and the individual transitional communities **seral stages,** or **seral communities.** In successive stages there is not only a change in the species of

Figure 50–15 The Dolly Sods area of West Virginia, a community subject to considerable natural stress. The branches of many of the coniferous trees are missing on one side, due mostly to the action of the prevailing winds and the development of ice at low temperatures. Very few plant species are able to grow in this area of sparse soil and harsh climate. (The stress is not entirely natural, for the area has been denuded of much of its original topsoil by disastrous fires in past years.)

Figure 50–16 Examples of succession. (*a*) Primary succession. This view shows small plants growing in a volcanic crater in Hawaii. (*b*) Old-field secondary succession, northern New York State. The conifers are replacing the pasture in the absence of grazing or cultivation. (*c*) Dune succession, Rose Island, Bahamas. Grass (*foreground*) gives way to palms (*background*), and in places, as shown in (*d*), the Australian *Casuarina* "pines" (*background*) and Brazilian "pepper" shrubs (*mid-foreground*) have displaced everything else to form a permanent disturbance climax. (*e*) An underground cavern in Florida suffered roof collapse when the water that formerly filled it subsided. The resulting hole in the ground filled gradually with silt, as shown in (*f*), and when it is viewed at ground level, a stable edaphic climax of cypress trees is found to be growing in the shallow water, as shown in (*g*). ((*a*), Charles Seaborn.)

organisms present but (at least in subclimax stages) usually also an increase in the number of species and in the total biomass.

The gradual process by which one community changes into another is called **succession.** Successions can be **primary,** in which a habitat lacking in preexisting life is colonized (as in the example given in Chapter 30), and **secondary,** in which a preexisting community whose succession has been retarded by stress is allowed to develop further. Ultimately, the two are thought to result in the same kind of climax community if they occur in the same geographic area.

The ultimate causes of these successions are complex. Climate and other physical factors play some role, but the succession is directed in part by the nature of the community itself, because the action of each seral community is to make the area less favorable for itself and more favorable for other species until the stable climax community is reached. Local physical factors such as the nature of the soil, the topography, and the amount of water may cause the succession of communities to stop short of or otherwise differ from the expected climax community in what is called an **edaphic climax.** Odd climax communities can also result from the importation of alien species, especially in areas that have been subject to ecological abuse. Such an exotic climax community is termed a **disturbance climax** (Fig. 50–16).

SOME EXAMPLES OF SUCCESSION

One of the classic studies of ecological succession was made on the shores of Lake Michigan (Fig. 50–17). As the lake has become smaller, it has left successively younger sand dunes, and the stages in ecological succession can be observed pro-

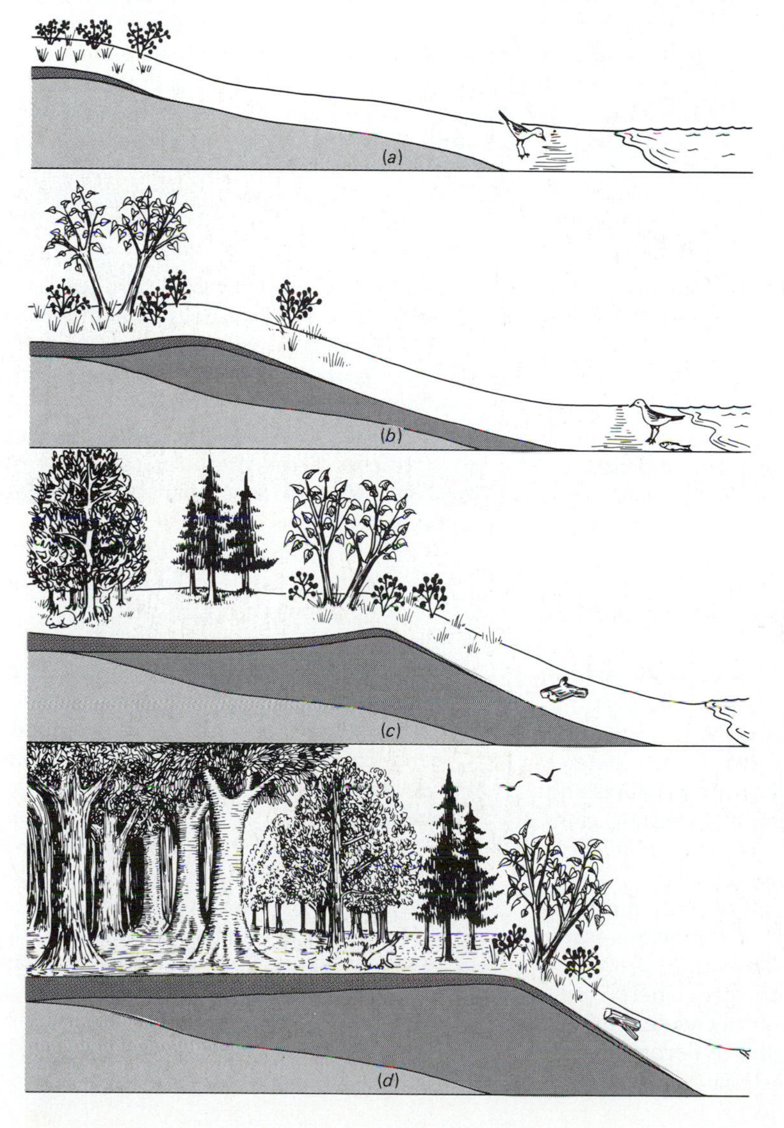

Figure 50–17 Stages in dune succession. Note the beach grass on the fore dunes, shrubs on older dunes, and ancient dunes with an established forest in the background.

gressively from the lake shore inland: The youngest dunes nearest the lake have only grasses and insects. The next older ones have shrubs such as cottonwoods. Evergreens occur next, and finally there is a beech–maple climax community, with deep rich soil full of earthworms and snails.

As the lake retreated it also left a series of ponds. The youngest of these contain little rooted vegetation and lots of bass and bluegills. Later the ponds become choked with vegetation and smaller in size as the basins fill. The ponds then become marshes and finally dry ground, invaded by shrubs and ending in the beech–maple climax forest. Human-made ponds, such as those behind dams, similarly tend to become filled up.

Another more dramatic example of community succession began on August 7, 1883, when a volcanic explosion occurred on the Indonesian island of Krakatoa, causing part of the island to disappear. The remainder was covered with hot volcanic debris to a depth of 60 m, and all life was obliterated. A year later some grass and a single spider were found. By 1908, 202 species of animals had taken up residence on the island. This increased to 621 species by 1919, and to 880 species by 1934, when there was a young forest on one part of the island. Similar studies at the site of the recent Mt. St. Helen's eruption have disclosed much more complex and rapid colonization, probably because it is a much less biologically isolated location.

WHY SUCCESSIONS SUCCEED

Why does ecological succession take place at all? Why don't communities of organisms spring into existence fullblown? Actually, in a few cases something very much like that does indeed take place, but as a rule a series of successor communities of varying lengths is necessary to prepare the habitat for the climax community. As each community modifies its environment (as, for instance when lichens break up rock to become soil), the habitat becomes less suited to its pioneers and more receptive to new organisms. Each community modifies its environment, so that it becomes less harsh and stressful. This permits the establishment of another community that is more complex and more specific in its requirements.

SUMMARY

I. Ecology is a natural science involving the study of interactions among living things, and between living things and their environment.

II. A community is an assemblage of diverse organisms that habitually associate with one another.
- *A.* An ecosystem is a fairly self-sufficient community consisting of producers, consumers, and decomposers, considered together with the non-living surroundings and environment, in which substances and energy are exchanged and flow among the components.
- *B.* An ecosystem may contain subsidiary communities.

III. An ecosystem usually contains producers, which practice autotrophic nutrition (for the most part this category is equivalent to the plants and protists), and consumers and decomposers, which are heterotrophs. The decomposers recycle waste products and dead bodies in the form of simple inorganic compounds for reuse by the producers. They may serve as the basis of certain food chains as well.
- *A.* The autotrophs include the photosynthetic autotrophs—certain bacteria, algae, and plants—and chemosynthetic autotrophs—bacteria.
- *B.* The consumers include carnivores, herbivores, omnivores, parasites, and detritus feeders (which have some of the characteristics of decomposers).
- *C.* The decomposers are the bacteria and fungi.

IV. Matter travels cyclically through ecosystems, but whereas matter has an essentially closed cycle so that it is recycled as well as reused, energy is not capable of being recycled because of the second law of thermodynamics.
- *A.* Energy enters most ecosystems via the photosynthesis of the producers and leaves ultimately in the form of heat radiation, which passes into outer space.
- *B.* By its influence on climate, solar energy has an additional major influence on the occurrence and adaptations of organisms.

V. Biomass is lost on each trophic level of a food chain. Food chains are usually not simple in nature, so that energy passes through most communities in the form of a network of paths—called a food web—which tend to render the community more stable than would be the case if all food chains were mutually independent.

VI. A limiting factor is that essential which is in the shortest supply. Potential limiting factors tend to interact to determine population density and size and distribution of individual organisms and of communities. By their interaction with specific adaptations, they define the ecological niche of each organism.

VII. The ecological niche is the complex of adaptations that determine an organism's life style and role in the community.

VIII. Organisms cooperate or compete. Among their relationships one may distinguish amensalism, commensalism, mutualism, and parasitism. Interspecific competition is a factor in the differentiation of ecological niches.

IX. By cooperative social behavior organisms are often able to increase their adaptive success.

X. Any kind of stress tends to simplify ecosystems.

XI. Communities progress through a series of stages to form a climax community, which has no successor.

POST-TEST

1. In the last analysis, energy is captured and incorporated into food only by the ______________ of biotic communities.
2. Decomposers ______________ materials for initial use by the ______________.
3. In both marine and terrestrial ecosystems, decomposing organic material tends to accumulate in a substratum known as the ______________ ______________.
4. The two major biomass circuits in all ecosystems are the ______________ circuit, based directly or indirectly upon the consumption of the primary producers, and the ______________ circuit, based upon dead materials.
5. A ______________ would be an example of a detritus feeder.
6. Some primary producers are not plantlike and do not utilize the sun as an energy source; these are the ______________ autotrophs.
7. Herbivores, carnivores, and omnivores must find and digest their food; their type of nutritional adaptation is known as the ______________ mode.
8. ______________ nutrition is employed by nonparasitic fungi and most decay bacteria.
9. Although matter moves through the biosphere by closed cycles, ______________ cannot be recycled, though it can be passed along to a limited extent through food chains.
10. Net productivity is the amount of ______________ that a given plant community is able to store in its body tissues over and above the demands of its own ______________.
11. Since energy is lost at each step of a food chain, there is a progressive decrease in ______________ that often limits the length of the food chain.
12. Limiting factors define a range of tolerance that determines both ______________ ______________ and ______________ of organisms.
13. A limiting factor is whatever essential is in ______________ supply.
14. ______________ is an important limiting factor in aquatic communities, for it penetrates water poorly.
15. Stress produced by overpopulation could be a ______________-dependent limiting factor.
16. The status or role or life style of an organism in a community is known as its ______________ ______________.
17. According to the ______________ ______________ principle (Gause's law), complete competitors cannot coexist in the same ecological niche.
18. The mutual inhibition of germination or growth by plants is termed ______________.
19. In general, stress tends to ______________ communities of organisms.
20. A ______________ ______________ is a community that is held to have no successor.
21. Early members of a ______________ of communities are usually more resistant to harsh environmental conditions than are their successors.

REVIEW QUESTIONS

1. What are some of the advantages to be gained by organisms living together in groups?
2. What is the difference between a sere and a climax community? between primary and secondary succession?
3. What is allelopathy? Why is it advantageous? In what kind of habitats do you think it would be most likely to occur?
4. Why is biomass lost on each successive trophic level of a food chain?
5. What are the various types of consumers that might be found in a typical ecosystem? List them and give their typical sources of food.
6. How does a detritus feeder differ from other consumers? Should it be considered a decomposer?
7. Describe or diagram the carbon cycle. Look up the nitrogen and phosphorus cycles in Chapter 30. How do they differ most significantly from the carbon cycle?
8. What is a limiting factor? What actual or potential limiting factors might govern the population of houseflies? of starlings? of lions? of people?
9. People and cockroaches in a sense occupy the same habitat, but not the same ecological niche. Describe the differences in niche between the two as completely as you can. Can the two organisms be said to be competitors? If so, how is it that they manage to coexist?
10. In parts of the American South, hardwoods are the usual climax vegetational type, but are economically much less valuable than the slash and yellow pine trees that characterize an earlier successional stage. Often, farmers periodically burn their woodlands (in controlled fashion), which if properly done does not harm the pines but eliminates the hardwoods. What ecological principle or principles does this practice illustrate?
11. Much meat is imported into the United States from developing countries. Is steak an ecological luxury in the United States? If so, why?

51
Population and Community Ecology

OUTLINE

I. Populations
  A. The characteristics of populations
  B. Population cycles
  C. Population dispersion
II. Biotic communities
  A. Terrestrial habitats
    1. The tundra biome
    2. The forest biomes
    3. The grassland biomes
    4. The chaparral biome
    5. The desert biomes
  B. Aquatic habitats
    1. Properties of water
    2. Marine life zones and life styles
      a. The edge of the sea: marshes and estuaries
      b. The intertidal zone
      c. The subtidal zone
      d. The neritic zone
      e. The oceanic region
    3. Freshwater life zones
  C. Succession in aquatic communities

LEARNING OBJECTIVES

After you have read this chapter you should be able to:

1. Give the principal characteristics or parameters of populations, including density, birth rate, death rate, age distribution, biotic potential, rate of dispersion, and growth form.
2. Identify the principal factors responsible for population limitation and dispersion.
3. Briefly describe the principal regional biotic communities (i.e., biomes) of the earth.
4. Describe at least three ways in which the aquatic habitat differs significantly from terrestrial habitats.
5. Describe such marine communities as the estuarine, intertidal, continental shelf, planktonic, and abyssal communities, summarizing (where applicable) their food web relationships, their zones of greatest productivity, their principal environmental constraints, and their general ecological significance.
6. Describe thermal and nutrient stratification in an aquatic habitat.
7. Describe the processes of eutrophication and ecological succession in a typical freshwater pond or lake.

In this chapter we will apply the basic principles of ecology to larger groupings of organisms. Ecology has been called the systems approach to nature. Here we will consider some of these systems and what makes them operate as they do.

Populations

A population may be defined as a group of organisms of the same species occupying a given area. It has characteristics that are a function of the group as a whole, and not of the individual members. Although, for instance, individuals are born and die, individuals do not have birth rates and death rates; these are characteristics of the population as a whole. Modern ecology deals especially with communities and populations, and the study of community organization is a particularly active field at present. Population and community relationships are often even more important in determining the occurrence and survival of organisms in nature than physical factors in the environment.

THE CHARACTERISTICS OF POPULATIONS

One important attribute of a population is **population density**—the number of individuals per unit area or volume, for example, human inhabitants per square mile, trees per acre in a forest, millions of diatoms per cubic meter of sea water. Population density is a measure of the population's success in a given region. Frequently in ecological studies it is important to know not only the population density, but also whether it is changing and, if so, what the rate of change is.

A graph in which the number of organisms or the logarithm of that number is plotted against time is a **population growth curve** (Fig. 51–1). Such curves are characteristic of populations rather than of a single species and are amazingly similar for populations of almost all organisms from bacteria to humans.

Population growth curves have a characteristic shape. When a few individuals enter a previously unoccupied area, growth is slow at first (the positive acceleration phase), and then becomes rapid and increases exponentially (the logarithmic phase). The growth rate eventually slows down as environmental resistance gradually increases (the negative acceleration phase) and finally reaches an equilibrium, or saturation level. This saturation level is termed the **carrying capacity** of the given environment.

The **birth rate,** or **natality,** of a population is simply the number of new individuals produced per unit time. The **maximum natality** is the largest number of organisms that could be produced per unit time under ideal conditions, when there are no limiting factors. This is a constant for a species, determined by physiological factors such as the number of eggs produced per female per unit time, the proportion of females in the species, and so on. The actual birth rate is usually considera-

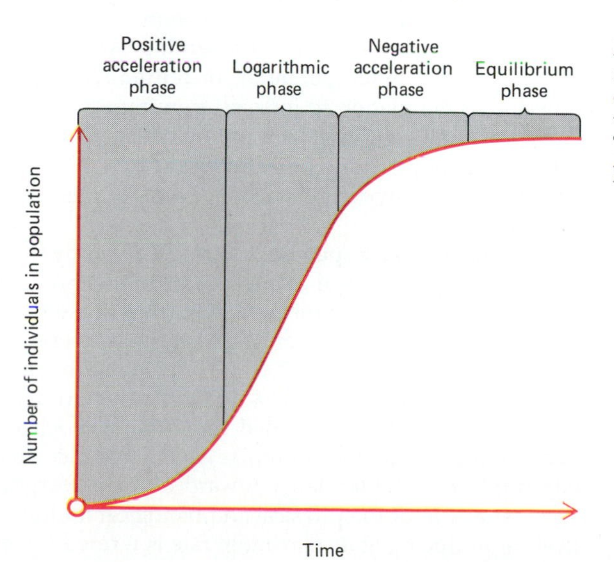

Figure 51–1 A typical sigmoid (S-shaped) growth curve of a population, one in which the total number of individuals is plotted against elapsed time. The shapes of the growth curves of all populations are highly similar.

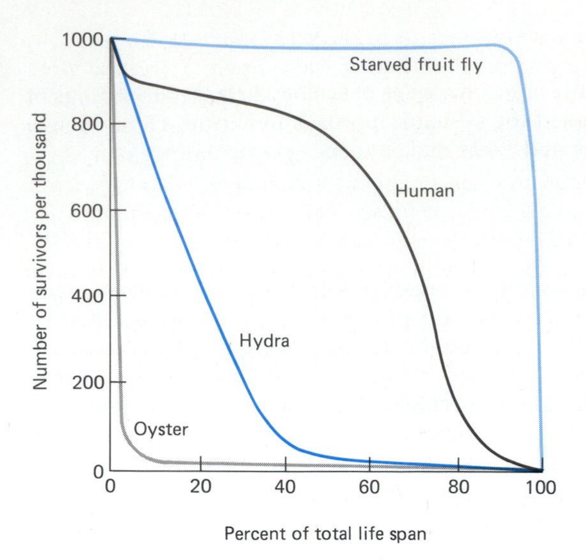

Figure 51–2 Survival curves of four different animals, plotted as number of survivors left at each fraction of the total life span characteristic of the species. The total life span of humans is about 100 years, with most, though by no means all, having died by the Biblical threescore and ten. Note that about 10% of all babies born die during the first few years of life. Only a small fraction of the human population dies between ages 5 and 45 years, but after age 45, the number of survivors decreases rapidly. Starved fruit flies live only about five days, but almost all the population lives the same length of time and most die at the same time. The vast majority of oyster larvae die almost immediately after hatching, but the few that do become attached to the proper sort of rock or to an old shell do survive. The survival curve of hydras is one that is more typical of most animals and plants than the others shown here. In this curve, a relatively constant fraction of the population dies off in each successive time period.

bly less than this, for not all the eggs laid hatch, not all the larvae or young survive, and so on. The size and composition of a population and a variety of environmental conditions affect actual birth rate.

The **mortality rate,** or **death rate,** of a population refers to the number of individuals dying per unit time. There is a theoretical **minimum mortality** (somewhat analogous to the maximum natality), which is the number of deaths that would occur under ideal conditions, deaths due simply to the physiological changes of old age. This minimum mortality rate is also a constant for a given population. The actual mortality rate will vary, depending upon physical factors and on size, composition, and density of population.

A **survival curve** may be obtained by plotting the number of survivors in a population against time (Fig. 51–2). If the units of the time axis are expressed as the percentage of total life span, the survival curves for organisms with very different total life spans can be compared. From such curves can be determined the stage in the life cycle at which a particular species is most vulnerable. Reducing or increasing mortality in this vulnerable period will have the greatest effect on the future size of the population. Since the death rate is more variable and affected to a greater extent by environmental factors than the birth rate, it has a primary role in controlling the size of a population.

It is obvious that populations which differ in the relative numbers of young and old will have different characteristics, different birth and death rates, and different prospects. Death rates typically vary with the ages of individuals in the population. Birth rates are usually proportional to the number of individuals able to reproduce. Rapidly growing populations have a high proportion of young forms. The age of fishes, for example, can be determined from the growth rings on their scales; studies of the age ratios of commercial fish catches are of great use in predicting future catches and in preventing overfishing of a region.

The term **biotic potential,** or **reproductive potential,** refers to the inherent power of a population to increase in numbers when the age ratio is stable and all environmental conditions are optimal. The biotic potential is defined mathematically as the slope of the population growth curve during the logarithmic phase of growth. When environmental conditions are less than optimal, the rate of population growth is less.

Even when a population is growing rapidly in number, each *individual* organism of reproductive age carries on reproduction at the same rate as at any other time; the increase in numbers is due to increased survival. At a conservative estimate, one man and one woman, with the cooperation of their children and grandchildren, could give rise to 200,000 progeny within a century, and a pair of fruit flies could multiply to give 3368×10^{52} offspring in a year. Since optimal conditions cannot possibly be maintained indefinitely, such biological catastrophes do not occur, but the situations in India, Africa, and elsewhere indicate the potential tragedy implicit in the tendency toward human overpopulation.

The sum of the physical and biological limiting factors that prevent a species from reproducing at its maximum rate is termed the **environmental resistance.** It is environmental resistance that saves the world from becoming converted into a solid mass of fruit flies, elephants, or people. Put another way, the difference between the potential ability of a population to increase and the actual change in the size of the population is a measure of environmental resistance. Environmental resistance is often low when a species is first introduced into a new territory, so that the species increases at an enormous rate, as when the rabbit was introduced into Australia, and the English sparrow and Japanese beetle were brought into the United States. However, as a species increases in number the environmental resistance to it also increases, in the form of both the organisms that prey upon or parasitize it, and in the competition among the members of the species for food and living space. Eventually, an equilibrium will be reached either by decreasing the birth rate or by increasing the death rate. This topic is discussed further in Chapter 52.

POPULATION CYCLES

Once a population has reached the equilibrium level, the numbers will vary up and down from year to year, depending on variations in environmental resistance or on factors intrinsic to the population. Some of these population variations are completely irregular, but others are regular and cyclic.

One of the best-known of these is the regular nine- to ten-year cycle of abundance and scarcity of the snowshoe hare and the lynx in Canada, which can be traced from the records of the number of pelts received by the Hudson Bay Company. The peak of the hare population comes about a year before the peak of the lynx population (Fig. 51–3). Since the lynx feeds on the hare, it is obvious that the lynx cycle is related to the hare cycle. More recent (1983) studies in Newfoundland demonstrate that predation by lynx also regulates the population of caribou on that island.

Lemmings and voles are small mouselike animals living in the northern tundra region. Every three or four years there is a great increase in the number of lemmings. They eat all the available food in the region and then migrate in vast numbers looking for food. They may invade villages in hordes, and finally many reach the sea and drown. The numbers of arctic foxes and snowy owls, which feed on lemmings, increase similarly. When the lemming population decreases, the foxes starve and the owls migrate south—accounting for the invasion of snowy owls observed in the United States every three or four years.

Although some cycles recur with great regularity, others do not. For example, in the carefully managed forests of Germany the populations of four species of moths whose caterpillars feed on pine needles were estimated from censuses made each year from 1880 to 1940. The numbers ranged from less than one to more than 10,000 per thousand square meters. The cycles of the four species were quite independent and irregular in frequency and duration. Ecologically speaking, each species was marching to its own tune.

Attempts to explain these vast oscillations in numbers on the basis of climatic changes have been unsuccessful. At one time it was believed that they were caused by sunspots, and the sunspot and lynx cycles do appear to have corresponded during the early part of the 19th century. However, the cycles are of slightly differ-

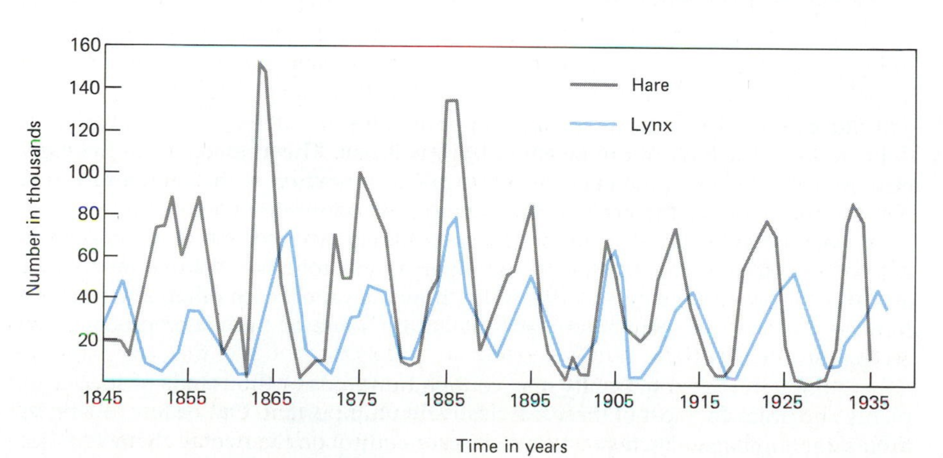

Figure 51–3 Changes in the abundance of the lynx and the snowshoe hare, as indicated by the number of pelts received by the Hudson's Bay Company. This is a classic case of cyclic oscillation in population density. (From MacLulich.)

ent lengths and by 1920 were completely out of phase, with sunspot maxima corresponding to lynx minima. Attempts to correlate these cycles with other periodic weather changes and with cycles of disease organisms have also been unsuccessful.

The snowshoe hares die off cyclically even in the absence of predators and known disease organisms. The animals apparently die of "shock," characterized by low blood sugar, exhaustion, convulsions, and finally death, symptoms that resemble the alarm response induced in laboratory animals subjected to physiological stress (Chapter 42).

This similarity led J. J. Christian to propose in 1950 that their death, like the alarm response, is the result of some upset in the adrenal–pituitary system. As population density increases, there is increasing physiological stress on individual hares, owing to crowding and competition for food. Some individuals are forced into poorer habitats, where food is less abundant and predators more abundant. The physiological stresses stimulate the hypothalamus to signal the adrenal medulla to secrete epinephrine. The pituitary secretes more ACTH, which stimulates the adrenal cortex to produce corticosteroids. The increased concentration of these hormones produces symptoms of stress, leading to physiological shock. In the latter part of the winter of a year of peak population, the stress of cold weather, lack of food, and the onset of the new reproductive season with its additional demands on the pituitary to secrete gonadotropic hormones may cause the adrenal–pituitary system to fail. The animal becomes unable to maintain its normal control of carbohydrate metabolism, and low blood sugar develops, leading to convulsions and death. This is a reasonable hypothesis, but the appropriate experiments and observations in the wild needed to test it have not yet been made.

POPULATION DISPERSION

Populations have a tendency to disperse, or spread out in all directions, until some physical barrier is reached. Within the area, the members of the population may occur at random (this is rarely found), they may be distributed more or less uniformly throughout the area (this occurs when there is competition or antagonism to keep them apart), or, most commonly, they may occur in small groups or clumps.

Group living may increase the competition between the members of the society for food or space, but this is more than counterbalanced by the greater survival power of the group during unfavorable periods (see Chapter 49.) Thus, a group of animals can have much greater resistance than a single individual to many adverse conditions and may be able to exploit food resources of the environment more effectively than they could if isolated.

As we saw in Chapter 49, in many animal species, individuals (or groups) are often found regularly spaced apart, so that each member (or social group of a population) tends to occupy a certain area or **territory,** which is defended against intrusion by other members of the same species and sex (or sometimes other ecologically competitive species). Recall that territoriality may have survival value for a species by ensuring an adequate amount of food, nesting materials, and cover for the young; in protecting the female and young against other males; and in limiting the population to a density that can be supported by the environment. Many species of birds and some mammals, fishes, crabs, and insects establish such territories, either as regions for gathering food or as nesting areas.

Biotic Communities

A **biotic community** is an assemblage of populations of different species living in a defined area or habitat. It can be either large or small. The interactions of the various kinds of organisms maintain the structure and function of the community and provide the basis for the ecological regulation of community succession.

Sometimes adjacent communities are sharply defined and separated from each other. More frequently they blend imperceptibly together. Why certain plants and animals constitute a given community, how they affect each other, and how we humans can control them to our advantage are some of the major problems of ecological research.

Although each community may contain hundreds or thousands of species of plants and animals, most of these are relatively unimportant. Only a few (owing to their size, numbers, or activities) exert a major control on the overall characteristics of the community. In land communities these major species are usually plants, for they both produce food and provide shelter for many other species. Many land communities are named for their dominant plants—sagebrush, oak–hickory, pine, and so on. Aquatic communities, containing no conspicuous large plants, are usually named for some physical characteristic—stream rapids community, mud flat community, or sandy beach community.

The population density of any organism that can be maintained in a particular habitat, the carrying capacity, depends not only upon the availability of food. Carrying capacity often depends upon subtler and not easily assessed factors such as the availability of adequate cover for protection against predators, the prevalence of diseases and natural enemies, and the presence of suitable nesting sites. (It was the lack of proper nesting sites that produced the apparent demise of the ivory-billed woodpecker in the Southeastern United States in the 1930s). It should be clear from these considerations that simply hatching quail or any other desired fish or game species in some artificial propagation program is not by itself an effective way to increase the population density. Usually most such newly released organisms simply fail to survive. The output of many an artifical game propagation program might be put to better use if simply packaged and sold in a supermarket.

Every species of animal and plant tends to produce more offspring than can survive within the normal range of the organism. Due to such factors as dispersal mechanisms and territoriality, there is a strong **population pressure** that tends to force the individuals of each species to spread out and become established in new territories. Competing species, predators, lack of food, adverse climate, and the unsuitability of the adjacent regions (perhaps owing to the lack of some requisite physical or chemical factor) all act to counterbalance the population pressure and to prevent the spread of the species. Since all these factors are subject to change, the range of a species tends to be dynamic rather than static and may change suddenly, often as a result of human activity. The spread of a species is prevented by **geographical barriers,** such as oceans, mountains, deserts, and large rivers, and is facilitated by **natural highways,** such as land connections between continents. The present distribution of plants and animals is determined by the barriers and highways that exist now and those that have existed in the geological past.

TERRESTRIAL HABITATS

Biogeographic realms are regions made up of whole continents or of large parts of a continent separated by major geographical barriers and characterized by the presence of certain unique animals and plants. Within these biogeographical realms, and established by a complex interaction of climate, physical factors, and biotic factors, are large, distinct, easily differentiated community units called **biomes.** A biome is a large community unit characterized by the kinds of plants and animals present.

Although a biome can be considered a very large ecosystem (or even a very large community), it differs substantially from such ecosystems as a small pond or patch of woodland in its complexity and internal variations. Thus, the coniferous taiga biome (discussed later on) contains lake, bog, and other ecosystems within itself, in addition to the overall evergreen forest ecosystem. The evergreen forest itself also varies considerably from one locality to the next within the biome.

In each biome the *kind* of climax vegetation is uniform—grasses, conifers, deciduous trees—but the particular *species* of plant may vary in different parts of the biome. The kind of climax vegetation depends upon the physical environment, and the two together determine the kinds of animals present. Although the climax community is the most important part of the character of a biome, the definition of a biome includes not only the actual climax community of a region but also the several intermediate communities that may precede the climax community.

There is usually no sharp line of demarcation between adjacent biomes. Instead, each blends with the next through a fairly broad transition region termed an **ecotone.** There is, for example, an extensive region in northern Canada where the tundra—another major biome—and coniferous forests blend to form the tundra–coniferous forest ecotone. The ecotonal community typically consists of some organisms from each of the adjacent biomes plus some that are characteristic of, and perhaps restricted to, the ecotone. The ecotone often contains both a greater number of species and a higher population density than in either adjacent biome; this is known as the **edge effect.**

TABLE 51–1
Some Major Biome Types

Biome	*Distribution*	*Typical Organisms*	*Characteristic Threats*	*Notes*
Tundra	Northernmost circumpolar community of any importance; essentially no equivalent in Southern Hemisphere; similar communities on high mountains of all latitudes	"Reindeer moss," sedges, heather, etc.; mosquitoes and some other insects in summer; owl, ptarmigan, grizzly and polar bears, arctic fox, musk ox, caribou, lemming, stoat, weasel, snowshoe hare, and similar forms; few or no reptiles or amphibians	Mechanical abrasion of slowly regenerating tundra; road building, oil pipelines	Very cold, usually low rainfall; kept wet by low rate of evaporation plus poor drainage caused by permafrost layer; extremely short growing season
Taiga	Northern Europe, Asia, and North America, but in areas of more moderate temperature than the tundra	Coniferous evergreens, some deciduous trees (chiefly in subclimax situations); small seed-eating birds plus their predators, such as hawks; fur-bearing carnivores such as mink, fisher, margin; elk, bear, puma, tiger (Siberia only), wolverine, wolves, etc.	Lumbering, unregulated hunting and trapping, and, in some areas, agricultural development	Thick blanket of dead needles overlying acidic soil poor in mineral nutrients
Temperate deciduous forests	Variable; some probably subclimax stages of evergreen forests; central and southern Europe, eastern North America, western China, Japan, New Zealand, etc.; tend to develop in temperate climates of moderate rainfall	Varied, familiar vertebrate and invertebrate animals; deciduous, hardwood trees and shrubs of many species, e.g., beech, oak, maple, cherry; most animals found near or on the forest floor	Plowing, which mixes top and subsoil, converting it to agricultural land; high human population densities, with spilling over of suburbs onto prime forest or agricultural land	Historically, industrial civilizations have tended to develop in this biome; in many areas, especially Europe, very little left of the original community at this time
Tropical rain forest	Tropical areas of high rainfall, e.g., Congo and Amazon Basins; grassy, open woodland (savanna) produced by more moderate rainfall or a pronounced dry season; some temperate rain forests, usually coniferous	Extremely rich diversity of life forms representing almost all phyla; habitat dominated by multiple stories of broad-leaved evergreen trees. Most animals and many epiphytic plants concentrated in the canopy or treetop zones; high temperatures result in very rapid decomposition of soil organic material.	Lumbering, ill-conceived agricultural development, depletion of animal populations (e.g., monkeys, leopards) by trophy hunting and collecting for the pet trade	Most soil nutrients tied up in living bodies of trees and other vegetation; when the natural vegetational cover is removed, a process of soil impoverishment and oxidation often turns the soil to a stony substance (laterite); original climax community slow to regenerate
Grasslands	Continental interiors, especially temperate situations with rather low rainfall, e.g., the North American Midwest and the Ukraine	Dominated by grasses; large herbivores such as bison, antelope, cattle; jackrabbits, rodents such as prairie dogs, wolves; rich, diverse array of ground-nesting birds	Original community destroyed by any and all agricultural development, for which this habitat is eminently suitable; now almost none of original community left; in places, massive erosion produced by overgrazing and overexploitation	Soil characteristically quite rich in mineral content, which may form a calcified layer some feet below the surface; useful for grazing and, where rainfall permits, for grain crops. Tropical savanna has widely dispersed trees as well as grasses
Deserts	Continental interiors; sparse rainfall; a few (e.g., the Gobi) temperate in climate; the majority, subtropical or tropical; tend to lie in rain shadows of mountains	Drought-resistant vegetation such as sagebrush, cacti, euphorbias, and even some kinds of algae; in moister deserts, animals may be numerous but tend to be nocturnal; numerous reptiles and mammals; some birds	Threatened in some places by irrigation and residential–industrial development; so far only local damage	One of the more variable biomes, the variation depending mainly on rainfall rates; some deserts are essentially devoid of life; irrigation frequently accentuates already high mineral content of soil to saline levels.

Where habitats become excessively fragmented (as in agricultural regions where woodland is divided into small woodlots) the genetic diversity is reduced. This is due probably to the difficulty of avoiding inbreeding within the small, isolated populations. Species diversity is also reduced because many species have minimum requirements for range or territory which may not be met in small plots of habitat isolated from one another by intervening agricultural or residential land use. Thus, despite the much greater edge produced between communities by such fragmentation, the edge effect may be absent or reduced.

The distribution of the biomes appears to be governed by the interaction of four main variables, which also interact to influence one another: growing season, annual temperatures (especially minimum and maximum temperatures), mineral availability, and rainfall (both average and minimum). Thus, temperature and growing season are important factors in the distribution of taiga and tundra, which have a circumpolar distribution. Elsewhere, rainfall and the closely related mineral content of the soil assume greater importance.

Some of the biomes recognized by ecologists are tundra, coniferous forest, deciduous forest, broad-leafed evergreen subtropical forest, grassland, desert, chaparral, and tropical rain forest (characteristics of some major biomes are summarized in Table 51–1). These biomes are distributed as belts around the world (Fig. 51–4). Thus, a person traveling from the equator to the North Pole may encounter tropical rain forests, grassland, desert, deciduous forest, and coniferous forest, and finally reach the tundra of northern Canada, Alaska, or Siberia. Traveling from the East Coast of North America westward, he or she might encounter deciduous forest first and then prairie, plains, desert, and coniferous forest.

Climatic conditions at higher altitudes are in many ways similar to those at higher latitudes. Therefore, the order in which biomes are encountered on the slopes of high mountains is somewhat similar to what a traveler would find on a poleward trip (Figs. 51–5 and 51–6).

The Tundra Biome

Between the Arctic Ocean and polar icecaps and forests to the south lies a circumpolar band of treeless, wet, arctic grassland called the **tundra** (Fig. 51–7). Some 5 million acres of tundra stretch across northern North America, northern Europe, and Siberia. The primary characteristics of this region are low temperatures and short growing season. The amount of precipitation is small, but that is usually not a limiting factor because the rate of evaporation is also low. In fact, the tundra is rather like a swamp in most areas. This is because of the poor drainage produced by the frozen soil in combination with very low topographical relief so that what stream flow does take place is sluggish at best. However, the tundra is to some extent physiologically dry since the low temperatures that prevail there can make it difficult or impossible for plants to absorb water effectively during much of the year. In contrast, mountain tundras tend to be literally dry both because the topography encourages drainage and because high-altitude winds tend to desiccate the habitat.

In typical tundra, the ground usually remains frozen except for the uppermost 10 or 20 cm, which thaws during the brief summer season. The permanently frozen deeper soil layer is called the **permafrost layer.** The rather thin carpet of vegetation includes lichens, mosses, grasses, sedges, and a few low shrubs. Many plants are strikingly adapted to the local conditions. For example, some arctic flowers follow the sun like solar power collectors, and have helically shaped corollas that focus the warm sun on the ovaries of the flower. The animals that are adapted to survive in the tundra are caribou and reindeer, arctic hare, arctic fox, polar bear, wolf, lemming, snowy owl, ptarmigan, and, during the summer, swarms of flies, mosquitoes, and a host of migratory birds (Fig. 51–8).

The caribou and reindeer are highly migratory because there is not enough vegetation produced in any one local area to support them. Although casual inspection might suggest that tundras are rather barren areas, a surprisingly large number of organisms are adapted to survive the cold. During the long daylight hours of the brief summer, the rate of primary production is quite high. The production from the vegetation on the land, from the plants in the many shallow ponds that dot the landscape, and from the phytoplankton in the adjacent Arctic Ocean provides enough food to support a variety of resident fish, mammals, and many kinds of migratory birds and insects. On warmer days, in parts of this habi-

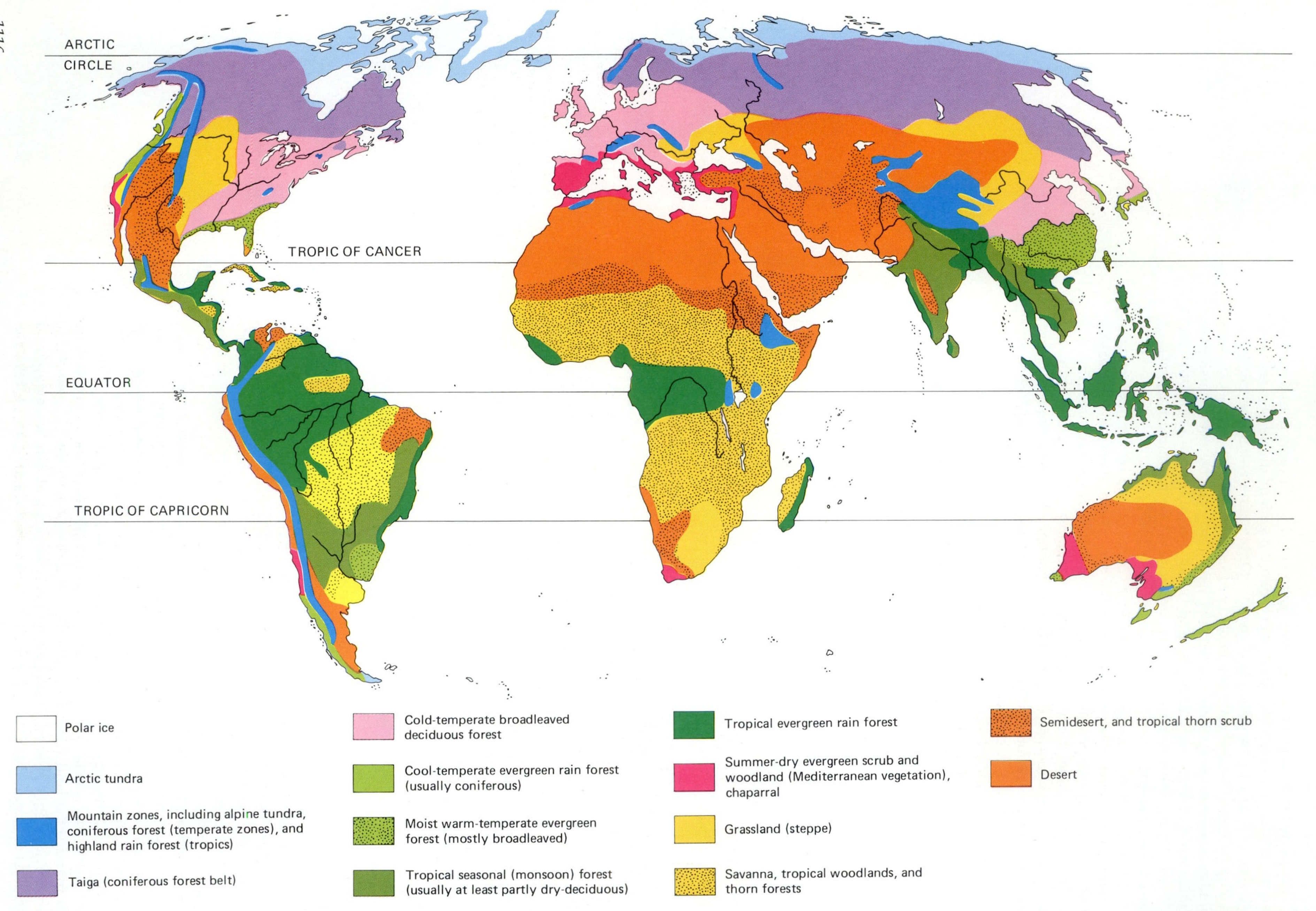

Figure 51–4 A map of the biomes of the earth. Note that only the tundra and the taiga are roughly circumpolar. Other biomes are influenced as much by rainfall as by growing season and therefore have a much less continuous distribution.

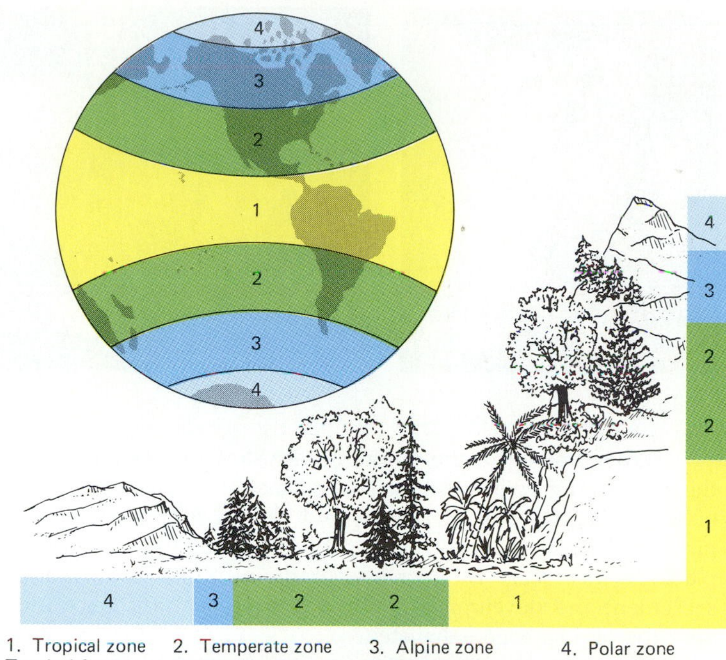

Figure 51–5 Diagram showing correspondence of life zones at successively higher altitudes at the same latitude (*1–4, right*) and at successively higher latitudes at the same altitude (*1–4, left; inset*).

Figure 51–6 Zonation at a northern temperate latitude. Notice the coniferous forest eventually giving way to bare rock and artic-type ice caps. This photo was taken in Glacier Park, Montana; the valley you see and the jagged faces of the rock are from movement of glaciers through the area. (E. R. Degginger.)

tat, some insects whose tissues contain antifreeze chemicals are able to be active even in the snow. (Similar cold-adapted insects, though inconspicuous, can be seen in winter in the temperate zone as well.)

Although there are tundra-like communities at high altitudes in the Southern Hemisphere, there is nothing to correspond to the tundra proper in Antarctic areas because of the lack of land at the proper latitudes.

Figure 51–7 An example of tundra vegetation, on Baylot Island in northeastern Canada. (E. R. Degginger.)

(*a*)

(*b*)

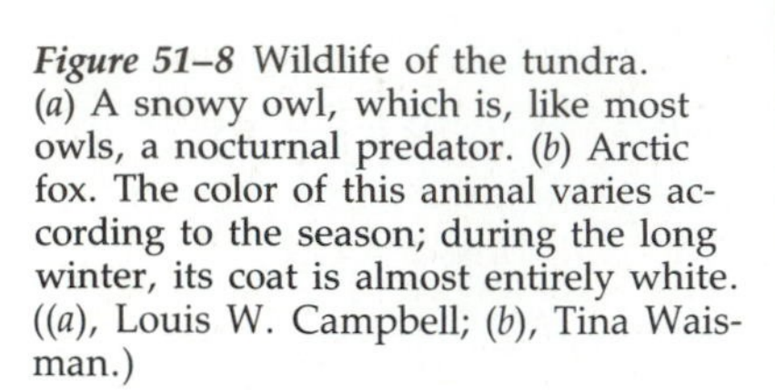

Figure 51–8 Wildlife of the tundra. (*a*) A snowy owl, which is, like most owls, a nocturnal predator. (*b*) Arctic fox. The color of this animal varies according to the season; during the long winter, its coat is almost entirely white. ((*a*), Louis W. Campbell; (*b*), Tina Waisman.)

When damaged, the tundra is very slow to regenerate, probably because of the very short growing season. This is a matter of great concern in the development of oil-producing areas of Alaska, and also in the maintenance of the Alaska oil pipeline. Damage done by military vehicles in the 1940s is still very evident in some places in the 1980s. The increased traffic resulting as an indirect effect of Arctic settlement and development may mar the tundra for generations to come, even after the oil has been exhausted.

The Forest Biomes

Several different types of forest biomes can be distinguished, and in the Northern Hemisphere these are arranged generally on a gradient from north to south or from high altitude to lower altitude, when precipitation permits. Adjacent to the tundra region either at high latitude or high altitude is the **northern coniferous forest,** or **taiga** (Fig. 51–9), which stretches across both North America and Eurasia just south of the tundra. This is characterized by spruce, fir, and pine trees, and by animals such as the snowshoe hare, lynx, Siberian tiger, and wolf. It is the southernmost of the circumpolar biomes.

The evergreen conifers provide dense shade throughout the year. This tends to inhibit development of shrubs and a herbaceous undergrowth (Fig. 51–10). Like the tundra, the taiga is a physiologically dry habitat, and the characteristic water-conserving adaptations of conifers serve them well in this habitat. Deciduous trees do occur in the taiga, but typically as subclimax vegetation. The continuous presence of green leaves permits photosynthesis to occur throughout much of the year despite the low temperature during the winter and results in a fairly high annual rate of primary production.

These coniferous forests are the major source of commercial lumber around the world. After they have fallen, the needles decay very slowly, and the soil devel-

Figure 51–9 An example of taiga, or northern coniferous forest, in Spruce Valley, McKinley Park, Alaska. (E. R. Degginger.)

Figure 51–10 A common inhabitant of the coniferous forest, a cow moose. Also seen in this photograph is the limited development of the underbrush typical of a dense coniferous forest. (J. N. A. Lott, McMaster University/BPS.)

Figure 51–11 A moist coniferous biome. This photograph shows conifers growing along the bank of the Hoh River in Olympic National Park, Washington State. (L. Egede-Nissen/BPS.)

ops a characteristic condition with a high organic content and high acidity, but with relatively little actual humus. The northern coniferous forest, like the tundra, shows a very marked seasonal periodicity, and the populations of animals tend to show marked peaks and depressions in numbers, as in the case of the snowshoe hare that was previously discussed.

Along the west coast of North America, from Alaska south to central California, is a region termed the **moist coniferous forest** biome, characterized by a much greater humidity, somewhat higher temperatures, and smaller seasonal temperature ranges than those of the classic coniferous forests farther north. There is high seasonal rainfall (from 75 to 375 cm per year) in addition to a great deal of moisture contributed by the frequent fogs. The seasonal distribution of the rainfall is, however, very uneven, so that the forest trees do have sharply defined growing seasons much like the growing seasons found (for different reasons) in more northerly, colder regions. There are forests of Sitka spruce in the northern section; western hemlock, arbor vitae, and Douglas fir in the Puget Sound area; and the coastal redwood, *Sequoia sempervirens,* in California. These forests typically have a luxuriant ground cover of ferns and other herbaceous plants. The potential timber production of this region is very great, and with careful foresting and replanting, the annual crop of lumber is very high, at least for the present (Fig. 51–11).

The **temperate deciduous forest** biome (Fig. 51–12) is found in areas with abundant, evenly distributed rainfall (75 to 150 cm annually) and moderate temperatures with distinct summers and winters. Temperate deciduous forest biomes originally covered eastern North America, all of Europe, parts of Japan and Australia, and the southern portion of South America.

The trees present—beech, maple, oak, hickory, and chestnut—lose their leaves during half the year; thus, the contrast between winter and summer is marked. The undergrowth of shrubs and herbs is generally well developed. The animals originally present in the North American forest included deer, bears, North American puma, moose, squirrels, gray foxes, bobcats, wild turkeys, and woodpeckers, among others (Fig. 51–13). Much of this forest region has now been replaced by cultivated fields and by cities.

(*a*)

(*b*)

Figure 51–12 Seasonal changes in a temperate deciduous forest. (*a*) Dense, green hardwood foliage during summer. (*b*) Color changes in foliage (see Chapter 27) during fall. ((*b*), E. R. Degginger.)

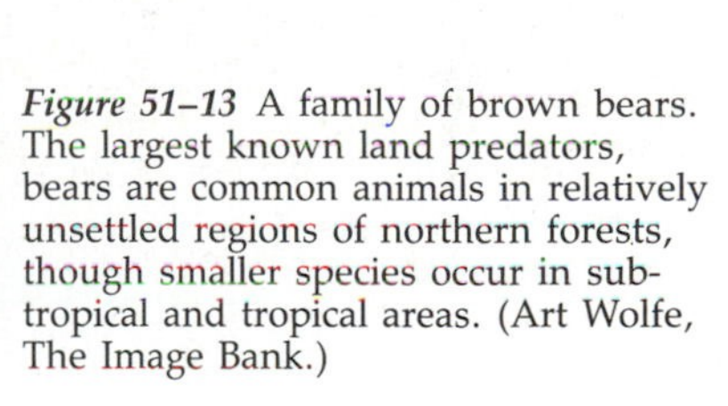

Figure 51–13 A family of brown bears. The largest known land predators, bears are common animals in relatively unsettled regions of northern forests, though smaller species occur in subtropical and tropical areas. (Art Wolfe, The Image Bank.)

Figure 51–14 A broad-leaved evergreen subtropical forest, near Miami, Florida.

In regions of fairly high rainfall but where temperature differences between winter and summer are less marked (as in parts of Florida) is the **broad-leaved evergreen subtropical forest** biome (Fig. 51–14). The vegetation includes live oaks, magnolias, tamarinds, and palm trees, with many vines and epiphytes such as orchids and Spanish "moss."

The variety of life reaches its maximum in the **tropical rain forests** (Fig. 51–15), which occupy low-lying areas near the equator with annual rainfalls of 200 cm or more. The thick rain forests, with a tremendous variety of plants and animals, are found in the valleys of the Amazon, Orinoco, Congo, and Zambesi rivers, and in parts of Central America, Malaya, Borneo, and New Guinea.

The extremely dense vegetation makes it difficult to study or even photograph the rain forest biome. The vegetation is vertically stratified, with tall trees often covered by vines, creepers, lianas, and epiphytes. Under the tall trees is a continuous evergreen carpet, the canopy layer, some 25 to 35 m in height. The lowest layer is an understory that becomes dense where there is a break in the canopy.

No single species of animal or plant is present in a tropical rain forest in large enough numbers to be dominant (Fig. 51–16). The diversity of species is remarkable; there may be more species of plants and insects in a few acres of tropical rain forest than in all of Europe. Part of this is due to the presence of epiphytes—commensal vegetation growing on the trees that produces extensive microhabitats of its own. Certain large epiphytic bromeliads, for instance, store as much as a gallon of rain water in their leaf cups, and absorb the water thus stored between rains by means of tiny scales on the bases of the leaves, which function a bit like root hairs. Mosquito larvae, other insects, and even specialized species of crabs and

Figure 51–15 The tropical rain forest. (*a*) A broad view of a rain forest on one of the Hawaiian Islands. It is possible that some trees you see in this photograph are species introduced by settlers and traders, who began to arrive in the mid-19th century. (*b*), (*c*) Tropical rain forest trees typically possess elaborate systems of buttress roots for support in the wet soil; in very high rainfall areas, their leaves may have elongated "drip tips" to drain water off rapidly so as to discourage the growth of tiny epiphytes, the epiphylls, which would otherwise develop on the leaves and cut off some sunlight from them. (*d*) Other, large epiphytes grow on trunks and branches and the vinelike lianas or even strangler figs (see Chapter 27). ((*a*), Charles Seaborn.)

frogs live in this odd aerial swamp. The trees of the tropical rain forest are usually evergreen (though not coniferous) and rather tall. Their roots are often shallow, forming a mat on the surface of the soil that can be as much as a meter thick, with swollen bases or flying buttresses that help to hold them upright despite their poor soil anchorage.

Since the temperature is high year around, decay organisms decompose organic litter before it really has time to become humus. At the same time, the usually high rate of rainfall leaches nutrients rapidly from the soil. However, highly developed mycorrhizae (Chapter 30) extract nutrients from decomposing material and transfer them to the roots of living plants before they have a chance to actually enter the soil itself. The absorptive mechanisms of the community are so efficient that there is often a lower mineral content in the runoff water than there was in the rain that fell there!

Low light intensity at the ground level may result in sparse herbaceous vegetation and actual bare spots in certain areas. Many of the animals live in the upper layers of the vegetation. Among the characteristic animals are monkeys, sloths, termites, ants, anteaters, many reptiles, and many brilliantly colored birds—parakeets, toucans, and birds of paradise. Actual jungle, with tangled low vegetation, develops only as an intermediate successional stage or on riverbanks where light intensity is high.

(a)

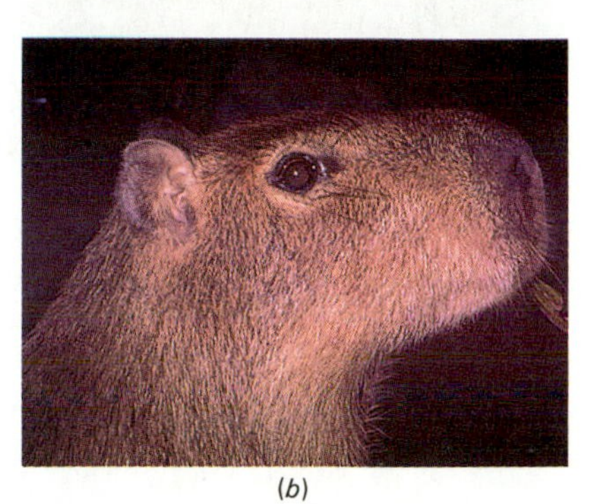

(b)

Figure 51–16 Animals of the rain forest. (*a*) An arrow-poison frog, from Columbia, living inside the cup of a bromeliad (an epiphyte). The bright colors of this frog are a warning that it is inedible. This protects the frog against almost all animals except humans, who use the frog's poison to dip the tips of the arrows they use for hunting. (*b*) A capibara, the largest rodent in South America. These animals are amphibious, spending large amounts of time in the water. The capibara is a staple food item in the diets of local people. ((*a*), Edmund D. Brodie, Jr., Adelphi University/BPS; (*b*), L. E. Gilbert, University of Texas at Austin/BPS.)

The Grassland Biomes

The **grassland biome** (Fig. 51–17) is found where rainfall is about 25 to 75 cm per year, not enough to support a forest, yet more than that of a true desert. Grasslands typically occur in the interiors of continents—the prairies of the western United States and those of Argentina, Australia, southern Russia, and Siberia. Grasslands provide natural pasture for grazing animals, and our principal agricultural food plants have been developed by artificial selection from the grasses.

The mammals of the grassland biome are either grazing or burrowing forms—bison, antelope, zebras, wild horses and asses, rabbits, ground squirrels, prairie dogs, and gophers. These typically aggregate into herds or colonies, which probably provides some protection against predators. The birds characteristic of grasslands are prairie chickens, meadowlarks, and rodent hawks.

The species of grasses present in any given grassland may range from tall species, 150 to 250 cm in height to short species of grass that do not exceed 15 cm in height. Some species of grass grow in clumps or bunches; others spread out and form sods with underground rhizomes. The roots of the several species of grass found in grasslands penetrate deeply into the soil, and the weight of the roots of a healthy plant in usually several times the weight of the shoot. In many grasslands, the above-ground parts of the grass are seasonally deciduous, dying in the cold or dry season, while the roots live on perenially.

Trees and shrubs may occur in grasslands either as scattered individuals or in belts along the streams and rivers. The soil of grasslands is very rich in humus because of the rapid—but not *too* rapid—growth and decay of the individual plants. It also usually is mineral-rich because the rainfall is relatively low, too low to wash the minerals out of the soil structure. In fact, a calcified layer of soil often

Figure 51–17 Western grassland of the North American plains. (E. R. Degginger.)

occurs just below the humus layer (horizon). The grassland soils are well suited for growing cultivated food plants such as corn and wheat, which are species of cultivated grasses. The grasslands are also well adapted to serve as natural pastures for cattle, sheep, and goats. However, when grasslands are subjected to consistent overgrazing and overplowing, they can be turned into deserts. Grassland itself can result from human disturbance of what would otherwise be forest. There is evidence that repeated burning of North American habitats by the original Indian inhabitants produced extensive prairies, which otherwise would have been forests.

Very little of the original grasslands is left in any temperate area. In the American Midwest, for instance, prairie vegetation mainly occurs in such relict areas as abandoned railroad rights of way.

There is a broad belt of **tropical grassland,** or **savanna,** in Africa lying between the Sahara desert and the tropical rain forest of the Congo basin. Other savannas are found in South America and Australia. Although the annual rainfall is high, as much as 125 cm, a distinct, prolonged dry season prevents the development of a forest. During the dry season there may be extensive fires, which play an important role in the ecology of the region. These tropical grasslands (Fig. 51–18) support great numbers and varieties of grazing animals such as antelopes and zebra, and hence predators such as lions.

How to make the best use of these African grasslands is a problem now facing the emerging nations of Africa as they work to raise the level of nutrition in their human populations. Most, if not virtually all, ecologists are of the opinion that it would be better to harvest the native herbivores—antelope, hippopotamuses, and wildebeests—on a sustained-yield basis rather than try to exterminate them completely and substitute cattle, as is usually (and unfortunately) the practice. The diversity of the natural population would mean broader use of all the resources of primary production, and the native species are immune to the many tropical parasites and diseases that plague the cattle that have been introduced. Moreover, introduced cattle, goats, and sheep are far more prone to produce ecologically ruinous overgrazing than the native animals are.

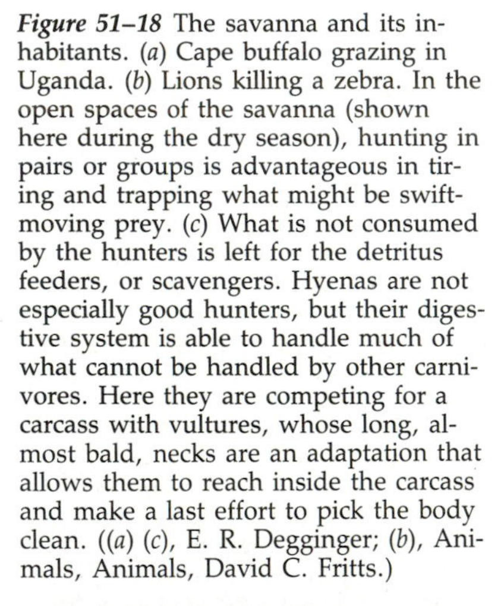

Figure 51–18 The savanna and its inhabitants. (*a*) Cape buffalo grazing in Uganda. (*b*) Lions killing a zebra. In the open spaces of the savanna (shown here during the dry season), hunting in pairs or groups is advantageous in tiring and trapping what might be swift-moving prey. (*c*) What is not consumed by the hunters is left for the detritus feeders, or scavengers. Hyenas are not especially good hunters, but their digestive system is able to handle much of what cannot be handled by other carnivores. Here they are competing for a carcass with vultures, whose long, almost bald, necks are an adaptation that allows them to reach inside the carcass and make a last effort to pick the body clean. ((*a*) (*c*), E. R. Degginger; (*b*), Animals, Animals, David C. Fritts.)

(*a*)

(*b*)

(*c*)

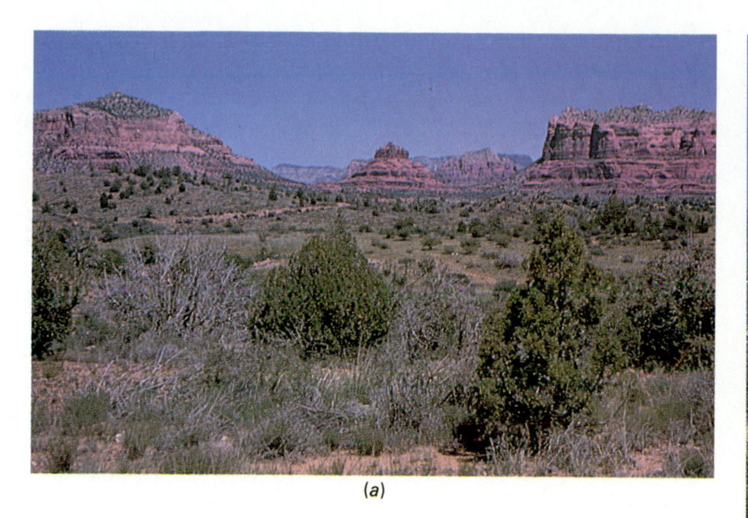

(a)

(b)

(c)

Figure 51–19 The chaparral biome. (*a*) A scrub juniper and oak community in Arizona. (*b*) The yucca is a plant that grows in both chaparral and moist desert environments. It spends a considerable part of its energy budget to create the large flowering stalk that you see; however, since vegetation is widely spaced, the stalk is necessary for it to be visible to its pollinator, the yucca moth. (*c*) A horned lizard, showing protective coloration (Chapter 45). This coloration is an important adaptation to life in a region where ground cover is scarce. ((*a*), J. Robert Waaland, University of Washington/BPS; (*b*) L. Egede-Nissen/BPS; (*c*), L. E. Gilbert, University of Texas at Austin/BPS.)

The Chaparral Biome

In mild, temperate regions of the world with relatively abundant rain in the winter but very dry summers, the climax community includes trees and shrubs with hard, thick evergreen leaves (Fig. 51–19). This type of vegetation is called **chaparral** in California and Mexico, macchie around the Mediterranean, and mellee scrub on Australia's south coast.

The trees and shrubs common in California's chaparral are chamiso and manzanita. Eucalyptus trees introduced from Australia's south coast into California's chaparral region have prospered and have to a considerable extent replaced the native woody vegetation in areas near cities.

Mule deer and many kinds of birds live in the chaparral during the rainy season but move north or to higher altitudes to escape the hot, dry summer. Brush rabbits, wood rats, chipmunks, lizards, wren-tits, and brown towhees are characteristic animals of the chaparral biome. During the hot, dry season there is an ever-present danger of fire, which can sweep rapidly over the chaparral slopes. Following a fire, shrubs sprout vigorously after the first rains and may reach maximum size within 20 years.

The Desert Biomes

A **desert** is defined as a region with less than 25 cm of rain per year, although deserts can also develop in certain hot regions where there may be more rainfall but an uneven distribution in the annual cycle. Vegetation is sparse and consists in North America of greasewood, sagebrush, or cactus, or their ecological equivalents on other continents. The individual plants in the desert are typically widely spaced, with large bare areas separating them. In the brief rainy season the California desert becomes carpeted with an amazing variety of wildflowers and grasses, most of which complete their life cycle from seed to seed in a few weeks. The animals present in the desert are reptiles, insects, and burrowing rodents such as the kangaroo rat and pocket mouse. These rodents are able to live without drinking water, depending instead on the water they are able to extract from the seeds and succulent cactus they eat, plus whatever water they are able to produce metabolically.

The small amount of rainfall may be due to a continued high barometric pressure, as in the Sahara and Australian deserts; to a geographical position directly downwind from high mountains, as in the western North American deserts; or to high altitude, as in the deserts of Tibet and Bolivia. The only absolute deserts, where little or no rain ever falls, are those of northern Chile and the central Sahara. These are almost as devoid of life as is the surface of the moon. Yet, where the soil is favorable an irrigated desert can be extremely productive because of the large amount of sunlight it may receive.

Two types of deserts can be distinguished on the basis of their average temperatures: "hot" deserts, such as those found in Arizona, characterized by the giant saguaro cactus, palo verde trees, and the creosote brush (Fig. 51–20); and "cool" deserts, such as those present in Idaho, dominated by sagebrush (Fig. 51–21).

(a)

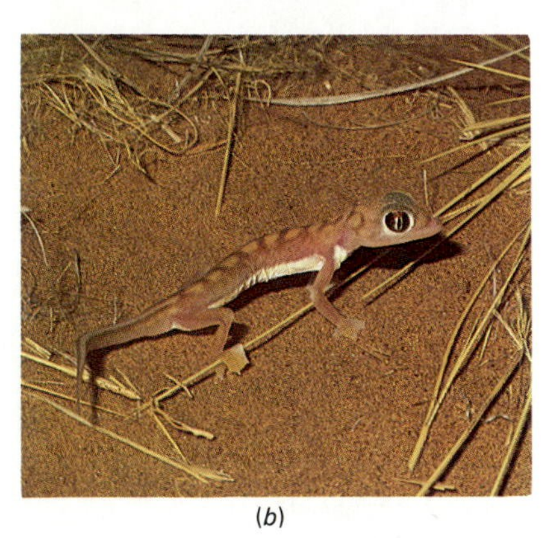

(b)

Figure 51–20 Two inhabitants of hot deserts. (*a*) The giant saguaro cactus of the southwestern United States. (*b*) A gecko, from the Central Namib desert (Africa). Its webbed feet act something like snowshoes so that the lizard can run rapidly on the surface of the sand. ((*a*), E. R. Degginger; (*b*), F. J. Odendaal, Stanford University/BPS.)

Certain reptiles and insects are well adapted for survival in deserts because of their thick, impervious integuments and the fact that they excrete dry waste matter. A few species of mammals are adapted to the desert and excrete very concentrated urine. They also avoid the sun by remaining in their burrows during the day. The camel and the desert birds must have an occasional drink of water but can go for long periods of time using the water stored in the body.

When deserts are irrigated, the large volume of water passing through the irrigation system may lead to the accumulation of salts in the soil as some of the water is evaporated; this will eventually limit the area's productivity. The water supply itself can fail if the watershed from which it is obtained is not cared for properly. In some places, such as parts of Egypt and Arizona, the underground water-storing geological formations **(aquifers)** are not recharged at all at the present day and contain fossil water remaining from past eras but no longer renewed. Irrigation mines this nonrenewable resource, which makes agriculture in those areas a strictly temporary undertaking. Even where more reasonable amounts of rainfall may occur, the ruins of old irrigation systems and of the civilizations they supported in the deserts of North Africa and the Near East remind us that the irrigated desert will retain its productivity only when the entire system is kept in appropriate balance.

AQUATIC HABITATS

To the casual observer, the aquatic habitat looks much the same everywhere, for its watery covering obscures it from view. Nevertheless, beneath and within this obscuring interface is a wide variety of communities and habitats, comparable in diversity to the terrestrial ecosystems, but different enough from them that the two might almost as well exist on separate planets.

Properties of Water

For a review of some of the special properties of water, you may want to refer to Chapter 2. Some important properties to remember are its high specific heat, which contributes to the temperature stability of large volumes of water; the lower density of ice as compared with liquid water, which permits the survival of many bottom-dwelling organisms during winter; and the fact that water is denser than air. This third property is important because organisms themselves are largely composed of water. This means that they have comparable density to that of the surrounding water, allowing the organisms to be buoyed up by it. The buoyancy afforded by water in turn permits a reduction in the need for heavy skeletal structures and allows a growth to a greater size. Aquatic animals of all phyla are able to attain much greater size in the aquatic realm. The blue whale, for instance, is the largest animal known to have existed at any time.

Last, but very significant, water does not transmit light nearly as well as air does. In part, this is a result of the suspended matter universally occurring in natural waters, but it is also a characteristic, to some extent, of pure water. Consequently, active photosynthetic organisms can occur only within about 100 m of the surface (Fig. 51–22). Furthermore, not all wavelengths of light are absorbed to equal

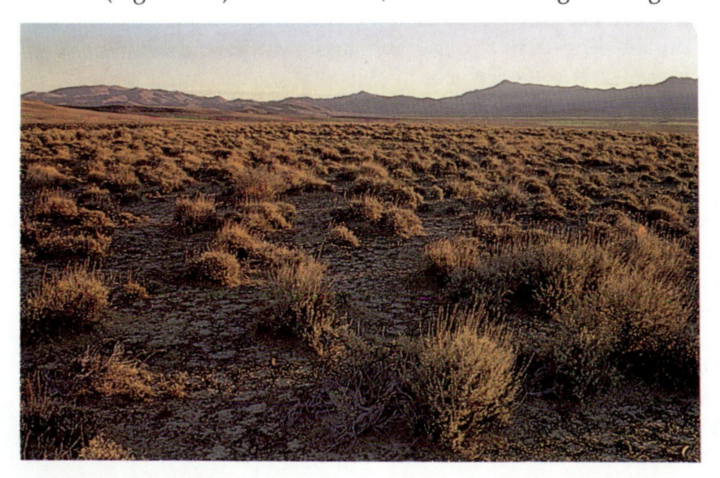

Figure 51–21 A sagebrush (*Artemesia tridentata*) environment in Nevada. This type of growth is typical of cold deserts. (Richard Weymouth Brooks, Photo Researchers, Inc.)

Figure 51–22 This offshore caye (or sand island) is surrounded by an area of shallow water. Notice that vegetation grows only in the shallows. That is because light penetrates water rather poorly; the deeper water becomes, the sparser is its bottom vegetation.

extents. The red end of the spectrum is absorbed rapidly, so that even in shallow depths, light is blue or green in color. These colors are least effective in photosynthesis, for which accessory photosynthetic pigments are only a partial compensation. In all, the aquatic habitat seems singularly unsuited to photosynthesis. Nevertheless, a very large portion of the world's photosynthesis occurs there—in the topmost meter or two, or sometimes only several centimeters.

Marine Life Zones and Life Styles

There has recently been a great upsurge of interest in oceanography in general and in marine ecology in particular as we have begun to appreciate that we have much to learn about the still-mysterious sea. The oceans, which cover 70% of the earth's surface, constitute one of the great reservoirs of living things and of the essential nutrients needed by both land and marine organisms. It is clear that the total biomass in the ocean far exceeds that of land and in fresh water.

The seas are continuous with one another, and marine organisms are restrained from spreading to all parts of the ocean only by factors such as temperature, salinity, and depth. However, the waters of the seas are continually moving in vast currents, such as the Gulf Stream, the North Pacific Current, and the Humboldt Current, which circle in a clockwise fashion in the Northern Hemisphere and counterclockwise in the Southern Hemisphere. These currents not only influence the distribution of marine forms but also have marked effects on the climates of the adjacent land masses. In addition, there are very slow currents of cold, dense water flowing at great depths from the polar regions toward the equator.

Where the wind consistently moves surface water away from steep coastal slopes, water from the deep is brought to the surface by a process termed **upwelling.** This water is cold and rich in nutrients that have accumulated in the depths. Regions of upwelling typically occur on the western coasts of continents, as in California, Peru, and Portugal, and are the most productive of all marine areas. The upwelling produced by the Peru Current has created one of the richest fisheries in the world and supports large populations of seabirds that deposit nitrate- and phosphate-rich **guano** on the headlands and adjacent coastal islands. These upwellings are very important in returning elements such as phosphorus to the surface for recycling.

Although the saltiness of the open ocean is relatively uniform, the concentrations of phosphates, nitrates, and other nutrients vary widely in different parts of the sea and at different times of the year and are usually the major factors limiting the biological productivity of the seas in a given region.

A gently sloping **continental shelf** usually extends some distance offshore. Beyond this, the ocean floor—the **continental slope**—drops steeply to the abyssal region. The region of shallow water over the continental shelf is called the **neritic zone.** It can be subdivided into **supratidal** (above the high tide mark), **intertidal** (between the high and low tide lines), and **subtidal** regions (Fig. 51–23).

The open sea beyond the edge of the continental shelf is the **oceanic region.** The upper layer of the ocean, into which enough light can penetrate to be effective

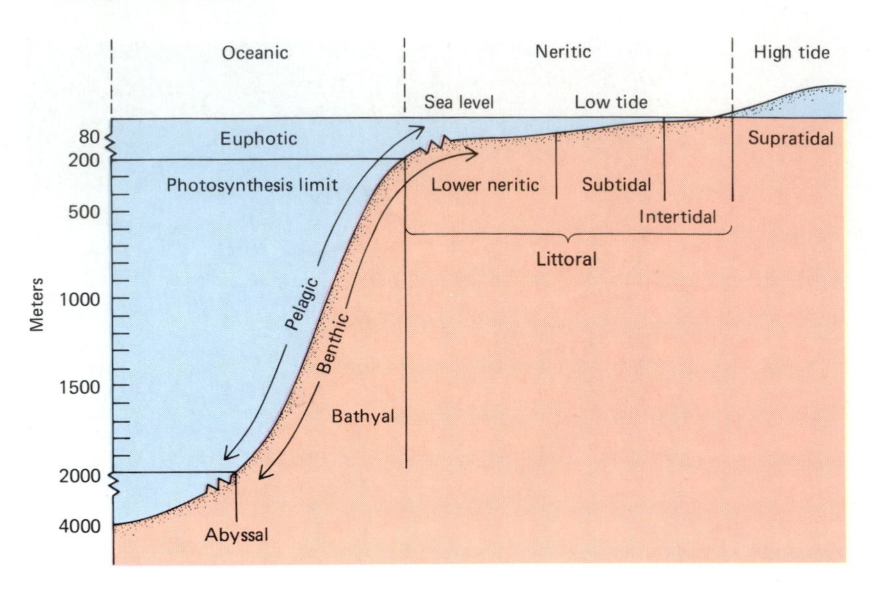

Figure 51–23 Zonation in the sea.

in photosynthesis, is known as the **euphotic zone.** The average lower limit of this zone is about 100 m, but in a few regions of clear tropical water it may extend to twice that depth. The regions of the ocean beneath the euphotic zone are called the **bathyal zone.** The bathyal zone extends over the continental slope to a depth of perhaps 2000 m. The depths of the ocean beyond this zone constitute the **abyssal zone.** Any bathyal or abyssal communities are entirely heterotrophic, except for those associated with hot springs.

The floor of the ocean is not flat but is thrown into gigantic ridges and valleys. Some of the ridges rise nearly to the surface (or above it where there are oceanic islands); some of the valleys lie 10,000 m below the surface of the sea. Huge underwater avalanches occur from time to time as parts of the ridges tumble into the valleys.

Some organisms are bottom-dwellers, collectively called **benthos,** and creep or crawl over the bottom or are **sessile** (attached to it). Others are **pelagic,** living in the open water, and are either **nekton** (active swimmers) or **plankton** (organisms that float with the current). The plankton includes algae, protozoans, small larval forms of a variety of animals, and a few worms. Recent evidence indicates that most of the biomass of the marine plankton consists of microorganisms of bacterial dimensions, the **picoplankton,** which had been missed in earlier plankton studies. The picoplankton contain chlorophyll and are presumably composed of cyanobacteria-like organisms, although surprisingly little is as yet known about them. The existence of the picoplankton will probably force considerable reinterpretation of the ecology of the open sea during the next few years. The nekton includes jellyfish, squid, fishes, turtles, seals, and whales. Some of the benthic animals—crabs, snails, sea stars, certain worms—crawl over the substrate. Clams and worms burrow into the sand, mud, or rock of the sea bottom. Sponges, sea anemones, corals, bryozoans, crinoids, oysters, barnacles, and tunicates are attached to the substratum.

THE EDGE OF THE SEA: MARSHES AND ESTUARIES. Where the sea meets the land there may be one of several kinds of ecosystems with distinctive characteristics: a rocky shore, a sandy beach, an intertidal mud flat, or a tidal estuary containing salt marshes (Fig. 51–24). An **estuary** is a coastal body of water, partly surrounded by land, with access to the open sea and usually a supply of fresh water from a river. Its salinity is intermediate between sea water and fresh water. Most estuaries, particularly those in temperate and arctic regions, undergo marked variations in temperature, salinity, and other physical properties in the course of a year. To survive there, estuarine organisms must have a wide range of tolerance to these changes.

The waters of estuaries are among the most naturally fertile in the world, frequently having a much greater productivity than the adjacent sea or the fresh water up the river. This high productivity is brought about by (1) the action of the tides, which promote a rapid circulation of nutrients and aid in the removal of waste products, (2) by the importation of nutrients into the estuarine ecosystem from the land drained by rivers and creeks that run into the estuary, and (3) by the

(a)

(b)

Figure 51–24 The estuarine environment. (*a*) Brigantine salt marsh estuary in New Jersey. Estuaries are the most biologically productive communities. They are the basis for many aquatic food chains; yet, they are subject to extensive destruction from land development and pollution from rivers that feed into them from populated areas. (*b*) The glasswort, *Salicornia virginia,* is an estuary plant that grows just above the high tide mark in marshy areas. It has adapted to salty soil conditions by its ability to accumulate comparable concentrations of salt in its own tissues. ((*a*), E. R. Degginger; (*b*) Roy R. Lewis, Coastal Creations.)

presence of many kinds of plants, which provide an extensive photosynthetic carpet, and whose roots and stems also mechanically trap much potential food material. Producers include the phytoplankton, the algae living in and on the mud, sand, rocks, or other hard surfaces, and the large attached plants—seaweeds, eel grasses, mangroves, and marsh grasses.

Some marsh grass is eaten by insects and other terrestrial herbivores, but most of it is converted to detritus and consumed by clams, crabs, and other marine detritus feeders. Estuaries are often able to support large populations of fish, oysters, shrimp, and other seafood, which can be tapped by **mariculture,** the marine equivalent of agriculture, as in the oyster farms of Japan where oysters are grown suspended on rafts hanging from floats. This is an excellent way of using the natural productivity of the estuaries to obtain a harvest of protein foods. The farms require considerable space and must be located well apart. They must also be protected from pollution—a very difficult task, since most estuarine areas have been heavily populated and developed by humans.

Estuaries and marshes are among the ecological regions of the world that are most seriously threatened by human activities. They were long considered to be worthless regions in which waste materials could be dumped with impunity. Many have been irretrievably lost by being drained, filled, and converted to housing developments or industrial sites. We are just beginning to appreciate that the best interests of all are served by maintaining estuaries in their natural state and protecting them from wastes and thermal and oil pollution.

THE INTERTIDAL ZONE. The gravitational pulls of the sun and moon each cause bulges on opposite sides of the earth, producing the high tides. The tides advance westward as the earth rotates eastward on its axis. Since the earth completes one rotation a day, there are two high tides and two low tides each day. The intertidal zone is the near-shore benthic zone that is periodically exposed by the tides. It is one of the most favorable of all habitats in the world, and many biologists believe that life may have originated there. The abundance of water, light, oxygen, carbon dioxide, and minerals makes it extremely salutary for plants, and the dense growth of plants, providing food and shelter, makes it an excellent habitat for animals. The autotrophs of the region include a wide variety of algae plus a few grasses. Many of the animals are sessile and are more or less permanently fixed to the sea bottom, though they may be pelagic at some stage of their life cycle. These sessile animals are usually restricted to certain depths of the intertidal zone (Fig. 51–25). There is keen competition among the animals for space and food, so the forms living there require special adaptations to survive.

Since the intertidal zone is exposed to air twice daily, its inhabitants must have some sort of protection against desiccation. Some animals avoid this by burrowing into the damp sand or rocks until the tide returns; others have shells that can be closed, retaining a supply of water inside. Many plants contain jelly-like substances such as agar, which absorb large quantities of water and retain it while the tide is out.

One of the outstanding characteristics of this region is the ever-present action of the waves. Organisms living on a sandy or rocky beach must have ways of

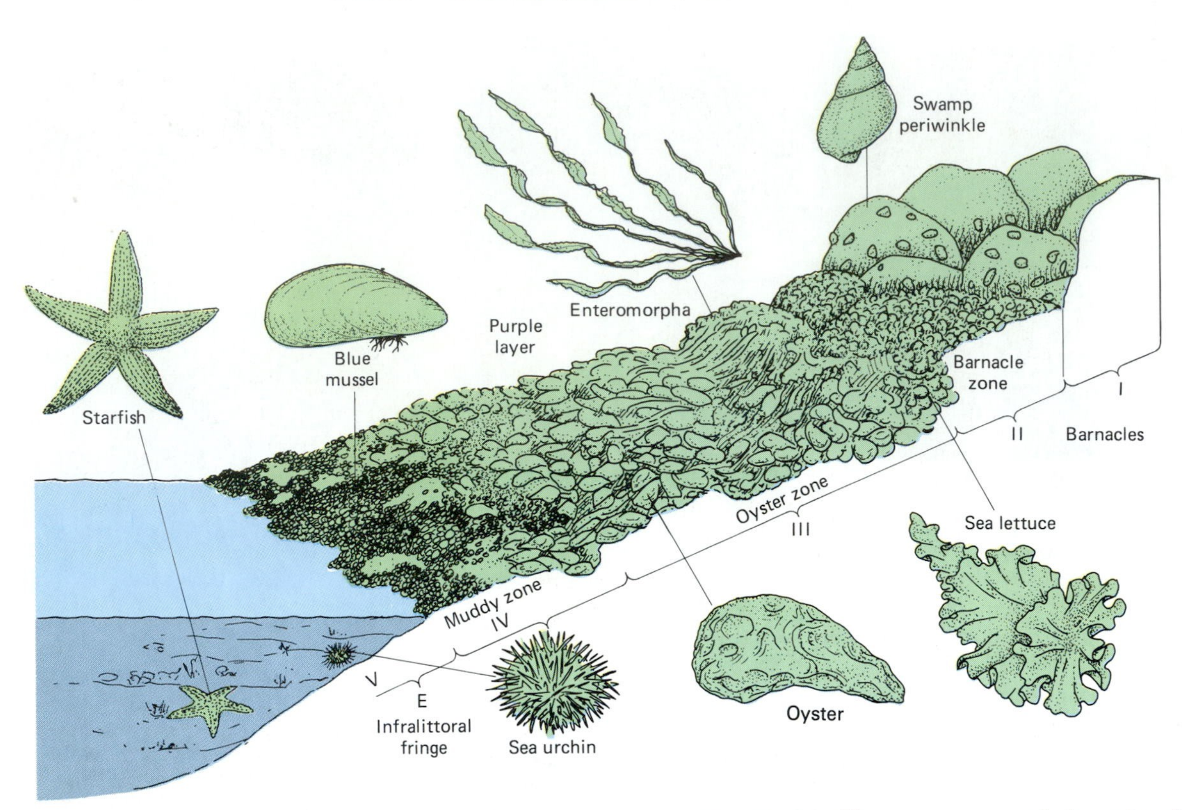

Figure 51–25 Zonation along a rocky shore of the Mid-Atlantic coastline. *I.* Bare rock with some black algae and swamp periwinkle. *II.* Barnacle zone. *III.* Oyster zone, which also includes sea lettuce and purple layer. *IV.* Muddy zone with mussel beds. *V.* Infralittoral zone, with seastars and other typical animals. (From R. Smith.)

resisting wave action (Fig. 51–26). The many seaweeds have tough pliable bodies, able to bend with the waves without breaking, while the animals either are encased in hard calcareous shells, such as those of mollusks, bryozoans, starfish, barnacles, and crabs, or are covered by a strong leathery skin that can bend without breaking, like that of the sea anemone and octopus.

The successive zones of the intertidal region can be seen clearly on most rocky shores (see Fig. 51–25). At the uppermost end is a zone of bare rock marking the transition between land and sea. Next is a spray zone with dark patches of algae on which periwinkles (*Littorina*) graze. Below this is the zone regularly covered by

(*a*)

(*b*)

(*c*)

Figure 51–26 Adaptations to life in the intertidal zone. (*a*) The seaweeds clinging to these rocks have little hard structure to their bodies, yet have strong holdfasts so they can remain in place against the onslaught of the tides. (*b*) These Bahamian snails shelter from the wave action by clinging tightly to a cavity in the rocks. Their shells are so shaped as to offer minimal resistance to the rushing seawater. (*c*) This crustacean is fairly well able to resist drying, and is motile. Thus it can forage actively for food and find hiding places in the rocks and seaweed, where it can stay sheltered during the extremes of high and low tide. ((*a*), (*c*), Charles Seaborn.)

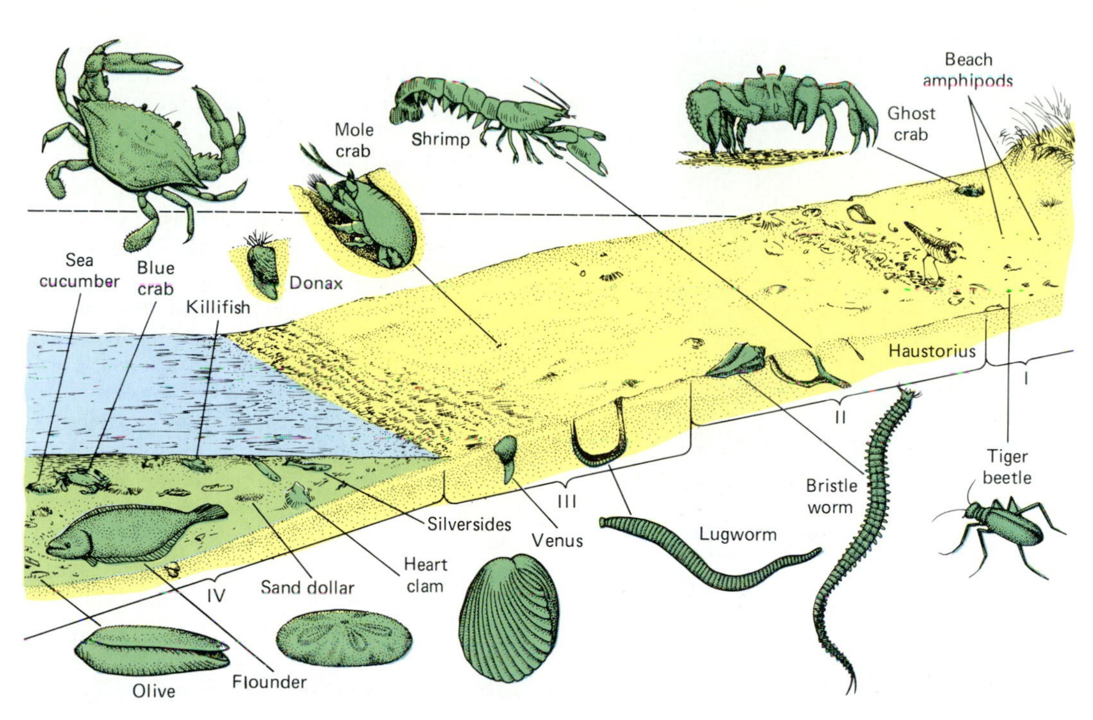

Figure 51–27 (*a*) Life zones on a sandy beach. The gradation, less marked than that of a rocky shore, is nevertheless evident. *I*. Supratidal zone: ghost crabs, and sand fleas. *II*. Flat beach zone: ghost shrimp, bristle worms, clams. *III*. Intertidal zone: clams, lugworms, sand crabs (which follow the retreating or advancing waters). *IV*. Subtidal zone.

the high tide. Rocks in this zone are encrusted with barnacles, limpets, and mussels. On the layer of rocks below this there is a cover of rock kelp and Irish moss containing small crabs and snails. Rocky coasts typically have tide pools that contain characteristic assemblages of animals and plants, whose members do, however, vary according to the depth, location, and other physical variables of the tide pools.

The sandy shore, despite its pleasant appearance, may be an even more harsh environment than the rocky shore. It is subject to all the extremes of the latter as well as the inconvenience of a constantly shifting substratum. This combination makes life on the surface almost impossible, so life has retreated below the surface. Zonation on a sandy beach is illustrated in Figure 51–27. It does not, however, conform to a universal pattern as does that of rocky shores. In the example depicted, the supralittoral zone is inhabited by ghost crabs and beach hoppers. These animals spend most of the daytime hidden in damp burrows and forage at night. Ghost crabs nightly go to the water to dampen their gill chambers. The intertidal zone is not as rich here as on the rocky shore, but it is the home of ghost shrimps, clams, and bristle worms. Lower down on the beach are lugworms, trumpet worms, and other species of clams. Two interesting inhabitants of this zone are the mole crab and the coquina clam. As waves roll up on the beach these small organisms emerge from the sand, ride the waves up the beach, and, as the velocity of the water decreases, burrow quickly into the sand as the waves retreat. Once settled, the crab extends its antennae, and the clam its siphon, to extract particulate food from the receding waves.

THE SUBTIDAL ZONE. The rather ill-defined subtidal zone is also thickly populated, for it has plenty of light and the nutrients required by plants. The absence of the periodic exposure to air and the diminished wave action permit many plants and animals to live here that could not survive in the tidal zone. Many species of fish and many single-celled algae occur in the subtidal zone. The larger seaweeds, which require a substratum for attachment, are found only in the shallower parts of the region. The subtidal zone is the home of cockles, razor clams, moon shells, and other mollusks. Still farther out are sea stars, sand dollars, sea cucumbers, killifish, silversides, and flounders.

In this area of the marine environment, as in all others, the principal **producer organisms,** equivalent to the flowering plants on land, are the **phytoplankton** (see Fig. 17–1), consisting principally of diatoms and dinoflagellates. It is difficult to appreciate their importance because they are so small. An absolutely minimal estimate would place their density at 375 million individuals per cubic meter. In temperate regions the phytoplankton undergoes two seasonal population explosions,

Figure 51–28 Mixed zooplankton from the offshore Atlantic. Seen in this photograph are various crustaceans (including copepods and *Calanus*), a jellyfish, and several worms from different phyla (at far right is an easily identifiable polychaete annelid). (George Whiteley, Photo Researchers, Inc.)

or **blooms,** one in the spring, and the other in late summer or fall. The mechanism is similar to that responsible for blooms in lakes. Not only **zooplankton** (Fig. 51–28), the tiny floating animals, but many other animals feed upon phytoplankton by filtering sea water.

In the open sea and large lakes, organisms die and sink to the bottom. Decomposition releases nutrients, but nutrients are unavailable to the producers up near the surface. In the fall, however, increased wave action, along with changes in temperature, carries nutrients upward—a phenomenon termed autumn **turnover**—which produces the autumn blooms of plankton. Other nutrients are carried to the poles or some other area of upwelling where they come to the surface and stimulate productive plankton communities. Neither phytoplankton nor any other photosynthetic autotroph can grow beneath the **compensation point**—the depth at which the demands of their metabolism exceed their ability to carry out photosynthesis.

In the winter, low temperatures and reduced light restrict photosynthesis to a low level. When spring brings higher temperatures and more light, photosynthesis accelerates, and there is an ample supply of nutrients. The nutrient supply is ample because the winter mixing of surface and deep water brings up nutrients that have fallen to and accumulated at the bottom. Within a fortnight, the diatoms multiply 10,000-fold. This prodigious growth accounts for the spring bloom. Soon, however, the nutrients are exhausted. Replacement from lower layers no longer occurs because warming of the surface water keeps it on top and prevents mixing, often forming a region of sharp temperature change, called a **thermocline,** with resulting **thermal stratification** between the deep, cold layers and the warmer upper layers. Nutrients are now once more locked in the bodies of animals that have eaten the phytoplankton or are slowly falling to the bottom in dead bodies. Whereas temperature and light were the limiting factors during the winter, nutrient level is the limiting factor during the summer, especially since existing phytoplankton are now being consumed by animals. Now nutrients begin to accumulate again in lower layers. As fall approaches, the upper layers of water begin to cool again. The accompanying density change, together with the autumn equinoctial gales, begins mixing the water again. Water rich in phosphates and nitrates is brought up from below. Other forms of phytoplankton, especially nitrogen-fixing cyanobacteria, now bloom until reduced nutrients or low temperature again intercedes (Fig. 51–29).

Another important population in the ocean is the zooplankton, composed of all the animals carried passively by moving water. Every major phylum of the animal kingdom is represented, if not as adults, then at least as eggs or larval stages. While small, the organisms of the zooplankton are visible through a 6× hand lens. The beauty and almost limitless variety of forms of these animals beg description. A sample haul near the Isle of Man yielded an average of 4500 planktonic animals per cubic meter. The same hauls gave about 727,000 planktonic plants per cubic meter. Despite the huge numbers, plankton biomass is not impressive when judged according to the vast volume of water that may be beneath it. It is estimated that the plankton of the open sea, if compressed, would form a layer no more than 3 mm thick.

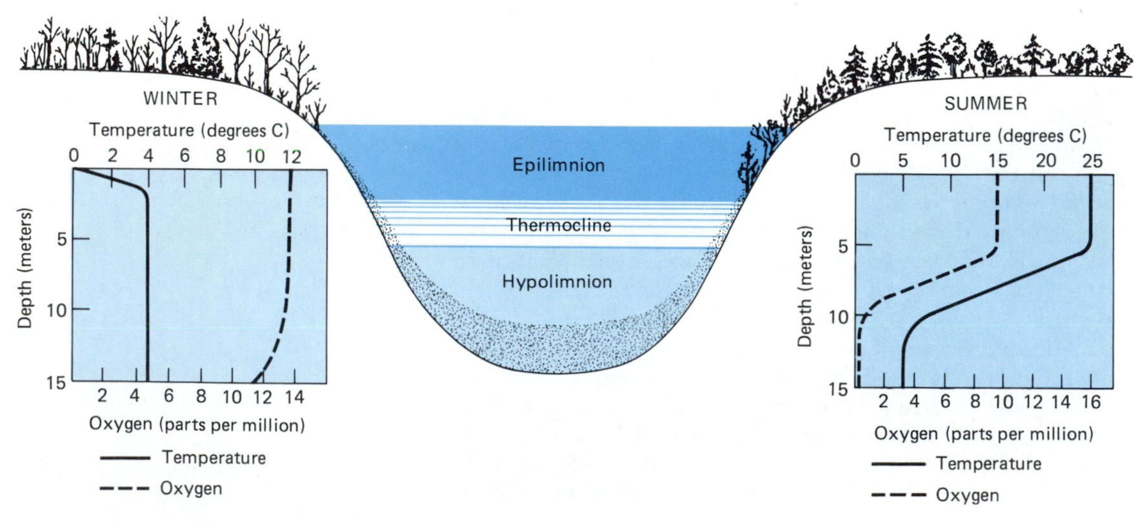

Figure 51–29 Thermal stratification in a north temperate zone lake, Linsley Pond, Connecticut. *Right,* Summer conditions; *left,* winter conditions. Note that in summer the oxygen-rich circulating layer of water (called the epilimnion) is separated from the cold, oxygen-poor water layer (the hypolimnion) by a broad zone, called the thermocline, which is characterized by a rapid change in temperature and oxygen with increasing depth. (From "Life in the Depths of a Pond," by E. S. Deevey, Jr., Copyright by Scientific American, Inc., all rights reserved.)

THE NERITIC ZONE. The active surface dwellers of the neritic zone (essentially the waters on the continental shelf) are many bony fishes, large crustaceans, turtles, seals, whales, and a host of sea birds. All these consumers are restricted in their distribution by temperature, salinity, and nutrients. There is horizontal zonation and vertical stratification. The greatest numbers of fishes (individuals, not species) are found in northern waters where there are cold upwellings. Only a few species make up the bulk of commercial fisheries. Three-fourths of the world's catch consists of herring, cod, haddock, pollock, salmon, flounder, sole, plaice, halibut, mackeral, tuna, and bonito groups. The cod and flounder groups are bottom fish. A great number of birds feed upon these and other marine organisms. Sandpipers, plovers, herons, curlews, and others search the supra- and intertidal zones; cormorants, sea ducks, and pelicans, the subtidal zones; and petrels and shearwaters, the lower neritic zone farther out to sea.

THE OCEANIC REGION. The **oceanic region** is less rich in species and numbers than the coastal areas, but it has its characteristic species. Many of these are transparent or bluish in color and since the sediment-free water of the open sea is marvelously transparent, these animals are nearly invisible. Animals that are too thick to be transparent frequently have smooth shiny and silvery bodies that almost make them invisible by mirroring the water in which they swim. Among the most characteristic animals of the open ocean are the baleen and toothed whales, the only mammals that are truly marine. (Porpoises are actually small toothed whales. Manatees or sea cows are partly freshwater in habitat.) The baleen, or whalebone, whales live on phytoplankton, which they strain from the water (Fig. 51–30). The toothed whales live on nekton, and giant squids comprise part of their diet. The distribution of these whales is correlated directly or indirectly with the distribution of phytoplankton.

The life forms of arctic seas differ from those of tropical seas. The boundaries between the two are not sharp and shift with the seasons. The major tropical oceans, the **Atlantic** and the **Indo-Pacific,** are separated from each other by regions of cold water. Each contains a large number of different species of animals and plants.

Above the surface of the open sea are the oceanic birds—the petrels, albatrosses, and frigate birds. They are not truly marine because they must come to land to breed. Apart from this requirement, though, they may spend all their lives at sea out of sight of land. As with the whales, their distribution may be worldwide but not uniform. They are restricted by the distribution of fishes, which in turn is dependent on the phytoplankton.

Eighty-eight percent of the ocean is more than one mile deep. The ocean is continuous throughout the world except for the deep water of the **Arctic Ocean,** which is cut off from the rest by a narrow submerged mountain range connecting Greenland, Iceland, and Europe. This is an area of great pressure and of perpetual night. Since no photosynthesis (and only a little chemosynthesis near the abyssal hot springs) is possible, the major source of energy is the constant rain of organic debris, the bodies and waste products of organisms in the surface layers, that falls toward the bottom. The other prerequisite for life, oxygen, gets to the bottom by means of the oceanic circulation discussed previously.

Figure 51–30 A baleen whale siphoning the surface of the open ocean for plankton. The bird nearby is hoping to capture some fish that has been stirred near the surface of the water by the motion of the whale. (David W. Hamilton, The Image Bank.)

Figure 51–31 A deep-sea fish. Most fish that live at great ocean depths have rather weak or vestigial eyes. Many have luminous organs for locating each other for mating. Inside the nearly transparent body of this particular fish you can see the remains of small crustaceans, which probably drifted down from the surface waters of the ocean. (Peter David, Photo Researchers, Inc.)

Little is known of life in the deep because of the enormous difficulty of observation. The pelagic animals are strong swimmers, not easily caught in nets. The scant knowledge available has been gleaned from studies of net hauls and by observation from special undersea craft or via underwater television. Most of the animals of the open sea are primarily detritus feeders. They are either **filter feeders,** which sieve out particles before they reach the bottom, or **grubbers,** detritus feeders which ingest sediment.

Since the number of members of any one species in these vast depths is small, finding a mate is more of a problem than in any other region, and some fish have a curious adaptation to ensure that reproduction will occur. At an early age the male becomes attached to and fuses with the head of the female, where he continues to live as a small (2.5-cm) parasite. In due course he becomes mature, and when the female lays her eggs, he releases his sperm into the water to fertilize them.

The bottom of the sea is a soft ooze made of the organic remains and shells of foraminiferans, radiolarians, and other animals and plants. Many invertebrates live on the ocean floor even at great depths. These are usually characterized by thin, almost transparent shells, whereas the related shallow-water forms, exposed to wave action, have thick, hard shells. Apparently even the greatest "deeps" are inhabited, for tube-dwelling worms have been dredged from depths of 8000 m, and sea urchins, sea stars, bryozoa, and brachiopods have been found at depths of 6000 m. Animals living on these bottom oozes typically have long, thin appendages and possess spines or stalks.

The productivity of the true oceanic habitat is so low, however, that it really amounts to a vast aquatic desert. Here there is no effective mixing of surface and abyssal waters because the great depth renders both the temperature changes and the wind action that occur above ineffective. Consequently the euphotic zone is impoverished in nutrients and cannot be highly productive. Opportunities to feed come rarely to the exclusively carnivorous fish of the abyss. Consequently, many deep sea fish have bizarre specializations (Fig. 51–31) connected with predation such as special jaw arrangements that may allow them to feed on victims that are physically larger than themselves! Since the bathyal and abyssal benthos ultimately depends on the photosynthesis of the euphotic zone, its biomass is very low (though somewhat higher than was previously believed). We will never feed the world's swelling population from the "limitless resources of the sea" because they are anything but limitless! (Fig. 51–32).

Freshwater Life Zones

Freshwater habitats can be divided into **standing water**—lakes, ponds, and swamps—and **running water**—rivers, creeks, and springs—each of which can be further subdivided. The biological communities of freshwater habitats are in general more familiar than the saltwater ones, and many of the animals used as specimens in biology classes are from fresh water—amoebas, hydras, planarians, crayfish, and frogs.

In much the same way as the zones of the ocean are distinguished, a body of standing water can be divided into the **littoral zone** (the shallow water near the shore), the **limnetic zone** (the surface waters away from the shore), and the **profundal zone** (the deep waters under the limnetic zone).

Aquatic life is probably most prolific in the littoral zone. Within this zone the plant communities form concentric rings around the pond or lake as the depth

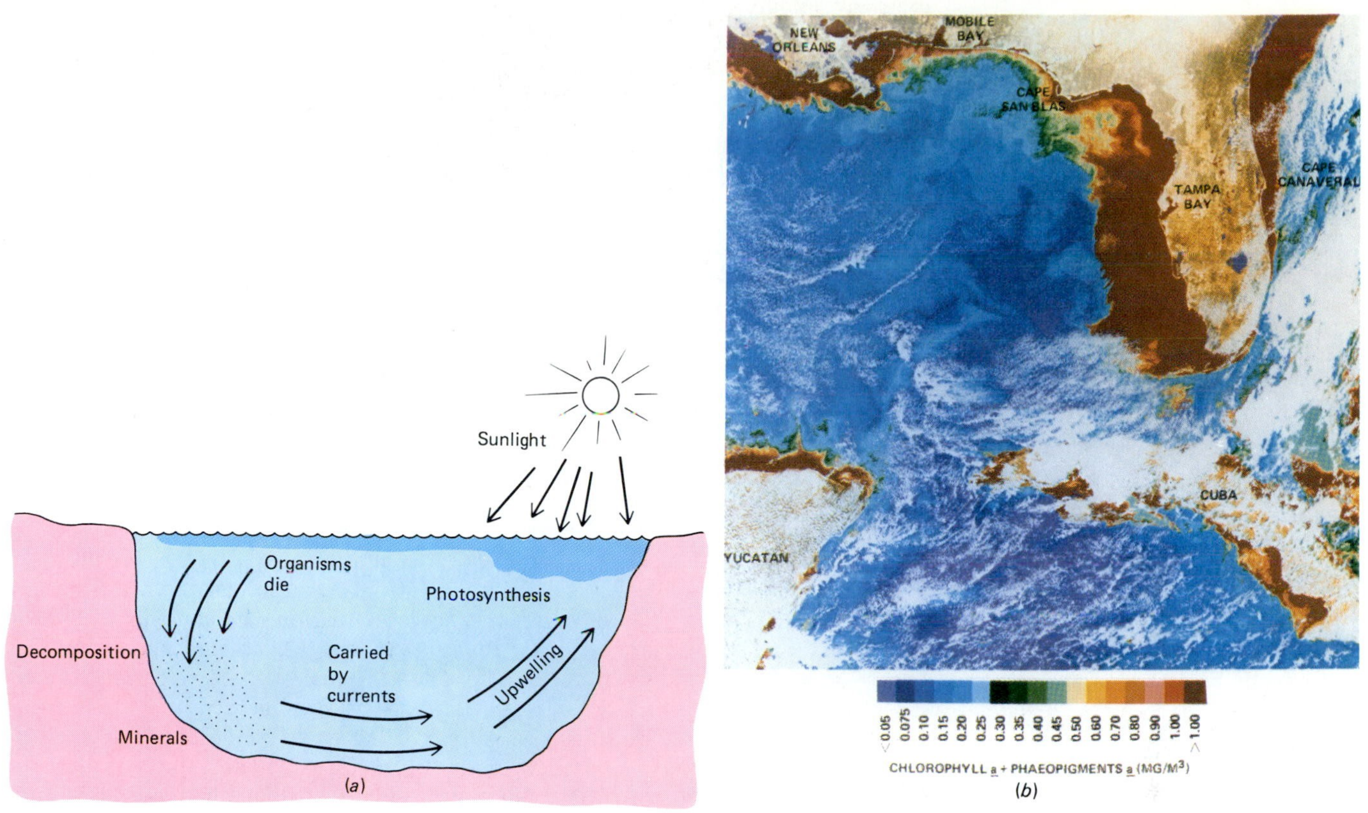

Figure 51–32 Mineral transport in the ocean. (*a*) In the open sea and large lakes, organisms die and sink to the bottom, forming the basis of the food chain that terminates in the detritus feeders. Decomposition releases nutrients, but these are located where they are of no use to the planktonic producers up near the surface. Nutrients are, however, carried to the poles or to some other area of upwelling, usually off the coast of a continent. There they come to the surface and stimulate the growth of productive planktonic communities. Phytoplankton cannot grow beneath the depth at which the demands of their metabolism exceed their ability to carry out photosynthesis (the compensation point). The role of upwelling in providing nutrients, plus the leaching of nutrients from continental land masses by freshwater drainage, results in a high concentration even of plankton in estuaries and continental shelf habitats. (*b*) The coastal enrichment of the ocean is graphically illustrated by this photograph, which was prepared using a special instrument that is sensitive to the chlorophyll content of the water, due mostly to the phytoplankton it contains. Note that plankton growth is confined mostly to continental margins; most of the open sea is therefore little more than an aquatic desert. ((*b*), NASA.)

increases (Fig. 51–33). At the shore proper are the cattails, bullrushes, arrowheads, and pickerelweeds—the emergent, firmly rooted vegetation linking water and land environments. Out slightly deeper are the rooted plants with floating leaves such as water lilies. Still deeper are the fragile, thin-stemmed water weeds, rooted but totally submerged. Here also are found diatoms, cyanobacteria, and green algae such as common green pond scum.

The littoral zone is also the scene of the greatest concentration of animals (Fig. 51–34), distributed in recognizable communities. In or on the bottom are various dragonfly nymphs, crayfish, isopods, worms, snails, and clams. Other animals live in or on plants and other objects projecting up from the bottom. These include the climbing dragonfly and damselfly nymphs, rotifers, flatworms, bryozoans, hydras, snails, and others. The zooplankton consists of water fleas such as *Daphnia,* rotifers, and ostracods. The larger freely swimming animals, collectively called the **nekton,** include diving beetles and bugs, dipterous larvae (e.g., mosquitoes), and large numbers of many other insects. Among the vertebrates are frogs, salamanders, snakes, and turtles. Floating members of the community, which together make up the **neuston,** include whirligig beetles, water striders, and numerous protozoans. Many pond fish (sunfish, top minnows, bass, pike, and gar) spend much of their time in the littoral zone.

The **limnetic,** or open-water zone, is occupied by many protists (dinoflagellates, diatoms, *Euglena*), many small crustaceans (copepods, cladocerans, and so on), and many fish. Its upper, warmer zone is known as the **epilimnion,** and its lower, colder zone is the **hypolimnion.** These zones are separated by a thermocline in the larger lakes.

Profundal (deep) life forms include bacteria, fungi, clams, bloodworms (larvae of midges), annelids, and other small animals capable of surviving in a region of little light and low oxygen. In contrast to ponds, where the littoral zone is large, the water usually shallow, and temperature stratification usually absent, lakes have large limnetic and profundal zones, a marked thermal stratification, and a seasonal cycle of heat and oxygen distribution.

Running water differs in three major aspects from lakes and ponds: (1) Current is a controlling and limiting factor; (2) land–water interchange is great because of the small size and depth of moving water systems as compared with lakes; and

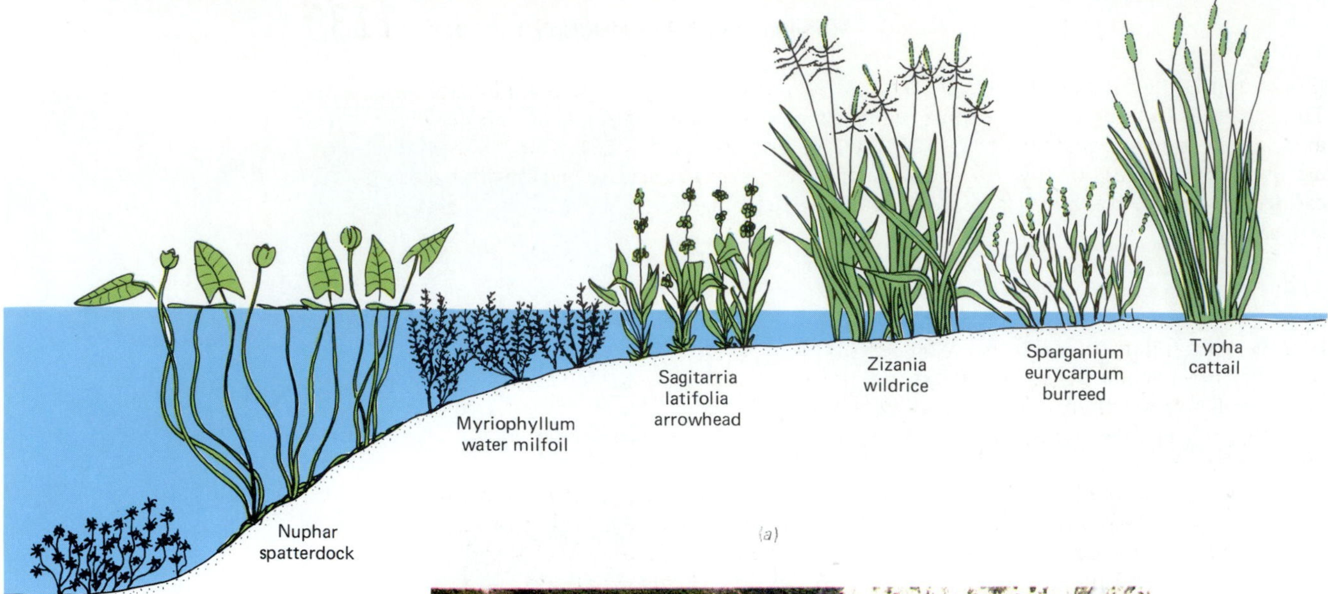

Figure 51–33 (*a*) Zonation of vegetation about ponds and along river banks. Note the changes in vegetation with water depth. (*b*) Bass spawning ground. Each circular crater is a bass "nest." These fish utilize the shallows of the littoral zone for breeding. Lakeside "development" could deplete a lake of bass without one of them being caught by a fisherman. (SELBYPIC.)

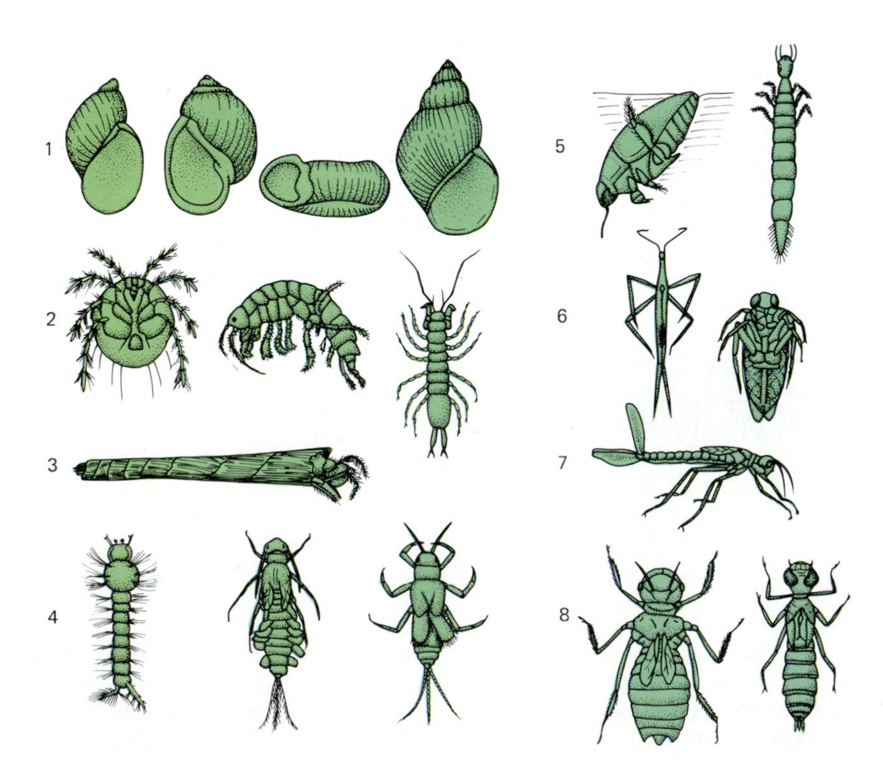

Figure 51–34 Some representative animals of the littoral zone of ponds and lakes. *1–4*, Mainly herbivores; *5–8*, predators. (After Pennak.)

(3) oxygen is almost always in abundant supply, except when there is pollution. The extremes of temperature tend to be greater than in standing water. Plants and animals living in streams are usually attached to surfaces; the nektonic animals that are present are exceptionally strong swimmers. Characteristic stream organisms are caddis fly larvae, blackfly larvae, attached green algae, encrusting diatoms, and aquatic mosses.

SUCCESSION IN AQUATIC COMMUNITIES

Freshwater habitats undergo succession much more rapidly than do most other life zones. Ponds become swamps, swamps become filled in and converted to dry land, and streams erode their banks and change their course. The kinds of plants and animals present may change markedly and show ecological successions similar to those on land.

The benthic habitats of lakes, ponds, and rivers are much more subject to successional processes than those in marine situations. Such succession is largely dependent upon the deposition of silt and is, therefore, more of a geologically than biologically based phenomenon, in contrast to the usual terrestrial successions. Nevertheless, it is just as relentless, if not more so. The woodland pond has been described as a temporary wet spot on the floor of the forest. In the absence of intervention, a freshwater lake or pond will become progressively shallower until it becomes a part of the surrounding terrain. (This can also happen in some estuaries and bays of the marine habitat.) The pond's aquatic vegetation will change at the same time. Bladderwort and tape grass may be among the plants appearing in the earliest stages in this succession. Eventually, water lilies or similar plants may appear. With the establishment of a marsh, a cattail and bullrush community may develop. The climax of this aquatic succession is not aquatic at all, but will be simply woodland, prairie, or bog.

The associated animal life undergoes dramatic changes in keeping with the plant successions. The more "desirable" species of fish are usually associated with the **oligotrophic** (early and intermediate) stages of succession. Successional changes in fish species are chiefly associated with rising water temperatures, decreasing oxygen content, and bottom changes affecting egg laying and breeding. These **eutrophic** changes, as they are known, are even more directly associated with progressive nutritional enrichment of the water, and they promote the growth of aquatic plants. (See the next chapter for more on eutrophication.)

SUMMARY

I. Organisms usually are not uniformly distributed, but make up populations, which in turn make up communities and ecosystems.
  A. Populations interact in such a way that they typically fluctuate in cyclic fashion.
  B. Populations are limited by environmental resistance, factors such as food supply, predators, and limited nesting areas which limit increase in population.
  C. Populations disperse as a result of such behavioral adaptations as territoriality; dispersal, however, does not ensure survival.

II. The main communities of the earth are geographically distributed in accordance with rainfall and growing season variations. The principal terrestrial biomes are summarized in Table 51–1.

III. The physical properties of water—principally its ability to absorb light, dissolve polar substances, transport insoluble substances, resist temperature change, and solidify at a lighter weight—determine the characteristics of the aquatic habitat.
  A. Aquatic communities include the plankton, the nekton, the bottom-dwelling benthos, and the intertidal organisms.
  B. Freshwater habitats undergo succession more rapidly than other life zones. Plant nutrients tend to accumulate in many bodies of water, giving rise to successional changes. Deposition of silt causes the body of water to become shallower, warmer, and sunnier. Plant growth is encouraged by all these factors. In the end, a terrestrial community is established.

POST-TEST

1. A ______________ is a group of organisms of the same species that occupy the same area.

2. Population density is determined by the interaction of ______________ rate, ______________ rate, and indirectly, by the ______________ ______________ found within the habitat.

3. The sum of all limiting factors that determine population density is called the ______________ resistance, which prevents a species from overrunning the earth.

4. According to the records kept by the Hudson Bay Company in Canada, there is a nine- to ten-year cycle of population increase and decrease involving

_______________ _______________ and the _______________ that preys upon them.

5. It has been proposed that a high snowshoe hare population density may lead to _______________-related disease among the animals, which in turn could lead to population decline; this would represent a _______________-_______________ limiting factor for the hares.

6. A _______________ is a very large ecosystem containing certain characteristic organisms.

7. Areas of contact among very large ecosystems are known as _______________; their _______________ diversity and population densities may be particularly high, an effect known as the _______________ effect.

8. The major _______________ biome is subarctic and arctic, and circumpolar; for the most part it is underlain by a permanently frozen layer of soil, the _______________.

9. South of this community there is another circumpolar one, the _______________, which is characterized by the dominance of _______________ trees.

10. That community has a southern extension, the moist coniferous forest biome, which is located on the _______________ Coast of the _______________ _______________ continent.

11. The tropical rain forest is characterized by a very shallow layer of _______________ and extremely effective mechanisms of _______________ mineral nutrients.

12. High _______________ of the soil is a potential problem faced by irrigation agriculture in desert areas.

13. In deep aquatic, especially oceanic habitats, mineral nutrients tend to accumulate in the _______________ areas.

14. In lakes, there is often a seasonal mixture of water in the spring and fall known as the _______________; this produces _______________ of plankton.

15. Estuaries are noted for their _______________ (*high/low*) salinity and _______________ (*high/low*) biological productivity.

16. The very smallest of the marine producers, which may actually be responsible for most of the ocean's productivity, are the _______________.

17. The _______________ zone contains organisms that often must possess adaptations to resist both wave action and periodic _______________.

18. The portion of the ocean that overlies the continental shelf is known as the _______________ region.

19. Freshwater lakes are usually most productive in their _______________ zone.

20. An older, nutritionally enriched lake is said to be _______________.

REVIEW QUESTIONS

1. What are some characteristics of populations that are not displayed by organisms individually?
2. What is meant by a biotic community? Give examples.
3. Which varies more in different populations of the same species: birth rates or death rates? Why?
4. Explain why there is a tendency toward an orderly sequence of communities leading to a climax community.
5. Discuss the factors that tend to keep the size of a population of animals in the wild relatively constant. What factors may cause cyclic variations in the size of such a population?
6. What is a biome? How does it differ from a biotic community?
7. Differentiate between plankton and nekton. Give examples of each.
8. In what ways do northern tundras differ from mountain tundras?
9. What distinctive properties of water produce the differences between terrestrial and marine organisms and communities?
10. Why is the productivity of oceanic habitats usually low?
11. Why are estuaries highly productive habitats?
12. Describe the subdivisions of the marine habitat, and give examples of animals typically found in each.
13. Compare the life forms of a swiftly flowing stream with that of a quiet lake.
14. Define the term ecotone, and give an example.

52
Human Ecology

OUTLINE

I. The ecological impact of primitive societies
II. Urbanization and the industrial revolution
III. The ecology of agriculture
  A. Agricultural communities
  B. Insecticides and biocides
IV. Use of water and wetlands
  A. Pollution and waste disposal
  B. Cultural eutrophication
  C. Thermal pollution
V. Air pollution
  A. Major pollutants
  B. Ecological effects of air pollution
  C. Meteorology of air pollution
VI. Other pollution: radioactive materials
VII. Extinction
VIII. Human overpopulation
  A. When is a nation overpopulated?
  B. The future

LEARNING OBJECTIVES

After you have read this chapter you should be able to:

1. Review the development and impact of modern human life styles upon the communities and ecosystems of the earth.
2. Describe the typical sequence of pollutional events that results from dumping saturating quantities of organic wastes into a watercourse.
3. Describe the process of pollutional eutrophication, differentiating between natural and cultural processes.
4. Discuss air pollution with respect to its principal ecological effects.
5. Summarize the process of extinction, listing factors that may contribute to the decline and extinction of endangered species and providing an example of each.
6. Relate overpopulation to specific environmental problems.

The effects of our technological society on the ecosphere rank with any geological disaster one might care to name, and the record of human squandering of natural resources is a dark one. The slaughter of the bison that once roamed the Western plains of the United States, the decimation of the whales, the destruction of tropical rain forest, and the pollution of streams with sewage, industrial, and farm wastes are some of the more flagrant examples of natural resources wasted beyond hope of rapid reclamation (Fig. 52–1).

Human beings have an ecology just as do prairie dogs, ants, or any other organism. This chapter is devoted to examining the ecological impact of human society upon the ecosystems of which humanity forms an often unknowing part.

The Ecological Impact of Primitive Societies

The ecological impact of humanity varies widely with the kinds of human societies and their respective technologies. No kind of technology has had a more direct and indirect ecological impact than that of agriculture.

The simplest agriculture is an extension of gathering behavior, in which normally scattered food plants are encouraged to grow in definite localities and in unnatural population densities. As societies become increasingly dependent upon agriculture and the herding of domestic animals, a way of life termed **pastoralism,** their hunting and gathering economy becomes steadily less significant, although some of it may persist even in the most technologically advanced societies.

Machine technology plays an increasingly important role in the life of such developing societies. Initially, the agriculture still depends upon primitive slash-and-burn techniques (Fig. 52–2). Trees are felled and then burned, releasing minerals into the soil. Such practices permit only a modest increase in the human population. But where conditions are right, soil fertility may be maintained in the same places for a number of generations. This permits the development of settled human populations with specialists such as traders, blacksmiths, kings, priests, and potters, who are supported by agricultural surpluses extracted from the farmers by trade or taxation.

In its initial stages, an agricultural revolution disrupts natural communities little more than does the hunting and gathering that preceded it. When a patch of soil is exhausted, the cultivators move on, and successional processes eventually return it to whatever climax state it originally enjoyed. Though a forest may be a patchwork of such abandoned farmsteads, it differs little from an undisturbed community—provided that population pressures do not force too rapid reuse of land that has not yet had time to recover. Forms of agriculture encouraging a more

Figure 52–1 Though the damage we have inflicted upon our environment is great, this chapter is not meant to discourage people from trying to save what we have left. More important, it hopes to bring to light issues that can be tackled by a combination of commitment and a knowledge of ecological principles. (*a*) This destruction of a forest for a phosphate mine provides an ugly contrast to the forests we studied in Chapter 51. However, it is possible to turn some of the technology used here to the purpose of restoring land that we have plundered in the past. (*b*) A graded, restored, and replanted strip mine. Though it is not the exact habitat it once was, any reasonable attempt to repair the land is preferable to leaving large stretches of territory to permanent waste. (SELBYPIC.)

(*a*)

(*b*)

Figure 52–2 Primitive slash-and-burn agriculture. (*a*) The forest trees are felled and (*b*) then burned, which releases mineral nutrients into the soil. (*c*) Crops are planted, among which are the dietary staple tapioca roots. As shown in (*d*) through (*g*), these are harvested, grated, soaked, cooked, and rolled out and dried for future consumption. (*h*) A clearing becomes a village, which when the soil has become depleted, is abandoned, so that the rain forest regenerates. ((*a*), (*c*)–(*e*), (*f*), and (*g*), courtesy of Hajit Kaur; (*b*), courtesy of Don Draper; (*h*), courtesy of Bert Watson.)

settled existence tend to be embraced readily, however; natural ecosystems then become displaced and are replaced with new communities composed of few species. These require constant effort and energy input to prevent successional change. Other areas, not suitable for farming, can be exploited by hoofed domestic animals raised for their wool, meat, or milk. All competing life forms and potential predators are systematically exterminated.

Urbanization and the Industrial Revolution

As agricultural surplus accumulated and various nonagricultural classes such as merchants and craftspeople developed, a pattern of residential land use emerged that at first required little space. However, with the Industrial Revolution and recent burgeoning of human population, residents have become major occupiers of surface formerly given over to agriculture or to natural ecosystems. Industries themselves require land and, in addition, alter land indirectly—for example, by pollution. The transportation facilities necessary to serve residences, shops, and businesses grew along with them, so that in some urban areas 60% or more of the land surface is used for nothing else but the movement of goods and persons (Fig. 52–3).

However profoundly the new agricultural communities differed from the forests or grasslands that preceded them, still they had continuity with earlier life styles. But with the Industrial Revolution, all this changed. Previously, the greatest part of the human species had not only lived in rural areas but also ate almost all the produce they raised and purchased little other food; such a life style is called **subsistence agriculture.** Today, in industrialized countries, only a fraction of the population farms the soil. This would hardly be possible if agriculture were labor-intensive, but our society has found ways of substituting high technology and substantial energy input for much of the human labor that agriculture formerly required.

The Ecology of Agriculture

Since the development of agriculture has permitted and produced most of the distinctive features of modern human ecology, let us first examine the ecology of agriculture itself.

AGRICULTURAL COMMUNITIES

An agricultural community such as a field of corn or of cabbages usually has characteristics that set it apart from the original community that it displaced.

(*a*)

(*b*)

Figure 52–3 Urbanization and patterns of land use. (*a*) Night satellite photo showing the eastern half of the United States taken by the Defense Meteorological Satellite Program, February 1974. (*b*) The filling in of wetlands to make way for industrial and residential development. ((*a*), official U.S. Air Force photo; (*b*) SELBYPIC.)

Figure 52–4 Monoculture. This vast field of soybeans consists not only of a single species but even of a single variety of that species of plant. In the case of fruit trees, great areas of land can be covered by an asexually propagated clone. (U.S. Department of Agriculture Soil Conservation Service.)

First, *an agricultural community is unstable* because it can be maintained only by human intervention and energy input. Maize (corn), for example, can reproduce itself only with human aid because it lacks an efficient seed dispersal adaptation. Another way of looking at the matter is that humans are essential for the continuance of agricultural communities.

Second, *an agricultural community is simpler than any natural ecosystem.* The plants of an agricultural community may be of only a single species. This **monoculture** greatly simplifies the community; since the cultigen that is being monocultured is one that can be consumed by humans, from a human viewpoint the community that results is very highly and desirably productive. However, this very monoculture renders the agricultural community unstable and highly susceptible to insect pests or diseases, which can spread from one host organism to another with little hindrance. In the natural state, host plants (those attacked by pests) are interspersed with others that a particular pest species will not eat. In monocultural agriculture, however, unnatural pest control methods are often required, which further unbalance the community (Fig. 52–4).

Third, *the chemical cycles of an agricultural community are often incomplete.* The **cultigens** (cultivated organisms such as maize that differ considerably from their wild ancestors) are consumed at a spot remote from where they are grown, and their substance is not usually returned to the soil but escapes from the terrestrial ecosystem through some sewage disposal plant or garbage dump. This interruption of nutrient cycle causes the soil to deteriorate unless artificially fertilized—which brings with it problems of its own, as will be discussed later.

Fourth, *competition among species is greatly reduced.* Farmers go to great lengths to suppress species considered undesirable—called **weeds**—that compete with cultigens or interfere with their harvest.

It should be obvious by now that an agricultural community is not and cannot be the same as the natural community it has replaced. The demands of agriculture simply do not permit this. Yet some compromise is possible, and with enlightened management, agricultural lands can be made to yield a modest return of wild game, or merely to serve as homes for wild species (Fig. 52–5).

Unfortunately, even in the most affluent and enlightened nations, a knowledge of ecology sufficient to induce most farmers and other citizens to sacrifice even a small part of the potential food production of their land in exchange for restoring ecological balance is lacking. In the heavily populated and economically less developed countries, such as India and Africa, the situation is even worse. Hungry inhabitants encroach continually on wild game preserves and exterminate most surviving wild animals from the land; sometimes they exchange pelts and trophies for money provided by black-market profiteers from other countries. In any case, where the pressures to improve living standards are high, conservation is too frequently perceived by such nations as a luxury affordable only by affluent nations.

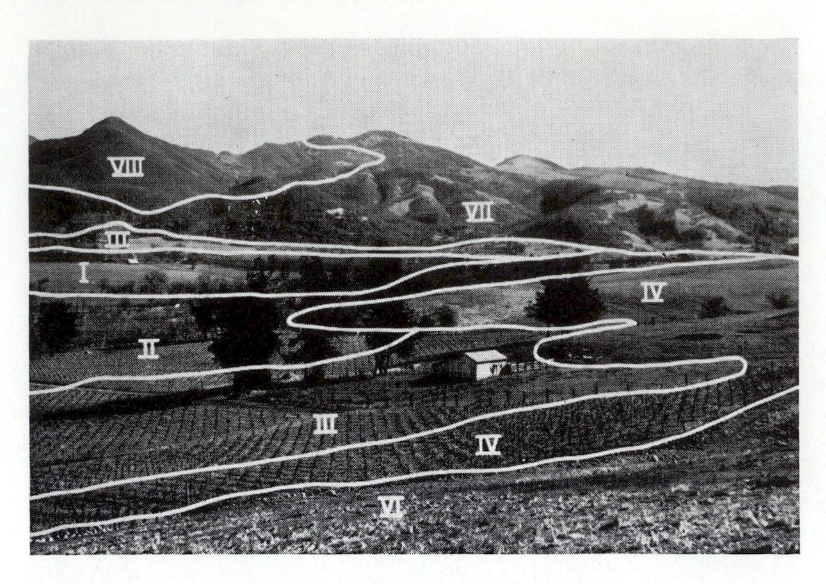

Figure 52–5 Classification of land according to its most appropriate use. Types I and II may be cultivated continuously; Types III and IV are subject to erosion and must be cultivated with great care; Types VI and VII are suitable for pasture or forests but not for cultivation; Type VIII is productive only as a habitat for game. This system of classification is supposedly a compromise between pressures for agricultural development and pressures for conservation. Yet, like many conservation issues, such planned use of land makes long-term economic sense. Conversion of much of the land to high-technology farming would require enormous expenditures of energy and might eventually prove unprofitable in light of the tendency of the topsoil at higher altitudes to erode. (United States Department of Agriculture Soil Conservation Service.)

INSECTICIDES AND BIOCIDES

Chemical pesticides, or biocides, have been in use since the 19th century. In recent years, a number of highly effective organic pesticides have been developed and put into widespread use by everyone from farmers and public health workers to suburban homeowners. But these substances have also created great ecological problems and have in some instances made pest control more, rather than less, difficult. Most of them fall into one of two main classes, the **chlorinated hydrocarbons** and the **anticholinesterases.** The chlorinated hydrocarbons interfere with nerve action by antagonizing the sodium pump, and the anticholinesterases act by rendering the enzyme cholinesterase incompetent (see Chapters 32 and 39). The anticholinesterases are, in general, more toxic than the chlorinated hydrocarbons, but they are also far less persistent in the environment. Humans are less susceptible than insects to the action of both classes of pesticides, partly because of their large body size, and partly because of differences in physiology. Pesticides that are resistant to environmental degradation are known as **hard pesticides,** or **persistent pesticides.** Excellent examples, the widely used chlorinated hydrocarbon biocides such as DDT (*d*ichloro*d*iphenyl*t*richloroethane) and dieldrin, pose a threat to terrestrial and even aquatic ecosystems, even though most of them are almost completely insoluble in water. Many of these compounds are not readily broken down by the metabolism of normal organisms and are not readily excreted either. They tend to accumulate in body fat (Fig. 52–6). By far most of the food metabolized by organisms is excreted as carbon dioxide, water, and nitrogenous waste products. The hard pesticides are not excreted, however, or at least not excreted as readily, even though they are taken in with food. A large portion of such a pesticide remains in the body of the animal that has consumed them. Thus, in a sprayed lawn the cricket that eats the grass has more biocide than the grass did; the toad that eats crickets has more than the crickets it eats; and so on. An ultimate predator such as a hawk (or human) concentrates in its body an appreciable portion of the biocides absorbed by many acres of vegetation. This process, called **biological magnification,** was first described in connection with the chlorinated hydrocarbon DDT. Although DDT is no longer much used in many enlightened countries (including the United States), almost everything said about DDT still applies to its persistent chemical successors, which are as widely used today as DDT ever was.

The harmful effects of DDT on predators and the natural enemies of pest organisms are not its only defects. Pesticide resistance occurs widely. DDT-resistant flies, mosquitoes, fish, and even mice now occur in some parts of the world. These animals require such high doses of pesticides to kill them that almost everything else in their habitat is exterminated, including their natural enemies. These enemies are more vulnerable to DDT, both because their populations are small and because of biological concentration of the toxicant. For instance, outbreaks of damaging plant mites have been traced to the elimination of predatory insects by the DDT that was intended to control another pest. DDT and almost all other pesticides

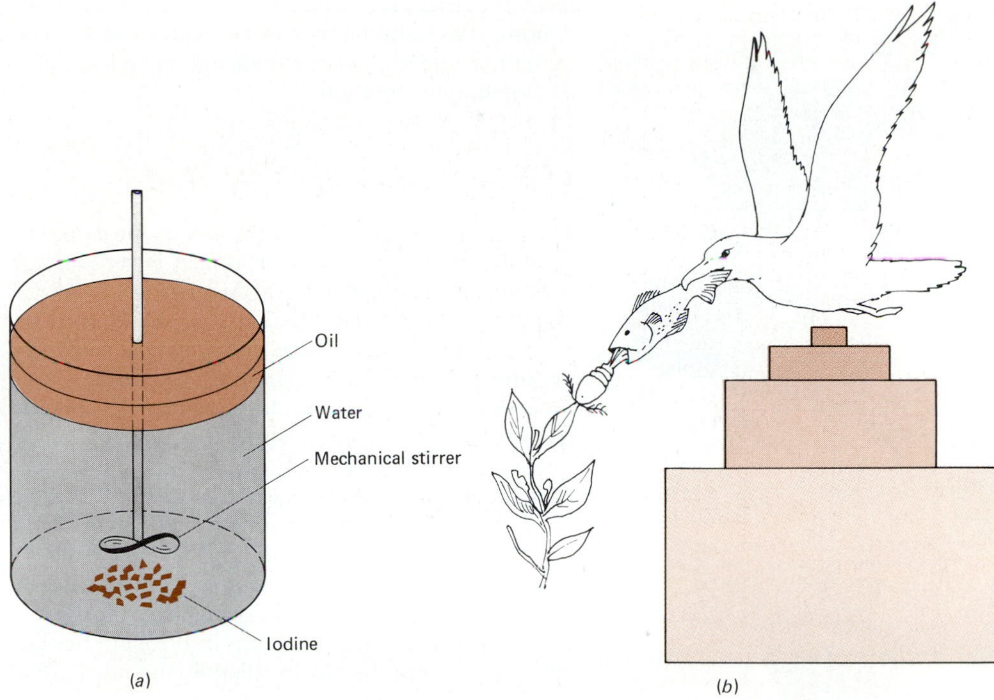

Figure 52–6 Biological magnification. (*a*) How a substance almost insoluble in water can become concentrated in another medium. Imagine a simple experiment with iodine, water, and mineral oil. The iodine is placed on the bottom of a vessel, and water is added. On top of the water is placed a layer of oil. The iodine is quite insoluble in water but very soluble in oil; it must pass through the water to get to the oil, however. Very soon, the water is saturated with iodine, which gives it a slight yellow tinge that never deepens. However, as hours pass, the iodine in the water enters the mineral oil, which absorbs it avidly. There is now less iodine in the water, and more dissolves to replace it. Still more is absorbed by the oil, so more dissolves in the water, and so on in a continuous process, which will proceed until almost all of the iodine is dissolved in the oil even though hardly any was dissolved in the intervening water at any time.

Like the iodine in the experiment, chlorinated hydrocarbons tend to be soluble in fat, although not in water, and most are very stable compounds. Thus, although they are present in only parts per billion or even trillion in surface water, they accumulate rapidly in the fats and oils of living things. This represents the first stage of biological magnification. (*b*) A toxicant pyramid, which provides further stages of concentration. Most of the biomass consumed on each level is excreted, but DDT and similar fat-soluble biocides are not excreted to anywhere near the same extent and therefore tend to become progressively concentrated in the higher members of the food chain. They are thus most dangerous to ultimate predators.

actually can *create* very serious pest problems in this way. The development of such resistance is not limited to DDT, but occurs with almost all pesticides. Even rats resistant to the rodenticide warfarin have been discovered.

It is true that the problems connected with persistent pesticides are reduced to some extent when degradable compounds such as the anticholinesterases, carbamates or organophosphates (which break down in nature) are used. Unfortunately, however, *there is no known pesticide that is free from undesirable ecological consequences.* The carbamate Sevin, for instance, is among the pesticides least toxic to mammals, yet is instant death for bees, and bees are vital for pollination of food and wild plants.

Since all pesticides kill more than the pests against which they are directed, in the end our pesticide problems can be solved only if we can avoid using them. Yet this now seems less likely to happen than it ever has. In fact, we currently use twice the amount of pesticides used in 1962, the year in which Rachel Carson's book *Silent Spring* was published.[1] Even hard (nonbiodegradable) pesticides continue in widespread use for special purposes or, in environmentally unenlightened countries, for any purpose. Moreover, DDT is illegally imported into the United States, or occurs as an impurity in other pesticides. We can avoid using pesticides, to be realistic, only if we are content to produce less food. We can produce less food only if we have fewer mouths to feed.

The World Health Organizaton has warned that an international ban of DDT before cheap, safe, and effective substitutes are found would be a disaster to world health. The agricultural production of the United States would decrease about 30% if all pesticides were banned, and the poorer nations of the world would suffer even more from decreased food supplies and increased incidences of insect-borne diseases, especially malaria. The ironic fact is, however, that the mosquitoes that carry malaria have in recent years developed behavioral and other mechanisms of resistance to DDT, so that while DDT remains damaging to everything else, it is steadily less effective against the pests for which it was intended!

The control of insect pests by chemical pesticides could be carried out more cautiously than it has in the past, and could in many cases be replaced with alternatives. Biological control methods do exist—although none of them permit as high a

[1] *Silent Spring* (see Readings at the end of this part) is often used to date the beginning of the modern environmental movement, although this movement certainly has roots extending much further back in time. In *Silent Spring*, Carson described the ecological damage done by pesticides, especially when carelessly applied without due thought of the consequences. Though not the first to do this, Carson attracted public attention to the pesticide problem in a way that no writer before her or since has been able to do.

level of agricultural production as our overpopulated world demands (Fig. 52–7). Among these alternatives to pesticides can be listed the use of natural enemies, insect hormones and sex attractants, the release of sterile males, and such practices as careful crop rotation.

(a)

(b)

(c)

Figure 52–7 Biological control, a pest control alternative to the use of biocides. (*a*) Male (*bottom*) and female (*top*) screwworm flies. The larvae produce tremendous damage in livestock. (*b*) Sterilized male screwworm flies are released. They mate with wild females, which cannot then reproduce since the females mate only once. (*c*) Biological control of the imported cabbage worm. Tiny wasps are released in cabbage plots. These parasitoids lay their eggs in the worms; when these hatch, the larvae devour the pests alive. (USDA photos.)

Use of Water and Wetlands

The primary productivity of the sea, as measured by the pounds of organic carbon produced per year per acre of surface, is very high. The productivity of the western Atlantic off the coast of North America is 2.5 to 3.5 tons of organic carbon per acre, and that of Long Island Sound is 2.5 to 4.5 tons per acre. The productivity of the average forest is about 1 ton per acre; most cultivated land fixes only about ¾ ton of organic carbon per acre; and only the fertile, intensively cultivated cornfields of Iowa produce as much as 4 tons per acre. Despite this high productivity, our actual harvest from the ocean, in terms of pounds of fish caught per acre of surface, is very low. Only the rich fishing grounds of the North Sea produce as much as 15 pounds of fish per acre. The ecological reasons for this are clear: The fish are secondary or tertiary consumers and are on top of a vast pyramid of producers. There are many organisms competing for the food energy fixed by the algae, in addition to the edible fish and crustaceans harvested from the sea.

The annual harvest of marine fish over the entire globe was 18 million tons in 1938 and reached 70 million tons in 1970, but has been declining since then. Almost 80% comes from the North Atlantic, the northern and western Pacific, and off the west coasts of Peru and Ecuador in South America. About half the total harvest is used as human food; the remainder is used as food for pets, poultry, and livestock.

Humans could undoubtedly recover for their use much more of the biological productivity of the sea. Although they might be reluctant to eat marine algae themselves, algae could be filtered from sea water and processed to become suitable as feed for cattle or other animals. This would, however, place us in direct competition with many of the very organisms (such as whales) that we would most like to harvest for food!

Actually, the history of our conservation of marine biotic resources to date is anything but encouraging. Several of the larger species of whales, for example, are endangered mostly as a direct result of excessive and unnecessary hunting (they do not form a major part of any nation's—including Japan's—diet, and yield no product for which there is no acceptable substitute). Even sardines, oysters, lobsters, mackerel, tuna and other traditional seafood organisms are becoming less common and therefore more expensive.This situation has resulted not only because of overfishing but also because of the destruction of wetlands by pollution or industrial development. Wetlands are ecologically vital because they form the basis of the food chains upon which many aquatic communities depend and serve as the refuges where marine organisms undergo the early stages of their life histories.

POLLUTION AND WASTE DISPOSAL

Let us begin our consideration of this topic by defining two important terms: waste and pollution. **Waste** may be defined as any product of our civilization that is usually discarded rather than used, or a formerly useful product that is no longer usable for its original purpose or for any other. **Pollution** exists when wastes or other substances have a significantly damaging effect upon public health, property, ecosystems, or esthetic values.

The classic form (but not the only form) of water pollution is that produced by industrial and municipal sewage, and indeed these wastes at present are probably of more biological and economic significance than any other kind. How do they produce their effects?

Organic wastes provide a rich source of nutrients for decay bacteria and fungi. Hence feces, blood from slaughterhouses, oxygen-demanding wastes from paper mills, and peelings from vegetable processing plants stimulate the growth of bacteria whose metabolism rapidly removes oxygen from the water. Such wastes contain large amounts of sediment, chemically combined nitrogen, phosphorus, carbon dioxide, methane, hydrogen sulfide, and smaller amounts of miscellaneous chemicals, heavy-metal ions, and pesticides. Industrial waste accounts for most of the water pollution of the United States. Usually far more concentrated than municipal sewage, industrial waste produces up to 12 times the amount of pollution per gallon of effluent than municipal wastes do. But what they lack in concentration municipal wastes somewhat make up in volume, especially when street drainage and surface runoff after rain and snow is taken into account.

A polluted environment is a demanding one. Not surprisingly, few organisms can tolerate it. Yet those that are able to exist in it often attain astronomical numbers and very large biomass. Most aquatic organisms, including, unfortunately, those we hold to be most desirable, are sensitive to the effects of pollution and are destroyed by it.

Near the source of the pollution, surprisingly, numbers of fish and other organisms may persist because the organic wastes have not had time to decay (Fig. 52–8). In severe instances, somewhat farther away, conditions are much worse, with bacteria and fungi that degrade organic sewage using up so much oxygen that actual anaerobic conditions exist. Still farther away, the polluted water has begun to purify itself so that the organic materials start to disappear, oxygen diffuses into the water from the air, and except for cultural eutrophication (as we will see shortly), something like a normal ecology is established.

CULTURAL EUTROPHICATION

Natural eutrophication proceeds as a result of the slow leaching and seepage of nutrient salts from the surrounding watershed. It is really a successional process. **Cultural eutrophication,** resulting from pollution, not only occurs far more rapidly than the natural process but differs in several other respects. In natural eutrophication, a slow accumulation of silt on the bottom usually also accompanies the nutritional enrichment, so that geologically old lakes are inclined to be shallow and warm. In cultural eutrophication, overstimulated plant growth and decay swiftly choke the body of water, leaving a nutrient-rich thick layer of sludge or organic ooze on the bottom (Fig. 52–9).

A nutrient-poor, or **oligotrophic,** lake may contain some of the most highly valued varieties of sport fish, but since it is poor in primary producers, its total productivity is low. Thus, from the human point of view, a moderate degree of eutrophication may be beneficial, and fishponds are often artificially fertilized for that very reason. But in the end, unrestrained cultural eutrophication almost always leads to ecological disaster and the end of the aquatic community's usefulness to humans. Newly dominant cyanobacteria are largely inedible by fish, and heavy plant growth of all types clogs the water and at night competes with fish for oxygen. When the cyanobacteria or plants are killed by winter frost or overcrowding (or anything else), their decomposition results in water pollution.

THERMAL POLLUTION

When a body of water becomes heated to such an extent that it is harmful to the natural inhabitants of that body of water, we speak of the condition as **thermal pollution.** Increasing water temperature has a number of biological effects. Oxygen is less soluble in warm water than in cold, so there is less available for the use of aquatic organisms. Despite this, increased temperature causes organic wastes to decompose more rapidly, hastening the local depletion of oxygen.

Most aquatic animals are poikilothermic; their body temperature increases when the water temperature increases. The warmer temperature increases the speed of their metabolic reactions, heightening their need for oxygen—which becomes less and less plentiful. As a result, fish may suffocate. In extreme instances of very high water temperatures, the organisms are simply cooked. In addition, thermal pollution has indirect effects on the migratory and other behavior of aquatic organisms and upon their seasonal temperature adjustments, which it can easily disrupt.

Thermal pollution arises from the discharge of water used for cooling power plant condensers and other industrial operations. Because of the universal demand for electric power, thermal pollution arising from this source is probably the most potentially threatening.

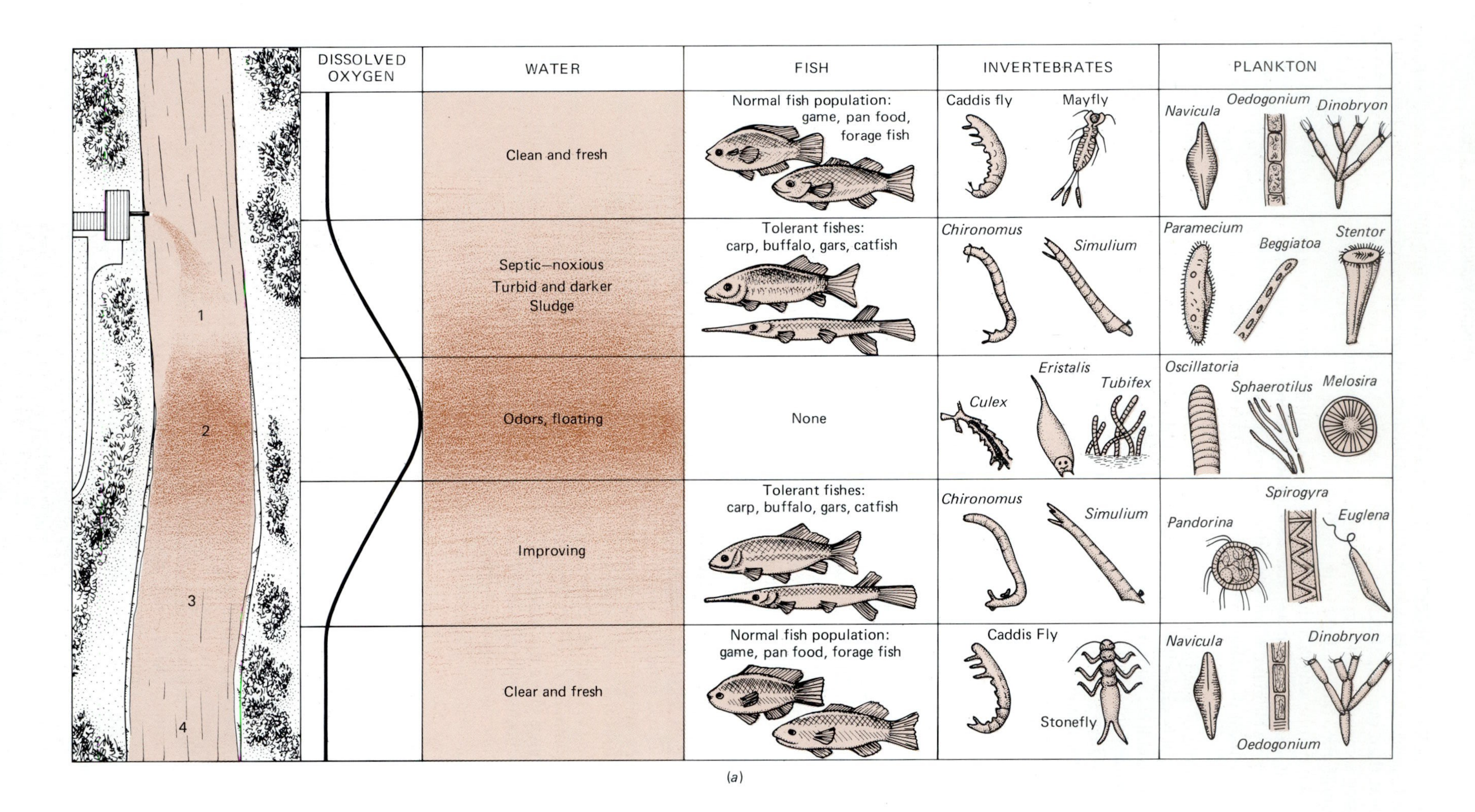

DISSOLVED OXYGEN
WATER
FISH
INVERTEBRATES
PLANKTON
1
2
3
4
Clean and fresh
Normal fish population: game, pan food, forage fish
Caddis fly
Mayfly
Navicula
Oedogonium
Dinobryon
Septic–noxious
Turbid and darker
Sludge
Tolerant fishes: carp, buffalo, gars, catfish
Chironomus
Simulium
Paramecium
Beggiatoa
Stentor
Odors, floating
None
Culex
Eristalis
Tubifex
Oscillatoria
Sphaerotilus
Melosira
Improving
Tolerant fishes: carp, buffalo, gars, catfish
Chironomus
Simulium
Pandorina
Spirogyra
Euglena
Clear and fresh
Normal fish population: game, pan food, forage fish
Caddis Fly
Stonefly
Navicula
Dinobryon
Oedogonium

(*a*)

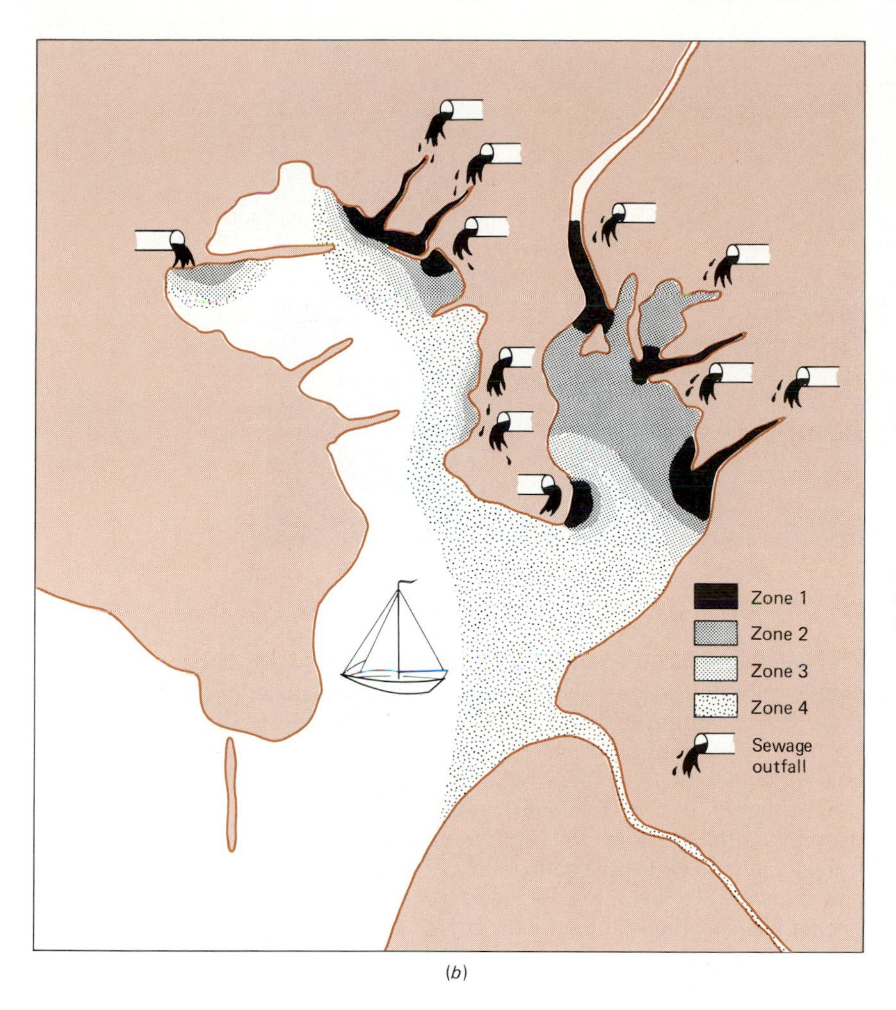

Figure 52–8 Pollution zonation. (*a*) Pollution zonation in a stream receiving untreated sewage. As the amount of oxygen dissolved in the water decreases, fishes disappear, and only organisms able to obtain oxygen from the surface, those that tolerate low oxygen tensions, or those that respire anaerobically are able to survive. When the sewage has been completely decomposed by bacteria, the species of plants and animals present in the stream return approximately to normal, although eutrophic changes may persist.
(*b*) Complex pollution zonation, as it occurs in an actual example. Note the influence of polluted rivers on this estuary. The occasional presence of zones *2* to *4* without zone *1* results from the presence of overloaded secondary treatment plants which did remove some waste. Many outlets, however, simply deposited raw sewage in the watercourses at the time this drawing was made. ((a) from Odum.)

Air Pollution

Have you ever considered the plight of a fish in a polluted lake? It is exposed to a continuous dose of toxic substances every moment, every day, from which, usually, it has no escape. Air pollution puts us in the position of the fish, for each of us must breathe about 20,000 times every day. With each breath we inhale a wide variety of toxic gases and particles spewed into the air by automobiles, power plants, industries (Fig. 52–10), and cigarette smokers. The United States alone dumps more than 220 million tons of pollutants into the air annually.

Figure 52–9 Experimentally induced eutrophication resulting from the addition of phosphate fertilizer to the upper portion of this lake. The resulting growth of cyanobacteria has given the water a whitish blue appearance in the photograph. The lower portion of the lake, which served as a control, is not affected. (Courtesy of Dr. D. W. Schindler, and *Science*. Copyright 1974 by the American Association for the Advancement of Science.)

Figure 52–10 Air pollution from a phosphate fertilizer processing plant. (SELBYPIC.)

MAJOR POLLUTANTS

The pollutants that account for most of air pollution are carbon monoxide, sulfur oxides, nitrogen oxides, hydrocarbons, and particulates. Their relative contributions to air pollution are indicated in Figure 52–11. These pollutants interact to form secondary pollutants and smog. Cigarette smoke, an important indoor air pollutant, contains several of these pollutants, as well as many others, and thus represents a highly concentrated source. Although the total volume of polluted air produced by smokers does not begin to compete with the amount spewed forth by industry, the health consequences may be more severe, for the air polluted by the smoker is the air closest to us—the very air we must breathe. Industrial pollution is generally somewhat diluted by the time it wafts our way. Nevertheless, the public health and ecological effects of industrial and other forms of gross air pollution are very substantial.

ECOLOGICAL EFFECTS OF AIR POLLUTION

Air pollution has many sources, ranging from the smokestacks of power plants to the exhaust pipe of the family car. Our consumptive life style ensures a multitude of sources of such air pollution, especially in manufacturing areas and the vicinity of cities. Air pollution at times reduces the amount of sunlight reaching Chicago by about 40%, and the amount reaching New York by 25%. When sunlight is reduced, photosynthesis is also diminished. In addition, trees and other plants absorb great quantities of pollutants directly from the air. For the most part, damage occurs in the photosynthetic tissue (mesophyll) of the leaf. The surfaces of mesophyll cells, which are moist to facilitate gas exchange in photosynthesis, are vulnerable to attack by toxic substances in the air. Fluoride particulates, photochemical smog (which is generated from automobile emissions by a chemically complex atmospheric process), sulfur dioxide, and perhaps even very fine soot can kill vegetation

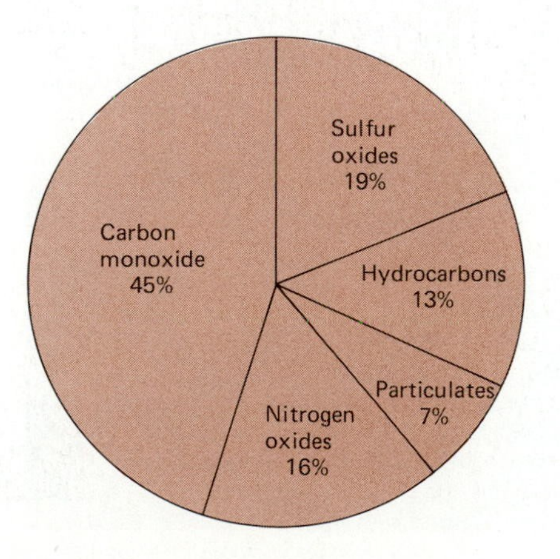

Figure 52–11 The principal of air pollutants.

in this way (Fig. 52–12). Some species of plants are more susceptible to certain pollutants than are other species. At a concentration of less than one part per million, the common pollutant nitrogen dioxide reduces the growth of tomato plants by 30%.

In Los Angeles County and areas up to 60 miles away, forests are being damaged by photochemical smog. Photosynthesis is reduced by 66% by smog concentrations of 0.25 part per million. Such decreases in photosynthesis slow the flow of resins under the bark, which in turn render the trees susceptible to plant disease and insect pests. With the increase in air pollution that present trends are producing, vast damage to vegetation may occur over the face of the North American continent, and even the world. This damage occurs not only directly, as described here, but also by means of the very threatening acid precipitation, discussed in the following section.

Animals, too, are exposed to air pollution, both by inhaling pollutants directly and by ingesting vegetation that has been contaminated by air pollutants. Such polluted areas are undesirable for grazing. The vegetation is of poor quality, and the animals inhale smoke and eat poisonous substances deposited on the grass. The acreage required to raise cattle increases in polluted areas, and illness and death in sheep and cattle have been traced to specific pollutants such as fluoride.

Figure 52–12 Effect of air pollution (sulfur dioxide) on potato leaves. (USDA photo.)

METEOROLOGY OF AIR POLLUTION

Atmospheric inversion is the cause of most reported air pollution episodes, though not of air pollution itself. In such cases, weather conditions form a lid of warm air above one of cooler polluted air. Though inversion layers do not actually increase pollution, they seal the pollutants below and prevent them from being dispersed (Fig. 52–13). In Los Angeles County, a noted example, inversion layers form when a layer of warm air moves in from the deserts to the east and lies over the mountains that surround the Los Angeles area. Beneath this warm air is a layer of cooler air that has moved in from the sea. Such inversions also form, though less regularly, in virtually every area of the country. Simple air stagnation, which can occur anywhere, is almost as bad, for with stagnation pollution is not dispersed by normal wind action.

It is that very wind action that results in **acid rain,** perhaps hundreds of miles from where the pollutants originated. It is common for acid rain to be caused by air

(*a*)

(*b*)

(*c*)

Figure 52–13 The meteorology of air pollution: three views of downtown Los Angeles. (*a*) In a normal pattern of air flow, warm air close to the ground rises, carrying with it most atmosphere pollutants. (*b*) A thermal inversion. At 75 meters a lid of warm air prevents the circulation of air from below. (*c*) Heavy pollution trapped under an inversion layer at 450 meters. (Courtesy of Los Angeles Air Pollution Control District.)

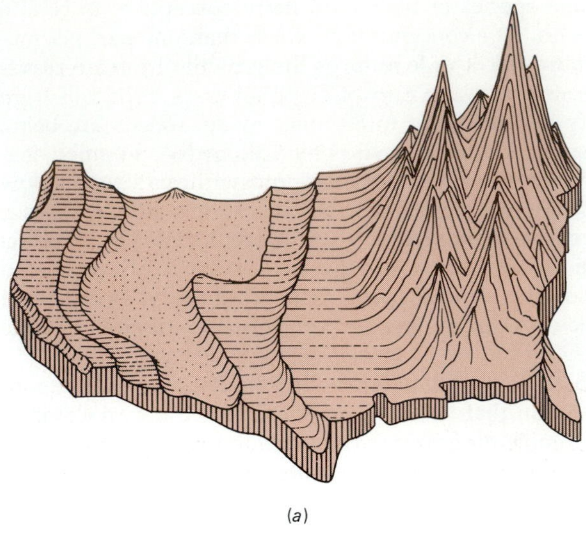

(a)

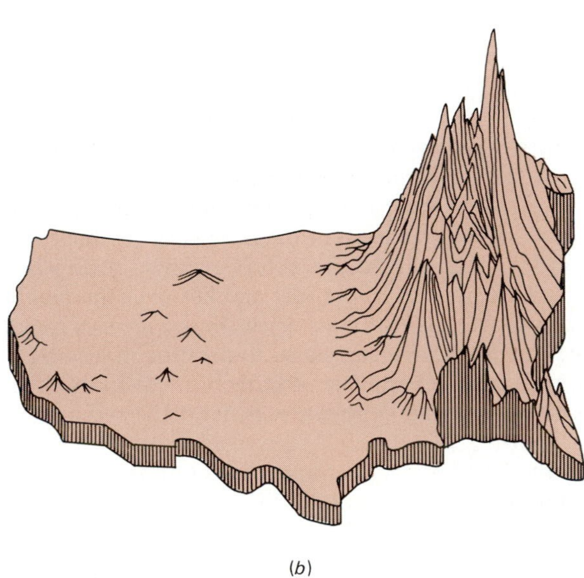

(b)

Figure 52–14 Air pollution is a serious environmental problem worldwide. (*a*) Rain acidity in the United States. The highest peaks represent the lowest pH of precipitation. Note that the problem is worst near the industrial areas of the East and Midwest, with secondary centers in portions of the heavily polluted West Coast. Ironically, however, because it is spread by winds, acid rain does not always affect the immediate area in which the pollution is produced. Instead, it can travel to nearby areas that might, in fact, be wilderness. Such is the problem in areas of the Adirondacks in New York and sections of southern Canada. (*b*) A map showing relative amounts of particulate pollution. Compare this with the map in (*a*).

pollution originating outside national boundaries, which has caused international disputes between such countries as Great Britain and Sweden, or the United States and Canada. Acid rain damages trees and other vegetation directly, probably leaches far more nutrients from the soil than it could possibly add, and where the surface waters are soft—that is, lack natural buffering capacity—kills fish and other aquatic life over vast areas, reaching into wild lakes and ponds seldom visited by people but despoiled by them nonetheless. (Fig. 52–14; see also Fig. 2–13).

Air pollution probably also affects climate on a global scale. Some scientists fear that if we continue to inject great quantities of particulates into the atmosphere we may bring on another ice age. Particulate matter could serve as nuclei for the condensation of high clouds, which would reflect sunlight away from the earth. One scientist has calculated that the addition of only 50 million tons of pollutant particles to the atmosphere could reduce the average surface temperature of our planet from its present 15°C (60°F) to about 7°C (40°F). Most forms of plant life could not survive in such a cold climate.

Other scientists fear that the vast amounts of carbon dioxide we are producing may lead to a kind of thermal pollution by means of a greenhouse effect. Excessive heat would be retained so that average temperatures would rise. Whatever the long-term effects on climate turn out to be, pollutants *are* likely to affect the delicate balance of the ecosphere. Our very inability to predict its effects should produce concern. It seems foolish to passively await a practical demonstration of just which prediction may turn out to be most correct.

Other Pollution: Radioactive Materials

Many kinds of pollution are possible. Their variety is limited only by our technology. One example of a thoroughly modern form of pollution is pollution from radioactive waste.

Strontium can serve as an excellent example of a radioactive pollutant. It is very similar to calcium in its chemical and physical properties and moves with calcium in its biogeochemical cycle. Calcium and strontium are washed out of rocks together and move down rivers to the ocean. One of the worst radioactive products of nuclear and thermonuclear weapons, radioactive strontium (^{90}Sr) has a half-life of 28 years—that is, 28 years after its initial production, half of it will still remain radioactive and virulent. Wherever it settles in the body, a radioactive material produces mutations and predisposes to cancer.

In the far North, radioactive fall-out of ^{90}Sr has been absorbed by lichens, and the caribou and reindeer that eat the lichens have the strontium concentrated in their flesh. Humans who eat the flesh of the animals or drink milk from them may then accumulate concentrated ^{90}Sr in their bones. In other parts of the world, ^{90}Sr has accumulated in the bones of both adults and children who obtain it in cow's milk. The cows obtained it by eating vegetation polluted with ^{90}Sr.

Little ^{90}Sr has been added to the environment since 1963, mostly because little atmospheric testing of nuclear weapons has been done since that date. If bans on nuclear testing continue, ^{90}Sr and the related ^{137}Cs (a radioactive isotope of cesium) will gradually be reduced in the environment. Both are by-products of fission reactions and enter the environment through fall-out or through faulty nuclear power plant waste disposal. They are especially hazardous because they accumulate in successive living organisms along the food chain (biological magnification), somewhat as DDT does. Studies in Great Britain showed that ^{90}Sr was 21 times more concentrated in grass than in soil and some 700 times more concentrated in sheep than in the grass.

Other radioactive materials such as ^{14}C or ^{3}H could become even more dangerous than ^{90}Sr following a nuclear war. The indirect ecological effects of such a conflict are beyond calculation (but see Focus on Human Ecology: The Last Winter), but it is likely that a large nuclear exchange directly involving only two nations would ultimately destroy all or almost all the major ecosystems of the earth.

Extinction

Extinction is the dying out of a variety, species, or any other taxonomic category of life. Extinction is a natural occurrence, something that has gone on since the first organic molecules might have formed in the primitive ocean. Yet the presence of human beings on the planet has brought about a rapid extinction of many types of organisms, an extinction that is often based on short-sighted decisions that might make a delicate woodland area expendable to airport runways, highways, or shopping centers. Once these areas are converted to human use, they go through successive changes, since human technology and land needs are always changing. However, the extinction of the organisms that depended on the original habitat for survival is permanent.

When an organism becomes extinct, the knowledge we could gain by observing its life style, studying its biochemistry, and finding its place in the diversity of life on the planet is forever lost. Scientifically, the loss is incalculable, for in the organism's blood or tissues there could have been some chemical, some enzyme or drug, that could have application toward improving the life of our own species. Esthetically, we feel revulsion by anything that takes away from the richness of life on this planet (Fig. 52–15).

The species most vulnerable to extinction appear to be those that are large, predatory, and migratory, such as the polar bear. However, almost every imaginable combination of characteristics occurs among the endangered species, for extinction also pursues animals that require large tracts of wilderness or solitude, that live in very specialized or restricted habitats, that compete with human beings in any respect, or that yield an economically valuable product (Fig. 52–16). Extinctions also result from importing alien species against which an endangered organism has no effective defense. Recently and ominously, environmental pollution has been

FOCUS ON

Human Ecology: The Last Winter

There has never been any doubt in the minds of informed people that nuclear war would be unimaginably disastrous, but until recently there has been surprisingly little attention paid to the probable long-term ecological consequences of nuclear exchange. Scientists assembled at a two-day conference in Washington, D.C., in the fall of 1983 agreed that of all possible ecological damage, the consequences of nuclear war would be by far the most serious, producing a horrendous climax to the environmental despoliations of our civilizations.

The biological concentration of such isotopes as strontium-90 has been studied since the 1950s, and it has been obvious that even a limited nuclear exchange would produce widespread environmental pollution by radioactive fallout. It has also been understood that the destruction of much of the protective atmospheric ozone layer by nuclear-generated nitrogen oxides would produce widespread crop damage, cancer, and disruption of the visual systems of animals whose sight is sensitive to light of very short wavelengths. What was new at the conference was an appreciation of what the smoke, soot, and pulverized earth produced by nuclear explosions would do to the terrestrial climate.

Such widespread dust has entered the earth's atmosphere before. There is much speculation that dust stirred up by asteroidal bombardment of the earth in the geological past may have been responsible for several widespread extinctions of life that are observable in the fossil record. Even within historical times, large volcanic eruptions have placed enough ash in the atmosphere to produce marked climatic changes. The eruption of the Indonesian volcano Tambura, which took place in 1815, ejected some 25 cubic *miles* of debris, much of which did not fall to earth immediately but remained in the atmosphere for a considerable time. In 1816 it produced changes in climate that resulted in a disastrous year for agriculture, the "year without a summer." There were no fewer than three killing frosts during the New England growing season and widespread hardship throughout the Northern Hemisphere.

It would appear that the dust and soot produced by even a "moderate" nuclear exchange could be expected to almost completely obscure sunlight over the entire Northern Hemisphere and probably the Southern Hemisphere as well. Even in summer, prevailing temperatures would immediately fall below freezing, dropping as low as perhaps −15°C to −20°C. The cold would persist, freezing natural bodies of water, in many cases to the bottom. Since the oceans would remain relatively warm for a considerable time, the resulting marked temperature difference between land and sea would produce storms of unprecedented violence. The darkness and cold, lasting for months on end, would cause most animals and plants to die, and probably most of them to become extinct, according to the conferees. This would be especially true of the tropical forms, but even temperate zone species would be decimated, particularly if the exchange occurred in the summer.

Although recovery of a sort could be expected within a few years, conventional agriculture would be impossible, not only because of climatic disruption but also because of the destruction of the industrial-based needs of agriculture such as fertilizer plants, fuel delivery, availability of agricultural machinery and spare parts, and the like. Starving people would hunt down any surviving animals that they could, and eventually would starve themselves. Moreover, radiation sickness and other disease states brought about by widespread chemical pollution resulting from the burning of synthetic material could be expected to weaken even the survivors. A report summarizing the conference's conclusions emphasized ". . . that survivors, at least in the Northern Hemisphere, would face extreme cold, water shortages, lack of food and fuel, heavy burdens of radiation and pollutants, disease and psychological stress—all in twilight or darkness."* Also predicted was the extinction of most tropical plants and animals, as well as of most temperate zone vertebrates. The report concluded on a particularly chilling note: Under these circumstances, the extinction of *Homo sapiens* cannot be excluded.

*Ehrlich, P., et al.: Long-term biological consequences of nuclear war. *Science,* 23 December 1983, pp. 1293–1300.

involved in some extinctions. Although it is difficult to be sure, the destruction of habitat appears to account, at least in part, for the bulk of extinctions and endangerments of species. The famous case of the now-extinct passenger pigeon probably resulted from the destruction of the beech forests in which it nested, as much as from the market hunting that is usually blamed. Another extinct species, the ivory-billed woodpecker, required large virgin tracts of cypress forest. Today there is little of that habitat left.

Figure 52–15 Some animal species faced with extinction. (*a*) Florida "panther," actually a type of puma. This last remnant of a population of predators that once ranged as far north as New York State is now confined by persecution and habitat destruction to a shrinking habitat in the Everglades. (*b*) The orangutan, once thought by some to be the living species closest to *Homo sapiens*. (*c*) The nene goose, a flightless bird from Hawaii. The nene has been preserved by a vigorous artificial propagation program. (*d*) Many species of the tortoises that Darwin observed in the Galapagos are nearing extinction. It is ironic that the very abundance of this species in earlier years helped Darwin to formulate some of his concepts of natural selection. (*e*) Predators are often persecuted out of proportion to the amount of damage they actually cause for ranchers. This photograph displays the attitude of whoever shot this coyote, left to hang on a ranch fence, presumably as a moral lesson to the other coyotes. (*f*) The chinese giant panda lives in a very restricted habitat and occupies a very specialized ecological niche. Zoos may be the only way that this species may survive. ((*a*), Florida Game and Fresh Water Fish Commission; (*b*), Ira Block, The Image Bank; (*c*); Charles Seaborn; (*d*); J. N. A. Lott, McMaster University/BPS; (*e*), Bob Martin; (*f*), National Zoological Park.)

Figure 52–16 Abundance is no guarantee against extinction. Conch mollusks are widely consumed as food throughout the West Indies and the Caribbean region. The shells are actually used as landfill or, as we see here, as a decorative building material. Despite its former abundance, the conch is becoming scarce in many of the locales in which it was formerly common because the animals are being consumed at a rate exceeding their natural replacement.

Human Overpopulation

Fifty years ago, many communities dumped their raw sewage into the nearest river or bay. They could depend upon natural bacterial decomposition to break it down with little negative effect upon the ecology of the waterway. Nature was able to do the job. As communities grew, however, additional numbers of people produced more sewage than nature could handle. As water became more and more polluted, communities had to invest in expensive sewage treatment systems. Technology, in other words, had to take over nature's functions.

In 1970, the average population density in the United States was one person to every ten acres of ice-free land. The average population density for the rest of the world was at the same level. Yet even if the current reduction in the birth rate continues, the population density of the United States will be double that number by the year 2000. (See Fig. 52–17 for other possibilities.)

More people mean more than just increased sewage, however. More people need more food, more clothing, more houses, schools, roads, automobiles, television sets, energy, and all the material goods that our society holds so dear. Each of these can be translated into increased pressure on the environment. For example, the need for more food results in the use of more biocides and more chemical fertilizers, which result in damage to the soil and increased land and water pollution. Production of chemical fertilizers requires use of more petroleum, contributing to energy shortages. At the same time, more people need more housing, more stores, schools, and roads, so that more land is taken out of agricultural production, leaving farmers to raise more food on less land.

Overpopulation is certainly one of the most pressing problems of our time—a problem from which we in the United States are *not* insulated! Aside from the moral issues involved, the National Security Council has described population increases around the world as a threat to our national security. How much greater a threat must it be to the national security of the emerging nations, in which population growth is now the greatest, and of their neighbors. In such countries, millions of people continue to move into overcrowded cities, contributing to social unrest and political upheaval, with world-wide effects.

The third of the world's population that lives in the developed countries consumes 85% of the earth's resources and is responsible for most of the stress placed upon the environment. Yet the other two thirds of humanity strive to consume at the level of the more affluent nations. It has been estimated that the maximum world population that could be supported at the United States level of affluence is about *one* billion. The environmental impact of enriching even this billion to current United States norms would involve increased industrial pollution, increased agricultural land erosion, further depletion of natural resources, and much more. However, aside from that, such a goal is totally unrealistic because at the current growth rate, world population will reach 8 billion by the year 2100. And inevitably the vast majority of human beings living then will subsist in unprecedented conditions of poverty.

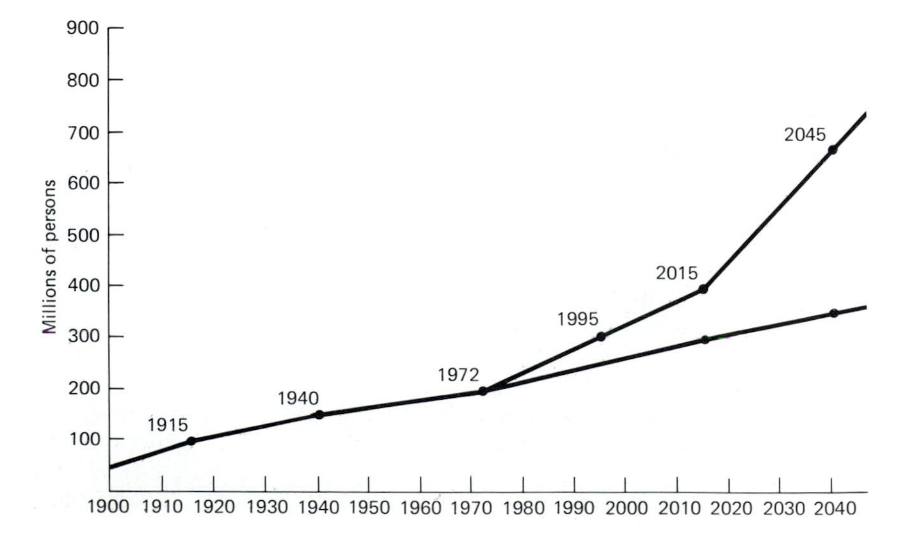

Figure 52–17 A comparison of the growth rates of the population of the United States based on projections of two children versus three children per family. By the year 2070, the difference in population is rather dramatic: one billion with a three-children birth rate; 350 million with a two-children birth rate.

In large part, overpopulation has resulted from advances in medicine. The conquest of such infectious diseases as malaria, yellow fever, smallpox, tuberculosis, and cholera has been responsible for lowering the death rate—especially in the less developed countries. Infant mortality in particular has been drastically reduced, and birth control has not begun to offset the increased numbers that survive. As a result, large families of children are now growing up and reproducing instead of dying in infancy, as was formerly the case.

WHEN IS A NATION OVERPOPULATED?

It is commonly understood in the United States that *other* countries, especially the less developed nations, are overpopulated, but surely not the United States itself! However, as suburban sprawl stretches out, merging one city into another, as farms are sold and the cost of food increases, as resources dwindle and we experience fuel and other shortages, as water pollution and air pollution produce increasing blight, many individuals no longer need to be convinced that overpopulation is a reality, a present reality, even in this country.

In the long run, the ideal population for a nation, continent, or planet is one that can be sustained indefinitely. Higher population densities than this generally result from temporary expansion of the carrying capacity of the habitat by technological means. These expansions thus far have always depended on the consumption of nonrenewable resources and on the passage of some of the costs to the ecosystem in the form of pollution. Since nonrenewable resources will dwindle and finally cease, and since the ability of the ecosystem to absorb pollution is limited, expansion cannot continue indefinitely. Populations dependent upon present-day technology eventually will be drastically reduced. Since more resources will by that time have been consumed, the resulting population is likely to be both materially poorer and fewer in number.

THE FUTURE

Mostafa K. Tolba, executive director of the United Nations environmental program, told delegates to the second United Nations Conference on the Global Environment on May 10, 1982, that if the nations of the world carry on with their present policies, by the turn of the century they would face "an environmental catastrophe which will witness devastation as complete, as irreversible, as any nuclear holocaust . . . ." Most of the delegates to the conference seemed to feel that this was exactly what is most likely to happen. Population pressures, technological and economic forces, and ingrained attitudes and prejudices oppose whatever measures have been attempted or might be, and it is tempting to view the ecological future of our planet in a very bleak light. However, many potential solutions do exist, which could do much to ameliorate our environmental crisis and could even buy us time in which permanent solutions could be found.

Clearly, the quality of all life on the planet depends on humanity's ability to limit its numbers. Hopefully, that goal can be accomplished through planning, and not through violence or disease. Although many, if not all, environmental problems stem in some way from the increasingly large human population, their severity has to do with our choices regarding our life style and how we go about supporting it. As biological beings, we share much in common with the fate of other life forms on the planet. We are also, however, beings with a unique capacity to reflect on and probe our biological identities. This talent is the key to ensuring our, and the planet's, continued survival.

SUMMARY

I. In primitive hunting and gathering societies, human beings have minimum environmental impact.

II. Agriculture permits a more settled and more thorough exploitation of the environment.

  A. Advanced agriculture, with substantial energy and technological input, has great environmental impact.

    1. In agriculture, people change ecosystems to benefit themselves. Generally they replace diverse natural communities with those consisting of one or two specially bred species. These artificial communities are unstable, simple, and usually lacking in some important ecological mineral cycles. Some wildlife can coexist with agricultural communities.

    2. Monoculture of agricultural species promotes

attack by pests and disease. This is typically combatted using pesticides.
   a. Persistent pesticides tend to destroy nontarget organisms and to become concentrated in predators by biological magnification.
   b. Less destructive pest control methods exist and should be employed where possible.

III. Pollution exists when the environment is degraded by wastes in its suitability for any of the organisms that normally would inhabit it, or when damage to property, esthetic values, or public health results.
   A. Severe aquatic pollution by organic substances produces a characteristic zonation. Eventually, much of the pollution is rendered harmless by natural processes, but only after a portion of the habitat has been badly degraded.
   B. Cultural eutrophication causes bodies of water to become filled with abnormal quantities of aquatic vegetation, often to become shallower, and in the end, to disintegrate as natural habitats.

IV. With every breath we inhale various pollutants deposited in the air by motor vehicles, industries, power plants, and other sources.
   A. An inversion layer may develop when a layer of warm air acts as a lid, sealing pollutants beneath so that they cannot be dispersed. Stagnation of air masses is of equal importance.
   B. Air pollution also affects climate on a global scale.

V. Species may become endangered or extinct by a variety of processes—overexploitation, persecution, environmental pollution, or the introduction of exotic species—but habitat destruction is the most important of these.

VI. Increases in population can be translated into increased needs for food, shelter, clothing, material goods, and educational, medical, and social services. These increases result in additional environmental stress.
   A. We have temporarily extended the carrying capacity of the earth for the human population by our technology. This expansion depends upon consumption of nonrenewable resources and the passage of some of the costs to the environment in the form of pollution.
   B. A nation may be considered overpopulated if its total population is too great to be sustained permanently by its resources. By this criterion, virtually all areas of the modern world must be judged to be overpopulated *at the present time*.

POST-TEST

1. Usually, hunting–gathering societies have ______________ impact on the ecosystems in which they are found.

2. Agriculturally modified communities are characterized by extremely ______________ species diversity; in addition, their ecological cycles are often ______________ within the community, and they can be perpetuated only by constant human intervention.

3. DDT and similar pesticides are very ______________ (*soluble/insoluble*) in water but very ______________ (*soluble/insoluble*) in body fat.

4. Due to ______________ ______________, persistent, DDT-like pesticides tend to accumulate in the highest level ______________ members of a food chain.

5. The primary productivity of the sea can be two to three times that of a typical ______________, and that of an ______________ can be as much as ten times that.

6. In the area closest to the source of pollution of a stream, the organic content of the water is ______________ (*high/low*) as is, surprisingly, the ______________ content.

7. Thermal pollution raises the ______________ of water, increases the ______________ requirements of the organisms that live there, and at the same time reduces the concentration of this substance in the water.

8. An important ecological effect of air pollution is ______________ rain, which appears to substantially damage many forests and kills fish in streams and lakes in areas where the water has little natural ______________ capacity, that is, where it is ______________.

9. An ______________ ______________ tends to trap air pollution beneath itself; such a layer occurs when ______________ (*cold/warm*) air overlies ______________ (*cold/warm*) air.

10. One cause of particular environmental concern is the possible ______________ effect that carbon dioxide, an unavoidable product of all combustion, is likely to produce; this effect could cause the climate to become considerably ______________.

11. About ______________ ______________ of the world's population lives in the developed countries; yet this fraction consumes about 85% of the world's resources and pollutes accordingly.

REVIEW QUESTIONS

1. Compare hunting and gathering societies with technological societies such as our own with respect to population density and to environmental impact.
2. Is slash-and-burn agriculture more ecologically acceptable than modern practices? Under what circumstances might slash-and-burn agriculture become ecologically damaging?
3. What are some ecological effects of air pollution?
4. What is an inversion layer? What are the effects of air pollution on the global climate?
5. List two types of water pollution. Describe the ecology of severe sewage pollution. How can damage from it be minimized?
6. When should a nation be considered overpopulated? Is the United States overpopulated?

Recommended Readings for Part VII

Books about Evolution

Ambrose, E. J. *The Nature and Origin of the Biological World.* New York, John Wiley, 1982. An up-to-date evolution textbook that discusses some of the current controversial views and treats them fairly.

Cairns-Smith, A. G. *Genetic Takeover and the Mineral Origins of Life.* Cambridge, Cambridge University Press, 1982. The chicken-and-egg riddle of the origin of life has always been one of replication: How can a cell without information reproduce? How can information reproduce without a cell? This author presents an ingenious and eccentric hypothesis.

Colbert, E. H. *Men and Dinosaurs.* New York, Dutton, 1968. The history not so much of dinosaurs as of their discovery. Other books by this author are also extremely worthwhile.

DeCamp, L. S. *The Great Monkey Trial.* New York, Doubleday, 1968. An entertaining, scholarly, but not always fair account of the Scopes Monkey Trial of Dayton, Tennessee, in 1925.

Desmond, A. J. *The Hot-blooded Dinosaurs.* New York, The Dial Press, 1978. A popular and readable—and fascinating—bit of biological advocacy to the effect that the dinosaurs were homeothermic.

Dobzhansky, T., F. G. Ayala, and G. L. Stebbins. *Evolution.* San Francisco, W. H. Freeman, 1977. One of the very best summaries of the action of the genetic mechanisms of evolution.

Greene, J. C. *Science, Ideology and World View.* Berkeley, University of California Press, 1982. How evolution influences the way we view the universe and our place in it.

———. *Darwin and the Modern World View.* New York, New American Library, 1963. Contains explicit discussion of the intellectual relationship between evolution and religion.

Margulis, L. *Symbiosis in Cell Evolution.* San Francisco, W. H. Freeman, 1981. The endosymbiotic theory of the origin of eukaryotes by a strong advocate.

Napier, J. R. *Primates and their Adaptations.* Burlington, N.C., Carolina Biological Supply Co., 1977 (booklet). A much more thorough discussion of the primates than we have had space to provide in this book.

Reader, J. *Missing Links, the Hunt for Earliest Man.* Boston, Little, Brown, 1981. An up to date and very readable recent history of the science of human paleontology.

Savage, J. M. *Evolution,* 3rd ed. New York, Holt, Rinehart and Winston, 1977. A brief college-level paperback text.

Stanfield, W. D. *The Science of Evolution.* New York, Macmillan, 1977. A typical and representative example of a modern evolution textbook.

Stanley, S. *Macroevolution: Pattern and Process.* San Francisco, W. H. Freeman, 1979. Does speciation and evolution of higher categories proceed in fits and starts? An exposition of punctuated equilibrium.

Stebbins, G. L. *Darwin to DNA, Molecules to Humanity.* San Francisco, W. H. Freeman, 1982.

Stone, I. *The Origin.* New York, Doubleday, 1980. A novelized but reasonably accurate and pleasant biography of Darwin.

Wickler, W. *Mimicry.* New York, McGraw-Hill, 1968. There is more to mimicry than meets the eye. The adaptive significance of this phenomenon is surprisingly complex.

Wilson, E. W. *Sociobiology, the New Synthesis.* Cambridge, N.H., The Belknap Press, 1975. Important (to say the least!) summary of Wilson's proposed mechanisms of the evolution of altruism and his views on kin selection. Equally relevent for behavior reading.

Magazine and Journal Articles about Evolution:

Ball, J. A. Extraterrestrial intelligence: where is everybody? *American Scientist* 68:656–663, 1980. Does evolution always proceed in what we would consider an upward way? If so, why do we seem to be alone in the universe?

Bambach, R. K. Responses to creationism. *Science* 220:851–853, 1983. Five anticreationist books and one creationist book are reviewed in this article.

Carson H. L., P. S. Nair, and F. M. Sene. *Drosophila* hybrids in nature: proof of gene exchange between sympatric species. *Science* 189:806–807, 1975. Speciation in progress?

Cloud, P. Evolution of ecosystems. *American Scientist* 62:54–66, 1974. Summary of the geologic timetable from an ecological point of view.

———, and M. F. Glaessner. The Ediacarian period and system: metazoa inherit the earth. *Science* 218:783–792, 1982. The naked metazoa of the upper precambrian, which now, according to the authors, deserve a new name and status.

Dobzhansky, T. Species of *Drosophila. Science* 177:664–670, 1972. The classic studies of speciation under laboratory conditions.

Eyde, R. H. The foliar theory of the flower. *American Scientist* 63:430–37, 1975. Did flowers originate from leaves? Maybe not.

Gilbert, L. E. The coevolution of a butterfly and a vine. *Scientific American,* August 1982, pp. 110–121. Some species of vines have coloration or even structures that visually mimic the eggs of certain tropical butterflies and inhibit the butterflies from depositing real eggs on them.

Goode, R. Darwinism: survival of a theory. *MD,* February 1983, pp. 116–172. The evolution of evolutionary theory.

Gould, S. J. Darwinism and the expansion of evolutionary theory. *Science* 216:380–387, 1982. Another, more complete but also somewhat more difficult discussion of the recent changes in evolutionary thinking.

Grivell, L. A. Mitochondrial DNA. *Scientific American,* March 1983, pp. 78–89. The nonstandard version of the genetic code of the mitochondria implies to this author that the organisms that originally gave rise to the mitochondria are now extinct.

Hiam, A. Airborne models and flying mimics. *Natural History* 91:42–49, 1982. Batesian (and other kinds of) mimicry.

Karp, L. E. The immortality of a cancer victim dead since 1951. *Smithsonian,* March 1976, pp. 51–56. The ubiquitous HeLa cell line, ideally adapted to laboratory conditions, has tended to contaminate cultures of other cells and displace them so that the most stringent measures must be taken to protect tissue cultures from these alien invaders. A bizarre example of natural selection in an unnatural environment.

Kirsch, J. A. W. The six-percent solution: second thoughts on the adaptedness of the Marsupalia. *American Scientist* 65:276–288, 1977. This author denies that marsupials are primitive, and along with it, the old notion that evolution is goal-directed.

Levinton, J. S. Charles Darwin and darwinism. *Bioscience* 32:495–500, 1982. A spirited defense of classical Darwinian evolution against modern challenge by evolutionary heresies.

Lewin, R. Adaptation can be a problem for evolutionists. *Science* 216:1212–1213, 1982. Did features now adaptive originate with their modern function?

Mayr, E. Darwin and natural selection. *American Scientist* 65:321–327, 1977. How did the concept of natural selection ever occur to Darwin?

Mossman, D. J., and W. A. S. Sarjeant. The footprints of extinct animals. *Scientific American,* January 1983, pp. 75–85. The bones of extinct animals are much rarer than their tracks.

Officer, C. B., and C. L. Drake. The Cretaceous-Tertiary transition. *Science* 219:1383–1390, 1983. These authors do not think that the dinosaurs and others were extraterrestrially expunged.

Rukang, W., and L. Shenlong. Peking man. *Scientific American,* June 1983, pp. 86–94. The culture of *Homo erectus.*

Russell, D. A. The mass extinctions of the late Mesozoic. *Scientific American,* January 1982, pp. 58–65. The new paleontological orthodoxy: The dinosaurs were extinguished by an asteroid.

Simon, C. Death star. *Science News* 125:250–252, 1984. Some scientists think that our sun has a companion star that periodically hurls lethal comets at the earth, leading to mass extinctions.

Trachtman, P. The search for life's origins–and a first "synthetic" cell. *Smithsonian,* June 1984, pp. 43–51.

Wiens, J. A. Competition or peaceful coexistence. *Natural History* 92:30–34, 1983. Sometimes there may be less competition between species than evolutionists often believe.

Woese, C. R. Archaebacteria. *Scientific American,* June 1981, pp. 98–122. If five kingdoms aren't enough for you, you might consider adding a sixth to accommodate these eccentric and possibly archaic bacteria.

Books about Behavior

Alcock, J. *Animal Behavior: An Evolutionary Approach.* 2nd edition. Sinauer, 1979.

Bonner, J. T. *The Evolution of Culture in Animals.* Princeton, Princeton University Press, 1980. A charming and scholarly little book.

Gould, J. L. *Ethology.* New York, Norton, 1982. A good general text.

Hinde, R. A. and J. S. Hinde. *Instinct and Intelligence.* Burlington, N.C., Carolina Biological Supply Co. (booklet). According to this author, the concepts of instinct and intelligence lack explanatory power.

Matthews, G. V. T. *Orientation and Position-finding by Birds.* Burlington, N.C., Carolina Biological Supply Co., 1974 (booklet). Though written before the magnetic senses of animals were at all understood, this is a good summary of research on the use of visual cues by birds in navigation.

Von Frisch, K. *Animal Architecture.* New York, Harcourt Brace Jovanovich, 1974. The construction projects of animals, from spider webs to anthills, not forgetting birds' nests.

Journal and Magazine Articles about Behavior

Alkon, D. L. Learning in a marine snail. *Scientific American* 249:70–84, 1983. Evidently for the first time, a cellular mechanism of learning has been defined.

Bodde, T. Killer bees: cause for alarm? *Bioscience* 32:648–649, 1982. Evolutionary and ecological implications of behavioral genetics in bees.

Bonner, J. T. Chemical signals of social amoebae. *Scientific American,* March 1983, pp. 114–123. Species-specific pheromones of organisms on the verge of multicellularity.

Eisner, T., and S. Nowicki. Spider web protection through visual advertisement: role of the stabilimentum. *Science* 219:185–186, 1983. About the heavy platforms of silk in the center of spider webs which warn birds not to fly into the web.

Gallup, G. G., Jr. Self-awareness in primates. *American Scientist* 67:417–421, 1979. Chimpanzees and the like may, in a sense, be persons.

Gould, S. J. The guano ring. *Natural History,* January 1982, pp. 12–19. An easily studied and easily interpreted example of territoriality.

Hölldobler, B. The wonderfully diverse ways of the ant. *National Geographic,* June 1984, pp. 774–813. Beautifully illustrated by John D. Dawson, this excellent and up-to-date article summarizes fascinating and little-known recent findings including, for example, harvester ant tournaments.

Kroodsma, D. E. The ecology of avian vocal learning. *Bioscience* 33:165–171, 1983. Learned and innate acquisition of birdsong, related to ecologic niche.

Maclean, G. L. Water transport by sand grouse. *Bioscience*

33:365–369, 1983. A better example of behavior as adaptation could scarcely be hoped for.

Partridge, B. L. The structure and function of fish schools. *Scientific American,* June, 1982. A review of the mechanisms by which fish organize themselves into schools.

Restak, R. M. Newborn Knowledge. *Science 82,* January-February 1982, pp. 58–65. Infants are behaviorally adapted for their role. The behavior is social, elaborate, and innate.

Roberts, L. Insights into the animal mind. *Bioscience* 33:362–364, 1983. Though perhaps animals do not think in the ordinary sense of the word, to view them as machines may be an unjustified industrial-age anthropomorphism.

Ryker, L. C. Acoustic and chemical signals in the life cycle of a beetle. *Scientific American,* June 1984, pp. 112–123. The interplay of acoustic and chemical signals enables the Douglas fir beetle to attract a mate, repel intruders, and regulate its population density.

Seeley, T. D. How honeybees find a home. *Scientific American,* October 1982. A discussion of how scout bees locate a suitable shelter for constructing a hive.

Shettleworth, S. J. Memory in food-hoarding birds. *Scientific American,* March 1983. A discussion of the ability of birds to remember where they have hidden food over a period of several months.

Thornhill, R. Sexual selection in the black-tipped hangingfly. *Scientific American,* June 1981, pp. 162–172. An example of classic Darwinian sexual selection with an odd twist.

Uetz, G. W., and G. E. Stratton. Communication in spiders. *Endeavor* 7:13–18, 1983. A summary and review of *Spider Communication,* by P. N. Witt and J. S. Rovner (eds.), which was published in 1982 by Princeton University Press. You might read this first.

Veit, P. G. Gorilla society. *Natural History,* March 1982, pp. 48–58. Once thought to be a uniquely human cultural universal, incest avoidance is also practiced among apes.

Books about Ecology

Ayensu, E. S. *Jungles.* New York, Crown Publishers, 1980. A handsome coffee table volume filled with fascinating information, this book is little short of a revelation.

Coker, R. E. *Streams, Lakes and Ponds.* New York, Harper and Row, 1968. Old, but never bettered, an extremely readable introduction to limnology.

Leigh, E. G., A. S. Rand, and D. M. Windsor. *The Ecology of a Tropical Rain Forest: Seasonal Rhythms and Long-term Changes.* Washington, D.C., Smithsonian Institution Press, 1983. By far the best study of a tropical ecosystem by a team of competent specialists, according to Ernst Mayr.

Whittaker, R. H. *Communities and Ecosystems,* 2nd ed. New York, Macmillan, 1975. An intermediate-level discussion of the community concept. Includes a summary of the biomes.

Magazine and Journal Articles about Ecology

Bergerud, A. T. Prey switching in a simple ecosystem. *Scientific American,* December 1983, pp. 130–141. How caribou enter into the classical lynx–snowshoe hare predator-prey relationship.

Boerner, R. E. J. Fire and nutrient cycling in temperate ecosystems. *Bioscience* 32:187–192, March 1982. By liberating nutrients stored in woody tissues, fire can increase ecosystem productivity.

Bryant, J. P. Hare trigger. *Natural History,* November 1981, pp. 46–52. The regulator of the snowshoe hare population, long a mystery, may result from toxin production by heavily browsed plants.

Deyrup, M. Deadwood decomposers. *Natural History,* March 1981, pp. 84–91. Decaying Douglas fir trunks support a community of organisms whose insects alone number more than 300 species.

Edmond, J. and K. Von Damm. Hot springs on the ocean floor. *Scientific American,* April 1983, pp. 78–93. A unique and extreme environment that supports the only known nonsolar ecosystems.

Karl, D. M., C. O. Wirsen, and H. W. Jannasch. Deep-sea primary production at the Galapagos hydrothermal vents. *Science* 207:1345–1347, 1980. The chemautotrophic basis of benthic hotspring communities.

Koehl, M. A. R. The interaction of moving water and sessile organisms. *Scientific American,* December 1982, pp. 124–134. Adaptations of intertidal and subtidal community organisms.

Lanyon, W. E. Fallow field guide to birds. *Natural History,* May 1982, pp. 61–67. As plant communities succeed, the composition of the animal species that inhabit them changes in a coordinated and regular fashion.

Mackoviak, P. A. Our microbial associates. *Natural History,* April 1983, pp. 80–87. Though not an ecosystem, the human body is a microcosmic microbial community. Are these little passengers of any use to us?

Marquis, R. E. Microbial barobiology. *Bioscience* 32:267–271, 1982. The biosphere may extend for miles underground.

MacMahon, J. A. Mount St. Helens revisited. *Natural History,* May 1982, pp. 14–24. Posteruption ecological succession.

Payne, W. J. Bacterial denitrification: asset or defect? *Bioscience* 33:319–324, 1983. The oceans may recycle 95 percent of their nitrogen each month. What ecological benefits, if any, accrue from the activities of denitrifying bacteria?

Savory, T. H. Hidden lives. *Scientific American,* July 1968, pp. 108–114. The community of the brown belt of the forest. Old, but still very good.

Smuts, G. L. Interrelations between predators, prey and their environment. *Bioscience* 28:316–319, 1978. The effect of predators upon populations of their prey is more likely to be marked at times and places where ecologic stress from other sources is significant.

Van Cleve, K., et al. Taiga ecosystems in interior Alaska. *Bioscience* 33:39–44, 1983. The ecology of the taiga.

Books about Human Ecology and Environmental Issues

Anonymous. *The International Book of the Forest.* New York, Simon and Schuster, 1981. Though tainted by the viewpoint of the commercial forest industries and to some extent an apologia for them, if its bias is kept in mind, this book is an excellent source of information not readily available elsewhere not only about forests but about their economic importance.

Ehrlich, P., and A. Ehrlich. *Extinction.* New York, Random

House, 1981. A powerful and compellingly reasoned compilation of arguments for the preservation of species. These authors have also written many other excellent titles in the environmental sciences, such as *The End of Affluence* and *The Population Bomb.*

Graham, F., Jr. *Since Silent Spring.* Boston, Houghton Mifflin, 1970. Rather than refer you to the original book, *Silent Spring,* by Rachel Carson, we thought it better to list a somewhat updated version.

Moran, J. M., M. D. Morgan, and J. H. Wiersma. *Introduction to Environmental Science.* San Francisco, W. H. Freeman, 1980. A fine introductory textbook.

Turk, J. *Introduction to Environmental Studies,* 2nd ed. Philadelphia, Saunders College Publishing, 1985. A fine general textbook.

Magazine and Journal Articles about Human Ecology and Environmental Issues

Beck, M., et al. The bitter politics of acid rain. *Newsweek,* April 25, 1983, pp. 36–37. A neat, brief summation of the political and economic reasons for acid rain.

Beddington, J. R., and R. M. May. The harvesting of interacting species in a natural ecosystem. *Scientific American* 247:62–70, 1982. If whales eat plankton and so do we, what will become of the whales?

Brown, L. R. Vanishing croplands. *Environment* 20:6–34, 1978. Mining, urbanization, erosion, and much else depletes the land now available for agriculture and encourages the despoliation of natural ecosystems.

Ehrlich, P. R., and H. A. Mooney. Extinction, substitution and ecosystem services. *Bioscience* 33:248–254, 1983. Why ecosystems should be indefinitely maintained in some approximation of their original, functional state.

Elias, T. S., and H. S. Irwin. Urban trees. *Scientific American,* November 1976, pp. 111–118. The kinds of trees that can grow in Brooklyn—an extreme example of ecosystem simplification by stress.

Finneran, K. Solar technology: A whether report. *Technology Review* 86:48–59, 1983. Is solar energy a truly practical and, if practical, widely applicable answer to our energy problems?

Grier, J. W. Ban of DDT and subsequent recovery of reproduction in bald eagles. *Science* 218:1232–1234, 1982. Decline and recovery of eagle populations is significantly related to the DDT-residue content of their eggs.

Grover, H. D. The climatic and biological consequences of nuclear war. *Environment* 26:7–38, 1984. Summary of the ecological effects of dust, haze, and other atmospheric insults of nuclear warfare.

Gwatkin, D. R., and S. K. Brandel. Life expectancy and population growth in the third world. *Scientific American,* May 1982, pp. 57–65. A decrease in death rate is the basic cause of population growth in underdeveloped countries.

Larson, W. E., F. J. Pierce, and R. H. Dowdy. The threat of soil erosion to long-term crop production. *Science* 219:458–465, 1983. One of the less publicized of the major ecological threats to our civilization.

Lockeretz, W. The lessons of the dust bowl. *American Scientist* 66:560–569, 1978. The principal lesson in this paper appears to be that we have not learned our lesson about "The most severe man-made environmental problem the United States has ever seen."

Pimental, D., and C. A. Edwards. Pesticides and ecosystems. *Bioscience* 32:595–600, 1982. Pesticide application as an ecological stress.

Radcliffe, B., and L. P. Gerlach. The ecology movement after ten years. *Natural History,* January 1981, pp. 12–18. Among educated, ecologically aware persons, environmental issues are still important, as to a lesser extent they are among the general population.

Reichert, W. Agriculture's diminishing diversity. *Environment* 24:6–11, 39–44, 1982. Genetic improvement and industrialization of agriculture have resulted in a very limited gene pool among plant and animal cultigens. This increases the likelihood that epidemics to which these strains are particularly vulnerable will produce widespread disaster.

Revelle, R. Carbon dioxide and world climate. *Scientific American* 247:35–43, 1982. As a result of fossil fuel use and the clearing and burning of forests, the global atmospheric content of carbon dioxide is increasing. What will this do to us?

Sagan, C. Nuclear war and climatic catastrophe: Some policy implications. *Foreign Affairs,* Winter 1984, 62:2, pp. 257–292. A definitive nuclear winter paper, written primarily for non-scientists.

Stommel, H., and E. Stommel. The year without a summer. *Scientific American,* June 1979, pp. 176–186. A natural environmental mishap, produced by the explosion of the volcano Krakatoa, illustrates the fragility of the heat balance of the earth's atmosphere.

Ward, G. M., T. M. Sutherland, and J. M. Sutherland. Animals as an energy source in third-world agriculture. *Science* 208:570–575, 1980. A model of clear thinking and careful discussion whose implications reach far beyond the issues immediately addressed by the authors. Appropriate technology is often traditional technology. But even traditional technology can be improved.

Westoff, C. F. Marriage and fertility in the developed countries. *Scientific American,* December 1978, pp. 51–55. What are the consequences of population growth changes in the developed countries, which may actually begin to decline before the year 2015?

Post-test Answers

CHAPTER 1

1. metabolism *2.* homeostasis *3.* stimulus *4.* cyclosis *5.* asexual reproduction *6.* respond *7.* c *8.* f *9.* b *10.* a *11.* g *12.* h *13.* d *14.* e *15.* plants and algae (some bacteria); animals; fungi and bacteria *16.* prokaryotes *17.* cyanobacteria and bacteria *18.* Protista *19.* hypothesis *20.* theory

CHAPTER 2

1. carbon, oxygen, hydrogen, nitrogen, phosphorus, calcium *2.* C; H; O *3.* trace elements *4.* electrons, protons, neutrons *5.* electrons *6.* atomic number *7.* mass number *8.* isotopes *9.* orbitals *10.* two *11.* positive *12.* helium, neon; noble gases *13.* chemical bond *14.* ions *15.* cations; anions *16.* f *17.* g *18.* e *19.* h *20.* b *21.* d *22.* c *23.* a *24.* capillary action *25.* specific heat *26.* pH *27.* donor; acceptor *28.* buffer

CHAPTER 3

1. organic compounds *2.* sugars, starches *3.* monosaccharides, hexose *4.* isomers *5.* ribose, deoxyribose *6.* glucose *7.* amino sugars *8.* vitamin A *9.* cholesterol *10.* a *11.* f *12.* d *13.* i *14.* j *15.* b *16.* e *17.* c *18.* h *19.* g *20.* buffers *21.* essential amino acids *22.* primary structure *23.* secondary structure *24.* hydrogen bonds *25.* tertiary structure *26.* denaturation *27.* enzymes, hormones, structural components *28.* DNA *29.* adenine, guanine, cytosine, thymine *30.* adenosine triphosphate (ATP) *31.* nicotinamide adenine dinucleotide

CHAPTER 4

1. a *2.* p *3.* k *4.* g *5.* h *6.* e *7.* f *8.* n *9.* o *10.* m *11.* j *12.* i *13.* q *14.* d *15.* c *16.* b *17.* l *18.* endoplasmic reticulum *19.* rough endoplasmic reticulum *20.* Golgi complex *21.* lysosomes *22.* peroxisomes *23.* cristae *24.* plastids *25.* microtubules *26.* cytoskeleton *27.* microtrabecular lattice *28.* cilia, flagella; microtubules; 2; 9 *29.* nuclei *30.* nucleoplasm; chromosomes *31.* diploid; haploid *32.* zygote *33.* cell wall *34.* cell cycle *35.* S phase *36.* prophase, metaphase, anaphase, telophase *37.* metaphase *38.* cytokinesis *39.* colchicine

CHAPTER 5

1. cell membrane *2.* receptor proteins *3.* lipid bilayer, proteins *4.* phospholipids, glycolipids *5.* hydrophilic; hydrophobic *6.* spontaneously *7.* assembling, sealing *8.* glycoproteins *9.* integral proteins *10.* microvilli *11.* cellulose *12.* peptidoglycan *13.* selectively permeable *14.* diffusion *15.* dialysis *16.* isotonic *17.* turgor pressure *18.* exocytosis *19.* endocytosis; phagocytosis *20.* desmosomes *21.* tight junctions *22.* gap junction

CHAPTER 6

1. epithelial tissue *2.* protection, absorption, secretion, sensation *3.* squamous, cuboidal, columnar *4.* pseudostratified ciliated epithelium *5.* stratified squamous epithelium *6.* glands *7.* stroma *8.* Reticular fibers *9.* 1. g; 2. f; 3. d; 4. e; 5. j; 6. b; 7. c; 8. i; 9. a; 10. h *10.* hemoglobin *11.* plasma *12.* 1. i; 2. c; 3. a; 4. b; 5. h; 6. e; 7. d; 8. g; 9. f *13.* root, shoot *14.* tracheids *15.* surface, fundamental, vascular

CHAPTER 7

1. energy *2.* kinetic energy *3.* photosynthesis, cellular respiration, cellular work *4.* Thermodynamics *5.* first law of thermodynamics *6.* exothermic *7.* endothermic *8.* entropy *9.* second law of thermodynamics *10.* free; temperature, pressure *11.* negative *12.* exergonic *13.* coupled *14.* additive *15.* activation energy *16.* catalyst *17.* enzymes *18.* enzyme–substrate complex *19.* active site *20.* temperature, enzyme, substrate, cofactors

CHAPTER 8

1. photons *2.* action spectrum *3.* more *4.* excited *5.* water *6.* NADPH, ATP *7.* oxidation *8.* grana *9.* thylakoid *10.* outer, inner, thylakoid *11.* intramembrane space, stroma, thylakoid space *12.* carotenoids, phycobilins *13.* photosystems *14.* reaction center; antenna *15.* H_2O *16.* NADPH *17.* electrons, protons; oxygen *18.* electron acceptors *19.* cyclic photophosphorylation *20.* protons *21.* ATP, stromal *22.* ribulose 1,5-bisphosphate, phosphoglycerate *23.* ribulose 1,5-bisphosphate; Calvin cycle (or C_3 pathway) *24.* phosphoenol pyruvate (PEP); CO_2 *25.* oxygen, phosphoglycolate

CHAPTER 9

1. cellular respiration *2.* glycolysis *3.* citric acid cycle *4.* electron transport system *5.* NAD, FAD *6.* activated; ATP *7.* deamination, citric acid cycle *8.* dehydrogenation *9.* decarboxylation *10.* proton; protons, ATP *11.* ATP *12.* inner membrane, mitochondria *13.* phosphorylation *14.* 36, 263 *15.* mitochondria, cytoplasm *16.* fermentation *17.* lactate

CHAPTER 10

1. heredity *2.* asexual reproduction *3.* zygote *4.* chromatin *5.* DNA, RNA, protein *6.* nucleosome *7.* centromere *8.* homologous chromosomes *9.* gene *10.* haploid, n; diploid, 2n *11.* polyploid *12.* synapsis *13.* crossing over *14.* 6 *15.* spermatogenesis *16.* d *17.* f *18.* h *19.* g *20.* j *21.* e *22.* c *23.* b *24.* a *25.* i

CHAPTER 11

1. locus *2.* alleles *3.* monohybrid cross *4.* dominant gene *5.* recessive gene *6.* homozygous; heterozygous *7.* first filial generation, F_1 generation *8.* phenotype *9.* genotype *10.* *a priori* probability *11.* product *12.* sum *13.* 0, 1 *14.* identical *15.* dihybrid cross *16.* complementary genes *17.* polygenes *18.* multiple alleles *19.* linked *20.* Barr body, genetic *21.* inbreeding *22.* hybrid vigor *23.* albinism *24.* phenylketonuria

CHAPTER 12

1. isogenic strains *2.* intelligence quotient (IQ) *3.* karyotype *4.* congenital defect *5.* aneuploidy *6.* trisomic *7.* monosomic *8.* nondisjunction *9.* translocation *10.* Down syndrome; trisomic; 21 *11.* Klinefelter syndrome; Turner syndrome *12.* metabolism *13.* hemoglobin *14.* cystic fibrosis *15.* amniocentesis *16.* negative eugenics

CHAPTER 13

1. phosphate, pentose sugar, nitrogenous base *2.* dATP, dGTP, dCTP, dTTP *3.* DNA polymerase *4.* nucleotides *5.* turns *6.* density *7.* base pairing *8.* complementary *9.* adenine, thymine; cytosine, guanine *10.* dead smooth; live rough *11.* DNA *12.* x-ray diffraction *13.* complementary *14.* S phase *15.* single *l6.* both directions *17.* bacteriophages *18.* semiconservative

CHAPTER 14

1. ribosomal, messenger, transfer *2.* mRNA *3.* proteins, nucleic acids *4.* codon *5.* anticodon *6.* transcription *7.* DNA-dependent RNA polymerase; ATP, GTP, CTP, UTP *8.* promoter *9.* processing *10.* introns *11.* exons *12.* translation *13.* activation, ATP *14.* 7-methyl guanosine *15.* codon; anticodon *16.* polyribosomes *17.* initiation, elongation, termination *18.* GTP *19.* antibiotics, streptomycin, tetracycline, erythromycin *20.* ribosomal cycle *21.* albinism *22.* chromosomal mutation *23.* frameshift mutation *24.* radiation, mutagens *25.* synonyms

CHAPTER 15

1. Inducer *2.* inducible enzymes *3.* repressed *4.* corepressor *5.* repressor; regulatory gene *6.* promoter *7.* operon *8.* mitochondria, chloroplasts *9.* chloroplast, nucleus *10.* maize (corn), McClintock *11.* halteres; control

CHAPTER 16

1. genetic engineering (or recombinant DNA) *2.* restriction endonucleases; bacteriophage *3.* recognition sites *4.* palindrome *5.* plasmids *6.* antibiotics *7.* replication *8.* hybridizing, DNA (or RNA) *9.* restriction endonuclease *10.* ligase; cloned (or amplified) *11.* fusing; β-galactosidase *12.* crown gall; bacterium; plasmid *13.* calcium phosphate *14.* self-replicating; protein (or product)

CHAPTER 17

1. taxonomy *2.* species *3.* reproductive isolation *4.* subspecies; genus *5.* genus, species *6.* splitter *7.* monophyletic *8.* Protista *9.* Prokaryota *10.* Animalia *11.* Eukaryota *12.* Animalia *13.* Chordata *14.* Vertebrata *15.* Mammalia *16.* Eutheria *17.* Primates *18.* Hominidae *19.* *Homo* *20.* (*Homo*) *sapiens.*

CHAPTER 18

1. DNA, RNA *2.* bacteria *3.* attachment; penetration; replication; assembly; release *4.* virus; cell wall *5.* nucleic acid *6.* lysogenic conversion *7.* lytic *8.* temperate *9.* transduction *10.* Interferons *11.* latent *12.* host (or normal); cancer cells *13.* Viroids; coat *14.* prophage *15.* lytic

CHAPTER 19

1. Monera *2.* nuclear membrane, organelles *3.* producers *4.* cell wall *5.* gram-positive *6.* mesosome *7.* Plasmids *8.* pili *9.* saprobes *10.* Facultative anaerobes *11.* conjugation *12.* spherical, rod-shaped, spiral *13.* cocci, bacilli, spirilla *14.* clostridia *15.* spirochete *16.* chlamydias *17.* Opportunistic *18.* Crown gall *19.* exotoxins *20.* cell wall

CHAPTER 20

1. single; colonial *2.* contractile vacuole *3.* protozoa; algal *4.* d *5.* b *6.* a *7.* c *8.* e. *9.* diatoms *10.* digestive system; insect *11.* dinoflagellates

CHAPTER 21

1. decomposers *2.* budding; fission *3.* hyphae; mycelium *4.* fruiting body *5.* true fungi; slime molds *6.* oospores *7.* sporangium; flagella *8.* zygospores *9.* mating types *10.* conidia; conidiophores *11.* ascospores; asci *12.* basidiospore *13.* gills *14.* sexual *15.* amoeba-like cells *16.* lichen *17.* roots; plants *18.* ergot *19.* penetrate plant cells to obtain nourishment *20.* barberry plant

CHAPTER 22

1. isogamy *2.* *Volvox* *3.* walls; cellulose *4.* haploid *5.* gamete *6.* identical; isomorphic *7.* Phaeophyta; algin *8.* phycocyanin; cyanobacteria *9.* mosses, liverworts; Bryophyta *10.* gametophyte; sporophyte *11.* haploid *12.* gemmae *13.* protonema; filamentous

CHAPTER 23

1. Equisetophyta; microphylls; silica *2.* sporophyte *3.* prothalli *4.* haploid *5.* haploid; triploid *6.* cycads (Cycadopsida) *7.* pistillate; pollen tube; generative nucleus *8.* leaves *9.* xylem; xylem vessels *10.* branches *11.* Monocots: parallel leaf veins; one cotyledon; endosperm; no secondary growth; flower parts in multiples of 3; separated vascular bundles. Dicots: branching leaf veins; two cotyledons; usually no endosperm; secondary growth, especially in woody dicots; flower parts in multiples of 4 or 5; in woody dicots, layers of vascular tissue (although the vascular bundles are discrete in herbaceous dicots).

CHAPTER 24

1. invertebrates *2.* sessile *3.* body cavity *4.* ectoderm *5.* mesoderm *6.* mouth *7.* cnidarians (also ctenophores and adult echinoderms) *8.* one quarter *9.* homologous *10.* spicules *11.* spongocoel; osculum *12.* polyp, medusa *13.* Cnidaria; Anthozoa *14.* flatworms *15.* smell *16.* eating poorly cooked pork that contains the larvae *17.* a complete digestive tract (tube-within-a-tube body plan), a separate circulatory system *18.* b *19.* c *20.* d *21.* e *22.* a *23.* c *24.* b *25.* e *26.* a *27.* d

CHAPTER 25

1. exoskeleton *2.* radula *3.* open *4.* trochophore *5.* metamerism *6.* lung *7.* shell *8.* locomotion *9.* present in the same animal *10.* increase the surface area of the intestine *11.* cerebral ganglia *12.* arthropods *13.* arthropods *14.* biting and grinding food *15.* excretory *16.* barnacles *17.* pinching claws *18.* excretory *19.* b *20.* c *21.* d *22.* a *23.* e *24.* d *25.* b

CHAPTER 26

1. echinoderms *2.* water vascular; tube *3.* arms (rays); test (shell) *4.* notochord; tubular nerve cord; gill slits *5.* Tunicates *6.* vertebral column; cranium *7.* Pisces, Tetrapoda *8.* Placoid *9.* cloaca *10.* food; digestion *11.* ray-finned; tetrapods (land vertebrates) *12.* gills *13.* tetrapods (land vertebrates) *14.* terrestrial; keeps the embryo moist and acts as a shock absorber *15.* birds, mammals *16.* lay eggs 0RaR *17.* c, d, e *18.* e *19.* d, e *20.* f *21.* c *22.* d *23.* a–e *24.* b *25.* c *26.* e *27.* d *28.* a

CHAPTER 27

1. anchorage; absorption *2.* guard cells *3.* stomata; water *4.* palisade parenchyma *5.* potassium (or K^+) *6.* sieve cells; companion cells; slime (or p-protein) *7.* source; sink; leaves *8.* water; minerals; cohesion *9.* food; heart *10.* hydrostatic pressure; guttation

CHAPTER 28

1. scutellum *2.* hypocotyl; radicle *3.* germination inhibitor *4.* maturation *5.* straightening; phytochrome *6.* apical meristems *7.* auxins *8.* leaves *9.* water *10.* primary thickening

CHAPTER 29

1. pulvinus; turgor pressure *2.* darker *3.* lowers; cellulose crosslinks *4.* ethylene *5.* amyloplasts; starch *6.* phytochrome; red *7.* red; infrared (or far red) *8.* gibberellins *9.* auxin; broad

CHAPTER 30

1. shortest *2.* molybdenum; nitrogen fixation *3.* gaseous *4.* potassium *5.* alga, cyanobacterium, fungus; pioneer *6.* minerals (plant mineral nutrients) *7.* ammonia; nitrate *8.* cyanobacteria *9.* symplast *10.* Casparian *11.* minerals; humus *12.* mycorrhizae *13.* field *14.* air

CHAPTER 31

1. genetic; identical *2.* new combinations *3.* propagules (or gemmae) *4.* scion; stock; cambium *5.* pollen; water *6.* staminate *7.* ovulate; pollen tube; archegonium; embryo *8.* petals *9.* male, female (staminate, pistillate) *10.* pollination *11.* simple *12.* fruit

CHAPTER 32

1. hydrostatic *2.* corneum *3.* molt *4.* appendicular *5.* origin; insertion *6.* sphincter *7.* maximally *8.* d; c; b; a; e; f; g (actually the last two steps alternate repeatedly) *9.* energy storage *10.* phosphocreatine

CHAPTER 33

1. digested; absorbed *2.* cecum; bacteria that digest cellulose *3.* rechewing partly digested food *4.* carnivores; omnivores *5.* ecto; endo *6.* mucosa; goblet *7.* peristalsis *8.* Incisors *9.* salivary glands *10.* esophagus; duodenum *11.* rugae *12.* gastric glands; stomach *13.* duodenum *14.* d *15.* a *16.* f *17.* e *18.* f, c *19.* b *20.* fat

CHAPTER 34

1. anabolism; catabolism *2.* basal metabolic *3.* energy used to carry on daily activities *4.* remains the same; increases *5.* sugars, starches *6.* fuel for cellular respiration *7.* glycogen synthesis (from glucose) *8.* reconverted to glucose *9.* gluconeogenesis *10.* liver; fat *11.* fats; acetyl coenzyme A *12.* acidic *13.* ingested in the diet *14.* all of the essential amino acids in sufficient quantity to support growth *15.* essential amino acids *16.* amino group; carbon *17.* A, D, E, K *18.* b *19.* d *20.* e *21.* a *22.* c

CHAPTER 35

1. c, e *2.* a, f *3.* b, d *4.* b *5.* oxygen *6.* serum *7.* gamma globulin *8.* anemia *9.* b *10.* e *11.* d *12.* c *13.* a *14.* K *15.* fibrin *16.* Arterioles *17.* capillaries *18.* pulmonary arteries *19.* semilunar *20.* systole; diastole *21.* volume of blood pumped by one ventricle in one minute *22.* the more blood the heart pumps *23.* blood pressure *24.* blood flow; resistance to blood flow *25.* aorta *26.* brain *27.* kidneys; liver *28.* heart attack *29.* blood pressure *30.* vasoconstrictors

CHAPTER 36

1. stimulating an immune response *2.* antibodies (or immunoglobulins) *3.* interferons *4.* redness, heat, swelling, pain *5.* bone marrow; thymus; lymph *6.* killer T cells *7.* memory *8.* plasma cells (differentiated B lymphocytes) *9.* antibody *10.* thymus *11.* antigen–antibody *12.* coat pathogens *13.* active *14.* passive *15.* HLA. *16.* one location to another in the same organism *17.* immune response; transplanted *18.* privileged *19.* allergen; allergic *20.* histamine; inflammation.

CHAPTER 37

1. thin; diffusion; moist; blood vessels *2.* tracheal; spiracles *3.* gills; bony fish *4.* lungs *5.* swim bladder *6.* air sacs *7.* trachea; bronchus *8.* alveoli *9.* diaphragm *10.* water loss *11.* e *12.* a *13.* b *14.* d *15.* c *16.* vital capacity *17.* oxygen *18.* oxyhemoglobin; Bohr effect *19.* bicarbonate ions *20.* oxygen *21.* inhaling dirty air *22.* emphysema *23.* cigarette smoking

CHAPTER 38

1. excretion *2.* uric acid *3.* urea *4.* protonephridia; flame *5.* metanephridia *6.* antennal (green) glands *7.* Malpighian tubules *8.* nephrons *9.* b *10.* d *11.* c *12.* a *13.* e *14.* a *15.* c *16.* e *17.* d *18.* b *19.* c *20.* capillaries; Bowman's capsule *21.* afferent arteriole; efferent arteriole *22.* glomerular filtrate *23.* reabsorbed *24.* excreted in the urine *25.* hypertonic *26.* collecting ducts; reabsorbed; decreased *27.* ADH; suffers tremendous fluid loss in the urine

CHAPTER 39

1. sense organs; central nervous system (CNS) *2.* synapses *3.* stimuli *4.* receptors; nerves *5.* neuron; neuroglia *6.* receive stimuli; transmit neural impulses; cell body; synapse *7.* regeneration of injured axons *8.* cellular sheath, myelin sheath *9.* cell bodies *10.* resting potential *11.* Excitatory *12.* depolarizes *13.* action potential *14.* all-or-none *15.* one node of Ranvier to the next *16.* resting potential; more *17.* neurotransmitter *18.* acetylcholine *19.* norepinephrine *20.* divergence *21.* reflex action *22.* afferent; spinal cord; integration

CHAPTER 40

1. nerve net *2.* Afferent; efferent *3.* brain *4.* visceral; pedal *5.* brain; spinal cord *6.* brain stem *7.* medulla; spinal canal *8.* c *9.* e *10.* b *11.* a *12.* d *13.* b *14.* d *15.* e *16.* a *17.* c *18.* c *19.* a *20.* e *21.* b *22.* beta *23.* REM *24.* vagus; internal organs of chest and upper abdomen *25.* energy; stress *26.* dorsal *27.* increasingly larger amounts are needed to obtain the desired effect.

CHAPTER 41

1. receptor; accessory *2.* receive stimuli from the outside world; position of body parts; orientation *3.* Photoreceptors; mechanoreceptors *4.* energy; electrical; receptor potential *5.* adaptation *6.* gravity; statolith *7.* vision; hairs; cupula *8.* muscle spindles; Golgi tendon; joint *9.* motion; stimulus *10.* Halteres *11.* equilibrium *12.* labyrinth; saccule, utricle; semicircular *13.* otoliths; gravity *14.* endolymph; ampulla *15.* inner; sound *16.* Corti; cochlear *17.* chemoreceptors *18.* rhodopsins *19.* c *20.* d *21.* b *22.* e *23.* a

CHAPTER 42

1. target tissues *2.* hormones *3.* chemical messenger produced by one type of cell that has a specific regulatory effect on the activity of another type of cell *4.* cell membrane; cyclic AMP *5.* negative feedback *6.* molting *7.* hypothalamus *8.* growth *9.* b *10.* c *11.* a *12.* e *13.* c *14.* e *15.* c *16.* cortex; cortisol *17.* hyper; GH *18.* thyroid *19.* glucose; protein *20.* anterior lobe of the pituitary; TSH

CHAPTER 43

1. fragmentation *2.* sexual and asexual generations *3.* develops into a new individual *4.* hermaphroditic *5.* gamete; zygote *6.* sterile *7.* a *8.* c *9.* d *10.* b *11.* e *12.* b *13.* d *14.* c *15.* a *16.* c *17.* b *18.* d *19.* e *20.* b *21.* a *22.* c *23.* d *24.* c *25.* d *26.* e *27.* b *28.* therapeutic *29.* female sterilization *30.* menopause

CHAPTER 44

1. morphogenesis; cellular differentiation *2.* cleavage *3.* inner cell mass; embryo *4.* gastrulation *5.* ectoderm; endoderm *6.* neural plate *7.* amniotic fluid *8.* placenta *9.* gestation *10.* newborn infant *11.* production of milk (for the nourishment of the young) *12.* determines the future differentiation of nearby cell groups *13.* frog *14.* development *15.* aging

CHAPTER 45

1. lost *2.* mutation *3.* natural selection (or genetic drift, differential immigration or emigration, founder effect, bottleneck effect) *4.* both; sickle cell anemia *5.* genetic drift; founder, bottleneck; small *6.* Industrial melanism *7.* Batesian *8.* interbreeding *9.* prezygotic *10.* plants; parthenogenesis *11.* stabilizing *12.* uniformitarianism *13.* acquired *14.* *Beagle* *15.* Malthus; starvation *16.* overproduction of offspring, variation among offspring, inheritability of that variation, natural selection for the variants.

CHAPTER 46

1. homology *2.* vestigial *3.* chimpanzees *4.* adult *5.* biogeography; evolution *6.* homologous; structural parallelism *7.* adaptive radiation *8.* radioactive decay; isotopes

CHAPTER 47

1. oxygen *2.* organic; proteins; nucleic acids *3.* heterotrophic; anaerobic *4.* reverse *5.* endosymbiotic; bacteria; chloroplasts *6.* Prokaryotes (or microorganisms) *7.* Cambrian *8.* Silurian *9.* Labyrinthodont; lobe-finned *10.* Mississippian, Pennsylvanian; Carboniferous *11.* mammal *12.* dinosaurs; asteroid *13.* Cenozoic; birds, insects; flowering plants *14.* *Australopithecus* *15.* *Homo sapiens* *16.* punctuated equilibrium; gaps

CHAPTER 48

1. signals from the environment *2.* Ethology *3.* geotropism *4.* positive chemotaxis *5.* circadian *6.* crepuscular *7.* releaser *8.* Instinctive; learned *9.* follow the parent about *10.* migratory restlessness

CHAPTER 49

1. social behavior *2.* society *3.* aggregation *4.* symbolic *5.* Pheromones *6.* dominance hierarchy *7.* home range *8.* conflict; population *9.* pair bond *10.* nurse; royal jelly *11.* genetic *12.* culture *13.* altruistic *14.* Kin *15.* propagating (making more copies of themselves).

CHAPTER 50

1. producers *2.* recycle; producers *3.* brown belt *4.* grazing; (organic) detritus *5.* buzzard, hyena, maggot, earthworm, etc. *6.* chemosynthetic *7.* holozoic *8.* Saprobic *9.* energy *10.* energy; metabolism (energy consumption) *11.* biomass *12.* population density; distribution *13.* shortest *14.* Light *15.* density *16.* ecological niche *17.* competitive exclusion *18.* allelopathy *19.* simplify *20.* climax community *21.* sere

CHAPTER 51

1. population *2.* birth, death; limiting factors *3.* environmental *4.* snowshoe hares; lynx *5.* stress; density-dependent *6.* biome *7.* ecotones; species; edge *8.* tundra; permafrost *9.* taiga; evergreen (coniferous) *10.* West; North American *11.* humus (topsoil); recycling *12.* salinity *13.* deepest (abyssal, profundal, etc.) *14.* turnover; blooms *15.* low; high *16.* picoplankton *17.* intertidal; drying (desiccation) *18.* neritic *19.* littoral *20.* eutrophic

CHAPTER 52

1. small (minimal) *2.* low; incomplete *3.* insoluble; soluble *4.* biological magnification; predatory *5.* forest; estuary *6.* high; oxygen *7.* temperature; oxygen *8.* acid; buffering; soft *9.* inversion layer; warm; cold 10. greenhouse; warmer *11.* one third

appendix A
The Classification of Organisms

The system of cataloging organisms used here is described in Chapter 1 and in Part IV. In the following synoptic survey the phyla are arranged according to the five-kingdom system advanced by R. H. Whittaker (*Science*, 163: 150–160, 1969). We have modified the system below the kingdom level, and also have omitted many groups (especially extinct ones), in order to simplify and clarify the vast number of diverse categories of living organisms and their relationships to one another. Groups mentioned that have not been discussed in Part IV are denoted by an asterisk. Note that we have omitted the viruses from this survey since they do not fit into any of the five kingdoms.

KINGDOM MONERA: Bacteria and cyanobacteria

Prokaryotic organisms that lack nuclear membranes, mitochondria, and other membranous organelles; flagella, when present, are solid (rather than the 9 + 2 type typical of eukaryotes). Typically unicellular, but some form colonies or filaments. The predominant mode of nutrition is absorptive, but some groups are photosynthetic or chemosynthetic. Reproduction is primarily asexual by fission, but genetic recombination does occur in some species. Monerans are nonmotile or move by the beating of flagella or by gliding.

SUBKINGDOM CYANOBACTERIA

Formerly known as blue-green algae. Photosynthetic; chlorophyll and accessory pigments on photosynthetic lamellae. Some species fix nitrogen. (About 2500 species.)

SUBKINGDOM SCHIZOMYCETES

Bacteria. Most are heterotrophs and free-living saprobes; others are symbionts. Some are photosynthetic or chemosynthetic. (About 3000 species.)

Eubacteria. True bacteria. A diverse group consisting of several parts (subgroups). Cocci, bacilli, or spirilla. Most are harmless saprobes ecologically important as decomposers, but a few can be pathogenic—for example, *Escherichia coli, Clostridium botulinum,* and *Staphylococcus aureus.*

Myxobacteria. Slime bacteria. Unicellular, short rods resembling eubacterial bacilli, but lacking a rigid cell wall. Gram-negative. Excrete slime and are capable of a slow gliding movement. Some species form fruiting bodies.

Spirochetes. Slender, flexible, spiral-shaped bacteria. Move by bending unique axial filament. Some are important pathogens. An example is *Treponema pallidum,* which causes syphilis.

Actinomycetes. Moldlike bacteria. Branching, filamentous bacteria; many produce moldlike spores. Most are saprobes that decompose organic material in soil. A few are pathogenic. For example, *Mycobacterium tuberculosis* causes human tuberculosis.

Mycoplasmas. Bacteria without cell walls. Also known as PPLO. Tiny bacteria bounded by pliable cell membrane.

Rickettsias. Obligate intracellular parasites; unable to carry on metabolism independently of host. Most parasitize certain arthropods.

Chlamydias. Obligate intracellular parasites; sometimes referred to as energy parasites. Infect almost every species of bird and mammal.

Archaebacteria. Obligate anaerobes that produce methane by producing carbon dioxide. Some taxonomists suggest that these organisms should be assigned to a separate kingdom.

KINGDOM PROTISTA

Primarily solitary unicellular or colonial eukaryotic organisms that do not form tissues and exhibit relatively little division of labor in colonial forms. Most groups contain predominantly unicellular representatives. All modes of nutrition occur in this kingdom, except chemosynthetic autotrophy. Life cycles may include both sexual and asexually reproducing phases and may be extremely complex, especially in parasitic forms. Locomotion is by cilia, flagella, ameboid movement or by other means. Flagella have 9 + 2 structure.

ANIMAL-LIKE PROTISTS: PROTOZOA

Phylum Mastigophora. Flagellates. Locomotion by means of flagella. Some freeliving; many symbiotic; some pathogenic (e.g., *Trypanosoma* causes sleeping sickness). Reproduction usually asexual by binary fission. Includes the euglenoid organisms, which have choloroplasts and pigmented "eyespot," but lack an outer cellulose cell wall. (About 2500 species.)

Phylum Sarcodina. Shelled or naked protozoa whose movement or feeding is associated with pseudopods. The amoebas belong to this group. (About 11,500 living species.)

Phylum Ciliata. Ciliates. Locomotion is by cilia. This group includes *Paramecium* and *Stentor.* Reproduction asexual by binary fission or sexual by conjugation. (About 7200 species.)

Phylum Sporozoa. Sporozoans. Parasitic protists that reproduce by spores and have no means of locomotion. Reproduce by an unusual type of multiple fission. *Plasmodium* causes malaria. (About 6000 species.)

THE ALGAL PROTISTS

Phylum Chrysophyta. The diatoms, golden-brown algae, and the yellow-green algae. Unicellular, photosynthetic. Cell walls composed of silica and manganese rather than cellulose. (About 10,000 species.)

Phylum Pyrrophyta. Includes the dinoflagellates. Unicellular, usually biflagellate; brown, red, or yellow; cell walls contain cellulose. (About 1100 species.)

KINGDOM FUNGI

All eukaryotes; mainly multicellular organisms. Cells often form a mycelium. All heterotrophs with saprobic or parasitic nutrition. Many multinucleate with nuclei dispersed in a walled and often septate mycelial syncytium. Most are nonmotile; live embedded in a medium or food supply. Reproduce by means of spores which may be produced sexually or asexually. Reproductive cycles typically include both sexual and asexual processes. Body usually haploid with brief diploid period following fertilization.

DIVISION EUMYCOPHYTA.

True fungi. Almost always have cell walls and hyphae and reproduce by means of spores.

Class Oomycetes. Water molds. Distinguished by their biflagellated asexual spores, which can swim to new locations. Some are unicellular; others consist of branched mycelia with coenocytic hyphae. Cellulose cell walls. Sexual stage includes oogonia within which eggs develop and slender antheridia that produce male gametes. Important parasites in this group include *Plasmopara,* which causes downy mildew of grapes, and *Saprolegnia,* which parasitizes fish.

Class Zygomycetes. Produce sexual resting spores called zygospores. Hyphae are coenocytic and cell walls consist mainly of chitin. Heterothallic (have two mating types). Example is black bread mold *Rhizopus nigricans.*

Class Ascomycetes. Sac fungi. Sexual reproduction involves formation of ascospores in little sacs called asci. Asexual reproduction involves production of spores called conidia, which pinch off from conidiophores. Hyphae usually have perforated septa. This class includes yeasts, powdery mildews, many molds that cause food spoilage, and edible truffles and morels. (About 30,000 species.)

Class Basidiomycetes. Club fungi. Includes mushrooms, toadstools, bracket fungi, puff balls, rusts, and smuts. Basidiospores develop from basidia. Heterothallic. Hyphae divided by perforated septa. (About 25,000 species.)

Class Deuteromyctes. Imperfect fungi. Sexual stage has not been observed. Most reproduce only by conidia. (About 25,000 species.)

DIVISION MYXOMYCOPHYTA

Slime molds. Organisms with life cycles that include separate cells, aggregations of cells, and spore-producing stages. During part of their life cycle they ingest food rather than absorb it.

Class Myxomycetes. Plasmodial slime molds. Spend part of their life cycle as a thin streaming, multinucleate mass of protoplasm that creeps along decaying leaves or wood.

Class Acrasiomycetes. Cellular slime molds. Amoebalike cells aggregate to form a multicellular pseudoplasmodium which eventually develops into a fruiting body that produces spores.

KINGDOM PLANTAE

Multicellular, eukaryotic, photosynthetic organisms with cell walls containing cellulose. Cells frequently contain vacuoles. Photosynthetic pigments in plastids. Main photosynthetic pigment is chlorophyll *a*; chlorophyll *b* and various carotenoids serve as accessory pigments. Nonmotile with a few exceptions. Differentiated tissues and organs. Both sexual and asexual reproduction, with alternation of gametophyte and sporophyte generations. The gametophyte generation is markedly reduced in the higher members of the kingdom.

ALGAL-LEVEL DIVISIONS

DIVISION CHLOROPHYTA. Green algae. Mainly aquatic. Unicellular, colonial, and multicellular forms. Some motile and flagellated. Examples are *Volvox* and *Ulva.* (About 7000 species.)

DIVISION PHAEOPHYTA. Brown algae multicellular, often large. Some with vascular tissues. Chlorophylls *a* and *c*, with fucoxanthin also present. Food stored in the form of laminarin (a carbohydrate). Zoospores with two lateral flagella. This group includes the giant kelps. (About 1500 species.)

DIVISION RHODOPHYTA. Red algae. Multicellular, usually marine. Some (coralline algae) have bodies impregnated with calcium carbonate. Chlorophyll *a* and (in some) *d*, with phycocyanin and phycoerythrin also present; food stored in the form of a chemically distinctive starch. No flagellae. (About 4000 species.)

TERRESTRIAL DIVISIONS

DIVISION BRYOPHYTA. Nonvascular terrestrial plants. Liverworts, hornworts, and mosses. Lack conducting tissues. Usually terrestrial, with a marked alternation of gametophyte and sporophyte generations. Of these, the more prominent is the gametophyte (sexual generation) on which the sporophyte is dependent. (About 25,000 species.)

Class Musci. Mosses. The gametophyte plant has an erect stem, and its leaves are arranged in a spiral. (About 15,000 species.)

Class Hepaticae. Liverworts. Usually simple, flat plants living in moist shady places although some tropical representatives depart from this description. (About 9000 species.)

**Class Anthoceratopsida.* Hornworts. Limited distribution. Haploid gametophyte is dominant; hornlike diploid sporophytes are often capable of photosynthesis and are independent. (About 100 species.)

DIVISION TRACHEOPHYTA: VASCULAR PLANTS

Class Rhynipsida. Leafless, rootless, vascular plants. All are extinct.

Class Lycopsida. The club mosses. Possess simple vascular tissues and spirally arranged true leaves called microphylls. Have biflagellate, motile sperm. (About 900 species.)

Class Filicopsida: Ferns and Horsetails.

Order Equisetophyta. The scouring rushes or horsetails. Simple vascular tissues, jointed stems, and reduced scalelike leaves. Modern representatives small, but some extinct types were treelike.

Order Filicophyta. Ferns. Ferns are generally homosporous (produce only one kind of spore), though water ferns are heterosporous (produce two distinct types of spores). Gametophyte is usually free-living and photosynthetic. Sporophyte generation highly vascular and much larger than gametophyte, with pronounced tendency to vegetative propagation as well as sporulation.

The Seed Plants: Gymnosperms

Conifers, cycads and many evergreen trees and shrubs, although not all representatives are evergreen. No true flowers or ovules are present and seeds are usually borne naked on the surface of the cone scales. However, fruitlike structures homologous to cones, or fruitlike cones are sometimes present. Endosperm-like tissue derived from gametophyte is haploid. (About 640 species.)

Class Coniferopsida. Conifers. Common, usually evergreen trees and shrubs, with needle-shaped leaves, plus some extinct, large-leaved evergreen trees, the *Cordaites* (Devonian-Permian). Ovules not enclosed. (About 575 species.)

Class Cycadopsida. Cycads. Palmlike plants with flagellated sperm. Found in tropical or subtropical regions. Considered the most primitive of modern seed plants. (About 120 species.)

Class Lyginopteridopsida. Seed ferns. Extinct plants that had fern-like leaves, but also had seeds.

Class Ginkgopsida. Broad-leaved deciduous gymnosperms. The *Ginkgo* tree is the only living representative.

Class Gnetopsida. Climbing shrubs or small trees found in tropical or semitropical regions with many characteristics similar to those of the angiosperms. (About 75 species.)

The Seed Plants: Angiosperms

Class Angiospermopsida. Flowering plants with seeds enclosed in ovary, which becomes (sometimes along with other flower parts) the fruit. (About 250,000 species.)

Subclass Dicotyledonae (or Magnoliopsida). Most flowering plants are dicots. Embryos with two cotyledons or seed leaves; vascular bundles in a ring in the stem, fused into layers in woody representatives. Leaves with netlike venation. Flower parts in multiples of five, four, or sometimes two. (About 175,000 species.)

Subclass Monocotyledonae (or Liliopsida). The bamboos, grasses, lilies, orchids, and palms. Leaves with parallel veins, stems with scattered vascular bundles, flower parts in multiples of three. Embryo with one cotyledon and usually a large endosperm. (About 75,000 species.)

KINGDOM ANIMALIA

Multicellular, heterotrophic organisms with eukaryotic cells lacking rigid cell walls, plastids, and photosynthetic pigments. Level of organization and tissue differentiation in complex forms far exceeds that of other kingdoms; development of sensory-neuromotor systems in most permits rapid response. Reproduction mainly sexual, with production of eggs and motile sperm by meiosis.

SUBKINGDOM PARAZOA

Cell differentiation present but tissue differentiation very limited. Adults sessile.

Phylum porifera. Sponges. Mainly marine. The body is perforated with many pores to admit water, from which food is filtered by choanocytes. Solitary, or form colonies. Asexual reproduction by budding; sexual reproduction in which sperm is released externally to swim to internal egg. Larva is motile. (About 10,000 species.)

Class Calcispongiae. Skeleton composed of calcium carbonate spicules.

Class Hexactinellida. Glass sponges. Skeleton of six-rayed siliceous spicules.

Class Demospongiae. Horny sponges. Skeleton of spongin fibers or of siliceous spicules that do not have six rays.

SUBKINGDOM EUMETAZOA

Tissue and organ system level of organization. Most have a mouth and digestive cavity.

BRANCH RADIATA

Animals with radial or biradial symmetry.

Phylum Cnidaria. Radially symmetrical, aquatic animals with a central gastrovascular cavity. The body wall consists of two layers of cells, in the outer of which are stinging cells called cnidocytes. Tentacles surround mouth. Polyp and medusa forms. Planula larva. (About 9000 species.)

Class Hydrozoa. Hydralike animals, either single or colonial. Typically, there is an alternation of polyp (asexual) and medusa (sexual) generations. Examples are *Hydra*, a freshwater form that lacks a medusa stage, and *Obelia*, a marine form.

Class Scyphozoa. True jellyfishes. Medusa is prominent body form; polyp stage restricted to small larval stage. Example is *Cyanea*, the largest jellyfish.

Class Anthozoa. Corals and sea anemones. No alternation of generations; polyps may be individual or colonial. Digestive cavity is partitioned by mesenteries, which increase effective surface area.

Phylum Ctenophora. Comb jellies. Biradial symmetry. Free-swimming; marine. Two tentacles and eight longitudinal rows of cilia resembling combs; animal moves by means of these bands of cilia. (About 100 species.)

BRANCH BILATERIA

Animals with bilateral symmetry.

Protostomes

Spiral, determinate cleavage; mouth develops from blastopore.

Acoelomates

No body cavity; region between body wall and internal organs filled with tissue (parenchyma).

Phylum Platyhelminthes. Flatworms. Acoelomate. Body dorsoventrally flattened; cephalization. Three tissue layers. Excretory organs are protonephridia with flame cells. Simple nervous system with ganglia in head region. (About 13,000 species.)

Class Turbellaria. Planarians. Free-living flatworms with ciliated epidermis.

Class Trematoda. Flukes. Parasitic, nonciliated flatworms with one or more suckers. Many are internal parasites with complicated life cycles requiring two or more hosts. Examples are the blood flukes (genus *Schistosoma*) and *Fasciola hepatica*, the liver fluke.

Class Cestoda. Tapeworms. Parasitic flatworms with no digestive tract; the body consists of a scolex and a long chain of proglottids in which eggs are produced. Life cycle includes more than one host. Examples are the beef tapeworm *Taenia saginata* and the pork tapeworm *Taenia solium*.

Phylum Nemertina (or Nemertea). Proboscis worms. Nonparasitic, acoelomate, marine animals with a complete digestive tract and a protrusible proboscis armed with a hook for capturing prey. The simplest animal with a circulatory system carrying blood. (About 650 species.)

Pseudocoelomates

Have body cavity that is not completely lined with mesoderm. Animals in this group also have a complete digestive tract extending from mouth to anus.

Phylum Nematoda. Roundworms. Slender, elongated, cylindrical worms; have pseudocoelom; complete digestive tract; covered with cuticle. Free-living and parasitic forms. Examples: *Ascaris*, hookworms, pinworms. (Very numerous; about 10,000 species.)

*Phylum Acanthocephala.** Small worms parasitic in arthropods. Head armed with many recurved hooks. No digestive tract. (About 500 species.)

*Phylum Nematomorpha.** Horsehair worms. Extremely thin, brown or black worms about 15 cm long. Adults are free-living, but the larvae are parasitic in insects. (About 200 species.)

Phylum Rotifera. Wheel animals. Microscopic, wormlike animals. Anterior end bears ciliated crown, the beating of which suggests a wheel. Posterior end tapers to a foot. Constant number of cells. (About 1800 species.)

*Phylum Gastrotricha.** Microscopic, wormlike animals resembling the rotifers but lacking the crownlike circle of cilia. (About 400 species.)

Schizocoelous Coelomates

These animals have a true coelom that develops as a schizocoel (i.e., the mesoderm splits to form the body cavity). A complete digestive tract is present.

*Phylum Sipuncula.** Cylindrical marine worms with retractable anterior end. Have tentacles or lobes around the mouth. (About 300 species.)

Phylum Mollusca. The mollusks. Unsegmented, soft-bodied animals usually covered by a dorsal shell. Have a ventral, muscular foot. Most of organs located above foot in visceral mass. A shell-secreting mantle covers the visceral mass and forms a mantle cavity, which contains gills. Trochophore and/or veliger larva. (About 100,000 species.)

Class Monoplacophora. *Neopilina* and its relatives. Single dome-like shell and multiple pairs of organs; considered primitive.

Class Polyplacophora. Chitons. Sluggish marine animals with a shell composed of eight transverse plates.

Class Gastropoda. Snails, slugs, whelks, abalones. Body twisted within spirally coiled shell; some lack shells. Well-developed head with tentacles.

Class Bivalvia. Clams, oysters, mussels, scallops. Lack a head and radula. Enclosed by two shells (valves) hinged dorsally and opening ventrally. Hatchet-shaped foot for burrowing.

Class Cephalopoda. Squids, octopods, cuttlefish. Marine animals with well-developed "head-foot," with 8 or 10 tentacles, and well-developed eyes and nervous system. Fast-swimming, predatory animals.

*Class Scaphopoda.** Tooth shells. Marine forms living in sand or mud. Tubular shell opens at both ends.

Phylum Annelida. Segmented worms. Both body wall and internal organs are segmented. Segments separated by septa. Some have nonjointed appendages. Setae aid in locomotion. Closed circulatory system, metanephridia, specialized regions of digestive tract. Trochophore larva. (About 10,000 species.)

Class Polychaeta. Most marine worms. Each segment of the body has a pair of paddlelike parapodia for swimming and many bristles (setae). Some burrow in sand and mudflats; some live in calcareous tubes, which they secrete; others swim freely in the ocean. Clamworms, tubeworms, fanworms.

Class Oligochaeta. Earthworms. Freshwater and terrestrial forms. Head indistinct. Few setae per segment; no parapodia. Hermaphroditic.

Class Hirudinea. Leeches. Predatory and/or ectoparasites. Suckers at both anterior and posterior ends. Setae and parapodia absent. Hermaphroditic.

*Phylum Pognophora.** Beard worms. Elongated marine animals. Live in chitinous tubes. Anterior end of body has one to many long tentacles. Digestive tract absent. (About 80 species.)

Phylum Onychopora. Rare, tropical animals, structurally intermediate between annelids and arthropods. Segmented internally like annelid, and has annelidlike excretory system. Open circulatory system as in arthropods, and insectlike respiratory system. Many pairs of short, conical legs with claws. (About 70 species.)

Phylum Arthropoda. Segmented animals with paired, jointed appendages and a hard, chitinous exoskeleton. Open circulatory system with dorsal heart. Hemocoel occupies most of body cavity and coelom is reduced. (About 900,000 species.)

Subphylum Trilobitomorpha. Trilobites.

Class Trilobita. Fossil trilobites. Extinct marine arthropods; segmented body covered by a hard, segmented shell; had head, thorax, and abdomen; each body segment had a pair of biramous appendages.

Subphylum Chelicerata. First pair of appendages are the chelicerae used to manipulate food; body consists of cephalothorax and abdomen. No mandibles; no antennae.

Class Merostomata. Horseshoe crabs. Long spikelike tail used in locomotion.

Class Arachnida. Spiders, scorpions, ticks, mites. Adults have no antennae; the first pair of appendages, the chelicerae, ends in pincers, the second pair (pedipalps) are used as jaws and the last four pairs are used for walking.

Class Pycnogonida. Sea spiders. Body greatly reduced in size.

Subphylum Crustacea. Two pairs of antennae; mandibles; biramous appendages; usually respire by means of gills.

Class Crustacea. Lobsters, crabs, shrimp, barnacles

Subphylum Uniramia. Uniramous (unbranched) appendages; single pair of antennae; mandibles.

Class Insecta. Insects. The largest class of animals. Mainly terrestrial. Body divided into a distinct head with four pairs of appendages; the thorax, with three pairs of legs and usually two pairs of wings; and the abdomen, which has no appendages. Respiration by means of trachea. Excretion through Malpighian tubules. About 28 orders. Some of the principal orders are described in the Focus on the Principal Orders of Insects on page 532 in the text.

Class Chilopoda. Centipedes. Each body segment, except the head and tail, has a pair of legs.

Class Diplopoda. Millipedes. Each external segment (actually two segments fused) bears two pairs of legs.

*Phylum Phoronida.** Wormlike marine forms that secrete and live in a chitinous tube. U-shaped digestive tract. Have lophophore, a crown of hollow tentacles surrounding the mouth. (About 10 species.)

*Phylum Bryozoa.** Moss animals. Mainly marine; sessile; colonial. Have lophophore. (About 4000 species.)

*Phylum Brachiopoda.** Lamp shells. Body attached by a stalk. Body enclosed within two calcareous shells. (About 300 species.)

*Phylum Entoprocta.** Body attached by stalk. Mouth and anus surrounded by lophophore (U-shaped circle of ciliated tentacles).

Deuterostomes

Radial indeterminate cleavage. Blastopore develops into anus; mouth forms from a second opening. Enterocoelous coelomates.

*Phylum Chaetognatha.** Arrow worms. Marine worms with dart-shaped bodies; part of plankton. (About 50 species.)

Phylum Echinodermata. Marine animals that have pentaradial symmetry as adults, but bilateral symmetry as larvae. Endoskeleton of small calcareous plates. Water vascular system; tube feet for locomotion. (About 6000 species.)

Class Crinoidea. Sea lilies, feather stars. Cup-shaped body with oral surface turned upward. Branched, feathery arms used to trap food.

Class Asteroidea. Sea stars. Body is central disk with broad rays (usually five) not sharply marked off from disk. Mouth on underside of disk.

Class Ophiuroidea. Brittle stars. Have long, narrow rays sharply differentiated from the central, disk-shaped body.

Class Echinoidea. Sea urchins and sand dollars. Spherical or flattened oval animals with many long spines; lack rays.

Class Holothuroidea. Sea cucumbers. Long, ovoid, soft-bodied; lack rays. Usually have a ring of tentacles around the mouth.

Phylum Hemichordata. Acorn worms. Marine animals with an anterior muscular proboscis, connected by a collar region to a long wormlike body. The larval form resembles an echinoderm larva. (About 80 species.)

Phylum Chordata. Notochord, pharyngeal gill slits, and dorsal, tubular nerve cord present at some time in life cycle.

Subphylum Urochordata. Tunicates. Marine animals. Adults are saclike, attached, filter-feeding animals with a tunic of cellulose. Larvae are free-swimming and have notochord in tail region. Notochord and nerve cord present only in larva.

Subphylum Cephalochordata. Lancelets. Marine animals with a segmented, elongated fishlike body. Burrow in sand and take in food by the beating of cilia on the anterior end. Notochord extends from tip of head to tip of tail. Best known example is *Amphioxus*.

Subphylum Vertebrata. Animals with a definite head and a trunk supported by a series of skeletal (bone or cartilage) structures called vertebrae surrounding or replacing the notochord in the adult. Closed circulatory system with ventral heart. Well-developed brain protected by cranium. A pair of well-developed eyes and other sense organs. Usually two pairs of limbs.

Superclass Pisces. Fish. Aquatic vertebrates.

Class Agnatha. Lamprey eels and hagfishes, and fossil ostracoderms. Lack true jaws. Lack paired appendages (fins).

Class Chondrichthyes. Sharks, rays, skates, and chimaeras. Fishes with a cartilaginous skeleton. Spiral valve present in intestine. Pelvic fins modified to form claspers in male.

Class Osteichthyes. Bony fishes. Swim bladder usually present. Gills covered by bony operculum.

Subclass Actinopterygii. Ray-finned fishes. Most modern Osteichthyes. Living members of this group include sturgeons, paddlefish, salmon, flounder, and most other familiar fishes.

Subclass Sarcopterygii. Lobe-finned fishes. Living members are lungfish.

Superclass Tetrapoda. Four-legged land vertebrates.

Class Amphibia. Frogs, toads, salamanders, and the extinct forms, labyrinthodonts. As larvae, these forms have gills; as adults, they have lungs. Skin usually scaleless, moist, and slimy.

Order Urodela. Salamanders, mud puppies, newts. Elongate body ending in a tail. Limbs not specialized for jumping.

Order Anura. Frogs and toads. Amphibians lacking a tail. Hind limbs modified for jumping.

Order Apoda. Caecilians. Amphibians with wormlike bodies that lack limbs. Eyes small or vestigial.

Class Reptilia. Lizards, snakes, reptiles, crocodiles, extinct dinosaurs. Poikilothermic tetrapods with lungs. Have amnion during development. Dry skin covered by epidermal scales. Three-chambered heart.

Order Chelonia. Turtles, terrapins, tortoises. Enclosed in a protective shell made up of bony plates. Lack teeth.

Order Squamata. Lizards, snakes, iguanas, geckos. Body covered by epidermal scales that overlap. Teeth present.

Order Crocodilia. Crocodiles, alligators, caimans, gavials. Elongate skull. Teeth present. Considered large, degenerate survivors of Triassic archosaurians.

Class Aves. Birds. Homeothermic animals covered with feathers. Anterior limbs modified as wings for flight. Present-day birds are toothless but the primitive ones had reptilian teeth. (More than 30 orders.)

Class Mammalia. Homeothermic animals whose skin is covered with hair. The females have mammary glands which secrete milk for the nourishment of the young.

Subclass Prototheria. Monotremes. Primitive egg-laying mammals that retain many reptilian features such as egg-laying and cloacae. Most extinct; only two species survive—the duckbilled platypus and the spiny anteater.

Subclass Metatheria. Marsupials. Young born alive, but in a very undeveloped state. They complete development within a pouch (marsupium) located on mother's abdomen. Teats open into marsupium. Includes opossums and a variety of forms found in Australia—kangaroos, wallabies, koala bears, wombats, and so on.

Subclass Eutheria. Placental mammals. Young develop within the uterus of the mother, obtaining nourishment by way of the placenta. About 17 living orders. (See Chapter 26, Focus on Some Orders of Living Placental Mammals.)

appendix B
Dissecting Terms

Your task of mastering new terms will be greatly simplified if you learn to dissect each new word. Many terms can be divided into a prefix, the part of the word that precedes the main root, the word root itself, and often a suffix, a word ending that may add to or modify the meaning of the root. As you progress in your study of biology, you will learn to recognize the more common prefixes, word roots, and suffixes. Such recognition will help you analyze new terms so that you can more readily determine their meaning, and will also help you remember them.

PREFIXES

a-, ab- from, away, apart (abduct, lead away, move away from the midline of the body)
a-, an- un-, -less, lack, not (asymmetrical, not symmetrical)
ad- (also **af-, ag-, an-, ap-**) to, toward (adduct, move toward the midline of the body)
ambi- both sides (ambidextrous, able to use either hand)
ante- forward, before (anteflexion, bending forward)
anti- against (anticoagulant, a substance that prevents coagulation of blood)

bi- two (biceps, a muscle with two heads of origin)
bio- life (biology, the study of life)
brady- slow (bradycardia, abnormally slow heart beat)

circum-, circ- around (circumcision, a cutting around)
co-, con- with, together (congenital, existing with or before birth)
contra- against (contraception, against conception)
crypt- hidden (cryptorchidism, undescended or hidden testes)
cyt- cell (cytology, the study of cells)

di- two (disaccharide, a compound made of two sugar molecules chemically combined)
dis- (also **di-** or **dif-**) apart, un-, not (dissect, cut apart)
dys- painful, difficult (dyspnea, difficult breathing)

end-, endo- within, inner (endoplasmic reticulum, a network of membranes found within the cytoplasm)
epi- on, upon (epidermis, upon the dermis)
eu- good, well (euphoria, a sense of well-being)
ex-, e-, ef- out from, out of (extension, a straightening out)
extra- outside, beyond (extraembryonic membrane, a membrane such as the amnion that protects the embryo)

hemi- half (cerebral hemisphere, lateral half of the cerebrum)
hetero- other, different (heterogeneous, made of different substances)
homo-, hom- same (homologous, corresponding in structure)
hyper- excessive, above normal (hypersecretion, excessive secretion)
hypo- under, below, deficient (hypodermic, below the skin; hypothyroidism, insufficiency of thyroid hormones)

in-, im- not (imbalance, condition in which there is no balance)
inter- between, among (interstitial, situated between parts)
intra- within (intracellular, within the cell)
iso- equal, like (isotonic, equal strength)

mal- bad, abnormal (malnutrition, poor nutrition)
mega- large, great (megakaryocyte, giant cell of bone marrow)
meta- after, beyond (metaphase, the stage of mitosis after prophase)

neo- new (neonatal, newborn during the first 4 weeks after birth)

oo- egg (oocyte, developing egg cell)
oligo- small, deficient (oliguria, abnormally small volume of urine)
Orth-, ortho- straight (orthodontist, one who straightens teeth)

para- near, beside, beyond (paracentral, near the center)
peri- around (pericardial membrane, membrane that surrounds the heart)
poly- many, much, multiple, complex (polysaccharide, a carbohydrate composed of many simple sugars)
post- after, behind (postnatal, after birth)
pre- before (prenatal, before birth)

retro- backward (retroperitoneal, located behind the peritoneum)

semi- half (semilunar, half-moon)
sub- under (subcutaneous tissue, tissue immediately under the skin)
super-, supra- above (suprarenal, above the kidney)
syn- with, together (syndrome, a group of symptoms which occur together and characterize a disease)

trans- across, beyond (transport, carry across)

SUFFIXES

-able, -ible able (viable, able to live)
-ac pertaining to (cardiac, pertaining to the heart)
-ad used in anatomy to form adverbs of direction (cephalad, toward the head)
-asis, -asia, -esis condition or state of (hemostasis, stopping of bleeding)

-cide kill, destroy (biocide, substance that kills living things)

-ectomy surgical removal (appendectomy, surgical removal of the appendix)
-emia condition of blood (anemia, without enough blood)

-gen something produced or generated or something that produces or generates (pathogen, something that can cause disease)
-gram record, write (electrocardiogram, a record of the electrical activity of the heart)

-graph record, write (electrocardiograph, an instrument for recording the electrical activity of the heart)

-ic adjective-forming suffix which means *of* or *pertaining to* (ophthalmic, of or pertaining to the eye)
-ist one who practices, deals with, or does (biologist, one who studies biology)
-itis inflammation of (appendicitis, inflammation of the appendix)

-logy study or science of (cytology, study of cells)

-oid like, in the form of (thyroid, in the form of a shield)
-oma tumor (carcinoma, a malignant tumor)
-osis indicates disease (psychosis, a mental disease)
-ous, -ose full of (poisonous, full of poison)

-pathy disease (dermopathy, disease of the skin)
-plasty reconstruction (rhinoplasty, reconstruction of the nose)

-scope instrument for viewing or observing (microscope, instrument for viewing small objects)
-stomy refers to a surgical procedure in which an artificial opening is made (colostomy, surgical formation of an artificial anus)

-tomy cut, incision into (appendectomy, incision into the appendix)

-uria refers to urine (polyuria, excessive production of urine)

SOME COMMON WORD ROOTS

aden gland, glandular (adenosis, a glandular disease)
alg pain (neuralgia, nerve pain)
angi, angio vessel, vascular (lymphangiogram, an x-ray of lymphatic vessels following injection of a radiopaque contrast media)
arthr joint (arthritis, inflammation of the joints)

bi, bio life (biology, study of life)
blast a formative cell, germ layer (osteoblast, cell that gives rise to bone cells)
brachi arm (brachial artery, blood vessel that supplies the arm)
bronch branch of the trachea (bronchitis, inflammation of the bronchi)
bry grow, swell (embryo, an organism in the early stages of development)

carcin cancer (carcinogenic, cancer-producing)
cardi heart (cardiac, pertaining to the heart)
cephal head (cephalad, toward the head)
cerebr brain (cerebral, pertaining to the brain)
cervic, cervix neck (cervical, pertaining to the neck)
chol bile (cholecystogram, an x-ray of the gallbladder)
chondr cartilage (chondrocyte, a cartilage cell)
chrom color (chromosome, deeply staining body in nucleus)
col, coli, colo colon (colitis, inflammation of the colon)
cran skull (cranial, pertaining to the skull)
cyst, cysti, cysto urinary bladder (cystitis, inflammation of the urinary bladder; cystogram, an x-ray of the urinary bladder)
cyt cell (cytology, study of the cells)

derm skin (dermatology, study of the skin)
duct, duc lead (duct, passageway)

ecol dwelling, house (ecology, the study of organisms in relation to their environment)
encephal, encephalo brain (encephalitis, inflammation of the brain)
enter intestine (enteritis, inflammation of the intestine)
evol to unroll (evolution, descent of complex organisms from simpler ancestors)

gastr stomach (gastritis, inflammation of the stomach)
gen generate, produce (gene, a hereditary factor)
glyc, glyco sweet, sugar (glycogen, storage form of glucose)
gon semen, seed (gonad, an organ producing gametes)

hem, em blood (hematology, the study of blood)
hepat, hepar liver (hepatitis, inflammation of the liver)
hist tissue (histology, study of tissues)
hom, homeo same, unchanging, steady (homeostasis, reaching a steady state)
hydr water (hydrolysis, a breakdown reaction involving water)
hyster, hystero uterus (hysterectomy, surgical removal of all or part of the uterus; hysterogram, an x-ray of the uterus)

lapar, laparo abdomen (laparotomy, incision into the abdomen)
laryng, laryngo larynx (laryngitis, inflammation of the larynx; otorhinolaryngology, study of the ear, nose, and larynx)
leuk white (leukocyte, white blood cell)
lith, litho stone or calculus (lithonephritis, inflammation of the kidney due to the presence of calculi)

macro large (macrophage, large phagocytic cell)
mamm breast (mammary glands, the glands that produce milk to nourish the young)
mening, meningo meninges (meningitis, inflammation of the membranes of the brain or spinal cord)
micro small (microscope, instrument for viewing small objects)
morph form (morphogenesis, development of body form)
my, myo muscle (myocardium, muscle layer of the heart)

nephr kidney (nephron, microscopic unit of the kidney)
neur, nerv nerve (neuralgia, pain associated with a nerve)

occiput back part of the head (occipital, back region of the head)
odont, odonto tooth (odontotomy, incision into a tooth)
ophthal, ophthalmo eye (ophthalmopathy, disease of the eye)
orchi, orchido, orchid testis (orchitis, inflammation of the testes; orchiectomy, surgical removal of the testis)
ost, oss bone (osteology, study of bones)
ot, oto ear (otitis, inflammation of the ear; otoscope, an instrument for examination of the ear)

path disease (pathologist, one who studies disease processes)
ped child (pediatrics, branch of medicine specializing in children)
ped, pod foot (biped, organism with two feet)
phag eat (phagocytosis, process by which certain cells ingest particles and foreign matter)
phil love (hydrophilic, a substance that attracts water)
phleb, phlebo vein (phlebitis, inflammation of a vein)
phren, phreno diaphragm (phrenocolic, of or pertaining to the diaphragm and colon)
proct anus (proctoscope, instrument for examining rectum and anal canal)
psych mind (psychology, study of the mind)
pyel, pyelo pelvis or kidney (pyelitis, inflammation of the renal pelvis)

rect, recto rectum (rectocolitis, inflammation of the rectum and colon)
rhin nose (rhinalgia, pain in the nose)

salping, salpingo uterine tube (salpingectomy, surgical removal of the uterine tube)
scler hard (atherosclerosis, hardening of the arterial wall)
som body (chromosome, deeply staining body in the nucleus)
splen, spleno spleen (splenectomy, surgical removal of the spleen)
stas, stat stand (stasis, condition in which blood stands, as opposed to flowing)

thromb clot (thrombus, a clot within the body)
thym, thymo thymus (thymectomy, surgical removal of the thymus)

ur urea, urine (urologist, a physician specializing in the urinary tract)

visc pertaining to an internal organ or body cavity (viscera, internal organs)

glossary/index*

*Bold face page numbers indicate pages on which the index term is defined; "il" following the page number indicates an illustration, "t" a table, "f" a focus, and "n" a footnote.

the metric system

Standard metric units		*Abbreviations*
Standard unit of mass	gram	g
Standard unit of length	meter	m
Standard unit of volume	liter	l

Some common prefixes		*Examples*
kilo	1,000	a kilogram is 1,000 grams
centi	0.01	a centimeter is 0.01 meter
milli	0.001	a milliliter is 0.001 liter
micro (μ)	one-millionth	a micrometer is 0.000001 (one-millionth) of a meter
[illegible]	one-billionth	[illegible]ogram is 10^{-9} (one-billionth) of a gram
pic[illegible]	one-trillion[illegible]	[illegible]m is 10^{-12} (one-trillionth) of a gram

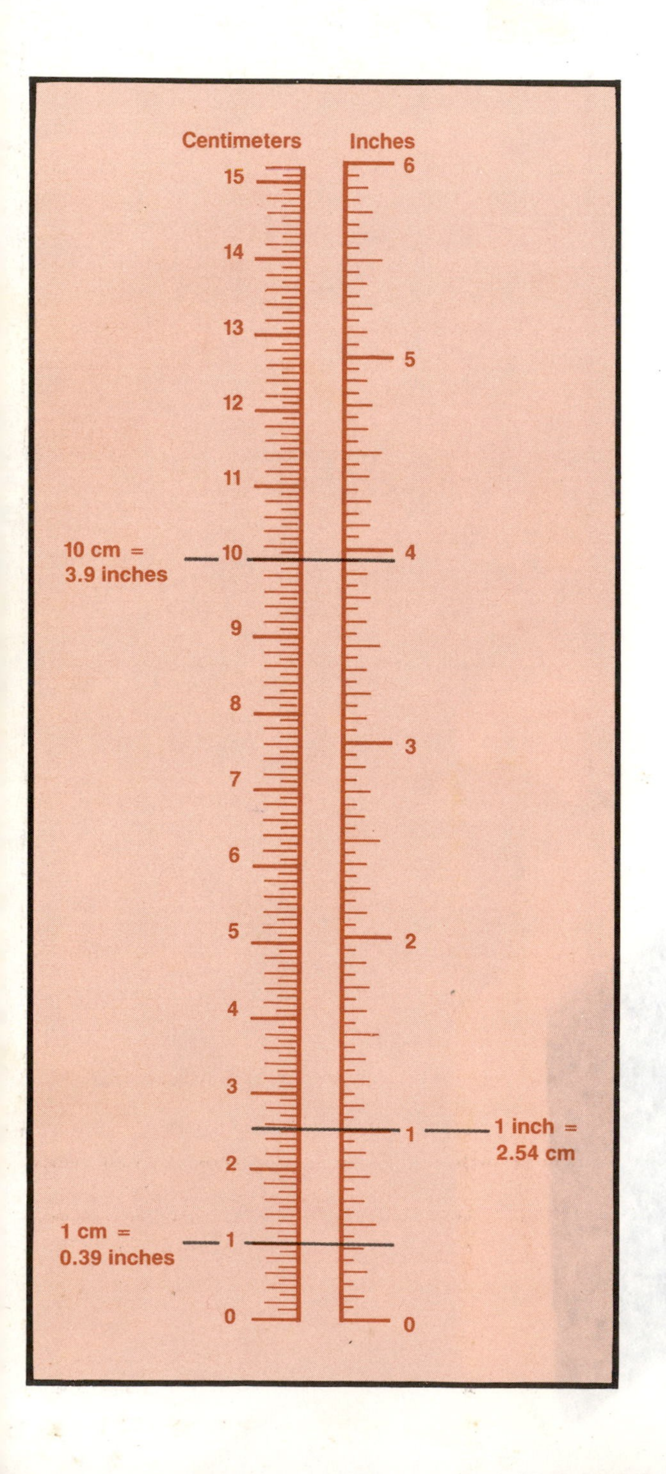

Think Metric!

A 154-lb person weighs 70 kilograms (kg).

A 5′ 6″ person is 165 cm long.

You are driving down the highway at 85.8 km per hour. That is the same speed as 55 mph.

A 70-kg human male has 5.6 liters of blood. That is about 6 quarts.

SOME COMMON UNITS OF LENGTH

Unit	*Abbreviation*	*Equivalent*
meter	m	approximately 39 in
centimeter	cm	10^{-2} m
millimeter	mm	10^{-3} m
micrometer	μm	10^{-6} m
nanometer	nm	10^{-9} m
angstrom	Å	10^{-10} m

Length Conversions

1 in	=	2.5 cm		1 mm	=	0.039 in
1 ft	=	30 cm		1 cm	=	0.39 in
1 yd	=	0.9 m		1 m	=	39 in
1 mi	=	1.6 km		1 m	=	1.094 yd
				1 km	=	0.6 mi

To convert	*Multiply by*	*To obtain*
inches	2.54	centimeters
feet	30	centimeters
centimeters	0.39	inches
millimeters	0.039	inches